MW01346711

PSYCHOTIC DISORDERS

PSYCHOTIC DISORDERS

COMPREHENSIVE CONCEPTUALIZATION AND TREATMENTS

EDITED BY

Carol A. Tamminga, MD

PROFESSOR AND CHAIR
DEPARTMENT OF PSYCHIATRY
UT SOUTHWESTERN MEDICAL CENTER
DALLAS, TX, USA

Elena I. Ivleva, MD, PhD

ASSOCIATE PROFESSOR
DEPARTMENT OF PSYCHIATRY
EARLY PSYCHOSIS PROGRAM, DIRECTOR
DIVISION OF TRANSLATIONAL NEUROSCIENCE RESEARCH IN SCHIZOPHRENIA
UT SOUTHWESTERN MEDICAL CENTER
DALLAS, TX, USA

Ulrich Reininghaus, PhD

HEISENBERG PROFESSOR AT THE DEPARTMENT OF PUBLIC MENTAL HEALTH
CENTRAL INSTITUTE OF MENTAL HEALTH (CIMH)
MANNHEIM, GERMANY
VISITING PROFESSOR
HEALTH SERVICE AND POPULATION RESEARCH DEPARTMENT
INSTITUTE OF PSYCHIATRY, PSYCHOLOGY, AND NEUROSCIENCE
KING'S COLLEGE LONDON
LONDON, UK

Jim van Os, MD, PhD

DEPARTMENT OF PSYCHIATRY, BRAIN CENTRE RUDOLF MAGNUS
UNIVERSITY MEDICAL CENTRE UTRECHT
UTRECHT, THE NETHERLANDS

OXFORD
UNIVERSITY PRESS

Oxford University Press is a department of the University of Oxford. It furthers the University's objective of excellence in research, scholarship, and education by publishing worldwide. Oxford is a registered trade mark of Oxford University Press in the UK and certain other countries.

Published in the United States of America by Oxford University Press
198 Madison Avenue, New York, NY 10016, United States of America.

Library of Congress Cataloging-in-Publication Data
Names: Tamminga, Carol A, editor. | Ivleva, Elena I., editor. |
Reininghaus, Ulrich, editor. | Os, J. van (Jim van), editor.
Title: Psychotic disorders : comprehensive conceptualization and treatments /
editors, Carol A. Tamminga, Elena I. Ivleva, Ulrich Reininghaus, Jim van Os.
Description: New York, NY : Oxford University Press, [2021] |
Includes bibliographical references and index.
Identifiers: LCCN 2019055816 (print) | LCCN 2019055817 (ebook) |
ISBN 9780190653279 (hardback) | ISBN 9780190653293 (epub) | ISBN 9780197501467
Subjects: MESH: Psychotic Disorders | Psychotic Disorders—therapy
Classification: LCC RC55 (print) | LCC RC55 (ebook) | NLM WM 200 |
DDC 616.89—dc23
LC record available at https://lccn.loc.gov/2019055816
LC ebook record available at https://lccn.loc.gov/2019055817

1 3 5 7 9 8 6 4 2

Printed by LSC Communications, United States of America

CONTENTS

SECTION 3
NEUROBIOLOGY OF PSYCHOSIS

HERITABILITY AND GENETICS

COGNITIVE BIOMARKERS OF PSYCHOSIS

NEUROPHYSIOLOGIC BIOMARKERS OF PSYCHOSIS

BRAIN IMAGING BIOMARKERS

PATHOPHYSIOLOGY OF PSYCHOSIS: NEUROTRANSMITTERS

PATHOPHYSIOLOGY: VOLTAGE-GATED ION CHANNELS IN PSYCHOSIS

PATHOPHYSIOLOGY: IMMUNE MECHANISMS

BRAIN CIRCUIT ALTERATIONS IN PSYCHOSIS

EARLY INTERVENTIONS

SECTION 6
FUTURE DIRECTIONS AND OPPORTUNITIES

PREFACE

The diagnosis of psychotic disorders in psychiatry has been systematized for decades, and described even longer, over millennia. We have lists of symptoms, criteria involving presentation, descriptions of illness course, and symptom constellations. The characteristics of schizophrenia, schizoaffective disorder, even bipolar disorder with psychosis and psychotic disorder NOS have been spelled out, subject to committee review as well as new formulations published at least once a decade. Who could question this process? These diagnostic specifications contribute remarkably to the practical communications around psychotic disorders, to reimbursement needs, to building up clinical care models, and to teaching. However, now that we regularly sample biological characteristics in people with psychosis, it is the biological heterogeneity within diagnoses that the field is encountering, and these challenge our standing ideas. Indeed, the scientific world has changed around these symptom-based diagnoses to require that biological distinctiveness be a part of the disease description, just as elsewhere in medicine. In psychotic disorders, biological testing began with genetic characteristics, seeking a genetic fingerprint of the disorders. We have been disappointed that genetics has not been more effective in constructing disease definitions with biological distinctiveness. Not that the genetic characterization has failed completely,[1–3] it is just that even genetic data demonstrate considerable biological heterogeneity within and across diagnoses. Many common psychiatric disorders overlap genetically with other diagnostically distinct disorders. There is also evidence that socioenvironmental risk (such as prenatal factors, childhood adversity, urbanicity, and ethnicity) is shared across psychiatric disorders, including the psychoses. Moreover, the genetic basis for a common disorder is cross-correlated with symptom determinants not at all in the diagnosis.[4] And, common diagnoses within a dimension (such as psychosis) have biological characteristics that are orthogonal to the disease phenomenology.[5] Sullivan[4] opines, "These results strongly suggest that our diagnostic categories do not define pathophysiological entities." Therefore, in order to create biologically based nomenclatures within psychiatry, we need to return to the drawing board with creative experiments, careful phenotyping, and biological goals.

This volume is an attempt to have the field address diagnostic heterogeneity and rich interconnections across syndromes within the dimension of psychosis and advance the work of identifying biological determinants. It is a very timely moment to do this in our field, because of the explosion in knowledge occurring in basic neuroscience, the unprecedented advances in genetics in other fields, and the parallel growth we see in clinically relevant understanding of human disease risk, interesting targets, and broad unexplored phenotypes for brain disorders, which we can use to test new biological targets. We are at a crossroads—an opportunistic threshold for biological advancement in the area of psychosis conceptualization and treatment.

Psychiatric disorders in the psychosis dimension such as schizophrenia, schizoaffective disorder, delusional disorder, schizophreniform disorder, brief psychotic disorder, psychosis NOS, psychotic bipolar disorder, and psychotic depression are all highly disabling disorders, with prominent psychotic features along with multifaceted symptoms (e.g., delusions, hallucinations, disorganized speech, and cognitive and affective disturbance). They are suspected to have complex etiologies, and all show heterogeneous outcomes. These are brain-mediated diseases that generate high financial, social, and personal burden for patients, caregivers, and society at large. Primary psychotic and primary mood disorders have predominantly been studied as separate entities since Emil Kraepelin proposed the dichotomy of "dementia praecox" and "manic depression" based on phenomenological characteristics. Despite this history, factor-analytic evidence on transdiagnostic dimensional phenotypes as well as many studies using biological outcomes suggest considerable overlap between conventional psychosis diagnoses and phenotypes of brain. But characteristically, the phenomenology of these psychotic disorders continues to be investigated in separate populations and clinics and are published in separate bodies of research, isolating these areas from each other. These psychosis "fields" have also been characterized by, and criticized for, too little crossover among researchers, clinicians, and service users. But, over the past decade or so, the phenomenology, mechanisms, etiology, and treatment of psychotic disorders have been increasingly investigated around biological models integrated around dimensions of psychopathology including psychosis.

This volume embraces a dimensional approach to psychosis and brings together international experts at the forefront of cross-disciplinary research into the etiology, mechanisms, and treatment of psychotic disorders. The ambition of this volume is to be integrative along three key dimensions, with the aim to:

1. Review, integrate, and critique research findings (phenomenology and biomarker) on the overlaps and interfaces between psychiatric disorders within the psychosis dimension, considering novel disease definitions and conventional disease groups.

2. Identify research findings on factors from all potential levels of mediation and causation (i.e., genetic, epigenetic, molecular and cellular, psychological, social, environmental, cultural) including how the biological and environmental factors interact and impact on each other and how they define pathophysiology and etiology.
3. Review existing treatment approaches, from psychopharmacology, to multiple modalities of psychotherapy, to neuromodulation and cognitive interventions. Consider the treatment, course, and outcome of psychiatric disorders in the psychosis dimension, using optimal biomarker characteristics as well as phenomenology to identify homogeneous disease groups, and bring together clinical, public health, and service user perspectives as well as predictors of treatment.

There is little evidence that, within the psychosis dimension, biologically relevant subgroups are specified by conventional diagnoses. Within conventional systems, even though nonschizophrenia psychotic disorders constitute 70% of the psychotic illness burden (only 30% of those with psychotic illness meet criteria for schizophrenia), the other groups comparatively receive little attention in the academic literature (which is more than 80% on schizophrenia and less than 20% on nonschizophrenia psychotic categories). The literature therefore tends to be biased toward the specific conventional diagnosis of schizophrenia, toward phenomenology as the specifier, and toward the poor outcome spectrum of psychosis.

In the meantime, the scientific evidence indicates that different conventional psychotic categories can be viewed as sharing some similar brain characteristics within the dimension of psychosis with a total lifetime prevalence of 3.5%, of which the diagnosis of schizophrenia represents less than a third. People with schizophrenia as a conventional diagnosis display substantial heterogeneity—both between- and within-course in persons—in psychopathology, treatment response, and outcome. And all conventional psychosis diagnoses extensively overlap biologically. Therefore, it is time to focus on the dimension of psychosis, study its components carefully, and use the heterogeneity within its broad boarders to segregate biologically based subgroups as they emerge as latent structures within the psychosis dimension which have common outcomes, treatment response, and biology.

No book has adopted the approach that we propose here. This volume will take an integrative, transdiagnostic, cross-disciplinary approach to the formulation and treatment of psychosis spectrum disorders. We start the volume by having a section around different conceptualizations of nosology active within the field, in organic as well as functional psychoses, and an articulation of the role of the recently proposed Research Diagnostic Criteria proposal in influencing the field. Also, several colleagues contribute chapters on the course of symptom manifestations in psychotic illnesses; here it is obvious that the early course of illness has become the target of intense interest and treatment focus. We recognize that psychiatry has always been intent on identifying the neurobiology of psychosis or, more likely, of the psychoses, suspecting that this would lead us to precision treatments. Interest in the genetics of psychosis has been central to this approach from the time that the human genome was sequenced. In addition to genetics, modern clinical scientists have hoped that specific phenotypes along with selective and critical biomarkers or biomarker batteries could make a contribution to the understanding of psychosis pathobiology. The biomarkers of interest include cognitive performance; neurophysiological markers including EEG and ocular motor measures; brain imaging especially contributes key biomarkers, many of which are functional; and animal models would be the critical instantiation of our ideas of psychosis pathobiology. Models of pathophysiology include target neurochemical systems with a full range of possible targets; inflammatory systems implicated in psychosis; and targeted brain systems where evidence indicates their role in psychopathology. The diversity of neurobiological targets may underlie disease heterogeneity and may eventually be the substrate for distinct psychosis diseases. Also, in these chapters, the broad area of environmental and psychological determinants of brain pathology is incorporated, including the effects of early life adversity. Treatment approaches for psychosis are reviewed as highly diverse, including known and postulated approaches. All of these areas are represented within this text. The book concludes with 3 perspectives conjecturing where we are going, all generalists arguing for novel ideas and change in conceptual systems, all in the context of effective psychosis understanding. We intend to have this text stimulate new issues, generate novel frameworks, carry new directions forward, and contribute to knowledge about psychotic disorders.

The opportunity is vast and promising, both at present and in the future, to investigate broadly across all psychoses; to find binding, explanatory, and effective biological principles; to define disease entities using biological characteristics; and to use all this to lead to improved treatment and basic disease understanding.

Carol A. Tamminga
Elena I. Ivleva
Ulrich Reininghaus
Jim van Os

REFERENCES

1. Ripke S. Biological insights from 108 schizophrenia-associated genetic loci. *Nature.* 2014;511(7510):421–427.
2. Genomic dissection of bipolar disorder and schizophrenia, including 28 subphenotypes. *Cell.* 2018;173(7):1705–1715.
3. Huckins LM, Dobbyn A, Ruderfer DM, et al. Gene expression imputation across multiple brain regions provides insights into schizophrenia risk. *Nat Genet.* 2019;51(4):659–674.
4. Sullivan PF, Agrawal A, Bulik CM, et al. Psychiatric genomics: an update and an agenda. *Am J Psychiatry.* 2018;175(1):15–27.
5. Clementz BA, Sweeney JA, Hamm JP, et al. Identification of distinct psychosis biotypes using brain-based biomarkers. *Am J Psychiatry.* 2016;173(4):373–384.

CONTRIBUTORS

Olesya Ajnakina, PhD
Department of Biostatistics and Health Informatics
King's College London
London, England, UK

Schahram Akbarian, MD, PhD
Department of Psychiatry and Friedman Brain Institute
Icahn School of Medicine at Mount Sinai
New York, NY, USA

Nikolai Albert, MD
Copenhagen Research Center for Mental Health CORE
Mental Health Center Copenhagen
Copenhagen University Hospital
Copenhagen, Denmark

Samuel J. Allen, MD
Lieber Institute for Brain Development
Baltimore, MD, USA

Kürşat Altınbaş, PhD
Department of Psychiatry
Selçuk University Medical Faculty
Konya, Turkey

Nancy Andreasen, MD, PhD
Department of Psychiatry
University of Iowa
Iowa City, IA, USA

Andrea M. Auther, PhD
Division of Psychiatry Research
The Zucker Hillside Hospital, Northwell Health
Glen Oaks, NY, USA
Department of Psychiatry
Zucker School of Medicine at Hofstra/Northwell
Hempstead, NY, USA

Ryan P. Balzan, BPsych (Hons), MPsych (Clin), PhD
College of Education, Psychology and Social Work
Flinders University
Adelaide, South Australia
Psychopathology and Early Intervention Lab II
Department of Psychiatry
The Medical University of Warsaw
Poland

MD Bauman, PhD,
Professor
Department of Psychiatry and Behavioral Sciences
University of California, Davis
Sacramento, CA, USA

Chyrell D. Bellamy, PhD
Department of Psychiatry
Yale School of Medicine
Program for Recovery and Community Health (PRCH)
New Haven, CT, USA

Iruma Bello, PhD
New York State Psychiatric Institute
Columbia University Medical Center
New York, NY, USA

Richard P. Bentall, PhD, FBA
Department of Psychology
University of Sheffield
South Yorkshire, England, UK

German E. Berrios, MD, DPM
Robinson College
University of Cambridge
Cambridge, UK, USA

Rahul Bharadwaj, PhD
Lieber Institute for Brain Development
Baltimore, MD, USA

Jeffrey Bishop, PharmD, MS, BCCPP, FCCP
Departments of Psychiatry and Experimental and Clinical Pharmacology
University of Minnesota
Minneapolis, MN, USA

Michael A. P. Bloomfield, PhD, BMBCh, MRCPsych, PGDip
Psychiatric Imaging Group
MRC London Institute of Medical Sciences
Imperial College London
Translational Psychiatry Research Group
Research Department of Mental Health Neuroscience
University College London
Department of Psychosis Studies
Institute of Psychiatry, Psychology and Neuroscience
King's College London
Clinical Psychopharmacology Unit
Research Department of Clinical, Educational and Health Psychology
University College London
London, England, UK

Jan Booij, MD, PhD
Department of Nuclear Medicine
Academic Medical Center
Amsterdam, The Netherlands

Denny Borsboom, PhD
Department of Psychological Methods
University of Amsterdam
Amsterdam, The Netherlands

Peter Bosanac, MBBS, MMed. (Psychiatry), MD, FRANZCP
Grad Dipl Mental Health Science (Clinical Hypnosis)
University of Melbourne
St Vincent's Hospital Melbourne
Melbourne, Australia

Lindy-Lou Boyette, PhD
Department of Clinical Psychology
University of Amsterdam
Amsterdam, The Netherlands

Alison Brabban
Consultant Clinical Psychologist
Tees, Esk and Wear Valleys National Health Service Foundation Trust
Darlington, England, UK

Alison Branitsky, MA
Research Assistant and Group Facilitator at Hearing Voices Network
Manchester, England, UK

Jérôme Brunelin, PhD, HDR
Lyon Neuroscience Research Center
University of Lyon
Lyon, France

Peter Buckley, MD
Virginia Commonwealth University School of Medicine
Richmond, VA, USA

Bauke Buwalda, PhD
Groningen Institute for Evolutionary Life Sciences (GELIFES)
University of Groningen
Groningen, The Netherlands

Tyrone D. Cannon, PhD
Department of Psychiatry
Yale University
New Haven, CT, USA

Mary Cannon, MD, PhD
Department of Psychiatry
Royal College of Surgeons in Ireland and Beaumont Hospital
Dublin, Ireland

William T. Carpenter, MD
Department of Psychiatry
University of Maryland School of Medicine
Baltimore, MD, USA

David J. Castle, MBChB, MSc, MD, DLSHTM, GCUT, FRCPsych, FRANZCP
University of Melbourne
St Vincent's Hospital Melbourne
School of Psychiatry and Neurosciences
The University of Western Australia
Department of Psychiatry
The University of Cape Town
St Vincent's Mental Health
Fitzroy, Victoria, Australia

Andrea Cipriani, MD, PhD
Department of Psychiatry
University of Oxford
Oxford, UK
Oxford Health NHS Foundation Trust
Oxford, UK

Mary Clarke, PhD
Department of Psychiatry
Royal College of Surgeons in Ireland, Education and Research Centre, Beaumont Hospital
Dublin, Ireland

Brett A. Clementz, PhD
Department of Psychology
University of Georgia
Athens, GA, USA

P. J. Conn, PhD
Lee Limbird Professor of Pharmacology, Director
Vanderbilt Center for Neuroscience Drug Discovery
Department of Pharmacology
Vanderbilt Kennedy Center
Vanderbilt University
Nashville, TN, USA

Barbara A. Cornblatt, PhD, MBA
Division of Psychiatry Research
The Zucker Hillside Hospital, Northwell Health
Glen Oaks, NY, USA
Department of Psychiatry
Zucker School of Medicine at Hofstra/Northwell
Hempstead, NY, USA
Center for Psychiatric Neuroscience
Feinstein Institute for Medical Research, Northwell Health
Manhasset, NY, USA

Xavier Cornejo
Department of Psychology
Wesleyan University
Middletown, CT, USA

Dirk Corstens, MD
Regional Institution for Ambulatory Mental Health Care (RIAGG Maastricht)
Maastricht, The Netherlands

Mark Costa, MD, MPH
Department of Psychiatry
Yale School of Medicine
Yale University
New Haven, CT, USA

Tom K. J. Craig, PhD
Department of Health Service and
Population Research
Institute of Psychiatry, Psychology and Neuroscience
King's College London
London, England, UK

Adam Crowther, BA (Hons), MA, (Oxon), MSc
Department of Psychology
Institute of Psychiatry, Psychology
and Neuroscience
King's College London
London, England, UK

Larry Davidson, PhD
Yale School of Medicine
Department of Psychiatry, Program for Recovery and
Community Health (PRCH)
New Haven, CT, USA

Philippe Delespaul, PhD
Department of Psychiatry and Neuropsychology
University of Maastricht
Maastricht, The Netherlands
Mondriaan Mental Health Trust
Maastricht/Heerlen, The Netherlands

Lisa Dixon, MD, MPH
New York State Psychiatric Institute
Columbia University Medical Center
New York, NY, USA

Graham M. L. Eglit, PhD
Department of Psychiatry and Sam and Rose
Stein Institute for Research on Aging
University of California
San Diego, CA, USA

Rana Elmaghraby, MD
University of Minnesota Medical School
Minneapolis, MN, USA

Sophie Erhardt, PhD
Department of Physiology and Pharmacology
Karolinska Institutet
Stockholm, Sweden

Anne-Kathrin J. Fett, PhD
Department of Clinical, Neuro and
Developmental Psychology, Faculty of Behavioral
and Movement Sciences, and Institute for Brain
and Behavior Amsterdam
Vrije Universiteit Amsterdam
Amsterdam, The Netherlands
Department of Psychology, City
University of London, Northampton Square
London, England, UK
Department of Psychosis Studies, Institute of Psychiatry,
Psychology and Neuroscience
King's College London
London, England, UK

Judith M. Ford, PhD
Mental Health Service
San Francisco VA Health Care System
San Francisco, CA, USA
Department of Psychiatry
University of California
San Francisco, CA, USA

Marta Di Forti, PhD
Social, Genetic and Developmental Psychiatry Centre
Institute of Psychiatry, Psychology and Neuroscience
King's College London
London, England, UK

Robert Freedman, MD
Department of Psychiatry F-546
University of Colorado Denver School of Medicine
Aurora, CO, USA

Toshi A. Furukawa, MD, PhD
Department of Health Promotion and Human Behavior
Kyoto University Graduate School of Medicine/School of
Public Health
Kyoto, Japan

Łukasz Gawęda, PhD
Department of Psychiatry and Psychotherapy
University Medical Center Hamburg-Eppendorf
Hamburg, Germany
College of Education, Psychology & Social Work
Flinders University
Adelaide, Australia

Felipe V. Gomes, PhD
Department of Pharmacology
Ribeirao Preto Medical School
University of Sao Paulo
Brazil

Marianne Goodman, MD
Department of Psychiatry
The Mount Sinai Hospital
New York, NY, USA

Anthony A. Grace, PhD
Department of Neuroscience
University of Pittsburgh
Pittsburgh, PA, USA

Rachael G. Grazioplene, PhD
Department of Psychology
University of Minnesota
Minneapolis, MN, USA

Sinan Guloksuz, MD, PhD
Department of Psychiatry and Neuropsychology, School
for Mental Health and Neuroscience, European Graduate
School of Neuroscience (EURON)
Maastricht University Medical Centre
Maastricht, The Netherlands
Department of Psychiatry
Yale School of Medicine
New Haven, CT, USA

Kimberly Guy
Yale Program for Recovery and Community Health,
Yale School of Medicine
Yale University
New Haven, CT, USA

Frédéric Haesebaert, MD, PhD
Lyon Neuroscience Research Center
Université Claude Bernard Lyon
Lyon, France

Holly K. Hamilton, PhD
Department of Veterans Affairs Sierra Pacific Mental Illness Research, Education, and Clinical Center (MIRECC)
San Francisco VA Health Care System
San Francisco, CA, USA

Marie C. Hansen, MA
Department of Psychiatry
Columbia University
New York City, NY, USA

Philip D. Harvey, PhD
Leonard M. Miller Professor of Psychiatry
University of Miami Miller School of Medicine
Miami, FL, USA
Research Service
Bruce W. Carter VA Medical Center
Miami, FL, USA

Takanori Hashimoto, MD, PhD
Department of Psychiatry and Behavioral Science
Kanazawa University
Kanazawa, Japan
Department of Psychiatry
University of Pittsburgh
Pittsburgh, PA, USA

Mark Hayward
Honorary Senior Research Fellow, School of Psychology
University of Sussex
Brighton, England, UK
Director of Research & Consultant Clinical Psychologist
Sussex Partnership NHS Foundation Trust
Worthing, West Sussex, UK

Dale C. Hesdorffer, PhD
Professor of Epidemiology
Gertrude H Sergievsky Center
Department of Epidemiology
Columbia University
New York, NY, USA

Kelsey Heslin, BA, MS
Department of Psychiatry
University of Iowa
Iowa City, IA, USA

S. Kristian Hill, PhD
Department of Psychology
College of Health Professions
Rosalind Franklin University
North Chicago, IL, USA

Carolin Hoffmann, PhD
Department of Psychiatry and Neuropsychology
School of Mental Health and Neuroscience
Maastricht University
Maastricht, The Netherlands

Oliver D. Howes, MRCPsych, PhD
Psychiatric Imaging Group, MRC London Institute of Medical Sciences
Imperial College London
Department of Psychosis Studies, Institute of Psychiatry, Psychology and Neuroscience
King's College London
London, England, UK

Thomas M. Hyde, MD, PhD
Lieber Institute for Brain Development
Baltimore, MD, USA
Department of Psychiatry and Behavioral Sciences
Johns Hopkins University School of Medicine
Baltimore, MD, USA
Department of Neurology
Johns Hopkins University School of Medicine
Baltimore, MD, USA

Matti Isohanni, MD, PhD
Psychiatrist, Psychotherapist, Professor of Psychiatry (emeritus)
University of Oulu
Oulu, Finland

Adela-Maria Isvoranu, MSc
Department of Psychological Methods
University of Amsterdam
Amsterdam, The Netherlands

Samantha Jankowski, BS
New York State Psychiatric Institute
New York, NY, USA

Behnam Javidfar, BS
Department of Psychiatry
Icahn School of Medicine at Mount Sinai
Manhattan, NY, USA

Daniel C. Javitt, MD, PhD
Department of Psychiatry
Columbia University
New York, NY, USA

Dilip V. Jeste, MD
Senior Associate Dean for Healthy Aging and Senior Care, Estelle and Edgar Levi Chair in Aging
Distinguished Professor of Psychiatry and Neurosciences
Director, Sam and Rose Stein Institute for Research on Aging
University of California
San Diego, CA, USA

Louise Johns, MA, DPhil, DClinPsy, PGCAP
Consultant Academic Clinical Psychologist
Oxford Health NHS Foundation Trust
Oxford, England, UK
Honorary Senior Research Fellow
Department of Psychiatry
University of Oxford
Oxford, England, UK

Kousuke Kanemoto, MD
Professor of Neuropsychiatry, Clinical, Personality and Abnormal Psychology
Aichi Medical University
Nagakute, Aichi, Japan

K. Nidhi Kapil-Pair, PhD
Mental Illness Research & Clinical Center
Icahn School of Medicine at Mount Sinai
Manhattan, NY, USA

Shitij Kapur, FRCPC, PhD, FMedSci
Department of Medicine, Dentistry and Health Sciences
University of Melbourne
Melbourne, Australia

Bibi S. Kassim
Icahn School of Medicine at Mount Sinai
Manhattan, NY, USA

Emma Sophia Kay, MSW
Doctoral Candidate, School of Social Work
University of Alabama
Tuscaloosa, AL, USA

Sarah K. Keedy, PhD
Department of Psychiatry and Behavioral Neuroscience
University of Chicago
Chicago, IL, USA

Richard S. E. Keefe, PhD
Department of Psychiatry and Behavioral Sciences
Duke University School of Medicine
Durham, NC, USA

Barrett Kern, PsyD
Department of Psychology
Roosevelt University
Chicago, IL, USA

Matcheri S. Keshavan, MD
Department of Psychiatry
Beth Israel Deaconess Medical Center
Boston, MA, USA

Alaptagin Khan, MD
Department of Psychiatry
Harvard Medical School
Boston, MA, USA
Developmental Biopsychiatry Research Program, McLean Hospital
Belmont, MA, USA

Tara Kingston, MD, MSc, MRCPsych
Molecular & Cellular Therapeutics
Royal College of Surgeons in Ireland
Dublin, Ireland
Cavan-Monaghan Mental Health Service
Cavan Hospital & St. Davnet's Hospital
Monaghan, Ireland

Anne S. Klee, PhD
VA Connecticut Healthcare System
Yale School of Medicine
Department of Psychiatry
New Haven, CT, USA

Joel E. Kleinman, MD PhD
Lieber Institute for Brain Development
Baltimore, MD, USA
Department of Psychiatry and Behavioral Sciences
Johns Hopkins University School of Medicine
Baltimore, MD, USA

Lydia Krabbendam, PhD
Department of Clinical, Neuro and Developmental Psychology
Faculty of Behavioral and Movement Sciences, and Institute for Brain and Behavior Amsterdam
Vrije Universiteit Amsterdam
Amsterdam, The Netherlands
Department of Psychosis Studies
King's College London
Institute of Psychiatry, Psychology and Neuroscience, London
London, England, UK

Nina V. Kraguljac, MD
Department of Psychiatry and Behavioral Neurobiology
University of Alabama at Birmingham
Birmingham, AL, USA

Marek Kubicki, MD, PhD
Psychiatry Neuroimaging Laboratory
Department of Psychiatry
Brigham and Women's Hospital, and Harvard Medical School
Boston, MA, USA
Department of Psychiatry
Massachusetts General Hospital, and Harvard Medical School
Boston, MA, USA
Department of Radiology
Brigham and Women's Hospital, and Harvard Medical School
Boston, MA, USA

Elizabeth Kuipers, PhD
Department of Psychology, Institute of Psychiatry
Institute of Psychiatry, Psychology and Neuroscience
King's College London
London, England, UK

Adrienne C. Lahti, MD
Department of Psychiatry and Behavioral Neurobiology
University of Alabama at Birmingham
Birmingham, AL, USA

Yulia Landa, PsyD, MS
Department of Psychiatry
Icahn School of Medicine at Mount Sinai
Manhattan, NY, USA

Charles H. Large, PhD
Autifony Therapeutics Limited
Stevenage Biosciences Catalyst
Stevenage, UK

Imke L. J. Lemmers-Jansen
Department of Educational and Family Studies
Faculty of Behavioral and Movement Sciences, and Institute for Brain and Behavior Amsterdam
Vrije Universiteit Amsterdam
Amsterdam, The Netherlands
Department of Clinical, Neuro and Developmental Psychology
Faculty of Behavioral and Movement Sciences, and Institute for Brain and Behavior Amsterdam
Vrije Universiteit Amsterdam
Amsterdam, The Netherlands

Rebekka Lencer, MD
Department of Psychiatry & Psychotherapy
Otto Creutzfeldt Center for Behavioral and Cognitive Neuroscience
University of Münster
Münster, Germany

Stefan Leucht, MD
Department of Psychiatry and Psychotherapy
Technische Universität München, Klinikum rechts der Isar
Munich, Germany

David A. Lewis, MD, PhD
Department of Psychiatry
University of Pittsburgh
Pittsburgh, PA, USA
Department of Neuroscience
University of Pittsburgh
Pittsburgh, PA, USA

Tania Lincoln, PhD
Department of Clinical Psychology and Psychotherapy
University of Hamburg
Hamburg, Germany

Katiah Llerena, PhD
Department of Veterans Affairs Sierra Pacific Mental Illness Research, Education, and Clinical Center (MIRECC)
San Francisco VA Health Care System
San Francisco, CA, USA

Eleanor Longden, PhD
Psychosis Research Unit, Greater Manchester Mental Health NHS Foundation Trust
Manchester, UK

Mario Losen, PhD
Department of Psychiatry and Neuropsychology
School of Mental Health and Neuroscience
Maastricht University
Maastricht, The Netherlands

Thies Lüdtke, MSc
Department of Psychiatry and Psychotherapy
University Medical Center Hamburg Eppendorf
Hamburg, Germany

Amanda E. Lyall, PhD
Psychiatry Neuroimaging Laboratory
Department of Psychiatry, Brigham and Women's Hospital, and Harvard Medical School
Boston, MA, USA
Department of Psychiatry, Massachusetts General Hospital, and Harvard Medical School
Boston, MA, USA

Maria Maier
Department of Psychology
Yale University
New Haven, CT, USA

Ashok Malla, MBBS, FRCPC, MRCPsych, DPM
Professor and Canada Research Chair in
Early Psychosis and Early Intervention in Youth Mental Health
Department of Psychiatry, McGill University,
Nominated Principal Investigator, ACCESS
Open Minds; Douglas Mental Health University Institute
Montréal, QC, Canada

Marina Mané-Damas, MS
Department of Psychiatry and Neuropsychology
School of Mental Health and Neuroscience
Maastricht University
Maastricht, The Netherlands

Ivana S. Marková, MBChB, MPhil, MD
Professor of Psychiatry
Hull York Medical School
University of Hull
Hull, UK
Hull York Medical School Allam Medical Building
University of Hull
Hull, UK

Pilar Martinez-Martínez, PhD
Department of Psychiatry and Neuropsychology
School of Mental Health and Neuroscience
Maastricht University
Maastricht, The Netherlands

Daniel H. Mathalon, MD, PhD
Mental Health Service, San Francisco VA Health Care System
San Francisco, CA, USA
Department of Psychiatry, University of California
San Francisco, CA, USA

Jennifer McDowell, PhD
Professor and Chair
Department of Psychology
University of Georgia
Athens, GA, USA

Thomas H. McGlashan, MD
Professor Emeritus
Department of Psychiatry
Yale University
New Haven, CT, USA

Patrick McGorry, AO, MD, PhD, FRC, FRANZ, FAA, FASSA
Executive Director, Orygen
Professor of Youth Mental Health, CYMH
University of Melbourne; NHMRC Senior Principal Research Fellow; President Society for Mental Health Research; President Schizophrenia International Research Society
Orygen, Parkville, VIC, Australia

Anna P. McLaughlin, MS
Department of Psychological Medicine
Institute of Psychiatry, Psychology and Neuroscience
King's College London
National Institute for Health Research (NIHR) Mental Health Biomedical Research Centre at South London Maudsley NHS Foundation Trust
King's College London
London, England, UK

Andreas Meyer-Lindenberg, MD, PhD
Department of Psychiatry and Psychotherapy, Central Institute of Mental Health
Mannheim, Germany

Jouko Miettunen, MSc, MPhil, PhD
Professor of Clinical Epidemiology
University of Oulu
Oulu, Finland

Jonathan Mill, PhD
University of Exeter Medical School
Complex Disease Epigenetics Group
Exeter, UK

Brian Miller, MD, MPH
Associate Professor
Department of Psychiatry and Health Behavior
Augusta University
Augusta, GA, USA

M. S. Moehle, PhD
Vanderbilt Center for Neuroscience Drug Discovery
Department of Pharmacology
Nashville, TN, USA

Peter C. Molenaar, PhD
Department of Psychiatry and Neuropsychology
School of Mental Health and Neuroscience
Maastricht University
Maastricht, The Netherlands

Valeria Mondelli, MD, PhD
Department of Psychological Medicine, Institute of Psychiatry, Psychology and Neuroscience
King's College London
National Institute for Health Research (NIHR) Mental Health Biomedical Research Centre at South London Maudsley NHS Foundation Trust
King's College London
Maurice Wohl Clinical Neuroscience Institute
London, England, UK

Marine Mondino, PhD
Neuroscience Research Center
University of Lyon
Lyon, France

Craig Morgan, PhD
Social Epidemiology Research Group,
Institute of Psychiatry, Psychology,
and Neuroscience
King's College London
London, England, UK

Steffen Moritz, PhD
Department of Psychiatry and Psychotherapy
University Medical Center Hamburg Eppendorf
Hamburg, Germany

Eric Morris, PhD
Senior Lecturer & Psychology Clinic Director, School of Psychology & Public Health
La Trobe University
Melbourne, Australia

Sarah E. Morris, PhD
Department of Psychiatry
Massachusetts General Hospital
Boston, MA, USA
Chief, Schizophrenia Spectrum Disorders
Research Program, Division of Adult
Translational Research, National Institute of Mental Health
Bethesda, MD, USA

Robin M. Murray, MD, FRS
Professor
Institute of Psychiatry, Psychology, and Neuroscience
King's College London
London, England, UK

Hong Ngo, PhD
Assistant Professor of Clinical Psychology
New York State Psychiatric Institute, Columbia University Medical Center
New York, NY, USA

Nnamdi Nkire, MD, MRCPsych, FRCPC
Molecular & Cellular Therapeutics, Royal College of Surgeons in Ireland
Dublin, Ireland
Cavan-Monaghan Mental Health Service, Cavan Hospital & St. Davnet's Hospital
Monaghan, Ireland

Merete Nordentoft, MD, MPH, PhD, DMSs
Copenhagen Research Center for Mental Health CORE, Mental Health Center Copenhagen
Copenhagen University Hospital, University of Copenhagen
Copenhagen, Denmark

Carol S. North, MD, MPE
Medical Director, The Altshuler Center for Education & Research at Metrocare Services and The Nancy and Ray L. Hunt Chair in Crisis Psychiatry and Professor of Psychiatry
The University of Texas Southwestern Medical Center
Dallas, TX, USA

Kyoko Ohashi, PhD
Department of Psychiatry, Harvard Medical School
Boston, MA, USA
Developmental Biopsychiatry Research Program and Brain Imaging Center, McLean Hospital
Belmont, MA, USA

Juliana Onwumere, PhD
Consultant Clinical Psychologist, National Psychosis Unit and Lecturer
Department of Psychology, Institute of Psychiatry
Institute of Psychiatry, Psychology and Neuroscience
King's College London
London, England, UK

Jennifer Pacheco, PhD
RDoC Unit, National Institute of Mental Health
Bethesda, MD, USA

Barton W. Palmer, PhD
Department of Psychiatry and Sam and Rose Stein Institute for Research on Aging
University of California
San Diego, CA, USA
Veterans Affairs San Diego Healthcare System
San Diego, CA, USA

Carmine M. Pariante, MD, PhD
Department of Psychological Medicine, Institute of Psychiatry, Psychology and Neuroscience
King's College London
National Institute for Health Research (NIHR) Mental Health Biomedical Research Centre at South London
Maudsley NHS Foundation Trust
King's College London
London, England, UK

Krystal Parker, PhD
Iowa Neuroscience Institute
Department of Psychiatry
University of Iowa
Iowa City, IA, USA

Godfrey Pearlson, MA, MBBS, MD
Professor of Psychiatry and Neuroscience
Yale University
New Haven, CT, USA

Matti Penttilä, MSc, MD, PhD
Institute of Medicine
Department of Psychiatry
University of Oulu
Oulu, Finland

Laura Pientka, MD
Department of Psychiatry
University of Minnesota
Minneapolis, MN, USA

Amy E. Pinkham, PhD
School of Behavioral and Brain Sciences
The University of Texas at Dallas
Richardson, TX, USA
Department of Psychiatry
The University of Texas Southwestern Medical School
Dallas, TX, USA

Ehsan Pishva, MD, PhD
Complex Disease Epigenetics Group
University of Exeter Medical School
Exeter, UK
Department of Psychiatry and Neuropsychology, School for Mental Health and Neuroscience (MHeNS)
Maastricht, The Netherlands

David E. Pollio, PhD
Distinguished Professor
Department of Social Work
University of Alabama
Birmingham, AL, USA

Albert R. Powers III, MD, PhD
Assistant Professor of Psychiatry
Yale University
New Haven, CT, USA

Lotta-Katrin Pries, PhD
Department of Psychiatry and Neuropsychology, School for Mental Health and Neuroscience, European Graduate School of Neuroscience (EURON)
Maastricht University Medical Centre
Maastricht, The Netherlands

J. D. Ragland, PhD
Professor, School of Medicine, Imaging Research Center
University of California at Davis
Sacramento, CA, USA

James L. Reilly, PhD
Department of Psychiatry & Behavioral Sciences
Northwestern University Feinberg School of Medicine
Chicago, IL, USA

Julie Repper, PhD
ImROC Programme, Nottinghamshire Healthcare NHS Foundation Trust
Nottingham, UK

Brian J. Roach, MS
Mental Health Service, San Francisco VA Health Care System
San Francisco, CA, USA

David L. Roberts, PhD
Department of Psychiatry
University of Texas Health San Antonio
San Antonio, TX, USA

Leah H. Rubin, PhD, MPH
Department of Neurology
Johns Hopkins University School of Medicine
Baltimore, MD, USA

Mar Rus-Calafell, PhD
Research Fellow and Senior Clinical Psychologist
Department of Psychiatry
University of Oxford, Warneford Hospital
Oxford, USA

Vincent Russell, MB, MSc, MRCPsych, FRCPC
Cavan-Monaghan Mental Health Service, Cavan Hospital & St. Davnet's Hospital
Monaghan, Ireland
Department of Psychiatry, Royal College of Surgeons in Ireland, Beaumont Hospital
Dublin, Ireland

Bart P. F. Rutten, MD, PhD
Professor and Head, Department of Psychiatry and Neuropsychology, School for Mental Health and Neuroscience (MHeNs), European Graduate School of Neuroscience (EURON), Faculty of Health, Medicine and Life Sciences
Maastricht University Medical Centre
Maastricht, The Netherlands

Elyn R. Saks, JD, PhD
Orrin B. Evans Distinguished Professor of Law, and Professor of Law, Psychology, and Psychiatry and the Behavioral Sciences, USC Gould School of Law
University of Southern California
Los Angeles, CA, USA

Marsal Sanches, MD, PhD
Department of Psychiatry and Behavioral Sciences
University of Texas Houston Health Sciences Center
Houston, TX, USA

Charles A. Sanislow, PhD
Department of Psychology and Program in Neuroscience and Behavior
Wesleyan University
Middletown, CT, USA

Jakob Scheunemann, MSc
Department of Psychiatry and Psychotherapy
University Medical Center Hamburg Eppendorf
Hamburg, Germany

C. M. Schumann, PhD
Department of Psychiatry and Behavioral Science
University of California Davis
Sacramento, CA, USA

Robert Schwarcz, PhD
Professor of Psychiatry
Maryland Psychiatric Research Center
Department of Psychiatry
University of Maryland School of Medicine, Maple and Locust Streets
Baltimore, MD, USA

Daniel Scott, PhD
Instructor
Department of Psychiatry
University of Texas Southwestern Medical Center
Dallas, TX, USA

Johanna Seitz, MD
Clinical Fellow in Psychiatry
Psychiatry Neuroimaging Laboratory
Department of Psychiatry, Brigham and Women's Hospital, and Harvard Medical School
Boston, MA, USA

Jean-Paul Selten, MD, PhD
Professor
Department of Psychiatry and Psychology
University of Maastricht
Leiden, The Netherlands
Rivierduinen, Institute for Mental Health Care
Leiden, The Netherlands

Joseph J. Shaffer, PhD
Post-Doctoral Research Scholar
Department of Radiology
University of Iowa
Iowa City, IA, USA

Juliet M. Silberstein
University of Miami Miller School of Medicine
Miami, FL, USA

Mike Slade, PhD
Professor of Mental Health Recovery and Social Inclusion
School of Health Sciences, Institute of Mental Health
University of Nottingham, UK

Jair C. Soares, MD, PhD
Professor and Chair
Department of Psychiatry and Behavioral Sciences
The University of Texas Health Science Center at Houston, UT Houston Medical School
Houston, TX, USA

Andreas Sprenger, PhD
Senior Researcher
Department of Neurology
University of Lübeck
Lübeck, Germany

Michael Stevens, PhD
Adjunct Professor of Psychiatry
Yale University
New Haven, CT, USA

Clara Strauss, PhD
Honorary Senior Lecturer, School of Psychology
University of Sussex
Brighton, UK
Clinical Research Fellow & Consultant Clinical Psychologist
Sussex Partnership NHS Foundation Trust
Brighton, UK

John A. Sweeney, PhD
Professor of Psychiatry
University of Cincinnati
Cincinnati, OH, USA

Rajiv Tandon, MD
Professor of Psychiatry
Department of Psychiatry
University of Florida
Gainsville, FL, USA

Martin H. Teicher, MD, PhD
Associate Professor of Psychiatry
Department of Psychiatry
Harvard Medical School
Boston, MA, USA
Developmental Biopsychiatry Research Program, McLean Hospital
Belmont, MA, USA

John Torous, MD
Instructor in Psychiatry
Beth Israel Deaconess Medical Center and Harvard Medical School
Boston, MA, USA

Michael R. Trimble, MD
Professor of Behavioral Neurology
Department of Behavioral Neurology
Institute of Neurology
London, UK

Samson Tse, PhD
Professor of Mental Health
Department of Social Work and Social Administration, Faculty of Social Sciences
University of Hong Kong
Hong Kong, China

Daniel H. Vaccaro, BA
Clinical Research Coordinator
Icahn School of Medicine
Mount Sinai Hospital
Medical Student
Columbia University Vagelos College of Physicians and Surgeons
New York, NY

Jim van Os, MD, PhD
Department of Psychiatry, Brain Centre Rudolf Magnus
University Medical Centre Utrecht
Utrecht, The Netherlands

Catherine van Zelst
Department of Psychiatry and Neuropsychology, Faculty of Health Medicine and Lifesciences
Maastricht University
Maastricht, Netherlands

Sophia Vinogradov, MD
Professor and Head
Department of Psychiatry
University of Minnesota
Minneapolis, MN, USA

John L. Waddington, PhD, DSc, MRIA
Molecular & Cellular Therapeutics, Royal College of Surgeons in Ireland
Dublin, Ireland
Jiangsu Key Laboratory of Translational Research & Therapy for Neuro-Psychiatric-Disorders and Department of Pharmacology, College of Pharmaceutical Sciences
Soochow University
Suzhou, China

Toby T. Winton-Brown, PhD
Research Fellow
Department of Neuroscience
Monash University
Melbourne, VIC, Australia

Scott W. Woods, MD
Professor of Psychiatry
Yale University
New Haven, CT, USA

Til Wykes, DBE
Department of Psychology
Institute of Psychiatry, Psychology and Neuroscience
King's College London
London, England, UK

S. E. Yohn, PhD
Post-Doctoral Fellow
Vanderbilt Center for Neuroscience Drug Discovery and Department of Pharmacology
Nashville, TN, USA

Xiang-Yang Zhang, MD, PhD
Associate Professor of Psychiatry and Behavioral Sciences
University of Texas Health Sciences Center
Houston, TX, USA

Sacha Zilkha, PhD
New York State Psychiatric Institute
Columbia University
New York, NY, USA

Eric C. Zimmerman, MD, PhD
Departments of Neuroscience, Psychiatry and Psychology (PGY-1)
University of Pittsburgh
Pittsburgh, PA, USA

Shenghua Zong, MD
Department of Psychiatry and Neuropsychology
School of Mental Health and Neuroscience
Maastricht University
Maastricht, The Netherlands

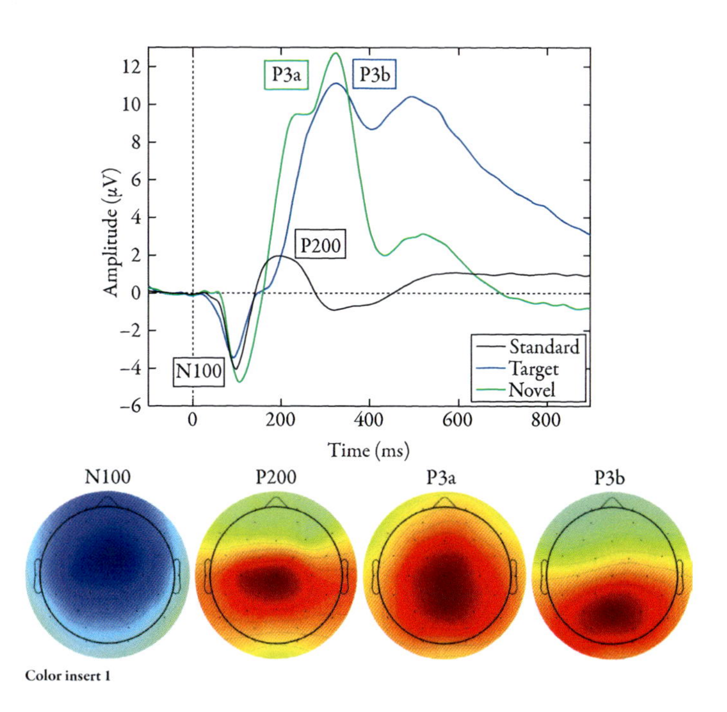

Color insert 1

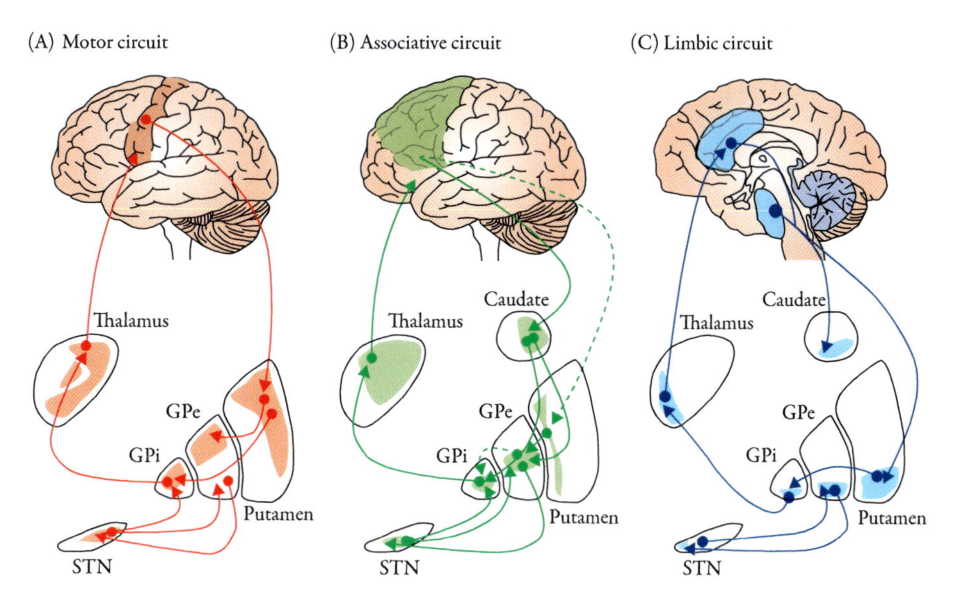

Color insert 2

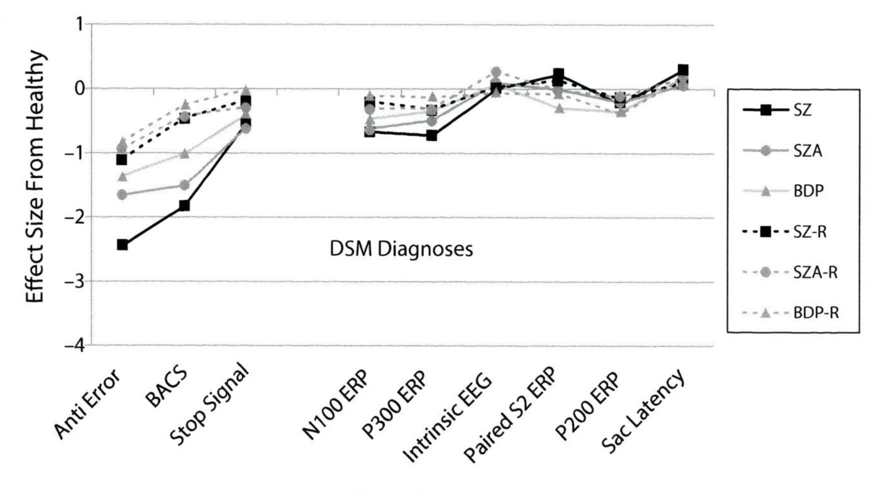

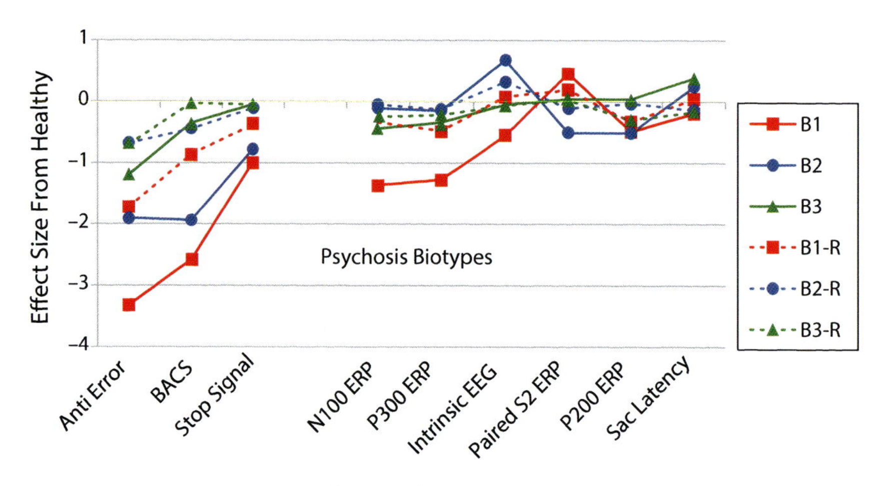

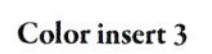
Color insert 3

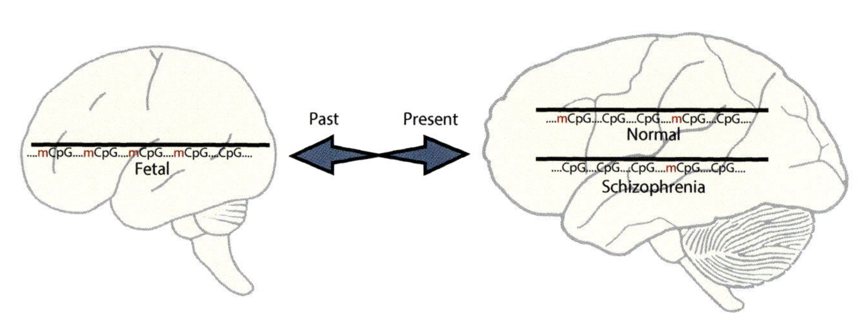

Color insert 4

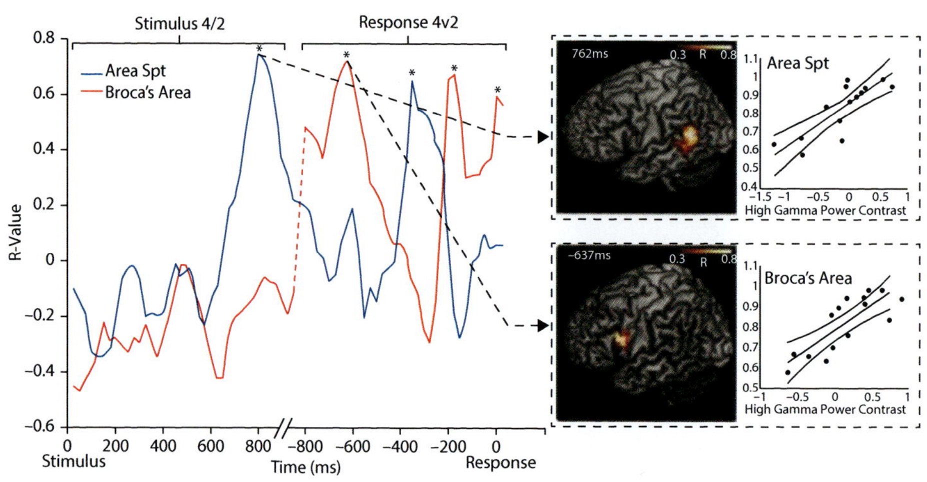

Color insert 5

SECTION 1

PHENOMENOLOGICAL CHARACTERISTICS AND DIMENSIONAL CONCEPTUALIZATION OF PSYCHOSIS

1

CONCEPTUALIZATION OF PSYCHOSIS IN PSYCHIATRIC NOSOLOGY

PAST, PRESENT, AND THE FUTURE

Matcheri S. Keshavan, John Torous, and Rajiv Tandon

INTRODUCTION

Growing evidence suggests that complex spectra of psychotic disorders are related to a large number of phenotypes associated with heterogeneous gene networks and environmental risk factors affecting diverse pathophysiological processes. However, the diagnostic concept of psychotic disorders remains largely as it was originally constructed over a century ago. Most of the criteria defining psychotic disorders continue to be based on clinicians' interpretation of the subjective reports of symptoms by patients. There continues to be uncertainty on what is precisely the core disturbance underlying schizophrenia and related psychotic disorders; the nature of the boundaries of this spectrum and those of the syndromes within the psychotic spectrum also remain a topic of debate.

The formulation of our current nosology of psychiatric disorders began in the late 19th century and culminated in publication of a section related to mental disorders (section V) in the sixth revision of the International Classification of Disease (ICD-6)[1] in 1949 and in the first edition of the American Diagnostic and Statistical Manual of Mental Disorders (DSM-I)[2] in 1952. In subsequent revisions of these texts (ICD 7–10 and DSM II–5), substantial changes in diagnostic criteria have been made, but the basic structure has been broadly preserved. Differences between the two systems have narrowed, and there is marked concordance between DSM-5[3] and ICD-10.[4] DSM-5 became operational in 2013; ICD-10 is in its final stages of revision and ICD-11 was released in June 2018. Schizophrenia has been one of the major diagnostic categories in all versions of both manuals, and the nosological approach to psychosis has received much attention. In this chapter, we outline the current conceptualization of psychotic disorders, explore the historical evolution of these concepts, and discuss future directions for psychiatric nosology.

THE PAST: EVOLUTION OF THE CONCEPT OF SCHIZOPHRENIA FROM THE 19TH CENTURY

Although schizophrenia has been studied as a specific disease entity for the past century,[5] its precise nature (i.e., core definition, boundaries, causes, and pathogenesis) remains undefined.[6] Since the differentiation between *dementia praecox* (a term originally coined by Morel), *manic-depressive insanity* (by Kraepelin [1856–1926]),[7] and *schizophrenia* (by Eugen Bleuler [1857–1939]),[8] its definitions have varied and its boundaries have expanded and receded over the past century. Thus, it is instructive to examine the varying definitions of schizophrenia over the past 150 years and trace its evolution to the present DSM-5[9] and ICD-11.[10]

The current construct of schizophrenia derives from Emil Kraepelin's formulation of *dementia praecox* in the late 19th century and his elaboration of this concept in the early part of the 20th century.[7] Until that time, there were two broad prevailing constructs of major psychiatric illness or *psychosis*. Guislain, Heinroth,[11] and Griesinger[12] postulated that there was only one basic form of psychosis with diverse manifestations attributable to endogenous and environmental factors ("*einheitpsychose*"), while others suggested that there were several distinct psychotic disorders (such as *catatonia*, *hebephrenia*, *folie circulaire*, *dementia paranoides*, *melancholia*, etc.).[13–15]

Kraepelin discerned two distinct patterns of course of illness among the various psychotic disorders and used these patterns to define and classify mental conditions. His approach was influenced by the contemporaneous delineation of *general paresis of insanity* (neurosyphilis or tertiary syphilis). The distinctive course and outcome among patients with this illness in psychiatric hospitals led to its definition, etiological identification, and development of the diagnostic Wasserman test, and later to specific treatment with penicillin. Kraepelin therefore concluded that course and outcome of illness were the best disease characteristics to use in defining and demarcating psychiatric disease entities. Based on his study of several hundred cases of hospitalized patients,[16] he delineated two distinct disorders: (1) *dementia praecox*, which included catatonia, hebephrenia, and paranoid states; and (2) *manic-depressive insanity*, which was comprised of folie circulaire and melancholia. Kraepelin[7] defined schizophrenia on the basis of its onset (in adolescence or early adulthood), course (deteriorating), and outcome (demence or "mental dullness"). He distinguished dementia praecox from manic-depressive insanity based on the chronicity and poor outcome of the former contrasted with the episodic nature and better outcome of the latter.

During the same time period, Eugen Bleuler[8] renamed the condition "schizophrenia" because he considered the splitting of different psychic functions to be its defining characteristic. He postulated that schizophrenia was defined by a set of basic or fundamental symptoms that were unique to schizophrenia and always present in those with this group of diseases, whereas its course and outcome were variable.[8] He called delusions and hallucinations *accessory symptoms* and considered them to be variable and nonspecific. He described fundamental symptoms (which would now be identified as negative symptoms) that included *autism*, *ambivalence*, *flat affect*, and *loosening of associations* ("the four As"). He further believed that many mild cases existed and considerably broadened the scope of the disease construct of schizophrenia. He included latent and simple forms as part of this entity that he called "the group of schizophrenias."

Influenced by the thinking of Karl Jaspers, Kurt Schneider (1887–1967)[17] postulated that impairment of empathic communication was the fundamental defect in schizophrenia. He considered "un-understandability" of the personal experience of the person with schizophrenia as pathognomonic. Based on this premise, he defined 11 *first-rank symptoms* which he considered to be diagnostic of schizophrenia.[17] Examples of these first-rank symptoms are likely familiar to readers today as positive symptoms of schizophrenia, including auditory hallucinations, thought withdrawal, thought broadcasting, thought insertion, and somatic hallucinations, among several others. Given central importance in the Present State Examination,[18] these symptoms were incorporated into both ICD-9 and DSM-III definitions of schizophrenia.

Definitions of schizophrenia over the past half-century from DSM-I through DSM-IV and ICD-6 through ICD-10 have all incorporated Kraepelinian chronicity, Bleulerian negative symptoms, and Schneiderian positive (first-rank) symptoms as part of their definitions, but the relative emphasis paid to these three roots has varied across versions. In the 1960s and 1970s, there was significant discordance between the DSM (I–II) and ICD (7–8) systems with regard to this issue. Whereas DSM-I[1] and DSM-II[19] highlighted the Bleulerian perspective (i.e., the emphasis on negative symptoms and very broad definition of the schizophrenias, including latent, pseudoneurotic, pseudopsychopathic, and residual subtypes), ICD-7 and ICD-8 stressed Kraepelinian chronicity. This overly broad definition of schizophrenia in these earlier DSM systems led to poor reliability in diagnosing schizophrenia and a marked discrepancy[20] between the diagnosis of schizophrenia in the United States and the rest of the world, which used the ICD system. In reaction to these discrepancies, the operationalized criteria of DSM-III[21] provided the narrowest definition of schizophrenia with an emphasis on Kraepelinian chronicity and Schneiderian positive (first-rank) symptoms. In fact, the imperative of improved reliability and the belief that positive symptoms could be more reliably diagnosed resulted in positive symptoms becoming the defining characteristic of schizophrenia in DSM-III and ICD-9. Furthermore, certain positive symptoms such as Schneiderian first-rank symptoms[22,23] were accorded special importance, and the presence of any single bizarre delusion or special auditory hallucinations (voices arguing or commenting) was sufficient for a diagnosis of schizophrenia in DSM-III–IV and ICD 9–10. Although there were modest changes in the definition of schizophrenia from DSM-III to DSM-IV (and from ICD-9 to ICD-10), the importance of positive symptoms with the special significance placed on bizarre delusions and specific hallucinations in the diagnostic criteria of schizophrenia remained.

THE PRESENT: DSM-5 AND ICD-11

Whereas DSM-III (introduced in 1980) substantially improved the reliability of the diagnosis of various psychotic disorders, limitations in their validity have become apparent over the past three decades.[24] Changes from DSM-III to DSM-III-R to DSM-IV were relatively modest, with the poor mapping of clinically defined psychotic syndromes to neurobiological findings (genes, structural and functional brain abnormalities, etc.), and the absence of major therapeutic advances in the treatment of psychotic disorders over this period necessitating a major reconsideration of the definition and description of psychotic disorders. Developers of DSM-5 and ICD-11 were confronted with the task of redressing the following shortcomings, and have made important modifications, reviewed herein.

THE PROBLEM OF CLINICAL HETEROGENEITY

Although the significant heterogeneity of schizophrenia has always been recognized,[25,26] with multiple etiological factors and pathophysiological processes, it has been treated as a singular entity.[6] Now it is almost certain, however, that our construct of schizophrenia encompasses not one but several diseases.[27] Until now, we tried to explain the heterogeneity of the illness by defining distinct subtypes: paranoid, disorganized, catatonic, simple, and undifferentiated. These classic subtypes have, however, been found to provide a very poor description of the enormous heterogeneity of this condition. Subtype stability is low, reliability of diagnosing them is weak, their validity is questionable, and only the paranoid and undifferentiated subtypes have been used with some frequency.[28] Because these subtypes have virtually no clinical or research utility,[29,30] they are being used infrequently,[30] and have been removed from DSM-5, with ICD-11 eliminating them as well.[10,31] Instead, the heterogeneity of schizophrenia is being formally described in terms of psychopathological dimensions (discussed later) that will be measured in all patients throughout the course of the illness.[32] Since the severity of different symptom domains varies in each patient over the illness course and in response to treatment, such dimensional assessment will be an invaluable tool to clinicians as they provide individualized, measurement-based care to their patients with schizophrenia.[33]

THE DIFFICULTIES OF CLINICAL STAGING

An important drawback in definitions of schizophrenia in DSM-IV and ICD-10 was the inability to characterize the distinct stages of the illness.[34] The classification system did not allow clinicians to adequately describe the course of their patient's illness (prodromal, or the clinical high risk state; first episode; chronic, etc.) or denote the patient's current clinical status (active symptoms; in partial or full remission). While the concept of staging has not formally been embraced by DSM-5 or ICD-11, provision of modified course specifiers in both classification systems have been added to help redress this anomaly.[10,31] Additionally, there was no provision in DSM-IV or ICD-10 to describe the early or late prodromal phases of the illness when there might be a possibility of arresting active pathology and preventing disease progression.[35,36] The addition of "attenuated psychosis syndrome" (discussed later) as a condition for further study in DSM-5[35,36] was intended to address this deficiency. It is believed that the still unsatisfactory outcome of schizophrenia in a significant proportion of individuals with the disorder is due to the identification of the illness and initiation of treatment late in the course of the illness after a substantial amount of damage has already occurred. The introduction of attenuated psychosis syndrome in DSM-5 supports the efforts of clinicians to recognize and monitor psychotic symptoms early in the course of their evolution, and, if necessary, intervene in these crucial early stages. These changes in DSM-5 should increase emphasis on early intervention in psychiatry, as in the rest of medicine. Although recognition of attenuated psychosis syndrome is important, the ability to reliably diagnose the condition in routine clinical practice has not been demonstrated[37] and its nosologic relationship to other psychiatric diagnostic entities has not been precisely defined. Consequently, attenuated psychosis syndrome is not in the main body (section 2) of the DSM-5 but in section 3 as a condition needing further study.[35] It should be emphasized that a diagnosis of attenuated psychosis syndrome is not an indication for routine antipsychotic treatment but should instead prompt a careful search for comorbidities (such as anxiety, depression, substance use disorder, etc.) and their appropriate treatment, along with close monitoring for a possible transition to psychosis.

THE PROBLEM OF BLURRED BOUNDARIES: SCHIZOAFFECTIVE DISORDER

An ongoing challenge has been the unsatisfactory classification of patients with concurrent psychotic and mood symptoms with imprecise boundaries between schizophrenia, schizoaffective disorder, and psychotic mood disorder. Additionally, the definition of schizoaffective disorder as a cross-sectional diagnosis led to its poor reliability, low stability, and limited clinical utility. This imprecise definition has also led to a more frequent diagnosis of this disorder than originally intended. Schizoaffective disorder in DSM-5 requires at least 2 weeks of nonaffective psychosis during the lifetime duration of the illness, and full mood episodes need to be present for the majority of the total active and residual course of illness, from the onset of psychotic symptoms up until the current diagnosis. The more circumscribed delineation of schizoaffective disorder in DSM-5 and its definition as a longitudinal disorder in both DSM-5 and ICD-11 should enhance its stability and clinical utility.[38]

THE PROBLEM OF CATEGORIES VERSUS DIMENSIONS OF PSYCHOTIC DISORDER

Schizophrenia is characterized by several psychopathological domains that have distinctive courses, patterns of treatment-response, and prognostic implications. The relative severity of these symptom dimensions varies across patients and within patients at different stages of their illness. Measuring the relative severity of these symptom dimensions through the course of illness in the context of treatment can provide useful information to the clinician about the nature of the illness in a particular patient and the specific impact of treatment on different aspects of the patient's illness (analogous to measuring pulse, temperature, blood pressure, respiratory rate, etc.). Thus DSM-5 offers a new scale, the Clinical-Rated Dimensions of Psychosis Symptoms Scale (C-RDPSS). As a simple rating scale (akin to a thermometer or a sphygmomanometer), it should encourage clinicians to explicitly assess and track changes in the severity of these dimensions and to use this information to guide treatment. The C-RDPSS is composed of six symptom domains with a total of 8 items with each item being scored on an anchored 0–4 scale.[32] Whereas positive symptoms and mood symptoms each include two items (delusions and hallucinations for positive symptoms, and depression and mania for mood symptoms), the other four dimensions each have one item. In addition to its clinical utility,[33,39] this dimensional measurement should prove useful from a research perspective and thereby permit studies on etiology and pathogenesis that cut across current diagnostic categories. Such approaches would be consistent with recent findings in genetics and neuroscience and the recent Research Domain Criteria (RDoC) project initiated by the National Institutes of Mental Health, which are discussed in greater detail in a subsequent section.[40]

DIFFICULTY IN USING CURRENT CLASSIFICATION TO GUIDE TREATMENTS

For the past century, schizophrenia was conceptualized as a singular disorder with a distinctive pathology, albeit with multiple etiological factors and diverse manifestations. DSM-5 attempts to move toward a more multifaceted and dimensional perspective, but the prior historical momentum of the single disorder paradigm has dictated a single specific treatment directed at the common pathology; thus far, the main treatment option has consisted of antipsychotic medications. Antipsychotics have, however, been found to be ineffective against some core features of the illness, such as negative symptoms and cognitive deficits.[41] Although researchers have assiduously searched for an alternative core pathology in schizophrenia, no single mechanism has been found and no single treatment found to be effective against all aspects of the illness. With DSM-5, schizophrenia is explicitly conceptualized as a multidimensional illness[42] with each dimension potentially explained by a distinct pathophysiology and the target of a unique treatment.

In DSM-5, the six distinct dimensions of schizophrenia that are defined and scored on an anchored 0–4 scale offer a further opportunity to guide treatment. Different schizophrenia patients exhibit varying admixtures of symptom severity across the six dimensions. Additionally, these dimensions respond differently to various treatments and different patients show dissimilar patterns of treatment response. Unique measurement of how each of these distinct dimensions responds to specific treatments in individual patients will enable monitoring of treatment response and appropriate treatment adjustments. The availability of a simple scale that provides measures of symptom severity across distinct dimensions of schizophrenia may aid in measurement-based, individualized treatment.

Currently, antipsychotic medications are particularly effective at treating positive symptoms and disorganization in schizophrenia. Specific treatments are in development to address negative symptoms and cognitive deficits in schizophrenia. Hopefully, some of these will be found to be effective and safe and will become available for clinical practice in the future. Currently, antidepressants and mood stabilizers are moderately effective against the mood symptoms in schizophrenia. Antipsychotic medications, benzodiazepines, and electroconvulsive therapy have demonstrated efficacy for the motor symptoms (including catatonia) in schizophrenia. As better treatments for various symptom dimensions are developed, more effective, "evidence-based" pharmacotherapy may become possible in schizophrenia. The dimensional approach of DSM-5 may be also useful for standardized reporting and monitoring of personalized prescribing patterns that go beyond antipsychotics.

THE CATATONIA CONUNDRUM

A significant limitation in the definition of psychotic disorders has been the variable definitions of catatonia and its discrepant treatment across the manual (e.g., subtype of schizophrenia but a specifier of mood disorders).[43,44] It is widely known that catatonia occurs across a variety of disorders, including schizophrenia, bipolar disorder and depression, and can also be secondary to general medical conditions. In DSM-5, a single definition of catatonia is used and catatonia is treated as a specifier across all conditions.[45] Catatonia not otherwise specified is added as a residual category to allow for the rapid diagnosis and specific treatment of catatonic symptoms in severely ill patients in whom the underlying diagnosis is not clear.

KEEPING PACE WITH CHANGE: DSM AS A LIVING DOCUMENT

Continuing flexibility for the classificatory systems is critical for keeping pace with the rapid conceptual changes and accumulation of knowledge in psychiatric disorders. Unlike previous editions of the DSM, DSM-5 has been designed as a "living document' with the provision for periodic updates (versions 5.1, 5.2, etc.)[3] and it is likely that the scale for measurement of the six psychotic dimensions will be elevated to the main body of the manual. Updates will also be made based on clinician feedback and new research findings. While DSM-5 has advanced the nosology of schizophrenia, the next section explores continuing challenges and future directions for the conceptualization and definition of schizophrenia.

CONTINUING CHALLENGES AND FUTURE DIRECTIONS

CHALLENGES IN CONCEPTUALIZATION OF PSYCHOTIC DISORDERS: TOWARD A SPECTRUM CONCEPT

Schizophrenia is clearly not one, but several disorders. As outlined in this section of this chapter, delineation of unique psychopathology dimensions in schizophrenia may facilitate the identification of distinct etiopathophysiological pathways to schizophrenia,[27,46] development of specific biological tests for the "different schizophrenias,"[47] elaboration of specific treatments for the different schizophrenias,[39] construction of an etiopathophysiological nosology of the schizophrenias, and most importantly, better outcome for patients with the disorder.

Family and epidemiological studies have shown that several different disorders tend to cluster among the biological relatives of patients with schizophrenia. Schizotypal personality disorder, characterized by psychopathological traits such as anhedonia, perceptual aberrations, and magical thinking, is seen frequently among relatives of schizophrenia patients, as shown by the Roscommon epidemiological study.[48] Related to schizotypal personality disorder is the concept of schizotaxia, a personality style characterized by introversion and cognitive "slippage" linked by Meehl to the polygenic etiology of schizophrenia.[49] It has been suggested that schizotypal personality disorder, paranoid personality disorder and other nonaffective psychotic disorders, such as schizophreniform disorder and atypical psychoses, form a continuum of liability, with schizophrenia at the more severe end of this spectrum. Classification approaches in psychiatry could usefully move toward a metastructure including partially overlapping spectra such as schizophrenia and affective and autism spectrum disorders. Such a framework may better map on to the overlapping psychopathology, gene, and brain circuitry alterations in these disorders.

RELIABILITY OF DIAGNOSIS CONTINUES TO BE LIMITED

Any successful classificatory system needs to have utility in having a common language of communication across clinicians, researchers, funding agencies, and the general public. Such a classification also needs to be reliable, so that any two clinicians assessing the same individuals need to come to the same conclusions. Diagnostic reliability is critically important for ensuring clear communication between clinicians, consistent treatment choices, and a steady path toward establishing validity. Diagnostic reliability was poor before DSM-III, largely because of the lack of specific diagnostic criteria. The development of operationalized criteria in DSM-III improved diagnostic reliability, leading many to believe that the problem of reliability was solved. However, the reliability of psychiatric diagnoses has not improved considerably between DSM-IV and DSM-5. DSM-5 field trials showed lower kappa reliability estimates than past field trials and the general research literature, leading to debate and criticism of the DSM approach.[50,51] The continuing challenges in reliability in these field trials have been attributed to methodological issues such as variable clinician training, small sample sizes, and high clinician attrition rates.[52]

THE VEXING PROBLEM OF VALIDITY

The successive iterations of classification of psychotic disorders have progressively demonstrated some clinical utility and progress that has been made toward improving reliability. However, a major question remains in regard to the validity of even the current classification. The continued reliance on subjectively rated symptoms and the lack of actionable biomarkers remain core roadblocks toward establishing validity. The field may continue to develop better means to reliably define and classify people with versus without certain clinical features, but how helpful will these efforts be toward the development of specific treatments or elucidation of etiology? Instead, the ability to use diagnostic categories to predict outcome and appropriately select differential treatment choices, as practiced in other fields of medicine, will be better served with a classification scheme that directly maps onto the underlying natural disease entities. Such a notion had its roots in the efforts of Robins and Guze[53] (1970), who proposed "external validators" such as course, brain pathology, and etiology as a way to develop psychiatric classification.[53]

The search for these biological ways to classify psychotic disorders is not new. Early in the last century, Leonhard[54] had proposed a classification of schizophrenia into endogenous and exogenous based on presumed etiology. Crow classified schizophrenia into type I (with predominant positive symptoms) and type II (with predominant negative symptoms).[55] In this model, type I is associated with dopaminergic dysfunction, while type II is associated with structural brain abnormalities. Crow postulated that alteration of language brain areas and lateralization of brain functions may be causative.[56] Andreasean et al. also proposed a distinction between negative and positive schizophrenia,[57] and developed rating scales for quantification of positive and negative symptoms;[58] Carpenter et al. proposed a classification of schizophrenia into those with versus without deficit symptoms.[59] All these approaches, however, assumed that schizophrenia is a distinct disease entity, and research on external validators supporting these distinctions has been limited.

THE CONCEPT OF ENDOPHENOTYPES

The imprecise boundaries of disorder categories defined by symptomatic measures underlie the challenge of identifying valid categories within the psychosis spectrum. This points to the need of identifying biomarkers that cut across syndromal categories. An important paradigm shift in the field in recent years is the concept of intermediate (or endo-)phenotypes, which provide footholds in the path from the phenome to the genome.[60] More specifically, an endophenotype can be defined by the characteristics of being (1) measurable; (2) heritable; (3) state-independent, that is, present in stable as well as acute phases of the illness; (4) and present in unaffected relatives and (5) cosegregating with the illness within families. Examples include the cerebral ventricular size and gray matter volumes, cognitive impairments, and electrophysiological measures such as amplitudes of the P300 event-related potentials.[61] These measures are often interrelated, so that independent "families" (or extended) intermediate phenotypes can potentially be identified, that link etiology, pathophysiology, and clinical expression (e.g., abnormal gamma synchrony, inefficient task-related prefrontal activation, and impaired working memory). This approach offers the promise of unscrambling/deconstruction as well as reconstructing the current concept of schizophrenia, thereby allowing a more straightforward and successful analysis of this biologically heterogeneous entity.[62] The examples of the RDoC and B-SNIP (Bipolar-Schizophrenia Network for Intermediate Phenotypes) projects offer respective useful examples of the deconstruction and reconstruction paradigms.

TOWARD DECONSTRUCTION OF DISEASE DOMAINS: THE RESEARCH DOMAIN CRITERIA (RDOC) PROJECT

The National Institute of Mental Health's RDoC project "seeks to organize information relevant to psychopathology based on dimensions of observable behavior and neurobiological measures . . . from genes to neural circuits to behaviors,"[40] across a range of psychiatric disorders, agnostic to symptom-based classification. The RDoC includes several domains such as negative valence systems, positive valance systems, cognitive systems, social processes, and arousal and regulatory systems.[63] The RDoC system proposes analysis of these domains across several units of analysis including genes, molecules, cells, circuits, physiology, behaviors, self-reports, and paradigms. The cells of this matrix represent a scaffold for enabling reproducible and validated psychiatric research. The RDoC takes on new significance when viewed through a historical lens. Applied to psychosis, the goal of splitting psychosis along biological lines is similar to the efforts in the late 19th century to search for and classify the illness by the brain lesions underlying unique types of psychosis. However, while the goal may be similar, the tools and theory are not. New approaches like genome-wide association studies, molecular makers of cellular signaling, and an array of neuroimaging techniques allow the promise of unparalleled windows into the pathophysiology of psychosis. Another difference from 19th-century efforts is that the focus is no longer on "lesions" but rather the recognition that psychosis is likely a network disorder involving widespread dysfunction in the brain.[64] However, RDoC is currently a research tool and not intended for clinical use. Still, in the few years since its promulgation, a plethora of new dimensional research on schizophrenia, as well as numerous other mental disorders has emerged.[65]

The B-SNIP approach offers an experimental strategy that may help in reconstructing diagnostic categories using intermediate phenotypes—a battery of biological measures of brain structure and function—in a manner agnostic to symptom-based categories.[66] This multisite study examined individuals with clinical diagnoses of schizophrenia, bipolar disorder, or schizoaffective disorder, as well as their biological relatives, with the goal of assessing how biological markers of illness such as brain imaging and EEG activity map onto current DSM-based diagnoses. The resulting biomarker-based categories, called "Biotypes" one, two, and three were able to better categorize subjects on biological characteristics and outperformed the DSM-based diagnoses.[67] Of further interest, the three resulting B-SNIP Biotypes had little resemblance to the clinical constructs of schizophrenia, schizoaffective disorder, or bipolar disorder. This suggests the biology of psychotic illness does not respect current clinical diagnoses or nosology, and offers the potential of further biologically based discoveries in coming years.

TOWARD A REVERSE NOSOLOGY

One reason for the rather limited biological distinction between current DSM categories may stem from the latter being based on clustering of symptoms and other clinical features alone, without any other external validators. Such an approach over time has become reified as the "gold standard" for validating any new biomarker data that emerges over time.[37,68] Imagine categorizing fever patients into those with versus without cough and breathlessness. How well would a sputum sample with microscopy distinguish between these groups and correctly diagnose bacterial pneumonia? Clearly, categorizing such patients based on biology (e.g., those with vs. without a finding on a chest X-ray) would have been a good start. We have recently suggested that a "reverse nosology" approach (see Figure 1.1) to reclassifying psychotic disorders based on a "bottom-up" analysis of genetic or neuroimaging findings may reveal more valid categories. In a recent analysis using B-SNIP data,[69] we examined the symptomatic correlates of schizophrenia polygene scores, using a large number of permutations of combinations of five symptom items selected from positive, negative, and mood symptom scales.[70] The top five clusters that correlated with progressive schizophrenia did not resemble any current ICD or DSM disease categories. The clinical symptom features that associated with genetic characteristics most highly were poor abstract thinking, lack of insight, disorientation, inner tension, and depression, that is, predominantly negative, cognitive, and affective characteristics. Strikingly, none of the PANSS positive (psychotic) symptoms appeared in the top five clusters. These results suggest the potential of a reverse nosology approach to dramatically alter current conceptualizations of schizophrenia, and warrant further investigation.

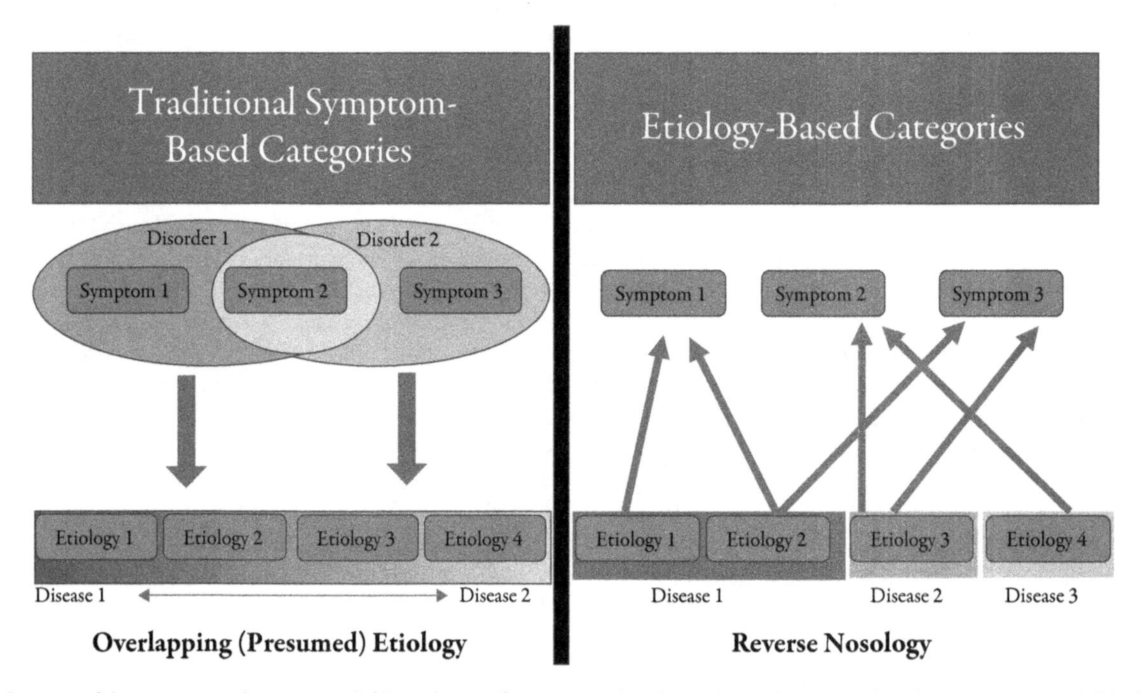

Figure 1.1 Schematic of the reverse nosology approach (the right panel) as compared to the traditional symptom-based category approach (the left panel).
Adapted from Tamminga et al., 2016.

In current clinical practice, biomarker and genetic data are not yet available to help clinicians. The proposals made thus far are currently research tools that allow researchers to identify biologically distinct, homogeneous psychosis groups in order to examine molecular, cellular, and system pathology as potential treatment targets. However, using a rich combination of clinical and physiological information together in clinical practice may offer classification approaches with better validity than those based only on clinical data. Recent developments in smartphone technology as well as widespread smartphone use among patients with psychotic disorders[71] offer an opportunity to obtain a richer phenotypic signature in individual patients on a moment-by-moment basis, including subjective, objective behavioral, and physiological data.[72] Accelerometers in the smartphones of subjects with schizophrenia can be used to infer sleep patterns, mobility data gathered from smartphone GPS in subjects with schizophrenia may augment relapse prediction,[73] and real-time surveys on smartphones in subjects with schizophrenia may help reveal cognitive biases.[74] Digital data derived from wearables and smartphones offers the potential of real-time, longitudinal, and objective data streams—and is increasingly available through open source smartphone platforms.

CONCLUSIONS

The evolving nosology of psychotic disorders, and especially schizophrenia, has been anything but linear. Reviewing the history of schizophrenia reveals the origins of current nosology and those forces that created the current clinical definitions of the illness. As this review reveals, classification of psychotic disorders has proceeded from categorical clusters of co-occurring symptoms and clinical features toward an increasing use of psychopathological dimensions. While these approaches have served the clinicians' needs of utility and reliability, the question of validity of these categories has been addressed only in recent years. Both dimensional (e.g., RDoC) and categorical (e.g., B-SNIP Biotypes) approaches are being used to develop valid ways to organize what we know about the biology of psychotic disorders. However, there is no reason why we need restrict ourselves to any of these approaches; it may well be that we use a combination of diagnostic schemes at more than one level to best describe a given patient's unique clinical signature. For example, in oncology to describe a lung cancer, we still use a clinical phenotypic classification (tumor, node, metastasis, TNM) to define the nature, extent and location of a tumor; a histopathological classification to describe what we see under the microscope (e.g., small cell, non–small cell); and a molecular classification (e.g., Herceptin receptor sensitive). A judicious combination of relevant clinical and biomarker information may best serve to address the diagnostic and prognostic questions posed by our patients.

Just over 100 years ago Emil Kraepelin used simple notecards to redefine what would become schizophrenia and bipolar disorders; today we have a myriad of new tools and methods and large datasets to advance understanding of these disorders. We can learn from both the successes and setbacks of prior efforts to define schizophrenia in order to best use the wealth of existing, new biological, and emerging digital data for increasing understanding and advancing care in psychotic disorders.

REFERENCES

1. World Health Organization. Manual of the International Statistical Classification of Diseases, Injuries and Causes of Death, Sixth Revision (ICD-6). World Health Organization, Geneva, 1949.
2. American Psychiatric Association. Diagnostic and Statistical Manual of Mental Disorders-1st edition (DSM-I). American Psychiatric Association, Washington D.C., 1952.
3. American Psychiatric Association. Diagnostic and Statistical Manual of Mental Disorders- 5th edition (DSM-5). American Psychiatric Association, Washington D.C., 2013.
4. World Health Organization. The ICD-10 Classification of Mental and Behavioral Disorders: Clinical Descriptions and Diagnostic Guidelines (CDDG). World Health Organization, Geneva, 1992.
5. Tandon R, Maj M. Nosological status and definition of schizophrenia: some considerations for DSM-V and ICD-11. Asian J Psychiatry 2008;1:22–27.
6. Tandon R, Keshavan MS, Nasrallah HA. Schizophrenia, "Just the facts": what we know in 2008. Part 1: overview. Schizophr Res. 2008;100:4–19.
7. Kraepelin E. Dementia Praecox and Paraphrenia, 1919, edited by GM Robertson. Krieger. New York, 1971.
8. Bleuler E. Dementia Praecox, or the Group of Schizophrenias, 1911, translated by J Zinkin. International University Press, New York, 1950.
9. Tandon R, Carpenter WT. DSM-5 status of psychotic disorders: 1-year pre-publication. Schizophr Bull. 2012;38:369–370.
10. Gaebel W, Zielasek J, Cleveland H-R. Psychotic disorders in ICD-11. Psychiatrie. 2013;10:11–17.
11. Heinroth JCA. Lehrbuch der Storungen des Seelenlebens oder der Seelenstorungen und ihrer Behandlung. Vogel, Leipzig, Germany, 1818.
12. Griesinger W. Die Pathologie und Therapie der psychischen Krankheiten. Krabbe, Stuttgart, Germany, 1845.
13. Hoenig J. The concept of schizophrenia: Kraepelin-Bleuler-Schneider. Br J Psychiatry. 1983;142:547–556.
14. Barnes MP, Saunders M, Walls TJ, Saunders I, Kirk CA. The syndrome of Karl Ludwig Kahlbaum. J Neurol Neurosurg Psychiatry. 1986;41:991–996.
15. Sedler MJ. Falret's discovery (1854): the origin of the concept of bipolar affective illness, translated by MJ Sedler and Eric C Dessain. Am J Psychiatry. 1983;140:1127–1133.
16. Jablensky A, Hugler H, von Cranach M, Kalinov K. Kraepelin revisited: a reassessment and statistical analysis of dementia praecox and manic-depressive insanity in 1908. Psychol Med. 1993;23:843–858.
17. Schneider K. Clinical Psychopathology, translated by MW Hamilton. Grune and Stratton, New York, 1959.
18. Wing JK, Cooper JE, Sartorius N. The Measurement and Classification of Psychiatric Symptoms. Cambridge University Press, Cambridge, 1974.
19. American Psychiatric Association. Diagnostic and Statistical Manual of Mental Disorders-2nd edition (DSM-II). American Psychiatric Association, Washington D.C., 1968.
20. Kendell RD, Cooper JR, Gourlay AJ, et al. Diagnostic criteria of American and British psychiatrists. Arch Gen Psychiatry. 1971;25:123–130.
21. American Psychiatric Association. Diagnostic and Statistical Manual of Mental Disorders-3rd edition (DSM-III). American Psychiatric Association, Washington D.C., 1980.
22. Mellor CS. First rank symptoms of schizophrenia. Br J Psychiatry. 1970;117:15–23.

23. Wing JK, Nixon J. Discriminating symptoms in schizophrenia: a report from the International Pilot Study of Schizophrenia. Arch Gen Psychiatry. 1975;32:953–959.
24. Tandon R. The nosology of schizophrenia: towards DSM-5 and ICD-11. Psychiatr. Clin North Am. 2012;35:555–569.
25. Wyatt RJ, Alexander RC, Egan MF, Kirch DG. Schizophrenia, "Just the facts": what do we know, how well do we know it. Schizophr Res. 1988;1:3–18.
26. Tandon R. Moving beyond findings: concepts and model-building in schizophrenia. J Psychiatr Res. 1999;29:255–260.
27. Keshavan MS, Nasrallah HA, Tandon R. Schizophrenia, "Just the facts" 6: moving ahead with the schizophrenia concept: from the elephant to the mouse. Schizophr Res. 2011;127:3–13.
28. Xu TY. The subtypes of schizophrenia. Shanghai Arch Psychiatr. 2011;23:106–108.
29. Tandon R, Nasrallah HA, Keshavan MS. Schizophrenia, "Just the facts." Part 4: clinical features and concept. Schizophr Res. 2009;110:1–23.
30. Braff DL, Ryan J, Rissling AJ, Carpenter WT. Lack of use in the literature in the last 20 years supports dropping traditional schizophrenia subtypes from DSM-5 and ICD-11. Schizophr Bull. 2013;39:751–753.
31. Tandon R, Gaebel W, Barch DM, et al. Definition and description of schizophrenia in DSM-5. Schizophr Res. 2013;150:3–9.
32. Barch DM, Bustillo J, Gaebel W, et al. Logic and justification for dimensional assessment of symptoms and related phenomena in psychosis: relevance to DSM-5. Schizophr Res. 2013;150:10–14.
33. Tandon R, Targum SD, Nasrallah HA, Ross R. Strategies for maximizing clinical effectiveness in the treatment of schizophrenia. J Psychiatr Practice. 2006;12:348–363.
34. McGorry PD. Risk syndromes, clinical staging, and DSM V: new diagnostic infrastructure for early intervention in psychiatry. Schizophr Res. 2010;120:49–53.
35. Tsuang MT, van Os J. Tandon R, et al. Attenuated psychosis syndrome in DSM-5. Schizophr Res. 2013;150:31–35.
36. Tandon N, Shah J, Keshavan MS, Tandon R. Attenuated psychosis and the schizophrenia prodrome: current status of risk identification and psychosis prevention. Neuropsychiatry. 2012; 2:345–353.
37. Insel TR, 2010. Rethinking schizophrenia. Nature. 2010;468: 187–193.
38. Malaspina D, Owen MJ, Heckers S, et al. Schizoaffective disorder in DSM-5. Schizophr Res. 2013;150:21–25.
39. Heckers S, Tandon R, Bustillo J. Catatonia in the DSM: shall we move or not? Schizophr Bull. 2010;36:205–207.
40. Insel T, Cuthbert B, Garvey M, et al. Research Domain Criteria (RDoC): toward a new classification framework n mental disorders. Am J Psychiatry. 2010;167:748–751.
41. Tandon R, Heckers S, Bustillo J, et al. Catatonia in the DSM-5. Schizophrenia Res. 2013;150:26–30.
42. Tandon R, Keshavan MS, Nasrallah HA. Schizophrenia, "Just the facts": what we know in 2008. Part 5: treatment and prevention. Schizophr Res. 2010;122:1–23.
43. Regier DA, Narrow WE, Clarke DE, et al. DSM-5 field trials in the United States and Canada, Part II: test-retest reliability of selected categorical diagnoses. Am J Psychiatry. 2013;170:59–70.
44. Tandon R. Antipsychotics in the treatment of schizophrenia: an overview. J Clin Psychiatry. 2011; 72(Suppl.1):4–8.
45. Carpenter WT., Tandon R. Psychotic disorders in DSM-5: summary of changes. Asian J Psychiatry. 2013;6:266–268.
46. Nasrallah HA, Tandon R, Keshavan MS. Beyond the facts in schizophrenia: closing the gaps in diagnosis, pathophysiology, and treatment. Epidemiol Psychiatr Sci. 2011;20:317–327.
47. Kapur S, Phillips AG, Insel TR. Why has it taken so long for biological psychiatry to develop clinical tests and what to do about it? Mol Psychiatry. 2012;17:1174–1179.
48. Kendler KS, Karkowski LM, Walsh D. The structure of psychosis: latent class analysis of probands from the Roscommon Family Study. Arch Gen Psychiatry. 1998;55(6):492–499.
49. Meehl PE. Schizotaxia revisited. Arch Gen Psychiatry. 1989;46(10):935–944.
50. Spitzer RL, Williams JB, Endicott J. Standards for DSM-5 reliability. Am J Psychiatry. 2012;169(5):537.
51. Chmielewski M, Clark LA, Bagby RM, Watson D. Method matters: understanding diagnostic reliability in DSM-IV and DSM-5. J Abnorm Psychol. 2015;124(3):764–769.
52. Jones KD. A critique of the DSM-5 field trials. J Nerv Ment Dis. 2012;200(6):517–519.
53. Robins E, Guze SB. Establishment of diagnostic validity in psychiatric illness: its application to schizophrenia. Am J Psychiatry. 1970;126(7):983–987.
54. Teichmann G. The influence of Karl Kleist on the nosology of Karl Leonhard. Psychopathology. 1990;23(4–6):267–276.
55. Crow TJ. The two-syndrome concept: origins and current status. Schizophr Bull. 1985;11(3):471.
56. Crow TJ. Schizophrenia as the price that Homo sapiens pays for language: a resolution of the central paradox in the origin of the species. Brain Res Rev. 2000;31(2):118–129.
57. Andreasen NC, Olsen S. Negative v positive schizophrenia: definition and validation. Archives Gen Psychiatry. 1982;39(7):789–794.
58. Andreasen NC. Positive vs. negative schizophrenia: a critical evaluation. Schizophr Bull. 1985;11(3):380.
59. Carpenter WT Jr, Heinrichs DW, Wagman AM. Deficit and non-deficit forms of schizophrenia: the concept. Am J Psychiatry. 1988;145:578–83.
60. Gottesman II, Gould TD. The endophenotype concept in psychiatry: etymology and strategic intentions. Am J Psychiatry. 2003;160(4):636–645.
61. Ivleva EI, Morris DW, Moates AF, Suppes T, Thaker GK, Tamminga CA. Genetics and intermediate phenotypes of the schizophrenia—bipolar disorder boundary. Neurosci Biobehav Rev. 2010;34:897–921.
62. Keshavan MS, Clementz BA, Pearlson GD, Sweeney JA, Tamminga CA. Reimagining psychoses: an agnostic approach to diagnosis. Schizophr Res. 2013;146(1):10–16.
63. Cuthbert BN, Insel TR. Toward the future of psychiatric diagnosis: the seven pillars of RDoC. BMC Med. 2013;11(1):126.
64. Silbersweig D, Loscalzo J. Precision psychiatry meets network medicine: network psychiatry. JAMA.
65. Carcone D, Ruocco AC. Six Years of Research on the National Institute of Mental Health's Research Domain Criteria (RDoC) Initiative: A Systematic Review. Front Cell Neurosci. 2017;11.
66. Keshavan MS, Morris DW, Sweeney JA, et al. A dimensional approach to the psychosis spectrum between bipolar disorder and schizophrenia: the Schizo-Bipolar Scale. Schizophr Res. 2011;133(1):250–254.
67. Clementz BA, Sweeney JA, Hamm JP, et al. Identification of distinct psychosis biotypes using brain-based biomarkers. Am J Psychiatry. 2015;173(4):373–384.
68. Keshavan MS, Clementz BA, Pearlson GD, Sweeney JA, Tamminga CA. Reimagining psychoses: an agnostic approach to diagnosis. Schizophr Res. 2013;146(1–3):10–16.
69. Tamminga CA, Pearlson GD, Stan AD, et al. Strategies for advancing disease definition using biomarkers and genetics: the bipolar and schizophrenia network for intermediate phenotypes. Biol Psychiatry: Cog Neurosci Neuroimaging. 2017;2(1): 20–27.
70. Tamminga CA, Pearlson GD, Stan AD, et al. Strategies for advancing disease definition using biomarkers and genetics: the bipolar and schizophrenia network for intermediate phenotypes. Biol Psychiatry: Cogn Neurosci Neuroimaging. 2017;2:20–27. doi:10.1016/j.bpsc.2016.07.005

71. Torous J, Chan SR, Tan SY, et al. Patient smartphone ownership and interest in mobile apps to monitor symptoms of mental health conditions: a survey in four geographically distinct psychiatric clinics. JMIR Mental Health. 2014;1(1).
72. Torous J, Onnela JP, Keshavan M. New dimensions and new tools to realize the potential of RDoC: digital phenotyping via smartphones and connected devices. Transl Psychiatry. 2017;7(3):e1053.
73. Torous J, Kiang MV, Lorme J, Onnela JP. New tools for new research in psychiatry: a scalable and customizable platform to empower data driven smartphone research. JMIR Mental Health. 2016;3(2).
74. Moran EK, Culbreth AJ, Barch DM. Ecological momentary assessment of negative symptoms in schizophrenia: relationships to effort-based decision making and reinforcement learning. J Ab Psychology. 2017;126(1):96.

2

HISTORICAL EPISTEMOLOGY OF THE "UNITARY PSYCHOSIS"

German E. Berrios and Ivana S. Marková

INTRODUCTION

The problem of whether madness was a unitary or manifold concept first appeared during the 18th century. At the time it reflected tensions between emerging concepts within medicine: (1) the view that diseases, like plants, were natural kinds and hence susceptible to being classified in the same way (*more botanico*);[1] (2) the proposal that diseases were composed of layers related to each other by flows of communication (e.g., causal links); and (3) the increasing application to the medical sciences of a form of Newtonian analytical methodology.[2] Given that subtler versions of these assumptions still govern the epistemology of medicine, it is not surprising that the problem of the "unitary psychosis," "monopsychosis," "psychotic continuum," etc., remains. Rather than repeating information already in the public domain,[3–8] this chapter focuses on the historical epistemology of the problem.

During the 17th century diseases were reconceptualized as "entities" exhibiting a stable ontology: "In the first place, it is necessary that all diseases be reduced to definite and certain species, and that, with the same care which we see exhibited by botanists in their phytologies; since it happens, at present, that many diseases, although included in the same genus, mentioned with a common nomenclature, and resembling one another in several symptoms, are, notwithstanding, different in their natures" (p13).[9]

Diseases were also believed to possess a multilayered structure, with (1) a bottom layer (organic or anatomical substratum) that contained the "pathological" changes that caused them, (2) a middle layer that served as a conduit for the transfer of information (chemicals or nervous fluids) from the bottom to the top layer (during the 19th century the function of this layer was reformulated into the new discipline of "physiology"), and (3) a surface or top layer that carried the "expression" of the disease and interacted with the environment. The three layers were connected by a bottom-to-top unidirectional causal link. This view culminates in the work of Morgagni.[10] At the end of the 18th century Lamarck[11] proposed that the causal arrow may be bidirectional (this view was resuscitated by epigenetics in the 20th century).[12]

A drive to classify all world objects started during the 17th century,[13] and to classify diseases during the following century. This led during the early 1700s to a debate as to which of the three layers should provide classificatory or taxonomic criteria. Interestingly, Linné and most of the other 18th-century classificators opted for the top layer and chose as criteria the sexual structure of flowers.[14] Wearing their nosologists' hats, the same men chose signs and symptoms in the case of diseases. One of the few thinkers that objected was Adanson,[15] who argued that all features be considered.[16] In the absence of machines with sufficient number-crunching power this proposal was utopian at the time. During the 20th century, it was rescued by numerical taxonomy.[17] A variant of this approach, under the name Research Domain Criteria[18] has recently been proposed: (1) consisting in the statistical analysis of proxy variables putatively representing the bottom and middle layer; (2) seeking to identify disease forms with more stable ontology than the formations provided by classical psychopathology that were mostly collected from observation of the top layer.[19]

The development during the 19th century of the "anatomo-clinical model of disease" was based on the multilayer proposal.[20] In the case of madness, the brain was selected to be the bottom layer. The accumulation of patients with similar complaints made possible by the new mental asylums allowed the search for correlations between post-mortem findings and clinical profile. *Ab initio* mental asylums mostly collected organic psychosis (neurosyphilis, traumatic epilepsy, alcoholism, chronic infectious diseases, etc.) and positive findings of this nature encouraged alienists to believe that it was time to use the bottom layer as the "true" source of taxonomic criteria.[21] Others believed that the said findings reflected organic diseases of the brain or were artifactual and hence the top layer should remain the source of criteria.[22]

The debate on whether mental disorders should be classified based on "etiology" (bottom layer) or "symptomatology" (top layer) has not yet been settled. It is unclear whether madness should be defined in terms of its complaints, that is, of the phenomenology of the suffering involved or of some experientially independent neurobiological event.[23]

The accumulation of patients with similar complaints also allowed for longitudinal observation. By the 1850s, "time" as a variable was incorporated into the definition and classification of madness (Kahlbaum played a role in this)[24] and led to differentiating acute and chronic forms of madness.[25] This in turn undermined the belief in its immutability and clinicians started to accept the view that the "same disease" could be successively expressed in various forms. Once again, this caused debate as to whether these forms should be considered as "different" types of madness. For example, the observation that during a long admission madness could change from melancholia, to mania to dementia led Griesinger to conclude that there was only one form of madness.[26]

By the 1850s, madness (*insanity, vesania, lunacy, folie, pazzia, irresein, wahnsinn*, etc.) started to be called "psychosis." This was more than a change in name as it required: (1) the detachment of madness from Cullen's old class of neuroses—defined until then as a condition associated with "lack of fever and absence of focal lesion," and (2) the creation of a group of new "diseases" defined as combinations of mental signs and symptoms (as identified in the top layer) kept together by some sort of a "glue" whose origins hopefully could be found in the bottom layer.

Since the 1850s the speculative "content" of the three layers of madness has been repeatedly replaced. Originally, the bottom layer only included speculative information concerning the anatomical (naked-eye) brain. By the time of Ramón y Cajal and Golgi, the information was being related to microscopic observation. From then on successive replacements have taken place with information from molecular, neuroimaging, and genetic research and new correlations have had to be sought. The middle layer has also undergone replacement and now includes the concept of endophenotype. Likewise, the top layer, which once included only the descriptions of mental symptoms (descriptive psychopathology) started to be called "phenomenological" by the 1930s, and more recently "phenotypical." Symptoms have become "criteria" and "items" in diagnostic and measuring instruments. There has been a shift from the qualitative to the quantitative and dimensional.[27] Epistemological changes of this nature are bound to affect how the concept of "unitary psychosis" is to be formulated and understood.

HISTORIOGRAPHY

Being a hybrid discipline, psychiatry and its objects present a special challenge to a conventional approach by either social or natural sciences alone.[28] "Historical epistemology"[29] is a method of study that seeks to identify the sociocultural systems that legitimize what is considered as "knowledge" within each historical period.[30] It thus historicizes the concept of madness and does not demand that its recognition require any form of ontological invariance.

Central to this historicist stance there is a concept-making mechanism called "convergence."[31] Psychiatric concepts result from the coming together (converging) during a given historical period of a word, a concept, and a referent. Words (newly coined or recycled) are used to name the convergence. Concepts (cognitive gadgets) are used to justify the convergence, explain the referent in question, and link the convergence to a *Weltanschauung*. Referents (that may be complaints, signs, notions, objects, institutions, etc.) can be considered as the content of the convergence. Only some convergences survive, and this may be due more to social convenience, aesthetics, economic and political value, etc., than to any correspondence with a hypothetical "truth" in the world.

Attending all convergences, central (explanatory, justificatory) and attending (supporting) concepts can be found. In the case of the "unitary psychosis," the central concepts are unity, psychosis, and stratification; the attending concepts are description, definition, causality, and classification. Brief information on the cultural biography of both types of concepts is provided in what follows.

THE CENTRAL CONCEPTS

As members of earlier convergences, the words "unitary" and "psychosis" were already in use in some European vernaculars. Before being recycled and built into a new convergence, both concepts needed to be refurbished.

THE TERM AND CONCEPT OF "UNITARY"

"Unity," "unicity," "unitary," "Einheit," "der Einzige," "único," "unique," etc.,[32] were all related to the ancient metaphor of the "monad," the "simple" (onefold) and the "indivisible" (atom). According to Aristotle, things were "one" by nature or by accident. "One" could be applied to the following situations, when: (1) a thing is continuous (like an organism), (2) the substratum of two things does not differ in kind, (3) a thing corresponds to one unit, and (4) two things fall within the same definition (Metaphysics 1016a 1–1017a).[33] Aristotle's choice of examples and language suggests that he had in mind a stratified model of things.

Ever since Aristotle the "one" (unity) has remained a central concept in Western culture, playing an important role in theology, mathematics, aesthetics, ethics, epistemology, logic, the law, and metaphysics. For example, Newton's "analytical" methodology consisted in the search for the simplest, indivisible components (unit, atom, monad) that constituted the structure of natural objects. Since Newton, this methodology has remained central to all disciplines, including psychology and psychiatry. The universal influence of Newtonian physics led to a common definition of "unity." Thus, the European vernaculars had little difficulty in translating *Einheitspsychose* into "unitary psychosis," "psychose unique,"[34] "psicosis única,"[35] etc., and believing that the definition of "unitary" in each case was the same. However, soon enough differences began to appear as to what did it actually mean for a "psychosis" (form of madness) to be considered as "unitary."[36] In some psychiatric cultures the "unitary" was considered as a feature of the bottom layer, in others of the top layer; yet in others the emphasis was cross-sectional observation and in others on longitudinal follow-up. Not taking these international differences into account has given rise to inaccurate historical accounts of the unitary psychosis.

STRATIFIED MODELS OF THE MIND

Stratified models of objects and of the mind, personality, self, disease, etc., were already available before the 19th century. After the 1870s, they became popular. First included in the metaphysical neurology of Hughlings Jackson,[37] they were soon to influence Ribot, Janet, Bleuler, Freud, Chaslin, Blondel, Ey, Scheler, Hoffmann,[38] etc. These writers adopted multilayer[39] or stratified[40] models to explain the putative functioning of the brain, mind, or spirit, and to explain fashionable clinical concepts such as "dissociation" where the planes of cleavage could be vertical or horizontal.[41] The multistrata model offers a tool to classify the meanings given to "unitary psychosis."

THE TERM, CONCEPT, AND STRUCTURE OF "PSYCHOSIS"

The term "psychosis"

The word "psychosis" was already in use during the 19th century to name the contents of normal consciousness: "name for mental affections as a class. Used very loosely for mental phenomena, states of consciousness, thoughts, ideas."[42] Baldwin iterated this definition: "a) used with regard to normal processes, psychosis is equivalent to the mental or psychical element in a psychophysical process, just as neurosis refers to that aspect of the process which belongs to the nervous system. The term simply designates this factor without implying any theory of relation of the mental to the physical; b) used as equivalent to the total state of consciousness existing at any one moment the term is a general designation for all concrete psychic facts" (p392).[43]

During the second half of the 19th century the term started to be used in Germany to refer to insanity[44] and Baldwin reflected this: "Used pathologically (and in this sense the usage is rapidly gaining ground both in foreign and in English literature), the term designates an abnormal mental condition, especially inasmuch as it is correlated with a specific disease-process (a disease-entity, if the term be allowed) with characteristic origin, course, and symptoms. The typical forms of insanity which can be scientifically differentiated would rank as psychoses in this sense" (p392).

Feuchtersleben's usage of "Psychosen" is, however, transitional:

> The second part, namely, the pathology of the mind, is again divided into the pathogeny, which seeks to trace the causality existing between the diseases of the mind and the body, and the nosography, which aims at exhibiting the phenomena, the natural history, and the so-called system of psychoses. ("*Der zweite Theil, die Pathologie der Seele, sondert sich weider in die Pathogenie, welche das Ursächliche zwischen Seelen -und Körperleiden,—und die Nosographie, welche die Erscheinungen [gleichsam die Naturgeschichte und das sogenannte System der Psychosen]) darzustellen sucht.*") (p5)[45]

Feuchtersleben is not yet referring to any disease or group of diseases but only offering a generic name for the contents of consciousness.

The concept of "psychosis"

The concept of insanity was transformed into that of "psychosis" sometime during the second half of the 19th century.[46] In Flemming's book "Psychose" is used 248 times, but the German alienist does not offer a definition, giving the impression that by the early 1860s everyone knew what it meant! To understand its new meaning,[47–49] the concept of psychosis must be mapped against three dichotomies: psychoses versus neuroses; functional versus organic; and exogenous versus endogenous.

Psychoses versus neuroses

Defined during the 18th century by Cullen as "clinical states without fever or localized lesions of the central nervous system," the "neuroses" constituted during his period a large class of diseases embracing all forms of insanity together with various neurological and medical disorders. This means that at the beginning of the 19th century this group was considered as representing (inter alia) "organic" disorders of the nervous system. On the other hand, the "psychoses" only named psychological, experiential states, unrelated to disease.

By 1900, the historian finds the "neuroses" named a small group of conditions: hysteria, hypochondria, secondary depression, Raynaud's disease, anxiety disorders, and obsessional-compulsive disease.[50] The attrition of the old neuroses resulted from the fact that anatomo-pathological research had throughout the century found that most of its members showed some "focalized" pathology and had to be removed into a new group which in the fullness of time gave rise to the new specialism of "neurology."

The small number of disorders left behind (where no focal brain changes had been found) was reinterpreted by Freud and others as resulting from "psychological" lesions and conflicts. Thus, a neat etiological crossover took place during the 19th century: the psychoses were to become "organic" while the "neuroses" were to be considered as "psychological" in origin.[51,52]

Functional versus organic

By the time of the First World War, the distinction between functional and organic had become fundamental to psychiatric description, explanation, and classification. The "functional" group of psychoses included dementia praecox, manic depressive insanity, paraphrenia, and paranoia. Thus Mendel wrote: "on the other hand, there is a great difference of opinion amongst authors as to how to divide those mental diseases in which no anatomical changes have hitherto been met and which do not belong under any of the forms named; they have been designated as functional psychoses, by which it is not said that anatomical changes do not exist, but only that we have so far been unable to verify them" (p160).[53] Mendel contrasted them with "organic psychoses" such as progressive paralysis of the insane, senile dementia, arteriosclerotic, and syphilitic psychoses.

By the end of World War I, Jaspers still listed three functional" psychoses, genuine epilepsy, schizophrenia, and manic-depressive illness:

> these three . . . have four points in common. In the first place their study gave rise to the concept of disease entity . . . in the second place the cases which belong to this group cannot be subsumed under the disorders of group I and Ill. One must however assume that many of these psychoses have a somatic base . . . in the third place (they) are not exogenous but endogenous psychoses. Heredity is an important cause . . . in the fourth place they all lack anatomical cerebral pathology. (pp607–608)[54]

Exogenous versus endogenous

Möbius[55] introduced de Candolle's[56] botanical terminology into neuropsychiatry. Endogenous disorders were those in which "the principal condition must lie in the individual, in a congenital disposition (*Anlage*), other factors being merely contingent and quantitative" (p140), for example, neurasthenia, hysteria, epilepsy, migraine, Huntington's chorea, Friedrich's disease. "Exogenous" disorders were supposedly "engendered from without" for example, trigeminal neuralgia, thyroid disease, multiple sclerosis, and Parkinson's disease. Möbius did not specify in relation to what space was "without" to be defined. Textual analysis shows that the boundary he considered was not the "skin," so it can be concluded that by "exogenous" Möbius did not mean "environmental." Kraepelin gave the terms a lease of life: "now the idea is gradually developing that the internal causes fall into two groups called by Möbius exogenous and endogenous. The former group showing uniform character and development; the latter being manifold and fluctuating" (p15).[57]

Historical analysis shows that Möbius misunderstood De Candolle. Describing tree-growth, the French botanist had written: "the older [layers] are at the centre and the younger ones at the circumference; from which it results that the plant hardens from inside out; I call these [vegetables] *Exogenes* (*exo* = outside; *genos* = to engender); . . . others organize themselves in a way in which the old, i.e. the harder [layers] are outside and growth occurs towards the centre; I call these the *Endogenes* (*endos* = inside; *genos* = to engender)" (p240). So, Möbius (and others since)[58] used this pair of concepts wrongly: "Endogenous" was not meant to signify "genetic" (as the term was created before such a concept existed) nor was "exogenous" meant to signify "environmental."

THE ATTENDING CONCEPTS

DESCRIPTION AND DEFINITION

Descriptio related to the capacity of a language to capture God's attributes.[59] By extension, it was used to name the perfect (complete) lists of attributes that language could generate in relation to any object in nature. Based on the "correspondence theory of truth," Condillac proposed during the 18th century that science itself should be defined as "well-made language, that is, as one capable of "perfect descriptions."[60,61]

Definitions can be thought of as cognitive and linguistic gadgets able to cut out fragments of reality or to construct ideal objects. They are used by all disciplines to render their objects stable.[62] More prescriptive than descriptive, psychiatric definitions tend to shape subjective reality. During the early 19th century, descriptive psychopathology was constructed as the psychiatric counterpart of medical semiology, and mostly included definitions of mental symptoms.[63] Ever since, there has been a debate on whether definitions carve madness at its joints.[64]

CAUSES AND REASONS

The concept of etiology, as used in medicine and psychiatry, is related to the philosophical concept of causality. The Latin "causa" meant: reason, motive, inducement, occasion. It translated the Greek words *α→σ←α* and *α°σιομ* (whence "aetiology"). Since Greek times, "*α→σ←α*" has been considered as a relational concept, i.e., that "without which another thing (called effect) cannot be."[65]

Aristotle's analysis of the concept lasted until the modern period:

> we call a cause: 1. that from which (as immanent material) a thing comes into being, e.g. the bronze of the statue and the silver of the saucer; 2. The form or pattern, i.e. the formula of the essence, and the classes which include this; 3. That from which the change or the freedom from change first begins, e.g. the man who has deliberated is a cause, and the father a cause of the child; and 4. The end, i.e. that for the sake of which a thing is. e.g. health is the cause of walking.[66]

Since then, four types have been recognized: material, formal, efficient, and final cause, and it has been debated whether the four are actually different and are of equal importance, and whether they work better in conjunction. The mechanistic revolution of the 17th century[67] focused on efficient causes although it has been periodically asked whether teleological (final) causality may play a role in biology.[68,69]

Around the same period, views on the etiology of disease became related to the notion of "efficient cause" (expressed via an "internal mechanism" residing in the bottom layer). During the 18th century, Hume attacked the concept of "efficient cause": "a cause is an object precedent and contiguous to another, and so united with it, that the idea of the one determines the mind to form the idea of the other, and the impression of the one to form a more lively idea of the other" (Section XIV; p172).[70]

Humean skepticism encouraged medicine to consider "cause" and "effect" as different entities and to require that in all cases it was shown that the former occurred before the latter. This "seriatim" view of causality can be found in the work of Pinel, Heinroth, Bayle, Prichard, Esquirol, Georget, Feuchtersleben, Morel, Griesinger, Guislain, etc., and has remained popular to this day. The conceptual complexities that characterize the current debate on causality in general philosophy[71] and philosophy of science[72] have barely touched psychiatry, where the old seriatim view remains the norm.[73]

Whether "reasons" can be considered as causes of behavior remains unresolved in philosophy[74] but is a popular view in the social sciences and psychoanalysis.[75] As a hybrid discipline,[76] psychiatry straddles the natural and social sciences and the causality of reasons in relation to mental sufferings and disorders remains an important question. Reasons in this sense need not be construed in material or mechanical terms but expressed in the language of folk psychology[77] (statements of the form "He went crazy because she abandoned him"). In common parlance it is well accepted that "reasons" provide adequate accounts for a range of human behaviors, even when the semantic engine that makes such process possible remains unknown.[78]

The main argument against reasons being causes is that they are considered as explanations "after the event" and hence do not have the form of seriatim, "proper," causal accounts. In medicine and psychiatry causality is still understood in terms of "general laws" (not of idiosyncratic explanations, i.e., related to specific situations). There is also the widespread belief that to be effective reasons must be "conscious" (intentional), that is, for person P to claim post facto that she did Y for reason X, she must have pre facto been fully aware of X and of the fact that X led to Y. However, after the psychoanalytical revolution, this view is likely to be considered as narrow. To many, the statement: "John gives many presents to Mary because he is in love with her, although he is yet not aware of it" would make perfect sense even if the issue of pre-fact awareness and conscious intentionality does not arise.[79]

LISTINGS, ORDERINGS, AND CLASSIFICATIONS

In the history of medicine, "nosology" and "nosography" have been used to name the classification and description of diseases, respectively.[80] In the philosophy of science, classification may mean: (1) the principles and rules according to which groupings are put together (taxonomy), (2) the act of grouping itself (the sortal act, which in the case of listings such as DSM-5 or ICD-10 is almost tantamount with "diagnosis"), and (3) the resulting groupings themselves.

Taxonomy is the normative discipline that governs the language, meta-language, and rules of classification.[81] The term "system" names the frame of a classification, and "domain" names its content, that is, the objects to be classified. The language of taxonomy also includes polarities such as "categorical vs. dimensional"; "monothetic vs. polythetic"; "natural vs. artificial," "top-to-bottom" vs. "bottom-up," "structured vs. listing," "hierarchical vs. nonhierarchical," "exhaustive vs partial," idiographic vs nomothetic," etc.

The "natural" vs. "artificial" polarity dates to the 18th century. In this usage, "natural" meant part of the natural world; and "artificial" meant man-made. Psychiatric classificators have since struggled with this distinction.[82] Those upholding an ontological view will argue that mental disorders are natural kinds (like plants or dogs) susceptible to full description and to a natural, stable classification. Those who believe them to be cultural objects will argue mental disorders result from the application of temporal narratives to human suffering and will consider all classifications as artificial. Currently, the former view is predominant and has led to efforts to replace qualitative psychopathological descriptions of human suffering by "proxy" variables that may capture the very ontology of "mental diseases."

The application of taxonomic rules to a given (open or closed) universe of objects gives rise to three types of groupings: classifications, orderings, and listings. Real classifications systematically organize the objects of a domain and have the power to extract new information and on this basis predict the existence of new facts, phenomena, or objects (e.g., the periodic table of elements). "Orderings" (e.g., the "scala naturae")[83] do not meet all taxonomic criteria, are epistemologically less powerful, and unable to extract new knowledge from their domain. "Listings" (e.g., DSM-5)[84] are the weakest form of grouping and have little or no epistemological power.[85]

THE VARIOUS MEANINGS OF "UNITARY PSYCHOSIS"

When the central and attending concepts briefly analyzed here are put together, it becomes apparent that the concept of "unitary psychosis" is multivocal. What determines which meaning is current or fashionable within each historical period is unclear. It is likely, however, that it will depend less on the advances of science than on their temporal economic and social value.

ILLUSTRATIONS FROM ITS HISTORY

The question of whether "madness" (insanity, lunacy, distraction, *folie*, *Wahnsinn*, *Pazzia*, *locura*, etc.) referred to one or various disturbances of behavior (e.g., lycanthropy, dementia, vesania, melancholia, fury, frenzy, mania, etc.) was already debated before the 19th century.[86] At the time, this was not known as the "unitary psychosis" problem, for neither of these concepts was available to the interested parties. Since the publication of "Unitary Psychosis: A conceptual history,"[87] efforts have been made to resolve the problem "empirically" by subjecting important clinical and genetic information to serious statistical analysis.[88] The fact that the problem has not yet been resolved suggests that its nature is conceptual rather than empirical. There is space in what follows only to describe two historical episodes illustrating unitarian views related to the bottom and the top layer.

A bottom-layer unitarian

Borrowing from Boissier, Linné, Vogel, Sagar, and McBride, Cullen[89] suggested that the insanities could be classified on anatomical, functional, symptomatic, and outcome criteria. An exponent of downward taxonomy,[90] he first evaluated the anatomical criteria: "there have occurred so many instances of this kind, that I believe physicians are generally disposed to suspect organic lesion of the brain to exist in almost every case of insanity" . . . "this, however, is probably a mistake; for we know that there have been many instances of insanity from which the persons have entirely recovered" (p139) . . . "Such transitory cases, indeed, render it probable, that a state of excitement, changeable by various causes had been the cause of such instances of insanity" (p140). Cullen seems to be suggesting that madness can also follow functional disturbances.

Cullen believed that this failure of the bottom layer to provide taxonomic criteria affected the classification of insanity in general: "having thus endeavoured to investigate the cause of insanity in general, it were to be wished that I could apply the doctrine to distinguishing the several species of it, according as they depend upon the different state and circumstances of the brain, and thereby to the establishing of scientific and accurately adapted method of cure. These purposes, however, appear to me to be extremely difficult to be attained" (p141). Based on this comment, Cullen criticizes those who have offered such classifications: "The ingenious Dr Arnold has been commendably employed in distinguishing the different species of insanity as they appear with respect to the mind; and his labours may hereafter prove useful, when we shall come to know something more of the different states of the brain corresponding to these different states of the mind; but at present I can make little application of his numerous distinctions" (p141).

Cullen concluded with a comment that some might wish to interpret as a forerunner of the "unitary" notion: "It appears to me that he [Dr Arnold] has chiefly pointed out and enumerated distinctions, that are mere varieties, which can lead to little or no variety of practice: and I am specially led to form the latter conclusion, because these varieties appear to me to be often combined together, and to be often changed into one another, in the same person; in whom we must therefore suppose a general cause of the disease" (p142).

Periodically discussed, the notion of a "psychosis continuum"[91,92] shares conceptual features with the "unitary psychosis." Its useful ambiguity allowed for the "continuum" to be related either to the bottom or the top layer. Debates in the 1970s and 1980s played on this double meaning.[93] For example, the statistical debate on whether a bimodality could be found between the clinical description of the psychosis was played at the top layer;[94] while the inference of a genetic continuum of sorts was clearly referred to the bottom layer.[95,96] This old debate has of late been resuscitated in relation to a putative genetic overlap between the major psychoses.[97–99]

Top-layer unitarians

Following Sydenham's injunction, Enlightenment physicians conceived of and classified diseases in the manner of a botanist.[100–102] It was a top-to-bottom mode of analysis based on three assumptions: (1) the objects to be classified were ontologically stable; (2) privileged features were present in all and were essential to their definition; and (3) the domain of objects to be classified resulted from the operation of a universal law or design. Before Darwin such a design was considered as resulting from a divine hand and hence "natural" objects reflected the natural order.

During the 19th century the debate on whether there were one or more forms of madness took place, per force, at the top layer. *Ab initio*, it concerned what was seen on cross-sectional observations. After "time" became included as a variable in the analysis of madness, observations turned longitudinal. In either case, whether comparing different forms of madness in the here and now or following up patient cohorts to see whether the nature of their madness had changed, the goal was to identify symptom-clusters that seemed to endure and breed through.

Cross-sectional observation had provided sufficient information to claim that symptom clusters such as phrensy, mania, melancholia, and dementia and idiocy were stable, different, and easily identifiable.[103] Longitudinal observation showing that over time the same patient could show melancholia, mania, and finally dementia led Griesinger,[104] Neumann,[105] and others to question the autonomy or independence of the static forms of madness and reignited the debate on the "one versus many."[106]

To explain the clinical stability of certain forms of madness, alienists used both the concept of pathognomonic symptom (which related to the top layer) and of "true-breeding" (which related to the bottom layer). The concept of the "pathognomonic" refers to "sign or symptom by which a disease may be known or distinguished; specifically characteristic or indicative of a particular disease."[107] First used in general medicine, by the 1820s it came into alienism and clinicians sought to identify them in the various forms of madness. By the second half of the century skepticism had set in and doctors were abandoning the idea that such signs could be identified.[108]

The concept of "true breeding" was part of the history of animal breeding and husbandry technologies which by the 18th century were fully developed. Dogs and horses could be bred not only for desirable physical features but also for temperament and psychological traits such as aggression or placidity, etc. It was not difficult to extrapolate from these time-tested observations that human behaviors, including madness, could also result from breeding.[109]

The introduction of so-called degeneration theory by Morel in 1857[110] not only related to his strong Roman Catholic beliefs concerning the original sin but also to his knowledge of natural history and husbandry.[111] Based on the transformism of Buffon and Lamarck, and on the old idea that human pedigrees may undergo drifts in their biological "quality," Morel proposed that successive generations of a tainted family could progressively show ever more serious forms of mental disorder until dementia ensued. Soon enough his views became popular in Europe[112] and confused the unitary psychosis debate further as the "breeding-true" criterion, an important explanation for the consistency for the forms of madness, did not seem to apply.

Influenced by Kahlbaum, Kraepelin started in the 1880s to reanalyze the top layer (he was not much interested in the bottom one, i.e., in etiology).[113] By the 8th edition of his *Lehrbuch*, just before World War I, he proposed two types of endogenous psychosis: dementia praecox and manic-depressive insanity. The dichotomy was criticized on theoretical[114,115] and clinical grounds.[116] By 1920 Kraepelin[117] had accepted the view that the top layer could not offer markers that differentiated the two forms of psychoses.

After World War II the "one versus many" debate was once again played out in the German Democratic Republic by Karl Leonhard and Helmut Rennert. The complex classification of the psychoses developed by the former[118] was repeatedly challenged by the latter, who instead proposed a "universal genesis of the endogenous psychoses."[119]

CONCLUSIONS

Information on the biography and views of those who have contributed to the debate on "unitary psychosis" has been iterated oftentimes (the required references are listed at the end). The fact that the problem keeps reappearing suggests that such iteration is no longer helpful. Given that the "one vs. many forms of madness" problem is likely to be conceptual rather than empirical, it is suggested here that the field is explored from a new perspective.

Following the methodological indications of historical epistemology, this chapter has started the task of identifying the central and supporting concepts that seem to have participated in the construction of the concept of unitary psychosis. It has also proposed that concepts be organized in terms of a simplified three-strata model that includes: (1) a bottom-layer whose nature is mostly organic and includes causality generators; (2) a middle-layer whose nature in the past was considered as "functional" but which now can be redefined as a sort of translator or compiler able to transfer information from the bottom layer to the top layer; and (3) a top or surface layer which contains the expressive forms and mechanisms required to communicate the signs and symptoms of disease.

The three layers are linked by flows of information, instructions, and values. In theory, these should be bidirectional, but in practice the 19th-century unidirectional view persists, and for all effects and purposes psychiatry still defines the essence of madness as resulting from the unidirectional flow of pathological signals originating in the bottom layer (brain) and emerging in the top layer as "mad behavior." This view encourages a facile reductionism.

Difficulties with the description and classification of what is currently called "mental disorders" can be traced back to conceptual tensions and incoherences affecting its epistemological frame. Hopefully, this chapter has provided sufficient purchase for the frame itself to be subject to analysis and renewal. Without understanding it, the concept of unitary psychosis will remain the mystery that currently it seems to be.

REFERENCES

1. López Piñero J M: *Historical Origins of the Concept of Neurosis*. Translated by D. Berrios. Cambridge: Cambridge University Press, 1983.
2. Guerlac H: Newton's method. In Wiener P P (ed) *Dictionary of the History of Ideas*. Vol 3. New York: Scribner's Sons, pp378–391, 1973.
3. Marcuse H: Zur Frage der Einheitspsychose. *Archiv für Psychiatrie und Nervenkrankheiten* 1926, 78: 682–683.
4. Vliegen J: *Die Einheitspsychose*. Stuttgart: Enke, 1980.
5. Llopis B: La psicosis única. *Archivos de Neurobiología* 1954, 17: 57–72.
6. Angst J: Historical aspects of the dichotomy between manic-depressive disorders and schizophrenia. *Schizophrenia Research* 2002, 57: 5–13.
7. Janzarik W: Nosographie und Einheitspsychose. In Huber G (ed) *Schizophrenie und Zyklothymie*. Stuttgart: Thieme, pp29–38, 1969.
8. Menninger K: The unitary concept of mental illness. *Bulletin of the Menninger Clinic* 1958, 22: 4–12.
9. *The Works of Thomas Sydenham*. Translated by Dr. Greenhill. Vol 1. London: Sydenham Society, 1858.
10. Morgagni J B: *The Seats and Causes of Diseases*. Translated by B Alexander. London: A Millar & T Cadell, 1769.
11. Gissis S B & Jablonka E (eds): *Transformations of Lamarckism: From Subtle Fluids to Molecular Biology*. Cambridge, MA: MIT Press, 2011.
12. Esteller M (ed): *Epigenetics in Biology and Medicine*. London: CRC Press, 2009.
13. Slaughter M M: *Universal Languages and Scientific Taxonomy in the Seventeenth Century*. Cambridge: Cambridge University Press, 1982.
14. Duris P: *Linné et la France: 1780–1850*. Genève: Droz, 1993.
15. Adanson M: *Familles des plantes*. Paris: Vincent, 1763.
16. Berrios G E: Classification in psychiatry: a conceptual history. *Australian and New Zealand Journal of Psychiatry* 1999, 33: 145–160.
17. Sokal R R & Sneath P H: *Principles of Numerical Taxonomy*. San Francisco: W H Freeman, 1963.
18. Cuthbert B N & Kozak M J: Constructing constructs for psychopathology: the NIMH Research Domain Criteria. *Journal of Abnormal Psychiatry* 2013, 122: 928–937.
19. Berrios G E: *The History of Mental Symptoms*. Cambridge: Cambridge University Press, 1996.
20. Ackerknecht E: *Medicine at the Paris Hospital 1794–1848*. Baltimore: Johns Hopkins Press, 1967.
21. Morel B A: *Traité de maladies mentales*. Paris: Masson, 1860.
22. Berrios G E: Baillarger and his essay on a classification of different general of insanity. *History of Psychiatry* 2008, 19: 358–373.
23. Berrios G E & Marková I S: Biological psychiatry: conceptual issues. In D'Haenen H, den Boer J A, & Willner P (eds) *Biological Psychiatry*. New York: John Wiley, pp3–24, 2002.
24. Kahlbaum K: *Die Gruppirung der psychischen Krankheiten und die Eintheilung der Seelenstörungen*. Danzig: A. W. Kafemann, 1863.
25. Lanteri Laura G: La chronicité dans la psychiatrie moderne française. *Annales* 1972, 3: 548–568.
26. Griesinger W: *Die Pathologie und Therapie der psychischen Krankheiten*. 2nd Edition. Stuttgart: Krabbe, 1861.
27. Marková I S & Berrios G E: Epistemology of mental symptoms. *Psychopathology* 2009, 42: 343–349.
28. Berrios G E & Marková I S: Towards a new epistemology of psychiatry. In Kirmayer L J, Lemeson R, & Cummings C A (eds) *Re-Visioning Psychiatry*. Cambridge: Cambridge University Press, pp41–64, 2015.
29. Max-Plank-Institut für Wissenschaftsgeschichte. Pre-Print 434. Conference: *Epistemology and History. From Bachelard and Canguilhem to Today's History of Science*. 2012.
30. Berrios G E: Historiography of mental systems and diseases. *History of Psychiatry* 1994, 5: 175–190.
31. Berrios G E: Convergences that are no more. *History of Psychiatry* 2011, 22: 133–136.
32. Dierse U: Einzige. In Ritter J (ed) *Historische Wörterbuch der Philosophie*. Vol 2. Darmstadt: Wissenschaftliche Buchgesellschaft, pp427–430, 1992.
33. *The Complete Works of Aristotle*. Edited by Jonathan Barnes. Princeton: Princeton University Press, pp3451–3455, 1984.
34. Grivois H (ed) *Psychose naissante, psychose unique?* Paris: Masson, 1991.
35. Valenciano L: La tesis de la psicosis única en la actualidad. In Llopis B. *Introducción dialéctica a la Psicopatología*. Madrid: Morata, pp 113–159, 1970.
36. Viotti-Daker M: *Die Kontinuität der Psychosen in den Werken Griesingers, Kahlbaums und Kraepelins und die Idee der Einheitspsychose*. Heidelberg University Dissertation, 1994.

37. Berrios G E: J H Jackson and "The factors of insanities." *History of Psychiatry* 2001, 12: 353–373.
38. Hoffmann H F: *Die Schichttheorie: Eine Anschauung von Natur und Leben*. Stuttgart: Enke, 1935.
39. Schiefele H: Schichtenlehre in der Psychiatrie un Psychologie. In Ritter J (ed) *Historische Wörterbuch der Philosophie*. Vol 8. Darmstadt: Wissenschaftliche Buchgesellschaft, pp1268–1271, 1992.
40. Rothacker E: *Die Schichten der Personlichkeit*. Bonn: H Bouvier, 1948.
41. Berrios G E: The concept of dissociation in psychiatry. *Rivista Sperimentale de Freniatria* 2018, 142: 29–50.
42. Tuke D H: *A Dictionary of Psychological Medicine*. 2 Vols. London: J and A Churchill, 1892.
43. Baldwin J M: *Dictionary of Philosophy and Psychology*. New York: Macmillan, 1901.
44. Flemming C F: *Pathologie und Therapie der Psychosen*. Berlin: August Hirschwald, 1859.
45. Feuchtersleben Baron von E: *Lehrbuch der ärztlichen Seelenkunde*. Wien: Carl Gerold, 1845.
46. Berrios G E: Historical aspects of the psychoses: 19th century issues. *British Medical Bulletin* 1987, 43: 484–498.
47. Bürgy M: Zur Geschichte und Phänomenologie des Psychose-Begriff. Eine Heidelberger Perspektive (1913–2008). *Nervenarzt* 2009, 80: 584–592.
48. Janzarik W: Der Psychose-Begriff. *Nervenarzt* 2003, 74: 3–11.
49. Haustgen T & Bourgeois M L: Cinquante ans d'histoire des psychoses a la Société médico-psychologique (1852–1902). *Annales Médico-Psychologiques* 2002, 160: 730–738.
50. López Piñero J M & Morales Meseguer J M: *Neurosis y psicoterapia: Un estudio histórico*. Madrid: Espasa-Calpe, 1970.
51. Beer D: The dichotomies psychosis/neurosis and functional/organic: a historical perspective. *History of Psychiatry* 1996, 7: 231–255.
52. Postel J: Les névroses. In Postel J & Quétel C (eds) *Nouvelle histoire de la psychiatrie*. Paris: Privat, pp357–366, 1983.
53. Mendel E: *Textbook of Psychiatry*. Translated by W Krauss. Philadelphia: Davis, 1907.
54. Jaspers K: *General Psychopathology*. 7th German Edition. Translated by J Hoenig & M W Hamilton. Manchester: Manchester University Press, 1963.
55. Möbius P J: *Abriss der Lehre von den Nervenkrankheiten*. Leipzig: Abel, 1893.
56. De Candolle A O: *Théorie élémentaire de la Botanique*. 2nd Edition. Paris: Deterville, 1819.
57. Kraepelin E: *Psychiatrie: Ein Lehrbuch für Studirende und Aerzte*. 5th Edition. Leipzig: J A Barth, 1896.
58. Lewis A: Endogenous and exogenous: a useful dichotomy. *Psychological Medicine* 1971, 1: 191–196.
59. Kaulbach F: Beschreibung. In Ritter J (ed) *Historische Wörterbuch der Philosophie*. Vol 1. Darmstadt: Wissenschaftliche Buchgesellschaft, pp 838–868, 1992.
60. Jahnke H N & Otte M: On "science as a language." In Jahnke H N & Otte M (eds) *Epistemological and Social Problems of the Sciences in the Early Nineteenth Century*. Dordrecht: Reidel, pp 75–89, 1981.
61. Berrios G E & Fuentenebro F: Chaslin and "Is psychiatry a well-made language?" *History of Psychiatry* 1995, 6: 387–406.
62. Robinson R: *Definition*. Oxford: Clarendon Press, 1950.
63. Berrios G E: Descriptive psychopathology: conceptual and historical aspects. *Psychological Medicine* 1984, 14: 303–313.
64. Campbell J K, O'Rourke M, & Slater M H (eds): *Carving Nature at Its Joints: Natural Kinds in Metaphysics and Science*. Cambridge, MA: MIT Press, 2011.
65. Berrios G E: Historical development of ideas about psychiatric aetiology. In Gelder M (ed) *New Oxford Textbook of Psychiatry*. Vol 1. Oxford: Oxford University Press, pp147–153, 2000.
66. Todd R B: The four causes: Aristotle's exposition and the ancients. *Journal of the History of Ideas* 1976, 37: 319–322.
67. Dijksterhuis E J: *The Mechanization of the World Picture*. Oxford: Clarendon Press, 1961.
68. Johnson M R: *Aristotle on Teleology*. Oxford: Clarendon Press, 2005.
69. Lenoir T: *The Strategy of Life: Teleology and Mechanics in Nineteenth Century German Biology*. Dordrecht: Reidel, 1982.
70. Hume D: *A Treatise of Human Nature*. L A Selby-Bigge edition. Oxford: Clarendon Press, 1967.
71. Clatterbaugh K: *The Causation Debate in Modern Philosophy 1637–1739*. New York: Routledge, 1999.
72. Wallace W A: *Causality and the Scientific Explanation. Vol II: Classical and Contemporary Science*. Ann Arbor: University of Michigan Press, 1974.
73. Riese W: *Le penseé causal en médecine*. Paris: Presses Universitaires de France, 1950.
74. Hutto D: A cause for concern: reasons, causes and explanations. *Philosophy and Phenomenological Research* 1999, 59: 381–401.
75. Axmacher N: Causation in psychoanalysis. *Frontiers in Psychology* 2013, 4: 1–4.
76. Berrios G E: Psychiatry and its objects. *Revista de Psiquiatría y Salud Mental (Barcelona)*, 2011, 4: 179–182.
77. Hutto D D: *Folk Psychological Narratives: The Sociocultural Basis of Understanding Reasons*. Cambridge, MA: MIT Press, 2008.
78. Enç B: *How We Act: Causes, Reasons and Intentions*. Oxford: Clarendon Press, 2003.
79. Farrell B: *The Standing of Psychoanalysis*. Oxford: Oxford University Press, 1981.
80. Berrios G E: The 19th century nosology of alienism: history and epistemology. In Kendler K & Parnas J (eds) *Philosophical Issues in Psychiatry II: Nosology*. Oxford: Oxford University Press, pp101–117, 2012.
81. Blackwelder R E: *Taxonomy*. New York: Wiley, 1967.
82. Diekmann A: *Klassifikation—System—"scala naturae."* Stuttgart: Wissenschaftliche Verlagsgesellschaft, 1992.
83. Formigari L: The chain of being. In Wiener P P (ed) *Dictionary of the History of Ideas*. Vol 1. New York: Scribner's Sons, pp325–335, 1973.
84. American Psychiatric Association: *Diagnostic and Statistical Manual of Mental Disorders*. 5th Edition. Arlington, VA: American Psychiatric Association, 2013.
85. Klosterkötter J: Traditionelle Klassificationssysteme psychischer Störungen. In Möller H J, Laux G, & Kapfhammer H P (eds) *Psychiatrie, Psychosomatik, Psychotherapie*. 5th Edition. Heidelberg: Springer, pp493–515, 2017.
86. Renynghe de Voxvrie G: Réactualisation du concept de psychose unique introduit par Joseph Guislain. *Acta Psychiatrica Belgica* 1993, 93: 203–219.
87. Berrios G E & Beer D: The notion of unitary psychosis: a conceptual history. *History of Psychiatry* 1994, 5: 13–36.
88. Möller H J: Systematic of psychiatric approaches between categorical and dimensional approaches: Kraepelin's dichotomy and beyond. *European Archives of Psychiatry and Clinical Neuroscience* 2008, 258 (Suppl. 2): 48–73.
89. Cullen W: *First Lines of the Practice of Physic*. New Edition, Volume IV. Edinburgh: Elliot, 1789.
90. Bowman I A: *William Cullen (1710–90) and the Primacy of the Nervous System*. Indiana University PhD Thesis, Michigan: Ann Arbor, 1975.
91. Marneros A, Andreasen N C, & Tsuang M T (eds): *Psychotic Continuum*. Berlin: Springer, 1995.
92. Müller N: Dichotomie oder Kontinuum? Gemeinsamkeiten und Besonderheiten in Neurobiologie und Therapie affectiver und schizophrener Psychosen. *Psychiatrie und Psychotherapie* 2010, 6: 139–143.
93. Berrios G E & Beer D: Unitary psychosis in English-speaking psychiatry: a conceptual history. In Mundt Ch & Saß H (eds) *Für und wider die Einheitspsychose*. Heidelberg: Springer, pp12–21, 1992.
94. Brockington I F, Kendell R E, Wainwright S, et al.: The distinction between the affective psychoses and schizophrenia. *British Journal of Psychiatry* 1979, 135: 243–248.
95. Crow T J: The continuum of psychosis and its implication for the structure of the gene. *British Journal of Psychiatry* 1986, 149: 419–429.
96. Crow T J: Psychosis as a continuum and the virogene concept. *British Medical Bulletin* 1987, 43: 754–767.

97. Lichtenstein P, Yip B H, Björk C, et al.: Common genetic influences for schizophrenia and bipolar disorder: a population-base study of 2 million nuclear families. *Lancet* 2009, 373: 1–12.
98. Cardno A G & Owen M J: Genetic relationships between schizophrenia, bipolar disorder and schizoaffective disorder. *Schizophrenia Bulletin* 2014, 40: 504–515.
99. Gandal M J, Haney J R, et al: Shared molecular neuropathology across major psychiatric disorders parallels polygenic overlap. *Science* 2018, 359: 693–697.
100. Larson J L: *Reason and Experience: The Representation of Natural Order in the Work of Carl Von Linné*. Berkeley: University of California Press, 1971.
101. Fischer-Homberger E: Eighteenth century nosology and its survivors. *Medical History* 1970, 14: 397–403.
102. Berg F: Linné et sauvages. *Lychnos* 1956, 16: 31–54.
103. Esquirol E: *Des maladies mentales considérées sous les rapports médical, hygiénique et médico-legal*. 2 Vols. Paris: Baillière, 1838.
104. Rennert H: Wilhelm Griesinger und die Einheitspsychose. *Wissenschaftliche Zeitschrift der Humboldt-Universität zu Berlin. Mathematisch-Naturwissenschaftliche Reihe* 1968, 17: 15–16.
105. Lanczik M: Heinrich Neumann und seine Lehre von der Einheitspsychose. *Fundamenta Psychiatrica* 1989, 3: 49–54.
106. Verwey G: *Psychiatry in an Anthropological and Biomedical Context: Philosophical Presuppositions and Implications of German Psychiatry, 1820–1870*, Dordrecht: Reidel, 1985.
107. *Oxford English Dictionary*. 2nd Edition. Oxford: Oxford University Press, 2009.
108. Hetch L: Pathognomonie. In Dechambre A (ed) *Dictionnaire Encyclopédique des Sciences Médicales*. Second Series, Vol 21. Paris: Masson, pp598–599, 1885.
109. Castle W E: *Heredity in Relation to Evolution and Animal Breeding*. London: Appleton and Company, 1911.
110. Morel B A: *Traité des dégénérescences physiques intellectuelles et morales de l'espèce humaine*. Paris: Baillière, 1857.
111. Genil-Perrin G: *Histoire des origines et de l'évolution de l'idée de dégénérescence en médicine mentale*. Paris: A Leclerc, 1913.
112. Bing F: La théorie de la dégénérescence. In Postel J & Quétel C (eds) *Nouvelle histoire de la psychiatrie*. Paris: Privat, pp351–356, 1983.
113. Berrios G E & Hauser R: The early development of Kraepelin's ideas on classification: a conceptual history. *Psychological Medicine* 1988, 18: 813–821.
114. Berrios G E, Dening R G, & Dening T R: Hoche and the significance of symptom complexes in psychiatry. *History of Psychiatry* 1991, 2: 329–343.
115. Kretschmer E: Gedanken über die Fortentwicklung der psychiatrischen Systematik: Bemerkungen zu vorstehender Abhandlung. *Zeitschrift für die gesamte Neurologie und Psychiatrie* 1919, 48: 370–377.
116. Mayer W: Über paraphrene Psychosen. *Zeitschrift für die gesamte Neurologie und Psychiatrie* 1921, 71: 187–206.
117. Kraepelin E: Die Erscheinungsformen des Irreseins. *Zeitschrift für die gesamte Neurologie und Psychiatrie* 1920, 62: 1–29.
118. Leonhard K: *Classification of Endogenous Psychosis*. Wien: Springer, 1999.
119. Kumbier E & Herpertz S C: Helmut Rennert's universal genesis of endogenous psychosis: the historical concept and its significance for today's discussion on unitary psychosis. *Psychopathology* 2010, 43: 335–344.

3

DIMENSIONAL CONCEPTUALIZATION OF PSYCHOSIS

Kürşat Altınbaş, Sinan Guloksuz, and Jim van Os

INTRODUCTION

In this chapter, we review the literature on the conceptualization of psychosis spectrum disorder. After a brief summary of historical formulations of major psychoses, we outline findings from clinical and general populations that point to a dimensional framework of psychosis, and discuss advantages and disadvantages for the purpose of clinical practice and research.

THE ORIGINS OF THE CONCEPTUALIZATION OF PSYCHOSIS

Historically, diagnostic differentiation between major forms of psychosis ("manifestations of insanity"), currently framed as bipolar disorders and schizophrenia, started around the 19th century. In 1863, Karl Kahlbaum reported his comprehensive analysis of diagnostic nosology, showing clinician bias in diagnosis. In this seminal monograph, Kahlbaum discussed the advantages and disadvantages of a strictly symptom-oriented classification of psychiatric syndromes.[1] In the late 1800s, another German psychiatrist, Ewald Hecker, pointed out the inadequacies of a psychiatric nosology that disregards the lack of consistency between disease forms and symptom complexes.[1] Both psychiatrists agreed that psychiatric syndromes were temporary conditions that could be observed across different psychiatric conditions. In parallel to conventions in general medicine that different syndromes have characteristic symptoms and prognosis, the nosology of psychiatric disorders shifted the focus from symptom-based syndromes to disease entities. Thereafter, Emil Kraepelin, mainly inspired by Hecker and Kahlbaum's clinical methodology, introduced his psychiatric classification. In 1896, he proposed that hebephrenia, catatonia, and paranoid psychoses are distinct disease categories.[1] In the sixth version of his textbook, he formulated the classical dichotomous approach that dementia praecox (schizophrenia) and manic-depressive insanity (bipolar disorder) have distinct symptoms and course.[2] In contrast to Kraepelin's dichotomy, Eugen Bleuler proposed a spectrum model from latent schizotypy and schizophrenia.[3] According to Bleuler's clinical conceptualization, a patient could be either predominantly "schizophrenic" or predominantly "manic-depressive," while most were somewhere in between these extremes of the spectrum.

The *Diagnostic and Statistical Manual for Mental Disorders* (DSM), published by the American Psychiatric Association, starting with the third version, adopted the categorical classification that is compatible with the medical model: Each diagnostic category represents a distinct disease phenotype with clear-cut boundaries. The neo-Kraepelinian movement, disappointed by the low reliability of psychiatric diagnoses, led the reform with a bottom-up approach, first introducing the Research Diagnostic Criteria and Feighner Criteria for research purposes, then adopting these principles in a clinical diagnostic classification system.[4] Interestingly, the taxonomy of psychotic disorders in these classification systems was influenced not only by Kraepelin's view but also—arguably to a greater degree—by Schneider's first rank symptoms and Langfeldt's concept of poor outcome schizophrenia. The classification of psychotic disorders, with a particular emphasis on schizophrenia, has largely survived in subsequent versions of DSM and the International Classification of Diseases (ICD).

Although the categorical approach has been favored by clinicians from a utilitarian perspective (e.g., increased reliability, compatibility with general medicine, adaptability to clinical judgment), categorical taxonomy has limitations, such as difficulty in fitting patients with atypical symptoms into a specific disease category.[5] Further, diagnostic validity and stability around a hypothesized diagnostic point of rarity are low.[6,7] Also, importantly, links between specific symptoms, needs for care and outcomes on the one hand and the specific psychotic disorder category on the other are weak, reducing clinical utility.[8]

The early nosological reform largely also failed to meet expectations in research. Biological psychiatry's attempts to "reverse-engineer" DSM and ICD psychotic disorder disease entities have been unsuccessful, resulting in inconsistent findings and "paradoxically" providing support for overlap between diagnoses and widespread "continuity."[9,10]

PSYCHOTIC SYMPTOMS ACROSS TRADITIONAL DIAGNOSTIC CATEGORIES

Evidence from experimental, psychopathological, neurobiological, and genetic studies indicate overlapping symptoms, treatments, outcomes, and biological and genetic markers between psychotic disorder categories, suggesting a multidimensional spectrum encompassing nonaffective and affective psychosis. Clinical research indicates that bipolar disorder and schizophrenia can be usefully conceptualized as the distant ends of a multidimenional severity continuum, in which schizoaffective disorder lies at the midpoint.[11] Both core positive symptoms of nonaffective psychotic disorders (delusions and hallucinations) and negative symptoms cross diagnostic categories.[12,13] Reversely, there is remarkable variation in premorbid course, symptom profile, treatment and outcome of patients diagnosed with the same categorical diagnosis.[14,15] Several psychometric studies have revealed that the spectrum of major psychotic disorders might be modeled using five main symptom dimensions: mania, positive symptoms, disorganization, depression, and negative symptoms.[16–18] To determine the organizational structure of symptom dimensions at onset and its concordance with categorical diagnoses, Russo and colleagues cross-sectionally analyzed data of 500 patients with first episode psychosis.[19] Factor analyses revealed six first order symptom dimensions including mania, negative, disorganization, depression, hallucinations, and delusions and also two higher order factors as affective and nonaffective psychoses. In a follow-up longitudinal design, they examined the stability of the organizational structure of symptom dimensions after 5 to 10 years of follow-up among 100 of these patients, and showed that the factorial structure of symptoms during the first episode remained stable over the follow-up period.[19] Recent symptom-focused studies indicate a bifactor model, encompassing both a general dimension as well as five symptom dimensions (positive and negative symptoms, mania, depression, and disorganization).[20–23] Boks and colleagues found that the proportion of manic syndrome expression, thought to be specific to the bipolar disorder, was 35% in paranoid schizophrenia, 54% in schizoaffective disorders, and 61% in bipolar disorder type I,[24] replicating the findings of the classic Northwick Park study.[25]

Consistent with the clinical data presented thus far, behavioral genetic studies have failed to provide evidence for the traditional diagnostic separation of mood and psychotic disorders.[26,27] For example, a large family study of Swedish population among patients with schizophrenia and bipolar disorder including more than two million nuclear families has demonstrated that major psychoses share a common genetic cause.[28] More recent molecular genetic studies confirm considerable genetic overlap between categories of psychotic and nonpsychotic mental disorders.[29,30] Evidence from neuroimaging studies, albeit inherently less precise and methodologically much more challenging, and therefore difficult to interpret, is also compatible with the notion of a transdiagnostic manifestation of psychosis with differences and similarities across diagnostic categories.[31–34] Finally, reviews indicate that environmental exposures (e.g., cannabis use, childhood adversity, minority position) also are nonspecifically associated with psychotic disorders.[35]

Clinical relevance

Arguably the most important argument for a dimensional conceptualization is that research shows that dimensions have added clinical value over and above a categorical representation. Thus, dimensional "diagnosis" allows for a flexible, personalized, and dynamic way of charting a patient's symptoms as depicted in Fig. 3.1,[22,36] which has advantages over static categories that

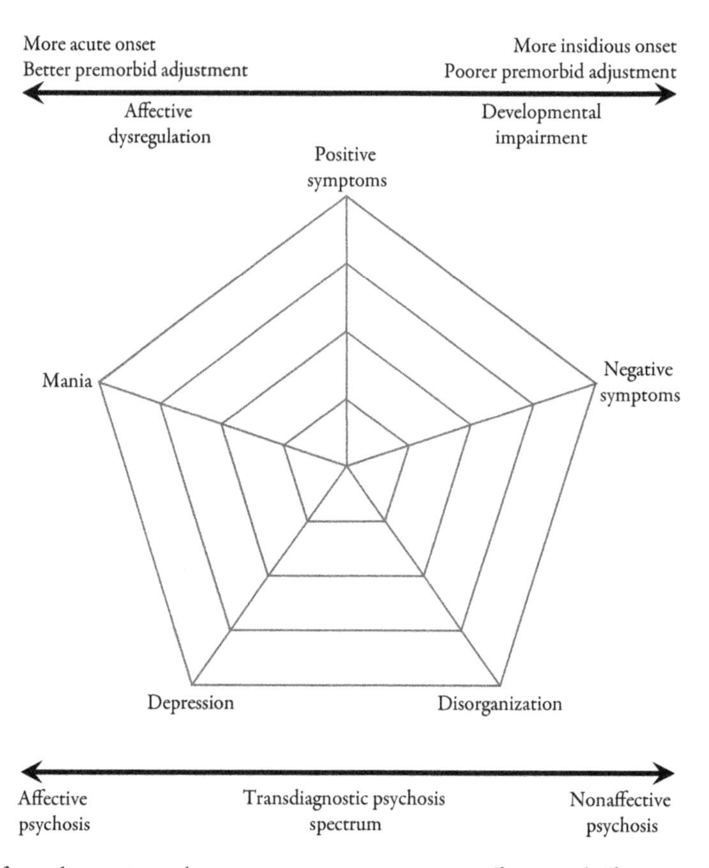

Figure 3.1 Schematic representation of transdiagnostic psychosis spectrum encompassing nonaffective and affective psychotic experiences.[22,36]

do not accurately convey personal psychopathology and cannot capture changes over time. Also, research indicates that dimensional information predicts course and outcome over and above the information conveyed by categorical diagnosis, which has to do with specific predictive associations of dimensions with onset, course, and other correlates, as summarized in Fig. 3.1.[37–40]

As outcome prediction represents one of the key functions of diagnosis in clinical medicine, this represents a cogent argument for the routine use of dimensional information in clinical practice in addition to categorical information.[8]

PSYCHOTIC PSYCHOPATHOLOGY ACROSS CLINICAL AND NONCLINICAL POPULATIONS

There is a growing body of data indicating psychotic experiences and associated codimensional expressions of affective, motivational, and cognitive variation[41–44] is also detectable at the level of the general population, compatible with the long-standing notion of "schizotypy" as a psychometric indicator of distributed psychosis proneness in the general population.[45] Meta-analyses of epidemiological data indicate that psychosis is phenomenologically, temporally, and etiologically continuous with the clinical syndrome, with prevalence rates of subthreshold psychotic symptoms varying between 5% and 15%, depending on age,[46–48] younger people having considerably higher rates than older cohorts.[49] Thus, expression of psychosis in the general population is associated with younger age, environmental risk factors for psychotic disorder, genetic risk associated with psychotic disorder,[50,51] and, over time, increased risk for transition to psychotic disorder.[46,47] Prevalence and incidence rates show considerable variation across sex, ethnic group, psychopathological context (higher rates being associated with more general psychopathology), and study methodology.[46,47] Similar to psychotic disorder, reviews indicate that the estimated median prevalence is much higher than median incidence, suggesting persistence in a considerable proportion of individuals. Nevertheless, for most, the experience of psychosis is transient and temporary, and likely includes hypnagogic or hypnopompic auditory perceptions, drug-induced states, misinterpretations of ambiguous noise, and periods of bereavement.[52–54] Persistence is not rare—systematic reviews estimate the yearly persistence rate at around 20%,[46,47] greater level of persistence predicting greater risk of transition to a clinical psychotic disorder.[55] The link between psychotic experiences and psychotic disorder can thus be summarized as a progression from psychosis-proneness to persistence to impairment and clinical transition. This relationship (risk persistence–clinical syndrome) is compatible with the observation that the rate of psychotic experiences reported from population-based studies is around 50–100 times higher than prevalence and incidence of schizophrenia spectrum disorders, thus likely representing the underlying expression of distributed genetic and nongenetic risk in the general population, confirming the notion of "schizotypy."[36]

Clinical relevance

Although framed as risk indicators, research indicates that psychotic experiences in the general adult population are not clinically neutral. In fact, most of the expression of psychotic experiences occurs, consistently across studies, in the context of a nonpsychotic disorder.[56–59] The coexpression of psychotic experiences with nonpsychotic psychopathology is not neutral, as copresence of a psychotic experience in the context of a nonpsychotic disorder is associated with clinical severity, copresence of other symptoms, functioning and outcome.[41,42,60–65] In line with this, greater level of psychotic experiences in nonpsychotic disorders is associated with more exposure to environmental risk factors like childhood adversity and more exposure to indicators of genetic risk.[66–68] Psychotic experiences, therefore, can be conceptualized as indicators of severity in nonpsychotic disorders.[36]

The occurrence and course of psychotic experiences can be summarized within a developmental framework as embedded in the psychosis proneness-persistence-impairment model, depicted in Fig. 3.2. Psychosis proneness represents distributed

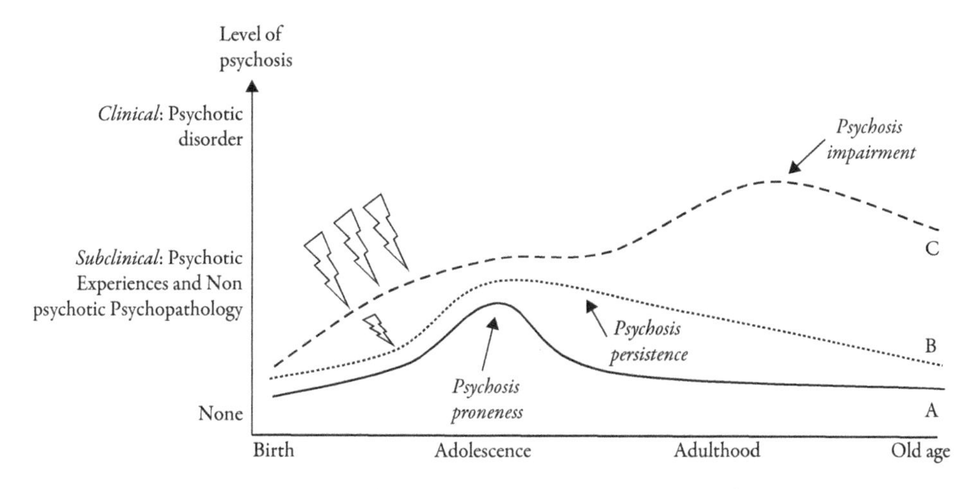

Figure 3.2 Psychosis proneness-persistence-impairment model of psychotic disorder. Person A has "normal" developmental expression of subclinical psychotic experiences (*psychosis proneness*) that are transient. Person B has similar expression, but longer *persistence* due to additional but mild environmental exposure and in the context of nonpsychotic psychopathology. Person C has longer persistence due to severe repeated environmental exposure and transition to clinical psychotic disorder with significant *impairment*.

genetic and nongenetic risk in the general population, associated with a degree of subthreshold expression of psychosis which, with increasing levels of developmental environmental exposure, takes on clinical relevance in, first, expression of psychotic experiences and multidimensional psychopathology at the level of nonpsychotic disorders and, finally, transition to full-blown psychotic disorder.

The psychosis proneness-persistence-impairment model implies an underlying mechanism of gene–environment interaction between distributed polygenic risk and developmental environmental exposures. This model was examined recently in a large 25-country study, funded by the European Union (EUGEI study[69]). The results showed independent and joint effects of molecular genetic liability (polygenic risk for schizophrenia) and environmental exposures (cannabis, childhood adversity) in psychosis spectrum disorder,[70] confirming the hypothesis.

A DIMENSIONAL PSYCHOSIS SPECTRUM SYNDROME: CURRENT APPROACHES

Taken together, clinical, genetic, epidemiological, and neurobiological findings indicate that there is overlap between mood and psychotic disorders, and overlap between clinical and nonclinical expressions of psychosis-associated psychopathology, challenging the classical categorization of major psychotic disorders represented in current diagnostic systems such as DSM and ICD. A spectrum approach in psychosis may be more productive. For example, a spectrum approach was introduced in DSM-5 for autism, such that "the symptoms of people with Autism Spectrum Disorder will fall on a continuum, with some individuals showing mild symptoms and others having much more severe symptoms."[71] Various alternative models have been proposed for psychotic disorders,[72–76] including a proposal for a psychosis spectrum disorder.[75,76] The combination of adjustable broad "spectrum" categories may facilitate a patient-centered transdiagnostic framework that embraces a dimensional approach, maximizing the usefulness of psychopathology for research and clinical practice.[77]

The Psychotic Disorders Workgroup of DSM-5 discussed introducing a dimensional approach embedded in the chapter of schizophrenia spectrum disorders by changing the order and structure of the psychotic disorders chapter using duration, severity, and number of psychotic symptoms to provide a more valid diagnostic classification.[78] Eventually, DSM-5 eliminated previously defined subtypes of schizophrenia and partially embraced a spectrum approach within a chapter titled "Schizophrenia Spectrum and Other Psychotic Disorders." However, the dimensional assessment of psychosis was only offered in Section III: "Emerging Measures and Models." The National Institute of Mental Health proposed the Research Domain Criteria (RDoC), allowing investigators to avoid having to direct research efforts along the lines of a restrictive categorical approach that might represent a pragmatic and viable option for clinical practice but might not align with underlying mechanisms.[79] Although RDoC was originally designed for research, a recent study extracting information from electronic health records showed that the RDoC may provide clinically relevant information that has already been documented by clinicians.[80] More recently, the Hierarchical Taxonomy of Psychopathology (HiTOP) consortium formulated a model that organizes psychopathology using six hierarchical layers of dimensional spectra based on evidence from statistical modeling of psychopathology.[81] Another possible representation of psychotic psychopathology is based on symptom networks, which has embedded in it the conceptually attractive and clinically valid paradigm that symptoms may cause other symptoms (for example social anxiety may cause paranoia, or low mood may cause nihilistic delusions).[82,83] The network model also can very easily accommodate spectrum features in the sense of clinical and subclinical severities of symptoms[67,84] and be extended to incorporate the impact of environmental[85] and genetic risks.[86] All these approaches are not mutually exclusive—the advantage of dimensional representations of psychopathology is that they can be flexibly combined with other representations, depending on the context in research or clinical practice.

In conclusion, there is evidence that both research and clinical practice will benefit from a transdiagnostic dimensional formulation that may lay the groundwork for an improved approach toward classification. Continuous dimensions allow for a more dynamic, personal, and precise assessment of psychopathology[87] and provide an improved framework to dissect cross-categorical pathoetiological processes. Similar to autism, a broad "umbrella" syndrome may be introduced, such as psychosis spectrum disorder,[76] which can serve as the starting point for a detailed multidimensional assessment in clinical practice and may also allow for the inclusion of clinically relevant subthreshold expressions of psychosis.

REFERENCES

1. Kendler KS, Engstrom EJ. Kahlbaum, Hecker, and Kraepelin and the transition from psychiatric symptom complexes to empirical disease forms. *Am J Psychiatry*. 2016;174(2):102–109.
2. Healy D, Harris M, Farquhar F, Tschinkel S, Le Noury J. Historical overview: Kraepelin's impact on psychiatry. *Eur Arch Psychiatry Clin Neurosci*. 2008;258(2):18.
3. Jablensky A. The diagnostic concept of schizophrenia: its history, evolution, and future prospects. *Dialogues Clin Neurosci*. 2010;12(3):271.
4. Dutta R, Greene T, Addington J, McKenzie K, Phillips M, Murray RM. Biological, life course, and cross-cultural studies all point toward the value of dimensional and developmental ratings in the classification of psychosis. *Schizophr Bull*. 2007;33(4):868–876.
5. Craddock N, Owen MJ. Rethinking psychosis: the disadvantages of a dichotomous classification now outweigh the advantages. *World Psychiatry*. 2007;6(2):84.
6. Frances AJ, Widiger T. Psychiatric diagnosis: lessons from the DSM-IV past and cautions for the DSM-5 future. *Annu Rev Clin Psychol*. 2012;8:109–130.
7. Fusar-Poli P, Cappucciati M, Rutigliano G, et al. Diagnostic stability of ICD/DSM first episode psychosis diagnoses: meta-analysis. *Schizophr Bull*. 2016;42(6):1395–1406.
8. van Os J, Kapur S. Schizophrenia. *Lancet*. 2009;374(9690):635–645.
9. Button KS, Ioannidis JP, Mokrysz C, et al. Power failure: why small sample size undermines the reliability of neuroscience. *Nat Rev Neuroscience*. 2013;14(5):365.

10. Kapur S, Phillips AG, Insel TR. Why has it taken so long for biological psychiatry to develop clinical tests and what to do about it? *Mol Psychiatry.* 2012;17(12):1174.
11. Mancuso SG, Morgan VA, Mitchell PB, Berk M, Young A, Castle DJ. A comparison of schizophrenia, schizoaffective disorder, and bipolar disorder: results from the second Australian national psychosis survey. *J Affect Disord.* 2015;172:30–37.
12. Van Os J, Gilvarry C, Bale R, et al. Diagnostic value of the DSM and ICD categories of psychosis: an evidence-based approach. *Soc Psychiatry Psychiatr Epidemiol.* 2000;35(7):305–311.
13. Strauss GP, Cohen AS. A transdiagnostic review of negative symptom phenomenology and etiology. *Schizophr Bull.* 2017;43(4):712–719.
14. Jablensky A. Subtyping schizophrenia: implications for genetic research. *Mol Psychiatry.* 2006;11(9):815.
15. Stroup TS. Heterogeneity of treatment effects in schizophrenia. *Am J Med.* 2007;120(4):S26–S31.
16. Allardyce J, Suppes T, Van Os J. Dimensions and the psychosis phenotype. *Int J Methods Psychiatr Res.* 2007;16(S1):S34–S40.
17. Tandon R. Schizophrenia and other psychotic disorders in DSM-5: clinical implications of revisions from DSM-IV. *Clin Schizophr Relat Psychoses.* 2013;7(1):16–19.
18. Potuzak M, Ravichandran C, Lewandowski KE, Ongür D, Cohen BM. Categorical vs dimensional classifications of psychotic disorders. *Compr Psychiatry.* 2012;53(8):1118–1129.
19. Russo M, Levine SZ, Demjaha A, et al. Association between symptom dimensions and categorical diagnoses of psychosis: a cross-sectional and longitudinal investigation. *Schizophr Bull.* 2013;40(1):111–119.
20. Quattrone D, Di Forti M, Gayer-Anderson C, et al. Transdiagnostic dimensions of psychopathology at first episode psychosis: findings from the multinational EU-GEI study. *Psychol Med.* 2018:1–14.
21. Reininghaus U, Böhnke JR, Chavez-Baldini U, et al. Transdiagnostic dimensions of psychosis in the Bipolar-Schizophrenia Network on Intermediate Phenotypes (B-SNIP). *World Psychiatry.* 2019;18(1):67–76.
22. Reininghaus U, Bohnke JR, Hosang G, et al. Evaluation of the validity and utility of a transdiagnostic psychosis dimension encompassing schizophrenia and bipolar disorder. *British J Psychiatry.* 2016;209(2):107–113.
23. Reininghaus U, Priebe S, Bentall RP. Testing the psychopathology of psychosis: evidence for a general psychosis dimension. *Schizophr Bull.* 2013;39(4):884–895.
24. Boks MP, Leask S, Vermunt JK, Kahn RS. The structure of psychosis revisited: the role of mood symptoms. *Schizophr Res.* 2007;93(1–3):178–185.
25. Johnstone EC, Crow TJ, Frith CD, Owens DG. The Northwick Park "functional" psychosis study: diagnosis and treatment response. *Lancet.* 1988;2(8603):119–125.
26. Kendler KS, Neale MC, Walsh D. Evaluating the spectrum concept of schizophrenia in the Roscommon Family Study. *Am J Psychiatry.* 1995;152(5):749–754.
27. Cardno AG, Rijsdijk FV, Sham PC, Murray RM, McGuffin P. A twin study of genetic relationships between psychotic symptoms. *Am J Psychiatry.* 2002;159(4):539–545.
28. Lichtenstein P, Yip BH, Björk C, et al. Common genetic determinants of schizophrenia and bipolar disorder in Swedish families: a population-based study. *Lancet.* 2009;373(9659):234–239.
29. Gandal MJ, Haney JR, Parikshak NN, et al. Shared molecular neuropathology across major psychiatric disorders parallels polygenic overlap. *Science.* 2018;359(6376):693–697.
30. Cross-Disorder Group of the Psychiatric Genomics Consortium, Lee SH, Ripke S, et al. Genetic relationship between five psychiatric disorders estimated from genome-wide SNPs. *Nat Genet.* 2013;45(9):984–994.
31. Pearlson GD, Clementz BA, Sweeney JA, Keshavan MS, Tamminga CA. Does biology transcend the symptom-based boundaries of psychosis? *Psychiatr Clin North Am.* 2016;39(2):165–174.
32. Ho NF, Li Hui Chong P, Lee DR, Chew QH, Chen G, Sim K. The amygdala in schizophrenia and bipolar disorder: a synthesis of structural MRI, diffusion tensor imaging, and resting-state functional connectivity findings. *Harv Rev Psychiatry.* 2019;27(3):150–164.
33. Haukvik UK, Tamnes CK, Soderman E, Agartz I. Neuroimaging hippocampal subfields in schizophrenia and bipolar disorder: a systematic review and meta-analysis. *J Psychiatr Res.* 2018;104:217–226.
34. Birur B, Kraguljac NV, Shelton RC, Lahti AC. Brain structure, function, and neurochemistry in schizophrenia and bipolar disorder-a systematic review of the magnetic resonance neuroimaging literature. *NPJ Schizophr.* 2017;3:15.
35. Guloksuz S, van Os J, Rutten BPF. The exposome paradigm and the complexities of environmental research in psychiatry. *JAMA Psychiatry.* 2018;75(10):985–986.
36. van Os J, Reininghaus U. Psychosis as a transdiagnostic and extended phenotype in the general population. *World Psychiatry.* 2016;15:118–124.
37. Van Os J, Fahy TA, Jones P, et al. Psychopathological syndromes in the functional psychoses: associations with course and outcome. *Psychol Med.* 1996;26(1):161–176.
38. Van Os J, Gilvarry C, Bale R, et al. A comparison of the utility of dimensional and categorical representations of psychosis. UK700 Group [In Process Citation]. *Psychol Med.* 1999;29(3):595–606.
39. Van Os J, Gilvarry C, Bale R, et al. To what extent does symptomatic improvement result in better outcome in psychotic illness? UK700 Group. *Psychol Med.* 1999;29(5):1183–1195.
40. Ajnakina O, Lally J, Di Forti M, et al. Utilising symptom dimensions with diagnostic categories improves prediction of time to first remission in first-episode psychosis. *Schizophr Res.* 2018;193:391–398.
41. Pries LK, Guloksuz S, Ten Have M, et al. Evidence that environmental and familial risks for psychosis additively impact a multidimensional subthreshold psychosis syndrome. *Schizophr Bull.* 2018.
42. van Rossum I, Dominguez MD, Lieb R, Wittchen HU, van Os J. Affective dysregulation and reality distortion: a 10-year prospective study of their association and clinical relevance. *Schizophr Bull.* 2011;37(3):561–571.
43. Dominguez MD, Saka MC, Lieb R, Wittchen HU, van Os J. Early expression of negative/disorganized symptoms predicting psychotic experiences and subsequent clinical psychosis: a 10-year study. *Am J Psychiatry.* 2010;167(9):1075–1082.
44. Werbeloff N, Dohrenwend BP, Yoffe R, van Os J, Davidson M, Weiser M. The association between negative symptoms, psychotic experiences and later schizophrenia: a population-based longitudinal study. *PloS One.* 2015;10(3):e0119852.
45. Kendler KS, McGuire M, Gruenberg AM, Walsh D. Schizotypal symptoms and signs in the Roscommon Family Study: their factor structure and familial relationship with psychotic and affective disorders. *Arch Gen Psychiatry.* 1995;52(4):296–303.
46. Van Os J, Linscott RJ, Myin-Germeys I, Delespaul P, Krabbendam L. A systematic review and meta-analysis of the psychosis continuum: evidence for a psychosis proneness–persistence–impairment model of psychotic disorder. *Psychol Med.* 2009;39(2):179–195.
47. Linscott R, Van Os J. An updated and conservative systematic review and meta-analysis of epidemiological evidence on psychotic experiences in children and adults: on the pathway from proneness to persistence to dimensional expression across mental disorders. *Psychol Med.* 2013;43(6):1133–1149.
48. Kelleher I, Connor D, Clarke MC, Devlin N, Harley M, Cannon M. Prevalence of psychotic symptoms in childhood and adolescence: a systematic review and meta-analysis of population-based studies. *Psychol Med.* 2012;42(9):1857–1863.
49. Verdoux H, Van Os J, Maurice-Tison S, Gay B, Salamon R, Bourgeois M. Is early adulthood a critical developmental stage for psychosis proneness? A survey of delusional ideation in normal subjects. *Schizophr Res.* 1998;29(3):247–254.
50. Jeppesen P, Larsen JT, Clemmensen L, et al. The CCC2000 birth cohort study of register-based family history of mental disorders and psychotic experiences in offspring. *Schizophr Bull.* 2015;41(5):1084–1094.
51. van Os J, van der Steen Y, Islam MA, et al. Evidence that polygenic risk for psychotic disorder is expressed in the domain of

neurodevelopment, emotion regulation and attribution of salience. *Psychol Med.* 2017;47(14):2421–2437.
52. Beavan V, Read J, Cartwright C. The prevalence of voice-hearers in the general population: a literature review. *Journal of Mental Health.* 2011;20(3):281–292.
53. Baumeister D, Sedgwick O, Howes O, Peters E. Auditory verbal hallucinations and continuum models of psychosis: a systematic review of the healthy voice-hearer literature. *Clin Psychol Rev.* 2017;51:125–141.
54. de Leede-Smith S, Barkus E. A comprehensive review of auditory verbal hallucinations: lifetime prevalence, correlates and mechanisms in healthy and clinical individuals. *Front Hum Neurosci.* 2013;7:367.
55. Dominguez MD, Wichers M, Lieb R, Wittchen HU, van Os J. Evidence that onset of clinical psychosis is an outcome of progressively more persistent subclinical psychotic experiences: an 8-year cohort study. *Schizophr Bull.* 2011;37(1):84–93.
56. Van Os J, Hanssen M, Bijl R, Ravelli A. Strauss (1969) revisited: a psychosis continuum in the general population? *Schizophr Res.* 2000;45(1–2):11–20.
57. Hanssen M, Peeters F, Krabbendam L, Radstake S, Verdoux H, van Os J. How psychotic are individuals with non-psychotic disorders? *Soc Psychiatry Psychiatr Epidemiol.* 2003;38(3):149–154.
58. Jeppesen P, Clemmensen L, Munkholm A, et al. Psychotic experiences co-occur with sleep problems, negative affect and mental disorders in preadolescence. *J Child Psychol Psychiatry.* 2015;56(5):558–565.
59. McGrath JJ, Saha S, Al-Hamzawi A, et al. Psychotic experiences in the general population: a cross-national analysis based on 31,261 respondents from 18 countries. *JAMA Psychiatry.* 2015;72(7):697–705.
60. Wigman JT, van Nierop M, Vollebergh WA, et al. Evidence that psychotic symptoms are prevalent in disorders of anxiety and depression, impacting on illness onset, risk, and severity—implications for diagnosis and ultra-high risk research. *Schizophr Bull.* 2012;38(2):247–257.
61. van Nierop M, Viechtbauer W, Gunther N, et al. Childhood trauma is associated with a specific admixture of affective, anxiety, and psychosis symptoms cutting across traditional diagnostic boundaries. *Psychol Med.* 2015;45(6):1277–1288.
62. Kaymaz N, van Os J, de Graaf R, Ten Have M, Nolen W, Krabbendam L. The impact of subclinical psychosis on the transition from subclinicial mania to bipolar disorder. *J Affect Disord.* 2007;98(1-2):55–64.
63. Wigman JT, van Os J, Abidi L, et al. Subclinical psychotic experiences and bipolar spectrum features in depression: association with outcome of psychotherapy. *Psychol Med.* 2014;44(2):325–336.
64. Perlis RH, Uher R, Ostacher M, et al. Association between bipolar spectrum features and treatment outcomes in outpatients with major depressive disorder. *Arch Gen Psychiatry.* 2011;68(4):351–360.
65. Saha S, Scott JG, Varghese D, McGrath JJ. The association between general psychological distress and delusional-like experiences: a large population-based study. *Schizophr Res.* 2011;127(1-3):246–251.
66. Guloksuz S, van Nierop M, Lieb R, van Winkel R, Wittchen HU, van Os J. Evidence that the presence of psychosis in non-psychotic disorder is environment-dependent and mediated by severity of non-psychotic psychopathology. *Psychol Med.* 2015;45(11):2389–2401.
67. Guloksuz S, van Nierop M, Bak M, et al. Exposure to environmental factors increases connectivity between symptom domains in the psychopathology network. *BMC Psychiatry.* 2016;16:223.
68. Smeets F, Lataster T, Viechtbauer W, Delespaul P, G.R.O.U.P. Evidence that environmental and genetic risks for psychotic disorder may operate by impacting on connections between core symptoms of perceptual alteration and delusional ideation. *Schizophr Bull.* 2015;41(3):687–697.
69. European Network of National Networks studying Gene–Environment Interactions in Schizophrenia, van Os J, Rutten BP, et al. Identifying gene–environment interactions in schizophrenia: contemporary challenges for integrated, large-scale investigations. *Schizophr Bull.* 2014;40(4):729–736.
70. Guloksuz S, Pries LK, Delespaul P, et al. Examining the independent and joint effects of molecular genetic liability and environmental exposures in schizophrenia: results from the EUGEI study. *World Psychiatry.* 2019;18(2):173–182.
71. American Psychiatric Association. *Diagnostic and statistical manual of mental disorders.* 5th ed. [computer program]. Arlington, VA: American Psychiatric Publishing, 2013.
72. Craddock N, Owen MJ. The Kraepelinian dichotomy—going, going . . . but still not gone. *Br J Psychiatry.* 2010;196:92–95.
73. Altinbas K, Smith DJ, Craddock N. Rediscovering the bipolar spectrum/Bipolar spektrumunun yeniden kesfi. *Noro-Psikyatri Arsivi.* 2011;48(3):167.
74. Van Os J. "Salience syndrome" replaces "schizophrenia" in DSM-V and ICD-11: psychiatry's evidence-based entry into the 21st century? *Acta Psychiatr Scand.* 2009;120(5):363–372.
75. Keshavan MS, Nasrallah HA, Tandon R. Schizophrenia, "Just the Facts" 6. Moving ahead with the schizophrenia concept: from the elephant to the mouse. *Schizophr Res.* 2011;127(1–3):3–13.
76. Guloksuz S, Van Os J. The slow death of the concept of schizophrenia and the painful birth of the psychosis spectrum. *Psychol Med.* 2018;48(2):229.
77. van Os J, Guloksuz S, Vijn TW, Hafkenscheid A, Delespaul P. The evidence-based group-level symptom-reduction model as the organizing principle for mental health care: time for change? *World Psychiatry.* 2019;18(1):88–96.
78. Heckers S, Barch DM, Bustillo J, et al. Structure of the psychotic disorders classification in DSM-5. *Schizophr Res.* 2013;150(1):11–14.
79. Cuthbert BN, Insel TR. Toward new approaches to psychotic disorders: the NIMH Research Domain Criteria project. In: Oxford University Press; 2010.
80. McCoy TH, Castro VM, Rosenfield HR, Cagan A, Kohane IS, Perlis RH. A clinical perspective on the relevance of Research Domain Criteria in electronic health records. *Am J Psychiatry.* 2015;172(4):316–320.
81. Kotov R, Krueger RF, Watson D, et al. The Hierarchical Taxonomy of Psychopathology (HiTOP): a dimensional alternative to traditional nosologies. *J Abnorm Psychol.* 2017;126(4):454–477.
82. Borsboom D, Cramer AO, Schmittmann VD, Epskamp S, Waldorp LJ. The small world of psychopathology. *PLoS One.* 2011;6(11):e27407.
83. Guloksuz S, Pries LK, van Os J. Application of network methods for understanding mental disorders: pitfalls and promise. *Psychol Med.* 2017:1–10.
84. van Os J, Lataster T, Delespaul P, Wichers M, Myin-Germeys I. Evidence that a psychopathology interactome has diagnostic value, predicting clinical needs: an experience sampling study. *PLoS One.* 2014;9(1):e86652.
85. Isvoranu AM, Borsboom D, van Os J, Guloksuz S. A network approach to environmental impact in psychotic disorder: brief theoretical framework. *Schizophr Bull.* 2016;42(4):870–873.
86. Isvoranu AM, Guloksuz S, Epskamp S, van Os J, Borsboom D, Investigators G. Toward incorporating genetic risk scores into symptom networks of psychosis. *Psychol Med.* 2019:1–8.
87. Kraemer HC. DSM categories and dimensions in clinical and research contexts. *Int J Methods Psychiatr Res.* 2007;16(Suppl 1):S8–S15.

TRANSDIAGNOSTIC DIMENSIONS OF PSYCHOSIS

4

APPLYING RESEARCH DOMAIN CRITERIA (RDOC) DIMENSIONS TO PSYCHOSIS

Sarah E. Morris, Jennifer Pacheco, and Charles A. Sanislow

INTRODUCTION

Debates over categorical versus dimensional approaches across all areas of psychopathology are not new but have received greater attention over the past few decades.[1,2] This attention has tended to focus more on certain DSM-defined disorders (e.g., personality disorders) rather than others (e.g., psychotic disorders).[3] In fact, in the lead-up to the publication of the most recent edition of the Diagnostic and Statistical Manual of Mental Disorders (DSM-5),[4] personality disorders were argued to be a test case for dimensional approaches that might one day be extended more broadly to clinical syndromes.[5] When confronted with the prospect of thinking about severe mental illness in terms of gradations, researchers and clinicians alike might find the task daunting at first blush. On the other hand, when studying or treating familiar components of clinical syndromes, the prospect might seem more inviting. Clinicians routinely measure treatment gains in dimensional terms, as researchers who focus on intermediate phenotypes are aware that dichotomies of normality and pathology are largely constructed.

For disorders such as depression and anxiety, it is easy for clinicians and researchers to imagine gradation of experiences such as sadness or worry. It may be that thinking about the symptoms of clinical syndromes such as "schizophrenia" collectively invites a categorical division of wellness and illness simply because the daunting nature of the clinical expression of the pathology is so far removed from usual and familiar human experiences. However, it has been the case for some time that dimensional conceptions of psychotic disorders have empirical merit.[6] We argue that the syndromal approach to defining research agendas for mental illness has constrained progress by obscuring dimensional properties of clinical symptoms and their associated internal mechanisms.

In this chapter, we detail the National Institute of Mental Health (NIMH) Research Domain Criteria (RDoC) framework, an approach that is dimensional by definition. After describing the development and framework of RDoC, we consider ways that it may help advance research into psychotic disorders. We argue that research based on intermediate phenotypes, rather than studies constrained by categorical syndromes that collectively encompass variants of psychopathological systems, recognizes the dimensional nature of many sorts of mechanisms that help to account for clinical suffering involved with psychotic-spectrum psychopathology (e.g., thought disorder, delusions, and hallucinations).

BACKGROUND OF RDOC

The National Institute of Mental Health launched the RDoC initiative in 2009 in response to a scientific objective in the 2008 NIMH Strategic Plan to develop new ways of classifying mental illnesses based on dimensions of observable behavioral and neurobiological measures. This goal emerged from the realization that, in contrast to most other medical specialties and despite enormous recent advances in neuroscientific methods, diagnostic tools in psychiatry remain restricted to subjective symptoms and observable signs, and that there were essentially no laboratory tests—biological or behavioral—to validate clinical diagnoses. Moreover, the current approach using diagnostic syndromes defined by lists of clinically observed symptoms invited the assumption that disorders were best seen as categorial. Despite decades of clinical neuroscience research, there have been few advances in our understanding of etiology, detection of reliable biomarkers, or development of novel therapies that relate to categorically defined clinical syndromes. Although diagnostic "categories" reflect commonly occurring symptom groupings, variation in the features of these diagnoses has hindered research advances.[7]

The inherent problems of diagnoses based on syndromes defined by symptoms include heterogeneity, comorbidity, and overspecification. These were part of the motivation for NIMH to unharness clinical research from a categorical framework that implied a bright line separating health from illness. Admittedly, syndromal diagnostic criteria provided an important tool for psychiatric treatment by clarifying "caseness" of collective disease processes. It also allowed reliable categorization of patients that led to advances over roughly the past four decades. It does seem, though, that it is among the most seemingly intractable forms of psychopathology that progress has been limited. In contrast to psychotic disorders, much progress has been in understanding the mechanisms and developing treatments for anxiety disorders, for instance. For the latter, the relative simplicity of psychopathological processes and the potential to model internal mechanisms with animal research no doubt offered a leg up relative to psychosis. Following that lead, RDoC provides a framework to parse dimensions of functioning and to clarify when a failure of such an identified mechanism corresponds to human suffering. From there, clinical problems can be redefined based on what is not working properly in the individual rather than manifest observation of a failed mechanism.

For those working in the competitive arena for funding, a need for an alternative approach was apparent. Diagnostic categories implied by clinical syndromes (i.e., DSM/ICD) had taken a strong hold on the research enterprise such that the entire system of grant applications, clinical trials, journals, pharmaceutical development, and regulations was harnessed to diagnostic categories. Studying psychopathology so defined resulted in restricting the range of variance in ways that obscured aspects of mechanisms that might more completely explain a clinical problem, or even obscure a compensatory process that interacted with health and illness. In peer review for scientific merit, there was little room to explore alternative approaches to mainstream classification, leading, in essence, to reification of clinically observed syndromes for which associated mechanisms were less clear.[8] This, in turn, hindered progress for treatment development, as evidenced by a noticeable decline in research for new pharmacologic agents.[9] The RDoC was introduced to permit investigators to think beyond the diagnostic boundaries and to encourage them to reach across categories by studying relevant constructs not necessarily specific to a single DSM disorder. In so doing, gains in integrative neuroscience could be brought to bear on improving the understanding and treatment of mental illness as the neurobiological mechanisms relevant to mental disorders could be better clarified when not constrained by a categorically defined syndrome.

The RDoC is a research framework and not intended to be used for clinical diagnosis. It was never intended to be an alternative or a replacement for the DSM. For clinical care and communication, and for billing and insurance, the DSM and ICD remain the diagnostic standards. The RDoC should be viewed as independent of existing diagnostic approaches; a helpful way to guide researchers when designing their studies. Ultimately, the new knowledge that emerges from careful and rigorous research—using the RDoC framework or other alternative classification approaches—will inform diagnostic practices and aid in developing more effective treatments, but whether those changes will be minor adjustments or significant reconceptualizations remains to be seen.

NIMH intends for the RDoC to serve as a framework for research in which fundamental neurobehavioral dimensions that cut across traditional disorder categories are used as the basis for characterizing and classifying patients in clinical studies. The RDoC framework is designed to focus on the primary neural circuits that support human behavior and to integrate many levels of measuring and understanding information (from genomics to self-report) to better understand the basic dimensions of functioning corresponding to the full range of human behavior. While diagnostic categories are needed in their current form for many things, RDoC is a focused research initiative with the goal of modifying and informing a new classification system in the future. Its success will be achieved through rigorous study and validation of transdiagnostic dimensional constructs.

PSYCHOSIS DIMENSIONS BEFORE RDOC

As is evident throughout this textbook, dimensional approaches for psychosis are not new. As far back as 1965, motivated by the opinion that "exclusive adherence to nosology has not served biological psychiatry well . . . (and) . . . has been a factor stunting its growth," researchers encouraged a "functional psychopathology," in which rigid categories were set aside and symptoms were considered in terms of their "elemental constituents," such as variation in disturbances in perception, memory, and motivation.[9] Other researchers focusing more squarely on clusters of symptoms on which clinical syndromes have been based used multivariate statistical methods that suggested various dimensional components ranging in number.[10]

Perhaps one of the more notable breaks from categories dates back to work by Strauss, Carpenter, and Bartko[6] in the lead-up to the DSM-III. They broke the schizophrenia construct into six domains of psychosis that were relevant to schizophrenia, but that could also relate to psychotic processes not necessarily connected to schizophrenia. Their work led to the positive-negative symptom distinction, and one might speculate that the fervor with which the positive-negative symptom dimension was embraced might have obscured the broader idea of cross-cutting dimensions. Even so, the idea of cross-cutting dimensions persisted to the DSM-IV,[11] which proposed three dimensions, *Psychotic*, *Disorganized*, and *Negative*, that were included in the "Alternative Dimensional Descriptors for Schizophrenia" (pp. 710–711).

However, one problem with these studies—as well as factor analytic work—is that the starting point is, as with DSM-III onward, based on symptoms that stem from syndromes defined by clinical observation. In part, it may be that the degree of granularity is not quite right. Especially for the kinds of complex behaviors seen in psychotic-spectrum disorders, which stem from behavioral observations, corresponding mechanisms are less likely to be proximally connected. That said, reorganizing symptom-level data using structural approaches have been helpful to break from the categories of the DSM, for instance internalizing-externalizing dimensions.[12] While these approaches may optimize the organization of syndrome-derived symptoms, there may still be difficulty connecting these to internal mechanisms.[13] The psychoneurometric approach offers one model to connect more complex phenotypes in the "Self-Report" units of the RDoC Matrix with neural mechanisms.[14] One of the motivating forces for RDoC was the need for an organizing structure that was not limited to any of these alternative approaches, but rather provided a means to incorporate the progress that has been made in these areas.

DEVELOPMENT OF THE RDOC FRAMEWORK

The crux of the RDoC initiative is the RDoC matrix, which is offered as a research framework intended to break down psychiatric illness into fundamental behavioral, psychological, and biological processes, and to encourage psychopathology research that incorporates advances in behavioral neuroscience.[15,16] The matrix is composed of higher-level *Domains* of functioning, reflecting contemporary knowledge about major systems of cognition, motivation, and social behavior. The *rows* of the matrix represent specified functional concepts summarizing data about a specified dimension of behavior (*Constructs*), which are characterized in aggregate by the *Units of Analysis* in *columns* that intersect with the row of constructs. In the current iteration of the RDoC matrix, there are five Domains: Negative Valence Systems, Positive Valence Systems, Cognitive Systems, Systems for Social Processes, and Arousal/Regulatory Systems. The Units of Analysis are genes, molecules, cells, circuits, physiology, behavior, and self-reports. The matrix also has a separate column to specify well-validated paradigms used in studying each construct.

The RDoC matrix is intended as a framework to help organize research hypotheses and findings as the field explores novel approaches to psychiatric classification. The RDoC approach is not prescriptive or mandatory for NIMH funding. Research that focuses on the diagnostic categories based on clinical description (i.e., DSM or ICD) continues to be programmatically supported and funded by NIMH. That said, there is an expectation that contemporary clinical research should account for heterogeneity of such diagnoses. The RDoC offers researchers one way to embrace and try to explain the heterogeneity, but it is not necessarily the only way (and not the only dimensional approach supported by NIMH funding). The Consortium of Genetics of Schizophrenia (COGS),[17] the Bipolar-Schizophrenia Network for Intermediate Phenotypes (B-SNIP),[18] and Systems Neuroscience of Psychosis (SyNoPsis)[19] projects provide prime examples of efforts focused on shifting away from existing diagnostic categories to explore other approaches to classification. Of note, these projects do not necessarily shy away from characterizing research participants using DSM diagnoses, but rather follow a fundamental tenet of the RDoC approach; namely, to not constrain dimensional variance by narrowly focusing on a category. The B-SNIP project, for instance, includes patients diagnosed with schizophrenia, schizoaffective disorder, and bipolar disorder.

UNPACKING COMPLEX PSYCHOSIS PHENOTYPES

As should be evident, the RDoC is a framework that allows researchers to unharness their hypotheses from the constraints imposed by clinical syndromes and categorical diagnoses, yet still provide a way to organize findings so that researchers can communicate with one another using a shared vocabulary and advance scientific discovery. Despite the growing pains of thinking outside the diagnostic box, an extant body of existing work related to dimensional psychopathology, intermediate phenotypes, and neural mechanisms existed before the advent of the RDoC. The RDoC merely offers a framework to organize findings from this kind of research, and to relate these findings to clinical impairment. For RdoC-framed research on psychosis dimensions, we offer the following principles.

Principle 1: Focus on narrow aspects of psychopathology, rather than polythetic diagnoses: The scientific challenges presented by diagnostic classifications based on collections of symptoms are well described in prior work (e.g.,[8,20,21]) and exemplified by Galatzer-Levy and Bryant's calculation that there are more than 636,000 ways to combine symptoms to meet the diagnostic criteria for post-traumatic stress disorder in the DSM-5.[22] Implicit to the categorical approach is an assumption that, other than the core symptoms required to meet criteria, all symptoms should be weighted equally in making the diagnosis. A test of these assumptions focused on DSM-IV criteria for Major Depressive Disorder (MDD) suggests that they are not empirically supported.[23] Specifically, the relationship between the number of symptoms and degree of functional impairment was linear, such that a thresholding approach was not justified. In addition, certain symptoms (sleep problems, energy loss, and psychomotor disturbance) were more strongly related to functioning than others, suggesting that they should not necessarily be considered equally relevant to disorder.

A prominent exemplar of this symptom-first approach can be seen in the B-SNIP study. In this ongoing symptom-first study, individuals who experience psychosis complete a battery of biomarker tests, including psychophysiological, cognitive, and self-report measures. Although the participants have been diagnosed with schizophrenia, schizoaffective disorder, or bipolar disorder, the data are analyzed independent of diagnosis to derive data-driven classifications. Using this approach, three clusters of patients, or "biotypes" emerged, based largely on differences in cognitive control and sensorimotor reactivity. When the biotypes were validated by examining group differences in measures that were not used to derive the clusters, the three groups—more so than the diagnostic groups—were differentiated by various characteristics, including level of impairment in functioning, brain structural features, and also by the rates of psychosis and level of psychosocial functioning in patients' relatives, suggesting that the biotypes are neurobiologically distinct.[18]

The B-SNIP project and its discoveries represent a major step toward personalized treatment for individuals with psychosis. Psychosis is itself a heterogeneous phenomenology, however, and future work will determine whether focusing on characteristics associated with a narrower—but still heterogeneous—aspect of psychopathology, such as auditory hallucinations,[24] might yield improved classification validity.

Principle 2: Dimensionality: The categorical approach to diagnosis—requiring a certain number of symptoms to make a diagnosis—implies that individuals who meet that threshold are qualitatively different from those who fall one or two symptoms short. In contrast, a central premise of RDoC is to identify a system (i.e., an internal mechanism) and show how it relates to functional behavior, and to provide this explanation in dimensional terms. Inherent to this principle is that investigators should assume dimensionality in psychopathology, biological systems, and behavioral processes, unless clusters, categories, "tipping points," and/or discontinuities are empirically detected.

The prospect of dimensionalizing psychotic illness might at first blush present an intellectual challenge when thinking clinically about thoughts and behaviors that are quite far removed from typical, such that they are difficult to place onto a dimension that includes health. In contrast to the kind of suffering seen in depression, where it is not a stretch to imagine an extreme state of sadness to infer what the extreme state of depression is like, there is something about psychosis that seems distinctly different. Among the symptoms of psychosis, reality distortion is perhaps most difficult to conceptualize within the RDoC framework.[25] This may arise perhaps due to the frightening quality of a complete departure from reality; however, when considered in the context of models of normative neurobehavioral processes such as source monitoring and cognitive control, dimensionality is evident.

The Source Monitoring Framework[26] offers a theory of cognition to describe a process of reality monitoring. Minor errors in ascribing the source of information, for example, misattributing from where you learned of a recent political fact, are likely familiar to all. Source monitoring offers a framework to explain the processes—cognitive and neural—between perceived and imagined events, and the ways that these different kinds of events might be confused. In this model, errors might be self-generated or perceptually derived: "Benign failures in this monitoring processes are called misperceptions, errors in memory, and unfounded or self-deceptive beliefs. Less benign failures lead to hallucinations and delusions" (p. 34).[27] These processes, well studied in healthy populations, have utility for understanding more extreme derailments seen in psychoses.

A review of work addressing brain mechanisms and reality monitoring in healthy groups and in individuals diagnosed with schizophrenia has demonstrated disruptions in areas of prefrontal cortex and other areas implicated in cognitive control as well as memory processes (hippocampus) and perspective taking/understanding of self (parietal areas).[28] These interrelated circuits appear to be more sensitive to self-generated memories than to externally derived information. These and other findings provide a foothold for RDoC-informed research. With an RDoC approach, in contrast to making comparisons between those diagnosed with schizophrenia and healthy controls, the magnitude of the neural dimension may be related to a particular behavioral or symptom dimension (or both). This approach breaks free from the constraints of the diagnosis which, as we have described, can either constrain the variance of a mechanism (resulting in overspecification), or introduce unwanted variation in the clinical sample (resulting in heterogeneity). Moreover, the dimensional approach captures the full range of variation in a mechanism, providing greater opportunity to probe when and how disruptions in a mechanism occur.

Integral to this work on dimensional psychopathology and to the RDoC itself is the need for careful interpretation of dimensionality and examination of assumptions about dimensions, and how to translate evidence of dimensionality to clinically actionable information. Baumeister and colleagues' examination[29] of the literature on auditory-verbal hallucinations in patients and healthy voice-hearers was informed by the distinction between quasi-dimensional models, in which psychotic experiences and distress are assumed to be part of the same dimension and resulting from the same mechanism, and fully dimensional models, in which the tendency to experience psychotic symptoms is assumed to be distributed in the general population and is independent of clinical distress.[30] This analysis highlights the importance of avoiding the assumption that dimensions are equivalent to continua and the difficulty of discriminating between a single illness with a range of outcomes and latent categorical structures that are obscured by the appearance of a continuum.[31] Relatedly, one must empirically test whether dimensionality of a symptom indicates dimensionality of a mechanism; in other words, clarifying the distinctions between "deviations of degree as well as deviations of kind" might point to meaningful differences between individuals and valid predictors of illness course.[32]

Principle 3: Apply neuroscience: One of the major motivations for the RDoC was to more effectively use integrative neuroscience to address psychiatric research questions. For that reason, the initial starting point for generating the domains and constructs in the matrix was to consider the major functional domains which the brain has evolved to accomplish and to establish evidence of an implementing circuit as a criterion for constructs. This is not to say that every RDoC study must include a measure of brain activity but framing of hypotheses and interpretation of data should—at a minimum—be informed by current knowledge about the structural and functional organization of the brain.

The B-SNIP study described earlier is an example of selecting a broad array of patients across diagnostic categories, and then using patterns of brain activity to identify participants based on dimensional mechanisms that cut across the standard diagnoses. Such "new" groupings then provide independent variables for further research using these mechanisms as a starting point for future research.[33] In another example, researchers took advantage of heterogeneous symptom patterns within the diagnostic category of depression. Rather than attempting to parse subtypes of depression based on co-occurring clinical symptoms, Drysdale and colleagues examined patterns of functional connectivity and used multivariate statistical approaches (principal components analysis and cluster analysis) to identify "low-dimensional representations" to identify patterns in resting-state connectivity, which were then related back to distinct profiles of clinical symptoms and differential responsiveness to transcranial magnetic stimulation.[34]

Another study examined cellular phenotypes among individuals diagnosed with the heterogeneous bipolar disorder category; in that study, researchers identified electrophysiological characteristics in induced pluripotent stem cells that identified individuals who exhibited clinical improvement following administration of lithium and those who did not with 92% accuracy.[35] These differential treatment effects would be missed if the heterogeneity within the diagnostic group were ignored. These studies used DSM diagnoses as a starting point, but future studies may explore whether physiological features such as these are useful for early detection of elevated risk of certain types of psychopathology or as phenotypes for transdiagnostic genomics studies. Indeed, putative mechanisms identified in one DSM disorder (or a select set of DSM disorders) may be relevant for other forms of psychopathology outside the clinical conceptualized grouping.[36]

Pioneering work by Ralph Hoffman examined the relation between auditory hallucinations and repetitive transcranial magnetic stimulation (rTMS) applied to temporoparietal cortex, including Wernicke's area.[37] In this approach, Hoffman and colleagues identified their patient groups based on the clinical phenomenon of persistent auditory hallucinations, and not on DSM diagnoses. Specific temporoparietal areas active during the time that each patient was "hearing voices" were identified during fMRI prior to treatment, and the spatial coordinates of those areas of activation were targeted with the rTMS treatment on an individual basis for each patient, illustrating a personalized medicine approach.

In sum, these different approaches illustrate the identification of an element of a biological system that, when dysfunctional, can be related back to a clinical problem. The range of these approaches also shows that the DSM is not "off limits" but rather is not center stage. Finally, in all of these approaches, there were dimensional properties to both the putative mechanism and the clinical expression of pathology.

Given that much of what is known about behavioral neuroscience has emerged from animal research, there are questions about how animal research relates to the RDoC framework. Just as research in humans has been constricted by an overly rigid focus on symptom-based syndromes, animal research has confronted the dilemma of studying complex psychiatric disorders which don't exist in animals.[38] Evidence from analog studies in nonhumans is apparent across domains, but perhaps especially in the Negative Valence Domain, for which the bench to bedside translational trajectory is, arguably, most advanced. The value and relevance of animal research in the age of the RDoC is clear; such studies are part of an iterative process of discovery such that mechanistic studies in animals can inform efforts to understand emerging phenotypes, validate novel classification schemes, and experimentally examine the role of environmental and genetic influences on the development of brain circuits related to psychopathological outcomes.[39] The RDoC domains find homology in animal systems, and the framework's focus on circuit-based explanations of phenotypes and integrative methodologies should facilitate synergy between animal and human research.[40]

Principle 4: Be integrative: The RDoC framework has its foundations in a modern understanding of brain–behavior relationships, but it is not only focused on neuroscience or circuit-based measures and RDoC research should be cautious about attributing to one unit of analysis more validity than another. The matrix recommends several units of analysis (the columns of the matrix), and the hope is that researchers will use them to inform and constrain each other in integrative ways. It is difficult to think about neural circuitry without also thinking about psychology, behavior, and our personal experiences. None of these systems work in isolation, and the RDoC provides a framework to articulate the dynamic interactions across units of analysis. It can be thought of as a more pluralistic system, looking across different measurement types to integrate information, rather than having one measurement be the foundation, or all-encompassing. It is a way for researchers whose work may be more or less focused on particular grouping of units or analysis (or domains) to relate their findings to one another.

The RDoC's emphasis on integrative measures should not be taken to suggest that all units of analysis will neatly converge within a construct or in association with a specific aspect of psychopathology. Within a single diagnostic group, and, presumably, across patient groups, some units of analysis may show a dimensional distribution and others may exhibit naturally occurring categories.[41] Spanning self-report, behavior, and biological measures is a challenge, and it is important to bear in mind the likelihood that "many, if not most, clinically observed symptoms are at a level of abstraction such that no statistical parsing of their co-occurrence will render them more neatly associated with underlying circuit function."[42] Novel analytic and computational methods, such as normative modeling approaches, hold promise for detecting various patterns of association between high dimension data across RDoC units of analysis.[43] Identifying associations across units of analysis is only a first step; identifying the mechanisms by which psychological and the biological processes converge is an essential next step.[44]

In considering the rows and columns of the RDoC matrix, it is notable that there are no structural features that depict development or various aspects of a person's environment. This should not be interpreted to mean that that these are unimportant to the RDoC approach to psychopathology and mental health.[45] It is important to keep in mind that the entirety of the RDoC approach is not reflected in the matrix, and although development and environment are not reflected in the physical representation of the matrix, they are known to be critically important factors in studying the full range of psychological functioning, including dimensions of psychosis,[46,47] and are appropriate for study using the RDoC approach. Many NIMH-funded RDoC research projects specifically examine developmental trajectories and explore the influence of environment on severity and outcomes of psychiatric symptoms. Quite the opposite of ignoring these aspects, the RDoC considers them to be crucial for helping to explain heterogeneity in psychopathology, for classifying people who may be at increased risk, and for identifying time periods that may be best suited for intervention. Not all measures/units are likely to converge, and certain units may be more informative for understanding or detecting a certain aspect of psychopathology, and this is likely to vary across developmental stages.[48] Accordingly, it is important for the RDoC to retain the various units of analysis on equal footing.

Development in the context of the RDoC framework does not only relate to neurodevelopment; the RDoC is consistent with a comprehensive and modern understanding of the developmental risk model, with intersecting brain changes and environmental impacts, such as adversity, urbanicity, and cannabis use,[49] including risk factors that do not specifically produce pathology but reduce resiliency to risk factors (e.g., risk genes that disrupt protection processes against the neural effects of hypoxia[50]).

Principle 5: Assume interactions between constructs: The structure of the matrix may implicitly convey a rigid structure of distinct domains and constructs, but, just as the brain is densely interconnected, RDoC research should consider interactions among constructs and across domains. Especially when considering complex clinical phenotypes such as delusions, no single construct or domain will be adequate to capture the breadth and complexity.[51]

Corlett and colleagues provide a strong example of a highly integrative approach to understanding delusions.[52] Informed by basic neuroscience, the model incorporates cognitive, affective, and perceptual constructs and proposes that delusions result from disrupted learning due to inappropriate signaling of overly precise prediction errors. The model is computationally based and incorporates a hierarchical organization of systems, associated with distinct neuromodulators, providing a rich foundation for testing of specific mechanistic hypotheses and exploring a range of delusion phenotypes across a dimension of symptomatic severity.

Related to this principle, a common misconception about the RDoC is that researchers must adhere to the specific domains and constructs listed in the RDoC matrix. To the contrary, RDoC was never intended to move the field from one set of classifications and constraints to another. While those constructs were proposed and rigorously vetted by groups of experts in the field to represent strong validity and promise for study, they are not the only ones. These are upheld as exemplars, which demonstrate types of constructs that cut across units of analysis and can be applied in a trans-diagnostic manner. A source of this confusion may have been the initial series of RDoC funding announcements which required that proposals focus on specific and existing domains of the RDoC matrix. In general, however, NIMH encourages studies that incorporate new, well-justified constructs. It is through this type of expansion of research that the RDoC matrix can grow and evolve as it was intended.

PROCESS FOR PROGRESS: HOW RDOC WILL CHANGE OVER TIME

From its beginning, the RDoC matrix was intended to be a flexible structure for which updates and changes are expected as new research emerges that broadens the understanding of psychopathology and guides future research. The RDoC Unit and Workgroup have grappled with developing a careful process for modifying the matrix while simultaneously emphasizing that the matrix is a collection of strong exemplars, not the definitive or complete set of domains and constructs.[53] In 2016, the National Advisory Mental Health Council (NAMHC) approved the formation of a Changes to the Matrix (CMAT) workgroup, which was given the charge of developing procedures for proposing new domains and constructs or other types of changes to the matrix and of convening experts to evaluate the proposed changes. With input from several experts, the CMAT workgroup proposed a reorganization of the Positive Valence Systems domain and revised construct definitions, which was approved by NAMHC in May 2018. The workgroup will next consider a similar reorganization of the Negative Valence System.

Beyond these changes to domains and constructs, there are several other possible updates to the matrix that are worthy of consideration. For example, the elements currently listed in the *Circuits* unit of analysis tend to be brain regions, rather than fully specified circuits.[54] Addressing this shortcoming will require an in-depth review of the behavioral neuroscience literature—including the results of the first cohort of studies supported by RDoC grants—and novel ways of integrating knowledge across units of analysis,[55] leading, perhaps, to an entirely new organizational structure for the RDoC; This is an important priority area for the future of the framework.

Additional future changes are indicated by emerging findings in genomics. Epigenetic changes and immune dysregulation are implicated in transdiagnostic psychopathology,[56] two areas that are not currently included in the matrix.

CONCLUDING COMMENTS: GOING FORWARD

The pending 10th anniversary of the introduction of the RDoC provides an opportunity to reflect on the next phase of scientific challenges and questions that will be the focus of future investigation.

What is the optimal "grain size" for RDoC constructs? Many researchers have embraced the idea that intermediary phenotypes may offer better purchase for clarifying cognitive and neural mechanisms in schizophrenia, and modern behavioral neuroscience has yielded highly detailed models and revealed fine-grained differentiations between interrelated neural and cellular processes which are useful for understanding brain function. More data are needed, however, to determine how narrow—or broad—neurobehavioral constructs should be to be maximally informative about psychopathology and its treatment. For example, a relatively narrowly defined construct such as "effort valuation" might be an appropriate grain size for detecting response to a specific intervention, but a more generalized assessment of reward responsivity might be more useful for identifying individuals at risk of mood disorder.

Balancing standardization and innovation in measurement: To detect complex patterns of relationships between constructs and psychopathology, studies with sample sizes in the thousands (and, for genetic discovery, tens or hundreds of thousands) must be compiled, via large coordinated data collection efforts or combining existing data. The benefits of coordinating research efforts in ways that allows "crosstalk" are relevant in times where research is costly, and full range of (dimensional) variation of phenomena can provide important information relative to limitations of focusing on narrow, discrete phenomena Explicitly addressing the costs and benefits of prioritizing efforts to standardize measurement and data units will help move the field forward practically while maintaining an awareness of limitations of "common denominators."

Developing valid and reliable behavioral assessments and laboratory tests. The RDoC matrix includes a list of paradigms that can be used to assess constructs. The elements in the *Paradigms* column of the matrix are largely based on recommendations from participants in the original five workshops in 2010–2011 at which the constructs were defined. More recently, the RDoC unit convened an NAMHC workgroup to critically review the Paradigms column of the matrix and conduct a thorough evaluation of candidate tasks and measures for each construct using a set of standardized criteria. The workgroup report[57] provides a comprehensive resource to help investigators select assessment tools and highlights areas in need of further research. The dynamic nature of the matrix accommodates innovation by leaving open the possibility to add new tasks/measures as they are developed.

Novel analytic approaches and computational psychiatry: The RDoC approach encourages analyses that explore dimensionality and integrate data from across units of analysis, which requires statistical methods that may be less familiar to investigators who are accustomed to between-groups approaches. The RDoC's dimensional approach should not be misunderstood, however, to preclude the existence of discontinuities or tipping points in neurobehavioral processes which may be informative for understanding the transition from health to illness or for identifying subgroups for the purpose of personalizing treatment.

Computational methods represent another emerging area with great promise of high priority to NIMH, both in developing highly specific and testable models of brain–behavior relationships and in suggesting novel "computational phenotypes."[58] Both data-driven and theory-driven approaches are highly suitable for RDoC-informed studies.[59]

Focus on symptoms or functioning? Symptoms of psychosis, depression, mania, and anxiety are often accompanied by impairment in everyday functioning, but sometimes the relationship between symptoms and functioning is inconsistent. Performance on even very important clinical/translational tasks sometimes does not correlate strongly with functioning.[60] As a field, psychiatry is concerned with alleviating suffering and, arguably, the assumption has been that if suffering is alleviated, then functioning will improve. It will be important to remain aware of that assumption and perhaps to shift to a dual focus on psychopathology and functioning, given the evidence of shared and unique contributors.[61]

Maj has argued that the multiple efforts to improve diagnosis and classification of mental disorders are a sign of a paradigm shift of Kuhnian proportions.[62] We do agree that change is afoot, though we wish to conclude in a cautionary manner with a modest outlook. For purposes of keeping the record clear, we also restate our vision for the role of RDoC as we move to the next phase.

The seemingly intractable psychotic spectrum disorders certainly motivate a "new approach." Yet, despite promises that a great cure is just around the corner, a "magic bullet" for psychosis has never been realized. We surmise that no such panacea is immediately in the offing, be it new technology or a new diagnosis. The RDoC workgroup was not the first to call attention to these problems.[63–65] However, our group was the first to propose a structure for the field that would serve as a nomological net on which research that integrates basic functioning as measured across multiple systems could be organized, and research could be cast in designs not harnessed to the DSM.

REFERENCES

1. Dutta R, Greene T, Addington J, McKenzie K, Phillips M, Murray RM. Biological, life course, and cross-cultural studies all point toward the value of dimensional and developmental ratings in the classification of psychosis. *Schizophr Bull.* 2007;33(4):868–876.
2. Lecrubier Y. Refinement of diagnosis and disease classification in psychiatry. *Eur Arch Psychiatry Clin Neurosci.* 2008;258(1):6–11.
3. Cuthbert BN. Dimensional models of psychopathology: research agenda and clinical utility. *J Abnorm Psychol.* 2005;114(4):565–569.
4. American Psychiatric Association. *Diagnostic and Statistical Manual of Mental Disorders.* 5th ed. Washington, DC: Author; 2013.
5. First MB, Bell CC, Cuthbert BN, et al. Personality disorders and relational disorders: a research agenda for addressing crucial gaps in DSM. In: Kupfer DJ, First MB, Regier DA, eds. *A Research Agenda for DSM-V.* Washington, DC: American Psychiatric Association; 2002:123–199.
6. Strauss JS, Carpenter WT Jr, Bartko JJ. The diagnosis and understanding of schizophrenia. Part III. Speculations on the processes that underlie schizophrenic symptoms and signs. *Schizophr Bull.* 1974;11:61–69.
7. Hyman SE. Revolution stalled. *Sci Transl Med.* 2012;4(155):155cm111–155cm111.
8. Hyman SE. The diagnosis of mental disorders: the problem of reification. *Annu Rev Clin Psychol.* 2010;6(1):155–179.
9. Van Praag HM, Asnis GM, Kahn RS, et al. Nosological tunnel vision in biological psychiatry. *Ann N Y Acad Sci.* 1990;600(1):501–510.
10. Kendler KS, Karkowski LM, Walsh D. The structure of psychosis: latent class analysis of probands from the Roscommon family study. *Arch Gen Psychiatry.* 1998;55(6):492–499.
11. American Psychiatric Association. *Diagnostic and Statistical Manual of Mental Disorders.* 4th ed. Washington, DC: Author; 1994.
12. Sanislow CA, Pine DS, Quinn KJ, et al. Developing constructs for psychopathology research: research domain criteria. *J Abnorm Psychol.* 2010;119(4):631–639.
13. Sanislow CA. Connecting psychopathology meta-structure and mechanisms. *J Abnorm Psychol.* 2016;125(8):1158–1165.
14. Yancey JR, Venables NC, Patrick CJ. Psychoneurometric operationalization of threat sensitivity: relations with clinical symptom and physiological response criteria. *Psychophysiology.* 2016;53(3):393–405.
15. Cuthbert BN. Research domain criteria: toward future psychiatric nosologies. *Dialogues Clin Neurosci.* 2015;17(1):89–97.
16. Kozak MJ, Cuthbert BN. The NIMH research domain criteria initiative: background, issues, and pragmatics. *Psychophysiology.* 2016;53(3):286–297.
17. Calkins ME, Dobie DJ, Cadenhead KS, et al. The Consortium on the Genetics of Endophenotypes in Schizophrenia: model recruitment, assessment, and endophenotyping methods for a multisite collaboration. *Schizophr Bull.* 2007;33(1):33–48.
18. Clementz BA, Sweeney JA, Hamm JP, et al. Identification of distinct psychosis biotypes using brain-based biomarkers. *Am J Psychiatry.* 2016;173(4), 373–384. https://doi.org/10.1176/appi.ajp.2015.14091200
19. Strik W, Stegmayer K, Walther S, Dierks T. Systems neuroscience of psychosis: mapping schizophrenia symptoms onto brain systems. *Neuropsychobiology.* 2017;75(3):100–116.

20. Kapur S, Phillips AG, Insel TR. Why has it taken so long for biological psychiatry to develop clinical tests and what to do about it[quest]. *Mol Psychiatry*. 2012;17(12):1174–1179.
21. Guloksuz S, van Os J. The slow death of the concept of schizophrenia and the painful birth of the psychosis spectrum. *Psychol Med*. 2017;48(2):229–244.
22. Galatzer-Levy IR, Bryant RR. 636,120 ways to have post-traumatic stress disorder. *Perspectives on Psychological Science*. 2013;8(6):651–662.
23. Sakashita C, Slade T, Andrews G. Empirical investigation of two assumptions in the diagnosis of DSM-IV major depressive episode. *Austral NZ J Psychiatry*. 2007;41(1):17–23.
24. Ford JM, Morris SE, Hoffman RE, et al. Studying hallucinations within the NIMH RDoC framework. *Schizophr Bull*. 2014;40(Suppl 4):S295–S304.
25. Carpenter WT. The RDoC Controversy: Alternate Paradigm or Dominant Paradigm? *Am J Psychiatry*. 2016;173(6):562–563.
26. Mitchell KJ, Johnson MK. Source monitoring 15 years later: what have we learned from fMRI about the neural mechanisms of source memory? *Psychological Bulletin*. 2009;135(4):638–677.
27. Johnson MK. Discriminating the origin of information. In: Oltmanns TF, Maher BA, eds. *Delusional beliefs*. Oxford, UK: John Wiley & Sons; 1988:34–65.
28. Simons JS, Garrison JR, Johnson MK. Brain mechanisms of reality monitoring. *Trends Cogn Sci*. 2017;21(6):462–473.
29. Baumeister D, Sedgwick O, Howes O, Peters E. Auditory verbal hallucinations and continuum models of psychosis: a systematic review of the healthy voice-hearer literature. *Clin Psychol Rev*. 2017;51:125–141.
30. Claridge G, Beech T. Fully and quasi-dimensional constructions of schizotypy. In: Raine A, Mednick SA, Lencz T, eds. *Schizotypal Personality*. Cambridge: Cambridge University Press; 1995:192–216.
31. Kaymaz N, van Os J. Extended psychosis phenotype—yes: single continuum—unlikely. *Psychol Med*. 2010;40(12):1963–1966.
32. Mittal VA, Wakschlag LS. Research domain criteria (RDoC) grows up: strengthening neurodevelopment investigation within the RDoC framework. *Journal of Affective Disorders*. 2017;216:30–35.
33. Insel TR, Cuthbert BN. Brain disorders? Precisely. *Science*. 2015;348(6234):499–500.
34. Drysdale AT, Grosenick L, Downar J, et al. Resting-state connectivity biomarkers define neurophysiological subtypes of depression. *Nat Med*. 2017;23(1):28–38.
35. Stern S, Santos R, Marchetto MC, et al. Neurons derived from patients with bipolar disorder divide into intrinsically different sub-populations of neurons, predicting the patients/' responsiveness to lithium. *Mol Psychiatry*. 2017.
36. Cross-Disorder Group of the Psychiatric Genomics Consortium. Genetic relationship between five psychiatric disorders estimated from genome-wide SNPs. *Nature Genetics*. 2013;45(9):984–994.
37. Hoffman RE, Wu K, Pittman B, et al. Transcranial magnetic stimulation of Wernicke's and right homologous sites to curtail "voices": a randomized trial. *Biol Psychiatry*. 2013;73(10):1008–1014.
38. Nestler EJ, Hyman SE. Animal models of neuropsychiatric disorders. *Nat Neurosci*. 2010;13:1161.
39. Kaffman A, Krystal JJ. New frontiers in animal research of psychiatric illness. In: Kobeissy FH, ed. *Psychiatric Disorders: Methods and Protocols*. Totowa, NJ: Humana Press; 2012:3–30.
40. Anderzhanova E, Kirmeier T, Wotjak CT. Animal models in psychiatric research: The RDoC system as a new framework for endophenotype-oriented translational neuroscience. *Neurobiol Stress*. 2017;7:47–56.
41. Jalbrzikowski M, Ahmed KH, Patel A, et al. Categorical versus dimensional approaches to autism-associated intermediate phenotypes in 22q11.2 microdeletion syndrome. *Biol Psychiatry: Cogn Neurosci Neuroimaging*. 2017;2(1):53–65.
42. Bilder RM. On the hierarchical organization of psychopathology and optimizing symptom assessments for biological psychiatry. *Biol Psychiatry: Cogn Neurosci Neuroimaging*. 2017;2(4):300–302.
43. Marquand AF, Wolfers T, Mennes M, Buitelaar J, Beckmann CF. Beyond lumping and splitting: a review of computational approaches for stratifying psychiatric disorders. *Biol Psychiatry: Cogn Neurosci Neuroimaging*. 2016;1(5):433–447.
44. Miller GA, Rockstroh BS, Hamilton HK, Yee CM. Psychophysiology as a core strategy in RDoC. *Psychophysiology*. 2016;53(3):410–414.
45. Garvey M, Avenevoli S, Anderson K. The National Institute of Mental Health research domain criteria and clinical research in child and adolescent psychiatry. *J Am Acad Child Adolesc Psychiatry*. 2016;55(2):93–98.
46. Zavos HS, Freeman D, Haworth CA, et al. Consistent etiology of severe, frequent psychotic experiences and milder, less frequent manifestations: a twin study of specific psychotic experiences in adolescence. *JAMA Psychiatry*. 2014;71(9):1049–1057.
47. Binbay T, Drukker M, Elbi H, et al. Testing the psychosis continuum: differential impact of genetic and nongenetic risk factors and comorbid psychopathology across the entire spectrum of psychosis. *Schizophr Bull*. 2012;38(5):992–1002.
48. Franklin JC, Jamieson JP, Glenn CR, Nock MK. How developmental psychopathology theory and research can inform the research domain criteria (RDoC) project. *J Clin Child Adolesc Psychol*. 2015;44(2):280–290.
49. Murray RM, Bhavsar V, Tripoli G, Howes O. 30 years on: How the neurodevelopmental hypothesis of schizophrenia morphed into the developmental risk factor model of psychosis. *Schizophr Bull*. 2017;43(6):1190–1196.
50. Schmidt-Kastner R, van Os J, Esquivel G, Steinbusch HWM, Rutten BPF. An environmental analysis of genes associated with schizophrenia: hypoxia and vascular factors as interacting elements in the neurodevelopmental model. *Mol Psychiatry*. 2012;17:1194.
51. MacDonald IAW. Studying delusions within research domain criteria: the challenge of configural traits when building a mechanistic foundation for abnormal beliefs. *Schizophr Bull*. 2017;43(2):260–262.
52. Feeney EJ, Groman SM, Taylor JR, Corlett PR. Explaining delusions: reducing uncertainty through basic and computational neuroscience. *Schizophr Bull*. 2017;43(2):263–272.
53. Garvey MA, Cuthbert BN. Developing a motor systems domain for the NIMH RDoC program. *Schizophr Bull*. 2017;43(5):935–936.
54. Elmer GI, Brown PL, Shepard PD. Engaging research domain criteria (RDoC): neurocircuitry in search of meaning. *Schizophr Bull*. 2016;42(5):1090–1095.
55. Bilder RM, Howe AG, Sabb FW. Multilevel models from biology to psychology: mission impossible? *J Abnorm Psychol*. 2013;122(3):917–927.
56. The Network Pathway Analysis Subgroup of the Psychiatric Genomics Consortium. Psychiatric genome-wide association study analyses implicate neuronal, immune and histone pathways. *Nat Neurosci*. 2015;18(2):199–209.
57. National Advisory Mental Health Council Workgroup on Tasks and Measures for RDoC. Behavioral assessment methods for RDoC constructs. 2016; https://www.nimh.nih.gov/about/advisory-boards-and-groups/namhc/reports/rdoc_council_workgroup_report_153440.pdf. Accessed 1/22/2018.
58. Ferrante M, Redish AD, Oquendo MA, Averbeck BB, Kinnane ME, Gordon JA. Computational psychiatry: a report from the 2017 NIMH workshop on opportunities and challenges. *Mol Psychiatry*. 2018.
59. Huys QJM, Maia TV, Frank MJ. Computational psychiatry as a bridge from neuroscience to clinical applications. *Nat Neurosci*. 2016;19(3):404–413.
60. Olbert CM, Penn DL, Kern RS, et al. Adapting social neuroscience measures for schizophrenia clinical trials, Part 3: Fathoming external validity. *Schizophr Bull*. 2013;39(6):1211–1218.
61. Iorfino F, Hickie IB, Lee RSC, Lagopoulos J, Hermens DF. The underlying neurobiology of key functional domains in young people with mood and anxiety disorders: a systematic review. *BMC Psychiatry*. 2016;16(1):156.

62. Maj M. Narrowing the gap between ICD/DSM and RDoC constructs: possible steps and caveats. *World Psychiatry*. 2016;15(3):193–194.
63. Weinberger DR. Biological phenotypes and genetic research on schizophrenia. *World Psychiatry*. 2002;1(1):2–6.
64. Lang PJ, McTeague LM. The anxiety disorder spectrum: fear imagery, physiological reactivity, and differential diagnosis. *Anxiety, Stress, Coping*. 2009;22(1):5–25.
65. Krueger RF. The structure of common mental disorders. *Arch Gen Psychiatry*. 1999;56(10):921–926.

5

SCHIZOPHRENIA, SCHIZOAFFECTIVE DISORDER, BIPOLAR DISORDER

Barrett Kern and Sarah K. Keedy

ROLE OF PSYCHOSIS FOR DSM-5 DIAGNOSIS OF SCHIZOPHRENIA, SCHIZOAFFECTIVE DISORDER, AND BIPOLAR DISORDER

Schizophrenia (SCZ), schizoaffective disorder (SAD), and bipolar disorder with psychotic features (BDP) are the likely diagnoses when individuals exhibit psychotic symptoms and other biological causes have been ruled out. These are, therefore, psychiatry's "flagship" disorders for prominent psychosis presentations. Hence, there are several ways to understand the construct of *psychosis* through the examination of these disorders. Initial conceptualizations emphasized that these are among the most severe forms of mental illness. The dominance of dysfunctional reality testing has been a key feature entrenched throughout iterations of the diagnostic criteria for these disorders. Psychosis is now most commonly understood to be present if there are hallucinations or delusions. These specific symptoms operationalize the notion of reality testing gone awry, with hallucinations signaling an inability to appreciate the lack of an external stimulus associated with a sensory perception; and delusions signaling inability to adjust one's beliefs in the face of contradictory evidence that would generally sway others. Hallucinations and delusions are now the only symptoms found in the current diagnostic criteria of all three diagnoses (SCZ, SAD, and BDP). By contrast, these disorders are distinguished from one another by their variation in mixture of such psychosis symptoms and mood episodes (whether manic, depressive, or mixed) presenting over time. In this chapter, we endeavor to provide a more detailed map of the relation of these primary psychotic disorders in terms of the way psychosis is or is not commonly characterized in them. A highlight of this examination is the progressive separation of psychotic and mood symptoms, and varying distinction of psychosis symptoms themselves within each disorder. Infused in this review are questions around the utility versus biological reality of boundaries between these primary psychotic disorders.

Current (DSM-5) BDP is the diagnosis given when there is history of a manic or mixed mood episode, with or without history of depressive mood episodes, and at some point during any type of mood episode, hallucinations or delusions are present. This is a relatively restricted psychosis symptom presentation with the requirement that hallucinations or delusions be accompanied by a mood episode. Notably, psychosis in BDP is restricted to hallucinations and/or delusions. Disorganized thought process or/and behavior, which are possible psychosis symptoms in SCZ, are not offered as possible symptoms for bipolar mood episodes to then be labeled as psychotic. In BPD, distinction is made as to whether the psychotic symptoms are mood-congruent or mood-incongruent, e.g., whether the hallucinations or delusions are related to content or emotional tone of the mood episode. For example, hallucinations that are critical voices occurring during a depressive episode or delusions of being supernaturally powerful during a manic episode are defined as mood-congruent. Examples of mood-incongruent symptoms include critical voices during an elated, grandiose mania. This distinction appears to have prognostic value, with mood-incongruent psychosis indicative of worse outcome and functioning,[1,2] though not all studies find this.[3] Frequency of mood-congruent vs. mood-incongruent psychotic symptoms in bipolar disorder is roughly equal, and commonly, both mood-congruent and mood-incongruent psychosis symptoms are present at once.[4,5] Notably, there is no specifier or other diagnostic implication for whether hallucinations or delusions are observed. Further, it is not diagnostically relevant whether psychotic symptoms occur during manic or depressive episodes, or both. Finally, in the bipolar disorders, an otherwise hypomanic episode is defined as manic with psychotic features if there are hallucinations or delusions present.

Schizophrenia is the diagnosis given when at least two of the following symptoms are present for a considerable portion of a 1-month period (Criterion A): hallucinations, delusions, disorganized speech, disorganized or catatonic behavior, and negative symptoms. There is an additional rule that at least one of the two symptoms be from among the first three listed. No mood disturbance is diagnostically relevant, although in recognition of the distress associated with living with schizophrenia, diagnosticians are encouraged to be sparing in adding comorbid depressive disorder diagnoses. As with BDP and SAD, a diagnosis of SCZ can be made with a variety of combinations of psychosis symptoms, and no specific symptom combination is considered to be pathognomonic to the disorder nor does any symptom profile render a diagnostic specifier. This is new in DSM-5 relative to its history. All

DSM editions prior to the 5th had SCZ subtype specifiers on the basis of the dominance of particular psychosis symptoms, such as a paranoid subtype, a disorganized subtype, and so on. DSM-5 dropped subtypes due to low stability and validity.[6] SCZ is now similar to BDP and SAD in that a specific type of psychotic symptoms is not diagnostically relevant.

For a diagnosis of SAD, Criterion A of SCZ must be met and mood episodes (depressive or manic) must be present for the majority of time during active and residual phases of illness. Hallucinations and/or delusions may or may not be present during such mood episodes, but the diagnosis of SAD requires at least 2 weeks of hallucinations and/or delusions (typically much more than that is observed) outside of any mood episodes. Type of mood episode in SAD is noted as a specifier to the diagnosis. There is a "depressive" specifier if only depressive episodes occur, and a "bipolar" specifier if manic episodes are observed and depressive episodes may or may not also occur. Hence, BDP and SAD are distinguished on allowance of mood and psychosis symptom overlap: in BDP, psychosis must *only* occur during mood episodes, but in SAD, psychosis must occur for 2 weeks *outside* of mood episodes. Another difference is that there is no specifier for mood congruency of psychosis symptoms that may occur during mood episodes in SAD. However, like BDP, both mood-congruent and mood-incongruent psychotic symptoms are equally likely in SAD.[7]

In sum, either hallucinations or delusions are required symptoms across BDP, SAD, and SCZ, and the distinguishing symptom among them is essentially prominence of mood episodes. While these are features of a categorical diagnostic system, it is of course not the only way to conceptualize the nature of the presentation of psychosis and mood symptoms among individuals with psychotic disorders. One apparent alternative view is that these disorders are points along a dimensional continuum. One exemplar operationalization of such a dimensional approach is the Schizo-Bipolar Scale,[8] which is a single rating given from 0 to 9, where bipolar disorder with psychosis and with regular, repeated manic episodes anchors one end (score of 0) and SCZ, where no mood episodes are apparent, anchors the other end (score of 9). More equivalent presentations of psychosis and mood episodes, as in SAD, define midpoint values (scores of 4 or 5). The remaining integers designate variation in prominence of mood or psychosis. The application of this scale in samples of individuals with bipolar and schizophrenia spectrum disorders shows that there is good representation throughout the available rating values 0–9, confirming what experienced clinicians observe, which is that there is much more heterogeneity among these disorders than is captured by categorical diagnostic systems. There are also many rating instruments where each psychosis and mood symptom is rated for its severity, better capturing heterogeneity. Examples are the Positive and Negative Symptom Rating Scale (PANSS), the Young Mania Rating Scale, and others. Notably, in the most recent DSM (5th edition), a scale called the Clinician-Rated Dimensions of Psychosis Symptom Severity was introduced. This scale is composed of 8 items, each a psychotic or mood symptom quantitatively rated for severity. This is an approach similar to the widely used rating scales in research like the PANSS. While not required for diagnosis per DSM-5, the inclusion of this rating scale in DSM-5 signals increased recognition of the utility of a dimensional approach to psychotic disorders.

A final point for this basic mapping of key similarities and differences across SCZ, SAD, and BDP in their current diagnostic conceptualization is to acknowledge that individuals can and do shift among them over time,[9,10] though some directional trajectories are not possible as necessitated by the diagnostic criteria. For example, BDP can become SAD, but not the other way around, as once psychosis is exhibited for 2 weeks outside of mood episodes, SAD is applicable and BDP is ruled out. Estimates are that about 20% of first episode diagnoses of SCZ change to bipolar,[11], 16% of SAD first episode diagnoses shift to SCZ, and 4% of SCZ first episode diagnoses shift to SAD.[12] Such observations, among others, lend support to considering whether these three syndromes are variations of a common underlying pathology.[13]

THE SHIFTING CONCEPTUALIZATION OF THE RELATEDNESS OF PSYCHOTIC AND AFFECTIVE SYMPTOMS

A feature of current (DSM-5) conceptualizations of BDP, SAD, and SCZ is the categorical separation of psychotic and mood disorders. However, this parallel treatment has not always been the case. Early in diagnostic systems, mood symptoms were subordinate to the construct of psychosis. Here, we trace this progression through DSM editions, a convenient, though incomplete and still unresolved, accounting of the iterations of thought on these common symptoms of psychotic disorders. We minimize attention to rationale and evolving schools of thought influencing the progression, addressed elsewhere in this volume. Instead we seek merely to highlight the shifts themselves to spotlight the challenge posed by the endeavor of distinguishing among psychotic disorders.

The precursor to DSM was the classification system of Emil Kraepelin, who made the key distinction between affective and nonaffective psychopathology, naming manic-depressive psychosis (now bipolar disorder) and dementia praecox (now SCZ). This distinction between affective and nonaffective psychotic disorders has since always been maintained. Throughout DSM editions, SCZ, SAD, and BDP with or without psychotic features have remained distinguished from one another in some manner. Though the diagnoses shift in terms of how closely or distantly they are portrayed as related, they always include some manner of a conceptualization of psychosis weaving through the diagnostic criteria.

DSM I and II presented versions of all three diagnoses under a "Psychotic Disorders" category. In these early editions, the term "psychotic" was used to indicate a level of severity of illness, rather than referring to a cluster of specific symptoms. DSM-I defined *psychosis* as a profound disturbance in one's existence, involving break from reality, and operationalized by impairments in thinking, behavior, *and* mood. Diagnostic terms in DSM-I were "schizophrenic reaction" and "manic-depressive reaction," reflecting psychodynamic conceptualization dominant at the time. There were multiple subtypes of schizophrenic reaction including a *schizoaffective* subtype. However, there was hedging regarding this subgroup, with notation stating that, over time, individuals with this diagnosis tended be "basically schizophrenic in nature" (APA, 1952, p. 27), suggesting mood symptoms to be less relevant or less likely to persist. With psychodynamically oriented etiology as a key organizing principle of DSM-I, there was another section where psychotic symptoms could be observed in a different and very specified relation to affective disturbance. This was the Psychophysiologic Autonomic and Visceral Disorders section, ailments located to any of a variety of specified organs, but in all cases, thought to be caused by repressed emotion, and distinguished from conversion disorders. One disorder in this section was *psychophysiologic reaction of organs of special sense*. This diagnosis could apply if an aberrant sensory experience, namely a reported hallucination, was complained of as a relatively isolated symptom, and it would be ascribed to a pathological affective state.

A final notable feature of DSM-I is that, while it refers to hallucinations and delusions as symptoms of psychotic disorders, it did not have detailed descriptions or examples defining them. What it did have, however, was a list of specifiers to use with diagnoses. Among the specifiers were some to indicate the sensory modality of hallucinations (APA, 1952, p. 128), a feature unique to this edition among all DSMs, although in DSM-5 some version of symptom-specific assessment is encouraged (e.g., the Clinician-Rated Dimensions of Psychosis scale, mentioned previously). On the other hand, "delusions" in DSM-I was a specifier but there was no demarcation of delusion type (grandiose, erotomanic, etc.).

While DSM-II ceased use of "reactions" as part of diagnostic terminology for psychotic disorders, it retained DSM-I's basic organization of a superordinate Psychoses category, with manic-depressive illness and SCZ under it. Schizoaffective remained a subtype of schizophrenia. However, for schizoaffective diagnosis, there was no clear guidance pertaining to when depression or elation become pronounced enough to warrant this specifier. The psychoanalytic influence of etiology continued to shape the concepts and language which were expanded somewhat relative to DSM-I. For example, individuals with the paranoid SCZ subtype were described as using the "mechanism of projection, which ascribes to others characteristics he cannot accept himself" (APA, 1968, p. 34). This was meant to explain why the paranoid subtype did not present with the same level of disorganization, another psychosis symptom, as seen in the catatonic and hebephrenic subtypes. Finally, DSM-II retained a Psychophysiologic Disorders category with a "disorder of organ of special sense" diagnosis, still distinguished from conversion disorders, specifying a primacy to affective pathology that could result in psychosis. This latter diagnosis still was placed outside the Psychotic Disorders section, thus indicating it to be a less severe mental illness.

In DSM-III, nosology shifted toward atheoretical descriptions of symptoms constituting syndromes (diagnoses). As such, the term "psychotic" ceased to refer primarily to a level of severity and also ceased to be a general category of disorders. In the same sense that DSM-III moved toward being lists of symptoms, *psychosis* moved similarly—most commonly operationalized as hallucinations and delusions, when specified. Aside from this conceptual shift, there was similarity to prior editions in that SCZ and affective disorders, inclusive of bipolar disorder, remained parallel to one another in organization. DSM-III contained enhanced details regarding each syndrome, including content of hallucinations and delusions as typically experienced in a variety of disorders. For example, hallucinations characteristic of SCZ were specified as most commonly auditory and not mood-related, and several types of delusions possibly present were specified. Only one such symptom was required for a diagnosis of SCZ. Delusions and hallucinations in bipolar disorder called for a "with psychotic features" specifier, along with noting whether they were mood congruent. This specifier became bipolar disorder's main link to psychosis as a construct, and in DSM-III, psychosis symptoms during mood episodes could be hallucinations, delusions, or grossly bizarre behavior (the latter symptom is uniquely presented as possible in bipolar disorder among DSM editions). The level of descriptive detail was an innovation of DSM-III. Yet in contrast to this improved detail describing SCZ and mood disorders as parallel groups and with clear distinguishing criteria, SAD was somewhat unmoored relative to its prior conceptualizations. This was apparent in two ways. First, it was more distally related to SCZ in becoming its own disorder, no longer a schizophrenia subtype. Second, it was located in the "Psychotic Disorders Not Elsewhere Classified" chapter, and instead of being given carefully defined diagnostic criteria as other disorders were, SAD was a disorder of exclusion for when neither SCZ nor bipolar disorder seemed clearly applicable.

DSM-IV (1994) brought back the term "psychotic" as a category, and merged it with SCZ in a chapter titled "Schizophrenia and Other Psychotic Disorders." Consistent with all editions, there remained a parallel treatment of affective disorders, titled "Mood Disorders," where manic-depressive presentations (bipolar disorder) were located. Diagnostic criteria themselves in DSM-IV became less descriptive relative to DSM-III, although the material written to clarify the criteria remained extensive in description. DSM-IV mandated that a diagnosis of SCZ would be made if two psychotic symptoms were observed, not just the one as in DSM-III (among hallucinations, delusions, disorganized behavior, or prominent negative symptoms), although certain versions of the symptoms, the so-called first rank symptoms, could be the only symptoms present and lead to a SCZ diagnosis. Hallucinations and delusions are reviewed extensively, and cultural background is encouraged as a consideration for determining presence or absence of psychotic symptoms. Affective disorders in DSM-IV could have a specifier of "with psychotic features" as was the case in DSM-III, but only if the symptoms were hallucinations or delusions. Schizoaffective disorder remained its own disorder but was now reunited with SCZ by being located with it in the same chapter. It received its own diagnostic criteria, but such criteria remained vague with respect to the extent to which affective symptomatology played a role in this diagnosis—noting only that mood episodes should be present "a significant amount."

DSM-5 continues the methodology of using detailed symptomatology to define diagnoses and describing the symptoms at length in text surrounding the criteria, and maintaining a general demarcation between SCZ, SAD, and BDP. Schizophrenia and SAD are each in the "Schizophrenia Spectrum and Other Psychotic Disorders" chapter, whereas as BDP appears in the "Bipolar and Related Disorders" chapter, which DSM-5 separates from the "Depressive Disorders" for the first time. Schizoaffective disorder in DSM-5 includes criteria that specify mood episodes must occur for a majority of the active and residual phases of illness. Although it appears in the "Schizophrenia and Other Psychotic Disorders" chapter, conceptually, this criterion of duration of mood episodes arguably pushes SAD, as a diagnosis, phenomenologically closer to affective disorders. This is in interesting contrast to DSM-I, where schizoaffective presentations were thought to be "essentially schizophrenic." All DSM editions include text acknowledging limitations to their nosological systems. Being no exception, DSM-5, like its predecessors, did not resolve the long-standing conceptual and practical conundrums accompanying a categorical classification system attempting to be imposed on dimensional life experiences.

BARRIERS TO UNDERSTANDING PSYCHOTIC SYMPTOMS IN SCHIZOAFFECTIVE DISORDER

Subsequent sections of this chapter will address psychosis symptom features across and within the psychotic disorders. However, a prefacing section is called for to attempt to directly address the conundrum of SAD, with the least validity and prognostic value of the three disorders discussed. One key reason for this problem is the lack of specificity for the extent of affective symptomatology required for a SAD diagnosis. This renders it a group that is more heterogeneous and unreliably identified than SCZ or BDP, reducing comparability and therefore progress across studies. While DSM-5 specifies "a majority" of time for affective symptoms, DSM-IV only stated that mood episodes be present "a substantial portion" of the total duration of illness, a directive with less clear operationalization. Thus, previous research on SAD using DSM-IV may have included individuals with both psychosis and mood symptoms who would not have been included using the DSM-5 criteria. Comparisons among other DSM editions are similarly discouraging.

Even for the apparently clearer SAD criterion of hallucinations or delusions being present for 2 weeks in the absence of a mood episode, there can easily be disagreement in interpreting this. While a delusional belief endorsed every day for 2 weeks may be fairly uncontroversial, hallucinations as discreet experiences have a less clear relation to the rule of being "present" for 2 weeks. Some individuals report voices with constant commentary on their actions and thoughts, likely crossing the threshold of "present for 2 weeks." However, what about hallucinations experienced as on and off throughout the day? Or hallucinations lasting only a few seconds each day for 2 weeks? What about hallucinations continuously occurring for 5 days for all waking hours but not for the other 11 days without intervention? In practical terms, this is the role of "clinical judgment," but this does not well-serve the research enterprise where the goal is that one sample of SAD is similar to another sample in another study.

Another issue is the time window needed for a schizoaffective diagnosis, so that the proportion of affective to psychotic symptoms over time can be evaluated. The actual proportion of mood versus psychotic symptoms is harder to determine when the person has only experienced a year or two of illness compared with 15 years of illness. For example, someone evaluated 2 years into their disorder may appear to be 90% mood symptoms with psychosis and 10% psychotic symptoms outside of mood symptoms, which warrants a diagnosis of SAD. However, 5 years later, when re-evaluated, this person may have only 30% mood symptoms during psychosis and 70% psychotic symptoms outside of mood, warranting a diagnosis of SCZ. Even later in life, this same person's symptom profile could shift again and cause the person to revert back to a diagnosis of SAD. While such fluctuation may accurately reflect nature for a proportion of individuals, it is poorly addressed in the usual categorical diagnosis framework.

The challenge of SAD is met in the literature in one of three ways: one is that it is excluded from study. Another is that it is lumped in with SCZ, constituting a "schizophrenia spectrum" group. A third is it is lumped in with bipolar disorder, and perhaps major depression with psychotic features, constituting an "affective psychosis" group for comparison with nonaffective psychosis (SCZ). For the remainder of this chapter, we specify which approach is used where possible, and acknowledge the net effect may be one of dissatisfaction in terms of providing clarity on questions of psychosis in SAD.

THE RELATEDNESS OF PSYCHOSIS SYMPTOMS IN THE PSYCHOTIC DISORDERS

While a distinction between affective and nonaffective psychotic disorders has been maintained in some manner in DSM history, the biological meaningfulness of such a distinction remains to be established. Research addressing this question occurs at multiple levels of inquiry, from genes to behavior. At the symptom level, the focus of this chapter, such work has included efforts to assess patterns of psychosis and mood symptom co-occurrence and covariance of severity. Commonly, this approach confirms some rationale for a separation of psychosis and mood symptoms, though such findings are largely conducted with SCZ, with or without SAD included. A specific finding that has held some sway is the five-factor model, with factors being positive symptoms (hallucinations and delusions), negative symptoms (anhedonia, avolition, etc.), disorganization, depression, and mania.[14] However, variations of the five-factor model have been reported following lack of replication. Such replication problems may be due to issues such as studying acutely ill vs. treated patients,[15] in which there may be different underlying structures of symptom covariance. However, most factor structures proposed as alternative to the five-factor model typically retain separation of mood and psychosis symptoms. This was the case for a study including all three major psychotic disorders (SCZ, SAD, and BDP).[16] While these findings offer no clarity into the spectrum of the co-occurrence of mood and psychosis symptoms, they do support the idea of a common underlying etiology *within* each factor. This supports an approach in which such symptoms groupings may be fruitful targets for research.

Other empirical efforts to understand the symptoms of SCZ, SAD, and BPD address more qualitative questions, such as whether certain psychosis symptoms are unique when experienced in the context of particular disorders. Historically, some symptoms were thought to be pathognomonic, e.g., Schneider's listing of "first rank" symptoms of schizophrenia. These are symptoms proposed as likely to lead most reliably to diagnosis due to not typically appearing in other disorders. These symptoms include auditory hallucinations of voices conversing with each other or commenting on the hearer's thoughts and behaviors, thought insertion, thought withdrawal, thought broadcasting, delusions of control, and bizarre delusions. These first-rank symptoms were incorporated into DSM editions to make a diagnosis of SCZ easier, rendering additional symptoms not necessary. However, by giving more weight to these types of symptoms, such criteria may have skewed the diagnostic landscape and led to more people being diagnosed with SCZ than would have been had the criteria not reflected this ideology. Over time, first-rank symptoms have not been supported as pathognomonic of SCZ versus SZA or BPD, but they are less frequent and less severe in BPD relative to SCZ.[17,18] This observation of quantitative rather than qualitative differences between SZ, SZA, and BP is echoed in many studies comparing the disorders on biomarkers such as cognition, neuroimaging, and electrophysiology.[19,20]

Beyond questions of pathognomonic psychosis symptoms for any particular psychotic disorder, some work has addressed whether features of psychosis symptoms are particular to any diagnoses. Waters and Fernyhough reviewed 43 studies to identify any characteristic of hallucinations that were specific to schizophrenia compared with other psychiatric diagnoses, and found only age of onset of hallucinations in late adolescence predicted SCZ. Features such as lack of control, interference, and occurring in more than one sensory modality did not differentiate SCZ from BPD.[21] For delusions, less work has been done, but one report of hospitalized patients with delusions found SCZ patients (n = 138) had a more severe pervasive quality to their delusions compared with BPD (n = 72), but no differences on other qualities such as conviction, preoccupation, and negative affect. This same study also reported more frequent body/mind control and thought broadcasting delusion subtypes in SCZ compared with BDP, but no difference for six other delusion subtypes.[22] A proposed cognitive mechanism for delusional propensity, reality monitoring deficits, was found in affective psychosis and SCZ groups with delusion histories relative to affective psychosis and SCZ groups who have never had delusions.[23] More work is needed comparing delusional features across psychotic disorders.

Finally, some research had addressed temporal or causal relationships between delusions and hallucinations. One such line of work proposes a mechanism in which hallucinations may be the first symptom to manifest, followed by the individual creating faulty explanations for the experience that subsequently crystalize and become delusions, and some evidence supports this possibility.[24] However, much more work on such ideas is needed. Certainly this proposal can only account for a portion of individuals, as first episode psychosis cases have been readily identified where delusions emerge first, or where delusions and hallucinations emerge concurrently.[25]

FEATURES OF AND CHALLENGES TO UNDERSTANDING PSYCHOTIC SYMPTOMATOLOGY WITHIN DIAGNOSES

Given agreement that key shared properties across SCZ, SAD, and BDP are hallucinations and delusions, we assess what is known regarding these symptoms within each disorder, circumventing the problem of a paucity of direct comparisons of the disorders. This sequential organization by diagnostic category highlights the problem of nonparallel sets of knowledge of phenomenology across the psychotic disorders.

SCHIZOPHRENIA AND SCHIZOAFFECTIVE DISORDER

Literature on hallucinations and delusions among SCZ and SAD is much richer than for bipolar disorder. Lifetime experience of hallucinations is estimated at about 80% in SCZ, and perhaps slightly less (about 75%) of schizoaffective cases. Frequency of types of hallucinations in SCZ/SAD are 64%–80% auditory, 23%–31% visual, 9%–19% tactile, and 6%–10% olfactory.[26] Gustatory hallucinations are estimated at about 4%–10% in one recent report,[18] and while not always the case, some ascribe these to nonpsychiatric etiology.[27] Studies tend to agree that auditory hallucinations are the most common modality for hallucinations in schizophrenia. While this is likely essentially true based on that consistency of report, the accounting for the magnitude of this dominance of auditory hallucinations in schizophrenia must be acknowledged as dependent to some degree on the role of the diagnostic criteria being used. The most obvious influence is DSM-III and IV, where a diagnosis of SCZ was possible via reporting two or more voices conversing or a voice keeping up a running commentary. No other single hallucination experience resulted in a SCZ diagnosis. In other words, someone with a severe type of visual hallucination would not be diagnosed with SCZ under these DSM editions, and therefore not be included in a study of SCZ hallucination subtypes. While this is not likely a large problem, it is a reminder of the effect diagnostic systems can have.

Reports of SCZ/SAD delusions, regardless of type, are somewhat less common than reports of hallucinations. Estimates of any delusion present in SCZ or SAD range from about 50% to 95%.[28,29] Delusions range widely in terms of theme, level of conviction, and functional impact, and there is more literature focused on specific types of delusions (paranoid, grandiose) in SCZ/SAD than there is on delusions generally. Sometimes there is distinction made as to whether delusions are "bizarre," meaning outside the realm of physical reality. For example, a bizarre delusion would involve believing one's thoughts are being removed via satellite technology. A nonbizarre delusion example would be a belief that other cars driving behind you contain individuals surveilling you. The former is not consistent with current technological capability; the latter is something that *can* happen. Similar to criteria for conversing or commenting voices, DSM-IV had a criterion for bizarre delusions serving as a single positive psychosis symptom that could be solely present and count toward a diagnosis of SCZ. This rule, too, has potentially influenced estimates of delusions in SCZ. Nonbizarre delusions are themselves a likely source of unreliability, as they can be difficult to accurately identify or differentiate from actual experience. For example, individuals with SCZ may report that everyone on the street seems to be looking at them or that people on a bus whisper about them. This may be delusional, or may be experienced over time due to other odd features of the person's behavior (or both). Similarly, some individuals who have not been adequately educated about the nature of their symptoms may develop an explanation for their experience that sounds delusional in nature but may simply be a reflection of accessible knowledge. For example, not understanding hallucinations as a possible symptom of a disorder may leave an individual to rely on other explanations, such as spiritual sources. There is some evidence to suggest that the erroneous attribution of hallucinatory experiences to external sources, in combination with other factors (e.g., impaired cognition, adverse childhood events), is one of the pathways from nonclinical hallucinatory experiences to full psychotic disorder.[30]

BIPOLAR DISORDER

The literature on psychosis in bipolar disorder tends to focus on the prevalence of *any* psychosis, regardless of whether the symptoms are hallucinations or delusions. This is consistent with the irrelevance of which symptom is present for a specifier of psychosis being added to a bipolar disorder diagnosis, but limits understanding of the two symptoms. Prevalence estimates of any psychotic symptom in bipolar disorder range widely. A review of DSM-III-based studies reported a range of 20%–50%[31]. A 2017 study with inpatient and outpatient recruitment found slightly higher lifetime psychosis history of 57.5% of bipolar patients.[32] However, limiting the individuals studied to those recruited from treatment settings may cause an overestimate of psychosis rates, especially given that individuals with bipolar disorder and no psychosis are less likely to seek treatment. An epidemiological study conducted in Australia reported 87.5% of individuals with bipolar disorder experienced delusions, typically for short durations, in their lifetime.[33] Hence, prevalence of psychosis in bipolar disorder remains to be established. There is some evidence suggesting that psychosis is slightly more common in manic relative to depressive episodes[34] when it does occur.

Psychosis during mood episodes of bipolar disorder has been thought to indicate a more severe form of bipolar illness, but evidence for this is equivocal. Some findings are that psychosis in bipolar disorder is associated with earlier age of onset of bipolar symptoms, greater prevalence of comorbid substance use disorders, or more psychiatric hospitalizations (both voluntary and involuntary),[35] but this is not always found.[36,37] Other evidence suggests psychosis in bipolar disorder is associated with more favorable disease features, e.g., shorter mood episodes and lower rates of comorbid disorders.[38] These latter observations could stem from faster identification of need for intervention, as more rapid access to care may reduce overall burden of illness (e.g., less interference with occupational and interpersonal functioning[32]; shorter duration of episodes).

A recent meta-analysis[34] reported the most common types of delusions in bipolar disorder are grandiose (60.7%), influence/control (50.9%), and persecutory (48.2%). Delusions of grandeur were noted to more frequently accompany mania relative to other delusion types. As grandiosity appears most prominent in BPD, we note that the veracity of the findings elevating grandiosity in BPD among others is questionable. Distinguishing between psychotic and nonpsychotic grandiosity is not clearly demarcated in most assessment instruments. For example, the PANSS Grandiosity item allows for "poorly formed delusions." This has unclear implication for making a binary diagnostic judgment of "with psychotic features" or not. Grandiosity is, at minimum, an exaggeration of characteristics or abilities. For example, musicians experiencing grandiosity may have thoughts that they possess much greater mastery of an instrument that they actually do. However, grandiose delusions may be present if something verifiably untrue is expressed, such as having been signed to a major record label. While this distinction between grandiosity and grandiose delusion is clear, a novice musician might report strongly held thoughts that they are destined to be famous and take steps such as relocating cities based on this belief. Gathering details about any functional impairment related to a belief like this, whether the person experiences this belief outside of mood episodes, and whether the individual is able to provide a logical explanation for why they hold this conviction may help to distinguish grandiosity from grandiose delusion. Yet, different clinicians' judgments of responses to such questioning may or may not agree with respect to deciding whether this is delusional. In this way, grandiosity is an exemplar symptom to conceptualize as occurring along a severity dimension. Yet, for practical reasons, this dimensional symptom must still somehow coexist alongside the need for a categorical cut-point in order to diagnose "with (or without) psychotic features."

Hallucinations in bipolar disorder are estimated to have prevalence rates ranging from 12% to 63%.[39] However, despite this wide range, it is fairly clear that when hallucinations are reported, auditory is the most common sensory modality and slightly more likely in manic than depressive episodes.[34]

CONCLUSION

While there are distinctions made between SCZ, SAD, and BDP, it is challenging to support these categorical diagnostic boundaries when examining them closely. The considerable overlap of psychosis symptomatology across SCZ, SAD, and BDP directly challenges the validity of such separation. This dilemma is nothing new, as it can be traced through all the iterations of definitions of these disorders in previous editions of the DSM. The existence of diagnostic systems categorizing primary psychosis and mood disorder presentations has been motivated in part by the need to promote similarity of groups across studies. While most would agree such an effort is worthwhile, the general state of knowledge regarding the neurobiology underlying the striking nature of these primary psychotic disorders remains inadequate, calling into question continued adherence to such a system, at least for research purposes. Recognition that features of psychosis within SCZ, SAD, and BDP are qualitatively similar, along with appreciation of the enduring instability within the field of conceptualizations of SCZ, SAD, and BPD phenomenologically, suggests that biological processes associated with psychotic symptoms, and disease etiology in general, in these disorders may be the same. Under such logic, researchers may be able to advance the current understanding of psychotic disorders by conducting transdiagnostic research. Further, routine detailed dimensional characterization and biological measurement of broadly included psychosis subjects in research studies may lead to exciting and novel advancements in our understanding of the perplexing presentation of major psychotic disorders.

REFERENCES

1. Leon AC, Solomon DA, Mueller TI, et al. A brief assessment of psychosocial functioning of subjects with bipolar I disorder: the LIFE-RIFT. Longitudinal Interval Follow-up Evaluation-Range Impaired Functioning Tool. *J Nerv Ment Dis*. 2000;188(12):805–812.
2. Strakowski SM, Williams JR, Sax KW, Fleck DE, DelBello MP, Bourne ML. Is impaired outcome following a first manic episode due to mood-incongruent psychosis? *J Affect Disord*. 2000;61(1–2):87–94.
3. Carlson GA, Kotov R, Chang S-W, Ruggero C, Bromet EJ. Early determinants of four-year clinical outcomes in bipolar disorder with psychosis. *Bipolar Disorders*. 2012;14(1):19–30.
4. Fennig S, Fennig-Naisberg S, Karant M. Mood-congruent vs. mood-incongruent psychotic symptoms in affective psychotic disorders. *Isr J Psychiatry Relat Sci*. 1996;33(4):238–245.
5. Tohen M, Tsuang MT, Goodwin DC. Prediction of outcome in mania by mood-congruent or mood-incongruent psychotic features. *Am J Psychiatry*. 1992;149(11):1580–1584.
6. Tandon R, Gaebel W, Barch DM, et al. Definition and description of schizophrenia in the DSM-5. *Schizophr Res*. 2013;150(1):3–10.
7. Winokur G, Scharfetter C, Angst J. The diagnostic value in assessing mood congruence in delusions and hallucinations and their relationship to the affective state. *Eur Arch Psychiatry Neurol Sci*. 1985;234(5):299–302.
8. Keshavan MS, Morris DW, Sweeney JA, et al. A dimensional approach to the psychosis spectrum between bipolar disorder and schizophrenia: the Schizo-Bipolar Scale. *Schizophr Res*. 2011;133(1–3):250–254.
9. Bromet EJ, Kotov R, Fochtmann LJ, et al. Diagnostic shifts during the decade following first admission for psychosis. *Am J Psychiatry*. 2011;168(11):1186–1194.
10. Salvatore P, Baldessarini RJ, Tohen M, et al. McLean-Harvard International First-Episode Project: two-year stability of DSM-IV diagnoses in 500 first-episode psychotic disorder patients. *J Clin Psychiatry*. 2009;70(4):458–466.
11. Kim JS, Baek JH, Choi JS, Lee D, Kwon JS, Hong KS. Diagnostic stability of first-episode psychosis and predictors of diagnostic shift from non-affective psychosis to bipolar disorder: a retrospective evaluation after recurrence. *Psychiatry Res*. 2011;188(1):29–33.
12. Fusar-Poli P, Cappucciati M, Rutigliano G, et al. Diagnostic stability of ICD/DSM first episode psychosis diagnoses: meta-analysis. *Schizophr Bull*. 2016;42(6):1395–1406.
13. Ivleva EI, Morris DW, Moates AF, Suppes T, Thaker GK, Tamminga CA. Genetics and intermediate phenotypes of the schizophrenia—bipolar disorder boundary. *Neurosci Biobehav Rev*. 2010;34(6):897–921.
14. <Altered white gray proportions in striatum of patients with schizophrenia Tamagaki 2005.pdf>.
15. Anderson A, Wilcox M, Savitz A, et al. Sparse factors for the positive and negative syndrome scale: which symptoms and stage of illness? *Psychiatry Res*. 2015;225(3):283–290.
16. Anderson AE, Mansolf M, Reise SP, et al. Measuring pathology using the PANSS across diagnoses: inconsistency of the positive symptom domain across schizophrenia, schizoaffective, and bipolar disorder. *Psychiatry Res*. 2017;258:207–216.
17. Rosen C, Grossman LS, Harrow M, Bonner-Jackson A, Faull R. Diagnostic and prognostic significance of Schneiderian first-rank symptoms: a 20-year longitudinal study of schizophrenia and bipolar disorder. *Comprehensive Psychiatry*. 2011;52(2):126–131.
18. Shinn AK, Pfaff D, Young S, Lewandowski KE, Cohen BM, Ongur D. Auditory hallucinations in a cross-diagnostic sample of psychotic disorder patients: a descriptive, cross-sectional study. *Comprehensive Psychiatry*. 2012;53(6):718–726.
19. Kempf L, Hussain N, Potash JB. Mood disorder with psychotic features, schizoaffective disorder, and schizophrenia with mood features: trouble at the borders. *International Review of Psychiatry*. 2005;17(1):9–19.
20. Pearlson GD, Clementz BA, Sweeney JA, Keshavan MS, Tamminga CA. Does biology transcend the symptom-based boundaries of psychosis? *Psychiat Clin N Am*. 2016;39(2):165–+.
21. Waters F, Fernyhough C. Hallucinations: a systematic review of points of similarity and difference across diagnostic classes. *Schizophr Bull*. 2017;43(1):32–43.
22. Appelbaum PS, Robbins PC, Roth LH. Dimensional approach to delusions: comparison across types and diagnoses. *American Journal of Psychiatry*. 1999;156(12):1938–1943.
23. Radaelli D, Benedetti F, Cavallaro R, Colombo C, Smeraldi E. The reality monitoring deficit as a common neuropsychological correlate of schizophrenic and affective psychosis. *Behav Sci (Basel)*. 2013;3(2):244–252.
24. De Loore E, Gunther N, Drukker M, et al. Persistence and outcome of auditory hallucinations in adolescence: a longitudinal general population study of 1800 individuals. *Schizophr Res*. 2011;127(1–3):252–256.
25. Compton MT, Potts AA, Wan CR, Ionescu DF. Which came first, delusions or hallucinations? An exploration of clinical differences among patients with first-episode psychosis based on patterns of emergence of positive symptoms. *Psychiatry Res*. 2012;200(2–3):702–707.
26. McCarthy-Jones S, Smailes D, Corvin A, et al. Occurrence and co-occurrence of hallucinations by modality in schizophrenia-spectrum disorders. *Psychiatry Res*. 2017;252:154–160.
27. Mueser KT, Bellack AS, Brady EU. Hallucinations in schizophrenia. *Acta Psychiatr Scand*. 1990;82(1):26–29.
28. Harrow M, MacDonald AW 3rd, Sands JR, Silverstein ML. Vulnerability to delusions over time in schizophrenia and affective disorders. *Schizophr Bull*. 1995;21(1):95–109.
29. Paolini E, Moretti P, Compton MT. Delusions in first-episode psychosis: Principal component analysis of twelve types of delusions and

demographic and clinical correlates of resulting domains. *Psychiatry Res.* 2016;243:5–13.
30. Krabbendam L, Myin-Germeys I, Hanssen M, et al. Hallucinatory experiences and onset of psychotic disorder: evidence that the risk is mediated by delusion formation. *Acta Psychiatr Scand.* 2004;110(4):264–272.
31. Pope HG Jr., Lipinski JF Jr. Diagnosis in schizophrenia and manic-depressive illness: a reassessment of the specificity of "schizophrenic" symptoms in the light of current research. *Arch Gen Psychiatry.* 1978;35(7):811–828.
32. Dell'Osso B, Camuri G, Cremaschi L, et al. Lifetime presence of psychotic symptoms in bipolar disorder is associated with less favorable socio-demographic and certain clinical features. *Compr Psychiatry.* 2017;76:169–176.
33. Morgan VA, Mitchell PB, Jablensky AV. The epidemiology of bipolar disorder: sociodemographic, disability and service utilization data from the Australian National Study of Low Prevalence (Psychotic) Disorders. *Bipolar Disord.* 2005;7(4):326–337.
34. Smith LM, Johns LC, Mitchell R. Characterizing the experience of auditory verbal hallucinations and accompanying delusions in individuals with a diagnosis of bipolar disorder: a systematic review. *Bipolar Disorders.* 2017;19(6):417–433.
35. Bora E, Yucel M, Pantelis C. Neurocognitive markers of psychosis in bipolar disorder: a meta-analytic study. *J Affect Disord.* 2010;127(1–3):1–9.
36. Keck PE Jr., McElroy SL, Havens JR, et al. Psychosis in bipolar disorder: phenomenology and impact on morbidity and course of illness. *Comprehensive Psychiatry.* 2003;44(4):263–269.
37. Burton CZ, Ryan KA, Kamali M, et al. Psychosis in bipolar disorder: does it represent a more "severe" illness? *Bipolar Disorders.* 2017.
38. Aminoff SR, Hellvin T, Lagerberg TV, Berg AO, Andreassen OA, Melle I. Neurocognitive features in subgroups of bipolar disorder. *Bipolar Disorders.* 2013;15(3):272–283.
39. Toh WL, Thomas N, Rossell SL. Auditory verbal hallucinations in bipolar disorder (BD) and major depressive disorder (MDD): a systematic review. *J Affect Disord.* 2015;184:18–28.

6

PSYCHOTIC SYMPTOMS IN BIPOLAR DISORDER

Marsal Sanches, Xiang-Yang Zhang, and Jair C. Soares

INTRODUCTION

Bipolar disorder (BD) is a severe, recurrent mental illness characterized by fluctuations in the affective state and energy level.[1] It includes two major types: BD type I, characterized by recurrent episodes of mania and, typically, depression, and BD type II, with recurrent major depression and hypomania. The latter has been traditionally considered a milder form of BD.[2] Bipolar disorder corresponds to one of the most prevalent severe mental disorders, affecting 2.6% of the US adult population in a given year, with a lifetime prevalence of 3.9%.[3,4] More than 90% of BD patients experience at least two lifetime acute affective episodes, with most patients experiencing multiple recurrences of mania and depression, even with treatments of proven efficacy.[5]

Moreover, BD is associated with significant functional impairment, as well as increased mortality secondary to suicide and later adverse outcomes of co-occurring medical illnesses.[6] It also imposes a significant burden on society, due to its association with increased economic and family burden, poor quality of life, increased absenteeism and disability as well as suicidality.[1] Nevertheless, despite clinical significance of BD, its pathophysiological mechanisms are not completely understood, and, to date, there are no valid, clinically applicable biomarkers for the disorder.

Psychotic features are frequent among BD patients, and may occur during both depressive and manic episodes.[7] In the present chapter, we discuss some of the epidemiological and clinical aspects of psychotic symptoms in bipolar patients.

PREVALENCE AND PHENOMENOLOGY OF PSYCHOTIC SYMPTOMS IN BD

Psychotic symptoms in BD include delusions,[8] hallucinations (including mood-incongruent and Schneiderian first-rank symptoms (e.g., third-person auditory hallucinations),[7,9] and formal thought disorder.[7] The prevalence of psychosis is high among BD patients.[10] Epidemiological data show that nearly 50% of BD patients exhibit a lifetime history of psychotic symptoms,[7] and a recent study reported a lifetime occurrence of psychotic symptoms in more than half of BD patients.[10] In a large, comprehensively characterized sample of 1,342 BD-I patients,[11] a high frequency of lifetime psychotic symptoms (73.8%) was reported, including delusions (68.9%), hallucinations (42.7%), mood incongruent symptoms (30.1%), Schneiderian symptoms (21.2%), and formal thought disorder (59.7%). Notably, the authors also reported that 28.6% of BD-I patients had visual hallucinations, which is comparable to the rate of visual hallucinations reported among patients with schizophrenia[12] and corresponds to a much higher proportion than the previously reported 14% prevalence of visual hallucinations in BD.[8]

Moreover, a high prevalence of psychotic symptoms among BD-II patients has also been reported and seems to vary broadly, from 3 to 45%.[13–16] A comparison between subgroups of BD-I and BD-II patients indicated that 95.7% of BD-I patients vs. 4.3% of BD-II patients experienced psychotic features in their lifetime, suggesting that such symptomatology is much more frequent in BD-I patients.[10]

DEMOGRAPHIC CHARACTERISTICS ASSOCIATED WITH PSYCHOTIC SYMPTOMS IN BD

Several studies demonstrated comparable rates of hallucinations in BD patients of both genders.[7,17,18] However, the study by Upthegrove et al. (2015)[8] reported a higher rate of psychotic symptoms among female BD patients than among males. Another recent study also reported that female patients with BD are more likely to suffer from hallucinations, compared to males.[11] It is worthy of mentioning that, in contrast to schizophrenia, sex ratios are nearly equal in BD.[19,20] Women with BD report higher level of childhood maltreatment than men,[11] and childhood trauma has been linked to risk for psychosis in affective disorders among females but not among males.[21] These findings support possible interactions between female gender and childhood mistreatment with respect to the risk of psychotic symptoms.

Other studies have also reported associations between early life events and a history of hallucinations in BD.[8,22] Moreover, it has been reported that the relationship between childhood adversity and psychosis in BD is particularly strong for auditory hallucinations.[11] It is noteworthy that this relationship is also reported in schizophrenia and is independent of specific types of childhood adversity,[23,24] indicating that this relationship likely spans diagnostic boundary of severe mental illness.

CLINICAL CORRELATES OF PSYCHOTIC SYMPTOMS IN BD

Numerous studies described associations between psychotic symptoms and earlier illness onset among BD patients,[8,17] as well as greater symptom severity,[18,25] more severe illness course/mood episodes, and frequent hospitalizations[8,17,18] Associations of psychosis with lower response to lithium,[26,27] and higher rates of comorbidity[25] have also been described. A more recent study indicates that BD patients with a history of psychotic symptoms have an earlier illness onset and more mania-related hospitalizations, as compared to BD patients without psychotic symptoms.[11,18] Interestingly, the most recent genome-wide association study (GWAS) showed that BD patients with psychotic features have significantly higher schizophrenia polygenic risk scores than those patients without psychotic features. Moreover, high polygenic risk scores for schizophrenia in BD patients are associated with more frequent hospitalizations and earlier onset of the disease.[28] These data suggest that, within the bipolar spectrum, BD patients with a history of psychosis may have different genetic basis. In addition, previous studies have shown more frequent depressive episodes in BD-I patients with mood-incongruent psychotic symptoms.[29] Nevertheless, this association has not been replicated by two recent studies.[8,11]

COGNITIVE FUNCTION IN PSYCHOTIC VS. NONPSYCHOTIC BD

Evidence suggests that a history of hallucinations and/or delusions in the course of BD-I is associated with severe cognitive deficits. A worse performance on the Stroop test (a measure of executive function) among psychotic vs. nonpsychotic BD has been previously reported.[30] A meta-analysis comparing cognitive performance of bipolar patients with and without a history of psychosis revealed that the patients with a history of psychosis performed significantly worse in four cognitive domains (planning and reasoning, working memory, verbal memory, and processing speed), with moderately greater impairment on these tasks (Cohen's $d = 0.30–0.55$).[17] However, several other studies reported no differences in cognitive deficits between BD patients with or without history of psychotic symptoms.[31–35] Interestingly, a recent study showed that patients with a history of delusions had a significantly higher educational level and premorbid IQ, but no significant differences in current IQ, compared with patients without psychotic symptoms.[11] Taken together, these findings indicate that the relationship between cognitive function and psychotic symptoms is still unclear and deserves further investigation.

SUBSTANCE USE IN PSYCHOTIC VS. NONPSYCHOTIC BD

Substance use disorders (SUDs) are common among bipolar patients.[36–38] A recent meta-analysis showed that the SUD with the highest prevalence in BD were alcohol use disorder (42%), followed by cannabis use disorder (20%). Moreover, males had a higher lifetime prevalence of SUD compared to females. Bipolar disorder with a comorbid SUD was associated with earlier age of onset and a higher number of hospitalizations than BD without SUD.[36] Another meta-analysis described associations between SUD and male gender, number of manic episodes, and history of suicidality.[38] These results demonstrate that SUDs may affect many aspects of BD relevant to clinical course, psychopathology and prognosis. Nevertheless, studies specifically addressing a possible relationship between SUD and psychotic symptoms in BD are rare. A recent study reported a lack of correlation between lifetime SUD and psychotic symptoms among BD patients. Moreover, current or lifetime alcohol and substance use were not found to be associated with delusions, hallucinations, mood incongruent symptoms, Schneiderian symptoms, and disorganized speech.[11]

SUMMARY

Many studies have demonstrated the relevance of psychosis in BD patients, especially in BD-I.[7,8,11,17] A history of psychotic symptoms in BD has been associated with several demographic and clinical characteristics, including earlier age of onset,[8] higher symptom severity,[11] higher frequency of mood episodes, especially manic episodes, and a higher number of hospital admissions, as well as possibly more severe cognitive impairment.[11,18,32,34,39] Moreover, certain modalities of psychotic symptoms (for example, hallucinations) seem to be particularly associated with a history of childhood maltreatment among bipolar patients.[11]

Since similar psychotic symptoms occur in both schizophrenia and BD, it has been hypothesized that these disorders may constitute a diagnostic continuum with shared neurobiological characteristics.[40] Some authors advocate that this psychosis continuum would extend from BD to schizoaffective disorder, with schizophrenia occupying the other end of the spectrum.[41] Considerable evidence from biomarker-focused studies supports highly overlapping clinical, cognitive, neurophysiologic, and brain-imaging characteristics across schizophrenia, schizoaffective disorder, and psychotic BD.[42–44]

Finally, it has been proposed that, within the bipolar spectrum, psychotic BD could represent a distinct disease subtype,[45] and psychotic BD may be associated with different patterns of genetic vulnerability.[28,46] Longitudinal studies focusing on distinct risk factors related to psychotic symptoms among bipolar patients, combined with recent genetic insights spanning diagnostic boundaries, may help clarify pathophysiological aspects and diagnostic and nosological implications of psychotic symptoms in BD.

REFERENCES

1. Grande I, Berk M, Birmaher B, Vieta E. Bipolar disorder. Lancet. 2016; 387(10027): 1561–1572.

2. Dell'Osso B, Holtzman JN, Goffin KC, et al. American tertiary clinic-referred bipolar II disorder compared to bipolar I disorder: more severe in multiple ways, but less severe in a few other ways. J Affect Disord. 2015; 188: 257–262.
3. Merikangas KR, Jin R, He JP, et al. Prevalence and correlations of bipolar spectrum disorder in the world mental health survey initiative. Arch Gen Psychiatry. 2011; 68(3): 241–251.
4. Soares JC. Recent advances in the treatment of bipolar mania, depression, mixed states, and rapid cycling. Int Clin Psychopharmacol. 2000; 15(4): 183–196.
5. Vázquez GH, Holtzman JN, Lolich M, Ketter TA, Baldessarini RJ. Recurrence rates in bipolar disorder: systematic comparison of long-term prospective, naturalistic studies versus randomized controlled trials. Eur Neuropsychopharmacol. 2015; 25(10): 1501–1512.
6. Hayes JF, Miles J, Walters K, King M, Osborn DP. A systematic review and meta-analysis of premature mortality in bipolar affective disorder. Acta Psychiatr Scand. 2015; 131(6): 417–425.
7. Keck PK Jr., McElroy SL, Havens JR, et al. Psychosis in bipolar disorder: phenomenology and impact on morbidity and course of illness. Compr Psychiatry. 2003; 44(4): 263–269.
8. Upthegrove R, Chard C, Jones L, et al. Adverse childhood events and psychosis in bipolar affective disorder. Br J Psychiatry. 2015; 206: 191–197.
9. Carlson GA, Kotov R, Chang SW, Ruggero C, Bromet EJ. Early determinants of four- year clinical outcomes in bipolar disorder with psychosis. Bipolar Disorders. 2012; 14: 19–30.
10. Dell'Osso B, Camuri G, Cremaschi L, et al. Lifetime presence of psychotic symptoms in bipolar disorder is associated with less favorable socio-demographic and certain clinical features. Compr Psychiatry. 2017; 76: 169–176.
11. van Bergen AH, Verkooijen S, Vreeker A, et al. The characteristics of psychotic features in bipolar disorder. Psychol Med, in press.
12. Bauer SM, Schanda H, Karakula H, et al. Culture and the prevalence of hallucinations in schizophrenia. Comprehensive Psychiatry. 2011; 52: 319–325.
13. Yildiz A, Sachs GS. Age onset of psychotic versus nonpsychotic bipolar illness in men and in women. J Affect Disord 2003; 74(2): 197–201.
14. Mantere O, Suominen K, Leppämäki S, Valtonen H, Arvilommi P, Isometsä E. The clinical characteristics of DSM-IV bipolar I and II disorders: baseline findings from the Jorvi Bipolar Study (JoB). Bipolar Disord. 2004; 6(5): 395–405.
15. Mazzarini L, Colom F, Pacchiarotti I, Nivoli AM, Murru A, Bonnin CM. Psychotic versus non-psychotic bipolar II disorder. J Affect Disord. 2010; 126(1–2): 55–60.
16. Goes FS, Sadler B, Toolan J, et al. Psychotic features in bipolar and unipolar depression. Bipolar Disord. 2007; 9(8): 901–906.
17. Bora E, Yücel M, Pantelis C. Neurocognitive markers of psychosis in bipolar disorder: a meta-analytic study. Journal of Affective Disorders. 2010; 127: 1–9.
18. Özyildirim I, Çakir S, Yazici O. Impact of psychotic features on morbidity and course of illness in patients with bipolar disorder. European Psychiatry. 2010; 25: 47–51.
19. Hendrick V, Altshuler LL, Gitlin MJ, Delrahim S, Hammen C. Gender and bipolar illness. J Clin Psychiatry. 2000; 61: 393–396.
20. Aleman A, Kahn RS, Selten JP. Sex differences in the risk of schizophrenia: evidence from meta-analysis. Arch Gen Psychiatry. 2003; 60: 565–571.
21. Fisher H, Morgan C, Dazzan P, et al. Gender differences in the association between childhood abuse and psychosis. Br J Psychiatry. 2009; 194: 319–325.
22. Hammersley P, Dias A, Todd G, Bowen-Jones K, Reilly B, Bentall RP. Childhood trauma and hallucinations in bipolar affective disorder: preliminary investigation. Br J Psychiatry. 2003; 182: 543–547.
23. Read J, van Os J, Morrison AP, Ross CA. Childhood trauma, psychosis and schizophrenia: a literature review with theoretical and clinical implications. Acta Psychiatr Scand. 2005; 112: 330–350.
24. Varese F, Smeets F, Drukker M, et al. Childhood adversities increase the risk of psychosis: a meta-analysis of patient-control, prospective-and cross-sectional cohort studies. Schizophr Bull. 2012; 38: 661–671.
25. Coryell W, Leon AC, Turvey C, Akiskal HS, Mueller T, Endicott J. The significance of psychotic features in manic episodes: a report from the NIMH collaborative study. J Affect Disord. 2001; 67(1–3): 79–88.
26. Maj M. The effect of lithium in bipolar disorder: a review of recent research evidence. Bipolar Disorders. 2003; 5: 180–188.
27. Maj M, Pirozzi R, Bartoli L, Magliano L. Long-term outcome of lithium prophylaxis in bipolar disorder with mood-incongruent psychotic features: a prospective study. J Affect Disord. 2002; 71: 195–198.
28. Bipolar Disorder and schizophrenia Working Group of the Psychiatric Genomics Consortium. Genomic dissection of bipolar disorder and schizophrenia, including 28 subphenotypes. Cell. 2018; 173: 1705–1715.
29. Tohen M, Zarate CA Jr., Hennen J, et al. The McLean-Harvard First-Episode Mania Study: prediction of recovery and first recurrence. Am J Psychiatry. 2003; 160(12): 2099–2107.
30. Selva G, Salazar J, Balanzá-Martínez V, et al. Bipolar I patients with and without a history of psychotic symptoms: do they differ in their cognitive functioning? J Psychiatric Res. 2007; 41: 265–272.
31. Glahn DC, Bearden CE, Cakir S, et al. Differential working memory impairment in bipolar disorder and schizophrenia: effects of lifetime history of psychosis. Bipolar Disorders. 2006; 8: 117–123.
32. Glahn DC, Bearden CE, Barguil M, et al. The neurocognitive signature of psychotic bipolar disorder. Biological Psychiatry. 2007; 62: 910–916.
33. Savitz J, van der Merwe L, Stein DJ, Solms M, Ramesar R. Neuropsychological status of bipolar I disorder: impact of psychosis. Br J Psychiatry. 2009; 194: 243–251.
34. Simonsen C, Sundet K, Vaskinn A, et al. Neurocognitive dysfunction in bipolar and schizophrenia spectrum disorders depends on history of psychosis rather than diagnostic group. Schizophr Bull. 2011; 37: 73–83.
35. Aminoff SR, Hellvin T, Lagerberg TV, Andreassen OA, Melle I. Neurocognitive features in subgroups of bipolar disorder. Bipolar Disorders. 2013; 15: 272–283.
36. Hunt GE, Malhi GS, Cleary M, Lai HM, Sitharthan T. Prevalence of comorbid bipolar and substance use disorders in clinical settings, 1990–2015: Systematic review and meta-analysis. J Affect Disord. 2016; 206: 331–349.
37. Hunt GE, Malhi GS, Cleary M, Lai HM, Sitharthan TJ. Comorbidity of bipolar and substance use disorders in national surveys of general populations, 1990–2015: Systematic review and meta-analysis. J Affect Disord. 2016; 206: 321–330.
38. Messer T, Lammers G, Müller-Siecheneder F, Schmidt RF, Latifi S. Substance abuse in patients with bipolar disorder: a systematic review and meta-analysis. Psychiatry Res. 2017; 253: 338–350.
39. Levy B, Medina AM, Weiss RD. Cognitive and psychosocial functioning in bipolar disorder with and without psychosis during early remission from an acute mood episode: a comparative longitudinal study. Comprehensive Psychiatry. 2013; 54: 618–626.
40. van Os J, Reininghaus U. Psychosis as a transdiagnostic and extended phenotype in the general population. World Psychiatry. 2016; 15: 118–124.
41. Craddock N, O'Donovan MC, Owen MJ. The genetics of schizophrenia and bipolar disorder: dissecting psychosis. J Med Genet. 2005; 42: 193–204.
42. Hill SK, Reilly JL, Keefe RS, et al. Neuropsychological impairments in schizophrenia and psychotic bipolar disorder: findings from the Bipolar-Schizophrenia Network on Intermediate Phenotypes (B-SNIP) study. Am J Psychiatry. 2013; 170(11): 1275–1284.
43. Clementz BA, Sweeney JA, Hamm JP, et al. Identification of distinct psychosis biotypes using brain-based biomarkers. Am J Psychiatry. 2016; 173(4): 373–384.

44. Ivleva EI, Clementz BA, Dutcher AM, et al. Brain structure biomarkers in the psychosis biotypes: findings from the Bipolar-Schizophrenia Network for Intermediate Phenotypes. Biol Psychiatry. 2017; 82(1): 26–39.
45. Toni A, Perugi G, Mata B, Madaro D, Maremmani I, Akiskal HS. Is mood-incongruent manic psychosis a distinct subtype? Eur Arch Psychiatry and Clin Neurosci. 2001; 251: 12–17.
46. Potash JB, Yen-Feng C, MacKinnon DF, et al. Familial aggregation of psychotic symptoms in a replication set of 69 bipolar disorder pedigrees. Am J Med Genet. Part B, Neuropsychiatr Genet. 2003; 116B: 90–97.

7

MAJOR DEPRESSIVE DISORDER WITH PSYCHOTIC FEATURES

CONFRONTING AND RESOLVING THE DIMENSIONAL CHALLENGE

John L. Waddington, Tara Kingston, Nnamdi Nkire, and Vincent Russell

INTRODUCTION

The autobiographical experiences of Rebecca Lawrence,[1] a consultant psychiatrist in the United Kingdom who is also a patient with an ICD-10 diagnosis of *F33.3 Major depressive disorder, recurrent, severe with psychotic symptoms*, represent one individual's publicly articulated, personal journey along a route with origins in ancient times: that a person with what would now be called a depressive illness can also show the delusional beliefs and/or experience hallucinatory phenomena that would now be called psychosis.[2,3] On this background the enduring debate regarding the "Kraepelinian dichotomy," i.e., the extent to which dementia praecox (schizophrenia) and psychotic disorders with a prominent mood component constitute distinct categories of serious mental illness,[4,5] is in reality only a relatively recent issue along a far more protracted route.

In 2013 we[6] re-emphasized that the conceptualization, classification and understanding of major depressive disorder (MDD) with psychotic features (MDDP) (also known as psychotic depression[7]) has long been both unsatisfactory and contentious. Nosology is embracing a dimensional approach that is complementary to or independent of contemporary diagnostic criteria that struggle to separate these features reliably into distinct categories. Thus, psychotic illness is posited to be best described in terms of varying severities along numerous dimensions of psychopathology,[4–12] rather than by categories that are "purely phenomenological, and largely arbitrary, and not based on valid etiological concepts or mechanisms of illness or genetic predispositions."[13]

However, while these formulations have offered several exemplars beyond schizophrenia, to encompass deficit syndrome schizophrenia, schizoaffective disorder, and bipolar disorder, MDDP is conspicuous in its absence. Persons experiencing psychosis on a background of MDD reflect starkly the co-occurrence of psychotic and affective diagnoses and typify robustly the intersection of psychotic and affective dimensions of psychopathology; thus, it remains enigmatic that MDDP has not received comparable systematic consideration.[6] The present review seeks to fill this void in terms of epidemiology, psychopathology, neuropsychology, pathobiology, treatment, and long-term outcome of MDDP vis-à-vis schizophrenia, schizoaffective disorder, and bipolar disorder. This is followed by a conceptual reanalysis of MDDP vis-à-vis MDD that leads to consideration of the trajectory of a broad milieu of psychosis that is but one element in a yet broader dimensional space of psychopathology and dysfunction.

EPIDEMIOLOGY

THE FIRST PSYCHOTIC EPISODE

Studies that have systematically ascertained and assessed large numbers of incident cases of MDDP at the first psychotic episode, together with other psychotic diagnoses, on a representative basis, and followed these prospectively over a prolonged period, in a manner satisfying the methodological criteria proposed[14] for the epidemiological study of any disease, are limited. Several studies have included MDDP and constitute important datasets for the study of first episode psychosis[10,15–20] However, they are less able to fully illuminate the epidemiology of MDDP vis-à-vis schizophrenia, schizoaffective disorder, bipolar disorder and other psychotic disorders due to restricted representativeness at a population level, diversity of ascertainment in terms of modes of presentation included, arbitrary upper age cut-offs, presentation of data in the absence of information on the population at risk, and/or combining MDDP with bipolar disorder into a composite of "affective psychosis"; additionally, they are, to varying extents, subject to the well-recognized variables of urbanicity[21,22] and migration.[23,24] The only available meta-analysis, involving four studies, all subject to the above limitations, indicates the incidence of "depressive psychoses" to be 5.3/100,000 person-years (95% confidence interval 3.7–7.6); comparative data: schizophrenia 15.2/100,000 (11.9–19.5); bipolar disorder 3.7/100,000 (3.0–4.5); schizoaffective disorder was not analyzed separately due to "volatility" in this diagnosis.[25]

Using an alternative approach, the Cavan-Monaghan First Episode Psychosis Study (CAMFEPS) has incepted 432 cases of first episode psychosis as ascertained on an epidemiological basis in two rural counties in Ireland, with DSM-IV diagnosis as a post hoc assessment rather than an inclusion/exclusion criterion. As described previously in detail,[26–28] this study involves a defined catchment area, case ascertainment via all routes to care (i.e., public, private and forensic; inpatient, outpatient and home-based), full diagnostic scope (i.e., all 12 DSM-IV psychotic diagnoses, including MDDP) and no arbitrary upper age cut-off (i.e., cases incepted throughout the adult lifespan); there is no urbanicity and minimal migration. On this basis, CAMFEPS has generated incidence data that include the following for diagnoses made 6 months following first presentation with psychosis: schizophrenia 6.4/100,000, 3.1-fold more common in men than in women; schizoaffective disorder 2.0/100,000, 1.9-fold more common in men than in women; bipolar disorder 6.9/100,000, indistinguishable between men and women; MDDP 7.3/100,000, indistinguishable between men and women; mean age at first presentation with psychosis, a proxy for age at onset, was indistinguishable between schizophrenia, schizoaffective disorder, and bipolar disorder, being younger in men than in women only for schizophrenia, but was almost 20 years older for MDDP[26,28] (and Nkire et al., in preparation). When juxtaposed with previous evidence that psychotic depression may be overrepresented in older persons,[29] this emphasizes how the application of arbitrary upper age cut-offs may render studies of MDDP seriously incomplete and unrepresentative.

OVERVIEW

In such terms, these four psychotic diagnoses may appear to be characterized by distinct epidemiological "signatures." However, it is essential that these "signatures" not be misinterpreted as in any way validating such diagnoses; these differences are quantitative but not qualitative, as each diagnosis, including MDDP, can occur in either sex, with the onset of psychosis occurring at any age over the adult life span, from the teens through to the ninth decade.[26,28]

PSYCHOPATHOLOGY

THE FIRST PSYCHOTIC EPISODE

On a background of factor analytic studies of psychopathology in first episode studies that have indicated diverse solutions,[30,31] few studies have evaluated the psychopathology of MDDP in systematic comparison with schizophrenia, schizoaffective disorder, and bipolar disorder within epidemiologically representative populations and using widely applied assessment instruments to allow such comparisons. Among cases of first episode psychosis ascertained via an urban university medical center, community agencies, private referrals, and state-operated facilities, those with MDDP evidenced lower positive symptom scores on the Positive and Negative Syndrome Scale (PANSS) than did those with schizophrenia (within which schizoaffective disorder and schizophreniform disorder were subsumed) or bipolar disorder, with schizophrenia and bipolar disorder being indistinguishable; those with bipolar disorder evidenced lower negative symptom scores than did those with schizophrenia or MDDP, with schizophrenia and MDDP being indistinguishable; depression scores were higher in MDDP than in schizophrenia or bipolar disorder, with schizophrenia and bipolar disorder being indistinguishable.[32] In CAMFEPS, schizophrenia, schizoaffective disorder, bipolar disorder, and MDDP evidenced indistinguishable PANSS positive symptom and general symptom scores; negative symptom scores were slightly lower in schizoaffective disorder and MDDP than in schizophrenia, and lower still in bipolar disorder[26,28] (and Nkire et al., in preparation).

OVERVIEW

While these four psychotic diagnoses may be characterized by some differences in psychopathological "signature," where present these are quantitative rather than qualitative (but see[33]), with considerable symptom overlap across these diagnostic groups, and each of these domains of psychopathology can occur in each diagnostic group, including MDDP, in either sex at any age over the adult life span. Also, it is important to note that, using most currently available instruments, depressive symptom and negative symptom scores in psychotic illness can be confounded.[28,34] The challenge for the field is whether this indicates some failure of current assessment instruments to distinguish intrinsically distinct domains of psychopathology, or that a symptom such as anhedonia can occur in both affective and psychotic disorders as a unitary domain of psychopathology (see "Conceptual re-analysis").

NEUROPSYCHOLOGY

THE FIRST PSYCHOTIC EPISODE

While cognitive impairment is a well-recognized feature of both nonaffective and affective psychosis that is associated with poor functional outcome, this relates primarily to comparisons between schizophrenia and bipolar disorder,[35] with MDDP having received less independent investigation. On meta-analysis of patients at any stage of illness, those with affective psychosis were impaired on global cognition and on each of six cognitive domains, with large effect sizes for most measures and no differences between bipolar disorder and MDDP.[36] In a subsequent study at the first psychotic episode, the schizophrenia group exhibited widespread impairment and the bipolar group only limited impairment across six cognitive domains; the MDDP group also demonstrated widespread impairment, though over a narrower range of measures than for schizophrenia.[37] It has been reported that negative but not positive symptoms are associated with impairment in verbal memory and executive function in MDDP, as reported also in schizophrenia, with this relationship enduring for verbal memory after controlling for severity of depression.[38]

In CAMFEPS, ascertainment of first episode psychosis cases was throughout the adult lifespan, up to the tenth decade, together with instances of psychosis due to a general medical condition. This necessitated initial assessment of general cognition using the Mini-Mental State Examination (MMSE), so as to identify any instances of marked cognitive impairment, followed by evaluation of executive dysfunction; neither MMSE scores nor executive dysfunction differed between schizophrenia, schizoaffective disorder, bipolar disorder, and MDDP[28] (and Nkire et al, in preparation). Regarding insight, a previous study reported this to be more impaired in bipolar patients experiencing a manic episode than in those experiencing a mixed or depressive episode or MDDP,[39] while in CAMFEPS insight did not differ between schizophrenia, schizoaffective disorder, bipolar disorder, and MDDP (Nkire et al., in preparation).

PREMORBID FEATURES

While systematic review and meta-analysis indicates premorbid IQ to be lower in individuals who go on to receive a diagnosis of schizophrenia,[40] to a greater extent than in those who go on to receive a diagnosis of bipolar disorder,[41] studies of premorbid IQ in MDDP have indicated no difference from schizophrenia.[37,42] In CAMFEPS, neither premorbid IQ nor premorbid functioning differed between schizophrenia, schizoaffective disorder, bipolar disorder, and MDDP[28] (and Nkire et al., in preparation).

OVERVIEW

While these four psychotic diagnoses may be characterized by some differences in neuropsychological "signature," where present these are quantitative but not qualitative, with considerable overlap in the extent of cognitive impairment across these diagnostic groups, and each of these domains of cognition can be impaired in each diagnostic group, including MDDP, in either sex at any age over the adult life span.

PATHOBIOLOGY

GENETICS

Recent insights from extensive genome-wide association studies (GWAS) studies indicate some shared genetic risk between schizophrenia and bipolar disorder, particularly in terms of common genes of small effect[43,44] and less so for rare copy number variations (CNVs) of large effect,[45,46] but have rarely included schizoaffective disorder.[47] While family studies indicate some risk for psychotic depression to be partly shared with both affective illness and psychotic disorders, linkage and association studies only variably suggest some risk loci shared across the psychosis spectrum.[48] Thus, in the absence of GWAS/CNV studies that include MDDP as an independent diagnostic group, there is, as yet, no reliable evidence base.

NEUROIMAGING

The multitude of structural and functional neuroimaging studies on schizophrenia and bipolar disorder[49] is in marked contrast to the paucity of studies on MDDP.[50,51] Thus, in the face of so few neuroimaging studies that include MDDP as an independent diagnostic group, there is, as yet, no reliable evidence base.

NEUROLOGY

As for neuroimaging, the numerous studies on neurological soft signs in schizophrenia and bipolar disorder,[52] while not wholly consistent,[53,54] are in marked contrast to the paucity of studies on MDDP. In CAMFEPS, the occurrence of neurological soft signs was similar in MDDP, schizophrenia, schizoaffective disorder, and bipolar disorder[28] (and Nkire et al., in preparation).

NEUROPHYSIOLOGY

On investigating oculomotor abnormalities, each of schizophrenia, bipolar disorder, and MDDP showed similarly elevated antisaccade error rates at the first psychotic episode; however, antisaccade latencies were elevated only in schizophrenia and reflexive saccades were hypometric only in MDDP.[55] In a related oculomotor paradigm, maintenance pursuit velocity was reduced in all three diagnostic groups, but motion perception during pursuit initiation was not impaired in any group; open loop pursuit velocity was reduced in schizophrenia and bipolar disorder but not in MDDP.[56]

OVERVIEW

While the evidence base in terms of genetics, neuroimaging, and neurological soft signs is too sparse to allow confident consideration of MDDP vis-à-vis schizophrenia, schizoaffective disorder, and bipolar disorder, juxtaposition of the available evidence with neurophysiological findings suggests quantitative more than qualitative differences. Systematic genetic, neuroimaging, and neurological studies across these four psychotic diagnoses are required, and these should include oculomotor processes.

TREATMENT

RANDOMIZED CONTROLLED TRIALS

The many challenges in treating MDDP are well recognized but remain unresolved.[57] In the face of such challenges, there are two main sources of information and opinion: meta-analysis of randomized controlled trials (RCTs) and distillation of consensus among multiple national and international guidelines.

As RCTs constitute the "gold standard" that should inform guidelines and resultant individual physician practice, the results of the most recent meta-analysis of pharmacological treatment,[58] involving 12 RCTs with a total of 929 participants and focusing on the main outcome of reduction more in severity of depression than of psychosis, suggests the following: (1) limited evidence for efficacy of either an antidepressant or an antipsychotic alone; (2) the combination of an antidepressant plus an antipsychotic is more effective than antidepressant monotherapy; (3) the combination of an antidepressant plus an antipsychotic is more effective than antipsychotic monotherapy; (4) the combination of an antidepressant plus an antipsychotic is more effective than placebo. However, the authors emphasize differences between studies with regard to diagnosis, uncertainty as to randomization and allocation concealment, differences in effectiveness between the various antidepressants and antipsychotics used, and diversity in outcome criteria; on this basis, they conclude "Psychotic depression is heavily understudied, limiting confidence in the conclusions drawn. Some evidence indicates that combination therapy with an antidepressant plus an antipsychotic is more effective than either treatment alone. Evidence is limited for treatment with an antidepressant alone or with an antipsychotic alone."[58]

NATIONAL AND INTERNATIONAL GUIDELINES

Among national and international guidelines and associated treatment algorithms, a recent systematic review and analysis of first-, second-, and third-line treatment recommendations[59] indicated the following: four guidelines suggest two first-line treatments, i.e., combination therapy with an antidepressant plus an antipsychotic or electroconvulsive therapy (ECT), with no second- or third-line alternatives; two suggest first-line combination therapy with an antidepressant plus an antipsychotic and second-line treatment with ECT; two suggest first-line treatment with an antidepressant, second-line combination therapy with an antidepressant plus an antipsychotic, and ECT as third-line treatment; one suggests two first-line treatments, i.e., an antidepressant or ECT, and second-line combination therapy with an antidepressant plus an antipsychotic. In three instances a tricyclic antidepressant is preferred and in one instance a selective serotonin or serotonin-noradrenaline reuptake inhibitor is preferred, there being an independent systematic review suggesting that tricyclic antidepressant treatment of MDDP may, occasionally, result in exacerbation of psychosis[60]; in three instances a second-generation antipsychotic is preferred and in one instance a first-generation antipsychotic is preferred; in three instances ECT is a second- or third-line choice where severe suicidality or a threatening somatic condition is present; in three instances, lithium is suggested where initial pharmacological treatment regimens fail to achieve full remission. The authors emphasize the high degree of heterogeneity found in these treatment guidelines; on this basis, they conclude, "The lack of consensus is most likely due to the relatively limited evidence base on this topic and emphasizes the need for further studies on the treatment of psychotic depression."[59]

OVERVIEW

Any consensus across these two approaches, i.e., meta-analysis of RCTs and distillation among multiple national and international guidelines is, in reality, limited; the most reliable finding/common recommendation of first-line combination therapy with an antidepressant plus an antipsychotic occurs among a broad range of alternative therapeutic approaches and treatment hierarchies. The greatest consensus is the need for further studies on treatment of MDDP.

LONG-TERM OUTCOME: DIAGNOSIS

DIAGNOSTIC STABILITY

While each of the diagnoses of schizophrenia, schizoaffective disorder, bipolar disorder, and MDDP differs operationally, they all share the emergence of psychotic symptoms at some point during the course of illness. Early in the course of psychotic illness, the maelstrom of symptoms can lead to considerable diagnostic fluidity. Thereafter, long-term (in)stability of diagnosis may yield important information: do initial diagnoses remain stable, diverge, or converge to a smaller number of diagnoses that might be considered more fundamental diagnostic nodes[27]? While the importance of these issues is widely recognized and has received considerable attention, MDDP has received only limited consideration, such that the most recent meta-analysis on the stability of first episode psychotic diagnoses pools MDDP into a composite of "affective spectrum psychoses."[61]

Prospective studies of MDDP indicate variable diagnostic stabilities and transitions across diverse populations and follow-up periods.[15–17,20] In CAMFEPS,[27] 55% of inception diagnoses of MDDP were retained at 6-year follow-up; 5% were revised to schizophrenia, 8% to schizoaffective disorder, 10% to bipolar I disorder, 2% to bipolar II disorder, depressed, with psychotic features, and 2% to substance-induced psychosis; 13% were deceased due to natural causes or accident and 5% due to suicide. There is evidence that diagnostic transition from MDDP, as for other diagnoses, occurs primarily in the early phase of illness following the first psychotic episode and is more stable in the longer term.[15,26,27] In counterpoint, diagnosis of MDDP at follow-up accrued modest (1%–10%) transitions of inception diagnoses from brief psychotic disorder, schizoaffective disorder, bipolar disorder, schizophrenia, delusional disorder, substance-induced disorder, and psychotic disorder not otherwise specified.[15–17,20,27]

OVERVIEW

Recent meta-analysis includes the following prospective stability values: schizophrenia 90%; affective spectrum psychoses (a composite of bipolar disorder and MDDP) 84%; schizoaffective disorder 72%; delusional disorder 59%; brief psychotic disorder 56%; psychotic disorder not otherwise specified 36%.[61] For comparison, a mean prospective stability value for MDDP across five studies[15–17,20,27] is 61%. Rather than indicating diagnostic instability, such transitions may reflect the "ebb-and-flow" of the fluid psychopathology of psychotic illness on which is imposed a nosology that is purely phenomenological and largely arbitrary, and not based on valid etiological concepts or mechanisms of illness or genetic predispositions.[13]

LONG-TERM OUTCOME: FUNCTIONING

CLINICAL AND FUNCTIONAL OUTCOME

There are a multitude of studies on the long-term clinical and functional outcome of schizophrenia[8,62] and bipolar disorder,[63,64] less so for schizoaffective disorder.[65,66] However, very few have systematically addressed the long-term outcome of MDDP and these utilize various indices to indicate variable extents of recovery, remission and relapse.[67–71] In CAMFEPS, at 6-year follow-up positive psychotic symptoms were less prominent in MDDP and bipolar disorder than in schizophrenia and schizoaffective disorder, while negative symptoms, impaired functioning and reduction in objectively determined quality of life were less prominent in MDDP and bipolar disorder, intermediate in schizoaffective disorder and most prominent in schizophrenia; however, subjectively determined quality of life in MDDP did not differ from other diagnoses. Service engagement was highest for MDDP, intermediate for schizoaffective disorder and bipolar disorder, and lowest for schizophrenia[27] (and Kingston et al., in preparation).

OVERVIEW

Findings to date, though limited, indicate that MDDP is characterized by poor long-term outcomes in terms of psychopathology, functionality, service engagement, and quality of life that, while somewhat less marked than for schizophrenia and schizoaffective disorder, are similar to those for bipolar disorder. However, these differences are quantitative rather than qualitative, with considerable overlap in outcomes across these diagnostic groups. While continued treatment for MDDP with a given medication regimen over a 4-month period can remain effective,[72] medication use over considerably longer periods may not be associated with either remission or relapse.

The most adverse of all outcomes is death. In CAMFEPS, at 6-year follow-up we noted[27] that 13% of patients with MDDP were deceased due to natural causes or accident and 5% due to suicide (see "Long-term outcome: diagnosis"), a higher rate of mortality than was evident for any other diagnosis. This may reflect, at least in part, CAMFEPS incepting cases over the entire adult life span without any arbitrary upper age-cut-off, such that the mean age of patients at the first psychotic episode for MDDP was 20 years older than for schizophrenia, schizoaffective disorder, or bipolar disorder. However, in AESOP a marginal increase in attempted suicide and self-harm in MDDP was noted at 10-year follow-up, even in the face of an arbitrary upper age cut-off.[71] These two studies are cautionary and emphasize a need for further consideration of, and attention to, both mental and physical health in MDDP.[27]

CONCEPTUAL REANALYSIS

MAJOR DEPRESSIVE DISORDER WITH VS WITHOUT PSYCHOTIC FEATURES

That there exists a diagnostic category of MDDP carries with it the presumption that there exists also a diagnostic category of MDD without psychotic features and that this distinction exists on an "either-or" basis, i.e., a binary distinction. How valid is that presumption? Numerous authoritative reviews and meta-analyses have described the epidemiology,[73] psychopathology,[73] neuropsychology,[74] and pathobiology[75–77] of MDD, variable aspects of which do and do not differ from the equally extensive literatures on schizophrenia and bipolar disorder (see "Epidemiology," "Psychopathology," "Neuropsychology," and "Pathobiology"). On this background, formal comparisons with MDD indicate MDDP to show, in addition to self-selection for greater severity of psychosis: indistinguishable premorbid IQ;[78–80] indistinguishable[78,81,82] or greater[80,83] severity of depression; greater neuropsychological impairment,[84] more abnormal brain activation during cognitive tasks,[80,82] greater reduction in amygdalar volume,[83] greater disruption in communication between the hypothalamus and subgenual cortex in association with cortisol dysregulation,[85] and indistinguishable risk for suicide.[86] Overall, the psychopathology of MDDP vis-à-vis MDD may be better measured with a composite scale that indicates severity of depression to be unrelated to severity of psychosis.[87]

However, where present, these differences are quantitative rather than qualitative, with considerable overlap between MDDP and MDD, and therefore they do not, in themselves, validate these diagnoses. As recently re-emphasized, "a good classification should have points of rarity between classes"[88] and few are present in relation to MDDP vs. MDD. Furthermore, "The identification of psychiatric symptoms generally involves the ascription of a cut-off point to what is essentially a set of continua,"[88] hence the criticality of whether "psychosis" is best conceptualized as a continuum or as a category. This challenge is relevant not only in the context of MDDP vs. MDD but also for bipolar disorder with vs. without psychotic features,[28] the presence of mood-congruent vs. mood-incongruent psychotic features in both MDDP and bipolar disorder, and indeed for the general population.

"PSYCHOTIC-LIKE" EXPERIENCES IN THE GENERAL POPULATION AND IN MAJOR DEPRESSIVE DISORDER

A large body of research now indicates that subclinical "psychotic-like" experiences can be identified in the general population (median annual incidence 2.5%, prevalence 7.2%), in whom it may represent an early behavioral manifestation of risk for psychosis; such psychotic experiences appear transitory in about 80% of individuals, while around 20% go on to develop more persistent psychotic experiences and 7% a transdiagnostic psychotic disorder.[89,90]

On this population basis, such "psychotic-like" experiences would be expected in MDD; for example, systematic review indicates auditory verbal hallucinations to be present in 5%–41% of individuals with MDD.[91] More specifically, recent prospective studies of adolescents and young adults have shown, after exclusion of cases of any psychotic disorder, that having MDD was associated with a 4-fold increase in scores on a delusional inventory and a 3-fold increase in either hallucinatory or delusional experiences on diagnostic interview, relative to persons without MDD;[92] the association between depression and psychotic experiences may weaken as depression resolves.[93] Similarly, in a recent study using a large population sample, 14% of those with MDD reported a hallucinatory experience, this being a 4-fold increase relative to those without MDD.[94] Earlier propositions that the psychosis of MDDP arises as a component of greater severity of depression in MDD, in association with endogenous depression/melancholia, are not wholly consistent with recent evidence.[81,87,95] Rather, it would appear that for some individuals with MDD, traversing a continuum of severity of "psychotic-like" experiences beyond an arbitrary cut-off results in the binary transition from a diagnosis of MDD to a diagnosis of MDDP.

PATHOBIOLOGY OF DEPRESSION "AND" RATHER THAN "VS." PSYCHOSIS

That "psychotic-like" experiences are more common in MDD than in the general population suggests some neurobiological relationship between depression and psychosis. A recent twin study has shown that depressive symptoms and "psychotic-like" experiences in the community during adolescence share common genetic influences,[96] and GWAS data indicate pleiotropic effects of genes associated with shared risk for schizophrenia, bipolar disorder, MDD, autism spectrum disorder, and attention-deficit/hyperactivity disorder.[43,97] The genetic relationship between MDD and psychotic disorder has been accentuated by two recent studies indicating that earlier onset of MDD and higher overall symptom severity is associated with greater genetic overlap with schizophrenia and bipolar disorder.[98,99]

At a putative mechanistic level, a recent meta-analysis of 193 voxel-based structural MRI studies involving 7,381 patients and 8,511 controls across 6 diagnostic groups (schizophrenia, bipolar disorder, MDD, substance use disorder, obsessive-compulsive disorder, and anxiety) revealed concordance across these diagnoses for reduction of gray matter volume in an anterior insula/dorsal anterior cingulate-based network, slightly more prominent in the psychotic vs. nonpsychotic groups, in association with deficits in executive function across diagnoses.[100]

Among individual domains of psychopathology, anhedonia is highly relevant to and informative on these issues, as it is recognized to be a fundamental component of MDD,[101] with or without psychotic features, and of the negative symptoms of schizophrenia.[102] The neural circuits serving the reward-related processes of anhedonia include the ventral striatum, prefrontal cortical regions, and their afferent and efferent projections.[103] More specifically, a recent meta-analysis of functional MRI studies on anhedonia in MDD vs. schizophrenia indicated anticipatory anhedonia to be associated with decreased activation in frontal-striatal networks that included the dorsal anterior cingulate, and both middle and medial frontal gyri, while consummatory anhedonia was associated with reduced activation in ventral striatal areas; critically, both MDD and SZ patients showed similar neurobiological impairments,[104] which overlap with those identified in the above meta-analysis of structural MRI studies.[100]

While risk factors for MDDP have received little previous attention, they have indicated a general overlap between familial and environmental risk factors vis-à-vis MDD; a parental history of bipolar disorder, but not of schizophrenia, may be a risk factor for MDDP.[105] A recent study has indicated MDDP and schizophrenia (but less so bipolar disorder) at 10-year follow-up to share the following baseline characteristics relative to a control population: family history of mental illness; family history of psychosis; childhood adversity; lower educational attainment; living alone; infrequent contact with friends; no close confidants; unemployment; more neurological soft signs.[106]

SYNTHESIS

The now substantive evidence that the arbitrary diagnoses of schizophrenia, schizoaffective disorder, bipolar disorder, MDD, and MDDP overlap in numerous domains of evaluation violates the long-recognized requirement that a good classification system should have points of rarity between classes; rather, such evidence is in accordance with the identification of psychiatric symptoms generally involving the imposition of arbitrary cutoff points to what are essentially continua.[88,107] There is increasing opinion and evidence that diagnostic categories such as schizophrenia, schizoaffective disorder and bipolar disorder reflect not discrete entities but, rather, domains characterized by variations in pathobiological processes, psychopathological dimensions, and functional characteristics, the boundaries of which are likely arbitrary and in continuity or intersection with other domains of mental illness, through to the limits of "normal" human experience and functioning;[89,90,108,109] these concepts are also influential in the area of translational neuroscience.[110] The present review indicates that the MDD-MDDP domain should join what we currently classify as schizophrenia, schizoaffective disorder, bipolar disorder, and likely other arbitrary diagnostic categories in what may be, in reality, a yet broader milieu of psychotic, affective, and other developmental abnormalities in an extended dimensional space of psychopathology and dysfunction.

ACKNOWLEDGMENTS

The authors' studies were supported by the Stanley Medical Research Institute and Cavan-Monaghan Mental Health Service.

REFERENCES

1. Lawrence R, Lawrie SM. Psychotic depression. *Br Med J* 2012;345:e6994.
2. Million T, Grossman S, Meagher S. *Masters of the mind: Exploring the story of mental illness from ancient times to the new millennium.* Hoboken, NJ: John Wiley, 2004.
3. Telles-Correia D, Marques JG. Melancholia before the twentieth century: fear and sorrow or partial insanity? *Front Psychol* 2015;6:81.
4. Craddock N, Owen MJ. The Kraepelinian dichotomy—going, going . . . but still not gone. *Br J Psychiat* 2010;196(2):92–95.
5. Kendler KS. Psychosis within vs outside of major mood episodes: a key prognostic and diagnostic criterion. *JAMA Psychiatry.* 2013;70(12):1263–1264.
6. Waddington JL, Buckley PF. Psychotic depression: an underappreciated window to explore the dimensionality and pathobiology of psychosis. *Schizophr Bull* 2013;39(4): 754–755.
7. Swartz CM, Shorter E. *Psychotic depression.* New York, NY: Cambridge University Press, 2007.
8. van Os J, Kapur S. Schizophrenia. *Lancet* 2009;374(9690): 635–645.
9. Heckers S, Barch DM, Bustillo J, et al. Structure of the psychotic disorders classification in DSM-5. *Schizophr Res* 2013;150(1):11–14.
10. Cotton S, Lambert M, Schimmelmann B, et al. Depressive symptoms in first episode schizophrenia spectrum disorder. *Schizophr Res* 2012;134(1):20–26.
11. Harvey PD, Twamley EW, Pinkham AE, Depp CA, Patterson TL. Depression in schizophrenia: associations with cognition, functional capacity, everyday functioning, and self-assessment. *Schizophr Bull* 2017;43(3):575–582.
12. Upthegrove R, Marwaha S, Birchwood M. Depression and schizophrenia: cause, consequence, or trans-diagnostic issue? *Schizophr Bull* 2017;43(2):240–244.
13. Weinberger DR, Glick ID, Klein DF. Whither Research Domain Criteria (RDoC)? The good, the bad, and the ugly. *JAMA Psychiatry* 2015;72(12):1161–1162.
14. Poser CM. The epidemiology of multiple sclerosis: a general overview. *Ann Neurol* 1994;36(Suppl2):S180–S193.
15. Bromet EJ, Kotov R, Fochtmann LJ, et al. Diagnostic shifts during the decade following first admission for psychosis. *Am J Psychiat* 2011;168(11):1186–1194.
16. Salvatore P, Baldessarini RJ, Tohen M, et al. McLean-Harvard International First-Episode Project: two-year stability of DSM-IV diagnoses in 500 first-episode psychotic disorder patients. *J Clinical Psychiat* 2009;70(4):458–466.
17. Whitty P, Clarke M, McTigue O, et al. Diagnostic stability four years after a first episode of psychosis. *Psychiat Serv* 2005;56(9): 1084–1088.
18. Kirkbride JB, Fearon P, Morgan C, et al. Heterogeneity in incidence rates of schizophrenia and other psychotic syndromes: findings from the 3-center AeSOP study. *Arch Gen Psychiat* 2006;63(3):250–258.
19. Reay R, Mitford E, McCabe K, Paxton R, Turkington D. Incidence and diagnostic diversity in first-episode psychosis. *Acta Psychiatr Scand* 2010;121(4):315–319.
20. Chang WC, Pang SLK, Chung DWS, Chan SSM. Five-year stability of ICD-10 diagnoses among Chinese patients presented with first-episode psychosis in Hong Kong. *Schizophr Res* 2009;115(2):351–357.
21. Kelly BD, O'Callaghan E, Waddington JL, et al. Schizophrenia and the city: A review of literature and prospective study of psychosis and urbanicity in Ireland. *Schizophr Res* 2010;116(1):75–89.
22. Vassos E, Pedersen CB, Murray RM, Collier DA, Lewis CM. Meta-analysis of the association of urbanicity with schizophrenia. *Schizophr Bull* 2012;38(6):1118–1123.
23. Cantor-Graae E, Selten JP. Schizophrenia and migration: a meta-analysis and review. *Am J Psychiat* 2005;162(1):12–24.
24. Bourque F, van der Ven E, Malla A. A meta-analysis of the risk for psychotic disorders among first-and second-generation immigrants. *Psychol Med* 2011;41(5):897–910.
25. Kirkbride JB, Errazuriz A, Croudace TJ, et al. Incidence of schizophrenia and other psychoses in England, 1950–2009: a systematic review and meta-analyses. *PloS One.* 2012;7(3):e31660.
26. Baldwin P, Browne D, Scully PJ, et al. Epidemiology of first-episode psychosis: illustrating the challenges across diagnostic boundaries through the Cavan-Monaghan study at 8 years. *Schizophr Bull* 2005;31(3):624–638.
27. Kingston T, Scully P, Browne D, et al. Diagnostic trajectory, interplay and convergence/divergence across all 12 DSM-IV psychotic diagnoses: 6-year follow-up of the Cavan-Monaghan First Episode Psychosis Study (CAMFEPS). *Psychol Med* 2013;43(12):2523–2533.
28. Owoeye O, Kingston T, Scully PJ, et al. Epidemiological and clinical characterization following a first psychotic episode in major depressive disorder: comparisons with schizophrenia and bipolar I disorder in the Cavan-Monaghan First Episode Psychosis Study (CAMFEPS). *Schizophr Bull* 2013;39(4):756–765.
29. Gournellis R, Oulis P, Howard R. Psychotic major depression in older people: a systematic review. *Int J Geriat Psychiat* 2014;29(8):784–796.
30. Salvatore P, Khalsa H, Hennen J, et al. Psychopathology factors in first-episode affective and non-affective psychotic disorders. *J Psychiat Res* 2007;41(9):724–736.
31. Russo M, Levine SZ, Demjaha A, et al. Association between symptom dimensions and categorical diagnoses of psychosis: a cross-sectional and longitudinal investigation. *Schizophr Bull* 2014;40(1):111–119.
32. Rosen C, Marvin R, Reilly JL, et al. Phenomenology of first-episode psychosis in schizophrenia, bipolar disorder, and unipolar depression: a comparative analysis. *Clin Schizophr Relat Psychoses* 2012;6(3):145–151.
33. Stanghellini G, Raballo A. Differential typology of delusions in major depression and schizophrenia. A critique to the unitary concept of "psychosis." *J Affect Disord* 2015;171:171–178.
34. Lako IM, Bruggeman R, Knegtering H, et al. A systematic review of instruments to measure depressive symptoms in patients with schizophrenia. *J Affect Disord* 2012;140(1):38–47.
35. Bora E, Pantelis C. Meta-analysis of cognitive impairment in first-episode bipolar disorder: comparison with first-episode schizophrenia and healthy controls. *Schizophr Bull* 2015;41(5):1095–1104.
36. Bora E, Yücel M, Pantelis C. Cognitive impairment in affective psychoses: a meta-analysis. *Schizophr Bull* 2010;36(1):112–125.
37. Zanelli J, Reichenberg A, Morgan K, et al. Specific and generalized neuropsychological deficits: a comparison of patients with various first-episode psychosis presentations. *Am J Psychiat* 2009;167(1):78–85.
38. Che AM, Gomez RG, Keller J, et al. The relationships of positive and negative symptoms with neuropsychological functioning and their ability to predict verbal memory in psychotic major depression. *Psychiat Res* 2012;198(1):34–38.
39. Dell'Osso L, Pini S, Cassano GB, et al. Insight into illness in patients with mania, mixed mania, bipolar depression and major depression with psychotic features. *Bipolar Disord* 2002;4(5):315–322.
40. Khandaker GM, Barnett JH, White IR, Jones PB. A quantitative meta-analysis of population-based studies of premorbid intelligence and schizophrenia. *Schizophr Res* 2011;132(2-3):220–227.
41. Martino DJ, Samamé C, Ibañez A, Strejilevich SA. Neurocognitive functioning in the premorbid stage and in the first episode of bipolar disorder: a systematic review. *Psychiat Res* 2015;226(1):23–30.
42. Hill SK, Reilly JL, Harris MSH, et al. A comparison of neuropsychological dysfunction in first-episode psychosis patients with unipolar depression, bipolar disorder, and schizophrenia. *Schizophr Res* 2009;113(2):167–175.
43. Cross-Disorder Group of the Psychiatric Genomics Consortium. Genetic relationship between five psychiatric disorders estimated from genome-wide SNPs. *Nat Genetics* 2013;45(9):984–994.
44. Schizophrenia Working Group of the Psychiatric Genomics Consortium. Biological insights from 108 schizophrenia-associated genetic loci. *Nature* 2014;511(7510):421–427.
45. Green EK, Rees E, Walters JTR, et al. Copy number variation in bipolar disorder. *Mol Psychiat* 2016;21(1):89–93.

46. CNV and Schizophrenia Working Groups of the Psychiatric Genomics Consortium; Psychosis Endophenotypes International Consortium. Contribution of copy number variants to schizophrenia from a genome-wide study of 41,321 subjects. *Nat Genetics* 2017;49(1):27–35.
47. Cardno AG, Owen MJ. Genetic relationships between schizophrenia, bipolar disorder, and schizoaffective disorder. *Schizophr Bull* 2014;40(3):504–515.
48. Domschke K. Clinical and molecular genetics of psychotic depression. *Schizophr Bull* 2013;39(4):766–775.
49. Birur B, Kraguljac NV, Shelton RC, Lahti AC. Brain structure, function, and neurochemistry in schizophrenia and bipolar disorder—a systematic review of the magnetic resonance neuroimaging literature. *NPJ Schizophrenia* 2017;3:15.
50. de Azevedo-Marques Périco C, Duran FL, Zanetti MV, et al. A population-based morphometric MRI study in patients with first-episode psychotic bipolar disorder: comparison with geographically matched healthy controls and major depressive disorder subjects. *Bipolar Disord* 2011;13(1):28–40.
51. Busatto GF. Structural and functional neuroimaging studies in major depressive disorder with psychotic features: a critical review. *Schizophr Bull* 2013;39(4):776–786.
52. Whitty PF, Owoeye O, Waddington JL. Neurological signs and involuntary movements in schizophrenia: intrinsic to and informative on systems pathobiology. *Schizophr Bull* 2009;35(2):415–424.
53. Rigucci S, Dimitri-Valente G, Mandarelli G, et al. Neurological soft signs discriminate schizophrenia from bipolar disorder. *J Psychiatr Pract* 2014;20(2):147–153.
54. Chrobak AA, Siwek GP, Siuda-Krzywicka K, et al. Neurological and cerebellar soft signs do not discriminate schizophrenia from bipolar disorder patients. *Prog Neuro-Psychopharmacol Biol Psychiat* 2016;64(1):96–101.
55. Harris MS, Reilly JL, Thase ME, Keshavan MS, Sweeney JA. Response suppression deficits in treatment-naive first-episode patients with schizophrenia, psychotic bipolar disorder and psychotic major depression. *Psychiat Res* 2009;170(2):150–156.
56. Lencer R, Reilly JL, Harris MS, Sprenger A, Keshavan MS, Sweeney JA. Sensorimotor transformation deficits for smooth pursuit in first-episode affective psychoses and schizophrenia. *Biol Psychiat* 2010;67(3):217–223.
57. Rothschild AJ. Challenges in the treatment of major depressive disorder with psychotic features. *Schizophr Bull* 2013;39(4):787–796.
58. Wijkstra J, Lijmer J, Burger H, Geddes J, Nolen WA. Pharmacological treatment of psychotic depression. Cochrane Database Syst Rev 2015;7:CD004044.
59. Leadholm AKK, Rothschild AJ, Nolen WA, Bech P, Munk-Jørgensen P, Østergaard SD. The treatment of psychotic depression: is there consensus among guidelines and psychiatrists? *J Affect Disord* 2013;145(2):214–220.
60. Kantrowitz JT, Tampi RR. Risk of psychosis exacerbation by tricyclic antidepressants in unipolar major depressive disorder with psychotic features. *J Affect Disord* 2008;106(3):279–284.
61. Fusar-Poli P, Cappucciati M, Rutigliano G, et al. Diagnostic stability of ICD/DSM first episode psychosis diagnoses: meta-analysis. *Schizophr Bull* 2016;42(6):1395–1406.
62. Owen MJ, Sawa A, Mortensen PB. Schizophrenia. *Lancet* 2016;388(10039):86–97.
63. Belmaker RH. Bipolar disorder. *N Engl J Med* 2004;351(5):476–486.
64. Grande I, Berk M, Birmaher B, Vieta E. Bipolar disorder. *Lancet* 2016;387(10027):1561–1572.
65. Pagel T, Franklin J, Baethge C. Schizoaffective disorder diagnosed according to different diagnostic criteria—systematic literature search and meta-analysis of key clinical characteristics and heterogeneity. *J Affect Disord* 2014;156:111–118.
66. Rink L, Pagel T, Franklin J, Baethge C. Characteristics and heterogeneity of schizoaffective disorder compared with unipolar depression and schizophrenia—a systematic literature review and meta-analysis. *J Affect Disord* 2016;191:8–14.
67. Craig TJ, Grossman S, Bromet EJ, Fochtmann LJ, Carlson GA. Medication use patterns and two-year outcome in first-admission patients with major depressive disorder with psychotic features. *Compr Psychiat* 2007;48(6):497–503.
68. Naz B, Craig TJ, Bromet EJ, Finch SJ, Fochtmann LJ, Carlson GA. Remission and relapse after the first hospital admission in psychotic depression: a 4-year naturalistic follow-up. *Psychol Med* 2007;37(8):1173–1181.
69. Haim R, Rabinowitz J, Bromet E. The relationship of premorbid functioning to illness course in schizophrenia and psychotic mood disorders during two years following first hospitalization. *J Nerv Mental Dis* 2006;194(10):791–795.
70. Tohen M, Khalsa H-MK, Salvatore P, Vieta E, Ravichandran C, Baldessarini RJ. Two-year outcomes in first-episode psychotic depression: The McLean–Harvard first-episode project. *J Affect Disord* 2012;136(1):1–8.
71. Heslin M, Lappin J, Donoghue K, et al. Ten-year outcomes in first episode psychotic major depression patients compared with schizophrenia and bipolar patients. *Schizophr Res* 2016;176(2-3):417–422.
72. Wijkstra J, Burger H, van den Broek WW, et al. Long-term response to successful acute pharmacological treatment of psychotic depression. *J Affect Disord* 2010;123(1):238–242.
73. Kupfer DJ, Frank E, Phillips ML. Major depressive disorder: new clinical, neurobiological, and treatment perspectives. *Lancet* 2012;379(9820):1045–1055.
74. Lee RS, Hermens DF, Porter MA, Redoblado-Hodge MA. A meta-analysis of cognitive deficits in first-episode major depressive disorder. *J Affect Disord* 2012;140(2):113–124.
75. Kempton MJ, Salvador Z, Munafò MR, et al. Structural neuroimaging studies in major depressive disorder: meta-analysis and comparison with bipolar disorder. *Arch Gen Psychiat* 2011;68(7):675–690.
76. Levinson DF, Mostafavi S, Milaneschi Y, et al. Genetic studies of major depressive disorder: why are there no GWAS findings, and what can we do about it? *Biol Psychiat* 2014;76(7):510–512.
77. Müller VI, Cieslik EC, Serbanescu I, Laird AR, Fox PT, Eickhoff SB. Altered brain activity in unipolar depression revisited: meta-analyses of neuroimaging studies. *JAMA Psychiat* 2017;74(1):47–55.
78. Hill SK, Keshavan MS, Thase ME, Sweeney JA. Neuropsychological dysfunction in antipsychotic-naive first-episode unipolar psychotic depression. *Am J Psychiat* 2004;161(6):996–1003.
79. Gomez RG, Fleming SH, Keller J, et al. The neuropsychological profile of psychotic major depression and its relation to cortisol. *Biol Psychiat* 2006;60(5):472–478.
80. Kelley R, Garrett A, Cohen J, et al. Altered brain function underlying verbal memory encoding and retrieval in psychotic major depression. *Psychiat Res: Neuroimaging* 2013;211(2):119–126.
81. Forty L, Jones L, Jones I, et al. Is depression severity the sole cause of psychotic symptoms during an episode of unipolar major depression? a study both between and within subjects. *J Affect Disord* 2009;114(1):103–109.
82. Garrett A, Kelly R, Gomez R, Keller J, Schatzberg AF, Reiss AL. Aberrant brain activation during a working memory task in psychotic major depression. *Am J Psychiat* 2011;168(2):173–182.
83. Keller J, Shen L, Gomez RG, et al. Hippocampal and amygdalar volumes in psychotic and nonpsychotic unipolar depression. *Am J Psychiat* 2008;165(7):872–880.
84. Zaninotto L, Guglielmo R, Calati R, et al. Cognitive markers of psychotic unipolar depression: a meta-analytic study. *J Affect Disord* 2015;174:580–588.
85. Sudheimer K, Keller J, Gomez R, et al. Decreased hypothalamic functional connectivity with subgenual cortex in psychotic major depression. *Neuropsychopharmacology* 2015;40(4):849–860.
86. Leadholm AKK, Rothschild AJ, Nielsen J, Bech P, Østergaard SD. Risk factors for suicide among 34,671 patients with psychotic and non-psychotic severe depression. *J Affect Disord* 2014; 156:119–125.
87. Østergaard SD, Meyers BS, Flint AJ, et al. Measuring psychotic depression. *Acta Psychiatr Scand* 2014;129(3):211–220.

88. Bebbington P, Freeman D. Transdiagnostic extension of delusions: schizophrenia and beyond. *Schizophr Bull* 2017;43(2):273–282.
89. Linscott R, van Os J. An updated and conservative systematic review and meta-analysis of epidemiological evidence on psychotic experiences in children and adults: on the pathway from proneness to persistence to dimensional expression across mental disorders. *Psychol Med* 2013;43(6):1133–1149.
90. van Os J, Reininghaus U. Psychosis as a transdiagnostic and extended phenotype in the general population. *World Psychiat* 2016;15(2):118–124.
91. Toh WL, Thomas N, Rossell SL. Auditory verbal hallucinations in bipolar disorder (BD) and major depressive disorder (MDD): a systematic review. *J Affect Disord* 2015;184:18–28.
92. Varghese D, Scott J, Welham J, et al. Psychotic-like experiences in major depression and anxiety disorders: a population-based survey in young adults. *Schizophr Bull* 2009;37(2):389–393.
93. Sullivan SA, Wiles N, Kounali D, et al. Longitudinal associations between adolescent psychotic experiences and depressive symptoms. *PloS One*. 2014;9(8):e105758.
94. Kelleher I, DeVylder JE. Hallucinations in borderline personality disorder and common mental disorders. *Br J Psychiat* 2017;210(3):230–231.
95. Østergaard SD, Bille J, Søltoft-Jensen H, Lauge N, Bech P. The validity of the severity–psychosis hypothesis in depression. *J Affect Disord* 2012;140(1):48–56.
96. Zavos HM, Eley TC, McGuire P, et al. Shared etiology of psychotic experiences and depressive symptoms in adolescence: a longitudinal twin study. *Schizophr Bull* 2016;42(5):1197–1206.
97. Cross-Disorder Group of the Psychiatric Genomics Consortium. Identification of risk loci with shared effects on five major psychiatric disorders: a genome-wide analysis. *Lancet* 2013;381(9875):1371–1379.
98. Power RA, Tansey KE, Buttenschøn HN, et al. Genome-wide association for major depression through age at onset stratification: Major Depressive Disorder Working Group of the Psychiatric Genomics Consortium. *Biol Psychiat* 2017;81(4):325–335.
99. Verduijn J, Milaneschi Y, Peyrot WJ, et al. Using clinical characteristics to identify which patients with major depressive disorder have a higher genetic load for three psychiatric disorders. *Biol Psychiat* 2017;81(4):316–324.
100. Goodkind M, Eickhoff SB, Oathes DJ, et al. Identification of a common neurobiological substrate for mental illness. *JAMA Psychiatry* 2015;72(4):305–315.
101. Rizvi SJ, Pizzagalli DA, Sproule BA, Kennedy SH. Assessing anhedonia in depression: Potentials and pitfalls. *Neurosci Biobehav Rev* 2016;65:21–35.
102. Marder SR, Galderisi S. The current conceptualization of negative symptoms in schizophrenia. *World Psychiatry* 2017; 16(1):14–24.
103. Der-Avakian A, Markou A. The neurobiology of anhedonia and other reward-related deficits. *Trends Neurosci* 2012;35(1):68–77.
104. Zhang B, Lin P, Shi H, et al. Mapping anhedonia-specific dysfunction in a transdiagnostic approach: an ALE meta-analysis. *Brain Imaging Behav* 2016;10(3):920–939.
105. Østergaard SD, Waltoft BL, Mortensen PB, Mors O. Environmental and familial risk factors for psychotic and non-psychotic severe depression. *J Affect Disord* 2013;147(1-3):232–240.
106. Heslin M, Desai R, Lappin J, et al. Biological and psychosocial risk factors for psychotic major depression. *Soc Psychiat Psychiatr Epidemiol* 2016;51(2):233–245.
107. Kendell R. Diagnosis and classification of functional psychoses. *Br Med Bull* 1987;43(3):499–513.
108. Waddington J, Hennessy R, O'Tuathaigh C, Owoeye O, Russell V. Schizophrenia and the lifetime trajectory of psychotic illness: developmental neuroscience and pathobiology, redux. In: Brown A, Patterson PH, eds. *The Origins of Schizophrenia*. New York, NY: Columbia University Press, 2012:3–21.
109. Owen MJ. New approaches to psychiatric diagnostic classification. *Neuron* 2014;84(3):564–571.
110. Pletnikov M, Waddington J. *Modeling the Psychopathological Dimensions of Schizophrenia: From Molecules to Behavior.* Amsterdam, The Netherlands: Elsevier, 2016.

8

PSYCHOSIS IN PERSONALITY DISORDERS

K. Nidhi Kapil-Pair, Yulia Landa, Marie C. Hansen, Daniel H. Vaccaro, and Marianne Goodman

INTRODUCTION

This chapter reviews psychotic symptoms and clinical presentations in the ten major personality disorders: paranoid, schizoid, schizotypal, antisocial, borderline, histrionic, narcissistic, avoidant, dependent, and obsessive-compulsive. For each personality disorder, prevalence, origin, and subtypes of psychosis are discussed along with descriptions of psychobiological underpinnings (where known).

Both Diagnostic and Statistical Manual of Mental Disorders (DSM) and the International Classification of Diseases (ICD) are evolving into dimensional approaches for research and clinical purposes.[1] Prior to DSM-5, personality disorders (PDs) were categorized by *clusters*.[2] "Cluster A" PDs are characterized by odd, eccentric thinking or behavior, and include paranoid, schizoid, and schizotypal PDs. Traditionally, Cluster A PDs were viewed as attenuated psychotic disorders for their early symptomatic presentations, revealing patterns in cognitions and behaviors that may fully remit over time and are amenable to change with early intervention.[3] "Cluster B" PDs are characterized by dramatic, overly emotional, or unpredictable thinking or behavior, and include antisocial, borderline, histrionic, and narcissistic PDs. "Cluster C" PDs are characterized by anxious, fearful thinking or behavior, and include avoidant, dependent, and obsessive-compulsive PDs. Personality disorders will be discussed by these clusters, as historically detailed.

Delusions are most common in paranoid PD, followed by schizoid PD in one population sample.[4] Schizoid and schizotypal PDs are most rare and severe, in terms of intersections with psychosis and greater proneness to schizophrenia, as evidenced by emergent biomarker-based research.[5] Other prevalent PDs associated with psychotic presentations are paranoid, schizoid, schizotypal, avoidant, and dependent.[6] Furthermore, various psychotic features were found in borderline PD, specifically characterized by ubiquitous odd thinking, unusual perceptual experiences, nondelusional paranoia, *quasi*-psychotic (atypical, circumscribed, transient psychotic experiences), and *typical* psychotic manifestations (Schneiderian first-rank, prolonged, widespread, bizarre psychotic symptoms).[7] These findings together illustrate psychotic presentations across clusters of PDs.

Schizophrenia spectrum PDs, which traditionally include Cluster A PDs, are thought to reflect phenotype expression associated with liability for schizophrenia. Growing evidence provides insight into how certain personality disturbances common to schizophrenia-spectrum PDs may co-occur with psychosis and vary in clinical severity.[7] High rates of comorbid PDs exist with schizophrenia (40%–60%), compared to the general population (6%–13%).[8,9] However, one of the limitations is that comorbid PDs are often examined within a single type of psychotic disorder, most commonly schizophrenia. Nevertheless, there is a substantial variance in rates of comorbid PDs, which warrants investigation of a broader psychosis dimension,[8] including schizophrenia, schizoaffective disorder, psychotic mood disorders, schizophreniform disorder, delusional disorder, brief psychotic disorder, and psychosis associated with substance use or medical conditions. The wide range of PD comorbidities across studies may be attributed to variability in conceptualization, diagnostic tools, methodological approaches and cultural factors.[8] However, there is a need for transdiagnostic approaches encompassing key constructs—such as phenotypes, biomarkers, and genetic markers—which may provide clues to the nature of overlap between various PDs and psychotic disorders.

To illustrate confluence, Table 8.1 compiles psychotic symptoms found across all PDs. There are two main types of psychotic symptoms: hallucinations and delusions. Hallucinations are characterized by particular sensory modality: visual, auditory, tactile, proprioceptive, gustatory, or olfactory. Kraepelin described six subtypes of delusions defined according to their content: ideas of reference, persecution, exalted ideas (grandiose delusions), sin (or jealousy), sexual ideas (erotomania), and influence (somatic or induced *folie à deux*, delusions in which people believe they are influenced or controlled).[3] As demonstrated in Table 8.1, delusional symptoms are far more prevalent than hallucinations across all PD clusters, with the greatest expression in Cluster A and borderline PDs.

In the following sections, psychosis—as it broadly and specifically presents in each PD—is discussed. Schizotypal, borderline and antisocial are the most studied PDs from a neurobiological point of view.[10] Relationships between PDs and psychosis are not well understood, partially due to

Table 8.1 ASSOCIATED PSYCHOTIC SYMPTOMS BY PERSONALITY DISORDER

PERSONALITY DISORDER		VISUAL	AUDITORY	TACTILE	PROPRIOCEPTIVE	OLFACTORY	REFERENCE	PERSECUTORY	GRANDEUR	JEALOUSY	EROTOMANIA	SOMATIC	INDUCED FOLIE À DEUX
PD Cluster		**Hallucinations***					**Delusions**						
	Paranoid	de Portugal et al. (2013)	de Portugal et al. (2013)	de Portugal et al. (2013)		de Portugal et al. (2013)	de Portugal et al. (2013); Sadock et al. (2014)	de Portugal et al. (2013); Freeman et al.(2016); Kinderman et al. (2003); Shapiro (1965); Verlag (2016)	de Portugal et al. (2013)	de Portugal et al. (2013)	de Portugal et al. (2013)	de Portugal et al. (2013)	
A	**Schizoid**	de Portugal et al. (2013)	de Portugal et al. (2013)	de Portugal et al. (2013)			de Portugal et al. (2013)	de Portugal et al. (2013)		de Portugal et al. (2013)	Vaknin (2007)	de Portugal et al. (2013)	
	Schizotypal	Hoerman et al. (2003); Sadock et al. (2014)	de Portugal et al.(2013); Siever et al. (1993)	de Portugal et al. (2013)	Hoerman et al. (2003)		de Portugal et al. (2013); Sadock et al. (2014)	de Portugal et al. (2013); Hoselle et al. (2014); Siever et al. (1993)	Hoerman et al. (2003)	de Portugal et al. (2013)			
	Antisocial											Bishop & Holt (1980); Sadock et al. (2014); Williams (2014)	Petrikis et al. (2013); Shah et al. (2016)
B	**Borderline**	Barnow et al. (2010); Schroeder et al.(2012; 2013); Suzuki et al. (1998); Yee et al. (2005)	Daalman et al.(2011); Gras et al.(2014); Glaser et al. (2010); Hepworth et al. (2010); Kingdon et al. (2010); Merrett et al. (2106); Pearse et al. (2014); Schroeder et al. (2012; 2013); Slotema et al.(2012); Suzuki et al. (1998); Tschoke et al. (2014); Yee et al. (2005)			Barnow et al. (2010); Yee et al. (2005)	Vaknin (2007)	Kingdon et al. (2010); Schroeder et al. (2013); Suzuki et al. (1998)		Costa et al. (2015)	Costa et al. (2015); Sansone & Sansone (2010); Suzuki et al. (1998)	Suzuki et al. (1998)	
	Histrionic											Ferrari et al. (2015); Fisher (1999); Sadock et al. (2014)	Petrikis et al. (2013); Shah et al. (2016)
	Narcissistic						Vaknin (2007)		Hoerman et al. (2013); Vaknin (2007)	Hoerman et al. (2013); Vaknin (2007)	Hoerman et al. (2013); Vaknin (2007)		

	Avoidant	de Portugal et al. (2013)		de Portugal et al. (2013)	de Portugal et al. (2013); Vaknin (2007)	de Portugal et al. (2013)	de Portugal et al. (2013)	de Portugal et al. (2013)		Fisher (1999); de Portugal et al. (2013)	
C	**Dependent**										Kaplan & Sadock (2015); Petrikis et al. (2013); Shah et al. (2016); Srivasta & Borkar (2010)
	Obsessive-Compulsive		de Portugal et al. (2013)	de Portugal et al. (2013)	de Portugal et al. (2013)		de Portugal et al. (2013)		de Portugal et al. (2013)		

*Gustatory hallucination, or *phantosmia*, is not represented in table due to lack of relevant findings related to PDs, to date.

limitations of study design, e.g., cross-sectional studies do not allow to distinguish between cause and effect of PDs and psychosis manifestations.[11] Studies employing longitudinal design and monitoring attenuated psychotic syndromes in subjects at risk, from adolescence to young adulthood, are likely to provide important clues and inform genetic, molecular, and neurocircuit underpinnings.

PSYCHOSIS IN PARANOID PERSONALITY DISORDER

Paranoid PD incorporates patterns of thought and behavior centered on pervasive and enduring distrust of others, which typically begin in early adulthood and occur across a range of situations. Features of the disorder include suspecting without sufficient basis that others are attempting to harm or deceive; a reluctance to confide in others or share personal information; unjustified suspicions about faithfulness of romantic partners; misperceptions that everyday events or speech are personal attacks; and hypersensitivity to interpersonal slights. Paranoid PD is a relatively common PD, found in about 2%–4% in the general population.[12,13] In clinical samples, this disorder is more commonly diagnosed in men than women.[13] Paranoid PD is often considered part of the broad category of "schizophrenia spectrum disorders," however genetic liability research has found inconclusive results regarding familial risk or linkages between schizophrenia and paranoid PD.[14,15] Due to the suspicious nature of individuals with this disorder, research that specifically focused on paranoid PD is limited.[14,16]

In longitudinal studies, Birkeland[17] found a clinical sample of patients with paranoid PD who manifested frank delusions. Paranoid PD may include periods of transient psychosis, possibly triggered by increased stress. Maladaptive information processing have been implicated in paranoid delusions: *abnormal attentional biases* to threat-related information;[18-19] *abnormal attributional biases*, or excessive tendency to make external-personal (other blaming) attributions for negative events; *jumping to conclusions*, or reasoning bias; *abnormal referential biases*, or excessive tendency to personalize others' statements or actions as having a special meaning for them; *diminished insight*, or poor capacity for self-reflection; and *theory of mind deficits*, or inability to understand others' thoughts and feelings.[18,20,21] Isolation and social avoidance are found to correlate with paranoia,[20] and isolation-induced ruminations increase the degree to which one's behavior is consistent with paranoid attitudes.[22,23] Tendencies toward these thinking styles, combined with social isolation, may play an important role in development of psychotic-level processing in paranoid PD.

PSYCHOSIS IN SCHIZOID PERSONALITY DISORDER

Schizoid PD is characterized by interpersonal coldness, disinterest in social relationships, and limited capacity to express emotions (i.e., alexithymia can co-occur but is not synonymous). In a study of individuals at risk for psychosis, conversion was best predicted by a diagnosis of schizoid PD,[9] suggesting interpersonal deficits, such as difficulty expressing emotions and engaging with others, as important indicators for later development of a psychotic disorder. Schizoid PD is thought to impact approximately 5% of the general population, although prevalence is not clearly established.[13] Some studies suggest a 2:1 male/female ratio, but exact sex ratio in this disorder is unknown.[13] In comparison to schizotypal PD, individuals with schizoid PD similarly lack close relationships with others, but do not often experience the cognitive-perceptual distortions of individuals with schizotypal PD.[24] In addition, although both schizoid PD and schizotypal PD are associated with social deficits, schizoid PD is defined more by an *indifference* to social relationships, rather than distrust or suspicion regarding others.[24] Beck[25] described how lack of contact with others may lead individuals with schizoid PD to engage in overabundance of fantasy, at times, leading to brief psychotic episodes.[26]

PSYCHOSIS IN SCHIZOTYPAL PERSONALITY DISORDER

Schizotypal PD is defined by criteria pertaining to social and interpersonal deficits, cognitive/perceptual disturbances, and eccentric and odd behavior. Similar to schizophrenia, cognitive deficits, including working memory impairment and context processing, are central to the clinical presentation of schizotypal PD. These cognitive deficits are associated with alterations in prefrontal regions, and are thought to involve a hypodopaminergic state, particularly with D1 receptor dysfunction in the dorsolateral prefrontal cortex.[27] Schizotypal PD occurs in approximately 3% of the population, with DSM-5 suggesting greater diagnosis rates among males.[13]

Schizotypal PD represents combinations of *both* cognitive-perceptual experiences (i.e., positive-like) and social and interpersonal deficits (i.e., negative-like).[17] Individuals with schizotypal PD exhibit long-standing interpersonal complications related to their extreme suspicious and paranoid beliefs.[7] Paranoid ideas and/or distorted perceptions coupled with periods of elevated psychosocial stress put individuals at risk for experiencing brief psychotic episodes.[17] Additionally, individuals with schizotypal PD exhibit magical thinking, odd beliefs, or overvalued ideas, and may have brief psychotic episodes, but not as frequent, prolonged, or intense as those displayed in individuals with schizophrenia.

Structural imaging studies in schizotypal PD indicate decrements in gray matter volumes in the temporal lobe, and less consistent findings in the frontal lobe and striatum. Temporal lobe volume reductions, particularly in the left hemisphere, are similar to those found in schizophrenia, but more limited in scope.[28] While temporal lobe volume reductions continue to progress over time in schizophrenia, this is not the case with schizotypal PD.[29] Smaller temporal lobe volume in schizotypal PD is associated with clinical features such as memory dysfunction and odd speech.[30] Ivleva et al. (2013) examined a large sample of first-degree relatives of psychosis probands, arranged by psychosis dimension, from the

Bipolar-Schizophrenia Network for Intermediate Phenotypes (B-SNIP).[31] The relatives who met criteria for Cluster A PDs (combined across schizoid, paranoid, and schizotypal) showed considerable gray matter volume reductions distributed over a broad range of regions, including frontal and temporal cortex. In contrast, some studies find no alterations in the frontal structure in individuals with schizotypal PD,[27] interpreted as a possible protective factor, i.e., lack of frontal lobe involvement in schizotypal PD may be protective from fully manifested schizophrenia-like psychosis.[27] In studies examining striatum volume in schizotypal PD, discrepant findings are reported for the putamen and caudate. Chemerinski et al. (2013)[32] found increased putamen volumes in schizotypal PD compared to schizophrenia, which correlated with reductions in paranoid symptoms, and no volume differences in the caudate. However, others report smaller caudate volumes in schizotypal PD, compared to controls, and no differences in the putamen volume.[33]

More recent studies have examined white matter tract integrity using diffusion tensor imaging (DTI). Nakamura et al. (2005)[34] found abnormalities in frontotemporal connectivity: specifically, bilateral reductions in degree of anisotropy in the uncinate fasciculus were present in schizotypal PD. Reduced anisotropy in the right uncinated fasciculus correlated with several clinical symptoms, including ideas of reference, restricted affect, social anxiety, and suspiciousness. The anterior limb of the internal capsule (a structure that connects fibers between the frontal lobe and thalamus) demonstrates abnormalities in schizotypal PD; specifically, in the dorsal section, fewer fiber tracts were found in schizotypal PD that correlated with heightened schizotypal PD symptom severity.[35]

To better understand the role of dopamine dysfunction in schizotypal PD, positron emission tomography (PET) and single photon emission computed tomography (SPECT) studies have been conducted to clarify receptor binding properties. Striatal presynaptic dopamine release was greater in schizotypal PD than in healthy controls,[36] and midway between the dopamine release observed in schizophrenia. Similar receptor binding studies in healthy individuals with schizotypy traits have also shown abnormalities in presynaptic striatal dopamine release and their correlations with symptom severity.[37]

PSYCHOSIS IN ANTISOCIAL PERSONALITY DISORDER

Antisocial PD is characterized by disregard for others' needs or feelings, lack of remorse, aggressive, often violent behavior, impulsivity, and persistent lying, stealing, or conning others. In DSM-5, 12-month prevalence rates of antisocial PD range from 0.2% to 3%, and more common in men than women.[38] Although research examining links between antisocial PD and psychosis is scarce, findings indicate that antisocial traits co-occur with psychotic symptoms. Antisocial PD has been linked to somatic delusions,[13,39,40] and *folie à deux*—shared delusions.[41,42] In contrast, in a study comparing psychotic/psychotic-like symptoms across different Cluster B PDs, Perry and Klerman (1980) found that borderline PD but not antisocial PD correlated with presence of psychotic or psychotic-like symptoms over a 1-year period.[43] Likewise, Moran et al. (2004) found no differences in course or symptomatology of schizophrenia among men with or without antisocial PD.[44] However, research suggests *antisocial traits* may be linked to psychotic symptomatology in borderline PD and early-onset psychosis. Goodman (1999) showed childhood antisocial behavior, sadism in particular, predicts psychotic symptoms in borderline PD.[45] Huber et al. (2016) reported that patients with early-onset psychosis had elevated scores of agitation and aggressive behavior, and these scores correlated with antisocial PD; moreover, earlier age of psychosis onset was correlated with antisocial PD.[46] In summary, while psychotic symptoms are generally not characteristic of antisocial PD, antisocial traits/behavior appear to show complex interactions with other disorders involving psychotic symptoms.

PSYCHOSIS IN BORDERLINE PERSONALITY DISORDER

Borderline PD is characterized by impulsivity, risky or self-injurious behaviors, unstable and intense relationships, intense fear of being alone or abandoned, and fluctuations in mood, often in reaction to interpersonal stress. Borderline PD occurs in 1%–2% of the population, but definitive prevalence studies are not yet available.[38] Borderline PD is twice as common in women than in men.

Of patients with borderline PD, 20%–50% experience psychotic symptoms, i.e., hallucinations and delusions.[47] Studies suggests that auditory verbal hallucinations in borderline PD are similar to those in psychotic disorders.[47] Although empirical evidence has yet to support treatment efficacy, some researchers recommend psychotherapy interventions, such as cognitive-behavioral therapy, for treating long-standing auditory verbal hallucinations in borderline PD.[48] Drug trials lend some support to the use of atypical antipsychotics for reducing borderline PD's psychotic or psychotic-like symptoms, but studies are yet to investigate borderline PD's hallucinations and delusions as targeted constructs.[49]

PHENOMENOLOGY

To differentiate psychotic symptoms in borderline PD from other psychotic disorders, Zanarini et al. (1990) drew distinctions between "disturbed," nonpsychotic symptoms; "quasi" psychotic symptoms; and "true" psychotic symptoms. Quasi-psychotic symptoms—transient, circumscribed, or atypical hallucinations and delusions—have been proposed as a marker of borderline PD.[50] Researchers later challenged this divide between "quasi" and "true" psychotic symptoms. Yee et al. (2005) found 29% of patients with borderline PD reported auditory verbal hallucinations. They also found such hallucinations were "persistent, longstanding, and a significant source of distress and disability," indicating that terms like "quasi-hallucination" trivialize the phenomenon, and should be avoided.[51] Two studies found auditory verbal hallucinations in borderline PD were similar in phenomenology to auditory verbal hallucinations in schizophrenia. Kingdon et al. (2010) compared the experience of auditory verbal hallucinations and paranoid delusions in patients with borderline PD to those with schizophrenia, as well as patients with comorbid borderline PD and schizophrenia.[52] They found 46% and 29% of borderline PD patients experienced auditory verbal hallucinations and paranoid delusions, respectively. All three groups were similar in experiences of voices: no significant differences in conviction, frequency, duration, or beliefs about location of the voices were found. However, borderline PD group reported higher distress and more negative content of voices. Slotema et al. (2012) also examined phenomenology of auditory verbal hallucinations in borderline PD, and found no differences, compared to psychotic disorders, except for the item "disruption of life" (which was rated higher in patients with psychotic disorders).[53] In borderline PD, auditory verbal hallucinations were experienced for a mean duration of 18 years, with a mean frequency of "at least daily," lasting several minutes or more, and with ensuing distress highly rated. The study concluded patients with borderline PD fulfill the criteria for hallucinations *proper*, suggesting clinicians use the term *auditory verbal hallucinations* rather than "*pseudo-*" or "*quasi-*" hallucinations, to prevent trivialization and promote adequate diagnosis and treatment.[53]

RISK FACTORS

Comorbid psychiatric disorders, daily stressors, and childhood trauma have been suggested as risk factors for borderline PD, although further exploration is required.[32,47,50,54–56,57,58–61] Some studies have found evidence for relationships between psychosis and depression, substance abuse, and comorbid manic features,[32,50,59–61] however, there is no conclusive evidence to link borderline PD's psychotic symptoms with comorbidities. More investigation is needed, especially given that post-traumatic stress disorder is present in 58%–79% of patients with borderline PD.[50,61]

Patients with borderline PD report strong reactions to daily life stress, with increased psychotic symptoms.[57] Previous studies in patients with psychotic disorders reported daily life stress was also associated with severity of psychosis. Researchers found patients with borderline PD displayed the strongest psychotic reactivity.[57] According to Barnow et al. (2012), this phenomenon may result from sensitization of the hypothalamo-pituitary axis.[58] Furthermore, studies have shown that patients with borderline PD with high levels of dissociation have a dysregulated cortisol response to stress.

For borderline PD, childhood trauma is highly prevalent: 40%–76% report experiencing childhood sexual abuse, and 92% report childhood emotional abuse.[47] In some studies, childhood rape and physical assault predicted visual and auditory hallucinations in adulthood in individuals with and without psychotic disorders.[54,55] While many studies have shown associations between childhood trauma and adult psychotic symptoms, these data are limited in borderline PD populations.[47]

CLINICAL MANAGEMENT

In light of the reviewed findings, researchers encourage routine screening and assessment of psychotic symptoms in borderline PD in order to provide proper treatment.[47] Treatment of symptoms may involve specialized psychotherapy interventions. Hepworth et al. (2013) suggests that cognitive-behavioral therapy for psychosis (CBTp), specifically focused on cognitive appraisals and the relationship with voices, may have clinical utility for those with borderline PD.[48] In a sample of 45 patients with distressing auditory verbal hallucinations (10 borderline PD-only, 23 schizophrenia-only, and 12 both), they found no group differences in beliefs about power or malevolence of voices. They did, however, differ in their affective responses: patients with borderline PD reported more emotional resistance, and schizophrenia-only patients reported more emotional engagement. Researchers argue an individual's relationship with voices should be a target of intervention, and suggest applying, adapting, and evaluating a wide range of cognitive models and methods, from classic cognitive therapy to newer techniques, such as mindfulness.[48] Additionally, Schroeder et al. (2013) recommend that interventions should address daily stressors, given that psychotic symptoms in borderline PD are often related to stress and acute emotional crisis. Future directions could explore the efficacy of psychotherapy interventions in treating hallucinations and delusions in borderline PD, such as CBTp.

Treatment may involve atypical antipsychotics, in addition to psychotherapeutic interventions. Only a small number of trials have examined efficacy of medications in reducing psychotic symptoms in borderline PD. These studies lend some support to use of certain atypical antipsychotics in patients with borderline PD experiencing psychotic symptoms. However, these trials used general instruments rather than detailed assessments of psychotic symptoms.[47] Available studies mostly assessed changes in "psychotic-like" symptoms, like paranoid ideation and dissociative symptoms, and "psychoticism," which includes items ranging from hallucinations to schizoid lifestyle. One open study of clozapine reported hallucinations as a specific endpoint.[62] To date, there have been only four randomized controlled trials assessing the efficacy of atypical antipsychotics for psychotic or psychotic-like symptoms in borderline PD (see Table 8.2).[62–76]

In a 12-week randomized clinical trial by Zanarini et al. (2011),[77] 451 outpatients were administered olanzapine at low and moderate doses. Results showed reduced paranoid ideation and dissociation items on the ZAN-BPD scale with the moderate dose of olanzapine (5–10mg/d, $N = 148$), compared to the placebo group ($N = 153$). The SCL-90-R total score was also reduced, but scores on psychoticism and paranoid ideation were not reported.[77] An open label extension study showed sustained and improved ZAN-BPD total scores. The report also presented data from the ZAN-BPD paranoid ideation subscale, but it was not powered to determine differences in this domain. While this trial demonstrates impressive reductions in paranoid ideation, more research is needed to further support preliminary evidence showing reductions in

Table 8.2 ATYPICAL ANTIPSYCHOTICS FOR PSYCHOTIC OR "PSYCHOTIC-LIKE" SYMPTOMS IN BPD

DRUG	STUDY	TYPE	GROUPS	DOSE (MG/D)	DURATION	RELEVANT OUTCOMES
Olanzapine (OLZ)	Schulz et al. (1999)[63]	Open	OLZ (*N* = 9) comorbid dysthymia	2.5–10	8 wk.	SCL-90 PSY ↓ SCL-90 PAR ↓ BPRS TD ↓
	Zanarini et al. (2001)[64]	RCT	OLZ (*N* = 19) Placebo (*N* = 9)	5.33 ± 2.43	6 mo.	SCL-90 PAR ↓
	Schulz et al. (2008)[65]	RCT	OLZ (*N* = 155) Placebo (*N* = 159)	2.5–20	12 wk.	ZAN-BPD TOT ↓ SCL-90-R TOT ↓
	Zanarini et al. (2011)[66]	RCT	Low OLZ (*N* = 150) Med OLZ (*N* = 148) Placebo (*N* = 153)	Low: 2.5 Med: 5–10	12 wk.	with medium dose: ZAN-BPD PID ↓ SCL-90-R TOT ↓
	Zanarini et al. (2012)[67]	Open ext.	OLZ (*N* = 320)	2.5–20	12 wk.	ZAN-BPD TOT ↓ sustained, improved
Aripiprazole (ARI)	Nickel et al. (2006)[68]	RCT	ARI (*N* = 26) Placebo (*N* = 26)	15	8 wk.	SCL-90-R PSY ↓ SCL-90-R PAR ↓
	Nickel et al. (2007)[69]	RCT	ARI (*N* = 26) Placebo (*N* = 26)	15	18 mo.	SCL-90-R PSY ↓ SCL-90-R PAR ↓
	Bellino et al. (2008)[70]	Open	ARI (*N* = 16) + sertraline	10–15	12 wk.	BPDSI PID ↓
Clozapine (CLZ)	Frankenburg et al. (1993)[62]	Open	CLZ (*N* = 15) 'atypical psychotic symptoms'	75–55	02-9 mo.	BPRS Halluc. ↓ BPRS Pos. Symp. ↓
	Benedetti et al. (1998)[71]	Open	CLZ (*N* = 15) severe "psychotic-like symptoms"	25–100	4 mo.	BPRS Psychotic Cluster ↓
Quetiapine (QTP)	Gruettert et al. (2005)[72]	Case Series	QTP (*N* = 12) psychosis	300–750	12 wk.	SCL-90 PSY ↓
	Mauri et al. (2007)[73]	Naturalistic	QTP (*N* = 13)	250–1000	2 wk.	BPRS TOT ↓ PANSS TOT ↓
Risperidone (RSP)	Friedel et al. (2008)[74]	Open	RSP (*N* = 18)	1.8 ± 0.37	8 wk.	BDRS CPI ↓ BSI PSY ↓
Paliperidone (PAL)	Bellino et al. (2011)[75]	Open	PAL (*N* = 18)	3–6	12 wk.	BPRS TOT ↓ BPDSI PID ↓
Ziprasidone (ZSD)	Pascual et al. (2008)[76]	RCT	ZSD (*N* = 30) Placebo (*N* = 30)	40–200	12 wk.	CGI-BPD: No significant effect

psychoticism, as well as emerging improvements in certain psychotic symptoms, like paranoid delusional thinking.[77]

PSYCHOSIS IN HISTRIONIC PERSONALITY DISORDER

Histrionic PD is characterized by attention seeking; shallow, rapidly changing, and excessive emotions; being easily influenced by others; and dramatic or sexually provocative behavior to gain attention; and may include excessive concern with physical appearance. Limited findings from general population studies suggest prevalence of approximately 1%–3% for this disorder.[13] In community samples, histrionic and dependent PDs had considerably lower prevalence when compared to other PDs.[78] Due to the lower prevalence of this disorder, literature on psychosis in histrionic PDs is limited.

Literature suggests that in histrionic PD associations between perception and behavior may be impacted by information processing biases.[79] Some core beliefs associated with distress in histrionic PD are: "I am inadequate and unable to handle life on my own" and "It is necessary to be loved by everyone, all the time."[79] Research has found associations between histrionic PD and delusional thinking, specifically somatization.[78,79,80]

While the term "histrionic" replaced "hysterical" in DSM-III, the term "hysterical psychosis" was historically applied to individuals with *hysterical* personalities. These patients were often found "acting out delusion-like fantasies, and were considered psychotic, but in fairly good contact with reality" (p. 95).[81] The literature pertaining to hysterical psychosis suggests the condition arose from delusional ideas, possible hallucinations, and bizarre behaviors, and marked the failure of a psychological defense.[82–84] Likewise, the delusional state that accompanies presentations of hysteria uniquely involves overt somatic disability combined with the lack of insight.[80]

Among psychotic symptoms, delusional thinking is thought to be associated the most with histrionic PD, specifically somatic and shared delusions, i.e., *folie à deux*.[41,42] Delusional manifestations are attributed to a combination of neurobiological, cognitive, and psychological factors, which need further examination in histrionic PD. More research across dimensions of psychopathology is required to draw further associations between histrionic PD and other co-occurring conditions.

PSYCHOSIS IN NARCISSISTIC PERSONALITY DISORDER

Narcissistic PD is characterized by beliefs that one is particularly special and more important than others; fantasies about power, success, attractiveness; failure to recognize needs and feelings of others; exaggeration of achievements or talents; and expectations of constant praise and attention. Literature suggests two types of narcissistic PD: "overt" (grandiose) and "covert" (vulnerable), both with excessive self-interest.[85] The DSM-5 estimates prevalence for narcissistic PD to range from less than 1% to 6% in community samples.[38]

While empirical research on psychosis in narcissistic PD is limited, theoretical literature on narcissistic PD is extensive, and presents psychosis (loss of reality testing) as a metric to understanding the extent of pathology. However, psychosis manifestations in narcissistic PD are often "psychotic-like," or more subthreshold than overt, representing transient disconnections with reality.[85] In comparison to individuals with psychotic disorders, individuals with narcissistic PD are aware of their external environment.[86] Individuals with narcissistic PD tend to develop aggrandizing narratives, or emotionally vest in personalized *myths*; these experiences may reach "psychotic-like" levels. Psychoanalytic literature suggests individuals with severe narcissistic pathology may use distortion, sadistic aggression, perverse logic, and seduction to ward against a weakened ego, and provide the individual with enough *ego strength* (i.e., cultivated resilience of the core sense of self) to defend against traumatic separation deemed imminent.[40] Pathology results in the individual maintaining conflated fantasies or confabulations.[40,86] While self-aggrandizing personal myths of people with narcissistic PD can trend toward psychosis (i.e., delusional thinking), they often remain grounded in consensual reality.[86]

PSYCHOSIS IN AVOIDANT PERSONALITY DISORDER

Avoidant PD is characterized by feeling inadequate, inferior, timid, isolated, or unattractive; hypersensitivity to criticism or rejection; avoidance of activities that require interpersonal contact; social inhibition; avoiding new activities or meeting strangers; extreme shyness in social situations and personal relationships; and fear of disapproval, embarrassment, or ridicule.[87] The DSM-5 estimates prevalence of this disorder to be 2%–3% of the general population.[38]

Perceptual disturbances (such as tactile and olfactory hallucinations) and a wide range of delusional thinking are common in avoidant PD (see Table 8.1).[4,80,86] Additionally, several key features of avoidant PD, such as social detachment and isolative interpersonal behaviors, are common in other psychotic disorders,[87] indicating areas of phenomenological and neurobiological overlap requiring further investigation.

PSYCHOSIS IN DEPENDENT PERSONALITY DISORDER

Dependent PD is characterized by excessive dependence on others and feeling the need to be taken care of, submissive or clingy behavior toward others, fear of having to provide self-care for oneself; lack of self-confidence, requiring excessive advice and reassurance from others to make even small decisions; and difficulty initiating or maintaining one's own projects due to lack of self-confidence. DSM-5 estimates prevalence of this disorder at 0.6%, more common in women than in men.[38]

Folie à deux, delusions shared by two individuals, is a relatively rare psychotic syndrome linked to "prepsychotic personality traits" found in histrionic, antisocial, and dependent PDs.[41,42] Delusions may be shared by more than two: *folie à trois*, *folie à quatre*, or *folie à plusieurs* ("madness of many"). The literature on psychotic features in dependent PD includes such variants as spousal psychosis, psychosis among parent–child dyads, as well as *folie à familie*, shared psychosis among the family. One study found predisposing factors in family members related to social isolation and dependent personality traits. Without implying causation, there is a link between dependent PD and shared psychotic disorders that necessitates further exploration. Individuals with dependent PD and aforementioned psychotic manifestations appear to respond well to psychotherapy and isolation from the proband (individual with endogenous psychosis), and usually do not require treatment with antipsychotics.[88]

PSYCHOSIS IN OBSESSIVE-COMPULSIVE PERSONALITY DISORDER

Obsessive-compulsive PD is a disorder consisting of impairments in self-functioning in either *identity* (i.e., sense of self derived predominantly from productivity, with constricted expression of emotions and experiences), or *self-direction* (i.e., difficulty completing tasks and goals due to extreme conscientiousness, moralistic attitudes, and rigid, high, inflexible standards), and impairments in interpersonal functioning, consisting of either difficulties with *empathy* (i.e., understanding others' perspectives and feelings), or *intimacy* (i.e., relationships are considered secondary to occupational productivity; marked by rigidity and negative affect). In order to meet criteria, an individual must possess pathological traits characterized by *compulsivity*, behaviors including perfectionism; inability to tolerate errors or imperfections; difficulty changing viewpoints; preoccupation with details/organization; and *negative affectivity* or perseveration at tasks despite failures or effectiveness.[38] Obsessive-compulsive PD has a prevalence of 3%–8% in the United States, and is the most common PD found in the general population.[89]

Emerging research suggests there is a more specific relationship between delusions and obsessional beliefs, possibly with a shared aspects of mechanism.[90] Direct links between obsessive-compulsive PD and psychotic symptoms are yet to be investigated; nevertheless, evidence suggests common co-occurrence of obsessive-compulsive PD and paranoid and schizotypal PDs.[91] For example, a study using the DSM-III-R criteria found elevated odds ratios for comorbid obsessive-compulsive PD and paranoid and schizoid PDs.[92] Similarly, Hummelen et al.(2008) found that obsessive-compulsive PD was associated with paranoid PD more frequently than with any other PD.[93] These overlaps suggest that obsessive-compulsive PD may have common psychological and/or neurobiological underpinnings with Cluster A PDs, which are not captured in its current categorization as a disorder of predominantly fearful/anxious affect. However, research comparing the comorbidity of obsessive-compulsive PD with Cluster A PDs has yet to be conducted with DSM-5 criteria.

CONCLUSIONS

In summary, when examining PDs across domains, links to psychosis spectrum presentations are quite specific and nuanced, and possibly share features with other comorbidities. Transdiagnostic frameworks, such as the National Institutes of Mental Health's Research Domain Criteria (RDoC)[1] initiative, more adequately inform major systems of emotion, cognition, motivation, and social behavior, and may potentially provide vital information for diagnostic and clinical purposes. Furthermore, ICD-10[94] distinguishes from acute, transient psychosis sometimes seen in developing countries (i.e., various cultures), which consider other domains in the trajectory of personality development, adaptation, and functioning—healthy, or otherwise. For instance, consideration of relationships between childhood trauma and psychotic symptoms can be better understood with comprehensive investigatory models that may also consider the sociocultural and transgenerational effects of trauma on the entire system of an individual. Personality disorder research is usually limited by small sample sizes, use of self-report measures, clinical interviews, inclusion criteria determined via telephone screens, and unrepresentative samples, such as students, psychiatric patients' relatives, and control groups from other studies.[95] Research indicates that personality and psychopathological dimensions are dynamic complexities that offer stronger evidence when examined under longitudinal approaches, not cross-sectional studies.[4] By approaching these conditions from multiple disciplines, such as across neurobiologic and cognitive behavioral domains, a greater range of correlates can potentially inform conditions and treatments of psychosis in personality disorders.

REFERENCES

1. Morris SE, Cuthbert BN. Research Domain Criteria: cognitive systems, neural circuits, and dimensions of behavior. Dialogues in clinical neuroscience. 2012;14(1):29.
2. Clinic M. Personality Disorders. 2017; Available from: http://www.mayoclinic.org/diseases-conditions/personality-disorders/symptoms-causes/dxc-20247656.
3. Bentall RP, Corcoran R, Howard R, Blackwood N, Kinderman P. Persecutory delusions: a review and theoretical integration. Clinical psychology review. 2001;21(8):1143–1192.
4. de Portugal E, Díaz-Caneja CM, González-Molinier M, de Castro MJ, del Amo V, Arango C, et al. Prevalence of premorbid personality disorder and its clinical correlates in patients with delusional disorder. Psychiatry research. 2013;210(3):986–993.
5. Via E, Orfila C, Pedreño C, Rovira A, Menchón JM, Cardoner N, et al. Structural alterations of the pyramidal pathway in schizoid and schizotypal cluster A personality disorders. International journal of psychophysiology. 2016;110:163–170.
6. Bolinskey PK, James AV, Cooper-Bolinskey D, Novi JH, Hunter HK, Hudak DV, et al. Revisiting the blurry boundaries of schizophrenia: spectrum disorders in psychometrically identified schizotypes. Psychiatry research. 2015;225(3):335–340.
7. Oliva F, Dalmotto M, Pirfo E, Furlan PM, Picci RL. A comparison of thought and perception disorders in borderline personality disorder and schizophrenia: psychotic experiences as a reaction to impaired social functioning. BMC psychiatry. 2014;14(1):239.
8. Moore EA, Green MJ, Carr VJ. Comorbid personality traits in schizophrenia: prevalence and clinical characteristics. Journal of psychiatric research. 2012;46(3):353–359.
9. Schultze-Lutter F, Klosterkotter J, Michel C, Winkler K, Ruhrmann S. Personality disorders and accentuations in at-risk persons with and without conversion to first-episode psychosis. Early Interv Psychiatry. 2012;6(4):389–398.
10. Perez-Rodriguez M, New A, Siever L. The neurobiology of personality disorders: the shift to DSM-5. In Charney DS, Nestler EJ (eds.), Neurobiology of mental illness, 4th edition. New York: Oxford University Press; 2013.
11. Wei Y, Zhang T, Chow A, Tang Y, Xu L, Dai Y, et al. Co-morbidity of personality disorder in schizophrenia among psychiatric outpatients in China: data from epidemiologic survey in a clinical population. BMC psychiatry. 2016;16(1):224.
12. Torgersen S, Kringlen E, Cramer V. The prevalence of personality disorders in a community sample. Arch Gen Psychiatry. 2001;58(6):590–596.
13. Sadock BJ, Sadock VA, Ruiz P. Kaplan and Sadock's synopsis of psychiatry: behavioral sciences/clinical psychiatry. Wolters Kluwer Health; 2014.
14. Webb CT, Levinson DF. Schizotypal and paranoid personality disorder in the relatives of patients with schizophrenia and affective disorders: a review. Schizophr Res. 1993;11(1):81–92.
15. Tienari P, Wynne LC, Laksy K, Moring J, Nieminen P, Sorri A, et al. Genetic boundaries of the schizophrenia spectrum: evidence from

the Finnish Adoptive Family Study of Schizophrenia. American journal of psychiatry. 2003;160(9):1587–1594.
16. Skodol AE, Bender DS, Morey LC, Clark LA, Oldham JM, Alarcon RD, et al. Personality disorder types proposed for DSM-5. J Pers Disord. 2011;25(2):136–169.
17. Birkeland SF. Delusional psychosis in individuals diagnosed with paranoid personality disorder: a qualitative study. Current Psychology. 2014;33(2):219–228.
18. Kinderman P, Prince S, Waller G, Peters E. Self-discrepancies, attentional bias and persecutory delusions. Br J Clin Psychol. 2003;42(Pt 1):1–12.
19. Freeman D, Waite F, Emsley R, Kingdon D, Davies L, Fitzpatrick R, et al. The efficacy of a new translational treatment for persecutory delusions: study protocol for a randomised controlled trial (The Feeling Safe Study). Trials. 2016;17(1):134.
20. Freeman D, Garety PA, Bebbington PE, Smith B, Rollinson R, Fowler D, et al. Psychological investigation of the structure of paranoia in a non-clinical population. Brit J Psychiat. 2005;186:427–435.
21. Freeman D, Garety PA, Kuipers E. Persecutory delusions: developing the understanding of belief maintenance and emotional distress. . Psychological medicine. 2001; 1293–1306.
22. Cooper J, Croyle RT. Attitudes and attitude change. Annu Rev Psychol. 1984;35:395–426.
23. Snyder M, Kendzierski, D. Acting on one's attitudes: procedures for linking attitude and behavior Journal of experimental social psychology. 1982;18(2):165–183.
24. Chemerinski E, Siever LJ. The schizophrenia spectrum personality disorders. In: Weinberger DR, Harrison PJ (eds.), Schizophrenia. Hoboken, NJ: John Wiley & Sons; 2011.
25. Beck ATF, A. Cognitive therapy of personality disorders. New York: Guilford Press; 1990.
26. Chadwick PK. Peer-professional first-person account: schizophrenia from the inside—phenomenology and the integration of causes and meanings. Schizophrenia bulletin. 2007;33(1):166–173.
27. Rosell DR, Futterman SE, McMaster A, Siever LJ. Schizotypal personality disorder: a current review. Curr Psychiatry Rep. 2014;16(7):452.
28. Hazlett EA, Buchsbaum MS, Haznedar MM, Newmark R, Goldstein KE, Zelmanova Y, et al. Cortical gray and white matter volume in unmedicated schizotypal and schizophrenia patients. Schizophr Res. 2008;101(1–3):111–123.
29. Takahashi T, Zhou SY, Nakamura K, Tanino R, Furuichi A, Kido M, et al. A follow-up MRI study of the fusiform gyrus and middle and inferior temporal gyri in schizophrenia spectrum. Progress in neuro-psychopharmacology and biological psychiatry. 2011;35(8):1957–1964.
30. Dickey CC, McCarley RW, Voglmaier MM, Niznikiewicz MA, Seidman LJ, Demeo S, et al. An MRI study of superior temporal gyrus volume in women with schizotypal personality disorder. American journal of psychiatry. 2003;160(12):2198–2201.
31. Ivleva EI, Bidesi AS, Keshavan MS, Pearlson GD, Meda SA, Dodig D, Schretlen, DJ. Gray matter volume as an intermediate phenotype for psychosis: Bipolar-Schizophrenia Network on Intermediate Phenotypes (B-SNIP). American journal of psychiatry. 2013;170(11):1285–1296.
32. Chemerinski E, Byne W, Kolaitis JC, Glanton CF, Canfield EL, Newmark RE, et al. Larger putamen size in antipsychotic-naive individuals with schizotypal personality disorder. Schizophr Res. 2013;143(1):158–164.
33. Koo MS, Levitt JJ, McCarley RW, Seidman LJ, Dickey CC, Niznikiewicz MA, et al. Reduction of caudate nucleus volumes in neuroleptic-naive female subjects with schizotypal personality disorder. Biol Psychiatry. 2006;60(1):40–48.
34. Nakamura M, McCarley RW, Kubicki M, Dickey CC, Niznikiewicz MA, Voglmaier MM, et al. Fronto-temporal disconnectivity in schizotypal personality disorder: a diffusion tensor imaging study. Biol Psychiatry. 2005;58(6):468–478.
35. Hazlett EA, Collazo T, Zelmanova Y, Entis JJ, Chu KW, Goldstein KE, et al. Anterior limb of the internal capsule in schizotypal personality disorder: fiber-tract counting, volume, and anisotropy. Schizophr Res. 2012;141(2–3):119–127.
36. Abi-Dargham A, Kegeles LS, Zea-Ponce Y, Mawlawi O, Martinez D, Mitropoulou V, et al. Striatal amphetamine-induced dopamine release in patients with schizotypal personality disorder studied with single photon emission computed tomography and [123I]iodobenzamide. Biol Psychiatry. 2004;55(10):1001–1006.
37. Woodward ND, Cowan RL, Park S, Ansari MS, Baldwin RM, Li R, et al. Correlation of individual differences in schizotypal personality traits with amphetamine-induced dopamine release in striatal and extrastriatal brain regions. American journal of psychiatry. 2011;168(4):418–426.
38. American Psychiatric Association. Diagnostic and statistical manual of mental disorders (DSM-5). American Psychiatric Pub; 2013.
39. Bishop ER, Holt AR. Pseudopsychosis: a reexamination of the concept of hysterical psychosis. Comprehensive psychiatry. 1980;21(2):150–161.
40. Williams P. Orientations of psychotic activity in defensive pathological organizations. International journal of psychoanalysis. 2014;95(3):423–440.
41. Petrikis P, Andreou C, Garyfallos G, Karavatos A. Incubus syndrome and folie à deux: a case report. European psychiatry. 2003;18(6):322.
42. Shah K, Breitinger S, Avari J, Francois D. Late-onset folie à deux in monozygotic twins. Schizophrenia research. 2017;182:142–143.
43. Perry C, Klerman, GL. Clinical features of the borderline personality disorder. American journal of psychiatry. 1980;37(2):165–173.
44. Moran P, Hodgins S. The correlates of comorbid antisocial personality disorder in schizophrenia. Schizophrenia bulletin. 2004;30(4):791–802.
45. Goodman G, Hull JW, Clarkin JF, Yeomans FE. Childhood antisocial behaviors as predictors of psychotic symptoms and DSM-III-R borderline criteria among inpatients with borderline personality disorder. Journal of personality disorders. 1999;13(1):35–46.
46. Huber CG, Hochstrasser L, Meister K, Schimmelmann BG, Lambert M. Evidence for an agitated-aggressive syndrome in early-onset psychosis correlated with antisocial personality disorder, forensic history, and substance use disorder. Schizophrenia research. 2016;175(1):198–203.
47. Schroeder K, Fisher HL, Schäfer I. Psychotic symptoms in patients with borderline personality disorder: prevalence and clinical management. Current opinion in psychiatry. 2013;26(1):113–119.
48. Hepworth CR, Ashcroft K, Kingdon D. Auditory hallucinations: a comparison of beliefs about voices in individuals with schizophrenia and borderline personality disorder. Clinical psychology and psychotherapy. 2013;20(3):239–245.
49. Pope HG Jr., Jonas JM, Hudson JI, Cohen BM, Tohen M. An empirical study of psychosis in borderline personality disorder. American journal of psychiatry. 1985;142(11):1285–1290.
50. Zanarini MC, Gunderson JG, Frankenburg FR. Cognitive features of borderline personality disorder. American journal of psychiatry. 1990;147(1):57.
51. Yee L, Korner AJ, McSwiggan S, Meares RA, Stevenson J. Persistent hallucinosis in borderline personality disorder. Comprehensive psychiatry. 2005;46(2):147–154.
52. Kingdon DG, Ashcroft K, Bhandari B, Gleeson S, Warikoo N, Symons M, et al. Schizophrenia and borderline personality disorder: similarities and differences in the experience of auditory hallucinations, paranoia, and childhood trauma. Journal of nervous and mental disease. 2010;198(6):399–403.
53. Slotema CW, Daalman K, Blom JD, Diederen KM, Hoek HW, Sommer I. Auditory verbal hallucinations in patients with borderline personality disorder are similar to those in schizophrenia. Psychological medicine. 2012;42(9):1873–1878.
54. Shevlin M, Murphy J, Read J, Mallett J, Adamson G, Houston JE. Childhood adversity and hallucinations: a community-based study using the National Comorbidity Survey Replication. Social psychiatry and psychiatric epidemiology. 2011;46(12):1203–1210.
55. Daalman K, Boks M, Diederen K, de Weijer AD, Blom JD, Kahn RS, et al. The same or different? A phenomenological comparison of

auditory verbal hallucinations in healthy and psychotic individuals. J Clin Psychiatry. 2011;72(3):320–325.
56. Berenbaum H, Thompson RJ, Milanak ME, Boden MT, Bredemeier K. Psychological trauma and schizotypal personality disorder. Journal of abnormal psychology. 2008;117(3):502.
57. Glaser J, Van Os J, Thewissen V, Myin-Germeys I. Psychotic reactivity in borderline personality disorder. Acta psychiatrica scandinavica. 2010;121(2):125–134.
58. Barnow S, Limberg A, Stopsack M, Spitzer C, Grabe H, Freyberger H, et al. Dissociation and emotion regulation in borderline personality disorder. Psychological medicine. 2012;42(4):783–794.
59. Nishizono-Maher A, Ikuta N, Ogiso Y, Moriya N, Miyake Y, Minakawa K. Psychotic symptoms in depression and borderline personality disorder. Journal of affective disorders. 1993;28(4):279–285.
60. Miller FT, Abrams T, Dulit R, Fyer M. Psychotic symptoms in patients with borderline personality disorder and concurrent axis I disorder. Psychiatric services. 1993;44(1):59–61.
61. Benvenuti A, Rucci P, Ravani L, Gonnelli C, Frank E, Balestrieri M, et al. Psychotic features in borderline patients: is there a connection to mood dysregulation. Bipolar disord. 2005;7(4):338–343.
62. Frankenburg FR, Zanarini MC. Clozapine treatment of borderline patients: a preliminary study. Comprehensive psychiatry. 1993;34(6):402–405.
63. Schulz SC, Camlin KL, Berry SA, Jesberger JA. Olanzapine safety and efficacy in patients with borderline personality disorder and comorbid dysthymia. Biological psychiatry. 1999;46(10):1429–1435.
64. Zanarini MC, Frankenburg FR. Olanzapine treatment of female borderline personality disorder patients: a double-blind, placebo-controlled pilot study. Journal of clinical psychiatry. 2001.
65. Schulz SC, Zanarini MC, Bateman A, Bohus M, Detke HC, Trzaskoma Q, et al. Olanzapine for the treatment of borderline personality disorder: variable dose 12-week randomised double-blind placebo-controlled study. British journal of psychiatry. 2008;193(6):485–492.
66. Zanarini MC, Frankenburg FR, Reich DB, Silk KR, Hudson JI, McSweeney LB. The subsyndromal phenomenology of borderline personality disorder: a 10-year follow-up study. American journal of psychiatry. 2007;164(6):929–935.
67. Zanarini M, Schulz S, Detke H, Zhao F, Lin D, Deberdt W, et al. Open-label treatment with olanzapine in patients with borderline personality disorder. European psychiatry. 2008;23:S93.
68. Nickel MK, Muehlbacher M, Nickel C, Kettler C, Gil FP, Bachler E, et al. Aripiprazole in the treatment of patients with borderline personality disorder: a double-blind, placebo-controlled study. American journal of psychiatry. 2006;163(5):833–838.
69. Nickel MK, Loew TH, Gil FP. Aripiprazole in treatment of borderline patients, part II: an 18-month follow-up. Psychopharmacology. 2007;191(4):1023–1026.
70. Bellino S, Paradiso E, Bogetto F. Efficacy and tolerability of aripiprazole augmentation in sertraline-resistant patients with borderline personality disorder. Psychiatry research. 2008;161(2):206–212.
71. Bendetti F, Sforzini L, Colombo C, Marrei C, Smeraldi E. Low-dose clozapine in acute and continuation treatment of severe borderline personality disorder. Journal of clinical psychiatry. 1998.
72. Gruettert T, Friege L. Quetiapine in patients with borderline personality disorder and psychosis: a case series. International journal of psychiatry in clinical practice. 2005;9(3):180–186.
73. Mauri MC, Volonteri LS, Fiorentini A, Pirola R, Bareggi SR. Two weeks' quetiapine treatment for schizophrenia, drug-induced psychosis and borderline personality disorder: a naturalistic study with drug plasma levels. Expert opinion on pharmacotherapy. 2007;8(14):2207–2213.
74. Friedel RO, Jackson WT, Huston CS, May RS, Kirby NL, Stoves A. Risperidone treatment of borderline personality disorder assessed by a borderline personality disorder-specific outcome measure: a pilot study. Journal of clinical psychopharmacology. 2008;28(3):345–347.
75. Bellino S, Bozzatello P, Rinaldi C, Bogetto F. Paliperidone ER in the treatment of borderline personality disorder: a pilot study of efficacy and tolerability. Depression research and treatment. 2011.
76. Pascual JC, Soler J, Puigdemont D, Perez-Egea R, Tiana T, Alvarez E, et al. Ziprasidone in the treatment of borderline personality disorder: a double-blind, placebo-controlled, randomized study. Journal of clinical psychiatry. 2008;69(4):603–608.
77. Zanarini MC, Schulz SC, Detke HC, Tanaka Y, Zhao F, Lin D, et al. A dose comparison of olanzapine for the treatment of borderline personality disorder: a 12-week randomized, double-blind, placebo-controlled study. Journal of clinical psychiatry. 2011;72(10):1353–1362.
78. Samuels J, Eaton WW, Bienvenu OJ, Brown CH, COSTA PT, Nestadt G. Prevalence and correlates of personality disorders in a community sample. British journal of psychiatry. 2002;180(6):536–542.
79. Novais F, Araujo A, Godinho P. Historical roots of histrionic personality disorder. Front psychol. 2015;6:1463.
80. Fisher C. Hysteria: a delusional state. Medical hypotheses. 1999;53(2):152–156.
81. Siomopoulos V. Hysterical psychosis: psychopathological aspects. British journal of medical psychology. 1971;44(2):95–100.
82. Gift TE, Strauss JS, Young Y. Hysterical psychosis: An empirical approach. American journal of psychiatry. 1985.
83. Poupart F. The hysterical organization. International journal of psychoanalysis. 2014;95(6):1109–1129.
84. Pattison E. Why no cases of hysterical psychosis? American journal of psychiatry. 1986;143(8):1070-a-1.
85. Kohut H. The analysis of the self. New York: Int. Univ. Press.; 1971.
86. Vaknin S. Malignant self love: narcissism revisited. Narcissus Publishing; 2007.
87. Sanislow CA, Bartolini EE, Zoloth EC. Avoidant personality disorder. In: Ramachandran VS, editor. Encyclopedia of human behavior. 2nd ed. San Diego: Academic Press; 2012. pp. 257–266.
88. Ohnuma T, Arai H. Genetic or psychogenic? A case study of "folie a quatre" including twins. Case reports in psychiatry. 2015.
89. Diedrich A, Voderholzer U. Obsessive–compulsive personality disorder: A current review. Current psychiatry reports. 2015; 17(2):1–10.
90. Bebbington PE, Freeman D. Transdiagnostic extension of delusions: schizophrenia and beyond. Schizophrenia bulletin. 2017;43(2):273–282.
91. Diedrich A, Voderholzer U. Obsessive-compulsive personality disorder: a current review. Curr Psychiatry Rep. 2015;17(2):2.
92. Rossi A, Marinangeli MG, Butti G, Kalyvoka A, Petruzzi C. Pattern of comorbidity among anxious and odd personality disorders: the case of obsessive-compulsive personality disorder. CNS spectrums. 2000;5(9):23–26.
93. Hummelen B, Wilberg T, Pedersen G, Karterud S. The quality of the DSM-IV obsessive-compulsive personality disorder construct as a prototype category. Journal of nervous and mental disorder. 2008;196:446–455.
94. World Health Organization. The ICD-10 classification of mental and behavioural disorders: diagnostic criteria for research. World Health Organization; 1993.
95. Coid J, Yang M, Tyrer P, Roberts A, Ullrich S. Prevalence and correlates of personality disorder in Great Britain. British journal of psychiatry. 2006;188(5):423–431.

9

SYMPTOM NETWORK MODELS OF PSYCHOSIS

Adela-Maria Isvoranu, Lindy-Lou Boyette, Sinan Guloksuz, and Denny Borsboom

Disorders within the psychosis spectrum are highly heterogeneous and multifactorial.[1,2] Despite intensive research over the past century, the causes and pathogenesis of psychosis are still unclear, and no genetic marker has been consistently linked to developing a psychotic disorder.[3–5] In recent years, in an attempt to overcome this conundrum, the conceptualization of mental disorders as networks of interacting symptoms has gained considerable ground.[4,6] This conceptualization aligns well with practitioners' viewpoints as it focuses on concrete symptoms and their interrelations, rather than on abstract latent disorders or syndromes.[7]

Even though direct influences of one symptom on another are routinely observed in clinical practice (e.g., if a patient shows social withdrawal, this may soon lead to the patient displaying paranoid ideation and vice versa), in classical (psychometric) approaches to psychosis, which underlie most common psychometric practices in research, such direct influences are not modeled. Instead, symptoms are treated as passive psychometric indicators of a (set of) latent variable(s)—thus, it is assumed that symptoms are simply a result of the underlying disorder, rather than influencing each other.[4,6,8] As a result, correlations between symptoms are in a nontrivial sense *spurious*: symptoms cluster together *because of their common dependence on the disorder*.

The assumption that correlations between symptoms arise from a common cause, which has been deemed problematic by both psychometricians and clinicians,[4,7] has spurred the development of alternative psychometric approaches to mental disorders, in which symptoms are viewed instead as networks of mutually interacting components. Collectively, these lines of research have become known as *the network approach* to mental disorders. The centerpiece of the network approach is the idea that symptoms are active causal agents in producing disorder states, and that the study of their causal interaction is central to progress in understanding and treating mental disorders.

This chapter aims to introduce the network approach to mental disorders in the context of psychotic symptomatology. We first discuss standard approaches to psychotic disorders, highlighting unresolved issues, and we then provide an introduction to network models of psychosis, with a focus on the general theoretical framework. We concentrate on how (environmental and genetic) risk factors can be understood from a network perspective and how they can be included in network models. We complete the chapter with a discussion on how network models can be integrated into treatment approaches.

PSYCHOSIS: DISEASE MODEL VERSUS NETWORK MODEL

For the past twenty years, we have been biding time and asserting that certain etiological information is just around the corner;[9] we still wait.[10] A reconsideration of our basic strategies and fundamental assumptions may be in order.[11]

The concept of schizophrenia, and generally the disorders within the psychosis spectrum, as a *disease entity* emerged over a century ago[2] and it soon became a central (theoretical and applied) framework in the fields of psychiatry and clinical psychology. This framework is now known in mental health research and psychiatry as the *disease model*.[4] The scientific terminology used by clinicians and researchers alike quickly aligned with this view, and to date idioms such as "suffering from schizophrenia" are used to indicate the presence of a mental disorder.[6] In addition, comprehensive research has argued in favor of using the term "disease" when describing schizophrenia, as the term "disease" implies a discrete entity with a specific etiology and therefore better aligns with the clarity and heuristic value of the *disease model*.[2] Today, the use of the term "schizophrenia" almost immediately *implies* disease entity and in spite of fuzzy and unreliable results, biological approaches still focus on reverse-engineering this hypothesized disease entity using case-controls paradigms.[12]

The disease model strongly relies on the assumption that the nature of symptoms and the interactions between symptoms are not themselves relevant. It is argued that their presence is simply a result of the underlying disorder and that both research and treatment should focus on the disorder per se (see Figure 9.1). One example of this line of thinking that was very influential in the past decades is the idea that mental disorders are in fact diseases of the brain—literally "brain disorders"—, the nature of which will be uncovered through neuroscientific investigation.[13] Notably, it is not neuroscience

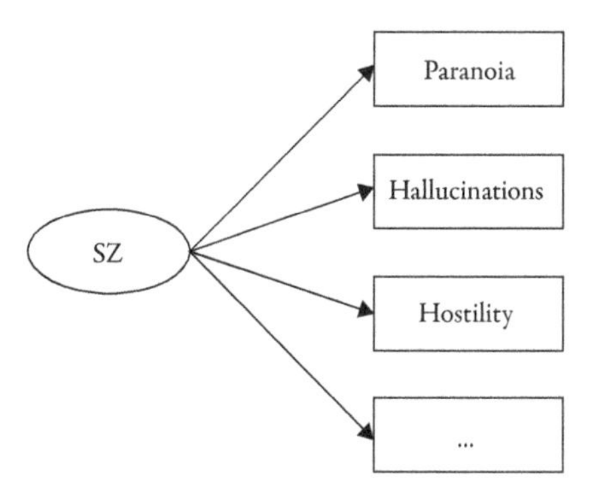

Figure 9.1 (Simplified) visualization of the disease model for schizophrenia. According to this model, schizophrenia (SZ) is the root cause of its symptoms (three symptoms examples as extracted from Positive and Negative Syndrome Scale[14]), which co-occur only because they are all caused by the same disorder.

per se that is the problem, but the reductionist taxonomy that current standard (psychometric) approaches are grounded on.

With this framework in mind, extensive research has focused on identifying psychological and biological essences of mental disorders[4] and on identifying *common causes* that give rise to symptomatology.[6] However, this quest has not been as successful as expected and perhaps unsurprisingly the underlying assumption that symptoms are interchangeable has proved problematic—to date few researchers and even fewer clinicians defend it.[4,5,7] Unlike in a medical context, where symptoms point to a disorder, which can then be confirmed with further testing (e.g., blood testing, X-ray), there is no diagnostic tool for any mental disorder.[15] The diagnosis is entirely based on the clinical manifestations, and the progress of a disorder is determined only by tracing the progress of the symptoms and global functioning. Within a clinical context the dynamics between the symptoms themselves therefore become of central interest, and a causal structure such as *Paranoid Ideation* → *Anxiety* → *Social Isolation* can often be observed. Given that the disease model does not allow (neither theoretically nor statistically) for such direct causal relations between symptoms, recent years have seen an emergence of network models of mental disorders, in which disorders are presumed to arise from this exact *causal interplay* between symptoms instead of being the result of a common cause. This approach, which turns the disease model on its head because it proposes that interactions between symptoms are themselves the causes of disorder states, aligns well with clinical intuition and current findings in the field; as a result, it currently spurs many novel research approaches to the old questions of what mental disorders are and where they come from.

Following these developments, the hypothesis has been advanced that *common cause* mechanisms for mental disorders cannot be found because no such mechanisms exist.[6] This hypothesis has spread quickly through the research fields of psychology and psychiatry. In particular, it has been argued that symptoms may cluster together not only because of a *shared origin*, but because they have the power to directly trigger and influence each other.[4,7] This novel theoretical framework became known as the *network approach to psychopathology*[4] and the fundamental assumption underlying it is that mental disorders arise from a complex interplay between symptoms and psychological, biological, and sociological components;[16] this interplay can be captured in a network model (see Figure 9.2). To date, network models have been applied to a wide range of constructs, including depression,[17–24] post-traumatic stress disorder,[25] psychosis,[26–31] autism,[32,33] social anxiety disorders,[34,35] substance abuse,[36] personality,[37,38] quality of life research,[39] and the more general structure of psychopathology.[40–42]

While networks models are based on innovative statistical methodology and offer radically new opportunities for theory formation and mathematical simulation of mental disorders, the notion that symptoms influence each other has been well known. In fact, such symptom–symptom interactions, transcending traditional diagnostic categories, have long been observed and acknowledged by clinical practitioners—network models just naturally accommodate it.[43] Importantly, in contrast to the mysterious latent entities posited in conventional statistical models of mental disorders, interactions between symptoms are often mundane and well understood. For instance, if a person thinks they are being followed by someone who would like to harm them (i.e., present signs of *paranoid ideation*), this person may soon become too *anxious* to walk outside and to freely carry out daily activities (i.e., present signs of *social isolation*). This behavior could then in turn reactivate the initial paranoia, as the patient is not provided with any counterevidence of the paranoid belief, allowing it to be sustained. In this way, the symptoms enter a feedback loop in which they reinforce each other; such feedback loops are often identified and are central in clinical practice. If symptom interactions are sufficiently strong, or if the symptom network is sufficiently large, the system can then enter into a stable state of sustained symptom activation. This arises when the interactions between symptoms are such that the symptom network is able to maintain its own activation—a state that we phenomenologically recognize as a mental disorder.

To wit, within this framework, disorders are *not* "mere labels" or social constructions: a stable state of problem activation is as real as anything, and as such in the network approach mental disorders are explained, but not explained away. Importantly, however, in the network approach a disorder such as schizophrenia is no longer understood as a *disease entity* that explains the symptoms. Instead, it is an alternative stable state of a network that is *constituted* by the *causal interactions* between symptoms. It can be illuminating to think of this in terms of an ecosystem analogy. It is well known that ecosystems, such as lakes, have alternative stable states (e.g., a lake can be clear or turbid).[44] These stable states are a function of a highly complex interplay between various components that together determine their overall state (e.g., the number of fish in the lake, the state of the algae population, the amount of nitrate in the soil, etc.). Importantly, the stable states of a lake are real: the lake is either clear or turbid as a matter of objective empirical fact. However, the stable state of the lake is *not* properly conceptualized as a latent disorder that "underlies" or "explains" the state of relevant components in the system: it refers to a global, emergent state of the lake that arises out of the interactions between the components in the system, rather than a mysterious unknown latent entity that is responsible for the aligned state of the system components as we may encounter it (e.g., turbid water, an increased algae population, fewer fish). In the same way, network approaches conceptualize disorders as emergent global properties of the symptom interactions: they *are* real, but do *not* signify a central underlying disorder in the form of a generally applicable pathogenic pathway.

Symptom interactions can be captured in a *network structure* (see Figure 9.2 for a hypothetical example of a network structure). Within such a network structure, symptoms are represented as *nodes*, while the relations between the symptoms are represented as *edges*.[38,45] Notably, the nodes within a network do not have to be limited to symptoms, but can also be, for instance, risk factors pertaining to the disorder (details on how the influence of the genes and environment can be included in symptom networks are discussed later), as well as nonsymptomatic variables or protective variables. It is, however, important for network dynamics that these elements/variables can both influence other variables and be influenced by other variables (e.g., gender would not qualify here as it is immutable).[43] A green edge indicates a positive relation between nodes, while a red edge indicates a negative relation between nodes.[45] For instance, if two nodes representing the typical symptoms *anxiety* and *paranoid ideation* are linked together by a green edge, this link indicates that the two nodes may *positively* activate each other (i.e., if a person is suffering from anxiety, the person is also most likely to present symptoms of paranoid ideation and/or vice versa). Likewise, if two nodes representing the typical symptoms *excitement* and *blunted affect* are linked together by a red edge, this link indicates that the two nodes may *negatively* activate each other (i.e., if a person is suffering from blunted affect, the person is also less likely to present symptoms of excitement and/or vice versa). Factors that impinge on the symptoms from outside together make up the *external field* of the symptoms. Note that in this case "from outside" means "from outside of the network" rather than "from outside of the person," so in addition to precipitating life events, drug use, or social circumstances, the external field can also include variable that relate to brain function and other bodily processes.[6]

In the absence of a strong theory on the structure of these interactions, networks are commonly constructed from empirical data in an explorative fashion, by searching for the network structure that best explains the associations present in a dataset.[4,38,45] Given the rapidly increasing popularity of these data-driven network models, it is important to note, however, that while there is a strong relation between network theory (which deals with causal interactions between symptoms) and statistical network models (which represent the statistical association between symptom measures and possibly other variables), that relation is by no means one-to-one. That is, many different statistical models could be used to inform a given network theory, and in some cases network theories are in fact constructed without any explicit statistical model; for example, by asking clinicians or respondents which symptoms activate each other.[33,47,48]

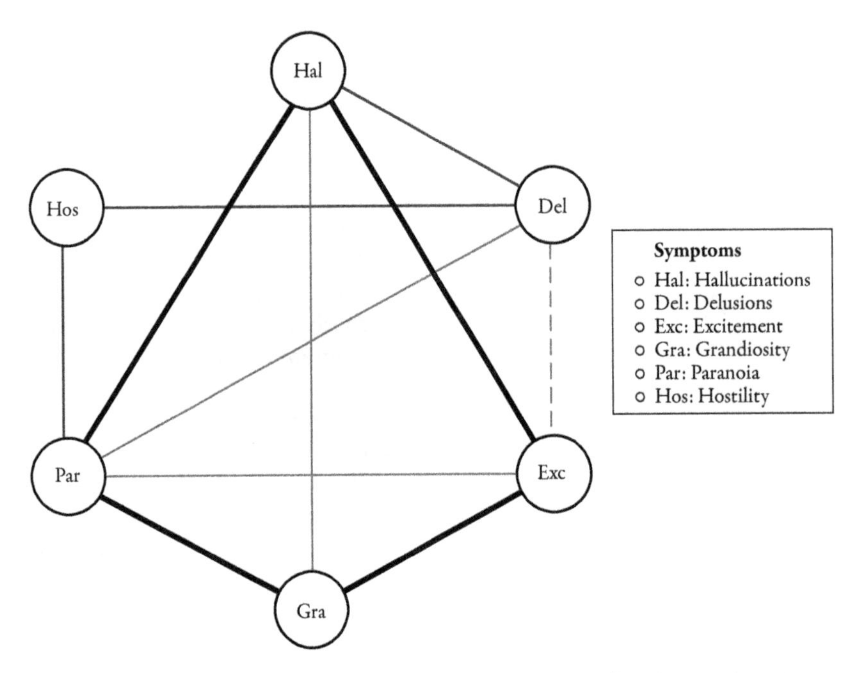

Figure 9.2 Visualization of a *hypothetical* network model of psychosis constructed using the *qgraph*[45] package in the *R Statistical Software*.[46] Full edges indicate positive associations, while dashed edges indicate negative associations between nodes. According to this model, the symptoms of schizophrenia (six symptom examples as extracted from Positive and Negative Syndrome Scale[14]) have causal power to trigger and influence each other.

TOWARD INCORPORATING ENVIRONMENTAL AND GENETIC RISK SCORES INTO NETWORK MODELS

1972: *"We are optimistically hopeful that the current mass of research on families of schizophrenics will discover an endophenotype, either biological or behavioral (psychometric pattern), which will not only discriminate schizophrenics from other psychotics, but will also be found in all the identical co-twins of schizophrenics whether concordant or discordant. All genetic theorizing will benefit from the development of such an indicator."*[49]

Widespread research has focused on identifying biological and environmental risk factors involved in the onset and progression of schizophrenia. Despite such efforts, risk factors that cause psychotic manifestations remain elusive.[50,51] Meta-analytical work suggests that substance abuse (especially cannabis), growing up in an urbanized area, and developmental trauma are among the most common environmental influences that increase the susceptibility to psychosis.[51,52] Other factors include ethnicity, season of birth, prenatal stress, and obstetric complications, but to a much lesser extent and with more inconsistent results.[50] In terms of a biological component, the look for biomarkers for schizophrenia has been extensive. However, to date, even though schizophrenia is known to be a highly heritable mental condition, with a rough heritability estimate up to 80%, the search for candidate genes for schizophrenia has showed no robust findings.[51] The diagnosis of schizophrenia is therefore based entirely on its clinical features and no other determinants have yet been incorporated into the diagnostic scheme.[53] Notably, if all psychotic symptoms were indeed caused by an underlying disease entity—as schizophrenia is argued to be—one would expect to identify that symptoms have the same (or at least similar) risk factors (see Figure 9.3A). That is because risk factors are expected to only influence the liability to develop a disorder, while the disorder would, in turn, cause the symptoms.[54] Current findings in the field of psychosis and depression suggest, however, that symptoms substantively differ in terms of their risk factors,[24,26,27] lending support to the importance of investigating the individual symptoms of a disorder and their causal associations—not only associations to each other, but also associations to the external field of influence.

In accordance, one can investigate the effects of the external field of influence (i.e., risk factors such as genes or environmental components) on a network structure in two ways. First, we can examine whether such factors have a *direct main* effect on the nodes (see Figure 9.3B, left panel). As highlighted earlier, nodes within a network are not restricted to symptoms only, but can be expanded to almost any variable—as long as the variable can both *influence* and *be influenced* by other variables.[43] As such, a node can represent any environmental,

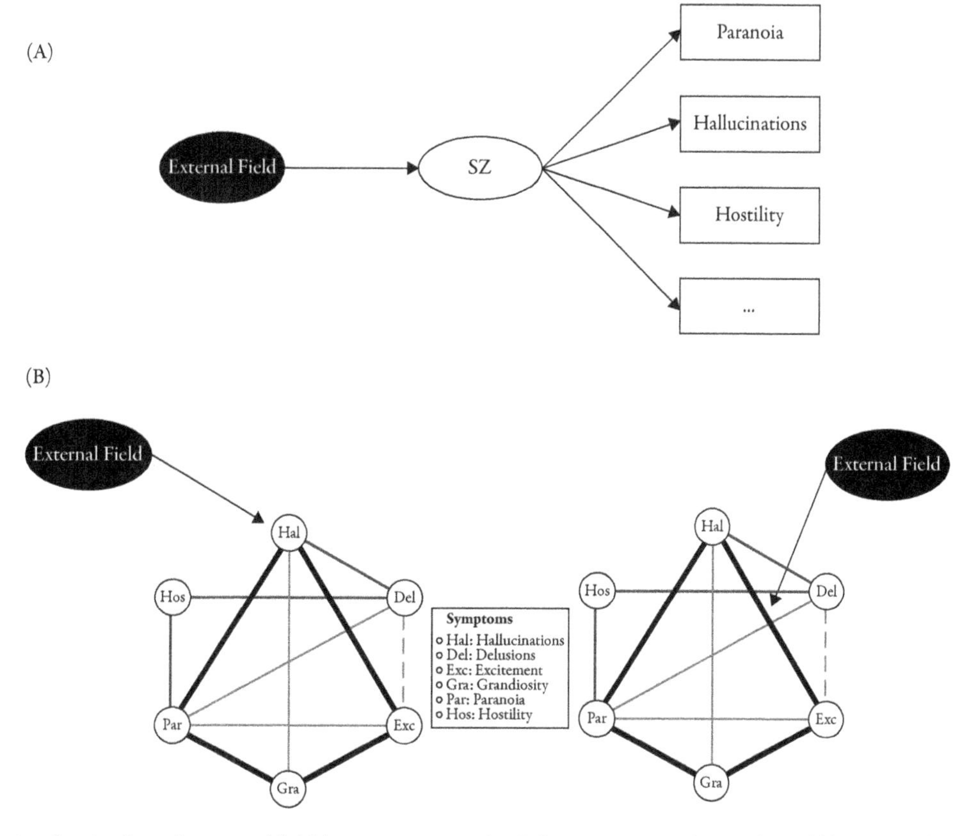

Figure 9.3 **A.** Visualization showing how the external field (e.g., environmental risk factors, genetic risk scores) would be expected to influence symptoms from a *common cause* perspective. The influence takes place via the latent variable (schizophrenia), which then causes the symptoms. **B.** Visualization showing how the external field (e.g., environmental risk factors, genetic risk scores) would be expected to influence symptoms in *network models*. The influence takes place either as a *main effect*, on the node itself or as an *indirect secondary effect*, on the symptom–symptom interactions.

biological, or cognitive component that is deemed relevant for the disorder. To assess the main effect of these components we can construct a network that incorporates (1) environmental risk factors and symptoms, (2) genetic risk factors—such as a polygenic risk score—and symptoms, (3) both environmental risk factors and genetic risk scores and symptoms, depending on the availability of the data. In this way, we do not only examine whether symptoms differ in terms of their risk factors, but we can also identify associations between the risk factors themselves, including potential gene-by-environment interrelations. As network models are still novel, to date, we are aware of only one study that has incorporated polygenic risk scores into symptom networks and found associations between the polygenic risk score and mainly positive psychotic symptoms and depressive symptoms;[55] we are however aware of work in progress and expect extensive research to follow this avenue soon.

In terms of investigating the associations between environmental risk scores and individual symptoms of mental disorders, several studies in the field of depression[20,24,54] and psychosis[26,27] have focused on assessing this interplay using network analysis. Isvoranu and colleagues[26,27] have identified that (1) when all symptoms of the Positive and Negative Syndrome Scale (PANSS)[14] are included in a network alongside childhood trauma, general psychopathology symptoms appear to mediate the associations between trauma and the positive and negative symptoms; (2) when using general population data to examine the interrelations between environmental exposure—as measured by urbanicity, developmental trauma, and cannabis use—and psychopathology expression, the three environmental factors distinctly affect the symptomatology, with the main connective pathway being through cannabis use. In addition, strong associations between cannabis use and urbanicity, as well as between developmental trauma and cannabis use were identified.[27] Even though this line of research is in its early stages, it shows high promise in potentially unraveling means by which the external field influences symptoms. In short, with a network framework in mind, we convert from trying to understand and identify risk factors for schizophrenia to trying to identify risk factors for *symptoms* of schizophrenia and for the *interaction* between these symptoms.

A second means (see Figure 3B, right panel) by which we can investigate the influence of the external field on symptomatology is through examining whether this has an *indirect secondary* effect on the network, by acting on the symptom–symptom connections. Thus, a stressor in the external field (e.g., grief, traumatic experiences, genetic components) cannot only be said to activate the network by either activating one or more symptoms (as detailed earlier), but also by *increasing the strength of interactions between the symptoms themselves.* Recent findings in large general population cohorts support this hypothesis—environmental exposure to risk factors such as childhood trauma, urbanicity, and cannabis use seem to alter the connectivity between the symptoms in a dose-response fashion, progressively triggering the transition toward a severe condition.[56–59] Isvoranu and colleagues[27] found that subjects exposed to cannabis showed increased connectivity between psychopathological dimensions such as *depression* and *anxiety* as well as between *interpersonal sensitivity* and *phobic anxiety*, compared to subjects not exposed to any high-risk environmental factors. Correspondingly, Guloksuz and colleagues[57] showed using two general population cohorts that environmental exposure increased the degree of symptom connectivity within all symptom dimensions and related diagnostic categories, and further argued that the emergence of psychosis may be related to severity of increasingly connected psychopathology rather than to a specific illness category.

With these findings in mind, the influence of the external field as expressed through adverse environmental determinants seems to lead to a more strongly connected and thus less resilient symptom network, facilitating the transition to an alternative stable state of mental disorder. In particular, it may further be possible that part of the missing heritability issues faced at the moment are due to the fact that genetic vulnerability may be expressed as increased connectivity of symptom interactions rather than as a main effect on the network (or on the disorder). In addition, the impact of the environment and/or genetics does not appear to be specific to particular diagnosis categories, but vulnerability factors seem to rather be spread between the traditional diagnostic categories (e.g., childhood trauma,[60–63] cannabis use[64–66]). Finally, thinking of this from a comorbidity perspective, if the external field has an influence on individual symptoms and the symptom–symptom interactions—which we know are often shared between disorders—the finding that vulnerability factors are similar and common between different diagnostic categories is no longer surprising.

CONCEPTUALIZING TREATMENT FROM A NETWORK PERSPECTIVE

If a mental disorder indeed refers to a self-sustaining alternative stable state of a symptom network, as network theory holds, then this has direct implications for how to conceptualize treatment. To a considerable extent, existing treatments can be re-interpreted from a network perspective, and we may perhaps better understand the treatment mechanisms by looking at them in this way. Below, we give an overview of some thoughts and observations that may fit within this framework. It is also possible that, in the future, novel treatment protocols may be partly based on network theory; although we are still far removed from having operational network treatment, individualized network modeling has already been implemented in proof-of-concept studies,[67] as we will also discuss.

As emphasized throughout this chapter, in a network model, there are basically three factors that determine its dynamics: the nodes in the network, the connections in the network, and the external field of factors outside of the network that influence it.[6] In accordance, we can consider three types of interventions from a network perspective; namely, those that target nodes in the network (henceforth: *node-interventions*), those that target connections in the network (henceforth: *edge-interventions*), and those that target the external field of factors outside of the network that influence it (henceforth: *field-interventions*).

Assuming, for the moment, that we take the network to be the collection of symptoms that a patient with a psychotic disorder endorses, a node-intervention would be an intervention that is directly aimed at the reduction of a specific symptom. For instance, the use of an antipsychotic drug targeting derogatory auditory hallucinations may be considered an example of a node-intervention. In a similar fashion, a wide array of cognitive behavioral techniques that are aimed directly at challenging the content of a cognition could be considered node-interventions. For instance, consider a patient who believes that the voice that he or she hears originates from an outside source, but is willing to examine this belief. A behavioral experiment, a common technique in cognitive-behavioral therapy, may be conducted, for instance asking the patient to tape the voice. Similarly, sleep reduction for reducing insomnia, response-prevention for compulsions, use of a drug with sedative qualities for agitation or interoceptive exposure for anxiety may all be considered examples of node-interventions.

Edge-interventions—interventions targeting relations between symptoms—may lie in various approaches in therapies that aim to teach a person how to improve cognitive, behavioral, and affective coping skills that may impact symptom–symptom relations. For instance, in the case of derogatory auditory hallucinations, an edge-intervention may involve being taught a new adequate coping strategy by a clinician or support group, such as the use of distraction, humming, or setting limits and structure to the contact, instead of an unhelpful coping strategy, such as loudly arguing with the voices whenever they occur (for instance leading to agitation and/or social isolation due to social rejection). Quite recently, several new therapies that are specifically designed to increase general cognitive, behavioral, and affective coping skills have gained ground for psychosis, such as metacognitive training (MCT)[68] and acceptance and commitment therapy (ACT).[69] Metacognitive training for psychosis aims to gain insight into unhelpful cognitive distortions putatively thought to be related to delusional beliefs (e.g., overconfidence in decisions, jumping to conclusions) and improve general metacognitive processes ("thinking about thinking"). Acceptance and commitment therapy aims, among other things, to increase psychological flexibility, such as more openness with unwanted thoughts and feelings that are unlikely to be amenable to direct change strategies, and encourages value-directed action. Instead of directly targeting symptoms when they occur, these types of interventions may decrease the connections, and therefore the disruptive impact of symptoms within a network.

Field-interventions target factors that are not directly part of the symptom network, but that do influence it from outside. For example, interventions include approaches that aim to restore or improve a person's community maintenance, such as entering a work project, which can reduce multiple symptoms and provide a buffer against different cascades of symptoms, and interventions that try to improve the physical environment of the person (e.g., by providing housing). Also interventions aimed at increasing the social support network of a person, or fighting the stigma of mental illness, may influence an individual's symptom network from the outside.

We recognize that the distinction between types of interventions is not always clear-cut and do not argue that a "direct" intervention cannot have multiple and indirect effects. The general aim of these examples is to provide an idea of how existing interventions could be interpreted in a network theoretical framework. Currently, there is no way to use this information, but in the future one could think of systematic intervention planning that uses network models for the selection of (sequences of) interventions. In such cases, one would start with a (partly) personalized network model that captures the person's most important symptoms and interactions between them. One then asks: assuming a given set of node-, edge-, and field-interventions, what would be the most effective way of changing and charging the symptom network to repel it from its disorder state and create an enduring healthy attractor state? Network models could, in such a methodology, play the role of navigation software, in that they can suggest a number of routes that may result in effective treatment. Clearly, we have not even begun yet with seriously engaging such approaches, but in time network-based intervention planning may offer an attractive extension of the treatment professional's toolkit.

One of the problems that will have to be solved in order for this type of approach to work is that we need ways to assess a given person's network structure. There are currently two ways of approaching this problem, both of which are promising but also need considerable work before they can be effectively put to use. First, Frewen and colleagues[46] have developed so-called Perceived Causal Relations (PCR) scales. These scales first query which symptoms a person has, and then query the extent to which combinations of symptoms influence each other. One can administer these scales to clients themselves (resulting in a self-reported symptom network) or to others (e.g., family, friends, healthcare professionals), to get a qualitative idea of the most important connections between symptoms. The advantage of this methodology is that it is relatively fast (one can administer a PCR scale in less than an hour) and that it is straightforward to combine different sources of evidence (e.g., one can easily combine networks resulting from self-assessment and from assessment by others). The downside is that, if one wants to use the methodology to get a view of the actual symptom network, one has to assume that people can actually assess causal relations between symptoms. Although there is some evidence that people naturally represent disorders in this way,[47] it is not clear that this representation is also accurate.

A second way of getting a glimpse of symptom-symptom interactions is by using ecological momentary assessment (EMA) protocols. In such approaches, which have seen tremendous growth in recent years due to the availability of smartphones, the person answers a short questionnaire multiple times a day. If sufficiently many data-points are available, time-series analyses can be used to estimate the network structure of the interactions between queried symptoms.[70,71] In a proof-of-concept study, Kroeze and colleagues[67] used this process to estimate a network structure for a number of clients in therapy; the network was subsequently shown to the client, and interventions were selected partly based on the network structure. Fisher and Boswell[72] have pioneered similar treatment approaches using time series analyses, although they did not explicitly use network structures but instead based treatment on dynamic factor analysis models. The main advantage of these approaches is that they can uncover symptom–symptom interactions that people may not be able to identify themselves.

A promising future extension is that it will become possible to combine subjective assessments with more complex and objective behavioral and physical measures (e.g., movement, location, physiological data, etc.). However, it should be noted that current approaches suffer from a number of serious drawbacks that need to be addressed in future work: (1) data-analysis methods are currently suboptimal as they do not allow to model mean trends but only deviations from these trends, (2) there is no way to currently integrate processes that operate at different time scales (moments, hours, days, weeks), (3) many symptoms and external factors simply cannot be assessed using EMA, either for practical or for principled reasons. Discussing such limitations of the EMA approaches—as well as more general limitations of the network approach—is beyond the scope of this chapter; for detailed theoretical and methodological considerations we direct the reader to the cited review articles.[43,73]

CONCLUDING REMARKS

The field of mental healthcare is in need of conceptualizations of pathology that are aligned with the complexity and multitude of symptoms and problems that most patients present with. Tools that could advance our understanding of psychiatric disorders, as well as provide substantiated intervention selection could be highly beneficial to clinical practice. In time, the network approach may provide such tools. Notably, it is unlikely that there is such a thing as a "one-size fits all treatment," especially given the inherent heterogeneity of psychosis. Thus, it is likely that intervention planning and monitoring will require personalized network models, which may vary from person to person. Endorsing integrative treatments approaches that are tailored to the individual, as well as embracing new avenues that allow us to combine subjective assessments with physical measures, will likely contribute to the advancement of the field and clinical practice. These possibilities are within reach, and in the next few decades we are bound to learn to what extent the promises of network theory in coordinating research and treatment will be fulfilled.

REFERENCES

1. Weinberger DR, Harrison PJ, eds. *Schizophrenia*. Oxford, UK: Wiley-Blackwell; 2010. doi:10.1002/9781444327298.
2. Tandon R, Keshavan MS, Nasrallah HA. Schizophrenia, "just the facts," what we know in 2008. Part 1. Overview. *Schizophr Res*. 2008;100(1–3):4–19. doi:10.1016/j.schres.2008.01.022.
3. Tandon R, Keshavan MS, Nasrallah HA. Schizophrenia, "just the facts," what we know in 2008. 2. Epidemiology and etiology. *Schizophr Res*. 2008;102(1–3):1–18. doi:10.1016/j.schres.2008.04.011.
4. Borsboom D, Cramer AOJ. Network analysis: an integrative approach to the structure of psychopathology. *Annu Rev Clin Psychol*. 2013;9:91–121. doi:10.1146/annurev-clinpsy-050212-185608.
5. Kendler KS, Zachar P, Craver C. What kinds of things are psychiatric disorders? Psychol Med. 2011;41(6):1143–1150. doi:10.1017/S0033291710001844.
6. Borsboom D. A network theory of mental disorders. *World Psychiatry*. 2017;16(1):5–13. doi:10.1002/wps.20375.
7. Fried EI. Problematic assumptions have slowed down depression research: why symptoms, not syndromes are the way forward. *Front Psychol*. 2015;6(March):309. doi:10.3389/fpsyg.2015.00309.
8. Schmittmann VD, Cramer AOJ, Waldorp LJ, Epskamp S, Kievit RA, Borsboom D. Deconstructing the construct: a network perspective on psychological phenomena. *New Ideas Psychol*. 2013;31(1):43–53. doi:10.1016/j.newideapsych.2011.02.007.
9. Gottesman II, McGuffin P, Farmer AE. Clinical genetics as clues to the "real" genetics of schizophrenia (a decade of modest gains while playing for time). *Schizophr Bull*. 1987;13(1):23–47. doi:10.1093/schbul/13.1.23.
10. Sullivan PF. Schizophrenia genetics: the search for a hard lead. *Curr Opin Psychiatry*. 2008;21(2):157–160. doi:10.1097/YCO.0b013e3282f4efde.
11. Tandon R, Keshavan MS, Nasrallah HA. Schizophrenia, "just the facts," what we know in 2008. 2. Epidemiology and etiology. *Schizophr Res*. 2008;102(1–3):1–18. doi:10.1016/j.schres.2008.04.011.
12. Guloksuz S, van Os J. The slow death of the concept of schizophrenia and the painful birth of the psychosis spectrum. 2017. doi:10.1017/S0033291717001775.
13. Insel TR, Cuthbert BN. Brain disorders? precisely. *Science (80-)*. 2015;348(6234):499–500. doi:10.1126/science.aab2358.
14. Kay SR, Fiszbein A, Opler LA. The Positive and Negative Syndrome Scale (PANSS) for schizophrenia. *Schizophr Bull*. 1987;13(2):261–276. doi:10.1093/schbul/13.2.261.
15. Kapur S, Phillips AG, Insel TR. Why has it taken so long for biological psychiatry to develop clinical tests and what to do about it? Mol Psychiatry. 2012;17(12):1174–1179. doi:10.1038/mp.2012.105.
16. Epskamp S, Waldorp LJ, Mõttus R, Borsboom D. Discovering psychological dynamics: the Gaussian graphical model in cross-sectional and time-series data. 2016:1–56. http://arxiv.org/abs/1609.04156.
17. Boschloo L, Van Borkulo CD, Borsboom D, Schoevers RA. A prospective study on how symptoms in a network predict the onset of depression. *Psychother Psychosom*. 2016;85(3):183–184. doi:10.1159/000442001.
18. Fried EI, Epskamp S, Nesse RM, Tuerlinckx F, Borsboom D. What are "good" depression symptoms? comparing the centrality of DSM and non-DSM symptoms of depression in a network analysis. *J Affect Disord*. 2015;189:314–320. doi:10.1016/j.jad.2015.09.005.
19. van de Leemput IA, Wichers M, Cramer AOJ, et al. Critical slowing down as early warning for the onset and termination of depression. *Proc Natl Acad Sci*. 2014;111(1):87–92. doi:10.1073/pnas.1312114110.
20. Fried EI, Bockting C, Arjadi R, et al. From loss to loneliness: the relationship between bereavement and depressive symptoms. *J Abnorm Psychol*. 2015;124(2):256–265. doi:10.1037/abn0000028.
21. Bringmann LF, Lemmens LHJM, Huibers MJH, Borsboom D, Tuerlinckx F. Revealing the dynamic network structure of the Beck Depression Inventory-II. *Psychol Med*. 2015;45(04):747–757. doi:10.1017/S0033291714001809.
22. van Borkulo CD, Boschloo L, Borsboom D, Brenda WJHP, Waldorp LJ, Schoevers RA. Association of symptom network structure with the course of depression. 2016;72(12):1219–1226. doi:10.1001/jamapsychiatry.2015.2079.
23. Cramer AOJ, Borsboom D, Aggen SH, Kendler KS. The pathoplasticity of dysphoric episodes: differential impact of stressful life events on the pattern of depressive symptom inter-correlations. *Psychol Med*. 2012;42(5):957–965. doi:10.1017/S003329171100211X.
24. Fried EI, Nesse RM. The impact of individual depressive symptoms on impairment of psychosocial functioning. *PLoS One*. 2014;9(2):e90311. doi:10.1371/journal.pone.0090311.

25. McNally RJ, Robinaugh DJ, Wu GWY, Wang L, Deserno M, Borsboom D. Mental disorders as causal systems: a network approach to posttraumatic stress disorder. *Clin Psychol Sci*. 2011:1–14. doi:10.1177/2167702614553230.
26. Isvoranu AM, van Borkulo CD, Boyette LL, Wigman JTW, Vinkers CH, Borsboom D. A network approach to psychosis: pathways between childhood trauma and psychotic symptoms. *Schizophr Bull*. 2016;(Advance Access):1–10. doi:10.1093/schbul/sbw055.
27. Isvoranu AM, Borsboom D, van Os J, Guloksuz S. A network approach to environmental impact in psychotic disorder: brief theoretical framework. *Schizophr Bull*. 2016;42(4):870–873. doi:10.1093/schbul/sbw049.
28. Bak M, Drukker M, Hasmi L, Van Jim OS. An n=1 clinical network analysis of symptoms and treatment in psychosis. *PLoS One*. 2016;11(9). doi:10.1371/journal.pone.0162811.
29. van Rooijen G, Isvoranu AM, Meijer CJ, van Borkulo CD, Ruhé HG, de Haan L. A symptom network structure of the psychosis spectrum. *Schizophrenia Research*. 2016.
30. Galderisi S, Rucci P, Kirkpatrick B, et al. Interplay among psychopathologic variables, personal resources, context-related factors, and real-life functioning in individuals with schizophrenia a network analysis. *JAMA Psychiatry*. 2018. doi:10.1001/jamapsychiatry.2017.4607.
31. Fonseca-Pedrero E, Ortuño J, Debbané M, et al. The network structure of schizotypal personality traits. *Schizophr Bull*. 2018;44(suppl_2):S468–S479. doi:10.1093/schbul/sby044.
32. Deserno MK, Borsboom D, Begeer S, Geurts HM. Multicausal systems ask for multicausal approaches: a network perspective on subjective well-being in individuals with autism spectrum disorder. *Autism*. 2016. doi:10.1177/1362361316660309.
33. Ruzzano L, Borsboom D, Geurts HM. Repetitive behaviors in autism and obsessive-compulsive disorder: new perspectives from a network analysis. *J Autism Dev Disord*. 2015;45(1):192–202. doi:10.1007/s10803-014-2204-9.
34. Beard C, Millner AJ, Forgeard MJC, et al. Network analysis of depression and anxiety symptom relations in a psychiatric sample. *Psychol Med*. 2016;(2016):1–11. doi:10.1017/S003329171 6002300.
35. Heeren A, McNally RJ. An integrative network approach to social anxiety disorder: The complex dynamic interplay among attentional bias for threat, attentional control, and symptoms. *J Anxiety Disord*. 2016;42:95–104. doi:10.1016/j.janxdis.2016.06.009.
36. Rhemtulla M, Fried EI, Aggen SH, Tuerlinckx F, Kendler KS, Borsboom D. Network analysis of substance abuse and dependence symptoms. *Drug Alcohol Depend*. 2016;161:230–237. doi:10.1016/j.drugalcdep.2016.02.005.
37. Cramer AOJ, van der Sluis S, Noordhof A, et al. Dimensions of normal personality as networks in search of equilibrium: you can't like parties if you don't like people. *Eur J Pers*. 2012;26(4):414–431. doi:10.1002/per.1866.
38. Costantini G, Epskamp S, Borsboom D, et al. State of the aRt personality research: a tutorial on network analysis of personality data in R. *J Res Pers*. 2015;54:13–29. doi:10.1016/j.jrp.2014.07.003.
39. Kossakowski JJ, Epskamp S, Kieffer JM, Borkulo CD, Rhemtulla M, Borsboom D. The application of a network approach to Health-Related Quality of Life (HRQoL): introducing a new method for assessing HRQoL in healthy adults and cancer patients. *Qual Life Res*. 2015;25(4):781–792. doi:10.1007/s11136-015-1127-z.
40. Tio P, Epskamp S, Noordhof A, Borsboom D. Mapping the manuals of madness: comparing the ICD-10 and DSM-IV-TR using a network approach. *Int J Methods Psychiatr Res*. 2016;25(4):267–276. doi:10.1002/mpr.1503.
41. Borsboom D, Cramer AOJ, Schmittmann VD, Epskamp S, Waldorp LJ. The small world of psychopathology. *PLoS One*. 2011;6(11). doi:10.1371/journal.pone.0027407.
42. Boschloo L, Schoevers RA, van Borkulo CD, Borsboom D, Oldehinkel AJ. The network structure of psychopathology in a community sample of preadolescents. *J Abnorm Psychol*. 2016; 125(4):599–606. doi:10.1037/abn0000150.
43. Fried EI, Cramer AOJ. Moving forward: challenges and direction for psychopathological network theory and methodology. *Perspect Psychol Sci*. 2017. doi:10.17605/OSF.IO/BNEK.
44. Scheffer M, Bascompte J, Brock WA, et al. Early-warning signals for critical transitions. *Nature*. 2009;461(7260):53–59. doi:10.1038/nature08227.
45. Epskamp S, Cramer AOJ, Waldorp LJ, Schmittmann VD, Borsboom D. qgraph: network visualizations of relationships in psychometric data. *J Stat Softw*. 2012;48(4):1–18. http://www.jstatsoft.org/v48/i04.
46. R Development Core Team. *R A Lang environ stat comput*. 2015; 55:275–286.
47. Frewen PA, Allen SL, Lanius R a, Neufeld RWJ. Perceived causal relations: novel methodology for assessing client attributions about causal associations between variables including symptoms and functional impairment. *Assessment*. 2012;19(4):480–493. doi:10.1177/1073191111418297.
48. Kim NS. Clinical psychologists' theory-based representations of mental disorders affect their diagnostic reasoning and memory. *Diss Abstr Int Sect B Sci Eng*. 2002;63(3-B):1565.
49. I.I. G. Schizophrenia and genetics: a twin study vantage point. *ACADPRESS, NEW YORK, NY*. 1972.
50. Leask SJ. Environmental influences in schizophrenia: the known and the unknown. *Adv Psychiatr Treat*. 2004;10(5):323–330. doi:10.1192/apt.10.5.323.
51. van Os J, Kapur S. Schizophrenia. *Lancet*. 2009;374(9690):635–645. doi:10.1016/S0140-6736(09)60995-8.
52. van Os J, Kenis G, Rutten BPF. The environment and schizophrenia. *Nature*. 2010;468(7321):203–212. doi:10.1038/nature09563.
53. Tsuang MT, Stone WS, Faraone SV. Genes, environment and schizophrenia. *Br J Psychiatry*. 2001;178(suppl. 40). doi:10.1192/bjp.178.40.s18.
54. Fried EI, Nesse RM, Zivin K, Guille C, Sen S. Depression is more than the sum score of its parts: individual DSM symptoms have different risk factors. *Psychol Med*. 2014;44(10):2067–2076. doi:10.1017/S0033291713002900.
55. Isvoranu A-M, Guloksuz S, Epskamp S, van Os J, Borsboom D, GROUP. Towards incorporating genetic risk scores into symptom networks of psychosis. *Psychol Med*. 2019.
56. Guloksuz S, Nierop M Van, Lieb R, Winkel R Van, Wittchen H, Os J Van. Evidence that the presence of psychosis in non-psychotic disorder is environment-dependent and mediated by severity of non-psychotic psychopathology. *Psychol Med*. 2015:1–13. doi:10.1017/S0033291715000380.
57. Guloksuz S, van Nierop M, Bak M, et al. Exposure to environmental factors increases connectivity between symptom domains in the psychopathology network. *BMC Psychiatry*. 2016;16(1):223. doi:10.1186/s12888-016-0935-1.
58. Fusar-Poli P, Nelson B, Valmaggia L, Yung AR, McGuire PK. Comorbid depressive and anxiety disorders in 509 individuals with an at-risk mental state: Impact on psychopathology and transition to psychosis. *Schizophr Bull*. 2014;40(1):120–131. doi:10.1093/schbul/sbs136.
59. Smeets F, Lataster T, Viechtbauer W, et al. Evidence that environmental and genetic risks for psychotic disorder may operate by impacting on connections between core symptoms of perceptual alteration and delusional ideation. *Schizophr Bull*. 2015;41(3):687–697. doi:10.1093/schbul/sbu122.
60. Wiersma JE, Hovens JGFM, Van Oppen P, et al. The importance of childhood trauma and childhood life events for chronicity of depression in adults. *J Clin Psychiatry*. 2009;70(7):983–989. doi:10.4088/JCP.08m04521.
61. Heim C, Nater UM, Maloney E, Boneva R, Jones JF, Reeves WC. Childhood trauma and risk for chronic fatigue syndrome: association with neuroendocrine dysfunction. *Arch Gen Psychiatry*. 2009;66(1):72–80. doi:10.1001/archgenpsychiatry.2008.508.

62. Dannlowski U, Stuhrmann A, Beutelmann V, et al. Limbic scars: long-term consequences of childhood maltreatment revealed by functional and structural magnetic resonance imaging. *Biol Psychiatry.* 2012;71(4):286–293. doi:10.1016/j.biopsych.2011.10.021.
63. Read J, Van Os J, Morrison AP, Ross CA. Childhood trauma, psychosis and schizophrenia: a literature review with theoretical and clinical implications. *Acta Psychiatr Scand.* 2005;112(5):330–350. doi:10.1111/j.1600-0447.2005.00634.x.
64. Patton GC, Coffey C, Carlin JB, Degenhardt L, Lynskey M, Hall W. Cannabis use and mental health in young people: cohort study. *BMJ.* 2002;325(7374):1195–1198. doi:10.1136/bmj.325.7374.1195.
65. Bovasso GB. Cannabis abuse as a risk factor for depressive symptoms. *Am J Psychiatry.* 2001;158(12):2033–2037. doi:10.1176/appi.ajp.158.12.2033.
66. Arseneault L, Cannon M, Poulton R, Murray R, Caspi A, Moffitt TE. Cannabis use in adolescence and risk for adult psychosis: longitudinal prospective study. *BMJ.* 2002;325(7374):1212–1213. doi:10.1136/bmj.325.7374.1212.
67. Kroeze R, van der Veen D, Servaas M, et al. Personalized feedback on symptom dynamics of psychopathology: a proof-of-principle study. *J Pers Res.* 2017;3(1):1–10. doi:10.17505/jpor.2017.01.
68. Moritz S, Woodward TS. Metacognitive training in schizophrenia: from basic research to knowledge translation and intervention. *Curr Opin Psychiatry.* 2007;20(6):619–625. doi:10.1097/YCO.0b013e3282f0b8ed.
69. Hayes SC, Strosahl KD, Wilson KG. *Acceptance and commitment therapy: an experiential approach to behavior change.* 1999.
70. Bringmann LF, Vissers N, Wichers M, et al. A network approach to psychopathology: new insights into clinical longitudinal data. *PLoS One.* 2013;8(4):e60188. doi:10.1371/journal.pone.0060188
71. Epskamp S, van Borkulo CD, van der Veen D, Isvoranu AM, Riese H, Cramer AOJ. Personalized network modeling in psychopathology: the importance of contemporaneous and temporal connections. *Clin Psychol Sci.* 2017;6(3):416–427.
72. Fisher AJ, Boswell JF. Enhancing the personalization of psychotherapy with dynamic assessment and modeling. *Assessment.* 2016;23(4):496–506. doi:10.1177/1073191116638735.
73. Guloksuz S, Pries L-K, van Os J. Application of network methods for understanding mental disorders: pitfalls and promise. *Psychol Med.* 2017:1–10. doi:10.1017/S0033291717001350.

PSYCHOSIS IN GENERAL MEDICAL CONDITIONS AND ORGANIC BRAIN DISORDERS

10

ORGANIC PSYCHOSIS

PHENOTYPIC DEVIANTS OR CLUES TO SCHIZOPHRENIA?

Peter Buckley and Brian Miller

Like epilepsy, schizophrenia is the name of a syndrome of which the idiopathic type is merely the commonest form among many.[1]

INTRODUCTION

Ms. Susannah Cahalan's story is illustrative and compelling.[2] As a US newspaper reporter, she dramatically brings to life the conundrum of organic vs. functional psychoses. Her book—provocatively titled *Brain on Fire*—tells the frightening story of her psychotic state that was misattributed as schizophrenia. She presented in a psychotic state, was carefully examined and with a thoughtful clinical assessment (including a brain scan), was found to have a psychosis without any organic cause. By chance, a blood sample was later taken to test for NMDA receptor antibodies and it tested positive—revealing the (heretofore obscure) diagnosis of autoimmune encephalitis. Did Ms. Cahalan have some inflammatory psychosis—equivalent to but just less demonstrable than a psychosis that might be associated with an infective encephalitis—or did she have some form of "neuroinflammatory schizophrenia," a most extreme example of immune dysfunction that is now being documented in some patients with schizophrenia? Are such distinctions even valid or heuristic?

THE ENIGMA OF DIAGNOSIS

Although the diagnostic enigma of "organic" versus "functional" psychosis might well be more "center stage" now with modern-day conceptualization of schizophrenia from nosological (e.g., Diagnostic and Statistical Manual, APA, 2013) and neurobiological[3] perspectives, it is clear that the relationship of schizophrenia to the schizophrenia-like organic psychoses has long been recognized as an issue of fundamental importance.[1,4–5] Current diagnostic criteria for schizophrenia are based in no small part, on the presence and persistence of psychotic phenomena (hallucinations, delusions, thought disorder, and/or disorganized behavior) in the absence of any organic pathology—either the lack of a brain disorder or the lack of a medical condition that could be causative. This is "ground zero" for schizophrenia. And yet, it is perhaps a false dichotomy from several vantage points. First, by stating what schizophrenia is not, it assumes that we know what it is. Second, the diagnosis of schizophrenia is based on what one scholar once said is "the quicksand of symptomatology." This is a real problem, especially if symptoms can occur from any of a myriad of conditions. Third, and relatedly, there is no "schizophrenic symptom." None of the features we ascribe to the diagnosis of schizophrenia are either sensitive or specific to this condition. Wilson's disease, an inborn error of copper metabolism that is characterized by hepatolenticular degeneration, is recognized clinically by Kayser-Fleischer (KF) rings—copper deposits on the cornea. Wilson's disease can—and does—present as psychosis. The KF rings are pathognomonic for Wilson's disease and when seen through the slit lamp on eye examinations, the diagnosis is assured. There are no such "KF rings" for schizophrenia. Fourth, as the evidence amply reviewed in this book attests to, there *are* brain changes in schizophrenia. Their existence belies the dichotomy of "functional" and "organic" psychosis. Fifth, while the presence of medical conditions could be considered exclusionary to the diagnosis of schizophrenia, there are many medical comorbidities that are commonly seen in people with schizophrenia. While these interrelationships may be complex and not necessarily causal, it is also short-sighted to conclude that medical conditions in people with schizophrenia occur independently and without influence on their psychotic symptoms.

HOW SIGNIFICANT ARE "BRAIN LESIONS" AND OTHER "ORGANIC" FINDINGS IN "FIRST-EPISODE PSYCHOSIS?"

Less than 5% of patients with a first episode of psychosis (FEP)—with a presentation that is indistinguishable from schizophrenia—will have some "organic" pathology. Examples on brain imaging include partial or complete agenesis of the corpus callosum, cavum septum pellucidum, and white matter lesions. When such findings are seen in individuals who present with an otherwise indistinguishable FEP, they are considered as incidental and unrelated to the psychosis. And yet, early on in the neuroimaging of schizophrenia, such anomalous "brain lesions" were viewed as evidence for a "neurodevelopmental hypothesis" of schizophrenia. Other examples include thyroid dysfunction or adrenal hyperplasia.

In other instances where lesions are more directly attributable to some known brain disease, the psychosis is attributed to that condition and any connotation of a schizophreniform psychosis is withheld. Stark examples are psychosis with meningioma and psychosis with neurosarcoidosis.

Johnston and colleagues[6] in their study of first-episode schizophrenia detected 23 patients with evident organic psychoses. Their symptoms, however, were similar to the 92 other patients with typical schizophrenia. In another study,[7] the symptom pattern for patients with organic psychoses (n = 53) was indistinguishable (apart from an excess of auditory hallucinations) from that of patients with regular schizophrenia. Cutting[8] compared the pattern of symptoms in organic psychoses (n = 74) and relapsed schizophrenia and found only minor differences in the quality of delusions and hallucinations.

Thus, it is perhaps of no surprise that, in a well-screened study of first-episode schizophrenia, Johnstone and collaborators[9] described a group whose psychosis was readily attributable to physical illness. Physical causes included alcohol and drug abuse, neurosarcoidosis, carcinoma, syphilis, head injury, and autoimmune and thyroid disease. The wide variety of reported conditions that are of etiological relevance is similar to findings of other studies.[4,7] Falkenberg and colleagues[10] evaluated radiological findings in brain MRI scans of 108 research study patients and 240 patients undergoing clinical assessment of FEP, with matched healthy volunteers. Brain abnormalities—overwhelmingly inconsequential—were noted in 6% of the research FEP group and in 15% of the clinical FEP group. The low "diagnostic yield" from brain scans of seemingly FEP populations have prompted economic considerations that brain imaging is an expensive, inefficient, and not cost-effective assessment tool.[11] However, use of brain imaging still prevails and is an integral part of the assessment of an FEP.

We now move on to consider some selective "organic" psychoses, recognizing of course that this is not an exhaustive list. Accordingly, we have chosen to focus on three illustrative conditions. Common causes of organic psychosis are also highlighted in Box 10.1.

Box 10.1 COMMON CAUSES OF ORGANIC PSYCHOSIS

- Autoimmune disorders (lupus, sarcoidosis, anti-NMDA-receptor encephalitis)
- Basal ganglia disorders (Huntington's disease, Wilson's disease)
- Brain tumors and paraneoplastic syndrome
- Endocrine disorders (thyroid disease, Cushing's disease, Cushing's syndrome)
- Epilepsy
- Head injury
- Infections (HIV, syphilis)
- Major neurocognitive disorders (Alzheimer's disease, Lewy body dementia, Parkinson's disease)
- Metabolic disorders (acute intermittent porphyria)
- Multiple sclerosis
- Narcolepsy
- Nutritional deficiency (vitamin B12 deficiency)
- Stroke
- Velocardiofacial syndrome (22q deletion syndrome)

HEAD TRAUMA AND PSYCHOSIS

In 1993, we published a case series of patients who had long-standing schizophrenia and whom prior clinicians had diagnosed and treated as schizophrenia.[12] When scanned as a supplemental project to a large MRI study in schizophrenia, their scans revealed marked brain changes, specifically and seemingly more selectively in frontal and temporal lobe regions. Had an MRI scan been available and/or ordered at the time of their initial presentation, it is likely their physician would have considered their psychosis as due to head trauma. The relationship between brain trauma and psychiatric illnesses is strongest for depression, post-traumatic stress disorder, and aggression.[13] Earlier reports[4,14] stressed the temporal relationship between the cerebral insult and onset of schizophrenia. They also stressed that these psychoses are generally associated with an absence of genetic risks for schizophrenia, thus lending greater weight to their etiological importance.[4,15]

In a Finnish study of brain injury among war veterans, Achte and colleagues[16] reported that 23% of cases with paranoid schizophrenia developed within the first year after injury. It is postulated that early trauma that precedes brain maturation may be of particular importance in the genesis of some forms of psychosis.[17] The literature on head trauma and psychosis stretches back, but is hard to interpret in the "pre-brain scan" era. It is clear, however, that head trauma is a risk for schizophrenia.[13,18,19] A study by Sachdev and colleagues[20] compared 45 patients with psychosis after head trauma with a matched sample of patients with traumatic brain injury who were not psychotic. The psychotic group had more brain damage, especially in temporo-parietal regions. Interestingly, the onset of psychosis was more often than not insidious and for the entire group, there was a latency of 55 months between the head trauma and onset of psychosis.

A recurrent theme in research on "organic phenocopies" of schizophrenia is whether the condition (on its own) caused the psychosis or whether it was the "straw that broke the camel's back" and was synergistic with another known risk factor for schizophrenia. Genetic liability is the most often studied association. In the Sachdev study cited earlier, a family history of psychosis was a risk factor. However, from available evidence, it appears that head trauma can also cause psychosis as an independent risk factor.

Major head trauma more often results in organic brain syndromes, of which psychosis can be and often is a part. Less severe head trauma can cause a psychosis indistinguishable from schizophrenia. More subtle head trauma—with minor concussion—might also be a risk factor for schizophrenia. We are presently witnessing a scientific reappraisal of "mild head injury," and further research might reveal an ever more common association with schizophrenia. Time will tell.

EPILEPSY AND PSYCHOSIS

Perhaps the most widely reviewed relationship between organic brain disease and schizophrenia concerns epilepsy and the schizophrenia-like psychoses.[21,22] Psychotic phenomena are common in epilepsy, in both pre- and postictal states. They are most commonly associated with temporal lobe epilepsy (TLE), a relationship described in the seminal paper by Davison and Bagley.[4] The literature upholds a laterality to this relationship, with left-sided TLE more apt to have psychosis. Indeed, this association, and particularly the observation of psychotic symptoms in patients with epilepsy, has contributed to the improved understanding of a fundamental issue in schizophrenia, i.e., the issue of laterality. In addition, greater understanding of neurophysiology that has been gained through the study of epilepsy has helped form a conceptual framework to the basis of some symptoms of schizophrenia and affective disorders.[23,24] This is particularly so in consideration of the process of kindling.[22]

Gold studied the cognitive deficits in patients with TLE in comparison with patients with schizophrenia.[24] They found a more selective pattern of memory loss in patients with TLE. In contrast, Allebone and colleagues[25] conducted a comprehensive review of psychosis and epilepsy—debunking the "conventional wisdom," they found more generalized brain changes beyond merely temporo-limbic changes. These sentiments are also echoed by epilepsy leaders who now view epilepsy as more of a network dysfunction rather than as the result of a select brain lesion.[26]

PSYCHOSIS IN OTHER KNOWN NEUROPSYCHIATRIC CONDITIONS

Psychosis is a concomitant of a variety of neuropsychiatric conditions, from neurosarcoidosis to systemic lupus erythematosus (SLE) and to other neuropsychiatric conditions like multiple sclerosis and autoantibody-mediated inflammatory conditions like NMDA receptor autoimmune encephalitis. As one such example, psychosis occurs in sarcoidosis, a widespread granulomatous condition that is thought to have an immune basis.[27] Psychosis is most common in the context of an adverse effect of the use of steroids to treat sarcoidosis. However, neurosarcoidosis may also of itself cause psychosis. Psychosis may even occasionally be a presenting feature of sarcoidosis. There does not appear to be laterality effect and/or selective distribution of sarcoid lesions in the brain that are associated with psychosis. Paradoxically, low-dose steroids are also a treatment for psychosis associated with neurosarcoidosis.

The relationship of psychosis to other immune-related neuropsychiatric conditions is less clear. In multiple sclerosis, depression is the most common psychiatric comorbidity.[28] Psychosis and bipolar disorder are less common, although speculatively these may be presenting features of multiple sclerosis.[7,29] Psychosis is uncommon, and this is more likely to be related to steroid treatments for multiple sclerosis.

The extent to which treatments for neuropsychiatric disorders (especially steroids) can cause psychosis is much more prevalent than the psychosis being the result of the underlying neuropsychiatric condition. Psychosis occurring in Parkinson's disease (PD) is an apt illustration. L-dopa-related psychosis is a common clinical scenario in PD, and it substantially complicates management. While clozapine and quetiapine have been found to be preferentially useful for treating psychosis in PD, our field is intrigued by the implication of pimavanserin, an inverse agonist for serotonin 5HT2A receptors, which now has an approved indication for treatment of psychosis in PD.[30]

In addition to the lay public attention from Ms. Cahalan's book *Brain on Fire*, there is substantial scientific interest in the topic of anti-NMDA receptor encephalitis. Gaughran and colleagues[31] found only one patient (and also one control) positive for anti-NMDA receptor antibodies in a 98-patient sample of FEP. In a German study that focused directly on patients with anti-NMDA receptor encephalitis, 72% had a normal MRI and in the remaining patients the abnormal pattern was diffuse.[32] Interest in this uncommon yet intriguing autoimmune-mediated psychosis is presently heightened, as it may also provide a window to understand neuroinflammation in schizophrenia, including advancing novel treatment approaches.[33]

TREATMENT CONSIDERATIONS

In considering the plethora of conditions in which organic psychoses can arise, there are a number of treatment principles (Box 10.2). First, as in all aspects of medicine, a thorough evaluation is needed to determine what is the underlying condition(s) that is (are) contributing to the psychosis. This will dictate the approach to treatment. Second, optimize the management of the primary neuropsychiatric condition. The treatment, if effective, may also resolve the psychosis if it is directly related to the condition (e.g., neurosarcoidosis). Third, use antipsychotic medications sparingly—if at all. Patients with neuropsychiatric conditions are exquisitely sensitive to the side effects of antipsychotic medications. This is most notable for first-generation antipsychotics, wherein the propensity for motor adverse effects and tardive dyskinesia is greatly heightened. In general, second-generation antipsychotics are preferred for such circumstances, particularly so for psychosis in PD (see earlier discussion). Fourth, reevaluate the need for treatment over time. Many organic psychoses resolve either spontaneously and/or with appropriate treatment to the underlying neuropsychiatric condition, and/or with judicious short-term use of antipsychotics. Since these circumstances are more variable than the intractable course of schizophrenia, we should not simply assume that antipsychotics are necessary into longer-term care. A reevaluation after a psychosis has subsided is prudent.

Box 10.2 **PRINCIPLES IN THE MANAGEMENT OF ORGANIC PSYCHOSES**

Evaluate comprehensively to determine underlying organic brain condition
Optimize treatment of this condition
Use antipsychotic medications sparingly—if at all
Reevaluate over time
Explain carefully to patients and family
Document well in medical records

Finally, since these psychoses differ tremendously in clinical manifestation and course from schizophrenia, it is important to provide patients and relatives alike with as much understanding of the reason for the psychosis as is available. Pointing out the distinction between organic psychosis and schizophrenia is helpful, as the patient might believe he/she has also "contracted" schizophrenia. Once these are well explained, patients will also better tolerate the ambiguity of associations between psychosis and neuropsychiatric conditions.

CONCLUDING REMARKS

The psychiatric literature is replete with intriguing examples of psychoses concomitant to conditions that can affect the brain and to medications that can induce psychosis as a side effect. This review illustrated the conceptual and clinical implications of these associations, especially with respect to our understanding of schizophrenia. Greater understanding, coupled with and driven by more refined investigations of neuropsychiatric conditions and of the neurobiology of schizophrenia itself, will likely lead to further convergence of the fundamental nosology of schizophrenia.

In a thoughtful reflection on the origins of schizophrenia, Sir Robin Murray provides a contemporary perspective on this dichotomy that has been central to our understanding and diagnostic approaches to schizophrenia.[34] He concludes: "I expect to see the end of the concept of schizophrenia soon. Already the evidence that it is a discrete entity rather than just the severe end of psychosis has been fatally undermined. Furthermore, the syndrome is already beginning to break down, for example, into those cases caused by copy number variants, drug abuse, social adversity, etc. Presumably, this process will accelerate and the term schizophrenia will be confined to history, like dropsy."

REFERENCES

1. Mapother E. Mental syndromes associate with head injury. *Br Med J* 1937;2:1055–1073.
2. Cahalan S. *Brain on Fire: My Month of Madness*. Free Press; 2012.
3. Gershon ES, Pearlson G, Keshavan MS, et al. Genetic analysis of deep phenotyping projects in common disorders. *Schizophr Res* 2017.
4. Davison K, Bagley CR. Schizophrenia-like psychoses associated with organic disorders of the nervous system—a review of the literature. In: Herrington RN (ed); *British Journal of Psychiatry, Special Publication No. 4: Current Problems in Neuropsychiatry*. Headley Bros., Ashford (N); 1969.
5. Davison K. Schizophrenia-like psychoses associated with organic cerebral disorders: a review. *Psychiatr Dev* 1983;1(1):1–33.
6. Johnstone E, Cooling J, Frith C, et al. Phenomenology of organic and functional psychoses and the overlap between them. *Br J Psychiatry* 1988;153:770–776.
7. Feinstein A, du Boulay G, Ron MA. Psychotic illness in multiple sclerosis. *Br J Psychiatry* 1992;161:680–685.
8. Cutting J. The phenomenology of acute organic psychosis: comparison with acute schizophrenia. *Br J Psychiatry* 1987;151:324–332.
9. Johnstone E, MacMillan F, Crow TJ. The occurrence of organic disease of aetiological significance in a population of 268 cases of first-episode schizophrenia. *Psychol Med* 1987;17:371–379.
10. Falkenberg I, Bennetti S, Raffin M, et al. Clinical utility of magnetic resonance imaging in first-episode psychosis. *Br J Psychiatry* 2017;211:231–237.
11. Borgwardt S, Schmidt A. Implementing magnetic resonance imaging into clinical routine screening in patients with psychosis? *Br J Psychiatry* 2017;211(4):192–193.
12. Buckley PF, Stack JP, Madigan C, O'Callaghan E, Larkin C, Redmond O, et al. Magnetic resonance imaging of the schizophrenia-like psychoses associated with cerebral trauma: clinicopathological correlates. *Am J Psychiatry* 1993;150:146–148.
13. Hesdorffer DC, Rauch SL, Tamminga CA. Long-term psychiatric outcomes following traumatic brain injury: a review of the literature. *J Head Trauma Rehab* 2009;24:452–459.
14. Hillbom E. Schizophrenia-like psychoses after brain trauma. *Acta Psychiatr Neurol Scand 60* 1951;36–47.
15. Lewis SW. The secondary schizophrenias. In: *Schizophrenia*. Blackwell Science Ltd. 1995.
16. Achte K, Jarho L, Kyykka T, Vesterinen E. Paranoid disorders following war brain damage. *Psychopath* 1991;24:309–315.
17. Lewis SW. Congenital risk factors for schizophrenia. *Psychol Med* 1989;19:5–13.
18. Kim E. Does traumatic brain injury predispose individuals to develop schizophrenia? *Curr Opin Psychiatry* 2008;21:286–289.
19. Fujii DE, Ahmed I. Psychotic disorder caused by traumatic brain injury. *Psychiatr Clin North Am* 2014;37:113–124.
20. Sachdev P, Smith JS, Cathcart S. Schizophrenia-like psychosis following traumatic brain injury: a chart-based descriptive and case-control study. *Psychol Med* 2001;31:231–239.
21. Slater E, Beard AW. The schizophrenia-like psychosis of epilepsy. *Proc Royal Soc Med* 1962;55:1–7.
22. Mace CJ. Epilepsy and schizophrenia. *Br J Psychiatry* 1993;163:439–445.
23. Roberts CW, Done DJ, Bruton C, Crow TJ. A "mock-up" of schizophrenia: temporal lobe epilepsy and schizophrenia-like psychosis. *Biol Psychiat* 1990;127–143.
24. Gold JM, Hermann BP, Wyler A, et al. Schizophrenia and temporal lobe epilepsy: a neuropsychological study. *Arch Gen Psychiatry* 1994;51:265–272.
25. Allebone J, Kanaan R, Wilson SJ. Systematic review of structural and functional brain abnormalities in psychosis of epilepsy. *J Neurol Neurosurg Psychiatry* 2017.
26. Scharfman HE, Kanner AM, Friedman A, et al. Epilepsy as a network disorder (2): what can we learn from other network disorders such as dementia and schizophrenia, and what are the implications for translational research? *Epilepsy Behav* 2018;78:302–312.
27. Greene JJ, Neumann IC, Poulik JM, et al. The protean neuropsychiatric and vestibuloauditory manifestations of neurosarcoidosis. *Audiol Neurootol* 2017;22:205–217.
28. Murphy R, O'Donoghue S, Counihan T, et al. Neuropsychiatric syndromes of multiple sclerosis. *J Neurology Neurosurg Psychiatry* 2017;88(8):697–708.
29. Hutchinson M, Stack JP, Buckley PF: Bipolar affective disorder prior to the onset of multiple sclerosis. *Acta Neurologica Scandinavica* 1993;6:388–393.
30. Sahli ZT, Tarazi FI. Pimavanserin: novel pharmacotherapy for Parkinson's disease psychosis. *Expert Opin Drug Discov* 2018;13(1):103–110.
31. Gaughran F, Lally J, Beck K, et al. Brain-relevant antibodies in first episode psychosis: a matched up case control study. *Psychol Medicine* 2017.
32. Peer M, Pruss H, Ben-Dayan I, et al. Functional connectivity of large scale brain networks in patients with anti NMDA receptor encephalitis: an observational study. *Lancet Psychiatry* 2017;4(10):768–774.
33. Miller BJ, Buckley P. Monoclonal antibody immunotherapy in psychiatric disorders. *Lancet Psychiatry* 2017;4(1):13–15.
34. Murray RM. Mistakes I have made in my research career. *Schizophrenia Bull* 2017;43(2):253–256.

11

EPILEPSY AND PSYCHOSIS

Michael R. Trimble, Kousuke Kanemoto, and Dale C. Hesdorffer

HISTORICAL BACKGROUND

Anyone interested in the dimensions of the psychoses must be well connected with the history of the comorbidities between epilepsy and psychopathology, and the intimate links between the psychoses and seizures, with a history going back well over 2,000 years. Descriptions of psychotic behaviors seemingly associated with seizures were represented on the Greek stage in the tragedies of Euripides, but Hippocrates (460–370 BC), in his text "The Sacred Disease," emphasized the importance of the brain as the source of madness and of epilepsy.[1] It was not however until the 19th century that the combination came under renewed scrutiny, in part with the establishment of asylums for chronic psychiatric disorders, and the increasing numbers of university departments with chairs of psychiatry (better termed "neuropsychiatry"), such as those occupied by Wilhelm Griesinger (1817–1868), Henry Maudsley (1835–1918), Emil Kraepelin (1856–1926), and many others who studied the interlinks between seizures and psychotic states. An immediate problem was that both the epilepsies and the psychoses are heterogeneous disorders, and with a growing split between the physicians who were somatically orientated and those adopting a more defined psychological approach (eventually becoming the neurologists and psychiatrists of today) the dialectical relationship between the two became confused and for a while quite lost.[1]

In this chapter we wish to draw some of the threads of this history together, and emphasize the key role that epilepsy has played in our exploration of the underlying etiology and pathogenesis of the psychoses, and indeed to the organic psychoses generally.

SEVERITY DIMENSION OF PSYCHOSIS: CLASSIFICATIONS OLD AND NEW

While the concept of "endogenous psychosis" has disappeared in the latest versions of the Diagnostic and Statistical Manual (DSM), a rule ultimately articulated by Schneider[2] still controls the basis of the currently most prevailing classification system of psychiatric illness. That is, psychotic illness is unobtrusively dichotomized into apparent organic psychosis and psychosis not attributable to known organic diseases, even in DSM-5.[3] The fact that the pages dedicated to the former are six times more than the latter demonstrates that the main focus of the DSM diagnostic system is clearly oriented toward the former. Organic psychoses are seemingly regarded as illegitimate children slipped into psychiatric diagnosis, waiting to be handed over to the proper hands of neurologists. Once organic psychoses are strongly suspected, the DSM classification system seemingly loses interest in further describing symptomatic details.

Thus, DSM-5 is devoid of the detailed description of the different phenomenological features of different organic psychoses, although they are fundamentally different. Thus, the delusory-hallucinatory experiences in schizophrenia show a far closer kinship with those in epilepsy than in diffuse Levy body disease or Alzheimer's disease. Among the organic psychoses, psychoses in patients with epilepsy have many features in common with "endogenous" psychoses.

In the current DSM system, "endogenous psychoses" or psychoses not attributable to definite organic causes, are subclassified into schizophrenia, schizoaffective disorder, delusional disorder, brief psychotic disorder, and schizophreniform disorder. If we are allowed to set aside postictal psychosis, psychoses in patients with epilepsy could be mostly allocated to one of these subgroups of psychotic disorders in the DSM system. In Box 11.1, the classification of psychotic illness in patients with epilepsy proposed by Landolt in 1963[4] is demonstrated. While the productive and psychotic twilight states seemingly correspond to acute interictal psychosis, chronic disorder agrees with that form studied by Slater (1963).[5] However, prior to the demarcation of postictal psychosis as an independent clinical entity, authors were not so clearly aware of ictal, postictal, and interictal states of psychosis in epilepsy.[4,6,7]

Since the rediscovery of postictal psychosis in the 1990s (it was well recognized in the 19th century), most classifications of psychoses in patients with epilepsy have been based on the chronological relationship to seizures. Transient psychotic events are divided into acute interictal and postictal psychosis, the former including forced normalization.

Less than 5% of psychoses in patients with epilepsy results as a manifestation of nonconvulsive status epilepticus, which often occurs in patients without a prior history of epilepsy. Psychosis as a manifestation of aura continua should be preferably exempted from nonconvulsive status epilepticus and discussed along with postictal psychosis, because nonconvulsive status epilepticus with continuous or semicontinuous

Box 11.1 CLASSIFICATION OF PSYCHOSES IN PATIENTS WITH EPILEPSY BEFORE CONCEPTUALIZATION OF POSTICTAL PSYCHOSIS[4]

A. Paroxysmal symptoms or seizures[1]

B. Episodic form

1. Psychotic episodes
    a. Postictal twilight states[2]
    b. Petit mal status[3]
    c. Twilight states due to nonepileptic organic causes
    d. Productive and psychotic twilight states (forced normalization)[4]
2. Episodic dysthymia (forced normalization)

C. Chronic disorder

[1]Psychosis as seizure equivalents; [2]Nonconvulsive status epilepticus with continous or semicontinuous spike-wave complex on EEG; [3]Postictal confusion; [4]Acute interictal psychosis

[4]Landolt H. [Semi-conscious and moody states in epilepsy and their electroencephalographic demonstration]. *Deutsche Zeitschrift fur Nervenheilkunde*. 1963;185:411–430.

spike-waves on EEG, or spike-wave-stupor, and aura continua are quite different physiological entities, being closely related to postictal psychosis. Symptomatically speaking, while peculiar subjective feelings and altered ways of experience stand out in aura continua, decline of drive and attention is noted in spike-wave stupor in most cases.[8]

LIFETIME PREVALENCE AND EPIDEMIOLOGY OF PSYCHOSIS IN THE GENERAL POPULATION

In a Finnish population-based study, a nationally representative group of 8,028 people aged 30 years and older were screened for psychotic disorders. Those with a positive screen were interviewed, using the SCID for psychosis and the CIDI for bipolar I disorder. The lifetime prevalence of any psychosis was 3.06%. The prevalence was 1.94% for nonaffective psychotic disorders (i.e., 0.87% for schizophrenia, 0.32% for schizoaffective disorder, 0.07% for schizophreniform disorder, 0.18% for delusional disorder, 0.05% for brief psychotic disorder, and 0.45% for psychotic disorder NOS); and the prevalence was 0.59% for affective psychoses (i.e., 0.24% for bipolar I disorders—0.12% with psychotic features, 0.12% without psychotic disorder—and 0.35% with major depressive disorder with psychotic features); the prevalence was 0.42% for substance-induced psychotic disorder (i.e., 0.41% alcohol-induced, 0.03% other-substance-induced); and 0.21% for psychotic disorders due to a general medical condition.[9]

In a population-based study of 7,076 people aged 18–64 years, in the Netherlands Mental Survey, the CIDI was used to determine the prevalence of psychosis for DSM-III-R, comparing urban to rural groups by population density.[10] There was an increasing trend in the prevalence odds ratio for any psychotic disorder ($p < 0.001$) across area density/kilogram^2 areas (<500 OR = 1.58 [0.64–3.89]; 500–999 OR = 2.55 [1.09–5.96]; 100–1499 OR = 2.55 [1.09–5.96]; 1500–2499 OR = 3.21 [1.40–7.37]; and ≥2500 OR = 4.74 [2.09–10.73]).

PSYCHOSIS IN EPILEPSY

Among psychiatric disorders, psychosis has the strongest association with epilepsy. In community-based studies of epilepsy examining the prevalence of psychosis, 3.1% to 9.2% have psychosis,[11–14] 0.7%–3.3% have affective psychosis,[13,15] and 0.7%–1.2% have schizophrenia.[11–14] These prevalence figures are higher than those in general population data.[9,10] A systemic study demonstrated that psychotic symptoms (hallucinations and delusions in full consciousness) are present in about 5% of patients with epilepsy.[16] When people with epilepsy are compared to controls, psychotic disorders are 3-fold or more common in epilepsy.[13,14] Among people with prevalent epilepsy, there is a 3.8-fold increased prevalence of schizophrenia (95% CI: 2.72–5.40), a 6.5-fold increased prevalence of organic psychosis (95% CI: 5.42–7.74) and a 3.9-fold increased prevalence of other psychoses (95% CI: 3.42–4.43).[14]

Studies examining the time order of the relationship between psychosis/schizophrenia and epilepsy have been conducted in large population-based registries because both disorders are relatively rare. Hesdorffer has shown that the cumulative data suggest a bidirectional relationship between incident epilepsy and incident psychosis. A bidirectional relationship was also seen in the UK general practice research database.[17] Psychosis was associated with a statistically significant increased risk for developing epilepsy in each of the 3 years before epilepsy onset (incident rate ratio [IRR] = 15.7, 8.5, and 7.7 in the 3rd-1st year before epilepsy onset), and incident epilepsy was associated with a statistically significant increased risk for developing a first-ever psychosis in each of the 3 years after epilepsy onset (IRR = 10.9, 4.8, and 4.0 respectively). Similar results were observed when epilepsy of unknown etiology was examined. In a Swedish study,[18] hospitalized psychosis was associated with an increased risk for developing epilepsy (OR = 2.7; 95% CI 1.6–4.8).

In an early Danish record-linkage study, the standardized incidence ratio (SIR) assessed the risk for nonorganic nonaffective psychoses, nonaffective psychosis, and schizophrenia in people with epilepsy compared to controls, excluding all people with a learning disability and/or substance misuse, as these are associated with epilepsy.[19] The standardized incidence ratio was significantly increased 2.15-fold for schizophrenia ($p < 0.001$) for men and women, 2.74-fold for nonaffective psychosis ($p < 0.0001$), and 2.75-fold for nonorganic nonaffective psychoses ($p < 0.001$). Psychomotor epilepsy was associated with a 5.07-old increased risk for nonorganic, non-affective psychosis (95% CI: 3.42–7.24) in males, greater than other epilepsy types. Data for females and for both females and males did not differ. In general, the risk for schizophrenia was increased most for psychomotor epilepsy. This was seen in males (SIR = 2.56; 95% CI 1.03–5.27), was not statistically significant in females, and was seen in both genders (SIR = 2.35; 95% CI: 1.17–4.21). There was overlap across epilepsy types for nonaffective psychosis.

A second Danish registry study[20] examined the association between epilepsy and risk for schizophrenia and schizophrenia-like psychosis in the presence and absence of a personal history of epilepsy, and a family history of psychoses, adjusting for age, sex, calendar year of diagnosis, place of birth, paternal and maternal age at birth, and birth order. Among those with no family history of psychosis, a personal history of epilepsy was associated with an increased risk for schizophrenia (adjusted RR = 2.61; 95% CI 2.29–2.99) and with an increased risk for schizophrenia-like psychosis(adjusted RR = 3.12; 95% CI 2.83–3.42). However, a family history of epilepsy was associated with a minor risk for schizophrenia (adjusted RR = 1.17; 95% CI 1.05–1.31) and for schizophrenia-like psychosis (adjusted RR = 1.26; 95% CI 1.15–1.37). A personal history of epilepsy was associated with a greater risk for schizophrenia and schizophrenia-like psychosis compared to a family history of epilepsy across a family history of schizophrenia, family history of schizophrenia-like psychosis, and family history of affective psychosis. There were no gender differences, but increasing age was associated with a statistically significant increased risk for developing schizophrenia in people with epilepsy, suggesting that the median age at onset of schizophrenia after epilepsy is greater than the onset of 22 years reported in the general population.[21]

INTERICTAL PSYCHIATRIC DISORDERS

The clinical picture of acute interictal psychosis corresponds relatively well with schizophreniform disorders in DSM-5. It develops typically within a couple of weeks or several months and lasts typically for more than a month until medical intervention is introduced. As previously mentioned, symptomatically there is a similarity between "endogenous" schizophreniform disorder (or schizophrenia) and interictal psychosis, being apparently far greater than that between interictal psychosis and other various organic psychoses, although bizarreness or extravagantness of the contents of delusory hallucinatory experiences such as observed in schizophrenia were mostly absent in interictal psychosis. Psychotic symptoms of most cases with forced normalization (see discussion later) and drug-induced ones are included in acute interictal psychosis. Generalized epilepsies are more often encountered in acute interictal psychoses than in postictal ones, although the ratio occupied by partial epilepsy with temporal or medial frontal origin still prevails in this category.

Chronic psychotic states in patients with epilepsy are even more heterogeneous than acute psychotic ones with regard to backgrounds as well as symptoms. First, a "psychodynamic hypothesis" proposed by Pond half a century ago[22] and continuously being advocated by recent studies (e.g., Adachi)[23] needs to be emphasized in this clinical setting. It postulates that epilepsy is nothing but an impetus, which triggers psychotic outbreaks in susceptible patients. "Psychodynamic" here has nothing to do with Freudian metapsychology but emphasizes interplay between individual susceptibility and a variety of social as well as physical stresses on brain function, which, irrespective of localization, are assumed to increase the risk of later development of psychosis. Second, as Slater and Beard[5] suggested, organic lesions leading ultimately to epilepsy could promote the risk of being psychotic independently from epileptic activity. He noted as others have found, that patients with lesions including limbic areas might be exposed to greater risk. Third, some parts of acute interictal psychosis and postictal psychosis eventually become chronic after repeated episodes.[24] The mechanisms are unclear, but Flor Henry (1969)[25] postulated that it was the seizures rather than an organic lesion per se that were important, and a "kindling" process and alteration of dopamine-related activity may be involved.[26] The series of psychoses studied by Slater and Beard (1963)[5] importantly noted that in comparison with schizophrenia, the schizophrenia-like psychoses, as they referred to them, had more affective warmth and were less likely to have premorbid schizoid personality traits, to undergo progressive personality deterioration, and to have a family history. The majority of their cases had a paranoid schizophrenia-like illness. Later studies have confirmed such findings.[27]

BIPOLAR DISORDER AND EPILEPSY

It used to be confidently stated that bipolar disorder was rare in patients with epilepsy,[28] and classical manic-depressive illness is in our experience far from being common in epilepsy. This statement applies to a time of classification prior to the use of standardized diagnostic manuals such as the DSM-III and onward, and were also based on clinical impression rather than being assessed by the use of invalidated rating scales. However in the postictal state, as part of a postictal psychosis, manic or hypomanic pictures could develop, although more generally the presentation was one of a mixed affective state with psychotic features. Kanner noted postictal hypomanic symptoms in 22% patients, often with associated psychotic phenomenology.[29] It is also the case that some patients with temporal lobe epilepsy show energetic dysthymic states, resembling bipolar II disorder, but the status of these to the bipolar spectrum is debatable. Another setting in which hypomanic or even manic-depressive psychoses are seen is after temporal lobectomy, especially of the nondominant temporal cortex).[30]

PARADOXES

The epidemiology reviewed thus far clearly establishes a close association between epilepsy and psychoses, but is largely concentrated on the interictal psychoses, in part because of their more ready identification. However, as we have pointed out, within this category is a veritable mixture of neuropsychiatric conditions. This should be perhaps least confusing with the postictal psychoses. However, the situation is far from clear, in part on account of the confusion between seizures, as identifiable clinical events and epilepsy. Thus, the seizure is but one manifestation of pathological and electrophysiological events in the brain, which are present in some, but not all forms of epilepsy, continuously. It is often naively thought that once a seizure has occurred, and ended, that the problem is, at least for a while resolved. Yet this belies that fact that there are continuing neuronal events well beyond the actual seizure that impinge on behavior. These may be quite long lasting, and hence even deciding which events are ictal and postictal is contentious. This is one reason why in clinical practice postictal psychoses are often missed with sometimes disastrous consequences.

ICTALLY RELATED PSYCHIATRIC DISORDERS

Postictal confusion occurs immediately following the seizure with disruption of the EEG, usually with diffuse slowing. Patients present with a variety of behaviours with lived episodes of disorientation. Rarely lasting more than a few hours, they can occasionally persist for days or longer especially in the learning disabled and in the elderly. Recurrent seizures, which do not allow for mental state recovery between attacks can also lead to prolonged states of obtunding. In some patients suicidal ideation is reported. Ictally driven psychoses may be schizophrenia-like, especially in complex partial seizure status, although usually present with a variety of hallucinatory and delusional experiences and cognitive disruption in association with a disturbed EEG.

These should not be confused with postictal psychoses, although the latter may emerge from the former. Logsdail and Toone[31] gave descriptions of postictal psychoses. Most have complex partial seizures with secondary generalization, with an increase in seizures prior to the onset of psychosis, usually as a cluster. A lucid interval, from 1 to 6 days was clearly described in 80% of cases. Thus, the immediate consequences of the seizures have cleared, and there is calm before the storm.

On follow-up several years later, 21% had become chronically psychotic. Notable features of this postictal state are a mixed affective picture, religious hallucinations and delusions, fear of impending death, and the relative absence of clouding of consciousness. The EEG often is not so abnormal, and may show an improved pattern on the interictal state suggesting links between this state and forced normalization (see later discussion). Since there can be well-preserved contact with the environment, paranoid states with command hallucinations to harm the self or others are well described, and may be obeyed, in contrast to any attempts at self-harm that occur in the immediate postictal phase of confusion, when such directed actions are not possible.

Kanemoto[32] noted 2% of patients attending an epilepsy center had a postictal psychosis, but of the subgroup with temporal lobe epilepsy, the figure was 11%. In their series, the "lucid interval" ranged from less than 24 hours to 7 days, and the psychosis itself from 12 hours to longer than 1 month. They noted a close link between psychic auras such as déjà vu and ictal fear.

Kanner has studied postictal disorders in some detail.[33] They reported in addition to the psychoses (7%), a spectrum that included anxiety (45%), dysphoric/depressive symptoms (45%), and hypomania (22%). The latter can be floridly psychotic and may combine with bizarre delusions very disturbed behavior.

The relation of postictal psychosis with psychotic aura continua is more complicated. While some psychotic episodes in patients with epilepsy can be shown to be manifestations of aura continua, that is, direct epileptic equivalents,[34–36] in others, ictal activities have not been confirmed during psychotic episodes as a sequel to clustered seizures with depth EEG studies.[37,38] In some exceptional cases with postictal psychosis, however, psychotic aura continua seemed to ensue soon after clustered seizures.[35]

Several existing data hint at a close link between postictal psychosis and bipolar mood disorder besides the affective signs and symptoms embedded within the postictal psychosis itself. A family study suggested that prevalence of postictal psychosis among patients with a positive family history of psychotic disorders was not increased, but was 3.49 times higher (p = 0.001) among patients who had a positive family history of mood disorder in their first or second relatives.[39] In the study of Kohler et al.,[40] the ratios of mood and anxiety disorder tended to increase after a temporal lobectomy in patients with ictal fear. This could be another argument for the link between postictal psychotic events and mood disorder, because ictal fear and postictal psychosis might be closely associated.[32,41] Further, a preoperative history of postictal psychosis or prolonged confusional states in those who have a temporal lobectomy leads to postoperative mood disorder in some cases.[42]

BIOLOGICAL ANTAGONISMS AND THE CONCEPT OF FORCED NORMALIZATION

We have dealt with comorbidities of epilepsy that are ictal, (directly related to the seizure), and interictal disorders unrelated in time to the seizure. Hinted at earlier was the suggestion that psychotic behaviors can be somehow linked with the aftermath of an electrophysiological seizure (postictal), but sometimes also may clinically occur in relatively clear consciousness, and with an improved EEG compared to the interictal state. The concept of forced normalization is in effect a seizure-related disorder, recognized in the 19th century, and rediscovered in the middle of the 20th. The latter was due to the introduction of the EEG and the ability to record from patients on a daily basis.

Heinrich Landolt (1917–1971) recorded changes in the EEG during preseizure dysphoric episodes and during limited periods of frank psychosis lasting days or weeks.[43] He noted improvement in EEG activity during such episodes and referred to this as "forced normalization" (*Forcierte Normalizierung).*[4] Although the classic clinical picture is of a psychosis, which can resemble schizophrenia, with delusions and hallucinations appearing in a state of unconfused consciousness, often a mixed picture is seen, with marked affective components sometimes as a paranoid or even manic psychosis. In lesser forms, the clinical picture may merely present as an exacerbation or precipitation of previously observed behavioral problems as seizures remit. Detection of the psychosis is very difficult in those with learning disability, and episodes of paranoid aggression mixed with dysphoria may be mistaken for more straightforward behavioral problems, unrelated to epilepsy. The disturbed behavior may last days or weeks. It is often terminated by a seizure, either brought about by a decrease in antiseizure medications or ECT, and the EEG abnormalities then return.

The importance of forced normalization has again become apparent with the introduction of the new generation of Antiepileptic drugs in the past 25 years. Many of these are given to patients with persistent seizures, and some have strong antiseizure properties. Observations of forced normalization have been reported with several newer anticonvulsant agents especially when intractable seizures are quite suddenly decreased of stopped.

Landolt defined forced normalization thus:[4] "Forced normalisation is the phenomenon characterised by the fact that, with the recurrence of psychotic states, the EEG becomes more normal, or entirely normal, as compared with previous and subsequent EEG findings." It was in effect an EEG phenomenon. He noted that similar "normalization" could be provoked by anticonvulsant drugs, and noted that, at the end of a psychotic episode, the EEG returned to being abnormal. It was mainly but not exclusively related to temporal lobe epilepsy.

Wolf gave this the name "paradoxical normalization."[44] Thus, as a general rule in epilepsy, if the behavior deteriorates, so does the EEG (e.g., in nonconvulsive status, or in encephalopathies), yet in forced normalization, the EEG improves but the behavior becomes worse. The clinical consequences of this disorder are so important, and it is so often missed or misunderstood. In essence, the object of treating epilepsy is to suppress seizures, and the idea that such management might induce severe behavioral syndromes (including a psychosis) is out of reach for many who consider that epilepsy is no more than recurrent seizures and the inconvenience they cause.

AFFINITY AND ANTAGONISM

Understanding the biological associations between epilepsy and the psychoses thus presents us with a seeming paradox. An "affinity" is supported by observations of an increased incidence of psychoses in people with epilepsy, especially with temporal lobe epilepsy and the occurrence of psychosis in the ictal and post-ictal periods. The "antagonism" hypothesis is supported by the syndrome of paradoxical normalization. The importance of the latter reflects on earlier epidemiological observations that seizures were observed infrequently in some psychiatric populations, and, in the presence of psychosis, epilepsy seemed to have a better prognosis. These, plus clinical observations going back to the 19th century that certain psychosis improved after a seizure led von Meduna to use convulsive therapy for the treatment of schizophrenia.[45] A plausible explanation of the relation of epilepsy to psychosis and vice versa must explain this paradox.

The question as to the validity of the concept of forced normalization has been by explored by Trimble and Schmitz,[43] where translations of the original papers from Landolt and others are given, with contemporary observations from several investigators. Landoldt's observations were made at a historical time of considerable division of neurology and psychiatry and before epileptology became almost a separate specialty. Further, the number of drugs available for the treatment of the condition, were distinctly limited.

Landolt pointed out how the underlying epilepsy in many of their cases was "psychomotor," implying a temporal lobe pathology underlying the seizures. At this time the concept of the limbic system and its anatomical connections, and relationship to the control of emotional behavior was barely on the horizon, and the ideas of neurotransmission were in their infancy. However, the observation that suppression of seizures of a temporal lobe origin could lead to an alternative clinical expression (psychosis) has been but one of the pieces of the jigsaw of unraveling the role of the limbic lobe of the brain in the regulation of behavior.

Electroconvulsive therapy emerged as an effective treatment for severe psychiatric disorders in the postwar period, and about the same time removal of one or the other of the temporal lobes to treat intractable seizures became possible. One observation of profound significance was that of Flor-Henry on operated patients.[25] He observed that those with a left-sided temporal lobe focus were more likely to have a schizophrenia-like psychosis, contrasting this with a right-sided abnormality linking with an affective disorder. This was such an important observation, now rarely commented on. However, the links between a left-sided hippocampal abnormality (the pathology was available at that time as the operations were lobectomies) and a schizophrenia-like psychosis directly led to the search for neuroanatomical underpinnings of schizophrenia, focusing on the medial temporal structures in the dominant hemisphere. Since studies supporting the laterality hypothesis in schizophrenia-like psychoses have been made in epilepsy and schizophrenia in the absence of epilepsy, using surface EEG, depth electrode recordings, computed tomography, neuropathology, neuropsychology, positron-emission tomography (PET), and more recently with MRI. The literature has been summarized by Trimble and George.[27] The findings do not of course simply relate to one hemisphere, but the epilepsy link led to the development of more biological approaches to understanding schizophrenia and away from purely sociological hypotheses. Trimble pointed out that a specific group of hallucinations and delusions, defined by Schneider, and referred to as First Rank Symptoms, which usually (but by no means exclusively) are relevant to the diagnosis of schizophrenia, are important in this story.[2] They may be signifiers of temporal lobe dysfunction, representing as they do disturbances of language and symbolic representation. In this sense, he equated them to a Babinski sign for a neurologist, i.e., pointing to a location and lateralization of an abnormality in the central nervous system.

With regard to the antagonism, Wolf and Trimble emphasized the antagonism was not between epilepsy and schizophrenia, but between seizures and the symptoms and signs of psychotic states.[46] In other words, within the *increased association* between epilepsy and psychoses was an *antagonism* between seizures and hallucinations and delusions. In chronic cases, this emerged when the pattern of the seizures sometimes changed in the course of epilepsy, seizures diminishing and psychotic symptoms emerging. However, acutely this could be revealed through states of forced normalization.

This potential, for the sudden transformation of an aberrant cerebral process from one state to another, helps explain a mutability of the clinical pictures. The latter emphasizes the changing underlying neurophysiological/neurochemical status of the brain, brought about by artificially altering the neuromodulators within the brain, more specifically by the enhancement of inhibitory factors, explaining why prescribing anticonvulsant drugs and suddenly stopping seizures can have such adverse effects. This makes heuristic sense, if the brain is viewed as a dynamic but self-contained energy system; sudden change does not allow for the resetting of a previously established homeostasis. Landolt referred to his phenomenon as based on a "supranormal inhibitory process,"[4] but he could not, in the absence of the neurochemical understanding which we now have, take this any further. Stevens,[47] uniting the neurology and psychiatry with an understanding not only of the neuroanatomy of the limbic forebrain, but also neurotransmitters, put forward the view that "all that spikes is not fits," and that inhibition by dopamine, GABA, and other transmitters are essential to prevent the normal brain from being in a continual state of status epilepticus. Excess inhibition however, in her model predisposes to psychosis (e.g., dopamine agonism being psychotogenic), the alternative to this being the stunning effect of the antipsychotic drug clozapine in intractable cases of psychosis, this drug being the antipsychotic most likely to provoke seizures.

The most important thing about forced normalization is to recognize it. Many of those dealing with epilepsy are reluctant to accept that stopping seizures, which must be that main aim of treatments, could in and of itself lead to alternative expressions of the underlying neuropathology and neurophysiology that also leads to the epileptic seizures. In that medial temporal structures are involved in many case of intractable epilepsy, and in disorders such as schizophrenia and mania, it is hardly surprising that such alternating clinical pictures can be observed.

The mechanisms of the switch from seizures to a floridly abnormal mental state remain unclear, yet the observations raise important questions about the biological relevance of seizures and the seizure threshold, and the close link between cerebral energetic distributions and underlying neuroanatomical circuitry. Thus, in a brain that has learned to have seizures beginning in certain defined areas of the cortex, the effects of quite suddenly altering the flow of electrochemical forces by the introduction of a compound that alters the local neurochemical environment must be taken into account. There are so few intracranial recordings of patients in the transition from one phase of forced normalization to another, so what is happening at say the medial temporal cortex when the clinical picture changes is unknown. What is known is that the surface EEG changes (normalizes). But it is known that a "limbic status" can last several days without any surface expression, and there is at least one case documented in whom a short-lived psychosis developed with disappearance of the surface EEG changes, following electrical depolarization of medial temporal structures).[34]

It should be emphasized that these phenomena represent only a small number of the psychoses seen in people with epilepsy, and complete cessation of seizures does not usually lead to psychosis. The populations at risk are most likely seen in tertiary epilepsy centers, on account of intractable epilepsy, or in psychiatric settings. However, these observations and theories, derived from neurological and psychiatric research have enhanced our understanding of the basic mechanisms of the psychoses.

TREATMENTS

Treatment of psychotic disorder in patients with epilepsy almost always involves two mutually supplementary pharmacotherapies: adjustment of antiepileptic treatment and administration of dopamine-antagonists or benzodiazepines. While it remains unclear which antiseizure agents are more contributory to a psychotic outbreak, carbamazepine, valproate, and lamotrigine are most often recommended as least psychosis-provoking agents. While some authors have incriminated levetiracetam as a psychosis-provoking agent,[48] others have emphasized less psychiatric problems with this drug.[49] Although topiramate, zonisamide, phenytoin, and ethosuximide have been also listed as provoking psychoses in patients with epilepsy, most studies are retrospective and often adopted a biased group as control subjects, if any were included. Further, it should be noted that the number of the patients given a particular antiepileptic drug varies greatly across different studies from which both the subjects and controls are selected.

In acute interictal psychosis, a relatively small dose of dopamine-antagonists is mostly effective.[50] Except for clozapine and possibly zotepine, dopamine-antagonists are safely applicable in patients with epilepsy within a therapeutic range. In severe cases of intractable psychoses, clozapine has been used, although EEG studies before treatment need to be taken to ensure that any deterioration after prescription is not due to a developing underlying nonconvulsive status.[51] In postictal psychosis, sedative measures should be preferably induced as expeditiously as practicable. Depending on the severity and response to the drugs, either benzodiazepines or dopamine-blockers with potent sedative property is often chosen.

While a history of acute interictal psychosis is a risk for recurrence of psychosis even after successful surgery, postictal psychoses disappear as a rule if seizures are controlled with surgical measures, although a history of postictal psychosis is associated with a slightly unfavorable surgical outcome. It should be noted, however, that this is not a contraindication to surgical intervention, because a substantial proportion of postsurgical psychotic episodes remit swiftly.

CONCLUSIONS

In this chapter, we have summarized the quite complicated subject of the relationship between the epilepsies and psychoses. There are different forms and presentations of both conditions, and unraveling the associations requires considerable neuropsychiatric expertise. The relative lack of interest in the organic foundations of psychiatric disorders by psychiatrists of the 20th century up to the present time remains a paradox, since the urgency has been for psychiatry not only to be allied to medicine and adopt a discipline based on scientific principles but also to use neuroscience-based methods to explore and expand understanding of brain based on principles of the discipline. Epilepsy has been central to this enterprise, from the time of the Greek physicians, to the significant interest and exploration of epilepsy with the clinical introduction of the EEG in the latter half of the last century. Epilepsy and the other organic psychoses receive most attention from neuropsychiatry, a discipline in its modern form becoming recognized worldwide, with a long historic tradition.[45]

REFERENCES

1. Trimble MR. *The psychoses of epilepsy*. New York: Raven Press; 1991.
2. Schneider K. *Klinische Psychopathologie (Siebente, verbesserte Auflage)*. Stuttgart: Thieme;1966.
3. American Psychiatric Association. *Diagnostic and statistical manual of mental disorders (5th ed.)*. Washington, DC: American Psychiatric Association; 2013.
4. Landolt H. [Semi-conscious and moody states in epilepsy and their electroencephalographic demonstration]. *Deutsche Zeitschrift fur Nervenheilkunde*. 1963;185:411–430.
5. Slater E, Beard AW, Glithero E. The schizophrenialike psychoses of epilepsy. *British journal of psychiatry: the journal of mental science*. 1963;109:95–150.
6. Kohler GK. [Concept determination and classification of the so-called epileptic psychoses]. *Schweizer Archiv fur Neurologie, Neurochirurgie und Psychiatrie = Archives suisses de neurologie, neurochirurgie et de psychiatrie*. 1977;120(2):261–281.
7. Bruens JH. *Psychoses in epilepsy*. New York: Wiley; 1969.
8. Guberman A, Cantu-Reyna G, Stuss D, Broughton R. Nonconvulsive generalized status epilepticus: clinical features, neuropsychological testing, and long-term follow-up. *Neurology*. 1986;36(10):1284–1291.
9. Perala J, Suvisaari J, Saarni SI, et al. Lifetime prevalence of psychotic and bipolar I disorders in a general population. *Archives of general psychiatry*. 2007;64(1):19–28.
10. van Os J, Hanssen M, Bijl RV, Vollebergh W. Prevalence of psychotic disorder and community level of psychotic symptoms: an urban-rural comparison. *Archives of general psychiatry*. 2001;58(7):663–668.
11. Edeh J, Toone B. Relationship between interictal psychopathology and the type of epilepsy: results of a survey in general practice. *British journal of psychiatry: the journal of mental science*. 1987;151:95–101.
12. Jalava M, Sillanpaa M. Concurrent illnesses in adults with childhood-onset epilepsy: a population-based 35-year follow-up study. *Epilepsia*. 1996;37(12):1155–1163.
13. Stefansson SB, Olafsson E, Hauser WA. Psychiatric morbidity in epilepsy: a case controlled study of adults receiving disability benefits. *Journal of neurology, neurosurgery, and psychiatry*. 1998;64(2):238–241.
14. Gaitatzis A, Carroll K, Majeed A, J WS. The epidemiology of the comorbidity of epilepsy in the general population. *Epilepsia*. 2004;45(12):1613–1622.
15. Forsgren L. Prevalence of epilepsy in adults in northern Sweden. *Epilepsia*. 1992;33(3):450–458.
16. Clancy MJ, Clarke MC, Connor DJ, Cannon M, Cotter DR. The prevalence of psychosis in epilepsy; a systematic review and meta-analysis. *BMC psychiatry*. 2014;14:75.
17. Hesdorffer DC, Ishihara L, Mynepalli L, Webb DJ, Weil J, Hauser WA. Epilepsy, suicidality, and psychiatric disorders: a bidirectional association. *Annals of neurology*. 2012;72(2):184–191.
18. Adelow C, Andersson T, Ahlbom A, Tomson T. Hospitalization for psychiatric disorders before and after onset of unprovoked seizures/epilepsy. *Neurology*. 2012;78(6):396–401.
19. Bredkjaer SR, Mortensen PB, Parnas J. Epilepsy and non-organic non-affective psychosis: national epidemiologic study. *British journal of psychiatry: the journal of mental science*. 1998;172:235–238.
20. Qin P, Xu H, Laursen TM, Vestergaard M, Mortensen PB. Risk for schizophrenia and schizophrenia-like psychosis among patients with epilepsy: population based cohort study. *BMJ*. 2005;331(7507):23.
21. Thorup A, Waltoft BL, Pedersen CB, Mortensen PB, Nordentoft M. Young males have a higher risk of developing schizophrenia: a Danish register study. *Psychological medicine*. 2007;37(4):479–484.
22. Pond DA. Psychiatric aspects of epilepsy in children. *Journal of mental science*. 1952;98(412):404–410.
23. Adachi N, Akanuma N, Ito M, et al. Epileptic, organic and genetic vulnerabilities for timing of the development of interictal psychosis. *British journal of psychiatry: the journal of mental science*. 2010;196(3):212–216.
24. Tarulli A, Devinsky O, Alper K. Progression of postictal to interictal psychosis. *Epilepsia*. 2001;42(11):1468–1471.
25. Flor-Henry P. Psychosis and temporal lobe epilepsy. A controlled investigation. *Epilepsia*. 1969;10(3):363–395.
26. Trimble MR. The relationship between epilepsy and schizophrenia: a biochemical hypothesis. *Biol Psychiatry*. 1977;12:299–304.
27. Trimble MR, George M. *Biological psychiatry*, 3rd edition. Chichester: Wiley; 2010.
28. Wolf P. Manic episodes in epilepsy. In Akimoto H, Kazamatsuri H, Ward SM, eds. *Advances in epilepsy: XIII epilepsy international symposium*. New York: Raven Press; 1982:237–240.
29. Kanner AM, Soto A, Gross-Kanner H. Prevalence and clinical characteristics of postictal psychiatric symptoms in partial epilepsy. *Neurology*. 2004;62(5):708–713.
30. Mace CJ, Trimble MR. Psychosis following temporal lobe surgery: a report of six cases. *Journal of neurology, neurosurgery, and psychiatry*. 1991;54(7):639–644.
31. Logsdail SJ, Toone BK. Post-ictal psychoses: a clinical and phenomenological description. *British journal of psychiatry: the journal of mental science*. 1988;152:246–252.
32. Kanemoto K. Post-ictal psychoses, revisited. In Trimble M, Schmitz B, eds. *The neuropsychiatry of epilepsy*. Cambridge, UK: Cambridge University Press; 2002:117–134.
33. Kanner AM. Recognition of the various expressions of anxiety, psychosis, and aggression in epilepsy. *Epilepsia*. 2004;45(Suppl 2):22–27.
34. Wieser H. Electrophysiological aspects of forced normalisation. In Trimble M, Schmitz B, eds. *Forced normalisation and alternative psychoses of epilepsy*. Petersfield, UK: Wrightson Biomedical Publishing; 1998:95–142.
35. Kanemoto K. Periictal Capgras syndrome after clustered ictal fear: depth-electroencephalogram study. *Epilepsia*. 1997;38(7):847–850.
36. Takeda Y, Inoue Y, Tottori T, Mihara T. Acute psychosis during intracranial EEG monitoring: close relationship between psychotic symptoms and discharges in amygdala. *Epilepsia*. 2001;42(6):719–724.
37. Mathern GW, Pretorius JK, Babb TL, Quinn B. Unilateral hippocampal mossy fiber sprouting and bilateral asymmetric neuron loss with episodic postictal psychosis. *Journal of neurosurgery*. 1995;82(2):228–233.
38. So NK, Savard G, Andermann F, Olivier A, Quesney LF. Acute postictal psychosis: a stereo EEG study. *Epilepsia*. 1990;31(2):188–193.

39. Alper K, Devinsky O, Westbrook L, et al. Premorbid psychiatric risk factors for postictal psychosis. *Journal of neuropsychiatry and clinical neurosciences.* 2001;13(4):492–499.
40. Kohler CG, Carran MA, Bilker W, O'Connor MJ, Sperling MR. Association of fear auras with mood and anxiety disorders after temporal lobectomy. *Epilepsia.* 2001;42(5):674–681.
41. Savard G, Andermann F, Olivier A, Remillard GM. Postictal psychosis after partial complex seizures: a multiple case study. *Epilepsia.* 1991;32(2):225–231.
42. Kanemoto K, Kim Y, Miyamoto T, Kawasaki J. Presurgical postictal and acute interictal psychoses are differentially associated with postoperative mood and psychotic disorders. *Journal of neuropsychiatry and clinical neurosciences.* 2001;13(2):243–247.
43. Trimble MR, Schmitz B, eds. *Forced normalisation and alternative psychoses of epilepsy.* Petersfield, UK: Biomedical Publishing; 1998.
44. Wolf P. Forced normalisation. In: Trimble MR, Bolwig TG, eds. *Aspects of Epilepsy and Psychiatry.* Chichester: Wiley and Sons; 1986:101–112.
45. Trimble MR. *The intentional brain: motion, emotion and the development of modern neuropsychiatry.* Baltimore: Johns Hopkins Press; 2016.
46. Wolf P, Trimble MR. Biological antagonism and epileptic psychosis. *British journal of psychiatry: the journal of mental science.* 1985;146:272–276.
47. Stevens JR. Psychopathology and brain dysfunction. In: Shagass C, Gershon SAF, eds. *Psychopathology and brain dysfunction.* New York: Raven Press; 1977.
48. Chen Z, Lusicic A, O'Brien TJ, Velakoulis D, Adams SJ, Kwan P. Psychotic disorders induced by antiepileptic drugs in people with epilepsy. *Brain: a journal of neurology.* 2016;139(Pt 10):2668–2678.
49. Mula M, Trimble MR, Sander JW. Are psychiatric adverse events of antiepileptic drugs a unique entity? a study on topiramate and levetiracetam. *Epilepsia.* 2007;48(12):2322–2326.
50. Tadokoro Y, Oshima T, Kanemoto K. Interictal psychoses in comparison with schizophrenia—a prospective study. *Epilepsia.* 2007;48(12):2345–2351.
51. Langosch JM, Trimble MR. Epilepsy, psychosis and clozapine. *Human psychopharmacology.* 2002;17(2):115–119.

12

UNDERSTANDING SEX DIFFERENCES IN PSYCHOSIS THROUGH THE EXPLORATION OF HORMONAL CONTRIBUTIONS

Leah H. Rubin

There is compelling evidence that sex modulates the risk, clinical presentation, and course of psychosis. Sex differences are present when examining each diagnostic entity separately or when using a transdiagnostic focus in individuals with or at high-risk for psychosis. In schizophrenia (SZ), females are at decreased risk of developing the disease[1] and the disease course is more benign in women than in men. Females on average have a later age of onset (25–30 females vs. 20–25 males) and a smaller peak of onset (45–54) coinciding with the menopausal transition, less pre- and perinatal complications, shorter and less frequent episodes, less severe negative and more affective symptoms, better premorbid functioning, and a better treatment response to antipsychotic medication versus males.[2–4] The genetic risks for schizophrenia are often either sex-specific (only in men or women) or sex-dependent (larger effect in men or women).[5] In bipolar disorder (BPD), females are particularly sensitive to the impact of reproductive life events as triggers for episode onset.[6,7] Females with BPD transitioning through the menopause are at increased risk for affective dysregulation compared to age-matched males and younger males and females with BPD.[8] Finally, studies using a transdiagnostic lens in individuals with or at high-risk for psychosis reflect findings reported in SZ including age of onset, illness duration, and functioning (e.g., maintaining relationships).[9–11] In the most recent large-scale study in psychosis, females younger than 45 had less previous hospitalizations versus same aged males.[10] However, the converse was demonstrated among patients greater than 55 years of age, suggesting a possible protection from psychotic episodes during the reproductive years for females.

Sex differences in cognition and social cognition are well documented in healthy individuals, and these normal cognitive sexual dimorphisms are preserved in psychosis. In healthy individuals, women excel at fine motor skills, psychomotor speed (digit symbols), verbal memory, and fluency, whereas men excel at visuospatial abilities and gross motor skills.[12–15] Women are also more emotionally perceptive and reactive to emotional stimuli and demonstrate enhanced emotional memory versus men.[16–18] Effect sizes (Cohen's *d*) for sex differences range from 0.30 to 0.90 with the smallest being verbal and the largest being visuospatial abilities.[19] In SZ, recent[20] but not earlier studies[21–24] also demonstrate these sex differences. One of the most important differences across studies was how tasks were grouped. Early studies grouped tasks as a function of domain (i.e., memory, motor skills) whereas recent studies[25,26] grouped tests according to the direction of sex differences—a psychoneuroendocrine approach. Using this approach, female patients performed better on verbal memory[24,26,27] and fluency[28] and worse on visuospatial abilities versus male patients.[24,26,28] In first-episode psychosis, female patients also cognitively rebound more rapidly than men on "female-dominant" tasks after initiation of antipsychotic treatment, and female patients decline more rapidly than males on "male-dominant" tasks (Figure 12.1).[25] This finding that women show greater cognitive improvement on tests that favor women (e.g., verbal memory) aligns with evidence suggesting that estrogen might protect against cognitive dysfunction in SZ. Antipsychotic treatment may have facilitated a normalization of estrogen's activational regulation of neurophysiology and cognition.

In healthy individuals, sex differences in emotion processing and cognition are paralleled by sex differences in both brain structure and function. Sex differences in human brain morphology[29] and in activation during behavioral tasks are apparent in networks related to emotion processing, language, and visuospatial abilities. Sex differences in emotion processing typically include differential activation of prefrontal, temporal, and limbic brain regions.[17,30] For language, the most common areas showing sex differences are in the inferior frontal and superior temporal gyrus whereas for visuospatial abilities differences are most common in posterior parietal cortex.[31–33] Sex differences in regional brain networks are also consistent with findings from structural and functional neuroimaging studies.[17,29–31,33,34] Typically, studies demonstrate differences in functional brain networks including amygdala, insula, cingulate (anterior, posterior), and prefrontal cortex (dorsolateral, medial) which are also important for emotion and cognition.[35–38] Notably, these normal brain sexual dimorphisms are often disrupted in SZ.[39–42]

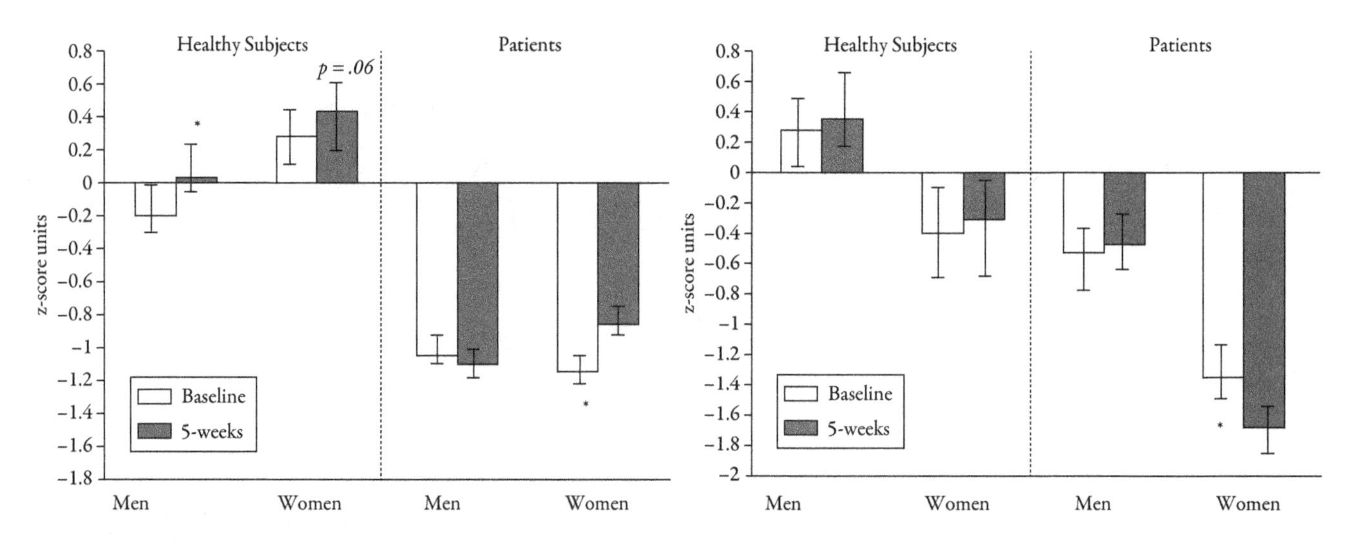

Figure 12.1 The magnitude in the change in cognitive performance depended on the combined influence of treatment, sex, and whether the tests favor men or women in the general population (and in this sample). NOTE. From Rubin LH, Haas GL, Keshavan MS, Sweeney JA, Maki PM. 2008. Sex difference in cognitive response to a in first episode schizophrenia. *Neuropsychopharmacology* 33(2):290–297. The post hoc analysis showed a significant Group x Sec x Time x Test Interaction, $F(1, 105) = 4.75, p = 0.04$. Asterisks denote significant pairwise difference in within-subject change over time. Female patients showed significant improvements on "female" tests and a decrease in "male" tests over time. Male patients did not show significant improvement on "male" or "female" tests. Composite scores were first created by computing within-test z-scores followed by between-test z-scores. Therefore, the composite scores will not sum to 0 as typically expected.

These striking sex differences in psychosis raise the hypothesis that the disorders are affected by neuroendocrine factors differing between men and women. The most promising endocrine candidates to play a role in SZ are sex hormones (i.e., estrogen) which have broad brain and behavioral effects. Additionally, two sexually dimorphic neurohormones that often act in a "ying/yang" fashion—oxytocin (OT) and arginine vasopressin (AVP)—also appear to play a strong role. Emerging evidence over the past 10 years from basic science, clinical, and epidemiological studies implicate neuroendocrine factors in the pathophysiology of SZ and psychosis. Focusing on neuroendocrine factors in part may provide insight into the etiology of psychosis and guide treatment development at least for a subset of patients. Here we discuss the role of both endogenous and exogenous hormones on symptomatology as well as cognition and emotion processing, given that deficits in these behaviors persist despite effective treatments for positive symptoms. There remains a pressing need for novel treatment strategies for these deficits, and such treatments may need to differ by sex.

Hormones are a focus in psychosis partially because alterations in peripheral hormone levels are often reported among chronic and treatment-naïve individuals with psychosis. Alterations are reported in estrogen, progesterone, testosterone, OT, and AVP.[43–47] In our own work, we demonstrated that: (1) SZ patients (lesser degree in BPD) and their relatives show decreased AVP levels versus controls; (2) in acutely ill untreated first-episode psychosis, patients show increased AVP levels versus controls; (3) relatives with a psychosis spectrum personality disorder show lower AVP levels than other relatives; and (4) AVP and OT levels are highly familial.[43,44] Resting AVP level in particular has several characteristics of a promising endophenotype in psychosis. Next, we aim to examine the effects of endogenous and exogenous hormones on clinical, cognitive, and emotion outcomes in psychosis.

Evidence from menstrual cycle studies demonstrate the important modulatory role of neuroendocrine factors in psychosis and offer a way to examine changes in endogenous hormone levels in relation to changes in clinical outcomes. Many,[45,48–52] but not all, menstrual cycle studies[53] demonstrate that symptomatology fluctuates across the cycle, with decreased severity in the midluteal (high estrogen/progesterone) versus the early follicular phase (low estrogen/progesterone). In one of the earliest studies, five treated outpatient women with SZ (aged 29–49) completed the Abbreviated Symptom Checklist.[48] Women reported less overall symptoms during the midluteal versus the early follicular phase. However, the generalizability of this finding was limited, based on a small sample and exclusiveness of self-report to determine cycle phase. A later study investigated changes in Brief Psychiatric Rating Scale (BPRS) symptom scores at five points across the cycle (2nd, 7th, 14th, 21th, 28th day) in 32 acutely ill premenopausal women with SZ (6 unmedicated).[50] Cross-correlations revealed negative associations between estradiol levels and the total BPRS score. In contrast, two subsequent studies (*n*'s = 39, 24) in treated premenopausal women with chronic SZ (aged 19–47) showed no changes on the total BPRS score (raters blind to menstrual status) between menstruation (low estrogen/progesterone) and 1 week prior to menstruation (high estrogen/progesterone).[51,52] Only Choi et al.[51] analyzed hormone levels (estradiol, progesterone), but they were unrelated to BPRS ratings. Similarly, Thompson et al.[53] reported no changes in the Positive and Negative Syndrome Scale (PANSS) scores (raters blind to menstrual status) across the cycle in 29 premenopausal women with psychosis (4 unmedicated). Although not significant, means were in the expected directions, with women experiencing decreased symptoms during the luteal versus follicular phase ($d = 0.28$ for total, 0.16 positive, 0.10 negative, 0.20 general symptoms). Although cycle phase was verified by hormone assays, phase was not tightly controlled. Sessions occurred once in the follicular (days 1–14) and once in the luteal phase (days 14–28), 14 days apart, allowing for a broad range of estradiol and progesterone levels. Moreover, symptom–hormone correlations were not reported.

In the largest study to date, 125 Caucasian premenopausal women with SZ were assessed on the BPRS and PANSS during the follicular (days 2–4), periovulatory (days 10–12), and midluteal (days 20–22) phases of the cycle (raters blind to cycle phase).[49] Symptom severity was the same during the follicular versus the periovulatory phase. However, there were decreased symptoms (total BPRS and all PANSS scores) during the midluteal versus periovulatory phase and during the midluteal versus the follicular phase. Moreover, higher peripheral estradiol levels (irrespective of phase) was associated with less overall symptomatology. Although this is the largest and most systematic menstrual cycle study to date, progesterone-symptom associations were not assessed. Progesterone is also low at the follicular and periovulatory phase and high at the midluteal phase. Thus, it is impossible to know whether progesterone levels were influencing the decrease in symptoms during the midluteal phase.

In the most recent study, 23 women with either SZ or schizoaffective disorder were assessed at two menstrual cycle phases, early follicular (days 2–4; low estrogen/progesterone) and midluteal (days 20–22; high estrogen/progesterone). Phase at first session was counterbalanced, and sessions were held during two separate cycles 42 days apart. Menstrual phase was initially estimated by counting back to Day 1 of the last menstrual cycle and then later validated with estradiol and progesterone levels.[45] Although we found a better symptom presentation (PANSS total, positive, general) during the midluteal versus the early follicular phase, symptom changes were not related to estradiol or progesterone changes. Rather, we demonstrated for the first time that endogenous OT levels, which were consistent across the cycle, were associated with symptoms in female patients. Higher OT levels were associated with lower scores (PANSS total, positive general, prosocial; Figure 12.2). We have since demonstrated associations between OT and positive symptoms, although in the opposite direction.[43,54] Taken together, these studies suggest that hormones can impact symptoms in SZ. Additionally, these studies demonstrate the importance of multiple neuroendocrine systems including OT. To our knowledge, no study to date has focused on AVP across the menstrual cycle among individuals with psychosis.

To date, there are only a few published menstrual cycle studies in psychosis that include cognitive and/or emotion processing.[45,53,55] In the first study, neither controls nor patients showed cognitive enhancements during the luteal phase (days 14–28), though both showed visuospatial ($d = 0.41$ patients; controls $d = 0.13$) and attention/perceptual speed ($d = 0.28$ patients; controls $d = 0.42$) enhancements during the follicular phase (days 1–14).[53] Additionally, only patients showed

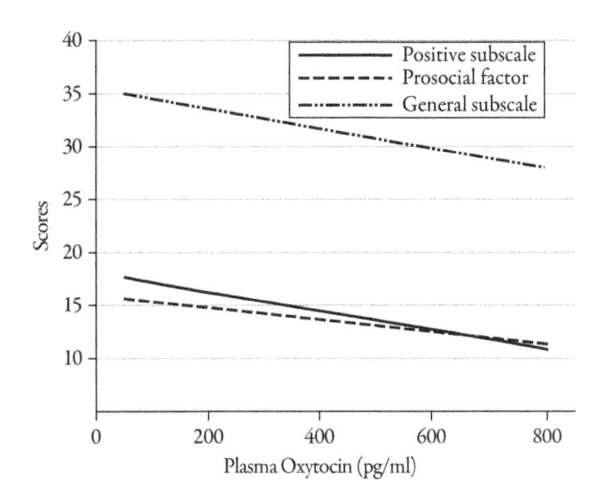

Figure 12.2 Average oxytocin levels relate to average clinical symptoms or on the PANSS in women with schizophrenia/schizoaffective disorder. NOTE. From Rubin LH, Carter CS, Drogos L, Poumajafi-Nazarloo H, Sweeney JA, Maki PM. Peripheral oxytocin is associated with reduced symptom severity in schizophrenia. *Schizophr Res* 2010;124(1–3):13–21. Fitted subscale scores on the Positive and Negative Syndrome Scale (PANSS) using multilevel random coefficient models as a functions of plasma oxytocin levels in female patients for three scales: positive subscale ($p < 0.001$), general psychopathology subscale ($p < 0.01$), and the prosocial factor ($p < 0.05$).

fine motor skill enhancements during the follicular versus the luteal phase ($d = 0.30$). Estrogen levels were not correlated with cognition within menstrual cycle phase. Progesterone-cognition associations were not examined. In our menstrual cycle study, which tightly controlled for cycle phase,[26,45,55] cognitive changes were not seen on "female-dominant" (verbal learning/memory, psychomotor speed, fine motor skills, fluency) or on a "male-dominant" test (visuospatial abilities).[26] With emotion processing, we did find female patients and controls to more accurately identify facial emotions during the follicular versus midluteal phase.[55] This shift in performance was unrelated to sex steroid hormones but rather higher OT levels were associated with perceiving faces as happier.

Critical support for the role of hormones in modulating the clinical course of SZ also comes from randomized clinical trials (RCTs) in psychosis. To date, there are 16 RCTs (2 cross-over trials) examining the effects of adjunctive estrogen or elective estrogen receptor modulators (SERMS; raloxifene) on symptoms (Table 12.1). SERMS are "designer" estrogenic agents that act as estrogen agonists in some tissue (bone) and estrogen antagonists in reproductive tissues and therefore have beneficial effects on certain tissues without increasing risk of certain reproductive cancers. Twelve out of 13 studies demonstrate a beneficial effect of short-term adjunctive hormone use on symptoms. Of the three studies demonstrating no symptom benefits, two studies had null findings.[56,57] The third reported raloxifene being less efficacious than placebo on symptoms in a large sample of 200 severely ill postmenopausal women with SZ or schizoaffective disorder.[58] A number of factors might explain discrepancies across studies including length of treatment, menopausal status, biological sex, type and dose of estradiol, use of progestogens, and sample size. Nevertheless, the majority of findings, as demonstrated in a recent meta-analysis,[59] is in support of adjunctive estrogen/SERMS showing symptom benefits.

Table 12.1 EFFECTS OF ADJUNCTIVE HORMONES IN 16 RANDOMIZED CLINICAL TRIALS ON SYMPTOMS AND/ OR COGNITION IN PSYCHOSIS

AUTHOR	*N*	DIAGNOSIS	MEAN &/OR AGE RANGE; % MALE	TREATMENT	DURATION (WEEKS)	OUTCOME
Kulkarni[60–62]	102	Schizophrenia spectrum (chronic)	33[ǀ]; 0%	Antipsychotic + high dose estradiol (100 mg transdermal every 4 days)	4	E ↓ PANSS total, positive, general
Akhondzad-eh[63]	32	Schizophrenia (chronic)	33[ǀ]; 0%	Haloperidol + ethinyl estradiol (.5 mg/ day oral)	8	E ↓ PANSS total, positive, general
Louza[64]	40	Schizophrenia (chronic)	32; 18–49; 0%	Haloperidol + CEE (.625 mg/day oral)	4	E ↓ BPRS total & positive (trends)
Bergemann[X 57,65]	46	Schizophrenia (acute/ paranoid/ hypo-estrogenic)	38; 16–67; 0%	Antipsychotic + three-phase estrogen-progestin combination oral • Follicular: 4 mg 17β-estradiol • Ovulation: 4 mg 17β-estradiol/1 mg norethisterone acetate • Luteal: 1 mg 17β-estradiol	8	E no effect on PANSS/BPRS; ↑ abstract priming; no cognitive improvements
Ko[*66]	28	Schizophrenia spectrum (chronic)	34[ǀ]; 0%	Antipsychotic + CEE(.625 mg/day oral)/ MPA (2.5 mg/day oral)	8	E ↓ SANS; ↑ verbal memory, fluency, & some measures of psychomotor speed/ visual scanning (TMT A, not DSST); no effect on visual recognition, mental flexibility
Kulkarni[67]	53	Schizophrenia spectrum (chronic)	32; 100%	Antipsychotic + estradiol valerate (2 mg/ day oral)	2	E ↓ PANSS general
Usall[68]	33	Schizophrenia (chronic, prominent negative symptoms)	62[ǂ]; 0%	Antipsychotic + raloxifene hydrochloride (60 mg/day oral)	12	Raloxifene ↓ PANSS total, positive, negative, general
Ghafari[69]	32	Schizophrenia (chronic)	34[ǀ]; 0%	Antipsychotic + CEE (0.625 mg/day oral)	4	E ↓ PANSS total, positive, negative, general
Kianimehr[70]	46	Schizophrenia (chronic)	60[ǂ]; 0%	Risperidone + raloxifene hydrochloride (120 mg/day oral)	8	Raloxifine ↓ PANSS positive
Huerta-Ramos[71]	35	Schizophrenia (chronic)	61[ǂ]; 0%	Antipsychotic + raloxifene hydrochloride (60 μg/day oral)	12	Raloxifine ↑ verbal learning, fluency; semantic clustering, psychomotor speed (trends)
Weikert[X† 56]	79	Schizophrenia spectrum (chronic)	35; 18–51; 62%	Antipsychotic + raloxifene hydrochloride (120 mg/day oral)	13	Raloxifine no effect on PANSS; 6 weeks ↑ verbal memory, fluency, attention/ processing speed 6 weeks; sex-specific analyses fluency benefit only in females; 13 weeks ↑ attention/ processing speed
Kulkarni[72]	183	Schizophrenia spectrum (treatment resistant)	35; 18–45; 0%	Antipsychotic + estradiol (200 μg transdermal or 100 μg transdermal)	8	E (100 & 200μg) ↓ PANSS total, positive, general; no effect on RBANS
Khodaie-Ardakani[73]	46	Schizophrenia (chronic)	31; 18--55; 100%	Risperidone + raloxifene hydrochloride (120 mg/day oral)	8	Raloxifene ↓ PANSS total, negative, general
Kulkarni[74,75]	56	Schizophrenia spectrum (chronic)	53; 40–70; 0%	Antipsychotic + raloxifene hydrochloride (120 mg/day oral)	12	Raloxifene ↓ PANSS total, general; no effect on RBANS
Usall[76]	70	Schizophrenia (chronic)	62; >45; 0%	Antipsychotic + raloxifene hydrochloride (60 mg/day oral)	24	Raloxifene ↓ PANSS total, negative, general; ↓ SANS alogia
Weiser[58]	200	Schizophrenia spectrum (chronic)	56; 45–65; 0%	Antipsychotic + raloxifene hydrochloride (120 mg/day oral)	16	Raloxifene ↑ PANSS total, positive, negative, general; no effect on BACS

E = estrogen. CEE = Conjugated equine estrogen. MPA = medroxyprogesterone acetate. PANSS= Positive and Negative Syndrome Scale. SANS = Scale for the Assessment of Negative Symptoms. BPRS=Brief Psychiatric Rating Scale. TMT = Trail Making Test. DSST = Digit Symbol Substitution Test. BACS = Brief Assessment of cognition in schizophrenia. *examined negative symptoms only. †Studied 79 individuals but only 30 were women. ǀchild-bearing age. ǂ postmenopausal. Xcross-over study.

To date, there are relatively fewer RCTs of adjunctive hormone therapy/estrogenic agents in SZ including cognitive endpoints.[65–66,77] To date, four of six studies show some cognitive benefits (Table 12.1). The first efficacious study was in 28 premenopausal women on the SZ spectrum.[66] Adjunctive conjugated estrogen with medroxyprogesterone acetate improved verbal memory, fluency, and processing speed but not on visual recognition or psychomotor speed/visual scanning. While in some ways the pattern of effects conforms to the expectation that estrogen would improve performance on "female-dominant" tests, the finding of improvements on sexually "neutral" tests suggest that estrogen effects may extend beyond cognitive domains that affect women. One possibility is that hormones may have a broader effect in this population.

The second study showing cognitive benefits was in a subset[57] of 19 treated hypoestrogenic women with paranoid SZ.[65] Adjuvant hormone treatment improved performance on a measure of comprehension of metaphoric speech and/or thought disturbance. Despite this improvement, no benefits were noted on a cognitive battery that included tests of verbal and spatial abilities and attention. Outcomes from this RCT differ from the previous study and there were study differences with respect to age and the type and dose of estrogen treatment. Nevertheless, this trial added to the literature suggesting that hormones may modulate cognition in SZ.

The last two studies demonstrating cognitive benefits used adjuvant raloxifene. First, 12 weeks of adjunctive raloxifene (60 μg/day) improved verbal learning and fluency in postmenopausal women with SZ. It showed trends on psychomotor speed.[71] Next, in a cross-over study, 13 weeks of a higher dose of raloxifene (120μg/day) in premenopausal women and men across the schizophrenia spectrum showed improvements on verbal memory, attention/processing speed, and fluency (driven by women) after 6 weeks.[56] After 13 weeks, benefits were only retained on attention/processing speed. Two neuroimaging substudies were conducted in subsets of patients from this larger study. Raloxifene increased prefrontal activity in 30 patients (8 women) during an emotional response inhibition task and the magnitude of the association was modulated by variation in estrogen receptor alpha (ESR-1).[78] Raloxifene also altered neural activation in the hippocampus and inferior frontal gyrus during an emotional face recognition task in 20 patients.[79]

Findings of beneficial effects of estrogen/ SERMS on symptoms and cognition may reflect any one of a number of mechanisms of action. There is a wealth of evidence from studies demonstrating sex differences in, and estrogen effects on, brain dopamine systems.[80] In nonhuman primates, estrogen affects the density of dopaminergic cells in the substantia nigra, suggesting a modulation of cell-packing density by expansion of synaptic refinements and connections.[81] There is also a sex difference in the effect of estrogen on striatal dopamine concentrations and release in rats, with significant effects observed in female but not male rats.[82] In human studies, women on average have a higher capacity for presynaptic dopamine synthesis versus men and may have more D2-family receptors than males.[83] Estrogen also upregulates dopaminergic function in postmenopausal women.[84] The beneficial effects of estrogen on hippocampal-dependent cognitive tests appear to be at least partially due to an improvement in basal forebrain cholinergic function, particularly cholinergic projections to the hippocampus.[85] The role of dopamine in this behavioral effect has been less widely studied, though one study suggested that estrogen withdrawal due to ovariectomy in rats produced impairments in hippocampal-dependent memory tasks and reductions in hippocampal dihydroxyphenylacetic acid (dopamine metabolite) and that these two effects were independent.[86] Estrogen may enhance postsynaptic dopamine activity, and by broadly enhancing fronto-striato-thalamo-cortical circuitry and intrinsic neocortical function, estrogen-dopamine interactions provide a plausible mechanism for some aspects of the sex differences we observed in cognitive response to antipsychotic medication over time.

While dopamine-estrogen interactions are one possibility, the full range of mechanisms by which antipsychotic treatment might differentially enhance symptoms and the magnitude of sex biases in patients' cognition is unknown. Another explanation is that female sex steroid hormones may interact with antipsychotic medications to potentiate treatment effects on neuronal structure and function. Studies suggest that estrogen alters responsivity to antipsychotic medication.[87–91] Chronic estrogen and standard antipsychotic treatments promote similar and additive effects on neural and behavioral measures of dopaminergic function. In rats, both estrogen and haloperidol increase the number of striatal dopamine receptors, and estrogen potentiates antipsychotic-induced catalepsy.[87] In monkeys and rats, haloperidol and estrogen produce similar and additive effects on dyskinesia.[88,89] Estrogen also helps to maintain plasma haloperidol levels[90] and affects the potency of haloperidol in rats.[91] As a result, estrogen enhancing medications may have a beneficial role in women with SZ.

In addition to estrogen, numerous studies demonstrate benefits of intranasal OT on symptoms, cognition, and emotion processing in SZ. To date, there have been nine RCTs examining intranasal OT in SZ (Table 12.2) of which eight show benefits on at least one outcome. These findings also nicely parallel single-dose studies of intranasal OT in SZ.[92–100] The primary study limitations are that they comprise small numbers of predominately men and therefore findings cannot necessarily be generalized to women.

Table 12.2 EFFECTS OF INTRANASAL OXYTOCIN (OT) IN NINE RANDOMIZED CLINICAL TRIALS ON SYMPTOMS, COGNITION, AND/OR EMOTION PROCESSING IN PSYCHOSIS

AUTHOR	*N*	DIAGNOSIS	MEAN &/OR AGE RANGE; % MALE	TREATMENT	DURATION (WEEKS)	OUTCOME
Feifel X [101,102]	15	Schizophrenia (chronic); moderate positive symptoms	48; 80%	Antipsychotic + OT (20 IU 2x/day week 1; 40 IU 2x/day week 2–3)	3	OT ↓ PANSS total, positive, negative; ↑ verbal learning/memory not working memory
Pedersen[103]	20	Schizophrenia (chronic); moderate positive symptoms	37; 18–55; 85%	Antipsychotic + OT (24 IU 2x/day for 2 weeks)	2	OT ↓ PANSS total, positive, general; ↑ theory of mind, trustworthiness
Lee[104]	28	Schizophrenia spectrum (chronic)	40; 80%	Antipsychotic + OT (20 IU 2x/day)	3	OT no effects on BPRS/SANS in overall sample; ↓ SANS total, avolition/anhedonia/asociality in inpatients only; ↑ odor identification
Modabbernia[105]	40	Schizophrenia (chronic)	32; 18–50; 83%	Risperidone + OT (20 IU 2x/day week 1; 40 IU 2x/day week 2–8)	8	OT ↓ PANSS total, positive, negative
Gibson[106]	14	Schizophrenia (chronic); moderate positive symptoms	36; 18–55; 79%	Antipsychotic + OT (24 IU 2x/day)	6	OT ↓ PANSS negative; ↑ fear recognition, perspective taking
Cacciotti-Saija[107]	52	Schizophrenia spectrum (early psychosis)	16–35	Antipsychotic + OT (24 IU 2x/day)	6	OT no benefit on PANSS or social cognition; ↓ SANS
Brambilla X [108]	31	Schizophrenia (chronic); PANSS ≥55 & CGI-S ≥4	30; 18–45; 84%	Antipsychotic + OT (40 IU 1x/day)	16	OT ↑ MSCEIT understanding emotions & ↑ speed on emotion priming task for facial affect recognition; no effect on implicit priming, theory of mind
Jarskog[109]	60	Schizophrenia spectrum (chronic); moderate positive symptoms	39; 18–65; 76%	Antipsychotic + OT (24 IU 2x a day)	12	OT ↑ interpersonal relationships; ↓ PANSS negative symptoms (trend)
Buchanan[110]	58	Schizophrenia spectrum (chronic)	46; 18–65; 24%	Antipsychotic + OT (24 IU 2x a day)	6	OT no effects on SANS or MATRICS

PANSS = Positive and Negative Syndrome Scale. SANS = Scale for the Assessment of Negative Symptoms. BPRS = Brief Psychiatric Rating Scale. CGI-S = Clinical Global Impressions-Severity scale score. MSCEIT = Mayer-Salovey-Caruso Emotional Intelligence Test. X cross-over study.

The mechanisms by which OT may have therapeutic effects on symptoms is unknown, but may involve modulation of dopamine and glutamate neurotransmission. Dopamine-neuropeptide interactions are of importance given our findings that OT relates to positive symptoms,[45] which reflect a hyperactivity of dopamine transmission. Mesolimbic dopamine projections are thought to be hyperactive in SZ, and animal studies demonstrate that OT administered peripherally inhibits dopamine transmission in the mesolimbic pathway.[111] In male rats, high doses of OT reverse the social deficits caused by chronic phencyclidine treatment, a noncompetitive NMDA antagonist.[112] In male rats, OT treatment was found to reverse deficits in prepulse inhibition (PPI) induced by dizocilpine, a noncompetitive NMDA antagonist.[113] However, OT had smaller effects following PPI induced-deficits caused by amphetamine and OT was unable to reverse apomorphine-induced deficits in PPI. Prepulse inhibition is a measure of sensorimotor gating used to test the efficacy of antipsychotics. Oxytocin knock-out mice also show PPI disruptions following administration of phencyclidine.[114] In rats, clozapine also increases secretion of plasma OT.[115] Taken together, these studies suggest that endogenous OT may influence the severity of psychotic symptoms in SZ and may also impact the therapeutic effects of antipsychotic drugs.

Our more recent work into the impact of epigenetic modification of the oxytocin receptor gene (*OXTR*) also provides insights into potential mechanisms of OT and emotion processing.[116] We found that greater methylation of *OXTR* was observed in individuals with prototypic SZ versus psychotic BPD. Additionally, it was only in female patients and controls that methylation was associated with emotion processing and in relation to structural brain regions subserving emotion processing. Interestingly we noted these patterns despite alterations in peripheral OT levels in any psychotic disorders suggesting that greater methylation may have been a homeostatic compensation. Thus, the OT pathway may be disrupted at the level of the receptor and not the level of neuropeptide synthesis, and that in psychotic disorders, this effect is relatively specific to women with SZ. This sex difference in the pattern of association between *OXTR* methylation and peripheral OT levels might help explain why beneficial effects of intranasal OT are seen in male patients. In males, intranasal OT might increase signaling at the *OXTR* to enhance emotional processing. However, as previously noted, studies of the consequences of intranasal OT in females are uncommon, and little is known regarding sex differences in response to OT therapies or their variability in relation to genotype.

Despite AVP and testosterone alterations in psychosis, few RCTs focus on these adjunctive hormones in relation to clinical and cognitive outcomes. With respect to AVP, three studies have focused on intranasal 1-desamino-8-D-arginine vasopressin (DDAVP), a synthetic analogue of AVP. Two of three RCTs show beneficial effects of DDAVP. In the first, DDAVP (0.3–0.6 to a max of 1μg/kg/day) reduced scores on negative symptoms r versus placebo among 12 chronic treatment-resistant patients with SZ (aged 18–37; 58% male).[117] Similarly, in 40 individuals (aged 18-50; 80% men) with chronic SZ, intranasal DDAVP (20 mcg/day) improved PANSS total, negative, and general symptoms.[118] However, in a cross-over study, 3 weeks of DDAVP (60 μg/day) did not improve symptoms on the BPRS, Scale for the Assessment of Negative Symptoms (SANS), or cognition in 21 patients with chronic SZ (aged 20–60; 81% male).[119] The role of AVP treatment in psychosis warrants further study as a clinical treatment.

To date, there are few studies on the treatment of psychosis with adjunctive testosterone. Short-term, 4 weeks, adjuvant testosterone (5 g of 1% gel) improved PANSS negative symptoms in 30 men with chronic SZ (aged 20–49) versus placebo.[120] The benefit of testosterone on symptoms in males may in part reflect the interplay between testosterone and brain dopamine systems, as animal studies demonstrate that testosterone increases molecular changes in dopamine signaling in adolescent male rats.[121,122]

In sum, there is strong evidence that neuroendocrine factors may play a critical role in both the clinical presentation, cognition, and emotion processing in psychosis. Taken together, neuroendocrine factors need to be taken into account in studies in developing neurobiological models of psychotic disorders, as well as considered as potential treatment options at least for a subset of individuals.

ACKNOWLEDGMENTS

I thank Drs. Pauline M. Maki, John A. Sweeney, and C. Sue Carter for their unwavering support and input on this body of work.

REFERENCES

1. Aleman A, Kahn RS, Selten JP. Sex differences in the risk of schizophrenia: evidence from meta-analysis. *Arch Gen Psychiatry*. 2003;60(6):565–571.
2. Grigoriadis S, Seeman MV. The role of estrogen in schizophrenia: implications for schizophrenia practice guidelines for women. *Can J Psychiatry*. 2002;47(5):437–442.
3. Hafner H. Gender differences in schizophrenia. *Psychoneuroendocrinology*. 2003;28(Suppl 2):17–54.
4. Goldstein JM, Walder DJ. Sex differences in schizophrenia: the case for developmental origins and etiological implications. In: Sharma T, Harvey P, eds. *The early course of schizophrenia*. Oxford, UK: Oxford University Press; 2006:147–173.
5. Goldstein JM, Cherkerzian S, Tsuang MT, Petryshen TL. Sex differences in the genetic risk for schizophrenia: history of the evidence for sex-specific and sex-dependent effects. *American journal of medical genetics Part B, Neuropsychiatric genetics: the official publication of the International Society of Psychiatric Genetics*. 2013;162B(7):698–710.
6. Freeman MP, Gelenberg AJ. Bipolar disorder in women: reproductive events and treatment considerations. *Acta Psychiatr Scand*. 2005;112(2):88–96.
7. Diflorio A, Jones I. Is sex important? Gender differences in bipolar disorder. *Int Rev Psychiatry*. 2010;22(5):437–452.
8. Marsh WK, Ketter TA, Rasgon NL. Increased depressive symptoms in menopausal age women with bipolar disorder: age and gender comparison. *J Psychiatr Res*. 2009;43(8):798–802.
9. Hanlon MC, Campbell LE, Single N, et al. Men and women with psychosis and the impact of illness-duration on sex-differences: the second Australian national survey of psychosis. *Psychiatry Res*. 2017;256:130–143.
10. Shlomi Polachek I, Manor A, Baumfeld Y, et al. Sex differences in psychiatric hospitalizations of individuals with psychotic disorders. *J Nerv Ment Dis*. 2017;205(4):313–317.
11. Barajas A, Ochoa S, Obiols JE, Lalucat-Jo L. Gender differences in individuals at high-risk of psychosis: a comprehensive literature review. *The Scientific World Journal*. 2015;2015:430735.
12. Andreano JM, Cahill L. Sex influences on the neurobiology of learning and memory. *Learn Mem*. 2009;16(4):248–266.
13. Gur RC, Gur RE. Memory in health and in schizophrenia. *Dialogues Clin Neurosci*. 2013;15(4):399–410.
14. Voyer D, Voyer S, Bryden MP. Magnitude of sex differences in spatial abilities: a meta-analysis and consideration of critical variables. *Psychol Bull*. 1995;117(2):250–270.
15. Kimura D. *Sex and cognition*. London: MIT Press; 1999.
16. Whittle S, Yucel M, Yap MB, Allen NB. Sex differences in the neural correlates of emotion: evidence from neuroimaging. *Biol Psychol*. 2011;87(3):319–333.
17. Stevens JS, Hamann S. Sex differences in brain activation to emotional stimuli: a meta-analysis of neuroimaging studies. *Neuropsychologia*. 2012;50(7):1578–1593.
18. Cahill L. Sex-related influences on the neurobiology of emotionally influenced memory. *Ann N Y Acad Sci*. 2003;985:163–173.
19. Halpern DF. *Sex differences in cognitive abilities*. Hillsdale, NJ: Erlbaum; 1986.
20. Leger M, Neill JC. A systematic review comparing sex differences in cognitive function in schizophrenia and in rodent models for schizophrenia, implications for improved therapeutic strategies. *Neurosci Biobehav Rev*. 2016;68:979–1000.

21. Hoff AL, Wieneke M, Faustman WO, et al. Sex differences in neuropsychological functioning of first-episode and chronically ill schizophrenic patients. *Am J Psychiatry*. 1998;155(10):1437–1439.
22. Goldstein JM, Seidman LJ, Goodman JM, et al. Are there sex differences in neuropsychological functions among patients with schizophrenia? *Am J Psychiatry*. 1998;155(10):1358–1364.
23. Goldberg TE, Gold JM, Torrey EF, Weinberger DR. Lack of sex differences in the neuropsychological performance of patients with schizophrenia. *Am J Psychiatry*. 1995;152(6):883–888.
24. Gur RC, Ragland JD, Moberg PJ, et al. Computerized neurocognitive scanning: II. The profile of schizophrenia. *Neuropsychopharmacology*. 2001;25(5):777–788.
25. Rubin LH, Haas GL, Keshavan MS, Sweeney JA, Maki PM. Sex difference in cognitive response to antipsychotic treatment in first episode schizophrenia. *Neuropsychopharmacology*. 2008;33(2):290–297.
26. Rubin LH, Carter CS, Drogos LL, Pournajafi-Nazarloo H, Sweeney JA, Maki PM. Effects of sex, menstrual cycle phase, and endogenous hormones on cognition in schizophrenia. *Schizophr Res*. 2015;166(1–3):269–275.
27. Fiszdon JM, Silverstein SM, Buchwald J, Hull JW, Smith TE. Verbal memory in schizophrenia: sex differences over repeated assessments. *Schizophr Res*. 2003;61(2–3):235–243.
28. Halari R, Mehrotra R, Sharma T, Ng V, Kumari V. Cognitive impairment but preservation of sexual dimorphism in cognitive abilities in chronic schizophrenia. *Psychiatry Research*. 2006;141:129–139.
29. Ruigrok AN, Salimi-Khorshidi G, Lai MC, et al. A meta-analysis of sex differences in human brain structure. *Neurosci Biobehav Rev*. 2014;39:34–50.
30. Sacher J, Neumann J, Okon-Singer H, Gotowiec S, Villringer A. Sexual dimorphism in the human brain: evidence from neuroimaging. *Magn Reson Imaging*. 2013;31(3):366–375.
31. Tomasino B, Gremese M. Effects of stimulus type and strategy on mental rotation network: an activation likelihood estimation meta-analysis. *Frontiers in human neuroscience*. 2015;9:693.
32. Costafreda SG, Fu CH, Lee L, Everitt B, Brammer MJ, David AS. A systematic review and quantitative appraisal of fMRI studies of verbal fluency: role of the left inferior frontal gyrus. *Hum Brain Mapp*. 2006;27(10):799–810.
33. Wagner S, Sebastian A, Lieb K, Tuscher O, Tadic A. A coordinate-based ALE functional MRI meta-analysis of brain activation during verbal fluency tasks in healthy control subjects. *BMC Neurosci*. 2014;15:19.
34. Rubin LH, Yao L, Keedy SK, et al. Sex differences in associations of arginine vasopressin and oxytocin with resting-state functional brain connectivity. *J Neurosci Res*. 2017;95(1–2):576–586.
35. Zuo XN, Kelly C, Di Martino A, et al. Growing together and growing apart: regional and sex differences in the lifespan developmental trajectories of functional homotopy. *J Neurosci*. 2010;30(45):15034–15043.
36. Filippi M, Valsasina P, Misci P, Falini A, Comi G, Rocca MA. The organization of intrinsic brain activity differs between genders: a resting-state fMRI study in a large cohort of young healthy subjects. *Hum Brain Mapp*. 2013;34(6):1330–1343.
37. Biswal BB, Mennes M, Zuo XN, et al. Toward discovery science of human brain function. *Proc Natl Acad Sci U S A*. 2010; 107(10):4734–4739.
38. Allen EA, Erhardt EB, Damaraju E, et al. A baseline for the multivariate comparison of resting-state networks. *Frontiers in systems neuroscience*. 2011;5:2.
39. Goldstein JM, Seidman LJ, O'Brien LM, et al. Impact of normal sexual dimorphisms on sex differences in structural brain abnormalities in schizophrenia assessed by magnetic resonance imaging. *Arch Gen Psychiatry*. 2002;59(2):154–164.
40. Walder DJ, Seidman LJ, Makris N, Tsuang MT, Kennedy DN, Goldstein JM. Neuroanatomic substrates of sex differences in language dysfunction in schizophrenia: a pilot study. *Schizophr Res*. 2007;90(1–3):295–301.
41. Abbs B, Liang L, Makris N, Tsuang M, Seidman LJ, Goldstein JM. Covariance modeling of MRI brain volumes in memory circuitry in schizophrenia: sex differences are critical. *Neuroimage*. 2011;56(4):1865–1874.
42. Gur RE, Kohler C, Turetsky BI, et al. A sexually dimorphic ratio of orbitofrontal to amygdala volume is altered in schizophrenia. *Biol Psychiatry*. 2004;55(5):512–517.
43. Rubin LH, Carter CS, Bishop JR, et al. Reduced levels of vasopressin and reduced behavioral modulation of oxytocin in psychotic disorders. *Schizophr Bull*. 2014;40(6):1374–1384.
44. Rubin LH, Carter CS, Bishop JR, et al. Peripheral vasopressin but not oxytocin relates to severity of acute psychosis in women with acutely-ill untreated first-episode psychosis. *Schizophr Res*. 2013;146(1–3):138–143.
45. Rubin LH, Carter CS, Drogos L, Pournajafi-Nazarloo H, Sweeney JA, Maki PM. Peripheral oxytocin is associated with reduced symptom severity in schizophrenia. *Schizophr Res*. 2010;124(1–3):13–21.
46. Turan. T, Uysal C, Asdemir A, Kılıç E. May oxytocin be a trait marker for bipolar disorder? *Psychoneuroendocrinology*. 2013.
47. Howes OD, Wheeler MJ, Pilowsky LS, Landau S, Murray RM, Smith S. Sexual function and gonadal hormones in patients taking antipsychotic treatment for schizophrenia or schizoaffective disorder. *J Clin Psychiatry*. 2007;68(3):361–367.
48. Hallonquist JD, Seeman MV, Lang M, Rector NA. Variation in symptom severity over the menstrual cycle of schizophrenics. *Biol Psychiatry*. 1993;33(3):207–209.
49. Bergemann N, Parzer P, Runnebaum B, Resch F, Mundt C. Estrogen, menstrual cycle phases, and psychopathology in women suffering from schizophrenia. *Psychol Med*. 2007;37(10):1427–1436.
50. Riecher-Rossler A, Hafner H, Stumbaum M, Maurer K, Schmidt R. Can estradiol modulate schizophrenic symptomatology? *Schizophr Bull*. 1994;20(1):203–214.
51. Choi SH, Kang SB, Joe SH. Changes in premenstrual symptoms in women with schizophrenia: a prospective study. *Psychosom Med*. 2001;63(5):822–829.
52. Harris AH. Menstrually related symptom changes in women with schizophrenia. *Schizophr Res*. 1997;27(1):93–99.
53. Thompson K, Sergejew A, Kulkarni J. Estrogen affects cognition in women with psychosis. *Psychiatry Res*. 2000;94(3):201–209.
54. Rubin LH, Wehring HJ, Demyanovich H, et al. Peripheral oxytocin and vasopressin are associated with clinical symptom severity and cognitive functioning in midlife women with chronic schizophrenia. submitted.
55. Rubin LH, Carter CS, Drogos L, et al. Sex-specific associations between peripheral oxytocin and emotion perception in schizophrenia. *Schizophr Res*. 2011;130(1–3):266–270.
56. Weickert TW, Weinberg D, Lenroot R, et al. Adjunctive raloxifene treatment improves attention and memory in men and women with schizophrenia. *Mol Psychiatry*. 2015;20(6):685–694.
57. Bergemann N, Mundt C, Parzer P, et al. Estrogen as an adjuvant therapy to antipsychotics does not prevent relapse in women suffering from schizophrenia: results of a placebo-controlled double-blind study. *Schizophr Res*. 2005;74(2–3):125–134.
58. Weiser M, Levi L, Burshtein S, et al. Raloxifene plus antipsychotics versus placebo plus antipsychotics in severely ill decompensated postmenopausal women with schizophrenia or schizoaffective disorder: a randomized controlled trial. *J Clin Psychiatry*. 2017.
59. Heringa SM, Begemann MJ, Goverde AJ, Sommer IE. Sex hormones and oxytocin augmentation strategies in schizophrenia: a quantitative review. *Schizophr Res*. 2015;168(3):603–613.
60. Kulkarni J, de Castella A, Smith D, Taffe J, Keks N, Copolov D. A clinical trial of the effects of estrogen in acutely psychotic women. *Schizophr Res*. 1996;20(3):247–252.
61. Kulkarni J, De Castalla A, Downey M, Taffe J, Fitzgerald P. Estrogen—a useful adjunct in the treatment of men with schizophrenia? *Schizophrenia Research*. 2002;53(3 Suppl 1):10.
62. Kulkarni J, de Castella A, Fitzgerald PB, et al. Estrogen in severe mental illness: a potential new treatment approach. *Arch Gen Psychiatry*. 2008;65(8):955–960.
63. Akhondzadeh S, Nejatisafa AA, Amini H, et al. Adjunctive estrogen treatment in women with chronic schizophrenia: a

double-blind, randomized, and placebo-controlled trial. *Prog Neuropsychopharmacol Biol Psychiatry*. 2003;27(6):1007–1012.
64. Louza MR, Marques AP, Elkis H, Bassitt D, Diegoli M, Gattaz WF. Conjugated estrogens as adjuvant therapy in the treatment of acute schizophrenia: a double-blind study. *Schizophr Res*. 2004;66(2–3):97–100.
65. Bergemann N, Parzer P, Jaggy S, Auler B, Mundt C, Maier-Braunleder S. Estrogen and comprehension of metaphoric speech in women suffering from schizophrenia: results of a double-blind, placebo-controlled trial. *Schizophr Bull*. 2008;34(6):1172–1181.
66. Ko YH, Joe SH, Cho W, et al. Effect of hormone replacement therapy on cognitive function in women with chronic schizophrenia. *International journal of psychiatry in clinical practice*. 2006;10(2):97–104.
67. Kulkarni J, de Castella A, Headey B, et al. Estrogens and men with schizophrenia: is there a case for adjunctive therapy? *Schizophr Res*. 2011;125(2–3):278–283.
68. Usall J, Huerta-Ramos E, Iniesta R, et al. Raloxifene as an adjunctive treatment for postmenopausal women with schizophrenia: a double-blind, randomized, placebo-controlled trial. *J Clin Psychiatry*. 2011;72(11):1552–1557.
69. Ghafari E, Fararouie M, Shirazi HG, Farhangfar A, Ghaderi F, Mohammadi A. Combination of estrogen and antipsychotics in the treatment of women with chronic schizophrenia. *Clinical schizophrenia & related psychoses*. 2013;6(4):172–176.
70. Kianimehr G, Fatehi F, Hashempoor S, et al. Raloxifene adjunctive therapy for postmenopausal women suffering from chronic schizophrenia: a randomized double-blind and placebo controlled trial. *Daru: journal of Faculty of Pharmacy, Tehran University of Medical Sciences*. 2014;22:55.
71. Huerta-Ramos E, Iniesta R, Ochoa S, et al. Effects of raloxifene on cognition in postmenopausal women with schizophrenia: a double-blind, randomized, placebo-controlled trial. *Eur Neuropsychopharmacol*. 2014;24(2):223–231.
72. Kulkarni J, Gavrilidis E, Wang W, et al. Estradiol for treatment-resistant schizophrenia: a large-scale randomized-controlled trial in women of child-bearing age. *Mol Psychiatry*. 2015;20(6):695–702.
73. Khodaie-Ardakani MR, Khosravi M, Zarinfard R, et al. A placebo-controlled study of raloxifene added to risperidone in men with chronic schizophrenia. *Acta medica Iranica*. 2015;53(6):337–345.
74. Kulkarni J, Gurvich C, Lee SJ, et al. Piloting the effective therapeutic dose of adjunctive selective estrogen receptor modulator treatment in postmenopausal women with schizophrenia. *Psychoneuroendocrinology*. 2010;35(8):1142–1147.
75. Kulkarni J, Gavrilidis E, Gwini SM, et al. Effect of adjunctive raloxifene therapy on severity of refractory schizophrenia in women: a randomized clinical trial. *JAMA Psychiatry*. 2016;73(9): 947–954.
76. Usall J, Huerta-Ramos E, Labad J, et al. Raloxifene as an adjunctive treatment for postmenopausal women with schizophrenia: a 24-week double-blind, randomized, parallel, placebo-controlled trial. *Schizophr Bull*. 2016;42(2):309–317.
77. Bergemann N, Parzer P, Kaiser D, Maier-Braunleder S, Mundt C, Klier C. Testosterone and gonadotropins but not estrogen associated with spatial ability in women suffering from schizophrenia: a double-blind, placebo-controlled study. *Psychoneuroendocrinology*. 2008;33(4):507–516.
78. Kindler J, Weickert CS, Schofield PR, Lenroot R, Weickert TW. Raloxifene increases prefrontal activity during emotional inhibition in schizophrenia based on estrogen receptor genotype. *Eur Neuropsychopharmacol*. 2016;26(12):1930–1940.
79. Ji E, Weickert CS, Lenroot R, et al. Adjunctive selective estrogen receptor modulator increases neural activity in the hippocampus and inferior frontal gyrus during emotional face recognition in schizophrenia. *Translational psychiatry*. 2016;6:e795.
80. Dluzen DE. Neuroprotective effects of estrogen upon the nigrostriatal dopaminergic system. *J Neurocytol*. 2000;29(5–6):387–399.
81. Leranth C, Roth RH, Elsworth JD, Naftolin F, Horvath TL, Redmond DE Jr. Estrogen is essential for maintaining nigrostriatal dopamine neurons in primates: implications for Parkinson's disease and memory. *J Neurosci*. 2000;20(23):8604–8609.
82. McDermott JL, Liu B, Dluzen DE. Sex differences and effects of estrogen on dopamine and DOPAC release from the striatum of male and female CD-1 mice. *Exp Neurol*. 1994;125(2):306–311.
83. Kaasinen V, Nagren K, Hietala J, Farde L, Rinne JO. Sex differences in extrastriatal dopamine d(2)-like receptors in the human brain. *Am J Psychiatry*. 2001;158(2):308–311.
84. Craig MC, Cutter WJ, Wickham H, et al. Effect of long-term estrogen therapy on dopaminergic responsivity in post-menopausal women—a preliminary study. *Psychoneuroendocrinology*. 2004;29(10):1309–1316.
85. Gibbs RB, Gabor R, Cox T, Johnson DA. Effects of raloxifene and estradiol on hippocampal acetylcholine release and spatial learning in the rat. *Psychoneuroendocrinology*. 2004;29(6):741–748.
86. Heikkinen T, Puolivali J, Liu L, Rissanen A, Tanila H. Effects of ovariectomy and estrogen treatment on learning and hippocampal neurotransmitters in mice. *Horm Behav*. 2002;41(1):22–32.
87. Di Paolo T, Poyet P, Labrie F. Effect of chronic estradiol and haloperidol treatment on striatal dopamine receptors. *Eur J Pharmacol*. 1981;73(1):105–106.
88. Bedard PJ, Boucher R, Daigle M, Di Paolo T. Similar effect of estradiol and haloperidol on experimental tardive dyskinesia in monkeys. *Psychoneuroendocrinology*. 1984;9(4):375–379.
89. Gordon JH, Borison RL, Diamond BI. Estrogen in experimental tardive dyskinesia. *Neurology*. 1980;30(5):551–554.
90. Grimm JW, Aravagiri M, See RE. Ovariectomy results in lower plasma haloperidol levels in rats following chronic administration. *Pharm Res*. 1998;15(10):1640–1642.
91. Roberts DC, Dalton JC, Vickers GJ. Increased self-administration of cocaine following haloperidol: effect of ovariectomy, estrogen replacement, and estrous cycle. *Pharmacol Biochem Behav*. 1987;26(1):37–43.
92. Goldman MB, Gomes AM, Carter CS, Lee R. Divergent effects of two different doses of intranasal oxytocin on facial affect discrimination in schizophrenic patients with and without polydipsia. *Psychopharmacology (Berl)*. 2011.
93. Averbeck BB, Bobin T, Evans S, Shergill SS. Emotion recognition and oxytocin in patients with schizophrenia. *Psychol Med*. 2011:1–8.
94. Fischer-Shofty M, Shamay-Tsoory SG, Levkovitz Y. Characterization of the effects of oxytocin on fear recognition in patients with schizophrenia and in healthy controls. *Frontiers in neuroscience*. 2013;7:127.
95. Fischer-Shofty M, Brune M, Ebert A, Shefet D, Levkovitz Y, Shamay-Tsoory SG. Improving social perception in schizophrenia: the role of oxytocin. *Schizophr Res*. 2013;146(1–3):357–362.
96. Davis MC, Lee J, Horan WP, et al. Effects of single dose intranasal oxytocin on social cognition in schizophrenia. *Schizophr Res*. 2013;147(2–3):393–397.
97. Davis MC, Green MF, Lee J, et al. Oxytocin-augmented social cognitive skills training in schizophrenia. *Neuropsychopharmacology*. 2014;39(9):2070–2077.
98. Woolley JD, Chuang B, Lam O, et al. Oxytocin administration enhances controlled social cognition in patients with schizophrenia. *Psychoneuroendocrinology*. 2014;47:116–125.
99. Guastella AJ, Ward PB, Hickie IB, et al. A single dose of oxytocin nasal spray improves higher-order social cognition in schizophrenia. *Schizophr Res*. 2015;168(3):628–633.
100. Michalopoulou PG, Averbeck BB, Kalpakidou AK, et al. The effects of a single dose of oxytocin on working memory in schizophrenia. *Schizophr Res*. 2015;162(1–3):62–63.
101. Feifel D, Macdonald K, Nguyen A, et al. Adjunctive intranasal oxytocin reduces symptoms in schizophrenia patients. *Biol Psychiatry*. 2010;68(7):678–680.
102. Feifel D, Macdonald K, Cobb P, Minassian A. Adjunctive intranasal oxytocin improves verbal memory in people with schizophrenia. *Schizophr Res*. 2012;139(1–3):207–210.
103. Pedersen CA, Gibson CM, Rau SW, et al. Intranasal oxytocin reduces psychotic symptoms and improves theory of mind and social perception in schizophrenia. *Schizophr Res*. 2011;132(1):50–53.

104. Lee MR, Wehring HJ, McMahon RP, et al. Effects of adjunctive intranasal oxytocin on olfactory identification and clinical symptoms in schizophrenia: results from a randomized double blind placebo controlled pilot study. *Schizophr Res*. 2013;145(1–3):110–115.
105. Modabbernia A, Rezaei F, Salehi B, et al. Intranasal oxytocin as an adjunct to risperidone in patients with schizophrenia: an 8-week, randomized, double-blind, placebo-controlled study. *CNS Drugs*. 2013;27(1):57–65.
106. Gibson CM, Penn DL, Smedley KL, Leserman J, Elliott T, Pedersen CA. A pilot six-week randomized controlled trial of oxytocin on social cognition and social skills in schizophrenia. *Schizophr Res*. 2014;156(2–3):261–265.
107. Cacciotti-Saija C, Langdon R, Ward PB, et al. A double-blind randomized controlled trial of oxytocin nasal spray and social cognition training for young people with early psychosis. *Schizophr Bull*. 2015;41(2):483–493.
108. Brambilla M, Cotelli M, Manenti R, et al. Oxytocin to modulate emotional processing in schizophrenia: a randomized, double-blind, cross-over clinical trial. *Eur Neuropsychopharmacol*. 2016;26(10):1619–1628.
109. Jarskog LF, Pedersen CA, Johnson JL, et al. A 12-week randomized controlled trial of twice-daily intranasal oxytocin for social cognitive deficits in people with schizophrenia. *Schizophr Res*. 2017;185:88–95.
110. Buchanan RW, Kelly DL, Weiner E, et al. A randomized clinical trial of oxytocin or galantamine for the treatment of negative symptoms and cognitive impairments in people with schizophrenia. *J Clin Psychopharmacol*. 2017;37(4):394–400.
111. Sarnyai Z, Kovacs GL. Role of oxytocin in the neuroadaptation to drugs of abuse. *Psychoneuroendocrinology*. 1994;19(1): 85–117.
112. Lee PR, Brady DL, Shapiro RA, Dorsa DM, Koenig JI. Social interaction deficits caused by chronic phencyclidine administration are reversed by oxytocin. *Neuropsychopharmacology*. 2005;30(10):1883–1894.
113. Feifel D, Reza T. Oxytocin modulates psychotomimetic-induced deficits in sensorimotor gating. *Psychopharmacology (Berl)*. 1999;141(1):93–98.
114. Caldwell HK, Stephens SL, Young WS, 3rd. Oxytocin as a natural antipsychotic: a study using oxytocin knockout mice. *Mol Psychiatry*. 2009;14(2):190–196.
115. Uvnas-Moberg K, Alster P, Svensson TH. Amperozide and clozapine but not haloperidol or raclopride increase the secretion of oxytocin in rats. *Psychopharmacology (Berl)*. 1992;109(4):473–476.
116. Rubin LH, Connelly JJ, Reilly JL, et al. Sex and diagnosis specific associations between DNA methylation of the oxytocin receptor gene with emotion processing and temporal-limbic and prefrontal brain volumes in psychotic disorders. *Biological psychiatry Cognitive neuroscience and neuroimaging*. 2016;1(2):141–151.
117. Iager AC, Kirch DG, Bigelow LB, Karson CN. Treatment of schizophrenia with a vasopressin analogue. *Am J Psychiatry*. 1986;143(3):375–377.
118. Hosseini SM, Farokhnia M, Rezaei F, et al. Intranasal desmopressin as an adjunct to risperidone for negative symptoms of schizophrenia: a randomized, double-blind, placebo-controlled, clinical trial. *Eur Neuropsychopharmacol*. 2014;24(6):846–855.
119. Stein D, Bannet J, Averbuch I, Landa L, Chazan S, Belmaker RH. Ineffectiveness of vasopressin in the treatment of memory impairment in chronic schizophrenia. *Psychopharmacology (Berl)*. 1984;84(4):566–568.
120. Ko YH, Lew YM, Jung SW, et al. Short-term testosterone augmentation in male schizophrenics: a randomized, double-blind, placebo-controlled trial. *J Clin Psychopharmacol*. 2008;28(4):375–383.
121. Purves-Tyson TD, Owens SJ, Double KL, Desai R, Handelsman DJ, Weickert CS. Testosterone induces molecular changes in dopamine signaling pathway molecules in the adolescent male rat nigrostriatal pathway. *PLoS One*. 2014;9(3):e91151.
122. Purves-Tyson TD, Handelsman DJ, Double KL, Owens SJ, Bustamante S, Weickert CS. Testosterone regulation of sex steroid-related mRNAs and dopamine-related mRNAs in adolescent male rat substantia nigra. *BMC Neurosci*. 2012;13:95.

SECTION 2

PSYCHOSIS COURSE AND LIFETIME MANIFESTATIONS

EARLY PSYCHOSIS

13

CLINICAL PHENOMENOLOGY OF THE PRODROME FOR PSYCHOSIS

Albert R. Powers III, Thomas H. McGlashan, and Scott W. Woods

HISTORY OF THE PRODROME CONCEPT

Schizophrenia is a heterogeneous disease entity. Its manifestations, clinical course, functioning, response to treatment, and ultimate outcomes vary widely across individuals. Because of this heterogeneity, the line between the normal spectrum and frank psychosis may be difficult to define. Furthermore, individuals who will go on to develop psychosis often experience significant signs and symptoms from an early age in a clinical picture that can overlap substantially with other neurodevelopmental disorders. These signs and symptoms may include social, cognitive, psychomotor, linguistic, attentional, and emotional disturbances in addition to attenuated psychotic symptoms such as increased suspiciousness, perceptual abnormalities, and thought disorganization. This heterogeneous combination of signs and symptoms constitutes a potential prodrome to schizophrenia. These may be accompanied by an increasingly large number of objective measures found to be abnormal prior to the onset of full psychosis, including brain structure, function, and chemistry, as well as behavioral, computational, electrophysiological, and metabolic abnormalities.

Despite the evident heterogeneity, it has been recognized for over a century that a prodrome for psychosis exists.[1] In his extensive work on early psychosis in the mid-20th century, Harry Stack Sullivan noted that attenuated forms of the Schneiderian first-rank symptoms (i.e., hallucinations, delusions, thought disorder) were present months or even years before a patient's first admission for outright psychotic symptoms.[2] Soon thereafter, the term "prodrome" was introduced to describe this constellation of attenuated psychotic symptoms,[3] which were characterized more extensively by Hafner and colleagues in the mid-1990s.[4–7] The validity of the "prodromal" symptoms concept was subsequently supported by data on the duration, manifestations, and accompanying features of this period gathered retrospectively from individuals who have experienced their first episode of psychosis (see discussion later). While not all participants recalled a period of attenuated psychotic symptoms, between 80% and 90% did.[5]

COMPARATIVE NOSOLOGY

Over the past 2–3 decades, novel approaches to the evaluation and treatment of psychosis have shifted to a preventive strategy that attempts to identify prospectively those who are at high risk of developing psychosis based on their clinical presentation. These approaches have variably identified those who qualify as being in: (1) an ultra high risk (UHR) state of developing psychosis as ascertained by the Comprehensive Assessment of At-Risk Mental States (CAARMS)[8] or (2) a clinical high risk (CHR) syndrome for developing psychosis as ascertained by the Structured Interview for Psychosis-Risk Syndromes (SIPS).[9] Drawing from pools of help-seeking participants referred for the presence of symptoms concerning for the subsequent development of psychosis, studies of these constructs have been deemed to possess acceptable reliability and validity (see discussion later) despite the use of distinct diagnostic tools.

Because the SIPS approach adapted an early version of the CAARMS, the UHR and CHR constructs exhibit significant overlap. Both identify similar symptoms characteristic of the at-risk state and both require that the patient has never before experienced a fully psychotic episode. Both approaches designate three at-risk syndromes: (1) attenuated psychotic symptoms (APSS) syndrome, (2) brief limited intermittent psychotic symptoms (BLIPS), and (3) genetic vulnerability for psychosis plus a marked decline in psychosocial functioning (Genetic Risk and Deterioration Syndrome, or GRD). Because of the similarities we refer hereafter to both approaches as "CHR."

There are differences in the SIPS and CAARMS approaches as well, however. To qualify as being at CHR on the SIPS, patients must have had the relevant symptoms in the past month, and have worsened significantly in the year prior to presentation,[10] whereas a significant decline in psychosocial functioning is a requirement for the CAARMS designation.[11] Despite these differences (and some others—see [12]), both instruments appear to identify patient populations that transition to psychosis at similar rates after ascertainment.[13]

One area of ongoing discrepancy between measures involves not the beginning but the end of the prodromal period—the transition to psychosis. To identify the onset of psychosis, the CAARMS requires the occurrence of at least one fully (positive) psychotic symptom several times a week for over 1 week. By contrast, the SIPS criteria require at least one fully psychotic symptom multiple times per week for at least 1 month, or for as little as 1 day if this symptom is seriously disorganizing or dangerous. It should be noted that both the CAARMS and SIPS definitions of psychosis onset are weighted heavily toward positive symptoms; thus, a patient who presents with severe functional impairment and/or negative symptoms but with relatively mild positive symptoms may not reach criteria for conversion to frank psychosis.[14]

METHODS OF PROSPECTIVE SYMPTOM CAPTURE

Attenuated positive symptoms and other symptoms characteristic of prodromal states are currently collected via three methods: structured interviews, screening tools, and self-report surveys. Structured interviews take about 2 hours to complete and require a clinically experienced interviewer who has undergone certification training[15] to administer. Both SIPS and CAARMS now feature rating scales for the severity of CHR symptoms. The SIPS severity scale is called the Scale of Psychosis-Risk Symptoms, or the SOPS. It contains five positive symptom items, six items for negative symptoms, four for disorganization, and four for general symptoms and generates a total score (0 to 114) and subscale scores for each domain. Screening instruments[16,17] are administered by self-report or interviewer-assisted self-report and are generally used to assist decisions regarding which individuals to invite to a structured interview. Self-report surveys have been employed mostly in epidemiologic studies.

A number of studies have found high rates of psychotic-like symptoms (PLEs) in the general or general clinical populations,[18,19] as assessed by self-report survey. It is important to emphasize the methodologic differences between self-report surveys and structured interviews. Studies comparing self-report versus SIPS interview methods consistently find that rates of meeting SIPS criteria for CHR are much lower than rates of endorsing PLEs by self-report.[20–23] These findings are supported by large differences between CHR patients and healthy subjects on interview-based symptom ratings. Our unpublished review found that the SOPS total in CHR (median across 40 published samples) was 35.9 and 2.3 in healthy subjects (21samples). Positive symptom subscale median scores were 9.9 and 0.8, respectively.

SYMPTOMS AT CHR PRESENTATION

Patients who meet CHR criteria are help-seeking, currently symptomatic, functionally and cognitively impaired, and at high risk of developing psychosis.[10] The vast majority of patients meeting CHR criteria qualify for APSS syndrome.[24]

There is variability in the frequency with which the different attenuated positive symptoms are endorsed by CHR patients. Among the five SOPS positive symptoms, the most commonly endorsed items are P1 (unusual thought content), P2 (suspiciousness), and P4 (perceptual abnormalities). P5 (disorganized communication) and P3 (grandiosity) especially are endorsed much less often. This ranking pattern was generally consistent across studies.[25,26] While these symptoms are often present on initial CHR assessment, it is common that comorbid psychopathology such as anxiety, depression, and panic disorder have driven patients to seek treatment before presentation as CHR[27] (see Figure 13.1).

There is also some evidence that symptom profiles at presentation have changed over time since the delineation of the CHR syndrome, with temporally earlier cohorts presenting with a larger array of attenuated psychotic symptoms, higher disorganization scores, and more individuals with genetic risk

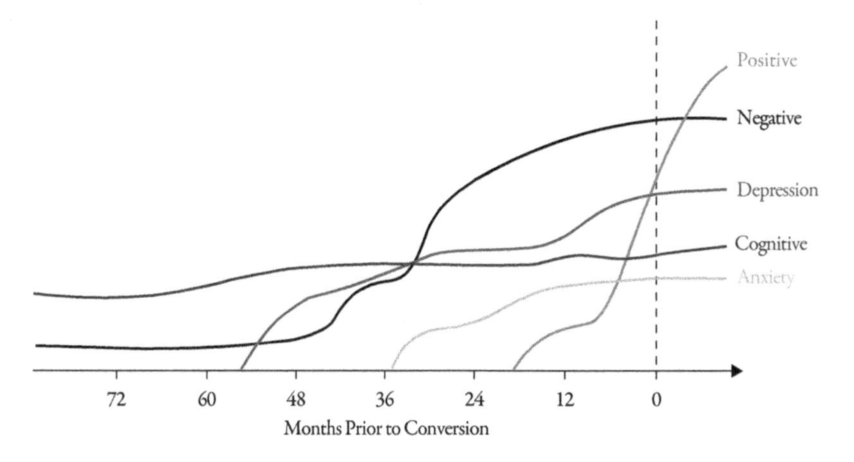

Figure 13.1 Course of symptoms during the prodromal period.
Adapted from Hafner et al.[28]

factors.[29] Additionally, recent work demonstrates that attenuated psychotic symptoms tend to improve over time after baseline assessment.[30] While most improvement tends to occur in the first 6 months after baseline assessment, continued improvement is evident as late as 24 months after baseline for most positive symptoms. These factors may contribute, along with potential treatment effects, length-of-time bias, and lead-time bias, to a well-documented decline in conversion rates in individuals with CHR over the past two decades.[29]

RESEARCH DIAGNOSIS OF CHR

GENERAL POPULATION STUDIES

In contrast to the high rates of endorsement of self-report PLEs in the general population, rates of CHR diagnosis based on structured interview in the general population are much lower. The two largest such epidemiologic studies employing SIPS interview methods recorded rates of 0.4% (from Switzerland, N = 1,229, ages 16–40)[31] and 1.6% (from China, N = 2,800, mean age 19).[32]

An age effect appears consistent in the interview-based epidemiologic studies, however. Three studies that focused on younger children and adolescents reported higher rates with the SIPS: 7.9% (from Switzerland, N = 76, ages 8–15),[33] 9.0% (from Ireland, N = 212, ages 11–13),[34] and 12.0% (from Israel, N = 2,800, ages 13–16).[35]

Rates in the population studies also appear to vary depending on whether the native criteria for CHR or modified criteria are used. Several studies reported that relaxation of the SIPS criteria requirements for minimum symptom frequency or onset/worsening yielded a higher proportion of subjects meeting criteria.[33,34,36,37]

The relatively low rates of CHR diagnosis in the general population, when using interview methods, do support the validity of the CHR diagnostic process in the research setting. However, the higher rates in children and younger adolescents suggest the need for additional research on the validity of CHR diagnoses in the younger ages and caution for now against application of the criteria in patients younger than 12.

CLINICAL EPIDEMIOLOGY STUDIES

Interestingly, in contrast to the generally low rates of CHR diagnosis in the general population, rates in clinical epidemiology studies of adolescents and young adults are substantially higher. Studies of consecutive inpatient admissions have reported rates ranging from 24%[38] to 38%[39] and, most studies of outpatient help-seekers, rates from 16%[40] to 36%.[41] One apparent outlier study from China recorded a rate at first mental health outpatient visit of 5.5%.[25]

These relatively high rates of CHR syndrome diagnosis suggest that CHR is a common and largely unrecognized clinical presentation in adolescent and young adult clinical settings. Future work should determine whether greater diagnostic focus on CHR as opposed to the traditional emphasis on the nonpsychotic conditions can improve outcomes. In addition, it has long been known that the presence of nonpsychotic disorder can predict future psychosis, albeit weakly.[42] Future studies should investigate whether patients whose nonpsychotic disorder is comorbid with CHR largely account for that risk.

DIAGNOSTIC RELIABILITY

DIAGNOSTIC RELIABILITY IN THE RESEARCH CONTEXT

Early reports using the SIPS and the CAARMS found high rates of interrater reliability on the CHR diagnosis in the research context.[11,15] Our group has subsequently located an additional 20 published reports of interrater reliability on the CHR diagnosis using the SIPS, as of June 2018. Reported diagnostic reliability was in the "excellent" range (kappa ≥ 0.80 or percent concordance ≥ 0.90) in 17 of 21 studies (81%). In the remaining four studies reliability was close to "excellent."

Recently, a critique of the CHR diagnostic work[43] has included a focus on the distinction between CHR and frank psychosis. While it is true that both CHR and the CHR structured interview definitions of psychosis are based in large part on the same attenuated positive symptom scales, it is not clear this is a fatal flaw. To rely on medical analogies, the risk syndromes of borderline hypertension and borderline diabetes are each based on the same scale as their frank diagnoses (blood pressure and blood glucose, respectively).

This criticism does, however, illuminate a relative weakness of CHR work to date: the reliability and validity of the CHR vs. psychosis distinction. Some of the reliability studies mentioned earlier included psychotic subjects in their "not CHR" groups, but we are aware of only three studies that have reported specifically on the reliability of the SIPS CHR vs. psychosis distinction. In the first report by the SIPS developers,[44] 2/11 non-CHR cases were rated as psychotic by one or more raters. Raw data from that study show perfect agreement on the psychosis judgment: both of two raters agreed on one case and three raters all agreed in the other. Another study[45] of 37 raters found a percent concordance of 0.93 and a kappa of 0.68. Presumably the wider-than-expected divergence between indices reflected an imbalance in numbers of CHR vs. psychotic subjects. The third study reported categorical agreement with the gold standard on whether SOPS positive symptoms were in the CHR vs. psychotic range.[46] Kappa was excellent at 0.90.

Even though the available data suggest that the CHR vs. psychosis judgment is reliable in the research context, considering the importance of the issue the database is sparse. Hopefully the van Os and Guloksuz critique will spur additional research on this point. Certainly CHR research groups have ample opportunity to do such work when screening previously undiagnosed psychosis out from CHR samples and when doing evaluations for possible conversion.

Studies on the validity of the structured interview distinction between CHR and frank psychosis are needed as well. It is not enough that no CHR subjects receive psychosis diagnoses at Structured Interview for DSM (SCID) or that all SIPS-defined psychosis subjects do,[26] because the SCID judgments have usually been made by the CHR evaluator and constrained by the CHR interview results. Moreover, although there are many studies showing that CHR patients are less ill than first episode psychosis patients, this group has generally been recruited by methods of convenience and not as a sample newly diagnosed as psychotic by the CHR structured interview.

We are aware of only six samples where psychotic patients defined by the SIPS were compared to CHR.[39,47–51] Three studies made comparison on the Global Assessment of Functioning, with all three showing that functioning was strongly and significantly lower in the SIPS-defined psychosis patients. On the symptom measures the findings were mixed, with some studies showing significantly higher symptoms in the SIPS-defined psychosis patients and others showing the groups did not differ significantly. Meta-analysis will be required to determine the synthesized conclusion, and again, more research is needed on this issue. Perhaps the ideal design would identify three groups (new psychosis discovered by CHR structured interview, CHR, neither CHR nor psychosis) and follow all groups over time on outcomes related to severity of illness and functioning.

DIAGNOSTIC RELIABILITY IN THE CLINICAL CONTEXT

While the evidence of diagnostic reliability of the CHR syndrome is strong in the research context, much less work has been done in the clinical context. DSM-5 taskforce included Attenuated Psychosis Syndrome (APS) in its field trial.[52–54] Results for APS, however, were inconclusive due to the small number of subjects recruited ($N = 17$). Kappa for diagnostic agreement was nevertheless in the middle of the range of the coded diagnoses assessed.[55] An additional study[56] provided 30 minutes of training in DSM-5 APS diagnosis to 11 community clinicians and evaluated 34 patients. Findings suggest that training sessions should be longer than 30 minutes.

DIAGNOSTIC COMORBIDITY

Early studies reported that rates of Axis I diagnoses comorbid with CHR at ascertainment are high.[57] Subsequently, high rates of comorbidity have been confirmed by meta-analysis,[58] with estimates of depressive disorder prevalence at 41% and anxiety disorder prevalence at 15%. The high rates of comorbidity were one of the reasons that the DSM-5 Psychotic Disorders Work Group gave for placing Attenuated Psychosis Syndrome in the appendix rather than as a uniquely coded diagnosis in the main text. Diagnostic comorbidity at baseline, however, is not unique to CHR, and comorbidity rates in fact are high in general when structured interviews are performed.[59] Epidemiology studies have reported lifetime comorbidity rates of 56%[60] and 60%[61] and 12-month rates of 45%[62] and 46%.[63] Comorbidity is also common in studies of schizophrenia, with rates from a systematic review[64] ranging from 6% to 57% for major depressive disorder and from 3% to 33% for panic disorder. A first episode psychosis study reported the rates of at least one SCID comorbid disorder as 37%.[65] In addition, the presence of baseline diagnostic comorbidity has not predicted emergence of psychosis.[66]

OTHER COMORBIDITIES—FUNCTIONAL AND COGNITIVE IMPAIRMENT

Neurocognitive impairments are often present prior to the onset of attenuated psychotic symptoms among those who are CHR.[67,68] These impairments are also often present in similar patterns in unaffected relatives of those with schizophrenia,[69] suggesting these deficits may serve as functional endophenotypes for psychotic illness.[70]

Additionally, strong evidence supports the idea that functional impairments presage illness onset in CHR. Difficulty maintaining social relationships and academic work have been consistently reported,[71,72] leading to recent emphasis on reliable measurement of social and role functioning in CHR[73,74] and on interventions focused on ameliorating these difficulties.[75]

Taken together, neurocognitive and functional declines have been shown to be significantly predictive of conversion to psychosis in CHR.[76] These symptoms have been most closely tied to impairments in quality of life and everyday functioning,[77] causing some to argue that, rather than conversion to frank psychosis, functional impairment should be the primary outcome measure in studies of CHR.[69]

COURSE OF CHR

Early work conceived of CHR as a *state* of risk, whereas more recent views tend to consider CHR, with not only its accompanying risk but also its associated symptoms, distress, and cognitive and functional impairment, as a syndrome.[78] The CHR syndrome does *not* always lead inevitably to psychosis; in fact psychosis is the outcome in only a minority of cases. In addition to conversion to psychosis, the CHR syndrome may follow a course characterized by remission or by continued subthreshold illness during the period of observation. For the SIPS, four current status outcomes beyond conversion have been specified: continued *progression*, *persistence* of stable symptoms, *partial remission*, and *full remission*.[79] The partial vs. full remission distinction is modeled after similar course specifiers for affective disorder instantiated into DSM.

DURATION OF CHR

Estimates of prodrome duration from retrospective studies have ranged from less than 1 to almost 8 years, largely due to differences in definitions of prodrome and psychosis onsets in these early studies.[80–82] For example, some (including Bleuler[83]) argued that only psychosis-like symptoms should be considered diagnostic of a prodrome, while others emphasized functional deterioration as a hallmark of the prodromal period.[84] Similarly, early definitions of psychosis onset varied widely, with some defining it as the start of disabling or "florid" psychotic symptoms,[6] and some favoring the onset of treatment as the hallmark of illness onset.[80,85] Subsequently, establishment of a standard set of criteria for prodome and psychosis onset[44,86,87] have allowed for a fuller characterization of prodromal duration, although this has primarily been undertaken as a way of standardizing dates of prodrome onset and conversion from retrospectively gathered data.[88] Large-scale longitudinal data gathering through consortia such as the North American Prodrome Longitudinal Study (NAPLS) has allowed for identification of prodrome and psychosis onset that is partly prospective and thus potentially more precise. This is because retrospectively gathered data may be influenced by "effort after meaning," when patients and families identify a single event that seems to have signified the major changes experienced over the course of an illness, when in fact this may not be an accurate reflection of events.[87,89] In data gathered as part of the second phase NAPLS, we found that the average duration of the prodromal period among those who convert to psychosis is about 22 months. Data approximated a "rule of thirds": one third of CHR patients who convert will do so by 1 year after CHR syndrome onset, another third between 1 and 2 years after onset, and the final third greater than 2 years after onset.[90]

CONVERSION FROM CHR

A meta-analysis concluded that 22% of CHR patients convert to psychosis within 1 year, 29% within 2 years, and 36% within 3 years.[91]

REMISSION FROM CHR

A similar meta-analysis reported that remission occurs in slightly under half of CHR patients who do not convert.[92]

DIAGNOSTIC PLURIPOTENTIALITY

One controversy in the CHR field has been whether clinical outcomes in CHR patients are specific to psychosis or whether they are diagnostically pluripotential, meaning that CHR "turns into" other, nonpsychotic, disorders.[93] The possibility of pluripotentiality has been supported by relatively high rates of affective and anxiety disorder diagnoses at follow-up.[94] Recently, however, studies including non-CHR comparison groups have found that neither emergence of nor persistence of nonpsychotic disorder is higher in CHR cases than in appropriate controls.[95–97] These findings have been supported by others using different methods.[98,99] It is possible that the original perception of high rates of follow-up nonpsychotic disorder comorbidity among CHR individuals is due to higher rates of baseline comorbid disorder than in non-CHR patients to begin with, as seen in one large study.[26] The relative frequency of baseline comorbid disorder in CHR vs. non-CHR groups appears to be another area ripe for meta-analysis.

PREDICTION OF COURSE

PREDICTIVE VALIDITY OF A CHR DIAGNOSIS FOR CONVERSION

The predictive validity of a CHR diagnosis for development of subsequent frank psychosis has been evaluated in a recent meta-analysis.[13] The meta-analysis located a total of 11 longitudinal studies comparing psychosis outcomes in CHR subjects with those from subjects who were referred for CHR evaluation but did not meet CHR criteria. These non-CHR patients appear to constitute the optimal ecological control group. Of the 11 studies, 5 used the SIPS,[26,36,40,48,51] 4 used the CAARMS, and 2 used other instruments. A similar but more recent analysis that focused exclusively on patients referred for CAARMS evaluation (total six studies) found that the risk of developing frank psychosis among those who meet CAARMS criteria was 5.3 times higher than among those who did not.[100]

We have subsequently located an additional four such studies employing the SIPS.[101–104] Meta-analysis of the nine studies found that the risk of developing frank psychosis among those who meet SIPS CHR criteria was 7.0 times higher than among those who did not, a highly significant effect (Figure 13.2).

PREDICTION OF OUTCOME WITHIN CHR SAMPLES

The NAPLS group first published an individualized risk calculator for psychosis among persons at CHR.[105] Factors that

	CHR		non-CHR			Risk Ratio
Study or Subgroup	Events	Total	Events	Total	Weight	M-H, Random, 95% CI
Addington 2012	14	111	2	71	14.1%	4.48 [1.05, 19.11]
Kline 2015	4	21	0	33	4.4%	13.91 [0.79, 245.83]
Kobayashi 2008	4	16	0	66	4.4%	35.47 [2.01, 627.38]
Lindgren 2014	2	51	1	97	6.2%	3.80 [0.35, 40.95]
Liy 2011	21	59	0	92	4.6%	66.65 [4.11, 1079.72]
Manninen 2014	1	7	3	45	7.6%	2.14 [0.26, 17.82]
Schultze-Lutter 2014	109	194	7	52	34.1%	4.17 [2.07, 8.41]
Simon 2012	10	73	0	49	4.5%	14.19 [0.85, 236.69]
Woods 2009	88	303	3	135	20.1%	13.07 [4.21, 40.57]
Total (95% CI)		**835**		**640**	**100.0%**	**6.98 [3.74, 13.02]**
Total events	253		16			

Heterogeneity: $Tau^2 = 0.17$; $Chi^2 = 9.94$, $df = 8$ ($P = 0.27$); $I^2 = 20\%$
Test for overall effect: $Z = 6.10$ ($P < 0.00001$)

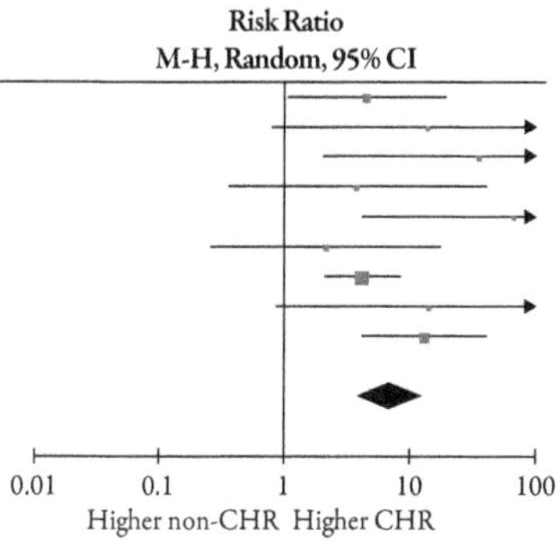

Risk Ratio
M-H, Random, 95% CI
0.01 0.1 1 10 100
Higher non-CHR Higher CHR

Figure 13.2 Predictive validity of the CHR diagnosis using the SIPS.
Random effects meta-analysis of studies using Structured Interview for Psychosis-Risk Syndromes (SIPS) for clinical high risk (CHR) diagnosis and reporting conversion rates in CHR and non-CHR groups. M-H, Mantel-Haenszel method of weighting studies. Studies are cited in bibliography.

contributed to prediction were symptom severity of SOPS unusual thought content and suspiciousness items, cognitive impairment, and functional decline. The calculator was initially tested in a new sample independent from those that generated the hypothesized predictors and has subsequently been independently replicated in two further samples.[106,107]

THE ROLE OF PERCEPTUAL ABNORMALITIES

Although subthreshold positive symptoms including unusual thought content, suspiciousness, and disorganized thinking[30] have been reliably shown to predict conversion to psychosis in CHR, perceptual abnormalities as a category have mixed evidence regarding predictive validity. Four studies have demonstrated that higher perceptual abnormality scores are associated with higher-than-average CHR risk, but 17 others failed to demonstrate significant prediction within CHR samples.[108]

Recently, two studies reported that perceptual abnormalities of different modalities may be differentially predictive of conversion,[109,110] although others have failed to replicate these findings.[108]

The poor predictive power of perceptual abnormalities within CHR samples has raised questions whether perceptual abnormalities should contribute to a CHR diagnosis.[111] The key analyses for the diagnostic criteria issue, however, should include evaluating predictive power of perceptual abnormalities for conversion within samples containing both CHR and non-CHR patients.

REFERENCES

1. Bleuler E. *Dementia Praecox or the Group of the Schizophrenias.* New York: International Universities Press; 1911.
2. Sullivan HS. The onset of schizophrenia: 1927. *Am J Psychiatry* 1994;151(6 Suppl):134–139.
3. Mayer-Gross W. Die Klinik. In: Beringer K, Bürger-Prinz H, Gruhle HW, et al., eds. *Spezieller Teil: Fünfter Teil die Schizophrenie.* Berlin, Heidelberg: Springer; 1932:293–578.
4. Hafner H, Riecher-Rossler A, Maurer K, Fatkenheuer B, Loffler W. First onset and early symptomatology of schizophrenia: a chapter of epidemiological and neurobiological research into age and sex differences. *Eur Arch Psychiatry Clin Neurosci* 1992;242(2–3):109–118.
5. Hafner H, Maurer K, Loffler W, et al. The epidemiology of early schizophrenia. Influence of age and gender on onset and early course. *Br J Psychiatry Suppl* 1994(23):29–38.
6. Häfner H, Maurer K, Löffler W, et al. Onset and early course of schizophrenia. *Search for the Causes of Schizophrenia*: Springer; 1995:43–66.
7. Hafner H, Maurer K, Loffler W, et al. The ABC Schizophrenia Study: a preliminary overview of the results. *Soc Psychiatry Psychiatr Epidemiol* 1998;33(8):380–386.
8. Yung A, Phillips L, McGorry PD. *Treating Schizophrenia in the Prodromal Phase.* London: Taylor & Francis; 2004.
9. McGlashan TH, Walsh BC, Woods SW. *The Psychosis-Risk Syndrome: Handbook for Diagnosis and Follow-Up.* New York: Oxford University Press; 2010.
10. Woods SW, Walsh BC, Saksa JR, McGlashan TH. The case for including Attenuated Psychotic Symptoms Syndrome in DSM-5 as a psychosis risk syndrome. *Schizophr Res* 2010;123(2–3):199–207.
11. Yung AR, Yuen HP, McGorry PD, et al. Mapping the onset of psychosis: the Comprehensive Assessment of At-Risk Mental States. *Aust N Z J Psychiatry* 2005;39(11–12):964–971.
12. Schultze-Lutter F, Schimmelmann BG, Ruhrmann S, Michel C. "A rose is a rose is a rose," but at-risk criteria differ. *Psychopathology* 2013;46(2):75–87.
13. Fusar-Poli P, Cappucciati M, Rutigliano G, et al. At risk or not at risk? Meta-analysis of the prognostic accuracy of psychometric interviews for psychosis prediction. *World Psychiatry* 2015;14:322–332.
14. Fusar-Poli P, Van Os J. Lost in transition: setting the psychosis threshold in prodromal research. *Acta Psychiatr Scand* 2013;127(3):248–252.
15. Miller TJ, McGlashan TH, Rosen JL, et al. Prodromal assessment with the structured interview for prodromal syndromes and the scale of prodromal symptoms: predictive validity, interrater reliability, and training to reliability. *Schizophr Bull* 2003;29(4):703–715.
16. Brodey BB, Addington J, First MB, et al. The Early Psychosis Screener (EPS): item development and qualitative validation. *Schizophr Res* 2018;197:504–518.
17. Kline E, Wilson C, Ereshefsky S, et al. Psychosis risk screening in youth: a validation study of three self-report measures of attenuated psychosis symptoms. *Schizophr Res* 2012;141(1):72–77.
18. Yung AR, Buckby JA, Cotton SM, et al. Psychotic-like experiences in nonpsychotic help-seekers: associations with distress, depression, and disability. *Schizophr Bull* 2006;32(2):352–359.
19. van Os J, Linscott RJ, Myin-Germeys I, Delespaul P, Krabbendam L. A systematic review and meta-analysis of the psychosis continuum: evidence for a psychosis proneness-persistence-impairment model of psychotic disorder. *Psychol Med* 2009;39(2):179–195.
20. Granö N, Karjalainen M, Edlund V, et al. Changes in depression, anxiety and hopelessness symptoms during family- and community-oriented intervention for help-seeking adolescents and adolescents at risk of psychosis. *Nordic J Psychiatry* 2014;68(2):93–99.
21. Schultze-Lutter F, Renner F, Paruch J, Julkowski D, Klosterkötter J, Ruhrmann S. Self-reported psychotic-like experiences are a poor estimate of clinician-rated attenuated and frank delusions and hallucinations. *Psychopathology* 2014;47(3):194–201.
22. Schultze-Lutter F, Klosterkötter J, Gaebel W, Schmidt SJ. Psychosis-risk criteria in the general population: frequent misinterpretations and current evidence. *World Psychiatry* 2018;17(1):107–108.
23. Fusar-Poli P, Raballo A, Parnas J. What is an attenuated psychotic symptom? on the importance of the context. *Schizophr Bull* 2017;43(4):687–692.
24. Fusar-Poli P, Howes OD, Allen P, et al. Abnormal prefrontal activation directly related to pre-synaptic striatal dopamine dysfunction in people at clinical high risk for psychosis. *Mol Psychiatry* 2011;16(1):67–75.
25. Zhang T, Li H, Woodberry KA, et al. Prodromal psychosis detection in a counseling center population in China: an epidemiological and clinical study. *Schizophr Res* 2014;152(2–3):391–399.
26. Woods SW, Addington J, Cadenhead KS, et al. Validity of the prodromal risk syndrome for first psychosis: findings from the North American prodrome longitudinal study. *Schizophr Bull* 2009;35(5): 894–908.
27. Hafner H, Maurer K, an der Heiden W. ABC Schizophrenia study: an overview of results since 1996. *Soc Psychiatry Psychiatr Epidemiol* 2013;48(7):1021–1031.
28. Hafner H, Maurer K, Trendler G, an der Heiden W, Schmidt M, Konnecke R. Schizophrenia and depression: challenging the paradigm of two separate diseases--a controlled study of schizophrenia, depression and healthy controls. *Schizophr Res* 2005;77(1):11–24.
29. Hartmann JA, Yuen HP, McGorry PD, et al. Declining transition rates to psychotic disorder in "ultra-high risk" clients: investigation of a dilution effect. *Schizophr Res* 2016;170(1):130–136.
30. Addington J, Liu L, Buchy L, et al. North American prodrome longitudinal study (NAPLS 2): the prodromal symptoms. *J Nerv Ment Dis* 2015;203(5):328–335.

31. Schultze-Lutter F, Michel C, Ruhrmann S, Schimmelmann BG. Prevalence and clinical significance of DSM-5-attenuated psychosis syndrome in adolescents and young adults in the general population: The Bern Epidemiological At-Risk (BEAR) Study. *Schizophr Bull* 2014;40(6):1499–1508.
32. Wang L, Shi J, Chen F, et al. Family perception and 6-month symptomatic and functioning outcomes in young adolescents at clinical high risk for psychosis in a general population in China. *PLoS ONE* 2015;10(9).
33. Schimmelmann BG, Michel C, Martz-Inrgartinger A, Linder C, Schultze-Lutter F. Age matters in the prevalence and clinical significance of ultra-high-risk symptoms and criteria in the general population: findings from the BEAR and BEARS-Kid studies. *World Psychiatry* 2015;14:189–197.
34. Kelleher I, Murtagh A, Molloy C, et al. Identification and characterization of prodromal risk syndromes in young adolescents in the community: a population-based clinical interview study. *Schizophr Bull* 2012;38(2):239–246.
35. Koren D, Lacoua L, Rothschild-Yakar L, Parnas J. Disturbances of the basic self and prodromal symptoms among young adolescents from the community: a pilot population-based study. *Schizophr Bull* 2016;42(5):1216–1224.
36. Schultze-Lutter F, Klosterkötter J, Ruhrmann S. Improving the clinical prediction of psychosis by combining ultra-high risk criteria and cognitive basic symptoms. *Schizophr Res* 2014;154(1–3):100–106.
37. Salokangas RKR, Heinimaa M, Ilonen T, et al. Vulnerability to and current risk of psychosis: Description, experiences and preliminary results of the detection of early psychosis or DEEP project. *Neurology Psychiatry Brain Res* 2004;11(1):37–44.
38. Gerstenberg M, Hauser M, Al-Jadiri A, et al. Frequency and correlates of DSM-5 attenuated psychosis syndrome in a sample of adolescent inpatients with nonpsychotic psychiatric disorders. *J Clin Psychiatry* 2015;76(11):e1449–e1458.
39. Raballo A, Monducci E, Ferrara M, Nastro PF, Dario C. Developmental vulnerability to psychosis: selective aggregation of basic self-disturbance in early onset schizophrenia. *Schizophr Res* 2018.
40. Kobayashi H, Nemoto T, Koshikawa H, et al. A self-reported instrument for prodromal symptoms of psychosis: testing the clinical validity of the PRIME Screen-Revised (PS-R) in a Japanese population. *Schizophr Res* 2008;106(2–3):356–362.
41. Comparelli A, Corigliano V, De Carolis A, et al. Anomalous self-experiences and their relationship with symptoms, neuro-cognition, and functioning in at-risk adolescents and young adults. *Comprehensive Psychiatry* 2016;65:44–49.
42. Weiser M, Reichenberg A, Rabinowitz J, et al. Association between nonpsychotic psychiatric diagnoses in adolescent males and subsequent onset of schizophrenia. *Arch Gen Psychiatry* 2001;58(10):959–964.
43. van Os J, Guloksuz S. A critique of the "ultra-high risk" and "transition" paradigm. *World Psychiatry* 2017;16(2):200–206.
44. Miller TJ, McGlashan TH, Rosen JL, et al. Prospective diagnosis of the initial prodrome for schizophrenia based on the Structured Interview for Prodromal Syndromes: preliminary evidence of interrater reliability and predictive validity. *Am J Psychiatry* 2002;159(5):863–865.
45. McFarlane WR, Levin B, Travis L, et al. Clinical and functional outcomes after 2 years in the early detection and intervention for the prevention of psychosis multisite effectiveness trial. *Schizophr Bull* 2015;41(1):30–43.
46. Addington J, Epstein I, Liu L, French P, Boydell KM, Zipursky RB. A randomized controlled trial of cognitive behavioral therapy for individuals at clinical high risk of psychosis. *Schizophr Res* 2011;125(1):54–61.
47. Koike S, Takano Y, Iwashiro N, et al. A multimodal approach to investigate biomarkers for psychosis in a clinical setting: The integrative neuroimaging studies in schizophrenia targeting for early intervention and prevention (IN-STEP) project. *Schizophr Res* 2013;143(1):116–124.
48. Simon AE, Grädel M, Cattapan-Ludewig K, et al. Cognitive functioning in at-risk mental states for psychosis and 2-year clinical outcome. *Schizophr Res* 2012;142(1–3):108–115.
49. Tso IF, Taylor SF, Grove TB, et al. Factor analysis of the Scale of Prodromal Symptoms: data from the Early Detection and Intervention for the Prevention of Psychosis Program. *Early Intervention Psychiatry* 2017;11(1):14–22.
50. Waltz JA, Demro C, Schiffman J, et al. Reinforcement learning performance and risk for psychosis in youth. *J Nervous Mental Dis* 2015;203(12):919–926.
51. Liu C-C, Lai M-C, Liu C-M, et al. Follow-up of subjects with suspected pre-psychotic state in Taiwan. *Schizophr Res* 2011;126(1–3):65–70.
52. Clarke DE, Narrow WE, Regier DA, et al. DSM-5 field trials in the United States and Canada, part I: study design, sampling strategy, implementation, and analytic approaches. *Am J Psychiatry* 2013;170(1):43–58.
53. Narrow WE, Clarke DE, Kuramoto SJ, et al. DSM-5 field trials in the United States and Canada, part III: development and reliability testing of a cross-cutting symptom assessment for DSM-5. *Am J Psychiatry* 2013;170(1):71–82.
54. Regier DA, Narrow WE, Clarke DE, et al. DSM-5 field trials in the United States and Canada, part II: test-retest reliability of selected categorical diagnoses. *Am J Psychiatry* 2013;170(1):59–70.
55. Fusar-Poli P, Carpenter W, Woods S, McGlashan T. Attenuated psychosis syndrome: ready for DSM-5.1? *Annual Rev Clin Psychology* 2014;10:155–192.
56. Woods S, Walsh B, McGlashan T. Diagnostic reliability and validity of the proposed DSM-5 attenuated psychosis syndrome (abstract). *Early Intervention Psychiatry* 2012;6(Suppl. 1):33.
57. Lencz T, Smith CW, Auther A, Correll CU, Cornblatt B. Nonspecific and attenuated negative symptoms in patients at clinical high-risk for schizophrenia. *Schizophr Res* 2004;68(1):37–48.
58. Fusar-Poli P, Nelson B, Valmaggia L, Yung AR, McGuire PK. Comorbid depressive and anxiety disorders in 509 individuals with an at-risk mental state: impact on psychopathology and transition to psychosis. *Schizophr Bull* 2014;40(1):120–131.
59. Rettew DC, Lynch AD, Achenbach TM, Dumenci L, Ivanova MY. Meta-analyses of agreement between diagnoses made from clinical evaluations and standardized diagnostic interviews. *International J Meth Psychiatric Res* 2009;18(3):169–184.
60. Kessler RC, McGonagle KA, Zhao S, et al. Lifetime and 12-month prevalence of DSM-III-R psychiatric disorders in the United States: results from the national comorbidity survey. *Arch Gen Psychiatry* 1994;51(1):8–19.
61. Robins LN, Locke BZ, Regier DA. An overview of psychiatric disorders in America. In: Robins LN, Regier DA, eds. *Psychiatric Disorders in America: The Epidemiologic Catchment Area Study.* New York: Maxwell Macmillan International; 1991.
62. Kessler RC, Wai TC, Demler O, Walters EE. Prevalence, severity, and comorbidity of 12-month DSM-IV disorders in the national comorbidity survey replication. *Arch Gen Psychiatry* 2005;62(6):617–627.
63. Kessler RC, Avenevoli S, Costello EJ, et al. Prevalence, persistence, and sociodemographic correlates of DSM-IV disorders in the national comorbidity survey replication adolescent supplement. *Arch Gen Psychiatry* 2012;69(4):372–380.
64. Buckley PF, Miller BJ, Lehrer DS, Castle DJ. Psychiatric comorbidities and schizophrenia. *Schizophr Bull* 2009;35(2):383–402.
65. Sim K, Swapna V, Mythily S, et al. Psychiatric comorbidity in first episode psychosis: the early psychosis intervention program (EPIP) experience. *Acta Psychiatr Scand* 2004;109(1):23–29.
66. Albert U, Tomassi S, Maina G, Tosato S. Prevalence of non-psychotic disorders in ultra-high risk individuals and transition to psychosis: A systematic review. *Psychiatry Res* 2018;270:1–12.
67. Lencz T, Smith CW, McLaughlin D, et al. Generalized and specific neurocognitive deficits in prodromal schizophrenia. *Biol Psychiatry* 2006;59(9):863–871.
68. Cannon TD, Bearden CE, Hollister JM, Rosso IM, Sanchez LE, Hadley T. Childhood cognitive functioning in schizophrenia

patients and their unaffected siblings: a prospective cohort study. *Schizophr Bull* 2000;26(2):379–393.
69. Sitskoorn MM, Aleman A, Ebisch SJ, Appels MC, Kahn RS. Cognitive deficits in relatives of patients with schizophrenia: a meta-analysis. *Schizophr Res* 2004;71(2–3):285–295.
70. Hill SK, Reilly JL, Keefe RS, et al. Neuropsychological impairments in schizophrenia and psychotic bipolar disorder: findings from the Bipolar-Schizophrenia Network on Intermediate Phenotypes (B-SNIP) study. *Am J Psychiatry* 2013;170(11):1275–1284.
71. Carrion RE, McLaughlin D, Goldberg TE, Olvet D, Auther A, Cornblatt BA. Prediction of functional outcome in individuals at clinical high-risk for psychosis. *Schizophr Res* 2012;136: S138–S139.
72. Addington J, Penn D, Woods SW, Addington D, Perkins DO. Social functioning in individuals at clinical high risk for psychosis. *Schizophr Res* 2008;99(1–3):119–124.
73. Cornblatt BA, Auther AM, Niendam T, et al. Preliminary findings for two new measures of social and role functioning in the prodromal phase of schizophrenia. *Schizophr Bull* 2007;33(3):688–702.
74. Cornblatt BA, Carrion RE, Addington J, et al. Risk factors for psychosis: impaired social and role functioning. *Schizophr Bull* 2012;38(6):1247–1257.
75. Niendam TA, Jalbrzikowski M, Bearden CE. Exploring predictors of outcome in the psychosis prodrome: implications for early identification and intervention. *Neuropsychol Rev* 2009;19(3):280–293.
76. Carrion RE, Demmin D, Auther AM, et al. Duration of attenuated positive and negative symptoms in individuals at clinical high risk: associations with risk of conversion to psychosis and functional outcome. *J Psychiatr Res* 2016;81:95–101.
77. Green MF, Kern RS, Braff DL, Mintz J. Neurocognitive deficits and functional outcome in schizophrenia: are we measuring the "right stuff"? *Schizophr Bull* 2000;26(1):119–136.
78. Fusar-Poli P, Rocchetti M, Sardella A, et al. Disorder, not just state of risk: meta-analysis of functioning and quality of life in people at high risk of psychosis. *Br J Psychiatry* 2015;207(3):198–206.
79. Woods SW, Walsh BC, Addington J, et al. Current status specifiers for patients at clinical high risk for psychosis. *Schizophr Res* 2014;158:69–75.
80. Mcglashan TH. The Chestnut-Lodge follow-up-study, 2: long-term outcome of schizophrenia and the affective-disorders. *Arch Gen Psychiatry* 1984;41(6):586–601.
81. Huber G, Gross G, Schuttler R, Linz M. Longitudinal studies of schizophrenic patients. *Schizophr Bull* 1980;6(4):592–605.
82. Yung AR, McGorry PD. The initial prodrome in psychosis: descriptive and qualitative aspects. *Aust N Z J Psychiatry* 1996;30(5):587–599.
83. Bleuler M. *The Schizophrenic Disorders: Long-Term Patient and Family Studies*. New Haven, CT: Yale University Press; 1978.
84. Wing JK. Comments on the long-term outcome of schizophrenia. *Schizophr Bull* 1988;14(4):669–673.
85. Gift TE, Strauss JS, Harder DW, Kokes RF, Ritzler BA. Established chronicity of psychotic symptoms in first-admission schizophrenic patients. *Am J Psychiatry* 1981;138(6):779–784.
86. Woods SW, Miller TJ, McGlashan TH. The "prodromal" patient: both symptomatic and at-risk. *CNS Spectr* 2001;6(3):223–232.
87. Yung AR, McGorry PD. The prodromal phase of first-episode psychosis: past and current conceptualizations. *Schizophr Bull* 1996;22(2):353–370.
88. Klosterkotter J, Hellmich M, Steinmeyer EM, Schultze-Lutter F. Diagnosing schizophrenia in the initial prodromal phase. *Arch Gen Psychiatry* 2001;58(2):158–164.
89. Hirsch S, Cramer P, Bowen J. The triggering hypothesis of the role of life events in schizophrenia. *Br J Psychiatry Suppl* 1992(18): 84–87.
90. Powers AR III, Mathalon DH, Addington J, et al. Duration of the psychosis prodrome. Submitted.
91. Fusar-Poli P, Bonoldi I, Yung AR, et al. Predicting psychosis: meta-analysis of transition outcomes in individuals at high clinical risk. *Arch Gen Psychiatry* 2012;69(3):220–229.
92. Simon AE, Borgwardt S, Riecher-Rössler A, Velthorst E, de Haan L, Fusar-Poli P. Moving beyond transition outcomes: meta-analysis of remission rates in individuals at high clinical risk for psychosis. *Psychiatry Res* 2013;209(3):266–272.
93. McGorry P, Van Os J. Redeeming diagnosis in psychiatry: timing versus specificity. *Lancet* 2013;381(9863):343–345.
94. Lin A, Wood SJ, Nelson B, Beavan A, McGorry P, Yung AR. Outcomes of nontransitioned cases in a sample at ultra-high risk for psychosis. *Am J Psychiatry* 2015;172(3):249–258.
95. Webb JR, Addington J, Perkins DO, et al. Specificity of incident diagnostic outcomes in patients at clinical high risk for psychosis. *Schizophr Bull* 2015;41:1066–1075.
96. Woods SW, Powers AR III, Taylor JH, et al. Lack of diagnostic pluripotentiality in patients at clinical high risk for psychosis: specificity of comorbidity persistence and search for pluripotential subgroups. *Schizophr Bull* 2018;44(2):254–263.
97. Fusar-Poli P, Rutigliano G, Stahl D, et al. Long-term validity of the At Risk Mental State (ARMS) for predicting psychotic and non-psychotic mental disorders. *European Psychiatry* 2017; 42:49–54.
98. Schultze-Lutter F, Schimmelmann BG, Klosterkötter J, Ruhrmann S. Comparing the prodrome of schizophrenia-spectrum psychoses and affective disorders with and without psychotic features. *Schizophr Res* 2012;138(2–3):218–222.
99. Lee TY, Lee J, Kim M, Choe E, Kwon JS. Can we predict psychosis outside the clinical high-risk state? a systematic review of non-psychotic risk syndromes for mental disorders. *Schizophr Bull* 2018;44(2):276–285.
100. Oliver D, Kotlicka-Antczak M, Minichino A, Spada G, McGuire P, Fusar-Poli P. Meta-analytical prognostic accuracy of the comprehensive assessment of at risk mental states (CAARMS): the need for refined prediction. *European Psychiatry* 2018;49: 62–68.
101. Kline E, Thompson E, Demro C, Bussell K, Reeves G, Schiffman J. Longitudinal validation of psychosis risk screening tools. *Schizophr Res* 2015;165(2–3):116–122.
102. Addington J, Piskulic D, Perkins D, Woods SW, Liu L, Penn DL. Affect recognition in people at clinical high risk of psychosis. *Schizophr Res* 2012;140(1–3):87–92.
103. Manninen M, Lindgren M, Therman S, et al. Clinical high-risk state does not predict later psychosis in a delinquent adolescent population. *Early Intervention Psychiatry* 2014;8(1):87–90.
104. Lindgren M, Manninen M, Kalska H, et al. Predicting psychosis in a general adolescent psychiatric sample. *Schizophr Res* 2014;158(1–3):1–6.
105. Cannon TD, Yu C, Addington J, et al. An individualized risk calculator for research in prodromal psychosis. *Am J Psychiatry* 2016;173(10):980–988.
106. Zhang T, Li HJ, Tang Y, et al. Validating the predictive accuracy of the NAPLS-2 psychosis risk calculator in a clinical high-risk sample from the SHARP (Shanghai at risk for psychosis) program. *Am J Psychiatry* 2018;175(9):906–908.
107. Carrión RE, Cornblatt BA, Burton CZ, et al. Personalized prediction of psychosis: external validation of the NAPLS-2 psychosis risk calculator with the EDIPPP project. *Am J Psychiatry* 2016;173(10):989–996.
108. Niles HF, Walsh BC, Woods SW, Powers AR III. Does hallucination perceptual modality impact psychosis risk? Submitted.
109. Ciarleglio AJ, Brucato G, Masucci MD, et al. A predictive model for conversion to psychosis in clinical high-risk patients. *Psychol Med* 2018:1–10.
110. Lehembre-Shiah E, Leong W, Brucato G, et al. Distinct relationships between visual and auditory perceptual abnormalities and conversion to psychosis in a clinical high-risk population. *Jama Psychiatry* 2017;74(1):104–106.
111. Perkins DO, Jeffries CD, Cornblatt BA, et al. Severity of thought disorder predicts psychosis in persons at clinical high-risk. *Schizophr Res* 2015;169(1–3):169–177.

14

PREDICTORS OF CONVERSION TO PSYCHOSIS

Rachael G. Grazioplene and Tyrone D. Cannon

The existence of a well-characterized prodromal period for psychosis,[1] paired with evidence that early intervention promotes better long-term outcomes,[2] has given rise to a body of literature that aims to identify specific, quantifiable risk factors that prospectively predict which individuals will experience a clinically defined psychosis. Broadly, research in this area is focused on the development and refinement of algorithms to predict psychosis, isolation of mechanisms of onset of psychosis, and the development and testing of interventions that may prevent onset of full psychosis and associated functional deterioration. The availability of well-performing algorithms for prediction of psychosis would enable their application in prevention study contexts, maximizing the number of cases for whom intervention is most critical and minimizing the number of "false positive" cases who will not have converted to full psychosis even without intervention. Isolation of potential mechanisms that promote worsening from an at-risk to a fully psychotic clinical state is critical for the development of novel interventions to prevent onset of psychosis and associated functional disability. Studies of the psychosis prodrome are thus highly focused on pragmatic aims with great public health significance.

There are two major branches of longitudinal research that characterize research in this area: the first branch encompasses studies that recruit from help-seeking samples based on clinical high risk for psychosis (CHR-P) criteria; the second branch measures theoretically risk-relevant traits in general population or cohort samples to determine whether traits measured/reported in non-help-seeking people predict later psychosis or psychosis-related functional impairment.

Compared to population-level sampling, the CHR-P approach benefits from enriched likelihood of psychotic outcomes, often involves more detailed clinical assessments, and typically collects more densely sampled time points. However, a major pitfall of this approach is that the individuals in these studies are generally already ill enough to have sought treatment (for emerging psychotic impairment and/or comorbid psychopathology), making the CHR-P paradigm more relevant to secondary prevention (with the goal of lessening the impact of existing illness) than to primary prevention (with the goal of preventing illness before it occurs) goals. Population-based sampling approaches ameliorate the major biases that are likely to be introduced via the use of help-seeking samples, making them "risk diluted" by nature, and introducing practical considerations that limit the feasibility of in depth assessments and frequent follow-up. These two approaches (represented schematically in Figure 14.1) thus have substantial trade-offs in terms of their respective limitations and the nature of the insights that can be gained.

In this chapter, we review findings from clinical high risk and general population-based sampling research related to prediction of psychosis, discussing what has been learned from these two approaches as well as theoretical and practical pitfalls of each. We conclude with a discussion of how to reconcile different approaches to research in psychosis prediction and incorporate them under a transdiagnostic framework.

CLINICAL HIGH RISK STUDIES

Samples ascertained based on the presence of prodromal features may be referred to as clinical high risk (CHR; also called clinical high risk-psychosis, or CHR-P), ultra high risk (UHR), and at risk mental state (ARMS). Although these terms emerged from different research programs and are associated with subtle differences in diagnostic criteria, the different labels all pertain to the goal of ascertaining a group of individuals enriched for risk for imminent onset of psychosis. The CHR samples are usually ascertained from help-seeking populations either via recruitment or referral. And CHR status is determined using instruments designed to assess genetic risk (i.e., by family history), measure prodromal symptoms, and operationalize clinical cutoffs for symptom severity, duration, and insight as well as global functioning. Commonly implemented diagnostic tools include the Structured Interview for Psychosis-Risk Syndrome (SIPS) and its associated Scale of Prodromal Symptoms (SOPS)[3] and the Comprehensive Assessment of At Risk Mental States (CAARMS).[4] Both of these instruments specify criteria for several risk syndromes: attenuated positive symptoms (APS; also called attenuated psychotic syndrome or attenuated positive symptom syndrome; APSS), brief limited intermittent psychotic symptoms (BLIPS), and/or genetic risk and deterioration (GRD). Participants meet criteria for APS if they exhibit at least one attenuated positive symptom (unusual thought content, suspiciousness, perceptual abnormalities, disorganized cognition, or grandiosity), but do not meet criteria for a fully psychotic form of mental illness.[5] In the SIPS system, these symptoms must have begun or deteriorated within the 12 months prior to ascertainment

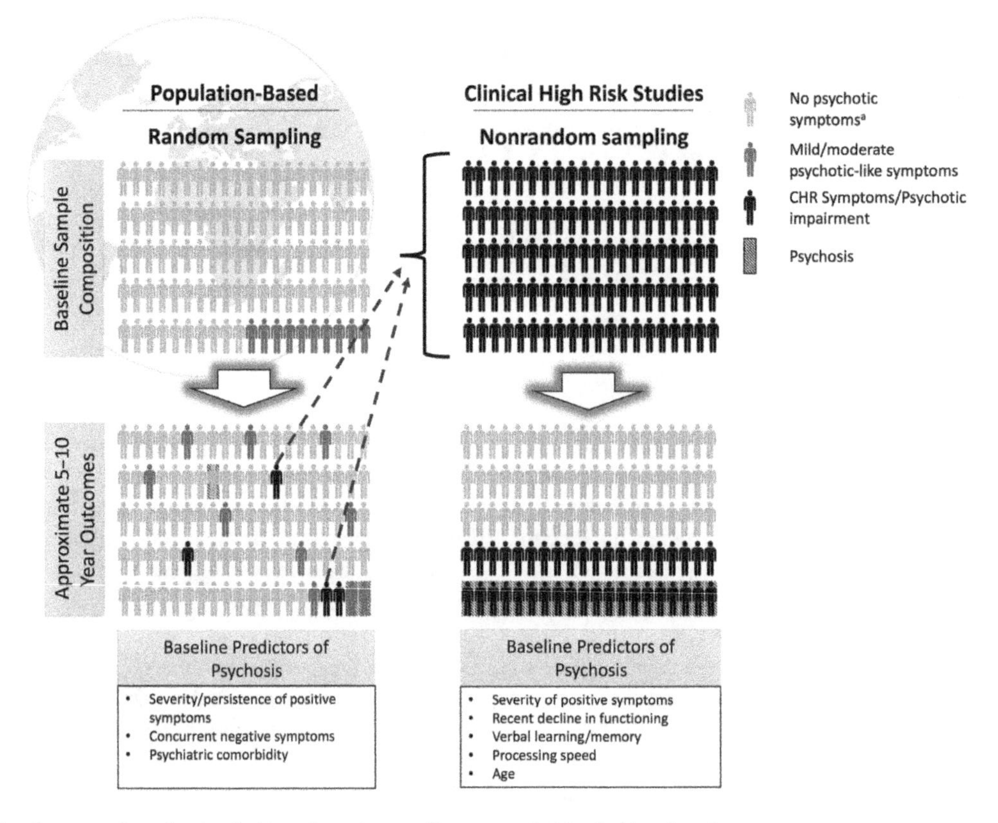

Figure 14.1 Proposed pathways and mechanism linking obstetric complications and risk of schizophrenia.

for an individual to qualify for prodromal status, while in the CAARMS system, more long-standing symptoms may be present. Attenuated positive symptoms accounts for the vast majority of cases in CHR research (85%).[6] And GRD, also called "trait and state risk factors," is defined as a recent decline in functioning paired with either genetic liability or schizotypal personality disorder.[7] Typically GRD makes up a small portion of these samples (5%) and is the only subgroup that may be categorized as high risk even if positive symptoms are not evident (though in most cases, GRD and APS co-occur). Those meeting BLIPS criteria constitute approximately 10% of high risk samples, and their baseline presentation is characterized by a history of clinically severe psychotic symptoms that do not meet diagnostic criteria for psychosis in terms of duration and/or frequency. Like GRD, BLIPS typically occurs against a backdrop of APS features, though it represents the most severe expression of prodromal pathology. Across samples ascertained using CHR criteria, between 15% and 35% of participants progress to meeting criteria for clinical psychosis within 2 years from study recruitment. Since 35% represents a dramatic (roughly 400x) increase compared to risk at the population level,[8,9] meeting CHR criteria clearly indexes a heightened risk of conversion in its own right.

Another approach to clinical risk ascertainment, used on its own or in combination with the SIPS and/or CAARMS, focuses on "basic symptoms," which include subjective perceptual and cognitive disturbances, impairment in social functioning, and some nonspecific symptoms not explicitly related to psychosis risk (e.g., somatic complaints, affective impairments, stress tolerance). Instruments for measuring basic symptoms include the Bonn Scale for the Assessment of Basic Symptoms (BSABS) and the Schizophrenia Proneness Instrument (SPI). Conversion risks for studies in which high risk status is determined by basic symptoms criteria are mixed, often much lower than the 15%–35% range associated with CHR criteria, but sometimes approaching that range, although over a much longer time interval.[10] One study that examined basic symptoms in 160 patients referred based on suspicion of psychosis risk (but not based on standard prodromal criteria) showed that only 4% of those who lacked basic symptoms at baseline transitioned to psychosis within 9.6 years (compared to a 70% transition rate in patients with at least one cognitive or perceptual basic symptom present at baseline).[11]

MAXIMIZING PREDICTION OF PSYCHOSIS IN CHR SAMPLES

Regardless of the criteria used for ascertaining clinical risk, it is important to keep in mind that most CHR participants do not go on to develop psychosis. Among the 65% to 85% who do not, roughly half remit the symptoms that indexed their heightened risk and the remainder continue to manifest stable levels of low-grade psychotic-like symptoms. Given this heterogeneity, factors other than CHR status must be used to increase the accuracy of risk prediction. The relevant factors could be related to the severity of attenuated symptoms (i.e., higher severity implying closer proximity to the psychosis threshold) as well as other psychopathological and/or environmental factors that could influence whether an at-risk individual will convert to psychosis. Clinical high risk studies typically collect several measures that could be used to improve prediction, including (but not limited to) social and/or cognitive functioning, clinical symptoms (e.g., Brief Psychiatric Rating Scale), global functioning, and demographic assessments.[5]

Several factors and traits have emerged as consistent predictors of psychosis within high risk samples. Across 5 CHR samples with $N > 100$ between 2001 and 2012, the presence and/or severity of virtually all prodromal symptoms emerged as predictors of imminent psychosis, revealing heterogeneous multivariate prediction models across studies.[8,9,11–13] The most consistent risk factors were severity of unusual thought content, suspiciousness/paranoia, social/functional impairment, cognitive disorganization, and negative symptoms (anhedonia/asociality). Other baseline predictors included duration of prodromal symptoms prior to recruitment, verbal impairment, family history of illness, poor global functioning, cognitive impairment, and drug use. More recently reported studies replicate the baseline risk prediction value of disorganized communication, poor social functioning, and verbal memory.[14,15] The baseline predictive utility of general cognitive ability and domain-specific neurocognitive deficits remains controversial. There is some consensus that, compared to nonconverters, converters are impaired on domain-specific neurocognitive functioning, demonstrating deficits in verbal learning/memory, working memory, and attention.[15] There is also evidence that these group differences are not accounted for by general cognitive ability, suggesting that general intellectual functioning deficits lack predictive validity for psychosis compared to domain-specific assessments.[16]

Some studies have examined whether combining CHR and basic symptoms criteria result in higher transition rates compared with either approach in isolation. In a relatively large sample ($N = 246$) and using a recruitment design that included SIPS determination of CHR as well as a partial assessment of basic symptoms (subjective cognitive disturbances only, based on nine items from the BSABS), Schultze-Lutter et al. (2014)[10] reported that those with both CHR status and basic symptoms at baseline were more than twice as likely to convert to psychosis within 2 years (66%) compared to those who met criteria for CHR only (28% converted) or basic symptoms only (23% converted). In contrast, another study that employed the SIPS and measured basic symptoms ($N = 183$) reported that the inclusion of a basic symptom measure (subjective cognitive disturbances) did not improve risk prediction: the transition rate in participants who met criteria for CHR and reported subjective cognitive disturbances was 21%, which was roughly equal to the transition rates in those who met criteria for CHR alone (20.6%) but still higher those who met basic symptoms criteria alone (4.5%).[9] More research is needed to determine the utility of basic symptoms for risk prediction.

Another approach to disentangling the heterogeneity in risk profiles is through a more fine-grained stratification of CHR samples into subgroups based on symptom configuration, temporal course, and/or severity, for example using latent class profiles[17] or clinical staging models.[18] Healey et al.[17] showed that a CHR subgroup with the highest negative symptoms and the lowest functioning scores (cognitive and social) were twice as likely to convert compared with a more moderate expression profile, while a subgroup with intact cognitive performance had the lowest psychosis risk. Carrión et al.[18] found that CHR individuals who maintained a negative symptom profile (without positive symptoms) throughout follow-up had the lowest conversion rate; among those who had positive symptoms, those with the highest positive symptoms at baseline transitioned most rapidly to psychosis, suggesting that these groups were further along in the prodromal course. Though staging models may be of clinical utility, it is unclear whether creating subcategories based on continuous measures provides the best approach to understanding multivariate risk in the prodrome.

The prediction models summarized earlier refer to group-averaged results. To be maximally clinically useful, prediction algorithms should be able to scale risk in an individual patient during their initial clinical contact. Such individualized risk calculation is possible when a large dataset on a reference population is available from which risks can be calculated based on one or more predictor variables. Well-performing risk calculators have been developed in numerous somatic disease contexts, including cardiovascular disease and cancer,[19–21] where they provide a rationale for clinicians to pursue more or less invasive intervention strategies, based on the level of risk implied by an individual's profile across a set of risk factors. They also inform patients and their family members to help make complex treatment decisions.

Using time-to-event regression in a sample of 596 CHR participants (84 converters; NAPLS-2 sample), Cannon et al.[22] created an individualized risk calculator that can be used to estimate risk of conversion in individual patients ascertained using SIPS/SOPS criteria. Predictor variables were selected a priori based on the most consistent predictors of conversion reported in non-NAPLS CHR publications. Of the eight selected predictor variables, severity of unusual thought content and suspiciousness (summed scores), recent social functioning decline, verbal learning/memory (summed scores), processing speed, and age distinguished converters from nonconverters, while stressful life events, family history of psychosis, and traumas did not predict conversion.[22] The overall model achieved a C-index of 0.71, which is in the range of those of established calculators currently in use for cardiovascular disease and cancer recurrence risk, with C-indices of 0.58 to 0.81.[19,21] The risk algorithm was able to generate 2-year conversion probability for each subject based on individual symptom profiles and demonstrated comparable sensitivity and specificity in external validation.[23] In both the NAPLS-2 and the replication dataset, severity of baseline positive symptoms was the strongest predictor of conversion.

The most immediate uses of the risk calculator are likely to be in the selection of individual subjects for participation in clinical (prevention) trials, given the desire to avoid exposing cases with lower transition risks to the potential adverse consequences of any interventions and given the potential to evaluate whether interventions differ in effectiveness based on initial risk levels and/or profiles across predictors. In terms of clinical practice outside the context of a prevention trial, at this point the most likely use is for the clinician to be able to communicate to the patient and family a scaling of risk that could help to recruit their cooperation with a monitoring and/or intervention plan.

MECHANISMS OF ONSET IN CHR SAMPLES

Many CHR studies also include assessments of biological factors that may be associated with risk for and progression to full psychosis, including structural, functional, metabolic, and neurochemical brain imaging and blood and/or saliva samples for genomics, proteomics, and hormone assays. As these domains of assessment are the focus of other chapters in this volume, findings related to these measures are mentioned only briefly here. Current models of psychosis emphasize disruptions in integrated synaptic activity and neuronal connectivity.[24,25] Given that the causes and consequences of psychosis are generally confounded in studies of diagnosed patients and that some of the contributing factors are likely to change in the transition to psychosis,[26] studies tracking potential biomarkers using a prospective, longitudinal design can be very informative.

Several such prospective longitudinal neuroimaging studies have now been published, encompassing multiple independent samples of CHR cases.[27–30] All of these studies found that CHR cases who converted to psychosis showed a steeper rate of cortical gray matter loss compared with nonconverters. Among the studies that also included a healthy comparison group, the rate of gray matter decline was also greater in converters compared with age- and gender-matched controls.[27,28,30] A recent study by the NAPLS consortium, with the largest CHR group reported to date ($N = 274$, including 35 converters), showed that converters experienced a steeper rate of gray matter loss in right superior frontal, middle frontal, and medial orbitofrontal cortex, even when a stringent multiple correction method was applied at the whole-brain level.[31]

While there is evidence suggesting that antipsychotic drug use may contribute to cortical thinning,[32,33] antipsychotics are not solely responsible for the progressive gray matter loss observed in psychosis. Patterns of cortical thinning similar to those reported in medicated patients have been reported in drug-naïve converters[31] and unmedicated first-episode schizophrenia patients,[34] indicating that at least some amount of gray matter reduction in schizophrenia is independent of antipsychotic drug exposure. In addition, longitudinal imaging studies of twins discordant for schizophrenia and other genetically informative samples have observed that a steeper rate of gray matter decline is associated with a genetic diathesis to schizophrenia.[35,36] Notably, in the NAPLS longitudinal MRI study, prefrontal gray matter decline was predicted by baseline levels of pro-inflammatory cytokines in plasma.[31] Inflammatory markers are elevated in postmortem neural tissue from patients with schizophrenia,[37–40] and these same markers are associated with microglial-mediated synaptic pruning and dendritic retraction in animal models,[41,42] thus providing a potential mechanism for the reduced neuropil seen in patients.[24,43,44] Gyrification abnormalities may also help to explain apparent cortical thinning effects in the prodromal period, as indicated by a recent report that used network-based gyrification indices to classify converters and nonconverters.[45] More research is needed to establish whether the risk relevance of abnormal cortical gyrification overlaps with that of cortical thinning. Finally, although gray matter decline appears likely to be a key feature of psychosis risk, recent evidence suggests that assessment of deviation from age-typical maturational trajectories may be of limited clinical use: in the NAPLS longitudinal study, the predictive value of the "brain age gap" in cortical thinning was highly collinear with age at ascertainment, and thus did not improve performance of the NAPLS-based risk calculator.[46]

Though accelerated decline in cortical gray matter among UHR cases who convert to psychosis may be one of the leading indicators of processes underlying the development of psychosis, theoretically it seems likely that changes in synaptic signaling and functional connectivity are the more proximal mechanisms underlying symptom expression. A recent rs-fMRI study of CHR cases in the NAPLS consortium observed a pattern highly consistent with effects reported in chronic psychosis.[47] Specifically, CHR converters to psychosis showed increased thalamic connectivity with sensorimotor cortices, but reduced thalamic connectivity with PFC, anterior cingulate cortex, and cerebellum.[48] In the same sample, alterations in cross-paradigm connectivity in cerebello-thalamo-cortical networks were linked to conversion risk and were correlated with disorganized symptoms.[49] However, a critical gap in knowledge remains; namely, whether disrupted functional connectivity progresses prior to onset of psychosis and in association with accelerated gray matter decline. A recent longitudinal analysis of NAPLS-2 consortium data suggests that conversion is preceded by progressive alterations in network-level functional properties at rest (network efficiency, network diversity), and some of these network-level changes are correlated with cortical thinning.[50] Although promising, this study was considered preliminary due to the small number of converters ($N = 8$). Exploring progressive functional-structural risk signatures comprehensively requires multiple assessments prior to onset, enabling application of time-lagged and growth-curve analytic methods capable of establishing temporal precedence and mediation among multiple cascading influences.[51]

CRITIQUE OF CHR APPROACH

Much has been learned about the psychosis risk syndrome via the CHR construct, yet there are several features of high risk sampling that raise important concerns about the degree to which the CHR risk criteria reflect the full range of manifestation of prospective psychosis risk markers that might exist in the general population. This is an important consideration, both from a basic research perspective and from the vantage point of creating effective early treatment interventions. As noted previously, CHR samples are nearly always recruited from populations that are distressed and help-seeking. Although this recruitment strategy generates psychosis conversion enrichment, it also signals that CHR participants are already experiencing changes in psychological functioning that merit attention, and therefore those included may not offer an optimal window for primary preventative intervention. Although the epidemiology of non-help-seekers in the population who convert to psychosis is not precisely known, recent retrospective studies indicate that only 30%–50% of first episode patients report experiencing prodromal symptoms.[52] Comorbidity in CHR samples (e.g., anxiety, depression) is the norm, not the exception,[53,54] suggesting that many patients initially seek treatment for symptoms other than those that qualify them for a CHR designation; although patients with psychosis often have high comorbidity, this raises important concerns about the extent to which CHR samples represent the prodromal course of psychosis in general. In the same vein, the conversion rate shifts dramatically depending on the baseline sampling criteria and years of follow-up, resulting in a lack of consensus about what exactly constitutes the high-risk window.

Another major issue limiting an accurate understanding of psychosis risk is that most CHR research conceptualizes the psychosis prodrome from the vantage of diagnostic criteria for schizophrenia, despite the known transdiagnostic significance of psychosis. Relatedly, the CHR model considers transition to psychosis tantamount to a harbinger of chronicity, while all other outcomes are treated as relatively more optimal. In fact, as noted previously, the first episode of psychosis in a high risk participant may be transient and not accompanied by long-term functional deficits, while many CHR individuals who do *not* experience a frank psychosis suffer from other psychopathologies and/or long-term functional deficits.[55] This issue is not unique to CHR research—reliance on clinical syndromes paired with a lack of transdiagnostic theoretical frameworks is a limiting factor in mental health science in general. For example, the presence of negative symptoms in CHR samples increases prospective risk for psychosis, but there is evidence that negative symptoms confer general psychopathology risk and are not specific to the psychosis prodrome.[56]

Concerns about representativeness are reinforced by two recent examinations of the performance of CHR criteria in the transdiagnostic prediction of psychosis. Across approximately two thousand individuals who made first contact with a community psychological assistance clinic, high risk presentation at baseline (as measured by the CAARMS) did not demonstrate prognostic accuracy for psychosis conversion compared to a reference group, regardless of whether follow-up was conducted at 2 years or at final contact (median = 87 months).[57] There was also no evidence that tailoring treatment based on CHR group membership resulted in more effective care.[57] In another sample of over 90,000 patients who were first assessed in the South London healthcare system on the basis of any nonpsychotic mental health problem (but including bipolar disorder and intermittent psychotic experiences), those meeting CHR criteria at baseline (assessed using CAARMS) accounted for only 5.19% (52/1001) and 1.19% (12/1010) of 6-year transitions to psychosis in the reference and validation samples, respectively.[58] Granted, these estimates based on CHR status are likely to be biased downward due to the relatively high sample age range (mean age = 32), the relative long follow-up latency compared to most CHR studies (6 years as opposed to 2 or fewer), and the enriched presence of many mental illnesses at baseline. Poor predictive validity of a CHR instrument (CAARMS) in general clinical populations does not necessarily damn the CHR construct, especially since there is still evidence that it has predictive utility in samples recruited explicitly based on CHR criteria. However, it remains to be demonstrated how well cases who transition to full psychosis when ascertained via CHR criteria correspond with all first episodes of psychosis in the population.

POPULATION-BASED AND COHORT STUDIES

One way to refine and contextualize the findings from CHR research is to conduct prospective population-level and/or cohort studies. Generally speaking, population sampling studies adopt the theoretical stance that risk for psychosis represents a continuum at the population level and that the expression of psychosis risk should be considered at least partly independent from diagnostic boundaries. Although there is some controversy about transdiagnostic/continuum frameworks of psychopathology in the clinical research community,[59] cumulative evidence from research in psychometrics,[60] epidemiology,[61] neurobiology,[62] and genetics[63] suggests that psychotic-spectrum illnesses are the result of continuous latent traits that probabilistically influence risk of clinically impairing psychopathology.

Lifetime prevalence rates indicate that psychotic disorders are relatively rare, affecting approximately 3.5% of the population.[64] Compared to clinically defined expressions of psychosis risk, more subtle/infrequent expressions of latent risk can be quantified at higher rates in population samples. For example, conservative estimates indicate that at least 7%–8% of people report some clinically severe psychotic experiences (PEs) in their lifetime (delusions and hallucinations), though the majority of these experiences are infrequent or remit with time, and do not appear to cause distress (~80%).[61] Milder expression of psychotic-like symptoms, including unusual beliefs and magical thinking, are even more common, as high as 20%–30% across cultures.[65,66] These subclinical PEs are operationalized using a variety of clinical and/or personality constructs (e.g., psychotic-like experiences [PLEs], positive schizotypy),[67] and associations between these traits and later psychosis can be examined.

Compared to CHR research efforts, relatively fewer of these large prospective studies have been conducted. A meta-analysis of six general population cohorts (seven total studies) that assessed the association between PEs at baseline and incidence of clinical psychosis over a 25 year time frame found that self-reported subclinical PEs were weakly associated with a higher rate of transition to psychosis, but not other mental illnesses, and reported a dose-response association between PE severity/frequency and risk of transition.[67] Sample sizes ranged from 534 to 4,762, mean baseline age ranged from 11 to 41 years, and follow-up latency ranged from 1 to 24 years.[68–74] Where they were measured at baseline, motivational impairments,[68] depression,[73] anxiety,[71] environmental risk (socioeconomic status, cannabis use),[68] psychiatric comorbidity and/or conduct problems in childhood,[74] and social impairments[71] increased the prospective risk of a psychotic outcome. Corroborating the findings from CHR studies, the addition of negative symptoms increased psychosis risk when positive symptoms were also present, however population studies underline the fact that individuals who report only negative symptoms do not appear to display elevated risk for later psychotic disorder.[68,75]

Only one study in the meta-analysis[67] (in a subsample of the Dominguez et al.[68] sample) collected more than two time points across an age range similar to CHR studies (ages 14–17, N = 845).[69] In this adolescent sample, prevalence of subclinical PEs was 22% across all time points. While most PEs did not recur (60%–70%), those with recurrent PEs had the highest rates of psychosis (16%–27%) compared to rates of transition in those never reporting a PE (3.7%) or those reporting only one PE time point (5.3%). Although the rate of transition was lowest in those never reporting a PE, the majority of subjects ultimately experiencing psychosis at the final follow-up were in this group (64%), raising the concern that the study design could not adequately capture prodromal symptom windows. A more recent report sampled from 9,498 youth aged 8–21, followed up approximately 2 years later in a subsample based on presence (N = 249) or absence (N = 254) of subthreshold psychotic symptoms at baseline.[76] Those with psychotic symptoms at baseline were more likely to have symptoms after 2 years (50%) compared to the rate of emergent symptoms in the control baseline group (16%). Participants in the baseline-positive group were more likely to have a diagnosable psychotic disorder at follow-up (6.8% vs. <1%), and had significantly higher comorbidity of nonpsychotic psychiatric symptoms at follow-up (76.8% vs. 48.6%). Adolescents with persistent symptoms endorsed higher agreement on most symptoms at baseline, though the only baseline measures that differed significantly between persistence and remission groups were endorsement of suspiciousness and perception of thought control. Although the developmental window in this sample is currently too short to approximate long-term psychosis rates, Calkins et al.[76] demonstrate the potential utility of a hybrid approach: using detailed CHR (and other) clinical assessments in a representative baseline sample makes it possible to identify cases based on subtler—and ostensibly earlier—expressions of CHR-like symptoms. Compared to the usual population study approach, restricting follow-up sample size using baseline criteria increases the feasibility of collecting detailed longitudinal psychometric and biometric data. Since the control sample was asymptomatic at baseline, but still demonstrated longitudinal incidence of CHR-like symptoms, this approach appears promising for capturing factors that predict the emergence and eventual course of new symptoms.

Incidence and severity of any mild/moderate psychotic experiences in prospective population-based samples, especially those that persist over time, do appear to have some predictive utility for psychosis risk. However, the vast majority of those who eventually convert to psychosis in these studies do not report any subthreshold PEs when measured at baseline. This does not necessarily mean that the majority of people who develop psychosis do not experience prodromal symptoms that include subclinical PEs, since it could also indicate that follow-ups in these studies were too infrequent to capture a symptomatic time frame. Though some studies demonstrate promise for the use of multidimensional trait assessments to assess prospective population-level risk (e.g., psychometric schizotypy), the multidimensional trait approach is rare.[77] The population-based studies that have implemented more comprehensive trait inventories to prospectively assess psychosis risk either did not assess differential psychopathology risk,[78] or reported based on samples for which long-term outcomes (e.g., transition, remission, or emergence) are not yet available.[76] As incorporation of biometric assessments becomes more commonplace in population and cohort studies, novel prospective risk signatures may be identified. For example, a recent study of over 1.1 million men in the Swedish Conscript Study demonstrated a link between poorer visual acuity at age 18 and lifetime risk for psychotic disorder (10,769 cases). This effect was independent of whether vision was correctable, and risk was more pronounced in cases with higher intraocular acuity differences. Surprisingly, there was no association between sibling visual acuity and proband psychosis risk, indicating the apparent absence of shared genetic or shared environmental influences. Potential explanations for this association include the insidious cumulative effects of social/occupational limitations resulting from any amount/duration of myopia.[79]

CRITIQUES OF POPULATION-BASED APPROACHES

Population-based studies can overcome some of the challenges introduced via the use of help-seeking samples (as summarized earlier), but they are difficult to compare directly with findings from the CHR literature. Population samples must be extremely large in order to have sufficient statistical power to model predictors of psychosis, since only approximately 1 in 200 people in general population samples will transition to psychosis over a period of 2 years.[9] This creates several practical challenges that ultimately limit the interpretability of findings: First, the administration of detailed interview-type and/or biometric assessments by trained clinicians is rarely feasible in very large samples. Second, it is almost always inefficient to study biological mechanisms in population samples, given the disproportionate cost, risk, and time demands posed by collecting neuroimaging and/or other biological assays. Third, postadolescent age ranges are typical in population studies, and time-to-follow-up is long compared to CHR designs (often 10 years[72] or longer[70,71]), which means that population studies will fail to capture the prodromal phase of many participants who ultimately experience psychosis. The population-based studies of adolescents identified several individuals experiencing new symptoms at first follow-up,[69,76] and found that persistence of symptoms mattered more than any incidence. Finally, there is questionable overlap between what is considered a psychotic outcome in CHR studies compared to what is labeled as psychosis in population-based samples.

Many population-based studies define psychotic outcomes more broadly compared to the strict requirements of CHR-studies, which means that psychotic outcomes like "psychotic impairment"[68,69] or distress caused by psychotic symptoms[73] are probably more comparable to baseline CHR symptoms than to outright psychosis as defined by instruments like the SIPS or CAARMS (see Figure 14.1).[1]

Though difficult to compare directly, some common themes emerge across CHR and population-based studies. Regardless of study design, the most consistent baseline predictor of clinically significant psychosis is the severity and persistence of subclinical positive symptoms. In cases where positive symptoms are present at baseline, risk for transition is amplified by a variety of factors that can be thought of as indexing poor functioning and/or risk for psychopathology in general, including comorbid psychiatric symptoms, social/motivational deficits (including negative symptoms), cognitive impairments, drug use, and environmental stressors. Preliminary findings from some CHR studies indicate that neural connectivity alterations are linked to psychosis etiology (see earlier section, "Mechanisms of onset in CHR samples"), but as yet there is no comprehensive account from either CHR nor population-based studies of what causes psychosis and its prodromal syndromes.

INTEGRATIVE AND TRANSDIAGNOSTIC APPROACHES

Both CHR and population-based approaches have contributed unique and complementary insights for the prediction of psychotic outcomes, but the evidence summarized so far points to the need for new standards of psychosis risk assessment. The hybrid approach taken by Calkins et al.[76] is an encouraging example of how the distinct advantages of different sampling approaches might be combined: this study used broad sampling for baseline ascertainment, yet also overcame some of the practical limitations inherent to population studies by restricting follow-up to more manageably sized subsamples. In addition to this type of integration, we believe that there are other study design choices that could enhance risk prediction in the future. Since all the studies reported herein are limited in the diversity of their psychometric tools, perhaps the most brute force solution for improving psychosis risk prediction is to simply add more diverse and detailed psychometric assessments to longitudinal population studies, with the goal in mind of capturing more information about the person-level context in which symptoms occur. These could include comprehensive measurement of personality traits (both normal- and clinical-range), characteristic adaptations (e.g., self-concept, attitudes, goals), intelligence testing, experience sampling methods, narrative assessments, and any other psychometrically rigorous assessments of behaviors/events relevant to adolescent and young adult development. Several factors other than those typically examined in either CHR or population studies have been associated with psychosis risk, including cannabis use,[80] sleep disturbances,[81] perceptions of dependability in close social relationships,[82] subtle alterations in language use,[83] and disturbance in a basic sense of self.[84] Self disturbances, once a core aspect of the psychosis construct, have been mostly overlooked in recent psychosis risk research. Anomalous self-awareness (for example feeling as if one's self is dissociated from conscious thought) is a major theme in the subjective reports of prodromal individuals,[85] and there is evidence that including self-disturbances in CHR studies enhances positive prediction.[84] Notably, the presence and severity of these symptoms in CHR adolescents appears to confer selective risk for psychotic spectrum disorders.[86] In general, more humanistic, appraisal-based measures have fallen out of favor in the psychosis research, despite evidence that "nonspecific" complaints from prodromal individuals (motivational deficits, concentration problems, fatigue) are often shorthand for subjective anomalous experiences that more specifically characterize impending psychosis.[87] For example, difficulty concentrating may be due to chaotic or racing thoughts, but this latter specificity will not be captured by most diagnostic checklists.[87] Using appraisal-based measures in combination with standard clinical interviews, trait/symptom assessments, and biological assays represents another promising opportunity for integrative research in psychosis risk prediction.

Almost without exception, risk prediction approaches conceptualize psychosis from the vantage of clinically defined schizophrenia. Given the mounting evidence that the fundamental features of psychosis transcend traditional psychopathology categories,[88] it is important to keep in mind that virtually any interpretation of the literature to date is limited by the absence of studies that can examine whether shared transdiagnostic psychosis phenomenology reflects shared etiological vulnerabilities. Reconciling CHR and population sampling approaches in combination with more detailed assessments that incorporate transdiagnostic theory and methodology is necessary in order to advance the science of psychosis prediction.

1 A notable exception to this trend is the study by Werbeloff et al.,[67] which used hospitalization for nonaffective psychosis as the psychotic outcome criterion. Arguably, this criterion is even more restrictive than those criteria defined by CHR studies.

REFERENCES

1. Yung AR, McGorry PD. The prodromal phase of first-episode psychosis: past and current conceptualizations. *Schizophrenia Bulletin.* 1996;22:353–370.
2. Penn DL, Waldheter EJ, Perkins DO, Mueser KT, Lieberman JA. Psychosocial treatment for first-episode psychosis: a research update. *American Journal of Psychiatry.* 2005;162:2220–2232.
3. McGlashan T, Walsh B, Woods S. The psychosis-risk syndrome: handbook for diagnosis and follow-up. 2010.
4. Yung AR, Yuen HP, McGorry PD, et al. Mapping the onset of psychosis: the Comprehensive Assessment of At-Risk Mental States. *Australian and New Zealand Journal of Psychiatry.* 2005;39:964–971.
5. Addington J, Heinssen R. Prediction and prevention of psychosis in youth at clinical high risk. *Annual Review of Clinical Psychology.* 2012;8:269–289.

6. Fusar-Poli P, Cappucciati M, Borgwardt S, et al. Heterogeneity of psychosis risk within individuals at clinical high risk. *JAMA Psychiatry*. 2016;73:113.
7. Nelson B, Yuen HP, Wood SJ, et al. Long-term follow-up of a group at ultra high risk ("prodromal") for psychosis: the PACE 400 study. *JAMA Psychiatry*. 2013;70:793–802.
8. Cannon TD, Cadenhead K, Cornblatt B, et al. Prediction of psychosis in youth at high clinical risk: a multisite longitudinal study in North America. *Archives of General Psychiatry*. 2008;65(1):28–37.
9. Ruhrmann S, Schultze-Lutter F, Salokangas RKR, et al. Prediction of psychosis in adolescents and young adults at high risk. *Archives of General Psychiatry*. 2010;67(3):241–251.
10. Schultze-Lutter F, Ruhrmann S, Fusar-Poli P, Bechdolf A, G Schimmelmann B, Klosterkotter J. Basic symptoms and the prediction of first-episode psychosis. *Current Pharmaceutical Design*. 2012;18:351–357.
11. Klosterkötter J, Hellmich M, Steinmeyer EM, Schultze-Lutter F. Diagnosing schizophrenia in the initial prodromal phase. *Archives of General Psychiatry*. 2001;58:158–164.
12. Demjaha A, Valmaggia L, Stahl D, Byrne M, McGuire P. Disorganization/cognitive and negative symptom dimensions in the at-risk mental state predict subsequent transition to psychosis. *Schizophrenia Bulletin*. 2012;38(2):351–359.
13. Yung AR, Phillips LJ, Yuen HP, McGorry PD. Risk factors for psychosis in an ultra high-risk group: psychopathology and clinical features. *Schizophrenia Research*. 2004;67(2–3):131–142.
14. Addington J, Liu L, Perkins DO, Carrion RE, Keefe RSE, Woods SW. The role of cognition and social functioning as predictors in the transition to psychosis for youth with attenuated psychotic symptoms. *Schizophrenia Bulletin*. 2017;43(1):57–63.
15. Cornblatt BA, Carrión RE, Auther A, et al. Psychosis prevention: A modified clinical high risk perspective from the recognition and prevention (RAP) Program. *American Journal of Psychiatry*. 2015;172(10):986–994.
16. Seidman LJ, Shapiro DI, Stone WS, et al. Association of Neurocognition With Transition to Psychosis. *JAMA Psychiatry*. 2016;73(12):1239–1248.
17. Healey KM, Penn DL, Perkins D, Woods SW, Keefe RSE, Addington J. Latent Profile Analysis and Conversion to Psychosis: Characterizing Subgroups to Enhance Risk Prediction. *Schizophrenia Bulletin*. 2017.
18. Carrión RE, Correll CU, Auther AM, Cornblatt BA. A severity-based clinical staging model for the psychosis prodrome: Longitudinal findings from the New York recognition and prevention program. *Schizophrenia Bulletin*. 2017;43(1):64–74.
19. Kattan MW, Yu C, Stephenson AJ, Sartor O, Tombal B. Clinicians versus nomogram: predicting future technetium-99m bone scan positivity in patients with rising prostate-specific antigen after radical prostatectomy for prostate cancer. *Urology*. 2013;81(5):956–961.
20. Lee TH, Marcantonio ER, Mangione CM, et al. Derivation and prospective validation of a simple index for prediction of cardiac risk of major noncardiac surgery. *Circulation*. 1999;100(10):1043–1049.
21. Pfeiffer RM, Park Y, Kreimer AR, et al. Risk prediction for breast, endometrial, and ovarian cancer in white women aged 50 y or older: derivation and validation from population-based cohort studies. *PLoS Medicine*. 2013;10(7):e1001492.
22. Cannon TD, Yu C, Addington J, et al. An individualized risk calculator for research in prodromal psychosis. *American Journal of Psychiatry*. 2016;173(10):980–988.
23. Carrión RE, Cornblatt BA, Burton CZ, et al. Personalized prediction of psychosis: external validation of the NAPLS-2 psychosis risk calculator with the EDIPPP project. *Schizophrenia Bulletin*. 2016;173:989–996.
24. Glausier JR, Lewis DA. Dendritic spine pathology in schizophrenia. *Neuroscience*. 2013;251:90–107.
25. Stephan KE, Friston KJ, Frith CD. Dysconnection in schizophrenia: from abnormal synaptic plasticity to failures of self-monitoring. *Schizophrenia Bulletin*. 2009;35(3):509–527.
26. Cannon TD. Neurodevelopment and the transition from schizophrenia prodrome to schizophrenia: research imperatives. *Biological Psychiatry*. 2008;64(9):737–738.
27. Borgwardt SJ, McGuire PK, Aston J, et al. Reductions in frontal, temporal and parietal volume associated with the onset of psychosis. *Schizophrenia Research*. 2008;106(2–3):108–114.
28. Pantelis C, Velakoulis D, McGorry PD, et al. Neuroanatomical abnormalities before and after onset of psychosis: a cross-sectional and longitudinal MRI comparison. *Lancet*. 2003;361:281–288.
29. Sun D, Phillips L, Velakoulis D, et al. Progressive brain structural changes mapped as psychosis develops in "at risk" individuals. *Schizophrenia Research*. 2009;108(1–3):85–92.
30. Ziermans TB, Schothorst PF, Schnack HG, et al. Progressive structural brain changes during development of psychosis. *Schizophrenia Bulletin*. 2012;38(3):519–530.
31. Cannon TD, Chung Y, He G, et al. Progressive reduction in cortical thickness as psychosis develops: a multisite longitudinal neuroimaging study of youth at elevated clinical risk. *Biological Psychiatry*. 2015;77(2):147–157.
32. Fusar-Poli P, Smieskova R, Kempton MJ, Ho BC, Andreasen NC, Borgwardt S. Progressive brain changes in schizophrenia related to antipsychotic treatment? A meta-analysis of longitudinal MRI studies. *Neuroscience and Biobehavioral Reviews*. 2013;37(8):1680–1691.
33. Navari S, Dazzan P. Do antipsychotic drugs affect brain structure? A systematic and critical review of MRI findings. *Psychological Medicine*. 2009;39(11):1763–1777.
34. Haijma SV, Van Haren N, Cahn W, Koolschijn PC, Hulshoff Pol HE, Kahn RS. Brain volumes in schizophrenia: a meta-analysis in over 18 000 subjects. *Schizophrenia Bulletin*. 2013;39(5):1129–1138.
35. McIntosh AM, Owens DC, Moorhead WJ, et al. Longitudinal volume reductions in people at high genetic risk of schizophrenia as they develop psychosis. *Biological Psychiatry*. 2011;69(10):953–958.
36. Brans RG, van Haren NE, van Baal GC, Schnack HG, Kahn RS, Hulshoff Pol HE. Heritability of changes in brain volume over time in twin pairs discordant for schizophrenia. *Archives of General Psychiatry*. 2008;65(11):1259–1268.
37. Rao JS, Kim HW, Harry GJ, Rapoport SI, Reese EA. Increased neuroinflammatory and arachidonic acid cascade markers, and reduced synaptic proteins, in the postmortem frontal cortex from schizophrenia patients. *Schizophrenia Research*. 2013;147(1):24–31.
38. Fillman SG, Cloonan N, Catts VS, et al. Increased inflammatory markers identified in the dorsolateral prefrontal cortex of individuals with schizophrenia. *Molecular Psychiatry*. 2013;18(2):206–214.
39. Fung SJ, Joshi D, Fillman SG, Weickert CS. High white matter neuron density with elevated cortical cytokine expression in schizophrenia. *Biological Psychiatry*. 2014;75(4):e5–7.
40. Catts VS, Wong J, Fillman SG, Fung SJ, Weickert CS. Increased expression of astrocyte markers in schizophrenia: association with neuroinflammation. *Australian and New Zealand Journal of Psychiatry*. 2014.
41. Milatovic D, Gupta RC, Yu Y, Zaja-Milatovic S, Aschner M. Protective effects of antioxidants and anti-inflammatory agents against manganese-induced oxidative damage and neuronal injury. *Toxicology and Applied Pharmacology*. 2011;256(3):219–226.
42. Schafer DP, Lehrman EK, Stevens B. The "quad-partite" synapse: microglia-synapse interactions in the developing and mature CNS. *Glia*. 2013;61(1):24–36.
43. Selemon LD, Goldman-Rakic PS. The reduced neuropil hypothesis: a circuit based model of schizophrenia. *Biological Psychiatry*. 1999;45(1):17–25.
44. Selemon LD, Rajkowska G, Goldman-Rakic PS. Elevated neuronal density in prefrontal area 46 in brains from schizophrenic patients: application of a three-dimensional, stereologic counting method. *Journal of Comparative Neurology*. 1998;392(3):402–412.
45. Das T, Borgwardt S, Hauke DJ, et al. Disorganized gyrification network properties during the transition to psychosis. *JAMA Psychiatry*. 2018;75(6):613–622.
46. Chung Y, Addington J, Bearden CE, et al. Adding a neuroanatomical biomarker to an individualized risk calculator for psychosis: a proof-of-concept study. *Schizophrenia Research*. 2019.
47. Anticevic A, Cole MW, Repovs G, et al. Characterizing thalamo-cortical disturbances in schizophrenia and bipolar illness. *Cerebral Cortex*. 2013;[Epub].

48. Anticevic A, Haut K, Murray JD, et al. Association of thalamic dysconnectivity and conversion to psychosis in youth and young adults at elevated clinical risk. *JAMA Psychiatry*. 2015.
49. Cao H, Chen OY, Chung Y, et al. Cerebello-thalamo-cortical hyperconnectivity as a state-independent functional neural signature for psychosis prediction and characterization. *Nature Communications*. 2018;9(1):3836.
50. Cao H, Chung Y, McEwen SC, et al. Progressive reconfiguration of resting-state brain networks as psychosis develops: preliminary results from the North American Prodrome Longitudinal Study (NAPLS) consortium. *Schizophrenia Research*. 2019.
51. McArdle JJ, Hamgami F, Jones K, et al. Structural modeling of dynamic changes in memory and brain structure using longitudinal data from the normative aging study. *Journals of Gerontology Series B, Psychological Sciences and Social Sciences*. 2004;59(6):P294–304.
52. Shah JL, Crawford A, Mustafa SS, Iyer SN, Joober R, Malla AK. Is the clinical high-risk state a valid concept? Retrospective examination in a first-episode psychosis sample. *Psychiatric Serv*. 2017;68(10):1046–1052.
53. Fusar-Poli P, Cappucciati M, Borgwardt S, et al. Heterogeneity of psychosis risk within individuals at clinical high risk. *JAMA Psychiatry*. 2016;73(2):113–120.
54. Rutigliano G, Valmaggia L, Landi P, et al. Persistence or recurrence of non-psychotic comorbid mental disorders associated with 6-year poor functional outcomes in patients at ultra high risk for psychosis. *Journal of Affective Disorders*. 2016;203:101–110.
55. Yung AR, Nelson B, Thompson A, Wood SJ. The psychosis threshold in ultra high risk (prodromal) research: is it valid? *Schizophrenia Research*. 2010;120:1–6.
56. Strauss GP, Cohen AS. A transdiagnostic review of negative symptom phenomenology and etiology. *Schizophrenia Bulletin*. 2017;43(4):712–719.
57. Conrad AM, Lewin TJ, Sly KA, et al. Utility of risk-status for predicting psychosis and related outcomes: evaluation of a 10-year cohort of presenters to a specialised early psychosis community mental health service. *Psychiatry Research*. 2017;247(November 2016):336–344.
58. Fusar-Poli P, Rutigliano G, Stahl D, et al. Development and validation of a clinically based risk calculator for the transdiagnostic prediction of psychosis. *JAMA Psychiatry*. 2017;74(5):493–500.
59. Lawrie SM. Whether psychosis is best conceptualized as a continuum or in categories is an empirical, practical and political question. *World Psychiatry*. 2016;15:125–126.
60. Krueger RF, Markon KE. Reinterpreting comorbidity: a model-based approach to understanding and classifying psychopathology. *Annual Review of Clinical Psychology*. 2006;2:111–133.
61. Linscott RJ, van Os J. An updated and conservative systematic review and meta-analysis of epidemiological evidence on psychotic experiences in children and adults: on the pathway from proneness to persistence to dimensional expression across mental disorders. *Psychological Medicine*. 2013;43(6):1133–1149.
62. Nelson MT, Seal M, Pantelis C, Phillips L. Evidence of a dimensional relationship between schizotypy and schizophrenia: a systematic review. *Neuroscience and Biobehavioral Reviews*. 2013;37:317–327.
63. Kendler KS. A joint history of the nature of genetic variation and the nature of schizophrenia. *Molecular Psychiatry*. 2014:1–7.
64. McGrath JJ, Saha S, Al-Hamzawi AO, et al. Age of onset and lifetime projected risk of psychotic experiences: cross-national data from the world mental health survey. *Schizophrenia Bulletin*. 2016;42:933–941.
65. Nuevo R, Chatterji S, Verdes E, Naidoo N, Arango C, Ayuso-Mateos JL. The continuum of psychotic symptoms in the general population: a cross-national study. *Schizophrenia Bulletin*. 2012;38:475–485.
66. Preti A, Sisti D, Rocchi MBL, et al. Prevalence and dimensionality of hallucination-like experiences in young adults. *Comprehensive Psychiatry*. 2014;55:826–836.
67. Kaymaz N, Drukker M, Lieb R, et al. Do subthreshold psychotic experiences predict clinical outcomes in unselected non-help-seeking population-based samples? A systematic review and meta-analysis, enriched with new results. *Psychological Medicine*. 2012;42(11):2239–2253.
68. Dominguez MDG, Saka MC, Lieb R, Wittchen HU, Van Os J. Early expression of negative/disorganized symptoms predicting psychotic experiences and subsequent clinical psychosis: A 10-year study. *American Journal of Psychiatry*. 2010;167(9):1075–1082.
69. Dominguez MDG, Wichers M, Lieb R, Wittchen HU, Van Os J. Evidence that onset of clinical psychosis is an outcome of progressively more persistent subclinical psychotic experiences: an 8-year cohort study. *Schizophrenia Bulletin*. 2011;37(1):84–93.
70. Poulton R, Caspi A, Moffitt TE, Cannon M, Murray RM, Harrington H. Children's self-reported psychotic symptoms and adult schizophreniform disorder: a 15-year longitudinal study. *Archives of General Psychiatry*. 2000;57:1053–1058.
71. Werbeloff N, Drukker M, Dohrenwend BP, et al. Self-reported attenuated psychotic symptoms as forerunners of severe mental disorders later in life. *Archives of General Psychiatry*. 2012;69(5):467–475.
72. Chapman LJ, Chapman JP, Kwapil TR, Eckblad M, Zinser MC. Putative psychosis-prone subjects 10 years later. *Journal of Abnormal Psychology*. 1994;103:171–183.
73. Hanssen M, Bak M, Bijl R, Vollebergh W, van Os J. The incidence and outcome of subclinical psychotic experiences in the general population. *British Journal of Clinical Psychology*. 2005;44:181–191.
74. Welham J, Scott J, Williams G, et al. Emotional and behavioural antecedents of young adults who screen positive for non-affective psychosis: a 21-year birth cohort study. *Psychological Medicine*. 2009;39:625–634.
75. Kwapil TR, Gross GM, Silvia PJ, Barrantes-Vidal N. Prediction of psychopathology and functional impairment by positive and negative schizotypy in the Chapmans' ten-year longitudinal study. *Journal of Abnormal Psychology*. 2013;122:807–815.
76. Calkins ME, Moore TM, Satterthwaite TD, et al. Persistence of psychosis spectrum symptoms in the Philadelphia Neurodevelopmental Cohort: a prospective two-year follow-up. *World Psychiatry*. 2017;16(1):62–76.
77. Debbané M, Eliez S, Badoud D, Conus P, Flückiger R, Schultze-Lutter F. Developing psychosis and its risk states through the lens of schizotypy. *Schizophrenia Bulletin*. 2015;41:S396–S407.
78. Hayes JF, Osborn DPJ, Lewis G, Dalman C, Lundin A. Association of late adolescent personality with risk for subsequent serious mental illness among men in a Swedish Nationwide Cohort Study. *JAMA Psychiatry*. 2017:1–9.
79. Hayes JF, Picot S, Osborn DPJ, Lewis G, Dalman C, Lundin A. Visual acuity in late adolescence and future psychosis risk in a cohort of 1 million men. *Schizophrenia Bulletin*. 2018.
80. Gage SH, Jones HJ, Burgess S, et al. Assessing causality in associations between cannabis use and schizophrenia risk: a two-sample Mendelian randomization study. *Psychological Medicine*. 2017;47:971–980.
81. Oh HY, Singh F, Koyanagi A, Jameson N, Schiffman J, DeVylder J. Sleep disturbances are associated with psychotic experiences: findings from the National Comorbidity Survey Replication. *Schizophrenia Research*. 2016;171(1–3):74–78.
82. Haidl T, Rosen M, Schultze-Lutter F, et al. Expressed emotion as a predictor of the first psychotic episode: results of the European prediction of psychosis study. *Schizophrenia Research*. 2018;199:346–352.
83. Corcoran CM, Carrillo F, Fernández-Slezak D, et al. Prediction of psychosis across protocols and risk cohorts using automated language analysis. *World Psychiatry*. 2018;17(1):67–75.
84. Nelson B, Thompson A, Yung AR. Basic self-disturbance predicts psychosis onset in the ultra high risk for psychosis "prodromal" population. *Schizophrenia Bulletin*. 2012;38(6):1277–1287.
85. Henriksen MG, Raballo A, Parnas J. The pathogenesis of auditory verbal hallucinations in schizophrenia: a clinical–phenomenological account. *Philosophy, Psychiatry, and Psychology*. 2015;22:165–181.
86. Raballo A, Monducci E, Ferrara M, Fiori Nastro P, Dario C. Developmental vulnerability to psychosis: selective aggregation of basic self-disturbance in early onset schizophrenia. *Schizophrenia Research*. 2018;201:367–372.
87. Parnas J, Henriksen MG. Disordered self in the schizophrenia spectrum. *Harvard Review of Psychiatry*. 2014;22:251–265.
88. Van Os J, Reininghaus U. Psychosis as a transdiagnostic and extended phenotype in the general population. *World Psychiatry*. 2016;15(2):118–124.

15

FIRST-EPISODE PSYCHOSIS

PHENOMENOLOGY, ONSET, COURSE, AND EARLY INTERVENTION (OPUS)

Merete Nordentoft and Nikolai Albert

FIRST-EPISODE PSYCHOSIS AND ITS CONSEQUENCES

Psychosis in the schizophrenia spectrum is among the most severe mental disorders. It has major influences in terms of the burden and cost for the people living with the consequences of symptoms and disabilities; for their relatives, including partners and children; and for society in terms of lost workforce and need for treatment.[1–4] Moreover, it is often associated with risk for comorbid substance abuse,[5] depression, and somatic illnesses; fatal and nonfatal suicidal behavior is frequent.[6]

Following the WHO Pilot Study of Schizophrenia,[7] the incidence of schizophrenia was believed to show rather modest variations across countries worldwide. However, a later comprehensive review demonstrated substantial variation in incidence of schizophrenia,[8] with an average rate of 15 per 100,000 inhabitants per year. Recent Danish studies have shown incidence rates for narrow schizophrenia (ICD 10 – F20.x) more than twice as high (40 per 100,000 for men and 31 per 100,000 for women)[9] and a cumulated life-time risk of schizophrenia spectrum disorders (ICD 10 – F2x) of 1.8% for the entire population.[10] Analyses of sex-specific incidence rates have consistently reported higher rates in males than in females.[8,10–12]

Thus, psychosis in the schizophrenia spectrum constitutes a major public health problem, and that is the background for the World Health Organization recently stating that early intervention should have a high priority.[13]

The onset of psychosis can be acute, but symptoms often develop gradually, and social consequences may already be manifest, such as loss of job or affiliation to school, social isolation, changed interests and habits, and diurnal rhythm. Sometimes, frequent use of substances such as cannabis or cocaine worsens the condition and impedes treatment response. In most cases, both the young person and the family do not understand the impact and the consequences of the illness and do not know about helpful and necessary precautions.

In the most severe instances, psychotic symptoms such as hallucinations and delusions can appear very intrusive to the young person and can therefore have serious consequences not only for the afflicted person but also for relatives. In rare cases it can even lead to tragic consequences for staff members in psychiatry or social services and people in the community. Thus, treatment of psychotic symptoms must have high priority. Staff members need good clinical and social skills in order to make treatment of psychotic symptoms an attractive option for the young person with psychosis.

Often, the daily life of a young person with psychosis is severely affected by the psychotic condition. It is therefore necessary to offer help with social issues, such as risk of being expelled from school, difficulties in interaction with social services, and unpaid bills and debt. In addition to being obviously helpful for the person, such initiatives form the basis for a good working alliance, which is necessary to engage the person in treatment.

Even though psychotic symptoms are often the most dramatic manifestation of a psychotic condition, negative symptoms (e.g., affective flattening, alogia, apathy, anhedonia) are more decisive for the outcome of the illness. The presence of severe negative symptoms is very debilitating and hinders participation in work or education. Psychopharmacological treatment has limited effect on negative symptoms.[14,15]

Early treatment is a prerogative, and many factors are likely to affect how soon after onset of symptoms patients with schizophrenia or schizophrenia-like psychosis seek treatment. A wide array of factors may interact to negatively influence treatment-seeking behavior. These might be psychopathological factors such as insidious onset, persecutory ideas, social withdrawal, level of insight into illness, and disruptive behavior, as well as sociocultural factors such as educational level of patient and family, beliefs and knowledge about mental illness, availability of and access to healthcare, and stigmatization associated with treatment for mental illness. The first barrier to treatment access is that somebody will have to recognize the symptoms as psychotic features that can be treated. This makes it very important that symptoms of psychosis are well known and that psychiatric treatment is seen as an attractive possibility for help. Relatives, teachers, peers, and the patients themselves should be able to recognize the symptoms and be aware that treatment is available. It is a widespread public opinion that psychiatric conditions are hopeless. It is not very well recognized that in many cases the symptoms can be treated successfully, that the prognosis for psychiatric conditions is diverse, and that many can live a normal life. In many cases, psychotic symptoms can be relieved. The public image of psychiatry should be positive, and psychiatry needs to deserve a positive picture. It should be easy to get access to treatment, and the treatment should be flexible, respectful, and of high quality.

THE BACKGROUND AND RATIONALE FOR THE EARLY INTERVENTION MODEL

In one of the first long term-studies of the course of schizophrenia, Manfred Bleuler found the illness to have a heterogeneous trajectory. Bleuler described seven typical schizophrenic trajectories in his seminal work *The Schizophrenic Disorder: Long-Term Patient and Family Studies.*[16] A few patients had a "catastrophic trajectory," going straight from acute onset to a chronic debilitated state; most experienced a more fluctuating trajectory. Some had only one psychotic episode, while others had several episodes but might be fully remitted between episodes. These findings were confirmed in other long-term follow-up studies published around the same time.[17,18] These studies were primarily prevalence studies, typically including all patients admitted to a given hospital in a given period, with several patients having had several hospitalizations prior to inclusion in the studies. True incidence cohorts of patients with schizophrenia were published in the 1980s and 1990s, allowing true estimates of the course of schizophrenia to be developed.

Several studies of first-episode schizophrenia or first-episode psychosis patients have been published since 1980.[19] Herein we address a few of these because their results have direct implication for the later development of the early intervention model.

One of the most thorough investigations of psychopathology over time was conducted in the Madras study, the 10-year results of which were published in the mid-1990s.[20,21] Using monthly assessments of psychotic and negative symptoms, the researchers found that symptomatology in the early years showed a high degree of changeability compared to the later years, where symptoms were stable over time. This finding indicated that, even if for many of the patients there was steady improvement over time, in most cases symptoms that remained after 6 years had to be considered chronic.

Other studies investigated the effects of early predictors on long-term outcome. Carpenter and Straus looked at 11-year outcomes of patients diagnosed with schizophrenia and found that, allowing for some changeability, the psychopathological outcome and number of social contacts were more or less the same at the 11-year follow-up as they had been at the 2-year examination.[22] The International Study of Schizophrenia gathered long-term data from several WHO studies initiated in the 1960s and 1970s and analyzed 15- to 25-year outcomes.[23] This large study found that the duration of psychotic symptoms within the first 2 years of illness was the strongest predictor of long-term outcome.

Most of the early incidence cohorts found that many of the patients had been ill for a long duration prior to initiation of treatment. One study[24] found that patients who were treated early in their course of illness had a better 20-year outcome than those who were diagnosed later in their course. In the early 1990s Wyatt[25] and Loebel et al.[26] published highly influential papers on the effect of late treatment on later outcome. Wyatt's review was primarily based on studies comparing patients from the preneuroleptic era with patients from the neuroleptic era. It was implied in the study that patients who were treated after the introduction of antipsychotic medication would experience shorter duration of psychosis. The author concluded that a majority of the included studies showed improvement in illness course after the introduction of antipsychotic medication. From this it was extrapolated that there might be a toxic biological effect of psychosis.[25] Loebel et al. showed in a cohort of 70 patients diagnosed with schizophrenia that the duration of psychosis prior to initiation of antipsychotic treatment had a negative effect on level and rate of remission after initiation of treatment. They also suggested that there could be a toxic effect of psychosis acting on the brain and causing long-term damage.[26] Even if the relationship between DUP and outcome is hotly debated,[27,28] several later studies and meta-analyses have concluded that DUP has an influence on later outcome of psychotic illnesses.[29–32]

Based on these findings, the goal of the early intervention model was to direct resources toward the early phase of illness. Based on the incidence cohorts, it was hypothesized that the early years of illness constituted a critical phase in which symptomatology showed a higher level of plasticity and that targeted interventions could be more successful in intervening in these early years than in the later years, when symptoms seemed to be more chronic. If intervention in these early years was successful in changing the illness burden, this might have long-term effect, given the early years' high predictive value on later outcome.[33,34] This thrust was twofold. First, patients had to be identified earlier so that treatment could start at an earlier stage of the illness. Second, treatment facilities targeting young patients in their early course of illness had to be developed.

OPUS: TREATMENT AND RESULTS

The OPUS I study was one of the first randomized clinical trials designed to test the early intervention hypothesis. It was initiated in 1998 and recruited participants with a first-episode schizophrenia spectrum disorder (ICD 10 – F2).[35] Participants were between 18 and 45 years old and had not received more than 12 weeks of antipsychotic medication prior to inclusion in the trial.[36] Totally, 547 participants were recruited and randomized to either 2 years of OPUS treatment or treatment as usual.

TREATMENT APPROACH

The name OPUS was taken from the world of music; it is not an acronym. It was meant to illustrate how all members of an orchestra must be synchronized and in tune if the individual musicians are to be part of a harmonized whole. In line with this thinking, the OPUS teams were designed as

Box 15.1 OPUS TREATMENT

1. Assertive community treatment
2. Social skill training
3. Family involvement

The treatment was provided by a multidisciplinary team with a maximal caseload of 10:1.

Aside from the organizational elements of the OPUS teams, the attitude toward the patients was an important part of the development of treatment. The attitude was describes as:

- A long-awaited guest, who you want to feel welcome and at home during a long visit.
- A collaborator, whose insights and attitudes are decisive for the outcome.
- An individual with personal preferences that should be taken into account in the treatment to the greatest extent possible.

multidisciplinary teams who held weekly team meetings to encourage a variety of professionals to seek guidance from their peers with different professional backgrounds. The teams consisted of a psychiatrist, psychologists, nurses, social workers, and a vocational therapist. All team members, with the exception of the psychiatrist, functioned as case managers. The average patient:case manager ratio in the trial was 10:1.

The treatment (Box 15.1) had three main pillars: modified assertive community treatment,[37] social skill training,[38] and family involvement.[39,40] The case manager was responsible for establishing and maintaining contact with the patient. Meetings with patients were held at patient homes, at the OPUS facilities, or at other places where the patient felt most comfortable. The office hours were 8 am to 5 pm, but the patient could leave a message on the case manager's cell phone and expect to be contacted the next day. The case manager was further responsible for establishing contact with at least one family member of the patient and inviting that person to be part of the treatment. The family involvement included psychoeducation and multifamily groups as developed by McFarlane.[40] The multifamily groups ran for 18 months with fortnightly meetings of four to six families, including the patient. The groups focused on practical problem solving and coping strategies. Patients in need also participated in social skill training, focusing on coping with symptoms, social skills, and problem solving. The groups included no more than six patients and two therapists.

The welcoming approach

The patient should be considered *a long-awaited guest who you want to feel welcome and at home during a long visit.*

Patients who were included in the study were all newly diagnosed with a schizophrenia spectrum disorder and had not received antipsychotic medication for more than 12 weeks. Given the long duration of untreated psychosis (median 48 weeks,[41] many patients had lived with symptoms, both psychotic and nonpsychotic, for a long duration prior to enrollment in the trial. Many had consulted several treatment facilities, professional and nonprofessional, prior to start of OPUS treatment. It was extremely important at the first visit to recognize this long journey and the accomplishment of actually attending the first consultation.

The welcoming approach had several practical implications. First, it should be easy to get an appointment, and the invitation letter should include the choice to call and change the time or location of the appointment. Treatment facilities should be welcoming. Upon arrival, the patient should be welcomed warmly by a staff member and offered something to drink. It was important to signal the informal aspect of the setting by ensuring that the waiting area was seen more as an area for the patient than for the clinical staff, for instance by placing a football table or other games there. A box for suggestions for improvements should be placed in the waiting area and the area should be accessible, without locked doors, to the widest extent possible. At the first consultation, both the psychiatrist and preferably the primary contact person should be present. The first consultation should include an introduction to the treatment facilities and the staff.

The welcoming approach also implied training of staff members. If staff members are overburdened or faced with unsolvable problems and the institution is not aware of the relationship between staff members and clients, staff members can develop an unfriendly attitude toward patients. Thus, the welcoming attitude required training and careful maintenance.

At the first visit, the patient might not tell the full story or might change or deny parts of the story. Details that might not be disclosed initially could include that symptoms had lasted longer, were more frequent, or were more severe or more bizarre than had first been reported. School or work attendance might have been lower, substance abuse might have been present or more severe, or crime, debt, or shameful or humiliating social relations might not have been revealed by the patient at the initial visit. When such things were not told, it could be difficult for the professional to understand what was going on, and part of the patient's behavior might be incomprehensible. Often, as patients started to trust, the story could be told more fully and accurately. Patients might reveal that they had been abused during childhood, had been beaten by their partner, had been drunk while responsible for a child, had committed several crimes, or owed large amounts of money to criminal groups. In such situations it was important to allow time to pass and to trust that a fuller picture would emerge. It was important not to condemn the patient because of untold or changed parts of their story. Actually, the extent to which patients reveal sensitive parts of their story to a complete stranger at the first psychiatric consultation was, and is, impressive.

It was even more important not to condemn patients when they failed to carry out planned changes in behavior (e.g., started using drugs again, failed to take medication as prescribed, or did not keep appointments). To remain welcoming, it was important that the staff be aware of this possibility and trained on how to handle told and untold details and unkept promises. When staff was not trained and kept aware of the importance of not condemning the patients, there was a danger that the staff might reject the patient due to occurrences that might otherwise be perceived as lies. To consider the patient as a long-awaited guest meant that rules such as "three strikes and you are out" were totally unacceptable. The attitude was that patients who did not come to treatment visits regularly were those in greatest need of treatment. Activities to get in contact and build an alliance with the patient were very important. In some cases this might involve contacting relatives to try to motivate the patient or help in identifying possibilities for meeting the patient. If the situation had come to a complete deadlock, compulsory admission could be considered. According to Danish legislation, this can occur if the patient is considered psychotic and either is endangering himself or herself or is likely to deteriorate further if treatment is not implemented, and if attempts at motivation have been fruitless. When such unfortunate instances occurred, it might have seemed hypocritical to speak of a welcoming approach; nonetheless, it was important to express openly that the steps that were taken were not intended to punish the patient or to demonstrate who held power, but to arrange help, support, and treatment and to avoid further damage.

The collaborating and personal approach

The patient should be considered as *a collaborator, whose insights and attitudes are decisive for the outcome and as an individual with personal preferences that should be taken into account in the treatment to the greatest extent possible.*

To view the patient as collaborator with insights and preferences that should be taken into account was considered more than just a positive attitude; it was a vital tool in harmonizing and reaching the treatment goal set by the clinician and the patient. It was considered unproductive simply to ensure that the patient was treated with medication, maybe even compulsory, if the treatment was not maintained for more than a short period. If a turbulent psychotic episode was to be followed by a more stable period in which establishment, or reestablishment, of the treatment alliance, solving of social problems, and planning rehabilitation could be addressed, it was necessary that the patient have the principal role in the play. The patient could be perceived not only as a guest but also as a collaborating partner whose preferences were key to successful completion of the treatment. In planning this process, it was essential to listen, be patient, and consider the patient's insight, maturity, capabilities, and wishes for the future to evolve over a period that could be longer than the staff had initially predicted.

A goal of the treatment was therefore, as much as initiation of medical treatment, development of a therapeutic environment where the patients felt safe and respected, so that objection about medication, for example, could be debated openly and frankly. Only through development of such trust could one hope that patients would confide to the clinician that they were planning to stop taking their medication. If such plans were expressed, it was important that the clinician listen and try to understand the patients' motivations. The clinician might advise the patients; however, patients who wanted to discontinue treatment would do so with or without approval. Therefore, it was more important to remain in contact with, and as best as possible guide, the patient rather than trying to force the patient to continue treatment.

RESULTS FROM THE OPUS I AND II TRIALS

The treatment just described was developed for the OPUS I trial and was later implemented in OPUS teams all over Denmark.[42] The interventional part of the OPUS I trial lasted for 2 years; participants were assessed after the first and second year. At the 1-year and 2-year assessments, the OPUS treatment showed positive effect on almost every outcome measure. Participants who had received the OPUS treatment had fewer negative and psychotic symptoms, higher functional levels, and greater user satisfaction. The effect on negative symptoms was especially notable, as no treatment prior had shown such remarkable influence on these symptoms. Also, fewer participants were diagnosed with a comorbid substance or alcohol disorder. The OPUS patients were treated with lower doses of antipsychotic medication, and user satisfaction was higher in the OPUS group.[43] The OPUS trial was the first evidence in a large randomized clinical trial that specialized early intervention could affect early outcomes after a first-episode psychosis. During recent decades, other randomized clinical trials, carried out by research groups around the world, have also demonstrated that early intervention services can improve treatment outcomes in first-episode psychosis and bipolar disorder while treatment is ongoing.[44–51]

The OPUS intervention is considered a complex medical intervention and there is no specific knowledge regarding which part of the intervention is most effective.

As negative symptoms are strong predictors of long-term outcome, we carried out detailed analyses of the effects of OPUS treatment on various domains of negative symptoms. These analyses showed that the treatment did not amend any one of the four negative domains (affective flattening, alogia, anhedonia, apathy) more than the others. Rather, there was a mean global improvement on all domains, indicating that the case manager identified individual possibilities for improvement in the individual patient. It is believed that, given the assertive approach and smaller caseload, flexibility and time give the case manager and the multidisciplinary team opportunities to recognize where the patient is most ready and able and to choose appropriate approaches to remedy.[52]

Following the end of the 2-year intervention, patients were referred to treatment as usual, in most cases at community health centers. When participants were re-interviewed at the 5-year follow-up (3 years after end of the OPUS treatment), patients who had received OPUS treatment seemed to have relapsed to the symptomatic and functional level of the treatment as usual (TAU) patients (Figure 15.1). However, patients from the intervention group had fewer bed days and lower use of supported housing.[53] The 10-year follow-up did not show any differences between the OPUS and the TAU groups, except that use of supported housing, and thus the ability

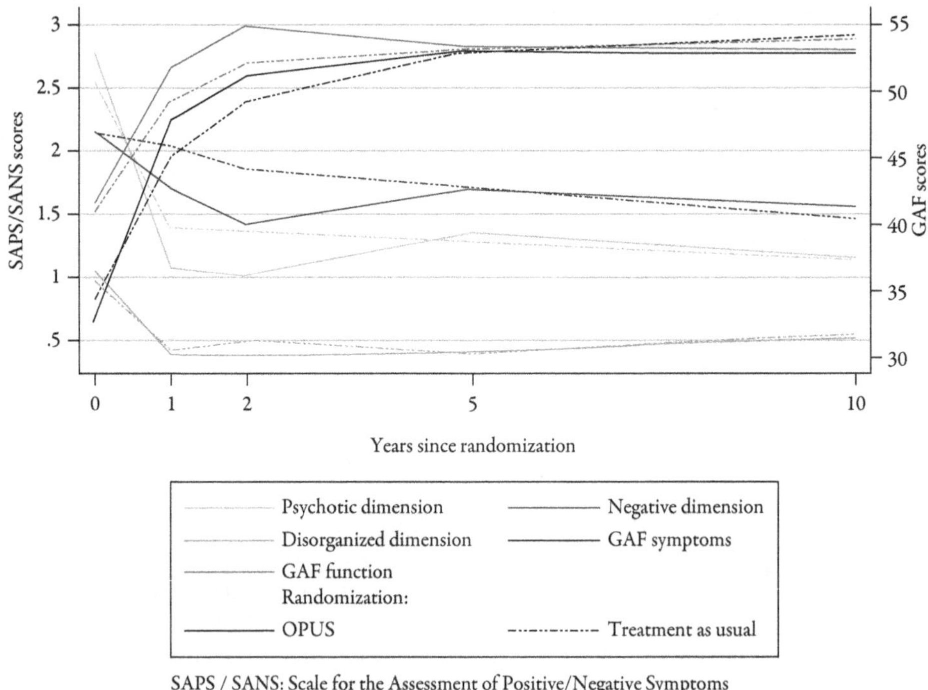

Figure 15.1 Psychopathological and functional outcomes at 1-, 2-, 5-, and 10-year follow-up.[54]

to live independently, was significantly higher in the OPUS group until the 7th year (5 years after the end of the intervention treatment.

Even with this partial positive long-term outcome, it was unanticipated that the intervention did not seem to have any long-term effects on the psychopathology or functional level of the patients when it had been so successful in remedying these outcomes during the intervention.

To investigate whether this relapse could be prevented, we initiated the OPUS II trial.[55,56] Based on the OPUS I trial, the OPUS treatment had been implemented all through Denmark and we therefore recruited participants to the OPUS II trial from the now-established OPUS teams.[42] In all, 400 participants were recruited and randomized to either the now-standard 2 years of OPUS treatment followed by TAU (in most cases, in community health centers) or the experimental 5-year OPUS treatment. Participants were recruited 18 months into their OPUS treatment, and follow-up interviews were conducted 3.5 years later (5 years after participants had started OPUS treatment). Our hypothesis was that participants randomized to 2-year OPUS treatment would relapse, as has occurred in the OPUS I trial, and that the good results could be maintained in the group that received 5 years of OPUS. Contrary to this prediction, at the 5-year follow-up we could not detect any differences between the two treatment groups on psychopathology, level of functioning, cognitive function, substance abuse, or hospitalizations. Nevertheless, participants in the 5-year OPUS group reported higher client satisfaction and a stronger working alliance with their contact person.

We did not find any sign of a relapse in either of the two treatment groups; rather, all participants seemed to improve on psychopathology, level of functioning, and cognitive function (Figure 15.2). We believe that this positive trajectory was due partly to developments in the TAU arm from the time of the OPUS I trial to the OPUS II trial. First, elements that were novel with the introduction of OPUS treatment, such as psychoeducation and family involvement, are now mandatory parts of the treatment provided in community health centers. Further, the use of assertive community treatment (ACT) is more widespread than at the time of the OPUS I trial. As many as 19% of the participants in the OPUS II trial, randomized to 2 years of OPUS treatment, had been in contact with an ACT team during the follow-up period. The ACT teams have a caseload of 10:1, and this means that for the most impaired participants we compared two interventions that, at least case-manger-wise, were equally intensive. We suggest as an interpretation of the results from the OPUS II trial that the positive effects of early intervention programs can be upheld either by prolonging treatment or by high-resourced TAU treatment with ACT for the most impaired patients.

LONG-TERM FOLLOW-UP OF THE OPUS I TRIAL

Studies of long-term outcomes are a key focus in schizophrenia research. There are rather few long-term studies, but several of them have identified a proportion of patients with psychotic illness with a good long-term outcome (e.g., recovery and remission and not all received antipsychotic medication).[1,23,57,58] These cohort studies provided valuable information on the heterogeneity of psychotic illness regarding course and outcome. As Figure 15.3 shows, at the 10-year

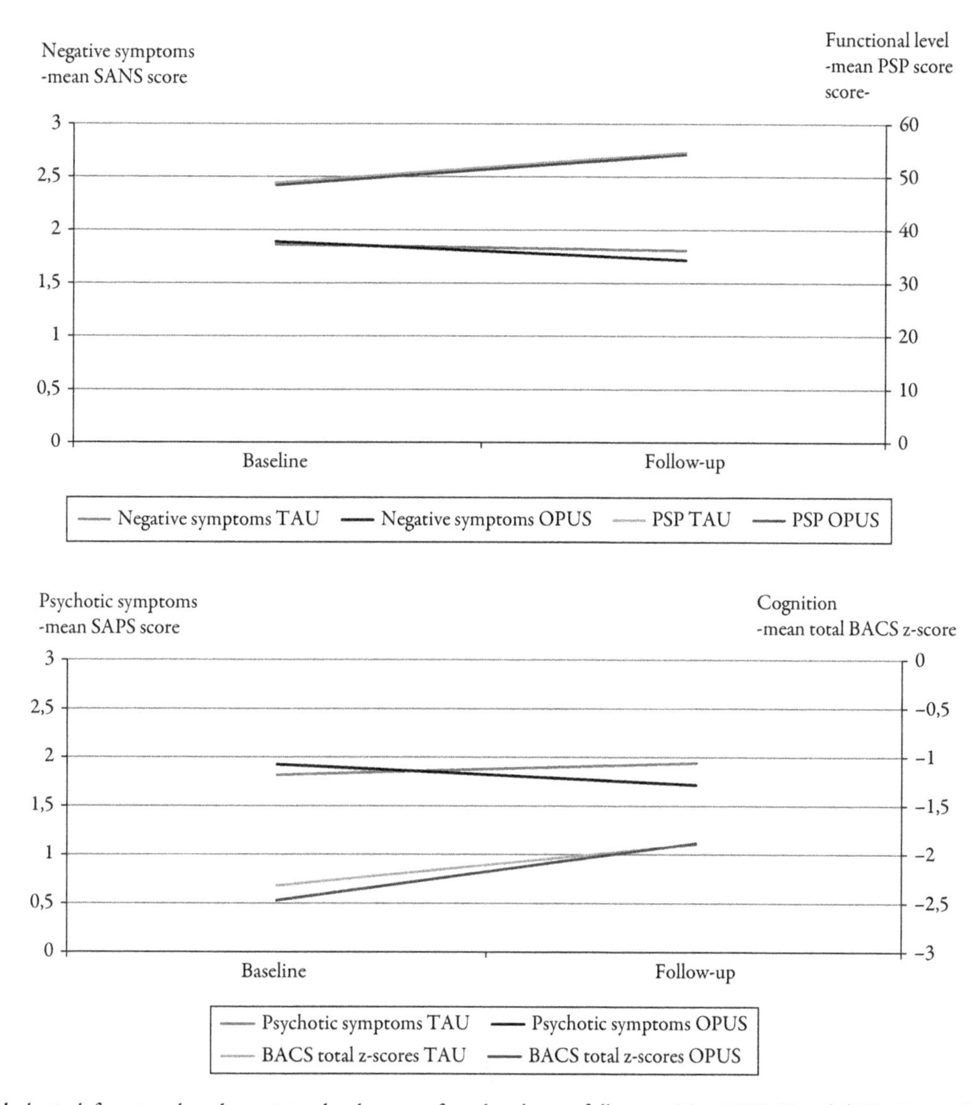

Figure 15.2 Psychopathological, functional, and cognitive development from baseline to follow-up. The OPUS II trial. (PSP: Personal and Social Performance scale, BACS: Brief Assessment of Cognition in Schizophrenia, TAU: Treatment as Usual)

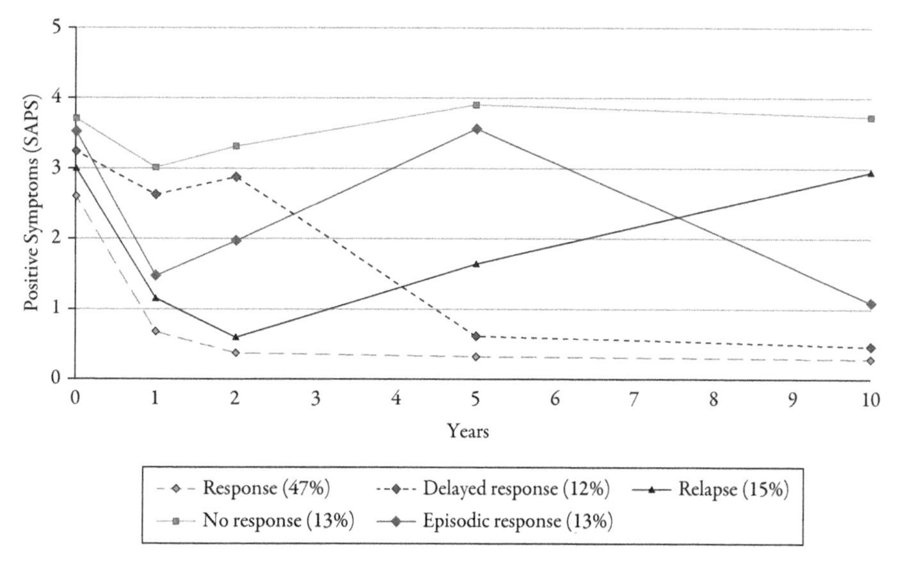

Figure 15.3 Psychotic symptom trajectories over 10 years.[59]

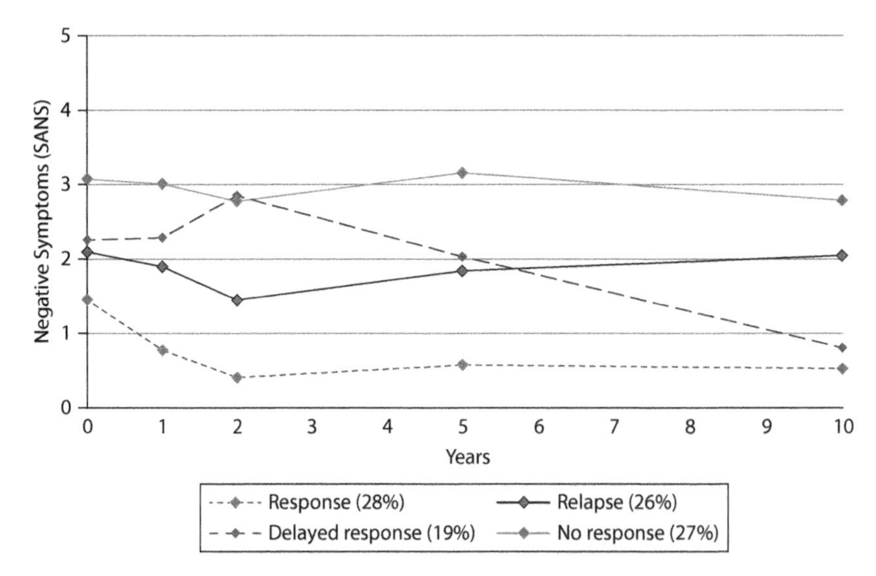

Figure 15.4 Negative symptom trajectories over 10 years.[59]

follow-up of the OPUS cohort, the majority of patients had achieved remission of psychotic symptoms.[59]

On the other hand, as Figure 15.4 shows, the 10-year follow-up of the OPUS cohort also showed that the trajectories of negative symptoms did not reveal the same optimistic picture, as a larger proportion still had marked negative symptoms after 10 years.

Negative symptoms have major influence on the daily level of functioning. Figure 15.5 shows the percentages of 400 participants in the OPUS II trial who were in competitive employment during the follow-up period.[60,61] The groups are categorized based on remission status of psychotic and/or negative symptoms at the baseline assessment. It is clear from the figure that remission of negative symptoms, compared to psychotic symptoms, was far more decisive for labor market affiliation.

In the 10-year follow-up study of the OPUS I cohort, patients were classified in four groups based on dichotomizing patients in remitted and non-remitted cases (based on any global SAPS score ≥3 versus all global SAPS scores < 3) and antipsychotic treatment (any/no) at time of follow-up (Table 15.1).[62,63]

These four categories were analyzed at several follow-ups, and a rather consistent pattern emerged, with approximately 25% who were in remission and had not received antipsychotic medications at each time of follow-up (Figure 15.6). Specific analysis demonstrated that the pattern was rather stable between 5- and 10-year follow-ups.[62,63]

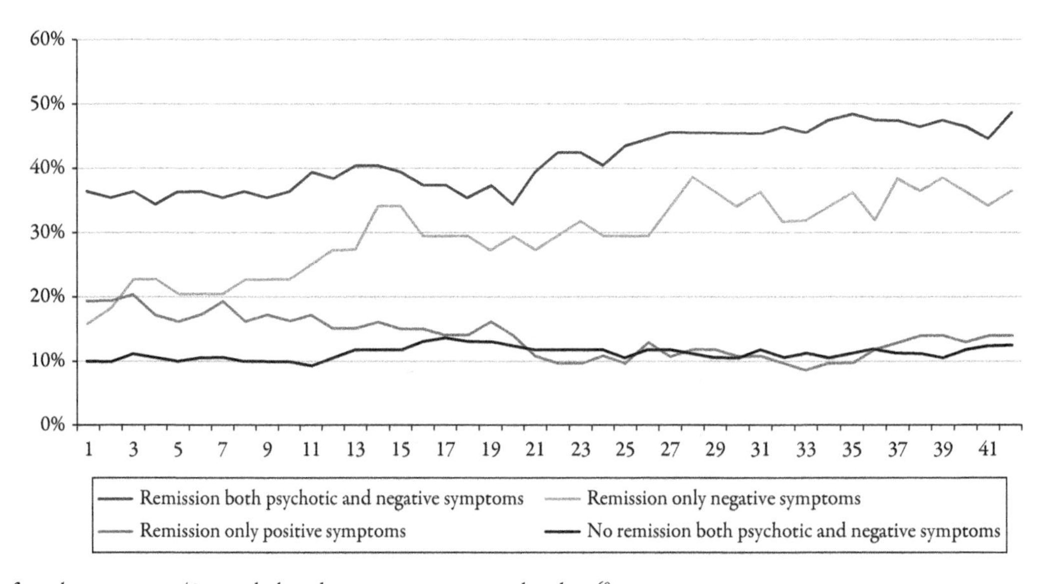

Figure 15.5 Rates of employment over 42 months based on remission status at baseline.[60]
Data sources: The OPUS II trial and Dream database.[61]

Table 15.1 CATEGORIZING OF PATIENTS 10 YEARS AFTER INITIATION OF TREATMENT BASED ON REMISSION OF PSYCHOTIC SYMPTOMS AND USE OF ANTIPSYCHOTIC MEDICATION.[62,63]

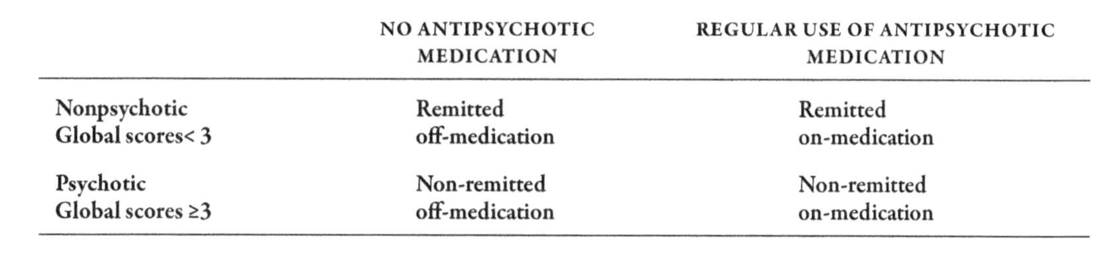

	NO ANTIPSYCHOTIC MEDICATION	REGULAR USE OF ANTIPSYCHOTIC MEDICATION
Nonpsychotic Global scores< 3	Remitted off-medication	Remitted on-medication
Psychotic Global scores ≥3	Non-remitted off-medication	Non-remitted on-medication

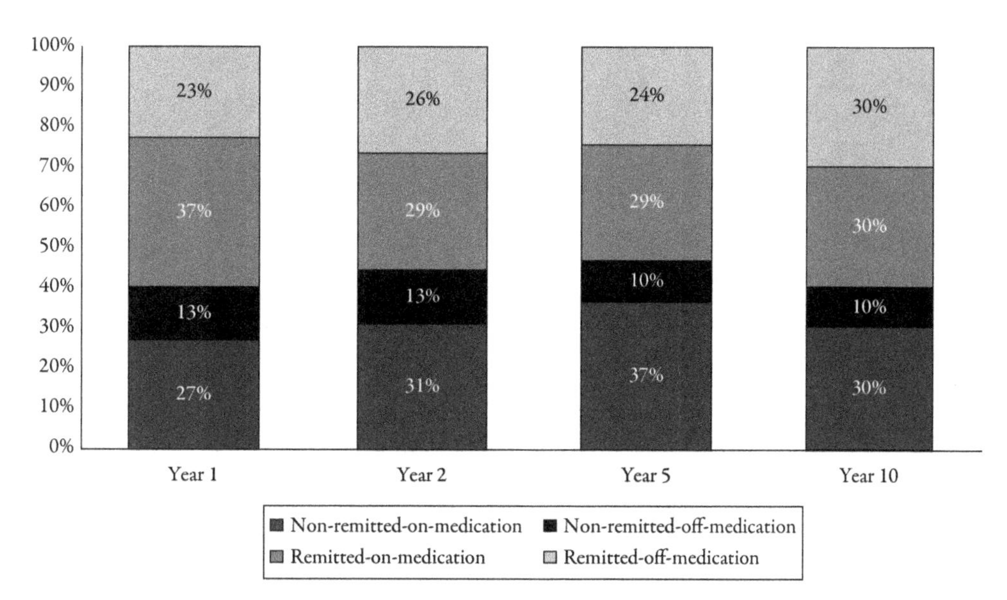

Figure 15.6 Distribution of patients remitted-off-medication, remitted-on-medication, non-remitted-off-medication, and non-remitted-on-medication by year of follow-up.[62]

REFERENCES

1. Morgan C, Lappin J, Heslin M, et al. Reappraising the long-term course and outcome of psychotic disorders: the AESOP-10 study. *Psychol Med.* 2014;44(13):2713–2726. doi:10.1017/S0033291714000282.
2. Global Burden of Disease Study 2013 Collaborators. Global, regional, and national incidence, prevalence, and years lived with disability for 301 acute and chronic diseases and injuries in 188 countries, 1990–2013: a systematic analysis for the Global Burden of Disease Study 2013. *Lancet.* 2015;386(9995):743–800. doi:10.1016/S0140-6736(15)60692-4.
3. Flachs E, Eriksen L, Koch M, et al. *[Sygdomsbyrden I Danmark], Burden of disease in Denmark.* Copenhagen; 2015.
4. Millier A, Schmidt U, Angermeyer MC, et al. Humanistic burden in schizophrenia: a literature review. *J Psychiatr Res.* 2014;54(1):85–93. doi:10.1016/j.jpsychires.2014.03.021.
5. Hjorthøj C, Østergaard MLD, Benros ME, et al. Association between alcohol and substance use disorders and all-cause and cause-specific mortality in schizophrenia, bipolar disorder, and unipolar depression: A nationwide, prospective, register-based study. *Lancet Psychiatry.* 2015;2(9):801–808. doi:10.1016/S2215-0366(15)00207-2.
6. Nordentoft M, Mortensen PB, Pedersen CB. Absolute Risk of Suicide After First Hospital Contact in Mental Disorder. *Arch Gen Psychiatry.* 2011;68(10):1058–1064. doi:10.1001/archgenpsychiatry.2011.113.
7. Jablensky A, Sartorius N, Ernberg G, et al. Schizophrenia: manifestations, incidence and course in different cultures: a World Health Organization ten-country study. *Psychol Med Monogr Suppl.* 1992;20:1–97. doi:10.1017/CBO9781107415324.004.
8. McGrath J, Saha S, Chant D, Welham J. Schizophrenia: a concise overview of incidence, prevalence, and mortality. *Epidemiol Rev.* 2008;30:7–11. doi:10.1093/epirev/mxn001.
9. Kühl JOG, Laursen TM, Thorup A, Nordentoft M. The incidence of schizophrenia and schizophrenia spectrum disorders in Denmark in the period 2000–2012: a register-based study. *Schizophr Res.* 2016. doi:10.1016/j.schres.2016.06.023.
10. Pedersen CB, Mors O, Bertelsen A, et al. A comprehensive nationwide study of the incidence rate and lifetime risk for treated mental disorders. *JAMA Psychiatry.* 2014;71(5):573. doi:10.1001/jamapsychiatry.2014.16.
11. Thorup A, Waltoft BL, Pedersen CB, Mortensen PB, Nordentoft M. Young males have a higher risk of developing schizophrenia: a Danish register study. *Psychol Med.* 2007;37(4):479–484. doi:10.1017/S0033291707009944.
12. van der Werf M, Hanssen M, Köhler S, et al. Systematic review and collaborative recalculation of 133,693 incident cases of schizophrenia. *Psychol Med.* 2014;44(1):9–16. doi:10.1017/S0033291712002796.
13. World Health Organization. Mental health action plan 2013–2020. *WHO Libr Cat Data.* 2013:1–44. doi:ISBN 978 92 4 150602 1.
14. Fusar-Poli P, Papanastasiou E, Stahl D, et al. Treatments of negative symptoms in schizophrenia: meta-analysis of 168 randomized placebo-controlled trials. *Schizophr Bull.* 2015;41(4):892–899. doi:10.1093/schbul/sbu170.
15. Rummel-Kluge C KWLS. Antidepressants for the negative symptoms of schizophrenia (Review). *Cochrane Database Syst Rev.* 2006;(3):1–55. doi:10.1002/14651858.CD005581.pub2.www.cochranelibrary.com.

16. Bleuler M. *The schizophrenic disorder: long term patient and family studies*. Westford, MA: Murray; 1978.
17. Ciompi L. Catamnestic long-term study on the course of life and aging of schizophrenics. *Schizophr Bull.* 1980;6(4):606–618. doi:10.1093/schbul/6.4.606.
18. Huber G, Gross G, Schiittler R, Linz M. Longitudinal studies of schizophrenic patients. *Schizophr Bull.* 1980;6(4):592–605.
19. Menezed NM, Arenovich T, Zipursky RB. A systematic review of longitudinal outcome studies of first-episode psychosis. *Psychol Med.* 2006;36:1349–1362. doi:10.1017/S0033291706007951.
20. Eaton WW, Thara R, Federman B, Melton B, Liang K. Structure and course of positive and negative symptomes in Schizophrenia. *Arch Gen Psychiatry.* 1995;52:127–134.
21. Thara R, Henrietta M, Joseph A, Rajkumar S, Eaton WW. Ten-year course of schizophrenia—the Madras longitudinal study. *Acta Psychiatr Scand.* 1994;90:329–336.
22. Carpenter WT, Strauss JS. The prediction of outcome in schizophrenia IV: eleven-year follow-up of the Washington IPSS cohort. *J Nerv Ment Dis.* 1991;179(9):517–525.
23. Harrison G, Hopper K, Craig T, et al. Recovery from psychotic illness: a 15-and 25-year international follow-up study. *Brit J Psychiat.* 2001;178:506–517.
24. Helgason L. Twenty years' follow-up of first psychiatric presentation for schizophrenia: what could have been prevented? *Acta Psychiatr Scand.* 1990;81:231–235.
25. Wyatt J. Neuroleptics and the natural course of schizophrenia. *Schizophr Bull.* 1991;17(2):325–351.
26. Loebel AD, Lieberman JA, Alvir JMJ, Mayerhoff DI, Geisler SH, Szymanski SR. Duration of psychosis and outcome in first-episode schizophrenia. *Am J Psychiatry.* 1992;149(9):1183–1188. doi:10.1176/ajp.149.9.1183.
27. Norman RMG, Malla AK. Duration of untreated psychosis: a critical examination of the concept and its importance. *Psychol Med.* 2001;31:381–400. doi:10.1017/S0033291701003488.
28. Ho BC, Andreasen NC, Flaum M, Nopoulos P, Miller D. Untreated initial psychosis: its relation to quality of life and symptom remission in first-episode schizophrenia. *Am J Psychiatry.* 2000;157(5):808–815. doi:10.1176/appi.ajp.157.5.808.
29. Marshall M, Lewis S, Lockwood A, Drake R, Jones P, Croudace T. Association between duration of untreated psychosis and outcome in cohorts of first-episode patients. *Arch Gen Psychiatry.* 2005;62(9):975–983. doi:10.1001/archpsyc.62.9.975.
30. Perkins DO, Gu H, Boteva K, Lieberman JA. Relationship between duration of untreated psychosis and outcome in first-episode schizophrenia: a critical review and meta-analysis. *Am J Psychiatry.* 2005;162(10):1785–1804. doi:10.1176/appi.ajp.162.10.1785.
31. Penttilä M, Jääskeläinen E, Hirvonen N, Isohanni M, Miettunen J. Duration of untreated psychosis as predictor of long-term outcome in schizophrenia: systematic review and meta-analysis. *Br J Psychiatry.* 2014;205(2):88–94. doi:10.1192/bjp.bp.113.127753.
32. ten Velden Hegelstad W, Larsen TK, Auestad B, et al. Long-term follow-up of the TIPS early detection in psychosis study: effects on 10-year outcome. *Am J Psychiatry.* 2012;169:374–380.
33. Birchwood M, Macmillan F. Early intervention in schizophrenia. *Br J Psychiatry.* 1993;170:2–5. doi:10.4103/0019-5545.42402.
34. Birchwood M. *Early intervention in psychosis: a guide to concepts, evidence and intervention.* (Birchwood M, Fowler D, Jackson C, eds.). Chichester, UK: John Wiley & Sons; 2002. www.wiley.com.
35. World Health Organization. *The ICD-10 classification of mental and behavioural disorders.* 1st edition. Geneva: World Health Organization; 1993.
36. Jorgensen P, Nordentoft M, Abel MB, Gouliaev G, Jeppesen P, Kassow P. Early detection and assertive community treatment of young psychotics: the Opus study; rationale and design of the trial. *Soc Psychiatry Psychiatr Epidemiol.* 2000;35(7):283–287. doi:10.1007/s001270050240.
37. Stein LI, Santos AB. *Community treatment of persons with severe mental illness.* New York, NY: Norton; 1998.
38. Liberman RP, Wallace CJ, Blackwell G, Kopelowicz A, Vaccaro J V., Mintz J. Skills training versus psychosocial occupational therapy for persons with persistent schizophrenia. *Am J Psychiatry.* 1998;155(8):1087–1091. doi:10.1176/ajp.155.8.1087.
39. Anderson CM, Reiss DJ, Hogarty GE. *Schizophrenia and the family: a practitioner's guide to psychoeducation and management.* 1986.
40. McFarlane WR. Multiple-family groups and psychoeducation in the treatment of schizophrenia. *Arch Gen Psychiatry.* 1995;52(8):679. doi:10.1001/archpsyc.1995.03950200069016.
41. Jeppesen P, Petersen L, Thorup A, et al. The association between premorbid adjustment, duration of untreated psychosis and outcome in first-episode psychosis. *Psychol Med.* 2008;38(8):1157–1166. doi:10.1017/S0033291708003449.
42. Nordentoft M, Melau M, Iversen T, et al. From research to practice: how OPUS treatment was accepted and implemented throughout Denmark. *Early Interv Psychiatry.* 2013(June 2010). doi:10.1111/eip.12108.
43. Petersen L, Jeppesen P, Thorup A, et al. A randomised multicentre trial of integrated versus standard treatment for patients with a first episode of psychotic illness. *BMJ.* 2005;331(7517):602. doi:10.1136/bmj.38565.415000.E01.
44. Craig TKJ. The Lambeth Early Onset (LEO) Team: randomised controlled trial of the effectiveness of specialised care for early psychosis. *BMJ.* 2004;329(7474):1067. doi:10.1136/bmj.38246.594873.7C.
45. Gafoor R, Nitsch D, McCrone P, et al. Effect of early intervention on 5-year outcome in non-affective psychosis. *Br J Psychiatry.* 2010;196(5):372–376. doi:10.1192/bjp.bp.109.066050.
46. Grawe RW, Falloon IRH, Widen JH, Skogvoll E. Two years of continued early treatment for recent-onset schizophrenia: a randomised controlled study. *Acta Psychiatr Scand.* 2006;114(5):328–336. doi:10.1111/j.1600-0447.2006.00799.x.
47. Kessing LV, Hansen HV, Hvenegaard A, et al. Treatment in a specialised out-patient mood disorder clinic v. standard out-patient treatment in the early course of bipolar disorder: randomised clinical trial. *Br J Psychiatry.* 2013;202(3):212–219. doi:10.1192/bjp.bp.112.113548.
48. Marshall M, Rathbone J. Early intervention for psychosis. *Cochrane Database Syst Rev.* 2011;(6):CD004718. doi:10.1002/14651858.CD004718.pub3.
49. Nordentoft M, Rasmussen JO, Melau M, Hjorthøj CR, Thorup A A E. How successful are first episode programs? A review of the evidence for specialized assertive early intervention. *Curr Opin Psychiatry.* 2014;27(3):167–172. doi:10.1097/YCO.0000000000000052.
50. Kane JM, Robinson DG, Schooler NR, et al. Comprehensive versus usual community care for first-episode psychosis: 2-year outcomes from the NIMH RAISE early treatment program. *Am J Psychiatry.* 2016;173(4):362–372. doi:10.1176/appi.ajp.2015.15050632.
51. Ruggeri M, Bonetto C, Lasalvia A, et al. Feasibility and effectiveness of a multi-element psychosocial intervention for first-episode psychosis: results from the cluster-randomized controlled GET UP PIANO trial in a catchment area of 10 million inhabitants. *Schizophr Bull.* 2015;41(5):1192–1203. doi:10.1093/schbul/sbv058.
52. Thorup A, Petersen L, Jeppesen P, et al. Integrated treatment ameliorates negative symptoms in first episode psychosis-results from the Danish OPUS trial. *Schizophr Res.* 2005;79(1):95–105. doi:10.1016/j.schres.2004.12.020.
53. Bertelsen M, Jeppesen P, Petersen L, et al. Five-year follow-up of a randomized multicenter trial of intensive early intervention vs standard treatment for patients with a first episode of psychotic illness: the OPUS trial. *Arch Gen Psychiatry.* 2008;65(7):762–771. doi:10.1001/archpsyc.65.7.762.
54. Secher RG, Hjorthoj CR, Austin SF, et al. Ten-year follow-up of the OPUS specialized early intervention trial for patients with a first episode of psychosis. *Schizophr Bull.* 2015;41(3):617–626. doi:10.1093/schbul/sbu155.
55. Melau M, Jeppesen P, Thorup A, et al. The effect of five years versus two years of specialised assertive intervention for first

episode psychosis—OPUS II: study protocol for a randomized controlled trial. *Trials*. 2011;12(1):72. doi:10.1186/1745-6215-12-72.
56. Albert N, Melau M, Jensen H, et al. Five years of specialised early intervention versus two years of specialised early intervention followed by three years of standard treatment for patients with a first episode psychosis: randomised, superiority, parallel group trial in Denmark (OPUS II). *BMJ*. 2017;356:i6681. doi:10.1136/bmj.i6681.
57. Harrow M, Jobe TH, Faull RN. Do all schizophrenia patients need antipsychotic treatment continuously throughout their lifetime? A 20-year longitudinal study. *Psychol Med*. 2012;42(10):2145–2155. doi:10.1017/S0033291712000220.
58. Moilanen J, Haapea M, Miettunen J, et al. Characteristics of subjects with schizophrenia spectrum disorder with and without antipsychotic medication: a 10-year follow-up of the northern Finland 1966 birth cohort study. *Eur Psychiatry*. 2013;28(1):53–58. doi:10.1016/j.eurpsy.2011.06.009.
59. Austin SF, Mors O, Budtz-Jørgensen E, et al. Long-term trajectories of positive and negative symptoms in first episode psychosis: a 10-year follow-up study in the OPUS cohort. *Schizophr Res*. 2015;168(1–2):84–91. doi:10.1016/j.schres.2015.07.021.
60. Albert N. Specialised early intervention in first episode schizophrenia spectrum disorders, the effect of prolonging the specialised treatment from two to five years compared to treatment as usual. The OPUS II trial. 2016.
61. Statistics Denmark. The DREAM database, Statistics Denmark. http://www.dst.dk/da/TilSalg/Forskningsservice/Data/Andre_Styrelser.aspx#.
62. Gotfredsen D, Wils RS, Hjorthøj C, et al. Stability and development of psychotic symptoms and the use of antipsychotic medication—long-term follow-up. *Psychol Med*. 2017. https://doi.org/10.1017/S0033291717000563.
63. Wils RS, Gotfredsen DR, Hjorthøj C, et al. Antipsychotic medication and remission of psychotic symptoms 10 years after a first-episode psychosis. *Schizophr Res*. 2016. doi:10.1016/j.schres.2016.10.030.

16

EVIDENCE-BASED TREATMENT AND IMPLEMENTATION FOR EARLY PSYCHOSIS

Sacha Zilkha, Iruma Bello, Hong Ngo, Samantha Jankowski, and Lisa Dixon

The last decade has witnessed enthusiasm and promise for early intervention services in psychotic disorders, particularly schizophrenia. It is hoped that providing the right services early can alter the course and trajectory of psychotic illness by increasing rates of recovery including employment, life satisfaction, and community tenure.[1–4] The scientific foundation for this hopefulness rests on two fundamental observations. First, that shorter duration of untreated psychosis (DUP) has been associated with better short- and long-term symptom and functional outcomes when compared to longer DUP.[5–8] While causation has not been established, the implication of this association is that shortening the DUP may improve outcomes. The second observation is that specific treatments and treatment packages, called coordinated specialty care (CSC) in the United States, improve outcomes when compared to usual care at least in the short run.[4] However, achieving better outcomes and recovery over the long term with early intervention services depends on the extent to which these benefits persist over time.

This chapter provides an overview of key studies addressing the promise of early intervention services, organized around the notion of "before," "during," and "after" CSC care. "Before" refers to the challenge and importance of reducing DUP and treating psychosis at its earliest phase. While we appreciate that psychotic illnesses likely start prior to the onset of psychotic symptoms, perhaps even before the clinical high risk period, or "prodrome," we focus here only on the DUP. "During" refers to CSC treatment and the evidence for improved outcomes. "After" refers to the evidence for long-term impact. Space precludes full examination of each of these issues given the proliferation of research on these topics. We then describe the challenges and opportunities of large-scale implementation of a multi-element CSC program in New York State, OnTrackNY, focusing on its attempts to optimize impact before, during, and after CSC treatment.

BEFORE COORDINATED SPECIALTY CARE: THE IMPORTANCE OF DURATION OF UNTREATED PSYCHOSIS

The importance of treating illnesses as close as possible to their onset is a guiding principle of the provision of medical care across a variety of disorders. Such treatment can provide secondary and tertiary prevention, and often prevent disease progression. The lag between the onset of psychosis and receipt of treatment is referred to as the "duration of untreated psychosis" (DUP). While the definition of DUP is straightforward, the construct is more complex. In most studies, treatment initiation is considered to be the start of antipsychotic medications.[9] However, other studies have defined treatment as including psychosocial interventions, hospital admission, or other contact with a treatment provider.[10–13]

Given the large number of individual studies that examine the association between DUP and outcomes, we will focus on important meta-analyses.[10,12,14] Marshall and colleagues[12] evaluated an array of outcomes including depression, anxiety, social functioning, overall functioning, quality of life, positive symptoms, negative symptoms, rates of remission, time to remission, and relapse at baseline, 6, 12, and 24 months. The study included 4,490 people over the age of 16 but under 60 with first episode psychosis (FEP) from 26 studies. These studies defined DUP as the time from psychosis onset to neuroleptic treatment or hospital admission. The mean DUP for all studies was 124 weeks, which declined to 103 weeks after removing one outlier from the final analyses. Correlations between DUP and outcomes at baseline were limited; DUP was significantly correlated with symptoms of depression and anxiety, and quality of life. However, at follow-up time points, there were consistent negative correlations between DUP and an array of outcomes. Specifically, at 6 months, longer DUP was significantly correlated with greater levels of positive symptoms, negative symptoms, depression, and anxiety as well as reduced overall and social functioning and lower rates of achieving remission. At 12 months, longer DUP was significantly correlated with more positive, negative, depression, and anxiety symptoms, lower quality of life, and lower overall functioning, and individuals with longer DUP were not as likely to be in remission and took longer to achieve remission. While only 2 studies followed patients for 24 months, the link between longer DUP and greater positive symptoms, poorer quality of life, and overall functioning persisted. Importantly, Marshall and colleagues[12] identified nine studies that controlled for premorbid functioning as this might confound results, and found that premorbid functioning did not explain the associations between DUP and outcomes.

Boonstra and colleagues[14] focused their meta-analysis on negative symptoms and on extending the follow-up period beyond 2 years. The authors identified 16 eligible studies that included 3,339 participants with firs-episode nonaffective psychosis with a mean DUP of 61.4 weeks. A definition of DUP was not included in the study inclusion/exclusion criteria. As predicted, shorter DUP was associated with lower severity of negative symptoms at baseline, short-term follow-up (1–2 years) and long-term follow-up (5–8 years). Of note, individuals with DUP less than 9 months had significantly greater improvements in negative symptoms than those with longer DUP.

Pentilla and colleagues[10] expanded consideration of the association of DUP to longer-term outcomes, focusing on studies with a minimum length of follow-up of 2 years. The authors included 39 articles from 33 samples. Sample size ranged from 23 to 776. This study did not solely focus on first episode samples but included studies in which the majority of participants had received a diagnosis of schizophrenia. Individual studies in this meta-analysis defined DUP as the time between onset of psychosis and treatment initiation, which included antipsychotic medication, hospital admission, psychosocial treatment, or contact with treatment services. The mean DUP was 61.3 weeks. Results indicated that longer DUP was significantly associated with poor overall symptom outcomes, higher severity of negative and positive symptoms, inability to reach remission, reduced social functioning, and worse global functioning based on indices such as the Global Assessment of Functioning or a combination of unspecified social and clinical factors.[10]

The results of the recent National Institute of Mental Health (NIMH) Recovery After an Initial Schizophrenia Episode Early Treatment Program (RAISE ETP) study are also informative regarding the potential impact of DUP.[4] In this study, DUP was defined as the time between psychosis onset and first time antipsychotic medication treatment.[9] The study included 404 individuals with first-episode nonaffective psychosis between 15 and 40 years of age with a mean DUP of 193.5 weeks and a median DUP of 74 weeks. Using a cluster randomized design, outcomes of 223 participants receiving "NAVIGATE," the CSC program, were compared to 181 individuals receiving usual care after 2 years of treatment. Notably, DUP moderated the effects of NAVIGATE such that individuals with a DUP less than 74 weeks benefited significantly more on quality of life and symptom measures compared to those with a DUP greater than 74 weeks.[4]

Overall, evidence for the association between DUP and outcome is robust. At the same time, a causal linkage has not been firmly established. An array of studies have attempted to identify biological correlates of DUP and outcomes that could provide a causal roadmap.[11] Some of the driving hypotheses being explored relate to the notion that psychosis is itself toxic to the brain and that taking antipsychotic medications or participating in other treatments could arrest that process. Although reviewing this specific body of literature is beyond this chapter, what remains clear is that more studies are necessary to fully understand the extent of any causal relationship between DUP and outcomes.

A logical corollary of the potential link between DUP and outcomes is that reducing DUP should improve outcomes. An expanding body of research has focused on reducing DUP. In a review of DUP reduction interventions, Lloyd-Evans and colleagues (2011) evaluated 11 DUP intervention studies, which included 8 discrete interventions aimed at increasing early psychosis detection and timely treatment. Three of the interventions included education campaigns for general practitioners to identify early signs of psychosis and encouraging timely referral to care; the remaining initiatives involved a multi-intervention approach. These multi-element interventions included large-scale public service announcements across various media outlets, outreach to schools, face-to-face and written contact with general practitioners and other healthcare providers, and a telephone line for the public to call for advice. The review surmised that educating general practitioners alone has little impact on reducing DUP and similarly establishing early intervention services in and of themselves probably had little impact on DUP. Rather, taking a multifocused approach, including interventions targeting multiple audiences through various modes of communication appeared to yield the greatest benefits.[15]

The Treatment Intervention in Psychosis Study (TIPS)[16] demonstrated how a successful early intervention service in Norway decreased DUP and improved outcomes for individuals with early psychosis over time. The TIPS implemented a targeted early psychosis detection program in two health sectors and compared individuals living in those sectors to individuals living in two other health sectors without a specialized early psychosis detection program. However, each health sector had equivalent early psychosis treatment centers. The early detection initiative included an educational campaign aimed at the general population focused on psychosis and a more targeted campaign directed at social workers, high school healthcare providers and general health practitioners. The campaigns included a telephone number that potential patients or other referral sources could call to connect with a specialized early detection team. TIPS initially enrolled 141 individuals in the early detection sectors and 140 individuals in non-early-detection sectors. The study included baseline, 2 year, 5-year, and 10-year follow-up points. At baseline, 2 years, and 5 years, individuals in the early detection group experienced fewer positive and negative symptoms, and at years 2 and 5, they had fewer cognitive and depressive symptoms. At 10 years, the early detection group was more likely to have achieved recovery as defined by a combination of symptomatic and functional outcomes including employment. While DUP itself was not a predictor of recovery, this study underscores the possible benefits to early detection in treating individuals with early psychosis.

In summary, the significant association between DUP and outcomes has challenged the field to focus on early detection prior to treatment with CSC. More work is needed to determine how to shorten DUP and understand how DUP may influence outcomes.

DURING COORDINATED SPECIALTY CARE

Research studies conducted globally have contributed to the evidence base supporting treatment approaches for early psychosis. Studies have focused on specific treatment components (i.e., single-element studies), as well as multi-element team-based approaches. Single-element studies have examined low dose atypical antipsychotic medication,[17–19] family therapy,[20,21] cognitive behaviorally oriented therapy,[22–24] group therapy,[25,26] and work and school supports.[27,28] While these evidence-based treatments have shown promise on their own and have been recommended to treat schizophrenia across the life span,[29] multi-element services that combine each of the single elements have demonstrated consistently better short-term outcomes.[1,2,4,30–33] The most recent studies conducted in the United States have led to the creation of the label "coordinated specialty care" (CSC) to represent these team-based multi-element programs. In what follows we briefly describe some of the key multi-element randomized trials.

The first randomized control trial (RCT) of multi-element care for early psychosis was the OPUS study, which took place in Denmark.[32] In this study, 547 individuals ages 18–45 with psychosis who had no more than 12 weeks exposure to antipsychotic medications were recruited between 1998 and 2000.[34] These individuals were randomly assigned to multi-element care (OPUS) or treatment as usual (TAU). Individuals randomized to OPUS received services for 2 years based on the assertive community treatment model, which included individualized case management, family groups, low-dose antipsychotic medications, and, when indicated, cognitive-behavioral therapy (CBT) and social skills training. OPUS clinicians preferred to see individuals at home, though the treatment team could see the individual at another location in the community or in the office, and the patient to staff ratio was 10:1. In contrast, individuals in the TAU group had monthly meetings with a psychiatric nurse in a community mental health center, consultations with a social worker, and medication when indicated. Home visits were infrequent, and participants consulted psychiatric emergency departments for care after office hours. The patient to staff ratio was 25:1.[34] OPUS treatment lasted for 2 years, after which participants received TAU. After 2 years of treatment, OPUS participants had lower levels of positive and negative symptoms of psychosis, reductions in substance use, and increased engagement in and satisfaction with treatment compared to individuals in the control group.[35]

The Lambeth Early Onset (LEO) study took place in the United Kingdom and was the second RCT to test the impact of multi-element care for early psychosis.[30] Participants were 144 individuals living in London, England, between the ages of 16 and 40 with nonaffective psychosis who did not have a primary substance use disorder or an organic psychosis; they were randomized to receive multi-element care or standard care. These individuals were eligible if this was their first or second time seeking mental health services.[36] In this study, multi-element care lasted 18 months and included atypical antipsychotic medications at low doses, CBT, family therapy, and vocational services. Standard care involved treatment by teams not trained in specialized services for early psychosis at a local community mental health center in the Lambeth section of London. Individuals who received the specialized intervention had fewer hospital readmissions, better medication adherence, better occupational functioning and quality of life compared to those who do not receive the specialized care, and were more likely to stay in the study.[30,36]

Grawe and colleagues[31] conducted a small-scale RCT in Norway for individuals between the ages of 18 and 35 who had schizophrenia for less than 2 years. Thirty participants were randomized to receive team-based, multi-element care, and 20 participants were randomized to receive standard treatment. Services in both the multi-element treatment team and standard care conditions were multimodal and intensive, with case management, antipsychotic medications, and low patient-to-staff ratios. Unlike the control group, the treatment group received specialized team-based care that included CBT skills, family psychoeducation, CBT for family communication, and crisis services in the home. After 2 years, individuals in the experimental condition demonstrated significantly better outcomes in terms of negative and positive symptoms compared to participants receiving standard care, however number of hospitalizations and major psychotic episodes did not differ significantly between groups.[31]

The largest multi-element study to date was conducted across 10 clinical sites in China by Guo and colleagues.[3] The study enrolled 1,268 individuals 16–50 years of age with an onset of psychosis within 5 years. Participants were randomly assigned to the control condition, which involved antipsychotic medications and monthly medication management visits, or to the experimental condition, which involved antipsychotic medications and monthly medication management visits in addition to 4 straight hours of psychosocial groups, which included psychoeducation, family psychoeducation and support, skills training, and CBT one time monthly for 12 months. Families were required to bring participants to the clinic for treatment in both conditions. By the end of the study, 406 individuals received the multi-element intervention and 338 individuals received medication alone. After 1 year, individuals in the experimental group had significantly greater improvement in insight, social functioning, obtaining employment or education, activities of daily living, and quality of life in addition to lower rates of "clinical relapse," which was defined by worsening symptoms, hospitalization, need for increased level of psychiatric treatment, self-harming behaviors, or violent behaviors.

The Recovery After an Initial Schizophrenia Episode (RAISE) projects and the Specialized Treatment Early in Psychosis (STEP) program[33] were the foundational US-based studies of CSC. The RAISE ETP study,[4] as mentioned earlier, compared NAVIGATE to TAU among individuals who were between 15 and 40 years of age, diagnosed with nonaffective psychosis, who had only experienced one episode of psychosis and were treated with antipsychotic medications for 6 months or less. NAVIGATE included evidence-based prescribing of antipsychotic medications, family psychoeducation, and supported employment and education services.

The individual therapy component, titled individual resiliency training, consists of CBT-based strategies for symptoms management, skills training, and substance abuse treatment using shared decision-making with a focus on promoting individual resilience, recovery, and goal attainment.[37] Treatment as usual included available community services as determined by clinicians within community clinics who did not receive specialized training in or had experience offering specialized early psychosis treatment. Individuals receiving NAVIGATE remained in treatment longer, had more improvement on quality of life measures, were more likely to have a job or be in school, and experienced greater symptom reduction compared to participants in the TAU clinics after 2 years of treatment.[4]

Srihari et al.[33] compared their comprehensive early psychosis program (STEP), which included antipsychotic medications, CBT, family education, and case management to help individuals access education and employment supports, to TAU. Treatment received in the TAU condition varied because it was determined by the participant's current provider or by an outside treatment provider to whom they were referred. The sample included 120 individuals who had an onset of psychosis of less than 5 years before entry into the study and fewer than 12 weeks of exposure to antipsychotic medications. After 1 year of treatment, STEP participants had fewer total hospital admissions, fewer hospital days and a greater likelihood of being employed or in school.

The study of CSC both affirms our knowledge of the utility of early specialized treatment and underscores gaps that still remain in the knowledge-base. While it is evident that there are improved outcomes associated with delivering multi-element treatment models, there is no evidence to confidently indicate which of these elements is most critical. More research is needed to determine the extent to which enhancing treatment components to current CSC models in areas such as cognitive impairment, substance use and suicide prevention, might further improve outcomes. Furthermore, it is not easy to compare CSC studies to generalize results more globally because the elements of CSC differ from one location to the next, patient inclusion criteria varies greatly, and there is heterogeneity in terms of outcomes measured. While the data are promising, many questions remain unanswered. CSC requires further elaboration of what, for whom and for which outcomes.

AFTER COORDINATED SPECIALTY CARE

If CSC is a time-limited service, its ultimate promise to change the long-term outcomes of schizophrenia rests on the extent to which the benefits of CSC extend beyond the duration of program participation. The key studies described here examine the persistence of benefits of CSC over time and emerging strategies to enhance long-term benefits.

Follow-up of both LEO and OPUS participants revealed that benefits of these models decayed over time. In 2005, a follow-up study to the LEO study assessed participants between three and a half and five years after initiating treatment.[30] Seventy percent of the original study participants were included in this follow-up. Despite significant differences in hospital admissions and readmissions obtained between the control group and those who received specialized treatment at 18-month follow-up, there were no significant differences in hospitalization rates or days in the hospital between groups 3.5 to 5 years after individuals initiated treatment. The OPUS study also examined outcomes post-specialized treatment at 5- and 10-year follow-up points after treatment initiation. After 5 years, many of the treatment benefits dropped off including differences in substance use, antipsychotic medication use, rates of remission, and positive or negative symptoms of psychosis.[38] However, individuals who received the specialized intervention were significantly less likely to be living in supported housing and, unlike in LEO, these individuals had fewer hospital days than individuals who received standard care.[38] At the 10-year mark, individuals who were in the experimental group were less likely to be in supportive housing and had fewer hospital days compared to those individuals who in standard care, but no other differences were observed.[34]

In addition to measuring long-term outcomes following CSC, several studies have looked at the impact of extending CSC. In 2011, Norman and colleagues published results from a nonexperimental study in Ontario Canada, which assessed 2- and 5-year outcomes for 132 individuals enrolled in the Prevention and Early Intervention Program for Psychoses (PEPP).[39] PEPP provided high-intensity care to individuals with nonaffective psychosis for 2 years with lower intensity services for an additional 3 years. During the initial two years of PEPP, participants received integrated medical and psychosocial treatment following an assertive case management model, which included family members, community resources, educational partners, and employers to support community integration.[40] After 2 years of treatment, the majority of participants demonstrated symptom stability and had their services reduced to only psychiatrist visits. Individuals were assessed every 3–6 months, and if there was a worsening of symptoms, service intensity would increase. Individuals receiving PEPP for 2 years, and lower-intensity care for 3 additional years, demonstrated symptom reduction and continued improvement in global functioning from years 3 to 5. Norman and colleagues compared PEPP outcomes with those of the OPUS trial. At 2 years there were no differences found between PEPP and OPUS participants in terms of social functioning or positive symptoms, however, PEPP participants and those OPUS participants in the control condition, had higher levels of negative symptoms than the OPUS participants receiving multi-element care. After 5 years, PEPP participants appeared to have lower severity of positive symptoms and improved social functioning compared to both the OPUS experimental and control conditions[40] suggesting that they benefited from continued services.

The first published RCT to test the effect of extending specialty care for early psychosis took place in Hong Kong and compared the impact of 2 years of the Early Assessment Service for Young People (EASY) and an additional year of specialized treatment to 2 years of EASY and a year of "step-down care."[41] EASY was established in 2001 and provides community-awareness efforts, an open system of referral and two years of specialty care for individuals ages 15 to 25 with FEP.[42] EASY includes medication services, psychoeducation, stress management, coping skills, and relapse prevention, in addition to family and caregiver support. Cognitive-behavioral therapy is provided when psychotic symptoms are deemed "treatment resistant." On average, in a 1-year period, individuals meet with the team for a total of 16 total sessions of the intervention. In this trial, 82 individuals were randomized to the extended specialty care group, receiving EASY for 2 years followed by an additional year of case management that aligned with EASY's model and focused on re-establishment of social networks and returning to work. Seventy-eight individuals were randomized to 2 years of EASY and 1 year of "step-down care," which consisted of follow-up with a psychiatrist, antipsychotic medications, and crisis services. At the end of treatment, individuals having received 3 years of specialty care had fewer negative and depressive symptoms, better global functioning, and attended outpatient appointments more frequently than those in the step-down group. Chang and colleagues conducted a follow-up to this study to evaluate long-term benefits of the specialized care extension, two years post the three years of treatment. This follow-up included 67 individuals from the step-down treatment group and 76 from the specialty care group. Despite initial post-treatment benefits, no significant differences on the previous outcomes measured were found at the 3-year follow-up.[42]

The OPUS II trial tested an extension of OPUS treatment, comparing 197 participants who received 5 years of OPUS versus 203 participants who received 2 years of OPUS in addition to 3 years of TAU.[43] After 5 years of OPUS treatment, individuals maintained contact with specialized services more often, had stronger alliances with providers, and had higher satisfaction with services compared to individuals who received 2 years of OPUS. However, negative symptoms, positive and negative symptom remission, substance use, medication adherence, suicidal ideation, work, and hospital days did not differ significantly between groups.[43]

While it is clear that CSC provides symptom, social, and occupational improvements for individuals with early psychosis, benefits after specialized care appear to drop off over time.[30,34,38] To better understand the maintenance of CSC treatment gains over time, researchers have tested specialized early intervention services that go beyond the standard 1 to 2 years of treatment typically offered.[40,41,43] However, it seems presumptive to draw conclusions, since only a few studies have examined extended treatment and results have been equivocal and inconsistent thus far.[40,41,43] This work will continue, and likely requires focusing on the heterogeneity of illness, defining subgroups of patients who benefit from extended treatments and predictors for successful treatment outcomes.

ONTRACKNY: FROM RESEARCH TO PRACTICE

BACKGROUND AND OVERVIEW

Research examining DUP, CSC, and post-CSC outcomes is continuing to evolve. Variants of early detection and delivery of CSC have become standard in many high-income countries. The United States has lagged behind, a situation that is now changing due to the availability of federal funding to support the development and evaluation of CSC.[44]

Early intervention programs across the United States have been expanding rapidly. This expansion has led to a better understanding of what it takes to implement CSC in real-world settings. In the following sections, we describe issues in the development and policies of OnTrackNY, New York State's CSC program, to demonstrate how this program has tried to meet the challenges of optimizing services before, during, and after CSC treatment. The name "OnTrackNY" was chosen to epitomize its positive and hopeful mission.

OnTrackNY evolved from the RAISE Implementation and Evaluation Study,[2] which developed and tested a CSC program called Connection. The research phase ended in June 2013, and RAISE Connection transitioned to OnTrackNY in June 2013. At that time, external review from leaders in the field was sought, and the program was modified based on expert feedback. OnTrackNY leadership presumed that while many CSC studies were positive, the evidence base did have significant limitations, and the model would evolve and encompass best practices in response to emerging science as well as local experience and adaptation. Initial state funding supported four teams. In 2014, House of Representatives Bill 3547 allocated 25 million dollars set aside to support evidence-based early psychosis programs as part of the Community Mental Health Block Grant program, an amount that was doubled two years later. This increased funding allowed OnTrackNY to expand to a total of 21 teams, which have enrolled 815 individuals thus far.

BEFORE ONTRACKNY

OnTrackNY teams use a robust outreach, recruitment, and evaluation strategy to reduce DUP and connect individuals to early intervention services quickly. Each OnTrackNY team has an outreach and recruitment coordinator (ORC) who builds and maintains collaborative relationships in the community, increasing the likelihood of referrals to the program, identifying gaps in early detection, and reducing stigma. The outreach strategy is systematic, comprehensive, and tier-based, so that outreach is first conducted in places that are likely to yield the highest number of referrals (e.g., inpatient units) and gradually expanded to other referral sources (e.g., schools). Strategic outreach and recruitment activities inform the community and potential referral sources (providers, service seekers, and family members) about the program's eligibility criteria and the treatment offered.

The ORC is also responsible for conducting screening and eligibility evaluations. We have found that having a designated individual with protected time to do these tasks typically facilitates rapid engagement and program enrollment. OnTrackNY teams respond to inquiries and referrals within 24 hours. As quickly as possible, the ORC schedules a meeting with the referred individual and family (separately or together) to establish rapport, describe program services, and if possible begin to conduct a formal eligibility evaluation in locations that are convenient for the individual and family.

The eligibility evaluation primarily consists of a timeline assessment to determine whether the referred individual has a primary nonaffective psychotic disorder and will be best served by the CSC treatment. This assessment allows for a broader understanding of when symptoms of psychosis were present and under what context, and to determine differential diagnosis and onset of psychotic symptoms. For each psychotic symptom, the level of intensity, frequency, and impact on behavior and/or functioning, and degree of conviction for delusions are determined. Eligible individuals are between the ages of 16 and 30, with psychotic symptoms present for at least 1 week and less than 2 years. Exclusionary criteria include affective mood disorders (e.g., bipolar disorder and depression with psychotic features), substance-induced psychosis, psychosis due to a general medical condition, intellectual disabilities or autism spectrum disorder, or serious or chronic medical illness significantly impairing function independent of psychosis.

Assessment challenges: eligibility determination

Eligibility determinations can be difficult given the low incidence of FEP and the challenge of accurately assessing the onset of psychotic symptoms. Clinicians sometimes struggle with differentiating FEP from other mental illnesses such as primary mood disorders with psychotic features, substance-induced psychosis, obsessive-compulsive disorder, posttraumatic stress disorder, personality disorders, and prodromal schizophrenia. In order to address this need, the ORCs within OnTrackNY receive enhanced training on differential diagnosis and DSM-5 diagnostic criteria for mental illnesses, and validated assessment tools, and they participate in frequent discussion using case examples to develop their expertise. Having an experienced clinician accessible to the team for discussing diagnostically complex incoming referrals also helps to increase the accuracy of eligibility determinations.

DURING ONTRACKNY: CLINICAL AND TREATMENT COMPONENTS

OnTrackNY treatment is person-centered, recovery-focused, and culturally competent, and uses a shared decision-making (SDM) framework with a primary goal of helping young individuals maintain and achieve their goals for school, work, and social relationships. Shared decision-making is a collaborative process that provides specific steps for individuals and their providers to make healthcare decisions together, taking into account the best scientific evidence available and the person's values and preferences.[45] The young person becomes the central member of the team's efforts, and the treatment plan is guided by the individual's life goals and ambitions. An important tenet for this service is the use of assertive outreach strategies that help engage and retain program participants and families over time. The team uses a flexible approach, which allows meetings to happen in the community as needed, with an emphasis on conveying hope for recovery.

The treatment components of OnTrackNY consist of *evidence-based prescribing* practices that prioritize the lowest effective doses of antipsychotic medications with the fewest side effects; a focus on *health and wellness/primary care coordination*, which allows for monitoring and treating cardiometabolic factors associated with antipsychotic medications; *case management* to help individuals and families with concrete needs; *cognitive-behavioral based therapy*, which provides a general supportive approach coupled with coping skills to promote resiliency; *family support and education* to promote family involvement; *supported education and employment* services based on the individual placement and support model; and *peer support* services to enhance engagement and provide additional support and advocacy for program participants.

Challenges with clinical concepts: engagement, recovery, SDM, and cultural competency

Building a sustained connection with participants and families has proven to be challenging in some cases. Sometimes participants do not want to meet with providers, decide to stop treatment as they start feeling better, or come to believe that continued treatment is not the best option. OnTrackNY teams are trained on several strategies to enhance engagement with participants and their supports. Teams are encouraged to remain person-centered across encounters to help the individuals feel that they are part of the program. Engagement is not measured by a minimum number of visits attended or participation in all treatment components. Instead, services are tailored to individual needs and preferences, and the team consistently conveys that they are there to support the individual to attain recovery. For individuals who believe that they do not need any help or support, team members might focus on working on other goals such as school or employment. Additionally, the team members emphasize the young person's resilience and strengths. Notwithstanding, it is recommended that the team continue doing assertive outreach for the small proportion of participants who disengage early on.

OnTrackNY teams work closely with participants and families using SDM to convey information, recommendations, and a sense of respect and inclusion in treatment decisions. Sometimes, however, individuals and families come to the team with competing goals or preferences. In these instances, the team has to navigate encouraging family participation in the decision-making process while respecting the individuals' preference to make independent decisions. Young people sometimes have a sense of invincibility, which influences the way they consume the information provided by team members, which must also be respected (e.g., deciding to stop taking antipsychotic medications as soon as they feel better, deciding to continue using drugs or alcohol). Thus, this requires clinicians to tolerate a certain level of risk by allowing young people to make decisions that may go against their recommendations. At the same time, teams are trained to understand that SDM is a process that usually leads to the individual feeling empowered and staying engaged with the team while trying out different options. Team members are reminded that there are continuous opportunities for the individual to decide differently if they realize the original decision did not work out as expected.

Providing culturally competent services is also a powerful way to promote engagement and person-centered care. Although teams are trained in assessing and integrating culture into the treatment approach, providers continue to grapple with specific situations where cultural factors are central. With the help of cultural competency and gender and sexuality experts and input from the teams, a guide on culture and FEP was developed, which details common themes that arise and best practices for delivering culturally competent services within OnTrackNY.

Challenges with clinical interventions: treating suicidality, violence, and cognitive health

Other areas that have come to require additional guidance and training include assessing and treating suicidality, violence, and cognitive health. Providers generally feel adept at screening for suicidality and implementing a safety plan when appropriate. However, further assistance is needed in developing a more systematic approach for assessing the extent of suicide risk and weighing risk and protective factors. This is particularly necessary when participants might be denying suicidal ideation and yet still engage in impulsive or dangerous behaviors. Similarly, providers require guidance in effectively assessing risk of violence and aggression and learning strategies for addressing these issues.

Given the evidence of cognitive remediation for improving outcomes in schizophrenia by targeting information-processing skills and thinking, teams will benefit from recently developed training materials to build providers' proficiency in the assessment and treatment of cognitive health using restorative and compensatory strategies. Both types of cognitive remediation can be conducted individually or in a group, have been manualized to facilitate dissemination and implementation, and are effectively paired with educational, vocational, and social interventions, and therefore can be integrated into the CSC model.

AFTER ONTRACKNY

Discharge planning is a key component of the OnTrackNY model. On average, OnTrackNY participants stay in the program for 2 years, and teams are encouraged to remind clients and families early on that the treatment is time-limited. To prepare clients and families psychologically and practically, discharge planning typically begins 6 months prior to an individual's transitioning out of the program. Primary clinicians play an integral role in transition planning, which includes educating clients and families about treatment options, ensuring clients and families have the needed coping skills to move forward and providing ample support to ensure successful transition to the next provider/s. Throughout the transition, there is an ongoing focus on recovery goals and ensuring a comprehensive aftercare plan that includes work and/or school, sufficient supply of medications and community supports. While this may be a difficult transition for clients and families, teams are encouraged to celebrate transitions and show clients and families that leaving OnTrackNY is a positive step forward in the young person's recovery.

OUTCOMES

As a condition of funding, all OnTrackNY teams submit client- and team-level data regarding key service components and client outcomes as measures of fidelity on a monthly basis to the New York State Office of Mental Health and receive monthly continuous quality improvement–focused reports based on this data. During OnTrackNY's tenure, data have indicated that program participants derive significant benefits from program participation that are consistent with findings of controlled research, including impact on school and work participation as well as rates of hospitalization and use of acute services.[46]

CONCLUSION

In conclusion, there is a large international evidence-based body of data supporting the effectiveness of early intervention services for FEP, which has led to the rapid dissemination of CSC programs across the United States. Studies support the benefits that program participants attain while receiving these services early in their illness including reduced hospitalization rates and involvement in education and employment goals. However, more studies are needed to determine whether these benefits are sustained after treatment ends. OnTrackNY is one of several CSC models, which have proven effective in improving outcomes. The widespread implementation of this program across the state of New York and several other states has provided valuable information regarding the strengths of the model and areas where it necessitates expansion and adaptation as it is implemented across diverse communities. Strategies for enhancing the CSC model are continuously being developed and informed by the emerging research in early intervention services for FEP and by the feedback of stakeholders implementing these programs in real-world settings. This creates an adaptable model that can accommodate and integrate new evidence-based interventions and treatment approaches to meet individual needs across communities.

REFERENCES

1. Breitborde NJ, Bell EK, Dawley D, et al. The Early Psychosis Intervention Center (EPICENTER): Development and six-month outcomes of an American first-episode psychosis clinical service. *BMC Psychiatry*. 2015;15(1):266. doi:10.1186/s12888-015-0650-3.
2. Dixon LB, Goldman HH, Bennett ME, et al. Implementing coordinated specialty care for early psychosis: The RAISE connection program. *Psychiat Serv*. 2015;66(7):691–698. doi:10.1176/appi.ps.201400281.
3. Guo X, Zhai J, Liu Z, et al. Effect of antipsychotic medication alone vs combined with psychosocial intervention on outcomes of early-stage schizophrenia: A randomized, 1-year study. *Arch Gen Psychiat*. 2010;67(9):895–904. doi:10.1001/archgenpsychiatry.2010.105.

4. Kane JM, Robinson DG, Schooler NR, et al. Comprehensive versus usual community care for first-episode psychosis: 2-year outcomes from the NIMH RAISE early treatment program. *AM J Psychiat.* 2015;173(4):362–372. doi:10.1176/appi.ajp.2015.15050632.
5. Chang WC, Tang JY, Hui CL, et al. Prediction of remission and recovery in young people presenting with first-episode psychosis in Hong Kong: A 3-year follow-up study. *Aust NZ J Psychiat.* 2012; 46(2): 100–108. doi: 10.1177/0004867411428015.
6. Tapfumaneyi A, Johnson S, Joyce J, et al. Predictors of vocational activity over the first year in inner-city early intervention in psychosis services. *Early Interv Psychia.* 2015;9(6):447–458. doi:10.1111/eip.12125.
7. Petersen L, Thorup A, Øqhlenschlæger J, et al. Predictors of remission and recovery in a first-episode schizophrenia spectrum disorder sample: 2-year follow-up of the OPUS trial. *Can J Psychiat.* 2008;53(10):660–670. doi:10.1177/070674370805301005.
8. Major BS, Hinton MF, Flint A, Chalmers-Brown A, McLoughlin K, Johnson S. Evidence of the effectiveness of a specialist vocational intervention following first episode psychosis: A naturalistic prospective cohort study. *Soc Psychiatry Psychiatr Epidemiol.* 2010;45(1):1–8. doi:10.1007/s00127-009-0034-4.
9. Addington J, Heinssen RK, Robinson DG, et al. Duration of untreated psychosis in community treatment settings in the United States. *Psychiat Serv.* 2015;66(7):753–756. doi: 10.1176/appi.ps.201400124
10. Penttila M, Jaaskelainen E, Hirvonen N, Isohanni M, Miettunen J. Duration of untreated psychosis as predictor of long-term outcome in schizophrenia: Systematic review and meta-analysis. *Brit J Psychiat.* 2014;205(2):88–94. doi:10.1192/bjp.bp.113.127753.
11. Anderson KK, Rodrigues M, Mann K, et al. Minimal evidence that untreated psychosis damages brain structures: A systematic review. *Schizophr Res.* 2015;162(1–3):222–233. doi:10.1016/j.schres.2015.01.021.
12. Marshall M, Lewis S, Lockwood A, Drake R, Jones P, Croudace T. Association between duration of untreated psychosis and outcome in cohorts of first-episode patients. *Arch Gen Psychiat.* 2005; 62(9):975–983. doi:10.1001/archpsyc.62.9.975.
13. Compton MT, Carter T, Bergner E, et al. Defining, operationalizing and measuring the duration of untreated psychosis: Advances, limitations and future directions. *Early Interv Psychia.* 2007;1(3):236–250. doi:10.1111/j.1751-7893.2007.00036.x
14. Boonstra N, Klaassen R, Sytema S, et al. Duration of untreated psychosis and negative symptoms: A systematic review and meta-analysis of individual patient data. *Schizophr Res.* 2012;142(1):12–19. Doi: 10.1016/j.schres.2012.08.017
15. Lloyd-Evans B, Crosby M, Stockton S, et al. Initiatives to shorten duration of untreated psychosis: Systematic review. *Brit J Psychiat.* 2011;198(4):256–263. doi:10.1192/bjp.bp.109.075622.
16. Hegelstad WTV, Larsen TK, Auestad B, et al. Long-term follow-up of the TIPS early detection in psychosis study: Effects on 10-year outcome. *Am J Psychiat.* 2012;*169*(4), 374–380. 10.1176/appi.ajp.2011.11030459.
17. Sanger TM, Lieberman JA, Tohen M, Grundy S, Beasley JC, Tollefson GD. Olanzapine versus haloperidol treatment in first-episode psychosis. *Am J Psychiat.* 1999;156(1):79–87. doi:10.1176/ajp.156.1.79.
18. Berger GE, Proffitt T-M, McConchie M, et al. Dosing quetiapine in drug-naive first-episode psychosis. *J Clin Psychiat.* 2008;69(11):1702–1714.
19. Robinson DG, Woerner MG, Delman HM, Kane JM. Pharmacological treatments for first-episode schizophrenia. *Schizophrenia Bull.* 2005;31(3):705–722. doi:10.1093/schbul/sbi032.
20. Goldstein MJ, Rodnick EH, Evans JR, May PR, Steinberg MR. Drug and family therapy in the aftercare of acute schizophrenics. *Arch Gen Psychiat.* 1978;35(10):1169–1177. doi:10.1001/archpsyc.1978.01770340019001.
21. Zhang M, Wang M, Li J, & Phillips MR (1994). Randomised-control trial of family intervention for 78 first-episode male schizophrenic patients. *Brit J Psychiat.* 1994;165(suppl 24):96–102.
22. Lewis S, Tarrier N, Haddock G, et al. Randomised controlled trial of cognitive-behavioural therapy in early schizophrenia: Acute-phase outcomes. *Brit J Psychiat.* 2002;181(43):s91–s97. doi:10.1192/bjp.181.43.s91.
23. Jackson HJ, McGorry PD, Killackey E, et al. Acute-phase and 1-year follow-up results of a randomized controlled trial of CBT versus Befriending for first-episode psychosis: The ACE project. *Psychol Med.* 2008;38(5):725–735. doi:10.1017/s0033291707002061.
24. Tarrier N, Lewis S, Haddock G, et al. Cognitive-behavioural therapy in first-episode and early schizophrenia: 18-month follow-up of a randomised controlled trial. *Brit J Psychiat.* 2004;184(3):231–239. doi:10.1192/bjp.184.3.231.
25. Miller R, Mason SE. Using group therapy to enhance treatment compliance in first episode schizophrenia. *Soc Work Groups.* 2002;24(1):37–51. doi:10.1300/j009v24n01_04.
26. Lecomte T, Leclerc C, Wykes T, Lecomte J. Group CBT for clients with a first episode of schizophrenia. *J Cogn Psychother.* 2003;17(4):375–383. doi: 10.1891/jcop.17.4.375.52538
27. Killackey E, Jackson HJ, Mcgorry PD. Vocational intervention in first-episode psychosis: Individual placement and support v. treatment as usual. *Brit J Psychiat.* 2008;193(2):114–120. doi:10.1192/bjp.bp.107.043109.
28. Nuechterlein KH, Subotnik KL, Turner LR, Ventura J, Becker DR, Drake RE. Individual placement and support for individuals with recent-onset schizophrenia: Integrating supported education and supported employment. *Psychiatr Rehabil J.* 2008;31(4):340–349. doi:10.2975/31.4.2008.340.349.
29. Dixon LB, Dickerson F, Bellack AS, et al. The 2009 schizophrenia PORT psychosocial treatment recommendations and summary statements. *Schizophrenia Bull.* 2010;36(1):48–70. doi:10.1093/schbul/sbp115.
30. Gafoor R, Nitsch D, McCrone P, et al. Effect of early intervention on 5-year outcome in non-affective psychosis. *Brit J Psychiat.* 2010;196(5):372–376. doi:10.1192/bjp.bp.109.066050.
31. Grawe RW, Falloon IRH, Widen JH, Skogvoll E. Two years of continued early treatment for recent-onset schizophrenia: A randomised controlled study. *Acta Psychiat Scand.* 2006;114(5):328–336. doi:10.1111/j.1600-0447.2006.00799.x.
32. Nordentoft M, Jeppesen P, Petersen L, et al. OPUS project: A randomised controlled trial of integrated psychiatric treatment in first episode psychosis. *Schizophr Res.* 2003;60(1):297. doi:10.1016/s0920-9964(03)80503-0.
33. Srihari VH, Tek C, Kucukgoncu S, et al. First-episode services for psychotic disorders in the U.S. public sector: A pragmatic randomized controlled trial. *Psychiat Serv.* 2015;66(7):705–712. doi:10.1176/appi.ps.201400236.
34. Secher RG, Hjorthoj CR, Austin SF, et al. Ten-year follow-up of the OPUS specialized early intervention trial for patients with a first episode of psychosis. *Schizophrenia Bull.* 2014;41(3):617–626. doi:10.1093/schbul/sbu155.
35. Petersen L, Jeppesen P, Thorup A. A randomised multicentre trial of integrated versus standard treatment for patients with a first episode of psychotic illness. *BMJ.* 2005;331(7517):602. doi:10.1136/bmj.38565.415000.e01.
36. Craig TK, Garety P, Power P, et al. The Lambeth Early Onset (LEO) Team: Randomised controlled trial of the effectiveness of specialised care for early psychosis. *BMJ.* 2004;329(7474):1067. doi:10.1136/bmj.38246.594873.7C.
37. Penn DL, Meyer PS, Gottlieb JD. Individual resiliency training (IRT). 2014 Available at http://www.navigateconsultants.org/wp-content/uploads/2017/05/IRT-Manual.pdf.
38. Bertelsen M, Jeppesen P, Petersen L, et al. Five-year follow-up of a randomized multicenter trial of intensive early intervention vs standard treatment for patients with a first episode of psychotic illness: The OPUS trial. *Arch Gen Psychiat.* 2008;65(7):762–771. doi:10.1001/archpsyc.65.7.762.
39. Norman RM, Manchanda R, Malla AK, Windell D, Harricharan R, Northcott S. Symptom and functional outcomes for a 5year early intervention program for psychoses. *Schizophr Res.* 2011;129(2):111–115. doi:10.1016/j.schres.2011.04.006.

40. Malla A, Norman R, McLean T, Scholten D, Townsend L. A Canadian programme for early intervention in non-affective psychotic disorders. *Aust NZ J Psychiat.* 2003;37(4):407–413. doi: 10.1046/j.1440-1614.2003.01194.x
41. Chang WC, Chan GHK, Jim OTT, et al. Optimal duration of an early intervention programme for first-episode psychosis: Randomised controlled trial. *Brit J Psychiat.* 2015;206(6):492–500. doi: 10.1192/bjp.bp.114.150144.
42. Chang WC, Kwong VWY, Lau ESK, et al. Sustainability of treatment effect of a 3-year early intervention programme for first-episode psychosis. *Brit J Psychiat.* 2017;1–8. doi: 10.1192/bjp.bp.117.198929.
43. Albert N, Melau M, Jensen H, et al. Five years of specialized early intervention versus two years of specialized early intervention followed by three years of standard treatment for patients with a first episode psychosis: Randomized, superiority, parallel group trial in Denmark (OPUS II). *BMJ.* 2017;356:i6681. doi: 10.1136/bmj.j1015
44. Heinssen RK, Goldstein AB, Azrin ST. Evidence-based treatment for first-episode psychosis: Components of coordinated specialty care. 2014. Available at http://www.nimh.nih.gov/health/topics/schizophrenia/raise/coordinated-specialty-care-for-first-episode-psychosis-resources.shtml.
45. Elwyn G, Frosch D, Thomson R, et al. Shared decision making: A model for clinical practice. *J Gen Intern Med.* 2012;27:1361–1367. doi: 10.1007/s11606-012-2077-6.
46. Nossel I, Bello I, Scodes J, Wall M, Smith T, Malinovsky I, Marino L, Dixon L. Outcomes of community based coordinated specialty care program in New York State (submitted for publication).

PSYCHOSIS OVER LIFE SPAN AND LATE-LIFE PSYCHOSIS

17

LIFE SPAN DEVELOPMENT OF SCHIZOPHRENIA

SYMPTOMS, CLINICAL COURSE, AND OUTCOMES

Matti Isohanni, Jouko Miettunen, and Matti Penttilä

INTRODUCTION

Schizophrenia is a serious illness characterized by a mixture of symptoms, varying through the life span. A life span epidemiology and life course approach offer an interdisciplinary framework for guiding research on health, disease, human development, and aging. Schizophrenia is rare before puberty. Its incidence increases rapidly until the mid-20s, before declining over the following decades.[1] The peak age of onset is between 20 and 29 years.[2]

In this chapter, our focus relies mainly on DSM-IV and ICD-10 schizophrenia. We will follow age and disease periods from life span, starting from the premorbid period, then to illness onset, midlife course, and ending with old age schizophrenia. We describe main symptom trajectories, clinical courses, medications, and outcomes. We also review main life span findings from our research project, the Northern Finland Birth Cohort 1966. Finally, we discuss the methodological challenges, and clinical and translational relevance of these studies.

PSYCHOSIS: KEY SYMPTOMATOLOGY

KEY SYMPTOM DOMAINS IN PSYCHOSES

Psychotic symptoms are challenging to define, due to their heterogeneous phenomenology. The main feature of psychosis is the distortion of reality. Psychotic symptoms have been traditionally categorized into positive and negative, as well as general. Other symptom domains include disorganization, excitement, depression, and cognitive impairment.[3]

Current understanding of the phenomenology of psychosis focuses on the combination of biology and psychology, considered as a continuum comparable with other phenotypes in psychiatry and medicine.[4,5] People may have psychotic experiences (PE) and mild psychotic-like symptoms for a short period of time without progressing to a full clinical syndrome. The DSM-5 includes attenuated psychosis syndrome. Many who do not have a confirmed psychotic disorder endorse items related to the presence of hallucinations and delusions. Mean lifetime prevalence of ever having a lifetime PE is 5.8%, with hallucinatory experiences being more common (5.2%) than delusional experiences (1.3%).[6] For most people PEs are transient and not disabling. PEs may be related to an increased risk of a range of mental disorders (e.g., depression, anxiety disorders) in addition to psychoses.[7] Severe psychotic disorders often include a prominent decline of functioning. Most studies have been conducted in severe psychoses, such as schizophrenia. The longitudinal course of less severe psychotic disorders is poorly understood.

SYMPTOM DOMAINS OF SCHIZOPHRENIA

Since early definitions, psychotic symptoms have been core to the diagnosis of schizophrenia. Positive symptoms, including hallucinations, particularly auditory, and delusions, have long been considered as definitive symptoms in schizophrenia. However, they also exist in other psychotic disorders. Negative symptoms include apathy, avolition, blunted affect, poverty of speech, asociality, and anhedonia. The decline of cognitive functioning, such as impairment of memory and speed of processing, is a characteristic domain, although cognitive deficits are also observed in affective and other psychoses.[8,9]

PSYCHOTIC FEATURES IN OTHER PSYCHIATRIC DISORDERS

In delusional disorders (persistent delusions and other deficits/symptoms), symptoms are primarily restricted to a single psychosis domain; often the only symptoms are delusions. Schizoaffective disorder includes the same phenomenology of symptoms as schizophrenia, along with prominent mood symptoms. When considering the continuum of psychotic symptoms, schizoaffective disorder is "located" between schizophrenia and bipolar disorder. Schizotypal and other schizophrenia-spectrum personality disorders (paranoid, schizoid) are associated with characteristic "psychosis-like" phenomena; and may include periods of transient, frank psychosis, ultimately leading to schizophrenia.[10]

Psychotic symptoms can also be a feature of mood disorders, especially in severe major depressive disorder or bipolar disorder. Our current understanding is that a continuum of psychotic symptoms is also shown in the course of mood disorders. Some psychotic symptoms can be present even in mild disorders, although the prevalence is higher in more severe forms of mood disorders.[11] The distinguishing difference in mood disorders is that the manifestation of psychotic symptoms is limited to the time periods of either depression or mania.

Anxiety disorders and post-traumatic stress disorder may include psychotic-like experiences or even frank psychotic symptoms for short periods of time.[11] In post-traumatic stress disorder, the differential diagnosis with dissociative states can be challenging. Personality disorders, particularly borderline personality disorder, are associated with an increased prevalence of psychotic symptoms, and 25%–50% of individuals with unstable personality disorder report psychotic symptoms.[10] Organic psychoses can manifest with symptoms similar to schizophrenia, but have an identifiable organic cause, and psychotic symptoms are usually associated with other symptoms of brain injury or illness. Similarly, psychotic symptoms associated with substance use disorders have different etiology and prognosis than schizophrenia. The onset of psychosis and initiation of treatment of schizophrenia may be affected by concurrent substance use disorder.

LIFE-COURSE OF SYMPTOMS IN SCHIZOPHRENIA

SYMPTOMS AT PRODROMAL PERIOD

Schizophrenia usually does not begin suddenly. We can see this more clearly by appreciating the events leading up to the diagnosis.[12] Subtle developmental deviances in motor, cognitive, emotional, and behavioral domains, although mostly within normal limits, are found in individuals who later develop schizophrenia. They follow a developmental trajectory that partly differs from that of the general population, supporting the neurodevelopmental hypothesis.[13] It is possible to identify risk factors for schizophrenia, but they have limited predictive value. Four dimensions relating to prodromal negative symptoms have been proposed with possible linkage to early signs of positive symptoms.[14] The disturbance of perception of self may be the core symptom dimension in the prodromal period.[15] Mild cognitive and negative symptoms are often dominant in the prodromal phase, with positive symptoms being either short or attenuated.

The beginning of the prodromal period is defined as the onset of the duration of untreated illness (DUI), which has not been studied as extensively as the duration of untreated psychosis (DUP). The criteria to define the onset of first psychotic symptoms have been easier to develop and use, although still challenging.[16] The DUP is often relatively short, with most people accessing treatment within a month after the onset of psychosis. However, many people have a remarkably longer interval between the illness onset and initiation of treatment. The mean of DUP has been in the approximate range of 6 and 12 months.[17] The association between long DUP and poor outcome, in both the short- and long-term course,[17] has provided rationale for early detection and intervention in psychoses.

Early detection of psychosis is highly relevant, both from an individual and public health point of view, although challenging. Studies indicate that changes in neurodevelopment may happen at very early stages, during the first years of life, including the fetal and perinatal periods. It remains unclear whether certain early risk factors are causal or rather that they are just the markers of deviant neurodevelopment. Our current understanding of the early phases of psychosis highlights the importance of functional impairment associated with symptom manifestations, especially negative symptoms and thought disorder.[18]

SYMPTOMS AT FIRST EPISODE AND FIRST YEARS OF ILLNESS

The peak age of onset for schizophrenia is in late adolescence and early adulthood.[2] The main hypothesis for this pattern of onset relates to altered neurodevelopmental trajectories.[19] The pathway to prominent symptoms from the "at risk" mental state remains unclear. Deviations in brain functioning and structure, as well as cognition, may become more pronounced. Some of those at risk recover, indicating that protective factors, resilience, and care may play an important role.

Hypotheses on the development of positive symptoms are still inconclusive. Delusions may follow auditory hallucinations, combined with negative symptoms. Positive symptoms can be treated quite effectively with antipsychotics, but commonly with side effects. The course of negative symptoms is rather stable during the first years of illness. Medication and other treatment interventions have had limited long-term efficacy on these, even in controlled settings.

All of these symptoms have a negative impact on the successful achievement of normal milestones in early adulthood, in the areas of education, employment, personal relations, and fecundity. The disruption to the development of life skills at this age has a long-lasting effect on social skills and adult identity.[20]

SYMPTOMS IN MIDLIFE AND OLD AGE

Symptoms frequently reach a plateau in midlife, 5–10 years after disease onset. Positive symptoms are less prominent in one-third of aging patients, but some may continue to suffer from severe positive symptoms until the end of their life. Negative symptoms may fluctuate, causing difficulties in social life and the ability to function. Negative symptoms may be aggravated by depression, or medication side effects.

LIFE-COURSE OF CLINICAL AND SOCIAL OUTCOMES OF SCHIZOPHRENIA

MIDLIFE COURSE OF SCHIZOPHRENIA

Midlife progression follows different course patterns.[21,22] The mechanisms behind disease progression—and targeted interventions to address them—remain largely unknown. Study populations might be selected toward more severe or milder cases, and epidemiologically sound midlife progression is difficult to assess. Many clinicians and scientists assumed some pessimism about the potential for full or even partial recovery. This dates back to Kraepelin's belief that most experience a poor long-term prognosis with a defect end state.

Most longitudinal studies suggest that a chronic, deteriorating course with relapses/exacerbations is a frequent, but not necessarily inevitable, outcome. Early age at onset is usually a poor prognostic factor.[2] Family history of psychosis, poor premorbid adjustment, cannabis use, poorer clinical and social outcomes, and larger deficits in cognition are associated with earlier age of onset.[2] A recent systematic review and meta-analysis[23] studied age of onset of psychosis and long-term (at least 2-year follow-up) outcomes. Younger age at onset predicted more hospitalizations, negative symptoms, and relapses, poorer social and occupational functioning, and poorer global outcome.

The course of schizophrenia may also be relatively benign. Clinical and research evidence indicates that there is a subgroup of individuals who reach symptomatic and functional recovery and have a nonprogressive course (13.5% in meta-analysis[22]), showing good overall outcomes (depending on methods and sample). Although full recovery is uncommon (as is the case for many somatic disorders), many effective interventions can provide major symptomatic relief, reduce disability, and optimize recovery.

Neurocognitive deficits are a central dimension in schizophrenia. The deficits emerge during deviant neurodevelopment, indicate existing premorbid cognitive dysfunction, and are evident both in early childhood and in adolescence. Cognitive alterations in several domains become established at the first psychotic episode and remain relatively stable during the illness course in midlife, for up to 10 years[24] and also in older persons with schizophrenia[25] In midlife, most studies do not show further cognitive deterioration.[26]

Somatic health is affected in many ways during midlife.[27] High medical comorbidity is common, and contributes to excess, premature mortality. Individuals with schizophrenia have a higher risk for diabetes, which is increased by both first- and second-generation antipsychotics. Metabolic deviances may start at an early age, and in the early phase of illness, but may be mild at onset. Besides genetic factors, diet, unhealthy habits, medications, and mobility limitations related to negative symptoms may also contribute to weight gain and obesity.

Premature mortality is found in all ages and both genders, largely due to unhealthy lifestyle factors and poor quality of somatic treatments. The standardized mortality ratio (SMR) in schizophrenia appears to decrease with age. Excess mortality seems to be the highest between the ages 30 and 34. Between 30 and 50 years of age, all-causes mortality is 5–10 times higher than in the general population in both genders. Over time, life expectancy in the general population is improving as also in psychotic disorders. There seems to exist a differential mortality cap for individuals with psychosis.[28]

Treatment-resistant schizophrenia is a severe and long-lasting form of schizophrenia.[29,30] One-fifth to one-third of people with schizophrenia have treatment-resistant illness. Healthcare costs of treatment-resistant patients are 3- to 11-fold higher than in schizophrenia in general, causing 60% to 80% of the total economic burden of schizophrenia. Minimal scientific understanding of treatment-resistant schizophrenia[31] contributes to the paucity of tailored effective treatments. Promising evidence is available for clozapine and some other second-generation antipsychotics, and for cognitive-behavioral therapy and electroconvulsive therapy. Early, active treatment and medication adherence after the first episode are crucial as it may prevent the treatment resistance.

In summary, the prognosis of schizophrenia is subject to individual variation. The illness course in midlife is often deteriorating, but recovery is also possible. Generally, greater brain matter loss, cognitive deterioration, excess somatic comorbidity, and mortality are seen in early and midlife phases of schizophrenia than are observed in controls. Cognitive impairment plateaus but is associated with poorer prognosis and functional outcome.

AGING IN SCHIZOPHRENIA

As with the general population, the proportion of people with schizophrenia over 60 years is increasing. Over one-fifth are diagnosed after the age of 40. Older people with schizophrenia are in danger of experiencing double stigmatization: one caused by schizophrenia itself and another, by general agism.

Epidemiology

Schizophrenia is most commonly a lifelong disease. The number of elderly individuals with schizophrenia will double in next few decades. Despite excess mortality, many survive into old age. The prevalence of schizophrenia in the population aged over 65 years is around 1%, and 3 out of 100 patients with schizophrenia received this diagnosis at the age of 60 years or later.[32]

Symptomatology and outcomes

The prognosis of schizophrenia in later life is variable. Few prospective longitudinal studies[33,34] have analyzed disease trajectories from early midlife into old age. Multiple courses exist, ranging from full or partial recovery with no or minimal residual functional deficits to complete disability due to psychiatric and somatic reasons.

Some patients are apparently symptom-free and have a favorable clinical course with a diminishment of psychotic symptoms and may stop receiving treatment.[35] This is partly explained by survival and selection biases: the severely symptomatic and unwell have died or are institutionalized. Recovered and untreated cases are often excluded in research. This may increase the risk of presenting a biased, overly pessimistic view of disease outcomes. Some have a cyclical course with exacerbations and/or continuous symptomatology.

Cognition

Older people with schizophrenia may experience the second "drop" in cognitive function during illness, or relatively greater age-associated declines in cognitive functioning than healthy individuals.[36] Several factors modify accelerated cognitive aging and cognitive decline: somatic comorbidity, gender, and care. In schizophrenia there are sizable and heterogeneous deficits in global cognitive functioning at age 65. A meta-analysis[25] concluded that large and generalized impairments exist in both global and domain-specific neuropsychological functions in schizophrenia, relative to their age-matched peers. There are strong associations between cognitive function and later functional outcomes.

Somatic comorbidities

Somatic comorbidities are common and may be linked to shared genetic factors as well as premature, accelerated physical aging, but also unhealthy life styles and habits, such as minimal physical activity or exercise, unhealthy diet, or smoking, and to antipsychotic medication. Barriers to adequate diagnostics and care for somatic problems (due to both the patient and medical care systems) may contribute to general health problems.

Excess mortality

In schizophrenia, the life span is reduced by an average of 10–20 years. Relapses and medication noncompliance increase the risk of death. Causes of premature death include excess cardiac and pulmonary problems, smoking, complications of diabetes, and accidents. The effects of schizophrenia on years of potential life lost and life expectancy are substantial and have not lessened over time.[37]

The risk for suicide in schizophrenia is elevated. The predictors of suicide include a greater number of hospitalizations, recent admission, recent discharge, comorbid mood disorders, substance abuse, personality disorders, and previous suicide attempts. Improving lifestyle, early diagnostics, adequate somatic and psychiatric care, optimal treatments, and suicide prevention are key targets in reducing mortality rates throughout the lifetime.

In summary, in old age, many people with schizophrenia achieve at least partial recovery, with diminished psychotic symptoms and attainment of moderate biopsychosocial balance, functional capacity, and quality of life. Nevertheless, a considerable proportion of patients continue to experience troublesome symptomatology, functional deficits, or complete disability, and require continuous psychiatric, social, and somatic screening and care.

SCHIZOPHRENIA FROM FETAL PERIOD TO MIDLIFE: THE CONTRIBUTION OF NORTHERN FINLAND BIRTH COHORT 1966

Birth cohort design is useful in studying developmental and disease trajectories in psychiatric and medical disorders. The Northern Finland Birth Cohort 1966 (NFBC1966) is an extensive (over 12,000 people) birth cohort. It has contributed close to 1,000 peer-review publications in various fields of medicine (http://www.oulu.fi/nfbc). The schizophrenia research arm in the NFBC1966 has been active for nearly 30 years, enabling several important findings. The key published articles (2018 over 100) were reviewed by Jääskeläinen et al.[38] The NFBC1966 results stress the neurodevelopmental[39] nature of disease.

The most essential risk factors for schizophrenia within the NFBC1966 were male gender, parental psychosis, unwanted pregnancy, perinatal brain damage, low birth weight, late age at learning developmental milestones, central nervous system infections by the age of 14 years, and not attending expected school grade.[38,40] Smoking at the age of 14, psychosocial stressors, and not being married predicted negative symptoms.

Psychiatric outcomes were heterogeneous, but relatively poor. Being single, having an early onset, insidious onset, suicidal ideations upon the first admission, a rehospitalization, and a high number of treatment days in the early stages of the illness predicted a poorer clinical outcome after a minimum follow-up of 10 years.[38,41]

In midlife, brain structure[38,42] and neurocognitive alterations[26,43] were observed. High lifetime doses of antipsychotics were associated with alterations in brain morphometry, neurocognition, and poorer clinical outcomes.[42–45] Clinical follow-up was inadequate.[46] Somatic comorbidity occurred early and was common.[38] The overall mortality rate in schizophrenia was 14% before the age of 27 years, with half of the deaths being suicides. The suicide rate in schizophrenia before the age of 39 was 7% with 71% of suicides occurring during the first 3 years after the illness onset.[38]

The results of NFBC1966 on delayed or deviant development in psychiatric, somatic, educational, cognitive, and occupational outcomes in midlife schizophrenia are in contrast with normative development in NFBC1966 nonpsychotic controls (even at-risk[47]) and also Finnish population. A Finnish conscript study[48] of men born between 1962 and 1976 (n = 419,523) found steady maturation in personality traits and increasing cognitive abilities.

Clinical and translational conclusions[49] highlight early and accurate diagnostics and care of psychiatric adversities, individualized antipsychotic medications using personalized, lower range of recommended doses, and activity and continuity of treatments. As physical illnesses are diagnosed late and treated insufficiently, their identification needs to be improved, and more coordinated and continuous care should be achieved between medical providers.

ANTIPSYCHOTICS DURING THE LIFE SPAN

Millions of people use antipsychotic medications; thousands of clinicians prescribe them every day. Antipsychotics are effective for the treatment of psychotic symptoms and for relapse prevention. Current antipsychotics diminish illness expression, but do not restore lost complex brain functions. Some patients stay free of relapse without antipsychotics but current diagnostic means or predictive models do not allow for a reliable identification of such individuals. Low and moderate exposure to antipsychotics reduces mortality[50] despite an increased risk of sudden death and excess mortality in schizophrenia in general. Antipsychotics have no, or minimal, efficacy for negative symptoms or cognitive impairment. The major clinical challenge of balancing the long-term risk-benefit ratio of antipsychotics was recently reviewed by Goff et al.[51] and Correll et al.[52]

THE LONGITUDINAL USE OF ANTIPSYCHOTICS

A life span view of antipsychotic medication practices stresses careful documentation of doses and responses, planning of medication as part of the whole treatment program, and individualized and tailored selection and dosing (dose as low as possible). Chlorpromazine equivalent dose year and defined daily dose (DDD) year are measures of cumulative exposure to medication.[43,53] Medication management or clinical follow-up and review aiming at effective and safe use is generally inadequate.[46,54,55]

The clinical use of antipsychotics becomes more stable after the first 5 years.[44] With age, doses and cumulative exposure to antipsychotics are usually smaller, and a considerable proportion of patients do not use any antipsychotics. In some studies, favorable outcome was associated with low and steady antipsychotic medication and was unfavorable with high long-term cumulative use or antipsychotic polypharmacy.[44,56,57]

Harms related to antipsychotics tend to accumulate through years.[58] The final (but not only) pathway to psychosis is dopamine dysregulation. Antipsychotics can upregulate D2 receptors producing receptor supersensitivity, and also affect brain morphology.[59] They may lead to serious neurological and metabolic side effects, neuroleptic-induced dysphoria, and negative symptoms. Side effects increase the risk of heart disease, diabetes, stroke, social functioning, and worsening cognition.[43]

IS IT POSSIBLE TO REDUCE AN ANTIPSYCHOTIC DOSE OR DISCONTINUE MEDICATION?

In most treatment recommendations, the antipsychotics dose range in maintenance treatment varies between 150 and 900 mg/day chlorpromazine equivalents, to be continued for at least 1 year after the first episode, and 5 years after multiple episodes. Maintenance antipsychotic therapy and second-generation antipsychotics are a standard in long-term treatment. However, a general shift has been to avoid maximal doses and polypharmacy, to favor the smallest effective doses and choose an appropriate antipsychotic agent that causes minimal side effects. In a therapeutic community study, intensive psychosocial interventions and medication management reduced the mean dose of antipsychotics on an acute psychosis ward from 370 mg/day chlorpromazine equivalents to 160 mg/day. Eight percent of patients in the therapeutic community ward had extrapyramidal symptoms, contrasted to 15%–21% in traditional psychosis wards.[60]

Current guidelines are rather nonspecific in regard to antipsychotic dose tapering and discontinuation. Guidelines allow for low doses and stopping, but do not encourage this strategy or suggest how to go about tapering, i.e., at what point in the clinical course of illness, over what time period, etc. Scientific evidence on dose reduction or medication discontinuation is currently based primarily on observational studies. Long-term (7-year) follow-up of a randomized early dose reduction/discontinuation[61] demonstrated superior long-term functional outcome in the dose reduction/discontinuation group, even though the initial relapse rate was about twofold higher. Two recent studies having very long follow-ups (10 and 20 years) demonstrate that long-term antipsychotic treatment is associated with improved outcomes until 3[62] or 8 years.[63] However, most schizophrenia individuals are in midlife, and duration of illness is 10–20 years. Thus, evidence for the content and cost-effectiveness of their antipsychotic medication is limited.

A subgroup of patients are well managed during the stabilized phase on illness, especially with regard to functional remission, with small doses of antipsychotics, and even without permanent antipsychotic medication.[44,56] Presently, there are no reliable predictors that could help identify patients who respond well to a dose reduction/discontinuation strategy. A long relapse-free period is a weak predictor.[56]

The smallest effective dose could be an advisable strategy to achieve effectiveness and tolerability. The following prerequisites may assist in this: accurate diagnostics; maximal use of careful follow-up and psychosocial interventions; individualized and personalized treatment program with additional nonpharmacological care and patient and family guidance. Team-based comprehensive care improves functional and clinical outcomes[64] by delivering not only medical but also psychosocial aspects of care. Ethical principles provide guidance, especially Hippocrates' principle of preventing harm: *primum non nocere* (first, do not harm). The risk of long term serious side effects is one reason to consider minimizing antipsychotic doses by maximizing structured medication management and psychosocial treatments.

In summary, antipsychotic medications are effective for acute psychosis and in prevention of relapses and excess mortality. Long-term use may include harmful effects, as also does inappropriate withholding or discontinuing medication. Limited scientific evidence for the dose reduction or discontinuation exists. When aiming for an optimal benefit-risk ratio and balancing symptomatic, functional, and somatic outcomes, lower ranges of effective dosing could be one goal. Joint psychosocial treatments and good medication management may help to achieve lowest effective doses and optimal medication selection and switching, dose escalation, reduction, or discontinuation, and may improve sustained adherence.

DESCRIPTIVE LIFE SPAN MODELS OF LIFE COURSE APPROACH

Current knowledge on the course of schizophrenia has remained fragmented. We have not succeeded in putting many pieces of the complex puzzle together. Many key etiological and prognostic factors remain unidentified. We have developed a descriptive, multilevel life span model of schizophrenia. This kind of model may help integrate different types of evidence, and thus provide a coherent framework to the didactic goals, and guide future research.

CHALLENGES IN LIFE COURSE STUDIES

REGISTER AND COHORT STUDIES

In longitudinal and life course studies, a lifelong health perspective is one aim: to produce relevant data for clinical decision-making of health policy, treatment organizations, clinicians, patients, and relatives. In Nordic countries, registers are particularly useful for studying health outcomes.[65]

The earliest cohorts started in the 1940s. For example, the Iowa Longitudinal Study of longitudinal brain development,[66] or the Chicago follow-up study of long-term antipsychotic use.[57] Welham et al.[67] reviewed birth cohort studies on antecedents of schizophrenia, finding 11 birth cohorts from 7 countries. These studies have provided prospective developmental history data about the development of adult psychosis.

POTENTIAL DIFFICULTIES IN LONGITUDINAL STUDIES

Many longitudinal studies of schizophrenia have produced divergent pictures of its risks, course, and outcomes. Heterogeneity of the study samples and follow-ups is one major reason. Most studies are performed at onset or during short (under 2 years) follow-ups—but most patients are sick for longer periods or are in midlife or old age.

Many potential sources of biases are linked to long follow-ups: e.g., attrition, selection, and loss. Many potential confounding factors may influence outcomes: time, period, aging, diagnostics, treatments, comorbidity, medication, etc. There is a risk of residual confounding or confounding by indication due to naturalistic, observational design. Randomized controlled trials (RCTs) in long follow-ups and life span studies are difficult to design.

Practical difficulties relate to long follow-ups. Scientists, scientific organizations, and study populations all get older, are lost to follow-ups, retire, or die. Instruments and methods tend to change over time. Technology (e.g., brain-imaging techniques and programs, cognitive testing batteries), ethical principles, funding, and data collection and protection may also change. Scientists doing longitudinal studies often report that they have lost their long-term funding, or key scientist(s), or a functioning team, or permission to use data, or personal health and ability to work, etc. On the other hand, long follow-ups allow to collect large and unique data bases, as in Nordic countries.[65]

The following quality criteria are often used in longitudinal and life span studies. What was the diagnostic validity and accuracy? Does diagnostic updating occur? Do different, or new, diagnostic concepts and criteria confound the results? How representative and epidemiologically sound was the study sample? Attrition and lost cases are reality; how selective are they? How were attrition and confounding factors taken into account, e.g. smoking, weight, aging, time period, antipsychotic and other medications, other treatments? In long follow-ups power issues are often challenging as the number of cases decreases. It is often necessary to estimate what kind of signal it is possible to detect using available sample sizes. Do the results concord with current scientific views and treatment recommendations? What translational impact, clinical relevance, and clinical recommendation can be drawn from the results? How is the research team constructed and managed?[68,69] Does it have enough continuity and critical mass of senior and junior researchers? Is the team able to identify, prepare for, and adjust to future challenges, partly also by junior scientists? Is there enough national and international collaboration and networking to pull knowledge or data or perform systematic reviews or meta-analyses?

MEDICATION STUDIES

Randomized controlled trials can help determine the short-term efficacy and adverse effects of a treatment. However, RCTs are difficult to conduct, particularly during very long follow-ups. They tend to be reductionistic when analyzing the complex interactions between brain, environment, and drug effect typical in severe mental disorders.[70] Also, RCTs often involve methodological problems: poor compliance, short follow-up periods, and even slow study processes. And RCTs tend not to detect the heterogeneity of effects and different individual/subgroup effect sizes.

Observational, naturalistic, nonexperimental settings may offer an optimal setting for investigating the long-term effects of treatment that often are impossible to study in RCTs. However, the patients are not treated randomly. This may cause residual, latent confounding or confounding by indication: very sick patients often get longer treatments and/or higher doses of antipsychotics. Medication status may, in part, be the consequence of whether patients are doing well or poorly rather than the cause (reverse causation). Residual confounding can often be controlled, if sufficient data on care, illness course and severity, and attrition exist.[71]

CONCLUDING REMARKS

The life span view of schizophrenia is not widely studied. The focus in most treatment algorithms and recommendations is on the acute phase and early years of illness. However, most patients are sick for longer periods, or are in midlife or old age, and evidence-based care is required. In clinical decision-making, these limitations must be acknowledged and more personalized medicine approaches applied.

No major breakthroughs in medication or psychosocial interventions for schizophrenia are in sight. Current treatments diminish illness expression, but do not usually restore lost function or age-specific adulthood opportunities. In the future, the integration in genetics, neurobiology, and environmental factors may provide more effective interventions,[72] as also may increasing pharmacoepidemiological data on long-term use of antipsychotics.[44,50,63,71]

Early detection of psychosis has been a key area of development in psychiatry in recent decades. Predicting the onset of psychosis is challenging: less than 50% of at-risk persons experience transition to psychotic disorder.[73] Advanced risk assessment tools that can ascertain the probability of conversion to psychosis in individual patients have been developed.[74,75]

Models showing promising success (e.g.,[76–78]) for early intervention in psychosis have been developed, but only a small proportion of at-risk persons receive these services, e.g., in prodromal clinics.[79,80] Evidence for the cost-effectiveness of early detection is limited, as is the long-term impact of early intervention. Psychotherapeutic interventions show the most promise in the early phases.[81,82]

Protective and preventive factors for schizophrenia are still understudied and underused. The following have been considered: good obstetric care, decreasing child abuse and bullying, improving treatment of migrants, better urban planning, early detection and care, physical activity, good education, refraining from substance use, relapse prevention, and good psychiatric and somatic care. Studies often focus on negative factors such as risks or poor outcomes with, perhaps, less attention paid to positive factors such as protection, resilience, risk cases with healthy outcomes,[47] human rights, happiness, and hope. Some studies also show an association between creativity and mental disorders.[83]

Experienced clinicians/scientists often treat complicated, high-risk, treatment-resistant patients and find limited success, while cases with good response and recovery are often lost to follow-up ("the clinician's illusion").[84] This "pessimism bias" or sampling bias may reduce clinicians' confidence to taper, change, or discontinue antipsychotics, or use psychotherapies, or continue treatments over years and decades.

Instilling hope improves treatment compliance and encourages the maintenance of social relationships.[85] Hope promotes relationships with relatives, peers, staff, and pets; increases success in daily life; and encourages patients to take an active role in managing their illness. In everyday life, hope increases the importance of care, cure, optimal medication, friends, relatives, home, food, activities, and nature. Despite many challenges, individuals with schizophrenia are not necessarily less happy than their peers.[86]

REFERENCES

1. Häfner H, Maurer K, Loffler W, et al. The epidemiology of early schizophrenia: influence of age and gender on onset and early course. *Br J Psychiatry* 1994; 164(suppl. 23): 29–38.
2. Miettunen J, Immonen J, McGrath J, Isohanni M, Jääskeläinen E. The age of onset of schizophrenia spectrum disorders. In: de Girolamo G, McGorry P, Sartorius N, eds. *The age of onset of mental disorders: ethiopathogenetic and treatment implications*. Springer; 2019:55–73. doi:10.1007/978-3-319-72619-9_4
3. Arango C, Carpenter WT. The schizophrenia construct: symptomatic presentation. In: Weinberger DR, Harrison PJ, eds. *Schizophrenia*. 3rd ed. Oxford, UK: Wiley-Blackwell; 2011:9–23.
4. Tamminga CA, Pearlson G, Keshavan M, Sweeney J, Clementz B, Thaker G. Bipolar and schizophrenia network for intermediate phenotypes: outcomes across the psychosis continuum. *Schizophr Bull*. 2014; 40: 131–137.
5. Pearlson GD, Clementz BA, Sweeney JA, Keshavan MS, Tamminga CA. Does biology transcend the symptom-based boundaries of psychosis? *Psychiatr Clin North Am*. 2016; 39: 165–174.
6. McGrath JJ, Saha S, Al-Hamzawi A, et al. Psychotic experiences in the general population: A cross-national analysis based on 31261 respondents from 18 countries. *JAMA Psychiatry* 2015; 72: 697–705.
7. McGrath JJ, Saha S, Al-Hamzawi A, et al. The bidirectional associations between psychotic experiences and DSM-IV mental disorders. *Am J Psychiatry* 2016; 173: 997–1006.
8. Hill SK, Reilly JL, Keefe RS, et al. Neuropsychological impairments in schizophrenia and psychotic bipolar disorder: findings from the Bipolar-Schizophrenia Network on Intermediate Phenotypes (B-SNIP) study. *Am J Psychiatry* 2013; 170: 1275–1284.
9. Zaninotto L, Guglielmo R, Calati R, et al. Cognitive markers of psychotic unipolar depression: a meta-analytic study. *J Affect Disord*. 2015; 174: 580–588.
10. Balaratnasingam S, Janca A. Normal personality, personality disorder and psychosis: current views and future perspectives. *Curr Opin Psychiatry* 2015; 28: 30–34.
11. Varghese D, Scott J, Welham J, et al. Psychotic-like experiences in major depression and anxiety disorders: a population-based survey in young adults. *Schizophr Bull*. 2011; 37: 389–393.
12. Jones PB. Adult mental health disorders and their age at onset. *Br J Psychiatry* 2013; 202: 5–10.
13. Murray RM, Lewis SW. Is schizophrenia a neurodevelopmental disorder? *Br Med J*. 1987; 295: 681–682.
14. Møller P, Husby R. The initial prodrome in schizophrenia: searching for naturalistic core dimensions of experience and behavior. *Schizophr Bull*. 2000; 26: 217–232.
15. Nelson B, Yung AR, Bechdolf A, McGorry PD. The phenomenological critique and self-disturbance: implications for ultra-high risk ("prodrome") research. *Schizophr Bull*. 2008; 34: 381–392.
16. Register-Brown K, Hong LE. Reliability and validity of methods for measuring the duration of untreated psychosis: a quantitative review and meta-analysis. *Schizophr Res*. 2014; 160: 20–26.
17. Penttilä M, Jääskeläinen E, Hirvonen N, Isohanni M, Miettunen J. Duration of untreated psychosis as predictor of long-term outcome in schizophrenia: systematic review and meta-analysis. *Br J Psychiatry* 2014; 205: 88–94.
18. Fulford D, Niendam TA, Floyd EG, et al. Symptom dimensions and functional impairment in early psychosis: more to the story than just negative symptoms. *Schizophr Res*. 2013; 147: 125–131.
19. Kochunov P, Hong LE. Neurodevelopmental and neurodegenerative models of schizophrenia: white matter at the center stage. *Schizophr Bull*. 2014; 40: 721–728.
20. Morgan VA, Waterreus A, Carr V. Responding to challenges for people with psychotic illness: updated evidence from the Survey of High Impact Psychosis. *Aust N Z J Psychiatry* 2017; 5: 124–140.
21. Isohanni M, Jääskeläinen E, Tanskanen P, et al. Midlife progression in schizophrenia. *Psychiatria Fennica* 2010; 41: 10–34.
22. Jääskeläinen E, Juola P, Hirvonen N, et al. A systematic review and meta-analysis of recovery in schizophrenia. *Schizophr Bull*. 2013; 39: 1296–1306.
23. Immonen J, Jääskeläinen E, Korpela H, Miettunen J. Age at onset and the outcomes of schizophrenia: systematic review and meta-analysis. *Early Interv Psychiatry* 2017; 11: 453–460.
24. Bozikas VP, Andreou C. Longitudinal studies of cognition in first episode psychosis: a systematic review of the literature. *Aust N Z J Psychiatry* 2011; 45: 93–108.
25. Irani F, Kalkstein S, Moberg EA, Moberg PJ. Neuropsychological performance in older patients with schizophrenia: a meta-analysis of cross-sectional and longitudinal studies. *Schizophr Bull*. 2011; 37: 1318–1326.
26. Rannikko I, Jääskeläinen E, Miettunen J, et al. Predictors of long-term change in adult cognitive performance: systematic review and data from the Northern Finland Birth Cohort 1966. *Clin Neuropsychol*. 2016; 30: 17–50.
27. Smith, DJ, Langan J, McLean G, Guthrie B, Mercer SW. Schizophrenia is associated with excess multiple physical-health

comorbidities but low levels of recorded cardiovascular disease in primary care: cross-sectional study. *BMJ Open* 2013; 3: pii: e002808.
28. Saha S, Chant D, McGrath JJ. A systematic review of mortality in schizophrenia: is the differential mortality gap worsening over time? *Arch Gen Psychiatry* 2007; 64: 1123–1131.
29. Seppälä A, Molins C, Miettunen J, et al. What do we know about treatment-resistant schizophrenia? A systematic review. *Psychiatria Fennica* 2016; 47: 95–127.
30. Howes OD, McCutcheon R, Agid O, et al. Treatment-resistant schizophrenia: treatment response and resistance in psychosis (TRRIP) working group consensus guidelines on diagnosis and terminology. *Am J Psychiatry* 2017; 174: 216–29.
31. Molins C, Roldán A, Corripio I, et al. Response to antipsychotic drugs in treatment-resistant schizophrenia: conclusions based on systematic review. *Schizophr Res.* 2016; 178: 64–67.
32. Howard R, Rabins PV, Seeman MV, Jeste DV. Late-onset schizophrenia and very-late-onset schizophrenia-like psychosis: an international consensus. The International Late Onset Schizophrenia group. *Am J Psychiatry* 2000; 157: 172–178.
33. Cowling D, Miettunen J, Jääskeläinen E, et al. Ageing in schizophrenia: a review. *Psychiatria Fennica* 2012; 43: 39–68.
34. Talaslahti T, Alanen H, Hakko H, Isohanni M, Häkkinen U, Leinonen E. Mortality and causes of death in elderly patients with schizophrenia. *Int J Geriatr Psychiatry* 2012; 27: 1131–1137.
35. Harding CM, Brooks GW, Ashikaga T, Strauss JS, Breier A. The Vermont longitudinal study of persons with severe mental illness, II: long-term outcome of subjects who retrospectively met DSM-III criteria for schizophrenia. *Am J Psychiatry* 1987; 144: 727–735.
36. Rajji TK, Miranda D, Mulsant BH. Cognition, function, and disability in patients with schizophrenia: a review of longitudinal studies. *Can J Psychiatr.* 2014; 59: 13–17.
37. Hjorthøj C, Stürup AE, McGrath JJ, Nordentoft M. Years of potential life lost and life expectancy in schizophrenia: a systematic review and meta-analysis. *Lancet Psychiatry* 2017; 4: 295–301.
38. Jääskeläinen E, Haapea M, Rautio N, et al. Twenty years of schizophrenia research in the Northern Finland Birth Cohort 1966: a systematic review. *Schizophr Res Treatm.* 2015; 2015: 524875.
39. Kobayashi H, Isohanni M, Jääskeläinen E, et al. Linking the developmental and degenerative theories of schizophrenia: association between infant development and adult cognitive decline. *Schizophr Bull.* 2014; 40: 1319–1327.
40. Isohanni M, Miettunen J, Mäki P, et al. Risk factors for schizophrenia: follow-up data from the Northern Finland 1966 Birth Cohort Study. *World Psychiatry* 2006; 5: 168–171.
41. Juola P, Miettunen J, Veijola J, Isohanni M, Jääskeläinen E. Predictors of short- and long-term clinical outcome in schizophrenic psychosis—the Northern Finland 1966 Birth Cohort study. *Eur Psychiatry* 2013; 28: 263–268.
42. Guo JY, Huhtaniska S, Miettunen J, et al. Longitudinal regional brain volume loss in schizophrenia: relationship to antipsychotic medication and change in social function. *Schizophr Res.* 2015; 168: 297–304.
43. Husa AP, Moilanen J, Murray GK, et al. Lifetime antipsychotic medication and cognitive performance in schizophrenia at age 43 years in a general population birth cohort. *Psychiatry Res.* 2017; 247: 130–138.
44. Moilanen JM, Haapea M, Jääskeläinen E, et al. Long-term antipsychotic use and its association with outcomes in schizophrenia—the Northern Finland Birth Cohort 1966. *Eur Psychiatry* 2016; 36: 7–14.
45. Veijola J, Guo JY, Moilanen JS, et al. Longitudinal changes in total brain volume in schizophrenia: relation to symptom severity, cognition and antipsychotic medication. *PLoS One* 2014; 9: e101689.
46. Nykänen S, Puska V, Tolonen J-P, et al. Use of psychiatric medications in schizophrenia and other psychoses in a general population sample. *Psychiatry Res.* 2016; 235: 160–168.
47. Keskinen E, Marttila R, Koivumaa-Honkanen H, et al. Search for protective factors for psychosis: a population based sample with special interest in unaffected individuals with parental psychosis. *Early Interv Psychiatry* 2016: Sep 13 [Epub ahead of print].
48. Jokela M, Pekkarinen T, Sarvimäki M, Terviö M, Uusitalo R. Secular rise in economically valuable personality traits. *Proc Natl Acad Sci U S A* 2017; 114: 6527–6532.
49. Isohanni M, Miller B, Koivumaa-Honkanen H, et al. Clinical relevance of Finnish population based studies in schizophrenia, with special reference to the Northern Finland 1966 Birth Cohort Study. *Psychiatria Fennica* 2011; 42: 65–86.
50. Taipale H, Mittendorfer-Rutz E, Alexanderson K, et al. Antipsychotics and mortality in a nationwide cohort of 29,823 patients with schizophrenia. *Schizophr Res.* 2017; doi: 10.1016/j.schres.2017.12.010 [Epub ahead of print].
51. Goff DC, Falkai P, Fleischhacker WW, et al. The long-term effects of antipsychotic medication on clinical course in schizophrenia. *Am J Psychiatry* 2017; 174: 840–849.
52. Correll C, Rubio J, Kane J. What is the risk-benefit ratio of long-term antipsychotic treatment in people with schizophrenia? *World Psychiatry* 2018;17: 149–160.
53. Leucht S, Samara M, Heres S, Davis JM. Dose equivalents for antipsychotic drugs: the DDD method. *Schizophr Bull.* 2016; 42: 90–94.
54. Ran MS, Weng X, Chan CL, et al. Different outcomes of never treated and treated patients with schizophrenia: 14-year follow-up study in rural China. *Br J Psychiatry* 2015; 207: 495–500.
55. Isohanni M, Miettunen J, Jääskeläinen E, Jani Moilanen J, Hulkko A, Huhtaniska S. Under-utilized opportunities to optimize medication management in long-term treatment of schizophrenia. *World Psychiatry* 2018; 17: 132–133.
56. Moilanen J, Haapea M, Miettunen J, et al. Characteristics of subjects with schizophrenia spectrum disorder with and without antipsychotic medication: a 10-year follow-up of the Northern Finland 1966 Birth Cohort study. *Eur Psychiatry* 2013; 28: 53–58.
57. Harrow M, Jobe TH, Faull RN, Yang J. A 20-year multi-follow up longitudinal study assessing whether antipsychotic medications contribute to work functioning in schizophrenia. *Psychiatry Res.* 2017; 256: 267–274.
58. Murray RM, Quattrone D, Natesan S, et al. Should psychiatrists be more cautious about the long-term prophylactic use of antipsychotics? *Br J Psychiatry* 2016: 209: 361–365.
59. Huhtaniska S, Jääskeläinen E, Hirvonen N, et al. Long-term antipsychotic use and brain changes in schizophrenia: a systematic review and meta-analysis. *Hum Psychopharmacol.* 2017; 32: e2574.
60. Isohanni, M. The psychiatric ward as a therapeutic community. Acta Universitatis Ouluensis Oulu: University of Oulu; 1983; D 111 n:o 5.
61. Wunderink L, Nieboer RM, Wiersma D, et al. Recovery in remitted first-episode psychosis at 7 years of follow-up of an early dose reduction/discontinuation or maintenance treatment strategy: long-term follow-up of a 2-year randomized clinical trial. *JAMA Psychiatry* 2013; 70: 913–920.
62. Hui C, Honer W, Lee E, et al. Long-term effects of discontinuation from antipsychotic maintenance following first-episode schizophrenia and related disorders: a 10 year follow-up of a randomised, double-blind trial. *Lancet Psychiatry* 2018; 5: 432–442.
63. Tiihonen J, Tanskanen A, Taipale H. 20-year nationwide follow-up study on discontinuation of antipsychotic treatment in first-episode schizophrenia. *Am J Psychiatry* 2018; 175: 765–773.
64. Kane JM, Robinson DG, Schooler NR, et al. Comprehensive versus usual community care for first-episode psychosis: 2-year outcomes from the NIMH RAISE Early Treatment Program. *Am J Psychiatry* 2016; 173: 362–372.
65. Miettunen J, Suvisaari J, Haukka J, Isohanni M. Use of register data for psychiatric epidemiology in the Nordic Countries. In: Tsuang M, Tohen M, Jones P, eds. *Textbook in psychiatric epidemiology*. 3rd ed. Chichester, West Sussex: Wiley-Blackwell; 2011:117–131.
66. Andreasen NC, Nopoulos P, Magnotta V, Pierson R, Ziebell S, Ho BC. Progressive brain change in schizophrenia: a prospective longitudinal study of first-episode schizophrenia. *Biol. Psychiatry* 2011; 70: 672–679.

67. Welham J, Isohanni M, Jones P, McGrath J. The antecedents of schizophrenia: a review of birth cohort studies. *Schizophr Bull.* 2009; 35: 603–623.
68. Isohanni, M. Administrative aspects in longitudinal studies: how to navigate on a stormy and dangerous ocean? *Acta Psychiatrica Scandinavica* 2001;104: 1–3.
69. Isohanni M, Isohanni I, Veijola J. How should a scientific team be effectively formed and managed? *Nord J Psychiatry* 2002; 56: 157–162.
70. Wang P, Brookhart A, Ulbricht C, Schneeweiss S. The pharmacoepidemiology of psychiatric medications. In: Tsuang M, Tohen M, Jones P, eds. *Textbook in psychiatric epidemiology*. 3rd ed. Chichester, West Sussex: Wiley-Blackwell, 2011; 155–165.
71. Tiihonen J, Mittendorfer-Rutz E, Majak M et al. Real-world effectiveness of antipsychotic treatments in a nationwide cohort of 29 823 patients with schizophrenia. *JAMA Psychiatry*. 2017; 74: 686–693.
72. Howes OD, McCutcheon R, Owen MJ, Murray RM. The role of genes, stress, and dopamine in the development of schizophrenia. *Biol Psychiatry* 2017; 81: 9–20.
73. Schultze-Lutter F, Michel C, Schmidt SJ, et al. EPA guidance on the early detection of clinical high risk states of psychoses. *Eur Psychiatry* 2015; 30: 405–416.
74. Cannon T, Yu C, Addington J, et al. An individualized risk calculator for research in prodromal psychosis. *Am J Psychiatry* 2016; 173: 980–988.
75. Savill M, D'Ambrosio J, Cannon T, Loewy R. Psychosis risk screening in different populations using the Prodromal Questionnaire: a systematic review. *Early Interv Psychiatry* 2018; 12: 3–14.
76. Nordentoft M, Jesper Rasmussen Ø, Melau M, et al. How successful are first episode programs? A review of the evidence for specialized assertive early intervention. *Curr Opin Psychiatry* 2014; 27: 167–172.
77. Bello I, Lee R, Malinovsky I, et al. OnTrackNY: The development of a coordinated specialty care program for individuals experiencing early psychosis. *Psychiatr Serv*. 2017; 68: 318–320.
78. Dixon L, Goldman H, Srihari V, Kane JM. Transforming the treatment of schizophrenia in the United States: the RAISE initiative. *Annu Rev Clin Psychol.* 2018; 14: 237–258.
79. Amos A. Assessing the cost of early intervention in psychosis: a systematic review. *Aust N Z J Psychiatry* 2012; 46: 719–734.
80. Clarke M, McDonough CM, Doyle R, Waddington JL. Are we really impacting duration of untreated psychosis and does it matter? Longitudinal perspectives on early intervention from the Irish Public Health Services. *Psychiatr Clin North Am.* 2016; 39: 175–186.
81. Miklowitz DJ, O'Brien MP, Schlosser DA, et al. Family-focused treatment for adolescents and young adults at high risk for psychosis: results of a randomized trial. *J Am Acad Child Adolesc Psychiatry* 2014; 53: 848–858.
82. Ising HK, Lokkerbol J, Rietdijk J, et al. M. Four-year cost-effectiveness of cognitive behavior therapy for preventing first-episode psychosis: the Dutch Early Detection Intervention Evaluation (EDIE-NL) trial. *Schizophr Bull.* 2017; 43: 365–374.
83. Lauronen E, Veijola J, Isohanni I, Jones PB, Nieminen P, Isohanni M. Links between creativity and mental disorder. *Psychiatry* 2004; 67: 81–98.
84. Cohen P, Cohen J. The clinician's illusion. *Arch Gen Psychiatry* 1984; 41: 1178–1182.
85. Kylmä J, Juvakka T, Nikkonen M, Korhonen T, Isohanni M. Hope and schizophrenia: an integrative review. *J Psych Mental Health Nursing* 2006; 13: 651–664.
86. Agid O, McDonald K, Fervaha G, et al. Values in first-episode schizophrenia. *Can J Psychiatry* 2015; 60: 507–514.

18

PHENOMENOLOGICAL CHARACTERISTICS OF PSYCHOSIS OF AGING

PSYCHOSIS AND DEMENTIA INTERPHASE

Graham M. L. Eglit, Barton W. Palmer, and Dilip V. Jeste

BRIEF HISTORICAL BACKGROUND

In his seminal case study of the disease that would later bear his name, Alois Alzheimer described a woman of previously normal functioning who presented with progressive changes in personality and cognition in her early 50s. This woman, Auguste Deter, was admitted in 1901 to the municipal mental hospital in Frankfurt, Germany, where, at the time, Alzheimer was researching the neuropathological correlates of psychiatric symptoms. In Alzheimer's descriptions, Deter's first symptoms consisted of striking delusions and behavior changes 8 months prior to her hospitalization. Deter was noted to voice extreme jealousy toward her husband, was often concerned that someone was trying to kill her, and was occasionally observed screaming wildly for hours at a time. Over time, she became increasingly forgetful and her speech became unintelligible. In her final days, she was profoundly apathetic and in a near vegetative state, spending most of her day in bed with her legs pulled up.[1]

Prior to Auguste Deter's death in 1906, Alzheimer had relocated to Munich to work with the most influential German psychiatrist at that time, Emil Kraepelin. In fact, it was Kraepelin who introduced the term "Alzheimer's disease" to distinguish the disease ascribed to Auguste Deter from the historically more familiar senile dementia. While Kraepelin published on the findings of Alzheimer, his main area of research was on another debilitating neuropsychiatric illness with prominent psychotic features, schizophrenia. Kraepelin referred to schizophrenia as *dementia praecox*, reflecting his emphasis on the course and age of onset of the disease. According to Kraepelin, schizophrenia was characterized by onset in late adolescence or early adulthood (*praecox*) and resulted in progressive deterioration in mental functioning (*dementia*). While Kraepelin later acknowledged that neither early onset nor progressive deterioration were universally characteristic of patients manifesting schizophrenia-like symptoms, the influence of his description of schizophrenia as *dementia praecox* persisted for many years.[1]

The study of psychosis in aging populations has challenged several of Kraepelin's early concepts regarding the course and age of onset of schizophrenia.[2] Consistent with Alzheimer's descriptions, though, recent research has demonstrated that psychosis is highly prevalent in Alzheimer's and other neurodegenerative diseases.[3] This chapter reviews current thinking on psychosis in late life, with a particular focus on schizophrenia and psychosis secondary to dementia. In particular, this chapter focuses on the prevalence of psychosis among older adults, the longitudinal course of schizophrenia into adulthood, the onset of schizophrenia-like psychosis in late life, and the nature of psychotic syndromes that emerge in the context of a dementing disorder. Special emphasis will be placed on the course and phenomenology of psychotic and cognitive symptoms in these two disorders in order to provide guidelines to assist in differential diagnosis.

LATE-LIFE PSYCHOSIS

Late-life psychosis is a common and frequently disabling occurrence among older adults. While definitions vary slightly, psychosis is typically conceptualized as consisting of the presence of hallucinations and delusions. The International Classification of Diseases and Health Related Problems, 10th Revision (ICD-10) additionally considers abnormalities of behavior as a feature of psychosis, while the Diagnostic and Statistical Manual of Mental Disorders, 5th Edition (DSM-5) includes formal thought disorder as an additional component of a psychotic syndrome,[4,5] Nonetheless, the presence of hallucinations and delusions represents the core defining feature of psychosis.

Psychosis is prevalent among older adults.[6] Cross-sectional studies have reported the presence of psychotic symptoms in 5%–15% of geropsychiatric inpatients, 10%–62% of nursing home patients, 10% of adults older than 85, and 27% of community-dwelling psychiatric outpatients.[7,8] Overall, the risk of developing psychotic symptoms in late life has been estimated to be as high as 23% and exceeds the rate of psychosis among younger adults.[6–8]

The etiology of psychosis often differs in older as opposed to younger adults. Psychosis can either be due to a psychiatric disorder (termed "primary psychosis") or can be secondary to a medical or neurological disorder (termed "secondary psychosis"). However, phrased this way, this dichotomy between primary and secondary psychosis may echo the historical misconception of "functional" versus "organic" psychoses,[9] i.e., the historical notion that the primary psychoses are best conceptualized as psychogenic rather than reflections of brain function. A more accurate and contemporary assertion is to view the primary psychoses as neurodevelopmental, whereas the secondary psychoses reflect acquired disruption of the central nervous system.[10] Primary psychosis is the more common etiology in younger adults, and schizophrenia is the most common cause of these symptoms in this age group,[11] although psychotic symptoms can also emerge in the context of mood disorders (major depressive disorder, bipolar disorder) and are found in schizoaffective disorder.[4] In contrast, approximately 60% of psychotic disorders in late life (i.e., after age 60) are secondary to another condition.[7,8] The most common conditions resulting in a secondary psychosis are delirium, substance use, general medical conditions (especially metabolic, infectious, neurological, and endocrine disorders), and dementia. Dementia accounts for nearly 40% of all psychotic disorders among the elderly and thus represents the most common cause of psychosis in late life.[12]

SCHIZOPHRENIA

PHENOMENOLOGY

Schizophrenia is a chronic neurodevelopmental disorder characterized by hallucinations, delusions, disorganized speech, grossly disorganized or catatonic behavior, and negative symptoms (diminished expression and/or avolition).[4] In order to meet criteria for a diagnosis of schizophrenia, DSM-5 requires the presence of two or more of these symptoms for at least 6 months' duration with an associated decline in social or occupational functioning. In a change from DSM-IV, DSM-5 stipulates that at least one of these symptoms must be a core psychotic symptom, i.e., hallucinations, delusions, or disorganized speech. This change is consistent with DSM-5's conceptualization of schizophrenia as a fundamentally psychotic disorder.[13]

Hallucinations in schizophrenia can occur in any sensory modality, although auditory hallucinations are most common. Typically, these hallucinations consist of voices talking among one another or directly to the patient. Voices may be single or multiple, clear or unintelligible, and may be identified as voices of family, friends, strangers, or God. In some cases, voices may be connected to the content of the patient's delusions. Perhaps most troubling for the patient are command auditory hallucinations, in which voices command the patient to perform specific actions.[14]

There is considerable heterogeneity in the content of delusions in schizophrenia. Delusions of control (the belief that one's mind or body are under the control of an external agent), thought insertion (the belief that one's internal thoughts are those of other people who are intruding into the patient's mind), thought withdrawal (the belief that thoughts have been removed from one's mind), and thought broadcasting (the belief that one's thoughts are being broadcast to others) have long been associated with schizophrenia. A variety of other delusions may occur, including delusions of grandiosity and mind reading. However, persecutory delusions and delusions of reference are the most common. Notably, delusions are often significantly impacted by a person's life and sociocultural context. For instance, delusions of grandiosity may involve the belief that one is Jesus Christ in a predominantly Christian culture as opposed to identification with the prophet Mohammad in a Muslim culture.[14,15]

AGE OF ONSET

Psychotic symptoms most often emerge in late adolescence to early adulthood in schizophrenia. In a small minority of individuals (20%–25%), psychotic symptoms first appear after the age of 40.[16,17] Psychotic symptoms can also emerge after the age of 60; however, in such cases, these psychotic symptoms often co-occur with declining cognitive functioning. In recognition of these distinct ages of onset, subtypes of schizophrenia have been proposed. These subtypes include (1) early-onset psychosis, with prodrome onset in adolescence and early adulthood; (2) late-onset psychosis, with prodrome onset in the 40s and 50s; and (3) very late-onset schizophrenia-like psychosis, with prodrome onset after age 60.[18]

Research has explored similarities and differences in clinical presentations among these subtypes. Comparisons between early-onset and late-onset subtypes have found that a family history of schizophrenia and early childhood maladjustment is common to both subtypes.[19] In addition, early- and late-onset subtypes of schizophrenia are characterized by similar severity of positive symptoms, presence of nonspecific structural abnormalities on brain imaging, and number of minor physical anomalies. Finally, these subtypes also have a similar overall pattern and stability of neuropsychological deficits and are both responsive to neuroleptic medication.[19–21]

There are also differences in clinical presentation between early- and late-onset schizophrenia subtypes. The gender composition of these subtypes is perhaps the most significant of these differences. While both men and women have the highest incidence of new-onset schizophrenia in their 20s and 30s, women have a second peak of incidence in their mid-late 40s, leading to a disproportionate number of women constituting the late-onset schizophrenia subtype. Recognition that this second peak co-occurs in the context of the hormonal changes of menopause has led to the estrogen hypothesis. This hypothesis maintains that estrogen is protective against the expression of schizophrenia in premenopausal women, and reductions in estrogen levels during menopause increase rates of schizophrenia among women by lessening this protective effect.[22] Late-onset schizophrenia also appears to be characterized by less severe general psychopathology, better everyday functioning, superior health-related quality of life, and less severe negative symptoms.[19,23]

In contrast, very late-onset schizophrenia-like psychosis differs from early- and late-onset schizophrenia in several substantial ways. Phenomenologically, very late-onset schizophrenia-like psychosis is associated with a greater likelihood of visual, olfactory, and tactile hallucinations than early- and late-onset schizophrenia, which are both more commonly associated with auditory hallucinations.[18,24] There is also a lower genetic load and less often a family history of schizophrenia in very late-onset schizophrenia-like psychosis.[19] Perhaps most significantly, very late-onset schizophrenia-like psychosis often co-occurs with progressive cognitive deterioration and specific brain structural abnormalities on MRI such as stroke, tumors, and regional atrophy.[25–28] Taken together, this research suggests that very-late onset schizophrenia-like psychosis is probably due to a different etiology than early- and late-onset schizophrenia in most cases. Vascular and neurodegenerative etiologies are most likely for this subtype, given the co-occurrence of progressive cognitive deterioration and neurological changes on brain imaging.

SCHIZOPHRENIA AND AGING

Given that typical age of onset in schizophrenia is in late adolescence to early adulthood, older adults with schizophrenia have likely been living with this illness for 30 or 40 years.[16] Research among these older adults has challenged many previously held assumptions about the progressive nature of schizophrenia. Contrary to these prior assumptions, much of this research has reported a benign course with some notable improvements in functioning in older age.[2]

One of the core assumptions of Kraepelin's conceptualization of schizophrenia as *dementia praecox* was that cognition declines over the course of this illness. However, the overwhelming majority of research on cognition in schizophrenia has not supported this assumption (reviewed in [29]). This research has shown that schizophrenia is associated with mild premorbid cognitive deficits of about 5 or 10 IQ points.[30] There is some evidence for a subtle cognitive decline of approximately an additional 5 IQ points from pre- to post-onset of psychotic symptoms.[31] However, following this initial onset period, the vast majority of community-dwelling persons with schizophrenia exhibit a remarkably stable level of cognitive functioning. For instance, Heaton et al.[32] followed 142 community-dwelling schizophrenia patients and 209 normal comparison participants for 2–10 years and found that the overall rate and pattern of cognitive changes did not differ across groups. This relative stability may not generalize to the small minority (approximately 10%) of schizophrenia patients who are chronically institutionalized, as these patients exhibit greater than age-expected cognitive decline and increased rates of dementia. Nonetheless, for the vast majority of community-dwelling schizophrenia patients, cognitive function does not decline at an accelerated rate relative to individuals without schizophrenia in older age (however, see [2,33]).

Perhaps even more strikingly, psychosocial functioning appears to improve over the course of the life span among persons with schizophrenia. Older adults with schizophrenia report improved mental-health-related quality of life, a measure of the impact of a chronic illness on an individual's overall well-being and functioning.[34,35] In addition, older schizophrenia patients report greater acceptance and self-management abilities with respect to their symptoms, and exhibit improved medication adherence, a reduction in positive symptoms, and decreased rates of relapse and hospitalization.[36,37] Moreover, a small percentage of patients (approximately 10%) exhibit sustained remission of symptoms.[38]

In contrast to spared cognitive functioning, physical aging appears to be accelerated in schizophrenia.[2] Schizophrenia has long been known to involve a reduced life span. Recent meta-analytic estimates suggest that the lives of individuals with schizophrenia are 14.5 years shorter than those of persons without schizophrenia.[39] There are several medical comorbidities in schizophrenia that may contribute to this shortened life span, including increased rates of metabolic syndrome and coronary heart disease.[40] In addition, persons with schizophrenia exhibit elevated levels of biomarkers of inflammation and high rates of insulin resistance.[41–45] Several causes of accelerated aging have been posited, including the effects of smoking, a sedentary lifestyle, antipsychotic medications (especially atypical antipsychotics), substance use, and limited access to healthcare. However, at least some of the comorbidities found in schizophrenia appear to be a consequence of the biology of schizophrenia itself.[2]

PSYCHOSIS ASSOCIATED WITH DEMENTIA

Psychosis is also a common symptom among individuals with dementia. Dementia, renamed "major neurocognitive disorder" in DSM-5, is characterized by impairment in one or more cognitive domains resulting in impaired independence in everyday functioning. Relevant cognitive domains include complex attention, executive function, learning/memory, perceptual motor, and social cognition. Several disease processes can cause dementia/major neurocognitive disorder. Many of these disease processes are also associated with high rates of psychotic symptoms, including Alzheimer's disease, Lewy body disease, and Parkinson's disease.[4]

ALZHEIMER'S DISEASE

Alzheimer's disease is the most common neurodegenerative disease and the leading cause of dementia among older adults, accounting for 60%–80% of all cases of dementia.[46–48] Alzheimer's disease is characterized by early and prominent memory impairments. As the disease progresses, additional cognitive abilities decline, including language, executive functioning, and visuospatial abilities.[49] The prevalence of Alzheimer's disease increases with age, from approximately 3% of those ages 65–74 to 32% of those ages 85 years and older.[50]

Psychotic symptoms are highly prevalent in Alzheimer's disease. Approximately 40% of Alzheimer's disease patients report psychotic symptoms. Delusions tend to be more common than hallucinations, with prevalence estimates of delusions ranging from 27% to 35% and hallucinations from 13% to 18%.[51] Psychotic symptoms appear to be particularly prevalent in the middle stage of the disease.[3]

Several risk factors for psychotic symptoms have been identified. Hallucinations tend to be associated with less education, more severe cognitive impairment, longer duration of disease, falls, and use of anxiolytics, and are more common among African Americans.[52,53] In contrast, delusions appear to be associated with a unique set of predictors, which include depression, aggression, poorer general health, use of antihypertensives, and older age.[52] Overall, psychotic symptoms more often occur among individuals with more severe cognitive impairment, older age, older age at onset, longer duration of illness, and who are of African American ethnic background.[3]

In contrast to schizophrenia, where auditory hallucinations predominate, visual hallucinations tend to be the most common type of hallucination in Alzheimer's disease. These visual hallucinations are often complex, fully formed, and three dimensional, and are most commonly of familiar people from the past, animals, or objects. Delusions are most frequently persecutory in nature in Alzheimer's disease. The most commonly reported themes of these delusions include delusions of theft, infidelity, abandonment, and believing that their house is not their home.[3] Clinically, it is important to distinguish primary delusions from false beliefs related to memory-loss and/or misidentifications.[54]

LEWY BODY DISEASE

Lewy body disease is the third leading cause of neurodegenerative dementia among the elderly.[55] In addition to cognitive impairment, symptoms in Lewy body disease (also termed "dementia with Lewy bodies") consist of difficulties maintaining alertness/arousal, dream enactment behavior, Parkinsonian motor signs (stooped posture, postural instability/gait disturbance, muscle rigidity, and tremor), and psychotic symptoms, especially visual hallucinations. In fact, visual hallucinations are considered a core feature of Lewy body disease in the DSM-5.[4]

Given that visual hallucinations are a core feature in diagnosis, it is perhaps unsurprising that psychotic symptoms are highly prevalent in Lewy body disease. In contrast to Alzheimer's disease, visual hallucinations tend to be more common than delusions in Lewy body disease and tend to occur early in the disease course. Nagahama et al.[56] report that 43% of individuals with Lewy body disease presented with visual hallucinations at diagnosis. Over the course of the disease, the prevalence of visual hallucinations may reach as high as 93%.[57] Auditory hallucinations occur in Lewy body disease as well, but at a much lower rate, and they rarely occur in patients without visual hallucinations.[58] Phenomenologically, visual hallucinations tend to be fully formed, detailed, three-dimensional objects, people, or animals.[59]

Delusions, while less common than visual hallucinations, are also frequently reported in Lewy body disease. Prevalence estimates suggest that delusions are present at diagnosis in up to 57% of patients.[57,60] Phenomenologically, delusions of misidentification, including Capgras syndrome and phantom boarder syndrome, are particularly common, occurring in over 80% of Lewy body disease cases with delusions.[60] As with Alzheimer's disease, clinicians should be mindful to distinguish between delusions and false beliefs resulting from memory impairment and other cognitive difficulties.[54]

PARKINSON'S DISEASE

Parkinson's disease is the second most common neurodegenerative disease following Alzheimer's disease.[61] However, only a subset of individuals with Parkinson's disease subsequently develop dementia. Overall, 3%–5% of dementia cases are attributable to Parkinson's disease.[62] Prevalence of Parkinson's disease is estimated at approximately 0.3% of the general population and 1% of persons over the age of 60.[63] Parkinson's disease is characterized by cardinal motor signs of bradykinesia (slowness of movement), rigidity, resting tremor, and postural instability and gait disturbance.[64] Subtle cognitive changes are often present at diagnosis, but significant cognitive impairment warranting a diagnosis of dementia typically does not develop until several years after motor symptom onset. In fact, a dementia syndrome within a year of onset of motor symptoms is an exclusionary criteria for Parkinson's disease.[65] Individuals presenting with both Parkinsonian motor signs and a dementia syndrome at onset would most likely meet criteria for dementia with Lewy bodies.[66] Nonetheless, diagnosis of Parkinson's disease confers a sixfold risk of developing dementia over time compared to people of the same age without Parkinson's disease. The average time from motor symptom onset to dementia in Parkinson's disease is approximately 10 years.[67,68]

Prevalence estimates of psychosis in Parkinson's disease range from approximately 20% to 60% and tend to increase with greater disease duration.[69–73] For instance, in a 12-year follow-up study, Fossa et al.[74] reported an 18% prevalence of psychosis at baseline, which increased to 60% at 12-year follow-up. Among traditional psychotic symptoms, hallucinations are more common than delusions.[70] More recently, a National Institute of Health NINDS-NIMH working group proposed revised criteria for psychosis in Parkinson's disease.[75] In addition to hallucinations and delusions, these revised criteria include minor forms of hallucinations, including illusions or misperceptions of real stimuli, presence hallucinations (the false sense that someone is nearby), and passage hallucinations (sensation that a person or animal is passing in the person's peripheral visual field). In a comparative study, applying these revised criteria increased the cross-sectional prevalence of psychosis from 43% to 60%, indicating that prior epidemiological studies may have underestimated the prevalence of these minor psychotic phenomena.[76]

Similar to Lewy body disease and Alzheimer' disease, hallucinations in Parkinson's disease are most commonly visual, although they can occur in auditory, tactile, olfactory, and gustatory modalities as well.[77,78] Hallucinations in nonvisual modalities tend to occur alongside visual hallucinations and often do not emerge until the advanced stage of the disease.[79,80] When auditory hallucinations occur, they tend to be vague and nonthreatening and rarely interact with the patient.[81] Visual hallucinations are important clinical variables in Parkinson's disease, as they predict an accelerated rate of cognitive decline and conversion to Parkinson's disease dementia.[67]

Visual hallucinations are most commonly of two types. Simple hallucinations refer to flashes, dots, shapes, lines, and pattern-like swirly grids or tiles. In contrast, complex hallucinations consist of fully-formed stable objects, most often appearing as familiar and unfamiliar people and faces, animals, or inanimate objects.[82,83] The majority of visual hallucinations in Parkinson's disease are nonthreatening and are typically brief, lasting seconds to minutes, with patients retaining insight. However, over time, these "benign" hallucinations may become more disconcerting to the patient, especially as insight and cognition diminish in advanced stages of the disease.[84,85]

Delusions in Parkinson's disease usually involve a single, consistent theme, such as jealousy, paranoia, infidelity, abandonment, or somatic illness.[86] Similar to Lewy body disease, misidentification syndromes are common in Parkinson's disease, with Capgras syndrome and Fregoli syndrome (the belief that familiar people are often malevolently disguised as strangers) being particularly prevalent. In one study, misidentification syndromes were reported in 16.7% of Parkinson's disease patients.[87] Overall, delusions are estimated to affect approximately 5%–10% of Parkinson's disease patients.[70] Comparisons between schizophrenia and Parkinson's disease have revealed that while persecutory delusions are common in both conditions, delusions of grandiosity, reference, and bizarre beliefs are more frequent in schizophrenia.[88]

COMPARISONS AMONG LATE-LIFE PSYCHOTIC DISORDERS

Table 18.1 presents a review of the course and phenomenology of psychosis in early- and late-onset schizophrenia, very late-onset schizophrenia-like psychosis, Alzheimer's disease, Lewy body disease, and Parkinson's disease. Inspection of this table reveals important differences across these conditions that can assist with differential diagnosis.

Typical age of onset differs widely across these conditions. While psychotic symptom onset typically occurs in the 20s–30s and 40s–50s in early- and late-onset schizophrenia, respectively, psychotic symptoms typically do not emerge until much later in life in very late-onset schizophrenia-like psychosis, Alzheimer's disease, Lewy body disease, and Parkinson's disease.[16,18] In these latter conditions, psychotic symptoms usually do not begin until at least the sixth decade of life, and often do not occur until even later than this.[18,46,49,55,61,63]

While hallucinations and delusions are common across each of these diagnoses, the phenomenology of these psychotic symptoms differs. The typical modality of hallucinations is perhaps the most discriminating phenomenological feature across these diagnoses: auditory hallucinations are most common in early- and late-onset schizophrenia, while visual hallucinations are more typical in very late-onset schizophrenia-like psychosis, Alzheimer's disease, Lewy body disease, and Parkinson's disease.[3,14,18,24,58,77,78] The predominant theme of delusions is less specific, however. This is because while the predominant theme of delusions differs somewhat across these conditions, there is a great deal of variability in the theme of delusions within each diagnosis and a moderate degree of overlap in these themes across diagnoses.[3,15,60,86] Similarly, the relative prevalence of delusions and hallucinations does not provide much assistance in differential diagnosis, as these symptoms may co-occur in each condition, although hallucinations are notably more common than delusions in Lewy body disease and Parkinson's disease.[57,77]

The presence of cognitive and neurological changes also helps discriminate among these conditions. In early- and late-onset schizophrenia, cognition may decline slightly at the time

Table 18.1 AGE OF ONSET, COURSE, AND PHENOMENOLOGY OF PSYCHOTIC SYMPTOMS IN EARLY-ONSET SCHIZOPHRENIA, LATE-ONSET SCHIZOPHRENIA, VERY LATE-ONSET SCHIZOPHRENIA-LIKE PSYCHOSIS, ALZHEIMER'S DISEASE, LEWY BODY DISEASE, AND PARKINSON'S DISEASE

	EARLY-ONSET SCHIZOPHRENIA	LATE-ONSET SCHIZOPHRENIA	VERY LATE-ONSET SCHIZOPHRENIA-LIKE PSYCHOSIS	ALZHEIMER'S DISEASE	LEWY BODY DISEASE	PARKINSON'S DISEASE
Typical Age of Onset	20s and 30s	40s and 50s	60+	65+	65+	65+
Most Prevalent Psychotic Symptom	Hallucinations	Hallucinations	Hallucinations	Delusions	Hallucinations	Hallucinations
Most Prevalent Modality of Hallucinations	Auditory	Auditory	Visual	Visual	Visual	Visual
Most Common Themes of Delusions	Persecution and Reference	Persecution and Reference	Variable (Depends on Etiology)	Persecution	Misidentification	Misidentification
Presence of Specific Abnormalities on Brain Imaging	No	No	Yes	Yes	Yes	Yes
Presence of Progressive Cognitive Decline	No	No	Yes	Yes	Yes	Yes

of psychotic symptom onset.[31] After this initial period, however, cognition remains stable in most cases.[29,32] Furthermore, brain imaging in early- and late-onset schizophrenia is unlikely to reveal specific neurological abnormalities.[19] In contrast, in very late-onset schizophrenia-like psychosis, Alzheimer's disease, Lewy body disease, and in some cases of Parkinson's disease, psychotic symptoms tend to emerge in the context of gradual and progressive declines in cognitive function. These declines are further accompanied by brain-imaging abnormalities suggestive of a specific etiology.[25,26,49,55,67]

IMPLICATIONS FOR TREATMENT/ MANAGEMENT

Pharmacological and nonpharmacological treatment of schizophrenia and other psychotic disorders is covered in other chapters in this book. In general, antipsychotics and other medications should be used with considerable caution because of the high risk of side effects and interactions with different medications that older adults are likely to be receiving. Use of antipsychotics in older persons with dementia carries FDA black box warnings regarding strokes and mortality. Psychosocial or behavioral treatments are essential for most of the older adults with psychotic disorders.

SUMMARY

Research exploring the trajectory of schizophrenia in older adulthood has called into question several previously held assumptions about the course of this illness. Contrary to prior thinking, most individuals with schizophrenia experience a relatively benign course in their later years, exhibiting an improvement in psychosocial functioning and a rate of cognitive decline consistent with normal aging. In contrast, biological aging appears to be accelerated in schizophrenia, leading to a significantly greater physical comorbidity and reduced life span.[2] Cross-sectional studies have found elevated levels of inflammation and insulin resistance in schizophrenia.[41–45] Further research is needed to explore the longitudinal course of these biological processes and their interactions over time. Development of interventions to target these biological changes will be important for extending the life span of individuals with schizophrenia.

While schizophrenia is the most common cause of psychotic symptoms among younger adults, psychotic symptoms most commonly emerge in the context of dementia among individuals over the age of 60.[12] Psychotic symptoms are important clinical phenomena in these cases. The presence of visual hallucinations is a core diagnostic feature of dementia with Lewy bodies. In addition, psychotic symptoms connote a more severe phenotype associated with a more rapid rate of cognitive decline in Alzheimer's disease[89] and place individuals with Parkinson's disease at greater risk of subsequently developing dementia.[90] Psychosis in dementia is also associated with increased nursing home placement,[91,92] caregiver stress,[93,94] and worsened quality of life.[95] Clinicians should thus be mindful of the possible presence of these symptoms and their implications for patients.

REFERENCES

1. Cipriani G, Dolciotti C, Picchi L, Bonuccelli U. Alzheimer and his disease: a brief history. *Neurol Sci*. 2011;32(2):275–279.
2. Jeste DV, Wolkowitz OM, Palmer BW. Divergent trajectories of physical, cognitive, and psychosocial aging in schizophrenia. *Schizophr Bull*. 2011;37(3):451–455.
3. Ropacki SA, Jeste DV. Epidemiology of and risk factors for psychosis of Alzheimer's disease: a review of 55 studies published from 1990 to 2003. *Am J Psychiatry*. 2005;162(11):2022–2030.
4. American Psychiatric Association. *Diagnostic and statistical manual of mental disorders (5th ed.)*. Washington, DC: American Psychiatric Association; 2013.
5. World Health Organization. *International statistical classification of diseases and related health problems, 10th revision (ICD-10)*. Geneva, Switzerland: World Health Organization; 1992.
6. Ostling S, Skoog I. Psychotic symptoms and paranoid ideation in a nondemented population-based sample of the very old. *Arch Gen Psychiatry*. 2002;59(1):53–59.
7. Holroyd S, Laurie S. Correlates of psychotic symptoms among elderly outpatients. *International journal of geriatric psychiatry*. 1999;14(5):379–384.
8. Webster J, Grossberg GT. Late-life onset of psychotic symptoms. *The American journal of geriatric psychiatry: official journal of the American Association for Geriatric Psychiatry*. 1998;6(3): 196–202.
9. Beer MD. The dichotomies: psychosis/neurosis and functional/ organic: a historical perspective. *History of psychiatry*. 1996;7(26 Pt 2):231–255.
10. Marenco S, Weinberger DR. The neurodevelopmental hypothesis of schizophrenia: following a trail of evidence from cradle to grave. *Development and psychopathology*. 2000;12(3):501–527.
11. Perala J, Suvisaari J, Saarni SI, et al. Lifetime prevalence of psychotic and bipolar I disorders in a general population. *Arch Gen Psychiatry*. 2007;64(1):19–28.
12. Reinhardt MM, Cohen CI. Late-life psychosis: diagnosis and treatment. *Curr Psychiatry Rep*. 2015;17(2):1.
13. Tandon R, Gaebel W, Barch DM, et al. Definition and description of schizophrenia in the DSM-5. *Schizophrenia research*. 2013;150(1):3–10.
14. Wong AHC, Van Tol HHM. Schizophrenia: from phenomenology to neurobiology. *Neuroscience & Biobehavioral Reviews*. 2003;27(3):269–306.
15. Tandon R, Nasrallah HA, Keshavan MS. Schizophrenia, "just the facts" 4: clinical features and conceptualization. *Schizophrenia research*. 2009;110(1–3):1–23.
16. Harris MJ, Jeste DV. Late-onset schizophrenia: an overview. *Schizophr Bull*. 1988;14(1):39–55.
17. Jeste DV, Lanouette N, Vahia IV. Schizophrenia and paranoid disorders. In: Blazer DG, Steffense DC, eds. *Textbook of geriatric psychiatry (4th ed.)*. Washington, DC: American Psychiatric Publishing; 2009:317–332.
18. Howard R, Rabins PV, Seeman MV, Jeste DV. Late-onset schizophrenia and very-late-onset schizophrenia-like psychosis: An international consensus. The International Late-Onset Schizophrenia Group. *Am J Psychiatry*. 2000;157(2):172–178.
19. Jeste DV, Symonds LL, Harris MJ, Paulsen JS, Palmer BW, Heaton RK. Nondementia nonpraecox dementia praecox? Late-onset schizophrenia. *The American Journal of Geriatric Psychiatry*. 1997;5(4):302–317.
20. Jeste DV, Harris MJ, Krull A, Kuck J, McAdams LA, Heaton R. Clinical and neuropsychological characteristics of patients with late-onset schizophrenia. *The American journal of psychiatry*. 1995;152(5):722–730.

21. Lohr JB, Alder M, Flynn K, Harris MJ, McAdams LA. Minor physical anomalies in older patients with late-onset schizophrenia, early-onset schizophrenia, depression, and Alzheimer's disease. *Am J Geriatr Psychiatry*. 1997;5(4):318–323.
22. Gogos A, Sbisa AM, Sun J, Gibbons A, Udawela M, Dean B. A role for estrogen in schizophrenia: clinical and preclinical findings. *International journal of endocrinology*. 2015;2015:615356.
23. Vahia IV, Palmer BW, Depp C, et al. Is late-onset schizophrenia a subtype of schizophrenia? *Acta Psychiatr Scand*. 2010;122(5):414–426.
24. Howard R, Almeida O, Levy R. Phenomenology, demography and diagnosis in late paraphrenia. *Psychol Med*. 1994;24(2):397–410.
25. Barak Y, Aizenberg D, Mirecki I, Mazeh D, Achiron A. Very late-onset schizophrenia-like psychosis: clinical and imaging characteristics in comparison with elderly patients with schizophrenia. *J Nerv Ment Dis*. 2002;190(11):733–736.
26. Palmer BW, McClure FS, Jeste DV. Schizophrenia in late life: findings challenge traditional concepts. *Harvard Review of Psychiatry*. 2001;9(2):51–58.
27. Mazeh D, Zemishlani C, Aizenberg D, Barak Y. Patients with very-late-onset schizophrenia-like psychosis: a follow-up study. *The American journal of geriatric psychiatry: official journal of the American Association for Geriatric Psychiatry*. 2005;13(5):417–419.
28. Girard C, Simard M. Elderly patients with very late-onset schizophrenia-like psychosis and early-onset schizophrenia: cross-sectional and retrospective clinical findings. *Open Journal of Psychiatry*. 2012;2(4):305.
29. Palmer BW, Dawes SE, Heaton RK. What do we know about neuropsychological aspects of schizophrenia? *Neuropsychology review*. 2009;19(3):365–384.
30. Bilder RM, Reiter G, Bates J, et al. Cognitive development in schizophrenia: follow-back from the first episode. *J Clin Exp Neuropsychol*. 2006;28(2):270–282.
31. Mesholam-Gately RI, Giuliano AJ, Goff KP, Faraone SV, Seidman LJ. Neurocognition in first-episode schizophrenia: a meta-analytic review. *Neuropsychology*. 2009;23(3):315–336.
32. Heaton RK, Gladsjo JA, Palmer BW, Kuck J, Marcotte TD, Jeste DV. Stability and course of neuropsychological deficits in schizophrenia. *Arch Gen Psychiatry*. 2001;58(1):24–32.
33. Thompson WK, Savla GN, Vahia IV, et al. Characterizing trajectories of cognitive functioning in older adults with schizophrenia: does method matter? *Schizophrenia research*. 2013;143(1):90–96.
34. Reine G, Simeoni MC, Auquier P, Loundou A, Aghababian V, Lancon C. Assessing health-related quality of life in patients suffering from schizophrenia: a comparison of instruments. *Eur Psychiatry*. 2005;20(7):510–519.
35. Folsom DP, Depp C, Palmer BW, et al. Physical and mental health-related quality of life among older people with schizophrenia. *Schizophrenia research*. 2009;108(1-3):207–213.
36. Shepherd S, Depp CA, Harris G, Halpain M, Palinkas LA, Jeste DV. Perspectives on schizophrenia over the lifespan: a qualitative study. *Schizophrenia bulletin*. 2012;38(2):295–303.
37. Post F. Schizo-affective symptomatology in late life. *Br J Psychiatry*. 1971;118(545):437–445.
38. Granholm E, McQuaid JR, McClure FS, et al. A randomized, controlled trial of cognitive behavioral social skills training for middle-aged and older outpatients with chronic schizophrenia. *Am J Psychiatry*. 2005;162(3):520–529.
39. Hjorthoj C, Sturup AE, McGrath JJ, Nordentoft M. Years of potential life lost and life expectancy in schizophrenia: a systematic review and meta-analysis. *Lancet Psychiatry*. 2017;4(4):295–301.
40. Jin H, Folsom D, Sasaki A, et al. Increased Framingham 10-year risk of coronary heart disease in middle-aged and older patients with psychotic symptoms. *Schizophrenia research*. 2011;125(2-3):295–299.
41. Hong S, Lee EE, Martin AS, et al. Abnormalities in chemokine levels in schizophrenia and their clinical correlates. *Schizophrenia research*. 2017;181:63–69.
42. Lee EE, Eyler LT, Wolkowitz OM, et al. Elevated plasma F2-isoprostane levels in schizophrenia. *Schizophrenia research*. 2016;176(2–3):320–326.
43. Lee EE, Hong S, Martin AS, Eyler LT, Jeste DV. Inflammation in schizophrenia: cytokine levels and their relationships to demographic and clinical variables. *The American journal of geriatric psychiatry: official journal of the American Association for Geriatric Psychiatry*. 2017;25(1):50–61.
44. Joseph J, Depp C, Martin AS, et al. Associations of high sensitivity C-reactive protein levels in schizophrenia and comparison groups. *Schizophrenia research*. 2015;168(1-2):456–460.
45. Fan X, Goff DC, Henderson DC. Inflammation and schizophrenia. *Expert Rev Neurother*. 2007;7(7):789–796.
46. Mayeux R, Stern Y. Epidemiology of Alzheimer's disease. *Cold Spring Harbor Perspectives in Medicine*. 2012;2(8):1–18.
47. Nowrangi MA, Rao V, Lyketsos CG. Epidemiology, assessment, and treatment of dementia. *Psychiatr Clin North Am*. 2011;34(2):275–294, vii.
48. Reitz C, Brayne C, Mayeux R. Epidemiology of Alzheimer disease. *Nat Rev Neurol*. 2011;7(3):137–152.
49. Smith GE, Bondi MW. Alzheimer's disease. In: Smith GE, Bondi MW, eds. *Mild cognitive impairment and dementia: definitions, diagnosis, and treatment*. New York, NY: Oxford University Press; 2013.
50. Hebert LE, Weuve J, Scherr PA, Evans DA. Alzheimer disease in the United States (2010–2050) estimated using the 2010 census. *Neurology*. 2013;80(19):1778–1783.
51. Zhao QF, Tan L, Wang HF, et al. The prevalence of neuropsychiatric symptoms in Alzheimer's disease: systematic review and meta-analysis. *Journal of affective disorders*. 2016;190:264–271.
52. Bassiony MM, Steinberg MS, Warren A, Rosenblatt A, Baker AS, Lyketsos CG. Delusions and hallucinations in Alzheimer's disease: Prevalence and clinical correlates. *International journal of geriatric psychiatry*. 2000;15:99–107.
53. Ostling S, Gustafson D, Blennow K, Borjesson-Hanson A, Waern M. Psychotic symptoms in a population-based sample of 85-year-old individuals with dementia. *Journal of geriatric psychiatry and neurology*. 2011;24(1):3–8.
54. Cantillon M, De La Puente AM, Palmer BW. Psychosis in Alzheimer's disease. *Seminars in clinical neuropsychiatry*. 1998;3(1):34–40.
55. Ferman TJ. Dementia with Lewy bodies. In: Smith GE, Bondi MW, eds. *Mild cognitive impairment and dementia: definitions, diagnosis, and treatment*. New York, NY: Oxford University Press; 2013.
56. Nagahama Y, Okina T, Suzuki N, Matsuda M, Fukao K, Murai T. Classification of psychotic symptoms in dementia with Lewy bodies. *The American journal of geriatric psychiatry: official journal of the American Association for Geriatric Psychiatry*. 2007;15(11): 961–967.
57. Ballard CG, O'Brien J, Lowery K, et al. A prospective study of dementia with Lewy bodies. *Age Ageing*. 1998;27(5):631–636.
58. Ferman TJ, Boeve B, Silber MH, et al. Hallucinations and delusions associated with the REM sleep behavior disorder/dementia syndrome. *Journal of Neuropsychiatry and Clinical Neuroscience*. 1997;9(692):66.
59. Mosimann UP, Rowan EN, Partington CE, et al. Characteristics of visual hallucinations in Parkinson disease dementia and dementia with Lewy bodies. *The American journal of geriatric psychiatry: official journal of the American Association for Geriatric Psychiatry*. 2006;14(2):153–160.
60. Ballard C, Holmes C, McKeith I, et al. Psychiatric morbidity in dementia with Lewy bodies: a prospective clinical and neuropathological comparative study with Alzheimer's disease. *The American journal of psychiatry*. 1999;156(7):1039–1045.
61. de Lau LM, Breteler MM. Epidemiology of Parkinson's disease. *Lancet Neurol*. 2006;5(6):525–535.
62. Aarsland D, Zaccai J, Brayne C. A systematic review of prevalence studies of dementia in Parkinson's disease. *Mov Disord*. 2005;20(10):1255–1263.
63. Nussbaum RL, Ellis CE. Alzheimer's disease and Parkinson's disease. *The New England journal of medicine*. 2003;348(14): 1356–1364.
64. Gelb DJ, Oliver E, Gilman S. Diagnostic criteria for Parkinson disease. *Arch Neurol*. 1999;56(1):33–39.

65. Aarsland D, Bronnick K, Larsen JP, Tysnes OB, Alves G. Cognitive impairment in incident, untreated Parkinson disease: the Norwegian ParkWest study. *Neurology*. 2009;72(13):1121–1126.
66. McKeith IG, Dickson DW, Lowe J, et al. Diagnosis and management of dementia with Lewy bodies: third report of the DLB Consortium. *Neurology*. 2005;65(12):1863–1872.
67. Aarsland D, Andersen K, Larsen JP, Lolk A, Kragh-Sorensen P. Prevalence and characteristics of dementia in Parkinson disease: an 8-year prospective study. *Arch Neurol*. 2003;60(3):387–392.
68. Aarsland D, Perry R, Brown A, Larsen JP, Ballard C. Neuropathology of dementia in Parkinson's disease: a prospective, community-based study. *Ann Neurol*. 2005;58(5):773–776.
69. Aarsland D, Larsen JP, Cummins JL, Laake K. Prevalence and clinical correlates of psychotic symptoms in Parkinson disease: a community-based study. *Arch Neurol*. 1999;56(5):595–601.
70. Fenelon G, Alves G. Epidemiology of psychosis in Parkinson's disease. *Journal of the neurological sciences*. 2010;289(1-2):12–17.
71. Graham JM, Grunewald RA, Sagar HJ. Hallucinosis in idiopathic Parkinson's disease. *Journal of neurology, neurosurgery, and psychiatry*. 1997;63(4):434–440.
72. Holroyd S, Currie L, Wooten GF. Prospective study of hallucinations and delusions in Parkinson's disease. *Journal of Neurology, Neurosurgery & Psychiatry*. 2001;70(6):734–738.
73. Sanchez-Ramos JR, Ortoll R, Paulson GW. Visual hallucinations associated with Parkinson's disease. *Archives of Neurology*. 1996;53(12):1265–1268.
74. Forsaa EB, Larsen JP, Wentzel-Larsen T, et al. A 12-year population-based study of psychosis in Parkinson disease. *Arch Neurol*. 2010;67(8):996–1001.
75. Ravina B, Marder K, Fernandez HH, et al. Diagnostic criteria for psychosis in Parkinson's disease: report of an NINDS, NIMH work group. *Movement disorders: official journal of the Movement Disorder Society*. 2007;22(8):1061–1068.
76. Fenelon G, Soulas T, Zenasni F, Cleret de Langavant L. The changing face of Parkinson's disease-associated psychosis: a cross-sectional study based on the new NINDS-NIMH criteria. *Movement disorders: official journal of the Movement Disorder Society*. 2010;25(6):763–766.
77. Tousi B, Frankel M. Olfactory and visual hallucinations in Parkinson's disease. *Parkinsonism & related disorders*. 2004;10(4):253–254.
78. Fenelon G, Thobois S, Bonnet AM, Broussolle E, Tison F. Tactile hallucinations in Parkinson's disease. *J Neurol*. 2002;249(12):1699–1703.
79. Fenelon G, Mahieux F, Huon R, Ziegler M. Hallucinations in Parkinson's disease: prevalence, phenomenology and risk factors. *Brain*. 2000;123 (Pt 4):733–745.
80. Goetz CG, Stebbins GT, Ouyang B. Visual plus nonvisual hallucinations in Parkinson's disease: development and evolution over 10 years. *Movement disorders: official journal of the Movement Disorder Society*. 2011;26(12):2196–2200.
81. Diederich NJ, Goetz CG, Stebbins GT. Repeated visual hallucinations in Parkinson's disease as disturbed external/internal perceptions: focused review and a new integrative model. *Movement disorders: official journal of the Movement Disorder Society*. 2005;20(2):130–140.
82. Barnes J, David AS. Visual hallucinations in Parkinson's disease: a review and phenomenological survey. *Journal of neurology, neurosurgery, and psychiatry*. 2001;70(6):727–733.
83. Friedman JH, Agarwal P, Alcalay R, et al. Clinical vignettes in Parkinson's disease: a collection of unusual medication-induced hallucinations, delusions, and compulsive behaviours. *Int J Neurosci*. 2011;121(8):472–476.
84. Goetz CG, Ouyang B, Negron A, Stebbins GT. Hallucinations and sleep disorders in PD: ten-year prospective longitudinal study. *Neurology*. 2010;75(20):1773–1779.
85. Papapetropoulos S, Mash DC. Psychotic symptoms in Parkinson's disease: from description to etiology. *J Neurol*. 2005;252(7): 753–764.
86. Marsh L. Psychosis in Parkinson's disease. *Primary Psychiatry*. 2005;12:56–62.
87. Pagonabarraga J, Llebaria G, Garcia-Sanchez C, Pascual-Sedano B, Gironell A, Kulisevsky J. A prospective study of delusional misidentification syndromes in Parkinson's disease with dementia. *Movement disorders: official journal of the Movement Disorder Society*. 2008;23(3):443–448.
88. Black DW, Boffeli TJ. Simple schizophrenia: past, present, and future. *The American journal of psychiatry*. 1989;146(10): 1267–1273.
89. Murray PS, Kumar S, Demichele-Sweet MA, Sweet RA. Psychosis in Alzheimer's disease. *Biological psychiatry*. 2014;75(7):542–552.
90. Santangelo G, Trojano L, Vitale C, et al. A neuropsychological longitudinal study in Parkinson's patients with and without hallucinations. *Movement disorders: official journal of the Movement Disorder Society*. 2007;22(16):2418–2425.
91. Goetz CG, Stebbins GT. Mortality and hallucinations in nursing home patients with advanced Parkinson's disease. *Neurology*. 1995;45(4):669–671.
92. Magni E, Binetti G, Bianchetti A, Trabucchi M. Risk of mortality and institutionalization in demented patients with delusions. *Journal of geriatric psychiatry and neurology*. 1996;9(3):123–126.
93. Stella F, Banzato CE, Quagliato EM, Viana MA, Christofoletti G. Psychopathological features in patients with Parkinson's disease and related caregivers' burden. *International journal of geriatric psychiatry*. 2009;24(10):1158–1165.
94. Kaufer DI, Cummings JL, Christine D, et al. Assessing the impact of neuropsychiatric symptoms in Alzheimer's disease: the Neuropsychiatric Inventory Caregiver Distress Scale. *Journal of the American Geriatrics Society*. 1998;46(2):210–215.
95. McKinlay A, Grace RC, Dalrymple-Alford JC, Anderson T, Fink J, Roger D. A profile of neuropsychiatric problems and their relationship to quality of life for Parkinson's disease patients without dementia. *Parkinsonism & related disorders*. 2008;14(1):37–42.

SECTION 3

NEUROBIOLOGY OF PSYCHOSIS

HERITABILITY AND GENETICS

19

GENETIC NEUROPATHOLOGY REVISITED

GENE EXPRESSION IN PSYCHOSIS

Samuel J. Allen, Rahul Bharadwaj, Thomas M. Hyde, and Joel E. Kleinman

INTRODUCTION

The discovery of decreased dopamine in the striatum of postmortem brains diagnosed with Parkinson's disease led to the development of medications that have helped to reduce some of the burden of the disease.[1–4] A second transformative finding that found a restriction fragment length polymorphism association at chromosome 4 with Huntington's disease, although not a postmortem brain study, opened the door to an improved understanding of genetic-based neurological disorders.[5] Similar findings have unfortunately not been demonstrated for the neuropathology and genetics of psychiatric disorders including schizophrenia and bipolar disorder.

Numerous studies of postmortem brains of patients diagnosed with schizophrenia or bipolar disorders have failed to identify pathognomonic lesions. Moreover, replicable findings have been relatively scarce as well. Genetic studies have fared much better, despite failing to identify any single gene that accounted for a psychiatric disorder. Studies of thousands of patients and cohorts, such as the Psychiatric Genomics Consortium II (PGCII), have identified a large number of genetic variations that confer risk for psychiatric disease, but each variant accounts for only a small portion of risk for these disorders.[6]

Nevertheless, postmortem brain studies of gene expression in schizophrenia and bipolar disorder still offer promise in elucidating molecular mechanisms that have the potential to identify novel drug targets. This review covers two broad areas related to genetic neuropathology: (1) postmortem differential gene expression comparisons between case (i.e., schizophrenia and bipolar disorder) and control brains and (2) studies that find associations between expression changes and genetic risk variants, expression quantitative loci (eQTLs) for schizophrenia and bipolar disorder.

DIFFERENTIAL EXPRESSION

Analysis of differential gene expression between cases and controls has long been by far the most popular strategy for molecular psychiatry research to identify genes and mechanisms to target pharmacologically (see Table 19.1). Studies investigating gene expression first used qRT-PCR and microarrays. More recently, RNA sequencing technology has allowed for thousands of comparisons of expression between patients and controls. Unfortunately, the vast majority of the many "findings" have not been replicated in either schizophrenia or bipolar disorder studies; this may be related to the differences between these different technologies.

SCHIZOPHRENIA

Two genes that have been studied extensively in schizophrenia and that have yielded reproducible results are *DRD2* (dopamine receptor D2) and *GAD1* (glutamic acid decarboxylase 1). The *DRD2* gene is highly expressed in multiple brain areas such as the basal ganglia, hippocampus, and the mesolimbic reward pathway. Multiple postmortem brain studies found increased D2 receptors in the striatum of patients with schizophrenia,[7–9] which have been confirmed by some, but not all, neuroimaging studies of drug-naïve patients.[10–12] The latter was crucial for interpreting the results of the postmortem studies insofar as antemortem treatment was one of the major confounds that might account for the previously observed increases.

The most replicated expression change, *reduced* GAD67 expression in prefrontal cortex (PFC) of patients with schizophrenia,[13–16] has also been shown to be associated with a putative risk allele for schizophrenia.[17] In so far as GAD67 is one of the major transcripts of *GAD1*, one of the two genes involved in the synthesis of GABA (the major inhibitory neurotransmitter in mammalian brain), this finding suggests that there is decreased GABA in the PFC of patients with schizophrenia.[18] Moreover, insofar as GABA plays a major role in early brain development, this finding lends support to the neurodevelopmental hypothesis of schizophrenia.

A third gene, *ErbB4* (erb-b2 receptor tyrosine kinase 4), a member of the Tyr protein kinase and epidermal growth factor receptor families, has specific slice variants/isoforms significantly elevated in dorsolateral prefrontal cortex (DLPFC) of patients with schizophrenia in two entirely separate samples/studies.[19,20] Moreover, this increased expression was also found in layer 4 of the DLPFC by a third independent laboratory with a separate cohort of patients with schizophrenia.[21]

A transmembrane receptor ligand gene in the ErbB4 pathway, *NRG1* (neuregulin 1) is also of interest, even though there

Table 19.1 SUMMARY OF GENES IMPLICATED IN SCHIZOPHRENIA BY DIFFERENTIAL EXPRESSION STUDIES

GENE	GENE PRODUCT	FUNCTION/ROLE
DRD2	D2 dopamine receptor	Modulatory neurotransmission and mesolimbic reward pathway transmission
GAD1	Glutamate decarboxylase 1	Catalyzes the production of GABA
ErbB4	ErbB4 tyrosine kinase receptor 4	Growth factor and cell differentiation
NRG1	Neuregulin 1	Cell-cell signaling and development
DISC1	Disrupted in schizophrenia 1	Neurite growth and cortex development
GRM3	Glutamate metabotropic receptor 3	G-protein coupled receptor for glutamate
TACR3	Tachykinin receptor 3	Neuropeptide signaling
IGF2	Insulin-like growth factor 2	Imprinted gene involved in general growth and development
CALB1	Calbindin 1	Calcium binding protein implicated in Huntington Disease
KCNA1	Voltage-gated potassium channel (subfamily A, member 1)	Ionotropic channels broadly involved in neurotransmission
KCNH2	Voltage-gated potassium channel (subfamily H, member 2)	
KCNC3	Voltage-gated potassium channel (subfamily C, member 3)	
KCNK1	Voltage-gated potassium channel (subfamily K, member 1)	
KCNN1	Potassium calcium-activated channel (subfamily N, member 1)	
SCN9A	Voltage-gated sodium channel alpha subunit 9	
SCN1B	Voltage-gated sodium channel, beta subunit 1	
GRIN3A	Glutamate ionotropic receptor (NMDA type, subunit 3A)	General NMDA-glutamate neurotransmission
GABRA5	Gamma-aminobutyric acid receptor type A, Alpha 5	Inhibitory neurotransmission
GABRB3	Gamma-aminobutyric acid receptor type A, beta 3	
KCNIP3	Voltage-gated potassium channel interacting protein 3	Auxiliary protein that modulates voltage-gated potassium channels

are not replicable expression studies to date. The transcript of interest is a novel alternative transcript that is associated with a putative schizophrenia risk allele.[22–24] This transcript is preferentially expressed in human fetal PFC and disappears by 4 years of age.[25] Insofar as it is not expressed in childhood, adolescence, or later years, it is not possible for there to be differential expression in the postmortem human PFC of patients with schizophrenia, illustrating the importance of temporal gene expression regulation in psychiatric disease.

Similarly, putative risk alleles in *KCNH2* (potassium voltage-gated channel subfamily H member 2), are associated with another novel, alternative transcript preferentially expressed in human fetal PFC and both PFC and hippocampus of patients with schizophrenia relative to controls.[26] Although this finding has not been replicated to date, genes such as *KCNH2* and *NRG1* have novel alternative transcripts preferentially expressed in human fetal brain, a characteristic that they share with transcripts associated with Psychiatric Genomics Consortium positive Single nucleotide polymorphisms (SNPs).[27]

Still another gene of interest is *DISC1* (disrupted in schizophrenia 1), a gene involved in neurite growth and cortical development that was first identified in a Scottish family of patients with schizophrenia and affective disorder patients where their diagnoses were associated with a translocation from chromosome 1 to 11.[28] Common genetic variants for *DISC1* (not the translocation) thought to increase risk for schizophrenia are also associated with alternative transcripts preferentially expressed in normal human fetal PFC and in hippocampus from patients with schizophrenia.[29] These transcripts are truncated and are thought to make truncated DISC1 proteins. Ordinarily, the DISC1 protein is a scaffolding protein that binds a number of other proteins, but truncated DISC1 proteins cannot bind a number of these other proteins as the truncated protein no longer allows the binding, given that these sites are gone. As a consequence, the DISC1 genetic variants that are associated with these DISC1 alternative transcripts are also associated with decreased expression in hippocampus (and to a lesser degree in DLPFC) of a family of other genes including *LIS1* (lissencephaly gene 1), *NUDEL* (nudE neurodevelopment protein 1 like 1) and *FEZ1* (fasciculation and elongation protein zeta 1), three other putative schizophrenia candidate genes.[30]

The metabotropic glutamate receptor *GRM3* (glutamate metabotropic receptor 3) is another gene of interest, which was first a candidate gene based on its possible role in phencyclindine-induced psychoses.[31] Common genetic variants associated with the risk for schizophrenia are associated with an alternative transcript expressed in DLPFC, which is preferentially expressed in fetal PFC.[32] Although there is a risk allele in GRM3 identified by the PGCII, that genetic variant is not clearly associated with the novel transcript identified in our earlier studies nor is there consistently differential expression in cohorts of schizophrenic patients relative to controls.[32]

Recently, one of the more comprehensive postmortem studies of gene expression was published using RNA sequencing in over 500 human DLPFC (schizophrenia n = 258 and controls n = 279).[33] In this study there were 239 isoforms that were differentially expressed, including 94 upregulated and 145 downregulated. The most upregulated gene was *TACR3* (tachykinin receptor 3), which codes for a tachykinin receptor, and the most downregulated gene was *IGF2* (insulin like growth factor 2), an important paternal imprinting gene involved in growth and development. In addition, *GABRA5* (gamma-aminobutyric acid type A receptor alpha5 subunit), which encodes a GABA chloride channel, and *CALB1* (calbindin 1), a gene belonging to the calcium-binding superfamily, were among the top differentially expressed genes.

A comparable similar large RNA sequencing study of 495 subjects, 175 with schizophrenia, has recently been published, finding differential expression for numerous genes including voltage-gated potassium and sodium channels (*KCNA1, KCNC3, KCNK1, KCNN1, SCN9A*), glutamate and GABA-gated ion channels (*GRIN3A, GABRA5, GABRB3*), and ion channel auxiliary subunits (*KCNIP3, SCN1B*).[34] The difference in expression between the two large RNA sequencing studies may be related to two potential confounds, age and RNA quality. While both studies have similar age ranges, the former study has a much larger group of elderly patients and controls, a factor that is well known to affect RNA expression.[35] Again, while both studies have high quality, comparable RNA, the latter has used a new, presumably improved method for conveying RNA quality with region specific RNA degradation data.[36]

BIPOLAR DISORDER

Differential gene expression studies of postmortem brains of bipolar disorder are far less in number compared to those investigating schizophrenia. A major effort to correct this deficit was made by the Stanley Medical Research Institute, with two series of postmortem brain studies involving 15 bipolar subjects in the first study and 35 in the second one.[37,38] Numerous other studies from the NIMH, Harvard Brain Bank, and others have failed to produce any consistent replicable findings, similar to what has occurred in schizophrenia. While the number of studies of bipolar disorder is less than that for schizophrenia, this is only one possible explanation for the failures to replicate. Inadequate numbers of studies and patients; mismatches for age, sex, and race; different assays that measure different alternative transcripts; antemortem treatments; substance abuse; and RNA quality are confounds that might contribute to failures to replicate. To be certain, none of these confounds are unique to bipolar disorder postmortem brain studies. If there is a unique confound, perhaps it relates to diagnosis, where both bipolar I and bipolar II subjects are lumped together as well as bipolar patients with and without psychosis. Exactly which, if any, of these confounds contributes to the failure to replicate, remains to be determined.

RISK VARIANTS

The paucity of reproducible postmortem brain studies comparing patients with schizophrenia and bipolar disorder with controls has led to a new approach. Rather than making comparisons between patients and controls, comparisons have been made between subjects with risk alleles relative to those without risk alleles (Table 19.2). Insofar as the risk and nonrisk alleles can occur in homozygotes or heterozygotes, there are three groups to be compared for expression in either normal controls or patients. Fortunately, for this approach the Psychiatric Genomics Consortium (PGC) has identified numerous loci for both schizophrenia and bipolar disorder.[6] Using PGC loci as a starting point, quantitative expression is measured in homozygotes with the risk allele, heterozygotes, and homozygotes of the nonrisk allele, resulting in eQTLs that are useful for a number of reasons.

First, eQTL associations can be identified in normal controls, which are not confounded by antemortem treatment or substance abuse. Second, the use of normal controls in determining an eQTL mitigates any uncertainties associated with psychiatric diagnosis. Third, the eQTL identifies genes/transcripts by which allelic variation increases risk for psychiatric illness. The eQTL may elucidate the molecular biological mechanism by which a genetic variation increases risk for a psychiatric illness, meaning the eQTL addresses etiology. Fourth, insofar as gene expression is one of the most proximate intermediate phenotypes for the risk allele of the illness, they allow for smaller numbers than the association between the risk allele and the illness with much greater statistical significance. Lastly, once the mechanism or specific transcript has been identified in the normal controls, the results then can be confirmed in patients and comparisons between patients and controls for the expression of that specific transcript can follow. Hopefully, this genetic-risk-driven targeted approach will lead to comparisons between patients and controls that can be replicated across different laboratories and studies.

SCHIZOPHRENIA

The previously mentioned Fromer et al. study[33] was not just about differential gene expression. It also focused on the PGCII genetic risk loci identifying five loci associated in each instance with a single gene (*FURIN, TSNARE1, CNTN4, CLCN3*, and *SNAP91*). Two of these associations have been

Table 19.2 SUMMARY OF GENES IMPLICATED IN RISK VARIANT STUDIES

GENE	PRODUCT	FUNCTION/ROLE
Schizophrenia		
FURIN	Furin	Ubiquitous protease activity for various cellular pathways
TSNARE1	T-SNARE domain containing 1	Endomembrane transport system and vesicle docking/fusion
CNTN4	Contactin 4	Axon-associated cell adhesion
CLCN3	Chloride voltage-gated channel 3	Antiporter of chloride ions and protons
SNAP91	Synaptosome associated protein 91	Endomembrane transport and clathrin-mediated endocytosis
BORCS7	BLOC-1 related complex subunit 7	Lysosome localization
AS3MT	Arsenite methyltransferase	Arsenic metabolism
SNX19	Sorting nexin 19	Vesicle trafficking
C4A	Complement C4A	Classical complement pathway
Bipolar Disorder		
CACNA1C	Calcium voltage-gated channel, subunit alpha 1C	Mediates calcium entry into cells
ANK3	Ankyrin 3	Links membrane to cytoskeleton
ZNF804A	Zinc finger protein 804A	Transcription factor associated with schizophrenia and bipolar disorder

confirmed in the Jaffe et al. data set,[34] *TSNARE1* and *CLCN3*, with another 40 genes in 29 other loci. Twelve of the other loci are associated with just one gene, but not necessarily one transcript.

The 10q24.32 locus, one of the most significantly positive loci, is associated with both *BORCS7* (BLOC-1 related complex subunit 7) and a novel isoform in *AS3MT* (AS3MT^{d2d3}).[39] This novel form of *AS3MT* (arsenite methyltransferase) is lacking in arsenic methyltransferase activity compared to the full AS3MT transcript. BORCS7 has been implicated in lysosome transport, while the novel AS3MT isoform's biological function remains unknown.[40] Apparently, different patients have an association to either *BORCS7* or *AS3MT*. Both genes have increased expression in DLPFC of patients with schizophrenia. Lastly, the true culprit is not the genetic variant identified in the PGCII study, but a variable number tandem repeat in linkage disequilibrium with the SNP only 269 base pairs away.

A second PGCII locus, 11q25, is also associated with another single gene, *SNX19* (sorting nexin 19), which plays a role in vesicle trafficking.[41] This transcript, although present in human fetal brain, is not preferentially expressed early in brain development, nor is there differential expression in the DLPFC or hippocampus of patients with schizophrenia.

Last, but certainly not the least, the major histocompatibility complex (MHC) locus, which spans several megabases of chromosome 6, is the most significant and perhaps challenging locus for schizophrenia risk. It has been found to be associated with *C4A* (complement C4A), a gene involved in the classical activation pathway of antibodies, the expression of which is increased in five brain regions of patients with schizophrenia including anterior cingulate cortex, orbital frontal cortex, parietal cortex, cerebellum, and corpus callosum.[42]

BIPOLAR

Not unlike what has been seen in expression studies, there are fewer patients with bipolar disorder in the GWAS and fewer risk loci identified to date. Nevertheless, there are at least loci with eQTLs that appear promising in shedding light on the etiology of bipolar disorder: *CACNA1C*, *ANK3*, and *ZNF804A*.

Comparisons between the Wellcome Trust Case Control Consortium and STEP-UCL have implicated *CACNA1C* as a viable risk gene for bipolar disorder.[43] *CACNA1C* codes for the alpha 1C subunit of voltage-gated calcium channels. Bigos et al. were able to demonstrate that one risk SNP (rs1006737) in *CACNA1C* is associated with increased expression of *CACNA1C* mRNA in human brain[44] and was determined to be not only a bipolar disorder risk variant but also a risk variant for schizophrenia. Interestingly, the researchers also found that the risk SNP was associated with increased hippocampal activity during emotional-processing tasks and increased prefrontal activity when performing executive cognitive tasks. More recently, as PGCII got larger, a different SNP in *CACNA1C* was identified as the "risk SNP." At this time, it is unclear which transcript in *CACNA1C* is associated with the "new risk SNP."

A second gene, *ANK3*, codes for ankyrin G, which serves to link various membrane proteins to the actin cytoskeleton, which is involved in a multitude of cellular processes. Genetic variants in *ANK3* that increased risk for bipolar disorder are associated with alternative isoforms of ANK3 in human brain.[45]

Lastly, *ZNF804A* was originally identified as a gene of interest in schizophrenia, but later studies further implicated in the gene in bipolar disorder as well. *ZNF804A* encodes zinc finger protein 804A, which acts as a transcription factor for various other genes. A psychosis risk allele associated with ZNF804A mRNA expression was identified in fetal human brain.[46] Subsequently, the risk SNP was associated with expression of a specific novel, truncated alternative transcript in fetal human PFC whose function has yet to be determined.[47] This alternative transcript is once again preferentially expressed in human fetal PFC and in DLPFC of patients with schizophrenia. Interestingly enough, this same transcript is reduced in DLPFC of patients with bipolar disorder.

DISCUSSION ON RISK VARIANTS

Although the initial decades of research on gene expression and psychosis were not terribly successful in producing either replicable results or improved diagnosis and/or treatment, the future looks more promising for at least three reasons. First and foremost, the initial studies of a dozen patients and controls were doomed by insufficient power in the face of confounds of antemortem treatment, substance abuse, degraded RNA, and mismatches of age, sex, and race. The field has now moved to studies that are 100–300 patients in number. Second, we now have the power of RNA sequencing so that we are no longer looking at one gene at a time. Third, we have improved methods of data analysis that do not use RINs or pH blindly to covary for RNA degradation. Last, but far from least, we have a reliable scientific entry point where we can use GWAS positive genetic variants and their associated transcripts to elucidate mechanisms for risk and query differential expression in a hypothesis-based way.

What have we found thus far? First, there are replicable findings. Second, the use of PGCII risk alleles to identify specific alternative transcripts gives us better targets to make our comparisons. Even when they are negative for differential expression, they can yield the molecular mechanisms by which risk alleles increase risk for schizophrenia or bipolar disorder. Hopefully, this new approach will yield replicable and more useful findings which will lead to improved diagnosis and treatment of psychoses, especially schizophrenia and bipolar disorder.

REFERENCES

1. Ehringer H and Hornykiewicz O. Verteilung von noradrenalin und dopamine (3-hydroxytyramin) im gehirn des menschen und ihr verhalten bei erkrankungen des extrapyramidelen systems. *Klin Wschr* 1960;38:1236–1239. Republished in English translation in *Parkinsonism and Related Disorders* 1998; 4:53–57
2. Hornykiewicz O. L-Dopa. *J Parkinsons Dis.* 2017;7(Suppl 1):S3–S10.
3. Cotzias G, Van Woert M, Schiffer L. Aromatic amino acids and modification of parkinsonism. *New England Journal of Medicine.* 1967;276(7):374–379. doi:10.1056/nejm196702162760703.
4. Cotzias G, Papavasiliou P, Gellene R. Modification of parkinsonism—chronic treatment with L-Dopa. *New England Journal of Medicine.* 1969;280(7):337–345. doi:10.1056/nejm196902132800701.
5. Gusella J, Wexler N, Conneally P, et al. A polymorphic DNA marker genetically linked to Huntington's disease. *Nature.* 1983; 306(5940):234–238.
6. Schizophrenia Working Group of the Psychiatric Genomics Consortium. Biological insights from 108 schizophrenia-associated genetic loci. *Nature.* 2014;511(7510):421–427.
7. Seeman P, Chau-Wong M, Tedesco J, Wong K. Brain receptors for antipsychotic drugs and dopamine: direct binding assays. *Proceedings of the National Academy of Sciences.* 1975;72(11):4376–4380.
8. Seeman P, Bzowej NH, Guan HC, et al. Human brain D1 and D2 dopamine receptors in schizophrenia, Alzheimer's, Parkinson's, and Huntington's diseases. *Neuropsychopharmacology.* 1987;1(1):5–15.
9. Cross A, Crow T, Ferrier I, et al. Dopamine receptor changes in schizophrenia in relation to the disease process and movement disorder. *J Neural Transm Suppl.* 1983;18(265):265–272.
10. Wong D, Wagner H, Tune L, et al. Positron emission tomography reveals elevated D2 dopamine receptors in drug-naive schizophrenics. *Science.* 1986;234(4783):1558–1563.
11. Jönsson E, Nöthen M, Grünhage F, et al. Polymorphisms in the dopamine D2 receptor gene and their relationships to striatal dopamine receptor density of healthy volunteers. *Molecular Psychiatry.* 1999; 4(3):290–296.
12. Farde L, Wiesel F, Stone-Elander S, et al. D2 dopamine receptors in neuroleptic-naive schizophrenic patients. *Archives of General Psychiatry.* 1990;47(3):213–219.
13. Akbarian S, Kim J, Potkin S, et al. Gene expression for glutamic acid decarboxylase is reduced without loss of neurons in prefrontal cortex of schizophrenics. *Archives of General Psychiatry.* 1995;52(4):258–266.
14. Volk D, Austin M, Pierri J, et al. Decreased glutamic acid decarboxylase67 messenger RNA expression in a subset of prefrontal cortical γ-aminobutyric Acid neurons in subjects with schizophrenia. *Archives of General Psychiatry.* 2000;57(3):237–245
15. Guidotti A, Auta J, Davis JM, et al. Decrease in reelin and glutamic acid decarboxylase67 (GAD67) expression in schizophrenia and bipolar disorder. *Archives of General Psychiatry.* 2000;57(11):1061–1069.
16. Mirnics K, Middleton FA, Marquez A, et al. Molecular characterization of schizophrenia viewed by microarray analysis of gene expression in prefrontal cortex. *Neuron.* 2000;28(1):53–67.
17. Straub R, Lipska B, Egan M, et al. Allelic variation in GAD1 (GAD67) is associated with schizophrenia and influences cortical function and gene expression. *Molecular Psychiatry.* 2007;12(9):854–869.
18. Benes FM, McSparren J, Bird E, SanGiovanni J, Vincent S. Deficits in small interneurons in prefrontal and cingulate cortices of schizophrenic and schizoaffective patients. *Archives of General Psychiatry.* 1991;48(11):996–1001.
19. Law A, Kleinman J, Weinberger D, et al. Disease-associated intronic variants in the ErbB4 gene are related to altered ErbB4 splice-variant expression in the brain in schizophrenia. *Human Molecular Genetics.* 2007;16(2):129–141.
20. Silberberg G, Darvasi A, Pinkas-Kramarski R, et al. The involvement of ErbB4 with schizophrenia: Association and expression studies. *American Journal of Medical Genetics.* 2006;141B(2):142–148.
21. Chung D, Volk D, Arion D, et al. Dysregulated ErbB4 splicing in schizophrenia: selective effects on parvalbumin expression. *American Journal of Psychiatry.* 2016;173(1):60–68.
22. Stefansson H, Sigurdsson E, Steinthorsdottir V, et al. Neuregulin 1 and susceptibility to schizophrenia. *Am J Hum Genet.* 2002;71(4): 877–892.
23. Law A, Lipska B, Weickert C, et al. Neuregulin 1 transcripts are differentially expressed in schizophrenia and regulated by 5′ SNPs associated with the disease. *Proc Natl Acad Sci USA.* 2006;103(17):6747–6752.
24. Paterson C, Wang Y, Kleinman J, Law A. Schizophrenia risk variation in the NRG1 gene exerts effects on NRG1-IV splicing during

fetal and early postnatal human neocortical development. *American Journal of Psychiatry*. 2014;171(9):979–989.

25. Petryshen T, Middleton F, Kirby A, et al. Support for involvement of neuregulin 1 in schizophrenia pathophysiology. *Molecular Psychiatry*. 2005;10(4):366–374.
26. Huffaker S, Chen J, Nicodemus K, et al. A primate-specific, brain isoform of KCNH2 affects cortical physiology, cognition, neuronal repolarization and risk of schizophrenia. *Nat Med*. 2009;15(5):509–518.
27. Birnbaum R, Jaffe A, Chen Q, Hyde T, Kleinman J, Weinberger D. Investigation of the prenatal expression patterns of 108 schizophrenia-associated genetic loci. *Biological Psychiatry*. 2015;77(11).
28. Blackwood D, Fordyce A, Walker M, et al. Schizophrenia and affective disorders—cosegregation with a translocation at chromosome 1q42 that directly disrupts brain-expressed genes: clinical and P300 findings in a family. *American Journal of Human Genetics*. 2001;69(2):428–433.
29. Nakata K, Lipska B, Hyde T, et al. DISC1 splice variants are upregulated in schizophrenia and associated with risk polymorphisms. *Proceedings of the National Academy of Sciences*. 2009;106(37):15873–15878.
30. Lipska B, Peters T, Hyde T, et al. Expression of DISC1 binding partners is reduced in schizophrenia and associated with DISC1 SNPs. *Human Molecular Genetics*. 2006;15(8):1245–1258.
31. Takahata R, Moghaddam B. Activation of glutamate neurotransmission in the prefrontal cortex sustains the motoric and dopaminergic effects of phencyclidine. *Neuropsychopharmacology*. 2003;28(6):1117–1124.
32. Sartorius L, Weinberger D, Hyde T, Harrison P, Kleinman J, Lipska B. Expression of a GRM3 splice variant is increased in the dorsolateral prefrontal cortex of individuals carrying a schizophrenia risk SNP. *Neuropsychopharmacology*. 2008;33(11):2626–2634.
33. Fromer M, Roussos P, Sieberts S, et al. Gene expression elucidates functional impact of polygenic risk for schizophrenia. *Nature Neuroscience*. 2016;19(11):1442–1453.
34. Jaffe A, Straub R, Shin J. Developmental and genetic regulation of the human cortex transcriptome in schizophrenia. *BioRxiv*. 2017.
35. Colantuoni C, Lipska B, Ye T, et al. Temporal dynamics and genetic control of transcription in the human prefrontal cortex. *Nature*. 2011;478(7370):519–523.
36. Jaffe A, Tao R, Norris A, et al. qSVA framework for RNA quality correction in differential expression analysis. *Proc Natl Acad Sci USA*. 2017;114(27):7130–7135.
37. Kim S, Webster M. The Stanley neuropathology consortium integrative database: a novel, web-based tool for exploring neuropathological markers in psychiatric disorders and the biological processes associated with abnormalities of those markers. *Neuropsychopharmacology*. 2009;35(2):473–482.
38. Kim S, Cho H, Lee D, Webster MJ. Association between SNPs and gene expression in multiple regions of the human brain. *Translational Psychiatry*. 2012;2(5):e113.
39. Li M, Jaffe AE, Straub RE, et al. A human-specific AS3MT isoform and BORCS7 are molecular risk factors in the 10q24.32 schizophrenia-associated locus. *Nature Medicine*. 2016;22(6):649–656.
40. Pu J, Schindler C, Jia R, Jarnik M, Backlund P, Bonifacino J. BORC, a multisubunit complex that regulates lysosome positioning. *Dev Cell*. 2015;33(2):176–188.
41. Ma L, Semick S, Jaffe A, et al. Sorting nexin 19 (SNX19) contributes to schizophrenia risk via overexpression of novel transcripts. Submitted to *Mol Cell Bio* for publication.
42. Sekar A, Bialas A, de Rivera H, et al. Schizophrenia risk from complex variation of complement component 4. *Nature*. 2016: 177–183.
43. Ferreira M, O'Donovan M, Meng Y, et al. Collaborative genome-wide association analysis supports a role for ANK3 and CACNA1C in bipolar disorder. *Nature Genetics*. 2008;40(9):1056–1058.
44. Bigos K, Mattay V, Callicott J, et al. Genetic variation in CACNA1C affects brain circuitries related to mental illness. *Archives of General Psychiatry*. 2010;67(9):939–945.
45. Rueckert E, Barker D, Ruderfer D, et al. Cis-acting regulation of brain-specific ANK3 gene expression by a genetic variant associated with bipolar disorder. *Molecular Psychiatry*. 2012;18(8): 922–929.
46. Hill MJ, Bray NJ. Evidence that schizophrenia risk variation in the ZNF804A gene exerts its effects during fetal brain development. *American Journal of Psychiatry*. 2012;169(12):1301–1308.
47. Tao R, Cousijn H, Jaffe A, et al. Expression of ZNF804A in human brain and alterations in schizophrenia, bipolar disorder, and major depressive disorder. *JAMA Psychiatry*. 2014; 71(10):1112–1120.

20

EPIGENOMIC REGULATION IN PSYCHOSIS

Bibi S. Kassim, Behnam Javidfar, and Schahram Akbarian

INTRODUCTION

Psychosis spectrum disorders lack a unifying molecular or cellular pathology, and most cases are believed to be of multifactorial etiology with numerous environmental and genetic components involved. Conventional psychopharmacology, including drugs targeting monoamine signaling (e.g., dopaminergic, serotonergic, and noradrenergic pathways) elicits insufficient therapeutic responses in one-half or less of patients diagnosed with schizophrenia and related illnesses such as depression with psychosis (Gerhard et al., 2017; Tondo et al., 2017). Thus, it is necessary to further explore the neurobiology and molecular pathology of psychosis spectrum disorders in order to develop novel treatment strategies of higher efficacy. In this chapter we discuss one potentially promising avenue to achieve progress, neuroepigenomics. Epi- (Greek for *over, above*) genetic decoration of the genome residing in each of our brain cells is a highly regulated cell-specific and life-long process involving multiple molecular layers, including (for example) DNA methylation, many different types of covalent and post-translational histone modifications, and complex nonrandom chromosomal conformations including "loopings" and self-folding "domains" extending across many hundreds of kilobases of linear genome. It is now generally accepted that these and many other epigenomic determinants remain "plastic" throughout all periods of development and aging, with ongoing dynamic regulation even in neurons and other highly differentiated brain cells. We explore in this chapter some of the implications resulting from this conclusion. This includes a discussion on how neuroepigenomics offers insights into biological function of underlying genetic risk architecture of psychosis, which includes a considerable portion of noncoding DNA offering no function by sequence analyses alone. Then, we summarize emerging insights and early evidence from preclinical studies that identified potential epigenetic drug targets and therapies in psychosis.

EPIGENOMIC IMPLICATIONS FOR NORMAL AND DISEASED HUMAN BRAIN DEVELOPMENT

DNA METHYLATION

Comprehensive exploration of the functional organization of the human genome has to go far beyond sequencing of the 6 billion base pairs that define genetic information in the cell nuclei of all somatic tissues, including brain. In vertebrates, including humans, two related but functionally very different types of DNA cytosine modifications, methylation (m) and hydroxymethylation (hm), provide the bulk of the epigenetic modifications in vertebrate DNA together with some of their chemical intermediates (Ito et al., 2011; Kriaucionis and Heintz, 2009). These are most encountered at the site of CpG dinucleotides, and there many other sites of non-CpG cytosines methylated in brain tissue (Xie et al., 2012). Curiously, while the largest amount (97%) of mC5s are found in intragenic and intergenic sequences and within DNA repeats (Maunakea et al., 2010), only a few of these studies have explored brain DNA methylation changes at repeat DNA, and the bulk of brain methylation work has focused on the gene repressive actions of this mark when positioned at gene promoters and other regulatory sequences that control transcription (Lomvardas and Maniatis, 2016). Finally, there is evidence that vertebrate epigenomes (including brain tissue) harbor very limited amounts of adenine methylation, and then, mostly in areas rich with a repeat DNA (Koziol et al., 2016). The functional significance of this type of DNA N(6) adenine methylation in brain has yet to be explored.

Developmental and disease-associated regulation of DNA methylation

There can be little doubt that DNA methylation landscapes show substantial reorganization during the course of normal development and aging of the human cerebral cortex. This includes a fast rise in methyl-cytosine levels at many promoters during the transition from peri- to postnatal age, that then continues at a slower pace into old age in conjunction with subtle changes (mostly a decline) in expression of transcripts originating from these promoters (Hernandez et al., 2011; Jaffe et al., 2016; Numata et al., 2012; Siegmund et al., 2007). Given the relative ease of surveying DNA methylation on a genome-wide scale with commercially available or custom-made kits, it is not surprising that there is already a considerable body of literature on the DNA methylome of subjects diagnosed with schizophrenia, bipolar disorder, and depression with and without suicide. One early study (Mill et al., 2008) examined approximately 12,000 regulatory regions in frontal cortices of psychotic patients and control subjects. They found differential DNA methylation patterns at numerous loci, including several genes involved in GABAergic and glutamatergic neurotransmission, and in brain development. Interestingly, this study included both schizophrenia and bipolar disorder patients, implying that some epigenetic signatures may not be specific to schizophrenia but rather may be associated more broadly with psychosis disorders. A second study, interrogating 450,000 (2%) of CpG sites genome-wide, reported that CpGs associated with close to 3,000 genes were differentially methylated in schizophrenia (Wockner et al., 2014). Furthermore, a recent study by Turecki and colleagues mapped DNA methylation landscapes in specific cell types of the anterior cingulate cortex of 49 depressed subjects who died by suicide with or without a severe history of childhood abuse, and reported an association between specific methylation alterations in the DNA of oligodendrocytes, at sites of myelin-regulating genes (Lutz et al., 2017). Importantly, these methylation changes were associated with changes in the myelin-thickness of small-diameter axons (Lutz et al., 2017), making this study one of the first to link specific types of epigenetic regulation to a disease-relevant histological alteration in human brain tissue. However, these early genome-scale DNA methylomics studies only allow limited conclusions because of the relatively small sample sizes. To this end, larger-scale studies with many hundreds of clinical cases and controls will allow for more firm assessment of whether or not there is a tractable epigenetic risk architecture in the brain tissue of individuals diagnosed with psychosis.

Of note, it is now generally accepted that schizophrenia and major mood disorders show substantial overlap in genetic (including risk-associated common DNA polymorphisms and haplotypes) and environmental disease factors (Uher and Zwicker, 2017). In the remaining portion of our DNA methylation subchapter, we discuss in more detail how two recently published studies provide an early glimpse of how brain DNA methylation mapping in normal and diseased brain could connect the genomic risk map for the heritable etiology of psychosis (built from common variants by genome-wide association [GWA]) from its neurodevelopmental origins. Jaffe and colleagues (2016) studied DNA methylation in tissue homogenates from prefrontal cortex using the popular Illumina 450 (genome scale) array to study over 520 postmortem cases collected across a wide age range, from the 14th week of gestation to 80 years of age, with 191 cases diagnosed with schizophrenia included in the adult cohort. The authors combined their large dataset with existing gene expression profiles for many samples of the same kind of cohort and with single nucleotide polymorphism (SNP)/common DNA polymorphism–based risk maps linked to schizophrenia by GWA. The study reported significant developmental remodeling of brain DNA methylation patterns, with >230,000 of their 456,000 autosomal CpG dinucleotides affected, most during the transition from the second fetal trimester to postnatal life. This finding (which resonates with some of the aforementioned studies conducted on smaller postmortem brain cohorts) may be explained in part based on normal developmental shifts in cell-type composition, attributed to decreasing progenitor pools and emergence of glia and neuronal differentiation in the more mature tissue samples. The authors then examined how their collection of age-regulated CpGs matched with GWA maps from the Psychiatric Genomics Consortium (PGC), which has so far discovered 108 risk haplotypes (small blocks of the genome in linkage disequilibrium) comprising thousands of SNPs (Schizophrenia Working Group of the Psychiatric Genomics, 2014). Notably, CpGs differentially methylated during the transition from fetal to postnatal ages were somewhat more likely (53% versus 50.7%) to fall in the vicinity of PGC SNPs than the remaining age-regulated CpGs, and this effect was primarily mediated by CpGs that were hypermethylated in the fetal PFC relative to postnatal and adult PFC. While the biological significance of such small (but nonetheless tractable) differences remain unclear, the authors report that 94% of the 2,104 CpGs that were differentially methylated in PFC of adult subjects with schizophrenia (as compared with adults without) showed significant methylation drifts during the transition from fetal to postnatal age (Jaffe et al., 2016). Strikingly, DNA methylation alterations in the adult PFC of subjects with the disease were ~65-fold more likely to match a CpG defined by an age-related drift during the fetal-to-postnatal transition as opposed to the shift from adolescence to old age. The authors conclude that the majority of DNA methylation changes encountered in the adult brain of a subject diagnosed with schizophrenia are less likely to reflect epigenetic (dys)regulation around the time when schizophrenia symptoms emerge, but instead could be the result of (not further defined) disease-relevant mechanisms that operate during much earlier phases of development (Figure 20.1).

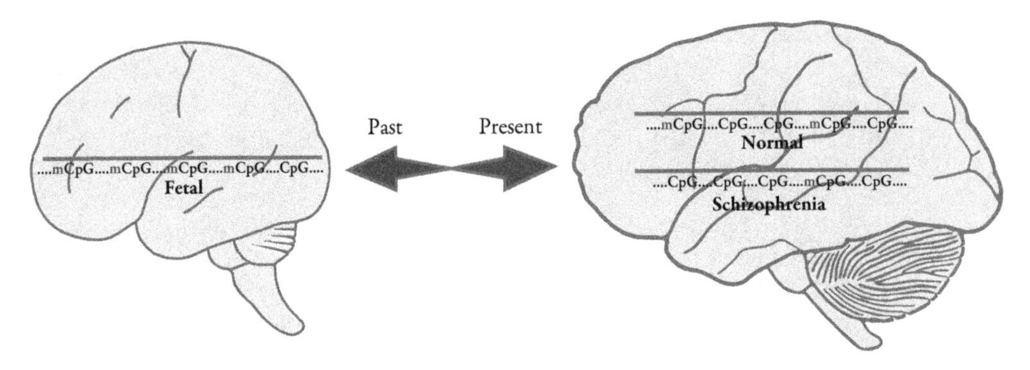

Figure 20.1 DNA methylation signatures in schizophrenia brain connect past and present. Left, during normal fetal development, a subset of genomic sequences are heavily methylated (red). Right, these become relatively hypomethylated in healthy adults, but excessively hypomethylated in PFC of subjects with schizophrenia. Many of the developmentally tagged methyl-CpG sequences (red) are positioned in the vicinity of risk-associated DNA polymorphisms for schizophrenia (see text). See color insert 4 in front of book for full colorized version.

This conclusion is further supported by the study of Hannon and colleagues who examined in detail the role of DNA sequence variation on the methylation landscape of the developing human brain (Hannon et al., 2016). Using a similar methylomics platform as Jaffe et al. (2016), they assayed 166 fetal postmortem brains coupled with high-density SNP profiling of the samples. The resulting methylation quantitative trait loci (mQTL) map identified ~17,000 SNPs that associate with DNA methylation levels (mQTLs), The authors then undertook a similar analysis individually of adult prefrontal cortex, striatum and cerebellum. Here they observed strong correlations between the two time points, implying that the vast majority of sequence-mediated effects on DNA methylation operate across the lifespan of the human brain. However, when comparing their mQTL map with GWA study (or GWAS) data for schizophrenia, they observed a specific and significant enrichment of schizophrenia risk loci with brain mQTLs. Therefore, the two DNA methylation studies (Hannon et al., 2016; Jaffe et al., 2016), taken together, provide strong evidence that the epigenetic architecture of schizophrenia, and possibly of other types of psychosis spectrum disorder, intersect with the underlying genetic architecture, which could impact prenatal development. Moreover, such genotype–epigenotype interactions maintain a "molecular echo" throughout the postnatal life span, potentially affecting the brain at-risk throughout all maturational periods in childhood and adolescence. However, it should be noted that, in both studies, overall methylation differences linked with schizophrenia-associated SNPs tended to quite subtle, showing just 6.7% difference per allele for the average mQTL 3 and 1.3% average methylation difference between diseased and control tissue. It is difficult to explore the functional consequences such as changes in gene expression, given such small differences in methylation levels. Nonetheless, the two studies discussed (Hannon et al., 2016; Jaffe et al., 2016), will likely offer a significant advancement in the field, by providing a testable hypothesis, namely that DNA polymorphisms contributing to the heritable risk for schizophrenia and related disorders could impact the epigenetic landscape of the immature brain, which in turn could be detrimental, affecting future steps of neurodevelopment. It is exciting that these studies in normal and diseased human brain resonate with recently reported cell type–specific DNA methylation mappings in mouse brain (Mo et al., 2015). In adult mouse cerebral cortex, the DNA methylome of various neuronal subtypes harbors hundreds of sequences; many of these are located upstream of gene transcription start sites, which maintain a CpG methylation profile reflecting past gene expression patterns from earlier, including prenatal, phases of development (Mo et al., 2015). If these principles could be confirmed for the neuronal epigenomes of the adult human cortex, then the reported alterations in the diseased postmortem brain tissue (Jaffe et al., 2016), integrated across many such loci, could link epigenomic dysregulation in individual disease cases to the developmental window of vulnerability. Such type of integrative DNA methylomics may indeed make it possible to link genome organization and function in fetal brain to psychotic illness in adolescent or adult subjects.

HISTONE MODIFICATIONS

The elementary unit of chromatin is the nucleosome, or 146 bp of genomic DNA wrapped around an octamer of core histones H3/H4/H2A/H2B, connected by linker DNA and linker histones. Histone modifications, including the methylation and acetylation of specific lysine residues, provide important building blocks for the "epigenome," and organize the genome into many tens of thousands of transcriptional units, clusters of condensed chromatin, and other features that are differentially regulated in different cell types and developmental stages of the organism (see Figure 20.2 for an overview). (For the reader interested in additional discussions and descriptions of various histone modification markings, we refer to Peter and Akbarian, 2011; Tessarz and Kouzarides, 2014; Zhang et al., 2015; Zhou et al., 2011.) However, only very few postmortem brain studies have embarked on histone modification mappings. This is extremely surprising, considering that a recent study with more than 60,000 participants identified regulatory pathways for methyl-histone H3-lysine 4 (H3K4me), a residue-specific histone mark which on a genome-wide scale is associated with gene expression and open chromatin (Akbarian and Huang, 2009; Eissenberg and Shilatifard, 2010; Liu et al., 2015; Tessarz and Kouzarides, 2014), as the top-ranking biological system by GWA across the three disorders, schizophrenia, bipolar illness, and depression, representing some of the most common psychosis and mood spectrum disorders (Network and Pathway Analysis Subgroup of Psychiatric Genomics, 2015). Until recently, genome-scale H3K4 methylation mappings were limited to other types of neuropsychiatric disease, including Huntington's and autism (Bai et al., 2015; Dong et al., 2015; Shulha et al., 2012; Vashishtha et al., 2013). These studies, taken together, imply that H3K4 methylation landscapes in cortical neurons are highly sensitive to functional perturbations and could reflect cortical dysfunction encountered in both degenerative and nondegenerative types of neuropsychiatric disease. Presently, to the best of our knowledge, only two studies embarked on genome-scale H3K4 methylation mapping in tissues from subjects with schizophrenia, and no data exist for other types of psychosis. One study explored the genome-wide distribution of the H3K4-methyl mark in neuronal nuclei from prefrontal cortex of 17 cases with schizophrenia compared to 17 controls (Mitchell et al., 2017). The study identified more than 1,311 genomic regions with differential H3K4 methylation in the disease cases, primarily due to excessive H3K4 methylation with significant enrichment for gene categories associated with synaptic function and neuronal connectivity and complex behaviors. Among the set of H3K4 hypermethylated sequences, there was prominent enrichment for specific transcription factor motifs, including *MYOCYTE-SPECIFIC ENHANCER FACTOR MEF2C* and related MEF proteins. Cell culture and in vivo mouse brain work indicates that the H3K4 hypermethylation as observed in the postmortem cortex of subjects with schizophrenia likely reflects decreased MEF2C function, while overexpression of *Mef2c* in prefrontal neurons improves working memory and other types of cognition, specifically memory, in the mouse (Mitchell et al.,

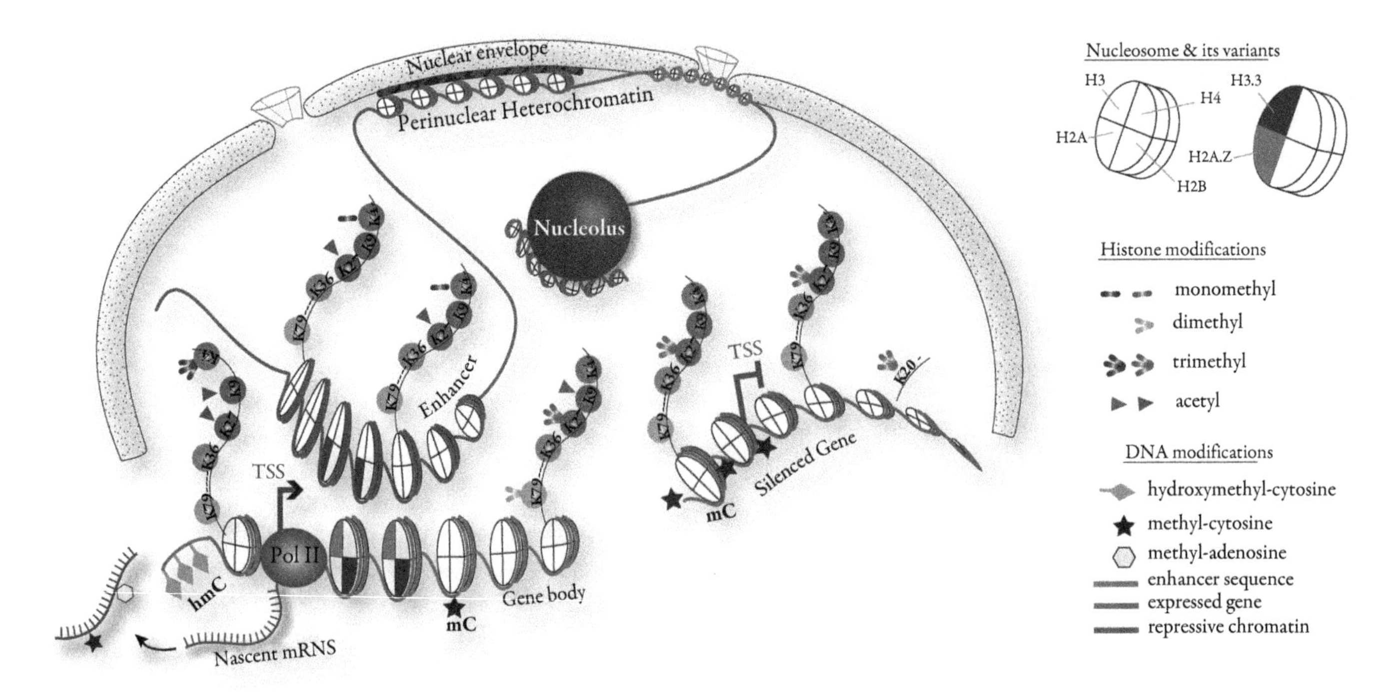

Figure 20.2 The epigenome, from nucleus to nucleosome. Illustration of gene poised for transcription by polymerase II (Pol II) initiation complex, with nucleosome free interval at transcription start site (TSS). distal enhancer sequence is, via chromatin loopings, moved in close proximity to active gene. heterochromatic portions of the genome, including silenced gene and heterochromatic structures bordering the nuclear envelope and pore complex, and also the nucleolar periphery. A small subset of representative histone variants and histone H3 site-specific lysine (K) residues at N-terminal tail (K4, K9, K27, K36) or core fold domain of the (histone) octamer (K79) and the H4K20 residue are shown as indicated, together with panel of mono- and trimethyl, or acetyl modifications that differentiate between active promoters, transcribed gene bodies, and repressive chromatin, as indicated. DNA cytosines that are hydroxymethylated at the C5 position are mostly found at active promoters, while methylated cytosines are positioned within body of actively transcribed genes and around repressed promoters and in constitutive heterochromatin. Finally, there is the "epitranscriptome," which involves a large number of RNA modifications such as, for example, methylation of cytosine and adenosine residues. In contrast to DNA and chromatin modifications, the investigation of RNA modification changes in the context of mood and psychosis spectrum disorders has barely begun.

2017). Dysregulated H3K4 methylation landscapes were also reported for olfactory neuroepithelium cultured from intranasal biopsies of a very small cohort of subjects with schizophrenia, with a significant enrichment for genes associated with oxidative stress pathways (Kano et al., 2013). Work conducted by consortia, including PsychENCODE (PsychENCODE consortium et al., 2015) is now providing sequence-specific histone modification mappings, including methyl-H3K4, on a genome-wide scale for much larger postmortem brain cohorts with hundreds of disease and control cases included, and already has provided new insights into the neurogenomic risk architectures of common psychiatric disease, including schizophrenia (Girdhar et al., 2018; Wang et al., 2018). Interestingly, H3K4 methylation undergoes developmental remodeling (mostly demethylation) at more than 1,000 loci in prefrontal neurons, a process that is most prominent in the perinatal period and early infancy, followed by a slow and progressive decrease (or, at some loci, increase) in H3K4 methylation throughout childhood and adolescence (Shulha et al., 2013). Such prolonged developmentally regulated "pruning" of excess H3K4 methylation markings in prefrontal neurons is an interesting observation, given that this cell population is, at least in some cases with schizophrenia, affected by increased levels of H3K4 methylation at specific promoter and other regulatory sequences (Mitchell et al., 2017).

EPIGENOMIC ALTERATIONS INDUCED BY DRUGS FREQUENTLY PRESCRIBED FOR PSYCHOSIS SPECTRUM DISORDERS

Studies in animals and cellular models have shown that many of the drugs prescribed for the treatment of psychosis spectrum disorders impact the epigenome of brain cells, including DNA methylation and histone modifications (Bredy et al., 2007; Cheng et al., 2008; Dong et al., 2008; Green et al., 2017; Kwon and Houpt, 2010; Li et al., 2004; Mill et al., 2008; Shimabukuro et al., 2006). These include antipsychotic drugs targeting dopamine and serotonin receptors and lithium and mood-stabilizers affecting the balance of neuronal excitation and inhibition. However, it is still not clear whether such types of medication-induced changes are linked to specific mechanisms of actions mediating the drugs' therapeutic effects in psychosis, or instead merely a reflection of indirect biological effects and of epiphenomena. Early studies exploring H3K4 methylation levels at specific promoter sequences of candidate genes such as *GAD1* (encoding a GABA synthesis enzyme) showed that the direction of methylation changes in diseased tissue are opposite to those observed after exposure to antipsychotic drugs (Huang et al., 2007). However, studies on the impact of commonly prescribed psychotropic medications, including (but not limited to) antipsychotics, mood-stabilizers, and antidepressants, on the reported alterations after genome-scale profiling of DNA methylation (Jaffe et al., 2016; Mill et al., 2008) (Lutz et al., 2017) and H3K4 methylation (Mitchell et al., 2017) in postmortem tissue from psychosis spectrum cases, show that this area is not fully understood.

EPIGENETIC DRUG TARGETS IN PSYCHOSIS?

REGULATORS OF H3K4 METHYLATION

As already reviewed, multiple regulators of H3K4 methylation show by GWA a surprisingly strong link to the genetic risk architecture of schizophrenia and related common psychiatric disorders (Network and Pathway Analysis Subgroup of Psychiatric Genomics, 2015). Furthermore, the H3K4-methyl mark shows strong neuron-specific regulation at genes involved in cerebral cortex in glutamatergic and dopaminergic signaling (Dincer et al., 2015), with significant gene-specific alterations in cells and brain tissue from subjects with schizophrenia (Huang et al., 2007; Kano et al., 2013; Mitchell et al., 2017). Importantly, mutations in various genes encoding H3K4-methyl regulators, including the histone H3-lysine 4 specific methyltransferase *SETD1A/KMT2F*, have been linked to rare monogenic forms of neurodevelopmental disease and even adult-onset schizophrenia (Singh et al., 2016; Takata et al., 2016; Takata et al., 2014; Vallianatos and Iwase, 2015). Furthermore, neuron-specific ablation of the H3K4 methyltransferase *Kmt2a/Mll1* in key nodes of the neural circuitry underlying psychosis, including prefrontal cortex and ventral striatum, was associated with defective synaptic plasticity, increased anxiety, altered response profiles to dopaminergic probes and impairments in working memory (Jakovcevski et al., 2015; Shen et al., 2016). In addition, genetic ablation of H3K4 methyl-transferase genes elicits robust defects in hippocampal learning and memory (Gupta et al., 2010; Kerimoglu et al., 2013). Therefore, it is timely to explore changes in cognition and behavior in preclinical models after drug-induced interference with the H3K4-methyl-regulome.

HISTONE ACETYLATION AND DNA METHYLATION

Drugs that interfere with transcriptional regulation hold promise as novel psychopharmacological treatments. Interestingly, sodium valproate, a frequently prescribed antiepileptic drug and mood stabilizer is a weak but broad-acting inhibitor of histone deacetylase enzymes (HDAC) (Guidotti et al., 2011). Histone acetylation facilitates gene expression while deacetylation and therefore HDAC activity is commonly associated with repressive chromation remodeling, as it cleaves off the acetyl-groups from the histone lysine residues (Sharma et al., 2006). However, the therapeutic range of valproate plasma levels (?) in treated clinical populations appears to be below HDAC inhibition levels (Hasan et al., 2013). HDAC inhibitors (HDACi) are thought to upregulate gene expression at some loci by shifting the balance of these processes toward acetylation (of promoter-bound histones). In preclinical model systems, HDACi improve learning and memory function in a variety of paradigms (Lopez-Atalaya et al., 2014) and, if the findings in small animal paradigms would apply to humans, indeed could exert therapeutic effects in a broad range of psychiatric and neurodegenerative disease (Baltan et al., 2011; Chuang et al., 2009; Covington et al., 2009; Fischer et al., 2010; Morris et al., 2010; Schroeder et al., 2007; Tsou et al., 2009).

Lastly, several DNA methylation inhibitors could affect synaptic plasticity and hippocampal learning and memory in mice and rats, which could be used to modulate reward and addiction-related behavior (Han et al., 2010; LaPlant et al., 2010; Levenson et al., 2006; Lubin et al., 2008; Miller et al., 2010; Miller and Sweatt, 2007). These promising findings resulting from preclinical and translational research are highly encouraging. However, whether or not there is indeed therapeutic potential of chromatin modifying drugs in the context of psychosis spectrum disorder remains to be determined.

TREATING PSYCHIATRIC DISEASE BY (EPI)GENOMIC EDITING?

In the previous paragraphs, we speculated about whether drug-induced interference with epigenomic pathways could be exploited to treat psychosis, or at least change cognition and behavior in the preclinical animal model. However, inhibition (or activation) of histone modification enzymes and many other chromatin regulatory proteins very likely will affect widespread areas of neuronal and glial genomes, thereby broadly affecting neurological function. This could be easily associated with an increased likelihood for nonspecific effects and unwanted side effects. However, many noncoding DNA regulating gene expression, such as promoters and enhancers, play an important role in the genetic risk architecture of schizophrenia and are enriched for H3K4 methylation as compared to the portion of the genome not associated with disease risk (Roussos et al., 2014). Therefore, as an alternative approach, easily testable in preclinical models, one could target chromatin in a highly locus- and sequence-selective manner. In theory, the molecular toolboxes for such types of preclinical experiments already exist. Specifically, genome editing strategies via RNA-guided nucleases, including the clustered regularly interspaced short palindromic repeat (CRISPR)-CRISPR-associated protein systems (CRISPR-Cas), introduced to the field only a few years ago (Doudna and Charpentier, 2014), have now been widely adopted in all areas of genomic medicine, including the neurosciences (Heidenreich and Zhang, 2016). Mutations and disruption of enhancer sequences, even in instances where the enhancer was positioned on the linear genome many kilobases apart from the designated target promoter, has been accomplished for several neuropsychiatric risk genes, including the NMDA receptor gene *GRIN2B* (Bharadwaj et al., 2014) and the *FOXG1* transcription factor (Won et al., 2016). However, until now, such experiments have been conducted primarily in cell culture systems in order to test for transcriptional changes after targeted mutations in the regulatory element. Therefore, the next phase of experiments should include in vivo genomic editing of risk-associated promoter and enhancer sequences, and then test for changes in cognition and behavior in the animal. However, such types of genomic editing may carry drawbacks given that mutagenic interventions are likely to be irreversible. However, CRISPR-Cas and other RNA-guided nuclease systems can easily be converted into epigenomic editing tools (by using mutant protein with inactivated nuclease function, fused to a transcriptional activator such as VP64 or P300(Hilton et al., 2015), or a repressor such as KRAB (Thakore et al., 2015), even with multilocus manipulation (Stricker et al., 2017). With the underlying DNA sequence left intact, it will be interesting to explore whether simultaneous epigenomic targeting of enhancer and promoter sequences within multiple risk haplotypes could offer a promising approach to effectively alter cognition and behavior. With the recent accomplishment of region-specific multiplex gene editing in adult mouse brain in vivo (Zetsche et al., 2017), we predict that the approaches proposed here will soon move center stage in preclinical research.

ACKNOWLEDGMENTS

Work conducted in the authors' laboratories is sponsored by the National Institutes of Health.

REFERENCES

Akbarian, S., and Huang, H.S. (2009). Epigenetic regulation in human brain-focus on histone lysine methylation. Biol Psychiatry *65*, 198–203.

Bai, G., Cheung, I., Shulha, H.P., Coelho, J.E., Li, P., Dong, X., Jakovcevski, M., Wang, Y., Grigorenko, A., Jiang, Y., *et al.* (2015). Epigenetic dysregulation of hairy and enhancer of split 4 (HES4) is associated with striatal degeneration in postmortem Huntington brains. Hum Mol Genet *24*, 1441–1456.

Baltan, S., Murphy, S.P., Danilov, C.A., Bachleda, A., and Morrison, R.S. (2011). Histone deacetylase inhibitors preserve white matter structure and function during ischemia by conserving ATP and reducing excitotoxicity. J Neurosci *31*, 3990–3999.

Bharadwaj, R., Peter, C.J., Jiang, Y., Roussos, P., Vogel-Ciernia, A., Shen, E.Y., Mitchell, A.C., Mao, W., Whittle, C., Dincer, A., *et al.* (2014). Conserved higher-order chromatin regulates NMDA receptor gene expression and cognition. Neuron *84*, 997–1008.

Bredy, T.W., Wu, H., Crego, C., Zellhoefer, J., Sun, Y.E., and Barad, M. (2007). Histone modifications around individual BDNF gene promoters in prefrontal cortex are associated with extinction of conditioned fear. Learn Mem *14*, 268–276.

Cheng, M.C., Liao, D.L., Hsiung, C.A., Chen, C.Y., Liao, Y.C., and Chen, C.H. (2008). Chronic treatment with aripiprazole induces differential gene expression in the rat frontal cortex. Int J Neuropsychopharmacol *11*, 207–216.

Chuang, D.M., Leng, Y., Marinova, Z., Kim, H.J., and Chiu, C.T. (2009). Multiple roles of HDAC inhibition in neurodegenerative conditions. Trends Neurosci *32*, 591–601.

Covington, H.E., 3rd, Maze, I., LaPlant, Q.C., Vialou, V.F., Ohnishi, Y.N., Berton, O., Fass, D.M., Renthal, W., Rush, A.J., 3rd, Wu, E.Y., *et al.* (2009). Antidepressant actions of histone deacetylase inhibitors. J Neurosci *29*, 11451–11460.

Dincer, A., Gavin, D.P., Xu, K., Zhang, B., Dudley, J.T., Schadt, E.E., and Akbarian, S. (2015). Deciphering H3K4me3 broad domains associated with gene-regulatory networks and conserved epigenomic landscapes in the human brain. Transl Psychiatry *5*, e679.

Dong, E., Nelson, M., Grayson, D.R., Costa, E., and Guidotti, A. (2008). Clozapine and sulpiride but not haloperidol or olanzapine activate brain DNA demethylation. Proc Nat Acad Sci U S A *105*, 13614–13619.

Dong, X., Tsuji, J., Labadorf, A., Roussos, P., Chen, J.F., Myers, R.H., Akbarian, S., and Weng, Z. (2015). The role of H3K4me3 in transcriptional regulation is altered in Huntington's disease. PLoS One *10*, e0144398.

Doudna, J.A., and Charpentier, E. (2014). Genome editing: the new frontier of genome engineering with CRISPR-Cas9. Science *346*, 1258096.

Eissenberg, J.C., and Shilatifard, A. (2010). Histone H3 lysine 4 (H3K4) methylation in development and differentiation. Developmental Biology *339*, 240–249.

Fischer, A., Sananbenesi, F., Mungenast, A., and Tsai, L.H. (2010). Targeting the correct HDAC(s) to treat cognitive disorders. Trends Pharmacol Sci *31*, 605–617.

Gerhard, T., Stroup, T.S., Correll, C.U., Huang, C., Tan, Z., Crystal, S., and Olfson, M. (2017). Antipsychotic medication treatment patterns in adult depression. J Clin Psychiatry.

Girdhar, K., Hoffman, G.E., Jiang, Y., Brown, L., Kundakovic, M., Hauberg, M.E., Francoeur, N.J., Wang, Y.C., Shah, H., Kavanagh, D.H., *et al.* (2018). Cell-specific histone modification maps in the human frontal lobe link schizophrenia risk to the neuronal epigenome. Nat Neurosci *21*, 1126–1136.

Green, A.L., Zhan, L., Eid, A., Zarbl, H., Guo, G.L., and Richardson, J.R. (2017). Valproate increases dopamine transporter expression through histone acetylation and enhanced promoter binding of Nurr1. Neuropharmacology *125*, 189–196.

Guidotti, A., Auta, J., Chen, Y., Davis, J.M., Dong, E., Gavin, D.P., Grayson, D.R., Matrisciano, F., Pinna, G., Satta, R., *et al.* (2011). Epigenetic GABAergic targets in schizophrenia and bipolar disorder. Neuropharmacology *60*, 1007–1016.

Gupta, S., Kim, S.Y., Artis, S., Molfese, D.L., Schumacher, A., Sweatt, J.D., Paylor, R.E., and Lubin, F.D. (2010). Histone methylation regulates memory formation. J Neurosci *30*, 3589–3599.

Han, J., Li, Y., Wang, D., Wei, C., Yang, X., and Sui, N. (2010). Effect of 5-aza-2-deoxycytidine microinjecting into hippocampus and prelimbic cortex on acquisition and retrieval of cocaine-induced place preference in C57BL/6 mice. Eur J Pharmacol *642*, 93–98.

Hannon, E., Spiers, H., Viana, J., Pidsley, R., Burrage, J., Murphy, T.M., Troakes, C., Turecki, G., O'Donovan, M.C., Schalkwyk, L.C., *et al.* (2016). Methylation QTLs in the developing brain and their enrichment in schizophrenia risk loci. Nat Neurosci *19*, 48–54.

Hasan, A., Mitchell, A., Schneider, A., Halene, T., and Akbarian, S. (2013). Epigenetic dysregulation in schizophrenia: molecular and clinical aspects of histone deacetylase inhibitors. Eur Arch Psychiatry Clin Neurosci *263*, 273–284.

Heidenreich, M., and Zhang, F. (2016). Applications of CRISPR-Cas systems in neuroscience. Nat Rev Neurosci *17*, 36–44.

Hernandez, D.G., Nalls, M.A., Gibbs, J.R., Arepalli, S., van der Brug, M., Chong, S., Moore, M., Longo, D.L., Cookson, M.R., Traynor, B.J., *et al.* (2011). Distinct DNA methylation changes highly correlated with chronological age in the human brain. Hum Mol Genet *20*, 1164–1172.

Hilton, I.B., D'Ippolito, A.M., Vockley, C.M., Thakore, P.I., Crawford, G.E., Reddy, T.E., and Gersbach, C.A. (2015). Epigenome editing by a CRISPR-Cas9-based acetyltransferase activates genes from promoters and enhancers. Nat Biotechnol *33*, 510–517.

Huang, H.S., Matevossian, A., Whittle, C., Kim, S.Y., Schumacher, A., Baker, S.P., and Akbarian, S. (2007). Prefrontal dysfunction in schizophrenia involves mixed-lineage leukemia 1-regulated histone methylation at GABAergic gene promoters. J Neurosci *27*, 11254–11262.

Ito, S., Shen, L., Dai, Q., Wu, S.C., Collins, L.B., Swenberg, J.A., He, C., and Zhang, Y. (2011). Tet proteins can convert 5-methylcytosine to 5-formylcytosine and 5-carboxylcytosine. Science *333*, 1300–1303.

Jaffe, A.E., Gao, Y., Deep-Soboslay, A., Tao, R., Hyde, T.M., Weinberger, D.R., and Kleinman, J.E. (2016). Mapping DNA methylation across development, genotype and schizophrenia in the human frontal cortex. Nat Neurosci *19*, 40–47.

Jakovcevski, M., Ruan, H., Shen, E.Y., Dincer, A., Javidfar, B., Ma, Q., Peter, C.J., Cheung, I., Mitchell, A.C., Jiang, Y., *et al.* (2015). Neuronal Kmt2a/Mll1 histone methyltransferase is essential for prefrontal synaptic plasticity and working memory. J Neurosci *35*, 5097–5108.

Kano, S., Colantuoni, C., Han, F., Zhou, Z., Yuan, Q., Wilson, A., Takayanagi, Y., Lee, Y., Rapoport, J., Eaton, W., *et al.* (2013). Genome-wide profiling of multiple histone methylations in olfactory cells: further implications for cellular susceptibility to oxidative stress in schizophrenia. Mol Psychiatry *18*, 740–742.

Kerimoglu, C., Agis-Balboa, R.C., Kranz, A., Stilling, R., Bahari-Javan, S., Benito-Garagorri, E., Halder, R., Burkhardt, S., Stewart, A.F., and Fischer, A. (2013). Histone-methyltransferase MLL2 (KMT2B) is required for memory formation in mice. J Neurosci *33*, 3452–3464.

Koziol, M.J., Bradshaw, C.R., Allen, G.E., Costa, A.S., Frezza, C., and Gurdon, J.B. (2016). Identification of methylated deoxyadenosines in vertebrates reveals diversity in DNA modifications. Nat Struct Mol Biol *23*, 24–30.

Kriaucionis, S., and Heintz, N. (2009). The nuclear DNA base 5-hydroxymethylcytosine is present in Purkinje neurons and the brain. Science *324*, 929–930.

Kwon, B., and Houpt, T.A. (2010). Phospho-acetylation of histone H3 in the amygdala after acute lithium chloride. Brain Res *1333*, 36–47.

LaPlant, Q., Vialou, V., Covington, H.E., 3rd, Dumitriu, D., Feng, J., Warren, B.L., Maze, I., Dietz, D.M., Watts, E.L., Iniguez, S.D., *et al.* (2010). Dnmt3a regulates emotional behavior and spine plasticity in the nucleus accumbens. Nat Neurosci *13*, 1137–1143.

Levenson, J.M., Roth, T.L., Lubin, F.D., Miller, C.A., Huang, I.C., Desai, P., Malone, L.M., and Sweatt, J.D. (2006). Evidence that DNA (cytosine-5) methyltransferase regulates synaptic plasticity in the hippocampus. J Biol Chem *281*, 15763–15773.

Li, J., Guo, Y., Schroeder, F.A., Youngs, R.M., Schmidt, T.W., Ferris, C., Konradi, C., and Akbarian, S. (2004). Dopamine D2-like antagonists induce chromatin remodeling in striatal neurons through cyclic AMP-protein kinase A and NMDA receptor signaling. J Neurochem *90*, 1117–1131.

Liu, L., Jin, G., and Zhou, X. (2015). Modeling the relationship of epigenetic modifications to transcription factor binding. Nucleic Acids Res *43*, 3873–3885.

Lomvardas, S., and Maniatis, T. (2016). Histone and DNA modifications as regulators of neuronal development and function. Cold Spring Harb Perspect Biol *8*.

Lopez-Atalaya, J.P., Valor, L.M., and Barco, A. (2014). Epigenetic factors in intellectual disability: the Rubinstein-Taybi syndrome as a paradigm of neurodevelopmental disorder with epigenetic origin. Prog Mol Biol Transl Sci *128*, 139–176.

Lubin, F.D., Roth, T.L., and Sweatt, J.D. (2008). Epigenetic regulation of BDNF gene transcription in the consolidation of fear memory. J Neurosci *28*, 10576–10586.

Lutz, P.E., Tanti, A., Gasecka, A., Barnett-Burns, S., Kim, J.J., Zhou, Y., Chen, G.G., Wakid, M., Shaw, M., Almeida, D., *et al.* (2017). Association of a history of child abuse with impaired myelination in the anterior cingulate cortex: convergent epigenetic, transcriptional, and morphological evidence. Am J Psychiatry, appiajp201716111286.

Maunakea, A.K., Nagarajan, R.P., Bilenky, M., Ballinger, T.J., D'Souza, C., Fouse, S.D., Johnson, B.E., Hong, C., Nielsen, C., Zhao, Y., *et al.* (2010). Conserved role of intragenic DNA methylation in regulating alternative promoters. Nature *466*, 253–257.

Mill, J., Tang, T., Kaminsky, Z., Khare, T., Yazdanpanah, S., Bouchard, L., Jia, P., Assadzadeh, A., Flanagan, J., Schumacher, A., *et al.* (2008). Epigenomic profiling reveals DNA-methylation changes associated with major psychosis. Am J Hum Genet *82*, 696–711.

Miller, C.A., Gavin, C.F., White, J.A., Parrish, R.R., Honasoge, A., Yancey, C.R., Rivera, I.M., Rubio, M.D., Rumbaugh, G., and Sweatt, J.D. (2010). Cortical DNA methylation maintains remote memory. Nat Neurosci *13*, 664–666.

Miller, C.A., and Sweatt, J.D. (2007). Covalent modification of DNA regulates memory formation. Neuron *53*, 857–869.

Mitchell, A.C., Javidfar, B., Pothula, V., Ibi, D., Shen, E.Y., Peter, C.J., Bicks, L.K., Fehr, T., Jiang, Y., Brennand, K.J., *et al.* (2017). MEF2C transcription factor is associated with the genetic and epigenetic risk architecture of schizophrenia and improves cognition in mice. Mol Psychiatry.

Mo, A., Mukamel, E.A., Davis, F.P., Luo, C., Henry, G.L., Picard, S., Urich, M.A., Nery, J.R., Sejnowski, T.J., Lister, R., *et al.* (2015). Epigenomic signatures of neuronal diversity in the mammalian brain. Neuron *86*, 1369–1384.

Morris, M.J., Karra, A.S., and Monteggia, L.M. (2010). Histone deacetylases govern cellular mechanisms underlying behavioral and synaptic plasticity in the developing and adult brain. Behav Pharmacol *21*, 409–419.

Network, and Pathway Analysis Subgroup of Psychiatric Genomics, C. (2015). Psychiatric genome-wide association study analyses implicate neuronal, immune and histone pathways. Nat Neurosci *18*, 199–209.

Numata, S., Ye, T., Hyde, T.M., Guitart-Navarro, X., Tao, R., Wininger, M., Colantuoni, C., Weinberger, D.R., Kleinman, J.E., and Lipska, B.K. (2012). DNA methylation signatures in development and aging of the human prefrontal cortex. Am J Hum Genet *90*, 260–272.

Peter, C.J., and Akbarian, S. (2011). Balancing histone methylation activities in psychiatric disorders. Trends Mol Med *17*, 372–379.

PsychENCODE consortium, Akbarian, S., Liu, C., Knowles, J.A., Vaccarino, F.M., Farnham, P.J., Crawford, G.E., Jaffe, A.E., Pinto, D., Dracheva, S., *et al.* (2015). The PsychENCODE project. Nat Neurosci *18*, 1707–1712.

Roussos, P., Mitchell, A.C., Voloudakis, G., Fullard, J.F., Pothula, V.M., Tsang, J., Stahl, E.A., Georgakopoulos, A., Ruderfer, D.M., Charney, A., *et al.* (2014). A role for noncoding variation in schizophrenia. Cell Rep *9*, 1417–1429.

Schizophrenia Working Group of the Psychiatric Genomics, C. (2014). Biological insights from 108 schizophrenia-associated genetic loci. Nature *511*, 421–427.

Schroeder, F.A., Lin, C.L., Crusio, W.E., and Akbarian, S. (2007). Antidepressant-like effects of the histone deacetylase inhibitor, sodium butyrate, in the mouse. Biol Psychiatry *62*, 55–64.

Sharma, R.P., Rosen, C., Kartan, S., Guidotti, A., Costa, E., Grayson, D.R., and Chase, K. (2006). Valproic acid and chromatin remodeling in schizophrenia and bipolar disorder: preliminary results from a clinical population. Schizophr Res *88*, 227–231.

Shen, E.Y., Jiang, Y., Javidfar, B., Kassim, B., Loh, Y.E., Ma, Q., Mitchell, A.C., Pothula, V., Stewart, A.F., Ernst, P., *et al.* (2016). Neuronal deletion of Kmt2a/Mll1 histone methyltransferase in ventral striatum is associated with defective spike-timing-dependent striatal synaptic plasticity, altered response to dopaminergic drugs, and increased anxiety. Neuropsychopharmacology *41*, 3103–3113.

Shimabukuro, M., Jinno, Y., Fuke, C., and Okazaki, Y. (2006). Haloperidol treatment induces tissue- and sex-specific changes in DNA methylation: a control study using rats. Behav Brain Funct *2*, 37.

Shulha, H.P., Cheung, I., Guo, Y., Akbarian, S., and Weng, Z. (2013). Coordinated cell type-specific epigenetic remodeling in prefrontal cortex begins before birth and continues into early adulthood. PLoS Genet *9*, e1003433.

Shulha, H.P., Cheung, I., Whittle, C., Wang, J., Virgil, D., Lin, C.L., Guo, Y., Lessard, A., Akbarian, S., and Weng, Z. (2012). Epigenetic signatures of autism: trimethylated H3K4 landscapes in prefrontal neurons. Arch Gen Psychiatry *69*, 314–324.

Siegmund, K.D., Connor, C.M., Campan, M., Long, T.I., Weisenberger, D.J., Biniszkiewicz, D., Jaenisch, R., Laird, P.W., and Akbarian, S. (2007). DNA methylation in the human cerebral cortex is dynamically regulated throughout the life span and involves differentiated neurons. PLoS One *2*, e895.

Singh, T., Kurki, M.I., Curtis, D., Purcell, S.M., Crooks, L., McRae, J., Suvisaari, J., Chheda, H., Blackwood, D., Breen, G., *et al.* (2016). Rare loss-of-function variants in SETD1A are associated with schizophrenia and developmental disorders. Nat Neurosci *19*, 571–577.

Stricker, S.H., Koferle, A., and Beck, S. (2017). From profiles to function in epigenomics. Nat Rev Genet *18*, 51–66.

Takata, A., Ionita-Laza, I., Gogos, J.A., Xu, B., and Karayiorgou, M. (2016). De novo synonymous mutations in regulatory elements contribute to the genetic etiology of autism and schizophrenia. Neuron *89*, 940–947.

Takata, A., Xu, B., Ionita-Laza, I., Roos, J.L., Gogos, J.A., and Karayiorgou, M. (2014). Loss-of-function variants in schizophrenia risk and SETD1A as a candidate susceptibility gene. Neuron *82*, 773–780.

Tessarz, P., and Kouzarides, T. (2014). Histone core modifications regulating nucleosome structure and dynamics. Nat Rev Mol Cell Biol *15*, 703–708.

Thakore, P.I., D'Ippolito, A.M., Song, L., Safi, A., Shivakumar, N.K., Kabadi, A.M., Reddy, T.E., Crawford, G.E., and Gersbach, C.A. (2015). Highly specific epigenome editing by CRISPR-Cas9 repressors for silencing of distal regulatory elements. Nat Methods *12*, 1143–1149.

Tondo, L., Vazquez, G.H., and Baldessarini, R.J. (2017). Depression and mania in bipolar disorder. Curr Neuropharmacol *15*, 353–358.

Tsou, A.Y., Friedman, L.S., Wilson, R.B., and Lynch, D.R. (2009). Pharmacotherapy for Friedreich ataxia. CNS Drugs *23*, 213–223.

Uher, R., and Zwicker, A. (2017). Etiology in psychiatry: embracing the reality of poly-gene-environmental causation of mental illness. World Psychiatry *16*, 121–129.

Vallianatos, C.N., and Iwase, S. (2015). Disrupted intricacy of histone H3K4 methylation in neurodevelopmental disorders. Epigenomics *7*, 503–519.

Vashishtha, M., Ng, C.W., Yildirim, F., Gipson, T.A., Kratter, I.H., Bodai, L., Song, W., Lau, A., Labadorf, A., Vogel-Ciernia, A., *et al.* (2013). Targeting H3K4 trimethylation in Huntington disease. Proc Natl Acad Sci U S A *110*, E3027–3036.

Wang, D., Liu, S., Warrell, J., Won, H., Shi, X., Navarro, F.C.P., Clarke, D., Gu, M., Emani, P., Yang, Y.T., *et al.* (2018). Comprehensive functional genomic resource and integrative model for the human brain. Science *362*.

Wockner, L.F., Noble, E.P., Lawford, B.R., Young, R.M., Morris, C.P., Whitehall, V.L., and Voisey, J. (2014). Genome-wide DNA methylation analysis of human brain tissue from schizophrenia patients. Transl Psychiatry *4*, e339.

Won, H., de la Torre-Ubieta, L., Stein, J.L., Parikshak, N.N., Huang, J., Opland, C.K., Gandal, M.J., Sutton, G.J., Hormozdiari, F., Lu, D., *et al.* (2016). Chromosome conformation elucidates regulatory relationships in developing human brain. Nature *538*, 523–527.

Xie, W., Barr, C.L., Kim, A., Yue, F., Lee, A.Y., Eubanks, J., Dempster, E.L., and Ren, B. (2012). Base-resolution analyses of sequence and parent-of-origin dependent DNA methylation in the mouse genome. Cell *148*, 816–831.

Zetsche, B., Heidenreich, M., Mohanraju, P., Fedorova, I., Kneppers, J., DeGennaro, E.M., Winblad, N., Choudhury, S.R., Abudayyeh, O.O., Gootenberg, J.S., *et al.* (2017). Multiplex gene editing by CRISPR-Cpf1 using a single crRNA array. Nat Biotechnol *35*, 31–34.

Zhang, T., Cooper, S., and Brockdorff, N. (2015). The interplay of histone modifications - writers that read. EMBO Rep *16*, 1467–1481.

Zhou, V.W., Goren, A., and Bernstein, B.E. (2011). Charting histone modifications and the functional organization of mammalian genomes. Nat Rev Genet *12*, 7–18.

21

DNA MODIFICATIONS IN SCHIZOPHRENIA

Ehsan Pishva and Bart P. F. Rutten

Epigenetic processes act to dynamically control gene expression independently of DNA sequence variation. The epigenetic regulation of gene expression, mediated by molecular processes including DNA modifications, histone modifications, and the action of noncoding RNAs (ncRNAs), is critical for cellular differentiation and development. Despite identical genomic content in every nucleated cell in the human body, the complex interplay of epigenetic mechanisms regulates the expression of genes and results in differentiated cells with distinct physiological functions.[1] Epigenetic changes also induce transcriptional reprogramming throughout the lifetime as a result of external environmental stimuli.[2] In this chapter, we discuss the dynamics of DNA modification processes in the central nervous system and their role in the regulation of gene expression and brain development. We provide an overview of the current evidence supporting the notion that alterations to DNA methylation—the most intensely studied epigenetic modification in human populations—plays a role in the etiology of schizophrenia (SCZ). Finally, we discuss the existing challenges and future directions for studies of epigenetic variation in SCZ.

DNA MODIFICATION: REGULATION OF TRANSCRIPTION IN THE CENTRAL NERVOUS SYSTEM

DNA methylation refers to the transfer of a methyl group (CH3) from S-adenosyl methionine (SAM) covalently to the 5-position of cytosine residues in DNA. Although 5mC primarily occurs in the context of a CpG dinucleotide,[3] recent findings also indicate the occurrence of 5mC in the context of cytosine-adenine (CpA), cytosine-cytosine (CpC), and cytosine-thymine (CpT) dinucleotides.[4] Interestingly, enrichment of non-CpG 5mC has been described in pluripotent stem cells, oocytes, and neurons,[5] with recent studies suggesting an important role specifically in the brain.[6]

DNA methyltransferases (DNMTs) catalyze the methylation of cytosine, with DNMT1 being responsible for maintenance, and DNMT3A and DNMT3B facilitating de novo DNA methylation. Methyl-CpG-binding domain proteins (MeCp2 and MBDs) can be recruited by methylated CpG dinucleotides and inhibit the histone deacetylase activities that facilitate the remodeling of chromatin. Altered chromatin structure can change the accessibility of the transcription factors to the regulatory regions of genes and ultimately alter the transcriptional activity of a given genomic region.

The frequency of CpG dinucleotides in the human genome is ~1% of total DNA bases. Nevertheless, more than 70%–80% of all CpG dinucleotides in the genome are methylated.[7] CpG islands (CGIs) are regions of the genome with relatively high frequency of CpG dinucleotides, being generally unmethylated and encompassing gene promoter regions. Methylated cytosines in CGIs are often associated with transcriptional repression,[8] but recent findings suggest that the relationship between 5-mC and transcriptional activity may be more complex. DNA methylation within the gene body, for example, often correlates with active transcription[9] and alternative splicing.[10]

The crucial role of DNA methylation in the normal development and functioning of the human central nervous system is exemplified by studies of the gene encoding methyl-CpG-binding protein (MeCP2), which is essential for the normal functioning of neurons. MeCP2 binds to methylated CpGs and silences gene expression by recruiting histone deacetylase (HDAC) activity and inducing chromatin remodeling. Mutations in the MeCP2 gene cause global gene silencing,[11] resulting in Rett syndrome, a progressive neurodevelopmental disorder.[12] Growing evidence implicates a critical role for cell-specific DNA modifications in key neurobiological and cognitive functions such as neuronal plasticity,[13] memory formation,[14] and circadian processes.[15] Furthermore, aberrations in the DNA modification levels are correlated with normal brain development[4,5,16] as well as with a range of neurodevelopmental and neuropsychiatric disorders including Parkinson's[17] and Huntington's disease,[18] amyotrophic lateral sclerosis,[19] Alzheimer's disease,[20] and major depression disorder.[21]

Although the methylation of cytosine is the most stable epigenetic modification in vertebrates and the best characterized in the context of human health and disease, there is increasing interest in the role of other modifications to DNA. Methyl groups can be either actively or passively removed from the cytosine, and the active demethylation process starts with conversion of 5-mC to 5-hydroxymethylation (5-hmC) catalyzed by ten-eleven translocation enzymes (TET1, TET2, and TET3), which also catalyze conversion of 5-hmC to other forms of cytosine, such as 5-formylcytosine (5-fC) and 5-carboxylcytosine (5-CaC).[22] 5hmC, 5fC, and 5caC modifications were initially considered to be intermediate byproducts of active 5mC demethylation, but the identification of relatively stable levels of 5-hmC in non-dividing cells suggest an independent role in the regulation of gene expression.[23] Importantly, 5hmC is present at relatively high levels in the central nervous system,[24] being particularly enriched in the vicinity of genes with synapse-related functions[25] and highly dynamic during human brain development.[26] Although initially hypothesized to represent an intermediate step of the DNA demethylation pathway, 5hmC is now assumed to have specific functional roles in the brain, with implications for the etiology of disorders such as SCZ. However, despite the accumulating evidence suggesting an important role for the other types of cytosine modifications in human neurodevelopment, no publication has yet reported on associations between 5hmC, 5fC, or 5-CaC and SCZ or SCZ-related phenotypes. Importantly, many of the standard methods used to quantify 5mC are unable to discriminate between 5mC and 5hmC.

DNA MODIFICATIONS AND SCHIZOPHRENIA: CURRENT STATE OF PLAY

Epidemiological and molecular evidence support both genetic and environmental contributions to the etiology of SCZ. The most replicated environmental risks for SCZ occur at critical periods early in development, a time of rapid cell replication when the epigenome is known to be particularly labile in response to external factors and the standard epigenetic signals driving development and tissue differentiation are being established.[27] Furthermore, despite sharing the same DNA sequence, the concordance rate between monozygotic (MZ) twins for SCZ is consistently estimated to be <65%, potentially reflecting environmentally or stochastically mediated epigenetic variation.[28] Finally, many of the currently prescribed treatments for SCZ are associated with epigenomic changes,[29] further implicating epigenetic processes in disease etiology.

The initial evidence linking DNA methylation processes to SCZ and other psychotic disorders focused on the analyses of DNA methylation levels in regulatory elements such as the promoters or enhancers regions proximal to a priori candidate genes. These analyses were mostly targeted at promoter regions of genes encoding neurotransmitter related gamma-aminobutyric acid (GABA)-ergic, glutamatergic,[30] serotonergic,[31] and dopaminergic systems.[32] While most of the candidate gene studies profiled (relatively) small numbers of samples that were obtained from a range of different (and often inappropriate) cell and tissue types, no consistent pattern of SCZ-associated methylation changes was identified.

The first epigenome-wide association study (EWAS) of SCZ investigated methylation differences in ~12K CpG islands across the genomic DNA derived from frontal-cortex postmortem brain tissue.[33] Rapid advances in genomic profiling technology over the following decade, along with the development of powerful bioinformatics tools, has facilitated the assessment of cytosine modifications at single base resolution across the whole genome. Whole-genome bisulfite sequencing is considered the gold standard method for robust quantification of DNA modifications. However, given the sufficient precision at an affordable cost, microarray platforms are now widely being adopted by the research community to measure methylation levels in large numbers of samples. The Illumina EPIC HumanMethylation array,[34] for instance, can interrogate over 850K CpG sites across the genome with a coverage up to 99% of RefSeq genes.[35]

Several EWAS analyses of SCZ have reported multiple differentially methylated CpG positions (DMPs) and regions (DMRs) annotated to genes involved in biological networks and pathways relevant to psychiatric disorders and neurodevelopment. An overview of the findings from studies of DNA methylation in SCZ is given in Table 21.1. These findings are described in detail in multiple review articles.[36,37] However, it should be noted that the findings of the first EWASs in SCZ have not been fully consistent with each other, or reproducible. Such discrepancies may be due to inadequate statistical power, variations in measurement tools in different studies, or a spectrum of confounding factors, including tissue and cell type, age, sex, medication exposure, and reverse causation. Furthermore, the majority of the current evidence is limited to the use of whole blood or bulk brain tissue, which is a mixture of many different cell types, and cell-type heterogeneity is likely to be an important confounding factor in existing EWAS analyses.[38]

Table 21.1 SUMMARY OF PUBLISHED DNA METHYLATION STUDIES IN SCZ USING POSTMORTEM BRAIN TISSUE AND BLOOD

REFERENCE	TISSUE SOURCE	SAMPLE NUMBER	METHOD	KEY FINDINGS
Mill et al., 2008[39]	PFC	35 SCZ, 35 CTRL	CpG island microarrays	Significant SCZ-associated differences in DNA methylation at multiple loci
Dempster et al., 2011[28]	Peripheral blood	22 MZ twin pairs discordant for SCZ	Illumina HM27K	Numerous loci with DNA methylation differences between discordant twins
Kinoshita et al., 2013[40]	Peripheral blood leukocytes	24 medication free-SCZ, 23 CTRL; 3 MZ twin pairs discordant for SCZ	Illumina HM27K	Methylation status at 10,747 CpG sites significantly associated with SCZ with an FDR threshold of 5%. 234 of these significantly different between discordant twins
Nishioka et al., 2013[41]	Whole blood	18 FESCZ, 15 CTRL	Illumina HM27K	Numerous SCZ-associated loci
Chen et al., 2014[42]	CER	39 SCZ, 43 CTRL	Illumina HM27K	Four CpG sites showing correlated differential gene expression and methylation associated with SCZ, with an FDR threshold of 20%.
Numata et al., 2014[43]	DPFC	106 SCZ, 110 CTRL	Illumina HM27K	Methylation status at 107 CpG sites significantly associated with SCZ after Bonferroni correction. Several cis-mQTLS identified, including SCZ-associated risk SNPs
Wockner et al., 2014[44]	PFC	24 SCZ, 24 CTRL	Illumina HM450K and pyrosequencing	Numerous SCZ-associated loci with some of them overlapping with Kinoshita et al., 2013[40]
Pidsley et al., 2014[45]	PFC, CER	Discovery cohort: 21 SCZ, 23 CTRL, Replication cohort: 18 SCZ, 15 CTRL	Illumina HM450K and pyrosequencing	4 SCZ-associated DMPs with an FDR threshold of 5% in the PFC. One SCZ-associated DMR in the PFC spanning the gene body of NRN1. Top SCZ-associated modules of comethylated sites in the PFC enriched for neuropsychiatric- and neurodevelopmental-linked genes SCZ. SCZ-associated DMPs in the PFC were enriched for neurodevelopmental-associated genes
Liu et al., 2014[46]	Whole blood	98 SCZ, 108 CTRL	Illumina HM450K	Numerous SCZ-associated loci
Aberg et al., 2014[47]	Whole blood	Discovery cohort: 760 SCZ, 738 CTRL Replication cohort 1: 178 SCZ, 182 CTRL Replication cohort 2: 561 SCZ, 582 CTRL	MBD-seq and pyrosequencing of bisulfite-PCR amplicons	Methylation status at 139 sites significantly associated with SCZ with an FDR threshold of 1%. One site at FAM63B replicated in the pyrosequencing assay with a P Value lower than the multiple testing thresholds. Several of the SCZ-associated sites were linked to hypoxia
Wockner et al., 2015[48]	PFC	24 SCZ, 24 CTRL + 2 publicly available cohorts from Pidsley et al., 2014[45]	Illumina HM450K	Identification of several SCZ associated DMRs including *CERS3*, *DPPA5*, *REC8*, *PRDM9*, *LY6G5C*, and *DDX43*
Montano et al., 2016[49]	Whole blood	Discovery cohort: 689 SCZ, 645 CTRL, Replication cohort: 247 SCZ and 250 CTRL	Illumina HM450K	923 SCZ-associated DMPs at an FDR threshold of 20%, 625 of which replicated. These included *RPS6KA1*, *MAD1L1*, *KLF13*, *SULT4A1*, *NPDC1*, *AUTS2*, *PIK3R1*, and *PNPO*
Jaffe et al., 2016[50]	DPFC	108 SCZ patients and 136 CTRL	Illumina HM450K	Identification of 2,104 SCZ associated CpGs (Bonferroni-adjusted $P < 0.05$). Enrichment of DMPs associated with prenatal-postnatal transition in SCZ GWAS regions.
Hannon et al., 2016[51]	PFC (BA 9), STR and CER	76 PFC (38 SCZ and 38 CTRL), 82 STR (37 SCZ, 45 CTRL) and 77 CER (37 SCZ and 40 CTRL)	Illumina HM450K	Significant enrichment of SCZ-associated GWAS variants in fetal brain mQTLs

INTEGRATING GENETIC AND EPIGENETIC APPROACHES TO UNDERSTAND THE ETIOLOGY OF SCHIZOPHRENIA

In recent years, considerable progress has been made in understanding the role of both common and rare genetic variation in SCZ using forward genetics approaches. The most recent genome-wide association study (GWAS) included >100,000 individuals, for example, and identified 145 loci harboring common risk variants robustly associated with SCZ, with evidence for a substantial polygenic component within signals falling below genome-wide levels of significance.[52] Despite this success, however, there remains uncertainty about the specific causal genes involved in SCZ, and how their function is regulated in development and disease. Many SCZ-associated GWAS variants reside in large regions of strong the untranslated region and do not directly index coding changes affecting protein structure; instead, they are hypothesized to influence gene regulation, a notion supported by their enrichment in regulatory domains, including enhancers and regions of open chromatin.[53] Because these regulatory signatures are both tissue- and developmental-stage-specific, it is critical that genetic effects on transcriptional regulation are mapped in the correct cell-types and at the critical stages of development. A recent study characterized >270,000 fetal brain DNA methylation quantitative trait loci (mQTLs);[54] although many of these were found to be developmentally stable, a subset displayed fetal-specific effects that are not present in adult brain. Notably, fetal brain mQTLs are enriched among SCZ risk loci, supporting the hypothesis that disease-associated variants have regulatory genomic consequences during brain development.[51]

DNA MODIFICATIONS AND ENVIRONMENTAL RISKS FOR SCHIZOPHRENIA

A wealth of epidemiologic evidence implicates pre- and postnatal environmental exposures in the risk of developing psychotic disorders in later life.[55] There are consistent findings that connect exposure to nutritional factors, childhood trauma, minority group position, drug abuse, and pre- and perinatal complications to an increased risk of psychotic disorders.[55] Although several of these environmental factors have been directly associated with altered DNA methylation profiles,[56] studies attempting to establish (mediating) links between DNA methylation, exposure to environmental risk factors, and psychosis are challenged by difficulties in prospectively capturing the exposure to the environmental factor, and by tissue-specificity of DNA methylation profiles, among other conceptual and methodological challenges.[57] Currently, there is no evidence for direct links between the risk of SCZ, environmental exposures (e.g., childhood trauma, obstetric complications, prenatal factors), and altered DNA methylation profiles.[58] However, there is strong evidence indicating changes in DNA methylation induced by environmental factors that have been epidemiologically associated with SCZ. For example, it has been shown that prenatal nutritional deficiency, which has been known as a risk factor for SCZ,[59] can also induce persistent changes in DNA methylation depending on sex and embryonic developmental stage.[60] Further efforts (e.g., using large twin populations or prospective epidemiological studies) are required to improve our understanding about the epigenetic changes induced by environmental factors and development of SCZ.

EXPERIMENTAL LIMITATIONS AND FUTURE DIRECTIONS

Contrary to analyses of DNA sequence variation (i.e., in GWAS), in which a large number of samples can be profiled together from various tissues, epigenetic marks show distinct cell and tissue specific patterns. Therefore, it is likely that profiling neural cells from patients will be crucial for studying epigenetic variations in neuropsychiatric disorders. Studying human brain tissue in vivo is not feasible; therefore, postmortem brains remain the only source for tissue specific investigation of epigenetic mechanisms in the etiology of SCZ. However, obtaining an adequate number of high-quality tissue samples from patients diagnosed with SCZ and individuals without a diagnosis of mental disorders while being matched for possible confounding factors such as age, sex, and smoking is a major challenge. Moreover, the influence of cellular heterogeneity is rarely controlled for in analyses of SCZ-associated DNA modifications. Even where brain tissue is used, analyses are primarily focused on 5mC, and restricted by their use of polyclonal or "bulk" postmortem tissue samples that comprise a complex mix of neuronal and non-neuronal cell-types.[61] Recent efforts by the PsychENCODE consortium and others aim to generate a catalog of non-coding elements and epigenomic variation in purified cell types from multiple brain regions of the human brain.[62] For example, methods have been developed to isolate and profile purified neuronal nuclei from postmortem brain tissue, using fluorescence-activated cell sorting (FACS) to gate and select NeuN+ and NeuN- immunolabeled populations of nuclei prior to the analysis of DNA and histone modifications.[63] Such efforts provide a basis for cell-specific profiling of DNA modifications and gene expression, which are required for fine-mapping SCZ-associated genetic variants, and improve our understanding about the functional consequences of human genomic variations.

Longitudinal study designs are optimal for establishing the direction of causality in epigenetic epidemiology, enabling an analysis of whether epigenetic profiles in patients are associated with disease risk or whether they represent secondary effect of the disease chronicity or antipsychotics. To date, only one study examined longitudinal methylome-wide profiles of blood derived DNA in people with an at risk mental state (ARMS) and compared the profiles of subjects who converted to psychosis, as compared to ARMS subjects who did not convert to psychosis.[64] A limitation of these longitudinal analyses is, however, the fact that they are limited to profiling epigenetic variation in peripheral tissues such as blood.

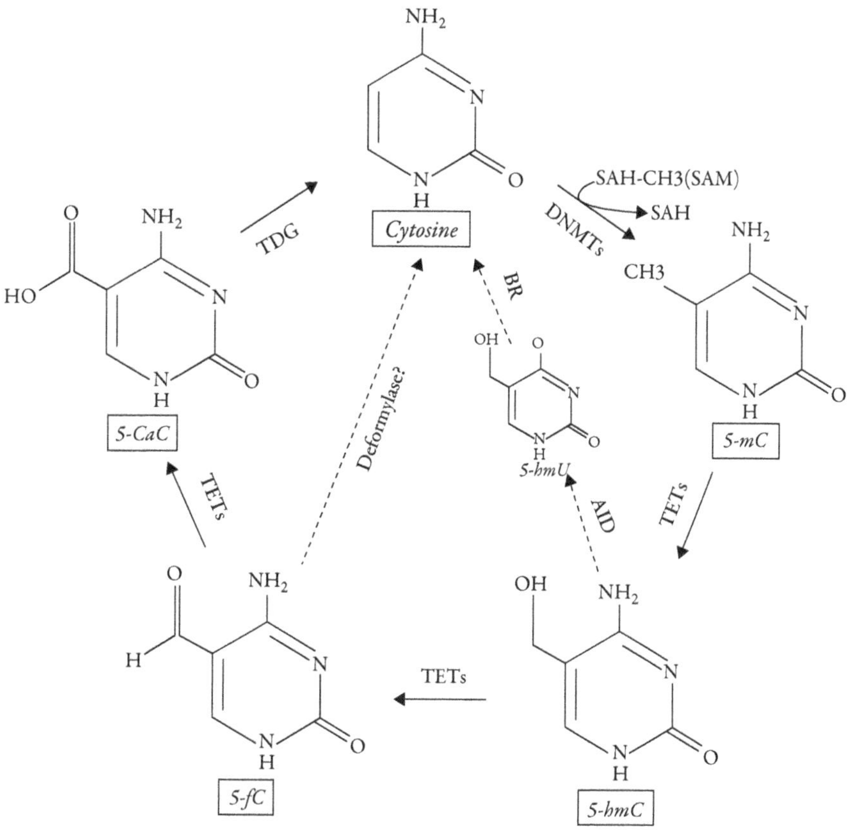

Figure 21.1 DNA methylation and demethylation process. Cytosine can be methylated to 5mC by DNMTs. 5mC can be oxidized to 5hmC, 5fC, and 5caC by ten-eleven-translocation (TET) proteins. 5caC can return to cytosine via base excision repair catalyzed by thymine DNA glycosylase (*TDG*). 5mC and 5hmC can also be demethylated via deamination by activation induced cytidine deaminase (AID)/apolipoprotein B mRNA editing enzyme, catalytic polypeptide (APOBEC).

Another noteworthy issue concerns the heterogeneous nature of neuropsychiatric disorders. SCZ is a complex disease with a broad range of symptoms including hallucination, delusion, thought abnormalities, cognitive impairment, mood dysorder, and a nxiety. Therefore, the accurate characterization of the phenotype heterogeneity and dynamicity may be important to unravel the genomic and epigenomic architecture of SCZ. The importance of phenotype definition and selection on GWAS findings has been recently highlighted.[65] Importantly, given the tissue and cell heterogeneity of the epigenetic marks, SCZ EWAS analyses are not feasible in sample sizes such as those recruited for genetic epidemiology. Therefore, it is perhaps even more crucial to overcome the challenges related to reduced statistical power resulting from phenotypic heterogeneity and dynamicity in epigenomic analyses of SCZ (see Figure 21.1).

CONCLUSION

In conclusion, DNA modifications play an important role in the development and functioning of the brain, although the current evidence for SCZ-associated epigenetic variation is limited, partly due to common methodological challenges in studying epigenetic modifications in mental disorders, including inadequate statistical power, reverse causation, and cellular heterogeneity. Future studies of epigenetic variation in SCZ should aim to (1) integrate multi-omic data across the regions and cell-types of the brain, and (2) incorporate dimensional clinical phenotyping of SCZ throughout the life span in conjunction with (preferably prospectively collected) data on environmental exposures across life as well as on (prospectively collected) epigenetic profiles on accessible tissue sources. This might identify regulatory genomic networks underlying onset, severity, and/or duration of clinical expression of SCZ, thereby providing novel insights into mediating and moderating biological processes that might inform future etiological studies, identify biomarkers, and nominate novel (drug, nutritional, or other) targets for interventions to prevent, attenuate, and/or treat SCZ.

REFERENCES

1. Reik W. Stability and flexibility of epigenetic gene regulation in mammalian development. *Nature.* 2007;447(7143):425–432.
2. Dolinoy DC, Weidman JR, Jirtle RL. Epigenetic gene regulation: linking early developmental environment to adult disease. *Reprod Toxicol.* 2007;23(3):297–307.
3. Ehrlich M, Wang RY. 5-Methylcytosine in eukaryotic DNA. *Science.* 1981;212(4501):1350–1357.
4. Lister R, Mukamel EA, Nery JR, et al. Global epigenomic reconfiguration during mammalian brain development. *Science.* 2013;341(6146):629–640.

5. Jang HS, Shin WJ, Lee JE, Do JT. CpG and non-CpG methylation in epigenetic gene regulation and brain function. *Genes (Basel).* 2017;8(6).
6. Kinde B, Gabel HW, Gilbert CS, Griffith EC, Greenberg ME. Reading the unique DNA methylation landscape of the brain: non-CpG methylation, hydroxymethylation, and MeCP2. *Proc Natl Acad Sci U S A.* 2015;112(22):6800–6806.
7. Ehrlich M, Gama-Sosa MA, Huang LH, et al. Amount and distribution of 5-methylcytosine in human DNA from different types of tissues of cells. *Nucleic Acids Res.* 1982;10(8):2709–2721.
8. Deaton AM, Bird A. CpG islands and the regulation of transcription. *Gene Dev.* 2011;25(10):1010–1022.
9. Yang XJ, Han H, De Carvalho DD, Lay FD, Jones PA, Liang GN. Gene body methylation can alter gene expression and is a therapeutic target in cancer. *Cancer Cell.* 2014;26(4):577–590.
10. Flores K, Wolschin F, Corneveaux JJ, Allen AN, Huentelman MJ, Amdam GV. Genome-wide association between DNA methylation and alternative splicing in an invertebrate. *BMC Genomics.* 2012;13.
11. Nan X, Campoy FJ, Bird A. MeCP2 is a transcriptional repressor with abundant binding sites in genomic chromatin. *Cell.* 1997; 88(4):471–481.
12. Amir RE, Van den Veyver IB, Wan M, Tran CQ, Francke U, Zoghbi HY. Rett syndrome is caused by mutations in X-linked MECP2, encoding methyl-CpG-binding protein 2. *Nature Genetics.* 1999;23(2):185–188.
13. Borrelli E, Nestler EJ, Allis CD, Sassone-Corsi P. Decoding the epigenetic language of neuronal plasticity. *Neuron.* 2008;60(6):961–974.
14. Day JJ, Sweatt JD. DNA methylation and memory formation. *Nat Neurosci.* 2010;13(11):1319–1323.
15. Azzi A, Dallmann R, Casserly A, et al. Circadian behavior is light-reprogrammed by plastic DNA methylation. *Nat Neurosci.* 2014;17(3):377–382.
16. Feng J, Chang H, Li E, Fan G. Dynamic expression of de novo DNA methyltransferases Dnmt3a and Dnmt3b in the central nervous system. *J Neurosci Res.* 2005;79(6):734–746.
17. Miranda-Morales E, Meier K, Sandoval-Carrillo A, Salas-Pacheco J, Vazquez-Cardenas P, Arias-Carrion O. Implications of DNA methylation in Parkinson's disease. *Front Mol Neurosci.* 2017;10:225.
18. De Souza RA, Islam SA, McEwen LM, et al. DNA methylation profiling in human Huntington's disease brain. *Hum Mol Genet.* 2016;25(10):2013–2030.
19. Martin LJ, Wong M. Aberrant regulation of DNA methylation in amyotrophic lateral sclerosis: a new target of disease mechanisms. *Neurotherapeutics.* 2013;10(4):722–733.
20. Lunnon K, Smith R, Hannon E, et al. Methylomic profiling implicates cortical deregulation of ANK1 in Alzheimer's disease. *Nature Neuroscience.* 2014;17(9):1164–1170.
21. Pishva E, Rutten BPF, van den Hove D. DNA methylation in major depressive disorder. *Adv Exp Med Biol.* 2017;978:185–196.
22. Guo JU, Su Y, Zhong C, Ming GL, Song H. Hydroxylation of 5-methylcytosine by TET1 promotes active DNA demethylation in the adult brain. *Cell.* 2011;145(3):423–434.
23. Irwin RE, Thakur A, O'Neill KM, Walsh CP. 5-Hydroxymethylation marks a class of neuronal gene regulated by intragenic methylcytosine levels. *Genomics.* 2014;104(5):383–392.
24. Kriaucionis S, Heintz N. The nuclear DNA base 5-hydroxymethylcytosine is present in Purkinje neurons and the brain. *Science.* 2009;324(5929):929–930.
25. Khare T, Pai S, Koncevicius K, et al. 5-hmC in the brain is abundant in synaptic genes and shows differences at the exon-intron boundary. *Nat Struct Mol Biol.* 2012;19(10):1037–U1094.
26. Spiers H, Hannon E, Schalkwyk LC, Bray NJ, Mill J. 5-hydroxymethylcytosine is highly dynamic across human fetal brain development. *BMC Genomics.* 2017;18.
27. Rutten BPF, Mill J. Epigenetic mediation of environmental influences in major psychotic disorders. *Schizophrenia Bull.* 2009;35(6):1045–1056.
28. Dempster EL, Pidsley R, Schalkwyk LC, et al. Disease-associated epigenetic changes in monozygotic twins discordant for schizophrenia and bipolar disorder. *Hum Mol Genet.* 2011;20(24): 4786–4796.
29. Boks MP, de Jong NM, Kas MJ, et al. Current status and future prospects for epigenetic psychopharmacology. *Epigenetics-US.* 2012;7(1):20–28.
30. Huang HS, Akbarian S. GAD1 mRNA expression and DNA methylation in prefrontal cortex of subjects with schizophrenia. *Plos One.* 2007;2(8).
31. Carrard A, Salzmann A, Malafosse A, Karege F. Increased DNA methylation status of the serotonin receptor 5HTR1A gene promoter in schizophrenia and bipolar disorder. *J Affect Disord.* 2011;132(3):450–453.
32. Zhang AP, Yu J, Liu JX, et al. The DNA methylation profile within the 5'-regulatory region of DRD2 in discordant sib pairs with schizophrenia. *Schizophrenia Research.* 2007;90(1–3): 97–103.
33. Mill J, Tang T, Kaminsky Z, et al. Epigenomic profiling reveals DNA-methylation changes associated with major psychosis. *Am J Hum Genet.* 2008;82(3):696–711.
34. Pidsley R, Zotenko E, Peters TJ, et al. Critical evaluation of the Illumina MethylationEPIC BeadChip microarray for whole-genome DNA methylation profiling. *Genome Biol.* 2016;17(1):208.
35. Pruitt KD, Tatusova T, Maglott DR. NCBI reference sequence (RefSeq): a curated non-redundant sequence database of genomes, transcripts and proteins. *Nucleic Acids Res.* 2005;33(Database issue):D501–504.
36. Dempster E, Viana J, Pidsley R, Mill J. Epigenetic studies of schizophrenia: progress, predicaments, and promises for the future. *Schizophrenia Bull.* 2013;39(1):11–16.
37. Nishioka M, Bundo M, Kasai K, Iwamoto K. DNA methylation in schizophrenia: progress and challenges of epigenetic studies. *Genome Med.* 2012;4(12):96.
38. Jeffries AR, Mill J. Profiling regulatory variation in the brain: methods for exploring the neuronal epigenome. *Biological Psychiatry.* 2017;81(2):90–91.
39. Mill J, Tang T, Kaminsky Z, et al. Epigenomic profiling reveals DNA-methylation changes associated with major psychosis. *Am J Hum Genet.* 2008;82(3):696–711.
40. Kinoshita M, Numata S, Tajima A, et al. DNA methylation signatures of peripheral leukocytes in schizophrenia. *Neuromolecular Med.* 2013;15(1):95–101.
41. Nishioka M, Bundo M, Koike S, et al. Comprehensive DNA methylation analysis of peripheral blood cells derived from patients with first-episode schizophrenia. *J Hum Genet.* 2013;58(2): 91–97.
42. Chen C, Zhang C, Cheng L, et al. Correlation between DNA methylation and gene expression in the brains of patients with bipolar disorder and schizophrenia. *Bipolar Disord.* 2014;16(8):790–799.
43. Numata S, Ye T, Herman M, Lipska BK. DNA methylation changes in the postmortem dorsolateral prefrontal cortex of patients with schizophrenia. *Front Genet.* 2014;5:280.
44. Wockner LF, Noble EP, Lawford BR, et al. Genome-wide DNA methylation analysis of human brain tissue from schizophrenia patients. *Transl Psychiatry.* 2014;4:e339.
45. Pidsley R, Viana J, Hannon E, et al. Methylomic profiling of human brain tissue supports a neurodevelopmental origin for schizophrenia. *Genome Biol.* 2014;15(10):483.
46. Liu J, Siyahhan Julnes P, Chen J, Ehrlich S, Walton E, Calhoun VD. The association of DNA methylation and brain volume in healthy individuals and schizophrenia patients. *Schizophr Res.* 2015;169(1–3):447–452.
47. Aberg KA, McClay JL, Nerella S, et al. Methylome-wide association study of schizophrenia: identifying blood biomarker signatures of environmental insults. *JAMA Psychiatry.* 2014;71(3):255–264.
48. Wockner LF, Morris CP, Noble EP, et al. Brain-specific epigenetic markers of schizophrenia. *Transl Psychiatry.* 2015;5:e680.
49. Montano C, Taub MA, Jaffe A, et al. Association of DNA methylation differences with schizophrenia in an epigenome-wide association study. *JAMA Psychiatry.* 2016;73(5):506–514.

50. Jaffe AE, Gao Y, Deep-Soboslay A, et al. Mapping DNA methylation across development, genotype and schizophrenia in the human frontal cortex. *Nat Neurosci.* 2016;19(1):40–47.
51. Hannon E, Spiers H, Viana J, et al. Methylation Quantitative Trait Loci (Mqtl) in the developing human brain and their enrichment in genomic regions associated with schizophrenia. *Eur Neuropsychopharm.* 2017;27:S307–S308.
52. Pardinas AF, Holmans P, Pocklington AJ, et al. Common schizophrenia alleles are enriched in mutation-intolerant genes and in regions under strong background selection. *Nat Genet.* 2018;50(3):381–389.
53. Schaub MA, Boyle AP, Kundaje A, Batzoglou S, Snyder M. Linking disease associations with regulatory information in the human genome. *Genome Res.* 2012;22(9):1748–1759.
54. Hannon E, Spiers H, Viana J, et al. Methylation QTLs in the developing brain and their enrichment in schizophrenia risk loci. *Nat Neurosci.* 2016;19(1):48–54.
55. van Os J, Kenis G, Rutten BP. The environment and schizophrenia. *Nature.* 2010;468(7321):203–212.
56. Rutten BPF, Vermetten E, Vinkers CH, et al. Longitudinal analyses of the DNA methylome in deployed military servicemen identify susceptibility loci for post-traumatic stress disorder. *Mol Psychiatry.* 2018;23(5):1145–1156.
57. Mill J, Heijmans BT. From promises to practical strategies in epigenetic epidemiology. *Nat Rev Genet.* 2013;14(8):585–594.
58. Pishva E, Kenis G, van den Hove D, et al. The epigenome and postnatal environmental influences in psychotic disorders. *Soc Psychiatry Psychiatr Epidemiol.* 2014;49(3):337–348.
59. Susser E, Neugebauer R, Hoek HW, et al. Schizophrenia after prenatal famine: further evidence. *Arch Gen Psychiatry.* 1996;53(1): 25–31.
60. Tobi EW, Lumey L, Talens RP, et al. DNA methylation differences after exposure to prenatal famine are common and timing- and sex-specific. *Hum Mol Genet.* 2009;18(21):4046–4053.
61. Jeffries AR, Mill J. Profiling regulatory variation in the brain: methods for exploring the neuronal epigenome. *Biol Psychiatry.* 2017;81(2):90–91.
62. PsychENCODE consortium, Akbarian S, Liu C, et al. The PsychENCODE project. *Nat Neurosci.* 2015;18(12):1707–1712.
63. Kundakovic M, Jiang Y, Kavanagh DH, et al. Practical guidelines for high-resolution epigenomic profiling of nucleosomal histones in postmortem human brain tissue. *Biol Psychiatry.* 2017;81(2): 162–170.
64. Kebir O, Chaumette B, Rivollier F, et al. Methylomic changes during conversion to psychosis. *Mol Psychiatry.* 2017;22(4):512–518.
65. Chaste P, Klei L, Sanders SJ, et al. A genome-wide association study of autism using the Simons simplex collection: does reducing phenotypic heterogeneity in autism increase genetic homogeneity? *Biological Psychiatry.* 2015;77(9):775–784.

22

ENDOPHENOTYPES

A WINDOW ON THE GENETICS OF SCHIZOPHRENIA

David Braff

The importance of endophenotype research in schizophrenia and psychosis studies is well established and rapidly evolving. This chapter focuses on schizophrenia endophenotypes but also explores the psychosis dimension. Endophenotypes are heritable biomarkers that inform us about key gene to phene pathways in psychosis. Since the endophenotype concept was introduced and then expanded by Gottesman and colleagues, endophenotypes have played a crucial role in advancing our understanding of the gene to phene pathways of schizophrenia.[1–4] Originally, Gottesman introduced the concept of endophenotypes as quantitative laboratory-based measures (neurocognitive, neurophysiological, neuroimaging, biochemical gene transcript measures, etc.) invisible to the naked eye but measures that augment and enhance the scope of clinical observation.

In this context endophenotypes are (1) heritable biomarkers which are quantitative and are observed in psychosis patients relative to healthy control subjects. In addition, (2) Gottesman posited that first-degree relatives of schizophrenia probands have quantitative endophenotype scores that are intermediate between the deficit scores of schizophrenia patients and normal values of unimpaired healthy control subjects. Interestingly, this second criteria is based on laboratory observation and is purely mathematical. It is not mechanistic in terms of why these intermediate values occur in relatives of schizophrenia patients but is a crucial quantitative observation. This intermediate score of relatives might be a result of incomplete penetrance of the genetic risk diathesis of schizophrenia, a lower polygenic risk profile, the result of high risk with protective genetic or environmental factors or many other interacting dynamic mammalian biological processes. And (3) these quantitative deficits in patients are primarily related to state rather than trait. Over time, it has appeared that there is a state (e.g., high levels of positive or negative symptoms) influence on endophenotype deficits, but there is also a strong stable trait related influence. In addition, (4) Gottesman thought that a reasonable criterion would also be cosegregation of the endophenotype with schizophrenia in families. This later component of the endophenotype definition is variably used in current research. One reason for this is that the boundary between schizophrenia and bipolar psychotic disorders is more fluid than previously thought. So cosegregation to a fuzzy diagnosis may be blurred since we may be dealing with a psychosis dimension where, for example, two diagnostic entities—schizophrenia and bipolar psychosis—are fused and confused in evaluating cosegregation (see BSNIP psychosis dimension discussion later). Thus, we are left primarily with endophenotypes as quantitative laboratory-based biomarkers, which have deficits in their quantitative scores in schizophrenia patients and also have values for first-degree relatives which are intermediate between the patients and normal controls. Relative state independency is also commonly observed. This is all quite salient since the use of quantitative measures rather than fuzzy case-control dichotomization yields up to 100 times greater statistical efficiency (associated with 10 times the power to detect), allowing for "better" gene finding via advanced statistical genetics methods.[5,6]

Gottesman's construct of endophenotypes has proved to be highly useful. Endophenotype research has increased dramatically since the concept was introduced and there are now thousands of endophenotype references in the literature, elucidating the neural and genomic substrates of schizophrenia and other neuropsychiatric disorders (e.g., see Figure 22.1). This growing harvest of endophenotype data and neurobiological studies is being produced in behavioral, clinical, and genomic domains.[7–11]

ON THE QUANTITATIVE NATURE OF ENDOPHENOTYPE RESEARCH

Endophenotypes tap into a plethora of neurocognitive, neurophysiological, neuroimaging, biochemical, gene transcript, and many other domains of function that produce quantitative measures where the psychotic patient has deficits and their first-degree relatives have intermediate values between those of patients and normal controls. This is a vast realm of quantitative endophenotypes that includes many of the most important dependent measures and biomarkers in psychosis research. It appears that the "knowledge to wisdom gap" in neuropsychiatric genetics is going to be narrowed by endophenotype research. We have many facts (Knowledge) but integrating them into a comprehensive picture of schizophrenia genetics and neural substrates (Wisdom) has proved profoundly challenging.[12,13] The endophenotype strategy is

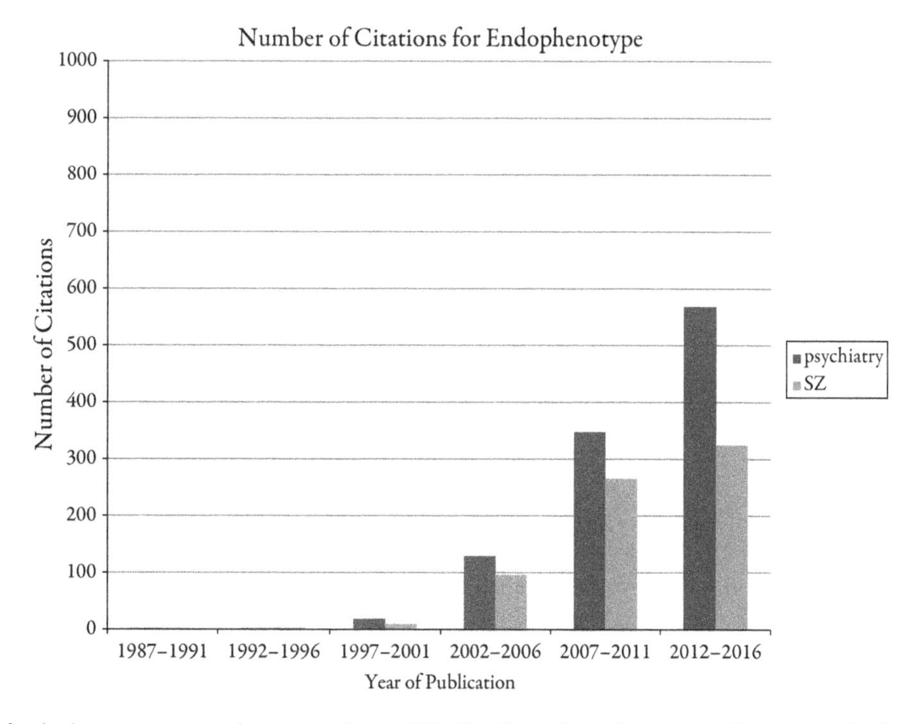

Figure 22.1 The number of endophenotype citations has increased since 1987. This figure shows the number of citations for "endophenotypes and psychiatry" and "endophenotypes and schizophrenia." As discussed in the text, many articles using endophenotypes (see PPI example in the text) are not included in this search, which is therefore an underestimate of the impact of the endophenotype strategy.

based on the strong inference that endophenotypes are genetically and behaviorally "simpler" than our broad diagnostic criteria and these laboratory-based observations occupy a crucial place in the gene to phene pathways of schizophrenia. It has been pointed out that anyone who has diagnostically interviewed "several hundred schizophrenia patients and then administered the letter number sequencing (LNS) working memory test to the same patients would undoubtedly say that yes endophenotypes are simpler than the disorder."[7] This is admittedly a face validity argument for the use of endophenotypes, and must be understood in a broad context, but it is salient. Because of their relative simplicity vs. the complexity of "whole human being DSM diagnosis," the ability to use quantitative endophenotypes to resolve many conundrums in schizophrenia research is reflected by a dramatic increase in citations from 1987 to the current time (see Figure 22.1).

This almost logarithmic-appearing increase in endophenotype research citations in Figure 22.1 is quite impressive. What is less apparent is that thousands of articles using endophenotypes do not use the term "endophenotype" and are lost in most literature searches for endophenotypes. This is reflected by the fact that for prepulse inhibition (PPI), a quantitative measure of sensorimotor gating, a well-known endophenotype, there are about 3,000 citations since the concept of sensorimotor gating was introduced into the neuropsychiatric literature.[14,15] But if you search for PPI endophenotype the number of citations is much, much smaller. Thus, even Figure 22.1 represents a significant underestimate of the use and influence of neurocognitive and neurophysiological, brain imaging, and other (e.g., inflammatory markers) endophenotypes.

ON INTERMEDIATE ENDOPHENOTYPE VALUES IN FIRST-DEGREE RELATIVES OF SCHIZOPHRENIA PATIENTS

As endophenotype research has expanded dramatically, some challenges have become apparent. For example when clinically unaffected first-degree relatives of schizophrenia patients have some level of endophenotype deficits, are these relatives really "normal" clinically? The answer, based on very well curated studies, is yes they are normal and do not have symptoms that meet the criteria for neuropsychiatric disorders in DSM 5.[16,17] It is important to stress this, lest family members of patients become "pathologized" needlessly based on quantitative biomarker scores. A second question is related to the lack of psychopathology and schizophrenia in so many of these relatives of schizophrenia patients. Is this intermediate endophenotype finding but the absence of diagnostic labels the result of a low level of cumulative endophenotypic deficits (see endophenotype ranking value [ERV] section later) or some other factor? Related to the question: could a "low dose" of endophenotypic loading actually increase genetic fitness and account for part of schizophrenia's persistence in the population over time? Ever since the interesting work by Jarvik and Chadwick on "The Odyssean Personality,"[18] it has been posited that perhaps carrying a "low level" of suspicion (paranoia) might be adaptive, especially in the long history of homo sapiens where in an ancient world populated by apex predators such as saber-toothed tigers, hypervigilance and high levels of possible harm detection might be adaptive fitness traits. One could posit that the saber-toothed tigers now are morphed into predatory human beings and that an increased "risk detector" might still be adaptive if one what is thinking of where one's children's might safely attend after school activities or where to invest money (cf. Bernie Madoff and the 2008 financial crisis of the US economy). In any case this seems to be a fascinating but hard to define research area of the possible fitness advantages of "disease genes" in neuropsychiatry. Certainly the fitness profile of other risk genes and endophenotypes of other medical disorders is simpler to understand (e.g., sickle cell anemia and its relationship to malaria resistance). Unfortunately our field of endeavor is much more complex than hematology. We are dealing with the brain: 75 billion neurons arranged in complex circuits housed in a protective cranial vault making in vivo studies and the detection of adaptive evolutionary advantages quite difficult to directly measure.

Another explanation for some level of endophenotype deficits in first-degree relatives is that perhaps the relatives of first-degree relatives of schizophrenia patients simply have a subthreshold cumulative level of endophenotype and/or risk gene and environmental (GXE) loading (see ERV section later). In addition, it is important to note that risk genes may interact with protective genes (G) and/or protective environmental (E) factors, which reflects the dynamic interplay of multiple risk and protective factors in the stress diathesis model proposed by Zubin many years ago,[19] now commonly invoked in the familiar Gene X Environment (GXE) paradigm of neuropsychiatric disease cumulative risk assessment.

ENDOPHENOTYPES ARE "CONSORTIA FRIENDLY" AND ARE PLATFORMS FOR COLLABORATIONS

A major advantage of the endophenotype strategy is that they can be employed both in individual laboratories and also across Consortia to increase sample size. In fact, currently there are many well-quality-assured and well-curated family and even case-control multisite studies of endophenotypes—to cite just a few: Multiplex Multigenerational Investigation of Schizophrenia (MGI), Project Among African-Americans to Explore Risks for Schizophrenia (PAARTNERS), Bipolar-Schizophrenia Network for Intermediate Phenotypes (BSNIP), Consortium on the Genetics of Schizophrenia (COGS), Cognitive Genomics Consortium (COGENT), Psychosis Endophenotype International Consortium (PEIC), the NIMH/Lieber group, Australian Schizophrenia Research Bank and many other projects reflecting the high level of interest in using the endophenotype strategy in many laboratories around the world.[20–25] These studies also afford the opportunity for key "spin off" collaborative efforts with outstanding laboratories in various populations (simplex, multiplex, African American families). For example, a COGS study done with MC King and J McClellan and colleagues at the University of Washington using COGS trios showed that there was a network of prefrontal early neurodevelopmental linked de novo mutations in genetic sequencing that are associated with schizophrenia.[26] This study was made possible since COGS had family trios so we could identify which endophenotype-deficit-laden probands had harmful mutations not found in their parents. In concert, similar COGS samples along with those from MGI and PAARTNERS provided subjects for methylation studies done with Andrew Feinberg's laboratory at Johns Hopkins and showed an interesting result indicating that endophenotype-deficit-laden individuals had an excess of methylation marks obtained via the CHARM analyses developed at Johns Hopkins School of Medicine (JHSM).[27] Environmentally (E) generated methyl markers are an important vantage point in studying the neurobiological effects of epigenetic events. These collaborations stimulated the organization of TOSCA: A Trio Of Schizophrenia ConsortiA consisting of COGS, MGI, and PAARTNERS. These important collaborative studies also reflect the fact that looking at DNA alone without examining endophenotypes, family, neurodevelopmental, and environmental factors is limited when one begins to explore the full range of genomic underpinnings of schizophrenia and neurocognition, neurophysiology, imaging, and even real-world function. This elegant tapestry of endophenotype-guided research presents a puzzle that is still being explored. We have discovered much but are still early in our journey of explore the genomics and neurobiology of schizophrenia as discussed in the conclusion of this chapter.[12,13]

TWO (OF MANY) CONSORTIA USING ENDOPHENOTYPES IN PSYCHOSIS

THE CONSORTIUM ON THE GENETICS OF SCHIZOPHRENIA (COGS)

COGS-1 is a family-based endophenotype study, and COGS-2 is a follow-up case-control study. For COGS-1, Braff, Freedman, Schork, and Gottesman described the strategy of the seven-site family study.[4] Families were ascertained by an affected schizophrenia proband who had at least one unaffected sibling and also both parents available and consenting for an extensive battery of neurocognitive and neurophysiological testing. Nested within the COGS-1 study of over 300 families were trios of both parents and at least one affected (and one unaffected) sibling. This allows investigators to parse de novo from common ancient mutations in schizophrenia, a distinction that has been discussed extensively in the literature. As Michael Owen et al. point out, both de novo and common mutations occur and contribute to the risk of schizophrenia.[28] Their insight into resolving this argument of de novo versus common mutations is very important and could easily be extended to other straw man debates in the world of neuropsychiatric genomics. The reality is that in this space which is full of conundrums and challenges, there is more than enough room for multiple approaches.

The association studies of the 12 primary COGS neurocognitive and neurophysiological endophenotypes were selected a priori based on the extant literature on found gene-endophenotype associations in a 42-gene glutamate-related network that has an ERBB4-NRGL hub.[16,17] These two genes are intimately related to glutamate neurotransmission as well as other synaptic and neurodevelopmental processes and have highly pleiotropic significant associations to 9 of 12 endophenotypes (see Figure 22.2). It is important to note that

Chr	Gene	Prepulse inhibition	P50 suppression	Antisaccade	Continuous performance	Letter-number span	Verbal learning	Abstraction	Face memory	Spatial memory	Spatial processing	Sensorimotor dexterity	Emotion recognition
1p34.3	GRIK3	*								*			*
1q21.3	SNAPAP				*								
1q23.3	**NOS1AP**	*					*			*		*	**
1q31.3	ASPM											*	
1q42.2	**DISC1**		**	*	*	*							
2p12	**CTNNA2**				*	*		*			*		
2q34	**ERBB4**		*	*		*	*	*	*	**		*	
3p25.3	SLC6A1						*		*				
3q13.31	DRD3												*
4q22.3	**GRID2**	**	*			*					*	*	*
5p15.3	SLC6A3								*		*	*	
5q12.3	HTR1A						*						
5q32	HTR4											**	
5q32	CAMK2A	**				**							
5q34	GABRB2							*					
6p21.3	GRM4										*		**
6q23.2	TAAR6				*								
6q24.3	**GRM1**				*		*	*	*	**	*	*	
6q25.1	ESR1								*		*		
7p15.3	SP4					*		*			*		
7q21.1	GRM3		**										
7q22.1	**RELN**			**	*				*		*	*	**
8p21.3	SLC18A1			*									
8p12	**NRG1**	*					**	**			***	*	
9q31.1	GRIN3A	*											
9q34.2	DBH	*											
9q34.3	GRIN1											**	
10q23.2	GRID1				*							**	
11p13	SLC1A2				**			*		*			
11q14.3	GRM5								*			**	
11q23.1	NCAM1	**											
11q23.2	DRD2			*									
11q23.3	GRIK4		*	***			*						
12p13.1	GRIN2B						*						**
12q24.1	DAO											**	
13q14.2	HTR2A			**		*					*		
14q32.3	AKT1								*				
15q13.3	CHRNA7	**											
16p13.2	GRIN2A												*
17p13.3	YWHAE												*
17p13.1	DLG4					*					*		
17q21.3	CRHR1			*									
20q11.2	SLC32A1							*					
20q13.3	CHRNA4										*	***	
22q11.2	PRODH			*									
22q11.2	COMT	**					*		*				

Figure 22.2 Genes associated significantly with COGS endophenotypes using the total significance test (TST) to account for covariates and family wise error. $^{*}p < 0.01$; $^{**}p < 0.001$; $^{***}p < 0.0001$

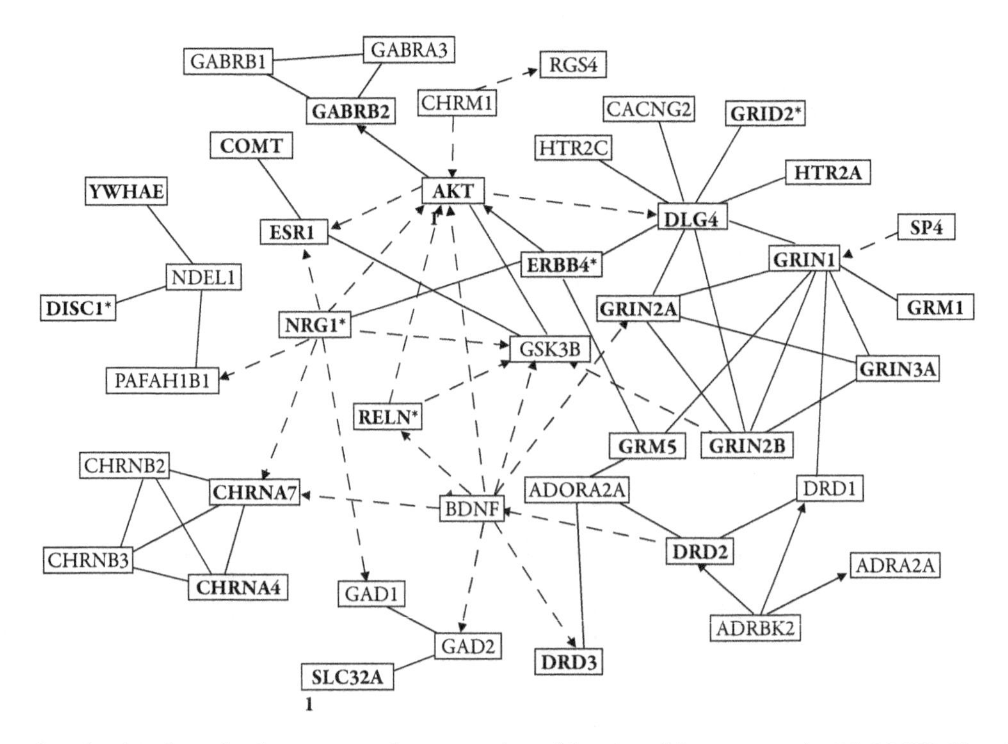

Figure 22.3 Pathway analysis detailing the molecular interactions between a subset of the 94 candidate genes on the COGS SNP Chip. Genes are represented as nodes, and the molecular interactions between nodes are represented by lines and arrows, with solid lines representing direct protein–protein or protein–DNA interactions, solid arrows representing phosphorylation, and dashed arrows representing indirect effects on expression, activation, or inhibition. Gene functions and relationships were determined by Ingenuity Pathway Analysis. Genes associated ($p < 0.01$) with at least one endophenotype are shown in bold, with an asterisk (*) indicating those genes associated with >3 endophenotypes.

understanding complex gene–gene networks and interactions in this complex database as reflected in Figure 22.2 is challenging, but that these challenges can be met. Laura Lazzeroni developed a novel total significance test (TST) for looking concurrently at these multiple genes and neurocognitive and neurophysiological endophenotypes and accounting simultaneously for key covariates. This allows for both family wise error corrections and the inclusion of important covariates such as age. The COGS initial results and related extended studies[16,17] allowed us to specify the 42-gene network underlying key schizophrenia related endophenotypic deficits. Figure 22.3 illustrates how we were able to identify a glutamate-related gene network underlying key endophenotypes in cognitive and information-processing deficits in schizophrenia patients.

THE BIPOLAR-SCHIZOPHRENIA NETWORK FOR INTERMEDIATE PHENOTYPES (BSNIP)

The Bipolar-Schizophrenia Network for Intermediate Phenotypes (BSNIP) Consortium has probed the "psychosis dimension" and examines both schizophrenia and bipolar psychosis patients. BSNIP is a five-site, meticulously curated study and explores the concept of "psychotic disorders" rather than separate schizophrenia and bipolar psychosis disorders. Conceptually this project uses a dimension of psychosis rather than individual diagnoses. BSNIP colleagues have pioneered this approach as being a good strategy for examining the genetics and endophenotypic profiles of multiple psychotic disorders.[29] Interestingly, supporting the dimensional view, the BSNIP family trees reflected virtually equal numbers of pure single diagnosis and mixed schizophrenia and bipolar diagnosis patients.[22] Using the BACS cognitive score, the BSNIP investigators showed that the cognition, ocular motor smooth pursuit and antisaccade performance,[30] and regional brain volume[31] were all very similar in distribution across all psychotic subjects, suggesting a severity continuum that is not based on distinct DSM-5 diagnosis but is based on quantitative endophenotype scores. Supporting the psychosis dimension interpretation, single endophenotypes only sorted in a weak way across conventional diagnoses. BSNIP scholars concluded that conventional diagnoses, strongly grounded in subjective phenomenology and qualitative impressions, are less compelling and generate less confidence in group separation compared with quantitative endophenotype-based categorization. The independent brain characteristics captured stronger quantifiable traits, which can be used in characterizing brain diseases and may override clinical phenomenology traits, which are qualitative.[29]

The BSNIP group organized disease units around endophenotypic characteristics of the probands and generated biologically informative outcomes that are now being tested in follow-up studies.[8,32] Thus, endophenotype-based criteria can be used to create categories that are informative and biologically homogeneous and can to be used in both molecular discovery and in clinical trials. In terms of BSNIP alone, it is possible that these endophenotypic observations and the identification of three distinct cluster-derived "biotypes" may be a first approximation that is needed to better describe the neurobiology of broad dimensional psychotic disorders. This represents a proven principle that neurobiological constructed groups of psychotic individuals can vary independently of their overall diagnosis and challenges conventional diagnostic labels. While this disease categorization is not ready for "primetime usage" for clinical diagnosis it does offer a window onto the future of how we might proceed with heuristic diagnostic entities that are linked more closely to underlying genomic, neural substrate and clinical outcome domains. As has been pointed out by Braff and Tamminga, "after all it is the quantification of hematocrit, not the level of patient fatigue which is the monitor for anemia pathophysiology—although fatigue is important for patient care."[7] Thus, the BSNIP group and others envision a neuropsychiatric field in which quantitative endophenotypic measures of the psychosis dimension might well be very useful, generative, and accurate measures of disease genetics, neurobiology, and severity, and also may determine treatment response.

ENDOPHENOTYPES AND THE NIMH RESEARCH DOMAIN CRITERIA (RDOC)

Endophenotypes are uniquely positioned to answer questions raised by the RDoC initiative of NIMH.[33] Before the RDoC was formalized, endophenotypes were tapping into crucial RDoC described domains. Fundamentally, endophenotypes offer a significant statistical genetics and conceptual clarity and behavioral assessment advantage over fuzzy DSM-based qualitative clinical diagnostic phenotypes in tapping into RDoC domains of function. This is because diagnosis is based on clinical impression and there are few strongly validated quantitative measures used to arrive at DSM diagnoses. It is important to point out that endophenotypes are strategically important quantitative measures as indicated by the title of the Insel and Cuthbert 2000 commentary, "Endophenotypes: Bridging Genomic Complexity and Disorder Heterogeneity."[33] A well-studied endophenotype, PPI, is the second reference in this paper, which proposes the use of endophenotypes in the very complex world of disorder gene finding in order to understand the heterogeneity of neuropsychiatric disorders. Also, PPI genetics has been studied extensively in rodents and humans and the genes and neural circuits underlying PPI deficits are fairly well known.[34] Thus, Gottesman's concept of endophenotypes preceded the RDoC criteria and offered a foundation for its conceptualization. The concept of using neurobiological research domains based on endophenotypes is useful and part of a continuing strategy for unraveling the mysteries of psychosis itself.

ENDOPHENOTYPES AND THE ENDOPHENOTYPE RANKING VALUE SYSTEM

Glahn, Almasy, Blangero, and colleagues have published a number of important papers outlining a strategy to create high dimensional endophenotype ranking for risk genes in major neuropsychiatric disorders.[35] Their premise is that gene finding is enhanced by using quantitative endophenotypes. In the family context, quantitative endophenotypes have up to 100 times the efficiency of case-control studies for gene finding for a disorder such as schizophrenia with a 1% prevalence. This translates into a 10-fold increase in efficiency (i.e., fewer subjects are needed) in sample size, so an N of 5,000 with quantitative endophenotype measures has the gene-finding power of an N of 50,000 DSM-identified case-control subjects. Thus endophenotypes offer a substantial statistical genetics advantage in power needed for gene finding vs. traditional diagnoses.

The ERV authors document success in utilizing the ERV technique for complex illnesses.[35,36] They convincingly show that family-based observations to assess both the heritability of endophenotypes and genetic correlation with disease liability is a practical and powerful strategy. This reinforces the idea that family studies combined with quantitative endophenotypes interact together to greatly increase our power to detect and rank the importance of neuropsychiatric endophenotypes. They essentially rank endophenotype values via the ERV, which is a measure of the genetic utility of endophenotypes for understanding the neurobiology of illnesses. It is interesting that the ERV endophenotype ranking system (wittily named after <u>Irv</u> Gottesman, the founder of the endophenotype concept in neuropsychiatry) provides an unbiased and empirically derived method that allows us to choose high-value endophenotypes in a manner that balances the strength of the genetic signal of endophenotypes and the strength of its relation to the risk genes of the disorder of interest. The authors have defined ERV by using the square root of the heritability of illness X, the square root of the heritability of its endophenotype, and their genetic correlation, and it is expressed by the formula that they have discussed in several papers ($ERV_{ie} = |\sqrt{h_i^2}\sqrt{h_e^2}\rho_g|$). These authors have applied their ERV analyses to high-dimensional sets of traits and have ranked significant endophenotypes for various neuropsychiatric disorders. In their 2012 paper the authors performed an automated high dimensional search for endophenotypes ranking 37 wide ranging behavioral/neurocognitive candidate endophenotypes for manic depressive disorder (MDD) including measures from neurocognition to gene transcripts, illustrating the power of the ERV formulation. ERV values vary between 0 and 1, where higher values indicate that the endophenotype and the illness are more strongly influenced by *shared* genetic factors.[35] Thus, ERV reflects that endophenotypes tap into an illness with significant pleiotropy and reduces the heterogeneity of the disease and can focus on the shared genetic factors that influence both the illness and the endophenotype. As described earlier, many endophenotypes can be assessed. This high-throughput ERV methodology has many advantages, and the authors indicate that it is similar to screening methods developed for drug discovery, although the differences are also clear. Specifically, drug discovery screening can be conducted in a totally agnostic way, since repurposing of drugs is possible in a high-throughput environment. In contrast, analyses of multiple variables at once including endophenotypes and heritable neuropsychiatric disorders entails significant a priori selection based on existing research. In addition, one could create a table of endophenotype risk based on the ERV and then create total endophenotype risk load, analogous to polygenic risk load. One could even relate endophenotype risk load to polygenic risk load and then see how they relate to one another. The ERV approach is well positioned to inform psychosis research and gene/risk finding in coming years.

ENDOPHENOTYPES AND THE MATRICS INITIATIVE

The Measurement and Treatment Research to Improve Cognition in Schizophrenia (MATRICS) initiative was designed to identify neurocognitive targets for new drug development.[37] The rationale was that since it is well established that neurocognitive impairment correlates with poor social and vocational function and outcome, identifying specific schizophrenia- related neurocognitive targets for new drug treatment that would improve functional outcome was feasible.[37,38] The MATRICS identified seven domains of neurocognitive function which are targets for new drug development in order to reverse cognitive deficits and improve both vocational and social functioning in schizophrenia patients. The goal was that the development of a consensus regarding the most promising neurocognitive targets would lead to new molecular targets from this project. Ultimately the MATRICS domains that are fundamentally impaired in schizophrenia and are also FDA-endorsed targets for new treatments include verbal learning and memory, visual learning and memory, working memory, attention and vigilance, processing speed, problem-solving, reasoning, and social cognition.

Endophenotypes used in large neurocognitive and endophenotype studies such as the COGS, PAARTNERS, MGI, and other consortia named earlier, assess some or all of the seven domains of neurocognitive functioning that are identified in the MATRICS initiative. Thus, one way of looking at the vast literature on endophenotypes is that they both identify MATRICS-related neurocognitive deficits and also, via candidate gene, Genome wide Association Study (GWAS), and other genetic platforms, identify potential molecular targets for remediation of these deficits. Thus, with genomic loci identified that are associated with and have control over the neurocognitive domains identified in the MATRICS initiative, we hypothesize that precision-based medicine might advance individualized treatments based on ERV risk profiles. In addition, although the MATRICS was designed to look at neuropharmacological treatments, it is clear that "rewiring the brain" treatments based on cognitive-behavior therapy and sensory training are also extremely important in advancing our armamentarium of treatments of neurocognitive deficits. For example, the work of Vinogradov, Merzenich, and colleagues[39] on sensory training and rewiring the brain is very important. In this context, sensory training is given in 40 sessions and even 6 months later results in improved cognition, which is identified as an FDA-endorsed MATRICS target. Drug treatments do not have this long-lasting effect after discontinuation, and it is achieved with a cognitive intervention that has no biological side effects—a very impressive accomplishment. This very important finding is sometimes overlooked in the treatment literature, which has a preponderance of drug-related studies. In addition, it is of note that endophenotypes, such as electrophysiological EEG-based mismatch negativity (MMN), where MMN deficits are associated with poor functional outcome, might also identify individuals who are prone to developing neurocognitive deficits.[40] The EEG MMN can even be measured very early in mammalian development, and in humans this MMN, an automated "difference detector" brainwave to salient stimuli, can actually be measured from fetuses in utero.[41] Obviously, this is an extreme example of the possible value of neurocognitive and neurophysiological endophenotypes early in mammalian development, since it is unlikely that in utero monitoring and measurement of high-risk fetuses will lead to any drug treatment interventions in the near or even moderately distant future. But high-risk children might well benefit from no side effect, neurocognitive interventions early in life. The whole area of risk detection via endophenotype deficits is rapidly expanding. The link between quantitative neurocognitive endophenotypes with their genetic associations and MATRICS identified neurocognitive treatment targets offers many possibilities for precision-based treatment pathways as discussed briefly in this chapter.

CONCLUSIONS AND FUTURE DIRECTIONS

Neurobiologically informed quantitative endophenotype strategy guided studies are complementary to GWAS and Whole Genome Sequencing (WGS) case control studies. They are rooted in the fertile soil of neurobiological measures that we know are related to disorder pathogenesis, risk genes, and etiology. In addition, heritable quantitative endophenotypes are able to be analyzed through the prism of quantitative statistical genetics methods as illustrated earlier.[5] Originally, Gottesman proposed the quantitative endophenotype strategy in order to resolve many challenges entailed in identifying risk genes verses the use of qualitative DSM diagnoses of psychosis disorders. The promise of the endophenotype strategy, has been partially realized, as quantitative endophenotypes have identified the genetic basis of such areas as glutamate neurotransmission deficits underlying neurocognitive and neurophysiological endophenotypes in schizophrenia.[7,16,17] Ultimately GWAS and sequencing studies of schizophrenia will to link up with laboratory-based quantitative endophenotype measures and clinical real-world function to "close the loop" on understanding clinically important genes, gene networks, and pathways.

Endophenotypes offer an important bridge between genomic heterogeneity and psychosis. Endophenotypes occupy an important position on the gene to phene pathway and are more proximal to risk genes than are traditional disorder diagnoses. In addition, quantitative endophenotypes offer significant advantages for psychosis gene finding versus fuzzy case-control designs.[13] Endophenotype–gene correlations and associations can be utilized in schizophrenia gene network construction as illustrated earlier (Figure 22.2 and Figure 22.3). These risk networks may provide molecular targets for precision-based drug and/or behavioral treatment interventions in the future. Future directions for neuropsychiatric genomics: What is the likely time course for the future of therapeutic advances based on endophenotypes and other genetic studies in neuropsychiatric disorders? Even for a genetically and neuronally simple neuropsychiatric disorder such as Huntington's disease (HD), where we know the simple Mendelizing mode of heritability, the exact neural substrates that are targeted by HD, and even how increased numbers of CAG trinucleotide repeats have a linear relationship to earlier age of onset (and more penetrant amine toxicity), we have no effective treatments, not to even mention cures. We have only the ability to have genetic counselors tell gene carriers that there is a 50–50 probability of passing the disease on to their offspring. Thus it is reasonable to advise the public that most common complex psychiatric disorders will take a relatively long time to reach truly new powerful "actionable" therapeutic strategies for schizophrenia and other psychotic patients seeking new therapies. But we must also remember that existing drug, cognitive, and psychosocial therapies, while not ideal, are quite helpful to our patients.

There are three likely pathways for the journey of dramatic new scientific discovery: (1) serendipity: an "aha" moment of novel intuition, e.g., the discovery of chlorpromazine as an antipsychotic; (2) a novel, lightning-strike "Kuhnian" insight that changes a field (e.g., Darwin and Mendel's work); and (3) relatively incremental advances, the path for most of the science advances now being conducted.[12] As Sam Barondes insightfully said about current gene discovery in neuropsychiatric genetics, "this is the first step in a journey of a thousand miles."[42] This view of the likely long time course facing us in gene and endophenotype discovery to therapeutics in schizophrenia research has been reinforced by Eric Landers and Francis Collins.[43,44] This fact should not discourage us: we will have many exciting opportunities to advance our field. One or another insight or finding may even have a unique, field-changing Kuhnian impact. The foundations being constructed now should have long-term rewards.

Final thoughts: The pathways from gene discovery to therapeutic advances in schizophrenia and psychosis are uniquely amenable to the endophenotype strategy as discussed earlier. The vulnerability to psychosis is highly polygenic and involves the neurobiological output of the 75 billion neurons and their circuits encased in our cranial vaults. The outputs of long-line neural circuits and shorter line micro circuits will have to be examined via many converging strategies in the coming years. Clearly, the use of quantitative endophenotypes is one powerful complementary strategy (other strategies for risk gene discovery are discussed in this comprehensive volume). The endophenotype strategy is one crucial methodology that needs to be fully utilized as we attempt to relieve the suffering of our patients and their families from the highly heterogeneous, somewhat mystifying and very challenging psychotic disorders that uniquely afflict human beings.

REFERENCES

1. Gottesman II, Shields J. Genetic theorizing and schizophrenia. *Br J Psychiatry.* 1973;122(566):15–30.
2. Gottesman II, Erlenmeyer-Kimling L. Family and twin strategies as a head start in defining prodromes and endophenotypes for hypothetical early-interventions in schizophrenia. *Schizophr Res.* 2001;51(1):93–102.
3. Gottesman II, Gould TD. The endophenotype concept in psychiatry: etymology and strategic intentions. *Am J Psychiatry.* 2003;160(4):636–645.
4. Braff DL, Freedman R, Schork NJ, Gottesman II. Deconstructing schizophrenia: an overview of the use of endophenotypes in order to understand a complex disorder. *Schizophr Bull.* 2007;33(1):21–32.
5. Blangero J, Williams JT, Almasy L. Robust LOD scores for variance component-based linkage analysis. *Genet Epidemiol.* 2000;19(Suppl 1):S8–S14.
6. Almasy L, Blangero J. Multipoint quantitative-trait linkage analysis in general pedigrees. *Am J Hum Genet.* 1998;62(5):1198–1211.
7. Braff DL, Tamminga CA. Endophenotypes, epigenetics, polygenicity and more: Irv Gottesman's dynamic legacy. *Schizophr Bull.* 2017;43(1):10–16.
8. Braff DL, ed. Endophenotypes in schizophrenia [Special issue]. *Schizophr Res.* 2015;163(1–3):1–80.
9. Gur RE, Nimgaonkar VL, Almasy L, et al. Neurocognitive endophenotypes in a multiplex multigenerational family study of schizophrenia. *Am J Psychiatry.* 2007;164:813–819.
10. Tan H-Y, Callicott JH, Weinberger DR. Intermediate phenotypes in schizophrenia genetics redux: is it a no brainer? *Mol Psychiatry.* 2008;13:233–238.
11. Ivleva EI, Morris DW, Moates AF, Suppes T, Thaker GK, Tamminga CA. Genetics and intermediate phenotypes of the schizophrenia—bipolar disorder boundary. *Neurosci Biobehav Rev.* 2010;34(6):897–921.
12. Braff L, Braff DL. The neuropsychiatric translational revolution: still very early and still very challenging. *JAMA Psychiatry.* 2013;70(8):777–779.
13. Braff DL. NIMH Neuropsychiatric genomics: crucial foundational accomplishments and the extensive challenges that remain. *Mol Psychiatry.* In press.
14. Braff DL, Stone C, Callaway E, Geyer M, Glick I, Bali L. Prestimulus effects on human startle reflex in normals and schizophrenics. *Psychophysiology.* 15:339–343, 1978.
15. Graham FK, Putnam LE, Leavitt LA. Lead-stimulation effects of human cardiac orienting and blink reflexes. *J Exp Psychol Hum Percept Perform.* 1975;104(2):175–182.
16. Greenwood TA, Lazzeroni LC, Murray SS, et al. Analysis of 94 candidate genes and 12 endophenotypes for schizophrenia from the Consortium on the Genetics of Schizophrenia. *Am J Psychiatry.* 2011;168:930–946.
17. Greenwood TA, Light GA, Swerdlow NR, Radant AD, Braff DL. Association analysis of 94 candidate genes and schizophrenia-related endophenotypes. *PLoS One.* 2012;7(1):e29630.
18. Jarvik LF, Chadwick SB. Schizophrenia and survival. In: Hammer, Salzinger, and Sutton, eds. *Psychopathology.* New York, NY: Wiley; 1972: 57–73.
19. Zubin J, Spring B. Vulnerability—a new view of schizophrenia. *J Abnor Psychol.* 1977;86(2):103–126.
20. Gur RE, Nimgaonkar VL, Almasy L, et al. Neurocognitive endophenotypes in a multiplex multigenerational family study of schizophrenia. *Am J Psychiatry.* 2007;164:813–819.
21. Aliyu MH, Calkins ME, Swanson CL Jr, et al. PAARTNERS Study Group. Project among African-Americans to explore risks for

schizophrenia (PAARTNERS): recruitment and assessment methods. *Schizophr Res.* 2006;87(1–3):3–44.

22. Tamminga CA, Ivleva EI, Keshavan MS, et al. Clinical phenotypes of psychosis in Bipolar-Schizophrenia Network on Intermediate Phenotypes (B-SNIP). *Am J Psychiatry.* 2013;170(11):1263–1274.
23. Braff DL, ed. The use of endophenotypes to deconstruct and understand the genetic architecture, neurobiology, and guide future treatments of the group of schizophrenias [Special issue]. *Schizophr Bull.* 2007;33(1).
24. Lencz T, Knowles E, Davies G, et al. Molecular genetic evidence for overlap between general cognitive ability and risk for schizophrenia: a report from the Cognitive Genomics consorTium (COGENT). *Mol Psychiatry.* 2014;19:168–174.
25. Loughland C, Draganic D, McCabe K, et al. Australian Schizophrenia Research Bank: a database of comprehensive clinical, endophenotypic, and genetic data for aetiological studies of schizophrenia. *Aust N Z J Psychiatry.* 2010;44(11):1029–1035.
26. Gulsuner S, Wash T, Watts AC, et al. Spatial and temporal mapping of de novo mutations in schizophrenia to a fetal profrontal cortical network. *Cell.* 2013;154(3):518–529.
27. Montano C, Taub M, Jaffe A, et al. Association of DNA methylation differences with schizophrenia in an epigenome-wide association study. *JAMA Psychiatry.* 2016;73(5):506–514.
28. Owen MJ, Craddock N, O'Donovan MC. Suggestion of roles for both common and rare risk variants in genome-wide studies of schizophrenia. *Arch Gen Psychiatry.* 2010;67(7):667–673.
29. Keshavan MS, Clementz BA, Pearlson GD, Sweeney JA, Tamminga CA. Reimagining psychoses: an agnostic approach to diagnosis. *Schizophr Res.* 2013;146(1–3):10–16.
30. Lencer R, Sprenger A, Reilly JL, et al. Pursuit eye movements as an intermediate phenotype across psychotic disorders: evidence from the B-SNIP study. *Schizophr Res.* 2015;169(1–3):326–333.
31. Ivleva EI, Bidesi AS, Keshavan MS, et al. Gray matter volume as an intermediate phenotype for psychosis: Bipolar-Schizophrenia Network on Intermediate Phenotypes (B-SNIP). *Am J Psychiatry.* 2013;170(11):1285–1296.
32. Tamminga CA, Pearlson GD, Stan AD, et al. Strategies for advancing disease definition using biomarkers and genetics: the Bipolar and Schizophrenia Network for Intermediate Phenotypes. *Biol Psychiatry.* In press.
33. Insel T, Cuthbert B. Endophenotypes: bridging genomic complexity and disorder heterogeneity. *Biol Psychiatry.* 2009;66(11):988–989.
34. Swerdlow NR, Braff DL, Geyer MA. Sensorimotor gating of the startle reflex: what we said 25 years ago, what has happened since then, and what comes next. *J Psychopharmacol.* 2016;30(11):1072–1081.
35. Glahn DC, Curran JE, Winkler AM, et al. High dimensional endophenotype ranking in the search for major depression risk genes. *Biol Psychiatry.* 2012;71:6–14.
36. Glahn DC, Williams JT, McKay DR, et al. Discovering schizophrenia endophenotypes in randomly ascertained pedigrees. *Biol Psychiatry.* 2015;77(1):75–83.
37. Green MF, Nuechterlein KH, Gold JM, et al. Approaching a consensus cognitive battery for clinical trials in schizophrenia: the NIMH-MATRICS conference to select cognitive domains and test criteria. *Biol Psychiatry.* 2004;56(5):301–307.
38. Green MF, Nuechterlein KH. The MATRICS initiative: developing a consensus cognitive battery for clinical trials. *Schizophr Res.* 2004;72(1):1–3.
39. Fisher M, Holland C, Merzenich MM, Vinogradov S. Using neuroplasticity-based auditory training to improve verbal memory in schizophrenia. *Am J Psychiatry.* 2009;166(7):805–811.
40. Light GA, Braff DL. Mismatch negativity deficits are associated with poor functioning in schizophrenia patients. *Arch Gen Psychiatry.* 2005;62:127–136.
41. Draganova R, Eswaran H, Murphy P, Huotilainen M, Lowery C, Preissl H. Sound frequency change detection in fetuses and newborns, a magnetoencephalographic study. *Neuroimage.* 2005;28: 354–361.
42. Barondes S, quoted in Carey B (2016, January 27). Scientists move closer to understanding schizophrenia's cause. *New York Times.* Retrieved from https://www.nytimes.com/2016/01/28/health/schizophrenia-cause-synaptic-pruning-brain-psychiatry.html
43. Lander E, quoted in Carey B (2016, January 27). Scientists move closer to understanding schizophrenia's cause. *New York Times.* Retrieved from https://www.nytimes.com/2016/01/28/health/schizophrenia-cause-synaptic-pruning-brain-psychiatry.html
44. Collins F, quoted in Fikes BJ (2015, March 6). Genomic medicine needs a long-term tack. *San Diego Union Tribune.* Retrieved from http://www.sandiegouniontribune.com/business/biotech/sdut-genomic-medicine-collins-nih-2015mar06-htmlstory.html

COGNITIVE BIOMARKERS OF PSYCHOSIS

23

COGNITIVE BIOMARKERS OF PSYCHOSIS

S. Kristian Hill, Richard S. E. Keefe, and John A. Sweeney

Generalized cognitive dysfunction is now recognized as a core feature of psychotic disorders and among the best predictors of functional disability and a wide range of long-term outcomes. Although the etiology and developmental course of cognitive dysfunction is not entirely clear, some form of cognitive dysfunction or deviation from cognitive expectation precedes psychotic symptoms by years. Thus, by the time clinical symptoms of acute (or subclinical) psychosis manifest, the disease process may be well underway and more resistant to intervention. By focusing on cognition, the early course of deviation from cognitive expectation, and cognitive biomarkers, there is excellent potential for early identification of the disease process at a point in development when interventions could be more efficacious. Progress in this area will require a refocus or reconceptualization of psychosis primarily as a cognitive disorder in which psychotic symptoms appear later in the course of the disorder and/or in more severe cases. Novel methodologies are needed for identifying cognitive decline or stagnation as well as novel classification systems that incorporate longitudinal cognitive trajectories. It will also be important to clarify base rates of deviation from cognitive expectation across the psychosis spectrum. Deficits in affective processes and social cognition may also be important for early identification of individuals who are unable to keep pace with peers and fall progressively further behind in a manner that is distinct from their parents and siblings. Greater research attention building on the heritability of cognitive abilities, family studies, and intermediate phenotyping may also be needed to simplify and inform the genetic architecture of psychotic disorders. In this manner, the field can begin to more effectively dissect the disorder(s) into discrete subgroups linked not only to genetics, but cognitive dysfunction and developmental cognitive profiles.

The clinical diagnosis of schizophrenia and other psychotic disorders has traditionally relied on clinical phenomenology and behavioral observation, and is dependent on subjective factors such as self-reports and clinical judgment. Additionally, clinical presentation may vary over time and in response to treatment, thereby leading to changes in diagnosis over the course of illness. For example, individuals initially diagnosed with bipolar disorder were 103 times more likely than controls to be diagnosed with schizoaffective disorder by 45 years of age.[1] Similarly, those with an initial diagnosis of schizophrenia are 80 times more likely to have a lifetime diagnosis of schizoaffective disorder.[1] This illustrates both the considerable overlap in clinical phenomenology across psychotic disorders and the complex diagnostic consequences of malleable symptom presentation.

There are several reasons to be interested in cognitive and neuropsychological biomarkers for psychotic disorders. First, cognitive impairments represent a major cause of persistent functional disability in these disorders.[2] Consequently, efforts to reduce functional disability are an active target of new pharmacological treatments.[3–5] Second, neuropsychological deficits are most pronounced in schizophrenia and thus have potential clinical value for differential diagnosis. Although cognitive dysfunction was not included as a diagnostic criteria in the DSM-5, the issue was given serious consideration and continues to be an active topic of discussion for future diagnostic classification systems.[6,7] Third, because cognitive deficits are predictive of clinical outcome, neuropsychological assessments can be a useful tool for planning long-term treatment and clinical management. Fourth, to the extent that cognitive deficits are familial in psychotic disorders, these deficits may provide useful endophenotypes for advancing gene discovery. Finally, consistent with the Research Domain Criteria (RDoC) approach to using biomarkers for developing novel biological classification systems for psychotic disorders, cognitive studies show promise in identifying a more schizophrenia-like subgroup with distinctive EEG alterations and more pronounced structural MRI abnormalities.[8]

This chapter reviews key findings regarding the nature and extent of cognitive neuropsychological impairment across the psychosis spectrum, the utility of neurocognitive measures in evaluating treatment studies, familial patterns of cognitive impairment, and prodromal studies that may inform understanding of neurodevelopmental cognitive trajectories.

COGNITION ACROSS THE PSYCHOSIS SPECTRUM

With respect to classification of major psychotic disorders, Kraepelin's diagnostic model for severe mental illness, especially in the early years of his writing, proposed discrete underlying disease processes for bipolar disorder and schizophrenia. Yet, in addition to overlapping clinical phenomena (i.e., psychosis and affective symptoms) cutting across diagnostic boundaries, neurophysiological, genetic, and neuroimaging markers also show considerable overlap.[9–13] These findings have raised questions about the boundaries between schizophrenia and related disorders with a mixture of psychotic and affective features (schizoaffective disorder, psychotic depression, bipolar disorder with psychosis) as well as whether dimensional models, or novel categorical models using biological rather than traditional phenomenological criteria, may better account for the overlapping relationship of currently defined clinical syndromes.[14,15]

In recent years cognitive and neuropsychological dysfunction have come into view as core features not only of schizophrenia[16–18] but also of a wide range of psychotic disorders.[19–21] The clinical neuropsychological approach is a promising approach to identifying objective biomarkers. Given the highly reliable procedures for assessing a wide range of cognitive abilities, neuropsychological measures are well suited for investigating neurocognitive dysfunction across the psychotic disorders. For schizophrenia, neuropsychological studies indicate a generalized cognitive deficit that is present at the first episode of psychosis, relatively stable over time, and largely independent of clinical status or antipsychotic treatment.[22,23] Neuropsychological disturbances have also been recognized as highly prevalent and the best-established causes of functional disability.[24–26] Objective impairments are present across a wide range of cognitive domains that are relatively stable over time (rather than progressive through the early course of illness), largely independent of clinical presentation, and unaffected by treatment with antipsychotics or mood stabilizers.[22,27,28]

Schizoaffective disorder is often considered part of the "schizophrenia spectrum," but it is unclear whether schizoaffective disorder is an or an independent disorder, an intermediate disorder between schizophrenia and affective disorders, or a form of bipolar disorder. Related debates abound as to whether schizophrenia and schizoaffective disorders are part of a continuum[29,30] or distinct conditions.[31] The few studies directly comparing schizoaffective disorder and schizophrenia on cognitive measures have been mixed. Schizoaffective patients tend to perform intermediate between schizophrenia and affective disorder group on cognitive, physiological, and functional outcome measures with some statistically significant group differences observed,[32–40] but such differences have not been observed reliably.[33,41–46]

Neuropsychological deficits now appear to be common and enduring trait-like features in bipolar disorder, especially in those with psychotic features, rather than state-like symptoms present only during acute episodes.[19] Studies comparing diagnostic groups across the psychosis spectrum indicate a similar profile of generalized impairment across psychotic disorders that differs primarily in level of dysfunction with schizophrenia groups showing more severe cognitive impairments than bipolar disorder or major depression with psychotic features.[21,47,48] Furthermore, patients with psychotic depression and bipolar disorder with psychosis tend to display significant deficits compared to their nonpsychotic counterparts.[49–54] This suggests that psychosis appears to be associated with a common profile of diffuse neurocognitive dysfunction with similar modest changes in cognition after antipsychotic treatment (and improved clinical symptom ratings),[19] regardless of whether psychotic symptoms occur in the context of schizophrenia or affective psychosis.

In comparison to bipolar disorder, investigation of the schizophrenia premorbid or prodromal phase has been studied more extensively. Early studies reported deficits in motor development and brain dysmaturation, during childhood and early adolescence, well before clinical onset of schizophrenia.[55–58] Premorbid or prodromal cognitive deficits have since been studied extensively, and meta-analytic studies estimate an IQ deficit of 0.5 standard deviations in prodromal patients who go on to develop schizophrenia compared to their typically developing peers.[59–61]

Cognitive and neuropsychological impairments are typically defined as performance significantly below that of an appropriately matched control group. However, this approach is biased toward patients with moderate to severe impairments and is overly exclusive of those with more mild deficits (who are more likely to benefit from pharmacological therapy). Furthermore, cross-sectional comparison with healthy control groups does not take into account premorbid ability level and may lead to misclassification of patients who perform in the normal range, but show a significant decline compared to premorbid abilities. Thus, future studies might benefit from considering the relationship between premorbid cognitive levels and current performance on cognitive measures as a useful tool for understanding disease related cognitive effects.

NEURODEVELOPMENTAL COGNITIVE BIOMARKERS

Higher rates of pre- and perinatal complications as well as observations of cognitive, motor, and language abnormalities before illness onset have stimulated neurodevelopmental theories of psychosis. In essence, neurodevelopmental theories propose that normal development is disrupted and begins a cascade of events unfolding over years (or even decades), eventually leading to an acute psychotic episode and diagnosis of a psychotic disorder.[62–65] The mechanisms through which neurodevelopmental disturbances disrupt brain maturation are unclear, but are thought to entail various genetic and environmental risk factors at critical developmental junctures. The nature and timing of developmental cognitive abnormalities remains unclear, however retrospective studies of psychosis patients before illness onset as well as studies of individuals at high risk (for a psychotic disorder) have indicated that some level of cognitive dysfunction is present before illness onset.

Historically, cognitive impairments were thought to evolve over the course of the illness either from disease processes or exposure to neuroleptics. However, studies of first-episode psychosis patients with little or no previous exposure to antipsychotics have indicated that generalized cognitive impairments are already present at first episode. Follow-up studies over the early years after illness onset generally indicate that cognitive deficits are relatively stable over time (even during acute psychotic episodes) in the early years after illness onset.[18,19,22,27,28,66,67] These findings suggest that the more severe cognitive impairment in chronic patients compared to first-episode patients[66] may result in part from available samples being more disabled, requiring more active long-term care and, thus, having a greater presence in medication management clinics from which chronic patient samples are recruited. However, as some accelerated age-related brain changes may occur in schizophrenia in mid- to late-life, more advanced later life cognitive changes may also occur.[68,69]

Early neurodevelopmental models of schizophrenia assumed a relatively normal period of development through late adolescence or early adulthood, at which time an environmental trigger would release the diathesis for onset of the disorder leading to acute psychotic symptoms and cognitive deterioration.[70–72] However, the notion that childhood and adolescent development prior to onset of psychosis is not normal is now generally accepted.[66,73] The results of neuropsychological studies are largely consistent with, and thus support, this shift in perspective.

DEVELOPMENTAL TRAJECTORIES

Although early life intellectual abilities predict a wide range of outcomes such as educational attainment, income, physical functioning, health outcomes, mortality rates, mental health, and hospitalization for psychiatric disorders,[74–78] schizophrenia has a uniquely diminished developmental trajectory. The timing and consistency of developmental cognitive dysfunction remains unclear, but retrospective findings suggest a divergence of individuals who eventually developed schizophrenia from age-related peers around the age of 13 (several years before the typical onset of psychotic symptoms).[79,80] Additionally, one prospective study reported poor cognitive performance in children who later developed schizophrenia as early as 7 years of age and thereafter.[81] The course of developmental progression in those at risk for adult onset bipolar disorder remains largely unstudied. Recent research has focused on understanding the developmental trajectory of cognitive function prior to onset of a psychotic illness. With respect to premorbid or prodromal cognitive trajectories, two developmental profiles seem most likely.[81] The developmental deterioration profile is characterized by (relatively) normal cognitive development followed by decline and deterioration. In contrast, the neurodevelopmental lag hypothesis predicts a rate of cognitive development that lags behind typically developing peers such that prepsychotic individuals fall further behind their typically developing peers due to a lack of normal progress rather than a loss of previously acquired cognitive skills. This pattern has also been seen in pediatric onset bipolar disorder.[82] In this profile, the developmental trajectory of pre-illness individuals slowly diverges from a typical development, as healthy individuals consistently take larger developmental steps forward. Although high-risk individuals are making steady developmental progress, the progress is reduced and they fall further behind their peers. The main distinction between the deterioration approach and the disruption of normal development is a lack of deterioration or loss of previously developed cognitive skills in the neurodevelopmental approach.

NEURODEGENERATION HYPOTHESIS

Premorbid cognitive deficits have been well established in schizophrenia and, although investigated less extensively, affective psychosis. The notion that cognitive measures may have utility as biomarkers of psychosis that may be evident during the prodromal phase led to numerous studies of high-risk samples in which subthreshold psychotic symptoms characterize the prodromal phase. Studies of high-risk individuals have consistently shown significant cognitive impairments compared to controls, but impairments were less severe than the deficits seen in schizophrenia patients.[83–85] Additionally, some studies reported significantly lower performance for those who later experienced psychosis,[86] and signs of cognitive decline during the transition to overt psychosis.[87,88] In a large longitudinal study focused on neurocognitive impairments that predicted the transition to psychosis, a profile of high premorbid verbal abilities combined with impaired declarative memory was a strong predictor of conversion to a psychotic disorder within a few years.[89] This pattern of high verbal premorbid abilities combined with later impairments predictive of transition to psychosis could be interpreted as consistent with a disease process characterized by ongoing deterioration.[90] Decline from estimated premorbid (verbal) abilities compared to observed neuropsychological abilities has been reported in schizophrenia[91–93] and across the psychosis spectrum.[94] Consistent with other indicators, deviation from cognitive expectation was significantly worse in schizophrenia[94] and may reflect a more severe disease process in schizophrenia. Additional studies are needed to replicate findings of cognitive deterioration in the prodromal phase, clarify the extent of meaningful cognitive decline in prodromal samples, and explore the possibility that prodromal subtypes exist and may be associated with different neurodevelopmental and cognitive profiles. Furthermore, although familiality of deviation from expectation was significant, the magnitude of this effect was small compared to both the familiality of cognitive abilities and the level of decline from expectation in unaffected family members.[21,94] Thus, profiles of cognitive decline may only be weakly heritable, potentially reflecting a disruption of heritable cognitive traits by disease-related processes.

Many prodromal studies have used follow-back or retrospective designs with data collection at just one time point. Birth-cohort studies, particularly those with assessments at multiple time points, may address some of these methodological issues. Growth charting from the Philadelphia Neurodevelopmental Cohort identified 2,321 individuals who endorsed psychotic symptoms between the ages of 8 and 21. Based on the Penn Computerized Neuropsychological Battery, those who endorsed psychotic symptoms had lower predicted age (based on cognitive performance) compared to peers who did not endorse psychotic features.[95] Participants from the Dunedin, New Zealand, longitudinal study on health and behavior who would later develop schizophrenia showed no evidence of developmental deterioration on WISC-R subtests from 7 to 13 years of age. However, the preschizophrenia group showed a profile consistent with a developmental lag for nonverbal problem-solving and freedom from distractibility subtests. Additionally, cognitive profiles for verbal reasoning and conceptualization were notable for developmental deficits which appeared early and remained stable through early adolescence.[81] These findings are consistent with the notion that prodromal deficits may be better conceptualized as a failure to acquire certain cognitive skills rather than a neurodegenerative process involving loss of previously acquired skills.

DEVELOPMENTAL LAG

There is a growing literature indicating that schizophrenia and bipolar disorder share a pattern of generalized and stable cognitive deficits. From a cognitive perspective, stable generalized impairments, even during euthymic states, have been well documented in bipolar disorder.[50,96] Because pediatric-onset bipolar disorder is more common than early-onset schizophrenia, investigating developmental cognitive trajectories in pediatric bipolar disorder may shed light on the premorbid or prodromal progression of cognitive dysfunction in bipolar disorder. In a 3-year follow-up study of primarily euthymic pediatric bipolar disorder patients, increasing deficits compared to normal age-related progression in executive function and verbal memory were reported.[97] Thus, cognitive profiles, over a 3-year period, were consistent with a developmental lag in pediatric bipolar disorder rather than a degenerative process (loss of function) characterized by loss of previously acquired cognitive abilities. This may reflect disease-related effects on the maturation of prefrontal and mesial temporal cortex perhaps involving deficient synaptogenesis and/or inefficient or excessive synaptic pruning. It is also possible that the illness may reduce interest and/or engagement in select academic activities leading to a more modest developmental trajectory. More recently, a 2-year follow-up indicated significant cognitive improvement in several cognitive domains in small sample of early-onset bipolar disorder. These gains were more substantial than improvement seen in controls,[98] and, although these findings need to be confirmed in larger studies, this opens the possibility for catch-up periods of cognitive development in some individuals.

One early approach to understanding the nature and progression of cognitive dysfunction in psychosis is to compare patients in different stages of illness. Cross-sectional studies and indirect comparisons across studies suggested that first-episode and chronic patients[99] were more similar than first-episode patients and high-risk individuals.[84,85,100] This pattern suggests potential progressive impairment in cognitive process during the transition to psychosis. However, longitudinal studies that are likely to be more informative about the course of cognitive dysfunction in psychosis might be confounded by low-functioning individuals who are more likely to convert. Meta-analysis of first-episode and high-risk follow-up studies found no evidence of cognitive deterioration during or after the transition to diagnosis with a psychotic disorder.[83] In fact, high-risk individuals, first-episode patients, and healthy controls all showed cognitive improvement over time and no evidence of loss of acquired cognitive skills near illness onset. Some may argue that the discrepancy between cross-sectional comparisons across stages and longitudinal studies of high-risk individuals lies with the high-risk participants who transition to onset of psychosis during the follow-up period. Specifically, should cognitive deterioration occur during the transition to psychosis, then high-risk individuals who transition will show declines compared to nontransitioning participants and controls. However, meta-analysis showed no differences in longitudinal studies between converters and nonconverters.[83] The lone study that supported the degeneration hypothesis noted that all participants in the transition group were treated with antipsychotic medications whereas just one in the nontransition group was treated with an antipsychotic.[101] This suggests that the transition group displayed more severe symptoms, which hastened treatment with antipsychotics. The early need for antipsychotic treatment may be related to more severe cognitive impairments[21,102] rather than a profile of deterioration or deficits exacerbated by pharmacologic treatment.

FAMILY STUDIES

The high heritability of neuropsychological abilities suggests that cognitive measures may be useful as cognitive biomarkers, as individuals with a high degree of genetic overlap or familial high risk appear to show a similar, albeit muted, pattern of cognitive deficits similar to their affected relative. The schizophrenia literature has indicated familial patterns of neurocognitive dysfunction for some time.[21,103–105] Deficits in episodic memory, working memory, attention, response inhibition, and reasoning have consistently been reported in offspring, monozygotic and dizygotic twins, and other unaffected relatives.[39,46,106–111] There are fewer studies evaluating relatives of individuals with bipolar disorder and even fewer directly comparing relatives of both schizophrenia and bipolar patients. In relatives of bipolar patients, neuropsychological deficits have been reported in a range of unaffected relatives and in several areas (i.e., verbal learning and memory, working memory, executive function, face memory, response inhibition),[112–114] but cognitive deficits have not been consistently reported.[115] Meta-analysis does little to clarify the familial pattern of cognitive dysfunction in bipolar disorder as small effect sizes are reported for select areas.[116] Overall, familial studies of cognitive dysfunction in bipolar disorder and other affective psychosis is relatively underdeveloped and has lagged behind similar work in schizophrenia.

Few studies have directly compared relatives of both individuals with schizophrenia and bipolar disorder to explore the potential for similar familial patterns. In a small Korean sample of unaffected first-degree relatives of bipolar patients with psychosis and schizophrenia probands both showed working memory dysfunction compared to controls, whereas only relatives of schizophrenia probands showed verbal fluency deficits.[117] Working memory impairments were interpreted as a shared endophenotype, while verbal fluency impairment was proposed to be a schizophrenia-related phenotype. Another small family study was pivotal in drawing attention to dimensional models of psychosis. Initial analyses were based on diagnostic category and failed to distinguish schizophrenia and psychotic bipolar probands. Additionally, when their respective first-degree relatives were compared there were no significant differences. However, when relatives were collapsed across diagnoses and classified as affected (positive lifetime history of psychotic symptoms) or unaffected (no history of psychosis symptoms), there were significant differences between the two relative groups on a range of neuropsychological domains, but probands and affected relatives did not differ.[118]

In a large sample study by the Bipolar-Schizophrenia Network for Intermediate Phenotypes (B-SNIP) consortium addressing these issues, a complex pattern of familial cognitive effects was observed when probands with schizophrenia and psychotic bipolar disorders and their unaffected relatives were examined. Specifically, relatives of schizophrenia probands displayed mild cognitive deficits regardless of the presence of personality disorder features. However, among relatives of psychotic bipolar probands mild cognitive dysfunction was only observed in relatives with traits of Cluster A or B personality disorders.[21] These findings suggest that schizophrenia risk is associated with cognitive disturbances regardless of schizotypal traits. In contrast, risk related to cognitive disturbances in bipolar disorder with psychosis was considerably more closely related to schizotypal and erratic personality disorder features.

Recent work with bipolar disorder suggests that unaffected relatives can be grouped based on proband subtype as globally impaired, selectively impaired, or intact.[119,120] This was consistent with the idea that neurocognitive alterations may represent a promising endophenotype in bipolar disorder, as appears to be the case in schizophrenia, but with differences in cognition and personality traits across disorders. Relatives of the globally impaired group performed in qualitatively similar ways to bipolar probands, and performed worse than the remaining relative groups.[120] This supports the notion that the threshold for familial cognitive dysfunction differs among discrete subgroups of psychosis.

REFOCUSING ON COGNITION

Kahn and Keefe recently reminded the field that both Kraepelin and Bleuler prioritized cognitive dysfunction over psychotic symptoms in the diagnosis of schizophrenia.[121] They suggested that conceptualizing schizophrenia as a cognitive illness and focusing on identifying the emergence of cognitive dysfunction earlier in development would better inform the disorder, and thus facilitate early diagnosis and intervention at a more favorable point in development. Reduction of psychotic symptoms is the primary target of most treatment regimens, yet treatment efficacy for schizophrenia has not improved substantially since the development of chlorpromazine (if at all—see [122,123]).

Persistent cognitive deficits are problematic in terms of educational and occupational function. More than a decade ago the Measurement and Treatment Research to Improve Cognition in Schizophrenia (MATRICS) initiative identified cognitive enhancement as a critical target for advancing treatment and improving quality of life in schizophrenia.[124] However, available antipsychotic drugs have not favorably tipped the cognitive scales in a meaningful way,[27,125–129] although there may be some cause for guarded optimism.[130,131]

One of the primary tenets of this call to refocus on cognition was based on findings that individuals who develop schizophrenia show evidence of cognitive dysfunction years before an acute psychotic episode. By shifting focus to cognition, onset of illness would shift a decade or more earlier than we currently conceptualize onset. In the current framework of onset being primarily related to acute psychosis, early diagnosis and treatment programs may be later than ideal, as the illness, and associated cognitive and neurobiological changes, may then be considerably advanced and less amenable to alteration. The key challenge to unraveling this illness is to identify individuals at the highest risk before onset and understanding the markers of cognitive change and their neurobiological mechanisms. This may require new methodological approaches that extend beyond correlational analyses and mean group changes and take into account patterns of variability in individual differences in change. That is, different individuals may show different levels of change at different ages, and this intraindividual variability may have utility in predicting interventions for individual patients. For example, increased within-person variability preceded and predicted mean level change on a variety of outcomes including anger, hostility, depression, and memory.[132] Critical changes in developmental trajectories may provide a signal of the onset of a dynamic process that is only later seen in terms of the onset of psychotic symptoms. A focus on this hypothesis requires a shift away from assessing cognition in adults with a history of acute psychotic episodes to evaluating cognitive underperformance compared to expectation in large samples of high-risk individuals with frequent longitudinal time-points—similar to the North American Prodrome Longitudinal Study (NAPLES).

SUMMARY

Generalized cognitive dysfunction is now recognized as a core feature of psychotic disorders and among the best predictors of functional disability and a wide range of long-term outcomes. Although the etiology and developmental course of cognitive dysfunction is not entirely clear, some form of cognitive dysfunction or deviation from cognitive expectation precedes psychotic symptoms by years. Thus, by the time clinical symptoms of acute (or subclinical) psychosis manifest, the disease process may be well underway and more resistant to intervention. By focusing on cognition, the early course of deviation from cognitive expectation, and cognitive biomarkers, there is excellent potential for early identification of the disease process at a point in development when interventions could be more efficacious. Progress in this area will require a refocus or reconceptualization of psychosis primarily as a cognitive disorder in which psychotic symptoms appear later in the course of the disorder and/or in more severe cases. Novel methodologies are needed for identifying cognitive decline or stagnation as well as novel classification systems that incorporate longitudinal cognitive trajectories. It will also be important to clarify base rates of deviation from cognitive expectation across the psychosis spectrum. Deficits in affective processes and social cognition may also be important for early identification of individuals who are unable to keep pace with peers and fall progressively further behind in a manner that is distinct from their parents and siblings. Greater research attention, building on the heritability of cognitive abilities, family studies, and intermediate phenotyping, may also be needed to simplify and inform the genetic architecture of psychotic disorders. In this manner, the field can begin to more effectively dissect the disorder(s) into discrete subgroups linked not only to genetics but also to cognitive dysfunction and developmental cognitive profiles.

REFERENCES

1. Laursen TM, Agerbo E, Pedersen CB. Bipolar disorder, schizoaffective disorder, and schizophrenia overlap: a new comorbidity index. *J Clin Psychiatry* 2009;70:1432–1438.
2. Green MF. What are the functional consequences of neurocognitive deficits in schizophrenia? *Am J Psychiatry* 1996;153:321–330.
3. Harvey PD, Cornblatt BA. Pharmacological treatment of cognition in schizophrenia: an idea whose method has come. *Am J Psychiatry* 2008;165:163–165.
4. Hill SK, Reilly JL, Harris MS, Khine T, Sweeney JA. Oculomotor and neuropsychological effects of antipsychotic treatment for schizophrenia. *Schizophr Bull* 2008;34:494–506.
5. Keefe RS, Buchanan RW, Marder SR et al. Clinical trials of potential cognitive-enhancing drugs in schizophrenia: what have we learned so far? *Schizophr Bull* 2011;39:417–435.
6. Keefe RS, Fenton WS. How should DSM-V criteria for schizophrenia include cognitive impairment? *Schizophr Bull* 2007;33: 912–920.
7. Bora E, Yucel M, Pantelis C. Cognitive impairment in schizophrenia and affective psychoses: implications for DSM-V criteria and beyond. *Schizophr Bull* 2010;36:36–42.
8. Clementz BA, Sweeney JA, Hamm JP et al. Identification of distinct psychosis biotypes using brain-based biomarkers. *Am J Psychiatry* 2016;173:373–384.
9. Berrettini WH. Are schizophrenic and bipolar disorders related? A review of family and molecular studies. *Biol Psychiatry* 2000;48:531–538.
10. Bramon E, Sham PC. The common genetic liability between schizophrenia and bipolar disorder: a review. *Curr Psychiatry Rep* 2001;3:332–337.
11. Badner JA, Gershon ES. Meta-analysis of whole-genome linkage scans of bipolar disorder and schizophrenia. *Mol Psychiatry* 2002;7:405–411.
12. Baron M. Genetics of schizophrenia and the new millennium: progress and pitfalls. *Am J Hum Genet* 2001;68:299–312.
13. Tamminga CA, Pearlson G, Keshavan M, Sweeney J, Clementz B, Thaker G. Bipolar and schizophrenia network for intermediate phenotypes: outcomes across the psychosis continuum. *Schizophr Bull* 2014;40 Suppl 2:S131–S137. doi: 10.1093/schbul/sbt179.:S131-S137.
14. Guloksuz S, van OJ. The slow death of the concept of schizophrenia and the painful birth of the psychosis spectrum. *Psychol Med* 2017;1–16.
15. Keshavan MS, Morris DW, Sweeney JA et al. A dimensional approach to the psychosis spectrum between bipolar disorder and schizophrenia: the Schizo-Bipolar Scale. *Schizophr Res* 2011;133:250–254.
16. Bilder RM, Goldman RS, Robinson D et al. Neuropsychology of first-episode schizophrenia: initial characterization and clinical correlates. *Am J Psychiatry* 2000;157:549–559.
17. Saykin AJ, Shtasel DL, Gur RE et al. Neuropsychological deficits in neuroleptic naive patients with first-episode schizophrenia. *Arch Gen Psychiatry* 1994;51:124–131.
18. Hoff AL, Sakuma M, Wieneke M, Horon R, Kushner M, DeLisi LE. Longitudinal neuropsychological follow-up study of patients with first-episode schizophrenia. *Am J Psychiatry* 1999;156:1336–1341.
19. Hill SK, Reilly JL, Harris MSH et al. A comparison of neuropsychological dysfunction in first-episode psychosis patients with unipolar depression, bipolar disorder, and schizophrenia. *Schizophr Res* 2009;113:167–175.
20. Reichenberg A, Harvey PD, Bowie CR et al. Neuropsychological function and dysfunction in schizophrenia and psychotic affective disorders. *Schizophr Res* 2009;35:1022–1029.
21. Hill SK, Reilly JL, Keefe RS et al. Neuropsychological impairments in schizophrenia and psychotic bipolar disorder: findings from the Bipolar and Schizophrenia Network on Intermediate Phenotypes (B-SNIP) study. *Am J Psychiatry* 2013;10.
22. Hill SK, Schuepbach D, Herbener ES, Keshavan MS, Sweeney JA. Pretreatment and longitudinal studies of neuropsychological deficits in antipsychotic-naive patients with schizophrenia. *Schizophr Res* 2004;68:49–63.
23. Reilly JL, Sweeney JA. Generalized and specific neurocognitive deficits in psychotic disorders: utility for evaluating pharmacological treatment effects and as intermediate phenotypes for gene discovery. *Schizophr Bull* 2014;40:516–522.
24. Harvey PD, Strassnig M. Predicting the severity of everyday functional disability in people with schizophrenia: cognitive deficits, functional capacity, symptoms, and health status. *World Psychiatry* 2012;11:73–79.
25. Green MF. Cognitive impairment and functional outcome in schizophrenia and bipolar disorder. *J Clin Psychiat* 2006;67:e12.
26. Harvey PD, Wingo AP, Burdick KE, Baldessarini RJ. Cognition and disability in bipolar disorder: lessons from schizophrenia research. *Bipolar Disord* 2010;12:364–375.
27. Keefe RS, Sweeney JA, Gu H et al. Effects of olanzapine, quetiapine, and risperidone on neurocognitive function in early psychosis: a randomized, double-blind 52-week comparison. *Am J Psychiatry* 2007;164:1061–1071.
28. Sweeney JA, Haas GL, Keilp JG, Long M. Evaluation of the stability of neuropsychological functioning after acute episodes of schizophrenia: one-year followup study. *Psychiatry Res* 1991;38:63–76.
29. Gooding DC, Tallent KA. Spatial working memory performance in patients with schizoaffective psychosis versus schizophrenia: a tale of two disorders? *Schizophr Res* 2002;53:209–218.
30. Lake CR, Hurwitz N. Schizoaffective disorders are psychotic mood disorders; there are no schizoaffective disorders. *Psychiatry Res* 2006;143:255–287.
31. Ketter TA, Wang PW, Becker OV, Nowakowska C, Yang Y. Psychotic bipolar disorders: dimensionally similar to or categorically different from schizophrenia? *J Psychiatr Res* 2004;38:47–61.
32. Beatty WW, Jocic Z, Monson N, Staton RD. Memory and frontal lobe dysfunction in schizophrenia and schizoaffective disorder. *J Nerv Ment Dis* 1993;181:448–453.
33. Hooper SR, Giuliano AJ, Youngstrom EA et al. Neurocognition in early-onset schizophrenia and schizoaffective disorders. *J Am Acad Child Adolesc Psychiatry* 2010;49:52–60.
34. Verdoux H, Liraud F. Neuropsychological function in subjects with psychotic and affective disorders. Relationship to diagnostic category and duration of illness. *Eur Psychiatry* 2000;15:236–243.
35. Szoke A, Meary A, Trandafir A et al. Executive deficits in psychotic and bipolar disorders—implications for our understanding of schizoaffective disorder. *Eur Psychiatry* 2008;23:20–25.
36. Soh P, Narayanan B, Khadka S et al. Joint coupling of awake EEG frequency activity and MRI gray matter volumes in the psychosis dimension: a BSNIP study. *Front Psychiatry* 2015;6:162. doi: 10.3389/fpsyt.2015.00162. eCollection@2015.:162.
37. Lencer R, Sprenger A, Reilly JL et al. Pursuit eye movements as an intermediate phenotype across psychotic disorders: evidence from the B-SNIP study. *Schizophr Res* 2015;169:326–333.

38. Hill SK, Buchholz A, Amsbaugh H et al. Working memory impairment in probands with schizoaffective disorder and first degree relatives of schizophrenia probands extend beyond deficits predicted by generalized neuropsychological impairment. *Schizophr Res* 2015;166:310–315.
39. Reilly JL, Frankovich K, Hill S et al. Elevated antisaccade error rate as an intermediate phenotype for psychosis across diagnostic categories. *Schizophr Bull* 2014;40:1011–1021.
40. Ruocco AC, Reilly JL, Rubin LH et al. Emotion recognition deficits in schizophrenia-spectrum disorders and psychotic bipolar disorder: findings from the Bipolar-Schizophrenia Network on Intermediate Phenotypes (B-SNIP) study. Schizophr Res. 158[1-3], 105–112. 2014.
41. Goldstein G, Shemansky WJ, Allen DN. Cognitive function in schizoaffective disorder and clinical subtypes of schizophrenia. *Arch Clin Neuropsychol* 2005;20:153–159.
42. Miller LS, Swanson-Green T, Moses JA, Jr., Faustman WO. Comparison of cognitive performance in RDC-diagnosed schizoaffective and schizophrenic patients with the Luria-Nebraska Neuropsychological Battery. *J Psychiatr Res* 1996;30:277–282.
43. Stip E, Sepehry AA, Prouteau A et al. Cognitive discernible factors between schizophrenia and schizoaffective disorder. *Brain Cogn* 2005;59:292–295.
44. Fiszdon JM, Richardson R, Greig T, Bell MD. A comparison of basic and social cognition between schizophrenia and schizoaffective disorder. *Schizophr Res* 2007;91:117–121.
45. Glahn DC, Bearden CE, Cakir S et al. Differential working memory impairment in bipolar disorder and schizophrenia: effects of lifetime history of psychosis. *Bipolar Disord* 2006;8:117–123.
46. Ethridge LE, Soilleux M, Nakonezny PA et al. Behavioral response inhibition in psychotic disorders: diagnostic specificity, familiality and relation to generalized cognitive deficit. *Schizophr Res* 2014;159:491–498.
47. Glahn DC, Almasy L, Blangero J et al. Adjudicating neurocognitive endophenotypes for schizophrenia. *Am J Med Genet B Neuropsychiatr Genet* 2007;144:242–249.
48. Bora E, Yucel M, Pantelis C. Neurocognitive markers of psychosis in bipolar disorder: a meta-analytic study. *J Affect Disord* 2010;127:1–9.
49. Goldberg TE, Torrey EF, Gold JM et al. Genetic risk of neuropsychological impairment in schizophrenia: a study of monozygotic twins discordant and concordant for the disorder. *Schizophr Res* 1995;17:77–84.
50. Hill SK, Keshavan MS, Thase ME, Sweeney JA. Neuropsychological dysfunction in antipsychotic-naive first-episode unipolar psychotic depression. *Am J Psychiatry* 2004;161:996–1003.
51. Jeste DV, Heaton SC, Paulsen JS, Ercoli L, Harris J, Heaton RK. Clinical and neuropsychological comparison of psychotic depression with nonpsychotic depression and schizophrenia. *Am J Psychiatry* 1996;153:490–496.
52. Mojtabai R, Bromet EJ, Harvey PD, Carlson GA, Craig TJ, Fennig S. Neuropsychological differences between first-admission schizophrenia and psychotic affective disorders. *Am J Psychiatry* 2000;157:1453–1460.
53. Reichenberg A, Weiser M, Rabinowitz J et al. A population-based cohort study of premorbid intellectual, language, and behavioral functioning in patients with schizophrenia, schizoaffective disorder, and nonpsychotic bipolar disorder. *Am J Psychiatry* 2002;159:2027–2035.
54. Zihl J, Gron G, Brunnauer A. Cognitive deficits in schizophrenia and affective disorders: evidence for a final common pathway disorder. *Acta Psychiatr Scand* 1998;97:351–357.
55. Keshavan MS, Hogarty GE. Brain maturational processes and delayed onset in schizophrenia. *Dev Psychopathol* 1999;11:525–543.
56. Walker EF, Trotman HD, Goulding SM et al. Developmental mechanisms in the prodrome to psychosis. *Dev Psychopathol* 2013;25:1585–1600.
57. Pruessner M, Cullen AE, Aas M, Walker EF. The neural diathesis-stress model of schizophrenia revisited: an update on recent findings considering illness stage and neurobiological and methodological complexities. *Neurosci Biobehav Rev* 2017;73:191–218. doi: 10.1016/j.neubiorev.2016.12.013. Epub; 2016 Dec 16:191–218.
58. Walker EF, Savoie T, Davis D. Neuromotor precursors of schizophrenia. *Schizophr Bull* 1994;20:441–451.
59. Woodberry KA, Giuliano AJ, Seidman LJ. Premorbid IQ in schizophrenia: a meta-analytic review. *Am J Psychiatry* 2008;165:579–587.
60. Lieberman JA, Perkins D, Belger A et al. The early stages of schizophrenia: speculations on pathogenesis, pathophysiology, and therapeutic approaches. *Biol Psychiatry* 2001;50:884–897.
61. Cannon TD, Yu C, Addington J et al. An individualized risk calculator for research in prodromal psychosis. *Am J Psychiatry* 2016;173:980–988.
62. Rapoport JL, Addington AM, Frangou S, Psych MR. The neurodevelopmental model of schizophrenia: update 2005. *Mol Psychiatry* 2005;10:434–449.
63. Rapoport JL, Giedd JN, Gogtay N. Neurodevelopmental model of schizophrenia: update 2012. *Mol Psychiatry* 2012;17:1228–1238.
64. Harrison PJ. Schizophrenia: a disorder of neurodevelopment? *Curr Opin Neurobiol* 1997;7:285–289.
65. Wolf SS, Weinberger DR. Schizophrenia: a new frontier in developmental neurobiology. *Isr J Med Sci* 1996;32:51–55.
66. Sweeney JA, Haas GL, Li S. Neuropsychological and eye movement abnormalities in first-episode and chronic schizophrenia. *Schizophr Bull* 1992;18:283–293.
67. Keefe RS, Silva SG, Perkins DO, Lieberman JA. The effects of atypical antipsychotic drugs on neurocognitive impairment in schizophrenia: a review and meta-analysis. *Schizophr Bull* 1999;25:201–222.
68. Harvey PD, Rosenthal JB. Cognitive and functional deficits in people with schizophrenia: evidence for accelerated or exaggerated aging? *Schizophr Res* 2017;10.
69. Zhang W, Deng W, Yao L et al. Brain structural abnormalities in a group of never-medicated patients with long-term schizophrenia. *Am J Psychiatry* 2015;172:995–1003.
70. Keshavan MS, Murray RM. Neurodevelopment and adult psychopathology. Cambridge, United Kingdom: Cambridge University Press: 1997.
71. Maynard TM, Sikich L, Lieberman JA, LaMantia AS. Neural development, cell-cell signaling, and the "two-hit" hypothesis of schizophrenia. *Schizophr Bull* 2001;27:457–476.
72. Weinberger DR. Implications of normal brain development for the pathogenesis of schizophrenia. *Arch Gen Psychiatry* 1987;44:660–669.
73. Haas GL, Sweeney JA. Premorbid and onset features of first-episode schizophrenia. *Schizophr Bull* 1992;18:373–386.
74. Gale CR, Batty GD, Tynelius P, Deary IJ, Rasmussen F. Intelligence in early adulthood and subsequent hospitalization for mental disorders. *Epidemiology* 2010;21:70–77.
75. Batty GD, Deary IJ, Gottfredson LS. Premorbid (early life) IQ and later mortality risk: systematic review. *Ann Epidemiol* 2007;17:278–288.
76. Jokela M, Batty GD, Deary IJ, Silventoinen K, Kivimaki M. Sibling analysis of adolescent intelligence and chronic diseases in older adulthood. *Ann Epidemiol* 2011;21:489–496.
77. von SS, Deary IJ, Kivimaki M, Jokela M, Clark H, Batty GD. Childhood behavior problems and health at midlife: 35-year follow-up of a Scottish birth cohort. *J Child Psychol Psychiatry* 2011;52:992–1001.
78. Deary IJ, Batty GD, Pattie A, Gale CR. More intelligent, more dependable children live longer: a 55-year longitudinal study of a representative sample of the Scottish nation. *Psychol Sci* 2008;19:874–880.
79. Fuller R, Nopoulos P, Arndt S, O'Leary D, Ho BC, Andreasen NC. Longitudinal assessment of premorbid cognitive functioning in patients with schizophrenia through examination of standardized scholastic test performance. *Am J Psychiatry* 2002;159:1183–1189.
80. van Oel CJ, Sitskoorn MM, Cremer MP, Kahn RS. School performance as a premorbid marker for schizophrenia: a twin study. *Schizophr Bull* 2002;28:401–414.

81. Reichenberg A, Caspi A, Harrington H et al. Static and dynamic cognitive deficits in childhood preceding adult schizophrenia: a 30-year study. *Am J Psychiatry* 2010;167:160–169.
82. Pavuluri MN, West A, Jindal K, Hill SK, Sweeney JA. Longitudinal study of neurocognitive function in pediatric bipolar disorder: three-year followup shows patients lagging behind healthy youth. *Am J Psychiatry*. In press.
83. Bora E, Murray RM. Meta-analysis of cognitive deficits in ultra-high risk to psychosis and first-episode psychosis: do the cognitive deficits progress over, or after, the onset of psychosis? *Schizophr Bull* 2014;40:744–755.
84. Giuliano AJ, Li H, Mesholam-Gately RI, Sorenson SM, Woodberry KA, Seidman LJ. Neurocognition in the psychosis risk syndrome: a quantitative and qualitative review. *Curr Pharm Des* 2012;18:399–415.
85. Fusar-Poli P, Deste G, Smieskova R et al. Cognitive functioning in prodromal psychosis: a meta-analysis. *Arch Gen Psychiatry* 2012;69:562–571.
86. Seidman LJ, Giuliano AJ, Meyer EC et al. Neuropsychology of the prodrome to psychosis in the NAPLS consortium: relationship to family history and conversion to psychosis. *Arch Gen Psychiatry* 2010;67:578–588.
87. Hawkins KA, Addington J, Keefe RS et al. Neuropsychological status of subjects at high risk for a first episode of psychosis. *Schizophr Res* 2004;67:115–122.
88. Keefe RS, Perkins DO, Gu H, Zipursky RB, Christensen BK, Lieberman JA. A longitudinal study of neurocognitive function in individuals at-risk for psychosis. *Schizophr Res* 2006;88:26–35.
89. Seidman LJ, Shapiro DI, Stone WS et al. Association of neurocognition with transition to psychosis: baseline functioning in the second phase of the North American Prodrome Longitudinal Study. *JAMA Psychiatry* 2016;73:1239–1248.
90. Lieberman JA. Is schizophrenia a neurodegenerative disorder? A clinical and neurobiological perspective. *Biol Psychiatry* 1999;46:729–739.
91. Keefe RS, Eesley CE, Poe MP. Defining a cognitive function decrement in schizophrenia. *Biol Psychiatry* 2005;57:688–691.
92. Woodward ND, Heckers S. Brain structure in neuropsychologically defined subgroups of schizophrenia and psychotic bipolar disorder. *Schizophr Bull* 2015;41:1349–1359.
93. Weickert TW, Goldberg TE, Gold JM, Bigelow LB, Egan MF, Weinberger DR. Cognitive impairments in patients with schizophrenia displaying preserved and compromised intellect. *Arch Gen Psychiatry* 2000;57:907–913.
94. Hochberger WC, Combs T, Reilly JL et al. Deviation from expected cognitive ability across psychotic disorders. *Schizophr Res* 2017;10.
95. Gur RC, Calkins ME, Satterthwaite TD et al. Neurocognitive growth charting in psychosis spectrum youths. *JAMA Psychiatry* 2014;71:366–374.
96. Pavuluri MN, Schenkel LS, Aryal S et al. Neurocognitive function in unmedicated manic and medicated euthymic pediatric bipolar patients. *Am J Psychiatry* 2006;163:286–293.
97. Pavuluri MN, West A, Hill SK, Jindal K, Sweeney JA. Neurocognitive function in pediatric bipolar disorder: 3-year follow-up shows cognitive development lagging behind healthy youths. *J Am Acad Child Adolesc Psychiatry* 2009;48:299–307.
98. Lera-Miguel S, Andres-Perpina S, Fatjo-Vilas M, Fananas L, Lazaro L. Two-year follow-up of treated adolescents with early-onset bipolar disorder: changes in neurocognition. *J Affect Disord* 2015;172:48–54. doi: 10.1016/j.jad.2014.09.041. Epub; 2014 Oct 17:48–54.
99. Mesholam-Gately RI, Giuliano AJ, Goff KP, Faraone SV, Seidman LJ. Neurocognition in first-episode schizophrenia: a meta-analytic review. *Neuropsychology* 2009;23:315–336.
100. Bang M, Kim KR, Song YY, Baek S, Lee E, An SK. Neurocognitive impairments in individuals at ultra-high risk for psychosis: who will really convert? *Aust N Z J Psychiatry* 2015;49:462–470.
101. Wood SJ, Brewer WJ, Koutsouradis P et al. Cognitive decline following psychosis onset: data from the PACE clinic. *Br J Psychiatry Suppl* 2007;51:s52–s57. doi: 10.1192/bjp.191.51.s52.:s52-s57.
102. Eum S, Hill SK, Rubin LH et al. Cognitive burden of anticholinergic medications in psychotic disorders. *Schizophr Res* 2017.
103. Kremen WS, Seidman LJ, Pepple JR, Lyons MJ, Tsuang MT, Faraone SV. Neuropsychological risk indicators for schizophrenia: a review of family studies. *Schizophr Bull* 1994;20:103–119.
104. Keefe RS, Silverman JM, Roitman SE et al. Performance of nonpsychotic relatives of schizophrenic patients on cognitive tests. *Psychiatry Res* 1994;53:1–12.
105. Scala S, Lasalvia A, Seidman LJ, Cristofalo D, Bonetto C, Ruggeri M. Executive functioning and psychopathological profile in relatives of individuals with deficit v. non-deficit schizophrenia: a pilot study. *Epidemiol Psychiatr Sci* 2013;1–13.
106. Byrne M, Hodges A, Grant E, Owens DC, Johnstone EC. Neuropsychological assessment of young people at high genetic risk for developing schizophrenia compared with controls: preliminary findings of the Edinburgh High Risk Study (EHRS). *Psychol Med* 1999;29:1161–1173.
107. van Erp TG, Saleh PA, Rosso IM et al. Contributions of genetic risk and fetal hypoxia to hippocampal volume in patients with schizophrenia or schizoaffective disorder, their unaffected siblings, and healthy unrelated volunteers. *Am J Psychiatry* 2002;159:1514–1520.
108. Conklin HM, Calkins ME, Anderson CW, Dinzeo TJ, Iacono WG. Recognition memory for faces in schizophrenia patients and their first-degree relatives. *Neuropsychologia* 2002;40:2314–2324.
109. Horan WP, Braff DL, Nuechterlein KH et al. Verbal working memory impairments in individuals with schizophrenia and their first-degree relatives: findings from the Consortium on the Genetics of Schizophrenia. *Schizophr Res* 2008;103:218–228.
110. Egan MF, Goldberg TE, Gscheidle T et al. Relative risk for cognitive impairments in siblings of patients with schizophrenia. *Biol Psychiatry* 2001;50:98–107.
111. Hoti F, Tuulio-Henriksson A, Haukka J, Partonen T, Holmstrom L, Lonnqvist J. Family-based clusters of cognitive test performance in familial schizophrenia. *BMC Psychiatry* 2004;4:20.
112. Gourovitch ML, Torrey EF, Gold JM, Randolph C, Weinberger DR, Goldberg TE. Neuropsychological performance of monozygotic twins discordant for bipolar disorder. *Biol Psychiatry* 1999;45:639–646.
113. Keri S, Kelemen O, Benedek G, Janka Z. Different trait markers for schizophrenia and bipolar disorder: a neurocognitive approach. *Psychol Med* 2001;31:915–922.
114. McIntosh AM, Harrison LK, Forrester K, Lawrie SM, Johnstone EC. Neuropsychological impairments in people with schizophrenia or bipolar disorder and their unaffected relatives. *Br J Psychiatry* 2005;186:378–385.
115. Zalla T, Joyce C, Szoke A et al. Executive dysfunctions as potential markers of familial vulnerability to bipolar disorder and schizophrenia. *Psychiatry Res* 2004;121:207–217.
116. Arts B, Jabben N, Krabbendam L, van Os J. Meta-analyses of cognitive functioning in euthymic bipolar patients and their first-degree relatives. *Psychol Med* 2007;1–15.
117. Kim D, Kim JW, Koo TH, Yun HR, Won SH. Shared and distinct neurocognitive endophenotypes of schizophrenia and psychotic bipolar disorder. *Clin Psychopharmacol Neurosci* 2015;13:94–102.
118. Ivleva EI, Morris DW, Osuji J et al. Cognitive endophenotypes of psychosis within dimension and diagnosis. *Psychiatry Res* 2012;196:38–44.
119. Burdick KE, Russo M, Frangou S et al. Empirical evidence for discrete neurocognitive subgroups in bipolar disorder: clinical implications. *Psychol Med* 2014;44:3083–3096.
120. Russo M, Van Rheenen TE, Shanahan M et al. Neurocognitive subtypes in patients with bipolar disorder and their unaffected siblings. *Psychol Med* 2017;1–14.
121. Kahn RS, Keefe RS. Schizophrenia is a cognitive illness: time for a change in focus. *JAMA Psychiatry* 2013;70:1107–1112.
122. Hegarty JD, Baldessarini RJ, Tohen M, Waternaux C, Oepen G. One hundred years of schizophrenia: a meta-analysis of the outcome literature. *Am J Psychiatry* 1994;151:1409–1416.

123. Lieberman JA. Effectiveness of antipsychotic drugs in patients with chronic schizophrenia: efficacy, safety and cost outcomes of CATIE and other trials. *J Clin Psychiatry* 2007;68:e04.
124. Buchanan RW, Davis M, Goff D et al. A summary of the FDA-NIMH-MATRICS workshop on clinical trial design for neurocognitive drugs for schizophrenia. *Schizophr Bull* 2005;31:5–19.
125. Jarskog LF, Lowy MT, Grove RA et al. A phase II study of a histamine H(3) receptor antagonist GSK239512 for cognitive impairment in stable schizophrenia subjects on antipsychotic therapy. *Schizophr Res* 2015;164:136–142.
126. Marx CE, Lee J, Subramaniam M et al. Proof-of-concept randomized controlled trial of pregnenolone in schizophrenia. *Psychopharmacology (Berl)* 2014;231:3647–3662.
127. Umbricht D, Keefe RS, Murray S et al. A randomized, placebo-controlled study investigating the nicotinic alpha7 agonist, RG3487, for cognitive deficits in schizophrenia. *Neuropsychopharmacology* 2014;39:1568–1577.
128. Weiser M, Heresco-Levy U, Davidson M et al. A multicenter, add-on randomized controlled trial of low-dose d-serine for negative and cognitive symptoms of schizophrenia. *J Clin Psychiatry* 2012;73:e728–e734.
129. Harvey PD, Ogasa M, Cucchiaro J, Loebel A, Keefe RS. Performance and interview-based assessments of cognitive change in a randomized, double-blind comparison of lurasidone vs. ziprasidone. *Schizophr Res* 2011;127:188–194.
130. Kreinin A, Bawakny N, Ritsner MS. Adjunctive pregnenolone ameliorates the cognitive deficits in recent-onset schizophrenia: an 8-week, randomized, double-blind, placebo-controlled trial. *Clin Schizophr Relat Psychoses* 2017;10:201–210.
131. Keefe RS, Meltzer HA, Dgetluck N et al. Randomized, double-blind, placebo-controlled study of encenicline, an alpha7 nicotinic acetylcholine receptor agonist, as a treatment for cognitive impairment in schizophrenia. *Neuropsychopharmacology* 2015;40:3053–3060.
132. Biesanz JC, West SG, Kwok OM. Personality over time: methodological approaches to the study of short-term and long-term development and change. *J Pers* 2003;71:905–941.

24

SOCIAL COGNITION IN PSYCHOSIS

Amy E. Pinkham and David L. Roberts

Social cognition has enjoyed a rapid rise in popularity within schizophrenia research over the last 20 years. Since publication of the first conceptual paper addressing social cognition in schizophrenia in 1997,[1] over 1,350 papers have included the linked terms "social cognition" and "schizophrenia," and 775 of these have appeared in the last 5 years (PsycINFO search conducted in May 2017). These studies span basic and applied research including animal models, human neuroimaging, human behavior, and clinical investigations ranging from behavioral to pharmacological interventions. Despite the majority of these studies focusing on schizophrenia, they also span diagnostic categories and include schizophrenia, schizoaffective disorder, clinical high-risk samples, schizotypy, and psychotic bipolar disorder. Thus, social cognition has emerged as a translational and transdiagnostic area of investigation.

Definitions of social cognition are numerous but can be broadly summarized as the perception, processing, and utilization of social information. Specific definitions include "the human ability and capacity to perceive the intentions and dispositions of others"[2, p. 28] and "the processes that subserve behavior in response to conspecifics, and, in particular, to those higher cognitive processes subserving the extreme, diverse, and flexible social behaviors."[3, p. 469] Combining several such definitions, an NIMH Workshop on Social Cognition in Schizophrenia characterized social cognition as "the mental operations that underlie social interactions, including perceiving, interpreting, and generating responses to the intentions, dispositions, and behaviors of others."[4, p. 1211] These definitions highlight the important link between social cognition and social behavior and indicate that social cognition may be critical for understanding the social dysfunction that is a hallmark feature of schizophrenia.

In this chapter, we review our current understanding of social cognition in psychosis by first discussing the domains typically included under the broader construct of social cognition and how individuals with psychosis perform within these domains. Next, we present information supporting the importance of social cognition with emphasis on work demonstrating that social cognition is a significant determinant of real-world functioning. We then review the growing body of literature addressing neurobiological bases of impaired social cognition. Finally, we examine behavioral and pharmacological treatment approaches and evaluate the efficacy of these interventions for improving social cognition and social functioning.

DOMAINS OF SOCIAL COGNITION

Social cognition has traditionally been parsed into domains representing specific abilities or skills such as emotion recognition or mentalizing. These domains are not empirically derived, however, and this has led to debate about which, and how many, domains should be included. In an attempt to provide consensus, Pinkham and colleagues conducted an extensive two-step survey of experts in the fields of schizophrenia, autism, and social psychology.[5] Step one solicited nominations for key domains of social cognition, and step two presented nominated domains back to the experts for validation. This process identified four core domains of social cognition:

1. *Emotion processing* is broadly defined as perceiving and using emotions.[4] At a lower perceptual level, it includes emotion perception/recognition (i.e., identifying and recognizing emotional displays from facial expressions and/or non-face cues such as voice), and at a higher level it includes understanding emotions and managing or regulating emotions.
2. *Social perception* refers to decoding and interpreting social cues in others.[6,7] It includes social context processing and social knowledge, which can be defined as knowing social rules and how these rules influence others' behaviors.[8]
3. *Mentalizing* is defined as the ability to represent the mental states of others, including the inference of intentions, dispositions, and/or beliefs.[9,10] Mentalizing is also referred to as theory of mind, mental state attribution, or cognitive empathy.[11]
4. *Attributional style/bias* describes the way in which individuals explain the causes, or make sense, of social events or interactions.[4,10]

While these domains represent expert consensus, they still lack empirical validation. Indeed, the few factor analyses available suggest social cognition may be best represented as a single factor[12] or parsed according to level of information processing (i.e., perception vs. inferential and regulatory processing) rather than domain of social information (e.g., emotion vs. mental state).[13,14] It is also important to note that these domains do not cover the entirety of social cognition. Topics such as self-perception, metacognition, and empathy all tap social cognition but have received relatively limited attention in psychosis. Nevertheless, the afore listed categories do highlight the multidimensional nature of social cognition and provide a useful framework for reviewing our current understanding of social cognitive abilities in individuals with psychosis.

EMOTION PROCESSING

Studies addressing the lower, perceptual level of emotion processing have largely focused on explicit recognition of emotion. While these studies have examined multiple modalities of communicating emotion such as prosody[15] and bodily expressions,[16] the majority have been devoted to emotion recognition from the faces of others. This work consistently demonstrates impairments in individuals with psychosis, and recent meta-analyses report large effect sizes (e.g., d = .91[17] and g = .89[18]) for the comparison of patients and healthy individuals.

Reviews of emotion recognition abilities in psychosis are numerous,[17,19] and several points are worth emphasis. First, emotion recognition deficits in psychosis are evident even when compared to individuals with other psychiatric disorders such as depression[20] and may exist on a continuum of increasing severity from psychotic bipolar disorder to schizoaffective disorder to schizophrenia.[21] Second, recognition impairments are greater for negative emotions, and particularly fear, than for positive emotions.[16,22] Patients also tend to misattribute negative emotions to neutral faces,[22] and this pattern is related to active paranoid ideation.[23] Third, emotion recognition abilities are also related to negative[24] and disorganized symptoms.[25] Fourth, individuals with schizophrenia show abnormal visual scanning of faces that includes reduced time viewing salient features (e.g., eyes and mouth),[26,27] which may contribute to emotion recognition impairments. Finally, despite thoroughly documented deficits in explicitly recognizing facial emotions, an emerging body of work suggests that implicit processing of facial emotion may be intact in psychosis.[28,29] These findings indicate that emotion recognition impairments may be the result of difficulties arising during contextual appraisal and integration of information rather than perception.[28,30]

Studies addressing the higher level skills associated with emotion processing such as emotion management have been less numerous, but they also show a large effect size (g = .88) for the difference in task performance between individuals with psychosis and healthy controls.[18] Likewise, investigation of emotion regulation strategies such as cognitive reappraisal or acceptance also demonstrates that individuals with schizophrenia engage in these processes less frequently than healthy controls.[31,32]

SOCIAL PERCEPTION

Social perception spans a number of abilities and skills related to the identification and utilization of social cues. These include biological motion detection, interpretation of voice and body cues (e.g., as measured by the Profile of Nonverbal Sensitivity),[33] and understanding typical relationships between individuals (e.g., as measured by the Relationships Across Domains Task).[34] Social knowledge, or the awareness of the procedures and goals associated with routine social situations (e.g., as measured by the Situational Feature Recognition Test),[35] is also included in social perception but is viewed as a more foundational skill that aids in social perception.

Work examining this range of skills in psychosis suggests that as skills become more complex, the level of impairment becomes more pronounced. For example, initial studies demonstrate that perception of abstract, rather than concrete, cues is more difficult for patients. Concrete social cues involve observations of an actor's behavior (e.g., "what is she doing?") and characteristics (e.g., "what is she wearing?"), whereas abstract cues consist of inferences of affect and goals. In a series of studies, Corrigan and colleagues found that individuals with schizophrenia were more sensitive to, and better able to recognize, concrete social cues rather than abstract ones.[35–37] More recently, a meta-analysis that separated social knowledge from social perception found that the effect size for social perception impairment (g = 1.04) was approximately twice as large as that for social knowledge (g = .54).[18] The effect size for biological motion detection (standardized mean difference = .66) also supports a pattern of incremental impairments across skills as this skill is often considered to be among the most basic social cognitive abilities.[38] Importantly however, measurement issues plague this area as many tasks show poor psychometric properties including floor and ceiling effects that might be related to findings of greater deficit on more complex tasks.[39,40]

MENTALIZING

Difficulties in understanding the mental states of others are wide-ranging in psychosis and occur in very basic skills such as identifying false beliefs[41] as well as highly nuanced skills such as recognizing faux pas.[42] Meta-analyses consistently report overall large effect sizes ranging from .99 to 1.25 when comparing the performance of patients and healthy individuals.[43,44] There is however, some debate about whether these are state- or trait-related impairments. Several studies report individuals in remission show more normative levels of performance,[45,46] therefore suggesting a state-dependent impairment. The previously mentioned meta-analyses however support a trait deficit by noting that mentalizing difficulties persist in remitted patients[43] and are not related to symptom severity.[44]

While severity of symptoms may not be critical, the specific type of symptoms one experiences may be related to the nature of mentalizing difficulties one displays. As first noted by Abu-Akel, poor performance on a mentalizing task could be due to either a lack of mentalizing (i.e., failure to infer the intentions of others) or to a "hyper-theory of mind" (i.e., generating too many possible inferences and then choosing the wrong one).[47] Frith later hypothesized that undermentalizing may be more likely in individuals with predominant negative symptoms whereas overmentalizing would be evident in individuals with prominent positive symptoms like paranoia.[48] These ideas have received relatively less attention due to the difficulties differentiating types of mentalizing errors; however, initial investigations support links between undermentalizing and negative symptoms and between overmentalizing and positive symptoms.[49,50]

ATTRIBUTIONAL STYLE/BIAS

The majority of early work addressing attributional style in psychosis highlighted two biases that are common in individuals with paranoia or persecutory ideation. The first, the externalizing bias, refers to the tendency to make external attributions for negative outcomes.[51,52] The second bias, the personalizing bias, can been seen largely as a clarification of the externalizing bias. Specifically, when making an external attribution, individuals can either attribute the event to a person (i.e., an external personal attribution) or to situational factors (i.e., an external situational attribution). Several studies suggest that individuals with more severe persecutory delusions commit a personalizing bias by most often attributing negative events to others.[52,53] Unfortunately, however, this literature is not without inconsistencies. A recent meta-analysis failed to find evidence for either an externalizing or a personalizing bias in individuals with psychosis relative to healthy individuals.[18] This lack of group differences persisted even when examining a subset of patients with persecutory delusions.

A promising new avenue for attribution work is investigation of a hostile attributional style. Individuals with schizophrenia make more hostile attributions than healthy controls,[40,54] and in direct group comparisons, patients with high paranoia show greater hostility bias than nonparanoid patients.[55] Correlational analyses also show links between hostile attributions and increased rates of paranoia in both patients[56] and healthy individuals.[57] The tendency to make hostile attributions is also highly related to the likelihood of engaging in aggressive or violent acts[58] and is among the best predictors of social functioning.[59] A hostile attributional style may therefore be an important component for understanding symptoms of psychosis as well as functional impairment.

WHY IS SOCIAL COGNITION IMPORTANT FOR PSYCHOSIS?

At the most basic level, social cognition represents a significant area of impairment for individuals with psychosis. As indicated earlier, effect sizes for performance differences between patients and controls fall into the medium-to-large range, and these impairments are evident across social cognitive domains. Evidence is also building in support of social cognition as an endophenotype.[60,61] For example, emotion-processing impairments are evident in first-episode individuals,[62] individuals at clinical high risk for developing psychosis,[63] and first-degree relatives of individuals with psychosis.[64] The same is true for mentalizing and social perception.[65,66] Additionally, both cross-sectional[67] and longitudinal studies[68,69] demonstrate that social cognitive impairment is stable across phases of illness and time. Thus, these are long-standing impairments that are unlikely to be the result of a degenerative process after illness onset.

Correlational studies,[70] factor analyses,[71] and differential deficit designs[72] all also indicate that social cognition is largely independent from neurocognition (e.g., attention, memory, etc.). Such findings highlight that social cognitive impairments are not simply secondary to difficulties in more core cognitive processes.[73]

Perhaps most importantly, social cognition contributes to real-world outcomes like community and social functioning.[74] Social cognition has a larger influence on outcomes than neurocognition[75] and mediates the relation between neurocognition and social outcomes.[76] Thus, there is evidence that social cognition serves as both an independent predictor and a mediator of outcome, which suggests that improving social cognition may generalize to better daily functioning. As explained in more detail later, this is in fact the case. Remediation of social cognitive impairment leads to better real-world social outcomes, including social adjustment, social functioning, social relationships, and social skills.[77,78]

A NEUROBIOLOGICAL BASIS FOR SOCIAL COGNITIVE IMPAIRMENTS

The strong links between social cognition and behavior have prompted explorations of the mechanisms that may underlie social cognitive impairments, and attention has been devoted to examining neural processing of social stimuli. This work is rooted in social neuroscience and the robust findings that some brain regions and networks appear to be specialized for processing social information. In first describing this "social brain," Brothers emphasized the role of the orbitofrontal cortex, superior temporal sulcus, and amygdala.[79] These regions remain key points of investigation; however, as knowledge has accumulated, the field has shifted from focusing on structures in isolation to emphasizing networks that involve multiple neural regions with reciprocal connections between them.[80] Interestingly, many of these networks map onto the social cognitive domains that have been of primary focus in schizophrenia.[81] For example, a "social perception" network, centered on the amygdala and including orbitofrontal cortex, striatum, nucleus accumbens, and visual cortices, has been implicated in processing emotion and socially salient stimuli.[82] Similarly, a "mentalizing" network comprising medial prefrontal cortex, superior temporal sulcus and surrounding superior temporal gyrus, and temporo-parietal junction[83,84] has been tied to thinking about others and reflecting on oneself. Identification of such networks raises the possibility that abnormal functioning of, or within, these networks may result in social cognitive impairment.

To date, a large body of literature has accrued to support this hypothesis by reporting that individuals with psychosis show both structural and functional abnormalities in social cognitive neural networks (reviewed in [85,86]). For example, meta-analyses of neural activation during tasks of facial emotion recognition demonstrate reduced activation for patients relative to controls in amygdala, fusiform gyrus, anterior cingulate cortex, and thalamus.[87,88] It should be noted though that results concerning amygdala are nuanced in that patients may actually show increased amygdala responses to neutral stimuli that yield small differences between neutral and emotional conditions. This lack of difference between conditions can be easily misinterpreted as reduced amygdala activation, which may confound many interpretations of group differences.[89] Reductions in activation in medial prefrontal cortex in patients are also widely reported across a variety of nonverbal mentalizing tasks.[90,91] Thus, abnormal functioning of these networks does appear to be a viable mechanism for social cognitive impairment in psychosis.

Importantly, several neuroimaging studies have also demonstrated that activation of these networks is predictive of functional outcomes. In our own work, we found that greater modulation of neural responses to social stimuli were significantly correlated with improved social functioning in both healthy individuals and individuals with schizophrenia.[92] In another study, we again found a relationship between activation and functioning in patients such that the amount of activation in the amygdala in response to direct gaze expressions of anger was significantly and positively correlated to levels of social and occupational functioning.[93] Such relationships have also been found for mentalizing,[94,95] which provides a strong argument for causal links between neural activation, social cognition, and functional outcome. These data also highlight the importance of treating social cognitive impairments and pursuing remediation strategies that will normalize neural processes.

SOCIAL COGNITIVE TREATMENT

Over the past 15 years, social cognitive treatment for schizophrenia has rapidly developed, and this work has been summarized in several recent meta-analyses and review papers.[96,97] One limitation to these reviews is that scant attention has been paid to the putative mechanisms of social cognitive treatment. The current section provides a brief critical analysis of social cognitive treatments with an emphasis on mechanisms of action.

The earliest intervention studies explicitly targeting social cognition applied established learning techniques to social stimuli. These techniques include drill-and-repeat practice, errorless learning, positive reinforcement, feature abstraction, verbalization, and self-instruction. Use of these techniques led to improved performance on measures of face emotion perception,[98] social perception,[99] and mentalizing.[100] Subsequent studies have continued to incorporate basic learning principles to varying degrees. Some approaches have adhered closely to traditional training principles, with efforts to link these techniques to treatment targets specified by cognitive neuroscience. We refer to these as "bottom-up" approaches. Other approaches, which we call "top-down," have departed somewhat from traditional learning techniques, and applied principles from social psychology and psychotherapeutic modalities.

"BOTTOM-UP" INTERVENTION APPROACHES

Treatment approaches influenced by cognitive neuroscience endeavor to strengthen the functioning of social cognitive brain circuitry by targeting narrow, clearly definable social cognitive abilities with massed trials of drill-and-repeat practice. This approach has been most successfully applied to lower-level domains of social cognition in which targets can be tightly defined and tasks narrowly operationalized. For example, emotion perception tasks require subjects to interpret a well-defined stimulus (a human face or voice) and select an emotion from a small set of response options (e.g., *happy*, *angry*, etc.). This enables use of drill-and-repeat training with hundreds of similar items. Early emotion perception training programs utilized this strategy paired with corrective feedback (e.g., [98]). Programs like the Micro-Expression Training Tool[101] extended this work by targeting more subtle and differentiated expressions of facial emotion. Most recently, approaches such as the SocialVille training program have further refined face emotion perception training by incorporating facial movement, facial emotion memory, and discrimination tasks as well as computer-based adaptive-design algorithms to deliver graded-difficulty and errorless-learning paradigms.[102]

Efforts to apply bottom-up techniques to higher-level social cognitive processes (e.g., mentalizing and attributional bias) have been challenged by developing narrowly operationalized behavioral tasks that lend themselves to drill-and-repeat learning. Unlike emotion perception, mentalizing offers neither a cleanly definable stimulus set nor a predefined set of response options, both of which are constrained only by the limits of what a person could be thinking or intending. Training paradigms for mentalizing have tended to follow the same structure as assessment measures in this area, using story-based paradigms in which the participant is told or shown the first portion of a social vignette, and then asked to use this information to infer the thoughts or intentions of a story character. This use of vignettes is problematic because it draws heavily on brain functions other than mentalizing—including sustained attention, working memory, verbal processing, and social knowledge[103]—in a manner that is not consistent or controlled across items. Additionally, each item in a story-based paradigm is relatively long and labor intensive, diminishing the number of trials that a participant can complete per training session, and thereby decreasing the ability to engage bottom-up learning mechanisms.

"TOP-DOWN" INTERVENTION APPROACHES

Top-down techniques target higher-order social cognition by teaching participants to explicitly apply interpretive strategies to complex social stimuli. This approach, seen in interventions such as Social Cognitive Skills Training[104] and Social Cognition and Interaction Training (SCIT),[105] typically uses a group training format in which participants analyze social pictures, videos, and scenarios as well as social challenges in their own day-to-day lives. Training is less time intensive than bottom-up approaches and used more diffuse, nonspecific social learning and interaction opportunities by providing a context in which to teach and practice applying social cognitive strategies. It is assumed however that top-down approaches do not provide sufficient rehearsal to retrain the automatic functioning of lower-level brain systems. Rather, limited rehearsal is used to help participants comprehend and practice controlled, effortful use of compensatory social cognitive strategies. For example, SCIT uses face emotion perception training paradigms like those used in bottom-up approaches, but with many fewer training trials. Instead, guided discussion is used to help participants: (1) understand links between facial clues and specific emotions (e.g., smile = happy), (2) learn to yoke one's confidence in judgments to the type and quantity of clues, (3) appreciate that it is impossible to be 100% certain of another person's emotional state, and (4) develop self-awareness of the role one's own emotional state plays in influencing interpretation of others' emotions.

To increase participants' self-awareness of their own interpretive biases, top-down approaches use techniques also employed in cognitive psychotherapy and in social psychological debiasing paradigms. For example, SCIT uses an approach similar to cognitive therapy's Antecedent-Belief-Consequence[106] exercise to help participants analyze difficult social experiences in their lives.

A key challenge faced by top-down approaches is that neurocognitive deficits in schizophrenia hinder participants' ability to deliberatively learn, recall, and apply new strategies—a challenge that is circumvented in bottom-up training approaches through use of repetition. Top-down approaches have taken on this challenge in several ways. First, top-down approaches maximize the salience of training content by focusing on the real-world level of social cognition. This creates frequent opportunities for humor, shared meaning, empathic attunement, and self-relevant experience during training sessions, which contributes both to high patient satisfaction and to top-down programs being widely adopted in routine clinical settings.[105]

Top-down approaches also achieve learning with minimal rehearsal by maximizing the simplicity, accessibility, and memorability of the strategic approaches that they teach. For example, *Mary/Eddie/Bill* is an approach to both mentalizing and attributional bias that teaches patients three basic attributional styles by linking each to a colorful character prototype. My-fault Mary exhibits an internalizing-personalizing style, Easy Eddie a situationalizing-style, and Blaming Bill an externalizing-personalizing style. Patients learn to flexibly interpret all social situations from each of these three perspectives, which may improve perspective-taking and decrease attributional bias. Use of these character prototypes enable patients to quickly learn this heuristic, and to retain the ability to use it for weeks after training has stopped.[107,108]

At this point, it is unclear whether bottom-up or top-down treatment mechanisms are more effective, or whether one or the other is more feasible or acceptable to patients. Top-down approaches delivered in a group format have the advantage of being more ecologically valid, whereas computer-aided bottom-up approaches have the potential advantage of improving brain efficiency without reliance on patients' effortfully deploying strategies in their day-to-day lives. As data accumulate from currently underway trials (reviewed later), and as outcome measurement becomes more specific and sensitive, it will be important to examine whether bottom-up or top-down approaches differ in efficacy, and if differences are moderated by patient level factors.

EFFICACY OF BEHAVIORALLY BASED SOCIAL COGNITIVE TREATMENTS

As noted earlier, several recent meta-analyses and review papers have summarized the efficacy of social cognitive treatments for schizophrenia. The most recent meta-analysis included 16 studies and found large weighted effect sizes for emotion perception (d = .87) and medium-to-large effect sizes for mentalizing (d = .70).[97] A smaller set of studies included measures of social perception (four studies) and attributional bias (seven studies), and yielded large (d = 1.29) and small-to-medium (d = .30–.52) effect sizes, respectively.[97] From a mechanistic standpoint, it is notable that two recent reviews[96,97] have concluded that neurocognitive training (e.g., memory and attention) is *not* a necessary "building block" for social cognitive improvement.

Generalization to social functioning arguably is the most important outcome in determining the value of social cognitive treatments. An early meta-analysis reported medium-to-large effect sizes on social functioning (d = .78).[109] However, Horan and Green[96] note that although six of nine studies in this meta-analysis showed treatment-related improvements in social functioning, these studies suffer from a range of methodological limitations, and they are particularly inconsistent in their use of functional outcome measures.

Indeed, at this relatively early point in the investigation of social cognitive treatments, methodological limitations considerably hinder the drawing of strong conclusions. Studies are limited by small sample size, use of quasi-experimental design, nonblinded assessment, insufficient monitoring of treatment fidelity, and use of a diverse set of outcome measures with inconsistent psychometric properties. Only two small studies have examined durability of social cognitive treatment effects after treatment termination.[110,111] It is also notable that most studies have compared social cognitive intervention to treatment as usual rather than an active control treatment. Although some studies have compared social cognitive treatment to neurocognitive remediation, no studies have compared two social cognitive treatments with different putative mechanisms of action, and no reviews have attempted to compare outcomes of studies that use bottom-up versus top-down approaches to treatment.

Finally, the potential effects of social cognitive treatment on neural systems are only beginning to be elucidated. One recent article[112] reviewed these effects across 11 studies. Although the diversity of methods and small sample sizes temper conclusions, the majority of studies registered positive intervention-related changes. This remains an exciting area for future investigation.

PHARMACOLOGICAL/ SOMATIC TREATMENTS

Several studies have established that antipsychotic medications do not appreciably improve social cognition in schizophrenia.[113,114] Very recently, there has been a surge of research on oxytocin (OT), based on evidence that OT increases the perceived salience of social cues.[115] Several small trials have found modest improvements in social cognition after acute OT administration,[116] and improvements in both social cognition and symptoms after 14 days of OT administration.[117] Trials combining OT with behavioral social cognitive treatment have registered mixed results.[118,119] Notably, evidence has emerged that OT's salience-mediated effects may have a negative impact in subgroups with social sensitivity, including patients with prominent paranoia.[120] This possibility suggests that symptom-based subgroups may have different mechanisms of social cognitive difficulty—such as undermentalizing or insufficient social sensitivity versus over-mentalizing or excessive social sensitivity—and that these differences may have important treatment implications. Overall, caution and more research are needed to fully understand the potential of OT as a treatment to improve social cognition in schizophrenia.

A final area of potential somatic treatments for social cognitive dysfunction is brain stimulation techniques, such as transcranial direct current stimulation (tDCS). This approach has shown some evidence of efficacy in remediation of basic neurocognitive functions, such as memory and attention.[121] A preliminary study in social cognition randomized 36 participants to single-session anodal (stimulating) tDCS, cathodal (inhibiting) tDCS, or sham tDCS and found significant improvement in the anodal condition on one of four social cognitive tasks (emotion perception),[122] warranting further investigation of this treatment approach.

CONCLUSIONS

From the information presented in this chapter, four primary conclusions come to the forefront. First, individuals with psychosis display deficits or biases in multiple domains of social cognition that appear to be largely independent of neurocognition. Second, social cognitive deficits are critically important for understanding impaired social and vocational functioning in individuals with psychosis. Third, abnormal functioning of the "social brain" may be a mechanism for impaired social cognition and functioning. Fourth, there is solid evidence that psychosocial treatments can be used to improve emotion perception and theory of mind in schizophrenia, and some evidence that doing so improves social functioning. The combination of these factors leaves little question about the importance of social cognition for furthering our understanding of psychosis and provides a solid foundation for continuing both basic and applied research programs that will move the field toward optimized interventions.

REFERENCES

1. Penn DL, Corrigan PW, Bentall RP, Racenstein J, Newman L. Social cognition in schizophrenia. *Psychological Bulletin.* 1997;121(1):114.
2. Brothers L. The social brain: a project for integrating primate behaviour and neuropsychology in a new domain. *Concepts in Neuroscience.* 1990;1:27–51.
3. Adolphs R. Social cognition and the human brain. *Trends in Cognitive Sciences.* 1999;3(12):469–479.
4. Green MF, Penn DL, Bentall R, et al. Social cognition in schizophrenia: an NIMH workshop on definitions, assessment, and research opportunities. *Schizophrenia Bulletin.* 2008;34(6):1211–1220.
5. Pinkham AE, Penn DL, Green MF, Buck B, Healey K, Harvey PD. The social cognition psychometric evaluation study: results of the expert survey and RAND panel. *Schizophr Bull.* 2014;40(4):813–823.
6. Penn DL, Ritchie M, Francis J, Combs D, Martin J. Social perception in schizophrenia: the role of context. *Psychiatry Res.* 2002;109(2):149–159.
7. Sergi MJ, Green MF. Social perception and early visual processing in schizophrenia. *Schizophrenia Research.* 2003;59(2–3):233–241.
8. Corrigan PW, Green MF. Schizophrenic patients' sensitivity to social cues: the role of abstraction. *The American Journal of Psychiatry.* 1993;150(4):589–594.
9. Frith CD. *The Cognitive Neuropsychology of Schizophrenia.* Psychology Press; 1992.
10. Penn DL, Addington J, Pinkham A. Social cognitive impairments. In: Lieberman JA, Stroup TS, Perkins DO, eds. *A Textbook of Schizophrenia.* Washington: American Psychiatry Press; 2006.
11. Shamay-Tsoory SG. The neural bases for empathy. *The Neuroscientist: A Review Journal Bringing Neurobiology, Neurology and Psychiatry.* 2011;17(1):18–24.
12. Browne J, Penn DL, Raykov T, et al. Social cognition in schizophrenia: factor structure of emotion processing and theory of mind. *Psychiatry Res.* 2016;242:150–156.
13. Mancuso F, Horan WP, Kern RS, Green MF. Social cognition in psychosis: multidimensional structure, clinical correlates, and relationship with functional outcome. *Schizophrenia Research.* 2011;125(2-3):143–151.
14. Lin YC, Wynn JK, Hellemann G, Green MF. Factor structure of emotional intelligence in schizophrenia. *Schizophrenia Research.* 2012;139(1–3):78–81.
15. Hoekert M, Kahn RS, Pijnenborg M, Aleman A. Impaired recognition and expression of emotional prosody in schizophrenia: Review and meta-analysis. *Schizophrenia Research.* 2007;96(1):135–145.
16. Bigelow NO, Paradiso S, Adolphs R, et al. Perception of socially relevant stimuli in schizophrenia. *Schizophrenia Research.* 2006;83(2–3):257–267.
17. Kohler CG, Walker JB, Martin EA, Healey KM, Moberg PJ. Facial emotion perception in schizophrenia: a meta-analytic review. *Schizophr Bull.* 2010;36(5):1009–1019.
18. Savla GN, Vella L, Armstrong CC, Penn DL, Twamley EW. Deficits in domains of social cognition in schizophrenia: a meta-analysis of the empirical evidence. *Schizophr Bull.* 2013;39(5):979–992.
19. Pinkham AE, Gur RE, Gur RC. Affect recognition deficits in schizophrenia: neural substrates and psychopharmacological implications. *Expert Review of Neurotherapeutics.* 2007;7(7):807–816.

20. Weniger G, Lange C, Ruther E, Irle E. Differential impairments of facial affect recognition in schizophrenia subtypes and major depression. *Psychiatry Res.* 2004;128(2):135–146.
21. Ruocco AC, Reilly JL, Rubin LH, et al. Emotion recognition deficits in schizophrenia-spectrum disorders and psychotic bipolar disorder: Findings from the Bipolar-Schizophrenia Network on Intermediate Phenotypes (B-SNIP) study. *Schizophrenia Research.* 2014;158(1–3):105–112.
22. Kohler CG, Turner TH, Bilker WB, et al. Facial emotion recognition in schizophrenia: intensity effects and error pattern. *Am J Psychiatry.* 2003;160(10):1768–1774.
23. Pinkham AE, Brensinger C, Kohler C, Gur RE, Gur RC. Actively paranoid patients with schizophrenia over attribute anger to neutral faces. *Schizophrenia Research.* 2011;125(2-3):174–178.
24. Sachs G, Steger-Wuchse D, Kryspin-Exner I, Gur RC, Katschnig H. Facial recognition deficits and cognition in schizophrenia. *Schizophrenia Research.* 2004;68(1):27–35.
25. Ventura J, Wood RC, Jimenez AM, Hellemann GS. Neurocognition and symptoms identify links between facial recognition and emotion processing in schizophrenia: meta-analytic findings. *Schizophrenia Research.* 2013;151(1–3):78–84.
26. Sasson N, Tsuchiya N, Hurley R, et al. Orienting to social stimuli differentiates social cognitive impairment in autism and schizophrenia. *Neuropsychologia.* 2007;45(11):2580–2588.
27. Simpson C, Pinkham AE, Kelsven S, Sasson NJ. Emotion recognition abilities across stimulus modalities in schizophrenia and the role of visual attention. *Schizophrenia Research.* 2013;151(1-3):102–106.
28. Kring AM, Siegel EH, Barrett LF. Unseen affective faces influence person perception judgments in schizophrenia. *Clinical Psychological Science: A Journal of the Association for Psychological Science.* 2014;2(4):443–454.
29. Shasteen JR, Pinkham AE, Kelsven S, Ludwig K, Payne BK, Penn DL. Intact implicit processing of facial threat cues in schizophrenia. *Schizophrenia Research.* 2016;170(1):150–155.
30. Sasson NJ, Pinkham AE, Weittenhiller LP, Faso DJ, Simpson C. Context effects on facial affect recognition in schizophrenia and autism: behavioral and eye-tracking evidence. *Schizophr Bull.* 2016;42(3):675–683.
31. Kimhy D, Vakhrusheva J, Jobson-Ahmed L, Tarrier N, Malaspina D, Gross JJ. Emotion awareness and regulation in individuals with schizophrenia: implications for social functioning. *Psychiatry Research.* 2012;200(2):193–201.
32. Perry Y, Henry JD, Grisham JR. The habitual use of emotion regulation strategies in schizophrenia. *British Journal of Clinical Psychology.* 2011;50(2):217–222.
33. Rosenthal R, DePaulo BM, Jall JA. *The PONS Test Manual: Profile of Nonverbal Sensitivity.* Irvington Pub; 1979.
34. Sergi MJ, Fiske AP, Horan WP, et al. Development of a measure of relationship perception in schizophrenia. *Psychiatry Research.* 2009;166(1):54–62.
35. Corrigan PW, Green MF. The Situational Feature Recognition Test: A measure of schema comprehension for schizophrenia. *International Journal of Methods in Psychiatric Research.* 1993.
36. Corrigan PW, Garman A, Nelson D. Situational feature recognition in schizophrenic outpatients. *Psychiatry Research.* 1996;62(3):251–257.
37. Corrigan PW, Nelson DR. Factors that affect social cue recognition in schizophrenia. *Psychiatry Research.* 1998;78(3):189–196.
38. Okruszek Ł, Pilecka I. Biological motion processing in schizophrenia—systematic review and meta-analysis. *Schizophrenia Research.* 2017.
39. Kern RS, Penn DL, Lee J, et al. Adapting social neuroscience measures for schizophrenia clinical trials, part 2: trolling the depths of psychometric properties. *Schizophr Bull.* 2013;39(6):1201–1210.
40. Pinkham AE, Penn DL, Green MF, Harvey PD. Social cognition psychometric evaluation: results of the initial psychometric study. *Schizophr Bull.* 2016;42(2):494–504.
41. Brune M, Abdel-Hamid M, Lehmkamper C, Sonntag C. Mental state attribution, neurocognitive functioning, and psychopathology: what predicts poor social competence in schizophrenia best? *Schizophrenia Research.* 2007;92(1–3):151–159.
42. Zhu CY, Lee TM, Li XS, Jing SC, Wang YG, Wang K. Impairments of social cues recognition and social functioning in Chinese people with schizophrenia. *Psychiatry and Clinical Neurosciences.* 2007;61(2):149–158.
43. Sprong M, Schothorst P, Vos E, Hox J, van Engeland H. Theory of mind in schizophrenia: meta-analysis. *Br J Psychiatry.* 2007;191:5–13.
44. Chung YS, Barch D, Strube M. A meta-analysis of mentalizing impairments in adults with schizophrenia and autism spectrum disorder. *Schizophrenia Bulletin.* 2014;40(3):602–616.
45. Drury VM, Robinson EJ, Birchwood M. "Theory of mind" skills during an acute episode of psychosis and following recovery. *Psychological Medicine.* 1998;28(5):1101–1112.
46. Pousa E, Duno R, Brebion G, David AS, Ruiz AI, Obiols JE. Theory of mind deficits in chronic schizophrenia: evidence for state dependence. *Psychiatry Res.* 2008;158(1):1–10.
47. Abu-Akel A. Impaired theory of mind in schizophrenia. *Pragmatics & Cognition.* 1999;7(2):247–282.
48. Frith CD. Schizophrenia and theory of mind. *Psychological Medicine.* 2004;34(03):385–389.
49. Fretland RA, Andersson S, Sundet K, Andreassen OA, Melle I, Vaskinn A. Theory of mind in schizophrenia: error types and associations with symptoms. *Schizophrenia Research.* 2015;162(1):42–46.
50. Montag C, Dziobek I, Richter IS, et al. Different aspects of theory of mind in paranoid schizophrenia: evidence from a video-based assessment. *Psychiatry Research.* 2011;186(2):203–209.
51. Janssen I, Versmissen D, Campo JA, Myin-Germeys I, Van Os J, Krabbendam L. Attribution style and psychosis: evidence for an externalizing bias in patients but not individuals at high risk. *Psychological Medicine.* 2006;36:771–778.
52. Langdon R, Ward PB, Coltheart M. Reasoning anomalies associated with delusions in schizophrenia. *Schizophr Bull.* 2010;36(2):321–330.
53. Mehl S, Landsberg MW, Schmidt A-C, et al. Why do bad things happen to me? Attributional style, depressed mood, and persecutory delusions in patients with schizophrenia. *Schizophrenia Bulletin.* 2014;40(6):1338–1346.
54. Buck B, Iwanski C, Healey KM, et al. Improving measurement of attributional style in schizophrenia; a psychometric evaluation of the Ambiguous Intentions Hostility Questionnaire (AIHQ). *Journal of Psychiatric Research.* 2017.
55. Pinkham AE, Harvey PD, Penn DL. Paranoid individuals with schizophrenia show greater social cognitive bias and worse social functioning than non-paranoid individuals with schizophrenia. *Schizophrenia Research: Cognition.* 2016;3:33–38.
56. Buck BE, Pinkham AE, Harvey PD, Penn DL. Revisiting the validity of measures of social cognitive bias in schizophrenia: Additional results from the Social Cognition Psychometric Evaluation (SCOPE) study. *British Journal of Clinical Psychology.* 2016;55(4):441–454.
57. Combs DR, Penn DL, Wicher M, Waldheter E. The Ambiguous Intentions Hostility Questionnaire (AIHQ): a new measure for evaluating hostile social-cognitive biases in paranoia. *Cognitive neuropsychiatry.* 2007;12(2):128–143.
58. Harris ST, Oakley C, Picchioni MM. A systematic review of the association between attributional bias/interpersonal style, and violence in schizophrenia/psychosis. *Aggression and Violent Behavior.* 2014;19(3):235–241.
59. Lahera G, Herrera S, Reinares M, et al. Hostile attributions in bipolar disorder and schizophrenia contribute to poor social functioning. *Acta Psychiatrica Scandinavica.* 2015;131(6):472–482.
60. Green MF, Horan WP, Lee J. Social cognition in schizophrenia. *Nature Reviews Neuroscience.* 2015.
61. Gur RC, Gur RE. Social cognition as an RDoC domain. *American Journal of Medical Genetics Part B: Neuropsychiatric Genetics.* 2016;171(1):132–141.
62. Barkl SJ, Lah S, Harris AW, Williams LM. Facial emotion identification in early-onset and first-episode psychosis: a systematic review with meta-analysis. *Schizophrenia Research.* 2014;159(1):62–69.

63. Kohler CG, Richard JA, Brensinger CM, et al. Facial emotion perception differs in Young persons at genetic and clinical high-risk for psychosis. *Psychiatry research.* 2014;216(2):206–212.
64. Allott KA, Rice S, Bartholomeusz CF, et al. Emotion recognition in unaffected first-degree relatives of individuals with first-episode schizophrenia. *Schizophrenia Research.* 2015;161(2):322–328.
65. Barbato M, Liu L, Cadenhead KS, et al. Theory of mind, emotion recognition and social perception in individuals at clinical high risk for psychosis: findings from the NAPLS-2 cohort. *Schizophrenia Research: Cognition.* 2015;2(3):133–139.
66. Bora E, Pantelis C. Theory of mind impairments in first-episode psychosis, individuals at ultra-high risk for psychosis and in first-degree relatives of schizophrenia: systematic review and meta-analysis. *Schizophrenia Research.* 2013;144(1):31–36.
67. Comparelli A, Corigliano V, De Carolis A, et al. Emotion recognition impairment is present early and is stable throughout the course of schizophrenia. *Schizophrenia Research.* 2013;143(1):65–69.
68. Horan WP, Green MF, DeGroot M, et al. Social cognition in schizophrenia, part 2: 12-month stability and prediction of functional outcome in first-episode patients. *Schizophrenia Bulletin.* 2012;38(4):865–872.
69. Piskulic D, Liu L, Cadenhead KS, et al. Social cognition over time in individuals at clinical high risk for psychosis: findings from the NAPLS-2 cohort. *Schizophrenia Research.* 2016;171(1):176–181.
70. Ventura J, Wood RC, Hellemann GS. Symptom domains and neurocognitive functioning can help differentiate social cognitive processes in schizophrenia: a meta-analysis. *Schizophr Bull.* 2011; advance access.
71. Sergi MJ, Rassovsky Y, Widmark C, et al. Social cognition in schizophrenia: relationships with neurocognition and negative symptoms. *Schizophr Res.* 2007;90(1–3):316–324.
72. Pinkham AE, Sasson NJ, Kelsven S, Simpson CE, Healey K, Kohler C. An intact threat superiority effect for nonsocial but not social stimuli in schizophrenia. *Journal of Abnormal Psychology.* 2014;123(1):168.
73. Pinkham AE. Social cognition and its relationship to neurocognition. *Cognitive Impairment in Schizophrenia: Characteristics, Assessment and Treatment.* 2013:126.
74. Couture SM, Penn DL, Roberts DL. The functional significance of social cognition in schizophrenia: a review. *Schizophrenia Bulletin.* 2006;32(1):S44–SS63.
75. Fett A-KJ, Viechtbauer W, Penn DL, van Os J, Krabbendam L. The relationship between neurocognition and social cognition with functional outcomes in schizophrenia: a meta-analysis. *Neuroscience & Biobehavioral Reviews.* 2011;35(3):573–588.
76. Schmidt SJ, Mueller DR, Roder V. Social cognition as a mediator variable between neurocognition and functional outcome in schizophrenia: empirical review and new results by structural equation modeling. *Schizophrenia Bulletin.* 2011;37(suppl 2):S41–S54.
77. Kurtz MM, Richardson CL. Social cognitive training for schizophrenia: a meta-analytic investigation of controlled research. *Schizophr Bull.* 2012;38(5):1092–1104.
78. Fiszdon JM. Introduction to social cognitive treatment approaches for schizophrenia. In: Roberts DL, Penn DL, eds. *Social Cognition in Schizophrenia.* New York, NY: Oxford University Press; 2013:285–310.
79. Brothers L. The social brain: a project for integrating primate behavior and neurophysiology in a new domain. *Concepts in Neuroscience.* 1990;1:27–51.
80. Stanley DA, Adolphs R. Toward a neural basis for social behavior. *Neuron.* 2013;80(3):816–826.
81. Kennedy DP, Adolphs R. The social brain in psychiatric and neurological disorders. *Trends in Cognitive Sciences.* 2012;16(11):559–572.
82. Adolphs R. What does the amygdala contribute to social cognition? *Annals of the New York Academy of Sciences.* 2010;1191:42–61.
83. Amodio DM, Frith CD. Meeting of minds: the medial frontal cortex and social cognition. *Nature Reviews.* 2006;7(4):268–277.
84. Saxe R. Uniquely human social cognition. *Current Opinion in Neurobiology.* 2006;16(2):235–239.
85. Pinkham AE, Penn DL, Perkins DO, Lieberman J. Implications for the neural basis of social cognition for the study of schizophrenia. *The American Journal of Psychiatry.* 2003;160(5):815–824.
86. Pinkham AE. The social cognitive neuroscience of schizophrenia. In: Roberts DL, Penn DL, eds. *Social Cognition in Schizophrenia: From Evidence to Treatment.* New York, NY: Oxford University Press; 2013:263–284.
87. Delvecchio G, Sugranyes G, Frangou S. Evidence of diagnostic specificity in the neural correlates of facial affect processing in bipolar disorder and schizophrenia: a meta-analysis of functional imaging studies. *Psychological Medicine.* 2013;43(03):553–569.
88. Taylor SF, Kang J, Brege IS, Tso IF, Hosanagar A, Johnson TD. Meta-analysis of functional neuroimaging studies of emotion perception and experience in schizophrenia. *Biological Psychiatry.* 2012;71(2):136–145.
89. Anticevic A, Van Snellenberg JX, Cohen RE, Repovs G, Dowd EC, Barch DM. Amygdala recruitment in schizophrenia in response to aversive emotional material: a meta-analysis of neuroimaging studies. *Schizophr Bull.* 2012;38(3):608–621.
90. Benedetti F, Bernasconi A, Bosia M, et al. Functional and structural brain correlates of theory of mind and empathy deficits in schizophrenia. *Schizophrenia Research.* 2009;114(1-3):154–160.
91. Brune M, Ozgurdal S, Ansorge N, et al. An fMRI study of "theory of mind" in at-risk states of psychosis: comparison with manifest schizophrenia and healthy controls. *NeuroImage.* 2011;55(1):329–337.
92. Pinkham AE, Hopfinger JB, Ruparel K, Penn DL. An investigation of the relationship between activation of a social cognitive neural network and social functioning. *Schizophrenia Bulletin.* 2008;34(4):688–697.
93. Pinkham AE, Loughead J, Ruparel K, Overton E, Gur RE, Gur RC. Abnormal modulation of amygdala activity in schizophrenia in response to direct-and averted-gaze threat-related facial expressions. *American Journal of Psychiatry.* 2011;168(3):293–301.
94. Dodell-Feder D, Tully LM, Lincoln SH, Hooker CI. The neural basis of theory of mind and its relationship to social functioning and social anhedonia in individuals with schizophrenia. *NeuroImage: Clinical.* 2014;4:154–163.
95. Smith MJ, Schroeder MP, Abram SV, et al. Alterations in brain activation during cognitive empathy are related to social functioning in schizophrenia. *Schizophrenia Bulletin.* 2014:sbu023.
96. Horan WP, Green MF. Treatment of social cognition in schizophrenia: current status and future directions. *Schizophrenia Research.* 2017.
97. Kurtz MM, Gagen E, Rocha NB, Machado S, Penn DL. Comprehensive treatments for social cognitive deficits in schizophrenia: a critical review and effect-size analysis of controlled studies. *Clinical Psychology Review.* 2016;43:80–89.
98. Silver H, Goodman C, Knoll G, Isakov V. Brief emotion training improves recognition of facial emotions in chronic schizophrenia: a pilot study. *Psychiatry Research.* 2004;128(2):147–154.
99. Corrigan PW, Hirschbeck JN, Wolfe M. Memory and vigilance training to improve social perception in schizophrenia. *Schizophrenia Research.* 1995;17(3):257–265.
100. Sarfati Y, Passerieux C, Hardy-Baylé M-C. Can verbalization remedy the theory of mind deficit in schizophrenia? *Psychopathology.* 2000;33(5):246–251.
101. Russell TA, Chu E, Phillips ML. A pilot study to investigate the effectiveness of emotion recognition remediation in schizophrenia using the micro-expression training tool. *British Journal of Clinical Psychology.* 2006;45(4):579–583.
102. Rose A, Vinogradov S, Fisher M, et al. Randomized controlled trial of computer-based treatment of social cognition in schizophrenia: the TRuSST trial protocol. *BMC Psychiatry.* 2015;15(1):142.
103. Greig TC, Bryson GJ, Bell MD. Theory of mind performance in schizophrenia: diagnostic, symptom, and neuropsychological correlates. *The Journal of Nervous and Mental Disease.* 2004;192(1):12–18.
104. Horan WP, Kern RS, Tripp C, et al. Efficacy and specificity of social cognitive skills training for outpatients with psychotic disorders. *Journal of Psychiatric Research.* 2011;45(8):1113–1122.

105. Roberts DL, Penn DL, Combs DR. *Social Cognition and Interaction Training (SCIT): Group Psychotherapy for Schizophrenia and Other Psychotic Disorders, Clinician Guide.* Oxford University Press; 2015.
106. Beck JS. *Cognitive Behavior Therapy: Basics and Beyond.* Guilford Press; 2011.
107. Roberts DL, Kleinlein P, Stevens B. An alternative to generating alternative interpretations in social cognitive therapy for psychosis. *Behavioural and Cognitive Psychotherapy.* 2012;40(4):491–495.
108. Roberts DL, Liu PY-T, Busanet H, Maples N, Velligan D. A tablet-based intervention to manipulate social cognitive bias in schizophrenia. *American Journal of Psychiatric Rehabilitation.* 2017;20(2):143–155.
109. Kurtz MM, Richardson CL. Social cognitive training for schizophrenia: a meta-analytic investigation of controlled research. *Schizophrenia Bulletin.* 2011;38(5):1092–1104.
110. Combs DR, Adams SD, Penn DL, Roberts D, Tiegreen J, Stem P. Social Cognition and Interaction Training (SCIT) for inpatients with schizophrenia spectrum disorders: preliminary findings. *Schizophrenia research.* 2007;91(1):112–116.
111. Roberts DL, Penn DL. Social cognition and interaction training (SCIT) for outpatients with schizophrenia: a preliminary study. *Psychiatry research.* 2009;166(2):141–147.
112. Campos C, Santos S, Gagen E, et al. Neuroplastic changes following social cognition training in schizophrenia: a systematic review. *Neuropsychology Review.* 2016;26(3):310–328.
113. Penn DL, Keefe RS, Davis SM, et al. The effects of antipsychotic medications on emotion perception in patients with chronic schizophrenia in the CATIE trial. *Schizophrenia Research.* 2009;115(1):17–23.
114. Roberts DL, Penn DL, Corrigan P, Lipkovich I, Kinon B, Black RA. Antipsychotic medication and social cue recognition in chronic schizophrenia. *Psychiatry Research.* 2010;178(1):46–50.
115. Rosenfeld AJ, Lieberman JA, Jarskog LF. Oxytocin, dopamine, and the amygdala: a neurofunctional model of social cognitive deficits in schizophrenia. *Schizophrenia Bulletin.* 2010;37(5):1077–1087.
116. Davis MC, Lee J, Horan WP, et al. Effects of single dose intranasal oxytocin on social cognition in schizophrenia. *Schizophrenia Research.* 2013;147(2):393–397.
117. Pedersen CA, Gibson CM, Rau SW, et al. Intranasal oxytocin reduces psychotic symptoms and improves Theory of Mind and social perception in schizophrenia. *Schizophrenia Research.* 2011;132(1):50–53.
118. Cacciotti-Saija C, Langdon R, Ward PB, et al. A double-blind randomized controlled trial of oxytocin nasal spray and social cognition training for young people with early psychosis. *Schizophrenia Bulletin.* 2014;41(2):483–493.
119. Davis MC, Green MF, Lee J, et al. Oxytocin-augmented social cognitive skills training in schizophrenia. *Neuropsychopharmacology.* 2014;39(9):2070.
120. Zik JB, Roberts DL. The many faces of oxytocin: implications for psychiatry. *Psychiatry Research.* 2015;226(1):31–37.
121. Nienow TM, MacDonald A 3rd, Lim KO. TDCS produces incremental gain when combined with working memory training in patients with schizophrenia: a proof of concept pilot study. *Schizophrenia Research.* 2016;172(1–3):218.
122. Rassovsky Y, Dunn W, Wynn J, et al. The effect of transcranial direct current stimulation on social cognition in schizophrenia: a preliminary study. *Schizophrenia Research.* 2015;165(2):171–174.

25

SELF-AWARENESS IN SCHIZOPHRENIA

AFFECTED DOMAINS AND THEIR IMPACT

Juliet M. Silberstein, Amy E. Pinkham, and Philip D. Harvey

The idea that individuals with schizophrenia exhibit a lack of self-awareness is not new. In fact, Kraepelin and Bleuler reported unawareness of illness as integral to the disease over 100 years ago. Then, insight was viewed as binary and dependent on whether individuals acknowledged their illness and understood that their psychotic symptoms were not veridical. More nuanced definitions of clinical insight have since emerged that consider the level of awareness individuals have regarding multiple aspects of their illness (e.g., symptoms and need for treatment).[1,2] Significantly, 50% to 80% of individuals exhibit some level of impaired clinical insight.[3–5] People in denial of their disease are more likely to have high rates of hospitalization and refusal of treatment, including medications.[6–8] This chapter will move beyond clinical insight to address the patient's capacity to self-evaluate across the multiple domains of functioning now seen to be important in people with schizophrenia.

Advances in the definitions of and research on metacognition have added a layer of complexity to the assessment of clinical insight. Historical emphasis on discrete positive and negative symptoms would place focus on a person's inaccurate belief that their delusion or hallucination is real (impaired clinical insight) without assessing that patient's ability to reflect on their thoughts or to become aware that their conclusion is incorrect.[1] This emphasis on processing one's own thoughts has been termed metacognition and is most simply defined as *thinking about thinking*. More complex conceptualizations of metacognition define it as encompassing a spectrum spanning from discrete acts (self-awareness of errors or identification of emotion) to much more complicated, synthetic acts (integration of experience into worldview).[9] Metacognition can be deployed in several ways, including the comparative processing of one's own thoughts as well as those of others (e.g., theory of mind [ToM]),[10,11] recognition of anomalous beliefs (e.g., cognitive insight),[1] awareness of neuropsychological dysfunction (neurocognitive insight),[12] the ability to evaluate the level of competence of functional skills, the attribution of agency to one's self,[13,14] and the ability to discriminate correct from incorrect decisions.[15] This chapter focuses on the last few of these areas, addressing how individuals evaluate their own abilities and performance;[16] we refer to this type of self-awareness as introspective accuracy (IA).

In our view, IA overlaps with metacognition but is distinct. Introspective accuracy can describe impairments that result from errors in the metacognitive process but can also be applied to many additional different domains, such as judgments of competence in social or other adaptively relevant situations.[17] Introspective accuracy is separable from other metacognitive constructs such as clinical and cognitive insight in that both constructs focus on understanding clinical phenomena such as anomalous experiences and erroneous inferences,[2,7] whereas IA focuses on the understanding of specific skills and ability levels, which are themselves dependent on the intactness of certain lower-level component skills and the absence of interfering symptoms. Consistent with the NIMH's Research Domain Criteria (RDoC) matrix for social processes that separates "self-knowledge" and "understanding mental states," IA is also distinct from ToM in that the latter is other-focused rather than self-focused and geared toward the content of thoughts rather than the correctness of those thoughts.[16] Introspective accuracy therefore maps onto the RDoC subconstruct of "Self-Knowledge," which is defined, in part, as the ability to make judgments about one's abilities both globally and momentarily.

Impairments in IA are evidenced by discrepancies between how one self-evaluates abilities and achievements and his/her actual performance across clinical, functional, social, and neurocognitive domains. For instance, a person who thinks he has a good memory but actually forgets frequently has poor neurocognitive IA and is likely to suffer real-world consequences because he will not engage in compensatory behaviors such as writing down directions. Other domains impacted by IA include functional skills and clinical outcomes. A person who has difficulty evaluating their ability to write a check or use an automatic teller machine might be unable to attain financial independence, and an individual who underestimates their disability is more likely to be noncompliant with rehabilitation suggestions. In the social realm, a person who incorrectly believes she is *not* good at reading others' emotions (i.e., lack of IA regarding social cognitive abilities) may avoid interactions with others. People who think that they lack social competence may choose to withdraw socially, even if their true limitation is anxiety and not competence. Introspective accuracy is not global, meaning a patient might have insight in one domain and lack it in another.[18] Functional outcomes themselves are clearly not unifactorial, which would require that any true predictor of functional outcomes not be a single generic deficit.

Additionally, as demonstrated earlier, the direction of IA impairments can reflect either the over- or underestimation of abilities, which we label introspective bias (IB). Thus, IA refers to correspondence between one's estimation of his/her abilities and his/her actual abilities and IB refers to the direction of inaccuracy. Importantly, the bidirectional nature of IB represents another point of contrast between IA and cognitive insight, which is geared only toward overconfidence in the validity of beliefs,[12] and the clinical insight concept, wherein unawareness of abnormal experiences is intrinsically unidimensional.

As detailed in this chapter, our initial work with individuals with schizophrenia has found IA impairments in the domains of neurocognitive performance, social cognitive performance, and functional abilities.[2,19–26] These studies have shown that patients' self-assessments correlate minimally with clinicians' assessments or objective performance.[21,22,26–28] Importantly, we, and others, have also found that IA deficits across various domains have negative consequences on morbidity and mortality as these impairments impact medication adherence, suicidality, everyday activities, vocational functioning, and social outcomes.[27–30]

INTROSPECTIVE ACCURACY IN HEALTHY INDIVIDUALS

The concept of IA is not unique to schizophrenia. A multitude of research studies have shown that healthy individuals most commonly claim they are "above average."[31,32] Overestimation of one's own ability has been shown among elderly individuals who reported overconfidence in their driving ability,[33] cashiers who reported their ability to spot underage customers better than their peers,[34] and chess players who overestimated their chances of winning.[35] Socially, people rate themselves above average on positive traits such as sophistication and sensibility while rating themselves below average on negative traits such as neuroticism and impracticality.[36] While healthy individuals generally tend to overestimate their strengths, task difficulty can thwart healthy individuals' overestimation bias. For example, when study participants are presented with an arduous task such as computer programming or given a deliberately dense article, individuals show negative IB and assess their performance as worse than average.[37–39] Healthy individuals' self-assessment, at best, modestly predicts their skills and character, and often overshoots their ability.[40]

Kruger and Dunning (1999, 2008) published a series of studies across numerous domains (e.g., college exams to humor tests), settings (e.g., laboratory and naturalistic), and subject populations (e.g., students to gun club members) showing the lowest quartile of performers in any one cohort greatly overestimated their ability. Individuals in the bottom 25% reported their performance above the 60th percentile and overestimated their raw score by 20% to 30%. Conversely, the highest quartile of performers slightly underestimated their raw score by 8.5%.[40,41] This pattern seen in low-skilled individuals has been further replicated among physicians evaluating their board examination performance,[42,43] medical students assessing their interview and test performance,[44,45] and lab technicians assessing their technical knowledge.[46] Poor-performers' misestimation of their neurocognitive capacities raises the questions of whether IA errors in individuals with and without severe mental illness occur due to similar neuropsychological pathways.

A direct pathway between low skill level and overestimation of ability has been frequently supported in healthy individuals. Often, among individuals without a severe mental illness, the skills needed to correctly respond are also the same skills needed to judge the response. For example, to be able answer a math problem you must know the mathematical rules. You must also understand those same rules to check for errors.[41] Expert musicians more accurately reported their strengths and weaknesses than novice musicians,[47] doctorates better ranked the difficulty of physics problems than undergraduates,[48] and students with high verbal abilities superiorly estimated their comprehension compared to students with lower abilities.[49] Kruger and Dunning (1999) found that objective performance was strongly correlated with the extent of IA deficits, where IA improved stepwise with each quartile increase. Importantly, the study showed that targeted skill development improved IA. Students who received logic training, aligned with the neurocognitive assessment, better differentiated their correct from incorrect answers. In fact, low-skilled students improved their IA abilities so greatly that their self-assessments were on par with those made by high-skilled students.[41] In sum, IA impairment was a direct effect of a cognitive skill gap and could be improved by neurocognitive interventions or feedback.

Evidence correlating IA with neurocognitive capacity becomes murkier when the concept of IA is broadened. Self-assessment of choices can occur prior, during, or after performance. Further, IA can measure absolute accuracy (e.g., assessment matches performance) or relative accuracy (e.g., assessments across a number of components are correlated).[49–51] While skill level mediates the precision of judgments following the completion of a performance-based assessment, few studies have shown an interaction between ability level and relative or predictive accuracy.[49] Further, IA has been shown to account for the variance in one's performance that cannot be explained by neurocognitive capacity alone.[52] Future research is needed to differentiate possible overlapping but distinct pathways comprising facets of IA and capacity levels.

Schizophrenia research has largely suggested neurocognitive capacity and IA impairment processes are independent. Repeated studies have shown negligible links between performance and self-assessment.[21,22,26–28] Further, impaired IA has been shown to have a stronger correlation to everyday outcomes than can be explained by skill deficits alone.[17,24,53] As we discuss later, people with schizophrenia may lack access to the opportunities to deploy and then evaluate their skills. Asking someone who has never worked about their vocational competence is like asking a healthy person who has never flown a plane whether a Boeing or an Airbus is easier to operate in an emergency.

INTROSPECTIVE ACCURACY AND BIAS ACROSS FUNCTIONAL DOMAINS IN SCHIZOPHRENIA

Impaired IA in schizophrenia has been demonstrated across a multidimensional continuum including the domains of neurocognition, social cognition, and functional abilities.[17] On average, patients overestimate their functional potential and underestimate their global disability in comparison to clinician ratings. This misestimation can predict impaired outcomes across domains.[17,26]

COGNITIVE PERFORMANCE

Cognitive deficits are a central component of schizophrenia with at least 70-75% of patients performing below normative standards. While cognitive impairment has been linked to reduced IA,[21] a recent meta-analysis of the relation between cognition and IA found a correlation of only $r = .16$. Further, the mean scores on performance-based neurocognitive tests did not differ across levels of IA ($p > .14$) for people with schizophrenia, and even those with high IA had composite scores 1.3 standard deviation (SD) below normative standards for healthy individuals.[54] Thus, global cognitive impairment on its own is unlikely to be a critical moderator of IA.

Introspective accuracy of cognitive capacity can be assessed in a few ways. Patients can take a modified performance-based cognitive assessment. For example, the Wisconsin Card Sorting Test (WCST) is an executive functioning test in which the person needs to solve the test through performance feedback. When errors are made, the strategy in use needs to be discarded; when responses are correct, the strategy is retained. In our VALERO II study, we used a modified WCST and asked participants to rate their confidence in the correctness of each response and to decide whether they wanted the response to count toward their total score. Thus, performance could be split into domains of accuracy (correct responses), appraisal (average confidence for correct and incorrect responses), and judgment (proportion of responses "offered" as correct as a function of actual accuracy). The second main way to assess cognitive IA is to compare interview-based assessments of cognitive abilities completed by the patient to those completed by an informant. The Cognitive Assessment Interview (CAI),[55] a 10-item instrument that assesses the cognitive domains that are diminished in schizophrenia, and the Schizophrenia Cognition Rating Scale (SCoRS),[56] a similar 20-question interview, are two validated cognitive ratings tools. Comparing self- and clinician- reports provides direct index of cognitive IA.[57]

Patients' self-reported cognitive skills consistently do not correlate with their performance on cognitive tasks or with clinician reports of their cognitive capacity.[58] Between 28% and 52% of individuals with schizophrenia who show cognitive deficits fail to indicate that they have any problems,[59–61] and our own work has demonstrated that over half of patients overestimate when compared to both clinician ratings and performance on neuropsychological (NP) tests.[21,26,62] Interestingly, ratings by high-contact informants (who were unaware of patients' test performance) were much more strongly related to patients' objective test performance, compared with patient self-reports. Shared variance estimates of 25% have been detected. The convergence of clinician ratings of cognitive performance with objective test data has been impressive, and suggests that clinician ratings may be a useful proxy for extensive neuropsychological testing.[22] These correlations also provide evidence that clinician ratings are useful as a standard against which to judge the accuracy of patients' self-assessments.

SOCIAL COGNITIVE PERFORMANCE

Consistent with the pattern of IA for cognition, patients with schizophrenia tend to make more high-confidence errors than healthy individuals on social cognitive tasks. To assess IA in social cognition, patients complete a typical social cognitive task, such as determining emotions from facial stimuli or examining the eye region of the face to determine the mental state of the depicted person. Immediately after responding to each stimulus, participants rate their confidence in the correctness of that response. When making social cognitive judgments of others' emotions or mental states, individuals with schizophrenia make more high confidence errors than healthy individuals.[53,63] These high-confidence errors also are more likely to occur for more difficult stimuli. Strikingly, these difficulties appear to be specific to self-assessment. When asked to determine if the social behaviors of another individual were socially appropriate, individuals with schizophrenia were as able as healthy individuals to recognize social mistakes.[64] This work suggests that, at least within the domain of social cognition, IA impairment is not due to generalized poor judgment or global cognitive limitations. This finding is consistent with extensive research suggesting that psychotic patients with impaired clinical insight can detect psychotic ideas on the part of others.

Individuals with schizophrenia's ability to accuracy and confidently judge the social behavior of others provides an opportunity to discuss a complementary process to IA, that of decision-making competence.[50,51,65–67] The goal of improving IA assumes individuals have the capacity to choose actions that match their self-assessment, however, people ultimately have *control* over whether or not to make the most appropriate choice. Consider an individual who wrongly believes she comprehends the directions for a heavy and dangerous piece of machinery (poor IA), and enthusiastically turns it on (good control). A second man realizes he does not know how to operate the machine (good IA) but starts the machine anyways (poor control). Their common decision puts both individuals at risk, and demonstrates that *introspective* and *control* accuracy influence one's performance, actions, and functioning.[50,51,65–67]

A study we conducted showed people with schizophrenia failed to adjust their choices or effort-level despite accurately answering hard and easy questions (e.g., good IA and poor control). We used the Bell-Lysaker Emotion Recognition Test (BLERT), which asks participants to identify emotion from short video clips, to evaluate patients' with schizophrenia (n = 57) and healthy controls' (n = 47) social cognitive performance and IA of that performance. Participants rated their confidence in the accuracy of their responses after each item. As a proxy for effort, the time to complete each item was recorded. Both patients and healthy controls were less confident in and spent more time on incorrect items. However, individuals with schizophrenia did not adjust their effort between correct and incorrect items to the same degree as healthy individuals (p < .001). While both patients and healthy controls accurately recognized when an item was more challenging to them, patients did not adjust their effort accordingly. These results link IA and metacognition in that even with accurate IA, patients failed to use their assessment to change strategy.[68]

EVERYDAY FUNCTIONING

The VALERO-II study quantified the direct impact of impairments in introspective accuracy on everyday functioning. We asked 214 individuals with schizophrenia to self-evaluate their cognitive ability on the CAI rating scale and to self-report their everyday functioning in social, vocational, and everyday activities domains on the specific level of function (SLOF). Concurrently, high-contact clinicians rated these same abilities with identical rating scales. Participants with schizophrenia took performance-based tests to assess their cognitive abilities and everyday functional skills. We then predicted everyday functioning, as rated by the clinicians, with the discrepancies between self-assessed and clinician-assessed functioning, and patients' scores on the performance-based measures.

We found no correlational relationship with patient reports and either performance-based assessments or high contact clinician reports,[26,62] nor did these reports relate to objective data regarding attaining functional milestones such as employment or residential status.[24] These findings are consistent with multiple meta-analyses reporting a lack of correlation between objective and subjective ratings of functioning and quality of life.[69]

Although overestimation was more common, the difference scores between patient and clinician ratings in the VALERO study were not unidirectional. We found that patients with schizophrenia were more likely to overestimate (60%) their functioning than underestimate; however, the distributions of discrepancy scores was normal, and at least 40% of patients saw themselves as performing equivalently to or more poorly than their clinicians rated them. These findings confirm that deficits in IA are bidirectional and highlight the need to consider IB when examining IA. Further, the findings suggest that discrepancy scores are not due to some global problems in understanding normative cognitive performance as referenced by the CAI. At present, it is unclear what contributes to underestimation in some patients and overestimation in others, but given that these tendencies are likely to directly impact behavior, understanding the moderators of IB is necessary.[57]

FUNCTIONAL OUTCOMES AND SIGNIFICANCE OF INTROSPECTIVE ACCURACY

To guide recovery efforts and develop better treatments that improve the everyday lives of patients with severe mental illness, a complete understanding of the factors influencing suboptimal functioning is necessary. To date, studies of the determinants of everyday functional deficits in schizophrenia[19,20,70,71] have stalled at accounting for 50% or less of the variance in real-world functioning. Although neurocognition, social cognition, and functional skills appear to be globally related to functioning and are more relevant than symptom severity,[72–74] many individual neurocognitive and social cognitive domains account for 2% or less of the variance in real world functional or social outcomes.[72,75] It is unlikely that some additional undiscovered subdomains of social cognition or neurocognition will lead to substantial gains in prediction. For example, results of the Cognitive Neuroscience Test Reliability and Clinical Applications for Schizophrenia (CNTRACS) Consortium study did not lead to substantial gains in predicting everyday disability rated by clinicians despite using highly selective neuroscience-oriented tests with clear neurobiological correlates.[23]

Likewise, most research on self-awareness has investigated clinical and cognitive insight, albeit critical in areas of nonadherence, relapse potential, and suicide, may not be associated with everyday functional outcomes. Cognitive insight, and awareness of anomalous beliefs in particular, has only been linked to global indices of function that are multiply-determined (i.e., living independently or subjective quality of life),[2,76,77] and even these results have been mixed, with one study showing a relation between cognitive insight and psychopathology but not functional outcome.[78] In our own preliminary data from a sample of 72 individuals with schizophrenia, correlations between informant-rated everyday functioning and cognitive insight were all minimal (r < 0.1, p > 0.2).

A handful of studies in schizophrenia[79,80] and HIV[81] however have shown that global indices of introspective accuracy relate to functional outcomes even when controlling for neurocognitive abilities and symptom severity. Using data again from VALERO phase II (n = 214), we examined the extent that global IA, momentary IA, skill level, and objective performance predicted functional outcomes. As previously described, we collected data that allowed us to examine IA on a real-time basis using item-by-item judgments of performance on the Wisconsin Card sorting test (WCST). Global IA, an individual's accuracy in estimating their current functioning and relative level of ability, was measured using the discrepancy between high-contact clinicians' and patients' self-reports of cognition through the CAI rating and functional outcomes through the SLOF rating. Neurocognitive abilities were evaluated by a modified MATRICS consensus cognitive battery (MCCB) test and functional capacity with the UCSD Performance-based skills assessment (UPSA-B), in which participants were asked to perform everyday tasks related to communication and finances.[82] See Table 25.1 for the results.

Table 25.1 DERIVED GLOBAL MISESTIMATION OF FUNCTIONING SCORE PREDICTING REAL-WORLD OUTCOMES AMONG PERSONS WITH SCHIZOPHRENIA[82]

SLOF INTERVIEWER RATINGS	STEP	VARIABLE(S)	*P*	*R2*
SLOF Interpersonal	1	Misestimation score	<.001	.29
	2	UPSA-B total score	<.001	.31
SLOF Activities	1	Misestimation score	<.001	.42
	2	Global NP *T* score	<.001	.45
SLOF Vocational	1	Misestimation score	<.001	.37
	2	UPSA-B total score	<.001	.42

Pearson correlations revealed that the discrepancies between self-assessed and clinician-rated neurocognitive performance and everyday functioning were strongly correlated with clinician-rated impairments in everyday functioning. Specifically, those whose introspections about their own functional performance were the least accurate were rated as functioning the worst by their clinicians. More critical was the finding that in a regression analysis, global impairments in IA regarding overall cognitive functioning (as assessed by CAI) and everyday functioning (as assessed by SLOF) were more strongly correlated with impairments in clinician-rated everyday functioning than performance-based ability measures including the MCCB and the UPSA-B. The variance in clinician-rated everyday functioning accounted for by misestimation of functioning was quite substantial. Further, impaired momentary IA, real-time assessment of performance on the WCST, also accounted for 8%variance in everyday functioning beyond the influences of impaired IA of global cognition and functioning.[82]

These reveal IA as a promising new lead in the search for transdiagnostic determinants of real-world functional outcome. To summarize, impaired introspective accuracy, as indexed by difference scores between clinician ratings and self-reports, was a more potent predictor of everyday functional deficits in social, vocational, and everyday activities domains than scores on performance-based measures of cognitive abilities and functional skills. Impaired introspective accuracy was the single best predictor of everyday functioning in all 3 domains, with actual abilities considerably less important.[57]

Data from our Social Cognition Psychometric Evaluation (SCOPE; n = 218) study also support the functional significance of social cognitive IA. Outpatients with schizophrenia (n = 179) and healthy controls (n = 104) completed a series of social cognitive performance-based assessments twice: once at baseline and again 2–4 weeks later. The Penn Emotion Recognition Task (ER-40) asked participants to choose the correct emotion label for 40 color photographs of static faces and rate the accuracy of their response for each picture. Preliminary data shows that IA for item-by-item responses significantly predicted variance in both social competence (β = 0.34, p = .004) and functional capacity (β = 0.34, p = .004). Importantly, clinical insight as assessed by the Positive and Negative Syndrome Scale was not related to outcomes (social competence or functional capacity), or IA, all $p > .18$. These results lead to the provocative conclusion that the ability to accurately self-assess one's ongoing performance in multiple domains (i.e., IA regarding cognition, social cognition, social competence, and functional capacity) may be the most important determinant of everyday functional success, even when traditionally important predictors of the abilities to perform functional skills are considered.[83]

MOOD, LIFE MILESTONES, AND PSYCHOTIC SYMPTOMS AS MODERATORS OF INTROSPECTIVE ACCURACY AND BIAS

Mild depression has been correlated with a more accurate self-evaluation of ability across both healthy and pathological populations. This phenomenon is called "depressive realism" and refers to the finding that healthy individuals with mild depression show more IA than those with no depression, whose IB is to overestimate their abilities. A meta-analysis of depressive realism found that healthy individuals overestimated their abilities by double the estimates made by individuals with mild depression.[84]

Our cross-sectional work supports a similar relationship in schizophrenia, in that higher levels of depression have been associated with reduced overestimation of real-world functioning (r = -.44, p < .001).[62] However, our most recent analyses have raised the possibility that the relationship between mood and IA/IB in schizophrenia may not be linear across all domains. We examined IA for three real-world outcomes as a function of depression using data from the VALERO study. As expected, IA impairments were most pronounced in the nondepressed group in that they overestimated their interpersonal function, everyday activities, and vocational functioning to the greatest extent when compared to clinician-reported functioning. However, IA did not improve steadily as depression increased for interpersonal function and everyday activities. Rather, IA impairments tracked the presence or absence of depression, regardless of severity. This was in contrast to the pattern for vocational function, where we did see better IA as the severity of depression increased (Table 25.2).[85]

Table 25.2 IA* FOR REAL-WORLD OUTCOMES AS A FUNCTION OF DEPRESSION[85]

	NO DEPRESSION MEAN (SD)	MILD DEPRESSION MEAN (SD)	MODERATE DEPRESSION MEAN (SD)
SLOF Interpersonal	0.4 (0.9)	0.1 (0.9)	0.1 (0.8)
SLOF Activities	0.5 (0.9)	0.1 (0.9)	0.1 (0.8)
SLOF Vocational	0.5 (1.0)	0.3 (0.8)	0.1 (0.9)

* Higher scores indicate greater impairments in IA

Additionally, employment history has been correlated with differing levels of self-assessment among people with schizophrenia. Unemployed patients with no employment history are most likely to overestimate their vocational potential. These patients have been shown to rate their vocational skills on par or above skill ratings from employed or previously employed patients with schizophrenia. Interestingly, unemployed patients with prior work experience reported the least vocational potential.[24,57] Some patients seem to overestimate their potential, engaging and failing at difficult tasks due to inadequate skill or ability to adjust effort. Conversely, other patients might have diminished motivation to attempt tasks within their ability.[17,68] Again, this highlights the clinical importance of accurately assessing introspective accuracy.

Increased psychotic symptoms have also been linked to poorer IA. Higher levels of paranoia among even nonclinical individuals have been associated with overestimation of abilities on tasks of visual perception[86] and general knowledge.[87] Among individuals with schizophrenia, our work shows that higher symptom ratings on suspiciousness, grandiosity, stereotyped thinking, and social withdrawal are all predict poorer IA regarding real-world functioning.[62]

ACCURACY OF INFORMANTS

Our research, as described in this chapter, has quantified IA as discrepancy scores between clinician and patients' assessment ratings, and patients' scores on performance-based assessments because clinicians' ratings are convergent with objective evidence. Clinician ratings have been found to be significantly correlated with patients' performance across neurocognitive, social, and functional tasks, while ratings generated by an informant who simply knows the patient and is willing to provide ratings are likely to be uninformative.[26]

In a systematic study of validity of reports of various informants (VALERO-I), we compared correlations between reports of everyday functioning with objective measures of cognitive test performance and tests of the ability to perform everyday functional skills. Correlations between ratings generated by friend or relative informants and other information were almost shocking in their lack of validity (Table 25.3). This is significant because if a friend or relative provides information of limited usefulness, the report could easily lead to clinical decisions or clinical trials outcome ratings with high potential for weak validity.[26]

Caregivers who had regular contact with patients had much more valid ratings when performance on functionally relevant objective measures was considered. Patients with caregivers had greater impairments in everyday outcomes, however, suggesting that this subset was more impaired than the overall sample. For patients without caregivers, other sources of information—including careful observation by high-contact clinicians—seem to be required to generate a valid assessment of functioning.[26]

IMPLICATIONS FOR TREATMENT

The emerging data that IA impairments both in self-estimates of global ability and in real-time decisions are significantly correlated to functional outcomes indicate that IA is an important avenue for clinical treatment. It has been well documented that poor self-awareness in the domain of illness can lead to reduced adherence to medication, followed by relapse, leading to emergency room treatments or acute admissions, and behavior associated with violence or self-harm.[6–8] Introspective accuracy adds a dimension to this existing knowledge of poor outcomes associated with insight, cognitive, and metacognitive deficits. To address awareness and insight, cognitive therapy has become a prominent adjunct to antipsychotics as an intervention in the treatment of schizophrenia,

Table 25.3 LOW VALIDITY IN RATINGS OF PATIENTS' COGNITION AND FUNCTION BY NONCAREGIVER INFORMANTS (FRIENDS AND FAMILY)[26]

SOURCES OF DATA COMPARED WITH RATINGS BY FRIENDS AND FAMILY	DATA FROM EACH SOURCE WITH WHICH INFORMANTS' RATINGS LACKED CORRELATION
Objective performance data	Neuropsychological test performance; performance-based measures of functional capacity
Clinician ratings	Social functioning; vocational functioning; ability to perform everyday functional skills
Patient self-reports	Social functioning; vocational functioning; ability to perform everyday functional skills

which has been shown to have a small to medium effect on positive symptoms beyond medication.[88,89]

"Metacognitive" behavior therapy (MCT) has emerged as the focus of some treatment efforts,[90–93] but only recently have interventions specifically targeting IA emerged.[89,94] Metacognitive deficits have been shown to be trait-like and existing on a continuum; thus, MCT interventions must be individualized.[9] Valid and reliable assessments of cognitive and metacognitive ability have become a focus in schizophrenia research. For example, the Measurement and Treatment Research to Improve Cognition in Schizophrenia (MATRICS) project,[95] was aimed at need for standardized methods for assessment in treatment studies aimed at cognitive impairment associated with schizophrenia. Our research projects, SCOPE and VALERO, aimed to find valid assessments of social cognition and functional outcomes, respectively. As the number of reliable assessment tools increases, clinicians will be more able to identify the areas of deficit causing poor functional outcomes. Beyond assessment tools, clinician's assessments have been shown to be highly accurate and aligned with performance measures.

Modifiers outlined in this chapter can help clinicians characterize impairments in introspective accuracy (Table 25.4). Subjective reports of depression have a bell-shaped relationship with introspective accuracy. A self-reported depression score of 0 by a disabled schizophrenia patient suggests some unawareness of an unfortunate life situation; mild to moderate scores are associated with more accurate self-assessment; and more severe scores, as seen in other conditions can predict overestimation of disability.[21] Psychosis and negative symptoms are associated with reduced introspective accuracy and global overreporting of functional competence.[62]

The work in systemizing MCT and to assess these interventions with randomized controlled trials is just beginning, however, initial studies look promising. Metacognitive behavior therapy has been shown to reduce overconfident errors and maintenance of false beliefs even when given compelling counterevidence. A randomized control trial ($n = 150$) showed that MCT had a reduction in severity of delusions ($p = .05$), and improved self-reported quality of life ($p = .08$) after 3 years.[89] While promising, MCT interventions have focused primarily on symptom reduction rather than functional improvement. Numerous studies have indicated that symptom reduction in schizophrenia does not typically produce sustained change in everyday functioning. It is likely that in order to maximize chances for recovery, interventions will need to focus broadly on the full range of IA deficits (i.e., neurocognition, social cognition, and functional abilities).

Table 25.4 PATIENT CHARACTERISTICS THAT PREDICT ACCURACY OF IA

CHARACTERISTIC	IB IS LIKELY . . .
Low depression	Overestimation
Moderate depression	Underestimation (Possible)
Psychosis	Accurate
Negative symptoms	Overestimation
Impaired cognition	Overestimation
Never employed	Overestimation
Unemployed with prior work history	Accurate

CONCLUSIONS

Inaccurate self-assessment is common in people with schizophrenia, much as in the general population. However, when inaccurate self-assessment is combined with objectively low scores on ability variables, the result can be significant incremental contributions to everyday disability. Inaccurate self-assessment occurs in domains of clinical symptoms, neurocognitive and social cognitive performance, and everyday functioning including social, vocational, and everyday activities. Inaccurate self-assessments tend to correlate with each other and there are several determinants of these self-assessments, including mood states. Treatment of these impairments has been attempted with some success, particularly in areas of social metacognition. This is a developing area of research and identification of the neurobiological correlates of these impairment domains is an important topic for future research.

REFERENCES

1. Beck AT, Rector NA, Stoler N, Grant PM. *Schizophrenia: Cognitive Theory, Research, and Therapy.* New York, NY: Guilford Press; 2009.
2. Riggs SE, Grant PM, Perivoliotis D, Beck AT. Assessment of cognitive insight: a qualitative review. *Schizophr Bull.* 2012;38(2):338–350.
3. Amador XF, Flaum M, Andreasen NC, et al. Awareness of illness in schizophrenia and schizoaffective and mood disorders. *Arch Gen Psychiatry.* 1994;51(10):826–836.
4. Amador XF, David AS. *Insight and Psychosis.* New York, NY: Oxford University Press; 1998.
5. Carpenter WT, Strauss JS, Bartko JJ. Flexible system for the diagnosis of schizophrenia: report from the WHO International pilot study of schizophrenia. *Science.* 1973;182(4118):1275–1278.
6. Lin IF, Spiga R, Fortsch W. Insight and adherence to medication in chronic schizophrenics. *J Clin Psychiatry.* 1979;40(10):430–432.
7. Beck AT, Baruch E, Balter JM, Steer RA, Warman DM. A new instrument for measuring insight: the Beck Cognitive Insight Scale. *Schizophr Res.* 2004;68(2–3):319–329.
8. Van Putten T, Crumpton E, Yale C. Drug refusal in schizophrenia and the wish to be crazy. *Arch Gen Psychiatry.* 1976;33(12):1443–1446.
9. Lysaker PH, Dimaggio G. Metacognitive capacities for reflection in schizophrenia: implications for developing treatments. *Schizophr Bull.* 2014;40(3):487–491.
10. Nelson TO, Stuart RB, Howard C, Crowley M. Metacognition and clinical psychology: a preliminary framework for research and practice. *Clinical Psychology and Psychotherapy.* 1999;6(2):73–79.
11. Dimaggio G, Lysaker P. *Metacognition and Severe Adult Mental Disorders: From Research to Treatment.* London: Routledge; 2010.
12. Medalia A, Thysen J. A comparison of insight into clinical symptoms versus insight into neuro-cognitive symptoms in schizophrenia. *Schizophr Res.* 2010;118(1-3):134–139.
13. Metcalfe J, Greene MJ. Metacognition of agency. *J Exp Psychol Gen.* 2007;136(2):184–199.

14. Hur JW, Kwon JS, Lee TY, Park S. The crisis of minimal self-awareness in schizophrenia: a meta-analytic review. *Schizophr Res.* 2014;152(1):58–64.
15. Fleming SM, Weil RS, Nagy Z, Dolan RJ, Rees G. Relating introspective accuracy to individual differences in brain structure. *Science.* 2010;329(5998):1541–1543.
16. Koren D, Seidman LJ, Goldsmith M, Harvey PD. Real-world cognitive—and metacognitive—dysfunction in schizophrenia: a new approach for measuring (and remediating) more "right stuff." *Schizophr Bull.* 2006;32(2):310–326.
17. Harvey P, Pinkham A. Impaired self-assessment in schizophrenia: why patients misjudge their cognition and functioning. *Current Psychiatry.* 2015;14:6.
18. Yahav T, Maimon T, Grossman E, Dahan I, Medalia O. Cryo-electron tomography: gaining insight into cellular processes by structural approaches. *Curr Opin Struct Biol.* 2011;21(5):670–677.
19. Bowie CR, Leung WW, Reichenberg A, et al. Predicting schizophrenia patients' real-world behavior with specific neuropsychological and functional capacity measures. *Biol Psychiatry.* 2008;63(5):505–511.
20. Bowie CR, Reichenberg A, Patterson TL, Heaton RK, Harvey PD. Determinants of real-world functional performance in schizophrenia subjects: correlations with cognition, functional capacity, and symptoms. *Am J Psychiatry.* 2006;163(3):418–425.
21. Bowie CR, Twamley EW, Anderson H, Halpern B, Patterson TL, Harvey PD. Self-assessment of functional status in schizophrenia. *J Psychiatr Res.* 2007;41(12):1012–1018.
22. Durand D, Strassnig M, Sabbag S, et al. Factors influencing self-assessment of cognition and functioning in schizophrenia: implications for treatment studies. *Eur Neuropsychopharmacol.* 2015;25(2):185–191.
23. Gold JM, Barch DM, Carter CS, et al. Clinical, functional, and intertask correlations of measures developed by the Cognitive Neuroscience Test Reliability and Clinical Applications for Schizophrenia Consortium. *Schizophr Bull.* 2012;38(1):144–152.
24. Harvey PD, Sabbag S, Prestia D, Durand D, Twamley EW, Patterson TL. Functional milestones and clinician ratings of everyday functioning in people with schizophrenia: overlap between milestones and specificity of ratings. *J Psychiatr Res.* 2012;46(12):1546–1552.
25. Keefe RS, Davis VG, Spagnola NB, et al. Reliability, validity and treatment sensitivity of the Schizophrenia Cognition Rating Scale. *Eur Neuropsychopharmacol.* 2015;25(2):176–184.
26. Sabbag S, Twamley EM, Vella L, Heaton RK, Patterson TL, Harvey PD. Assessing everyday functioning in schizophrenia: not all informants seem equally informative. *Schizophr Res.* 2011;131(1–3):250–255.
27. Patterson TL, Semple SJ, Shaw WS, et al. Self-reported social functioning among older patients with schizophrenia. *Schizophr Res.* 1997;27(2–3):199–210.
28. McKibbin C, Patterson TL, Jeste DV. Assessing disability in older patients with schizophrenia: results from the WHODAS-II. *J Nerv Ment Dis.* 2004;192(6):405–413.
29. Green MF, Schooler NR, Kern RS, et al. Evaluation of functionally meaningful measures for clinical trials of cognition enhancement in schizophrenia. *Am J Psychiatry.* 2011;168(4):400–407.
30. Holshausen K, Bowie CR, Mausbach BT, Patterson TL, Harvey PD. Neurocognition, functional capacity, and functional outcomes: the cost of inexperience. *Schizophr Res.* 2014;152(2–3):430–434.
31. Alicke MD A. Global self-evaluation as determined by the desirability and controllability of trait adjectives. *J Pers Soc Psychol.* 1985;49:(6):1621–1630.
32. Dunning D, McElwee RO. Idiosyncratic trait definitions: implications for self-description and social judgment. *J Pers Soc Psychol.* 1995;68(5):936–946.
33. Marottoli RA, Richardson ED. Confidence in, and self-rating of, driving ability among older drivers. *Accid Anal Prev.* 1998;30(3):331–336.
34. McCall M, Nattrass K. Carding for the purchase of alcohol: I'm tougher than other clerks are. *Journal of Applied Social Psychology.* 2001;31(10):2184–2194.
35. Park YJ, Santos-Pinto L. Overconfidence in tournaments: evidence from the field. *Theory and Decision.* 2010;69(1):143–166.
36. Dunning D, Meyerowitz JA, Holzberg AD. Ambiguity and self-evaluation: the role of idiosyncratic trait definitions in self-serving assessments of ability. *J Pers Soc Psychol.* 1989;57(6):1082–1090.
37. Kruger J. Lake Wobegon be gone! The "below-average effect" and the egocentric nature of comparative ability judgments. *J Pers Soc Psychol.* 1999;77(2):221–232.
38. Miller DT, McFarland C. Pluralistic ignorance: When similarity is interpreted as dissimilarity. *J Pers Soc Psychol.* 1987;53(2):298–305.
39. Burson KA, Larrick RP, Klayman J. Skilled or Unskilled, but Still Unaware of It: How Perceptions of Difficulty Drive Miscalibration in Relative Comparisons. *J Pers Soc Psychol.* 2006;90(1):60–77.
40. Ehrlinger J, Johnson K, Banner M, Dunning D, Kruger J. Why the unskilled are unaware: further explorations of (absent) self-insight among the incompetent. *Organ Behav Hum Decis Process.* 2008;105(1):98–121.
41. Kruger J, Dunning D. Unskilled and unaware of it: how difficulties in recognizing one's own incompetence lead to inflated self-assessments. *J Pers Soc Psychol.* 1999;77(6):1121–1134.
42. Didwania A, Kriss M, Cohen ER, McGaghie WC, Wayne DB. Internal medicine postgraduate training and assessment of patient handoff skills. *J Grad Med Educ.* 2013;5(3):394–398.
43. Mehdizadeh L, Sturrock A, Myers G, Khatib Y, Dacre J. How well do doctors think they perform on the General Medical Council's Tests of Competence pilot examinations? A cross-sectional study. *BMJ Open.* 2014;4(2):e004131.
44. Hodges B, Regehr G, Martin D. Difficulties in recognizing one's own incompetence: novice physicians who are unskilled and unaware of it. *Acad Med.* 2001;76(10 Suppl):S87–S89.
45. Sawdon M, Finn G. The "unskilled and unaware" effect is linear in a real-world setting. *J Anat.* 2014;224(3):279–285.
46. Haun DE, Leach A, Vivero R, Foley A. Are they microcytes or macrocytes? Can we do a better job? *MLO Med Lab Obs.* 2002;34(1):40–42.
47. Hallam S. The development of expertise in young musicians: strategy use, knowledge acquisition and individual diversity. *Music Education Research.* 2001;3(1):7–23.
48. Chi MTH, Feltovich PJ, Glaser R. Categorization and representation of physics problems by experts and novices. *Cognitive Science.* 1981;5(2):121–152.
49. Maki RH, Shields M, Wheeler AE, Zacchilli TL. Individual differences in absolute and relative metacomprehension accuracy. *Journal of Educational Psychology.* 2005;97(4):723–731.
50. Schraw G. A conceptual analysis of five measures of metacognitive monitoring. *Metacognition and Learning.* 2008;4(1):33–45.
51. Veenman MVJ, Van Hout-Wolters BHAM, Afflerbach P. Metacognition and learning: conceptual and methodological considerations. *Metacognition and Learning.* 2006;1(1):3–14.
52. Veenman MVJ, Wilhelm P, Beishuizen JJ. The relation between intellectual and metacognitive skills from a developmental perspective. *Learning and Instruction.* 2004;14(1):89–109.
53. Moritz S, Woznica A, Andreou C, Kother U. Response confidence for emotion perception in schizophrenia using a continuous facial sequence task. *Psychiatry Res.* 2012;200(2-3):202–207.
54. Nair A, Aleman A, Davis A. Cognitive Functioning and awareness of illness in schizophrenia: a review and meta-analysis. In: Harvey P, ed. *Cognitive Impairment in Schizophrenia: Characteristics, Assessment, and Treatment.* New York: Cambridge University Press; 2013:18.
55. Ventura J, Subotnik KL, Ered A, Hellemann GS, Nuechterlein KH. Cognitive Assessment Interview (CAI): validity as a co-primary measure of cognition across phases of schizophrenia. *Schizophr Res.* 2016;172(1–3):137–142.
56. Keefe RS, Poe M, Walker TM, Kang JW, Harvey PD. The Schizophrenia Cognition Rating Scale: an interview-based assessment and its relationship to cognition, real-world functioning, and functional capacity. *Am J Psychiatry.* 2006;163(3):426–432.
57. Gould F, Sabbag S, Durand D, Patterson TL, Harvey PD. Self-assessment of functional ability in schizophrenia: milestone achievement

and its relationship to accuracy of self-evaluation. *Psychiatry Res.* 2013;207(1–2):19–24.
58. Moritz S, Balzan RP, Bohn F, et al. Subjective versus objective cognition: evidence for poor metacognitive monitoring in schizophrenia. *Schizophr Res.* 2016;178(1–3):74–79.
59. Medalia A, Lim RW. Self-awareness of cognitive functioning in schizophrenia. *Schizophr Res.* 2004;71(2–3):331–338.
60. Medalia A, Thysen J, Freilich B. Do people with schizophrenia who have objective cognitive impairment identify cognitive deficits on a self report measure? *Schizophr Res.* 2008;105(1–3):156–164.
61. Medalia A, Thysen J. Insight into neurocognitive dysfunction in schizophrenia. *Schizophr Bull.* 2008;34(6):1221–1230.
62. Sabbag S, Twamley EW, Vella L, Heaton RK, Patterson TL, Harvey PD. Predictors of the accuracy of self assessment of everyday functioning in people with schizophrenia. *Schizophr Res.* 2012;137(1–3):190–195.
63. Kother U, Veckenstedt R, Vitzthum F, et al. "Don't give me that look"—overconfidence in false mental state perception in schizophrenia. *Psychiatry Res.* 2012;196(1):1–8.
64. Langdon R, Connors MH, Connaughton E. Social cognition and social judgment in schizophrenia. *Schizophrenia Research: Cognition.* 2014;1(4):171–174.
65. Koriat A, Goldsmith M. Monitoring and control processes in the strategic regulation of memory accuracy. *Psychol Rev.* 1996;103(3):490–517.
66. Koren D, Poyurovsky M, Seidman LJ, Goldsmith M, Wenger S, Klein EM. The neuropsychological basis of competence to consent in first-episode schizophrenia: a pilot metacognitive study. *Biol Psychiatry.* 2005;57(6):609–616.
67. Koren D, Seidman LJ, Poyurovsky M, et al. The neuropsychological basis of insight in first-episode schizophrenia: a pilot metacognitive study. *Schizophr Res.* 2004;70(2–3):195–202.
68. Cornacchio D, Pinkham AE, Penn DL, Harvey PD. Self-assessment of social cognitive ability in individuals with schizophrenia: appraising task difficulty and allocation of effort. 2017;179:85–90.
69. Tolman AW, Kurtz MM. Neurocognitive predictors of objective and subjective quality of life in individuals with schizophrenia: a meta-analytic investigation. *Schizophr Bull.* 2012;38(2):304–315.
70. Bowie CR, Depp C, McGrath JA, et al. Prediction of real-world functional disability in chronic mental disorders: a comparison of schizophrenia and bipolar disorder. *Am J Psychiatry.* 2010;167(9):1116–1124.
71. Harvey PD, Raykov T, Twamley EW, Vella L, Heaton RK, Patterson TL. Validating the measurement of real-world functional outcomes: phase I results of the VALERO study. *Am J Psychiatry.* 2011;168(11):1195–1201.
72. Fett AK, Viechtbauer W, Dominguez MD, Penn DL, van Os J, Krabbendam L. The relationship between neurocognition and social cognition with functional outcomes in schizophrenia: a meta-analysis. *Neurosci Biobehav Rev.* 2011;35(3):573–588.
73. Green MF, Kern RS, Braff DL, Mintz J. Neurocognitive deficits and functional outcome in schizophrenia: are we measuring the "right stuff"? *Schizophr Bull.* 2000;26(1):119–136.
74. Tabares-Seisdedos R, Balanza-Martinez V, Sanchez-Moreno J, et al. Neurocognitive and clinical predictors of functional outcome in patients with schizophrenia and bipolar I disorder at one-year follow-up. *J Affect Disord.* 2008;109(3):286–299.
75. Depp CA, Mausbach BT, Harmell AL, et al. Meta-analysis of the association between cognitive abilities and everyday functioning in bipolar disorder. *Bipolar Disord.* 2012;14(3):217–226.
76. Kim JH, Lee S, Han AY, Kim K, Lee J. Relationship between cognitive insight and subjective quality of life in outpatients with schizophrenia. *Neuropsychiatr Dis Treat.* 2015;11:2041–2048.
77. Phalen PL, Viswanadhan K, Lysaker PH, Warman DM. The relationship between cognitive insight and quality of life in schizophrenia spectrum disorders: symptom severity as potential moderator. *Psychiatry Res.* 2015;230(3):839–845.
78. O'Connor JA, Wiffen B, Diforti M, et al. Neuropsychological, clinical and cognitive insight predictors of outcome in a first episode psychosis study. *Schizophr Res.* 2013;149(1–3):70–76.
79. Lysaker PH, Erickson MA, Buck B, et al. Metacognition and social function in schizophrenia: associations over a period of five months. *Cogn Neuropsychiatry.* 2011;16(3):241–255.
80. Lysaker PH, Dimaggio G, Carcione A, et al. Metacognition and schizophrenia: the capacity for self-reflectivity as a predictor for prospective assessments of work performance over six months. *Schizophr Res.* 2010;122(1–3):124–130.
81. Casaletto KB, Moore DJ, Woods SP, Umlauf A, Scott JC, Heaton RK. Abbreviated goal management training shows preliminary evidence as a neurorehabilitation tool for HIV-associated neurocognitive disorders among substance users. *Clin Neuropsychol.* 2016;30(1):107–130.
82. Gould F, McGuire LS, Durand D, et al. Self-assessment in schizophrenia: accuracy of evaluation of cognition and everyday functioning. *Neuropsychology.* 2015;29(5):675–682.
83. Pinkham AE, Penn DL, Green MF, Harvey PD. Social cognition psychometric evaluation: results of the initial psychometric study. *Schizophr Bull.* 2016;42(2):494–504.
84. Moore MT, Fresco DM. Depressive realism: a meta-analytic review. *Clin Psychol Rev.* 2012;32(6):496–509.
85. Harvey PD, Twamley EW, Pinkham AE, Depp CA, Patterson TL. Depression in schizophrenia: associations with cognition, functional capacity, everyday functioning, and self-assessment. *Schizophr Bull.* 2017;43(3):575–582.
86. Moritz S, Goritz AS, Van Quaquebeke N, Andreou C, Jungclaussen D, Peters MJ. Knowledge corruption for visual perception in individuals high on paranoia. *Psychiatry Res.* 2014;215(3):700–705.
87. Moritz S, Goritz AS, Gallinat J, et al. Subjective competence breeds overconfidence in errors in psychosis: a hubris account of paranoia. *J Behav Ther Exp Psychiatry.* 2015;48:118–124.
88. Hofmann SG, Asnaani A, Vonk IJ, Sawyer AT, Fang A. The efficacy of cognitive behavioral therapy: a review of meta-analyses. *Cognit Ther Res.* 2012;36(5):427–440.
89. Moritz S, Veckenstedt R, Andreou C, et al. Sustained and "sleeper" effects of group metacognitive training for schizophrenia: a randomized clinical trial. *JAMA Psychiatry.* 2014;71(10):1103–1111.
90. Moritz S, Andreou C, Schneider BC, et al. Sowing the seeds of doubt: a narrative review on metacognitive training in schizophrenia. *Clin Psychol Rev.* 2014;34(4):358–366.
91. Moritz S, Veckenstedt R, Bohn F, et al. Complementary group metacognitive training (MCT) reduces delusional ideation in schizophrenia. *Schizophr Res.* 2013;151(1-3):61–69.
92. Moritz S, Veckenstedt R, Randjbar S, Vitzthum F, Woodward TS. Antipsychotic treatment beyond antipsychotics: metacognitive intervention for schizophrenia patients improves delusional symptoms. *Psychol Med.* 2011;41(9):1823–1832.
93. Mazza M, Lucci G, Pacitti F, et al. Could schizophrenic subjects improve their social cognition abilities only with observation and imitation of social situations? *Neuropsychol Rehabil.* 2010;20(5):675–703.
94. Moritz S, Kerstan A, Veckenstedt R, et al. Further evidence for the efficacy of a metacognitive group training in schizophrenia. *Behav Res Ther.* 2011;49(3):151–157.
95. Nuechterlein KH, Green MF, Kern RS, et al. The MATRICS Consensus Cognitive Battery, part 1: test selection, reliability, and validity. *Am J Psychiatry.* 2008;165(2):203–213.

NEUROPHYSIOLOGIC BIOMARKERS OF PSYCHOSIS

26

NEUROPHYSIOLOGIC BIOMARKERS OF PSYCHOSIS

EVENT-RELATED POTENTIAL BIOMARKERS

Judith M. Ford, Holly K. Hamilton, Katiah Llerena, Brian J. Roach, and Daniel H. Mathalon

When EEG was first used to study mental illness in the 1960s, the hope was that it would someday provide a diagnostic "laboratory test" and a treatment target. Although Hans Berger (1), the father of EEG, was himself a psychiatrist, his interest in brain waves was in their potential to enable paranormal communication. While the latter has not been realized, we are getting closer to realizing the potential of the former, as we discuss in this chapter. Here, we discuss EEG-based event-related potential (ERP) biomarkers of schizophrenia, as well as bipolar disorder and youth at clinical high-risk for psychosis. We limit our scope to auditory responses.

EVENT-RELATED POTENTIALS (ERPS)

EEG is a coarse measure of brain activity that represents a mixture of activity from hundreds of different neural sources. Neural responses associated with sensory, cognitive, and motor events can be extracted from the EEG by averaging the EEG, time locked to an event, over many presentations of the event (2). These ERPs provide a millisecond-to-millisecond read-out of the brain's responses, from the initial 8th nerve response to clicks to later cortical responses associated with deviance detection, response categorization, and motor response preparation and execution.

P50

The P50 is typically studied in schizophrenia using a paired stimulus or "conditioning-testing" paradigm with two identical "click"-like stimuli (S1 and S2) separated by 500 ms. P50 is a positive component, is largest at the central midline site Cz, and peaks at 50 ms after the stimulus. P50 amplitude to S2 is attenuated relative to S1, and this S2/S1 ratio reflects suppression of P50 to S2 relative to S1. P50 suppression is thought to reflect inhibitory control, or sensory gating, with S1 activating inhibitory neural mechanisms that minimize interference from S2(3). A schematic of P50 to S1 and S2 is shown in Figure 26.1.

Schizophrenia patients generally exhibit significantly poorer suppression than healthy individuals, with meta-analyses confirming robust effect sizes for the P50 suppression deficit that exceed many other cognitive and neurobiological abnormalities (4–6). However, effects across individual studies have varied, possibly due to methodological or participant differences (7). Although the suppression deficit is commonly believed to result from a failure of response inhibition to S2, some suggest it may result from an attenuated S1 amplitude without reduction of the S2 amplitude (e.g., see 8).

Nevertheless, P50 suppression abnormalities are seen across the spectrum of psychosis: acutely ill and clinically stable schizophrenia patients (4, 6, 7), schizophrenia patients early in their illness (9), schizotypal personality disorder (10), psychotic bipolar disorder (11, 12), post-traumatic stress disorder (13), and clinical high-risk samples (14, 15). Moreover, family studies indicate that the P50 suppression deficit is heritable (16–18). While P50 suppression does appear to be highly sensitive to schizophrenia, it may not be specific to the disorder (e.g., 13, 19).

P50 suppression abnormalities may reflect a fundamental neural deficit that contributes to attentional difficulties associated with schizophrenia, including difficulties in filtering out responses to irrelevant or redundant sensory perceptual information and internally generated stimuli (20), as well as the sense of sensory inundation and feeling overwhelmed. Indeed, self-reported experience of perceptual disturbances that are presumably related to sensory gating have been shown to be associated with P50 suppression abnormalities in schizophrenia (21). However, evidence relating abnormal P50 suppression to observable clinical symptoms and cognitive impairments in schizophrenia has been mixed. Some studies demonstrate clear associations between P50 abnormalities and negative symptoms as well as attention, working memory, and speeded information processing in schizophrenia (22, 23), while others have not replicated these findings (24, 25).

A distributed network involving hippocampal, temporal, prefrontal, and thalamic regions has been implicated in the generation of P50 and its suppression (26–29). Pharmacological and animal studies have demonstrated the importance of the cholinergic system in sensory gating, with stimulation of alpha7-nicotinic receptors regulating response suppression to redundant stimuli (30). Furthermore, P50 suppression abnormalities in schizophrenia can transiently

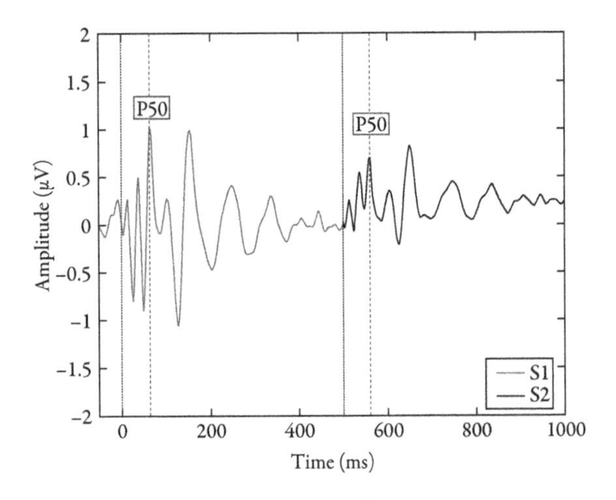

Figure 26.1 Prototypic ERPs to S1 and S2 "click"-like sounds, with a 500 ms interstimulus interval (ISI), and filtered between 10 and 50 Hz. In this and all figures, voltage in microvolts is on the y-axis, time in milliseconds is on the x-axis, negative is plotted down, EEG at scalp sites is referenced to mastoids.

be normalized by nicotine or nicotine agonists (e.g., 31, 32). P50 suppression abnormalities have been linked to a subunit of the CHRNA7 cholinergic receptor (33, 34) as well as to a catechol-O-methyltransferase polymorphism (35).

N100/P200

Fifty years ago, auditory ERPs were used in "evoked response audiometry" to assess hearing in people whose behavioral reports could not be obtained or trusted. Specifically, the N100 (N1), a negative going component of the ERP peaking 100 ms after stimulus onset, was used as an index of hearing because of its sensitivity to loudness (36). It was also sensitive to other physical variables, which all are under experimenter control: stimulus rise time (37), interstimulus interval (ISI) (38), sound duration (39), and the degree to which a tone differs in frequency from the preceding tone(s) (40). In the 1970s, its sensitivity to variables not under experimenter control, such as attention (41) and arousal (42), shifted N100 out the realm of audiometry and into cognitive neuroscience. Consequently, in addition to being recorded to sounds in a series of unchanging sounds, it is more often recorded in series of standard (frequent) and deviant (infrequent) sounds, in selective attention paradigms to assess the amount of attention a person pays to an "unattended" stimulus, and in paradigms involving motor actions that produce the sounds.

Regardless of the eliciting paradigm, N100 amplitude is often, but not always, reduced in schizophrenia, as we discussed in a literature review (43). In that review, we asked if N100 could be used as a biomarker of schizophrenia, and, if so, under what conditions. We noted that N100 reduction in schizophrenia depends on a number of factors: The degree of its reduction in schizophrenia is greater with long (>1 sec) compared to short ISIs; it is greater with greater task demands; its relationship to medications is complex; and it is not sensitive to clinical symptoms. Nevertheless, abnormalities in early auditory processing, reflected in N100 reduction, might serve as a biomarker of psychosis spectrum, as it is reduced in first-degree relatives of schizophrenia patients (44, 45), bipolar disorder patients (46), and in youth at clinical high risk for psychosis (47). In the years since that review appeared, subsequent papers are generally consistent with our conclusions.

N100 is followed by P2 or P200, a positive going potential, peaking at about 200 ms. For years, it was measured as the peak-to-peak difference between N100 and P200 (i.e., "N100–P200" or "N1–P2"). Although P200 invariably follows N100, they can be distinguished experimentally (48–51), topographically on the scalp (48, 52, 53), and functionally (54). Yet, N100 and P200 often covary (55). Thus, although cerebral structures that generate them may overlap to some extent, N100 and P200 are unlikely to reflect a single underlying neural process, and therefore, are best measured and studied independently of each other.

The functional significance of P200 is poorly understood (56, 57). P200 may reflect an attention-modulated process required for the performance of an auditory discrimination task (58), or when elicited by a nontarget stimulus in an oddball paradigm, it may reflect an attentional shift toward the stimulus, and reflect some aspects of the classification process (59). Thus, although both are obligatory responses to tones, N100 and P200 might be considered reflections of different perceptual stages in the auditory processing stream.

Like N100 amplitude, P200 reductions are sometimes (60–63), but not always (64), reported in schizophrenia. In an fMRI/ERP fusion analysis (54), we found a consistent pattern of covariation between increased P200 amplitude, activation of bilateral superior temporal gyrus/middle temporal gyrus, activation of frontal and parietal regions, and suppression of visual cortex and subdivisions of the default mode network (medial prefrontal cortex and posterior cingulate cortex). To the extent that P200 reflects the allocation of perceptual resources (59), our data show that people who expend more perceptual resources to the tone during passive listening have greater suppression of the default mode network and visual cortex. Schizophrenia patients with smaller joint-P200 components tended to be more amotivated and apathetic.

A prototypic ERP to a standard tone is shown in Figure 26.2 (black line), where the N100 and P200 and their scalp topographies can be seen.

MISMATCH NEGATIVITY (MMN)

Mismatch negativity (MMN) is elicited by an infrequent deviant sound presented within a series of repeated standard sounds (65). It is an index of sensory "echoic" memory, since the detection of auditory deviance depends on the short-term formation of a memory trace of the standard sounds present in the immediately preceding time window (66, 67). However, MMN also reflects longer term (minutes to hours) synaptic plasticity (68) and predictive coding, with the MMN signaling a prediction error when the auditory expectancy is violated by a deviant stimulus (69–71). Mismatch negativity can be elicited by deviance in one or more dimensions of auditory

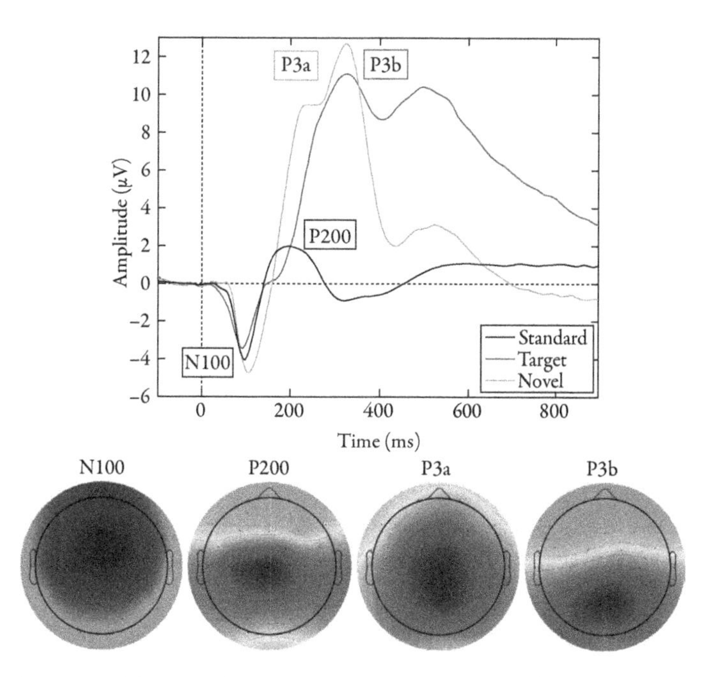

Figure 26.2 Prototypic ERPs to standard tones (black) recorded at Cz, novel sounds recorded at Fz (green), and target tones recorded at Pz (blue), all low pass filtered at 30 Hz. N100 and P200 are shown in the ERP to standard tones (black), P3a to novel sounds (green), and P3b to target sounds (blue). The scalp distributions of these components at their peak latencies are shown for N100, P200, P3a, and P3b. Nose is pointing upward. Note they are on different voltage scales in order to maximize visualization of the topographic, rather than voltage, differences. See color insert 1 in front of book for full colorized version.

stimuli, including pitch, duration, intensity and location (72). A schematic of ERPs to a standard tone, a duration deviant, and the difference waveform (deviant—standard) is shown in Figure 26.3.

Because MMN is elicited preattentively (65, 73, 74) it can assess auditory processing deficits in neuropsychiatric disorders without the confounding influences of motivation and attention associated with higher order cognitive tasks(75). Mismatch negativity amplitude reduction in psychosis is well documented (76, 77), including in chronic (78–94), early illness (80, 85, 93, 95–97), and first-episode schizophrenia (86, 87, 98-101) (but see 91, 94, 100, 102, 103). Reductions are seen in individuals at clinical high risk for psychosis (79, 85, 95), with greater reductions in those who subsequently transition to psychosis (97, 100, 104), adolescents with psychotic experiences (105), and, to a lesser extent, bipolar patients (106, 107). Its reduction in schizophrenia is not due to medication and is seen in unmedicated patients (79, 81, 100) (but see 102, 108, 109). Its sensitivity to schizophrenia does not depend on deviant type (101). While MMN is generally not sensitive to clinical symptoms, it is sensitive to functional status, being normal in schizophrenia patients who have significant role function in their communities (110). Whether a significant improvement in functional status could rescue MMN is not known.

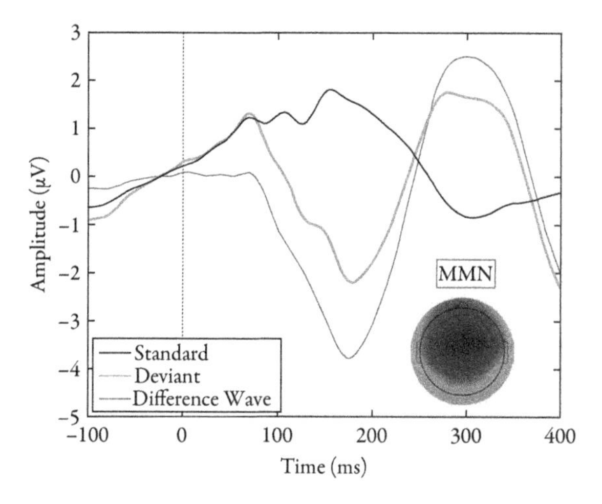

Figure 26.3 Prototypic ERPs to 50 ms duration standard tones (black) and 100 ms duration deviants. The difference waveform (deviant–standard) is shown in pink. ERPs were recorded at Cz, and low pass filtered at 30 Hz. The scalp distribution of MMN is shown at its peak latency.

P300

P300 amplitude reduction in schizophrenia is arguably the most replicated biological finding in the illness (111, 112), making it a likely electrophysiological marker of psychosis risk. Although the cognitive significance of P300 is still debated (113), prevailing views consider it to reflect attentional resource allocation (114), phasic attentional shifts (115), working memory updating of stimulus context (116), or stimulus salience (117). Its latency reflects processing speed or efficiency during stimulus evaluation (118) independent of motor preparation time (119).

Most P300 research uses the oddball paradigm, in which an infrequent deviant stimulus presented in a stream of standard stimuli elicits a P300 (120). There are two varieties of P300, P3a and P3b, with different psychological antecedents, neural generators, scalp topographies, and latencies. P3b is elicited when the infrequent stimulus is a "target," requiring a voluntary or "top-down" shift of attention (pressing a button to targets) or an updating of memory (running count of targets). It has a midline parietal maximum peaking about 300 ms following stimulus onset (121). Comprehensive reviews and meta-analyses indicate that P3b amplitude reduction and latency delay are among the most consistently replicated functional brain abnormalities reported in schizophrenia (122, 123). Prototypic ERPs to target tones (P3b) and novel sounds (P3a) are shown in Figure 26.2 with their distinctly different scalp topographies.

P3a is elicited by an unexpected, task-irrelevant salient stimulus in the oddball paradigm. It reflects involuntary, phasic "bottom-up" attention necessary for rapid detection, evaluation, and adaptation to unexpected and potentially important changes in the environment (124). P3a has a frontocentral scalp maximum and habituates rapidly (125, 126). Relatively few studies have examined P3a in schizophrenia, but most of these studies show amplitude reductions in response to novel or salient sounds (127–129).

Reductions in P3b (130, 131) and P3a (132, 133) are heritable and associated with disrupted genes (134, 135) implicated in the neurodevelopmental dysfunction that underlies schizophrenia (136), consistent with their potential roles as genetic endophenotypic markers of schizophrenia risk (137). Furthermore, P3b and P3a are reduced in individuals at clinical high risk for psychosis (47, 85, 138–142), particularly those who subsequently transition to psychosis (143, 144). P3b (145) and P3a (146) are also reduced in bipolar patients to the same extent as in schizophrenia patients.

P300 amplitude deficits reflect a stable trait factor, but also show clinical state fluctuations. Patients with worse psychosis (47), disorganization (129), negative symptoms (129, 147), and social and occupational functioning (129, 148) have smaller P3b amplitudes. Reduced P3a is associated with more severe negative symptoms in patients (149), but results have been mixed (129). In a single study of both state and trait effects, P3a and P3b both tracked clinical state over time independent of medication status (128). Importantly, even at their best clinical state, patients still had smaller P3a and P3b amplitudes compared to controls. This is consistent with greater P300 amplitude reduction in institutionalized inpatients relative to outpatients (150). Although neuroleptic medications increase P3b amplitude (151, 152), deficits persist after antipsychotic medications are withdrawn (153). P3b reduction and delay is greater with longer illness durations, suggesting sensitivity to a progressive pathophysiological process (154). P3a is related to frontal gray matter reduction, and target P3b is related to temporal gray matter reduction (155), suggesting reductions in both reflect structural brain deficits associated with the illness.

AUDITORY STEADY STATE RESPONSE

Recently, investigators have taken advantage of new algorithms and greater computing power to decompose EEG oscillations into power and phase components. Gamma band (30–80 Hz) oscillations are especially relevant to schizophrenia research because of their role in coding and binding information required for perception (156, 157) and their dependence on gamma-aminobutyric acid (GABAA) (158–164) and glutamatergic N-methyl-D-aspartate (NMDA) (165–168) receptors that are affected by schizophrenia (169).

The auditory steady state response (ASSR) is elicited by an auditory stimulus repeated at a fixed rate or frequency (170). Across a wide range of tested frequencies, the ASSR has its peak power with a 40 Hz repetition rate (170–172). This could be due to the phase-synchronized overlap of individual GBRs that span 100 ms and linearly summate to produce a peak ASSR amplitude when stimuli are presented every 25 ms (i.e., at a 40 Hz frequency) (173). The ASSR takes 200–300 ms to reach a stable magnitude. A recent meta-analysis determined that 40 Hz ASSR reductions in schizophrenia were robust across 20 different EEG and magnetoencephalograpy studies, and neither the type of time-frequency measure (phase vs. power) nor the physical stimulus characteristics influenced this (174). There are no robust relationships between clinical symptoms and ASSR (174), although there is a modest relationship with cognition (175).

Both 40 Hz ASSR and GBR abnormalities have been found in the first-degree relatives of schizophrenia patients (176, 177) and clinically high-risk youth (178, 179). ASSR deficits are also reported in bipolar disorder (180).

CONCLUSIONS

EEG has kept its promise as a biomarker of mental illness; each component discussed is compromised in people with schizophrenia and to varying degrees, across the psychosis spectrum. Thus, they seem to be stable biomarkers of psychosis vulnerability. However, they tend to be relatively insensitive to the specific clinical symptoms that define psychosis, such as hallucinations, delusions, and thought disorder. This could be due to several reasons. First, none of the task conditions that elicit these ERP components are designed to elicit those symptoms or their underpinnings. Second, valid assessment of hallucinations and delusions relies on clinicians' understanding of a patient's self-report of psychotic experiences that are also difficult for the patient to describe and interpret. However, negative symptoms do seem to correlate with P300, perhaps because amotivation and inability to maintain focus contribute to both negative symptoms and smaller P300s. Also, negative symptom assessment relies on observation by trained clinicians and on relatively objective reports by the patients, who can more easily report details of their daily activities and living situations than the hallucinations and delusions they experience. While this might suggest that negative symptom ratings would be more valid and thereby correlate better with ERP components, this does not seem to be borne out in the literature. Third, because of the motivational and cognitive limitations of some people with schizophrenia, most of the components described here can be elicited passively and minimally tax motivation and higher order processes. Thus, they, in turn, are limited in what they can reveal about some symptoms.

An early hurdle in our effort to use ERPs as biomarkers of different mental illnesses was the need to establish "specificity" of an ERP abnormality to a specific DSM diagnosis. With the promulgation of the Research Domain Criteria, we are now less concerned about the specificity of an ERP abnormality to a single diagnosis, and more concerned about what deficits in the different components might reveal about dysfunctions that cut across diagnostic boundaries. For example, P50 suppression deficits may reflect deficits in neural inhibition; N100/P200 deficits might reflect deficits in attentional resource availability and allocation; MMN deficits might reflect deficits in neural plasticity; and P300 deficits might reflect deficits in updating of working memory (P3b) and orienting to changes in the environment (P3a). When acquired together, functional MRI and ERPs may point to specific neuroanatomical targets for noninvasive brain stimulation.

REFERENCES

1. Berger H. Uber das Elektroenkephalogramm des Menschen. Arch Psychiatr Nervenkr. 1929;87:527–570.
2. Luck SJ. An introduction to the event-related potential technique. Cambridge, MA: The MIT Press; 2005.
3. Nagamoto HT, Adler LE, Waldo MC, Freedman R. Sensory gating in schizophrenics and normal controls: effects of changing stimulation interval. Biol Psychiatry. 1989;25:549–561.
4. Bramon E, Rabe-Hesketh S, Sham P, Murray RM, Frangou S. Meta-analysis of the P300 and P50 waveforms in schizophrenia. Schizophr Res. 2004;70:315–329.
5. Heinrichs RW. In search of madness: schizophrenia and neuroscience. Oxford University Press; 2001.
6. Patterson JV, Hetrick WP, Boutros NN, Jin Y, Sandman C, Stern H, Potkin S, Bunney WE Jr. P50 sensory gating ratios in schizophrenics and controls: a review and data analysis. Psychiatry Res. 2008;158:226–247.
7. de Wilde OM, Bour LJ, Dingemans PM, Koelman JHTM, Linszen DH. A meta-analysis of P50 studies in patients with schizophrenia and relatives: differences in methodology between research groups. Schizophr Res. 2007;97:137–151.
8. Johannesen JK, Kieffaber PD, O'Donnell BF, Shekhar A, Evans JD, Hetrick WP. Contributions of subtype and spectral frequency analyses to the study of P50 ERP amplitude and suppression in schizophrenia. Schizophr Res. 2005;78:269–284.
9. Yee CM, Nuechterlein KH, Morris SE, White PM. P50 suppression in recent-onset schizophrenia: clinical correlates and risperidone effects. J Abnorm Psychol. 1998;107:691–698.
10. Cadenhead KS, Light GA, Geyer MA, Braff DL. Sensory gating deficits assessed by the P50 event-related potential in subjects with schizotypal personality disorder. Am J Psychiatry. 2000;157:55–59.
11. Schulze KK, Hall MH, McDonald C, Marshall N, Walshe M, Murray RM, Bramon E. P50 auditory evoked potential suppression in bipolar disorder patients with psychotic features and their unaffected relatives. Biol Psychiatry. 2007;62:121–128.
12. Olincy A, Martin L. Diminished suppression of the P50 auditory evoked potential in bipolar disorder subjects with a history of psychosis. Am J Psychiatry. 2005;162:43–49.
13. Neylan TC, Fletcher DJ, Lenoci M, McCallin K, Weiss DS, Schoenfeld FB, Marmar CR, Fein G. Sensory gating in chronic posttraumatic stress disorder: reduced auditory P50 suppression in combat veterans. Biol Psychiatry. 1999;46:1656–1664.
14. Brockhaus-Dumke A, Schultze-Lutter F, Mueller R, Tendolkar I, Bechdolf A, Pukrop R, Klosterkoetter J, Ruhrmann S. Sensory gating in schizophrenia: P50 and N100 gating in antipsychotic-free subjects at risk, first-episode, and chronic patients. Biol Psychiatry. 2008;64:376–384.
15. Myles-Worsley M, Ord L, Blailes F, Ngiralmau H, Freedman R. P50 sensory gating in adolescents from a pacific island isolate with elevated risk for schizophrenia. Biol Psychiatry. 2004;55:663–667.
16. Young DA, Waldo M, Rutledge JH, 3rd, Freedman R. Heritability of inhibitory gating of the P50 auditory-evoked potential in monozygotic and dizygotic twins. Neuropsychobiology. 1996;33:113–117.
17. Hall MH, Schulze K, Rijsdijk F, Picchioni M, Ettinger U, Bramon E, Freedman R, Murray RM, Sham P. Heritability and reliability of P300, P50 and duration mismatch negativity. Behav Genet. 2006;36:845–857.
18. Greenwood TA, Braff DL, Light GA, Cadenhead KS, Calkins ME, Dobie DJ, Freedman R, Green MF, Gur RE, Gur RC, et al. Initial heritability analyses of endophenotypic measures for schizophrenia: the consortium on the genetics of schizophrenia. Arch Gen Psychiatry. 2007;64:1242–1250.
19. Boutros NN, Gooding D, Sundaresan K, Burroughs S, Johanson CE. Cocaine-dependence and cocaine-induced paranoia and midlatency auditory evoked responses and sensory gating. Psychiatry Res. 2006;145:147–154.
20. Braff DL, Grillon C, Geyer MA. Gating and habituation of the startle reflex in schizophrenic patients. Arch Gen Psychiatry. 1992;49:206–215.
21. Johannesen JK, Bodkins M, O'Donnell BF, Shekhar A, Hetrick WP. Perceptual anomalies in schizophrenia co-occur with selective impairments in the gamma frequency component of midlatency auditory ERPs. J Abnorm Psychol. 2008;117:106–118.
22. Smith AK, Edgar JC, Huang M, Lu BY, Thoma RJ, Hanlon FM, McHaffie G, Jones AP, Paz RD, Miller GA, et al. Cognitive abilities and 50- and 100-msec paired-click processes in schizophrenia. Am J Psychiatry. 2010;167:1264–1275.
23. Hamilton HK, Williams TJ, Ventura J, Jasperse LJ, Owens EM, Miller GA, Subotnik KL, Neuchterlein KH, Yee CM. Clinical and cognitive significance of auditory sensory processing deficits in schizophrenia. Am J Psychiatry. 2018;175:275–283.
24. Potter D, Summerfelt A, Gold J, Buchanan RW. Review of clinical correlates of P50 sensory gating abnormalities in patients with schizophrenia. Schizophr Bull. 2006;32:692–700.
25. Sánchez-Morla EM, Santos JL, Aparicio A, García-Jiménez MÁ, Soria C, Arango C. Neuropsychological correlates of P50 sensory gating in patients with schizophrenia. Schizophr Res. 2013;143:102–106.
26. Tregellas JR, Davalos DB, Rojas DC, Waldo MC, Gibson L, Wylie K, Du YP, Freedman R. Increased hemodynamic response in the hippocampus, thalamus and prefrontal cortex during abnormal sensory gating in schizophrenia. Schizophr Res. 2007;92:262–272.
27. Williams TJ, Nuechterlein KH, Subotnik KL, Yee CM. Distinct neural generators of sensory gating in schizophrenia. Psychophysiology. 2011;48:470–478.
28. Grunwald T, Boutros NN, Pezer N, von Oertzen J, Fernández G, Schaller C, Elger CE. Neuronal substrates of sensory gating within the human brain. Biol Psychiatry. 2003;53:511–519.
29. Mayer AR, Hanlon FM, Franco AR, Teshiba TM, Thoma RJ, Clark VP, Canive JM. The neural networks underlying auditory sensory gating. Neuroimage. 2009;44:182–189.
30. Adler LE, Olincy A, Waldo M, Harris JG, Griffith J, Stevens K, Flach K, Nagamoto H, Bickford P, Leonard S, et al. Schizophrenia, sensory gating, and nicotinic receptors. Schizophr Bull. 1998;24:189–202.
31. Adler LE, Hoffer LJ, Griffith J, Waldo MC, Freedman R. Normalization by nicotine of deficient auditory sensory gating in the relatives of schizophrenics. Biol Psychiatry. 1992;32:607–616.
32. Martin LF, Kem WR, Freedman R. Alpha-7 nicotinic receptor agonists: potential new candidates for the treatment of schizophrenia. Psychopharmacology (Berl). 2004;174:54–64.
33. Freedman R, Coon H, Myles-Worsley M, Orr-Urtreger A, Olincy A, Davis A, Polymeropoulos M, Holik J, Hopkins J, Hoff M, et al. Linkage of a neurophysiological deficit in schizophrenia to a chromosome 15 locus. Proc Natl Acad Sci U S A. 1997;94:587–592.

34. Leonard S, Gault J, Hopkins J, Logel J, Vianzon R, Short M, Drebing C, Berger R, Venn D, Sirota P, et al. Association of promoter variants in the alpha7 nicotinic acetylcholine receptor subunit gene with an inhibitory deficit found in schizophrenia. Arch Gen Psychiatry. 2002;59:1085–1096.
35. Lu BY, Martin KE, Edgar JC, Smith AK, Lewis SF, Escamilla MA, Miller GA, Canive JM. Effect of catechol O-methyltransferase val(158)met polymorphism on the p50 gating endophenotype in schizophrenia. Biol Psychiatry. 2007;62:822–825.
36. Keidel WD, Spreng M. Neurophysiological evidence for the Stevens power function in man. J Acoust Soc Am. 1965;38:191–195.
37. Putnam LE, Roth WT. Effects of stimulus repetition, duration, and rise time on startle blink and automatically elicited P300. Psychophysiology. 1990;27:275–297.
38. Davis H, Mast T, Yoshie N, Zerlin S. The slow response of the human cortex to auditory stimuli: recovery process. Electroencephalogr Clin Neurophysiol. 1966;21:105–113.
39. Onishi S, Davis H. Effects of duration and rise time of tone bursts on evoked V-potentials. J Acoust Soc Am. 1968;44:572–591.
40. Butler RA. Effect of changes in stimulus frequency and intensity on habituation of the human vertex potential. J Acoust Soc Am. 1968;44:945–950.
41. Hillyard SA, Hink RF, Schwent VL, Picton TW. Electrical signs of selective attention in the human brain. Science. 1973;182:177–180.
42. Naatanen R, Michie PT. Early selective-attention effects on the evoked potential: a critical review and reinterpretation. Biol Psychol. 1979;8:81–136.
43. Rosburg T, Boutros NN, Ford JM. Reduced auditory evoked potential component N100 in schizophrenia—a critical review. Psychiatry Res. 2008;161:259–274.
44. Ford JM, Mathalon DH, Roach BJ, Keedy SK, Reilly JL, Gershon ES, Sweeney JA. Neurophysiological evidence of corollary discharge function during vocalization in psychotic patients and their nonpsychotic first-degree relatives. Schizophr Bull. 2013;39:1272–1280.
45. Turetsky BI, Greenwood TA, Olincy A, Radant AD, Braff DL, Cadenhead KS, Dobie DJ, Freedman R, Green MF, Gur RE, et al. Abnormal auditory N100 amplitude: a heritable endophenotype in first-degree relatives of schizophrenia probands. Biol Psychiatry. 2008;64:1051–1059.
46. Umbricht D, Koller R, Schmid L, Skrabo A, Grubel C, Huber T, Stassen H. How specific are deficits in mismatch negativity generation to schizophrenia? Biol Psychiatry. 2003;53:1120–1131.
47. del Re EC, Spencer KM, Oribe N, Mesholam-Gately RI, Goldstein J, Shenton ME, Petryshen T, Seidman LJ, McCarley RW, Niznikiewicz MA. Clinical high risk and first episode schizophrenia: auditory event-related potentials. Psychiatry Res. 2015;231:126–133.
48. Wang J, Mathalon DH, Roach BJ, Reilly J, Keedy SK, Sweeney JA, Ford JM. Action planning and predictive coding when speaking. Neuroimage. 2014;91:91–98.
49. Ford JM, Roth WT, Kopell BS. Attention effects on auditory evoked potentials to infrequent events. Biological Psychology. 1976;4:65–77.
50. Oades R, Dittmann-Balcar A, Zerbin D. Development and topography of auditory event-related potentials (ERPs): mismatch and processing negativity in individuals 8–22 years of age. Psychophysiology. 1997;34:677–693.
51. Ford JM, Roth WT, Menon V, Pfefferbaum A. Failures of automatic and strategic processing in schizophrenia: comparisons of event-related brain potential and startle blink modification. Schizophr Res. 1999;37:149–163.
52. Vaughan HG, Jr., Ritter W, Simson R. Topographic analysis of auditory event-related potentials. Prog Brain Res. 1980;54:279–285.
53. Verkindt C, Bertrand O, Thevenet M, Pernier J. Two auditory components in the 130–230 ms range disclosed by their stimulus frequency dependence. Neuroreport. 1994;5:1189–1192.
54. Ford JM, Roach BJ, Palzes VA, Mathalon DH. Using concurrent EEG and fMRI to probe the state of the brain in schizophrenia. Neuroimage Clin. 2016;12:429–441.
55. Paiva TO, Almeida PR, Ferreira-Santos F, Vieira JB, Silveira C, Chaves PL, Barbosa F, Marques-Teixeira J. Similar sound intensity dependence of the N1 and P2 components of the auditory ERP: Averaged and single trial evidence. Clin Neurophysiol. 2016;127:499–508.
56. Woodman GF. A brief introduction to the use of event-related potentials in studies of perception and attention. Atten Percept Psychophys. 2010;72:2031–2046.
57. Crowley KE, Colrain IM. A review of the evidence for P2 being an independent component process: age, sleep and modality. Clin Neurophysiol. 2004;115:732–744.
58. Novak G, Ritter W, Vaughan HG, Jr. Mismatch detection and the latency of temporal judgements. Psychophysiology. 1992;29:398–411.
59. Garcia-Larrea L, Lukaszewicz AC, Mauguiere F. Revisiting the oddball paradigm: non-target vs neutral stimuli and the evaluation of ERP attentional effects. Neuropsychologia. 1992;30:723–741.
60. Roth WT, Goodale J, Pfefferbaum A. Auditory event-related potentials and electrodermal activity in medicated and unmedicated schizophrenics. Biol Psychiatry. 1991;29:585–599.
61. Roth WT, Horvath TB, Pfefferbaum A, Kopell BS. Event related potentials in schizophrenics. Electroencephalogr Clin Neurophysiol. 1980;48:l27–l39.
62. Salisbury DF, Collins KC, McCarley RW. Reductions in the N1 and P2 auditory event-related potentials in first-hospitalized and chronic schizophrenia. Schizophr Bull. 2010;36:991–1000.
63. Ethridge LE, Hamm JP, Pearlson GD, Tamminga CA, Sweeney JA, Keshavan MS, Clementz BA. Event-related potential and time-frequency endophenotypes for schizophrenia and psychotic bipolar disorder. Biol Psychiatry. 2015;77:127–136.
64. Potts GF, Hirayasu Y, O'Donnell BF, Shenton ME, McCarley RW. High-density recording and topographic analysis of the auditory oddball event-related potential in patients with schizophrenia. Biol Psychiatry. 1998;44:982–989.
65. Naatanen R, Teder W, Alho K, Lavikainen J. Auditory attention and selective input modulation: a topographical ERP study. Neuroreport. 1992;3:493–496.
66. Naatanen R, Jacobsen T, Winkler I. Memory-based or afferent processes in mismatch negativity (MMN): a review of the evidence. Psychophysiology. 2005;42:25–32.
67. Naatanen R. Mismatch negativity (MMN): perspectives for application. International journal of psychophysiology: official journal of the International Organization of Psychophysiology. 2000;37:3–10.
68. Stephan KE, Baldeweg T, Friston KJ. Synaptic plasticity and dysconnection in schizophrenia. Biol Psychiatry. 2006;59:929–939.
69. Garrido MI, Kilner JM, Stephan KE, Friston KJ. The mismatch negativity: a review of underlying mechanisms. Clin Neurophysiol. 2009;120:453–463.
70. Friston K. A theory of cortical responses. Phil Trans R Soc B. 2005;360:815–836.
71. Todd J, Michie PT, Schall U, Ward PB, Catts SV. Mismatch negativity (MMN) reduction in schizophrenia—impaired prediction-error generation, estimation or salience? International journal of psychophysiology: official journal of the International Organization of Psychophysiology. 2012;83:222–231.
72. Naatanen R, Pakarinen S, Rinne T, Takegata R. The mismatch negativity (MMN): towards the optimal paradigm. Clin Neurophysiol. 2004;115:140–144.
73. Fischer C, Morlet D, Bouchet P, Luaute J, Jourdan C, Salord F. Mismatch negativity and late auditory evoked potentials in comatose patients. Clin Neurophysiol. 1999;110:1601–1610.
74. Naatanen R, Alho K. Generators of electrical and magnetic mismatch responses in humans. Brain Topogr. 1995;7:315–320.
75. Mathalon DH, Ford JM. Divergent approaches converge on frontal lobe dysfunction in schizophrenia. The American journal of psychiatry. 2008;165:944–948.
76. Umbricht D, Krljes S. Mismatch negativity in schizophrenia: a meta-analysis. Schizophr Res. 2005;76:1–23.

77. Naatanen R, Kahkonen S. Central auditory dysfunction in schizophrenia as revealed by the mismatch negativity (MMN) and its magnetic equivalent MMNm: a review. Int J Neuropsychopharmacol. 2009;12:125–135.
78. Javitt DC, Grochowski S, Shelley AM, Ritter W. Impaired mismatch negativity (MMN) generation in schizophrenia as a function of stimulus deviance, probability, and interstimulus/interdeviant interval. Electroencephalogr Clin Neurophysiol. 1998;108:143–153.
79. Brockhaus-Dumke A, Tendolkar I, Pukrop R, Schultze-Lutter F, Klosterkotter J, Ruhrmann S. Impaired mismatch negativity generation in prodromal subjects and patients with schizophrenia. Schizophr Res. 2005;73:297–310.
80. Umbricht DS, Bates JA, Lieberman JA, Kane JM, Javitt DC. Electrophysiological indices of automatic and controlled auditory information processing in first-episode, recent-onset and chronic schizophrenia. Biol Psychiatry. 2006;59:762–772.
81. Catts SV, Shelley AM, Ward PB, Liebert B, McConaghy N, Andrews S, Michie PT. Brain potential evidence for an auditory sensory memory deficit in schizophrenia. Am J Psychiatry. 1995;152:213–219.
82. Shelley AM, Ward PB, Catts SV, Michie PT, Andrews S, McConaghy N. Mismatch negativity: an index of a preattentive processing deficit in schizophrenia. Biol Psychiatry. 1991;30:1059–1062.
83. Javitt DC, Doneshka P, Zylberman I, Ritter W, Vaughan HG, Jr. Impairment of early cortical processing in schizophrenia: an event-related potential confirmation study. Biol Psychiatry. 1993;33:513–519.
84. Michie PT, Budd TW, Todd J, Rock D, Wichmann H, Box J, Jablensky AV. Duration and frequency mismatch negativity in schizophrenia. Clin Neurophysiol. 2000;111:1054–1065.
85. Jahshan C, Cadenhead KS, Rissling AJ, Kirihara K, Braff DL, Light GA. Automatic sensory information processing abnormalities across the illness course of schizophrenia. Psychol Med. 2012;42:85–97.
86. Oades RD, Wild-Wall N, Juran SA, Sachsse J, Oknina LB, Ropcke B. Auditory change detection in schizophrenia: sources of activity, related neuropsychological function and symptoms in patients with a first episode in adolescence, and patients 14 years after an adolescent illness-onset. BMC Psychiatry. 2006;6:7.
87. Oknina LB, Wild-Wall N, Oades RD, Juran SA, Ropcke B, Pfueller U, Weisbrod M, Chan E, Chen EY. Frontal and temporal sources of mismatch negativity in healthy controls, patients at onset of schizophrenia in adolescence and others at 15 years after onset. Schizophr Res. 2005;76:25–41.
88. Rasser PE, Schall U, Todd J, Michie PT, Ward PB, Johnston P, Helmbold K, Case V, Soyland A, Tooney PA, et al. Gray matter deficits, mismatch negativity, and outcomes in schizophrenia. Schizophr Bull. 2011;37:131–140.
89. Light GA, Braff DL. Mismatch negativity deficits are associated with poor functioning in schizophrenia patients. Arch Gen Psychiatry. 2005;62:127–136.
90. Light GA, Braff DL. Stability of mismatch negativity deficits and their relationship to functional impairments in chronic schizophrenia. Am J Psychiatry. 2005;162:1741–1743.
91. Magno E, Yeap S, Thakore JH, Garavan H, De Sanctis P, Foxe JJ. Are auditory-evoked frequency and duration mismatch negativity deficits endophenotypic for schizophrenia? High-density electrical mapping in clinically unaffected first-degree relatives and first-episode and chronic schizophrenia. Biol Psychiatry. 2008;64:385–391.
92. Kiang M, Braff DL, Sprock J, Light GA. The relationship between preattentive sensory processing deficits and age in schizophrenia patients. Clin Neurophysiol. 2009;120:1949–1957.
93. Javitt DC, Shelley AM, Silipo G, Lieberman JA. Deficits in auditory and visual context-dependent processing in schizophrenia: defining the pattern. Arch Gen Psychiatry. 2000;57:1131–1137.
94. Salisbury DF, Shenton ME, Griggs CB, Bonner-Jackson A, McCarley RW. Mismatch negativity in chronic schizophrenia and first-episode schizophrenia. Arch Gen Psychiatry. 2002;59:686–694.
95. Atkinson RJ, Michie PT, Schall U. Duration mismatch negativity and P3a in first-episode psychosis and individuals at ultra-high risk of psychosis. Biol Psychiatry. 2012;71:98–104.
96. Todd J, Michie PT, Schall U, Karayanidis F, Yabe H, Naatanen R. Deviant matters: duration, frequency, and intensity deviants reveal different patterns of mismatch negativity reduction in early and late schizophrenia. Biol Psychiatry. 2008;63:58–64.
97. Perez VB, Woods SW, Roach BJ, Ford JM, McGlashan TH, Srihari VH, Mathalon DH. Automatic auditory processing deficits in schizophrenia and clinical high-risk patients: forecasting psychosis risk with mismatch negativity. Biol Psychiatry. 2013.
98. Hermens DF, Ward PB, Hodge MA, Kaur M, Naismith SL, Hickie IB. Impaired MMN/P3a complex in first-episode psychosis: cognitive and psychosocial associations. Prog Neuropsychopharmacol Biol Psychiatry. 2010;34:822–829.
99. Kaur M, Battisti RA, Ward PB, Ahmed A, Hickie IB, Hermens DF. MMN/P3a deficits in first episode psychosis: comparing schizophrenia-spectrum and affective-spectrum subgroups. Schizophr Res. 2011;130:203–209.
100. Bodatsch M, Ruhrmann S, Wagner M, Muller R, Schultze-Lutter F, Frommann I, Brinkmeyer J, Gaebel W, Maier W, Klosterkotter J, et al. Prediction of psychosis by mismatch negativity. Biol Psychiatry. 2011;69:959–966.
101. Hay RA, Roach BJ, Srihari VH, Woods SW, Ford JM, Mathalon DH. Equivalent mismatch negativity deficits across deviant types in early illness schizophrenia-spectrum patients. Biol Psychol. 2015;105:130–137.
102. Devrim-Ucok M, Keskin-Ergen HY, Ucok A. Mismatch negativity at acute and post-acute phases of first-episode schizophrenia. Eur Arch Psychiatry Clin Neurosci. 2008;258:179–185.
103. Valkonen-Korhonen M, Purhonen M, Tarkka IM, Sipila P, Partanen J, Karhu J, Lehtonen J. Altered auditory processing in acutely psychotic never-medicated first-episode patients. Brain Res Cogn Brain Res. 2003;17:747–758.
104. Shaikh M, Valmaggia L, Broome MR, Dutt A, Lappin J, Day F, Woolley J, Tabraham P, Walshe M, Johns L, et al. Reduced mismatch negativity predates the onset of psychosis. Schizophr Res. 2012;134:42–48.
105. Murphy JR, Rawdon C, Kelleher I, Twomey D, Markey PS, Cannon M, Roche RA. Reduced duration mismatch negativity in adolescents with psychotic symptoms: further evidence for mismatch negativity as a possible biomarker for vulnerability to psychosis. BMC Psychiatry. 2013;13:45.
106. Hermens DF, Chitty KM, Kaur M. Mismatch negativity in bipolar disorder: A neurophysiological biomarker of intermediate effect? Schizophr Res. 2018;191:132–139.
107. Jahshan C, Cadenhead KS, Rissling AJ, Kirihara K, Braff DL, Light GA. Automatic sensory information processing abnormalities across the illness course of schizophrenia. Psychol Med. 2011:1–13.
108. Kirino E, Inoue R. The relationship of mismatch negativity to quantitative EEG and morphological findings in schizophrenia. J Psychiatr Res. 1999;33:445–456.
109. Korostenskaja M, Dapsys K, Siurkute A, Maciulis V, Ruksenas O, Kahkonen S. Effects of olanzapine on auditory P300 and mismatch negativity (MMN) in schizophrenia spectrum disorders. Progress in neuro-psychopharmacology and biological psychiatry. 2005;29:543–548.
110. Hamilton HK, Perez VB, Ford JM, Roach BJ, Jaeger J, Mathalon DH. The relationship between auditory processing deficits and independent role functioning in schizophrenia. Schizophrenia Bulletin. 2018;44:492–504.
111. Jeon YW, Polich J. Meta-analysis of P300 and schizophrenia: patients, paradigms, and practical implications. Psychophysiology. 2003;40:684–701.
112. Ford JM. Schizophrenia: the broken P300 and beyond. Psychophysiology. 1999;36:667–682.
113. Rasmusson DX, Allen JD. Intertarget interval does not affect P300 within oddball series. Int J Psychophysiol. 1994;17:57–63.
114. Polich J. Habituation of P300 from auditory stimuli. Psychobiology. 1989;17:19–28.
115. Knight RT. Evoked potential studies of attention capacity in human frontal lobe lesions. In: Levin HS, Eisenberg HM, Benton AL,

editors. Frontal lobe function and dysfunction. New York: Oxford University Press; 1991: pp. 139–153.
116. Donchin E, Coles M. Is the P300 component a manifestation of context updating? (Commentary on Verleger's critique of the context updating model). Behavioral and Brain Science. 1988;11:357–374.
117. Sutton S, Tueting P, Zubin J, John ER. Information delivery and the sensory evoked potential. Science. 1967;155:1436–1439.
118. Duncan-Johnson CC, Donchin E. On quantifying surprise: the variation in event-related potentials with subjective probability. Psychophysiology. 1977;14:456–467.
119. McCarthy G, Donchin E. A metric for thought: a comparison of P300 latency and reaction time. Science. 1981;221:79–89.
120. Polich J. Updating P300: an integrative theory of P3a and P3b. Clin Neurophysiol. 2007;118:2128–2148.
121. Polich J. P300, probability, and interstimulus interval. Psychophysiology. 1990;27:396–403.
122. Ford JM, Roth WT, Pfefferbaum A. P3 and schizophrenia. Annals of the NY Academy of Sciences. 1992;658:146–162.
123. Bramon E, Rabe-Hesketh S, Sham P, Murray RM, Frangou S. Meta-analysis of the P300 and P50 waveforms in schizophrenia. Schizophr Res. 2004;70:315–329.
124. Daffner KR, Scinto LF, Calvo V, Faust R, Mesulam MM, West WC, Holcomb PJ. The influence of stimulus deviance on electrophysiologic and behavioral responses to novel events. J Cogn Neurosci. 2000;12:393–406.
125. Iacono WG, Carlson SR, Malone SM. Identifying a multivariate endophenotype for substance use disorders using psychophysiological measures. Int J Psychophysiol. 2000;38:81–96.
126. Almasy L, Blangero J. Endophenotypes as quantitative risk factors for psychiatric disease: rationale and study design. American Journal of Medical Genetics. 2001;105:42–44.
127. Ford JM, Roth WT, Menon V, Pfefferbaum A. Failures of automatic and strategic processing in schizophrenia: Comparisons of event-related potential and startle blink modification. Schizophr Res. 1999;37:149–163.
128. Mathalon DH, Ford JM, Pfefferbaum A. Trait and state aspects of P300 amplitude reduction in schizophrenia: a retrospective longitudinal study. Biol Psychiatry. 2000;47:434–449.
129. Perlman G, Foti D, Jackson F, Kotov R, Constantino E, Hajcak G. Clinical significance of auditory target P300 subcomponents in psychosis: differential diagnosis, symptom profiles, and course. Schizophr Res. 2015;165:145–151.
130. Groom MJ, Bates AT, Jackson GM, Calton TG, Liddle PF, Hollis C. Event-related potentials in adolescents with schizophrenia and their siblings: a comparison with attention-deficit/hyperactivity disorder. Biol Psychiatry. 2008;63:784–792.
131. Bestelmeyer PEG, Phillips LH, Crombie C, Benson P, St.Clair D. The P300 as a possible endophenotype for schizophrenia and bipolar disorder: evidence from twin and patient studies. Psychiatry Res. 2009;169:212–219.
132. Turetsky BI, Cannon TD, Gur RE. P300 subcomponent abnormalities in schizophrenia: III. Deficits in unaffected siblings of schizophrenic probands. Biol Psychiatry. 2000;47:380–390.
133. Turetsky BI, Bilker WB, Siegel SJ, Kohler CG, Gur RE. Profile of auditory information-processing deficits in schizophrenia. Psychiatry Res. 2009;165:27–37.
134. Blackwood DH, Fordyce A, Walker MT, St Clair DM, Porteous DJ, Muir WJ. Schizophrenia and affective disorders—cosegregation with a translocation at chromosome 1q42 that directly disrupts brain-expressed genes: clinical and P300 findings in a family. Am J Hum Genet. 2001;69:428–433.
135. Hennah W, Thomson P, Peltonen L, Porteous D. Genes and schizophrenia: beyond schizophrenia: the role of DISC1 in major mental illness. Schizophr Bull. 2006;32:409–416.
136. Shaikh M, Hall MH, Schulze K, Dutt A, Li K, Williams I, Walshe M, Constante M, Broome M, Picchioni M, et al. Effect of DISC1 on the P300 waveform in psychosis. Schizophr Bull. 2013;39:161–167.
137. Turetsky BI, Dress EM, Braff DL, Calkins ME, Green MF, Greenwood TA, Gur RE, Gur RC, Lazzeroni LC, Nuechterlein KH, et al. The utility of P300 as a schizophrenia endophenotype and predictive biomarker: clinical and socio-demographic modulators in COGS-2. Schizophr Res. 2015;163:53–62.
138. Bramon E, Shaikh M, Broome M, Lappin J, Berge D, Day F, Woolley J, Tabraham P, Madre M, Johns L, et al. Abnormal P300 in people with high risk of developing psychosis. Neuroimage. 2008;41:553–560.
139. Frommann I, Brinkmeyer J, Ruhrmann S, Hack E, Brockhaus-Dumke A, Bechdolf A, Wolwer W, Klosterkotter J, Maier W, Wagner M. Auditory P300 in individuals clinically at risk for psychosis. Int J Psychophysiol. 2008;70:192–205.
140. Mondragon-Maya A, Solis-Vivanco R, Leon-Ortiz P, Rodriguez-Agudelo Y, Yanez-Tellez G, Bernal-Hernandez J, Cadenhead KS, de la Fuente-Sandoval C. Reduced P3a amplitudes in antipsychotic naive first-episode psychosis patients and individuals at clinical high-risk for psychosis. J Psychiatr Res. 2013;47:755–761.
141. Ozgurdal S, Gudlowski Y, Witthaus H, Kawohl W, Uhl I, Hauser M, Gorynia I, Gallinat J, Heinze M, Heinz A, et al. Reduction of auditory event-related P300 amplitude in subjects with at-risk mental state for schizophrenia. Schizophr Res. 2008;105:272–278.
142. van der Stelt O, Lieberman JA, Belger A. Auditory P300 in high-risk, recent-onset and chronic schizophrenia. Schizophr Res. 2005;77:309–320.
143. van Tricht MJ, Nieman DH, Koelman JH, van der Meer JN, Bour LJ, de Haan L, Linszen DH. Reduced parietal P300 amplitude is associated with an increased risk for a first psychotic episode. Biol Psychiatry. 2010;68:642–648.
144. Hamilton HK, Woods SW, Roach BJ, Llerena K, McGlashan TH, Srihari VH, Ford JM, Mathalon DH. Auditory and visual oddball stimulus processing deficits in schizophrenia and the psychosis risk syndrome: forecasting psychosis risk with P300. In preparation.
145. Bestelmeyer PE, Phillips LH, Crombie C, Benson P, St Clair D. The P300 as a possible endophenotype for schizophrenia and bipolar disorder: evidence from twin and patient studies. Psychiatry Res. 2009;169:212–219.
146. Jahshan C, Wynn JK, Altshuler LL, Glahn DC, Green MF. Cross-diagnostic comparison of duration mismatch negativity and P3a in bipolar disorder and schizophrenia. Bipolar Disorder 2012;14:239–248.
147. Bruder GE, Kayser J, Tenke CE, Friedman M, Malaspina D, Gorman JM. Event-related potentials in schizophrenia during tonal and phonetic oddball tasks: relations to diagnostic subtype, symptom features and verbal memory. Biol Psychiatry. 2001;50:447–452.
148. Light GA, Swerdlow NR, Thomas ML, Calkins ME, Green MF, Greenwood TA, Gur RE, Gur RC, Lazzeroni LC, Nuechterlein KH, et al. Validation of mismatch negativity and P3a for use in multi-site studies of schizophrenia: characterization of demographic, clinical, cognitive, and functional correlates in COGS-2. Schizophr Res. 2015;163:63–72.
149. Merrin EL, Floyd TC. P300 responses to novel auditory stimuli in hospitalized schizophrenic patients. Biol Psychiatry. 1994;36:527–542.
150. Ford JM, Mathalon DH, Marsh L, Faustman WO, Harris D, Hoff AL, Beal M, Pfefferbaum A. P300 amplitude is related to clinical state in severely and moderately ill patients with schizophrenia. Biol Psychiatry. 1999;46:94–101.
151. Asato N, Hirayau Y, Ofura C, Hokama H, Ohta H, Arakaki H, Hirayasu A, Nakamoto H, Yamamoto K, Randall M. Are event-related potential abnormalities in schizophrenics trait or state dependent? In: Ogura C, Koga Y, Shimokochi M, editors. Recent advances in event-related brain potential research. Amsterdam: Elsevier; 1996: pp. 564–567.
152. Coburn KL, Shillcutt SD, Tucker KA, Estes KM, Brin FB, Merai P, Moore NC. P300 delay and attenuation in schizophrenia: reversal by neuroleptic medication. Biol Psychiatry. 1998;44:466–474.
153. Faux SF, McCarley RW, Nestor PG, Shenton ME, Pollak SD, Penhune V, Mondrow E, Marcy B, Peterson A, Horvath T, et al. P300 topographic asymmetries are present in unmedicated schizophrenics. Electroencephalogr Clin Neurophysiol. 1993;88:32–41.

154. Mathalon DH, Ford JM, Rosenbloom M, Pfefferbaum A. P300 reduction and prolongation with illness duration in schizophrenia. Biol Psychiatry. 2000;47:413–427.
155. Ford JM, Sullivan EV, Marsh L, White PM, Lim KO, Pfefferbaum A. The relationship between P300 amplitude and regional gray matter volumes depends upon the attentional system engaged. Electroencephalogr Clin Neurophysiol. 1994;90:214–228.
156. Buzsaki G, Draguhn A. Neuronal oscillations in cortical networks. Science. 2004;304:1926–1929.
157. Tallon-Baudry C, Bertrand O, Delpuech C, Permier J. Oscillatory gamma-band (30–70 Hz) activity induced by a visual search task in humans. J Neurosci. 1997;17:722–734.
158. Sohal VS, Zhang F, Yizhar O, Deisseroth K. Parvalbumin neurons and gamma rhythms enhance cortical circuit performance. Nature. 2009;459:698–702.
159. Hashimoto T, Volk DW, Eggan SM, Mirnics K, Pierri JN, Sun Z, Sampson AR, Lewis DA. Gene expression deficits in a subclass of GABA neurons in the prefrontal cortex of subjects with schizophrenia. J Neurosci. 2003;23:6315–6626.
160. Lewis DA, Hashimoto T, Volk DW. Cortical inhibitory neurons and schizophrenia. Nat Rev Neurosci. 2005;6:312–324.
161. Deng C, Huang XF. Increased density of GABAA receptors in the superior temporal gyrus in schizophrenia. Exp Brain Res. 2006;168:587–590.
162. Impagnatiello F, Guidotti AR, Pesold C, Dwivedi Y, Caruncho H, Pisu MG, Uzunov DP, Smalheiser NR, Davis JM, Pandey GN, et al. A decrease of reelin expression as a putative vulnerability factor in schizophrenia. Proc Natl Acad Sci U S A. 1998;95: 15718–15723.
163. Lewis DA, Cho RY, Carter CS, Eklund K, Forster S, Kelly MA, Montrose D. Subunit-selective modulation of GABA type A receptor neurotransmission and cognition in schizophrenia. Am J Psychiatry. 2008;165:1585–1593.
164. Gonzalez-Burgos G, Lewis DA. GABA neurons and the mechanisms of network oscillations: implications for understanding cortical dysfunction in schizophrenia. Schizophr Bull. 2008;34:944–961.
165. Roopun AK, Cunningham MO, Racca C, Alter K, Traub RD, Whittington MA. Region-specific changes in gamma and beta2 rhythms in NMDA receptor dysfunction models of schizophrenia. Schizophr Bull. 2008;34:962–973.
166. Coyle JT, Tsai G, Goff D. Converging evidence of NMDA receptor hypofunction in the pathophysiology of schizophrenia. Ann N Y Acad Sci. 2003;1003:318–327.
167. Doheny HC, Faulkner HJ, Gruzelier JH, Baldeweg T, Whittington MA. Pathway-specific habituation of induced gamma oscillations in the hippocampal slice. Neuroreport. 2000;11:2629–2633.
168. Krystal JH, Anand A, Moghaddam B. Effects of NMDA receptor antagonists: implications for the pathophysiology of schizophrenia. Arch Gen Psychiatry. 2002;59:663–664.
169. Mathalon DH, Sohal VS. Neural oscillations and synchrony in brain dysfunction and neuropsychiatric disorders: it's about time. JAMA Psychiatry. 2015;72:840–844.
170. Galambos R, Makeig S, Talmachoff PJ. A 40-Hz auditory potential recorded from the human scalp. Proceedings of the National Academy of Sciences of the United States of America. 1981;78:2643–2647.
171. Pastor MA, Artieda J, Arbizu J, Marti-Climent JM, Penuelas I, Masdeu JC. Activation of human cerebral and cerebellar cortex by auditory stimulation at 40 Hz. J Neurosci. 2002;22:10501–10506.
172. O'Donnell BF, Hetrick WP, Vohs JL, Krishnan GP, Carroll CA, Shekhar A. Neural synchronization deficits to auditory stimulation in bipolar disorder. Neuroreport. 2004;15:1369–1372.
173. Bohorquez J, Ozdamar O. Generation of the 40-Hz auditory steady-state response (ASSR) explained using convolution. Clin Neurophysiol. 2008;119:2598–2607.
174. Thune H, Recasens M, Uhlhaas PJ. The 40-Hz auditory steady-state response in patients with schizophrenia: a meta-analysis. JAMA Psychiatry. 2016;73:1145–1153.
175. Light GA, Hsu JL, Hsieh MH, Meyer-Gomes K, Sprock J, Swerdlow NR, Braff DL. Gamma band oscillations reveal neural network cortical coherence dysfunction in schizophrenia patients. Biol Psychiatry. 2006;60:1231–1240.
176. Hong LE, Summerfelt A, McMahon R, Adami H, Francis G, Elliott A, Buchanan RW, Thaker GK. Evoked gamma band synchronization and the liability for schizophrenia. Schizophr Res. 2004;70:293–302.
177. Leicht G, Karch S, Karamatskos E, Giegling I, Moller HJ, Hegerl U, Pogarell O, Rujescu D, Mulert C. Alterations of the early auditory evoked gamma-band response in first-degree relatives of patients with schizophrenia: hints to a new intermediate phenotype. Journal of Psychiatric Research. 2011;45:699–705.
178. Perez VB, Roach BJ, Woods SW, Srihari VH, McGlashan TH, Ford JM, Mathalon DH. Early auditory gamma-band responses in patients at clinical high risk for schizophrenia. Supplements to Clinical Neurophysiology. 2013;62:147–162.
179. Tada M, Nagai T, Kirihara K, Koike S, Suga M, Araki T, Kobayashi T, Kasai K. Differential alterations of auditory gamma oscillatory responses between pre-onset high-risk individuals and first-episode schizophrenia. Cereb Cortex. 2016;26:1027–1035.
180. Rass O, Krishnan G, Brenner CA, Hetrick WP, Merrill CC, Shekhar A, O'Donnell BF. Auditory steady state response in bipolar disorder: relation to clinical state, cognitive performance, medication status, and substance disorders. Bipolar Disord. 2010;12:793–803.

27

OCULOMOTOR BIOMARKERS OF ILLNESS, RISK, AND PHARMACOGENETIC TREATMENT EFFECTS ACROSS THE PSYCHOSIS SPECTRUM

James L. Reilly, Jennifer McDowell, Jeffrey Bishop, Andreas Sprenger, and Rebekka Lencer

INTRODUCTION

Oculomotor, or eye movement, measures have been extensively used to assess the functional brain systems involved in cognitive and sensorimotor processes that are selectively intact or disturbed in various clinical conditions, including psychosis.[1–3] Indeed, the first report of eye movement dysfunction in schizophrenia was reported over a century ago.[4] Eye movements include (1) saccades, or rapid movements that shift point of gaze reflexively or under volitional control, and (2) smooth pursuit eye movements (SPEM), or movements in which the eyes track a moving target requiring matching of eye velocity to target velocity. Increasingly, oculomotor measures have been considered as potential biomarkers to evaluate dysfunction of functional brain systems thought to be shared across disorders that are perhaps more proximal to genetic and other neurobiologic risk factors than clinical signs or symptoms. They have also been used to evaluate genetic and pharmacogenetic effects on these functional brain systems.

Eye movement paradigms offer particular advantages as biomarkers of illness, risk for illness, and of genetic and pharmacogenetic variation on select cognitive and sensorimotor processes. First, the neurophysiologic and neurochemical basis of eye movement control has been well characterized from both single unit recording studies of nonhuman primates and functional imaging studies in humans.[5–8] Second, oculomotor performance is reliably measured and quantified in both behavioral and neuroimaging lab settings, and paradigms can be experimentally manipulated to better understand differences in performance between groups or changes associated with treatment.[9] Lastly, due to the relative ease of performance on eye movement tasks and limited subject burden, they can be used across a range of ages and clinical severity levels.[10]

In this chapter, we summarize findings from saccadic and SPEM studies across proband groups with psychotic disorders and their relatives. Emphasis is given to studies that compare oculomotor performance between psychosis groups, including schizophrenia, schizoaffective disorder, and psychotic bipolar proband groups when possible. The review concludes with a summary of recent findings demonstrating the utility of eye movement measures for demonstrating associations to specific genetic variants and differential genetic effects of antipsychotic treatment on select cognitive and sensorimotor processes.

FINDINGS FROM SACCADIC EYE MOVEMENT TASKS

Saccades are fast eye movements that redirect visual gaze and exist in a hierarchy from simple, reflexive types of behavior such as prosaccades, to more cognitively complex behaviors such as antisaccades or memory guided saccades[11] (see Figure 27.1A). A prosaccade is a visually guided glance made to an external stimulus or target (Figure 27.1Bi). Correct performance requires stimulus-response mapping whereby the eyes are directed toward the peripheral cue accurately and quickly (typically taking about 150 msec). More cognitively complex saccades are volitionally driven, generated in response to instructions or internal goals and which typically take 50 to 100 msec longer to initiate than prosaccades. An antisaccade requires inhibition of a glance toward a cue and the generation of a glance to the mirror image (opposite direction, same distance) location (see Figure 27.1Bii). A glance toward the cue constitutes an error and is considered a failure of inhibition.[12] Finally, a memory guided saccade requires inhibition of a glance toward the brief appearance of a cue (typically 50 msec) and during a delay interval, and the subsequent generation of a saccade after a delay interval to the remembered location of the cue guided only by the internalized representation of the cue's location (Figure 27.1Biii). As such, memory guided saccades require both inhibition and working memory.[3]

Saccadic behavior is easily recorded in the laboratory. Performance is most often characterized by the following variables:(1) *error rate* (the number of incorrect trials / the number of total scoreable trials), (2) *latency* (reaction time; time in msec from the cue onset to the initiation of a saccade), and (3) *gain* (accuracy of a response; [eye position / target position]). The brain networks underlying saccadic behavior are relatively well understood based on human imaging studies[7,13] (see Figure 27.1A). Saccadic performance has been

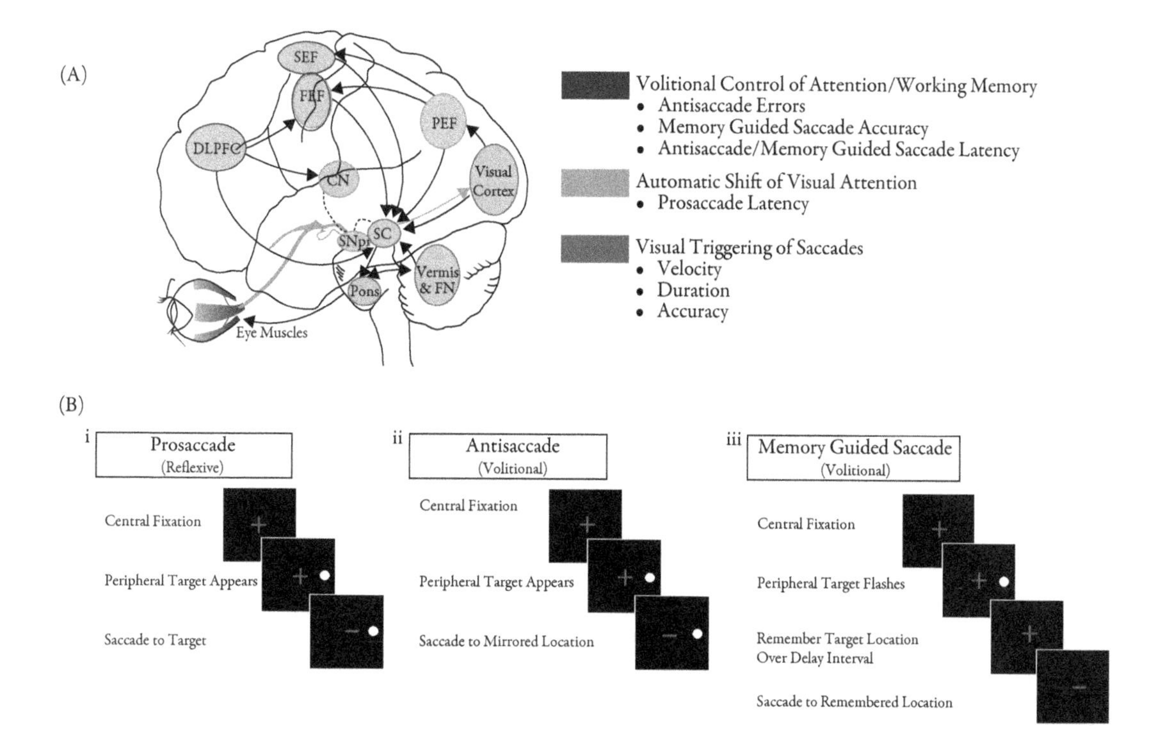

Figure 27.1 **A. Hypothetical summary of hierarchical regions controlling reflexive and volitional saccades and corresponding behavioral measures.** Regions implicated in the executive or volitional control of saccades are identified in blue. These regions sit atop regions in green that reflect control of more reflexive saccadic movements, which in turn sits atop a system processing visual input that drives saccades. Saccadic measures theoretically under control of these hierarchically organized regions are listed. CN = caudate nucleus; DLPFC = dorsolateral prefrontal cortex; FN = fastigial nucleus; FEF = frontal eye field; PEF = parietal eye field; SC = superior colliculus; SEF = supplementary eye field; SNpr = substantia nigra pars reticulata. **B. Example of prosaccade, antisaccade and memory guided saccade paradigms. i.** Prosaccade tasks require a visually guided glance to a peripheral cue as accurately and quickly as possible. **ii.** Antisaccade tasks require inhibition of the reflexive response to glance to a peripheral cue and instead generate a saccade toward a mirrored image location. **iii.** Memory guided saccade tasks require inhibition of the reflexive response to glance to a brief peripheral target and instead remember where the target appeared and generate a saccade to the remembered location after a delay. Red arrows reflect the correct direction of gaze for each task.

investigated in patients with psychotic disorders and their relatives, most often in schizophrenia. While prosaccade and memory guided saccade performance has been less extensively studied, antisaccade performance represents one of the most widely evaluated and promising neurophysiological biomarkers for psychosis.[14–16]

PROSACCADES

Behavioral findings in probands

Error rate

Prosaccade tasks require programming an eye movement to a visual target and result in low error rates across all relevant groups, including psychosis probands, relatives and other at-risk populations, and healthy people (see [15,17,18]).

Latency

Overall, people with schizophrenia show normal range prosaccade latencies and respond to manipulations (like changes in fixation characteristics) that result in well-documented increases or decreases in reaction times among healthy participants.[15,19] People with bipolar disorder also generally show preserved prosaccade performance. In the largest data collection effort of its kind, the Bipolar-Schizophrenia Network for Intermediate Phenotypes (B-SNIP) study[15] reported on a large sample of probands with schizophrenia (n = 267), schizoaffective disorder (n = 150) and bipolar disorder (n = 202). There were no group effects on prosaccade latency. An earlier study demonstrated that prosaccade latency in first-episode treatment-naïve probands with schizophrenia, bipolar disorder, major depression with psychosis, and major depression without psychosis did not differ from healthy subjects.[20] While prosaccade latencies seem to be unaffected in people who have received antipsychotic medication chronically, there is evidence of a specific medication effect. First-episode schizophrenia patients have prosaccades characterized by faster reaction times, which are then normalized or even lengthened after antipsychotic treatment.[21]

Gain

Many fewer studies report on prosaccade amplitudes than reaction time. No overall group difference was reported among the B-SNIP proband groups.[15] The study by Harris et al. (2009)[20] showed that schizophrenia, bipolar, and major depression without psychosis groups had normal gain. The group with major depressive disorder with psychosis, however, had hypometric saccade amplitudes (decreased gain). While most studies suggest that prosaccade gain is fairly accurate in people with schizophrenia, slight hypometria is consistent with an antipsychotic medication effect.[22]

Behavioral findings in relatives

The literature on family studies of people with schizophrenia (particularly) is large. A meta-analysis designed to compare effects of various eye movement measures across studies demonstrated that prosaccade latency and gain did not differ between relatives of people with schizophrenia and control subjects.[1] Normal prosaccade metrics have been reported in siblings specifically,[23] as they provide a more appropriate age match than parent control groups. The B-SNIP data collection effort included relatives of probands with: (2) schizophrenia (n = 304), (2) schizoaffective disorder (n = 193), and (3) bipolar disorder (n = 242). All groups showed performance that did not differ from healthy participants, with the exception of modestly reduced prosaccade latency in the relatives of the schizoaffective probands.[15]

ANTISACCADES

Behavioral findings in probands

Error rate

Antisaccade tasks require programming a saccade to the opposite location of a visual cue, with an initial glance toward the cue (instead of opposite) constituting an error. Increased antisaccade errors in people with schizophrenia is one of the most reliable and replicable behavioral measures in the field. The Consortium on the Genetics of Schizophrenia (COGS) study reported that antisaccade error rates (as well as antisaccade latency and gain) are stable measures of performance[24] across more than a thousand subjects (N = 1,280 people with schizophrenia over multiple sites and two studies). The data on bipolar disorder are historically less consistent, suggesting that poor antisaccade performance among such cases may be associated with mood state[25] and/or the presence of psychosis.[26] The B-SNIP data[15] show that all psychosis proband groups (schizophrenia, schizoaffective, and bipolar with psychosis) had elevated error rates compared to controls and that these elevations are unrelated to symptoms and to antispsychotic treatment. There is a graded response deficit with schizophrenia probands demonstrating the greatest error rate, followed by schizoaffective and bipolar probands (who do not differ from each other). Similarly, Harris et al. (2009) showed that individuals with first -episode psychosis made more antisaccade errors than healthy individuals, with no significant differences across the three psychosis groups (schizophrenia, bipolar, and major depression with psychosis).[20]

Latency

Increased latencies are frequently observed for correct antisaccades in people with schizophrenia.[3] The B-SNIP data showed that schizophrenia probands had the longest latencies for correct responses, followed by schizoaffective probands, followed by bipolar probands, who did not differ from the healthy group.[15]

Gain

Typically, hypometric antisaccade gain is observed in schizophrenia when compared to healthy people.[18,19]

Behavioral findings in relatives

The majority of data in relatives arises from studies of schizophrenia. In an early study of over 360 subjects, antisaccade error rate was titrated with relatedness to disease such that probands had the highest rate, followed by first-degree relatives, followed by second-degree relatives, who approached the performance of healthy people.[19] This is a typical result, as summarized by a meta-analysis of antisaccade data showing that both error rates and correct latencies are reliably increased in relatives of people with schizophrenia.[14] A study of monozygotic twins discordant for schizophrenia showed increased antisaccade latency and degreased gain in both twins,[23] suggesting that these may be the risk markers, and also that those variables do not arise as a function of medication.[18] In the B-SNIP study,[15] the relatives of all the psychosis proband groups showed increased antisaccade error rates compared to controls, although that did not differ across diagnostic categories (and each proband group had higher error rates than their respective relative group). Of note, relatives with elevated psychosis spectrum personality traits had greatest error rates; yet, even those relatives without these personality traits had greater error rates compared to controls. Error rate was modestly and comparably familial across schizophrenia, schizoaffective, and bipolar pedigrees. In terms of correct antisaccade latency, none of the B-SNIP relative groups differed from the control group. Finally, these data revealed an interesting relationship between prosaccade latency and antisaccade error rate. In all proband and relative groups, people who made faster prosaccades also made more antisaccade errors with the one exception—this correlation was not upheld in the schizophrenia group. In sum, antisaccade performance, most consistently increased error rate in psychosis probands and their relatives, suggest that it indexes a risk for psychosis, rather than for a specific diagnosis.

MEMORY GUIDED SACCADES

Behavioral findings in probands

Error rate

Far fewer studies have been conducted of memory-guided saccades than prosaccades or antisaccades, and these are nearly exclusively among individuals with schizophrenia. Memory guided saccade, or oculomotor delayed response, tasks require inhibition of the tendency to look to a cue when it appears and instead make a delayed response after a period guided by internally represented spatial location information. Thus, it is a task that requests both inhibition and spatial working memory.[3] Errors occur when subjects incorrectly look to the target location during the delay, thus indicating a failure of inhibition. Individuals with schizophrenia commit an elevated number of inhibitory errors.[27–33] Higher error rates also occur among subjects considered ultra-high risk for schizophrenia, particularly those with elevated neurologic soft signs.[34]

Latency

Similar to findings on the antisaccade task, increased latency of memory guided saccades compared to prosaccades is typically observed in schizophrenia patients,[31,33,35–37] representing greater time to plan and initiate a volitional saccade to mentally represented locations.

Gain

Reduced gain of the primary memory guided saccade is typically observed among patients with schizophrenia.[35,36,38] This is particularly true under paradigm conditions that place greater demands on working memory circuitry such as an increase in delay period duration or provision of interfering stimuli during the delay period.[33,36,37] In addition, accuracy of the final resting eye position after small correct saccades are made to the "decided upon" location of the cue is also reduced indicating an imprecision or decay of the spatial location information stored in working memory vs. inaccuracy of the motor plan to direct the initial saccade.[32,33,37] As reviewed in what follows, antipsychotic medications appear to exacerbate pretreatment deficits in memory guided saccade accuracy,[33,37] an effect that appears dependent on genetic factors.[39]

Behavioral findings in relatives

Compared to prosaccade and antisaccade tasks, far fewer studies characterizing performance of schizophrenia relatives on memory guided saccade tasks have been conducted and many had smaller sample sizes. Increased inhibitory errors during the delay period have been reported in some[31,32] but not all studies;[40,41] positive findings of increased error rate are smaller than those reported on antisaccade tasks. To date no studies have reported differences in memory guided response latency among relatives,[40,41] which is in contrast to findings of probands. Finally, the majority of studies have reported no differences in the gain of memory guided saccades among schizophrenia relatives.[40–42] Overall, despite also being a volitional eye movement task, findings among relatives on memory guided saccade tasks appear far less robust than those reported on antisaccade tasks.

NEUROIMAGING FINDINGS IN SACCADIC EYE MOVEMENT STUDIES

In healthy individuals, saccades are supported by widespread cortical and subcortical regions, including occipital cortex, posterior parietal cortex, frontal and supplementary eye fields, superior colliculus, thalamus, striatum/basal ganglia, and cerebellum[7,12,13] (see Figure 27.1A). With the transition from simple to more complex behavior (as in prosaccades to antisaccades or memory guided saccades), neural circuitry may show stronger, more extensive activation within this basic circuitry and/or recruitment of new neural regions to support task performance. Structural measures arise from magnetic resonance imaging (MRI) studies and include quantification of gray and white matter volume globally or regionally, or quantify white matter tracts (via diffusion tensor imaging [DTI]). Functional measures arise from functional MRI (fMRI), electroencephalography (EEG) or magnetoencephalography (MEG) studies that allow for analyses of resting or task-related states.

Broadly, there is long-standing evidence of structural brain abnormalities in schizophrenia. B-SNIP data showed widespread reductions in gray matter volume in schizophrenia (n = 146), schizoaffective disorder (n = 90) and psychotic bipolar disorder (n = 115) that overlap between schizophrenia and schizoaffective disorder, and are more focal and localized to frontotemporal cortex in the bipolar group.[43] In a similar vein, there is evidence of widespread disconnectivity between anatomical regions in schizophrenia (as shown by DTI studies).[44] A recent study is among the first to assess the integrity of connections specific to saccadic circuitry in people with schizophrenia. White matter fibers were identified and tracked between saccade-generating nodes of frontal eye fields and putamen in basal ganglia. People with schizophrenia demonstrated lower white matter integrity than healthy people with good antisaccade performance,[45] providing a direct link between saccadic performance and the fiber tracts underlying its circuitry.

The majority of fMRI studies using saccade tasks have been conducted in individuals with schizophrenia, and generally demonstrate reduced activation in regions underlying control of eye movements compared to healthy individuals, particularly on volitional saccade tasks. Among first-episode schizophrenia patients performing a prosaccade task prior to treatment, reduced activation was observed in dorsal prefrontal regions which increased after 4 to 6 weeks of antipsychotic treatment and patients were no longer distinguishable from controls.[46] On antisaccade and memory guided saccade tasks that require greater volitional control, there is evidence in schizophrenia of broadly reduced activation across the oculomotor circuitry, particularly in striatal and prefrontal regions.[27,47,48] One fMRI study including a proband group and their siblings showed three distinct patterns of activation during performance on antisaccade and memory guided saccade tasks: (1) regions that were unchanged across all three groups, (2) regions that showed decreased activation only in the patients, and (3) a set of regions that differentiated between probands and family members, the latter of which may reflect a risk for developing the disorder.[49]

A recent study recorded EEG during steady state stimuli during prosaccade and antisaccade tasks[17] in probands with schizophrenia (n = 33), schizoaffective disorder (n = 20), and bipolar disorder with psychosis (n = 45). Because there is a growing awareness that DSM-type clinical diagnoses may not be neurobiologically distinct, these data were analyzed twice. First, they were analyzed grouped into their respective diagnostic category. Second, they were analyzed using a set of brain-based biomarkers (called "Biotypes") that were independent of clinical characteristics.[50] When based on clinical diagnosis, all psychosis probands showed poor antisaccade performance and diminished baseline oscillatory phase synchrony, with few distinctions between psychosis groups. When categorized by Biotypes, group differentiation was observed. For instance, Biotype 1 showed baseline and stimulus activity that was diminished, whereas it was accentuated in Biotype 2.[17] These data (and others that directly compare between clinical diagnoses and brain-based Biotypes) suggest that the Biotypes approach captures a distinction between subgroups that has proved to be elusive in DSM-type diagnoses of psychotic disorders.

FINDINGS FROM STUDIES OF SMOOTH PURSUIT EYE MOVEMENTS

Smooth pursuit eye movements (SPEM) provide optimal visual acuity of small moving objects by matching eye velocity to object/target velocity.[5] SPEMs are always driven by a combination of visual and nonvisual components that have to be integrated in various brain networks. The velocity of the object image on the retina, i.e., retinal slip velocity, represents the visual input signal for pursuit. SPEM control requires sensorimotor transformation of the visual motion signal into an oculomotor command and its integration with nonvisual mechanisms such as feedback about performance quality, integration of an efference copy signal of the motor command and prediction of the target movement. Two main SPEM subfunctions include: (1) SPEM initiation, which mainly relies on retinal input for fast visual motion information and early sensorimotor processing; and (2) SPEM maintenance, which more heavily depends on nonvisual, cognitive factors that become more important the longer SPEM is maintained.[51] Both, SPEM initiation and SPEM maintenance have been studied in patients with psychotic disorders and their relatives, first and most often in schizophrenia and schizoaffective disorder.[2,14,52] More recent studies indicate that pursuit deficits are also present in probands with psychotic mood disorders and their relatives,[53–58] suggesting that SPEM dysfunctions represent a common neurophysiological intermediate phenotype across psychotic disorders.

SPEM INITIATION

Pursuit initiation comprises the first 100 ms of pursuit that do not rely on visual feedback of the target movement ("open loop"). To avoid disturbing catch-up saccades during pursuit initiation, the classical foveo-petal step-ramp stimulus was designed, in which movement of a pursuit target in the intended direction ("ramp") is preceded by a target "step" in the opposite direction. Parameters of interest are *pursuit initiation latency* and *initial eye acceleration or mean eye velocity*, respectively[59] (Figure 27.2B).

Stimuli with step and ramp in the same direction are called foveo-fugal step-ramp and allow for measuring the saccadic *eye position error* of the initial catch-up saccade during the pursuit initiation phase. This measure indicates how correctly the system takes into account target velocity when calculating the landing position of the catch-up saccade on the moving target indicating integrity of visual motion processing during the open loop phase.

Behavioral findings of pursuit initiation in probands

Initial eye acceleration and initial eye velocity gain

The majority of studies that have investigated pursuit initiation in patients with psychotic disorders reported similarly reduced initial eye acceleration and initial eye velocity gain, in both first-episode and chronically ill patients with schizophrenia spectrum and mood disorders, specifically psychotic bipolar disorders.[55,56,60–63] Provided that initial eye acceleration represents the most direct measure of the ability to use visual motion information for pursuit drive, these findings imply a diminished capacity for using motion information for sensorimotor transformations in the pursuit system across psychotic disorders.

Pursuit latency

Studies in patients with schizophrenia and psychotic bipolar disorder including studies with first-episode, unmedicated patients, do not support altered reaction time until the eyes start following a moving target, making it unlikely that gross attentional impairment could cause pursuit initiation deficits.[55,56,62]

Saccadic accuracy during pursuit initiation

The few studies that have investigated the saccadic position error on moving targets in foveo-fugal ramp tasks in patients with schizophrenia and bipolar disorder did not observe impairment in any group.[55,56,64] Since saccade accuracy requires a precise analysis of target speed, this finding suggests sufficient integrity of visual motion perception to guide eye movements to moving targets.[65]

Together, these findings on pursuit initiation in probands imply that while visual motion signals are used effectively to guide saccades on a moving target and to drive the onset of pursuit responses, reduced initial eye acceleration and velocity gain provide evidence of a diminished capacity for using this motion information for sensorimotor transformations in the pursuit system.

Behavioral findings of pursuit initiation in relatives

Few studies have focused on alterations of pursuit initiation in relatives of probands with psychotic disorders, with the majority of studies reporting reduced initial eye acceleration and initial eye velocity compared to healthy control.[61,66] Initiation impairments observed in relatives were less severe than those in probands, and did not differ by proband diagnosis. Recent results from the B-SNIP study including relatives of probands with schizophrenia (n = 306), schizoaffective disorders (n = 217), and psychotic bipolar disorder (n = 273) showed that pursuit initiation was impaired not only in subgroups of relatives with elevated psychosis spectrum personality traits but also in those without these traits.[61] Thus, impaired motion information processing for sensorimotor transformations during pursuit initiation may indicate increased susceptibility to psychosis even in the absence of clinical symptoms.

SPEM MAINTENANCE

Pursuit maintenance occurs about 300 ms after pursuit onset, and is referred to as the "closed loop" phase, reflecting that the system receives feedback of its own actions. Note that during optimal pursuit when a velocity gain of 1 is achieved, retinal slip velocity is close to zero, thus pursuit maintenance has to rely on predictive signals derived from nonvisual SPEM components.

Specific aspects of pursuit maintenance must be distinguished. First, *early maintenance gain* determined with ramp or step-ramp tasks indicates early sensorimotor integration of visual feedback about performance accuracy in intervals 350 to 500 ms after pursuit onset (Figure 27.2B). Second, *predictive maintenance gain* is measured later during closed-loop pursuit, when the movement is mostly driven by nonvisual components. Predictive maintenance gain is best assessed with targets continuously moving from left to right and vice versa in the horizontal plane, e.g., triangular wave or sinusoidally moving targets. Typically, early sensorimotor maintenance gain is lower than predictive maintenance reflecting the important impact of nonvisual components driving pursuit maintenance.[5] This is most obvious with fully predictable oscillating tasks without abrupt target reversals. Here, high gains can be achieved even with relatively high target speeds up to 32°/s.

Another experimental approach for studying nonvisual input to pursuit control is to blank the visual target during ongoing pursuit movements.[51] Parameters of interest are *eye deceleration* after target extinction and *residual eye velocity/gain* when pursuing the imagined, nonvisible target. Eye velocity will increase again upon reappearance of the visible target.

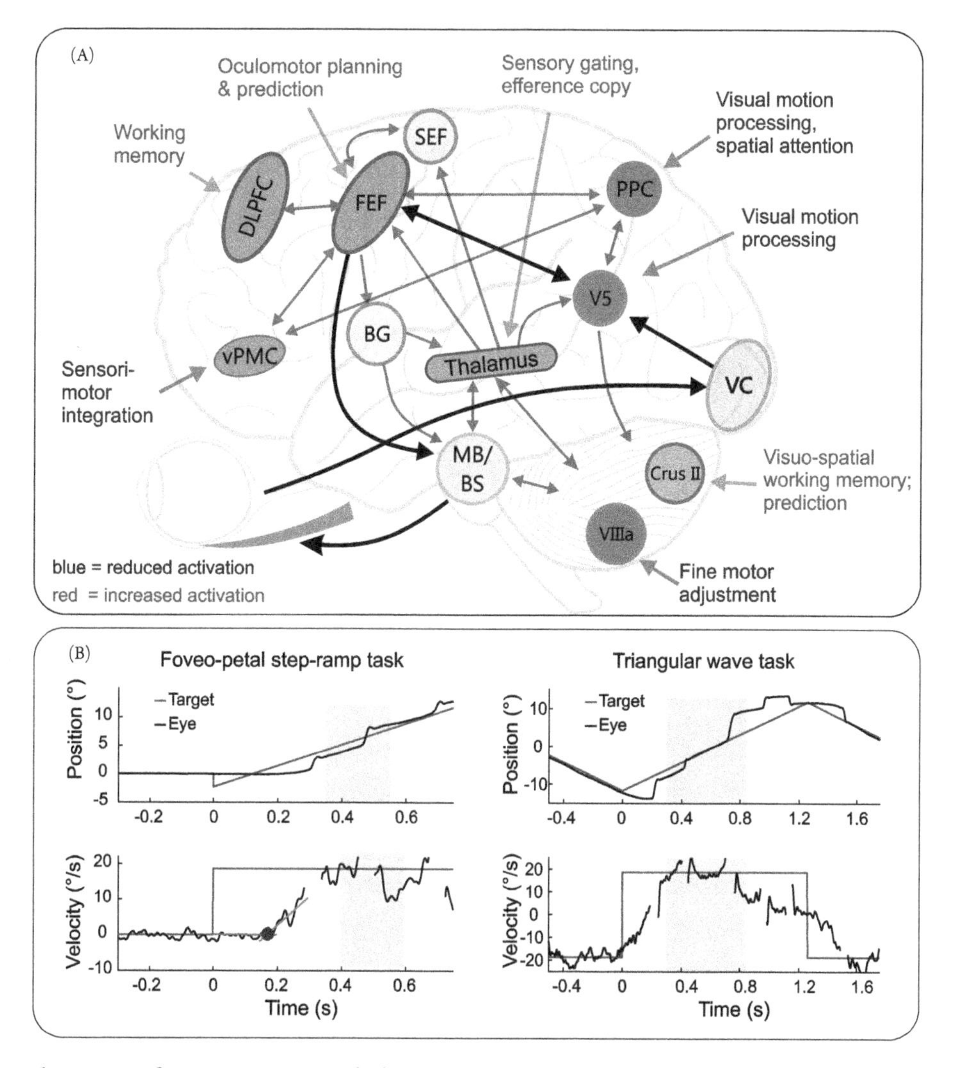

Figure 27.2 **A. Hypothetical summary of main pursuit network alterations in psychotic disorders from fMRI studies.** During SPEM, input from visual sensory motion area VS and its projections to posterior parietal association cortex (PPC) for perceptual analysis is reduced. At the same time, increased activation of a cortico-thalamo-cerebellar network including the frontal eye fields (FEF) as well as dorsolateral prefrontal cortex (DLPFC) indicates enhanced recruitment of nonvisual components for oculomotor planning, prediction, and use of working memory resources. Additionally, reduced activation in ventral premotor cortex (vPMC) and cerebellar area VIII suggests disturbed sensorimotor integration and fine motor adjustment. Main pursuit network connections are indicated in black. VC = primary visual cortex; SEF = supplementary eye field; BG = basal ganglia, MB/BS = mid-brain/brainstem.

B. Examples of pursuit stimuli with pursuit recordings (eye position and eye velocity). Foveopetal step-ramp tasks are used to measure saccade free pursuit initiation. Variables of interest are pursuit latency (time between target step and dot), initial eye acceleration, and early sensorimotor gain (measured in gray shaded interval). Triangular wave tasks are used to measure predictive maintenance gain in predefined intervals (gray shaded) excluding artifacts induced by target reversals.

Behavioral findings of pursuit maintenance in probands

Early sensorimotor maintenance gain

Reduced early maintenance gain has been reported predominantly among probands with schizophrenia, both unmedicated first-episode and chronically ill patients, but also among patients with psychotic mood disorders, specifically bipolar disorder.[55,56,60,61] Differences in gain impairment severity between patients with schizophrenia and bipolar disorder, both first-episode and chronically ill, imply that gain impairments may increase over the course of illness more severely in patients with schizophrenia, perhaps due to specific illness-related factors. In light of the earlier-mentioned impairments on pursuit initiation, findings on early maintenance gain suggest that across psychotic disorders diminished capacity for visual motion processing are not compensated by visual feedback control under early closed loop conditions.

Predictive maintenance gain

This measure is most commonly used for characterizing smooth pursuit ability in probands, predominantly schizophrenia, and represents one of the most robust findings.[2] Diefendorf and Dodge were the first to measure SPEM in a range of patients with different psychiatric disorders.[4] In 1973, Holzman et al. rediscovered these findings and were the first to report SPEM deficits in probands' relatives, suggesting SPEM as genetic susceptibility marker to schizophrenia.[67] They also established the hypothesis that SPEM deficits are due to impaired motion signal integration.[52]

While earlier reports claimed that pursuit maintenance deficits are specific to schizophrenia, more recent studies, including the large-scale B-SNIP study, showed that predictive maintenance gain was similarly impaired across probands with schizophrenia, schizoaffective disorder, and psychotic bipolar disorder.[61] As seen with early maintenance gain, predictive maintenance gain impairments were most severe in probands with schizophrenia compared to psychotic bipolar disorder and schizoaffective disorder pointing to disease-related factors modulating maintenance performance.[61]

In contrast to this, using fully predictable oscillating tasks that avoid pursuit initiation at target reversals did not reveal any maintenance gain deficits across first-episode psychosis probands.[56] This finding underlines the potential compensatory use of cognitive, predictive components to pursuit drive in probands. This explanation is in line with more detailed analyses of pursuit maintenance from target blanking tasks that showed that both probands with schizophrenia-spectrum disorders and those with bipolar disorder were able to follow a nonvisible imagined target as well as healthy participants.[62,68] Others reported reduced eye velocity if the blanking interval occurred at the beginning of a ramp, reflecting impaired pursuit initiation than pursuit maintenance.[69] Together, these findings support a model of smooth pursuit disturbances in psychotic disorders not being a pure failure in adding nonvisual, cognitive contributions to pursuit drive in higher-order cortical circuits but rather being a consequence of deficits in sensorimotor transformation of visual motion information as seen with pursuit initiation and early sensorimotor maintenance pursuit.

Behavioral findings of pursuit maintenance in relatives

The majority of studies that investigated pursuit performance in relatives reported predictive maintenance gain impairments in relatives that were less severe than those seen in probands, mostly with schizophrenia. This includes studies of complete pedigrees as required for genetic linkage studies,[70] group comparisons of first-degree relatives[14] using also tasks with target blanking,[71] and defining eye tracking dysfunction (ETD) on an individual basis for use in genetic studies.[14] When comparing ETD rates between family members of pedigrees with multiple and single cases of schizophrenia, no differences were observed.[72] Few studies included relatives of probands with psychotic bipolar disorder reporting similar alterations of maintenance gain compared to relatives of schizophrenia probands.[53,54,57,58] In contrast to impaired pursuit initiation seen across relatives, the large-scale B-SNIP study did not yield evidence for impairments in early or predictive maintenance gain among any relative groups.[61] However, familiality estimates, which reflect within-family relatedness of pursuit measures between probands and their first-degree relatives, were highest for early sensorimotor maintenance gain suggesting that this measure may represent a valuable intermediate phenotype for risk to psychosis although no direct pursuit maintenance impairment was seen in relatives. This pattern of altered pursuit initiation but rather unimpaired pursuit maintenance in relatives across psychotic disorders suggests that in relatives preserved nonvisual components of pursuit may still support pursuit maintenance despite impairments in sensorimotor processing.

NEUROIMAGING FINDINGS IN SMOOTH PURSUIT EYE MOVEMENT STUDIES

Research in nonhuman primates and human functional neuroimaging studies provide a strong translational framework for clinical research of pursuit eye movement deficits[5] (Figure 2A). Findings from fMRI studies support the model of disturbed visual motion processing underlying impaired SPEM in probands. During SPEM, unilateral V5 activation is decreased[73] and that pursuit eye and target velocity, respectively, are less strongly correlated to activity in V5, lateral intraparietal cortex, supplementary eye field and putamen in probands with schizophrenia than in healthy subjects.[74,75] This implies that V5 output may be less available to sensorimotor systems for pursuit control in probands. Another study showed that during SPEM, activation of the right ventral premotor cortex was reduced in probands while activations in the left dorsolateral prefrontal cortex, the right thalamus, and the left cerebellar hemisphere were increased.[76] This network of increased brain activation during SPEM supports a model of compensatory recruitment of nonvisual input from thalamo-prefrontal-cerebellar networks for pursuit maintenance in patients. In line with this enhanced activation was observed in bilateral frontal eye fields, anterior cingulate, the right superior temporal cortex in patients with schizophrenia compared to controls during pursuit with target blanking.[76]

Only one study directly compared SPEM-related brain activation between unmedicated first-episode probands with schizophrenia-spectrum disorders and psychotic bipolar disorder.[77] During passive visual motion processing both proband groups showed reduced activation in the posterior parietal projection fields of motion-sensitive area V5 in line with reduced bottom-up transfer of visual motion information from V5 to parietal association cortex. During SPEM, neural activation was enhanced in anterior intraparietal sulcus and insula across probands, and in dorsolateral prefrontal cortex and dorsomedial thalamus in schizophrenia patients, presumably reflecting compensatory mechanisms provided by a thalamo-prefrontal network. Direct comparison could not identify any differences in SPEM-related brain areas between proband groups suggesting a common systems-level mechanism of pursuit deficits involving reduced transfer of motion signals to parietal cortex. Follow-up fMRI studies after 6 weeks of antipsychotic treatment did not reveal differences in SPEM-related brain activation in either proband group suggesting a stability of alterations in motion processing and pursuit systems after short-term treatment with antipsychotic medication and clinical stabilization.[7] One limitation was a smaller sample size of the bipolar group requiring larger studies to confirm the similarity in findings across probands with schizophrenia-spectrum disorders and bipolar disorder.

One PET-study that reported reduced frontal eye field activation in relatives of schizophrenia probands, who showed ETD which was in contrast to relatives without ETD,[78] however, samples sizes were very small.

GENETIC ALTERATIONS AND PHARMACOGENETIC EFFECTS OF ANTIPSYCHOTICS ON EYE MOVEMENTS

In this final section, we review recent findings that identify genetic alterations and pharmacogenetic effects of antipsychotic treatment on memory guided saccade and SPEM. As reported earlier, reduced memory guided saccade gain observed among individuals with first episode schizophrenia prior to treatment was demonstrated to worsen after 6 weeks of second-generation antipsychotic treatment despite clinical stabilization.[33,37] In a subsequent study including some of these patients Bishop et al. (2015) evaluated whether polymorphisms of the Type-3 metabotropic glutamate receptor gene (*GRM3*) and selected variants in candidate dopamine genes (i.e., dopamine-D2 receptor gene; *DRD2* and catecohol-o-methyltransferase gene; *COMT*) were associated with pretreatment memory guided saccade performance and change observed after antipsychotic treatment.[39] Polymorphisms within these candidate genes were selected based on functional association studies implicating them in working memory function. At baseline prior to treatment, individuals with the *COMT* Met158Met genotype had better working memory performance than Val allele carriers, which is consistent with prior studies. Worsening of memory guided saccade accuracy observed after antipsychotic treatment was associated with variation in *GRM3* polymorphisms, such that those patients with the rs1468412_TT genotype exhibited a substantial worsening saccade accuracy compared to those with the rs1468412_AA genotype. No influence of *COMT* or *DRD2* variants was seen on treatment effects. Thus, the *GRM3* findings identified a potential subgroup of patients with cognitive sensitivity to treatment, suggesting that a genetic predisposition toward inefficient cortical processing supporting working memory performance that is exacerbated by antipsychotics.

First hints of genetic linkage between SPEM predictive maintenance gain and microsatellite markers on the short arm of chromosome 6 (6p21-23) were reported from families multiply affected with schizophrenia[79] and were later replicated independently.[80] Since then, genetic association studies using eye movement phenotypes have focused on single nucleotide polymorphisms (SNPs) in candidate genes for schizophrenia including COMT and neuregulin-1 (NRG1) besides others (e.g., [81–83]). Following a model of different genes modulating specific SPEM subfunctions one previous study in first-episode psychosis patients including schizophrenia spectrum disorders and psychotic bipolar disorders reported an association of pursuit initiation impairments with a dopamine D2 receptor gene (DRD2) while pursuit maintenance was associated to candidate SNPs in metabotropic glutamate receptor 3 protein (GRM3).[84] These findings suggest that genetic alterations of striatal D2 receptor activity could contribute to pursuit initiation deficits in psychotic disorders consistent with the role of dopaminergic neurotransmission in the striatum regulating speed of motor response initiation. Alterations in GRM3 coding for the mGluR3 protein may impair pursuit maintenance by compromising higher perceptual and cognitive processes that depend on optimal glutamate signaling in corticocortical circuits. Additionally, DRD2 and GRM3 genotypes selectively modulated the severity of impairments in pursuit initiation and predictive maintenance resulting from antipsychotic treatment.

A very recent genome-wide association study (GWAS) including 849 participants from the B-SNIP consortium reported associations of pursuit initiation with genes involved in nuclear trafficking and gene silencing (*IPO8*), fast axonal guidance and synaptic specificity (*PCDH12*), transduction of nerve signals (*NRSN1*) and retinal degeneration (*LMO7*), while predictive pursuit maintenance showed suggestive association with genes regulating synaptic glutamate release (*SH3GL2*).[85] Additionally, antisaccade error rate was significantly associated with a noncoding region at chromosome 7 in participants of predominantly African ancestry. Exploratory pathway analyses revealed pursuit initiation and predictive maintenance to be associated with genes implicated in broader nervous system development and function. Collectively, these findings highlight the importance of genes related to disease risk alongside other unique genetic contributions to eye movement phenotypes associated with psychotic disorders.

CONCLUSIONS

Evidence is accumulating that eye movements are sensitive biomarkers to assess alterations in select cognitive and sensorimotor processes among individuals with or at risk for psychosis spectrum disorders, and how such processes may be differentially impacted by genetic factors and antipsychotic treatments. In recent years, this work has been advanced by large-scale studies such as COGS and B-SNIP, which have assembled substantial samples to begin to address the promise of eye movements as putative endophenotypes or biomarkers of illness. Inhibitory errors on the antisaccade task appear to represent a robust and graded deficit across the psychosis spectrum from schizophrenia to psychotic bipolar disorder, while elevated yet comparable error rates among relatives suggests this marks familial risk for psychosis in general. Specific abnormalities in the use of early visual motion information for sensorimotor transformation on smooth pursuit are apparent among both probands and relatives, while visual feedback and cognitive components for dynamical pursuit adjustments appears restricted to probands. While much of this work has been conducted among individuals with schizophrenia, findings in other psychotic disorders appear qualitatively similar. Further research is needed across a range of categorical diagnoses and dimensional symptom and neurobiological characterizations to further support the biomarker utility of eye movements in psychosis research.

REFERENCES

1. Klein C, Ettinger U. A hundred years of eye movement research in psychiatry. *Brain and Cognition*. 2008;68(3):215–218. doi:10.1016/j.bandc.2008.08.012.
2. O'Driscoll GA, Callahan BL. Smooth pursuit in schizophrenia: a meta-analytic review of research since 1993. *Brain and Cognition*. 2008;68(3):359–370. doi:10.1016/j.bandc.2008.08.023.
3. Gooding DC, Basso MA. The tell-tale tasks: a review of saccadic research in psychiatric patient populations. *Brain and Cognition*. 2008;68(3):371–390. doi:10.1016/j.bandc.2008.08.024.
4. Diefendorf AR, Dodge R. An experimental study of the ocular reactions of the insane from photographic records. *Brain*. 1908.
5. Lencer R, Trillenberg P. Neurophysiology and neuroanatomy of smooth pursuit in humans. *Brain and Cognition*. 2008;68(3):219–228. doi:10.1016/j.bandc.2008.08.013.
6. Ilg UJ, Thier P. The neural basis of smooth pursuit eye movements in the rhesus monkey brain. *Brain and Cognition*. 2008;68(3):229–240. doi:10.1016/j.bandc.2008.08.014.
7. McDowell JE, Dyckman KA, Austin BP, Clementz BA. Brain and cognition. *Brain and Cognition*. 2008;68(3):255–270. doi:10.1016/j.bandc.2008.08.016.
8. Johnston K, Everling S. Neurophysiology and neuroanatomy of reflexive and voluntary saccades in non-human primates. *Brain and Cognition*. 2008;68(3):271–283. doi:10.1016/j.bandc.2008.08.017.
9. Smyrnis N. Metric issues in the study of eye movements in psychiatry. *Brain and Cognition*. 2008;68(3):341–358. doi:10.1016/j.bandc.2008.08.022.
10. Luna B, Velanova K, Geier CF. Development of eye-movement control. *Brain and Cognition*. 2008;68(3):293–308. doi:10.1016/j.bandc.2008.08.019.
11. Hutton SB. Cognitive control of saccadic eye movements. *Brain and Cognition*. 2008;68(3):327–340. doi:10.1016/j.bandc.2008.08.021.
12. Munoz DP, Everling S. Look away: the anti-saccade task and the voluntary control of eye movement. *Nat Rev Neurosci*. 2004;5(3):218–228. doi:10.1038/nrn1345.
13. Sweeney JA, Luna B, Keedy SK, McDowell JE, Clementz BA. fMRI studies of eye movement control: Investigating the interaction of cognitive and sensorimotor brain systems. *NeuroImage*. 2007;36:T54–T60. doi:10.1016/j.neuroimage.2007.03.018.
14. Calkins ME, Iacono WG, Ones DS. Brain and cognition. *Brain and Cognition*. 2008;68(3):436–461. doi:10.1016/j.bandc.2008.09.001.
15. Reilly JL, Frankovich K, Hill S, et al. Elevated antisaccade error rate as an intermediate phenotype for psychosis across diagnostic categories. *Schizophrenia Bulletin*. 2014;40(5):1011–1021. doi:10.1093/schbul/sbt132.
16. Millard SP, Shofer J, Braff D, et al. Prioritizing schizophrenia endophenotypes for future genetic studies: An example using data from the COGS-1 family study. *Schizophrenia Research*. 2016;174(1–3):1–9. doi:10.1016/j.schres.2016.04.011.
17. Hudgens-Haney ME, Ethridge LE, Knight JB, et al. Intrinsic neural activity differences among psychotic illnesses. *Psychophysiology*. 2017;54(8):1223–1238. doi:10.1111/psyp.12875.
18. Ettinger U, Aichert DS, Wöstmann N, Dehning S, Riedel M, Kumari V. Response inhibition and interference control: effects of schizophrenia, genetic risk, and schizotypy. *J Neuropsychol*. 2017;34(1–3):1016. doi:10.1111/jnp.12126.
19. McDowell JE, Myles-Worsley M, Coon H, Byerley W, Clementz BA. Measuring liability for schizophrenia using optimized antisaccade stimulus parameters. *Psychophysiology*. 1999;36(1):138–141.
20. Harris MSH, Reilly JL, Thase ME, Keshavan MS, Sweeney JA. Psychiatry Research. *Psychiatry Research*. 2009;170(2–3):150–156. doi:10.1016/j.psychres.2008.10.031.
21. Reilly JL, Harris MSH, Keshavan MS, Sweeney JA. Abnormalities in visually guided saccades suggest corticofugal dysregulation in never-treated schizophrenia. *Biological Psychiatry*. 2005;57(2):145–154. doi:10.1016/j.biopsych.2004.10.024.
22. Reilly JL, Lencer R, Bishop JR, Keedy S, Sweeney JA. Pharmacological treatment effects on eye movement control. *Brain and Cognition*. 2008;68(3):415–435. doi:10.1016/j.bandc.2008.08.026.
23. Ettinger U, Kumari V, Crawford TJ, et al. Smooth pursuit and antisaccade eye movements in siblings discordant for schizophrenia. *Journal of Psychiatric Research*. 2004;38(2):177–184.
24. Radant AD, Millard SP, Braff DL, et al. Robust differences in antisaccade performance exist between COGS schizophrenia cases and controls regardless of recruitment strategies. *Schizophrenia Research*. 2015;163(1–3):47–52. doi:10.1016/j.schres.2014.12.016.
25. Gooding DC, Mohapatra L, Shea HB. Temporal stability of saccadic task performance in schizophrenia and bipolar patients. *Psychol Med*. 2004;34(5):921–932.
26. Curtis CE, Calkins ME, Grove WM, Feil KJ, Iacono WG. Saccadic disinhibition in patients with acute and remitted schizophrenia and their first-degree biological relatives. *Am J Psychiatry*. 2001;158(1):100–106. doi:10.1176/appi.ajp.158.1.100.
27. Camchong J, Dyckman KA, Chapman CE, Yanasak NE, McDowell JE. Basal ganglia-thalamocortical circuitry disruptions in schizophrenia during delayed response tasks. *BPS*. 2006;60(3):235–241. doi:10.1016/j.biopsych.2005.11.014.
28. Everling S, Krappmann P, Preuss S, Brand A, Flohr H. Hypometric primary saccades of schizophrenics in a delayed-response task. *Exp Brain Res*. 1996;111(2):289–295.
29. Park S. Association of an oculomotor delayed response task and the Wisconsin Card Sort Test in schizophrenic patients. *Int J Psychophysiol*. 1997;27(2):147–151.
30. Park S, Holzman PS. Schizophrenics show spatial working memory deficits. *Arch Gen Psychiatry*. 1992;49(12):975–982.
31. McDowell JE, Brenner CA, Myles-Worsley M, Coon H, Byerley W, Clementz BA. Ocular motor delayed-response task performance among patients with schizophrenia and their biological relatives. *Psychophysiology*. 2001;38(1):153–156.
32. Landgraf S, Amado I, Bourdel M-C, Leonardi S, Krebs M-O. Memory-guided saccade abnormalities in schizophrenic patients and

their healthy, full biological siblings. *Psychol Med.* 2008;38(6):861–870. doi:10.1017/S0033291707001912.
33. Reilly JL, Harris MSH, Keshavan MS, Sweeney JA. Adverse effects of risperidone on spatial working memory in first-episode schizophrenia. *Arch Gen Psychiatry.* 2006;63(11):1189–1197. doi:10.1001/archpsyc.63.11.1189.
34. Caldani S, Bucci MP, Lamy J-C, et al. Saccadic eye movements as markers of schizophrenia spectrum: exploration in at-risk mental states. *Schizophrenia Research.* 2017;181:30–37. doi:10.1016/j.schres.2016.09.003.
35. McDowell JE, Clementz BA. Ocular-motor delayed-response task performance among schizophrenia patients. *Neuropsychobiology.* 1996;34(2):67–71.
36. Park S, Holzman PS, Goldman-Rakic PS. Spatial working memory deficits in the relatives of schizophrenic patients. *Arch Gen Psychiatry.* 1995;52(10):821–828.
37. Reilly JL, Harris MSH, Khine TT, Keshavan MS, Sweeney JA. Antipsychotic drugs exacerbate impairment on a working memory task in first-episode schizophrenia. *Biological Psychiatry.* 2007;62(7):818–821. doi:10.1016/j.biopsych.2006.10.031.
38. Karoumi B, Ventre-Dominey J, Vighetto A, Dalery J, d'Amato T. Saccadic eye movements in schizophrenic patients. *Psychiatry Research.* 1998;77(1):9–19.
39. Bishop JR, Reilly JL, Harris MSH, et al. Pharmacogenetic associations of the type-3 metabotropic glutamate receptor (GRM3) gene with working memory and clinical symptom response to antipsychotics in first-episode schizophrenia. *Psychopharmacology.* 2015;232(1):145–154. doi:10.1007/s00213-014-3649-4.
40. Ross RG, Harris JG, Olincy A, Radant A, Adler LE, Freedman R. Familial transmission of two independent saccadic abnormalities in schizophrenia. *Schizophrenia Research.* 1998;30(1):59–70.
41. Ross RG, Heinlein S, Zerbe GO, Radant A. Saccadic eye movement task identifies cognitive deficits in children with schizophrenia, but not in unaffected child relatives. *J Child Psychol Psychiatry.* 2005;46(12):1354–1362. doi:10.1111/j.1469-7610.2005.01437.x.
42. Diwadkar VA, Sweeney JA, Boarts D, Montrose DM, Keshavan MS. Oculomotor delayed response abnormalities in young offspring and siblings at risk for schizophrenia. *CNS Spectr.* 2001;6(11):899–903.
43. Ivleva EI, Bidesi AS, Keshavan MS, et al. Gray matter volume as an intermediate phenotype for psychosis: Bipolar-Schizophrenia Network on Intermediate Phenotypes (B-SNIP). *Am J Psychiatry.* 2013;170(11):1285–1296. doi:10.1176/appi.ajp.2013.13010126.
44. Kubicki M, McCarley R, Westin C-F, et al. A review of diffusion tensor imaging studies in schizophrenia. *Journal of Psychiatric Research.* 2007;41(1–2):15–30. doi:10.1016/j.jpsychires.2005.05.005.
45. Schaeffer DJ, Rodrigue AL, Burton CR, et al. White matter structural integrity differs between people with schizophrenia and healthy groups as a function of cognitive control. *Schizophrenia Research.* 2015;169(1-3):62–68. doi:10.1016/j.schres.2015.11.001.
46. Keedy SK, Reilly JL, Bishop JR, Weiden PJ, Sweeney JA. Impact of antipsychotic treatment on attention and motor learning systems in first-episode schizophrenia. *Schizophrenia Bulletin.* 2015;41(2):355–365. doi:10.1093/schbul/sbu071.
47. Raemaekers M, Jansma JM, Cahn W, et al. Neuronal substrate of the saccadic inhibition deficit in schizophrenia investigated with 3-dimensional event-related functional magnetic resonance imaging. *Arch Gen Psychiatry.* 2002;59(4):313–320.
48. McDowell JE, Brown GG, Paulus M, et al. Neural correlates of refixation saccades and antisaccades in normal and schizophrenia subjects. *BPS.* 2002;51(3):216–223.
49. Camchong J, Dyckman KA, Austin BP, Clementz BA, McDowell JE. Common neural circuitry supporting volitional saccades and its disruption in schizophrenia patients and relatives. *BPS.* 2008;64(12):1042–1050. doi:10.1016/j.biopsych.2008.06.015.
50. Clementz BA, Sweeney JA, Hamm JP, et al. Identification of distinct psychosis biotypes using brain-based biomarkers. *Am J Psychiatry.* 2016;173(4):373–384. doi:10.1176/appi.ajp.2015.14091200.
51. Barnes GR. Cognitive processes involved in smooth pursuit eye movements. *Brain and Cognition.* 2008;68(3):309–326. doi:10.1016/j.bandc.2008.08.020.
52. Levy DL, Sereno AB, Gooding DC, O'Driscoll GA. Eye tracking dysfunction in schizophrenia: characterization and pathophysiology. *Curr Top Behav Neurosci.* 2010;4:311–347.
53. Ivleva EI, Moates AF, Hamm JP, et al. Smooth pursuit eye movement, prepulse inhibition, and auditory paired stimuli processing endophenotypes across the schizophrenia-bipolar disorder psychosis dimension. *Schizophrenia Bulletin.* 2014;40(3):642–652. doi:10.1093/schbul/sbt047.
54. Kathmann N, Hochrein A, Uwer R, Bondy B. Deficits in gain of smooth pursuit eye movements in schizophrenia and affective disorder patients and their unaffected relatives. *Am J Psychiatry.* 2003;160(4):696–702. doi:10.1176/appi.ajp.160.4.696.
55. Sweeney JA, Luna B, Haas GL, Keshavan MS, Mann JJ, Thase ME. Pursuit tracking impairments in schizophrenia and mood disorders: step-ramp studies with unmedicated patients. *BPS.* 1999;46(5):671–680.
56. Lencer R, Reilly JL, Harris MS, Sprenger A, Keshavan MS, Sweeney JA. Sensorimotor transformation deficits for smooth pursuit in first-episode affective psychoses and schizophrenia. *BPS.* 2010;67(3):217–223. doi:10.1016/j.biopsych.2009.08.005.
57. Blackwood DH, Sharp CW, Walker MT, Doody GA, Glabus MF, Muir WJ. Implications of comorbidity for genetic studies of bipolar disorder: P300 and eye tracking as biological markers for illness. *Br J Psychiatry Suppl.* 1996;(30):85–92.
58. Rosenberg DR, Sweeney JA, Squires-Wheeler E, Keshavan MS, Cornblatt BA, Erlenmeyer-Kimling L. Eye-tracking dysfunction in offspring from the New York High-Risk Project: diagnostic specificity and the role of attention. *Psychiatry Research.* 1997;66(2–3):121–130.
59. Carl JR, Gellman RS. Human smooth pursuit: stimulus-dependent responses. *J Neurophysiol.* 1987;57(5):1446–1463.
60. Clementz BA, McDowell JE. Smooth pursuit in schizophrenia: abnormalities of open- and closed-loop responses. *Psychophysiology.* 1994;31(1):79–86.
61. Lencer R, Sprenger A, Reilly JL, et al. Pursuit eye movements as an intermediate phenotype across psychotic disorders: Evidence from the B-SNIP study. *Schizophrenia Research.* 2015;169(1–3):326–333. doi:10.1016/j.schres.2015.09.032.
62. Trillenberg P, Sprenger A, Talamo S, et al. Visual and non-visual motion information processing during pursuit eye tracking in schizophrenia and bipolar disorder. *Eur Arch Psychiatry Clin Neurosci.* 2017;267(3):225–235. doi:10.1007/s00406-016-0671-z.
63. Farber RH, Clementz BA, Swerdlow NR. Characteristics of open- and closed-loop smooth pursuit responses among obsessive-compulsive disorder, schizophrenia, and nonpsychiatric individuals. *Psychophysiology.* 1997;34(2):157–162.
64. Thaker GK, Ross DE, Buchanan RW, et al. Does pursuit abnormality in schizophrenia represent a deficit in the predictive mechanism? *Psychiatry Research.* 1996;59(3):221–237.
65. Newsome WT, Wurtz RH, Komatsu H. Relation of cortical areas MT and MST to pursuit eye movements. II. Differentiation of retinal from extraretinal inputs. *J Neurophysiol.* 1988;60(2):604–620.
66. Clementz BA, Sweeney JA, Hirt M, Haas G. Pursuit gain and saccadic intrusions in first-degree relatives of probands with schizophrenia. *Journal of Abnormal Psychology.* 1990;99(4):327–335.
67. Holzman PS, Proctor LR, Hughes DW. Eye-tracking patterns in schizophrenia. *Science.* 1973;181(4095):179–181.
68. Sprenger A, Trillenberg P, Nagel M, Sweeney JA, Lencer R. Enhanced top-down control during pursuit eye tracking in schizophrenia. *Eur Arch Psychiatry Clin Neurosci.* 2013;263(3):223–231. doi:10.1007/s00406-012-0332-9.
69. Hong LE, Turano KA, O'Neill H, et al. Refining the predictive pursuit endophenotype in schizophrenia. *Biological Psychiatry.* 2008;63(5):458–464. doi:10.1016/j.biopsych.2007.06.004.

70. Lencer R, Malchow CP, Trillenberg-Krecker K, Schwinger E, Arolt V. Eye-tracking dysfunction (ETD) in families with sporadic and familial schizophrenia. *BPS*. 2000;47(5):391–401.
71. Thaker GK. Neurophysiological endophenotypes across bipolar and schizophrenia psychosis. *Schizophrenia Bulletin*. 2007;34(4):760–773. doi:10.1093/schbul/sbn049.
72. Lencer R, Trillenberg-Krecker K, Schwinger E, Arolt V. Schizophrenia spectrum disorders and eye tracking dysfunction in singleton and multiplex schizophrenia families. *Schizophrenia Research*. 2003;60(1):33–45.
73. Hong LE, Tagamets M, Avila M, Wonodi I, Holcomb H, Thaker GK. Specific motion processing pathway deficit during eye tracking in schizophrenia: a performance-matched functional magnetic resonance imaging study. *BPS*. 2005;57(7):726–732. doi:10.1016/j.biopsych.2004.12.015.
74. Lencer R, Nagel M, Sprenger A, Heide W, Binkofski F. Reduced neuronal activity in the V5 complex underlies smooth-pursuit deficit in schizophrenia: evidence from an fMRI study. *NeuroImage*. 2005;24(4):1256–1259. doi:10.1016/j.neuroimage.2004.11.013.
75. Nagel M, Sprenger A, Steinlechner S, Binkofski F, Lencer R. Altered velocity processing in schizophrenia during pursuit eye tracking. Lappe M, ed. *PLoS ONE*. 2012;7(6):e38494. doi:10.1371/journal.pone.0038494.
76. Nagel M, Sprenger A, Nitschke M, et al. Different extraretinal neuronal mechanisms of smooth pursuit eye movements in schizophrenia: an fMRI study. *NeuroImage*. 2007;34(1):300–309. doi:10.1016/j.neuroimage.2006.08.025.
77. Lencer R, Keedy SK, Reilly JL, et al. Psychiatry research: neuroimaging. *Psychiatry Research: Neuroimaging*. 2011;194(1):30–38. doi:10.1016/j.pscychresns.2011.06.011.
78. O'Driscoll GA, Benkelfat C, Florencio PS, et al. Neural correlates of eye tracking deficits in first-degree relatives of schizophrenic patients: a positron emission tomography study. *Arch Gen Psychiatry*. 1999;56(12):1127–1134.
79. Arolt V, Lencer R, Purmann S, et al. Testing for linkage of eye tracking dysfunction and schizophrenia to markers on chromosomes 6, 8, 9, 20, and 22 in families multiply affected with schizophrenia. *Am J Med Genet*. 1999;88(6):603–606.
80. Matthysse S, Holzman PS, Gusella JF, et al. Linkage of eye movement dysfunction to chromosome 6p in schizophrenia: additional evidence. *Am J Med Genet B Neuropsychiatr Genet*. 2004;128B(1):30–36. doi:10.1002/ajmg.b.30030.
81. Haraldsson HM, Ettinger U, Magnusdottir BB, et al. COMT val(158)met genotype and smooth pursuit eye movements in schizophrenia. *Psychiatry Research*. 2009;169(2):173–175. doi:10.1016/j.psychres.2008.10.003.
82. Haraldsson HM, Ettinger U, Magnusdottir BB, et al. Neuregulin-1 genotypes and eye movements in schizophrenia. *Eur Arch Psychiatry Clin Neurosci*. 2010;260(1):77–85. doi:10.1007/s00406-009-0032-2.
83. Smyrnis N, Kattoulas E, Stefanis NC, Avramopoulos D, Stefanis CN, Evdokimidis I. Schizophrenia-related neuregulin-1 single-nucleotide polymorphisms lead to deficient smooth eye pursuit in a large sample of young men. *Schizophrenia Bulletin*. 2011;37(4):822–831. doi:10.1093/schbul/sbp150.
84. Lencer R, Bishop JR, Harris MSH, et al. Association of variants in DRD2 and GRM3 with motor and cognitive function in first-episode psychosis. *Eur Arch Psychiatry Clin Neurosci*. 2014;264(4):345–355. doi:10.1007/s00406-013-0464-6.
85. Lencer R, Mills LJ, Alliey-Rodriguez N, et al. Genome-wide association studies of smooth pursuit and antisaccade eye movements in psychotic disorders: findings from the B-SNIP study. *Translational Psychiatry*. 2017. doi:10.1038/tp.2017.210.

BRAIN IMAGING BIOMARKERS

28

STRUCTURAL CONNECTIVITY IN PSYCHOSIS

Amanda E. Lyall, Johanna Seitz, and Marek Kubicki

INTRODUCTION

STRUCTURAL CONNECTIVITY IN PSYCHOSIS

Aberrations in functional connectivity have been frequently reported in many psychiatric disorders and were hypothesized to arise from altered communication within, as well as abnormal integration between, different brain systems. The idea that psychosis is characterized as a disconnection syndrome, rather than a disturbance of a single region or network is known as the dysconnectivity hypothesis.[1] Since functional circuits are associated with an array of anatomical structures (i.e., axons grouped into tracts or bundles) providing communication and connectivity both within and between those circuits, the dysconnectivity hypothesis further implies that those anatomical connections might be disrupted in psychosis.[2] In this chapter, we summarize and highlight the state of the field for structural connectivity research in psychotic disorders.

WHAT IS STRUCTURAL CONNECTIVITY?

Structural connectivity, at its simplest, can be defined as communication between different anatomical parts of the brain. Traditionally, structural connectivity is thereby represented by anatomical structures physically connecting two brain areas, which can range in size from a single axon to large, densely packed bundles of myelinated axons, also called white matter tracts or fasciculi. While a large portion of the computational power of the human brain is generated in the cortico-cortical neuronal connections found within gray matter, the coordinated communication between distinct regions of the brain is supported by long-range myelinated white matter tracts. Based on their anatomy and function, tracts can be grouped in three general categories: (1) Projection tracts, which connect cortical areas with subcortical and lower-order regions like the thalamus, striatum, cerebellum, and the brainstem (e.g., thalamic radiations); (2) association tracts, which connect different areas within the same hemisphere (e.g., arcuate fasciculus), and (3) commissural tracts that connect opposite hemispheres with each other (e.g., corpus callosum).

STRUCTURAL CONNECTIVITY IN NEUROIMAGING

When studying structural connectivity with neuroimaging techniques, it is important to understand the many ways that anatomical correlates can be represented. Structural connectivity studies employ an array of neuroimaging methods that aim to *model* and *analyze* different components of anatomical connections. However, the interpretation of results must be considered carefully, because neuroimaging methodologies only provide indirect measurements of underlying biological features.

In general, there are three primary sets of methods with three distinct aims (see Figure 28.1): (1) investigate the "health" of white matter across the entire brain (Figure 28.1A); (2) investigate anatomical white matter structures that connect specific brain regions (Figure 28.1B); (3) investigate coordinated and/or correlated changes within gray or white matter brain regions using statistical approaches (Figure 28.1C). The first two groups of methods focus almost entirely on the study of white matter, whereas the last group can focus on both gray and white matter.

METHODS TO MEASURE STRUCTURAL CONNECTIVITY

WHOLE BRAIN APPROACHES

As stated in the introduction, local measurements of white matter "integrity" introduced above can be further processed, combined, and analyzed in various ways, to gain information about structural connectivity. One of the most common methods is to use parametric mapping techniques to study and compare between groups the properties of all the voxels in the brain at once. In general, these are best used when there are no a priori hypotheses regarding specific anatomical connections. In this approach, also referred to as voxel-based morphometry (VBM), all parametric contrasts can be registered to a template image, which can be either an atlas or representative scan from the study population. This provides a one-to-one correspondence for all voxels in the image across the study population. After ensuring proper registration of images to the template, the respective values can be statistically compared

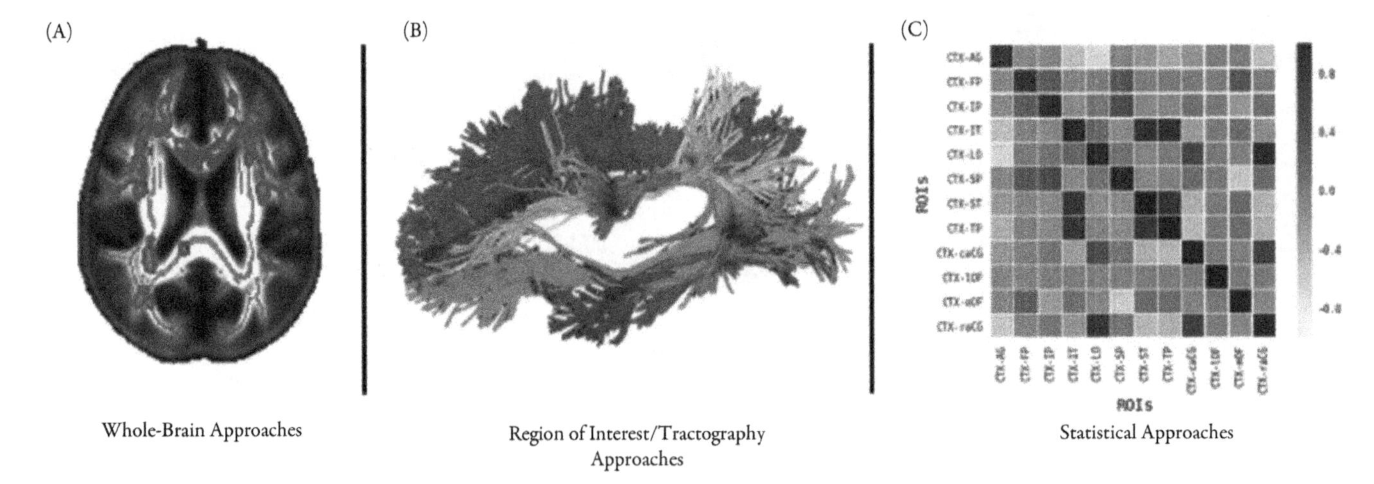

Figure 28.1 **Panel A** shows a representative image of the outcome of tract-based spatial statistics,[3] one example of whole-brain approaches to the study of structural connectivity. In this image, voxels along the skeleton that are statistically significant after correction for multiple comparisons ($p < 0.05$) are shown in the clusters. **Panel B** shows an example of a tractography approach, whereby specific white matter fiber tracts serve as anatomical connections between discrete cortical brain regions. In this image, there are five left hemispheric tracts defined: the cingulum bundle, the arcuate fasciculus, the superior longitudinal fasciculus, the inferior fronto-occipital fasciculus and the inferior longitudinal fasciculus. **Panel C** shows an example of a statistical approach to study structural connectivity. This image is a correlation matrix of 12 cortical regions of interest and shows the correlative relationship between the measures within each region of interest (e.g., cortical thickness) with the measures of every other region of interest included in the analysis. The strength of the relationship is defined by the color bar to the right.

voxel by voxel across populations. As this approach requires many comparisons to be carried out, it is necessary to employ stringent corrections for Type I error. The VBM methodology has been evolving, with more recent development being tract-based spatial statistics (TBSS).[3] In TBSS, all the registered datasets are utilized to create a *mean fractional anisotropy (FA) skeleton* (whole-brain skeletonized average of all FA values), which is used to represent the principal white matter tracts within the brain (see Figure 28.1A). The FA values in voxels proximal to the skeleton are then projected onto the skeleton and used for nonparametric statistical analyses either across the entire skeleton or averaged within predefined regions of interest. This method improves brain-to-brain registration, which then allows for enhanced statistical power.

REGION OF INTEREST/ TRACTOGRAPHY APPROACHES

Another way to study anatomical connections between brain regions is through analyzing diffusion properties in specific predefined anatomical connections, i.e., tracts, using the three-dimensional modeling method called "tractography" (see Figure 28.1B). This approach uses the diffusion tensor within each voxel to consider the probability that two adjacent voxels are connected via white matter fiber tracts based on both the degree of anisotropy and orientation of the tensors. White matter tracts are extracted from an initial seeding region, which is either manually or automatically defined. Parameters, such as FA, can be extracted for the identified fiber tracts and either averaged over each tract or sampled along tracts. Tractography is the most anatomically accurate neuroimaging method for defining entire trajectories of white matter tracts. It is therefore extremely useful for the identification of subtle differences in structural connectivity between healthy and clinical populations. However, this level of accuracy requires high resolution data, profound anatomical knowledge, and a great deal of time and computational power.

STATISTICAL APPROACHES TO STRUCTURAL CONNECTIVITY

The third most popular approach to the study of structural connectivity is investigating coordinated growth patterns of cortical areas (see Figure 28.1C). As mentioned earlier, this definition of structural connectivity is based on the concept that areas with a high degree of functional or maturational connectivity will increase or decrease in size together.[4] These synchronized patterns of structural growth are suggested to be indirect reflections of interregional associations, or the relative "connectedness," between cortical areas within the brain. In these approaches, regional values for any structural measures of gray matter, such as cortical thickness or rate of volume growth over time, are correlated with other regions of the brain which constructs structural covariance matrices,[4] which then can be compared across different populations. It is important to note that these approaches do not directly comment on anatomic structures connecting brain areas, they demonstrate how different brain areas act as a structural system.

STRUCTURAL CONNECTIVITY IN PSYCHOSIS

Studies of structural connectivity in populations of individuals with psychotic disorders have clearly demonstrated that connectivity abnormalities are not only a consequence of psychosis but also may be related to increased vulnerability for psychosis. In what follows, we review the findings of the most prominent of the psychotic disorders: schizophrenia.

SCHIZOPHRENIA

Chronic schizophrenia

Many of the initial structural connectivity studies of schizophrenia chose to focus on patients in the chronic stage of illness. In these studies, the most common index evaluated in both whole brain and tract-based methods was FA. Whole-brain voxel-based studies tend to report significant reductions in FA in chronic patients when compared to healthy control populations,[5–8] though a few studies exist that report no differences.[9,10] A meta-analysis of 15 voxel-based diffusion tensor imaging studies in chronic schizophrenia patients determined that significant reductions were found in only two regions, left frontal deep white matter and left temporal deep white matter.[11] However, this was followed-up with a larger voxel-wise meta-analysis and meta-regression on 79 VBM studies with chronic schizophrenia patients showing FA reductions in interhemispheric fibers, anterior thalamic radiation, inferior longitudinal fasciculi, inferior frontal occipital fasciculi, cingulum bundle, and fornix.[12] In addition, magnetization transfer imaging studies have shown that patients with schizophrenia have a reduced magnetization transfer ratio (MTR, an indirect measure of myelin content), in the white matter of several brain regions, including the prefrontal areas, the left temporal lobe, areas adjacent to visual processing centers, the corpus callosum, the fornix, and the internal capsule.[13–15] Moreover, lower MTR values in certain white matter bundles are associated with worse quality of life and impaired processing speed in chronic patients.[14,16] Yet, white matter pathology was shown to be not consistent across VBM studies, as a study by Melanakos and colleagues (2011) found that even the most consistently reported pathological regions in schizophrenia were only reported independently in less than 35% of the datasets studied.[17]

Multiple studies using tractography approaches also found alterations in the structural connectivity in chronic schizophrenia in specific association, projection, and commissural white matter tracts. Significant differences have been reported in intrahemispheric fronto-temporal connections, including the cingulum bundle,[18,19] uncinate fasciculus[20] and arcuate fasciculus, which is often delineated as part of the superior longitudinal fasciculus.[21] Other tracts that experience structural abnormalities in patients with chronic schizophrenia are those terminating in the frontal and prefrontal cortices,[22,23] including the internal capsule[8] and thalamic radiations[24] as well as the corpus callosum[25–27] and connections between the medial orbitofrontal and anterior cingulate cortices.[28] These findings also echo those found with voxel-based approaches, such that the majority of these studies show a reduction in FA in each of the tracts studied (for reviews see [29]), though there are studies in which increased FA have been found, for instance in the arcuate fasciculus, in schizophrenia patients.[30,31]

Finally, investigations that employ statistical approaches to the study of structural connectivity deficits in chronic schizophrenia show that patients with schizophrenia exhibit less connected and less hierarchically organized networks.[32,33] Once again, these deficits are most prominent within fronto-temporal regions.[34–36] Interestingly, van den Heuvel and colleagues (2013) found that patients with schizophrenia exhibited a significantly lower number of connections within midline frontal, parietal, and insular regions when compared to healthy controls.[34] These findings were also echoed by a another study by Klauser and others (2016), with an even larger group ($n = 326$ patients, 197 controls), which showed that 52% of the white matter connections were affected in chronic schizophrenia.[24] A comprehensive summary of all structural covariance studies, particularly earlier analyses investigating interregional gray and white matter volume correlations, within chronic schizophrenia patients and controls can be found within the review by Wheeler and colleagues (2014).[29]

Taken together, the above studies indicate that the chronic stage of schizophrenia is a disorder affecting the "brain connections." The hope for future directions would be to develop more individualized approaches to separate subgroups based on the underlying structural connectivity characteristics. Research in chronic schizophrenia would therefore benefit from a greater focus on underrepresented patient groups, such as treatment-resistant[37,38] or late-onset schizophrenia[39,40] to further tease out distinguishing factors between subgroups, and isolate more specific aberrations in structural connectivity that bring us closer to an understanding of the etiological factors influencing and shaping the progression of schizophrenia into its chronic form.

First-episode psychosis populations

Complementary to studies in chronic populations, structural connectivity research focusing on patients exhibiting their first episode of psychosis can help differentiate between primary and secondary markers of illness. Investigations into first-episode or "recent-onset" patients tend to recruit patients that have experienced transition to psychosis within the past 3 to 5 years. Even more important are studies that can recruit first-episode antipsychotic-naïve populations, which can help attest to the possible contribution that medication may have on structural connectivity.

Several studies employing whole-brain methodologies report widespread reductions in FA in first-episode schizophrenia patients, even in patients who have minimal exposure to antipsychotic medications.[41–43] Alterations in structural connectivity are principally reported in the frontal, temporal, and parietal lobes.[41–43] Interestingly, a study by Melicher and colleagues (2015) found widespread reductions in FA in the early stages of schizophrenia using TBSS but, more importantly, showed that the extent of the finding was dependent on the size of the population studied.[44] In this study, the authors suggested that underpowered studies may ultimately result in poor or false spatial resolution, whereby studies only report localized differences when in fact the whole brain may be affected.[44]

Within tractography studies, there exists some evidence for tract-specific alterations in structural connectivity that are akin to those reported in chronic populations. Reductions in FA within discrete white matter tracts have been found in first-episode populations in the uncinate fasciculus,[45–47] the arcuate fasciculus/superior longitudinal fasciculus,[45,47] the fornix,[48,49] and the occipital-frontal fasciculus.[50,51] However, study results are also inconsistent,[46,52–55] which might be again related to differences in methodology and sample size (for review see [29]).

Relative to chronic populations, there are fewer studies that have employed statistical approaches to investigate structural connectivity in first-episode populations. However, of the few studies that have employed this approach, the findings share similar themes to previous work conducted in chronic populations. One study showed that 86% of the disrupted connections in recent-onset patients overlapped with the disruptions in chronic patients.[56] These focal reductions were primarily found within frontal and parietal interhemispheric connections.[56] Interestingly, a study by di Biase and colleagues (2017) shows that recent-onset patients exhibit an intermediate amount of connectivity reduction compared to chronic patients, suggesting that the degree of dysconnectivity is less severe and may progress with illness duration.[56] Aberrant network connectivity has also been reported in two independent populations of medication-naïve first-episode patients, suggesting that the altered "connectome pathology" is likely not a product of medication effects.[57,58]

Distinct from first-episode populations are a rare clinical population of patients with early-onset schizophrenia. These individuals experience their first psychotic episode prior to the age of 18 years (or 13 years for childhood onset schizophrenia) and tend to present with more severe symptoms than patients with adult-onset schizophrenia.[59] A recent systematic review by Tamnes and colleagues (2016) reported that 21 diffusion studies were conducted in early-onset populations.[60] Overall, similar to other schizophrenia populations, early-onset patients exhibit reductions in FA relative to healthy control populations with large variability in the anatomical locations, which include frontal, parietal, and temporal white matter, as well as the corpus callosum, cingulum, superior longitudinal fasciculus and a few others (for comprehensive review see [60]).

In summary, it appears that structural connectivity differences are present in the early phases of schizophrenia, but possibly to a lesser degree than those that are found in chronic populations. As stated previously, the results exhibit a high degree of heterogeneity, such that no specific white matter connection has been found to be reliably identified across all studies focusing on first-episode or early-onset populations.

High-risk and prodromal populations

In recent decades, there has been a significant shift in focus toward the study of abnormalities that may have occurred prior to the onset of illness, also known as the prodromal period, for schizophrenia. The central goal of these studies is to identify structural connectivity deficits that are found in subsets of individuals who are at risk for developing a psychotic disorder. These studies make it possible to learn more about the trajectories of structural connectivity abnormalities in early stage psychosis patients as well as identify specific structural markers or signatures that predispose an individual to develop a specific disorder.

There are a few whole-brain TBSS studies that have reported significant differences in FA within focal regions in clinical high and ultra-high-risk subjects. For instance, four whole-brain TBSS studies found reductions in FA in clinical high risk subjects,[61–64] with specific reductions in FA found within the superior longitudinal fasciculus,[61,62] the corpus callosum,[62,63] left inferior longitudinal fasciculus,[62] the right external and internal capsule,[62] and within the superior frontal gyrus.[65] Longitudinal studies investigating structural connectivity differences in high-risk subjects who later convert versus those who do not convert to psychosis also find mixed results. Carletti and colleagues (2012) found significant reductions in the FA within left frontal white matter in converters that was not present in those who did not convert[62] while Bloemen and colleagues (2009) showed reductions in FA in the right putamen and left superior temporal lobe but increased FA in left medial temporal lobe in converters.[66] Another group reported that while FA within the corpus callosum was reduced at baseline in all at-risk subjects, improvements in positive symptoms directly correlated with longitudinal increases in FA within the corpus callosum in patients who ultimately did not convert to psychosis.[67] Still others find no significant differences in converters versus nonconverters.[43] Magnetization transfer imaging was also recently used to study subjects at ultra-high risk for psychosis, revealing MTR changes in the cingulate gyrus, the stria terminalis, the entorhinal cortex, and some areas of the frontal lobe.[68]

The most reported tractography differences identified in clinical high and ultra-high risk subjects are reductions in FA within the frontal and temporal white matter connections (for review see [69,70]). Tomyshev and colleagues (2017) investigated 18 different white matter tracts in a cohort of 27 ultra-high risk male patients and found reductions in FA within the left anterior thalamic radiation, the anterior forceps, and the left inferior longitudinal fasciculus.[71] In two more targeted analyses, Bernard and colleagues (2015 and 2017) found altered developmental trajectories in hippocampal-thalamic and cerebello-thalamo-cortical white matter connections in individuals that were at ultra-high risk for psychosis, both of which were found to be related to disease progression.[72,73]

There are relatively fewer studies that have employed statistical approaches to clinical high-risk populations, though the popularity of these methods promises more studies will emerge. In a study by Drakesmith and others (2015) of 123 subjects with psychotic experiences and 127 healthy controls, the authors found that individuals with attenuated psychotic symptoms also exhibit reduced global and local efficiency, specifically within the default mode network.[74] Similarly, Schmidt and colleagues (2017), applied a graph theoretical framework to a population of 24 clinical high-risk subjects and report a disruption in organization that was linked to severity of negative symptoms,[75] echoing what has been previously reported in first-episode and chronic populations.

Whole-brain approaches in genetic risk populations have reported mild reductions in FA compared to healthy controls that are intermediate to those reported in schizophrenia patients,[31,64,76,77] though some studies report no differences.[78] The differences found were primarily in regions such as the superior longitudinal fasciculus,[76,77] corpus callosum,[64] cingulum bundle,[31] and fornix,[77] all regions that have been previously identified in first-episode or chronic populations.

Tractography studies have also reported differences in genetic risk populations. In a study by Boos and colleagues (2013), tractography data from 126 patients and 123 of their unaffected relatives showed that there was a significant increase in FA in the left and right arcuate fasciculus in the nonpsychotic individuals.[79] Meanwhile, Kubicki and colleagues (2013) found no significant differences in FA between genetic risk individuals and healthy controls within the arcuate, uncinate, and inferior fronto-occipital fasciculi, though a difference in axial diffusivity (AD) was found in high-risk patients for all tracts.[80] De Leeuw and colleagues (2017) recently found that FA in the fronto-striatal white matter tract connecting the nucleus accumbens to the dorsolateral prefrontal cortex was increased in genetic high-risk patients relative to controls.[81] Moreover, while FA showed age-related increases in the control subjects, the same increase was not observed in the genetic high-risk subjects,[81] potentially indicating a genetically mediated alteration in white matter structural connectivity.

Finally, several studies have been conducted using statistical approaches in genetic high-risk populations. Altered structural connectome network properties, namely reduced global and local efficiency are similar, but not as pronounced, as what is observed in first-episode and chronic populations, in genetic high-risk populations,[82,83] even in infancy.[84] These alterations in structural connectivity were primarily found in frontal, striatal, and thalamic areas.[83]

Combined, despite the heterogeneity of findings within both the clinical and genetic high-risk literature, there is converging evidence that alterations in structural connectivity may precede symptom onset. These studies also provide an avenue for understanding what structural connectivity deficits may be more related to genetic or neurodevelopmental mechanisms versus alterations that are more likely attributed to burgeoning clinical pathology.

LIMITATIONS AND FUTURE DIRECTIONS

Despite the significant advances that have been made toward establishing a relationship between structural connectivity and psychosis, the etiology of psychosis symptomology and illness onset are still unclear. The heterogeneity of the findings provide evidence that psychosis exists on a spectrum and that structural connectivity abnormalities may be more related to specific clinical phenotypes rather than distinct phenomenological disease entities. With the advent of the Research Domain Criteria approach, as put forth by the National Institutes of Mental Health,[85] new investigations have started to search for possible reliable biomarkers that could distinguish between subtypes of clinical populations and better inform treatment. Additionally, many of the studies in psychosis-related populations share similar limitations: (1) small sample sizes leading to a lack of statistical power, (2) possible confounding effects of both differential and/or long-term medication, (3) a lack of well-powered longitudinal studies, and (4) a presently incomplete understanding of the impacts of various demographics variables, including, but not limited to, age and sex.

Of all the limitations currently facing the field of structural connectivity, the largest is likely sample size. Unlike the field of genetics, recruiting and acquiring imaging data from psychotic populations is a difficult task. Ultimately, small sample sizes lead to a lack of power and restricted generalizability. While meta-analyses can serve as useful tools for understanding general trends within a field, the likely reality is that most of the studies included within meta-analyses are underpowered themselves, leading to potentially limited or aberrant conclusions.[44] As such, it is important as the field moves forward to find ways to collaborate, similar in nature to those types of analyses done within the field of genetics. One such example of this is the ENIGMA consortium.[86] The ENIGMA consortium recently published the results of their meta-analysis which consisted of 4,322 individuals from 29 independent sites. In this study, they found that effect sizes varied by region, with the anterior corona radiata as well as the body and genu of the corpus callosum showing the greatest differences.[86] Additionally, as the field begins to embrace the spectrum hypothesis, as well as universal data acquisition protocols (e.g., Human Connectome Project [HCP][87]), studies collecting larger samples with a potentially wider definition of psychosis and more uniform imaging acquisition protocols across individual studies may begin to emerge. Through these larger-scale analyses, it may be possible to identify specific biotypes that would be related to more specific biological subtypes of illness.

Medication effects are another persistent confound within many studies. Though there are presently significant efforts to recruit antipsychotic-naïve populations in early stages of psychosis, these are difficult to collect and, once again, usually result in small sample sizes or attrition from the study. Disentangling the effects of the illness versus the medication is especially important in understanding disease progression from acute to chronic psychosis. Prolonged medication has previously been associated with a reduction in glial cell number in adult mice.[88] In a longitudinal study by Ho and colleagues (2011), prolonged or more intense antipsychotic treatment was associated with smaller gray matter tissue volumes and a progressive decline in white matter volume.[88] However, there are some studies that suggest that antipsychotics have differential effects on tissue dependent on the type of medication[89] or a potentially ameliorative effect on tissue.[90,91] It is important for future studies to attempt to understand the extent of the impact that various types of medications, whether it be antipsychotics or mood stabilizers, have on structural connectivity abnormalities within patients experiencing psychosis. A more complete understanding of this effect will isolate which alterations in structural connectivity are primary versus secondary markers of illness.

One way that we may be able to accomplish this goal is through the execution of effective and well-powered longitudinal studies. The present lack of long-term studies investigating structural connectivity hinders the field's ability to understand the natural progression of disease states from illness onset to chronic stages as well as what factors contribute to underlying structural pathologies. Potentially more importantly, these studies can also provide a clearer picture of the features that predict positive outcomes in psychotic patients. However, longitudinal studies, despite their importance, face similar difficulties as cross-sectional studies: medication effects, clinical heterogeneity, and attrition.

Additionally, there are many demographics variables that have a significant influence on the structural connectivity phenotypes observed in psychotic patients. For instance, multiple studies have suggested that schizophrenia is a disorder classified by accelerated aging, whereby patients experience faster age-related decline than healthy individuals.[92–94] Similarly, a few studies have linked aberrations in structural connectivity with illness duration,[95] duration of untreated symptoms,[96] and age of onset,[97] though an equal number of studies showing no relationship between structural connectivity and these factors also exist (e.g., [98]). In a recent study by Bijanki and colleagues (2015), the authors showed that age was strongly mediating an observed relationship between FA and schizophrenia patients' Scale for the Assessment of Negative Symptoms (SANS) scores,[99] suggesting that possible previous correlations reported in the literature that were not factoring in the effect of age could be confounded.

Another key demographic factor that has gained considerable traction in the last few years is the effect of sex. Research trends have begun to posit that underlying biological factors mediating or contributing to the psychosis may be different between males and females. In a recent systematic review of diffusion tensor imaging studies in schizophrenia by Shahab and others (2017), of the 75 studies that met the inclusion criteria, most showed no effect of sex, however those that did find an effect showed reductions in FA and a more pronounced effect of the disorder in female patients.[100] As many studies tend to collect fewer female patients than male patients, they might be not powered enough to detect sex differences, despite the fact that incidence, age of onset, clinical manifestation, prognosis, and treatment response differ between the two sexes.[100] Therefore, future studies should aim to advance the understanding of potential sex-specific effects as they may help inform the development of novel treatments or improve pre-existing interventions.

In summary, these demographic variables, as well as several others not discussed here have a potentially substantial impact on the structural connectivity of the brain in both healthy and clinical populations. Therefore, it is important that future studies be cognizant of these effects to properly interpret the results of clinical investigations.

As the definition of psychosis expands, so must our understanding of the myriad of underlying biological factors that may contribute to symptom presentation and illness severity. To accomplish this goal, research inquiries may require a wider perspective of the possible biological processes ongoing within the central nervous system, but also to influencing factors outside of the central nervous system, that may interact with, or act upon, the structural connections within the brain.

Moreover, imaging technology is also evolving to provide more anatomically accurate data. More complex acquisition schemes are being developed, some to increase directional resolution of diffusion data (i.e., multi-shell diffusion imaging[101] or diffusion spectrum imaging [DSI][102]), while others aim to increase spatial resolution (i.e., protocols developed by the HCP,[87] compressed sensing[103]). Data processing and image analysis methods are also becoming more sophisticated and have already acquired the ability to up-sample low-resolution legacy data (track density imaging [TDI][104]), or resolve the fiber crossing limitations of tractography algorithms (multitensor tractography,[105] spectrum harmonics[106]). The application of all these methods will provide a better way to extract more refined, biologically relevant structures that may lead to advances in our understanding of structural connectivity differences in psychotic populations.

SUMMARY

In summary, the current state of the field suggests that despite the relative heterogeneity of the findings, aberrations in structural connectivity are likely a key component of psychosis pathology. With the increasing recognition and application of the psychosis spectrum model, paired with forward progress toward improvements in data resolution and quality, and larger-scale collaborative efforts leveraging cutting-edge imaging paradigms, the next decade will yield considerable advances in knowledge about the biological bases of psychosis and its relationship to observed structural connectivity abnormalities. These innovative approaches will hopefully help to establish reliable imaging biomarkers of psychosis and encourage (by providing noninvasive in vivo monitoring tools) the development of new therapeutics options and interventions that may help to address more specific aspects of psychosis symptomology or provide alternative treatments for those currently treatment-resistant.

REFERENCES

1. Friston KJ. The disconnection hypothesis. *Schizophr Res.* 1998;30(2):115–125.
2. Canu E, Agosta F, Filippi M. A selective review of structural connectivity abnormalities of schizophrenic patients at different stages of the disease. *Schizophrenia Research.* 2014. doi:10.1016/j.schres.2014.05.020.
3. Smith SM, Jenkinson M, Johansen-Berg H, et al. Tract-based spatial statistics: voxelwise analysis of multi-subject diffusion data. *NeuroImage.* 2006;31(4):1487–1505. doi:10.1016/j.neuroimage.2006.02.024.
4. Alexander-Bloch A, Raznahan A, Bullmore E, Giedd J. The convergence of maturational change and structural covariance in human cortical networks. *Journal of Neuroscience.* 2013;33(7):2889–2899. doi:10.1523/JNEUROSCI.3554-12.2013.
5. Scheel M, Prokscha T, Bayerl M, Gallinat J, Montag C. Myelination deficits in schizophrenia: evidence from diffusion tensor imaging. *Brain Struct Funct.* 2013;218(1):151–156. doi:10.1007/s00429-012-0389-2.
6. Roalf DR, Ruparel K, Verma R, Elliott MA, Gur RE, Gur RC. White matter organization and neurocognitive performance variability in schizophrenia. *Schizophrenia Research.* 2013;143(1):172–178. doi:10.1016/j.schres.2012.10.014.

7. Fujino J, Takahashi H, Miyata J, et al. Impaired empathic abilities and reduced white matter integrity in schizophrenia. *Prog Neuropsychopharmacol Biol Psychiatry*. 2014;48:117–123. doi:10.1016/j.pnpbp.2013.09.018.
8. Ellison-Wright I, Nathan PJ, Bullmore ET, et al. Distribution of tract deficits in schizophrenia. *BMC Psychiatry*. 2014;14(1):1–13. doi:10.1186/1471-244X-14-99.
9. Steel RM, Bastin ME, McConnell S, et al. Diffusion tensor imaging (DTI) and proton magnetic resonance spectroscopy (1H MRS) in schizophrenic subjects and normal controls. *Psychiatry Res*. 2001;106(3):161–170.
10. Foong J, Symms MR, Barker GJ, Maier M, Miller DH, Ron MA. Investigating regional white matter in schizophrenia using diffusion tensor imaging. *Neuroreport*. 2002;13(3):333–336.
11. Ellison-Wright I, Bullmore E. Meta-analysis of diffusion tensor imaging studies in schizophrenia. *Schizophr Res*. 2009;108(1–3):3–10. doi:10.1016/j.schres.2008.11.021.
12. Bora E, Fornito A, Radua J, et al. Neuroanatomical abnormalities in schizophrenia: a multimodal voxelwise meta-analysis and meta-regression analysis. *Schizophrenia Research*. 2011;127(1–3):46–57. doi:10.1016/j.schres.2010.12.020.
13. Kubicki M, PARK H, Westin C-F, et al. DTI and MTR abnormalities in schizophrenia: analysis of white matter integrity. *NeuroImage*. 2005;26(4):1109–1118. doi:10.1016/j.neuroimage.2005.03.026.
14. Palaniyappan L, Al-Radaideh A, Mougin O, Gowland P, Liddle PF. Combined white matter imaging suggests myelination defects in visual processing regions in schizophrenia. *Neuropsychopharmacology*. 2013;38(9):1808–1815. doi:10.1038/npp.2013.80.
15. Wei Y, Collin G, Mandl RCW, et al. Cortical magnetization transfer abnormalities and connectome dysconnectivity in schizophrenia. *Schizophrenia Research*. 2017. doi:10.1016/j.schres.2017.05.029.
16. Faget-Agius C, Boyer L, Wirsich J, et al. Neural substrate of quality of life in patients with schizophrenia: a magnetisation transfer imaging study. *Sci Rep*. 2015;5(1):67. doi:10.1038/srep17650.
17. Melonakos ED, Shenton ME, Rathi Y, Terry DP, Bouix S, Kubicki M. Voxel-based morphometry (VBM) studies in schizophrenia—can white matter changes be reliably detected with VBM? *Psychiatry Res*. 2011;193(2):65–70. doi:10.1016/j.pscychresns.2011.01.009.
18. Kubicki M, Westin C-F, Nestor PG, et al. Cingulate fasciculus integrity disruption in schizophrenia: a magnetic resonance diffusion tensor imaging study. *BPS*. 2003;54(11):1171–1180.
19. Sun Z, Wang F, Cui L, et al. Abnormal anterior cingulum in patients with schizophrenia: a diffusion tensor imaging study. *Neuroreport*. 2003;14(14):1833–1836. doi:10.1097/01.wnr.0000094529.75712.48.
20. Kubicki M, Westin C-F, Maier SE, et al. Uncinate fasciculus findings in schizophrenia: a magnetic resonance diffusion tensor imaging study. *Am J Psychiatry*. 2002;159(5):813–820. doi:10.1176/appi.ajp.159.5.813.
21. Nakamura K, Kawasaki Y, Takahashi T, et al. Reduced white matter fractional anisotropy and clinical symptoms in schizophrenia: a voxel-based diffusion tensor imaging study. *Psychiatry Res*. 2012;202(3):233–238. doi:10.1016/j.pscychresns.2011.09.006.
22. Molina V, Lubeiro A, Soto O, et al. Alterations in prefrontal connectivity in schizophrenia assessed using diffusion magnetic resonance imaging. *Prog Neuropsychopharmacol Biol Psychiatry*. 2017;76:107–115. doi:10.1016/j.pnpbp.2017.03.001.
23. Zhou Y, Fan L, Qiu C, Jiang T. Prefrontal cortex and the dysconnectivity hypothesis of schizophrenia. *Neurosci Bull*. 2015;31(2):207–219. doi:10.1007/s12264-014-1502-8.
24. Klauser P, Baker ST, Cropley VL, et al. White matter disruptions in schizophrenia are spatially widespread and topologically converge on brain network hubs. *Schizophrenia Bulletin*. August 2016:sbw100. doi:10.1093/schbul/sbw100.
25. Knöchel C, Oertel-Knöchel V, Schönmeyer R, et al. Interhemispheric hypoconnectivity in schizophrenia: fiber integrity and volume differences of the corpus callosum in patients and unaffected relatives. *NeuroImage*. 2012;59(2):926–934. doi:10.1016/j.neuroimage.2011.07.088.
26. Whitford TJ, Savadjiev P, Kubicki M, et al. Fiber geometry in the corpus callosum in schizophrenia: evidence for transcallosal misconnection. *Schizophrenia Research*. 2011;132(1):69–74. doi:10.1016/j.schres.2011.07.010.
27. Zhuo C, Liu M, Wang L, Tian H, Tang J. Diffusion tensor MR imaging evaluation of callosal abnormalities in schizophrenia: a meta-analysis. van Amelsvoort T, ed. *PLoS ONE*. 2016;11(8):e0161406. doi:10.1371/journal.pone.0161406.
28. Ohtani T, Bouix S, Lyall AE, et al. Abnormal white matter connections between medial frontal regions predict symptoms in patients with first episode schizophrenia. *Cortex*. 2015;71:264–276. doi:10.1016/j.cortex.2015.05.028.
29. Wheeler AL, Voineskos AN. A review of structural neuroimaging in schizophrenia: from connectivity to connectomics. *Front Hum Neurosci*. 2014;8:653. doi:10.3389/fnhum.2014.00653.
30. Rotarska-Jagiela A, Oertel-Knoechel V, DeMartino F, et al. Anatomical brain connectivity and positive symptoms of schizophrenia: a diffusion tensor imaging study. *Psychiatry Res*. 2009;174(1):9–16. doi:10.1016/j.pscychresns.2009.03.002.
31. Knöchel C, O'Dwyer L, Alves G, et al. Association between white matter fiber integrity and subclinical psychotic symptoms in schizophrenia patients and unaffected relatives. *Schizophrenia Research*. 2012;140(1–3):129–135. doi:10.1016/j.schres.2012.06.001.
32. Micheloyannis S. Graph-based network analysis in schizophrenia. *World J Psychiatry*. 2012;2(1):1–12. doi:10.5498/wjp.v2.i1.1.
33. van Dellen E, Bohlken MM, Draaisma L, et al. Structural brain network disturbances in the psychosis spectrum. *Schizophrenia Bulletin*. 2016;42(3):782–789. doi:10.1093/schbul/sbv178.
34. van den Heuvel MP, Sporns O, Collin G, et al. Abnormal rich club organization and functional brain dynamics in schizophrenia. *JAMA Psychiatry*. 2013;70(8):783–792. doi:10.1001/jamapsychiatry.2013.1328.
35. Sun Y, Dai Z, Li J, Collinson SL, Sim K. Modular-level alterations of structure-function coupling in schizophrenia connectome. *Hum Brain Mapp*. 2017;38(4):2008–2025. doi:10.1002/hbm.23501.
36. Zhang W, Deng W, Yao L, et al. Brain structural abnormalities in a group of never-medicated patients with long-term schizophrenia. *Am J Psychiatry*. 2015;172(10):995–1003. doi:10.1176/appi.ajp.2015.14091108.
37. Holleran L, Ahmed M, Anderson-Schmidt H, et al. Altered interhemispheric and temporal lobe white matter microstructural organization in severe chronic schizophrenia. 2014;39(4):944–954. doi:10.1038/npp.2013.294.
38. Nakajima S, Takeuchi H, Plitman E, et al. Neuroimaging findings in treatment-resistant schizophrenia: a systematic review: lack of neuroimaging correlates of treatment-resistant schizophrenia. *Schizophrenia Research*. 2015;164(1–3):164–175. doi:10.1016/j.schres.2015.01.043.
39. Jones DK, Catani M, Pierpaoli C, et al. A diffusion tensor magnetic resonance imaging study of frontal cortex connections in very-late-onset schizophrenia-like psychosis. *Am J Geriatr Psychiatry*. 2005;13(12):1092–1099. doi:10.1176/appi.ajgp.13.12.1092.
40. Hahn C, Lim HK, Lee C-U. Neuroimaging findings in late-onset schizophrenia and bipolar disorder. *J Geriatr Psychiatry Neurol*. 2014;27(1):56–62. doi:10.1177/0891988713516544.
41. Szeszko PR, Ardekani BA, Ashtari M, et al. White matter abnormalities in first-episode schizophrenia or schizoaffective disorder: a diffusion tensor imaging study. *Am J Psychiatry*. 2005;162(3):602–605. doi:10.1176/appi.ajp.162.3.602.
42. Federspiel A, Begre S, Kiefer C, Schroth G, Strik WK, Dierks T. Alterations of white matter connectivity in first episode schizophrenia. *Neurobiol Dis*. 2006;22(3):702–709.
43. Peters BD, Blaas J, de Haan L. Diffusion tensor imaging in the early phase of schizophrenia: what have we learned? *Journal of Psychiatric Research*. 2010;44(15):993–1004. doi:10.1016/j.jpsychires.2010.05.003.
44. Melicher T, Horacek J, Hlinka J, et al. White matter changes in first episode psychosis and their relation to the size of sample

studied: a DTI study. *Schizophrenia Research*. 2015;162(1–3):22–28. doi:10.1016/j.schres.2015.01.029.
45. Price G, Cercignani M, Parker GJM, et al. White matter tracts in first-episode psychosis: a DTI tractography study of the uncinate fasciculus. *NeuroImage*. 2008;39(3):949–955. doi:10.1016/j.neuroimage.2007.09.012.
46. Kawashima T, Nakamura M, Bouix S, et al. Uncinate fasciculus abnormalities in recent onset schizophrenia and affective psychosis: a diffusion tensor imaging study. *Schizophr Res*. 2009;110(1–3):119–126. doi:10.1016/j.schres.2009.01.014.
47. Seitz J, Zuo JX, Lyall AE, et al. Tractography analysis of 5 white matter bundles and their clinical and cognitive correlates in early-course schizophrenia. *Schizophrenia Bulletin*. 2016;42(3):762–771. doi:10.1093/schbul/sbv171.
48. Luck D, Malla AK, Joober R, Lepage M. Disrupted integrity of the fornix in first-episode schizophrenia. *Schizophrenia Research*. 2010;119(1–3):61–64. doi:10.1016/j.schres.2010.03.027.
49. Fitzsimmons J, Hamoda HM, Swisher T, et al. Diffusion tensor imaging study of the fornix in first episode schizophrenia and in healthy controls. *Schizophrenia Research*. 2014;156(2–3):157–160. doi:10.1016/j.schres.2014.04.022.
50. Szeszko PR, Robinson DG, Ashtari M, et al. Clinical and neuropsychological correlates of white matter abnormalities in recent onset schizophrenia. *Neuropsychopharmacology*. 2008;33(5):976–984. doi:10.1038/sj.npp.1301480.
51. Lee SH, Kubicki M, Asami T, et al. Extensive white matter abnormalities in patients with first-episode schizophrenia: a diffusion tensor imaging (DTI) study. *Schizophr Res*. 2013;143(2–3):231–238. doi:10.1016/j.schres.2012.11.029.
52. Fitzsimmons J, Schneiderman JS, Whitford TJ, et al. Cingulum bundle diffusivity and delusions of reference in first episode and chronic schizophrenia. *Psychiatry Res*. 2014;224(2):124–132. doi:10.1016/j.pscychresns.2014.08.002.
53. Whitford TJ, Kubicki M, Pelavin PE, et al. Cingulum bundle integrity associated with delusions of control in schizophrenia: preliminary evidence from diffusion-tensor tractography. *Schizophrenia Research*. 2015;161(1):36–41. doi:10.1016/j.schres.2014.08.033.
54. Peters BD, de Haan L, Dekker N, et al. White matter fiber tracking in first-episode schizophrenia, schizoaffective patients and subjects at ultra-high risk of psychosis. *Neuropsychobiology*. 2008;58(1):19–28. doi:10.1159/000154476.
55. Luck D, Buchy L, Czechowska Y, et al. Fronto-temporal disconnectivity and clinical short-term outcome in first episode psychosis: a DTI-tractography study. *Journal of Psychiatric Research*. 2011;45(3):369–377. doi:10.1016/j.jpsychires.2010.07.007.
56. Di Biase MA, Cropley VL, Baune BT, et al. White matter connectivity disruptions in early and chronic schizophrenia. *Psychol Med*. 2017;12:1–14. doi:10.1017/S0033291717001313.
57. Zhang R, Wei Q, Kang Z, et al. Disrupted brain anatomical connectivity in medication-naïve patients with first-episode schizophrenia. *Brain Struct Funct*. 2014. doi:10.1007/s00429-014-0706-z.
58. Li F, Lui S, Yao L, et al. Altered white matter connectivity within and between networks in antipsychotic-naive first-episode schizophrenia. *Schizophrenia Bulletin*. 2017. doi:10.1093/schbul/sbx048.
59. Clemmensen L, Vernal DL, Steinhausen H-C. A systematic review of the long-term outcome of early onset schizophrenia. *BMC Psychiatry*. 2012;12(1):150. doi:10.1186/1471-244X-12-150.
60. Tamnes CK, Agartz I. White matter microstructure in early-onset schizophrenia: a systematic review of diffusion tensor imaging studies. *J Am Acad Child Adolesc Psychiatry*. 2016;55(4):269–279. doi:10.1016/j.jaac.2016.01.004.
61. Karlsgodt KH, Niendam TA, Bearden CE, Cannon TD. White matter integrity and prediction of social and role functioning in subjects at ultra-high risk for psychosis. *Biol Psychiatry*. 2009;66(6):562–569. doi:10.1016/j.biopsych.2009.03.013.
62. Carletti F, Woolley JB, Bhattacharyya S, et al. Alterations in white matter evident before the onset of psychosis. *Schizophrenia Bulletin*. 2012;38(6):1170–1179. doi:10.1093/schbul/sbs053.
63. Clemm von Hohenberg C, Pasternak O, Kubicki M, et al. White matter microstructure in individuals at clinical high risk of psychosis: a whole-brain diffusion tensor imaging study. *Schizophrenia Bulletin*. 2014;40(4):895–903. doi:10.1093/schbul/sbt079.
64. Skudlarski P, Schretlen DJ, Thaker GK, et al. Diffusion tensor imaging white matter endophenotypes in patients with schizophrenia or psychotic bipolar disorder and their relatives. *Am J Psychiatry*. 2013;170(8):886–898. doi:10.1176/appi.ajp.2013.12111448.
65. Peters BD, Schmitz N, Dingemans PM, et al. Preliminary evidence for reduced frontal white matter integrity in subjects at ultra-high-risk for psychosis. *Schizophrenia Research*. 2009;111(1–3):192–193. doi:10.1016/j.schres.2009.03.018.
66. Bloemen OJN, de Koning MB, Schmitz N, et al. White-matter markers for psychosis in a prospective ultra-high-risk cohort. *Psychol Med*. 2010;40(8):1297–1304. doi:10.1017/S0033291709991711.
67. Katagiri N, Pantelis C, Nemoto T, et al. A longitudinal study investigating sub-threshold symptoms and white matter changes in individuals with an "at risk mental state" (ARMS). *Schizophrenia Research*. 2015;162(1–3):7–13. doi:10.1016/j.schres.2015.01.002.
68. Bohner G, Milakara D, Witthaus H, et al. MTR abnormalities in subjects at ultra-high risk for schizophrenia and first-episode schizophrenic patients compared to healthy controls. *Schizophrenia Research*. 2012;137(1–3):85–90. doi:10.1016/j.schres.2012.01.020.
69. Samartzis L, Dima D, Fusar Poli P, Kyriakopoulos M. White matter alterations in early stages of schizophrenia: a systematic review of diffusion tensor imaging studies. *J Neuroimaging*. 2014;24(2):101–110. doi:10.1111/j.1552-6569.2012.00779.x.
70. Canu E, Agosta F, Filippi M. A selective review of structural connectivity abnormalities of schizophrenic patients at different stages of the disease. *Schizophrenia Research*. 2015;161(1):19–28. doi:10.1016/j.schres.2014.05.020.
71. Tomyshev AS, Lebedeva IS, Akhadov TA, et al. MRI study for the features of brain conduction pathways in patients with an ultra-high risk of endogenous psychoses. *Bull Exp Biol Med*. 2017;162(4):425–429. doi:10.1007/s10517-017-3631-3.
72. Bernard JA, Orr JM, Mittal VA. Abnormal hippocampal-thalamic white matter tract development and positive symptom course in individuals at ultra-high risk for psychosis. *NPJ Schizophr*. 2015;1(1):npjschz20159. doi:10.1038/npjschz.2015.9.
73. Bernard JA, Orr JM, Mittal VA. Cerebello-thalamo-cortical networks predict positive symptom progression in individuals at ultra-high risk for psychosis. *NeuroImage: Clinical*. 2017;14:622–628. doi:10.1016/j.nicl.2017.03.001.
74. Drakesmith M, Caeyenberghs K, Dutt A, et al. Schizophrenia-like topological changes in the structural connectome of individuals with subclinical psychotic experiences. *Hum Brain Mapp*. 2015. doi:10.1002/hbm.22796.
75. Schmidt A, Crossley NA, Harrisberger F, et al. Structural network disorganization in subjects at clinical high risk for psychosis. *Schizophrenia Bulletin*. 2017;43(3):583–591. doi:10.1093/schbul/sbw110.
76. Prasad KM, Upton CH, Schirda CS, Nimgaonkar VL, Keshavan MS. White matter diffusivity and microarchitecture among schizophrenia subjects and first-degree relatives. *Schizophrenia Research*. 2015;161(1):70–75. doi:10.1016/j.schres.2014.09.045.
77. Zhou Y, Liu J, Driesen N, et al. White matter integrity in genetic high-risk individuals and first-episode schizophrenia patients: similarities and disassociations. *Biomed Res Int*. 2017;2017:3107845–3107849. doi:10.1155/2017/3107845.
78. Koivukangas J, Björnholm L, Tervonen O, et al. White matter structure in young adults with familial risk for psychosis—the Oulu Brain and Mind Study. *Psychiatry Res*. 2015;233(3):388–393. doi:10.1016/j.pscychresns.2015.06.015.
79. Boos HBM, Mandl RCW, van Haren NEM, et al. Tract-based diffusion tensor imaging in patients with schizophrenia and their non-psychotic siblings. *Eur Neuropsychopharmacol*. 2013;23(4):295–304. doi:10.1016/j.euroneuro.2012.05.015.
80. Kubicki M, Shenton ME, Maciejewski PK, et al. Decreased axial diffusivity within language connections: a possible biomarker of

schizophrenia risk. *Schizophrenia Research*. 2013;148(1–3):67–73. doi:10.1016/j.schres.2013.06.014.
81. de Leeuw M, Bohlken MM, Mandl RC, Hillegers MH, Kahn RS, Vink M. Changes in white matter organization in adolescent offspring of schizophrenia patients. *Neuropsychopharmacology*. 2017;42(2):495–501. doi:10.1038/npp.2016.130.
82. Collin G, Kahn RS, de Reus MA, Cahn W, van den Heuvel MP. Impaired rich club connectivity in unaffected siblings of schizophrenia patients. *Schizophrenia Bulletin*. 2014;40(2):438–448. doi:10.1093/schbul/sbt162.
83. Bohlken MM, Brouwer RM, Mandl RCW, et al. Structural brain connectivity as a genetic marker for schizophrenia. *JAMA Psychiatry*. 2016;73(1):11–19. doi:10.1001/jamapsychiatry.2015.1925.
84. Shi F, Yap P-T, Gao W, Lin W, Gilmore JH, Shen D. Altered structural connectivity in neonates at genetic risk for schizophrenia: a combined study using morphological and white matter networks. *NeuroImage*. 2012;62(3):1622–1633. doi:10.1016/j.neuroimage.2012.05.026.
85. Insel TR. The NIMH Research Domain Criteria (RDoC) project: precision medicine for psychiatry. *Am J Psychiatry*. 2014;171(4):395–397. doi:10.1176/appi.ajp.2014.14020138.
86. Thompson PM, Stein JL, Medland SE, et al. The ENIGMA Consortium: large-scale collaborative analyses of neuroimaging and genetic data. *Brain Imaging Behav*. 2014;8(2):153–182. doi:10.1007/s11682-013-9269-5.
87. Van Essen DC, Smith SM, Barch DM, Behrens TEJ, Yacoub E, Ugurbil K. The WU-Minn Human Connectome project: an overview. *NeuroImage*. 2013;80:62–79. doi:10.1016/j.neuroimage.2013.05.041.
88. Ho B-C, Andreasen NC, Ziebell S, Pierson R, Magnotta V. Long-term antipsychotic treatment and brain volumes: a longitudinal study of first-episode schizophrenia. *Arch Gen Psychiatry*. 2011;68(2):128–137. doi:10.1001/archgenpsychiatry.2010.199.
89. Bartzokis G, Lu PH, Nuechterlein KH, et al. Differential effects of typical and atypical antipsychotics on brain myelination in schizophrenia. *Schizophr Res*. 2007;93(1–3):13–22. doi:10.1016/j.schres.2007.02.011.
90. Ebdrup BH, Raghava JM, Nielsen MØ, Rostrup E, Glenthøj B. Frontal fasciculi and psychotic symptoms in antipsychotic-naive patients with schizophrenia before and after 6 weeks of selective dopamine D2/3 receptor blockade. *J Psychiatry Neurosci*. 2016;41(2):133–141. doi:10.1503/jpn.150030.
91. Reis Marques T, Taylor H, Chaddock C, et al. White matter integrity as a predictor of response to treatment in first episode psychosis. *Brain*. 2014;137(Pt 1):172–182. doi:10.1093/brain/awt310.
92. Cropley VL, Klauser P, Lenroot RK, et al. Accelerated gray and white matter deterioration with age in schizophrenia. *Am J Psychiatry*. 2017;174(3):286–295. doi:10.1176/appi.ajp.2016.16050610.
93. Koutsouleris N, Davatzikos C, Borgwardt S, et al. Accelerated brain aging in schizophrenia and beyond: a neuroanatomical marker of psychiatric disorders. *Schizophrenia Bulletin*. 2014;40(5):1140–1153. doi:10.1093/schbul/sbt142.
94. Kochunov P, Glahn DC, Rowland LM, et al. Testing the hypothesis of accelerated cerebral white matter aging in schizophrenia and major depression. *Biol Psychiatry*. 2013;73(5):482–491. doi:10.1016/j.biopsych.2012.10.002.
95. Wu C-H, Hwang T-J, Chen Y-J, et al. Primary and secondary alterations of white matter connectivity in schizophrenia: a study on first-episode and chronic patients using whole-brain tractography-based analysis. *Schizophrenia Research*. 2015;169(1–3):54–61. doi:10.1016/j.schres.2015.09.023.
96. Filippi M, Canu E, Gasparotti R, et al. Patterns of brain structural changes in first-contact, antipsychotic drug-naive patients with schizophrenia. *American Journal of Neuroradiology*. 2014;35(1):30–37. doi:10.3174/ajnr.A3583.
97. Kyriakopoulos M, Perez-Iglesias R, Woolley JB, et al. Effect of age at onset of schizophrenia on white matter abnormalities. *Br J Psychiatry*. 2009;195(4):346–353. doi:10.1192/bjp.bp.108.055376.
98. Kanaan RAA, Borgwardt S, McGuire PK, et al. Microstructural organization of cerebellar tracts in schizophrenia. *Biol Psychiatry*. 2009;66(11):1067–1069. doi:10.1016/j.biopsych.2009.07.028.
99. Bijanki KR, Hodis B, Magnotta VA, Zeien E, Andreasen NC. Effects of age on white matter integrity and negative symptoms in schizophrenia. *Schizophrenia Research*. 2015;161(1):29–35. doi:10.1016/j.schres.2014.05.031.
100. Shahab S, Stefanik L, Foussias G, Lai M-C, Anderson KK, Voineskos AN. Sex and diffusion tensor imaging of white matter in schizophrenia: a systematic review plus meta-analysis of the corpus callosum. *Schizophrenia Bulletin*. 2017. doi:10.1093/schbul/sbx049.
101. Jbabdi S, Sotiropoulos SN, Savio AM, Graña M, Behrens TEJ. Model-based analysis of multishell diffusion MR data for tractography: how to get over fitting problems. *Magn Reson Med*. 2012;68(6):1846–1855. doi:10.1002/mrm.24204.
102. Wedeen VJ, Hagmann P, Tseng W-YI, Reese TG, Weisskoff RM. Mapping complex tissue architecture with diffusion spectrum magnetic resonance imaging. *Magnetic Resonance in Medicine*. 2005;54(6):1377–1386. doi:10.1002/mrm.20642.
103. Lingala SG, DiBella E, Jacob M. Deformation corrected compressed sensing (DC-CS): a novel framework for accelerated dynamic MRI. *IEEE Trans Med Imaging*. 2015;34(1):72–85. doi:10.1109/TMI.2014.2343953.
104. Calamante F, Tournier J-D, Heidemann RM, Anwander A, Jackson GD, Connelly A. Track density imaging (TDI): validation of super resolution property. *NeuroImage*. 2011;56(3):1259–1266. doi:10.1016/j.neuroimage.2011.02.059.
105. Malcolm JG, Shenton ME, Rathi Y. Filtered multitensor tractography. *IEEE Trans Med Imaging*. 2010;29(9):1664–1675. doi:10.1109/TMI.2010.2048121.
106. Zhan W, Stein EA, Yang Y. Mapping the orientation of intravoxel crossing fibers based on the phase information of diffusion circular spectrum. *NeuroImage*. 2004;23(4):1358–1369. doi:10.1016/j.neuroimage.2004.07.062.

29

FUNCTIONAL CONNECTIVITY BIOMARKERS OF PSYCHOSIS

Godfrey Pearlson and Michael Stevens

INTRODUCTION

In general, psychiatry is plagued by heterogeneity within and overlap across conventional diagnostic categories and currently lacks valid biological diagnostic tests. More specifically among psychotic syndromes there is well-documented multilevel overlap (e.g., for treatment response, clinical symptomatology, familial liability, risk genes). Only recently have researchers begun to examine brain-related biological findings across different diagnoses with psychosis to address classification-related questions. At the present time, the majority of studies in psychosis has compared schizophrenia patients to healthy controls and does not yet extend across the spectrum of psychosis.

An important focus of neuroscientific inquiry in psychosis involves examining distributed brain networks. This includes functional brain connectivity measures derived from functional MRI (fMRI)—both task-based and resting-state, which provide distinct but usually complementary information about how the brain regions that form neural systems are configured, engage together, and influence each other in the service of different types of information processing. In step with emerging recognition of the importance of network connectivity to cognition, a predominant working hypothesis is that disrupted connectivity in specific, identifiable functional brain circuits underpins psychosis phenomenology. This hypothesis has been articulated in various ways by different researchers. Although most theoretical formulations emphasize the importance of synaptic efficacy, neuroplasticity, or modulation on neural system development or dysfunction,[1,2] few studies of the much broader psychosis dysconnectivity hypothesis actually focus on those levels of inquiry. Instead, over the past decade the idea of dysconnectivity (i.e., abnormal connectivity, as opposed to disconnectivity, a term implying absent connections), has been tested primarily with fMRI-based neuroimaging tools and fMRI time-series analysis methods. An obvious question in psychosis is whether presumed abnormal functional connectivity applies equally to all cortical networks as suggested by Paul Meehl[3] or is localized to particular functional circuits or networks. Examples could include localization to either the default mode network (DMN) or other task-negative circuits that are characteristically most active when the brain is not engaged in performance of goal-directed tasks, versus the task-positive networks that characteristically increase activity during cognitive processing. These latter neocortical circuits include the fronto-parietal and cingulo-opercular networks, whose activity is typically anticorrelated with that of DMN.

Currently, dysconnection models generally tend to be rather nonspecific or not fully articulated. Many studies have a "black box" aspect, because they focus on the systems level without empirical or direct conceptual reference to the underlying factors that might give rise to any profile of abnormal dysconnection measured by fMRI. Nor do such studies typically account for complexities such as that any given neuron can be influenced by multiple neurotransmitters in ways difficult to model through conventional approaches.

But using fMRI to describe the profile of distributed circuit dysconnectivity has not been unproductive. Indeed, the key point of this chapter is to argue that MRI-based research to date has been essential to move the field to a point where it is just now becoming feasible to address the more salient issues of relating fMRI-measured dysconnection to their underlying physiological, cellular, and genetic etiologies. Indeed, in a recent, promising line of inquiry, Stephan and others have employed computational modeling approaches to explore inputs and outputs to yield testable predictions to both MRI levels of inquiry and other neural measurements after administering neuropharmacologic probes.[4,5] However, in order to know the most important questions to ask, the field first has grappled with isolating and defining what aspects of complex brain network functional connectivity relate to key factors in psychosis. Relevant issues include whether different DSM-based psychotic illnesses show similar or different abnormal functional connection patterns (syndrome specificity), the state versus trait nature of these markers (i.e., status as illness- versus illness-risk markers) and their possible heritability (i.e., are they endophenotypes) across diagnoses. Additional issues are their relationship to various drug treatments (i.e., are they treatment artifacts) that often show differential association with diagnosis.[6] Thus, the most promising research questions explored to date have involved understanding how neural system connectivity as measured by fMRI relates to other observations in across psychosis-related clinical phenomenology.

Arguably it is important to better understand how the developmental nature of brain dysconnectivity in psychotic illnesses may relate to typical symptoms in a way that is not characteristic of other processes that also alter structural and functional connectivity with subsequent problems in functional network inefficiency, for example those following adult traumatic brain injury.[7] From this foundation, neuroscientists can begin to disentangle how alterations at the gene level might lead to changed microcircuit and cell-to-cell communication, how such alterations are reflected in the large-scale neural system differences assessed through fMRI (as discussed here) or EEG, and how these physiological differences translate ultimately into psychiatric symptoms and syndromes.

As important as these questions are to current psychosis neuroscience, their impact on our understanding of psychosis as a dysconnectivity syndrome must be viewed in context of the limitations of neuroimaging functional connectivity methods. Before summarizing and synthesizing fMRI-generated evidence for psychosis and schizophrenia dysconnectivity, we start with a strong caveat of interpretation. Above all, it is important to always recall that the statistical relationships and dependencies observed with fMRI data are only a functional "snapshot" of something we believe to represent fundamental properties about how the brain is wired, and how various network nodes communicate with each other. In reality, few of those questions yet have detailed answers. While we know that there are canonical, commonly found brain networks such as frontoparietal systems that engage when cognitive control is needed, reward system regions that engage when learning takes place, or even the much examined "default mode network" whose functional relevance and place within neurocognitive models of various psychiatric disorders is the subject of active, intense debate. But our accumulated understanding of this ever-growing body of connectivity research starts and stops where we can describe various ways in which various patient groups with psychosis show connectivity-related abnormalities in these putative functional systems. Only recently have researchers studied the relationship between connectivity profiles and factors that directly influence human brain synaptic function (e.g., glutamate challenge via ketamine versus whole-brain connectivity). Ketamine induces a robust whole-brain connectivity pattern that can be differentially modulated by drugs of different mechanism and clinical profile.[8] Overall, commonly used fMRI connectivity measurements have no direct relationship to any such factors and have to be determined through painstaking experimentation. Similarly, Schmidt and colleagues[6] discuss how abnormal frontoparietal connectivity may derive from abnormal N-methyl-D-acetylase (NMDA) receptor-mediated plasticity, while other reviews, e.g., Khalili-Mahani,[9] discuss more general principles of pharmacological manipulations of resting-state fMRI. Research also has shown default mode and cingulo-opercular networks are both probably influenced by dopamine. So if central dopamine abnormalities are dysfunction, this would be a general feature that affects the whole brain. Despite these piecemeal advances in our understanding, fMRI can best be described as providing a useful "keyhole" through which to observe a limited, if highly informative, measurement of brain network function with restricted biological specificity and unknown underpinnings. Nevertheless, it is a well-suited vehicle for synthesizing what we know about the relationship of distributed network function to clinically relevant phenomenology.

The MRI methods that we discuss in what follows have all been applied to dysconnectivity in psychiatric disorders. Therefore we briefly review the basis of fMRI connectivity measurements and describe the most frequently used approaches applied to psychosis before detailing some key findings. Brain activity is correlated in both space and time and occurs on a millisecond scale. Temporally, there are recurrent, predictable fluctuations ranging from thousandths to hundreds of Hz that can be linked either between different sources within a specified frequency or across different frequencies (cross-frequency coupling). Spatially, activities are correlated across disparate brain regions that "switch" on or off synchronously to form independent networks such as the default mode, cingulo-opercular, etc., networks. Both temporal and spatial abnormalities have been shown repeatedly in schizophrenia and other psychotic disorders. Together, these complex spatiotemporal oscillations can be parameterized to characterize either highly localized or widely distributed brain connectivity profiles depending on the technique used. For example, summary measurements can be derived within circuits (functional connectivity; FC), pairwise across different circuits (functional network connectivity; FNC), or moment-to-moment across circuits (dynamic functional connectivity). FMRI operates on a scale of seconds, assessing an indirect proxy of brain function tied to blood-oxygenation levels. As such, it is removed in both time (a thousand times more coarse than actual brain activity) and in terms of biological mechanism. However, it remains of interest because convincing evidence has been accumulated to link neuronal firing patterns in distal regions and hemodynamic connectivity. Moreover, the same networks are conserved across the brain in terms of task activity coactivation, fMRI timeseries cross-correlation and related methods, and even structural covariance to convince researchers that canonical network exist that can be tied to specific cognitive functions.

To avoid redundancy in describing many of these technical and conceptual issues in detail, the reader is referred to a recent overview of resting-state fMRI in neuropsychiatric disorders[10] to read alongside this chapter. That paper summarizes and tabulates major issues, concepts, and fundamentals, including definitions of major regions and circuits involved, the utility of resting-state fMRI imaging studies, optimal methodologic acquisition and analysis approaches, associated problems and pitfalls of such studies in neuropsychiatric populations, relationships between structural and functional connectivity, and recommended key features that should be employed to evaluate scientific reports in this topic area. This can be usefully read in conjunction with a second review article examining resting-state and task-based FC from a developmental perspective in the context of adolescent brain network maturation.[11]

RESTING-STATE CONNECTIVITY

The term "resting state" is most commonly used to describe a task-free fMRI timeseries, collected typically when patients lie awake and alert to the MRI, but are not instructed to engage cognitively. However, examining fMRI signal's spatial and temporal coherence across brain regions during both resting state and during task activation reveals topologically similar circuits that display functional coherence (i.e., correlations in spontaneous activity, usually at low frequencies). Thus while FC varies between task and rest, it reorganizes in a predictable manner between the two states, while keeping identifiable, flexible distributed network characteristics, allowing efficient responses to a changing cognitive environment. Of note, "resting state" is often mistakenly interpreted as representing a pure task-absent baseline. While one can gain useful connectivity information from both resting state and active tasks, and while contrasting them directly can be useful in some contexts, it should always be kept in mind that "resting state" is merely an uncontrolled state of active consciousness and subject to numerous interpretive caveats. Resting-state studies are especially sensitive to intrinsic and extrinsic noise from such factors as cardiac and respiration rates, subject head movement, test-related anxiety, tobacco and caffeine consumption, prescribed medication, and recreational drug use. Since all of these are likely to vary systematically between psychosis patients and controls, there is a risk that not only will such confounds interfere with data interpretation but also they may be misinterpreted as actual primary disease-related brain differences. Strategies to minimize these issues include studying unaffected close relatives sharing genetic risk but not taking antipsychotic or mood stabilizing medications, assessing at-high-risk unmedicated subjects. Many of these were captured in a point-counterpoint discussion initiated by Weinberger and Radulescu's[12] paper, and responded to by Calhoun, Glahn, and Pearlson.[13] We spend more time discussing resting state here because there are far more studies of rs-FC MRI than of task-based FC. This is attributable in part to the brevity and ease of collection of resting-state measures, their straightforward analysis (e.g., through seed-based or independent component analysis-based approaches) and avoidance of task administration in a population that is frequently cognitively impaired (and thus experiences difficulties in learning cognitive paradigms), distractible, and inclined to move more in the scanner. Most resting-state studies of psychosis examined the DMN in this context among the various intrinsic networks that could be chosen, in part due to its relative ease of identification and study. Similar considerations apply to the study of DMN during task-based functional studies. As a caveat, because there are multivariate associations between multimodal imaging phenotypes and behavioral, clinical lifestyle and health-related measures including age, alcohol use, and body mass index independent of psychiatric diagnosis,[14] such factors need to be modeled, along with in-scanner movement to avoid confounding measures that may vary systematically between psychiatric and healthy control samples.

MAJOR RESTING-STATE FINDINGS: FUNCTIONAL CONNECTIVITY AND FUNCTIONAL NETWORK CONNECTIVITY

Assessment of functional connectivity patterns in the resting state yields temporally coherent activations arrayed in highly reproducible resting-state networks (RSNs). The 2015 Lo paper[15] is important because it compared classic resting-state FC measures of brain networks to graph theoretical analyses of network connections (using small-world topology measures such as clustering, efficiency homogeneity, and randomness). Compared to controls, schizophrenia patients had significantly reduced FC as well as alterations in all global topology metrics. Unaffected relatives of schizophrenia patients showed similar, lesser findings than patients, suggesting that these abnormalities represent familial risk not explained by antipsychotic treatment. A recent critical comprehensive review of such resting-state studies in schizophrenia[16] concludes that resting-state disruptions are related to positive and negative symptoms and cognitive manifestations of the disorder.

Other papers have looked across multiple resting state networks. Abnormalities in both resting-state FC and FNC in psychosis subjects in the Bipolar-Schizophrenia Network for Intermediate Phenotypes (B-SNIP) study have been demonstrated.[17–20] Khadka[20] reported that multiple resting-state components in schizophrenia and psychotic bipolar subjects displayed aberrant FC compared to healthy controls. Some of these were unique to schizophrenia probands, some were shared between the illnesses, and also abnormal in unaffected relatives of probands, as previously identified in DMN.[21] The study thus revealed both unique and overlapping within-network resting-state FC brain connectivity abnormalities in these disorders. A related report from the same group[17] examined resting-state FNC in many of the same subjects, reporting that schizophrenia and psychotic bipolar probands shared several abnormal RSN connections, some of which were correlated with key symptoms, but networks additionally displayed unique abnormalities that distinguished the disorders. Two of the aforementioned network pairs were also abnormal in unaffected relatives of bipolar probands. Notably a paralimbic network that was abnormal only in bipolar probands contained multiple mood-related regions including amygdala, parahippocampal gyrus, hippocampus, mesial temporal cortex, insula, subgenual cingulate, and ventrolateral prefrontal cortex. Overall, as described by many others for schizophrenia,[16,22] abnormal relationships between the DMN and task-positive networks characterize psychosis more broadly.

Heritability of synchronous activity in brain networks and of DMN activity specifically[24] have been shown in healthy individuals. In a study of over 1,300 individuals consisting of patients, their unaffected relatives and controls from the B-SNIP consortium,[18] analysis of resting-state data identified three DMNs, all of which manifested reduced connectivity in patient groups, that correlated with symptom severity, and one of these networks was similarly abnormal in relatives of schizophrenia patients. Further analysis in probands and controls only revealed five sub-DMNs that correlated significantly with an equal number of multi-single nucleotide polymorphism genetic networks governing such processes as in NMDA-related long-term potentiation, axon guidance, synaptogenesis, protein kinase A, and immune response signaling. One interpretation of dynamic FC assessments and psychosis is that the clinical relevance of FC organization is best exemplified by altered brain state dynamics, as emphasized by Cohen.[25]

A number of additional resting-state experimental designs and analyses have been employed in psychotic illnesses.

REGIONAL HOMOGENEITY MEASURES

Regional homogeneity (ReHo) is a voxel-based brain measure assessing local (often voxel-by-voxel) functional brain connectivity and synchronization of a given voxel and its nearest neighbors, using Kendall's coefficient of concordance as an index of similarity, and is presumed to represent the degree of local synchronization of fMRI time courses or regional integration of information processing.[26]

As such it has been presumed to yield more specific information about the similarity of brain activity between more discrete regional brain regions or clusters than either ICA- or region-of-interest/seed-based assessments. Viewed from another perspective, altered ReHo reflects in imbalance in local functional connectivity. ReHo abnormalities have been reported in both schizophrenia and bipolar illness[26,27] as well as in other neuropsychiatric conditions, and as potential biomarkers for psychosis risk[28] and adolescent-onset schizophrenia.[26] Adolescent-onset, drug-naïve schizophrenia patients compared to healthy controls exhibited significantly increased ReHo values in both superior medial prefrontal cortices, and decreases in left superior temporal gyrus and right precentral and inferior parietal regions. Support vector machine analysis using a combination of ReHo value selected from three regions accurately discriminated patients and controls with high sensitivity and specificity.[26]

THE "CHRONNECTOME"—QUANTIFYING DYNAMIC FUNCTIONAL CONNECTIVITY

Because current analytic approaches average across the scan acquisition, they ignore potentially important short-term, dynamic connectivity properties that may vary temporally in predictable ways during the assessment period. Calhoun and colleagues[29] published methods to quantify mutually informed or statistically correlated regional brain activity across all RSNs that change or evolve over time, using sliding time windowing or wavelet-based approaches.[25] Preliminary evidence suggests that there exist not only short-lived time-varying connection patterns in the realm of tens of seconds,[30] but also various dynamic functional connectivity (DFC) "states" that are stable and reproducible, and whose metrics are significantly different in schizophrenia compared to healthy controls.[30–33] Du[30] derived whole-brain, region-of-interest-based DFC measures in psychosis patients including those with schizophrenia, bipolar, and schizoaffective disorders as well as healthy controls, finding that DFC measures in general were unrelated to current medication status and were more numerous and informative, better discriminated among diagnostic groups than did traditional static FC measures, and were also more closely related to active clinical symptoms. Many of these states varied widely among diagnostic groups and on several measures DSM psychotic disorders were arrayed across a severity spectrum with schizophrenia being most abnormal and, bipolar disorder, least. Lottman[32] examined DFC measures in schizophrenia patients in the unmedicated state and following 1 and 6 weeks of risperidone treatment, compared to matched healthy controls. Using traditional static connectivity measures unmedicated schizophrenia patients showed the expected increased and decreased connectivity between several different independent components compared to controls. With dynamic measures, patients were characterized by significantly increased connectivity between thalamus and sensorimotor network in one state, and by short-term mean dwell times and fraction of time spent in the sparsely connected state, but the reverse for both of these through all measures in the intermediately connected state. Antipsychotic treatment normalized mean dwell times. The authors concluded that patients' static FC differences may result less from persisting functional dysconnections than from partially modifiable alterations in temporal dynamics of brain networks. Finally, Miller,[33] in a large (N > 300) dataset, detected several differences between schizophrenia patients and controls in both temporally variant joint functional-domain brain connectivity and in cross joint functional-domain intertemporal information flow.

Finally, slow fluctuations in intrinsic activity (low-frequency oscillations) characterize the resting brain, and can be quantified by measuring either the *amplitude of low frequency fluctuations (ALFF)* or their *fractional amplitude (fALFF)* of resting state. These measures do not represent true connectivity estimates, although conceptually connected. They have been explored across psychosis, e.g., Meda,[19] to compare traditional across-DSM psychotic disorders as well as empirically biologically based transdiagnostic psychosis subtypes, termed "Biotypes," from the B-SNIP sample.

RESTING-STATE VERSUS TASK-BASED FUNCTIONAL CONNECTIVITY

While FC differs between rest and task (e.g., alterations in within-network connectivity and across communication hubs, reflected in overall global integration) it is currently unclear whether data derived from these two assessment approaches reflect the same types of short-term underlying events.[34] Trade-offs between task- versus non-task-based paradigms for the purpose of studying psychotic disorders have been previously reviewed in depth in Pearlson and Calhoun.[35] The majority of task-based studies challenge psychosis subjects by employing "cognitive stress tests" on which these patients perform poorly, and that may relate to underlying theoretical construct, e.g., dopaminergic "tuning" of prefrontal circuitry. Most brain imaging researchers have chosen among cognitive domains known to be impaired in patients with psychosis, such as working memory or attention. Classical fMRI cognitive tasks and resting state exist on a continuum; such tasks as auditory oddball paradigms are sufficiently straightforward and show accuracy rates that are comparable between psychotic patients and controls, yet with markedly different brain responses. Patients can differ from healthy controls when switching between task-related and nontask paradigms, suggesting strategies using information from both designs simultaneously.

Problems with task-based connectivity approaches include the fact that patients may be unmotivated, sedated and fatigued, distracted by illness symptoms such as hallucinations, and challenged in retaining complex instruction sets, and may have their performance further undermined by impaired sustained attention. As task difficulty levels become more demanding, psychotic patients may become increasingly unmotivated and demoralized. This combination inevitably leads to problems in disentangling impaired performance from the associated abnormal task-related BOLD response. Various solutions to these inherent problems have included selecting healthy controls that perform behaviorally as poorly as patients on the study paradigm, and choosing minimal-effort tasks such as oddball discrimination where patient/control performance discrepancies are minimal. Some investigators select paradigms that lend themselves readily to parametric designs (e.g., Sternberg working memory task), thus allowing group behavioral and BOLD comparisons across a range of task difficulty. Sheffield and Barch[22] suggest that future studies could minimize much of the ambiguity that plagues current task-related studies in schizophrenia both by adopting standardized cognitive tasks and by assessing multiple cognitive domains within individual studies. In addition, fMRI connectivity studies using active task probes suffer from the same analytic contamination found in resting-state data (e.g., physiological confounds from cardiac and respiratory signals, head movement, context-specific anxiety, or recent nicotine and caffeine consumption or medications). While all these issues need to be controlled and/or explicitly examined in each study, many of these issues can best be addressed or circumvented by assessing patients with multiple psychosis diagnoses within the same study.

MAJOR FINDINGS FROM TASK-BASED FUNCTIONAL CONNECTIVITY STUDIES OF PSYCHOSIS

Birur and collaborators[36] reviewed 13 task-based fMRI papers in chronically medicated schizophrenia compared to bipolar patients. Three of those studies specifically denoted their bipolar subjects as belonging to the psychotic subtype, and one examined both psychotic and nonpsychotic bipolar patients. Due in part to the wide variety of task paradigms employed in these studies, it is difficult to draw general conclusions, e.g., even the studies that used the N-back task reported few similar findings. The same review examined connectivity across 16 resting-state fMRI studies; 9 of these included psychotic bipolar individuals of which 4 included both psychotic and nonpsychotic subjects. A number of these papers are particularly informative. Anticevic and colleagues[37] addressed the issue of specificity of the presence of psychosis and in bipolar disorder. They analyzed resting-state fMRI data using a ventral anterior cingulate seed to reveal connectivity of the dorsal medial prefrontal region. Bipolar patients lacking a history of psychosis showed increased coupling of these regions compared to healthy controls, while an opposite finding of reduced regional connectivity characterized both schizophrenia and psychotic bipolar patients (who were not significantly different from each other).

INTEGRATION: A CRITICAL SUMMARY OF FUNCTIONAL CONNECTIVITY PUBLICATIONS IN SCHIZOPHRENIA AND PSYCHOTIC MOOD DISORDERS

Sheffield and Barch[22] review both the resting-state and cognitive task-based FC literatures in schizophrenia comprehensively; this is also done on a smaller scale by Mwansisya.[38] They conclude that schizophrenia patients display both functional connectivity and functional network connectivity abnormalities across two major groups of circuits: cortical task-negative and task-positive networks and the cortico-cerebellar-striatal-thalamic loop. They further conclude that these two types of functional connectivity abnormalities are not differentially expressed relative to particular cognitive abilities, but are rather nonspecific and generalized in nature, consistent with the more globally expressed cognitive deficits detected in schizophrenia. In turn, work from the B-SNIP group[39] has shown a relative lack of diagnostic specificity for cognitive deficits across different DSM psychosis diagnoses, consistent with the broad similarities of FC and FNC abnormalities across the same syndromes. Sheffield and Barch[22] raise the interesting point that cognitive impairment in schizophrenia, including abnormal attention, working memory, processing speed, and executive functioning, may be related to abnormal FC within and between task-positive and task-negative circuits (also see Repovs,[40] Unschuld[41]). They argue that this in turn is part of a larger-scale, dynamic communication breakdown involving disrupted FC between multiple cortical areas (in particular prefrontal cortex), thalamus, cerebellum, and striatum. Mwansisya[38] places more emphasis on overlapping task and resting-state fMRI abnormalities in prefrontal regions (including dorsolateral and orbitofrontal regions) and in the left superior temporal gyrus, consistent with earlier work by Pearlson and colleagues in heteromodal association cortex.[42]

The preceding discussion leads us to another broad question—that of the relationship of circuit connectivity abnormalities to psychosis symptoms. This is an especially difficult area to study because of confounds related to widely different methodologic approaches, treatment effects, and heterogeneous patient populations. As noted by Karbasforoushan and Woodward,[43] obtaining general inferences from the FC literature is difficult due to heterogeneity in methodologic approaches of study design and analysis, sample sizes, and definitions of clinical groups (e.g., psychotic versus nonpsychotic bipolar distinctions). Nevertheless, a number of research groups have made valiant attempts to explore the area of symptomatic relationships. For example Northoff et al.[44,45] review abnormalities in FC (spatial) and coupling between different frequency fluctuations (temporal connectivity) in this context. They argue that the poor performance of schizophrenia patients in temporal prediction and judgments of temporal order and simultaneity are likely linked to abnormal temporal connectivity such as disturbed ALFF. In addition they posit links between early sensory processing in schizophrenia and altered cross-frequency coupling, as well as the latter and cognitive symptoms. For example, global temporal coordination abnormalities could be associated with generalized cognitive deficits. Finally they argue that typical psychotic symptoms might also be explained by an alteration of spontaneous activity altering the direction of the relationship between the DMN and central executive network, ultimately leading to "a mixing of internally and externally oriented thoughts." Addressing the question of auditory verbal hallucinations more specifically, Ćurčić-Blake[46] review FC both during occurrence of active auditory hallucinations in the resting state as well as task-based differences in individuals who do or do not have such experiences. They conclude that the former type of studies ('symptom capture') generally report FC increases among auditory, language and basal ganglia regions, while the latter reveal variable, illness phase-related, altered inter-hemispheric connectivity between language association areas. Northoff[47] explored possible relationships between auditory hallucinations and abnormally increased resting-state activity within auditory cortex as further modulated (again abnormally) by the DMN, resulting in confusion between intrinsic activity versus that normally evoked by external auditory stimuli. In addition, Alderson-Day[48] linked disruption of resting connectivity in the left superior temporal gyrus to the occurrence of auditory verbal hallucinations, although with a modest replication across studies. This localization is consistent with earlier reports of gray matter volume reductions in the same region, first reported by Barta.[49] Finally Robinson[50] explored relationships between sense of agency in schizophrenia and resting-state alterations in the brain's hypothesized self-referential regions.

ARE THERE KEY CIRCUITS?

ROLE OF THE THALAMUS

Several groups have hypothesized that the thalamus may play a central role in psychosis-related dysconnectivity. Its many heterogeneous, topographically arrayed nuclei have diverse inputs and multiple connections to cortex. These diverse circuits subserve a wide variety of neural functions, from motor output to emotion processing, suggesting that thalamic pathology could provoke widespread disarray in functional connectivity.[51–53] A recent review[54] reevaluates the role of thalamus from one limited primarily to sensory transmission to the cortex, to include nonsensory, complex, ongoing, bidirectional interactions between frontal cortex and thalamus expressed in varied behavioral contexts. Among other purposes, this fronto-thalamic connectivity serves to underpin, maintain, and update persisting frontal activity in a manner consistent with representing the external environment as is needed for performing such diverse tasks as working memory and motor planning. Giraldo-Chica and Woodward[55] critically review the schizophrenia resting-state fMRI literature, also paying attention to bipolar disorder. Because the thalamus is relatively small and distinctly variegated in nature, accurate, reproducible parcellation has proven challenging. Woodward[56] reported reduced connectivity between prefrontal cortex and thalamus, but increased thalamic connectivity to somatosensory and motor cortex. Anticevic[57] used a panthalamic, rather than a cortical seed, reproducing above connectivity finding, and extending it to observations of thalamic hyperconnection to additional cortical sensory regions. Overall, thalamo-cortical FC abnormalities have been demonstrated in psychosis patients at all stages, and occur in both schizophrenia and bipolar disorder,[58,57] but are not clearly related to symptoms.[55] Many studies have focused on one or another single thalamic nucleus, making problematic any generalizable statements about the thalamus or agreement along findings to date. Murray and Anticevic[59] review this literature in general, and advocate for a computational model-based approach that generates predictions for results of specific thalamic disorganization in terms of larger-scale effects on reciprocal sensory versus associative thalamo-cortical connections. The latter group[53] builds on this idea in experiments using ketamine, the NMDA receptor antagonist, in a glutamatergic challenge paradigm.

ROLE OF FRONTO-PARIETAL CONNECTIVITY

An alternative hypothesis with much supporting evidence is that fronto-parietal network connectivity is disturbed both in the resting state and in task-related fMRI, e.g., during working memory processing. Schmidt et al.[6] review much of the recent evidence for fronto-parietal network dysconnectivity across patients with chronic psychoses, first-episode patients, individuals at clinical or genetic high-risk for psychosis, and non-clinically identified populations with psychosis-like experiences (e.g., "voice-hearers"). They produce a convincing argument that fronto-parietal connectivity is disturbed during both resting states and working memory processing (with an emphasis on graph theory-based analyses) in all of the earlier-mentioned individuals, thus representing a psychosis risk endophenotype.

WHERE WE ARE AND WHERE DO WE GO FROM HERE?

Are psychoses diseases of connectivity, or is disrupted connectivity a conceptual or observational waypoint best viewed as an epiphenomenon removed from the actual disease process, and thus a macroscopic consequence of more fundamental underlying molecular or cellular pathology? Romme et al.[60] review how in schizophrenia white matter disconnections are especially prevalent in those cortical regions with notable expression of schizophrenia-related risk genes. This gene-related white matter connectivity presumably underpins the structurally related functional dysconnectivity reported by Skudlarski and others. [61] As noted by Arikkath and Mirnics,[62] the genetic liability plausibly interferes with early brain development via changed gene expression during crucial developmental periods. Or, circuit-specific dysfunction might be related to other, as yet unknown features of how microcircuits are formed in the process of building a brain from a combination of gene-guided and experiential influences over time. For example, Richmond et al.[63]review how from a graph-theoretical perspective, genetic and environmental factors influence the development of functional and structural brain networks from infancy through adolescence. Ultimately, micro-level elements at the cellular level coordinate dynamically to produce the macro-level BOLD signals that constitute RS-fMRI data. Biophysical investigations in non-human primates, e.g. Chen et al.,[64] suggest that low-frequency correlations are a widespread emergent property of neural systems that exist even in the primate spinal cord. Resting-state connectivity spatial patterns derived from such experiments genuinely represent FC between brain regions and are strongly associated with other FC data-types (e.g., as assessed by electrophysiology), underpinning the validity of resting-state fMRI as representing FC.

Currently, the field of FC biomarkers in psychosis is essentially descriptive, seeking patterns derived from fMRI, but lacking a basic understanding of the underlying pathophysiology. Investigators will need to move beyond these simpler paradigms toward a full integration with connectivity-based neuroscience, perhaps through biological manipulation (e.g., paradigms involving behavioral training or pharmacologic manipulation) to probe brain systems more directly. In addition, studies can assess average task performance across a diversity of cognitive tasks. In parallel, these investigations need to move beyond schizophrenia to encompass other psychotic disorders. As noted in the introduction to this chapter, the field's ultimate aim is to trace the path from psychosis risk genes, to developmentally altered molecular and neurobiological systems such as cerebral cell migration patterns, via an understanding of cerebral micro-circuits, through physiological connectivity measurements such as those depicted by fMRI, and finally to clinical and cognitive observations and to allied subjective patient self-reports. Elucidating these relationships will ultimately help in our understanding the underlying basis of dysconnectivity—whether this is structural in nature (e.g., Skudlarski[61]), and/or based in a biochemical/neurotransmitter abnormality, such as the NMDA-receptor or glutamate release.[6,16] Therefore we need more studies that assess clinical phenotypes together with biological variables from different domains, including both structural and functional brain imaging together with genetic risk factors in the context of childhood and adolescent brain system development[11,65]. Obviously, such investigations are necessarily large-scale and expensive, so that funding them devolves in part to national policy-related budgetary priorities. Another important question is whether certain critical brain circuits or important nodes are specifically implicated in psychotic disorders, or whether dysfunction is widely and equally distributed. Some of these questions are explored in a recent editorial by Alexander-Bloch[66] in relationship to connectomic measures within brain hub or nodal regions. A connectome-based approach to functional abnormalities in schizophrenia and bipolar disorder has been particularly helpful in summarizing how network abnormalities occur at both the large and local scales, and static and dynamic contexts.[67]

The role of genetic influences on brain development and maintenance of neural circuits needs further explication[24] as does further understanding of the endophenotype status of specific dysconnectivity patterns. Thompson et al.[68] suggest an approach for elucidating how genetic factors influence brain connectivity, considered both structurally (using diffusion MRI) and at the level of functional networks. Several of the studies reviewed earlier[17–20],[21,69] show that some of these resting-state measures are present in unaffected first-degree relatives, either within or across psychotic disorders, suggesting that at least a proportion of FC and FNC abnormalities represent predispositions to psychotic disorders, rather than only biomarkers of disease status, i.e., state versus trait phenomena.[35] Such studies also help rule out potential disease-related confounds, such as secondary medication effects on brain connectivity or even dyskinesia-related in-scanner movement. Medication effects on connectivity patterns, both in terms of disease-related confounds and as indices of successful treatment, can be better understood via the longitudinal study of unmedicated patients before and following treatment initiation and of unaffected first-degree relatives. Some investigations are beginning to examine dysconnectivity associated with psychosis in specific genetic deletion syndromes such as 22q11.2.[70]

In our opinion, computational modeling approaches allied with neuropharmacologic probes, such as those used to address connectional changes associated with ketamine challenge, have the potential to yield significant information.[71] The Human Connectome Project (HCP) is an ongoing set of consortium-based studies endeavoring to improve and standardize brain imaging acquisition and analysis tools for fMRI (and structural) studies. Such standardization, if widely adopted, will ultimately prove to be extremely helpful in terms of comparing dysconnection findings from different labs more directly. In addition, the related HCP Lifespan project, together with emerging data from the NIH-funded Adolescent Brain Cognitive Development (ABCD)[72] study supplemented by information from the United Kingdom Biobank investigations,[73] will yield important information on the normal development of brain network patterns and the temporal relationship of the appearance of fMRI abnormalities to first emergence of symptoms. These observations should help identify individuals at risk for psychosis at an earlier stage than currently, with important implications for more timely treatment interventions. Already, some groups have begun projects examining resting state connectivity associated with the onset and progression of psychotic illnesses.[74]

At the moment then, the field is best summarized as having gathered diverse cross-disciplinary data that bear on the question of dysconnection, but as noted elsewhere,[62] linking these isolated findings into a coherent story remains a work in progress. The field awaits the paradigm shift that will inform us both which dysconnectivity patterns are truly psychosis-specific as well as their underlying pathophysiological basis.

REFERENCES

1. Friston KJ. Schizophrenia and the disconnection hypothesis. *Acta Psychiatrica Scandinavica. Supplementum*. 1999;395:68–79.
2. Coyle JT, Balu DT, Puhl MD, Konopaske GT. History of the concept of disconnectivity in schizophrenia. *Harvard Review of Psychiatry*. 2016;24(2):80–86.
3. Lenzenweger MF. Schizotaxia, schizotypy, and schizophrenia: Paul E. Meehl's blueprint for the experimental psychopathology and genetics of schizophrenia. *J Abnorm Psychol.* 2006;115(2):195–200.
4. Stephan KE, Iglesias S, Heinzle J, Diaconescu AO. Translational perspectives for computational neuroimaging. *Neuron*. 2015;87(4):716–732.
5. Anticevic A, Murray JD, Barch DM. Bridging levels of understanding in schizophrenia through computational modeling. *Clin Psychol Sci.* 2015;3(3):433–459.
6. Schmidt A, Diwadkar VA, Smieskova R, et al. Approaching a network connectivity-driven classification of the psychosis continuum: a selective review and suggestions for future research. *Frontiers in Human Neuroscience*. 2015;8:1047.
7. Hayes JP, Bigler ED, Verfaellie M. Traumatic brain injury as a disorder of brain connectivity. *J Int Neuropsychol Soc.* 2016;22(2):120–137.
8. Joules R, Doyle OM, Schwarz AJ, et al. Ketamine induces a robust whole-brain connectivity pattern that can be differentially modulated by drugs of different mechanism and clinical profile. *Psychopharmacology*. 2015;232(21–22):4205–4218.
9. Khalili-Mahani N, Rombouts SARB, van Osch MJP, et al. Biomarkers, designs, and interpretations of resting-state fMRI in translational pharmacological research: a review of state-of-the-art, challenges, and opportunities for studying brain chemistry. *Human Brain Mapping.* 2017;38:2276–2325.
10. Pearlson GD. Applications of resting state functional MR imaging to neuropsychiatric diseases. *Neuroimaging Clin N Am.* 2017;27(4):709–723.
11. Stevens MC. The contributions of resting state and task-based functional connectivity studies to our understanding of adolescent brain network maturation. *Neurosci Biobehav Rev.* 2016;70:13–32.
12. Weinberger DR, Radulescu E. Finding the elusive psychiatric "lesion" with 21st-century neuroanatomy: a note of caution. *Am J Psychiatry*. 2016;173(1):27–33.
13. Calhoun V, Glahn D, Pearlson G. Finding the elusive psychiatric "lesion" with 21st-century neuroanatomy: a note of caution. In: Weinberger DR, Radulescu E, eds. Online comment ed. *Schizophrenia Research Forum* 2015.
14. Moser DA, PhD, Doucet GE, PhD, Lee WH, PhD, et al. Multivariate associations among behavioral, clinical, and multimodal imaging phenotypes in patients with psychosis *JAMA Psychiatry*. Mar 7 2018.
15. Zac Lo CY, Su TW, Huang CC, et al. Randomization and resilience of brain functional networks as systems-level endophenotypes of schizophrenia. *Proc Natl Acad Sci USA*. 2015;112(29):9123–9128.
16. Hu ML, Zong XF, Mann JJ, et al. A Review of the functional and anatomical default mode network in schizophrenia. *Neuroscience Bulletin*. 2017;33(1):73–84.
17. Meda SA, Gill, A., Stevens, M.C., Lorenzoni, R.P., Glahn, D.C., Calhoun, V.D., Sweeney, J.A., Tamminga, C.A., Keshavan, M.S., Thaker, G., Pearlson, G.D. Differences in resting-state fMRI functional network connectivity between schizophrenia and psychotic bipolar probands and their unaffected first-degree relatives. *Biol Psychiatry*. 2012;71(10):881–889.
18. Meda SA, Ruano G, Windemuth A, et al. Multivariate analysis reveals genetic associations of the resting default mode network in psychotic bipolar disorder and schizophrenia. *Proc Natl Acad Sci USA*. 2014;111(19):E2066–2075.
19. Meda SA, Wang Z, Ivleva EI, et al. Frequency-specific neural signatures of spontaneous low-frequency resting state fluctuations in psychosis: evidence from bipolar-schizophrenia network on intermediate phenotypes (B-SNIP) Consortium. *Schizophr Bull.* 2015;41(6):1336–1348.
20. Khadka S, Meda SA, Stevens MC, et al. Is aberrant functional connectivity a psychosis endophenotype? A resting state functional magnetic resonance imaging study. *Biol Psychiatry*. 2013;74(6):458–466.
21. Liu H, Kaneko Y, Ouyang X, et al. Schizophrenic patients and their unaffected siblings share increased resting-state connectivity in the task-negative network but not its anticorrelated task-positive network. *Schizophr Bull.* 2012;38(2):285–294.
22. Sheffield JM, Barch DM. Cognition and resting-state functional connectivity in schizophrenia. *Neurosci Biobehav Rev.* 2016;61:108–120.
23. Richiardi J, Altmann A, Milazzo AC, et al. Correlated gene expression supports synchronous activity in brain networks. *Science.* 2015;348(6240):1241–1244.
24. Glahn DC, Winkler AM, Kochunov P, et al. Genetic control over the resting brain. *Proc Natl Acad Sci USA.* 2010;107(3):1223–1228.
25. Cohen JR. The behavioral and cognitive relevance of time-varying, dynamic changes in functional connectivity. *NeuroImage.* 2017:1–11.
26. Wang S, Zhang Y, Lv L, et al. Abnormal regional homogeneity as a potential imaging biomarker for adolescent-onset schizophrenia: a resting-state fMRI study and support vector machine analysis. *Schizophr Res.* 2017.
27. Liang MJ, Zhou Q, Yang KR, et al. Identify changes of brain regional homogeneity in bipolar disorder and unipolar depression using resting-state FMRI. *Public Library of Science One.* 2013;8(12):e79999.
28. Wang S, Wang G, Lv H, Wu R, Zhao J, Guo W. Abnormal regional homogeneity as potential imaging biomarker for psychosis risk syndrome: a resting-state fMRI study and support vector machine analysis. *Scientific Reports.* 2016;6:27619.
29. Calhoun VD, Miller R, Pearlson G, Adali T. The chronnectome: time-varying connectivity networks as the next frontier in fMRI data discovery. *Neuron.* 2014;84(2):262–274.
30. Du Y, Pearlson GD, Lin D, et al. Identifying dynamic functional connectivity biomarkers using GIG-ICA: application to schizophrenia, schizoaffective disorder, and psychotic bipolar disorder. *Human Brain Mapping.* 2017;38(5):2683–2708.
31. Yaesoubi M, Miller RL, Bustillo J, Lim KO, Vaidya J, Calhoun VD. A joint time-frequency analysis of resting-state functional connectivity reveals novel patterns of connectivity shared between or unique to schizophrenia patients and healthy controls. *NeuroImage Clinical.* 2017;15:761–768.
32. Lottman KK, Kraguljac NV, White DM, et al. Risperidone effects on brain dynamic connectivity-a prospective resting-state fMRI study in schizophrenia. *Frontiers in Psychiatry.* 2017;8.
33. Miller RL, Vergara VM, Keator DB, Calhoun VD. A method for intertemporal functional-domain connectivity analysis: application to schizophrenia reveals distorted directional information flow. *IEEE Trans Biomed Eng.* 2016;63(12):2525–2539.
34. Gonzalez-Castillo J, Bandettini PA. Task-based dynamic functional connectivity: recent findings and open questions. *NeuroImage.* 2017.
35. Pearlson GD, Calhoun VD. Convergent approaches for defining functional imaging endophenotypes in schizophrenia. *Frontiers in Human Neuroscience.* 2009;3:37.
36. Birur B, Kraguljac NV, Shelton RC, Lahti AC. Brain structure, function, and neurochemistry in schizophrenia and bipolar disorder—a systematic review of the magnetic resonance neuroimaging literature. *NPJ Schizophrenia.* 2017;3:15.
37. Anticevic A, Savic A, Repovs G, et al. Ventral anterior cingulate connectivity distinguished nonpsychotic bipolar illness from psychotic bipolar disorder and schizophrenia. *Schizophr Bull.* 2015;41(1):133–143.
38. Mwansisya TE, Hu A, Li Y, et al. Task and resting-state fMRI studies in first-episode schizophrenia: a systematic review. *Schizophr Res.* 2017;189:9–18.

39. Tamminga CA, Pearlson G, Keshavan M, Sweeney J, Clementz B, Thaker G. Bipolar and schizophrenia network for intermediate phenotypes: outcomes across the psychosis continuum. *Schizophr Bull.* 2014;40 Suppl 2:S131–137.
40. Repovs G, Csernansky JG, Barch DM. Brain network connectivity in individuals with schizophrenia and their siblings. *Biol Psychiatry.* 2011;69(10):967–973.
41. Unschuld PG, Buchholz AS, Varvaris M, et al. Prefrontal brain network connectivity indicates degree of both schizophrenia risk and cognitive dysfunction. *Schizophr Bull.* 2014;40(3):653–664.
42. Pearlson GD, Petty RG, Ross CA, Tien AY. Schizophrenia: a disease of heteromodal association cortex? *Neuropsychopharmacology: Offici al Publication of the American College of Neuropsychopharmacology.* 1996;14(1):1–17.
43. Karbasforoushan H, Woodward ND. Resting-state networks in schizophrenia. *Current Topics in Medicinal Chemistry.* 2012;12(21):2404–2414.
44. Northoff G, Duncan NW. How do abnormalities in the brain's spontaneous activity translate into symptoms in schizophrenia? From an overview of resting state activity findings to a proposed spatiotemporal psychopathology. *Prog Neurobiol.* 2016;145–146:26–45.
45. Northoff G. The brain's spontaneous activity and its psychopathological symptoms—"spatiotemporal binding and integration." *Prog Neuropsychopharmacol Biol Psychiatry.* 2018;80(Pt B):81–90.
46. Ćurčić-Blake B, Ford JM, Hubl D, et al. Interaction of language, auditory and memory brain networks in auditory verbal hallucinations. *Prog Neurobiol.* 2017;148:1–20.
47. Northoff G. Are auditory hallucinations related to the brain's resting state activity? A "neurophenomenal resting state hypothesis." *Clinical Psychopharmacology and Neuroscience: The Official Scientific Journal of the Korean College of Neuropsychopharmacology.* 2014;12(3):189–195.
48. Alderson-Day B, McCarthy-Jones S, Fernyhough C. Hearing voices in the resting brain: a review of intrinsic functional connectivity research on auditory verbal hallucinations. *Neurosci Biobehav Revs.* 2015;55:78–87.
49. Barta PE, Pearlson GD, Powers RE, Richards SS, Tune LE. Auditory hallucinations and smaller superior temporal gyral volume in schizophrenia. *Am J Psychiatry.* 1990;147(11):1457–1462.
50. Robinson JD, Wagner NF, Northoff G. Is the sense of agency in schizophrenia influenced by resting-state variation in self-referential regions of the brain? *Schizophr Bull.* 2016;42(2):270–276.
51. Pergola G, Selvaggi P, Trizio S, Bertolino A, Blasi G. The role of the thalamus in schizophrenia from a neuroimaging perspective. *Neurosci Biobehav Revs.* 2015;54:57–75.
52. Glahn DC, Laird AR, Ellison-Wright I, et al. Meta-analysis of gray matter anomalies in schizophrenia: application of anatomic likelihood estimation and network analysis. *Biol Psychiatry.* 2008;64(9):774–781.
53. Anticevic A, Cole MW, Repovs G, et al. Connectivity, pharmacology, and computation: toward a mechanistic understanding of neural system dysfunction in schizophrenia. *Frontiers in Psychiatry.* 2013;4:169.
54. Acsady L. The thalamic paradox. *Nature Neuroscience.* Jun 27 2017;20(7):901–902.
55. Giraldo-Chica M, Woodward ND. Review of thalamocortical resting-state fMRI studies in schizophrenia. *Schizophr Res.* 2017;180:58–63.
56. Woodward ND, Karbasforoushan H, Heckers S. Thalamocortical dysconnectivity in schizophrenia. *Am J Psychiatry.* 2012;169(10):1092–1099.
57. Anticevic A, Cole MW, Repovs G, et al. Characterizing thalamocortical disturbances in schizophrenia and bipolar illness. *Cerebral Cortex.* 2014;24(12):3116–3130.
58. Woodward ND, Heckers S. Mapping thalamocortical functional connectivity in chronic and early stages of psychotic disorders. *Biol Psychiatry.* 2016;79(12):1016–1025.
59. Murray JD, Anticevic A. Toward understanding thalamocortical dysfunction in schizophrenia through computational models of neural circuit dynamics. *Schizophr Res.* 2017;180:70–77.
60. Romme IAC, de Reus MA, Ophoff RA, Kahn RS, van den Heuvel MP. Connectome disconnectivity and cortical gene expression in patients with schizophrenia. *Biol Psychiatry.* 2017;81(6):495–502.
61. Skudlarski P, Jagannathan K, Anderson K, et al. Brain connectivity is not only lower but different in schizophrenia: a combined anatomical and functional approach. *Biol Psychiatry.* 2010;68(1):61–69.
62. Arikkath J, Mirnics K. Connecting the dots. *Society of Biological Psychiatry.* 2017;81(6):463–464.
63. Richmond S, Johnson KA, Seal ML, Allen NB, Whittle S. Development of brain networks and relevance of environmental and genetic factors: a systematic review. *Neurosci Biobehav Rev.* 2016;71:215–239.
64. Chen LM, Yang PF, Wang F, Mishra A, Shi Z, Wu R, Wu TL, Wilson GH 3rd, Ding Z, Gore JC. Biophysical and neural basis of resting state functional connectivity: Evidence from non-human primates. *Magn Reson Imaging.* 2017;39:71–81. doi:10.1016/j.mri.2017.01.020. Epub 2017 Feb 2. Review.PMID:28161319
65. Roalf DR. Progress toward elucidating commonalities in mental disorders using brain imaging and publicly available data. *JAMA Psychiatry.* 2018;75(3):295–296.
66. Alexander-Bloch AF. Disconnectionism in biological psychiatry. *Biol Psychiatry.* 2017;82(10):e75–e77.
67. Narr KL, Leaver AM. Connectome and schizophrenia. *Current Opinion in Psychiatry.* 2015;28(3):229–235.
68. Thompson PM, Ge T, Glahn DC, Jahanshad N, Nichols TE. Genetics of the connectome. *NeuroImage.* 2013;80:475–488.
69. Whitfield-Gabrieli S, Thermenos HW, Milanovic S, et al. Hyperactivity and hyperconnectivity of the default network in schizophrenia and in first-degree relatives of persons with schizophrenia. *Proc Natl Acad Sci USA.* 2009;106(4):1279–1284.
70. Scariati E, Padula MC, Schaer M, Eliez S. Long-range dysconnectivity in frontal and midline structures is associated to psychosis in 22q11.2 deletion syndrome. *J Neural Transm.* 2016;123(8):823–839.
71. Driesen NR, McCarthy G, Bhagwagar Z, et al. Relationship of resting brain hyperconnectivity and schizophrenia-like symptoms produced by the NMDA receptor antagonist ketamine in humans. *Mol Psychiatry.* 2013;18(11):1199–1204.
72. Bjork JM, Straub LK, Provost RG, Neale MC. The ABCD study of neurodevelopment: Identifying neurocircuit targets for prevention and treatment of adolescent substance abuse. *Curr Treat Options Psychiatry.* 2017;4(2):196–209.
73. UK Biobank data on 500,000 people paves way to precision medicine—editorial. *Nature.* 2018;562(7726):163–164.
74. Chung Y, Cannon TD. Brain imaging during the transition from psychosis prodrome to schizophrenia. *J Nerv Ment Dis.* 2015;203(5):336–341.

30

MR SPECTROSCOPY

Adrienne C. Lahti and Nina V. Kraguljac

INTRODUCTION

Magnetic resonance spectroscopy (MRS) allows the noninvasive measurement of the chemical composition of tissues, energy metabolism, and neurotransmitter levels in vivo in the human brain.[1] It detects magnetic resonance signals produced by atomic nuclei in the tissue.

Data can either be acquired from one area of interest with single-voxel spectroscopy, or across a slab of smaller areas of interest with multi-voxel chemical shift imaging (also known as magnetic resonance spectroscopic imaging [MRSI]). A number of different acquisition sequences are available to use. In its basic form, single-voxel techniques use either a spin-echo (point resolved spectroscopy [PRESS]) or simulated echo (simulated echo acquisition mode [STEAM]) pulse sequence. More recently developed fast spectroscopic imaging techniques focus on k-space sampling strategies (e.g., echo planar spectroscopic imaging [EPSI]), which mitigate artifacts from eddy currents and spectral aliasing effects.[1] A spectral editing method, MEshcher-GArwood Point RESolved Spectroscopy [MEGA-PRESS], has been developed to allow a separation of the GABA signal from the stronger overlying signals, by collecting two interleaved datasets that differ in their editing pulses, differentially affecting the GABA spin system.[2,3] The two datasets are then subtracted, leaving only those peaks that are affected by the editing pulses. In MRSI, phase-encoding gradients are used in one, two, or three dimensions to sample k-space. Here, the spectra of all voxels are acquired simultaneously,[4] but the spectrum of a voxel may be contaminated by signals originating from adjacent voxels. Generally, single voxel spectroscopy is used when accurate quantification is paramount, and MRSI is used when information on spatial distribution is of interest.[5]

For data acquisition, a number of steps are taken:

1. Shimming is the process of homogenizing the magnetic field and is performed to enhance the sensitivity and resolution of the metabolite signal. Shimming is easier in smaller regions acquired by single-voxel spectroscopy, compared to MRSI.
2. Setting the power for applying the pulses in the localization and suppression sequences.
3. Suppression of signals from water, often with CHESS[6] or VAPOR[7] pulses.
4. Quantification of the signal amplitude of the spectra that will provide an estimate of the amount of signal generating molecules.
5. Acquisition of a reference signal to account for a number of confounding factors. Most commonly, this is done by including an additional scan measuring the unsuppressed voxel water peak, using another metabolite such as creatine from the originally acquired spectrum, or including an external phantom that contains known quantities of the metabolites of interest. (For an in-depth review, see.[8])

The general quality of data depends on magnetic field strength, size and composition of the brain region of interest, acquisition time, and concentration of metabolites.[9] Spectral quality is typically reported with measures of the signal-to-noise ratio (SNR), line width of the water resonance or the N-acetyl-aspartate (NAA) peak, and Cramér Rao Lower Bounds (CRLB), which is a measure of uncertainty. Because metabolites are differentially distributed in gray matter, white matter, and cerebrospinal fluid, it is also important to correct for differences in tissue composition. For rigorous reporting, all those measures should be provided.

The most widely used type of spectroscopy leverages the signal produced by protons to quantify different metabolites (^{1}H-MRS) (Figure 30.1). Metabolites that can be measured include NAA, a putative marker of neuronal integrity; choline (Cho), a marker of cellular turnover and cell membrane breakdown; myoinositol (mI), a metabolite that plays an important role in osmoregulation; creatine (Cr), a signal that is related to phosphate metabolism; and the amino acids GABA and the combination of glutamate (Glu) + glutamine (Gln), often expressed as Glx (glutamate+glutamine).[10] At field strengths greater than 3 Tesla is also possible for ^{1}H-MRS to reliably separate the glutamine peak from the glutamate peak. Other spectroscopy techniques include ^{31}phosphorus spectroscopy, which provides a wide range of information on energy metabolism, and ^{13}Carbon spectroscopy, which use cerebral glucose metabolism to assess glutamine synthesis and glutamatergic transmission.

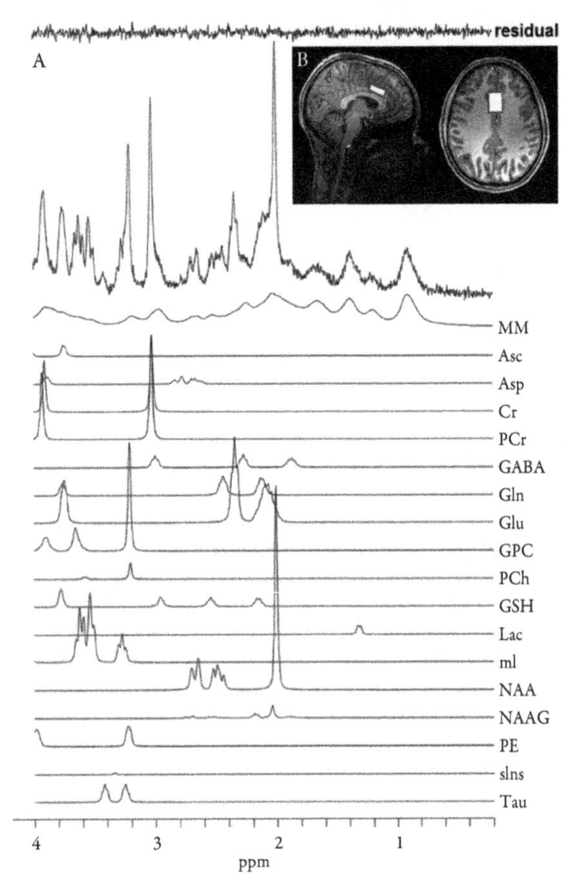

Figure 30.1 A. Representative spectrum obtained at 7T with a STEAM sequence. The spectral fit is shown in red over the acquired spectrum. The individual metabolite fits are shown below. The difference between the spectrum and fit (residual) is shown at the top of the figure. Abbreviations: Asc: ascorbate, Asp: aspartate, Cr: creatine, GABA: gamma-aminobutyric acid, Gln: glutamine, Glu: glutamate, GPC: glycerophosphocholine, GSH: glutathione, Lac: lactate, mI: *myo*-Inositol, MM: macromolecules, NAA: *N*-acetylaspartate, NAAG: *N*acetylaspartylglutamate, PCh: phosphocholine, PCr: phosphocreatine, PE: phosphorylethanolamine, sIns: *scyllo*-Inositol, Tau: taurine. B. MRS voxel placement in the dorsal anterior cingulate cortex.

NAA, CHOLINE, AND OTHER METABOLITES

Abnormalities in schizophrenia extend to a number of neurometabolites varying across different brain regions and illness stages. Investigating neurometabolite levels in antipsychotic-naïve and antipsychotic-free subjects, a recent meta-analysis reports consistent decreases in NAA, suggestive of reduced neuronal integrity, in the thalamus and frontal white matter, but not in temporal lobe areas.[11] Meta-analytic evidence also suggests a decrease in NAA in the frontal lobe as well as the hippocampus and thalamus in chronic schizophrenia patients.[12] Importantly, alterations may occur in the transition between the at-risk phase and first psychotic episode,[13] and may become more prominent with illness progression.[14] Other metabolites including Cho and mI are less consistently reported to be abnormal, but may reflect a neuro-inflammatory process in the early illness stage.[15]

The literature is also inconsistent in demonstrating relationships between the severity of clinical symptoms across dimensions and specific neurometabolites. While the majority of studies fail to establish such relationships, a number of reports find positive correlations between frontal lobe NAA and cognitive performance[16–18] as well as negative correlations between frontal lobe NAA and negative symptom severity.[16,19–21] In addition, choline alterations in the basal ganglia have been linked to drug-induced parkinsonism[22] and tardive dyskinesia.[23] It is important to note that these relationships were reported more often in the earlier literature, suggesting that methodological aspects such as magnetic field strength, stringent data quality control, and correction for partial volume effects may reduce the probability of spurious findings.

Taking advantage of newer MRS acquisition techniques, a number of other metabolites have now been studied in the context of symptom severity in schizophrenia. Implicating altered anterior cingulate cortex bioenergetics in the illness, a recent high field study reported elevated lactate levels associated with poorer cognitive performance and functional capacity in medicated patients with schizophrenia.[24] Glutathione, a metabolite that acts as an antioxidant in the brain, is reported to be negatively correlated with negative symptom severity in chronic patients, suggesting oxidative stress as potential target for new drug development.[25] There is still much knowledge to be gained from these historically difficult-to-detect metabolites, but preliminary results hold promise for detecting mechanistically relevant biomarkers that can serve as targets for novel intervention.[10]

GLUTAMATE, GLUTAMINE, AND GABA

Together, the excitatory glutamatergic neuron and the inhibitory GABAergic interneuron represent the basic processing unit throughout the cerebral cortex.[26] After release from presynaptic terminals, glutamate and GABA are taken up by glial cells, and converted to glutamine that is then returned to neurons and converted back to the original neurotransmitters. Over 75% of the brain's total energy consumption is coupled to the cycling of these neurotransmitters.[27] Studies of glutamate and GABA have considerable interest because recent theories have implicated abnormalities in the excitation/inhibition balance as a result of N-methyl-D-aspartate (NMDA) receptor hypofunction on GABA interneurons in the pathophysiology of schizophrenia.[28–31] Postmortem studies have reported glutamatergic and GABAergic abnormalities in schizophrenia.[32] In addition, in humans, NMDA receptor antagonists induce a behavioral phenotype that mirrors the symptoms of schizophrenia,[33] including cognitive impairments.[34–36] MRS studies measuring glutamate or Glx in schizophrenia have consistently reported abnormal glutamatergic measurements,[37–42] although there is great heterogeneity of results owing to differences in voxel location, stage of illness (high risk, first episode, and chronic populations), medication status (medication-naïve, unmedicated, and medicated), magnet strength, sequence acquisitions, and data post processing.[43] In addition, it is important to keep in mind that glutamate measurements obtained with MRS reflect the total amount of glutamate in the voxel, and cannot be equated with glutamatergic transmission. Measurements of GABA in schizophrenia have been less frequent, but abnormalities have been identified as well.[44–47] Glutamine and especially the glutamine/glutamate ratio are of interest because they are thought to index the rate of glutamatergic synaptic activity better than glutamate.[48,49]

A number of recent meta-analyses and comprehensive reviews have attempted to summarize the growing number of studies measuring glutamatergic metabolites.[10,11,37,38,50] The most recent meta-analysis (1,686 cases and 1,451 controls) of cross-sectional studies[37] found evidence of significant elevations in glutamate in the basal ganglia, glutamine in the thalamus, and Glx in the basal ganglia and medial temporal lobe (MTL). There was also an indication that levels of glutamatergic metabolites were sensitive to illness stage: Glx was elevated in medial frontal cortex in a high-risk population, but not in first-episode psychosis (FEP) or chronic schizophrenia, while elevated Glx in MTL was seen with chronic schizophrenia but not in the high risk or FEP populations. A few words of caution when it comes to high risk and FEP populations: about two-thirds of the high-risk population will not convert to psychosis, leaving the interpretation of high-risk MRS studies complicated to say the least; the definition of FEP population typically includes patients who can be up to 2 and sometimes 4 years from the start of treatment, which makes this population a heterogeneous one. The aforementioned meta-analysis failed to find an association between glutamatergic measures and symptom severity. While associations with positive or negative symptoms have been difficult to identify (see review in [51]), some studies have suggested that Glx levels are elevated in patients whose positive symptoms are not responding to antipsychotic treatment.[52–54] Although no association was found between glutamatergic metabolites and antipsychotic doses in this meta-analysis, this question is better addressed by prospective longitudinal studies.

Only one meta-analysis summarized the GABA measurements (538 controls, 526 cases).[44] These measurements have been most commonly obtained in the medial prefrontal cortex (mPFC) compared to other areas. No significant group differences were found in the mPFC, parietal/occipital lobe and striatum.

MEDICATION-NAÏVE STUDIES

The study of medication-naïve patients is of critical importance to understand the pathophysiology of the illness without the confounding effect of medication. However, only a few studies have enrolled truly medication naïve subjects (MNS). In a large sample (60 MNS) that included participants enrolled in prior studies,[41,55] Plitman,[15] extending the findings of de la Fuente-Sandoval, reported elevated glutamate levels in the associative striatum. In a sample of 15 MNS, Wood[56] reported no group difference in Glx levels in the medial temporal cortex. In a sample of 21 MNS, Theberge[57] reported elevated glutamine levels in the anterior cingulate cortex (ACC) and thalamus; groups that included part of this initial cohort were reported later: these studies confirmed findings of elevated glutamine in the thalamus[58,59] and the ACC.[59] In a sample of 28 MNS, Kelemen[60] reported reduced GABA levels in the occipital cortex. Clearly it is difficult to derive any conclusion from these limited samples studied in nonoverlapping regions.

HIGH-FIELD (7T) STUDIES

High-field (7T) MRS offers improved detection of distinct glutamate and glutamine signals. To date, only five studies of schizophrenia using 7T MRS have been published; they all report measurements of glutamate, glutamine, and GABA.[61–65] Rowland[63] found an elevated glutamine/glutamate ratio in the ACC in medicated patients, Thakkar[64] observed reduced glutamate and GABA in the occipital cortex, but no abnormalities in the basal ganglia (also in medicated patients), and Brandt[61] found that ACC glutamate decreased with age in medicated patients but not controls. Marsman[62] observed lower GABA in medial prefrontal or parieto-occipital cortices in medicated patients. Reid[65] found ACC glutamate to be significantly lower in a group of FEP subjects. With the exception of Reid,[65] these 7T studies used mixed or older samples with relatively long illness durations. There is an urgent need to enroll medication-naïve cohorts that are followed longitudinally.

LONGITUDINAL STUDIES OF GLUTAMATE AND GABA

There are a limited number of longitudinal studies evaluating the effect of treatment with antipsychotic medication on glutamatergic and GABAergic metabolites. Short-term studies are better suited to capture specific effect of treatment, as longer studies are more likely to be influenced by environmental factors. In chronic patients washed out of medications, Szulc reported a decrease in temporal lobe Glx following 4 weeks of treatment with a variety of antipsychotic medications.[66] In medication-naïve FEP subjects compared to healthy controls, de la Fuente-Sandoval observed higher baseline striatal glutamate and a significant reduction in striatal glutamate after four weeks of risperidone treatment.[55] Using an edited sequence to measure GABA in the dorsal caudate and mPFC in a large group of medication-naïve FEP patients and healthy controls, the same group reported higher baseline GABA levels in the two regions and a significant reduction in mPFC GABA after 4 weeks of risperidone treatment.[67]

THE KETAMINE MODEL OF PSYCHOSIS, SPECTROSCOPY FINDINGS

Subanesthetic doses of ketamine, a noncompetitive NMDAR blocker, are found to transiently induce a behavioral phenotype similar to that seen in schizophrenia,[33–36,68,69] making it a popular pharmacological model for psychosis. In response to ketamine infusion in healthy subjects, a number of spectroscopy studies reported increases of glutamatergic indices in the ACC,[70,71] mPFC,[72] and hippocampus.[31] However, two studies found no change in the mPFC[73] and the occipital cortex.[74] Potentially reconciling these discrepancies, others demonstrated that ketamine's effects on Glx are transient,[75] and may be dose-dependent. Ketamine studies do not report consistent relationships between changes in glutamatergic indices and symptom severity. Only one study reported a positive correlation between post-ketamine glutamate levels and positive symptom severity,[71] whereas another found a trend level correlation between increase in glutamine and poorer performance on a cognitive control task.[70] It is possible that clinical symptoms emerge from a disruption in functional networks secondary to glutamatergic excess which may obscure the link between clinical symptom severity and neurometabolite alterations. In other words, it is possible that the glutamate system acts as a moderating variable between functional network level dysfunction and symptom severity.[31]

RELATIONSHIP BETWEEN GLUTAMATE, GLUTAMINE, AND GABA AND COGNITION

Behavioral pharmacology has provided the bulk of the evidence for the involvement of ionotropic glutamate receptors in some of the cognitive processes known to be impaired in schizophrenia.[76] In addition, in humans, ketamine has been shown to worsen cognitive performance on a variety of tests.[35,36,77,78] Because there is no available pharmacological treatment for cognitive dysfunction in schizophrenia, there is a strong interest in characterizing the relationship between glutamate or GABA and impaired cognition with the ultimate goal of identifying a target for remediation. A number of such studies have been published and are summarized in Table 30.1 according to whether glutamate metabolites were found elevated, reduced, or not different compared to controls (limited to studies performed at 3T or higher). Five studies reported elevated glutamate metabolites in schizophrenia;[18,24,39,40,79] two studies found prefrontal Gln/Glu to be negatively associated with cognition in the combined group (schizophrenia and control);[24,79] another one,[18] by far the largest, found white matter Glx to be positively and negatively associated with overall cognition in controls and schizophrenia, respectively. Four studies failed to find a difference in glutamate metabolites;[80–83] one reported white matter Glx measured in a slice above the lateral ventricles to be associated with cognitive performance in patients, but not in controls,[82] and another one reported substantia nigra (SN) Glx to be positively associated with cognition in controls, but not in schizophrenia.[83] Finally, two studies reported reduced glutamate metabolites;[65,84] one of them reported a negative association between ACC glutamate and a measure of immediate memory in controls, but not in schizophrenia.[65]

Five studies in schizophrenia have evaluated the relationship between GABA levels and cognitive assessments (Table 30.2).[39,46,65,84,86] In the context of no group difference in GABA levels, a negative association was identified between ACC GABA and overall cognitive performance in patients, but not in controls.[65] In the presence of reduction in GABA levels, one study reported mPFC GABA to be negatively associated with cognitive performance in patients, but not in controls.[86] Finally the other study found a positive association between ACC GABA and a measure of attention in the combined group.[84]

Cautiously because most of these studies were small, whether or not patients presented glutamatergic or GABA abnormalities, altered relationships between these metabolites and cognition were observed.

GLUTAMATE, GLUTAMINE, GABA, AND OTHER IMAGING MODALITIES

Not surprisingly considering their role in neuroenergetics, glutamate and GABA have been found to be key components of the fMRI blood oxygen level dependent (BOLD) response.[87] Shedding light into how differences in glutamate levels could influence the way the brain implement cognitive processes, Falkenberg[88] showed that the BOLD response can be differentially modulated based on resting-state (rs) glutamate levels: in individuals with low rs glutamate the BOLD response increased as a cognitive task became more difficult, whereas in those with high rs glutamate levels the BOLD response increased when the task was easier. A recent and comprehensive review of combined MRS/fMRI studies performed in controls reported evidence for a negative correlation between GABA and stimulus-induced BOLD responses within the measured region, while glutamate was positively correlated with the stimulus-induced response within the measured regions, as well as within regions distant from the origin. This suggests that glutamate is also related to long-range connections between regions.[89] In schizophrenia, only a few studies, including our own,[81,90–92] have evaluated the relationships between glutamate and the BOLD response. In healthy controls, we reported a robust positive correlation between glutamate measured in the hippocampus and inferior frontal activation during a memory task; this association was not present in patients with schizophrenia.[90] Likewise, Falkenberg reported an altered relationship between ACC glutamate and the BOLD response in the inferior parietal cortex during a cognitive control task in schizophrenia.[93] Interestingly, Overbeek identified a similar altered relationship between ACC glutamate and the BOLD response in the posterior default mode network during cognitive performance in schizophrenia.[92]

A number of studies have now started to combine MRS with other modalities. Because schizophrenia is a heterogeneous disorder likely involving multiple underlying pathological mechanisms,[94] using several modalities has the potential to interrogate different neurobiological aspects of the illness. Combining MRS and PET, Stone reported a negative association between hippocampal glutamate and striatal [18F] DOPA uptake in a population of high-risk subjects, but not in controls, and a suggestion that this opposite association was more pronounced in those who later transition to psychosis.[95]

Table 30.1 STUDIES EVALUATING THE RELATIONSHIP BETWEEN GLUTAMATERGIC METABOLITES AND COGNITION IN SCHIZOPHRENIA

AUTHORS	STAGE OF ILLNESS	REGION	COGNITIVE DOMAINS	SCANNER AND METHOD	PARTICIPANTS: SCHIZOPHRENIA (SZ)/HEALTHY CONTROLS (HC)	APD	GLUTAMATE METABOLITES (SZ VS. HC)	GLUTAMATE METABOLITES/ COGNITION
Shirayama et al. (2010)[79]	Chronic	Medial PFC (mPFC)	Verbal fluency, Wisconsin card sorting test (WCST), Trail making, Digit span distraction test (DSDT), Stroop test, Iowa gambling test	3 T PRESS	19/18	Yes	↑glutamine/glutamate (Gln/Glu)	mPFC Gln/Glu negatively associated with the WCST and DSDT in combined SZ -HC
Kegeles et al. (2012)[39]	Mixed	mPFC dorsolateral PFC (dlPFC)	N-Back working memory test	3 T J-edited	9 medication-naïve/ 5 unmedicated, and 16 medicated SZ/22	Yes & No	↑ Glx in mPFC in unmedicated vs. HC; no difference in the dlPFC	No associations
Kraguljac et al. (2013)[40]	Chronic	Hippocampus	Repeatable battery for assessment of neuropsycho-logical status (RBANS)	3 T PRESS		No	↑Glx	No associations
Rowland et al. (2016)[46]	Chronic	Anterior cingulate cortex (ACC)	General cognitive performance using the MATRICS Consensus Cognitive Battery (MCCB)	7 T STEAM	27/29	yes	↑Gln and Gln/Glu	ACC Gln/Glu negatively associated with MCCB in combined SZ -HC
Bustillo et al. (2017)[18]	Chronic Broad age range	Axial supraventri-cular slab of gray (GM) and white matter (WM)	MCCB	3 T MRSI	104/97	yes	↑Glx in GM and WM	WM Glx positively and negatively associated with the MCCB Total score in HC and SZ, respectively
Rowland et al. (2009)[80]	Chronic	Middle frontal Inferior frontal	RBANS	3 T PRESS	10 deficit/10 nondeficit SZ/11	Yes	Glx: No difference	No associations
Reid et al. (2010)[81]	Chronic	ACC	RBANS	3 T PRESS	26/23	yes	Glx: No difference	No associations
Bustillo et al. (2011)[82]	Early (<30 years) and later (>30 years)	Slice superior to the lateral ventricles	Broad neuropsychological battery	4 T Proton echoplanar spectroscopic imaging (PEPSI)	Young: 12/10 Old: 18/18	Yes	Glx: No difference	WM Glx positively associated with cognitive performance in SZ, but not HC
Reid et al. (2013)[83]	Chronic	Substantia nigra (SN)	RBANS	3 T PRESS	35/22	Yes	Glx: No difference	SN Glx positively associated with the RBANS total score in HC, but not SZ
Rowland et al. (2013)[84]	Chronic	ACC Centrum semiovale (CSO)	RBANS	3 T MEGA-PRESS	11 young SZ (YSZ)-10 old SZ (OSZ) /10 YHC-10 OHC	Yes	↓ Glx in ACC and CSO	No associations
Reid et al. (2018)[65]	FEP	ACC	RBANS	7 T STEAM	21/21	yes	↓glutamate	Negative association between Gln and RBANS Immediate Memory in HC, but not in SZ
Dempster et al. (2015)[85]	FEP	ACC Thalamus	Paced Auditory Serial Addition Task (PASAT) WCST, Trails	4 T STEAM	13 medication-naïve/3 medicated/No HC	Yes & No	No HC for comparison	ACC Gln positively and thalamus Gln negatively associated with the PASAT

Table 30.2 STUDIES EVALUATING THE RELATIONSHIP BETWEEN GABA AND COGNITION IN SCHIZOPHRENIA

AUTHORS	STAGE OF ILLNESS	REGION	COGNITIVE DOMAINS	SCANNER AND METHOD	PARTICIPANTS: SCHIZOPHRENIA (SZ)/ HEALTHY CONTROLS (HC)	APD	GABA (SZ VS. HC)	GABA/COGNITION
Kegeles et al. (2012)[39]	Chronic	Medial PFC (mPFC), dorsolateral PFC	N-Back working memory test	3 T J-edited	9 medication-naïve/5 unmedicated, and 16 medicated SZ/22	Yes & No	↑ GABA in mPFC in unmedicated vs HC; no difference in the dlPFC	No associations
Reid et al. (2018)[65]	FEP	ACC	Repeatable battery for assessment of neuropsychological status (RBANS)	T STEAM	21/21	Yes	GABA: No difference	Negative association between GABA and RBANS Total score in SZ, but not in HC
Rowland et al. (2016)[46]	Chronic	Anterior cingulate cortex (ACC)	General cognitive performance using the MATRICS Consensus Cognitive Battery (MCCB)	7 T STEAM	27/29	Yes	GABA: No difference	No associations
Rowland et al. (2013)[84]	Chronic	ACC Centrum semiovale (CSO)	RBANS	3 T MEGA-PRESS	11 young SZ (YSZ)- 10 old SZ (OSZ) /10 YHC-10 OHC	Yes	↓ ACC GABA in OSZ vs OHC; no GABA difference in CSO	ACC GABA positively associated with RBANS Attention score in combined SZ-HC
Marsman et al. (2014)[85]	Chronic	mPFC Parieto-occipital	Wechler Adult Intelligence Scale (WAIS)	7 T MEGA-sLASER	17/23	Yes	↓GABA in mPFC	mPFC GABA negatively associated with WAIS scores in SZ, but not HC

Two studies have combined MRS with diffusion tensor imaging.[80,96] To further the understanding of the deficit syndrome in schizophrenia, Rowland evaluated measures of MRS in prefrontal and parietal cortices and of fractional anisotropy (FA) in the track that connect these regions; reduced FA, but no differences in MRS metabolites were reported.[80] Studying FA of the cingulum bundle, Reid reported a negative association between ACC Glx and white matter integrity, suggesting that glutamate excitotoxicity could lead to white matter alterations.[96] Two studies combined MRS with gray matter structural measures.[40,97] Also suggestive of glutamate excitotoxicity, Kraguljac[40] reported a significant negative association between hippocampal Glx and hippocampal volumetric measures in patients, but not in controls; and Plitman[97] a significant negative association between striatal Glx and precommisural caudate volume in FEP subjects. Combining 7T MRS with magnetoencephalography in FEP, Gawne failed to find an association between ACC GABA and gamma band activity evoked by a 40 Hz auditory stimulus.[98] Rowland[48] reported an association between smaller auditory mismatch negativity (MMN) amplitude and lower GABA, lower glutamate and higher ratio of glutamine to glutamate in patients, but not in controls.

FUNCTIONAL MRS

All the aforementioned studies measured neurometabolites during a resting state, thus reflecting steady state concentrations. The recent development of functional MRS (fMRS), where changes in glutamate and GABA are measured during task performance, might provide a more fine-grained understanding of the link between metabolites and cognitive processes. An advantage of this technique is that, in contrast to the fMRI BOLD signal, fMRS has the potential to differentiate between changes in inhibitory and excitatory activity. (For a comprehensive review of the state of fMRS in cognitive neuroscience and psychiatric research, see [99].) So far, only two studies have reported fMRS experiment contrasting patients with schizophrenia with controls; both are suggestive of an impaired glutamatergic modulation, either during a Stroop task[100] or during a heat pain stress.[101]

PHOSPHOROUS MRS

Phosphorus MRS (^{31}P MRS) allows the quantification of phosphorus metabolites falling into two categories: cell membrane phospholipid (phosphomonoesters [PME] and phosphodiesters [PDE]), and energy related metabolites (adenosine triphosphate [ATP], phosphocreatine [PCr], inorganic phosphate [Pi]). The studies of these metabolites have the potential to interrogate fundamental neurobiological mechanisms, including membrane turnover and bioenergetics, which have been implicated in the pathology of schizophrenia. However, this technique requires expertise and special hardware. In addition, higher field strength is critical to provide good SNR and spectral resolution.

PME are membrane phospholipid precursors and PDE are breakdown products of membrane phospholipids. A recent review of ^{31}P MRS studies in schizophrenia[102] concluded that one relatively consistent pattern seen in schizophrenia was that of decreased PME in subcortical regions.

In the mitochondria, ATP, the high energy reservoir, is formed from adenosine diphosphate (ADP) and Pi. This process is coupled to the reaction which transfers the energy moiety of ATP to Cr to generate PCr. PCr is shuttled to the cytoplasm, where the reversible reaction can generate ATP from ADP and PCr. The shuttle of these phosphate moieties is fundamental for brain function; while ATP levels are maintained within a narrow range, PCr levels can fluctuate and act as an energy buffer. The comprehensive review of Yuksel[102] concluded that there were no consistent abnormalities in energy-related phosphorus metabolites between patients and controls. As with other MRS studies, heterogeneity of methods (field strength, voxel placement, acquisition, and post-processing methods), patient populations (high risk, FEP, chronic, medication naïve, free, or medicated), and small sample sizes have all contributed to muddy the results.

One promising technique is dynamic ^{31}P MRS, where changes in these bioenergetic markers between a rest and an activation state are measured, such as during visual stimulation. Such a study has already been published in a population of bipolar patients[103] studied at rest and during a photic stimulation paradigm.

^{13}CARBON SPECTROSCOPY

^{13}Carbon spectroscopy traces ^{13}Carbon in the glial and neuronal departments to directly determine the in vivo rates of the Krebs cycle, glutamate-glutamine cycle, and oxidative pathways.[104] But because it is a very technically complex technique, only a few studies have been conducted in human subjects, and direct quantification of altered cycling fluxes has yet to be performed in schizophrenia. Perhaps the most relevant study conducted in human subjects to date shows that subanaesthetic ketamine doses in healthy subjects significantly increase labeling of glutamate-C4, GABA-C2, and glutamine-C4 in the medial prefrontal cortex, and numerical increase labeling of these metabolites in the hippocampus both with [1-^{13}C]glucose and [2-^{13}C]acetate infusions,[105] suggesting a link between altered glutamate cycling and psychosis.

CONCLUDING REMARKS

Spectroscopy is a flexible imaging method allowing interrogating various pathophysiological pathways and thus ideally suited for the study of schizophrenia. The availability of higher field strength magnets, improved hardware, new optimized pulse sequences, and better spectral processing all have contributed to improve the quality of the data. Functional and dynamic MRS hold great promise for the future.

In schizophrenia, decreased NAA and altered glutamatergic/GABA measurements have been identified relatively consistently. However, because of the vast heterogeneity of published studies, little is still known about the trajectory of those abnormalities over the course of the illness. There is an urgent need for multicenter studies to characterize large and well-defined cohorts of patients, using harmonized imaging methods.

ACKNOWLEDGMENTS

This work was supported by the National Institute of Mental Health (R01MH102951 & MH113800, ACL; K23MH106683, NVK).

REFERENCES

1. Zhu H, Barker PB. MR spectroscopy and spectroscopic imaging of the brain. *Methods Mol Biol* 2011;711:203–226.
2. Mullins PG, McGonigle DJ, O'Gorman RL, Puts NA, Vidyasagar R, Evans CJ, Cardiff Symposium on MRSoG, Edden RA. Current practice in the use of MEGA-PRESS spectroscopy for the detection of GABA. *Neuroimage* 2014;86:43–52.
3. O'Gorman RL, Michels L, Edden RA, Murdoch JB, Martin E. In vivo detection of GABA and glutamate with MEGA-PRESS: reproducibility and gender effects. *J Magn Reson Imaging* 2011;33(5):1262–1267.
4. Skoch A, Jiru F, Bunke J. Spectroscopic imaging: basic principles. *Eur J Radiol* 2008;67(2):230–239.
5. van der Graaf M. In vivo magnetic resonance spectroscopy: basic methodology and clinical applications. *Eur Biophys J* 2010;39(4):527–540.
6. Haase A, Frahm J, Hanicke W, Matthaei D. 1H NMR chemical shift selective (CHESS) imaging. *Phys Med Biol* 1985;30(4):341–344.
7. Tkac I, Starcuk Z, Choi IY, Gruetter R. In vivo 1H NMR spectroscopy of rat brain at 1 ms echo time. *Magn Reson Med* 1999;41(4):649–656.
8. Cecil KM. Proton magnetic resonance spectroscopy: technique for the neuroradiologist. *Neuroimaging Clin N Am* 2013;23(3):381–392.
9. Schwerk A, Alves FD, Pouwels PJ, van Amelsvoort T. Metabolic alterations associated with schizophrenia: a critical evaluation of proton magnetic resonance spectroscopy studies. *J Neurochem* 2014;128(1):1–87.
10. Wijtenburg SA, Yang S, Fischer BA, Rowland LM. In vivo assessment of neurotransmitters and modulators with magnetic resonance spectroscopy: application to schizophrenia. *Neurosci Biobehav Rev* 2015;51:276–295.
11. Iwata Y, Nakajima S, Plitman E, et al. Neurometabolite levels in antipsychotic-naive/free patients with schizophrenia: a systematic review and meta-analysis of (1)H-MRS studies. *Prog Neuropsychopharmacol Biol Psychiatry* Mar 23 2018.
12. Kraguljac NV, Reid M, White D, Jones R, den Hollander J, Lowman D, Lahti AC. Neurometabolites in schizophrenia and bipolar disorder—a systematic review and meta-analysis. *Psychiatry Res* 2012;203(2-3):111–125.
13. Brugger S, Davis JM, Leucht S, Stone JM. Proton magnetic resonance spectroscopy and illness stage in schizophrenia—a systematic review and meta-analysis. *Biol Psychiatry* 2011;69(5):495–503.
14. Liemburg E, Sibeijn-Kuiper A, Bais L, et al. Prefrontal NAA and Glx levels in different stages of psychotic disorders: a 3T 1H-MRS study. *Sci Rep* 2016;6:21873.
15. Plitman E, de la Fuente-Sandoval C, Reyes-Madrigal F, Chavez S, Gomez-Cruz G, Leon-Ortiz P, Graff-Guerrero A. Elevated myo-inositol, choline, and glutamate levels in the associative striatum of antipsychotic-naive patients with first-episode psychosis: a proton magnetic resonance spectroscopy study with implications for glial dysfunction. *Schizophr Bull* 2016;42(2):415–424.
16. Bertolino A, Sciota D, Brudaglio F, et al. Working memory deficits and levels of N-acetylaspartate in patients with schizophreniform disorder. *Am J Psychiatry* 2003;160(3):483–489.
17. Ohrmann P, Siegmund A, Suslow T, et al. Cognitive impairment and in vivo metabolites in first-episode neuroleptic-naive and chronic medicated schizophrenic patients: a proton magnetic resonance spectroscopy study. *J Psychiatr Res* 2007;41(8):625–634.
18. Bustillo JR, Jones T, Chen H, et al. Glutamatergic and neuronal dysfunction in gray and white matter: a spectroscopic imaging study in a large schizophrenia sample. *Schizophr Bull* 2017;43(3): 611–619.
19. Callicott JH, Bertolino A, Egan MF, Mattay VS, Langheim FJ, Weinberger DR. Selective relationship between prefrontal N-acetylaspartate measures and negative symptoms in schizophrenia. *Am J Psychiatry* 2000;157(10):1646–1651.
20. Aydin K, Ucok A, Guler J. Altered metabolic integrity of corpus callosum among individuals at ultra high risk of schizophrenia and first-episode patients. *Biol Psychiatry* 2008;64(9):750–757.
21. He ZL, Deng W, Li ML, Chen ZF, Collier DA, Ma X, Li T. Detection of metabolites in the white matter of frontal lobes and hippocampus with proton in first-episode treatment-naive schizophrenia patients. *Early Interv Psychiatry* 2012;6(2):166–175.
22. Yamasue H, Fukui T, Fukuda R, Kasai K, Iwanami A, Kato N, Kato T. Drug-induced parkinsonism in relation to choline-containing compounds measured by 1H-MR spectroscopy in putamen of chronically medicated patients with schizophrenia. *Int J Neuropsychopharmacol* 2003;6(4):353–360.
23. Ando K, Takei N, Matsumoto H, Iyo M, Isoda H, Mori N. Neural damage in the lenticular nucleus linked with tardive dyskinesia in schizophrenia: a preliminary study using proton magnetic resonance spectroscopy. *Schizophr Res* 2002;57(2–3):273–279.
24. Rowland LM, Pradhan S, Korenic S, Wijtenburg SA, Hong LE, Edden RA, Barker PB. Elevated brain lactate in schizophrenia: a 7 T magnetic resonance spectroscopy study. *Transl Psychiatry* 2016;6(11):e967.
25. Matsuzawa D, Obata T, Shirayama Y, et al. Negative correlation between brain glutathione level and negative symptoms in schizophrenia: a 3T 1H-MRS study. *PLoS One* 2008;3(4):e1944.
26. Logothetis NK. What we can do and what we cannot do with fMRI. *Nature* 2008;453(7197):869–878.
27. Rothman DL, Behar KL, Hyder F, Shulman RG. In vivo NMR studies of the glutamate neurotransmitter flux and neuroenergetics: implications for brain function. *Annu Rev Physiol* 2003;65:401–427.
28. Moghaddam B, Javitt D. From revolution to evolution: the glutamate hypothesis of schizophrenia and its implication for treatment. *Neuropsychopharmacologyy* 2012;37(1):4–15.
29. Olney JW, Farber NB. Glutamate receptor dysfunction and schizophrenia. *Arch Gen Psychiatry* 1995;52(12):998–1007.
30. Javitt DC. Twenty-five years of glutamate in schizophrenia: are we there yet? *Schizophrenia bulletin* 2012;38(5):911–913.
31. Kraguljac NV, Frolich MA, Tran S, et al. Ketamine modulates hippocampal neurochemistry and functional connectivity: a combined magnetic resonance spectroscopy and resting-state fMRI study in healthy volunteers. *Mol Psychiatry* 2017;22(4):562–569.
32. Cohen SM, Tsien RW, Goff DC, Halassa MM. The impact of NMDA receptor hypofunction on GABAergic neurons in the pathophysiology of schizophrenia. *Schizophr Res* 2015;167(1-3):98–107.
33. Lahti AC, Koffel B, LaPorte D, Tamminga CA. Subanesthetic doses of ketamine stimulate psychosis in schizophrenia. *Neuropsychopharmacology* 1995;13(1):9–19.
34. Lahti AC, Weiler MA, Tamara Michaelidis BA, Parwani A, Tamminga CA. Effects of ketamine in normal and schizophrenic volunteers. *Neuropsychopharmacology* 2001;25(4):455–467.
35. Parwani A, Weiler MA, Blaxton TA, Warfel D, Hardin M, Frey K, Lahti AC. The effects of a subanesthetic dose of ketamine on verbal memory in normal volunteers. *Psychopharmacology (Berl)* 2005;183(3):265–274.

36. Kraguljac NV, Carle M, Froelich MA, Tran S, Yassa MA, White DM, Reddy A, Lahti AC. Mnemonic discrimination deficits in first-episode psychosis and a ketamine model suggests dentate gyrus pathology linked to N-methyl-D-aspartate receptor hypofunction *Biol Psychiatry: CNNI* 2017; in press.
37. Merritt K, Egerton A, Kempton MJ, Taylor MJ, McGuire PK. Nature of glutamate alterations in schizophrenia: a meta-analysis of proton magnetic resonance spectroscopy studies. *JAMA Psychiatry* 2016;73(7):665–674.
38. Marsman A, van den Heuvel MP, Klomp DW, Kahn RS, Luijten PR, Hulshoff Pol HE. Glutamate in schizophrenia: a focused review and meta-analysis of (1)H-MRS studies. *Schizophr Bull* 2013;39(1):120–129.
39. Kegeles LS, Mao X, Stanford AD, et al. Elevated prefrontal cortex gamma-aminobutyric acid and glutamate-glutamine levels in schizophrenia measured in vivo with proton magnetic resonance spectroscopy. *Arch Gen Psychiatry* 2012;69(5):449–459.
40. Kraguljac NV, White DM, Reid MA, Lahti AC. Increased hippocampal glutamate and volumetric deficits in unmedicated patients with schizophrenia. *JAMA Psychiatry* 2013;70(12):1294–1302.
41. de la Fuente-Sandoval C, Leon-Ortiz P, Favila R, Stephano S, Mamo D, Ramirez-Bermudez J, Graff-Guerrero A. Higher levels of glutamate in the associative-striatum of subjects with prodromal symptoms of schizophrenia and patients with first-episode psychosis. *Neuropsychopharmacology* 2011;36(9):1781–1791.
42. Kraguljac NV, White DM, Hadley J, Reid MA, Lahti AC. Hippocampal-parietal dysconnectivity and glutamate abnormalities in unmedicated patients with schizophrenia. *Hippocampus* 2014;24(12):1524–1532.
43. Egerton A, Bhachu A, Merritt K, McQueen G, Szulc A, McGuire P. Effects of antipsychotic administration on brain glutamate in schizophrenia: a systematic review of longitudinal 1H-MRS studies. *Front Psychiatry* 2017;8:66.
44. Egerton A, Modinos G, Ferrera D, McGuire P. Neuroimaging studies of GABA in schizophrenia: a systematic review with meta-analysis. *Transl Psychiatry* 2017;7(6):e1147.
45. Chiapponi C, Piras F, Piras F, Caltagirone C, Spalletta G. GABA system in schizophrenia and mood disorders: a mini review on third-generation imaging studies. *Front Psychiatry* 2016;7:61.
46. Rowland LM, Krause BW, Wijtenburg SA, et al. Medial frontal GABA is lower in older schizophrenia: a MEGA-PRESS with macromolecule suppression study. *Mol Psychiatry* 2016;21(2):198–204.
47. Marenco S, Meyer C, Kuo S, et al. Prefrontal GABA levels measured with magnetic resonance spectroscopy in patients with psychosis and unaffected siblings. *Am J Psychiatry* 2016;173(5):527–534.
48. Rowland LM, Summerfelt A, Wijtenburg SA, et al. Frontal glutamate and gamma-aminobutyric acid levels and their associations with mismatch negativity and digit sequencing task performance in schizophrenia. *JAMA Psychiatry* Feb 2016;73(2):166–174.
49. Ongur D, Haddad S, Prescot AP, Jensen JE, Siburian R, Cohen BM, Renshaw PF, Smoller JW. Relationship between genetic variation in the glutaminase gene GLS1 and brain glutamine/glutamate ratio measured in vivo. *Biol Psychiatry* 2011;70(2):169–174.
50. Poels EM, Kegeles LS, Kantrowitz JT, Javitt DC, Lieberman JA, Abi-Dargham A, Girgis RR. Glutamatergic abnormalities in schizophrenia: a review of proton MRS findings. *Schizophr Res* 2014;152(2-3):325–332.
51. Merritt K, McGuire P, Egerton A. Relationship between glutamate dysfunction and symptoms and cognitive function in psychosis. *Front Psychiatry* 2013;4:151.
52. Ota M, Ishikawa M, Sato N, et al. Glutamatergic changes in the cerebral white matter associated with schizophrenic exacerbation. *Acta Psychiatr Scand* 2012;126(1):72–78.
53. Egerton A, Brugger S, Raffin M, Barker GJ, Lythgoe DJ, McGuire PK, Stone JM. Anterior cingulate glutamate levels related to clinical status following treatment in first-episode schizophrenia. *Neuropsychopharmacology* 2012;37(11):2515–2521.
54. Demjaha A, Egerton A, Murray RM, Kapur S, Howes OD, Stone JM, McGuire PK. Antipsychotic treatment resistance in schizophrenia associated with elevated glutamate levels but normal dopamine function. *Biol Psychiatry* 2014;75(5):e11–13.
55. de la Fuente-Sandoval C, Leon-Ortiz P, Azcarraga M, et al. Glutamate levels in the associative striatum before and after 4 weeks of antipsychotic treatment in first-episode psychosis: a longitudinal proton magnetic resonance spectroscopy study. *JAMA Psychiatry* 2013;70(10):1057–1066.
56. Wood SJ, Berger GE, Wellard RM, Proffitt T, McConchie M, Velakoulis D, McGorry PD, Pantelis C. A 1H-MRS investigation of the medial temporal lobe in antipsychotic-naive and early-treated first episode psychosis. *Schizophr Res* 2008;102(1-3):163–170.
57. Theberge J, Bartha R, Drost DJ, et al. Glutamate and glutamine measured with 4.0 T proton MRS in never-treated patients with schizophrenia and healthy volunteers. *Am J Psychiatry* 2002;159(11):1944–1946.
58. Aoyama N, Theberge J, Drost DJ, et al. Grey matter and social functioning correlates of glutamatergic metabolite loss in schizophrenia. *Br J Psychiatry* 2011;198(6):448–456.
59. Theberge J, Williamson KE, Aoyama N, et al. Longitudinal grey-matter and glutamatergic losses in first-episode schizophrenia. *Br J Psychiatry* 2007;191:325–334.
60. Kelemen O, Kiss I, Benedek G, Keri S. Perceptual and cognitive effects of antipsychotics in first-episode schizophrenia: the potential impact of GABA concentration in the visual cortex. *Prog Neuropsychopharmacol Biol Psychiatry* 2013;47:13–19.
61. Brandt AS, Unschuld PG, Pradhan S, et al. Age-related changes in anterior cingulate cortex glutamate in schizophrenia: A (1)H MRS Study at 7 Tesla. *Schizophr Res* 2016;172(1-3):101–105.
62. Marsman A, Mandl RC, Klomp DW, et al. GABA and glutamate in schizophrenia: a 7 T (1)H-MRS study. *Neuroimage Clin* 2014;6:398–407.
63. Rowland LM, Pradhan S, Korenic S, Wijtenburg SA, Hong LE, Edden RA, Barker PB. Elevated brain lactate in schizophrenia: a 7 T magnetic resonance spectroscopy study. *Transl Psychiatry* 2016;6(11):e967.
64. Thakkar KN, Rosler L, Wijnen JP, Boer VO, Klomp DW, Cahn W, Kahn RS, Neggers SF. 7T proton magnetic resonance spectroscopy of gamma-aminobutyric acid, glutamate, and glutamine reveals altered concentrations in patients with schizophrenia and healthy siblings. *Biol Psychiatry* 2017;81(6):525–535.
65. Reid MA, Salibi N, White DM, Gawne TJ, Denney TS, Lahti AC. 7T proton magnetic resonance spectroscopy of the anterior cingulate cortex in first-episode schizophrenia. *Schizophr Bull* Jan 29 2018.
66. Szulc A, Galinska B, Tarasow E, et al. Proton magnetic resonance spectroscopy study of brain metabolite changes after antipsychotic treatment. *Pharmacopsychiatry* 2011;44(4):148–157.
67. de la Fuente-Sandoval C, Reyes-Madrigal F, Mao X, et al. Prefrontal and striatal gamma-aminobutyric acid levels and the effect of antipsychotic treatment in first-episode psychosis patients. *Biol Psychiatry* 2018;83(6):475–483.
68. Lahti AC, Holcomb HH. Schizophrenia, VIII: pharmacologic models. *The Am J Psychiatry* 2003;160(12):2091.
69. Weiler MA, Thaker GK, Lahti AC, Tamminga CA. Ketamine effects on eye movements. *Neuropsychopharmacology* 2000;23(6):645–653.
70. Rowland LM, Bustillo JR, Mullins PG, et al. Effects of ketamine on anterior cingulate glutamate metabolism in healthy humans: a 4-T proton MRS study. *Am J Psychiatry* 2005;162(2):394–396.
71. Stone JM, Dietrich C, Edden R, et al. Ketamine effects on brain GABA and glutamate levels with 1H-MRS: relationship to ketamine-induced psychopathology. *Mol Psychiatry* 2012;17(7):664–665.
72. Javitt DC, Carter CS, Krystal JH, et al. Utility of imaging-based biomarkers for glutamate-targeted drug development in psychotic disorders: a randomized clinical trial. *JAMA Psychiatry* 2018;75(1):11–19.
73. Taylor MJ, Tiangga ER, Mhuircheartaigh RN, Cowen PJ. Lack of effect of ketamine on cortical glutamate and glutamine in healthy volunteers: a proton magnetic resonance spectroscopy study. *J Psychopharmacol* 2012;26(5):733–737.

74. Valentine GW, Mason GF, Gomez R, Fasula M, Watzl J, Pittman B, Krystal JH, Sanacora G. The antidepressant effect of ketamine is not associated with changes in occipital amino acid neurotransmitter content as measured by [(1)H]-MRS. *Psychiatry Res*2011;191(2):122–127.
75. Kegeles LS, Mao X, Ojeil N, et al. J-editing/MEGA-PRESS time-course study of the neurochemical effects of ketamine administration in healthy humans. *Proc Intl Soc Mag Reson Med* 2013;21(2013):1206.
76. Robbins TW, Murphy ER. Behavioural pharmacology: 40+ years of progress, with a focus on glutamate receptors and cognition. *Trends Pharmacol Sci* 2006;27(3):141–148.
77. Newcomer JW, Farber NB, Jevtovic-Todorovic V, Selke G, Melson AK, Hershey T, Craft S, Olney JW. Ketamine-induced NMDA receptor hypofunction as a model of memory impairment and psychosis. *Neuropsychopharmacology* 1999;20(2):106–118.
78. Krystal JH, D'Souza DC, Karper LP, et al. Interactive effects of subanesthetic ketamine and haloperidol in healthy humans. *Psychopharmacology (Berl)* 1999;145(2):193–204.
79. Shirayama Y, Obata T, Matsuzawa D, et al. Specific metabolites in the medial prefrontal cortex are associated with the neurocognitive deficits in schizophrenia: a preliminary study. *Neuroimage* 2010;49(3):2783–2790.
80. Rowland LM, Spieker EA, Francis A, Barker PB, Carpenter WT, Buchanan RW. White matter alterations in deficit schizophrenia. *Neuropsychopharmacology* 2009;34(6):1514–1522.
81. Reid MA, Stoeckel LE, White DM, et al. Assessments of function and biochemistry of the anterior cingulate cortex in schizophrenia. *Biol Psychiatry* 2010;68(7):625–633.
82. Bustillo JR, Chen H, Gasparovic C, et al. Glutamate as a marker of cognitive function in schizophrenia: a proton spectroscopic imaging study at 4 Tesla. *Biol Psychiatry* 2011;69(1):19–27.
83. Reid MA, Kraguljac NV, Avsar KB, White DM, den Hollander JA, Lahti AC. Proton magnetic resonance spectroscopy of the substantia nigra in schizophrenia. *Schizophr Res* 2013;147(2-3):348–354.
84. Rowland LM, Kontson K, West J, Edden RA, Zhu H, Wijtenburg SA, Holcomb HH, Barker PB. In vivo measurements of glutamate, GABA, and NAAG in schizophrenia. *Schizophr Bull* 2013;39(5):1096–1104.
85. Dempster K, Norman R, Theberge J, Densmore M, Schaefer B, Williamson P. Glutamatergic metabolite correlations with neuropsychological tests in first episode schizophrenia. *Psychiatry Res* 2015;233(2):180–185.
86. Marsman A, Mandl RC, Klomp DW, et al. GABA and glutamate in schizophrenia: a 7 T (1)H-MRS study. *Neuroimage Clin* 2014;6:398–407.
87. Shulman RG, Hyder F, Rothman DL. Insights from neuroenergetics into the interpretation of functional neuroimaging: an alternative empirical model for studying the brain's support of behavior. *J Cereb Blood Flow Metab* 2014;34(11):1721–1735.
88. Falkenberg LE, Westerhausen R, Specht K, Hugdahl K. Resting-state glutamate level in the anterior cingulate predicts blood-oxygen level-dependent response to cognitive control. *Proc Natl Acad Sci U S A* 2012;109(13):5069–5073.
89. Duncan NW, Wiebking C, Northoff G. Associations of regional GABA and glutamate with intrinsic and extrinsic neural activity in humans-a review of multimodal imaging studies. *Neurosci Biobehav Rev* 2014;47:36–52.
90. Hutcheson NL, Reid MA, White DM, et al. Multimodal analysis of the hippocampus in schizophrenia using proton magnetic resonance spectroscopy and functional magnetic resonance imaging. *Schizophr Res* 2012;140(1–3):136–142.
91. White DM, Kraguljac NV, Reid MA, Lahti AC. Contribution of substantia nigra glutamate to prediction error signals in schizophrenia: a combined magnetic resonance spectroscopy/functional imaging study. *NPJ Schizophr* 2015;1:14001.
92. Overbeek G, Gawne TJ, Reid MA, Salibi N, Kraguljac NV, White DM, Lahti AC. Relationship between cortical excitation and inhibition and task-induced activation and deactivation: A combined MR Spectroscopy and functional MRI study at 7T in first episode psychosis. *Biological Psychiatry: CNNI* 2018: in press.
93. Falkenberg LE, Westerhausen R, Craven AR, et al. Impact of glutamate levels on neuronal response and cognitive abilities in schizophrenia. *Neuroimage Clin* 2014;4:576–584.
94. Joyce EM, Roiser JP. Cognitive heterogeneity in schizophrenia. *Curr Opin Psychiatry* 2007;20(3):268–272.
95. Stone JM, Howes OD, Egerton A, et al. Altered relationship between hippocampal glutamate levels and striatal dopamine function in subjects at ultra high risk of psychosis. *Biol Psychiatry* 2010;68(7):599–602.
96. Reid MA, White DM, Kraguljac NV, Lahti AC. A combined diffusion tensor imaging and magnetic resonance spectroscopy study of patients with schizophrenia. *Schizophr Res* 2016;170(2-3):341–350.
97. Plitman E, Patel R, Chung JK, et al. Glutamatergic metabolites, volume and cortical thickness in antipsychotic-naive patients with first-episode psychosis: implications for excitotoxicity. *Neuropsychopharmacology* 2016;41(10):2606–2613.
98. Gawne TJ, Overbeek GJ, Killen JF, White DM, Reid MA, Lahti AC. A combined magnetoencephalography and 7T proton magnetic resonance spectroscopy study in first episode schizophrenia. *71st Annual Meeting of the Society of Biological Psychiatry*. Atlanta, GA; 2016.
99. Stanley JA, Raz N. Functional Magnetic Resonance Spectroscopy: The "New" MRS for Cognitive Neuroscience and Psychiatry Research. *Front Psychiatry* 2018;9:76.
100. Taylor R, Neufeld RW, Schaefer B, Densmore M, Rajakumar N, Osuch EA, Williamson PC, Theberge J. Functional magnetic resonance spectroscopy of glutamate in schizophrenia and major depressive disorder: anterior cingulate activity during a color-word Stroop task. *NPJ Schizophr* 2015;1:15028.
101. Chiappelli J, Shi Q, Wijtenburg SA, et al. Glutamatergic Response to Heat Pain Stress in Schizophrenia. *Schizophr Bull* Sep 23 2017.
102. Yuksel C, Tegin C, O'Connor L, Du F, Ahat E, Cohen BM, Ongur D. Phosphorus magnetic resonance spectroscopy studies in schizophrenia. *J Psychiatr Res* 2015;68:157–166.
103. Yuksel C, Du F, Ravichandran C, et al. Abnormal high-energy phosphate molecule metabolism during regional brain activation in patients with bipolar disorder. *Mol Psychiatry* 2015;20(9):1079–1084.
104. Moreno A, Ross BD, Bluml S. Direct determination of the N-acetyl-L-aspartate synthesis rate in the human brain by (13)C MRS and [1-(13)C]glucose infusion. *J Neurochem* 2001;77(1):347–350.
105. Chowdhury GM, Behar KL, Cho W, Thomas MA, Rothman DL, Sanacora G. (1)H-[(1)(3)C]-nuclear magnetic resonance spectroscopy measures of ketamine's effect on amino acid neurotransmitter metabolism. *Biol Psychiatry* 2012;71(11):1022–1025.

PATHOPHYSIOLOGY OF PSYCHOSIS

NEUROTRANSMITTERS

31

DOPAMINERGIC MECHANISMS UNDERLYING PSYCHOSIS

Michael A. P. Bloomfield and Oliver D. Howes

INTRODUCTION

The theory that dopaminergic mechanisms play a role in psychosis has developed since the mid-twentieth century. Here we review the dopamine hypothesis, the main lines of evidence in support of it, the relationship of dopaminergic function to symptoms of psychosis, current research and its limitations. A PubMed search with the terms "dopamine AND schizophrenia" elicits over 9,500 articles (search conducted June 2019). Inevitably space constraints mean we will not be able to consider all this work and we apologize in advance to colleagues where this includes their work. Instead we aim to provide an overview of the main lines of evidence as currently considered by the field.

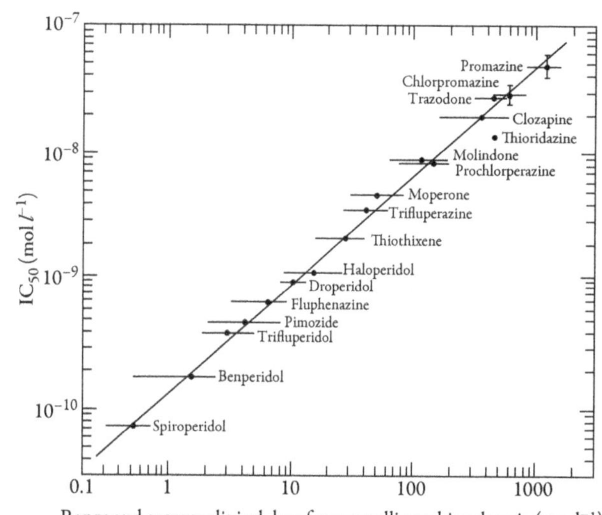

Figure 31.1 The relationship between dopamine receptor half maximal inhibitory concentration and the clinical dose of antipsychotics required for treating schizophrenia.
Seeman et al. (1976).[4]

EVIDENCE OF DOPAMINERGIC DYSFUNCTION IN PSYCHOSIS

THE ORIGINS OF THE DOPAMINE HYPOTHESIS

The dopamine hypothesis was initially proposed by van Rossum[1] (and modified several times[2] since), and has been a dominant pathoetiological theory of psychosis for decades following seminal research which found that chlorpromazine and haloperidol, medicines used to treat schizophrenia, affected dopaminergic function[3] and that the clinical potency of antipsychotics was directly related to their affinity for dopamine receptors[4] (Figure 31.1). In parallel, the stimulant drug amphetamine, which causes dopamine release, was found to cause psychotic symptoms in healthy individuals[5] and worsen psychosis in people with schizophrenia.[6]

POSTMORTEM STUDIES

Elevated levels of brain dopamine and dopamine receptors in people with schizophrenia were initially reported in postmortem studies.[7,8] Increased striatal dopamine was unrelated to antipsychotic medication, whereas increased dopamine receptor binding sites were only present for patients in whom antipsychotic medication had been continued until death. This was interpreted as reflecting an iatrogenic increase in dopamine receptor availability. However, an earlier postmortem study[9] also found increased dopamine receptor availability in antipsychotic-naïve patients which would not support drug-induced changes in receptor availability.

PSYCHOSIS AS SUBCORTICAL HYPERDOPAMINERGIA AND FRONTAL HYPODOPAMINERGIA

The dopamine hypothesis was later modified[10] to propose that subcortical hyperdopaminergia gave rise to positive symptoms, and frontal hypodopaminergia gave rise to negative symptoms. Around the same time, other versions of the dopamine hypothesis incorporated other neurotransmitters, such as serotonin.[11]

THE DOPAMINE HYPOTHESIS IN THE AGE OF MOLECULAR IMAGING

Molecular imaging including positron emission tomography (PET) has provided evidence that elevated striatal presynaptic dopamine synthesis leads to psychotic symptoms.[12] Presynaptic dopaminergic function can be indexed using radiolabeled levo-dihydroxyphenylalanine ([^{18}F]-DOPA) to measure dopamine synthesis capacity or by measuring the change in radiotracer binding to dopamine type 2 and 3 ($D_{2/3}$) receptors after experimentally inducing dopamine release. Although not replicated in all studies,[13] elevated striatal dopamine synthesis capacity has been found in people with schizophrenia taking medication and those that were medication-naïve,[14] people at risk of psychosis,[12] and first-degree relatives of people with schizophrenia.[15] Importantly, dopamine elevation is positively correlated with severity of prodromal psychotic symptoms.[12] Subsequently, two separate meta-analyses[2,16] have reported that, compared to healthy controls, people with schizophrenia have increased striatal dopamine synthesis capacity with a large effect size of $d = .79$.

Other aspects of dopaminergic function have also been investigated. In a study[17] following dopamine depletion with alpha-methyl-p-tyrosine (α-MPT) administration, people with schizophrenia exhibited increased dopamine type 2 (D_2) receptor availability compared to controls. A meta-analysis[2] of studies comparing baseline D_2 receptor availability reported a small increase in patients with psychosis compared to controls, however this is not reliably seen in patients who are antipsychotic-naïve. Studies of the vesicular monoamine transporter[18] and dopamine transporter availability[2,19] found no significant difference between patients and controls. Elevated dopamine synthesis and release capacity is therefore the most widely replicated in vivo neurochemical finding in psychosis[2] and predicts the subsequent development of psychosis in at-risk individuals[20] (Figure 31.2). Basal $D_{2/3}$ receptor occupancy by dopamine is elevated in schizophrenia, indicative of higher baseline synaptic dopamine levels.[21] Since striatal baseline synaptic dopamine levels and dopamine release are closely correlated in psychosis,[22] the same abnormality likely underlies both.

There has been ongoing interest in the role of extra-striatal brain regions especially frontal regions in light of proposed hypofrontality.[10,23] While cortical dopamine receptor levels appear unaltered,[24] people with psychosis appear to have reduced amphetamine-induced dopamine release in the dorsolateral prefrontal cortex (DLPFC) and this correlated with DLPFC activation during working memory functional magnetic resonance imaging (fMRI),[25] providing evidence for reduced cortical dopamine levels. Furthermore, there is recent evidence of reduced dopamine type 1 (D_1) receptor in the

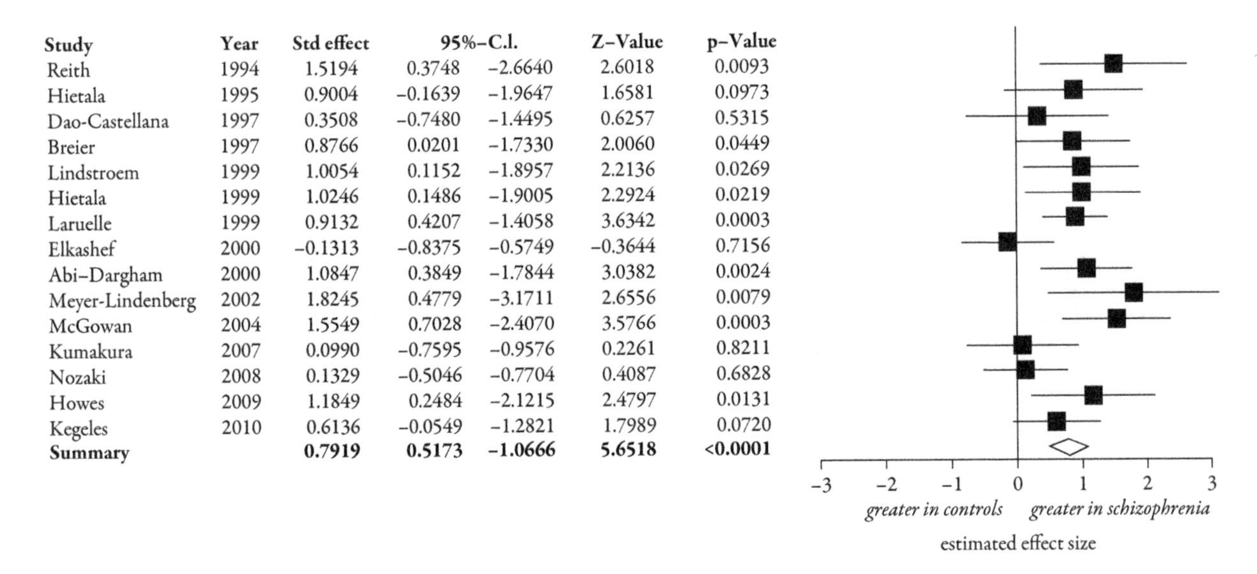

Study	Year	Std effect	95%-C.I.		Z-Value	p-Value
Reith	1994	1.5194	0.3748	−2.6640	2.6018	0.0093
Hietala	1995	0.9004	−0.1639	−1.9647	1.6581	0.0973
Dao-Castellana	1997	0.3508	−0.7480	−1.4495	0.6257	0.5315
Breier	1997	0.8766	0.0201	−1.7330	2.0060	0.0449
Lindstroem	1999	1.0054	0.1152	−1.8957	2.2136	0.0269
Hietala	1999	1.0246	0.1486	−1.9005	2.2924	0.0219
Laruelle	1999	0.9132	0.4207	−1.4058	3.6342	0.0003
Elkashef	2000	−0.1313	−0.8375	−0.5749	−0.3644	0.7156
Abi-Dargham	2000	1.0847	0.3849	−1.7844	3.0382	0.0024
Meyer-Lindenberg	2002	1.8245	0.4779	−3.1711	2.6556	0.0079
McGowan	2004	1.5549	0.7028	−2.4070	3.5766	0.0003
Kumakura	2007	0.0990	−0.7595	−0.9576	0.2261	0.8211
Nozaki	2008	0.1329	−0.5046	−0.7704	0.4087	0.6828
Howes	2009	1.1849	0.2484	−2.1215	2.4797	0.0131
Kegeles	2010	0.6136	−0.0549	−1.2821	1.7989	0.0720
Summary		**0.7919**	**0.5173**	**−1.0666**	**5.6518**	**<0.0001**

Figure 31.2 Studies of presynaptic dopaminergic function: Forrest plot showing the effect size and 95% confidence intervals of the difference between patients with schizophrenia and controls by study. There was evidence of a significant elevation in schizophrenia with a summary effect size of $d = .79$.
Reproduced from Howes et al. (2012).[2]

DLPFC in people with first-episode psychosis who are medication naïve.[26]

CLINICAL HIGH-RISK GROUPS

The development of structured clinical interviews has enabled the identification of people with attenuated psychotic symptoms, some of whom are in the prodrome to a psychotic disorder. Evidence of dopaminergic dysfunction in these clinical high-risk groups includes increased peripheral dopamine metabolites,[27] the efficacy of dopamine antagonists in reducing symptom severity,[28] and increased dopamine synthesis capacity[12] which was positively associated with the severity of prodromal symptoms. These findings have been extended to increased dopamine release (discussed later),[29] replicated,[30] and found to be specific to those individuals who later progress to psychosis.[20] Likewise, in people in the psychosis prodrome, dopamine synthesis capacity increases further with the development of acute psychosis.[20] Recently, reduced striatal dopamine synthesis capacity has been reported in patients with schizophrenia during remission of positive symptoms.[31] Taken together, these findings suggest there are dynamic changes in the dopamine system linked to psychosis.

DOPAMINE DYSFUNCTION IN PSYCHOSIS ACROSS DIAGNOSES

There is evidence that dopaminergic dysfunction is a transdiagnostic feature of psychosis. Even with comorbid conditions where the dopamine system is blunted, such as addiction,[32] there is a relationship between dopamine release and increase in psychotic symptoms. Increased striatal dopamine synthesis capacity has been observed in other syndromes of psychosis including bipolar affective disorder[33] whereby positive psychotic symptoms correlated with dopamine synthesis capacity when combined with a first episode schizophrenia sample, irrespective of diagnosis,[33] and temporal lobe epilepsy.[34]

THE ROLE OF DOPAMINE IN COGNITION—MOTIVATION AND SALIENCE

To understand how dopamine dysfunction may underlie psychosis, we first describe the role of dopamine in cognition. The striatum is a key structure of the dopamine system and a component of the basal ganglia. Based on the nigrostriatal hypodopaminergia observed in Parkinson's disease, these brain systems were originally described as being primarily motor in function. However, there is mounting evidence in favor of key roles for them in cognition.

Dopamine is a monoamine neurotransmitter that is most abundant in the striatum (Figure 31.3), limbic system, frontal cortex, and hypothalamus. The dopamine synthesis pathway involves conversion of tyrosine to DOPA via tyrosine hydroxylase, followed by conversion of DOPA to dopamine via dopamine decarboxylase. The striatum projects topographically to the pallidal complex, the ventral tegmental area, and the substantia nigra.[35] The outputs from the globus pallidus, internal segment (GPi)/SN, then project back to the cortex via the thalamus, completing the basic striato-thalamo-cortical circuit. This is known as the direct pathway. The side loop, from the striatum via the globus pallidus, external segment (GPe), passes through the subthalamic nucleus to the GPi, and is referred to as the indirect pathway.

There are four major dopaminergic projections in the human brain: the nigrostriatal pathway (from the substantia nigra to the striatum), the mesolimbic pathway (from the

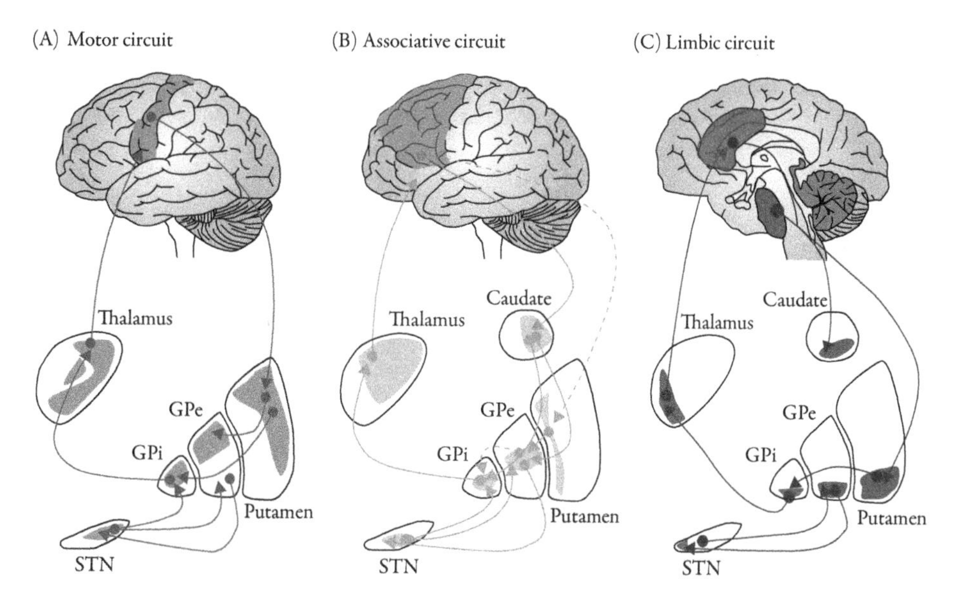

Figure 31.3 The major circuits of the striatum. The caudate nucleus, putamen, and subthalamic nucleus (STN) receive cortical and subcortical inputs. The internal segment of the globus pallidus (GPi) and the substantia nigra pars reticulata (not shown) are the main output nuclei. The striatum is involved in motor (A), cognitive (B) and emotional (C) processing, and these functions show a topographical organization.[36] See color insert 2 in front of book for full colorized version

ventral tegmental area [VTA] to the nucleus accumbens and amygdala via the medial forebrain bundle), the mesocortical pathway (from the VTA to the frontal cortex via the median forebrain bundle), and the tuberoinfundibular pathway (from the hypothalamus to the pituitary gland and median eminence).

Selective damage to dopaminergic fibers causes feeding and drinking deficits.[37] Mesolimbic lesions reduce forward locomotion for reward-seeking behavior,[38] while nigrostriatal lesions disrupt appetitive behavior.[39] Likewise, dopamine antagonism reduces the rewarding effects of food,[40] water,[41] stimulants,[42] and lateral hypothalamic electrical stimulation.[43] Whereas dopamine-antagonist-treated rodents can perform operant conditioning tasks, they display disrupted reward-related learned behaviors.[44,45] These findings gave rise to a range theories of dopaminergic function in behavior relating to hedonia,[46] reward,[40] reinforcement,[47] and motivational salience.[48] Here, reinforcement refers to the "stamping-in" of the association between a stimulus and a reward[49] and dopamine antagonism results in a progressive decline in reinforced behavior.[46]

Rewarding and reward-associated stimuli increase motivational arousal.[50] The motivating effect of rewards obtained before a behaviour is known as *priming*. There is evidence that priming is, to some extent, dopamine-dependent[51] and dopaminergic potentiation can prime food and drug-seeking behavior.[52,53] However, dopaminergic mechanisms are not necessary for prereward motivation, but rather they amplify this process.[54] In other words, there is "enhancement of reward '*wanting*' without enhanced '*liking*,'" as wanting and liking have been proposed to represent two separate dimensions of reward function.[48] There is evidence from human in vivo imaging that stimulant-induced euphoria is correlated with drug-induced dopamine release.[55] Importantly, in addition to processing *positively* rewarding events, dopamine is also involved in "*negatively* rewarding" events, i.e., aversive stimuli.[56]

The priming (drive-like) effects of an encounter with an otherwise neutral stimulus that has acquired motivational importance through prior association with a reward is termed incentive motivation. Whilst dopamine has been proposed to be central for incentive motivation,[57] it has been argued that the main role of dopamine in incentive motivation lies in the "stamping in" of the reward value of a stimulus.[58] This is based on the finding that dopamine blockade severely impedes Pavlovian conditioning,[45] while established incentive motivational stimuli can result in conditioned behaviors under dopaminergic blockade.[59]

Electrophysiological studies demonstrate that the midbrain dopaminergic system signals reward prediction errors.[60] Initially unpredictable rewards elicit dopamine neuron activations. As the reward becomes predicted by the conditioned stimulus with continued experience, the reward-elicited activations decrease. If the predicted reward does not occur (e.g., because of an incorrect response or omission of reward to a correct response) then a depression in dopamine neuron activity occurs at the same time as an increase in activity would have occurred had the reward been obtained (Figure 31.4).

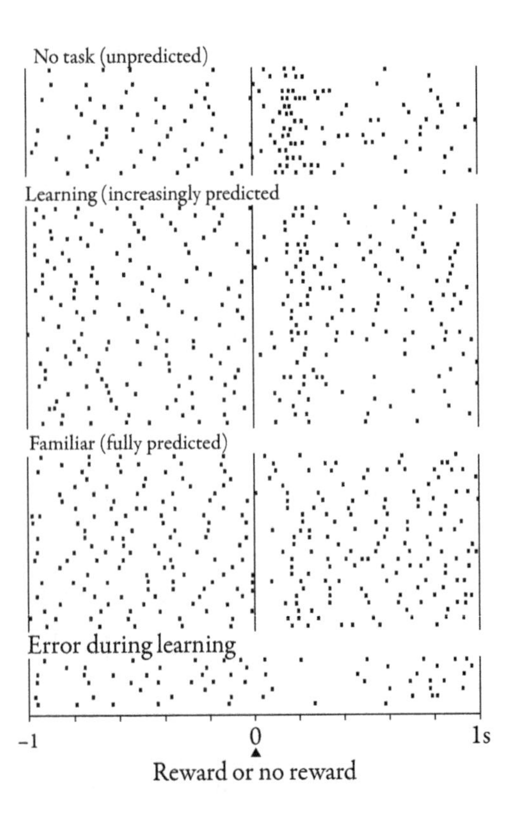

Figure 31.4 Coding of reward-prediction error during learning by dopamine neurons. (*Dots*) Neuronal impulses, each horizontal line is showing the firing of one dopamine neuron, with the chronological sequence in each panel being from *top* to *bottom*. Rewards were small quantities of apple juice delivered to the mouth of a monkey. No task: The temporally unpredicted occurrence of reward outside of any task induces reliable neuronal activation to reward. Learning: The presentation of a novel picture pair in a two-picture discrimination task leads to uncertain behavioral performance with unpredictable occurrence of reward and dopamine response. (*Top* to *bottom*) Response decreases with increasing picture acquisition (only correct trials shown). Familiar: Presentation of known pictures in same task leads to predictable occurrence of reward and no dopamine response. Error during learning: Error performance with novel pictures leads to omission of reward.
Hollerman and Schultz (1998).[60]

In addition to encoding the motivational value for positive outcomes, there are populations of neurons that respond to salient but nonrewarding (e.g., aversive) stimuli thus encoding a motivational salience.[61]

LINKING DOPAMINE DYSFUNCTION TO SYMPTOMS: THE ABERRANT SALIENCE HYPOTHESIS

In light of the findings linking dopaminergic dysfunction to schizophrenia it was suggested that striatal dopamine dysregulation drives the development of psychosis.[62] Based on animal evidence that striatal dopamine release underlies the attribution of motivational salience to stimuli[48] and evidence in humans, a mechanism to explain how elevated striatal dopamine could lead to psychotic symptoms was proposed[63] where dopamine dysregulation leads to the aberrant assignment of salience to internal and external stimuli, and psychotic symptoms are secondary to cognitive rationalization of these experiences. Please see the chapter by Kapur in this book for a full account of this hypothesis.

LINKING THE STRESS RESPONSE AND DOPAMINE

Given the implications for an organism's survival, the dopamine system's functions in salience processing and learning mean that it plays key roles in the brain's response to stressful experiences. Acute and chronic stressors differentially alter dopaminergic function in a region-specific manner. Acute stressors lead to cortical dopamine release which subsequently can attenuate striatal dopamine release[64] potentially via cortical glutamatergic projections to the midbrain and striatum[65] Chronic stress alters dopaminergic responses to later activating stimuli,[66] whilst cross-sensitizing the response to subsequent stimulant administration.[67] Dopaminergic dysfunction caused by early life stress appears to be complex and can be long-standing, including persistence into adulthood.[68–72] Divergent directionality of effects in the different components of the dopamine system may underlie the nature of the stress paradigm deployed in different studies.

Studies in humans have confirmed that stress alters dopamine function. PET studies with healthy volunteers have demonstrated widespread stress-induced cortical dopamine release[73] and stress-induced cortisol levels were positively associated with amphetamine-induced dopamine release in the striatum.[74] There is evidence of a relationship between levels of childhood adversity and the spatial extent of cortical dopamine release.[75] Elevated urinary dopamine metabolite levels have been reported in girls with a history of sexual abuse compared to those without.[76] Ventral striatal dopamine release was increased in response to psychosocial stress in humans who reported insufficient early life maternal care.[77]

Given that people with psychosis report high levels of psychosocial stressors, the effects of stress on the dopamine system and the findings of increased dopamine synthesis capacity in psychosis,[78] a number of studies have investigated the dopamine response to stressors in people with psychosis. People with schizophrenia, those at clinical high risk (CHR) and first-degree relatives have a potentiated response of the peripheral dopamine metabolite, homovanillic acid, to acute stress.[79] Compared to controls, first-degree relatives of people with schizophrenia show less widespread cortical dopaminergic response to stress,[80] nonetheless this was related to subjective stress and psychotic experiences caused by stress.[81] Mizrahi et al. found that antipsychotic-naïve people with schizophrenia and those at clinical high-risk have sensitized psychosocial stress-induced dopamine release in the associative and sensorimotor striatal subdivisions of (Figure 31.5).[29] Taken together with positive associations between childhood adversity and dopamine release,[82] these findings support the view that stress-induced dopaminergic dysfunction is involved in the pathophysiology of psychosis.

THE FINAL COMMON PATHWAY THEORY

The hypothesis that dopaminergic mechanisms are central to psychosis has become highly influential. This followed the discovery of the dopaminergic action of antipsychotic medicines[3] and the theory proposed by Davis et al.[10] relating positive symptoms to subcortical hyperdopaminergia and negative symptoms to frontal hypodopaminergia. These theories and subsequent research were synthesized into a model in which various risk factors for psychosis converge on dopaminergic pathways to cause psychosis through creating a state of aberrant salience as the "Final Common Pathway."[83]

A consistent finding in schizophrenia imaging research[2] has been that patients with psychosis have elevated striatal

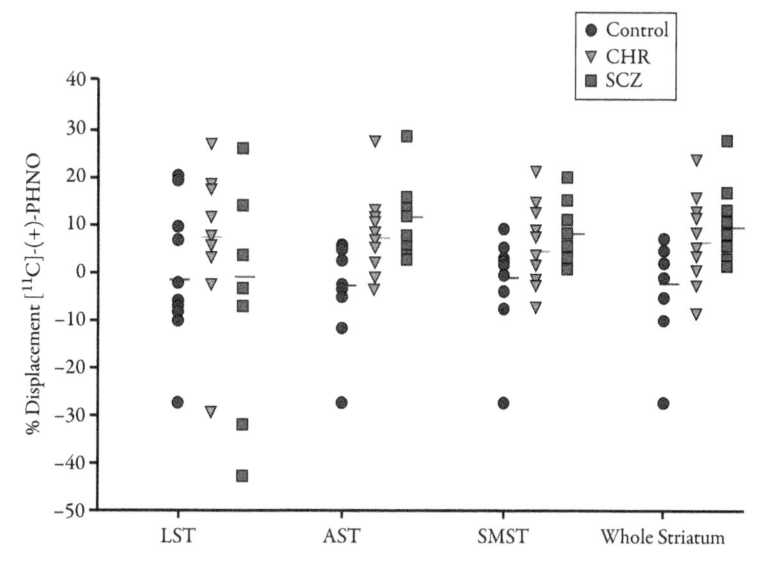

Figure 31.5 [^{11}C]-PHNO positron emission tomography response to stress in the corpus striatum and its functional subdivisions. AST, associative striatum; CHR, clinical high-risk group; LST, limbic striatum; SCZ, schizophrenia group; SMST, sensorimotor striatum.
Figure taken from Mizrahi et al. (2012).[29]

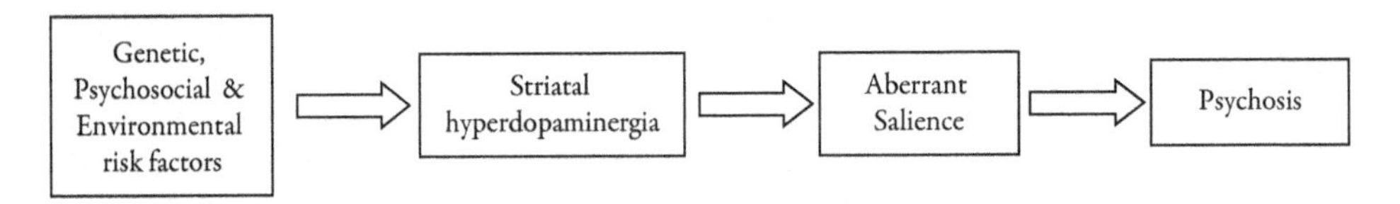

Figure 31.6 Multiple hits interact to result in striatal dopamine dysregulation to alter the appraisal of stimuli and resulting in psychosis, while current antipsychotic drugs act downstream of the primary dopamine dysregulation.

dopamine synthesis capacity, striatal dopamine release,[84] and baseline occupancy of D_2 receptors by dopamine.[21] Furthermore, all currently licensed antipsychotic drugs block D_2 receptors.[85] The emergent genetic evidence is also in keeping with a role for dopaminergic dysfunction in psychosis. While other genes are also significant at the genome-wide level, a significant contributor to the polygenic risk of schizophrenia is likely to be the gene encoding the D2 receptor (DRD2 gene), as identified by the Schizophrenia Working Group of the Psychiatric Genomics Consortium[86] genome-wide association study.

A variety of environmental psychosocial risk factors increase the risk of psychosis. Many of these share the common factor of being psychosocial stressors. These risk factors include migration, urban upbringing and abuse during childhood, and these relate to experiences of social isolation or subordination.[87] Evidence from animal studies indicate that both social isolation and subordination are associated with hyperdopaminergia,[68,69] as are models of maternal separation.[88] Other risk factors including obstetric complications have also been linked to dopaminergic dysfunction in animal models[89] and humans.[90]

Dysfunction of other neurotransmitter systems likely converges on the dopamine system to give rise to psychosis. For example, using the methylazoxymethanol acetate mouse model,[91] deficits in hippocampal gamma-aminobutyric acid (GABA) interneurons leads to glutamate dysfunction and disinhibited modulation of midbrain dopamine cells. Likewise, acute administration of ketamine, an N-methyl-D-aspartate (NMDA) receptor antagonist which produces an acute psychotic state, results in increased cortical and striatal dopamine levels.[92] Thus bridging a major theory of how glutamatergic dysfunction gives rise to psychosis, the NMDA hypofunction model, with the dopamine hypothesis.

Therefore, the Final Common Pathway hypothesis suggests that the multiple genetic and environmental risk factors for psychosis interact with each other in terms of their effects on the dopamine system (Figure 31.6). This has also extended to the prodrome and the extended (schizotypal) phenotype.[20]

PREDICTION ERRORS AND COMPUTATIONAL MODELS

Contemporary computational models of brain function have informed recent permutations of the dopamine hypothesis to emphasize the role of cortical-subcortical interactions in integrating existing internal models of the external world with incoming sensory information. Within this framework, sensory information is salient when it violates cortical predictive models of the environment. It is these mismatches between anticipated and actual sensory information that drive adaptive changes to the brain's models of the world.[93] Since this process is finely tuned by subcortical dopaminergic signaling, subtle changes in dopaminergic transmission could result in dysfunction of the brain's internal models resulting in psychotic experiences.[93] Given the role of the dopamine system in prediction error signaling, recent computational models[94] have proposed that chronic mismatches between events and dopamine signals can lead to alterations within the go and no-go pathways, which could explain how dopaminergic dysfunction could result in both positive and negative symptoms of psychosis.

LINKING GENES TO NEUROBIOLOGY

There is growing evidence that genes associated with increased risk of psychosis converge on a number of neurobiological pathways that may shed light on the underlying pathophysiology of the disorder. Examples of these relevant to dopaminergic mechanisms underlying psychosis include the dopamine D2 receptor,[86] glutamatergic dysfunction,[95] and the dopaminergic-related neuregulin 1–erbB2 receptor tyrosine kinase 4–phosphoinositide 3 kinase–protein kinase B (NRG1-ERBB4-PI3K-AKT1) pathway.[96] Further evidence comes from a transgenic murine model that overexpresses striatal D2 receptors during development (the D2OE mouse) resulting in diffuse effects on the rest of the brain.[97] This results in permanent alterations in prefrontal cognition which may be mediated via aberrant cortical D_1 receptor signaling and dopamine turnover. Plastic changes within pallidal go/no-go pathways[98] and attenuated VTA responsivity[99] may underlie the observed reward deficits in this model. Together with work in prodromal psychosis, the D2OE findings raise the possibility that developmental D_2 receptor dysfunction may precede symptoms and may lead to additional pathological consequences beyond positive psychotic experiences including negative symptoms and cognitive deficits.

DUAL HIT

Taking the effects of stress exposure on the dopamine system with the genetic evidence gives rise to a "dual hit" model of dopaminergic dysfunction in psychosis.[100] The first "hit" refers to the genetic vulnerability involving dopaminergic and related genes (Schizophrenia Working Group 2014).

These may alter the brain's developmental trajectory to induce genetic vulnerability through direct and diffuse effects on circuitry as demonstrated by the D2OE mouse. The second "hit" refers to environmental stressors that aggravate the existing genetic vulnerability. The majority of environmental risk factors for psychosis affect dopaminergic transmission (e.g., drugs, stress and urbanicity).[78] Together these then result in further dopamine-mediated network dysfunction[101] and clinical psychosis.

LIMITATIONS OF THE DOPAMINE HYPOTHESIS

The dopamine hypothesis of psychosis is not without criticism.[102] The main tenets of this relate to cause and effect in terms of dopamine and psychosis, and the mode of action of antipsychotics. Moncrieff argues that the question of whether altered dopamine synthesis necessarily *causes* psychosis still needs to be addressed. In particular, she argues that if patients are more aroused, as would occur in psychosis, then this may account for findings of hyperdopaminergia.[12] It remains controversial whether there are differences in the clinical phenomenology of amphetamine-induced psychosis compared to schizophreniform psychosis.[103,104] Amphetamine-induced psychosis results in abnormal stereotyped movements and posturing in animals,[105] and this is likewise observed in untreated chronic psychosis, albeit without the external motor hyperactivity.[104]

Further limitations lie in the methodological challenges that preclude the demonstration of aberrant phasic dopaminergic activity in people with psychosis. The tools available to image the dopamine system in vivo, PET and fMRI, lack the temporal and spatial resolution of electrophysiology and microdialysis used in animal research and on which these hypotheses are based. Furthermore, studies of dopamine metabolite cerebrospinal fluid (CSF) levels have produced conflicting results, which may reflect complicated interactions between antipsychotics and dopamine metabolite levels.[106] However, a likely confounding factor in the interpretation of these results is that CSF levels provide a marker of whole brain dopamine metabolism and therefore lack the spatial specificity of chemical imaging studies.

An additional complicating factor in the dopamine hypothesis is that, clozapine, which is the treatment of choice for refractory schizophrenia,[107] has relatively weak affinity for the D_2 receptor[108] and it has been hypothesized that its therapeutic superiority to the other antipsychotics is due to antagonism of the serotonin type 2a ($5HT_{2A}$) receptor.[11] This is particularly interesting given that $5HT_{2A}$ agonism results in marked hallucinations, albeit predominantly visual, as per the mode of action of the classic psychedelics such as lysergic acid diethylamide (LSD), psilocybin, and mescaline.[109] Likewise, not all patients with schizophrenia appear to exhibit increases in striatal dopamine synthesis capacity: those who do not respond to standard antipsychotic treatment have no significant elevation in dopamine synthesis capacity, compared to controls, while patients who do respond to antipsychotics do exhibit increased dopamine synthesis capacity.[110] The idea that patients who remain symptomatic despite antidopaminergic treatment may not have a primary dopaminergic elevation gave rise to the hypothesis that schizophrenia as a syndrome may arise from multiple neurochemical abnormalities. It has been argued, therefore, that schizophrenia could be classified according to the presence or absence of hyperdopaminergia, type A and type B respectively.[111]

A further limitation is that the aberrant salience hypothesis does not fully explain all the phenomenology of psychosis. For example, while it is capable of accounting for delusion formation, it is less clear how this extends to hallucinations, thought alienation, and the emotional valence of paranoia. By the same token, it needs to be recognized that many other neurochemical and neurophysiological alterations have been reported in schizophrenia and other psychotic disorders. As discussed earlier, striatal dopamine dysfunction is generally seen as the final step in the development of psychosis, but it remains to be determined which alterations a central to leading to the dopamine dysfunction, and how this occurs. Determining this is central to efforts to develop preventive interventions for people who may be at risk of psychosis, as well as developing better treatments for those with established psychoses.

CONCLUSION

The dopamine hypothesis has been highly influential in our understanding of the neurobiology of psychosis. This is based on its position, unusual in the field of psychopharmacology, whereby there are coherent and converging lines of evidence linking a neurochemical alteration (striatal hyperdopaminergia) with a condition (psychosis) with evidence that this is transdiagnostic.[33,34] Furthermore, inducing striatal dopamine release, a model of hyperdopaminergia, in otherwise healthy individuals induces psychotic experiences,[5] there is a relationship between severity of psychotic experience and dopamine synthesis and release capacity[12,62] and medicines used to reduce psychotic symptoms block dopamine receptors.[4] Many of the key findings supporting the dopamine hypothesis have been widely, although not universally, replicated and are supported by meta-analyses with large effect sizes. Given the effect sizes in these meta-analyses, it would take a large number of new negative studies to overturn these findings.

While striatal dopamine dysfunction may not be the case for everyone with psychosis, it remains likely that dopaminergic mechanisms underlie psychotic experience for a significant proportion of individuals. Future research should address why dopamine antagonists are not helpful for everyone, how to reduce treatment side-effects, and the role of interactions between dopaminergic and nondopaminergic mechanisms in psychosis and should provide a coherent mechanistic explanation of the varied phenomenology of psychosis. In addition, we must develop new interventions to help people experiencing distress associated with psychosis and understand which treatments work best for whom.

Overall, it seems that the dopamine hypothesis is standing the test of time. It is also clear that it has evolved with increasing understanding of the role of the dopaminergic system in healthy brain function together with the nature of the dopaminergic alterations in psychosis. Finally, it is likely that the dopamine hypothesis of psychosis will continue to evolve further.

REFERENCES

1. van Rossum JM. The significance of dopamine-receptor blockade for the mechanism of action of Neuroleptic drugs. *Archives Internationales de Pharmacodynamie et de therapie.* 1966;160(2):492–494.
2. Howes OD, Kambeitz J, Kim E, et al. The nature of dopamine dysfunction in schizophrenia and what this means for treatment. *Archives of General Psychiatry.* 2012;69(8):776–786.
3. Carlsson A, Lindqvist M. Effect of chlorpromazine or haloperidol on formation of 3-methoxytyramine and normetanephrine in mouse brain. *Acta Pharmacol Toxicol (Copenh).* 1963;20:140–144.
4. Seeman P, Lee T, Chau-Wong M, Wong K. Antipsychotic drug doses and neuroleptic/dopamine receptors. *Nature.* 1976;261(5562):717–719.
5. Connell P. Amphetamine psychosis.. *Br Med J* 1957;5018:582.
6. Lieberman JA, Kane JM, Alvir J. Provocative tests with psychostimulant drugs in schizophrenia. *Psychopharmacology.* 1987;91(4):415–433.
7. Bird ED, Spokes EG, Iversen LL. Increased dopamine concentration in limbic areas of brain from patients dying with schizophrenia. *Brain.* 1979;102(2):347–360.
8. Mackay AV, Iversen LL, Rossor M, et al. Increased brain dopamine and dopamine receptors in schizophrenia. *Archives of General Psychiatry.* 1982;39(9):991–997.
9. Lee T, Seeman P. Elevation of brain neuroleptic/dopamine receptors in schizophrenia. *American Journal of Psychiatry.* 1980; 137(2):191–197.
10. Davis KL, Kahn RS, Ko G, Davidson M. Dopamine in schizophrenia: a review and reconceptualization. *American Journal of Psychiatry.* 1991;148(11):1474–1486.
11. Huttunen M. The evolution of the serotonin-dopamine antagonist concept. *Journal of Clinical Psychopharmacology.* 1995;15(1 Suppl 1):4S–10S.
12. Howes OD, Montgomery AJ, Asselin MC, et al. Elevated striatal dopamine function linked to prodromal signs of schizophrenia. *Arch Gen Psychiatry.* 2009;66(1):13–20.
13. Elkashef AM, Doudet D, Bryant T, Cohen RM, Li SH, Wyatt RJ. 6-(18)F-DOPA PET study in patients with schizophrenia: positron emission tomography. *Psychiatry Res.* 2000;100(1):1–11.
14. Howes OD, Montgomery AJ, Asselin MC, Murray RM, Grasby PM, McGuire PK. Molecular imaging studies of the striatal dopaminergic system in psychosis and predictions for the prodromal phase of psychosis. *British Journal of Psychiatry Supplement.* 2007;51:s13–18.
15. Huttunen J, Heinimaa M, Svirskis T, et al. Striatal dopamine synthesis in first-degree relatives of patients with schizophrenia. *Biological Psychiatry.* 2008;63(1):114–117.
16. Fusar-Poli P, Meyer-Lindenberg A. Striatal presynaptic dopamine in schizophrenia, part II: meta-analysis of [(18)F/(11)C]-DOPA PET studies. *Schizophr Bull.* 2013;39(1):33–42.
17. Voruganti L, Slomka P, Zabel P, et al. Subjective effects of AMPT-induced dopamine depletion in schizophrenia: correlation between dysphoric responses and striatal D(2) binding ratios on SPECT imaging. *Neuropsychopharmacology.* 2001;25(5):642–650.
18. Taylor SF, Koeppe RA, Tandon R, Zubieta JK, Frey KA. In vivo measurement of the vesicular monoamine transporter in schizophrenia. *Neuropsychopharmacology.* 2000;23(6):667–675.
19. Chen KC, Yang YK, Howes O, et al. Striatal dopamine transporter availability in drug-naive patients with schizophrenia: a case-control SPECT study with [(99m)Tc]-TRODAT-1 and a meta-analysis. *Schizophr Bull.* 2013;39(2):378–386.
20. Howes OD, Bose SK, Turkheimer F, et al. Dopamine synthesis capacity before onset of psychosis: a prospective [18F]-DOPA PET imaging study. *American Journal of Psychiatry.* 2011;168(12):1311–1317.
21. Abi-Dargham A, Rodenhiser J, Printz D, et al. Increased baseline occupancy of D2 receptors by dopamine in schizophrenia. *Proceedings of the National Academy of Sciences of the United States of America.* 2000;97(14):8104–8109.
22. Abi-Dargham A, van de Giessen E, Slifstein M, Kegeles LS, Laruelle M. Baseline and amphetamine-stimulated dopamine activity are related in drug-naive schizophrenic subjects. *Biological Psychiatry.* 2009;65(12):1091–1093.
23. Weinberger DR, Berman KF. Prefrontal function in schizophrenia: confounds and controversies. *Philosophical Transactions of the Royal Society of London Series B, Biological Sciences.* 1996;351(1346):1495–1503.
24. Kambeitz J, Abi-Dargham A, Kapur S, Howes OD. Alterations in cortical and extrastriatal subcortical dopamine function in schizophrenia: systematic review and meta-analysis of imaging studies. *British Journal of Psychiatry.* 2014;204(6):420–429.
25. Slifstein M, van de Giessen E, Van Snellenberg J, et al. Deficits in prefrontal cortical and extrastriatal dopamine release in schizophrenia: a positron emission tomographic functional magnetic resonance imaging study. *JAMA Psychiatry.* 2015;72(4):316–324.
26. Stenkrona P, Matheson GJ, Halldin C, Cervenka S, Farde L. D1-dopamine receptor availability in first-episode neuroleptic naive psychosis patients. *Int J Neuropsychopharmacol.* 2019.
27. Sumiyoshi T, Kurachi M, Kurokawa K, et al. Plasma homovanillic acid in the prodromal phase of schizophrenia. *Biological Psychiatry.* 2000;47(5):428–433.
28. Ruhrmann S, Bechdolf A, Kuhn KU, et al. Acute effects of treatment for prodromal symptoms for people putatively in a late initial prodromal state of psychosis. *British Journal of Psychiatry Supplement.* 2007;51:s88–95.
29. Mizrahi R, Addington J, Rusjan PM, et al. Increased stress-induced dopamine release in psychosis. *Biological Psychiatry.* 2012;71(6):561–567.
30. Egerton A, Chaddock CA, Winton-Brown TT, et al. Presynaptic striatal dopamine dysfunction in people at ultra-high risk for psychosis: findings in a second cohort. *Biological Psychiatry.* 2013;74(2):106–112.
31. Avram M, Brandl F, Cabello J, et al. Reduced striatal dopamine synthesis capacity in patients with schizophrenia during remission of positive symptoms. *Brain.* 2019;142(6):1813–1826.
32. Thompson JL, Urban N, Slifstein M, et al. Striatal dopamine release in schizophrenia comorbid with substance dependence. *Mol Psychiatry.* 2013;18(8):909–915.
33. Jauhar S, Nour MM, Veronese M, et al. A test of the transdiagnostic dopamine hypothesis of psychosis using positron emission tomographic imaging in bipolar affective disorder and schizophrenia. *JAMA Psychiatry.* 2017;74(12):1206–1213.
34. Reith J, Benkelfat C, Sherwin A, et al. Elevated dopa decarboxylase activity in living brain of patients with psychosis. *Proceedings of the National Academy of Sciences of the United States of America.* 1994;91(24):11651–11654.
35. Haber SN. The basal ganglia. In: Jurgen K, Mai GP, eds. *The human nervous system.* Academic Press; 2012:680–740.
36. Jahanshahi M, Obeso I, Rothwell JC, Obeso JA. A fronto-striato-subthalamic-pallidal network for goal-directed and habitual inhibition. *Nature Reviews Neuroscience.* 2015;16(12):719–732.
37. Ungerstedt U. Adipsia and aphagia after 6-hydroxydopamine induced degeneration of the nigro-striatal dopamine system. *Acta Physiologica Scandinavica Supplementum.* 1971;367:95–122.
38. Smith GP. The arousal function of central catecholamine neurons. *Annals of the New York Academy of Sciences.* 1976;270:45–56.
39. Ervin GN, Fink JS, Young RC, Smith GP. Different behavioral responses to L-DOPA after anterolateral or posterolateral hypothalamic injections of 6-hydroxydopamine. *Brain Research.* 1977;132(3):507–520.

40. Wise RA, Spindler J, deWit H, Gerberg GJ. Neuroleptic-induced "anhedonia" in rats: pimozide blocks reward quality of food. *Science.* 1978;201(4352):262–264.
41. Gerber GJ, Sing J, Wise RA. Pimozide attenuates lever pressing for water reinforcement in rats. *Pharmacology, Biochemistry, and Behavior.* 1981;14(2):201–205.
42. De Wit H, Wise RA. Blockade of cocaine reinforcement in rats with the dopamine receptor blocker pimozide, but not with the noradrenergic blockers phentolamine or phenoxybenzamine. *Canadian Journal of Psychology.* 1977;31(4):195–203.
43. Fouriezos G, Wise RA. Pimozide-induced extinction of intracranial self-stimulation: response patterns rule out motor or performance deficits. *Brain Research.* 1976;103(2):377–380.
44. Wise RA, Schwartz HV. Pimozide attenuates acquisition of lever-pressing for food in rats. *Pharmacology, Biochemistry, and Behavior.* 1981;15(4):655–656.
45. Spyraki C, Fibiger HC, Phillips AG. Cocaine-induced place preference conditioning: lack of effects of neuroleptics and 6-hydroxydopamine lesions. *Brain Research.* 1982;253(1–2):195–203.
46. Wise RA. Neuroleptics and operant behavior: The anhedonia hypothesis. *Behavioral and Brain Sciences.* 1982;5(1):39–53.
47. Fibiger HC, Mason ST. The effects of dorsal bundle injections of 6-hydroxydopamine on avoidance responding in rats. *British Journal of Pharmacology.* 1978;64(4):601–605.
48. Berridge KC, Robinson TE. What is the role of dopamine in reward: hedonic impact, reward learning, or incentive salience? *Brain Research Brain Research Reviews.* 1998;28(3):309–369.
49. Wise RA, Rompre PP. Brain dopamine and reward. *Annual Review of Psychology.* 1989;40:191–225.
50. Bindra D. Neuropsychological interpretation of the effects of drive and incentive-motivation on general and instrumental behavior. *Psychological Review.* 1968;75:1–22.
51. Esposito RU, Faulkner W, Kornetsky C. Specific modulation of brain stimulation reward by haloperidol. *Pharmacology, Biochemistry, and Behavior.* 1979;10(6):937–940.
52. Roitman MF, Stuber GD, Phillips PE, Wightman RM, Carelli RM. Dopamine operates as a subsecond modulator of food seeking. *Journal of Neuroscience: the Official Journal of the Society for Neuroscience.* 2004;24(6):1265–1271.
53. de Wit H, Stewart J. Reinstatement of cocaine-reinforced responding in the rat. *Psychopharmacology.* 1981;75(2):134–143.
54. Wyvell CL, Berridge KC. Intra-accumbens amphetamine increases the conditioned incentive salience of sucrose reward: enhancement of reward "wanting" without enhanced "liking" or response reinforcement. *Journal of Neuroscience* 2000;20(21):8122–8130.
55. Laruelle M, Abi-Dargham A, van Dyck CH, et al. SPECT imaging of striatal dopamine release after amphetamine challenge. *Journal of Nuclear Medicine.* 1995;36(7):1182–1190.
56. Salamone JD. The involvement of nucleus accumbens dopamine in appetitive and aversive motivation. *Behavioural Brain Research.* 1994;61(2):117–133.
57. Stewart J, de Wit H, Eikelboom R. Role of unconditioned and conditioned drug effects in the self-administration of opiates and stimulants. *Psychol Rev.* 1984;91(2):251–268.
58. Wise RA. Dopamine, learning and motivation. *Nature Reviews Neuroscience.* 2004;5(6):483–494.
59. McFarland K, Ettenberg A. Haloperidol does not affect motivational processes in an operant runway model of food-seeking behavior. *Behavioral Neuroscience.* 1998;112(3):630–635.
60. Hollerman JR, Schultz W. Dopamine neurons report an error in the temporal prediction of reward during learning. *Nature Neuroscience.* 1998;1(4):304–309.
61. Bromberg-Martin ES, Matsumoto M, Hikosaka O. Dopamine in motivational control: rewarding, aversive, and alerting. *Neuron.* 2010;68(5):815–834.
62. Laruelle M, Abi-Dargham A. Dopamine as the wind of the psychotic fire: new evidence from brain imaging studies. *Journal of Psychopharmacology.* 1999;13(4):358–371.
63. Kapur S. Psychosis as a state of aberrant salience: a framework linking biology, phenomenology, and pharmacology in schizophrenia. *American Journal of Psychiatry.* 2003;160(1):13–23.
64. Pani L, Porcella A, Gessa GL. The role of stress in the pathophysiology of the dopaminergic system. *Molecular Psychiatry.* 2000;5(1):14–21.
65. Del Arco A, Mora F. Prefrontal cortex-nucleus accumbens interaction: in vivo modulation by dopamine and glutamate in the prefrontal cortex. *Pharmacology, Biochemistry, and Behavior.* 2008;90(2):226–235.
66. Cabib S, Puglisi-Allegra S. Stress, depression and the mesolimbic dopamine system. *Psychopharmacology.* 1996;128(4):331–342.
67. Antelman SM, Eichler AJ, Black CA, Kocan D. Interchangeability of stress and amphetamine in sensitization. *Science.* 1980;207(4428):329–331.
68. Tidey JW, Miczek KA. Social defeat stress selectively alters mesocorticolimbic dopamine release: an in vivo microdialysis study. *Brain Res.* 1996;721(1--2):140–149.
69. Hall FS, Wilkinson LS, Humby T, Robbins TW. Maternal deprivation of neonatal rats produces enduring changes in dopamine function. *Synapse.* 1999;32(1):37–43.
70. Brake WG, Zhang TY, Diorio J, Meaney MJ, Gratton A. Influence of early postnatal rearing conditions on mesocorticolimbic dopamine and behavioural responses to psychostimulants and stressors in adult rats. *Eur J Neurosci.* 2004;19(7):1863–1874.
71. Mangiavacchi S, Masi F, Scheggi S, Leggio B, De Montis MG, Gambarana C. Long-term behavioral and neurochemical effects of chronic stress exposure in rats. *J Neurochem.* 2001;79(6):1113–1121.
72. Shimamoto A, Debold JF, Holly EN, Miczek KA. Blunted accumbal dopamine response to cocaine following chronic social stress in female rats: exploring a link between depression and drug abuse. *Psychopharmacology (Berl).* 2011;218(1):271–279.
73. Lataster J, Collip D, Ceccarini J, et al. Psychosocial stress is associated with in vivo dopamine release in human ventromedial prefrontal cortex: a positron emission tomography study using [(1)(8)F] fallypride. *NeuroImage.* 2011;58(4):1081–1089.
74. Wand GS, Oswald LM, McCaul ME, et al. Association of amphetamine-induced striatal dopamine release and cortisol responses to psychological stress. *Neuropsychopharmacology.* 2007;32(11):2310–2320.
75. Kasanova Z, Hernaus D, Vaessen T, et al. Early-life stress affects stress-related prefrontal dopamine activity in healthy adults, but not in individuals with psychotic disorder. *PloS one.* 2016;11(3):e0150746.
76. De Bellis MD, Lefter L, Trickett PK, Putnam FW, Jr. Urinary catecholamine excretion in sexually abused girls. *J Am Acad Child Adolesc Psychiatry.* 1994;33(3):320–327.
77. Pruessner JC, Champagne F, Meaney MJ, Dagher A. Dopamine release in response to a psychological stress in humans and its relationship to early life maternal care: a positron emission tomography study using [11C]raclopride. *Journal of Neuroscience.* 2004;24(11):2825–2831.
78. Howes OD, Murray RM. Schizophrenia: an integrated sociodevelopmental-cognitive model. *Lancet.* 2014;383(9929):1677–1687.
79. Brunelin J, d'Amato T, van Os J, Cochet A, Suaud-Chagny MF, Saoud M. Effects of acute metabolic stress on the dopaminergic and pituitary-adrenal axis activity in patients with schizophrenia, their unaffected siblings and controls. *Schizophrenia Research.* 2008;100(1–3):206–211.
80. Lataster J, Collip D, Ceccarini J, et al. Familial liability to psychosis is associated with attenuated dopamine stress signaling in ventromedial prefrontal cortex. *Schizophr Bull.* 2014;40(1):66–77.
81. Hernaus D, Collip D, Lataster J, et al. Psychotic reactivity to daily life stress and the dopamine system : a study combining experience sampling and [18F]fallypride positron emission tomography. *J Abnorm Psychol.* 2015;124:27–37.
82. Oswald LM, Wand GS, Kuwabara H, Wong DF, Zhu S, Brasic JR. History of childhood adversity is positively associated with ventral striatal dopamine responses to amphetamine. *Psychopharmacology (Berl).* 2014;231(12):2417–2433.

83. Howes OD, Kapur S. The dopamine hypothesis of schizophrenia: version III—the final common pathway. *Schizophr Bull.* 2009;35(3):549–562.
84. Abi-Dargham A, Gil R, Krystal J, et al. Increased striatal dopamine transmission in schizophrenia: confirmation in a second cohort. *American Journal of Psychiatry.* 1998;155(6):761–767.
85. Frankle WG, Laruelle M. Neuroreceptor imaging in psychiatric disorders. *Annals of Nuclear Medicine.* 2002;16(7):437–446.
86. Schizophrenia Working Group of the Psychiatric Genomics Consortium. Biological insights from 108 schizophrenia-associated genetic loci. *Nature.* 2014;511(7510):421–427.
87. van Winkel R, Stefanis NC, Myin-Germeys I. Psychosocial stress and psychosis: a review of the neurobiological mechanisms and the evidence for gene-stress interaction. *Schizophr Bull.* 2008;34(6):1095–1105.
88. Kehoe P, Shoemaker WJ, Triano L, Hoffman J, Arons C. Repeated isolation in the neonatal rat produces alterations in behavior and ventral striatal dopamine release in the juvenile following amphetamine challenge. *Behavioral Neuroscience.* 1996;110:1435–1444.
89. Boksa P, El-Khodor BF. Birth insult interacts with stress at adulthood to alter dopaminergic function in animal models: possible implications for schizophrenia and other disorders. *Neuroscience and Biobehavioral Reviews.* 2003;27(1–2):91–101.
90. Froudist-Walsh S, Bloomfield MA, Veronese M, et al. The effect of perinatal brain injury on dopaminergic function and hippocampal volume in adult life. *eLife.* 2017;6.
91. Grace AA. Dopamine system dysregulation and the pathophysiology of schizophrenia: insights from the methylazoxymethanol acetate model. *Biological Psychiatry.* 2017;81(1):5–8.
92. Kokkinou M, Ashok AH, Howes OD. The effects of ketamine on dopaminergic function: meta-analysis and review of the implications for neuropsychiatric disorders. *Molecular Psychiatry.* 2018;23(1):59–69.
93. Fletcher PC, Frith CD. Perceiving is believing: a Bayesian approach to explaining the positive symptoms of schizophrenia. *Nature Reviews Neuroscience.* 2009;10(1):48–58.
94. Maia TV, Frank MJ. An Integrative perspective on the role of dopamine in schizophrenia. *Biological Psychiatry.* 2017;81(1):52–66.
95. Corlett PR, Honey GD, Krystal JH, Fletcher PC. Glutamatergic model psychoses: prediction error, learning, and inference. *Neuropsychopharmacology.* 2011;36(1):294–315.
96. Hatzimanolis A, McGrath JA, Wang R, et al. Multiple variants aggregate in the neuregulin signaling pathway in a subset of schizophrenia patients. *Translational Psychiatry.* 2013;3:e264.
97. Simpson EH, Kellendonk C. Insights about striatal circuit function and schizophrenia from a mouse model of dopamine d2 receptor upregulation. *Biological Psychiatry.* 2017;81(1):21–30.
98. Cazorla M, de Carvalho FD, Chohan MO, et al. Dopamine D2 receptors regulate the anatomical and functional balance of basal ganglia circuitry. *Neuron.* 2014;81(1):153–164.
99. Krabbe S, Duda J, Schiemann J, et al. Increased dopamine D2 receptor activity in the striatum alters the firing pattern of dopamine neurons in the ventral tegmental area. *Proceedings of the National Academy of Sciences of the United States of America.* 2015;112(12):E1498–1506.
100. Howes OD, McCutcheon R, Owen MJ, Murray RM. The role of genes, stress, and dopamine in the development of schizophrenia. *Biological Psychiatry.* 2017;81(1):9–20.
101. Horga G, Cassidy CM, Xu X, et al. Dopamine-related disruption of functional topography of striatal connections in unmedicated patients with schizophrenia. *JAMA Psychiatry.* 2016;73(8):862–870.
102. Moncrieff J. A critique of the dopamine hypothesis of schizophrenia and psychosis. *Harvard Review of Psychiatry.* 2009;17(3): 214–225.
103. Snyder SH, Aghajanian GK, Matthysse S. Prospects for research on schizophrenia. V. Pharmacological observations, drug-induced psychoses. *Neurosciences Research Program Bulletin.* 1972;10(4):430–445.
104. Batki SL, Harris DS. Quantitative drug levels in stimulant psychosis: relationship to symptom severity, catecholamines and hyperkinesia. *American Journal on Addictions.* 2004;13(5):461–470.
105. Van Rossum J, Hurkmans AT. Mechanism of Action of Psychomotor Stimulant Drugs. Significance of Dopamine in Locomotor Stimulant Action. *International Journal of Neuropharmacology.* 1964;3:227–239.
106. Widerlov E. A critical appraisal of CSF monoamine metabolite studies in schizophrenia. *Annals of the New York Academy of Sciences.* 1988;537:309–323.
107. Taylor DM, Douglas-Hall P, Olofinjana B, Whiskey E, Thomas A. Reasons for discontinuing clozapine: matched, case-control comparison with risperidone long-acting injection. *British Journal of Psychiatry.* 2009;194(2):165–167.
108. Seeman P. Atypical antipsychotics: mechanism of action. *Canadian Journal of Psychiatry/Revue Canadienne de Psychiatrie.* 2002;47(1):27–38.
109. Nichols DE. Hallucinogens. *Pharmacol Ther.* 2004;101(2):131–181.
110. Demjaha A, Murray RM, McGuire PK, Kapur S, Howes OD. Dopamine synthesis capacity in patients with treatment-resistant schizophrenia. *American Journal of Psychiatry.* 2012;169(11):1203–1210.
111. Howes OD, Kapur S. A neurobiological hypothesis for the classification of schizophrenia: type A (hyperdopaminergic) and type B (normodopaminergic). *British Journal of Psychiatry.* 2014;205(1):1–3.

32

GLUTAMATE IN THE PATHOPHYSIOLOGY OF SCHIZOPHRENIA

Daniel C. Javitt

INTRODUCTION

All current medications for schizophrenia, including both typical and atypical antipsychotics, function primarily by blocking dopamine D_2 receptors (D_2R), supporting the development of dopamine models of schizophrenia (review in [1]). Nevertheless, only ~65% of individuals with schizophrenia show a significant clinical response to antipsychotics, with the remainder showing either incomplete or even no appreciable response. Furthermore, antipsychotics have limited effectiveness against schizophrenia-associated negative symptoms and neurocognitive impairments.[2] Thus, the majority of individuals with schizophrenia show marked reductions in social and role function despite best available treatment.

Glutamatergic models of schizophrenia were first proposed approximately 30 years ago[3–5] and continue to enable new research conceptualizations of schizophrenia, and to catalyze new treatment development approaches.[6] Over the past 30 years, such models have been increasingly supported by genetic, brain imaging, and neurochemistry-based research, and have led to updated conceptualizations regarding causes of poor functional outcome in schizophrenia.

Glutamatergic models are based on the early observation that phencyclidine (PCP), ketamine, and related drugs induced schizophrenia-like psychotic effects,[7] followed somewhat later by the observation that these compounds induce their unique behavioral effects by blocking neurotransmission at N-methyl-D-aspartate-type glutamate receptors (NMDARs).[4] The present chapter reviews the alterations in the glutamate system, relative to the pathophysiological bases of schizophrenia.

GLUTAMATERGIC SYSTEMS IN BRAIN

Glutamate is the most abundant neurotransmitter in mammalian brain, and accounts for ~60% of all neurons and ~80% of all synapses. All cortical pyramidal neurons utilize glutamate as their primary neurotransmitter. Glutamatergic neurons mediate virtually all inputs to cortex, as well as the majority of cortico-cortical and corticofugal transmission. Glutamatergic efferents from cortex and thalamus regulate subcortical systems implicated in schizophrenia, such as the mesolimibic and mesocortical dopamine systems.[4,8] Glutamatergic projection neurons from cochlea, brainstem, and retina mediate auditory and visual sensory afferents into cortex, through structures such as medial (MGN) and lateral (LGN) geniculate nuclei. Glutamatergic models thus inherently predict disruptions of information processing throughout distributed brain regions.

Glutamatergic neurotransmission is regulated by local circuit GABAergic inhibitory neurons, which are divided into three main subtypes based on expressed markers. These include parvalbumin (PV)-positive interneurons, which are associated primarily with modulation of high-frequency (gamma) activity in cortex; somatostatin-positive (SOM) interneurons, which are considered to provide a blanket of inhibition within cortex and regulate slower frequency oscillations in cortex; and vasoactive intestinal peptide (VIP)-positive interneurons, which poke "holes" within the inhibitory layer, permitting local disinhibition.[9]

At the presynaptic level, glutamate is released from presynaptic terminals following neuronal excitation, and then cleared from the synaptic cleft by glutamate transporters (EAAC1-3) located on astrocytes. In astrocytes, glutamate is converted to glutamine, which is then released into the extracellular via "System N" small neutral amino acid (SNAT) transporters, reaccumulated in the presynaptic terminal by "system A" SNAT transporters and recycled into glutamate by phosphatase-activated glutaminase.[10]

During normal function, recycling of glutamate released from the neurotransmitter pool accounts for ~50% of brain energy expenditure. Thus, imaging techniques such as functional MRI (fMRI) primarily image glutamatergic transmission in the brain. Furthermore, drug-induced increases in regional blood flow ("pharmacoBOLD") can be used to imaging glutamate-related processes.[11] In addition, reuptake of glutamate by astrocytes stimulates de novo glutamine synthesis beyond that produced by recycling.[10] As a result, neuronal activity results in increased total concentrations of glutamate+glutamine (termed "Glx") and can be detected with magnetic resonance spectroscopy (MRS).[11]

As with other neurotransmitters, glutamate acts through multiple receptor types, which are divided into ionotropic and metabotropic subtypes. Ionotropic receptors are divided into alpha-amino-3-hydroxy-5-methylisoxazole-4-propionate (AMPAR), NMDAR, and kainite receptor subtypes. Of these, AMPAR mediate primarily fast glutamatergic neurotransmission (activation duration <10 ms) and are primarily permeable to Na^+, although Ca^{2+}-permeable AMPAR channels may be present on specific neuronal types, such as PV and SOM interneurons.[12]

By contrast, NMDARs are permeable to both Na^+ and Ca^{2+}. Entry of Ca^{2+} via NMDAR serves as the primary trigger for initiation of plasticity-related processes including long-term potentiation (LTP) and long-term depression (LTD) via a cascade of second messenger systems. NMDAR-dependent LTP, in turn, increases AMPAR insertion into the postsynaptic membrane, leading to potentiation of transmission through both NMDAR and AMPAR in the potentiated synapse.[13]

NMDARs are hetero-oligomers consisting of variable distributions of NR1, NR2A-D, and NR3A,B subunits (also termed GluN1, GluN2, and GluN3) that are encoded by *GRIN1*, *GRIN2A-D*, and *GRIN3A-B* genes, respectively. These subunits confer distinct properties to the NMDAR. Thus, for example, NMDAR containing NR2B subunits have higher Ca^{2+} conductance than NR2A, and so are better able to initiate LTP. During development there is a shift from NR2B to NR2A following the critical period, presumably in service of reducing neuronal plasticity.[14]

NMDAR have additional features that enable them to subserve highly specific roles in the CNS. Thus, NMDAR produce excitatory potentials that persist for up to ~250 ms, which enable recurrent NMDAR circuits to maintain memory traces of up to ~2 sec.[15] These may form the basis for several forms of cortical short-term memory. In addition, NMDAR require binding of multiple glutamate molecules for activation,[16] allowing them to produce nonlinear gain effects.[17]

Finally, NMDAR are blocked by Mg^{2+} at resting membrane potential, so that even following glutamate binding and channel opening, no net current flow occurs. The sensitivity of NMDAR both to neurotransmitter (glutamate) and membrane potential enables them to function in a "Hebbian" manner to form connections between distinct brain pathways.[4]

Although glutamate serves as the main endogenous agonist at NMDAR, the NMDAR complex also contains allosteric binding sites for (1) the small, monocarboxylic amino acids glycine and D-serine, and (2) glutathione/redox potential (Figure 32.1A). Both glycine and D-serine are present at high concentrations in brain, and appear to play a cooperative role in NMDAR modulation.[18] Glycine is synthesized de novo in the brain from serine via serine hydroxymethyltransferase (SHMT). Glycine levels in brain are regulated by glycine specific transporters (GlyT1, GlyT2) as well as small neutral amino acid (SNAT) transporter.[19] Glycine also regulates functional properties of brain microglia, which regulate presynaptic glutamatergic neurons in brain at least in part through release of glycine.[20]

D-Serine is synthesized in brain from L-serine via serine racemase (SR) and degraded by D-amino acid oxidase (DAAO, also called DAO).[21] DAAO, in turn, is modulated by a gene product termed G72 or DAO-Activator (DAOA).[14] In addition, novel alanine-insensitive D-serine transporters are localized in presynaptic nerve terminals and may play a

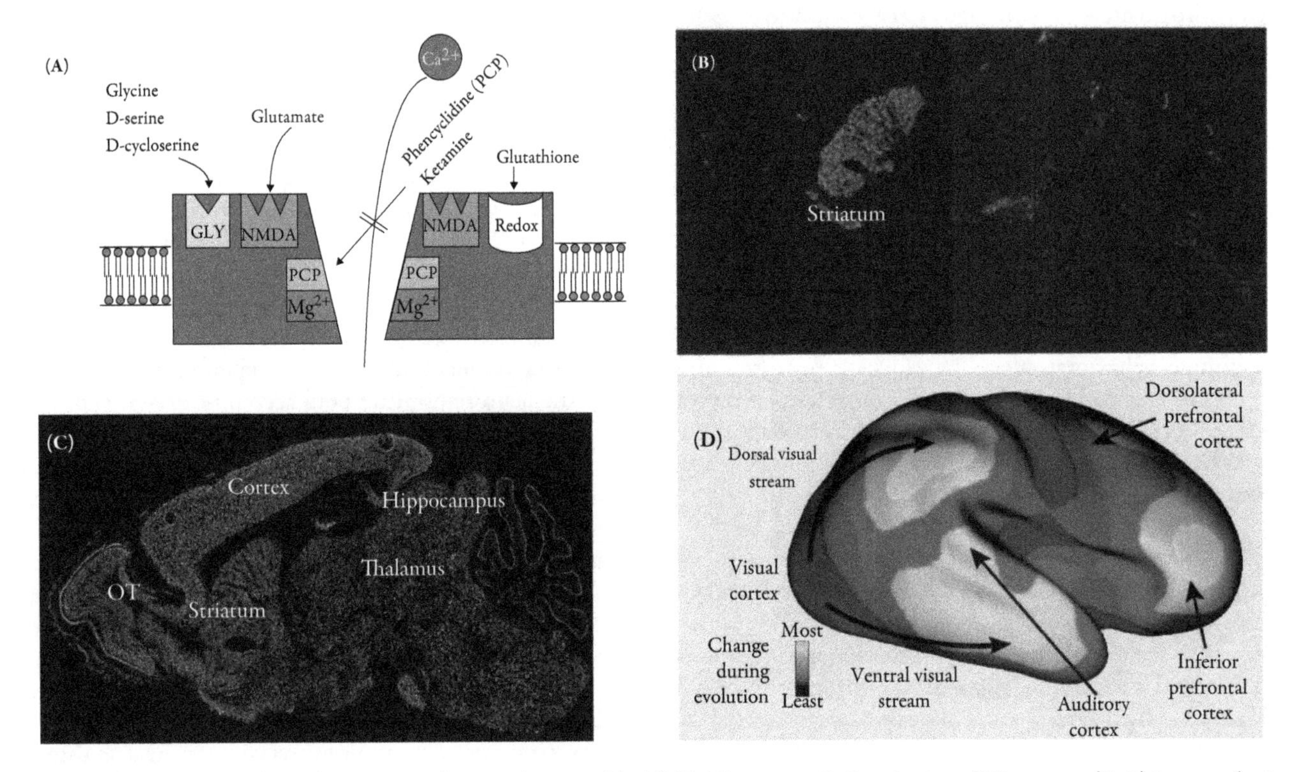

Figure 32.1 NMDA receptors in brain function. *A.* Schematic diagram of the NMDAR receptor. *B.* Distribution of D2 receptor (D_2R) in mouse brain, showing distribution almost entirely in striatum. *C.* Distribution of NMDAR in brain showing localization throughout brain, including cortex, hippocampus, thalamus, and subcortical structures. *D.* Map of cortical expansion during evolution, showing enlargement of sensory as well as frontal brain regions.
Javitt and Zukin (1991).[4]

role in D-serine regulation.[22] Whether D-serine in brain is localized primarily to astrocytes ("glia-transmission") or presynaptic nerve terminals is a topic of active debate.[23,24]

Metabotropic glutamate receptors are divided into three groups. Group I consists of mGluR1 and mGluR5, which in general increase NMDAR activation, while Group II consists of mGluR2 and mGluR3, which inhibit presynaptic glutamate release. Group III consists of mGluR4, mGluR6, mGluR7, and mGluR8 and in general inhibit NMDAR function.[25]

CONVERGENCES OF GLUTAMATE AND DOPAMINE MODELS

Since the initial development of dopaminergic theories of schizophrenia, hyperactivity of dopaminergic systems has been extensively documented. Thus, for example, dopamine levels in brain may be assessed indirectly using radio-receptor imaging, in which displacement of D_2R-targeted SPECT (e.g., ^{123}I-IBZM)[26] or PET (e.g., ^{11}C-raclopride)[27,28] ligands can be used to assess amphetamine-induced dopamine release. Consistent with dopaminergic theories, acutely psychotic individuals show increased basal dopamine levels, increased amphetamine-induced dopamine release,[27] increased presynaptic striatal dopaminergic turnover,[29] and altered resting state connectivity within frontostriatal pathways.[30]

Although these studies document the presence of dopaminergic hyperactivity, they do not indicate the underlying causes. Intrinsic abnormalities of the dopamine system may contribute in some individuals. However, in others primary dysfunction of the glutamatergic system may play a critical role. Thus, acute administration of the NMDAR antagonist ketamine leads to acute displacement of D_2R ligands such as ^{11}C-raclopride in striatum[28,31] and cingulate cortices[32] in healthy volunteers, suggestive of increased presynaptic dopamine release, and induces a schizophrenia-like potentiation of amphetamine-induced dopamine release.[27]

In rodents, as in humans, PCP treatment leads to enhanced amphetamine-induced dopamine release in frontal cortex and dorsal-, but not ventral-, striatum,[19,33,34] consistent with findings in schizophrenia.[27] Effects in rodents are reversed by simultaneous treatment with NMDAR/glycine-site agonists,[35,36] supporting the role of NMDAR in dopaminergic regulation.

Dopaminergic hyperactivity, when present, serves as an excellent predictor of response to antipsychotic medication.[37,38] Nevertheless, dopaminergic hyperactivity, of itself, does not correlate with baseline levels of psychotic symptoms. Furthermore, treatment nonresponders have normal dopamine levels but may show elevations in glutamate,[39] consistent with underlying NMDAR dysfunction.[11]

In striatum, D_2R and NMDAR exert offsetting effects on GABAergic medium spiny neurons in the indirect pathway, which mediates inhibitory outflow from the striatum.[4,40] These neurons receive both glutamatergic axons arising from cortex and thalamus, and dopamine projections from midbrain dopaminergic nuclei. Furthermore, NMDAR and D_2R produce offsetting effects on activation of these neurons, with NMDAR leading to net neuronal activation, and D_2R neurons reducing NMDAR-mediated excitations.[41] Thus both dopaminergic hyperactivity and NMDAR hypoactivity reduce inhibitory outflow in the indirect pathway, leading to locomotor hyperactivity in rodents,[40] and potentially to psychosis in humans.

GLUTAMATERGIC MODELS AND COGNITIVE DYSFUNCTION IN SCHIZOPHRENIA

As opposed to dopamine receptors, which are highly localized to frontostriatal brain systems (Figure 32.1B), NMDAR show a widespread distribution in brain with high density in subcortical as well as cortical brain regions. In cortex, NMDAR are localized to sensory as well as associative brain regions (Figure 32.1C). Thus, as opposed to dopaminergic models that predict localized cognitive dysfunction, NMDAR models of schizophrenia predict a widespread pattern of cortical dysfunction, with deficits that are regionally diffuse, but restricted to NMDAR-mediated processes.

Symptomatic and cognitive deficits resembling those of schizophrenia, including impairments in processes such as executive function and working memory are induced by NMDAR antagonists including PCP[7] or ketamine.[14,42,43] By contrast, to ketamine, dopaminergic agents such as amphetamine induce primarily positive, but not negative symptoms, and do not induce schizophrenia-like cognitive impairments.[14] Thus, glutamatergic models are especially relevant to forms of schizophrenia associated with prominent negative symptoms and sensory-level impairments.

Dissociations between glutamate and dopamine models are seen most clearly in relationship to sensory processing abnormalities associated with schizophrenia.[44] Sensory cortices, in general, receive limited dopaminergic innervation, and thus historically have been little studied in schizophrenia research. For example, Blueler referred to sensory processes as "intact simple functions."[45]

By contrast, NMDAR play a prominent role in both cortical and subcortical sensory pathways, suggesting that NMDAR-dysfunction in schizophrenia, if it exists, should result in NMDAR-related sensory abnormalities. Moreover, auditory sensory and dorsal stream visual regions have shown extensive elaboration during primate development, equivalent to that observed in frontal regions such as inferior frontal gyrus[46] (Figure 32.1D), suggesting that deficits in sensory processing may be highly relevant to human-specific behaviors. Two sensory biomarkers have proven particularly useful as indices of NMDAR dysfunction in schizophrenia.

AUDITORY MISMATCH NEGATIVITY (MMN)

Auditory mismatch negativity (MMN) is an event-related potential (ERP) that reflects comparison between stimuli at

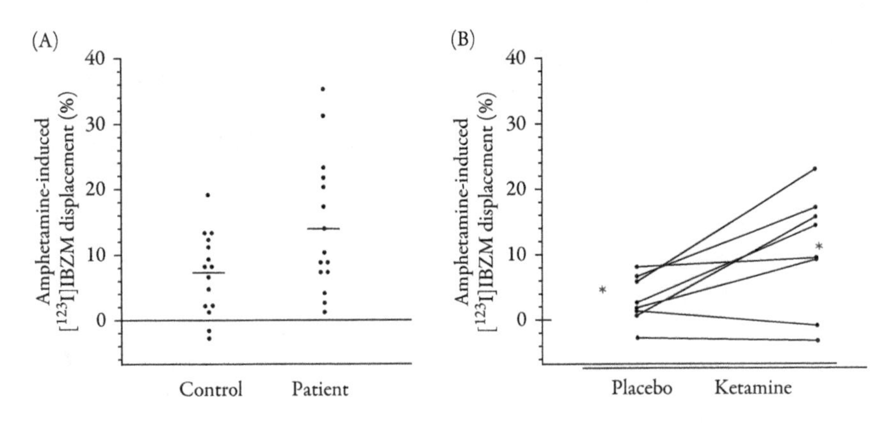

Figure 32.2 Schematic representation of NMDAR involvement in MMN generation. *A.* MMN paradigm showing standard and deviant stimuli. *B.* Representative illustration of the deficit in schizophrenia, consistent with generators primarily in auditory sensory cortex. *C.* Postulated role of NMDAR in which presentation of standard stimuli depolarizes the postsynaptic membrane, and subsequent presentation of the deviant activates current flow through open, unblocked channels.
From Kantrowitz and Javitt (2010)[46].

the level of auditory sensory cortex (rev. in [47]) (Figure 32.2A). Mismatch negativity generation has been shown to be sensitive to NMDAR antagonists in monkey,[48,49] rodent,[50,51] and human[52,53] studies. Mismatch negativity also correlates with plasma[54] and frontal brain[55] glutamate levels, supporting NMDAR involvement. Pre-ketamine MMN amplitude correlates with the degree of psychotomimetic effect observed following ketamine administration,[56] further suggesting that it may serve as an index of baseline NMDAR function at the level of auditory cortex.

Deficits in MMN have been extensively documented in schizophrenia with large effect-sizes ($d = .8$) across studies.[57] Moreover, MMN deficits correlate highly with impaired outcome[58,59] and differentiate poor from good outcome subjects.[60] Mismatch negativity deficits precede illness onset in individuals at clinical high risk and contribute to risk for conversion,[61–63] as well as poor current social and role function.[64] Mismatch negativity deficits are relatively selective for schizophrenia vs. other psychotic or neurodegenerative disorders.[65]

Conversely, MMN generation is not affected by dopaminergic[66–68] or serotonergic[68–70] modulation in healthy volunteers, is not associated with catecholaminergic indices in schizophrenia,[71] and is neither improved nor worsened by treatment with typical or atypical antipsychotics.[72–74]

As opposed to the differential sensitivity of MMN to NMDAR vs. modulatory inputs, more complex cognitive processes in schizophrenia, such as performance on the "AX"-type visual continuous performance task (AX-CPT) is affected by both NMDAR antagonists and psychotomimetic serotonergic 5-HT2A receptor agonists (e.g. psilocybin, dimethyltryptamine). Thus, whereas complex, attention-dependent processing may be disrupted at multiple levels of processing and are therefore sensitive to effects of both glutamatergic and modulatory neurotransmitters, specific aspects of sensory processing such as MMN generation appear relatively selective to glutamatergic dysfunction.

VISUAL P1

Visual P1 reflects activation of subcortical and early cortical visual pathways. As with MMN, visual P1 generation particularly to low contrast, low spatial frequency stimuli is sensitive to NMDAR dysfunction in animal models, and may therefore be used as an in vivo index of NMDAR dysfunction. As in the auditory system, schizophrenia patients have significant early visual processing deficits that were largely overlooked until recently, and that likely reflect NMDAR dysfunction within the early visual system.[46] In visual neurons, NMDAR mediate nonlinear gain processes that are specifically relevant to functioning of the magnocellular visual system. In individuals with schizophrenia, reductions in contrast gain are observed to both steady-state (ssVEP) and transient and visual ERP resemble the pattern induced by NMDAR antagonists in rodents.[44]

As with auditory MMN, deficits in visual P1 are associated with behavioral impairments and poor functional outcome.[75] Specifically, visual deficits interfere with several processes required for everyday function, including perceptual closure, face emotion recognition, and reading.[44] Deficits in these early stages of processing significantly contribute to the personal experience of schizophrenia,[46] while also providing effective biomarkers for research into neural mechanisms underlying impaired glutamatergic function.

PATHOPHYSIOLOGICAL IMPLICATIONS OF TREATMENT TRIALS

The role of glutamate in the pathophysiology of schizophrenia may also be explored pharmacologically through treatment studies. In general, glutamatergic models suggest that treatments that stimulate NMDAR receptor function should be therapeutically beneficial. One test of this hypothesis comes from the studies targeting the glycine/D-serine modulatory site of the NMDAR.

Initial controlled clinical studies were performed in the early 1990s using glycine at doses of up to 8000 mg/kg/d (~60 g/d)[76] Subsequent studies were done with D-serine, which showed similar levels of benefit but at significantly lower doses of 30–120 mg/kg (2–8 g/d).[77] A concern regarding use of D-serine is a potential for nephrotoxicity at high doses (e.g., >4 g/d).[77]

Recent meta-analyses support use of NMDAR agonists in combination with non-clozapine antipsychotics with moderate effect size across studies[78,79] (Figure 32.3A). D-Serine has

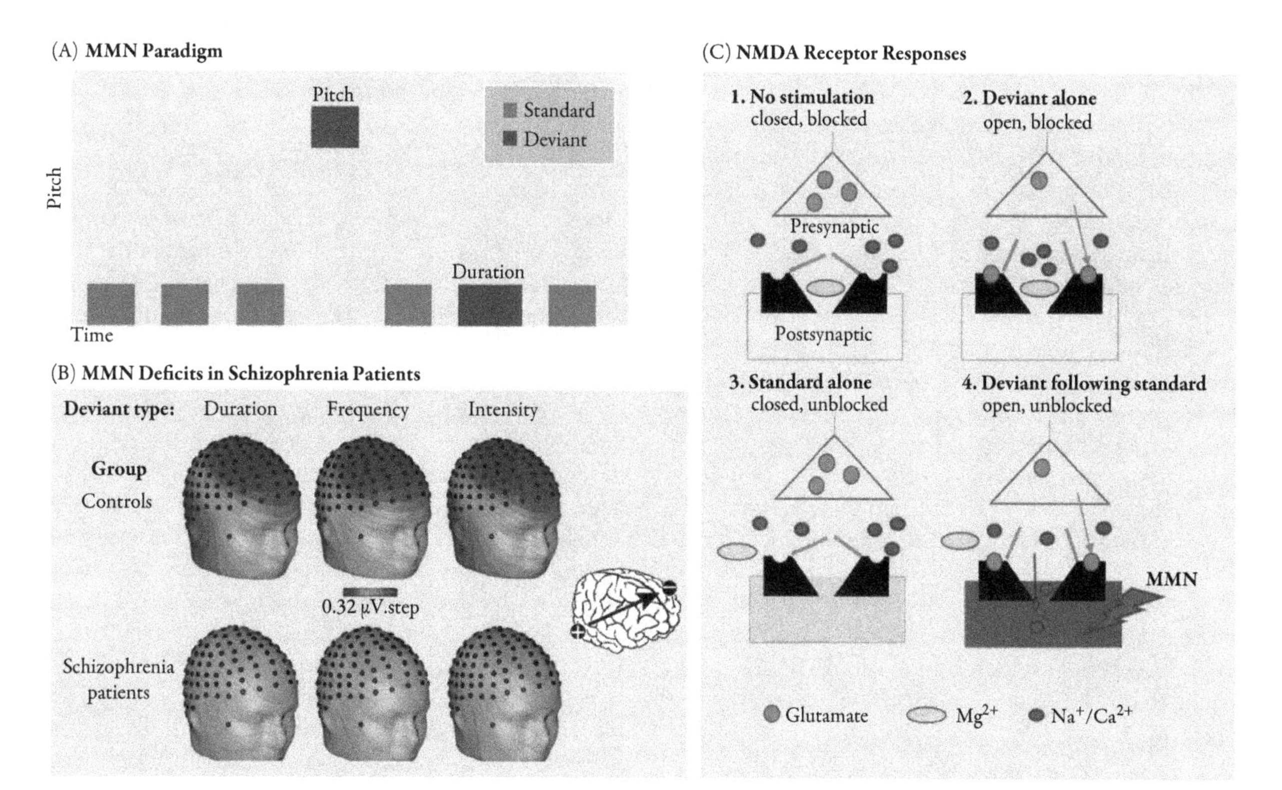

Figure 32.3 Effect of D-serine in schizophrenia. *A.* Meta-analysis showing significant, moderate-large effect (d = .7) as adjunctive treatment to antipsychotics across all studies in individuals with established schizophrenia. *B.* Change in negative symptoms during D-serine vs. placebo treatment in individuals at clinical high risk for schizophrenia.
From [79].

also been shown to be effective against negative symptoms in clinical high risk individuals[79] (Figure 32.3B).

Negative results have been reported for both glycine[80] and D-serine.[81] However, these effects have occurred primarily in the context of multi-center trials with large placebo effects (>20% change in symptoms). D-Cycloserine, an NMDAR glycine-site partial agonist when used at doses of ~50 mg/d, has in general proven less effective than glycine and D-serine,[82] suggesting the need for full-, rather than partial-, agonists.

Most recently, NMDAR agonists such as glycine and D-serine have been shown to reverse NMDAR antagonist effects on MMN in rodents,[51] and to reverse MMN[83,84] and visual P1[84] reductions in schizophrenia. Intermittent D-Serine treatment has also been shown to enhance auditory cortical plasticity.[85] Thus, although glycine or D-serine are not well suited for widespread therapeutic use, their beneficial effects on both symptoms and neurophysiological deficits support glutamatergic models.

Although most studies with NMDAR agonists use continuous dosing, we have also recently observed that repeated intermittent treatment with D-serine, combined with plasticity training, may help reverse cognitive deficits in schizophrenia. In that study, plasticity was assessed using a repeated auditory discrimination task. D-Serine administration both improved plasticity during the learning task, and increased MMN amplitudes to trained tones. Thus, both continuous and intermittent dosing strategies need to be explored further.[85]

GLYCINE TRANSPORT INHIBITORS

Extracellular glycine levels in brain are above the level needed to fully saturate the glycine binding site of the NMDA receptor.[86] NMDAR are protected from these circulating levels by GlyT1 and SNAT transporters that are colocalized with NMDAR and serve to maintain low subsaturating glycine levels within the local region of the NMDAR synapse.[19,87]

Initial studies were performed with the low affinity glycine derivative glycyldodecylamide (GDA), and demonstrated the ability of the compound to reverse NMDAR antagonist-induced behavioral effects in rodents.[88] Subsequent studies confirmed the effectiveness of high affinity glycine transport inhibitors in a range of rodent[36] and primate[89] models relevant to schizophrenia.

Initial clinical studies were performed with sarcosine (N-methylglycine), a naturally occurring compound that serves as an antagonist of both GlyT1 and SNAT5 transporters.[19] Across trials, sarcosine has had robust clinical effects with no failures to replicate,[90–92] supporting glutamatergic models. Nevertheless, definitive multicenter trials have not yet been conducted.

Clinical trials have been conducted with high-affinity GlyT1 transport inhibitors, of which bitopertin (RG1678, Roche) has been tested most extensively. In an initial phase II trial, bitopertinGlyT1 occupancy associated with beneficial treatment response was in the range of 30%–50%, with loss of efficacy at higher occupancies.[93] Of subsequent phase III studies, those targeting persistent predominant negative symptoms showed no significant between-group difference.[94] However, in all studies the placebo response was >30%-.

By contrast, in studies targeting "suboptimally treated" symptoms of schizophrenia characterized by persistent positive and negative symptoms despite adequate antipsychotic treatment, 1 of 3 studies was strongly positive (p = .0028),[95] and the positive result remained significant even following correction for multiple comparisons (threshold p = .017).[96] The one positive study differed from the negative studies only in level of placebo-response, not in the level of response to the experimental agent (bitopertin). These findings thus highlight the need for improved clinical trials methodology in studying treatment nonresponsive schizophrenia subjects.

Nevertheless, across studies bitopertin showed a prominent inverted-U shaped response curve and narrow therapeutic window, suggesting loss of benefit at higher doses.[96] Reasons for the loss of benefit need to be determined. Bitopertin also did not induced glycine-like potentiation of MMN.[97] Overall, these findings suggest that targeting the GlyT1 transporter exclusively may be insufficient and that more balanced inhibition of glycine transport (as in the case of sarcosine) may be needed.

METABOTROPIC RECEPTORS

In addition to compounds that target NMDAR directly, it has been proposed that reversal of NMDAR-induced increases in brain glutamate levels may be therapeutically beneficial.[98] The most advanced compound, pomaglumetad (LY2140023), was tested in several phase II and phase III studies. In an initial phase II study, pomaglumetad showed significant beneficial effects vs. placebo.[99] However, subsequent phase III studies were negative.[100]

An open question remains whether doses used in the clinical studies were sufficient to adequately engage central mGluR2/3 receptors, and therefore whether the clinical program adequately tested the underlying hypothesis. In addition, a post hoc analysis of the data suggested that effects might be greatest in early stage patients.[101] Studies to investigate the utility of Pomaglumetad in more acute schizophrenia populations, as well as in prodromal subjects, therefore remain underway. In addition, novel mGluR2/3 positive allosteric modulators with potentially higher potency for presynaptic mGluR2/3 receptors and greater CNS penetrance are currently underway.[102]

In addition, agents that stimulate mGluR5 will theoretically stimulate NMDAR and thus potentially could be used as treatments for schizophrenia.[25] To date, however, such compounds have been associated with preclinical neurotoxicity profiles that limit ability of candidate compounds to be advanced to clinical studies.

IN VIVO EVIDENCE FOR GLUTAMATERGIC DYSFUNCTION

As opposed to other neurotransmitter systems, glutamate binds with relatively low affinity to glutamate receptors. Thus, radioreceptor imaging approaches like those used with dopamine cannot be used for study of the glutamatergic system. Alternative approaches are therefore required.

MRS IMAGING

Glutamate is a dicarboxylic acid with resonant centers that allow its detection with ^{1}H-MRS imaging. It is known that acute NMDAR blockade significantly increases MRS glutamate+glutamine (Glx) levels, suggesting that such measures can be used as an indirect readout of NMDAR dysfunction.[11]

A recent meta-analysis concluded that glutamate and Glx elevations in basal ganglia and Glx in medial temporal lobe are statistically reliable across all stages of the illness, and that elevated medial frontal glutamate levels were reliable in individuals at high risk for schizophrenia but not in first-episode psychosis or chronic schizophrenia.[103]

Highlighting the potential importance of the striatal glutamate increase, a complementary relationship between glutamate and striatal dopamine release has recently been observed in which presynaptic dopamine levels in striatum correlate positively with striatal glutamate but negatively with medial prefrontal glutamate. Thus, an increase in striatal glutamate may contribute to dopaminergic excesses in schizophrenia.[8] By contrast, higher prefrontal glutamate levels have been observed in treatment-resistant vs. treatment-responsive patients.[39]

POSTMORTEM STUDIES

Further supporting evidence for glutamatergic dysfunction comes from the study of NMDAR density in postmortem brain. Recent meta-analyses have found significant reductions in both NR1 and NR2C subunits in dorsolateral prefrontal cortex in schizophrenia,[104,105] consistent with glutamatergic theories. Differential alterations have also been observed in schizophrenia and bipolar disorder.[106]

BLOOD/CSF MEASURES

Reduction in brain levels of glycine and D-serine may also contribute to NMDAR dysfunction in schizophrenia. A recent meta-analysis demonstrated significant reductions of D-serine in both brain tissue/CSF (SMD~.4) and blood (SMD ~.7).[107] By contrast, no consistent changes in brain or plasma glycine levels have been observed.[108]

Concentrations of endogenous NMDAR antagonists may also be increased in schizophrenia. For example, kynurenic acid is a tryptophan metabolite that blocks both NMDAR and α_7 nicotinic receptors. Elevated levels of kynurenic acid have been observed in blood and CSF in schizophrenia, suggesting that targeting metabolic pathways leading to kynurenic acid synthesis may be therapeutically beneficial.[109,110]

GENETIC FACTORS

It is increasingly realized that schizophrenia is a highly polygenic disorder with numerous mutations conferring vulnerability in any given individual.[111] Theoretically, these mutations could affect presynaptic glutamate release, postsynaptic NMDAR, or metabolism of endogenous NMDAR modulators, such as glycine or D-serine. To date, evidence has accumulated suggesting that all three factors may be etiological in specific subsets of individuals. In general, both copy number variants (CNV) and single-nucleotide changes as detected in genome-wide association studies (GWAS) may contribute.[111]

PRESYNAPTIC GLUTAMATE TERMINALS

Both GWAS[111] and CNV[112–115] studies have consistently shown involvement of genes affecting glutamatergic neurotransmission. Although the number of genes included in these networks is large, several genes have been investigated individually. These include DISC1,[116] dysbindin (DTNBP1),[117] neuregulin (NRG1),[118] and presynaptic glutamate transporters.[112–114] Familial segregation in a mutation of the mGluR5 receptor, which modulates NMDAR function, has also been reported[115]

An unexpected finding from GWAS studies has been involvement of genes involved in immune function, such as molecules in the human leukocyte antigen (HLA) locus (also call major histopathology complex, MHC),[119] and complement component 4.[120,121] While this is sometimes interpreted as supporting autoimmunity hypotheses, a more likely explanation is that several of the MHCI molecules initially evolved in the CNS to regulate pruning of presynaptic terminals, and only later were incorporated into the immune system.[122,123] Similarly, complement C4 plays a critical role in microglia-mediated pruning of presynaptic terminals.[124] The finding of MHC-related risk thus converge with theories of excessive presynaptic glutamatergic pruning during late adolescence in schizophrenia.[125]

NMDAR-RELATED PATHOLOGY

Genetic polymorphisms of NR1, NR2A and NR2B receptors have all been associated with schizophrenia.[105] Moreover, polymorphisms of the *GRIN2B* subunit may lead to NR1 downregulation, contributing to postmortem findings.[105] Rare variants have also been reported that implicate proteins related to NMDAR, including mGluR5.[115] Finally, adequate function of the NMDAR requires anchoring and regulation within the postsynaptic density. Abnormalities within the postsynaptic interactome, especially involving PSD-95, may thus also contribute to glutamatergic dysfunction in schizophrenia.[126]

MODULATORY GENES

NMDAR are regulated by D-serine and glycine. Genetic abnormalities have been reported in genes regulating D-serine synthesis including SR, DAAO,[127] and DAOA/G72.[128] In general, these abnormalities would be predicted to contribute to the reduced levels of D-serine observed in schizophrenia.

Glycine in the brain is synthesized primarily via serine hydroxymethyltransferase (SHMT) and degraded via glycine dehydrogenase (GLDC). Although these enzymes have not been linked to schizophrenia through either GWAS- or CNV-type studies, one informative family has been described in which psychotic illness (bipolar disorder, schizophrenia) was associated with duplications/triplications of the GLDC gene. In these individuals, symptoms improved markedly following supplementation with glycine.[129]

SUMMARY

Just as dopaminergic hyperactivity may be considered the final common pathway to paranoid psychoses across disorders, glutamatergic dysfunction may be viewed as the final common pathway to a form of schizophrenia that encompasses global cognitive impairment and prominent negative symptoms, along with positive symptoms.[14]

As with other neurochemical abnormalities, glutamatergic dysfunction in schizophrenia may be caused by a variety of factors that accumulate in any one individual to reduce integrity of glutamatergic function in general, as well as NMDAR function in particular. Regardless of cause, glutamatergic deficits undermine neural processing throughout the brain, leading to impaired function at both subcortical and cortical levels.

Glutamatergic deficits uniquely account for deficits in sensory processing in schizophrenia, such as auditory MMN and visual P1, and may contribute along with dysfunction of other neurotransmitter systems in impairments in higher level function. Although no medications are clinically available that target NMDAR dysfunction in schizophrenia, meta-analyses suggest beneficial effects of compounds such as glycine and D-serine that serve as positive NMDAR modulators. Other glutamatergic targets such as presynaptic mGluR2/3 and postsynaptic mGluR5 receptors are also targets of ongoing research.

Overall, glutamatergic models of schizophrenia are strongly supported by pathophysiological data. Whether these findings can be translated into improved clinical treatments, however, remains to be determined.

REFERENCES

1. Howes OD, Kapur S. A neurobiological hypothesis for the classification of schizophrenia: type A (hyperdopaminergic) and type B (normodopaminergic). *Br J Psychiatry.* 2014;205(1):1–3.
2. Javitt DC. Current and emergent treatments for symptoms and neurocognitive impairment in schizophrenia. *Current Treatment Options in Psychiatry.* 2015;1(2):107–120.
3. Javitt DC. Negative schizophrenic symptomatology and the PCP (phencyclidine) model of schizophrenia. *Hillside J Clin Psychiatry.* 1987;9(1):12–35.
4. Javitt DC, Zukin SR. Recent advances in the phencyclidine model of schizophrenia. *American Journal of Psychiatry.* 1991;148(10):1301–1308.
5. Coyle JT. The glutamatergic dysfunction hypothesis for schizophrenia. *Harv Rev Psychiatry.* 1996;3(5):241–253.
6. Moghaddam B, Javitt D. From revolution to evolution: the glutamate hypothesis of schizophrenia and its implication for treatment. *Neuropsychopharmacology.* 2012;37(1):4–15.

7. Domino E, Luby E. Abnormal mental states induced by phencyclidine as a model of schizophrenia. In: Domino E, ed. *PCP (Phencyclidine): Historical and Current Perspectives.* Ann Arbor, Mich: NPP Books; 1981:401–418.
8. Gleich T, Deserno L, Lorenz RC, et al. Prefrontal and striatal glutamate differently relate to striatal dopamine: potential regulatory mechanisms of striatal presynaptic dopamine function? *J Neurosci.* 2015;35(26):9615–9621.
9. Karnani MM, Jackson J, Ayzenshtat I, et al. Opening holes in the blanket of inhibition: localized lateral disinhibition by VIP interneurons. *J Neurosci.* 2016;36(12):3471–3480.
10. Overstreet-Wadiche L, Wadiche JI. Good housekeeping. *Neuron.* 2014;81(4):715–717.
11. Javitt DC, Carter CS, Krystal J, et al. Multicenter validation study of biomarkers for glutamate-targeted drug development in psychotic disorders: a randomized clinical trial. *JAMA Psychiatry.* in press.
12. Matta JA, Pelkey KA, Craig MT, Chittajallu R, Jeffries BW, McBain CJ. Developmental origin dictates interneuron AMPA and NMDA receptor subunit composition and plasticity. *Nat Neurosci.* 2013;16(8):1032–1041.
13. Granger AJ, Nicoll RA. Expression mechanisms underlying long-term potentiation: a postsynaptic view, 10 years on. *Philos Trans R Soc Lond B Biol Sci.* 2014;369(1633):20130136.
14. Kantrowitz JT, Javitt DC. N-methyl-D-aspartate (NMDA) receptor dysfunction or dysregulation: the final common pathway on the road to schizophrenia? *Brain Res Bull.* 2010;83(3–4):108–121.
15. Wong KF, Wang XJ. A recurrent network mechanism of time integration in perceptual decisions. *J Neurosci.* 2006;26(4):1314–1328.
16. Javitt DC, Frusciante MJ, Zukin SR. Rat brain N-methyl-D-aspartate receptors require multiple molecules of agonist for activation. *Mol Pharmacol.* 1990;37(5):603–607.
17. Fox K, Daw N. A model for the action of NMDA conductances in the visual cortex. *Neural Comput.* 1992;4:59–83.
18. Javitt DC, Zukin SR. Interaction of [3H]MK-801 with multiple states of the N-methyl-D-aspartate receptor complex of rat brain. *Proc Natl Acad Sci U S A.* 1989;86(2):740–744.
19. Javitt DC, Duncan L, Balla A, Sershen H. Inhibition of system A-mediated glycine transport in cortical synaptosomes by therapeutic concentrations of clozapine: implications for mechanisms of action. *Mol Psychiatry.* 2005;10(3):275–287.
20. Komm B, Beyreis M, Kittl M, Jakab M, Ritter M, Kerschbaum HH. Glycine modulates membrane potential, cell volume, and phagocytosis in murine microglia. *Amino Acids.* 2014;46(8): 1907–1917.
21. Sershen H, Hashim A, Dunlop DS, Suckow RF, Cooper TB, Javitt DC. Modulating NMDA receptor function with D-amino acid oxidase inhibitors: understanding functional activity in PCP-treated mouse model. *Neurochem Res.* 2016;41(1–2):398–408.
22. Javitt DC, Balla A, Sershen H. A novel alanine-insensitive D-serine transporter in rat brain synaptosomal membranes. *Brain Res.* 2002;941(1–2):146–149.
23. Wolosker H, Balu DT, Coyle JT. The rise and fall of the D-serine-mediated gliotransmission hypothesis. *Trends Neurosci.* 2016;39(11):712–721.
24. Papouin T, Henneberger C, Rusakov DA, Oliet SHR. Astroglial versus Neuronal D-Serine: Fact Checking. *Trends Neurosci.* 2017.
25. Herman EJ, Bubser M, Conn PJ, Jones CK. Metabotropic glutamate receptors for new treatments in schizophrenia. *Handbook of Experimental Pharmacology.* 2012(213):297–365.
26. Abi-Dargham A, van de Giessen E, Slifstein M, Kegeles LS, Laruelle M. Baseline and amphetamine-stimulated dopamine activity are related in drug-naive schizophrenic subjects. *Biol Psychiatry.* 2009;65(12):1091–1093.
27. Kegeles LS, Abi-Dargham A, Frankle WG, et al. Increased synaptic dopamine function in associative regions of the striatum in schizophrenia. *Arch Gen Psychiatry.* 2010;67(3):231–239.
28. Vollenweider FX, Vontobel P, Oye I, Hell D, Leenders KL. Effects of (S)-ketamine on striatal dopamine: a [11C]raclopride PET study of a model psychosis in humans. *J Psychiatr Res.* 2000;34(1):35–43.
29. Howes OD, Kapur S. The dopamine hypothesis of schizophrenia: version III—the final common pathway. *Schizophr Bull.* 2009;35(3):549–562.
30. Sarpal DK, Robinson DG, Lencz T, et al. Antipsychotic treatment and functional connectivity of the striatum in first-episode schizophrenia. *JAMA Psychiatry.* 2015;72(1):5–13.
31. Breier A, Adler CM, Weisenfeld N, et al. Effects of NMDA antagonism on striatal dopamine release in healthy subjects: application of a novel PET approach. *Synapse.* 1998;29(2):142–147.
32. Aalto S, Ihalainen J, Hirvonen J, et al. Cortical glutamate-dopamine interaction and ketamine-induced psychotic symptoms in man. *Psychopharmacology (Berl).* 2005;182(3):375–383.
33. Balla A, Sershen H, Serra M, Koneru R, Javitt DC. Subchronic continuous phencyclidine administration potentiates amphetamine-induced frontal cortex dopamine release. *Neuropsychopharmacology.* 2003;28(1):34–44.
34. Sershen H, Balla A, Aspromonte J, Xie S, Cooper T, Javitt D. Characterization of interactions between phencyclidine and amphetamine in rodent prefrontal cortex and striatum: implications in NMDA/glycine-site-mediated dopaminergic dysregulation and dopamine transporter function. *Neurochemistry International.* 2008;52(1–2):119–129.
35. Javitt DC, Balla A, Burch S, Suckow R, Xie S, Sershen H. Reversal of phencyclidine-induced dopaminergic dysregulation by N-methyl-D-aspartate receptor/glycine-site agonists. *Neuropsychopharmacology.* 2004;29(2):300–307.
36. Balla A, Schneider S, Sershen H, Javitt DC. Effects of novel, high affinity glycine transport inhibitors on frontostriatal dopamine release in a rodent model of schizophrenia. *European Neuropsychopharmacology.* 2012;22(12):902–910.
37. Sarpal DK, Argyelan M, Robinson DG, et al. Baseline striatal functional connectivity as a predictor of response to antipsychotic drug treatment. *American Journal of Psychiatry.* 2016;173(1): 69–77.
38. Weinstein JJ, Chohan MO, Slifstein M, Kegeles LS, Moore H, Abi-Dargham A. Pathway-specific dopamine abnormalities in schizophrenia. *Biol Psychiatry.* 2017;81(1):31–42.
39. Mouchlianitis E, Bloomfield MA, Law V, et al. Treatment-resistant schizophrenia patients show elevated anterior cingulate cortex glutamate compared to treatment-responsive. *Schizophr Bull.* 2016;42(3):744–752.
40. Roseberry TK, Lee AM, Lalive AL, Wilbrecht L, Bonci A, Kreitzer AC. Cell-type-specific control of brainstem locomotor circuits by basal ganglia. *Cell.* 2016;164(3):526–537.
41. Gerfen CR, Surmeier DJ. Modulation of striatal projection systems by dopamine. *Annu Rev Neurosci.* 2011;34:441–466.
42. Krystal JH, Bennett A, Abi-Saab D, et al. Dissociation of ketamine effects on rule acquisition and rule implementation: possible relevance to NMDA receptor contributions to executive cognitive functions. *Biol Psychiatry.* 2000;47(2):137–143.
43. Kantrowitz JT, Javitt DC. Thinking glutamatergically: changing concepts of schizophrenia based upon changing neurochemical models. *Clinical Schizophrenia and Related Psychoses.* 2010;4(3): 189–200.
44. Javitt DC. When doors of perception close: bottom-up models of disrupted cognition in schizophrenia. *Annu Rev Clin Psychol.* 2009;5:249–275.
45. Javitt DC. Sensory processing in schizophrenia: neither simple nor intact. *Schizophr Bull.* 2009;35(6):1059–1064.
46. Javitt DC, Freedman R. Sensory processing dysfunction in the personal experience and neuronal machinery of schizophrenia. *American Journal of Psychiatry.* 2015;172(1):17–31.
47. Javitt DC, Sweet RA. Auditory dysfunction in schizophrenia: integrating clinical and basic features. *Nat Rev Neurosci.* 2015;16(9):535–550.
48. Javitt DC, Steinschneider M, Schroeder CE, Vaughan HG, Jr., Arezzo JC. Detection of stimulus deviance within primate primary auditory cortex: intracortical mechanisms of mismatch negativity (MMN) generation. *Brain Res.* 1994;667(2):192–200.

49. Gil-da-Costa R, Stoner GR, Fung R, Albright TD. Nonhuman primate model of schizophrenia using a noninvasive EEG method. *Proc Natl Acad Sci U S A.* 2013;110(38):15425–15430.
50. Ehrlichman RS, Maxwell CR, Majumdar S, Siegel SJ. Deviance-elicited changes in event-related potentials are attenuated by ketamine in mice. *J Cogn Neurosci.* 2008;20(8):1403–1414.
51. Lee M, Balla A, Sershen H, Sehatpour P, Lakatos P, Javitt DC. Rodent mismatch negativity (MMN)/theta neuro-oscillatory response as a translational neurophysiological biomarker for N-methyl-D-aspartate receptor-based new treatment development in schizophrenia. *Neuropsychopharmacology.* 2017.
52. Umbricht D, Schmid L, Koller R, Vollenweider FX, Hell D, Javitt DC. Ketamine-induced deficits in auditory and visual context-dependent processing in healthy volunteers: implications for models of cognitive deficits in schizophrenia. *Arch Gen Psychiatry.* 2000;57(12):1139–1147.
53. Rosburg T, Kreitschmann-Andermahr I. The effects of ketamine on the mismatch negativity (MMN) in humans—a meta-analysis. *Clin Neurophysiol.* 2016;127(2):1387–1394.
54. Nagai T, Kirihara K, Tada M, et al. Reduced mismatch negativity is associated with increased plasma level of glutamate in first-episode psychosis. *Sci Rep.* 2017;7(1):2258.
55. Rowland LM, Summerfelt A, Wijtenburg SA, et al. Frontal glutamate and gamma-aminobutyric acid levels and their associations with mismatch negativity and digit sequencing task performance in schizophrenia. *JAMA Psychiatry.* 2016;73(2):166–174.
56. Umbricht D, Koller R, Vollenweider FX, Schmid L. Mismatch negativity predicts psychotic experiences induced by NMDA receptor antagonist in healthy volunteers. *Biol Psychiatry.* 2002;51(5):400–406.
57. Avissar M, Xie S, Vail B, Lopez-Calderon J, Wang Y, Javitt DC. Meta-analysis of mismatch negativity to simple versus complex deviants in schizophrenia. *Schizophr Res.* 2017.
58. Kim M, Kim SN, Lee S, et al. Impaired mismatch negativity is associated with current functional status rather than genetic vulnerability to schizophrenia. *Psychiatry Res.* 2014;222(1-2):100–106.
59. Thomas ML, Green MF, Hellemann G, et al. Modeling deficits from early auditory information processing to psychosocial functioning in schizophrenia. *JAMA Psychiatry.* 2017;74(1):37–46.
60. Lee M, Sehatpour P, Dias EC, et al. A tale of two sites: differential impairment of frequency and duration mismatch negativity across a primarily inpatient versus a primarily outpatient site in schizophrenia. *Schizophr Res.* 2017.
61. Bodatsch M, Ruhrmann S, Wagner M, et al. Prediction of psychosis by mismatch negativity. *Biol Psychiatry.* 2011;69(10):959–966.
62. Perez VB, Woods SW, Roach BJ, et al. Automatic auditory processing deficits in schizophrenia and clinical high-risk patients: forecasting psychosis risk with mismatch negativity. *Biol Psychiatry.* 2014;75(6):459–469.
63. Naatanen R, Todd J, Schall U. Mismatch negativity (MMN) as biomarker predicting psychosis in clinically at-risk individuals. *Biol Psychol.* 2016;116:36–40.
64. Carrion RE, Cornblatt BA, McLaughlin D, et al. Contributions of early cortical processing and reading ability to functional status in individuals at clinical high risk for psychosis. *Schizophr Res.* 2015;164(1–3):1–7.
65. Baldeweg T, Hirsch SR. Mismatch negativity indexes illness-specific impairments of cortical plasticity in schizophrenia: a comparison with bipolar disorder and Alzheimer's disease. *Int J Psychophysiol.* 2015;95(2):145–155.
66. Korostenskaja M, Kicic D, Kahkonen S. The effect of methylphenidate on auditory information processing in healthy volunteers: a combined EEG/MEG study. *Psychopharmacology (Berl).* 2008;197(3):475–486.
67. Leung S, Croft RJ, Baldeweg T, Nathan PJ. Acute dopamine D(1) and D(2) receptor stimulation does not modulate mismatch negativity (MMN) in healthy human subjects. *Psychopharmacology (Berl).* 2007;194(4):443–451.
68. Leung S, Croft RJ, Guille V, et al. Acute dopamine and/or serotonin depletion does not modulate mismatch negativity (MMN) in healthy human participants. *Psychopharmacology (Berl).* 2010;208(2):233–244.
69. Umbricht D, Vollenweider FX, Schmid L, et al. Effects of the 5-HT2A agonist psilocybin on mismatch negativity generation and AX-continuous performance task: implications for the neuropharmacology of cognitive deficits in schizophrenia. *Neuropsychopharmacology.* 2003;28(1):170–181.
70. Heekeren K, Daumann J, Neukirch A, et al. Mismatch negativity generation in the human 5HT2A agonist and NMDA antagonist model of psychosis. *Psychopharmacology (Berl).* 2008;199(1):77–88.
71. Hansenne M, Pinto E, Scantamburlo G, et al. Mismatch negativity is not correlated with neuroendocrine indicators of catecholaminergic activity in healthy subjects. *Hum Psychopharmacol.* 2003;18(3):201–205.
72. Korostenskaja M, Kahkonen S. What do ERPs and ERFs reveal about the effect of antipsychotic treatment on cognition in schizophrenia? *Curr Pharm Des.* 2009;15(22):2573–2593.
73. Umbricht D, Javitt D, Novak G, et al. Effects of clozapine on auditory event-related potentials in schizophrenia. *Biol Psychiatry.* 1998;44(8):716–725.
74. Umbricht D, Javitt D, Novak G, et al. Effects of risperidone on auditory event-related potentials in schizophrenia. *The International Journal of Neuropsychopharmacology.* 1999;2(4):299–304.
75. Martinez A, Hillyard SA, Bickel S, Dias EC, Butler PD, Javitt DC. Consequences of magnocellular dysfunction on processing attended information in schizophrenia. *Cereb Cortex.* 2012;22(6):1282–1293.
76. Javitt DC. Glutamate as a therapeutic target in psychiatric disorders. *Mol Psychiatry.* 2004;9(11):984–997, 979.
77. Kantrowitz JT, Malhotra AK, Cornblatt B, et al. High dose D-serine in the treatment of schizophrenia. *Schizophr Res.* 2010;121(1–3):125–130.
78. Singh SP, Singh V. Meta-analysis of the efficacy of adjunctive NMDA receptor modulators in chronic schizophrenia. *CNS Drugs.* 2011;25(10):859–885.
79. Kantrowitz JT, Woods SW, Petkova E, et al. D-serine for the treatment of negative symptoms in individuals at clinical high risk of schizophrenia: a pilot, double-blind, placebo-controlled, randomised parallel group mechanistic proof-of-concept trial. *Lancet Psychiatry.* 2015;2(5):403–412.
80. Buchanan RW, Javitt DC, Marder SR, et al. The Cognitive and Negative Symptoms in Schizophrenia Trial (CONSIST): the efficacy of glutamatergic agents for negative symptoms and cognitive impairments. *American Journal of Psychiatry.* 2007;164(10):1593–1602.
81. Weiser M, Heresco-Levy U, Davidson M, et al. A multicenter, add-on randomized controlled trial of low-dose d-serine for negative and cognitive symptoms of schizophrenia. *J Clin Psychiatry.* 2012;73(6):e728–734.
82. Heresco-Levy U, Javitt DC. Comparative effects of glycine and D-cycloserine on persistent negative symptoms in schizophrenia: a retrospective analysis. *Schizophr Res.* 2004;66(2–3):89–96.
83. Greenwood LM, Leung S, Michie PT, et al. The effects of glycine on auditory mismatch negativity in schizophrenia. *Schizophr Res.* 2017.
84. Kantrowitz JT, Epstein ML, Lee M, et al. Improvement in mismatch negativity generation during d-serine treatment in schizophrenia: correlation with symptoms. *Schizophr Res.* 2017.
85. Kantrowitz JT, Epstein ML, Beggel O, et al. Neurophysiological mechanisms of cortical plasticity impairments in schizophrenia and modulation by the NMDA receptor agonist D-serine. *Brain.* 2016;139(Pt 12):3281–3295.
86. Javitt DC, Heresco-Levy U. Are glycine sites saturated in vivo? *Arch Gen Psychiatry.* 2000;57(12):1181–1183.
87. Javitt DC. Glycine transport inhibitors for the treatment of schizophrenia: symptom and disease modification. *Curr Opin Drug Discov Devel.* 2009;12(4):468–478.
88. Javitt DC, Balla A, Sershen H, Lajtha A. A.E. Bennett Research Award. Reversal of phencyclidine-induced effects by glycine and glycine transport inhibitors. *Biol Psychiatry.* 1999;45(6):668–679.
89. Roberts BM, Shaffer CL, Seymour PA, Schmidt CJ, Williams GV, Castner SA. Glycine transporter inhibition reverses ketamine-induced working memory deficits. *Neuroreport.* 2010;21(5):390–394.

90. Tsai GE, Lin PY. Strategies to enhance N-methyl-D-aspartate receptor-mediated neurotransmission in schizophrenia, a critical review and meta-analysis. *Curr Pharm Des.* 2010;16(5):522–537.
91. Amiaz R, Kent I, Rubinstein K, Sela BA, Javitt D, Weiser M. Safety, tolerability and pharmacokinetics of open label sarcosine added on to anti-psychotic treatment in schizophrenia—preliminary study. *Isr J Psychiatry Relat Sci.* 2015;52(1):12–15.
92. Strzelecki D, Kaluzynska O, Szyburska J, Wlazlo A, Wysokinski A. No changes of cardiometabolic and body composition parameters after 6-month add-on treatment with sarcosine in patients with schizophrenia. *Psychiatry Res.* 2015;230(2):200–204.
93. Umbricht D, Yoo K, Youssef E, et al. Glycine transporter type 1 (GLYT1) inhibitor RG1678: positive results of the proof-of-concept study for the treatment of negative symptoms in schizophrenia. *Neuropsychopharmacol.* 2010;35:s320–321.
94. Bugarski-Kirola D, Blaettler T, Arango C, et al. Bitopertin in negative symptoms of schizophrenia-results from the phase III FlashLyte and DayLyte studies. *Biol Psychiatry.* 2017;82(1):8–16.
95. Bugarski-Kirola D, Iwata N, Sameljak S, et al. Efficacy and safety of adjunctive bitopertin versus placebo in patients with suboptimally controlled symptoms of schizophrenia treated with antipsychotics: results from three phase 3, randomised, double-blind, parallel-group, placebo-controlled, multicentre studies in the SearchLyte clinical trial programme. *Lancet Psychiatry.* 2016;3(12):1115–1128.
96. Javitt DC. Bitopertin in schizophrenia: glass half full? *Lancet Psychiatry.* 2016;3(12):1092–1093.
97. Kantrowitz JT, Nolan KA, Epstein ML, et al. Neurophysiological effects of bitopertin in schizophrenia. *J Clin Psychopharmacol.* 2017;37(4):447–451.
98. Moghaddam B, Krystal JH. Capturing the angel in "angel dust": twenty years of translational neuroscience studies of NMDA receptor antagonists in animals and humans. *Schizophr Bull.* 2012;38(5):942–949.
99. Patil ST, Zhang L, Martenyi F, et al. Activation of mGlu2/3 receptors as a new approach to treat schizophrenia: a randomized phase 2 clinical trial. *Nat Med.* 2007;13(9):1102–1107.
100. Downing AM, Kinon BJ, Millen BA, et al. A double-blind, placebo-controlled comparator study of LY2140023 monohydrate in patients with schizophrenia. *BMC Psychiatry.* 2014;14:351.
101. Kinon BJ, Millen BA, Zhang L, McKinzie DL. Exploratory analysis for a targeted patient population responsive to the metabotropic glutamate 2/3 receptor agonist pomaglumetad methionil in schizophrenia. *Biol Psychiatry.* 2015;78(11):754–762.
102. Ellaithy A, Younkin J, Gonzalez-Maeso J, Logothetis DE. Positive allosteric modulators of metabotropic glutamate 2 receptors in schizophrenia treatment. *Trends Neurosci.* 2015;38(8):506–516.
103. Merritt K, Egerton A, Kempton MJ, Taylor MJ, McGuire PK. Nature of glutamate alterations in schizophrenia: a meta-analysis of proton magnetic resonance spectroscopy studies. *JAMA Psychiatry.* 2016;73(7):665–674.
104. Catts VS, Lai YL, Weickert CS, Weickert TW, Catts SV. A quantitative review of the postmortem evidence for decreased cortical N-methyl-D-aspartate receptor expression levels in schizophrenia: how can we link molecular abnormalities to mismatch negativity deficits? *Biol Psychol.* 2016;116:57–67.
105. Weickert CS, Fung SJ, Catts VS, et al. Molecular evidence of N-methyl-D-aspartate receptor hypofunction in schizophrenia. *Mol Psychiatry.* 2013;18(11):1185–1192.
106. Beneyto M, Meador-Woodruff JH. Lamina-specific abnormalities of NMDA receptor-associated postsynaptic protein transcripts in the prefrontal cortex in schizophrenia and bipolar disorder. *Neuropsychopharmacology.* 2008;33(9):2175–2186.
107. Cho SE, Na KS, Cho SJ, Kang SG. Low d-serine levels in schizophrenia: a systematic review and meta-analysis. *Neurosci Lett.* 2016;634:42–51.
108. Brouwer A, Luykx JJ, van Boxmeer L, Bakker SC, Kahn RS. NMDA-receptor coagonists in serum, plasma, and cerebrospinal fluid of schizophrenia patients: a meta-analysis of case-control studies. *Neurosci Biobehav Rev.* 2013;37(8):1587–1596.
109. Erhardt S, Schwieler L, Imbeault S, Engberg G. The kynurenine pathway in schizophrenia and bipolar disorder. *Neuropharmacology.* 2017;112(Pt B):297–306.
110. Wonodi I, Schwarcz R. Cortical kynurenine pathway metabolism: a novel target for cognitive enhancement in Schizophrenia. *Schizophr Bull.* 2010;36(2):211–218.
111. Schork AJ, Wang Y, Thompson WK, Dale AM, Andreassen OA. New statistical approaches exploit the polygenic architecture of schizophrenia—implications for the underlying neurobiology. *Curr Opin Neurobiol.* 2016;36:89–98.
112. Horiuchi Y, Iida S, Koga M, et al. Association of SNPs linked to increased expression of SLC1A1 with schizophrenia. *Am J Med Genet B Neuropsychiatr Genet.* 2012;159B(1):30–37.
113. Myles-Worsley M, Tiobech J, Browning SR, et al. Deletion at the SLC1A1 glutamate transporter gene co-segregates with schizophrenia and bipolar schizoaffective disorder in a 5-generation family. *Am J Med Genet B Neuropsychiatr Genet.* 2013;162B(2):87–95.
114. Rees E, Walters JT, Chambert KD, et al. CNV analysis in a large schizophrenia sample implicates deletions at 16p12.1 and SLC1A1 and duplications at 1p36.33 and CGNL1. *Hum Mol Genet.* 2014;23(6):1669–1676.
115. Timms AE, Dorschner MO, Wechsler J, et al. Support for the N-methyl-D-aspartate receptor hypofunction hypothesis of schizophrenia from exome sequencing in multiplex families. *JAMA Psychiatry.* 2013;70(6):582–590.
116. Maher BJ, LoTurco JJ. Disrupted-in-schizophrenia (DISC1) functions presynaptically at glutamatergic synapses. *PLoS One.* 2012;7(3):e34053.
117. Karlsgodt KH, Robleto K, Trantham-Davidson H, et al. Reduced dysbindin expression mediates N-methyl-D-aspartate receptor hypofunction and impaired working memory performance. *Biol Psychiatry.* 2011;69(1):28–34.
118. Hahn CG, Wang HY, Cho DS, et al. Altered neuregulin 1-erbB4 signaling contributes to NMDA receptor hypofunction in schizophrenia. *Nat Med.* 2006;12(7):824–828.
119. Corvin A, Morris DW. Genome-wide association studies: findings at the major histocompatibility complex locus in psychosis. *Biol Psychiatry.* 2014;75(4):276–283.
120. Sekar A, Bialas AR, de Rivera H, et al. Schizophrenia risk from complex variation of complement component 4. *Nature.* 2016;530(7589):177–183.
121. Presumey J, Bialas AR, Carroll MC. Complement system in neural synapse elimination in development and disease. *Adv Immunol.* 2017;135:53–79.
122. Adelson JD, Sapp RW, Brott BK, et al. Developmental sculpting of intracortical circuits by MHC Class I H2-Db and H2-Kb. *Cereb Cortex.* 2016;26(4):1453–1463.
123. Fourgeaud L, Davenport CM, Tyler CM, Cheng TT, Spencer MB, Boulanger LM. MHC class I modulates NMDA receptor function and AMPA receptor trafficking. *Proc Natl Acad Sci U S A.* 2010;107(51):22278–22283.
124. Sellgren CM, Sheridan SD, Gracias J, Xuan D, Fu T, Perlis RH. Patient-specific models of microglia-mediated engulfment of synapses and neural progenitors. *Mol Psychiatry.* 2017;22(2):170–177.
125. Feinberg I. Cortical pruning and the development of schizophrenia. *Schizophr Bull.* 1990;16(4):567–570.
126. Li J, Zhang W, Yang H, et al. Spatiotemporal profile of postsynaptic interactomes integrates components of complex brain disorders. *Nat Neurosci.* 2017;20(8):1150–1161.
127. Shinkai T, De Luca V, Hwang R, et al. Association analyses of the DAOA/G30 and D-amino-acid oxidase genes in schizophrenia: further evidence for a role in schizophrenia. *Neuromolecular Medicine.* 2007;9(2):169–177.
128. Abi-Saab WM, D'Souza DC, Moghaddam B, Krystal JH. The NMDA antagonist model for schizophrenia: promise and pitfalls. *Pharmacopsychiatry.* 1998;31 Suppl 2:104–109.
129. Levy D, Coleman M, Godfrey L, et al. Successfully targeted treatment of a medically actionable mutation in psychotic disorders. *Schizophr Bull.* 2015;41(Suppl 1):S207.

33

GABAERGIC MECHANISMS IN PSYCHOSIS

Takanori Hashimoto and David A. Lewis

INTRODUCTION

Cognitive processes mediated by cerebral cortical circuits are the emergent properties of complex interactions between glutamatergic excitatory pyramidal neurons and GABAergic inhibitory interneurons. Although the latter constitute only 20%–25% of cortical neurons, they play critical roles not only in keeping excitatory activity in check but also in regulating the timing and synchrony of cortical circuit function.[1] These roles are implemented by a diverse complement of GABA neurons, which differ in multiple properties. For example, cortical GABA neurons can be classified into three major subsets based on the expression of parvalbumin (PV), somatostatin (SST), or vasoactive intestinal peptide (VIP)[2–4] that make distinct contributions to cortical information processing. PV neurons, through their perisomatic innervation of pyramidal neurons, regulate coincident input detection, response gain and synchronization.[4] In contrast, SST neurons, which provide inhibitory inputs to the distal dendrites of pyramidal neurons, modulate dendritic integration and contribute to processing and encoding of specific information.[4–6] Finally, via their inhibitory synapses on SST neurons, and to a lesser extent on PV neurons, VIP neurons disinhibit pyramidal neurons.[2,4,7]

Given the frequency and severity of cognitive impairments in schizophrenia and bipolar disorder,[8] it is not surprising that both in vivo and postmortem studies have found evidence of altered cortical GABA neurotransmission in these illnesses. In this chapter, we review this evidence that has been obtained through modern postmortem studies with quantitative, molecular biological methods. We first focus on schizophrenia, a representative psychotic disorder for which the largest body of literature is available on cortical and hippocampal GABA systems. Most postmortem studies included subjects with schizoaffective disorder, so the cited findings apply to both schizophrenia and schizoaffective disorder unless otherwise specified. Then, we review the postmortem literature on the GABA system in bipolar disorder, a mood disorder that is often associated with psychotic symptoms, to explore the extent to which alterations in the GABA system differ across diagnoses or are associated with the pathophysiology of psychosis. All findings reviewed in this chapter refer to studies of total gray matter (GM) unless otherwise specified.

GABA NEURONS IN SCHIZOPHRENIA

The density or total number of cortical GABA neurons has been assessed by many studies in which small, nonpyramidal, putative GABA neurons were differentiated from pyramidal neurons and glial cells using Nissl staining. An early study reported lower two-dimensional densities of nonpyramidal neurons in layer 2 of the dorsolateral prefrontal cortex (DLPFC) and anterior cingulate cortex (ACC) from subjects with schizophrenia.[9] However, the same group found that the two-dimensional density of nonpyramidal neurons was significantly lower in layer 2 in the ACC of subjects with bipolar disorder, but not those with schizophrenia.[10] Although the discrepancy in findings was attributed to differences in illness severity and chronicity between the schizophrenia cohorts, subsequent studies reported that the overall density of putative GABA neurons was increased[11,12] or unchanged[13] in the DLPFC of subjects with schizophrenia. Furthermore, stereological assessments of total neuron numbers in the entire cerebral cortex[14] and the prefrontal cortex[15] did not detect a significant effect of diagnosis, suggesting that the number GABA neurons is unlikely to be altered substantially in the cortex of subjects with schizophrenia.

Expression of the 67 kDa isoform of glutamic acid decarboxylase (GAD67), the enzyme responsible for most cortical GABA synthesis, has been analyzed in multiple postmortem studies of schizophrenia. Most studies using in situ hybridization (ISH) or polymerase chain reaction (PCR) found lower GAD67 mRNA levels across many cortical regions, including the DLPFC,[13,16–21] orbitofrontal cortex (OFC),[22,23] ACC,[22,24] superior temporal gyrus (STG),[22] primary motor cortex (M1),[24] and primary (V1) visual cortices[24] of subjects with schizophrenia compared with unaffected comparison subjects. Western blot analyses confirmed lower GAD67 protein levels in the DLPFC and temporal cortex (TC).[16,20,25]

At the cellular level, the densities of GAD67 mRNA-positive neurons were reduced by 18%–48% in the DLPFC and by 28%–53% in the ACC without a change in the mRNA level per positive neuron.[13,26–30] The reductions of GAD67 mRNA-positive neurons were prominently observed in layers 2–5 of the DLPFC[13,26,27] and layer 2 in the ACC.[29,30] Together, these findings indicate that GAD67 mRNA expression is markedly reduced in a subset of GABA neurons, whereas it is relatively unaffected in other GABA neurons.

Furthermore, a recent study using quantitative immunohistochemistry (IHC) revealed that the density of GAD67-positive axon terminals that co-express the vesicular GABA transporter (vGAT), a specific marker of GABA neuron terminals, was decreased by 16% without a change in the total density of vGAT-positive terminals in the DLPFC,[31] consistent with a marked reduction of GAD67 in a subset of cortical GABA neurons in schizophrenia.

On the other hand, expression of the 65kDa isoform of the enzyme, GAD65, appears to be unaffected at both mRNA and protein levels in the DLPFC and TC.[16,25,32,33] The density of axon terminals of GABA neurons that express GAD65 was also reported to be unaltered in the DLPFC and ACC.[32] Although one study reported slightly lower levels of GAD65 mRNA across the DLPFC, ACC, M1 and V1,[24] this finding could reflect the inclusion of subjects with schizoaffective disorder in the schizophrenia cohort, as a subsequent study reported that GAD65 mRNA and protein levels were significantly lower in the DLPFC of subjects with schizoaffective disorder, but not those with schizophrenia.[34]

Across all subfields of the hippocampus, neither GAD67 nor GAD65 appears to be altered in schizophrenia as assessed by the density of positive neurons, mRNA levels per positive neuron,[35] or the density of GAD65-positive axon terminals,[36] although tissue mRNA levels of both enzymes were reported to be decreased in some subfields.[37,38]

Axon terminals of GABA neurons contain GABA transporter-1 (GAT1) and vGAT, responsible for reuptake of synaptically released GABA and storage of GABA into presynaptic vesicles, respectively. Early ISH studies reported lower GAT1 mRNA levels[39] and a reduced density of GAT1 mRNA-positive neurons without a change in GAT1 mRNA levels per positive neuron[40] in the DLPFC of subjects with schizophrenia. A subsequent PCR study demonstrated lower GAT1 mRNA levels across the DLPFC, ACC, M1, and V1.[24] Although these studies reported small-to-moderate reductions in GAT1 mRNA levels in schizophrenia based on the analyses of small cohorts of schizophrenia and comparison subjects ($n \leq 12$ in each group), neither GAT1 protein immunoreactivity nor the overall density of GAT1-positive synaptic boutons was altered in the DLPFC[41] (See later discussion of GAT1 protein alterations in a specific class of GABA neurons). Furthermore, recent studies using larger subject cohorts did not replicate the lower DLPFC GAT1 mRNA levels in schizophrenia.[33,42] For vGAT, studies are consistent in reporting unaltered[43] or only slightly reduced[42] mRNA levels, and no alterations in the density of vGAT protein-positive terminals or vGAT protein levels per positive terminal, in the DLPFC of schizophrenia subjects.[31]

The activity of GABA neurons is regulated by excitatory synaptic inputs from cortical pyramidal neurons and subcortical projection neurons. In the ACC and DLPFC of schizophrenia subjects, the density of GAD67 mRNA-positive neurons that also expressed the mRNA for the excitatory GluN2A N-methyl-D-aspartate (NMDA) receptor subunit were significantly reduced, whereas GluN2A mRNA levels per GAD67-containing neuron were not significantly altered.[27,29] Similarly, in the ACC, the density of GAD67 mRNA-positive neurons that co-expressed GluK1 kainate receptor subunit mRNA was reduced, whereas GluK1 mRNA levels per GAD67-containing neuron were unchanged or increased.[30,33] These findings suggest that a subset of cortical GABA neurons has a deficiency in glutamate neurotransmission via NMDA and kainate receptors in schizophrenia.

PV NEURONS IN SCHIZOPHRENIA

Multiple postmortem studies have analyzed cortical expression of PV in schizophrenia. Levels of PV mRNA have consistently been reported to be lower in schizophrenia in multiple cortical regions, including the DLPFC,[44–48] ACC,[24,49] OFC,[50] M1,[24] and V1.[24] In contrast, most studies did not detect a significant change in the density or total number of neurons that contain PV mRNA or protein across cortical regions, including the DLPFC,[44,51–55] ACC,[56] entorhinal cortex (ERC),[57] planum temporale (PT),[58] and V1,[51] suggesting that levels of PV expression per neuron are reduced, an interpretation supported by findings of lower PV mRNA[44] and protein[54,55] per neuron (Figure 33.1), whereas the complement of PV neurons is not altered.[44,54,55] The lower densities of PV-immunoreactive neurons reported in some studies[59,60] likely reflects a reduced ability to detect lower levels of the protein due to tissue fixation conditions.[61]

In contrast to cortical areas, both the density and total number of PV neurons were decreased in the hippocampus of subjects with schizophrenia.[38,62] Considering that both the densities of GAD67 mRNA- and GAD65 mRNA-positive neurons were unchanged in the hippocampus,[35] PV protein levels in the hippocampus, as in the neocortex, appear to be reduced below a detectable level in a subset of PV neurons.

In the DLPFC of subjects with schizophrenia, mRNAs encoding PV,[44] GAD67,[13,26,44] GAT1,[53] and KCNS3 voltage-gated potassium channel subunit[63] were all shown to be lower in individual PV neurons (Figure 33.1). These findings likely indicate altered GABA neurotransmission by PV neurons for the reasons summarized in the following paragraphs.

First, the expression of PV, GAD67, and KCNS3 mRNAs are regulated by neuronal activity,[64,65] indicating that lower levels of these transcripts reflect reduced activity of PV neurons. This idea was supported by findings reviewed earlier that a subset of GABA neurons have lower mRNA levels for excitatory glutamate NMDA and kainate receptor subunits.[27,29,30,33] This subset of GABA neurons appears to include PV neurons as the density of DLPFC PV neurons that coexpress GluN2A mRNA was decreased in layers 3 and 4 without a change in GluN2A mRNA levels per PV neuron[66] (Figure 33.1).

Second, Erb-B2 receptor tyrosine kinase 4 (ErbB4), a receptor tyrosine kinase of neuregulin-1 that regulates the formation of excitatory synapses on PV neurons,[67] appears to be dysregulated in cortical PV neurons in schizophrenia. Although mRNA levels of total ErbB4 were unaltered in the GM tissue, relative mRNA levels of the minor splice variants (JM-a and CYT-1) to those of the major splice variants (JM-b and CYT-2) were elevated in the DLPFC of schizophrenia subjects[48,68,69] (Figure 33.1). This shift to the minor variants occurs in layer 4,[48] where PV neurons are greatly enriched,[4,70] but not in layer 2,[48] where another subset of calretinin (CR)-containing

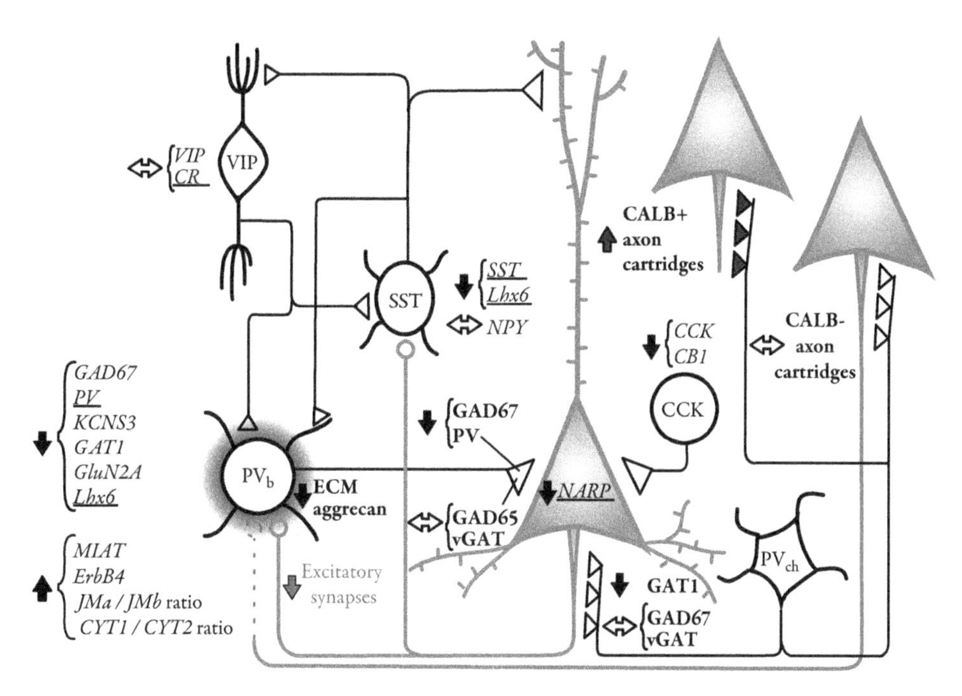

Figure 33.1 Summary of expression changes in mRNAs and proteins in the soma and axon terminals, respectively, of GABA neurons in the DLPFC of subjects with schizophrenia. The directions of alterations are indicated by arrows at the left of the names of mRNAs (italics) and proteins (bolds). Among PV neurons, basket cells are much more numerous than chandelier neurons and most abundant in the middle layer 4.[4,70,80] Therefore, some alterations associated with PV neurons, especially those detected in the middle layers, are illustrated within basket cells, although they might also be present in chandelier cells. Molecules found to be altered in the same manner in the cortex of subjects with bipolar disorder are underlined. Acronyms: CALB, calbindin; CR, calretinin; ECM, extra cellular matrix; ErbB4, Erb-B2 receptor tyrosine kinase 4; GAD65, glutamic acid decarboxylase 65kDa; GAD67, glutamic acid decarboxylase 67kDa; GAT1, GABA transporter 1; Lhx6, LIM homeobox 6; MIAT, myocardial infarction associated transcript; NPY, neuropeptide Y; PNN, perineuronal net; PV, parvalbumin; PVb, PV-positive basket cell, PVch, PV-positive chandelier cell; SST, somatostatin; vGAT, vesicular GABA transporter; VIP, vasoactive intestinal peptide.

GABA neurons that express ErbB4[71] are preferentially localized.[4,70] Furthermore, a regulator of ErbB4 splicing, myocardial infarction associated transcript (MIAT), is increased in putative PV neurons in the DLPFC of schizophrenia subjects[48] (Figure 33.1). Relative to the major splice variants, the signaling through the minor variants could impair the localization of the postsynaptic density (PSD) to excitatory synapses,[72] reducing the number of excitatory synaptic inputs onto PV neurons. Indeed, a recent study demonstrated a lower density of excitatory synapses, which were dual-labeled by vesicular glutamate transporter 1 and postsynaptic density protein 95 (PSD95), on PV neuron soma in DLPFC layer 4 of schizophrenia subjects[54] (Figure 33.1). Furthermore, the density of excitatory synapses on PV neuron soma was positively correlated with mRNA levels of PV and GAD67 in schizophrenia,[54] supporting the idea that fewer excitatory inputs to PV neurons resulted in an activity-dependent reduction in the expression of these transcripts.

Third, excitatory neurotransmission onto PV neurons is also regulated by neuronal activity-regulated pentraxin (NARP), which is mainly expressed in pyramidal neurons in response to neuronal activation[73] and secreted at their presynaptic terminals that contact PV neurons.[74] NARP binding to excitatory synapses promotes the clustering of α-amino-3-hydroxy-5-methyl-4-isoxazolepropionic acid (AMPA) receptors and enhances glutamate neurotransmission on PV neurons.[74] In the DLPFC of schizophrenia subjects, NARP mRNA levels were decreased[75] (Figure 33.1). Furthermore, NARP mRNA levels were positively correlated with GAD67 mRNA levels,[75] suggesting that a deficit in glutamate neurotransmission on PV neurons contributes to lower activity-dependent GAD67 levels in PV neurons in schizophrenia.

Fourth, most cortical PV neurons are surrounded by perineuronal nets (PNNs), an extracellular matrix (ECM) that influences activity-dependent neuronal plasticity by regulating localization of potassium channels and AMPA receptors in PV neurons.[76] The density of PNNs labeled with *Wisteria floribunda* agglutinin (WFA), a lectin that binds to the major proteoglycan components of PNNs, was reported to be reduced in the DLPFC[77] and ERC[78] of subjects with schizophrenia. These changes appear to be due to lower levels of the constituent molecules of PNNs as a recent study using high-resolution confocal microscopy demonstrated that the intensities of PNN labeling by WFA or by an antibody against aggrecan, the major proteoglycan in PNNs, were both reduced without a change in the density of PNNs in the DLPFC of subjects with schizophrenia[55] (Figure 33.1). Together, these findings suggest that activity-dependent plasticity of cortical PV neurons is altered in schizophrenia.

Two major subtypes of PV neurons are present in the cortex and hippocampus: basket cells and chandelier cells. Basket cells, which constitute the majority of PV neurons,[4,79,80] furnish inhibitory synapses on the soma and proximal dendrites of pyramidal neurons and other PV neurons, whereas chandelier cells form vertical arrays of synaptic terminals (termed axon cartridges) exclusively on the axon initial segment of pyramidal neurons. In the primate cortex, both GAD67 and GAD65 proteins are detectable in basket cell terminals, whereas only GAD67 was detected in axon cartridges of chandelier cells.[81,82] In schizophrenia, the densities of basket cell terminals that contain PV or GAD65 appear to be unchanged in the DLPFC[32,82] and hippocampus.[36] However, in these terminals, PV and GAD67 protein levels were reduced without a change in GAD65 protein levels[20,82] (Figure 33.1). In contrast, chandelier cell axon cartridges have normal levels of GAD67,[31] but appear to have lower levels of GAT1 protein as reflected in the lower densities of GAT1-positive axon cartridges across layers 2–4 of the DLPFC,[41,83] an interpretation supported by findings from a recent study demonstrating that the densities of axon cartridges were actually higher in layer 2 and unchanged across layers 3–6 in schizophrenia subjects.[84] Interestingly, among all layer 2 axon cartridges, only those that contain calbindin (CALB) were increased and CALB-negative axon cartridges were not affected. The greater density of chandelier cell axon cartridges could explain, at least in layer 2, the postsynaptic increase in $GABA_A$ alpha2-positive pyramidal neuron axon initial segments.[85] Taken together, these findings suggest that (1) chandelier cells furnish more CALB-positive axon cartridges on layer 2 pyramidal neurons, (2) some axon cartridges across layers 2–4 have markedly reduced levels of GAT1, which is below the limit of detection, and (3) GAD67 levels are unaffected in axon cartridges in the DLPFC of schizophrenia subjects (Figure 33.1).

SST NEURONS IN SCHIZOPHRENIA

SST mRNA levels were reported to be lower across the DLPFC,[24,46,47,86–88] ACC,[24] OFC,[50,88] M1,[24] V1,[24] and hippocampus,[38] and SST peptide levels were lower in the ACC, inferior frontal gyrus (IFG), STG, and parahippocampal gyrus (PHG), but not in the DLPFC or V1, in an earlier radioimmunoassay.[89] In addition, the majority of postmortem studies reported lower densities or total numbers of SST- and SST mRNA-positive neurons in the DLPFC,[33,90] ERC,[91] and hippocampus[38] of subjects with schizophrenia. As a lower number of GABA neurons is unlikely in the cortex[11–15] and hippocampus,[35] these findings suggest that SST mRNA and protein levels were reduced below the limits of detection in some SST neurons (Figure 33.1). In other SST neurons, SST expression appear to be reduced, but still detectable, as indicated by lower SST mRNA levels per positive neuron in the DLPFC of subjects with schizophrenia[90] (Figure 33.1). Interestingly, NARP was also detected in excitatory synapses on SST neurons and associated with activity-dependent formation of excitatory synapses on these neurons.[92] Therefore, given that SST expression is activity-dependent,[93] lower SST mRNA levels might be caused by a decrease in the density of excitatory synapses on SST neurons due to reduced NARP expression in pyramidal neurons.[75]

In the cortex, a subpopulation of SST neurons corresponds to the majority of Neuropeptide Y (NPY) neurons.[3,4,94] However, in schizophrenia, NPY mRNA levels were reported to be unaltered[46,95] or lower[45,86,88] in the DLPFC. Furthermore, lower NPY mRNA and protein levels might be selective to some cortical regions,[88,89] although the affected regions do not have corresponding effects on mRNA and protein levels. Both SST neurons and NPY neurons exist in the superficial white matter (sWM)[90,96] and NPY was detected in a subset of sWM SST neurons.[97] The discrepancy among the studies of cortical NPY expression could be due to contamination of sWM tissues into cortical homogenates and/or inclusion of subjects with schizoaffective disorder in the schizophrenia cohorts, as NPY mRNA levels were found to be decreased in the sWM, but not in the GM, of the DLPFC of subjects with schizoaffective disorder[96] (Figure 33.1). Interestingly, in schizophrenia subjects, the density of SST neurons in the sWM was reported to be higher in the DLPFC[97] and OFC,[50] although SST mRNA levels in the sWM were higher only in the OFC[50] and unaltered in the DLPFC.[90] The higher SST neuron density and SST mRNA levels might reflect the higher densities of interstitial WM neurons, the majority of which were shown to be GABA neurons, in schizophrenia.[23,97]

COMMON CHANGES IN PV NEURONS AND SST NEURONS IN SCHIZOPHRENIA

Several changes appear to be common to both PV and SST neurons and thus might reflect shared pathogenic mechanisms. Both PV neurons and SST neurons are derived from progenitors in the medial ganglionic eminence of the subcortical telencephalon.[98] LIM homeobox 6 (Lhx6), a transcription factor that regulates migration and differentiation of both PV and SST neurons, is selectively expressed by these neurons during development and adulthood.[99,100] Lhx6 mRNA levels were lower in the DLPFC of subjects with schizophrenia[47] (Figure 33.1). Furthermore, subjects with schizophrenia or schizoaffective disorder have been classified into two groups by unbiased clustering based on the expression levels of four transcripts that were shown to be altered in PV and SST neurons (i.e., GAD67, PV, SST and Lhx6) in the DLPFC.[47,87] These studies demonstrated that about 50% of subjects with either schizophrenia or schizoaffective disorder had a low GABA marker (LGM) phenotype (i.e., lower levels of GAD67, PV, SST, and Lhx6 transcripts), whereas in the non-LGM subjects, the mRNA levels for all these GABA markers were unaltered or only slightly decreased. The LGM phenotype shared by the same set of schizophrenia subjects suggests a pathogenic mechanism operating early in development when both subsets of neurons are under the same developmental regulation.

VIP NEURONS IN SCHIZOPHRENIA

Most cortical VIP neurons, a separate interneuron subgroup from PV and SST neurons, express CR.[3,4,94] Converging lines of evidence indicate that VIP/CR neurons are unaffected in the cortex of subjects with schizophrenia. First, the density of CR-positive neurons was reported to be unaltered across the DLPFC,[52,54,59,60,101] ACC,[56] A1, and PT.[58] Second, CR protein levels per positive neuron were unaltered in the DLPFC,[54] A1, and PT.[58] Third, analyses of protein and/or mRNA levels consistently reported no alterations in CR or VIP expression across multiple cortical regions, including the

DLPFC, ACC, OFC, M1, and V1[24,33,44,46,48,86,88,89] (Figure 33.1). Finally, the density of excitatory synapses on the soma of CR neurons was unaltered in the DLPFC of subjects with schizophrenia.[54]

OTHER GABA NEURONS IN SCHIZOPHRENIA

In the cortex, cholecystokinin (CCK) peptide is present in the soma of small VIP-positive neurons that do not express CR and of large basket cells that do not express PV, SST, CR, or VIP.[3,4,102] CCK mRNA is expressed heavily by these GABA neurons, although it was also detected at lower levels in a subgroup of pyramidal neurons.[103] Analyses of cortical CCK mRNA expression provided mixed results: reduced levels in the DLPFC,[86,104] STG,[104] and ERC,[105] and unaltered levels in the DLPFC,[88] OFC,[88] and ACC.[105] Interestingly, cannabinoid 1 receptor (CB1R), the principal cannabinoid receptor in the brain, is predominantly expressed in CCK-positive large basket cells.[106–108] In the DLPFC of subjects with schizophrenia, although the binding of selective ligands to CB1R was reported to be increased,[109–112] presumably reflecting a greater amount and/or a higher affinity of CB1R in the membrane,[113] both the mRNA and protein levels of CB1R were found to be reduced[114] (Figure 33.1). The reductions in CB1R mRNA were correlated with the changes in CCK mRNA levels in the DLPFC of the same schizophrenia subjects,[114] suggesting that among CCK neurons, large basket cells that coexpress CB1 might be preferentially affected in subjects with schizophrenia (Figure 33.1).

COMPARISONS TO BIPOLAR DISORDER

In the neocortex, the change in cortical GAD67 expression appears to be more modest and less common in bipolar disorder relative to schizophrenia. In bipolar disorder, the density of GAD67 mRNA-positive neurons was reported be lower in the DLPFC[27,28] and ACC.[29,30] Similar to schizophrenia, the lower GAD67 mRNA-positive neuron density was restricted to layer 2 in the ACC,[29,30] which might be associated with a reduced density of small, nonpyramidal neurons in layer 2 of the ACC.[10] GAD67 mRNA levels in the total GM were reported to be lower[16] or unchanged[115] in the DLPFC, and unchanged in the ACC.[22] These differences could reflect the smaller percentage of bipolar disorder subjects (30%) that exhibited the LGM phenotype compared to schizophrenia subjects (50%).[87] Cortical GAD65 levels were found to be unaltered at both mRNA and protein levels in the DLPFC of subjects with bipolar disorder.[16,115] In the hippocampus of bipolar disorder subjects, studies consistently reported reduced expression of GAD67 and GAD65, including lower densities of neurons that express GAD67 or GAD65 mRNAs,[35] lower tissue levels of GAD67 mRNA,[37,116] and lower tissue and cellular levels of GAD65 mRNAs.[35,37]

Similar to schizophrenia, the densities of GAD67 mRNA-positive neurons that coexpressed GluN2A NMDA receptor subunit mRNA were significantly reduced, without a change in GluN2A mRNA levels per GAD67-containing neuron, across layers 2-6 of the ACC of subjects with bipolar disorder.[29] However, this pattern of change in GluN2A mRNA expression was not detected in the DLPFC of bipolar disorder subjects.[27] The density of GluK1 mRNA-positive neurons that co-expressed GAD67 mRNA was also lower, without a change in GluK1 mRNA levels per GAD67-containing neuron, in the ACC of subjects with bipolar disorder.[30] Glutamate neurotransmission via NMDA and kainate receptors might be deficient in a subset of GABA neurons in the ACC, but not in the DLPFC, of subjects with bipolar disorder.

Similar to observations in schizophrenia subjects, PV mRNA levels were decreased,[87,115] without a change in the density of PV neurons,[59,60] in the DLPFC of bipolar disorder subjects (Figure 33.1). However, in contrast to schizophrenia, the density and total number of PV neurons were decreased in the ERC.[57,91] Similar to schizophrenia subjects, the density and total number of PV neurons and PV mRNA levels were found to be decreased in the hippocampus of bipolar disorder subjects.[62,116] However, it is not clear whether this reflects the loss of PV neurons or markedly reduced PV expression in individual neurons, as the densities of GAD67 mRNA-positive and GAD65 mRNA-positive neurons were reported to be decreased in the hippocampus of subjects with bipolar disorder.[35] Interestingly, mRNA levels for NARP were also found to be decreased in the DLPFC of subjects with bipolar disorder, especially those with psychotic features[75] (Figure 33.1), suggesting that a deficit in glutamic neurotransmission to PV neurons might be a common mechanism for alterations in PV neurons, and possibly SST neurons, in psychotic disorders. In contrast to schizophrenia, the densities of basket cell terminals that express GAD65 were found to be decreased in the DLPFC and ACC of subjects with bipolar disorder,[32] whereas the density of GAT1-immunoreactive chandelier cell axon cartridges was unchanged in the DLPFC of non-schizophrenia psychiatric subjects including those with psychotic features.[41,83]

Similar to schizophrenia subjects, subjects with bipolar disorder exhibited decreases in the density and total number of SST neurons in the ERC[91] and hippocampus,[116] although it is unclear whether these deficits reflect the loss of SST neurons or markedly reduced SST expression in individual neurons. On the other hand, studies of cortical SST expression in subjects with bipolar disorder are inconsistent, reporting increased,[117] nonsignificantly decreased,[115] and decreased[88] mRNA levels in the DLPFC. As in the case of GAD67 mRNA levels, the smaller percentage of LGM subjects with bipolar disorder[87] might explain these inconsistent results of cortical SST mRNA levels.

Similar to schizophrenia, VIP neurons appear to be relatively spared in subjects with bipolar disorder, as studies failed to detect a significant change in the density of CR neurons[56,59,60,62] or CR mRNA levels[88,115] in the DLPFC,[88,115] ACC,[56] OFC,[88] or hippocampus,[62] although one study reported lower VIP mRNA levels in the DLPFC and OCF of subjects with bipolar disorder.[88]

CONCLUSIONS

The findings reviewed in this chapter suggest that the cortical GABA system has several key features in subjects with schizophrenia (Figure 33.1).

First, the number of GABA neurons does not appear to be altered, but a subset of GABA neurons have markedly lower GAD67 mRNA levels across cortical regions, with the remaining GABA neurons expressing normal levels of GAD67 mRNA.[11–13,15,26,29–31]

Second, the GABA neurons with deficient GAD67 mRNA include PV neurons.[44] The density of PV neurons is not altered in schizophrenia,[44,51–58] but in addition to lower GAD67 mRNA, they also exhibit lower levels of PV, KCNS3, and LHX6 mRNAs.[24,44–48,63,86] Furthermore, the inhibitory synapses formed by two PV neuron subtypes, basket cells and chandelier cells, are differentially altered. The axon terminals of basket cells contain lower levels of GAD67 and PV proteins but are not changed in density.[20,82] In contrast, GAD67 protein levels are not altered in the axon cartridges of chandelier cells,[31] but GAT1 protein levels are reduced below detection in a subset of axon cartridges.[41,83]

Third, in addition to PV neurons, SST, but not VIP, neurons are affected in schizophrenia, as indicated by lower levels of SST and LHX6 mRNAs, whereas VIP and CR mRNA levels are relatively unaffected.[24,33,46–48,86,118]

Fourth, excitatory inputs to PV and SST neurons appear to be lower in schizophrenia. For example, gene expression for NARP, which regulates AMPA receptor clustering in excitatory synapses on PV neurons and possibly SST neurons,[74] is lower,[75] suggesting deficient glutamate transmission in these neurons. In addition, ErbB4, which promotes the formation of excitatory synapses on PV neurons, exhibited a dysregulation in alternative splicing with an increase in the ratio of minor to major splice variants.[48,68,69] This splicing shift was predicted to result in fewer excitatory synapses on PV neurons, which was confirmed in the DLPFC of schizophrenia subjects.[54]

Finally, these changes in transcripts and their cognate proteins in PV and SST neurons are unlikely to be the consequence of treatment with antipsychotics, because monkeys chronically exposed to antipsychotics at clinically relevant dosages did not exhibit a similar change.[26,44,47,54,63,75,83,86]

REFERENCES

1. Wamsley B, Fishell G. Genetic and activity-dependent mechanisms underlying interneuron diversity. *Nat Rev Neurosci.* 2017;18(5):299–309.
2. Pfeffer CK, Xue M, He M, Huang ZJ, Scanziani M. Inhibition of inhibition in visual cortex: the logic of connections between molecularly distinct interneurons. *Nat Neurosci.* 2013;16(8):1068–1076.
3. Lake BB, Ai R, Kaeser GE, et al. Neuronal subtypes and diversity revealed by single-nucleus RNA sequencing of the human brain. *Science.* 2016;352(6293):1586–1590.
4. Tremblay R, Lee S, Rudy B. GABAergic interneurons in the neocortex: from cellular properties to circuits. *Neuron.* 2016;91(2):260–292.
5. Kim D, Jeong H, Lee J, et al. Distinct roles of parvalbumin- and somatostatin-expressing interneurons in working memory. *Neuron.* 2016;92(4):902–915.
6. Kamigaki T, Dan Y. Delay activity of specific prefrontal interneuron subtypes modulates memory-guided behavior. *Nat Neurosci.* 2017;20(6):854–863.
7. Pi HJ, Hangya B, Kvitsiani D, Sanders JI, Huang ZJ, Kepecs A. Cortical interneurons that specialize in disinhibitory control. *Nature.* 2013;503(7477):521–524.
8. Barch DM, Ceaser A. Cognition in schizophrenia: core psychological and neural mechanisms. *Trends Cogn Sci.* 2012;16(1):27–34.
9. Benes FM, McSparren J, Bird ED, SanGiovanni JP, Vincent SL. Deficits in small interneurons in prefrontal and cingulate cortices of schizophrenic and schizoaffective patients. *Arch Gen Psychiatry.* 1991;48(11):996–1001.
10. Benes FM, Vincent SL, Todtenkopf M. The density of pyramidal and nonpyramidal neurons in anterior cingulate cortex of schizophrenic and bipolar subjects. *Biol Psychiatry.* 2001;50(6):395–406.
11. Selemon LD, Rajkowska G, Goldman-Rakic PS. Abnormally high neuronal density in the schizophrenic cortex. A morphometric analysis of prefrontal area 9 and occipital area 17. *Arch Gen Psychiatry.* 1995;52(10):805–818; discussion 819–820.
12. Selemon LD, Rajkowska G, Goldman-Rakic PS. Elevated neuronal density in prefrontal area 46 in brains from schizophrenic patients: application of a three-dimensional, stereologic counting method. *J Comp Neurol.* 1998;392(3):402–412.
13. Akbarian S, Kim JJ, Potkin SG, et al. Gene expression for glutamic acid decarboxylase is reduced without loss of neurons in prefrontal cortex of schizophrenics. *Arch Gen Psychiatry.* 1995;52(4):258–266.
14. Pakkenberg B. Total nerve cell number in neocortex in chronic schizophrenics and controls estimated using optical disectors. *Biol Psychiatry.* 1993;34(11):768–772.
15. Thune JJ, Uylings HB, Pakkenberg B. No deficit in total number of neurons in the prefrontal cortex in schizophrenics. *J Psychiatr Res.* 2001;35(1):15–21.
16. Guidotti A, Auta J, Davis JM, et al. Decrease in reelin and glutamic acid decarboxylase67 (GAD67) expression in schizophrenia and bipolar disorder: a postmortem brain study. *Arch Gen Psychiatry.* 2000;57(11):1061–1069.
17. Hashimoto T, Bergen SE, Nguyen QL, et al. Relationship of brain-derived neurotrophic factor and its receptor TrkB to altered inhibitory prefrontal circuitry in schizophrenia. *J Neurosci.* 2005;25(2):372–383.
18. Straub RE, Lipska BK, Egan MF, et al. Allelic variation in GAD1 (GAD67) is associated with schizophrenia and influences cortical function and gene expression. *Mol Psychiatry.* 2007;12(9):854–869.
19. Duncan CE, Webster MJ, Rothmond DA, Bahn S, Elashoff M, Shannon Weickert C. Prefrontal GABA(A) receptor alpha-subunit expression in normal postnatal human development and schizophrenia. *J Psychiatr Res.* 2010;44(10):673–681.
20. Curley AA, Arion D, Volk DW, et al. Cortical deficits of glutamic acid decarboxylase 67 expression in schizophrenia: clinical, protein, and cell type-specific features. *Am J Psychiatry.* 2011;168(9):921–929.
21. Kimoto S, Bazmi HH, Lewis DA. Lower expression of glutamic acid decarboxylase 67 in the prefrontal cortex in schizophrenia: contribution of altered regulation by Zif268. *Am J Psychiatry.* 2014;171(9):969–978.
22. Thompson M, Weickert CS, Wyatt E, Webster MJ. Decreased glutamic acid decarboxylase(67) mRNA expression in multiple brain areas of patients with schizophrenia and mood disorders. *J Psychiatr Res.* 2009;43(11):970–977.
23. Joshi D, Fung SJ, Rothwell A, Weickert CS. Higher gamma-aminobutyric acid neuron density in the white matter of orbital frontal cortex in schizophrenia. *Biol Psychiatry.* 2012;72(9):725–733.
24. Hashimoto T, Bazmi HH, Mirnics K, Wu Q, Sampson AR, Lewis DA. Conserved regional patterns of GABA-related transcript expression in the neocortex of subjects with schizophrenia. *Am J Psychiatry.* 2008;165(4):479–489.
25. Impagnatiello F, Guidotti AR, Pesold C, et al. A decrease of reelin expression as a putative vulnerability factor in schizophrenia. *Proc Natl Acad Sci U S A.* 1998;95(26):15718–15723.

26. Volk DW, Austin MC, Pierri JN, Sampson AR, Lewis DA. Decreased glutamic acid decarboxylase67 messenger RNA expression in a subset of prefrontal cortical gamma-aminobutyric acid neurons in subjects with schizophrenia. *Arch Gen Psychiatry.* 2000;57(3):237–245.
27. Woo TU, Kim AM, Viscidi E. Disease-specific alterations in glutamatergic neurotransmission on inhibitory interneurons in the prefrontal cortex in schizophrenia. *Brain Res.* 2008;1218:267–277.
28. Veldic M, Guidotti A, Maloku E, Davis JM, Costa E. In psychosis, cortical interneurons overexpress DNA-methyltransferase 1. *Proc Natl Acad Sci U S A.* 2005;102(6):2152–2157.
29. Woo TU, Walsh JP, Benes FM. Density of glutamic acid decarboxylase 67 messenger RNA-containing neurons that express the N-methyl-D-aspartate receptor subunit NR2A in the anterior cingulate cortex in schizophrenia and bipolar disorder. *Arch Gen Psychiatry.* 2004;61(7):649–657.
30. Woo TU, Shrestha K, Amstrong C, Minns MM, Walsh JP, Benes FM. Differential alterations of kainate receptor subunits in inhibitory interneurons in the anterior cingulate cortex in schizophrenia and bipolar disorder. *Schizophr Res.* 2007;96(1–3):46–61.
31. Rocco BR, Lewis DA, Fish KN. Markedly lower glutamic acid decarboxylase 67 protein levels in a subset of boutons in schizophrenia. *Biol Psychiatry.* 2016;79(12):1006–1015.
32. Benes FM, Todtenkopf MS, Logiotatos P, Williams M. Glutamate decarboxylase(65)-immunoreactive terminals in cingulate and prefrontal cortices of schizophrenic and bipolar brain. *J Chem Neuroanat.* 2000;20(3–4):259–269.
33. Guillozet-Bongaarts AL, Hyde TM, Dalley RA, et al. Altered gene expression in the dorsolateral prefrontal cortex of individuals with schizophrenia. *Mol Psychiatry.* 2014;19(4):478–485.
34. Glausier JR, Kimoto S, Fish KN, Lewis DA. Lower glutamic acid decarboxylase 65-kDa isoform messenger RNA and protein levels in the prefrontal cortex in schizoaffective disorder but not schizophrenia. *Biol Psychiatry.* 2015;77(2):167–176.
35. Heckers S, Stone D, Walsh J, Shick J, Koul P, Benes FM. Differential hippocampal expression of glutamic acid decarboxylase 65 and 67 messenger RNA in bipolar disorder and schizophrenia. *Arch Gen Psychiatry.* 2002;59(6):521–529.
36. Todtenkopf MS, Benes FM. Distribution of glutamate decarboxylase65 immunoreactive puncta on pyramidal and nonpyramidal neurons in hippocampus of schizophrenic brain. *Synapse.* 1998;29(4):323–332.
37. Benes FM, Lim B, Matzilevich D, Subburaju S, Walsh JP. Circuitry-based gene expression profiles in GABA cells of the trisynaptic pathway in schizophrenics versus bipolars. *Proc Natl Acad Sci U S A.* 2008;105(52):20935–20940.
38. Konradi C, Yang CK, Zimmerman EI, et al. Hippocampal interneurons are abnormal in schizophrenia. *Schizophr Res.* 2011;131(1–3):165–173.
39. Ohnuma T, Augood SJ, Arai H, McKenna PJ, Emson PC. Measurement of GABAergic parameters in the prefrontal cortex in schizophrenia: focus on GABA content, GABA(A) receptor alpha-1 subunit messenger RNA and human GABA transporter-1 (HGAT-1) messenger RNA expression. *Neuroscience.* 1999;93(2):441–448.
40. Volk D, Austin M, Pierri J, Sampson A, Lewis D. GABA transporter-1 mRNA in the prefrontal cortex in schizophrenia: decreased expression in a subset of neurons. *Am J Psychiatry.* 2001;158(2):256–265.
41. Woo TU, Whitehead RE, Melchitzky DS, Lewis DA. A subclass of prefrontal gamma-aminobutyric acid axon terminals are selectively altered in schizophrenia. *Proc Natl Acad Sci U S A.* 1998;95(9):5341–5346.
42. Hoftman GD, Volk DW, Bazmi HH, Li S, Sampson AR, Lewis DA. Altered cortical expression of GABA-related genes in schizophrenia: illness progression vs developmental disturbance. *Schizophr Bull.* 2015;41(1):180–191.
43. Fung SJ, Sivagnanasundaram S, Weickert CS. Lack of change in markers of presynaptic terminal abundance alongside subtle reductions in markers of presynaptic terminal plasticity in prefrontal cortex of schizophrenia patients. *Biol Psychiatry.* 2011;69(1):71–79.
44. Hashimoto T, Volk DW, Eggan SM, et al. Gene expression deficits in a subclass of GABA neurons in the prefrontal cortex of subjects with schizophrenia. *J Neurosci.* 2003;23(15):6315–6326.
45. Mellios N, Huang HS, Baker SP, Galdzicka M, Ginns E, Akbarian S. Molecular determinants of dysregulated GABAergic gene expression in the prefrontal cortex of subjects with schizophrenia. *Biol Psychiatry.* 2009;65(12):1006–1014.
46. Fung SJ, Webster MJ, Sivagnanasundaram S, Duncan C, Elashoff M, Weickert CS. Expression of interneuron markers in the dorsolateral prefrontal cortex of the developing human and in schizophrenia. *Am J Psychiatry.* 2010;167(12):1479–1488.
47. Volk DW, Matsubara T, Li S, et al. Deficits in transcriptional regulators of cortical parvalbumin neurons in schizophrenia. *Am J Psychiatry.* 2012;169(10):1082–1091.
48. Chung DW, Volk DW, Arion D, Zhang Y, Sampson AR, Lewis DA. Dysregulated ErbB4 splicing in schizophrenia: selective effects on parvalbumin expression. *Am J Psychiatry.* 2016;173(1):60–68.
49. McMeekin LJ, Lucas EK, Meador-Woodruff JH, et al. Cortical PGC-1alpha-dependent transcripts are reduced in postmortem tissue from patients with schizophrenia. *Schizophr Bull.* 2016;42(4):1009–1017.
50. Joshi D, Catts VS, Olaya JC, Shannon Weickert C. Relationship between somatostatin and death receptor expression in the orbital frontal cortex in schizophrenia: a postmortem brain mRNA study. *NPJ Schizophr.* 2015;1:14004.
51. Woo TU, Miller JL, Lewis DA. Schizophrenia and the parvalbumin-containing class of cortical local circuit neurons. *Am J Psychiatry.* 1997;154(7):1013–1015.
52. Tooney PA, Chahl LA. Neurons expressing calcium-binding proteins in the prefrontal cortex in schizophrenia. *Prog Neuropsychopharmacol Biol Psychiatry.* 2004;28(2):273–278.
53. Bitanihirwe BK, Woo TU. Transcriptional dysregulation of gamma-aminobutyric acid transporter in parvalbumin-containing inhibitory neurons in the prefrontal cortex in schizophrenia. *Psychiatry Res.* 2014;220(3):1155–1159.
54. Chung DW, Fish KN, Lewis DA. Pathological basis for deficient excitatory drive to cortical parvalbumin interneurons in schizophrenia. *Am J Psychiatry.* 2016;173(11):1131–1139.
55. Enwright JF, Sanapala S, Foglio A, Berry R, Fish KN, Lewis DA. Reduced labeling of parvalbumin neurons and perineuronal nets in the dorsolateral prefrontal cortex of subjects with schizophrenia. *Neuropsychopharmacology.* 2016;41(9):2206–2214.
56. Cotter D, Landau S, Beasley C, et al. The density and spatial distribution of GABAergic neurons, labelled using calcium binding proteins, in the anterior cingulate cortex in major depressive disorder, bipolar disorder, and schizophrenia. *Biol Psychiatry.* 2002;51(5):377–386.
57. Pantazopoulos H, Lange N, Baldessarini RJ, Berretta S. Parvalbumin neurons in the entorhinal cortex of subjects diagnosed with bipolar disorder or schizophrenia. *Biol Psychiatry.* 2007;61(5):640–652.
58. Smiley JF, Hackett TA, Bleiwas C, et al. Reduced GABA neuron density in auditory cerebral cortex of subjects with major depressive disorder. *J Chem Neuroanat.* 2016;76(Pt B):108–121.
59. Beasley CL, Zhang ZJ, Patten I, Reynolds GP. Selective deficits in prefrontal cortical GABAergic neurons in schizophrenia defined by the presence of calcium-binding proteins. *Biol Psychiatry.* 2002;52(7):708–715.
60. Sakai T, Oshima A, Nozaki Y, et al. Changes in density of calcium-binding-protein-immunoreactive GABAergic neurons in prefrontal cortex in schizophrenia and bipolar disorder. *Neuropathology.* 2008;28(2):143–150.
61. Stan AD, Lewis DA. Altered cortical GABA neurotransmission in schizophrenia: insights into novel therapeutic strategies. *Curr Pharm Biotechnol.* 2012;13(8):1557–1562.
62. Zhang ZJ, Reynolds GP. A selective decrease in the relative density of parvalbumin-immunoreactive neurons in the hippocampus in schizophrenia. *Schizophr Res.* 2002;55(1–2):1–10.
63. Georgiev D, Arion D, Enwright JF, et al. Lower gene expression for KCNS3 potassium channel subunit in parvalbumin-containing neurons in the prefrontal cortex in schizophrenia. *Am J Psychiatry.* 2014;171(1):62–71.

64. Lee KY, Royston SE, Vest MO, et al. N-methyl-D-aspartate receptors mediate activity-dependent down-regulation of potassium channel genes during the expression of homeostatic intrinsic plasticity. *Mol Brain.* 2015;8:4.
65. Cohen SM, Ma H, Kuchibhotla KV, et al. Excitation-transcription coupling in parvalbumin-positive interneurons employs a novel CaM kinase-dependent pathway distinct from excitatory neurons. *Neuron.* 2016;90(2):292–307.
66. Bitanihirwe BK, Lim MP, Kelley JF, Kaneko T, Woo TU. Glutamatergic deficits and parvalbumin-containing inhibitory neurons in the prefrontal cortex in schizophrenia. *BMC Psychiatry.* 2009;9:71.
67. Del Pino I, Garcia-Frigola C, Dehorter N, et al. Erbb4 deletion from fast-spiking interneurons causes schizophrenia-like phenotypes. *Neuron.* 2013;79(6):1152–1168.
68. Law AJ, Kleinman JE, Weinberger DR, Weickert CS. Disease-associated intronic variants in the ErbB4 gene are related to altered ErbB4 splice-variant expression in the brain in schizophrenia. *Hum Mol Genet.* 2007;16(2):129–141.
69. Joshi D, Fullerton JM, Weickert CS. Elevated ErbB4 mRNA is related to interneuron deficit in prefrontal cortex in schizophrenia. *J Psychiatr Res.* 2014;53:125–132.
70. Conde F, Lund JS, Jacobowitz DM, Baimbridge KG, Lewis DA. Local circuit neurons immunoreactive for calretinin, calbindin D-28k or parvalbumin in monkey prefrontal cortex: distribution and morphology. *J Comp Neurol.* 1994;341(1):95–116.
71. Neddens J, Fish KN, Tricoire L, et al. Conserved interneuron-specific ErbB4 expression in frontal cortex of rodents, monkeys, and humans: implications for schizophrenia. *Biol Psychiatry.* 2011;70(7):636–645.
72. Veikkolainen V, Vaparanta K, Halkilahti K, Iljin K, Sundvall M, Elenius K. Function of ERBB4 is determined by alternative splicing. *Cell Cycle.* 2011;10(16):2647–2657.
73. Tsui CC, Copeland NG, Gilbert DJ, Jenkins NA, Barnes C, Worley PF. NARP, a novel member of the pentraxin family, promotes neurite outgrowth and is dynamically regulated by neuronal activity. *J Neurosci.* 1996;16(8):2463–2478.
74. Chang MC, Park JM, Pelkey KA, et al. NARP regulates homeostatic scaling of excitatory synapses on parvalbumin-expressing interneurons. *Nat Neurosci.* 2010;13(9):1090–1097.
75. Kimoto S, Zaki MM, Bazmi HH, Lewis DA. Altered markers of cortical gamma-aminobutyric acid neuronal activity in schizophrenia: role of the NARP gene. *JAMA Psychiatry.* 2015;72(8):747–756.
76. Favuzzi E, Marques-Smith A, Deogracias R, et al. Activity-dependent gating of parvalbumin interneuron function by the perineuronal net protein brevican. *Neuron.* 2017;95(3):639–655 e610.
77. Mauney SA, Athanas KM, Pantazopoulos H, et al. Developmental pattern of perineuronal nets in the human prefrontal cortex and their deficit in schizophrenia. *Biol Psychiatry.* 2013;74(6):427–435.
78. Pantazopoulos H, Woo TU, Lim MP, Lange N, Berretta S. Extracellular matrix-glial abnormalities in the amygdala and entorhinal cortex of subjects diagnosed with schizophrenia. *Arch Gen Psychiatry.* 2010;67(2):155–166.
79. Krimer LS, Zaitsev AV, Czanner G, et al. Cluster analysis-based physiological classification and morphological properties of inhibitory neurons in layers 2-3 of monkey dorsolateral prefrontal cortex. *J Neurophysiol.* 2005;94(5):3009–3022.
80. Zaitsev AV, Gonzalez-Burgos G, Povysheva NV, Kroner S, Lewis DA, Krimer LS. Localization of calcium-binding proteins in physiologically and morphologically characterized interneurons of monkey dorsolateral prefrontal cortex. *Cereb Cortex.* 2005;15(8):1178–1186.
81. Fish KN, Sweet RA, Lewis DA. Differential distribution of proteins regulating GABA synthesis and reuptake in axon boutons of subpopulations of cortical interneurons. *Cereb Cortex.* 2011;21(11):2450–2460.
82. Glausier JR, Fish KN, Lewis DA. Altered parvalbumin basket cell inputs in the dorsolateral prefrontal cortex of schizophrenia subjects. *Mol Psychiatry.* 2014;19(1):30–36.
83. Pierri JN, Chaudry AS, Woo TU, Lewis DA. Alterations in chandelier neuron axon terminals in the prefrontal cortex of schizophrenic subjects. *Am J Psychiatry.* 1999;156(11):1709–1719.
84. Rocco BR, DeDionisio AM, Lewis DA, Fish KN. Alterations in a unique class of cortical chandelier cell axon cartridges in schizophrenia. *Biol Psychiatry.* 2016;82(1):40–48.
85. Volk DW, Pierri JN, Fritschy JM, Auh S, Sampson AR, Lewis DA. Reciprocal alterations in pre- and postsynaptic inhibitory markers at chandelier cell inputs to pyramidal neurons in schizophrenia. *Cereb Cortex.* 2002;12(10):1063–1070.
86. Hashimoto T, Arion D, Unger T, et al. Alterations in GABA-related transcriptome in the dorsolateral prefrontal cortex of subjects with schizophrenia. *Mol Psychiatry.* 2008;13(2):147–161.
87. Volk DW, Sampson AR, Zhang Y, Edelson JR, Lewis DA. Cortical GABA markers identify a molecular subtype of psychotic and bipolar disorders. *Psychol Med.* 2016;46(12):2501–2512.
88. Fung SJ, Fillman SG, Webster MJ, Shannon Weickert C. Schizophrenia and bipolar disorder show both common and distinct changes in cortical interneuron markers. *Schizophr Res.* 2014;155(1–3):26–30.
89. Gabriel SM, Davidson M, Haroutunian V, et al. Neuropeptide deficits in schizophrenia vs. Alzheimer's disease cerebral cortex. *Biol Psychiatry.* 1996;39(2):82–91.
90. Morris HM, Hashimoto T, Lewis DA. Alterations in somatostatin mRNA expression in the dorsolateral prefrontal cortex of subjects with schizophrenia or schizoaffective disorder. *Cereb Cortex.* 2008;18(7):1575–1587.
91. Wang AY, Lohmann KM, Yang CK, et al. Bipolar disorder type 1 and schizophrenia are accompanied by decreased density of parvalbumin- and somatostatin-positive interneurons in the parahippocampal region. *Acta Neuropathol.* 2011;122(5):615–626.
92. Spiegel I, Mardinly AR, Gabel HW, et al. Npas4 regulates excitatory-inhibitory balance within neural circuits through cell-type-specific gene programs. *Cell.* 2014;157(5):1216–1229.
93. Marty S, Onteniente B. The expression pattern of somatostatin and calretinin by postnatal hippocampal interneurons is regulated by activity-dependent and -independent determinants. *Neuroscience.* 1997;80(1):79–88.
94. Kubota Y, Hattori R, Yui Y. Three distinct subpopulations of GABAergic neurons in rat frontal agranular cortex. *Brain Res.* 1994;649(1–2):159–173.
95. Caberlotto L, Hurd YL. Reduced neuropeptide Y mRNA expression in the prefrontal cortex of subjects with bipolar disorder. *Neuroreport.* 1999;10(8):1747–1750.
96. Morris HM, Stopczynski RE, Lewis DA. NPY mRNA expression in the prefrontal cortex: selective reduction in the superficial white matter of subjects with schizoaffective disorder. *Schizophr Res.* 2009;115(2–3):261–269.
97. Yang Y, Fung SJ, Rothwell A, Tianmei S, Weickert CS. Increased interstitial white matter neuron density in the dorsolateral prefrontal cortex of people with schizophrenia. *Biol Psychiatry.* 2011;69(1):63–70.
98. Xu Q, Cobos I, De La Cruz E, Rubenstein JL, Anderson SA. Origins of cortical interneuron subtypes. *J Neurosci.* 2004;24(11):2612–2622.
99. Liodis P, Denaxa M, Grigoriou M, Akufo-Addo C, Yanagawa Y, Pachnis V. Lhx6 activity is required for the normal migration and specification of cortical interneuron subtypes. *J Neurosci.* 2007;27(12):3078–3089.
100. Georgiev D, Gonzalez-Burgos G, Kikuchi M, Minabe Y, Lewis DA, Hashimoto T. Selective expression of KCNS3 potassium channel alpha-subunit in parvalbumin-containing GABA neurons in the human prefrontal cortex. *PLoS One.* 2012;7(8):e43904.
101. Daviss SR, Lewis DA. Local circuit neurons of the prefrontal cortex in schizophrenia: selective increase in the density of calbindin-immunoreactive neurons. *Psychiatry Res.* 1995;59(1–2):81–96.
102. Kubota Y, Kawaguchi Y. Two distinct subgroups of cholecystokinin-immunoreactive cortical interneurons. *Brain Res.* 1997;752(1–2):175–183.

103. Schiffmann SN, Vanderhaeghen JJ. Distribution of cells containing mRNA encoding cholecystokinin in the rat central nervous system. *J Comp Neurol.* 1991;304(2):219–233.
104. Virgo L, Humphries C, Mortimer A, Barnes T, Hirsch S, de Belleroche J. Cholecystokinin messenger RNA deficit in frontal and temporal cerebral cortex in schizophrenia. *Biol Psychiatry.* 1995;37(10):694–701.
105. Bachus SE, Hyde TM, Herman MM, Egan MF, Kleinman JE. Abnormal cholecystokinin mRNA levels in entorhinal cortex of schizophrenics. *J Psychiatr Res.* 1997;31(2):233–256.
106. Marsicano G, Lutz B. Expression of the cannabinoid receptor CB1 in distinct neuronal subpopulations in the adult mouse forebrain. *Eur J Neurosci.* 1999;11(12):4213–4225.
107. Bodor AL, Katona I, Nyiri G, et al. Endocannabinoid signaling in rat somatosensory cortex: laminar differences and involvement of specific interneuron types. *J Neurosci.* 2005;25(29):6845–6856.
108. Eggan SM, Lewis DA. Immunocytochemical distribution of the cannabinoid CB1 receptor in the primate neocortex: a regional and laminar analysis. *Cereb Cortex.* 2007;17(1):175–191.
109. Dean B, Sundram S, Bradbury R, Scarr E, Copolov D. Studies on [3H] CP-55940 binding in the human central nervous system: regional specific changes in density of cannabinoid-1 receptors associated with schizophrenia and cannabis use. *Neuroscience.* 2001;103(1):9–15.
110. Zavitsanou K, Garrick T, Huang XF. Selective antagonist [3H]SR141716A binding to cannabinoid CB1 receptors is increased in the anterior cingulate cortex in schizophrenia. *Prog Neuropsychopharmacol Biol Psychiatry.* 2004;28(2):355–360.
111. Dalton VS, Long LE, Weickert CS, Zavitsanou K. Paranoid schizophrenia is characterized by increased CB1 receptor binding in the dorsolateral prefrontal cortex. *Neuropsychopharmacology.* 2011;36(8):1620–1630.
112. Jenko KJ, Hirvonen J, Henter ID, et al. Binding of a tritiated inverse agonist to cannabinoid CB1 receptors is increased in patients with schizophrenia. *Schizophr Res.* 2012;141(2–3):185–188.
113. Volk DW, Eggan SM, Horti AG, Wong DF, Lewis DA. Reciprocal alterations in cortical cannabinoid receptor 1 binding relative to protein immunoreactivity and transcript levels in schizophrenia. *Schizophr Res.* 2014;159(1):124–129.
114. Eggan SM, Hashimoto T, Lewis DA. Reduced cortical cannabinoid 1 receptor messenger RNA and protein expression in schizophrenia. *Arch Gen Psychiatry.* 2008;65(7):772–784.
115. Sibille E, Morris HM, Kota RS, Lewis DA. GABA-related transcripts in the dorsolateral prefrontal cortex in mood disorders. *Int J Neuropsychopharmacol.* 2011;14(6):721–734.
116. Konradi C, Zimmerman EI, Yang CK, et al. Hippocampal interneurons in bipolar disorder. *Arch Gen Psychiatry.* 2011;68(4): 340–350.
117. Nakatani N, Hattori E, Ohnishi T, et al. Genome-wide expression analysis detects eight genes with robust alterations specific to bipolar I disorder: relevance to neuronal network perturbation. *Hum Mol Genet.* 2006;15(12):1949–1962.
118. Volk DW, Radchenkova PV, Walker EM, Sengupta EJ, Lewis DA. Cortical opioid markers in schizophrenia and across postnatal development. *Cereb Cortex.* 2012;22(5):1215–1223.

34

ALTERATION IN NICOTINIC RECEPTORS IN PSYCHOTIC DISORDERS

MOLECULAR NEUROBIOLOGY AND CLINICAL RELEVANCE

Robert Freedman

NICOTINIC ACETYLCHOLINE RECEPTORS

Nicotinic acetylcholine receptors are a subclass of acetylcholine receptors, originally identified by their responsiveness to nicotine. The first molecular cloning came from the torpedo fish electric organ, which is composed of post synaptic elements of the neuromuscular junction linked in electrical series to generate a high potential capable of delivering a shock. The organ's acetylcholinesterase activity led to this identification. Subsequently, its amino acid series was sequenced and the DNA sequence was deduced. The receptor contains two alpha subunits and two beta subunits and either a gamma or delta subunit. It forms a 5-membered ring that can admit cations into the muscle when activated by two molecules of acetylcholine that bind in the cleft between alpha and beta subunits. It then twists slightly to open its cation channel.[1] Acetylcholine is removed quickly by acetylcholinesterase, but nicotine is not. The receptor quickly twists shut with the nicotine, or acetylcholine if acetylcholinesterase is inhibited, still attached. This desensitized or inactivated form of the receptor cannot be reactivated until the acetylcholine or nicotine is removed. The receptor then returns to its native state, where it can be reactivated.[2] The neuromuscular junction rarely loses function in this process, because it has a large reserve supply of receptors in the cycle. The exception is when over 90% of the receptors are destroyed by an immune reaction in myasthenia gravis. Then overuse of acetylcholinesterase leads to acetylcholine itself desensitizing the receptor and blocking neuromuscular contraction, producing the paralysis termed a cholinergic crisis.[3] This molecular pharmacology of activation and desensitizing inactivation also occurs with the central nervous system nicotinic acetylcholine receptors.

Other types of nicotinic receptors were cloned because of their nucleic acid similarity to the neuromuscular junction and numbered in order of their discovery, with the neuromuscular receptor subunits numbered 1. The most common configuration, two alpha subunits, alpha3, 4, 5, or 6, and three beta subunits, beta2, 3, or 4, is also common in the brain. The nervous system nicotinic acetylcholine receptors were identified long after the muscarinic acetylcholine receptors. They were initially overlooked because they rapidly desensitize to nicotine. Unlike the neuromuscular junction, there are far fewer receptors at each synapse, and thus desensitization to nicotine results in rapid loss of function, making them difficult to detect.[1]

The alpha7 receptor is unique because it alone has 5 identical subunits. It is thought to be activated by two molecules of acetylcholine, but how the subunits are configured to form two active sites is unknown. A chaperone molecule such as neuregulin is required for its assembly. The alpha7 receptor is difficult to detect because it desensitizes most rapidly of all receptors. It was cloned in chickens and eventually in humans.[4] Alpha4 subunits form receptors than admit only sodium ions, but alpha7 subunits form receptors that admit all cations, including calcium.[5]

The nicotinic receptors have diverse roles, but two are most prominent in human brain function. For the alpha-beta type receptors, the presynaptic release of neurotransmitters, particularly dopamine, is responsible for many of their behavioral effects.[6] Release of other neurotransmitters has been linked to their analgesic effects. Because of the rapid desensitization of nicotinic receptors, cigarette smoking is an ideal delivery method. The heated nicotine-containing vapor is rapidly absorbed into the pulmonary circulation, which activates the receptors maximally. Levels fall quickly, which minimizes desensitization.[7] Smokers feel alerted and are motivated to repeat the experience, consistent with other activators of dopaminergic neurotransmission. As smoking continues chronically, however, blood levels of nicotine become consistently higher and more persistent. The consequence is that the depth and frequency of smoking increase, as the smoker attempts to recapture the initial effects. A polymorphism in the *CHRNA5* subunit that decreases the sensitivity of receptors has been linked to smoking addiction to the point that it is also associated with lung cancer.[8,9]

The alpha7-nicotinic receptor has been primarily associated with sensory gating, the normal decrease in the brain's activation when stimuli are repeated.[10] A nicotinic receptor, then not identified, was suspected based on early studies in animal models that demonstrated that sensory gating was blocked by alpha-bungarotoxin, a snake venom that binds to both alpha1 and alpha7-based nicotinic receptors. None of the alpha-beta receptors are antagonized by alpha-bungarotoxin. Subsequently, the cloning of the alpha7-nicotinic receptor enabled its identification.[4]

Alpha7-nicotinic receptors are found both pre- and postsynaptically, including sites within both cholinergic and glutamate receptors. Their postsynaptic distribution is almost exclusively on inhibitory interneurons. Two prominent sites in the brain are the inhibitory neurons of the hippocampus and inhibitory neurons that form the nucleus reticularis thalamus. In the periphery, alpha7-nicotinic receptors are found on many neuroendocrine tissues including adrenal medulla and the small cells of the lung. They also mediate the vagal influence on the immediate immune cytokine response that is part of the inflammatory reaction.[11]

CHRNA7 is on chromosome 15, at 15q14. It is a 10-exon gene with a GC-rich promoter.[12] This part of chromosome 15 is imprinted with a parent of origin effect. The imprinting is controlled by a gene at 15q12. Imprinting errors, which can affect *CHRNA7*, are responsible for Prader-Willi syndrome, a childhood onset illness characterized by insatiable eating to the point of morbidity and mortality. Prader-Willi patients with imprinting errors often become psychotic as well.[13] More distally is the gene for Marfan's syndrome, which has been reported to cosegregate with schizophrenia in isolated populations, suggesting that the mutation in *FBN1*, the fibrillin gene, and a nearby second gene, perhaps *CHRNA7*, are cosegregrating because of their proximity.[14]

CHRNA7 is flanked by repeated sequences unrelated to *CHRNA7* itself, but which have caused some issues with it. These repeated sequences sometimes bind to each other during meiosis, and the results include inversion and duplication of *CHRNA7* or its deletion. At some point in human evolution, after separation from other primates, there was a partial duplication of *CHRNA7* and fusion to another gene forming a new gene *CHRFAM7A*.[12] This gene contains exons 5 to 10 of the *CHRNA7*. While it does not make a functional receptor gene itself, it inserts into the pentameric ring, which decreases the current that flows through the receptor. *CHRFAM7A*'s expression increases during early development and also during infection. Deletion of *CHRNA7* on one of an individual's two chromosomes often results in psychosis.[15] *CHRNA7* deletion is the second most likely copy number variation (CNV) likely to result in psychosis, after chromosome 22q. The deletion can span several genes on chromosome 15q14, but not *CHRFAM7A*. However, the critical region appears to be narrowing to *CHRNA7*, based on many case reports. *CHRNA7* deletions are not specific for schizophrenia but are found in autism spectrum disorder as well.[16] In ADHD there are both deletions and duplications, which generally also result in activation of the gene.[17]

CHRNA7 has not been prominent in most large genome-wide association studies, which indicates that there are no common polymorphisms near it that are linked to schizophrenia. Several studies have shown linkage to schizophrenia using other markers.[18] Our group and now others have shown association and linkage of *CHRNA7* to the P50 inhibitory gating deficit, consistent with its pathophysiological role in the deficit.[19,20]

Perhaps the strongest piece of evidence for *CHRNA7*'s role in schizophrenia is the survey performed at the Lieber Institute of the mRNA expression of 58 genetic candidate genes for schizophrenia. *CHRNA7* was the most extensively affected mRNA, with significant decrease over most of the frontal cortex in postmortem specimens from persons who had schizophrenia, compared to controls.[21]

Nicotinic receptors appear early in fetal brain development. When the neural epithelium is first formed, waves of electrical impulses travel across it. Early synapses are formed using components of the immune system. Then the cells begin to secrete acetylcholine, which activates nicotinic receptors. As development progresses, both GABA and N-methyl-D-aspartate (NMDA) glutamate receptors mediate transmission over the developing synaptic connections. GABA is an excitatory neurotransmitter throughout most of fetal life, because the chloride gradient in the neurons is too small for the GABA chloride channel to lower membrane voltage and instead it raises membrane voltage to cause excitation. Activation of alpha7-nicotinic receptors induces more release of GABA, which stimulates the development of the mature form of the chloride membrane transporter KCC2, which replaces the embryonic and ineffective NKCC1 in the last stages of fetal development. The result is that the chloride gradient is increased and GABA's chloride channel becomes inhibitory, its mature function. In animals with null mutations of *CHRNA7*, this transition does not fully occur.[22] The alpha7-nicotinic receptor is also required for a similar transition between immature NMDA glutamate receptors to mature, faster acting AMPA/kainate glutamate receptors. Neither of these transitions occur fully in schizophrenia, as assessed in post mortem brain samples.[23,24] A special fetal alpha7-nicotinic neurobiology supports this transition. The endogenous ligand that activates the alpha7-nicotinic receptors is not acetylcholine from synapses, because the transition occurs before cholinergic fibers have reached the forebrain from their midbrain cell bodies. The likely ligand is choline, which is found in the amniotic fluid. Millimolar concentrations of choline in the fluid are sufficient to activate alpha7-nicotinic receptors. The effect is pharmacologically specific as shown by blockade of the effect by antagonists.[25] The expression of *CHRNA7* and alpha7-nicotinic receptors is increased in fetal life, compared to after birth.[26] The increase in receptor expression in human fetuses is nearly 10-fold.[27]

NICOTINIC RECEPTORS AS DRUG TARGETS FOR THE TREATMENT OF SCHIZOPHRENIA

Nicotinic drugs for the treatment of schizophrenia have been primarily targeted to alpha7-nicotinic receptors. Nicotine itself has been explored as a cognitive treatment with mixed results, because patients with schizophrenia often smoke quite heavily, more than any other diagnostic group.[28] There has been a commonly accepted hypothesis that their smoking is an attempt at self-medication. Patients report some soothing effect, but no consistent clinical effect has ever been reported. We found that nicotine acutely administered has an effect on a neurocognitive measure of attention in patients with schizophrenia who were nonsmokers.[29] Patients who were smokers did not have any effect, but their chronically high levels of nicotine likely keep their nicotinic receptors largely in a desensitized state.

The first study of a specific alpha7-nicotinic agonist used 3-(2,4-dimethoxybenzylidene) anabaseine (DMXB-A). Anabaseine is a marine toxin from nemertine worms, a small worm found in the Puget Sound.[30] The addition of the benzylidine ring made the compound a more specific alpha7-nicotinic receptor, with 1,000-fold selectivity compared to other nicotinic receptors. The addition of the methoxygroups result in a compound that is a relatively weak partial agonist, compared to acetylcholine and nicotine. Removal of one or both methoxy groups and replacement by hydroxyl groups, which can occur in the brain, increase its agonist effect. The compound's absorption is rapid, less than an hour after oral doses, and it is also rapidly eliminated, with a half-life of about an hour.[31] DMXB-A had some cognitive effects immediately, which continued during a 5-day trial in normals. The side effect data also appeared favorable.

The first trial in schizophrenia occurred in 12 patients who were treated with two different doses with a placebo comparator in a crossover design. Each treatment was given for one day.[32] Because of the drug's short half-life, a second half dose was given 2 hours after the first to attempt to produce a more prolonged drug exposure. There were no significant adverse effects. During each treatment arm, the patients' P50 evoked potential was recorded in a sensory gating, conditioning-testing paradigm to assess inhibition, which is measured as the decrease in amplitude of the response to the second (S2) or testing stimulus because of recurrent inhibition and other inhibitory mechanisms engaged during the first (S1) or conditioning stimulus. The effect is expressed as the ratio S2 amplitude/S1 amplitude; lower values indicate increased inhibition. As noted earlier, patients with schizophrenia have deficits in this inhibition that are genetically linked to the *CHRNA7* locus and associated with polymorphisms in the gene. Therefore, this paradigm was used as a neurobiological measure of target engagement. At the time of this experiment, there was no PET ligand available. One has become available and it is displaceable by DMXB-A in animal models.[33] The P50 amplitude ratio (S2/S1) after DMXB-A 75 mg was 0.30 ± 0.22, compared to 0.71 ± 0.51 after placebo treatment, which was a highly significant difference. The P50 ratio had decreased to the level generally observe in normal subjects.

The subjects were also assessed using the Repeatable Battery for the Assessment of Neuropsychological Status (RBANS). DMXB-A 75 mg increased the Total Score, with a specific effect on the Attention Score. Positive but less significant effects were seen on all scales. The effects of a higher dose of DMXB-A 150 mg were also positive for both P50 ratio and RBANS, but the effects were less than DMXB-A 75mg. However, the difference between the two doses was not significant. Effects on both measures were not correlated with the plasma levels of DMXB-A or the hydroxyl metabolites.

Based on these Phase 1 results, a second Phase 2 clinical trial was initiated with DMXB-A 75 mg or 150 mg twice daily or placebo, again using a crossover design. Each treatment lasted 1 month.[34] Several fMRI measures were used as biomarker of engagement, principally limbic activity during both smooth pursuit eye movements and the resting state. Increased activity in this region is correlated with decreased P50 inhibition in schizophrenia. DMXB-A decreased the activity, consistent with its previously demonstrated action to increase inhibition. The effect was dependent on the patient's *CHRNA7* genotype. A polymorphism in the 3' promoter, whose minor allele is associated with schizophrenia, was examined.[35] The patients with the minor allele responded well to 150 mg DMXB-A, but their response to 75 mg DMXB-A was decreased, compared to patients with two major alleles. The homozygous minor allele carriers were the most affected. This *CHRNA7* genomic effect is consistent with the hypothesis that the minor allele associated with schizophrenia causes decreased *CHRNA7* expression.

A significant clinical effect on the Scale for Assessment of Negative Symptoms over all three arms of the study was found. The effect was greater for the DMXB-A 150 mg dose, particularly on the anhedonia and alogia subscales. There was no significant effect over the three arms on the neurocognitive measure used, the MATRICS Consensus Clinical Battery. Practice effects over time were significant and therefore only the first arm was secondarily analyzed, compared to the patient's pretrial baseline. There was a significant effect for

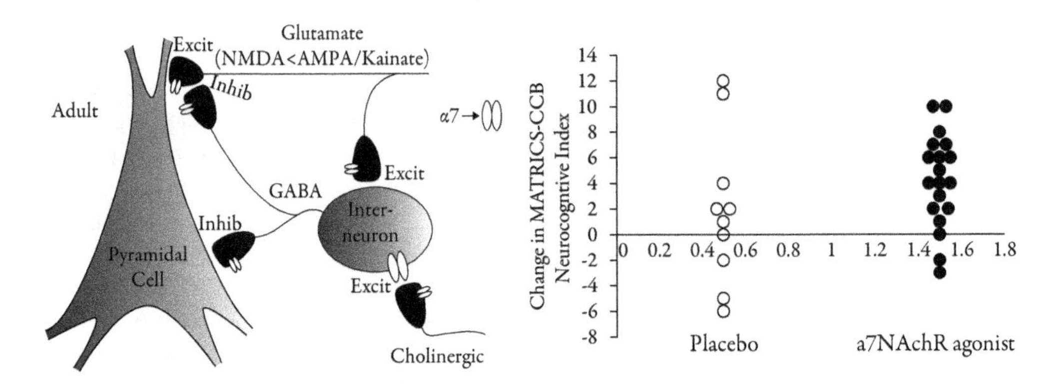

Figure 34.1 Alpha-7 nicotinic receptors in adult. Alpha7-nicotinic receptors are expressed postsynaptically on interneurons in the hippocampus. Their activation by a nicotinic agonist in patients with schizophrenia increases neurocognitive performance, compared to placebo.[34]

DMXB-A at both doses, compared to baseline, particularly for attention-vigilance and executive function indices and no significant effect for placebo. The effects were not related to plasma levels, but were correlated with resting state limbic fMRI activity. The trial was a two-site trial with the University of Maryland Psychiatric Research Center. There was no MATRICS CCB composite measure available, because the Social Cognition test had not been released. Later, the MATRICS CCB Neurocognitive Index was released, which does not require the Social Cognition test. It also showed positive effects of DMXB-A in the first arm of the study.

A broadly based industry effort to find compounds similar to DMXB-A ensued. Initially, a number of companies led by Memory Pharmaceuticals concentrated on making long-acting full agonists, because DMXB-A is a short-acting, weak partial agonist. These compounds did not do well even in initial preliminary trials. A second round of effort shifted emphasis to long-acting partial agonists like DMXB-A with much longer half-lives, one greater than 24 hours.[36] Several companies reported successful initial Phase 2 trials.[36–38] The trials showed effects on executive function and attention-vigilance similar to those shown for DMXB-A. However, none of these candidates has shown significant cognitive effects in schizophrenia in larger trials.[39–41] Similar problems were observed in companion trials for cognition in Alzheimer's dementia. Some observers have suggested that problems in the conduct of Phase 3 trials are partially responsible, with as many as 100 sites competing for patients. Most sites have too few patients to become proficient on the MATRICS CCB or comparable neurocognitive assessments.

However, there also were pharmacological problems with the industry development programs. Full agonists of alpha7-nicotinic receptors are highly desensitizing with chronic administration, as is required for Phase 3. For example, varenicline, which blocks the effect of nicotine from cigarette smoke by desensitizing receptors, is a full agonist at alpha7-nicotinic receptors.[42] The long duration of action of the industrial agonists exacerbates the problem, because longer exposure to the receptors also increases desensitization. For industry drug, severe constipation developed, a well-known side effect of 5HT-3 serotonin receptor antagonists such as ondansetron. The 5HT-3 receptor and the alpha7-nicotinic receptor have considerable molecular similarity. Tropisteron, a compound available in Asia as a 5HT-3 antagonist, has considerable alpha7-nicotinic receptor activity and normalized P50 inhibition in clinical trials in both Japan and China.[43,44] Because of these issues in Phase 3, in part reflecting pharmaceutical marketing efforts that require psychotropic compounds to have once daily administration, industry activity has subsided.

Based on the initial success of longer acting industrial compounds, we decided to reformulate DMXB-A to be longer acting to see if its effects would increase. The compound was reformulated with hypromellose, which delays release from the stomach by forming a gel-like matrix around the drug. We administered DMXB-A extended release (DMXBA-ER 150 mg) four times per day for one month. DMXB-A levels rose over several hours and remained elevated until about 10 pm. Plasma levels reached with DMXBA-ER 150 mg were comparable to those found with the immediate release formulation. With DMXBA-ER there were no effects on the P50 ratio, indicating that the anticipated biological effect did not occur. Performance on the MATRICS-CCB and SANS ratings showed no effects as well. This failure to see effects of DMXBX-ER were compared to the effects of the immediate release DMXB-A For example DMXB-A immediate release improved the MATRICS CCB Attention-Vigilance Index by 7.80 ± 8.85 T-score units, which was significantly greater than the effect of DMXBA-ER, -1.83 ± 6.46 T-score units.[45]

The value of the DMXBA-ER trial is that it demonstrated, using the same molecule, the pharmacokinetic limitations to the activation of alpha7-nicotinic receptors by agonists. The immediate release formulation's ability to reach a high brain concentration quickly and then to be eliminated rapidly may be particularly effective because DMXB-A activates a nitric-oxide-mediated second messenger response in neurons that appears to extend its effect after brief receptor stimulation, whereas prolonged exposure results in loss of effect.[46]

With the immediate release formulation, the increase of 4.20 ± 3.17 T-score units in the MATRICS CCB Neurocognitive Index is equivalent to the 4.8 ± 1.8 increase associated with employment in patients with schizophrenia, compared to patients who are not employed.[47] This pharmacokinetic limitation on drugs that target alpha7-nicotinic receptors as agonists, which likely applies to all the failed industry Phase 2 trials, means that a potentially valuable therapeutic intervention is unlikely to be developed.

Because of the limitation of the agonist strategy, there have been attempts to discover positive allosteric modulators (PAMs), which change the properties of the alpha7-nicotinic receptor either to increase its activation by acetylcholine (Type I PAM), or to decrease its desensitization (Type II PAM). Diazepam, a Type I PAM at GABA receptors is a model. The type II PAM PNU-120596 blocks desensitization of the channel and opens channels that have already been desensitized by the agonist nicotine.[48] Type II PAMs possibly could increase toxicity from nicotine by keeping open all receptor channels activated by nicotine. The type II PAM JNJ39393406 was not effective in its initial trial in schizophrenia.[49] The desensitization of the alpha7-nicotinic receptor is believed to be a use-dependent, readily reversible type of signal plasticity that may shape synaptic efficacy in various brain regions and protects neurons from excessive excitation by tight control of cytosolic calcium homeostasis.[2] Thus, any disruption of desensitization may also cloud the observation of any potential therapeutic response to alpha7-nicotinic receptor modulation. The type I PAM, AVL-3288, on the other hand, increased the effectiveness of acetylcholine and does not reverse effects of desensitization.[50] AVL-3288 increased cognition modestly in normal subjects and was safe in normal smokers.[51] Its first Phase 1b trial in schizophrenia has been initiated.[52]

Another drug that activates alpha7-nicotinic receptors is clozapine. Clozapine is not a direct agonist. However, it releases large amounts of acetylcholine, particularly in the hippocampus. The mechanism is unknown but could involve effects mediated by presynaptic serotonergic receptors on the cholinergic synapses. As a result, alpha7-nicotinic receptors are activated. P50 sensory gating is normalized in patients who receive clozapine, whereas generally does not occur with other antipsychotics, and many of the patients decrease or even cease smoking.[53,54]

All the trials have added alpha7-nicotinic agonists to the patient's current antipsychotics, with the rationale that, except for clozapine, the nicotinic mechanism of action is distinct from their primarily antidopaminergic effect. There is no reason to believe that the two drugs would either enhance or interfere with each other, and the trials suggest that they do not. Cigarette smoking is more likely to interfere, as the chronically high nicotine levels of most smokers interfere with the effects of nicotine itself.[29] The initial two trials of DMXB-A studied only patients who were nonsmokers.[32,34] DMXB-A is extensively metabolized by the CYP enzymes induced by nicotine. The one trial that studied smoking subjects failed in part because plasma levels of DMXB-A were too low.[45] Most industrial trials have studied smokers, but none has attempted to investigate possible changes in smoking behavior. One trial of tropisetron found that effects observed in nonsmokers did not occur in smokers.[42] An open question is whether a nicotinic agonist would stop these patients from smoking, as clozapine often does.

Because *CHRNA7* is not a major genetic factor in schizophrenia, it is reasonable to question whether alpha7-nicotinic treatments are likely to have effects for the many patients in whom this mechanism appears to be normal already. The trial data with DMXB-A find that it is actually effective at lower doses in patients who do not have a *CHRNA7* polymorphism associated with schizophrenia, and the postmortem studies indicate that *CHRNA7* mRNA is decreased in many patients, more than would be expected to have *CHRNA7* polymorphisms.[21,35] One study found no effect of *CHRNA7* polymorphisms on the generally altered level of *CHRNA7* mRNA expression, which is consistent with the hypothesis that other genes also affect *CHRNA7* mRNA expression, notably *NRG1*, the gene for neuregulin.[55–57] In drug development more generally, genes with small effect may not explain heritability well, but they often are useful drug targets. For example, in adult onset diabetes, *PPARgamma* has only modest influence on heritability, yet it is the target for the glitazones, a widely used class of drugs for this illness.[58]

Even in the most responsive patients, the addition of an alpha7-nicotinic agonist enhances cognition substantially, but it does not fully restore the patient's mental well-being. That prompted us to examine further the developmental role of *CHRNA7* in the prenatal risk for schizophrenia.

PRENATAL STUDIES OF CHRNA7 ACTIVATION

As described earlier, alpha7-nicotinic receptors are expressed much more widely and at higher levels in the fetal brain than at any other time in life cycle.[26,27] Their endogenous agonist is choline from the amniotic fluid throughout most of fetal development.[25] Fetal development is also the time when neurobiological defects associated with later schizophrenia are formed, as judged by their appearance as motor deficiencies in babies who later in adult life are diagnosed with schizophrenia. Therefore we investigated in mouse models whether dietary supplementation with choline of the dams from conception through weaning would enhance neurobiological development.[59,60] The offspring were then fed normal choline diets in adulthood. The marker used, a murine analogue of human P50 sensory gating, was normalized in adults who had received this treatment.

An observational study had previously showed positive effects of higher maternal choline diets on cognitive function in the children at 7 years of age.[61] The first double-blind controlled trial used 5400 mg phosphatidylcholine capsules daily, containing 750 mg choline, a 200% increase in their amount of choline, compared to the 360 mg estimated from diet.[62] Phosphatidylcholine is freely interchangeable with choline in the blood. It is the preferred supplement, because it is impervious to most colonic bacteria. Choline itself is metabolized by colonic bacteria to trimethylurea, which imparts a fishy odor, or trimethylamineoxide, which is atherogenic.[63] These metabolites are not increased by phosphatidylcholine.[64] The maternal supplement was begun by 18 weeks of gestation and continued through the first month of nursing. Plasma choline levels were twice the mean for the placebo-treated women. The trial found no effect on cognition as measured at 1 year of age by the infant's ability to find an object concealed by the investigator after different delays, up to 24 hours. The strength of the trial is the complete measurement of choline and its metabolites in the mother's plasma. A weakness of the trial is that the primary cognitive task undergoes a linear phase of improvement from 6 to 24 months, which means that the test was quite challenging for the 10- to 12-month-old infants in this study and their performance was not at a steady state level at the time of testing. Seventy-one percent of the enrolled women had their infants assessed, and of these, 46% could not complete the entire testing protocol.[65]

In a second double-blind clinical trial, which is from our group, 50 women took phosphatidylcholine 6300 mg daily from week 17 of gestation though delivery, a 250% increase in choline levels over their normal diet. After birth, the newborns received 100 mg of phosphatidylcholine drops from 2 weeks post birth until 3 months of age. Fifty women and their newborns were given placebo. Because of the newborns' limited behavioral repertoire, the principal outcome measure was P50 auditory evoked potential inhibitory sensory gating at 1 month of age. This measure was previously shown to be abnormal in infants whose parents had psychosis or whose mothers smoked or were depressed, all risk factors associated with later schizophrenia in the offspring.[66] The phosphatidylcholine-supplemented group had significantly fewer infants with abnormal P50 inhibitory sensory gating.[67]

The same children were reexamined at 3.5 years of age with the Childhood Behavioral Checklist (CBCL). Children from phosphatidylcholine-supplemented pregnancies, now on regular diets, had significantly fewer problems noted in the CBCL in attention and social interaction. This effect was related to their P50 sensory gating at 1 month of age. The magnitude of difference in CBCL ratings at 3.5 years between

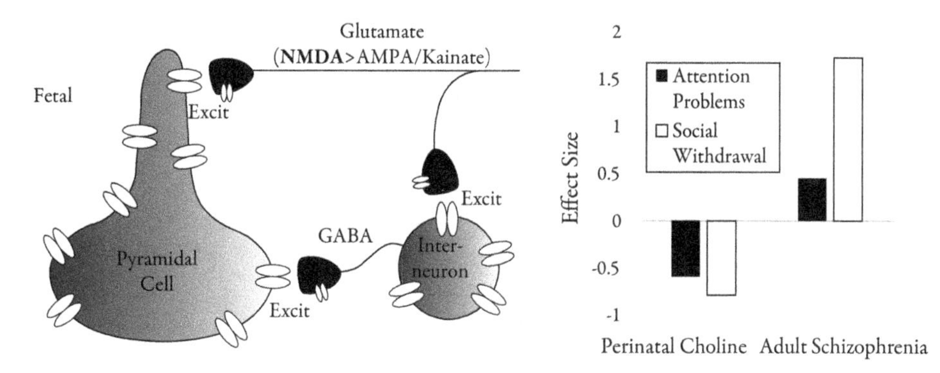

Figure 34.2 Alpha7-nicotinic receptors in fetuses. Alpha7-nicotinic receptors are expressed on both interneurons and pyramidal cells in the hippocampus at overall higher levels than in adults. Their activation by choline during gestation decreased social and attention problems in the children at 3.5 years of age.[68] As a comparison, parents of children who later developed schizophrenia as adults report that their children had a similar magnitude of social and attention problems.[69]

phosphatidylcholine-supplemented children and placebo control children for attention problems (d' = 0.59) and social problems (d' = 0.79) was similar to the magnitude of difference found in a retrospective comparison of adults with schizophrenia and a control group, whose parents had rated them on the CBCL based on their recall when their child was 3-4 years old (Figure 34.2).[68,69]

Strengths of this study are the double-blind placebo-controlled design, the relatively long-term assessment of behavior and its relation to an early neurobiological marker of effect, and a *CHRNA7* genotypic effect in the infants. Children with the *CHRNA7* genotype associated with schizophrenia were more likely to have social abnormalities, but this genetic effect disappeared in those children whose mothers had been treated with phosphatidylcholine during gestation.[68] A similar *CHRNA7* genotype effect with P50 sensory gating had been shown in their infancy.[67] The intervention was also found to be equally effective for infants who have a mother with schizophrenia.[68] Weaknesses are its small size, the 50% attrition of the groups by 3.5 years as mothers were lost to contact, and the lack of full follow-up assessments to adulthood, when schizophrenia would generally appear. Prospective trials to show significant effects on the incidence of schizophrenia would require treatment of thousands of mothers with decades of follow-up. Investigation of the early developmental course to assess the presence or absence of childhood symptoms later associated with illness would seem the only practical strategy and is consistent with RDoC, which includes developmental studies of discrete elements of a clinical illness.[70] P50 sensory gating is in the Perceptual Construct, Auditory Subconstruct, whose clinical manifestation is auditory hallucinations.

A recent survey showed that fewer than half of pregnant women reach an adequate level through normal diet, approximately 2450 mg phosphatidylcholine, equivalent to 450 mg of choline daily.[71] A case report reported very low levels of choline in a mother with bipolar disorder treated with lithium, but this finding has not been extended to other women.[72] The Institute of Medicine, National Academy of Sciences, recommends an upper limit of for phosphatidylcholine of 24,000 mg/day, well below the amounts used in clinical trials.[73] No safety concerns have been identified at or below this level.

Although epidemiologists have long recognized that prenatal maternal infection, smoking, and depression increase risk of schizophrenia in the offspring, there have been no attempts to ameliorate that risk. The molecular biology of schizophrenia also points to fetal brain development as a first period when genes can exert their pathogenic effects. While other risks occur as developmental to adulthood progresses, there is no evidence that what occurs in fetal life can later be fully remedied. Therefore, it is a uniquely important time for interventions. Phosphatidylcholine supplementation would not be expected to prevent all schizophrenia, but it might safely ameliorate symptoms or even decrease incidence. There is current debate about whether maternal supplements to improve the child's health should be or should not be discussed with inquiring mothers based on positive, but limited data from trials.[74] However, because many months to years are required for gestation and then for the child to become old enough to evaluate outcome, more data will not be available in time for a woman who is currently contemplating pregnancy to decide. Just as fetal development is a unique period of risk for illness, so too is pregnancy a unique opportunity to alter each child's developmental course toward later illness.

ACKNOWLEDGMENT

This work was supported by the Anschutz Foundation and the Institute for Children's Mental Disorders. The infant work was performed by the late Dr. Randal Ross. There are no conflicts of interest to disclose.

REFERENCES

1. Changeux J-P, Edelstein SJ. *Nicotinic Acetylcholine Receptors: From Molecular Biology to Cognition.* New York, New York: Odile Publishing, 2005.
2. Quick MW, Lester HA. Desensitization of neuronal nicotinic receptors. *J Neurobiology* 2002;53:457–478.
3. Osserman KE, Kaplan LI. Studies in myasthenia gravis: use of edrophronium chloride (Tensilon) in differentiating myasthenic from cholinergic weakness. *Arch Neurol Psychiat* 1953;70:385–392.

4. Peng X, Katz M, Gerzanich V, Anand R, Lindstrom J. Human alpha 7 acetylcholine receptor: cloning of the alpha 7 subunit from the SH-SY5Y cell line and determination of pharmacological properties of native receptors and functional alpha 7 homomers expressed in Xenopus oocytes. *Mol. Pharmacol* 1994;45:546–554.
5. Mulle C, Coquet D, Korn H, Changeux J-P. Calcium influx through nicotinic receptor in rat central neurons: Its relevance to cellular regulation. *Neuron* 1992;8:135–143.
6. de Kloet SF, Mansvelder HD, De Vries TJ. Cholinergic modulation of dopamine pathways through nicotinic acetylcholine receptors. *Biochem Pharmacol* 2015;97:425–438.
7. Benowitz NL. Pharmacology of nicotine: addiction and therapeutics. *Annu Rev Pharmacol Toxicol* 1996;36:597–613.
8. Beirut LJ, Stitzel JA, Wang JC et al. Variants in the nicotinic receptors alter the risk for nicotine dependence. *Am J Psychiat* 2008; 165:1163–1171.
9. Thorgeirsson TE, Geller F, Sulem P, et al. A variant associated with nicotine dependence, lung cancer and peripheral arterial disease. *Nature* 2008;452:638–642.
10. Freedman R. Alpha7-nicotinic acetylcholine receptor agonists for cognitive enhancement in schizophrenia. *Ann Rev Med* 2014;65:245–261.
11. Wang H, Yu M, Ochani M, et al. Nicotinic acetylcholine receptor alpha7 subunit is an essential regulator of inflammation. *Nature* 2003;421:384–388.
12. Gault J, Robinson M, Berger, R et al. Genomic organization and partial duplication of the human α7 neuronal nicotinic acetylcholine receptor gene. *Genomics* 1998;52:173–185.
13. Vogels A, Matthijs G, Legius E, et al. Chromosome 15 maternal uniparental disomy and psychosis in Prader-Willi syndrome 2003;40:72–73.
14. Sirota P, Frydman M, Sirota L. Schizophrenia and Marfan syndrome. *Brit J Psychiat*1990;157:433–436.
15. Stefansson H, Rujescu D, Cichon S et al. Large recurrent microdeletions associated with schizophrenia. *Nature* 2008;455:232–236.
16. Bacchelli E, Battaglia A, Cameil C et al. Analysis of *CHRNA7* rare variants in autism spectrum disorder susceptibility. *Am J Med Genetics Part A* 2015; 167A:715–723.
17. Williams NM, Franke B, Mick E et al. Genome-wide analysis of copy number variants in attention deficit hyperactivity disorder: the role of rare variants and duplications at 15q13.3. *Am J Psychiatry* 2012;169:195–204.
18. Leonard, S, Gault J, Moore T et al. Further Investigation of a chromosome 15 locus in schizophrenia: Analysis of affected sibpairs from the NIMH Genetics Initiative. *Neuropsychiatric Genetics* 1998;81:308–312.
19. Freedman, R, Coon H, Myles-Worsley M. et al. Linkage of a neurophysiological deficit in schizophrenia to a chromosome 15 locus. *Proc Nat Acad Sci USA* 1997;94:587–592.
20. Houy E, Raux G, Thibaut F et al. The promoter -194 C polymorphism of the nicotinic alpha 7 receptor gene has a protective effect against the P50 sensory gating deficit. *Mol. Psychiatry* 2004;9:320–322.
21. Guillozet-Bongaarts AL, Hyde TM, Dalley RA et al. Altered gene expression in the dorsolateral prefrontal cortex of individuals with schizophrenia. *Mol Psychiatry* 2014; 19:478–485.
22. Liu Z, Neff RA, Berg DK. Sequential interplay of nicotinic and GABAergic signaling guides neuronal development. Science 2006;314:1610–1613.
23. Hyde TM, Lipska BK, Ali T et al. Expression of GABA signaling molecules KCC2, NKCC1, and GAD1 in cortical development and schizophrenia. *J Neurosci* 2011;31: 11088–11095.
24. Kerwin R, Patel S, Meldrim N. Quantitative audioradiographic analysis of glutamate binding sites in the hippocampal formation in normal and schizophrenic brain. *Neurosci* 1990;39: 25–32.
25. Frazier CJ, Rollins YD, Breese CR et al. Acetylcholine activates an alpha-bungarotoxin sensitive nicotinic current in rat hippocampal interneurons, but not pyramidal cells. *J Neurosci* 1997;18: 1187–1195.
26. Birnbaum R, Jaffe AE, Hyde TM, et al. Prenatal expression patterns of genes associated with neuropsychiatric disorders. *Am J Psychiatry* 2014;171:758–767.
27. Court JA, Lloyd S, Johnson M et al. Nicotinic and muscarinic cholinergic receptor binding in the human hippocampal formation during development and aging. *Brain Res. Dev. Brain Res.* 1997;101:93–101.
28. de Leon J, Dadvand M, Canuso C et al Schizophrenia and smoking: an epidemiological survey in a state hospital. *Am J Psychiatry* 1995;152: 453–455.
29. Harris JG, Kongs S, Allensworth D A, Sullivan B, Zerbe G, Freedman, R. Effects of nicotine on cognitive deficits in schizophrenia. *Neuropsychopharmacol* 2004;29:1378–1385.
30. de Fiebre CM, Meyer EM, Henry JC, Muraskin SI, Kem WR, Papke RL. Characterization of a series of anabaseine-derived compounds reveals that the 3-(4)- dimethylaminocinnamylidine derivative is a selective agonist at neuronal nicotinic alpha-7/125 I-alpha-bungarotoxin receptor subtypes. *Mol Pharmacol* 1995;47:164–171.
31. Kitagawa H, Takenouchi T, Azuma R et al. Safety, pharmacokinetics, and effects on cognitive function of multiple doses of GTS-21 in healthy, male volunteers. *Neuropsychopharmacol* 2003;28:542–551.
32. Olincy A, Harris JG, Johnson L et al. Proof-of-concept trial of an a7- nicotinic agonist in schizophrenia. *Arch Gen Psychiat* 2006:63:630–638.
33. Wong DF, Kuwabara H, Pomper M et al. Human brain imaging of α7 nAChR with [(18)F]ASEM: a new PET radiotracer for neuropsychiatry and determination of drug occupancy. *Mol Imaging Biol* 2014;16:730–738.
34. Freedman R, Olincy A, Buchanan RW et al. Initial phase 2 trial of a nicotinic agonist in schizophrenia. *Am J Psychiat* 2008;165:1040–1047.
35. Tregellas JR, Tanabe J, Roja, DC et al. Effects of an alpha 7-nicotinic agonist on default network activity in schizophrenia. *Biol Psychiat* 2011:69:7–11.
36. Haig G, Bain E, Robieson W et al. A randomized, double-blind trial to assess the efficacy and safety of ABT-126, a selective α7 nicotinic acetylcholine receptor agonist in the treatment of cognitive Impairment in subjects with schizophrenia. *Am J Psychiatry* 2016;173:827–835.
37. Keefe RSE, Meltzer HA Dgetluck N et al. Randomized, double-blind, placebo-controlled study of encenicline, an α7 nicotinic acetylcholine receptor agonist, as a treatment for cognitive impairment in schizophrenia. *Neuropsychopharmacol* 2015;40:3053–3060.
38. Lieberman JA, Dunbar G, Segreti AC et al. A randomized exploratory trial of an alpha-7 nicotinic receptor agonist (TC-5619) for cognitive enhancement in schizophrenia. *Neuropsychopharmacol* 2013;38:968–975.
39. Haig G, Wang D, Othman AA, Zhao J. The α7 nicotinic agonist ABT-126 in the treatment of cognitive impairment associated with schizophrenia in nonsmokers: Results from a randomized controlled Phase 2b study. *Neuropsychopharmacology* 2016;41:2893–2902.
40. Umbricht D, Keefe RS, Murray S et al. A randomized, placebo-controlled study investigating the nicotinic alpha7 agonist, RG3487, for cognitive deficits in schizophrenia. *Neuropsychopharmacol* 2014;39:1568–1577.
41. Walling D, Marder SR, Kane J et al. Phase 2 trial of an alpha-7 nicotinic receptor agonist (TC-5619) in negative and cognitive symptoms of schizophrenia. *Schizophr Bull* 2016;42:335–343.
42. Picciotto MR, Addy NA, Mineur YS, Brunzell DH. It's not "either/or" activation and desensitization of nicotinic acetylcholine receptors: both contribute to behaviors related to nicotine addiction and mood. *Prog Neurobiol* 2008;84:329–342.
43. Koike K, Hashimoto K, Takai N et al. Tropisetron improves deficits in auditory sensory P50 suppression in schizophrenic patients. *Schizophr Res* 2005;76:67–72.
44. Zhang XY, Liu L, Liu S et al. Short-term tropisetron treatment and cognitive and P50 auditory gating deficits in schizophrenia. *Am J Psychiat* 2012;169:974–981.

45. Kem WR, Olincy A, Johnson L et al. Pharmacokinetic limitations on effects of an alpha7 nicotinic receptor agonist in schizophrenia: Randomized trial with an extended release formulation. *Neuropsychopharmacol* in press.
46. Adams CE, Stevens KE, Kem WR, Freedman R. Inhibition of nitric oxide synthase prevents alpha 7 nicotinic receptor-mediated restoration of inhibitory auditory gating in rat hippocampus. *Brain Res* 200;877:235–244.
47. August SM, Kiwanuka JN, McMahon RP, Gold JM. The MATRICS Consensus Cognitive Battery (MCCB): clinical and cognitive correlates. *Schizophr Res* 2012;134:76–82.
48. Hurst RS, Hajós M, Raggenbass M et al. A novel positive allosteric modulator of the alpha7 neuronal nicotinic acetylcholine receptor: In vitro and in vivo characterization. *J Neurosci* 2005;25:4396–4405.
49. Winterer G, Gallinat J, Brinkmeyer J et al. Allosteric alpha-7 nicotinic receptor modulation and P50 sensory gating in schizophrenia: a proof-of-mechanism study. *Neuropharmacol* 2013;64:197–204.
50. Ng HJ, Whittemore ER, Tran MB et al. 2007. Nootropic alpha7 nicotinic receptor allosteric modulator derived from GABAA receptor modulators. *Proc Natl Acad Sci USA* 2007;104:18059–18064.
51. Gee KW, Olincy A, Kanner R et al. First in human trial of a type I positive allosteric modulator of alpha7-nicotinic acetylcholine receptors: Pharmacokinetics, safety, and evidence for neurocognitive effect of AVL-3288. J Psychopharmacol 2017;31:434–441 52. New York State Psychiatric Institute. Clinical Trial of AVL-3288 in Schizophrenia Patients, Clinical Trials.gov 2016;NCT02978599.
53. Nagamoto HT, Adler LE, Hea RA, Griffith JM, McRae KA, Freedman R. Gating of auditory P50 in schizophrenics: unique effects of clozapine. *Biol Psychiat* 1996;40:181–188.
54. George TP, Serynak MJ, Ziedonis DM, Woods SW. Effects of clozapine on smoking in chronic schizophrenic outpatients. *J Clin Psychiat* 1997;56:344–346.
55. Stephens S, Logel J, Barton A et al. Association of the 5 -upstream regulatory region of the 7 nicotinic acetylcholine receptor subunit gene (*CHRNA7*) with schizophrenia. *Schizophr Res* 2009;109:102–112.
56. Kunii Y, Zhang W, Xu Q et al. *CHRNA7* and *CHRFAM7A* mRNAs: Co-Localized and their expression levels altered in the postmortem dorsolateral prefrontal cortex in major psychiatric disorders. *Am J Psychiat* 2015;172:1122–1130.
57. Mathew SV, Law AJ, Lipska BK et al. α7 nicotinic acetylcholine receptor mRNA expression and binding in postmortem human brain are associated with genetic variation in neuregulin 1. *Hum Mol Genet* 2007;16: 2921–2932.
58. Altshuler D, Hirschhorn JN, Klannemark M et al. The common PPARgamma Pro12Ala polymorphism is associated with decreased risk of type 2 diabetes. *Nat Genet* 2000;26:76–80.
59. Walker EF, Savoie T, Davis D. Neuromotor precursors of schizophrenia. *Schizophr Bull* 1994;20:441–451.
60. Stevens KE, Adams CE, Yonchek J et al. Permanent improvement in deficient sensory inhibition in DBA/2 mice with increased perinatal choline. *Psychopharmacol (Berl)* 2008;198:413–420.
61. Boeke CE, Gillman MW, Hughes MD et al. Choline intake during pregnancy and child cognition at age 7 years. *Am J Epidemiol* 2013;177:1338–1347.
62. Cheatham CL, Goldman BD, Fischer LM et al. Phosphatidylcholine supplementation in pregnant women consuming moderate-choline diets does not enhance infant cognitive function: a randomized, placebo-controlled trial. *Am J Clin Nutr* 2012;96:1465–1472.
63. Tang WHT, Zeneng W. Levison BS et al. Intestinal microbial metabolism of phosphatidylcholine and cardiovascular risk. *N Engl J Med* 2013;368:1575–1584.
64. Roe AJ, Zhang S, Bhadelia RA et al. Choline and its metabolites are differently associated with cardiometabolic risk factors, history of cardiovascular disease, and MRI-documented cerebrovascular disease in older adults. *Am J Clin Nutr* 2017;105:1283–1290.
65. Barr R, Dowden A, Hayne H. Developmental changes in deferred imitation by 6- to 24-month-old infants. *Infant Behav Dev* 1996;19:159–170.
66. Hunter SK, Kisley MA, McCarthy L, Freedman R, Ross RG. Diminished cerebral inhibition in neonates associated with risk factors for schizophrenia: parental psychosis, maternal depression, and nicotine use. *Schizophr Bull* 2011;37:1200–1208.
67. Ross RG, Hunter SK, McCarthy L et al. Perinatal choline effects on neonatal pathophysiology related to later schizophrenia risk. *Am J Psychiat* 2013;170:290–298.
68. Ross R, Hunter SK, Hoffman MC et al. Perinatal phosphatidylcholine supplementation and early childhood behavior problems: Evidence for *CHRNA7* moderation. *Am J Psychiat* 2016;173: 509–516.
69. Rossi A, Pollice R, Daneluzzo E et al. Behavioral neurodevelopmental abnormalities and schizophrenic disorder: a retrospective evaluation with the Child Behavior Checklist (CBCL). *Schizophr Res* 2000;44:121–128.
70. Casey BJ, Oliveri ME, Insel T. A neurodevelopmental perspective on the research domain criteria (RDoC) framework. *Biol Psychiat* 2014;76:350–353.
71. Masih SP, Plumpire L, Ly A et al. Pregnancy Canadian women achieve recommended intakes of one-carbon nutrients through prenatal supplementation but the supplement concentration, including choline, requires reconsideration. *J Nutr* 2015; 145: 1824–1834.
72. Gossell-William M, Fletcher H, Zeisel S. Unexpected depletion in plasma choline and phosphatidylcholine concentration in a pregnant women with bipolar affective disorder being treated with lithium, haloperidol, and benztropine: a case report. *J Med Case Reports* 2008;20:55.
73. Institute of Medicine of the National Academies of Science of the United States. *Dietary Reference Intakes for Thiamin, Riboflavin, Niacin, Vitamin B6, Folate, Vitamin B12, Pantothenic Acid, Biotin, and Choline.* Washington DC, 2010.
74. Ramaswami R, Serhan CN, Levy BD, Makrides, M. Fish oil supplementation in pregnancy. *N Engl J Med* 2016;375:2599–2601.

35

MUSCARINIC ACETYLCHOLINE RECEPTORS IN THE ETIOLOGY AND TREATMENT OF SCHIZOPHRENIA

M. S. Moehle, S. E. Yohn, and P. J. Conn

INTRODUCTION

Schizophrenia is a debilitating psychiatric disorder that effects ~1% of the population[1] and is characterized by three main clusters of symptoms; positive, negative, and cognitive.[2] The positive symptoms are associated with an excess or distortion of normal functions and often include hallucinations, delusions, and paranoia.[2] Negative symptoms are associated with a loss of normal functions and include social withdrawal, apathy, anhedonia, and blunted affect.[3] Cognitive symptoms include deficits in working memory and cognitive flexibility.[4,5] Patients with schizophrenia may exhibit any combination of these symptoms, and no single symptom is a hallmark of disease.[2]

Currently available medications mostly treat the positive symptom cluster mainly through antagonism of either the dopamine receptor subtype 2 (D_2) or serotonin receptor subtype 2A ($5\text{-}HT_{2a}$). Although antipsychotics can effectively treat the hallucinations and paranoia associated with the positive symptom domain, these medications either are partially effective or worsen expression of negative and cognitive symptoms of schizophrenia.[3,6–9] Moreover, these medications can even be sufficient to induce symptoms that mirror the negative and cognitive symptoms of schizophrenia in otherwise healthy individuals.[10] Additionally, these medications often have severe side effects, including impaired motor function, weight gain, and several peripheral adverse effects.[11–13] Taken together these limitations highlight a major unmet clinical need both to treat the cognitive and negative symptoms and to find better treatments for the positive symptoms. The major failings in the treatment of schizophrenia may in part result from insufficient understanding of its disease process.

While the causes of schizophrenia are likely complex, changes in mesolimbic and mesocortical dopamine (DA) are thought to be a major driver of these symptom clusters.[14] However, the pathological mechanisms behind these changes in DA are not well understood, and the limitation of D_2 antagonists warrants examining other neurotransmitters and signaling mechanisms outside of DA. Indeed, diverse lines of evidence now support multiple disturbances in glutamate, GABA, and acetylcholine (ACh) as possible drivers or modifiers of schizophrenia.[15–19]

Of particular interest is the cholinergic system. ACh regulates multiple brain regions and circuits thought to be involved in schizophrenia, including the basal ganglia (BG), prefrontal cortex (PFC), and hippocampus (HPC).[13,19] Cholinergic projections from several discrete brain regions innervate these regions. ACh from the hindbrain cholinergic nuclei of the pedunculopontine nucleus (PPN) and laterodorsal tegmental nucleus (LDT) release onto midbrain DA neurons as well as BG nuclei, which regulate locomotor and reward-related behaviors.[20–22] Basal forebrain cholinergic nuclei are a major source of cholinergic projections to the PFC, HPC, and other cortical structures and regulate sleep, cognition, and memory.[23] Additionally, the striatum includes local cholinergic interneurons that do not project outside of the striatum. These cholinergic interneurons regulate movement, reward, and other aspects of BG output.[24,25] Between these distinct cholinergic nuclei and interneurons, ACh has a major potential to regulate or modify several of the symptom clusters of schizophrenia.

ACh signals through two distinct classes of receptors, the nicotinic and muscarinic ACh receptors (nAChR and mAChR, respectively). The nAChRs are a class of ionotropic ACh receptors that are nonselective cation channels. The nAChRs are widely expressed and have widespread actions throughout the brain, including profound effects on DA signaling.[12] There are also diverse actions of nAChRs outside of DA regulation that are extensively discussed in other chapters of this book. In addition to fast excitatory neurotransmission through the ionotropic nAChRs, ACh acts on mAChRs. The mAChRs have diverse actions throughout the central nervous system (CNS), and play a pivotal role in the regulation of cognition, movement, reward, and sleep.[13,26] Additionally, multiple lines of evidence from genetic association studies, clinical studies, and preclinical studies point to a role for mAChRs in the pathophysiology and as potential targets for schizophrenia therapeutics.

MUSCARINIC ACETYLCHOLINE RECEPTORS

The mAChRs belong to the superfamily of G protein coupled receptors (GPCRs), and are prototypical family A or class 1 GPCRs.[27] The mAChR class of ACh receptors possess five unique subtypes termed M_1–M_5. These subtypes differ in the G proteins to which they couple.[28] M_1, M_3, and M_5 couple to G_q/G_{11} proteins, which lead to the activation of phospholipase C and formation of the second messenger inositol triphosphate and other second messengers. These G_q-coupled mAChR subtypes induce mobilization of intracellular calcium, and typically increase the excitability of neurons through closure of potassium channels and activation of cation channels.[27,28] In contrast, M_2 and M_4 couple to $G_{i/o}$ proteins, and often decrease neuronal excitability and synaptic transmission.[27] The actions of ACh through these distinct classes of mAChRs have powerful, and sometimes opposing, effects on the output of diverse neuronal populations.[29] Because of this, mAChRs have been viewed as promising targets for multiple CNS disorders, and major efforts from both academic labs and pharmaceutical companies are focusing on understanding the roles of mAChRs in CNS disorders as well as the development of novel ligands to target these receptors.[13,26,30]

Interestingly, broad spectrum agonists and antagonists of mAChRs have been used clinically and lend insights into the potential utility of mAChR ligands as pharmaceutical agents. Clinical utility of mAChR agonists has been somewhat limited, but mAChR agonists have proven useful in treatment of glaucoma and some autoimmune disorders.[31] However, the utility of mAChR agonists is severely limited due to the side of effects of these compounds; including salivation, lacrimation, urination, diarrhea, gastrointestinal (GI) distress, and emesis (termed SLUDGE syndrome).[32] These side effects are thought to be primarily mediated by M_2 and M_3 mAChRs in the peripheral nervous system.[13,30] Nonspecific antagonists of mAChRs also have clinical utility in the treatment of a number of peripheral disorders, such as overactive bladder, pulmonary disorders, and GI disturbances.[33] Antagonists of mAChRs also provided the first treatments for Parkinson's Disease and other movement disorders.[34] However, mAChR antagonists also induce adverse effects, notably memory dysfunction, through actions on both peripheral and central mAChRs. Due to this, if individual mAChRs can be linked to specific CNS disorders, highly subtype-selective mAChR ligands may have greater potential as therapeutic agents.

MUSCARINIC ACETYLCHOLINE RECEPTORS CAN REGULATE BRAIN CIRCUITS INVOLVED IN SYMPTOMS OF SCHIZOPHRENIA

Broad-spectrum antagonists of mAChRs offered some of the first evidence for the potential involvement of mAChRs in the pathophysiology of schizophrenia. Administration of mAChR antagonists induces behavioral responses in rodents that have been viewed as correlates of some of the positive and cognitive symptoms observed in schizophrenia patients. These include hyperlocomotion, disrupted prepulse inhibition (PPI), reduced latent inhibition (LI), and multiple cognitive deficits.[35,36] These data suggest that blockade of mAChR signaling is sufficient to disrupt activity of brain circuits that may be involved in the positive and cognitive symptom clusters in schizophrenia patients. Consistent with this, mAChR antagonists induce cognitive deficits and sensory hallucinations in healthy humans,[37,38] and worsen the cognitive and positive symptoms in schizophrenia patients.[4,39] Conversely, increasing cholinergic signaling through blocking the breakdown of ACh through acetylcholinesterase inhibitors improves cognitive function and reduces behaviors associated with positive symptoms in preclinical animal models and in some open-label trials of these medications from other CNS disorders.[40–42] These preclinical and clinical findings suggest that activation or potentiation of mAChR signaling could be beneficial for schizophrenia. However, the clinical utility of broad spectrum agonists is limited due to severe adverse side effects. This necessitates the understanding of the roles of individual mAChR subtypes in modulating circuits that are disrupted in schizophrenia patients.

EVIDENCE FOR ROLES OF SPECIFIC MACHR SUBTYPES IN CIRCUITS THOUGHT TO BE INVOLVED IN SCHIZOPHRENIA

The expression patterns of the individual mAChR subtypes point to a potential role for specific mAChR subtypes in modulating the brain circuits underlying symptom manifestation in schizophrenia. M_1 and M_4 are highly expressed in brain areas thought to be involved in schizophrenia.[43–45] M_1 is expressed throughout the striatum, PFC, and HPC, giving it the potential to modify the underlying cognitive disruptions in schizophrenia.[43,44,46] M_4 is highly expressed in multiple structures throughout the striatum, including on direct pathway spiny projection neurons (SPNs), presynaptically on cortical inputs onto SPNs, and in cholinergic interneurons. This gives the potential for M_4 to have a powerful role in regulating BG function.[13,30,47] The M_2 and M_3 mAChR subtypes are largely limited to the periphery, although both are expressed at low levels in the CNS, including expression of M_2 in striatal cholinergic interneurons.[44,48] Additionally, M_5 can be found on midbrain dopaminergic neurons although relatively little is known about its role on DA neurons other than it potentially has opposing effects on DA release at cell bodies and DA terminals in ex vivo slice electrophysiology experiments.[44,49] Based on the expression patterns of mAChR subtypes, the M_1 and M_4 subtypes are prime candidates as mAChRs that may play important roles in the actions of mAChR agonists and antagonists that could be relevant for schizophrenia.

Outside of the expression patterns of M_1 and M_4, binding studies also suggest that there are changes in expression of mAChRs in schizophrenia. Radioligand binding studies revealed decreased levels of mAChRs in the HPC, BG, PFC, and cingulate cortex in schizophrenia patients,[50–53] and positron emission tomography (PET) imaging studies suggest that M_1 and M_4 are decreased in unmedicated patients.[54] Additionally, imaging studies in patients and controls suggest that patients with the highest negative symptoms showed the lowest binding potentials of M_1 and M_4 PET ligands, suggesting that at least in some patient populations, potentiation of mAChRs could be beneficial.[55]

Genetic studies also point to the M_1 and M_4 receptors as being particularly important in schizophrenia. Analysis of *CHRM4*, encoding M_4, has revealed at least two single nucleotide polymorphisms that increase risk for schizophrenia and are associated with a decreased response to pharmacotherapy.[56,57] Additionally, *CHRM1*, encoding M_1, has been shown to have decreased promoter methylation and mRNA levels suggesting altered transcriptional regulation of *CHRM1* in schizophrenia, and patients with the lowest M_1 correlate with the worse cognitive scores.[55,58] The genes for the other mAChR subtypes, *CHRM2*, *CHRM3*, and *CHRM5*, have no known alterations at the variant, transcriptional, or expression level, again suggesting a unique role for M_1 and M_4 receptors in the pathophysiology of schizophrenia.

Over the last three decades, there has been a concerted effort to develop selective agonists of M_1 and M_4 mAChRs. One of these, xanomeline, an M_1/M_4-preferring agonist, advanced through phase III clinical trials. Xanomeline was first evaluated for potential efficacy in the treatment of cognitive deficits observed in Alzheimer's disease (AD) patient populations. In a large multicenter trial, xanomeline failed to reach its endpoint for providing significant efficacy in treatment of cognitive symptoms in AD patients. However, xanomeline had robust efficacy in reducing behavioral disturbances in AD patients that are similar to positive symptoms observed in schizophrenia (i.e., vocalizations, hallucinations, and delusions).[59–61] This exciting finding in AD patients provided the basis for initiating a small double-blind, placebo-controlled clinical trial in schizophrenia patients. Interestingly, xanomeline produced significant improvement in all three major symptom clusters in schizophrenia patients within the first week after initiating treatment. Further, the efficacy of xanomeline surpassed that of the benchmark antipsychotics.[62] Consistent with this, xanomeline has robust efficacy in animal models that predict antipsychotic and procognitive efficacy.[63–65] Thus, xanomeline reduced amphetamine-induced hyperlocomotion as well as amphetamine or MK-801, an NMDA antagonist, induced disruptions in PPI and LI.[36,63,64,66–68] Additionally xanomeline was found to be procognitive in novel object recognition and could attenuate cortical pathologies in genetic models of schizophrenia.[66,69] Although these results from animal models and clinical trials were very exciting, further clinical development of xanomeline was eventually discontinued due to side effects that were likely due to the effects on peripheral M_2 or M_3 receptors.[59,60] This provides direct evidence in humans that mAChR agonists that are more highly selective at M_1 and M_4 relative to M_2 and M_3 may provide great clinical benefit in schizophrenia patients.

ORTHOSTERIC VERSUS ALLOSTERIC MODULATORS

The GPCRs can be targeted by small molecules that act at a number of sites on the receptor. The orthosteric site is where the neurotransmitter or other endogenous ligands (hormone, peptide, etc.) typically bind. For the mAChRs, this is where ACh would normally bind.[28] Drugs targeting this site can either activate (agonist) or block (antagonist) the activity of the receptor. Unsurprisingly, this site is highly conserved among members of the cholinergic receptors, and this has made designing drugs that target individual mAChR subtypes at the orthosteric site challenging.[13,26]

In addition to acting at orthosteric sites, small molecules can also bind to allosteric sites to modulate activity of the receptor.[70] These sites are physically distinct from the orthosteric site, and a single GPCR can have multiple allosteric sites. Binding of a molecule to allosteric sites can either positively (positive allosteric modulator; PAM) or negatively modulate (negative allosteric modulator; NAM) the response of the receptor to its endogenous ligand. The mechanisms by which PAMs and NAMs exert their effects can be complex and have been extensively reviewed.[71,72] These physically distinct sites from the orthosteric site offer a unique opportunity to develop molecules that can target a single GPCR and not broadly target an entire group of GPCRs (i.e., mAChRs). Recent drug discovery efforts have optimized allosteric modulators for M_1 and M_4 that have unprecedented selectivity for a single mAChR relative to other GPCR subtypes.[47,73] These allosteric molecules have allowed for the interrogation of specific receptors in disease states and basic physiology, and could provide major breakthroughs in the treatment of CNS disorders.

M_1 PAMS AS A POTENTIAL TREATMENT FOR COGNITIVE AND NEGATIVE SYMPTOMS IN SCHIZOPHRENIA

Recent breakthroughs in the discovery of highly selective M_1 PAMs, along with the availability of knockout mice, has provided an unprecedented opportunity to evaluate the functional roles of M_1 in brain circuits that are thought to be disrupted in schizophrenia patients. M_1 may be a critical modulator of circuits controlling memory and attention due to the postsynaptic localization of M_1 on cholinergic projections to the PFC as well as colocalization of M_1 with NMDA glutamate receptors in the HPC.[44] This localization pattern raises the possibility that these receptors could provide a target for the treatment of cognitive deficits in schizophrenia as well as memory disorders such as AD.[12] Recent clinical evidence also points to a critical role for M_1 in cognition. Administration of an mAChR antagonist, biperiden, reduced both visual and verbal learning in healthy controls and patients with psychosis.[74] Animal studies of known memory-related circuits and cellular correlates of memory point to a critical role for M_1 in memory and cognitive function. Potentiation of M_1 in the medial PFC increases synaptic excitation of pyramidal cells and can potentiate CA1 hippocampal pyramidal cell firing.[75,76] Additionally, M_1 knockout mice lack cholinergic-dependent long-term potentiation (LTP) at the CA3 to CA1 synapse in the HPC, and potentiation of M_1 can induce LTD at this synapse through stimulating endocytosis of GluA2-containing AMPA receptors.[76–79] Outside of the PFC and

HPC, M_1 potentiation can also increase the intrinsic excitability of SPNs.[80] Consistent with these electrophysiological findings, behavioral studies of global M_1 knockout mice show deficits in mPFC and HPC-dependent memory tasks such as nonmatch to sample.[78,80,81]

In addition to modulating major hippocampal and cortical glutamatergic pathways, M_1 can also modulate DA signaling and release. Global M_1 knockout mice have elevated levels of extracellular DA in the striatum and increased spontaneous locomotion compared to littermate controls. These animals also have exaggerated locomotor responses to the psychostimulant amphetamine.[82] Taken together, these data suggest that M_1 activation may reduce subcortical DA transmission and related motor behaviors. M_1 receptors are also highly expressed in both direct and indirect pathway striatal SPNs, and M_1 activation can alter firing patterns and synaptic integration, likely through modulation of potassium and calcium channels.[80,83,84] While these data suggest that M_1 activation may reduce DA signaling in the BG, both nonselective agonists and the M_1 muscarinic agonist AC260584 can acutely stimulate DA efflux in the PFC and striatum.[85] In schizophrenia, DA hypofunction in cortical and extrastriatal areas may contribute to cognitive and negative symptom manifestation,[86] and in healthy controls, frontal cortex and striatal DA release is critical for working memory and behavioral flexibility.[87] Therefore, agents that enhance DA transmission, especially in the cortex, may provide treatment for cognitive and negative symptoms of schizophrenia. While the relative importance of M_1 actions in basal ganglia and extrastriatal dopaminergic circuits is not entirely clear, it is possible that actions of M_1 activation on both DA transmission and on glutamatergic projections neurons in cortex and HPC are important for M_1-mediated effects on memory and cognitive function.

Because of the challenges presented by nonselective muscarinic agonists, as well as data suggesting the potential for M_1 selective pharmacological agents to treat schizophrenia; a great deal of effort has been placed on developing M_1-selective PAMs. Recently, there has been remarkable progress in the development of highly selective M_1 PAMs. These agents have provided tools to further understand the ability of M_1 to modulate brain circuits altered by schizophrenia. Studies with an early M_1 PAM, BQCA, showed that administration of M_1-selective PAMs enhance memory function and increase spontaneous PFC brain activity.[75,81,88] Additionally, BQCA attenuates MK-801 induced cognitive deficits in HPC-dependent memory tasks.[89] BQCA can also potentiate the effects of atypical antipsychotics in MK-801-induced disruptions in memory tasks. PQCA, an analogue of BQCA, has robust efficacy in cognition assays of several species.[90] These results suggest that M_1 PAMs may have clinical efficacy for the treatment of schizophrenia, and necessitate further development of M_1 PAMs.

Newer generations of M_1 PAMs, such as VU0453595, PF-06767832, and VU6004256, with increased selectivity and specificity and better pharmacokinetic properties, have led to an increased understanding of the physiological role of M_1 in brain circuits that are relevant for schizophrenia. Acute administration of NMDA receptor antagonists, such as MK-801 and PCP, causes tonic excitation of PFC neurons, deregulates plasticity, and leads to social and cognitive deficits in animals after treatment.[91,92] Furthermore, repeated administration of NMDA receptor antagonists to rodents for 1 week during juvenile development leads to lasting deficits in cortical and cognitive function that persist long after administration of NMDA receptor antagonists has been discontinued and can be observed in adult animals. Interestingly, the M_1 PAM VU0453595 can reverse behavioral and social deficits in young adult mice that received transient repeated administration of PCP during juvenile development.[93] Furthermore, VU0453595 restores an M_1-dependent form of LTD (mLTD) at the hippocampal-PFC synapse that is observed in these animals.[93] This also provides evidence that loss of mLTD at the hippocampal-PFC synapse contributes to the cognitive impairments present in animal models of schizophrenia.[93] Interestingly, this restoration of mLTD may be endocannabinoid (eCB)-dependent. Recent evidence suggests that activation of M_1 triggers release of eCBs, presynaptic CB_1 receptor activation, and depression of glutamatergic transmission at this synapse.[94] Beyond eCBs, loss of mLTD could be related to a dysfunction in the muscarinic regulation of GABAergic transmission in the PFC.[95] However, at present, the mechanism by which M_1 activation induces mLTD or by which M_1 PAMs reverse plasticity and behavioral deficits in this model are not entirely known.

Other second-generation PAMs have also shown efficacy in preclinical models of schizophrenia in both cognitive tasks and correlates of the positive symptom domain. PF-06767832 can attenuate learning and memory deficits induced by the anticholinergic scopolamine in the Morris water maze assay. In preclinical models of antipsychotic efficacy, PF-06767832 could also block amphetamine-induced disruptions in PPI.[73] Another M_1 PAM, VU6004256, was shown to ameliorate cognitive disruption in a genetic model of schizophrenia where the NR1 subunit of NMDA receptor is knocked down.[96] Additionally in models of antipsychotic efficacy in this NR1 model, VU6004256 was found to attenuate increased spontaneous locomotion present in these mice.[96] Together these studies support the idea that M_1 activation may have a critical role in modulating cognitive functions and positive symptoms, dependent on both the PFC and mesocortical DA pathways, and suggest that M_1 allosteric activators may serve as a novel approach for the treatment of cognitive, negative, and positive symptom domains observed in schizophrenic patients. This hypothesis is beginning to be tested in phase 1 and 2 clinical trials, and early results indicate that MK-7622 could attenuate muscarinic antagonist–induced deficits in cognition and enhance quantitative electroencephalography correlated cognitive measures in healthy human volunteers.[97]

Interestingly, the mode of action of M_1 targeting compounds may be of critical importance. As already discussed, allosteric modulators have allowed for the development of molecules that have unprecedented specificity and selectivity in targeting a single GPCR over the orthosteric site. However, some compounds may exhibit mixed modalities and do not function as pure PAMs that only potentiate actions of the endogenous orthosteric agonist, ACh. While prototypical GPCR PAMs only potentiate responses to orthosteric agonists, ago-PAMs have been identified for multiple GPCRs that both potentiate responses and have allosteric agonist activity in the absence of orthosteric agonists.[72] Interestingly, two recently reported M_1 PAMs, PF-06767832 and MK-7622, have robust ago-PAM activity at M_1, induce mLTD in the HPC, and induce robust seizures and behavioral convulsions in mice.[73,98–100] However pure PAMs such as VU0453595 and VU6004256 do not display many of these properties.[98–100] These pure M_1 PAMs do not possess intrinsic agonist activity but only potentiate responses to orthosteric agonists and do not have a seizure liability.[50,98,100] Taken together, these findings highlight the potential utility of M_1 PAMs in reducing cognitive and possibly negative and positive symptoms in schizophrenia patients, but also highlight the importance of optimizing pure PAM to avoid unwanted liabilities.

M_4 PAMS MAY PROVIDE A NOVEL APPROACH FOR TREATMENT OF PSYCHOSIS

Like M_1, the development of M_4 knockout mice, as well as truly selective pharmacological tools has significantly expanded our understanding of M_4 in brain circuits that are involved in schizophrenia. M_4 receptors are abundantly expressed on several structures throughout the striatum including presynaptically on cortico-striatal projections as well as direct pathway SPNs, suggesting that M_4 could be a key regulator of striatal function.[43] In addition, M_4 is highly expressed in PFC, HPC, and other forebrain regions where it could regulate multiple circuits important for different symptom domains in schizophrenia.[44] Studies using nonselective pharmacological agents combined with genetic deletion of M_4, have suggested that this receptor may play a unique role in modulating brain circuits relevant to schizophrenia. In preclinical studies nonselective mAChR agonists, such as xanomeline, exert multiple antipsychotic-like effects in animal models, and are absent or greatly reduced in M_4 knockout mice.[63,64,101] This suggests that M_4 plays a dominant role in the antipsychotic-like effects of mAChR agonists. Interestingly, the effects of atypical antipsychotics also are reduced in M_4 knockout mice.[101] In addition to the absence of effect in global M_4 knockout mice, the antipsychotic-like effects of xanomeline are largely absent in mice in which M_4 is selectively deleted from striatal SPNs that express the D_1 subtype of DA receptor (D_1-SPNs) which constitute the direct pathway of the basal ganglia.[45,63] These data suggest a major role for M_4, and especially M_4 expressed in D_1-SPNs in mediating the antipsychotic effects of xanomeline.

Discovery of the first generation of M_4-selective PAMs, namely VU0010010 and LY2033298, provided the first compounds that are highly selective for M_4 relative to other mAChR subtypes and allowed studies that increase our understanding of brain circuits and behaviors modulated by M_4.[102–104] In ex vivo electrophysiology studies, VU0010010 selectively potentiated reductions in excitatory, but not inhibitory, synapses in hippocampal neurons, indicating a key role for M_4 in regulating hippocampal function.[103] LY2033298 is a relatively weak PAM for rodent M_4, but possessed better pharmacokinetic and pharmacodynamic properties suitable for in vivo dosing. This compound potentiated the behavioral effects of the nonselective mAChR agonist oxotremorine in reversing deficits in PPI and in decreasing DA levels in the striatum, and these effects were absent in M_4 knockout mice.[104,105] These early studies with first-generation M_4 PAMs provided evidence that selective potentiation of M_4 signaling may provide a novel approach in the modulation of brain circuits and behaviors related to schizophrenia.

Subsequent generations of M_4 selective PAMs, such as VU0467154, have overcome initial challenges with potency, selectivity, and pharmacokinetic properties of earlier compounds, and have allowed for greater interrogation of M_4 in behavioral assays.[47] These compounds have robust efficacy in multiple rodent models that are similar to those of atypical antipsychotics, including reversal of amphetamine and MK-801-induced disruptions in PPI and hyperlocomotion.[47,102,106,107] M_4 PAMs also reverse deficits in genetic models of schizophrenia, such as NR1 knockdown mice.[108] In accordance with the observed antipsychotic-like effects, M_4 PAMs attenuate amphetamine-induced increases in extracellular DA in the striatum and nucleus accumbens.[106] Functional magnetic resonance imaging studies in rodents also give anatomical and functional evidence to support the notion that M_4 PAMs have antipsychotic efficacy. Administration of the M_4 PAMs VU0467154 or VU1052100 block amphetamine D_1-agonist induced changes in cerebral blood volume, an indirect correlate of neuronal activity, in several brain regions including the striatum.[106,109] These data give robust preclinical evidence to suggest antipsychotic effects of M_4 PAMs.

The mechanisms underlying the efficacy of M_4 PAMs in brain circuits that are relevant for schizophrenia are likely complex. M_4 is expressed in several structures in the striatum including on direct pathway SPNs, cortical inputs into the striatum, as well as on cholinergic interneurons.[44] M_4 potentiation can directly inhibit DA-mediated responses in direct pathway SPN cell bodies,[110] SPN terminals in the substantia nigra pars reticulata,[109] and inhibits excitatory transmission via induction of LTD in direct pathway SPNs.[110,111] Additionally, even though M_4 is not expressed on midbrain DA neurons or their terminals in the striatum,[44] M_4 potentiation induces a sustained inhibition of DA release as measured through ex vivo fast scanning cyclic voltammetry studies.[107] This inhibition is dependent on the release of an eCB from direct pathway SPNs acting on CB_2 receptors on DA terminals.[107] Interestingly, typical antipsychotics exert their effects by inhibiting postsynaptic DA receptors, and these mechanistic studies suggest that M_4 PAMs may also act by reducing dopaminergic signaling. However, unlike DA receptor antagonists, the mechanism by which M_4 acts would allow selective inhibition of DA release in the striatum. This is interesting in light of recent imaging studies suggesting that excessive activity of dopaminergic projections to the dorsal striatum are especially important for schizophrenia and that there is a hypofunction of extrastriatal dopaminergic pathways in schizophrenia patients.[86] Based on these combined clinical and preclinical studies, it is possible that M_4 PAMs allow highly selective inhibition of DA signaling in striatal circuits that are important for schizophrenia, while sparing extrastriatal DA pathways that may contribute to the adverse effects of currently available antipsychotics.

In contrast to DA antagonists, recent data suggest that M_4 PAMs may also have pro-cognitive effects. Recent reports suggest that M_4 PAMs reverse amphetamine or MK-801-induced disruptions in cognitive tasks and enhance some aspects of cognitive function in wild type animals.[47,96,112] VU0467154 reverses deficits in fear conditioning induced by amphetamine or MK-801 as well as pairwise discrimination tasks.[47] In the HPC and at cortico-striatal synapses, M_4 can decrease glutamate release, suggesting that M_4 may be able to normalize aberrant glutamatergic transmission and firing patterns in these preclinical models of schizophrenia.[75,103,111] Additionally, administration of an M_4 PAM can increase acquisition time of pairwise discrimination tasks and conditioned freezing paradigms compared to vehicle-treated control animals.[112] Based on these exciting new data, it is possible that M_4 PAMs could exert robust antipsychotic efficacy and also improve cognitive function in schizophrenia patients.

CONCLUSION

mAChRs have long been viewed as potential targets for the treatment of schizophrenia. Recent advances that have produced several novel small molecule scaffolds with incredible selectivity and specificity for individual mAChR subtypes. Use of these compounds in combination with genetic, behavioral, imaging, and electrophysiological techniques have highlighted that M_1 and M_4 mAChRs have immense potential to treat the positive, negative, and cognitive symptoms of schizophrenia and may potentially drive some of the pathological changes in schizophrenia. Pure PAMs of the M_1 and M_4 mAChR may provide the selectivity and specificity necessary to achieve the potential clinical benefit of these agents while avoiding the unwanted side effects presented by activation of M_2 and M_3 mAChRs. Highly selective M_1 and M_4 PAMs are rapidly advancing for testing in schizophrenia patients, and the first M_1 PAMs are now in early stages of clinical development. Ultimately, advancing small-molecule PAMs of the M_1 and M_4 mAChR into clinical evaluation in schizophrenia patients will allow us to directly test the hypothesis that these agents will provide effective new treatments for this disorder.

REFERENCES

1. McGrath J, Saha S, Chant D, Welham J. Schizophrenia: a concise overview of incidence, prevalence, and mortality. *Epidemiologic Reviews.* 2008;30(1):67–76.
2. Patel KR, Cherian J, Gohil K, Atkinson D. Schizophrenia: overview and treatment options. *Pharmacy and Therapeutics.* 2014;39(9):638–645.
3. Makinen J, Miettunen J, Isohanni M, Koponen H. Negative symptoms in schizophrenia: a review. *Nordic Journal of Psychiatry.* 2008;62(5):334–341.
4. Bowie CR, Harvey PD. Cognitive deficits and functional outcome in schizophrenia. *Neuropsychiatric Disease and Treatment.* 2006;2(4):531–536.
5. Goff DC, Hill M, Barch D. The treatment of cognitive impairment in schizophrenia. *Pharmacology, Biochemistry, and Behavior.* 2011;99(2):245–253.
6. Leucht S, Corves C, Arbter D, Engel RR, Li C, Davis JM. Second-generation versus first-generation antipsychotic drugs for schizophrenia: a meta-analysis. *Lancet.* 2009;373(9657):31–41.
7. Green MF. What are the functional consequences of neurocognitive deficits in schizophrenia? *American Journal of Psychiatry.* 1996;153(3):321–330.
8. Greenwood KE, Landau S, Wykes T. Negative symptoms and specific cognitive impairments as combined targets for improved functional outcome within cognitive remediation therapy. *Schizophr Bull.* 2005;31(4):910–921.
9. Remington G, Foussias G, Fervaha G, et al. Treating negative symptoms in schizophrenia: an update. *Current Treatment Options in Psychiatry.* 2016;3:133–150.
10. Buchanan RW. Persistent negative symptoms in schizophrenia: an overview. *Schizophrenia Bulletin.* 2007;33(4):1013–1022.
11. Parsons B, Allison DB, Loebel A, et al. Weight effects associated with antipsychotics: a comprehensive database analysis. *Schizophrenia Research.* 2009;110(1-3):103–110.
12. Jones CK, Byun N, Bubser M. Muscarinic and nicotinic acetylcholine receptor agonists and allosteric modulators for the treatment of schizophrenia. *Neuropsychopharm.* 2012;37(1):16–42.
13. Conn PJ, Jones CK, Lindsley CW. Subtype-selective allosteric modulators of muscarinic receptors for the treatment of CNS disorders. *Trends in Pharmacological Sciences.* 2009;30(3):148–155.
14. Guillin O, Abi-Dargham A, Laruelle M. Neurobiology of dopamine in schizophrenia. *International Review of Neurobiology.* 2007;78:1–39.
15. Javitt DC. Glutamatergic theories of schizophrenia. *Israel Journal of Psychiatry and Related Sciences.* 2010;47(1):4–16.
16. Marsman A, van den Heuvel MP, Klomp DW, Kahn RS, Luijten PR, Hulshoff Pol HE. Glutamate in schizophrenia: a focused review and meta-analysis of (1)H-MRS studies. *Schizophr Bull.* 2013;39(1):120–129.
17. Wassef A, Baker J, Kochan LD. GABA and schizophrenia: a review of basic science and clinical studies. *Journal of Clinical Psychopharmacology.* 2003;23(6):601–640.
18. Taylor SF, Tso IF. GABA abnormalities in schizophrenia: a methodological review of in vivo studies. *Schizophrenia Research.* 2015;167(1–3):84–90.
19. Kruse AC, Kobilka BK, Gautam D, Sexton PM, Christopoulos A, Wess J. Muscarinic acetylcholine receptors: novel opportunities for drug development. *Nat Rev Drug Discov.* 2014;13(7):549–560.
20. Beninato M, Spencer RF. A cholinergic projection to the rat substantia nigra from the pedunculopontine tegmental nucleus. *Brain Research.* 1987;412(1):169–174.
21. Dautan D, Huerta-Ocampo I, Witten IB, et al. A major external source of cholinergic innervation of the striatum and nucleus accumbens originates in the brainstem. *Journal of Neuroscience.* 2014;34(13):4509–4518.
22. Xiao C, Cho Jounhong R, Zhou C, et al. Cholinergic mesopontine signals govern locomotion and reward through dissociable midbrain pathways. *Neuron.* 2015;90(2):333–347.
23. Heimer L. Basal forebrain in the context of schizophrenia. *Brain Research. Brain Research Reviews.* 2000;31(2–3):205–235.
24. Bonsi P, Cuomo D, Martella G, et al. Centrality of striatal cholinergic transmission in basal ganglia function. *Frontiers in Neuroanatomy.* 2011;5:6.
25. Pisani A, Bernardi G, Ding J, Surmeier DJ. Re-emergence of striatal cholinergic interneurons in movement disorders. *Trends in Neurosciences.* 2007;30(10):545–553.
26. Conn PJ, Christopoulos A, Lindsley CW. Allosteric modulators of GPCRs: a novel approach for the treatment of CNS disorders. *Nat Rev Drug Discov.* 2009;8(1):41–54.

27. Caulfield MP, Birdsall NJ. International Union of Pharmacology. XVII. Classification of muscarinic acetylcholine receptors. *Pharmacological Reviews.* 1998;50(2):279–290.
28. Ishii M, Kurachi Y. Muscarinic acetylcholine receptors. *Current Pharmaceutical Design.* 2006;12(28):3573–3581.
29. Xiang Z, Thompson AD, Jones CK, Lindsley CW, Conn PJ. Roles of the M1 muscarinic acetylcholine receptor subtype in the regulation of basal ganglia function and implications for the treatment of Parkinson's disease. *Journal of Pharmacology and Experimental Therapeutics.* 2012;340(3):595–603.
30. Foster DJ, Conn PJ. Allosteric modulation of GPCRs: new insights and potential utility for treatment of schizophrenia and other CNS disorders. *Neuron.*2017;94(3):431–446.
31. Gil D, Spalding T, Kharlamb A, et al. Exploring the potential for subtype-selective muscarinic agonists in glaucoma. *Life Sciences.* 2001;68(22):2601–2604.
32. Peters NL. Snipping the thread of life: ANTIMUSCARINIC side effects of medications in the elderly. *Archives of Internal Medicine.* 1989;149(11):2414–2420.
33. Chapple CR. Muscarinic receptor antagonists in the treatment of overactive bladder. *Urology.* 2000;55(5):33–46.
34. Vernier VG, Unna KR. The experimental evaluation of antiparkinsonian compounds. *Annals of the New York Academy of Sciences.* 1956;64(4):690–704.
35. Barak S, Weiner I. Scopolamine induces disruption of latent inhibition which is prevented by antipsychotic drugs and an acetylcholinesterase inhibitor. *Neuropsychopharmacology.* 2007;32(5):989–999.
36. Shannon HE, Hart JC, Bymaster FP, et al. Muscarinic receptor agonists, like dopamine receptor antagonist antipsychotics, inhibit conditioned avoidance response in rats. *J Pharmacol Exp Ther.* 1999;290(2):901–907.
37. MCEVOY JP. The clinical use of anticholinergic drugs as treatment for extrapyramidal side effects of neuroleptic drugs. *Journal of Clinical Psychopharmacology.* 1983;3(5):288–302.
38. Perry EK, Perry RH. Acetylcholine and hallucinations: disease-related compared to drug-induced alterations in human consciousness. *Brain and Cognition.* 995;28(3):240–258.
39. Buchanan RW, Freedman R, Javitt DC, Abi-Dargham A, Lieberman JA. Recent advances in the development of novel pharmacological agents for the treatment of cognitive impairments in schizophrenia. *Schizophrenia Bulletin.* 2007;33(5):1120–1130.
40. Ferreri F, Agbokou C, Gauthier S. Cognitive dysfunctions in schizophrenia: potential benefits of cholinesterase inhibitor adjunctive therapy. *Journal of Psychiatry and Neuroscience.* 2006;31(6):369–376.
41. Ribeiz SR, Bassitt DP, Arrais JA, Avila R, Steffens DC, Bottino CM. Cholinesterase inhibitors as adjunctive therapy in patients with schizophrenia and schizoaffective disorder: a review and meta-analysis of the literature. *CNS drugs.* 2010;24(4):303–317.
42. Friedman JI, Adler DN, Howanitz E, et al. A double blind placebo controlled trial of donepezil adjunctive treatment to risperidone for the cognitive impairment of schizophrenia. *Biol Psychiatry.* 2002;51(5):349–357.
43. Hersch SM, Gutekunst CA, Rees HD, Heilman CJ, Levey AI. Distribution of M1–M4 muscarinic receptor proteins in the rat striatum: light and electron microscopic immunocytochemistry using subtype-specific antibodies. *Journal of Neuroscience.* 1994;14(5 Pt 2):3351–3363.
44. Levey AI, Kitt CA, Simonds WF, Price DL, Brann MR. Identification and localization of muscarinic acetylcholine receptor proteins in brain with subtype-specific antibodies. *Journal of Neuroscience.* 1991;11(10):3218–3226.
45. Jeon J, Dencker D, Wortwein G, et al. A subpopulation of neuronal M(4) muscarinic acetylcholine receptors plays a critical role in modulating dopamine-dependent behaviors. *Journal of Neuroscience.* 2010;30(6):2396–2405.
46. Lebois EP, Bridges TM, Lewis LM, et al. Discovery and characterization of novel subtype-selective allosteric agonists for the investigation of M1 receptor function in the central nervous system. *ACS Chemical Neuroscience.* 2010;1(2):104–121.
47. Bubser M, Bridges TM, Dencker D, et al. Selective activation of M4 muscarinic acetylcholine receptors reverses MK-801-induced behavioral impairments and enhances associative learning in rodents. *ACS Chemical Neuroscience.* 2014;5(10):920–942.
48. Takeuchi T, Fujinami K, Goto H, et al. Roles of M2 and M4 muscarinic receptors in regulating acetylcholine release from myenteric neurons of mouse ileum. *Journal of Neurophysiology.* 2005;93(5):2841–2848.
49. Foster DJ, Gentry PR, Lizardi-Ortiz JE, et al. M5 receptor activation produces opposing physiological outcomes in dopamine neurons depending on the receptor's location. *he Journal of Neuroscience.* 2014;34(9):3253–3262.
50. Crook JM, Tomaskovic-Crook E, Copolov DL, Dean B. Low muscarinic receptor binding in prefrontal cortex from subjects with schizophrenia: a study of Brodmann's areas 8, 9, 10, and 46 and the effects of neuroleptic drug treatment. *American Journal of Psychiatry.* 2001;158(6):918–925.
51. Crook JM, Tomaskovic-Crook E, Copolov DL, Dean B. Decreased muscarinic receptor binding in subjects with schizophrenia: a study of the human hippocampal formation. *Biol Psychiatry.* 2000;48(5):381–388.
52. Dean B, McLeod M, Keriakous D, McKenzie J, Scarr E. Decreased muscarinic1 receptors in the dorsolateral prefrontal cortex of subjects with schizophrenia. *Mol Psychiatry.* 2002;7(10):1083–1091.
53. Dean B, Soulby A, Evin GM, Scarr E. Levels of [(3)H]pirenzepine binding in Brodmann's area 6 from subjects with schizophrenia is not associated with changes in the transcription factor SP1 or BACE1. *Schizophrenia Research.* 2008;106(2–3):229–236.
54. Raedler TJ, Knable MB, Jones DW, et al. In vivo determination of muscarinic acetylcholine receptor availability in schizophrenia. *American Journal of Psychiatry.* 2003;160(1):118–127.
55. Scarr E, Cowie TF, Kanellakis S, Sundram S, Pantelis C, Dean B. Decreased cortical muscarinic receptors define a subgroup of subjects with schizophrenia. *Mol Psychiatry.* 2009;14(11):1017–1023.
56. Scarr E, Sundram S, Keriakous D, Dean B. Altered hippocampal muscarinic M4, but Not M1, receptor expression from subjects with schizophrenia. *Biological Psychiatry.* 2007;61(10):1161–1170.
57. Scarr E, Um JY, Cowie TF, Dean B. Cholinergic muscarinic M4 receptor gene polymorphisms: a potential risk factor and pharmacogenomic marker for schizophrenia. *Schizophrenia Research.* 2013;146(1–3):279–284.
58. Scarr E, Craig JM, Cairns MJ, et al. Decreased cortical muscarinic M1 receptors in schizophrenia are associated with changes in gene promoter methylation, mRNA and gene targeting microRNA. *Translational Psychiatry.* 2013;3:e230.
59. Bodick NC, Offen WW, Levey AI, et al. Effects of xanomeline, a selective muscarinic receptor agonist, on cognitive function and behavioral symptoms in Alzheimer disease. *Archives of Neurology.* 1997;54(4):465–473.
60. Bodick NC, Offen WW, Shannon HE, et al. The selective muscarinic agonist xanomeline improves both the cognitive deficits and behavioral symptoms of Alzheimer disease. *Alzheimer Disease and Associated Disorders.* 1997;11(Suppl 4):S16–22.
61. Veroff AE, Bodick NC, Offen WW, Sramek JJ, Cutler NR. Efficacy of xanomeline in Alzheimer disease: cognitive improvement measured using the Computerized Neuropsychological Test Battery (CNTB). *Alzheimer Disease and Associated Disorders.* 1998;12(4):304–312.
62. Shekhar A, Potter WZ, Lightfoot J, et al. Selective muscarinic receptor agonist xanomeline as a novel treatment approach for schizophrenia. *American Journal of Psychiatry.* 2008;165(8):1033–1039.
63. Dencker D, Wortwein G, Weikop P, et al. Involvement of a subpopulation of neuronal M4 muscarinic acetylcholine receptors in the antipsychotic-like effects of the M1/M4 preferring muscarinic receptor agonist xanomeline. *Journal of Neuroscience.* 2011;31(16):5905–5908.
64. Shannon HE, Rasmussen K, Bymaster FP, et al. Xanomeline, an M1/M4 preferring muscarinic cholinergic receptor agonist, produces antipsychotic-like activity in rats and mice. *Schizophrenia Research.* 2000;42(3):249–259.

65. Woolley ML, Carter HJ, Gartlon JE, Watson JM, Dawson LA. Attenuation of amphetamine-induced activity by the non-selective muscarinic receptor agonist, xanomeline, is absent in muscarinic M4 receptor knockout mice and attenuated in muscarinic M1 receptor knockout mice. *European Journal of Pharmacology.* 2009;603(1-3):147–149.
66. Barak S, Weiner I. The M(1)/M(4) preferring agonist xanomeline reverses amphetamine-, MK801- and scopolamine-induced abnormalities of latent inhibition: putative efficacy against positive, negative and cognitive symptoms in schizophrenia. *International Journal of Neuropsychopharmacology.* 2011;14(9):1233–1246.
67. Perry KW, Nisenbaum LK, George CA, Shannon HE, Felder CC, Bymaster FP. The muscarinic agonist xanomeline increases monoamine release and immediate early gene expression in the rat prefrontal cortex. *Biol Psychiatry.* 2001;49(8):716–725.
68. Stanhope KJ, Mirza NR, Bickerdike MJ, et al. The muscarinic receptor agonist xanomeline has an antipsychotic-like profile in the rat. *Journal of Pharmacology and Experimental Therapeutics.* 2001;299(2):782–792.
69. Brown JW, Rueter LE, Zhang M. Predictive validity of a MK-801-induced cognitive impairment model in mice: implications on the potential limitations and challenges of modeling cognitive impairment associated with schizophrenia preclinically. *Progress in Neuro-Psychopharmacology and Biological Psychiatry.* 2014;49:53–62.
70. May LT, Leach K, Sexton PM, Christopoulos A. Allosteric modulation of G protein–coupled receptors. *Annual Review of Pharmacology and Toxicology.* 2007;47(1):1–51.
71. Gregory KJ, Sexton PM, Christopoulos A. Overview of receptor allosterism. *Current Protocols in Pharmacology.* 2010;Chapter 1:Unit 1.21.
72. Conn PJ, Lindsley CW, Meiler J, Niswender CM. Opportunities and challenges in the discovery of allosteric modulators of GPCRs for treating CNS disorders. *Nature reviews. Drug discovery.* 2014;13(9):692–708.
73. Davoren JE, Lee CW, Garnsey M, et al. Discovery of the potent and selective M1 PAM-agonist N-[(3R,4S)-3-hydroxytetrahydro-2H-pyran-4-yl]-5-methyl-4-[4-(1,3-thiazol-4-yl)ben zyl]pyridine-2-carboxamide (PF-06767832): evaluation of efficacy and cholinergic side effects. *J Med Chem.* 2016;59(13):6313–6328.
74. Vingerhoets C, Bakker G, van Dijk J, et al. The effect of the muscarinic M_1 receptor antagonist biperiden on cognition in medication free subjects with psychosis. *European Neuropsychopharmacology.* 27(9):854–864.
75. Shirey JK, Brady AE, Jones PJ, et al. A selective allosteric potentiator of the M1 muscarinic acetylcholine receptor increases activity of medial prefrontal cortical neurons and restores impairments in reversal learning. *Journal of Neuroscience.* 2009;29(45):14271–14286.
76. Buchanan KA, Petrovic MM, Chamberlain SE, Marrion NV, Mellor JR. Facilitation of long-term potentiation by muscarinic M(1) receptors is mediated by inhibition of SK channels. *Neuron.* 2010;68(5):948–963.
77. Hamilton SE, Nathanson NM. The M1 receptor is required for muscarinic activation of mitogen-activated protein (MAP) kinase in murine cerebral cortical neurons. *Journal of Biological Chemistry.* 2001;276(19):15850–15853.
78. Anagnostaras SG, Murphy GG, Hamilton SE, et al. Selective cognitive dysfunction in acetylcholine M1 muscarinic receptor mutant mice. *Nature Neuroscience.* 2003;6(1):51–58.
79. Xiong C-H, Liu M-G, Zhao L-X, et al. M1 muscarinic receptors facilitate hippocampus-dependent cognitive flexibility via modulating GluA2 subunit of AMPA receptors. *Neuropharmacology.* 2019;146:242–251.
80. Lv X, Dickerson JW, Rook JM, Lindsley CW, Conn PJ, Xiang Z. M1 muscarinic activation induces long-lasting increase in intrinsic excitability of striatal projection neurons. *Neuropharmacology.* 2017;118:209–222.
81. Gould RW, Dencker D, Grannan M, et al. Role for the M1 muscarinic acetylcholine receptor in top-down cognitive processing using a touchscreen visual discrimination task in mice. *ACS Chem Neurosci.* 2015;6(10):1683–1695.
82. Gerber DJ, Sotnikova TD, Gainetdinov RR, Huang SY, Caron MG, Tonegawa S. Hyperactivity, elevated dopaminergic transmission, and response to amphetamine in M1 muscarinic acetylcholine receptor-deficient mice. *Proc Natl Acad Sci U S A.* 2001;98(26):15312–15317.
83. Jeong J-Y, Kweon H-J, Suh B-C. Dual regulation of R-type Ca(V)2.3 channels by M(1) muscarinic receptors. *Molecules and Cells.* 2016;39(4):322–329.
84. Ben-Ari Y, Aniksztejn L, Bregestovski P. Protein kinase C modulation of NMDA currents: an important link for LTP induction. *Trends in Neurosciences.* 1992;15(9):333–339.
85. Li Z, Bonhaus DW, Huang M, Prus AJ, Dai J, Meltzer HY. AC260584 (4-[3-(4-butylpiperidin-1-yl)-propyl]-7-fluoro-4H-benzo[1,4]oxazin-3-one), a selective muscarinic M1 receptor agonist, increases acetylcholine and dopamine release in rat medial prefrontal cortex and hippocampus. *European Journal of Pharmacology.* 2007;572(2-3):129–137.
86. Weinstein JJ, Chohan MO, Slifstein M, Kegeles LS, Moore H, Abi-Dargham A. Pathway-specific dopamine abnormalities in schizophrenia. *Biol Psychiatry.* 2017;81(1):31–42.
87. Cools R, D'Esposito M. Inverted-U shaped dopamine actions on human working memory and cognitive control. *Biological Psychiatry.* 2011;69(12):e113–e125.
88. Ma L, Seager MA, Wittmann M, et al. Selective activation of the M1 muscarinic acetylcholine receptor achieved by allosteric potentiation. *Proceedings of the National Academy of Sciences.* 2009;106(37):15950–15955.
89. Choy KH, Shackleford DM, Malone DT, et al. Positive allosteric modulation of the muscarinic M1 receptor improves efficacy of antipsychotics in mouse glutamatergic deficit models of behavior. *J Pharmacol Exp Ther.* 2016;359(2):354–365.
90. Lange HS, Cannon CE, Drott JT, Kuduk SD, Uslaner JM. The M1 muscarinic positive allosteric modulator PQCA improves performance on translatable tests of memory and attention in rhesus monkeys. *J Pharmacol Exp Ther.* 2015;355(3):442–450.
91. Homayoun H, Moghaddam B. NMDA receptor hypofunction produces opposite effects on prefrontal cortex interneurons and pyramidal neurons. *he Journal of Neuroscience.* 2007;27(43): 11496–11500.
92. Ninan I, Jardemark KE, Wang RY. Differential effects of atypical and typical antipsychotic drugs on N-methyl-D-aspartate- and electrically evoked responses in the pyramidal cells of the rat medial prefrontal cortex. *Synapse.* 2003;48(2):66–79.
93. Ghoshal A, Rook JM, Dickerson JW, et al. Potentiation of M1 muscarinic receptor reverses plasticity deficits and negative and cognitive symptoms in a schizophrenia mouse model. *Neuropsychopharmacology.* 2016;41(2):598–610.
94. Martin HGS, Bernabeu A, Lassalle O, et al. Endocannabinoids mediate muscarinic acetylcholine receptor-dependent long-term depression in the adult medial prefrontal cortex. *Frontiers in Cellular Neuroscience.* 2015;9:457.
95. Ghoshal A, Moran SP, Dickerson JW, et al. Role of mGlu5 receptors and inhibitory neurotransmission in M1 dependent muscarinic LTD in the prefrontal cortex: Implications in Schizophrenia. *ACS Chemical Neuroscience.* 2017;8(10):2254–2265.
96. Grannan MD, Mielnik CA, Moran SP, et al. Prefrontal cortex-mediated impairments in a genetic model of NMDA receptor hypofunction are reversed by the novel M1 PAM VU6004256. *ACS Chem Neurosci.* 2016;7(12):1706–1716.
97. Uslaner JM, Kuduk SD, Wittmann M, et al. Preclinical to human translational pharmacology of the novel M_1 positive allosteric modulator MK-7622. *Journal of Pharmacology and Experimental Therapeutics.* 2018;365(3):556–566.
98. Moran SP, Doyle CA, Xiang Z, et al. M1 Allosteric modulators without agonist activity in the medial prefrontal cortex may provide the optimal profile for cognition enhancement. *Neuroscience* 2017; 2017; Washington, D.C.

99. Rook JM, Abe M, Cho HP, et al. Diverse effects on M1 signaling and adverse effect liability within a series of M1 ago-PAMs. 2017;8(4):866–883.
100. Moran SP, Dickerson JW, Cho HP, Xiang Z. M1-positive allosteric modulators lacking agonist activity provide the optimal profile for enhancing cognition. 2018;43(8):1763–1771.
101. Watt ML, Rorick-Kehn L, Shaw DB, et al. The muscarinic acetylcholine receptor agonist BuTAC mediates antipsychotic-like effects via the M4 subtype. *Neuropsychopharmacology.* 2013;38(13):2717–2726.
102. Brady AE, Jones CK, Bridges TM, et al. Centrally active allosteric potentiators of the M(4) muscarinic acetylcholine receptor reverse amphetamine-induced hyperlocomotor activity in rats. *Journal of Pharmacology and Experimental Therapeutics.* 2008;327(3):941–953.
103. Shirey JK, Xiang Z, Orton D, et al. An allosteric potentiator of M4 mAChR modulates hippocampal synaptic transmission. *Nature Chemical Biology.* 2008;4(1):42–50.
104. Chan WY, McKinzie DL, Bose S, et al. Allosteric modulation of the muscarinic M4 receptor as an approach to treating schizophrenia. *Proceedings of the National Academy of Sciences.* 2008;105(31):10978–10983.
105. Leach K, Loiacono RE, Felder CC, et al. Molecular mechanisms of action and in vivo validation of an M4 muscarinic acetylcholine receptor allosteric modulator with potential antipsychotic properties. *Neuropsychopharmacology.* 2010;35(4):855–869.
106. Byun NE, Grannan M, Bubser M, et al. Antipsychotic drug-like effects of the selective M4 muscarinic acetylcholine receptor positive allosteric modulator VU0152100. *Neuropsychopharmacology.* 2014;39(7):1578–1593.
107. Foster DJ, Wilson JM, Remke DH, et al. Antipsychotic-like effects of M4 positive allosteric modulators are mediated by CB2 receptor-dependent inhibition of dopamine release. *Neuron.* 2016;91(6):1244–1252.
108. Grannan M, Bubser M, Bridges T, et al. Effects of the M4 muscarinic receptor positive allosteric modulator VU0467154 on cognition and pyramidal cell firing properties in layer V of the mPFC (845.9). *FASEB Journal.* 2014;28(1 Supplement).
109. Moehle, MS, Pancani, T, Byun, N, Yohn, SE, Wilson, GH, et al. Cholinergic projections to the substantia nigra pars reticulata inhibit dopamine modulation of basal ganglia through the M4 muscarinic receptor. *Neuron.* 2017;96(6):1358–1372.e4.
110. Shen W, Plotkin JL, Francardo V, et al. M4 muscarinic receptor signaling ameliorates striatal plasticity deficits in models of L-DOPA-induced dyskinesia. *Neuron.* 2015;88(4):762–773.
111. Pancani T, Foster DJ, Moehle MS, et al. Allosteric activation of M4 muscarinic receptors improve behavioral and physiological alterations in early symptomatic YAC128 mice. *Proceedings of the National Academy of Sciences.* 2015;112(45):14078–14083.
112. Gould RW, Grannan MD, Gunter BW, et al. Cognitive enhancement and antipsychotic-like activity following repeated dosing with the selective M4 PAM VU0467154. *Neuropharmacology.* 2017.

36

KYNURENIC ACID IN BRAIN FUNCTION AND DYSFUNCTION

FOCUS ON THE PATHOPHYSIOLOGY AND TREATMENT OF SCHIZOPHRENIA

Robert Schwarcz and Sophie Erhardt

THE KYNURENINE PATHWAY OF TRYPTOPHAN DEGRADATION IN THE MAMMALIAN BRAIN

In mammals, the essential amino acid tryptophan is degraded primarily by the kynurenine pathway (KP), a cascade of enzymatic steps leading to the generation of several biologically active compounds. KP metabolism is initiated by the oxidative ring opening of tryptophan by two distinct enzymes, indoleamine-2,3-dioxygenase (IDO), which exists in two isoforms (IDO1 and IDO2) and tryptophan-2,3-dioxygenase (TDO2). After downstream formation of the pivotal metabolite—and pathway namesake—kynurenine, the KP segregates into various branches. The main arm of the pathway, involving several intermediates, leads to NAD^+, and several side chains terminate in acidic metabolites, which cannot be degraded further enzymatically (Figure 36.1).

KP metabolites are collectively termed "kynurenines" and have been shown to be involved in many diverse physiological and pathological processes.[1] However, although a century has passed since kynurenines were first recognized as major catabolic products of tryptophan, they have generally been viewed as the "poor relatives" of the prominent tryptophan metabolites serotonin and melatonin. Very little attention was paid to their possible involvement in biological processes in the mammalian brain until the 1980s. Discovery of the neuroexcitatory and excitotoxic properties of quinolinic acid (QUIN), a selective but rather weak N-methyl-D-aspartate receptor (NMDAR) agonist, first suggested that endogenous kynurenines may participate in brain function—possibly as modulators of glutamatergic neurotransmission. This idea soon gained further momentum by the realization that another KP metabolite, kynurenic acid (KYNA), is a broad-spectrum, competitive antagonist of ionotropic excitatory amino acid receptors.[2] During the following years, identification of several kynurenines and KP enzymes in the mammalian brain, and characterization of the mechanisms that regulate the formation and disposition of KP metabolites, provided increasing evidence that kynurenines may play distinct roles in neuronal functions.[3] However, even though neurons do contain KP enzymes,[4] kynurenines were found to be preferentially produced in nonneuronal cells and were therefore not considered to be classic neurotransmitters. Moreover, it turned out that the two main KP branches may be physically segregated in the brain. Thus, astrocytes, which contain kynurenine aminotransferases (KATs), readily produce KYNA from kynurenine but lack kynurenine 3-monooxygenase (KMO) and therefore cannot convert kynurenine to 3-hydroxykynurenine (3-HK) (Figure 36.1). In contrast, the de novo synthesis of 3-HK, a pro-excitotoxic KP metabolite that is closely linked to redox processes, and of its numerous downstream metabolites is believed to take place preferentially in microglial cells.[5]

KMO is a major gatekeeper of the KP. Because of its low activity in the brain and relatively low K_m for kynurenine (~20 μM), this enzyme is more rapidly saturated by rising brain kynurenine concentrations than KATs (K_m values: ~1 mM). It follows that cerebral KMO exerts special control over the fate of kynurenine within the brain. In situations where kynurenine influx from the blood increases, for example after pharmacological inhibition of KMO in the periphery kynurenine levels within the brain will eventually exceed the catabolic capacity of KMO, resulting in disproportionally large KYNA production.[6]

KYNURENINES AND SCHIZOPHRENIA: FOCUS ON KYNURENIC ACID

The idea that kynurenines may be associated with the pathophysiology of schizophrenia (SZ) was first explored several decades ago, albeit without a convincing conceptual rationale. These studies revealed no consistent abnormalities in KP metabolism in either urine or postmortem brain tissue of patients. The scenario changed in the 1980s, when the focus of SZ research began to shift from dopamine and other monoamines to glutamate, which was beginning to be recognized as the major excitatory neurotransmitter in the brain. Of special relevance in this context, NMDA receptors were shown to play a critical role in cognitive processes, which are impaired in SZ patients,[7] and psychotomimetic drugs such

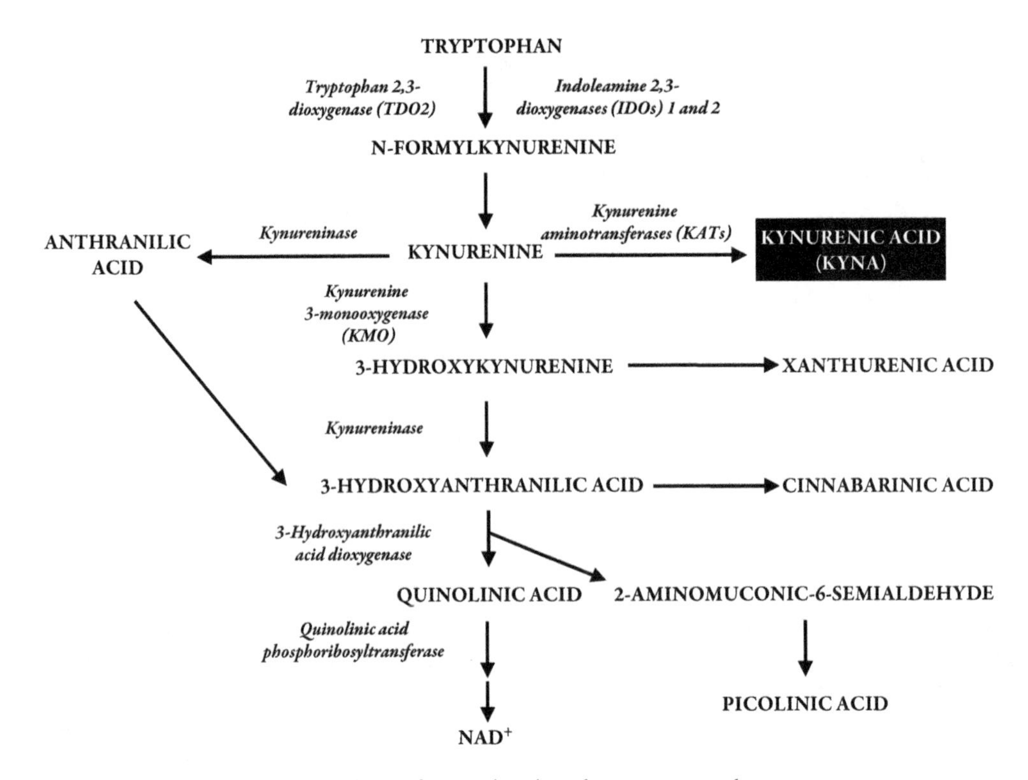

Figure 36.1 Schematic representation of the kynurenine pathway of tryptophan degradation in mammals.

as phencyclidine and ketamine, which produce a spectrum of symptoms reminiscent of SZ in healthy individuals were found to be effective NMDA receptor antagonists.[8] Based on these seminal discoveries, investigators began to consider the possibility that a reduction in glutamatergic neurotransmission may underlie the etiology of SZ. Especially following the demonstration that KYNA inhibits the obligatory glycine co-agonist site (the "glycine$_B$" receptor) of the NMDA receptor competitively with an IC_{50} of ~10 μM,[9] a participation of endogenous kynurenines in SZ pathology became plausible and triggered interest in a thorough re-evaluation of the KP in body fluids and tissues of patients.

So far, studies of KP metabolism in cerebrospinal fluid (CSF) and postmortem brain in SZ have focused mostly on analytes that are either neuroactive themselves or critically involved in the regulation of pathway function. As QUIN is found at normal levels in the CSF, and 3-HK levels are unaltered in the postmortem brain,[5] the activity in the neurotoxic branch of the pathway is supposedly unaffected in the disease. In contrast, the concentrations of both kynurenine and KYNA have been repeatedly found to be significantly elevated both in the cerebral cortex and in the CSF of SZ patients as compared to control subjects, and these increases were unrelated to antipsychotic medication.[5,6] Notably, analyses of peripheral kynurenines have so far provided an inconclusive picture.[10] Thus, some studies found no difference in serum 3-HK levels between first-episode neuroleptic-naïve patients and controls, whereas others suggested that serum 3-HK is decreased following antipsychotic therapy and may predict the severity of clinical symptoms in neuroleptic-naïve, first-episode SZ patients. Yet another study found elevated levels of both 3-HK and KYNA in dermal fibroblasts of individuals diagnosed with SZ or bipolar disorder. With regard to KYNA, an early study reported increased plasma levels in patients with SZ, whereas subsequent studies found reduced levels in patients with either SZ or bipolar disorder and an increased KYNA/3-HK ratio that was associated with cognitive impairments. Interestingly, following a psychological stress challenge, the salivary concentration of KYNA increases more in SZ patients who do not tolerate stressful tasks, and correlates with more severe psychiatric symptoms.[11]

Measurements of other KP metabolites in the serum of patients with SZ have revealed a marked increase in the levels of anthranilic acid,[12] whereas xanthurenic acid levels were found to be reduced.[13] Taken together, the relationship between the concentration of kynurenines in plasma, CSF, saliva, and brain has not been comprehensively evaluated so far, and studies in animals indicate that impairments in KP metabolism in the CSF and in brain tissue are probably more relevant for the clinical manifestations of SZ than peripheral anomalies.

Several explanations, which are by no means mutually exclusive, have been proposed to account for the higher central levels of KYNA seen in people with SZ. Thus, the elevation may be a direct consequence of enhanced synthesis upstream in the KP, caused by increased activity of either TDO2 or IDO1 (see Figure 36.1). Indeed, in individuals with SZ, as well as in bipolar disorder patients with psychosis, gene expression of *TDO2* and the density of TDO-immunopositive cells are significantly elevated in two pivotal brain regions, i.e., the prefrontal and the anterior cingulate cortex.[14] Notably, TDO2 and IDO1 activities are stimulated in stressful situations[15] and/or during infections or other challenges to the immune system,[16] causing these enzymes to generate more kynurenine in circumstances that are frequently linked to the pathophysiology of SZ (discussed later). This mechanism is not restricted to intracerebral events since kynurenine, in contrast to KYNA, is actively transported to the brain from the circulation.[17] Increased kynurenine production in the periphery therefore causes a prompt, corresponding surge in cerebral kynurenine levels, which, in turn, stimulates KYNA formation in the brain.[6]

Another mechanism, namely a reduction in KMO activity related to genetic, environmental, or other causes and shifting KP metabolism toward enhanced kynurenine and KYNA production (Figure 36.1), too, may account for the abnormally high central kynurenine and KYNA levels seen in SZ. In fact, postmortem analyses revealed significant and highly correlated decreases in cortical *KMO* gene expression (−33%) and KMO enzyme activity (−30%) in individuals with SZ.[18] Like the increases in brain KYNA levels, which were confirmed in the same tissues, these changes were unrelated to antipsychotic medication. Of note, independent analyses identified distinct single nucleotide polymorphisms (SNPs) in the *Kmo* gene in SZ patients in Japan, the United States, and Sweden, although these findings could not always be replicated in larger cohorts.[10] Furthermore, a nonsynonymous SNP (rs1053230) in the *Kmo* gene has been shown to affect the CSF KYNA levels in patients with SZ and bipolar disorder.[19] Interestingly, the C allele of rs1053230 is also inversely associated with KMO in epilepsy, as lymphoblastoid cell lines and hippocampal biopsies from patients with at least one C allele show lower KMO expression than patients with no C allele. Direct evidence of causality in this context comes from studies in experimental animals, as substantial increases in KYNA levels are seen in the brain of mice with a genomic elimination of *Kmo*.[20–22]

Of note, when examining rare mutations constituting a polygenic burden in SZ, the lowest identified nominal P-value for disruptive mutations was seen for another KP enzyme, kynureninase, which converts 3-HK to 3-hydroxyanthranilic acid (Figure 36.1).[23] The possible functional significance of this phenomenon has so far not been examined, however.

THE ROLE OF IMMUNE ACTIVATION

Activation of the immune system, which is increasingly believed to play a role in the pathophysiology of SZ too, may underlie the observed increase in brain kynurenine and KYNA levels in the disease. In fact, a meta-analysis shows highly significant effect sizes for serum IL-1β, sIL-2R, IL-6, and TNF-α in first-episode psychosis.[24] Maybe more importantly, increased levels of IL-6 and IL-1β are seen in the CSF of patients, and mRNAs of IL-1β, IL-6, and IL-8 are increased in the frontal cortex (Brodmann area 10) and dorsolateral prefrontal cortex, respectively, in patients with SZ. Similarly, bipolar individuals show upregulated brain mRNA and protein levels of IL-1β and increased CSF IL-1β concentrations especially when they have a history of psychosis.[10] It is therefore tempting to assume that the increased CSF levels of kynurenine and downstream catabolites, which have been repeatedly observed in humans during various infections of the central nervous system (CNS),[25] may be causally related to those immunological factors. Indeed, the KP is readily inducible by neuroinflammatory stimuli, in particular by the cytokines interferon (IFN)-γ, IL-1β, and IL-6. Thus, IFN-γ is a potent inducer of IDO1, whereas IL-1β increases KYNA production in human cortical astrocytes[26] and in endometrioma stromal cells by inducing TDO2. The strong correlation between IL-1β and TDO2 expression in prefrontal gray matter of healthy donors further supports the proposed mechanistic connection between immune factors and KP metabolism in the brain.[26]

KP genes do not have to be directly involved in KP impairments and, in particular, in the elevations in brain KYNA seen in individuals with SZ and bipolar disorder. For example, a recent study concluded that a genetic variation that activates a glial signaling pathway may cause an increase in KYNA synthesis and thereby induce psychosis and cognitive impairment in patients with bipolar disorder.[26] Specifically, the investigators correlated the results of a genome-wide association study (GWAS) with CSF KYNA levels in patients with bipolar disorder, and discovered a significant association with a SNP (rs10158645) that is located within chromosome 1p21.3. While the allelic distribution in this SNP did not differ when all bipolar disorder patients were compared with healthy controls, the minor allele was significantly overrepresented in bipolar persons with a history of psychosis. After replication in an independent cohort, this genetic variant, which was associated with reduced SNX7 expression, was also linked to deficits in executive function in the disease. The diminished cognitive flexibility of carriers of the minor rs10158645 allele was especially noteworthy because of the causal connection between KYNA and cognitive functions (discussed later).

A separate study using fibroblasts and lymphoblastoid cell lines revealed a strong association between the minor allele in rs10158645 and decreased expression of a nearby gene, sorting nexin 7 (SNX7). In hepatocytes, downregulation of SNX7 increases the levels of caspase-8 which in turn cleaves pro-IL-1β into its active form. Indeed, a strong inverse coexpression of SNX7 and caspase-8, as well as between SNX7 and TDO2 expression, is observed in prefrontal gray matter from healthy donors.[26] In brain tissue samples obtained postmortem, caspase-8 mRNA expression was increased specifically in bipolar patients with a history of psychotic features. Interestingly, these studies also revealed a link to dopaminergic dysfunction. Thus, the minor allele in rs10158645 was associated not only with CSF KYNA but also with CSF levels of the dopamine metabolite homovanillic acid (HVA). The statistical model applied suggested that this increase in HVA was a consequence of elevated KYNA levels. Further supporting causality, a strong correlation between CSF KYNA and CSF HVA had previously been found in patients with SZ and healthy controls. Taken together, these studies suggest that decreased SNX7 expression, through caspase-8-driven activation of IL-1β, is linked to increased concentrations of CSF KYNA, which in turn increase dopamine activity, leading to psychosis and cognitive inflexibility in patients.[26]

INCREASED BRAIN KYNA CAUSES COGNITIVE DEFICITS

As described and discussed in a number of articles in this volume, cognitive impairments have emerged as primary determinants of the long-term functional disability characteristic of SZ. These deficits occur in the majority of individuals with SZ, originate in early adolescence preceding psychotic symptoms of young adulthood, are often orthogonal to psychotic symptoms, and persist even after pharmacological control of psychosis.[7] Additionally, the severity of cognitive deficits predicts symptom relapse and poor treatment adherence in first-episode patients. Cognitive impairments in SZ are therefore hypothesized to be due to primary neuronal dysfunction rather than chronicity or neurodegeneration. With the increased recognition of the fact that reducing cognitive impairments in SZ would significantly improve functional outcomes and quality of life, the NIMH initiated the large-scale Measurement and Treatment Research to Improve Cognition in Schizophrenia (MATRICS) project. The first MATRICS committee identified seven primary domains critical for developing targeted procognitive treatments in the disease: working memory, verbal learning and memory, visual learning and memory, attention and vigilance, reasoning and problem-solving, speed of processing, and social cognition.

A second MATRICS committee, the CNTRICS, focusing on pharmacological strategies for cognitive treatment, prominently highlighted the therapeutic potential of approaches that would improve alpha7-nicotinic acetylcholine receptor (α7nAChR) and NMDAR function, both of which are critically involved in physiological processes underlying learning, memory, and other manifestations of synaptic plasticity. Although not conceptualized with KYNA in mind, this viewpoint concurred with suggestions that elevated KYNA levels might be especially pertinent to the cognitive deficits seen in persons with SZ.[5] Thus, KYNA not only competes with glutamate at the NMDAR but, at lower concentrations, preferentially inhibits the obligatory glycine co-agonist ("glycineB") site of the receptor as well as the α7nAChR (Figure 36.2). Of special interest in this context, KYNA acts as a *noncompetitive* antagonist of an allosteric site located in the extracellular domain of the α7nAChR, which overlaps with a site that is activated by the cognition-enhancing drug galantamine.[5] The α7nAChR may therefore be the preferred target of endogenous KYNA in at least part of the mammalian brain, resulting in a series of functionally relevant effects downstream.

In the mammalian brain, KYNA is present in low nanomolar (mouse, rat) to low micromolar (human) concentrations.[27] In line with its ability to inhibit α7nAChR and NMDAR function, even relatively modest increases in endogenous KYNA levels in the brain cause deficits in visuospatial working memory, contextual learning and memory, and prepulse inhibition and habituation of auditory-evoked potentials in rodents.[10,28] Furthermore, in both rats and humans, increased levels of KYNA in brain and CSF are associated with impaired cognitive flexibility,[10] and a recent study revealed that focal infusions of nanomolar concentrations of KYNA into the rat prefrontal cortex selectively attenuate the inhibitory component of local field potential responses, biasing the excitatory-inhibitory balance of prefrontal synaptic activity toward a state of disinhibition. This disruption, which occurred primarily at local GABAergic synapses, resulted from the blockade of presynaptic α7nAChRs and could contribute to the cognitive deficits observed in SZ.[29] Finally, in further support of the idea that elevated brain KYNA levels are *causally* involved in cognitive impairments, cognitive abnormalities are also seen in *Kmo* knockout mice,[21] and several of these dysfunctional states are reminiscent of the cognitive domains listed by the MATRICS committee. They may therefore be translationally

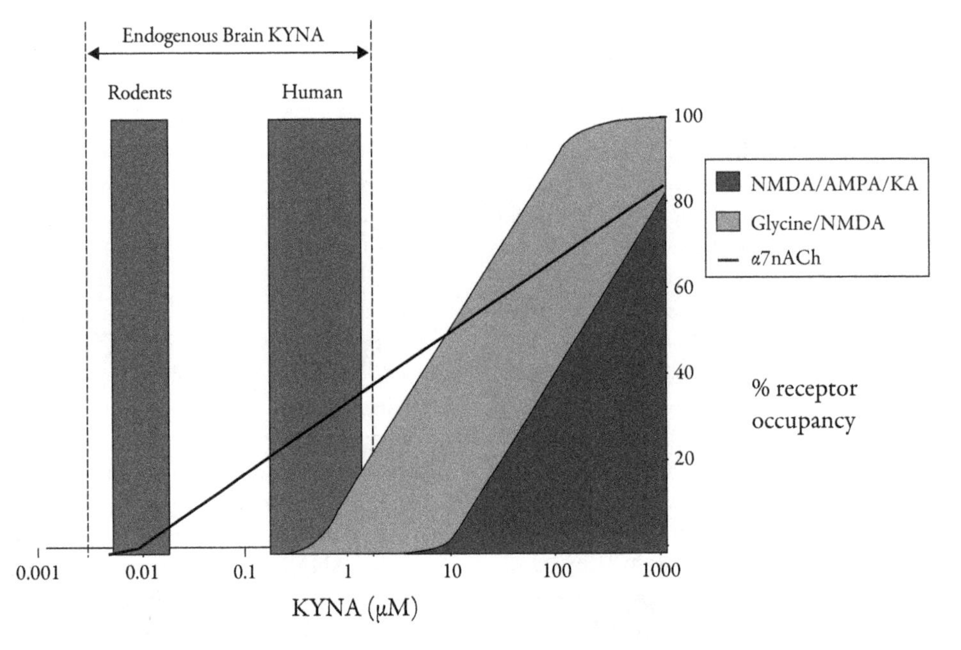

Figure 36.2 Cholinergic and glutamatergic receptor targets of KYNA at endogenous and pharmacological concentrations.

relevant, specifically supporting a role of KYNA in the cognitive deficits seen in persons with SZ.

PERINATAL MANIPULATION OF BRAIN KYNURENINES

Epidemiological studies leave little doubt that stress or bacterial or viral infections in utero constitute a significant risk factor for the development of SZ. This insight triggered a large number of studies in both rodents and nonhuman primates, which were designed to examine the consequences of stress or challenges to the immune system during the prenatal or early postnatal period. These experiments consistently showed adverse effects on the developing brain, leading to structural anomalies, neurochemical changes, and cognitive impairments reminiscent of SZ during adolescence and in adulthood. Probably initiated by the activation of IDO1 or TDO2 by pro-inflammatory cytokines or stimulation by corticosterone, maternal infections or stress in the perinatal period elevate brain KYNA levels in the offspring,[5,10] providing a conceptual link between early KP impairments and the emergence of SZ-like phenomena later in life.

The proposed connection between these adversities, stimulation of KP metabolism, and the pathophysiology of SZ is supported by studies in which KYNA levels were raised perinatally in the rat brain. In one experimental approach, pregnant rats were continuously fed KYNA's brain-penetrant bioprecursor kynurenine during the last week of gestation, i.e., at a time when ambient cerebral KYNA levels in the mammalian brain are already exceptionally high. Starting immediately after birth and until biochemical and behavioral testing in adulthood, the offspring then received normal rodent chow. Remarkably, and for reasons that are still only poorly understood, animals that had hyperphysiological brain KYNA levels during fetal development again presented with higher than normal levels of extracellular KYNA in the prefrontal cortex and in the hippocampus (the two brain areas tested so far) starting approximately 1 month after birth, i.e., during early adolescence. These animals also had pronounced deficits in hippocampal memory, learning and spatial memory, contextual memory, and executive function.[30] Moreover, they showed decreased expression of α7nAChRs, reductions in dendritic spine density, and subsensitivity to mesolimbic stimulation of glutamate release in the prefrontal cortex.[31]

Repeated systemic administration of the KMO inhibitor Ro 61-8048 to the pregnant dam can also be used to raise KYNA levels experimentally in the fetal rat brain. This treatment, too, causes delayed structural and functional abnormalities in the brain of the offspring, including an increase in dendritic complexity, a reduction in hippocampal long-term potentiation, and several distinct neurochemical changes in hippocampus and cortex.[32,33] These studies again highlight the critical importance of maintaining normal KP homeostasis in the maturing brain in utero. Further indirect support for the idea that abnormally high KYNA levels in the fetal brain have adverse long-term consequences comes from the demonstration that prenatal activation of α7nAChRs, possibly by counteracting the elevation in KYNA, reduces experimentally induced inflammation in the fetal brain and ameliorates the detrimental effects of maternal infection in the offspring.[34]

KP manipulations during the early postnatal period or in adolescence, too, influence behavioral performances in adult animals. Thus, in mice, repeated systemic kynurenine injections on postnatal days (PD) 7–16 or PD 7–10 enhance sensitivity to amphetamine-induced increases in locomotor activity and deficits in social behavior. In rats, intermittent exposure to kynurenine during adolescence causes impairments in two hippocampus-dependent cognitive processes, i.e., novel object recognition and contextual fear memory, as well as a deficit in social behavior, in adulthood. Moreover, kynurenine-treated rats show increased incentive salience of cues associated with reward, perhaps informing about the heightened sensitivity to drug-related cues seen in persons with SZ. Very interestingly, early exposure to kynurenine also prevents long-term potentiation after a burst of high-frequency stimulation that is sufficient to produce a robust effect in vehicle-treated rats. Jointly, all these results strengthen the hypothesis that developmental events resulting in the increased presence and function of KYNA in the brain have significant construct validity for the study of cognitive deficits in persons with SZ.[35]

THE NEUROBIOLOGY OF KYNA: AN EVOLVING STORY

It took the realization of a plausible causal connection between KYNA and the cognitive impairments seen in SZ to generate genuine interest in the mechanisms that control KP metabolism—and especially KYNA function—in the mammalian brain. Although important questions remain to be answered, many of the basic characteristics of KYNA neurobiology are now understood. As KYNA does not cross the blood-brain barrier to a significant extent,[17] and as the activity of both TDO2 and IDO1 in the brain is very low under physiological conditions, the neosynthesis of KYNA in the brain is normally mainly driven by peripherally derived kynurenine, which readily enters the brain from the circulation using the large neutral amino acid transporter.[17] Blood-derived kynurenine is then rapidly accumulated by astrocytes, where it is irreversibly transaminated to produce KYNA. The synthesis of KYNA in astrocytes is catalyzed by several enzymes, which were originally named for various natural substrates (glutamine, α-aminoadipate, and aspartate), but were renamed KATs in the context of KP biology. While the relative contributions of these KATs to cerebral KYNA biosynthesis, as well as their respective roles under various physiological and pathological conditions, still need to be elaborated in detail, several lines of evidence indicate that KAT II (= α-aminoadipate aminotransferase) deserves special attention. This was initially proposed on theoretical grounds, i.e., based on the fact that the enzyme's only other known endogenous substrate, α-aminoadipate, is present in the brain at much lower concentrations than the abundant amino acids glutamine (the classic substrate of KAT I and KAT III) and aspartate (the classic substrate of KAT IV).[36] The competitive substrate kynurenine may therefore be more readily transaminated in the presence of endogenous α-aminoadipate than in the presence of either glutamine or aspartate. A more mundane argument can be made based on the fact that the optimal pH for KAT II is 7.4, whereas the optimal pH for the other enzymes is in the alkaline range.[36]

Direct support for a functionally significant role of KAT II in cerebral KYNA formation has come from mice with a targeted deletion of KAT II. These knockout mice display a variety of phenotypic and molecular alterations as a result of decreased KYNA formation in the brain, including increased α7nAChR function in the hippocampus, enhanced cognition, and increased vulnerability to quinolinic acid-induced excitotoxicity.[5,37] These findings, together with the more recent development of selective KAT II inhibitors (discussed later), show that functional disinhibition occurs in the brain when KAT II activity is compromised.

Although highly relevant, fluctuations in kynurenine levels are not the only mechanism controlling the status of endogenous KYNA in the brain. While cerebral KYNA is not degraded enzymatically, its steady-state levels are also determined by the effectiveness of an unspecific, probenecid-sensitive transporter, which controls the efflux of a number of acidic metabolites from the brain.[38] In addition, recent studies have demonstrated that the organic acidic amino acid transporters OAT1 and OAT3 may be responsible for the active reuptake of extracellular KYNA levels in the brain,[39] though the role of these proteins remains to be confirmed in vivo.

Following its production from kynurenine in the brain, newly formed KYNA enters the extracellular milieu very rapidly, and its release does not appear to be controlled by Ca^{2+} or other conventional mechanisms. However, several mechanisms are in place to regulate the formation of endogenous KYNA from kynurenine. Although none of these effects are likely to influence KYNA levels specifically, their ability to control KYNA neosynthesis may have significant ramifications for brain physiology and pathology. Examples include 2-oxoacids, such as pyruvate and 2-oxoglutarate, which function as cosubstrates/aminoacceptors of the aminotransferase reaction and endogenous pro-oxidants (e.g., peroxynitrite and hydroxyl radicals), which promote the nonenzymatic conversion of kynurenine to KYNA. In the case of KAT II, regulation may involve the lysine metabolite α-aminoadipate, which is present in the mammalian brain in micromolar concentrations, competes with kynurenine for enzymatic transamination, and thus interferes readily with the production of KYNA. As noted, all these processes are readily understood and can operate in every tissue, though they may have special effects on brain function and dysfunction.

Other regulatory mechanisms are brain-specific and are less well understood. For example, as discovered using tissue slices in vitro and then verified by in vivo microdialysis, KYNA neosynthesis is substantially reduced under depolarizing conditions. This effect, as well as the substantial reduction in KYNA neosynthesis in the absence of glucose, is dependent on the integrity of neuron-glia interactions. These and related findings, including the decline in extracellular KYNA levels following selective astrocytic poisoning, and the essentially exclusive presence of KAT II expression and immunoreactivity in astrocytes, indicate that astrocytes play a central role in cerebral KYNA production. On the other hand, these studies also demonstrate that neuronal activity participates actively in glial KYNA synthesis in the brain and suggest that impairments in these intercellular mechanisms may be responsible for the very rapid increases in extracellular KYNA, which are seen in response to an acute excitotoxic insult or seizure activity. Conceivably, they may also underlie the—presumably slower—dysregulation of cerebral KYNA levels in SZ and other chronic brain diseases.[5]

Fluctuations in brain KYNA levels have remarkable consequences on classic neurotransmitters, supporting the classification of the metabolite as an astrocyte-derived neuromodulator (Figure 36.3). Thus, even modest elevations in KYNA rapidly reduce the extracellular concentration of glutamate. This effect, first described in the rat striatum,[40] has been studied in considerable detail, mainly using in vivo microdialysis in unanesthetized rodents. As reviewed elsewhere,[41] KYNA concentrations in the mid-nanomolar range consistently—and reversibly—decrease glutamate levels by 30%–40% in every brain region studied so far. Pharmacological studies indicate that this effect, which occurs very rapidly and is also achieved by applying KYNA's immediate bioprecursor kynurenine, is mediated by KYNA's inhibition of α7nAChRs, which are prominently situated on glutamatergic nerve terminals in the mammalian brain.[5] Selective inhibition of NMDARs, a possible alternative mechanism, is unlikely to contribute because the effect of KYNA on extracellular glutamate is not duplicated or influenced by the specific and potent NMDA receptor antagonist 7-chlorokynurenic acid, and is not affected by coperfusion with D-serine, a selective endogenous agonist of the $glycine_B$ site of the NMDAR. Interestingly, the influence of KYNA on glutamate is bidirectional, i.e., extracellular levels of the neurotransmitter *rise* rapidly when KYNA synthesis is compromised by the local application of a KAT II inhibitor. Specificity was confirmed in experiments which showed that this effect of KAT II inhibition is readily prevented by the coadministration of very low (nanomolar) concentrations of KYNA.

Experimentally induced up- or downregulation of brain KYNA also inversely affects the extracellular levels of dopamine and GABA in the rat brain, and KAT II inhibition causes a substantial increase in the extracellular levels of acetylcholine in the medial prefrontal cortex of rats. Like the acute effects on extracellular glutamate described earlier, these bidirectional effects of KYNA occur rapidly and do not appear to be brain-region-specific.[41] However, though so far mainly examined with regard to hippocampal glutamate levels in KAT II knockout mice, prolonged reductions in brain KYNA result in quantitatively very similar, *chronic* increases in extracellular neurotransmitter concentrations.[37] Notably, where tested, the acute results of KYNA fluctuations, again like the effects of KYNA on glutamate, were consistently found to be closely linked to α7nAChR function. Thus, the KYNA-induced reductions in dopamine and GABA are readily prevented by coadministration of galantamine or, in the case of striatal dopamine, by the α7nAChR agonist choline. In further support of the involvement of α7nAChRs, the

Figure 36.3 Astrocyte-derived KYNA controls the extracellular levels of several classic neurotransmitters. See text for details.

receptor antagonist methyllycaconitine duplicates, but is not additive with, the effects of KYNA.[41]

Interestingly, extracellular KYNA levels in the rat brain in turn decrease rapidly in response to glutamate and after pharmacological stimulation of dopaminergic function, and are also dysregulated by continuous exposure to nicotine.[41] Possible functional ramifications of these apparent feedback mechanisms, which are presumably mediated by neurotransmitter receptors on astrocytes, as well as their potential relevance for brain physiology and pathology, have not been examined to date.

MANIPULATION OF BRAIN KYNA: THERAPEUTIC OPPORTUNITIES

Inhibitors of KAT II allow investigators a direct route to reduce the levels of KYNA in the brain and, in particular, to target the KYNA pool that appears to be most relevant for the rapid mobilization of this neuromodulator. The first selective inhibitor, (*S*)-4-(ethylsulfonyl)benzoylalanine (ESBA), was described in 2006 and has been used successfully as an experimental tool in the rodent brain. This compound, which must be applied intracerebrally, and newer, systemically active agents, reliably decrease the extracellular concentration of KYNA by 30%–40%, irrespective of brain region.[5]

In view of the repeated demonstration of KYNA-induced cognitive deterioration in experimental animals and the well-established evidence suggesting a pathophysiologically relevant role of elevated brain KYNA in SZ (discussed earlier), KAT II inhibitors were very soon considered for possible use in humans. During recent years, this idea received strong support from convergent studies in rodents and nonhuman primates, showing that KAT II inhibition causes remarkable cognitive enhancement in several relevant paradigms.[42] This was first demonstrated in KAT II knockout mice, which show significantly improved performance in three cognitive paradigms as well as increased long-term potentiation in vitro.[37] Subsequent studies revealed that ESBA administration improves the performance of rats in the Morris water maze and, importantly, documented remarkable pro-cognitive effects of two orally active enzyme inhibitors, namely PF-04859989[43] and BFF-816.[44] Thus, PF-04859989 counteracts drug-induced memory impairments and disruption of auditory gating, and improves performance in a sustained attention task in rodents, and also effectively antagonizes ketamine-induced working memory deficits in nonhuman primates. In excellent agreement, BFF-816 administration enhances spatial and reference memory in rats and attenuates the contextual memory deficit in the offspring of kynurenine-treated dams. Notably, and in accordance with the proposed central role of α7nAChRs in the neurochemical effects of KYNA (discussed earlier), galantamine effectively prevents KYNA-induced deficits in attentional set-shifting and working memory. Taken together, all these preclinical results favor the likely translational relevance of KAT II inhibitors, especially supporting the idea that specific pharmacological reduction of brain KYNA levels may improve cognitive deficits in SZ and, possibly, other psychiatric diseases.[27]

OUTLOOK

In spite of the strong arguments that can be made for the development of KAT II inhibitors for cognitive improvement in humans, and especially in persons with SZ, several challenges remain. On theoretical grounds, these include resolution of the possible roles of the ever expanding number of KYNA targets in brain physiology and pathology, including the G protein-coupled receptor GPR 35, which affects Ca^{2+} fluxes in astrocytes and secondarily reduces glutamatergic neurotransmission, and the aryl hydrocarbon receptor. Elucidation of the functional, bidirectional interactions between oxidative processes and brain KYNA, which has recently garnered substantial attention, is also highly relevant but may be even more difficult.[1,38]

In light of recent findings indicating a prominent role of KYNA in both normal and abnormal brain development, future investigative efforts ought to be directed at a better understanding of the ontogenetic trajectory of the KP, and in particular the dynamics of the KP between mother, placenta, and fetus.[35] In this context, it also seems timely to investigate the possibility that tryptophan produced by the gut microbiome affects KYNA function and dysfunction in the brain at various stages of development, as well as in adulthood. Finally, emphasis must be placed on the thorough examination of possible sex differences and, circadian and seasonal phenomena, which are just beginning to be investigated with regard to brain KYNA.[27]

Although these fundamental questions regarding KYNA neurobiology have obvious practical ramifications, some additional unresolved issues have particular relevance in the clinical realm. Thus, from both diagnostic and therapeutic perspectives, genetic causes of cerebral KYNA malfunction will need to be examined in depth, and environmental risk factors must be carefully assessed longitudinally in human populations. Furthermore, as there is remarkably little known about the possible *functional* relationship between the KP—and KYNA formation in particular—and the serotonin branches of tryptophan degradation, future clinical studies ought to include parallel measurements of both catabolic routes in individual probands. It will be imperative, however, to clarify if and how measurements of peripheral serotonin and KP metabolism can be used as indicators, i.e., biomarkers, of central events, and especially of KYNA formation in the brain. Finally, from the perspective of drug design and development, as well as the need to verify target engagement in the brain noninvasively in vivo, it will also be important to develop novel imaging techniques so that the effects of pharmacological KYNA manipulations in the brain can be monitored unequivocally. Challenges in this regard include the distinct structure of human KAT II, which complicates tagging of the enzyme with a radiolabeled tracer,[36] and the multitude of potential receptor sites (discussed earlier). Resolution of these and related issues is expected to provide clinical researchers with the means to carefully assess the KYNA hypothesis in persons with SZ and, it is hoped, allow them to successfully use a fundamentally new pharmacological approach to attenuate cognitive deficits associated with the disease.

ACKNOWLEDGMENTS

The authors gratefully acknowledge the many contributions of postdoctoral fellows, graduate and undergraduate students, and research assistants in their laboratories. Over the years, the work reviewed in the present article was supported by grants from governmental institutions and non-profit Research Foundations in the United States and in Sweden.

REFERENCES

1. Stone TW, Stoy N, Darlington LG. An expanding range of targets for kynurenine metabolites of tryptophan. *Trends Pharmacol Sci.* 2013;34(2):136–143.
2. Perkins MN, Stone TW. An iontophoretic investigation of the actions of convulsant kynurenines and their interaction with the endogenous excitant quinolinic acid. *Brain Res.* 1982;247(1):184–187.
3. Schwarcz R. Metabolism and function of brain kynurenines. *Biochem Soc Trans.* 1993;21(1):77–82.
4. Guillemin GJ, Cullen KM, Lim CK, et al. Characterization of the kynurenine pathway in human neurons. *J Neurosci.* 2007;27(47):12884–12892.
5. Schwarcz R, Bruno JP, Muchowski PJ, Wu H-Q. Kynurenines in the mammalian brain: when physiology meets pathology. *Nat Rev Neurosci.* 2012;13(7):465–477.
6. Erhardt S, Olsson SK, Engberg G. Pharmacological manipulation of kynurenic acid: potential in the treatment of psychiatric disorders. *CNS Drugs.* 2009;23(2):91–101.
7. Schaefer J, Giangrande E, Weinberger DR, Dickinson D. The global cognitive impairment in schizophrenia: consistent over decades and around the world. *Schizophr Res.* 2013;150(1):42–50.
8. Jentsch JD, Roth RH. The neuropsychopharmacology of phencyclidine: from NMDA receptor hypofunction to the dopamine hypothesis of schizophrenia. *Neuropsychopharmacology.* 1999;20(3):201–225.
9. Birch PJ, Grossman CJ, Hayes AG. Kynurenic acid antagonises responses to NMDA via an action at the strychnine-insensitive glycine receptor. *Eur J Pharmacol.* 1988;154(1):85–87.
10. Erhardt S, Schwieler L, Imbeault S, Engberg G. The kynurenine pathway in schizophrenia and bipolar disorder. *Neuropharmacology.* 2017;112 (Pt B):297–306.
11. Chiappelli J, Pocivavsek A, Nugent KL, et al. Stress-induced increase in kynurenic acid as a potential biomarker for patients with schizophrenia and distress intolerance. *JAMA Psychiatry.* 2014;71(7):761–768.
12. Oxenkrug G, van der Hart M, Roeser J, Summergrad P. Anthranilic acid: a potential biomarker and treatment target for schizophrenia. *Ann Psychiatry Ment Health.* 2016;4(2):1–8.

13. Fazio F, Lionetto L, Curto M, et al. Xanthurenic acid activates mGlu2/3 metabotropic glutamate receptors and is a potential trait marker for schizophrenia. *Sci Rep*. 2015;5(1):17799.
14. Miller CL, Llenos IC, Dulay JR, Weis S. Upregulation of the initiating step of the kynurenine pathway in postmortem anterior cingulate cortex from individuals with schizophrenia and bipolar disorder. *Brain Res*. 2006;1073–1074:25–37.
15. Badawy AA-B, Namboodiri AMA, Moffett JR. The end of the road for the tryptophan depletion concept in pregnancy and infection. *Clin Sci*. 2016;130(15):1327–1333.
16. Strasser B, Becker K, Fuchs D, Gostner JM. Kynurenine pathway metabolism and immune activation: peripheral measurements in psychiatric and co-morbid conditions. *Neuropharmacology*. 2017;112(Pt B):286–296.
17. Fukui S, Schwarcz R, Rapoport SI, Takada Y, Smith QR. Blood-brain barrier transport of kynurenines: implications for brain synthesis and metabolism. *J Neurochem*. 1991;56(6):2007–2017.
18. Wonodi I, Stine OC, Sathyasaikumar K V, et al. Downregulated kynurenine 3-monooxygenase gene expression and enzyme activity in schizophrenia and genetic association with schizophrenia endophenotypes. *Arch Gen Psychiatry*. 2011;68(7):665–674.
19. Lavebratt C, Olsson S, Backlund L, et al. The KMO allele encoding Arg452 is associated with psychotic features in bipolar disorder type 1 and with increased CSF KYNA level and reduced KMO expression. *Mol Psychiatry*. 2014;19(3):334–341.
20. Parrott JM, Redus L, Santana-Coelho D, Morales J, Gao X, O'Connor JC. Neurotoxic kynurenine metabolism is increased in the dorsal hippocampus and drives distinct depressive behaviors during inflammation. *Transl Psychiatry*. 2016;6(10):e918.
21. Erhardt S, Pocivavsek A, Repici M, et al. Adaptive and behavioral changes in kynurenine 3-monooxygenase knockout mice: relevance to psychotic disorders. *Biol Psychiatry*. 2017;82(10):756–765.
22. Tashiro T, Murakami Y, Mouri A, et al. Kynurenine 3-monooxygenase is implicated in antidepressants-responsive depressive-like behaviors and monoaminergic dysfunctions. *Behav Brain Res*. 2017;317:279–285.
23. Purcell SM, Moran JL, Fromer M, et al. A polygenic burden of rare disruptive mutations in schizophrenia. *Nature*. 2014;506(7487):185–190.
24. Upthegrove R, Manzanares-Teson N, Barnes NM. Cytokine function in medication-naive first episode psychosis: a systematic review and meta-analysis. *Schizophr Res*. 2014;155(1-3):101–108.
25. Saito K, Heyes MP. Kynurenine pathway enzymes in brain: properties of enzymes and regulation of quinolinic acid synthesis. *Adv Exp Med Biol*. 1996;398:485–492.
26. Sellgren CM, Kegel ME, Bergen SE, et al. A genome-wide association study of kynurenic acid in cerebrospinal fluid: implications for psychosis and cognitive impairment in bipolar disorder. *Mol Psychiatry*. 2016;21(10):1342–1350.
27. Schwarcz R, Stone TW. The kynurenine pathway and the brain: challenges, controversies and promises. *Neuropharmacology*. 2017;112(Pt B):237–247.
28. Chess AC, Simoni MK, Alling TE, Bucci DJ. Elevations of endogenous kynurenic acid produce spatial working memory deficits. *Schizophr Bull*. 2007;33(3):797–804.
29. Flores-Barrera E, Thomases DR, Cass DK, et al. Preferential disruption of prefrontal GABAergic function by nanomolar concentrations of the α7nACh negative modulator kynurenic acid. *J Neurosci*. 2017;37(33):7921–7929.
30. Pocivavsek A, Thomas MAR, Elmer GI, Bruno JP, Schwarcz R. Continuous kynurenine administration during the prenatal period, but not during adolescence, causes learning and memory deficits in adult rats. *Psychopharmacology (Berl)*. 2014;231(14):2799–2809.
31. Pershing ML, Bortz DM, Pocivavsek A, et al. Elevated levels of kynurenic acid during gestation produce neurochemical, morphological, and cognitive deficits in adulthood: implications for schizophrenia. *Neuropharmacology*. 2015;90:33–41.
32. Khalil OS, Forrest CM, Pisar M, Smith RA, Darlington LG, Stone TW. Prenatal activation of maternal TLR3 receptors by viral-mimetic poly(I:C) modifies GluN2B expression in embryos and sonic hedgehog in offspring in the absence of kynurenine pathway activation. *Immunopharmacol Immunotoxicol*. 2013;35(5):581–593.
33. Pisar M, Forrest CM, Khalil OS, et al. Modified neocortical and cerebellar protein expression and morphology in adult rats following prenatal inhibition of the kynurenine pathway. *Brain Res*. 2014;1576:1–17.
34. Wu W-L, Adams CE, Stevens KE, Chow K-H, Freedman R, Patterson PH. The interaction between maternal immune activation and alpha 7 nicotinic acetylcholine receptor in regulating behaviors in the offspring. *Brain Behav Immun*. 2015;46:192–202.
35. Notarangelo FM, Pocivavsek A. Elevated kynurenine pathway metabolism during neurodevelopment: implications for brain and behavior. *Neuropharmacology*. 2017;112(Pt B):275–285.
36. Rossi F, Schwarcz R, Rizzi M. Curiosity to kill the KAT (kynurenine aminotransferase): structural insights into brain kynurenic acid synthesis. *Curr Opin Struct Biol*. 2008;18(6):748–755.
37. Potter MC, Elmer GI, Bergeron R, et al. Reduction of endogenous kynurenic acid formation enhances extracellular glutamate, hippocampal plasticity and cognitive behavior. *Neuropsychopharmacology*. 2010;35(8):1734–1742.
38. Moroni F, Cozzi A, Sili M, Mannaioni G. Kynurenic acid: a metabolite with multiple actions and multiple targets in brain and periphery. *J Neural Transm*. 2012;119(2):133–139.
39. Uwai Y, Honjo H, Iwamoto K. Interaction and transport of kynurenic acid via human organic anion transporters hOAT1 and hOAT3. *Pharmacol Res*. 2012;65(2):254–260.
40. Carpenedo R, Pittaluga A, Cozzi A, et al. Presynaptic kynurenate-sensitive receptors inhibit glutamate release. *Eur J Neurosci*. 2001;13(11):2141–2147.
41. Pocivavsek A, Notarangelo FM, Wu HQ, Bruno JP SR. Astrocytes as pharmacological targets in the treatment of schizophrenia: focus on kynurenic acid. In: *Modeling the Psychopathological Dimensions of Schizophrenia: Handbook of Behavioral Neuroscience*. 2015:423–443.
42. Jayawickrama GS, Sadig RR, Sun G, et al. Kynurenine aminotransferases and the prospects of inhibitors for the treatment of schizophrenia. *Curr Med Chem*. 2015;22(24):2902–2918.
43. Kozak R, Campbell BM, Strick CA, et al. Reduction of brain kynurenic acid improves cognitive function. *J Neurosci*. 2014;34(32):10592–10602.
44. Wu H-Q, Okuyama M, Kajii Y, Pocivavsek A, Bruno JP, Schwarcz R. Targeting kynurenine aminotransferase II in psychiatric diseases: promising effects of an orally active enzyme inhibitor. *Schizophr Bull*. 2014;40 (Suppl 2):S152–158.

PATHOPHYSIOLOGY

VOLTAGE-GATED ION CHANNELS IN PSYCHOSIS

37

GENETIC ASSOCIATION OF VOLTAGE-GATED ION CHANNELS WITH PSYCHOTIC DISORDERS

Charles H. Large

INTRODUCTION: THE ROLE OF VOLTAGE-GATED ION CHANNELS IN CNS FUNCTION

A goal for the treatment of psychotic disorders, such as schizophrenia and bipolar disorder, is to correct the dysfunction of neural circuits that have been implicated in a range of symptoms from altered sensory perception to impaired cognition. However, this goal is complicated by the absence of an obvious lesion or clear disease process. The brains of people with psychotic illness just appear to work differently. Dopamine receptor blocking strategies repress the florid symptoms of psychosis, but do little to influence the underlying dysfunction of brain circuits, and provide little or no improvement in the patient's cognitive function.[1] Voltage-gated ion channels (VGICs) are proteins that define neurons: they are required for all aspects of neural processing and signaling. Could modulation of these highly dynamic proteins be useful targets in the treatment of psychosis? This chapter reviews examples of VGICs that have been genetically linked to psychosis and which may take us toward the goal of better treatments for these devastating disorders.

Voltage-gated ion channel (VGIC) proteins form an aqueous pore across cellular lipid membranes through which charged ions (Na^+, Ca^{2+}, K^+, and Cl^-) can pass. Ion selectivity is set by specific amino acids within the channel pore. VGICs are sensitive to changes in the electrostatic charge across the membrane, which drives conformational changes in the protein complex, allowing the channels to open and close. VGICs are highly dynamic.

Eleven ion channel families have been identified,[2,3,4] for which structural information is increasingly available (e.g., [5,6]). The structure and function of the voltage-sensor domain, ion selectivity filter, as well as gating and inactivation mechanisms have been worked out in detail using molecular genetics and structural imaging techniques. Most recently, the development of cryogenic electron microscopy, which freezes the channel protein in its physiological conformation prior to imaging, has successfully resolved the structure of several VGICs at angstrom resolution.[7,8,9] The structural elements of VGICs combine to confer a range of biophysical properties that characterize individual channels and channel families. These properties affect the voltage range over which channels open and close, the speed with which this happens, whether or not the channels inactivate after opening, and how quickly they recover from inactivation.[10] Databases, such as Channelpedia (http://channelpedia.net)[11] provide a useful resource to navigate among the 140+ ion channel proteins that have been cloned.

VGIC channels sculpt the electrical dynamics of excitable cells, and in the case of neurons, orchestrate their characteristic ability to integrate synaptic input, fire action potentials, and release neurotransmitters. Knowledge of the biophysical properties of a specific neuronal channel, coupled with information about its subcellular location (somatic, dendritic, or axonal) permits prediction of its physiological role, and how the channel might contribute to human disease.[12] Given the frequent association between VGICs and cellular excitability, drugs targeting VGICs are most commonly directed toward the treatment of neurological disorders such as epilepsy and neuropathic pain, as well as cardiac disorders such as arrhythmias. This chapter considers examples of VGICs that have been linked to psychotic disorders by genetic association.

Genome-wide association studies (GWAS) continue to identify new genetic variants (single nucleotide polymorphisms, SNPs) that influence the risk of psychosis. Schizophrenia, for example, has been associated with more than 100 SNPs,[13] many of which have now been replicated across independent studies, and several of which are linked to multiple psychiatric phenotypes.[14,15] DSM-V and ICD-10 provide symptom-based classifications that remain useful to the clinician to provide a psychiatric diagnosis to the patient, but it seems increasingly likely that these diagnostic structures will be replaced eventually by parameters that capture genetic risk together with objective measures of brain dysfunction. In this future, ion channels may provide interesting opportunities for therapeutic intervention because they can allow targeted regulation of neuronal dysfunction.

CACNA1C (CAV1.2)

Calcium signaling and the calcium channels that regulate the flow of calcium ions into and within cells have been linked to a range of neuropsychiatric disorders.[16] The CACNA1C gene, which codes for the L-type calcium channel, Cav1.2

α1C subunit,[17,18,19] was identified in a human GWAS as a risk locus for bipolar disorder, schizophrenia, major depressive disorder, autism spectrum disorder, and attention deficit hyperactivity disorder.[20] Similarly, in a GWAS for psychosis proneness[21] and schizophrenia,[22] genome-wide significant results for CACNA1C and CACNB2 (a beta subunit that influences the properties of calcium channel alpha subunits) were obtained, strongly implicating calcium signaling in the etiology of the disease. This latter study has been supported by genetic association studies combining independent samples,[23] meta-analyses across multiple studies,[24,25] and case-control studies.[26] Association with bipolar disorder has also been replicated across multiple studies.[27] Cav1.2 is highly expressed in the brain and accounts for more than 80% of L-type calcium current in neurons.[28,29] The channels are expressed on dendrites and the cell soma, and couple neuronal excitation with Ca2+ signaling that modulates gene transcription, synaptic plasticity, and neuronal survival.[30 31]

In order to understand how Cav1.2 channel expression or function might be altered with the disease-associated allele, Yoshimizu et al.[32] generated human neurons from fibroblasts from individuals carrying risk genotypes or nonrisk genotypes at the rs1006737 locus within intron 3 of the CACNA1C gene. Electrophysiology and quantitative PCR characterization found an increase in L-type channel current density, and increased mRNA expression of the channel in neurons derived from subjects with the homozygous risk genotype compared with nonrisk genotypes. Similarly, Bigos et al.[33] found that a risk-associated SNP predicted increased expression of CACNA1C mRNA in postmortem human brain, and in an fMRI study they observed increased hippocampal and prefrontal cortical BOLD activity evoked by emotional processing or executive function tasks, respectively, further suggestive of increased calcium channel activity. Relevant to these findings, a gain of function mutation in the S6 transmembrane domain of CACNA1C (G406R) reduces voltage-dependent inactivation,[34] and gives rise to Timothy Syndrome, an autosomal dominant disorder characterized by significant cardiac and psychiatric symptoms.[35] Taken together, these studies suggest that increased activity of Cav1.2 channels increases the risk of psychosis. Drugs that inhibit the channels may therefore be useful in the treatment of these disorders.

Cav1.2 channels are critical during brain development, and have been linked to the development of parvalbumin-positive GABAergic interneurons,[36] which are implicated in the pathology of schizophrenia[37] (discussed later); however, in adult brain, expression of the channels in these interneurons appears to be low.[38] Two hypotheses can be put forward: (1) the risk genotypes for CACNA1C increase Cav1.2 function during brain development leading to alterations in brain architecture that predispose to psychosis, and/or (2) increased Cav1.2 function in adult brain contributes to risk of psychosis. The implications of these hypotheses for therapeutic intervention are clear: if increased risk of psychosis is due only to the altered Cav1.2 activity during brain development, drugs that target this channel may have little impact in the adult patient. However, if increased Cav1.2 function continues to contribute to psychosis risk into adulthood, then L-type channel blocking drugs might be useful.

Knockout of Cav1.2 expression is lethal, but conditional knockout of forebrain expression gives rise to anxiety-like behavior,[39] reduced spatial memory,[40] and reduced hippocampal neurogenesis.[41] Reduced expression or pharmacological block of Cav1.2 has been shown to affect presynaptic dopamine function in the VTA, characterized by a loss of sensitivity of dopamine release to blockade of the dopamine transporter (DAT), and reduced locomotor sensitization to the DAT inhibitor, GBR12909.[42] These rodent knockout studies presumably do not model the association between CACNA1C and psychiatric illness, which appears to be characterized by increased Cav1.2 expression;[43] however, they do confirm the importance of the channels within circuits involved in psychosis and cognition, such as the mesolimbic dopamine system, frontal cortex, and hippocampus.

Drugs that inhibit L-type calcium channels are used to control blood pressure and treat cardiac arrhythmias.[44] Several have been evaluated in the treatment of psychiatric illness. Evidence suggests possible utility in the control of rapid-cycling bipolar disorder, but results have been mixed.[45] No studies of the effects of L-type calcium channel inhibitors in schizophrenia have reported to date, although a study of isradipine,[46] a dihydropyridine, for cognitive enhancement in patients diagnosed with schizophrenia and schizoaffective disorder is listed as recruiting at www.clinicaltrials.gov (NCT01658150).

KCNH7 (ERG3, KV11.3)

A GWAS of a Taiwanese bipolar 1 patient population found an association with a locus (rs6736615) in the KCNH7 gene, which codes for subtype 3 of the ether-à-go-go (ERG) family of potassium channels, Kv11.3.[47] In a separate study, Strauss et al.[48] conducted a whole-exome search for low-frequency alleles shared among patients with bipolar spectrum disorders in an Amish population. They identified a different variant of KCNH7 (c.1181G>A; p.Arg394His; rs78247304) as a risk allele for bipolar disorder in this group. The association with bipolar disorder is considered preliminary, since neither of these two genetic studies reached genome-wide significance. When expressed in a neuroblastoma cell line, the Arg394His variant of the Kv11.3 channel showed a rightward shift in its voltage-dependence of activation (greater depolarization is required to open the channels), and relatively slowed activation kinetics compared to the channel encoded by the common allele.[49] Kv11.3 is thought to be mainly expressed in the central nervous system,[50] unlike Kv11.1 which is the well-known cardiac "hERG" channel associated with cardiac arrhythmias and Long-QT syndrome.[51] ERG3 (Kv11.3) mRNA was detected in cortex, hippocampus, and basal ganglia in the rat and was shown to be expressed by Purkinje cells in the cerebellum.[52] Kv11.3 channels activate relatively slowly in response to depolarization to potentials above -20mV and thus act as a progressive brake to slow down repetitive action potential firing, an effect known as spike frequency adaptation.[53] Knockdown of ERG3 channels enhances the excitability of hippocampal CA1 pyramidal neurons and dentate gyrus granule cells, and is associated with altered seizure thresholds.[54] Furthermore, expression of ERG3 is reduced in hippocampal epileptogenic tissue in both humans and mouse models of epilepsy.

Reduced activation of the Arg394His variant of Kv11.3 channels would be predicted to reduce spike frequency adaptation under conditions of high neuronal excitation and contribute to an increase excitability of neurons in hippocampus and other brain areas. Thus the Kv11.3 variant appears to confer risk of psychosis through reduced channel function leading to increased excitability of the nervous system. Therefore, a positive modulator of Kv11.3 channels would be indicated for the treatment of psychosis.

Several drugs, including a number of antipsychotic agents, are known to block Kv11 channels, causing a lengthening of the cardiac QT interval, which can cause life-threatening arrhythmias.[55] Aside from this cardiovascular effect, it might be interesting to explore the relationship between antipsychotic efficacy and Kv11.3 inhibition. One might postulate that significant Kv11.3 would be contraindicated for the treatment of bipolar disorder; at least in those patients carrying the Arg394His variant. Small molecule activators of Kv11 channels, such as NS1643[56] and ICA-105574 have been identified and shown to have antiarrhythmic efficacy in cardiac systems in vitro.[57] The potential of these or related drugs to influence neural activity has not been explored in humans, mainly due to their lack of selectivity between brain Kv11 channels and cardiac Kv11.1. However, studies have shown that NS1643 can reduce sensitivity to epileptogenic stimuli in mice.[58] There is a clear rationale to invest in the discovery and development of novel, selective Kv11.3 channel modulators.

CLCN3 (CLC-3)

Schizophrenia and other forms of psychosis are well known to be polygenic,[59] with genetic association studies showing that there are at least 100 gene loci where variants confer an increased risk of these disorders. The study of postmortem brain tissue from patients with a psychiatric diagnosis compared to controls allows the detection of SNPs that affect gene expression (expression quantitative trait loci, eQTLs). Recently, a study by the CommonMind Consortium (www.synapse.org/CMC) undertook to sequence RNA obtained from dorsolateral prefrontal cortex of a large number of subjects with schizophrenia (n = 258) and control subjects (n = 279) in order to identify gene loci where variation is associated with altered gene expression.[60] Of the loci identified, eight involved only a single gene. One such locus included the CLCN3 gene, where the risk allele (rs10520163) was associated with upregulation of expression. CLCN3, which codes for the brain-specific voltage-gated chloride channel 3 (ClC-3), is expressed at the synapse where it regulates glutamatergic transmission.[61] A later study, using proton magnetic resonance spectroscopy (1H-MRS), found that levels of glutamate in the brains of 56 patients with schizophrenia and 67 healthy subjects were correlated with glutamate-related risk genes, including the CLCN3 risk variant (rs10520163).[62] The effect of increased ClC-3 expression associated with the risk allele is unclear; knock-out of the channels leads to increased glutamate release from hippocampal synapses in mice,[63] but ClC-3 is also colocalized with the vesicular GABA transporter (VGAT) in the CA1 region of the hippocampus where channel knockdown also impacts GABA transmission.[64] The net effect of knockout of the channels in the mouse is to cause hippocampal neurodegeneration, but the effect of increased ClC-3 expression in the adult brain has not been studied. Furthermore, it is possible that the risk allele might have its main impact during neurodevelopment, since ClC-3 channels are known to be involved in cellular migration and cell cycle control; for example, in the lung,[65] heart,[66] and in certain cancers.[67]

Given the broad expression of ClC-3 channels across multiple organs, it may prove difficult to administer a ClC-3 inhibitor safely, without risking cardiac or respiratory side-effects. A better understanding of the relationship between the risk allele in the CLCN3 gene locus and the expression of psychosis will be important in order to determine whether these channels might offer a valid treatment approach.

SECONDARY ASSOCIATION BETWEEN VGICS AND PSYCHOSIS

VGICs are a final determinant of neuronal function, affecting excitability, dendritic integration, action potential firing, synaptic release, and plasticity. Expression of a given channel in a given neuron defines the latter's phenotype; however, other factors operate at different timescales to modify the relative contribution of VGICs to fine-tune neural activity and underpin their plasticity. These factors include transcriptional regulation,[68] trafficking or localization to different subcellular compartments (e.g., [69,70], and see what follows), colocalization with accessory subunits that modify function (e.g., [71,72]), and C-terminal-protein interactions that couple the ion channel to intracellular pathways controlling a range of cellular activities (e.g., [73]). These factors affect neuronal function on a timescale of hours to days. In addition, direct phosphorylation of the VGIC protein can modulate gating kinetics or inactivation on a timescale of seconds to minutes, and couples both intrinsic neural status and neurotransmitter receptor activation to electrophysiological activity.[74,75,76] For example, neuromodulators such as 5-HT, noradrenaline, dopamine, acetylcholine, and histamine can indirectly alter VGIC function in the target neuron.[77,78] Dopamine, thought to be critical to the expression of psychosis, is linked to the modulation of VGICs in a number of target neurons,[79,80] and raises the possibility that pharmaceutical intervention via these channels might offer an alternative to antidopaminergic drugs to treat symptoms of psychosis.

As discussed in the earlier part of this chapter, there are few direct associations between VGIC genes and psychosis. However, ANK3, the gene that codes for Ankyrin-G, has been implicated in bipolar disorder by multiple GWAS in a range of patient populations and is involved in the subcellular localization of sodium channels and the Kv7.2/3 potassium channel.[81]

The variant of the ANK3 gene associated with bipolar disorder causes reduced expression of exon 1b of Ankyrin-G in PV interneurons, leading to fewer channels located at the axonal initial segment. The consequence is an increased firing threshold and reduced maximal firing frequency of these inhibitory interneurons, contributing to altered E/I balance and reduced gamma synchrony within cortical circuits.

Finally, epigenetic factors are likely to play a part in altering VGIC expression and function in ways that have yet to be elucidated. Precedence for this comes from the recent observation that N-methyl-D-aspartate (NMDA) receptor hypofunction in the neurodevelopmental methylazoxymethanol (MAM) model for schizophrenia may arise from the excessive activity of a transcriptional repressor, RE1-silencing transcription factor, and the repressive histone marker H3K27me3 at the promoter on Grin2b, which codes for the NR2B subunit of the ligand-gated NMDA receptor ion channel. [82] It is highly likely that similar mechanisms might alter the balance of VGIC expression and function during development or in the adult brain in a way that might predispose to psychosis. Understanding these mechanisms will be important and may lead to new rational approaches to the treatment of psychotic disorders.

DISCUSSION

This chapter has considered specific examples of ion channels that have been implicated in the pathophysiology of psychotic disorders by genetic association.

It is perhaps surprising that so few VGIC genes have been associated with psychosis. However, VGICs are very highly conserved across species, with functional mutations tending to cause severe or life-threating disorders, such as epilepsy or cardiac arrhythmias. The examples discussed in this chapter appear to involve quite subtle changes in channel expression or function, and it remains unclear whether these changes predominantly affect neurodevelopment, leaving the brain susceptible to later insult, or whether they contribute directly to the emergence of psychotic symptoms in the mature brain. Targeting these channels in adult patients risks significant neurological and cardiac side effects for possibly little benefit if the channels play only a neurodevelopmental role in the disorder.

Few of these channels have been targeted by drug discovery programs to date, despite an increase in understanding of these complex, dynamic proteins. VGICs are not easy targets for medicinal chemistry: there are no traditional ligand binding sites, functional assays are essential for screening (reducing the practicality for high-throughput screening), and useful compounds tend to have only subtle effects on the channel protein (strong effects tend to be lethal; cf. certain toxins in nature). Given the typically low affinity of VGIC modulators for their targets, relatively high (micromolar) concentrations of drug are required for efficacy. This brings a range of issues for human drug development: high oral dose requirements and limitations of human plasma concentrations that can be achieved. Perhaps not surprisingly, relatively few pharmaceutical or biotech companies have yet taken up the VGIC challenge. However, with over 140 of channels still to explore, there remains a great deal of opportunity for new medicines for complex CNS disorders, such as psychosis.

REFERENCES

1. Clissold M, Crowe SF. Comparing the effect of the subcategories of atypical antipsychotic medications on cognition in schizophrenia using a meta-analytic approach. J Clin Exp Neuropsychol. 2018;20:1–17.
2. Alexander SPH, Kelly E, Marrion N, et al. The concise guide to pharmacology 2015/16: Overview. Br J Pharmacol. 2015;172:5729–5743.
3. Catterall WA, Wisedchaisri G, Zheng N. The chemical basis for electrical signalling. Nat Chem Biol. 2017;13(5):455–463.
4. Armstrong CM, Hille B. Voltage-gated ion channels and electrical excitability. Neuron. 1998;20(3):371–380.
5. Catterall WA. Voltage-gated sodium channels at 60: structure, function and pathophysiology. J Physiol. 2012;590(11):2577–2589.
6. Kuang Q, Purhonen P, Hebert H. Structure of potassium channels. Cell Mol Life Sci. 2015;72(19):3677–3693.
7. Shen H, Zhou Q, Pan X, et al. Structure of a eukaryotic voltage-gated sodium channel at near-atomic resolution. Science. 2017;355(6328).
8. Lee CH, MacKinnon R. Structures of the human HCN1 hyperpolarization-activated channel. Cell. 2017;168(1–2):111–120.
9. Henderson R. Overview and future of single particle electron cryomicroscopy. Arch Biochem Biophys. 2015;581:19–24.
10. Hille B. Ion channels in excitable membranes. 3rd ed. Sinauer Associates. 2001.
11. Ranjan R, Khazen G, Gambazzi L, et al. Channelpedia: an integrative and interactive database for ion channels. Front Neuroinform. 2011;5:36.
12. Noebels J. Precision physiology and rescue of brain ion channel disorders. J Gen Physiol. 2017;149(5):533–546.
13. Schizophrenia Working Group of the Psychiatric Genomics, C. Biological insights from 108 schizophrenia-associated genetic loci. Nature. 2014;511:421–427.
14. Purcell SM, et al. Common polygenic variation contributes to risk of schizophrenia and bipolar disorder. Nature. 2009;460:748–752.
15. Horwitz T, Lam K, Chen Y, et al. A decade in psychiatric GWAS research. Mol Psychiatry. 2018 Jun 25. doi: 10.1038/s41380-018-0055-z.
16. Nanou E, Catterall WA. Calcium channels, synaptic plasticity, and neuropsychiatric disease. Neuron. 2018;98(3):466–481.
17. Catterall WA, Perez-Reyes E, Snutch TP, Striessnig J. Nomenclature and structure-function relationships of voltage-gated calcium channels. Pharmacol Rev 2005;57:411–425.
18. Tanabe T, Takeshima H, Mikami A, et al. Primary structure of the receptor for calcium channel blockers from skeletal muscle. Nature. 1987;328:313–318.
19. Dolphin AC. Calcium channel diversity: multiple roles of calcium channel subunits. Curr Opin Neurobiol. 2009; 19:237–244.
20. Cross-Disorder Group of the Psychiatric Genomics C. Identification of risk loci with shared effects on five major psychiatric disorders: a genome-wide analysis. Lancet 2013;381:1371–1379.
21. Ortega-Alonso A, Ekelund J, Sarin AP et al. Genome-wide association study of psychosis proneness in the Finnish population. Schizophr Bull. 2017. doi: 10.1093/schbul/sbx006.
22. Ripke S, O'Dushlaine C, Chambert K, Moran JL, Kahler AK, Akterin S, Bergen SE, Collins AL, Crowley JJ, Fromer M, et al. Genome-wide association analysis identifies 13 new risk loci for schizophrenia. Nat Genet. 2013;45(10):1150–1159.
23. Takahashi S, Glatt SJ, Uchiyama M, et al. Meta-analysis of data from the Psychiatric Genomics Consortium and additional samples supports association of CACNA1C with risk for schizophrenia. Schizophr Res. 2015;168(1–2):429–433.

24. Nie F, Wang X, Zhao P, Yang H, et al. Genetic analysis of SNPs in CACNA1C and ANK3 gene with schizophrenia: A comprehensive meta-analysis. Am J Med Genet B Neuropsychiatr Genet. 2015;168(8):637–648.
25. Jiang H, Qiao F, Li Z, et al. Evaluating the association between CACNA1C rs1006737 and schizophrenia risk: a meta-analysis. Asia Pac Psychiatry. 2015;7(3):260–267.
26. Porcelli S, Lee SJ, Han C, et al. CACNA1C gene and schizophrenia: a case-control and pharmacogenetic study. Psychiatr Genet. 2015;25(4):163–167.
27. Bhat S, Dao DT, Terrillion CE et al. CACNA1C (Cav1.2) in the pathophysiology of psychiatric disease. Prog Neurobiol. 2012;99(1):1–14.
28. Sinnegger-Brauns MJ, Huber IG, Koschak A, et al. Expression and 1,4-dihydropyridine-binding properties of brain L-type calcium channel isoforms. Mol Pharmacol 2009;75:407–414.
29. Bhat S, Dao DT, Terrillion CE et al. CACNA1C (Cav1.2) in the pathophysiology of psychiatric disease. Prog Neurobiol. 2012;99(1):1–14.
30. Berger SM, Bartsch D. The role of L-type voltage-gated calcium channels Cav1.2 and Cav1.3 in normal and pathological brain function. Cell Tissue Res. 2014;357(2):463–476.
31. Bhat S, Dao DT, Terrillion CE et al. CACNA1C (Cav1.2) in the pathophysiology of psychiatric disease. Prog Neurobiol. 2012;99(1):1–14.
32. Yoshimizu T, Pan JQ, Mungenast AE et al. Functional implications of a psychiatric risk variant within CACNA1C in induced human neurons. Mol Psychiatry. 2015;20(2):162–169.
33. Bigos KL, Mattay VS, Callicott JH. Genetic variation in CACNA1C affects brain circuitries related to mental illness. Arch GenPsychiatry. 2010;67:939–945.
34. Barrett CF, Tsien RW. The Timothy syndrome mutation differentially affects voltage- and calcium dependent inactivation of Cav1.2 L-type calcium channels. Proc Nat Acad Sci USA. 2008; 105:2157–2162.
35. Dixon RE, Cheng EP, Mercado JL, Santana LF. L-type Ca2+ channel function during Timothy syndrome. Trends Cardiovasc Med. 2012;22(3):72–76.
36. Jiang M, Swann JW. A role for L-type calcium channels in the maturation of parvalbumin-containing hippocampal interneurons. Neuroscience. 2005; 135:839–850.
37. Lewis DA. Inhibitory neurons in human cortical circuits: substrate for cognitive dysfunction in schizophrenia. Curr Opin Neurobiol. 2014;26:22–26.
38. Vinet J, Sík A. Expression pattern of voltage-dependent calcium channel subunits in hippocampal inhibitory neurons in mice. Neuroscience. 2006;143(1):189–212.
39. Lee AS, Ra S, Rajadhyaksha AM, Britt JK et al. Forebrain elimination of cacna1c mediates anxiety-like behavior in mice. Mol Psychiatry. 2012; 17:1054–1055.
40. White JA, McKinney BC, John MC et al. Conditional forebrain deletion of the Ltype calcium channel Ca V 1.2 disrupts remote spatial memories in mice. Learning Memory. 2008;15:1–5.
41. De Jesús-Cortés H, Rajadhyaksha AM, Pieper AA. Cacna1c: Protecting young hippocampal neurons in the adult brain. Neurogenesis. 2016;3(1):e1231160.
42. Terrillion CE, Dao DT, Cachope R, et al. Reduced levels of Cacna1c attenuate mesolimbic dopamine system function. Genes Brain Behav. 2017 Feb 10. doi: 10.1111/gbb.12371.
43. Bigos KL, Mattay VS, Callicott JH. Genetic variation in CACNA1C affects brain circuitries related to mental illness. Arch Gen Psychiatry. 2010;67:939–945.
44. Liao P, Soong TW. CaV1.2 channelopathies: from arrhythmias to autism, bipolar disorder, and immunodeficiency. Pflugers Arch. 2010;460(2):353–359.
45. Casamassima F, Hay AC, Benedetti A, et al. L-type calcium channels and psychiatric disorders: a brief review. Am J Med Genet B Neuropsychiatr Genet. 2010;153B(8):1373–1390.
46. Anekonda TS, Quinn JF. Calcium channel blocking as a therapeutic strategy for Alzheimer's disease: the case for isradipine. Biochim Biophys Acta. 2011;1812(12):1584–1590.
47. Kuo PH, Chuang LC, Liu JR et al. Identification of novel loci for bipolar I disorder in a multi-stage genome-wide association study. Prog. Neuropsychopharm. Biol. Psychiat. 2014;51:58–64.
48. Strauss KA, Markx S, Georgi B et al. A population-based study of KCNH7 p.Arg394His and bipolar spectrum disorder. Hum Mol Genet. 2014;23:6395–6406.
49. Strauss KA, Markx S, Georgi B et al. A population-based study of KCNH7 p.Arg394His and bipolar spectrum disorder. Hum Mol Genet. 2014;23:6395–406.
50. Papa M, Boscia F, Canitano A et al. Expression pattern of the ether-a-gogo-related (ERG) K+ channel-encoding genes ERG1, ERG2, and ERG3 in the adult rat central nervous system. J Comp Neurol. 2003;466:119–135.
51. Thomas D, Karle CA, Kiehn J. The cardiac hERG/IKr potassium channel as pharmacological target: structure, function, regulation, and clinical applications. Curr Pharm Des. 2006;12:2271–2283.
52. Papa M, Boscia F, Canitano A et al. Expression pattern of the ether-a-gogo-related (ERG) K+ channel-encoding genes ERG1, ERG2, and ERG3 in the adult rat central nervous system. J Comp Neurol. 2003;466:119–135.
53. Chiesa N, Rosati B, Arcangeli A et al. A novel role for HERG K+ channels: spike-frequency adaptation. J Physiol. 1997;501(2): 313–318.
54. Xiao K, Sun Z, Jin X, et al. ERG3 potassium channel-mediated suppression of neuronal intrinsic excitability and prevention of seizure generation in mice. J Physiol. 2018 Jul 17. doi: 10.1113/JP275970.
55. Polcwiartek C, Kragholm K, Schjerning O et al. Cardiovascular safety of antipsychotics: a clinical overview. Expert Opin Drug Saf. 2016;15(5):679–688.
56. Bilet A, Bauer CK. Effects of the small molecule HERG activator NS1643 on Kv11.3 channels. PLoS One. 2012;7:e50886.
57. Meng J, Shi C, Li L et al. Compound ICA-105574 prevents arrhythmias induced by cardiac delayed repolarization. Eur J Pharmacol. 2013;718(1–3):87–97.
58. Xiao K, Sun Z, Jin X, et al. ERG3 potassium channel-mediated suppression of neuronal intrinsic excitability and prevention of seizure generation in mice. J Physiol. 2018 Jul 17. doi: 10.1113/JP275970.
59. Purcell SM, et al. Common polygenic variation contributes to risk of schizophrenia and bipolar disorder. Nature. 2009;460:748–752.
60. Fromer M, Roussos P, Sieberts SK et al. Gene expression elucidates functional impact of polygenic risk for schizophrenia. Nat Neurosci. 2016;19(11):1442–1453.
61. Guzman RE, Alekov AK, Filippov M, Hegermann J, Fahlke C. Involvement of ClC-3 chloride/ proton exchangers in controlling glutamatergic synaptic strength in cultured hippocampal neurons. Front Cell Neurosci. 2014;8:143.
62. Bustillo JR, Patel V, Jones T et al. Risk-conferring glutamatergic genes and brain glutamate plus glutamine in schizophrenia. Front Psychiatry. 2017;8:79.
63. Guzman RE, Alekov AK, Filippov M, Hegermann J, Fahlke C. Involvement of ClC-3 chloride/ proton exchangers in controlling glutamatergic synaptic strength in cultured hippocampal neurons. Front Cell Neurosci. 2014;8:143.
64. Riazanski V, Deriy LV, Shevchenko PD et al. Presynaptic CLC-3 determines quantal size of inhibitory transmission in the hippocampus. Nat Neurosci. 2011;14(4):487–494.
65. Lamb FS, Graeff RW, Clayton GH et al. Ontogeny of CLCN3 chloride channel gene expression in human pulmonary epithelium. Am J Respir Cell Mol Biol. 2001;24(4):376–381.
66. Duan DD. The ClC-3 chloride channels in cardiovascular disease. Acta Pharmacol Sin. 2011;32(6):675–684.
67. Arcangeli A, Becchetti A. Novel perspectives in cancer therapy: targeting ion channels. Drug Resist Updat. 2015;21–22:11–19.

68. Dehorter N, Ciceri G, Bartolini G et al. Tuning of fast-spiking interneuron properties by an activity-dependent transcriptional switch. Science. 2015;349(6253):1216–1220.
69. Shah MM, Hammond RS, Hoffman DA. Dendritic ion channel trafficking and plasticity. Trends Neurosci. 2017;33(2010): 307–316.
70. Wang S, Stanika RI, Wang X et al. Densin-180 controls the trafficking and signaling of L-type voltage-gated Cav1.2 Ca2+ channels at excitatory synapses. J Neurosci. 2017;37(18):4679–4691.
71. Goldfarb M. Voltage-gated sodium channel-associated proteins and alternative mechanisms of inactivation and block. Cell Mol Life Sci. 2012;69(7):1067–1076.
72. Norris AJ, Foeger NC, Nerbonne JM. Neuronal voltage-gated K+ (Kv) channels function in macromolecular complexes. Neurosci Lett. 2010;486(2):73–77.
73. Zhang Y, Zhang XF, Fleming MR et al. Kv3.3 channels bind Hax-1 and Arp2/3 to assemble a stable local actin network that regulates channel gating. Cell. 2016;165(2):434–448.
74. Yang S, Ben-Shalom R, Ahn M, et al. β-arrestin-dependent dopaminergic regulation of calcium channel activity in the axon initial segment. Cell Rep. 2016;16(6):1518–1526.
75. Kitai ST, Surmeier DJ. Cholinergic and dopaminergic modulation of potassium conductances in neostriatal neurons. Adv Neurol. 1993;60:40–52.
76. Greif GJ, Lin YJ, Liu JC, Freedman JE. Dopamine-modulated potassium channels on rat striatal neurons: specific activation and cellular expression. J Neurosci. 1995;15(6):4533–4544.
77. D'Adamo MC, Servettini I, Guglielmi L et al. 5-HT2 receptors-mediated modulation of voltage-gated K+ channels and neurophysiopathological correlates. Exp Brain Res. 2013;230(4): 453–462.
78. Atzori M, Lau D, Tansey EP et al. H2 histamine receptor-phosphorylation of Kv3.2 modulates interneuron fast spiking. Nat Neurosci. 2000;3(8):791–798.
79. Surmeier DJ, Kitai ST. D1 and D2 dopamine receptor modulation of sodium and potassium currents in rat neostriatal neurons. Prog Brain Res. 1993;99:309–324.
80. Lacey MG, Mercuri NB, North RA. Dopamine acts on D2 receptors to increase potassium conductance in neurones of the rat substantia nigra zona compacta. J Physiol. 1987;392:397–416.
81. Lopez AY, Wang X, Xu M, et al. Ankyrin-G isoform imbalance and interneuronopathy link epilepsy and bipolar disorder. Mol Psychiatry. 2017;22(10):1464–1472.
82. Gulchina Y, Xu SJ, Snyder MA, et al. Epigenetic mechanisms underlying NMDA receptor hypofunction in the prefrontal cortex of juvenile animals in the MAM model for schizophrenia. J Neurochem. 2017;143(3):320–333.

38

VOLTAGE-GATED ION CHANNELS IN NEURAL CIRCUITS IMPLICATED IN PSYCHOTIC DISORDERS

Charles H. Large

INTRODUCTION: UNDERSTANDING NETWORK DYSFUNCTION IN PSYCHOSIS

Debate about the nature of the pathology underlying psychotic disorders has continued since their first description. The apparent absence of organic brain changes fueled the concept of schizophrenia as a "functional" disorder. However, during the 1980s brain imaging techniques began to find consistent evidence for differences in the structure of the schizophrenic brain,[1,2,3] although there has been little evidence for a progressive degeneration aside from that linked to chronic antipsychotic drug use.[4] More recent theories of schizophrenia have considered "abnormalities in neural circuits and a fundamental cognitive process"[5] that are most likely developmental in origin. Molecular targets such as voltage-gated ion channels (VGICs) have a direct role in regulating the activity of neurons and neural circuits, and thus their involvement in "abnormalities in neural circuits" merits consideration.

Our current understanding of these abnormalities comes from many years of careful measurement of postmortem brain tissue from patients with schizophrenia and bipolar disorder (reviewed elsewhere in this book). Briefly, reduced spine density of cortical glutamatergic pyramidal neurons has been a consistent finding,[6,7,8,9] and has been proposed to give rise to reduced recurrent excitation within cortical circuits, which in turn leads to a perhaps compensatory reduction in the activation of local circuit GABAergic inhibitory neurons, in particular parvalbumin-positive (PV+) interneurons.[10] Alterations in markers for PV+ interneuron function, including PV itself, support this view.[11,12,13,14] There is a consensus of opinion that reduced PV+ interneuron function is an early feature of the pathology of schizophrenia which is associated with cognitive and perhaps early negative symptoms of the disorder.[15,16] Recent postmortem data confirm that PV+ interneurons are not lost, with the observed reduction in levels of PV, a calcium-binding protein, implying reduced activity.[17]

A second line of evidence comes from the study of N-methyl-D-aspartate (NMDA) antagonist drugs, such as ketamine or phencylidine, which can mimic some symptoms of psychosis and can induce behavioral abnormalities and cognitive deficits in animals.[18] Many years of research in both animals and humans has provided a reasonable understanding of how this specific pharmacological disruption of neuronal network activity can generate psychosis, cognitive, and perhaps also negative symptoms of schizophrenia. It also turns out that it is important to distinguish between acute and chronic effects of NMDA antagonist administration:

Acute treatment with an NMDA antagonist increases frontal cortical glutamate and dopamine transmission,[19] and increases spontaneous gamma frequency network synchronization.[20,21] This hyperactivity state is presumed to swamp task-related signals leading to an acute disruption of behavior and psychosis-like symptoms. The acute NMDA antagonist model may mimic the increase in resting-state gamma frequency EEG activity in patients with psychosis, and is in line with the concept of a "noisy brain"[22] in patients with psychotic disorders.

Reduced PV+ interneuron activity was initially thought to contribute to symptoms produced by acute NMDA receptor antagonists, with the suggestion that the drugs directly block NMDA receptors required for the normal activation of interneurons by glutamatergic pyramidal cell collaterals. However, activation of PV+ interneurons by pyramidal neurons, which is critical to their ability to synchronize cortical circuits at gamma frequencies, involves fast AMPA-mediated glutamatergic activation, not the slow excitatory responses mediated by NMDA receptors.[23,24] Thus, it seems likely that acute treatment with an NMDA antagonist does not immediately cause a reduction in PV+ interneuron activity, and this would be consistent with the observed enhancement of spontaneous network synchronization.

In contrast to acute NMDA antagonist challenge, repeated (subchronic) dosing with NMDA antagonists does lead to reduced levels of PV in frontal cortex and hippocampus of experimental animals.[25,26] The hypothesis is that prolonged network hyperactivity caused by repeated dosing with NMDA antagonists permanently damages sensitive PV+ interneurons possibly through direct excitotoxic effects.[27]

Repeated NMDA antagonist treatment is associated with reduced power and phase locking of *evoked* or task-related gamma-frequency electroencephalographic activity in rodents; these deficits are also consistently observed in patients with schizophrenia.[28,29,30,31,32] Reduced gamma EEG power is consistent with reduced PV+ interneuron function, and is observed in first-episode patients,[33] as well as subjects at high risk of developing schizophrenia, prior to the emergence of psychotic symptoms.[34] This suggests that neural circuit abnormalities involving PV+ interneurons are an early, perhaps fundamental, feature of the emergence of schizophrenia. Alterations in gamma-related biomarkers can also be observed in patients with psychotic bipolar disorder[35] and Fragile X syndrome.[36]

In conclusion, postmortem analysis of human schizophrenic and bipolar brains, coupled with observation of the effects of acute or repeated treatment with NMDA antagonists, provides the basis from which to propose certain consistent abnormalities in neural circuitry in patients with psychotic disorders. In particular, reduced function of PV+ interneurons may be an early feature, perhaps connecting multiple psychiatric diagnoses. This emerging hypothesis raises the possibility that modulation of PV+ interneurons could be a fruitful approach to the treatment of certain psychotic disorders. Several of the novel pharmacological approaches currently in clinical development target PV+ neurons (although not exclusively). These include muscarinic M1 receptor modulators and alpha7-nicotinic acetylcholine receptor modulators. However, a defining feature of PV+ interneurons is their fast firing ability, which is a requirement to coordinate the activity of cortical networks. VGICs, which confer this ability, may be an appropriate way to target PV+ interneuron function in patients with psychotic disorders. In this chapter, specific VGICs that have this potential are reviewed.

As a final note of introduction, deficits in PV+ interneuron function may be only one of many pathologies that can give rise to psychosis. Schizophrenia and bipolar disorder are almost certainly heterogeneous conditions. Thus, new drugs focused on these specific pathologies should be aimed at patients with symptoms or features that are consistent with the target hypothesis. For example, drugs intended to rescue PV+ interneuron pathology should be evaluated initially in patients who show evidence of reduced gamma frequency EEG activity. Initiatives to map out the heterogeneity of psychosis will be critical to the development of targeted treatments.[37]

SODIUM CHANNELS (NAV1.2)

Interest in voltage-gated sodium channels[38] as targets for the treatment for psychosis initially arose from small, investigator-led clinical studies of the effects of lamotrigine in patients with schizophrenia.[39] Lamotrigine was first developed as a broad-spectrum anticonvulsant, and was subsequently found to be a use-dependent inhibitor of voltage-gated sodium channels.[40] Voltage-gated sodium channels are essential to neural function, underpinning the depolarizing phase of the action potential. The channels show characteristic rapid activation and inactivation, followed by a slower recovery from inactivation that shapes the firing pattern of the neuron.[41] Lamotrigine, and other use-dependent sodium channel blockers, stabilize the inactivated state of the channels and thus reduce high-frequency firing, useful as a means to brake seizure activity. Lamotrigine later proved to be an effective mood-stabilizer for patients with bipolar disorder, with a unique ability to prevent depressive episodes.[42] The efficacy of lamotrigine in bipolar disorder has been linked to its sodium channel activity, although other pharmacological effects cannot be ruled out.[43,44] Lamotrigine has been shown to modulate glutamate release in vitro,[45] and this contributed to interest in the drug to control the perceived hyper-glutamatergic state thought to occur in schizophrenia.[46] As described earlier, this state can be modeled by acute administration of low doses of NMDA receptor antagonists, such as ketamine, phencyclidine, or MK-801.[47] Consistent with expectation, prior treatment with lamotrigine could prevent the acute effects of NMDA antagonists in rodents and humans.[48,49,50,51,52] However, subsequently, two double-blind placebo-controlled trials of lamotrigine as add-on therapy for the treatment of schizophrenia failed to show significant efficacy versus positive or negative symptoms, although one of the two studies did show a small, but significant effect on some measures of cognition.[53] In retrospect, it is possible that these clinical studies focused on the wrong population of patients. Preclinical efficacy of lamotrigine was observed in rodent models based on acute treatment with NMDA antagonists in otherwise normal animals. A recent human imaging study suggests that acute ketamine or PCP may model early episodes of psychosis,[54] associated with increased basal neural circuit activity. Intuitively, a sodium channel blocker might be expected to dampen this hyperactivity. However, the two GSK-sponsored trials recruited relatively chronic patients (time since illness onset is not reported, but the average age of subjects was 40 years[55]). In the later phase of the illness, and perhaps due to chronic antipsychotic treatment, other neural circuit changes may have occurred which are less susceptible to sodium channel inhibition. Future clinical studies might wish to evaluate the efficacy of lamotrigine to treat first-episode psychosis.

Recently, a genome-wide significant association has been reported between cognitive function and polymorphisms in the sodium channel gene SCN2A. An association was observed in both schizophrenia patients and healthy subjects, the latter showing a trend for reversed allelic directionality.[56] Sequencing of brain RNA in the same study revealed reduced expression of two of three SCN2A alternative transcripts in the schizophrenia group compared to the control samples. Mapping of the expression of Nav1.2 channels in human cortex and hippocampus[57] suggests coexpression with parvalbumin (PV), a marker for the fast-spiking GABAergic interneurons that coordinate gamma frequency network synchronization that has been strongly linked to cognitive function[58] and dysfunction in schizophrenia.[59,60] However, a wider distribution of Nav1.2 across other neuron types is probable,[61] and mutations in SCN2A that give rise to both gain of function or loss of function of Nav1.2 channels are associated with a range of severe epilepsy syndromes and developmental disorders.[62] Mutations in the SCN2A gene that give rise to reduced channel function or expression have been associated with autism.[63,64] We can hypothesize that a subtle reduction in PV interneuron sodium channel expression in schizophrenia could lead to cognitive impairment through a disruption of network synchronization; however, it is not obvious that a sodium channel inhibitor would correct this situation. Studies with lamotrigine do show counterintuitive positive effects on frontal cortical activity that are opposite to inhibitory effects of the drug on motor circuits.[65,66] Furthermore, clinical experience with the drug does point to positive effects on cognition in epilepsy populations that cannot be completely explained by reduced seizure burden.[67] Thus, Nav1.2 remains a target of potential interest for novel treatments for schizophrenia. Understanding the effects of sodium channel blocking drugs on biomarkers linked to specific circuit dysfunction in patients at different stages of illness will be an important next step to determine where and when these drugs might be best used.

KCNS3 (KV9.3)

The potassium channel alpha subunit, Kv9.3, coded by the KCNS3 gene[68] is a modulatory subunit that forms a heteromer with the Kv2.1 channel in brain and some peripheral tissues.[69,70] Kv2.1 channels mediate a delayed rectifier current that contributes to action potential repolarization, regulation of firing frequency, and resting membrane potential.[71] The channels have a somato-dendritic expression in the rat hippocampus and are present on both pyramidal cells and interneurons. In pyramidal neurons, knockdown of Kv2.1 expression using antisense RNA results in a broadening of the somatic action potential, but only under conditions of rapid firing or during ictal activity,[72] suggesting that the channels assist repolarization during periods of excessive firing. Kv9.3 subunits are expressed exclusively on PV+ interneurons,[73] where they modify Kv2.1 function by accelerating closed-state inactivation and inhibiting open-state inactivation.[74] The presence of the Kv9.3 subunit thus ensures a robust Kv2.1-mediated delayed rectifier current in response to excitatory input. The dendritic, postsynaptic location of Kv2.1/9.3 channels on PV+ interneurons suggests that they impact the shape of AMPA-mediated excitatory postsynaptic potentials (EPSPs), ensuring that these have a short duration. This is important to maintain the temporal accuracy of PV+ interneuron activation by glutamatergic inputs from pyramidal neurons in the network, allowing for precisely timed inhibitory feedback at gamma frequencies.[75,76] Given the high rates of activation at the pyramidal neuron-PV+ interneuron synapse, the presence of the Kv9.3 subunit helps to maintain the Kv2.1-mediated repolarizing current, which would otherwise tend to inactivate. A reduction in the expression of KCNS3 mRNA has been observed in postmortem tissue from patients with schizophrenia.[77] Messenger RNA levels were 23% lower in schizophrenia brain tissue overall, and 40% lower when assessed in individual PV+ interneurons. In this same study, no changes in KCNS3 mRNA levels were observed in macaque monkeys exposed to antipsychotic drugs for up to 27 months, suggesting that the reduction observed in human postmortem schizophrenia brain was not due to prior antipsychotic treatment. The relevance of these changes in KCNS3 mRNA expression to network function and the symptoms of schizophrenia is set out in detail by Giorgiev et al.,[78] who propose that drugs that enhance the function of Kv9.3 subunits could be useful to restore network activity and would be predicted to treat cognitive symptoms in patients with schizophrenia. To date, compounds that selectively modulate Kv9.3 channels have not been described. Drugs that target Kv2.1 channels, could also be considered, but these are likely to have effects beyond PV+ interneurons, and in particular may affect cardiac or pulmonary function.

KCNC (KV3.1, KV3.2)

Kv3 voltage-gated potassium channels (Kv3.1–4) are activated by depolarization of the neuronal plasma membrane to potentials above -20 mV; they open rapidly during the depolarizing phase of the neuronal action potential to initiate repolarization and prevent sodium channel inactivation. As the neuron begins to repolarize, the channels deactivate quickly and so do not contribute significantly to the after-hyperpolarization.[79,80] These distinct properties allow the channels to terminate the action potential rapidly without compromising action potential threshold, rise time, or magnitude, and without increasing the duration of the refractory period that follows. Consequently, neurons expressing Kv3 channels can sustain action potential firing at high frequencies.[81] Kv3.1–3 subtypes are expressed mainly in the central nervous system, whereas Kv3.4 channels are also found in skeletal muscle and sympathetic neurons.[82] Kv3.1–3 channel subtypes are differentially expressed in corticolimbic brain areas,[83,84,85] thalamus,[86] and cerebellum.[87] Kv3.1 and Kv3.3 channels are also expressed at high levels in auditory brainstem nuclei.[88,89]

Kv3.1 and Kv3.2 channels are particularly expressed by fast-spiking, PV+ interneurons in corticolimbic circuits. The importance of these interneurons for higher brain function is summarized earlier in this chapter. Preliminary evidence suggests reduced expression of Kv3.1, but not Kv3.2 in the dorsolateral prefrontal cortex of patients with schizophrenia who had not been taking antipsychotic drugs for at least 2 months before death.[90] Reductions were seen in levels of both mRNA and protein. Subjects that were on antipsychotic drugs at the time of death did not show a reduction in Kv3.1 expression. In the same study, chronic treatment of rats with haloperidol or risperidone led to an increase in Kv3.1 protein in prefrontal cortex. Reduced Kv3.1 expression has also been observed in an animal model of cortical dysfunction that recapitulates aspects of the cognitive and social deficits observed in schizophrenia, and which shows reduced PV+ interneuron function.[91,92] A 20%–30% reduction in Kv3.1 message in this model was accompanied by a reduction in PV mRNA, a calcium-binding protein whose expression reflects levels of activity of the interneurons, suggesting an association between the expression of Kv3.1 channels and firing activity. Studies of mice with a targeted deletion of Kv3.1 or Kv3.2 channels confirm their importance for the ability of PV+ interneurons to synchronize neural activity.[93,94]

Compounds that enhance the function of Kv3.1 and Kv3.2 channels have been described.[95,96] Rosato-Siri et al. showed that one of these, AUT1, caused a leftward shift in the voltage-dependence of activation of human recombinant Kv3.1 and Kv3.2 channels. The compound also restored the ability of somatosensory cortex PV+ interneurons to fire at high frequency.[97] Related compounds have been evaluated in adult rodents that previously received PCP for 7 days (scPCP), followed by 6 weeks' washout. As described earlier, subchronic dosing with PCP produces an enduring cognitive and social behavioral deficit in the rats, with evidence of reduced PV expression in prefrontal cortex and hippocampus.[98] These deficits were rescued by acute treatment with AUT00206, an analogue of AUT1.[99] In this same study, 3 weeks' chronic treatment of the scPCP rats with AUT00206 consistently rescued their cognitive deficit, and postmortem analysis showed that the drug increased the PV cell count in infralimbic cortex and hippocampus, consistent with enhanced PV+ interneuron activity. These data using novel modulators of Kv3 channels suggest that the drugs could restore the activity of PV+ interneurons, restore network function, and ameliorate cognitive and negative symptoms in patients with schizophrenia. A first clinical trial of AUT00206 in patients with schizophrenia is in progress www.clinicaltrials.gov (NCT03164876).

DISCUSSION

This chapter has considered specific examples of ion channels that have been implicated in the pathophysiology of psychotic disorders or else are considered credible targets to correct abnormalities of neural circuit dysfunction deduced from post-mortem pathology or animal models. This review is not exhaustive, since it does not cover alternative pathologies that have been proposed to underlie psychotic illness; for example, neuroinflammatory mechanisms,[100] and this review does not cover the potential opportunity to target ion channels to regulate dopamine neurotransmission, the core target of all current marketed antipsychotic drugs. With respect to this latter, a number of ion channel targets with links to dopamine were investigated in the 1980s and 1990s. For example, Kv1 channels regulating dopamine release,[101] and the coupling of dopamine receptors to sodium and potassium channels on striatal neurons.[102,103] However, this early research does not appear to have progressed toward clinical development of new therapeutics.

Improving understanding of the neural circuits that likely underpin the development and expression of psychosis has allowed the rational selection of novel VGIC targets. The VGIC targets described in this chapter were highlighted because of their specific pattern of expression on relevant neural elements (Kv9.3 and Kv3.1/Kv3.2), and/or because "use-dependent" mechanisms allow for relatively safe (i.e., subtle) levels of channel modulation (NaV1.2). For example, analysis of the efficacy of sodium channel blocking drugs in the treatment of epilepsy suggests that just 10%–20% channel modulation is required for clinical effect, which is just as well, since higher levels of modulation are associated with a strong neural depression.[104] Thus, in practice, successful targeting of VGICs for the treatment of psychotic disorders requires both a link to the underlying disease mechanism and a means to modulate the channels without disrupting normal neural (or cardiac) function. Sodium channel blockers, such as lamotrigine, are well tolerated at doses that are used in the treatment of bipolar disorder to stabilize mood. A good balance between efficacy and CNS side-effects may also be possible with Kv3.1/Kv3.2 modulators, given the benign side-effect profile observed so far in animals and humans (Autifony Therapeutics Ltd, unpublished results). It is too early to judge the potential for Kv9.3 modulator compounds, which do not yet exist, and as described earlier, the therapeutic window available for Kv7 modulators may not be consistent with the treatment of psychosis.

In conclusion, VGICs offer means to modulate directly the activity of neural circuits implicated in psychosis. As discussed in the previous chapter, they are not easy proteins to target for drug discovery, lacking traditional ligand binding sites and requiring often complex functional assays to screen for suitable modulators. However, VGICs have a singular advantage over other drug targets: their modulation tends to alter neural function in real time and thus they lend themselves to use of noninvasive translational techniques, such as electroencephalography, magnetoelectroencephalography, and functional brain imaging. These techniques can be used to confirm CNS effect in early clinical trials. Furthermore, there are a range of specific electrophysiological biomarkers that have been shown to be altered in psychosis that can be used to assess novel VGIC drug effect (e.g., evoked potentials such as mismatch negativity[105]). Given the enduring belief that psychosis is a "functional" disorder, it seems reasonable to expect that drugs targeting VGICs will emerge in the future and find a niche in the treatment of these devastating illnesses.

REFERENCES

1. Andreasen NC, Smith MR, Jacoby CG et al. Ventricular enlargement in schizophrenia: definition and prevalence. Am J Psychiatry. 1982. 139(3):292–296.
2. Owens DG, Johnstone EC, Crow TJ et al. Lateral ventricular size in schizophrenia: relationship to the disease process and its clinical manifestations. Psychol Med. 1985. 15(1):27–41.
3. Tyrera P, Mackay A. Schizophrenia: no longer a functional psychosis. TiNs. 1986. 9:537–538.
4. Lewis DA. Antipsychotic medications and brain volume: do we have cause for concern? Arch Gen Psychiatry. 2011. 68(2):126–127.
5. Andreasen NC. Schizophrenia: the fundamental questions. Brain Res Brain Res Rev. 2000. 31(2–3):106–112.
6. Garey LJ, Ong WY, Patel TS et al. Reduced dendritic spine density on cerebral cortical pyramidal neurons in schizophrenia. J Neurol Neurosurg Psychiatry. 1998. 65:446–453.
7. Sweet RA, Henteleff RA, Zhang W et al. Reduced dendritic spine density in auditory cortex of subjects with schizophrenia. Neuropsychopharmacology. 2009. 34:374–389.
8. Glantz LA, Lewis DA (2000) Decreased dendritic spine density on prefrontal cortical pyramidal neurons in schizophrenia. Arch Gen Psychiatry. 57:65–73.
9. Konopaske GT, Lange N, Coyle JT, Benes FM Prefrontal cortical dendritic spine pathology in schizophrenia and bipolar disorder. JAMA Psychiatry. 2014. 71(12):1323–1331.
10. Glausier JR, Lewis DA. Dendritic spine pathology in schizophrenia. Neuroscience. 2013. 251, 90–107.
11. Hashimoto T, Volk DW, Eggan SM et al. Gene expression deficits in a subclass of GABA neurons in the prefrontal cortex of subjects with schizophrenia. J Neurosci. 2003. 23(15):6315–6326.
12. Benes FM, Lim B, Matzilevich D et al. Circuitry-based gene expression profiles in GABA cells of the trisynaptic pathway in

schizophrenics versus bipolars. Proc Natl Acad Sci U S A. 2008. 105(52):20935–20940.
13. Volk DW, Radchenkova PV, Walker EM et al. Cortical opioid markers in schizophrenia and across postnatal development. Cereb Cortex. 2012. 22(5):1215–1223.
14. Lewis DA, Curley AA, Glausier JR, Volk DW. Cortical parvalbumin interneurons and cognitive dysfunction in schizophrenia. Trends Neurosci. 2012. 35(1):57–67.
15. Lisman J. Excitation, inhibition, local oscillations, or large-scale loops: what causes the symptoms of schizophrenia? Curr Opin Neurobiol. 2012. 22(3):537–544.
16. Millan MJ, Andrieux A, Bartzokis G et al. Altering the course of schizophrenia: progress and perspectives. Nat Rev Drug Discov. 2016. 15(7):485–515.
17. Enwright JF, Sanapala S, Foglio A, Berry R, Fish KN, Lewis DA. Reduced labeling of parvalbumin neurons and perineuronal nets in the dorsolateral prefrontal cortex of subjects with schizophrenia. Neuropsychopharmacology. 2016. 41(9):2206–2214.
18. Large CH. Do NMDA receptor antagonist models of schizophrenia predict the clinical efficacy of antipsychotic drugs? J Psychopharmacol. 2007. 21(3):283–301.
19. Adams B, Moghaddam B. Corticolimbic dopamine neurotransmission is temporally dissociated from the cognitive and locomotor effects of phencyclidine. J Neurosci. 1998. 18(14):5545–5554.
20. Saunders JA, Gandal MJ, Siegel SJ. NMDA antagonists recreate signal-to-noise ratio and timing perturbations present in schizophrenia. Neurobiol Dis. 2012. 46(1):93–100.
21. Hiyoshi T, Kambe D, Karasawa J, Chaki S. Differential effects of NMDA receptor antagonists at lower and higher doses on basal gamma band oscillation power in rat cortical electroencephalograms. Neuropharmacology. 2014. 85:384–396.
22. White RS, Siegel SJ. Cellular and circuit models of increased resting-state network gamma activity in schizophrenia. Neuroscience. 2016. 321:66–76.
23. Hu H, Martina M, Jonas P. Dendritic mechanisms underlying rapid synaptic activation of fast-spiking hippocampal interneurons. Science. 2010. 327(5961):52–58.
24. Rotaru DC, Yoshino H, Lewis DA et al. Glutamate receptor subtypes mediating synaptic activation of prefrontal cortex neurons: relevance for schizophrenia. J Neuroscience. 2011. 31(1):142–156.
25. Pratt JA, Winchester C, Egerton A et al. Modelling prefrontal cortex deficits in schizophrenia: implications for treatment. Br J Pharmacol. 2008. 153(Suppl 1):S465–S470.
26. Neill JC, Barnes S, Cook S et al. Animal models of cognitive dysfunction and negative symptoms of schizophrenia: focus on NMDA receptor antagonism. Pharmacol Ther. 2010. 128(3):419–432.
27. Farber NB, Wozniak DF, Price MT et al. Age-specific neurotoxicity in the rat associated with NMDA receptor blockade: potential relevance to schizophrenia? Biol Psychiatry. 1995. 38(12):788–796.
28. Kwon JS, O'Donnell BF, Wallenstein GV et al. Gamma frequency-range abnormalities to auditory stimulation in schizophrenia. Arch Gen Psychiatry. 1999. 56:1001–1005.
29. Woo T-UW, Spencer K, McCarley RW. Gamma oscillation deficits and the onset and early progression of schizophrenia. Harv Rev Psychiatry. 2010. 18(3):173–189.
30. Hirano Y, Oribe N, Kanba S et al. Spontaneous gamma activity in schizophrenia. JAMA Psychiatry. 2015. 72(8):813–821.
31. Spencer KM, Salisbury DF, Shenton ME, McCarley RW. Gamma-band auditory steady-state responses are impaired in first episode psychosis. Biol Psychiatry. 2008. 64:369–375.
32. Gonzalez-Burgos G, Cho RY, Lewis DA. Alterations in cortical network oscillations and parvalbumin neurons in schizophrenia. Biol Psychiatry. 2015. 77(12):1031–1040.
33. Haenschel C, Bittner RA, Waltz J et al. Cortical oscillatory activity is critical for working memory as revealed by deficits in early-onset schizophrenia. J Neuroscience. 2009. 29(30): 9481–9489.
34. Perez VB, Roach BJ, Woods SW et al. Early auditory gamma-band responses in patients at clinical high risk for schizophrenia. Suppl Clin Neurophysiol. 2013. 62:147–162.
35. Ethridge LE, Hamm JP, Pearlson GD et al: Event-related potential and time-frequency endophenotypes for schizophrenia and psychotic bipolar disorder. Biol Psychiatry. 2015. 77:127–136.
36. Ethridge LE, White SP, Mosconi MW et al. Reduced habituation of auditory evoked potentials indicate cortical hyper-excitability in Fragile X Syndrome. Translational Psychiatry. 2016. 6:e787.
37. Clementz BA, Sweeney JA, Hamm JP et al. Identification of distinct psychosis biotypes using brain-based biomarkers. Am J Psychiatry. 2016. 173(4):373–384.
38. Catterall WA, Goldin AL, Waxman SG. Nomenclature and structure-function relationships of voltage-gated sodium channels. Pharmacol Rev. 2005. 57:397–409.
39. Large CH, Webster EL, Goff DC. The potential role of lamotrigine in schizophrenia. Psychopharmacology. 2005. 181(3):415–436.
40. Xie X, Lancaster B, Peakman T, Garthwaite J. Interaction of the antiepileptic drug lamotrigine with recombinant rat brain type IIA Na+ channels and with native Na+ channels in rat hippocampal neurones. Pflugers Arch. 1995. 430(3):437–446.
41. Xie X, Dale TJ, John VH et al. Electrophysiological and pharmacological properties of the human brain type IIA Na+ channel expressed in a stable mammalian cell line. Pflugers Arch. 2001. 441(4):425–433.
42. Bowden CL, Singh V. Lamotrigine (Lamictal IR) for the treatment of bipolar disorder. Expert Opin Pharmacother. 2012. 13(17):2565–2571.
43. Xie X, Hagan RM. Cellular and molecular actions of lamotrigine: Possible mechanisms of efficacy in bipolar disorder. Neuropsychobiology. 1998. 38(3):119–130.
44. Large CH, Di Daniel E, Li X, George MS. Neural network dysfunction in bipolar depression: clues from the efficacy of lamotrigine. Biochem Soc Trans. 2009. 37(Pt 5):1080–1084.
45. Leach MJ, Marden CM, Miller AA. Pharmacological studies on lamotrigine, a novel potential antiepileptic drug: II. Neurochemical studies on the mechanism of action. Epilepsia. 1986. 27(5):490–497.
46. Javitt DC. Glutamatergic theories of schizophrenia. Isr J Psychiatry Relat Sci. 2010. 47(1):4–16.
47. Large CH. Do NMDA receptor antagonist models of schizophrenia predict the clinical efficacy of antipsychotic drugs? J Psychopharmacol. 2007. 21(3):283–301.
48. Large CH, Bison S, Sartori I et al. The efficacy of sodium channel blockers to prevent phencyclidine-induced cognitive dysfunction in the rat: potential for novel treatments for schizophrenia. J Pharmacol Exp Ther. 2011. 338(1):100–113.
49. Quarta D, Large CH. Effects of lamotrigine on PCP-evoked elevations in monoamine levels in the medial prefrontal cortex of freely moving rats. J Psychopharmacol. 2011. 25(12):1703–1711.
50. Hunt MJ, Garcia R, Large CH, Kasicki S. Modulation of high-frequency oscillations associated with NMDA receptor hypofunction in the rodent nucleus accumbens by lamotrigine. Prog Neuropsychopharmacol Biol Psychiatry. 2008. 32(5):1312–1319.
51. Gozzi A, Large CH, Schwarz A et al. Differential effects of antipsychotic and glutamatergic agents on the phMRI response to phencyclidine. Neuropsychopharmacology. 2008. 33(7):1690–1703.
52. Doyle OM, De Simoni S, Schwarz AJ et al. Quantifying the attenuation of the ketamine pharmacological magnetic resonance imaging response in humans: a validation using antipsychotic and glutamatergic agents. J Pharmacol Exp Ther. 2013. 345(1):151–160.
53. Goff DC, Keefe R, Citrome L et al. Lamotrigine as add-on therapy in schizophrenia: results of 2 placebo-controlled trials. J Clin Psychopharmacol. 2007. 27(6):582–589.
54. Anticevic A, Corlett PR, Cole MW et al. N-methyl-D-aspartate receptor antagonist effects on prefrontal cortical connectivity better model early than chronic schizophrenia. Biol Psychiatry. 2015. 77(6):569–580.
55. Goff DC, Keefe R, Citrome L et al. Lamotrigine as add-on therapy in schizophrenia: results of 2 placebo-controlled trials. J Clin Psychopharmacol. 2007. 27(6):582–589.
56. Dickinson D, Straub RE, Trampush JW et al. Differential effects of common variants in SCN2A on general cognitive ability, brain

physiology, and messenger RNA expression in schizophrenia cases and control individuals. JAMA Psychiatry. 2014. 71(6):647–656.
57. Wang W, Takashima S, Segawa Y et al. The developmental changes of Na(v)1.1 and Na(v)1.2 expression in the human hippocampus and temporal lobe. Brain Res. 2011. 1389:61–70.
58. Cannon J, McCarthy MM, Lee S et al. Neurosystems: brain rhythms and cognitive processing. Eur J Neurosci. 2014. 39(5):705–719.
59. Gonzalez-Burgos G, Cho RY, Lewis DA. Alterations in cortical network oscillations and parvalbumin neurons in schizophrenia. Biol Psychiatry. 2015. 77(12):1031–1040.
60. Uhlhaas PJ, Singer W. High-frequency oscillations and the neurobiology of schizophrenia. Dialogues Clin Neurosci. 2013. 15(3):301–313.
61. Whitaker WR, Clare JJ, Powell AJ et al. Distribution of voltage-gated sodium channel alpha-subunit and beta-subunit mRNAs in human hippocampal formation, cortex, and cerebellum. J Comp Neurol. 2000. 422(1):123–139.
62. Wolff M, Johannesen KM, Hedrich UB et al. Genetic and phenotypic heterogeneity suggest therapeutic implications in SCN2A-related disorders. Brain. 2017 Mar 4. doi: 10.1093/brain/awx054.
63. Sanders SJ, Murtha MT, Gupta AR et al. De novo mutations revealed by whole-exome sequencing are strongly associated with autism. Nature. 2012. 485(7397):237–241.
64. Tavassoli T, Kolevzon A, Wang AT et al. De novo SCN2A splice site mutation in a boy with Autism spectrum disorder. BMC Med Genet. 2014. 15:35.
65. Li X, Ricci R, Large CH et al. Interleaved transcranial magnetic stimulation and fMRI suggests that lamotrigine and valproic acid have different effects on corticolimbic activity. Psychopharmacology. 2010. 209(3):233–244.
66. Li X, Large CH, Ricci R et al. Using interleaved transcranial magnetic stimulation/functional magnetic resonance imaging (fMRI) and dynamic causal modeling to understand the discrete circuit specific changes of medications: lamotrigine and valproic acid changes in motor or prefrontal effective connectivity. Psychiatry Res. 2011. 194(2):141–148.
67. Meador KJ, Baker GA. Behavioral and cognitive effects of lamotrigine. J Child Neurol. 1997. 12(Suppl 1):S44–S47.
68. Chandy KG, Grissmer S, Gutman GA et al. Voltage-gated potassium channels: Kv9.3. IUPHAR/BPS Guide to Pharmacology, http://www.guidetopharmacology.org/GRAC/ObjectDisplayForward?objectId=569.
69. Stocker M, Kerschensteiner D. Cloning and tissue distribution of two new potassium channel alpha-subunits from rat brain. Biochem Biophys Res Commun. 1998. 248(3):927–934.
70. Kerschensteiner D, Soto F, Stocker M. Fluorescence measurements reveal stoichiometry of K+ channels formed by modulatory and delayed rectifier alpha-subunits. Proc Natl Acad Sci U S A. 2005. 102(17):6160–6165.
71. Du J, Haak LL, Phillips-Tansey E et al. Frequency-dependent regulation of rat hippocampal somato-dendritic excitability by the K+ channel subunit Kv2.1. J Physiol. 2000. 522(Pt 1):19–31.
72. Du J, Haak LL, Phillips-Tansey E et al. Frequency-dependent regulation of rat hippocampal somato-dendritic excitability by the K+ channel subunit Kv2.1. J Physiol. 2000. 522(Pt 1):19–31.
73. Georgiev D, González-Burgos G, Kikuchi M et al. Selective expression of KCNS3 potassium channel α-subunit in parvalbumin-containing GABA neurons in the human prefrontal cortex. PLoS One. 2012. 7(8):e43904.
74. Kerschensteiner D, Monje F, Stocker M. Structural determinants of the regulation of the voltage-gated potassium channel Kv2.1 by the modulatory α-subunit Kv9.3. J Biol Chem. 2003. 278(20):18154–18161.
75. Ainsworth M, Lee S, Cunningham MO et al. Rates and rhythms: a synergistic view of frequency and temporal coding in neuronal networks. Neuron. 2012. 75(4):572–583.
76. Rotaru DC, Lewis DA, Gonzalez-Burgos G. The role of glutamatergic inputs onto parvalbumin-positive interneurons: relevance for schizophrenia. Rev Neurosci. 2012. 23(1):97–109.
77. Georgiev D, Arion D, Enwright JF et al. Lower gene expression for KCNS3 potassium channel subunit in parvalbumin-containing neurons in the prefrontal cortex in schizophrenia. Am J Psychiatry. 2014. 171(1):62–71.
78. Georgiev D, Arion D, Enwright JF et al. Lower gene expression for KCNS3 potassium channel subunit in parvalbumin-containing neurons in the prefrontal cortex in schizophrenia. Am J Psychiatry. 2014. 171(1):62–71.
79. Rudy B, Chow A, Lau D et al. Contributions of Kv3 channels to neuronal excitability. Ann N Y Acad Sci. 1999. 868:304–343.
80. Rudy B, McBain CJ. Kv3 channels: voltage-gated K+ channels designed for high-frequency repetitive firing. Trends Neurosci. 2001. 24(9):517–526.
81. Martina M, Schultz JH, Ehmke H et al. Functional and molecular differences between voltage-gated K+ channels of fast-spiking interneurons and pyramidal neurons of rat hippocampus. J Neurosci. 1998. 18(20):8111–8125.
82. Weiser M, Vega-Saenz de Miera E, Kentros C et al. Differential expression of Shaw-related K+ channels in the rat central nervous system. J Neurosci. 1994. 14(3 Pt 1):949–972.
83. Chow A, Erisir A, Farb C et al. K(+) channel expression distinguishes subpopulations of parvalbumin- and somatostatin-containing neocortical interneurons. J Neurosci. 1999. 19(21):9332–9345.
84. McDonald AJ, Mascagni F. Differential expression of Kv3.1b and Kv3.2 potassium channel subunits in interneurons of the basolateral amygdala. Neuroscience. 2006. 138(2):537–547.
85. Chang SY, Zagha E, Kwon ES et al. Distribution of Kv3.3 potassium channel subunits in distinct neuronal populations of mouse brain. J Comp Neurol. 2007. 502(6):953–972.
86. Kasten MR, Rudy B, Anderson MP. Differential regulation of action potential firing in adult murine thalamocortical neurons by Kv3.2, Kv1, and SK potassium and N-type calcium channels. J Physiol. 2007. 584(Pt 2):565–582.
87. Sacco T, De Luca A, Tempia F. Properties and expression of Kv3 channels in cerebellar Purkinje cells. Mol Cell Neurosci. 2006. 33(2):170–179.
88. Grigg JJ, Brew HM, Tempel BL. Differential expression of voltage-gated potassium channel genes in auditory nuclei of the mouse brainstem. Hear Res. 2000. 140(1–2):77–90.
89. Li W, Kaczmarek LK, Perney TM. Localization of two high-threshold potassium channel subunits in the rat central auditory system. J Comp Neurol. 2001. 437(2):196–218.
90. Yanagi M, Joho RH, Southcott SA et al. Kv3.1-containing K(+) channels are reduced in untreated schizophrenia and normalized with antipsychotic drugs. Mol Psychiatry. 2014. 19(5):573–579.
91. Neill JC, Barnes S, Cook S et al. Animal models of cognitive dysfunction and negative symptoms of schizophrenia: focus on NMDA receptor antagonism. Pharmacol Ther. 2010. 128(3):419–432.
92. Pratt JA, Winchester C, Egerton A et al. Modelling prefrontal cortex deficits in schizophrenia: implications for treatment. Br J Pharmacol. 2008. 153(Suppl 1):S465–S470.
93. Joho RH, Ho CS, Marks GA. Increased gamma- and decreased delta-oscillations in a mouse deficient for a potassium channel expressed in fast-spiking interneurons. J Neurophysiol. 1999. 82(4):1855–1864.
94. Harvey M, Lau D, Civillico E et al. Impaired long-range synchronization of gamma oscillations in the neocortex of a mouse lacking Kv3.2 potassium channels. J Neurophysiol. 2012. 108(3):827–833.
95. Rosato-Siri MD, Zambello E, Mutinelli C et al. A novel modulator of Kv3 potassium channels regulates the firing of parvalbumin-positive cortical interneurons. J Pharmacol Exp Ther. 2015. 354(3):251–260.
96. Brown MR, El-Hassar L, Zhang Y et al. Physiological modulators of Kv3.1 channels adjust firing patterns of auditory brain stem neurons. J Neurophysiol. 2016. 116(1):106–121.
97. Rosato-Siri MD, Zambello E, Mutinelli C et al. A novel modulator of Kv3 potassium channels regulates the firing of parvalbumin-positive cortical interneurons. J Pharmacol Exp Ther. 2015. 354(3):251–260.
98. Neill JC, Barnes S, Cook S et al. Animal models of cognitive dysfunction and negative symptoms of schizophrenia: focus on NMDA receptor antagonism. Pharmacol Ther. 2010. 128(3):419–432.
99. Leger M, Alvaro G, Large CH et al. Neurobiological dysfunction in a sub-chronic phencyclidine rat model of schizophrenia symptomatology. J Psychopharmacol. 2015. 29(8) suppl. D05.

100. Howes OD, McCutcheon R. Inflammation and the neural diathesis-stress hypothesis of schizophrenia: a reconceptualization. Transl Psychiatry. 2017. 7(2):e1024.
101. Lacey MG, Mercuri NB, North RA. Dopamine acts on D2 receptors to increase potassium conductance in neurones of the rat substantia nigra zona compacta. J Physiol. 1987. 392:397–416.
102. Kitai ST, Surmeier DJ. Cholinergic and dopaminergic modulation of potassium conductances in neostriatal neurons. Adv Neurol. 1993. 60:40–52.
103. Greif GJ, Lin YJ, Liu JC, Freedman JE. Dopamine-modulated potassium channels on rat striatal neurons: specific activation and cellular expression. J Neurosci. 1995. 15(6):4533–4544.
104. Large CH, Kalinichev M, Lucas A et al. The relationship between sodium channel inhibition and anticonvulsant activity in a model of generalised seizure in the rat. Epilepsy Res. 2009. 85(1):96–106.
105. Javitt DC. Neurophysiological models for new treatment development in schizophrenia: early sensory approaches. Ann N Y Acad Sci. 2015. 1344:92–104.

PATHOPHYSIOLOGY

IMMUNE MECHANISMS

39

INFLAMMATORY MECHANISMS IN PSYCHOSIS

Anna P. McLaughlin, Carmine M. Pariante, and Valeria Mondelli

INTRODUCTION

The immune system is responsible for defending the body against injury, infections, and pathogens as well as protecting it from the detrimental effects of stress and environmental change. Accordingly, the immune system can influence and alter a wide variety of biological processes in order to protect itself, with intricate and far-reaching effects on mood, cognition, brain function, and physical health. When a threat is identified the immune system initially responds with acute inflammation, by recruiting immune cells and increasing blood flow, experienced as redness, warmth, and swelling. However, inflammation in the context of psychiatric disorders is not typically experienced or observed in such extremes, as it usually occurs systemically rather than being localized to a specific site of injury or infection. This type of chronic inflammation does not follow a clear biological trajectory, as it persists over long periods of time and is less obvious than acute inflammation. Nevertheless, chronic inflammation is a maladaptive process characterized by increased inflammatory markers in both the periphery and central nervous system (CNS). Indeed, chronic inflammation is believed to contribute to the development of many debilitating conditions such as depression,[1] diabetes,[2] and Alzheimer's,[3] and may explain the high rates of comorbid physical health problems and neurodegenerative processes that exist in patients with psychosis.

It is still unclear what causes the increased immune system activity in patients with psychosis, but epidemiological studies show a potential link between early life infection or trauma and increased risk of psychotic disorders.[4–6] Perinatal infection or trauma is hypothesized to result in heightened immune activation, where the immune system becomes persistently activated in a state of mild, chronic inflammation. Another hypothesis is that the immune system becomes "primed" in response to perinatal infection or trauma, and therefore exerts a greater inflammatory response to any event deemed as a threat or attack on the body, including social stress. This increased inflammation, primarily mediated by pro-inflammatory cytokines and microglia, could have a detrimental effect on neurotransmission, neurodevelopment, and cognitive processes, ultimately leading to the development of psychotic symptoms.[7,8] For other inflammatory mechanisms that are not discussed in this chapter, but which are also likely to be involved in the development of psychosis, the reader is referred to reviews discussing the role of the microbiome,[9] blood-brain barrier,[10] auto-immune antibodies[11] and genetic variability.[12]

CYTOKINES AND PSYCHOSIS

Activation of the immune system triggers immune cells to release small proteins known as cytokines, which are key to mediating inflammatory and anti-inflammatory responses in the body and CNS. Cytokines released by immune cells typically include interleukins (IL), soluble interleukins (sIL), interferons (IFNs), and tumor necrosis factors (TNFs). Cytokines are pleiotropic in nature and play an important role in cell signaling, with downstream effects that can alter neurotransmission, synaptic plasticity, behavior, and mood. Increased concentrations of pro-inflammatory cytokines in plasma, serum, and within the CNS are indicative of immune system activation. The presence of high levels of pro-inflammatory cytokines in patients with psychotic disorders have supported the hypothesis that immune activation, and more specifically abnormalities or upregulation of specific cytokine networks, may play a role in the pathogenesis of psychosis.

EVIDENCE FROM META-ANALYSES ON CYTOKINE LEVELS IN PSYCHOSIS

A considerable amount of evidence has accumulated from cross-sectional studies investigating cytokine abnormalities in the blood of patients with schizophrenia and bipolar disorder with mania.[13–15] However, the most compelling evidence of cytokine abnormalities in the pathogenesis of psychosis comes from studies of first-episode psychosis (FEP) patients, who are free from many of the confounds associated with antipsychotic medication and illness chronicity. Our group conducted a systematic review focusing exclusively on FEP patients, and found IL-6, TNF-α, and IL-1β to be consistently elevated across studies.[16] These findings were confirmed by a meta-analysis of medication-naïve FEP patients that also found increased levels of the same pro-inflammatory cytokines.[17] In an individual study that investigated both serum levels and leukocyte gene expression of cytokines, an increase in IL-6 and TNF-α was found in FEP patients compared with controls, suggesting that immune cells could be responsible for releasing increased levels of these circulating cytokines.[18]

Another meta-analysis evaluating the effect of antipsychotic medication in unmedicated FEP patient found that IL-6, IL-2, and IL-1β decreased significantly after antipsychotic treatment was started, but TNF-α, IL-17, and IFN-γ remained stable after commencing antipsychotic treatment.[19] These results are in line with that of Miller and colleagues,[15] suggesting that IL-6 and IL-1β seem to be consistently elevated during acute psychosis. The most recent and thorough meta-analysis of peripheral immune activity in unmedicated FEP patients found elevations in IL-6, IL-7, and IFN-γ that remained significant after accounting for data skew, study quality, and matching participants on both their physiology and environment.[20] Elevations in TNF-α, IL-1β, and IL-8, among others, were also observed, but these findings were not as robust to sensitivity analyses. Interestingly, the results of this study did not support immune activation only being present in a subgroup of patients with psychosis, but rather that elevated immune activity is a general feature of psychosis.

To understand whether pro and anti-inflammatory cytokines circulating in the periphery have an impact on CNS inflammation, studies sampling the cerebrospinal fluid (CSF) of participants can be used to measure concentrations of cytokines in the CNS. Two large meta-analyses of patients with schizophrenia found increases in IL-6 and IL-8 in both the blood and CSF of patients with schizophrenia.[21] IL-6 was increased regardless of whether patients were on medication, but levels were higher in recent-onset patients compared with chronic schizophrenia patients.[22] The concurrent increases of these cytokines support that increased inflammation in the periphery is accompanied by increased inflammation within the CNS. The consistent elevations of IL-6 in both the periphery and CNS warrants further investigation, as this pleiotropic cytokine is capable of dysregulating metabolism,[23] influencing neurogenesis,[24] and activating microglia,[25] which are all common features of schizophrenia.

Abnormal levels of many inflammatory cytokines may not, however, be specific to psychosis, but rather may be associated with the presence or severity of psychiatric symptoms. A recent-meta analysis found elevations of several pro-inflammatory cytokines, including IL-6, in schizophrenia, bipolar, and major depressive disorder patients during acute stages of illness.[26] Pharmacological treatment led to a reduction in pro-inflammatory cytokines and an increase in anti-inflammatory cytokines, but again this was not specific to schizophrenia or bipolar disorder with mania. The increase of pro-inflammatory cytokines during acute psychiatric episodes may be more indicative of the stress that these patients experience, rather than a specific set of cytokines being associated with a specific disorder. This may explain why many patients suffering with psychiatric disorders suffer from overlapping symptoms, such as cognitive decline and physical health issues.

EVIDENCE FROM LONGITUDINAL AND EPIDEMIOLOGICAL STUDIES ON CYTOKINES LEVELS AND PSYCHOSIS

Longitudinal studies investigating the causal role of cytokines in psychosis are rare, but they provide valuable insight as to whether cytokine abnormalities precede the confounding effects of antipsychotic medication, psychological stress, and cardio-metabolic risk factors associated with psychosis. The Avon Longitudinal Study of Parents and Children found a causal, dose-response association between higher IL-6 concentrations at age 9 and subclinical psychotic experiences at age 18, as well as psychotic disorder at age 18.[27] This association was independent of the effects of early childhood psychological and behavioral problems, and was still significant after adjusting for BMI, gender, ethnicity, and maternal depression. A more recent study found that higher serum C-reactive protein (CRP) levels in adolescence was associated with the development of schizophrenia by 27 years of age, with higher CRP being associated with earlier-onset schizophrenia at trend-level.[28] This linear, dose-response association persisted after controlling for age, sex, BMI, maternal education, and tobacco and alcohol use. While this study did not directly measure cytokine levels, CRP is a widely recognized marker of peripheral inflammation, which is induced at the molecular level by pro-inflammatory cytokines IL-1, IL-6, and TNF-α.[29] This was supported by a large meta-analysis investigating the effect of prenatal immune activation, showing that maternal elevations in CRP, IL-8, and IL-10 during pregnancy were associated with an increased risk of developing schizophrenia in offspring.[30]

Epidemiological studies have shown that viral infections, respiratory infections, and *Toxoplasma gondii* during pregnancy all increase the risk of offspring developing schizophrenia later in life.[5] However, it has been suggested that the infection is not necessarily the risk factor, but that the increase in maternal inflammatory cytokines may have a detrimental, long-lasting effect on the neurodevelopment of the fetus.[8] Specifically, increased levels of pro-inflammatory cytokines IL-8 (also known as CXCL8) and TNF-α during pregnancy have been associated with increased risk for schizophrenia in offspring,[4] while increased levels of anti-inflammatory cytokines during pregnancy have been associated with a lower risk of offspring developing psychosis.[31] Ellman and colleagues[32] found that exposure to IL-8 in utero was associated with decreased brain volume in adulthood, in patients with schizophrenia compared with healthy controls. This is also strongly supported by experimental animal models, showing that a disruption in the balance of maternal pro- and anti-inflammatory cytokines during the prenatal period leads to disturbed neurotransmission and reduced brain volume in offspring, as well as multiple schizophrenia-related behavioral abnormalities later in life.[33]

MECHANISMS LINKING INFLAMMATORY CYTOKINES TO DEVELOPMENT OF PSYCHOSIS

Considerable evidence now shows that disturbances in dopamine, glutamate, and gamma-aminobutyric acid (GABA) are involved in the pathogenesis of psychotic symptoms,[34] and evidence from animal models suggests that altered levels of cytokines may be an upstream trigger for these disturbances. Animal models of maternal immune activation can invasively measure cytokine and neurotransmitter levels, demonstrating how cytokines can influence synthesis and release of neurotransmitters, and how these changes are associated with psychotic-like behaviors. Dopamine hyperactivity has been strongly associated with the presence of acute psychotic symptoms, such as delusions and hallucinations,[35] and cytokines such as IL-6 can have a profound effect on dopaminergic activity. In rodent models of maternal immune activation, increased maternal IL-6 in rodents was associated with hyperactivity of the mesolimbic dopaminergic system and psychotic-like behaviors in rodent offspring.[36] Similarly, increased sensitivity to amphetamines increased in vitro striatal dopamine release and schizophrenia-like behaviors in attention processing, which manifested in adult offspring after prenatal immune activation.[37] Furthermore, these behaviors were reversed after administration of antipsychotics haloperidol and clozapine, which both act by blocking dopamine receptors.

Altered levels of cytokines may also be an upstream cause of disturbances in excitatory glutamate and inhibitory GABAergic neurotransmission. Increased levels of pro-inflammatory cytokines, IL-6 and IL-1β, have been shown to increase production of kynurenic acid, which is the only known, naturally occurring antagonist of the *N*-methyl-D-aspartate (NMDA) receptor.[38] NMDA receptor antagonism leads to increased glutamate release, which disrupts the balance between glutamate and GABA.[39] This imbalance then overstimulates the mesolimbic dopamine pathway increasing dopamine release, generating acute psychotic and delusional symptoms.[40] Similarly, dysfunctional NMDA receptors in the prefrontal cortex may inhibit the mesocortical dopamine pathway, reducing dopamine neurotransmission and generating the negative and affective symptoms typically observed in patients with schizophrenia.[40] Supporting this, in a mouse model of psychosis IL-6 was shown to be key in mediating the effect of ketamine, which is a potent NMDA antagonist and capable of inducing psychotic symptoms in healthy humans.[41] Again, both haloperidol and clozapine significantly reduce levels of kynurenic acid, which may contribute to their antipsychotic properties.[41] Accordingly, higher levels of kynurenic acid have been found in the CSF of medication-naïve FEP patients compared with healthy controls.[42]

Immune dysfunction as mediated through cytokines has also been suggested to contribute to reductions in brain structure, function, and neurogenesis. The hippocampus is one of the most affected brain regions, and hippocampal abnormalities are considered a trademark of psychosis even in individuals at risk for developing the disorder. Interestingly, the hippocampus is also sensitive to peripheral inflammation, as administration of a typhoid injection to healthy volunteers increased plasma levels of inflammatory cytokines (IL-6, IL-1, and TNF-α) and led to impaired hippocampal glucose metabolism and subsequent spatial memory deficits.[43] Increased IL-6 concentrations have also been associated with decreased hippocampal volume in FEP patients.[44] In UHR individuals who developed psychosis elevated levels of cytokines (TNF-α, IFN-γ, and IL-2) have been found to correlate with prefrontal gray matter tissue loss.[45] Similarly, in a sample of patients with chronic schizophrenia, patients with increased levels of inflammatory cytokines showed a 17% reduction in the volume of Broca's area and a 20% deficit in verbal fluency, compared with patients that had low levels of inflammatory cytokines.[46] Regarding neurogenesis, cytokines such as IL-1β have been found to exert a negative effect on neurogenesis through activation of the kynurenine pathway.[47] In a review investigating pro-inflammatory cytokines in vivo, both beneficial and detrimental effects of cytokines on neurogenesis were observed, suggesting that cytokines may play a highly complex role in the cognitive processes of neuropsychiatric disorders.[48] While clear associations have been observed between pro-inflammatory cytokines and deficits in brain structure and function, further research is warranted to investigate the exact mechanisms through which this occurs, particularly in the context of neurogenesis.

Considering the evidence from meta-analyses, longitudinal and epidemiological studies, and animal models, increased levels of pro-inflammatory cytokines appear to play a key role in the pathophysiology of psychosis. The strongest evidence comes from consistent observations of increased levels of pro-inflammatory cytokines during acute psychotic episodes and in FEP patients. The exact neurobiological mechanisms through which altered levels of cytokines influence mood and behavior in humans still requires further clarification, but it is likely that pro-inflammatory cytokines exert downstream effects on dopamine, glutamate, and GABA neurotransmission.[7] However, cytokines are not solely responsible for the immune abnormalities observed in patients with psychosis, as they themselves are actually messengers sent out from the CNS and they act as mediators between the peripheral immune system and the CNS. Exactly how and why pro-inflammatory cytokines are released from the CNS remains to be explored, and the answer may lie with the resident immune cells of the CNS, also known as microglia.

MICROGLIAL ACTIVATION AND PSYCHOSIS

Microglia are immune cells located within the CNS, also known as the macrophages of the brain. Microglia extend and contract to monitor the CNS for infections, pathogens, and neuronal injury, covering both networks of synaptic connections and glial signaling networks, to preserve efficient immune homeostasis within the CNS. When microglia identify a pathogen or injury in the brain they become "activated," whereby they change morphology, proliferate, and migrate to the site of injury to destroy the pathogen or remove dead cells.

In a healthy brain this process is an efficient immune response to prevent damage from insult or injury, however if microglia become chronically activated over extended periods of time they may remove or destroy cells that are essential for neuronal function.[49]

Microglia are typically characterized depending on their activity state; when microglia are monitoring or maintaining homeostasis within the CNS this is referred to as a quiescent or resting state and when they are responding to injury or inflammation this is considered an activated state. The activated state of microglia has also been further classified into two distinct phenotypes, depending on whether they are acting in a pro-inflammatory or anti-inflammatory manner within the CNS, referred to as M1 or M2 microglial activation respectively.[50] The M1 phenotype is associated with pro-inflammatory cytokines and cytotoxic effects, while the M2 phenotype is associated with anti-inflammatory cytokines and neuroprotective effects. It was originally believed that the M1 and M2 phenotypes could be differentiated through morphological analysis, however recent research indicates that these distinct states only seem to appear in vitro, but not in vivo.[51] Instead, it seems that microglia are constantly in flux responding to the environmental changes within the CNS, so the M1 and M2 phenotypes are more likely to be present in a continuum.

EVIDENCE FROM POSTMORTEM STUDIES

Trepanier and colleagues,[52] conducted the first systematic review on 22 studies assessing microglial markers (HLA, CD11b, CD68, and calprotectin) in the postmortem brain samples of schizophrenia patients. Of these, 11 studies reported an increase, 3 reported a decrease and 8 reported no difference in microglial density, however considerable heterogeneity was observed between brain regions and layers within regions being analyzed. Another factor that further complicated interpretation of findings was that many of the studies did not consider the effect of suicide on microglial activity. This is relevant because a study by Schneider et al.,[53] found that suicide, rather than psychiatric diagnosis, predicted increased microglial activation. Furthermore, in many postmortem studies a higher proportion of the patient group have committed suicide compared with the control group,[54] calling into question whether these changes are specifically associated with psychosis or more broadly associated with suicidality in psychiatric disorders. This demonstrates the need to consider the effect of suicide on microglial activation, in both patient and control groups. While there does seem to be a change of microglial activity in patients with psychosis, it is presently unclear whether stress and inflammation are common underlying pathways leading to a variety of psychiatric disorders, or whether specific disorders and symptoms are associated with specific inflammatory phenotypes.

More recently, van Kesteren and colleagues[55] conducted a meta-analysis of 41 studies and found a significant increase in microglial density in postmortem brain samples of schizophrenia patients compared with controls ($p = .003$). Variability between studies in the brain regions analyzed contributed to significant heterogeneity, and the authors cautioned despite finding a significant, positive result, that this could be due to the limited number of studies rather than a consistent effect. HLA and CD68 were the most commonly used markers of microglial density, however the authors pointed out that these markers cannot distinguish between microglia and macrophages. Although macrophages are unlikely to be present within brain tissue, new markers that are specific for microglia or macrophages would allow for greater confidence in results.

It may be that different markers of inflammation are associated with specific psychiatric symptoms. To date, only one group investigating neuroinflammation in psychosis have stratified their sample depending on individual inflammatory phenotype. Fillman et al.[56] found increased mRNA expression of microglial markers in the dorsolateral prefrontal cortex of patients with psychosis compared with controls ($p = .002$). The authors also defined a highly altered inflammatory group within sample, characterized by higher levels of IL-1b, IL-6, IL-8, and SERPINA3 mRNA expression (all at $p < .0001$), which constituted 19% psychotic patients and 5% of controls. Remarkably, this highly inflamed group did not include any subjects with schizoaffective disorder, patients with a depressive subtype of schizophrenia, nor any patients who had committed suicide, suggesting that these multiple markers of brain inflammation were highly associated with a "pure" diagnosis of schizophrenia. The same group then later went on to investigate inflammatory phenotype transdiagnostically and found that a significantly higher proportion of schizophrenia patients had a highly inflammatory profile (43%), compared to an almost-significant proportion of bipolar patients (32%) and a small proportion of controls (17%).[56] Had the authors further characterized bipolar patients by the presence of mania, this could have provided insight into whether these multiple markers of brain inflammation were predominantly associated with a diagnosis of schizophrenia, or psychotic and manic symptoms more broadly.

One study by Busse and colleagues[54] investigated hippocampal microglial activation in relation to paranoid or negative symptoms and found increased activation in schizophrenia patients with paranoid symptoms ($n = 10$) compared with controls ($n = 11$). In contrast, schizophrenia patients with predominantly negative symptoms ($n = 7$), demonstrated higher densities of hippocampal immune cells compared with controls. Unfortunately, due to the small sample size it is not possible to draw conclusions on whether paranoid schizophrenia represents a different inflammatory phenotype of schizophrenia associated with microglial activation, but this hypothesis does require further investigation.

Postmortem studies can only typically investigate the later stages of the disorder, but immune system activation is most likely to occur during the early stages, which is when the most brain tissue loss occurs. Studies that can investigate microglial activation in vivo in individuals at risk for psychosis, or shortly after they develop it, may be more representative of the immune system activation that is likely to occur in the prodromal stages of psychosis.

EVIDENCE FROM PET STUDIES

Microglial activation can be measured noninvasively using positron emission tomography (PET), a neuroimaging technique that involves injecting a radiotracer that binds to 18-kDa translocator protein (TSPO), which is an outer mitochondrial membrane expressed by activated microglia. TSPO expression is therefore used as a surrogate marker of microglial activation, and increased TSPO has been found in conditions where neuroinflammation is present such as stroke, neurodegeneration, and traumatic brain injury.[57] Providing further support of the association between TSPO and inflammation, injection of LPS also increases TSPO expression in humans as measured with PET.[58] Early PET studies investigating TSPO expression have typically used the first-generation radiotracer 11C-(R)-PK11195, however this tracer is now considered to have poor signal-to-noise ratio and nonspecific binding to TSPO. Studies using second-generation radiotracers have the benefit of improved signal-to-noise ratio for TSPO binding, but they also have high affinity to plasma proteins that are simultaneously increased during peripheral inflammation,[59] which is not always accounted for in analyses.

To date, PET studies investigating TSPO binding in patients with schizophrenia and psychosis have demonstrated highly conflicting results. Several early studies using first-generation radio-tracers found an increase in TSPO binding in patients compared with controls.[60–62] More recent studies using second-generation radiotracers have also found an increase in TSPO binding in patients compared with controls.[63] However, this has been contradicted by a similar amount of early and later studies using both first and second-generation radiotracers, which found no differences between groups.[64–69] In fact, some studies have even found a decrease in TSPO binding in patients compared with controls.[70–72] The numerous methodological issues associated with TSPO measurement in PET studies could somewhat account for the contradicting results, as one of the most pertinent issues is that TSPO is expressed also in endothelial cells and the levels of expression may not always completely correspond to expression from microglial cells. Furthermore, individuals have varying genetic affinity to TSPO radiotracers, which can lead to an approximate difference of 30% in TSPO binding between high-affinity and mixed-affinity individuals.[73]

Accordingly, meta-analyses of PET studies are faced with similar issues, as pooling data from studies that have used different first- and second-generation radiotracers and different outcome measures makes combined analysis difficult. Furthermore, even meta-analyses in this field can still have relatively low power to detect an effect once studies are grouped by the radiotracer used or analysis technique employed, or once patient groups are divided by illness stage. Despite these issues, two high-quality meta-analyses have been carried out in this field so far. Plaven-Sigray et al.[74] conducted the first and concluded that patients demonstrated decreased binding compared with controls, particularly in the hippocampus, frontal cortex, and temporal cortex. In contrast, the second meta-analysis by Marques et al.[75] came to a different conclusion depending on how they split the studies. When Marques et al., grouped studies that used binding potential (BP) as the measure of TSPO binding, they observed an overall increase in patients compared with controls. When the authors grouped studies using volume of distribution (VT) to as a measure of TSPO binding, they found no difference between groups. However, it should be noted that studies using BP as an outcome measure were mostly using first-generation radiotracers, while studies using VT were mostly using second-generation radiotracers, giving more weight to the conclusion that current evidence does not point to a difference in TSPO binding in patients with psychosis compared with controls.

Recent studies have sought to investigate the intricacies that may underlie the relationship between psychotic symptoms and TSPO binding. Some of the new hypotheses being put forward include TSPO binding being associated with GABA levels in the prefrontal cortex,[76] or that TSPO binding is associated with detrimental changes in hippocampal morphology as psychotic illness progresses.[70] One of the most recent studies, published after both meta-analyses, observed differential TSPO uptake in patients between acute psychotic episodes and post-treatment periods. The effect was differential depending on the age of patients, suggesting an age-related increase in TSPO binding in psychotic patients compared with controls.[77] These results are also related to the findings of Ottoy et al.,[73] who noticed that microglial activation was present in patients with schizophrenia, dependent on age. Although the sample sizes are again too small to draw conclusions from, the hypothesis of accelerated aging in patients with psychosis, driven by overactive microglia, has gained support from a number of different sources.[78,79] Such studies provide exciting new avenues to explore how microglial activity may drive the changes in brain structure, volume loss, and neurodegeneration that are observed in longitudinal studies of patients with psychosis.[80]

In summary, it is presently unclear whether TSPO binding is indeed increased or decreased in patients with psychosis, or whether is it associated with antipsychotic medication, or more broadly related to the stress they experience.[81] Changes in microglial activity in psychosis are likely to be altered on quite a small scale compared to acute inflammatory disorders, and it is also possible that increased TSPO binding may represent upregulated neuroprotective and anti-inflammatory activity of microglia and other immune cells. The future use of more specific radiotracers for microglia,[82] and larger sample sizes will prove useful in accurately identifying any potential differences in microglial activation that may exist in patients with psychosis.

EVIDENCE FROM ANIMAL STUDIES

Understanding whether changes in microglial activity or morphology is causally related to the development of psychotic symptoms is currently a critical area of research; however in vivo measurement of microglial activation in humans is currently limited to PET. Although animal models of psychosis cannot model many of the sociodemographic factors involved in the development of psychotic disorders, they do allow for the measurement of microglial changes to be studied in vivo. Changes in microglial activity and morphology preceding the development of psychotic-like behaviors of animals can be carefully extrapolated, to aid us in understanding how microglial activity may influence neurodevelopmental changes preceding the onset of psychosis.

Animal models investigating microglial changes in schizophrenia are lacking, but a recent study by Notter et al.,[83] measured TSPO expression in a mouse model of schizophrenia as well as a mouse model of acute, induced neurodegeneration. The authors found decreased TSPO expression in the prefrontal cortex, associated with increased inflammatory cytokines and schizophrenia-like behaviors, but in the neurodegenerative model they found strongly increased TSPO expression. Altered TSPO expression was not restricted to microglia but was also found in astrocytes and vascular endothelial cells. These findings highlight the need for studies to quantify microglial number, shape, and phenotype as well as peripheral inflammatory biomarkers when measuring TSPO expression, to understand how TSPO binding alters after stress or trauma.

Current techniques face many obstacles in accurately quantifying and classifying microglial activation, but evidence continues to emerge that microglial activity is linked to the pathophysiology of psychotic symptoms. However, it is possible that microglial activation is nonspecific to psychosis and more closely represents a state of severe stress closely linked to psychiatric symptoms and suicidal behaviors. Indeed, the topic of stress-induced microglial activation in the context of psychiatric disorders has recently been reviewed comprehensively,[81] with a consensus that while little is currently known and evidence is conflicting, there seems to be a mechanistic effect of microglia either in the development or increasing severity of psychiatric symptoms. Research into microglia may well continue to surprise us as we discover to what extent microglia contribute to healthy immune and neurodevelopmental function. Indeed, a new subtype of "dark microglia" was recently discovered in rodents after exposure to stress,[84] demonstrating how limited our current knowledge of microglia is.

MECHANISMS LINKING MICROGLIA ACTIVATION TO DEVELOPMENT OF PSYCHOSIS

In a healthy brain microglia respond to sudden injury or acute infection with pro-inflammatory activation, preventing further tissue from being damaged and removing the pathogen. Once removed, microglia quickly shift to anti-inflammatory, neuroprotective activity, so that debris clearance and tissue repair can take place. If pro-inflammatory microglial activity persists without a subsequent shift to anti-inflammatory activity, such as in the event of chronic stress, then this may lead to a state of chronic, low-grade inflammation within the CNS. Persistent pro-inflammatory microglial activation would lead to the removal or destruction of cells that are essential to neuronal function and could result in irreversible impairments in synaptic plasticity and neurotransmission, which has been hypothesized as a potential pathway to the development of psychotic symptoms.

Additionally, microglia are also believed to be responsible for synaptic pruning, which is a process that removes or "prunes" unnecessary or inefficient synapses. Immune activation during critical periods of neurodevelopment, such as in utero or adolescence, may induce aberrant microglial pruning, leading to disturbances in temporal, cognitive, and emotional processes. Both excessive and insufficient synaptic pruning can have deleterious consequences, as reduced activity in neurodevelopment can lead to deficits in synaptic connectivity, whereas overactivity has been linked to cognitive decline.[79,85,86] Paolicelli and colleagues,[86] were the first to demonstrate that microglia prune synapses during critical neurodevelopmental periods in mice. Interestingly, microglial pruning itself was necessary for the mice to develop healthy and mature synapses. A more recent review by the same group emphasized that the time window during which microglial pruning occurs greatly influences how detrimental aberrant pruning is on neural circuit formation and how long lasting the effects are.[87] The exact molecular mechanisms through which microglia and synapses communicate are still unclear, but it is possible that genetic factors could be mediating this process.[88]

ANTI-INFLAMMATORY TREATMENTS FOR PSYCHOSIS

The accumulating evidence describing immune activation in psychotic disorders opens exciting new possibilities for novel immune-modulating treatment strategies, which may prove more effective than current antipsychotic medications that target dopamine receptors. The tetracycline antibiotic minocycline has received a particularly great deal of attention in this regard, due to its broad anti-inflammatory properties and ability to reduce microglial activity after stress or immune activation.[89] When used as an add-on therapy with antipsychotic medication, minocycline has been found to ameliorate the negative symptoms of schizophrenia,[90–93] albeit with some inconsistencies between studies. Several meta-analyses have also confirmed the efficacy of minocycline for improving schizophrenia symptoms compared with placebo, when used as an add-on therapy.[94–96] A recently published study found not only that adjunctive minocycline improved symptoms but also that this improvement correlated with a reduction in levels of pro-inflammatory cytokines IL-1β and IL-6.[97] Unfortunately, minocycline does not seem be effective in first-episode psychosis, as a large clinical trial found no benefit in addition to routine clinical care,[98] Perhaps the efficacy of minocycline could be improved if patients are stratified by their baseline inflammation, as this would allow the drug to be targeted toward patients who are the most likely to benefit from it. At this early stage it seems minocycline has some benefit for patients already showing marked negative symptoms, but further understanding of how minocycline inhibits the immune response to improve such symptoms requires more detailed investigation (see Figure 39.1).

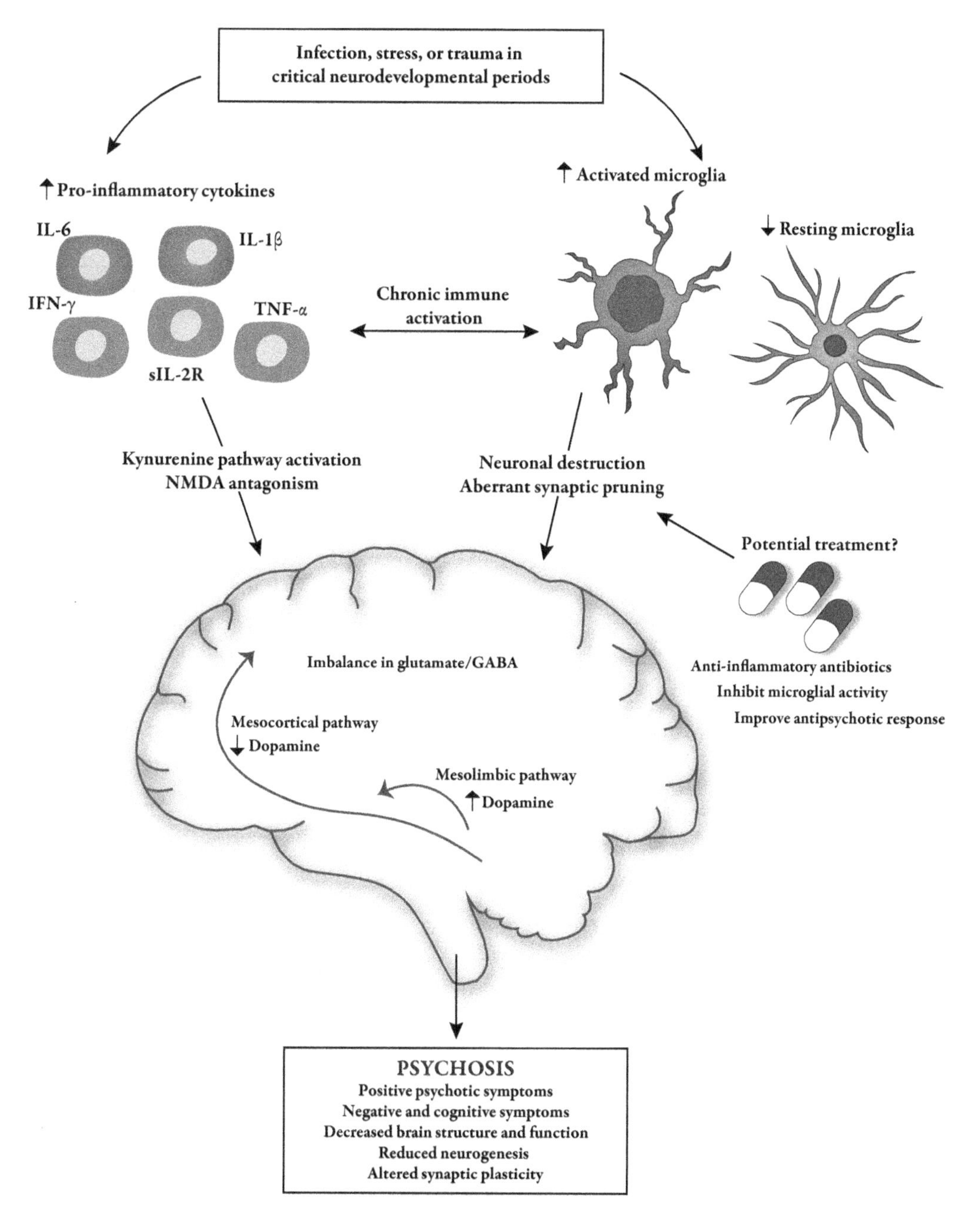

Figure 39.1 Infection, stress, or trauma during critical neurodevelopmental periods may lead to chronic inflammation, where increased pro-inflammatory cytokines and microglial activation persist. Pro-inflammatory cytokines could activate the kynurenine pathway, causing NMDA antagonism, leading to an imbalance of dopamine, glutamate, and GABA neurotransmission, while microglial activation could cause neuronal destruction and aberrant synaptic pruning, leading to the development of psychosis. Anti-inflammatory treatments that inhibit microglial activity, may be effective in ameliorating psychotic symptoms when used as an add-on treatment to antipsychotics.

CONCLUSION

We have discussed the evidence demonstrating a role for increased inflammation in psychotic disorders, specifically focusing on the mechanisms of pro-inflammatory cytokines and activated microglia. Epidemiological studies and animal models strongly implicate nonspecific inflammation as a causal factor, with early life infection and stress both strongly associated with an increased risk of developing psychosis. Although much of the mechanistic evidence has been derived from animal models, these models are consistent with epidemiological evidence and their findings accurately reflect the neurodevelopmental trajectory of psychosis, whereby psychotic symptoms and cognitive deficits typically manifest in early adulthood. However, it should be noted that while there is growing evidence for increased inflammation in patients with psychotic disorders, there is also a considerable amount of data showing increased inflammation in patients with depression, and more research is emerging in patients with other psychiatric disorders. It may be that increased inflammation broadly increases risk for developing a psychiatric disorder, and individual variability in gene risk variants may predispose an individual toward developing a specific disorder, in the event they are exposed to infection or trauma.

ACKNOWLEDGMENTS

This research has been supported by the National Institute for Health Research (NIHR) Mental Health Biomedical Research Centre at South London and Maudsley NHS Foundation Trust and King's College London. The views expressed are those of the authors and not necessarily those of the NHS, the NIHR or the Department of Health.

REFERENCES

1. Dantzer, R., O'Connor, J. C., Freund, G. G., Johnson, R. W., & Kelley, K. W. From inflammation to sickness and depression: when the immune system subjugates the brain. *Nat Rev Neurosci* **9**, nrn2297 (2008).
2. Xu, H. *et al.* Chronic inflammation in fat plays a crucial role in the development of obesity-related insulin resistance. *J Clin Invest* **112**, 1821–1830 (2003).
3. Blasko, I. *et al.* How chronic inflammation can affect the brain and support the development of Alzheimer's disease in old age: the role of microglia and astrocytes. *Aging Cell* **3**, 169–176 (2004).
4. Buka, S. L. *et al.* Maternal cytokine levels during pregnancy and adult psychosis. *Brain Behav Immun* **15**, 411–420 (2001).
5. Brown, A. S., & Derkits, E. J. Prenatal infection and schizophrenia: a review of epidemiologic and translational studies. *Am J Psychiatry* **167**, 261–280 (2010).
6. Nettis, M., Pariante, C. M., & Mondelli, V. Early-life adversity, systemic inflammation and comorbid physical and psychiatric illnesses of adult life. *Curr Top Behav Neurosci* (2019). doi:10.1007/7854_2019_89
7. Girgis, R. R., Kumar, S. S., & Brown, A. S. The cytokine model of schizophrenia: emerging therapeutic strategies. *Biol Psychiat* **75**, 292–299 (2014).
8. Müller, N., Weidinger, E., Leitner, B., & Schwarz, M. J. The role of inflammation in schizophrenia. *Front Neurosci* **9**, 372 (2015).
9. Severance, E. G., & Yolken, R. H. From infection to the microbiome: an evolving role of microbes in schizophrenia. *Curr Top Behav Neurosci* (2019). doi:10.1007/7854_2018_84
10. Pollak, T. A. *et al.* The blood–brain barrier in psychosis. *Lancet Psychiatry* **5**, 79–92 (2018).
11. Jeppesen, R., & Benros, M. Autoimmune diseases and psychotic disorders. *Front Psychiatry* **10**, 131 (2019).
12. Sekar, A. *et al.* Schizophrenia risk from complex variation of complement component 4. *Nature* **530**, 177–83 (2016).
13. Potvin, S. *et al.* Inflammatory cytokine alterations in schizophrenia: a systematic quantitative review. *Biol Psychiat* **63**, 801–808 (2008).
14. Modabbernia, A., Taslimi, S., Brietzke, E., & Ashrafi, M. Cytokine alterations in bipolar disorder: a meta-analysis of 30 studies. *Biol Psychiat* **74**, 15–25 (2013).
15. Miller, B. J., Buckley, P., Seabolt, W., Mellor, A., & Kirkpatrick, B. Meta-analysis of cytokine alterations in schizophrenia: clinical status and antipsychotic effects. *Biological Psychiat* **70**, 663–71 (2011).
16. Zajkowska, Z., & Mondelli, V. First-episode psychosis: an inflammatory state? *Neuroimmunomodulat* **21**, 102–108 (2014).
17. Upthegrove, R., Manzanares-Teson, N., & Barnes, N. M. Cytokine function in medication-naive first episode psychosis: a systematic review and meta-analysis. *Schizophr Res* **155**, 101–8 (2014).
18. Nicola, M. *et al.* Serum and gene expression profile of cytokines in first-episode psychosis. *Brain Behav Immun* **31**, 90–95 (2013).
19. Capuzzi, E., Bartoli, F., Crocamo, C., Clerici, M., & Carrà, G. Acute variations of cytokine levels after antipsychotic treatment in drug-naïve subjects with a first-episode psychosis: A meta-analysis. *Neurosci Biobehav Rev* **77**, 122–128 (2017).
20. Pillinger, T. *et al.* A meta-analysis of immune parameters, variability, and assessment of modal distribution in psychosis and test of the immune subgroup hypothesis. *Schizophrenia Bull* (2018). doi:10.1093/schbul/sby160
21. Wang, A. K., & Miller, B. J. Meta-analysis of cerebrospinal fluid cytokine and tryptophan catabolite alterations in psychiatric patients: comparisons between schizophrenia, bipolar disorder, and depression. *Schizophr Bull* (2017). doi:10.1093/schbul/sbx035
22. Gallego, J. A. *et al.* Cytokines in cerebrospinal fluid of patients with schizophrenia spectrum disorders: New data and an updated meta-analysis. *Schizophr Res* (2018). doi:10.1016/j.schres.2018.07.019
23. Borovcanin, M. *et al.* Interleukin-6 in schizophrenia—is there a therapeutic relevance? *Front Psychiatry* **8**, 221 (2017).
24. Erta, M., Quintana, A., & Hidalgo, J. Interleukin-6, a major cytokine in the central nervous system. *Int J Biol Sci* **8**, 1254–1266 (2012).
25. Streit, W. J., Hurley, S. D., McGraw, T. S., & Semple-Rowland, S. L. Comparative evaluation of cytokine profiles and reactive gliosis supports a critical role for interleukin-6 in neuron-glia signaling during regeneration. *J Neurosci Res* **61**, 10–20 (2000).
26. Baumeister, D., Akhtar, R., Ciufolini, S., Pariante, C., & Mondelli, V. Childhood trauma and adulthood inflammation: a meta-analysis of peripheral C-reactive protein, interleukin-6 and tumour necrosis factor-α. *Mol Psychiatr* **21**, 642 (2016).
27. Khandaker, G. M., Pearson, R. M., Zammit, S., Lewis, G., & Jones, P. B. Association of serum interleukin 6 and C-reactive protein in childhood with depression and psychosis in young adult life: a population-based longitudinal study. *JAMA Psychiatry* **71**, 1121–1128 (2014).
28. Metcalf, S. *et al.* Serum C-reactive protein in adolescence and risk of schizophrenia in adulthood: a prospective birth cohort study from Finland. *Brain Behav Immun* **66**, e2–e3 (2017).
29. Castell, J. V. *et al.* Acute-phase response of human hepatocytes: Regulation of acute-phase protein synthesis by interleukin-6. *Hepatology* **12**, 1179–1186 (1990).
30. Zhang, J., Luo, W., Huang, P., Peng, L., & Huang, Q. Maternal C-reactive protein and cytokine levels during pregnancy and the risk of selected neuropsychiatric disorders in offspring: a systematic review and meta-analysis. *J Psychiatr Res* **105**, 86–94 (2018).
31. Allswede, D. M., Buka, S. L., Yolken, R. H., Torrey, F. E., & Cannon, T. D. Elevated maternal cytokine levels at birth and risk for psychosis in adult offspring. *Schizophr Res* **172**, 41–45 (2016).
32. Ellman, L. M. *et al.* Structural brain alterations in schizophrenia following fetal exposure to the inflammatory cytokine interleukin-8. *Schizophr Res* **121**, 46–54 (2010).
33. Meyer, U., Feldon, J., & Yee, B. K. A review of the fetal brain cytokine imbalance hypothesis of schizophrenia. *Schizophr Bull* **35**, 959–972 (2009).
34. Howes, O., McCutcheon, R., & Stone, J. Glutamate and dopamine in schizophrenia: an update for the 21st century. *J Psychopharmacol* **29**, 97–115 (2015).
35. Kapur, S., Mizrahi, R., & Li, M. From dopamine to salience to psychosis—linking biology, pharmacology and phenomenology of psychosis. *Schizophr Res* **79**, 59–68 (2005).
36. Aguilar-Valles, A., Jung, S., Poole, S., Flores, C., & Luheshi, G. N. Leptin and interleukin-6 alter the function of mesolimbic dopamine neurons in a rodent model of prenatal inflammation. *Psychoneuroendocrino* **37**, 956–969 (2012).
37. Zuckerman, L., Rehavi, M., Nachman, R., & Weiner, I. Immune activation during pregnancy in rats leads to a postpubertal emergence of disrupted latent inhibition, dopaminergic hyperfunction, and altered limbic morphology in the offspring: a novel neurodevelopmental model of schizophrenia. *Neuropsychopharmacol* **28**, 1300248 (2003).
38. Campbell, B. M., Charych, E., Lee, A. W., & Möller, T. Kynurenines in CNS disease: regulation by inflammatory cytokines. *Front Neurosci-Switz* **8**, 12 (2014).

39. Lisman, J. E. *et al.* Circuit-based framework for understanding neurotransmitter and risk gene interactions in schizophrenia. *Trends Neurosci* **31**, 234–242 (2008).
40. Schwartz, T. L., Sachdeva, S., & Stahl, S. M. Glutamate neurocircuitry: theoretical underpinnings in schizophrenia. *Front Pharmacol* **3**, 195 (2012).
41. Ceresoli-Borroni, G., Rassoulpour, A., Wu, H.-Q., Guidetti, P., & Schwarcz, R. Chronic neuroleptic treatment reduces endogenous kynurenic acid levels in rat brain. *J Neural Transm* **113**, 1355–1365 (2006).
42. Erhardt, S. *et al.* Kynurenic acid levels are elevated in the cerebrospinal fluid of patients with schizophrenia. *Neurosci Lett* **313**, 96–98 (2001).
43. Harrison, N. A., Doeller, C. F., Voon, V., Burgess, N., & Critchley, H. D. Peripheral inflammation acutely impairs human spatial memory via actions on medial temporal lobe glucose metabolism. *Biol Psychiat* **76**, 585–593 (2014).
44. Mondelli, V. *et al.* Higher cortisol levels are associated with smaller left hippocampal volume in first-episode psychosis. *Schizophr Res* **119**, 75–78 (2010).
45. Cannon, T. D. Brain biomarkers of vulnerability and progression to psychosis. *Schizophrenia Bull* **42**, S127–S132 (2016).
46. Fillman, S. *et al.* Elevated peripheral cytokines characterize a subgroup of people with schizophrenia displaying poor verbal fluency and reduced Broca's area volume. *Mol Psychiatr* **21**, 1090 (2016).
47. Zunszain, P. A. *et al.* Interleukin-1β: a new regulator of the kynurenine pathway affecting human hippocampal neurogenesis. *Neuropsychopharmacol* **37**, npp2011277 (2011).
48. Borsini, A., Zunszain, P. A., Thuret, S., & Pariante, C. M. The role of inflammatory cytokines as key modulators of neurogenesis. *Trends Neurosci* **38**, 145–157 (2015).
49. Réus, G. Z. *et al.* The role of inflammation and microglial activation in the pathophysiology of psychiatric disorders. *Neuroscience* **300**, 141–154 (2015).
50. Tang, Y., & Le, W. Differential roles of M1 and M2 microglia in neurodegenerative diseases. *Mol Neurobiol* **53**, 1181–1194 (2016).
51. Ransohoff, R. M. A polarizing question: do M1 and M2 microglia exist? *Nat Neurosci* **19**, 987–991 (2016).
52. Trépanier, M., Hopperton, K., Mizrahi, R., Mechawar, N., & Bazinet, R. Postmortem evidence of cerebral inflammation in schizophrenia: a systematic review. *Mol Psychiatr* **21**, 1009 (2016).
53. Schnieder, T. P. *et al.* Microglia of prefrontal white matter in suicide. *J Neuropathology Exp Neurology* **73**, 880–890 (2014).
54. Busse, S. *et al.* Different distribution patterns of lymphocytes and microglia in the hippocampus of patients with residual versus paranoid schizophrenia: Further evidence for disease course-related immune alterations? *Brain Behav Immun* **26**, 1273–1279 (2012).
55. van Kesteren, C. *et al.* Immune involvement in the pathogenesis of schizophrenia: a meta-analysis on postmortem brain studies. *Transl Psychiat* 7, e1075 (2017).
56. Fillman, S. *et al.* Increased inflammatory markers identified in the dorsolateral prefrontal cortex of individuals with schizophrenia. *Mol Psychiatr* **18**, 206 (2013).
57. Jacobs, A. H., Tavitian, B., & INM Consortium. Noninvasive molecular imaging of neuroinflammation. *J Cereb Blood Flow Metabolism* **32**, 1393–1415 (2012).
58. Sandiego, C. M. *et al.* Imaging robust microglial activation after lipopolysaccharide administration in humans with PET. *Proc National Acad Sci* **112**, 12468–12473 (2015).
59. Turkheimer, F. E. *et al.* The methodology of TSPO imaging with positron emission tomography. *Biochem Soc T* **43**, 586–592 (2015).
60. van Berckel, B. N. *et al.* Microglia activation in recent-onset schizophrenia: a quantitative (R)-[11C]PK11195 positron emission tomography study. *Biol Psychiatr* **64**, 820–2 (2008).
61. Holmes, S. *et al.* In vivo imaging of brain microglial activity in antipsychotic-free and medicated schizophrenia: a [11C](R)-PK11195 positron emission tomography study. *Mol Psychiatr* **21**, 1672 (2016).
62. Doorduin, J. *et al.* Neuroinflammation in schizophrenia-related psychosis: a PET study. *J Nucl Med* **50**, 1801–1807 (2009).
63. Bloomfield, P. S. *et al.* Microglial activity in people at ultra high risk of psychosis and in schizophrenia: an [11C]PBR28 PET brain imaging study. *Am J Psychiatry* **173**, 44–52 (2016).
64. Takano, A. *et al.* Peripheral benzodiazepine receptors in patients with chronic schizophrenia: a PET study with [11C]DAA1106. *Int J Neuropsychopharmacology* **13**, 943–950 (2010).
65. Kenk, M. *et al.* Imaging neuroinflammation in gray and white matter in schizophrenia: an in-vivo pet study with [18F]-FEPPA. *Schizophr Bull* **41**, 85–93 (2014).
66. Coughlin, J. *et al.* In vivo markers of inflammatory response in recent-onset schizophrenia: a combined study using [11C]DPA-713 PET and analysis of CSF and plasma. *Transl Psychiatr* **6**, e777 (2016).
67. van der Doef, T. F. *et al.* In vivo (R)-[11C]PK11195 PET imaging of 18kDa translocator protein in recent onset psychosis. *NPJ Schizophrenia* **2**, 16031 (2016).
68. Hafizi, S. *et al.* Imaging microglial activation in untreated first-episode psychosis: a PET study with [18F]FEPPA. *Am J Psychiat* **174**, 118–124 (2017).
69. Biase, D. M. *et al.* PET imaging of putative microglial activation in individuals at ultra-high risk for psychosis, recently diagnosed and chronically ill with schizophrenia. *Transl Psychiat* 7, e1225 (2017).
70. Hafizi, S. *et al.* TSPO expression and brain structure in the psychosis spectrum. *Brain Behav Immun* (2018). doi:10.1016/j.bbi.2018.06.009
71. Collste, K. *et al.* Lower levels of the glial cell marker TSPO in drug-naive first-episode psychosis patients as measured using PET and [11C]PBR28. *Mol Psychiatr* **22**, 850–856 (2017).
72. Notter, T. *et al.* Translational evaluation of translocator protein as a marker of neuroinflammation in schizophrenia. *Mol Psychiatr* **23**, 323 (2017).
73. Ottoy, J. *et al.* [18F]PBR111 PET imaging in healthy controls and schizophrenia: test—retest reproducibility and quantification of neuroinflammation. *J Nucl Medicine Official Publ Soc Nucl Medicine* **59**, 1267–1274 (2018).
74. Plavén-Sigray, P. *et al.* Positron emission tomography studies of the glial cell marker translocator protein in patients with psychosis: a meta-analysis using individual participant data. *Biol Psychiat* **84**, 433–442 (2018).
75. Marques, T. *et al.* Neuroinflammation in schizophrenia: meta-analysis of in vivo microglial imaging studies. *Psychol Med* 1–11 (2018). doi:10.1017/s0033291718003057
76. Silva, T. *et al.* GABA levels and TSPO expression in people at clinical high risk for psychosis and healthy volunteers: a PET-MRS study. *J Psychiatry Neurosci Jpn* **44**, 111–119 (2019).
77. Picker, L. *et al.* State-associated changes in longitudinal [18F]-PBR111 TSPO PET Imaging of psychosis patients: evidence for the accelerated ageing hypothesis? *Brain Behav Immun* (2018). doi:10.1016/j.bbi.2018.11.318
78. Picker, L. J., Morrens, M., Chance, S. A., & Boche, D. Microglia and brain plasticity in acute psychosis and schizophrenia illness course: a meta-review. *Frontiers Psychiatry* **8**, 238 (2017).
79. Howes, O., & McCutcheon, R. Inflammation and the neural diathesis-stress hypothesis of schizophrenia: a reconceptualization. *Transl Psychiat* 7, e1024 (2017).
80. Mallya, A. P., & Deutch, A. Y. (Micro)glia as effectors of cortical volume loss in schizophrenia. *Schizophrenia Bull* (2018). doi:10.1093/schbul/sby088
81. Mondelli, V., Vernon, A. C., Turkheimer, F., Dazzan, P., & Pariante, C. M. Brain microglia in psychiatric disorders. *Lancet Psychiatry* **4**, 563–572 (2017).
82. Narayanaswami, V. *et al.* Emerging PET radiotracers and targets for imaging of neuroinflammation in neurodegenerative diseases: outlook beyond TSPO. *Mol Imaging* **17**, 1536012118792317 (2018).
83. Notter, T., & Meyer, U. Microglia and schizophrenia: where next? *Mol Psychiatr* **22**, 788–789 (2017).
84. Bisht, K. *et al.* Dark microglia: A new phenotype predominantly associated with pathological states. *Glia* **64**, 826–839 (2016).
85. Schafer, D. P. *et al.* Microglia sculpt postnatal neural circuits in an activity and complement-dependent manner. *Neuron* **74**, 691–705 (2012).

86. Paolicelli, R. C. *et al.* Synaptic pruning by microglia is necessary for normal brain development. *Science* **333**, 1456–1458 (2011).
87. Paolicelli, R. C. *et al.* TDP-43 depletion in microglia promotes amyloid clearance but also induces synapse loss. *Neuron* **95**, 297–308.e6 (2017).
88. Mokhtari, R., & Lachman, H. M. The major histocompatibility complex (MHC) in schizophrenia: a review. *J Clin Cell Immunol* 7, 1–7 (2016).
89. Hinwood, M., Morandini, J., Day, T., & Walker, F. Evidence that microglia mediate the neurobiological effects of chronic psychological stress on the medial prefrontal cortex. *Cereb Cortex* **22**, 1442–1454 (2012).
90. Chaudhry, I. B. *et al.* Minocycline benefits negative symptoms in early schizophrenia: a randomised double-blind placebo-controlled clinical trial in patients on standard treatment. *J Psychopharmacol* **26**, 1185–1193 (2012).
91. Kelly, D. L. *et al.* Adjunctive minocycline in clozapine-treated schizophrenia patients with persistent symptoms. *J Clin Psychopharm* **35**, 374–381 (2015).
92. Liu, F. *et al.* Minocycline supplementation for treatment of negative symptoms in early-phase schizophrenia: a double blind, randomized, controlled trial. *Schizophr Res* **153**, 169–176 (2014).
93. Khodaie-Ardakani, M.-R. *et al.* Minocycline add-on to risperidone for treatment of negative symptoms in patients with stable schizophrenia: randomized double-blind placebo-controlled study. *Psychiat Res* **215**, 540–546 (2014).
94. Solmi, M. *et al.* Systematic review and meta-analysis of the efficacy and safety of minocycline in schizophrenia. *CNS Spectrums* **22**, 415–426 (2017).
95. Xiang, Y.-Q. *et al.* Adjunctive minocycline for schizophrenia: a meta-analysis of randomized controlled trials. *Eur Neuropsychopharm* **27**, 8–18 (2017).
96. Cho, M. *et al.* Adjunctive use of anti-inflammatory drugs for schizophrenia: a meta-analytic investigation of randomized controlled trials. *Australian New Zealand J Psychiatry* 000486741983502 (2019). doi:10.1177/0004867419835028
97. Zhang, L. *et al.* Minocycline adjunctive treatment to risperidone for negative symptoms in schizophrenia: association with pro-inflammatory cytokine levels. *Prog Neuro-psychopharmacology Biological Psychiatry* **85**, 69–76 (2018).
98. Deakin, B. *et al.* The benefit of minocycline on negative symptoms of schizophrenia in patients with recent-onset psychosis (BeneMin): a randomised, double-blind, placebo-controlled trial. *Lancet Psychiatry* **5**, 885–894 (2018).

40

AUTOIMMUNE PROCESSES IN MENTAL DISORDERS

Marina Mané-Damas, Carolin Hoffmann, Shenghua Zong, Peter C. Molenaar, Mario Losen, and Pilar Martinez-Martínez

INTRODUCTION

Mental disorders, such as schizophrenia, still have poorly understood pathophysiology. Evidence suggests that hypofunction of glutamatergic and dopaminergic neurotransmission, and dysfunction of potassium channels play central roles [1–3].

Genetic studies in large cohorts have indicated that certain common and rare genetic variants increase the risk for neuropsychiatric disorders, including genes encoding ion channel components and synaptic transmission proteins. For example, a genome-wide analysis identified risk loci in L-type calcium channel genes with shared effects on five major neuropsychiatric disorders [4]. Additionally, genetic modifications in proteins involved in synaptic differentiation and transmission have been associated with higher risk of developing schizophrenia [5, 6] and autism [5, 7]. Already 20 years ago, there was speculation about the existence of a subgroup of schizophrenia patients with immunological abnormalities including the presence of pro-inflammatory factors and antinuclear antibodies, and a higher prevalence suffering from autoimmunity [8]. Specific variants in genes encoding human leukocyte antigen (HLA) regions (major histocompatibility complex [MHC] molecules), important for antigen presentation and immune system regulation, have been associated with schizophrenia risk [9, 10]. These findings may support the notion for a role of neuro-inflammation and possibly autoimmunity in neuropsychiatric disorders.

The role of antibodies in neuropsychiatric disorders attracted interest when, two decades ago, Dalmau described autoimmune encephalitis for the first time. Patients developed psychiatric and severe neurological manifestations coincidental with the presence of an ovarian teratoma. Both serum and cerebrospinal fluid (CSF) showed reactivity to neuronal tissue, present also in the tumor. Autoantibodies specifically target the subunit 1 (NR1 o GluR1) of the N-methyl-D-aspartate receptor (NMDAR). Tumor resection and immunosuppressive treatment resulted in a complete recovery, with the return of the normal social behavior and executive functions [11, 12].

In perspective, it is entirely conceivable that antibodies targeting neurotransmitter receptors and voltage- and ligand-activated ion channels could cause psychiatric symptoms. Neurotransmitter receptors blockers such as phencyclidine, ketamine, quinuclidinyl benzilate, and lysergic acid diethylamide (LSD), affecting NMDAR and muscarinic acetylcholine and serotonin (5HT) receptors, are all potent hallucinogens. Autoantibodies might reduce or impair synaptic transmission and therefore reduce functioning of neurons in a neural network, ultimately creating severe mental manifestations.

This book chapter aims to review what is currently known about the occurrence of autoantibodies against neuronal receptors and voltage-gated ion channels in the context of mental disorders and its correlation and co-occurrence with neurological manifestations (Figure 40.1).

AUTOIMMUNITY

The immune system protects the body against the attack of virus particles, bacteria, and other pathogens. A fast but unspecific response is provided by the innate or humoral immune

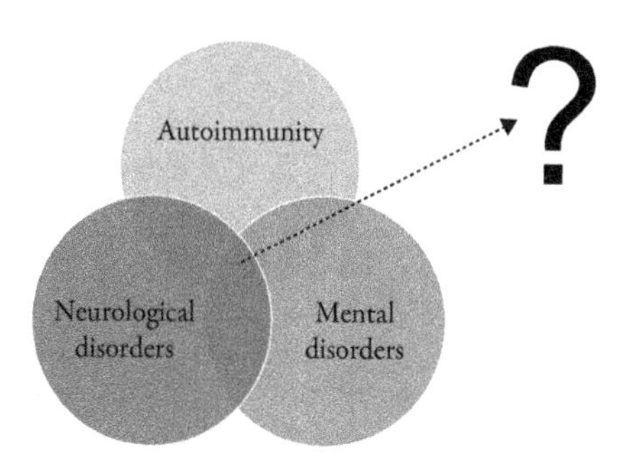

Figure 40.1 Venn diagram of relationship between mental and neurological disorders and autoimmunity.
In this chapter, we describe the clear overlap of mental and neurological manifestations in certain autoimmune disorders. Autoimmune encephalitis exemplifies this phenomenon, where autoantibodies against neuronal surface antigens can result in psychiatric and neurological manifestations. This chapter aims to give insights into the common area of the three disciplines, with strong attention to those cases where neurological symptoms are subtle and can be overlooked. In some cases that could result in misdiagnoses and administration of antipsychotic medication instead of immunosuppressants.

system, followed by a late and more specific reaction involving lymphocytes and other immune cells, known as the adaptive or cell-mediated response. The generation of specialized cells, T and B lymphocytes, requires a maturation process with several checkpoints to detect and delete defective cells. One of these mechanisms is the acquisition of tolerance, which is generated both centrally and peripherally in the lymphoid organs and where lymphocytes are screened for binding to self and non-self-antigens. However, failure of this mechanism may result in autoimmunity: the attack of the immune system against self-antigens, leading to tissue damage. Several cellular mechanisms are involved in this process such as the activation of the cytotoxic response and the generation of autoantibodies by B cells and plasma cells. Autoimmune diseases can be classified depending on the organ that is affected, e.g., endocrine, musculoskeletal, nervous system, etc.

Autoimmune diseases are estimated to affect a substantial 7%–9% of the general population [13]. In many autoimmune diseases of the peripheral nervous system, antibodies are directed against antigens on the surface of nerve or muscle membranes, like ion channels, such as the voltage-dependent calcium channel in Lambert-Eaton myasthenic syndrome [14], the potassium channel in neuromyotonia [15], or neurotransmitter ion channel receptors such as the nicotinic acetylcholine receptor (AChR) in myasthenia gravis [16].

Diseases where such ion channels are affected by autoantibodies or genetic causes have been named channelopathies. Therefore, autoantibodies targeting neurotransmitter receptors in the central nervous system (CNS) are also channelopathies in the sense that impaired or altered functioning of ion channels or brain receptors cause neuropsychiatric symptoms. Alternative mechanisms, such as complement activation, could contribute to the degeneration of the membrane around the targeted antigen, potentially including glia, astrocytes, or oligodendrocytes (see refs. [17–19] for a review on antibody effector mechanisms). Successful immunosuppressive therapy is dependent on the type of cells that cause the attack, their location, and whether the antibody-induced damage is reversible.

BLOOD-BRAIN BARRIER

The blood-brain barrier (BBB) is an impermeable barrier of hydrophilic substances and mostly large molecules, between the blood circulation and the CNS, protecting against possible toxic influences. Movement of antibodies or plasma cells through the BBB is still poorly characterized. Antibodies can be generated on both sides of that membrane, intrathecally or peripherally, from where they access the brain by crossing the BBB through different mechanisms. However, antibodies are known to cross the human BBB to a limited extent [20], and it has been found that pathogenic autoantibodies can reach sites in the CNS and exert immunologic effects there [21]. Such diseases include paraneoplastic neurological syndromes [22], neuromyelitis optica [23], epilepsy [24, 25], Morvan's syndrome, and limbic encephalitis [26]. The mechanisms by which antibodies pass the BBB under normal conditions are still not fully defined. The BBB becomes more permeable after local inflammation, e.g., in multiple sclerosis the permeability is increased sixfold [27]. Moreover, neuroinflammation may lead to local production of antibodies by B- cells and or plasma cells in the brain itself [28]. High antibody levels in the CSF might indicate intrathecal synthesis of antibodies by migrated antibody-producing cells in the brain, in line with evidence showing that antibody levels in CSF are more closely related with relapses in NMDAR encephalitis than in serum [29]. CSF concentration of B-cell-attracting chemokines have been correlated with intrathecal production of NMDAR antibodies, relapses, and limited response to treatment [30].

Subsequently, it is unlikely that the BBB is equally permeable to antibodies or antibody-producing cells throughout the brain, factors affecting the BBB integrity with respect to transport of antibodies may show regional specificity. This would explain that the "phenotype" of the antibody effect may well depend on the location where the antibody enters the brain and therefore justify how the same autoantibody might have substantially different pathological effects in different groups of patients. For instance, antibodies against contactin-associated protein-like 2 (CASPR2), a voltage-gated potassium channel complex (VGKCC) associated protein, are involved in limbic encephalitis and Morvan's syndrome [26].

CHANNELOPATHIES IN NEUROPSYCHIATRIC DISORDERS

Autoimmune encephalitis is estimated to have an annual incidence of 1.2 cases/100,000 [31]. After the development of prodromal symptoms, more than 50% of patients present severe psychotic manifestations, resembling those seen in schizophrenia and other mental disorders [32], including delusional and delirious thinking, anxiety, insomnia, paranoid thoughts, maniac and aggressive behavior, depression, hallucinations, cognitive impairment, and catatonic movement disorder [33–39]. In absence of obvious neurological manifestations, some patients have first been admitted into psychiatric units [40, 41].

Sixteen neuronal surface antigens have recently been identified in autoimmune encephalitis (Table 40.1). These antibodies seem to be responsible for the pathophysiology, since they are rarely found in healthy individuals and in the majority of patients with other neurological diseases and patients recover after immunosuppressive therapy [42, 43]. Due to the overlap of clinical manifestations between autoimmune encephalitis and mental disorders, the investigation of autoantibodies in mental disorder cohorts has gained interest. Primarily, NMDAR and VGKCC autoantibodies, in patients with acute schizophrenia and major depression, have been studied. However, the characteristics of the detection methods (Figure 40.2) as well as the variability among the cohorts analyzed have resulted in very inconsistent results, and led to controversy in the field.

Increased serological prevalence of autoantibodies, against NMDAR and VGKCC, have been described in different cohorts of patients with first episode schizophrenia, catatonic

Table 40.1 AUTOIMMUNE TARGETS IN NEUROPSYCHIATRIC DISORDERS

TARGETED ANTIGEN	DISEASE	SUBUNIT/ASSOCIATED PROTEIN	REFERENCES
AMPAR	Limbic encephalitis Psychiatric symptoms	GluR1/2 GluR3B	[50] [51]
DR	Basal ganglia encephalitis Acute psychosis	D2	[42] [52]
$GABA_BR$	Limbic encephalitis		[50]
GAD	Limbic encephalitis Stiff person syndrome	GAD65, GAD67	[50] [53]
GlyR	Progressive encephalomyelopathy		[54]
mGluR5	Ophelia's syndrome		[55, 56]
Nicotinic AChR	Schizophrenia	α7	[57, 58]
NMDAR	Encephalitis Schizophrenia	NR1	[12] [44, 46]
VGKCC (LGI1 and CASPR2 [26, 59])	Limbic encephalitis Morvan's syndrome Schizophrenia	LGI1 or CASPR2 CASPR2	[26, 37, 44, 59] [26, 37, 44, 45, 59]

and disorganized schizophrenia, depression, and bipolar disorder. However, no differences were detected between diseased and healthy individuals and higher antibody prevalences were associated with IgA or IgM isotypes [44–49]. The relevance of other isotypes than IgG is still not clear, as indicated in a recent study that investigated all antibody isotypes against the NMDAR, and found that only IgG autoantibodies had clinical and pathological relevance [60]. Most of these studies used live CBA as a detection method. Although the epitope conformation is preserved in a better manner, this technique has shown to report lower specificity and higher rates of positivity compared to fixed techniques, which are regularly used as a diagnostic method [29, 61].

Contrastingly, NMDAR autoantibodies were not detected in a cohort of 80 schizophrenia patients. In this study the authors analyzed IgG by different techniques including fixed CBA, brain IHC, and primary neuronal cell culture to validate the results [62]. Commercial and in-house CBAs, together with IHC were used to analyze the presence of NMDAR autoantibodies in a cohort of 475 patients with schizophrenia spectrum disorder. None of the patients had IgG autoantibodies [63]. In line with that, IgG antibodies, directed to NMDAR and other receptors, were undetectable in serum from a cohort of 70 first-episode psychosis [64] and in a group of 78 first-episode schizophrenia and 234 chronic schizophrenia patients [65]. Nevertheless, 2 out of 61 patients with first-episode psychosis presented serum IgG autoantibodies to the NMDAR, analyzed by commercial assays [66]. In a cohort of 925 patients admitted to the acute psychiatric care unit, serum IgG autoantibodies against NMDAR ($n = 5$), CASPR2 ($n = 7$), and AMPAR ($n = 1$) were identified by commercial CBA [67].

Psychiatric examinations rarely include CSF analysis for any clinical purpose. Antibody testing is performed mainly in serum or plasma, however, sensitivity between serum/plasma and CSF differs. Serum sensitivity is lower compared to CSF for NMDAR autoantibodies and in some cases autoantibodies are only detectable in CSF [29]. However, that varies depending on the antigen.

Studies analyzing the prevalence of autoantibodies in CSF are scarce in these cohorts, due to the low percentage of patients undergoing lumbar puncture. Autoantibodies to NMDAR and VGKCC were identified in CSF in 1 and 2 individuals respectively in a psychotic cohort of 125 patients (serum or plasma analysis were not included) [68]. Absence of IgG autoantibodies against several neuronal antigens in CSF was reported in a cohort of 119 patients with mental disorders and absence of neurological symptoms (not all were tested for all antigens). However, low serological antibody levels against CASPR2 were found in 3 cases (only 81 paired serum samples were available) [69].

Despite the fact that autoantibodies are rarely found in psychotic disorder cohorts, higher prevalence of IgG autoantibodies have been found in specific patient groups. Specifically, serum autoantibodies against the NMDAR and D2DR were found respectively in 5 and 3 out of 43 children with first-episode acute psychosis by flow cytometry, live CBA, and primary neuronal cell culture [52]. Schizophrenia or schizoaffective refractory patients, with more than 15 years of disease history on average, and absence of MRI abnormalities and neurological signs, including seizures, is another attention-grabbing group. NMDAR autoantibodies were found in 3 out of 43 cases using live CBA [70]. Ultimately, in a postpartum psychosis cohort, 2 out of 96 patients presented NMDAR autoantibodies by CBA. Yet, patients recovered without immunotherapy [71].

Even though ultimately the identification of the antigen is essential to gain insight into the pathogenic mechanisms underlying these disorders, some studies reported the presence of autoantibodies against unidentified neuronal antigens.

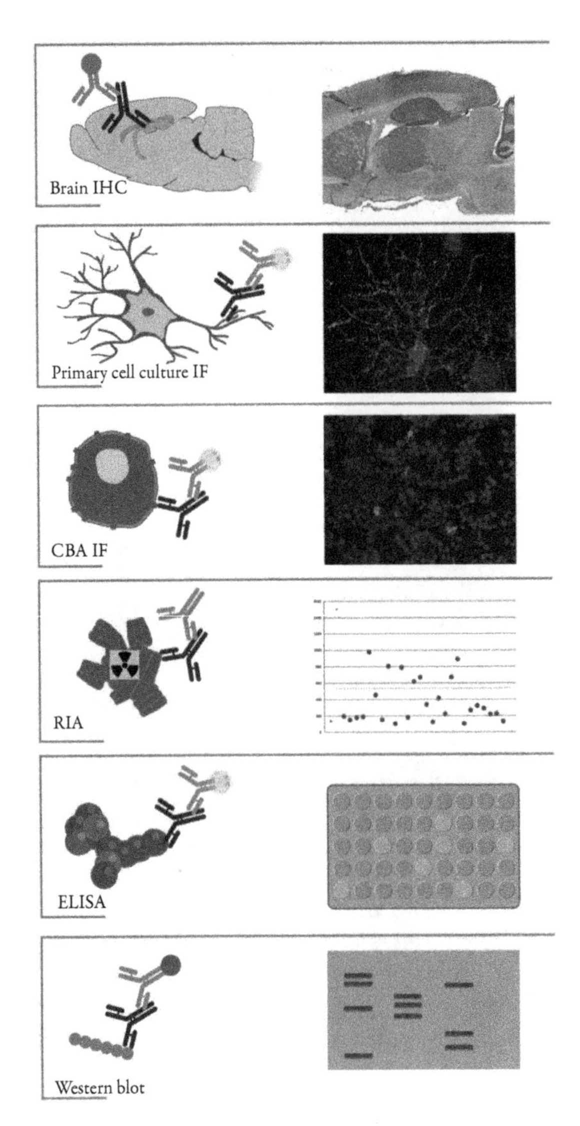

Figure 40.2 Antibody-mediated diagnostic assays.
Brain IHC is a tissue-based assay using rat brain slices that contain brain antigens; Primary cell culture IF are hippocampal neurons sourced from rat embryos, presenting neuronal antigens and used in live CBA IF; CBA IF can be performed live or fixed and is based on the overexpression of a specific antigen in a cell line; RIA assay uses a radiolabeled antigen (e.g., protein or membrane extract) which is precipitated after formation of immunocomplexes; ELISA is a quantitative technique where the purified antigen or a specific region of it (for example the pathogenic epitope) is coated in the plate; and Western blot uses pure antigen or protein extract; proteins are then separated depending on molecular weight and thereafter transferred to a membrane. In all methods, serum, plasma, or CSF from patient/control individuals is incubated with the different sources of antigen defined by the different methods. Human IgG-specific antibodies (brown) bound to the antigens present are recognized by secondary antibodies (green), labeled with different visualizing methods such as color deposit (brain IHC, WB, ELISA) and fluorescent dyes (primary cell culture IF and CBA IF). In the case of the RIA, the antigen is labeled with a radioactive component, and incubated with patient sera, plasma, or CSF, and a secondary antibody against human IgG is used for the immunoprecipitation, if the patient's antibodies bind to the antigen, a pellet will be generated that will be radioactive. All methods vary in terms of sensitivity-specificity, quantitative-qualitative terms. Combination of some of these methods is essential to properly diagnose patients and avoid false (positive and negative) results. IHC: immunohistochemistry, IF: immunofluorescence, CBA: cell-based assay, RIA: radioimmunoassay, ELISA: enzyme-linked immunosorbent assay, CSF: cerebrospinal fluid, IgG: immunoglobulin G.

For example, in the postpartum psychosis cohort mentioned earlier, two additional patients presented unknown reactivity in the hippocampus by brain IHC and on live hippocampal neurons [52]. In a cohort of 925 patients admitted to the acute psychiatric care unit, 17 had IgG unknown reactivity on rat and monkey brain sections commercially available [67]. A large study revealed robust hippocampal IgG staining by rat brain IHC and positive immune reactivity to primary hippocampal neurons in 5 out 1,739 patients from the Netherlands Study of Depression and Anxiety (NESDA). In contrast, no immune reactivity was detected in any of the 492 nondiseased controls (Zong et al., in preparation). In line with these results, in another study investigating psychotic disorders patients, 14 out of 621 patients with psychosis, 1 out of 70 patients with affective disorders, and 5 out of 259 nondiseased controls showed IgG unknown reactivity in the hippocampus by rat brain IHC (Hoffmann et al., in preparation).

ASSAYS FOR ANTIBODIES—FALSE RESULTS AND OTHER PITFALLS

False results pose a considerable problem for the interpretation of the data, especially if probability of specific neuronal antibodies in the test group is low. This is because the prevalence and the number of false positives together affect the reliability of a positive test outcome: the greater the number of false positives and the lower the prevalence, the lower the reliability of the assay. The proportion of individuals without a disease (healthy individuals or patients with other disorders related to the studied disease) that an immunological assay can predict correctly is known as clinical specificity. Sensitivity is defined as the fraction of disease cases that an assay correctly predicts [72, 73]. In a group of patients with psychosis, it is likely that the prevalence of these antibodies will be < 1% per specific neuronal target, making the reliability of a positive outcome rather small. There are several reasons why a test might give false results, for example:

- Low or undetectable autoantibody levels (could be as a result of the high affinity of the antibody to the antigen in vivo, leaving a small proportion of antibodies in circulation [74])
- Low affinity of the autoantibody to the antigen in vitro
- Coexistence of different autoantibodies in the same specimen
- Unspecific binding of the autoantibodies to different components of the assay or to a "related" antigen which cross-reacts with the antigen used in the test (e.g., a similar receptor or ion channel)
- Inaccessibility of the autoantibodies as a result of the assay (for example, when antibodies are internalized)
- Conformational changes of the antigen compared to in vivo (for example live versus fixed CBAs [61] or the use

of denaturalized proteins, peptides, or cell membrane extracts)

- Protein homology variation between human and the species used in the assay (usually rodents), especially when changes are located in the extracellular regions

Thus, the outcome of individual tests would be a poor guidance for counseling immunotherapy and certainly would make simple screening for antibodies in psychosis unfeasible. The use of two different detection assays in tandem should be implemented in order to decrease the likelihood of false positive results. Combining CBAs, IHC, and/or primary hippocampal cell culture assays and/or comparing the test outcomes between different laboratories might help to obtain robust and reproducible results. For example, the use of a single technique could explain the different prevalence of autoantibodies to the $\alpha7$ subunit of the AChR in schizophrenia. In the first study, 5 out of 21 patients presented autoantibodies to the recombinant protein by ELISA [57] while only 1 patient was found positive in a recent study with a similar but larger cohort, including 566 patients diagnosed with schizophrenia, bipolar disorders or other mental disorders, and 110 healthy controls. In the second study, membrane extracts from cells overexpressing the antigen were incubated with patient sera and the binding detected by radioimmunoassay [75] (see Figure 40.2 for autoantibody detection methods). The inclusion of control samples aids in the reduction of "false positives" to negligible levels. For a case in point, consider a study that analyzed the presence of NMDAR autoantibodies in different cohorts including controls. Using different techniques, the authors showed that pathogenic autoantibodies were restricted to autoimmune encephalitis cases and not relevant in the other groups, including schizophrenia, neurodegenerative disorders, and healthy individuals [60]. Additionally, screening of neuronal autoantibodies should be done in an antigen-combined panel rather than for individual antigens due to potential coexistence of different autoantibodies in the same patient [76, 77].

Extensive training of researchers is as important as the quality and standardization of the assays, since the correct interpretation of the results directly depends on it. Lastly, it is essential to perform an exhaustive psychiatric and neurological examination of the patients to help their preselection, most importantly by recognizing specific symptoms that can be associated with autoimmune variants of the disease. Discovering such relevant clinical details is therefore a number one challenge before we can start treating patients on the outcome of individual test assays.

EFFECT OF IMMUNOSUPPRESSION AND NEW THERAPEUTIC APPROACHES

The symptoms of patients with pure psychosis with autoimmune encephalitis are treatable and reversible. However, a fast intervention in the diagnosis and treatment might avoid the prolonged exposure of the target antigen and the corresponding possible brain damage. High-dose intravenous immunoglobulin (IVIG), plasma exchange or plasmapheresis, and intravenous steroids compose the first-line immunotherapy. A relevant proportion of patients present a poor improvement or no response to first-line immunosuppressants and have to follow second-line immunosuppressants, which involve cell therapy, including rituximab and cyclophosphamide [12].

NMDAR encephalitis is treatable, and the majority of patients who presented with isolated psychiatric symptoms at the first episode responded well to first-line immunotherapy [33, 78–83], with early treatment predicting better outcome [12]. It should be noted that upon recovery, the CSF samples still tested positive in most cases (i.e., in 24 out of 28 patients) [29]. However, a combination of intravenous steroids, IVIG together with cyclophosphamide and rituximab was needed to treat some patients with an acute psychotic episode, after NMDAR autoantibodies were detected, which resulted in a slow progressive recovery [84].

VGKCC antibody-associated encephalopathy treated with first-line therapies shows a reduction of specific antibody levels and an improvement of the cognitive functions. However, memory deficits are usually not fully reversible, suggesting that there is some damage in the hippocampus [85–93].

Novel therapeutic strategies such as proteasome inhibitors, targeting plasma cells—responsible for the synthesis of great majority of the secreted antibodies—have been used to treat autoimmune diseases [94]. A small group of patients with treatment-resistant NMDAR encephalitis recovered after bortezomib treatment, a first-generation proteasome inhibitor [95]. Bortezomib successfully depleted plasma cells in experimental animal models of autoimmune diseases [94] and is therefore a potential promising therapeutic for refractory patients.

The shortage of cases where patients with purely psychiatric manifestations have recovered after immunomodulatory therapies makes it difficult to define whether these treatments are equally beneficial and efficient in this group of patients.

CONCLUSIONS

Emergent recognition by clinicians of neuronal autoantibodies with psychotic symptomatology, although rare, has led to increased successful diagnosis. Furthermore, in some cases, an overlap of phenotypes between autoimmune encephalitis and mental disorders is evident. Therefore, it is conceivable that an antibody against an ion channel or a neurotransmitter receptor is always likely to evoke, to a certain extent, recognizable neurological symptoms. However, in some cases these symptoms are subtle or may be overlooked when patients are first seen in psychiatry, rather than in neurology, departments.

The prevalence of autoantibodies in patients without any neurological involvement remains difficult to determine. Limiting factors include the methodological differences of the detection assays, their sensitivity, and the fact that in many cases it is difficult or impossible to analyze CSF. There is also a need to study the presence of autoantibodies targeting other novel antigens that are not detectable by currently used methods, e.g., those that are not homologous in rat and human or antigens on astrocytes and glial cells.

Although cases of autoimmune encephalitis with isolated psychiatric symptoms are expected to be rare, we would like to highlight the importance of considering the diagnosis of autoimmune encephalitis in patients presenting with psychosis. This is especially vital for patients with no previous history of psychotic disorders, a rapid progression of the disease, or resistance to antipsychotic medication. Many case reports describing patients with pronounced psychotic manifestations provide an unclear methodology for autoantibody detection, making it very difficult to interpret results. Early diagnosis is crucial, since the majority of patients responds well to readily available immunotherapy.

ACKNOWLEDGMENTS

We thank Tanya Mohile for providing language and writing assistance and Geertjan van Zonneveld for technical assistance on the design of the figures.

REFERENCES

1. Vukadinovic, Z., and I. Rosenzweig, *Abnormalities in thalamic neurophysiology in schizophrenia: could psychosis be a result of potassium channel dysfunction?* Neurosci Biobehav Rev, 2012. **36**(2): p. 960–968.
2. Corti, C., et al., *Altered levels of glutamatergic receptors and Na+/K+ ATPase-alpha1 in the prefrontal cortex of subjects with schizophrenia.* Schizophr Res, 2011. **128**(1–3): p. 7–14.
3. Stephan, K.E., T. Baldeweg, and K.J. Friston, *Synaptic plasticity and dysconnection in schizophrenia.* Biol Psychiatry, 2006. **59**(10): p. 929–939.
4. Smoller, J.W., et al., *Identification of risk loci with shared effects on five major psychiatric disorders: a genome-wide analysis.* Lancet, 2013. **381**(9875): p. 1371–1379.
5. Gauthier, J., et al., *Truncating mutations in NRXN2 and NRXN1 in autism spectrum disorders and schizophrenia.* Hum Genet, 2011. **130**(4): p. 563–573.
6. Rujescu, D., et al., *Disruption of the neurexin 1 gene is associated with schizophrenia.* Hum Mol Genet, 2009. **18**(5): p. 988–996.
7. Vaags, Andrea K., et al., *Rare deletions at the neurexin 3 locus in autism spectrum disorder.* Am J Hum Genet, 2012. **90**(1): p. 133–141.
8. Ganguli, R., et al., *Autoimmunity in schizophrenia: a review of recent findings.* Ann Med, 1993. **25**(5): p. 489–496.
9. Wright, P., et al., *Schizophrenia and HLA: a review.* Schizophr Res, 2001. **47**(1): p. 1–12.
10. Crespi, B.J., and D.L. Thiselton, *Comparative immunogenetics of autism and schizophrenia.* Genes Brain Behav, 2011. **10**(7): p. 689–701.
11. Dalmau, J., et al., *Paraneoplastic anti-N-methyl-D-aspartate receptor encephalitis associated with ovarian teratoma.* Ann Neurol, 2007. **61**(1): p. 25–36.
12. Dalmau, J., et al., *Clinical experience and laboratory investigations in patients with anti-NMDAR encephalitis.* Lancet Neurol, 2011. **10**(1): p. 63–74.
13. Cooper, G.S., M.L. Bynum, and E.C. Somers, *Recent insights in the epidemiology of autoimmune diseases: improved prevalence estimates and understanding of clustering of diseases.* J Autoimmun, 2009. **33**(3–4): p. 197–207.
14. Fukunaga, H., et al., *Passive transfer of Lambert-Eaton myasthenic syndrome with IgG from man to mouse depletes the presynaptic membrane active zones.* Proc Natl Acad Sci U S A, 1983. **80**(24): p. 7636–7640.
15. Shillito, P., et al., *Acquired neuromyotonia: evidence for autoantibodies directed against K+ channels of peripheral nerves.* Ann Neurol, 1995. **38**(5): p. 714–722.
16. Lindstrom, J.M., *Antibody to acetylcholine receptor in myasthenia gravis: prevalence, clinical correlates, and diagnostic value.* Neurology, 1998. **51**(4): p. 933.
17. Gomez, A.M., et al., *Antibody effector mechanisms in myasthenia gravis-pathogenesis at the neuromuscular junction.* Autoimmunity, 2010. **43**(5–6): p. 353–370.
18. Zong, S., et al., *Neuronal surface autoantibodies in neuropsychiatric disorders: are there implications for depression?* Front Immunol, 2017. **8**(752).
19. Hoffmann, C., et al., *Autoantibodies in neuropsychiatric disorders.* Antibodies, 2016. **5**(2).
20. Poduslo, J.F., G.L. Curran, and C.T. Berg, *Macromolecular permeability across the blood-nerve and blood-brain barriers.* Proc Natl Acad Sci U S A, 1994. **91**(12): p. 5705–5709.
21. Diamond, B., et al., *Losing your nerves? Maybe it's the antibodies.* Nat Rev Immunol, 2009. **9**(6): p. 449–456.
22. Honnorat, J., and J.C. Antoine, *Paraneoplastic neurological syndromes.* Orphanet J Rare Dis, 2007. **2**: p. 22.
23. Lennon, V.A., et al., *A serum autoantibody marker of neuromyelitis optica: distinction from multiple sclerosis.* Lancet, 2004. **364**(9451): p. 2106–2112.
24. Solimena, M., et al., *Autoantibodies to glutamic acid decarboxylase in a patient with stiff-man syndrome, epilepsy, and type I diabetes mellitus.* N Engl J Med, 1988. **318**(16): p. 1012–1020.
25. Errichiello, L., et al., *Autoantibodies to glutamic acid decarboxylase (GAD) in focal and generalized epilepsy: a study on 233 patients.* J Neuroimmunol, 2009. **211**(1–2): p. 120–123.
26. Irani, S.R., et al., *Antibodies to Kv1 potassium channel-complex proteins leucine-rich, glioma inactivated 1 protein and contactin-associated protein-2 in limbic encephalitis, Morvan's syndrome and acquired neuromyotonia.* Brain, 2010. **133**(9): p. 2734–2748.
27. Cutler, R.W., G.V. Watters, and J.P. Hammerstad, *The origin and turnover rates of cerebrospinal fluid albumin and gamma-globulin in man.* J Neurol Sci, 1970. **10**(3): p. 259–268.
28. Corcione, A., et al., *Recapitulation of B cell differentiation in the central nervous system of patients with multiple sclerosis.* Proc Natl Acad Sci U S A, 2004. **101**(30): p. 11064–11069.
29. Gresa-Arribas, N., et al., *Antibody titres at diagnosis and during follow-up of anti-NMDA receptor encephalitis: a retrospective study.* Lancet Neurol, 2014. **13**(2): p. 167–177.
30. Leypoldt, F., et al., *Investigations on CXCL13 in Anti–N-methyl-D-aspartate receptor encephalitis: a potential biomarker of treatment response.* JAMA neurology, 2015. **72**(2): p. 180–186.
31. Divyanshu, D., et al., *Autoimmune encephalitis epidemiology and a comparison to infectious encephalitis.* Ann Neurol, 2018. **83**(1): p. 166–177.
32. Al-Diwani, A., et al., *Synaptic and neuronal autoantibody-associated psychiatric syndromes: controversies and hypotheses.* Front Psychiatry, 2017. **8**: p. 13.
33. Kayser, M.S., et al., *Frequency and characteristics of isolated psychiatric episodes in anti-N-methyl-d-aspartate receptor encephalitis.* JAMA Neurol, 2013. **70**(9): p. 1133–1139.
34. Dalmau, J., et al., *Anti-NMDA-receptor encephalitis: case series and analysis of the effects of antibodies.* Lancet Neurol, 2008. 7(12): p. 1091–1098.
35. Hacohen, Y., et al., *Paediatric autoimmune encephalopathies: clinical features, laboratory investigations and outcomes in patients with or without antibodies to known central nervous system autoantigens.* J Neurol Neurosurg Psychiatry, 2013. **84**(7): p. 748–755.
36. Vincent, A., et al., *Autoantibodies associated with diseases of the CNS: new developments and future challenges.* Lancet Neurol, 2011. **10**(8): p. 759–772.

37. Pruss, H., and B.R. Lennox, *Emerging psychiatric syndromes associated with antivoltage-gated potassium channel complex antibodies.* J Neurol Neurosurg Psychiatry, 2016. **87**(11): p. 1242–1247.
38. Al-Diwani, A., et al., *The psychopathology of NMDAR-antibody encephalitis in adults: a systematic review and phenotypic analysis of individual patient data.* Lancet Psychiatry, 2019. **6**(3): p. 235–246.
39. Rani A. Sarkis, M.D., M.Sc., et al., *Anti-N-methyl-D-aspartate receptor encephalitis: a review of psychiatric phenotypes and management considerations: a report of the American Neuropsychiatric Association committee on research.* J Neuropsychiatry Clin Neurosci, 2019. **31**(2): p. 137–142.
40. Bost, C., O. Pascual, and J. Honnorat, *Autoimmune encephalitis in psychiatric institutions: current perspectives.* Neuropsychiatr Dis Treat, 2016. **12**: p. 2775–2787.
41. Maat, P., et al., *Psychiatric phenomena as initial manifestation of encephalitis by anti-NMDAR antibodies.* Acta Neuropsychiatrica, 2013. **25**(3): p. 128–136.
42. Dalmau, J., C. Geis, and F. Graus, *Autoantibodies to synaptic receptors and neuronal cell surface proteins in autoimmune diseases of the central nervous system.* Physiological Reviews, 2017. **97**(2): p. 839–887.
43. Kreye, J., et al., *Human cerebrospinal fluid monoclonal N -methyl-D-aspartate receptor autoantibodies are sufficient for encephalitis pathogenesis.* Brain, 2016. **139**(10): p. 2641–2652.
44. Zandi, M.S., et al., *Disease-relevant autoantibodies in first episode schizophrenia.* J Neurol, 2011. **258**(4): p. 686–688.
45. Somers, K.J., et al., *Psychiatric manifestations of voltage-gated potassium-channel complex autoimmunity.* J Neuropsychiatry Clin Neurosci, 2011. **23**(4): p. 425–433.
46. Steiner, J., et al., *Increased prevalence of diverse N-methyl-D-aspartate glutamate receptor antibodies in patients with an initial diagnosis of schizophrenia: specific relevance of IgG NR1a antibodies for distinction from N -methyl-D-aspartate glutamate receptor encephalitis.* JAMA Psychiatry, 2013: p. 1–8.
47. Hammer, C., et al., *Neuropsychiatric disease relevance of circulating anti-NMDA receptor autoantibodies depends on blood-brain barrier integrity.* Mol Psychiatry, 2013.
48. Dahm, L., et al., *Seroprevalence of autoantibodies against brain antigens in health and disease.* Ann Neurol, 2014. **76**(1): p. 82–94.
49. Lennox, B.R., et al., *Prevalence and clinical characteristics of serum neuronal cell surface antibodies in first-episode psychosis: a case-control study.* Lancet Psychiatry, 2017. **4**(1): p. 42–48.
50. Vincent, A., et al., *Autoantibodies associated with diseases of the CNS: new developments and future challenges.* Lancet Neurol, 2011. **10**(8): p. 759–772.
51. Goldberg-Stern, H., et al., *Glutamate receptor antibodies directed against AMPA receptors subunit 3 peptide B (GluR3B) associate with some cognitive/psychiatric/behavioral abnormalities in epilepsy patients.* Psychoneuroendocrinology, 2014. **40**: p. 221–231.
52. Pathmanandavel, K., et al., *Antibodies to surface dopamine-2 receptor and N-methyl-D-aspartate receptor in the first episode of acute psychosis in children.* Biol Psychiatry, 2015. 77(6): p. 537–547.
53. Graus, F., A. Saiz, and J. Dalmau, *Antibodies and neuronal autoimmune disorders of the CNS.* J Neurol, 2010. **257**(4): p. 509–517.
54. Hutchinson, M., et al., *Progressive encephalomyelitis, rigidity, and myoclonus: a novel glycine receptor antibody.* Neurology, 2008. **71**(16): p. 1291–1292.
55. Lancaster, E., et al., *Antibodies to metabotropic glutamate receptor 5 in the Ophelia syndrome.* Neurology, 2011. 77(18): p. 1698–1701.
56. Mat, A., et al., *Ophelia syndrome with metabotropic glutamate receptor 5 antibodies in CSF.* Neurology, 2013. **80**(14): p. 1349–1350.
57. Chandley, M.J., et al., *Increased antibodies for the alpha7 subunit of the nicotinic receptor in schizophrenia.* Schizophr Res, 2009. **109**(1–3): p. 98–101.
58. Hoffmann, C., et al., *Alpha7 acetylcholine receptor autoantibodies are rare in sera of patients diagnosed with schizophrenia or bipolar disorder.* PLoS One, under revision.
59. Lai, M., et al., *Investigation of LGI1 as the antigen in limbic encephalitis previously attributed to potassium channels: a case series.* Lancet Neurol, 2010. **9**(8): p. 776–785.
60. Hara, M., et al., *Clinical and pathogenic significance of IgG, IgA, and IgM antibodies against the NMDA receptor.* Neurology, 2018. **90**(16): p. e1386–e1394.
61. Jézéquel, J., et al., *Cell- and single molecule-based methods to detect anti-N-methyl-D-aspartate receptor autoantibodies in patients with first-episode psychosis from the OPTiMiSE project.* Biol Psychiatry, 2017. **82**(10): p. 766–772.
62. Masdeu, J.C., et al., *Serum IgG antibodies against the NR1 subunit of the NMDA receptor not detected in schizophrenia.* Am J Psychiatry, 2012. **169**(10): p. 1120–1121.
63. de Witte, L.D., et al., *Absence of N-methyl-D-aspartate receptor IgG autoantibodies in schizophrenia: the importance of cross-validation studies.* JAMA Psychiatry, 2015. **72**(7): p. 731–733.
64. Mantere, O., et al., *Anti-neuronal anti-bodies in patients with early psychosis.* Schizophrenia Research, 2017.
65. Chen, C.-H., et al., *Seroprevalence survey of selective anti-neuronal autoantibodies in patients with first-episode schizophrenia and chronic schizophrenia.* Schizophr Res, 2017. **190**: p. 28–31.
66. Arboleya, S., et al., *Anti-NMDAR antibodies in new-onset psychosis. Positive results in an HIV-infected patient.* Brain, Behav Immun, 2016. **56**: p. 56–60.
67. Schou, M., et al., *Prevalence of serum anti-neuronal autoantibodies in patients admitted to acute psychiatric care.* Psychol Med, 2016. **46**(16): p. 3303–3313.
68. Endres, D., et al., *Immunological findings in psychotic syndromes: a tertiary care hospital's CSF sample of 180 patients.* Front Hum Neurosci, 2015. **9**: p. 476.
69. Oviedo-Salcedo, T., et al., *Absence of cerebrospinal fluid antineuronal antibodies in schizophrenia spectrum disorders.* Br J Psychiatry, 2018: p. 1–3.
70. Beck, K., et al., *Prevalence of serum N-methyl-D-aspartate receptor autoantibodies in refractory psychosis.* Br J Psychiatry, 2015. **206**(2): p. 164–165.
71. Bergink, V., et al., *Autoimmune encephalitis in postpartum psychosis.* Am J Psychiatry, 2015. **172**(9): p. 901–908.
72. Bossuyt, X., *Clinical performance characteristics of a laboratory test: a practical approach in the autoimmune laboratory.* Autoimmun Rev, 2009. **8**(7): p. 543–548.
73. E.K., S., ed. Clinical interpretation of laboratory procedures. Tietz fundamentals of clinical chemistry, ed. A.E.R. Burtis C.A. 1996, W.B. Saunders Company: Philadelphia. 192–199.
74. Castillo-Gomez, E., et al., *The brain as immunoprecipitator of serum autoantibodies against N-methyl-D-aspartate receptor subunit NR1.* Ann Neurol, 2016. **79**(1): p. 144–151.
75. Hoffmann, C., et al., *Alpha7 acetylcholine receptor autoantibodies are rare in sera of patients diagnosed with schizophrenia or bipolar disorder.* PLoS One, 2018. **13**(12): p. e0208412.
76. Gresa-Arribas, N., et al., *Antibodies to inhibitory synaptic proteins in neurological syndromes associated with glutamic acid decarboxylase autoimmunity.* PLoS One, 2015. **10**(3): p. e0121364.
77. Ohkawa, T., et al., *Identification and characterization of $GABA_A$ receptor autoantibodies in autoimmune encephalitis.* J Neurosci, 2014. **34**(24): p. 8151–8163.
78. Shimoyama, Y., et al., *Anti-NMDA receptor encephalitis presenting as an acute psychotic episode misdiagnosed as dissociative disorder: a case report.* J Clin Rep, 2016. **2**(1): p. 22.
79. Mariotto, S., et al., *Persistence of anti-NMDAR antibodies in CSF after recovery from autoimmune encephalitis.* Neurol Sci, 2017. **38**(8): p. 1523–1524.
80. Hermans, T., et al., *Anti-NMDA receptor encephalitis: still unknown and underdiagnosed by physicians and especially by psychiatrists?* Acta Clinica Belgica, 2017: p. 1–4.
81. Simabukuro, M.M., C.H.d.A. Freitas, and L.H.M. Castro, *A patient with a long history of relapsing psychosis and mania presenting with anti-NMDA receptor encephalitis ten years after first episode.* Dement Neuropsychol, 2015. **9**(3): p. 311–314.
82. Mangalwedhe S.B., et al., *Anti–N-methyl-d-aspartate receptor encephalitis presenting with psychiatric symptoms.* J Neuropsychiatry Clin Neurosci, 2015. **27**(2): p. e152–e153.

83. Heekin, R.D., et al., *Anti-NMDA receptor encephalitis in a patient with previous psychosis and neurological abnormalities: a diagnostic challenge*. Case Rep Psychiatry, 2015. **2015**: p. 253891.
84. Ryan, S.A.M.C., et al., *Anti-NMDA receptor encephalitis: a cause of acute psychosis and catatonia*. J Psychiatr Pract, 2013. **19**(2): p. 157–161.
85. Buckley, C., et al., *Potassium channel antibodies in two patients with reversible limbic encephalitis*. Ann Neurol, 2001. **50**(1): p. 73–78.
86. Pruss, H., *Postviral autoimmune encephalitis: manifestations in children and adults*. Curr Opin Neurol, 2017. **30**(3): p. 327–333.
87. Schott, J.M., et al., *Amnesia, cerebral atrophy, and autoimmunity*. Lancet, 2003. **361**(9365): p. 1266.
88. Vincent, A., et al., *Potassium channel antibody-associated encephalopathy: a potentially immunotherapy-responsive form of limbic encephalitis*. Brain, 2004. **127**(Pt 3): p. 701–712.
89. Amitava Ganguli, et al., *Voltage-gated, potassium-channel antibody-associated limbic encephalitis presenting as acute psychosis*. J Neuropsychiatry Clin Neurosci, 2011. **23**(2): p. E32–E34.
90. Parthasarathi, U.D., et al., *Psychiatric presentation of voltage-gated potassium channel antibody-associated encephalopathy*. Br J Psychiatry, 2006. **189**: p. 10.1192/bjp.bp.105.012864.
91. Gotkine, M., et al., *Limbic encephalitis presenting as a post-partum psychiatric condition*. J Neurol Sci, 2011. **308**(1): p. 152–154.
92. Anand, I., et al., *VGKC-complex antibody mediated encephalitis presenting with psychiatric features and neuroleptic malignant syndrome—further expanding the phenotype*. Devel Med Child Neurol, 2012. **54**(6): p. 575–576.
93. Kruse, J.L., et al., *Psychiatric autoimmunity: N-methyl-d-aspartate receptor IgG and beyond*. Psychosomatics, 2015. **56**(3): p. 227–241.
94. Gomez, A.M., et al., *Proteasome inhibition with bortezomib depletes plasma cells and specific autoantibody production in primary thymic cell cultures from early-onset myasthenia gravis patients*. J immunol (Baltimore, Md.: 1950), 2014. **193**(3): p. 1055–1063.
95. Scheibe, F., et al., *Bortezomib for treatment of therapy-refractory anti-NMDA receptor encephalitis*. Neurology, 2017. **88**(4): p. 366–370.

BRAIN CIRCUIT ALTERATIONS IN PSYCHOSIS

41

THE CIRCUITRY OF MIDBRAIN DOPAMINE SYSTEM DYSREGULATION IN SCHIZOPHRENIA

Felipe V. Gomes, Eric C. Zimmerman, and Anthony A. Grace

INTRODUCTION

There is substantial evidence that the dopamine system is hyperresponsive in schizophrenia patients. This includes the fact that all currently available antipsychotic medications have the ability to block dopamine D2 receptors,[1] and that their clinical potency correlates directly with their affinity for these receptors.[2,3] Moreover, drugs that induce dopamine release or increase dopaminergic transmission, such as amphetamine and L-DOPA, can induce psychotic symptoms in healthy individuals, and furthermore, acute exposure to these drugs at doses that are insufficient to induce psychotic symptoms in healthy subjects has been shown to exacerbate psychosis in schizophrenia patients.[4,5] These findings are complemented by more recent imaging studies in patients showing increased presynaptic markers of dopamine system function[6,7] and exaggerated dopamine release in response to amphetamine. Strikingly, the amplitude of this increase correlates with a worsening of psychotic symptoms.[8] Indeed, studies have shown that both the onset of the disorder[9] and response to treatment[10] correlate with dopamine system activity.

Although the dopamine system is hyperresponsive in schizophrenia, there is little evidence for dysfunction within dopaminergic neurons themselves.[11,12] This has led to the idea that aberrant dopamine transmission and associated symptoms may actually be driven by a disruption of afferent brain regions that regulate dopamine neuron activity. In particular, frontal cortical and hippocampal regions, which are known to display progressive changes in schizophrenia that correlate with illness phase,[13,14] have been proposed to play a prominent role in dopamine system dysfunction. In addition, recent evidence from animal studies has also implicated subregions of the thalamus in regulating the dopamine system via gating of corticohippocampal interactions.[15]

REGULATION OF THE ACTIVITY STATES OF MIDBRAIN DOPAMINE NEURONS

The dopamine system is unique among the modulatory systems in the brain in that it has discrete projections to brain regions involved in a diverse array of functions ranging from motor behavior to affect, reward, and cognition.[12] Anatomical and behavioral studies have shown that the dopamine system can be subdivided into distinct neuronal populations with respect to their location, projections, and behavioral function.[16,17] Rodent studies have shown that differentiation of dopamine system function is most prominent along the mediolateral axis.[18] In particular, the more medial dopamine neurons of the ventral tegmental area (VTA) innervate more ventromedial nucleus accumbens regions that are involved in reward processing. In contrast, VTA dopamine neurons that are located at the transition between the lateral VTA and the substantia nigra preferentially innervate the associative striatum. Finally, dopamine neurons in the substantia nigra project mainly to the habit-formation and motor-related dorsomedial and dorsolateral striatum, respectively.[18] In primates the relative positions of dopamine neurons shift, but their projection- and function-specific differentiation is retained. In lieu of a prominent stand-alone VTA, as in the rodent, primate limbic VTA neurons exist dorsal to the dorsally projecting substantia nigra neurons, forming a dorsal tier limbic/cortical dopamine system and a ventral tier motor/habit formation dopamine system within the substantia nigra.[19,20] Thus, distinct subgroups of dopamine neurons that have distinct projection sites and afferent drive are capable of independently regulating activity states and modulating information flow in separate circuits with distinct functions.[21]

Midbrain dopamine neurons exhibit specific activity states that are regulated by multiple afferent systems and have unique functional implications. Using in vivo electrophysiological recordings, we have been able to gain insight into the dopamine neuron activity states underlying many pathophysiological conditions.[12] Dopamine neurons are known to generate their own activity through a pacemaker conductance.[22] However, in vivo studies in normal rats have shown that approximately half of midbrain dopamine neurons are not spontaneously active, and instead exist in a hyperpolarized, quiescent state (i.e., not firing spontaneously).[23–25] Of those dopamine neurons that are spontaneously active, their firing can take the form of two distinct discharge patterns: single-spike firing and burst firing. Single-spike firing is characterized by a slow, irregular tonic firing pattern (i.e., tonic activity).[25,26] In contrast, when an organism experiences a behaviorally salient stimulus, such as the presentation of a potential threat or an unexpected reward,[27,28] there is a transition into a rapid burst-firing mode (i.e., phasic firing).[29] Thus, burst firing is the behaviorally relevant response of dopamine neurons to stimuli. Our studies have shown that tonic population activity and phasic burst firing combine to modulate the functional output of the dopamine system.[12]

Tonic and phasic dopamine neuron activity are regulated by different afferent systems. Tonic firing is modulated by a powerful GABAergic input from the ventral pallidum, which is capable of hyperpolarizing VTA dopamine neurons below the threshold for firing. It is via this projection that the ventral pallidum controls the proportion of dopamine neurons in the VTA that are firing spontaneously,[30] which is referred to as population activity. Inactivation of ventral pallidum releases VTA dopamine neurons from inhibition and causes them to fire spontaneously, thus enhancing population activity. The ventral pallidum in turn is potently regulated by a pathway that arises from the ventral subiculum of the hippocampus. When the ventral subiculum is activated, it provides a robust glutamatergic drive of GABAergic projection neurons in the nucleus accumbens, providing feedforward inhibition to the ventral pallidum and releasing the VTA dopamine neurons from inhibition (Figure 41.1). The result is an increase in dopamine neuron population activity. In this manner, the ventral subiculum-ventral striatal-ventral pallidal pathway controls the tonic baseline activity of the dopamine system.[11,21] Importantly, artificially activating the ventral subiculum is sufficient to increase the population activity of dopamine neurons in the VTA[31] and to release dopamine in the nucleus accumbens.[32]

Burst firing in dopamine neurons is driven by a different glutamatergic input, which arises from the brainstem pedunculopontine tegmentum (PPTg; Figure 41.1). This glutamatergic input acts on dopamine neurons via activation of NMDA receptors.[11,30] NMDA receptors are unique in that they exhibit magnesium block at more hyperpolarized membrane potentials,[33] and can only be activated by glutamate in neurons that are depolarized and firing spontaneously. In the case of dopamine neurons, this means that burst firing can only occur in neurons that are already in a depolarized, spontaneously active state, i.e., tonically active. Therefore, the proportion of dopamine neurons that are spontaneously active prior to PPTg activation determines the amplitude of burst-firing-driven phasic dopamine release. Collectively, these findings indicate that tonic activity is a critical parameter with respect to the functional output of the dopamine system; it sets the baseline level of responsivity of the system to behaviorally relevant stimuli.[34] For this reason, population activity sets the "gain" or the level of amplification of the phasic response,[11] i.e., the more dopamine neurons firing, the bigger the phasic response to stimuli.[21]

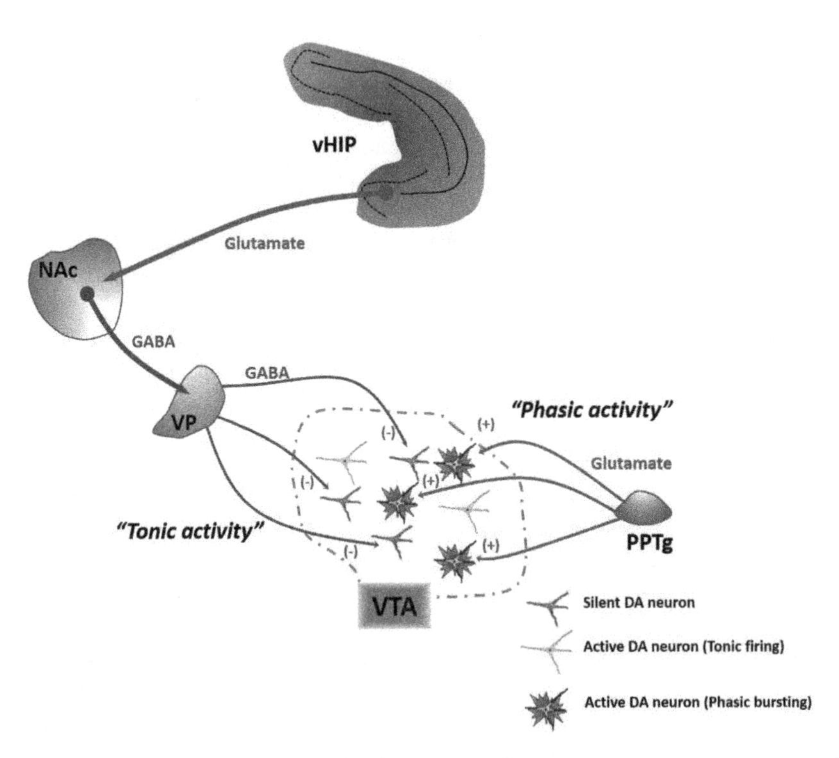

Figure 41.1 Tonic and phasic dopamine (DA) neuron activity are regulated by distinct afferent systems. DA neurons generate their own activity through a pacemaker conductance. However, a substantial population of DA neurons is not firing spontaneously, being held in a hyperpolarized state by a GABA-mediated inhibitory input from the ventral pallidum (VP). The VP, in turn, is controlled by a pathway originating in the ventral hippocampus (vHIP). The vHIP projects to the nucleus accumbens (NAc), which inhibits the VP. By contrast, phasic burst firing is driven by glutamatergic inputs arising from several areas, primary among these being the pedunculopontine tegmentum (PPTg). This afferent system works to regulate activity states within the population of spontaneously active DA neurons, because only neurons that are firing spontaneously can burst fire—NMDA channels on hyperpolarized ("silent") DA neurons are under magnesium block and will not change state. Therefore, the PPTg provides the rapid, behaviorally salient phasic signal, whereas the VP, by controlling the number of DA neurons firing, determines the gain of the phasic signal.

MIDBRAIN DOPAMINE SYSTEM OVERDRIVE IN SCHIZOPHRENIA

Although abnormalities in the dopamine system cannot account for the broad array of schizophrenia symptoms, there is abundant evidence that subcortical hyperdopaminergia is highly correlated with psychosis.[6] Kapur and others have theorized that psychotic symptoms, particularly delusions and hallucinations, emerge as a state of aberrant salience.[35,36] It is well established that in healthy subjects subcortical dopamine tone contributes to the attribution of incentive salience, a process by which a stimulus grabs attention and motivates goal-directed behavior due to associations with reward or punishment.[37,38] Animal studies have shown that stimuli become motivationally salient when the release of dopamine in the striatum coincides with their perception.[35,39] In schizophrenia, it has been proposed that an overresponsive midbrain dopamine system could result in an abnormal assignment of salience to innocuous stimuli. This is believed to be reflected in the schizophrenia patient as massively altered information processing, leading to flooding of the sensorium with unfiltered input and inappropriate attribution of salience.[36] In fact, midbrain dopamine dysfunction is associated with abnormalities in the ability of psychotic patients to discriminate between motivationally salient and neutral stimuli.[40] Notably, high-risk subjects for psychosis also seem to be more likely to attribute salience to irrelevant stimuli. [41,42]

In humans, subcortical dopamine tone can be measured using positron emission tomography (PET) of the radioligands fluorodopa and raclopride. Fluorodopa is taken up into active presynaptic dopaminergic terminals and therefore is thought to reflect the population activity of dopamine neurons—greater dopamine neuron population activity would correspond to a higher number of active terminals and hence greater fluorodopa uptake.[43] In contrast, raclopride acts as an antagonist of D2 receptors and is displaced from binding by endogenous dopamine. Raclopride displacement is believed to reflect phasic release since D2 receptors are highly concentrated in the synapse, where a phasic response will cause massive dopamine release (on the order of hundreds of micromolar to millimolar concentrations) driven by burst events, e.g., amphetamine-induced striatal dopamine release.[12] Performed in awake subjects, amphetamine-stimulated phasic events are likely a combination of the ability of this drug to increase attention to stimuli, combined with the attenuation of uptake into terminals. Therefore, fluorodopa uptake is thought to measure tonic dopamine neuron activity, which we posit reflects the responsivity of the system, whereas amphetamine-stimulated raclopride displacement reflects the phasic, behaviorally relevant impact of dopamine.

Increased striatal fluorodopa uptake has been consistently observed in schizophrenia patients, as well as increased striatal synaptic dopamine levels at rest or induced by amphetamine and psychosocial stress, as measured by raclopride displacement.[6,7,44,45] In addition to changes in the striatum, increased dopamine synthesis capacity in the midbrain in schizophrenia has also identified in imaging studies.[46] These findings indicate that schizophrenia is associated with elevations in striatal and midbrain dopamine synthesis capacity and elevated baseline and stimulated striatal dopamine levels, pointing to a critical role for presynaptic dopamine dysfunction in the disease. Importantly, it has also been observed that hyperdopaminergia predates the first psychotic break in ultra-high-risk individuals experiencing prodromal symptoms.[47] Additionally, ultra-high-risk individuals who later transition to psychosis have greater dopamine synthesis capacity than those who do not.[9,48,49] Furthermore, in the transition group, there was a direct relationship between the magnitude of dopamine alteration and the severity of prodromal psychotic symptoms and neurocognitive dysfunction.[47] These findings suggest that presynaptic dopaminergic dysfunction may start before psychosis onset, possibly leading to increased stimulus responsivity prior to the emergence of overstimulation and misinterpretation.

Altered dopamine synthesis capacity can also be used to predict antipsychotic efficacy. Studies have shown that dopamine synthesis capacity is significantly lower in patients with treatment-resistant schizophrenia.[10,50] We have published previously that one potential therapeutic action of antipsychotic drugs is the induction of dopamine neuron depolarization block, thereby attenuating dopamine neuron activity and responsivity.[51–53] Moreover, the ability of antipsychotic drugs to induce depolarization block is dependent on the baseline activity state of the dopamine neurons, with a more activated dopamine system leading to a more rapid induction of depolarization block.[53] These findings suggest that antipsychotic treatment may be ineffective in treatment-resistant patients with lower dopamine synthesis capacity because these patients do not exhibit a sufficiently activated dopamine system. What could be responsible for psychosis in such individuals? A study by Abi-Dargham and colleagues showed that schizophrenia patients who are also diagnosed with comorbid substance use disorders exhibit diminished baseline dopamine tone. However, amphetamine administration was still able to induce dopamine release, which was associated with an acute and transient increase in positive symptoms.[54] Therefore, it may be that the dopamine system in treatment-resistant patients is not sufficiently activated for antipsychotic drug-induced depolarization block to be efficacious in treating symptoms.

Importantly, evidence from neuroimaging studies has also pointed toward regional specificity in dopamine alterations in schizophrenia. Specifically, the associative striatum, which contains the rostral caudate, rostral putamen, and postcomissural caudate, exhibits the most pronounced enhancement of fluorodopa uptake in prodromal patients with psychotic symptoms and the largest increase in dopamine release in schizophrenia patients.[47,55] This region is functionally linked to associative cortical regions such as the dorsolateral prefrontal cortex and is thought to play a role in the attentional salience of stimuli, suggesting that these alterations may be involved in the development of psychosis.[56,57]

HIPPOCAMPUS HYPERACTIVITY DRIVES DOPAMINE SYSTEM DYSREGULATION IN SCHIZOPHRENIA

One of the most influential hypotheses of schizophrenia proposes that a hypofunction of the cortical dopamine system may underlie the subcortical hyperdopaminergic state and psychosis.[58,59] Although such a deficit in cortex may exist and contribute to other schizophrenia symptoms (e.g., negative or cognitive symptoms), the evidence for a causal link between cortical dopamine and psychosis is lacking. However, abnormalities in the midbrain dopamine system do seem to be a secondary consequence of disturbances in brain areas other than the dopamine neurons themselves.[60] In fact, data from schizophrenia patients have indicated that hyperactivity in the anterior hippocampus (ventral hippocampus in rodents) is strongly correlated with psychosis in schizophrenia. Furthermore, when the ventral hippocampus output is artificially stimulated in rats, both the VTA dopamine population activity and the amount of dopamine released in the nucleus accumbens are increased.[30–32] This evidence fits with the observation that the ventral subiculum of the hippocampus-nucleus accumbens-ventral pallidum pathway controls the level of tonic activity (gain) of the dopamine system. Therefore, hyperdopaminergia in schizophrenia might result from a pathological enhancement of this critical gain of function.[12]

In schizophrenia, alterations in hippocampal anatomy, perfusion, and activity are consistently reported.[61] Hippocampal volume, which may be genetically determined,[62] has been found to be reduced at the onset of schizophrenia and continues to decrease as the disease progresses.[14,63] Altered hippocampal function has been demonstrated to contribute to impaired information processing and to be responsible for some of the cognitive deficits in patients with schizophrenia. Moreover, the anterior hippocampus seems to be hyperactive in schizophrenia.[64] This hyperactivity correlates with psychosis and can also predict conversion to schizophrenia.[65–67] Interestingly, in ultra-high-risk individuals this hyperactivity seems to begin in anterior CA1 and spreads to the subiculum after the onset of psychosis.[68] This glutamatergic hyperactivity may drive for the loss of volume at disease onset. Therefore, the coupling between hippocampal hyperactivity and subcortical hyperdopaminergic activity could serve as a potential indicator of risk that may help to predict which subjects will later develop psychosis.[69]

The excessive hippocampal activity observed in schizophrenia patients is accompanied by a substantial reduction of the calcium-binding protein parvalbumin in GABAergic interneurons of the hippocampus.[70–72] This parvalbumin loss is thought to disrupt the activity of parvalbumin-expressing interneurons, resulting in disinhibition of hippocampal glutamatergic pyramidal neurons, and thereby promoting a hyperdopaminergic state as discussed earlier. Given the evidence correlating hippocampal dysfunction with psychosis in schizophrenia, we propose that aberrant hippocampal activity may underlie the dopamine hyperresponsiveness observed in this disorder (Figure 41.2).

Parallel observations in animal models of schizophrenia suggest that hippocampal overdrive leads to a hyperresponsive dopamine system. The methylazoxymethanol acetate (MAM) neurodevelopmental disruption model of schizophrenia is based on the administration of MAM (a mitotoxin) to pregnant rats at gestational day 17. This prenatal treatment produces a state that resembles schizophrenia along pharmacological, anatomical, physiological, and behavioral dimensions. These alterations become evident in MAM-treated offspring only after puberty.[72,73] Among these changes is hippocampal hyperactivity, manifested as increased pyramidal cell firing rate and disrupted slow rhythmic activity.[74,75] This pathologically enhanced drive from the ventral subiculum of

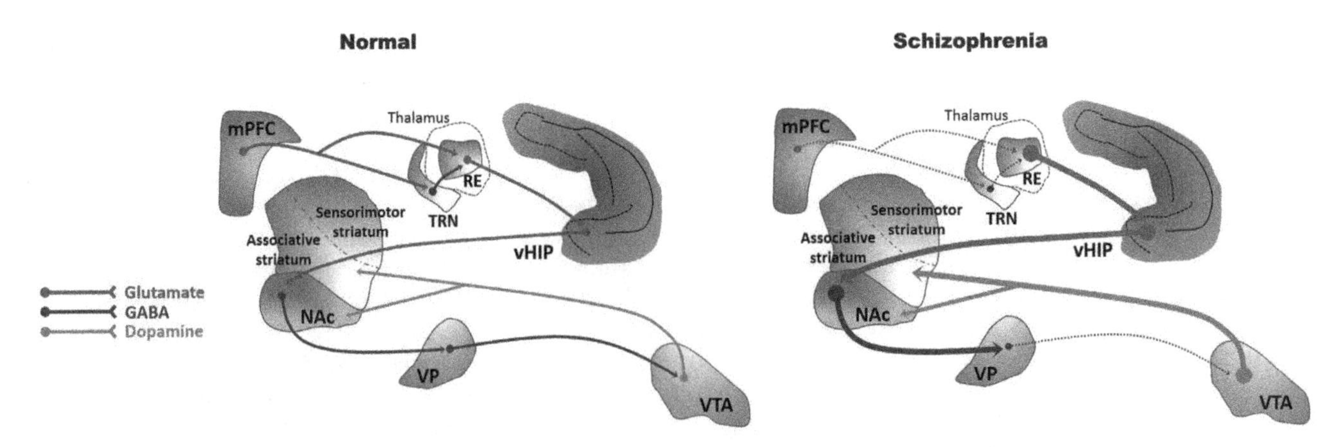

Figure 41.2 The ventral hippocampus (vHIP) regulates midbrain dopamine system activity via a polysynaptic circuit. The vHIP excites neurons in the nucleus accumbens (NAc) that, in turn, inhibits ventral pallidal (VP) activity. Given that the VP provides an inhibitory tone to VTA dopamine neurons, activation of the vHIP will result in an increase in dopamine neuron activity. In schizophrenia, hippocampal hyperactivity, which has been suggested to result from a loss of parvalbumin interneurons, leads to an increase in active dopamine neurons projecting to the associative striatum. Recently, it was demonstrated that disruption of corticothalamic-hippocampal interactions can induce a hyperdopaminergic state. These findings suggest that, in schizophrenia, dysfunction of the medial prefrontal cortex (mPFC) could disinhibit the nucleus reuniens (RE), possibly via loss of feed-forward inhibition from the reticular nucleus of the thalamus (TRN). This would in turn drive the vHIP and enhance VTA DA neuron population activity. Therefore disruptions in either the mPFC-RE projection or the vHIP interneuron control could have parallel or additive effects on disrupting DA neuron activity.

the hippocampus results in a doubling in the number of dopamine neurons firing spontaneously (i.e., increased tonic population activity) in the VTA.[75] The inactivation of the ventral subiculum completely normalizes the enhanced dopamine population activity and reverses enhanced locomotor hyperresponsivity to amphetamine. These findings suggest a causal link between excessive hippocampal activity and dopamine system hyperresponsivity in MAM rats. Interestingly, neurons constituting the lateral VTA projection to the associative striatal regions are the most severely affected in MAM animals, paralleling findings in schizophrenia patients discussed earlier.[75] Also similar to schizophrenia patients,[70–72] MAM rats show a reduction in parvalbumin-expressing interneurons within the ventral subiculum.[74,76,77] Altogether, these findings indicate that a functional loss of parvalbumin-expressing interneurons in the ventral subiculum could lead to alterations in its output that drive a hyperresponsive dopamine system underlying psychosis.[12]

POTENTIAL INVOLVEMENT OF THALAMIC DISRUPTION IN DOPAMINE SYSTEM ABNORMALITIES IN SCHIZOPHRENIA

The hippocampus is a well-established site of dysfunction in schizophrenia, but it resides within an extended network of several other brain regions. Many functional imaging studies have demonstrated aberrant functional and structural connectivity between the hippocampus and other brain regions implicated in schizophrenia pathophysiology,[78–80] suggesting that circuit dysfunction in schizophrenia is likely present in distributed networks. One region that has long been considered as a potential node of dysfunction in the disease is the thalamus,[81] but only recently has significant progress been made in describing the neurobiology that may underlie this role. Specifically, the anterior, midline, and mediodorsal nuclei of the thalamus have garnered increased attention due to their highly interconnected nature with respect to the hippocampus and prefrontal cortex across species.[82–84]

Initial postmortem studies in patients with schizophrenia reported a decrease in both thalamic size and cell number in the whole thalamus, as well as the anterior and mediodorsal nuclei specifically.[85,86] However, results from further studies proved inconsistent.[87] More consistent are structural imaging studies of the thalamus, the majority of which report reduced whole thalamus gray matter volume, as well as volume reductions in the anterior, midline, and mediodorsal nuclei.[80] These reductions are seen in individuals with both ultra-high risk and familial high risk for schizophrenia.[88,89] Unlike the case for the hippocampus, most functional imaging studies report no differences in intrinsic, resting-state activity of the thalamus in schizophrenia patients.[89] However, differences in resting-state functional connectivity between thalamus and hippocampus/prefrontal cortex have been reported in the disease. Reduced connectivity between the whole thalamus and anterior hippocampus has been reported.[78] Schizophrenia patients have been found to exhibit hyperconnectivity between anterior, midline, and mediodorsal nuclei and sensory cortices, but hypoconnectivity with prefrontal regions. This finding is present at both the chronic and early stages of schizophrenia as well as in ultra-high-risk individuals and first-episode subjects, and predicts conversion to psychosis in these groups.[78,90–95] In addition, studies using ketamine to model psychosis in healthy human subjects have demonstrated enhanced gamma band oscillations in the thalamus and hippocampus, as well as thalamic dysconnectivity patterns consistent with those seen in schizophrenia.[96] Finally, a reduction in sleep spindles (non–rapid eye movement sleep oscillations generated by the thalamic reticular nucleus) has been consistently reported in patients with schizophrenia, and the magnitude of this reduction has been shown to be inversely correlated with the severity of psychotic symptoms.[97,98]

Recent preclinical work in rodents has demonstrated possible circuit abnormalities underlying and resulting from the thalamic dysfunction observed in patients with schizophrenia. Our group and others have focused on the nucleus reuniens of the midline thalamus because of its heavy interconnections with the hippocampus and prefrontal cortex across species.[82,83,99] In rodents, the reuniens forms the primary route of communication between the prefrontal cortex and ventral hippocampus and is necessary for behaviors requiring coordinated activity of these two regions, e.g., spatial working memory and fear learning.[100–105] Recently, our group reported that activation of the reuniens can enhance population activity in VTA dopamine neurons via its excitatory projection to the ventral hippocampus. In addition, the reuniens appears to mediate the regulation of VTA dopamine neurons by the ventromedial prefrontal cortex.[15,106] The ventromedial prefrontal cortex of the rat, specifically the prelimbic prefrontal cortex, is also important for attenuation of stress responses via inhibition of the amygdala.[107] Indeed, lesions of this prefrontal region early in life will enable stressful stimuli administered prepubertally to lead to a condition in normal rats that is similar to that observed after MAM treatment.[108] These findings and others suggest that loss of top-down prefrontal regulation via disruption of corticothalamic communication (Figure 41.2), as has been observed in schizophrenia patients, could contribute to hippocampal overdrive, parvalbumin neuron loss, and a hyperdopaminergic state characteristic of this disorder.

CONCLUSION

Progress in understanding the pathophysiology of schizophrenia has been slow in coming over the past several decades, due to the complex multifactorial nature of the symptomatology, the delayed onset of psychosis, and the multiple neural systems that appear to play a role in the disease. The observations that antipsychotic drugs to block dopamine receptors and that dopamine releasing agents can mimic some aspects of schizophrenia have certainly led to insights into the disorder, but may have also obscured identification of the site of pathology. Recently, there has been a convergence of information from both clinical and preclinical studies that has provided a unique window into this complex disorder: the dopamine system itself may be normal, but is being dysregulated by other structures. Notable among these structures is the hippocampus, a region that controls the responsivity of the dopamine system, is hyperactive in patients with schizophrenia, and exhibits loss of a specific stress-sensitive interneuron type that may drive this hyperactivity.[109]

The focus on hippocampal hyperactivity and disruption of normal rhythmic activity provides a unique insight into the disorder; i.e., schizophrenia is not due to a loss of activity in a specific brain region. Indeed, loss of activity is more typically associated with a focused abnormality, and the ability of the brain to compensate for the disruption. However, it is unlikely that the hippocampus is acting in isolation to produce the myriad symptoms observed in patients with schizophrenia. The hippocampus itself is highly interconnected with other brain regions, and therefore a hippocampus that is both hyperactive and dysrhythmic would be expected to lead to widespread disturbances across cortical and subcortical structures that could contribute to cognitive and negative symptoms as well.

The possibility that this is due to interneuron loss also raises interesting questions. During development, excitatory circuits develop first, with inhibitory interneurons entering later to stabilize these circuits. Interneurons introduce rhythmic oscillatory activity in circuits, allowing for coherent information flow among interconnected regions.[12,110] Given that parvalbumin-containing interneurons are generated relatively late in development and are stress sensitive, they could represent a common site of pathology across neurodevelopmental psychiatric disorders.[12] Our knowledge about interneuron deficits in schizophrenia may provide unique insights into developing treatments for schizophrenia that leverage the potential to intervene at a sufficiently early time point in the pathological process to prevent transition to psychosis in susceptible individuals.

FUNDING

Research activity of the authors is supported by grants from US National Institutes of Health (MH 109199 to ECZ, MH57440 to A.A.G.). A.A.G. also has received funds from Johnson & Johnson, Lundbeck, Pfizer, GlaxoSmithKline, Merck, Takeda, Dainippon Sumitomo, Otsuka, Lilly, Roche, Asubio, Abbott, Autofony, Janssen, and Alkermes.

REFERENCES

1. Kapur S, Remington G. Dopamine D(2) receptors and their role in atypical antipsychotic action: still necessary and may even be sufficient. *Biol Psychiatry.* 2001;50(11):873–883.
2. Seeman P, Lee T. Antipsychotic drugs: direct correlation between clinical potency and presynaptic action on dopamine neurons. *Science.* 1975;188(4194):1217–1219.
3. Creese I, Burt DR, Snyder SH. Dopamine receptor binding predicts clinical and pharmacological potencies of antischizophrenic drugs. *Science.* 1976;192(4238):481–483.
4. Janowsky DS, el-Yousel MK, Davis JM, Sekerke HJ. Provocation of schizophrenic symptoms by intravenous administration of methylphenidate. *Arch Gen Psychiatry.* 1973;28(2):185–191.
5. Angrist B, Sathananthan G, Wilk S, Gershon S. Amphetamine psychosis: behavioral and biochemical aspects. *J Psychiatr Res.* 1974;11:13–23.
6. Kambeitz J, Abi-Dargham A, Kapur S, Howes OD. Alterations in cortical and extrastriatal subcortical dopamine function in schizophrenia: systematic review and meta-analysis of imaging studies. *Br J Psychiatry.* 2014;204(6):420–429.
7. Howes OD, Kambeitz J, Kim E, et al. The nature of dopamine dysfunction in schizophrenia and what this means for treatment. *Arch Gen Psychiatry.* 2012;69(8):776–786.
8. Laruelle M, Abi-Dargham A. Dopamine as the wind of the psychotic fire: new evidence from brain imaging studies. *J Psychopharmacol.* 1999;13(4):358–371.
9. Howes OD, Bose SK, Turkheimer F, et al. Dopamine synthesis capacity before onset of psychosis: a prospective [18F]-DOPA PET imaging study. *Am J Psychiatry.* 2011;168(12):1311–1317.
10. Demjaha A, Murray RM, McGuire PK, Kapur S, Howes OD. Dopamine synthesis capacity in patients with treatment-resistant schizophrenia. *Am J Psychiatry.* 2012;169(11):1203–1210.
11. Grace AA. Dopamine system dysregulation by the hippocampus: implications for the pathophysiology and treatment of schizophrenia. *Neuropharmacology.* 2012;62(3):1342–1348.
12. Grace AA. Dysregulation of the dopamine system in the pathophysiology of schizophrenia and depression. *Nat Rev Neurosci.* 2016;17(8):524–532.
13. Cannon TD, Chung Y, He G, et al. Progressive reduction in cortical thickness as psychosis develops: a multisite longitudinal neuroimaging study of youth at elevated clinical risk. *Biol Psychiatry.* 2015;77(2):147–157.
14. Chakos MH, Schobel SA, Gu H, et al. Duration of illness and treatment effects on hippocampal volume in male patients with schizophrenia. *Br J Psychiatry.* 2005;186:26–31.
15. Zimmerman EC, Grace AA. The nucleus reuniens of the midline thalamus gates prefrontal-hippocampal modulation of ventral tegmental area dopamine neuron activity. *J Neurosci.* 2016;36(34):8977–8984.
16. Bjorklund A, Dunnett SB. Dopamine neuron systems in the brain: an update. *Trends Neurosci.* 2007;30(5):194–202.
17. Roeper J. Dissecting the diversity of midbrain dopamine neurons. *Trends Neurosci.* 2013;36(6):336–342.
18. Ikemoto S. Dopamine reward circuitry: two projection systems from the ventral midbrain to the nucleus accumbens-olfactory tubercle complex. *Brain Res Rev.* 2007;56(1):27–78.
19. Szabo J. Organization of the ascending striatal afferents in monkeys. *J Comp Neurol.* 1980;189(2):307–321.
20. Lynd-Balta E, Haber SN. The organization of midbrain projections to the striatum in the primate: sensorimotor-related striatum versus ventral striatum. *Neuroscience.* 1994;59(3):625–640.
21. Grace AA, Floresco SB, Goto Y, Lodge DJ. Regulation of firing of dopaminergic neurons and control of goal-directed behaviors. *Trends Neurosci.* 2007;30(5):220–227.
22. Grace AA, Bunney BS. Intracellular and extracellular electrophysiology of nigral dopaminergic neurons—1. Identification and characterization. *Neuroscience.* 1983;10(2):301–315.
23. Freeman AS, Bunney BS. Activity of A9 and A10 dopaminergic neurons in unrestrained rats: further characterization and effects of apomorphine and cholecystokinin. *Brain Res.* 1987;405(1):46–55.
24. Freeman AS, Meltzer LT, Bunney BS. Firing properties of substantia nigra dopaminergic neurons in freely moving rats. *Life Sci.* 1985;36(20):1983–1994.
25. Grace AA, Bunney BS. The control of firing pattern in nigral dopamine neurons: single spike firing. *J Neurosci.* 1984;4(11):2866–2876.
26. Grace AA, Bunney BS. Opposing effects of striatonigral feedback pathways on midbrain dopamine cell activity. *Brain Res.* 1985;333(2):271–284.
27. Schultz W. Reward functions of the basal ganglia. *J Neural Transm.* 2016;123(7):679–693.

28. Lammel S, Ion DI, Roeper J, Malenka RC. Projection-specific modulation of dopamine neuron synapses by aversive and rewarding stimuli. *Neuron.* 2011;70(5):855–862.
29. Grace AA, Bunney BS. The control of firing pattern in nigral dopamine neurons: burst firing. *J Neurosci.* 1984;4(11):2877–2890.
30. Floresco SB, West AR, Ash B, Moore H, Grace AA. Afferent modulation of dopamine neuron firing differentially regulates tonic and phasic dopamine transmission. *Nat Neurosci.* 2003;6(9):968–973.
31. Floresco SB, Todd CL, Grace AA. Glutamatergic afferents from the hippocampus to the nucleus accumbens regulate activity of ventral tegmental area dopamine neurons. *J Neurosci.* 2001;21(13):4915–4922.
32. Legault M, Rompre PP, Wise RA. Chemical stimulation of the ventral hippocampus elevates nucleus accumbens dopamine by activating dopaminergic neurons of the ventral tegmental area. *J Neurosci.* 2000;20(4):1635–1642.
33. Mayer ML, Westbrook GL, Guthrie PB. Voltage-dependent block by Mg2+ of NMDA responses in spinal cord neurones. *Nature.* 1984;309(5965):261–263.
34. Lodge DJ, Grace AA. The hippocampus modulates dopamine neuron responsivity by regulating the intensity of phasic neuron activation. *Neuropsychopharmacology.* 2006;31(7):1356–1361.
35. Kapur S. Psychosis as a state of aberrant salience: a framework linking biology, phenomenology, and pharmacology in schizophrenia. *Am J Psychiatry.* 2003;160(1):13–23.
36. Howes OD, Kapur S. The dopamine hypothesis of schizophrenia: version III—the final common pathway. *Schizophr Bull.* 2009;35(3):549–562.
37. Berridge KC. The debate over dopamine's role in reward: the case for incentive salience. *Psychopharmacology.* 2007;191(3):391–431.
38. Berridge KC, Robinson TE. What is the role of dopamine in reward: hedonic impact, reward learning, or incentive salience? *Brain Res Rev.* 1998;28(3):309–369.
39. Schultz W, Dayan P, Montague PR. A neural substrate of prediction and reward. *Science.* 1997;275(5306):1593–1599.
40. Murray GK, Corlett PR, Clark L, et al. Substantia nigra/ventral tegmental reward prediction error disruption in psychosis. *Mol Psychiatry.* 2008;13(3):239, 267–276.
41. Schmidt A, Antoniades M, Allen P, et al. Longitudinal alterations in motivational salience processing in ultra-high-risk subjects for psychosis. *Psychol Med.* 2017;47(2):243–254.
42. Roiser JP, Howes OD, Chaddock CA, Joyce EM, McGuire P. Neural and behavioral correlates of aberrant salience in individuals at risk for psychosis. *Schizophr Bull.* 2013;39(6):1328–1336.
43. Modinos G, Allen P, Grace AA, McGuire P. Translating the MAM model of psychosis to humans. *Trends Neurosci.* 2015;38(3):129–138.
44. Abi-Dargham A, Rodenhiser J, Printz D, et al. Increased baseline occupancy of D2 receptors by dopamine in schizophrenia. *Proc Natl Acad Sci U S A.* 2000;97(14):8104–8109.
45. Mizrahi R, Addington J, Rusjan PM, et al. Increased stress-induced dopamine release in psychosis. *Biol Psychiatry.* 2012;71(6):561–567.
46. Howes OD, Williams M, Ibrahim K, et al. Midbrain dopamine function in schizophrenia and depression: a post-mortem and positron emission tomographic imaging study. *Brain.* 2013;136(Pt 11):3242–3251.
47. Howes OD, Montgomery AJ, Asselin MC, et al. Elevated striatal dopamine function linked to prodromal signs of schizophrenia. *Arch Gen Psychiatry.* 2009;66(1):13–20.
48. Egerton A, Chaddock CA, Winton-Brown TT, et al. Presynaptic striatal dopamine dysfunction in people at ultra-high risk for psychosis: findings in a second cohort. *Biol Psychiatry.* 2013;74(2):106–112.
49. Allen P, Luigjes J, Howes OD, et al. Transition to psychosis associated with prefrontal and subcortical dysfunction in ultra high-risk individuals. *Schizophr Bull.* 2012;38(6):1268–1276.
50. Kim E, Howes OD, Veronese M, et al. Presynaptic dopamine capacity in patients with treatment-resistant schizophrenia taking clozapine: an [18F]DOPA PET study. *Neuropsychopharmacology.* 2017;42(4):941–950.
51. Grace AA. The depolarization block hypothesis of neuroleptic action: implications for the etiology and treatment of schizophrenia. *J Neural Transm.* 1992;36:91–131.
52. Grace AA, Bunney BS. Induction of depolarization block in midbrain dopamine neurons by repeated administration of haloperidol: analysis using in vivo intracellular recording. *J Pharmacol Exp Ther.* 1986;238(3):1092–1100.
53. Valenti O, Cifelli P, Gill KM, Grace AA. Antipsychotic drugs rapidly induce dopamine neuron depolarization block in a developmental rat model of schizophrenia. *J Neurosci.* 2011;31(34):12330–12338.
54. Thompson JL, Urban N, Slifstein M, et al. Striatal dopamine release in schizophrenia comorbid with substance dependence. *Mol Psychiatry.* 2013;18(8):909–915.
55. Kegeles LS, Abi-Dargham A, Frankle WG, et al. Increased synaptic dopamine function in associative regions of the striatum in schizophrenia. *Arch Gen Psychiatry.* 2010;67(3):231–239.
56. Winton-Brown TT, Fusar-Poli P, Ungless MA, Howes OD. Dopaminergic basis of salience dysregulation in psychosis. *Trends Neurosci.* 2014;37(2):85–94.
57. Haber SN. The primate basal ganglia: parallel and integrative networks. *J Chem Neuroanat.* 2003;26(4):317–330.
58. Davis KL, Kahn RS, Ko G, Davidson M. Dopamine in schizophrenia: a review and reconceptualization. *Am J Psychiatry.* 1991;148(11):1474–1486.
59. Weinberger DR. Implications of normal brain development for the pathogenesis of schizophrenia. *Arch Gen Psychiatry.* 1987;44(7):660–669.
60. Grace AA, Gomes FV. The circuitry of dopamine system regulation and its disruption in schizophrenia: insights into treatment and prevention. *Schizophr Bull.* 2019;45(1):148–157.
61. Heckers S, Konradi C. Hippocampal pathology in schizophrenia. *Curr Top Behav Neurosci.* 2010;4:529–553.
62. Hibar DP, Adams HH, Jahanshad N, et al. Novel genetic loci associated with hippocampal volume. *Nat Commun.* 2017;8:13624.
63. Pantelis C, Velakoulis D, McGorry PD, et al. Neuroanatomical abnormalities before and after onset of psychosis: a cross-sectional and longitudinal MRI comparison. *Lancet.* 2003;361(9354):281–288.
64. Heckers S, Konradi C. GABAergic mechanisms of hippocampal hyperactivity in schizophrenia. *Schizophr Res.* 2015;167(1-3):4–11.
65. Silbersweig DA, Stern E, Frith C, et al. A functional neuroanatomy of hallucinations in schizophrenia. *Nature.* 1995;378(6553):176–179.
66. Stone JM, Howes OD, Egerton A, et al. Altered relationship between hippocampal glutamate levels and striatal dopamine function in subjects at ultra high risk of psychosis. *Biol Psychiatry.* 2010;68(7):599–602.
67. Schobel SA, Lewandowski NM, Corcoran CM, et al. Differential targeting of the CA1 subfield of the hippocampal formation by schizophrenia and related psychotic disorders. *Arch Gen Psychiatry.* 2009;66(9):938–946.
68. Schobel SA, Chaudhury NH, Khan UA, et al. Imaging patients with psychosis and a mouse model establishes a spreading pattern of hippocampal dysfunction and implicates glutamate as a driver. *Neuron.* 2013;78(1):81–93.
69. Abi-Dargham A, Horga G. The search for imaging biomarkers in psychiatric disorders. *Nat Med.* 2016;22(11):1248–1255.
70. Zhang ZJ, Reynolds GP. A selective decrease in the relative density of parvalbumin-immunoreactive neurons in the hippocampus in schizophrenia. *Schizophr Res.* 2002;55(1-2):1–10.
71. Benes FM, Lim B, Matzilevich D, Walsh JP, Subburaju S, Minns M. Regulation of the GABA cell phenotype in hippocampus of schizophrenics and bipolars. *Proc Natl Acad Sci U S A.* 2007;104(24):10164–10169.
72. Konradi C, Yang CK, Zimmerman EI, et al. Hippocampal interneurons are abnormal in schizophrenia. *Schizophr Res.* 2011; 131(1-3):165–173.
73. Gomes FV, Rincon-Cortes M, Grace AA. Adolescence as a period of vulnerability and intervention in schizophrenia: Insights from the MAM model. *Neurosci Biobehav Rev.* 2016;70:260–270.

74. Lodge DJ, Behrens MM, Grace AA. A loss of parvalbumin-containing interneurons is associated with diminished oscillatory activity in an animal model of schizophrenia. *J Neurosci.* 2009;29(8):2344–2354.
75. Lodge DJ, Grace AA. Aberrant hippocampal activity underlies the dopamine dysregulation in an animal model of schizophrenia. *J Neurosci.* 2007;27(42):11424–11430.
76. Lodge DJ, Grace AA. Divergent activation of ventromedial and ventrolateral dopamine systems in animal models of amphetamine sensitization and schizophrenia. *Int J Neuropsychopharmacol.* 2012;15(1):69–76.
77. Gill KM, Grace AA. Corresponding decrease in neuronal markers signals progressive parvalbumin neuron loss in MAM schizophrenia model. *Int J Neuropsychopharmacol.* 2014;17(10):1609–1619.
78. Samudra N, Ivleva EI, Hubbard NA, et al. Alterations in hippocampal connectivity across the psychosis dimension. *Psychiatry Res.* 2015;233(2):148–157.
79. Bernard JA, Orr JM, Mittal VA. Abnormal hippocampal-thalamic white matter tract development and positive symptom course in individuals at ultra-high risk for psychosis. *NPJ Schizophrenia.* 2015;1.
80. Pergola G, Selvaggi P, Trizio S, Bertolino A, Blasi G. The role of the thalamus in schizophrenia from a neuroimaging perspective. *Neurosci Biobehav Rev.* 2015;54:57–75.
81. McGhie A, Chapman J. Disorders of attention and perception in early schizophrenia. *Br J Med Psychol.* 1961;34:103–116.
82. Amaral DG, Cowan WM. Subcortical afferents to the hippocampal formation in the monkey. *J Comp Neurol.* 1980;189(4): 573–591.
83. Vertes RP, Linley SB, Hoover WB. Limbic circuitry of the midline thalamus. *Neurosci Biobehav Rev.* 2015;54:89–107.
84. Mitchell AS, Chakraborty S. What does the mediodorsal thalamus do? *Front Syst Neurosci.* 2013;7:37.
85. Young KA, Manaye KF, Liang C, Hicks PB, German DC. Reduced number of mediodorsal and anterior thalamic neurons in schizophrenia. *Biol Psychiatry.* 2000;47(11):944–953.
86. Pakkenberg B. Pronounced reduction of total neuron number in mediodorsal thalamic nucleus and nucleus accumbens in schizophrenics. *Arch Gen Psychiatry.* 1990;47(11):1023–1028.
87. Dorph-Petersen KA, Lewis DA. Postmortem structural studies of the thalamus in schizophrenia. *Schizophr Res.* 2017;180:28–35.
88. Harrisberger F, Buechler R, Smieskova R, et al. Alterations in the hippocampus and thalamus in individuals at high risk for psychosis. *NPJ Schizophrenia.* 2016;2:16033.
89. Pergola G, Trizio S, Di Carlo P, et al. Grey matter volume patterns in thalamic nuclei are associated with familial risk for schizophrenia. *Schizophr Res.* 2017;180:13–20.
90. Kuhn S, Gallinat J. Resting-state brain activity in schizophrenia and major depression: a quantitative meta-analysis. *Schizophr Bull.* 2013;39(2):358–365.
91. Woodward ND, Karbasforoushan H, Heckers S. Thalamocortical dysconnectivity in schizophrenia. *Am J Psychiatry.* 2012;169(10):1092–1099.
92. Woodward ND, Heckers S. Mapping thalamocortical functional connectivity in chronic and early stages of psychotic disorders. *Biol Psychiatry.* 2016;79(12):1016–1025.
93. Anticevic A, Haut K, Murray JD, et al. Association of thalamic dysconnectivity and conversion to psychosis in youth and young adults at elevated clinical risk. *JAMA Psychiatry.* 2015;72(9):882–891.
94. Anticevic A, Cole MW, Repovs G, et al. Characterizing thalamo-cortical disturbances in schizophrenia and bipolar illness. *Cereb Cortex.* 2014;24(12):3116–3130.
95. Anticevic A, Yang G, Savic A, et al. Mediodorsal and visual thalamic connectivity differ in schizophrenia and bipolar disorder with and without psychosis history. *Schizophr Bull.* 2014;40(6):1227–1243.
96. Rivolta D, Heidegger T, Scheller B, et al. Ketamine dysregulates the amplitude and connectivity of high-frequency oscillations in cortical-subcortical networks in humans: evidence from resting-state magnetoencephalography-recordings. *Schizophr Bull.* 2015;41(5):1105–1114.
97. Ferrarelli F, Tononi G. The thalamic reticular nucleus and schizophrenia. *Schizophr Bull.* 2011;37(2):306–315.
98. Ferrarelli F, Tononi G. Reduced sleep spindle activity point to a TRN-MD thalamus-PFC circuit dysfunction in schizophrenia. *Schizophr Res.* 2017;180:36–43.
99. Reagh ZM, Murray EA, Yassa MA. Repetition reveals ups and downs of hippocampal, thalamic, and neocortical engagement during mnemonic decisions. *Hippocampus.* 2017;27(2):169–183.
100. Ito HT, Zhang SJ, Witter MP, Moser EI, Moser MB. A prefrontal-thalamo-hippocampal circuit for goal-directed spatial navigation. *Nature.* 2015;522(7554):50–55.
101. Xu W, Sudhof TC. A neural circuit for memory specificity and generalization. *Science.* 2013;339(6125):1290–1295.
102. Cassel JC, Pereira de Vasconcelos A, Loureiro M, Cholvin T, Dalrymple-Alford JC, Vertes RP. The reuniens and rhomboid nuclei: neuroanatomy, electrophysiological characteristics and behavioral implications. *Prog Neurobiol.* 2013;111:34–52.
103. Vetere G, Kenney JW, Tran LM, et al. Chemogenetic interrogation of a brain-wide fear memory network in mice. *Neuron.* 2017;94(2):363–374 e364.
104. Griffin AL. Role of the thalamic nucleus reuniens in mediating interactions between the hippocampus and medial prefrontal cortex during spatial working memory. *Front Syst Neurosci.* 2015;9:29.
105. Hallock HL, Wang A, Shaw CL, Griffin AL. Transient inactivation of the thalamic nucleus reuniens and rhomboid nucleus produces deficits of a working-memory dependent tactile-visual conditional discrimination task. *Behav Neurosci.* 2013;127(6):860–866.
106. Patton MH, Bizup BT, Grace AA. The infralimbic cortex bidirectionally modulates mesolimbic dopamine neuron activity via distinct neural pathways. *J Neurosci.* 2013;33(43):16865–16873.
107. Rosenkranz JA, Moore H, Grace AA. The prefrontal cortex regulates lateral amygdala neuronal plasticity and responses to previously conditioned stimuli. *J Neurosci.* 2003;23(35):11054–11064.
108. Gomes FV, Grace AA. Prefrontal cortex dysfunction increases susceptibility to schizophrenia-like changes induced by adolescent stress exposure. *Schizophr Bull.* 2017;43(3):592–600.
109. Gomes FV, Zhu X, Grace AA. Stress during critical periods of development and risk for schizophrenia. *Schizophr Res.* 2019; doi: 10.1016/j.schres.2019.01.030.
110. Sultan KT, Brown KN, Shi SH. Production and organization of neocortical interneurons. *Front Cell Neurosci.* 2013;7:221.

42

FEELING AND REMEMBERING

EFFECTS OF PSYCHOSIS ON THE STRUCTURE AND FUNCTION OF THE AMYGDALA AND HIPPOCAMPUS

M. D. Bauman, J. D. Ragland, and C. M. Schumann

OVERVIEW

The amygdala and hippocampus are central components of a system in the medial temporal lobe (MTL) historically referred to as the "limbic system" and thought to form a boundary (or limbus) between the cerebral hemispheres and the brainstem (Figure 42.1). This system receives afferent connections from sensory association cortices, and provides efferent feedback to much of the cortex via thalamic nuclei. This led MacLean to hypothesize that the structures serves as, "a visceral brain that interprets and gives expression to its incoming information in terms of feeling" and that it might be thought of as, "the 'gut' component of memory" (p. 425, [1]). Current literature suggests that the highly interconnected amygdala and hippocampus play a central role in the formation of emotionally salient memories that may be disrupted in psychotic disorders.

In this chapter, we first provide a foundational understanding of hippocampus and amygdala structure and function in typically developing individuals, followed by a description of how psychotic disorders can disrupt structure and function, contributing to cognitive deficits and positive and negative symptom dimensions.

THE HIPPOCAMPUS

Understanding the importance of the hippocampus to long-term memory (LTM) became apparent in the late 1950s through case studies of Henry Molaison (patient H.M.) [2]. Following a bilateral anterior temporal lobectomy for intractable epilepsy, patient H.M. became densely amnesic, prompting extensive human and animal neuroscience research into the structure and function of the hippocampus.

STRUCTURE OF THE HIPPOCAMPAL FORMATION

The hippocampal formation (Figure 42.2) includes the hippocampal cortex (comprisingCA1–4), dentate gyrus, subiculum, and the parahippocampal gyrus (subdivided into anterior—perirhinal, and posterior—parahippocampal cortices). Not illustrated is the entorhinal cortex, which underlies

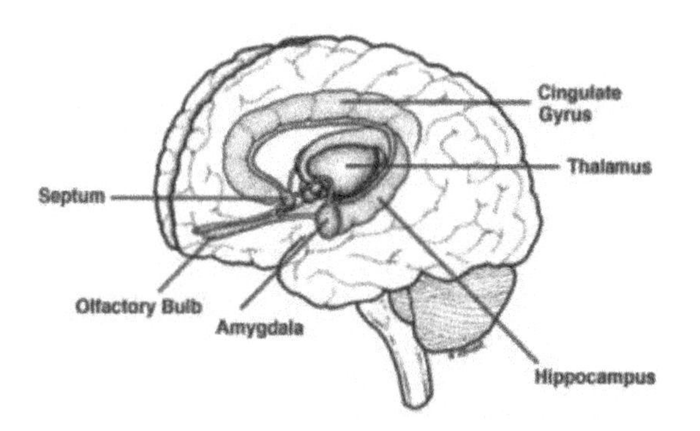

Figure 42.1 Location of the amygdala and hippocampus within the limbic system.

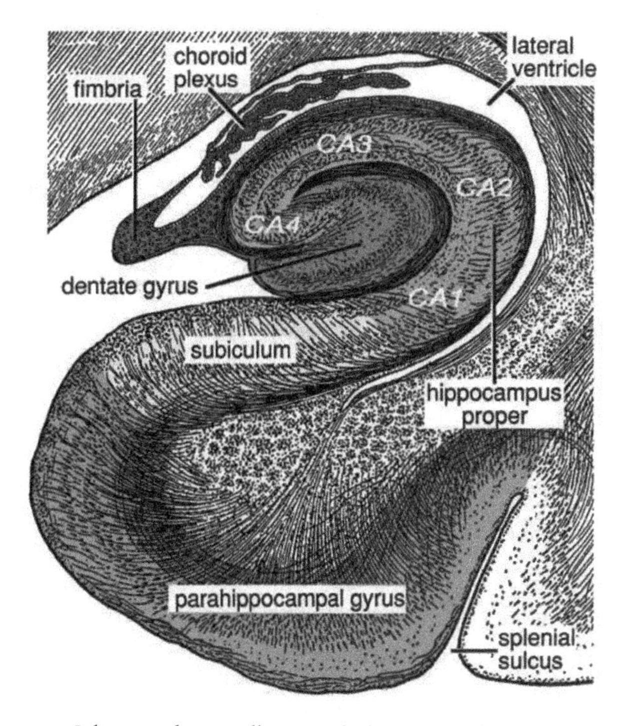

Figure 42.2 Schematic drawing illustrates the hippocampal cortex—termed the cornu ammonis (including CA1 . . . CA4); the dentate gyrus; subiculum; and parahippocampal gyrus—with entorhinal cortex (not pictured) underlying the anterior portion of the parahippocampal gyrus.

the boundary between the subiculum and perirhinal cortex. This highly complex assembly allows the hippocampus to hierarchically organize and create memory representations of multisensory information coming from association areas distributed throughout the brain. Sensory and perceptual information from association cortices is directed to perirhinal and parahippocampal cortices, which project to the entorhinal cortex, which, in turn, provides the majority of hippocampal input [3].

Organization occurs along an anterior/posterior gradient, with anterior inputs to the hippocampus providing detailed information about item features promoting episodic memories of *object identity*, and posterior inputs providing integrated information about *contextual features* of the learning environment [4]. This anterior/posterior division is preserved through a topographical organization of perirhinal and parahippocampal inputs into entorhinal cortex. Finally, at the top of this hierarchy, the hippocampal cortex receives convergent input in a topographically organized manner [5], with anterior hippocampus showing preferential connectivity with the perirhinal cortex, and posterior hippocampus showing preferential connectivity with parahippocampal cortex [6].

FUNCTION OF THE HIPPOCAMPAL FORMATION

Just as the hippocampal formation is an assembly of integrated, anatomically specific regions, LTM is composed of functionally unique episodic encoding and retrieval processes. A unique feature of episodic memory is that it engages autonoetic consciousness—or a type of "mental time travel" in which retrieval is accompanied by a sense of self-recollection or mental re-enactment of the original encoding event [7]. This capacity for self-recollection provides a key to understanding the division of labor within different components of the hippocampal formation.

Multicomponent models of episodic memory [8, 9] make a distinction between familiarity—in which something is correctly recognized as previously encountered based on the strength of the memory trace without retrieval of qualitative details about the encoding event and recollection, in which recognition is accompanied by retrieval of specific contextual details. This distinction can be illustrated by the difference between meeting a colleague at a conference who "you know you know," but don't know why and cannot recall their name (i.e., a sense of familiarity), versus the experience of remembering their name and institution, and recalling that they spilled red wine on your white linen dress at this same meeting 3 years ago (i.e., a sense of recollection).

Human and animal lesion studies and functional magnetic resonance imaging (fMRI) has consistently demonstrated that separate components of the hippocampal formation are differentially engaged by recollection and familiarity. fMRI studies revealed a double-dissociation, in which familiarity was found to specifically engage the perirhinal cortex, and recollection was accompanied by selective activation of the hippocampus and parahippocampal cortex [9]. These results supported three-component models, such as the binding of item and context (BIC) model [8, 9], which states that perirhinal and parahippocampal cortices encode item and context information, respectively, and the hippocampus forms a relational conjunction of item-context associations. Because familiarity depends on the strength of the item representation, it can be accomplished with the perirhinal cortex. However, because recollection requires retrieval of contextual details, it requires both the hippocampus and parahippocampal gyrus. As discussed in what follows, people with psychotic disorders appear to be disproportionately impaired in recollection versus familiarity-based retrieval [10, 11].

Another key functional dissociation describes how different subfields of the hippocampal cortex contribute to episodic encoding and retrieval. "Pattern separation" —or the ability to form distinct representations of events even when those events are quite similar, is a unique requirement for accurate encoding. Computational models and simulation studies [12, 13] demonstrate how high levels of lateral inhibition within the dentate gyrus results in sparse excitatory activity, promoting formation of discrete memories of events sharing overlapping features. Conversely, "pattern completion" refers to the ability of partial cues to lead to reinstatement of aspects of the original encoding event that are not currently present, which can foster recollection. Some computational models and electrophysiological studies suggest that this process is promoted by recurrent connections within the CA3 [14], while others stress the importance of CA1, subiculum, entorhinal, and parahippocampal cortices [15].

IMPACT OF PSYCHOTIC DISORDERS ON EPISODIC MEMORY PERFORMANCE

People with psychotic disorders such as schizophrenia suffer from a disproportionate episodic memory deficit [16, 17], contributing to loss of quality of life and poor functional outcomes [18]. However, not all components of episodic memory are equally impaired. People with schizophrenia have less consistent performance impairments when recognition is based on a sense of familiarity, and stronger and more consistent memory deficits when required to recollect some contextual or associative feature of the encoding event [10]. For example, a recent meta-analysis [11] demonstrated that familiarity deficits were smaller and more variable than consistently larger recollection deficits. Manipulation of encoding and retrieval conditions using the Relational and Item Specific Encoding task (RiSE [19]) revealed a three-way interaction in which patients with schizophrenia had unimpaired familiarity when instructed to encode visual object information based on item features, but were severely impaired when familiarity was tested following relational encoding. Recollection was impaired regardless of encoding process (Figure 42.3). In sum, patients appear most severely and consistently impaired when required to form relational memory representations during encoding to support a sense of either familiarity or recollection during retrieval.

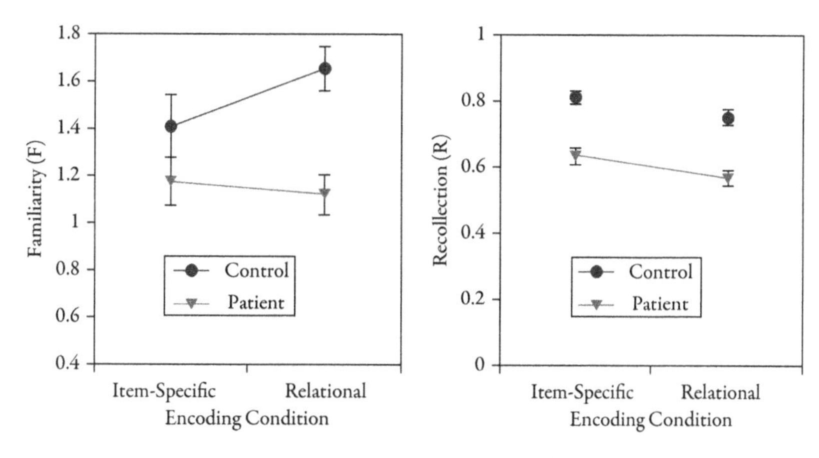

Figure 42.3 Performance of healthy controls and patients with schizophrenia on the Relational and Item Specific Encoding (RiSE) task. Familiarity performance (left panel) was only impaired in patients following relational encoding, whereas recollection (right panel) was impaired regardless of encoding condition.

IMPACT OF PSYCHOTIC DISORDERS ON HIPPOCAMPAL STRUCTURE

Early computerized tomography (CT) studies documented enlarged lateral and third ventricles in people with schizophrenia [20], and speculation arose as to whether this reflected shrinkage in adjacent structures, including hippocampus. A subsequent postmortem study [21] found that the amygdala, hippocampal formation, and parahippocampal gyrus were smaller in patients, whereas regions such as the caudate and putamen did not show volume reductions. A later review [22], concluded that there was evidence across studies for small (about 4%) but consistent bilateral hippocampal volume reductions in schizophrenia patients that could be observed early in the disease process, with little progression—arguing for a neurodevelopmental process. In the largest study to date, Van Erp and colleagues [23] analyzed MRI volumetric data from a multisite sample of 2028 schizophrenia patients and 2540 healthy controls, and confirmed a medium-sized decrease (Cohen's d = -0.46) in overall hippocampal volume in patients.

Efforts to identify an anterior/posterior gradient or specific subregions of hippocampal volume loss yielded less consistent findings. In an early study, Csernansky and colleagues [24] found that despite the lack of an overall volume reduction, shape deformations were localized to the head and body versus the tail of the hippocampal cortex. A higher resolution study [25] again found the most prominent shape abnormalities in anterior and midbody portions of the hippocampal cortex, corresponding to CA1 and CA2 subregions. However, for every study finding selective volumetric deficits in anterior hippocampus, an equal number found posterior deficits, or no evidence of an anterior/posterior gradient [26, 27].

Nevertheless, there have been some interesting functional and clinical associations with different portions of the hippocampus. Rametti and colleagues [28] found that although volume reductions were most prominent in posterior hippocampus, better memory was associated with larger anterior hippocampal volumes. Similarly, Wang and colleagues [29] found that better recollection was associated with larger volumes in the hippocampus and parahippocampal cortex, but not in the perirhinal cortex, consistent with the BIC model.

Although reduced hippocampal volumes are observed in antipsychotic-naïve patients [30], antipsychotics have also been associated with regionally specific decreases in dentate gyrus/CA4 volumes, and increases in subiculum volume [31]. Smaller anterior hippocampal volumes (primarily CA1) were also linked to more severe positive symptoms [30]. Finally, a review of postmortem structural and functional imaging studies [32] concluded that there was little evidence of hippocampal volume decrease in people with bipolar disorder, unless they also had an accompanying psychosis or were unmedicated.

IMPACT OF PSYCHOTIC DISORDERS ON HIPPOCAMPAL FUNCTION

Resting-state studies have yielded the most consistent pattern of regional hippocampal abnormalities in patients with schizophrenia. For example, a series of resting-state studies of cerebral blood volume (CBV) found CBV increases in patients relative to healthy controls within anterior hippocampus [33]. This was subsequently replicated using positron emission tomography (PET), which showed that increased blood flow in left anterior MTL was associated with more severe overall psychopathology [34], and greater auditory hallucinations [35].

Initial task-based PET studies produced less consistent results. For example, Heckers and colleagues [36] found that patients had decreased hippocampal activation and increased dorsolateral prefrontal cortex (DLPFC) activation during word retrieval. However, this was secondary to abnormally increased hippocampal rCBF in all task conditions. This same pattern of a generalized elevation in hippocampal rCBF in patients was also observed in several subsequent PET studies administering auditory recognition or word recall tasks [37]. Initial task-based fMRI studies of novelty detection [38] found evidence of reduced MTL activation in patients even though their task performance was unimpaired. Although several subsequent studies did link impaired word retrieval with reduced hippocampal activation in patients [39], a recent meta-analysis [40] found that the most consistent patient fMRI deficits during encoding and retrieval were in prefrontal cortex. The only consistent group difference in the MTL was abnormally increased activation of parahippocampal gyrus during encoding and retrieval, with no evidence of hippocampal group differences.

Rather than concluding that there are no memory-related functional deficits in the hippocampal formation in people with psychotic disorders, these mixed PET and fMRI findings point to the importance of using carefully designed memory paradigms to control encoding and retrieval demands related to differential memory impairments and hippocampal engagement [41]. For example, Achim and colleagues [42] found that MTL activation was unimpaired when patients successfully encoded visual object pairs that were semantically related, but was reduced when patients were required to encode arbitrary pairs of objects that were unrelated and associated with performance impairments. Similarly, a recent fMRI study of scene memory using eye-tracking methods [43] found that hippocampal activation was reduced only in the posterior hippocampus and only during a relational memory condition associated with specific eye-movement memory deficits.

INTEGRATED MODELS OF HIPPOCAMPAL DYSFUNCTION IN PSYCHOSIS

Despite this variability, several broad and well-supported conclusions can be drawn. First, there are small to medium-sized bilateral reductions in hippocampal volume in people with schizophrenia and bipolar psychosis, which appear neurodevelopmental rather than neurodegenerative in nature—although there may be some progressive change once medication is started. Second, anterior hippocampal blood volume, blood flow, and metabolism are abnormally increased in people with schizophrenia when they are at rest or engaged in memory tasks that do not produce performance impairments. This anterior hippocampal hyperactivity appears related to the severity of positive symptoms. Finally, episodic memory impairments are most striking when patients are required to encode relational memory representations to support recollection. These impairments are often accompanied by reduced hippocampal and DLPFC activation, with preliminary evidence of greater reductions in posterior hippocampus. These convergent results create a foundation for several integrative models.

Regarding clinical symptoms, one model [44] argues that aberrant dopamine signaling in striatum may be secondary to anterior hippocampal hyperactivity. The model proposes that efferent connections from a hyperactive anterior hippocampus increase excitatory glutamatergic input to the nucleus accumbens (NAc). An overactive NAc increases inhibitory GABAergic input to the ventral pallidum (VP). Because the VP is responsible for inhibiting dopamine neuron activity in the ventral tegmental area (VTA), increased VP inhibition results in a paradoxical increase in dopamine activity in the striatum. As previously noted, the link between anterior hippocampal hyperactivity and positive symptoms has been well established, providing support to the model.

An alternative model [45] proposes that reduced glutamatergic transmission in the dentate gyrus disrupts signaling in the CA3 subfield of the hippocampus, resulting in reduced pattern separation and increased pattern completion. The model proposes that this "shift toward pattern completion" could disrupt memory and increase the likelihood of delusions by strengthening inappropriate associations—leading to cognitive errors and formation of memories with psychotic content. Again, this model is consistent with evidence linking anterior hippocampal hyperactivity to positive symptoms and postmortem evidence of dysregulated glutamatergic transmission in patients' dentate gyrus [46]. However, as previously noted, memory studies indicate that recollection is differentially impaired in patients with schizophrenia, suggesting *reduced* rather than increased pattern completion.

A final model proposes that disrupted fronto-temporal integration may best explain the pattern of memory deficits and accompanying fMRI abnormalities commonly seen in patients with schizophrenia [41]. Consistent with evidence for a differential relational versus item memory impairment in schizophrenia, fMRI studies demonstrate that patients can successfully engage the ventrolateral prefrontal cortex (VLPFC) to promote item encoding, but show consistent DLPFC impairments associated with relational encoding deficits [19, 47]. Moreover, reduced hippocampal activation only appears during relational memory conditions, and is not present when patients are retrieving memories based on item-specific representations [47]. Finally, fMRI functional connectivity studies [48, 49] reveal that hippocampal to DLPFC connectivity is decreased, whereas hippocampal to VLPFC connectivity is abnormally increased in schizophrenia. This increase in hippocampal to VLPFC connectivity may reflect enhanced engagement of an item-specific and familiarity-based memory network in an attempt to compensate for a dysfunctional hippocampal to DLPFC relational and recollective memory network.

THE AMYGDALA

The amygdala is an almond-shaped brain region located in the anterior portion of the temporal lobe that plays a critical role in behaviors related to fear, emotion, memory, and social cognition [50]. Over a century of neuroscience research, beginning with nonhuman primate temporal lobe lesion studies carried out by Brown and Shaffer [51] and later by Kluver and Bucy [52], has revealed that macaque monkeys with temporal lobe and amygdala lesions develop a compulsive interest in objects, abnormal food preferences, and changes in tameness and sexual behavior. These changes, termed "Kluver-Bucy Syndrome," provided an initial link between the amygdala and socioemotional behavior. The amygdala is currently conceptualized as a "detector of perceptual saliency and biological relevance" [53], playing an essential role in survival and emotional well-being.

STRUCTURE OF THE AMYGDALA

The amygdala (Figure 42.4) can be subdivided into 13 discrete nuclei in nonhuman primate and human brains [54]. Studies utilizing neuroanatomical tracers in nonhuman primates reveal widespread connectivity between amygdala and cerebral cortex, including the frontal, insular, cingulate, temporal, parietal, and occipital cortices [55]. Most of these cortical structures generate projections back to the amygdala with the exception of parietal and occipital cortices, where the flow of information is thought to be unidirectional. The amygdala is also interconnected with many subcortical regions including the striatum, bed nucleus of the stria terminalis, basal forebrain, thalamus, hypothalamus, midbrain, and hindbrain.

The primate amygdala develops early in the prenatal period, forming 13 well-defined nuclei by birth [56]. There is also rapid enlargement of primate amygdala in the early postnatal period [57], which continues throughout childhood and well into adolescence [58], coinciding with ongoing functional maturational processes [59]. This structural and functional plasticity allows the amygdala to adapt to an ever-changing environment, and may also render the amygdala more vulnerable to neurodevelopmental disorders such as schizophrenia.

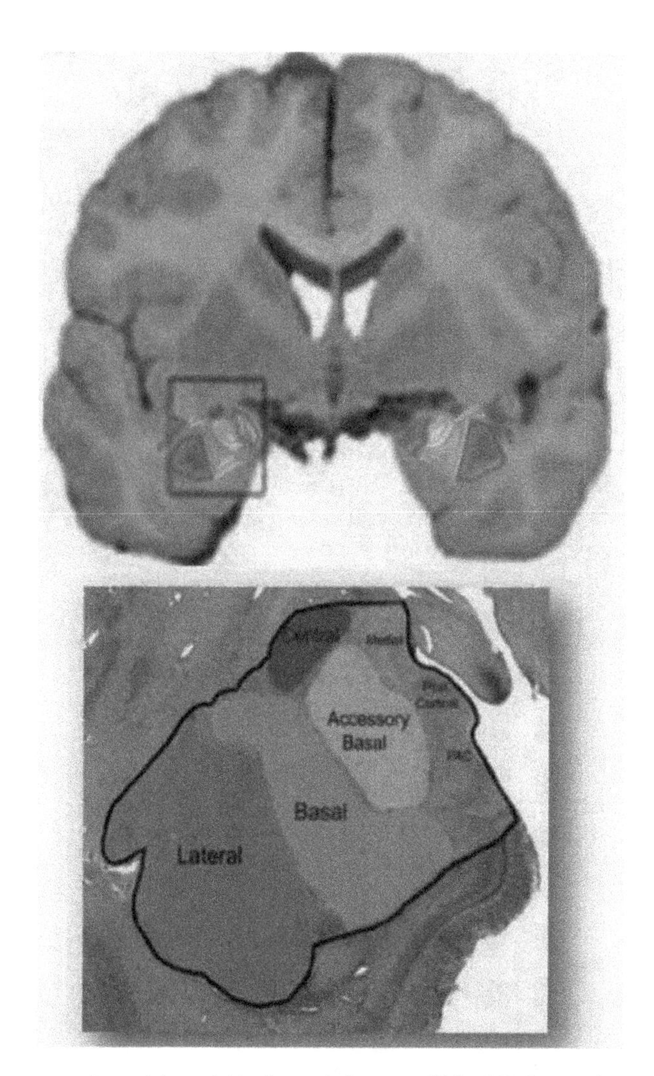

Figure 42.4 Amygdala nuclei in the medial temporal lobe. Nissl-stained image labeling lateral, basal, accessory basal, central, and medial/cortical nuclei.

FUNCTION OF THE AMYGDALA

Animal lesion studies have demonstrated that damage to the amygdala profoundly alters their ability to evaluate a wide array of environmental stimuli and adapt their behavior accordingly [60]. Although experimental lesion research is often the first approach in identifying brain regions essential for particular behaviors, there are obvious limitations, as the majority of amygdala lesion studies prior to the late 1990s also caused unintended collateral damage to surrounding cortex [61]. An alternative to animal lesions is to study humans with amygdala lesions secondary to disease, injury, or surgical procedures. For example, the bilateral MTL resection described for patient H.M. earlier in the chapter also included nearly complete bilateral removal of amygdala [62]. One of the most extensively studied patients with bilateral amygdala damage is patient S.M. [63], who developed difficulties attending to salient environmental stimuli, identifying potentially dangerous situations, and responding appropriately. Her lack of attention to the eye region of fearful faces makes it challenging to correctly identify the emotion of fear [64]. Patient S.M. is also unable to perceive deviations in societal norms [65] and has difficulty avoiding risk-taking behaviors [66].

Although changes in amygdala structure and function are reported in many psychotic disorders, we do not know if amygdala dysfunction plays a role in the initial pathophysiology or if the experience of living with a psychotic disorder alters the amygdala. This knowledge gap is important to bear in mind when reviewing the literature and we, therefore, place special emphasis on studies in high-risk populations and/or unaffected family members.

IMPACT OF PSYCHOTIC DISORDERS ON AMYGDALA STRUCTURE

Global changes in amygdala structure in schizophrenia

Volumetric studies of the amygdala using structural MRI of patients with schizophrenia have yielded inconsistent results, with the majority finding no difference in amygdala volume at any age or stage of the disorder. There is a trend for patients at high-risk for childhood or early-onset schizophrenia to have an enlarged amygdala during adolescence [67, 68]. Velakoulis and colleagues [69] also reported amygdala enlargement in first-episode but not in chronic patients who had been taking neuroleptic medications for several years. However, other research reported reduced amygdala volume in patients following an initial episode of psychosis [70].

These studies raise critical developmental issues and it is important to note that the amygdala undergoes a protracted growth trajectory that continues well beyond most other regions of the brain into adulthood [71]. It is therefore possible that, as in autism spectrum disorder [72], the amygdala may undergo an atypical growth trajectory in schizophrenia rather than remain consistently larger or smaller than in healthy participants. This extended maturational process may contribute to symptoms that typically present in late adolescence and early adulthood and may potentially play a role in "triggering" disease onset [73]. It is also possible that excitotoxicity associated with the disease process could contribute to reduced amygdala volume in some cases [74]. In summary, several factors, including age of first episode, type of psychotic symptoms, age of subjects at time of study, and treatment exposure make it difficult to draw definitive conclusions from cross-sectional structural MRI studies on amygdala neuropathology in schizophrenia.

Microscopic cellular changes in amygdala structure in schizophrenia

Although reductions in GABA levels and increases in dopamine have been linked to psychosis [75] quantitative studies of the postmortem amygdala have only recently emerged in the schizophrenia literature. Early postmortem studies examined a wide age range of subjects with mostly adult-onset schizophrenia and did not find any group differences [76]. Conversely, two studies employing modern stereological methods found patient decreases in the total number of neurons in the lateral nucleus of the amygdala [77, 78]. However, many of these patients were medicated or had a seizure disorder, and Beretta and colleagues [78] found that group differences were no longer significant after accounting for antipsychotics taken in the last 6 months before death. Finally, neurons expressing the peptide hormone somatostatin were found to be decreased in the lateral nucleus of the amygdala of both bipolar and schizophrenia cases [79]. Interactions between cellular and extracellular matrices that regulate neural functions may also play a role. For example, chondroitin sulfate proteoglycans (CSPG), one of the main organizational components of the extracellular matrix, is increased by more than 500% in deep nuclei of the amygdala of patients with schizophrenia but not bipolar disorder [80]. Although postmortem tissue studies face challenges related to small samples, comorbidity, pharmacological treatments, and potential substance abuse, the availability of increasingly sophisticated cellular and molecular technologies will continue to make substantial contributions to our understanding of the neurobiology of complex brain diseases.

IMPACT OF PSYCHOTIC DISORDERS ON AMYGDALA FUNCTION

Because amygdala dysfunction has been implicated in a wide range of neurodevelopmental and neuropsychiatric diseases we will use a dimensional approach to explore commonalities in amygdala dysfunction across psychotic disorders. As revealed by three recent meta-analyses of fMRI studies of emotional processing deficits in schizophrenia, the amygdala is consistently an area that responds differently in patients compared to healthy controls. The first meta-analysis by Li and colleagues [81] examined differences in activation peaks across studies of facial emotions and reported that bilateral amygdala activation is reduced in individuals with schizophrenia compared to healthy controls when processing facial emotions. A subsequent meta-analysis [82] evaluated both emotion perception and experience and found that individuals with schizophrenia show reduced bilateral amygdala activation in emotional perception paradigms, particularly in implicit studies of emotions. Finally, an analysis by Anticevic and colleagues [83] found that schizophrenia was associated with small reductions in amygdala activity in response to negatively valenced emotional images. However, this reduction was only found in studies using a "negative versus neutral" comparison and not in a "negative emotion versus baseline" comparison, suggesting that the aberrant amygdala response may have been due to elevated amygdala response to emotionally neutral stimuli.

This subtle, yet important, difference between responses to neutral emotional expression and baseline conditions is emerging as one of the more consistent findings in the field. There is increasing evidence that individuals with schizophrenia tend to experience emotionally neutral stimuli negatively [84], and that this behavioral response coincides with hyperactivation of the amygdala [85]. Increased amygdala activation in schizophrenia patients in response to neutral facial expressions has also been documented in a number of emotion recognition paradigms [86, 87].

Comparison of schizophrenia subgroups can also be informative. For example, amygdala activation patterns of paranoid individuals differ from both nonparanoid individuals and controls during emotion recognition tasks [88, 89], which may correlate with social functioning [90]. Individuals with schizophrenia also demonstrate a disconnect between arousal (as indexed by skin conductance) and amygdala activity, which was more pronounced in paranoid patients, compared to both nonparanoid patients and healthy controls [91]. Taken together, functional imaging studies provide ample evidence of aberrant amygdala activity in individuals with schizophrenia, though the nature of amygdala dysfunction may be influenced by the experimental paradigm, the developmental time course of illness, symptom defined subgroupings and the use of antipsychotic medications.

In addition to these focal deficits, individuals with schizophrenia demonstrate different patterns of amygdala connectivity as indexed by diffusional anisotropy [92, 93], resting state connectivity [94] and task-based functional connectivity [95]. Alterations in amygdala connectivity appear to be modulated by disease state and may correlate with symptom severity [96]. Bipolar patients with a history of psychosis also demonstrate abnormal amygdala-frontal connectivity, suggesting existence of common mechanisms underlying psychotic symptoms in the two disorders [97], though other studies detected differences in connectivity between the two patient populations [98]. Increasingly sophisticated techniques to quantify structural and functional connectivity will undoubtedly provide further insight into amygdala dysfunction within subtypes of schizophrenia and across the psychosis spectrum.

INTEGRATED MODELS OF AMYGDALA DYSFUNCTION IN PSYCHOSIS

Given that many neuropsychiatric disorders impair our ability to appraise and monitor environmental stimuli, it is not surprising that amygdala dysfunction has been consistently implicated [99]. However, it is unclear what features of amygdala pathology cut across psychotic disorders and what features may be unique. It also still needs to be determined whether structural and functional differences in individuals with schizophrenia are associated with underlying pathophysiology related to risk, or secondary to disease progression. Studies of amygdala pathology in unaffected relatives of individuals with schizophrenia have produced mixed findings [100, 101]. Medication effects also remain an important concern given antipsychotic medication effects on amygdala function in rodent models [102].

In spite of these challenges, promising models are beginning to take shape. For example, in a model of emotional hyperarousal, Grace and colleagues [103] proposed that the amygdala, though its connections with nucleus accumbens and prefrontal cortex, plays a key role in the socioemotional deficits in schizophrenia. Hyperresponse to neutral stimuli is also consistent with the aberrant salience hypothesis of schizophrenia, which suggests that subcortical dopamine dysfunction leads to aberrant assignment of salience [104, 105]. Although this salience model initially focused on striatal dopamine dysfunction, emerging evidence suggests that dopamine dysfunction in the amygdala may contribute to schizophrenia symptoms [106]. Indeed, an integrated two-hit model proposed that an amygdala lesion paired with a dopamine imbalance can produce intra-amygdala processing and related socioemotional deficits [107].

CONCLUDING REMARKS

In sum, a large body of research clearly establishes that psychotic disorders disrupt both the remembering and feeling aspects of limbic function, reflected by deficits in hippocampal and amygdala structure and function. Dysfunction in both structures contributes to negative and positive symptoms, and extends to dysfunctional reciprocal connections with prefrontal and striatal networks that are key to fully understanding these subcortical impairments. However, the literature also reveals a number of inconsistencies regarding the exact location and nature of these structural and functional deficits. Research on childhood onset neurodevelopmental disorders such as autism spectrum disorder demonstrates how structural differences change over time, and point to the importance of performing prospective developmental studies in high-risk individuals to better understand these dynamic changes and link them both to diagnostic and functional outcomes in social and role behavior. This improved developmental perspective will hopefully lead to increasingly effective early interventions designed to improve memory and socioemotional function and reduce illness burden.

ACKNOWLEDGMENTS

We thank Emily Lufburrow for her assistance in preparing this chapter.

REFERENCES

1. Maclean, P.D., *Some psychiatric implications of physiological studies on frontotemporal portion of limbic system (visceral brain).* Electroencephalogr Clin Neurophysiol, 1952. **4**(4): p. 407–418.
2. Corkin, S., *Lasting consequences of bilateral medial temporal lobectomy: clinical course and experimental findings in H.M.* Seminars in Neurology, 1984. **4**: p. 249–259.
3. Witter, M.P., et al., *Anatomical organization of the parahippocampal-hippocampal network.* Ann N Y Acad Sci, 2000. **911**: p. 1–24.
4. Eichenbaum, H., *Remembering: functional organization of the declarative memory system.* Curr Biol, 2006. **16**(16): p. R643–5.
5. Witter, M.P., G.W. Van Hoesen, and D.G. Amaral, *Topographical organization of the entorhinal projection to the dentate gyrus of the monkey.* J Neurosci, 1989. **9**(1): p. 216–228.
6. Libby, L.A., et al., *Differential connectivity of perirhinal and parahippocampal cortices within human hippocampal subregions revealed by high-resolution functional imaging.* J Neurosci, 2012. **32**(19): p. 6550–6560.
7. Tulving, E., *Episodic and semantic memory*, in *Organization of Memory*, E. Tulving and W. Donaldson, Editors. 1972, Academic Press: New York. p. 382–402.
8. Eichenbaum, H., A.P. Yonelinas, and C. Ranganath, *The medial temporal lobe and recognition memory.* Annu Rev Neurosci, 2007. **30**: p. 123–152.
9. Diana, R.A., A.P. Yonelinas, and C. Ranganath, *Imaging recollection and familiarity in the medial temporal lobe: a three-component model.* Trends Cogn Sci, 2007. **11**(9): p. 379–386.
10. Lepage, M., et al., *Associative memory encoding and recognition in schizophrenia: an event-related fMRI study.* Biol Psychiatry, 2006. **60**(11): p. 1215–1223.
11. Libby, L.A., et al., *Recollection and familiarity in schizophrenia: a quantitative review.* Biol Psychiatry, 2013. **73**(10): p. 944–950.
12. McClelland, J.L., B.L. McNaughton, and R.C. O'Reilly, *Why there are complementary learning systems in the hippocampus and neocortex: insights from the successes and failures of connectionist models of learning and memory.* Psychol Rev, 1995. **102**(3): p. 419–457.
13. Norman, K.A., and R.C. O'Reilly, *Modeling hippocampal and neocortical contributions to recognition memory: a complementary-learning-systems approach.* Psychol Rev, 2003. **110**(4): p. 611–646.
14. Rudy, J.W. and R.C. O'Reilly, *Contextual fear conditioning, conjunctive representations, pattern completion, and the hippocampus.* Behav Neurosci, 1999. **113**(5): p. 867–880.
15. Bakker, A., et al., *Pattern separation in the human hippocampal CA3 and dentate gyrus.* Science, 2008. **319**(5870): p. 1640–1642.
16. Aleman, A., et al., *Memory impairment in schizophrenia: a meta-analysis.* Am J Psychiatry, 1999. **156**(9): p. 1358–1366.
17. Saykin, A.J., et al., *Neuropsychological function in schizophrenia: selective impairment in memory and learning.* Arch Gen Psychiatry, 1991. **48**(7): p. 618–624.
18. Green, M.F., et al., *Neurocognitive deficits and functional outcome in schizophrenia: are we measuring the "right stuff"?* Schizophr Bull, 2000. **26**(1): p. 119–136.
19. Ragland, J.D., et al., *Relational and Item-Specific Encoding (RISE): task development and psychometric characteristics.* Schizophr Bull, 2012. **38**: p. 114–124.
20. Weinberger, D.R., et al., *Lateral cerebral ventricular enlargement in chronic schizophrenia.* Arch Gen Psychiatry, 1979. **36**(7): p. 735–739.
21. Bogerts, B., E. Meertz, and R. Schonfeldt-Bausch, *Basal ganglia and limbic system pathology in schizophrenia: a morphometric study of brain volume and shrinkage.* Arch Gen Psychiatry, 1985. **42**(8): p. 784–791.
22. Heckers, S., *Neuroimaging studies of the hippocampus in schizophrenia.* Hippocampus, 2001. **11**(5): p. 520–528.
23. van Erp, T.G., et al., *Subcortical brain volume abnormalities in 2028 individuals with schizophrenia and 2540 healthy controls via the ENIGMA consortium.* Mol Psychiatry, 2016. **21**(4): p. 585.

24. Csernansky, J.G., et al., *Hippocampal morphometry in schizophrenia by high dimensional brain mapping*. Proc Natl Acad Sci U S A, 1998. **95**(19): p. 11406–11.
25. Narr, K.L., et al., *Regional specificity of hippocampal volume reductions in first-episode schizophrenia*. Neuroimage, 2004. **21**(4): p. 1563–1575.
26. Rajarethinam, R., et al., *Hippocampus and amygdala in schizophrenia: assessment of the relationship of neuroanatomy to psychopathology*. Psychiatry Res, 2001. **108**(2): p. 79–87.
27. Weiss, A.P., et al., *Anterior and posterior hippocampal volumes in schizophrenia*. Schizophr Res, 2005. **73**(1): p. 103–112.
28. Rametti, G., et al., *Left posterior hippocampal density reduction using VBM and stereological MRI procedures in schizophrenia*. Schizophr Res, 2007. **96**(1–3): p. 62–71.
29. Wang, W.C., et al., *Hippocampal and parahippocampal cortex volume predicts recollection in schizophrenia*. Schizophr Res, 2014. **157**(1–3): p. 319–320.
30. Kalmady, S.V., et al., *Clinical correlates of hippocampus volume and shape in antipsychotic-naive schizophrenia*. Psychiatry Res, 2017. **263**: p. 93–102.
31. Rhindress, K., et al., *Hippocampal subregion volume changes associated with antipsychotic treatment in first-episode psychosis*. Psychol Med, 2017. **47**(10): p. 1706–1718.
32. Frey, B.N., et al., *The role of hippocampus in the pathophysiology of bipolar disorder*. Behav Pharmacol, 2007. **18**(5–6): p. 419–430.
33. Heckers, S. and C. Konradi, *GABAergic mechanisms of hippocampal hyperactivity in schizophrenia*. Schizophr Res, 2015. **167**(1–3): p. 4–11.
34. Friston, K.J., et al., *The left medial temporal region and schizophrenia. A PET study*. Brain, 1992. **115** (**Pt 2**): p. 367–382.
35. Silbersweig, D.A., et al., *A functional neuroanatomy of hallucinations in schizophrenia*. Nature, 1995. **378**(6553): p. 176–179.
36. Heckers, S., et al., *Impaired recruitment of the hippocampus during conscious recollection in schizophrenia*. Nat Neurosci, 1998. **1**(4): p. 318–323.
37. Heckers, S., et al., *Functional imaging of memory retrieval in deficit vs nondeficit schizophrenia*. Arch Gen Psychiatry, 1999. **56**(12): p. 1117–1123.
36. Zorrilla, L.T., D.V. Jeste, and G.G. Brown, *Functional MRI and novel picture-learning among older patients with chronic schizophrenia: abnormal correlations between recognition memory and medial temporal brain response*. Am J Geriatr Psychiatry, 2002. **10**(1): p. 52–61.
39. Weiss, A.P., et al., *Impaired hippocampal function during the detection of novel words in schizophrenia*. Biol Psychiatry, 2004. **55**(7): p. 668–675.
40. Ragland, J.D., et al., *Prefrontal activation deficits during episodic memory in schizophrenia*. Am J Psychiatry, 2009. **166**(8): p. 863–874.
41. Ranganath, C., M.J. Minzenberg, and J.D. Ragland, *The cognitive neuroscience of memory function and dysfunction in schizophrenia*. Biol Psychiatry, 2008. **64**(1): p. 18–25.
42. Achim, A.M., et al., *Selective abnormal modulation of hippocampal activity during memory formation in first-episode psychosis*. Arch Gen Psychiatry, 2007. **64**(9): p. 999–1014.
43. Ragland, J.D., et al., *Impact of schizophrenia on anterior and posterior hippocampus during memory for complex scenes*. Neuroimage Clin, 2017. **13**: p. 82–88.
44. Lodge, D.J., and A.A. Grace, *Hippocampal dysregulation of dopamine system function and the pathophysiology of schizophrenia*. Trends Pharmacol Sci, 2011. **32**(9): p. 507–513.
45. Tamminga, C.A., A.D. Stan, and A.D. Wagner, *The hippocampal formation in schizophrenia*. Am J Psychiatry, 2010. **167**(10): p. 1178–1193.
46. Gao, X.M., et al., *Ionotropic glutamate receptors and expression of N-methyl-D-aspartate receptor subunits in subregions of human hippocampus: effects of schizophrenia*. Am J Psychiatry, 2000. **157**(7): p. 1141–1149.
47. Ragland, J.D., et al., *Functional and neuroanatomic specificity of episodic memory dysfunction in schizophrenia: a functional magnetic resonance imaging study of the relational and item-specific encoding task*. JAMA Psychiatry, 2015. **72**(9): p. 909–916.
48. Meyer-Lindenberg, A.S., et al., *Regionally specific disturbance of dorsolateral prefrontal-hippocampal functional connectivity in schizophrenia*. Arch Gen Psychiatry, 2005. **62**(4): p. 379–386.
49. Wolf, D.H., et al., *Alterations of fronto-temporal connectivity during word encoding in schizophrenia*. Psychiatry Res, 2007. **154**(3): p. 221–232.
50. Murray, E.A., *The amygdala, reward and emotion*. Trends Cogn Sci, 2007. **11**(11): p. 489–497.
51. Brown, S., and E.A. Shafer, *An investigation into the functions of the occipital and temporal lobes of the monkey's brain*. Philosophical Transactions of the Royal Society of London: Biological Sciences, 1888. **179**: p. 303–327.
52. Kluver, H., and P. Bucy, *Preliminary analysis of functions of the temporal lobes in monkeys*. Archives of Neurology and Psychiatry, 1939. **42**: p. 979–1000.
53. Wang, S., et al., *Neurons in the human amygdala selective for perceived emotion*. Proc Natl Acad Sci U S A, 2014. **111**(30): p. E3110–9.
54. Amaral, D.G., et al., *Anatomical organization of the primate amygdaloid complex*, in *The Amygdala: Neurobiological Aspects of Emotion, Memory, and Mental Dysfunction,* J.P. Aggleton, Editor. 1992, Wiley-Liss: New York.
55. Freese, J.L., and D.G. Amaral, *Neuroanatomy of the primate amygdala*, in *The Human Amygdala* P.J. Whalen and E.A. Phelps, Editors. 2009, Guilford Press: US. p. 3–42.
56. Ulfig, N., M. Setzer, and J. Bohl, *Ontogeny of the human amygdala*. Ann N Y Acad Sci, 2003. **985**: p. 22–33.
57. Chareyron, L.J., et al., *Postnatal development of the amygdala: a stereological study in macaque monkeys*. J Comp Neurol, 2012. **520**(9): p. 1965–1984.
58. Giedd, J.N., *Normal development*. Child and Adolescent Psychiatric Clinics of Northern America, 1997. **6**(2): p. 281–288.
59. Qin, S., et al., *Immature integration and segregation of emotion-related brain circuitry in young children*. Proc Natl Acad Sci U S A, 2012. **109**(20): p. 7941–7946.
60. Emery, N.J., and D.G. Amaral, *The role of the primate amygdala in social cognition*, in *Cognitive Neuroscience of Emotion. Series in Affective Science.*, R.D. Lane and L. Nadel, Editors. 2000, Oxford University Press: New York. p. 156–191.
61. Meunier, M., et al., *Effects of aspiration versus neurotoxic lesions of the amygdala on emotional responses in monkeys*. Eur J Neurosci, 1999. **11**(12): p. 4403–4418.
62. Augustinack, J.C., et al., *H.M.'s contributions to neuroscience: a review and autopsy studies*. Hippocampus, 2014. **24**(11): p. 1267–1286.
63. Feinstein, J.S., R. Adolphs, and D. Tranel, *A tale of survival from the world of patient S. M.*, in *Living without an Amygdala*, D.G. Amaral and R. Adolphs, Editors. 2016, Guilford Press: US. p. 1–38.
64. Adolphs, R., et al., *A mechanism for impaired fear recognition after amygdala damage*. Nature, 2005. **433**(7021): p. 68–72.
65. Kennedy, D.P., et al., *Personal space regulation by the human amygdala*. Nat Neurosci, 2009. **12**(10): p. 1226–1227.
66. De Martino, B., C.F. Camerer, and R. Adolphs, *Amygdala damage eliminates monetary loss aversion*. Proc Natl Acad Sci U S A, 2010. **107**(8): p. 3788–3792.
67. Jacobsen, L.K., et al., *Progressive reduction of temporal lobe structures in childhood-onset schizophrenia*. Am J Psychiatry, 1998. **155**(5): p. 678–685.
68. Welch, K.A., et al., *Amygdala volume in a population with special educational needs at high risk of schizophrenia*. Psychol Med, 2010. **40**(6): p. 945–954.
69. Velakoulis, D., et al., *Hippocampal and amygdala volumes according to psychosis stage and diagnosis: a magnetic resonance imaging study of chronic schizophrenia, first-episode psychosis, and ultra-high-risk individuals*. Arch Gen Psychiatry, 2006. **63**(2): p. 139–149.
70. Joyal, C.C., et al., *The amygdala and schizophrenia: a volumetric magnetic resonance imaging study in first-episode, neuroleptic-naive patients*. Biol Psychiatry, 2003. **54**(11): p. 1302–1304.

71. Schumann, C.M., M.D. Bauman, and D.G. Amaral, *Abnormal structure or function of the amygdala is a common component of neurodevelopmental disorders*. Neuropsychologia, 2011. **49**(4): p. 745–759.
72. Amaral, D.G., C.M. Schumann, and C.W. Nordahl, *Neuroanatomy of autism*. Trends Neurosci, 2008. **31**(3): p. 137–145.
73. Benes, F.M., *Amygdalocortical circuitry in schizophrenia: from circuits to molecules*. Neuropsychopharmacology, 2010. **35**(1): p. 239–257.
74. Benes, F.M., *Emerging principles of altered neural circuitry in schizophrenia*. Brain Res Brain Res Rev, 2000. **31**(2–3): p. 251–269.
75. Haber, S.N., and J.L. Fudge, *The interface between dopamine neurons and the amygdala: implications for schizophrenia*. Schizophr Bull, 1997. **23**(3): p. 471–482.
76. Chance, S.A., M.M. Esiri, and T.J. Crow, *Amygdala volume in schizophrenia: post-mortem study and review of magnetic resonance imaging findings*. Br J Psychiatry, 2002. **180**: p. 331–338.
77. Kreczmanski, P., et al., *Volume, neuron density and total neuron number in five subcortical regions in schizophrenia*. Brain, 2007. **130**(Pt 3): p. 678–692.
78. Berretta, S., H. Pantazopoulos, and N. Lange, *Neuron numbers and volume of the amygdala in subjects diagnosed with bipolar disorder or schizophrenia*. Biol Psychiatry, 2007. **62**(8): p. 884–893.
79. Pantazopoulos, H., et al., *Decreased numbers of somatostatin-expressing neurons in the amygdala of subjects with bipolar disorder or schizophrenia: relationship to circadian rhythms*. Biol Psychiatry, 2017. **81**(6): p. 536–547.
80. Pantazopoulos, H., et al., *Extracellular matrix-glial abnormalities in the amygdala and entorhinal cortex of subjects diagnosed with schizophrenia*. Arch Gen Psychiatry, 2010. **67**(2): p. 155–166.
81. Li, H., et al., *Facial emotion processing in schizophrenia: a meta-analysis of functional neuroimaging data*. Schizophr Bull, 2010. **36**(5): p. 1029–39.
82. Taylor, S.F., et al., *Meta-analysis of functional neuroimaging studies of emotion perception and experience in schizophrenia*. Biol Psychiatry, 2012. **71**(2): p. 136–145.
83. Anticevic, A., et al., *Amygdala recruitment in schizophrenia in response to aversive emotional material: a meta-analysis of neuroimaging studies*. Schizophr Bull, 2012. **38**(3): p. 608–21.
84. Llerena, K., G.P. Strauss, and A.S. Cohen, *Looking at the other side of the coin: a meta-analysis of self-reported emotional arousal in people with schizophrenia*. Schizophr Res, 2012. **142**(1-3): p. 65–70.
85. Potvin, S., A. Tikasz, and A. Mendrek, *Emotionally neutral stimuli are not neutral in schizophrenia: a mini review of functional neuroimaging studies*. Front Psychiatry, 2016. **7**: p. 115.
86. Holt, D.J., et al., *Increased medial temporal lobe activation during the passive viewing of emotional and neutral facial expressions in schizophrenia*. Schizophr Res, 2006. **82**(2–3): p. 153–62.
87. Hall, J., et al., *Overactivation of fear systems to neutral faces in schizophrenia*. Biol Psychiatry, 2008. **64**(1): p. 70–73.
88. Williams, L.M., et al., *Fronto-limbic and autonomic disjunctions to negative emotion distinguish schizophrenia subtypes*. Psychiatry Res, 2007. **155**(1): p. 29–44.
89. Russell, T.A., et al., *Neural responses to dynamic expressions of fear in schizophrenia*. Neuropsychologia, 2007. **45**(1): p. 107–23.
90. Pinkham, A.E., et al., *An investigation of the relationship between activation of a social cognitive neural network and social functioning*. Schizophr Bull, 2008. **34**(4): p. 688–697.
91. Williams, L.M., et al., *Dysregulation of arousal and amygdala-prefrontal systems in paranoid schizophrenia*. Am J Psychiatry, 2004. **161**(3): p. 480–489.
92. Kalus, P., et al., *Volumetry and diffusion tensor imaging of hippocampal subregions in schizophrenia*. Neuroreport, 2004. **15**(5): p. 867–871.
93. Kalus, P., et al., *The amygdala in schizophrenia: a trimodal magnetic resonance imaging study*. Neurosci Lett, 2005. **375**(3): p. 151–156.
94. Peters, H., et al., *More consistently altered connectivity patterns for cerebellum and medial temporal lobes than for amygdala and striatum in schizophrenia*. Front Hum Neurosci, 2016. **10**: p. 55.
95. Bjorkquist, O.A., et al., *Altered amygdala-prefrontal connectivity during emotion perception in schizophrenia*. Schizophr Res, 2016. **175**(1–3): p. 35–41.
96. Anticevic, A., et al., *Amygdala connectivity differs among chronic, early course, and individuals at risk for developing schizophrenia*. Schizophr Bull, 2014. **40**(5): p. 1105–1116.
97. Anticevic, A., et al., *Global prefrontal and fronto-amygdala dysconnectivity in bipolar I disorder with psychosis history*. Biol Psychiatry, 2013. **73**(6): p. 565–573.
98. Mukherjee, P., et al., *Disconnection between amygdala and medial prefrontal cortex in psychotic disorders*. Schizophr Bull, 2016. **42**(4): p. 1056–1067.
99. Holt, D.J., and M.L. Phillips, *The human amygdala in schizophrenia*, in *The Human Amygdala*, P.J. Whalen and E.A. Phelps, Editors. 2009, Guilford Press: New York.
100. Rasetti, R., et al., *Evidence that altered amygdala activity in schizophrenia is related to clinical state and not genetic risk*. Am J Psychiatry, 2009. **166**(2): p. 216–225.
101. Habel, U., et al., *Genetic load on amygdala hypofunction during sadness in nonaffected brothers of schizophrenia patients*. Am J Psychiatry, 2004. **161**(10): p. 1806–1813.
102. Rosenkranz, J.A., and A.A. Grace, *Dopamine-mediated modulation of odour-evoked amygdala potentials during Pavlovian conditioning*. Nature, 2002. **417**(6886): p. 282–287.
103. Grace, A.A., *Gating of information flow within the limbic system and the pathophysiology of schizophrenia*. Brain Res Brain Res Rev, 2000. **31**(2–3): p. 330–41.
104. Gray, J.A., *Integrating schizophrenia*. Schizophr Bull, 1998. **24**(2): p. 249–266.
105. Kapur, S., R. Mizrahi, and M. Li, *From dopamine to salience to psychosis—linking biology, pharmacology and phenomenology of psychosis*. Schizophr Res, 2005. **79**(1): p. 59–68.
106. Pankow, A., et al., *Neurobiological correlates of delusion: beyond the salience attribution hypothesis*. Neuropsychobiology, 2012. **66**(1): p. 33–43.
107. Aleman, A. and R.S. Kahn, *Strange feelings: do amygdala abnormalities dysregulate the emotional brain in schizophrenia?* Prog Neurobiol, 2005. 77(5): p. 283–298.

43

THE CEREBELLUM IN PSYCHOSIS

Kelsey Heslin, Joseph J. Shaffer, Albert Powers, Nancy Andreasen, and Krystal Parker

INTRODUCTION

Psychosis involves a mismatch between thoughts and reality, and interpretations of the external environment that can involve commanding or commenting voices (auditory hallucinations), seeing objects that are not present (visual hallucinations) or being strongly committed to an idea or belief even though incontrovertible evidence indicates it is false (delusion). These breaks with reality are most commonly reported in patients with schizophrenia, bipolar disorder, and following drug use, but more recently similar symptoms have been reported in Parkinson's disease and autism. Although these diseases are vastly different in symptom presentation and diagnosis, the psychotic features are comparable. Recently, the cerebellum and its relayed connections to the frontal cortex have been implicated in the pathophysiology of psychosis (Figure 43.1).

For many years, damage to the cerebellum was associated exclusively with motor deficits like ataxia. However, case reports of the early 20th century described instances of individuals with progressive cerebellar degeneration or atrophy accompanied not only by ataxias but also by the appearance of psychoses including auditory hallucinations and delusions of persecution and grandeur (Keddie, 1969; Kutty and Prendes, 1981; Jurjus et al., 1994). Motor coordination impairment and neuropsychiatric deterioration appeared at similar times in these individuals. However, these cases lacked the anatomical detail necessary to directly implicate the cerebellum alone in the onset of these dramatic changes in thought and personality. Fortunately, modern advances in functional and structural imaging have allowed researchers to associate specific abnormalities in the cerebellum and its projections with neuropsychiatric disorders. For example, in Dandy-Walker Malformation, a condition characterized by incomplete development of the cerebellar vermis and cystic dilation of the fourth ventricle, some patients display a range of psychotic symptoms including hallucinations and delusions (Gan et al., 2012). In these cases, visualizing the extent of cerebellar hypoplasia with structural imaging is crucial in providing a differential diagnosis from other neuropsychiatric conditions (Williams et al., 2016).

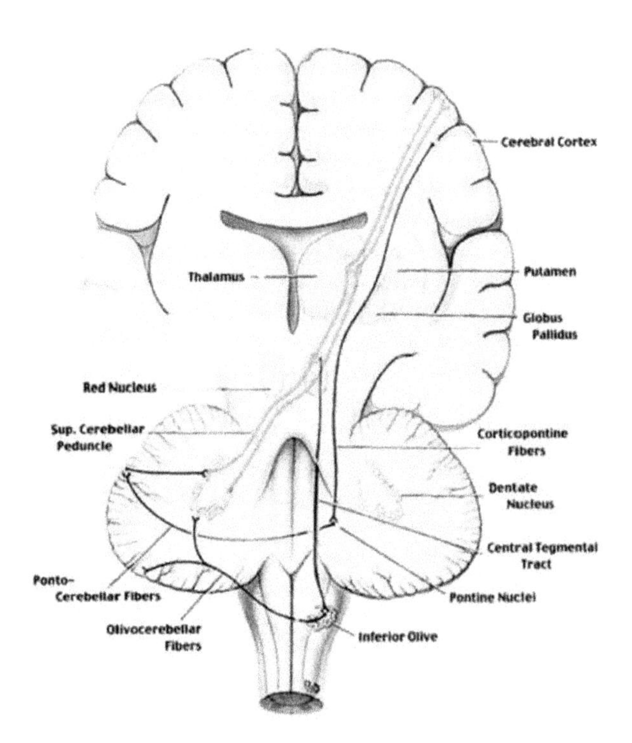

Figure 43.1 Cerebellar-thalamic-cortico-cerebellar circuitry.
Figure adapted from Andreasen et al., 2008.

Furthermore, modern approaches have also begun to reveal the full extent of cerebellar influence outside classically associated realms of motor control and balance in healthy individuals. The detection of a variety of cerebellocortical relays to areas outside the primary motor cortex, such as prefrontal and posterior parietal cortices, and limbic structures have fueled hypotheses that the cerebellum is necessary for information processing in a range of cognitive, sensorimotor, and affective functions. Cerebellar influence on these distributed regions is a putative basis for its implication in neuropsychiatric disease.

SCHIZOPHRENIA

The cerebellum has long been associated with control of motor movement, posture, and balance. Because of this, one of the earliest indications that the cerebellum is functioning abnormally in schizophrenia was the observation of motor disturbances (Shakow and Huston, 1936) or neurological "soft signs" (i.e., deficits in motor coordination, balance, and rapid movement sequencing) in patients that have experienced psychosis but remain naïve to neuroleptics (Gupta et al., 1995; Dazzan and Murray, 2002; Ho et al., 2004; van Harten et al.,

2017). Rather than appearing as the result of disease progression or pharmacological intervention, neurological soft signs are present at first onset of psychosis at a rate much higher than in the normal population (Gupta et al., 1995; Ho et al., 2004; Mouchet-Mages et al., 2011). Interestingly, longitudinal studies have also shown that children who go on to develop schizophrenia later in life score significantly lower on measures of motor coordination as early as 7 years of age (Crow et al., 1995; Cannon et al., 1999). The predictive nature of these early motor impairments hints at a disturbance in neurological structure and/or function of regions like the cerebellum prior to full onset of psychotic symptoms. Beyond being linked to motor coordination deficits, changes in cerebellar connectivity, structure, function, and gene expression can be distinguished in populations with schizophrenia and may help to reveal the neurological bases for psychotic symptoms in this disorder.

ABNORMALITIES OF CEREBELLAR STRUCTURE IN SCHIZOPHRENIA

Postmortem and neural imaging studies have demonstrated that there are observable structural differences in the cerebellum of patients with schizophrenia. In early postmortem studies, patients with schizophrenia were shown to have smaller anterior cerebellar vermis (lobules I–V) (Weinberger et al., 1980), lower linear density of Purkinje cells (Reyes and Gordon, 1981), and fewer Purkinje cells (Stevens, 1982) compared to subjects without a schizophrenia diagnosis. However, the prevalence of neuroleptic treatment and electroconvulsive therapy made the direct cause of these postmortem anatomical abnormalities challenging to interpret. The development of computerized tomography (CT) and magnetic resonance imaging (MRI) made it possible to estimate anatomical differences in vivo while taking disease time course into account. For example, Heath and colleagues (1979) demonstrated pathology of the cerebellar vermis in a substantial portion (40%) of CT scans from 85 patients diagnosed with schizophrenia (Heath et al., 1979). In line with this outcome, abnormal vermal anatomy has been one of the most consistent findings from structural imaging research on cerebellar changes in schizophrenia. A study comparing neuroleptic-naïve men diagnosed with schizophrenia to age-, sex-, and handedness-matched comparisons found smaller vermal volume in the patient group, which was negatively correlated with symptom severity scores in depression and paranoia (Ichimiya et al., 2001). These findings are congruent with reports that neuroleptic-naïve or neuroleptic-free male patients with schizophrenia have smaller anterior lobe and vermis area compared to age- and sex-matched healthy controls (Nopoulos et al., 1999). On the contrary, male patients with a history of chronic schizophrenia have increased vermal volume, which correlates with severity of positive symptoms and disordered thinking (Levitt et al., 1999). Contradictory findings of vermal volume in patients with schizophrenia may be due to dose-dependent effects of long-term neuroleptic treatment (Levitt et al., 1999; Ichimiya et al., 2001) as administration of neuroleptic medication has been shown to foster increased axonal sprouting (Benes et al., 1983), but higher doses in patients is associated with decreased general white matter over time (Ho et al., 2011).

While many studies find volumetric differences limited to cerebellar vermis, some have found significantly decreased gray matter volume in the cerebellar hemispheres (Marcelis et al., 2003; McDonald, 2005; Laidi et al., 2015) or entire cerebellum (Moberget et al., 2017) of patients with schizophrenia. However, findings of decreased gray matter in patients is certainly not limited to the cerebellum, but observed in other regions like the frontal and temporal lobes, and thalamus as well (Volz et al., 2000).

Importantly, positive symptoms, such as auditory verbal hallucinations, have been associated with decreased gray matter volume in cerebellar lobules VIIb and VIIIa in treatment resistant patients (Cierpka et al., 2017). Left lobule VI and X volumes are also significantly decreased in first-episode patients compared to healthy matched subjects (Kim et al., 2018). A meta-analysis of MRI data from over 900 patients and 1,000 healthy comparisons revealed lower cerebellar gray matter volume in the patient group, regardless of age, suggesting that this feature may be neurodevelopmental in nature, rather than an outcome of aging or disease progression (Moberget et al., 2017).

Furthermore, early structural abnormalities in patients with schizophrenia have been shown to be predictive of long-term disease outcomes. Wassink and colleagues (1999) longitudinally followed a group of 63 patients diagnosed with schizophrenia spectrum disorders for an average of 7 years (Wassink et al., 1999). MRI scans at 6-month intervals revealed that initial gross cerebellar volume was significantly negatively correlated with future duration of psychotic and negative symptoms, as well as psychosocial outcomes (i.e., quality of relationships and occupational performance). In contrast, initial cerebrum, temporal lobe, and ventricular volumes were not predictive of any outcome variables measured (Wassink et al., 1999). These data indicate that smaller initial cerebellar volume is associated with increased severity of psychopathology and poorer quality of life in patients diagnosed with schizophrenia spectrum disorders.

Genetic vulnerability may contribute to cerebellar volumetric differences in patients with schizophrenia spectrum disorders. Marcelis et al. (2003) compared whole-brain MRI scans from patients with a diagnosis of schizophrenia or schizoaffective disorder and psychosis to their first-degree relatives. They found that both patients and first-degree relatives had decreased cerebellar hemisphere gray matter and that patients had additional gray matter deficits in the thalamus and frontal lobes (Marcelis et al., 2003). These findings suggest abnormal cerebellar morphology as a possible heritable risk factor for psychosis. Moreover, first-degree relatives had higher global white matter volumes than patients and healthy comparison groups, which may represent a protective endophenotype absent in patients with psychosis (Marcelis et al., 2003).

While changes in gray matter volume can indicate dysfunction in a given area, the integrity of white matter tracts connecting the nodes of an extended circuit is also vital for function. As an essential node in a cortico-cerebellar-thalamic-cortical circuit (CCTCC) cerebellum is theorized to contribute to symptoms of schizophrenia, including cognitive dysfunction and psychosis (Andreasen et al., 1998; Andreasen and Pierson, 2008). When this information-processing loop functions suboptimally in schizophrenia, the result is a "cognitive dysmetria," or a disorganization of cognitive procedures and prioritization (Andreasen et al., 1998; Schmahmann, 1998). Diffusion tensor imaging (DTI) has revealed weaker white matter connections in the superior cerebellar peduncles in schizophrenia (Okugawa et al., 2006; Magnotta et al., 2008; Liu et al., 2011). The superior cerebellar peduncles carry the major efferent axon tracts from the cerebellum to the cerebral cortex via synaptic relays in areas like the thalamus and basal ganglia (Middleton and Strick, 1998; Bostan and Strick, 2018). However, not all studies report differences in superior cerebellar peduncle integrity between patients and healthy comparisons (Wang et al., 2003) and others have found that white matter integrity is decreased globally in patients with schizophrenia (Kanaan et al., 2009), and is not specific to the CCTCC.

FUNCTIONAL ABNORMALITIES IN SCHIZOPHRENIA

A body of functional imaging work has confirmed a disturbance in cortico-thalamic-cerebellar functional connectivity and loss of white matter integrity in this circuit in schizophrenia (Magnotta et al., 2008; Liu et al., 2011; Du et al., 2017). Resting-state functional connectivity within the cerebellum, and between the cerebellum and thalamus, the thalamus and frontal cortex, and the cerebellum and cerebral cortices (e.g., frontal and temporal) is also abnormal in patients with schizophrenia (Collin et al., 2011; Li et al., 2016; Woodward and Heckers, 2016; Ferri et al., 2018). Interestingly, individuals at clinical high risk for developing schizophrenia (i.e., individuals with prodromal attenuated psychotic symptoms) show an intermediate phenotype of cerebellar-thalamo-prefrontal resting state functional connectivity between those observed in first-episode patients and healthy comparisons (Du et al., 2017; Bang et al., 2018). These findings are in general agreement with reports of more severe cerebellar-thalamic-cortical dysconnectivity in clinical high-risk individuals who go on to develop psychosis compared to those who do not (Anticevic et al., 2015). However, others have reported that adolescent subjects at ultra-high risk for psychosis had a positive correlation between cerebellar-thalamic-cortical resting-state functional connectivity and positive symptom progression over the following year (Bernard et al., 2017). And still, other teams have only observed differences in resting state frontal lobe connectivity in first-episode patients, with thalamocortical connectivity differences emerging later in chronic stage subjects (Li et al., 2016). Even so, reduced connectivity in the cortico-thalamo-cerebellar circuit has been positively correlated with increased neurological soft signs (Mouchet-Mages et al., 2011), and greater impairment in cognitive and disorganization measures in schizophrenia patients (Repovs et al., 2011), while lower thalamo-cerebellar connectivity has correlated with delusions and bizarre behavior (Ferri et al., 2018). These general findings in decreased white matter and functional connectivity between the frontal cortex, thalamus, and cerebellum support the cognitive dysmetria hypothesis of schizophrenia as decreased connectivity between these areas may contribute to an anatomical basis for disorganized thought. However, connectivity abnormalities in schizophrenia are not exclusive to the CCTCC pathway.

Beyond abnormalities in resting-state functional connectivity, patients with schizophrenia also display altered patterns of activation during task performance. For example, patients with schizophrenia show decreased cerebellar activation in tasks requiring theory of mind, word and story recall, emotion attribution, and eyeblink conditioning (Koziol et al., 2013). Similar to findings of decreased functional connectivity between the frontal cortex, thalamus, and cerebellum in individuals at high genetic risk for schizophrenia (i.e., having two or more first- or second-degree relatives with schizophrenia), activation in these areas during performance of behavioral tasks reveals a striking pattern. During a verbal initiation task, found significantly lower activation in medial prefrontal, thalamic, and posterior cerebellar regions, with no difference in task performance, in young adults at high genetic risk for developing schizophrenia compared to matched normal-risk subjects (Whalley, 2003). This pattern of results again suggests aberrant CCTCC pathway activity as an existing anatomical feature associated with risk for developing the disease, not a factor relating to advanced disease progression.

Computationally, the cerebellum has been associated with production and updating of internal predictive models of the environment, pertaining to functions as wide-ranging as motor control (Ito, 2008; Shergill et al., 2014) and higher-level language processing (Guediche et al., 2015; Moberget et al., 2017). Temporary functional cerebellar lesions produced with transcranial magnetic stimulation induce consistent impairments in predicting sensory consequences of self-generated action (Miall et al., 2007; Lesage et al., 2012), a state of affairs observed in schizophrenia.

Furthermore, the involvement of the cerebellum in extended neural circuits that provide error feedback and adjustment during information processing (Dosenbach et al., 2008) may indicate an essential role in detecting erroneous percepts like auditory and visual hallucinations. For example, Powers and colleagues sought to delineate the role of sensory predictions in auditory-verbal hallucinations and psychosis (Powers et al., 2017). The authors adopted a predictive coding-based framework for perception, defined as the Bayesian combination of incoming sensory input with prior beliefs about the causes of sensory events (Friston, 2005). Within this framework, hallucinations (i.e., percepts in the absence of stimulus) may be viewed as a dominance of prior beliefs over incoming sensory evidence, thus producing percepts without any corresponding sensory event (Friston, 2005; Powers et al., 2016). The authors produced hallucinations via Pavlovian conditioning mechanisms (Ellson, 1941; Kot and Serper, 2002): a salient light and threshold-intensity tone were presented together repeatedly. Subsequently, participants reported hearing the tone despite its absence, contingent on the presence of the light. Although all participants experienced these *conditioned hallucinations*, those with a history of auditory hallucinations (with and without a diagnosable psychotic illness) experienced them more frequently compared to those without prior hallucinations. A network of regions classically identified during symptom-capture-based studies of hallucinations (i.e., superior temporal sulcus, anterior insula, auditory cortex, anterior cingulate) were active during conditioned hallucinations. Participants with past hallucinations were more likely to overtrust prior beliefs in comparison to incoming sensory evidence, with heightened activation of anterior insula and superior temporal sulcus in encoding these beliefs. By contrast, those with psychosis (regardless of whether they had hallucinations or not) were slower to update prior beliefs, with less corresponding activation in parahippocampal gyrus and cerebellum. The fact that cerebellar function was identified as being altered specifically in psychosis, regardless of symptomatology, supports its role as being central to psychotic illness (Andreasen, Paradiso, et al., 1998). Further elucidation of the network dynamics underlying interactions between these regions and the computational functions they subserve will help to clarify the role of cerebellar models in hallucinations and perception at large.

CELLULAR ABNORMALITIES

Schizophrenia diagnosis has been associated with abnormal expression of a number of genes and proteins in the cerebellum. Patients with schizophrenia have decreased expression of reelin and glutamic acid decarboxylase $(GAD)_{67}$ and GAD_{65} protein in the cerebellum and prefrontal cortex postmortem (Guidotti et al., 2000; Fatemi et al., 2005). Reelin, an extracellular matrix protein, is critical for neuronal migration patterning during embryonic development and synaptic plasticity in adulthood. Notably, reelin plays a crucial role in Purkinje cell migration and placement during development (Miyata et al., 2010). In line with observations of decreased cerebellar reelin expression, patients with schizophrenia also display decreased linear density of Purkinje cells, possibly indicating disrupted migration during development (Maloku et al., 2010). Furthermore, GAD_{65} and GAD_{67} are essential for gamma-aminobutyric acid (GABA) production, and are therefore critical for cell signaling and function of inhibitory interneurons and Purkinje cells. Purkinje cells are integral to the microcircuitry of the cerebellum and critically shape its output. Disruption in Purkinje cell placement and function may contribute to observed abnormalities in cerebellar function in patients with schizophrenia.

Abnormal expression of several synaptic proteins (i.e., proteins present on or near cell surfaces at synapses that play important roles in cell to cell communication) in the cerebellum of patients suggests that cell signaling is atypical. For example, a synaptic protein shown to be underexpressed in cerebellar tissue of schizophrenia patients is synaptosomal-associated protein (SNAP)-25, a SNARE protein necessary for synaptic vesicle docking (Mukaetova-Ladinska et al., 2002). Additionally, messenger RNA for synaptic proteins synaptophysin and complexin II are reduced in postmortem patient cerebellar tissue, with lower protein expression for complexin II as well (Eastwood et al., 2001). These findings contribute to the range of cerebellar abnormalities found in the schizophrenia population, but the relationship between cellular-level abnormalities and circuit-level disruptions is still incompletely understood.

BIPOLAR DISORDER

Roughly half of patients with bipolar disorder experience psychotic symptoms, the vast majority of those being patients diagnosed with bipolar I (Yurgelun-Todd, 2007). While cerebellar abnormalities in bipolar disorder have been examined less extensively relative to those in schizophrenia, some differences in bipolar populations have been noted.

ABNORMALITIES OF CEREBELLAR STRUCTURE BIPOLAR DISORDER

Decreased cerebellar cortex volume has been observed in patients diagnosed with bipolar I compared to healthy controls (Baldaçara et al., 2011), with similar volume loss to patients with schizophrenia (Rimol et al., 2010). However, not all studies have found decreased cerebellar cortical gray matter volume in bipolar patients (Loeber et al., 1999; McDonald et al., 2005), even when comparing subgroups with and without psychotic features (Laidi et al., 2015). Some studies have specifically reported decreased cerebellar vermal volume in bipolar disorder (Kim et al., 2013), similar to the level of atrophy seen in schizophrenia (Lippmann et al., 1982). However, it is unknown how disease progression or chronic medications play a role in volumetric differences. Bipolar patients that have experienced multiple manic episodes, but not first-episode patients, have smaller V3 vermis volume when compared with healthy controls (DelBello et al., 1999). This finding was replicated, and reduced V2 vermis volume was additionally found in both first-episode and multiepisode bipolar patients compared with healthy subjects (Mills et al., 2005). While a study by Monkul et al. (2008) observed no significant overall volume differences between participants with bipolar disorder and healthy controls, they did find an inverse relationship between V2 vermis volume and the number of previous affective episodes in males with bipolar disorder (Monkul et al., 2008). Cerebellar posterior vermal gray matter density has also been reported to progressively decrease based on illness duration in bipolar patients that were not treated with mood stabilizers or antipsychotic medications when compared to medicated patients and healthy volunteers (Kim et al., 2013). Additional reports also suggest that cerebellar atrophy may be related to familial risk. Brambilla et al. (2001) found that cerebellar cortex and vermis volume was reduced only in bipolar participants with a family history of the disorder (Brambilla et al., 2001). Both in participants with bipolar disorder and in healthy first-degree relatives, reduced cerebellar gray matter volume has been observed (Sarıçiçek et al., 2015). However, *increased* cerebellar gray matter volume has been found in populations at high genetic risk for bipolar that have not developed the disease or subsyndromic symptoms, which may be correlated with resistance (Kempton et al., 2009; Lin et al., 2018).

In terms of structural connectivity, a study by McDonald and colleagues did not observe any gray matter differences in cerebellar volume in bipolar patients, but did find a similar pattern of distributed white matter deficits to those observed in schizophrenia (McDonald, 2005). Using diffusion-weighted MRI, reduced axial and radial diffusivity has been shown in the cerebellum in participants with bipolar I compared with healthy controls (Ambrosi et al., 2016). Participants with bipolar II also show reduced axial diffusivity in the cerebellum, but no significant differences in cerebellar fractional anisotropy in either bipolar disorder group (Ambrosi et al., 2016).

FUNCTIONAL ABNORMALITIES IN BIPOLAR DISORDER

Decreased cerebro-cerebellar functional connectivity has been found in a range of functional networks (e.g., ventral attention and salience control) in bipolar patients with a history of psychosis (Shinn et al., 2017). These functional disruptions may help to explain disconnections between external stimuli and internal perceptions during psychotic or manic states.

In contrast, resting-state fMRI studies of bipolar II patients found increased functional connectivity in the left anterior lobe of the cerebellum (Wang et al., 2016), decreased functional connectivity in bilateral lobules VII and VIII, and decreased connectivity between the right posterior cerebellum and default mode network (Wang et al., 2017). However, relatively few studies have investigated altered cerebellar connectivity in bipolar disorder, and many have instead focused on seed-based analyses, which may or may not ultimately reveal altered cerebellar connectivity (Brady et al., 2016) or do not report on whether their findings included cerebellum (Argyelan et al., 2014).

Several studies have reported functional deficits in behavioral tasks that are thought to rely on the cerebellum in bipolar disorder. Chrobak and colleagues found that participants with bipolar disorder scored worse on the International Cooperative Ataxia Rating Scale (ICARS) with specific deficits in kinetic and oculomotor function (Chrobak et al., 2015). Likewise, participants with bipolar disorder show deficits in conditioned response acquisition and timing during eyeblink conditioning (Bolbecker et al., 2009). Furthermore, patients have increased timing variability in a finger-tapping task, likely due to disruptions in internal clock function (Bolbecker et al., 2011a), and deficits in postural control that include increased sway area and limited ability to correct posture that are worsened in the absence of visual indicators (Bolbecker et al., 2011b).

Altered cerebellar function has also been identified in bipolar disorder using functional neuroimaging. In an early study by Loeber and colleagues, decreased cerebellar blood volume was found in bipolar disorder using dynamic susceptibility contrast fMRI (Loeber et al., 1999). In contrast, Ketter et al. found glucose hypermetabolism in posterior brain structures, including the cerebellum, using positron-emission tomography (PET) that was independent of mood state, in a relatively large sample of treatment-resistant bipolar disorder patients (Ketter et al., 2001a). Strakowski and colleagues found reduced blood-oxygen level dependent (BOLD) response in the vermis in bipolar participants during the completion of the Stroop task (Strakowski et al., 2005). Reduced functional activity in the cerebellum has also been associated with a bipolar disorder and schizophrenia risk gene. Reduced BOLD activity during the Hayling sentence completion task has been associated with the DISC1 variant single nucleotide polymorphism (SNP) rs821633 in participants with bipolar disorder (Chakirova et al., 2011). Differences in cerebellar functional activity have also been identified between mood states. Functional activity in the cerebellum is increased in participants experiencing a mixed mood episode when compared with participants experiencing a depressed episode during a go-no-go task (Fleck et al., 2011). Likewise, decreased functional activity in cerebellar vermis has been reported during a flashing checkerboard task in both depressed and manic mood states compared with healthy subjects (Shaffer et al., 2017a).

CELLULAR ABNORMALITIES

Postmortem studies of the cerebellum in bipolar disorder have identified differences in gene expression and epigenetic markers. Postmortem cerebellum samples from individuals with bipolar disorder and healthy controls, show increased protein expression of tyrosine kinase B (TrKB), an important factor in synaptic plasticity (Soontornniyomkij et al., 2011). Likewise, in a pair of studies, Pinacho and colleagues found decreased protein expression of transcription factor specificity protein 4 (SP4) and 1 (SP1), which are known risk factors for bipolar disorder, in prefrontal cortex and cerebellum (Pinacho et al., 2011) and also increased phosphorylation of SP4 in the cerebellum (Pinacho et al., 2015).

Several studies suggest that cellular metabolism, particularly oxidative metabolism, may be disrupted in the cerebellum in bipolar disorder. Ben-Shachar and colleagues found increased expression of mitochondrial complex I genes in parieto-occipital cortex and decreased expression in the cerebellum (Ben-Shachar and Karry, 2008). Mitochondrial complex I is involved in electron transport during oxidative metabolism. This disruption may underlie the glucose hypermetabolism found by Ketter and colleagues using PET (Ketter et al., 2001a). This may also be supported by studies using a novel neuroimaging called T1 relaxation in the rotating frame ($T1\rho$). T1rho is sensitive to proton exchange and may indicate both fluctuations in local brain metabolism (possibly pH) and neuronal activity (Magnotta et al., 2012; Johnson et al., 2015). This method reveals significantly increased $T1\rho$ relaxation times in the cerebellum of bipolar patients in the euthymic and depressed or manic states (Johnson et al., 2015, 2018) and that lithium usage may prevent these differences (Johnson et al., 2015). These findings may suggest an increased reliance on glycolysis for cellular metabolism due to disruptions in electron transport in the mitochondria which may explain both the reported glucose hypermetabolism (Ketter et al., 2001b) and reduced pH (Johnson et al., 2015; Shaffer et al., 2017b; Johnson et al., 2018).

CONCLUSIONS

The evidence summarized here indicates that cerebellar morphology, function, and connectivity is abnormal in some cases of psychosis in schizophrenia and bipolar disorder. These abnormalities are often observed before diagnosis or early in disease onset, and therefore likely reflect some component of disease etiology rather than outcomes of chronic illness and pharmacological treatment. Undoubtedly, cerebellar irregularities occur among a number of other disrupted brain areas, transmitter systems, and functional networks during psychosis. How these distributed processes interact will be the topic of future research, along with the possibility of modulating cerebellar activity to aid in symptom reduction (Demirtas-Tatlidede et al., 2010; Parker et al., 2017).

REFERENCES

Ambrosi E, Chiapponi C, Sani G, Manfredi G, Piras F, Caltagirone C, Spalletta G (2016) White matter microstructural characteristics in bipolar I and bipolar II disorder: a diffusion tensor imaging study. J Affect Disord 189:176–183.

Andreasen NC, Paradiso S, O'Leary DS (1998) "Cognitive dysmetria" as an integrative theory of schizophrenia. Schizophr Bull 24:203–218.

Andreasen NC, Pierson R (2008) The role of the cerebellum in schizophrenia. Biol Psychiatry 64:81–88.

Anticevic A et al. (2015) Association of thalamic dysconnectivity and conversion to psychosis in youth and young adults at elevated clinical risk. JAMA Psychiatry 72:882.

Argyelan M, Ikuta T, Derosse P, Braga RJ, Burdick KE, John M, Kingsley PB, Malhotra AK, Szeszko PR (2014) Resting-state fMRI connectivity impairment in schizophrenia and bipolar disorder. Schizophr Bull 40:100–110.

Baldaçara L, Nery-Fernandes F, Rocha M, Quarantini LC, Rocha GGL, Guimarães JL, Araújo C, Oliveira I, Miranda-Scippa A, Jackowski A (2011) Is cerebellar volume related to bipolar disorder? J Affect Disord 135:305–309.

Bang M, Park H-J, Pae C, Park K, Lee E, Lee S-K, An SK (2018) Aberrant cerebro-cerebellar functional connectivity and minimal self-disturbance in individuals at ultra-high risk for psychosis and with first-episode schizophrenia. Schizophr Res

Benes F, Paskevich P, Domesick V (1983) Haloperidol-induced plasticity of axon terminals in rat substantia nigra. Science 221:969–971.

Ben-Shachar D, Karry R (2008) Neuroanatomical pattern of mitochondrial complex I pathology varies between schizophrenia, bipolar disorder and major depression. PLoS ONE 3:e3676.

Bernard JA, Orr JM, Mittal VA (2017) Cerebello-thalamo-cortical networks predict positive symptom progression in individuals at ultra-high risk for psychosis. NeuroImage Clin 14:622–628.

Bolbecker AR, Hong SL, Kent JS, Forsyth JK, Klaunig MJ, Lazar EK, O'Donnell BF, Hetrick WP (2011a) Paced finger-tapping abnormalities in bipolar disorder indicate timing dysfunction. Bipolar Disord 13:99–110.

Bolbecker AR, Hong SL, Kent JS, Klaunig MJ, O'Donnell BF, Hetrick WP (2011b) Postural control in bipolar disorder: Increased sway area and decreased dynamical complexity. PLoS ONE 6.

Bolbecker AR, Mehta C, Johannesen JK, Edwards CR, O'Donnell BF, Shekhar A, Nurnberger JI, Steinmetz JE, Hetrick WP (2009) Eyeblink conditioning anomalies in bipolar disorder suggest cerebellar dysfunction. Bipolar Disord 11:19–32.

Bostan AC, Strick PL (2018) The basal ganglia and the cerebellum: nodes in an integrated network. Nat Rev Neurosci 19:338–350.

Brady RO, Masters GA, Mathew IT, Margolis A, Cohen BM, Öngür D, Keshavan M (2016) State dependent cortico-amygdala circuit dysfunction in bipolar disorder. J Affect Disord 201: 79–87.

Brambilla P, Harenski K, Nicoletti M, Mallinger AG, Frank E, Kupfer DJ, Keshavan MS, Soares JC (2001) MRI study of posterior fossa structures and brain ventricles in bipolar patients. J Psychiatr Res 35:313–322.

Cannon M, Jones P, Huttunen MO, Tanskanen A, Murray RM (1999) Motor co-ordination deficits as predictors of schizophrenia among Finnish school children. Hum Psychopharmacol Clin Exp 14:491–497.

Chakirova G, Whalley HC, Thomson P a, Hennah W, Moorhead TW, Welch K a, Giles S, Hall J, Johnstone EC, Lawrie SM, Porteous DJ, Brown VJ, McIntosh a M (2011) The effects of DISC1 risk variants on brain activation in controls, patients with bipolar disorder and patients with schizophrenia. Psychiatry Res 192:20–28.

Chrobak AA, Siwek G, Tereszko A, Jeziorko S, Siuda-Krzywicka K, Arciszewska A, Siwek M, Dudek D (2015) Ataxia signs in bipolar disorder. Bipolar Disord 17:79.

Cierpka M, Wolf ND, Kubera KM, Schmitgen MM, Vasic N, Frasch K, Wolf RC (2017) Cerebellar contributions to persistent auditory verbal hallucinations in patients with schizophrenia. The Cerebellum 16:964–972.

Collin G, Hulshoff Pol H, Haijma S, Cahn W, Kahn R, Van Den Heuvel M (2011) Impaired cerebellar functional connectivity in schizophrenia patients and their healthy siblings. Front Psychiatry 2.

Crow TJ, Done DJ, Sacker A (1995) Childhood precursors of psychosis as clues to its evolutionary origins. Eur Arch Psychiatry Clin Neurosci 245:61–69.

Dazzan P, Murray R (2002) Neurological soft signs in first-episode psychosis: a systematic review. Br J Psychiatry 181:50s–57.

DelBello MP, Strakowski SM, Zimmerman ME, Hawkins JM, Sax KW (1999) MRI analysis of the cerebellum in bipolar disorder: a pilot study. Neuropsychopharmacol Off Publ Am Coll Neuropsychopharmacol 21:63–68.

Demirtas-Tatlidede A, Freitas C, Cromer JR, Safar L, Ongur D, Stone WS, Seidman LJ, Schmahmann JD, Pascual-Leone A (2010) Safety and proof of principle study of cerebellar vermal theta burst stimulation in refractory schizophrenia. Schizophr Res 124:91–100.

Dosenbach NUF, Fair DA, Cohen AL, Schlaggar BL, Petersen SE (2008) A dual-networks architecture of top-down control. Trends Cogn Sci 12:99–105.

Du Y, Fryer SL, Fu Z, Lin D, Sui J, Chen J, Damaraju E, Mennigen E, Stuart B, Loewy RL, Mathalon DH, Calhoun VD (2017) Dynamic functional connectivity impairments in early schizophrenia and clinical high-risk for psychosis. NeuroImage.

Eastwood S., Cotter D, Harrison P. (2001) Cerebellar synaptic protein expression in schizophrenia. Neuroscience 105:219–229.

Ellson DG (1941) Hallucinations produced by sensory conditioning. J Exp Psychol 28:1–20.

Fatemi SH, Stary JM, Earle JA, Araghi-Niknam M, Eagan E (2005) GABAergic dysfunction in schizophrenia and mood disorders as reflected by decreased levels of glutamic acid decarboxylase 65 and 67 kDa and Reelin proteins in cerebellum. Schizophr Res 72: 109–122.

Ferri J et al. (2018) Resting-state thalamic dysconnectivity in schizophrenia and relationships with symptoms. Psychol Med:1–8.

Fleck DE, Kotwal R, Eliassen JC, Lamy M, Delbello MP, Adler CM, Durling M, Cerullo MA, Strakowski SM (2011) Preliminary evidence for increased frontosubcortical activation on a motor impulsivity task in mixed episode bipolar disorder. J Affect Disord 133:333–339.

Friston KJ (2005) Hallucinations and perceptual inference. Behav Brain Sci 28:764–766.

Gan Z, Diao F, Han Z, Li K, Zheng L, Guan N, Kang Z, Wu X, Wei Q, Cheng M, Zhang M, Zhang J (2012) Psychosis and Dandy–Walker complex: report of four cases. Gen Hosp Psychiatry 34:102.e7–102.e11.

Guediche S, Holt LL, Laurent P, Lim S-J, Fiez JA (2015) Evidence for cerebellar contributions to adaptive plasticity in speech perception. Cereb Cortex N Y N 1991 25:1867–1877.
Guidotti A, Auta J, Davis JM, Gerevini VD, Dwivedi Y, Grayson DR, Impagnatiello F, Pandey G, Pesold C, Sharma R, Uzunov D, Costa E (2000) Decrease in reelin and glutamic acid decarboxylase67 (GAD67) expression in schizophrenia and bipolar disorder: a postmortem brain study. Arch Gen Psychiatry 57:1061.
Gupta S, Andreasen NC, Arndt S, Flaum M, Schultz SK, Hubbard WC, Smith M (1995) Neurological soft signs in neuroleptic-naive and neuroleptic-treated schizophrenic patients and in normal comparison subjects. Am J Psychiatry 152:191–196.
Heath RG, Franklin DE, Shraberg D (1979) Gross pathology of the cerebellum in patients diagnosed and treated as functional psychiatric disorders. J Nerv Ment Dis 167:585–592.
Ho B-C, Andreasen NC, Ziebell S, Pierson R, Magnotta V (2011) Long-term antipsychotic treatment and brain volumes: a longitudinal study of first-episode schizophrenia. Arch Gen Psychiatry 68:128–137.
Ho B-C, Mola C, Andreasen NC (2004) Cerebellar dysfunction in neuroleptic naive schizophrenia patients: clinical, cognitive, and neuroanatomic correlates of cerebellar neurologic signs. Biol Psychiatry 55:1146–1153.
Ichimiya T, Okubo Y, Suhara T, Sudo Y (2001) Reduced volume of the cerebellar vermis in neuroleptic-naive schizophrenia. Biol Psychiatry 49:20–27.
Ito M (2008) Control of mental activities by internal models in the cerebellum. Nat Rev Neurosci 9:304–313.
Johnson CP, Christensen GE, Fiedorowicz JG, Mani M, Shaffer JJ, Magnotta VA, Wemmie JA (2018) Alterations of the cerebellum and basal ganglia in bipolar disorder mood states detected by quantitative T1ρ mapping. Bipolar Disord.
Johnson CP, Follmer RL, Oguz I, Warren LA, Christensen GE, Fiedorowicz JG, Magnotta VA, Wemmie JA (2015) Brain abnormalities in bipolar disorder detected by quantitative T1ρ mapping. Mol Psychiatry 20:201–206.
Jurjus GJ, Weiss KM, Jaskiw GE (1994) Schizophrenia-like psychosis and cerebellar degeneration. Schizophr Res 12:183–184.
Kanaan RAA, Borgwardt S, McGuire PK, Craig MC, Murphy DGM, Picchioni M, Shergill SS, Jones DK, Catani M (2009) Microstructural organization of cerebellar tracts in schizophrenia. Biol Psychiatry 66:1067–1069.
Keddie KM (1969) Hereditary ataxia, presumed to be of the Menzel type, complicated by paranoid psychosis, in a mother and two sons. J Neurol Neurosurg Psychiatry 32:82–87.
Kempton MJ, Haldane M, Jogia J, Grasby PM, Collier D, Frangou S (2009) Dissociable brain structural changes associated with predisposition, resilience, and disease expression in bipolar disorder. J Neurosci 29:10863–10868.
Ketter TA, Kimbrell TA, George MS, Dunn RT, Speer AM, Benson BE, Willis MW, Danielson A, Frye MA, Herscovitch P, Post RM (2001a) Effects of mood and subtype on cerebral glucose metabolism in treatment-resistant bipolar disorder. Biol Psychiatry 49:97–109.
Ketter TA, Kimbrell TA, George MS, Dunn RT, Speer AM, Benson BE, Willis MW, Danielson A, Frye MA, Herscovitch P, Post RM (2001b) Effects of mood and subtype on cerebral glucose metabolism in treatment-resistant bipolar disorder. Biol Psychiatry 49:97–109.
Kim D, Byul Cho H, Dager SR, Yurgelun-Todd DA, Yoon S, Lee JH, Hea Lee S, Lee S, Renshaw PF, Kyoon Lyoo I (2013) Posterior cerebellar vermal deficits in bipolar disorder. J Affect Disord 150:499–506.
Kim T, Lee K-H, Oh H, Lee TY, Cho KIK, Lee J, Kwon JS (2018) Cerebellar structural abnormalities associated with cognitive function in patients with first-episode psychosis. Front Psychiatry 9.
Kot T, Serper M (2002) Increased susceptibility to auditory conditioning in hallucinating schizophrenic patients: a preliminary investigation. J Nerv Ment Dis 190:282–288.
Koziol LF, Budding D, Andreasen N, D'Arrigo S, Bulgheroni S, Imamizu H, Ito M, Manto M, Marvel C, Parker K, Pezzulo G, Ramnani N, Riva D, Schmahmann J, Vandervert L, Yamazaki T (2013) Consensus paper: the cerebellum's role in movement and cognition. Cerebellum Lond Engl.
Kutty IN, Prendes JL (1981) Psychosis and cerebellar degeneration. J Nerv Ment Dis 169:390–391.
Laidi C, d'Albis M-A, Wessa M, Linke J, Phillips ML, Delavest M, Bellivier F, Versace A, Almeida J, Sarrazin S, Poupon C, Le Dudal K, Daban C, Hamdani N, Leboyer M, Houenou J (2015) Cerebellar volume in schizophrenia and bipolar I disorder with and without psychotic features. Acta Psychiatr Scand 131:223–233.
Lesage E, Morgan BE, Olson AC, Meyer AS, Miall RC (2012) Cerebellar rTMS disrupts predictive language processing. Curr Biol CB 22:R794–R795.
Levitt JJ, McCarley RW, Nestor PG, Petrescu C, Donnino R, Hirayasu Y, Kikinis R, Jolesz FA, Shenton ME (1999) Quantitative volumetric MRI study of the cerebellum and vermis in schizophrenia: clinical and cognitive correlates. Am J Psychiatry 156:1105–1107.
Li T et al. (2016) Brain-wide analysis of functional connectivity in first-episode and chronic stages of schizophrenia. Schizophr Bull:sbw099.
Lin K, Shao R, Geng X, Chen K, Lu R, Gao Y, Bi Y, Lu W, Guan L, Kong J, Xu G, So K-F (2018) Illness, at-risk and resilience neural markers of early-stage bipolar disorder. J Affect Disord 238:16–23.
Lippmann S, Manshadi M, Baldwin H, Drasin G, Rice J, Alrajeh S (1982) Cerebellar vermis dimensions on computerized tomographic scans of schizophrenic and bipolar patients. Am J Psychiatry 139:667–668.
Liu H, Fan G, Xu K, Wang F (2011) Changes in cerebellar functional connectivity and anatomical connectivity in schizophrenia: a combined resting-state functional MRI and diffusion tensor imaging study. J Magn Reson Imaging 34:1430–1438.
Loeber RT, Sherwood AR, Renshaw PF, Cohen BM, Yurgelun-Todd DA (1999) Differences in cerebellar blood volume in schizophrenia and bipolar disorder. Schizophr Res 37:81–89.
Magnotta VA, Adix ML, Caprahan A, Lim K, Gollub R, Andreasen NC (2008) Investigating connectivity between the cerebellum and thalamus in schizophrenia using diffusion tensor tractography: a pilot study. Psychiatry Res 163:193–200.
Magnotta VA, Heo H-Y, Dlouhy BJ, Dahdaleh NS, Follmer RL, Thedens DR, Welsh MJ, Wemmie JA (2012) Detecting activity-evoked pH changes in human brain. Proc Natl Acad Sci U S A 109:8270–8273.
Maloku E, Covelo IR, Hanbauer I, Guidotti A, Kadriu B, Hu Q, Davis JM, Costa E (2010) Lower number of cerebellar Purkinje neurons in psychosis is associated with reduced reelin expression. Proc Natl Acad Sci U S A 107:4407–4411.
Marcelis M, Suckling J, Woodruff P, Hofman P, Bullmore E, van Os J (2003) Searching for a structural endophenotype in psychosis using computational morphometry. Psychiatry Res Neuroimaging 122:153–167.
McDonald C (2005) Regional volume deviations of brain structure in schizophrenia and psychotic bipolar disorder: computational morphometry study. Br J Psychiatry 186:369–377.
McDonald C, Bullmore E, Sham P, Chitnis X, Suckling J, MacCabe J, Walshe M, Murray RM (2005) Regional volume deviations of brain structure in schizophrenia and psychotic bipolar disorder: Computational morphometry study. Br J Psychiatry 186:369–377.
Miall RC, Christensen LOD, Cain O, Stanley J (2007) Disruption of state estimation in the human lateral cerebellum. PLoS Biol 5:e316.
Middleton FA, Strick PL (1998) Cerebellar output: motor and cognitive channels. Trends Cogn Sci 2:348–354.
Mills NP, Delbello MP, Adler CM, Strakowski SM (2005) MRI analysis of cerebellar vermal abnormalities in bipolar disorder. Am J Psychiatry 162:1530–1532.
Miyata T, Ono Y, Okamoto M, Masaoka M, Sakakibara A, Kawaguchi A, Hashimoto M, Ogawa M (2010) Migration, early axonogenesis, and Reelin-dependent layer-forming behavior of early/posterior-born Purkinje cells in the developing mouse lateral cerebellum. Neural Develop 5:23.
Moberget T et al. (2017) Cerebellar volume and cerebellocerebral structural covariance in schizophrenia: a multisite mega-analysis of 983 patients and 1349 healthy controls. Mol Psychiatry

Monkul ES, Hatch JP, Sassi RB, Axelson D, Brambilla P, Nicoletti MA, Keshavan MS, Ryan ND, Birmaher B, Soares JC (2008) MRI study of the cerebellum in young bipolar patients. Prog Neuropsychopharmacol Biol Psychiatry 32:613–619.

Mouchet-Mages S, Rodrigo S, Cachia A, Mouaffak F, Olie JP, Meder JF, Oppenheim C, Krebs MO (2011) Correlations of cerebello-thalamo-prefrontal structure and neurological soft signs in patients with first-episode psychosis. Acta Psychiatr Scand 123:451–458.

Mukaetova-Ladinska EB, Hurt J, Honer WG, Harrington CR, Wischik CM (2002) Loss of synaptic but not cytoskeletal proteins in the cerebellum of chronic schizophrenics. Neurosci Lett 317:161–165.

Nopoulos PC, Ceilley JW, Gailis EA, Andreasen NC (1999) An MRI study of cerebellar vermis morphology in patients with schizophrenia: evidence in support of the cognitive dysmetria concept. Biol Psychiatry 46:703–711.

Okugawa G, Nobuhara K, Minami T, Takase K, Sugimoto T, Saito Y, Yoshimura M, Kinoshita T (2006) Neural disorganization in the superior cerebellar peduncle and cognitive abnormality in patients with schizophrenia: A diffusion tensor imaging study. Prog Neuropsychopharmacol Biol Psychiatry 30:1408–1412.

Parker KL, Kim YC, Kelley RM, Nessler AJ, Chen K-H, Muller-Ewald VA, Andreasen NC, Narayanan NS (2017) Delta-frequency stimulation of cerebellar projections can compensate for schizophrenia-related medial frontal dysfunction. Mol Psychiatry 22:647–655.

Pinacho R, Saia G, Meana JJ, Gill G, Ramos B (2015) Transcription factor SP4 phosphorylation is altered in the postmortem cerebellum of bipolar disorder and schizophrenia subjects. Eur Neuropsychopharmacol 25:1650–1660.

Pinacho R, Villalmanzo N, Lalonde J, Haro JM, Meana JJ, Gill G, Ramos B (2011) The transcription factor SP4 is reduced in postmortem cerebellum of bipolar disorder subjects: Control by depolarization and lithium. Bipolar Disord 13:474–485.

Powers AR, Kelley M, Corlett PR (2016) Hallucinations as top-down effects on perception. Biol Psychiatry Cogn Neurosci Neuroimaging 1:393–400.

Powers AR, Mathys C, Corlett PR (2017) Pavlovian conditioning-induced hallucinations result from overweighting of perceptual priors. Science 357:596–600.

Repovs G, Csernansky JG, Barch DM (2011) Brain network connectivity in individuals with schizophrenia and their siblings. Biol Psychiatry 69:967–973.

Reyes M, Gordon A (1981) Cerebellar vermis in schizophrenia. Lancet 318:700–701.

Rimol LM, Hartberg CB, Nesvåg R, Fennema-Notestine C, Hagler DJ, Pung CJ, Jennings RG, Haukvik UK, Lange E, Nakstad PH, Melle I, Andreassen OA, Dale AM, Agartz I (2010) Cortical thickness and subcortical volumes in schizophrenia and bipolar disorder. Biol Psychiatry 68:41–50.

Sarıçiçek A, Yalın N, Hıdıroğlu C, Çavuşoğlu B, Taş C, Ceylan D, Zorlu N, Ada E, Tunca Z, Özerdem A (2015) Neuroanatomical correlates of genetic risk for bipolar disorder: a voxel-based morphometry study in bipolar type I patients and healthy first degree relatives. J Affect Disord 186:110–118.

Schmahmann JD (1998) Dysmetria of thought: clinical consequences of cerebellar dysfunction on cognition and affect. Trends Cogn Sci 2:362–371.

Shaffer JJ, Johnson CP, Fiedorowicz JG, Christensen GE, Wemmie JA, Magnotta VA (2017a) Impaired sensory processing measured by functional MRI in Bipolar disorder manic and depressed mood states. Brain Imaging Behav.

Shaffer JJ, Johnson CP, Long JD, Fiedorowicz JG, Christensen GE, Wemmie JA, Magnotta VA (2017b) Relationship altered between functional T1ρ and BOLD signals in bipolar disorder. Brain Behav 7.

Shakow D, Huston PE (1936) Studies of motor function in schizophrenia: I. Speed of Tapping. J Gen Psychol 15:63–106.

Shergill SS, White TP, Joyce DW, Bays PM, Wolpert DM, Frith CD (2014) Functional magnetic resonance imaging of impaired sensory prediction in schizophrenia. JAMA Psychiatry 71:28–35.

Shinn AK, Roh YS, Ravichandran CT, Baker JT, Öngür D, Cohen BM (2017) Aberrant cerebellar connectivity in bipolar disorder with psychosis. Biol Psychiatry Cogn Neurosci Neuroimaging 2:438–448.

Soontornniyomkij B, Everall IP, Chana G, Tsuang MT, Achim CL, Soontornniyomkij V (2011) Tyrosine kinase B protein expression is reduced in the cerebellum of patients with bipolar disorder. J Affect Disord 133:646–654.

Stevens JR (1982) Neuropathology of schizophrenia. Arch Gen Psychiatry 39:1131.

Strakowski SM, Adler CM, Holland SK, Mills NP, DelBello MP, Eliassen JC (2005) Abnormal fMRI brain activation in euthymic bipolar disorder patients during a counting Stroop interference task. Am J Psychiatry 162:1697–1705.

van Harten PN, Walther S, Kent JS, Sponheim SR, Mittal VA (2017) The clinical and prognostic value of motor abnormalities in psychosis, and the importance of instrumental assessment. Neurosci Biobehav Rev 80:476–487.

Volz H, Gaser C, Sauer H (2000) Supporting evidence for the model of cognitive dysmetria in schizophrenia—a structural magnetic resonance imaging study using deformation-based morphometry. Schizophr Res 46:45–56.

Wang F, Sun Z, Du X, Wang X, Cong Z, Zhang H, Zhang D, Hong N (2003) A diffusion tensor imaging study of middle and superior cerebellar peduncle in male patients with schizophrenia. Neurosci Lett 348:135–138.

Wang Y, Zhong S, Chen G, Liu T, Zhao L, Sun Y, Jia Y, Huang L (2017) Altered cerebellar functional connectivity in remitted bipolar disorder: A resting-state functional magnetic resonance imaging study. Aust N Z J Psychiatry:000486741774599.

Wang Y, Zhong S, Jia Y, Sun Y, Wang B, Liu T, Pan J, Huang L (2016) Disrupted resting-state functional connectivity in nonmedicated bipolar disorder. Radiology 000:151641.

Wassink TH, Andreasen NC, Nopoulos P, Flaum M (1999) Cerebellar morphology as a predictor of symptom and psychosocial outcome in schizophrenia. Biol Psychiatry 45:41–48.

Weinberger DR, Kleinman JE, Luchins DJ, Bigelow LB, Wyatt RJ (1980) Cerebellar pathology in schizophrenia: a controlled postmortem study. Am J Psychiatry 137:359–361.

Whalley HC (2003) fMRI correlates of state and trait effects in subjects at genetically enhanced risk of schizophrenia. Brain 127:478–490.

Williams AJ, Wang Z, Taylor SF (2016) Atypical psychotic symptoms and Dandy–Walker variant. Neurocase 22:472–475.

Woodward ND, Heckers S (2016) Mapping thalamocortical functional connectivity in chronic and early stages of psychotic disorders. Biol Psychiatry 79:1016–1025.

Yurgelun-Todd D (2007) Psychosis in bipolar disorder. In: The Spectrum of Psychotic Disorders (Fujii D, Ahmed I, eds), pp 137–155. Cambridge: Cambridge University Press.

SECTION 4

SOCIOENVIRONMENTAL MECHANISTIC FACTORS IN PSYCHOSIS

EARLY LIFE ADVERSITY

44

PERINATAL FACTORS IN PSYCHOSIS

Mary Clarke and Mary Cannon

INTRODUCTION

In this chapter, the perinatal risk factors for psychosis that we focus on are obstetric complications. Other perinatal exposures that may increase risk for psychosis, such as infection, are reviewed elsewhere in this book. The term "obstetric complications" refers to somatic deviations from the normal course of events over the pregnancy, delivery, and early neonatal periods.[1] Obstetric complications (OCs) have a long and varied history as one of the most-intensively investigated categories of environmental risk factor for psychotic disorders.[2,3] The psychosis-risk-increasing effects of OCs have been studied in many ways over the last 6 decades, first in high-risk studies where the rates of OCs experienced by the offspring of mothers with schizophrenia were examined, then in case-control studies that mainly employed maternal recall and case notes to document the perinatal history of people with schizophrenia, and finally in large population-based samples that took advantage of prospectively collected data, large sample sizes and the ability to control for confounders. In this chapter, we critically review the evidence for the association between OCs and schizophrenia and other psychotic disorders. However, the conclusion that can be reached from all lines of evidence is that OCs have a modest but consistently found association with schizophrenia and other psychotic disorders. Odds ratios (ORs) of 1.5 to 2.0 have been replicated in several population-based cohort studies using prospective data. As we document, there is a large amount of heterogeneity in the findings when individual OCs are examined but, as a category of risk factor, OCs and other events during the perinatal period represent a promising line of future inquiry and a possible area for preventive intervention. We have now come full circle, and the most recent work on OCs is again employing high-risk designs in an effort to elucidate the mechanisms underlying the association. We explore candidate mechanisms and set OCs in the context of other risk-increasing events in the developmental trajectory to psychosis.

HIGH-RISK STUDIES

Evidence from large population-based samples from Australia, Sweden, and Denmark show consistently that women with schizophrenia have increased rates of a range of pregnancy, birth, and delivery complications. Jablensky et al.[4] found increased rates of placental abruption, low birth weight, and cardiovascular congenital anomalies in the obstetric records of women with schizophrenia. Bennedsen et al.[5] found a higher rate of congenital malformations, stillbirths, and infant deaths among the offspring of women with schizophrenia. A series of studies from a large Danish register-linkage sample showed that many of these complications (in particular stillbirth, neonatal death, and sudden infant death) occur in the birth histories of women with other mental disorders[6–7] although rates of fatal birth defect were highest among women with a diagnosis of schizophrenia.[8] A recent study from the Helsinki High-Risk Study found that there were very few differences in the occurrence of OCs in offspring of mothers with schizophrenia spectrum disorders and controls.[9] However, most of the mothers in this sample developed psychosis after the birth of their offspring and the rates of OCs were higher among mothers in the sample with preexisting psychotic disorder. This is consistent with the Jablensky study,[4] and shows that the lack of consideration in some previous studies regarding the timing of maternal psychotic disorder may explain some of the heterogeneity in findings in the high-risk (studies. Maternal behaviors may account for the increased prevalence of OCs among high-risk offspring whose mothers have preexisting psychosis. The Helsinki High Risk Study found that three OCs predicted the development of psychosis in the offspring: placental abnormalities, prenatal infection, and maternal hypertension.[9] Given that most of the mothers in this sample developed psychosis after the birth of their offspring, maternal behaviors and characteristics are not likely to account for these associations, although the study was unable to examine this.

Recent high-risk studies have been able to take advantage of longitudinal designs to prospectively follow-up young people at familial high-risk (FHR) of psychosis. For example, the Edinburgh High Risk Study has recently shown that obstetric complications as part of a "polyenviromic risk score"; an aggregate score of exposure to environmental risk factors; could predict conversion to psychosis.[10] While OCs on their own were not predictive of conversion to psychosis in this FHR cohort,[11] this work shows that even in the absence of an independent association, OCs are important as an additive component in the complex, multifactorial pathway to psychotic disorder.

Evidence from population-based cohort studies

These studies have the following characteristics:[12] (1) large and psychiatrically well-defined samples of schizophrenia cases drawn from population-based registers or cohorts; (2) use of standardized, prospectively collected obstetric information from birth records or registers; and (3) control subjects drawn from the general population with OC information from the same source and context. Investigators did not use OC rating scales and usually reported odds ratios for all individual OCs recorded on the birth records and controlled for demographic confounders by matching or statistical adjustment. However, despite this seemingly optimal methodology, the results from these large population-based studies are far from consistent. Cannon et al.[13] carried out a meta-analysis covering population-based studies published up to and including 2001.

The results of the meta-analysis[13]can be summarized as follows. First, there was no significant heterogeneity between the studies and no evidence for publication bias. Second, the effect sizes for the relationship between obstetric risk factors and later schizophrenia were generally small, with odds ratios of less than two. Specifically, significant differences between cases and controls were found for the following nine variables in order of significance (OR:95% CI): uterine atony (2.2:1.5–3.5), bleeding in pregnancy (1.7:1.1–2.5), asphyxia (1.7:1.2–2.6), emergency caesarian section (3.2:1.4–7.5), birthweight <2000 g (3.9:1.4–10.8), birthweight <2500 g (1.7:1,2–2.3), congenital malformations (2.3:1.2–4.6), maternal diabetes in pregnancy (7.8:1.4–43.9); rhesus variables (2.0:1.01–3.96) (which includes rhesus incompatibility, rhesus negative mother, rhesus antibodies); and pre-eclampsia (1.4:0.99–1.85). Two complications nearly missed formal statistical significance: placental abruption (4.02:0.9–18.1) and head circumference <32 cm (1.3:0.8–2.06). Cannon et al.[13] grouped these variables into three categories: complications of pregnancy; abnormal fetal growth and development and complications of delivery. These effect sizes are similar in magnitude to those reported for the relationship between passive smoking and lung cancer or the risk of breast cancer among users of oral contraceptives. Given these small effects, it would be premature to draw causal inferences. Rather, the study of individual obstetric risk factors for schizophrenia can be conceptualized as the search for uncommon to rare risk factors for a relatively rare disorder, posing challenges for both the case-control and the classical cohort study design.

In the last decade, there have been few studies conducted with a similar design to those in the meta-analysis above. The only comparable study that reports on a broad range of obstetric complications is that of Byrne et al.[14] This large nested case-control study using Danish register-based data found significant, approximately twofold associations in the rate of schizophrenia among those who had experienced any one of the following complications: maternal non-attendance at antenatal appointments, gestational age of 37 weeks or below, pre-eclampsia, threatened premature delivery, hemorrhage during delivery, manual extraction of the baby, and maternal sepsis of childbirth and the puerperium. The findings were not appreciably altered by adjustment for the confounding effects of family psychiatric history, socioeconomic, and other demographic factors.

As noted earlier, Cannon et al.[13] summarized their findings by grouping the variables associated with later schizophrenia into three categories. These categories can be updated by incorporating the results from the Byrne et al. study[14] (in italics): (1) Complications of pregnancy: bleeding, pre-eclampsia, maternal diabetes, rhesus incompatibility, *maternal nonattendance at antenatal appointment, threatened premature delivery*; (2) Abnormalities of fetal growth and development: low birth weight, congenital malformations, small head circumference, *gestational age < 37weeks*; (3) Complications of delivery: asphyxia, uterine atony and emergency caesarean section, manual extraction of the baby, maternal sepsis of childbirth and puerperium; hemorrhage during delivery.

Investigation of specific individual complications

Some population-based studies have applied the power of the population-based methods to a more focused examination of certain complications of interest. They do not adopt the "atheoretical" stance of the prior population–based studies; rather, they have aimed to test a priori hypotheses. This approach builds on the findings and the methodological expertise of the previous cohort studies but adds a focus on causative mechanisms. For example, in a series of papers from a Danish Cohort using national population-based register data, Sorensen et al.[15,16] looked at the association between prenatal exposure to maternal hypertension/diuretic treatment and schizophrenia and at the association between prenatal exposure to analgesics and schizophrenia. The authors found that maternal hypertension and diuretic treatment during pregnancy are obstetric complications of note—experience of either one in the third trimester independently increases the odds of the exposed offspring developing schizophrenia twofold and experience of both increases the odds fourfold. Prenatal exposure to analgesics is an obstetric complication equally worthy of note with exposure increasing the odds of schizophrenia four-old, after adjustment for the confounding effects of parental history of schizophrenia, prenatal infection exposure, exposure to other drugs in the prenatal period, other pregnancy complications, parental social status, and parental age. And Insel et al.[17] reported that maternal–fetal blood incompatibility leads to a twofold increase in the rate of development of schizophrenia in adulthood among the exposed offspring in the Prenatal Determinants of Schizophrenia Study—a population-based birth cohort study from California. This finding replicates the findings of Hollister et al.[18] Also, in in the Prenatal Determinants of Schizophrenia Study, Insel et al.[19] reported that maternal mean haemoglobin of 10 g/dl or less was associated with a nearly fourfold increased rate of schizophrenia spectrum disorders.

The examination of specific "candidate" complications is likely to bring us closer to elucidating causal mechanisms in psychosis. For instance, low birthweight or poor fetal growth per se are not causes of psychosis but are a general indicator of a deviant neurodevelopmental process caused by other risk factors and genetic mutations that are more proximal to the underlying pathology. For example, a recent nested case-control study from the Philadelphia cohort of the Collaborative Perinatal Project has shown that low birth weight is a risk factor for psychosis only in the presence of prenatal exposure to influenza or hypoxia.[20]

WHAT IS THE POSSIBLE MECHANISM FOR THE ASSOCIATION BETWEEN OCS AND PSYCHOSIS?

OBSTETRIC COMPLICATIONS AND FAMILIAL RISK FOR SCHIZOPHRENIA

Ellman et al,[21] using Finnish register-based prospective data, has shown that while women with schizophrenia experienced significantly more OCs while pregnant compared to both mothers with a first-degree relative with schizophrenia and controls, there was no difference in the rate of OCs between those with a first-degree relative and controls. The lack of familial liability to OCs in this data indicates that familial liability to schizophrenia does not underlie the increased incidence of OCs among individuals with schizophrenia. Further evidence that OCs are not a proxy of familial risk was found in a number of other recent studies.[4,22–25]

OCS AS ONE "HIT" ON THE DEVELOPMENTAL TRAJECTORY TO PSYCHOSIS

Schlotz and Phillips[26] set forward various models by which pre- and perinatal complications can give rise to mental and behavioral disorders. They emphasize exposure to fetal adversity during sensitive periods of neurodevelopment and interactions between prenatal and postnatal factors as well as between prenatal and genetic factors as potentially having a causal role in the etiology of mental health outcomes in later life. In our own work, we have shown in a nest case-control study that OCs can interact with developmental delay to synergistically increase risk for schizophrenia spectrum disorders.[27] In Figure 44.1, we illustrate potential interrelationships between pre- and perinatal complications, genetic vulnerability, and environmental adversity and how these interrelationships may ultimately lead to an increased risk of psychotic disorders such as schizophrenia.

It is possible that OCs have a direct effect on fetal neurodevelopment[27]—for example, evidence suggests that perinatal hypoxia may have lasting effects on dopaminergic function. However, the same data also suggests a further indirect effect of OCs such that birth insults alter the manner, in which dopamine function is regulated by stress in adulthood.[28] Cannon et al.[29] have found that neurotrophic factors, perhaps stimulated in response to fetal distress, may be important in the etiology of schizophrenia. In this nested case-control study, assays from cord and maternal blood samples taken at delivery showed that, among schizophrenia patients, birth hypoxia was associated with a 20% decrease in brain derived neurotrophic factor (BDNF), while among the matched healthy controls birth hypoxia was associated with a 10% increase in BDNF. The deleterious effects of OCs, such as hypoxia and hyperbilirubinemia, on NMDA receptors has also been proposed as a potential mechanism.[30] Hypoxia itself has been proposed to mediate the effects of other OCs. Cannon et al.[31] found a linear relationship between the number of hypoxia causing OCs, such as abnormal fetal heart rate, third trimester heart rate, placental hemorrhaging, and risk of schizophrenia, suggesting that any association between these specific OCs and

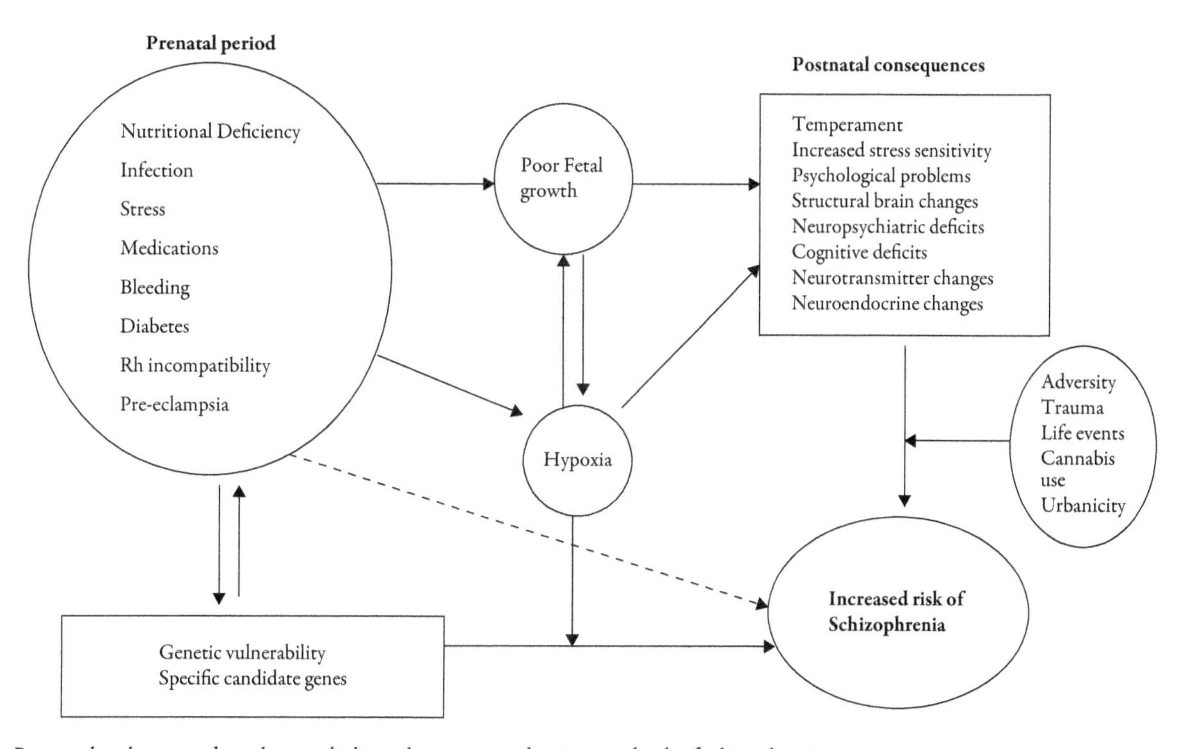

Figure 44.1 Proposed pathways and mechanism linking obstetric complications and risk of schizophrenia.

schizophrenia is accounted for by the effect of hypoxia on risk for schizophrenia. These data also suggest that hypoxia interacts with genetic susceptibility to further increase risk. The risk of schizophrenia increased with the number of hypoxia related OCs within families—given genetic vulnerability to schizophrenia, those who were exposed to hypoxia were more likely to develop schizophrenia than family members not exposed to hypoxia.

THE ROLE OF OCS IN GENE-ENVIRONMENT INTERACTIONS

The possibility of OCs interacting with underlying genetic vulnerability has been raised by several other lines of evidence.[32–35] In particular, recent findings show that many of the susceptibility genes identified for schizophrenia are affected by hypoxia.[36,37] Schmidt-Kastner et al.[36] have published a thought-provoking discussion of the possible role of hypoxia in gene regulation relevant to schizophrenia. The authors hypothesize that although some genes important for neurodevelopment are influenced by hypoxia in a physiological manner, excessive hypoxia could result in dysregulation of such genes and an abnormality of gene expression at a critical point of development. The authors propose that, since hypoxia is an element of ischemia and maternal factors may cause circulatory problems, both hypoxia and ischaemia should be considered jointly. Schmidt-Kastner et al.[36] report that about 3.5% of all genes are influenced by hypoxia and/or have vascular functions and devised a simple experiment to investigate the association of such genes with schizophrenia. They searched the literature and devised a list of currently hypothesized candidate genes for schizophrenia. They initially found that 71% of their list of candidate susceptibility genes for schizophrenia was influenced by ischemia-hypoxia and/or vascular expression. In a more stringent analysis, they examined only those genes that were mentioned in the reviews of Harrison and Weinberger (2005)[38] and Kirov et al.[39] The genes covered in these reviews were only those genes for which there was the most robust evidence. Schmidt-Kastner et al.[36] found that 8/11 of these genes (81%) met their criteria for ischemia-hypoxia regulation and/or vascular expression. In their discussion the authors note the findings of Prabakaran et al.[40] that almost 50% of the altered proteins identified by proteomics in prefrontal postmortem tissue in schizophrenia were associated with mitochondrial function and oxidative stress responses.

Nicodemus et al.[37] carried out a gene–environment interaction analysis in a family study of schizophrenia probands, siblings, and controls. Obstetric complication data was obtained by maternal recall and their severity was classified according to a standardized scale. Psychiatric diagnoses were made using a structured clinical interview. The authors then examined whether a set of schizophrenia candidate genes affected by hypoxia or involved in vascular function in the brain interacted with serious OCs to influence risk for schizophrenia. They found that 4 of the 13 genes examined showed evidence for significant serious OC-by-gene interaction. All of the OCs rated as serious had the potential to cause hypoxia—the most common were bleeding during pregnancy, extended labor, delivery problems, and respiratory distress at birth. The four genes found to interact with these OCs, namely AKT_1, BDNF, GRM3, and $DTNBP_1$, showed previous evidence of an association with schizophrenia and were affected by hypoxia. In keeping with the epidemiological findings reported earlier, the OCs identified in this study were diverse and occurred during both the prenatal and perinatal periods. This study provides evidence that serious OCs interact with genetic risk to increase risk for schizophrenia. However, any conclusions from this paper must be tempered by the fact that the study had limited power to detect interactions, tested a large number of SNPs thereby raising the possibility of false positives due to multiple testing, and used maternal recall of OCs decades after birth. One possibility of furthering this line of investigation is by enriching future samples for obstetric complications.[37]

Thus, evidence is starting to emerge that OCs may not only exert an independent effect on psychosis risk but may also exert a combined effect with underlying genetic vulnerability to psychosis. This raises the question of the nature of this combination—whether it is merely *additive*, whereby risk of illness results from a simple addition of accumulated independent risk factors, or whether it is *synergistic*, whereby the combined effects of these are greater than the sum of their individual effects on risk for psychosis Current evidence points to both additive and synergistic influences of OCs on psychosis risk.[34] However, the use of indirect measures of genetic effects, such as family history of psychotic disorder, in epidemiological samples makes finding strong evidence for a particular model of combined risk factor effects very difficult. However, both the additive and the synergistic models predict that a reduction in the incidence of OCs would have the effect of reducing overall incidence of schizophrenia.

OCS AND THE NEURODEVELOPMENTAL HYPOTHESIS OF SCHIZOPHRENIA

The concept of early life adverse events that affect neurodevelopment and interact with underlying genetic risk is central to the neurodevelopmental hypothesis of the etiology of schizophrenia. If OCs increase risk for schizophrenia by having long-lasting effects on neurodevelopment, perhaps in combination with other risk factors operating at different time points during development, then if and how OCs are associated with other markers of suboptimal neurodevelopment is of interest. Two main strands of evidence used in support of the neurodevelopmental hypothesis of schizophrenia are the presence of structural brain abnormalities preceding the onset of schizophrenia and the finding of an increased incidence of minor physical anomalies and neurological soft signs in childhood in those who later develop schizophrenia. Some but not all studies have demonstrated an association between OCs and either structural brain abnormalities or neurological abnormalities/minor physical anomalies in schizophrenia patients.

Structural brain abnormalities

Cannon et al.[41] found evidence that fetal hypoxia was associated with cortical gray matter reduction and cerebrospinal fluid increase in both schizophrenia patients and their siblings compared to healthy controls who did not have any family history of psychosis. This finding points to a gene–environment interaction in the etiology of schizophrenia. This study also reported evidence of a more complex interaction—the gene–environment effect was three times greater in those who were born small for their gestational age—indicating that OCs can interact both with underlying genetic risk and with each other to increase risk for schizophrenia. Cannon et al.[41] also found that fetal hypoxia correlated with ventricular enlargement but only within cases and not within their siblings or controls. Similarly, Falkai et al.[42] found that neonatal OCs were associated with the size of the ventricle-to brain ratio in a case-control study of schizophrenia patients, their relatives, and healthy controls with no psychiatric history. Ebner et al.[43] have found hippocampal volume reduction in those who had experienced OCs among schizophrenia patients, their first-degree relatives, and healthy controls, suggesting that OCs have an association with structural brain abnormalities that is independent of genetic susceptibility to schizophrenia. This finding replicates that of an association between OCs and hippocampal volume in sporadic schizophrenia.[44] Fearon et al.[45] have found that adolescents born very preterm or with very low birth weight showed many of the same structural brain abnormalities that are found in schizophrenia patients such as lateral ventricular enlargement and reduction in hippocampal volume.

Minor physical anomalies

In a case control study, Bramon et al.[46] showed that only schizophrenia patients who had a history of OCs had dermatoglyphic abnormalities. Fatjo-Vilas et al.[47] found in a case-control study that while a main group analysis did not show any differences between schizophrenia patients and controls in the incidence of dermatoglyphic anomalies, patients with a very low birth weight had an increased incidence of such anomalies. Similarly, Rosa et al.[48] found that among siblings discordant for schizophrenia, the affected siblings had a greater number of dermatoglyphic anomalies.

Dopamine model of psychosis

A recent model of the development of psychosis integrates features of the neurodevelopmental model with those of other models such as dopamine and cognitive models of psychosis.[49] According to this model, early life events from the prenatal period onward act on an underlying biological vulnerability of dopamine dysregulation. This model implies that prenatal and early life interventions that could reduce the incidence of exposure to environmental risk factors at this time could be of benefit in reducing risk for psychosis.

DO OCS INCREASE RISK FOR SCHIZOPHRENIA SPECTRUM DISORDERS ONLY?

Initial evidence suggested that the association between OCs and psychotic disorder is specific to schizophrenia and that no association is present between OCs and bipolar disorder.[50] Ogendhal et al.[51] using register-based data on a large Danish cohort found no evidence that OCs were related to bipolar disorder. Scott et al. [52] carried out a systematic review and meta-analysis of the effect of exposure to OCs on the risk of later developing bipolar disorder and found a pooled odds ratio of 1.1 (95% 0.76–1.35). However, OCs as a risk factor has been investigated much more extensively in relation to schizophrenia spectrum disorders compared to affective psychosis and very few studies present results on both outcomes. A recent meta-analysis from Laurens et al.[53] shows that there is evidence for OCs collectively as a risk factor for both schizophrenia spectrum disorders and affective psychosis and that there is limited evidence for any individual OC being associated with either outcome.

ARE OCS ASSOCIATED WITH PSYCHOTIC SYMPTOMS?

Psychotic symptoms are experienced not just by patients with psychiatric disorders but also by a substantial proportion of the general healthy population. In the absence of illness, these symptoms are also referred to as psychotic experiences (PEs) or subclinical psychotic symptoms. A meta-analysis by van Os et al.[54] reported a median prevalence of 5%–8% for these symptoms in the general population. Data on adolescents suggest that rates of PLEs are even higher among this age group with a meta-analysis from our group showing that the median prevalence of psychotic symptoms is 17% among children aged 9–12 years and 7.5% among adolescents aged 13–18 years.[55] Studying these nonpsychotic symptoms may therefore increase understanding of how psychotic experiences develop and potentially help elucidate etiological mechanisms underlying psychosis. Three studies have examined associations between adolescent psychotic symptoms and obstetric complications in 6,356 children from the Avon Longitudinal Study of Parents and Children.[56–58] The results are remarkably similar to findings for individuals with a diagnosis of schizophrenia. Significant associations were reported between risk of experiencing psychotic symptoms in adolescence and the following obstetric variables: maternal tobacco use and maternal alcohol use;[57] low birth weight and ponderal index;[56] maternal infection during pregnancy; maternal diabetes; need for resuscitation and 5-min Apgar score.[58] These results indicate that the same pre- and perinatal complications increase the risk both for psychotic disorder and for the risk of experiencing psychotic symptoms in childhood and adolescence. These results also provide further evidence that childhood and adolescent psychotic symptoms may lie on the causal pathway to psychotic.

ARE OCS ASSOCIATED WITH PRODROMAL SYNDROMES?

In a recent meta-analysis of environmental risk factors for psychosis in individuals at ultra-high risk for psychotic disorder due to having both psychotic symptoms and genetic risk for psychosis, there was a strong association between OCs and UHR status (odds ratio = 3.09). However, this association was for a general category of obstetric complications and no individual complication was important in the analysis[59]

CONCLUSION

The continued study of obstetric complications has the potential to inform us about the neurobiological mechanisms and pathways involved in psychosis. It is possible that focusing on those who have notable perinatal insults will help identify etiological homogeneous subgroups within the heterogeneous psychosis phenotype. It also seems likely that the genes influenced by certain obstetric complications are good candidates for the study of gene–environment interactions in psychosis. Obstetric complications may represent either the first or second "hit," in combination with underlying genetic vulnerability, for the onset and progression of a psychosis developmental trajectory. It is therefore possible that increased prenatal care would reduce risk for psychosis by reducing obstetrical risks, particularly for women with a history of psychosis whom we know are less likely to receive such care. Animal models will likely play an increasingly central role in elucidating the role of obstetric complications in the etiology of psychosis. These models have the ability to precisely determine both the nature and the timing of exposure to insult. It is only through sharpening measures of environmental insults and through gaining insight into critical windows of vulnerability that the complex etiopathogeneisis of psychosis can be elucidated.

REFERENCES

1. McNeil T. Perinatal influences in the development of schizophrenia. In: H Helmchen & FA Henn ed, Biological Perspectives of Schizophrenia, New York: John Wiley 1987:125–138.
2. Clarke MC, Harley M, Cannon M. The role of obstetric events in schizophrenia. Schizophr Bull, 2006;32:3–8.
3. Clarke MC, Kelleher I, Clancy M, Cannon M. Predicting risk and the emergence of schizophrenia. Psychiatr Clin North Am, 2012;35(3):585–612.
4. Jablensky AV, Morgan V, Zubrick SR, Bower C, Yellachich L. Pregnancy, delivery, and neonatal complications in a population cohort of women with schizophrenia and major affective disorders. Am J Psychiatry, 2005;162:79–91.
5. Bennedsen BE, Mortensen PB, Olesen AV, Henriksen TB. Congenital malformations, stillbirths, and infant deaths among children of women with schizophrenia. Arch Gen Psychiatry, 2001;58:674–679.
6. King-Hele S, Webb RT, Mortensen PB, et al. Risk of stillbirth and neonatal death linked with maternal mental illness: a national cohort study. Arch Dis Child Fetal Neonatal Ed, 2009;94(2):105–110.
7. Webb RT, Wicks S, Dalman C, et al. Influence of environmental factors in higher risk of sudden infant death syndrome linked with parental mental illness. Arch Gen Psychiatry, 2010;67:69–77.
8. Webb RT, Pickles AR, King-Hele SA, Appleby L, Mortensen PB, Abel KM. Parental mental illness and fatal birth defects in a national birth cohort. Psychol Med, 2008;38:1495–1503.
9. Suvisaari JM, Taxell-Lassas V, Pankakoski M, Haukka JK, Lönnqvist JK, Häkkinen LT. Obstetric complications as risk factors for schizophrenia spectrum psychoses in offspring of mothers with psychotic disorder. Schizophr Bull, 2013;39(5):1056–1066.
10. Padmanabhan JL, Shah JL, Tandon N, Keshavan MS. The "polyenviromic risk score": aggregating environmental risk factors predicts conversion to psychosis in familial high-risk subjects. Schizophr Res, 2017;181:17–22.
11. Johnstone EC, Ebmeier KP, Miller P, Owens DG, Lawrie SM. Predicting schizophrenia: findings from the Edinburgh High-Risk Study. Br J Psychiatry, 2005;186:18–25.
12. McNeil TF. Perinatal risk factors and schizophrenia: selective review and methodological concerns. Epidemiol Rev, 1995;17, 107–112.
13. Cannon M, Jones, P. B, Murray RM Obstetric complications and schizophrenia: Historical and meta-analytic review. Am J Psychiatry, 2002;159: 1080–1092.
14. Byrne M, Agerbo E, Bennedsen B, Eaton WW, Mortensen PB. Obstetric conditions and risk of first admission with schizophrenia: a Danish national register based study. Schizophr Res, 2007;97: 51–59.
15. Sorensen, H. J, Mortensen EL, Reinisch JM, Mednick SA. Do hypertensions and diuretic treatment in pregnancy increase the risk of schizophrenia in offspring? Am J Psychiatry, 2003;160:464–468.
16. Sorensen HJ, Mortensen EL, Reinisch JM, Mednick SA. Association between prenatal exposure to analgesics and risk of schizophrenia. British J Psychiatry, 2004;160:366–371.
17. Insel BJ, Brown AS, Bresnahan MA, Schaefer CA, Susser ES. Maternal-fetal blood incompatibility and the risk of schizophrenia in offspring. Schizophr Res, 2005;80:331–342.
18. Hollister JM, Laing P, Mednick SA. Rhesus incompatibility as a risk factor for schizophrenia in male adults. Arch Gen Psychiatry., 1996;53(1):19–24.
19. Insel BJ, Schaefer CA, McKeague IW, Susser ES, Brown AS. Maternal iron deficiency and the risk of schizophrenia in offspring. Arch Gen Psychiatry, 2008;10:1136–1144.
20. Fineberg AM, Ellman LM, Buka S, Yolken R, Cannon TD. Decreased birth weight in psychosis: influence of prenatal exposure to serologically determined influenza and hypoxia. Schizophr Bull, 2013:39(5):1037–1044.
21. Ellman LM, Hutttunen M, Lonnqvist J, Cannon TD. The effects of genetic liability for schizophrenia and maternal smoking during pregnancy on obstetric complications. Schizophr Res, 2007;93:229–236.
22. van Erp TG, Saleh PA, Rosso IM, et al. Contributions of genetic risk and fetal hypoxia to hippocampal volume in patients with schizophrenia or schizoaffective disorder, their unaffected siblings, and healthy unrelated volunteers. Am J Psychiatry, 2002;159:1514–1520.
23. Bersani G, Taddei I, Manuali G, et al. Severity of obstetric complications and risk of adult schizophrenia in male patients: a case-control study. J Matern Fetal Neonatal Med, 2003;14:35–38.
24. Nilsson E, Stalberg G, Lichtenstein P, Cnattingius S, Olausson PO, Hultman CM. Fetal growth restriction and schizophrenia: a Swedish twin study. Twin Res Hum Genet, 2005;8:402–408.
25. Boks MP, Selten JP, Leask S, Catelein S, van den Bosch RJ. Negative association between a history of obstetric complications and the number of neurological soft signs in first-episode schizophrenic disorder. Psychiatry Res, 2007;149:273–277.
26. Clarke MC, Tanskanen A, Huttunen M, et al. Increased risk of schizophrenia from additive interaction between infant motor developmental delay and obstetric complications: evidence from a population-based longitudinal study. Am J Psychiatry, 2011; 1295–1302.
27. Boog G. Obstetrical complications and further schizophrenia of the infant: A new methodological threat to the obstetrician? J Gynecol Obstet Biol Reprod (Paris), 2003;32:720–727.
28. Boksa P, El-Khodor BF. Birth insult interacts with stress at adulthood to alter dopaminergic function in animal models: possible

implications for schizophrenia and other disorders. Neurosci Biobehav Rev. 2003;27(1–2):91–101.
29. Cannon TD, Yolken R, Buka S, Torrey EF. Collaborative study group on the perinatal origins of severe psychiatric disorders. Decreased neurotrophic response to birth hypoxia in the etiology of schizophrenia. Biol Psychiatry, 2008;64:797–802.
30. Dalman C. Obstetric complications and risk of schizophrenia: an association appears undisputed, yet mechanisms are still unknown. Lakartidningen, 2003;100:1974–1979.
31. Cannon, T. D, Rosso IM, Hollister JM, Bearden CE, Sanchez HT. A prospective cohort study of genetic and perinatal influences in the etiology of schizophrenia. Schizophr Bull, 2000;26:249–256.
32. Mueser KT, McGurk SR. Schizophrenia. Lancet, 2004;19:2063–2072.
33. Devlin B, Klei L, Myles-Worsely M, et al. Genetic liability to schizophrenia Oceanic Palau: a search in the affected and maternal generation. Hum Genet, 2007;121:675–684.
34. Mittal VA, Ellman LM, Cannon TD. Gene–environment interaction and co-variation in schizophrenia: The role of obstetric complications. Schizophr Bull, 2008;34:1083–1094.
35. Fatemi SH, Folsom TD. The neurodevelopmental hypothesis of schizophrenia, revisited. Schizophr Bull, 2009;35:528–548.
36. Schmidt-Kastner R, van Os J, Steinbusch H, Schmitz C. Gene regulation by hypoxia and the neurodevelopmental origin of schizophrenia. Schizophr Res, 2006;84(2–3):253–271.
37. Nicodemus KK, Marenci S, Batten AJ, et al. Serious obstetric complications interact with hypoxia-regulated vascular-expression genes to influence schizophrenia risk. Mol Psychiatry, 2008;18:180–184.
38. Harrison PJ, Weinberger DR. Schizophrenia gene, gene expression and neuropathology: On the matter of their convergence. Mol Psychiatry, 2005;10:804.
39. Kirov G, O'Donovan MC, Owen MJ. Finding schizophrenia genes. J Clin Invest, 2005;115:1440–1448.
40. Prabakaran S, Swatton JE, Ryan MM, et al. Mitochondrial dysfunction in schizophrenia: Evidence for compromised brain metabolism and oxidative stress. Mol Psychiatry, 2004;9: 684–697.
41. Cannon TD, van Erp TG, Rosso IM, et al. Fetal hypoxia and structural brain abnormalities in schizophrenic patients, their siblings, and controls Arch Gen Psychiatry, 2002;59(1):35–41.
42. Falkai P, Schneider-Axmann T, Honer WG, et al. Influence of genetic loading, obstetric complications and premorbid adjustment on brain morphology in schizophrenia: an MRI study. Eur Arch Psychiatry Clin Neurosci, 2003;523:92–99.
43. Ebner F, Tepest R, Dani I, et al. The hippocampus in families with schizophrenia in relation to obstetric complications. Schizophr Res, 2008;104:71–78.
44. Stefanis N, Frabgou S, Yakeley J, et al. Hippocampal volume reduction in schizophrenia: effects of genetic risk and pregnancy and birth complications. Biol Psychiatry, 1999;46: 697–702.
45. Fearon P, O'Connell P, Frangou S, et al. Brain volumes in adult survivors of very low birth weight: a sibling-controlled study. Pediatrics, 2004;114:367–371.
46. Bramon E, Walshe M, McDonald C, et al. Dermatoglyphics and schizophrenia: a meta-analysis and investigation of the impact of obstetric complications upon a-b ridge count. Schizophr Res, 2005;75:399–404.
47. Fatjo-Vilas M, Gourion D, Campanera S, et al. New evidences of gene and environment interactions affecting prenatal neurodevelopment in schizophrenia-spectrum disorders: a family dermatologlyphic study. Schizophr Res, 2008;103:209–217.
48. Rosa A, Cuesta MJ, Peralta V, et al. Dermatoglyphic anomalies and neurocognitive deficits in sibling pairs discordant for schizophrenia spectrum disorders. Psychiatry Res, 2005;137, 215–221.
49. Howes OD, Murray RM. Schizophrenia: an integrated sociodevelopmental-cognitive model. Lancet, 2014;10;383(9929):1677–1687
50. Murray RM, Sham P, van Os J, Zanelli J, Cannon M, McDonald CA. A developmental model for similarities and dissimilarities between schizophrenia and bipolar disorder. Schizophr Res, 2004;71:405–416.
51. Ogendahl BK, Agerbo EM, Byrne M, Licgt RW, Eatin WW, Mortensen, PB. Indicators of fetal growth and bipolar disorder: a Danish national register-based study. Psychol Med, 2006;36: 1219–1224.
52. Scott J, McNeill Y, Cavanagh J, Cannon M, Murray R. Exposure to obstetric complications and subsequent development of bipolar disorder: systematic review. Br J Psychiatry, 2006;189: 3–11.
53. Laurens KR, Luo L, Matheson SL, et al. Common or distinct pathways to psychosis? A systematic review of evidence from prospective studies for developmental risk factors and antecedents of the schizophrenia spectrum disorders and affective psychoses. BMC Psychiatry, 2015;25;15:205.
54. van Os J, Linscott RJ, Myin-Germeys I, Delespaul P, Krabbendam, L. A systematic review and meta-analysis of the psychosis continuum: evidence for a psychosis proneness-persistence-impairment model of psychotic disorder. Psychol Med, 2009;39:179–195.
55. Kelleher I, Connor D, Clarke MC, Devlin N, Harley M, Cannon M. Prevalence of psychotic symptoms in childhood and adolescence: a systematic review and meta-analysis of population-based studies. Psychol Med, 2012;42(9):1857–1863.
56. Thomas K, Harrison G, Zammit S, et al. Association of measures of fetal and childhood growth with non-clinical psychotic symptoms in 12-year-olds: the ALSPAC cohort. Br J Psychiatry, 2009;194:521–526.
57. Zammit S, Odd D, Horwood A, et al. Investigating whether adverse prenatal and perinatal events are associated with non-clinical psychotic symptoms at age 12 years in the ALSPAC birth cohort. Psychol Med, 2009;39:1457–1467.
58. Zammit S, Thomas K, Thompson A, et al. Maternal tobacco, cannabis, and alcohol use during pregnancy and risk of adolescent psychotic symptoms in offspring. Br J Psychiatry,2009;195:294–300.
59. Fusar-Poli P, Tantardini M, De Simone S, et al. Deconstructing vulnerability for psychosis: meta-analysis of environmental risk factors for psychosis in subjects at ultra high-risk. Eur Psychiatry, 2017;40:65–75.

45

THE ROLE OF EARLY LIFE EXPERIENCE IN PSYCHOSIS

Richard P. Bentall

Socioenvironmental risk factors for psychosis can be broadly divided into two main groups: those that impact on populations and those that impact on the individual. The former group includes poverty,[1] economic inequalities,[2] urban environments,[3] migration[4] and membership of a minority ethnic group.[5] In the case of many of these risk factors, robust findings have been supported by meta-analyses, and the relevant literatures are reviewed in other chapters in this volume.

The second group of socioenvironmental factors, which are the focus of this chapter, have their impact at a personal level. These include nonoptimal characteristics of the family environment and adverse experiences during childhood such as maltreatment by caregivers, peers, or others. Historically, the role of these risk factors has been highly contested. In this chapter, I review what is known about their association with psychosis and consider whether these associations reflect true causal effects. As an advance caveat, in much of what follows I discuss outcomes using terms such as "psychosis" and "schizophrenia," but these concepts are problematic. There is currently considerable debate about the utility of these broad diagnostic categories;[6–9] although these debates are beyond the scope of the current review, I return to them briefly at the end.

METHODOLOGICAL CONSIDERATIONS

Before proceeding, it will be helpful to quickly consider two important methodological issues.

CAUSALITY

Philosophical treatments of causation since the days of David Hume have usually emphasized the regular and necessary relationship between putative causes and effects (constant conjunction) together with the implications counterfactual conditions (if X had not happened, would Y have occurred?).[10] Modern researchers have attempted to operationalize these concepts statistically but mindful that correlation does not prove causality and that observed relationships could be spurious, have often avoided talk of causation altogether in favor of muted language about "associations" and "risk factors."

Of course, the very concept of spurious correlation acknowledges that nonspurious correlations reflect genuine causal relationships.[11] In the case of the social environment, the task of sorting the nonspurious from the spurious is challenging because environmental variables can rarely be manipulated experimentally. Researchers must therefore use observational data to make inferences about counterfactual conditions (what would have happened had individuals not been exposed to the putative risk factor). Prospective data are usually considered to be stronger than retrospective data on the grounds that cause must precede effect and because of doubts about the accuracy of respondents' memories of what happened to them in childhood. In fact, cross-sectional and prospective designs are equally vulnerable to confounding by variables that influence both exposure to adversity and mental health. Moreover, studies in which psychiatric patients' retrospective reports have been checked against other sources of information suggest that they are usually accurate.[12]

Mathematical strategies for inferring causation from observational data have evolved considerably over recent decades[13] but both traditional and more modern approaches involve statistical adjustment for potential confounding variables. It is important to recognize that the selection of confounders to be controlled for should be based on an understanding of the likely causal relationships between risk factor, outcome and the confounders.[14] Of course, it is rare that all potential confounders can be identified and, therefore, some degree of uncertainty always pertains to any estimate of causality. A pragmatic approach, developed during the debates about the effects of smoking during the middle years of the last century, involves devising causal checklists, the most famous of which was proposed by the British statistician Austin Bradford Hill[15] (see Box 45.1). Hill was clear that no one criterion on his list trumps any of the others, but our confidence that a true causal relationship has been detected increases the more that the criteria are met.

HERITABILITY AND ENVIRONMENT

Heritability estimates of around 80% have been reported in psychosis research[16] and have often been wrongly interpreted as implying that there is insufficient room in the remaining variance for environmental factors to play a crucial role.[17] For this reason, it has sometimes been assumed that any apparently strong association between traumatic life events and mental illness can probably be attributed to genes influencing both exposure to adversity and the development of symptoms.[18]

In fact, heritability estimates are partial correlation coefficients that apply to populations and not to individuals, and

Box 45.1 HILL'S CRITERIA FOR INFERRING CAUSALITY FROM EPIDEMIOLOGICAL DATA

1. The association between a risk factor and the disease should be strong.
2. The data should be consistent across studies.
3. There should be a specific association between the risk factor and type of disease.
4. The risk factor should precede the disease.
5. There should be what Hill described as a biological gradient but which is now more often described as a dose-response relationship (the higher the dose—for example the more cigarettes smoked—the greater the risk of disease).
6. There should be plausible mechanisms to explain the association (in the case of lung cancer, the chemicals in tobacco smoke damage cells in the lung).
7. The epidemiological data should be consistent with other kinds of evidence (e.g., experiments with animals).
8. The effects should be reversible (if you stop smoking the risk of lung cancer decreases).
9. There should not be more plausible explanations (as far as possible, confounders have been eliminated).

so it is wrong to interpret them as estimates of the magnitude of causation. They assume an additive model in which the total variance in a trait is the sum of the variance attributable to genes and the variance attributable to the environment. In this kind of model, heritability is inevitably high in populations in which environmental variance is low (in a world in which everyone smokes exactly 20 cigarettes a day the heritability of lung cancer would approach 100% and yet it would still be reasonable to single out cigarettes a major cause of lung cancer). This artifact may help to explain why the heritabilities of IQ[19] and common psychiatric disorders[20] are greater in middle-class populations (where environmental variance is likely smaller) than in working-class populations.

Gene × environment interactions are also unaccounted for in standard heritability estimates. In the simplest case, in which genes increase the probability of exposure to an important environmental factor, it is possible for traits to be substantially influenced by environmental factors even if heritability approaches 100%.[21] In these kinds of scenarios, the decision about which effect is causal may reflect broader theoretical considerations; genes regularly preceding illness (constant conjunction) may be taken to imply a causal role, but if illness is contingent on the environmental exposure (without exposure the illness would not have occurred) a different interpretation might be appropriate.

Ideally, all studies of the antecedents of psychosis would consider both genes and environment. Unfortunately, with a few notable exceptions to be considered later, researchers have rarely measured both types of variables at the same samples.

INDIVIDUAL SOCIAL-ENVIRONMENTAL RISK FACTORS FOR PSYCHOSIS

FAMILY ENVIRONMENT

Over the past half century, studies of the influence of the family environment in psychosis have mostly focused on the effects of familial expressed emotion (EE) on risk of relapse once illness has occurred. Expressed emotion has been defined in terms of critical, hostile, or overcontrolling attitudes and behavior of family members toward the individual who is ill,[22] and there is substantial evidence from prospective studies that these kinds of attitudes are associated with a high risk of relapse.[23,24]

Despite this robust finding, the possible role of the family environment as an antecedent factor in psychosis has been neglected in recent years. This neglect has been in part a reaction to simplistic and highly stigmatizing theories of an earlier era, which attributed schizophrenia in offspring to victimization by "schizogenic mothers."[25] However, another deterrent has been the complexity of family relationships, in which the influences of family members on each other are often bidirectional and change with time. This complexity is evident in the findings of the UCLA high-risk study,[26] in which EE assessed by interviewing parents, affective style (an analogous measure) observed in family interactions, and communication deviance (discussed later) were measured in the parents of children attending a child guidance clinic for nonpsychotic problems (64 families, mean age of child = 15 years) who were followed-up for a minimum of 15 years (54 families were retained at the endpoint). Although all three family variables individually and jointly predicted a DSM-III diagnosis of schizophrenia at follow-up, subsequent analyses of family interactions found that the negativity of the child toward the parent was a strong predictor of the parent's negative affective style,[27] suggesting that at least some of the association between child and parental behavior could be explained by the evocative effect of the child's challenging behavior on the parent.

More recent researchers have attempted to determine the impact of high-EE in the family members of individuals with an at-risk mental state (ARMS) for psychosis. One study found lower EE in the families of ARMS patients compared to the families of offspring with active psychosis,[28] suggesting that EE develops in families in response to illness. However, two later studies found no difference between these two groups[29,30] and a longitudinal study found that ARMS patients living in high EE households exhibited worsening symptoms over time compared to those living in low-EE households.[31]

The case for a causal family influence is stronger for one of the other family variables measured in the UCLA high-risk study: communication deviance (CD). This term was introduced by Lyman Wynne and Margaret Singer[32] to describe a form of interfamilial communication that is vague, fragmented, and contradictory. The concept is multidimensional and includes contorted language (e.g., "This man is in the process of thinking of the process of becoming a doctor"), ambiguous referents (e.g., "Kid's stuff that's one thing, but something else is different too"), and problems at the level of pragmatics such as nonverbal behavior that disrupts communication. Early studies used a scoring system developed by Singer and Wynne[33] to classify speech collected from parents while they completed projective tests such as the Rorschach; hence, speech was recorded in the absence of the affected child. Later studies have used different coding methods, sometimes scoring parental speech during actual interactions with their children, e.g.,[34] sometimes using concepts closely aligned with CD such as communication disturbance (defined in terms of vague or missing linguistic references, ambiguous word meanings, and structural unclarities),[35] and sometimes prospective designs.[26]

The most consistent finding is that parents of offspring with psychosis show a high prevalence of CD. In the largest case-control study in the field, Wynne and colleagues[36] assessed 228 parents of mentally ill and healthy children, stratified according the severity of offspring illness (healthy, neurotic disorders, personality disorders, remitting schizophrenia, and unremitting schizophrenia). Parental CD predicted the severity of offspring diagnosis, which was interpreted as a dose-response effect. Many replications have been reported by research teams that were not involved in the development of the concept[35,37] and in countries other than the United States, where the earliest studies were conducted.[38–40] A recent meta-analysis of studies comparing CD in the parents of psychotic offspring and the parents of nonpsychotic controls,[41] retrieved 19 case-control studies and 1 prospective study (n = 1,753 parents), finding a large pooled effect size of Hedge's $g = 1.44$ (SE = 0.27; 95% CI: 0.92–1.97; $p < .001$) with significant heterogeneity; exclusion of one study by Wynne with an unusually large effects size reduced the overall effect to a still large $g = 0.97$ (SE = 0.11; 95% CI: 0.76–1.18; $p < .001$). Subgroup analyses selecting studies with different kinds of controls (e.g., parents of children with depression vs. parents of healthy children), or according to the methods used to elicit speech from parents did not materially influence the findings.

Interestingly, the effect size for mothers ($g = 0.89$; k = 7; SE = 0.18; 95% CI: 0.54–1.24) was higher than the effect size for fathers ($g = 0.39$; k = 6; SE = 0.16; 95% CI: 0.07–0.7), suggesting an environmental effect involving the quality of caregiving. Wynne[42] proposed that children learn to sustain foci of attention and to derive meaning from the world through communication with their caregivers. Hence, in children who are genetically vulnerable to schizophrenia, he argued, CD would impair cognitive and affective development.

Studies conducted by Wynne and his team found associations between parental CD and anxiety[43] and poor social competence[44] in 7- to 10-year-old children. In a recent British study, CD measured in healthy mothers expecting their first child was found to predict maternal responsivity to the child's distress at age 6 months.[45] On the one hand, the fact that CD was measured in advance of childbirth excludes any possibility of an evocative affect. On the other, the fact that CD was associated with mother's emotional responsiveness during a preverbal period of the child's life suggests that the construct may index a wider range of maternal traits than Wynne originally conceived.

ADVERSE CHILDHOOD EXPERIENCES

Outside of the domain of psychosis, considerable efforts have been made to understand the impact of traumatic childhood experiences on physical and mental health. Many of these studies, too numerous to review here, have focused on specific types of adversity, for example sexual abuse, and specific outcomes, for example depression. A notable exception is the Adverse Childhood Experiences (ACEs) Study, conducted by Vincent Felitti and Robert Anda, which examined a very large primary care population (N ≈ 19,000 in San Diego, California, and used a brief but subsequently widely used screening questionnaire which attempted to capture exposure to psychological abuse (e.g., "Did a parent or other adult in the household often or very often swear at, insult, or put you down?"), physical abuse ("Did a parent or other adult in the household often or very often push, grab, shove, or slap you?"), sexual abuse (e.g., "Did an adult or person at least 5 years older ever attempt oral, anal, or vaginal intercourse with you?") and various indices of less than optimal childrearing experiences. Consistent with previous research, the investigators found that ACEs were common, with about one in five reporting sexual abuse. Importantly, ACEs often co-occurred; about two-thirds reported at least one form of adversity and 12.5% reported at least four. Cumulative ACEs scores were associated with depression and suicide, with factors associated with reduced lifespan such as smoking and obesity, and with actual risk of diabetes, heart disease, stroke, and cancer.[46] These findings have been replicated elsewhere, for example in the United Kingdom[47] and a recent meta-analysis of 37 studies confirmed that multiple ACEs are associated with a wide range of negative physical and mental health outcomes.[48]

It is only recently that researchers have focused on the possible association between ACEs and psychosis. An influential 2005 review by John Read and his colleagues[9] compiled data from a large number of mostly uncontrolled studies. The lifetime prevalences of sexual abuse in patients were estimated to be 47.7% for females and 28.3% for males, with corresponding figures of 47.8% and 50.1% for physical abuse.

Subsequent studies have included cross-sectional epidemiological studies, longitudinal studies in which children experiencing adversity have been identified and then followed-up, and case (patient) vs. control studies. For example, a British epidemiological study[50] found an approximately 10-fold increased risk of psychosis following child sexual abuse and a longitudinal follow-up of the mental health records of Australian children known to legal services because of experiences of abuse found an approximately doubling of the risk of psychosis compared to age and sex-matched unabused controls.[51]

Some studies have examined bullying by peers as a potential risk factor. For example, researchers in London[52] compared first-episode psychosis patients and matched controls for reports of bullying occurring more than 5 years before entry into the study, and reported significantly more reports from the patients. In a study that uniquely examined potential bidirectional relationships between trauma and symptoms as well as the reversibility of the effects of trauma, Kelleher and his colleagues in Ireland[53] assessed psychotic experiences and experiences of bullying in a large cohort of adolescents who were followed-up one year later; there was a significant effect of psychosis at baseline on subsequent bullying (suggesting that children with preexisting psychotic traits are at increased risk of victimization) but this did not negate the larger predictive effect of bullying on later psychotic experiences, for which a dose-response effect was observed. In children who reported bullying at baseline but not at follow-up, there was an associated reduction in psychotic experiences.

Finally, some studies have considered loss of a parent at an early age, but these have been less consistent. For example, Morgan and his colleagues[54] found a threefold increased risk of early separation in first-episode psychotic patients in the United Kingdom compared to healthy controls but a case-control study in Japan[55] was unable to replicate this finding.

In a 2012 meta-analysis, Varese and colleagues[56] synthesized data on sexual abuse, physical abuse, emotional abuse, neglect, death of a parent, and bullying by peers from 18 case vs. control studies (n = 2,048), 10 prospective studies (n = 41,803) and 8 epidemiological studies (n = 35,546; see Figure 45.1). Overall, childhood trauma was associated with a significantly increased risk of psychosis in adulthood (OR = 2.78; 95% CI = 2.34–3.31). The effects were remarkably similar for the different study designs (case vs. control: OR = 2.72; 95% CI: 1.90–3.88; prospective studies: OR = 2.75; 95% CI: 2.17–3.47; epidemiological studies: OR = 2.99; 95% CI: 2.12–4.20) and there was an overall population attributable fraction (an estimate of

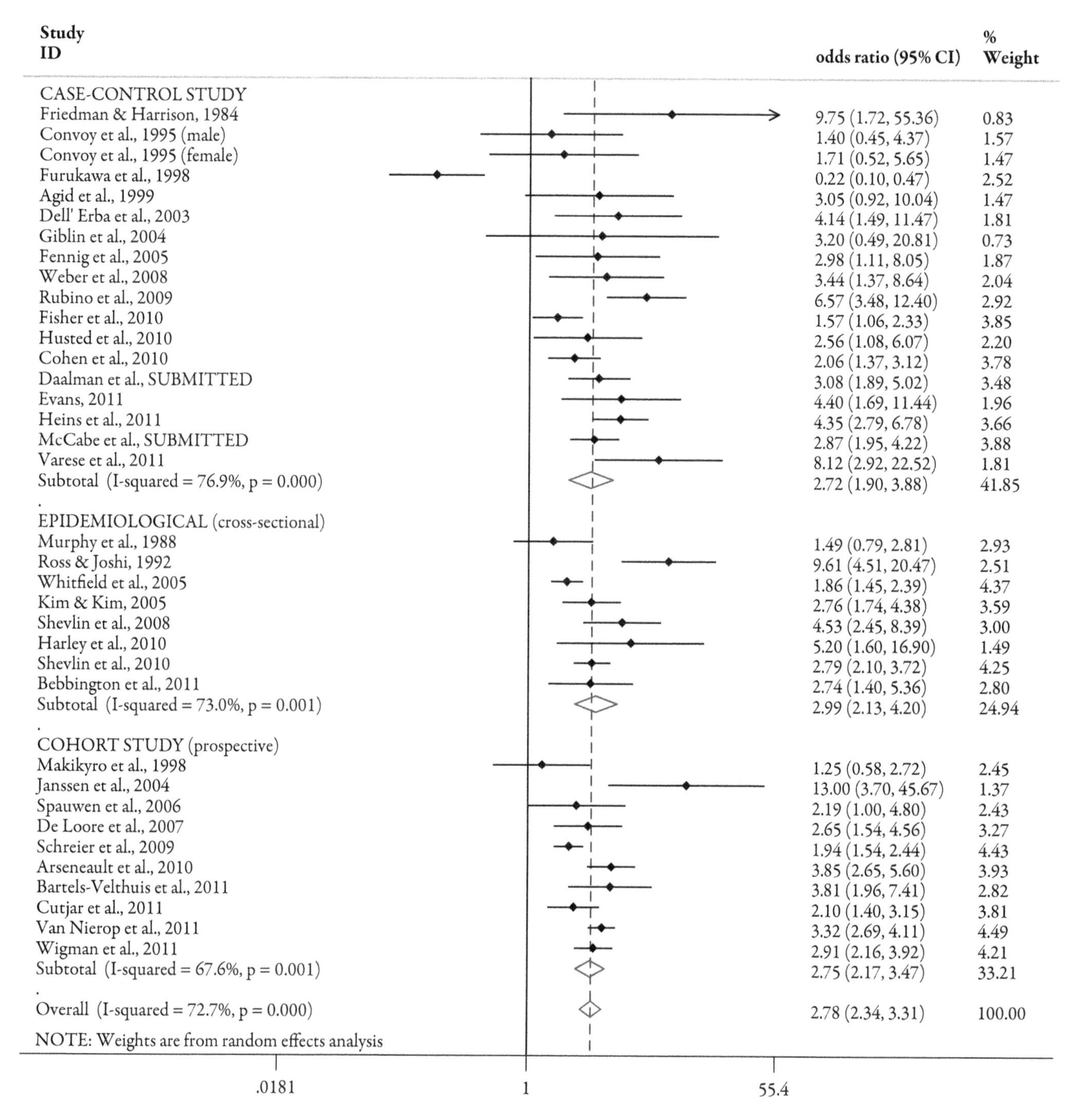

Figure 45.1 Forrest plot showing the relationship between childhood trauma and psychosis.
From Varese et al. 2012.[56]

the proportion of patients who would not have developed psychosis without exposure to the risk factor) of 33%.

Although there was statistically significant heterogeneity in the data, all the studies but one showed an effect in the expected direction. The exception was the Japanese study of the association between death of a parent and psychosis mentioned earlier[55] (including or excluding this study had little impact on the overall effect). Interestingly, when separate odds ratios were calculated for each of the different kinds of trauma, parental death was the only one that was nonsignificant. Seven epidemiological studies (including one from the ACEs series,[57] two longitudinal cohorts and one patient vs. control study had been interrogated for dose-response effects by the original researchers and a relationship had been found between cumulative trauma and psychosis risk in all but one. The exception was the only case vs. control study in this group, which used data collected from first episode patients and controls in the United Kingdom[58] and that may have been less sensitive to dose-response effects than the larger population-based and longitudinal studies.

A meta-analysis by Matheson and colleagues[59] published at about the same time, compared the strength of the association between ACEs (sexual abuse, physical abuse, and neglect) and illness in patients diagnosed with schizophrenia spectrum disorder, healthy controls, and patients with a range of other diagnoses. A significant difference was found in the seven studies that compared psychotic patients with healthy controls (n = 798 vs. 883; OR = 3.60, 95% CI: 2.08–6.23). Schizophrenia spectrum disorder patients were more likely to report childhood trauma than patients diagnosed with anxiety disorders (7 studies; n = 199 vs. 580; OR = 2.54; 95% CI: 1.29–5.01) and were also considerably less likely to report trauma than patients with a dissociative disorder diagnosis (n = 59 vs. 76; OR = 0.03: 95% CI: 0.01–0.15). However, no differences in reporting rates were found when comparisons were made with patients with affective psychoses, depressive disorder, or other psychoses.

Several studies that also have shown an association between traumatic experiences and psychosis have been published since these meta-analyses, and two based on the World Mental Health Surveys are worth noting. McGrath and colleagues[60] reported an analysis of the relationship between traumatic experiences in both childhood and adulthood and hallucinations and delusions in 24,464 respondents from 16 countries. The overall association of trauma exposure with probable psychosis was OR = 3.1, (95% CI: 2.7–3.7), and there was a dose-response effect between cumulative trauma so that, in those experiencing five or more exposures to trauma, the risk was OR = 7.6 (95%CI: 6.1–0.5). When individual traumas were examined, all types of personal trauma (e.g., rape, witnessing a death, being attacked) were associated with psychosis but most collective traumas (e.g., living in a war zone) were not. In a parallel analysis that focused only on ACEs,[61] it was reported that traumas reflecting maladaptive family functioning, including abuse and neglect, were most strongly associated with psychosis, and sexual abuse was especially strongly associated with the onset of psychotic experiences during childhood (OR = 8.5, 95% CI: 3.6–20.2). The population attributable risk associated with ACEs was estimated to be about a third. Overall, the findings from these reports are remarkably consistent with the findings from the Varese et al. meta-analysis.

A further related meta-analyses is also worth mentioning here. Palmier-Claus and colleagues[62] examined 13 case vs. control studies (n = 1,259 vs. 1,118) and 6 epidemiological studies (n > 2 million) of the association between childhood trauma (neglect, abuse, bullying, or parental loss) and bipolar disorder. There was an overall significant effect (OR = 2.63; 95% CI: 2.00–3.47). As in the earlier meta-analysis by Varese et al. (2012), similar effect sizes were observed for the two study designs (patients vs. control OR = 2.88; 95% CI: 2.04–4.06; epidemiological studies OR = 2.24; 95% CI: 1.40–3.57) and, as in Matheson et al.'s[59] meta-analysis, no difference was observed between the effect size for bipolar patients and that for schizophrenia spectrum patients.

Taken as a whole, the evidence, then, suggests a strong and consistent association between childhood trauma and psychosis broadly defined. Given this finding, it might be expected that some patients with psychosis will also meet the criteria for post-traumatic stress disorder (PTSD). This question is not as simple as it might at first appear because it is possible that PTSD and psychosis represent different and to some extent mutually exclusive outcomes of trauma exposure (perhaps depending on other facts such as genetic risk). It should also be noted that trauma in adulthood (not reviewed in this chapter), has also been linked to psychosis (although the evidence is not as extensive and robust as that on ACEs[63]) and so might also contribute to comorbidity between psychosis and PTSD. In a systematic review of 34 studies, Seow and colleagues[64] found a wide range of estimates of the prevalence of PTSD diagnoses in patients diagnosed with schizophrenia; however, the majority reported prevalence rates in the 20%–30% range.

GENETIC CONFOUNDING

As discussed earlier, it has often been assumed that the lion's share of causation in the development of psychosis must be attributed to genes and that apparent associations between psychosis and adversity may be inflated by genetic confounding.[18]

The possible role of genes in the association between parental communication deviance and psychosis was addressed in a study carried out in Finland, in which the adopted away offspring of mothers diagnosed with schizophrenia and of mothers without a diagnoses were followed-up and CD was measured in the adopting families (in most cases the adoptions had taken place before the child was 5 years old). The adoptees varied considerably in their age at the time of entry into the study, the youngest being 11 years and the oldest being 56 years.[65] Numerous reports with different subsamples and at different follow-up points have been published from this cohort. However, a consistent finding has been that adoptive family functioning interacts with genetic risk. For example, in a comparison of 128 index children matched with 128 controls, it was found that, as expected, the genetically at-risk offspring had a higher risk than the controls of both psychosis (10 cases vs. 1) and nonpsychotic disorders, but psychosis was only observed in genetically at-risk children who were raised by parents with CD.[66] In a later study with a wider sample, with the children followed up to a mean age of 44 years and employing a broader, summary measure of family functioning, evidence of the same kind of interaction between genetic and environmental risk was also found, leading the researchers to conclude not only that children at genetic risk of psychosis are especially sensitive to dysfunctional family relationships but also that healthy family relationships are protective.[67]

A strikingly different picture has emerged from studies that have examined the interplay between genetic risk for psychosis and adverse childhood experiences. Arseneault and colleagues[68] assessed psychotic experiences at age 12 in a large (n = 2,232) longitudinal cohort of twin pairs whose parents had been earlier interviewed about adverse events; psychotic experiences were related to trauma in a dose-response fashion and this effect was independent of genetic liability. In a smaller study of adult twins carried out in Spain,[69] Alemany and colleagues found that discordance in psychotic experiences between twin pairs was entirely accounted for by discordance in trauma experiences. Very similar findings were also obtained in a study of adolescent and adult twins in Belgium.[70]

This picture has been replicated in studies of patients. In a study carried out in the Netherlands, Heins and colleagues[71] assessed childhood trauma in psychotic patients and their siblings, finding that discordance for psychosis was related to trauma history. In a sample of first episode psychotic patients and healthy controls in London, Trotta and colleagues[72] found that the association between trauma experiences and psychosis was not confounded by genetic liability as assessed from family psychiatric history. Finally, the same researchers[73] assessed genetic liability to psychosis in first-episode patients and controls directly from DNA using polygenic load scores (a summary score based on counting known genetic markers of risk[74]) and again, risk attributable to genes was found to be independent of risk associated with childhood adversity.

Hence, the available evidence suggests that the interplay between genes and environmental factors in psychosis depends on the specific environmental factors. On the basis of a single but impressive study, family communication moderates genetic risk, so that it is protective when communication is optimal but increases the likelihood that genetic liability will be translated into symptoms when it is not. The effects of traumatic childhood experiences, on the other hand, appear to contribute to the likelihood of severe mental illness independently of genes.

EVALUATION AGAINST CAUSALITY CRITERIA

It is clear from the evidence reviewed earlier that most of Hill's nine criteria for causality are met by both the data on family environment and the data on childhood adverse experiences. The associations are strong (criterion 1); the findings are consistent (2); both types of risk factors clearly precede the onset of illness (4); there is evidence of a dose-response effect in both cases (5); the data are coherent with other evidence, namely the prevalence of PTSD symptoms in psychotic patients (7); the effects appear to be at least partially reversible (8); and other factors (genes) have been, for the most part, eliminated as an explanation (9). In the concluding sections of this chapter I briefly address the remaining two criteria—specificity of association (criterion 3) and the nature of the mechanisms linking adversity to psychosis (6). These two issues are intertwined because different symptoms might be expected to be linked to different kinds of adversity only if different mechanisms are responsible for the symptoms. Before proceeding further, it is worth noting that specificity is arguably the least important of Hill's causal criteria; after all, it is widely accepted that smoking tobacco causes not only lung cancer but also a variety of other ailments including bronchitis and heart disease.

In the case of psychosis, this issue is complicated because of persisting doubts about the validity of diagnoses such as "schizophrenia." As noted at the outset, there are currently a plethora of proposals about how to replace these diagnostic categories. Some researchers have attempted to identify broad empirically defined spectra of disorders, specifically the internalizing (mood), externalizing (aberrant and impulsive behavior) and psychosis (which groups all forms of psychosis into a single category) spectra.[75] Other researchers, on the other hand, have argued that there may be specific mechanisms underlying symptoms such as hallucinations and delusions,[6] and that symptoms may coalesce into syndromes not because they have common underlying causes but because symptoms can causally influence each other.[8] A possible unifying framework attempts to capture all forms of psychopathology within a hierarchical structure with broad spectra near the apex, syndromes occupying a middle level and symptoms at the bottom.[9] A very recent proposal, based on analyses of how symptoms cluster together within this kind of structure, is that there is a general disposition to psychopathology underlying all forms of mental illness, sometimes referred to simply as "p."[76]

Within this kind of hierarchical framework, it is possible to ask whether specific types of adversity are more associated with symptoms or the general factor. In this context, it is important to note that, even if there is an expectation of some effects at the symptom level, it is unlikely that there will be a simple lock-and-key type relationship between a specific type of adversity (say, sexual abuse) and a specific kind of symptom (say, hallucinations) because the various types of adversity are likely to have multiple and overlapping psychological effects.[77]

Some studies have pointed to specific associations at the symptom level. For example, several studies have reported associations between communication deviance in parents and thought disorder in children,[34,7}] and others, mainly conducted by myself and colleagues, have reported a particularly strong association between sexual abuse and hallucinations in population samples,[79,80] prisoners,[81] schizophrenia patients,[82] and patients with a diagnosis of bipolar disorder.[83] Some of the same studies have suggested that indices of disrupted early attachment relationships, for example neglect[84] or being raised in a children's home,[80,81] are more closely linked to paranoia. These associations can potentially be explained by the potency that these different types of adversities have to influence different psychological mechanisms.

On the other hand, the fact that childhood adversity is equally or near-equally associated with a wide range of diagnoses raises the possibility that they have their affect by influencing mechanisms that play a role in all forms of mental illness, perhaps emotional regulation difficulties linked with the "p" construct. Isvouranu and colleagues[85] used a novel statistical strategy to construct network models of the associations between five different kinds of childhood adversity and symptoms of psychosis in a large group of patients ($n = 552$). In this analysis, there were no direct associations between childhood trauma and positive symptoms and the pathways from all types of adversity to symptoms were mediated by symptoms of general psychopathology.

While the importance of early life experience for psychosis is now beyond doubt, much more research is required to map the causal pathways that explain why children who grow up in less than optimal environments are at a high risk of severe mental illness. Although beyond the scope of this review, these pathways may involve specific psychological mechanisms but also the effects of adversity on brain structure[86] and neurochemistry.[87] I began this review by noting that genetic studies of psychosis had been compromised by the failure to take into account life adversity; it is likely that investigations into the neurobiological mechanisms responsible for psychosis that have so far been conducted have been similarly limited by the failure to consider how the social world impacts on the human brain.

REFERENCES

1. Wicks S, Hjern A, Daman C. Social risk or genetic liability for psychosis? A study of children born in Sweden and reared by adoptive parents. American Journal of Psychiatry. 2010;167:1240–1246.
2. Johnson SL, Wibbels E, Wilkinson R. Economic inequality is related to cross-national prevalence of psychotic symptoms. Social Psychiatry and Psychiatric Epidemiology. 2015;50:1799–1807.
3. Vassos E, Pedersen CB, Murray RM, Collier DA, Lewis CM. Meta-analysis of the association of urbanicity with schizophrenia. Schizophrenia Bulletin. 2012;38:1118–1123.
4. Cantor-Graee E, Selten JP. Schizophrenia and migration: A meta-analysis and review. American Journal of Psychiatry. 2005;163:478–487.
5. Bécares L, Dewey ME, Das-Munshi J. Ethnic density effects for adult mental health: Systematic review and meta-analysis of international studies. Psychological Medicine. 2018;48:2054–2072.
6. Bentall RP. Madness explained: Psychosis and human nature. London: Penguin; 2003.
7. Insel T, Cuthbert B, Garvey M, et al. Research Domain Criteria (RDoC): Toward a new classification framework for research on mental disorders. American Journal of Psychiatry. 2010;167:748–751.
8. Borsboom D, Cramer AOJ. Network analysis: An integrative approach to the structure of psychopathology. Annual Review of Clinical Psychology. 2013;9:91–121.
9. Kotov R, Krueger RF, Watson D, et al. The Hierarchical Taxonomy of Psychopathology (HiTOP): A dimensional alternative to traditional nosologies. Journal of Abnormal Psychology. 2017;126:454–477.
10. Mumford S, Anjum RL. Causation: A very short introduction. Oxford: Oxford University Press; 2013.
11. Pearl J, Mackenzie D. The book of why: The new science of cause and effect. London: Penguin; 2018.
12. Fisher HL, Criag TK, Fearon P, et al. Reliability and comparability of psychosis patients' retrospective reports of childhood abuse. Schizophrenia Bulletin. 2011;37:546–553.
13. Pearl J, Glymour M, Jewell NP. Causal inference in statistics: A primer. Chichester: Wiley; 2016.
14. Glymour MM. Using causal diagrams to understand common problems in epidemiology. In: Oakes JM, Kaufman JS, eds. Methods in social epidemiology. San Francisco, CA: Jossey-Bass; 2006:387–422.
15. Hill AB. The environment and disease: Association or causation? Proceedings of the Royal Society for Medicine. 1965;58:295–300.
16. Sullivan PF, Kendler KS, Neale MC. Schizophrenia as a complex trait: Evidence from a meta-analysis of twin studies. Archives of General Psychiatry. 2003;60:1187–1192.
17. Gottesman II, Shields J. Genetic theorizing and schizophrenia. British Journal of Psychiatry. 1973;122:15–30.
18. Sideli L, Mule A, La Barbera D, Murray RM. Do child abuse and maltreatment increase the risk of schizophrenia? Psychiatry Investigations. 2012;9:87–99.
19. Turkheimer E, Haley A, Waldron M, D'Onofio B, Gottesman II. Socioeconomic status modifies heritability of IQ in young children. Psychological Science. 2003;14:623–628.
20. South SC, Krueger R. Genetic and environmental influences on internalizing psychopathology vary as a function of economic status. Psychological Medicine. 2011;41:107–117.
21. Dickins WT, Flynn JR. Heritability estimates versus large environmental effects: The IQ paradox resolved. Psychological Review. 2001;108:346–369.
22. Brown GW, Rutter M. The measurement of family activities and relationships: A methodological study. Human Relations. 1966;19:241–263.
23. Butzlaff RL, Hooley JM. Expressed emotion and psychiatric relapse: A meta-analysis. Archives of General Psychiatry. 1998;55:547–552.
24. Hooley JM. Expressed emotion and relapse of psychopathology. Annual Review of Clinical Psychology. 2007;3:329–352.
25. Sedgwick P. Psychopolitics. London: Pluto Press; 1982.
26. Goldstein MJ. The UCLA high-risk project. Schizophrenia Bulletin. 1987;13:505–514.
27. Cook WL, Strachan AM, Goldstein MJ, Miklowitz DJ. Expressed emotion and reciprocal affective relationships in disturbed adolescents. Family Process. 1989;28:337–348.
28. McFarlane WR, Cook WL. Family expressed emotion prior to onset of psychosis. 2007;46:185–197.
29. Meneghelli A, Alpi A, Pafumi N, Patelli G, Cocchi A. Expressed emotion in first-episode schizophrenia and in ultra high-risk patients: Results from the Programma2000 (Milan, Italy). Psychiatry Research. 2011;189:331–338.
30. Hamaie Y, Ohmuro N, Katsura M, et al. Criticism and depression among the caregivers of At-Risk Mental State and First-Episode Psychosis. Plos One. 2016;11:e0149875.
31. Schlosser DA, Zinberg JL, Lowey RL, et al. Predicting the longitudinal effects of the family environment on prodromal symptoms and functioning in patients at-risk for psychosis. Schizophrenia Research. 2010;118:69–75.
32. Wynne LC, Singer MT. Thought disorder and family relations of schizophrenics: I. A research strategy. Archives of General Psychiatry. 1963;9:191–198.
33. Singer MT, Wynne LC. Thought disorder and family relations of schizophrenics III. Methodology using projective techniques. Archives of General Psychiatry. 1965;12:187–200.
34. Velligan DI, Funderburg LG, Giesecke SL, Miller AL. Longitudinal analysis of communication deviance in the families of schizophrenic patients. Psychiatry. 1995;58:6–19.
35. Docherty NM, Gordinier S, Hall MJ, Cutting LP. Communication disturbances in relatives beyond the age of risk for schizophrenia and their associations with symptoms in patients. Schizophrenia Bulletin. 1999;25:851–862.
36. Wynne LC, Singer MT, Bartko J, Toohey M. Schizophrenics and their families: Research on parental communication. In: Tanner

JM, ed. Developments in psychiatric research. London: Hodder & Stoughton; 1977:254–286.
37. Miklowitz D, Velligan DI, Goldstein MJ, et al. Communication deviance in families of schizophrenic and manic patients. Journal of Abnormal Psychology. 1991;100:163–173.
38. Rund BR. Communication deviances in parents of schizophrenics. Family Process. 1986;25:133–147.
39. Holte A, Wickstrøm L. Relationships between personality development, interpersonal perception and communication in parents of schizophrenics, psychiatric controls and normal subjects. Acta Psychiatrica Scandinavica. 1991;84:46–57.
40. Tienari P, Wynne LC, Sorri A, et al. Genotype–environment interaction in schizophrenia-spectrum disorder: Long-term follow-up study of Finnish adoptees. British Journal of Psychiatry. 2004;184:216–222.
41. de Sousa P, Varese F, Sellwood W, Bentall RP. Parental communication deviance and psychosis: A meta-analysis. Schizophrenia Bulletin. 2014;40:756–768.
42. Wynne LC. The epigenesis of relational systems: A model for understanding family development. Family Process. 1984;23:297–318.
43. Wichstrøm L, Holte A, Wynne LC. Disqualifying family communication and anxiety in offspring at risk for psychopathology. Acta Psychiatrica Scandinavica. 1993;88:74–79.
44. Wichstrøm L, Holte A, Husby R, Wynne LC. Competence in children at risk for psychopathology predicted from confirmatory and disconfirmatory family communication. Family Process. 1993;32:203–220.
45. de Sousa P, Sellwood W, Fien K, et al. Mapping early environment using communication deviance: A longitudinal study of maternal sensitivity toward 6-month-old children. Development and Psychopathology. 2018.
46. Felitti VJ, Anda RF, Nordenberg D, et al. Relationship of childhood abuse and household dysfunction to many of the leading causes of death in adults: The Adverse Childhood Experiences (ACE) Study. American Journal of Preventive Medicine. 1998;14:245–258.
47. Bellis MA, Lowey H, Leckenby N, Hughes K, Harrison D. Adverse childhood experiences: Retrospective study to determine their impact on adult health behaviours and health outcomes in a UK population. Journal of Public Health. 2013;36:81–91.
48. Hughes K, Bellis MA, Hardcastle KA, et al. The effect of multiple adverse childhood experiences on health: A systematic review and meta-analysis. Lancet Public Health. 2017;2:e356–e366.
49. Read J, van Os J, Morrison AP, Ross CA. Childhood trauma, psychosis and schizophrenia: A literature review and clinical implications. Acta Psychiatrica Scandinavica. 2005;112:330–350.
50. Bebbington P, Jonas S, Kuipers E, et al. Childhood sexual abuse and psychosis: Data from a cross-sectional national psychiatric survey in England. British Journal of Psychiatry. 2011;199:29–37.
51. Cutajar MC, Mullen PE, Ogloff JRP, Thomas SD, Wells DL, Spataro J. Schizophrenia and other psychotic disorders in a cohort of sexually abused children. Archives of General Psychiatry. 2010;67:1112–1119.
52. Trotta A, Di Forti M, Mondelli V, et al. Prevalence of bullying victimisation amongst first-episode psychosis patients and unaffected controls. Schizophrenia Research. 2013;150:169–175.
53. Kelleher I, Keeley H, Corcoran P, et al. Childhood trauma and psychosis in a prospective cohort study: Cause, effect and directionality. American Journal of Psychiatry. 2013;170:734–741.
54. Morgan C, Kirkbride J, Leff J, et al. Parental separation, loss and psychosis in different ethnic groups: A case-control study. Psychological Medicine. 2007;37:495–503.
55. Furukawa T, Mizukawa R, Hirai T, Fujihara S, Takahashi K. Childhood parental loss and schizophrenia: Evidence against pathogenic but for some pathoplastic effects. Psychiatry Research. 1998;81:353–362.
56. Varese F, Smeets F, Drukker M, et al. Childhood adversities increase the risk of psychosis: A meta-analysis of patient-control, prospective and cross-sectional cohort studies. Schizophrenia Bulletin. 2012;38:661–671.
57. Whitfield CL, Dube SR, Felitti VJ, Anda RF. Adverse childhood experiences and hallucinations. Child Abuse and Neglect. 2005;29:797–810.
58. Fisher H, Jones PB, Fearon P, et al. The varying impact of type, timing and frequency of exposure to childhood adversity on its association with adult psychotic disorder. Psychological Medicine. 2010;40:1967–1978.
59. Matheson SL, Shepherd AM, Pinchbeck RM, Laurens KR, Carr VJ. Childhood adversity in schizophrenia: A systematic meta-analysis. Psychological Medicine. 2012;43:225–238.
60. McGrath JJ, Saha S, Lim C, et al. Trauma and psychotic experiences: Transnational data from the World Mental Health Survey. British Journal of Psychiatry. 2017;211:373–380.
61. McGrath JJ, McLaughlin KA, Saha S, et al. The association between childhood adversities and subsequent first onset of psychotic experiences: A cross-national analysis of 23 998 respondents from 17 countries. Psychological Medicine. 2017.
62. Palmier-Claus JE, Berry K, Bucci S, Mansell W, Varese F. Relationship between childhood adversity and bipolar affective disorder: systematic review and meta-analysis. British Journal of Psychiatry. 2016;209:454–459.
63. Beards S, Gayer-Anderson C, Borges S, Dewey ME, Fisher HL, Morgan C. Life events and psychosis: A review and meta-analysis. Schizophrenia Bulletin. 2013;39(4):740–747.
64. Seow LSE, Ong C, Mahesh MV, et al. A systematic review of comorbid post-traumatic stress disorder in schizophrenia. Schizophrenia Research. 2016;176:441–451.
65. Tienari P, Wynne LC, Moring J, et al. Finnish adoptive family study: sample selection and adoptee DSM-III-R diagnoses. Acta Psychiatrica Scandinavica. 2000;101:433–443.
66. Tienari P, Lahti I, Sorri A, Naarala M, Moring J, Wahlbeck K-E. The Finnish Adoption Family Study of Schizophrenia: Possible joint effects of genetic vulnerability and family environment. British Journal of Psychiatry. 1989;155, suppl 5:29–32.
67. Tienari P, Wynne LC, Sorri A, et al. Long term follow-up study of Finnish adoptees. British Journal of Psychiatry. 2004;184: 214–222.
68. Arseneault L, Cannon M, Fisher HL, Polanczyk G, Moffitt TE, Caspi A. Childhood trauma and children's emerging psychotic symptoms: A genetically sensitive longitudinal cohort study. American Journal of Psychiatry. 2011;168:65–72.
69. Alemany S, Goldberg X, van Winkel R, Gastro C, Peralta V, Fananas L. Childhood adversity and psychosis: Examining whether the association is due to genetic confounding using a monozygotic twin differences approach. European Psychiatry. 2013;28:207–212.
70. Lecei A, Decoster J, De Hert M, et al. Evidence that the association of childhood trauma with psychosis and related psychopathology is not explained by gene-environment correlation: A monozygotic twin differences approach. Schizophrenia Research. 2018.
71. Heins M, Simons C, Lataster T, et al. Childhood trauma and psychosis: A case-control and case-sibling comparison across different levels of genetic liability, psychopathology, and type of trauma. American Journal of Psychiatry. 2011;168:1286–1294.
72. Trotta A, Di Forti M, Iyegbe C, et al. Familial risk and childhood adversity interplay in the onset of psychosis. British Journal of Psychiatry Open. 2015;1:6–13.
73. Trotta A, Iyegbe C, di Forti M, et al. Interplay between schizophrenia polygenic risk score and childhood adversity in first presentation psychotic disorder: A pilot study. Plos One. 2016;11:e0163319.
74. Iyegbe C, Campbell D, Butler A, Ajnakina O, P. S. The emerging molecular architecture of schizophrenia, polygenic risk scores and the clinical implications for GxE research. Social Psychiatry and Psychiatric Epidemiology. 2014;49:169–182.
75. Kotov R, Chang SW, Fochtmann LJ, et al. Schizophrenia in the internalizing-externalizing framework: a third dimension? Schizophr Bull. 2011;37(6):1168–1178.
76. Lahey BB, Applegate B, Hakes JK, Zald DH, Hariri AR, Rathouz PJ. Is there a general factor of prevalent psychopathology during adulthood? Journal of Abnormal Psychology. 2012;121:971–977.

77. Bentall RP, de Sousa P, Varese F, et al. From adversity to psychosis: Pathways and mechanisms from specific adversities to specific symptoms. Social Psychiatry and Psychiatric Epidemiology. 2015;49:1011–1022.
78. Sass LA, Gunderson JG, Singer MT, Wynne LC. Parental communication deviance and forms of thinking in male schizophrenic offspring. Journal of Nervous and Mental Disease. 1984;172: 513–520.
79. Shevlin M, Dorahy M, Adamson G. Childhood traumas and hallucinations: An analysis of the National Comorbidity Survey. Journal of Psychiatric Research. 2007;41:222–228.
80. Bentall RP, Wickham S, Shevlin M, Varese F. Do specific early life adversities lead to specific symptoms of psychosis? A study from the 2007 The Adult Psychiatric Morbidity Survey. Schizophrenia Bulletin. 2012;38:734–740.
81. Shevlin M, McAnee G, Bentall RP, Murphy K. Specificity of association between adversities and the occurrence and co-occurrence paranoia and hallucinations: Evaluating the stability of risk in an adverse adult environment. Psychosis. 2015;7:206–216.
82. Wickham S, Bentall RP. Are specific early-life experiences associated with specific symptoms of psychosis: A patients study considering just world beliefs as a mediator. Journal of Nervous and Mental Disease. 2016;204:606–613.
83. Hammersley P, Dias A, Todd G, Bowen-Jones K, Reilly B, Bentall RP. Childhood trauma and hallucinations in bipolar affective disorder: A preliminary investigation. British Journal of Psychiatry. 2003;182:543–547.
84. Sitko K, Bentall RP, Shevlin M, O'Sullivan N, Sellwood W. Associations between specific psychotic symptoms and specific childhood adversities are mediated by attachment styles: An analysis of the National Comorbidity Survey. Psychiatry Research. 2014;217:202–209.
85. Isvoranu A-M, van Borkulo CD, Boyette L-L, Wigman JTW, Vinkers CH, Borsboom D. A network approach to psychosis: Pathways between childhood trauma and psychotic symptoms. Schizophrenia Bulletin. 2017;43:187–196.
86. Sheffield JM, Williams LF, Woodward ND, Heckers S. Reduced gray matter volume in psychotic disorder patients with a history of childhood sexual abuse. Schizophrenia Research. 2013;143:185–191.
87. Howes OD, Murray RM. Schizophrenia: An integrated sociodevelopmental-cognitive model. Lancet. 2014;383:1677–1687. http://dx.doi.org/10.1016/S0140-6736(13)62036-X.

46

SOCIOENVIRONMENTAL ADVERSITY ACROSS THE LIFE SPAN

Peter Bosanac and David J. Castle

INTRODUCTION

Cannabis sativa is the most widely used illicit drug among youth and is used more by males than females, with about half starting use prior to 18 years of age.[1] The lifetime prevalence of cannabis use disorder in people with established schizophrenia ranges from 13% to 64%.[1] Exposure to cannabis prior to the onset of frank psychosis carries a heightened risk for developing schizophrenia, as demonstrated in multiple longitudinal studies.[2]

Although cannabis use has been considered, albeit more contentiously in the past,[2] a causal factor in some cases of schizophrenia,[3,4] between 8% and 24% would not have developed the illness in the absence of cannabis use, contingent on the regional prevalence of use.[5] In addition, one long-term study using consistent diagnostic criteria for schizophrenia spanning over 35 years found that the proportion of those with a history of cannabis use had increased by a factor of two.[6] Also, the potency of cannabis has increased in recent decades,[2] with a greater risk of earlier onset, on average 6 years earlier than nonusers.[7]

Notwithstanding that the overwhelming majority of cannabis users will not develop a psychotic disorder, the relationship between cannabis, the developing brain, and psychosis is a complex one. One of the purported mechanisms in this relationship is that cannabis exacerbates subthreshold psychotic symptoms, that in the absence of such use, would have been transient.[8–10]

While the heritability of schizophrenia is marked at around 80% of the variance,[11] including those with a family history of schizophrenia,[12] the shared genetic predisposition of cannabis use and predisposition to schizophrenia is less clear.[13] Moreover, continued cannabis use in established schizophrenia worsens outcome, rather than mitigates the severity of symptoms, counter to any expectations of "self-medication."[14]

EARLY EXPOSURE AND RISK OF PSYCHOTIC DISORDER

In a seminal study of the risk of cannabis exposure for psychosis, military conscripts were six times likelier to develop schizophrenia over an ensuing 15-year period if they had used cannabis in excess of 50 times compared to those who had never used.[15] Subsequently, a dozen large-size, longitudinal (ranging from 1 to 35 years in duration) and predominantly cohort- or population-based studies[16,17] have attempted to clarify the causal role of cannabis in psychotic disorder. Three quarters of these studies have demonstrated a significantly heightened likelihood of psychotic symptoms or disorder[4,14,18–26] and a quarter have shown trend toward greater risk of developing psychosis.[16,27–29]

The divergence in the strength of these findings should be seen be in context of the confounding impact of other substances being used, and the diversity within the sampled clinical populations, namely first episode psychotic disorder compared with established schizophrenia. Also, older people with first-episode psychosis are less likely to use cannabis than their younger counterparts[30] and the greater proportion of male users of cannabis may account to some extent for the earlier onset of psychotic disorder in dual-diagnosis population, irrespective of cannabis use, thus confounding the purported association between cannabis and earlier onset of psychotic illness.[1]

Furthermore, a study comparing 49 first-episode people with schizophrenia and cannabis use disorder and 51 first-episode counterparts without cannabis use, found that lifetime exposure to cannabis was associated with greater severity of positive symptoms of schizophrenia, lower socioeconomic status, and male gender, but better social adjustment in childhood. The authors of this study concluded that the association between cannabis use disorder and onset of psychosis, was explicable in terms of the aforementioned demographic and clinical factors, irrespective of exposure to cannabis.[1] Another study of 785 individuals with nonaffective psychotic disorder found that age of illness onset was 1.8 years earlier in those with cannabis use in comparison to nonusers, when other variables, including gender were controlled for. Also, males in this study experienced onset of illness, on average, 1.3 years earlier than females. In 63.5% of cannabis-using participants, age at most intense cannabis use preceded the age of onset of psychosis.[31] Another multisite study of 143 first-episode psychotic disorder participants found a similarly earlier onset of illness in males and females who used cannabis. Postulated mechanisms included the severity of cannabis use, psychosocial functioning, and neurotoxicity.[32]

The age of commencement of cannabis use and the frequency of use in adolescence is associated with heightened risk of psychotic symptoms and an early onset of psychotic disorder.[33] In terms of cannabis exposure during potentially critical developmental periods of adolescence, those who had been using cannabis by 15 years of age had a four times greater risk of developing a psychotic disorder than those who commenced at 18 years of age or older.[4] Also, in the former birth cohort study of young people, in which those at heightened risk of developing a psychotic disorder were excluded at 11 years of age, cannabis use was associated with increased risk of developing schizophrenia, irrespective of other substances.[4] Continuing cannabis use in earlier, rather later in adolescence appears to confer a greater risk for the development of psychotic disorders.[33–39] A meta-analysis of 83 studies concluded that the age of onset of psychotic disorder was 2.7 years earlier than in nonusers.[40] This effect on earlier onset of psychosis in cannabis users was confirmed in a birth cohort twin study, which addressed potential, unknown confounders.[41] Furthermore, the frequency of cannabis use impacts on the age of onset of psychotic disorders.[42] A meta-analysis of 10 studies with a total of 66,816 participants found there was an odds ratio of 3.9 of developing psychosis in those using greater quantities of cannabis compared to those who did not use.[43] Having said this, cumulative exposure to cannabis may have a stronger impact on the development, rather than on earlier age of onset, in psychotic disorders.[44] Synthetic cannabinoids are also frequently used[45] and have emerged as a perpetuating factor for psychosis.[46]

There is evidence that cannabis exposure in adolescence, particularly in earlier phases, adversely affects the endocannabinoid system (which has a significant role in neurodevelopment) and other aspects of neurotransmitter (including the dopaminergic system) and central nervous system development. The evidence includes: animal models of early cannabis exposure; neuropsychological impairment and persistence following abstinence contingent on age of initial use, extent of use, and duration of abstinence; structural changes (frontal, amygdala, hippocampal, and cerebellar) with exposure in brain-imaging studies (albeit these findings are rather inconsistent); altered neural connectivity (hippocampal, corpus callosal, and commissural); and altered connectivity in the frontal-temporal regions and between hemispheres on functional neuroimaging.[47] It is possible that some of the differences in people exposed to cannabis and nonusers may be due to preexisting differences in these populations, including those who develop psychotic disorders, as well as differences among individuals that are not discernable to cross-sectional evaluation, as opposed to longitudinal studies.[47]

People experiencing psychosis in the setting of cannabis use have been found to have higher premorbid intelligence and social functioning, better cognitive functioning, and less in the way of neurological soft signs[48–51] than nonusers. These findings may, in part at least, be accounted for by a proportion of people with schizophrenia who do not use cannabis having preexisting neurodevelopment diatheses, and in turn impaired social and cognitive functioning in comparison to people using cannabis who have better premorbid functioning in these domains.[16]

NEUROBIOLOGY

The endocannabinoid system plays an important role in brain development, neuromodulation and neurogenesis.[50–53] Of the greater than 100 cannabinoids in the cannabis plant, delta-9-tetrahydrocannabinol (THC) and cannabidiol (CBD) are the most psychotropically important.[54] The cannabinoids act on the cannabinoid 1 (CB1) receptor, which is widely present in the central nervous system and act presynaptically on GABA and glutamatergic neurons, in particular cortical areas and the basal ganglia and hippocampus. Cannabinoid receptor type-2 (CB2) is mostly expressed peripherally, in the immune system, as well as cerebellum and brainstem.[16]

Synthetic cannabinoids have a greater potency than THC as a full agonist at the CB1 receptor, and they do not contain CBD to offset the effects of THC.[55–57]

Exposure to cannabis during adolescence, a key time of active remodeling and synaptic pruning in the brain,[58] may result in impaired maturation of prefrontal GABA interneurons by increased activation of cannabinoid-1 receptors in the dorsolateral prefrontal cortex[59,60] and consequent hindrance of necessary glutamatergic activity for the maturation to take place; this may in turn translate into a heightened predisposition to psychosis.[61] In addition, chronic cannabis use hampers axonal connectivity, particularly in people who commence at a younger age.[62] But, the critical age range in adolescence for the adverse impact of cannabis use, including prefrontal cortex maturation, has not been unequivocally clarified.[63] It has been postulated that cannabis may hasten the onset of schizophrenia via interaction with the latter's endophenotypes such as cognition or neurological soft signs.[8,64]

MECHANISM

While CBD content in cannabis may be protective against psychotic symptoms,[65] psychotic disorders may be precipitated by THC. The latter occurs via modulation of neurotransmission (presynaptic neurotransmitter release is inhibited by activation of cannabinoid 1(CB1) receptors by cannabis including dopaminergic activity in the brain, central to the onset and salient positive, negative, and cognitive symptoms in schizophrenia,[10] as well as impacting on noradrenergic and glutamatergic function.[8] Also, neurotoxicity or sensitization to excess striatal dopamine release mediated by cannabis activation of CB1 receptors[64,66] may precipitate psychosis following cannabis exposure. While reduced CB1 receptor levels in schizophrenia might mitigate against reduced GABA interneuron activity, cannabis use could oppose this protective factor.[67,68]

The potency of cannabis is itself a risk factor for psychosis.[16] A study of 410 first-episode psychosis patients compared with people who had not used showed an increased likelihood of approximately 3 to 5 times (odds ratio ranged from 2.9 to 5.4) in high-potency cannabis users, with the upper range affected by frequency of use being on a daily basis.[5] Moreover, people who use synthetic cannabinoids are also at increased risk of psychotic disorders.[46]

While people who use cannabis have a dampened capacity to produce and mobilize dopamine in the striatum on positron emission tomography studies, they are differentiated from other substance users by not having changes to dopamine D2 and D3 receptors in the striatum. Moreover, people using cannabis have a purported supersensitivity of postsynaptic dopamine receptors, despite only small release of dopamine (even with a challenge with amphetamines), which in combination with other predispositions may be a diathesis for psychosis.[16,69,70]

In a study of 102 people at heightened risk of developing a psychotic illness, the participants used cannabis predominantly to enhance mood in the setting of being less motivated to engage in pleasurable activities for mood enhancement, especially in the context of decreased motivation to seek pleasure otherwise. Hence, it was concluded that the presence of negative symptoms predisposed to cannabis use in individuals at high risk of developing psychosis.[71]

GENETIC DIATHESIS AND INTERFACE WITH ENVIRONMENT

Polygenic scoring is a means of measuring the association between cannabis exposure and alleles for increased risk of developing schizophrenia, with the latter having been determined from genome-wide association. In a study of 2,082 people who were well, comparing those who used cannabis and nonusers, polygenic scoring predicted a small degree of the variance in the use of cannabis. This was suggestive of a partially shared genetic predisposition for cannabis use and schizophrenia.[13] On the other hand, polygenic scoring has not differentiated the genetic risk for psychotic disorder in people using greater potency (THC) cannabis,[7] suggesting that some people are more predisposed to developing a psychotic disorder following cannabis use in comparison to others, possibly including those with copy number variants (the variable number among individuals of repeated sections of genome) with lower threshold diathesis for psychosis.[16]

The genetic risk of psychotic disorder in predisposed individuals has been linked with variants of genes that influence the dopaminergic system, including AKT1,[16,72–78] DRD2, the D2 receptor gene magnifying risk in the AKT1 polymorphism[75] and possibly Catechol-O-methyltransferase (COMT).[16,76–78] Although people who use cannabis and have the Val/Val rather than Met/Met or Val/Met COMT alleles[79] are thought to be at increased risk of experiencing a psychotic disorder, as well as that of a younger age of onset[8,12,79,80] other studies have either not replicated these findings[16] or have found that this polymorphism reduces working memory.[81]

NEUROLOGICAL AND COGNITIVE IMPACT

Studies of the impact of cannabis exposure on central nervous system structure in psychotic disorders have been fraught with methodological problems, and findings have been inconsistent. While an initial volumetric study of gray and white matter volumes of people with early course schizophrenia with cannabis use disorder compared to those without cannabis use found no significant difference in these,[82] this study was limited by an absence of evaluation of regions within the frontal lobes. Another volumetric magnetic resonance imaging study looking at different regions of gray and white matter volumes of 20 people with early course schizophrenia with cannabis use disorder, 31 people also with early course schizophrenia but without cannabis use, and 56 normal controls found differences in people with comorbid cannabis use in comparison to the other groups. In this study, people with the aforementioned comorbidity had less gray matter in their anterior cingulate regions compared with nonusers of cannabis, irrespective of the presence of schizophrenia.[83] Other studies in people with psychotic disorders and cannabis exposure have shown[84–87] increased striatal gray matter; possible reduction of cerebellar, left hippocampal, and right posterior cingulate volumes; increased loss of cortical gray matter volume; and increased third and lateral ventricular size in the first 5 years of psychosis; and significant loss of cerebellar white matter and changes of hippocampus morphology.

A systematic review of 21 neuroimaging studies of people with schizophrenia and substance use found that in some, but not all of these studies, cannabis (and alcohol) use increased brain structural abnormalities, particularly cerebellar. However, in keeping with the complex relationship of psychotic disorders and cannabis exposure, medial prefrontal cortex function appears to be preserved in functional imaging studies to date in this clinical population.[88]

In an observational analysis combining 1,574 male and female adolescents from three studies, males with high polygenic risk scores for schizophrenia who had cannabis use in early adolescence were found to have reduced thickness of the cortices.[89] This finding suggested that the mechanisms underpinning the development and maturation of the cortices are involved in conferring the association between heightened predisposition to schizophrenia and cannabis exposure.[89]

A two-hit neurodevelopmental diathesis, whereby an initial adverse impact on neurodevelopment renders the central nervous system vulnerable to another "hit," has been considered relevant to the predisposition for the development of a psychotic disorder in people with cannabis use. Such impacts on the left superior longitudinal fasciculus, a key bundle of white matter within the frontal lobe, has been implicated individually in asymptomatic users of cannabis as well as those with schizophrenia. What is not known is what the impact of cannabis use is on those with established schizophrenia, as well as those at risk of psychosis.[90]

A PET study of eight cannabis users with psychotic disorder compared with seven first-degree family members looking at dopamine release by THC found ligand displacement in the striatum consistent with the release of dopamine in both groups, particularly in the caudate nuclei. This was suggestive of a heightened sensitivity to THC and greater release of dopamine in people at risk of psychotic disorder.[91]

In a systematic review of 17 studies of early course schizophrenia, two studies found abnormal white matter integrity, mostly in projection, limbic, and corpus callosal tracts, but another found such abnormalities in all tracts except in limbic tracts.[92]

IMPLICATIONS FOR TREATMENT

Education for those at heightened risk of developing a psychotic disorder (e.g., genetic predisposition/family history,[93,94] ultra-high risk/prodrome of schizophrenia) about the exacerbated risk of developing a psychotic illness with cannabis use, is essential at an individual and population health level.

Alcohol intake, cigarette smoking, homelessness, financial and legal difficulties, and suboptimal treatment response are suggestive of heightened risk of cannabis use in psychotic disorders.[95] Screening for cannabis use and the evaluation of the nature, extent, and reasons for cannabis use is essential in people with a predisposition to, or established psychotic disorders, in parallel to evaluating readiness to change cannabis use. In people with psychotic disorders who use cannabis, evaluating the link between cannabis use and specific symptoms and the person's reasons for using cannabis is also necessary in understanding their developmental and illness trajectory, as well as in formulating treatment.

In a meta-analysis of people with psychotic disorders and ongoing cannabis use, there were higher rates of psychotic relapse, longer episodes of inpatient care, and greater severity of psychotic symptoms than in those who had ceased cannabis use or had never used.[96] Hence, a recommendation of reduced intake and cessation in individuals with psychotic disorder and cannabis use is essential in addressing and hopefully mitigating the severity of psychosis, in combination with education, motivational interviewing and cognitive-behavioral therapy.[16]

In terms of antipsychotic choice in people with schizophrenia who use cannabis, clozapine might have a particular role. A trial of clozapine (flexibly dosed to a range of 400 to 500 milligrams per day versus continuation of preexisting antipsychotic medication in which 31 people with schizophrenia and intercurrent cannabis use disorder were randomized to either group over 12 weeks, with open-label treatment but blinded longitudinal assessment demonstrated a moderate dampening of cannabis craving and reduced use in the group treated with clozapine.[97] While this study suggested a clinically significant benefit for clozapine compared to other antipsychotics in this comorbid population, the study was limited by the small number of participants, consecutive referrals for study entry by treating clinicians, and the non-blinded prescribing of medication. In the Australian Survey of High Impact Psychosis (SHIP) study, individuals who were prescribed clozapine had significantly lower odds of cannabis use, notwithstanding a similar lifetime odds to those not prescribed clozapine.[98] While there may be a potential benefit for clozapine over other antipsychotics in people with schizophrenia and significant cannabis use, the use of clozapine requires analysis of the individual patient's benefit versus risk, in, including hematological, cardiac, and metabolic monitoring in a specialist setting.

CONCLUSIONS

Although the relationship between cannabis, the developing brain, and psychosis is a complex one, cannabis exposure prior to the onset of frank psychosis carries a heightened risk for developing schizophrenia, as demonstrated in multiple longitudinal studies. While the age of initial cannabis use and the frequency of use in adolescence is associated with an increased risk of psychotic symptoms and early onset of illness, cumulative cannabis exposure may have a greater impact on the development, rather than age of onset in psychotic disorders. Additionally, in established schizophrenia, continued cannabis use worsens outcome rather than mitigating the severity of psychotic symptoms.

Cannabis exposure in adolescence adversely affects the endocannabinoid system's role in neurodevelopment and other aspects of neurotransmitter and central nervous system development. Some individuals are also more at risk of developing a psychotic disorder following cannabis use than others, possibly including those with copy number variants. In people with psychotic disorders, studies of the effects of cannabis exposure on central nervous system structure have been inconsistent.

At an individual and population health level, education is essential for those at heightened risk of developing a psychotic disorder with early and cumulative cannabis exposure. Hence, screening for cannabis use and the evaluation of the nature, extent, and reasons for cannabis use is also essential in people predisposed to, or with established psychotic disorders, in tandem with evaluating readiness to change cannabis use.

REFERENCES

1. Sevy S, Robinson DG, Napolitano B, et al. Are cannabis use disorders associated with an earlier age at onset of psychosis? A study in first episode schizophrenia. *Schizophr Res* 2010;120(1–3):101–107.
2. Murray RM, Quigley H, Quattrone D, Englund A, Di Forti M. 2016. Traditional marijuana, high-potency cannabis and synthetic cannabinoids: increasing risk for psychosis. *World Psychiatry* 2016;15(3):195–204.
3. Radhakrishnan R, Wilkinson ST, D'Souza DC. Gone to pot—a review of the association between cannabis and psychosis. *Front Psychiatry* 2014;5:54.
4. Arseneault L, Cannon M, Poulton R, Murray R, Caspi A, Moffitt TE. Cannabis use in adolescence and risk for adult psychosis: longitudinal prospective study. *BMJ* 2002;325:1212–1213.
5. Di Forti M, Marconi A, Carra E, et al. Proportion of patients inn South London with first episode psychosis attributable to use of high-potency cannabis: a case-control study. *Lancet Psychiatry* 2015;2(3):233–238.
6. Boydell J, van Os J, Caspi A, et al. Trends in cannabis use prior to first presentation with schizophrenia, in South-East London between 1965 and 1999. *Psychol Med* 2006;36:1441–1446.
7. Di Forti M, Salys H, Allegri F, et al. Daily use, especially at high potency cannabis drives the earlier onset of psychosis in cannabis users. *Schizophr Bull* 2014;40(6):1509–1517.
8. Dervaux A, Krebs M-O, Laqueille X. Is cannabis responsible for early onset psychotic illnesses? *Neuropsychiatry* 2011;1(3):203–207.

9. Kuepper R, van Os J, Lieb R, Wittchen HU, Höfler M, Henquet C. Continued cannabis use and risk of incidence and persistence of psychotic symptoms: 10 year follow-up cohort study. *BMJ* 2011;342:1–8.
10. Kuepper R, Morrison PD, van Os J, Murray RM, Kenis G, Henquet C. Does dopamine mediate the psychosis-inducing effects of cannabis? A review and integration of findings across disciplines. *Schizophr Res* 2010;121(1–3): 107–117.
11. Sullivan PF, Kendler KS, Neale MC. Schizophrenia as a complex trait—evidence from a meta-analysis of twin studies. *Arch Gen Psychiatry* 2003;60:1187–1192.
12. Henquet C, Rosa A, Delespaul P, et al. COMT ValMet moderation of cannabis-induced psychosis: a momentary assessment study of "switching on" hallucinations in the flow of daily life. *Acta Psychiatr Scand* 2009;119(2):156–160.
13. Power RA, Verweij KJ, Zuhair M, et al. Genetic predisposition to schizophrenia associated with increased use of cannabis. *Mol Psychiatry* 2014;19:1201–1204.
14. Murray R. Appraising the risks of reefer madness. *Cerebrum.* 2015:1.
15. Andreasson S, Allebeck P, Engstrom A, Rydberg U. Cannabis and schizophrenia: a longitudinal study of Swedish conscripts. *Lancet* 1987;2(8574):1483–1486.
16. Murray RM, Di Forti M. Cannabis and psychosis: what degree of proof do we require? *Biol Psychiatry* 2016;79:514–515.
17. Gage SH, Hickman M, Zammit S. Association between cannabis and psychosis: epidemiologic evidence. *Biol Psychiatry* 2016;79:549–556.
18. Rognli EB, Berge J, Håkansson A, Bramness JG. Long-term risk factors for substance-induced and primary psychosis after release from prison: a longitudinal study of substance users. *Schizophr Res* 2015;168:185–190.
19. Manrique-Garcia E, Zammit S, Dalman C, Hemmingsson T, Andreasson S, Allebeck P. Cannabis, schizophrenia and other non-affective psychoses: 35 years of follow-up of a population-based cohort. *Psychol Med* 2012;42:1321–1328.
20. Ferdinand RF, Sondeijker F, van der Ende J, Selten J-P, Huizink A, Verhulst FC. Cannabis use predicts future psychotic symptoms, and vice versa. *Addiction* 2005;100:612–618.
21. Henquet C, Krabbendam L, Spauwen J, et al. Prospective cohort study of cannabis use, predisposition for psychosis, and psychotic symptoms in young people. *BMJ* 2005;330:11.
22. Fergusson DM, Horwood LJ, Swain-Campbell NR. Cannabis dependence and psychotic symptoms in young people. *Psychol Med* 2003;33:15–21.
23. Zammit S, Allebeck P, Andréasson S, Lundberg I, Lewis G. Self reported cannabis use as a risk factor for schizophrenia in Swedish conscripts of 1969: historical cohort study. *BMJ* 2002;325(7374): 1199.
24. van Os J, Bak M, Hanssen M, Bijl RV, de Graaf R, Verdoux H. Cannabis use and psychosis: a longitudinal population-based study. *Am J Epidemiol* 2002;156:319–327.
25. Weiser M, Knobler HY, Noy S, Kaplan Z. Clinical characteristics of adolescents later hospitalized for schizophrenia. *Am J Med Genet* 2002;114:949–955.
26. Tien AY, Anthony JC. Epidemiological analysis of alcohol and drug use as risk factors for psychotic experiences. *J Nerv Ment Dis* 1990;178:473–480.
27. Gage SH, Hickman M, Heron J, et al. Associations of cannabis and cigarette use with psychotic experiences at age 18: findings from the Avon Longitudinal Study of Parents and Children. *Psychol Med* 2014;44:3435–3444.
28. Rössler W, Hengartner MP, Angst J, Ajdacic-Gross V. Linking substance use with symptoms of subclinical psychosis in a community cohort over 30 years. *Addiction* 2012;107:1174–1184.
29. Wiles NJ, Zammit S, Bebbington P, Singleton N, Meltzer H, Lewis G. Self-reported psychotic symptoms in the general population: results from the longitudinal study of the British National Psychiatric Morbidity Survey. *Br J Psychiatry* 2006;188: 519–526.
30. Wade D, Harrigan S, Edwards J, Burgess PM, Whelan G, McGorry PD. Substance misuse in first-episode psychosis: 15-month prospective follow-up study. *Br J Psychiatry* 2006;189:229–234.
31. Dekker N, Meijer J, Koeter M, et al. Age at onset of non-affective psychosis in relation to cannabis use, other drug use and gender. *Psychol Med* 2012;42(9):1903–11
32. Donoghue K, Doody GA, Murray RM, et al. Cannabis use, gender and age of onset of schizophrenia: data from the ÆSOP study. *Psychiatry Res* 2014; 30;215(3):528–532.
33. Bechtold J, Simpson T, White HR, Pardini D. Chronic adolescent marijuana use as a risk factor for physical and mental health problems in young adult men. *Psychol Addict Behav* 2015;29(3):552–563.
34. Wilkinson, ST, Radhakrishnan R, D'Souza DC. Impact of cannabis use on the development of psychotic disorders. *Curr Addict Rep* 2014;1:115–128.
35. Decoster J, van Os J., Myin-Germeys I, De Hert M, van Winkel R. Genetic variation underlying psychosis-inducing effects of cannabis: critical review and future directions. *Curr Pharm Des* 2012;18:5015–5023.
36. Casandro P, Fernandes C, Murray RM, Forti M. Cannabis use in young people: the risk for schizophrenia. Neuroscience and Biobehavioural Reviews 2011;35:1779–1787.
37. Moore TH, Zammit S, Lingford-Hughes S, et al. Cannabis use and risk of psychotic or affective mental health outcomes: a systematic review. *Lancet* 2007;28;370(9584):319–328.
38. Semple DM, McIntosh AM, Lawrie SM. Cannabis as a risk factor for psychosis: systematic review. *J Psychopharmacol* 2005;19:187–194.
39. Hall W, Degenhardt L. Cannabis use and psychosis: a review of clinical and epidemiological evidence. *Aust N Z J Psychiatry* 2000;34:26–34.
40. Large M, Sharma S, Compton MT, Slade T, Nielssen O. Cannabis use and earlier onset of psychosis: a systematic meta-analysis. *Arch Gen Psychiatry* 2011;68:555–561.
41. Callaghan RC, Cunningham JK, Allebeck P, et al. Methamphetamine use and schizophrenia: a population-based cohort study in California. *Am J Psychiatry* 2012;169(4):389–396.
42. Stefanis NC, Delespaul P, Henquet C, Bakoula C, Stefanis CN, Van Os J. Early adolescent cannabis exposure and positive and negative dimensions of psychosis. *Addiction* 2004;99(10):1333–1341.
43. Marconi A, Di Forti M, Lewis CM, Murray RM, Vassos E. Meta-analysis of the association between the level of cannabis use and risk of psychosis. *Schizophr Bull* 2016;42:1262–1269.
44. Stefanis NC, Dragovic M, Power BD, Jablensky A, Castle D, Morgan VA. Age at initiation of cannabis use predicts age at onset of psychosis: the 7- to 8-year trend. *Schizophr Bull* 2013;39:251–254.
45. Spaderna M, Addy PH, D'Souza DC. Spicing things up: synthetic cannabinoids. *Psychopharmacol* 2013;228(4):525–540.
46. Fattore L. Synthetic cannabinoids: further evidence supporting the relationship between cannabinoids and psychosis. *Biol Psychiatry* 2016;79:539–5448.
47. Volkow ND, Swanson JM, Evins AE, et al. Effects of cannabis use on human behavior, including cognition, motivation, and psychosis: a review. *JAMA Psychiatry* 2016;73:292–297.
48. Arnold C, Allott K, Farhall J, Killacky E, Cotton S, et al. Neurocognitive and social cognitive predictors of cannabis use in first-episode psychosis. *Schizophr Res* 2015;168:231–237.
49. Loberg EM, Helle S, Nygard M, Berle JO, Kroken RA, Johnsen E. The cannabis pathway to non-affective psychosis may reflect less neurobiological vulnerability. *Front Psychiatry* 2014;5:159.
50. Ferraro L, Russo M, O'Connor J et al. Cannabis users have higher premorbid IQ than other patients with first onset psychosis. *Schizophr Res* 2013;150:129–135.
51. Ruiz-Veguilla M, Callado LF, Ferrin M. Neurological soft signs in patients with psychosis and cannabis abuse: a systematic review and meta-analysis of paradox. *Curr Pharm Des* 2012;18:5156–5164.
52. Parakh P, Basu D. Cannabis and psychosis: have we found the missing links? *Asian J Psychiatr* 2013; 6: 281–287.
53. Yücel M, Bora E, Lubman DI, et al. The impact of cannabis use on cognitive functioning in patients with schizophrenia: a meta-analysis

of existing findings and new data in a first-episode sample. *Schizophr Bull* 2012;38(2):316–330.

54. Trezza V, Cuomo V, Vanderschuren LJ. Cannabis and the developing brain: insights from behavior. *Eur J Pharmacol* 2008;585(2–3):441–452.
55. GalveRoperh I, Aguado T, Palazuelos J, Guzman M. The endocannabinoid system and neurogenesis in health and disease. *Neuroscientist* 2007;13:109–114.
56. Hanus LO. Pharmacological and therapeutic secrets of plant and brain (endo)cannabinoids. *Med Res Rev* 2009;29:213–271.
57. Altinas M, Inanc L, Akcay Oruc G, Arpacioglu S, Gulec H. Clinical characteristics of synthetic cannabinoid-induced psychosis in relation to schizophrenia: a single-center cross-sectional analysis of concurrently hospitalized patients. *Neuropsychiat Dis Treat* 2016;12:1893–1900.
58. Lubman DI, Cheetham A, Yücel M. Cannabis and adolescent brain development. *Pharmacol Ther* 2015;148:1–16.
59. Dean B, Sundram S, Bradbury R, Scarr E, Copolov D. Studies on [^{3}H] CP-55940 binding in the human central nervous system: regional specific changes in density of cannabinoid-1 receptors associated with schizophrenia and cannabis use. *Neuroscience* 2001;103:9–15.
60. Tanda G, Pontieri F E, Di Chiara G. Cannabinoid and heroin activation of mesolimbic dopamine transmission by a common mu1 opioid receptor mechanism. *Science* 1997;276:2048–2050.
61. Caballero A, Tseng KY. Association of cannabis use during adolescence, prefrontal CB1 receptor signaling, and schizophrenia. *Front Pharmacol* 2012;3:101.
62. Zalesky A, Solowij N, Yucel M, et al. 2012. Effect of long-term cannabis use on axonal fibre connectivity. *Brain* 2012;135:2245–2255.
63. Schimmelmann BG, Conus P, Cotton SM, et al. Cannabis use disorder and age at onset of psychosis—a study in first-episode patients. *Schizophr Res* 2011;129(1):52–56.
64. D'Souza DC, Sewell RA, Ranganathan M. Cannabis and psychosis/schizophrenia: human studies. *Eur Arch Psychiatry Clin Neurosci* 2009;259(7):413–431.
65. Schubart CD, Sommer IE, van Gastel WA, Goetgebuer RL, Kahn RS, Boks MP. Cannabis with high cannabidiol content is associated with fewer psychotic experiences. *Schizophr Res* 2011;130:216–221.
66. Bossong MG, van Berckel BN, Boellaard R, et al. D9-tetrahydrocannabinol induces dopamine release in the human striatum. *Neuropsychopharmacology* 2009:34(3):759–766.
67. Eggan SM, Stoyak SR, Verrico CD, Lewis DA. Cannabinoid CB1 receptor immunoreactivity in the prefrontal cortex: Comparison of schizophrenia and major depressive disorder. *Neuropsychopharmacology* 2010;35(10):2060–2071.
68. Eggan SM, Hashimoto T, Lewis DA. Reduced cortical cannabinoid 1 receptor messenger RNA and protein expression in schizophrenia. *Arch Gen Psychiatry* 2008;65(7):772–784.
69. Volkow ND, Wang GJ, Telang F, *et al.* Decreased dopamine brain reactivity in marijuana abusers is associated with negative emotionality and addiction severity. *Proc Natl Acad Sci USA* 2014;111:E3149–56.
70. Murray RM, Mehta M, Di Forti M. Different dopaminergic abnormalities underlie cannabis dependence and cannabis-induced psychosis. *Biol Psychiatry* 2014;75:430–431.
71. Gill KE, Poe L, Azimov N, *et al.* Reasons for cannabis use among youths at ultra high risk for psychosis. *Early Interv Psychiatry* 2015;9:207–210.
72. van Winkel R, van Beveren NJ, Simons C. AKT1 moderation of cannabis-induced cognitive alterations in psychotic disorder. *Neuropsychopharmacology* 2011;36:2529–2537.
73. Di Forti M, Iyegbe C, Sallis H, et al. Confirmation that the AKT1 (rs2494732) genotype influences the risk of psychosis in cannabis users. *Biol Psychiatry* 2012;72:811–816.
74. Morgan CJ, Freeman TP, Powell J, Curran HV. AKT1 genotype moderates the acute psychotomimetic effects of naturalistically smoked cannabis in young cannabis smokers. *Transl Psychiatry* 2016;6:e738.
75. Colizzi M, Iyegbe C, Powell J, et al. Interaction between functional genetic variation of DRD2 and cannabis use on risk of psychosis. *Schizophr Bull* 2015;41:1171–1182.
76. McPartland JM, Duncan M, Di Marzo V, Pertwee RG. Are cannabidiol and Delta(9)- tetrahydrocannabivarin negative modulators of the endocannabinoid system? A systematic review. *Br J Pharmacol* 2015;172:737–753.
77. Thomas A, Baillie GL, Phillips AM, Razdan RK, Ross RA, Pertwee RG. Cannabidiol displays unexpectedly high potency as an antagonist of CB1 and CB2 receptor agonists in vitro. *Br J Pharmacol* 2007;150:613–623.
78. Karniol IG, Shirakawa I, Kasinski N, Pfeferman A, Carlini EA. Cannabidiol interferes with the effects of delta9-tetrahydrocannabinol in man. *Eur J Pharmacol* 1974;28:172–177.
79. Caspi A, Moffitt TE, Cannon M, et al. Moderation of the effect of adolescent-onset cannabis use on adult psychosis by a functional polymorphism in the catechol-O-methyltransferase gene: longitudinal evidence of a gene x environment interaction. *Biol Psychiatry* 2005;57:1117–1127.
80. Estrada G, Fatjó-Vilas M, Muñoz MJ, et al. Cannabis use and age at onset of psychosis: further evidence of interaction with COMT Val158Met polymorphism. *Acta Psychiatr Scand* 2011;123(6):485–492.
81. Tunbridge EM, Dunn G, Murray RM, et al. Genetic moderation of the effects of cannabis: catechol-O-methyltransferase (COMT) affects the impact of Delta9-tetrahydrocannabinol (THC) on working memory performance but not on the occurrence of psychotic experiences. *J Psychopharmacol* 2015;29:1146–1151.
82. Cahn W, Hulshoff Pol HE, Caspers E, van Haren NE, Schnack HG, Kahn RS. Cannabis and brain morphology in recent-onset schizophrenia. *Schizophr Res* 2004;67:305–307.
83. Szesko PR, Robinson DG, Sevy S, et al. Anterior cingulate grey-matter deficits and cannabis use in first-episode schizophrenia. *Br J Psychiatry* 2007;190:230–236.
84. Solowij N, Yücel M, Lorenzetti V, Lubman DI. Structural brain alterations in cannabis users: Association with cognitive deficits and psychiatric symptoms. In: Ritsner MS, ed. *The Handbook of Neuropsychiatric Biomarkers, Endophenotypes and Genes. Vol II Neuroanatomical and Neuroimaging Endophenotypes and Biomarkers.* Dordrecht, Germany: Springer; 2009:215–225.
85. Potvin S. Neuroimaging studies of substance abuse in schizophrenia. *Curr Psychiatry Rev* 2009;5(4):250.
86. Bangalore SS, Prasad KMR, Montrose DM, Goradia DD, Diwadkar VA, Keshaven MS. Cannabis use and brain structural alterations in first episode schizophrenia: a region of interest voxel based morphometric study. *Schizophr Res* 2008;99:1–6.
87. Rais M, Cahn W, Van Haren N, et al. Excessive brain volume loss over time in cannabis-using first-episode schizophrenia patients. *Am J Psychiatry* 2008;165:490–496.
88. Potvin SA, Bourgene J, Durand M. The neural correlates of mental rotation abilities in cannabis abusing patients with schizophrenia: an fMRI study. *Schizophr Res Treatment* 2013;543842.
89. French L, Gray C, Leonard G et al. Early cannabis use, polygenic risk score for schizophrenia and brain maturation in adolescence. *JAMA Psychiatry* 2015;72:1002–1011.
90. Bernier D, Cookey J, McAlindon D, et al. Multimodal neuroimaging of frontal white matter microstructure in early phase schizophrenia: the impact of early adolescent cannabis use. *BMC Psychiatry* 2013;13:264.
91. Kuepper R, Ceccarini J, Lataster J, et al. Delta-9- tetrahydracannabinol induced dopamine release as a function of

psychosis risk: 18 F fattypiride positron emission tomography study. PLoS One 2013;8(7):e 70378. Doi: 10.1371/journal.pone.0070378.
92. Cookey J, Bernier D, Tibbo PG. White matter changes in early phase schizophrenia and cannabis use: an update and systematic review of diffusion tensor imaging studies. *Schizophr Res* 2014;156(2–3):137–142.
93. Giordano GN, Ohlsson H, Sundquist K, Sundquist J, Kendler KS. The association between cannabis abuse and subsequent schizophrenia: a Swedish national co-relative control study. *Psychol Med* 2015;45(2):407–414.
94. Proal AC, Fleming J, Galvez-Buccollini JA, Delisi LE. A controlled family study of cannabis users with and without psychosis. *Schizophr Res* 2014;152(1):283–288.
95. Wynne J, Castle D. (2012). Addressing cannabis use in people with psychosis. In: Marijuana and Madness. Castle D, Murray RM, D'Souza CD, eds. 2nd ed. Cambridge, England: Cambridge University Press, 2012: 225–233.
96. Schoeler T, Monk A, Sami MB et al. Continued versus discontinued cannabis use in patients with psychosis: a systematic review and meta-analysis. *Lancet Psychiatry* 2016;3:215–225.
97. Brunette MF, Dawson R, O'Keefe CD, et al. A randomized trial of clozapine vs. other antipsychotics for cannabis use disorder in patients with schizophrenia. *J Dual Diagn* 2011;7:50–63.
98. Siskind DJ, Harris M, Phillipou A, et al. Clozapine users in Australia: their characteristics and experiences of care based on data from the 2010 National Survey of High Impact Psychosis. *Epidemiol Psychiatr Sci* 2016. doi: 10.1017/S2045796016000305.

47

MIGRATION, ETHNICITY, AND PSYCHOSES

Craig Morgan

INTRODUCTION

One of the most replicated findings in the epidemiology of psychotic disorders is the relatively high incidence rates among some migrant and minority ethnic groups. In a context of increasing levels of migration globally and growing ethnic and cultural diversity in many settings, understanding why rates of serious mental disorders are high in some minority populations is important. However, our knowledge of what factors drive these disparities is still limited. In this chapter, I provide an overview of what is known, of what gaps remain, and of the implications for our understanding of both the nature and etiology of psychoses and for public health and service responses.

RATES OF PSYCHOSIS IN MIGRANT AND MINORITY ETHNIC POPULATIONS

There is an extensive literature reporting high rates of psychotic disorders among several migrant and minority ethnic populations in high-income countries. Several systematic reviews and meta-analyses have sought to summarize these findings.[1–3] In one of the earliest, Cantor-Graae and Selten[1] reported a weighted mean relative risk (RR) for migrants (including second-generation) vs. nonmigrants, based on 18 studies, of around 2.9 (95% CI 2.5–3.4). In a later review, Bourque et al.[2] found very similar mean-weighted incidence rate ratios for first- (IRR 2.3, 95% CI 2.0–2.7) and second- (IRR 2.1, 95% CI 1.8–2.5) generation migrants, based on 21 studies. These reviews are useful in highlighting the consistency with which, overall, the literature reports high rates. However, reporting overall summary relative risks and rates creates the impression that there is a single, universal migrant effect. This obscures important variations between and among migrant and minority ethnic groups, and the literature, in fact, suggests a much more complex picture than is often assumed.

The variations in relative risks are striking and more interesting. This is hinted at in the reviews already mentioned. For example, Cantor-Graae and Selten[1] found that the highest relative risk was for those from developing countries (RR 3.3, 95% CI 2.8–3.9) and for those from areas where the majority population is black (RR 4.8, 95% CI 3.7–6.2). However, even these stratified analyses may still be too broad. In the United Kingdom the highest rates (and relative risks) have been reported for the black Caribbean population. Tortelli et al, in a meta-analysis of data from 18 studies, found a pooled IRR for black Caribbean vs. white British of 4.7 (95% 3.9–5.7).[3] Studies suggest rates for the black African population are also elevated relative to the white British population, but, in each relevant study, these tend to be lower than for black Caribbean.[4] The evidence for other minority ethnic groups in the United Kingdom suggests rates are either not increased or only modestly so (e.g., around 1.5, at most, for white non-British and Asian populations).[4–6] This should not be surprising. These are populations with different migratory histories and cultural heritages, living in diverse social contexts, and occupying varying social positions. This is true for the diverse migrant and minority ethnic populations in other countries and it is similarly no surprise that rates vary in these settings. For example, in the Netherlands, rates are particularly for Moroccan and Surinamese populations but less so for Turkish.[7,8] Further, there may be variations in relative risks by gender. In a study in east London, Kirkbride et al.[5] found evidence that rates may be specifically elevated among Pakistani (IRR 3.1) and Bangladeshi (IRR 2.3) women. In the Netherlands, there is strong evidence from several studies that the incidence is substantially higher among men from migrant groups compared with women, especially among those from the Maghreb (Morocco, Algeria, Lybia, Tunisia), with a ratio as high as 5:1.[9] More recent studies have reinforced the complexity of patterns of risk among migrant and minority groups. For example, a recent study using Swedish register data suggests the incidence of psychotic disorders may be particularly high among refugees relative to other migrants (i.e., IRR 2.9 for refugees vs. 1.7 for other migrants), a difference that was more pronounced among men.[10] In short, there is no single, universal, time-invariant migrant effect—as is often implied.

Further, using the language of migration to frame these findings is misleading. As noted later, most studies are of settled minority ethnic populations (including both first-generation migrants and their children, grandchildren, and so on) and, even for populations of recent migrants, the increased risk emerges only after a prolonged period in the new country. For this reason, I predominantly refer, from here on, to minority ethnic populations.

METHODOLOGICAL ARTIFACT AND MISDIAGNOSIS

As a first step, it is important to consider whether the findings of high rates in some migrant and minority groups are valid or an artifact of the way in which this question is studied. Research in this area has become increasingly robust: more comprehensive case finding, more accurate denominator data, standardized procedures for assessing psychopathology and assigning diagnoses, and so on. This, together with the fact that studies using different methods and approaches continue to report ethnic disparities, suggests most practical methodological issues—such as selection bias, incomplete case finding, accuracy of denominator data—are unlikely to account for these findings.[11] For brevity (and completeness), I include selective migration (i.e., the suggestion that those predisposed to psychosis are somehow more likely to migrate) among the artifactual explanations that have been convincingly discounted.

However, the thornier issue of misdiagnosis remains. For many, the problem revealed by these data is the mis- and consequent overdiagnosis of schizophrenia and other psychoses in minority populations. The essence of this is that clinicians and researchers wrongly diagnose schizophrenia or other psychotic disorders among those from minority ethnic groups when they are, in fact, experiencing a mood disorder or, more broadly, expressing distress in response to difficult conditions and experiences in culturally appropriate ways. From this perspective, psychiatry misattributes culturally appropriate expressions of distress among those from minority ethnic populations to symptoms of psychosis, thereby spuriously creating high rates of disorder. Debates on this issue are highly polarized.[12–14]

Where this has been directly assessed, the evidence suggests misdiagnosis is unlikely to fully explain the high rates. For instance, Hickling et al.,[15] in a study that compared diagnoses made by British and Jamaican psychiatrists on the same patients, found no differences in the percentage of black patients diagnosed with schizophrenia. Other studies, designed to investigate racial stereotyping, have found no evidence that psychiatrists are more likely to diagnose schizophrenia when the ethnicity of individuals described in case vignettes is black.[16,17] This, however, is one side of the story. As recent evidence and trends re-highlight (see what follows), the nature of psychoses is clearly such that diagnosis is often challenging, especially across diverse cultural groups. Further, given that low-level psychotic experiences are common and frequently co-occur with symptoms of depression, anxiety, and PTSD (e.g., [18,19]), it is plausible that predominantly affective disorders are sometimes misdiagnosed as psychotic disorders. What is more, there is some direct evidence that misdiagnosis does occur in relation some minority populations (e.g., in the United States).[20] This may not be enough to fully explain the high rates in many minority ethnic groups. But it is an important clinical issue, with profound consequences for management and treatment, and warrants much more research.

THE NATURE AND CAUSES OF PSYCHOSES

Before turning to substantive explanations of why rates are high in some groups, it is useful to briefly sketch three developments that have shaped current understandings of psychoses in general in recent years.

First, substantial evidence has emerged that low level, psychotic—or anomalous and unusual—experiences (such as fleeting and nondistressing hallucinations, suspiciousness, magical thinking, etc.) are common in the general population.[21,22] These experiences are associated with an increased risk of subsequent psychotic disorder and with similar risk factors to those for psychotic disorders.[19,23–26] These findings have led to renewed interest in continuum models of psychosis and stimulated much research on the basis that investigating anomalous experiences may tell us something about psychotic disorder. However, this is complicated by more recent findings that show these experiences are also associated with depression, anxiety, PTSD, and suicide.[18,19] It seems they may be a more general marker of distress. Of particular relevance, there are now several studies that have found relatively high prevalences of these experiences among migrant and minority populations, both in samples of adults and children and adolescents.[27–29] If psychosis does exist on a continuum, then these findings are what would be expected. In populations with high rates of disorder (i.e., with relatively large numbers at the extreme end of the continuum), we would expect to observe high proportions with varying expressions of the phenomena at each point along the continuum. The findings in relation to young people are particularly noteworthy, as it may be that disparities in experiences in childhood and adolescence foreshadow disparities in disorder in adulthood and, as such, provide an opportunity both to understand developmental trajectories toward psychoses and to intervene. This said, caution is warranted. All of the issues concerning misdiagnosis, i.e., cultural misinterpretation, are amplified many-fold in relation to these more nebulous and often crudely measured experiences.

Second, various trends in research have converged to re-highlight the marked heterogeneity of psychoses, both in terms of their manifestations and symptomatology and their course and outcome.[30] This—along with evidence that many risk factors are common across diagnoses—has further and increasingly challenged current diagnostic categories and boundaries. There is, for example, broad consensus that schizophrenia is, at best, a syndrome comprising multiple underlying disorders and, at worst, a spurious category that should be abandoned. This heterogeneity is important because much research is founded on the assumption that our diagnostic categories meaningfully distinguish individuals with distinct disorders. If this is not the case, much research is confounded. This underpins a trend to study psychoses broadly and, more radically, to dispense with diagnoses and focus on particular symptoms and processes (e.g., the Research Domain Criteria [RDoC] initiative).

Third, our understanding of the nature and aetiological architecture of psychotic disorders in general has increased substantially. There remain gaps, but there is some consensus on the broad contours. To begin with, it is clear that an array of factors, that are neither necessary nor sufficient in themselves to cause disorder, are associated with an increased risk of psychoses, spanning genetic, neurobiological, substance use, psychological, and social domains. For example, several social factors have been implicated at area (urbanicity, social fragmentation, ethnic density, etc.) and individual (bullying, abuse, life events, discrimination, etc.) levels.[31] That none are sufficient or necessary means that multiple factors must coparticipate over time—no doubt in various combinations—to push individuals along a developmental pathway to psychosis. Further, evidence is converging on interrelated psychological and biological mechanisms through which these factors increase risk, notably via effects on affective and cognitive processes and on physiological stress response and the dopamine system.[32–35] This evidence has recently been synthesized by Howes and Murray,[36] who draw from our socio-developmental model[11] to propose an integrated socio-developmental-cognitive model of schizophrenia.

CANDIDATE EXPLANATIONS

The preceding discussion provides a basis for considering the substantive factors that may contribute to high rates of psychotic disorder among some minority ethnic populations. What is the evidence that factors in each of the established domains of risk—genetic, neurodevelopmental, substance use, and social—may contribute to the elevated rates in some minority groups?

GENES AND NEURODEVELOPMENT

In an earlier review,[11] we concluded that there was no evidence that genetic or neurodevelopmental factors contributed to the high rates of psychoses observed in some minority groups. For example, there are at least three family studies that suggest similar degrees of parental (genetic) risk across ethnic groups.[37–39] Incidentally, these studies, in also finding evidence of elevated risk among siblings, more strongly hint at a role for environmental factors. Further, despite initial speculation about whether higher rates of prenatal infections or obstetric complications—both implicated in general in the etiology of psychoses—in some populations may underpin higher rates of disorder, there is no evidence to support this (albeit, the number of relevant studies is small [e.g., [40]]). More recently, there has been some speculation about a possible role for vitamin D deficiency, the hypothesis being that low prenatal vitamin D, as a consequence of reduced exposure to sunlight in moving from warm to colder northern European climates, may impact on brain development and increase risk for later psychoses in children of migrants.[41] No study has directly investigated this. However, that rates are relatively low among Asian populations in the United Kingdom, in which rates of vitamin D deficiency are high, suggests this is unlikely to be a prominent explanation. Similarly, our own work on direct measures of neurodevelopmental markers (e.g., minor physical anomalies) does not suggest these are contributory factors.[42] In the past 10 years or so, there has been no further research that has directly sought to investigate the potential contribution of these factors. The conclusion of our previous review consequently remains the same.

SUBSTANCE USE

There is now strong evidence that use of certain substances, particularly cannabis, is associated with an increased risk of psychosis, especially high potency forms with high concentrations of tetrahydrocannabinol (e.g., skunk).[43,44] This is particularly relevant here because cannabis use was one of the earliest and most controversial explanations proposed for the high rates of psychosis observed among the black Caribbean population in the United Kingdom. Previous work, however, has not provided any strong evidence to support this.[11,45] More recent work has been slow to emerge. However, it is likely that current interest in cannabis will produce samples that are sufficiently ethnically diverse to allow this question to be considered more fully and rigorously than before.

MIGRATION

A commonly proposed explanation—embedded in the framing of this issue in the language of migration—implicates the stresses of migration. However, this is unlikely to provide a full explanation for two reasons. The time between migration and onset is typically many years, and the disparities extend to those born to migrants in countries of settlement (i.e., not migrants at all). Holland et al., for example, found that the time from migration to first diagnosis was around 3 years for nonrefugee migrants.[10] This is in line with earlier reports. For refugees, it was shorter at around 2 years.[10] Veling et al.,[46] in a study of incident cases in the Netherlands found that earlier age of migration was associated with a greater increased risk. It may be that migration at an earlier age has a particularly pernicious impact on risk of psychosis, but earlier migration also means longer living, during formative years of childhood and adolescence, in the host society. It is likely, then, that these patterns are more a reflection of living in societies as part of a minority ethnic population than a consequence of the experiences and processes of migration. This is not to dismiss entirely migration and negative premigratory experiences, especially of persecution and trauma among refugees, and these may compound the effects of negative experiences postmigration. Indeed, this may explain the particularly high rates among refugees and those migrating from countries with especially high levels of poverty and violence. The major effects, however, are likely to arise from the adverse social conditions and experiences that many from minority ethnic groups encounter in the countries of settlement.

SOCIAL CONDITIONS AND EXPERIENCES

Perhaps the most striking change in the past 20 years in our models of the etiology of psychoses is that exposures over the life course to challenging social conditions and adverse experiences feature prominently.[36] In part, this shift has occurred because of research on migrant and minority populations. That rates of disorder vary by ethnic group and that well-established genetic and neurodevelopmental risk factors cannot plausibly account for these disparities strongly implicates socio-environmental exposures. The social exposures implicated in general include, at area level, population density,[47] social fragmentation and deprivation,[48] and ethnic density[49] and, at individual level, childhood adversities[50] such as bullying, family breakdown, neglect and abuse, adult life events,[51] discrimination,[52] and broad indicators of socioeconomic disadvantage.[53] Further, there is emerging evidence that these social adversities affect psychological and biological pathways implicated in the etiology of psychoses; that is, there are plausible mechanisms through which external social conditions and experiences may impact on biological and psychological processes to become embodied and manifest in experiences of paranoia, hallucinations, and disturbances of thought (e.g., [32,34]). What, then, are the specific putative social factors that may increase risk among some minority populations and what is the relevant evidence?

AREA LEVEL

Those from minority ethnic groups are more likely to live in densely populated and relatively disadvantaged urban areas. Living in these types of areas is—in general—associated with an increased risk of psychosis. Does this explain the high rates in some minority ethnic populations? Studies that have considered this do not find any strong evidence that this is the explanation. In an early study, Harrison et al.[54] found no evidence that area of residence could account for observed differences in rates between white British and black Caribbean populations in Nottingham, UK, and our subsequent study in three UK centers (AESOP) found similarly elevated rates by ethnic groups in all centers, despite there being of varying degrees of urbanicity (population density).[55] Further, in a more recent study in the United Kingdom, Kirkbride et al.[56] found similar ethnic disparities were evident in both rural and urban areas in a region in the east of England (i.e., two- to fourfold increased rates in black Caribbean, black African, and Pakistani populations, compared with white British).

At an area level, the most striking and consistent finding is that rates of psychosis are highest among migrant and minority ethnic groups where they form a smaller proportion of the local population.[49,57,58] This ethnic density effect stems back to the seminal study by Faris and Dunham[59] and has been replicated in many studies since, including several recent reports. Interpreting these findings, however, is difficult. When set alongside individual level data that suggests repeated exposure to discrimination is important (see next section) it is possible that living in areas of low ethnic density may increase risk because of exposure to more discrimination and hostility. Conversely, living in areas of high ethnic density may mitigate risks because of fewer negative experiences, more social supports, and greater cultural congruity. Das Munshi et al.,[60] in analyses of data from a UK national survey, did find some evidence to support this, e.g., those in areas of low own ethnic group density did report more experiences of racism, discrimination, and fewer social supports. However, no studies have directly investigated these possibilities in relation to psychotic disorder. Further, the effects for psychosis may not be uniform. Schofield et al.,[61] in a study using Danish register data, found evidence of ethnic density effects for second-, but not first-, generation migrants.

INDIVIDUAL LEVEL

At an individual level, there is some evidence implicating markers of childhood and adult disadvantage and, more specifically, discrimination. For example, our work from the AESOP study found that family breakdown (indexed by separation from parents) was both associated with increased odds of psychosis and was more common among black Caribbean populations in the United Kingdom.[62] This is in line with earlier work[63] and, more generally, with work in other contexts that suggests family breakdown in childhood is associated with a subsequent increased risk of psychosis.[64] However, parental separation is a crude marker that, at most, indexes exposures that occur before, during, or after that specifically impact on risk. We do not, though, know what these specific experiences are that increase risk within the context of family breakdown. There is a wider literature that provides strong evidence that household discord, neglect, and various forms of abuse increase risk of psychosis. However, these exposures have not, as far as we are aware, been considered directly in relation to minority ethnic groups and psychosis and we do not know whether they are more common or have stronger effects in these groups. Further, in AESOP we found similar patterns for markers of adult disadvantage and isolation: i.e., increased odds of psychosis in general; more common among black Caribbean and black African populations.[65] If these markers of disadvantage do index exposures that increase risk, then their greater prevalence in minority ethnic groups may partly underlie the reported high rates. However, this data relates to groups in the United Kingdom and the extent to which they relate to other contexts is unclear. Plus, the variables are crude and, at best, provide a hint, but they do little to help us understand the relevant exposures and mechanisms in a way that could inform strategies for prevention or intervention.

More specifically, the potential role of discrimination and perceived disadvantaged has been studied. Two early analyses of prevalence samples provided some initial evidence that experiences of racism and discrimination may be important.[66,67] Of particular note, Karlsen and Nazroo,[67] in an analysis of data from the Fourth National Survey of Ethnic Minorities in the United Kingdom, found an association between the estimated annual prevalence of psychosis and reports of perceived employer racism (OR 1.6), verbal abuse (OR 2.9), and racial attacks (OR 4.8). It is notable that the strongest effect was for experiences involving physical threat and violence (racial attacks). The limited relevant data from first-episode samples broadly supports these findings, albeit the approaches and measures used to capture discrimination and disadvantage vary. In AESOP we found that perceptions of disadvantage partly explained the high rates in black Caribbean and black African groups.[68] Veling et al.[69] in the Netherlands reported that the highest incidence rates were among those populations known to have the highest levels of perceived discrimination (i.e., Moroccan, IRR 4.8). In other analyses of case-sibling-control data on non-Western migrants, Veling et al.[70] found that cases were more likely to have a negative ethnic identity, compared with their matched controls. Together, these findings point to discrimination and perceived disadvantage, stemming from minority position, as potentially important factors.

SOCIAL DEFEAT

A prominent unifying hypothesis that seeks to bring the evidence together is that the high rates in some migrant and minority ethnic populations is a consequence of social defeat, i.e., the negative experience of being excluded from the majority group.[71,72] This hypothesis arose from analogy with animal studies that show rodents subject to threatening and intimidating behavior by other rodents become passive and submissive (i.e., defeated) and that this is associated with sensitization of the mesolimbic dopaminergic system, which has been implicated—in humans—in the underlying biology of psychoses. This hypothesis, that social defeat underpins the high rates of psychosis in some migrant and minority groups, has gained some traction.

However, the social defeat hypothesis of psychosis, as formulated, is problematic. First, it is largely tautological. It posits minority status (i.e., being excluded from a majority group) to account for high rates of disorder among those who occupy minority statuses. Second, in the animal studies that were the basis for the original formulation, social defeat is the outcome not the exposure. It is prolonged intimidating and threatening behavior (not outsider status) that produces the outcome—passivity and submission (social defeat). In other words, it is excessive and repeated threat that is associated with—or leads to—sensitization of the mesolimbic dopaminergic system in this model. Further, the end state of defeat that characterizes the rodents is reminiscent of a state of helplessness. This is why the social defeat paradigm is usually considered a model of depression.[73] This conceptual and theoretical confusion is reflected in the most recent formulation that reframes a disparate range of factors that have previously been associated with psychosis as indicators of social defeat, including—inexplicably—low IQ.

Unfortunately, the high rates of psychosis in some minority groups are unlikely to be explained so simply by a single exposure.[72] The range of factors involved and the mechanisms through which they impact on risk in minority groups are likely to be much more complex—and we need to embrace and seek to understand this complexity.

A SOCIODEVELOPMENTAL PATHWAY TO PSYCHOSIS

In previously synthesizing the evidence on rates of psychosis in minority ethnic populations, we proposed a sociodevelopmental pathway to account for the increases seen in some.[11] That is, we hypothesized a developmental pathway in which exposure to adversity and trauma (particularly in childhood and/or prior to and during migration)—in the absence of buffers and protective factors—interacts with underlying genetic risk and impacts on neurobiological development (in particular the stress response and dopamine systems) to create an enduring liability to psychosis (reflected in, for example, expression of anomalous and unusual experiences). This liability then becomes manifest as disorder in the event of further cumulative stressors and/or prolonged substance use, especially high potency cannabis. As noted, this proposal has been incorporated into broader models of psychosis.[36] Our purpose in highlighting a sociodevelopmental pathway is to draw attention to the possibility that there are some for whom adverse social conditions and experiences are the primary factors in the development of psychoses: that is, in the absence of these exposures, psychotic disorders would not have developed. It further follows from this that, in populations where adverse social conditions and experiences are more common, rates of psychosis will be higher. Our hypothesis is that this underpins the high rates in many migrant and minority ethnic populations.

The evidence that has accumulated in recent years—albeit fragmented and sporadic—both fits with this model and suggests refinements. It is, for example, possible to speculate more precisely about the kinds of social conditions and experiences that may be relevant, both in general and for migrant and minority groups in particular. There is growing evidence in general that contexts and experiences that involve high levels of interpersonal threat, hostility, and violence specifically increase risk of psychoses. For example, using data from the E-Risk study, Arseneault et al.[74] found that bullying and maltreatment, but not accidents, during childhood were associated with later psychotic (anomalous) experiences at age 12. This specifically implicates negative experiences involving intention to harm. In relation to migrant and minority groups, this fits more precisely with the evidence, e.g., of particularly high rates among refugees (exposed, by definition, to extreme threat) and of effects for discrimination, especially involving violence, as detailed earlier. This points to a more specific formulation of the relevant social exposures: that is, exposure—over the life course—to threat, hostility (including discrimination), and violence, especially in contexts of poverty, disadvantage, and isolation (e.g., in areas of low ethnic density).

SUMMARY: IMPLICATIONS AND GAPS

A core conclusion from this overview is that the patterns—and determinants—of variations in the occurrence of psychoses by ethnic group are more complex than commonly assumed.

We should not shy away from this. Acknowledging this complexity opens up new avenues for investigating, understanding, and ultimately addressing this critically important public health issue. What we know at present is that rates of psychotic disorder are high in some minority ethnic groups and that the most likely set of explanations for this center around adverse social conditions and experiences, especially those involving threat and violence. However, this conclusion on explanations is derived from a relatively small body of research in which the social is, at most, crudely characterized. Further, we do not know why rates are high in some minority ethnic groups in some settings and not others. This means it is not currently possible to formulate tangible, evidence-based proposals for feasible public health strategies and interventions to address this issue. In this respect, we are not much further forward than we were 10 years ago.

REFERENCES

1. Cantor-Graae E, Selten JP. Schizophrenia and migration: a meta-analysis and review. *Am J Psychiatry*. 2005;162:12–24.
2. Bourque F, van der Ven E, Malla A. A meta-analysis of the risk for psychotic disorders among first- and second-generation immigrants. *Psychol Med*. 2011;41(5):897–910.
3. Tortelli A, Errazuriz A, Croudace T, et al. Schizophrenia and other psychotic disorders in Caribbean-born migrants and their descendants in England: systematic review and meta-analysis of incidence rates, 1950–2013. *Soc Psychiatry Psychiatr Epidemiol*. 2015;50(7):1039–1055.
4. Fearon P, Kirkbride JB, Morgan C, et al. Incidence of schizophrenia and other psychoses in ethnic minority groups: results from the MRC AESOP Study. *Psychol Med*. 2006;36(11):1541–1550.
5. Kirkbride JB, Barker D, Cowden F, et al. Psychoses, ethnicity and socio-economic status. *Br J Psychiatry*. 2008;193(1):18–24.
6. Kirkbride JB, Errazuriz A, Croudace TJ, et al. Incidence of schizophrenia and other psychoses in England, 1950–2009: a systematic review and meta-analyses. *PLoS One*. 2012;7(3):e31660.
7. Selten JP, Veen N, Feller W, et al. Incidence of psychotic disorders in immigrant groups to the Netherlands. *Br J Psychiatry*. 2001;178:367–372.
8. Veling W, Selten JP, Veen N, Laan W, Blom JD, Hoek HW. Incidence of schizophrenia among ethnic minorities in the Netherlands: a four-year first-contact study. *Schizophr Res*. 2006;86(1–3):189–193.
9. van der Ven E, Veling W, Tortelli A, et al. Evidence of an excessive gender gap in the risk of psychotic disorder among North African immigrants in Europe: a systematic review and meta-analysis. *Soc Psychiatry Psychiatr Epidemiol*. 2016;51(12):1603–1613.
10. Hollander AC, Dal H, Lewis G, Magnusson C, Kirkbride JB, Dalman C. Refugee migration and risk of schizophrenia and other non-affective psychoses: cohort study of 1.3 million people in Sweden. *BMJ*. 2016;352.
11. Morgan C, Charalambides M, Hutchinson G, Murray RM. Migration, ethnicity, and psychosis: toward a sociodevelopmental model. *Schizophr Bull*. 2010;36(4):655–664.
12. Fernando S. *Mental Health, Race and Culture*. London: Macmillan; 1991.
13. McKenzie K. Moving the misdiagnosis debate forward. *International Review of Psychiatry*. 1999;11:153–161.
14. Selten JP, Hoek H. Does misdiagnosis explain the schizophrenia epidemic among immigrants from developing countries to Western Europe? *Soc Psychiatry Psychiatr Epidemiol*. 2008;43(12):937–939.
15. Hickling FW, McKenzie, K., Mullen, R., & Murray, R. A Jamaican psychiatrist evaluates diagnoses at a London psychiatric hospital. *Br J Psychiatry*. 1999;175:283–286.
16. Lewis G, Croft-Jeffreys C, David A. Are British psychiatrists racist? *Br J Psychiatry*. 1990;157:410–415.
17. Minnis H, McMillan A, Gillies M, Smith S. Racial stereotyping: survey of psychiatrists in the United Kingdom. *BMJ*. 2001;323(7318):905–906.
18. Kelleher I, Lynch F, Harley M, et al. Psychotic symptoms in adolescence index risk for suicidal behavior: findings from 2 population-based case-control clinical interview studies. *Arch Gen Psychiatry*. 2012;69(12):1277–1283.
19. Fisher HL, Caspi A, Poulton R, et al. Specificity of childhood psychotic symptoms for predicting schizophrenia by 38 years of age: a birth cohort study. *Psychol Med*. 2013;43(10):2077–2086.
20. Garb HN. Race bias, social class bias, and gender bias in clinical judgement. *Clinical Psychology: Science and Practice*. 1997;4(2):99–120.
21. Linscott RJ, van Os J. An updated and conservative systematic review and meta-analysis of epidemiological evidence on psychotic experiences in children and adults: on the pathway from proneness to persistence to dimensional expression across mental disorders. *Psychol Med*. 2013;43(6):1133–1149.
22. Kelleher I, Connor D, Clarke MC, Devlin N, Harley M, Cannon M. Prevalence of psychotic symptoms in childhood and adolescence: a systematic review and meta-analysis of population-based studies. *Psychol Med*. 2012;42(9):1857–1863.
23. Bentall RP, Fernyhough C. Social predictors of psychotic experiences: specificity and psychological mechanisms. *Schizophr Bull*. 2008;34(6):1012–1020.
24. Mackie C, Castellanos-Ryan N, Conrod P. Developmental trajectories of psychotic-like experiences across adolescence: Impact of victimization and substance use. *Psychol Med*. 2011;41(1):47–58.
25. Mackie C, O'Leary-Barrett M, Al-Khudhairy N, et al. Adolescent bullying, cannabis use and emerging psychotic experiences: A longitudinal general population study. *Psychol Med*. 2013;43(5):1033–1044.
26. Morgan C, Reininghaus U, Reichenberg A, et al. Adversity, cannabis use and psychotic experiences: evidence of cumulative and synergistic effects. *Br J Psychiatry*. 2014;204:346–353.
27. Morgan C, Fisher H, Hutchinson G, et al. Ethnicity, social disadvantage and psychotic-like experiences in a healthy population based sample. *Acta Psychiatr Scand*. 2009;119(3):226–235.
28. Adriaanse M, van Domburgh L, Hoek HW, Susser E, Doreleijers TAH, Veling W. Prevalence, impact and cultural context of psychotic experiences among ethnic minority youth. *Psychol Med*. 2015;45(3):637–646.
29. Laurens KR, West SA, Murray RM, Hodgins S. Psychotic-like experiences and other antecedents of schizophrenia in children aged 9-12 years: a comparison of ethnic and migrant groups in the United Kingdom. *Psychol Med*. 2008;38(8):1103–1111.
30. Silverstein SM, Moghaddam B, Wykes T. *Schizophrenia: Evolution and Synthesis*. Cambridge, MA: MIT Press; 2013.
31. Morgan C, McKenzie K, Fearon P. *Society and Psychosis*. Cambridge: Cambridge University Press; 2008.
32. Garety PA, Kuipers E, Fowler D, Freeman D, Bebbington PE. A cognitive model of the positive symptoms of psychosis. *Psychol Med*. 2001;31(2):189–195.
33. Garety PA, Bebbington P, Fowler D, Freeman D, Kuipers E. Implications for neurobiological research of cognitive models of psychosis: a theoretical paper. *Psychol Med*. 2007;37(10):1377–1391.
34. Borges S, Gayer-Anderson C, Mondelli V. A systematic review of the activity of the hypothalamic–pituitary–adrenal axis in first episode psychosis. *Psychoneuroendocrinology*. 2013;38(5):603–611.
35. Howes OD, Kapur S. The dopamine hypothesis of schizophrenia: version III: the final common pathway. *Schizophr Bull*. 2009;35(3):549–562.
36. Howes OD, Murray RM. Schizophrenia: an integrated sociodevelopmental-cognitive model. *Lancet*. 2014;383(9929):1677–1687.
37. Selten JP, Blom JD, van der Tweel I, Veling W, Leliefeld B, Hoek HW. Psychosis risk for parents and siblings of Dutch and Moroccan-Dutch patients with non-affective psychotic disorder. *Schizophr Res*. 2008;104(1–3):274–278.

38. Hutchinson G, Takei N, Fahy TA, et al. Morbid risk of schizophrenia in first-degree relatives of white and African-Caribbean patients with psychosis. *Br J Psychiatry*. 1996;169:776–780.
39. Sugarman PA, Craufurd D. Schizophrenia in the Afro-Caribbean community. *Br J Psychiatry*. 1994;164(4):474–480.
40. Hutchinson G, Takei N, Bhugra D, et al. The increased rate of psychosis among African-Caribbeans in Britain is not due to an excess of pregnancy and birth complications. *Br J Psychiatry*. 1997;171:145–147.
41. McGrath J. Migrant status, vitamin D and risk of schizophrenia. *Psychol Med*. 2011;41(4):892–893; author reply 894.
42. Dean K, Dazzan P, Lloyd T, et al. Minor physical anomalies across ethnic groups in a first episode psychosis sample. *Schizophr Res*. 2007;89(1–3):86–90.
43. Di Forti M, Morgan C, Dazzan P, et al. High-potency cannabis and the risk of psychosis. *Br J Psychiatry*. 2009;195(6):488–491.
44. Di Forti M, Quattrone D, Freeman TP, et al. The contribution of cannabis use to variation in the incidence of psychotic disorder across Europe (EU-GEI): a multicentre case-control study. *Lancet Psychiatry*. 2019.
45. Veen N, Selten JP, Hoek HW, Feller W, van der Graaf Y, Kahn R. Use of illicit substances in a psychosis incidence cohort: a comparison among different ethnic groups in the Netherlands. *Acta Psychiatr Scand*. 2002;105(6):440–443.
46. Veling W, Hoek HW, Selten JP, Susser E. Age at migration and future risk of psychotic disorders among immigrants in the Netherlands: a 7-year incidence study. *Am J Psychiatry*. 2011;168(12): 1278–1285.
47. Vassos E, Agerbo E, Mors O, Pedersen CB. Urban-rural differences in incidence rates of psychiatric disorders in Denmark. *Br J Psychiatry*. 2016;208(5):435–440.
48. Allardyce J, Gilmour H, Atkinson J, Rapson T, Bishop J, McCreadie RG. Social fragmentation, deprivation and urbanicity: relation to first-admission rates for psychoses. *Br J Psychiatry*. 2005;187:401–406.
49. Schofield P, Thygesen M, Das-Munshi J, et al. Ethnic density, urbanicity and psychosis risk for migrant groups—a population cohort study. *Schizophr Res*. 2017;190:82–87.
50. Morgan C, Gayer-Anderson C. Childhood adversities and psychosis: evidence, challenges, implications. *World Psychiatry*. 2016;15(2):93–102.
51. Beards S, Gayer-Anderson C, Borges S, Dewey ME, Fisher HL, Morgan C. Life events and psychosis: a review and meta-analysis. *Schizophr Bull*. 2013;39(4):740–747.
52. Veling W, Hoek HW, Mackenbach JP. Perceived discrimination and the risk of schizophrenia in ethnic minorities: a case-control study. *Soc Psychiatry Psychiatr Epidemiol*. 2008;43(12):953–959.
53. Stilo S, Di Forti M, Mondelli V, et al. Social disadvantage: cause or consequence of impending psychosis? *Schizophr Bull*. 2012.
54. Harrison G, Holton A, Neilson D, Owens D, Boot D, Cooper B. Severe mental disorder in Afro-Caribbean patients: some social, demographic and service factors. *Psychol Med*. 1989;19: 683–696.
55. Kirkbride JB, Fearon P, Morgan C, et al. Heterogeneity in incidence rates of schizophrenia and other psychotic syndromes: findings from the 3-center AeSOP study. *Arch Gen Psychiatry*. 2006;63(3):250–258.
56. Kirkbride JB, Hameed Y, Ioannidis K, et al. Ethnic minority status, age-at-immigration and psychosis risk in rural environments: evidence from the SEPEA study. *Schizophr Bull*. 2017;43(6):1251–1261.
57. Kirkbride JB, Morgan C, Fearon P, Dazzan P, Murray RM, Jones PB. Neighbourhood-level effects on psychoses: re-examining the role of context. *Psychol Med*. 2007;37(10):1413–1425.
58. Boydell J, van Os J, McKenzie K, et al. Incidence of schizophrenia in ethnic minorities in London: ecological study into interactions with environment. *BMJ*. 2001;323:1336–1338.
59. Faris R, Dunham H. *Mental Disorders in Urban Areas*. Chicago: University of Chicago Press; 1939.
60. Das-Munshi J, Becares L, Boydell JE, et al. Ethnic density as a buffer for psychotic experiences: findings from a national survey (EMPIRIC). *Br J Psychiatry*. 2012;201(4):282–290.
61. Schofield P, Thygesen M, Das-Munshi J, et al. Neighbourhood ethnic density and psychosis—is there a difference according to generation? *Schizophr Res*. 2018;195:501–505.
62. Morgan C, Kirkbride J, Leff J, et al. Parental separation, loss and psychosis in different ethnic groups: a case-control study. *Psychol Med*. 2007;37(4):495–503.
63. Mallett R, Leff J, Bhugra D, Pang D, Zhao JH. Social environment, ethnicity and schizophrenia. *Soc Psychiatry Psychiatr Epidemiol*. 2002;37:329–335.
64. Wicks S, Hjern A, Gunnell D, Lewis G, Dalman C. Social adversity in childhood and the risk of developing psychosis: a national cohort study. *Am J Psychiatry*. 2005;162:1652–1657.
65. Morgan C, Kirkbride J, Hutchinson G, et al. Cumulative social disadvantage, ethnicity and first-episode psychosis: a case-control study. *Psychol Med*. 2008;38(12):1701–1715.
66. Karlsen S, Nazroo JY. Relation between racial discrimination, social class, and health among ethnic minority groups. *Am J Public Health*. 2002;92(4):624–631.
67. Karlsen S, Nazroo J. Relation between racial discrimination, social class and health among ethnic minority groups. *Am J Public Health*. 2002;92:624–631.
68. Cooper C MC, Byrne M, Dazzan P, Morgan K, Hutchinson G, Doody G, Harrison G, Leff J, Jones P, Ismail K, Murray R, Bebbington P, Fearon P. Perceptions of disadvantage, ethnicity and psychosis: results from the ÆSOP study. *Br J Psychiatry*. 2008;192:185–190.
69. Veling W, Selten JP, Susser E, Laan W, Mackenbach JP, Hoek HW. Discrimination and the incidence of psychotic disorders among ethnic minorities in The Netherlands. *Int J Epidemiol*. 2007;36(4):761–768.
70. Veling W, Hoek HW, Mackenbach JP. Perceived discrimination and the risk of schizophrenia in ethnic minorities. *Soc Psychiatry Psychiatr Epidemiol*. 2008;43(12):953–959.
71. Selten JP, Cantor-Graae E. Social defeat: risk factor for schizophrenia? *Br J Psychiatry*. 2005;187:101–102.
72. Selten JP, van der Ven E, Rutten BPF, Cantor-Graae E. The social defeat hypothesis of schizophrenia: an update. *Schizophr Bull*. 2013;39(6):1180–1186.
73. Hollis F, Kabbaj M. Social defeat as an animal model for depression. *ILAR J*. 2014;55(2):221–232.
74. Arseneault L, Cannon M, Fisher HL, Polanczyk G, Moffitt TE, Caspi A. Childhood trauma and children's emerging psychotic symptoms: A genetically sensitive longitudinal cohort study. *Am J Psychiatry*. 2011;168(1):65–72.

PSYCHOLOGICAL MECHANISMS AND PSYCHOSIS

48

COGNITIVE AND EMOTIONAL PROCESSES IN PSYCHOSIS

Steffen Moritz, Thies Lüdtke, Łukasz Gawęda, Jakob Scheunemann, and Ryan P. Balzan

Research into the pathogenesis of schizophrenia and other psychiatric disorders is traditionally divided into two major streams; biogenetic and psychological. Whereas the former is primarily concerned with premorbid/ingrained impairment such as genetic aberrations, neurocognitive deficits, and cortical dysfunction, the latter focuses on emotional factors and cognitive biases (i.e., distortions in the processing, weighting, and retrieval of information). The two perspectives are not mutually exclusive,[1] though, and there is indeed an emerging literature on their interactions (Broyd et al., 2017; Howes and Murray, 2014). Yet, it remains useful for heuristic purposes to make a distinction whether psychosis is primarily viewed as an *abnormality*, or even a disability, implicating the need for biological treatments to correct for an alleged deficit (e.g., antipsychotic medication against a presumed dysregulated dopaminergic state) or else a *deviation from normality* amenable to psychological understanding and thus psychotherapy. These models also impact how patients and the public view the disorder; while a biological perspective may reduce blame, it seems to induce pessimism and promote the urge for social distance toward people with mental illness (Kvaale et al., 2013; Speerforck et al., 2014). This present review addresses the psychological stream of research into the underpinnings of schizophrenia, particularly for the formation and maintenance of positive symptoms. In a very provocative paper, Read, Bentall and Fosse (2009) have noted this line of research is both underfunded and under-represented in psychosis research (see also Goldfried, 2016). To some degree, this has historical reasons, as the pioneers of modern psychosis research, particularly Kraepelin (1899), whose work remains influential to the present day, put forward that schizophrenia (then called dementia praecox) reflects a severe brain disorder, without much empirical evidence and none that would be accepted today as unequivocal support. Kraepelin assumed a neurodegenerative course, while Jaspers (1913) posited that the hallmark symptoms of psychosis, delusional ideas, are not psychologically accessible. While another influential figure of psychiatry, Sigmund Freud (1963), disputed the latter assumption, he was pessimistic about the treatability of psychosis. As we summarize later, a psychological understanding of core psychotic symptoms has come a long way but many empirical gaps remain. Our review is primarily concerned with delusions, a core feature of psychosis. Without this symptom any diagnosis of schizophrenia or psychosis remains at least questionable, even though DSM-5 views delusions as only one out of five core symptoms (Association Psychiatric Association, 2013). Clinically significant delusions are a common symptom of schizophrenia, and may be experienced by over 70% of people during psychotic episodes (Lewis et al., 2009).

We divide this chapter into three sections: The first two sections highlight two recent models that—to a different degree—ascribe emotional and cognitive factors a causal role in the pathogenesis of (persecutory) delusions. We start with the model by Freeman and Garety (Freeman, 2016; Freeman and Garety, 2014; Garety and Freeman, 2013) which outlines six causal mechanisms for persecutory delusions, four of which are emotional in essence. This is followed by our own two-stage model (Moritz et al., 2016c). In the last section, we highlight areas that warrant further empirical research as well as unresolved issues. Our chapter leaves aside neurocognitive deficits, although these are intertwined with cognitive biases (Garety et al., 2013) and emotions (Moritz et al., 2017) in complex ways (e.g., awareness of cognitive deficits may compromise emotional well-being; emotional problems such as anxiety and stress can, in turn, impair cognitive performance). Neurocognitive deficits are addressed in greater detail in other chapters of this book. For reasons of constraint we also leave out societal aspects such as the impact of life adversities and urbanicity on later psychosis (Bentall et al., 2014; Heinz et al., 2013).

THE PSYCHOLOGICAL MODEL BY GARETY AND FREEMAN

A theoretical model by Freeman and Garety (Freeman, 2016; Freeman and Garety, 2014; Garety and Freeman, 2013) regards three dimensions (involving six core elements) as proximal and potentially causal for the formation of persecutory delusions/paranoia, the most prevalent subtype of delusions (for their original model see Figure 48.1):

1 Many authors start their articles highlighting the multicausality of psychosis and credit the respective other stream. Yet, in the remainder of the text one perspective is usually prevalent.

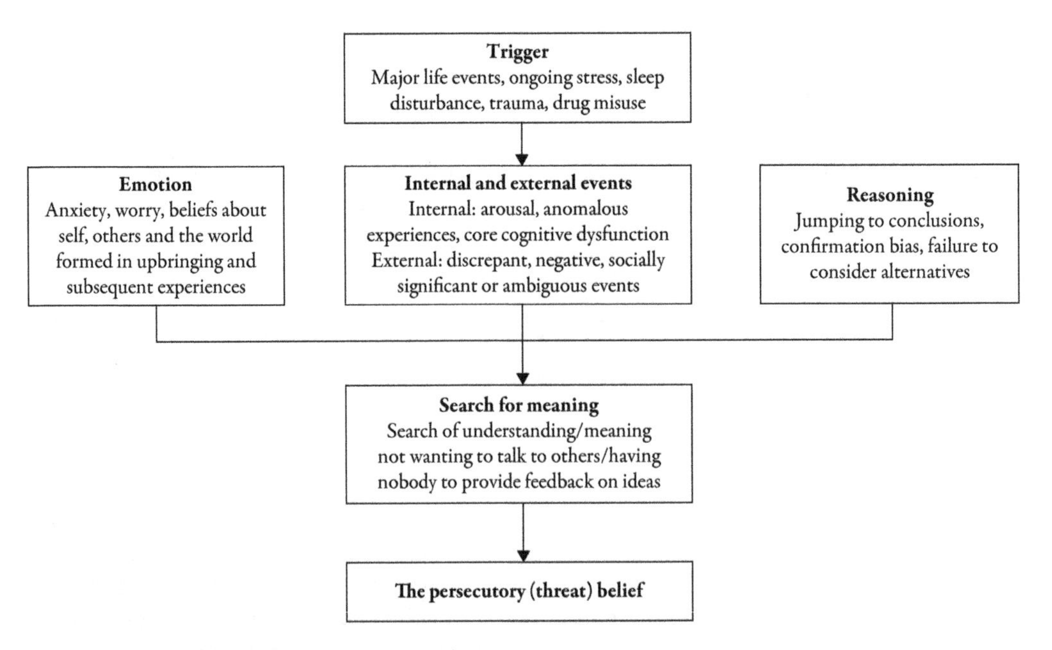

Figure 48.1 Model by Freeman and Garety (Garety and Freeman, 2013).

1. Emotional symptoms: these encompass a worry thinking style, negative beliefs about the self, interpersonal sensitivity, and sleep disturbance. Recently, the potential role of safety-seeking behaviors/avoidance for the pathogenesis of persecutory delusions has been stressed (Freeman, 2016).
2. Reasoning bias(es): These encompass belief inflexibility and particularly jumping to conclusions (JTC).
3. Anomalous internal experiences: These are essentially hallucinatory experiences and sensory irritations that an individual misattributes to an external source or agent. Anomalous internal experiences are also central to the delusion theories by Kapur ("salience theory") (Kapur, 2003), Corlett (Corlett et al., 2010), and Maher (2006), which posit that "surprising experiences demand surprising explanations" (Corlett et al., 2010, p. 360).

Given that reasoning biases are addressed in the next section when describing our own two-factor model, and the role of anomalous internal experiences are described in other chapters of this book, this section focuses on the role that emotional symptoms play in the development of persecutory delusions.

On first sight, it may come to a surprise that emotional problems and not cognitive deficits are dominant in the model, as affective psychosis (i.e., severe depression and bipolar disorder) versus nonaffective psychosis (i.e., schizophrenia, delusional disorder) have long been seen as independent disorders with different etiologies, again dating back to the days of Kraepelin. However, this is increasingly regarded as obsolete. Meta-analyses assert that affective disorders are common in psychosis (Buckley et al., 2009) with a prevalence of depressive mood (≥2 weeks) at first admission of 83%. According to Upthegrove (2009), the prevalence of depression is 45%–83% in the prodrome, 29%–75% in the acute phase, and 5%–54% post-psychosis. The notion that affective disorders may not only precede but also perhaps cause paranoia is supported by both basic research studies and treatment trials. Depressive symptoms are common precursors of psychosis and part of the (unspecific) schizophrenia prodrome (Häfner et al., 2013; Upthegrove, 2009), long before core psychosis symptoms surface. The majority of studies confirm a (predictive) relationship of negative affect with paranoia in both clinical and nonclinical populations (Freeman et al., 2012, 2015a; Jaya et al., 2018; Kramer et al., 2014; Morrison et al., 2015; So et al., 2018; Tone et al., 2011), or its mediating role (Freeman et al., 2014a; Lincoln et al., 2009).

The bulk of evidence for the four emotional components (i.e., worry thinking style, negative beliefs about the self, interpersonal sensitivity, and sleep disturbance) is extensive, however *negative beliefs about the self* is perhaps the best-established emotional factor in the model. Reviews on self-esteem and self-schemas converge on the conclusion that patients with persecutory delusions show low and fluctuating self-esteem as well as negative self-evaluation (Kesting and Lincoln, 2013; Tiernan et al., 2014), particularly those with "bad me" paranoia (i.e., higher deservedness of the paranoia).[2] The review by Kesting and Lincoln (2013) suggests that positive self-evaluations are spared or are at least less compromised (however see Collett et al., 2016).

Yet, not all studies have found strong ties between depression and paranoia (Moritz et al., 2017b; Sullivan et al., 2014), which might be owing to the differences in statistical procedures (Moritz et al., 2017b). There is also reason to believe

2 For press of space, our overview does not address implicit self-esteem. The notion that implicit self-esteem is lower than explicit self-esteem in psychosis, and that persecutory delusions enhance self-esteem through a personalization bias for negative events is currently not well-supported, partly owing to methodological problems to capture implicit self-esteem and symptomatic heterogeneity among those with persecutory delusions. Current reviews therefore do not entirely dismiss this hypothesis (Kesting and Lincoln, 2013; Tiernan et al., 2014).

that the relationship at times runs in the opposite direction: paranoid symptoms may prompt feelings of despair (e.g., job loss because of breakdown), humiliation and isolation (close friends and relatives are alienated by the experience of psychosis), which can culminate in post-psychosis depression (Upthegrove, 2009). Even if a strong causal relationship with paranoia is not replicated, depression and anxiety should constitute core treatment targets, as patients with psychosis consider working on their emotional problems a priority (Kuhnigk et al., 2012; Moritz et al., 2016a) and because of the high prevalence of emotional symptoms in the disorder (Buckley et al., 2009; Galletly et al., 2016).

Studies on depressive symptoms and their contribution to psychosis are complemented by an emerging research on coping or emotion regulation in psychosis (O'Driscoll et al., 2014) showing that patients with schizophrenia adopt dysfunctional strategies (e.g., rumination, thought suppression) rather than functional ones (e.g., acceptance, reappraisal) to deal with emotions. Yet, the specificity awaits to be established. Some studies indicate a relationship with depressive symptoms (Moritz et al., 2016b; Westermann and Lincoln, 2011) rather than positive symptoms with the possible exception of (experiential) avoidance (Freeman et al., 2005; Moutoussis et al., 2007; Udachina et al., 2014) and thought suppression (de Leede-Smith and Barkus, 2013; Jones and Fernyhough, 2006; Moritz et al., 2010; Morrison and Wells, 2003).

The trump card of the model by Garety and Freeman is a large body of investigations showing that the amelioration of emotional problems via specialized cognitive-behavioral therapy (CBT) and third-wave variants decrease positive symptoms, particularly delusions (Bullock et al., 2016; Ellett, 2013; Freeman et al., 2015b, 2014b; Myers et al., 2011; Visceglia and Lewis, 2011). While our own study could not find an effect of online CBT on positive symptoms, depressive symptoms were effectively reduced at a medium to strong effect size (Moritz et al., 2016b).

TWO-STAGE MODEL OF POSITIVE SYMPTOMS

Building on and integrating the work of other researchers, we have devised a theoretical two-stage model of positive symptoms that shares a number of elements with the aforementioned model but assembles these differently. Our theory does not necessarily contradict prior cognitive theories (Coltheart et al., 2011; Kapur, 2003) that view aberrant perceptions as the driving process of delusion formation. Stage 1 is at the heart of our heuristic model, which aims to explain the emergence of delusions. Stage 2 of the account tries to explain how delusional ideas may evolve into incorrigible convictions.

DELUSION FORMATION (STAGE 1)

We regard liberal acceptance as a key cognitive aberration in psychosis. To put it more plainly, patients reason like "bad statisticians," that is, that they assign meaning and momentum to weakly supported evidence. Our central claim is that the decision threshold for accepting hypotheses is lowered in psychosis; ideas and hypotheses that a healthy or nonpsychotic patient would reject, or put on hold until further verification checks are carried out, are accepted as possible or even plausible (Moritz et al., 2009, 2008, 2007; Moritz and Woodward, 2004). As shown by many authors, "strange" thoughts (e.g., people are making remarks about me; feelings of being looked at; believe to be destined to be someone very important) are common in the general population, too (van Os and Reininghaus, 2016), but they receive fewer and less severe reactions in nonpsychotic participants (Lincoln et al., 2014).

The idea that patients have liberal acceptance is borne out in two lines of research. First, it has been shown that patients with schizophrenia assign higher plausibility to interpretations that nonpsychotic individuals reject as absurd (McLean et al., 2017). This has been shown for both delusional (LaRocco and Warman, 2009) and nondelusional scenarios (Moritz and Woodward, 2004). Further, we have shown that patients with schizophrenia have a lower decision threshold than controls. Our group found that patients based decisions on much lower probability estimates than controls, for example 82% relative to 93% in controls (Moritz et al., 2012b, 2009, 2006; Moritz et al., 2016a; Veckenstedt et al., 2011). This parameter discriminated patients with psychosis from controls better than the conventional JTC bias (quick decision-making on the basis of little evidence) or draws to decision estimates (Moritz et al., 2012b; Moritz et al., 2017a). In fact, liberal acceptance may delay decision-making under some conditions, particularly when two or more hypotheses are simultaneously considered plausible (Moritz et al., 2007), and may even emerge in the absence of differences on JTC (Moritz et al., 2016a). As we have shown elsewhere in greater detail (Moritz et al., 2017b), liberal acceptance prompts overconfidence in errors while confidence in correct responses is at times even lower, which leads to "knowledge corruption" (defined as the proportion of high-confident responses that are false). This pattern of results has been confirmed in several studies on patients and analogue populations (Balzan, 2016; Eisenacher and Zink, 2017; Moritz and Woodward, 2006a). Subjective competence seems to moderate the effect which ties in well with the clinical observation that delusional ideas relate to topics for which the patient has some (subjective) expertise (Moritz et al., 2015a).

Patients in stage 1 of the process usually do not differ from nonpsychotic individuals in terms of the subjective probability of their initial ideas, including implausible ideas that later become delusions (see Figure 48.1, upper panel). However, they assign more weight and validity to their hypotheses. In this early stage of delusion formation, patients usually have some doubt (Jaspers, 1913; Klosterkötter, 1992). Accordingly, delusions at stage 1 are also the ones that are usually not acted on and patients often behave inconsequentially (Jaspers, 1913). The delusional idea is like a working hypothesis and increasingly absorbs the individual, a process that can take many weeks and evolves over time (Kapur, 2003; Klosterkötter, 1992).

DELUSION MAINTENANCE (STAGE 2)

Stage 2 of our model explains why delusional conviction is maintained, even in the presence of overwhelming counterevidence. Stage 2 can be subdivided into two stages: consolidation of primary delusional belief (stage 2a) and neglect of competing hypotheses (stage 2b) (see Figure 48.2).

Among the cognitive factors of stage 2 is the confirmation bias, which is also seen in nonclinical individuals (Fugelsang et al., 2004), but is suggested to be aggravated in patients (Balzan et al., 2013) (see Figure 48.2, lower panel). The confirmation bias denotes the tendency to look for proof or confirming evidence for one's hypotheses (Nickerson,

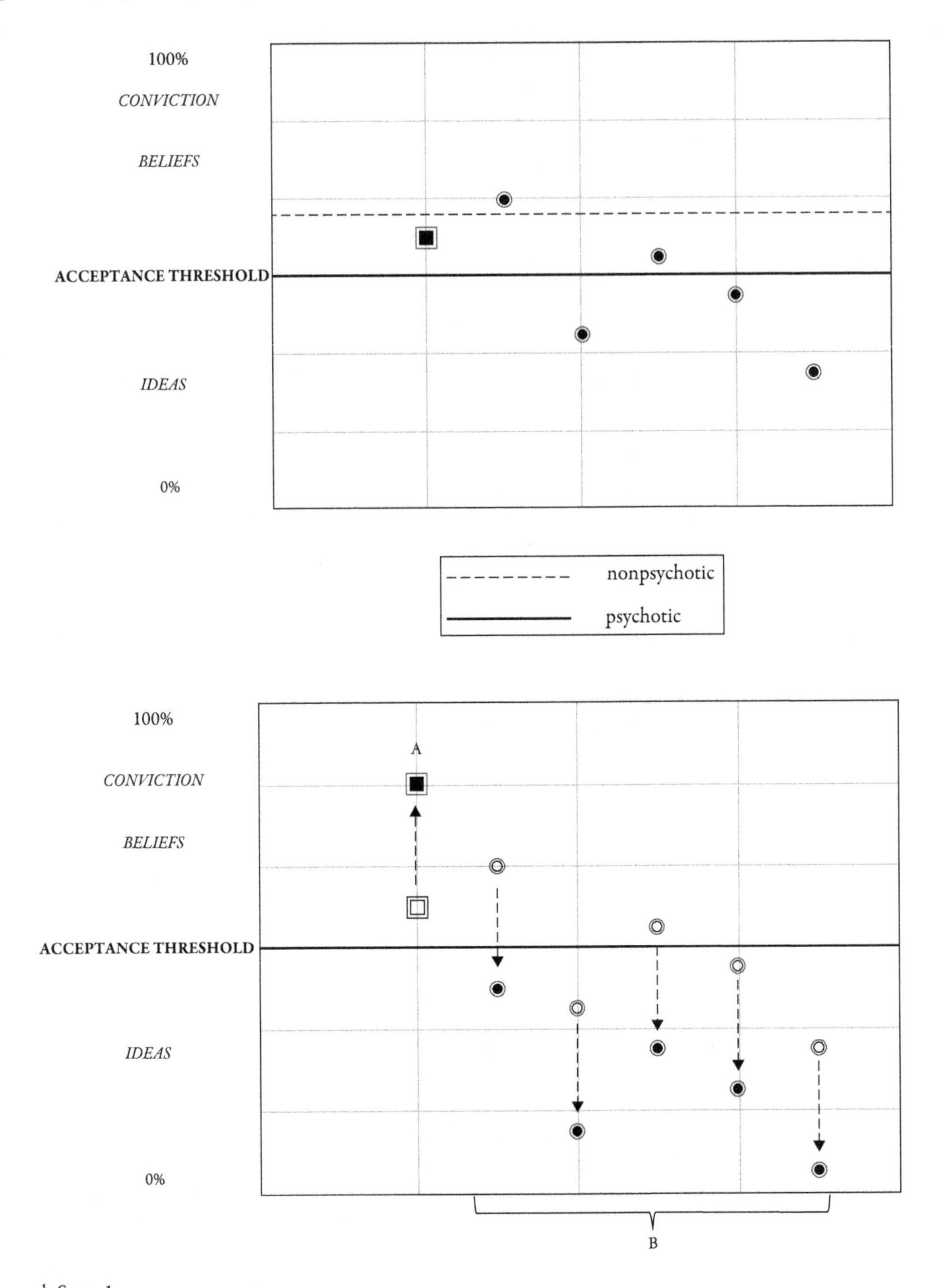

Figure 48.2 Upper panel: Stage 1
In individuals with psychosis, initial ideas are more easily accepted due to a lowered decision threshold compared to controls (dotted line: threshold of nonpsychotic individuals; full line: threshold of individuals with psychosis). In the example, the nonpsychotic individual pursues further only one (nondelusional) hypothesis.
Lower panel: Stage 2
Stage 2a: The delusional belief is augmented to a conviction by means of a confirmation bias and/or emotional factors (e.g., insight relief; see text) [A]
Stage 2b: Alternative hypotheses decrease in strength by means of a bias against disconfirmatory evidence (BADE) and, for example, withdrawal/avoidance (lack of corrective experiences) [B]
Square: delusional idea; circles: nondelusional idea(s)

1998; Oswald and Grosjean, 2004). Its counterpart is the *bias against disconfirmatory evidence* (BADE) which seems to distinguish patients with psychosis from psychiatric and nonpsychiatric controls (McLean et al., 2017) and has also been observed in high-risk "delusion-prone" populations (Buchy et al., 2007; Eisenacher et al., 2016; Woodward et al., 2008). The BADE may contribute to the systematization of the delusional idea because (challenging) alternative scenarios are ignored and receive less weight, receptively (stage 2b). This process is further intensified by isolation/loneliness and active social withdrawal/avoidance, as this limits opportunities for critical social feedback and corrective experiences (Fett et al., 2012; Freeman et al., 2002; Zimbardo, 1999). As described, the latter process also plays an important role in the model by Freeman and Garety (2013), who implicate belief inflexibility in their model (Garety and Freeman, 2013).

Apart from cognitive processes, emotional processes are essential as well. Otherwise, one may not explain why liberal acceptance should lead to delusional and not neutral ideas. Based on research on anxiety disorders (Soares et al., 2009), we argue that an emotionally charged explanation will receive more weight than a neutral one (de Jong, 2015). In other words, hypotheses that evoke fear are considered first and individuals may get stuck on these, particularly as initially generated hypotheses are judged as more plausible and therefore have a higher probability to prevail (Moritz and Woodward, 2004). In line with this, patients with psychosis and nonclinical individuals having delusion-like experiences prioritize processing of interpersonal threat (which is a central theme in persecutory delusions) (Freeman et al., 2013; Reininghaus et al., 2016), which may result in severe (inadequate) anxiety and increase the likelihood for accepting beliefs of being persecuted.

Another important emotional factor for maintenance is "insight relief" (Jaspers, 1913; Kapur, 2003; Kraepelin, 1919) as delusions often provide their holder with an explanation for his/her problems. Patients sometimes feel elated because the enemies are individuals or institutions usually of high rank or power imparting the patient with importance in return (Moritz et al., 2013b; Moritz et al., 2015c; Sundag et al., 2015). Indeed, most patients are ambivalent toward their symptoms (i.e., have both negative and positive attitudes toward symptoms) and a "gain from illness" can be observed (Moritz et al., 2015c).

Our model has been extended to hallucinations. While these are traditionally characterized as "psychotic," they occur in many nonpsychotic individuals, too (Laroi et al., 2014; Vellante et al., 2012). Loud, alien, vivid, and persistent intrusive thoughts are common in depression (Moritz et al., 2014c) and obsessive-compulsive disorder (OCD) (Moritz et al., 2014b; Röhlinger et al., 2015). These patients can "hear" their inner critic or see upcoming catastrophes. The nonpsychotic individual may also be stressed and intimidated by these experiences but usually knows they are self-generated and hold such ideas in check by doubt.

In our view, patients with psychosis liberally accept a bizarre explanation that externalizes self-generated phenomena such as loud thoughts and vivid images and equips it with a delusional superstructure. In patients with other disorders the aberrant experience often rests at the stage of an "as if" feeling (e.g., "it is *as if* there is someone is yelling at me") due to their higher threshold for accepting hypotheses.

According to our account a core element of treatment is to "plant the seeds of doubt," that is, to get patients to reflect on their judgment, decrease overconfidence in errors and look for further evidence before making momentous and firm judgments (Moritz et al., 2014a). Indeed, a new line of research shows that metacognitive (Eichner and Berna, 2016; Moritz et al., 2014a; Philipp et al., 2018) and reasoning training (Waller et al., 2015) decrease delusions and other positive symptoms at a small to medium effect size (Eichner and Berna, 2016). Preliminary evidence suggests that this might be owing to decreasing overconfidence (Koether, 2017) and raising the individual decision threshold (Andreou et al., 2016) but the mechanism of action is not fully elucidated and our hypothesis remains speculative at this point. Moreover, the group training seems less effective in highly acute patients (van Oosterhout et al., 2014) who may be too distracted to attend and fully grasp the concepts. In view of findings on emotional aspects (see earlier discussion) we have complemented MCT by modules addressing emotional problems and somewhat expanded on the initial rationale that only targeted cognitive biases. Recent programs also investigate the potential to target reasoning biases via computerized or Internet-based interventions (Garety et al., 2015; Moritz et al., 2015; Rüegg et al., 2018; Ward and Garety, 2017).

LIMITATIONS

Our aim was to demonstrate that emotional and reasoning processes play a vital role in psychosis. Apart from a rather consistent body of basic research showing that patients with psychosis differ from controls on a number of emotional and cognitive aspects, treatment studies assert that the amelioration of these factors leads to an improvement of symptoms. While some parts of the theories are well supported by evidence, other pieces of the puzzle are missing and many aspects require (independent) replication. Several problems that we summarize here need to be tackled in the future:

1. Comorbid diagnoses/specificity: Patients with schizophrenia suffer from a number of comorbid diagnoses (Buckley et al., 2009; Galletly et al., 2016). While drug and alcohol dependence as well as neurological disorders represent frequent exclusion criteria, it is virtually impossible to recruit a sample of patients with pure psychosis. It is therefore difficult to attribute group differences relative to healthy controls to the primary disorder. This caveat, of course, also applies to other fields of research and cannot be fully circumvented by correlational analyses as different syndromes—even seemingly contrary features like depression and paranoia—are often intertwined. We must acknowledge that with few exceptions (Moritz and Woodward, 2006b, 2006c) studies on BADE, overconfidence, and liberal acceptance have been conducted against nonclinical controls only. More studies involving psychiatric controls are

needed to pinpoint effects specific to psychosis. Studies should also actively consider the effects of medication. Medication is often treated as a confounding variable and as if it has a negligible impact on performance. However, there is amassing evidence that medication affects cognitive biases (Andreou et al., 2013), emotional well-being (Mizrahi et al., 2007; Moritz et al., 2013a), and neurocognition (Faber et al., 2012). Although antipsychotic medication is not as effective as formerly thought (Leucht et al., 2009; Murray et al., 2016) and its efficacy perhaps even declines over time against placebo (Rutherford et al., 2014), it is beyond doubt that it improves positive symptoms. More insight into how this effect is achieved via cognitive and/or emotional pathways is considered fruitful for garnering new models and treatment options for schizophrenia (including psychotherapy).

2. Intervention research: As shown, intervention studies targeting reasoning and emotional problems in psychosis show promise to ameliorate symptoms in addition to pharmacotherapy. Yet, longitudinal studies with long-term follow-ups are rare and we know very little about moderators of outcome, that is, which subgroup benefits from which particular intervention. In view of cost pressure in the healthcare system and adverse effects of failed interventions (e.g., relapse, fatalism due to false hope) it is pivotal to offer patients a promising treatment tailored to their symptoms, needs and resources. In addition, treatment effects are rather small and often not sustained at follow-up. Here, electronic reminders (e.g., smartphone Application) may be useful.
3. Causality: We do not fully know to which extent some emotional and cognitive factors are true risk factors (or a vulnerability to psychological problems as such), noncausal antecedents of psychosis, correlates, or consequences. We thus cannot be sure if intervention studies on these aspects ameliorate causal factors of psychosis or attenuate the consequences of psychosis. While high-risk studies can circumvent some of the problems inherent to patient trials, they also face major problems as they often involve individuals with in fact lower risk for vulnerability (e.g., highly educated student populations or individuals beyond the typical age of first exacerbation) and may therefore be more informative about resilience than risk factors for psychosis! Nonclinical controls scoring high on psychosis proneness scales sometimes show different response patterns than individuals who fulfill diagnostic criteria for psychosis requiring treatment. For example, associations at times differ between delusions and JTC (Ross et al., 2015) and coping styles (Kesting and Lincoln, 2013) for analogous population versus clinical samples. Of note, many participants with clinical ultra-high risk for psychosis showing the presence of cognitive and biological abnormalities (Egerton et al., 2013; Eisenacher et al., 2015; Johns et al., 2010; Thompson et al., 2012) will not develop psychosis (Fusar-Poli et al., 2016). We need to look both at resilience factors and the interaction between dysfunctional cognitive and emotional processes in order to improve our predictive power about the emergence of psychotic symptoms. Recent studies have suggested that the interaction between traumatic life events and dysfunctional cognitive appraisals may help to better predict psychosis (Appiah-Kusi et al., 2017; Gawęda et al., 2018b, 2018a; Hardy et al., 2016).
4. Latent variables: As shown, a set of emotional problems such as sleep disturbances, low self-esteem and worry may be relevant for the development of psychosis. Yet, we do not know if these factors are elementary or epiphenomena of hidden/latent variables. For example, Garety and Freeman (2013) note that sleep disturbances are perhaps causing paranoia via negative affect and anomalous experiences. Experimental studies in nonclinical populations showed that restricting sleep increases psychotic experiences and improving sleep reduces psychotic experiences, which is suggested to be in fact mediated by negative affect (Freeman et al., 2017; Reeve et al., 2017). Other depressive symptoms, such feeling slowed down or restless (Moritz et al., 2017b), which are not considered in (Freeman and Garety, 2014)'s model, also warrant further investigation.
5. Linkage between emotion and reasoning: In some places the review might have evoked the impression as if emotional and cognitive processes in psychosis present nonoverlapping routes to psychosis. It is too early to make firm conclusions, but some studies suggest that emotional factors and stress trigger or aggravate cognitive biases (Lincoln et al., 2010, 2011; Moritz et al., 2011; Moritz et al., 2015b). For example, a number of studies found that the JTC bias is exaggerated by stress and negative emotional states (Moritz et al., 2015b). Stress is clearly also associated with emotion processing (Myin-Germeys and van Os, 2007). Participants with psychosis reported a greater increase in stress in response to a stressor relative to healthy controls, which was predicted by a reduced awareness of and tolerance for distressing emotions (Lincoln et al., 2015). With respect to other biases, this is simply unknown because of a dearth of studies.
6. Dimensionality of psychotic symptoms: While it is now well established that there is a continuum of psychotic experiences in the general population (van Os et al., 2009; van Os and Reininghaus, 2016; Verdoux and Van Os, 2002), we need to further elucidate whether this is true for all psychotic symptoms. Factor analyses suggest that delusional symptoms can be segregated into a general (more benign?) subtype encompassing symptoms such as "There might be negative comments being circulated about me" and a more bizarre (more malign) subtype of symptoms such as "I can detect coded messages about me in the press/TV/radio" (Moritz et al., 2012b; Schlier et al., 2015). Whether these symptoms are governed by the same emotional and cognitive underpinnings is presently unknown.

7. Cognitive biases: As shown, research into cognitive biases has been very fruitful as to both our understanding and treatment. While much research has been devoted to JTC, source memory biases and overconfidence, other cognitive biases have received little attention and a number of findings on, for example, illusory correlation (Balzan et al., 2013), illusion of control or an excessive "truth effect" (Moritz et al., 2012a) await independent replication. Also, JTC has been barely studied in contexts other than probabilistic reasoning (Ziegler et al., 2008).

CONCLUSION

The present chapter aimed to provide an overview of the promising but still fragmentary knowledge we have accumulated over the past 20 years about emotional problems and cognitive biases in psychosis. It has been *made probable* that a subgroup of patients with schizophrenia have reasoning biases, most likely in addition to other (albeit less established) biases such as the confirmation bias and the bias against disconfirmatory evidence. More recently, emotional factors have been highlighted. Patients suffer from low self-esteem, have negative self-schemas, and display dysfunctional emotion regulation. Intervention studies highlight these aspects as potential mechanisms for improvement of the positive syndrome. To alleviate psychotic symptoms indirectly is not only an elegant way of treatment but also one that may raise treatment adherence as it matches preferences of patients.

While it is now beyond doubt that there are emotional and cognitive precursors to psychosis and that treatments targeting these aspects improve symptoms, the limitations and gaps in our knowledge represented a large part of this chapter. This was done very deliberately. Biases are not specific to psychosis (Kahneman and Tversky, 1996) and occur in nonclinical subjects, too, including researchers. We must be aware of file drawer effects/publication biases (Ioannidis et al., 2014) and a confirmatory bias in science (Fugelsang et al., 2004) that may obstruct progress in the field.

REFERENCES

Andreou, C., Moritz, S., Veith, K., Veckenstedt, R., Naber, D., 2013. Dopaminergic modulation of probabilistic reasoning and overconfidence in errors: a double-blind study. Schizophr. Bull. 40, 558–565. doi:10.1093/schbul/sbt064

Andreou, C., Wittekind, C., Fieker, M., Heitz, U., Veckenstedt, R., Bohn, F., Moritz, S., 2016. Individualized metacognitive therapy accelerates improvement of delusions in patients with psychosis: a randomized controlled rater-blind study. Schizophr. Res., submitted.

American Psychiatric Association, 2013. Diagnostic and statistical manual of mental disorders (DSM-5). American Psychiatric Pub.

Appiah-Kusi, E., Fisher, H.L., Petros, N., Wilson, R., Mondelli, V., Garety, P.A., Mcguire, P., Bhattacharyya, S., 2017. Do cognitive schema mediate the association between childhood trauma and being at ultra-high risk for psychosis? J. Psychiatr. Res. 88, 89–96. doi:10.1016/j.jpsychires.2017.01.003

Balzan, R., Delfabbro, P., Galletly, C., Woodward, T., 2013. Confirmation biases across the psychosis continuum: the contribution of hypersalient evidence-hypothesis matches. Br. J. Clin. Psychol. 52, 53–69. doi:10.1111/bjc.12000

Balzan, R.P., 2016. Overconfidence in psychosis: the foundation of delusional conviction? Cogent Psychol. 3, 1135855.

Bentall, R.P., de Sousa, P., Varese, F., Wickham, S., Sitko, K., Haarmans, M., Read, J., 2014. From adversity to psychosis: pathways and mechanisms from specific adversities to specific symptoms. Soc. Psychiatry Psychiatr. Epidemiol. 49, 1011–1022. doi:10.1007/s00127-014-0914-0

Broyd, A., Balzan, R.P., Woodward, T.S., Allen, P., 2017. Dopamine, cognitive biases and assessment of certainty: a neurocognitive model of delusions. Clin. Psychol. Rev. doi:10.1016/j.cpr.2017.04.006

Buchy, L., Woodward, T.S., Liotti, M., 2007. A cognitive bias against disconfirmatory evidence (BADE) is associated with schizotypy. Schizophr. Res. 90, 334–337. doi:10.1016/j.schres.2006.11.012

Buckley, P.F., Miller, B.J., Lehrer, D.S., Castle, D.J., 2009. Psychiatric comorbidities and schizophrenia. Schizophr. Bull. 35, 383–402. doi:10.1093/schbul/sbn135

Bullock, G., Newman-Taylor, K., Stopa, L., 2016. The role of mental imagery in non-clinical paranoia. J. Behav. Ther. Exp. Psychiatry 50, 264–268. doi:10.1016/j.jbtep.2015.10.002

Collett, N., Pugh, K., Waite, F., Freeman, D., 2016. Negative cognitions about the self in patients with persecutory delusions: an empirical study of self-compassion, self-stigma, schematic beliefs, self-esteem, fear of madness, and suicidal ideation. Psychiatry Res. 239, 79–84. doi:10.1016/j.psychres.2016.02.043

Coltheart, M., Langdon, R., McKay, R., 2011. Delusional belief. Annu. Rev. Psychol. 62, 271–298. doi:10.1146/annurev.psych.121208.131622

Corlett, P.R., Taylor, J.R., Wang, X.J., Fletcher, P.C., Krystal, J.H., 2010. Toward a neurobiology of delusions. Prog. Neurobiol. 92, 345–369. doi:10.1016/j.pneurobio.2010.06.007

de Jong, P.J., 2015. Danger-confirming reasoning and the persistence of phobic beliefs, in: Galbraith, N. (Ed.), Aberrant Beliefs and Reasoning. Psychology Press, Hove, United Kingdom, pp. 132–153.

de Leede-Smith, S., Barkus, E., 2013. A comprehensive review of auditory verbal hallucinations: lifetime prevalence, correlates and mechanisms in healthy and clinical individuals. Front. Hum. Neurosci. 7, 367. doi:10.3389/fnhum.2013.00367

Egerton, A., Chaddock, C.A., Winton-Brown, T.T., Bloomfield, M.A.P., Bhattacharyya, S., Allen, P., McGuire, P.K., Howes, O.D., 2013. Presynaptic striatal dopamine dysfunction in people at ultra-high risk for psychosis: findings in a second cohort. Biol. Psychiatry 74, 106–112. doi:10.1016/j.biopsych.2012.11.017

Eichner, C., Berna, F., 2016. Acceptance and efficacy of metacognitive training (MCT) on positive symptoms and delusions in patients with schizophrenia: a meta-analysis taking into account important moderators. Schizophr. Bull. 42, 952–962. doi:10.1093/schbul/sbv225

Eisenacher, S., Rausch, F., Ainser, F., Mier, D., Veckenstedt, R., Schirmbeck, F., Lewien, A., Englisch, S., Andreou, C., Moritz, S., Meyer-Lindenberg, A., Kirsch, P., Zink, M., 2015. Investigation of metamemory functioning in the at-risk mental state for psychosis. Psychol. Med. 1–12. doi:10.1017/S0033291715001373

Eisenacher, S., Rausch, F., Mier, D., Fenske, S., Veckenstedt, R., Englisch, S., Becker, A., Andreou, C., Moritz, S., Meyer-Lindenberg, A., Zink, M., 2016. Bias against disconfirmatory evidence in the "at-risk mental state" and during psychosis. Psychiatry Res. 238, 242–250. doi:http://dx.doi.org/10.1016/j.psychres.2016.02.028

Eisenacher, S., Zink, M., 2017. The importance of metamemory functioning to the pathogenesis of psychosis. Front. Psychol. 8, 304. doi:10.3389/fpsyg.2017.00304

Ellett, L., 2013. Mindfulness for paranoid beliefs: evidence from two case studies. Behav. Cogn. Psychother. 41, 238–242. doi:10.1017/S1352465812000586

Faber, G., Smid, H.G., Van Gool, A.R., Wiersma, D., Van Den Bosch, R.J., 2012. The effects of guided discontinuation of antipsychotics on neurocognition in first onset psychosis. Eur. Psychiatry 27, 275–280. doi:10.1016/j.eurpsy.2011.02.003

Fett, A.-K.J., Shergill, S.S., Joyce, D.W., Riedl, A., Strobel, M., Gromann, P.M., Krabbendam, L., 2012. To trust or not to trust: the dynamics of social interaction in psychosis. Brain 135, 976–984. doi:10.1093/brain/awr359

Freeman, D., 2016. Persecutory delusions: a cognitive perspective on understanding and treatment. Lancet Psychiatry 3, 685–692. doi:10.1016/S2215-0366(16)00066-3

Freeman, D., Dunn, G., Fowler, D., Bebbington, P., Kuipers, E., Emsley, R., Jolley, S., Garety, P., 2013. Current paranoid thinking in patients with delusions: the presence of cognitive-affective biases. Schizophr. Bull. 39, 1281–1287. doi:10.1093/schbul/sbs145

Freeman, D., Dunn, G., Murray, R.M., Evans, N., Lister, R., Antley, A., Slater, M., Godlewska, B., Cornish, R., Williams, J., Di Simplicio, M., Igoumenou, A., Brenneisen, R., Tunbridge, E.M., Harrison, P.J., Harmer, C.J., Cowen, P., Morrison, P.D., 2015a. How cannabis causes paranoia: using the intravenous administration of Δ9-tetrahydrocannabinol (THC) to identify key cognitive mechanisms leading to paranoia. Schizophr. Bull. 41, 391–399. doi:10.1093/schbul/sbu098

Freeman, D., Dunn, G., Startup, H., Pugh, K., Cordwell, J., Mander, H., Černis, E., Wingham, G., Shirvell, K., Kingdon, D., 2015b. Effects of cognitive behaviour therapy for worry on persecutory delusions in patients with psychosis (WIT): a parallel, single-blind, randomised controlled trial with a mediation analysis. Lancet Psychiatry 2, 305–313. doi:10.1016/S2215-0366(15)00039-5

Freeman, D., Emsley, R., Dunn, G., Fowler, D., Bebbington, P., Kuipers, E., Jolley, S., Waller, H., Hardy, A., Garety, P., 2014a. The stress of the street for patients with persecutory delusions: a test of the symptomatic and psychological effects of going outside into a busy urban area. Schizophr. Bull. 41, 1–9. doi:10.1093/schbul/sbu173

Freeman, D., Garety, P., 2014. Advances in understanding and treating persecutory delusions: a review. Soc. Psychiatry Psychiatr. Epidemiol. 49, 1179–1189. doi:10.1007/s00127-014-0928-7

Freeman, D., Garety, P.A., Bebbington, P.E., Smith, B., Rollinson, R., Fowler, D., Kuipers, E., Ray, K., Dunn, G., 2005. Psychological investigation of the structure of paranoia in a non-clinical population. Br. J. Psychiatry 186, 427–435. doi:10.1192/bjp.186.5.427

Freeman, D., Garety, P.A., Kuipers, E., Fowler, D., Bebbington, P.E., 2002. A cognitive model of persecutory delusions. Br. J. Clin. Psychol. 41, 331–347. doi:10.1348/014466502760387461

Freeman, D., Pugh, K., Dunn, G., Evans, N., Sheaves, B., Waite, F., Cernis, E., Lister, R., Fowler, D., 2014b. An early Phase II randomised controlled trial testing the effect on persecutory delusions of using CBT to reduce negative cognitions about the self: the potential benefits of enhancing self confidence. Schizophr. Res. 160, 186–192. doi:10.1016/j.schres.2014.10.038

Freeman, D., Sheaves, B., Goodwin, G.M., Yu, L.-M., Nickless, A., Harrison, P.J., Emsley, R., Luik, A.I., Foster, R.G., Wadekar, V., others, 2017. The effects of improving sleep on mental health (OASIS): a randomised controlled trial with mediation analysis. Lancet Psychiatry 4, 749–758.

Freeman, D., Stahl, D., McManus, S., Meltzer, H., Brugha, T., Wiles, N., Bebbington, P., 2012. Insomnia, worry, anxiety and depression as predictors of the occurrence and persistence of paranoid thinking. Soc. Psychiatry Psychiatr. Epidemiol. 47, 1195–1203. doi:10.1007/s00127-011-0433-1

Freud, S., 1963. Three case histories: The '"wolf man"', the '"rat man"' and the psychotic Doctor Schreber. Collier Books, New York, NY.

Fugelsang, J.A., Stein, C.B., Green, A.E., Dunbar, K.N., 2004. Theory and data interactions of the scientific mind: evidence from the molecular and the cognitive laboratory. Can. J. Exp. Psychol. 58, 86–95. doi:10.1037/h0085799

Fusar-Poli, P., Cappucciati, M., Borgwardt, S., Woods, S.W., Addington, J., Nelson, B., Nieman, D.H., Stahl, D.R., Rutigliano, G., Riecher-Rössler, A., Simon, A.E., Mizuno, M., Lee, T.Y., Kwon, J.S., Lam, M.M.L., Perez, J., Keri, S., Amminger, P., Metzler, S., Kawohl, W., Rössler, W., Lee, J., Labad, J., Ziermans, T., An, S.K., Liu, C.-C., Woodberry, K.A., Braham, A., Corcoran, C., McGorry, P., Yung, A.R., McGuire, P.K., 2016. Heterogeneity of psychosis risk within individuals at clinical high risk: a meta-analytical stratification. JAMA Psychiatry 73, 113–120. doi:10.1001/jamapsychiatry.2015.2324

Galletly, C., Castle, D., Dark, F., Humberstone, V., Jablensky, A., Killackey, E., Kulkarni, J., McGorry, P., Nielssen, O., Tran, N., 2016. Royal Australian and New Zealand College of Psychiatrists clinical practice guidelines for the management of schizophrenia and related disorders. Aust. N. Z. J. Psychiatry 50, 410–472. doi:10.1177/0004867416641195

Garety, P., Joyce, E., Jolley, S., Emsley, R., Waller, H., Kuipers, E., Bebbington, P., Fowler, D., Dunn, G., Freeman, D., 2013. Neuropsychological functioning and jumping to conclusions in delusions. Schizophr. Res. 150, 570–574. doi:10.1016/j.schres.2013.08.035

Garety, P., Waller, H., Emsley, R., Jolley, S., Kuipers, E., Bebbington, P., Dunn, G., Fowler, D., Hardy, A., Freeman, D., 2015. Cognitive mechanisms of change in delusions: an experimental investigation targeting reasoning to effect change in paranoia. Schizophr. Bull. 41, 400–410. doi:10.1093/schbul/sbu103

Garety, P.A., Freeman, D., 2013. The past and future of delusions research: from the inexplicable to the treatable. Br. J. Psychiatry 203, 327–333. doi:10.1192/bjp.bp.113.126953

Gawęda, Ł., Göritz, A.S., Moritz, S., 2018a. Mediating role of aberrant salience and self-disturbances for the relationship between childhood trauma and psychotic-like experiences in the general population. Schizophr. Res.

Gawęda, Ł., Prochwicz, K., Adamczyk, P., Frydecka, D., Misiak, B., Kotowicz, K., Szczepanowski, R., Florkowski, M., Nelson, B., 2018b. The role of self-disturbances and cognitive biases in the relationship between traumatic life events and psychosis proneness in a non-clinical sample. Schizophr. Res. 193, 218–224.

Goldfried, M.R., 2016. On possible consequences of National Institute of Mental Health funding for psychotherapy research and training. Prof. Psychol. 47, 77–83. doi:10.1037/pro0000034

Häfner, H., Maurer, K., An Der Heiden, W., 2013. ABC Schizophrenia study: an overview of results since 1996. Soc. Psychiatry Psychiatr. Epidemiol. 48, 1021–1031. doi:10.1007/s00127-013-0700-4

Hardy, A., Emsley, R., Freeman, D., Bebbington, P., Garety, P.A., Kuipers, E.E., Dunn, G., Fowler, D., 2016. Psychological mechanisms mediating effects between trauma and psychotic symptoms: The role of affect regulation, intrusive trauma memory, beliefs, and depression. Schizophr. Bull. 42, S34–S43. doi:10.1093/schbul/sbv175

Heinz, A., Deserno, L., Reininghaus, U., 2013. Urbanicity, social adversity and psychosis. World Psychiatry 12, 187–197. doi:10.1002/wps.20056

Howes, O.D., Murray, R.M., 2014. Schizophrenia: an integrated sociodevelopmental-cognitive model. Lancet. doi:10.1016/S0140-6736(13)62036-X

Ioannidis, J.P.A., Munafò, M.R., Fusar-Poli, P., Nosek, B.A., David, S.P., 2014. Publication and other reporting biases in cognitive sciences: Detection, prevalence, and prevention. Trends Cogn. Sci. doi:10.1016/j.tics.2014.02.010

Jaspers, K., 1913. Allgemeine Psychopathologie [General psychopathology]. 4. Auflage. Springer, Berlin (Germany). doi:10.1007/978-3-642-52895-8

Jaya, E.S., Ascone, L., Lincoln, T.M., 2018. A longitudinal mediation analysis of the effect of negative-self-schemas on positive symptoms via negative affect. Psychol. Med. 48, 1299–1307.

Johns, L.C., Allen, P., Valli, I., Winton-Brown, T., Broome, M., Woolley, J., Tabraham, P., Day, F., Howes, O., Wykes, T., McGuire, P., 2010. Impaired verbal self-monitoring in individuals at high risk of psychosis. Psychol. Med. 40, 1433–1442. doi:10.1017/S0033291709991991

Jones, S.R., Fernyhough, C., 2006. The roles of thought suppression and metacognitive beliefs in proneness to auditory verbal hallucinations in a non-clinical sample. Pers. Individ. Dif. 41, 1421–1432. doi:10.1016/j.paid.2006.06.003

Kahneman, D., Tversky, A., 1996. On the reality of cognitive illusions. Psychol. Rev. 103, 582–591; discussion 592–596. doi:10.1037/0033-295X.103.3.582

Kapur, S., 2003. Psychosis as a state of aberrant salience: a framework linking biology, phenomenology, and pharmacology in schizophrenia. Am. J. Psychiatry 160, 13–23. doi:10.1176/appi.ajp.160.1.13

Kesting, M.L., Lincoln, T.M., 2013. The relevance of self-esteem and self-schemas to persecutory delusions: A systematic review. Compr. Psychiatry 54, 766–789. doi:10.1016/j.comppsych.2013.03.002
Klosterkötter, J., 1992. The meaning of basic symptoms for the genesis of the schizophrenic nuclear syndrome. Jpn. J. Psychiatry Neurol. 46, 609–630. doi:10.1111/j.1440-1819.1992.tb00535.x
Koether, U., 2017. Bayesian analyses of the effect of metacognitive training on social cognition deficits and overconfidence in errors. J. Exp. Psychopathol.
Kraepelin, E., 1919. Dementia praecox and paraphrenia. Chicago Medical Books, Chicago.
Kraepelin, E., 1899. Psychiatrie. Ein Lehrbuch für Studierende und Aerzte [Psychiatry: a textbook for students and physicians]. J. A. Barth, Leipzig.
Kramer, I., Simons, C.J.P., Wigman, J.T., Collip, D., Jacobs, N., Derom, C., Thiery, E., Van Os, J., Myin-Germeys, I., Wichers, M., 2014. Time-lagged moment-to-moment interplay between negative affect and paranoia: New insights in the affective pathway to psychosis. Schizophr. Bull. 40, 278–286. doi:10.1093/schbul/sbs194
Kuhnigk, O., Slawik, L., Meyer, J., Naber, D., Reimer, J., 2012. Valuation and attainment of treatment goals in schizophrenia: perspectives of patients, relatives, physicians, and payers. J. Psychiatr. Pract. 18, 321–328. doi:10.1097/01.pra.0000419816.75752.65
Kvaale, E.P., Haslam, N., Gottdiener, W.H., 2013. The "side effects" of medicalization: A meta-analytic review of how biogenetic explanations affect stigma. Clin. Psychol. Rev. doi:10.1016/j.cpr.2013.06.002
LaRocco, V.A., Warman, D.M., 2009. Probability estimations and delusion-proneness. Pers. Individ. Dif. 47, 197–202. doi:10.1016/j.paid.2009.02.021
Laroi, F., Luhrmann, T.M., Bell, V., Christian, W.A., Deshpande, S., Fernyhough, C., Jenkins, J., Woods, A., 2014. Culture and hallucinations: overview and future directions. Schizophr. Bull. 40. doi:10.1093/schbul/sbu012
Leucht, S., Arbter, D., Engel, R.R., Kissling, W., Davis, J.M., 2009. How effective are second-generation antipsychotic drugs? A meta-analysis of placebo-controlled trials. Mol. Psychiatry 14, 429–447. doi:4002136 [pii] 10.1038/sj.mp.4002136
Lewis, S., Escalona, P., Keith, S., 2009. Phenomenology of schizophrenia, in: Sadock, B., Sadock, V., Ruiz, P. (Eds.), Kaplan and Sadock's comprehensive textbook of psychiatry. Lippincott Williams and Wilkins, Philadelphia, pp. 1433–1451.
Lincoln, T.M., Hartmann, M., Köther, U., Moritz, S., 2015. Dealing with feeling: specific emotion regulation skills predict responses to stress in psychosis. Psychiatry Res. 228, 216–22. doi:10.1016/j.psychres.2015.04.003
Lincoln, T.M., Möbius, C., Huber, M.T., Nagel, M., Moritz, S., 2014. Frequency and correlates of maladaptive responses to paranoid thoughts in patients with psychosis compared to a population sample. Cogn. Neuropsychiatry 19, 509–526. doi:10.1080/13546805.2014.931220
Lincoln, T.M., Peter, N., Schäfer, M., Moritz, S., 2010. From stress to paranoia: an experimental investigation of the moderating and mediating role of reasoning biases. Psychol. Med. 40, 169–171. doi:10.1017/S003329170999095X
Lincoln, T.M., Peter, N., Schäfer, M., Moritz, S., 2009. Impact of stress on paranoia: an experimental investigation of moderators and mediators. Psychol. Med. 39, 1129–1139. doi:10.1017/S0033291708004613
Lincoln, T.M., Salzmann, S., Ziegler, M., Westermann, S., 2011. When does jumping-to-conclusions reach its peak? The interaction of vulnerability and situation-characteristics in social reasoning. J. Behav. Ther. Exp. Psychiatry 42, 185–191. doi:10.1016/j.jbtep.2010.09.005
Maher, B.A., 2006. The relationship between delusions and hallucinations. Curr. Psychiatry Rep. 8, 179–183. doi:10.1007/s11920-006-0021-3
McLean, B.F., Mattiske, J.K., Balzan, R.P., 2017. Association of the jumping to conclusions and evidence integration biases with delusions in psychosis: a detailed meta-analytic approach. Schizophr. Bull. 43, 344–354.
Mizrahi, R., Rusjan, P., Agid, O., Graff, A., Mamo, D.C., Zipursky, R.B., Kapur, S., 2007. Adverse subjective experience with antipsychotics and its relationship to striatal and extrastriatal D2 receptors: A PET study in schizophrenia. Am. J. Psychiatry 164, 630–637. doi:10.1176/appi.ajp.164.4.630
Moritz, S., Andreou, C., Klingberg, S., Thoering, T., Peters, M.J. V, 2013a. Assessment of subjective cognitive and emotional effects of antipsychotic drugs: effect by defect? Neuropharmacology 72, 179–186. doi:10.1016/j.neuropharm.2013.04.039
Moritz, S., Andreou, C., Schneider, B.C., Wittekind, C.E., Menon, M., Balzan, R.P., Woodward, T.S., 2014a. Sowing the seeds of doubt: a narrative review on metacognitive training in schizophrenia. Clin. Psychol. Rev. 34, 358–366. doi:10.1016/j.cpr.2014.04.004
Moritz, S., Berna, F., Jaeger, S., Westermann, S., Nagel, M., 2016a. The customer is always right? Subjective target symptoms and treatment preferences in patients with psychosis. Eur. Arch. Psychiatry Clin. Neurosci. 1–5. doi:10.1007/s00406-016-0694-5
Moritz, S., Burnette, P., Sperber, S., Köther, U., Hagemann-Goebel, M., Hartmann, M., Lincoln, T.M., 2011. Elucidating the black box from stress to paranoia. Schizophr. Bull. 37, 1311–1317. doi:10.1093/schbul/sbq055
Moritz, S., Claussen, M., Hauschildt, M., Kellner, M., 2014b. Perceptual properties of obsessive thoughts are associated with low insight in obsessive-compulsive disorder. J. Nerv. Ment. Dis. 202, 562–565. doi:10.1097/NMD.0000000000000156
Moritz, S., Favrod, J., Andreou, C., Morrison, A.P., Bohn, F., Veckenstedt, R., Tonn, P., Karow, A., 2013b. Beyond the usual suspects: Positive attitudes towards positive symptoms is associated with medication noncompliance in psychosis. Schizophr. Bull. 39, 917–922. doi:10.1093/schbul/sbs005
Moritz, S., Göritz, A.S., Balzan, R.P., Gawęda, Ł., Kulagin, S.C., Andreou, C., 2017a. A new paradigm to measure probabilistic reasoning and a possible answer to the question why psychosis-prone individuals jump to conclusions. J. Abnorm. Psychol. 126, 406–415. doi:10.1037/abn0000262
Moritz, S., Göritz, A.S., Gallinat, J., Schafschetzy, M., Van Quaquebeke, N., Peters, M.J. V, Andreou, C., 2015a. Subjective competence breeds overconfidence in errors in psychosis: a hubris account of paranoia. J. Behav. Ther. Exp. Psychiatry 48, 118–124. doi:10.1016/j.jbtep.2015.02.011
Moritz, S., Göritz, A.S., McLean, B., Westermann, S., Brodbeck, J., 2017b. Do depressive symptoms predict paranoia or vice versa? J. Behav. Ther. Exp. Psychiatry 56, 113–121. doi:10.1016/j.jbtep.2016.10.002
Moritz, S., Hörmann, C.C., Schröder, J., Berger, T., Jacob, G.A., Meyer, B., Holmes, E. a, Späth, C., Hautzinger, M., Lutz, W., Rose, M., Klein, J.P., 2014c. Beyond words: sensory properties of depressive thoughts. Cogn. Emot. 28, 1047–1056. doi:10.1080/02699931.2013.868342
Moritz, S., Klein, J.P., Desler, T., Lill, H., Gallinat, J., Schneider, B.C., 2017a. Neurocognitive deficits in schizophrenia: are we making mountains out of molehills? Psychol. Med. 47, 2602–2612. doi:10.1017/S0033291717000939
Moritz, S., Köther, U., Hartmann, M., Lincoln, T.M., 2015b. Stress is a bad advisor. Stress primes poor decision making in deluded psychotic patients. Eur. Arch. Psychiatry Clin. Neurosci. 265, 461–469. doi:10.1007/s00406-015-0585-1
Moritz, S., Köther, U., Woodward, T.S., Veckenstedt, R., Dechêne, A., Stahl, C., 2012a. Repetition is good? An Internet trial on the illusory truth effect in schizophrenia and nonclinical participants. J. Behav. Ther. Exp. Psychiatry 43, 1058–1063. doi:10.1016/j.jbtep.2012.04.004
Moritz, S., Lüdtke, T., Pfuhl, G., Balzan, R., Andreou, C., 2017b. Liberal acceptance as a cognitive mechanism in psychosis: a 2-step-therory on the pathogenesis of positive symptoms in schizophrenia. Verhaltenstherapie 27, 108–118. doi:10.1159/000464256
Moritz, S., Lüdtke, T., Westermann, S., Hermeneit, J., Watroba, J., Lincoln, T.M., 2016b. Dysfunctional coping with stress in psychosis: an investigation with the Maladaptive and Adaptive Coping Styles (MAX) questionnaire. Schizophr. Res. 175, 129–135. doi:10.1016/j.schres.2016.04.025
Moritz, S., Mayer-Stassfurth, H., Endlich, L., Andreou, C., Ramdani, N., Petermann, F., Balzan, R.P., 2015. The benefits of doubt: cognitive bias correction reduces hasty decision-making in schizophrenia. Cognit. Ther. Res. 39, 627–635. doi:10.1007/s10608-015-9690-8

Moritz, S., Peters, M.J. V, Larøi, F., Lincoln, T.M., 2010. Metacognitive beliefs in obsessive-compulsive patients: a comparison with healthy and schizophrenia participants. Cogn. Neuropsychiatry 15, 531–548. doi:10.1080/13546801003783508

Moritz, S., Pfuhl, G., Lüdtke, T., Menon, M., Balzan, R.P., Andreou, C., 2016c. A two-stage cognitive theory of the positive symptoms of psychosis: highlighting the role of lowered decision thresholds. J. Behav. Ther. Exp. Psychiatry. doi:10.1016/j.jbtep.2016.07.004

Moritz, S., Rietschel, L., Veckenstedt, R., Bohn, F., Schneider, B.C., Lincoln, T.M., Karow, A., 2015c. The other side of "madness": frequencies of positive and ambivalent attitudes towards prominent positive symptoms in psychosis. Psychosis 7, 14–24. doi:10.1080/17522439.2013.865137

Moritz, S., Scheu, F., Andreou, C., Pfueller, U., Weisbrod, M., Roesch-Ely, D., 2016a. Reasoning in psychosis: risky but not necessarily hasty. Cogn. Neuropsychiatry 21, 91–106. doi:10.1080/13546805.2015.1136611

Moritz, S., Schröder, J., Klein, J.P., Andreou, C., Fischer, A., Arlt, S., 2016b. Effects of online intervention for depression on mood and positive symptoms in schizophrenia. Schizophr. Res. 175, 216–222.

Moritz, S., Van Quaquebeke, N., Lincoln, T.M., 2012b. Jumping to conclusions is associated with paranoia but not general suspiciousness: a comparison of two versions of the probabilistic reasoning paradigm. Schizophr. Res. Treatment 2012, 384039. doi:10.1155/2012/384039

Moritz, S., Veckenstedt, R., Randjbar, S., Hottenrott, B., Woodward, T.S., von Eckstaedt, F. V, Schmidt, C., Jelinek, L., Lincoln, T.M., 2009. Decision making under uncertainty and mood induction: further evidence for liberal acceptance in schizophrenia. Psychol. Med. 39, 1821–1829. doi:10.1017/S0033291709005923

Moritz, S., Woodward, T., 2004. Plausibility judgment in schizophrenic patients: evidence for a liberal acceptance bias. Ger. J. Psychiatry 7, 66–74.

Moritz, S., Woodward, T.S., 2006a. Metacognitive control over false memories: A key determinant of delusional thinking. Curr. Psychiatry Rep. 8, 184–190. doi:10.1007/s11920-006-0022-2

Moritz, S., Woodward, T.S., 2006b. A generalized bias against disconfirmatory evidence in schizophrenia. Psychiatry Res. 142, 157–165. doi:10.1016/j.psychres.2005.08.016

Moritz, S., Woodward, T.S., 2006c. The contribution of metamemory deficits to schizophrenia. J. Abnorm. Psychol. 115, 15–25. doi:10.1037/0021-843X.15.1.15

Moritz, S., Woodward, T.S., Hausmann, D., 2006. Incautious reasoning as a pathogenetic factor for the development of psychotic symptoms in schizophrenia. Schizophr. Bull. 32, 327–331. doi:10.1093/schbul/sbj034

Moritz, S., Woodward, T.S., Jelinek, L., Klinge, R., 2008. Memory and metamemory in schizophrenia: a liberal acceptance account of psychosis. Psychol. Med. 38, 825–32. doi:10.1017/S0033291707002553

Moritz, S., Woodward, T.S., Lambert, M., 2007. Under what circumstances do patients with schizophrenia jump to conclusions? A liberal acceptance account. Br. J. Clin. Psychol. 46, 127–137. doi:10.1348/014466506X129862

Morrison, A.P., Shryane, N., Fowler, D., Birchwood, M., Gumley, A.I., Taylor, H.E., French, P., Stewart, S.L.K., Jones, P.B., Lewis, S.W., Bentall, R., 2015. Negative cognition, affect, metacognition and dimensions of paranoia in people at ultra-high risk of psychosis: a multi-level modelling analysis. Psychol. Med. 45, 2675–2684. doi:10.1017/S0033291715000689

Morrison, A.P., Wells, A., 2003. A comparison of metacognitions in patients with hallucinations, delusions, panic disorder, and non-patient controls. Behav. Res. Ther. 41, 251–256. doi:10.1016/S0005-7967(02)00095-5

Moutoussis, M., Williams, J., Dayan, P., Bentall, R.P., 2007. Persecutory delusions and the conditioned avoidance paradigm: towards an integration of the psychology and biology of paranoia. Cogn. Neuropsychiatry 12, 495–510. doi:10.1080/13546800701566686

Murray, R.M., Quattrone, D., Natesan, S., Van Os, J., Nordentoft, M., Howes, O., Di Forti, M., Taylor, D., 2016. Should psychiatrists be more cautious about the long-term prophylactic use of antipsychotics. Br. J. Psychiatry. doi:10.1192/bjp.bp.116182683

Myers, E., Startup, H., Freeman, D., 2011. Cognitive behavioural treatment of insomnia in individuals with persistent persecutory delusions: a pilot trial. J. Behav. Ther. Exp. Psychiatry 42, 330–336. doi:10.1016/j.jbtep.2011.02.004

Myin-Germeys, I., van Os, J., 2007. Stress-reactivity in psychosis: evidence for an affective pathway to psychosis. Clin. Psychol. Rev. 27, 409–424. doi:10.1016/j.cpr.2006.09.005

Nickerson, R.S., 1998. Confirmation bias: A ubiquitous phenomenon in many guises. Rev. Gen. Psychol. 2, 175–220. doi:10.1037/1089-2680.2.2.175

O'Driscoll, C., Laing, J., Mason, O., 2014. Cognitive emotion regulation strategies, alexithymia and dissociation in schizophrenia, a review and meta-analysis. Clin. Psychol. Rev. 34, 482–495. doi:10.1016/j.cpr.2014.07.002

Oswald, M.E., Grosjean, S., 2004. Confirmation bias, in: Pohl, R.F. (Ed.), Cognitive illusions: a handbook on fallacies and biases in thinking, judgement and memory. Psychology Press, East Sussex, pp. 79–96. doi:10.13140/2.1.2068.0641

Philipp, R., Kriston, L., Lanio, J., Kühne, F., Härter, M., Moritz, S., Meister, R., 2018. Effectiveness of metacognitive interventions for mental disorders in adults—a systematic review and meta-analysis (METACOG). Clin. Psychol. Psychother. doi:10.1002/cpp.2345

Read, J., Bentall, R.P., Fosse, R., 2009. Time to abandon the bio-bio-bio model of psychosis: Exploring the epigenetic and psychological mechanisms by which adverse life events lead to psychotic symptoms. Epidemiol. Psichiatr. Soc. 18, 299–310. doi:10.1017/S1121189X00000257

Reeve, S., Emsley, R., Sheaves, B., Freeman, D., 2017. Disrupting sleep: the effects of sleep loss on psychotic experiences tested in an experimental study with mediation analysis. Schizophr. Bull. 44, 662–671.

Reininghaus, U., Kempton, M.J., Valmaggia, L., Craig, T.K.J., Garety, P., Onyejiaka, A., Gayer-Anderson, C., So, S.H., Hubbard, K., Beards, S., Dazzan, P., Pariante, C., Mondelli, V., Fisher, H.L., Mills, J.G., Viechtbauer, W., Mcguire, P., Van Os, J., Murray, R.M., Wykes, T., Myin-Germeys, I., Morgan, C., 2016. Stress sensitivity, aberrant salience, and threat anticipation in early psychosis: An experience sampling study. Schizophr. Bull. 42, 712–722. doi:10.1093/schbul/sbv190

Röhlinger, J., Wulf, F., Fieker, M., Moritz, S., 2015. Sensory properties of obsessive thoughts in OCD and the relationship to psychopathology. Psychiatry Res. 230, 592–596. doi:10.1016/j.psychres.2015.10.009

Ross, R.M., McKay, R., Coltheart, M., Langdon, R., 2015. Jumping to conclusions about the beads task? A meta-analysis of delusional ideation and data-gathering. Schizophr. Bull. 41, 1183–1191. doi:10.1093/schbul/sbu187

Rüegg, N., Moritz, S., Westermann, S., 2018. Metacognitive training online: a pilot study of an Internet-based intervention for people with schizophrenia. Zeitschrift fur Neuropsychol. 29. doi:10.1024/1016-264X/a000213

Rutherford, B.R., Pott, E., Tandler, J.M., Wall, M.M., Roose, S.P., Lieberman, J.A., 2014. Placebo response in antipsychotic clinical trials: A meta-analysis. JAMA Psychiatry 71, 1409–1421. doi:10.1001/jamapsychiatry.2014.1319

Schlier, B., Jaya, E.S., Moritz, S., Lincoln, T.M., 2015. The Community Assessment of Psychic Experiences measures nine clusters of psychosis-like experiences: a validation of the German version of the CAPE. Schizophr. Res. doi:10.1016/j.schres.2015.10.034

So, S.H.-W., Chau, A.K.C., Peters, E.R., Swendsen, J., Garety, P.A., Kapur, S., 2018. Moment-to-moment associations between negative affect, aberrant salience, and paranoia. Cogn. Neuropsychiatry 23, 299–306. doi:10.1080/13546805.2018.1503080

Soares, S.C., Esteves, F., Lundqvist, D., Öhman, A., 2009. Some animal specific fears are more specific than others: evidence from attention and emotion measures. Behav. Res. Ther. 47, 1032–1042. doi:10.1016/j.brat.2009.07.022

Speerforck, S., Schomerus, G., Pruess, S., Angermeyer, M.C., 2014. Different biogenetic causal explanations and attitudes towards persons with major depression, schizophrenia and alcohol dependence: is the concept of a chemical imbalance beneficial? J. Affect. Disord. 168, 224–228. doi:10.1016/j.jad.2014.06.013

Sullivan, S.A., Wiles, N., Kounali, D., Lewis, G., Heron, J., Cannon, M., Mahedy, L., Jones, P.B., Stochl, J., Zammit, S., 2014. Longitudinal associations between adolescent psychotic experiences and depressive symptoms. PLoS One 9, e105758. doi:10.1371/journal.pone.0105758

Sundag, J., Lincoln, T.M., Hartmann, M.M., Moritz, S., 2015. Is the content of persecutory delusions relevant to self-esteem? Psychosis 7, 237–248. doi:10.1080/17522439.2014.947616

Thompson, A., Papas, A., Bartholomeusz, C., Nelson, B., Yung, A., 2012. Externalized attributional bias in the ultra high risk (UHR) for psychosis population. Psychiatry Res. doi:10.1016/j.psychres.2012.10.017

Tiernan, B., Tracey, R., Shannon, C., 2014. Paranoia and self-concepts in psychosis: a systematic review of the literature. Psychiatry Res. doi:10.1016/j.psychres.2014.02.003

Tone, E.B., Goulding, S.M., Compton, M.T., 2011. Associations among perceptual anomalies, social anxiety, and paranoia in a college student sample. Psychiatry Res. 188, 258–263. doi:10.1016/j.psychres.2011.03.023

Udachina, A., Varese, F., Myin-Germeys, I., Bentall, R.P., 2014. The role of experiential avoidance in paranoid delusions: an experience sampling study. Br. J. Clin. Psychol. 53, 422–432. doi:10.1111/bjc.12054

Upthegrove, R., 2009. Depression in schizophrenia and early psychosis: implications for assessment and treatment. Adv. Psychiatr. Treat. 15, 372–379. doi:10.1192/apt.bp.108.005629

van Oosterhout, B., Krabbendam, L., de Boer, K., Ferwerda, J., van der Helm, M., Stant, A.D., van der Gaag, M., 2014. Metacognitive group training for schizophrenia spectrum patients with delusions: a randomized controlled trial. Psychol. Med. 44, 3025–35. doi:10.1017/S0033291714000555

van Os, J., Linscott, R.J., Myin-Germeys, I., Delespaul, P., Krabbendam, L., 2009. A systematic review and meta-analysis of the psychosis continuum: evidence for a psychosis proneness-persistence-impairment model of psychotic disorder. Psychol. Med. 39, 179–195. doi:10.1017/S0033291708003814

van Os, J., Reininghaus, U., 2016. Psychosis as a transdiagnostic and extended phenotype in the general population. World Psychiatry 15, 118–124.

Veckenstedt, R., Randjbar, S., Vitzthum, F., Hottenrott, B., Woodward, T.S., Moritz, S., 2011. Incorrigibility, jumping to conclusions, and decision threshold in schizophrenia. Cogn. Neuropsychiatry 16, 174–192. doi:10.1080/13546805.2010.536084

Vellante, M., Larøi, F., Cella, M., Raballo, A., Petretto, D.R., Preti, A., 2012. Hallucination-like experiences in the nonclinical population. J. Nerv. Ment. Dis. 200, 310–315. doi:10.1097/NMD.0b013e31824cb2ba

Verdoux, H., Van Os, J., 2002. Psychotic symptoms in non-clinical populations and the continuum of psychosis, in: Schizophrenia Research. pp. 59–65. doi:10.1016/S0920-9964(01)00352-8

Visceglia, E., Lewis, S., 2011. Yoga therapy as an adjunctive treatment for schizophrenia: a randomized, controlled pilot study. J. Altern. Complement. Med. 17, 601–607. doi:10.1089/acm.2010.0075

Waller, H., Emsley, R., Freeman, D., Bebbington, P., Dunn, G., Fowler, D., Hardy, A., Kuipers, E., Garety, P., 2015. Thinking Well: a randomised controlled feasibility study of a new CBT therapy targeting reasoning biases in people with distressing persecutory delusional beliefs. J. Behav. Ther. Exp. Psychiatry 48, 82–89. doi:10.1016/j.jbtep.2015.02.007

Ward, T., Garety, P.A., 2017. Fast and slow thinking in distressing delusions: A review of the literature and implications for targeted therapy. Schizophr. Res. 203, 80–87.

Westermann, S., Lincoln, T.M., 2011. Emotion regulation difficulties are relevant to persecutory ideation. Psychol. Psychother. 84, 273–287. doi:10.1348/147608310X523019

Woodward, T.S., Moritz, S., Menon, M., Klinge, R., 2008. Belief inflexibility in schizophrenia. Cogn. Neuropsychiatry 13, 267–277. doi:10.1080/13546800802099033

Ziegler, M., Rief, W., Werner, S.-M., Mehl, S., Lincoln, T.M., 2008. Hasty decision-making in a variety of tasks: does it contribute to the development of delusions? Psychol. Psychother. 81, 237–245. doi:10.1348/147608308X297104

Zimbardo, P.G., 1999. Discontinuity theory: cognitive and social searches for rationality and normality may lead to madness. Adv. Exp. Soc. Psychol. 31, 345–486. doi:10.1016/S0065-2601(08)60276-2

49

ABERRANT SALIENCE ATTRIBUTION AND PSYCHOSIS

Toby T. Winton-Brown and Shitij Kapur

Nearly two decades ago, Berridge and Robinson conceptualized a role for dopamine in motivational salience,[1] and in 2003 it was invoked as an explanation for psychosis in schizophrenia.[2] In the years since, research developing and testing the "aberrant salience hypothesis" has grown steadily. However even basic questions remain: How is salience calculated and attributed in the brain, and how can we measure it? How does it involve dopamine signaling and how does it go wrong? How does it change with the course of psychosis and with treatment? Does abnormal salience attribution underlie all psychoses? In this chapter, we review 15 years of research into aberrant salience processing in psychosis. We compare the different operationalizations of salience processing in online and offline tasks in controls and in various patient groups. We provide an up to date picture of the state of the art, and suggest directions for future research.

OVERVIEW OF THE ABERRANT SALIENCE MODEL OF PSYCHOSIS

The most profound distinction in psychic life seems to be that between what is meaningful and allows empathy and what in its particular way is ununderstandable, "mad" in the literal sense, schizophrenic psychic life

Karl Jaspers, General Psychopathology, 1913

The experience of psychosis, with its distortion of reality, unusual beliefs, altered perception, and related behavior, has historically confounded ordinary empathetic attempts to understand it. New models of understanding the development of psychotic symptoms aim to bridge this *ununderstandable* divide.

Models linking cognitive and biological levels are key[3]—understanding clinical syndromes in terms of alteration of normal information processing systems (memory, perception, attention, emotion), and their neural substrates (lesions, neurochemical changes, network disturbances).

The "aberrant salience" model of psychosis articulated nearly 15 years ago[2,4] posited that underlying psychosis was abnormal functioning of a "salience system" linked to abnormal dopamine function. At the time such a system was not well established, but the model provided a resonant heuristic for further work.

It resonated in our view for a number of reasons. First was this attempt to bridge this mind–brain divide. It proposed that a mediating link between dopamine dysfunction and psychotic symptoms is the aberrant assignment of incentive salience to innocuous stimuli. These stimuli now "mean something"—their unusual prominence is so compelling and anxiogenic they demand explanation, and the explanations they require may take the form of delusions, filtered through an individual's personal and cultural lens. Where the aberrantly salient stimuli are internal—thoughts, verbal images and memories, somatic sensations—the experiences are of hallucinations or passivity phenomena.

This link between brain and mind proved a useful model in explaining how psychotic symptoms may have arisen, and also how dopamine-blocking medications may help—not by erasing the delusion and hallucinations themselves but by quieting down the aberrantly salient experiences that are "fanning the flames."[5,6] It framed otherwise difficult-to-understand experiences of delusions and hallucinations in terms of the breakdown of a normal system, akin to the breakdown of other homeostatic systems like energy regulation in diabetes, or immune self-recognition in rheumatoid arthritis.

The model appeared to strike home with clinicians and patients in its capturing of the experience of early psychosis, particularly delusional mood and delusion formation. Phenomenologists of over a century ago described similar phenomena. For Klaus Conrad, "*Trema*" precedes the onset of delusions by a period of days, months, or even years.[7] The feeling at this time is of expectancy, that something is about to happen, a "marked change in motivational and emotional state" that eventually spreads to pervade the patient's entire experiential field, "imbued with the theme of the incipient delusion."[7]

> "I was in a higher and higher state of exhilaration and awareness. Things people said had hidden meaning... my senses were sharpened ... I became fascinated by the little insignificant things around me."[8]
>
> "[E]verything—objects, sound, events—took on special meaning for me."[8]

With "*apophany*,"[9] delusions appear as a relieving explanation for what had been a series of perplexing and disturbing experiences, and often involve the self as a central reference point of the universe ("*anastrophe*").

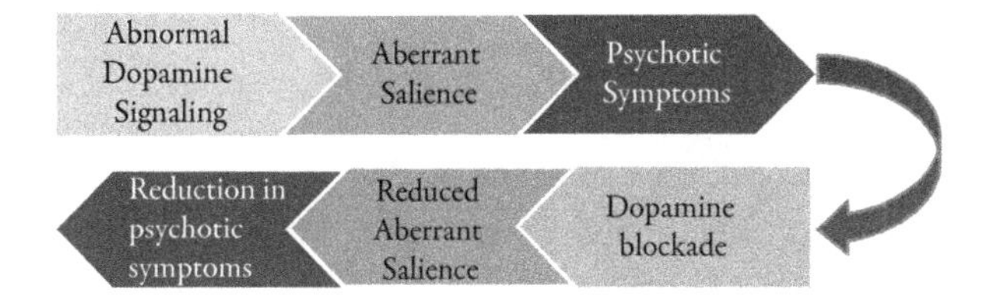

Figure 49.1 Key steps in model.

> "I felt like I was putting the pieces of a puzzle together . . . like mystical revelations . . . things began to fall together and make sense."[7]

Although delusional explanations serve to relieve the tension resulting from aberrant salient experiences, they can also lead to a distancing from reality and disturbance in behavior, now driven by preoccupying and unusual beliefs. Disturbed behavior leads to clinical attention, diagnosis, and treatment, usually incorporating dopamine blockade to reduce the drive to maintain these beliefs, which then slowly extinguish. Following dopamine blockade the normal experience of incentive salience may also be reduced, which may relate to avolitional and other negative symptoms, and lead to treatment discontinuation. Dopaminergic aberrant salience experiences may then recur and delusions may arise again.[2] When internally arising processes—visual imagery, verbal imagery, somatic sensations, and so on—are experienced as aberrantly salient, they are experienced as external, and experienced as hallucinations and passivity phenomena.

The aberrant salience model evolved from key prior work on the normal function of dopamine and its role in addictive behaviors, in particular reward anticipation, sensitization, learning, and motivation,[1] work that continues to evolve, as discussed later. A series of prior cognitive models share an emphasis on anomalous experiences and their appraisal. Gray and colleagues described *intrusions into conscious awareness* that arise from *deficits in integration of new input with stored memories*;[10] Frith related these to deficits in self-monitoring,[11] while Heinz explicitly posited a link between abnormal dopamine function and altered incentive salience in both psychosis and alcoholism.[12] Garety and colleagues suggest that prior reasoning and information-processing biases, schema, and emotional and social factors facilitate appraisals of the origins of these anomalous experiences as external, and beliefs and hallucinations becoming symptomatic.[13] There is overlap also with "one-step" models that regard perception and belief as indistinguishable, but emphasize dopamine dysfunction driven aberrant associative learning underlying the formation of psychotic symptoms.[14,15]

TESTING THE MODEL—PRELIMINARY CONSIDERATIONS

What essentially does the aberrant salience model claim, and how can these claims be tested? The model implies two main links in the *formation* of psychotic symptoms: first, that abnormal dopamine signaling leads to aberrant salience attribution, and second that this aberrant salience attribution leads to psychotic symptoms. A third step in the model relates to the *resolution* of psychotic symptoms, and the mechanism by which dopamine blocking antipsychotics have their rapid effect—not by acting directly on the delusional beliefs themselves, but by "dampening down" the aberrant salience experiences driving the beliefs. This allows delusions to extinguish, until antipsychotics are ceased, when the aberrant salience experiences reemerge and delusions become reactivated.

In psychosis research the aberrant salience model has had some degree of influence. However, for the model to be validated each of these steps requires interrogation and empirical testing in clinical psychosis populations. Progress to this end has been steady but limited in part from the ambiguity inherent in some concepts within the framework and by some basic questions that as yet remain unresolved.

Central to these is what exactly is meant by aberrant salience in the sense intended by the model, and therefore how it should be captured experimentally. Of course, in order to establish this, we need to first agree on what we mean by "salience": which aspects are most critically linked to dopamine function and which of these are likely to be altered as a result of dysregulated dopamine in psychosis. We need also to agree on what is meaningful "aberration"—which alterations in which salience processing are most likely to lead to psychotic symptoms? And which symptoms?

How to best capture these various aspects of salience behaviorally and neurally in humans is not clear, and fMRI studies outnumber behavioral and other modalities. In earlier experimental tasks a contrast is made on a single dimension between salient and nonsalient variants. As fMRI often relies on this subtraction it can be difficult to disentangle whether the observed differences are based on a reduced response to salient stimuli or increased response to neutral stimuli.

Interpretation of some these studies is also complicated by variations in illness stage and duration and by the inclusion of participants taking dopamine-blocking antipsychotic medication.

WHAT IS SALIENCE?

It can be helpful to understand function prior to dysfunction.[3] "Salience" in common usage is a nontechnical term that also has specific connotations in several technical fields. A consequent potential for ambiguity has confounded some of the empiric investigations of aberrant salience processing in psychosis—it can be unclear exactly what salience is, and what it is not.

Broadly speaking, salience refers to prominence within a context. This is most easily understood in terms of the sensory processing of stimulus features. For vision, certain physical features tend to be reliably salient: luminance, movement, color, contrast, and orientation.[16] These features can be combined in a summative "saliency map"[17] to guide the allocation of visual attention, used for example in robotics to enable successful navigation.[18] Highly salient stimuli, such as a loud bang or a flash of light, are attended and responded to largely independent of the organism and context. Mostly however, stimulus-driven processing interacts with internal state and trait factors to determine the most salient stimulus at a given point in time and place for a given organism,[19] pulling attention and cognition, and driving behavior to assist that organism in sorting between stimuli. This so-called selection problem, common to many organisms, also extends to internal stimuli: emotions, thoughts, memories, action plans, and movements are represented across the functional loops of the basal ganglia and are prioritized and selected on the basis of a "common currency"—their salience—and allocated cognitive and motor resources.[20]

Salience may also be defined in terms of this allocation. The first aspect of this is attention—that narrow beamed spotlight that brings certain stimuli to awareness and extracts and processes more information from them while allowing others fade into the background.[21] While attention itself is necessarily conscious, processes such as sensory filtering and sensory and behavioral orientation and priming may not be,[22] leading to the question of whether salient and hence "aberrantly salient" experiences must themselves be conscious ones—this remains unknown, and may bear on explanations for symptoms (such as hallucinations or primary delusions) where there is not a clear conscious anomalous experience preceding the development of a psychotic symptom.

More than just attentional diversion, a salient stimulus should lead to a behavioral response or "switch."[20] Here the influence of increased dopamine transmission transforms "cold" neural representations into "hot" objects to be actively sought, or avoided.[23] This aspect is crucial for the aberrant salience model of psychosis, which holds that altered dopamine function changes salience processing in a way that leads to psychotic symptoms. But just how does it do so?

WHAT DOES DOPAMINE DO?

INCENTIVE SALIENCE—REWARD

A key influence on behavior is the active pursuit of reward and the avoidance of punishment. "Reward" here refers to the positive value given to an object, a behavioral act or an internal state; rewards reinforce the behavior that led to them.[24] The role of dopamine has received particular attention in this context. While initial studies focused on the role of dopamine in the actual experience of pleasure [24] this was subsequently separated into "wanting" (*incentive salience*), "liking" (*hedonic impact*), and "learning" (*reward learning*).[1] Berridge and Robinson (1998) argued that dopamine systems mediate the motivational significance of rewards—the willingness to work for them—rather than either the pleasure they provide or the associative learning that results.[1,25]

Dopamine as a signal of incentive salience was the basis for the links with aberrant salience processing psychosis in the original formulation.[2] As well as being established experimentally, it seemed to account for the compelling power of the anomalous experiences as described by patients in the early phases of psychosis.

There is a good deal of evidence for reward-processing abnormalities in psychotic subjects. Initial studies used classical conditioning reward paradigms such as the monetary incentive delay (MID) task[26] that evolved from animal studies.[27] MID fMRI studies demonstrate robust ventral striatal activation in anticipation of reward,[28] and abnormalities in ventral striatal reward processing in psychosis have now been demonstrated in meta-analysis.[29] Initial application of the MID in subjects with schizophrenia revealed reductions in the ventral striatum (VS) signal to reward predicting relative to neutral cues,[30,31] although this was inconsistent,[32,33] and it was not clear whether these differences were due to reduced reward activations or increased neutral cue activations, or both. Activation differences were also in extrastriatal regions[32] and in the connectivity between VS and PFC.[34,35] In several of these studies, reduced reward-cue-related activations were correlated with the degree of negative symptoms[30,32,36,37]—evidence that reduced incentive-processing underlies volition and motivation deficits, but not toward the aberrant salience model, which predicts increased neutral activations would relate to positive symptoms.

Addressing this were two studies that recruited unmedicated first-episode psychosis (FEP) patients. In an early small sample reduction in VS activation to reward predicting cues correlated with positive symptoms at trend level,[36] confirmed in a later more definitive study.[31] Nielsen et al. (2012) recruited a moderate sized sample and showed that never-medicated FEP patients had reduced reward-cue-related activation relative to controls in the ventral striatum, ventral tegmental area (VTA), and ACC, and that these reductions were correlated with the extent of positive psychotic symptoms at the time of scanning. In contrast to the earlier studies this seemed to fit with the central prediction of the aberrant salience model.

INCENTIVE SALIENCE—AVERSION

However, the anomalous experiences of early psychosis are usually not rewarding—indeed they are often perplexing, threatening, and frightening. If aberrant salience experiences are just misplaced reward signals to neutral stimuli why it is that many of the experiences of aberrant salience in psychosis are imbued with a sense of fear and foreboding? It has been increasingly recognized that dopamine neurons may also function to encode the motivation to avoid aversive stimuli,[38] and so dopamine-mediated incentive salience may be bivalenced. In animal studies aversive events, such as footshocks, airpuffs, and restraint increase dopamine release in projection target regions and increase firing in some putative dopamine neurons.[39]

Moreover, it is now recognized that dopamine cell populations are functionally heterogeneous—VTA dopamine neurons can be either excited or inhibited by aversive events[40] and by stimuli predicting aversive events;[41] these may vary also according to projection targets.[42] In humans, dopaminergic regions activate in response to anticipation of both aversion[43] and monetary loss.[44]

If dopamine also encodes aversion, aberrant aversive coding of neutral stimuli may help explain why many of the anomalous experiences in early psychosis are imbued with a sense of threat and fear and lead to persecutory delusions. Two studies speak to this: following their demonstration of VS activation during the anticipation of aversive stimuli in controls,[43] Jensen and colleagues used an aversive Pavlovian conditioning task in medicated subjects with schizophrenia and demonstrated increased behavioral, physiological, and VS activation to neutral cues.[45] Romaniuk et al. used a similar aversive Pavlovian conditioning task and showed increased amygdala, VS, and midbrain activation to neutral cues; interestingly in the midbrain these activations correlated with delusional symptoms scores.[46]

SALIENCE AS NOVELTY

Midbrain dopamine cells have also been long known to respond to novel events such as lights and tones, regardless of incentive value,[47] and in humans there is considerable evidence for the involvement of dopamine in novelty processing.[48,49] Phenomenological accounts of psychosis frequently refer to an unusual sense of novelty,[2,9] and yet there have been relatively few studies of novelty-processing in subjects with psychosis. Most commonly studied are oddball paradigms that feature "novel" stimuli among repeated standard stimuli and "target" stimuli. Kiehl et al., for example, used an auditory oddball paradigm and showed reduced activation in bilateral frontal and temporal regions, but no relationship with symptoms.[50] More recently fMRI responses to novelty when studied alongside aversion and reward were not different in subjects at high clinical risk for psychosis relative to controls.[51]

SALIENCE AS PREDICTION ERROR

A further insight into dopamine neuron function came from Schultz and colleagues' recording individual dopamine neuron output in primates, highlighting the intimacy of prediction and reward.[27] While phasic dopamine responses do occur to unpredicted rewards, they are absent when predicted by a conditioned stimulus for that reward. Learning is driven by the mismatch between prediction and outcome, or prediction error (PE).[52] Computational models of prediction error account for the uncertainty, magnitude, and timing of rewarding outcomes[53] and can in be incorporated into statistical modeling of behavior and functional brain imaging. Large prediction errors, represented by large phasic dopamine signals,[53] are thus highly "salient," leading the organism to adjust behavior and cognition at ascending hierarchical scales.[15] Dysregulated dopamine signaling leading to phasic output uncoupled from context would be then aberrantly salient, and require significant behavioral and cognitive accommodation, even perhaps to the level of delusion.

As PEs closely follow phasic dopamine neuron responses[27] they are an appealing candidate construct for aberrant salience in the context of psychosis, and have been incorporated in a number of innovative ways into reward learning tasks in subjects with psychosis. Results in patient groups have varied but overall supported this construct, but a detailed synthesis in terms of aberrant salience is elusive.

Murray et al. used a reward learning paradigm in FEP subjects and controls and a Q-learning model to calculate individual trial to trial prediction errors that were incorporated into fMRI analyses.[54] Medicated and unmedicated subjects with psychosis responded faster to neutral cues, and had significantly reduced PE-correlated activation in the midbrain and ventral striatum to outcome in rewarded vs. neutral trials. In the midbrain this difference was driven both by reduced activation to reward and increased activation to neutral PEs, but not correlated with symptoms. Gradin et al. used a similar task in medicated subjects with schizophrenia[55] and found reduced extrastriatal reward PE-related activations in caudate, insula, thalamus, and amygdala-hippocampus that did correlate with positive psychotic symptoms. Similarly, Waltz et al.[56] manipulated the timing of a juice reward delivery following a predictive cue. Medicated subjects with schizophrenia show reduced responses in the putamen and frontal operculum to unexpected reward delivery, but no relationship with positive symptoms; negative symptoms related to reduced responses in the putamen to expected receipt of reward.[56]

Does this associative learning effect extend beyond reward? Corlett et al.[57] used a nonreward learning task to explore responses to learned associations between food and allergy. Subjects with psychosis demonstrated reduced VS signals during acquisition of associations, while differences in PE related activation in the PFC correlated with delusion scores,[57] suggesting that the PFC is updated with irrelevant information during delusion formation.

In order to disentangle reward and prediction error, Morris et al. compared activation patterns in the striatum and midbrain to expected and unexpected rewards and omissions in a Pavlovian learning game.[58] They found that in controls ventral striatal responses corresponded best to biphasic prediction errors—increased in response to unexpected reward and reduced in response to unexpected reward omissions. In subjects with schizophrenia this was altered—ventral striatal differences related to both increased activation to expected rewards and reduced responses to unexpected rewards, the former correlating with total and negative psychotic symptoms.[58] How might this be seen in terms of the aberrant salience model?

ABERRANT LEARNING AND THE SALIENCE ATTRIBUTION TASK

Are increased behavioral or neural responses to neutral stimuli sufficiently equivalent to aberrant salience? Some think perhaps not and have worked to create more specific probes. By adding additional stimulus dimensions irrelevant to reward outcome, behavioral and neural aspects of aberrant reward learning can be studied, analogous if not synonymous with aberrant salience. In an interesting variant on an instrumental learning game, Roiser and colleagues developed the salience attribution task (SAT) to measure responses to cues with reward relevant and irrelevant dimensions.[59] Implicit (reaction time differences) and explicit (visual scale) measures are taken of each subject's attribution of the significance of either form or color of picture cues in predicting monetary reward, which varies from block to block. Responses to reward-relevant and -irrelevant dimensions are considered adaptive and aberrant salience respectively; abnormalities are hypothesized to underlie negative and positive symptoms.

The SAT has been used both offline and during fMRI by several groups attempting to capture the dimension of aberrant salience processing in psychosis, and results have been promising if variable. In a series of studies, Roiser and colleagues demonstrate first that medicated FEP patients exhibit reduced adaptive salience but similar aberrant salience to controls.[60] Within the psychosis group those with delusional symptoms, however, showed greater explicit aberrant salience than those without, which correlated with delusion scores; there was no difference between patients who did and did not have hallucinations. Patients with negative symptoms also showed greater explicit aberrant salience, and negative symptoms also correlated with this measure.[60]

Using the same task alongside fMRI and 18-F-DOPA PET scanning in a group of clinical high risk (CHR) subjects, the same group then demonstrated increased aberrant salience relative to controls that was correlated with the severity of attenuated delusional ideation. fMRI activation to irrelevant stimulus features in the ventral striatum correlated with delusional ideation, and in the hippocampus related to striatal dopamine synthesis capacity[61]—the first time that dopamine function and aberrant salience were related directly in the same subjects.

In a subsequent larger study in three clinical groups—CHR subjects, FEP subjects, and controls—these differences in aberrant salience were not replicated, although there were differences in adaptive salience between controls and both patient groups.[62]

Nevertheless this is a promising strategy—Morris et al.[63] similarly tested the learning of relevant and irrelevant dimensions in a simple associative learning task that importantly controlled for general learning deficits. Medicated subjects with schizophrenia showed abnormal learning, with an attentional and learning bias toward irrelevant cues that correlated with positive psychotic symptoms.[63]

SALIENCE AND AGENCY

Contra Schultz et al.,[27] Redgrave and colleagues pointed out that the timing of phasic dopamine output following a sensory event (50–100 ms) precedes the attentive sensory processing, including gaze-shifting, required to make an accurate reward prediction, which therefore must remain unknown at the time of dopamine signaling.[64] They suggest that a key function of phasic dopamine output is instead to reinforce behaviors associated with salient sensory input, promoting a sense of agency.[64–66] That is, phasic dopamine signaling enables the organism to sort between stimuli for which it was and was not causally responsible. To us this also seems an attractive conception of dopamine function as dysfunction may then account for the misplaced sense of agency inherent in many psychotic symptoms, from paranoia through to passivity phenomena[11] and hallucinations;[67] while the concepts are not identical, there are clear links between notions of salience and agency.

Studies of agency in psychosis

However agency, like salience, is not an easy concept to operationalize and there are relatively few studies relating agency and salience in psychosis. Pankow et al.[68] compared responses of controls, subjects with subclinical delusions, and subjects with psychosis in schizophrenia while they applied neutral trait words to themselves or to a public figure. In the same three groups they performed an offline version of the SAT, exploring the relationship between these measures. They demonstrated a gradient of increasing aberrant salience attribution with increasing psychosis across the three groups. In schizophrenia subjects, reduced ventromedial prefrontal cortex (vmPFC) activation during self-referential processing correlated with increased aberrant salience attribution, suggesting a direct relationship between these processes.[68]

ABERRANT SALIENCE, APOPHENIA, AND PARIEDOLIA

Like responding to irrelevant stimuli in reward tasks, a tendency to find patterns ("pariedolia") and meaning ("apophenia") in noise has also been advanced as aberrant salience. Galdos et al. studied subjects with psychosis, their relatives and controls and found a dose-response relationship between the rate of speech illusion extracted from white noise and the degree of familial psychosis risk; this was accentuated for affectively salient speech.[69] In a similar vein Hoffman demonstrated that subjects at high clinical risk of psychosis extracted words from meaningless multispeaker babble—in this study the length of the speech illusion predicted later transition to psychosis.[70]

A SCALE FOR ABERRANT SALIENCE?

Outside the mostly cross-sectional studies discussed thus far, there is scant literature on the phenomenon of aberrant salience from a psychological standpoint, with one notable exception. The Aberrant Salience Inventory (ASI)[71] is a carefully validated 29-item trait measure that attempts to capture the experience of aberrant salience across the life span. It offers binary response choice to a breadth of different aberrant salience experiences, and so high scores reflect this breadth rather than indexing the extent or impact of any particular experience. To our knowledge there is no corresponding trait measure, which would be of great interest given the proposed waxing and waning of aberrant salience experiences with episodes of psychosis. The ASI has been applied in a large study of healthy controls oversampled for schizotypy: those with low self-concept clarity and high aberrant salience experiences had the highest levels of psychotic like experiences, particularly delusions.[72] The application of this scale in clinical psychosis populations is awaited with interest.

FUTURE DIRECTIONS

RECONCILING CONSTRUCTS WITH MULTIDIMENSIONAL AND MULTIMODAL STUDIES

As illustrated, salience and hence aberrant salience is a construct that may be viewed from a number of perspectives, even when limiting those to within a putatively "dopaminergic" framework.[73] Several of these when examined in isolation have shown promising cross-sectional alterations in subjects with psychosis relative to controls; some have been replicated and correlated with psychotic symptoms.

Reconciling these findings toward a harmonized conception and measure of aberrant salience is not straightforward, as few studies have compared these different aspects in the same subjects. The study by Pankow and colleagues is a commendable exception, along with a recent study by Winton-Brown et al.[51] that measured responses to reward, aversion, and novelty in the same CHR for psychosis subjects. They found alterations in striatal activation and connectivity relative to controls were largely confined to the reward domain, the latter relating to positive psychotic symptoms.[51]

A third multidimensional study by Boehme et al. tests the relationship of reward prediction error signaling and aberrant salience.[74] In a group of healthy controls using an fMRI reward learning task and the SAT offline, they demonstrate reduced reward PE signals in the striatum and prefrontal cortex related to increased aberrant salience, providing evidence toward a relationship between these constructs. Application in psychotic clinical groups is awaited, but in a subset of these subjects the researchers also performed 18-F-DOPA PET scans. They found that implicit behavioral aberrant salience also related to ventral striatal dopamine levels taken from a sphere centered on the peak PE signaling in the right VS.[74]

While tentative, this is of course of great interest. Incorporating direct measures of brain dopamine function into behavioral and functional neuroimaging studies is an innovation that adds another important layer of validation to the aberrant salience model. Although so far real-time phasic dopamine output is difficult to capture, 18-F-DOPA positron emission tomograpy (PET) has shown altered presynaptic dopamine availability in psychosis, probably reflecting altered tonic dopamine activity.[5,75] What this means for models that reference altered phasic dopamine signaling remains to be established, but multimodal studies provide intriguing clues.[74,76]

Similarly, few studies of aberrant salience in psychosis so far have attempted to incorporate measures of other psychological or social factors likely relevant to psychosis onset, such as reasoning biases, attributional styles, emotional influences and stress response sensitivity.[13] These studies, although also complex, are needed in order to validate the rich models that inform them. A recent study by Reininghaus and colleagues makes an innovative attempt. Using the Experience Sampling Methodology (ESM) in FEP, CHR, and control groups they ask whether experiences of stress sensitivity, aberrant novelty and salience, and threat anticipation contribute to psychotic experiences in daily life.[77] All three measures were associated with increased psychotic experiences in each of the three groups, with a particular role of aberrant salience evident in the CHR group and sensitivity to stressful events and threat anticipation playing a greater role in the FEP group.[77]

EXPERIMENTAL MANIPULATIONS OF SALIENCE ATTRIBUTION

Inducing psychotic symptoms via direct dopaminergic manipulation of salience processing would provide significant evidence toward the aberrant salience model, though ethically proscribed in clinical subjects. Two areas give analogies of this approach—the use of optogenetic technology in animal models, and observing the effects of dopaminergic therapy in Parkinson's disease (PD).

In mice, optogenetic activation of dopamine cells paired with reward delivery, in a sense mimicking an aberrantly salient event, leads to significantly altered behavior—overcoming associative blocking and preventing extinction learning of reward cues.[78] Similarly, optogenetic activation of dopamine neurons projecting to the nucleus accumbens can induce conditioned place preference, whereas activation of those projecting to the prefrontal cortex induces conditioned place aversion.[79] Extending this work to animal genetic disease model lines and animal psychosis phenotypes would be of great interest.

An opportunistic related framework may be provided by observing the effects of dopamine replacement therapy in PD.[80] In support of this is a behavioral study using the SAT in PD subjects before and after initiation of dopaminergic therapy showing increases in both adaptive and aberrant salience,[81] and a pilot study of the Aberrant Salience Inventory in PD patients that showed higher doses of dopamine replacement therapy related to increased ASI scores.[82]

DOES ABERRANT SALIENCE ATTRIBUTION UNDERLIE ALL PSYCHOSES?

A final important and largely unanswered question is that of the specificity of altered salience processing in psychosis. Does aberrant salience underlie all psychoses, or just psychosis in schizophrenia? Many of the studies described in this chapter limit recruitment to subjects meeting Diagnostic and Statistical Manual (DSM) or International Statistical Classification of Diseases and Related Health Problems (ICD) criteria for schizophreniform and/or affective psychoses and few are powered to detect differences by subgroup.

Might we speculate that aberrant salience underlies "final common pathway" psychoses[5] where dopamine transmission is altered? Where does this then leave treatment resistant psychoses where dopamine may be in fact normal?[83] Or affective psychoses where it seems plausible that salience processing may be affected differently, such as demonstrated by Gradin and colleagues comparing reward processing abnormalities in depression and schizophrenia and finding substantial overlap[55]—reward processing abnormalities may in fact be common to mania, depression, and schizophrenia.[84] We have little evidence yet to guide these questions.

Where do drug-induced or organic psychoses fit in this model? Bloomfield et al. find no difference in aberrant salience on the SAT in a small group of cannabis users compared to controls, but in users explicit (but not implicit) aberrant salience related to cannabis induced psychotic symptoms—the suggestion is that psychosis in cannabis users, rather than cannabis use per se, is mediated by aberrant salience.[85] Much further research in this area is needed.

Similarly we have a limited though evolving understanding of the progression of alterations in salience processing in relation to the development of psychotic symptoms from prodrome to first episode, remission, relapse, and chronic psychosis. Two studies mentioned earlier demonstrate abnormalities in clinical high risk samples,[51,61] the former has published longitudinal follow-up data that has showed resolution of the increased behavioral aberrant salience attribution relative to controls that had been evident at baseline.[86]

CONCLUSION AND FUTURE DIRECTIONS

Since the original formulation of a possible link between altered dopamine transmission, aberrant salience processing, and the development of psychotic symptoms there has been significant progress in the testing, refinement and validation of the model. It remains a promising model, but for this promise to be fulfilled several advancements need to be made. In groups of psychotic patients relative to controls there is now good cross-sectional evidence of altered reward anticipation, altered prediction error signaling and some evidence of altered aversion processing and aberrant reward learning, and in a number of studies this has been correlated with clinical measures of psychotic symptoms, particularly delusions. As yet we have neither a gold standard measure of aberrant salience nor a useful behavioral or bedside proxy measure, although both the SAT and the ASI are first steps, albeit ones with limitations. Remarkably missing are longitudinal data that have followed measures of salience across treatment as patients are getting better—this would be a most critical study.

Moving the field forward are studies that integrate different psychological constructs and imaging modalities, including direct and indirect measures of dopamine function and with dopaminergic manipulation. By applying these complex studies in different patient groups with longitudinal designs we will further elucidate the intriguing question of whether psychosis is after all a single unified transdiagnostic syndrome—resulting from a range of pathologies but ending in a final common pathway, and if and where aberrant salience lies on this pathway.

REFERENCES

1. Berridge KC, Robinson TE. What is the role of dopamine in reward: hedonic impact, reward learning, or incentive salience? *Brain Research Brain Research Reviews*. 1998;28(3):309–369.
2. Kapur S. Psychosis as a state of aberrant salience: a framework linking biology, phenomenology, and pharmacology in schizophrenia. *Am J Psychiatry*. 2003;160(1):13–23.
3. David AS. A method for studies of madness. *Cortex*. 2006;42(6):921–925. doi:10.1016/S0010-9452(08)70436-2.
4. Kapur S, Mizrahi R, Li M. From dopamine to salience to psychosis—linking biology, pharmacology and phenomenology of psychosis. *Schizophr Res*. 2005;79(1):59–68. doi:10.1016/j.schres.2005.01.003.
5. Howes OD, Kapur S. The dopamine hypothesis of schizophrenia: version III—the final common pathway. *Schizophr Bull*. 2009;35(3):549–562. doi:10.1093/schbul/sbp006.
6. Laruelle M, Abi-Dargham A. Dopamine as the wind of the psychotic fire: new evidence from brain imaging studies. *Journal of Psychopharmacology*. 1999;13(4):358–371. doi:10.1177/026988119901300405.
7. Mishara AL. Klaus Conrad (1905–1961): delusional mood, psychosis, and beginning schizophrenia. *Schizophr Bull*. 2010;36:9–13. doi:10.1093/schbul/sbp144.
8. Stanton B. First-person accounts of delusions. *Psychiatric Bulletin*. 2000;24(9):333–336. doi:10.1192/pb.24.9.333.
9. Mishara AL. Klaus Conrad (1905–1961): delusional mood, psychosis, and beginning schizophrenia. *Schizophr Bull*. 2010;36(1):9–13. doi:10.1093/schbul/sbp144.
10. Hemsley DR. A simple (or simplistic?) cognitive model for schizophrenia. *Behaviour Research and Therapy*. 1993;31(7):633–645. doi:10.1016/0005-7967(93)90116-C.
11. Frith CD, Blakemore S, Wolpert DM. Explaining the symptoms of schizophrenia: abnormalities in the awareness of action. *Brain Research Brain Research Reviews*. 2000;31(2–3):357–363.
12. Heinz A. Dopaminergic dysfunction in alcoholism and schizophrenia--psychopathological and behavioral correlates. *Eur Psychiatry*. 2002;17(1):9–16.
13. Garety PA, Kuipers E, Fowler D, Freeman D, Bebbington PE. A cognitive model of the positive symptoms of psychosis. *Psychol Med*. 2001;31(2):189–195. doi:10.1017/S0033291701003312.
14. Corlett PR, Frith CD, Fletcher PC. From drugs to deprivation: a Bayesian framework for understanding models of psychosis. *Psychopharmacology*. 2009;206(4):515–530. doi:10.1007/s00213-009-1561-0.
15. Fletcher PC, Frith CD. Perceiving is believing: a Bayesian approach to explaining the positive symptoms of schizophrenia. *Nat Rev Neurosci*. 2009;10(1):48–58. doi:10.1038/nrn2536.
16. Nothdurft H. Salience from feature contrast: additivity across dimensions. *Vision Research*. 2000;40(10–12):1183–1201.
17. Li Z. A saliency map in primary visual cortex. *Trends in Cognitive Sciences*. 2002;6(1):9–16.
18. Meger D, Forssén P-E, Lai K, et al. Curious George: an attentive semantic robot. *Robotics and Autonomous Systems*. 2008;56(6):503–511. doi:10.1016/j.robot.2008.03.008.
19. Corbetta M, Shulman GL. Control of goal-directed and stimulus-driven attention in the brain. *Nat Rev Neurosci*. 2002;3(3):215–229. doi:10.1038/nrn755.
20. Redgrave P, Prescott TJ, Gurney K. The basal ganglia: a vertebrate solution to the selection problem? *Neuroscience*. 1999;89(4):1009–1023.
21. Crick F. Function of the thalamic reticular complex: the searchlight hypothesis. *Proc Natl Acad Sci U S A*. 1984;81(14):4586–4590.
22. Sheeran P, Gollwitzer PM, Bargh JA. Nonconscious processes and health. *Health Psychol*. 2013;32(5):460–473. doi:10.1037/a0029203.
23. Iversen SD, Iversen LL. Dopamine: 50 years in perspective. *Trends Neurosci*. 2007;30(5):188–193. doi:10.1016/j.tins.2007.03.002.
24. Wise RA, Rompre PP. Brain dopamine and reward. *Annu Rev Psychol*. 1989;40(1):191–225. doi:10.1146/annurev.ps.40.020189.001203.
25. Flagel SB, Clark JJ, Robinson TE, et al. A selective role for dopamine in stimulus-reward learning. *Nature*. 2011;469(7328):53–57. doi:10.1038/nature09588.

26. Knutson B, Westdorp A, Kaiser E, Hommer D. FMRI visualization of brain activity during a monetary incentive delay task. 2000;12(1):20–27. doi:10.1006/nimg.2000.0593.
27. Schultz W. A neural substrate of prediction and reward. *Science*. 1997;275(5306):1593–1599. doi:10.1126/science.275.5306.1593.
28. Knutson B, Adams CM, Fong GW, Hommer D. Anticipation of increasing monetary reward selectively recruits nucleus accumbens. *J Neurosci*. 2001;21(16):RC159.
29. Radua J, Schmidt A, Borgwardt S, et al. Ventral striatal activation during reward processing in psychosis: a neurofunctional meta-analysis. *JAMA Psychiatry*. 2015;72(12):1243–1251. doi:10.1001/jamapsychiatry.2015.2196.
30. Juckel G, Schlagenhauf F, Koslowski M, et al. Dysfunction of ventral striatal reward prediction in schizophrenic patients treated with typical, not atypical, neuroleptics. *Psychopharmacology*. 2006;187(2):222–228. doi:10.1007/s00213-006-0405-4.
31. Nielsen MØ, Rostrup E, Wulff S, et al. Alterations of the brain reward system in antipsychotic naïve schizophrenia patients. *Biol Psychiatry*. 2012;71(10):898–905. doi:10.1016/j.biopsych.2012.02.007.
32. Waltz JA, Schweitzer JB, Ross TJ, et al. Abnormal responses to monetary outcomes in cortex, but not in the basal ganglia, in schizophrenia. *Neuropsychopharmacology*. 2010;35(12):2427–2439. doi:10.1038/npp.2010.126.
33. Walter H, Kammerer H, Frasch K, Spitzer M, Abler B. Altered reward functions in patients on atypical antipsychotic medication in line with the revised dopamine hypothesis of schizophrenia. *Psychopharmacology*. 2009;206(1):121–132. doi:10.1007/s00213-009-1586-4.
34. Diaconescu AO, Jensen J, Wang H, et al. Aberrant effective connectivity in schizophrenia patients during appetitive conditioning. *Front Hum Neurosci*. 2011;4:239. doi:10.3389/fnhum.2010.00239.
35. Schlagenhauf F, Sterzer P, Schmack K, et al. Reward feedback alterations in unmedicated schizophrenia patients: relevance for delusions. *Biol Psychiatry*. 2009;65(12):1032–1039. doi:10.1016/j.biopsych.2008.12.016.
36. Juckel G, Schlagenhauf F, Koslowski M, et al. Dysfunction of ventral striatal reward prediction in schizophrenia. *Neuroimage*. 2006;29(2):409–416. doi:10.1016/j.neuroimage.2005.07.051.
37. Simon JJ, Biller A, Walther S, et al. Neural correlates of reward processing in schizophrenia--relationship to apathy and depression. *Schizophr Res*. 2010;118(1–3):154–161. doi:10.1016/j.schres.2009.11.007.
38. Ungless MA, Argilli E, Bonci A. Effects of stress and aversion on dopamine neurons: implications for addiction. *Neurosci Biobehav Rev*. 2010;35(2):151–156. doi:10.1016/j.neubiorev.2010.04.006.
39. Anstrom KKK, Woodward DJD. Restraint increases dopaminergic burst firing in awake rats. *Neuropsychopharmacology*. 2005;30(10):1832–1840. doi:10.1038/sj.npp.1300730.
40. Brischoux F, Chakraborty S, Brierley DI, Ungless MA. Phasic excitation of dopamine neurons in ventral VTA by noxious stimuli. *Proc Natl Acad Sci U S A*. 2009;106(12):4894–4899. doi:10.1073/pnas.0811507106.
41. Mileykovskiy B, Morales M. Duration of inhibition of ventral tegmental area dopamine neurons encodes a level of conditioned fear. *J Neurosci*. 2011;31(20):7471–7476. doi:10.1523/JNEUROSCI.5731-10.2011.
42. Roitman MF, Wheeler RA, Wightman RM, Carelli RM. Real-time chemical responses in the nucleus accumbens differentiate rewarding and aversive stimuli. *Nat Neurosci*. 2008;11(12):1376–1377. doi:10.1038/nn.2219.
43. Jensen J, McIntosh AR, Crawley AP, Mikulis DJ, Remington G, Kapur S. Direct activation of the ventral striatum in anticipation of aversive stimuli. *Neuron*. 2003;40(6):1251–1257. doi:10.1016/S0896-6273(03)00724-4.
44. Carter RM, Macinnes JJ, Huettel SA, Adcock RA. Activation in the VTA and nucleus accumbens increases in anticipation of both gains and losses. *Frontiers in Behavioral Neuroscience*. 2009;3(August):21. doi:10.3389/neuro.08.021.2009.
45. Jensen J, Willeit M, Zipursky RB, et al. The formation of abnormal associations in schizophrenia: neural and behavioral evidence. *Neuropsychopharmacology*. 2008;33(3):473–479. doi:10.1038/sj.npp.1301437.
46. Romaniuk L, Honey GD, King JRL, et al. Midbrain activation during Pavlovian conditioning and delusional symptoms in schizophrenia. *Arch Gen Psychiatry*. 2010;67(12):1246–1254. doi:10.1001/archgenpsychiatry.2010.169.
47. Laurent PA. The emergence of saliency and novelty responses from reinforcement learning principles. *Neural Networks*. 2008;21(10):1493–1499. doi:10.1016/j.neunet.2008.09.004.
48. Bunzeck N, Duzel E. Absolute coding of stimulus novelty in the human substantia nigra/VTA. *Neuron*. 2006;51(3):369–379. doi:10.1016/j.neuron.2006.06.021.
49. Wittmann BC, Bunzeck N, Dolan RJ, Duzel E. Anticipation of novelty recruits reward system and hippocampus while promoting recollection. 2007;38(1):194–202. doi:10.1016/j.neuroimage.2007.06.038.
50. Kiehl KA, Stevens MC, Celone K, Kurtz M, Krystal JH. abnormal hemodynamics in schizophrenia during an auditory oddball task. *Biol Psychiatry*. 2005;57:1029–1040. doi:10.1016/j.biopsych.2005.01.035.
51. Winton-Brown T, Roiser JP, Schmidt A, et al. Altered activation and connectivity in a hippocampal-basal ganglia-midbrain circuit in subjects at ultra high risk for psychosis. *Transl Psychiatry*. 2017;7.
52. Ben Seymour, O'Doherty JP, Dayan P, et al. Temporal difference models describe higher-order learning in humans. *Nature*. 2004;429(6992):664–667. doi:10.1038/nature02581.
53. Tobler PN, Fiorillo CD, Schultz W. Adaptive coding of reward value by dopamine neurons. *Science*. 2005;307(5715):1642–1645.
54. Murray GK, Corlett PR, Clark L, et al. Substantia nigra/ventral tegmental reward prediction error disruption in psychosis. *Mol Psychiatry*. 2008;13(3):239, 267–276. doi:10.1038/sj.mp.4002058.
55. Gradin VB, Kumar P, Waiter G, et al. Expected value and prediction error abnormalities in depression and schizophrenia. *Brain*. 2011;134(6):1751–1764. doi:10.1093/brain/awr059.
56. Waltz JA, Schweitzer JB, Gold JM, et al. Patients with schizophrenia have a reduced neural response to both unpredictable and predictable primary reinforcers. *Neuropsychopharmacology*. 2009;34(6):1567–1577. doi:10.1038/npp.2008.214.
57. Corlett PR, Murray GK, Honey GD, et al. Disrupted prediction-error signal in psychosis: evidence for an associative account of delusions. *Brain*. 2007;130(Pt 9):2387–2400. doi:10.1093/brain/awm173.
58. Morris RW, Vercammen A, Lenroot R, et al. Disambiguating ventral striatum fMRI-related bold signal during reward prediction in schizophrenia. *Mol Psychiatry*. 2011;17(3):280–289. doi:10.1038/mp.2011.75.
59. Roiser JP, Stephan KE, Ouden den HEM, Friston KJ, Joyce EM. Adaptive and aberrant reward prediction signals in the human brain. 2010;50(2):657–664. doi:10.1016/j.neuroimage.2009.11.075.
60. Roiser JP, Stephan KE, Ouden den HEM, Barnes TRE, Friston KJ, Joyce EM. Do patients with schizophrenia exhibit aberrant salience? *Psychol Med*. 2009;39(2):199–209. doi:10.1017/S0033291708003863.
61. Roiser JP, Howes OD, Chaddock CA, Joyce EM, McGuire P. Neural and behavioral correlates of aberrant salience in individuals at risk for psychosis. *Schizophr Bull*. 2012;39(6):1328–1336. doi:10.1093/schbul/sbs147.
62. Smieskova R, Roiser JP, Chaddock CA, et al. Modulation of motivational salience processing during the early stages of psychosis. *Schizophr Res*. 2015;166:17–23. doi:10.1016/j.schres.2015.04.036.
63. Morris R, Griffiths O, Le Pelley ME, Weickert TW. Attention to irrelevant cues is related to positive symptoms in schizophrenia. *Schizophr Bull*. 2013;39(3):575–582. doi:10.1093/schbul/sbr192.
64. Redgrave P, Gurney K. The short-latency dopamine signal: a role in discovering novel actions? *Nat Rev Neurosci*. 2006;7(12):967–975. doi:10.1038/nrn2022.

65. Redgrave P, Gurney K, Reynolds J. What is reinforced by phasic dopamine signals? *Brain Research Reviews*. 2008;58(2):322–339. doi:10.1016/j.brainresrev.2007.10.007.
66. Redgrave P, Vautrelle N, Reynolds JNJ. Functional properties of the basal ganglia's re-entrant loop architecture: selection and reinforcement. *Neuroscience*. 2011;198:138–151. doi:10.1016/j.neuroscience.2011.07.060.
67. Allen P, Aleman A, McGuire PK. Inner speech models of auditory verbal hallucinations: evidence from behavioural and neuroimaging studies. *Int Rev Psychiatry*. 2007;19(4):407–415.
68. Pankow A, Katthagen T, Diner S, et al. Aberrant salience is related to dysfunctional self-referential processing in psychosis. *Schizophr Bull*. 2016;42(1):67–76. doi:10.1093/schbul/sbv098.
69. Galdos M, Simons C, Fernandez-Rivas A, et al. Affectively salient meaning in random noise: a task sensitive to psychosis liability. *Schizophr Bull*. 2011;37(6):1179–1186. doi:10.1093/schbul/sbq029.
70. E Hoffman R, Woods SW, Hawkins KA, et al. Extracting spurious messages from noise and risk of schizophrenia-spectrum disorders in a prodromal population. *B J Psychiatry*. 2007;191:355–356. doi:10.1192/bjp.bp.106.031195.
71. Cicero DC, Kerns JG, McCarthy DM. The Aberrant Salience Inventory: a new measure of psychosis proneness. *Psychol Assess*. 2010;22(3):688–701. doi:10.1037/a0019913.
72. Cicero DC, Docherty AR, Becker TM, Martin EA, Kerns JG. Aberrant salience, self-concept clarity, and interview-rated psychotic-like experiences. *Journal of Personality Disorders*. 2015;29(1):79–99. doi:10.1521/pedi_2014_28_150.
73. Winton-Brown TT, Fusar-Poli P, Ungless MA, Howes OD. Dopaminergic basis of salience dysregulation in psychosis. *Trends Neurosci*. 2014;37(2):85–94. doi:10.1016/j.tins.2013.11.003.
74. Boehme R, Deserno L, Gleich T, et al. Aberrant salience is related to reduced reinforcement learning signals and elevated dopamine synthesis capacity in healthy adults. *J Neurosci*. 2015;35(28):10103–10111. doi:10.1523/JNEUROSCI.0805-15.2015.
75. Howes OD, Kambeitz J, Kim E, et al. The nature of dopamine dysfunction in schizophrenia and what this means for treatment: meta-analysis of imaging studies. *Arch Gen Psychiatry*. 2012;69(8):776–786. doi:10.1001/archgenpsychiatry.2012.169.
76. Roiser JP, Howes OD, Chaddock CA, Joyce EM, McGuire P. Neural and behavioral correlates of aberrant salience in individuals at risk for psychosis. *Schizophr Bull*. 2013;39(6):1328–1336. doi:10.1093/schbul/sbs147.
77. Reininghaus U, Kempton MJ, Valmaggia L, et al. Stress sensitivity, aberrant salience, and threat anticipation in early psychosis: an experience sampling study. *Schizophr Bull*. 2016;42(3):712–722. doi:10.1093/schbul/sbv190.
78. Steinberg EE, Keiflin R, Boivin JR, Witten IB, Deisseroth K, Janak PH. A causal link between prediction errors, dopamine neurons and learning. *Nat Neurosci*. 2013;16(7):966–973. doi:10.1038/nn.3413.
79. Lammel S, Lim BK, Ran C, et al. Input-specific control of reward and aversion in the ventral tegmental area. *Nature*. 2012;491(7423):212–217. doi:10.1038/nature11527.
80. Poletti M. The dark side of dopaminergic therapies in Parkinson's disease: shedding light on aberrant salience. *CNS Spectr*. 2017;7:1–5. doi:10.1017/S1092852917000219.
81. Nagy H, Levy-Gigi E, Somlai Z, Takáts A, Bereczki D, Kéri S. The effect of dopamine agonists on adaptive and aberrant salience in Parkinson's disease. *Neuropsychopharmacology*. 2012;37(4):950–958. doi:10.1038/npp.2011.278.
82. Poletti M, Frosini D, Pagni C, et al. A pilot psychometric study of aberrant salience state in patients with Parkinson's disease and its association with dopamine replacement therapy. *Neurol Sci*. 2014;35(10):1603–1605. doi:10.1007/s10072-014-1874-6.
83. Demjaha A, Murray RM, McGuire PK, Kapur S, Howes OD. Dopamine synthesis capacity in patients with treatment-resistant schizophrenia. *Am J Psychiatry*. 2012;169(11):1203–1210. doi:10.1176/appi.ajp.2012.12010144.
84. Whitton AE, Treadway MT, Pizzagalli DA. Reward processing dysfunction in major depression, bipolar disorder and schizophrenia. *Curr Opin Psychiatry*. 2015;28(1):7–12. doi:10.1097/YCO.0000000000000122.
85. Bloomfield MAP, Mouchlianitis E, Morgan CJA, et al. Salience attribution and its relationship to cannabis-induced psychotic symptoms. *Psychol Med*. 2016;46(16):3383–3395. doi:10.1017/S0033291716002051.
86. Schmidt A, Antoniades M, Allen P, et al. Longitudinal alterations in motivational salience processing in ultra-high-risk subjects for psychosis. *Psychol Med*. 2017;47(2):243–254. doi:10.1017/S0033291716002439.

NEURAL CORRELATES OF SOCIOENVIRONMENTAL RISK AND PSYCHOSIS

50

NEURAL CORRELATES OF CHILDHOOD TRAUMA

Alaptagin Khan, Kyoko Ohashi, Maria Maier, and Martin H. Teicher

Childhood maltreatment is recognized as the most important preventable risk factor for a host of psychiatric disorders.[1] According to the Adverse Childhood Experiences (ACE) study, ACEs in the form of abuse and household dysfunction account for about 54%, 64%, and 67% of the population attributable risk (PAR) for depression,[2] addiction to illicit drugs,[3] and suicide attempts respectively.[4] Similarly, Green et al.[5] reported that childhood adversity was associated with 45% of the PAR for childhood onset psychiatric disorders. Maltreatment has also been found to hasten the onset and exacerbate course of bipolar disorder,[6–8] and to substantially increase risk for psychosis.[9–13] In fact, exposure to five or more ACEs has been reported to prospectively increase risk of receiving a prescription for an anxiolytic, antidepressant, antipsychotic, or mood stabilizer by 2-, 3-, 10- and 17-fold, respectively.[14]

Understanding how maltreatment increases risk for psychiatric and medical disorders is of crucial importance to prevent, preempt or treat these consequences of abuse and neglect.[15] Hence, in this section, we provide an overview of the neural correlates of childhood maltreatment focusing on potential effects on morphometry, function, connectivity, and network architecture. We also focus on the complex association between brain abnormalities and psychopathology, differences between maltreated and nonmaltreated individuals with the same primary psychiatric disorder and the role of maltreatment as a confounding factor in psychiatric neuroimaging studies.

Studying the effects of maltreatment on the brain is a complex undertaking, and published reports have numerous limitations. Of note is the fundamental problem of establishing a causal relationship between maltreatment and brain differences as individuals cannot be randomly assigned to experience abuse or neglect. Hence, findings on the effects of familial maltreatment can be interpreted in many ways, including the possibility that there are preexisting abnormalities that run in families and increase risk of being abused. This is remedied to some extent by observing parallel findings in individuals maltreated by unrelated individuals (neighbors, authority figures, peers), though the strongest evidence to date for causal relationships derive from translational studies in which remarkably similar brain changes have been observed in rodents and nonhuman primates exposed to early life stress.[16,17] Similarly, most studies rely on retrospective self-report of maltreatment, though as we have recently discussed[16,18] this limitation is not as severe as some might expect. It is particularly important to note that retrospective reports of childhood maltreatment have impressive test-retest reliability (e.g., Childhood Trauma Questionnaire $r = 0.88$,[19] Maltreatment and Abuse Chronology of Exposure scale[20] $r = 0.91$) and that retrospective reports are stable even in psychotic individuals and do not vary with the severity of their psychosis or depressive symptoms.[21]

MORPHOMETRY

Brain regions that appear to be especially vulnerable to the effects of childhood abuse typically have one or more of the following features: (1) a protracted postnatal development, (2) a high density of glucocorticoid receptors, (3) some degree of postnatal neurogenesis, or (4) being components of sensory systems that convey the adverse experience to conscious awareness.[15,16,22] Research on the neurobiological effects of childhood abuse began with studies assessing alterations in electrophysiology,[23–26] but was quickly eclipsed by MRI studies assessing structural and functional differences. Most studies use a region of interest (ROI) approach and have focused primarily on hippocampus, amygdala, and cerebral cortex. A significant number of studies have also used global approaches to identify alterations without preselection of ROIs.

HIPPOCAMPUS

The hippocampus, a key limbic structure involved in the formation and retrieval of memories, including autobiographical memories, is the most obvious target for the effects of childhood maltreatment. Portions of the hippocampal complex experience persistent postnatal neurogenesis and hippocampal pyramidal cells are densely populated with glucocorticoid receptors.[27] This renders the hippocampus highly susceptible to influence of excessive levels of corticosteroids, which suppress neurogenesis[28] and can markedly alter pyramidal cell morphology and even produce pyramidal cell death.[29]

There is compelling evidence that adults with histories of maltreatment have smaller hippocampi than nonmaltreated comparison subjects (see supplement[16]), though this is not necessarily true in children.[16] In two previous preliminary studies we found evidence that adult hippocampal vulnerability to effects of maltreatment differed based on developmental stage when maltreatment occurred.[30,31] We have now, in a more definitive study involving N = 336 participants, found that adult hippocampal volume in males was most significantly predicted by exposure to neglect, but not abuse, during first 7 years of development, while adult female hippocampal volume was most significantly predicted by abuse, but not neglect, at 10–11 and 15–16 years of age.[18] Further, we found that while dentate gyrus volume was affected by maltreatment in both genders, CA1 was much more strongly influenced by neglect than CA3 in males whereas CA3 was much more vulnerable than CA1 to abuse in females. Overall, this study suggests that neglect fosters inadequate hippocampal development in males while abuse produces a stress-related deficit in females.

The consistent finding of reduced hippocampal volume in adults with histories of childhood maltreatment but not in maltreated children suggests that there may be a significant delay between exposure and adverse outcome. This hypothesis is consistent with results from a preclinical study we conducted where developing rats were exposed to early life stress through repeated episodes of maternal separation. No differences were discernible in hippocampal synaptic density at weaning or at onset of puberty. However, by early adulthood there was a marked reduction in synaptic density in CA1 and CA3 which endured.[32]

The idea that there may be a delay or "silent period" between exposure and outcome is also seen clinically. We previously reported that there was, on average, a 9-year gap between first exposure to childhood sexual abuse in females with early exposure and full emergence of major depression or PTSD.[33] Similarly, childhood maltreatment markedly increases risk for binge drinking, with this phenomenon emerging most clearly in early adulthood.[34] It remains to be determined whether delayed neurobiological effects of maltreatment play a causal role in the emergence of psychopathology and whether this temporal delay may have relevance to the strong tendency of psychotic disorders to emerge between adolescence and early adulthood.[35]

AMYGDALA

The amygdala is another key limbic structure, critically involved in encoding of implicit emotional memories[36] and a core component of the threat detection and response system. Like the hippocampus, the amygdala has a high density of glucocorticoid receptors on pyramidal cells[37] and a postnatal developmental trajectory characterized by rapid initial growth followed by more sustained growth to peak volumes between 9 and 11 years with gradual pruning thereafter.[38] Preclinical studies show that the amygdala is highly vulnerable to exposure to early stress, which stimulates dendritic arborization and new spine formation on pyramidal cells leading to an increase in volume,[39,40] which is opposite to the effects of stress on the hippocampus. Further, stress-induced amygdala hypertrophy, unlike hippocampal hypotrophy, endures long after cessation of the stressor.[41]

One might assume that maltreatment-related alterations in amygdala volume should then be readily discernible, but this is not the case. Of the 43 studies reporting amygdala volume findings in subjects with maltreatment histories, or who were reared for periods of time in an institution, or by chronically depressed mothers, 24 reported a significant or nonsignificant reduction, 10 reported no significant difference, and 9 reported a significant or nonsignificant increase (see supplement in [16]). Studies reporting increased volume typically involved children reared in orphanages or adults with histories of disrupted attachments. Studies reporting significant reductions typically focused on adults with exposure to severe and persistent abuse with two studies reporting a graded reduction in amygdala volume in maltreated individuals following subsequent exposure to abuse or combat trauma.[42,43] This led us to propose that early (e.g., prepubertal) exposure to maltreatment leads to a significant increase in amygdala volume that persists in the absence of subsequent exposure, but that early exposure also sensitized the amygdala to undergo volume loss with postpubertal exposure, which does not occur without this prior sensitization.[15]

We also found in a sample of adults carefully assessed for quality of attachment as infants that volume of the right amygdala was predicted by degree of exposure to maltreatment at 10–11 years of age[31] while volume of left amygdala was predicted by degree of disrupted maternal communication and attachment at 18 months of age.[44] A key function of the amygdala is to serve as the center of a circuit involved in the detection and response to threat.[45] We have hypothesized that the left amygdala may be specialized to detect and respond to early developmental threat in the form of abandonment or neglect while the right amygdala may be specialized to detect later threats related to abuse. This makes sense from a hemispheric perspective as approach-related emotions are lateralized toward the left hemisphere and may serve an essential survival function when confronted by abandonment, whereas withdrawal-related emotions leading to flight responses are lateralized toward the right hemisphere.[46]

CEREBRAL CORTEX

The cerebral cortex, like the hippocampus and amygdala, possesses a population of stress-susceptible pyramidal cells with a high density of glucocorticoid receptors that peak during late adolescence–early adulthood.[47] There is also a population of glucocorticoid receptors on glial cells that are most densely distributed during the neonatal period and gradually decline. This suggests that the cerebral cortex may have two periods of heightened stress sensitivity: one during the period from infancy to early childhood, and the other from late adolescence to early adulthood, which also coincide with developmental periods of overproduction and pruning of dendrites, synapses, and receptors.[48] Overall, neglect and abuse are associated with significant reductions in total gray and white matter[49,50] and widespread reduction in cortical thickness.[51]

Sensory systems and pathways

Maltreatment appears to have most prominent effects on sensory cortex and association cortex in frontal and temporal regions. Sensory systems and pathways are the brain's first filters of information from the outside world and are known to manifest some of the most dramatic experience-dependent plastic responses.[52] Briefly, we found using unbiased, whole-brain analytical methods, that repeated exposure to parental verbal abuse was selectively associated with alterations in gray matter volume (GMV) in left superior temporal gyrus[53] and reduced integrity of the left arcuate fasciculus, which interconnects Broca's area and the surrounding frontal cortex with Wernicke's area and the superior temporal gyrus.[54] Similarly, visually witnessing interparental violence during childhood was associated with attenuated GMV or thickness in portions of visual cortex including right lingual gyrus, left occipital pole, and bilateral secondary visual cortex[55] and reduced integrity of the inferior longitudinal fasciculus, which interconnects visual cortex and limbic systems and helps determine our memory and emotional response to things we see.[56] We found that repeated episodes of sexual abuse by nonparental adults was associated with a substantial bilateral reduction in GMV in the primary visual cortex and visual association cortices,[57] while Heim et al.[58] found that repeated episodes of childhood sexual abuse in females was associated with thinning of somatosensory cortex, specifically the portion representing the clitoris and surrounding genital area. Overall, these differences can be explained as modifications to sensory systems and pathways that convey the aversive experience to consciousness, as a means of attenuating the effects of repeated exposures and thus reducing distress. Further, these modifications may shift how an individual responds to traumatic reminders, by altering conscious perception but leaving intact the subcortical pathways that provide a nonconscious route to circuits that can generate a rapid behavioral or emotional response to threats.[16]

Prefrontal cortex

The most consistent differences in higher level association cortex have been observed in morphometry of anterior cingulate cortex (ACC)[59–62] and dorsolateral[62,63] and ventromedial or orbital prefrontal cortex.[63–65] Consistent with the overall prolonged development of prefrontal cortex we provided preliminary evidence that this region, as a whole, was most susceptible to childhood abuse between ages 14–16 years.[30] However, there is also compelling evidence from the Avon Longitudinal Study that the ACC is particularly vulnerable to maltreatment during the first 6 years of life,[66] consistent with the idea of both early and late stages of cortical vulnerability.

CORPUS CALLOSUM

The corpus callosum is the largest white matter tract and plays a critically important role in interhemispheric communication, particularly between contralateral cortical regions. Motor, somatosensory, and parietal association cortices are interconnected through segments IV and V. Superior and inferior temporal cortices, posterior parietal cortex, and occipital cortices are interconnected through segments VI and VII. Myelinated regions such as the corpus callosum are potentially vulnerable to the impact of early exposure to excessive levels of stress hormones, which suppress glial cell division critical for myelination. One of the earliest and most consistent finding in maltreated children[67–69] and adults[30,70] is reduced area or integrity of the corpus callosum. Reductions in corpus callosum area were most frequently observed in segments IV and V (also known as the anterior midbody and the posterior midbody, respectively) and, next most frequently, in segments VI and VII (also known as the isthmus and the splenium, respectively). There seem to be marked sex differences in vulnerability as multiple studies reported a twofold greater reduction in corpus callosum area in maltreated males than in maltreated females. We also observed that, in males, corpus callosum area was most affected by exposure to neglect, whereas, in females, it was most affected by exposure to sexual abuse,[69] which parallels gender differences recently observed in hippocampal vulnerability.[18] Mid-portions of the corpus callosum may be particularly vulnerable to maltreatment between 9 and 10 years of age.[30]

FUNCTIONAL SYSTEMS

THREAT DETECTION AND RESPONSE SYSTEM

Abusive experiences can be construed as threats to survival, body integrity, or sense of self. The most reliable functional imaging finding in maltreated individuals is increased amygdala response to emotional faces, particularly those seen as threatening. This phenomenon has been observed in maltreated individuals during childhood[71,72] and adulthood[73] and even in childhood-maltreated adults without psychopathology.[59] Increased amygdala activation has been observed in orphans who experienced caregiver deprivation,[72,74] as well as in children exposed to either family violence,[71] stressful life events,[75] or physical or sexual abuse with symptoms of posttraumatic stress.[76] Interestingly, brain regions and pathways involved in regulating response to threatening stimuli[77–79] overlap extensively with regions found to differ structurally in maltreated individuals.[1,15,16] These include thalamus, auditory and visual cortex, anterior cingulate cortex, ventromedial prefrontal cortex, amygdala, and hippocampus. Further, diminished integrity has also been observed in the fiber tracts that interconnect these regions in maltreated individuals. Affected pathways in this circuit include the inferior longitudinal fasciculus, superior longitudinal fasciculus/arcuate fasciculus, uncinate fasciculus, cingulum bundle, and fornix.

REWARD ANTICIPATION AND RESPONSE SYSTEM

Another consistent functional imaging finding reported in at least six studies is diminished striatal BOLD response in maltreated individuals to anticipated or received reward in the monetary incentive delay task. This finding has been observed in both children and adults and includes orphans experiencing early deprivation;[80] children with reactive attachment disorder,[81] maltreated children at high risk for depression,[82] a birth cohort studied as adults who experienced early family adversity,[83] and adults reporting exposure to physical, sexual, or emotional abuse.[84] The principal regions regulating this response are key components of the reward system and consist of the mesolimbic and striatal target territories of the midbrain dopamine neurons.[85] These include the ACC, orbital prefrontal cortex, ventral striatum, and ventral pallidum. Other structures regulating this circuit include dorsal prefrontal cortex, amygdala, hippocampus, thalamus, lateral habenular, and pedunculopontine and raphe nuclei in the brainstem.[85] As noted earlier, maltreatment-related morphological differences have been observed in many of these regions. Diminished ventral striatal response to reward anticipation may be an important risk factor for, or component of, depression,[82,84] and may increase risk for drug addiction.[86]

NETWORK ARCHITECTURE

The brain is organized into networks, and alterations in network architecture may underlie many forms of psychopathology. We assessed cortical network architecture in 142 maltreated and 123 nonmaltreated individuals by delineating between subject intraregional correlations in cortical thickness—as regions that correlate in size tend to be functionally or structurally interconnected—and evaluated a 112-node network encompassing the entire cerebral cortex.[87] Graph theory was used to determine central importance of each node. Centrality was markedly reduced in the left ACC, temporal pole, and middle frontal gyrus, and was increased in right anterior insula and precuneus of maltreated individuals. Regions with greater centrality in the control group play an important role in emotional regulation, attention, and social cognition. The ACC is involved in regulating emotions, and in monitoring cognitive and motor responses during potential conflict situations. Both the temporal pole and middle frontal gyri are involved in aspects of social cognition such as theory of mind, person perception, and mentalizing. By contrast, regions with greater centrality in maltreated individuals seem to be primarily linked to self-awareness. The precuneus may be a critical region for self-centered mental imagery and self-referential thinking. The anterior insula is a locus for interoception, providing the substrate for internal body sensations such as thirst, hunger, and the need to void or eliminate. The anterior insula and ACC function together in a manner analogous to sensory and motor cortices to give rise to the feelings (insula) and motivations (ACC) underlying emotions. The anterior insula may also have a crucial role in self-awareness. Hence, maltreated individuals may be at heightened risk for psychopathology as a consequence of diminished ability to regulate emotions coupled with enhanced focus on internal bodily sensations and self-perception.[87]

More recently we used diffusion tensor imaging to assess differences in global network architecture in new groups of participants with moderate-to-severe (N = 140) versus no-to-low (N = 122) exposure to maltreatment.[88] Briefly, graph theory analysis revealed lower degree, strength, global efficiency, and higher pathlength and small-worldness in subjects with moderate-to-severe exposure. Local clustering was similar in both groups, but the different clusters were more strongly interconnected in the no-to-low exposure group. Brain network architecture needs to balance the opposing demands of integration and segregation in order to combine the presence of functionally specialized and segregated modules with a meaningful number of interconnecting links between the modules. This trade-off is reflected in the small-worldness of the network. The greater small-worldness in the moderate-to-severe maltreatment group is a consequence of a slightly sparser network with preserved local clustering but with a smaller number of interconnections between clusters.

We have speculated that alterations in global network architecture could serve as a major risk factor for psychopathology, as a weaker degree of integration between clusters may make it harder to effectively compensate for abnormalities that might occur within a node, cluster, or functional network.[88] This fits with the idea that psychopathology arises from multiple "hits"[89] such as genetic predisposition or exposure to traumatic events during different developmental stages. Waxing and waning of symptoms may reflect efforts of a less integrated network to compensate, shifting between periods of health and pathology as quasi-stable states.

ECOPHENOTYPES

We have recently proposed that maltreated and nonmaltreated individuals with the same primary psychiatric diagnosis are clinically, neurobiologically, and genetically distinct and that the maltreated subgroup constitutes a unique *ecophenotype* (i.e., a modified phenotype resulting from environmental influences).[1,15] From the clinical perspective dozens of reports show that maltreated subjects with psychopathology have an earlier age of onset, more comorbid symptoms, greater symptom severity, and poorer response to treatment[90] than nonmaltreated individuals with the same primary diagnosis.[1]

The hypothesis that they are distinctly different disorders is most directly supported by neuroimaging studies that identify abnormalities in the ecophenotype not observed in nonmaltreated individuals with the same diagnosis. The most researched example is hippocampal volume. Reduced hippocampal volume has been thought to play an important role in major depression (MDD).[91] However, three studies now show that reduced hippocampal volume is restricted to the maltreated ecophenotype.[92–94] Similarly, hyperactive amygdala response to sad faces has been proposed to play a significant role in MDD,[95,96] and reduced ACC volume to be critically involved in MDD,[97,98] anxiety,[99] psychosis[100] and substance abuse.[101] Nevertheless, evidence is emerging that these findings may only be relevant to the maltreated ecophenotype.[102–104] Additional differences include network architecture in subjects with MDD,[105] corpus callosum, and white matter abnormalities in bipolar disorder,[106,107] dorsolateral PFC and thalamic volume in schizophrenia,[108,109] and frontal, temporal, and parietal GMV in subjects with psychotic disorders.[104] In these studies, the ecophenotype differed markedly from controls, whereas nonmaltreated individuals with the same diagnoses did not. The most compelling example was recently published by Poletti et al.[110] They used voxel-based morphometry to compare subjects with schizophrenia and bipolar depression to controls in a large sample (N = 438). Participants were separated into those with maltreated (harsh parenting with overt family conflict and deficient nurturing) versus those without. Patients with maltreated differed from controls in inferior frontal gyrus GMV (both groups) and in insula and superior temporal and middle frontal gyri in schizophrenia. In contrast, there were no discernible differences in GMV between controls and subjects with schizophrenia or bipolar disorder without maltreated. These findings have led us to propose that morphometric abnormalities in stress-susceptible structures may only be discernible in the ecophenotype and that abnormalities in the nonmaltreated subtype may be limited to functional or neurochemical differences. Combining both maltreated and nonmaltreated subtypes in the same group may lead to nonreplicable findings and impede progress in delineating underlying neurobiology.

RESILIENCE

The most counterintuitive set of findings emerge from studies of maltreated individuals without psychopathology. Our initial hypothesis was that maltreatment-related brain changes and psychopathology would go hand-in-hand, such that maltreatment-related brain differences would be seen primarily in individuals with psychopathology, whereas brains of resilient subjects would resemble brains of unexposed healthy controls. Some studies comparing maltreated individuals with and without PTSD support this hypothesis. However, we identified many more studies, across various disorders that suggest that maltreatment-associated abnormalities were, by and large, independent of the presence or absence of current or lifetime psychopathology.[15,16] This implies that there is a subgroup of maltreated individuals who are more susceptible to the neurobiological consequences of maltreatment than to the psychiatric consequences. The most plausible explanation is that there are additional differences in the brains of the resilient subjects (either preexisting or adaptive) that enable them to compensate for abnormalities in stress-susceptible structures. The likelihood that resilient subjects are compensating, rather than unaffected, is consistent with the finding that maltreated individuals without psychopathology differed from healthy controls in the way they regulated their mood from hour to hour even though there were no differences in their average mood scores.[111] Specifically, resilient maltreated individuals showed greater variability in ratings of positive affect and heightened persistence of negative affect.

Our ongoing efforts suggest that resilient participants differ from susceptible and controls participants due to reduced centrality (nodal efficiency) in a handful of nodes including right amygdala and left inferior frontal gyrus. We suspect that these nodes may be functioning abnormally in maltreated individuals and creating problems in susceptible subjects. Resilient participants, on the other hand, may be able to effectively compensate despite their sparse networks, as these nodes in resilient subjects have a diminished ability to propagate information throughout the network. Identifying neurobiological differences between psychiatrically susceptible and resilient individuals is of paramount importance, as such information may provide crucial therapeutic insights. Although the obvious therapeutic strategy is to endeavor to reverse maltreatment-related abnormalities associated with psychopathology, an effective alternative may be to foster or strengthen compensatory brain adaptations.

CONCLUSIONS

Imaging studies provide a remarkable view of the potential effects of childhood abuse on brain structure, function, and connectivity. With few exceptions, consistent evidence has emerged for maltreatment-associated structural deficits in the adult hippocampus, corpus callosum, ACC, orbital frontal cortex and dorsolateral PFC. Consistent functional deficits have been observed in the amygdala when viewing emotional faces and in the striatum when anticipating award. Alterations in sensory systems and pathways that convey the adverse experience have been reported in individuals who experienced specific forms of childhood abuse. These observations are concordant with an experience-dependent adaptation hypothesis that suggests that such alterations promote avoidance and diminish approach responses. The similarity among findings is compelling, as maltreated children in these studies come from remarkably different sample populations. They may have been Romanian orphans, children without psychopathology or members of a birth cohort, or individuals selected because they had reactive attachment disorder, depressed mothers, were at high-risk for depression or had PTSD. Likewise, similar neuroimaging findings have been reported in adults with histories of childhood maltreatment, including subjects selected from the community without regard to psychopathology or specifically selected because they had no psychopathology, or had major depression, chronic PTSD, bipolar disorder, psychotic disorders, or personality disorders.

Almost all of the most significant morphometric differences reported in schizophrenia (e.g., reduced total, prefrontal, inferior frontal, temporal lobe, fusiform, insular, and hippocampal GMV[112]) have been reported in nonpsychotic (and often nonsymptomatic) individuals with childhood maltreatment,[15,16,49,113] calling into question their specific relevance to psychosis. The only readily apparent morphometric difference is in superior temporal gyrus GMV, which is substantially decreased in schizophrenia[112] but significantly increased in some maltreatment studies.[53,114] Overall, previously reported findings of structural and functional differences between psychiatric groups and healthy controls need to be re-evaluated to take into account the possible confounding influence of maltreatment. We also probably know little about neurobiological abnormalities in nonmaltreated individuals with psychiatric disorders, and this knowledge will only be gained by specifically comparing nonmaltreated clinical groups to nonmaltreated healthy controls. Attention to the developmental neurobiological consequences of childhood maltreatment/

early life stress may help to sharpen our understanding of the neurobiology of psychotic disorders and also result in more targeted approaches to treatment.

REFERENCES

1. Teicher, M.H., Samson, J.A. (2013) Childhood maltreatment and psychopathology: a case for ecophenotypic variants as clinically and neurobiologically distinct subtypes. Am J Psychiatry 170, 1114–1133.
2. Dube, S.R., Felitti, V.J., Dong, M., Giles, W.H., Anda, R.F. (2003) The impact of adverse childhood experiences on health problems: evidence from four birth cohorts dating back to 1900. Prev Med 37, 268–277.
3. Dube, S.R., Felitti, V.J., Dong, M., Chapman, D.P., Giles, W.H., *et al.* (2003) Childhood abuse, neglect, and household dysfunction and the risk of illicit drug use: the adverse childhood experiences study. Pediatrics 111, 564–572.
4. Dube, S.R., Anda, R.F., Felitti, V.J., Chapman, D.P., Williamson, D.F., *et al.* (2001) Childhood abuse, household dysfunction, and the risk of attempted suicide throughout the life span: findings from the Adverse Childhood Experiences Study. Jama 286, 3089–3096.
5. Green, J.G., McLaughlin, K.A., Berglund, P.A., Gruber, M.J., Sampson, N.A., *et al.* (2010) Childhood adversities and adult psychiatric disorders in the national comorbidity survey replication I: associations with first onset of DSM-IV disorders. Arch Gen Psychiatry 67, 113–123.
6. Daruy-Filho, L., Brietzke, E., Lafer, B., Grassi-Oliveira, R. (2011) Childhood maltreatment and clinical outcomes of bipolar disorder. Acta Psychiatr Scand 124, 427–434.
7. Leverich, G.S., McElroy, S.L., Suppes, T., Keck, P.E., Jr., Denicoff, K.D., *et al.* (2002) Early physical and sexual abuse associated with an adverse course of bipolar illness. Biol Psychiatry 51, 288–297.
8. Post, R.M., Leverich, G.S., Xing, G., Weiss, R.B. (2001) Developmental vulnerabilities to the onset and course of bipolar disorder. Dev Psychopathol 13, 581–598.
9. Arseneault, L., Cannon, M., Fisher, H.L., Polanczyk, G., Moffitt, T.E., *et al.* (2011) Childhood trauma and children's emerging psychotic symptoms: a genetically sensitive longitudinal cohort study. Am J Psychiatry 168, 65–72.
10. Bebbington, P., Jonas, S., Kuipers, E., King, M., Cooper, C., *et al.* (2011) Childhood sexual abuse and psychosis: data from a cross-sectional national psychiatric survey in England. Br J Psychiatry 199, 29–37.
11. Bendall, S., Jackson, H.J., Hulbert, C.A., McGorry, P.D. (2008) Childhood trauma and psychotic disorders: a systematic, critical review of the evidence. Schizophr Bull 34, 568–579.
12. Cutajar, M.C., Mullen, P.E., Ogloff, J.R., Thomas, S.D., Wells, D.L., *et al.* (2010) Psychopathology in a large cohort of sexually abused children followed up to 43 years. Child Abuse Negl 34, 813–822.
13. Fisher, H.L., Jones, P.B., Fearon, P., Craig, T.K., Dazzan, P., *et al.* (2010) The varying impact of type, timing and frequency of exposure to childhood adversity on its association with adult psychotic disorder. Psychol Med 40, 1967–1978.
14. Anda, R.F., Brown, D.W., Felitti, V.J., Bremner, J.D., Dube, S.R., *et al.* (2007) Adverse childhood experiences and prescribed psychotropic medications in adults. Am J Prev Med 32, 389–394.
15. Teicher, M.H., Samson, J.A. (2016) Annual research review: enduring neurobiological effects of childhood abuse and neglect. J Child Psychol Psychiatry 57, 241–266.
16. Teicher, M.H., Samson, J.A., Anderson, C.M., Ohashi, K. (2016) The effects of childhood maltreatment on brain structure, function and connectivity. Nat Rev Neurosci 17, 652–666.
17. Teicher, M.H., Tomoda, A., Andersen, S.L. (2006) Neurobiological consequences of early stress and childhood maltreatment: are results from human and animal studies comparable? Ann N Y Acad Sci 1071, 313–323.
18. Teicher, M.H., Anderson, C.M., Ohashi, K., Khan, A., McGreenery, C.E., *et al.* (2018) Differential effects of childhood neglect and abuse during sensitive exposure periods on male and female hippocampus. NeuroImage doi: 10.1016/j.neuroimage.2017.12.055.
19. Bernstein, D.P., Fink, L. (1998) Childhood Trauma Questionnaire Manual. The Psychological Corporation.
20. Teicher, M.H., Parigger, A. (2015) The "Maltreatment and Abuse Chronology of Exposure" (MACE) scale for the retrospective assessment of abuse and neglect during development. PLoS One 10, e0117423.
21. Fisher, H.L., Craig, T.K., Fearon, P., Morgan, K., Dazzan, P., *et al.* (2011) Reliability and comparability of psychosis patients' retrospective reports of childhood abuse. Schizophr Bull 37, 546–553.
22. Teicher, M.H., Andersen, S.L., Polcari, A., Anderson, C.M., Navalta, C.P., *et al.* (2003) The neurobiological consequences of early stress and childhood maltreatment. Neurosci Biobehav Rev 27, 33–44.
23. Ito, Y., Teicher, M.H., Glod, C.A., Ackerman, E. (1998) Preliminary evidence for aberrant cortical development in abused children: a quantitative EEG study. J Neuropsychiatry Clin Neurosci 10, 298–307.
24. Ito, Y., Teicher, M.H., Glod, C.A., Harper, D., Magnus, E., *et al.* (1993) Increased prevalence of electrophysiological abnormalities in children with psychological, physical, and sexual abuse. J Neuropsychiatry Clin Neurosci 5, 401–408.
25. Schiffer, F., Teicher, M.H., Papanicolaou, A.C. (1995) Evoked potential evidence for right brain activity during the recall of traumatic memories. J Neuropsychiatry Clin Neurosci 7, 169–175.
26. Teicher, M.H., Ito, Y., Glod, C.A., Andersen, S.L., Dumont, N., *et al.* (1997) Preliminary evidence for abnormal cortical development in physically and sexually abused children using EEG coherence and MRI. Ann N Y Acad Sci 821, 160–175.
27. Morimoto, M., Morita, N., Ozawa, H., Yokoyama, K., Kawata, M. (1996) Distribution of glucocorticoid receptor immunoreactivity and mRNA in the rat brain: an immunohistochemical and in situ hybridization study. Neurosci Res 26, 235–269.
28. Brown, E.S., Rush, A.J., McEwen, B.S. (1999) Hippocampal remodeling and damage by corticosteroids: implications for mood disorders. Neuropsychopharmacology 21, 474–484.
29. Sapolsky, R.M., Krey, L.C., McEwen, B.S. (1985) Prolonged glucocorticoid exposure reduces hippocampal neuron number: implications for aging. J Neurosci 5, 1222–1227.
30. Andersen, S.L., Tomoda, A., Vincow, E.S., Valente, E., Polcari, A., *et al.* (2008) Preliminary evidence for sensitive periods in the effect of childhood sexual abuse on regional brain development. J Neuropsychiatry Clin Neurosci 20, 292–301.
31. Pechtel, P., Lyons-Ruth, K., Anderson, C.M., Teicher, M.H. (2014) Sensitive periods of amygdala development: the role of maltreatment in preadolescence. Neuroimage 97, 236–244.
32. Andersen, S.L., Teicher, M.H. (2004) Delayed effects of early stress on hippocampal development. Neuropsychopharmacology 29, 1988–1993.
33. Teicher, M.H., Samson, J.A., Polcari, A., Andersen, S.L. (2009) Length of time between onset of childhood sexual abuse and emergence of depression in a young adult sample: a retrospective clinical report. J Clin Psychiatry 70, 684–691.
34. Shin, S.H., Miller, D.P., Teicher, M.H. (2013) Exposure to childhood neglect and physical abuse and developmental trajectories of heavy episodic drinking from early adolescence into young adulthood. Drug Alcohol Depend 127, 31–38.
35. Trotman, H.D., Holtzman, C.W., Ryan, A.T., Shapiro, D.I., MacDonald, A.N., *et al.* (2013) The development of psychotic disorders in adolescence: a potential role for hormones. Horm Behav 64, 411–419.
36. LeDoux, J.E. (1993) Emotional memory: in search of systems and synapses. Ann N Y Acad Sci 702, 149–157.
37. Sarrieau, A., Dussaillant, M., Agid, F., Philibert, D., Agid, Y., *et al.* (1986) Autoradiographic localization of glucocorticosteroid and progesterone binding sites in the human post-mortem brain. J Steroid Biochem 25, 717–721.

38. Uematsu, A., Matsui, M., Tanaka, C., Takahashi, T., Noguchi, K., *et al.* (2012) Developmental trajectories of amygdala and hippocampus from infancy to early adulthood in healthy individuals. PLoS One 7, e46970.
39. Mitra, R., Jadhav, S., McEwen, B.S., Vyas, A., Chattarji, S. (2005) Stress duration modulates the spatiotemporal patterns of spine formation in the basolateral amygdala. Proc Natl Acad Sci U S A 102, 9371–9376.
40. Vyas, A., Jadhav, S., Chattarji, S. (2006) Prolonged behavioral stress enhances synaptic connectivity in the basolateral amygdala. Neuroscience 143, 387–393.
41. Vyas, A., Pillai, A.G., Chattarji, S. (2004) Recovery after chronic stress fails to reverse amygdaloid neuronal hypertrophy and enhanced anxiety-like behavior. Neuroscience 128, 667–673.
42. Kuo, J.R., Kaloupek, D.G., Woodward, S.H. (2012) Amygdala volume in combat-exposed veterans with and without posttraumatic stress disorder: a cross-sectional study. Arch Gen Psychiatry 69, 1080–1086.
43. Whittle, S., Dennison, M., Vijayakumar, N., Simmons, J.G., Yucel, M., *et al.* (2013) Childhood maltreatment and psychopathology affect brain development during adolescence. J Am Acad Child Adolesc Psychiatry 52, 940–952 e941.
44. Lyons-Ruth, K., Pechtel, P., Yoon, S.A., Anderson, C.M., Teicher, M.H. (2016) Disorganized attachment in infancy predicts greater amygdala volume in adulthood. Behav Brain Res 308, 83–93.
45. Ohman, A. (2005) The role of the amygdala in human fear: automatic detection of threat. Psychoneuroendocrinology 30, 953–958.
46. Canli, T., Desmond, J.E., Zhao, Z., Glover, G., Gabrieli, J.D. (1998) Hemispheric asymmetry for emotional stimuli detected with fMRI. Neuroreport 9, 3233–3239.
47. Sinclair, D., Webster, M.J., Wong, J., Weickert, C.S. (2011) Dynamic molecular and anatomical changes in the glucocorticoid receptor in human cortical development. Mol Psychiatry 16, 504–515.
48. Tierney, A.L., Nelson, C.A., 3rd (2009) Brain development and the role of experience in the early years. Zero Three 30, 9–13.
49. De Bellis, M.D., Keshavan, M.S., Shifflett, H., Iyengar, S., Beers, S.R., *et al.* (2002) Brain structures in pediatric maltreatment-related posttraumatic stress disorder: a sociodemographically matched study. Biol Psychiatry 52, 1066–1078.
50. Sheridan, M.A., Fox, N.A., Zeanah, C.H., McLaughlin, K.A., Nelson, C.A., 3rd (2012) Variation in neural development as a result of exposure to institutionalization early in childhood. Proc Natl Acad Sci U S A 109, 12927–12932.
51. McLaughlin, K.A., Sheridan, M.A., Winter, W., Fox, N.A., Zeanah, C.H., *et al.* (2014) Widespread reductions in cortical thickness following severe early-life deprivation: a neurodevelopmental pathway to attention-deficit/hyperactivity disorder. Biol Psychiatry 76, 629–638.
52. Takesian, A.E., Hensch, T.K. (2013) Balancing plasticity/stability across brain development. Prog Brain Res 207, 3–34.
53. Tomoda, A., Sheu, Y.S., Rabi, K., Suzuki, H., Navalta, C.P., *et al.* (2011) Exposure to parental verbal abuse is associated with increased gray matter volume in superior temporal gyrus. Neuroimage 54 Suppl 1, S280–286.
54. Choi, J., Jeong, B., Rohan, M.L., Polcari, A.M., Teicher, M.H. (2009) Preliminary evidence for white matter tract abnormalities in young adults exposed to parental verbal abuse. Biol Psychiatry 65, 227–234.
55. Tomoda, A., Polcari, A., Anderson, C.M., Teicher, M.H. (2012) Reduced visual cortex gray matter volume and thickness in young adults who witnessed domestic violence during childhood. PLoS One 7, e52528.
56. Choi, J., Jeong, B., Polcari, A., Rohan, M.L., Teicher, M.H. (2012) Reduced fractional anisotropy in the visual limbic pathway of young adults witnessing domestic violence in childhood. Neuroimage 59, 1071–1079.
57. Tomoda, A., Navalta, C.P., Polcari, A., Sadato, N., Teicher, M.H. (2009) Childhood sexual abuse is associated with reduced gray matter volume in visual cortex of young women. Biol Psychiatry 66, 642–648.
58. Heim, C.M., Mayberg, H.S., Mletzko, T., Nemeroff, C.B., Pruessner, J.C. (2013) Decreased cortical representation of genital somatosensory field after childhood sexual abuse. Am J Psychiatry 170, 616–623.
59. Dannlowski, U., Stuhrmann, A., Beutelmann, V., Zwanzger, P., Lenzen, T., *et al.* (2012) Limbic scars: long-term consequences of childhood maltreatment revealed by functional and structural magnetic resonance imaging. Biol Psychiatry 71, 286–293.
60. De Bellis, M.D., Keshavan, M.S., Spencer, S., Hall, J. (2000) N-Acetylaspartate concentration in the anterior cingulate of maltreated children and adolescents with PTSD. Am J Psychiatry 157, 1175–1177.
61. Kelly, P.A., Viding, E., Wallace, G.L., Schaer, M., De Brito, S.A., *et al.* (2013) Cortical thickness, surface area, and gyrification abnormalities in children exposed to maltreatment: neural markers of vulnerability? Biol Psychiatry 74, 845–852.
62. Tomoda, A., Suzuki, H., Rabi, K., Sheu, Y.S., Polcari, A., *et al.* (2009) Reduced prefrontal cortical gray matter volume in young adults exposed to harsh corporal punishment. Neuroimage 47 Suppl 2, T66–71.
63. Edmiston, E.E., Wang, F., Mazure, C.M., Guiney, J., Sinha, R., *et al.* (2011) Corticostriatal-limbic gray matter morphology in adolescents with self-reported exposure to childhood maltreatment. Arch Pediatr Adolesc Med 165, 1069–1077.
64. De Brito, S.A., Viding, E., Sebastian, C.L., Kelly, P.A., Mechelli, A., *et al.* (2013) Reduced orbitofrontal and temporal grey matter in a community sample of maltreated children. J Child Psychol Psychiatry 54, 105–112.
65. Hanson, J.L., Chung, M.K., Avants, B.B., Shirtcliff, E.A., Gee, J.C., *et al.* (2010) Early stress is associated with alterations in the orbitofrontal cortex: a tensor-based morphometry investigation of brain structure and behavioral risk. J Neurosci 30, 7466–7472.
66. Jensen, S.K., Dickie, E.W., Schwartz, D.H., Evans, C.J., Dumontheil, I., *et al.* (2015) Effect of early adversity and childhood internalizing symptoms on brain structure in young men. JAMA Pediatr 169, 938–946.
67. De Bellis, M.D. (2002) Developmental traumatology: a contributory mechanism for alcohol and substance use disorders. Psychoneuroendocrinology 27, 155–170.
68. De Bellis, M.D., Baum, A.S., Birmaher, B., Keshavan, M.S., Eccard, C.H., *et al.* (1999) Developmental traumatology. Part I: Biological stress systems. Biol Psychiatry 45, 1259–1270.
69. Teicher, M.H., Dumont, N.L., Ito, Y., Vaituzis, C., Giedd, J.N., *et al.* (2004) Childhood neglect is associated with reduced corpus callosum area. Biol Psychiatry 56, 80–85.
70. Teicher, M.H., Samson, J.A., Sheu, Y.S., Polcari, A., McGreenery, C.E. (2010) Hurtful words: association of exposure to peer verbal abuse with elevated psychiatric symptom scores and corpus callosum abnormalities. Am J Psychiatry 167, 1464–1471.
71. McCrory, E.J., De Brito, S.A., Sebastian, C.L., Mechelli, A., Bird, G., *et al.* (2011) Heightened neural reactivity to threat in child victims of family violence. Current Biology 21, R947–R948.
72. Tottenham, N., Hare, T.A., Millner, A., Gilhooly, T., Zevin, J.D., *et al.* (2011) Elevated amygdala response to faces following early deprivation. Dev Sci 14, 190–204.
73. van Harmelen, A.L., van Tol, M.J., Demenescu, L.R., van der Wee, N.J., Veltman, D.J., *et al.* (2013) Enhanced amygdala reactivity to emotional faces in adults reporting childhood emotional maltreatment. Soc Cogn Affect Neurosci 8, 362–369.
74. Maheu, F.S., Dozier, M., Guyer, A.E., Mandell, D., Peloso, E., *et al.* (2010) A preliminary study of medial temporal lobe function in youths with a history of caregiver deprivation and emotional neglect. Cogn Affect Behav Neurosci 10, 34–49.
75. Suzuki, H., Luby, J.L., Botteron, K.N., Dietrich, R., McAvoy, M.P., *et al.* (2014) Early life stress and trauma and enhanced limbic activation to emotionally valenced faces in depressed and healthy children. J Am Acad Child Adolesc Psychiatry 53, 800–813 e810.
76. Garrett, A.S., Carrion, V., Kletter, H., Karchemskiy, A., Weems, C.F., *et al.* (2012) Brain activation to facial expressions in youth with PTSD symptoms. Depress Anxiety 29, 449–459.

77. LeDoux, J. (1996) Emotional networks and motor control: a fearful view. Prog Brain Res 107, 437–446.
78. LeDoux, J.E. (2002) Synaptic self: how our brains become who we are. Viking Penguin.
79. Maren, S., Phan, K.L., Liberzon, I. (2013) The contextual brain: implications for fear conditioning, extinction and psychopathology. Nat Rev Neurosci 14, 417–428.
80. Mehta, M.A., Gore-Langton, E., Golembo, N., Colvert, E., Williams, S.C., *et al.* (2010) Hyporesponsive reward anticipation in the basal ganglia following severe institutional deprivation early in life. J Cogn Neurosci 22, 2316–2325.
81. Takiguchi, S., Fujisawa, T.X., Mizushima, S., Saito, D.N., Okamoto, Y., *et al.* (2015) Ventral striatum dysfunction in children and adolescents with reactive attachment disorder: A functional MRI Study. BJPsych Open 1, 121–128.
82. Hanson, J.L., Hariri, A.R., Williamson, D.E. (2015) Blunted ventral striatum development in adolescence reflects emotional neglect and predicts depressive symptoms. Biol Psychiatry 78, 598–605.
83. Boecker, R., Holz, N.E., Buchmann, A.F., Blomeyer, D., Plichta, M.M., *et al.* (2014) Impact of early life adversity on reward processing in young adults: EEG-fMRI results from a prospective study over 25 years. PLoS One 9, e104185.
84. Dillon, D.G., Holmes, A.J., Birk, J.L., Brooks, N., Lyons-Ruth, K., *et al.* (2009) Childhood adversity is associated with left basal ganglia dysfunction during reward anticipation in adulthood. Biol Psychiatry 66, 206–213.
85. Haber, S.N., Knutson, B. (2010) The reward circuit: linking primate anatomy and human imaging. Neuropsychopharmacology 35, 4–26.
86. Balodis, I.M., Potenza, M.N. (2015) Anticipatory reward processing in addicted populations: a focus on the monetary incentive delay task. Biol Psychiatry 77, 434–444.
87. Teicher, M.H., Anderson, C.M., Ohashi, K., Polcari, A. (2014) Childhood maltreatment: altered network centrality of cingulate, precuneus, temporal pole and insula. Biol Psychiatry 76, 297–305.
88. Ohashi, K., Anderson, C.M., Bolger, E.A., Khan, A., McGreenery, C.E., *et al.* (2017) Childhood maltreatment is associated with alteration in global network fiber-tract architecture independent of history of depression and anxiety. Neuroimage 150, 50–59.
89. Bayer, T.A., Falkai, P., Maier, W. (1999) Genetic and non-genetic vulnerability factors in schizophrenia: the basis of the "two hit hypothesis." J Psychiatr Res 33, 543–548.
90. Nanni, V., Uher, R., Danese, A. (2012) Childhood maltreatment predicts unfavorable course of illness and treatment outcome in depression: a meta-analysis. Am J Psychiatry 169, 141–151.
91. Cole, J., Costafreda, S.G., McGuffin, P., Fu, C.H. (2011) Hippocampal atrophy in first episode depression: a meta-analysis of magnetic resonance imaging studies. J Affect Disord 134, 483–487.
92. Chaney, A., Carballedo, A., Amico, F., Fagan, A., Skokauskas, N., *et al.* (2014) Effect of childhood maltreatment on brain structure in adult patients with major depressive disorder and healthy participants. J Psychiatry Neurosci 39, 50–59.
93. Opel, N., Redlich, R., Zwanzger, P., Grotegerd, D., Arolt, V., *et al.* (2014) Hippocampal atrophy in major depression: a function of childhood maltreatment rather than diagnosis? Neuropsychopharmacology 39, 2723–2731.
94. Vythilingam, M., Heim, C., Newport, J., Miller, A.H., Anderson, E., *et al.* (2002) Childhood trauma associated with smaller hippocampal volume in women with major depression. Am J Psychiatry 159, 2072–2080.
95. Lee, B.T., Seong Whi, C., Hyung Soo, K., Lee, B.C., Choi, I.G., *et al.* (2007) The neural substrates of affective processing toward positive and negative affective pictures in patients with major depressive disorder. Prog Neuropsychopharmacol Biol Psychiatry 31, 1487–1492.
96. Suslow, T., Konrad, C., Kugel, H., Rumstadt, D., Zwitserlood, P., *et al.* (2010) Automatic mood-congruent amygdala responses to masked facial expressions in major depression. Biol Psychiatry 67, 155–160.
97. Lichenstein, S.D., Verstynen, T., Forbes, E.E. (2016) Adolescent brain development and depression: A case for the importance of connectivity of the anterior cingulate cortex. Neurosci Biobehav Rev.
98. Redlich, R., Almeida, J.J., Grotegerd, D., Opel, N., Kugel, H., *et al.* (2014) Brain morphometric biomarkers distinguishing unipolar and bipolar depression: a voxel-based morphometry-pattern classification approach. JAMA Psychiatry 71, 1222–1230.
99. Shang, J., Fu, Y., Ren, Z., Zhang, T., Du, M., *et al.* (2014) The common traits of the ACC and PFC in anxiety disorders in the DSM-5: meta-analysis of voxel-based morphometry studies. PLoS One 9, e93432.
100. Witthaus, H., Kaufmann, C., Bohner, G., Ozgurdal, S., Gudlowski, Y., *et al.* (2009) Gray matter abnormalities in subjects at ultra-high risk for schizophrenia and first-episode schizophrenic patients compared to healthy controls. Psychiatry Res 173, 163–169.
101. Koob, G.F., Volkow, N.D. (2010) Neurocircuitry of addiction. Neuropsychopharmacology 35, 217–238.
102. Grant, M.M., Cannistraci, C., Hollon, S.D., Gore, J., Shelton, R. (2011) Childhood trauma history differentiates amygdala response to sad faces within MDD. J Psychiatr Res 45, 886–895.
103. Malykhin, N.V., Carter, R., Hegadoren, K.M., Seres, P., Coupland, N.J. (2012) Fronto-limbic volumetric changes in major depressive disorder. J Affect Disord 136, 1104–1113.
104. Sheffield, J.M., Williams, L.E., Woodward, N.D., Heckers, S. (2013) Reduced gray matter volume in psychotic disorder patients with a history of childhood sexual abuse. Schizophr Res 143, 185–191.
105. Wang, L., Dai, Z., Peng, H., Tan, L., Ding, Y., *et al.* (2014) Overlapping and segregated resting-state functional connectivity in patients with major depressive disorder with and without childhood neglect. Hum Brain Mapp 35, 1154–1166.
106. Benedetti, F., Bollettini, I., Radaelli, D., Poletti, S., Locatelli, C., *et al.* (2014) Adverse childhood experiences influence white matter microstructure in patients with bipolar disorder. Psychol Med 44, 3069–3082.
107. Bucker, J., Muralidharan, K., Torres, I.J., Su, W., Kozicky, J., *et al.* (2014) Childhood maltreatment and corpus callosum volume in recently diagnosed patients with bipolar I disorder: data from the Systematic Treatment Optimization Program for Early Mania (STOP-EM). J Psychiatr Res 48, 65–72.
108. Kumari, V., Gudjonsson, G.H., Raghuvanshi, S., Barkataki, I., Taylor, P., *et al.* (2013) Reduced thalamic volume in men with antisocial personality disorder or schizophrenia and a history of serious violence and childhood abuse. Eur Psychiatry 28, 225–234.
109. Kumari, V., Uddin, S., Premkumar, P., Young, S., Gudjonsson, G.H., *et al.* (2014) Lower anterior cingulate volume in seriously violent men with antisocial personality disorder or schizophrenia and a history of childhood abuse. Aust N Z J Psychiatry 48, 153–161.
110. Poletti, S., Vai, B., Smeraldi, E., Cavallaro, R., Colombo, C., *et al.* (2016) Adverse childhood experiences influence the detrimental effect of bipolar disorder and schizophrenia on cortico-limbic grey matter volumes. J Affect Disord 189, 290–297.
111. Teicher, M.H., Ohashi, K., Lowen, S.B., Polcari, A., Fitzmaurice, G.M. (2015) Mood dysregulation and affective instability in emerging adults with childhood maltreatment: an ecological momentary assessment study. J Psychiatr Res 70, 1–8.
112. Haijma, S.V., Van Haren, N., Cahn, W., Koolschijn, P.C., Hulshoff Pol, H.E., *et al.* (2013) Brain volumes in schizophrenia: a meta-analysis in over 18 000 subjects. Schizophr Bull 39, 1129–1138.
113. De Bellis, M.D., Keshavan, M.S., Clark, D.B., Casey, B.J., Giedd, J.N., *et al.* (1999) Developmental traumatology. Part II: Brain development. Biol Psychiatry 45, 1271–1284.
114. De Bellis, M.D., Keshavan, M.S., Frustaci, K., Shifflett, H., Iyengar, S., *et al.* (2002) Superior temporal gyrus volumes in maltreated children and adolescents with PTSD. Biol Psychiatry 51, 544–552.

51

NEURAL CORRELATES OF URBAN RISK ENVIRONMENTS

Imke L. J. Lemmers-Jansen, Anne-Kathrin J. Fett, and Lydia Krabbendam

EPIDEMIOLOGY

The association between urbanicity and nonaffective psychosis has been supported by many epidemiological studies, which show elevated rates of psychotic disorders in densely populated areas.[1,2] Urban birth,[3,4] early life urbanicity, and current city living[5,6] are all associated with risk for psychotic disorder. However, exposure to urbanicity during early life has most effect on mental health outcomes, possibly through impact on a particularly sensitive period of human development, and there is no clear evidence that current city living has any effect over and above early urban exposure, representing a separate exposure of interest.[7–10] Of course, not all urban residents develop psychosis, and not all individuals who are affected by psychosis were brought up in urban areas, suggesting that multiple risk factors need to come together to render the urban environment "toxic." Nevertheless, higher rates of psychotic disorder in cities are accompanied by higher levels of psychosis proneness in the non-ill urban population, indicating that the impact of urban environments shifts the distribution of risk to higher levels in the entire population.[11]

The link between the urban environment and psychotic disorder may in part be explained by reverse causality, or a scenario that individuals or families with elevated liability for psychosis might be drawn toward city living. However, such a mechanism of "urban drift" cannot fully account for the observed associations between urbanicity and psychosis,[12,13] indicating that urban factors likely also impact causally.[14]

Recent studies show heterogeneous urban effects that vary depending on individual and contextual factors at neighborhood and country level.[15] For instance, different results have been found for different ethnic minority groups (depending on social neighborhood characteristics) and in relatively homogeneous Western, high-income countries (for a review see [16]). In addition, there is also variation in the incidence of psychotic disorder within the rural population.[17] North-European findings support long-established urbanicity-psychosis associations,[10,18–20] but South-European findings do not show such associations or even opposite effects.[21–23] Heterogeneity in urbanicity-psychosis risk associations has also been found in low- and middle-income countries worldwide.[24] A fine-grained and cross-country investigation is therefore required to understand national and international patterns of urbanicity variation impacting risk for or resilience against psychosis,[25] while taking into account periods of exposure (e.g., childhood or adulthood), multiple individual (e.g., polygenic risk and aspects of social cognition), and contextual factors (e.g., at neighborhood level or country).[23]

The urban environment may also be salutogenic in some aspects, moderating any risk-increasing impact. Thus, living in a city is generally associated with benefits like better (access to) healthcare, more employment opportunities, higher wages, and better schools.[26] Positive urbanicity factors may lie in the availability of resources, which may counteract negative urbanicity effects in some countries. However, besides these benefits, city life also has disadvantages that can affect physical well-being, mental health, and/or cognitive functioning.[27–30]

In addition to social factors that may function as stressors, lack of green space, environmental pollution, and toxin exposure have been suggested as urbanicity-related factors impacting psychosis risk,[29,31,32] as also have the frequency and intensity of social contacts differ between rural and urban dwellers.[26,33–36] This may be particularly problematic for patients with psychotic disorders because of alterations in social cognition and social interactions.[37,38] Epidemiological studies of urban impact in psychosis have suggested that the risk for psychosis associated with the urban environment interacts synergistically with genetic predisposition, a mechanism referred to as gene–environment interaction.[39] Thus, it could be hypothesized that encounters with strangers in urban environments overstrain social capacities of at-risk individuals, suggesting that the association between urbanicity and psychosis could partly originate in the cumulative effects of taxing social interactions acting synergistically with genetic risk (see also [40]).

Characteristics of the urban environment that reflect socio-environmental adversity and increased (social) stress have been investigated as possible explanations of the urbanicity-psychosis link.[25] In this work, urbanicity is defined as a proxy for other social stressors, such as social deprivation,[41] low levels of social capital impacting cohesion and trust,[42] disintegration of family networks, increased competition for resources,[43] minority group and ethnic density effects, perceived group discrimination,[43–46] or feelings of being inferior (or different) to another person. A common source of risk associated with all these factors is stress associated with perceived inequality, especially when the situation seems unchangeable, in line with the social defeat hypothesis.[47,48] Other variables associated with risk in the urban environment, such as lower socio-economic status (SES), a major predictor of both somatic and

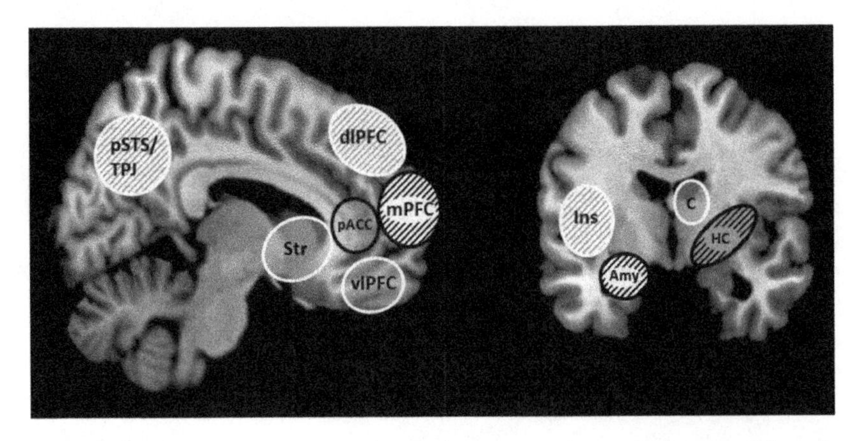

Figure 51.1 Brain areas implicated in the three possible mechanisms explaining the association between urbanicity and psychosis. In white the social brain, in black the stress processing network, and in gray areas associated with sensory gating deficits. Striped patterns indicate that the area is implicated in the two mechanisms with corresponding colors. Amy = amygdala; C = caudate; dlPFC = dorsolateral prefrontal cortex; HC = hippocampus; Ins = insula; mPFC = medial prefrontal cortex; pACC = perigenual anterior cingulate cortex; pSTS/TPJ: posterior superior temporal sulcus/temporo-parietal junction; Str = striatum; vlPFC = ventrolateral prefrontal cortex.

mental outcomes, particularly during early development, may also tap into the broader domain of social defeat.[49]

The human brain and its physiology are likely shaped by urban risk experiences, and effects may be larger in those with a genetic susceptibility to psychosis. Risk for psychosis, for example, could increase due to an altered stress response of the hypothalamic-pituitary-adrenal (HPA) axis and augmented dopamine signaling (DA).[50] Changes in these systems are thought to arise through sensitization, a process causing increased dopamine release in response to repeated stressors.[51–53] In sensitized individuals, even exposure to moderate levels of stress is associated with an excessive DA response.[45,51,54,55] This can influence cognitions and neural processing. Urbanicity may be associated with altered prefrontal activation, possibly in interaction with genetic variation in dopamine genes associated with psychotic illness.[56] Investigating the association between urbanicity and psychosis at a neural level can therefore give insights into hypotheses on the nature of these altered processes. Yet to date, few studies have systematically studied the neural mechanisms of urbanicity in patient populations with psychotic disorders.

In this chapter, we first review the literature on functional brain activity associated with urbanicity. Second, we discuss the association of urbanicity and brain structure and functional connectivity. We then review neuroimaging studies on the effects of air and noise pollution and lack of green space, followed by the neural signature of possible social factors underlying the association between urbanicity and psychosis (Figure 51.1). Some of the reviewed studies are derived from the animal literature, others were conducted in individuals with psychotic disorders. The majority however, were conducted in the general population. We conclude with a summary of the presented data and suggestions for future research.

URBANICITY AND BRAIN FUNCTION

Stress-related dysregulation of the HPA axis and the mesocorticolimbic dopamine system may play an important role in the development of psychosis. Stress stimulates the HPA axis, altering hippocampal activity, which in turn affects the mesolimbic dopamine system. This is thought to lead to aberrant salience, where patients attribute meaning to otherwise irrelevant stimuli.[52] The limbic system and dopamine are key to reward processes that are necessary for motivated behavior and learning, for example in social interactions. These processes may thus be relevant for understanding the urban effect.

A recent study investigated the association between urbanicity and stress in healthy individuals, at whole brain level, and within regions of interest (ROI), using the Montreal Stress Imaging Task (MIST) task that induces stress with arithmetical calculations.[57] Current city living was associated with higher activation of the amygdala during the stress paradigm, suggesting greater sensitivity to threat and negative emotions in city dwellers. Urban upbringing on the other hand, was associated with increased activity of the perigenual anterior cingulate cortex (pACC). The pACC is part of the limbic stress regulation system that is implicated in processing chronic social stressors, such as social defeat,[58] and modulates amygdala and HPA axis activity. These data were linked to schizophrenia, through previous findings of reduced cingulate gray matter, and connectivity abnormalities of the pACC with the amygdala during processing of affectively negative stimuli.

Krämer et al.[59] used the desire-reason-dilemma (DRD) paradigm, a task that distinguishes between conditioned reward processing and top-down suppressing of reward signals to investigate reward and urbanicity in healthy participants. In this paradigm, participants first learn that specific stimuli are associated with reward. Subsequently, they have to suppress the reward signal to achieve a superordinate goal. ROI analyses showed that during both processing and suppression of reward, current city living was associated with reduced activation in the left ventral tegmental area (VTA), a dopaminergic region modulating the midbrain DA system. Furthermore, city living was associated with increased activity in the amygdala, medial orbital cortex, and pACC during reward processing. This work shows that not only in stress processing but also in reward processing, urbanicity is associated with differences in

DA related brain areas,[59] supporting the possible role of urbanicity and DA dysregulation in the pathogenesis of psychosis.

Only one study has directly investigated brain function in psychosis in relation to urbanicity during upbringing. This study used the trust game (TG), tapping into real-time social behavior and social reward processing.[60] Trust is an important factor when it comes to building social capital and cohesion, and the reduction of social stress in healthy individuals.[61] Reduced baseline trust in an anonymous other was unrelated to urbanicity during upbringing. However, during repeated interactions with the same counterpart, lower urbanicity exposure during upbringing was associated with increased learning from positive feedback in patients, but not in controls. This tentatively suggests that learning from social feedback in individuals with psychosis may be sensitive to the effects of urbanicity exposure.[60] High urbanicity during upbringing was associated with differential activation of the amygdala bilaterally in patients compared to controls, mainly showing deactivation. No associations with urbanicity were found in mentalizing and reward-related areas.

Variation in brain activation during functional MRI with urbanicity exposure in healthy subjects has also been linked to specific genetic variation, although such studies crucially await replication. The neuropeptide S receptor 1 variant rs324981 interacted with urban upbringing on the amygdala stress response during an acute stress task,[62] showing that amygdala responses are influenced by urban upbringing, via a gene–environment interaction. A similar interaction was found during the n-back working memory task, where the catechol-O-methyltransferase (COMT) gene showed an interaction with urbanicity on prefrontal functioning. Low urban COMT-Met individuals showed efficient, COMT-Val individuals inefficient prefrontal activation. High urban individuals showed the reversed pattern, suggesting that urbanicity interacts with DA pathways, affecting cortical function.[56,63]

Summarizing, only a handful of studies has investigated the effects of urbanicity with functional MRI. The studies yielded heterogeneous results, stressing the need for more studies and, crucially, efforts aimed at exact replication.

URBANICITY AND BRAIN STRUCTURE

Following up on the Lederbogen study,[57] Haddad and colleagues[64] studied possible morphological correlates of urban upbringing in healthy subjects. Urbanicity was defined as urban upbringing, with three categories (rural, town, city). With increasing levels of urbanicity during upbringing, dlPFC and pACC volumes were reduced, the latter only in men.[64] The findings, while awaiting replication, tentatively suggest that urbanicity could affect brain structures, some of the effects on brain structures are specific to males. However, it remains unclear which urban factors would have a stronger influence on males and why.

Recently, initial steps were taken to investigate associations between urban upbringing and brain structure in patients with schizophrenia.[65–68] These studies have used the Dutch Genetic Risk and Outcome in Psychosis (GROUP) dataset, which includes a large sample of more than 1,100 nonaffective psychosis patients, 1,000 nonaffected siblings, 900 nonaffected parents, and 600 healthy controls. Gray matter volume was reduced in male patients with higher urbanicity exposure during development, whereas there was no such association in controls, siblings, and in female patients.[68] Another study by the same group[67] found that cortical thickness was reduced in patients compared to siblings and controls, however, urban upbringing did not have significant associations with cortical thickness. Also, no sex effects were found with respect to urban upbringing, contradicting Haddad et al.,[64] but it should be noted that the structural outcomes differed between studies. Resting-state functional connectivity of the PCC, a seed region of the default mode network (DMN), and of the nucleus accumbens (NAcc), a seed region for dopamine regulation within the mesocorticolimbic (MCL) system, did not reveal significant associations with urban upbringing either.[65,66] The DMN consists of the mPFC, the posterior cingulate cortex and the precuneus, the lateral parietal and temporal cortex, the hippocampus, and the parahippocampal gyrus, and is active during rest and deactivated during goal-directed behavior. The DMN is associated with many different processes, mainly with episodic and autobiographical memory, self-monitoring, and social cognitive processes regarding the self and other.[69] In the MCL system, dopamine transmission is regulated in top-down and bottom-up processes. Both siblings and patients showed reduced connectivity between the NaCC and MCL compared to controls, but again there was no evidence for a differential impact of urban upbringing and all these results require exact replication.[65,66]

NEURAL CORRELATES OF POSSIBLE URBANICITY-RELATED ENVIRONMENTAL AND SOCIAL MECHANISMS

In this section, we discuss neuroimaging studies that focus on environmental factors embedded in urban areas, namely pollution, noise, and lack of green space. We then review the available literature on neural correlates of social factors that have been associated with urban life, such as social capital, social exclusion, and social deprivation.

POLLUTION

Being exposed to pollution during early development may explain in part the association between urbanicity and psychosis.[70] City living is associated with a higher exposure to toxic substances in the air and sound pollution. Exposure to nitrogen dioxide and nitrogen oxides together statistically explained 60% of the association between urbanicity and adolescent psychotic experiences in one study.[71] However, another epidemiological study has suggested that exposure to air pollution is less likely to explain the association with urbanicity and schizophrenia risk, as the association with population density was much stronger than the distance to the nearest main road.[72] One study assessed functional connectivity, brain structure, and brain activation during a sensory task, where the children were presented with visual–auditory stimulation (viewing faces and listening to fast music) were assessed in a large sample of school children in Barcelona (8–12 years).[73] Higher level of pollution (i.e., amount of elemental carbon and nitrogen dioxide inside and outside the classroom) was associated with weaker functional DMN connectivity, between mPFC and angular gyrus (TPJ), and stronger connectivity with the frontal operculum at the lateral boundary of the DMN, indicating weaker segregation of network boundaries. Structural brain changes were not significantly associated with pollution. Research in children in Mexico City,[74,75] where the air quality is among the worst in the world, showed increased white matter hyperintensities (WMH). These WMH indicate white matter damage that can be linked to vascular oxidative stress, older age, dementia, risk for stroke, and inflammation.[76] Also, WMH have been associated with schizophrenia, with loss of connectivity and reduced information processing efficiency.[77] Rodent studies support the correlational evidence in humans, by showing a causal effect of air pollution on altered midbrain functioning and hippocampal structure, resulting in changed affective responses, and cognitive impairment.[78,79]

Sound pollution is exposure to noise. The intensity of the sound is of influence on perceived negative associations, but mainly the persistence (long-term exposure) and the information conveyed by the noise are of relevance.[32] Chronic noise induces stress and related neurochemical alterations in the brain,[80] and is associated with mental health problems like anxiety, depression, hostility, and a reduced quality of life.[81,82] However, not all people are equally susceptible to the stress-increasing effects of noise. Noise-sensitive individuals are more likely to experience negative emotions from unwanted sounds and show greater susceptibility to adverse effects of noise on health. Noise sensitivity is associated with sensory gating processes.[83] The inability to ignore unimportant noise, and sensory filtering or gating problems are known features of psychotic disorders.[84] Functional MRI research in patients with schizophrenia has shown that during restful states, hearing noise induced greater activation of the hippocampus in patients compared to controls. In healthy controls, it has been found that during a spatial selective attention task, where sounds deviating from the regular presented sound (250 Hz higher) had to be identified, noise was associated with reduced activation in the bilateral insula, cingulate gyrus, right TPJ, left dlPFC, and cerebellum, representing problems in the recruitment of task-related networks in response to the distracting noise. Patients in turn showed increased activity in the TPJ and ACC during task performance, and increased activity in the left hippocampus in response to noise compared to silence,[84] possibly indicating patients' problems to engage the ventral attention network. Together, this suggests a greater negative impact of noise on attention and task performance in patients with schizophrenia.

In sum, emerging and nonreplicated neuroimaging research has highlighted that several brain structures may be sensitive to the effect of different, prolonged environmental pollution factors. These may be relevant in explaining the association between urbanicity and psychosis that have been reported by some epidemiological studies. The limited correlational evidence in humans has been supported by experimental animal studies that suggest a negative causal effect of air pollution. It has also been suggested that patients with psychosis may be especially sensitive to the effects of prolonged noise, which may affect cognitive and neurochemical processes.

LACK OF GREEN SPACE

Growing up with low ambient availability of green space was found to be associated with a 1.5-fold increased risk of developing schizophrenia compared to growing up in areas rich in green space, even when controlling for urbanization and socioeconomic status.[85] The short-term behavioral effects of being in nature have been experimentally investigated, with many studies reporting a positive effect on well-being, which may last several hours after contact with nature.[86] Viewing nature, directly and in photographs also has benefits on several cognitive processes, including attention and temporal discounting, regardless of whether people grew up or currently lived in a rural or urban environment.[87,88] An early study used electroencephalography (EEG) during the comparison between urban versus nature pictures. Results showed that viewing green natural (vegetation) scenes elicited alpha activity in the central parietal cortex. A lower alpha amplitude while viewing urban pictures was associated with high arousal and feelings of anxiety.[89] The author concluded that "subjects felt more 'wakefully relaxed' while viewing the vegetation, as opposed to urban scenes," suggesting that nature reduces arousal in the brain. The first MRI study on the effects of viewing green space on the brain, conducted in healthy individuals, found increased activation of temporal and parietal lobes in the nature condition, while the frontal and occipital lobes were more active during urban scenery viewing. Rural scenes elicited positive affect, whereas urban pictures elicited negative affect, and this was accompanied by increased basal ganglia versus (para) hippocampal activity, respectively.[90] Subjective ratings of the scenes matched the neural outcomes, however, direct correlations between brain activation and subjective ratings of the images were not assessed in this study. These findings suggest a protective, regenerative characteristic of green, natural environments,[91] which is also supported by physical indices such as lower levels of cortisol; lower heart rate, muscle tension, and blood pressure; less arousal;[92,93] and activation of protective hormones and of the immune system.[94] A recent fMRI study investigated the immediate effects of a 90-minute walk in a natural as opposed to an urban environment, and showed that the walk in nature led to reduced rumination and subgenual prefrontal cortex (sgPFC) activation. The sgPFC is linked to self-focused behavioral withdrawal and rumination, supporting a restorative effect of nature, positively by distracting participants from depressed mood.[88] Furthermore, a structural MRI study investigating the impact of the current living environment on the human brain reported that the amount of forest within a 1 km radius of the home address was positively associated with amygdala integrity.[95] No associations of brain integrity were found with urban green, water, and wasteland around the home address. This could indicate that forests have positive effects on the integrity of the amygdala, or, reversely, that individuals with high structural amygdala integrity choose to live closer to forests.

In summary, epidemiological studies have linked the presence of green space during development to the risk for psychosis. Experimental approaches using short-term exposure to natural versus urban scenes, while awaiting a replication effort, have supported the notion that nature has beneficial effects on cognition and well-being, and may also impact on brain functioning in areas associated with stress processing. There is some evidence that not all green environments have the same effects and there is likely considerably heterogeneity between individuals. Future studies should focus on replication and investigating the specific characteristics of the green environment that elicit most positive effects on (mental) health in general, and psychosis risk in particular.

SOCIAL EXCLUSION

To investigate the immediate effect of social exclusion in humans, interactive games have been developed. The most frequently used paradigm is a ball-tossing game, Cyberball, where the participant throws a ball with two other players, and at a certain point is excluded from the game, while the other two players continue to throw the ball to each other. Studies in healthy participants showed that social exclusion elicits feelings of negative mood, and loss of control, belonging, and self-esteem, even when participants knew that the other players are computerized.[96] These behavioral effects are accompanied by activation in several regions associated with negative emotions, such as the subgenual ACC, middle temporal gyrus (MTG), and regions associated with emotion regulation,[97] such as the bilateral anterior insula, dorso-medial and ventral (lateral) prefrontal cortex (PFC), cuneus, and in the reward-related ventral striatum (VS).[98–100] During exclusion, parts of the DMN have shown increased activity, suggesting questioning the motives of other players, or rumination on the situation.[101] Only one study used this paradigm comparing patients with schizophrenia and healthy controls.[102] Healthy controls showed increased medial prefrontal cortex (mPFC) activation during exclusion, whereas patients exhibited a reduced mPFC response. Many studies have associated mPFC functioning with emotional and social information processing, and monitoring and updating of social values to plan future behavior. The authors suggested that differences in mPFC activation as a result of social exclusion could indicate a dysfunction of social valuation processes in schizophrenia. The magnitude of abnormality in mPFC responses correlated with the severity of positive symptoms, and not with antipsychotic medication, showing more impairment with increased illness severity.[102]

SOCIAL CAPITAL

Social capital refers to networks of relationships among people who live and work in a particular society. For effective social interactions trust, reciprocity, and cooperation are essential. These constructs can be operationalized in interactive neuroeconomics games. Participants are to make decisions about allocation of money between themselves and an unknown other. In order to make these social decisions, social cognition is required, including mentalizing abilities and empathy. Furthermore, trusting behavior is influenced by the rewarding effects of social cooperation. Patients, their first-degree relatives, and patients at clinical high-risk for psychosis encounter deficits in social cognition and reward processing, as reflected in altered activation of the social brain, and the related DMN.[103] Reduced trust could be explained by aberrant sensitivity to the rewarding propensity of social contact, and by impaired mentalizing skills.[104–106] These deficits might add to the instantiation and maintenance of psychotic symptoms. Initial research suggests that, despite the association between reward and urbanicity in healthy individuals,[59] urbanicity is not associated with the neural mechanisms of trust (social reward processing and learning, and mentalizing), but instead with social stress processing (see also X.2).[60]

SOCIAL DEPRIVATION

Due to ethical restrictions in humans and primates, research on social deprivation is mainly performed in rodents. Rodent studies provide evidence that social deprivation, operationalized as postweaning isolation, is associated with morphological changes, e.g., reduced dendritic length in the PFC and NAcc, areas associated with (conditioned) learning and motivated behavior.[107] Rats normally live in social groups, and young rats frequently engage in social behavior. Social isolation produces aberrant behavior such as hyperactivity in response to a novel environment. This behavior is also found in rats with hippocampal lesions,[108] suggesting that social isolation and hippocampal lesions may involve similar mechanisms. Isolation rearing resulted in elevated DA levels and psychotic-like features such as hyperreactivity to novel environments, cognitive impairment, and deficits in sensorimotor function.[109] Other research validated isolation rearing as a neurodevelopmental model of a "schizophrenia-like" state, further showing deficits in sensorimotor gating and reduced (m)PFC volume, both often found in schizophrenia.[110,111] Isolation and gating deficits might possibly explain social dysfunction in schizophrenia.

Summarizing, interpersonal social factors appear to be associated with aberrant activation of the DMN during social exclusion and reduced activation in the mPFC, dlPFC, and mentalizing areas in patients. The neural results are consistent with the notion of increased social stress impacting in predictable ways. Alternative or even complementary mechanisms may involve processes of mentalizing and sensory gating. Many patients with a psychotic disorder show deficits in mentalizing, that might be more pronounced when brought up in urban areas, as suggested by one epidemiological study, using a hearing impairment paradigm of reduced mentalization ability.[112] If a patient has problems identifying other people's emotions and intentions, the persistent confrontation with many unknown people in the environment could induce positive symptoms. This confrontation might also increase amygdala activity, reducing HPA activity, which in turn may induce maladaptive behavior resulting in social conflict.[50]

CONCLUSION AND FUTURE DIRECTIONS

While city living has been associated with greater health, possibly due to the availability of resources and economic factors for some,[26,113] in the area of mental health and psychosis, the effects are heterogeneous, but often appear to be adverse.[114] The existing literature on urbanicity, psychosis, and the brain is still scarce and awaiting replication, despite a recent surge of interest in the neural correlates of urban exposure.[86,94,114] Various mechanisms may explain reported effects of the city (for a review see [94]). Yet, the primary idea is that urban upbringing and/or city living increases sensitivity to stress, with a possible downstream impact on mesolimbic dopamine pathways,[115,116] through exposure to risk-attributes of densely populated urban environments. Factors that could explain the association between urbanicity and psychosis, for example, are air and noise pollution, lack of green space, social capital, and/or sensory gating deficits (see Figure 51.1). So far, the limited available evidence from fMRI studies assessing urbanicity in conjunction with (social) reward or stress tasks in healthy controls and patients highlighted the amygdala, mPFC, insula, and pACC as areas where effects of the urban environment may be mediated.[57,59,60] These results tentatively support the urbanicity-stress sensitization hypothesis, although some key regions that are affected by changes in the HPA axis, such as the hippocampus, have not shown differential activation. Findings are contradictory with regard to dopamine-related reward processing, leaving the hypothesis of sensitization of the reward systems as a result of prolonged city exposure a matter of further research.

With regard to functional connectivity, urbanicity could not account for differences found between patients with a psychotic disorder and healthy controls.[65,66] The existing studies investigating brain structure yielded contradictory results, with some evidence for urbanicity effects on some regions of stress processing and mentalizing,[64] and white matter volume, but no significant effect on cortical thickness.[67,68] Future research comparing methodology, setting, and definition of urban-rural categories will be necessary to generate further evidence on whether, where and how urbanicity influences brain structure.

In cities, access to green space is limited and air quality is often poor.[91,117,118] Exposure to green space appears to reduce arousal and rumination-related brain activation,[88,89,93] whereas urban stimuli increase activation of stress-related brain areas.[90] Although the evidence from neuroimaging studies is still scarce, there is some support for the urbanicity-stress hypothesis with respect to current city living. However, how current city living represents a separate exposure from urban upbringing remains unclear and remains to be validated in epidemiological studies. Some of these effects might be partly due to the impact of pollutants on the central nervous system, for example through neurochemical changes or neuroinflammation.[119–121] Animal studies associated pollution with reduced cognitive performance and increased negative mood symptoms, as well as changes in hippocampal morphology,[78] and with structural changes in the brain. Initial evidence supports that notion that in healthy individuals, urbanicity-related exposure to toxins is associated with DMN connectivity and white matter changes. Replication and further research is needed to elucidate the potential role of exposure to toxic vs. nonexposure to salutogenic (nature) environments in psychosis.

Urbanicity may be a proxy for a range of social risk attributes of the city, which may differ across geographical locations and countries in various ways. We discussed several imaging studies that used paradigms probing social tasks that may be of interest with regard to these social mechanisms. However, only one study examined such mechanisms in relation to urbanicity in individuals with psychosis, making it an important area for future investigations. In addition, it remains unknown whether critical periods exist during the human life span where risk associates of urbanicity may be particularly impactful. Together, this highlights the need for comprehensive, replicative, and systematic assessment of the risk-increasing and salutogenic attributes of urban environments in future neuroimaging research, elucidating how these attributes act on, and reinforce, the vulnerabilities for psychosis.

Urban environments represent a demanding environment for those living there. Limited space has to be shared with many others. Physically, the constant exposure to environmental and social stimuli means crowded spaces, noise, polluted air, possibly resulting in stress—although this is likely to differ substantially from person to person. Urban residents are challenged with many social encounters, the majority with unknown or slightly familiar others. For individuals with a vulnerability for psychosis this poses daily challenges, and it might overstrain their ability, causing (social) stress, possibly partly mediated by sensory gating deficits. Many of the research supports reviewed here support the social stress hypothesis, but the mechanisms causing this stress remain a topic for further investigation.

Finally, some initial research on gene–environment interactions, based on epidemiological observations, shows that effects of the city on the brain may be modulated by genes that are related to dopamine and stress expression.[56,62,63] Although this work requires careful replication, it is compatible with the hypothesis that the city may impact the brain of certain predisposed individuals. Researchers in the field will need to combine measures of vulnerability with fine-grained measures of developmental and, possibly, current urbanicity exposure and multifactorial assessments of social and environmental factors. Novel experimental methods such as virtual reality, in conjunction with skin conductance or heart rate measures, for example, could offer a way to assess the effect of urbanicity on (stress)physiology and cognitive functioning in a more controlled fashion. Epidemiological studies and animal models may help to clarify the effects of environmental pollution. Multimethod neuroimaging investigations of the effects of specific components of urbanicity on the neural mechanisms of risk for mental, neurodevelopmental, and neurodegenerative diseases in general, and schizophrenia in particular, have promising potential to develop strategies for pre- and intervention.

REFERENCES

1. Radua J, Ramella-Cravaro V, Ioannidis JPA, et al. What causes psychosis? An umbrella review of risk and protective factors. *World Psychiatry*. 2018;17(1):49–66.
2. van Os J, Pedersen CB, Mortensen PB. Confirmation of synergy between urbanicity and familial liability in the causation of psychosis. *American Journal of Psychiatry*. 2004;161(12):2312–2314.
3. Mortensen PB, Pedersen CB, Westergaard T, et al. Effects of family history and place and season of birth on the risk of schizophrenia. *New England Journal of Medicine*. 1999;340(8):603–608.
4. Laursen TM, Munk-Olsen T, Nordentoft M, Bo MP. A comparison of selected risk factors for unipolar depressive disorder, bipolar affective disorder, schizoaffective disorder, and schizophrenia from a Danish population-based cohort. *Journal of Clinical psychiatry*. 2007;68(11):1673–1681.
5. Sundquist K, Frank G, Sundquist J. Urbanisation and incidence of psychosis and depression. *British Journal of Psychiatry*. 2004;184(4):293–298.
6. McKenzie K, Murray A, Booth T. Do urban environments increase the risk of anxiety, depression and psychosis? An epidemiological study. *Journal of Affective Disorders*. 2013;150(3):1019–1024.
7. Pedersen CB, Mortensen PB. Evidence of a dose-response relationship between urbanicity during upbringing and schizophrenia risk. *Archives of General Psychiatry*. 2001;58(11):1039–1046.
8. Marcelis M, Takei N, van Os J. Urbanization and risk for schizophrenia: does the effect operate before or around the time of illness onset? *Psychological Medicine*. 1999;29(05):1197–1203.
9. Krabbendam L, Van Os J. Schizophrenia and urbanicity: a major environmental influence—conditional on genetic risk. *Schizophrenia Bulletin*. 2005;31(4):795–799.
10. Toulopoulou T, Picchioni M, Mortensen PB, Petersen L. IQ, the urban environment, and their impact on future schizophrenia risk in men. *Schizophrenia Bulletin*. 2017;43(5):1056–1063.
11. Van Os J, Hanssen M, Bijl RV, Vollebergh W. Prevalence of psychotic disorder and community level of psychotic symptoms: an urban-rural comparison. *Arch Gen Psychiatry*. 2001;58(7):663–668.
12. Selten J-P, Cantor-Graae E, Kahn RS. Migration and schizophrenia. *Current Opinion in Psychiatry*. 2007;20(2):111–115.
13. Colodro-Conde L, Couvy-Duchesne B, Whitfield JB, et al. Association between population density and genetic risk for schizophrenia. *JAMA Psychiatry*. 2018.
14. Paksarian D, Trabjerg BB, Merikangas KR, et al. The role of genetic liability in the association of urbanicity at birth and during upbringing with schizophrenia in Denmark. *Psychological Medicine*. 2018;48(2):305–314.
15. Manning N. Sociology, biology and mechanisms in urban mental health. *Social Theory and Health*. 2019;17(1):1–22.
16. Fett A-K, Lemmers-Jansen I, Krabbendam L. Psychosis and urbanicity—a review of the recent literature from epidemiology to neurourbanism. *Current Opinion in Psychiatry*. 2019.
17. Richardson L, Hameed Y, Perez J, Jones PB, Kirkbride JB. Association of environment with the risk of developing psychotic disorders in rural populations: findings from the Social Epidemiology of Psychoses in East Anglia study. *JAMA Psychiatry*. 2018;75(1):75–83.
18. Vassos E, Pedersen CB, Murray RM, Collier DA, Lewis CM. Meta-analysis of the association of urbanicity with schizophrenia. *Schizophrenia Bulletin*. 2012;38(6):1118–1123.
19. March D, Hatch SL, Morgan C, et al. Psychosis and place. *Epidemiologic Reviews*. 2008;30:84–100.
20. Kirkbride JB, Hameed Y, Ioannidis K, et al. Ethnic minority status, age-at-immigration and psychosis risk in rural environments: evidence from the SEPEA study. *Schizophr Bull*. 2017;43(6):1251–1261.
21. Quattrone D, Di Forti M, Gayer-Anderson C, et al. Transdiagnostic dimensions of psychopathology at first episode psychosis: findings from the multinational EU-GEI study. *Psychol Med*. 2018:1–14.
22. Bocchetta A, Traccis F. The Sardinian puzzle: concentration of major psychoses and suicide in the same sub-regions across one century. *Clinical Practice and Epidemiology in Mental Health*. 2017;13:246–254.
23. Jongsma HE, Gayer-Anderson C, Lasalvia A, et al. Treated incidence of psychotic disorders in the multinational EU-GEI study. *JAMA Psychiatry*. 2018;75(1):36–46.
24. DeVylder JE, Kelleher I, Lalane M, Oh H, Link BG, Koyanagi A. Association of urbanicity with psychosis in low- and middle-income countries. *JAMA Psychiatry*. 2018;75(7):678–686.
25. Heinz A, Deserno L, Reininghaus U. Urbanicity, social adversity and psychosis. *World Psychiatry*. 2013;12(3):187–197.
26. Glaeser E. Cities, productivity, and quality of life. *Science*. 2011;333(6042):592–594.
27. Gouin M, Flamant C, Gascoin G, et al. The association of urbanicity with cognitive development at five years of age in preterm children. *PloS One*. 2015;10(7):e0131749.
28. Stansfeld SA, Clark C. Health effects of noise exposure in children. *Current Environmental Health Reports*. 2015;2(2):171–178.
29. Attademo L, Bernardini F, Garinella R, Compton MT. Environmental pollution and risk of psychotic disorders: a review of the science to date. *Schizophrenia Research*. 2017;181:55–59.
30. Lambert KG, Nelson RJ, Jovanovic T, Cerdá M. Brains in the city: neurobiological effects of urbanization. *Neuroscience and Biobehavioral Reviews*. 2015;58:107–122.
31. van den Berg M, Wendel-Vos W, van Poppel M, Kemper H, van Mechelen W, Maas J. Health benefits of green spaces in the living environment: a systematic review of epidemiological studies. *Urban Forestry and Urban Greening*. 2015;14(4):806–816.
32. Savale P. Effect of noise pollution on human being: its prevention and control. *European Journal of Environment Research and Development*. 2014;8(4).
33. White KJC, Guest AM. Community lost or transformed? Urbanization and social ties. *City & Community*. 2003;2(3):239–259.
34. Witt LA. Urban-nonurban differences in social cognition: locus of control and perceptions of a just world. *The Journal of Social Psychology*. 1989;129(5):715–717.
35. Korte C. Urban-nonurban differences in social behavior and social psychological models of urban impact. *Journal of Social Issues*. 1980;36(3):29–51.
36. Söderström O, Söderström D, Codeluppi Z, Empson LA, Conus P. Emplacing recovery: how persons diagnosed with psychosis handle stress in cities. *Psychosis*. 2017;9(4):322–329.
37. Couture SM, Penn DL, Roberts DL. The functional significance of social cognition in schizophrenia: a review. *Schizophrenia Bulletin*. 2006;32(suppl 1):S44–S63.
38. Fett A-K, Viechtbauer W, Penn DL, van Os J, Krabbendam L. The relationship between neurocognition and social cognition with functional outcomes in schizophrenia: a meta-analysis. *Neuroscience and Biobehavioral Reviews*. 2011;35(3):573–588.
39. van Os J, Kenis G, Rutten BP. The environment and schizophrenia. *Nature*. 2010;468(7321):203–212.
40. Weiser M, Van Os J, Reichenberg A, et al. Social and cognitive functioning, urbanicity and risk for schizophrenia. *British Journal of Psychiatry*. 2007;191(4):320–324.
41. O'Donoghue B, Lyne J, Renwick L, et al. Neighbourhood characteristics and the incidence of first-episode psychosis and duration of untreated psychosis. *Psychological Medicine*. 2016;46(7):1367–1378.
42. Drukker M, Krabbendam L, Driessen G, van Os J. Social disadvantage and schizophrenia. *Social Psychiatry and Psychiatric Epidemiology*. 2006;41(8):595–604.
43. Zammit S, Lewis G, Rasbash J, Dalman C, Gustafsson J-E, Allebeck P. Individuals, schools, and neighborhood: a multilevel longitudinal study of variation in incidence of psychotic disorders. *Archives of General Psychiatry*. 2010;67(9):914–922.
44. Cantor-Graae E, Selten J-P. Schizophrenia and migration: a meta-analysis and review. *American Journal of Psychiatry*. 2005;162(1):12–24.
45. Kirkbride JB, Morgan C, Fearon P, Dazzan P, Murray RM, Jones PB. Neighbourhood-level effects on psychoses: re-examining the role of context. *Psychological Medicine*. 2007;37(10):1413–1425.

46. Veling W, Susser E, Van Os J, Mackenbach JP, Selten J-P, Hoek HW. Ethnic density of neighborhoods and incidence of psychotic disorders among immigrants. *American Journal of Psychiatry*. 2008;165(1):66–73.
47. Selten J-P, Cantor-Graae E. Social defeat: risk factor for schizophrenia? *British Journal of Psychiatry*. 2005;187(2):101–102.
48. Selten J-P, van der Ven E, Rutten BP, Cantor-Graae E. The social defeat hypothesis of schizophrenia: an update. *Schizophrenia Bulletin*. 2013;39(6):1180–1186.
49. Chiao JY. Neural basis of social status hierarchy across species. *Current Opinion in neurobiology*. 2010;20(6):803–809.
50. Tost H, Champagne FA, Meyer-Lindenberg A. Environmental influence in the brain, human welfare and mental health. *Nature Neuroscience*. 2015;18(10):1421–1431.
51. Myin-Germeys I, Delespaul P, Van Os J. Behavioural sensitization to daily life stress in psychosis. *Psychological Medicine*. 2005;35(5):733–741.
52. Heinz A, Schlagenhauf F. Dopaminergic dysfunction in schizophrenia: salience attribution revisited. *Schizophrenia Bulletin*. 2010;36(3):472–485.
53. Van Winkel R, Stefanis NC, Myin-Germeys I. Psychosocial stress and psychosis. A review of the neurobiological mechanisms and the evidence for gene-stress interaction. *Schizophrenia Bulletin*. 2008;34(6):1095–1105.
54. Kapur S. Psychosis as a state of aberrant salience: a framework linking biology, phenomenology, and pharmacology in schizophrenia. *American Journal of Psychiatry*. 2003;160(1):13–23.
55. Meyer-Lindenberg A, Tost H. Neural mechanisms of social risk for psychiatric disorders. *Nature Neuroscience*. 2012;15(5):663–668.
56. Reed JL, D'Ambrosio E, Marenco S, et al. Interaction of childhood urbanicity and variation in dopamine genes alters adult prefrontal function as measured by functional magnetic resonance imaging (fMRI). *PloS One*. 2018;13(4):e0195189.
57. Lederbogen F, Kirsch P, Haddad L, et al. City living and urban upbringing affect neural social stress processing in humans. *Nature*. 2011;474(7352):498–501.
58. LeDoux JE. Emotion circuits in the brain. *Annual Review of Neuroscience*. 2000;23(1):155–184.
59. Krämer B, Diekhof EK, Gruber O. Effects of city living on the mesolimbic reward system—an fMRI study. *Human Brain Mapping*. 2017;38(7):3444–3453.
60. Lemmers-Jansen IL, Fett A-KJ, Veltman DJ, Krabbendam L. Trust and the city—linking urban upbringing to neural mechanisms of trust in psychosis. *Australian and New Zealand Journal of Psychiatry*. Accepted for publication.
61. Takahashi T, Ikeda K, Ishikawa M, et al. Interpersonal trust and social stress-induced cortisol elevation. *Neuroreport*. 2005;16(2):197–199.
62. Streit F, Haddad L, Paul T, et al. A functional variant in the neuropeptide S receptor 1 gene moderates the influence of urban upbringing on stress processing in the amygdala. *Stress*. 2014;17(4):352–361.
63. Callicott J, Ihne J, Ursini G, et al. S. 11.01 Gene-environment interaction and prefrontal cortical function as measured by fMRI. *European Neuropsychopharmacology*. 2015;25:S125–S126.
64. Haddad L, Schäfer A, Streit F, et al. Brain structure correlates of urban upbringing, an environmental risk factor for schizophrenia. *Schizophrenia Bulletin*. 2014:sbu072.
65. Peeters SC, Gronenschild E, van de Ven V, et al. Altered mesocorticolimbic functional connectivity in psychotic disorder: an analysis of proxy genetic and environmental effects. *Psychological Medicine*. 2015;45(10):2157–2169.
66. Peeters SC, van de Ven V, Gronenschild EHM, et al. Default mode network connectivity as a function of familial and environmental risk for psychotic disorder. *PloS One*. 2015;10(3):e0120030.
67. Frissen A, van Os J, Lieverse R, Habets P, Gronenschild E, Marcelis M. No evidence of association between childhood urban environment and cortical thinning in psychotic disorder. *PloS One*. 2017;12(1):e0166651.
68. Frissen A, van Os J, Peeters S, Gronenschild E, Marcelis M. Evidence that reduced gray matter volume in psychotic disorder is associated with exposure to environmental risk factors. *Psychiatry Research: Neuroimaging*. 2018;271:100–110.
69. Supekar K, Uddin LQ, Prater K, Amin H, Greicius MD, Menon V. Development of functional and structural connectivity within the default mode network in young children. *Neuroimage*. 2010;52(1):290–301.
70. Fuller-Thomson E, Munro AP. Importance of considering the role of tetraethyl lead when examining the relationship between environmental pollution and psychotic disorders. *Schizophrenia Research*. 2018;195:570–571.
71. Newbury JB, Arseneault L, Beevers S, et al. Association of air pollution exposure with psychotic experiences during adolescence. *JAMA Psychiatry*. 2019.
72. Pedersen CB, Mortensen PB. Urbanization and traffic related exposures as risk factors for schizophrenia. *BMC Psychiatry*. 2006;6(1):2.
73. Pujol J, Martínez-Vilavella G, Macià D, et al. Traffic pollution exposure is associated with altered brain connectivity in school children. *Neuroimage*. 2016;129:175–184.
74. Calderón-Garcidueñas L, Mora-Tiscareño A, Ontiveros E, et al. Air pollution, cognitive deficits and brain abnormalities: a pilot study with children and dogs. *Brain and Cognition*. 2008;68(2):117–127.
75. Calderón-Garcidueñas L, Mora-Tiscareño A, Styner M, et al. White matter hyperintensities, systemic inflammation, brain growth, and cognitive functions in children exposed to air pollution. *Journal of Alzheimer's Disease*. 2012;31(1):183–191.
76. Debette S, Markus H. The clinical importance of white matter hyperintensities on brain magnetic resonance imaging: systematic review and meta-analysis. *BMJ*. 2010;341:c3666.
77. Hofman P, Krabbendam L, Vuurman E, Honig A, Jolles J. Schizophrenic patients are characterized by white matter hyperintensities, a controlled study. *Brain imaging in mild traumatic brain injury and neuropsychiatric disorders: a quantitative MRI study*. 2000:113.
78. Fonken LK, Xu X, Weil ZM, et al. Air pollution impairs cognition, provokes depressive-like behaviors and alters hippocampal cytokine expression and morphology. *Molecular Psychiatry*. 2011;16(10):987–995.
79. Levesque S, Surace MJ, McDonald J, Block ML. Air pollution and the brain: subchronic diesel exhaust exposure causes neuroinflammation and elevates early markers of neurodegenerative disease. *Journal of Neuroinflammation*. 2011;8(1):105.
80. Ravindran R, Devi RS, Samson J, Senthilvelan M. Noise-stress-induced brain neurotransmitter changes and the effect of Ocimum sanctum (Linn) treatment in albino rats. *Journal of Pharmacological Sciences*. 2005;98(4):354–360.
81. Akan Z, Yilmaz A, Özdemir O, Korpinar MA. Noise pollution, psychiatric symptoms and quality of life: noise problem in the east region of Turkey. *Journal of Inonu University Medical Faculty*. 2012;19(2):75–81.
82. Stansfeld SA. Noise, noise sensitivity and psychiatric disorder: epidemiological and psychophysiological studies. *Psychological Medicine Monograph Supplement*. 1992;22:1–44.
83. Kliuchko M, Heinonen-Guzejev M, Vuust P, Tervaniemi M, Brattico E. A window into the brain mechanisms associated with noise sensitivity. *Scientific Reports*. 2016;6:39236.
84. Tregellas JR, Smucny J, Eichman L, Rojas DC. The effect of distracting noise on the neuronal mechanisms of attention in schizophrenia. *Schizophrenia Research*. 2012;142(1):230–236.
85. Engemann K, Pedersen CB, Arge L, Tsirogiannis C, Mortensen PB, Svenning J-C. Childhood exposure to green space—a novel risk-decreasing mechanism for schizophrenia? *Schizophrenia Research*. 2018;199:142–148.
86. Bakolis I, Hammoud R, Smythe M, et al. Urban mind: using smartphone technologies to investigate the impact of nature on mental well-being in real time. *BioScience*. 2018.
87. Van der Wal AJ, Schade HM, Krabbendam L, Van Vugt M. Do natural landscapes reduce future discounting in humans? *Proceedings of the Royal Society of London B: Biological Sciences*. 2013;280(1773):20132295.

88. Bratman GN, Hamilton JP, Hahn KS, Daily GC, Gross JJ. Nature experience reduces rumination and subgenual prefrontal cortex activation. *Proceedings of the National Academy of Sciences.* 2015;112(28):8567–8572.
89. Ulrich RS. Natural versus urban scenes: some psychophysiological effects. *Environment and Behavior.* 1981;13(5):523–556.
90. Kim T-H, Jeong G-W, Baek H-S, et al. Human brain activation in response to visual stimulation with rural and urban scenery pictures: a functional magnetic resonance imaging study. *Science of the Total Environment.* 2010;408(12):2600–2607.
91. Maas J, Verheij RA, Groenewegen PP, De Vries S, Spreeuwenberg P. Green space, urbanity, and health: how strong is the relation? *Journal of Epidemiology and Community Health.* 2006;60(7):587–592.
92. Roe JJ, Aspinall PA, Mavros P, Coyne R. Engaging the brain: the impact of natural versus urban scenes using novel EEG methods in an experimental setting. *Environ Sci.* 2013;1(2):93–104.
93. Ulrich RS, Simons RF, Losito BD, Fiorito E, Miles MA, Zelson M. Stress recovery during exposure to natural and urban environments. *Journal of Environmental Psychology.* 1991;11(3):201–230.
94. Kuo M. How might contact with nature promote human health? Promising mechanisms and a possible central pathway. *Frontiers in Psychology.* 2015;6.
95. Kühn S, Düzel S, Eibich P, et al. In search of features that constitute an "enriched environment" in humans: associations between geographical properties and brain structure. *Scientific Reports.* 2017;7(1):11920.
96. Zadro L, Williams KD, Richardson R. How low can you go? Ostracism by a computer is sufficient to lower self-reported levels of belonging, control, self-esteem, and meaningful existence. *Journal of Experimental Social Psychology.* 2004;40(4):560–567.
97. Goldin PR, McRae K, Ramel W, Gross JJ. The neural bases of emotion regulation: reappraisal and suppression of negative emotion. *Biological Psychiatry.* 2008;63(6):577–586.
98. Falk EB, Cascio CN, O'Donnell MB, et al. Neural responses to exclusion predict susceptibility to social influence. *Journal of Adolescent Health.* 2014;54(5):S22–S31.
99. Eisenberger NI, Lieberman MD, Williams KD. Does rejection hurt? An fMRI study of social exclusion. *Science.* 2003;302(5643):290–292.
100. Masten CL, Eisenberger NI, Borofsky LA, et al. Neural correlates of social exclusion during adolescence: understanding the distress of peer rejection. *Social Cognitive and Affective Neuroscience.* 2009;4(2):143–157.
101. Bolling DZ, Pitskel NB, Deen B, et al. Dissociable brain mechanisms for processing social exclusion and rule violation. *NeuroImage.* 2011;54(3):2462–2471.
102. Gradin VB, Waiter G, Kumar P, et al. Abnormal neural responses to social exclusion in schizophrenia. *PloS One.* 2012;7(8):e42608.
103. Fett A-K, Shergill S, Krabbendam L. Social neuroscience in psychiatry: unravelling the neural mechanisms of social dysfunction. *Psychological Medicine.* 2015;45(6):1145–1165.
104. Fett A-K, Shergill S, Joyce D, et al. To trust or not to trust: the dynamics of social interaction in psychosis. *Brain.* 2012;135(3):976–984.
105. Gromann P, Shergill S, de Haan L, et al. Reduced brain reward response during cooperation in first-degree relatives of patients with psychosis: an fMRI study. *Psychological Medicine.* 2014;44(16):3445–3454.
106. Lemmers-Jansen IL, Fett A-KJ, Hanssen E, Veltman DJ, Krabbendam L. Learning to trust: social feedback normalizes trust behavior in first episode psychosis and clinical high risk. *Psychological Medicine.* 2018:1–11.
107. McGinty VB, Grace AA. Selective activation of medial prefrontal-to-accumbens projection neurons by amygdala stimulation and Pavlovian conditioned stimuli. *Cerebral Cortex.* 2007;18(8):1961–1972.
108. Alquicer G, Morales-Medina JC, Quirion R, Flores G. Postweaning social isolation enhances morphological changes in the neonatal ventral hippocampal lesion rat model of psychosis. *Journal of Chemical Neuroanatomy.* 2008;35(2):179–187.
109. King MV, Seeman P, Marsden CA, Fone KC. Increased dopamine D 2High receptors in rats reared in social isolation. *Synapse.* 2009;63(6):476–483.
110. Day-Wilson K, Jones D, Southam E, Cilia J, Totterdell S. Medial prefrontal cortex volume loss in rats with isolation rearing-induced deficits in prepulse inhibition of acoustic startle. *Neuroscience.* 2006;141(3):1113–1121.
111. Schubert M, Porkess M, Dashdorj N, Fone K, Auer D. Effects of social isolation rearing on the limbic brain: a combined behavioral and magnetic resonance imaging volumetry study in rats. *Neuroscience.* 2009;159(1):21–30.
112. van der Werf M, van Winkel R, van Boxtel M, van Os J. Evidence that the impact of hearing impairment on psychosis risk is moderated by the level of complexity of the social environment. *Schizophr Res.* 2010.
113. Dye C. Health and urban living. *Science.* 2008;319(5864):766–769.
114. Amodio DM. Can neuroscience advance social psychological theory? Social neuroscience for the behavioral social psychologist. *Social Cognition.* 2010;28(6):695.
115. Tost H, Meyer-Lindenberg A. Puzzling over schizophrenia: schizophrenia, social environment and the brain. *Nature Medicine.* 2012;18(2):211–213.
116. Lieberman JA, Sheitman BB, Kinon BJ. Neurochemical sensitization in the pathophysiology of schizophrenia: deficits and dysfunction in neuronal regulation and plasticity. *Neuropsychopharmacology.* 1997;17(4):205–229.
117. Van den Berg AE, Hartig T, Staats H. Preference for nature in urbanized societies: Stress, restoration, and the pursuit of sustainability. *Journal of Social Issues.* 2007;63(1):79–96.
118. Van den Berg AE, Jorgensen A, Wilson ER. Evaluating restoration in urban green spaces: does setting type make a difference? *Landscape and Urban Planning.* 2014;127:173–181.
119. Grandjean P, Landrigan PJ. Neurobehavioural effects of developmental toxicity. *Lancet Neurology.* 2014;13(3):330–338.
120. Block ML, Elder A, Auten RL, et al. The outdoor air pollution and brain health workshop. *Neurotoxicology.* 2012;33(5):972–984.
121. Brockmeyer S, D'Angiulli A. How air pollution alters brain development: the role of neuroinflammation. Vol 72016.

52

NEURAL CORRELATES OF ETHNIC MINORITY POSITION AND RISK FOR PSYCHOSIS

Jean-Paul Selten, Jan Booij, Bauke Buwalda, and Andreas Meyer-Lindenberg

INTRODUCTION

While there have been consistent reports in the literature of an increased risk of developing psychosis among members of ethnic minorities,[1–2] one may rightly ask the question as to *how* ethnic minority status could lead to an increased psychosis risk. In this chapter we discuss underlying mechanisms based on a recent review.[3]

Although there is no generally accepted theory of how minority position increases risk for psychosis, the pattern of risk factors for ethnic groups may provide an important clue. This pattern, for the larger part based on epidemiological studies within Europe, shows a higher risk for individuals with a black skin color than for those with other skin colors and a higher risk for individuals from developing countries than for those from developed countries. The highest risks have been reported for ethnic minorities who are least successful, such as African Caribbeans in the United Kingdom, Moroccan men in the Netherlands, and the Inuit in Denmark.[4–6] This observation strongly suggests that social defeat or social exclusion plays a role in contributing to the increased risk.

This idea is supported by the report of a *reduced* risk of developing psychosis in successful ethnic or religious minorities with a high level of social capital, such as the Swedish-speaking Finns in Finland or the Hutterites in Canada and the United States. Swedish-speaking Finns in Finland have a higher socioeconomic position, a lower divorce rate, and a longer life expectancy than the Finnish-speaking majority.[7] The Hutterites in Canada or the United States are economically successful and are known for their strong social cohesion.[8]

Indeed, the consistent reports in the literature of an increased risk for psychosis in subjects raised in urban areas, individuals with a low IQ, a hearing impairment, a nonheterosexual orientation, or a history of victimization in childhood,[9–11] suggest that social exclusion or social defeat is a coparticipating cause in the etiology of psychosis. The first attempt to find a common denominator for these findings was the social defeat hypothesis of psychosis.[12–15] This hypothesis posits that social defeat, defined as "subordinate position or outsider status"[12–13] or as "the negative experience of being excluded from the majority group,"[14–15] leads to an increased baseline activity and/or sensitization of the mesolimbic dopamine system and that these dopamine changes, in turn, place the individual at an increased risk of developing the disorder. Since a prevailing definition of social exclusion is "an enforced lack of social participation," the concepts of social exclusion and social defeat overlap.[16] The authors of the social defeat hypothesis emphasized that the variable of interest in their hypothesis is the subjective experience, which depends on the individual's interpretation of events.

An important model for social defeat stress in animals is the resident-intruder paradigm, whereby a male rodent (the intruder) is placed into the home cage of another male (the resident). The intruder is aggressively attacked and forced to display submissive behavior. This experience is not equivalent to the experience of social defeat in humans, because the defeat of the intruder is aimed at establishing a hierarchy, not at social exclusion. By social isolation of the animal after defeat, the experience of the animal would resemble more the experience of social exclusion in humans. Nonetheless, the resident-intruder paradigm is relevant for our understanding of the experience of social exclusion in humans, because humans experience social exclusion as defeating.

The principal question for this review is whether there are any plausible neurobiological mechanisms whereby social exclusion could lead to the development of a psychosis. Since this is a question about pathogenesis, not etiology, one could reformulate the question as to whether a contribution of social exclusion is compatible with the existing knowledge about pathogenesis. The pathogenic mechanisms of psychoses, however, are uncertain, and a large number of hypotheses have been proposed. Consequently, for practical purposes, this review examines whether a contribution of social exclusion is compatible with three important hypotheses on pathogenesis discussed in authoritative reviews: the social-cognition hypothesis, the dopamine hypothesis, and the neurodevelopmental hypothesis.[9,17]

Briefly, the *social-cognition hypothesis* posits that an impaired capacity to mentalize, i.e., to understand one's own and others' behavior in terms of mental states, such as intentions, wishes, beliefs, and emotions, plays an important role in the development of psychosis.[9] The *dopamine hypothesis*, version III, postulates that multiple "hits" interact to result in dopamine dysregulation, the final common pathway to psychosis.[18] Finally, the *neurodevelopmental hypothesis* posits that nonaffective psychotic disorder is due to an abnormal brain development, already manifest in childhood and youth, on account of delays in motor, social, and intellectual development, and later, at disease onset, in structural brain deficits.

As for this chapter, we included important studies that used concepts related to social exclusion or social defeat, such as discrimination, negative social evaluation, social adversity, social fragmentation, or social disadvantage.

IMPAIRED SOCIAL COGNITION

The concepts of social cognition, theory of mind, and mentalizing capacity are closely related. Since positive symptoms often involve misinterpretations of behavior observed in others, it is conceivable that processes that interfere with the normal acquisition of mentalizing ability increase the risk for psychosis. Since the big start in the development of mentalizing is made in relationship with the mother and during preschool years, neglect and abuse during these years are particularly harmful.[19] However, the development of mentalizing capacity continues in later years, in relation with peers, teachers, and other members of society. The evidence to support the notion that social exclusion interferes with this development comes from studies showing delays in the development of mentalizing in children with a serious hearing impairment.[20] Furthermore, it is conceivable that growing up as a member of a discriminated ethnic minority is damaging to the capacity to correctly infer the intentions of others, in particular those from the dominant group.[21]

DOPAMINE DYSREGULATION

There are consistent reports of increased dopamine synthesis capacity, increased dopamine release, and increased baseline synaptic dopamine concentrations.[22] An important topic here is sensitization, a process whereby repeated exposure to a given stimulus results in an enhanced response at subsequent exposure, in this example excess release of dopamine. Although there is a conspicuous lack of longitudinal studies in humans, the findings suggest that the mesolimbic dopamine system is sensitized. The question here is whether social exclusion, or exposures to stimuli related to social exclusion, lead to increased dopamine synthesis capacity, increased dopamine release, increased baseline synaptic dopamine concentrations, and/or sensitization of the dopamine system.

Animal studies provide ample evidence for this. First, after one or more episodes of defeat within the resident–intruder paradigm, the defeated animal shows evidence of an increased sensitivity to amphetamine (i.e., increased locomotor activity), increased dopamine release in the nucleus accumbens (part of the ventral striatum [VS]) and prefrontal cortex (homologous to the medial prefrontal cortex [mPFC] in humans), and increased firing of dopaminergic neurons in the ventral tegmental area (VTA).[23–26] Lengthy social isolation after the defeat amplifies the changes in dopamine activity, whereas return to the group mitigates them.[27] Second, an interesting experiment in cynomolgus macaques showed that a return to the social group after individual housing produced an increase in the amount or availability of dopamine D_2 receptors in the dominant monkeys, not in the subordinate ones.[28] These findings suggest that place in hierarchy could influence the dopaminergic system. Third, rat pups reared in isolation following weaning, develop, in adulthood, increased striatal synaptic dopamine concentrations and increased striatal dopamine release in response to cocaine or amphetamine.[29,30]

As for humans, a recent publication reported the results of two positron emission tomography (PET) substudies of dopamine function in first- and second-generation migrants.[31] The groups of migrants and nonmigrants included healthy volunteers, clinical high-risk subjects and antipsychotic-naïve patients with schizophrenia. The first substudy, using the dopamine $D_{2/3}$ receptor agonist tracer [^{11}C]-(+) -PHNO (which is very sensitive to detect changes in synaptic dopamine concentrations), showed that the psychological stress of exposure to the Montreal Imaging Stress Task induced a greater striatal dopamine release in migrants than in nonmigrants. The second study, using [^{18}F]-DOPA, showed a greater striatal dopamine synthesis capacity in migrants than in nonmigrants. These results provide the first evidence that the effect of ethnic minority status on the risk of developing psychosis may be mediated by an increase in dopamine function.

Other neuroreceptor imaging studies assessed the impact of hearing impairment and childhood trauma in nonpsychotic individuals. Gevonden et al. used single photon emission computed tomography (SPECT) to compare dopamine function in young adults with a serious acquired hearing impairment (SHI) to that in healthy controls.[32] The participants underwent two SPECT-scans with the dopamine $D_{2/3}$ antagonist [^{123}I]iodobenzamide, which is, like [^{11}C]raclopride, sensitive to detect changes in synaptic dopamine concentrations after an amphetamine challenge. There were no significant differences in baseline striatal $D_{2/3}$ receptor binding. However, amphetamine-induced striatal dopamine release was significantly greater among the participants with SHI than among the healthy controls. Reports of social exclusion were not associated with dopamine release after amphetamine, perhaps because self-reports of this phenomenon are biased by a tendency to give socially desirable replies.

Oswald et al. (2014) exposed a general population sample of young adults to two [^{11}C]raclopride PET-scans.[33] The first scan was preceded by saline, the second by amphetamine. The results showed a positive association between reports of childhood trauma and baseline $D_{2/3}$ receptor availability in the VS of men, not in women. Further, there was a positive association between trauma and amphetamine-induced dopamine release in the VS.

Egerton et al. (2016) used [^{18}F]-DOPA PET to investigate the impact of childhood adversity on individuals at ultra-high risk for psychosis and healthy volunteers.[34] The results showed that sexual and physical abuse (Cohen's $d = 0.75$) and unstable family arrangements ($d = 0.86$) were associated with increased dopamine synthesis capacity in the associative striatum. Interestingly, there was no relationship between dopamine synthesis and events (i.e., parental loss or separation) that do not necessarily involve intentional harm to the child.

To summarize, the findings of the two studies on baseline $D_{2/3}$ receptor binding were mixed. Further, although the use of a cross-sectional design does not permit definitive conclusions, the results on dopamine release and synthesis are compatible with the idea of dopamine sensitization. Future studies could examine groups with a history of exposure to social exclusion (e.g., immigrants from developing countries or subjects with a nonheterosexual orientation), preferably using a longitudinal design. Ideally, the first assessment is made before the exposure (e.g., when the migrant arrives in the new country), the second thereafter (e.g., after the migrant has spent several years in the country of destination). Without the use of a longitudinal design it remains uncertain when the sensitization has developed.

NEURAL MECHANISMS IN HUMANS LINKING SOCIAL EXCLUSION AND STRESS TO DOPAMINE DYSFUNCTION

The dopaminergic VS receives inputs from the VTA, the mPFC, anterior cingulate cortex (ACC), amygdala, and hippocampus, which are critical for salience and reward signaling. These circuits are strongly implicated in social cognition.[9]

Three networks center on subregions of the amygdala,[35] a key structure for the integration of emotion and social processing. A *social-perceptive network* connects ventrolateral amygdala (lateral nucleus, which receives rapid sensory inputs across all modalities) to sensory association areas of the temporal cortex and the orbitofrontal cortex (OFC). It has been implicated in decoding and interpreting social signals from others in the context of past experience and current goals. A *social-affiliative network* is anchored in the medial sector of the amygdala, which contains nuclei (especially the basal nucleus) that share anatomical connections with mesolimbic, reward-related areas of the ventromedial prefrontal cortex, medial temporal lobe, ventromedial striatum, and hypothalamus. This network relates to appetitive, prosocial, and trusting interactions. A *social aversion network* is centered on the dorsal sector of the amygdala, which contains nuclei (such as the central nucleus) that project to dorsal ACC, insula, ventrolateral striatum, hypothalamus, and brainstem. These regions are implicated in fear, fright or flight, and avoidant behavior.

Two other networks are relevant. The *mentalizing network* connects mPFC with the temporoparietal junction, superior temporal sulcus, precuneus and anterior temporal lobe (ATL). It serves social reasoning, social knowledge, actively thinking about others, reflecting on oneself, and theory of mind. It overlaps with the so-called default mode or resting-state network. Finally, the *empathy and mirror network* is engaged when vicariously experiencing states (such as pain) of others or in action observation and includes parts of the dorsal cingulate and anterior insula.

Recent work has shown that social risk factors such as socioeconomic status,[36] urban upbringing and living,[37] and ethnic minority status,[38] have convergent effects on social stress processing in a neural system centered on perigenual anterior cingulate cortex (pgACC)[39] (Figure 52.1). Social stress in members of ethnic minorities not only impacted activation of the pgACC differentially but also uncovered abnormalities in the VS linked to the degree of perceived discrimination against the ethnic minority group, further underlining the posited link to social exclusion.

We have also implicated the pgACC in the regulation of the human hypothalamic–pituitary–adrenal stress response system itself.[40] Furthermore, genome-wide significant common[41] and rare[42] susceptibility genes for psychosis impact the same network. Taken together, these data suggest a central role for a stress-associated convergent "risk circuit" for psychosis:[43] at the center of this network is pgACC, which regulates key limbic structures such as amygdala, hippocampus, and VS and in turn participates in regulatory interactions with higher-order lateral and mPFC structures such as Brodmann's areas 46 and 10. Taken together with the animal experimental evidence, this would suggest that striatal dopaminergic dysfunction is a consequence of prefrontal dysregulation of that region, in part through the strong bidirectional interactions between pgACC and VS, a hypothesis that could be tested using hybrid PET-fMRI.

Recent advances in social neuroscience link these networks up more specifically with processes linked to social exclusion

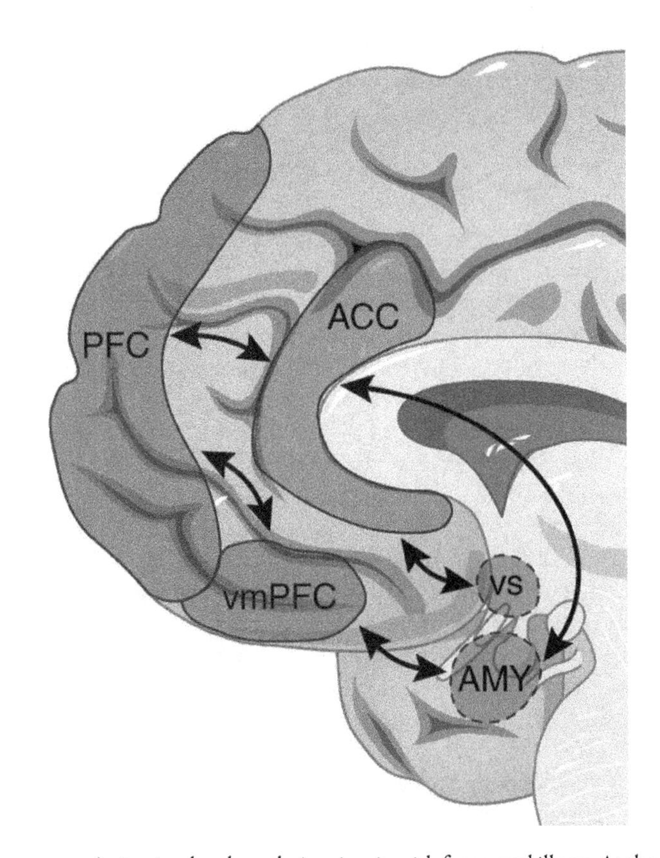

Figure 52.1 A circuit related to ethnic minority risk for mental illness. At the core of the proposed mechanism is pgACC, which regulates subcortical structures such as ventral striatum (VS) and amygdala (AMY) and is in turn modulated by ventromedial (vmPFC) and medial prefrontal (PFC) regions.

From Meyer-Lindenberg and Tost, 2012,[43] with permission. Meyer-Lindenberg, A. & Tost, H. 2012. Neural mechanisms of social risk for psychiatric disorders. Nat Neurosci 15: 663–668. doi: 10.1038/nn.3083.

in humans: prejudice and stereotype.[44,45] The social neuroscience of *prejudice*, or (usually negative) attitudes and emotional reactions to individuals based on (out-)group membership, maps across several networks. Prominently implicated is the social aversion network, where amygdala supports threat-based associations and anterior insula signals negative affect (which often accompanies a prejudiced response). In the social-perceptive network amygdala signals initial responses to salient positive or negative cues, including cues regarding group membership. The mPFC, a prominent component of both the social-affiliative and mentalizing networks, is engaged more strongly toward in-group than out-group members. Finally, appetitive responses such as positive attitudes and approach-related behavioral tendencies, which are often expressed toward in-group members, map on the striatum. *Stereotypes* (characteristics ascribed to a social group through [over]generalization) are likely represented in the ATL, which feeds into the mPFC. In this way, social stereotypes are thought to influence social attributions in dorsal mPFC activity. The interaction of these networks is critical for integrated social behavior. Neural projections from the amygdala and insula to the mPFC may support the integration of affective responses with mentalizing and empathy processes. Amygdala and OFC connect to the ATL via the uncinate fasciculus to support bidirectional interactions between affective responses and stereotype concepts. Signals from amygdala, insula, striatum, OFC, and ATL converge in regions of the mPFC, where information seems to be integrated in support of elaborate person representations. Finally, the joint influences of prejudiced affect and stereotype concepts on behavior are likely to converge in the dopaminergic striatum, which receives inputs from the amygdala, OFC, mPFC, lateral PFC, rostral ACC, ATL, and midbrain.

ABNORMAL NEURODEVELOPMENT

The central question, here, is whether social exclusion contributes to (1) delays in motor, social and intellectual development, or (2) structural brain deficits at the onset of psychosis.

As for the first question, the relationship between social exclusion and these developmental delays may be circular. While nobody will doubt that motor, social, and cognitive delays may lead to social exclusion and that exposure to social exclusion contributes to an impaired development of social skills, there is more debate on the issue as to whether the experience of social exclusion contributes to an impaired motor and intellectual development.[46] A consideration of this interesting discussion, however, is beyond the scope of this paper.

An answer to the second question, as to whether the human experience of social exclusion contributes to the gray matter reduction, white matter disruption, or ventricular enlargement observed at the onset of schizophrenia, is also uncertain. There have been reports of a decreased pgACC volume in men raised in cities[47] and in male second-generation immigrants,[48] but no studies of the relationship between (proxies for) social exclusion and widespread gray matter deficits, ventricular enlargement, or white matter disruption.

Some animal studies, however, did examine whether long-term social isolation or exposure to social defeat lead to the aforementioned brain changes. Fabricius et al. (2010) applied stereological volume estimates to rats isolated after weaning and found that isolated males had significantly smaller brains and smaller hippocampi than group-housed controls and larger ventricles than controls.[49] However, this was not seen in female rats. Schubert et al. (2009) applied magnetic resonance volumetry to the limbic system of isolated rats and observed no volume reduction in the hippocampi, but a 5% reduction in the volume of the mPFC, a region that is strongly directly and indirectly (through ACC) connected to the ventral striatum.[50]

Although it is uncertain whether decreased neurogenesis plays a role in the pathogenesis of psychosis, it is of interest that social defeat affects this phenomenon. Czeh et al. (2007) exposed adult rats to 5 weeks of daily social defeat and found that this led to decreased gliogenesis in the mPFC and to decreased neurogenesis in the gyrus dentatus, while there was only a minor impact on nonlimbic structures.[51] Other studies showed that acute social defeat stress suppressed hippocampal cell proliferation transiently up to 50%–75% ,[52] while chronic defeat resulted in a more subtle decrease of only 29%–33%.[53,54] Lack of neutrophil support and impaired neuronal vascular supply have been offered as explanations.[55] Interestingly, several postmortem studies have described decreased numbers of hippocampal neurons in schizophrenia patients.[56]

Taken together, these results indicate that it is important to examine whether social exclusion contributes to the development of the anatomic changes already present at the time of the first psychotic episode. For example, one could compare the development of the brain over years between excluded and nonexcluded adolescents. Adolescents can be excluded on account of various features, such as an ethnic or sexual minority status, a hearing impairment, or an odd appearance.

CONCLUSION

There is evidence to suggest that social exclusion impacts on human dopaminergic functioning and thereby influences the risk of developing psychosis among members of ethnic minorities. Studies of animals suggest that it is important to examine whether social exclusion contributes to the abnormal brain development in nonaffective psychotic disorder. Consequently, future studies, preferentially with a longitudinal design, should examine dopaminergic functioning and structural brain development in various socially excluded groups.

REFERENCES

1. Cantor-Graae E, Selten JP. Schizophrenia and migration: a meta-analysis and review. *Am J Psychiatry* 2005;162(1):12–24.
2. Bourque F, van der Ven E, Malla A. A meta-analysis of the risk for psychotic disorders among first- and second-generation immigrants. *Psychol Med* 2011;41(5):897–910.

3. Selten JP, Booij J. Buwalda B, Meyer-Lindernberg A. Biological mechanisms whereby social exclusion may contribute to the etiology of psychosis: a narrative review. *Schizophr Bull* 2017;43(2):287–292.
4. Tortelli A, Errazuriz A, Croudace T et al. Schizophrenia and other psychotic disorders in Caribbean-born migrants and their descendants in England: systematic review and meta-analysis of incidence rates, 1950–2013. *Soc Psychiatry Psychiatr Epidem* 2015;50(7):1039–1055.
5. van der Ven, E, Veling W, Tortelli A et al. Evidence of an excessive gender gap in the risk of psychotic disorder among North African immigrants in Europe: a systematic review and meta-analysis. *Soc Psychiatry Psychiatr Epidem* 2016;51(12):1603–1613.
6. Cantor-Graae E, Pedersen CB. Risk of schizophrenia in second-generation immigrants; a Danish population-based cohort study. *Psychol Med* 2007;37(4):485–494.
7. Suvisaari J, Opler M, Lindbohm ML, Sallmén M. Risk of schizophrenia and minority status: a comparison of the Swedish-speaking minority and the Finnish-speaking majority in Finland. *Schizophr Res* 2014;159(2-3):303–308.
8. Nimgaonkar VL, Fujiwara TM, Dutta M et al. Low prevalence of psychoses among the Hutterites, an isolated religious community. *Am J Psychiatry* 2000;157(7):1065–1070.
9. van Os J, Kenis G, Rutten BPF. The environment and schizophrenia. *Nature* 2010;468 (7321):203–212.
10. Gevonden MJ, Selten JP, Myin-Germeys I, de Graaf R, Ten Have M, van Dorsselaer S, van Os J, Veling W. Sexual minority status and psychotic symptoms: findings from the Netherlands Mental Health Survey and Incidence Studies (NEMESIS). *Psychol Med* 2014;44(2):421–433.
11. Varese F, Smeets F, Drukker M et al. Childhood adversities increase the risk of psychosis: a meta-analysis of patient-control, prospective and cross-sectional cohort studies. *Schizophr Bull* 2012;38(4):661–671.
12. Selten JP, Cantor-Graae E. Social defeat: risk factor for schizophrenia? *Br J Psychiatry* 2005;187:101–102.
13. Selten JP, Cantor-Graae E. Hypothesis: social defeat is a risk factor for schizophrenia. *BrJ Psychiatry* 2007;191(Suppl. 51):s9–s12.
14. Selten JP, van der Ven E, Rutten B, Cantor-Graae E. The social defeat hypothesis of schizophrenia: an update. *Schizophr Bull* 2013;39 (6):1180–1186.
15. Selten JP, van Os J, Cantor-Graae E. The social defeat hypothesis of schizophrenia: issues of measurement and reverse causality. *World Psychiatry* 2016;15(3):294–295.
16. Morgan C, Burns T, Fitzpatrick R, Pinfold V, Priebe S. Social exclusion and mental health: conceptual and methodological review. *Br J Psychiatry* 2007;191:477–483.
17. Howes OD, Murray RM. Schizophrenia: an integrated sociodevelopmental-cognitive model. *Lancet* 2014;383(9929):1677–1687.
18. Howes OD, Kapur S. The dopamine hypothesis of schizophrenia: version III—the final common pathway. *Schizophr Bull* 2009;35(3):549–562.
19. Colvert E, Rutter M, Kreppner J, et al. Do theory of mind and executive function deficits underlie the adverse outcomes associated with profound early deprivation? Findings from the English and Romanian adoptees study. *J Abnorm Child Psychol* 2008;36(7):1057–1068.
20. Peterson CC, Siegal M. Deafness, conversation and theory of mind. *J Child Psychol Psychiatry* 1995;36(3):459–474.
21. Branscombe NR, Schmitt MT. Perceiving pervasive discrimination among African Americans: implications for group identification and well-being. *J Personality Soc Psychol* 1999;77:135–149.
22. Howes OD, Kambeitz J, Kim E, et al. The nature of dopamine dysfunction in schizophrenia and what this means for treatment. *Arch Gen Psychiatry* 2012;69(8):776–786.
23. Tidey JW, Miczek KA. Social defeat stress selectively alters mesocorticolimbic dopamine release: an in vivo microdialysis study. *Brain Res* 1996;721(1–2):140–149.
24. de Jong JG, Wasilewski M, Van der Vegt BJ, Buwalda B, Koolhaas JM. A single social defeat induces short-lasting behavioral sensitization to amphetamine. *Physiol Behav.* 2005;83(5):805–811.
25. Trainor BC. Stress responses and the mesolimbic dopamine system: social contexts and sex differences. *Horm Behav* 2011;60(5):457–469.
26. Anstrom KK, Miczek KA, Budygin EA. Increased phasic dopamine signaling in the mesolimbic pathway during social defeat in rats. *Neuroscience* 2009;161(1):3–12.
27. Isovich E, Engelmann M, Landgraf R, Fuchs E. Social isolation after a single defeat reduces striatal dopamine transporter binding in rats. *Eur J Neurosci* 2001;13:1254–1256.
28. Morgan D, Grant KA, Gage HD, et al. Social dominance in monkeys; dopamine D2 receptors and cocaine self-administration. *Nat Neurosci* 2002;5(2):169–174.
29. Lapiz MD, Fulford A, Muchimapura S, Mason R, Parker T, Marsden CA. Influence of postweaning social isolation in the rat on brain development, conditioned behavior, and neurotransmission. *Neurosci Behav Physiol* 2003;33(1):13–29.
30. Kosten TA, Zhang XY, Kehoe P. Chronic neonatal isolation stress enhances cocaine-induced increases in ventral striatal dopamine levels in rat pups. *Brain Res Dev Brain Res* 2003;141(1–2): 109–116.
31. Egerton A, Howes OD Houle S et al. Elevated striatal dopamine function in immigrants and their children: a risk mechanism for psychosis. *Schizophr Bull* 2017;43(12):293–301.
32. Gevonden MJ, Booij J, van den Brink W, Heijtel D, van Os J. Selten JP. Increased release of dopamine in the striata of young adults with hearing impairment and its relevance for the social defeat hypothesis of schizophrenia. *JAMA Psychiatry* 2014;71(12):1364–1372.
33. Oswald LM, Wand GS, Kuwabara H, Wong DF, Zhu S, Brasic JR. History of childhood adversity is positively associated with ventral striatal dopamine responses to amphetamine. *Psychopharmacol Berl.* 2014;231(12):2417–2433.
34. Egerton A, Valmaggia L, Howes OD et al. Adversity in childhood linked to elevated striatal dopamine function in adulthood. *Schizophr Res* 2016;176(2–3): 171–176.
35. Bickart KC, Hollenbeck MC, Barrett LF, Dickerson BC. Intrinsic amygdala-cortical functional connectivity predicts social network size in humans. *J Neurosci* 2012;32(42):14729–14741.
36. Zink CF, Tong Y, Chen Q, Bassett DS, Stein JL, Meyer-Lindenberg A. Know your place: neural processing of social hierarchy in humans. *Neuron* 2008;58(2):273–283.
37. Lederbogen F, Kirsch P, Haddad L, et al. City living and urban upbringing affect neural social stress processing in humans. *Nature* 2011;474(7352):498–501.
38. Akdeniz C, Tost H, Streit F, et al. Neuroimaging evidence for a role of neural social stress processing in ethnic minority-associated environmental risk. *JAMA Psychiatry* 2014;71(6):672–680.
39. Tost H, Champagne FA, Meyer-Lindenberg A. Environmental influence in the brain, human welfare and mental health. *Nature Neurosci* 2015;18(10):1421–1431.
40. Boehringer A, Tost H, Haddad L, et al. Neural correlates of the cortisol awakening response in humans. *Neuropsychopharmacol* 2015;40(9):2278–2285.
41. Erk S, Meyer-Lindenberg A, Schnell K, et al. Brain function in carriers of a genome-wide supported bipolar disorder variant. *Arch Gen Psychiatry* 2010;67(8):803–811.
42. Stefansson H, Meyer-Lindenberg A, Steinberg S, et al. CNVs conferring risk of autism or schizophrenia affect cognition in controls. *Nature* 2014;505(7483):361–366.
43. Meyer-Lindenberg A, Tost H. Neural mechanisms of social risk for psychiatric disorders. *Nature Neuroscience* 2012;15(5):663–668.
44. Amodio DM. The neuroscience of prejudice and stereotyping. *Nature Rev Neurosci* 2014;15(10):670–682.
45. Kubota JT, Banaji MR, Phelps EA. The neuroscience of race. *Nature Neurosci* 2012;15(7):940–948.
46. Baumeister RF, Twenge JM, Nuss CK. Effects of social exclusion on cognitive processes: anticipated aloneness reduces intelligent thought. *J Personal Soc Psychol* 2002;83(4):817–827.

47. Haddad L, Schäfer A, Streit F, Lederbogen F, Grimm O, Wüst S, et al. Brain structure correlates of urban upbringing, an environmental risk factor for schizophrenia. *Schizophr Bull* 2015;41(1):115–122.
48. Akdeniz C, Schäfer, Streit, Haller L, Wüst S, Kirsch P et al. Sex/dependent association of perigenual anterior cingulate cortex volume and migration background, an environmental risk factor for schizophrenia. *Schizophr Bull* 2016; ePub ahead of print.
49. Fabricius K, Helboe L, Steiniger-Brach B, Fink-Jensen A, Pakkenberg B. Stereological brain volume changes in post-weaned socially isolated rats. *Brain Res* 2010;1345:233–239.
50. Schubert MI, Porkess MV, Dashdorj N, Fone KC, Auer DP. Effects of social isolation rearing on the l4mbic brain: a combined behavioural and magnetic resonance imaging volumetry study in rats. *Neuroscience* 2009;159(1): 21–30.
51. Czéh B, Müller-Keuker JI, Rygula R, et al. Chronic social stress inhibits cell proliferation in the adult medial prefrontal cortex; hemispheric asymmetry and reversal by fluoxetine treatment. *Neuropsychopharmacol* 2007;32(7):1490–1503.
52. Lagace DC, Donovan MH, DeCarolis NA, et al. Adult hippocampal neurogenesis is functionally important for stress-induced social avoidance. *Proc Natl Acad Sci USA* 2010;107(9):4436–4441.
53. Czéh B, Michaelis T, Watanabe T, et al. Stress-induced changes in cerebral metabolites, hippocampal volume, and cell proliferation are prevented by antidepressant treatment with tianeptine. *Proc Natl Acad Sci USA* 2001;98(22):12796–12801.
54. Czéh B, Welt T, Fischer AK, et al. Chronic psychosocial stress and concomitant repetitive transcranial magnetic stimulation: effect on stress hormone levels and adult hippocampal neurogenesis. *Biol Psychiatry* 2002;52(11)1057–1065.
55. Hammels C, Pishva E, De Vry J, et al. Defeat stress in rodents: from behavior to molecules. *Neurosci Biobehav Rev* 2015;59: 111–140.
56. Falkai P, Malchow B, Wetzestein K, et al. Decreased oligodendrocyte and neuron number in anterior hippocampal areas and the entire hippocampus in schizophrenia: a stereological postmortem study. *Schizophr Bull* 2016;42 Suppl 1;S4–S12.

53

RESILIENCE IN PSYCHOSIS SPECTRUM DISORDER

Lotta-Katrin Pries, Sinan Guloksuz, and Bart P. F. Rutten

INTRODUCTION

Resilience refers to an umbrella concept encompassing both differential susceptibility and differential recovery from severe adversities and episodes of mental ill-health.[1,2] The concept emerged in the 1970s, in the context of developmental psychopathology: Children who despite being exposed to significant adversities showed relative normal development and functioning later in life.[3] In the recent decade, the concept of resilience has received increased attention as it is suggested that understanding resilience may uncover opportunities for preventative and therapeutic interventions.

Resilience has been widely acknowledged as a dynamic and adaptive process, which can cause decreased vulnerability, promote quick recovery, and even enhance mental health after adversity (e.g., post-traumatic growth; PTG).[1,2,4,5] Similarly, the recent health concept moves away from explaining health as the absence of any illness toward conceptualizing it as "the ability to adapt and to self-manage."[6] The new mental health-care model encourages the strengthening of connectedness and resilience.[7] However, research on "resilience" has retained a sense of elusiveness, which is likely related—at least partly—to the varying approaches and conceptualizations attempting to measure aspects of resilience.[2]

Resilience has been investigated using a range of approaches focusing on different levels, including individual protective factors, and factors related to family or the society at large.[2,8] Furthermore, resilience has been investigated in respect to a single-incidence or multiple stressors,[9] while the working model in studies may aim to investigate monocausal (psychological, biomedical, or sociocultural) or complex multicausal (biopsychosocial) aspects.[2] Other recent theories on resilience are based on lifelong learning, such as the mismatch theory, which proposes that differential vulnerability to psychopathology may depend on the mismatch between early-life experiences and the environment encountered in later life.[10]

Research on resilience in psychosis spectrum disorder (PSD) may spark important new endeavors in focusing on positive factors at the biological, social, and psychological levels at different stages of life. For instance, early life adversities were associated with PSD in adulthood. However, many of those with a history of early adversity do not develop PSD in later life—a smaller fraction does not even manifest psychopathology.[11] In this regard, genetic and environmental factors (e.g., secure attachment, high self-esteem, coping mechanisms) play a critical role in shaping resilience to the effects of early adversities.

Further, protective factors might be especially important in early stages of illness. Individuals may use dynamic adaptive resources throughout the process of coping with traumatic-stress related to psychotic experiences or involuntary admission. Eventually, adaptive responses may also result in PTG, where individuals develop resilience (e.g., through relating to others, personal strength, spiritual change) against future traumatic events and psychotic experiences.[12] Resilience and protective personal resources have been associated with recovery[13] and improved daily functioning and global well-being,[14,15] which makes resilience a promising avenue of research in PSD.

In this chapter, we review and discuss the resilience concept in PSD. First, we summarize applied approaches to measure resilience. Second, we provide a synopsis of important psychological, social, and biological resilience factors with a specific attention on the different stages of psychopathology. Finally, we aim to give a comprehensive overview of the role of the interplay between resilience factors in PSD and contemplate future research directions.

HOW TO MEASURE "RESILIENCE"?

Qualitative and quantitative approaches to measure resilience are: (1) questionnaires,[16] (2) structured and semistructured interviews with subsequent systematized evaluation,[17] and (3) operant constructions assessing resilience (e.g., positive outcomes in the face of adversity).[18] However, the field lacks a consensus or a "gold standard" as a result of difficulties in assessing resilience and variation across operant constructions. Therefore, it is important to specify the way of operationalizing resilience by specifying: (1) study population, (2) exposure and assessment method, (3) the method for assessing "resilience," and (4) confounders.

Questionnaires on resilience have been applied at different stages of psychosis.[16,19] There are various questionnaires with moderate psychometric properties: e.g., (1) the Resilience Scale for Adults (RSA),[20] which assesses protective factors (personal competence, social competence, family coherence, social support and personal structure); (2) the Connor–Davidson Resilience Scale (CD-RISC),[21] which focuses on stress-coping abilities on five factors (personal competence, high standards, and tenacity; trust in one's instincts, tolerance of negative affect, and strengthening effects of stress; positive acceptance of change and secure relationships; control; spiritual influences); (3) the Brief Resilience Scale (BRS),[22] which investigates the ability to bounce back and recover; and (4) the Resilience Scale (RS),[23] which assesses personal competence and acceptance of self and life (e.g., "I feel that I can handle many things at a time"), have the strongest theoretical and empirical support in the literature.[24–26]

Such questionnaire-based studies have reported statistically significant differences between patient and control populations.[16] However, it should be noted that resilience questionnaires do not directly assess resilience but resilience-promoting aspects.[24] For instance, other measurement tools such as the five-factor model of personality might be more suitable in predicting adaptive functioning.[27]

Resilience can also be conceptualized as an operant construct, examining individual's resilient outcomes in the face of adversities.[8] Both diagnostic and average-level approaches are used.[9] The diagnostic approach constructs resilience within the binary model: presence *vs.* absence of illnesses.[9] The average-level approaches construct resilience on a continuum as average-level differences between individual outcomes.[9] Researchers have argued that the development of a valid resilience operant relies heavily on the adversity (e.g., first-episode psychosis) and outcome variables (e.g., psychosocial functioning), which may distinguish between resilient and nonresilient individuals.[5,8] Furthermore, researchers need to consider the multidimensional impact of stressors and the continuity of resilience outcomes.[8]

Finally, qualitative studies may enhance our understanding of the perspective of those individuals who may benefit from resilience-based treatment approaches. Researchers using qualitative methods to appraise individual characteristics have analyzed resilience from different angles, e.g., different stages of pathology,[17,28] symptoms,[29] and factors affecting treatment success.[30] As both patients and clinicians need to agree on the meaning of resilience, it is important to understand resilience not only through quantitative studies but also from patient's perspective and experience.[17,31]

Regarding the multidimensionality of resilience and limitations inherent to assessment techniques, a combination of quantitative (questionnaire, interview, operant-construct) and qualitative approaches would provide the optimal solution.

PSYCHOLOGICAL AND COGNITIVE MECHANISMS

SELF-ESTEEM

Elevated self-esteem is associated with increased well-being, psychological and physical health,[32] affect adaptation, and coping mechanisms.[33] Individuals with PSD show low self-esteem,[34,35] which is associated with increased severity and related distress of negative and positive psychotic experiences.[19,36] In a longitudinal study, positive self-esteem, negative emotions, and hopelessness as well as symptomatology and functioning at baseline predicted higher scores on subjective recovery 6-month later.[37] Furthermore, higher scores on self-esteem scales correlate with higher levels of quality of life (QoL)[35] and stigma-resilience,[38] and predict time to relapse.[39] Patients with higher self-esteem reported lower stigma awareness (of patients and family)[40] and higher resilience scores (on the resilience scale [RS]).[16]

In general, studies show robust associations between self-esteem and resilience, and thereby suggest that interventions aimed to boost self-esteem might be beneficial for enhancing resilience in PSD.

INSIGHT

It has been reported that the majority of patients with PSD (50%–80%) lacks appropriate insight in having a mental illness, albeit to varying degrees.[41] Higher levels of insight predict remission and better occupational functioning,[42–44] while insight into one's own illness and life situation may encourage patients to seek help and adhere to treatment.[45–47] A meta-analysis found that the level of overall insight (i.e., insight in having a mental disorder, symptoms and attribution of symptoms, the social consequences of the disorder and need for treatment) in schizophrenia was negatively associated with global, positive, and negative symptoms, and that this association was moderated by age at onset and acute patient status.[48] On the other hand, higher levels of insight may—possibly through the mechanisms of self-stigma and demoralization[49]—also lead to poor outcome such as higher hopelessness, depression, poor QoL, and increased suicidality.[47,50–52]

Overall, insight is important to reduce time to treatment at onset and increase treatment adherence over the course. Furthermore, insight can positively influence symptomatology and recovery, especially when combined with hope.[53] However, it is important to intervene the effects of demoralization and self-stigma.

OPTIMISM AND HOPE

Evidence suggests that levels of optimism, positive emotions, and hope may be positively associated with resilience-related factors such as coping abilities, self-esteem,[33,54–56] and recovery,[57] while oppositional behavior and expression of low optimism might increase risk for psychosis among individuals with psychotic-like experiences.[58] In patients diagnosed with schizophrenia, hope and optimism were correlated with positive outcomes, such as reduced symptoms,[59] higher functioning, higher self-esteem, social support,[54] treatment activity,[60,61] and QoL.[62] Furthermore, a recent cross-sectional study found that scores on a positive psychological index, including perceived stress, happiness, resilience (on the CD-RISC), and optimism were highly heterogeneous among patients diagnosed with schizophrenia (with some patients even displaying "normative" scores) and that the positive psychological index was positively associated with mental and physical health variables.

COPING

Coping styles describe pallets of responses benefiting adjustment to current and future stressors.[63,64] Compared to healthy controls, patients with psychosis displayed increased use of maladaptive (e.g., avoidance) and decreased use of adaptive (e.g., approaching) coping styles,[65] a pattern linked to declined functioning.[66] On a more positive note, however, increased use of adaptive coping styles was positively associated with psychosocial functioning and negatively with symptom severity in PSD.[67–69] On a meta-cognitive level, a positive self-appraisal of one's coping ability might also be beneficial, e.g., by decreasing the association between hopelessness and suicidality.[70]

Effective coping may reflect the ability to use various context-related styles and coping mechanisms that vary along recovery.[71] For instance, a qualitative study found that patients with psychosis, in response to significant life stressors, initially applied avoidance coping styles and eventually more problem-oriented styles.[72] Hereby, several factors may influence the use of coping styles: personality, sleep quality, neurocognition, and symptomatology.[71]

In line with an earlier perspective by Schwarzer and Taubert on coping,[73] researchers[63] theorize that coping is a dynamic process and that individuals use various coping mechanisms (reactive, anticipatory, preventive, and proactive) at different stages of psychosis. Reactive coping, (oriented toward the problems or associated emotions) may be important at psychosis onset, while anticipatory coping (learning to discover warning signs for psychosis) may help a patient develop strategies for relapse. Thus, preventive coping mechanisms may enhance resilience by decreasing vulnerabilities and developing buffers (e.g., increasing social contact and exercising).

Proactive coping is described as going beyond the response to direct stressors or challenges; the individual may engage in "tenacious goal striving" by seeking out new challenges and creating new opportunities.[63,73] At this step, individuals may learn to reappraise the emotional meaning of these stressors. Furthermore, it was argued that patients may regain control, change their view of their "self" and the illness, find hope and purpose, and eventually achieve mastery by reevaluating stressors and the "self."[63] Tait and colleagues[74] likewise suggest that strong and active identities may help individuals develop more adaptive recovery styles.

Few studies examined coping mechanisms and factors that go beyond direct stress responses.[63] Qualitative studies demonstrated that patients emphasized the role of connectedness, hope and optimism, identity, meaning in life, and empowerment for their recovery,[28,31,75] which are also important factors within the context of PTG. Davis and Brekke[76] found that positive reappraisal and intrinsic motivation partially mediated the connection between social support and role-functioning performance. However, at follow-up (6-month) the mediation effect was not found.[76] In another study, it was indicated that motivation was associated with vocational functioning, but not with illness severity or specific symptoms.[77]

Taken together, coping is an integral component of resilience framework, with adaptive coping explaining heterogeneity within clinical populations to some degree. Future research efforts may significantly benefit from examining coping at different stages of psychopathology, with a particular focus on proactive coping mechanism.

RELATIONSHIP AND SOCIAL SUPPORT

Increased social support—probably by means of self-esteem and stress buffer[78]—has been linked to resilience and mental health.[2,79] Social support is associated with better functioning, higher QoL,[80,81] reduced hospitalization, and decreased positive symptoms.[82,83] Perceived and actual support positively impact outcome through aspects such as connectedness, empowerment, and positive identity formation.[84]

Social interactions and support play important roles at different time points. Secure attachment during childhood is protective against the development of mental disorders by establishing positive views of the self and others (embodied mentalization) and thereby shaping future relationships.[85] Furthermore, secure attachment may be related to early help-seeking, better therapeutic alliance, and improved adherence to treatment in individuals with mental disorders.[86,87] On the contrary, childhood adversities are associated with insecure attachment that eventually may lead to more severe positive symptoms in PSD.[88]

Further, social connections and familial support may also encourage patients to seek clinical help and thereby reduce time to treatment substantially.[89–91] It was found that the relationship between employment status and longer duration of untreated psychosis (DUP) was moderated by social contact, with low social contact being associated with longer DUP in unemployed people.[92] Social support also increases medication adherence[93] and accelerates recovery.[94] An 18-month follow-up study showed that perceived emotional support mediated the relationship between family contact as well as network size and time to remission.[95] Finally, family members may help prevent relapses as they learn to attend subthreshold signs and take actions promptly.[96]

Individual characteristics of both patient and caregiver (sex, age, and personality) influence social support and social functioning.[68,71,97] While remitted patients display better social functioning than nonremitters,[98] the personality trait of extraversion was associated with increased help-seeking behavior.[71] Furthermore, caregivers may experience adjustment difficulties, display mental problems, and become overwhelmed by patient interaction. Meanwhile, unhealthy social relations can be detrimental to the recovery process. Perceived stress in caregivers was associated with expressed emotions (EEs)[99]—a family atmosphere characterized by high levels of hostility and criticism (aspects of EE) contributes to frequent relapses.[100,101]

Given the strong positive aspects of social support and relationships, a dynamic approach to increase social functioning is necessary to fully account for individual needs: e.g., more intense support and involvement of caregivers at onset followed by a reciprocal and equal relationship.[102]

NEUROBIOLOGY OF PSYCHOSIS RESILIENCE

STRESS REACTIVITY AND THE HPA-AXIS

The impact of environmental stressors is regulated by and impacts on biological systems, such as the hypothalamus-pituitary-adrenal axis (HPA-axis).[1,103] Similar to psychological responses, biological responses can be adaptive or maladaptive depending on the individual's coping ability, the timing, and the duration.[1,103] Prolonged biological responses may cause long-lasting changes in sequential stress-responses, e.g., sensitization.[103] For instance, childhood adversities and insecure attachment may increase psychosis proneness possibly through HPA-axis and dopamine dysregulations, neuroinflammation, reduced oxytocin, and oxidative stress damage.[85] HPA-axis dysregulations might eventually lead to functional brain changes,[85] likely mediated by epigenetic regulation of gene expression.[104]

HPA-axis dysregulation appears to play an important role at psychosis onset[105] with studies reporting that increased glucocorticoid levels (especially cortisol)[85,106] may impact psychosis proneness by increasing dopaminergic activity.[107,108]

GENETICS AND EPIGENETICS

A few studies have focused on possible protective effects of genes in psychosis. In a genome-wide association study,[109] researchers evaluated genetic modifiers of functioning measured with the short-form health survey (SF-12),[110] while controlling for symptom severity, age, and sex in patients with bipolar disorder, depression, or schizophrenia: (1) genome-wide significant associations between variations in genetic make-up and functional impairment were not found; (2) as symptom severity and covariates could only predict a small portion of functioning,[109] the authors speculate that resilience may relate to functioning, irrespective of severity of symptoms. While no genome-wide significant hit was found, the authors observed that genetic variants in ADAMST16 showed the strongest correlation with physical health–related quality of life.

In contrast to genes conferring risk for PSD,[111] protective genetic markers have not received much attention, and they are often evaluated in relation to environmental factors.[112,113] For instance, the impact of childhood adversities on psychotic-like experiences was found to interact with a *FKBP5* haplotype,[114] the *BDNF* Val66Met,[115] and the COMT Val158Met[88] polymorphisms.

Accumulating evidence, mainly from animal studies, has pinpointed associations between distinct epigenetic profiles and differential susceptibility to the impact of severe stress on behavioral and biochemical phenotypes.[116,117] Epigenetic modifications entail mechanisms (e.g., histones acetylation/methylation and DNA methylation) that influence gene expression through chemical modification of the DNA without altering the DNA sequence.[118] Experimental animal studies have yielded the first wave of evidence proposing epigenetic regulation of distinct genes and biological pathways in mediating differential susceptibility.[116,117] While animal studies have the advantage of directly analyzing epigenetic profiles in the brain, human studies on the impact of severe stress and trauma are limited to epigenetic measurements of accessible tissue types such as blood or saliva.

The first longitudinal epigenetic study on the impact of traumatic stress on mental health in relation to DNA methylation suggested that alterations of mental health phenotypes in response to trauma exposure might indeed be traced in epigenetic profiles of blood samples. This prospective genome-wide study reported that longitudinal changes in the level of post-traumatic stress symptoms 1 month before and 6 months after a deployment period were significantly associated with longitudinal changes in DNA methylation profiles at 17 positions and 12 regions.[119] Another interesting research field for resilience markers is the evaluation of telomere length (i.e., repetitive DNA sequences at the end of a chromosome).[120]

Although growing evidence provides support for investing in research into understanding genetic and epigenetic correlates of stress resilience mechanisms, far less attention has been directed to resilience research in PSD. There is thus an urgent need to investigate resilience in prospective human studies to trace the dynamics of mental functioning over time as well as in response to environmental exposure in PSD.

NEUROIMAGING

Comparing functionally resilient and nonresilient individuals (measured with the modified Global Assessment of Functioning [mGAF] scale[121]) with increased risk for PSD, volume differences in frontal, temporal, and parietal cortex at baseline and throughout development were observed.[122] Resilient and nonresilient individuals also demonstrated dissimilar development of frontal cortical regions (e.g., cingulate gyrus).[122] It is suggested that resilient individuals may display compensatory neuronal mechanisms, with better functioning leading to decreased tissue loss.[122] Furthermore, lower white matter (WM) integrity and decreased volume in corpus callosum (CC) regions were found in a high-risk group compared to a control group.[123,124] At follow-up, increased striatal volume, WM integrity in the CC, and central CC volume was correlated with decreased nonclinical positive[123,125] and negative symptoms[124] in nonconverters, which may signal resilience. Eventually, in a cross-sectional study on patients with schizophrenia, unaffected first-degree relatives, and healthy controls, resilience in unaffected relatives was suggested to be associated with local brain network efficiency and resting-state functional brain connectivity in temporal and subcortical regions.[126]

Patients with childhood onset schizophrenia (COS) and their healthy siblings show common abnormalities (WM growth deficits and cortical thinning).[127] However, resilience might be indicated by siblings who ultimately reach normal levels during late adolescence,[127–129] while the development of psychosis might be related to delayed development of brain areas.[127,128] Progression of psychopathology appears to be associated with an increase in striatal volume: It was found that at 8–9 years old, the striatum volume is similar in COS, controls, and unaffected siblings.[130] However, in comparison to controls, striatum volume increases in COS but not in siblings at later ages. However, both COS and unaffected siblings showed some subregional shape abnormalities.[130]

Eventually, it should be noted that treatment success or relapses may depend on neurobiological variations which may indicate a further level of resilience.[131–133] In this regard, it was argued that treatment focusing on cognitive abilities might be important in early stages of psychopathology where dynamic neurobiological changes occur.[134] Furthermore, it was emphasized that attention needs to be directed to the neurobiology of individual transdiagnostic symptoms in order to achieve advances.[134] In summary, structural and functional abnormalities of the brain appear to be diverse within and between individuals with PSD: some are already present in early stages, others develop with disease progression. However, these findings should be interpreted cautiously given methodological limitations such as small sample sizes and procedural and analytical issues.

RESILIENCE DURING DIFFERENT STAGES OF THE PATHOLOGY

Resilience is a multidimensional and dynamic concept that is inherently linked to both reduced vulnerability for PSD and swift recovery.[1] Furthermore, through PTG, resilience may even enhance the ability to better deal with subsequent struggles.[1] However, the nature of resilience has yet to be completely understood, as it includes many interconnected factors altering resilience in a dynamic way. Nonetheless, looking at the different phases of the pathology, it becomes apparent that some resilience factors play more important roles at the onset of psychopathology, while others are more influential during the course (see Figure 53.1). Similar to earlier work in the broader field of resilience research and mental health,[1] the upper part of Figure 53.1 shows possible theoretical pathways for PSD: the illness course (subclinical symptoms, onset, and recovery). Individuals may experience different courses: For instance, some individuals may have late-onset psychosis and show PTG (nonintermitted line), while others may have an enduring psychopathology (dotted line), or show an onset and recovery with differently severe symptoms (dashed lines).

The lower part of the Figure 53.1 displays social, psychological, and neurobiological resilience factors. Naturally, the factors within each of these levels interact and may act as moderators and mediators to each other. The resilience factors are semichronologically ordered depending on the time point when they emerge and are most influential. However, this should not be interpreted as if those factors are only effective at one time point. For example, stress-resistance genes are logically placed at the left corner of the figure, the chronological beginning. Nonetheless, gene–environment interaction effects can be observed throughout the life span, which may

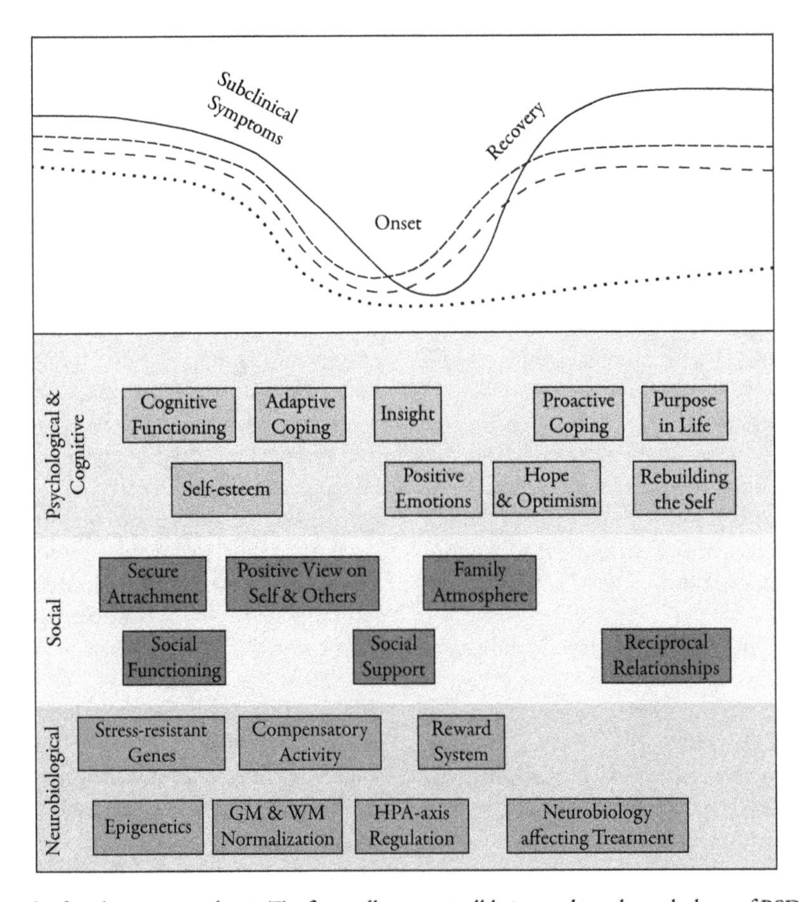

Figure 53.1 A conceptual framework of resilience in psychosis. The figure illustrates well-being and psychopathology of PSD over time. The upper part illustrates individual trajectories. The lower part summarizes the approximate timing of several factors relevant for resilience.

also relate to epigenetic changes. Similarly, although attachment styles are mainly formed in the early development, they influence an individual's resilience status throughout their life (e.g., affecting their view of themselves and others).[85]

Looking at the different phases of the pathology, it is apparent that before the first observed subclinical symptoms, different resilience factors can help individuals to develop strong foundations against current and future stressors. For instance, the development of good social and cognitive functioning,[68,135] as well as secure attachment styles[87] and self-esteem[136] may reduce vulnerability to mental disorders. Likewise, neurobiological background (e.g., HPA-axis regulation, GM and WM normalization) and environmental factors (e.g., secure attachment, self-esteem, coping mechanisms, and adaptation skills) may help an individual deal with early childhood adversities and related biological changes. The social environment, especially the family, plays a pivotal role in both reducing the time of untreated psychosis and instilling hope and optimism, which may facilitate remission and recovery.[90,95] Other factors that can impact the course of illness are positive emotions, adaptive coping, social support, and related neurobiological mechanisms (e.g., mesolimbic reward regulations).[1] Eventually, beyond recovery, patients may start using proactive coping mechanisms, finding a new purpose in life and rebuilding their perception of themselves beyond the diagnosis that was attached to them.[63]

It needs to be stressed that the aforementioned resilience factors describe only some of the aspects innate to the phenomenon; and there are different factors that were not discussed due to the scope of this chapter (e.g., demographic aspects such as age, sex, and education level, or life style aspects such as drug abuse and physical activity). Furthermore, it is increasingly accepted that resilience is a dynamic and active phenomenon with complex interactions of a range of factors.[4,5] In a qualitative study, it was indicated that patients with psychosis show resilience by using internal and external resources during different steps along and beyond recovery; acknowledgment of challenge, help-seeking behavior, acceptance, and commitment to face, deal with, and/or accept the challenges, ultimately leading to a personal sense of regaining control over their lives and becoming "resile."[28] Thus, resilience cannot be easily understood by measuring one or a few protective factors at one time point.

In practice, when studying resilience, researchers often use different resilience questionnaires in comparison to protective factors or outcomes.[16] However, resilience questionnaires do not directly assess resilience, but indirectly measure resilience-promoting aspects.[24] This method is not sufficient to assess the full concept of resilience. Another approach involves contrasting resilience to the development of a mental illness within a binary model.[9] Researchers may compare psychopathology versus nonpsychopathology and associate the status with protective variables (e.g., social support). However, inherently, this approach does not provide further information on individual differences along the continuum as a heterogeneous group is combined into one unit.[9] This approach also does not increase our understanding of resilience-enhancing variables: variables associated with the psychopathology will also affect the absence of the psychopathology, i.e., presence versus absence of social support may be a resilience or a risk factor, respectively. Furthermore, it is possible that protective factors differently impact on the different psychosis liability phenotypes, such as positive symptoms, cognitive, and affective symptoms[137]—a notion that cannot be sufficiently appraised by a simple binary model.

A few longitudinal studies investigated the interaction of diverse protective factors in PSD.[13,76,138] However, given the complexity, different approaches such as network analyses and the implementation of the exposome concept may contribute to further advances in the field.[139,140] Furthermore, research so far focused more on the onset rather than the course of psychosis. To improve our conceptual picture of resilience throughout the course of psychosis, future research efforts should also make use of innovative methods such as momentary assessment technologies of experience in the flow of daily life and identify the interactions and networks of experiences as the building blocks of the dynamic structure of resilience.[141]

CONCLUSION

Resilience is a complex dynamic network of interacting factors that include psychological, environmental, and biological aspects. Incorporation of a resilience framework in research efforts—a largely neglected area in PSD thus far—may leverage our efforts to improve the multidimensional PSD concept, develop prevention strategies, reduce the impact of early psychopathology, and improve outcome over the course.

REFERENCES

1. Rutten BP, Hammels C, Geschwind N, et al. Resilience in mental health: linking psychological and neurobiological perspectives. *Acta Psychiatr Scand*. 2013;128(1):3–20.
2. Davydov DM, Stewart R, Ritchie K, Chaudieu I. Resilience and mental health. *Clin Psychol Rev*. 2010;30(5):479–495.
3. Garmezy N. The study of competence in children at risk for severe psychopathology. 1974.
4. Snijders C, Pries LK, Sgammeglia N, et al. Resilience against traumatic stress: current developments and future directions. *Frontiers in Psychiatry*. 2018;9:676.
5. Kalisch R, Baker DG, Basten U, et al. The resilience framework as a strategy to combat stress-related disorders. *Nat Hum Behav*. 2017;1(11):784–790.
6. Huber M, Knottnerus JA, Green L, et al. How should we define health? *BMJ*. 2011;343.
7. van Os J, Guloksuz S, Vijn TW, Hafkenscheid A, Delespaul P. The evidence-based group-level symptom-reduction model as the organizing principle for mental health care: time for change? *World Psychiatry*. 2019;18(1):88–96.
8. Luthar SS, Cicchetti D, Becker B. The construct of resilience: a critical evaluation and guidelines for future work. *Child Dev*. 2000;71(3):543–562.
9. Bonanno GA, Diminich ED. Annual research review: positive adjustment to adversity–trajectories of minimal–impact resilience and emergent resilience. *Journal of Child Psychology and Psychiatry*. 2013;54(4):378–401.
10. Daskalakis NP, Oitzl MS, Schachinger H, Champagne DL, de Kloet ER. Testing the cumulative stress and mismatch hypotheses of psychopathology in a rat model of early-life adversity. *Physiol Behav*. 2012;106(5):707–721.

11. Sapienza JK, Masten AS. Understanding and promoting resilience in children and youth. *Curr Opin Psychiatry*. 2011;24(4):267–273.
12. Tedeschi RG, Calhoun LG. The Posttraumatic Growth Inventory: measuring the positive legacy of trauma. *J Trauma Stress*. 1996;9(3):455–471.
13. Torgalsbøen A-K. Sustaining full recovery in schizophrenia after 15 years: does resilience matter? *Clinical schizophrenia and related psychoses*. 2011;5(4):193–200.
14. Malla A, Payne J. First-episode psychosis: psychopathology, quality of life, and functional outcome. *Schizophr Bull*. 2005;31(3):650–671.
15. Gorwood P, Burns T, Juckel G, et al. Psychiatrists' perceptions of the clinical importance, assessment and management of patient functioning in schizophrenia in Europe, the Middle East and Africa. *Ann Gen Psychiatry*. 2013;12(1):8.
16. Mizuno Y, Hofer A, Suzuki T, et al. Clinical and biological correlates of resilience in patients with schizophrenia and bipolar disorder: a cross-sectional study. *Schizophr Res*. 2016;175(1-3):148–153.
17. Gooding PA, Littlewood D, Owen R, Johnson J, Tarrier N. Psychological resilience in people experiencing schizophrenia and suicidal thoughts and behaviours. *J Ment Health*. 2017:1–7.
18. Isaacs K, Mota NP, Tsai J, et al. Psychological resilience in U.S. military veterans: a 2-year, nationally representative prospective cohort study. *J Psychiatr Res*. 2017;84:301–309.
19. Shi J, Wang L, Yao Y, et al. Protective factors in Chinese university students at clinical high risk for psychosis. *Psychiat Res*. 2016;239:239–244.
20. Friborg O, Hjemdal O, Rosenvinge JH, Martinussen M. A new rating scale for adult resilience: what are the central protective resources behind healthy adjustment? *International Journal of Methods in Psychiatric Research*. 2003;12(2):65–76.
21. Connor KM, Davidson JR. Development of a new resilience scale: the Connor-Davidson Resilience Scale (CD-RISC). *Depress Anxiety*. 2003;18(2):76–82.
22. Smith BW, Dalen J, Wiggins K, Tooley E, Christopher P, Bernard J. The brief resilience scale: assessing the ability to bounce back. *Int J Behav Med*. 2008;15(3):194–200.
23. Wagnild GM, Young HM. Development and psychometric evaluation of the Resilience Scale. *J Nurs Meas*. 1993;1(2):165–178.
24. Windle G, Bennett KM, Noyes J. A methodological review of resilience measurement scales. *Health Qual Life Outcomes*. 2011;9(1):8.
25. Cosco TD, Kaushal A, Richards M, Kuh D, Stafford M. Resilience measurement in later life: a systematic review and psychometric analysis. *Health Qual Life Outcomes*. 2016;14(1):16.
26. Ahern NR, Kiehl EM, Sole ML, Byers J. A review of instruments measuring resilience. *Issues Compr Pediatr Nurs*. 2006;29(2):103–125.
27. Waaktaar T, Torgersen S. How resilient are resilience scales? The Big Five scales outperform resilience scales in predicting adjustment in adolescents. *Scandinavian Journal of Psychology*. 2010;51(2):157–163.
28. Henderson AR, Cock A. The responses of young people to their experiences of first-episode psychosis: harnessing resilience. *Community Ment Health J*. 2015;51(3):322–328.
29. Campbell MM, Sibeko G, Mall S, et al. The content of delusions in a sample of South African Xhosa people with schizophrenia. *BMC Psychiatry*. 2017;17(1):41.
30. Ruddle A. Qualitative evaluation of a cognitive behaviour therapy Hearing Voices Group with a service user co-facilitator. *Psychosis*. 2016;9(1):25–37.
31. Leamy M, Bird V, Le Boutillier C, Williams J, Slade M. Conceptual framework for personal recovery in mental health: systematic review and narrative synthesis. *Br J Psychiatry*. 2011;199(6):445–452.
32. Orth U, Robins RW, Widaman KF. Life-span development of self-esteem and its effects on important life outcomes. *J Pers Soc Psychol*. 2012;102(6):1271–1288.
33. Taylor SE, Stanton AL. Coping resources, coping processes, and mental health. *Annu Rev Clin Psychol*. 2007;3:377–401.
34. Pruessner M, Iyer SN, Faridi K, Joober R, Malla AK. Stress and protective factors in individuals at ultra-high risk for psychosis, first episode psychosis and healthy controls. *Schizophr Res*. 2011;129(1):29–35.
35. Wartelsteiner F, Mizuno Y, Frajo-Apor B, et al. Quality of life in stabilized patients with schizophrenia is mainly associated with resilience and self-esteem. *Acta Psychiatrica Scandinavica*. 2016;134(4):360–367.
36. Smith B, Fowler DG, Freeman D, et al. Emotion and psychosis: links between depression, self-esteem, negative schematic beliefs and delusions and hallucinations. *Schizophr Res*. 2006;86(1-3):181–188.
37. Law H, Shryane N, Bentall RP, Morrison AP. Longitudinal predictors of subjective recovery in psychosis. *Br J Psychiatry*. 2016;209(1):48–53.
38. Sibitz I, Unger A, Woppmann A, Zidek T, Amering M. Stigma resistance in patients with schizophrenia. *Schizophr Bull*. 2011;37(2):316–323.
39. Holding JC, Tarrier N, Gregg L, Barrowclough C. Self-esteem and relapse in schizophrenia: a 5-year follow-up study. *J Nerv Ment Dis*. 2013;201(8):653–658.
40. van Zelst C, van Nierop M, Oorschot M, et al. Stereotype awareness, self-esteem and psychopathology in people with psychosis. *PLoS One*. 2014;9(2):e88586.
41. Amador XF, Gorman JM. Psychopathologic domains and insight in schizophrenia. *Psychiatr Clin North Am*. 1998;21(1):27–42.
42. De Hert M, van Winkel R, Wampers M, Kane J, van Os J, Peuskens J. Remission criteria for schizophrenia: evaluation in a large naturalistic cohort. *Schizophr Res*. 2007;92(1-3):68–73.
43. Eberhard J, Levander S, Lindstrom E. Remission in schizophrenia: analysis in a naturalistic setting. *Compr Psychiatry*. 2009;50(3):200–208.
44. Chong CS, Siu MW, Kwan CH, et al. Predictors of functioning in people suffering from first-episode psychosis 1 year into entering early intervention service in Hong Kong. *Early Interv Psychiatry*. 2018;12(5):828–838.
45. Lincoln TM, Lullmann E, Rief W. Correlates and long-term consequences of poor insight in patients with schizophrenia. A systematic review. *Schizophr Bull*. 2007;33(6):1324–1342.
46. Velligan DI, Sajatovic M, Hatch A, Kramata P, Docherty JP. Why do psychiatric patients stop antipsychotic medication? A systematic review of reasons for nonadherence to medication in patients with serious mental illness. 2017;11:449.
47. Lysaker PH, Pattison ML, Leonhardt BL, Phelps S, Vohs JL. Insight in schizophrenia spectrum disorders: relationship with behavior, mood and perceived quality of life, underlying causes and emerging treatments. *World Psychiatry*. 2018;17(1):12–23.
48. Mintz AR, Dobson KS, Romney DM. Insight in schizophrenia: a meta-analysis. *Schizophr Res*. 2003;61(1):75–88.
49. Clarke DM, Kissane DW. Demoralization: its phenomenology and importance. *Aust N Z J Psychiatry*. 2002;36(6):733–742.
50. Hasson-Ohayon I, Kravetz S, Meir T, Rozencwaig S. Insight into severe mental illness, hope, and quality of life of persons with schizophrenia and schizoaffective disorders. *Psychiat Res*. 2009;167(3):231–238.
51. Drake RJ, Pickles A, Bentall RP, et al. The evolution of insight, paranoia and depression during early schizophrenia. *Psychol Med*. 2004;34(2):285–292.
52. Restifo K, Harkavy-Friedman JM, Shrout PE. Suicidal behavior in schizophrenia: a test of the demoralization hypothesis. *J Nerv Ment Dis*. 2009;197(3):147–153.
53. Lysaker PH, Campbell K, Johannesen JK. Hope, awareness of illness, and coping in schizophrenia spectrum disorders: evidence of an interaction. *J Nerv Ment Dis*. 2005;193(5):287–292.
54. Lecomte T, Corbière M, Théroux L. Correlates and predictors of optimism in individuals with early psychosis or severe mental illness. *Psychosis*. 2010;2(2):122–133.
55. Schrank B, Stanghellini G, Slade M. Hope in psychiatry: a review of the literature. *Acta Psychiatr Scand*. 2008;118(6):421–433.
56. Tugade MM, Fredrickson BL. Resilient individuals use positive emotions to bounce back from negative emotional experiences. *J Pers Soc Psychol*. 2004;86(2):320–333.
57. Resnick SG, Fontana A, Lehman AF, Rosenheck RA. An empirical conceptualization of the recovery orientation. *Schizophr Res*. 2005;75(1):119–128.

58. Dolphin L, Dooley B, Fitzgerald A. Prevalence and correlates of psychotic like experiences in a nationally representative community sample of adolescents in Ireland. *Schizophr Res.* 2015;169(1–3):241–247.
59. Lysaker PH, Salyers MP, Tsai J, Spurrier LY, Davis LW. Clinical and psychological correlates of two domains of hopelessness in schizophrenia. *J Rehabil Res Dev.* 2008;45(6):911–919.
60. Kukla M, Salyers MP, Lysaker PH. Levels of patient activation among adults with schizophrenia: associations with hope, symptoms, medication adherence, and recovery attitudes. *J Nerv Ment Dis.* 2013;201(4):339–344.
61. Oles SK, Fukui S, Rand KL, Salyers MP. The relationship between hope and patient activation in consumers with schizophrenia: results from longitudinal analyses. *Psychiat Res.* 2015;228(3):272–276.
62. Mashiach-Eizenberg M, Hasson-Ohayon I, Yanos PT, Lysaker PH, Roe D. Internalized stigma and quality of life among persons with severe mental illness: the mediating roles of self-esteem and hope. *Psychiat Res.* 2013;208(1):15–20.
63. Roe D, Yanos PT, Lysaker PH. Coping with psychosis: an integrative developmental framework. *J Nerv Ment Dis.* 2006;194(12):917–924.
64. Lazarus RS. Psychological stress and the coping process. 1966.
65. Horan WP, Ventura J, Mintz J, et al. Stress and coping responses to a natural disaster in people with schizophrenia. *Psychiat Res.* 2007;151(1–2):77–86.
66. Yanos PT, Moos RH. Determinants of functioning and well-being among individuals with schizophrenia: an integrated model. *Clin Psychol Rev.* 2007;27(1):58–77.
67. Kim KR, Song YY, Park JY, et al. The relationship between psychosocial functioning and resilience and negative symptoms in individuals at ultra-high risk for psychosis. *Aust N Z J Psychiatry.* 2013;47(8):762–771.
68. Rossi A, Galderisi S, Rocca P, et al. The relationships of personal resources with symptom severity and psychosocial functioning in persons with schizophrenia: results from the Italian Network for Research on Psychoses study. *Eur Arch Psychiatry Clin Neurosci.* 2017;267(4):285–294.
69. Mazor Y, Gelkopf M, Roe D. Posttraumatic growth among people with serious mental illness, psychosis and posttraumatic stress symptoms. *Compr Psychiatry.* 2018;81:1–9.
70. Johnson J, Gooding PA, Wood AM, Taylor PJ, Pratt D, Tarrier N. Resilience to suicidal ideation in psychosis: Positive self-appraisals buffer the impact of hopelessness. *Behav Res Ther.* 2010;48(9):883–889.
71. Phillips LJ, Francey SM, Edwards J, McMurray N. Strategies used by psychotic individuals to cope with life stress and symptoms of illness: a systematic review. *Anxiety Stress Coping.* 2009;22(4):371–410.
72. Robilotta S, Cueto E, Yanos PT. An examination of stress and coping among adults diagnosed with severe mental illness. *Israel Journal of Psychiatry and Related Sciences.* 2010;47(3):222–231.
73. Schwarzer R, Taubert S. Tenacious goal pursuits and striving toward personal growth: Proactive coping. *Beyond Coping: Meeting Goals, Visions and Challenges.* 2002:19–35.
74. Tait L, Birchwood M, Trower P. Adapting to the challenge of psychosis: personal resilience and the use of sealing-over (avoidant) coping strategies. *Br J Psychiatry.* 2004;185(5):410–415.
75. Pitt L, Kilbride M, Nothard S, Welford M, Morrison AP. Researching recovery from psychosis: a user-led project. *The Psychiatrist.* 2007;31(2):55–60.
76. Davis L, Brekke J. Social support and functional outcome in severe mental illness: the mediating role of proactive coping. *Psychiat Res.* 2014;215(1):39–45.
77. Fervaha G, Takeuchi H, Foussias G, Hahn MK, Agid O, Remington G. Achievement motivation in early schizophrenia: relationship with symptoms, cognition and functional outcome. *Early Interv Psychiatry.* 2018;12(6):1038–1044.
78. Brugha TS. Social Support. In: Morgan C, Bhugra D, Morgan C, Bhugra D, eds. *Principles of Social Psychiatry.* Wiley-Blackwell; 2010:461–476.
79. Cohen S, Wills TA. Stress, social support, and the buffering hypothesis. *Psychol Bull.* 1985;98(2):310–357.
80. Sanchez J, Rosenthal DA, Tansey TN, Frain MP, Bezyak JL. Predicting quality of life in adults with severe mental illness: extending the International Classification of Functioning, Disability, and Health. *Rehabil Psychol.* 2016;61(1):19–31.
81. Howard L, Leese M, Thornicroft G. Social networks and functional status in patients with psychosis. *Acta Psychiatr Scand.* 2000;102(5):376–385.
82. Norman RM, Malla AK, Manchanda R, Harricharan R, Takhar J, Northcott S. Social support and three-year symptom and admission outcomes for first episode psychosis. *Schizophr Res.* 2005;80(2-3):227–234.
83. Crush E, Arseneault L, Moffitt TE, et al. Protective factors for psychotic experiences amongst adolescents exposed to multiple forms of victimization. *J Psychiatr Res.* 2018;104:32–38.
84. Tew J, Ramon S, Slade M, Bird V, Melton J, Le Boutillier C. Social factors and recovery from mental health difficulties: a review of the evidence. *British Journal of Social Work.* 2011;42(3):443–460.
85. Debbane M, Salaminios G, Luyten P, et al. Attachment, neurobiology, and mentalizing along the psychosis continuum. *Front Hum Neurosci.* 2016;10:406.
86. Berry K, Barrowclough C, Wearden A. A review of the role of adult attachment style in psychosis: unexplored issues and questions for further research. *Clin Psychol Rev.* 2007;27(4):458–475.
87. Gumley AI, Taylor HE, Schwannauer M, MacBeth A. A systematic review of attachment and psychosis: measurement, construct validity and outcomes. *Acta Psychiatr Scand.* 2014;129(4):257–274.
88. Misiak B, Krefft M, Bielawski T, Moustafa AA, Sasiadek MM, Frydecka D. Toward a unified theory of childhood trauma and psychosis: a comprehensive review of epidemiological, clinical, neuropsychological and biological findings. *Neurosci Biobehav Rev.* 2017;75:393–406.
89. Gayer-Anderson C, Morgan C. Social networks, support and early psychosis: a systematic review. *Epidemiol Psychiatr Sci.* 2013;22(2):131–146.
90. Morgan C, Abdul-Al R, Lappin JM, et al. Clinical and social determinants of duration of untreated psychosis in the AESOP first-episode psychosis study. *Br J Psychiatry.* 2006;189(5):446–452.
91. Norman RM, Malla AK, Verdi MB, Hassall LD, Fazekas C. Understanding delay in treatment for first-episode psychosis. *Psychol Med.* 2004;34(2):255–266.
92. Reininghaus UA, Morgan C, Simpson J, et al. Unemployment, social isolation, achievement-expectation mismatch and psychosis: findings from the AESOP Study. *Soc Psychiatry Psychiatr Epidemiol.* 2008;43(9):743–751.
93. Ramirez Garcia JI, Chang CL, Young JS, Lopez SR, Jenkins JH. Family support predicts psychiatric medication usage among Mexican American individuals with schizophrenia. *Soc Psychiatry Psychiatr Epidemiol.* 2006;41(8):624–631.
94. Soundy A, Stubbs B, Roskell C, Williams SE, Fox A, Vancampfort D. Identifying the facilitators and processes which influence recovery in individuals with schizophrenia: a systematic review and thematic synthesis. *J Ment Health.* 2015;24(2):103–110.
95. Tempier R, Balbuena L, Lepnurm M, Craig TK. Perceived emotional support in remission: results from an 18-month follow-up of patients with early episode psychosis. *Soc Psychiatry Psychiatr Epidemiol.* 2013;48(12):1897–1904.
96. Herz MI, Lamberti JS, Mintz J, et al. A program for relapse prevention in schizophrenia: a controlled study. *Arch Gen Psychiatry.* 2000;57(3):277–283.
97. Thorup A, Albert N, Bertelsen M, et al. Gender differences in first-episode psychosis at 5-year follow-up–two different courses of disease? Results from the OPUS study at 5-year follow-up. *European Psychiatry.* 2014;29(1):44–51.
98. Brissos S, Dias VV, Balanza-Martinez V, Carita AI, Figueira ML. Symptomatic remission in schizophrenia patients: relationship with social functioning, quality of life, and neurocognitive performance. *Schizophr Res.* 2011;129(2–3):133–136.
99. Sadath A, Muralidhar D, Varambally S, Gangadhar B, Jose JP. Do stress and support matter for caring? The role of perceived stress and social support on expressed emotion of carers of persons with first episode psychosis. *Asian Journal of Psychiatry.* 2017;25:163–168.

100. Caqueo-Urizar A, Miranda-Castillo C, Lemos Giraldez S, Lee Maturana SL, Ramirez Perez M, Mascayano Tapia F. An updated review on burden on caregivers of schizophrenia patients. *Psicothema.* 2014;26(2):235–243.
101. Alvarez-Jimenez M, Priede A, Hetrick SE, et al. Risk factors for relapse following treatment for first episode psychosis: a systematic review and meta-analysis of longitudinal studies. *Schizophr Res.* 2012;139(1-3):116–128.
102. Topor A, Borg M, Mezzina R, Sells D, Marin I, Davidson L. Others: the role of family, friends, and professionals in the recovery process. *Archives of Andrology.* 2006;9(1):17–37.
103. de Kloet ER, Joels M, Holsboer F. Stress and the brain: from adaptation to disease. *Nat Rev Neurosci.* 2005;6(6):463–475.
104. McGowan PO, Sasaki A, D'Alessio AC, et al. Epigenetic regulation of the glucocorticoid receptor in human brain associates with childhood abuse. *Nat Neurosci.* 2009;12(3):342–348.
105. Corcoran C, Walker E, Huot R, et al. The stress cascade and schizophrenia: etiology and onset. *Schizophr Bull.* 2003;29(4):671–692.
106. Woodberry KA, Shapiro DI, Bryant C, Seidman LJ. Progress and future directions in research on the psychosis prodrome: a review for clinicians. *Harv Rev Psychiatry.* 2016;24(2):87–103.
107. Read J, Perry BD, Moskowitz A, Connolly J. The contribution of early traumatic events to schizophrenia in some patients: a traumagenic neurodevelopmental model. *Psychiatry: Interpersonal and Biological Processes.* 2001;64(4):319–345.
108. Walker EF, Diforio D. Schizophrenia: a neural diathesis-stress model. *Psychol Rev.* 1997;104(4):667–685.
109. McGrath LM, Cornelis MC, Lee PH, et al. Genetic predictors of risk and resilience in psychiatric disorders: a cross-disorder genome-wide association study of functional impairment in major depressive disorder, bipolar disorder, and schizophrenia. *Am J Med Genet B Neuropsychiatr Genet.* 2013;162B(8):779–788.
110. Ware J, Kosinski M, Keller S. SF-12 physical and mental health summary scales. Boston, MA: New England Medical Center. *The Health Institute.* 1995.
111. Ripke S, Neale B, Corvin A, et al. Biological insights from 108 schizophrenia-associated genetic loci. *Nature.* 2014;511(7510):421–427.
112. Wermter A-K, Laucht M, Schimmelmann BG, et al. From nature versus nurture, via nature and nurture, to gene× environment interaction in mental disorders. *European Child and Adolescent Psychiatry.* 2010;19(3):199–210.
113. van Os J, Rutten BP, Poulton R. Gene-environment interactions in schizophrenia: review of epidemiological findings and future directions. *Schizophr Bull.* 2008;34(6):1066–1082.
114. Cristobal-Narvaez P, Sheinbaum T, Rosa A, et al. The interaction between childhood bullying and the FKBP5 gene on psychotic-like experiences and stress reactivity in real life. *PLoS One.* 2016;11(7):e0158809.
115. de Castro-Catala M, van Nierop M, Barrantes-Vidal N, et al. Childhood trauma, BDNF Val66Met and subclinical psychotic experiences: attempt at replication in two independent samples. *Journal of Psychiatric Research.* 2016;83:121–129.
116. Feder A, Nestler EJ, Charney DS. Psychobiology and molecular genetics of resilience. *Nat Rev Neurosci.* 2009;10(6):446–457.
117. Russo SJ, Murrough JW, Han MH, Charney DS, Nestler EJ. Neurobiology of resilience. *Nat Neurosci.* 2012;15(11):1475–1484.
118. Jaenisch R, Bird A. Epigenetic regulation of gene expression: how the genome integrates intrinsic and environmental signals. *Nat Genet.* 2003;33:245–254.
119. Rutten BPF, Vermetten E, Vinkers CH, et al. Longitudinal analyses of the DNA methylome in deployed military servicemen identify susceptibility loci for post-traumatic stress disorder. *Mol Psychiatry.* 2018;23(5):1145–1156.
120. Çevik B, Mançe-Çalışır Ö, Atbaşoğlu EC, et al. Psychometric liability to psychosis and childhood adversities are associated with shorter telomere length: A study on schizophrenia patients, unaffected siblings, and non-clinical controls. 2019;111:169–185.
121. Hall RCW. Global assessment of functioning. *Psychosomatics.* 1995;36(3):267–275.
122. de Wit S, Wierenga LM, Oranje B, et al. Brain development in adolescents at ultra-high risk for psychosis: longitudinal changes related to resilience. *NeuroImage: Clinical.* 2016;12:542–549.
123. Katagiri N, Pantelis C, Nemoto T, et al. A longitudinal study investigating sub-threshold symptoms and white matter changes in individuals with an "at risk mental state"(ARMS). *Schizophrenia Research.* 2015;162(1):7–13.
124. Katagiri N, Pantelis C, Nemoto T, et al. Symptom recovery and relationship to structure of corpus callosum in individuals with an "at risk mental state." *Psychiatry Research Neuroimaging.* 2018;272:1–6.
125. Katagiri N, Pantelis C, Nemoto T, et al. Longitudinal changes in striatum and sub-threshold positive symptoms in individuals with an "at risk mental state" (ARMS). 2019;285:25–30.
126. Ganella EP, Seguin C, Bartholomeusz CF, et al. Risk and resilience brain networks in treatment-resistant schizophrenia. *Schizophr Res.* 2018;193:284–292.
127. Ordonez AE, Luscher ZI, Gogtay N. Neuroimaging findings from childhood onset schizophrenia patients and their non-psychotic siblings. *Schizophr Res.* 2016;173(3):124–131.
128. Zalesky A, Pantelis C, Cropley V, et al. Delayed development of brain connectivity in adolescents with schizophrenia and their unaffected siblings. *JAMA Psychiatry.* 2015;72(9):900–908.
129. Gogtay N, Hua X, Stidd R, et al. Delayed white matter growth trajectory in young nonpsychotic siblings of patients with childhood-onset schizophrenia. *Arch Gen Psychiatry.* 2012;69(9):875–884.
130. Chakravarty MM, Rapoport JL, Giedd JN, et al. Striatal shape abnormalities as novel neurodevelopmental endophenotypes in schizophrenia: a longitudinal study. *Hum Brain Mapp.* 2015;36(4):1458–1469.
131. Palaniyappan L, Marques TR, Taylor H, et al. Globally efficient brain organization and treatment response in psychosis: a connectomic study of gyrification. *Schizophr Bull.* 2016;42(6):1446–1456.
132. Remington G, Foussias G, Agid O, Fervaha G, Takeuchi H, Hahn M. The neurobiology of relapse in schizophrenia. *Schizophr Res.* 2014;152(2–3):381–390.
133. Mason L, Peters E, Williams SC, Kumari V. Brain connectivity changes occurring following cognitive behavioural therapy for psychosis predict long-term recovery. *Translational Psychiatry.* 2017;7(1):e1001.
134. Bartholomeusz CF, Cropley VL, Wannan C, Di Biase M, McGorry PD, Pantelis C. Structural neuroimaging across early-stage psychosis: Aberrations in neurobiological trajectories and implications for the staging model. *Aust N Z J Psychiatry.* 2017;51(5):455–476.
135. Gur RC, Calkins ME, Satterthwaite TD, et al. Neurocognitive growth charting in psychosis spectrum youths. *JAMA Psychiatry.* 2014;71(4):366–374.
136. Thoits PA. Self, identity, stress, and mental health. In: *Handbook of the sociology of mental health.* Springer; 2013:357–377.
137. Juckel G, Schaub D, Fuchs N, et al. Validation of the Personal and Social Performance (PSP) scale in a German sample of acutely ill patients with schizophrenia. *Schizophr Res.* 2008;104(1–3):287–293.
138. Harrow M, Jobe TH. Factors involved in outcome and recovery in schizophrenia patients not on antipsychotic medications: a 15-year multifollow-up study. *Journal of Nervous and Mental Disease.* 2007;195(5):406–414.
139. Guloksuz S, Pries LK, van Os J. Application of network methods for understanding mental disorders: pitfalls and promise. *Psychol Med.* 2017;47(16):2743–2752.
140. Guloksuz S, van Os J, Rutten BPF. The exposome paradigm and the complexities of environmental research in psychiatry. *JAMA Psychiatry.* 2018;75(10):985–986.
141. Verhagen SJ, Hasmi L, Drukker M, van Os J, Delespaul PA. Use of the experience sampling method in the context of clinical trials. *Evid Based Ment Health.* 2016;19(3):86–89.

SECTION 5

TREATMENT OF PSYCHOTIC DISORDERS

PHARMACOLOGICAL TREATMENTS

54

PHARMACOLOGICAL APPROACHES TO TREATMENT

Stefan Leucht, Andrea Cipriani, and Toshi A. Furukawa

INTRODUCTION

In this chapter we aim to cover treatment aspects of psychosis. The overarching concept is that if it were appropriate to speak about different dimensions of psychiatric disorders, psychosis being one of them, then antipsychotic drugs should be efficacious for all psychotic symptoms, whether they occur in schizophrenia, depression, mania, or any other psychotic disorder. The importance of this approach is that many psychiatrists believe that antipsychotics are primarily the treatment for schizophrenia, that antidepressants are the main option for depression, and that mood stabilizers are the main treatment for acute mania, even if patients with the latter two diagnoses present with psychotic symptoms. We therefore address the efficacy of antipsychotics in various psychotic disorders defined in this book.

SCHIZOPHRENIA

Schizophrenia is the prototypal psychotic disorder on which most evidence for the efficacy of antipsychotic drugs is available. Discussion of this disorder therefore constitutes the largest part of the chapter. We first discuss the efficacy of antipsychotics for the acute treatment of schizophrenia, and then we present the efficacy of antipsychotics for relapse prevention.

ANTIPSYCHOTIC DRUGS VERSUS PLACEBO—ACUTE TREATMENT

Efficacy on overall symptoms

In a meta-analysis that summarized all placebo-controlled antipsychotic drug trials since the introduction of chlorpromazine in the 1950s (105 studies [N] with 22,741 participants), the mean effect size for overall symptoms (mainly positive and negative symptom scale [PANSS] and Brief psychiatric rating scale [BPRS] total score) of all studies combined was 0.47 (95% CrI 0.42,0.51) (1). This translates into a difference of 9.6 PANSS points between drug and placebo, which corresponds to less than minimally improved (15 PANSS points) in analyses linking the PANSS/BPRS with the Clinical Global Inventory (CGI) in linking analyses (2, 3). However, as effect sizes are difficult to interpret, the authors also conducted an analysis of responder rates. Here, 51% of the antipsychotic-treated patients compared to 30% on placebo had at least a "minimal" response (N = 46, n = 8,918, relative risk [RR] 1.75 [1.59,1.97]), while 23% versus 14% had a "good" response (N = 38, n = 8,403, RR 1.96[1.65,2.44]) (1). While these data clearly show that antipsychotics are effective, the numbers are not very impressive. However, an analysis of response predictors revealed that effect sizes have decreased over the decades. On the average, a study published in 1970 would have had an effect size of 0.74, while the average effect size in 2015 was 0.38 (1). This decline of efficacy also becomes clear when comparing the overall responder rates found in the meta-analysis with those found in classical studies published in the 1960s. Cole 1964 reported the results of a National Institute of Mental Health (NIMH) study in which 61% of patients under drug vs. 22% under placebo showed marked or moderate improvement. In this study approximately 50% of patients experienced their first episode (4), while none of the studies in the meta-analysis that provided data on scale derived overall symptoms was a first-episode study and on the average the participants were rather chronic, but Rabinowitz et al. (5) showed that chronic patients respond worse to antipsychotics than first episode patients. Various analyses have shown that an increase of placebo-response over the years is a major factor explaining the decreasing drug-placebo differences (6–8).

Efficacy for positive symptoms and other symptom domains

Leucht et al. (1) also analyzed the effects of antipsychotics compared to placebo in terms of specific symptoms. The most important result in the context of this volume on psychosis was that the effect size for positive symptoms was similar to that of overall symptoms (N = 64, n = 18,174, standardized mean difference [SMD] 0.45 [0.40,0.50]), while the effect sizes for other domains were lower. More specifically, antipsychotics also significantly reduced negative symptoms and depressive symptoms associated with schizophrenia, but the mean standardized mean differences was only 0.35 (N = 69, n = 18,632, SMD 0.35[0.31,0.40]) for negative symptoms and 0.27 (N = 33, n = 9,658, SMD = 0.27 [0.20,0.34]) for depression (see Figure 54.1). This means that—as already their name suggests—the primary target of antipsychotics is still psychosis. It needs to be noted that all currently marketed antipsychotic drugs are dopamine antagonists or partial agonists,

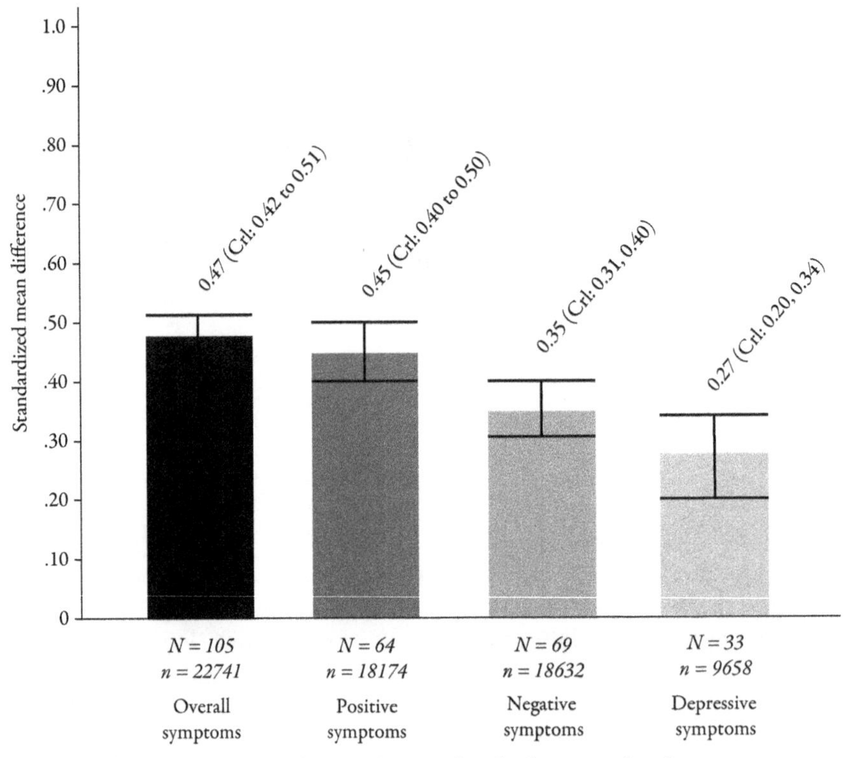

Figure 54.1 Example of an individualized case formulation of persecutory delusions.

making the role of dopamine blockade in this context the most likely mechanism of action.

Whether antipsychotic drugs also have a true, intrinsic effect on negative symptoms and depression is debated and cannot be answered based on the trials included by Leucht et al. (1) This meta-analysis focused on patients with acute exacerbations of schizophrenia who had positive symptoms. In such populations any effects on negative symptoms and depression can just be secondary to those of positive symptoms. Imagine patients who suffer a lot from hallucinations and delusions, such patients often withdraw from social relationships and do not interact much with others. This behavior can then look like negative symptoms, but they are secondary to the positive symptoms, and if positive symptoms are effectively treated, negative symptoms "secondarily" improve as well. Similar effects on depressive symptoms may occur if antipsychotic drugs reduce the stress associated with psychosis. Therefore, to understand whether some antipsychotics are also effective for primary negative symptoms, studies in relevant populations must be conducted. Ideally, these are patients who have prevailing, stable negative symptoms after treatment, but no or only a low degree of positive symptoms. During the literature search conducted for the aforementioned meta-analysis we found only a few studies (9–14) which overall found a superiority of the antipsychotics analyzed compared to placebo (N = 7, n = 817, SMD 0.42 [0.14,0.71], I2 56%). This result was mainly driven by five studies on low doses of amisulpride (9–12), a selective D2/D3 dopamine antagonist and 5-HT-7 X antagonist (Figure 54.2). Nevertheless, Figure 54.2 shows that there were also significant effects on depressive symptoms, again questioning the specificity of the findings. The best study from a methodological point of view of patients with predominant negative symptoms conducted to date might therefore be a recent large trial (n = 461) showing a superiority of the partial dopamine agonist cariprazine over risperidone in this population. In this trial, sponsored by the manufacturers of cariprazine, secondary effects due to reduction of positive symptoms, depressive symptoms, and extrapyramidal side effects had been well controlled for (15). That amisulpride and cariprazine are the drugs with the best documented efficacy for predominant negative symptoms was confirmed by a recent meta-analysis (16).

Finally, antipsychotic drugs seem to have a positive effect on quality of life (QOL; N = 6, n = 1,900, SMD 0.35 [0.16,0.51]) and social functioning compared to placebo (N = 10, n = 3,077, SMD 0.34 [0.21,0.47]) (1). As the included studies were short-term, improvement in these domains appear to arise already quite early on. But the low number of trials included must also be noted. These low numbers could raise suspicion of publication bias. On the other hand, it has become customary to measure QOL and functioning in short-term, placebo-controlled trials only recently, explaining the low number of trials currently available. There is a movement to emphasize these domains much more than previously. The rationale is that in the end QOL and functioning may be more important outcomes for patients than the mere reduction of positive symptoms. Some patients may be able to cope with a certain degree of positive symptoms, and QOL may present a proxy composite of the efficacy and side effects of antipsychotics.

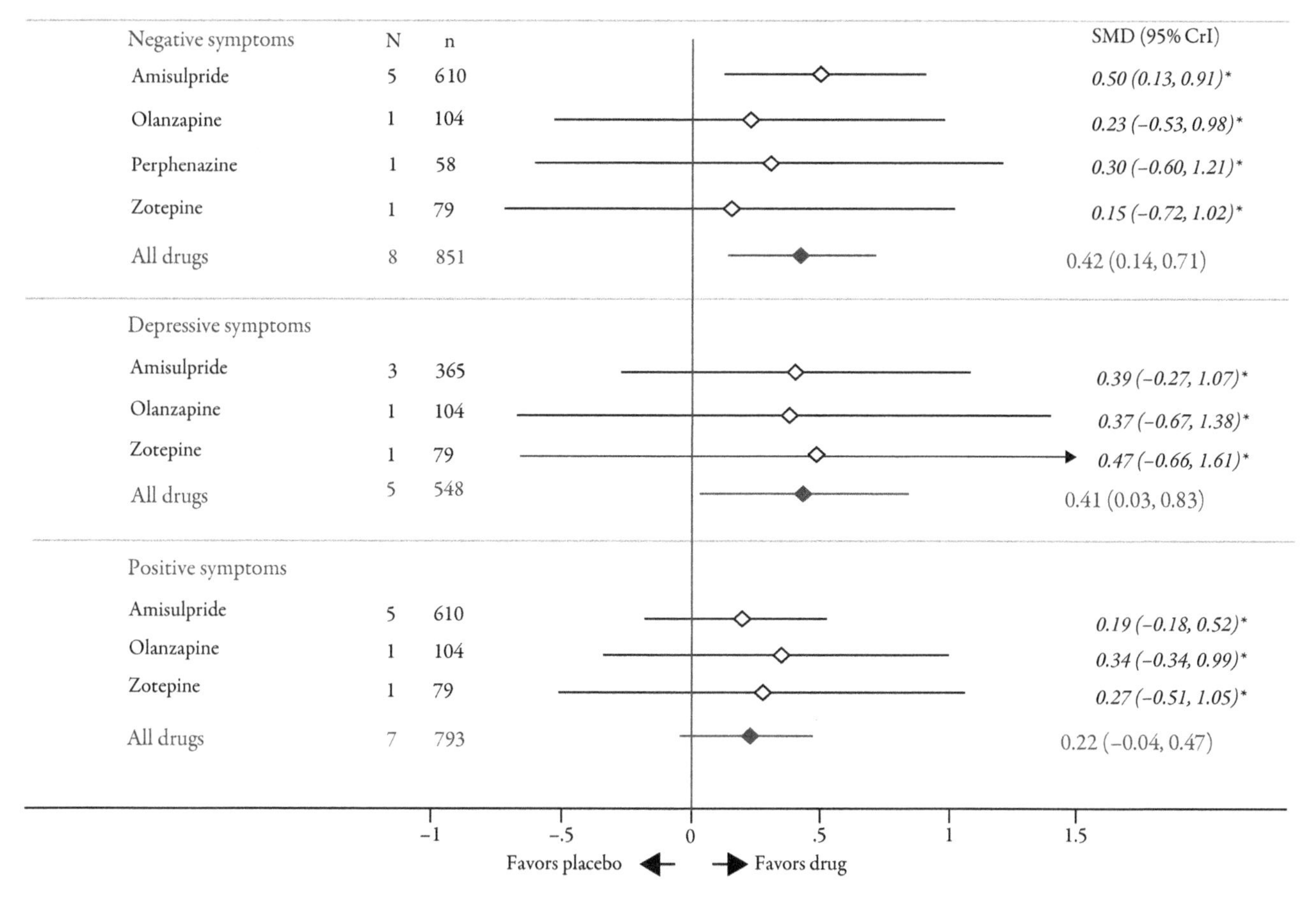

Figure 54.2 Vicious circles of voice maintenance.

Efficacy differences between compounds

It had for a long time been a psychopharmacological dogma that there are no efficacy differences between antipsychotics, except clozapine, which has been considered a superior drug since a pivotal trial compared to chlorpromazine published in 1988 (17). Indeed, two meta-analyses did not find convincing efficacy differences between the standard first-generation antipsychotics chlorpromazine and haloperidol compared to other first-generation antipsychotics (18, 19). However, the authors argued that the evidence was underpowered, because few randomized controlled trials (RCTs) and participants were available for each comparison. In contrast, a network meta-analysis composed by 212 RCTs with 43,049 found significant differences between some second-generation antipsychotics, haloperidol and chlorpromazine. These findings have been criticized in the light of the increasing placebo response described earlier, because the most recently introduced antipsychotics were also the least efficacious ones. But a sensitivity analysis in which placebo-controlled trials were excluded yielded similar results (20). The superiority of clozapine has recently been challenged by a meta-analysis of randomized, blinded trials in treatment resistant patients which did not find an efficacy superiority compared to other second-generation antipsychotics, clozapine was only superior to first-generation antipsychotics (22). However, a conventional pairwise meta-analysis by Siskind et al. (23) contradicted this result and found clozapine superior. Methodological issues have been discussed, but a reanalysis appears to be necessary to resolve the conflict (21).

Side effects

A detailed discussion of the side effects of antipsychotic drugs would go beyond the scope of this chapter on the efficacy of antipsychotics. Similar hierarchies as those for general efficacy have been established for weight gain, extrapyramidal side effects (with use of antiparkinsonian medication at least once in the studies as a proxy), prolactin increase, QTc prolongation, and sedation (20). But this side-effect analysis only included the second-generation antipsychotics available in 2013, haloperidol and chlorpromazine. Furthermore, there are many more side effects such as specific movement disorders (e.g., dystonias, akathisia, tardive dyskinesia), hyper- and hypotension, reduction of white blood cell count (in particular, agranulocytosis associated with clozapine), sexual side effects, delirium, and others. Some of these side effects can be severe and are the major downside of antipsychotic drugs.

Other pharmacological or biological treatments effective as monotherapies for schizophrenia

Antipsychotic drugs are currently the only treatment effective in monotherapy for schizophrenia, except eventually ECT in very old studies such as the classical one by May et al. (24), and this efficacy was confirmed by a Cochrane review (25). Multiple pharmacological augmentation strategies have been tested in RCTs over the last decades, and Correll et al. (26) summarized these findings in an overview of 29 meta-analyses testing 42 combination strategies. Due to quality concerns about the included studies rather than the about the quality of the meta-analyses, the authors concluded, "no single strategy can be recommended for patients with schizophrenia based on the current meta-analytic literature" (26). In terms of psychotherapeutic interventions, specific CBT programs have been developed and tested in multiple RCTs. Added to antipsychotics, CBT has been found effective for alleviation of psychosis (27). A trial suggested that CBT might also work as a monotherapy in people with schizophrenia who were not ready to take antipsychotic drugs, but a replication is needed (28).

ANTIPSYCHOTIC DRUGS VERSUS PLACEBO—MAINTENANCE TREATMENT

Overall efficacy

The major aim of maintenance treatment with antipsychotics of schizophrenia is the prevention of relapse. In a meta-analysis of all randomized, placebo-controlled antipsychotic drug trials in stable patients (65 RCTs with 6,493 participants), 64% of placebo-treated patients compared to 27% of drug treated patients relapsed within 7–12 months (29). In an overview of meta-analyses of commonly used medications, this was one of the largest differences between drug and placebo in medicine (30). As relapse is a relatively "soft" outcome, because it is usually measured by subjective rating scales, the authors also analyzed the number of patients who had to be rehospitalized and found that this was the case for 10% of the drug-treated patients compared to 26% of the placebo-treated patients (29). Rehospitalization is a "harder" outcome that has major cost implications, because in many countries the cost of hospitalization is much higher than that of medication. While antipsychotics reduced both the percentage of patients relapsed and the number of patients rehospitalized by approximately 60%, these numbers also suggest that not every relapse is so severe that it requires a hospitalization. Despite the established efficacy of antipsychotics for relapse prevention, there is a debate for how long antipsychotics should be prescribed. This debate goes beyond the scope of this chapter with a focus on efficacy, but it is driven by the side effects of antipsychotics including supersensitivity effects and a possible dose-related brain volume loss.

Efficacy in subgroups with a potentially better prognosis—first-episode patients and patients in remission

It is quite well established that up to 20% of patients with a first episode of schizophrenia will not have a second one in the next 5 years (31). Therefore, these patients may have a better prognosis. In the meta-analysis the reduction of relapse risk of first-episode patients was, however, similar to that of patients with more than one episode (29). Zipursky et al. (32) even estimated that after 2 years without medication 90% of first-episode patients would relapse, in contrast to only 3% of those on drugs. From this perspective, first-episode patients benefit as much from antipsychotic relapse prevention as multiple episode patients.

Efficacy differences between antipsychotic drugs in relapse prevention

A debate exists around whether there are efficacy differences between antipsychotic drugs in relapse prevention and this debate revolves around two questions. The first question is whether there are efficacy differences between drugs in relapse prevention. One meta-analysis compared first-generation with second-generation antipsychotics and found the latter—as a group—to be more efficacious, while there were few efficacy superiorities of individual second-generation antipsychotics compared to first-generation ones (33). In a meta-analysis comparing only second-generation antipsychotics Kishimoto et al. reported that no consistent superiority of any second-generation antipsychotic across efficacy and tolerability outcomes emerged (34). The second question is whether long-acting injectable (LAI) depot antipsychotics are superior to oral antipsychotics, because of safeguarded compliance. While the most up-to-date meta-analyses of RCTs found no difference between LAIs and oral formulations (35, 36), a superiority has been demonstrated in meta-analyses of mirror-image studies and of observational studies (37, 38). The hypothesis is that those patients who consent to RCTs are compliant per se, diluting the compliance superiority of LAIs. In contrast, mirror-image studies maybe biased due to expectancy effects, and in observational studies it is difficult to control for potential confounders.

SCHIZOAFFECTIVE DISORDER

The diagnostic criteria of schizoaffective disorder have changed several times over the past and diagnostic systems such as ICD-10 and DSM-V differ. Therefore, relatively few placebo-controlled trials are available that are restricted to schizoaffective disorder. Paliperidone was more efficacious than placebo in two trials (39, 40), aripiprazole outperformed placebo in a post hoc analysis of the schizoaffective subgroup of two studies (41), and paliperidone palmitate was superior to placebo in relapse prevention (42). Other studies found olanzapine to be superior to haloperidol (43); and risperidone was superior to haloperidol for depressive symptoms in some analyses (44). As schizoaffective disorder lies between schizophrenia and mood disorders, it can be debated whether afflicted patients should be treated with classical mood-stabilizers such as lithium, or antipsychotics or a combination of both. As several new antipsychotics have demonstrated mood-stabilizing properties, this question has been rarely addressed in the literature in recent years.

DELUSIONAL DISORDER

Delusional disorder is traditionally considered to be difficult to treat. A Cochrane review (45) found only one RCT comparing CBT with supportive therapy in 17 patients. Munro and Mok (46) presented a review of 209 case reports published since the 1960s. The average improvement rate in this literature of mostly case reports (limiting interpretability) was approximately 50%. Currently, more patients with delusional disorder are now treated with second-generation antipsychotics. Munoz-Negro and Cervilla (47) published a more recent review of 385 cases with delusional disorder and found a good response to antipsychotic treatment in 33.6% of the cases. The response rate to first-generation antipsychotics was higher (39%) than that of second-generation antipsychotics (28%), but this was an indirect comparison, without a controlled trial. It is possible that these response rates are exaggerations of the true antipsychotic effects, because clinicians tend to publish successful case reports rather than unsuccessful ones. The literature remains limited on the use of antipsychotic drugs to treat delusional disorders.

BRIEF PSYCHOTIC DISORDER

Brief psychotic disorder is a heterogeneous group of psychotic conditions that are of short duration, by definition. Their prognosis seems to be overall better than that of a first manifestation of schizophrenia (48), but relapses and diagnostic changes are frequent (49). Controlled studies are not available, therefore truly evidence-based statements on its treatment are not possible. In practice, treatment is usually symptom oriented and includes antipsychotics but also sedating agents such as benzodiazepines or mood-stabilizers. Whether antipsychotics should be used longer-term needs to weigh the by-definition self-limiting nature of the disorder on the one hand and the frequent relapses on the other (49).

PSYCHOTIC MANIA

Mania defines bipolar I disorder. DSM-5 (50) has modified DSM-IV criteria for mania by adding the requirement for increased activity/energy as a core symptom of mood elevation. This represents an effort to reduce the overdiagnosis of bipolar disorder driven by subjective report, and increases specificity (51). Psychotic symptoms in mania are described as mood congruent and represent an extension of grandiose interpretations, paranoid ideation, or heightened awareness. They are relatively common (52), however, in a study of 576 patients with acute mania, only 20% had a presentation dominated by psychosis (53). Presence of psychotic features may influence outcome but is also poorly characterized in relation to treatment response. The relative efficacy and acceptability of the treatments for mania have been analyzed using network meta-analysis, generally considered the highest level of evidence in treatment guidelines (54). All the antipsychotics showed superiority to placebo in terms of efficacy and the available evidence strongly supports the validity of the overall recommendation to use dopamine antagonist/partial agonists in mania. The individual rankings of drugs showed considerable overlap in confidence intervals, however, if efficacy and acceptability are considered together, risperidone and olanzapine should probably be considered the first-line treatment. Clozapine may also play a role by extrapolation from its superiority in schizophrenia (20) and limited observational data in treatment-resistant mania (55, 56). Many factors influence the choice of drug, including properties such as sedation (usually desirable in the short term but not in the long term), and the route of administration. The availability of parenteral formulations is valuable in emergencies and should form part of any local protocol for treating highly agitated patients (57). In the past, patients were habitually treated with high doses of antipsychotics, but this can produce marked extrapyramidal symptoms, which should be avoided, even when managing an emergency (51). Psychotic mania can be difficult to treat, so a combination of medications is often necessary. Although continuation of atypical antipsychotic adjunctive therapy after mania remission reduces relapse of mood episodes, the optimal duration is unknown. A 52-week double-blind placebo-controlled discontinuation trial recruited 159 patients with bipolar I disorder who recently remitted from a manic episode during treatment with risperidone or olanzapine adjunctive therapy to lithium or valproate (58). Patients were randomized to one of three conditions: discontinuation of risperidone or olanzapine and substitution with placebo at entry or at 24 weeks after entry or continuation of risperidone or olanzapine for the full duration of the study (52 weeks). The primary outcome measure was time to relapse of any mood episode. Risperidone or olanzapine adjunctive therapy for 24 weeks was beneficial, but continuation of risperidone beyond this period did not reduce the risk of relapse. Whether continuation of olanzapine beyond this period can reduce relapse risk remained unclear, but the potential benefit needs to be weighed against an increased risk of weight gain (average weight gain was 3.2 kg over 52 weeks).

PSYCHOTIC DEPRESSION

Major depression with psychotic features, called psychotic depression, is a distinct subtype of major depression (50). A general population survey reported that 18.5% of major depressive episodes are accompanied by psychotic features (59).

The treatment of choice recommended in the majority of the international guidelines is a combination of antidepressant (AD) and antipsychotic (AP) drugs (denoted here as COMB). The UK-NICE guideline is cautious in its wording by including the following: "For people who have depression with psychotic symptoms, consider augmenting the current treatment plan with antipsychotic medication (although the optimum dose and duration of treatment are unknown)." (60) They even add a caveat "There is no good quality evidence for pharmacological treatments of psychotic depression. However, there are practical problems in recruiting sufficient numbers of patients with psychotic depression and, therefore, practitioners may wish to consider lower levels of evidence." The Dutch guideline is also another exception, as they consider "starting with an AD and adding an AP if the patients does not respond" a reasonable option (61).

The body of evidence informing the treatment of psychotic depression has evolved in the past decade. A high-quality systematic review published in 2006 (62) identified 10 trials, with two trials (total $n = 77$) contributing to the COMB (AD + AP) vs. AD comparison and three (total $n = 188$) to the COMB (AD + AP) vs. AP comparison. The responses in this study showed that the RR was 1.44 (95%CI: 0.86 to 2.41) for the comparison between AD + AP vs AD only comparison, and was 1.92 (1.32 to 2.80) for the comparison of the AD + AP combination vs the AP alone. The authors therefore concluded that the best treatment for psychotic depression was AD monotherapy with the addition of AP for poor response, or starting with COMB (AD + AP); both regimens appear to be appropriate options, measured by the balance between risks and benefits.

Since then, the same group of researchers conducted their own trial and updated the systematic review in 2013 (63). They were now able to include five studies (total $n = 245$) for the comparison AD + AP vs. AD and four studies (total $n = 447$) for the comparison AD + AP vs. AP. Both comparisons now reached statistical significance, with RR of 1.42 (1.11 to 1.80) and 1.83 (1.40 to 2.38), respectively, for the clinical response. The authors now concluded that this evidence indicates that AD + AP is more effective than either treatment alone for psychotic depression. However, the literature is not convincing in this area and further research is needed.

Specifically, first, the preceding summary concerns the global response. Only a subset of the available studies examined the effects of COMB or AD or AP in terms of depression or psychosis, separately. Farahani et al. (64) report that COMB did not beat AD in terms of depression (SMD = -0.20, -0.44 to 0.03; 5 studies, total $n = 324$) or psychosis (SMD = -0.24, -0.85 to 0.38; 3 studies, total $n = 161$), whereas COMB beat AP in terms of depression (SMD=-0.49, -0.75 to -0.23; 4 studies, total $n = 428$) but not psychosis (SMD = -0.35, -0.70 to 0.01, 4 studies, total $n = 429$).

Second, the influence of mood congruence of the psychotic symptoms has not been studied. It is intuitively possible that patients with mood-congruent psychotic features may only need AD, while mood-incongruent psychotic features may necessitate the use of AP. However, there is no randomized evidence to support this suspicion. Rather, the available data suggest that the distinctions between mood-congruent and mood-incongruent psychotic features may not be as robust as suggested in the official diagnostic criteria (65, 66).

Third, there is no evidence to suggest how long the once-initiated COMB need be continued for a major depressive episode with psychotic features, or which drug should be withdrawn first. Studies of psychotic depression may offer a paradigm to examine the specificity of AD or AP for depressive or psychotic symptoms, whereas the current dimensional concept of mental illness might be inadequate for psychotic depression.

ORGANIC PSYCHOSES

Evidence for the treatment of most organic causes of psychoses is mainly based on case reports. However, psychosis associated with Parkinson's disease, dementia and delirium are examples for which meta-analyses of randomized-controlled trials have been published. Parkinson's disease psychosis can be caused by the disorder itself or by its treatment with prodopaminergic agents, which increase dopamine levels in the brain (67). Therefore, dose reduction of the antiparkinsonian drug must be considered. The trade-off in adding antipsychotics is that these patients are particularly vulnerable for extrapyramidal side effects. Therefore, second-generation antipsychotics are more appropriate than high-potency first-generation antipsychotics, but the former also differ in their risk for EPS (20). In terms of individual antipsychotics, a meta-analysis found that the randomized evidence is only positive for clozapine and pimavanserin, which according to these authors should be the treatment of choice (68).

The mainstays of pharmacological treatment of Alzheimer's disease are acetylcholinesterase inhibitors and memantine. Many studies have examined whether antipsychotic drugs are effective in reducing behavioral disturbances and psychosis associated with dementia, psychosis being the question in the context of this chapter. Most trial programs on the various second-generation antipsychotics did not provide proof of efficacy (69), and due to increased risk for cerebrovascular events and increased mortality (70) antipsychotics have received a black-boxed warning for the treatment of dementia from the Food and Drug Administration (FDA). According to the German dementia guideline, risperidone is the drug with the best evidence for the treatment of psychosis (71).

The circumstances under which a delirium can develop are multiple and therefore treating the underlying cause is the primary treatment goal. According to the current version (2010) of the National Institute of Clinical Excellence (NICE) guideline, verbal and nonverbal techniques to de-escalate the situation should be tried before pharmacological interventions are considered for the treatment of delirium (72). NICE recommends the short-term use of either haloperidol or olanzapine, if these are not effective and there if a person is distressed or considered a risk to themselves or others (72). However, a more recent guideline on the treatment of delirium in intensive care gave a low-quality recommendation that "atypical antipsychotics may reduce delirium duration," but that there is no evidence for the efficacy of haloperidol (73). Recent systematic reviews have also called the efficacy of haloperidol, and of antipsychotic drugs in general into question (74).

CONCLUSIONS

There is a remarkable imbalance as to how well the pharmacological treatment of the various psychotic disorders has been addressed by the literature, with schizophrenia having the most decisive literature. While hundreds of RCTs on the pharmacological treatment of schizophrenia are available, schizoaffective disorder, psychotic depression, and psychotic mania are far less well studied, and for delusional disorder, brief psychotic disorder, and most organic psychoses (with few exceptions such as dementia, Parkinson's disease, and delirium) the evidence seems to be mainly based only on case reports.

REFERENCES

1. Leucht S, Leucht C, Huhn M, Chaimani A, Mavridis D, Helfer B, Samara M, Rabaioli M, Bacher S, Cipriani A, Geddes JR, Salanti G, Davis JM. Sixty years of placebo-controlled antipsychotic drug trials in acute schizophrenia: systematic review, Bayesian meta-analysis, and meta-regression of efficacy predictors. Am J Psychiatry. 2017:appiajp201716121358.
2. Leucht S, Kane JM, Kissling W, Hamann J, Etschel E, Engel RR. Clinical implications of BPRS scores. Br J Psychiatry. 2005;187:363–371.
3. Leucht S, Kane JM, Kissling W, Hamann J, Etschel E, Engel RR. What does the PANSS mean? Schizophr Res. 2005;79:231–238.
4. Cole JO. Phenothiazine treatment in acute schizophrenia. Arch Gen Psychiatry. 1964;10:246–261.
5. Rabinowitz J, Werbeloff N, Caers I, Mandel FS, Stauffer V, Menard F, Kinon BJ, Kapur S. Determinants of antipsychotic response in schizophrenia: implications for practice and future clinical trials. J Clin Psychiatry. 2014;75:e308–316.
6. Agid O, Siu CO, Potkin SG, Kapur S, Watsky E, Vanderburg D, Zipursky RB, Remington G. Meta-regression analysis of placebo response in antipsychotic trials, 1970–2010. Am J Psychiatry. 2013;170:1335–1344.
7. Rutherford BR, Pott E, Tandler JM, Wall MM, Roose SP, Lieberman JA. Placebo response in antipsychotic clinical trials: a meta-analysis. JAMA Psychiatry. 2014;71:1409–1421.
8. Leucht S, Chaimani A, Leucht C, Huhn M, Mavridis D, Helfer B, Samara M, Cipriani A, Geddes JR, Salanti G, Davis JM. 60years of placebo-controlled antipsychotic drug trials in acute schizophrenia: meta-regression of predictors of placebo response. Schizophr Res. 2018;201:315–323.
9. Boyer P, Lecrubier Y, Puech AJ, Dewailly J, Aubin F. Treatment of negative symptoms in schizophrenia with amisulpride. Br J Psychiatry. 1995;166:68–72.
10. Lecrubier Y, Quintin P, Bouhassira M, Perrin E, Lancrenon S. The treatment of negative symptoms and deficit states of chronic schizophrenia: olanzapine compared to amisulpride and placebo in a 6-month double-blind controlled clinical trial. Acta Psychiatr Scand. 2006;114:319–327.
11. Danion JM, Rein W, Fleurot O. Improvement of schizophrenic patients with primary negative symptoms treated with amisulpride. Am J Psychiatry. 1999;156:610–616.
12. Loo H, Poirier-Littre MF, Theron M, Rein W, Fleurot O. Amisulpride versus placebo in the medium-term treatment of the negative symptoms of schizophrenia. Br J Psychiatry. 1997;170:18–22.
13. Collins AD, Dundas J. A double-blind trial of amitriptyline-perphenazine, perphenazine and placebo in chronic withdrawn inert schizophrenics. British Journal of Psychiatry. 1967;113:1425–1429.
14. Möller HJ, Riedel M, Müller N, Fischer W, Kohnen R. Zotepine versus placebo in the treatment of schizophrenic patients with stable primary negative symptoms: a randomized double-blind multicenter trial. Pharmacopsychiatry. 2004;37:270–278.
15. Nemeth G, Laszlovszky I, Czobor P, Szalai E, Szatmari B, Harsanyi J, Barabassy A, Debelle M, Durgam S, Bitter I, Marder S, Fleischhacker WW. Cariprazine versus risperidone monotherapy for treatment of predominant negative symptoms in patients with schizophrenia: a randomised, double-blind, controlled trial. Lancet. 2017;389:1103–1113.
16. Krause M, Zhu Y, Huhn M, Schneider-Thoma J, Bighelli I, Nikolakopoulou A, Leucht S. Antipsychotic drugs for patients with schizophrenia and predominant or prominent negative symptoms: a systematic review and meta-analysis. Eur Arch Psychiatry Clin Neurosci. 2018;268:625–639.
17. Kane JM, Honigfeld G, Singer J, Meltzer H, group at CCs. Clozapine for the treatment-resistant schizophrenic: a double-blind comparison with chlorpromazine. Archives of General Psychiatry. 1988;45:789–796.
18. Samara MT, Cao H, Helfer B, Davis JM, Leucht S. Chlorpromazine versus every other antipsychotic for schizophrenia: a systematic review and meta-analysis challenging the dogma of equal efficacy of antipsychotic drugs. European Neuropsychopharmacology. 2014;24:1046–1055.
19. Dold M, Tardy M, Samara MT, Li C, Kasper S, Leucht S. Are all first-generation antipsychotics equally effective in treating schizophrenia? A meta-analysis of randomised, haloperidol-controlled trials. World Journal of Biological Psychiatry. 2016;17:210–220.
20. Leucht S, Cipriani A, Spineli L, Mavridis D, Orey D, Richter F, Samara M, Barbui C, Engel RR, Geddes JR, Kissling W, Stapf MP, Lassig B, Salanti G, Davis JM. Comparative efficacy and tolerability of 15 antipsychotic drugs in schizophrenia: a multiple-treatments meta-analysis. Lancet. 2013;382:951–962.
21. Samara M, Leucht S. Clozapine in treatment-resistant schizophrenia. Br J Psychiatry. 2017;210:299.
22. Samara MT, Leucht C, Leeflang MM, Anghelescu IG, Chung YC, Crespo-Facorro B, Elkis H, Hatta K, Giegling I, Kane JM, Kayo M, Lambert M, Lin CH, Moller HJ, Pelayo-Teran JM, Riedel M, Rujescu D, Schimmelmann BG, Serretti A, Correll CU, Leucht S. Early improvement as a predictor of later response to antipsychotics in schizophrenia: a diagnostic test review. Am J Psychiatry. 2015;172:617–629.
23. Siskind D, McCartney L, Goldschlager R, Kisely S. Clozapine v. first- and second-generation antipsychotics in treatment-refractory schizophrenia: systematic review and meta-analysis. Br J Psychiatry. 2016.
24. May PR, Tuma AH, Yale C, Potepan P, Dixon WJ. Schizophrenia: a follow-up study of results of treatment. Arch Gen Psychiatry. 1976;33:481–486.
25. Tharyan P, Adams CE. Electroconvulsive therapy for schizophrenia. Cochrane Database Syst Rev. 2005:CD000076.
26. Correll CU, Rubio JM, Inczedy-Farkas G, Birnbaum ML, Kane JM, Leucht S. Efficacy of 42 pharmacologic cotreatment strategies added to antipsychotic monotherapy in schizophrenia: systematic overview and quality appraisal of the meta-analytic evidence. JAMA Psychiatry. 2017;74:675–684.
27. Bighelli I, Salanti G, Huhn M, Schneider-Thoma J, Krause M, Reitmeir C, Wallis S, Schwermann F, Pitschel-Walz G, Barbui C, Furukawa TA, Leucht S. Psychological interventions to reduce positive symptoms in schizophrenia: systematic review and network meta-analysis. World Psychiatry. 2018;17:316–329.
28. Morrison AP, Dunn G, Turkington D, Pyle M, Hutton P. Cognitive therapy for patients with schizophrenia—authors' reply. Lancet. 2014;384:401–402.
29. Leucht S, Tardy M, Komossa K, Heres S, Kissling W, Salanti G, Davis JM. Antipsychotic drugs versus placebo for relapse prevention in schizophrenia: a systematic review and meta-analysis. Lancet. 2012;379:2063–2071.
30. Leucht S, Hierl S, Dold M, Kissling W, Davis JM. Putting the efficacy of psychiatric and general medicine medication in perspective: a review of meta-analyses. Br J Psychiatry. 2012;200:97–106.
31. Robinson D, Woerner MG, Alvir JM, Bilder R, Goldman R, Geisler S, Koreen A, Sheitman B, Chakos M, Mayerhoff D, Lieberman JA. Predictors of relapse following response from a first episode of schizophrenia or schizoaffective disorder. Arch Gen Psychiatry. 1999;56:241–247.
32. Zipursky RB, Menezes NM, Streiner DL. Risk of symptom recurrence with medication discontinuation in first-episode psychosis: a systematic review. Schizophr Res. 2014;152:408–414.
33. Kishimoto T, Agarwal V, Kishi T, Leucht S, Kane JM, Correll CU. Relapse prevention in schizophrenia: a systematic review and

meta-analysis of second-generation antipsychotics versus first-generation antipsychotics. Mol Psychiatry. 2013;18:53–66.
34. Kishimoto T, Hagi K, Nitta M, Kane JM, Correll CU. Long-term effectiveness of oral second-generation antipsychotics in patients with schizophrenia and related disorders: a systematic review and meta-analysis of direct head-to-head comparisons. World Psychiatry. 2019;18:208–224.
35. Kishimoto T, Robenzadeh A, Leucht C, Leucht S, Watanabe K, Mimura M, Borenstein M, Kane JM, Correll CU. Long-acting injectable vs oral antipsychotics for relapse prevention in schizophrenia: a meta-analysis of randomized trials. Schizophr Bull. 2014;40:192–213.
36. Ostuzzi G, Bighelli I, So R, Furukawa TA, Barbui C. Does formulation matter? A systematic review and meta-analysis of oral versus long-acting antipsychotic studies. Schizophr Res. 2017;183:10–21.
37. Kishimoto T, Nitta M, Borenstein M, Kane JM, Correll CU. Long-acting injectable versus oral antipsychotics in schizophrenia: a systematic review and meta-analysis of mirror-image studies. J Clin Psychiatry. 2013;74:957–965.
38. Kirson NY, Weiden PJ, Yermakov S, Huang W, Samuelson T, Offord SJ, Greenberg PE, Wong BJ. Efficacy and effectiveness of depot versus oral antipsychotics in schizophrenia: synthesizing results across different research designs. J Clin Psychiatry. 2013;74:568–575.
39. Canuso CM, Schooler N, Carothers J, Turkoz I, Kosik-Gonzalez C, Bossie CA, Walling D, Lindenmayer JP. Paliperidone extended-release in schizoaffective disorder: a randomized, controlled study comparing a flexible dose with placebo in patients treated with and without antidepressants and/or mood stabilizers. J Clin Psychopharmacol. 2010;30:487–495.
40. Canuso CM, Lindenmayer JP, Kosik-Gonzalez C, Turkoz I, Carothers J, Bossie CA, Schooler NR. A randomized, double-blind, placebo-controlled study of 2 dose ranges of paliperidone extended-release in the treatment of subjects with schizoaffective disorder. J Clin Psychiatry. 2010;71:587–598.
41. Glick ID, Mankoski R, Eudicone JM, Marcus RN, Tran QV, Assuncao-Talbott S. The efficacy, safety, and tolerability of aripiprazole for the treatment of schizoaffective disorder: results from a pooled analysis of a sub-population of subjects from two randomized, double-blind, placebo-controlled, pivotal trials. Journal of Affective Disorders. 2009;115:18–26.
42. Fu DJ, Turkoz I, Simonson RB, Walling DP, Schooler NR, Lindenmayer JP, Canuso CM, Alphs L. Paliperidone palmitate once-monthly reduces risk of relapse of psychotic, depressive, and manic symptoms and maintains functioning in a double-blind, randomized study of schizoaffective disorder. J Clin Psychiatry. 2015;76:253–262.
43. Tran PV, Tollefson GD, Sanger TM, Lu Y, Berg PH, Beasley CM, Jr. Olanzapine versus haloperidol in the treatment of schizoaffective disorder: acute and long-term therapy. Br J Psychiatry. 1999;174:15–22.
44. Janicak PG, Keck PE, Davis JM, Kasckow JW, Tugrul K, Dowd SM, Strong J, Sharma RP, Strakowski SM. A double-blind, randomized, prospective evaluation of the efficacy and safety of risperidone versus haloperidol in the treatment of schizoaffective disorder. J Clin Psychopharmacol. 2001;21:360–368.
45. Skelton M, Khokhar WA, Thacker SP. Treatments for delusional disorder. Cochrane Database Syst Rev. 2015:CD009785.
46. Munro A, Mok H. An overview of treatment in paranoia/delusional disorder. Canadian Journal of Psychiatry Revue Canadienne de Psychiatrie. 1995;40:616–622.
47. Munoz-Negro JE, Cervilla JA. A systematic review on the pharmacological treatment of delusional disorder. J Clin Psychopharmacol. 2016;36:684–690.
48. Fusar-Poli P, Cappucciati M, Bonoldi I, Hui LM, Rutigliano G, Stahl DR, Borgwardt S, Politi P, Mishara AL, Lawrie SM, Carpenter WT, Jr., McGuire PK. Prognosis of brief psychotic episodes: a meta-analysis. JAMA Psychiatry. 2016;73:211–220.
49. Jager M, Riedel M, Moller HJ. [Acute and transient psychotic disorders (ICD-10: F23). Empirical data and implications for therapy]. Nervenarzt. 2007;78:745–746, 749–752.
50. American Psychiatric Association: Diagnostic and Statistical Manual of Mental Disorders, Fifth Edition. Arlington, VA, American Psychiatric Association; 2013.
51. Goodwin GM, Haddad PM, Ferrier IN, Aronson JK, Barnes T, Cipriani A, Coghill DR, Fazel S, Geddes JR, Grunze H, Holmes EA, Howes O, Hudson S, Hunt N, Jones I, Macmillan IC, McAllister-Williams H, Miklowitz DR, Morriss R, Munafo M, Paton C, Saharkian BJ, Saunders K, Sinclair J, Taylor D, Vieta E, Young AH. Evidence-based guidelines for treating bipolar disorder: Revised third edition recommendations from the British Association for Psychopharmacology. J Psychopharmacol. 2016;30:495–553.
52. Dunayevich E, Keck PE, Jr. Prevalence and description of psychotic features in bipolar mania. Current Psychiatry Reports. 2000;2:286–290.
53. Sato T, Bottlender R, Kleindienst N, Moller HJ. Syndromes and phenomenological subtypes underlying acute mania: a factor analytic study of 576 manic patients. Am J Psychiatry. 2002;159:968–974.
54. Leucht S, Chaimani A, Cipriani AS, Davis JM, Furukawa TA, Salanti G. Network meta-analyses should be the highest level of evidence in treatment guidelines. Eur Arch Psychiatry Clin Neurosci. 2016;266:477–480.
55. Green AI, Tohen M, Patel JK, Banov M, DuRand C, Berman I, Chang H, Zarate C, Jr., Posener J, Lee H, Dawson R, Richards C, Cole JO, Schatzberg AF. Clozapine in the treatment of refractory psychotic mania. Am J Psychiatry. 2000;157:982–986.
56. Li XB, Tang YL, Wang CY, de Leon J. Clozapine for treatment-resistant bipolar disorder: a systematic review. Bipolar Disorders. 2015;17:235–247.
57. Wilson MP, MacDonald K, Vilke GM, Feifel D. A comparison of the safety of olanzapine and haloperidol in combination with benzodiazepines in emergency department patients with acute agitation. Journal of Emergency Medicine. 2012;43:790–797.
58. Yatham LN, Beaulieu S, Schaffer A, Kauer-Sant'Anna M, Kapczinski F, Lafer B, Sharma V, Parikh SV, Daigneault A, Qian H, Bond DJ, Silverstone PH, Walji N, Milev R, Baruch P, da Cunha A, Quevedo J, Dias R, Kunz M, Young LT, Lam RW, Wong H. Optimal duration of risperidone or olanzapine adjunctive therapy to mood stabilizer following remission of a manic episode: A CANMAT randomized double-blind trial. Mol Psychiatry. 2016;21:1050–1056.
59. Ohayon MM, Schatzberg AF. Prevalence of depressive episodes with psychotic features in the general population. Am J Psychiatry. 2002;159:1855–1861.
60. NICE: Depression: the treatment and management of depression in adults (partial update of NICE clinical guideline 23). London, National Institute for Clinical Excellence; 2009.
61. Wijkstra J, Schubart CD, Nolen WA. Treatment of unipolar psychotic depression: the use of evidence in practice guidelines. The World Journal of Biological Psychiatry. 2009;10:409–415.
62. Wijkstra J, Lijmer J, Balk FJ, Geddes JR, Nolen WA. Pharmacological treatment for unipolar psychotic depression: systematic review and meta-analysis. Br J Psychiatry. 2006;188:410–415.
63. Wijkstra J, Lijmer J, Burger H, Cipriani A, Geddes J, Nolen WA. Pharmacological treatment for psychotic depression. Cochrane Database Syst Rev. 2015:Cd004044.
64. Farahani A, Correll CU. Are antipsychotics or antidepressants needed for psychotic depression? A systematic review and meta-analysis of trials comparing antidepressant or antipsychotic monotherapy with combination treatment. J Clin Psychiatry. 2012;73: 486–496.
65. Burch EA, Jr., Anton RF, Carson WH. Mood congruent and incongruent psychotic depressions: are they the same? Journal of Affective Disorders. 1994;31:275–280.
66. Fennig S, Fennig-Naisberg S, Karant M. Mood-congruent vs. mood-incongruent psychotic symptoms in affective psychotic disorders. The Israel Journal of Psychiatry and Related Sciences. 1996;33: 238–245.
67. Combs BL, Cox AG. Update on the treatment of Parkinson's disease psychosis: role of pimavanserin. Neuropsychiatric Disease and Treatment. 2017;13:737–744.

68. Wilby KJ, Johnson EG, Johnson HE, Ensom MHH. Evidence-based review of pharmacotherapy used for Parkinson's disease psychosis. Ann Pharmacother. 2017;51:682–695.
69. Schneider LS, Dagerman K, Insel PS. Efficacy and adverse effects of atypical antipsychotics for dementia: meta-analysis of randomized, placebo-controlled trials. American Journal of Geriatric Psychiatry. 2006;14:191–210.
70. Schneider LS, Dagerman KS, Insel P. Risk of death with atypical antipsychotic drug treatment for dementia: meta-analysis of randomized placebo-controlled trials. JAMA. 2005;294: 1934–1943.
71. Deuschl G, Maier W, et al.: S3 Leitlinie Demenzen. In: Leitlinien für Diagnostik und Therapie in der Neurologie. Edited by Neurologie DGf. www.dgn.org/leitlinien (last accessed 19.08.2017) 2016.
72. Excellence NNIoC. Delirium: prevention, diagnosis and management (clinical guideline). niceorguk/guidance/cg103 (last accessed 19072017). 2010.
73. Barr J, Fraser GL, Puntillo K, Ely EW, Gelinas C, Dasta JF, Davidson JE, Devlin JW, Kress JP, Joffe AM, Coursin DB, Herr DL, Tung A, Robinson BR, Fontaine DK, Ramsay MA, Riker RR, Sessler CN, Pun B, Skrobik Y, Jaeschke R, American College of Critical Care M. Clinical practice guidelines for the management of pain, agitation, and delirium in adult patients in the intensive care unit. Critical Care Medicine. 2013;41:263–306.
74. Neufeld KJ, Yue J, Robinson TN, Inouye SK, Needham DM. Antipsychotic medication for prevention and treatment of delirium in hospitalized adults: a systematic review and meta-analysis. Journal of the American Geriatrics Society. 2016;64:705–714.

55

ANIMAL MODELS OF PSYCHOSIS

APPROACHES AND VALIDITY

Daniel Scott

INTRODUCTION

Creation of a valid animal model of psychosis remains a challenge in psychiatric research for the simple reason that psychosis, by its nature, is a disorder detected by human thought and perception. As such, we are forced to rely on quantifiable, albeit indirect, behavioral analyses. The question of whether animal models for psychiatric conditions can be valid, and how to determine such validity has long been debated. Seminal work by Willner,[1] originally devised for the study of depression, proposed the validation of animal models through satisfaction of distinct types of validity. Validity of an animal model is achieved through the satisfaction of *face validity*, whether the model resembles the human condition; *predictive validity*, whether treatments used to treat the condition in humans similarly affects the animal model; and *construct validity*, whether the human and animal conditions share a common biological mechanism. Since that work was published in 1984, our understanding of biological psychiatry has evolved, and researchers have refined validation criteria. While Willner's pillars of validity remain widely accepted, more recent formulations have expanded on this framework. One formulation that may be appropriate for the modeling of psychosis in particular was proposed by Belzung and Lemoine in 2011.[2] When studying psychosis specifically, it is important to ensure not only that the animal preparations display validity but also that the specific behavioral tests used satisfy similar criteria. In this chapter, we examine the behavioral readouts of psychosis in animals and how these are affected by drugs, and assess the scientific relevance of a number of preparations used to create animal models of psychosis.

FACE VALIDITY/ETHOLOGICAL VALIDITY/BIOMARKER VALIDITY

Face validity posits that an animal model has validity if the behavioral phenotype resembles that of the human condition. Face validity has been defined as "the degree of phenomenological similarity between the model and the disorder to be modeled."[3,4] Belzung and Lemoine further refine this construct into both ethological validity, modeling the symptoms (and etiology, which will be discussed further in regard to construct validity), and biomarker validity, examining the dimension to be modeled on a more pathological level. Due to the unique nature of psychosis, animal models of face validity may be more relevant and appropriate for study of other psychiatric diseases and dimensions; facets of depression, such as anhedonia or learned helplessness and sociability symptoms of autism have obvious animal parallels. The same is hard to say for psychosis. However, there exists debate as to whether face validity refers to the *diagnosis* or to the specific *dimensions/symptoms* being studied. For the majority of psychiatric illnesses appearing in the *Diagnostic and Statistical Manual* (DSM-5), without a known distinct pathophysiology or definitive diagnostic tests, diagnosis is phenomenological. That being said, psychosis is a symptom that transcends many diagnoses, most notably schizophrenia, but also, but not limited to, brief psychotic disorder, schizotypal personality disorder, delusional disorder, schizoaffective disorder, catatonia, postpartum psychosis, post-traumatic stress disorder, Alzheimer's disease, Parkinson's, bipolar disorder, major depressive disorder, metabolic diseases, and epilepsy.[5] That being said, there may be some dimensional features common to psychosis across diagnoses, and those features form the bases of face validity.

ETHOLOGICAL VALIDITY

Most animal studies related to psychotic illnesses attempt to model schizophrenia generally, and as such, any behavioral phenotype revealed may share more homology with the negative or cognitive symptoms than with psychosis per se. Moreover, psychosis is, by definition, an internal, idiosyncratic experience. Determining how a quantifiable behavioral measure corresponds with such an internal process may be beyond the current state of animal behavioral research, and therefore, it may not be possible to objectively determine when a model presents sufficient face validity. With that in mind, there do exist a few animal behavioral paradigms which some scientists believe represent behavioral features of psychosis, or at least the positive symptoms of schizophrenia, and therefore may satisfy ethological validity. These can broadly be categorized into three distinct classes: locomotor, gating, and social behaviors.

Locomotor behaviors

Behavioral paradigms often used to model psychotic-like behaviors include heightened locomotor activity and stereotypy, particularly in response to pharmacological treatment; traits thought to resemble psychotic agitation and disorganized thought.[6] The types of treatments most often used include dopaminergic agonists or N-methyl-D-aspartate (NMDA) receptor antagonists; this feature of ethological validity overlaps significantly with induction validity, described in what follows. It can further be argued that models that assess amphetamine-induced hyperactivity additionally provide some face validity in that some research shows that individuals suffering from psychosis display excessive striatal dopamine in response to amphetamine administration relative to nonaffected individuals, which is associated with a worsening of psychosis.[7]

Stereotypy, or repetitive behavior, is distinct from perseverative behavior: stereotypy is defined strictly by repetitive motor behaviors, while perseveration represents a cognitive process wherein the animal is purposely repeating a specific behavioral response.[8] It must be cautioned, though, that the positive symptoms of schizophrenia are often believed to be composed of two separate modalities: the psychosis dimension, which includes hallucinations and delusions, and the disorganization dimension, which may be closer to what is being measuring via stereotypy.[9] However, it has been demonstrated that in humans experiencing first-episode psychosis, stereotypy was observed, and the degree of stereotypy correlates with the degree of positive symptoms as determined by Positive and Negative symptom Scales (PANSS) score, while not correlating with negative or cognitive symptoms of schizophrenia, lending some credibility to the idea that stereotypy may indeed be a readout of psychosis.[10]

Gating

A second category of behavioral tasks that display some face validity to psychosis are tests of gating. Gating refers to the automatic filtering of extraneous thoughts and sensory stimuli, thereby blocking their conscious recognition. A deficit in some form of gating has been documented in people suffering from psychosis.[11,12] represented by deficient sensorimotor gating or prediction error–dependent learning. Behavioral assessment of gating in animals can take various forms to resemble these human symptoms. One commonly used paradigm is a decrease in prepulse inhibition (PPI) of acoustic startle response. Prepulse inhibition is a measure of the degree to which the physical response to an acoustic startle is inhibited when preceded by an acoustic stimulus slightly louder than background. While in unaffected people this prepulse will inhibit the startle response, it has been shown in both patients with psychotic disorders and those at high risk for developing psychosis that the prepulse has a diminished effect on the startle response.[13–15] This phenotype suggests an impairment in sensorimotor gating, a biological process that can prevent sensory overload and disorganization of cognitive processing.[6,16] This may not be a model of psychosis per se, but instead modeling the "interface of psychosis and cognition."[17] Regardless, due to the homology between the human and animal paradigms, this task clearly fulfills face validity.

Another form of gating that can be used to model psychosis is Kamin blocking. This task relies on the formation of an association between a conditioned stimulus (CS) and an unconditioned stimulus (US), most often a mild shock or reward. If the CS is learned to predict the US, then pairing the CS (CS1) with another CS (CS2), using the same CS-US association, healthy people and control animals will ignore and therefore not learn the CS2-US association, while affected animals will respond to CS1 and CS2 similarly. Such an association can readily be applied to animal models, and is disrupted in models of psychosis which themselves show other forms of validity.[18]

Social cognition

A deficit in social cognition is a common feature associated with psychotic disorders, especially those developing late in life. Because this emerges most often in the elderly, along with paranoia, dementia, and mood disorders, there is some question about the overlap between psychosis, social cognition, and general cognitive function. However, because deficits in social cognition often coincide with the development of psychosis, it is widely accepted that social cognition is a dimension of psychosis.[19] Social cognition can be viewed as an emotional response based on a previous meeting, or familiarity. It has been demonstrated that such a response, based on measurements of skin conductance, is disrupted in schizophrenia.[20] Modeling social cognition in rodent studies is, fortunately, a fairly straightforward task. An animal will readily interact with a novel stranger. When re-exposed to the same social partner, a normal animal will display less time interacting, evidencing a decrease in social novelty, or normal social cognition. A failure for the animal to display this decrease in interaction time is representative of a deficit in social cognition. Importantly, this measure is distinct from social withdrawal, a negative symptom of schizophrenia, which may co-occur with psychosis. Therefore, to avoid this confound, in tests of social cognition, it is important to note no deficiency in the amount of social interaction on the first testing session, indicative of no social withdrawal.

It should be noted that numerous other behavioral paradigms are used in models of psychotic illness, including cognitive tasks assessing episodic or working memory, executive function, social interaction tasks, as described earlier, anxiety and depressive-like assessments, and others. However, as psychosis is most often a single dimension among many within various psychiatric diseases, models that display phenotypes related to these tasks must be viewed carefully; is the symptom modeled one of psychosis, or a separate feature of the disease? It is this complication that makes ethological validity of psychosis itself so difficult.

BIOMARKER VALIDITY

One further aspect of face validity, put forth by Belzung and Lemoine, is biomarker validity, where biomarkers, quantitative variables associated with the disease, are present in the animal model. It has been shown that humans with schizophrenia display increased hippocampal activity, which correlates with the degree of psychosis.[21,22] Indeed, animal preparations that display hippocampal hyperactivity often also show elevated amphetamine-induced locomotor activity,[23] but analysis of this biomarker is secondary to the development of the animal model, and hippocampal hyperactivity until recently has otherwise rarely been associated with analysis of psychosis in animal model systems. Additionally, reduced hippocampal volume has been observed in psychosis.[24] Further, it has been demonstrated that humans with schizophrenia demonstrate disruptions in synchrony between the hippocampus and prefrontal cortex, and this is associated with psychotic delusions.[25] While this frontotemporal synchrony is not often measured in animal models, models that do show this feature may satisfy biomarker validity as well. Importantly, unlike the inferential judgment needed for behavioral analysis, these biomarkers may be particularly specific and tractable, allowing for more objective and quantifiable modeling of psychosis.

Ultimately, when developing animal models for psychosis, there is little in the way of direct face validity connecting the human condition and the animal preparation. It is therefore imperative that these tasks, which may satisfy some degree of face validity, are paired with preparations using construct validity and are further validated with predictive validation.

PREDICTIVE VALIDITY/REMISSION VALIDITY/INDUCTION VALIDITY

REMISSION VALIDITY

Although few behavioral tests can satisfy face validity, animal studies of psychosis often validate their procedures by using predictive validity. Willner describes this generally as what Belzung and Lemoine refer to as "remission validity," where a treatment used to treat the symptom in humans is used to rescue a behavioral phenotype. In this regard, serendipity has long offered researchers opportunities to study psychosis in animal preparations. It is fortunate that pharmacological therapy used in the treatment of schizophrenia is most efficacious in regard to psychosis, therefore lending more relative credence to models satisfying predictive validity. However, we are still faced with the following dilemma: what behavioral phenotypes, affected by antipsychotic medication, represent the therapeutic effects of the medication, versus off-target effects, which are common in all antipsychotic pharmacological treatment? The behavioral paradigms described earlier, which satisfy face validity, do also exhibit remission validity on exposure to typical and atypical antipsychotic pharmacotherapy. Increased locomotor activity can be effectively blocked with such treatment,[26,27] though antipsychotic treatments that produce motor sedative effects could result in false positives during assessment. Deficits in PPI, likewise, are reversed by a range of typical and atypical antipsychotic medications,[28] and has further been suggested to enable the differentiation of these classes of drugs, depending on how the observed deficit is induced.[29]

One behavioral paradigm that can be used to assess predictive/remission validity, but that does not fulfill face validity, is the conditioned avoidance response. In this task, the animal is conditioned to avoid an aversive stimulus via a specific behavioral response, i.e., moving to a specific location. This natural behavior is blocked with antipsychotic treatment, at doses that otherwise do not affect escape behavior and that correspond to therapeutic levels of dosing and activity of the drug within the brain.[30] Importantly, because the conditioned response is a specific, rather than a general behavioral response, this task is unlikely to yield false positive results, as a general motor effect would similarly impact the conditioned avoidance as well as escape from the aversive stimulus itself. However, due to its lack of ethological validity, this paradigm may have limited utility in modeling psychosis, unless coupled with relevant biomarkers to further validate the model.

INDUCTION VALIDITY

Induction validity represents the converse of remission validity; that is, a treatment that can mimic psychosis in humans is used to induce a disease-like phenotype in animals. This is achieved through administration of dopaminergic drugs, such as cocaine, amphetamine, or methamphetamine,[31,32] glutamatergic NMDA receptor antagonists, such as ketamine or phencyclidine,[33] or serotonergic drugs, such as LSD,[34] all of which can be psychotomimetic. Indeed, the specific behavioral paradigms which display remission validity, in that they are sensitive to antipsychotic treatment, also display induction validity. Stereotypies can be induced by treatment with NMDA receptor antagonists,[35] and dopaminergic agonists, and serotonergic drugs can disrupt PPI.[28] Furthermore, treatment with such psychosis-inducing drugs can provoke other behavioral phenotypes as well, including behavioral sensitization, deficits in episodic and working memory, and effects on social interaction.[36,37] Moreover, often the behavioral effects induced by these drugs, most notably phencyclidine and LSD, can be attenuated with administration of antipsychotic medication,[34,38] lending further remission validity to these specific models. Still the question applies: what can we ascribe to psychosis specifically, apart from other, non-psychosis-like effects of these drugs? While some phenotypes can surely point toward nonpsychotic states, like sociability deficits or depressive-like phenotypes, the ambiguous nature of psychosis may include behavioral features that can plausibly be connected with psychosis, such as episodic memory deficits.[39] Indeed, such behavioral phenotypes must be further validated if they are to be considered appropriate for the animal study of psychosis. Moreover, due to the fact that psychoses can be both induced and treated by drugs of which the neural mechanism is understood, we can generate hypotheses regarding the specific biological mechanisms of psychosis, which lead us to Willner's final dimension of validity, construct validity.

CONSTRUCT VALIDITY

Construct validity relies on replicating a biological mechanism known to be involved in the disease pathology. Willner defined construct validity as an alignment of the theoretical accounts of the disease and modeled behavior.[40] Similarly, Sarter and Bruno[4] define construct validity as "a theory-driven, experimental substantiation of the behavioral and/or neuronal components of the model," which translates the hypothesized mechanisms of the human disease to analysis of animal behavior. It is this formulation of construct validity which Sarter and Bruno believed to be the most important criteria for validation of animal models. Much is known about the genes, environmental insults, and other factors that increase this risk of psychosis; construct validity can therefore provide the basis of an animal preparation. However, this must be further verified by behavioral or biomarker analysis and validated by sensitivity to antipsychotic medication in order to demonstrate causality and specificity. Moreover, the complexity of this form of validity should not be overlooked. Thus, Belzung and Lemoine proposed several criteria of validity that can be grouped under the general idea of construct validity. These forms of validity break down the notion of construct validity into distinct methods of developing the constructs themselves.

PATHOGENIC VALIDITY

One such criteria is pathogenic validity, which examines the specific methods by which a healthy, or otherwise unaffected animal can be transformed to a vulnerable or pathological animal.[2] This is broadly similar to what some behaviorists refer to as "etiological validity," where the causes of the human condition are applied to the animal, and the resultant behavioral phenotype is assumed to be disease-related.[3,41] Such constructs which may satisfy pathogenic validity include environmental stressors, physical brain manipulation, and genetic alteration.

Models with pathogenic validity

It is well established that environmental insults, be they prenatal, neonatal, during childhood, or during adolescence, can increase the risk of developing a psychotic illness.[42] Because of this relationship, a number of animal models have been developed that use such environmental stressors to induce behavioral effects relevant to psychosis. It must be cautioned, though, in many cases the particular stressors used do not necessarily resemble the environmental insults suffered by the humans who would develop a psychotic illness, nor do these stressors increase the risk of psychosis to the exclusion of other psychiatric illnesses. Thus, these models, though possessing some construct/pathogenic validity, must be further validated with face and/or predictive validity to more accurately model psychosis.

Prenatal

When administered to pregnant rats at gestational day 17 (GD17), MAM, an antimitotic agent[43] induces a number of traits in pups similar to those seen in human schizophrenics. These include, among others, reduced hippocampal volume, disorganization of pyramidal cells in CA3, and increased CA1 excitability.[44–47] Moreover, pups display heightened amphetamine- and MK-801-induced locomotor activity, as well as decreased PPI.[44,47,48] Furthermore, antipsychotic treatment effectively attenuates this increased sensitivity to MK-801.[48] These features provide this model with some additional face and predictive validity.

It has also been long hypothesized that maternal infection during pregnancy can increase the risk of psychotic illness in offspring.[49] This is believed to be due to the immune response induced in the mother as a result of viral infection. As such, administration of poly(I:C), a synthetic double-stranded RNA, to pregnant dams, which mimics viral infection and induces a robust immune response, is used to model the developmental insult, and produces a robust immune response in the pregnant mother. This model possesses more pathogenic validity than the GD17 MAM model, in that the immune response is more biologically relevant than the retardation of cell division caused by MAM. Moreover, this model produces many behavioral effects relevant to psychosis, including among others, impaired PPI and a heightened response to amphetamine,[50] again lending this model some face validity. Moreover, antipsychotic treatment of the offspring of dams treated with poly(I:C) can reverse or prevent many behavioral phenotypes, providing this model with some predictive validity as well.[51]

Neonatal

Childhood adversity has long been associated with risk of developing psychotic illness later in life. Such adversity can take many forms, including parental neglect, physical, emotional, or sexual abuse, and bullying, among others.[52] Such trauma can more than triple the likelihood of developing psychosis.[53] Moreover, rodents display robust maternal behavior, and pups rely on maternal care for normal neurodevelopment.[54] Thus, disruption of maternal care, through separation of the pups from their mother (maternal deprivation) provides a homologous model for neonatal adversity. There is generally a critical period at which this isolation exerts in long-term effects, with some researchers exposing pups to a long (~24h) separation prior to P14, while others will expose pups to shorter, daily bouts of isolation over several days or weeks, prior to weaning age (~P21).[55] It has further been demonstrated that maternal deprivation can induce behavioral phenotypes relevant to psychosis, including attenuated PPI and enhanced amphetamine-induced locomotor activity,[56,57] providing this model with some face validity as well. However, this procedure more often induces phenotypes relevant to anxiety and depression, introducing confounding factors to the etiology of psychosis.[58,59] Moreover, there is little evidence that the behavioral phenotype emerging in rodents exposed to maternal deprivation can be attenuated with antipsychotic treatment. Thus, though this paradigm does have solid etiological/pathogenic validity, it fails at providing remission/predictive validity.

Adolescent

It is well known that abuse of cannabis during adolescence can increase the risk of developing psychotic illness.[60] It is thought this is due to a psychoactive component in cannabis, THC, disrupting the endogenous cannabinoid system, which is involved in normal brain development,[61] including the maturation of neurotransmitter systems and cortical development.[62,63] This can be modeled in rodents, as adolescence occurs from P28–P50 in mice and rats, suggesting a critical period for THC or synthetic cannabinoid exposure, and behavioral analysis occurring during adulthood (after P75). This schedule has demonstrated a variety of behavioral effects, including reduced cognition, impaired PPI, hyperlocomotion, and decreased social interaction.[64] Thus, this model seems to provide robust face validity. As adolescent cannabis use seems to merely increase risk, as opposed to being decisively causative of psychosis in humans, this experimental paradigm is often combined with other models with pathogenic validity, or models of strain validity to isolate these developmental effects on a vulnerable population. While the combination of various models of construct validity will increase the overall strength of the model, it can be only through validation with face and/or predictive (remission) validity that one can be confident in the specificity of that which is being analyzed.

MECHANISTIC VALIDITY

Mechanistic validity posits that biological mechanisms thought to underlie the disease, when expressed in the animal, produce symptoms of the disease. This type of validity may overlap with induction validity, assuming the mechanism of the inducing agent resembles a theoretical biological mechanism of the condition.

Models with mechanistic validity

As previously discussed, much of our knowledge of the biological constructs underlying psychosis come from the serendipitous discoveries of first-generation antipsychotic drugs, as well as the psychotomimetic properties of NMDA receptor antagonists especially, but also dopaminergic and serotonergic drugs of abuse. These findings helped to formulate the dopamine and glutamate hypotheses of psychosis, and paved the way for the use of animal model systems employing these constructs.

Alternatively, due to the pathological implications of hippocampal dysfunction in people with psychosis, a model of neonatal ventral hippocampal lesion has also been employed. An advantage of this model is that it directly manipulates the physical properties of the brain in order to more resemble the decreased hippocampal volume present in psychosis.[24] And though this model is more reminiscent of schizophrenia as a whole, rather than psychosis as a dimension, it does provide face validity for psychosis both in reference to behavioral abnormalities, which emerge in adulthood (increased amphetamine-induced locomotor activity, decreased PPI), and in terms of brain pathology, as this lesion model induces numerous changes in the brain, especially in the prefrontal cortex and striatum, resembling pathology seen in human schizophrenics; this model also satisfies some degree of predictive validity, in that antipsychotic medication can reverse the amphetamine sensitivity.[65] However, while the pathology present in psychotic illness may present as only a small change from what is seen in healthy controls, these lesions present an extensive pathology, well beyond anything seen even in the most severe cases of psychosis.

Genetic animal model systems can be designed to mimic the proposed mechanism underlying psychosis. As the hypothesized mechanisms of psychosis generally rest on the idea of either reduced glutamate signaling or hyperdopaminergia, the genetic models that aim to recreate the mechanistic deficits seen in psychosis are most often based on those specific premises.

Postmortem analysis of brains of humans with psychotic schizophrenia have demonstrated a reduction of the NMDA receptor subunit GluN1 specifically in granule cells in the dentate gyrus of the hippocampus, which is further associated with increased basal activity within the hippocampus.[66] A mouse with a selective knockout of GluN1 within the dentate gyrus has been examined as a putative animal model of this feature of schizophrenia; this animal displays some face validity, with impaired PPI and social cognition, along with increased hippocampal activity in CA3.[67] Whether this animal preparation presents the underlying cause of psychosis, though, is still debatable. As it is unclear which specific mechanisms underlie the behavioral phenomenon of psychosis generally, more work with this animal model would be needed to further validate it as a model system.

In order to model a hypothesized hyperdopaminergic state associated with psychosis, mouse models with reduced expression or function of the dopamine transporter (DAT) knockdown mouse were developed.[68,69] Reduction in DAT function results in reduced uptake of extracellular dopamine levels, inducing the predicted hyperdopaminergic state. It must be noted, that although this mutation is used solely to mimic the functional changes in the brain associated with psychosis, there is some evidence that some humans with psychotic illness may have mutations in this gene as well, giving it a translational quality and strain validity.[70] These mouse models display spontaneous and increased amphetamine-induced hyperactivity and disorganized behavior. Moreover, the spontaneous hyperactivity is reversed with administration of typical and atypical antipsychotic drugs,[71] which, though unsurprising due to the known pharmacological action of those medications, still provide this model with predictive validity. However, since continued study has garnered such weak support for hyperdopaminergia in schizophrenia, many would view this as an outdated idea, without much overall validity.

HOMOLOGICAL VALIDITY

Finally, there is the idea of species and strain validity, which fall under the general category of homological validity. Species validity seeks to determine how a specific animal behavior corresponds to aspects of human cognition.[72] When dealing with psychosis, this can be quite problematic. As psychosis is such an individual experience, determining appropriate homologous measures can be difficult. To address this, satisfaction of species validity relies on the known correspondence between human and animal brain structure, function, and development. When dealing with higher-order tasks that tax cognitive function, some argue that a rodent behavioral model will be insufficient to provide sufficient species validity, as the cortical function necessary for such tasks is not homologous between the two systems. In such a case, primate studies may be used to assess such tasks. As it stands, it has been demonstrated that there exists dysfunction in the dorsolateral prefrontal cortex in those with psychosis or at high risk for developing psychosis,[73,74] with reduced dorsolateral prefrontal cortex (DLPFC) activity being associated with increased hippocampal activity, especially during tasks utilizing working memory.[75] Rodents do not have a "proper" DLPFC, thus necessitating the use of primate models to assess the functional role of this cortical region in the manifestation of psychosis.[76]

Strain validity represents the genetic predisposition to psychosis shared between human and animal preparations. As genetic engineering of mice has become a routine experimental technique, there exist a large number of animal model systems that fulfill strain validity. While a number of models of strain validity are described next, it must be cautioned that although a number of genetic mutations have been found to be associated with psychotic illnesses,[77] these mutations are far from fully penetrant, and may only slightly increase the risk of developing psychosis.

Models with strain validity

There have been numerous studies attempting to link specific genetic mutations with psychotic illnesses, as psychoses have substantial heritability, suggesting a genetic etiology. A few models utilizing known risk genes are discussed in this section. Although the specific genetic mutations that have been associated with psychotic illness tend to have minor effects on the individual, animal studies that look at sufficient numbers of animals with these human mutations, or other mutations in homologous genes, can provide some insight into the mechanisms and phenotypes associated with psychosis.

DISC1

In 1970, a chromosomal translocation was observed in a Scottish family, and this translocation, which truncated a specific gene, was associated with high incidence of psychiatric illness, including schizophrenia and bipolar disorders. In 2000, this gene, DISC1, was characterized.[78] As many people with this specific translocation, or other mutations in this gene, develop psychosis,[79] mouse models with mutations in this gene have long been studied for their supposed strain validity. Mice expressing a dominant negative form of DISC1 show hyperactivity and reduced PPI.[80] Mice with point mutations in this gene show, depending on the specific mutation, some combination of impaired PPI, social interaction, and working memory, increased depressive-like behavior, and hyperactivity.[81] However, because DISC1 mutations are associated with such a wide array of psychiatric conditions, it is unclear whether this model provides strong validity for psychosis per se, as opposed to other psychiatric illnesses.

Dysbindin/neuregulin-1

Susceptibility variants for psychotic diseases have been identified in a number of genes, including dysbindin (*DTNBP1*) and neuregulin-1 (*NRG1*).[82] *DTNBP1* knockout mice display increased spontaneous and amphetamine-induced locomotor activity and deficits in social interaction.[83] The *NRG1*-deficient mice show increased spontaneous locomotor activity, impaired PPI and social interaction, and reduced cortical NMDA receptors, suggesting some face validity.[83] Moreover this hyperactive phenotype can be reversed with clozapine, suggesting some remission validity as well.[84] Though mutations in these genes may provide the basis for some behavioral models of psychosis, these mutant animals may be best suited to provide insight into the details of neurochemical signaling in relation to psychosis, the mechanisms underlying the effects of antipsychotic medication, or interactions between putative genetic models and models with pathogenic validity.[85]

22q11.2

One large copy number variant (CNV) that increases the risk of developing psychosis is deletion of chromosome 22q11.2, associated with velocardiofacial syndrome (VCFS), or DiGeorge syndrome.[86] Patients with this condition face a number of symptoms, including peripheral defects, such as cardiac and facial abnormalities as well as mental and cognitive symptoms, including mental retardation, autism, and most notably, a 20- to 30-fold increase in the risk of developing schizophrenia.[87] Much of the critical portion of this chromosomal region is found on mouse chromosome 16,[88] enabling deletion of this chromosomal region to generate a mouse line with significant strain validity. Interestingly, mice with the targeted deletion show only a small range of behavioral effects, including impaired social memory [89] and sensorimotor gating.[90] These mice also display impaired hippocampal-prefrontal synchrony during a working memory task, indicating further biomarker validity in this model,[91] and supporting these behavioral outputs in mice as being markers of psychosis.

One gene in this region that has independently been associated with psychosis is COMT. Mice lacking only membrane-bound COMT show reduced PPI as well as increased aggressive and depressive-like phenotypes. These mice also show a hyperdopaminergic state in the striatum, suggesting some biomarker validity.[92] Moreover, mice lacking one copy of this gene display altered exploratory behavior without changes in activity and impaired working memory.[93,94] Meanwhile, there exists a psychiatric-disease-related point mutation at codon 158 in the COMT gene. Studies suggest that this polymorphism is associated with psychotic illness, including bipolar disorder and early onset of schizophrenia,[95] though some evidence suggests this gene may be more involved with the negative symptoms associated with these disorders, as

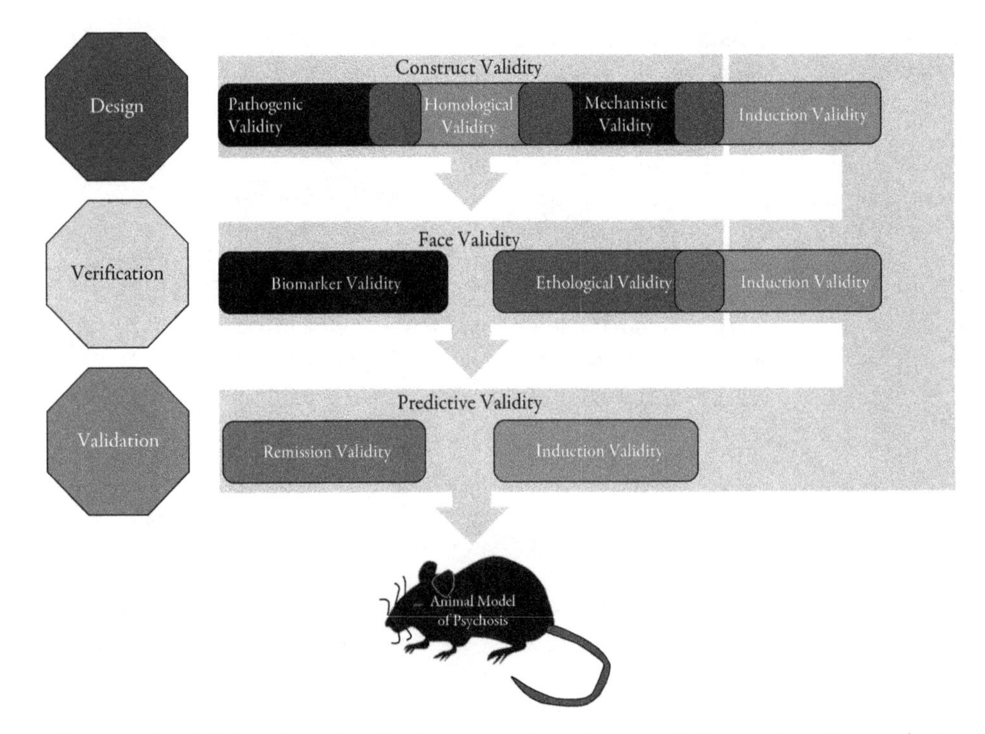

Figure 55.1 Development of an animal model of psychosis. Based on known genetic, environmental, pharmacological, or mechanistic causes of disease, a putative animal model can be designed (design). This model is then verified by demonstrating a behavioral phenotype or biomarkers associated with psychosis (verification). If these phenotypes can be attenuated with treatment shown to be effective at mitigating psychosis in humans (validation), then the animal preparation may be sufficient to effectively model psychosis. The specific types of validation used to generate each step in this process are indicated, as is where these types of validity may overlap.

opposed to psychosis per se.[96] Mice with this polymorphism display sexually dimorphic effects, with females displaying impaired PPI and impaired contextual fear conditioning.[97]

A lesson that can be taken from the review of animal models with mutations in historical psychosis candidate genes is that they fall short in yielding a definitive picture into the genetic bases of psychosis, but may shed some light on specific aspects of the mechanisms underlying this dimension.

SYNTHESIS AND CONCLUSIONS

Designing animal models of psychosis before disease pathophysiology has been defined is a difficult task, due both to the range of diseases in which psychosis is a dimension, and the complexity, heterogeneity, and internal nature of psychosis itself. However, we may be able to develop a system by which an appropriate animal model of psychosis can be produced (Figure 55.1). With the known risk factors and pathological changes observed in human psychosis, a putative preparation, which fulfills construct validity, can be produced. Alternatively, agents known to induce a psychotic-like state can be used to induce a phenotype in animals (induction validity). Upon the development of this animal preparation, the behavioral phenotypes and biomarkers associated with psychosis, which fulfill face validity, can be used to verify the effects of this preparation. If these symptoms can be reversed with antipsychotic treatment (remission validity), this preparation can be thoroughly validated.

Going forward, it is imperative that distinct, replicable biomarkers for psychosis, specifically, be determined in humans such that the confounds of nonspecific symptomatology and species-specific ambiguity be minimized. Only once the pathophysiology of psychosis is more fully understood, can the difficulties in modeling psychosis in animals be attenuated. Such progress can foster more complete and productive animal models for study, and open the door to generate new avenues for treatments. It must be the goal of the preclinical researcher to be able not only to parse out psychosis as a dimension to be studied but also to differentiate distinct manifestations of psychosis across disorders in order to advance this vital medical need.

REFERENCES

1. Willner P. The validity of animal models of depression. *Psychopharmacology (Berl).* 1984;83(1):1–16.
2. Belzung C, Lemoine M. Criteria of validity for animal models of psychiatric disorders: focus on anxiety disorders and depression. *Biol Mood Anxiety Disord.* 2011;1(1):9.
3. Geyer MA, Markou A. Animal models of psychiatric disorders. *Psychopharmacology: Fourth Generation of Progress*: Raven Press; 1995:787–798.
4. Sarter M, Bruno JP. Animal models in biological psychiatry. *Biological Psychiatry.* 2002:37–44.
5. Freudenreich O. Differential diagnosis of psychotic symptoms: medical mimics. *Psychiatric Times.* 2010;27(12):56–61.
6. Powell SB, Zhou X, Geyer MA. Prepulse inhibition and genetic mouse models of schizophrenia. *Behav Brain Res.* 2009;204(2):282–294.

7. Abi-Dargham A, Gil R, Krystal J, et al. Increased striatal dopamine transmission in schizophrenia: confirmation in a second cohort. *Am J Psychiatry*. 1998;155(6):761–767.
8. Morrens M, Hulstijn W, Lewi PJ, De Hert M, Sabbe BG. Stereotypy in schizophrenia. *Schizophr Res*. 2006;84(2–3):397–404.
9. Tordjman S, Drapier D, Bonnot O, et al. Animal models relevant to schizophrenia and autism: validity and limitations. *Behav Genet*. 2007;37(1):61–78.
10. Compton MT, Fantes F, Wan CR, Johnson S, Walker EF. Abnormal movements in first-episode, nonaffective psychosis: dyskinesias, stereotypies, and catatonic-like signs. *Psychiatry Res*. 2015;226(1):192–197.
11. Braff DL, Geyer MA. Sensorimotor gating and schizophrenia. Human and animal model studies. *Arch Gen Psychiatry*. 1990;47(2):181–188.
12. Corlett PR, Honey GD, Fletcher PC. Prediction error, ketamine and psychosis: an updated model. *J Psychopharmacol*. 2016;30(11):1145–1155.
13. Cadenhead KS, Geyer MA, Braff DL. Impaired startle prepulse inhibition and habituation in patients with schizotypal personality disorder. *Am J Psychiatry*. 1993;150(12):1862–1867.
14. Kumari V, Antonova E, Geyer MA. Prepulse inhibition and "psychosis-proneness" in healthy individuals: an fMRI study. *Eur Psychiatry*. 2008;23(4):274–280.
15. Ziermans TB, Schothorst PF, Sprong M, Magnee MJ, van Engeland H, Kemner C. Reduced prepulse inhibition as an early vulnerability marker of the psychosis prodrome in adolescence. *Schizophr Res*. 2012;134(1):10–15.
16. van den Buuse M, Garner B, Gogos A, Kusljic S. Importance of animal models in schizophrenia research. *Aust N Z J Psychiatry*. 2005;39(7):550–557.
17. Desbonnet L, Waddington JL, O'Tuathaigh CM. Mutant models for genes associated with schizophrenia. *Biochem Soc Trans*. 2009;37(Pt 1):308–312.
18. O'Tuathaigh CM, Salum C, Young AM, Pickering AD, Joseph MH, Moran PM. The effect of amphetamine on Kamin blocking and overshadowing. *Behav Pharmacol*. 2003;14(4):315–322.
19. La Salvia E, Chemali Z. A perspective on psychosis in late life and deficits in social cognition. *Harv Rev Psychiatry*. 2011;19(4):190–197.
20. Ameller A, Picard A, D'Hondt F, Vaiva G, Thomas P, Pins D. Implicit recognition of familiar and unfamiliar faces in schizophrenia: a study of the skin conductance response in familiarity disorders. *Front Psychiatry*. 2017;8:181.
21. Schobel SA, Lewandowski NM, Corcoran CM, et al. Differential targeting of the CA1 subfield of the hippocampal formation by schizophrenia and related psychotic disorders. *Arch Gen Psychiatry*. 2009;66(9):938–946.
22. Molina V, Reig S, Pascau J, et al. Anatomical and functional cerebral variables associated with basal symptoms but not risperidone response in minimally treated schizophrenia. *Psychiatry Res*. 2003;124(3):163–175.
23. Lodge DJ, Grace AA. Aberrant hippocampal activity underlies the dopamine dysregulation in an animal model of schizophrenia. *J Neurosci*. 2007;27(42):11424–11430.
24. Weinberger DR. Cell biology of the hippocampal formation in schizophrenia. *Biol Psychiatry*. 1999;45(4):395–402.
25. Lawrie SM, Buechel C, Whalley HC, Frith CD, Friston KJ, Johnstone EC. Reduced frontotemporal functional connectivity in schizophrenia associated with auditory hallucinations. *Biol Psychiatry*. 2002;51(12):1008–1011.
26. Castagne V, Moser PC, Porsolt RD. Preclinical behavioral models for predicting antipsychotic activity. *Adv Pharmacol*. 2009;57:381–418.
27. Leite JV, Guimaraes FS, Moreira FA. Aripiprazole, an atypical antipsychotic, prevents the motor hyperactivity induced by psychotomimetics and psychostimulants in mice. *Eur J Pharmacol*. 2008;578(2-3):222–227.
28. Geyer MA, Krebs-Thomson K, Braff DL, Swerdlow NR. Pharmacological studies of prepulse inhibition models of sensorimotor gating deficits in schizophrenia: a decade in review. *Psychopharmacology (Berl)*. 2001;156(2-3):117–154.
29. Porsolt RD, Moser PC, Castagne V. Behavioral indices in antipsychotic drug discovery. *J Pharmacol Exp Ther*. 2010;333(3):632–638.
30. Wadenberg ML. Conditioned avoidance response in the development of new antipsychotics. *Curr Pharm Des*. 2010;16(3):358–370.
31. Brady KT, Lydiard RB, Malcolm R, Ballenger JC. Cocaine-induced psychosis. *J Clin Psychiatry*. 1991;52(12):509–512.
32. Mahoney JJ, 3rd, Kalechstein AD, De La Garza R, 2nd, Newton TF. Presence and persistence of psychotic symptoms in cocaine-versus methamphetamine-dependent participants. *Am J Addict*. 2008;17(2):83–98.
33. Bubenikova-Valesova V, Horacek J, Vrajova M, Hoschl C. Models of schizophrenia in humans and animals based on inhibition of NMDA receptors. *Neurosci Biobehav Rev*. 2008;32(5):1014–1023.
34. De Gregorio D, Comai S, Posa L, Gobbi G. d-Lysergic acid diethylamide (LSD) as a model of psychosis: mechanism of action and pharmacology. *Int J Mol Sci*. 2016;17(11).
35. Hitri A, O'Connor DA, Cohen JM, Keuler DJ, Deutsch SI. Differentiation between MK-801- and apomorphine-induced stereotyped behaviors in mice. *Clin Neuropharmacol*. 1993;16(3):220–236.
36. Robinson TE, Becker JB. Enduring changes in brain and behavior produced by chronic amphetamine administration: a review and evaluation of animal models of amphetamine psychosis. *Brain Res*. 1986;396(2):157–198.
37. Amann LC, Gandal MJ, Halene TB, et al. Mouse behavioral endophenotypes for schizophrenia. *Brain Res Bull*. 2010;83(3-4):147–161.
38. Sams-Dodd F. Phencyclidine in the social interaction test: an animal model of schizophrenia with face and predictive validity. *Rev Neurosci*. 1999;10(1):59–90.
39. Tamminga CA. Psychosis is emerging as a learning and memory disorder. *Neuropsychopharmacology*. 2013;38(1):247.
40. Willner P, Mitchell PJ. Animal models of depression: a diathesis/stress approach. *Biological Psychiatry*. 2002:701–726.
41. Robbins TW, Sahakian BJ. "Paradoxical" effects of psychomotor stimulant drugs in hyperactive children from the standpoint of behavioural pharmacology. *Neuropharmacology*. 1979;18(12):931–950.
42. Laurens KR, Luo L, Matheson SL, et al. Common or distinct pathways to psychosis? A systematic review of evidence from prospective studies for developmental risk factors and antecedents of the schizophrenia spectrum disorders and affective psychoses. *BMC Psychiatry*. 2015;15:205.
43. Matsumoto H, Higa HH. Studies on methylazoxymethanol, the aglycone of cycasin: methylation of nucleic acids in vitro. *Biochem J*. 1966;98(2):20C–22C.
44. Moore H, Jentsch JD, Ghajarnia M, Geyer MA, Grace AA. A neurobehavioral systems analysis of adult rats exposed to methylazoxymethanol acetate on E17: implications for the neuropathology of schizophrenia. *Biol Psychiatry*. 2006;60(3):253–264.
45. Matricon J, Bellon A, Frieling H, et al. Neuropathological and reelin deficiencies in the hippocampal formation of rats exposed to MAM; differences and similarities with schizophrenia. *PLoS One*. 2010;5(4):e10291.
46. Sanderson TM, Cotel MC, O'Neill MJ, Tricklebank MD, Collingridge GL, Sher E. Alterations in hippocampal excitability, synaptic transmission and synaptic plasticity in a neurodevelopmental model of schizophrenia. *Neuropharmacology*. 2012;62(3):1349–1358.
47. Kallai V, Toth A, Galosi R, et al. The MAM-E17 schizophrenia rat model: comprehensive behavioral analysis of pre-pubertal, pubertal and adult rats. *Behav Brain Res*. 2017;332:75–83.
48. Le Pen G, Jay TM, Krebs MO. Effect of antipsychotics on spontaneous hyperactivity and hypersensitivity to MK-801-induced hyperactivity in rats prenatally exposed to methylazoxymethanol. *J Psychopharmacol*. 2011;25(6):822–835.
49. Mednick SA, Machon RA, Huttunen MO, Bonett D. Adult schizophrenia following prenatal exposure to an influenza epidemic. *Arch Gen Psychiatry*. 1988;45(2):189–192.

50. Meyer U, Feldon J, Schedlowski M, Yee BK. Towards an immunoprecipitated neurodevelopmental animal model of schizophrenia. *Neurosci Biobehav Rev.* 2005;29(6):913–947.
51. Reisinger S, Khan D, Kong E, Berger A, Pollak A, Pollak DD. The poly(I:C)-induced maternal immune activation model in preclinical neuropsychiatric drug discovery. *Pharmacol Ther.* 2015;149:213–226.
52. Morgan C, Gayer-Anderson C. Childhood adversities and psychosis: evidence, challenges, implications. *World Psychiatry.* 2016;15(2):93–102.
53. Matheson SL, Shepherd AM, Pinchbeck RM, Laurens KR, Carr VJ. Childhood adversity in schizophrenia: a systematic meta-analysis. *Psychol Med.* 2013;43(2):225–238.
54. Kaffman A, Meaney MJ. Neurodevelopmental sequelae of postnatal maternal care in rodents: clinical and research implications of molecular insights. *J Child Psychol Psychiatry.* 2007;48(3-4):224–244.
55. Niwa M, Matsumoto Y, Mouri A, Ozaki N, Nabeshima T. Vulnerability in early life to changes in the rearing environment plays a crucial role in the aetiopathology of psychiatric disorders. *Int J Neuropsychopharmacol.* 2011;14(4):459–477.
56. Ellenbroek BA, Riva MA. Early maternal deprivation as an animal model for schizophrenia. *Clinical Neuroscience Research.* 2003;3(4):297–302.
57. Matthews K, Hall FS, Wilkinson LS, Robbins TW. Retarded acquisition and reduced expression of conditioned locomotor activity in adult rats following repeated early maternal separation: effects of prefeeding, d-amphetamine, dopamine antagonists and clonidine. *Psychopharmacology (Berl).* 1996;126(1):75–84.
58. Macri S, Laviola G. Single episode of maternal deprivation and adult depressive profile in mice: interaction with cannabinoid exposure during adolescence. *Behav Brain Res.* 2004;154(1):231–238.
59. Barna I, Balint E, Baranyi J, Bakos N, Makara GB, Haller J. Gender-specific effect of maternal deprivation on anxiety and corticotropin-releasing hormone mRNA expression in rats. *Brain Res Bull.* 2003;62(2):85–91.
60. Stefanis NC, Delespaul P, Henquet C, Bakoula C, Stefanis CN, Van Os J. Early adolescent cannabis exposure and positive and negative dimensions of psychosis. *Addiction.* 2004;99(10):1333–1341.
61. Berghuis P, Rajnicek AM, Morozov YM, et al. Hardwiring the brain: endocannabinoids shape neuronal connectivity. *Science.* 2007;316(5828):1212–1216.
62. Wilson W, Mathew R, Turkington T, Hawk T, Coleman RE, Provenzale J. Brain morphological changes and early marijuana use: a magnetic resonance and positron emission tomography study. *J Addict Dis.* 2000;19(1):1–22.
63. Trezza V, Cuomo V, Vanderschuren LJ. Cannabis and the developing brain: insights from behavior. *Eur J Pharmacol.* 2008;585(2–3):441–452.
64. Rubino T, Parolaro D. Cannabis abuse in adolescence and the risk of psychosis: a brief review of the preclinical evidence. *Prog Neuropsychopharmacol Biol Psychiatry.* 2014;52:41–44.
65. Lipska BK. Using animal models to test a neurodevelopmental hypothesis of schizophrenia. *J Psychiatry Neurosci.* 2004;29(4):282–286.
66. Stan AD, Ghose S, Zhao C, et al. Magnetic resonance spectroscopy and tissue protein concentrations together suggest lower glutamate signaling in dentate gyrus in schizophrenia. *Mol Psychiatry.* 2015;20(4):433–439.
67. Segev A, Yanagi M, Scott D, et al. Reduced GluN1 in mouse dentate gyrus is associated with CA3 hyperactivity and psychosis-like behaviors. *Mol Psychiatry.* 2018.
68. van Enkhuizen J, Geyer MA, Halberstadt AL, Zhuang X, Young JW. Dopamine depletion attenuates some behavioral abnormalities in a hyperdopaminergic mouse model of bipolar disorder. *J Affect Disord.* 2014;155:247–254.
69. O'Neill B, Gu HH. Amphetamine-induced locomotion in a hyperdopaminergic ADHD mouse model depends on genetic background. *Pharmacol Biochem Behav.* 2013;103(3):455–459.
70. Greenwood TA, Schork NJ, Eskin E, Kelsoe JR. Identification of additional variants within the human dopamine transporter gene provides further evidence for an association with bipolar disorder in two independent samples. *Mol Psychiatry.* 2006;11(2):125–133, 115.
71. Spielewoy C, Roubert C, Hamon M, Nosten-Bertrand M, Betancur C, Giros B. Behavioural disturbances associated with hyperdopaminergia in dopamine-transporter knockout mice. *Behav Pharmacol.* 2000;11(3-4):279–290.
72. Robbins TW. Homology in behavioural pharmacology: an approach to animal models of human cognition. *Behav Pharmacol.* 1998;9(7):509–519.
73. Woodward ND, Waldie B, Rogers B, Tibbo P, Seres P, Purdon SE. Abnormal prefrontal cortical activity and connectivity during response selection in first episode psychosis, chronic schizophrenia, and unaffected siblings of individuals with schizophrenia. *Schizophr Res.* 2009;109(1–3):182–190.
74. Colibazzi T, Horga G, Wang Z, et al. Neural dysfunction in cognitive control circuits in persons at clinical high-risk for psychosis. *Neuropsychopharmacology.* 2016;41(5):1241–1250.
75. Meyer-Lindenberg AS, Olsen RK, Kohn PD, et al. Regionally specific disturbance of dorsolateral prefrontal-hippocampal functional connectivity in schizophrenia. *Arch Gen Psychiatry.* 2005;62(4):379–386.
76. Simen AA, DiLeone R, Arnsten AF. Primate models of schizophrenia: future possibilities. *Prog Brain Res.* 2009;179:117–125.
77. Biological insights from 108 schizophrenia-associated genetic loci. *Nature.* 2014;511(7510):421–427.
78. Millar JK, Wilson-Annan JC, Anderson S, et al. Disruption of two novel genes by a translocation co-segregating with schizophrenia. *Hum Mol Genet.* 2000;9(9):1415–1423.
79. Green EK, Grozeva D, Sims R, et al. DISC1 exon 11 rare variants found more commonly in schizoaffective spectrum cases than controls. *Am J Med Genet B Neuropsychiatr Genet.* 2011;156B(4):490–492.
80. Hikida T, Jaaro-Peled H, Seshadri S, et al. Dominant-negative DISC1 transgenic mice display schizophrenia-associated phenotypes detected by measures translatable to humans. *Proc Natl Acad Sci U S A.* 2007;104(36):14501–14506.
81. Clapcote SJ, Lipina TV, Millar JK, et al. Behavioral phenotypes of Disc1 missense mutations in mice. *Neuron.* 2007;54(3):387–402.
82. Gill M, Donohoe G, Corvin A. What have the genomics ever done for the psychoses? *Psychol Med.* 2010;40(4):529–540.
83. Jones CA, Watson DJ, Fone KC. Animal models of schizophrenia. *Br J Pharmacol.* 2011;164(4):1162–1194.
84. Pan B, Huang XF, Deng C. Antipsychotic treatment and neuregulin 1-ErbB4 signalling in schizophrenia. *Prog Neuropsychopharmacol Biol Psychiatry.* 2011;35(4):924–930.
85. Moran PM, O'Tuathaigh C, Papaleo F, Waddington JL. Dopaminergic function in relation to genes associated with risk for schizophrenia: translational mutant mouse models. *Prog Brain Res.* 2014;211:79–112.
86. Sullivan KE. The clinical, immunological, and molecular spectrum of chromosome 22q11.2 deletion syndrome and DiGeorge syndrome. *Curr Opin Allergy Clin Immunol.* 2004;4(6):505–512.
87. Bassett AS, Chow EW, AbdelMalik P, Gheorghiu M, Husted J, Weksberg R. The schizophrenia phenotype in 22q11 deletion syndrome. *Am J Psychiatry.* 2003;160(9):1580–1586.
88. Puech A, Saint-Jore B, Funke B, et al. Comparative mapping of the human 22q11 chromosomal region and the orthologous region in mice reveals complex changes in gene organization. *Proc Natl Acad Sci U S A.* 1997;94(26):14608–14613.
89. Piskorowski RA, Nasrallah K, Diamantopoulou A, et al. Age-dependent specific changes in area CA2 of the hippocampus and social memory deficit in a mouse model of the 22q11.2 deletion syndrome. *Neuron.* 2016;89(1):163–176.
90. Long JM, LaPorte P, Merscher S, et al. Behavior of mice with mutations in the conserved region deleted in velocardiofacial/DiGeorge syndrome. *Neurogenetics.* 2006;7(4):247–257.

91. Sigurdsson T, Stark KL, Karayiorgou M, Gogos JA, Gordon JA. Impaired hippocampal-prefrontal synchrony in a genetic mouse model of schizophrenia. *Nature*. 2010;464(7289):763–767.
92. Tammimaki A, Aonurm-Helm A, Zhang FP, et al. Generation of membrane-bound catechol-O-methyl transferase deficient mice with distinct sex dependent behavioral phenotype. *J Physiol Pharmacol*. 2016;67(6):827–842.
93. Babovic D, O'Tuathaigh CM, O'Sullivan GJ, et al. Exploratory and habituation phenotype of heterozygous and homozygous COMT knockout mice. *Behav Brain Res*. 2007;183(2):236–239.
94. Babovic D, O'Tuathaigh CM, O'Connor AM, et al. Phenotypic characterization of cognition and social behavior in mice with heterozygous versus homozygous deletion of catechol-O-methyl-transferase. *Neuroscience*. 2008;155(4):1021–1029.
95. Taylor S. Association between COMT Val158Met and psychiatric disorders: a comprehensive meta-analysis. *Am J Med Genet B Neuropsychiatr Genet*. 2017.
96. Clelland CL, Drouet V, Rilett KC, et al. Evidence that COMT genotype and proline interact on negative-symptom outcomes in schizophrenia and bipolar disorder. *Transl Psychiatry*. 2016;6(9):e891.
97. Risbrough V, Ji B, Hauger R, Zhou X. Generation and characterization of humanized mice carrying COMT158 Met/Val alleles. *Neuropsychopharmacology*. 2014;39(8):1823–1832.

PSYCHOLOGICAL TREATMENTS IN PSYCHOSIS

56

PSYCHOANALYTIC TREATMENT OF PSYCHOSIS

Elyn R. Saks

Psychodynamic treatment of psychosis, particularly schizophrenia, is controversial. Freud himself felt that psychotic people were too narcissistic, too inward looking, to engage in psychoanalysis; there could be no transference.[1] Others point out that analysis promotes regression, and psychotic people are already too regressed.[2] Still others point out that psychotic people do not have a functioning observing ego, which is necessary for surveying and assessing what is going on in one's mind.[3]

In this chapter I focus on psychoanalysis and long-term psychoanalytic psychotherapy. That is, I am not going to look at short-term psychodynamic treatment. I also am skeptical that there is a meaningful distinction between psychoanalysis and psychoanalytic psychotherapy, provided the latter involves at least 4-day-a-week treatment. I will use these terms interchangeably.

Despite Freud's pessimism about analytic treatment of the psychoses, early in the history of psychoanalysis, strong claims were made that psychosis could be treated psychoanalytically. This has been referred to as the "widening scope of psychoanalysis."[4] Analysts came to see this as overclaiming as to what psychoanalysis could do. The view that psychoanalysis was not helpful for people with psychosis became the conventional wisdom.[1]

More recently, there has been a return to thinking that, if conjoined with medication, psychoanalysis can help with psychosis.[5]

Certain schools of psychoanalysis, of course, are more open to this than others. Kleinians, for example, or Bionians, do treat people with psychosis, to good effect.[6]

In this chapter I review some of the more contemporary literature on psychodynamic treatment of psychosis, limiting myself to articles from 2016 and 2017. I then talk about some issues implicated by using psychoanalysis with psychosis. And finally I speculate on ways psychoanalysis can help with psychosis. I base this largely, in fact, in my own experience as a psychotic person benefiting from analysis.

Using one's first-person experience in discussions of psychoanalysis has a long history, beginning with Freud himself[7] and including others, like Kohut, who presumably was writing about himself in the classic paper, *The Two Analyses of Mr. Z.*[8]

In this 2016–2017 review I make some choices. I count therapies denoted as "mentalization" therapies as analytic. I count therapies described as "insight" as analytic. And I include "integrative" approaches for which analytic therapy is a component.

Hamm and Firmin[9] discuss a therapy with the acronym MERIT: meta-cognition reflection and insight therapy. They show MERIT at work with a disorganized schizophrenic man. He experienced improved coherence of speech as a component of his treatment and recovery.

De Jong et al.[10,11] also discuss meta-cognitive reflection and insight therapy with a patient with schizophrenia. This treatment approach can entail the use of cognitive behavioral, humanistic, or psychodynamic techniques. These would be interlaced in accordance with their relevance for promoting a synthesis of an integrated sense of the self and others.

Debbane et al.[12] discuss mentalization-based therapy in people at clinical high risk for psychosis (CHR). The authors distinguish between this and meta-cognition, while recognizing a role for both for preventative therapeutic treatment for CHR.

Holm-Hadulla et al.[13] describe a model of integrative psychotherapy for schizophrenia with cognitive-behavioral, psychodynamic, and existential elements.

Carr et al.[14] recommend integration of cognitive behavioral and psychodynamic approaches for people with psychosis. Also, they direct attention to recovery-oriented care concepts, cultural implications, and the personal narratives of each individual.

Koritar[15] identifies a shift in the analytic world between seeing therapeutic efficacy as deriving from making the unconscious conscious to seeing it as deriving from repeating one's past experience in the transference/countertransference. An extended discussion looks at the analysis of a psychotic patient. The analyst's ability to experience the patient's tortured inner world and the analyst's ability to symbolize this experience resulted in a newfound ability in the patient to reflect on and analyze his internal object relations.

In his book review of Garfield and Steinman[16] Bacal discusses the authors' self-psychology approach to a patient with psychosis. The reviewer questions whether the therapists were somewhat unique in their ability to work with this patient.

Hamm et al. (2017)[17] argue that self-direction is essential to recovery. There are complicating factors to this. Integrative psychotherapy has a potential role in optimizing self-direction and self-determination in the treatment of schizophrenia. They allege there is a particular value of therapy that is tied to its potential to address the barriers to self-directed recovery.

Hamm and Leonhardt[18] describe the role of interpersonal connection, personal narrative, and metacognition in integrative psychotherapy for schizophrenia. This approach combines cognitive, psychodynamic, and existential and dialogic models of self-experience.

Hamm et al.[19] argue that fragmentation of self-experience is central to psychosis and should be attended to in therapy. They suggest analytic therapy might be useful. They also identify the problem of the therapist colluding with the fragmentation. Therapists should relate to the whole person rather than a constellation of fragments.

Hasson-Ohayon et al. (2017)[20] propose a meta-cognitive intersubjective model for understanding challenges while working with psychotic patients. They identify three possible barriers to understanding challenges when working with such patients: differing client and therapist narratives about the role of the mental health system; different levels of client insight and therapist theoretical perspectives; and stigmatizing beliefs held by client and therapist alike.

In short, the idea of psychodynamic treatment for psychosis is alive and well. Many contemporary advocates for this recommend conjoint treatment, i.e., analytic therapy *and* medications.[21] Historically the norm was to think analysis could not happen if someone was on medications; we no longer think that.[22]

Having reviewed some of the literature, I now discuss some of the theoretical issues that psychodynamic treatment of psychosis raises.

First, at least with classical theory, conflict is emphasized versus deficit.[23] But perhaps psychosis is on a continuum with neurosis and therefore possibly treatable by a conflict approach. Many do indeed think that mental illness is dimensional rather than categorical.[24] So schizophrenia is on a spectrum with bipolar disorder, and bipolar disorder with psychotic depression, and psychotic depression with nonpsychotic depression.

The fact that these conditions may share symptoms, and that many of the same treatments work well for them, adds further fuel to the fire.[25]

That said, many if not most people also believe that the more serious illnesses do involve a biochemical dimension—a literal physical "defect."[26] Still, psychological approaches may help address such physical problems. If one is in a happier state of mind, for example, the pain of one's broken arm may be less. Of course, while the treatment may lead to more happiness, it does not cure the break itself. But some studies actually show that the same brain changes appear at the end of psychotherapy for depression and medication for depression.[27] Talking can in fact alter brain states.

And even if the "physical" problem can't be addressed by therapy, an ill person will have many issues on top of the biological issues; and these may involve conflict and may be successfully addressed by a conflict approach. For example, an anxious or angry person may have certain neurons firing or not; but may also be angry or anxious because she has authority issues as a result of her childhood; and these may impair her work life and be treatable by psychotherapy. And in general, even people with such serious disorders as schizophrenia can have work and relationship issues, just like everyone else; and these can be addressed in therapy, just like with everyone else.

Another important point, briefly acknowledged earlier, is that different schools of psychoanalysis have different theories of what psychosis consists in; and what needs to be done to effectively address it. For example, a self-psychologist will see deficits in healthy narcissism and think the patient needs a mirroring, empathic therapist—a good self-object.[28] Kleinians see some severe psychopathology as involving pre-Oedipal issues like envy or greed.[29] They may also see people with such pathology as never having passed through the paranoid/schizoid position into the depressive position.[29] Kleinians also speak in the language of the unconscious and think that getting to the patient's most profound anxiety is what will help him or her.[29]

In my own experience, I have had analysts with different theoretical orientations: a Kleinian, two classical analysts, an eclectic analyst, and a mentalization analyst. My experience, I think, provides good evidence that Freud was wrong about psychotic patients being unable to form a transference. I formed a very strong transference reaction to my therapists. And my hunch is that most psychotic people do.

The transference itself, of course, can be psychotic. I joke around that as I got better my view of my analyst went from thinking he or she was a dangerous monster wanting to kill me to thinking that he or she just didn't like me.

The regression point makes sense but shouldn't be dispositive. Further regression perhaps doesn't matter—it's malignant regression that does. And this can be managed in the therapeutic relationship.[30] As an example, the patient can be asked to sit up if the anonymity of the situation feels toxic to her and she starts decompensating.[31] Or certain questions can be answered if silence and passivity feels too threatening.[31]

Other schools may also welcome further regression of the patient.[32] It is only when the patient experiences the most infantile experiences, as for example Kleinians may think, that he or she can emerge free of his or her conflicts and other challenges.[32]

I have been in intensive psychoanalytic psychotherapy for thirty-eight years. I know the theory—that psychoanalysis is supposed to end. Termination is an important, even critical, part of the process.[33] But I am a lifer—I don't want to risk decompensation, which is what I experienced the last time I tried to stop treatment.

How might psychoanalytic treatment for psychosis work? Analytic therapy does a variety of things; not all are specific to psychoanalytic treatment, but all are part of why it helps. And these are things, I think, that occur across the different schools of psychoanalysis.

I note again that these ideas derive from my own experience. But there are reasons to think, again, this can be useful. In the end, studies will be needed, especially controlled studies, to see if these are really mechanisms that contribute to the efficacy of the process.

First, stress is bad for all illnesses, particularly mental illnesses.[34] Psychoanalytic treatment can help the patient identify her stressors and either figure out a way to avoid them or a way to cope with them.[35] As an example, I find traveling and giving talks stressful. So I try to give most talks via video. And when I do travel, I tend to travel with my husband, which can also help.

Second, psychoanalytic treatment fortifies the patient's observing ego.[36] He becomes better able to step back, observe what he is thinking and feeling, and identify problems with his thinking. He becomes much more in touch with his own mind.

For me, developing an observing ego has been critical. It was not until I got into analytic treatment that I began to identify what I was thinking and feeling—to identify the contents of my mind.

I came from a family that was not psychologically minded; we didn't learn to identify the things we were thinking or feeling, particularly "bad" or "negative" feelings. One of the biggest changes that resulted from my treatment was being able to mentalize my own mental states. And this extended to understanding others' minds too. My psychological world, so to speak, opened up to me.

An observing ego is also extremely important for people who are tempted—or tormented—by magical or psychotic thinking.[37] "I need to dismiss this thought; it's likely my illness acting up." Developing the ability to recognize when one's illness is acting up is a very important skill.[38] One can then take steps to address this before one's mind spins wildly out of control.

Third, psychodynamic treatment provides a safe place for the patient to bring her own feelings.[35] To talk about killing people with one's thoughts is not something you casually bring up with friends. Being able to say out loud the crazy and scary things the patient is feeling means he doesn't have to say them on the "outside." It serves as a sort of steam valve releasing steam.[38]

Fourth, psychodynamic interpretations can help detoxify frightening thoughts.[39] I remember one day saying a lot of violent things and my New Haven analyst, Dr. White, said that he thought I was saying violent things because I was really scared: the violence was my defense against fear. This made a lot of sense to me and my thinking quieted down a lot.

That said, therapists have mixed feelings about interpreting psychotic thoughts. One view—held by no dynamic therapist, I would think—is that psychotic ramblings are random firings of neurons with no meaning.[40] A second group thinks psychotic thoughts are meaningful—that essentially they tell the truth about one's psychic reality—but that it is not useful for the thoughts to be interpreted; the patient can't hear the interpretation.[41] The last view is that the thoughts tell the truth about psychic reality; and that it is often meaningful and helpful to interpret them.[42]

I am in the third school. I, say, have a thought that a classmate was trying to kill me. My analyst might relate that thought to my envy that she got a better grade than I did; my envy turned into anger and then was projected onto the classmate. An interpretation like this would be (has been) very helpful to me.

Last, in psychodynamic treatment one is working (one hopes) with a therapist who accepts one. A kind, and smart, and well-meaning person values you not only for the good, but also the bad and ugly. This is incredibly empowering and helps defuse a sense of shame about oneself.[43]

Of course, how analytic therapy helps is the $60,000 question, both with psychotic and with neurotic patients. Some analysts focus on insight, some on relationship.[44,45] I think both are equally important.

Another way to think of dynamic treatment of psychosis, as I have mentioned earlier, is that therapy can help with people's issues around work and relationship; and people with psychotic disorders can be helped on these issues too. We who struggle with mental illness want what Freud noted that everyone wants—to work and to love.

Still another idea is that people think today in terms of recovery.[46,47] We want to focus not just on reduction and remission of symptoms but also on quality of life. And, importantly, it is for the patient herself to determine what that is for her. Psychodynamic treatment can help people identify what quality of life is for them; and how best to achieve it.

These thoughts about how analysis can help are of course completely atheoretic. Studying how analysis works both with psychosis and the more garden-variety reasons people seek care is an important project.

I would like to make one final point: I think many of the other psychotherapeutic approaches, both to psychosis and to neurosis, borrow from psychoanalysis. For example, "underlying cognitions" in cognitive-behavioral therapy are the "unconscious fantasies" that the analyst uncovers. The techniques are of course different—e.g., we don't give analytic patients "homework"—but the approaches share a lot in common.

Indeed, analysis has affected ordinary MDs too. I hugged my doctor after hearing I was OK in a cancer context, and said words to the effect that "I don't really know you, but I really love you." He walked away, and then turned around and said "it's all transference!" For an ace cancer surgeon to say this was quite stunning.

Finally, it has lately been recognized that a combination of therapy and medications is the treatment of choice even for schizophrenia.[48] For many years the conventional wisdom was that medications were enough for schizophrenia and that therapy didn't help anyway.[49,50] The gold standard today, as reported even in the *New York Times*,[51] is that combined medication and therapy works best, for schizophrenia as for many other psychiatric disorders—and many other of the psychological challenges that people face.[52–54]

So, perhaps for some people, psychosis can go hand-in-hand with psychodynamic treatment in addition to medication. Perhaps the "widening scope," if it includes medications as well, was not so mistaken as once thought.

REFERENCES

1. Knapp PH, Levin S, Mccarter RH, Wermer H, Zetzel E. Suitability for psychoanalysis: a review of one hundred supervised analytic cases. *Psychoanal Q* 1960;29:459–477.
2. Thomas Z. Breaking through to the other side: a resident explores the benefits of time-limited psychodynamic therapy for patients with schizophrenia. *Psychodyn Psychiatry* 2017;45(1):59–77.

3. Garfield DA. Psychoanalytically informed psychotherapy of psychosis: the influences of American psychoanalysis. Panel report. *J Am Psychoanal Assoc* 2011;59(3):603–620.
4. Stone L. The widening scope of indications for psychoanalysis. *J Am Psychoanal Assoc* 1954;2(4):567–594.
5. Leichsenring F. Are psychodynamic and psychoanalytic therapies effective? A review of empirical data. *Int J Psychoanal* 2005;86(Pt 3):841–868.
6. Aguayo J. On understanding projective identification in the treatment of psychotic states of mind: the publishing cohort of H. Rosenfeld, H. Segal and W. Bion (1946–1957). *Int J Psychoanal* 2009;90(1):69–92.
7. Chessick RD. What hath Freud wrought? Current confusion and controversies about the clinical practice of psychoanalysis and psychodynamic psychotherapy. *Psychodyn Psychiatry* 2014;42(4):553–583.
8. Kohut H. The two analyses of Mr. Z. *Int J Psychoanal* 1979;60(1):3–27.
9. Hamm JA, Firmin RL. Disorganization and individual psychotherapy for schizophrenia: a case report of metacognitive reflection and insight therapy. *Journal of Contemporary Psychotherapy* 2016;46(4):227–234.
10. de JS, van DR, Pijnenborg GH, Lysaker PH. Metacognitive reflection and insight therapy (MERIT) with a patient with severe symptoms of disorganization. *J Clin Psychol* 2016;72(2):164–174.
11. de JS, van Donkersgoed RJ, Aleman A et al. Practical implications of metacognitively oriented psychotherapy in psychosis: findings from a pilot study. *J Nerv Ment Dis* 2016;204(9):713–716.
12. Debbané M, Benmiloud J, Salaminios G et al. Mentalization-based treatment in clinical high-risk for psychosis: a rationale and clinical illustration. *Journal of Contemporary Psychotherapy* 2016;46(4):217–225.
13. Holm-Hadulla R, Koutsoukou-Argyraki A. *Integrative psychotherapy of patients with schizophrenic spectrum disorders: the case of a musician suffering from psychotic episodes*. 26 ed. 2016.
14. Carr ER, McKernan LC, Hillbrand M, Hamlett N. Expanding traditional paradigms: an integrative approach to the psychotherapeutic treatment of psychosis. *Journal of Psychotherapy Integration* 2017;No.
15. Koritar E. Experience in the third space with a psychotic patient. *Canadian Journal of Psychoanalysis* 2015;23(1):114–123.
16. Bacal H. Self psychology and psychosis: the development of the self during intensive psychotherapy of schizophrenia and other psychoses. By David Garfield and Ira Steinman. *Psychoanal Q* 2016;85(4):1008–1017.
17. Hamm JA, Buck KD, Leonhardt BL, Luther L, Lysaker PH. Self-directed recovery in schizophrenia: attending to clients' agendas in psychotherapy. *Journal of Psychotherapy Integration* 2017;No.
18. Hamm JA, Leonhardt BL. The role of interpersonal connection, personal narrative, and metacognition in integrative psychotherapy for schizophrenia: a case report. *J Clin Psychol* 2016;72(2):132–141.
19. Hamm JA, Buck B, Leonhardt BL, Wasmuth S, Lysaker JT, Lysaker PH. Overcoming fragmentation in the treatment of persons with schizophrenia. *Journal of Theoretical and Philosophical Psychology* 2017;37(1):21–33.
20. Hasson-Ohayon I, Kravetz S, Lysaker PH. The special challenges of psychotherapy with persons with psychosis: intersubjective metacognitive model of agreement and shared meaning. *Clin Psychol Psychother* 2017;24(2):428–440.
21. Fonagy P. The effectiveness of psychodynamic psychotherapies: an update. *World Psychiatry* 2015;14(2):137–150.
22. Keshavan MS, Kaneko Y. Secondary psychoses: an update. *World Psychiatry* 2013;12(1):4–15.
23. Crichton JHM. Psychodynamic perspectives on staff response to inpatient misdemeanour. *Criminal Behav Ment Health* 1998;8(4):266–274.
24. Kessler RC. The categorical versus dimensional assessment controversy in the sociology of mental illness. *J Health Soc Behav* 2002;43(2):171–188.
25. Crow TJ. The continuum of psychosis and its implication for the structure of the gene. *Br J Psychiatry* 1986;149:419–429.
26. Jalava JV. *Science of conscience: Metaphysics, morality, and rhetoric in psychopathy research*. Simon Fraser University; 2007.
27. DeRubeis RJ, Siegle GJ, Hollon SD. Cognitive therapy versus medication for depression: treatment outcomes and neural mechanisms. *Nat Rev Neurosci* 2008;9(10):788–796.
28. McLean J. Psychotherapy with a narcissistic patient using Kohut's self psychology model. *Psychiatry (Edgmont)* 2007;4(10):40–47.
29. Trust TMK. *Envy and gratitude and other works 1946–1963*. Random House; 2011.
30. Furlan PM, Benedetti G. The individual psychoanalytic psychotherapy of schizophrenia: scientific and clinical approach through a clinical discussion group. *Yale J Biol Med* 1985;58(4):337–348.
31. Keats CJ, McGlashan TH. Intensive psychotherapy of schizophrenia. *Yale J Biol Med* 1985;58(3):239–254.
32. Gabbard GO. Classic article: introduction. *J Psychother Pract Res* 1999;8(1):64–65.
33. Picardi A, Gaetano P. Psychotherapy of mood disorders. *Clin Pract Epidemiol Ment Health* 2014;10:140–158.
34. Herbert J. Fortnightly review: stress, the brain, and mental illness. *BMJ* 1997;315(7107):530–535.
35. Manetta CT, Gentile JP, Gillig PM. Examining the therapeutic relationship and confronting resistances in psychodynamic psychotherapy: a certified public accountant case. *Innov Clin Neurosci* 2011;8(5):35–40.
36. Ridenour JM. Psychodynamic model and treatment of schizotypal personality disorder. *Psychoanalytic Psychology* 2016;33(1):129–146.
37. Peterkin AD, Dworkind M. Comparing psychotherapies for primary care: differences between short-term supportive and insight-oriented psychotherapies. *Can Fam Physician* 1991;37:719–725.
38. Wilson WH, Diamond RJ, Factor RM. A psychotherapeutic approach to task-oriented groups of severely ill patients. *Yale J Biol Med* 1985;58(4):363–372.
39. Craciun M. Time, knowledge, and power in psychotherapy: a comparison of psychodynamic and cognitive behavioral practices. *Qualitative Sociology* 2017;40(2):165–190.
40. Murphy D. Philosophy of psychiatry. In: Edward N. Zalta, editor. *The Stanford encyclopedia of philosophy*. Spring 2017 ed. Metaphysics Research Lab, Stanford University; 2017.
41. Murphy D. Explanation in Psychiatry. *Philosophy Compass* 2010;5(7):602–610.
42. Bentall RP, Beck AT. *Madness explained: psychosis and human nature*. Penguin Adult; 2004.
43. Kvrgic S, Cavelti M, Beck EM, Rusch N, Vauth R. Therapeutic alliance in schizophrenia: the role of recovery orientation, self-stigma, and insight. *Psychiatry Res* 2013;209(1):15–20.
44. Shedler J. The efficacy of psychodynamic psychotherapy. *Am Psychol* 2010;65(2):98–109.
45. Lilliengren P, Werbart A. *A model of therapeutic action grounded in the patients' view of curative and hindering factors in psychoanalytic psychotherapy*. 42 ed. 2005.
46. Cavelti M, Kvrgic S, Beck EM, Kossowsky J, Vauth R. Assessing recovery from schizophrenia as an individual process: a review of self-report instruments. *Eur Psychiatry* 2012;27(1):19–32.
47. Binder PE, Holgersen H, Nielsen GH. What is a "good outcome" in psychotherapy? A qualitative exploration of former patients' point of view. *Psychother Res* 2010;20(3):285–294.
48. Kane JM, Robinson DG, Schooler NR et al. Comprehensive versus usual community care for first-episode psychosis: 2-year outcomes from the NIMH RAISE early treatment program. *Am J Psychiatry* 2016;173(4):362–372.
49. Lieberman JA, Stroup TS, McEvoy JP et al. Effectiveness of antipsychotic drugs in patients with chronic schizophrenia. *N Engl J Med* 2005;353(12):1209–1223.
50. Miyamoto S, Duncan GE, Marx CE, Lieberman JA. Treatments for schizophrenia: a critical review of pharmacology and mechanisms of action of antipsychotic drugs. *Mol Psychiatry* 2005;10(1):79–104.

51. Carey B. New approach may alleviate schizophrenia. *New York Times* 2015 Oct 20;A1.
52. Cuijpers P, Sijbrandij M, Koole SL, Andersson G, Beekman AT, Reynolds CF, III. Adding psychotherapy to antidepressant medication in depression and anxiety disorders: a meta-analysis. *World Psychiatry* 2014;13(1):56–67.
53. Huhn M, Tardy M, Spineli LM et al. Efficacy of pharmacotherapy and psychotherapy for adult psychiatric disorders: a systematic overview of meta-analyses. *JAMA Psychiatry* 2014;71(6):706–715.
54. Pampallona S, Bollini P, Tibaldi G, Kupelnick B, Munizza C. Combined pharmacotherapy and psychological treatment for depression: a systematic review. *Arch Gen Psychiatry* 2004;61(7):714–719.

57

COGNITIVE-BEHAVIORAL THERAPY

Tania Lincoln and Alison Brabban

INTRODUCTION

The difficulties that patients with psychosis face are diverse and complex. On the one hand, there are the symptoms per se, such as persecutory delusions, threatening voices, or if negative symptoms dominate, the loss of drive and motivation. These experiences tend to be associated with anxiety, anger, or shame and with concerns related to the meaning or the consequences of the experiences. Psychotic symptoms tend to be associated with an array of interpersonal problems if patients believe they are being threatened by others, are misunderstood, cut off, or have difficulties engaging in meaningful communication. Ultimately, these problems can lead to increased social isolation. Moreover, the experience of an acute episode that might have involved voluntary or involuntary hospitalization can in itself be traumatizing. A psychotic episode may also leave a person feeling that there is something fundamentally wrong with him or her, which can be associated with a sense of hopelessness.

Reasons for coming into treatment

Andrew,[a] 26 years, student

Andrew sought outpatient therapy after being in hospital with his fourth psychotic episode. Inpatient treatment involving medication had considerably decreased his acute psychotic symptoms. However, he still reported low mood, lack of drive, and feelings of insecurity. He felt incapable at present of regaining morale and pursuing his studies. Furthermore, Andrew reported constantly worrying about relapse. Despite all of his efforts and support provided by his parents he had not been able to prevent relapse so far and he hoped that therapy would help him to "get a grip" on his psychosis.

Evelyn, 52, retired

Evelyn came to our outpatient unit without explicitly registering for the psychosis unit. She described problems with her family, her colleagues, and advisors in different psychosocial institutions. She reported to have been harassed, taken advantage of, betrayed, and laughed at by them. The disrespect and maltreatment by the family members was getting to her in a way that she felt she could no longer endure. Consequently, she had moved from the town she had been living in with them. She hoped that therapy would help to stabilize her mood. She also reported that the "schizophrenia" diagnosis in the report issued by her general practitioner made her angry as she felt that it did not apply to her. She also refused to take the medication he had prescribed.

Joanne, 26, student

Joanne was concerned about her next-door neighbors, who she felt might be abusing her in some way. She could hear them talking about her in the night, which caused her to lie awake feeling tense and anxious. She also felt distressed because she kept hearing people on the street saying nasty things about her. This was interfering with her academic endeavors at university. According to her medical report, she had developed paranoid schizophrenia a few years back. She acknowledged this and reported being afraid that all the stress she was having with the neighbors might cause a new episode.

a Note. All examples are from patients participating in two outpatient intervention studies modified in names and personal details and have also been provided in Lincoln and Beck.[1]

COGNITIVE-BEHAVIORAL INTERVENTIONS FOR PSYCHOTIC SYMPTOMS—A FEASIBLE APPROACH?

Most of the earlier clinical textbooks that describe cognitive-behavioral interventions for psychological disorders note that these interventions are contraindicated when it comes to schizophrenia or psychotic disorders, the concern being that targeting symptoms directly was likely to make matters worse. At the root of this concern was the assumption that psychotic symptoms were qualitatively different from normal experiences, purely biologically determined and therefore not amenable to reason and unlikely to be explicable or changeable by normal mechanisms of learning. Meanwhile, the idea that there is a qualitative difference between delusional and normal beliefs as well as hallucinations and normal perceptions has been strongly questioned by epidemiological studies that find high rates of delusion-like beliefs and hallucinations in healthy populations.[2,3] This demonstrates that it is difficult to draw a clear line between what we consider to be normal experiences and psychotic experiences. This continuum from healthy to clinical states of psychosis questions the belief that psychotic experiences are fundamentally different from normal ones and indicates that normal psychological mechanisms are involved in the formation and maintenance of psychotic symptoms.

Moreover, findings from research on cognitive and emotional correlates of psychotic symptoms suggest that psychotic experiences result from normal, though exaggerated, mechanisms of perception and reasoning and not merely from cognitive deficits. These findings have formed the basis for cognitive models of psychosis. As one of the most influential models in this area, Garety et al.[4] explain the development and maintenance of positive symptoms. This model draws on existing vulnerability-stress models of psychosis[5] and postulates that psychotic symptoms develop when stressors overload a person, causing them to have unusual experiences, such as increased arousal, bodily sensations, or hallucinations. According to this model, it is not the unusual experience itself that is crucial but the appraisal of the experience. For example, a person might attribute auditory hallucinations or other "strange" physical experiences to external factors (e.g., "Someone is trying to manipulate me") rather than perceiving the experience as a result of increased stress exposure (e.g., "I must be hearing voices because I'm so stressed and haven't slept for days"). The appraisal is assumed to be influenced by two core processes: the emotional condition of the person (e.g., someone already feeling anxious would be more likely to appraise the experience as threatening); and reasoning biases, such as a jumping-to-conclusions bias or attributional biases (see Moritz et al. in this book). The observation that delusional beliefs and other psychotic symptoms are at the extreme end of the "normal" spectrum of beliefs and appear to result from normal mechanisms of reasoning and perception has paved the way to modify CBT interventions originally developed by A.T. Beck for depression[6] to work with people experiencing psychosis.

THE PROMINENT TECHNIQUES

The following descriptions of the intervention used for positive symptoms are based on manuals of influential British researchers in this area,[7–10] a German treatment manual[11] and clinical experience. It needs noting though that CBTp is continually evolving and there is debate about which of the many new psychological interventions used in the treatment of psychosis may or may not be subsumed under the CBTp label.[12]

Most descriptions of CBTp, however, do converge on a number of key characteristics: One is stressing the importance of building a stable therapeutic relationship as well as on the development of an individual case formulation to develop a shared understanding of how symptoms might have arisen and are being maintained. Another is that distressing symptoms are conceptualized as part of a chain of preceding and resulting thoughts, behaviors, and feelings. The use of cognitive and behavioral interventions for working with psychotic symptoms as well as for changing dysfunctional beliefs about the self, other persons, and interventions to prevent relapse, is also an essential element. "Dysfunctional beliefs" might be the delusional beliefs per se, but could also be thoughts *about* the symptoms or could be thoughts related to the self or others. Ultimately, CBTp is geared toward helping patients work toward or achieve their personal goal(s). Psychotic symptoms are often a key obstacle to achieving these goals and as such will often but not always need to be dealt with as a core focus of the therapy.

CBTp is a collaborative process, sharing knowledge and skills with the patients to help them to manage their symptoms and distress going forward. Through the process of listening, validating, and educating, the therapy should aim to provide hope and to improve the patient's sense of self-efficacy and self-esteem.[13,14]

BUILDING A STABLE RELATIONSHIP

The essence of establishing a stable relationship is listening to the person and trying to understand their experiences regardless of how bizarre or unreasonable some of their beliefs might seem at first. Thus, the therapist strives toward seeing the world through the eyes of the patient. To build rapport, the therapist conveys understanding and empathy toward the reactions and feelings that arise from symptoms (e.g., "I can understand that you were really upset when you were worrying that you were being watched, even in your own house. It makes sense that you thought you needed to withdraw or hide."). Patients are more likely not only to be open about their unusual experiences but also to question their interpretations at a later point in therapy if they feel validated and understood. Some patients with delusions tend to be defensive because their experiences have been dismissed as irrational by friends and relatives or as delusional or insane by mental health professionals in the past. Indeed we have heard several patients say that they never went back to a therapist they felt was not taking them seriously. For building rapport and progressing in therapy it is essential that the therapist keeps in mind that CBTp is not about finding out what is "really" going on, but about finding out which beliefs are unhelpful and distressing for a patient. Nevertheless it can be helpful for a therapist to be aware that the prevalence of experiences of physical and sexual abuse and various forms of discrimination and social exclusion[15] are frequent in people with psychosis and "real" adverse experiences are likely to be at the bottom of many delusional beliefs.

To reduce symptom-related distress, the therapist uses a normalizing approach, emphasizing that psychotic experiences can be considered as fairly common (e.g., "Hearing voices does not necessarily mean you are crazy, because many people who are not considered to have mental health difficulties report hearing voices and still lead normal lives."). The effect this can have on patients is expressed by a former patient who was interviewed for a local magazine about his experience of CBTp: "It was like an enormous weight lifted when my therapist explained that that my thoughts and feelings were just an extreme variation of normal phenomena that can occur when someone is very stressed" (*Berliner Zeitung*, July 15, 2016).

Finally, an important aim in the first part of therapy is to convey hope for recovery and thereby enhance the patient's motivation to engage actively in promoting change. Focusing on the individual's personal goals as the intended outcomes delivers an intrinsic message that these are achievable and life can improve.[13]

DEVELOPING AND WORKING WITH FORMULATIONS

Another part of therapy is the collaborative development of a formulation, which refers to psychological mechanisms that explain how a person's problems might have arisen and are being maintained. Formulations use models derived from cognitive-behavioral research linking theory to the information provided by the individual patient. These formulations constitute working models that hold plausible explanations of experiences for the patient and build a foundation for individualized problem-orientated approach in therapy. Figure 57.1 illustrates a prototype formulation including an individual case example. This particular formulation postulates psychological mechanisms to explain anxiety and suspiciousness leading up to persecutory delusions and builds on the model published by Freeman and colleagues.[16]

Moreover, all of the described therapeutic approaches are derived from the assumption that symptoms, whether they are hallucinations, delusions, or negative symptoms, appear in a certain context characterized by emotional, cognitive, physiological, and behavioral factors. The therapist identifies situations, in which the "symptom" occurred and then elicits the triggers and short- and long-term cognitive, emotional, and behavioral consequences. This analysis of typical triggers and maintaining factors can then be used to derive targets for therapy.

INTERVENTIONS FOR AUDITORY HALLUCINATIONS

One of the first types of interventions developed for hallucinations focused mostly on the behavior that followed hallucinations by enhancing the use of coping strategies.[17] In this approach, which aims to increase a sense of control over hallucinations, patients are guided to write baseline-reports about the voices—their volume and duration—and their reaction to the voices. Then, both existing and new coping strategies, such as social communication, relaxation, music, withdrawal, etc., are evaluated in regard to the influence they have on the frequency and duration of the hallucination as well as the distress caused by it. The aim is to adopt the strategies that prove to be the most helpful. By helping patients to develop and use effective coping strategies, therapists may also be able to explore a number of key beliefs, such as "I have no control over my voices." Similarly, beliefs about the identity of a voice might be questioned once individuals realize they have some control over their experiences (e.g., would you be able to make your voices go away if it was the Devil?)

Subsequent approaches have focused more strongly on cognitive aspects, identifying and exploring certain beliefs

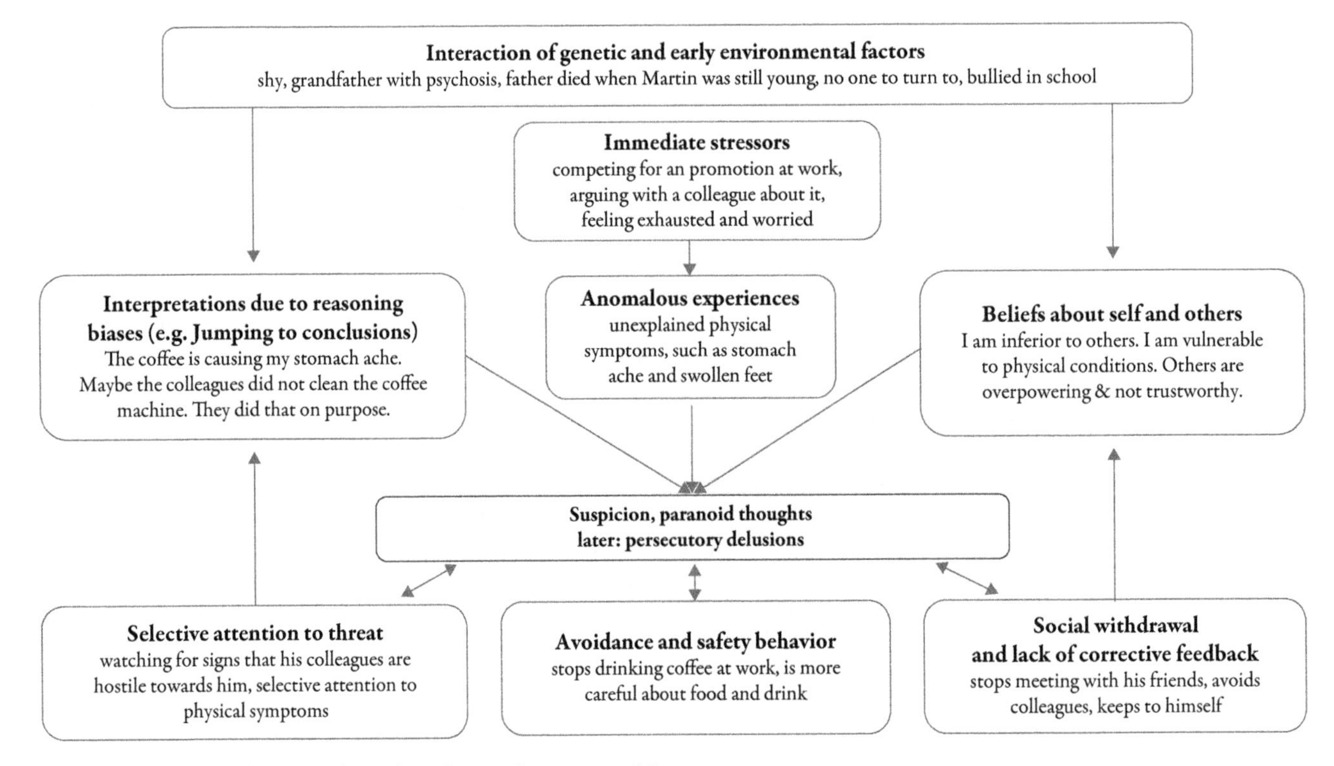

Figure 57.1 Example of an individualized case formulation of persecutory delusions

about the hallucinations that were considered to be linked to distress.[18] These types of so called "dysfunctional beliefs" may be catastrophizing (e.g., "Hearing voices means that I am crazy, need to go to hospital, and will never a live normal life again.") or considered delusional ("Hearing voices means that aliens must have implanted a chip in my brain," or "the neighbors are talking to me through the walls to frighten me."). They may also be related to the power or omniscience of the voices ("My voice has the power to harm me," "My voice knows everything about me") or the consequences of noncompliance with a voice ("I will be punished if I don't do what the voice says"). Figure 57.2 provides an illustration that could be used in therapy of how these types of beliefs can lead to further arousal and tension and thereby to the maintenance of voices and voice-related distress.

To reduce voice-related distress, CBTp uses methods and techniques found in traditional cognitive therapy to modify the beliefs seen as causing the distress. The therapist explores the patient's beliefs using gentle and empathic Socratic questioning (e.g., "What is the evidence that your voice has power over you? Is there anything that seems to challenge the power of your voices?") and guided discovery (e.g., "You told me last week that you did not respond to your voice during your conversation with Karen and that this helped you to concentrate better on what she was saying. What do you make of that in terms of the power of your voice?"). Furthermore, behavioral experiments can be helpful. For example, the belief that the voice is uncontrollable may be challenged by an exercise in which the patient discovers that he or she can suppress the voice by using coping strategies such as reading out loudly or by talking to others. Catastrophic interpretations of voices can be challenged by information about healthy voice hearers and normalizing.[7] Some interesting websites now provide numerous examples of how voice hearers cope with their voices (e.g., www.intervoiceonline.org). Another approach is to make links between low self-esteem and related beliefs about self and the content of voices and to focus on changing beliefs about self.

Working with auditory hallucinations with Joanne

To reduce the distress related to Joanne's auditory hallucinations, her therapist began by supporting Joanne to use coping-strategies in a more systematic manner. To set out, they used self-monitoring reports in which Joanne noted the appearance of hallucinations, the subjective sense of distress associated with them, and the coping strategies she was employing. By keeping record of the precipitating events, it became evident that Joanne's voices occurred most frequently when she was worrying about her upcoming exams and ruminating over the exhausting internship she was currently enrolled in. The protocols also demonstrated that not all of the coping strategies she was currently using were being helpful on a longer-term basis. For example, in order to cope with voices bothering her in the evening and the resulting difficulties in falling to sleep Joanne often drank wine. To escape from the voices on the street she withdrew into her room and listened to loud music. Joanne identified the "drinking wine strategy" as unhelpful, since it was only effective short-term and she often felt hungover in the morning. Similarly, she identified social withdrawal as not helpful, because it lowered her mood. She tested out several new strategies, of which two ("talking to roommates" and "distraction through chores or learning") turned out to be helpful, leading to reductions in distress and volume of voices. However, Joanne's conviction that a conspiracy against her was in progress made it difficult for her to adopt the new strategies, especially the strategy of talking to roommates, and caused an increase in focus on the voices. Thus, the therapist then began to intensify the work on delusions and the delusional beliefs related to the hallucinations.

There are also several novel approaches currently being developed that may also be classified under the umbrella of CBTp to varying degrees.[12] One of these approaches (relating therapy[19]) aims to change subordinate or aggressive styles of relating to the voices through assertiveness training, in relation to both voices and other individuals in the person's social environment. Another approach (avatar therapy[20]) uses

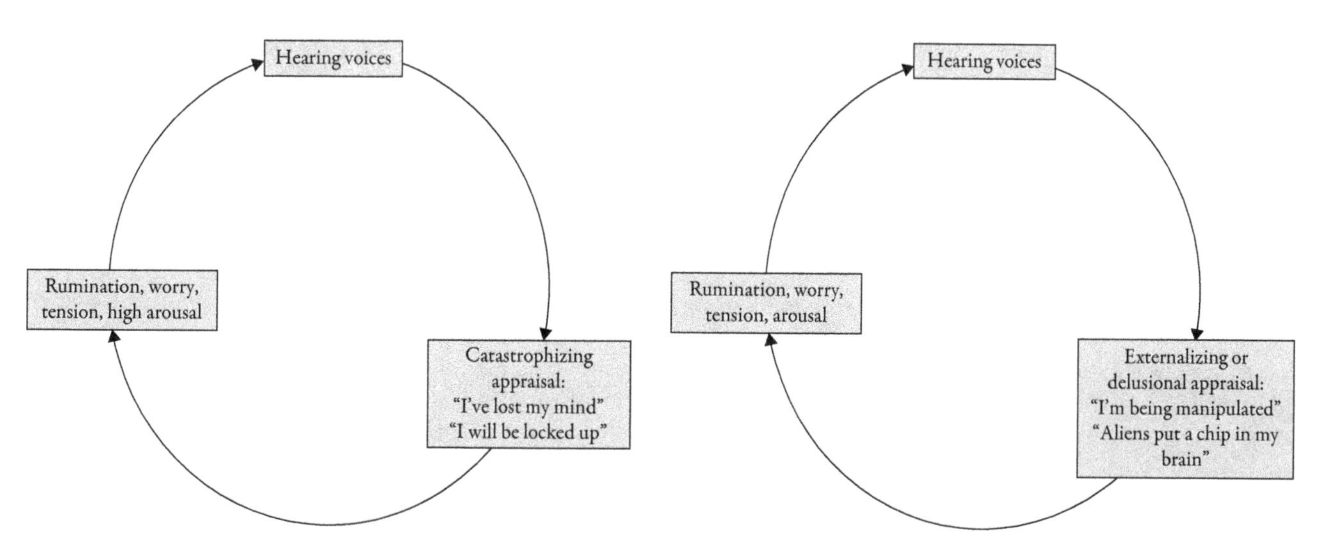

Figure 57.2 Vicious circles of voice maintenance

computer technology to create an avatar of each individual's voice, which is then controlled by the therapist in interaction with the patient. The avatar enables a powerful but safe "exposure" to the voice, leading to an increased sense of control over the voice as the avatar changes from being abusive to supportive in the course of therapy. In competitive memory training (COMET[21]), memories associated with positive self-esteem are retrieved and strengthened. The activation of this positive self-image is then used to weaken the negative content of voices and enhance self-esteem. Here too, the aim is to change the submissive relationship to the voices.

WORKING WITH DELUSIONS

It is essential to prepare the work on delusions thoroughly. Most importantly, and as described in the paragraph on therapeutic relationship, a premise is that the therapist is able to understand the origins of the delusional beliefs and how they are linked to the patient's background. Often, the beginnings of a delusion are easier to follow and validate than a complex conspiracy system that patients may present.

The beginning of delusional beliefs in Evelyn

Evelyn developed the idea that other family members were conspiring against her in a phase that was characterized by intense family conflicts. One day, she came back home and overheard her brother talking to her mother on the phone, mentioning her name. The first thought that occurred to her was that they were probably saying something negative about her.

Another step that can be helpful before beginning to challenge delusional beliefs is to look at what may motivate the patient to question the delusion belief as well as what may be motivating him or her to hold on to it. This can be done by collaboratively treating the core delusional beliefs as a hypothesis that can be correct, noncorrect, or partly correct. The therapist and patient explore the positive and negative consequences that holding on to or giving up the delusional beliefs might have. From the point of view of the patient, two options are particularly relevant. First, there is the option of giving up the belief. Evelyn could adopt the belief that she was stressed and feeling insecure and therefore misinterpreted the behavior of other family members as a conspiracy. However, if she takes this stance when, in fact, she interpreted the situation accurately (i.e., the family members were indeed talking about her in a negative way and planning how to get rid of her), then she would run the risk of trusting others mistakenly and becoming a naïve victim. Alternatively, Evelyn could make the mistake of holding on to the delusional belief even though it is unsubstantiated. In this case, she would risk of investing time and effort (e.g., moving town, giving up social contacts, etc.) for a false conviction. Moreover, she would miss the chance to take care of other important areas in life, such as making or obtaining meaningful work or seeking professional help.

Identifying possible consequences of abandoning or sticking with the delusional beliefs not only is relevant to enhancing the motivation to change but also provides the therapist with a clearer picture of why a patient might not (yet) be ready to question the validity of his or her beliefs. Even though there is no scientific evidence as yet to show that working with delusions can have possible adverse effects, exploring the validity of delusions tends to be unpromising if there is little to gain for the patient but a lot to be lost. This might be especially likely for delusions of grandeur. Some patients might find it difficult to let go of a delusional belief because they believe that confessing to having been mistaken comes equal to confessing to being crazy and needing help. To counter this concern, the therapist can discuss the normal mechanisms of how beliefs are formed by using examples such as religious or political beliefs, which make clear that it is not uncommon to adhere to beliefs with strong conviction even in the absence of convincing evidence.

Once a sufficient motivational basis is established, the therapist can proceed to a more direct albeit gentle exploration of delusional appraisals. Note that this part of therapy is not about who is right or wrong, but about gently supporting the patient to modify beliefs in a way that reduces distress. Moreover, it is about helping the patient to develop alternative explanations, even in the absence of definite evidence for or against any of the interpretations. Research by Freeman et al.[22] demonstrated that those with paranoid beliefs often struggle to generate alternative explanations for events, therefore the therapist may need to take a more proactive approach and suggest possibilities. Patients can then be encouraged to look at evidence in favor and not in favor of each alternative. Having alternative explanations for events can make it easier for a person to let go of the original belief in favor of another viewpoint.

In order to visualize the impact that dysfunctional and functional interpretations of events have on feelings and actions, the therapist can use the ABC-model of cognitive therapy, using diagrams to show the links between activating events (A), beliefs (B), and consequences (C). In working with delusions it is advisable to begin with a belief that is recent and less determined. For example, the therapist could ask: "Martin, you described that you were feeling upset at work yesterday when you thought that someone from work might have put something in your drink. You were thinking that these people might be part of the conspiracy against you. In the previous sessions we talked about why you might be more likely to believe that people are out to harm you. Would it be ok with you if we have a closer look at the evidence for and against your interpretation of what happened?" If the patient agrees, the therapist and patient go on to collaboratively note the evidence for the belief (e.g., coffee tasted strange, colleagues exchanged glances, stomachache in the evening, etc.) and against the belief (still alive today, coffee never tastes the same, etc.). The therapist may then discuss advantages of staying open-minded about the situation even in the absence of certainty of whether the belief is true or not. The effect this can have on patients' well-being is nicely expressed in an interview with a patient who received CBTp: "I learned that I don't have to surrender to my feelings—this was totally new for me. The other day I was in a restaurant and wasn't served at once. After a while, I started to think that the waiter didn't want me to be in there. Before therapy I would have left the restaurant frustrated or would have got worked up and complained. Now I can tolerate situations like that and consider the possibility that the waiter maybe just hasn't seen me or is stressed" (*Ihre Gesundheitsprofis Magazin*, June 14, 2016).

As an alternative to challenging the delusional beliefs per se, several more recently developed techniques focus on meta-cognitive beliefs (beliefs about beliefs). These approaches identify and work on thoughts that tend to be associated with the presence of delusional beliefs (e.g., "better to be distrustful than naive" or "because of my paranoia, I am on the safe side") and are hypothesized to be a maintaining factor.[23] Other approaches focus on the styles of information processing related to delusions. For example, the meta-cognitive training[24] teaches patients to modify their cognitive processing styles, such as to collect more information before drawing a conclusion. Another recent line of research has focused on targeting "causal" psychological factors, such as worry, sleep, or self-esteem.[25]

WORKING WITH NEGATIVE SYMPTOMS

Patients with negative symptoms are often made to feel that their problems are a result of weakness or laziness, or an unchangeable fate caused by the disorder. To correct these misconceptions and reduce pressure, the therapist can emphasize that withdrawal of the patient is understandable, and probably even a sensible response or survival strategy to protect himself from further distress.[26] For example, the therapist may say: "Everyone has problems staying motivated at times. This may be especially hard to do when we are under stress and experiencing a range of problems that make us feel vulnerable."[26(p270)] The next steps then aim at gradually increasing motivation to achieve personally meaningful goals and to support behavioral change. To begin with, the therapist and patient work together to identify attractive short- and long-term goals. To arrive at a shared understanding of what is preventing a patient from pursuing these goals, the therapist and patient collaboratively explore some of the recent situations in which the patient did or did not act in accordance with these goals. Beginning with the triggering situations, the responses of the patient are assessed on a cognitive, emotional, and behavioral level, as are the short- and long-term consequences of his or her behavior:

Example of not *reaching goals:*

Situation: Mate gave me a ring yesterday. I was lying in bed, feeling numb and slightly drowsy.

Response: Thoughts: Can't be bothered to get up. It will be awkward talking to someone now; Behavior: stayed in bed.

Consequences, short-term: Did nothing for next couple of hours, stayed alone, felt guilty; **long-term:** mate might give up asking me in future).

Example of reaching goals:

Situation: Left my room to make a cup of tea and noticed the pile of washing waiting to be done.

Response: Thoughts: I would feel better if I had some clean clothes to wear; Behavior: Put washing in machine and turned it on.

Consequences, short-term: felt brief sense of achievement; **long-term:** wore a clean shirt when mum came round the next day, which made me feel slightly more confident.

Building on the shared formulation derived from these individual examples, interventions for negative symptoms can be both cognitive and behavioral. On the cognitive level, the therapist attempts to elicit and challenge some of the core negative expectancy appraisals that tend to prevent patients from pursuing personally meaningful goals, such as "I am not going to have any fun anyway," and "I won't have anything to say."[26] These beliefs can be gently challenged (see also next section on schema work). Behavioral interventions can include activity monitoring and behavioral activation. They might also involve closer collaboration with the treating physician to change the medication status of a patient or encouraging the patient to increase his or her physical health condition. Social skills training can also be integrated into therapy if a lack of skills is seen as a maintaining factor for symptoms.

WORKING ON DYSFUNCTIONAL SELF-RELATED SCHEMAS

Working on improving self-esteem, self-acceptance, and acceptance of others is a vital aspect of CBTp. Core beliefs about the self and others and the circumstances that were associated with their development can be assessed using "life charts" (linking beliefs about self to previous and early experiences) and downward-arrow techniques (focused questions to assess self-schema). For a detailed description of cognitive intervention techniques see Leahy.[27] In a next step, these beliefs are linked to the delusional beliefs, the content of distressing voices or negative symptoms. The aim of this part of therapy is to get patients to question and abandon beliefs that are causing extreme distress or preventing them from achieving valued goals. This is achieved by exploring these assumptions using Socratic dialogue, reality testing, and discussing the implications of the beliefs. Because some patients find it difficult to describe thoughts and feelings, it can be helpful to work with imagery. In our work we found that schema therapy approaches[28] or techniques from compassion-focused-therapy[29] can also be useful. However, further evaluation of these approaches for psychotic disorders is needed. Finally, in order to achieve a stable growth in self-esteem and self-acceptance, patients should be encouraged to reengage in social interactions and support them in finding meaningful work.

Self-esteem focused interventions with Andrew

The therapist used imagery work that focused on childhood memories to help Andrew to understand where his beliefs of personal helplessness and inferiority arose. In the imagery exercises Andrew imagined scenes in which his father was looking down on him, shouting, and threatening to beat him up. In doing so, Andrew was able to describe the anxiety and feelings of defenselessness that went along with these types of experiences he had had with his father. Andrew realized that the extreme fear of others he experienced in his psychotic phases as well as the fact that he generally is easily intimidated by other people might originate in the feelings of helplessness toward the outbursts of his father. In therapy, the self-schema of the "defenseless boy" could be revised to match with many current experiences and a more functional self-schema could be established. In the end, Andrew was able to verbalize the fact that he is now a grown-up, taller in fact than his father, and able to stick up for himself. He began to feel more confident about being able to stand up to his father in future disagreements as well as to other persons if they crossed his boundaries. In order to further strengthen his sense of being able to defend himself he also took up more sports and enrolled for a karate course.

Relapse prevention with Andrew

Andrew and the therapist determined the preconditions, thoughts, feelings, and reactions that occurred prior to several previous psychotic episodes. He also asked his friends and parents about what they remembered. Andrew and the therapist were able to identify some similarities in the time prior to the episodes. For example, it became evident that exams and relationship conflicts tended to have a triggering effect. Also, Andrew discovered that his tendency to attach greater importance to minor things (e.g., interpreting things as special cues) might be a warning signal or an early symptom. Talking about these phases was also helpful in identifying some reactions that had helped Andrew in the past, such as relaxing activities, meeting friends that he had known for a long time and felt safe with and strategies that had turned out to be unhelpful or even increased tensions, such as certain complicated social relationships. Andrew wrote down these "early warning signals" as well as a number of helpful strategies for handling these. He involved his closest friends by giving them advice about what types of help he would prefer should a new episode occur. This also included the permission to contact his doctor.

RELAPSE PREVENTION

One aim of relapse prevention is to prepare patients to recognize early signs of relapse and respond to these signs. The therapist and patient collaboratively review what happened in the lead-up to previously experienced psychotic episodes. This involves a fairly detailed reconstruction of events, thoughts, and feelings occurring prior to previous psychotic episodes. It is helpful to use calendars or diaries, and note important markers, such as birthdays, travels, social events, having visitors, etc., that the patient can remember. As patients sometimes find it difficult to reconstruct these periods in their memory, it can be helpful to encourage them to involve friends or relatives. In a next step, the therapist gently probes for the more subtle experiences, thoughts and feelings. Although this kind of reconstruction can be time-consuming, it is an extremely helpful way of increasing the patient's awareness of individual risk-factors, early signals, and the development of symptoms. The patient is then encouraged to use this knowledge to recognize the signals and apply the therapeutic strategies (e.g., "I know that when I'm very stressed in my studies and several assignments are due in one go, I tend to hear my name or too easily get the impression that other people are talking about me behind my back. Therefore, at those times when I get the impression that other people are whispering about me, I need to remember that this is probably a stress symptom and it could be helpful to take things easier for a while.").

At the same time, it is crucial to reduce catastrophic appraisals of relapse, which are easily triggered by the awareness of early signals (e.g., "I'm sleeping poorly, this means I'm going to relapse.") and are likely to increase anxiety and thereby make relapse more, rather than less likely. The manual "Staying well after psychosis"[30] offers many valuable insights into the mechanisms of relapse and its prevention.

THE EMPIRICAL EVIDENCE

Today, the effectiveness of CBTp has been shown in many randomized controlled trials (RCTs). These RCTs typically compare CBTp combined with antipsychotic medication to antipsychotic medication alone (treatment as usual, TAU) or to medication combined with another psychological intervention (e.g., psychoeducation, supportive therapy). Numerous meta-analyses that differ in focus and methods summarize the findings from these studies. A detailed meta-analysis was conducted for the NICE-guidelines.[31] This meta-analysis included 31 RCTs, of which 19 compared CBTp to TAU. Compared to TAU, general psychopathology was found to be significantly reduced by the end of therapy (standardized mean difference [SMD] = -0.27 [95% CI: -0.45, -0.10]) and by the end of the follow-up period of 6 months (SMD = -0.23 [95% CI: -0.42, 0.04]) and 12 months (SMD = -0.40 [95% CI: -0.65, -0.15]). Thus, the effects for general psychopathology were found to be stable and in a small to moderate range. In regard to positive symptoms there were small significant effects at post-treatment and at a 12-month follow-up, but these were no longer significant after excluding some of the less methodologically rigorous studies. For negative symptoms, the comparison to TAU only became significant in the follow-up period up to 12 months, but remained stable at 12 to 24 months. There were also effects on rehospitalization and time spent in hospital in favor of CBTp, with CBTp patients spending an average of 8 days less in hospital in the 12–18 months after therapy. Based on these findings, the NICE guideline recommends CBTp be offered to patients in all phases. The authors of the guideline also recommend using a CBTp treatment manual and to offer at least 16 sessions of CBTp, while having regular supervision.

CBTP IN COMPARISON TO OTHER INTERVENTIONS

A meta-analysis by Turner et al.[32] investigated the effectiveness of CBTp compared to active control groups in a pool of 17 studies. They found CBTp to be more effective for the reduction of overall psychopathology and positive symptoms than control conditions (Hedges' g = 0.16, 95% CI: 0.04, 0.28). However, there was no significant advantage of CBTp for negative symptoms. In this meta-analysis, methodological differences between studies were controlled for and the findings remained robust. These findings are largely backed up by a Cochrane analysis[33] that found short-term advantages of CBTp over active control groups for overall symptoms, positive symptoms, and functioning—but not for negative symptoms. A recent network meta-analysis also approached the question of the effectiveness of different psychological approaches to psychosis.[34] This analysis aggregated data on the level of individual trials on various psychological interventions (e.g., CBTp, metacognitive training, mindfulness approaches, etc.). CBTp was the most strongly represented approach and was found to have significant effects compared to treatment as usual for positive, overall, and negative symptoms and functioning and compared to inactive control conditions for positive symptoms, which were not found for the other treatment approaches. However, neither of these meta-analyses included family interventions.

GENERALIZABILITY TO CLINICAL PRACTICE

Several studies have also investigated how well the effects from randomized controlled trials hold up in clinical practice. In one such study in Germany, 80 patients with psychosis were randomized to CBTp or a waiting list control condition.[35] Patients receiving CBTp showed a significantly stronger improvement in positive symptoms and general symptoms (Cohens d: 0.66) compared to those in the waiting group. However, there was no effect for negative symptoms. The positive effects of the therapy could be maintained at a one-year follow-up. This study and other effectiveness studies[36,37] demonstrate that the efficacy of CBTp can be generalized to clinical practice, despite the difference in patients, therapists and deliverance.

EFFICACY FOR SPECIFIC TARGET GROUPS

PATIENTS WHO REFUSE ANTIPSYCHOTIC MEDICATION

After a promising pilot study, Morrison and colleagues[38] conducted an RCT that included 72 patients with positive symptoms who had been refusing medication for at least 6 months. These patients were randomized to CBTp or TAU condition. The general symptomatology was assessed at 3- month periods between 3 and 18 months after therapy. Symptoms in the CBTp group were consistently lower compared to the TAU group with a medium effect in the moderate range (Cohen's d = 0.46). This study indicates that CBTp might be effective for outpatients who refuse antipsychotic medication. This initial trial is currently being followed up by several multicenter trials.

PATIENTS IN THE PRODROMAL PHASE

Several studies have demonstrated that CBT on its own, without additional medication, can be helpful for people exhibiting an at-risk mental state (ARMS) for psychosis, in order to prevent transition to psychosis. These studies have shown that CBT can reduce the risk of transition by more than half.[39]

PATIENTS WITH NEGATIVE SYMPTOMS

Because most of the CBT manuals focus strongly on positive symptoms (many of the studies even excluded patients without persistent positive symptoms), there is less evidence for CBTp for those people experiencing negative symptoms (e.g., [32]). Few studies have specifically investigated different variants of CBTp for negative symptoms. Grant et al.[40] used a cognitive approach to deal with negative symptoms in a sample of low-functioning patients with psychosis. Sixty patients were randomized to CBTp plus medication compared to medication alone. The authors found a significant improvement in functioning at the end of the 9-month period. They also found improvements in particular domains of negative symptoms, such as apathy and avolition, but not in others, such as anhedonia, affective expression, and alogia. The findings were confirmed in a smaller, uncontrolled trial of a briefer variant of the approach.[41] However, a German RCT compared a manualized variant of CBT for negative symptoms to cognitive remediation training and found no evidence of a superiority of CBTp.[42] It seems that although negative symptoms are not a contraindication for therapy, and CBTp is a promising approach, especially in regard to the motivational component of negative symptoms,[43] it needs to be developed further to increase its efficacy in this area.

SUMMARY AND OUTLOOK

Research on CBT for psychotic disorders has come a long way since the first case studies in the early nineties by UK pioneers. We are now faced with numerous treatment manuals offering detailed descriptions of how to deal with psychotic symptoms and consistent evidence demonstrating their effectiveness. Now, more effort needs to be undertaken to incorporate CBT for psychosis into regular training programs for clinical psychologists and for psychiatrists to make it widely available to patients. Finally, it will be exciting to see whether attempts that are underway to refine cognitive interventions by tailoring them to more specific symptoms[12] will be successful in improving the overall effectiveness and efficiency of cognitive interventions in the future.

ACKNOWLEDGMENTS

We thank Stephanie Mehl, Esther Jung, and Martin Wiesjahn for providing some of the case examples.

REFERENCES

1. Lincoln TM, Beck AT (2014). Psychosis. In: Hofmann SG, Rief W, eds. *The Wiley Handbook of Cognitive Behavioral Therapy*. Chichester: Wiley; 2014:437–461.
2. McGovern J, Turkington D. "Seeing the wood from the trees": A continuum model of psychopathology advocating cognitive behaviour therapy for schizophrenia. *Clin Psychol Psychother*. 2001;8:149–175.
3. van Os J, Linscott RJ, Myin-Germeys I, Delespaul P, Krabbendam L. A systematic review and meta-analysis of the psychosis continuum: Evidence for a psychosis proneness-persistence-impairment model of psychotic disorder. *Psychol Med*. 2009;39(2):179–195.
4. Garety PA, Kuipers L, Fowler D, Freeman D, Bebbington P. A cognitive model of the positive symptoms of psychosis. *Psychol Med*. 2001;31(2):189–195.
5. Zubin J, Spring B. Vulnerability—a new view of schizophrenia. *J Abnorm Psychol*. 1977;86:103–126.
6. Beck AT. The current state of cognitive therapy: A 40-year retrospective. *Arch Gen Psychiatry*. 2005;62:953–959.
7. Chadwick P, Birchwood M, Trower P. *Cognitive Therapy for Delusions, Voices and Paranoia*. Chichester: Wiley; 1996.
8. Fowler D, Garety P, Kuipers E. *Cognitive Behaviour Therapy for Psychosis. Theory and Practice*. Chichester: Wiley; 1995.
9. Kingdon DG, Turkington D. *Cognitive Therapy of Schizophrenia*. New York: Guilford Press; 2004.
10. Morrison AP, Renton JC, Dunn H, Williams S, Bentall RP. *Cognitive Therapy for Psychosis: A Formulation-Based Approach*. New York: Brunner-Routledge; 2004.
11. Lincoln TM. *Kognitive Verhaltenstherapie Der Schizophrenie: Ein Individuenzentrierter Ansatz*. 2nd ed. Göttingen: Hogrefe; 2014.
12. Lincoln TM, Peters E. A systematic review and discussion of symptom specific cognitive behavioural approaches to delusions and hallucinations. submitted.
13. Brabban A, Byrne R, Longden E, Morrison AP. The importance of human relationships, ethics and recovery-orientated values in the delivery of CBT for people with psychosis. *Psychosis*. 2017;9(2):157–166. doi:10.1080/17522439.2016.1259648
14. Morrison AP, Barratt S. What are the components of CBT for psychosis? A Delphi study. *Schizophr Bull*. 2010;36(1):136–142.
15. Van Os J, Kenis G, Rutten BPF. The environment and schizophrenia. *Nature*. 2010;468:203–212.
16. Freeman D, Garety PA, Kuipers E, Fowler D, Bebbington PE. A cognitive model of persecutory delusions. *Br J Clin Psychol*. 2002;41(4):331–347.
17. Tarrier N, Beckett N, Harwood S, Baker A, Yusupoff L, Ugarteburu I. A trial of two cognitive-behavioural methods of treating drug-resistant residual symptoms in schizophrenic patients: I. Outcome. *Br J Psychiatry*. 1993;162:524–532.
18. Chadwick P, Birchwood M. The omnipotence of voices: A cognitive approach to auditory hallucinations. *Br J Psychiatry*. 1994;165:190–201.
19. Hayward M, Berry K, McCarthy-Jones S, Strauss C, Thomas N. Beyond the omnipotence of voices: further developing a relational approach to auditory hallucinations. *Psychosis*. 2014;6(3):242–252. doi:10.1080/17522439.2013.839735
20. Leff J, Williams G, Huckvale M, Arbuthnot M, Leff AP. Avatar therapy for persecutory auditory hallucinations: What is it and how does it work? *Psychosis*. 2014;6(2):166–176. doi:10.1080/17522439.2013.773457
21. van der Gaag M, van Oosterhout B, Daalman K, Sommer IE, Korrelboom K. Initial evaluation of the effects of competitive memory training (COMET) on depression in schizophrenia-spectrum patients with persistent auditory verbal hallucinations: A randomized controlled trial. *Br J Clin Psychol*. 2012;51(2):158–171. doi:10.1111/j.2044-8260.2011.02025.x
22. Freeman D, Garety P, Fowler D, Kuipers E, Bebbington P, Dunn G. Why do people with delusions fail to choose more realistic explanations for their experiences? An empirical investigation. *J Clin Consult Psychol*. 2004;72(4):671–680.
23. Morrison AP, French P, Wells A. Metacognitive beliefs across the continuum of psychosis: Comparisons between patients with psychotic disorders, patients at ultra-high risk and non-patients. *Behav Res Ther*. 2007;45(9):2241–2246.
24. Moritz S, Woodward TS. Metacognitive training in schizophrenia: From basic research to knowledge translation and intervention. *Curr Opin Psychiatry*. 2007;20:619–625.
25. Freeman D. Persecutory delusions: A cognitive perspective on understanding and treatment. *Lancet Psychiatry*. 2016;3(7):685–692. doi:10.1016/S2215-0366(16)00066-3
26. Beck AT, Rector NA, Stolar N, Grant PM. *Schizophrenia: Cognitive Theory, Research and Therapy*. New York: Guilford Press; 2009.
27. Leahy RL. *Cognitive Therapy Techniques: A Practitioner's Guide*. New York: Guilford Press; 2003.
28. Young JE, Klosko JS, Weishaar ME. *Schema Therapy: A Practitioner Guide*. New York: Guilford Press; 2003.
29. Gilbert P. *Compassion Focused Therapy*. New York: Routledge Chapman & Hall; 2010.
30. Gumley A, Schwannauer M. *Staying Well After Psychosis: A Cognitive Interpersonal Approach to Recovery and Relapse Prevention*. Chichester: Wiley and Sons; 2006.
31. NCCMH. *Schizophrenia: Core Interventions in the Treatment and Management of Schizophrenia in Adults in Primary and Secondary Care*. Vol NICE Clinical Guideline 82. London: NICE; 2009.
32. Turner DT, van der Gaag M, Karyotaki E, Cuijpers P. Psychological interventions for psychosis: A meta-analysis of comparative outcome studies. *Am J Psychiatry*. 2014;171(5):523–538. doi:10.1176/appi.ajp.2013.13081159
33. Jones C, Hacker D, Cormac I, Meaden A, Irvine CB. Cognitive behaviour therapy versus other psychosocial treatments for schizophrenia. *Cochrane Database Syst Rev*. 2014;(4).//WOS:000303012300031.
34. Bighelli I, Salanti G, Huhn M, et al. Psychological interventions to reduce positive symptoms in schizophrenia: systematic review and network meta-analysis. *World Psychiatry*. 2018;17(3):316–329. doi:10.1002/wps.20577
35. Lincoln TM, Ziegler M, Mehl S, et al. Moving from efficacy to effectiveness in CBT for psychosis: A randomized-controlled clinical practice trial. *J Consult Clin Psychol*. 2012;80(4):674–686.
36. Farhall J, Freeman NC, Shawyer F, Trauer T. An effectiveness trial of cognitive behaviour therapy in a representative sample of outpatients with psychosis. *Br J Clin Psychol*. 2009;48:47–62.
37. Peters E, Landau S, McCrone P, et al. A randomised controlled trial of cognitive behaviour therapy for psychosis in a routine clinical service. *Acta Psychiatr Scand*. 2010;122(4):302–318. doi: 10.1111/j.1600-0447.2010.01572.x
38. Morrison AP, Turkington D, Pyle M, et al. Cognitive therapy for people with schizophrenia spectrum disorders not taking antipsychotic drugs: A single-blind randomised controlled trial. *Lancet*. 2014;383:1395–1403.
39. Hutton P, Taylor PJ. Cognitive behavioural therapy for psychosis prevention: A systematic review and meta-analysis. *Psychol Med*. 2013. doi:10.1017/S0033291713000354
40. Grant PM, Huh GA, Perivoliotis D, Stolar NM, Beck AT. Randomized trial to evaluate the efficacy of cognitive therapy for low-functioning patients with schizophrenia. *Arch Gen Psychiatry*. 2012;69(2):121–127.
41. Staring AB, Ter Huurne MA, van der Gaag M. Cognitive behavioral therapy for negative symptoms (CBT-n) in psychotic disorders: A pilot study. *J Behav Ther Exp Psychiatry*. 2013;44(3):300–306. doi:10.1016/j.jbtep.2013.01.004

42. Klingberg S, Wittdorf A, Meisner C, et al. Cognitive behavioral therapy versus supportive therapy for persistent positive symptoms in psychotic disorders: major results of the POSITIVE study. *Eur Arch Psychiatry Clin Neurosci.* 2011; Suppl 1:13.

43. Riehle M, Pillny M, Lincoln TM. Ist Negativsymptomatik bei Schizophrenie überhaupt behandelbar? Ein systematisches Literaturreview zur Wirksamkeit psychotherapeutischer Interventionen für Negativsymptomatik. *Verhaltenstherapie.* 2017;27(3):199–208. doi:10.1159/000478534

58

PSYCHOEDUCATION AS AN APPROACH TO TREATMENT OF SEVERE MENTAL ILLNESS

Emma Sophia Kay, David E. Pollio, and Carol S. North

Psychoeducation is a well-established approach to patient care for individuals coping with severe mental illness. Originally developed for persons coping with schizophrenia, psychoeducation practices have generalized to other psychiatric conditions, most notably bipolar disorders.[1] More recently, psychoeducation has been translated to a variety of other medical conditions. Psychoeducation has been carefully examined as an evidence-based practice[1] with in-depth discussion and reviews of the plethora of models and research related to this approach to patient-care. The purpose of this chapter is not to revisit in great detail the findings and conclusions from these studies, but rather to present an overview for readers not familiar with this approach, describe new developments in psychoeducation models, and address issues of increasing importance moving forward. This chapter begins with the definition of psychoeducation and then briefly describes the history and theories of the approach. Building on this foundation, the chapter proceeds to examine existing major psychoeducation models and subsequent extensions of the approach, review research conducted on the efficacy and effectiveness of psychoeducation, and address several important considerations for the implementation of psychoeducation. The chapter concludes with a discussion of promising future directions for psychoeducation approaches.

PSYCHOEDUCATION: DEFINITION AND HISTORY

Psychoeducation is best understood as an umbrella term for a group of intervention models designed to provide information, guidance, and support to individuals with severe mental illness and their families. Education and support are fundamental elements of psychoeducation. Psychoeducation models are not therapy per se, but some include therapeutic techniques or approaches. Many psychoeducation models incorporate patients' families in single-family or multiple-family group settings.[2,3] The family psychoeducation (FPE) model is, in fact, considered "one of the most effective psychosocial treatments ever developed"[4(p460)] and is endorsed by the US Department of Health and Human Services' Substance Abuse and Mental Health Services Administration (SAMHSA) as an evidence-based practice. The success of FPE may largely be attributable to current realities of mental health systems that rely heavily on family and/or primary caregiver support for successful long-term treatment. In tandem with pharmacotherapy, psychoeducation can foster coping skills[4–6] and self-efficacy[7,8] among patients and their families.

In the United States, the emergence of the practice of psychoeducation occurred during the deinstitutionalization movement, which began in 1955 with the closure of state mental hospitals.[9] By 1980, nearly half a million patients had been discharged from mental institutions to community care, and many families of these patients found themselves having to function in the role of primary caregiver. With no educational foundation to inform their caretaking responsibilities, families were left to develop their own means of coping in the face of seemingly inexplicable behavior from their family member with the illness.[10] Around this time, some psychiatrists began to advocate for family involvement, conferring benefits to both patients and families. In the late 1960s, the American psychiatrist Robert Liberman at the UCLA Clinical Research Center for Schizophrenia and Psychiatric Rehabilitation began researching family-based interventions for individuals with psychotic disorders, eventually codeveloping behavioral family therapy along with Ian Falloon, a New Zealander psychiatrist.[11] Several renowned psychiatrists received behavioral family therapy training at the UCLA Clinical Research Center, including William McFarlane, who would later develop the first formal family psychoeducation model.[11]

To understand how and why psychoeducation emerged as an accepted intervention for psychotic disorders, it is helpful to provide a brief discussion of its theoretical underpinnings. Because schizophrenia is well recognized as an illness of the brain with strong supporting biological evidence, established interventions necessarily draw on its biological underpinnings. McFarlane[12] was one of the first researchers to recognize that psychoeducation models must address biological causes, with the appreciation that social and familial processes are equally important, core components of this intervention for schizophrenia. McFarlane emphasized the need for a "biosocial" synthesis of biological and social factors in responding to the illness: "The state of the individual with schizophrenia is determined by a continuing interaction of specific biological dysfunctions of the brain with social processes."[12(p75)] Thus, although antipsychotic medication is a vital component of formal treatment for schizophrenia, successful long-term remission necessitates the ability of individuals with the illness to mitigate environmental stressors. McFarlane further emphasized that social feedback—especially through family involvement—shapes cognitive health and may act as a barrier, or even conversely as a facilitator, to relapse, underscoring the importance of helping families learn to provide effective social responses to their loved one with the illness.

PSYCHOEDUCATION MODELS

Originally developed to help families and patients with schizophrenia and other serious mental illness, psychoeducation was subsequently broadened for adaptations to other mental illness (e.g., bipolar disorder and eating disorders).[13] The flexibility of psychoeducation is well suited to the distinct and individual needs and strengths of each unique patient and family. Several distinct approaches merit further examination and review: single-patient psychoeducation, single-family psychoeducation, multiple-family psychoeducation, psychoeducation for first-episode psychosis, and smartphone-based psychoeducation. Each of these is discussed in the following sections.

ONE-ON-ONE PSYCHOEDUCATION

One-on-one psychoeducation, involving a single patient and the treating mental health professional, is not often mentioned in the literature, which focuses on single- and multiple-group family interventions. Several studies have investigated the efficacy of the individual psychoeducation model. Bossema et al. found that patients with a psychotic disorder had improved coping skills after receiving psychoeducation compared to before, but the gains were short-lived.[5] Preliminary findings from an 18-month pilot study of psychoeducation for psychosis found that the intervention group improved significantly more than a matched control group receiving treatment as usual between baseline and 6-month posttest in knowledge and insight.[14]

A randomized trial comparing an individual- to a group-based intervention in patients with bipolar disorder found that the group-based intervention had significantly fewer hospitalizations over an 8-year period.[15] In contrast, another found that bipolar patients preferred individual psychoeducation over group psychoeducation, providing an important reminder that patients' preferences have potential to shape an intervention's success.[16]

FAMILY PSYCHOEDUCATION (FPE)

Family psychoeducation is considered an evidence-based practice by SAMHSA. A SAMHSA toolkit documents the efficacy of FPE and provides an implementation guide for it.[17] FPE interventions are conducted by a collaborative team consisting of the patient, one or more family members or other caregivers, and treatment professionals. Family members are integral to this model and are considered "indispensable colleague(s) with differing expertise and potential skills."[4(p461)] Family members caring for their members during the early stages of illness in particular are thought to benefit from FPE.

The US Schizophrenia Patient Outcomes Research Team (PORT) provides guidelines for psychosocial treatment of schizophrenia.[18] For family-based interventions, PORT guidelines recommended that psychoeducation should continue at least 6 to 9 months to reduce risk of relapse and/or rehospitalization, especially for patients with a recent exacerbation of illness. If only a briefer intervention is possible or feasible, PORT advises that at least four sessions are needed.

MULTIFAMILY GROUP PSYCHOEDUCATION

A major application of the family psychoeducation model is multiple-family group (MFG) psychoeducation, which facilitates dialogue among groups of families sharing the strategies they have developed for caring for a family member with schizophrenia. McFarlane recommended groups of 4–8 families meeting on a biweekly basis for at least 2 years led by a clinician.[3] In these sessions, group members converse with one another to provide feedback on and suggestions for specific caregiving problems presented by the families.[3] In resource-limited settings that do not support such intensive intervention, however, an intensive 1-day workshop can benefit families.[6] There is also evidence from a Cochrane review that "brief" interventions comprising 10 or fewer sessions may prevent relapse, for at least for up to a year after their completion.[19]

MFG psychoeducation interventions have consistently been associated with positive outcomes for patients with schizophrenia and their families, including lower risk of relapse,[20–22] reduced hospitalization,[22,23] and improved medication adherence.[11] MFG psychoeducation has been demonstrated to be more effective than single-family psychoeducation.[21,22] One randomized study[22] found that MFG psychoeducation participants had significantly less relapse and rehospitalization compared to single-family group participants, and these gains were retained over 4 years.[21] Although the exact mechanism of change with psychoeducation is unclear, social support provided in a multiple-family format may help by countering the isolation that burdens so many people with psychotic disorders and their primary caregivers.[4,24] Consistent with this supposition, a study of the primary needs of participants in MFG revealed that chief among families' concerns was social isolation.[25]

Research suggests that MFG psychoeducation is cost-effective in real-world settings. McFarlane et al. noted that MFG psychoeducation models "require exactly one half of the staff" of single-family psychoeducation models.[22] They estimated that the cost-saving ratio for multiple- versus single-family psychoeducation group interventions was 34:1.[22(p685)] Another study applied this 34:1 ratio to a cost/burden analysis of MFG psychoeducation, finding that cost savings from MFG psychoeducation would be realized even if clinical benefits were reduced by 90%.[26] This estimated amount of cost savings was largely attributed to reduced hospitalization rates.

SMARTPHONE APPLICATIONS

In recent years, some researchers have set out to examine the feasibility, acceptability, and effectiveness of smartphone-based psychoeducation interventions for individuals with schizophrenia or bipolar disorder. Technology-based interventions offer flexible, convenient, and affordable interventions to patients and families for whom a face-to-face format is not feasible.

A team of Spanish researchers conducted a feasibility study of a smartphone application known as "SIMPLe" for adult bipolar disorder patients.[27] This application solicits regular user feedback for the purpose of providing appropriate and timely prompts for situations of risk for relapse. Users receive short psychoeducational messages based on their own data input related to mood, medication adherence, and other health-related categories. A pilot test of the application with 51 adult participants found that SIMPLe was generally found to be both feasible and acceptable, with a participant satisfaction rating of 86% for the application.[27]

Another smartphone-based intervention for adults with bipolar disorder is Personalized Real-Time Intervention for Stabilizing Mood (PRISM).[28] The PRISM intervention guides users to develop personalized action strategies based on the real-time information they provide. A randomized controlled trial of PRISM delivered by smartphone or face-to-face found no significant clinical outcome differences in the two conditions, but satisfaction ratings were higher in the intervention group.[28]

Smartphone interventions have also been developed for schizophrenia. In 2014, Ben-Zeev et al. conducted a feasibility study using a "FOCUS" smartphone application that they developed.[29] This application was aimed to prevent high-stress episodes that may precipitate relapse. A study of 33 adult participants in Chicago community-based programs using the application for 1 month found that more than 90% of participants found the FOCUS application to be feasible and acceptable.[29]

FIRST-EPISODE PSYCHOSIS

First-episode psychosis generally refers to the initial stages of illness in schizophrenia, but it lacks a standard definition.[30] The initial onset of schizophrenia is a critical period for the initiation of treatment, because untreated psychosis raises the risk of poor long-term outcomes.[31] Interventions tailored for first-episode psychosis have recently been developed and tested. Mueser et al. noted that such programs are more common in countries with single-payer medical systems such as Australia and the United Kingdom.[32] Recently, however, the National Institute of Mental Health (NIMH) developed a holistic program known as NAVIGATE that aims to help individuals in their first episode of psychosis to fully engage in mental health care and regain healthy functioning.[32] NAVIGATE draws on a strengths-based and motivational enhancement skillset and incorporates several psychoeducational components, including provision of patients and their families with educational materials about their illness and assistance in working together as a collaborative unit.[32]

The NAVIGATE program has produced positive clinical, functional, and quality-of-life- related outcomes.[33] A multisite trial involved 34 community treatment centers, participants randomly assigned to either the NAVIGATE program or treatment as usual in the community.[33] At the end of 2 years, NAVIGATE participants achieved greater improvement on the Positive and Negative Syndrome Scale (PANSS).[34] Although the two groups did not significantly differ on post-treatment scores on the Scales of Psychological Well-Being (SPWB),[35] duration of untreated psychosis (DUP) acted as a moderating variable for several components of the SPWB scale.[36] Specifically, for the Positive Relationships subscale, a shorter DUP was associated with significantly greater improvements in both groups. A shorter DUP was associated with significantly greater environmental mastery for participants in the NAVIGATE group only.[36] These results emphasize the importance of prompt and timely treatment early in the course of psychotic illness.

In 2013, an extensive systematic review of treatment for first-episode psychosis was conducted for the purpose of identifying essential components of care.[37] An international panel of experts assembled for this study used the consensus-building Delphi method to identify 32 critical components of evidence-based treatment services for early psychosis that ranged from tailored application of antipsychotic medication to social, group, and public level interventions.[37] A four-tier evidence-ranking system was used to provide a grade to each component from "A" to "D." MFG psychoeducation received an "A" grade with "strong evidence" to support it; individual and single-family group models received a "B" grade with "supportive evidence."[37]

Expanding on the work of Addington et al.,[37] White et al. conducted a study of 31 US programs that treat early psychosis through interviews of the directors of these programs.[38] Only two of the evidence-based components identified in the Addington et al. study were found to be used in all 31 programs: individual psychoeducation and outcomes/process tracking. Family therapy was featured in 30 of the programs.[38] These findings indicate that psychoeducation is widely considered to be an essential intervention for early psychosis.

PSYCHOEDUCATION AS AN EVIDENCE-BASED PRACTICE

Lefley[1] and others have concluded that psychoeducation deserves recognition as an evidence-based approach to treatment of mental illness. A number of psychoeducation models meet evidence-based standards.

A sizable body of literature informed by randomized controlled trials indicates that psychoeducation can reduce relapse in schizophrenia.[4] A recent Cochrane review[39] identified 44 randomized controlled trials conducted between 1988 and 2009 that compared psychoeducation plus standard care for schizophrenia to standard care alone. Primary outcomes examined were treatment adherence and relapse, because treatment adherence is necessary for relapse prevention, and adherence requires patients to understand and accept their illness—a focus of psychoeducation.[39] Overwhelmingly, these studies demonstrated that patients randomized to psychoeducation were more treatment adherent and less likely to relapse than patients randomized to standard care alone. The authors concluded that psychoeducation appears to improve short- and long-term (as long as 5 years) outcomes for adults with schizophrenia.[39]

The literature on psychoeducation is much smaller for bipolar disorder than for schizophrenia, consistent with the fact that psychoeducation was historically first developed as an intervention for schizophrenia. The evidence base for bipolar disorder is also strong: a meta-analysis of randomized trials conducted between 1991 and 2010 found that patients randomized to psychoeducation experienced improved medication adherence in 9 of 13 studies.[40] A more recent systematic review of randomized controlled trials for bipolar disorder concluded that psychoeducation improved medication adherence.[41] Additionally, group-focused interventions were found to be more effective for relapse prevention than one-on-one interventions.[41]

Although the ultimate goal of most psychoeducation programs is to prevent relapse and rehospitalization, there are broader outcomes to be considered. In the NAVIGATE study, Kane et al.[33] found that randomization to NAVIGATE resulted in significantly greater improvement in quality of life, as measured by the Heinrichs-Carpenter Quality of Life Scale (QLS).[42] Other areas of superior improvement in NAVIGATE were in subscales measuring social relationships, emotional engagement, and object/activity engagement.[33]

Most studies examining quality of life outcomes have been conducted outside the United States. A group of Italian researchers used the QLS scale in their study to identify levels of "real-world functioning" in a cross-sectional study of adults with schizophrenia who received psychoeducation.[43] High-functioning participants were more likely to be employed and to be in long-term relationships.[43] In Vietnam, Ngoc et al. found that quality of life increased significantly more among patients with schizophrenia randomly assigned to a family psychoeducation program compared to patients receiving treatment as usual.[44] In Brazil, researchers found that patients with bipolar disorder who received a psychoeducation intervention improved in several areas of quality of life, including social, emotional, and physical health, although the gain was not significantly greater in the intervention group than in the control group.[45]

American research has demonstrated that psychoeducation is associated with improved outcomes in education and employment. Participants randomized to the NAVIGATE intervention, who received targeted assistance from supported employment and education (SEE) specialists, had a greater increase in their educational and employment involvement over the 2-year study period compared to patients randomized to treatment as usual in the community.[46] The SEE component of the program was found to be a significant mediating factor of these outcomes.[46] This study's findings suggest that people with schizophrenia who are interested in attending school or obtaining employment are likely to benefit from regular, ongoing support of their educational and employment efforts.

McFarlane et al. investigated employment outcomes of MFG psychoeducation for schizophrenia.[47] Both the intervention and control groups received assistance in finding some type of employment, but only the intervention group involved family members in the effort. This family involvement resulted in the successful securement of jobs that were "potentially less stressful, since they involved social connections and thus carried a degree of familiarity for the participants.[47(p206)] After 18 months, participants receiving the MFG psychoeducation intervention were significantly more likely to be employed. The authors recommended that patients, families, clinicians, and employment specialists work synergistically toward attainment of optimal employment outcomes.[47]

Schizophrenia and bipolar disorder psychoeducation programs may also offer direct benefits to patients' families, including perceived reduction in caregiver burden,[48,49,50] decreased stigma toward schizophrenia,[44] reduced psychological distress,[13] and higher family functioning.[51] The importance of positive family outcomes cannot be overstated: greater levels of caregiver burden and family criticism of the family member with schizophrenia have been shown to be associated with higher risk of long-term relapse.[52]

CONSIDERATIONS FOR IMPLEMENTING PSYCHOEDUCATION

APPROACHES TO CURRICULA

There are two fundamental approaches to the development of educational materials in psychoeducation: structured and flexible curriculum strategies. Although a small number of structured curricula exist, psychoeducational approaches generally incorporate some element of flexibility. Among more structured approaches to curricula, perhaps the two best known are the National Alliance on Mental Illness (NAMI) 12-week "Family-to-Family Program"[53] and SAMHSA's evidence-based practice toolkit.[17] Both of these curricula use a set group of educational materials, presented in structured formats.

The NAVIGATE program for patients with first-episode psychosis has provided freely available online implementation materials (http://navigateconsultants.org/), which include psychoeducation curriculum files. The psychoeducation files consist of six separate program manuals for the program director, family clinician, patients, medication prescribers, and education and employment assistance providers. The website also provides a separate manual for the treatment team as a unit that emphasizes collaboration and coordination among all team members. A 1-hour, free consultation with a NAVIGATE specialist is available for program directors to gauge the program's potential fit for any agency.

Several primary psychoeducation components of the NAVIGATE program are designed to address different needs of persons with first-episode psychosis. Broadly, these include shared family-patient decision-making; strengths and resiliency focus; motivational enhancement; psychoeducation skills; family support; individualized medication management; family education; individual resiliency training; and supported education and employment.[32] The purpose of the NAVIGATE psychoeducation program is to shepherd recently diagnosed patients early in the course of psychosis and their families through a difficult and overwhelming period. The NAVIGATE program generally requires at least 2 years to complete, but because individual needs for quantity and duration of services may differ, the curriculum has a flexible time frame built into it.[32]

An MFG psychoeducation intervention developed by North at al., known as Psychoeducation Responsive to Families (PERF), is unique through its systematic involvement of patients and their families in the planning of their own specific course curriculum.[25] PERF can be tailored to any psychiatric disorder and groups of all ages and backgrounds. Accordingly, PERF has been used for youth with behavioral disorders,[54] runaway and homeless youth,[55] and military veterans.[56] PERF is thus an exemplar of work translating research efficacy to community settings that is adaptable to address the needs of many types of populations, even including recent application to patients infected with hepatitis C to help prepare them for antiviral treatment.[57]

Psychoeducation curricula have been tailored to distinct cultures, norms, and health payer systems internationally. In Japan, for example, a psychoeducation practitioner training program was developed for psychiatric hospital nurses.[58] Nurses who participated in the 2-day workshop significantly increased their psychoeducation self-efficacy and improved their attitudes toward psychoeducation.[58] A psychiatric nursing curriculum developed in Australia provides inpatient family psychoeducation services for first-episode psychosis. Called the "Journey to Recovery," this 5-week program provides psychoeducation to patients and their family caregivers and has been shown to increase caregivers' knowledge of psychosis.[59] In China, a randomized trial of a 10-session psychoeducation intervention for outpatients with schizophrenia and their primary caregivers demonstrated that patients in the intervention group improved their medication adherence, understanding of illness, and mental condition, and their caregivers demonstrated gains in self-efficacy and self-support and experienced reduced family burden.[60]

Canadian researchers have developed a brief group psychoeducation program aimed to reduce symptoms of psychosis and distress among outpatients with schizophrenia.[61] This 8-week intervention relied primary on cognitive-behavioral therapy (CBT) techniques. This small study (N = 24) found that patients who received the intervention showed a significant decrease in distress and symptoms of schizophrenia on the PANSS and Symptom Checklist-90.[62]

TRANSLATION/IMPLEMENTATION INTO COMMUNITY SETTINGS

Because most psychoeducation models allow a certain amount of flexibility in curriculum, psychoeducation can readily translate into community settings.[63] The long history of translation of psychoeducation to apply to families coping with psychiatric and chronic physical illnesses that has unfolded across more than two decades and resulted in the development of six distinct PERF models suggests the potential for further translation of family-responsive psychoeducation for use with other conditions.[63] Three primary principles are involved in establishment of new applications: establishing the need for the translation, building community partnerships, and adjusting the model to fit the identified population. Developing new applications of psychoeducation models requires inclusion of stakeholders at all levels. Stakeholders to be included are the families to be served, the practitioners who will be providing the intervention, and the organization hosting the intervention.

Another set of issues for translating psychoeducation models into community settings relates to the creation of specific groups. It has been learned that the implementation process for specific psychoeducation models must consider the structure and context of the group sessions, including the number of sessions, time of the meeting, the setting, likely barriers to individual attendance, facilitator roles, and pragmatic concerns such as transportation and whether some sort of food or drink might be served.[54,55,64] For example, a psychoeducation model was developed for children with serious emotional disturbance in a school setting, with the length of the intervention following the school year (9 months, with no sessions during holiday break).[54] Meetings were held at the school after school hours. To accommodate parent schedules, dinner was incorporated as part of the beginning of each session. Further, to facilitate the meetings of parents only, skilled babysitting was provided for all the children in a separate space, resulting in greater enthusiasm of parents for the group and greater desire to attend.

CONCLUSION/FUTURE DIRECTIONS

As an established evidence-based model for families coping with a member with schizophrenia, bipolar disorder, or other psychiatric illness, psychoeducation is a well-established part of the treatment of these disorders in conjunction with pharmacotherapy and psychosocial interventions. Further, the successful translation of psychoeducation from its original focus on schizophrenia to many other applications suggests the potential for the utility of psychoeducation across a broad range of other medical conditions, as evidenced by work in patients with hepatitis C.[57] Given the increasing importance of patient education as part of the hospital discharge process and demonstrated cost savings associated with multifamily psychoeducation groups, it is likely that psychoeducation will grow in importance not only for application to psychiatric illness but also to other medical conditions.

A particularly promising future direction is in the access to psychoeducation through the use of technology, as demonstrated by smartphone applications developed and tested in schizophrenia and bipolar disorder. Additionally, the potential for social media to deliver psychoeducation is underdeveloped and holds promise for delivering accurate information and support to families who do not have the time or resources to attend meetings for formal interventions in person. Further research is needed to assure the effectiveness of online models.

Psychoeducation as an intervention needs further establishment of rigorously empirical evidence to confirm the effectiveness of this intervention. Evidence for the efficacy of psychoeducation for families coping with schizophrenia and bipolar disorder has been clearly established, but evidence for the effectiveness of translations to other disorders and for models conducted outside of academic research settings is much less clear.

Further research is needed to deconstruct which elements of psychoeducation drive patient and family improvement. Questions remaining are what education is necessary to provide to families, the importance of group-level constructs such as cohesion or engagement, and whether there is any specific dose effect. These questions have all received limited attention in the literature, but each is important in further understanding what makes psychoeducation effective and pointing the way to effective dissemination.

REFERENCES

1. Lefley HP. *Family psychoeducation for serious mental illness.* New York: Oxford University Press; 2009.
2. Lukens EP, McFarlane WR. Psychoeducation as evidence-based practice: Considerations for practice, research, and policy. *Brief Treat Crisis Interv.* 2004;4(3):205–225.
3. McFarlane WR. Multiple-family groups and psychoeducation in the treatment of schizophrenia. *New Dir Ment Health Serv.* 1994;62:13–22.
4. McFarlane WR. Family interventions for schizophrenia and the psychoses: A review. *Fam Process.* 2016;55(3):460–482.
5. Bossema ER, de Haar CA, Westerhuis W, et al. Psychoeducation for patients with a psychotic disorder: Effects on knowledge and coping. *Prim Care Companion CNS Disord.* 2011;13(4).
6. Pollio DE, North CS, Reid DL, Miletic MM, McClendon JR. Living with severe mental illness—what families and friends must know: Evaluation of a one-day psychoeducation workshop. *Soc Work.* 2006;51(1):31–38.
7. Grabski B. Group psychoeducation in the bipolar disorder—its influence on the cognitive representation of illness and the basic personality dimensions: Study with control group. *Archives of Psychiatry and Psychotherapy.* 2016;18(3):18–26.
8. Solomon P, Draine, J, Mannion, E, Meisel, M. Impact of brief family psychoeducation on self-efficacy. *Schizophr Bull.* 1996;22(1):41–50.
9. Learning from history: Deinstitutionalization of people with mental illness as precursor to long-term care reform. Kaiser Family Foundation website. https://kaiserfamilyfoundation.files.wordpress.com/2013/01/7684.pdf. Accessed June 15, 2017.
10. Chafetz L, Barnes L. Issues in psychiatric caregiving. *Arch Psychiatr Nurs.* 1989;3(2):61–68.
11. Kopelowicz A, Zarate R, Wallace CJ, Liberman RP, Lopez SR, Mintz J. Using the theory of planned behavior to improve treatment adherence in Mexican Americans with schizophrenia. *J Consult Clin Psychol.* 2015;83(5):985–993.
12. McFarlane WR. *Multifamily groups in the treatment of severe psychiatric disorders.* New York: Guilford Press; 2002.
13. Kolostoumpis D, Bergiannaki JD, Peppou LE, et al. Effectiveness of relatives' psychoeducation on family outcomes in bipolar disorder. *Int J Emerg Ment Health.* 2015;44(4):290–302.
14. Walker H, Connaughton J, Wilson I, Martin CR. Improving outcomes for psychoses through the use of psycho-education: Preliminary findings. *J Psychiatr Ment Health Nurs.* 2012;19(10):881–890.
15. Kallestad H, Wullum E, Scott J, Stiles TC, Morken G. The long-term outcomes of an effectiveness trial of group versus individual psychoeducation for bipolar disorders. *J Affect Disord.* 2016;202:32–38.
16. Cakir S, Camuz Gümüş F. Individual or group psychoeducation: Motivation and continuation of patients with bipolar disorders. *International Journal of Mental Health.* 2015;44(4):263–268.
17. Family psychoeducation evidence-based practices (EBP) kit. Substance Abuse and Mental Health Services Administration website. https://store.samhsa.gov/product/Family-Psychoeducation-Evidence-Based-Practices-EBP-KIT/SMA09-4423. Accessed June 15, 2017.
18. Dixon LB, Dickerson F, Bellack AS, et al. The 2009 schizophrenia PORT psychosocial treatment recommendations and summary statements. *Schizophr Bull.* 2010;36(1):48–70.
19. Zhao S, Sampson S, Xia J, Jayaram MB. Psychoeducation (brief) for people with serious mental illness. Cochrane Database Syst Rev. 2015;(4):CD010823.
20. Hogarty GE, Schooler NR, Ulrich, RF. (1979). Fluphenazine and social therapy in the aftercare of schizophrenic patients. *Arch Gen Psychiatry.* 1979;36:1283–1294.
21. McFarlane WR, Link B, Dushay, R, Marchal, J, Crilly J. Psychoeducational multiple family groups: Four-year relapse outcomes in schizophrenia. *Fam Process.* 1995;34:127–144.
22. McFarlane WR., Lukens, EP, Link B, et al. Multiple-family groups and psychoeducation in the treatment of schizophrenia. *Arch Gen Psychiatry.* 1995;52:679–687.
23. Chien WT, Chan SW. The effectiveness of mutual support group intervention for Chinese families of people with schizophrenia: A randomised controlled trial with 24-month follow-up. *Int J Nurs Stud.* 2013;50(10):1326–1340.
24. Jewell TC, Downing D, McFarlane WR. Partnering with families: Multiple family group psychoeducation for schizophrenia. *J Clin Psychol.* 2009;65(8):868–878.
25. North CS, Pollio DE, Sachar B, Hong B, Isenberg K, Bufe G. The family as caregiver: A group psychoeducation model for schizophrenia. *Am J Orthopsychiatr.* 1998 68(1):39–46.
26. Breitborde NJK, Woods SW, Srihari VH. Multifamily psychoeducation for first-episode psychosis: A cost-effectiveness analysis. *Psychiatr Serv.* 2009;60(11):1477–1483.
27. Hidalgo-Mazzei D, Mateu A, Reinares M, et al. Psychoeducation in bipolar disorder with a SIMPLe smartphone application: Feasibility, acceptability and satisfaction. *J Affect Disord.* 2016;200:58–66.
28. Depp CA, Ceglowski J, Wang VC, et al. Augmenting psychoeducation with a mobile intervention for bipolar disorder: A randomized controlled trial. *J Affect Disord.* 2015;174:23–30.
29. Ben-Zeev D, Brenner CJ, Begale M, Duffecy J, Mohr DC, Mueser KT. Feasibility, acceptability, and preliminary efficacy of a smartphone intervention for schizophrenia. *Schizophr Bull.* 2014;40(6):1244–1253.
30. Breitborde NJ, Bell EK, Dawley D, et al. The Early Psychosis Intervention Center (EPICENTER):Development and six-month outcomes of an American first-episode psychosis clinical service. *BMC Psychiatry.* 2015;15:266.
31. Friis S, Melle I, Johannessen JO, et al. Early predictors of ten-year course in first-episode psychosis. *Psychiatr Serv.* 2016;67(4):438–443.
32. Mueser KT, Penn DL, Addington J, et al. The NAVIGATE program for first-episode psychosis: Rationale, overview, and description of psychosocial components. *Psychiatr Serv.* 2015;66(7):680–690.
33. Kane JM, Robinson DG, Schooler NR, et al. Comprehensive versus usual community care for first-episode psychosis: 2-year outcomes from the NIMH RAISE early treatment program. *Am J Psychiatry.* 2016;173(4):362–372.
34. Kay SR, Fiszbein A, Opler LA. The positive and negative syndrome scale (PANSS) for schizophrenia. *Schizophr Bull.* 1987;13(2):261–276.
35. Ryff CD. Happiness is everything, or is it? Explorations on the meaning of psychological well-being. *J Pers Soc Psychol.* 1989;57:1069–1081.

36. Browne J, Penn DL, Meyer-Kalos PS, et al. Psychological well-being and mental health recovery in the NIMH RAISE early treatment program. *Schizophr Res.* 2017;185:167–172.
37. Addington DE, McKenzie E, Norman R, Wang J, Bond GR. Essential evidence-based components of first-episode psychosis services. *Psychiatr Serv.* 2013;64(5):452–457.
38. White DA, Luther L, Bonfils KA, Salyers MP. Essential components of early intervention programs for psychosis: Available intervention services in the United States. *Schizophr Res.* 2015;168(1–2):79–83.
39. Xia J, Merinder LB, Belgamwar MR. Psychoeducation for schizophrenia. *Cochrane Database Syst Rev.* 2011;6:CD002831.
40. Batista TA, Von Werne Baes C, Juruena MF. Efficacy of psychoeducation in bipolar patients: Systematic review of randomized trials. *Psychol Neurosci.* 2011;4(3):409–416.
41. Bond K, Anderson IM. Psychoeducation for relapse prevention in bipolar disorder: A systematic review of efficacy in randomized controlled trials. *Bipolar Disord.* 2015;17(4):349–362.
42. Heinrichs DW, Hanlon TE, Carpenter WT Jr. The Quality of Life Scale: An instrument for rating the schizophrenic deficit syndrome. *Schizophr Bull.* 1984;10(3):388–398.
43. Rocca P, Montemagni C, Mingrone C, Crivelli B, Sigaudo M, Bogetto F. A cluster-analytical approach toward real-world outcome in outpatients with stable schizophrenia. *Eur Psychiatry.* 2016;32:48–54.
44. Ngoc TN, Weiss B, Trung LT. Effects of the family schizophrenia psychoeducation program for individuals with recent onset schizophrenia in Viet Nam. *Asian J Psychiatr.* 2016;22:162–166.
45. Cardoso TA, Farias CA, Mondin TC, et al. Brief psychoeducation for bipolar disorder: Impact on quality of life in young adults in a 6-month follow-up of a randomized controlled trial. *Psychiatry Res.* 2014;220(3):896–902.
46. Rosenheck R, Mueser KT, Sint K, et al. Supported employment and education in comprehensive, integrated care for first episode psychosis: Effects on work, school, and disability income. *Schizophr Res.* 2017;182:120–128.
47. McFarlane WR, Dushay RA, Deakins SM, et al. Employment outcomes in family-aided assertive community treatment. *Am J Orthopsychiatr.* 70(2):203–214.
48. Bulut M, Arslantas H, Ferhan Dereboy I. Effects of psychoeducation given to caregivers of people with a diagnosis of schizophrenia. *Issues in Mental Health Nursing.* 2016;37(11):800–810.
49. Fallahi Khoshknab M, Sheikhona M, Rahgouy A, Rahgozar M, Sodagari F. The effects of group psychoeducational programme on family burden in caregivers of Iranian patients with schizophrenia. *J Psychiatr Ment Health Nurs.* 2014;21(5):438–446.
50. Palli A, Kontoangelos K, Richardson C, Economou MP. Effects of group psychoeducational intervention for family members of people with schizophrenia spectrum disorders: Results on family cohesion, caregiver burden, and caregiver depressive symptoms. *Int J Ment Health.* 2015;44(4):277–289.
51. Tsiouri I, Gena A, Economou MP, Bonotis KS, Mouzas O. Does long-term group psychoeducation of parents of individuals with schizophrenia help the family as a system? A quasi-experimental study. *Int J Ment Health.* 2015;44(4):316–331.
52. Koutra K, Triliva S, Roumeliotaki T, et al. Impaired family functioning in psychosis and its relevance to relapse: A two-year follow-up study. *Compr Psychiatry.* 2015;62:1–12.
53. NAMI Family-to-Family. National Mental Health Alliance website. https://www.nami.org/Find-Support/NAMI-Programs/NAMI-Family-to-Family. Accessed June 14, 2017.
54. Pollio DE, McClendon JB, North CS, Reid DL, Jonson-Reid, M. The promise of school-based multifamily psychoeducation groups for families with a child coping with severe emotional or behavioral disorders. *Children and Schools.* 2005;27(2):111–116.
55. Pollio DE, Thompson SJ, Tobias L, Reid D, Spitznagel E. Longitudinal outcomes for youth receiving runaway/homeless shelter services. *J Youth Adolesc.* 2006;35:859–866.
56. Cohen AN, Glynn SM, Murray-Swank AB, et al. The family forum: Directions for the implementation of family psychoeducation for severe mental illness. *Psychiatr Serv.* 2008;59(1):40–88.
57. North CS, Pollio DE, Sims OT, et al. An effectiveness study of group psychoeducation for hepatitis C patients in community clinics. *Eur J Gastroenterol Hepatol.* 2017;29(6):679–685.
58. Matsuda M, Kono A. Development and evaluation of a psychoeducation practitioner training program (PPTP). *Arch Psychiatr Nurs.* 2015;29(4):217–222.
59. Petrakis M, Oxley J, Bloom H. Carer psychoeducation in first-episode psychosis: Evaluation outcomes from a structured group programme. *Int J Soc Psychiatry.* 2013;59(4):391–397.
60. Chan SW, Yip B, Tso S, Cheng BS, Tam W. Evaluation of a psychoeducation program for Chinese clients with schizophrenia and their family caregivers. *Patient Educ Couns.* 2009;75(1):67–76.
61. Goldberg JO, Wheeler H, Lubinsky T, Van Exan J. Cognitive coping tool kit for psychosis: Development of a group-based curriculum. *Cogn and Behav Pract.* 2007;14(1):98–106.
62. Derogatis LR, Cleary, PA. Factorial invariance across gender for the primary symptom dimensions of the SCL-90. *Br J Clin Psychol.* 1977;16:347–356.
63. Pollio DE, North CS, Hudson AM, Hong BA, Osborne VA, McClendon JB. Psychoeducation Responsive to Families (PERF): Translation of a multifamily group model. *Psychiatr Ann.* 2012;42(6):226–235.
64. McClendon J, Pollio DE, North CS, Reid D, Jonson-Reid, M. (2007). School-based groups for parents of children with emotional and behavioral disorders: Pilot results. *Fam Soc.* 2007;88(1): 124–129.

59

FAMILY INTERVENTIONS IN PSYCHOSIS

Juliana Onwumere and Elizabeth Kuipers

BACKGROUND

The deleterious impact of schizophrenia and related disorders on mortality rates and years of potential life lost are well documented (Hjorthoj et al., 2017; Jayatilleke et al., 2017). The social networks of individuals living with psychosis are typically small and adversely affected by their mental health problems (Sundenmann et al., 2012). In addition, many of those affected by psychosis report feeling lonely (Michalska da Rocha et al). Deficits in social functioning skills (Velthorst et al., 2016) and illness related stigma (Schulze & Angermeyer, 2003) are two of several factors that can make it difficult for people with psychosis to establish and maintain peer relationships. Nevertheless, many will be receiving support and care from close family and friends. These informal (unpaid) carers are typically the parents and partners of patients although other groups such as siblings and children fulfil these roles. Few people in these supportive roles would have made an active decision to take on the role and become a carer (Kuipers, 1992); for most people it was something they did because the situation demanded action. Given the long-term and fluctuating course of many psychotic disorders, becoming a caregiver can be a life-changing and a lifelong endeavor. Nevertheless, the valuable contributions to patient outcomes made by informal carers are increasingly acknowledged. Carers often respond to unmet needs within clinical, social, and psychological domains, and by doing so, help to facilitate patient progress and recovery. Recent data suggest that patients with carers, when compared to peers without, have lower mortality rates (Revier et al., 2015) and are less likely to experience a relapse or require a hospital admission (Norman et al., 2005).

THE DEMANDS OF CAREGIVING

The onset of psychosis in a relative is associated with a broad range of emotional and behavioral reactions. Though it receives minimal research attention, carers can report positive experiences from their role that include reports of increased self-esteem, an awareness of inner strengths and their priorities, and an improved relationship with the person they care for (Chen & Greenberg, 2005; Kulhara et al., 2012). The caregiving role is also associated with negative experiences (commonly reported as burden) (Nordstroem et al., 2017), which is recorded across different carer groups (e.g., as parents, siblings, partners) (Bowman et al., 2017). The wider caregiving literature that integrates different health conditions, confirms that the caregiving role can adversely affect a carer's quality of life, is positively linked to subjective experiences of loneliness (Vasileiou et al., 2017), and is associated with significantly higher rates of common mental disorders and poorer well-being, when compared to those in noncaregiving roles (Smith et al., 2014). The elevated levels of common mental disorders are especially observable in those undertaking caregiving tasks of at least 10 hours per week or more (Smith et al., 2014).

In psychotic disorders, the evidence indicates that carer reports of depression are high (Derajew et al., 2017); approximately 40% are likely to experience clinical depression (Hamaie et al., 2016; Sadath et al., 2017), with similar rates reported for other stress-related conditions (Guptha et al., 2015). Recent studies have indicated that carers can experience trauma (Kingston et al., 2016), burnout (Onwumere et al., 2017), sleep disturbance (Smith et al., 2018), and isolation (Hayley et al., 2015). For example, those in caregiving roles are recorded as being 10 times more socially isolated than noncaregiving peers (Hayes et al., 2015) and significantly more isolated than caregivers of patients with other health disorders with similar morbidity (Magliano et al., 2005). Carers also report experiencing high levels of worry about the relatives they care for. The worries are diverse but can relate to ongoing concerns about patient well-being; what their future holds, particularly regarding the future risk of relapse (Lai et al., 2017), risk of harm to self or others including to carers themselves (Katz et al., 2015; Onwumere et al., 2018), and a lack of supportive networks and friendships.

CAREGIVING RELATIONSHIPS AND OUTCOMES

Caregiving relationships can play an instrumental role in patient outcomes. Expressed emotion (EE) provides a measure of the quality of family interactions and relationships and is measured across dimensions comprising criticism, hostility, emotional over involvement, warmth, and positive comments (Vaughn & Leff, 1976). High EE reflects above-threshold levels of carer reports of criticism and/or hostility and/or emotional overinvolvement toward the person they provide care for (Vaughn & Leff, 1976). Low EE ratings reflect carers whose scores fell below the threshold on these scales. Though levels vary across studies and across the different measurement tools employed, high EE is not uncommon and has an approximate range of 40%–60% in early psychosis populations (Linzen et al., 1996; Heikkila et al., 2002). EE ratings are not a stable characteristic; carers can transition from low to high ratings and vice versa depending on several factors including carer reports of loss (Patterson et al., 2005).

Caregiving relationships characterized by high-EE compared to low-EE relationships are commonly predictive of poorer patient outcomes. Poorer outcomes are defined in terms of elevated rates of relapse, hospitalization (Bebbington & Kuipers, 1994), longer hospital admission (Marom et al., 2005; Koutra et al., 2015; Sadiq et al., 2017), cannabis use (Gonzalez-Blanch et al., 2015), and greater expression of psychosis symptoms, including voice loudness, preoccupation, and anger (Finnegan et al., 2014). Rates of patient relapse within high-EE households are at least twice the rate recorded in low-EE homes (Bebbington & Kuipers, 1994). The predictive relationship between high-EE and poorer patient outcomes is particularly observable in relationships characterized by high levels of criticism and can be a long-term predictor of poorer outcomes over several years (Cechnicki et al., 2013). The negative impact of high-EE relationships has also been evaluated at a neural level. Data from functional magnetic resonance assessments, undertaken with a psychosis population, indicated different patterns of brain activation following an individual's exposure to familial high EE. The affected regions (e.g., rostral anterior cingulate, middle superior frontal gyrus) were also those involved in the processing of aversive social information (Rylands et al., 2011). Difficult (high-EE) caregiving relationships are also linked with less optimal levels of carer functioning. Carers who report high levels of burden in their caregiving role and engage in less adaptive and more avoidant coping strategies, are also those more likely to be in high-EE caregiving relationships (Jansen et al., 2014; Raune et al., 2004; Kuipers et al., 2006). Further, carer criticism is a significant predictor of carer burden (Villalobos et al., 2017).

Though high EE remains a robust predictor of patient relapse in many studies worldwide (Bebbington & Kuipers, 1994; Sadiq et al., 2017), the last decade has also witnessed a growing body of research that highlights how high EE is not consistently or uniformly predictive of poorer outcomes in some ethnic groups. For example, Hashemi (1997) assessed British Pakistani and Sikh families alongside white British families. Though observations of high EE were found across all three groups (e.g., 80% in British Pakistani, 30% in British Sikh, 45% in white British), its predictive links with poorer patient outcomes were limited to the white British group in that high carer EE was predictive of greater patient relapse.

In a systematic review of 34 studies investigating emotional overinvolvement (EOI) across different cultural groups, the authors recorded an inconsistent relationship between high EOI and patient outcomes, which were often moderated by the presence/absence of additional factors such as carer reports of warmth (Singh et al., 2013). Rosenfarb et al. (2006) assessed family relationships in an ethnically mixed group of African American and white American carer patient pairs. The authors found that high levels of carer criticism and intrusiveness were not predictive of poorer outcomes in the African American but instead predicted reduced levels of patient relapse. In contrast, the traditional predictive high EE–patient relapse relationship was observed in their white peers. Similar findings have been observed in EE comparisons between Mexican American and white American samples. Thus, while the link between high EE and patient relapse were identified in the white American families, these links were not observed in the Mexican American group. Interestingly, an absence of carer warmth, however, was linked to poorer patient outcomes in the Mexican sample though not in their white peers (Kopelowicz et al., 2002).

Carers with high EE ratings can be differentiated from their low-EE peers by the specific beliefs they report about the illness (Dominguez-Marinez et al., 2017; Barrowclough & Hooley, 2003; Hooley & Campbell, 2002; Cherry et al., 2017). To illustrate, carers with high EE ratings attributable to elevated criticism and hostility are less likely to perceive their relative's problems as forming part of a recognizable illness. Instead, their relative's behaviors and overall presentation are conceptualized as something they have control over and are doing deliberately. High EE ratings due to EOI are more likely to be found in carers expressing high levels of self-blame (Barrowclough & Hooley, 2003), shame, and guilt (Cherry et al., 2017), and in those who attribute the relative's illness and suffering to something they did or failed to do. These carers perceive their relative as having no control of the illness and symptoms (Barrowclough & Hooley, 2003).

Patient perspectives of carer behaviors and the caregiving relationship are increasingly being investigated (Tsai et al., 2015). The literature confirms that patients are able to perceive negative affect from carers (Onwumere et al., 2009; Renshaw et al., 2008) and their perceptions can impact outcomes (Tompson et al., 2009; Hesse et al., 2015). In an early psychosis sample, patient perceptions of carer criticism toward them were positively correlated with self-reports of greater anxiety and depression (Tomlinson et al., 2014). Similarly, in patients with a diagnosis of schizophrenia, a negative perception of their family atmosphere was linked to the subsequent development of paranoia (Hesse et al., 2015).

Though it receives less research attention, the importance and effect of positive family communications and relations are increasingly acknowledged in the caregiver literature. Carer warmth and positive comments; though not included in the computation of low/high EE ratings, are part of the overall EE index and can have a measurable impact on the illness course and patient outcomes (Bebbington & Kuipers, 1994; Brietborde et al., 2007). Significantly lower rates of patient relapse are recorded in families reporting higher rates of carer warmth (Bertrando et al., 1992); a finding particularly observable in some ethnic groups (Lopez et al., 2004; Singh et al., 2013). Lee and colleagues recorded significantly lower rates of patient relapse at a 6- and 12-month follow period, when carers were rated highly on warmth or patients perceived greater levels of positive affect from carers (Lee et al., 2014). Recent work confirms a negative cycle between patient negative perceptions of family atmosphere and the later development of negative interpersonal self-concepts (Hesse et al., 2015).

FAMILY INTERVENTIONS IN PSYCHOSIS

Families report a number of unmet needs for help with dealing with problems, emotional support, communication, and preparing for and managing crises (Lal et al., 2017; Askey, 2009). Evidence-based family interventions have developed through integrating advancements in our understanding of psychosis and outcomes, the impact of caregiving, and carer and patient appraisals about the caregiving relationships. Family work provision in psychosis is recommended in treatment guidance in several global regions including Europe, the United States, Canada, and Australia (Norman et al., 2017; Galletly et al., 2016; NICE, 2014, Kreyenbuhl et al., 2010). In the United Kingdom (UK), for example, the National Institute for Health and Care Excellence (NICE) for England, Wales and Northern Ireland recommend the provision of family interventions to families who remain in close contact with patients. Recommended interventions are expected to run between 3 months and 1 year and for a minimum 10 ten sessions.

The evidence-based family interventions can have different formats in terms of their delivery style. Thus, interventions can be held in the family home or clinic, with one family or multiple groups of families seen, and can vary on inclusion of patients in the sessions (Kuipers et al., 2002; McFarlane, 2002; Barrowclough & Tarrier, 1992). Notwithstanding these differences, the interventions share several core features, which include the explicit recognition that a family member has psychosis, hence the offer of the intervention, and a conceptualization of psychosis within a bio-psycho-social framework as it applies to onset, maintenance and relapse. The interventions are neither predicated on nor offered with theories about underlying family dysfunction. The interventions remain focused on current issues that are affecting the family—in the "here and now," with an overarching aim of reducing the risk of patient relapse in the future. The interventions are based on a collaborative approach that purposively joins with family members and seeks to use their skills as active partners in optimizing family outcomes. Typically, interventions are offered alongside medication and other psychosocial treatments available within standard care, and are facilitated by trained clinicians. The key therapeutic strategies of family-based interventions are based on provision of information about psychosis, skill development in areas of problem-solving, positive styles of communication, adaptive coping, stress management, and facilitating support (Gracio, Goncalves-Pereira, & Leff, 2016; McFarlane, 2016; Sin & Norman, 2013; Kuipers et al., 2002; Barrowclough & Tarrier, 1992).

THE EVIDENCE BASE FOR FAMILY INTERVENTIONS

Efficacy data and family feedback attest to the positive impact of family interventions on patient and carer outcomes, across the illness course (Claxton et al., 2017; Gregory, 2009; Sin et al., 2017; Pharoah et al., 2010). The perceived benefits of the interventions, as identified by patients and their relatives, are broad but include notable improvements in skills related to communication, problem-solving and psychoeducation, alongside relapse prevention (Nilsen et al., 2016). Robust findings from several reviews and meta-analyses (e.g., Pharoah et al., 2010, Cochrane Review of 53 randomized controlled trials [RCTs]) confirm the superiority of family interventions in significantly reducing rates of patient relapse and hospitalization in longer-term groups. For reductions in relapse rates, the fixed effect relative risk (RR) was 0.55 (95% CI 0.5–0.6) and for admission to hospital it was 0.78 (95% CI 0.6–1.0). Family interventions also yield significant reductions in poor medication compliance with the relative risk recorded as 0.60 (95% CI 0.5–0.7) (Pharoah et al., 2010).

These key findings are also observable in first-episode and early course populations (Bird et al., 2010; Claxton et al., 2017; Onwumere et al., 2011; Ruggeri et al., 2017) and in ultra-high-risk psychosis groups (O'Brien et al., 2014). In their recent meta-analytic review of 17 papers from 14 separate studies, Claxton and colleagues (2017) confirmed favorable effects of family interventions on relapse at the end of treatment (RR = 0.58, 95% CI 0.34–1.00) but not follow up, and a positive effect on patient symptoms at follow up (d = -0.85, 95% CI -1.05- -0.20) but not end of treatment. Recent publication of long-term (i.e., 14 years) follow-up data found that in households where family interventions were delivered, patients had significantly better levels of engagement with medication treatments and social functioning when compared to those from the control group (Ran et al., 2016). In high-EE households, family interventions can significantly reduce high EE ratings to low EE (Claxton et al., 2017; Pharoah et al., 2010) and reduce levels of carer burden (Giron et al., 2010; Claxton et al., 2017). The efficacy of family intervention is also linked with improvements in patient social functioning (Pfammater et al., 2006). The cost-effective outcomes on the delivery of family interventions are good (Mihalopoulos et al., 2004; NICE, 2009), and their application to and implementation within routine services have proven efficacy (Kelly & Newstead, 2004; Ruggeri et al., 2015).

Carer functioning and outcomes, including willingness to continue in the caregiving role, are positively impacted by family interventions (Pharoah et al., 2010; Berglund et al., 2003). Data from a recent systematic review and meta-analysis of 32 RCTs and 2,858 carer participants highlighted significant improvements to carer global morbidities, reports of burden and expressed emotion (Sin et al., 2017). A similar pattern of findings noting positive improvements to carer well-being were also confirmed from a review of family interventions for early psychosis populations (Claxton et al., 2017).

FAMILY INTERVENTION IMPLEMENTATION

Interestingly, the robust evidence base of family interventions has not facilitated its wider provision in routine services. Ongoing difficulties are often observed in patient access to and provision of family interventions (Onwumere et al., 2016; Eassom et al., 2014; Selick et al., 2017; Addington et al., 2012). Recent studies suggest that access to family interventions can range between 0 to 53% across different and geographical areas (Ince et al., 2015; Onwumere et al., 2014; Prytys et al., 2011). Explanations for poor access have been attributed to a broad range of factors that include inadequate pathways for training practitioners and supervising and supporting their practice; difficulties experienced by clinicians with identifying and engaging families; organizational factors that adversely impact on the time available for clinicians to prioritize and complete the work, and pessimism among frontline staff about the value and efficacy of family interventions (Onwumere et al., 2016; Eassom et al., 2014; Durbin et al., 2014; Prytys et al., 2011; Fadden, 2006). Family factors (i.e., patient and relative) adversely affecting implementation rates for family intervention include a lack of understanding about the role and format of the interventions and perceived relevance to their needs, feeling uncomfortable with discussing family matters with others, and competing time demands (Selick et al., 2017; Fadden, 2006). Many carers, for example, are in paid employment and can have additional caregiving roles for other relatives that leave them with little time or flexibility for attending appointments (Onwumere et al., 2017).

MODELS OF FAMILY WORK FOR THE FUTURE

Against a background of the needs of every family being individual and variable, there is growing acknowledgment that not all families with a relative with psychosis will have need for, want, or indeed be best suited for, the full comprehensive evidence-based family interventions, despite treatment guidance (Cohen et al., 2008; Slade et al., 2003). A movement toward a triaged approach to intervention delivery has been mooted by some researchers, whereby comprehensive interventions are limited to the smaller group of families who have the more complex presentations and highest levels of clinical need (Onwumere et al., 2016; Cohen et al., 2008; Selwick et al., 2017; Mottaghipour & Bickerton, 2005). These may be, for example, families where the relative with psychosis frequently relapses, where the police are often called to manage family disputes and where the pattern of family relationships is complex (Grice, 2014; Onwumere et al., 2016). For the remainder of families, it has been argued that offering a range of interventions (Leggatt & Woodhead, 2016) and the least intensive components of family interventions required (e.g., psychoeducation, problem-solving skills, relapse prevention work) to deliver reductions in stress and promote adaptive coping, should be prioritized (e.g., Cohen et al., 2008). Though the approach seems sensible and has face validity, further empirical investigations are required to determine its value and added benefits in positively impacting the needs of families.

CONCLUSION

The impact of psychosis extends beyond the individual with diagnosis; it also affects their extended family and the quality of family relationships. The provision of evidence-based family interventions has been designed to promote positive family outcomes and mitigate the deleterious effects psychosis can have on family functioning and well-being. Though the evidence base is long and robust, access issues continue to prevail with many families not receiving the intervention. Further work is required on identifying the key mechanisms for positive change and optimal pathways for widening access across groups (Onwumere et al., 2016).

REFERENCES

1. Addington DE, Mckenzie E, Wang J, Smith HP, Adams B, Ismail, Z. (2012). Development of a core set of performance measures for evaluating schizophrenia treatment services. *Psychiatric Services*, 63(6), 584–591.
2. Askey R, Holmshaw J, Gamble C, Gray R. (2009). What do carers of people with psychosis need from mental health services? Exploring the views of carers, service users and professionals. *Journal of Family Therapy*, 31, 310–331.
3. Barrowclough C, Hooley J. M. (2003). Attributions and expressed emotion: A review. *Clinical Psychology Review*, 23(6), 849–880.
4. Barrowclough C, Tarrier N. (1992). Families of schizophrenia patients: cognitive behavioural intervention. Chapman and Hall, London.
5. Bebbington P, Kuipers L. (1994). The clinical utility of expressed emotion in schizophrenia. *Acta Psychiatrica Scandinavica Suppl*, 382, 46–53.
6. Berglund N, Vahlne JO, Edman A. (2003). Family intervention in schizophrenia: impact on family burden and attitude. *Social Psychiatry and Psychiatric Epidemiology*, 38(3), 116–121. doi: 10.1007/s00127-003-0615-6
7. Bowman S, Alvarez-Jimenez M, Wade D, Howie L, McGorry P. (2017). The positive and negative experiences of caregiving for siblings of young people with first episode psychosis. *Frontiers in Psychology*, 8, 730. doi: 10.3389/fpsyg.2017.00730.
8. Bird V, Premkumar P, Kendall T, Whittington C, Mitchell J, Kuipers E. (2010). Early intervention services, cognitive-behavioural therapy and family intervention in early psychosis: systematic review. *British Journal of Psychiatry*, 197(5), 350–356.
9. Cechnicki A, Bielanska A, Hanuszkiewicz I, Daren A. (2013). The predictive validity of expressed emotions (EE) in schizophrenia: a 20-year prospective study. *Journal of Psychiatric Research*, 47(2), 208–214. doi: 10.1016/j.jpsychires.2012.10.004
10. Chen FP, Greenberg JS. (2004). A positive aspect of caregiving: the influence of social support on caregiving gains for family members of relatives with schizophrenia. *Community Mental Health Journal*, 40(5), 423–435.
11. Cherry MG, Taylor PJ, Brown SL, Rigby JW, Sellwood W. (2017). Guilt, shame and expressed emotion in carers of people with long term mental health difficulties: a systematic review. *Psychiatry Research*. doi.org/10.1016/j.psychres.2016.12.056
12. Claxton M, Onwumere J, Fornells-Ambrojo M. (2017). Do family interventions improve outcomes in early psychosis? A systematic review and meta-analysis. *Frontiers in Psychology*, 8, 371. doi: 10.3389/fpsyg.2017.00371
13. Cohen AN, Glynn SM, Murray-Swank AB, Barrio C, Fischer EP, McCutcheon SJ, Dixon LB. (2008). The family forum: directions for the implementation of family psychoeducation for severe mental illness. *Psychiatric Services*, 59(1), 40–48. doi: 10.1176/appi.ps.59.1.40
14. Derajew H, Tolessa D, Tolu Feyissa G, Addisu F, Soboka M. (2017). Prevalence of depression and its associated factors among primary caregivers of patients with severe mental illness in southwest Ethiopia. *BMC Psychiatry*, 17:88. doi 10.1186/s12888-017-1249-7
15. Dominguez-Marinez T, Medina-Pradas C, Kwapil TR Barrantes-Vidal N. (2017). Relatives expressed emotion, distress and

attributions in clinical high risk and recent onset psychosis. *Psychiatry Research*, 247, 323–329.
16. Eassom E, Giacco D, Dirik A, Priebe S. (2014). Implementing family involvement in the treatment of patients with psychosis: a systematic review of facilitating and hindering factors. *British Medical Journal Open*, 4. doi: 10.1136/bmjopen-201400610810.1136/bmjopen-2014-006108
17. Fadden G. (2006). Training and disseminating family interventions for schizophrenia: developing family intervention skills with multi-disciplinary groups. *Journal of Family Therapy*, 28(1), 23–38. doi: doi 10.1111/j.1467-6427.2006.00335.x
18. Finnegan D, Onwumere J, Green C, Freeman D, Garety P, Kuipers E. (2014). Negative communication in psychosis: understanding pathways to poorer patient outcomes. *Journal of Nervous and Mental Disease*, 202(11), 829–832.
19. Galletly C, Castle D, Dark F, Humberstone V, Jablensky A, Killackey E, Kulkarni J, McGorry P, Nielssen O, Tran N. (2016). Royal Australian and New Zealand College of Psychiatrists clinical practice guidelines for the management of schizophrenia and related disorders. *Australian and New Zealand Journal of Psychiatry*, 50(5), 410–472
20. González-Blanch C, Gleeson JF Cotton SM, Crisp K, McGorry PD, Alvarez-Jimenez M. (2015). Longitudinal relationship between expressed emotion and cannabis misuse in young people with first-episode psychosis. *European Psychiatry*, 30(1):20–25. doi: 10.1016/j.eurpsy.2014.07.002
21. Gracio J, Gonclaves-Pereira M. Leff. (2016). Key elements of a family intervention for schizophrenia: a qualitative analysis of an RCT. *Family Process*. doi:10.1111/famp.12271.
22. Grice S. (2014) Southwark family intervention for psychosis service audit and review (South London and Maudsley NHS Foundation Trust internal document).
23. Giron M, Fernandez-Yanez A, Mana-Alvarenga S, Molina-Habas A, Nolasco A, Gomez-Beneyto M. (2010). Efficacy and effectiveness of individual family intervention on social and clinical functioning and family burden in severe schizophrenia: a 2-year randomized controlled study. *Psychological Medicine*, 40(1), 73–84.
24. Gregory M. (2009). Why are family interventions important? A family member perspective. In Lobban F, Barrowclough C (eds.): *A casebook of family interventions in psychosis*. London: John Wiley and Sons.
25. Gupta S, Isherwood G, Jones K, Van Impe K. (2015) Assessing health status in informal schizophrenia caregivers compared with health status in non-caregivers and caregivers of other conditions. *BMC Psychiatry*, 15, 162.
26. Hashemi AH. (1997). Schizophrenia, expressed emotion and ethnicity: a British Asian study. Unpublished PhD thesis, University of Birmingham.
27. Hayes L, Hawthorne G, Farhall J, O'Hanlon B, Harvey C. (2015). Quality of life and social isolation among caregivers of adults with schizophrenia: policy and outcomes. *Community Mental Health Journal* 51(5), 591–597. doi: 10.1007/s10597-015-9848-
28. Hooley JM, Campbell C. (2002). Control and controllability: beliefs and behaviour in high and low expressed emotion relatives. *Psychological Medicine*, 32(6), 1091–1099
29. Jayatilleke N, Hayes RD, Dutta R, Shetty H, Hotopf M, Chang CK, et al. (2017). Contributions of specific causes of death to lost life expectancy in severe mental illness. *European Psychiatry*, 43, 109–115.
30. Gupta S, Isherwood G, Jones K, Van Impe K. (2015). Assessing health status in informal schizophrenia caregivers compared with health status in non-caregivers and caregivers of other conditions. *BMC Psychiatry*. Advance online publication. doi 10.1186/s12888-015-0547-1
31. Hamaie Y, Ohmuro N, Katsura M, Obara C, Kikuchi T, Ito F, Miyakoshi T, Matsuoka H, Matsumoto K. (2016). Criticism and depression among the caregivers of at risk mental state and first episode psychosis patients. PLoS One 11, e0149875.
32. Hayes L, Hawthorne G, Farhall J, O'Hanlon B, Harvey C. (2015). Quality of life and social isolation among caregivers of adults with schizophrenia: Policy and outcomes. *Community Mental Health Journal*, 51(5), 591–597. doi: 10.1007/s10597-015-9848-
33. Hjorthøj C, Stürup A. M, McGrath J. J, Nordent M. (2017). Years of potential life lost and life expectancy in schizophrenia: a systematic review and meta-analysis. *Lancet Psychiatry*, 4, 295–301.
34. Ince P, Tai S, Haddock G. (2015). Using plain English and behaviourally specific language to increase the implementation of clinical guidelines for psychological treatments in schizophrenia. *Journal of Mental Health*, 24(3), 129–133. doi: 10.3109/09638237.2014.958213
35. Jansen JE, Lysaker PH, Harder S, Haahr UH, Lyse HG, Pedersen MB, Trauelsen AM, Simonsen E. (2014). Positive and negative caregiver experiences in first-episode psychosis: emotional overinvolvement, wellbeing and metacognition. *Psychology and Psychotherapy. Theory Research and Practice*, 87(3), 298–310.
36. Kelly M, Newstead L. (2004). Family intervention in routine practice: it is possible! *Journal of Psychiatric and Mental Health Nursing*, 11, 64–72.
37. Kingston C, Onwumere J, Keen N, Ruffell T, Kuipers E. (2016). Post traumatic symptoms in caregivers of people with psychosis and associations with caregiving experiences. *Journal of Trauma Dissociation*, 17(3), 307–321. doi.org/10.1080/15299732.2015.1089969
38. Koutra K, Triliva S, Roumeliotaki T, Stefanakis Z, Basta M, Lionis C, Vgontzas AN. (2015). Family functioning in first episode and chronic psychosis: the role of patient's symptoms severity and psychosocial functioning. *Community Mental Health Journal*, 52(6), 710–723.
39. Kuipers L. (1992). Needs of relatives of long-term psychiatric patients. In Thornicroft G, Brewin C, Wing C (eds.), *Measuring mental health needs* (pp. 291–307). Dorchester: Gaskell.
40. Kuipers E, Leff J, Lam D. (2002). *Family work for schizophrenia: a practical guide* (2nd ed.). London: Gaskell.
41. Kulhara P, Kate N, Grover S, Nehra R. (2012). Positive aspects of caregiving in schizophrenia: a review. *World Journal of Psychiatry*, 2(3), 43–48. doi: 10.5498/wjp.v2.i3.43
42. Kreyenbuhl J, Buchanan RW, Dickerson FB, Dixon LB. (2010) The Schizophrenia Patient Outcomes Research Team (PORT): updated treatment recommendations 2009. *Schizophrenia Bulletin*. 36(1): 94–103.
43. Lal S, Malla A, Marandola G, Thériault J, Tibbo P, Manchanda R, et al. (2017). "Worried about relapse": Family members' experiences and perspectives if relapse in first episode psychosis. *Early Intervention in Psychiatry*. doi: 10.1111/eip.12440
44. Lee G, Barrowclough C, Lobban F. (2014). Positive affect in the family environment protects against relapse in first-episode psychosis. *Social Psychiatry Psychiatric Epidemiology*, 49(3), 367–376. doi: 10.1007/s00127-013-0768-x
45. Lopez SR, Hipke KN, Polo AJ, Jenkins JH, Karno M, Vaughn C. (2004). Ethnicity, expressed emotion, attributions, and course of schizophrenia: family warmth matters. *Journal of Abnormal Psychology*, 113(3), 428–439.
46. Magliano L, Fiorillo A, Malangone C, De Rosa C, Maj M, National Mental Health Working Group. (2006) Social network in long-term diseases: a comparative study in relatives of persons with schizophrenia and physical illnesses versus a sample from the general population. *Social Science and Medicine*, 62(6), 1392–1402.
47. McFarlane WR. (2002). Multifamily groups in the treatment of severe psychiatric disorders. New York and London: Guildford Press.
48. McFarlane WR. (2016). Family interventions for schizophrenia and the psychoses: a review. *Family Process*, 55(3), 460–482.
49. Mihalopoulos C, Magnus A, Carter R, Vos T. (2004). Assessing cost effectiveness in mental health: family interventions for schizophrenia and related conditions. *Australian and New Zealand*, 38(7): 511–519.
50. Michalska da Rocha B, Rhodes S, Vasilopoulou E, Hutton P. (2017). Loneliness in psychosis: a meta-analytical review. *Schizophrenia Bulletin*. doi: 10.1093/schbul/sbx036

51. Mottaghipour Y, Bickerton A. (2005). The pyramid of family care: a framework for family involvement with adult mental health services. *Australian e-Journal for the Advancement of Mental Health*, 4(3), 1–8.
52. National Institute for Health and Care Excellence (NICE) (2009). Schizophrenia—core interventions in the treatment and management of schizophrenia in adults in primary and secondary. Clinical guideline 82.
53. National Institute for Health and Care Excellence (2014). *National Institute of Health and Clinical Excellence: core interventions in the treatment and management of schizophrenia in primary and secondary care*. London, UK: NICE.
54. Nilsen L, Frich JC, Friis S, Røssberg JI. (2014). Patients' and family members' experiences of a psychoeducational family intervention after a first episode psychosis: a qualitative study. *Issues in Mental Health Nursing*, 35(1), 58–68.
55. Nordstroem AL, Talbot D, Bernasconi C, Galani Berardo C, Lalonde J. (2017). Burden of illness of people with persistent symptoms of schizophrenia: a multinational cross sectional study. *International Journal of Social Psychiatry*, 63(2), 139–150.
56. Norman R, Lecomte T, Addington D, Anderson E. (2017). CPA treatment guidelines on psychosocial treatment of schizophrenia in adults. *Canadian Journal of Psychiatry/ La Revue Canadienne de Psychiatrie*, 1–7. doi: 10.1177/0706743717719894
57. Norman RMG, Malla AK, Manchanda R, Harricharan R, Takhar J, Northcott S. (2005). Social support and three-year symptom and admission outcomes for first episode psychosis. *Schizophrenia Research*, 80, 227–234.
58. O'Brien MP, Miklowitz DJ, Candan KA, Marshall C, Domingues I, Walsh BC, Cannon TD. (2014). A randomized trial of family focused therapy with populations at clinical high risk for psychosis: effects on interactional behavior. *Journal of Consulting and Clinical Psychology*, 82(1), 90–101. doi: 10.1037/a0034667
59. Onwumere J, Bebbington P, Kuipers E. (2011). Family interventions in early psychosis: specificity and effectiveness. *Epidemiology and Psychiatric Sciences*, 20(2), 113–119. doi: 10.1017/S2045796011000187
60. Onwumere J, Chung A, Boddington S, Little A, Kuipers E. (2014). Older adults with psychosis: a case for family interventions. *Psychosis, Psychological, Social and Integrative Approaches*, 6(2), 181–183. doi: 10.1080/17522439.2013.774436
61. Onwumere J, Grice S, Kuipers E, (2016). Delivering cognitive-behavioural family interventions for schizophrenia. *Australian Psychologist*, 51, 52–61
62. Onwumere J, Lotey G, Schulz J, James G, Afsharzadegan R, Harvey R, et al. (2017). Burnout in early course psychosis caregivers: the role of illness beliefs and coping styles. *Early Intervention in Psychiatry*, 11, 237–243.
63. Onwumere J, Kuipers E, Bebbington P, Dunn G, Freeman D, Fowler D, Garety P. (2009). Patient perceptions of caregiver criticism in psychosis: links with patient and caregiver functioning. *Journal of Nervous and Mental Disease*, 197(2), 85–91.
64. Onwumere J, Parkyn G, Learmonth S, Kuipers E. (2019). The last taboo: the experience of violence in first episode psychosis caregiving relationships. *Psychology and Psychotherapy Theory Research Practice*, 92, 1–9. DOI:10.1111/papt.12173
65. Patterson P, Birchwood M, Cochrane R. (2005). Expressed emotion as an adaptation to loss: prospective study in first-episode psychosis. *British Journal of Psychiatry*, 187(Suppl. 48), s59–s64.
66. Pfammatter M, Junghan UM, Brenner HD. (2006). Efficacy of psychological therapy in schizophrenia: conclusions from meta-analyses. *Schizophrenia Bulletin*, 32(Suppl 1), S64–80. doi: 10.1093/schbul/sbl030
67. Pharoah F, Mari J, Rathbone J, Wong W. (2010). Family intervention for schizophrenia. *Cochrane Database Systematic Review*, (12), CD000088. doi: 10.1002/14651858.CD000088.pub2
68. Prytys M, Garety P.A, Jolley S, Onwumere J, Craig T. (2011). Implementing the NICE guideline for schizophrenia recommendations for psychological therapies: a qualitative analysis of the attitudes of CMHT staff. *Clinical Psychology and Psychotherapy*, 18(1):48–59.
69. Ran MS, Chui CHK, Wong IY-L, Mao WJ, Lin FR, Liu B, Chan CL. (2016). Family caregivers and outcome of people with schizophrenia in rural china: 14 year follow up study. *Social Psychiatry and Psychiatric Epidemiology*. doi 10.1007/s00127-015-1169-0
70. Revier CJ, Reininghaus U, Dutta R, Fearon P, Murray RM, Doody GA, et al. (2015). Ten-year outcomes of first-episode psychoses in the MRC AESOP-10 study. *Journal of Nervous and Mental Disease*, 203(5), 379–386. doi: 10.1097/NMD.0000000000000295
71. Ruggeri M, Bonett C, Lasalvia A, Fioritti A, de Girolamo G, Santonastaso P, et al. (2015). Feasibility and effectiveness of a multi-element psychosocial intervention for first-episode psychosis: results from the cluster-randomized controlled get up piano trial in a catchment area of 10 million inhabitants. *Schizophrenia Bulletin*, 41(5):1192–1203.doi: 10.1093/schbul/sbv058
72. Ruggeri M, Lasalvia A, Santonastaso P, Pileggi F, Leuci E, Miceli M, et al. (2017). Family burden, emotional distress and service satisfaction in first episode psychosis. Data from the get up trial. *Frontiers in Psychology*, 8, 721. doi.org/10.3389/fpsyg.2017.00721
73. Rylands AJ, Mckie S, Elliot R, Deakin JFW, Tarrier N. (2011). A functional magnetic resonance imaging paradigm of expressed emotion in schizophrenia. *Journal of Nervous and Mental Disease*, 199(1), 25–29.
74. Sadiq S, Suhail K, Gleeson J, Alvarez-Jimenez M. (2017). Expressed emotion and the course of schizophrenia in Pakistan. *Social Psychiatry and Psychiatric Epidemiology*, 52(5), 587–593. doi 10.1007/s00127-017-1357-1
75. Sadath A, Muralidhar D, Varambally S, Gangadhar BN, Rose JR. (2017). Do stress and support matter for caring? The role of perceived stress and social support on expressed emotion of persons with first episode psychosis. *Asian Journal of Psychiatry*, 25, 163–168.
76. Schulze B, Angermeyer MC. (2003). Subjective experiences of stigma: a focus group study of schizophrenic patients, their relatives and mental health professionals. *Social Science and Medicine*, 56, 299–312.
77. Selick A, Durbin J, Vu N, O'Connor K, Volpe T, Lin E. (2016). Barriers and facilitators to implementing family support and education in early psychosis intervention programmes: a systematic review. *Early Intervention in Psychiatry*, 11(5), 365–374. doi: 10.1111/eip
78. Sin J, Gillard S, Spain D, Cornelius V, Chen T, Henderson C. (2017). Effectiveness of psychoeducational interventions for family carers of people with psychosis: a systematic review and meta-analysis. *Clinical Psychology Review*, 56, 13–24.
79. Singh SP, Harley K, Suhail K. (2013). Cultural specificity of emotional over involvement: a systematic review. *Schizophrenia Bulletin*, 39(2), 449–463. doi:10.1093/schbul/sbrl70.
80. Smith L, Onwumere J, Craig T, McManus S, Bebbington P, Kuipers E. (2014) Mental and physical illness in caregivers: results from an English national survey sample. *British Journal of Psychiatry*, 205(3):197–203.
81. Smith L, Onwumere J, Craig T Kuipers E. (2018). A role for poor sleep in determining distress in caregivers of individuals with early psychosis. *Early Intervention in Psychiatry*. doi: 10.1111/eip.12538
82. Slade M, Holloway F, Kuipers E. (2003). Skills development and family interventions in an early psychosis service. *Journal of Mental Health*, 12(4), 405–415.
83. Stain HJ, Galletly CA, Clark S, Wilson J, Killen EA, Anthes L, et al. (2012). Understanding the social costs of psychosis: the experience of adults affected by psychosis identified within the second Australian National Survey of Psychosis. *Australian and New Zealand Journal of Psychiatry*, 46, 879–889.
84. Sündermann O, Onwumere J, Kane F, Morgan C, Kuipers E. (2014). Social networks and support in first episode psychosis: exploring the role of loneliness and anxiety. *Social Psychiatry and Psychiatric Epidemiology*, 49(3), 359–366. doi.10.1007/s00127-013-0754-3
85. Tomlinson E, Onwumere J, Kuipers E. (2013). Distress and negative experiences of the care-giving relationship in early psychosis—does

social cognition play a role? *Early Intervention in Psychiatry*, 8(3), 253–260. doi: 10.1111/eip.12040
86. Vaughn CE, Leff JP. (1976) The measurement of expressed emotion in families of psychiatric patients. *British Journal of Social and Clinical Psychology*, 15(2), 157–165.
87. Vasileiou K, Barnett J, Barreto M, Vines J, Atkinson M, Lawson S, et al. Experiences of loneliness associated with being an informal caregiver: A qualitative investigation. *Frontiers in Psychology*, 8, 585. doi: 10.3389/fpsyg.2017.00585.
88. Velthorst E, Fett AKJ, Reichenberg A, Perlman G, van Os J, Bromet EJ, Kotov R. (2016).The 20-year longitudinal trajectories of social functioning in individuals with psychotic disorders. *American Journal of Psychiatry*. doi.org/10.1176/appi.ajp.2016.15111419
89. Villalobos BT, Ullman J, Wang Krick T, Alcantara D, Kopelowicz A, Lopez SR. (2017). Caregiver criticism, help giving and the burden of schizophrenia among Mexican American families. *British Journal of Clinical Psychology*, 56(3), 273–285.

60

PEER SUPPORT FOR PEOPLE WITH PSYCHIATRIC ILLNESS

A COMPREHENSIVE REVIEW

Chyrell D. Bellamy, Anne S. Klee, Xavier Cornejo, Kimberly Guy, Mark Costa, and Larry Davidson

"When I was in the psychiatric unit, we, the patients, were the biggest support to each other. We helped each other out as the nurses sat behind the glass bubble."
(Guy, as cited in, Bellamy et al.[1])

Many former patients or individuals with lived experience of mental illness might say that peer support—individuals with lived experience of mental illness providing support to others with similar lived experiences—is nothing new, that it has been happening informally throughout history, as described in the preceding quote by one of the authors of this chapter. Some might say that it is natural for humans to relate to individuals with similar lived experiences and for mutual support to occur because of those shared experiences. In fact, when we introduce peer support in presentations we often share the example of new parents who seek or are given support and guidance by more experienced parents to address daily childcare questions or concerns. What is new, since the 1980s, is the hiring of people with lived experiences as peer supporters in the behavioral health field to provide support to people with mental illnesses in formal mental health outpatient and inpatient facilities. Yale's Program for Recovery and Community Health (PRCH) has been at the forefront of the development and evaluation of peer support since the early 1990s, demonstrating that the involvement of those who have "been there" can be effective in engaging people, connecting them to needed services, and decreasing their use of substances and acute care while instilling hope, providing positive role models, and promoting engagement in self-care.[1–8] Peers can be instrumental in facilitating positive lifestyle and behavior changes and, by sharing their own "lived experiences," can help with motivation, coping, and illness self-management. While the use of peer support staff is still a relatively recent development, the US Centers for Medicare and Medicaid Services have added peer support to the menu of evidence-based and reimbursable services and peer health navigators are a key part of the person-centered behavioral health home model. Most states in the United States and some internationally, have developed a peer certification process to incorporate this relatively new and promising approach. This chapter provides an overview of peer support in formal roles within the behavioral and health workforce; it first provides an overview on what peer support is, why it is needed, and what it does to effect change; it then reviews the effectiveness research on peer support, specific duties/roles of peer supporters in practice, and ways to more effectively partner with peer supporters in the behavioral health field.

WHAT IS PEER SUPPORT?

Peer support generally falls in three broad categories: self-help groups, consumer operated services, and peer support providers.[3] In self-help, mutual support groups such as 12-step meetings, peer support is marked by a mutuality where members who share similar mental health problems voluntarily offer and receive support from one another. Consumer-operated organizations offer services that are developed and run for and by people with mental illness; examples include clubhouses and drop-in centers, peer-run centers.[9,10] They serve as a forum for shared experiences, foster community, and promote group pride.[11] Meanwhile, peer support employees are individuals with lived mental health experience hired by organizations to work directly with individuals who may be seen as harder to engage or those needing more recovery-related supports promoting self-management strategies. In both mutual support groups and consumer-run organizations, relationships are viewed as reciprocal in nature and everyone benefits, providing the opportunities to give back and get something in return from a fellow peer.[2] However, mutual support and consumer-operated or peer-run groups or services have not been fostered in agencies in the same way.

Relationships between peer support employees and the agency's clients/patients tend to be more formal and primarily unidirectional in nature, where the support is intended to help the client and not the peer provider specifically[4] (see Figure 60.1: Davidson graphic from 2010). In these relationships, the peer support provider typically shares less detail about his or her daily life and does not actively seek the client's support and friendship. When peer support providers do

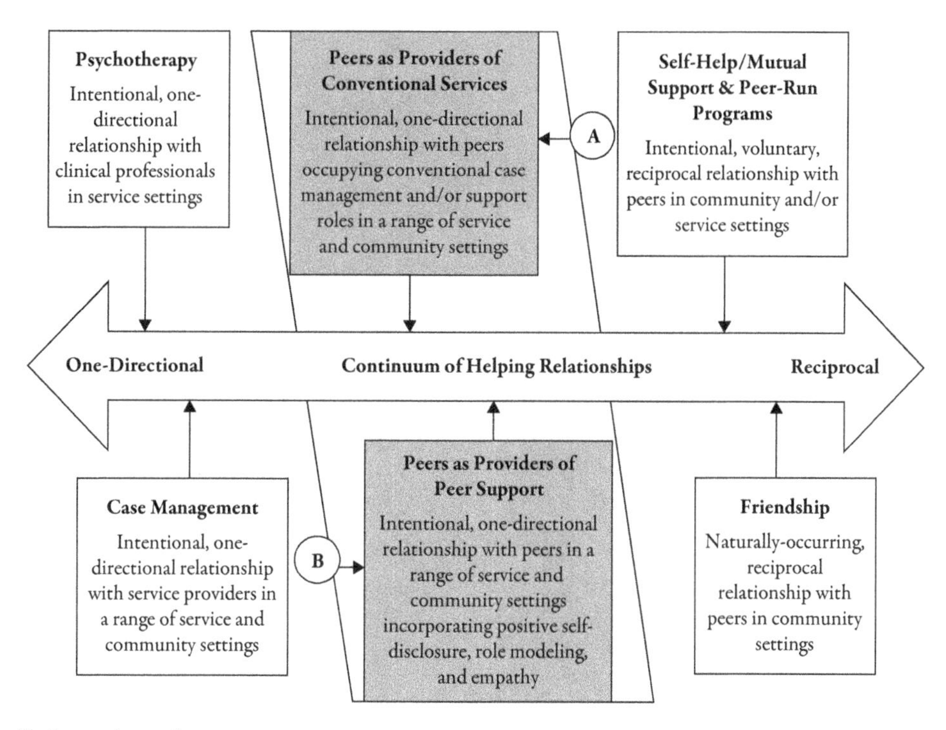

Figure 60.1 Continuum of helping relationships.
Davidson (2010).[6]

share information, they tend to focus on their own recovery stories, pitfalls they experienced, and the active steps they took to achieve their goals.

Despite the one-sided direction of these relationships, the peer support providers themselves still benefit from their work, experiencing an increased sense of self-worth and purpose in life.[12] These effects are aligned with Frank Riessman's[13] helper therapy principle, which contends that when an individual helps another person, the helper benefits as well.

WHY IS PEER SUPPORT NEEDED?

Unfortunately, many people with psychiatric diagnoses have experienced social isolation, poverty, stigmatization, discrimination, and demoralization, due to a variety of reasons including negative societal views on mental health in the Western world.[14] This stigmatization and discrimination is even more pronounced for people with diagnostic labels of schizophrenia and psychosis.[15] While stigma campaigns have helped a little,[16] for the most part, stigma and discrimination persists toward people with mental illness.[17] In the United States, for instance, there remains an association between mental illness and violence even though research indicates that people with mental illness are at a higher risk of being victims of violence rather than perpetrators.[15,16] This is in part due to the public's perceptions of mental illness that often are reinforced through how mental illness is portrayed in the media.[14,18] Against the backdrop of misinformation, pessimistic prognoses, and destructive stereotypes, peer staff provide invaluable, concrete proof of the reality of recovery. They instill hope and support that recovery is possible and demonstrate to persons with mental health conditions, their loved ones, and mental health practitioners alike that it is decent and caring people who experience psychiatric and substance use conditions and who, with sufficient and appropriate support, recover from them as well. Peer supporters can be beneficial in assisting individuals because of their experiences dealing with similar experiences of stigma, discrimination, and demoralization as former patients of the mental health system and as survivors residing in the community.

IS PEER SUPPORT EFFECTIVE?

Studies on the use of peers in the mental health field have demonstrated that peer staff have an ability to reach people who have been otherwise seen as difficult to engage. In research conducted by PRCH, peer interventions have been associated with fewer hospitalizations, fewer days in the hospital, longer community tenure after hospitalization, increased hope, improvements in self-care, enhanced sense of well-being in patients, decreased drug and alcohol use, and improvements in quality of life.[1–8] Similarly, individual studies have shown that program participants interacting with peers experience longer community tenures with fewer hospital stays,[19–22] shorter hospital stays,[19,22,23] and fewer emergency department visits[23] than those in treatment as usual. Studies also show no differences in rates of hospital admissions provided by peer supporters than care as usual[24,25] demonstrating outcomes that parallel those of professionally trained staff. Some studies also showed improved psychiatric symptomatology[26] and social functioning,[26–28] improved quality of life,[21] and increased treatment engagement.[8] Research studies also indicate that peer support contributes to improvements in empowerment among clients[29,30] and an increased sense of independence and empowerment for both the clients and peer supporters.[31] Furthermore, peer supporters themselves experience increases in self-esteem and self-confidence through their work supporting others[32–34] and their non-peer colleagues develop increased empathy and understanding of their clients by working alongside peer providers.[35] While several studies demonstrate these positive outcomes, meta-analytic[36–38] and systematic reviews;[39] describe a more complicated picture.

Pitt and colleagues[38] searched various registries and included 11 randomized control studies of consumers of mental health services employed as peer providers in mental healthcare settings from 1979 to 2012. They determined these studies were moderate to low in quality, citing various issues of research bias. They found that quality of life, symptom outcomes, satisfaction with treatment, number of hospital admissions, length of hospital stay, and other service use patterns for clients of teams with consumer-providers "were no better or worse" than outcomes achieved by professionals working in similar roles.[38] Lloyd-Evans and colleagues[40] conducted a systematic review and meta-analysis of 18 randomized trials of peer support interventions in nonresidential settings dating from 1982 to 2013. They determined there was little or no evidence that peer support was associated with positive outcomes for hospitalization, overall symptoms, or service satisfaction. Peer support was somewhat associated positively with hope, recovery, and empowerment, but not consistently within or across the different types of peer support. Fuhr and colleagues[37] conducted a meta-analysis of 14 randomized control studies from 1995 to 2012 .While three of the studies revealed positive results on quality of life and hope, for the majority of the studies no effects were found that indicated peer services enhanced any of the outcomes studied.

Chinman and colleagues[36] identified 20 studies conducted from 1995 to 2012 focusing on peer support for adults with serious mental illness. Taking into consideration the range of methodological rigor, the most positive impacts were found among studies with peers added to services and with peers delivering structured curricula. Across studies peer-delivered services brought reduced inpatient stays, increased levels of empowerment, improved engagement with care, and increased hopefulness, among other recovery outcomes with one exception.[36] A quasi-experimental study conducted by van Vught and colleagues[26] found several positive outcomes with the exception of increased days hospitalized among clients working with peers on an Assertive Community Treatment team. Cabassa and colleagues[41] conducted a systematic review of 18 studies dating from 1990 to 2015 focusing on health interventions for people with severe mental illness (SMI) involving peers. There were positive health outcomes related to self-management, dietary habits and communications with doctors, limited findings for physical activity, smoking, medication adherence, weight-related and cardiometabolic outcomes and mixed findings for quality of life, use of services, and self-related health outcomes. They found several methodological issues limiting the strength of the evidence and preventing the identification of the most beneficial interventions.

Most recently, Bellamy and colleagues[1] conducted an updated review based on meta-analyses and systematic reviews along with additional studies not included in previous reviews. The authors acknowledge the methodological limitations with a number of the studies. However, with the exception of the van Vugt study noted by Chinman and colleagues,[36] they contend that the evidence still supports that peer services results in outcomes "at least equivalent" to traditional care and/or services provided just by non-peer providers.[1] They further maintain that the evidence is strongest for peer support impacting recovery-oriented outcomes such as hope, quality of life, and empowerment than traditional mental health outcomes such as symptom severity and hospitalization rates.[1]

HOW DOES PEER SUPPORT WORK? MECHANISMS OF CHANGE

This question turns out to be complicated by a number of issues. Not the least of these complications is the fact that defining someone as a "peer"—which in this case means identifying someone as having a personal history of serious mental illness—tells us little about the person except for this one facet of his or her prior experience. As the idea of peer-provided services has spread, it has become more common for more experienced and established mental health professionals to disclose their own histories of mental illness; histories that in the past would have been kept private. While we have learned that peers can provide some conventional services as effectively as non-peers, what are the actual mechanisms at work?

Peer supporters help promote self-efficacy or belief in one's own abilities by sharing experiential knowledge and by modeling recovery and coping strategies.[32] Unlike traditional mental health practitioners, peer supporters disclose aspects of their illness and recovery stories and then draw on these experiences when working with their clients. They instill hope that recovery is possible and demonstrate that others, too, can and should assume more control over their lives.[4,9,42] While the intention is to assist the agency's clients and not themselves, the process of self-disclosure can lead to decreases in self-stigmatization and related improvements in quality of life and setting and achieving goals.[43]

Peer supporters are role models for recovery; they offer a view of the paths and steps others can take to achieve it themselves.[4,9] With role modeling, behaviors are learned through observation. Albert Bandura's (1977) social learning theory is based on the premise that a person identifies with a role model, and observes and then adopts the role model's behaviors, values, beliefs, and attitudes. Rosenthal and Bandura (1978) further describe observational learning effects as new patterns that one can learn and adopt after watching a model demonstrate novel responses that are not part of the observer's repertoire. Leon's Festinger's[44] social comparison theory offers additional insight into the psychological underpinnings of role modeling. It is based on the premise that individuals evaluate their beliefs and abilities by comparing themselves to others. Program participants can see themselves in peer supporters, compare themselves to them, and importantly see it is possible for them to achieve similar recovery goals. Anecdotally, peer supporters will often say they began working toward becoming a peer supporter after interacting with a peer supporter themselves.

Beyond being role models, peer supporters can further help service recipients to adopt new behaviors by encouraging them and helping them practice and refine these new skills. They offer real-world advice with managing symptoms, navigating complex service systems, living in poverty, and overcoming discrimination and stigma.[2,9] However, this guidance goes beyond practical advice; peer supporters help people get through difficult periods by providing support; relating to them; offering genuine empathy, trust, and acceptance; and communicating hope.[2,42,45] As their connections deepen, they encourage taking personal responsibility for one's actions and reinforce their rights to self-determination.[46]

Supported socialization activities allow individuals with serious mental illness to engage in their communities where they can experience caring, reciprocal relationships in which they themselves have something to offer others.[5] In their roles in mental health programs, peer supporters are the ideal team members to offer structured opportunities and supports that assist people with psychiatric disabilities to actively participate in community life. Having lived experience, peer supporters can share stories about their own journeys to engage in their communities explaining the steps they took and lessons they learned along the way. They can assist their clients in identifying social goals and then explore community opportunities such as connecting with interest groups or joining a religious congregation, among others. Because they themselves have worked through obstacles, peer supporters can assist individuals in identifying emotional barriers such as anxiety they may experience about being accepted by these groups and structural barriers such as difficulties navigating public transportation. By accompanying them at these early stages and supporting them as they transition into their new roles in the community, peer supporters can help their clients engage in meaningful community life.

In summary, what, then, might be considered the unique contributions, or active ingredients, of peers as providers of peer support? As described in Figure 60.2, these contributions fall into three basic categories. The first is the instillation of *hope* through positive *self-disclosure*, demonstrating to the service recipient that it is possible to go from being controlled by the illness to gaining some control over the illness, from being a victim to being the hero of one's own life journey.[47,48] The second expands this *role model*ing function to include self-care of one's illness and exploring new ways of using experiential knowledge, or "*street smarts*," in negotiating day-to-day life, not only with the illness but also with the social and human service systems, with having little to no income, with being unstably housed, with overcoming stigma, discrimination, and other trauma, etc.[48,49] The third focuses on the nature of the relationship between peer supporter and recipient, which is thought to be essential for the first two components to be effective. This relationship is characterized by trust, acceptance, understanding, and the use of *empathy*; empathy which in this case is paired with "*conditional regard*"—otherwise described as a peer supporter's ability to "read" a client based on having been in the same shoes he or she is in now. Their ability to empathize directly and immediately with their clients can be used in this particular way by peer supporters because they may have higher expectations and may place more demands on their clients, knowing that it is possible to recover, but also that it takes hard work to do so (e.g., "I know how you feel now, but I also know that you can have a better life").[41,49] This may at times lead to conflict and confrontation, but also is just as likely, if not more so, to lead to encouragement and inspiration.[3,41,49–53]

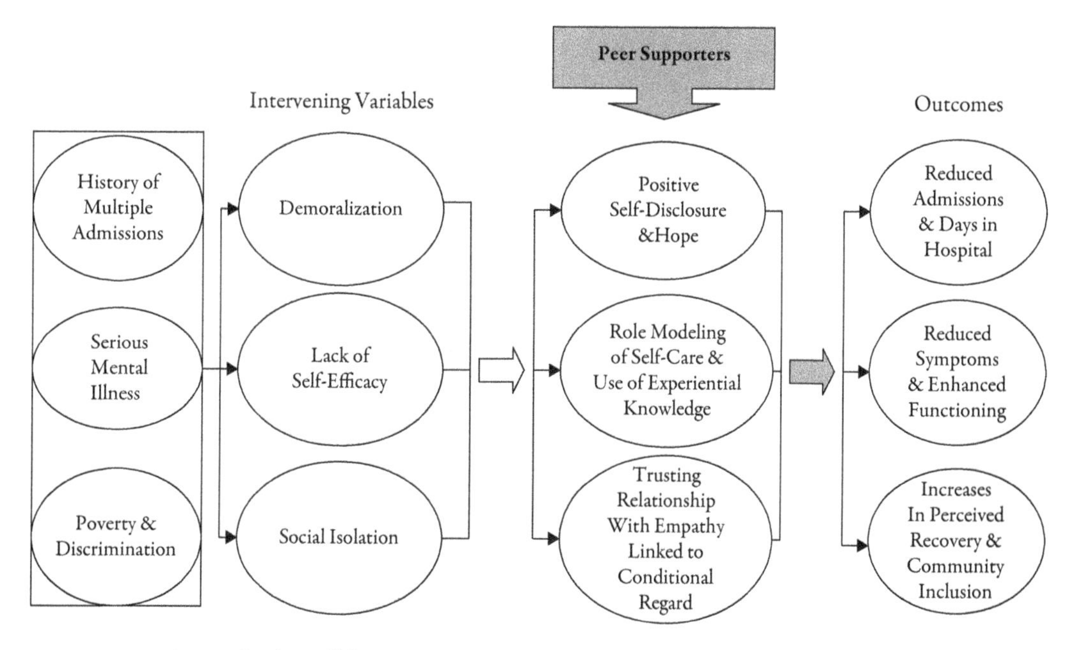

Figure 60.2 How peer supporters assist: mechanisms of change.

WHAT DO PEER SUPPORTERS DO IN PRACTICE?

At the root of peer support is the peer supporter's ability to connect and develop relationships with other individuals in recovery. The connection is often established because of the feeling of commonality—we are in this together—that often comes from having had similar lived experiences. It's a feeling of, "this person gets me." Peer supporters are also taught to enhance their relational skills. Intentional peer support (IPS) is one mechanism that is currently being used with a focus on relational skills.[45,54] In PRCH's previous studies on peer support the duties of the peer supporter (or as we defined the role, recovery mentor), were as follows:

Peer support is a combination of the theoretical foundation offered by IPS (focused on the principles and tasks: connection, worldview, mutuality, and moving toward[54] and the concrete interventions developed by the research team). In PRCH's earlier work on this model, we suggested that peer supporters are guided by the following five principles: (1) The initial focus of care is on the person's own understanding of his or her predicament (i.e., not necessarily the difficulties that brought him or her to the mentor's attention), and on the ways in which the mentor can be helpful in addressing this predicament, regardless of how the person understands it at the time; (2) Regardless of whether or not the person sought help, the mentor recognizes that the mentee had already embarked on his or her own life journey before meeting him or her; (3) Rather than dwelling on the mentee's distant past or worrying about his or her long-term future, mentors focus on the next several steps of the journey; (4) A mentor's credibility and effectiveness are enhanced to the degree that he or she is familiar with, and can anticipate interesting sites, common destinations, and important landmarks along the person's way; and (5) Mentors prepare for their role by acquiring tools that will be effective in addressing or compensating for symptoms and other sequelae of the illness that act as barriers to recovery.[55] It was within this framework that we developed the interventions of person-centered care planning and community inclusion that were evaluated in the "Culturally-Responsive, Person-Centered Care for Psychosis" project.[6] We combined these principles with that of Mead's work on IPS, which extends this model in several important ways. IPS is different from traditional service relationships because it does not start with the assumption of "a problem" that has brought the person to care.[45] Instead, mentors are taught to listen for how and why each person has learned to make sense of their experiences, and then to use the relationship to explore new ways of seeing, thinking, and doing. IPS also is sensitive to the prevalence of trauma in the lives of persons with mental illnesses, avoiding the implication that the person has done something "wrong" to require treatment, and instead asking the person about "what happened?"[54] and exploring alternative ways of dealing with what has happened. Instead of a focus on what the person needs to stop or avoid doing, mentors also encourage mentees to move toward what and where they would like to be. For one of PRCH's studies, we combined our earlier work with Mead's framework to train recovery mentors in the following three key areas described earlier, here listed in reverse order: (1) cultivating trusting relationships based on shared experiences, and characterized by acceptance, understanding, and empathy linked to conditional regard (incorporating Mead's concept of cultivating connections); (2) role modeling of self-care and use of experiential knowledge to explore new ways for the mentee to manage the mental illness and pursue personal goals (incorporating Mead's concepts of exploring worldviews and moving toward them); and (3) using positive self-disclosure to instill hope and restore the person's sense that he or she can make changes that will improve his or her life (incorporating Mead's concept of mutuality).

What happens at first contact is essential to developing intentional relationships between the peer supporter and the individual they are assisting. At the initial contact, the peer supporter explains his or her role and briefly shares his or her first-hand experience of recovery. This first meeting provides an opportunity for participants to connect with a "similar other" about "shared experiences." During this introduction, participants are also encouraged to talk about their experiences and to identify some of their expectations now that they are or have returned to the community. The peer supporter then offers participants assistance with "moving toward" their goals. This involves discussion of issues related to self-care and managing the person's illness, and role modeling of different strategies to minimize the disruptions associated with the illness. Just as importantly, though, this also involves assisting the person to identify those interests and goals he or she would like to pursue in the short-term, and providing the in vivo assistance needed for the person to do so. For these two components of the peer supporter's role, they are trained in person-centered care planning and community inclusion described in detail later. Throughout these activities, peer supporters are careful to avoid portraying themselves as the "experts" or conveying the message that the mentees are novices. Rather, the supporter's role is to explore with the mentee new ways of dealing with their situations and new ways of pursuing his or her interests and goals, with the mentee deciding on the destinations and selecting among the different pathways with which the mentor may be familiar.

Facilitation of person-centered planning involves organizing and conducting a series of planning meetings in collaboration with a participant and his or her primary clinician, that bring together the person with his or her network of professional and natural supports. The goal of these meetings is to discover a vision for a desirable future and to help the person develop an action plan to achieve that vision. Core principles of person-centered planning include: (1) primary direction in the planning process coming from the individual; (2) involvement of significant others and reliance on personal relationships as the primary source of support; (3) focus on capacities and assets rather than on limitations and deficits; and (4) an acceptance of uncertainty and setbacks as natural elements in the path to enhanced self-determination.[56]

Facilitation of community inclusion is based on the principle of person-centered planning, with emphasis on identifying, and promoting access to, integrated community settings rather than segregated settings designed for people with disabilities. To facilitate community inclusion, mentors help clients develop and use resource maps of their local communities based on their goals and interests.[57] Developing resource maps involves discovering existing, but untapped, resources and other potentially hospitable places and organizations in which the contributions of people with serious mental illnesses are welcomed and valued. Once identified, participants then receive in vivo, interactive training in joining community activities and organizations of their choice. In addition, participants are encouraged to and assisted in keeping a journal of their community outing experiences with entries such as words and stories, drawings, souvenirs, and photographs to use as a mechanism for remembering and celebrating their efforts and successes.

TRAINING PEER SUPPORTERS

Earlier we shared work being done to clarify the role of peer supporters. New York was one of the first states to train, certify, and hire peer support specialists.[58] In 2001, Georgia became the first state to fund peer support as a Medicaid-fundable service,[59] and on December 1, 2001, 35 mental health consumers had an opportunity to participate in a training and exam to become Georgia's first class of "Certified Peer Specialists," which allowed them to bill for Medicaid services.[60,61] Since 2001, Georgia's program has trained over 1,600 CPSs.[60]

As peer support became Medicaid fundable in many states, there has been a proliferation of state peer certification training programs (Connecticut and a few other US states do not directly receive Medicaid funding). Through data compiled by the Texas Institute for Excellence in Mental Health at the University of Texas at Austin as of July 2016, 41 states and the District of Columbia had established peer certification training programs and 2 states were in the process of developing certification programs.[62] Training programs vary in length and content with most occurring over 4 to 10 days. Topics taught generally include concepts of hope and recovery, interpersonal communication, telling your recovery story, identification and treatment of mental health disorders, wellness management, legal issues in mental health, stigma/discrimination issues in mental health, boundary issues, ethics/confidentiality/HIPAA, cultural competence and linguistic issues, overview of evidence-based practices, and workplace considerations. Modalities used to train peer supporters range but generally include didactics, discussions, role plays, and group exercises. Following training, most states administer an examination (written and/or oral) before certifying individuals. Criteria for eligibility varies across states. Most states require that applicants first and foremost identify as a person with a psychiatric and/or addiction history and then vary when it comes to other requirements. Some have a minimum age requirement, others expect applicants be out of the hospital for a defined period of time (we do not advocate for this requirement), while others want applicants to have a high school diploma among other criteria. In addition, most states do not offer the training to agencies or supervisors of peer supporters, which poses a challenge if the supervisors do not have lived experience of mental illness and recovery.[62]

PARTNERING WITH PEER SUPPORTERS IN THE WORKPLACE

The hiring and retention of peer supporters has been met with a few challenges, such as lack of role clarity—what is it that peer supporters will do; organizational culture—readiness to promote recovery-oriented care within the organization; and lack of preparation for supervisors of those in peer support roles. Part of the challenges in recognizing and including peer supporters in the behavioral health field, stems from beliefs about recovery. It has only been a short time since people received messages that recovery from mental illness was not possible. To go from a system that believed this to one that now embraces people with lived experiences working in behavioral health is still a far leap for some providers who have yet to embrace that people with lived experiences have a unique set of skills to add to the work of others on clinical teams.

PRCH is currently working on an approach to develop learning collaboratives for agencies that are interested in partnering with peer supporters in the workforce. The goal of this initiative is to help agencies establish the organizational culture and administrative infrastructure as well as to provide education on role clarity and supervision of peer supporters. Some of the challenges that peer supporters have faced in the workforce include: lack of role clarity, transition into employment in the behavioral health field, lack of a career ladder as a peer supporter, equitable pay for services provided, lack of effective supervision (educational, supportive, and administrative), microaggressions in the workplace by non-peer staff, and co-optation (staying a peer supporter versus taking on duties of case managers, etc.). Based on our experiences, we recommend that agencies plan for these challenges rather than simply hiring peer supporters. We provide an example below of how peer support came about in the US Department of Veterans Affairs and include workforce challenges and lessons learned.

CASE EXAMPLE OF PEER SUPPORT IN THE VETERANS HEALTH ADMINISTRATION

"Who then can so softly bind up the wound of another as he who has felt the same wound himself?"—Thomas Jefferson

Beginning in 2005, Veterans Health Administration (VA) medical centers began hiring peer support technicians (PSTs, later changed to peer support specialists) on their mental health clinical teams.[63] The 2008 *Handbook on Uniform Mental Health Services in VA Medical Centers and Clinics*[64] was developed in response to the goals of 2003 President's New Freedom Commission on Mental Health.[65] The handbook focused on a range of recovery-oriented services and required each VA medical center to offer peer support services to veterans with serious mental illnesses including the hiring of PSTs.[65] U.S. Public Law 110-387, established the following legislative requirements for peer support specialists working in the VA: a peer support specialist must be a veteran with an other than dishonorable discharge, be in recovery from a mental health condition for at least 1 year; and be trained and certified by a VA approved or State-approved not-for-profit certification organization.[66]

In the initial years of employment of PSTs in the VA, there was much staff resistance and confusion about the roles and responsibilities of PSTs.[67] As part of a randomized trial evaluating the implementation of PSTs in VA Mental Health Intensive Case Management teams, Chinman and colleagues used the Simpson transfer model to assess the teams' needs and preferences to help guide the implementation of PSTs on their teams. The researchers engaged with the teams to seek their input on desired characteristics of candidates, processes of the hiring, establishing roles and responsibilities, onboarding, training, and supervising PSTs.[63] Overall, the researchers found that using an organizational change model was useful for formalized planning. "We believe this approach can be helpful to facilities who have decided to deploy PSTs and useful to engage leadership who have not been predisposed to the idea."[63]

As of August 9, 2017, the Veterans Health Administration has 1,079 peer support specialists working in VA mental health programs across the nation (personal communication with Daniel O'Brien Mazza). In the VA, peer support specialists now work in various programs (intensive case management, homeless outreach and engagement, vocational, and justice outreach) and settings (residential, inpatient, outpatient) across mental health service lines. Peer support specialists experience a range of mental health conditions including psychotic, mood, anxiety, and substance use disorders with many having co-occurring disorders and a number having faced homelessness and/or unemployment at times in their lives. Often a peer support specialist with a particular disorder such as schizophrenia or post-traumatic stress disorder will work on a team that serves veterans with similar conditions and/or who faced similar life challenges. In the VA, peer support specialists are considered full clinical team members attending rounds, conducting individual and group sessions, educating veterans about setting goals, developing skills to manage illness, and documenting in the electronic health record. While they do not conduct formal assessments nor diagnose, the descriptive information that peer support specialists provide about their encounters with veterans is valued by team members. Teams value the perspectives of peer support specialists, see firsthand that recovery is possible, and use language that is more recovery-oriented. Similarly, research conducted outside the VA suggests that having peer support providers on clinical teams reduces stigma by shifting negative attitudes of mental health providers.[68,69]

CONCLUSION

In conclusion, peer support is the one of the fastest-growing occupations within the behavioral health workforce. Overall, the evidence suggests that peer support has benefits, and while more research is needed, current evidence suggests that adding peer support will more than likely enhance services for people with mental illness. Research in this area is challenging because definitions of peer support and the roles of peer supporters vary across sites and programs, making it difficult to assess what happens in practice. In addition, integrating or partnering effectively with peer supporters in the field is difficult for the same reasons. Much work needs to be done to further advance the practice and research of peer support, but as with any new or enhanced innovation, we must work with organizations and clinical providers to plan more effectively how to better partner with peer supporters so they can effectively deliver peer support services.

REFERENCES

1. Bellamy C, McDonald HE, Hawkins D, Guy K, Herring Y, Davidson L. Input from stakeholders: Hiring and retaining people in recovery in the behavioral health workforce. *Cadernos Brasileiros de Saúde Mental/Brazilian Journal of Mental Health*. 2017;9(21):66–78.
2. Bellamy CD, Rowe M, Benedict P, Davidson L. Giving back and getting something back: The role of mutual-aid groups for individuals in recovery from incarceration, addiction, and mental illness. *Journal of Groups in Addiction and Recovery*. 2012;7:223–236.
3. Davidson L, Chinman M, Kloos B, Weingarten R, Stayner D, Tebes JK. Peer support among individuals with severe mental illness: A review of the evidence. *Clinical Psychology: Science and Practice*. 1999;6:165–187.
4. Davidson L, Chinman M, Sells D, Rowe M. Peer support among adults with serious mental illness: a report from the field. *Schizophrenia Bulletin*. 2006;2005;32:443–450.
5. Davidson L, Shahar G, Stayner DA, Chinman MJ, Rakfeldt J, Tebes JK. Supported socialization for people with psychiatric disabilities: Lessons from a randomized controlled trial. *Journal of Community Psychology*. 2004;32:453–477.
6. Davidson L. *Effectiveness and cost-effectiveness of peer mentors in reducing hospital use*. National Institute of Mental Health application, R01 #MH091453L, L. Davidson, Principal Investigator. 2010.
7. Davidson L, Tondora JS, Staeheli MR, O'Connell MJ, Frey J, Chinman MJ. Recovery guides: An emerging model of community-based care for adults with psychiatric disabilities. In Lightburn A, Sessions P, eds., *Community based clinical practice*. London: Oxford University Press, 2006, pp. 476–501.
8. Sells D, Black R, Davidson L, Rowe M. Beyond generic support: Incidence and impact of invalidation in peer services for clients with severe mental illness. Psychiatric Services. 2008;59(11):1322–1327.
9. Solomon P. Peer support/peer provided services underlying processes, benefits, and critical ingredients. *Psychiatric Rehabilitation Journal*. 2004;27(4):392.
10. Stroul B. Rehabilitation in community support systems. In: Flexer R, Solomon P, eds. *Psychiatric rehabilitation in practice*. Boston, MA: Andover Medical, 1993.
11. Clay S, ed. *On our own, together: Peer programs for people with mental illness*. Nashville, TN: Vanderbilt University Press, 2005.
12. Mourra S, Sledge W, Sells D, Lawless M, Davidson L. Pushing, patience, and persistence: Peer providers' perspectives on supportive relationships. *American Journal of Psychiatric Rehabilitation*. 2014;17:307–328.
13. Riessman F. The "helper" therapy principle. *Social Work*. 1965:27–32.
14. Corrigan PW, Watson AC. Understanding the impact of stigma on people with mental illness. *World Psychiatry*. 2002;1(1):16–20.
15. Walsh E, Buchanan A, Fahy T. Violence and schizophrenia: Examining the evidence. *British Journal of Psychiatry*. 2002;180:490–495.

16. Evans-Lacko S, Corker E, Williams P, Henderson C, Thornicroft G. Effect of the Time to Change anti-stigma campaign on trends in mental-illness-related public stigma among the English population in 2003–13: an analysis of survey data. *Lancet Psychiatry.* 2014;1:121–128.
17. Pescosolido BA, Medina TR, Martin JK, Long JS. The "backbone" of stigma: Identifying the global core of public prejudice associated with mental illness. *American Journal of Public Health.* 2013;103:853–860.
18. Walsh E, Moran P, Scott C, et al. Prevalence of violent victimisation in severe mental illness. *British Journal of Psychiatry.* 2003;183:233–238.
19. Chinman MJ, Weingarten R, Stayner D, Davidson L. Chronicity reconsidered: Improving person-environment fit through a consumer-run service. *Community Mental Health Journal.* 2001;37:215–229.
20. Sledge WH, Lawless M, Sells D, Wieland M, O'Connell MJ, Davidson L. Effectiveness of peer support in reducing readmissions of persons with multiple psychiatric hospitalizations. *Psychiatric Services.* 2011;62:541–544.
21. Klein AR, Cnaan RA, Whitecraft J. Significance of peer social support with dually diagnosed clients: Findings from a pilot study. *Research on Social Work Practice.* 1998;8:529–551.
22. Min S, Whitecraft J, Rothbard AB, Salzer MS. Peer support for persons with co-occurring disorders and community tenure: A survival analysis. *Psychiatric Rehabilitation Journal.* 2007;30:207–213.
23. Clarke GN, Herinckx HA, Kinney RF, et al. Psychiatric hospitalizations, arrests, emergency room visits, and homelessness of clients with serious and persistent mental illness: Findings from a randomized trial of two ACT programs vs. usual care. *Mental Health Services Research.* 2000;2:155–164.
24. Solomon P, Draine J. One-year outcomes of a randomized trial of consumer case management. *Evaluation and Program Planning.* 1995;18:117–127.
25. O'Donnell M, Parker G, Proberts M, et al. A study of client-focused case management and consumer advocacy: The Community and Consumer Service Project. *Australian and New Zealand Journal of Psychiatry.* 1999;33:684–693.
26. van Vugt MD, Kroon H, Delespaul, Philippe AEG, Mulder CL. Consumer-providers in assertive community treatment programs: Associations with client outcomes. *Psychiatric Services.* 2012;63:477–481.
27. Forchuk C, Martin M-, Chan YL, Jensen E. Therapeutic relationships: From psychiatric hospital to community. *Journal of Psychiatric and Mental Health Nursing.* 2005;12:556–564.
28. Yanos PT, Primavera LH, Knight EL. Consumer-run service participation, recovery of social functioning, and the mediating role of psychological factors. *Psychiatric Services.* 2001;52:493–500.
29. Corrigan PW. Impact of consumer-operated services on empowerment and recovery of people with psychiatric disabilities. *Psychiatric Services.* 2006;57:1493–1496.
30. Dumont J, Jones K. Findings from a consumer/survivor defined alternative to psychiatric hospitalization. *Outlook.* 2002;3(Spring):4–6.
31. Ochocka J, Nelson G, Janzen R, Trainor J. A longitudinal study of mental health consumer/survivor initiatives: Part 3—A qualitative study of impacts of participation on new members. *Journal of Community Psychology.* 2006;34:273–283.
32. Salzer MS, Shear SL. Identifying consumer-provider benefits in evaluations of consumer-delivered services. *Psychiatric Rehabilitation Journal.* 2002;25:281–288.
33. Walker G, Bryant W. Peer support in adult mental health services: A metasynthesis of qualitative findings. *Psychiatric Rehabilitation Journal.* 2013;36:28–34.
34. Gidugu V, Rogers ES, Harrington S, et al. Individual peer support: A qualitative study of mechanisms of its effectiveness. *Community Mental Health Journal.* 2015;51:445–452.
35. Collins R, Firth L, Shakespeare T. "Very much evolving": A qualitative study of the views of psychiatrists about peer support workers. *Journal of Mental Health.* 2016;25:278–283.
36. Chinman M, George P, Dougherty RH, et al. Peer support services for individuals with serious mental illnesses: Assessing the evidence. *Psychiatric Services.* 2014;65:429–441.
37. Fuhr DC, Salisbury TT, De Silva MJ, et al. Effectiveness of peer-delivered interventions for severe mental illness and depression on clinical and psychosocial outcomes: A systematic review and meta-analysis. *Social Psychiatry and Psychiatric Epidemiology.* 2014;49:1691–1702.
38. Pitt V, Lowe D, Hill S, et al. Consumer-providers of care for adult clients of statutory mental health services. *Cochrane Database of Systematic Reviews.* 2013;3:CD004807.
39. Repper J, Carter T. A review of the literature on peer support in mental health services. *Journal of Mental Health.* 2011;20:362.
40. Lloyd-Evans B, Mayo-Wilson E, Harrison B, Istead H, Brown E, Pilling S, et al. A systematic review and meta-analysis of randomised controlled trials of peer support for people with severe mental illness. *BMC Psychiatry.* 2014;14(1):39.
41. Cabassa LJ, Camacho D, Vélez-Grau CM, Stefancic A. Peer-based health interventions for people with serious mental illness: A systematic literature review. *Journal of Psychiatric Research.* 2017;84:80–89.
42. Davidson L, Bellamy C, Guy K, Miller R. Peer support among persons with severe mental illnesses: A review of evidence and experience. *World Psychiatry.* 2012;11(2):123–128.
43. Corrigan PW, Morris S, Larson J, et al. Self-stigma and coming out about one's mental illness. *Journal of Community Psychology.* 2010;38:259–275.
44. Festinger L. A theory of social comparison processes. *Human Relations.* 1954;7(2):117–140.
45. Mead S, MacNeil C. Peer support: What makes it unique. *International Journal of Psychosocial Rehabilitation.* 2006;10(2):29–37.
46. Salzer MS, Brusilovskiy E, Schwenk E. Certified peer specialist roles and activities: Results from a national survey. *Psychiatric Services.* 2010;61:520–523.
47. Croft B, İsvan N. Impact of the 2nd story peer respite program on use of inpatient and emergency services. *Psychiatric Services.* 2015;66(6):632–637.
48. Pfeiffer PN, Heisler M, Piette JD, Rogers MA, Valenstein M. Efficacy of peer support interventions for depression: A meta-analysis. *Gen Hosp Psychiatry.* 2011;33(1):29–36.
49. Bryan AE, Arkowitz H. Meta-analysis of the effects of peer-administered psychosocial interventions on symptoms of depression. *American Journal of Community Psychology.* 2015;55(3–4): 455–471.
50. Mowbray CT, Moxley DP, Thrasher S, Bybee D, Harris S. Consumers as community support providers: Issues created by role innovation. *Community Mental Health Journal.* 1996;32(1):47–67.
51. Hibbard JH, Mahoney ER, Stockard J, Tusler M. Development and testing of a short form of the patient activation measure. *Health Serv Res.* 2005;40(6 Pt 1):1918–1930.
52. Hibbard JH, Greene J. What the evidence shows about patient activation: Better health outcomes and care experiences; fewer data on costs. *Health Affairs.* 2013;32(2):207–214.
53. Druss BG, Zhao L, Von Esenwein S, Morrato EH, Marcus SC. Understanding excess mortality in persons with mental illness: 17-year follow up of a nationally representative US survey. *Medical Care.* 2011;49(6):599–604.
54. Intentional Peer Support. 2017. http://www.intentionalpeersupport.org/
55. Kaplan R, Ganiats T, Rosen P, Sieber W, Anderson J. Development of a self-administered Quality of Well-Being scale (QWB-SA): Initial studies. *Quality of Life Research.* 1995:443–444.
56. Tondora J, Miller R, Slade M, Davidson L. *Partnering for recovery in mental health: A practical guide to person-centered planning.* Hoboken, NJ: John Wiley & Sons; 2014.
57. Bromage B, Kriegel L, Williamson B, Maclean K, Rowe M. Project Connect: A community intervention for individuals with mental illness. *American Journal of Psychiatric Rehabilitation.* 2017;20(3):218–233.
58. Grant EA, Swink N, Reinhart C, Wituk S. The development and implementation of a statewide certified peer specialist program.

In: Brown LD, Wituk S, eds. *Mental health self-help*. New York, NY: Springer, 2010: 193–209.

59. Sabin JE, Daniels N. Managed care: Strengthening the consumer voice in managed care: VII. The Georgia peer specialist program. *Psychiatric Services*. 2003;54(4):497–498.
60. Georgia Certified Peer Specialist Project. http://www.gacps.org/Home.html
61. Salzer MS. Certified peer specialists in the united states behavioral health system: An emerging workforce. In: Brown L, Wituk S, eds. *Mental health self-help*. New York, NY: Springer, 2010: 169–191.
62. Kaufman L, Kuhn W, Stevens Manser S. *Peer specialist training and certification programs: A national overview*. Texas Institute for Excellence in Mental Health, School of Social Work, University of Texas at Austin. 2016.
63. Chinman M, Shoai R, Cohen A. Using organizational change strategies to guide peer support technician implementation in the Veterans Administration. *Psychiatric Rehabilitation Journal*. 2010;33:269–277.
64. US Department of Veterans Affairs. VHA Handbook 1160.01 Uniform mental health services in VA medical centers and clinics. Washington, DC. 2008.
65. US Dept. of Health and Human Services, Substance Abuse and Mental Health Services Administration. *Achieving the promise: Transforming mental health care in America: final report, July 2003*. Rockville, MD: Presidents New Freedom Commission on Mental Health.
66. Veterans' Mental Health and Other Care Improvement Act of 2008, Public Law110-387, Section 405 modifies USC 740. 2008.
67. Gresen R, Chinman M, Davis M, et al. Early experiences of employing consumer-providers in the VA. *Psychiatric Services*. 2008;59:1315–1321.
68. Cook JA, Jonikas JA, Razzano L. A randomized evaluation of consumer versus nonconsumer training of state mental health service providers. *Community Mental Health Journal*. 1995;31:229–238
69. Dixon L, Hackman A, Lehman A. Consumers as staff in assertive community treatment programs. *Administration and Policy in Mental Health*. 1997;25:199–208.

61

MINDFULNESS AND ACCEPTANCE BASED THERAPIES FOR PSYCHOSIS

Louise Johns, Mark Hayward, Clara Strauss, and Eric Morris

INTRODUCTION

This chapter describes two approaches—person-based cognitive therapy (PBCT) and acceptance and commitment therapy (ACT)—that belong to the group of "third-wave" or contextual cognitive-behavioral therapies (CBTs). "Third-wave" cognitive and behavioral therapies unite approaches that emphasize clients' relationship with their symptoms.[1] Rather than targeting particular appraisals, as in traditional CBT, these interventions use mindfulness and acceptance to alter the way people relate to their internal experiences. In addition, ACT uses mindfulness and other practices to promote values-based living.[1] Some third-wave therapies (e.g., PBCT) are based on information processing accounts of human cognition, while others are fundamentally behavioral in perspective (e.g., ACT).

LINKS WITH COGNITIVE-BEHAVIORAL THERAPY FOR PSYCHOSIS (CBTP)

In PBCT, there is direct relationship with CBTp, as the therapy was developed from a cognitive model of psychotic symptoms, where the adoption of mindfulness aims to promote metacognitive awareness. For ACT, the influence of CBTp was indirect, as ACT emerged from a functional contextual perspective rather than a cognitive account of psychosis.[2] The development of ACT for psychosis extended a broad psychological flexibility model to the problems of psychosis, with a focus on direct behavior change based on personal values.

MINDFULNESS

Mindfulness can help individuals to learn to notice passing thoughts, feelings, or images, in order to develop a more decentered stance toward internal experiences. This may strengthen flexible responses to these experiences, and a sense of self as an experiencer (in both PBCT and ACT); additionally, mindfulness in ACT is used to support engagement with personal values as guides to action. Mindfulness is described as "paying attention in a particular way: on purpose, in the present moment, and non-judgmentally,"[3] where a person intentionally focuses their attention on present-moment experiences in a nonjudgmental and accepting way, and also with compassion and curiosity toward these experiences.[4] This state of mind can be contrasted with engaging in cognitive processes such as rumination, worry, planning, fantasizing; or behaving automatically without awareness (being on "auto-pilot").[5] PBCT is an example of a mindfulness-based intervention (MBI) because it foregrounds mindfulness practices as a means of promoting change. It is debatable whether ACT falls within this grouping. While mindfulness *as a process* is central to ACT, there are additional processes, such as flexible perspective taking and values clarification.

This chapter outlines the rationale and approach of PBCT and ACT for psychosis, present the current evidence base, and consider the implementation of these therapies in routine practice.

PERSON-BASED COGNITIVE THERAPY

The principles of PBCT[6] can be applied to working with psychosis experiences more broadly. However, PBCT has been explored most extensively in relation to the experience of hearing distressing voices. The cognitive model of voices[7] suggested that the distress caused by hearing voices is not solely attributable to the critical or commanding content of these experiences; in addition, distress can be influenced by the client's appraisals or beliefs about the power, control, and knowledge of the voices. PBCT builds on the cognitive model of voices in three ways: first, by foregrounding beliefs about the self—a variable that can maintain voice distress[8]—and exploring their accuracy and flexibility; second, by introducing the teaching of mindfulness skills that can facilitate the slower, more deliberate processing of a broader array of information to facilitate the re-evaluation of appraisals of both self and voices; and third, by reconceptualizing the therapeutic relationship.

Chadwick[6] describes the PBCT model as consisting of four "zones of proximal development," within which greater learning can occur within a therapeutic relationship as opposed to working by oneself.[9] These zones are (1) beliefs about voices—"symptomatic meaning"; (2) the accuracy of beliefs about self—"schemata"; (3) the flexibility of beliefs about self—"symbolic self"; and (4) mindfulness—"relationship to internal experience." Each of these zones is outlined in what follows, with reference to their use within PBCT groups for people experiencing distressing voices.[10] Initially, we describe the model's conceptualization of the therapeutic relationship as a "radical collaboration."

RADICAL COLLABORATION

PBCT draws on the well-established Rogerian principles[11] of warmth, genuineness, and unconditional positive regard to place the client at the heart of the therapeutic process. In addition, the therapist is encouraged to "meet the person" rather than their problem or symptoms. This requires the practice of PBCT to be supported by several positive assumptions about people with psychosis and the process of therapy (e.g., "psychotic experience is continuous with ordinary experience"; "effective therapy depends on understanding sources of distress, not sources of psychosis"; and "therapists aim to be themselves more fully with clients"). Furthermore, Chadwick[6] identifies a set of "anti-collaborative modes" that can inhibit a radically collaborative relationship (e.g., "It is my responsibility to keep the client in therapy" and "In order for me to be a competent therapist the client must change"), and which can be overcome through awareness, acceptance, and open discussion in supervision.

SYMPTOMATIC MEANING

Therapeutic work within this zone of proximal development (ZPD) will be familiar to therapists who have delivered CBTp, as it has a traditional focus on seeking to enhance clients' awareness of the beliefs they hold about the perceived power and control of their voices, and the extent to which these beliefs may be maintaining distress. In the context of cognitive biases that can restrict the array of information and evidence used to support beliefs (e.g., by paying attention primarily to confirmatory evidence[12]), clients are invited to deliberately seek and consider a broader array of information that may include evidence not supporting their beliefs about voice power and control. A balanced consideration that includes evidence supporting existing beliefs about voices is important in order to validate the client's current perspective. The invitation in therapy is not to discount evidence within one's usual gaze, but to look beyond the usual limits of that gaze and re-evaluate beliefs in the light of a broader array of evidence. There is no attempt to "persuade" clients that their beliefs are in any way false or wrong; clients are merely encouraged to work beyond the normal heuristics of information processing to ensure that their beliefs are based on more of the available evidence and are as accurate and helpful as possible. The work within this zone begins with a review of the evidence for and against beliefs about the power and control of voices (e.g., "voices have complete control"), and then moves to a consideration of evidence to support a linked belief about the self, e.g., "I have some personal power/control, even when voices are around."

SCHEMATA

Work within this zone builds on the insights developed about the restricted nature of information processing and its consequences for maintaining unhelpful and inaccurate beliefs. Having previously re-evaluated beliefs about the self in relation to voices, clients are invited to reflect on their beliefs about themselves more broadly. Negative core beliefs are commonly held by clients who hear distressing voices, yet the emphasis here is not on identifying and challenging negative self-schema. The rationale for this is that strongly held, fact-like, negative core beliefs about the self can be difficult to shift, particularly in a group setting with relatively few sessions. Participants are instead invited to identify and strengthen an existing but weakly held positive belief. This is facilitated by using an adaptation of Greenberg's two-chair method,[13] whereby clients are invited to sit in a "positive chair" and be interviewed by a therapist about a positive experience of their choosing. Clients are encouraged to recollect the experience by bringing to mind thoughts, feelings, and bodily sensations at that time. The primary question is how participants saw themselves as a person in that moment. What is being sought here is a positive core belief statement about the self, such as "I am okay as I am," "I am likable," or "I am capable." Clients often express a sense of pride and receive positive feedback from other group members. During the exercise, clients are invited to pay deliberate attention to and "soak up" their positive experience in a way that might strengthen these memories and enable them to be more readily available at other times. Clients are also invited to bring mindfulness to and soak-up any positive experiences between sessions.

SYMBOLIC SELF

Having acknowledged the existence of positive experiences within the schemata zone, clients are encouraged to accept both negative and positive experiences as part of their broader experience—and to notice the way they can move between these poles of experience from moment-to-moment, day-to-day, etc. In this way, clients are encouraged to view themselves as not defined entirely by negative experiences, including voices;[14] rather, they can hold a more balanced, flexible and sometimes contradictory view of themselves. Furthermore, when enveloped by negative experience, they may be able to recall a recent positive experience and consider the possibility that "it won't always be like this!"

RELATIONSHIP TO INTERNAL EXPERIENCE

The final ZPD facilitates the development of mindfulness skills that can be deployed to assist learning within each of the other zones. Mindfulness practices are offered during all sessions and have been adapted by Chadwick[6] in three ways to meet the needs of clients with psychosis. First, practice time is limited to 10 minutes maximum, as most clients find this is the most they can manage. Second, extended silences during practices are avoided, and therapists provide guidance throughout the practice. This is an important grounding method, and helps clients to decenter from intense internal experiences (voices, thoughts, feelings), and to reconnect with present-moment experience with clearer awareness. Third, practice outside sessions is not an essential requirement of the therapy—recordings of 10-minute guided practices are provided, and practice is encouraged.

The practice itself involves initially bringing awareness to the body and then to the breath. As is usual in mindfulness practice, clients are encouraged to use the breath as an anchor to the present moment. When the mind wanders (which it will) to thoughts, feelings, and voices, clients are invited to notice where the mind has wandered to, and to gently bring their awareness back to the sensations of breathing. The instruction to notice when the mind wanders to voices is threaded throughout the middle and end of the practice; therefore, voices are explicitly brought into the practice and are treated as any other experience is treated in mindfulness practice. For some clients, this guidance to sit with and observe voice experiences can be unsettling at first. Participants are usually well practiced in a range of distraction techniques, often recommended by other mental health professionals, and these can work to a greater or lesser extent to reduce distress from hearing voices.[15] Thus, clients can strongly believe that the only way to manage distress is through distraction. The first few attempts at sitting mindfully with voices can result in some increased distress for some people, which therapists can anticipate and normalize. These experiences and reactions can be drawn out during the inquiry following each mindfulness practice.

PBCT—AN INTEGRATED MODEL

Although the ZPDs are divided into a focus on mindfulness practice, and a focus on beliefs about voices and self, it is important to emphasize the conceptual links between the two. Rather than being conceptually distinct, a core aspect of PBCT is continually making links between what is learned through mindfulness practice and what is learned through evaluating beliefs about voices and self. The acquisition of mindfulness skills can enable clients to decenter (or "step back" from) from voices in a manner that provides an alternative response to typical patterns of getting absorbed and lost in voice comments and associated meanings. If clients are able to take a decentered perspective, even when voices are active, this can be used as evidence to support the belief that the client has some control. Furthermore, from this decentered perspective, clients are well placed to pay attention to and process information and evidence that may be beyond their usual gaze, thereby making new information available that can be incorporated into re-evaluations of beliefs about self and voices.

ACCEPTANCE AND COMMITMENT THERAPY (ACT)

ACT helps people to increase their psychological flexibility and engage more in values-based actions rather than behaviors that maintain distress and limit functioning. ACT is underpinned by a behavioral analytic account of language, relational frame theory (RFT[16]), and aims to reduce the direct impact of thoughts and language on behavior. It encourages clients to respond to internal experiences as "events in the mind" rather than literal content, and helps clients to develop a perspective of mindful acceptance toward these experiences. The intervention can be helpful when clients are struggling with experiences that cannot be controlled or when trying to control them leads to problems in everyday living. The approach helps clients to notice when attempts at "making sense" of experiences functions as an unhelpful form of control that maintains difficulties, facilitating a shift from trying to control internal events to focusing more on behavior changes that can lead to recovery.[17]

THE PSYCHOLOGICAL FLEXIBILITY MODEL

The six theoretical processes of ACT are shown in the "hexaflex" (Figure 61.1), and work together to increase psychological flexibility. These processes can be grouped into three sets of response style: open, aware, and active.[18]

Open

Acceptance and defusion skills promote openness toward internal experiences. Acceptance involves "making room for" these experiences without trying to avoid or suppress them (termed "experiential avoidance"). This acceptance is *not* a passive process of tolerance or resignation, but a full willingness to step toward difficult experiences, including psychotic symptoms, without struggling against them. Defusion skills help clients to "step back" from internal experiences, and see them for what they are (experiences), rather than what they say they are (guides to action and choices), thereby reducing unhelpful literal, rule-based responding to thoughts. Defusion works to undermine entanglement with thoughts and beliefs that promote restricted behaviors or avoidance.

Aware

Self as context refers to the sense of self (I, Here, Now) from which internal experiences are observed and contained. An awareness of this perspective, cultivated through mindful contact with the present moment, can loosen attachment to distressing thoughts, images, beliefs or hallucinations. The

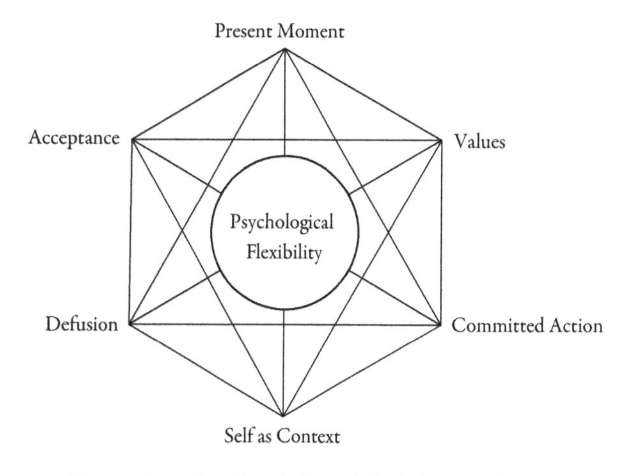

Figure 61.1 The ACT model of psychological flexibility. Used with Permission of S. C. Hayes.

rationale for mindfulness within ACT focuses on how it can promote flexible responding and taking action based on personal values. Mindfulness uses four processes from the psychological flexibility model: present-moment awareness, acceptance, defusion, and self as context.[19] This functional definition of mindfulness in ACT means that there is no linkage with particular mindfulness exercises, so that any method that changes these processes is relevant.[20]

Active

The heart of ACT involves assisting clients to become more active in their lives in their chosen ways. This happens through identifying personal values and using these values to inform steps toward meaningful goals and specific action plans. Small, achievable goals are slowly built on to form larger patterns of committed action.

RATIONALE FOR ACT FOR PSYCHOSIS (ACTP)

In addition to enhancing well-being and values-based living, ACT may benefit the particular problems and symptoms of psychosis. Some of the qualities (intrusive, uncontrollable, negative, frightening) of voices and delusional beliefs increase the likelihood that people respond to them with suppression or avoidance.[21] Conversely, psychotic experiences have high personal meaning, and some can be very engaging and interesting, especially in the context of little meaningful activity and social connection. Engagement with these experiences may be used to escape a mundane life, although with the potential for high personal cost in the long term.

Responding to voices

Hallucinations tend to be compelling experiences, and clients might resist or engage with voices depending on their appraisal of them. Resistance is an attempt to suppress or control voices, and can be divided into "fight" (attempts to confront) or "flight" (attempts to escape or avoid) responses.[22] However, these responses tend to be ineffective forms of coping in the long run, and can maintain voices and associated beliefs. "Fight" strategies such as shouting back or arguing with voices have been associated with poorer emotion control,[23] and "flight" responses, such as trying to block out voices, have been associated with depression.[24] Active engagement with voices is defined as "elective listening, willing compliance, and doing things to bring on the voices";[25] it involves listening to voices and directly accepting what they say. Passive engagement may incorporate submissive responses to voices, such as complying with commands.[26] Resistance and active and passive engagement may all inadvertently reinforce the experience of hearing voices, and continued preoccupation with voices can come at the cost of pursuing valued goals.

Delusional thinking

Experiences associated with delusional thinking, such as anxiety, shame, or humiliation can lead to avoidance of these experiences and of triggering situations (so-called passive avoidance). Some delusions, however, can be understood as "active" forms of experiential avoidance,[27] whereby the delusional symptom itself becomes a means of avoiding some other matter, for example, low self-esteem, guilt, or depression. The person verbally constructs an alternative reality, which they become preoccupied with via the processes of worry and rumination. Although, a person's delusions may not always start this way, these processes become maintaining factors and can have a negative impact on the person's valued life direction.

AN ACT APPROACH TO PSYCHOSIS

Given the negative impact of suppression and avoidance strategies in the long-term, acceptance is a potentially more adaptive response to psychotic experiences. ACT highlights a form of acceptance that fosters skills that can be applied as the psychotic experiences occur. This mindful acceptance is neither a specific coping strategy nor a process of providing meaning, but a particular style of relating to psychological events. It involves the skills of nonjudgmental awareness—deliberately observing mental events as they occur without judging them and without reacting to them; and disengagement (detachment) from the literal meaning of the content of voices and delusions, that is, distinguishing the actual experience (sounds/words) from what it represents (literal reality). The therapeutic focus of ACT can reduce the impact of psychotic symptoms and help the person to focus more on valued actions.[28] There is an emphasis on the *workability* of the individual's behavior, with greater choice, flexibility, and expansion of response options.[29] For example, a person who rigidly responds to hearing voices by social isolation and arguing with the voices may, through acceptance work, develop a broader range of behavioral responses to voice hearing. These might include activities such as going out of the house, having a conversation with another person, deliberately noticing the acoustic properties of the voices, or engaging in a valued activity, as well as their existing responses to try to control the voices. The clinical focus is to enhance the functional perspective-taking stance (self as experiencer), and promote flexible responding to the experience of hearing voices.[30]

In general, therapy proceeds by addressing the ACT processes in parallel, with each therapy session incorporating various elements of the model, and helping the person to make links between them. For example, a therapist might introduce a defusion exercise, discuss how this could be applied to reduce the client's struggle with voices or preoccupation with distressing beliefs, and then consider how the person could use defusion while engaging in a valued activity. Therapy may start by noting the costs of struggle with voices and thoughts, and introducing the idea of acceptance as an alternative response, with an increasing focus on values, willingness, and action as therapy progresses. Some adaptations to the practice of ACT are necessary to suit clients with psychosis. In particular, an emphasis on learning by addition (rather than replacement) is important for this client group, who often remain attached to non-acceptance-based coping strategies that can be effective in the short term. It can be helpful to introduce acceptance and willingness as additions "to the toolkit," and to encourage clients to "try willingness out" and see if this approach can result in more valued actions. Identifying and clarifying values in this client group, particularly those with established psychosis, can bring up themes of loss and missed opportunities. Some clients might be unclear about what they value, especially if they have personal histories of invalidation or trauma, or clients may have lost touch with their values through channeling their efforts on managing the psychosis. ACT focuses on the idea of "constructing" a meaningful life, starting from today. Connecting with values is seen as work in progress, and trying out new things. Importantly, compassion to self and others can be highlighted as a valued life direction.[31]

The therapeutic relationship is a key part of any psychological intervention for people with psychosis. Within ACT, the therapeutic relationship is validating, normalizing, and collaborative. It creates a context that teaches the limits of literal language for problem-solving, and encourages experiential learning of different ways of relating to private experiences while expanding values-based behaviors. The social context of the relationship involves "radical acceptance," an appreciation of the whole person. In addition to acceptance by the therapist, it includes acceptance by clients of themselves, including their unwanted experiences, and of other people. An important component of ACTp can be therapists' self-disclosures about the ways in which they struggle, which encourages a sense of universality and perspective taking. As with CBTp, ACTp is matched to the client's pace, and has a gentle, conversational style. Experiential and physicalizing exercises are used to bring the therapy alive, and to make the learning points more memorable.

In addition to individual formulation-based therapy, ACT has been developed as a group-based intervention for people with psychosis.[32] Particular aspects of the intervention lend themselves to application in a group format: many ACT metaphors are interactive and benefit from more people; observing others being present and willing can promote these processes in oneself; and making commitments in a social context is likely to strengthen action.

EVIDENCE BASE

A key question when considering third-wave approaches for psychosis is "are they effective?" Novel interventions can seem intuitively appealing, however, this appeal does not always translate into benefits for clients, and the various ways in which interventions are offered may not meet client needs in terms of format (e.g., individual versus group therapy) or duration (e.g., too few/too many sessions). Here, we review the research evidence for third-wave approaches for psychosis. We foreground randomized controlled trials (RCTs), as these are regarded as the "gold standard" research design when evaluating health interventions. Patients are assigned at random either to receiving the third-wave intervention or to receiving treatment as usual (TAU) or a control intervention, and the trial examines effects on outcomes such as relapse rates, symptoms of psychosis, and functioning. Any differences between these groups can be attributed to the intervention, as other possible differences between the groups should have been removed through the randomization process.

The first published RCT in this field focused on rehospitalization rates. Bach and Hayes[33] randomly allocated 80 inpatients with positive psychotic symptoms to TAU or to four individual sessions of ACT plus TAU. The ACT participants had significantly lower rehospitalization rates than TAU participants at 4-month follow-up, and this group difference was maintained 1 year post discharge.[34] Since this seminal study, there have been a number of RCTs in the area. Louise, Fitzpatrick, Strauss, Rossell, and Thomas[35] conducted a systematic review and meta-analysis of all RCTs measuring effects on psychosis symptoms (so not including the Bach and Hayes study,[33] which did not focus on symptoms). This review[35] took a systematic approach to identifying all published RCTs, meaning that the risk of bias is reduced as studies showing both positive and negative outcomes are included. The meta-analytic aspect of the review summarizes outcomes from all included RCTs, giving more weight to studies with larger sample sizes. The review identified 10 RCTs evaluating third-wave approaches for psychosis, with a total of 624 participants. Five studies evaluated group-based MBIs,[10,36–39] four studies evaluated individually administered ACT,[40–43] and one study evaluated compassion focused therapy (CFT).[44] Six of the RCTs had an inactive control condition, while four included an active control condition: psychoeducation,[37] integrated rehabilitation treatment,[39] and befriending.[41,42]

The meta-analysis showed significant postintervention between-group benefits of the third-wave approaches on measures of psychotic symptoms compared with the control conditions, with a small effect size (Hedges $g = 0.29$, 95% CI 0.04,0.54, $p = .021$). This demonstrates that people receiving the third-wave approaches reported significantly lower levels of psychotic symptoms than people not receiving these interventions, although the degree of benefit was modest. Significant postintervention between-group benefits were also found for symptoms of depression with a small-medium effect (Hedges $g = 0.39$, 95% CI 0.09,0.69, $p = .011$), so that people receiving these approaches were significantly less depressed following the intervention than people in the control conditions. In addition, significant benefits were shown on measures of mindfulness (Hedges $g = 0.56$, 95% CI 0.15,0.97, $p = .007$). This last finding is important as it suggests that the benefits of these third-wave approaches on symptoms of psychosis and depression are not simply due to nonspecific effects, but that benefits may be due to improvements in the targeted mindfulness processes (although this suggestion requires closer examination using mediation analyses). Effects were not statistically significant on all outcomes however. No differences were found between people who did and did not receive a third-wave intervention on measures of voice-related distress intensity (Hedges $g = -0.07$, 95% CI -0.37,0.23, $p = .65$), functioning (Hedges $g = 0.09$, 95% CI -0.23,0.42, $p = .57$), negative symptoms (Hedges $g = 0.09$, 95% CI -0.22,0.40, p = .56), or acceptance (Hedges $g = 0.08$, 95% CI -0.23,0.38, $p = .63$). Lack of significant effects on these outcomes (hypothesized to improve with third-wave approaches) may in part be due to therapeutic benefits of active control interventions. For example, befriending was offered in two studies as a control intervention,[41,42] and befriending may include active therapeutic ingredients by focusing on recovery-related discussions. What is therefore needed are RCTs comparing third-wave approaches to control conditions that do not contain therapeutic elements, such as a group-based but noninteractive activity (e.g., watching movies).

To summarize, RCTs to date demonstrate that, taken as a whole, third-wave approaches for psychosis lead to measurable benefits in rehospitalization rates, symptoms of psychosis, and depression, particularly when comparing these approaches to inactive control conditions. Benefits to symptoms of psychosis were in the medium range when comparing to TAU (Hedges $g = 0.46$), which suggests the degree of benefit may be similar to CBTp.[45] The MBIs were all delivered in groups (rather than individually), and the ACT intervention in the Bach and Hayes[33] study was brief (just four individual sessions), and so these approaches provide efficient ways of increasing access to evidence-based psychological approaches for psychosis. This is key given that access to individual CBTp is poor in public health settings,[46] despite its being recommended in treatment guidelines.[47,48]

IMPLEMENTATION IN PRACTICE

Delivery of individual CBTp has been restricted by low numbers of therapists, limited access to adequate training and supervision for staff, and a lack of protected time to deliver the interventions.[49] Briefer or group-based variants of CBTp have the potential to improve therapy access and dissemination. Targeting common processes that contribute to well-being is an effective approach to increasing therapy impact and access. PBCT and ACT approaches both aim to increase the process of psychological flexibility, which is a core component of mental well-being.[50] The focus on the common processes of mindfulness, acceptance, distancing, and values-based action also make PBCT and ACT interventions typically brief and well suited to group delivery and to the diverse presentations of psychosis.[2]

Brief group formats offer the potential for cost-effective, wide-scale delivery. As well as being offered to more clients at a time, they offer opportunities for staff development through cofacilitation, thereby disseminating expertise and increasing the scope of their delivery. Research supports the acceptability and feasibility of delivering group mindfulness and ACT interventions to people with psychosis, together with positive and clinically relevant outcomes.[10,51,52] There are standardized protocols available,[53–56] which range from 4 to 12 sessions. Mindfulness and acceptance concepts and skills can be taught and modeled within a group format, plus groups offer additional benefits of social support, normalizing, and access to other perspectives.[57,58] Group approaches can also appeal to clients who are unable or reluctant to engage with lengthy individual treatments.

In addition to delivery in community services, mindfulness-based and ACT interventions have been piloted and evaluated within in-patient settings. Delivery has been both individual and group-based, with the focus ranging from brief crisis management to longer term integrated treatment and therapeutic milieu.[59–61] These interventions are feasible on psychiatric ward, and their approach is well suited to this environment and the range of acute and chronic presenting problems. Recent protocols have been published for evaluation in larger scale RCTs, focusing on crisis intervention[62] and routine in-patient care.[63]

TRAINING AND SUPERVISION

Given that mindfulness and acceptance based therapies are relatively new, and not yet widely implemented, there are no clear-cut training guidelines for their delivery. Based on the RCT evidence, where we can be confident about therapy quality, these interventions should be delivered by psychological therapists, many of whom will be trained in CBTp (for which there are guidelines in the United Kingdom[64]). Teachers of mindfulness-based courses within the United Kingdom are expected to follow the good practice guidelines from the UK Network of Mindfulness-Based Teacher Training Organizations,[65] which involve completion of a teacher-training program or supervised pathway over at least 12 months. PBCT groups are facilitated by two psychological therapists, at least one of whom is CBT trained and at least one of whom is trained in mindfulness-based cognitive therapy (MBCT). Both of these skills could reside in one of the two facilitators, in which case criteria for the second facilitator can be flexible.

There is currently no formal training and accreditation process to deliver ACT, although most practitioners will be accredited cognitive or behavior therapists. As ACT is not a recognized MBI, there are no formal good practice guidelines for training to deliver mindfulness in ACT. Training may involve completion by facilitators of an 8-week MBCT course plus regular ongoing mindfulness practice. As mentioned earlier, it is possible to train a range of mental health professionals to deliver a brief manualized group protocol. Our group ACTp sessions were facilitated by a lead psychological therapist accompanied by one or two cofacilitators, who were mental health practitioners experienced in working with people with psychosis, and who had attended a bespoke two-day training event. It is also possible, and beneficial, to train peer supporters (people with personal experience of psychosis and of mental health services) as cofacilitators.[56] Peer support facilitators can offer group participants the opportunity to learn from their recovery journey and practice of employing psychological flexibility skills.

As with any psychological intervention, regular clinical supervision is essential. Supervisors of PBCT should be accredited CBTp and MBI practitioners; supervisors of ACT should also be CBTp trained, and have training or experience in delivering ACT and mindfulness. Supervision sessions include a mixture of support, problem-solving, and experiential learning.[66] Supervision also supports the facilitators to deliver the intervention with fidelity. This involves both adherence to the intervention (doing the key tasks involved, and avoiding doing things that are incompatible with the intervention), and competence (doing the intervention in a way that is impactful, with well-timed and sensitive use of techniques). For example, the debrief inquiry following a mindfulness exercise can be challenging, and facilitators' responses to participants may not always be consistent with the therapeutic approach. Morris[56] has developed the "ACTs of ACT" rating scale to track therapy fidelity, which comprises a checklist of seven ACT-consistent items and eight ACT-inconsistent items, rated for how much they are present in each session. The scale is rated by facilitators or by supervisors from direct observation or recordings of the group sessions. A similar adherence checklist was used in the RCT of PBCT for psychosis,[10] and the participants rated session content for elements being present/not present.

SUMMARY

Third-wave approaches focus on how people relate to their experiences, rather than the experiences themselves. PBCT and ACT are two of the most researched third-wave therapies for psychosis. Both approaches use mindfulness practice to help individuals learn to notice passing thoughts, voices, feelings or images, and develop a more decentered stance toward internal experiences. This can strengthen flexible responding to these experiences. PBCT integrates a mindfulness-based approach with a traditional CBTp, and ACT focuses more explicitly on values-based actions. Evidence suggests that these approaches can be as effective as CBTp, plus they can be delivered efficiently, either in a group-format or just a few individual sessions, and so offer a way to increase access to therapies for this client group. More research is needed to evaluate the effectiveness of these therapies, both in relation to their intended outcomes and in comparison with placebo control conditions.

REFERENCES

1. Hayes SC. Acceptance and commitment therapy, relational frame theory, and the third wave of behavioral and cognitive therapies. *Behav Ther*. 2004;35(4):639–665. doi:10.1016/S0005-7894(04)80013-3
2. Morris EMJ, Johns LC, Oliver JE. *Acceptance and Commitment Therapy and Mindfulness for Psychosis.* Wiley-Blackwell; 2013.
3. Kabat-Zinn J. *Wherever You Go, There You Are: Mindfulness Meditation in Everyday Life.* Hyperion; 1994.
4. Kabat-Zinn J. Mindfulness-based interventions in context: past, present, and future. *Clin Psychol Sci Pract.* 2006;10(2):144–156. doi:10.1093/clipsy.bpg016
5. Baer RA, Smith GT, Allen KB. Assessment of mindfulness by self-report: the Kentucky inventory of mindfulness skills. *Assessment.* 2004;11(3):191–206. doi:10.1177/1073191104268029
6. Chadwick P. *Person-Based Cognitive Therapy for Distressing Psychosis.* John Wiley & Sons; 2006.
7. Chadwick P, Birchwood M, Trower P. *Cognitive Therapy for Delusions, Voices and Paranoia.* Chichester: Wiley & Sons; 1996.
8. Fannon D, Hayward P, Thompson N, Green N, Surguladze S, Wykes T. The self or the voice? Relative contributions of self-esteem and voice appraisal in persistent auditory hallucinations. *Schizophr Res.* 2009;112(1–3):174–180. doi:10.1016/j.schres.2009.03.031
9. Vygotsky LS. *Mind in Society: The Development of Higher Psychological Processes.* Cambridge, MA: Harvard University Press; 1978.
10. Chadwick P, Strauss C, Jones AM, et al. Group mindfulness-based intervention for distressing voices: A pragmatic randomised controlled trial. *Schizophr Res.* 2016;175(1–3):168–173. doi:10.1016/j.schres.2016.04.001
11. Rogers CR. *On Becoming a Person.* London: Constable; 1961.
12. Maher BA. Delusional thinking and perceptual disorder. *J Individ Psychol.* 1974;30(1):98–113.
13. Greenberg LS, Rice LN, Elliott R. *Facilitating Emotional Change: The Moment-by-Moment Process.* New York: Guilford Press; 1993.
14. Goodliffe L, Hayward M, Brown D, Turton W, Dannahy L. Group person-based cognitive therapy for distressing voices: Views from the hearers. *Psychother Res.* 2010;20(4):447–461. doi:10.1080/10503301003671305
15. Hayward M, Edgecumbe R, Jones A-M, Berry C, Strauss C. Brief coping strategy enhancement for distressing voices: an evaluation in routine clinical practice. *Behav Cogn Psychother.* 2018;46(2):226–237. doi:10.1017/S1352465817000388
16. Hayes SC, Barnes-Holmes D, Roche B. *Relational Frame Theory: A Post-Skinnerian Account of Human Language and Cognition.* Springer; 2001.
17. Hayes SC, Strosahl K, Wilson KG. *Acceptance and Commitment Therapy: The Process and Practice of Mindful Change.* New York: Guilford Press; 2012.
18. Hayes SC, Villatte M, Levin M, Hildebrandt M. Open, aware and active: Contextual approaches as an emerging trend in the behavioral and cognitive therapies. *Annu Rev Clin Psychol.* 2011;7(1):141–168. doi:10.1146/annurev-clinpsy-032210-104449
19. Fletcher L, Hayes SC. Relational frame theory, acceptance and commitment therapy, and a functional analytic definition of mindfulness. *J Ration Cogn Ther.* 2005;23:315–336.
20. Hayes SC, Shenk C. Operationalizing mindfulness without unnecessary attachments. *Clin Psychol Sci Pract.* 2004;11:249–254. doi:10.1093/clipsy/bph079
21. Morris EM, Garety P, Peters E. Psychological flexibility and nonjudgemental acceptance in voice hearers: relationships with omnipotence and distress. *Aust New Zeal J Psychiatry.* 2014;48(12):1150–1162. doi:10.1177/0004867414535671
22. Gilbert P, Birchwood M, Gilbert J, et al. An exploration of evolved mental mechanisms for dominant and subordinate behavior in relation to auditory hallucinations in schizophrenia and critical thoughts in depression. *Psychol Med.* 2001;31(6):1117–1127.
23. Farhall J, Greenwood KM, Jackson HJ. Coping with hallucinated voices in schizophrenia: A review of self-initiated strategies and therapeutic interventions. *Clin Psychol Rev.* 2007. doi:10.1016/j.cpr.2006.12.002
24. Escher S, Delespaul P, Romme M, Buiks A, van Os J. Coping defence and depression in adolescents hearing voices. *J Ment Heal.* 2003;12:91–99.
25. Chadwick P, Birchwood M. The omnipotence of voices: A cognitive approach to auditory hallucinations. *Br J Psychiatry.* 1994;164(2):190–201. http://www.ncbi.nlm.nih.gov/pubmed/8173822. Accessed March 31, 2019.
26. Shawyer F, Mackinnon A, Farhall J, et al. Acting on harmful command hallucinations in psychotic disorders: an integrative approach. *J Nerv Ment Dis.* 2008;196(5):390–398. doi:10.1097/NMD.0b013e318171093b
27. García-Montes JM, Pérez-Álvarez M, Perona-Garcelán S. Acceptance and commitment therapy for delusions. In: Morris EM, Johns LC, Oliver JE, eds. *Acceptance and Commitment Therapy and Mindfulness for Psychosis.* Chichester, UK: Wiley-Blackwell; 2013.
28. Pérez-Álvarez M, García-Montes JM, Perona-Garcelán S, Vallina-Fernández O. Changing relationship with voices: New therapeutic perspectives for treating hallucinations. *Clin Psychol Psychother.* 2008;15(2):75–85. doi:10.1002/cpp.563
29. Pankey J, Hayes SC. Acceptance and commitment therapy for psychosis. *Int J Psychol Psychol Ther.* 2003;3(2):311–328.

30. Thomas N, Morris EMJ, Shawyer F, Farhall J. Acceptance and commitment therapy for voices. In: Morris EMJ, Johns LC, Oliver JE, eds. *Acceptance and Commitment Therapy and Mindfulness for Psychosis*. Chichester, UK: Wiley-Blackwell; 2013:95–111.
31. White RG. Treating depression in psychosis: Self-compassion as a valued life direction. In: Guadiano B, ed. *Incorporating Acceptance and Mindfulness into the Treatment of Psychosis: Current Trends and Future Directions*. Oxford: Oxford University Press; 2015:81–107.
32. McArthur A, Mitchell G, Johns LC. Developing acceptance and commitment therapy for psychosis as a group-based intervention. In: Morris EMJ, Johns LC, Oliver JE, eds. *Acceptance and Commitment Therapy and Mindfulness for Psychosis*. Chichester, UK: Wiley-Blackwell; 2013:219–239.
33. Bach P, Hayes SC. The use of acceptance and commitment therapy to prevent the rehospitalization of psychotic patients: A randomized controlled trial. *J Consult Clin Psychol*. 2002;70(5):1129–1139.
34. Bach P, Hayes SC, Gallop R. Long-term effects of brief acceptance and commitment therapy for psychosis. *Behav Modif*. 2012;36(2):165–181. doi:10.1177/0145445511427193
35. Louise S, Fitzpatrick M, Strauss C, Rossell SL, Thomas N. Mindfulness- and acceptance-based interventions for psychosis: Our current understanding and a meta-analysis. *Schizophr Res*. 2018;192:57–63. doi:10.1016/j.schres.2017.05.023
36. Chien WT, Lee IYM. The mindfulness-based psychoeducation program for Chinese patients with schizophrenia. *Psychiatr Serv*. 2013;64(4):376–379. doi:10.1176/appi.ps.002092012
37. Chien W., Thompson DR. Effects of a mindfulness-based psychoeducation programme for Chinese patients with schizophrenia: 2-year follow-up. *Br J Psychiatry*. 2014;205(1):52–59.
38. Langer ÁI, Cangas AJ, Salcedo E, Fuentes B. Applying mindfulness therapy in a group of psychotic individuals: A controlled study. *Behav Cogn Psychother*. 2012;40(1):105–109. doi:10.1017/S1352465811000464
39. López-Navarro E, Del Canto C, Belber M, et al. Mindfulness improves psychological quality of life in community-based patients with severe mental health problems: A pilot randomized clinical trial. *Schizophr Res*. 2015;168(1–2):530–536. doi:10.1016/j.schres.2015.08.016
40. Gaudiano BA, Herbert JD. Acute treatment of inpatients with psychotic symptoms using acceptance and commitment therapy: Pilot results. *Behav Res Ther*. 2006;44(3):415–437. doi:10.1016/j.brat.2005.02.007
41. Shawyer F, Farhall J, Mackinnon A, et al. A randomised controlled trial of acceptance-based cognitive behavioral therapy for command hallucinations in psychotic disorders. *Behav Res Ther*. 2012;50(2):110–121. doi:10.1016/j.brat.2011.11.007
42. Shawyer F, Farhall J, Thomas N, et al. Acceptance and commitment therapy for psychosis: Randomised controlled trial. *Br J Psychiatry*. 2017;210(2):140–148. doi:10.1192/bjp.bp.116.182865
43. White R, Gumley A, McTaggart J, et al. A feasibility study of acceptance and commitment therapy for emotional dysfunction following psychosis. *Behav Res Ther*. 2011;49(12):901–907. doi:10.1016/j.brat.2011.09.003
44. Braehler C, Gumley A, Harper J, Wallace S, Norrie J, Gilbert P. Exploring change processes in compassion focused therapy in psychosis: Results of a feasibility randomized controlled trial. *Br J Clin Psychol*. 2013;52(2):199–214. doi:10.1111/bjc.12009
45. Van der Gaag M, Valmaggia LR, Smit F. The effects of individually tailored formulation-based cognitive behavioral therapy in auditory hallucinations and delusions: A meta-analysis. *Schizophr Res*. 2014;156(1):30–37. doi:10.1016/j.schres.2014.03.016
46. The Schizophrenia Commission. *The Abandoned Illness: A Report from the Schizophrenia Commission*. London: Rethink Mental Illness; 2012.
47. Gaebel W, Riesbeck M, Wobrock T. Schizophrenia guidelines across the world: A selective review and comparison. *Int Rev Psychiatry*. 2011;23(4):379–387. doi:10.3109/09540261.2011.606801
48. National Institute for Health and Care Excellence. *Psychosis and Schizophrenia in Adults: Treatment and Management (CG178)*. London: National Institute for Health and Care Excellence; 2014.
49. Ince P, Haddock G, Tai S. A systematic review of the implementation of recommended psychological interventions for schizophrenia: Rates, barriers, and improvement strategies. *Psychol Psychother Theory, Res Pract*. 2016;89(3):324–350. doi:10.1111/papt.12084
50. Kashdan TB, Rottenberg J. Psychological flexibility as a fundamental aspect of health. *Clin Psychol Rev*. 2010;30(7):865–878. doi:10.1016/j.cpr.2010.03.001
51. Johns LC, Oliver JE, Khondoker M, et al. The feasibility and acceptability of a brief Acceptance and Commitment Therapy (ACT) group intervention for people with psychosis : The ""ACT for life" study. *J Behav Ther Exp Psychiatry*. 2016;50:257–263. doi:10.1016/j.jbtep.2015.10.001
52. Jacobsen P, Richardson M, Harding E, Chadwick P. Mindfulness for psychosis groups: Within-session effects on stress and symptom-related distress in routine community care. *Behav Cogn Psychother*. January 2019:1–10. doi:10.1017/S1352465818000723
53. Strauss C, Hayward M. Group person-based cognitive therapy for distressing psychosis. In: Morris EMJ, Johns LC, Oliver JE, eds. *Acceptance and Commitment Therapy and Mindfulness for Psychosis*. Chichester, UK: Wiley-Blackwell; 2013:240–255.
54. https://contextualscience.org/treatment_protocols.
55. Butler L, Johns LC, Byrne M, et al. Running acceptance and commitment therapy groups for psychosis in community settings. *J Context Behav Sci*. 2016;5(1):33–38. doi:10.1016/j.jcbs.2015.12.001
56. O'Donoghue EK, Morris EMJ, Oliver JE, Johns LC. *ACT for Psychosis Recovery: A Practical Manual for Group-Based Interventions Using Acceptance and Commitment Therapy*. California: New Harbinger; 2018.
57. Abba N, Chadwick P, Stevenson C. Responding mindfully to distressing psychosis: A grounded theory analysis. *Psychother Res*. 2008;18(1):77–87. doi:10.1080/10503300701367992
58. Walser RD, Pistorello J. ACT in group format. In: Hayes SC, Strosahl K, eds., *A Practical Guide to Acceptance and Commitment Therapy*. New York: Springer; 2004:347–372.
59. Jacobsen P, Morris E, Johns L, Hodkinson K. Mindfulness groups for psychosis: Key issues for implementation on an inpatient unit. *Behav Cogn Psychother*. 2011;39(3):349–353. doi:10.1017/S1352465810000639
60. Tyrberg MJ, Carlbring P, Lundgren T. Brief acceptance and commitment therapy for psychotic inpatients: A randomized controlled feasibility trial in Sweden. *Nord Psychol*. 2017;69(2):110–125. doi:10.1080/19012276.2016.1198271
61. Taylor R, Bremner G, Williams C, McDonald C, Morris E. Facilitating an acceptance and commitment therapy group in an acute inpatient setting: A pilot feasibility study. In preparation.
62. Jacobsen P, Peters E, Chadwick P. Mindfulness-based crisis interventions for patients with psychotic symptoms on acute psychiatric ward (amBITION study): Protocol for a feasibility randomised controlled trial. *Pilot Feasibility Stud*. 2016;2. doi:10.1186/s40814-016-0082-y
63. Gaudiano B, Davis C, Epstein-Lubow G, Johnson J, Mueser K, Miller I. Acceptance and commitment therapy for inpatients with psychosis (the REACH Study): Protocol for treatment development and pilot testing. *Healthcare*. 2017;5(2):23. doi:10.3390/healthcare5020023
64. Chandra A, Patterson E, Hodge S. *Standards for Early Intervention in Psychosis Services*. 2018. www.rcpsych.ac.uk/eipn.
65. UK Network for Mindfulness-Based Teachers Good practice guidelines for teaching mindfulness-based courses. https://www.mindfulnessteachersuk.org.uk/pdf/teacher-guidelines.pdf.
66. Morris EMJ, Bilich-Eric L. A Framework to Support Experiential Learning and Psychological Flexibility in Supervision: SHAPE. *Aust Psychol*. 2017;52(2):104–113. doi:10.1111/ap.12267

62

HEARING VOICES GROUPS

Alison Branitsky, Eleanor Longden, and Dirk Corstens

Hearing voices self-help and peer-support groups are an integral part of the work of the international Hearing Voices Movement (HVM), an influential psychiatric service-user/survivor organization that promotes a person-centered, recovery-oriented approach to voice hearing. The HVM was founded in 1988 and has its origins in the work of the social psychiatrist Marius Romme and researcher Sandra Escher,[1–2] who, in collaboration with voice hearer Patsy Hage, interviewed numerous voice hearers to gain insight on how they coped with and made sense of their experiences. Among other findings, Romme and Escher reported that not only did participants have diverse interpretations for their voices (e.g., psychodynamic, biomedical, spiritual, and paranormal) but also many were able to cope well with them outside of mainstream psychiatric services. More recent research has supported this contention, with cross-sectional studies estimating that between 5% and 16% of the adult population hears voices at least once in their lives,[3–5] many of whom likewise never require contact with psychiatric services.[4–6] Correspondingly, the HVM rejects the construction of voices as a psychologically meaningless symptom of illness and instead locates it as a normal human variation that arises within relevant social contexts; effectively that the presence of voices per se is less predictive of distress than the ways people interpret and cope with them, and that this in turn is something that may be strongly influenced by adverse life events.[7–10]

As such, a central part of the HVM approach is supporting voice hearers to find ways of accepting and making sense of their experiences and to provide frameworks for coping and recovery. As part of this endeavor individuals are encouraged to take ownership of their voices and to define the experience for themselves, with each subjective explanation accorded equal respect and consideration. Thus while the HVM emphasizes the documented link between voices, trauma, loss, and stress,[11–13] psychosocial models are not privileged over other accounts and a person is not told they are "wrong" for believing the origins of their voices are, for example, spiritual or paranormal but are instead supported to explore what this belief means to them. In this respect the HVM does not represent a particular therapeutic model, although it does emphasize the benefit of understanding voice hearing as an expression of emotionally significant information—in effect, as "messengers" that communicate unresolved conflict in the person's life.[2,6] Among other strategies, the HVM has therefore promoted the use of "voice profiling" (also known as "the construct"), which is an assessment tool devised by Romme and Escher (2000)[6] as a systematic way to formulate links between voices content/characteristics and adversity in the voice hearer's life.[8,14–15]

Consistent with models of "personal" as opposed to "clinical" recovery,[16] the HVM likewise does not see voice cessation as the only outcome of success, instead arguing that individuals can lead positive and fulfilling lives as voice hearers.[10] Correspondingly, the movement also advocates the value of accepting, understanding, and learning to cope and live with voices as a more constructive strategy than trying to constantly suppress or avoid them on the grounds that the latter may risk increasing voice frequency, negative content, and associated distress.[2,6,14,17–18] According to this perspective, avoidance and suppression, at least in some cases, may exacerbate dissociative divisions and limit opportunities for emotional exploration, whereas exploring and accepting voices—and by extension accepting oneself as a voice hearer—can provide a means of honoring subjective experience, devising a more coherent account of one's difficulties, and identifying relevant conflicts and issues to address in therapy.[14] In this regard, deconstructing links between symptom content and psychosocial conflict is likewise increasingly emphasized in models like cognitive-behavioral therapy for psychosis (CBTp) on the grounds that "[s]uch links often provide indications of . . . unresolved difficulties and associated negative self-evaluations . . . which may be closely intertwined with processes maintaining delusional beliefs and voices and may underpin aspects of the emotional reaction."[19(p.127);20–21]

The popularity of the HVM approach is most clearly evidenced in its global dissemination, with national Hearing Voices Networks currently established in 32 countries across Europe, North America, and Australasia, and additional emerging initiatives in Africa, Asia, and South America. Within these networks, whose activities are supported and coordinated via the charity and organizational body Intervoice (www.intervoiceonline.org), coalitions of voice hearers and their allies (e.g., friends, family members, academics, and mental health workers) have used numerous platforms and mediums to critique reductive biomedical views of voice hearing, challenge stigma, and promote messages of pride, empowerment, and recovery. Taken together, it is suggested that the movement's success offers "an attractive alternative for voice hearers who have not been fully helped by traditional approaches, who are searching for greater understanding and acceptance of their experiences, or who feel that their stories have not been heard or acknowledged.[9(p.S285)]

HEARING VOICES GROUPS

Since the earliest days of the HVM, a central tenet of both its practice and philosophy is that connecting and collaborating with peers can be an important aspect of recovery. As such a major part of the movement's work is supporting the development and successful running of hearing voices self-help groups: initiatives that provide a source of mutual understanding, companionship, and respect where individuals can gather together to explore distressing and overwhelming experiences.[22] Given that engaging with and/or discussing one's voices may not always be encouraged within mainstream services,[9,23–24] particularly for those with a diagnosis of schizophrenia, groups can therefore provide a powerful alternative space to talk openly about one's voices; explore their meaning, purpose, and origins; exchange insights and coping strategies; and share in both the struggle and successes of others without the threat of censorship or forced medication or hospitalization that may be perceived within some clinical settings.[22]

Such groups are becoming increasingly widespread, the majority of which are established within the community in addition to a growing number within acute and secure psychiatric and prison settings.[25] While groups must adhere to the principles described previously to qualify as HVM-affiliated, variations often exist in the day-to-day policies and running of different groups: for example, some may operate on a drop-in basis while others have fixed membership[22] and some discuss voice hearing specifically whereas others are open to individuals with other kinds of anomalous sensory experiences, nonshared beliefs, or other extreme states.[26] In accordance with the ethos of the HVM, which strives to create equity and alliance between experts by experience (voice hearers) and experts by profession (clinicians, academics),[27] groups are often facilitated or cofacilitated by mental health professionals. However, given that the personal ownership and interpretation of one's voices is privileged over medical knowledge, in cases where the group is run in conjunction with psychiatric services the role of the latter is to be purely advisory, or to assist with facilitation, resources, and infrastructure, rather than dispensing a medicalized interpretation of distress.[25]

Despite this variety, several key themes are inherent to all groups: that they are open to voice hearers regardless of psychiatric diagnosis, or lack thereof; that members shape the group and make decisions on how it is run; and that each member is deemed an expert by experience as someone who can both contribute valuable insight and wisdom and learn from the experience of others. As such, hearing voices groups are less structured than 12-step groups like Alcoholics Anonymous, and are markedly different from therapy groups in which one person sets the goals and agenda.[22] In contrast, the role of the facilitator in HVM groups is simply to guide the conversation and maintain the safety and confidentiality of the group: they have no more power or authority than any other member and all members are encouraged to ask questions of and engage with one another.

Consistent with the founding principles of the movement, an overarching goal of HVM groups is to help voice hearers better understand and articulate their individual experiences and to learn to cope with them—and in some instances learn and grow from them—under their own terms and in their own way.[22] For example, this might be achieved by members asking each other questions such as: What do the voices say? How many different voices are there? Have the voices changed over time? Do the voices resemble anyone you have encountered in real life? Do the voices usually occur alongside a particular emotion, or following a particular event? How do you feel when the voices are there? What do you think their purpose is?[22] Members often share if they have had similar experiences, as well as exchange strategies for coping and alleviating distress.

IMPACT OF HEARING VOICES GROUPS

The value of peer support in both statutory and voluntary sector psychiatric services is gaining increased recognition.[28] For example, a narrative review of 37 studies by Davidson et al. (2012)[29] found that peer workers were able to provide unique support to service-users on the basis of having "been there" themselves. Specifically, self-disclosure on the part of peer workers instilled hope that it was possible to live a fulfilling life despite encountering mental health difficulties. Many service-users reported that peer workers served as role models who could empathize with their experience while encouraging active striving toward recovery. In addition to emotional support, peer workers also provided practical help, such as assistance with navigating housing and employment issues. In turn, the presence of peer providers within mental health services was also found to improve social engagement, self-efficacy, life satisfaction, and feelings of control, hopefulness, and agency.

While there have been mixed results on the effectiveness of peer-delivered interventions within statutory services on clinical outcomes like hospitalization rates,[30–32] existing literature shows that such interventions can aid in personal and social recovery and are associated with increased empowerment,[33–35] self-esteem,[36] and social support.[37–38] In turn, benefits of peer-support are not unique to service-users; peer workers themselves may often report increased self-esteem and that being able to aid other people promoted their own recovery.[39–40]

Outside of statutory mental health services, community-based peer-run support groups for various complex mental health problems, including psychosis, evidence numerous social and clinical benefits. A randomized controlled trial by Castelein et al.[41] found that participation in a peer-support group helped enhance social networks through fostering mutual relationships; an important finding given that many people diagnosed with psychosis report impoverished social networks and limited avenues for support. Doughty and Tse's[42] meta-analysis likewise found that participating in entirely peer-run services or organizations resulted in fewer and shorter hospital admissions and enhanced overall psychological well-being. Qualitative studies of community-based peer-support groups similarly indicate that participants may benefit from increased self-awareness, increased self-advocacy and decision-making in regard to medication, and more meaningful interactions with traditional mental health service providers.[43] These benefits may partly be attributed to the fact that many psychiatric service-user/survivor led groups emphasize strengths over deficits[44] and position recovery as a dynamic, ongoing journey rather than a fixed end state.[45]

Table 62.1 ASSESSMENTS OF THE IMPACT AND EFFECTIVENESS OF HVM PEER-SUPPORT GROUPS

QUALITATIVE INVESTIGATIONS					
AUTHOR	COUNTRY	SAMPLE	GROUPS ASSESSED	METHODOLOGY	SOURCE
Beavan et al.[46]	Australia	N = 29 Mean age = 41 Female = 43%	7 community groups	Self-report feedback forms; thematic analysis	Peer reviewed journal article
Bowyer[47]	Australia	N = 37 Demographic details not reported	7 community groups	Self-report feedback forms	Service evaluation
Dos Santos & Beavan[48]	Australia	N = 4 Female = 50% 3—schizophrenia 1—bipolar disorder	3 community groups	Semi-structured interviews; interpretative phenomenological analysis	Peer reviewed journal article
Hendry[49]	UK	N = 7 Mean age = 39.14 Female = 42.86%	1 community group	Semi-structured interviews; interpretative phenomenological analysis	Unpublished D.Clin.Psych dissertation
Meddings et al.[50–51]	UK	N = 29 Mean age = 40.74 Female = 50%	1 community group	Semi-structured interviews; thematic analysis (17—limited evaluation 12—in-depth evaluation)	Service evaluation
Mind in Camden's Hearing Voices Prisons & Secure Units Project[52]	UK	N = 18 Mean age = 40 Female = 5.56% 7—schizophrenia 4—depression 2—schizoaffective disorder 1—bipolar disorder 1—drug-induced psychosis 2—no diagnosis 1—data missing	6 prison groups	Self-report feedback forms	Service evaluation
Oakland & Berry[53]	UK	N = 11 Mean age = 46 Female = 27%	1 day center (open to individuals not using the center); 1 community (based within health services); 1 day center (individuals can self-refer)	Semi-structured interviews; thematic analysis	Peer reviewed journal article

(*continued*)

Table 62.1 CONTINUED

QUALITATIVE INVESTIGATIONS					
AUTHOR	COUNTRY	SAMPLE	GROUPS ASSESSED	METHODOLOGY	SOURCE
Payne et al.[54]	UK	N = 8 Mean age = 47.88 Female = 50% 4—schizophrenia 4—data missing	2 community groups	Semi-structured interviews; interpretative phenomenological analysis	Peer reviewed journal article
Romme et al.[10]	International	N = 17 Mean age = 40.5 Female = 58.82%	17 community groups	Semi-structured interviews; self-report written narratives	Book chapter
Slater[55]	UK	N = 7 Female = 0%	Secure inpatient unit	Semi-structured interviews; thematic analysis	Service evaluation
Sørensen[56]	Denmark	N = 13 Mean age = 40.82 Female = 61.54%	1 community group	Interpretative phenomenological analysis	Unpublished MSc thesis
Longden et al.[57]	UK	N = 101 Mean age = 44.54 Female = 47.6% 90—psychotic disorder	62 community groups	Self-report questionnaire: HVGS[57]	Peer reviewed journal article
Meddings et al.[50–51]	UK	N = 29 Mean age = 40.74 Female = 50%	1 community group	Self-report questionnaires: BAVQ[59] CCES[61] PCS[63] PSYRATS-AH[60] RSES[62] TVRS[58] 17—limited evaluation 12—in-depth evaluation	Service evaluation

Note. BAVQ = Beliefs About Voices Questionnaire[59]; CCES = Consumer Constructed Empowerment Scale[61]; HVGS = Hearing Voices Groups Survey[57]; PCS = Personal Constructs Scale[63]; PSYRATS-AH = Psychotic Symptoms Rating Scale—Auditory Hallucinations[60]; RSES = Rosenberg's Self-Esteem Scale[62]; TVRS = Topography of Voice Rating Scale[58].

Although this literature supports the value of self-help and peer-support per se, very few studies have looked at the specific impact of HVM groups on voice hearers' overall quality of life. According to Corstens et al.,[9] this is partly a result of the "uneasy relationship" (p.S289) that exists between social science and medical research agendas and the HVM, in that the latter locates itself as a reformative social movement that favors survivor testimony as a primary evidence source and likewise rejects more traditional clinical outcomes, like voice cessation, as a key index of success. Nevertheless, while controlled efficacy studies are lacking, a number of small-scale cross-sectional studies and surveys (Table 62.1) have reported on individuals' experiences of participating in groups, as well as the group's wider impact on their life and well-being.

Romme and colleagues (2009)[10] were some of the first authors to systematically examine first-person accounts of the impact of HVM groups. They compiled recovery narratives from 50 voice hearers living in Europe, the United States, and New Zealand, wherein 17 individuals spoke specifically about the positive impact of groups on their ability to understand and cope with their experiences. Being in an accepting environment with fellow voice hearers proved to "kick-start" recovery for several participants (p.78), after which membership facilitated several aspects of their journeys to healing. For example, once acquainted with the group, voice hearers supported one another to explore the possible meanings of their voices, which in turn fostered a greater sense of control and acceptance. Participants also frequently reported that the groups allowed them to connect with people and feel less isolated; as one member expressed it, they were "creating a 'fellowship' around the voice hearing experience" (p. 82), which in turn promoted self-esteem and emphasized the inherent value of every individual's experience.

Subsequent qualitative work by Hendry[49] reported on seven members of an English HVM group who had been attending consistently for over 6 months. The interviews focused on participants' experiences within the group itself, most of whom reported positive themes and outcomes. Specifically, the group was deemed a space to come together with others in the same situation and delve into difficult experiences that could not always be freely discussed outside the group. Group consistency proved to be very useful, and members came to rely both on the other members and the group itself as a "secure base" on which to rely. In turn the experiences of other members helped to normalize individuals' own experiences: "it is nice to be able to talk about it and not be the freak in the room" (p.76). This normalization gave way to increased acceptance of oneself as a voice hearer, as reflected by the comment "[It] made me feel a bit more normal about what I was experiencing, made me accept it a bit more" (p.76). The group also appeared to act as a catalyst for change and personal growth, with several participants reporting increased self-esteem and confidence as well as a sense of purpose and achievement as a result of their attendance.

A further qualitative study was conducted by Oakland and Berry,[53] who interviewed 11 members from several groups in the United Kingdom about attendance, experiences within the group, and the impact of the group on quality of life. When discussing the group structure one participant, who attended a group that was facilitated by voice hearers only, noted that there was an equal balance of power between the facilitator and all the members, which may not have been the case in a group with a professional facilitator. However, another participant who attended a group that was cofacilitated by a mental health professional reported that the arrangement provided a supportive framework and the professional did not usurp power from the members. Other participants noted that they were pleased that the group set ground rules, such having no time-limit and no mandatory participation or attendance. These rules put participants at ease, as they knew there was no pressure to participate until they were ready and that they could continue coming to the group for as long as they liked. Two major themes subsequently emerged from the interviews: acceptance and hope. Participants reported having their experiences accepted by other people, often for the first time. Being part of the group also normalized the experience of voice hearing and facilitated connecting and opening up to others; as one participant stated, "I felt like I belonged somewhere so I could take off my mask and feel safe" (p. 124). Several people further observed that being part of the group was inspiring, in that seeing other people live happily with their voices instilled hope that they could too. In addition, members found that the skills and insights gained in the group could also benefit their lives outside the group. Several participants reported feeling a greater sense of empowerment and control, viewing themselves in a more positive light, and feeling more comfortable with their identities. One participant summarized such findings in the following terms: "It's started to help me build my self-esteem; from me hating myself to me being happy with the man I am today" (p. 125).

Similar findings are reported by dos Santos and Beavan,[48] who interviewed four participants from three different groups in Australia. All described being secretive about their experience of voice hearing prior to starting the group, particularly out of fear of hurting or burdening friends and family. In turn, although all participants reported anxiety during their first meeting, they subsequently expressed a feeling of connection and openness with the other members. Other identified themes included an appreciation of the care and compassion participants felt they received from other members, the way the group provided a constant space to rely on each week, and advice around coping strategies relayed by other members. Beyond the group, participants felt more comfortable and confident talking about their experiences with friends, family, and coworkers and felt an overall improvement in their self-esteem. Participants also felt as though they were better able to cope with their voices and conceptualized recovery as coexisting peacefully with their voices, rather than trying to eliminate them. One participant described this journey in the following way: "I'm not at all scared that I hear voices anymore, and they don't worry me in the slightest" (p. 33).

A later study by Beavan, de Jager, and dos Santos[46] provided somewhat similar findings, this time with a larger sample of 29 voice hearers from 7 different groups in Australia. Participants were asked to respond to the open-ended question, "What is especially important for you in participating in the group?" Sharing and feedback within the group was noted to be particularly useful, both for being able to confide in others, as well as being able to learn from other members. As such, the group was valued as a place to get support from people who had similar experiences and foster a sense of solidarity and belonging. Several participants also stated that they felt more in control and had acquired more strategies to cope with and relate to distressing voices. Talking to other voice hearers also served to normalize the experience, leading one member to acknowledge "hearing voices as part of normal human experience" (p. 62).

Similar themes were reported by Payne, Allen, and Lavender,[54] who conducted qualitative interviews with eight voice hearers from two peer-led UK-based groups. Participants commented on the nurturing environment of the group, citing a sense of compassion, acceptance, and camaraderie while noting it was a place where everyone shared a common identity but individual experiences were still acknowledged and respected. In turn, this commonality of experience could help inspire new friendships and the group as a whole was credited with fulfilling a need for belonging and connection—a need that may have been denied to participants at previous times in their lives and then reinforced through stigmatizing diagnoses. Consistent with other studies, the participants viewed the group as an "emotional container" and a place to safely unload feelings without the threat of any consequences. Being heard by other members of the group helped several participants to see different perspectives on their experiences, which in turn helped reduce anxiety as well as inspire a sense of trust in others. One member explicitly expressed the value of the consistency of the group: "I think for me, especially in my childhood, I didn't really have people I could rely on or trust to sort of be there in a positive way for me and so, I think it's something that's quite important to me" (p. 6). In dealing with voice hearing specifically, members reported gaining insight into the "purpose" and "message" of voices. While only one participant explicitly reported decreased frequency in voice hearing, all participants identified considerable improvement in other aspects of their functioning, such as returning to work or starting new relationships.

Several qualitative service evaluations have also been conducted assessing the effectiveness of HVM groups in various populations. These studies elucidate similar themes to those previously mentioned in the published studies, with a particular emphasis on the social aspects of the group, the positive impact on self-esteem, as well as the group being a place to gain novel insight into distressing experiences. A full summary of frequently endorsed themes across these different sources can be found in Table 62.2.

Meddings et al.[50–51] were the first to quantitatively investigate the effectiveness of HVM groups by assessing the experiences of 29 participants at baseline, after attending the group for 6 months, and after attending the group for 18 months using a battery of self-report measures comprising the Topography of Voices Rating Scale,[58] Beliefs About Voices Questionnaire,[59] selected items from the Psychotic Symptom Ratings Scales,[60] the Consumer Constructed Empowerment Scale,[61] Rosenberg's Self-Esteem Scale,[62] and the Personal Constructs Scale,[63] in which participants indicated what they hoped to gain from attending the group.

Given that HVM groups are nonclinical in nature, it is noteworthy that many members experienced clinically relevant improvements. For example, participants spent significantly fewer days in hospital after joining the group: an average of 39 days per year compared to an average of 8 days per year after joining. Furthermore, there was a borderline significant trend toward fewer hospital admissions overall and fewer compulsory hospital admissions, with four participants showing a clinically significant reduction in hospital admissions. Participants also used significantly more coping strategies when distressed by their voices, and felt significantly more able to talk about their experiences to others. Seven participants showed a clinically significant improvement in the number of coping strategies used. Participants also reported hearing voices less frequently, although volume and clarity were unaffected. In turn, while there was no change in how distressing, distracting, or controlling voices were, participants felt more equipped to cope with these experiences and more powerful relative to their voices. In terms of personal recovery-related outcomes, participants reported a significant increase in their sense of empowerment and self-esteem after attending the group. Furthermore, significantly more participants were employed, engaged in voluntary work, or attending college during the 18-month period than at baseline. In terms of personal constructs, the most commonly stated goals were to hear voices less often, to feel more normal and less isolated, to foster a sense of belonging by meeting others who hear voices, to cope better with the voices, to feel less anxious, and to be able to move forward in one's life. Participants reported significant improvement in domains encapsulated by their personal constructs, indicating that the group had the capacity to meet a diverse range of needs.

A more recent quantitative study was conducted by Longden, Read, and Dillon[57] who distributed the Hearing Voices Group Survey (HVGS), a 40-item instrument specifically created to assess groups' impact and effectiveness, to 62 English HVM groups. In addition to demographics details and group format, the HVGS contained 11 questions concerning participant experiences within the group, 12 assessing the impact of membership on social, occupational, and clinical aspects of participants' lives outside the group, and five questions relating to the group's impact on emotional well-being, all of which were scored on a five-point Likert scale. Responses from 101 individuals demonstrated that mean scores were significantly higher than the neutral midpoint across all three of these domains. For the category Experiences Within The Group, the questions with the highest scores above the neutral midpoint were: (1) "the group gives me support around voice hearing that I couldn't get elsewhere," (2) I have found it useful to meet other voice hearers in the group," and (3) "the group feels like a safe and confidential place to talk about difficult things." All other positive statements were rated closest to "agree," and all negative statements were rated closest to "disagree." The one exception was "I found the group distressing at times," which was ranked closest to "neither agree nor disagree."

Table 62.2 COMMONLY ENDORSED THEMES ABOUT THE PERCEIVED BENEFITS OF ATTENDING HVM GROUPS

THEME	DESCRIPTION	SOURCES	ENDORSEMENTS ACROSS STUDIES N (%)
Accepting and supportive environment	Groups provide a place where an individual's experiences are welcomed and can be shared without judgement.	Beavan et al.[46] Bowyer[47] Dos Santos & Beavan[48] Hendry[49] Meddings et al.[50–51] Mind in Camden[52] Oakland & Berry[53] Payne et al.[54] Romme et al.[10] Slater[55] Sørensen[56] Longden et al.[57]	12 (100%)
Improving self-confidence and self-esteem	Group attendance can facilitate positive feelings about oneself, which may also be transferred into other areas of one's life.	Beavan et al.[46] Bowyer[47] Dos Santos & Beavan[48] Hendry[49] Longden et al.[57] Meddings et al.[50–51] Mind in Camden[52] Oakland & Berry[53] Payne et al.[54] Romme et al.[10] Slater[55] Sørensen[56]	12 (100%)
Solidarity and shared experiences	Groups are a place to connect with people with similar experiences, and to recognise one's own experience in the stories of others.	Beavan et al.[46] Bowyer[47] Dos Santos & Beavan[48] Hendry[49] Longden et al.[57] Meddings et al.[50–51] Mind in Camden[52] Oakland & Berry[53] Payne et al.[54] Romme et al.[10] Slater[55] Sørensen[56]	12 (100%)
A place to talk about experiences	Groups provide the space for in-depth explorations of voice hearing and other distressing experiences that may be difficult to discuss elsewhere.	Beavan et al.[46] Bowyer[47] Dos Santos & Beavan[48] Hendry[49] Longden et al.[57] Meddings et al.[50–51] Mind in Camden[52] Oakland & Berry[53] Payne et al.[54] Romme et al.[10] Slater[55] Sørensen[56]	12 (100%)
Exchanging coping strategies	Members can share constructive strategies on how to cope and engage with voices and other distressing experiences.	Beavan et al.[46] Bowyer[47] Dos Santos & Beavan[48] Hendry[49] Longden et al.[57] Meddings et al.[50–51] Mind in Camden[52] Oakland & Berry[53] Payne et al.[54] Romme et al.[10] Slater[55]	11 (91.67%)

(continued)

Table 62.2 CONTINUED

THEME	DESCRIPTION	SOURCES	ENDORSEMENTS ACROSS STUDIES N (%)
Sense of agency in recovery	Members express optimism about the future, recovery potential, and of living a successful and meaningful life as someone who hears voices.	Beavan et al.[46] Dos Santos & Beavan[48] Hendry[49] Longden et al.[57] Meddings et al.[50–51] Mind in Camden[52] Oakland & Berry[53] Payne et al.[54] Romme et al.[10] Slater[55] Sørensen[56]	11 (91.67%)
Understanding one's experiences	Members may develop new knowledge and insights about their voices, which in turn can improve the ability to cope with and relate to them.	Beavan et al.[46] Dos Santos & Beavan[48] Hendry[49] Longden et al.[57] Meddings et al.[50–51] Mind in Camden[52] Oakland & Berry[53] Payne et al.[54] Romme et al.[10] Slater[55] Sørensen[56]	11 (91.67%)
Social connection	Groups provide the opportunity to socialize, meet new friends, and develop confidence to expand one's social networks both within and beyond the group.	Beavan et al.[46] Bowyer[47] Dos Santos & Beavan[48] Hendry[49] Longden et al.[57] Meddings et al.[50–51] Payne et al.[54] Romme et al.[10]	8 (66.67%)
Disclosing experiences with non-voice hearers	Members gain confidence in sharing their experiences with people outside the group.	Beavan et al.[46] Bowyer[47] Dos Santos & Beavan[48] Longden et al.[57] Meddings et al.[50–51] Romme et al.[10] Slater[55]	7 (58.33%)
Ownership of the group	Group members feel a sense of empowerment in the way the groups are facilitated or co-facilitated by fellow voice hearers and members can make decisions on how groups are run.	Hendry[49] Oakland & Berry[53] Payne et al.[54] Romme et al.[10]	4 (33.33%)

Positive experiences within the group were found to be correlated with numerous beneficial outcomes beyond the group itself. For example, the items (1) "the group has helped me feel less distressed by my voices," (2) "the group gives me support around voice hearing that I couldn't get elsewhere," (3) "the group has given me helpful information about making sense of my voice hearing experiences," and (4) "the group gives me positive messages about recovering from mental health problems," were all positively associated with social outcomes, such as making more friends, feeling more confident in social situations or asking for help, and developing stronger relationships with one's family. Furthermore, positive experiences within the group were positively correlated with emotional well-being variables, such as feeling better about oneself and having more hope for the future.

In addition to reported benefits, it should also be noted that participants across studies also described several challenges occurring within HVM groups. For example, several individuals reported vicarious emotional responses in which the low mood of another member impacted their own sense of well-being. This could also cause additional confusion and stress in terms of determining whether the change in mood should be attributed to others in the group or to an internal conflict within oneself.[49,54] Some participants additionally reported a "backlash" effect in terms of the voices becoming more hostile after being discussed with other people.[49] Other participants described negative experiences in terms of other members being reluctant to engage in certain interpretations for their voices, such as spiritual or technological explanations.[54] However, Longden et al.[57] demonstrated that even though slightly over a third of their participants felt distress during meetings, they still reported an extremely positive overall impact for the group. One speculative explanation for this finding is that while discussing issues related to voice hearing may be inherently painful, having an opportunity to acknowledge and explore difficult events with a compassionate audience of peers may alleviate emotional distress in the long-term.

Taken together, the overall themes that emerge from these studies can be seen to reflect the core benefits identified by Escher[64] of talking about one's voices-hearing experiences in a safe and supported environment. These benefits include identifying links between voice hearing experiences, life events, and emotions; reducing feelings of isolation and powerlessness by sharing one's experiences; exploring possible meanings of voices, which in turn may improve the hearer–voice relationship; increasing feelings of validation and affirmation by recognizing one's story in the experience of others; and cultivating a positive and healthy identity as a voice hearer. In turn, it is important to acknowledge that these themes are not only the province of those with complex mental health difficulties; rather they can be seen as reflecting the common human desire to feel connected with others, experience a sense of belonging, and have one's experiences witnessed by others who are able to openly share their own hardships and successes.

CONCLUSION

The fellowship, compassion, and mutual understanding that is often developed within HVM groups challenges conceptualizations of individuals with schizophrenia spectrum diagnoses as lacking insight and empathy and being more suited as passive recipients of support rather than providers of it.[22,25] In turn, the alliances formed in HVM groups between experts by experience and experts by profession have positive implications in that the groups may be a means of facilitating both clinical and social benefits for individuals who are unwilling to engage in psychiatric services or for whom mainstream psychotherapeutic and/or pharmacologic interventions have been unhelpful or unacceptable. Substantial survivor testimony, in addition to a small but persuasive body of research, attest to the valuable role of HVM groups and demonstrates that their impact can often extend beyond coping with voices to include improved functioning in other facets of their member's lives and a chance to develop meaningful connections with other voice hearers, as well as increasing emotional well-being and providing a sense of empowerment and hope for the future.

REFERENCES

1. Romme M, Escher S. Hearing voices. *Schizophr Bull.* 1989; 15(2): 209–216.
2. Romme M, Escher S. *Accepting Voices.* London, UK: Mind Publications; 1993.
3. van Os J, Linscott R, Myin-Germeys I, Krabbendam L. A systematic review and meta-analysis of the psychosis continuum: evidence for a psychosis-proneness-persistence-impairment model of psychotic disorder. *Psychol Med.* 2009; 39(2): 179–195.
4. Beavan V, Read J, Cartwright C. The prevalence of voice-hearers in the general population: a literature review. *J Ment Health.* 2011; 20(3): 281–292.
5. Tien AY. Distribution of hallucinations in the population. *Soc Psychiatry Psychiatr Epidemiol.* 1991; 26(2): 287–292.
6. Romme M, Escher S. *Making Sense of Voices.* London: Mind Publications; 2000.
7. Johnstone L. Voice hearers are people with problems, not patients with illnesses. In: Romme M, Escher S, eds. *Psychosis as a Personal Crisis: An Experience-Based Approach.* London: Routledge; 2012:27–36.
8. Corstens D, Longden E. The origins of voices: links between life history and voice hearing in a survey of 100 cases. *Psychosis.* 2013; 5(3): 270–285.
9. Corstens D, Longden E, McCarthy-Jones S, Waddingham R, Thomas N. Emerging perspectives from the Hearing Voices Movement: implications for research and practice. *Schizophr Bull.* 2014; 40(S4): S285–S294.
10. Romme M, Escher E, Dillon J, Corstens D, Morris M. *Living with Voices: 50 Stories of Recovery.* Ross-on-Wye: PCCS; 2009.
11. Bentall RP, Wickham S, Shevlin M, Varese F. Do specific early-life adversities lead to specific symptoms of psychosis? A study from the 2007 The Adult Psychiatric Morbidity Survey. *Schizophr Bull.* 2012; 38(4): 734–740.
12. Read J, van Os J, Morrison AP, Ross CA. Childhood trauma, psychosis and schizophrenia: a literature review with theoretical and clinical implications. *Acta Psychiat Scand.* 2005; 112(5): 330–350.
13. Varese F, Smeets F, Drukker M, et al. Childhood adversities increase the risk of psychosis: a meta-analysis of patient-control, prospective- and cross-sectional cohort studies. *Schizophr Bull.* 2012; 38(4): 661–671.
14. Corstens D, Escher S, Romme M. Accepting and working with voices: the Maastricht approach. In: Moskowitz A, Shäfer I, Dorahy MJ, eds. *Psychosis, Trauma, and Dissociation: Emerging Perspectives on Severe Pyschopathology.* Oxford: Wiley-Blackwell; 2008: 319–331.
15. Longden E, Corstens D, Escher S, Romme M. Voice hearing in a biographical context: a model for formulating the relationship between voices and life history. *Psychosis.* 2012; 4(3): 224–234.
16. Slade M, Longden E. Empirical evidence about recovery and mental health. *BMC Psychiatry.* 2015; 15(1).
17. Morrison AP. A cognitive analysis of auditory hallucinations: are voices to schizophrenia what bodily sensations are to panic? Behav Cogn Psychother. 1998; 26: 289–302.
18. Viega-Martinez C, Perez-Alvarez M, Garcia-Montes JM. Acceptance and commitment therapy applied to treatment of auditory hallucinations. *Clin Case Stud.* 2008; 7(2): 118–135.
19. Fowler D, Garety P, Kuipers E. Cognitive therapy for psychosis: formulation, treatment, effects and services implications. *J Ment Health.* 1998; 7(2): 123–133.
20. British Psychological Society: Division of Clinical Psychology. *Good Practice Guidelines on the Use of Psychological Formulation.* Leicester: British Psychological Society; 2011.
21. British Psychological Society: Division of Clinical Psychology. *Understanding Psychosis and Schizophrenia. A Report by the Division of Clinical Psychology.* Leicester: British Psychological Society; 2014.
22. Dillon J, Hornstein G. Hearing voices peer support groups: a powerful alternative for people in distress. *Psychosis.* 2013; 5(3): 286–295.
23. Kapur S. Psychosis as a state of aberrant salience: a framework linking biology, phenomenology, and pharmacology in schizophrenia. *Am J Psychiatry.* 2003; 160(1): 13–23.
24. McCabe R, Heath C, Burns T, Priebe S. Engagement of patients with psychosis in the consultation: conversation analytic study. *BMJ.* 2002; 325: 1148–1151.
25. Dillon J, Longden E. Hearing voices groups: creating safe spaces to share taboo experiences. In: Romme M, Escher S, eds. *Psychosis as*

a Personal Crisis: An Evidence-Based Approach. London: Routledge; 2012: 129–139.

26. Jones N, Marino C, Hansen MC. The Hearing Voices Movement in the United States: findings from a national survey of group facilitators. *Psychosis*. 2016; 8(2): 106–117.
27. Longden E, Corstens D, Dillon J. Recovery, discovery and revolution: the work of Intervoice and the Hearing Voices Movement. In: Cole S, Keenan S, Diamond B, eds. *Madness Contested: Power and Practice*. Ross-on-Wye: PCCS; 2013:161–180.
28. Davison L, Raakfeldt J, Strauss JS. *The Roots of the Recovery Movement in Psychiatry: Lessons Learned*. London: Wiley-Blackwell; 2010.
29. Davidson L, Bellamy C, Guy K, Miller R. Peer support among persons with severe mental illnesses: a review of evidence and experience. *World Psychiatry*. 2012; 11(2): 123–128.
30. Fuhr DC, Salisbury TT, De Silva MJ, et al. Effectiveness of peer-delivered interventions for severe mental illness and depression on clinical and psychosocial outcomes: a systematic review and meta-analysis. *Soc Psychiatry Psychiatr Epidemiol*. 2014; 49(11): 1691–1702.
31. O'Donnell M, Parker G, Proberts M. A study of client-focused case management and consumer advocacy: the community and consumer service project. *Aust N J Z Psychiatry*. 1999; 33(5): 684–693.
32. Chinman MJ, Weingarten R, Stayner D, Davidson L. Chronicity reconsidered: improving person-environment fit through a consumer-run service. *Community Ment Health J*. 2001; 37(3): 215–229.
33. Corrigan PW. Impact of consumer-operated services on empowerment and recovery of people with psychiatric disabilities. *Psych Serv*. 2006; 57(10): 1493–1496.
34. Resnick SG, Rosenheck RA. Integrating peer-provided services: a quasi-experimental study of recovery orientation, confidence, and empowerment. *Psych Serv*. 2008; 59(11): 1307–1317.
35. Ockocka J, Nelson G, Janzen R, Trainor J. A longitudinal study of mental health consumer/survivor initiative: part 3—a qualitative study of impact of participation on new members. *J Community Psychol*. 2006; 34(3): 273–283.
36. Davidson L, Chinman M, Kloos B, Weingarten R, Stayner D, Tebes JK. Peer support among individuals with severe mental illness: a review of the evidence. *Clin Psychol Sci Pract*. 1999; 6: 165–187.
37. Yanos TP, Primavera LH, Knight EL. Consumer-run service and participation, recovery of social functioning, and the mediating role of psychological factors. *Psych Serv*. 2001; 52(4): 493–500.
38. Trainor J, Shepherd M, Boydell KM, Leff A, Crawford E. Beyond the service paradigm: the impact and implications of consumer/survivor initiatives. *Psychiatr Rehabil J*. 1997; 21: 132–140.
39. Salzer MS, Shear SL. Identifying consumer-provider benefits in evaluations of consumer-delivered services. *Psychiatr Rehabil J*. 2002; 25(3): 281–288.
40. Ratzlaff S, McDiarmid D, Marty D, Rapp C. The Kansas consumer as provider program: measuring the effects of a supported education initiative. *Psychiatr Rehabil J*. 2006; 29(3): 174–182.
41. Castelein S, Bruggeman R, van Busschbach J, et al. The effectiveness of peer support groups in psychosis. *Acta Psychiatr Scand*. 2008; 118(1): 64–72.
42. Doughty C, Tse S. Can consumer-led mental health services be equally effective? an integrative review of CLMH services in high-income countries. *Community Ment Health J*. 2011; 47: 252–266.
43. Jones N, Corrigan PW, James D, Parker J, Larson N. Peer support, self-determination, and treatment engagement: a qualitative investigation. *Psychiatr Rehabil J*. 2013; 36(3): 209–214.
44. Ahmed AO, Doane NJ, Mabe PA, Buckley PF, Birgenheir D, Goodrum NM. Peers and peer-led interventions for people with schizophrenia. *Psychiatr Clin North Am*. 2012; 35(3): 699–715.
45. Deegan PE, Drake RE. Shared decisions making and medication management in the recovery process. *Psychiatr Serv*. 2006; 57(11): 1636–1639.
46. Beavan V, de Jager A, dos Santos B. Do peer-support groups for voice-hearers work? A small scale study of Hearing Voices Network support groups in Australia. *Psychosis*. 2017; 9(1): 57–66.
47. Bowyer R. Hearing Voices Network Australia Evaluation 2010. Unpublished Report. 2010.
48. dos Santos B, Beavan V. Qualitatively exploring Hearing Voices Network support groups. *J Ment Health Train Educ Pract*. 2015; 10(1): 26–38.
49. Hendry GL. What are the experiences of those attending a self-help hearing voices group: an interpretative phenomenological approach [dissertation]. University of Leeds; 2011.
50. Meddings S, Walley L, Collins T, Tullet F, McEwan B, Owen K. Are Hearing Voices groups effective? A preliminary evaluation. Unpublished Report. 2004.
51. Meddings S, Walley L, Collins T, Tullet F, McEwan B. The voices don't like it. *Mental Health Today*. 2006; 26–30.
52. Mind in Camden's Hearing Voices: Prisons and Secure Units Project. Evaluation Report. 2016.
53. Oakland L, Berry K. "Lifting the veil": a qualitative analysis of experiences in Hearing Voices Network groups. *Psychosis*. 2015; 7(2): 119–129.
54. Payne T, Allen J, Lavender T. Hearing Voices Network groups: experiences of eight voice hearers and the connection to group processes and recovery. *Psychosis*. 2017: 1–11.
55. Slater J. Piloting a Hearing Voices Group in a high secure psychiatric setting. Presented at: World Hearing Voices Congress 2010.
56. Sørensen TH. The Aarhus voice hearing program—an exploration of a recovery-oriented program [thesis]. Aarhus University; 2013.
57. Longden E, Read J, Dillon J. Assessing the impact and effectiveness of Hearing Voices Networks self-help groups. *Community Ment Health J*. 2017: in press.
58. Hustig HH, Hafner RJ. Persistent auditory hallucinations and their relationship to delusions and mood. *J Nerv Ment Dis*. 1990; 178: 264–267.
59. Chadwick P, Birchwood M. The omnipotence of voices II: the beliefs about voices questionnaire (BAVQ). *Br J Psychiatry*. 1995; 166: 773–776.
60. Haddock G, McCarron J, Tarrier N, Faragher EB. Scale to measure dimensions of hallucinations and delusions: the psychotic symptom rating scale (PSYRATS). *Psychol Med*. 1999; 29: 879–889.
61. Rogers ES, Chamberlin J, Ellison ML, Crean T. A consumer-constructed scale to measure empowerment among users of mental health services. *Psych Serv*. 1997; 48: 1042–1047.
62. Rosenberg M. *Society and the Adolescent Self-Image*. Princeton, NJ: Princeton University Press; 1965.
63. Kelly GA. *The Psychology of Personal Constructs*, vols. 1–2. New York, NY: Norton; 1955. (Reprinted by Routledge; 1991).
64. Escher S. Talking about voices. In: Romme M, Escher S, eds. *Accepting Voices*. London: Mind Publications; 1993: 50–59.

63

AVATAR THERAPY

A NEW DIGITAL THERAPY FOR AUDITORY VERBAL HALLUCINATIONS

Tom K. J. Craig and Mar Rus-Calafell

UNDERPINNINGS OF AVATAR THERAPY

AVATAR therapy (Audio-Visual Assisted Therapy Aid for Refractory auditory hallucinations) is a newly developed treatment for auditory verbal hallucinations (AVH). It involves a three-way interaction between client, therapist, and a digital representation of the image and voice of the entity (the avatar) that the person believes is the source of their voices. Using specially designed computer software, the client creates a visual representation of the entity (human or nonhuman) that they believe is talking to them. Additional software is used to transform the voice of the therapist to match the pitch and tone of the voice heard by the person, the two processes finally being combined to produce a computer simulation (a virtual agent or "avatar") through which the therapist can dialogue with the person. The therapist supports people to interact with the voice and challenge the negative things it says in order to change the relationship between the individual and their voice; from a distressing experience to a more positive or neutral interaction.

AVATAR therapy is part of a new and exciting wave of therapies that adopt an explicitly relational and dialectic approach to working with distressing voices. These approaches focus on the relationship between the voice hearer and the voices, and seek not necessarily to erase the experience of hearing the voice but to target specific aspects of that relationship and to improve the person's appraisal of control within the relationship.

VIRTUAL REALITY AS A VEHICLE FOR DELIVERING TREATMENT

Virtual reality (VR) refers to an artificial environment created by digital technology that gives users the impression of immersion in a realistic world in which they can interact with objects and animated avatars. The most complex implementations involve full immersion in 3D with visual, auditory, and touch sensations, but even simpler 2D implementations such as used in AVATAR therapy can also be highly interactive and experienced as very realistic.

Virtual reality offers researchers and clinicians the possibility not only of observing the user's real-time behavior when interacting with virtual agents but also of controlling and modifying the environment and the responses of the avatars or simulated stimuli and tasks. For psychosis, as noted in a recent systematic review by Rus-Calafell and colleagues (2017), VR is a promising method to be used in the assessment of neurocognitive deficits and the study of relevant clinical symptoms.[1]

The majority of studies have focused on using VR as an assessment tool to explore cognitive impairments and symptom assessment (mainly paranoid thoughts), with several studies using VR to explore the person's social functioning. For example, using daily environments (such as a virtual supermarket) to assess and demonstrate impaired executive functioning in schizophrenia[2] or the person's ability to plan grocery shopping,[3] or real-time interaction with virtual agents to assess aspects of social cognition.[4,5] The ability to program and so alter the environment and the appearance of the virtual agents that populate it, has also produced some very interesting insights into paranoid ideation and hallucinations.[6–8] There are, however, fewer studies using VR as a modality in which to deliver or supplement therapy. In their review, Rus-Calafell et al.[1] identified just eight studies including using VR as an adjunct to cognitive remediation,[9,10] rehearsing and improving job interview skills.[11] in social skills training,[12,13] in cognitive behavior therapy targeting paranoia,[14–16] and in AVATAR therapy.[16] From these and other studies, it appears that the use of VR is a safe and well-tolerated tool for people experiencing psychosis and people at high risk for psychosis.

The approach used in AVATAR therapy can be considered an implementation of partially immersive VR. It provides a *virtual embodiment* of the AVH experience by creating a physical representation of the agent behind the personified but usually disembodied voice. The approach to the creation of the avatar therefore attempts to get a close match to the facial appearance of what the person imagines the agent behind the AVH looks like. Additional matching of the speech of the avatar to the usual tone and content of the AVH provides even more realism to the experience and this combined realism seems to be a key aspect of therapy. This visualization of the AVH may facilitate two essential processes in the AVATAR therapy: (1) the validation of the experience and (2) the flow of dialogue with the voice through the sessions while modifying the type of relationship between the voice and the participant. By creating the virtual representation of the voice and sharing it with the therapist, the experience becomes something more tangible and less abstract while the sharing this avatar with the therapist helps to reassure the person that their experience is taken seriously and facilitates the therapy alliance. The person also understands that the conversation with the avatar will happen in a safe space, always supported by the therapist allowing them to take risks in confronting their voice that that they would previously have avoided. This standing up to the voice in a safe space over several sessions facilitates the generalization of this behavior to the persecuting voice and to others when they occur outside of the clinic setting.

The virtual reality-based technology used in the AVATAR therapy offers a unique way for clinicians to directly observe clients interacting with a representation of their voices and for working and modulating many aspects of the clients' relationships with these "voices."[17]

RELATIONAL APPROACHES TO VOICES

Auditory verbal hallucinations (AVH) are the most commonly reported form of auditory hallucination,[18,19] experienced by as many as 70% of individuals with a diagnosis of schizophrenia.[20] Typically, AVH have personal identities and are experienced as profoundly "real."[21] As many as three quarters of people with psychosis with AVH have mental images of the source of the voice when they hear it.[22] Some voice hearers use these images as evidence to support their beliefs about the voice (e.g., that it is omnipotent and omniscient because they have a concurrent image of God or the Devil), so there is some speculation that altering the content or meaning of the image, could result in a reduction of the associated distress and increase the sense of control over the experience.[23]

Whether or not heard with associated images, voices are often experienced as being in a superior or dominant relationship to the sufferer, for example, being members of public agencies such as the police or secret services, celebrities, religious leaders, or supernatural powers.[19,21,24,25] While some voices may be benign or even positive, the majority are hostile and distressing, exerting a severe impact on everyday functioning especially where they appear to come from entities who know the person's every movement and threaten dire consequences if their commands are not obeyed.

It has been suggested that social isolation and deficits in attribution of social meaning are also potential contributors for the development of AVHs.[26–28] Voices appraised as omnipotent and malevolent are associated with high levels of distress,[18] and this association is mediated by underlying social schemata that may have originated earlier in life. For example, how an individual perceives and responds to their voices may well reflect their experiences of other key relationships currently or in the past. Birchwood and colleagues[29,30] suggest that people who have experienced inferiority and powerlessness in key social relationships may establish social schema that make them more likely to experience similar feelings of powerlessness, distress, and depression in relation to their AVHs. Recent reviews have provided support for this hypothesis with the implication that therapies could benefit from targeting social and interpersonal variables.[18,31,32]

This relational perspective has contributed to new approaches of psychological therapy for AVH, that focus on the interpersonal relationship between the voice hearer and the voice.[33–35] In "Talking with Voices" Dirk Corstens and colleagues[33] draw on a therapy technique more commonly associated with psychoanalytic approaches in which one's personality is conceived as comprising a number of alternative selves each with their own perceptions and reactions to the world.[33] The voices heard by clients represent one of these selves that wants to be heard and may have important things to say to or about the client perhaps relating to some emotional crisis in their life.[33] The therapy involves asking the person to concentrate on the chosen voice and to locate it in the room. The facilitator (i.e., therapist) then engages in a dialogue with the voice, which is spoken by the voice hearer as an intermediary. There is no attempt to eradicate the AVH nor to target cognitive mechanisms or use specific techniques such as assertiveness training. Instead, the aim is to explore and change the relationship, so that the person and voice come to a better mutual understanding, the individual regains control, and the voice transitions from "a tormentor to an encouraging companion."[33] The approach has been reported through case series but has not yet been evaluated through a controlled trial.

Relating therapy (RT[35,36]), takes a somewhat different approach. It applies Birtchnell's (1996) interpersonal model to the voice-hearer relationship, identifying key interpersonal dimensions of power and proximity in how the individual relates to people and how this is reflected in how they relate to their AVHs.[37] Based on this initial awareness building stage, sessions (16 in total) expand into three phases: in phase 1, relationships in terms of power and proximity are discussed, linking this discussion to participant's experiences of relating to people and distressing voices; in phase 2, the therapist encourages the person to explore themes within the relational history of the hearer and their experience of relationships with voices, and interpersonal relating within the family and social environment; and finally, in phase 3, the therapy moves on to explore different ways of relating to the voice using a variety of techniques including assertiveness training (including role play and empty chair work) with the aim of increasing the person's sense of control within the relationship. A recent pilot study, including 29 people reporting distressing auditory hallucinations, showed RT was superior to treatment as usual for reducing distress associated with AVH.[35]

ESSENTIAL COMPONENTS OF AVATAR THERAPY

AVATAR therapy was first described by Julian Leff and colleagues[16] as a development of the relational or "dialogic" therapy approach to voices. It draws on the work of Dirk Corstens and colleagues described earlier.[33] The key insight was that the usual "empty chair" and role-play approach might be enhanced if the person could have a real-time conversation with the source of their AVH and that this could be effected by a digital representation (avatar) of their persecutor that spoke in a voice that was a close match in terms of the vocal characteristics and verbal content of the AVH. This digital representation was achieved by using a combination of commercially available software to construct an animated face coupled with bespoke software developed by Mark Huckvale and colleagues at UCL to provide real-time transformations of the therapist's voice (see Leff et al. 2013 for details[16]).

The avatar created for use in therapy is characterized in detail and is specific to the individual who will use it in therapy. The first step involves a detailed assessment of the AVH experience in the context of their current and previous significant relationships.[29,30] This includes information about the source of the voices (e.g., male or female, age, ethnicity, whether someone they have met or know), whether the voices seem to be part of a network (e.g., criminal gang, member of a former work team), whether they reflect unresolved social and emotional issues, and how the experience of AVH is influenced by the person's cultural background and experiences earlier in life such as childhood maltreatment, bullying and other traumatic events, low self-esteem, and comorbid depression. The nature of the relationship as it varies along dimensions of interpersonal power and proximity[37] also influences the evolving dialogue during therapy sessions.

As the therapy can only deal with one voice at a time, all this information is used to come to a decision about which voice to target (typically the most distressing but sometimes the lead character when voices arise from several entities that appear to know each other). Using this information about the AVH, the person assisted by the therapist selects one of a number of prerecorded transformations of the therapist's voice as an approximate match to the pitch and tone of their AVH. Each aspect of the selected voice is then further adjusted using on-screen controls until the client reports the closest possible match to what they usually hear. Successful achievement of the match is considered when the participants scores 75 or above on a "goodness of match" scale (ranging from 0 to 100). A similar process is used to create the visual image of the entity behind the voice, and the two are then combined to produce the final avatar (i.e., an animated talking virtual agent with enhanced lip-sync speech and eye blinking).

Therapy consists of approximately six sessions delivered at weekly intervals. In these, the therapist and participant sit in separate rooms both facing interconnected computers. The therapist is able to see the participant using a video link and to speak as himself or in the voice of the avatar, attending to the dialogue as it unfolds (Figure 63.1). The aim is to steer the dialogue over the therapy sessions so that the participant experiences a change in the relationship as the avatar cedes power and control.[16] The sessions are audiorecorded and provided on an MP3 player to the person with instructions to listen to the session daily and when troubled by their voices.

Each of the six sessions has three components taking about 50 minutes altogether: (1) an initial review of the previous week including use of the MP3 player and an account of any changes to the voices followed by a discussion of the planned targets for that day's session, (2) dialogue with the avatar (typically less than 15 minutes) and, finally (3) a brief discussion of that day's session and the provision of the MP3 recording to take away. The overall course of the therapy consists of two overlapping phases. In the first of these (typically sessions 1–3) the aim is to create a realistic simulation of the AVH experience, with the avatar saying what the person usually hears (mainly using verbatim statements from the voice captured during the initial assessment) in the hostile or intimidating tone typical of their AVH. This element of exposure is coupled with instructions to the person to look the avatar "in the eye," to be firm but polite in standing up to its comments,

THERAPIST

PARTICIPANT

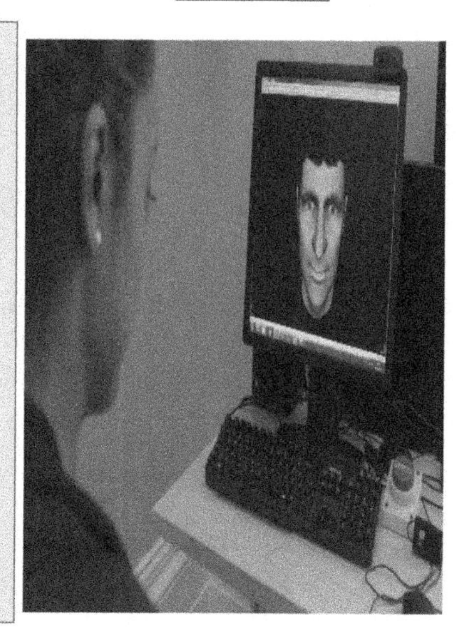

Figure 63.1 AVATAR therapy set-up. *Note*: Video Clips of AVATAR therapy.
Courtesy of Chris Chapman and the Wellcome Trust.

challenging its hostility and confidently asserting the intention to take control and lessen its power. The therapist encourages and praises assertive responses and makes subtle changes to the avatar reflecting recognition of the change in the client's behavior. While the most common initial response to the avatar is one of fearful submissiveness, other responses such as angry outbursts require strategies to help the person avoid getting "drawn in" to the taunts of the avatar.

The second phase (typically sessions 4 to 6) focuses on issues emerging from the assessment and the earlier sessions that seem to lie behind the AVH experience. Often these are experiences earlier in life of abuse or bullying. For example, one man experienced a highly critical and abusive voice of his father, who had died when the patient was still an adolescent. The voice repeatedly told him he was a failure, a weakling and a "waste of space" and compared him unfavorably to other children. As an adult reflecting on his childhood, he saw that his father was not only a very angry man but also one who had no friends and had fallen out with all his own family. If anything it was he who was the "failure." In the second phase of therapy, the client was encouraged to confront the avatar with these insights and the avatar (speaking as his deceased father) conceded that he had been a lonely and angry man. He expressed regret that he had taken out his own failure on his son and acknowledged the harm this may have done. He also expressed surprise and some pride when the person recounted recent successes and the fact that he continued to keep in touch with friends he had made before his illness.

Other strategies include techniques to bolster self-esteem, involving, say, the person reading a prepared list of their strengths and positive attributes, sometimes reported from an informant (e.g., friend or relative), and the avatar responding appropriately to hearing this alternative view of the individual. The avatar can also offer practical problem-solving advice by asking the person to try out alternative actions and or challenging avoidant behavior. For example, an avatar of a female voice challenged a man to the effect that he ought to go out and meet a "real" woman rather than spend so much time talking to her. He took the advice and in the final session reported that he had been out on a "date" with one of his sister's friends and had asked her out again. Both he and the avatar expressed delight at this success and agreed he really did not need the voice anymore.

The final session of AVATAR therapy often involves discussion around hopes and plans for the future. The avatar by this stage is encouraging and supportive, particularly of any steps to deal with the consequences of illness, as for example undertaking everyday tasks (shopping, socializing, returning to work). In some instances, when the goal of the therapy is to reduce engagement with the voice, the person and the avatar will say goodbye to each other.

AVATAR RESEARCH FINDINGS

MAIN RESULTS OF CONTROLLED STUDIES

AVATAR therapy has now been investigated in one pilot study and in a large randomized controlled trial. In the first,[16] 26 participants were randomized to therapy ($n = 14$) or a waiting list control group ($n = 12$). At the end of therapy, those in the waiting list were offered therapy. While there were no significant changes in outcome scores for the waiting list control participants, the AVATAR therapy group reported an average reduction of 8.7 points ($p = 0.003$) in the total score of the PSYRATS-AH rating scale for auditory hallucinations with three participants reporting a complete cessation of voices. Participants in the therapy arm also reported an average 5.9 point ($p = 0.004$) reduction in scores on the omnipotence and malevolence subscales of the revised Beliefs About Voices Questionnaire (BAVQ-R[38]). These results were essentially confirmed in an analysis comparing all those who received avatar at any point with those who did not. Although providing impressive results, this early pilot study also had a number of limitations. The therapy was delivered by one highly skilled therapist, the masking of the assessor was not reported, and the waiting list control condition does not take account of therapist time and attention.

The second, recently completed clinical trial takes the work one step further. A total of 150 participants, all of whom suffered from long-standing psychosis and had heard distressing voices for at least 12 months, were randomized 1:1 to either AVATAR therapy or a supportive counseling intervention. The primary outcome was the total score on the PSYRATS-AH scale at 12 weeks.[39] A detailed therapy manual was developed and a random selection of sessions rated for fidelity and competence by Julian Leff. Fidelity and competence was high and did not differ between therapists: overall adherence to the manual M = 18.9 (SD 2.3) out of a maximum of 21 (range 15–21) and average skill rating of 28.2 (SD 1.7) out of a maximum of 30 (range 25–30). The supportive counseling control condition was also manualized, and adherence to this manual was also independently evaluated by a senior counseling psychologist (M = 15.2; SD = 1.2, maximum 16). It focused predominantly on an individual's unique, personal potential to explore creativity, growth and psychological understanding by creating a nonjudgmental, nondirective, supportive, and safe environment, while avoiding the interpersonal treatment targets of AVATAR therapy (i.e., shift in power/control in the relationship with the voice[s]). It was delivered over the same number and duration of sessions as AVATAR therapy. At the end of each session participants recorded a weekly positive message onto an MP3 player to listen to during the week (thus matching the use of MP3 recordings in the AVATAR therapy arm).

One of the early concerns of the research team had been the high rate of dropout from therapy in the pilot study, and there is no doubt that the creation of the avatar and the initial sessions are challenging for both participant and therapist. In the current study however, therapy was completed by more than two-thirds of participants in both arms with rather higher dropout among those in the supportive therapy control arm (AVATAR therapy = 53 completers, 70% of the inclusions; supportive counseling = 50 completers, 64% of the inclusions). The reasons given were comparable in both and concerned logistical issues, problems with physical health, and statements that the approach was not what they wanted or was felt not relevant to their condition. Importantly, no dropout was thought to be the result of adverse effects of therapy, and across both groups, no adverse events occurred during or in the 6-month follow-up that were attributed to therapy.

In an intention to treat analysis, AVATAR therapy was significantly superior to supportive counseling at 12 weeks in terms of the total PSYRATS-AH score (estimated mean difference −3.82, SE 1.47, 95% CI −6.70 to −0.94; $p = .01$) and to subscores of frequency of voices (estimated mean difference −1.22. SE 0.38, 95% CI −1.97 to −0.84; $p < .001$) and associated distress (estimated mean difference −2.34 SE 0.98, 95%CI −4.26 to −.42; $p = .02$). There were also significant improvements in measures of the omnipotence of voices (estimated mean difference −2·07 SE 0.99, 95% CI −4.01 to −0.12; $p = .04$) their acceptance (estimated mean difference 3.0 SE 1.78, 95%CI 0.30 to 7.29; $p = .03$) measured by the Beliefs about Voices Questionnaire (BAVQ-R[38]) and the extent to which the participant had a sense of control over the experience, measured by the subscales "acceptance-based attitudes of the voices" and "commitment to effective action in response to the voice" of the Voice Acceptance and Action Scale (VAAS[40]). The reduction in total PSYRATS-AH at 12 weeks has a between group effect size (Cohen's *d*) of 0.8 that is both larger than anticipated and indeed larger than the mean effect size reported for voices ($d = 0.46$) in the most relevant recent meta-analysis of cognitive-behavioral therapy for psychosis.[41] At the 6-month follow-up, the numerical difference between the two therapy arms was maintained but the supportive counseling group continued to improve such that there was no longer a significant difference between the two arms at this point. No significant differences between the two groups were observed for any of the other secondary outcomes at either 12 or 24 weeks.

The lack of a treatment as usual control condition complicates the interpretation of the lack of a statistically significant difference between the two arms at 24 weeks. It seems unlikely that this reflects a regression to the mean in both groups, as participants were selected for persistent symptoms and were not recruited in a crisis so the baseline state would be expected to be relatively stable. Also, the early improvement in the avatar arm is maintained so the puzzle centers on what is happening in the supportive counseling. It is probable that supportive counseling is also beneficial for voices, with most studies finding small positive effects for supportive counseling and befriending in psychosis in contrast to smaller to zero effects of treatment as usual. Furthermore, supportive counseling in this trial went beyond simple attentional control and addressed practical concerns about living with psychosis, finding ways to improve current quality of life, coming to terms with past trauma, and identifying personal resources and qualities as well as providing audiotaped recordings of the most relevant 10 minutes of each session and encouraging participants to reflect on these as an attempt to control for the "homework" of AVATAR therapy.

WITHIN-THERAPY CHANGES

The changing nature of the avatar across the six sessions was reflected in how participants perceived the interaction. In a subsample of people included in the AVATAR therapy arm ($n = 20$ of 75), most reported that the avatar was a good representation of their voices: goodness of the match for the image (ranging from 0 to 100) was M = 75.95 (SD = 10.77) and for the voice was M = 78.50 (SD = 10.71). Participants also reported high levels of presence, understood as the degree of verisimilitude of the experience of the participant's sense of "hearing the voice" and the participant's perception of the avatar as a "voice talking to me," assessed by the adapted version of the Sense of Presence Questionnaire (SUS[42]) (M = 10.53 [SD = 3.08], ranging from 1 to 15) and this was maintained across all six sessions: $M_{s1} = 11.10$ SD = 2.24; $M_{s6} = 9.55$ SD = 3.61 $t(19) = 1.74$; $p > .05$. Reports of in-session anxiety and perceived hostility from the avatar also decreased in line with the changes to the avatar across the different sessions: anxiety levels when facing the avatar: $M_{s1} = 3.45$ SD = 1.27; $M_{s6} = 2.10$ SD = 1.25 $t(19) = 3.63$; $p = .002$, and perceived levels of hostility: $M_{s1} = 4.45$ SD = 0.605; $M_{s6} = 1.65$ SD = 0.93 $t(19) = 10.87$; $p = .00$ (see Figures 63.2 and 63.3).

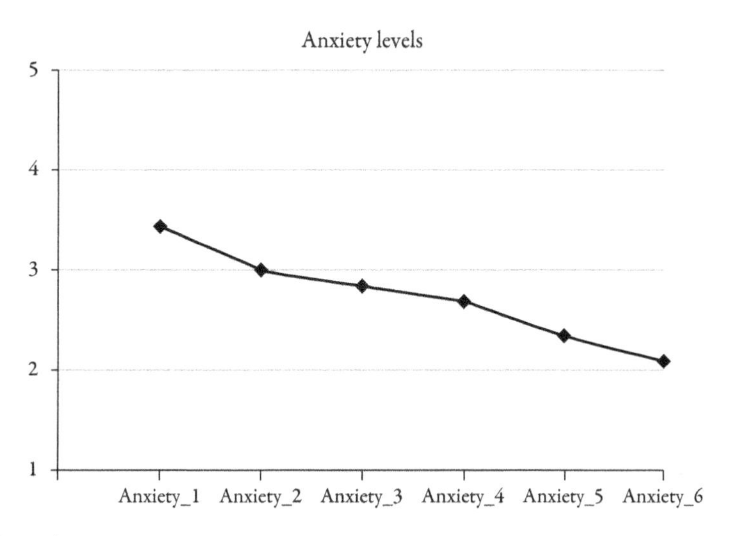

Figure 63.2 Levels of anxiety when facing the avatar.

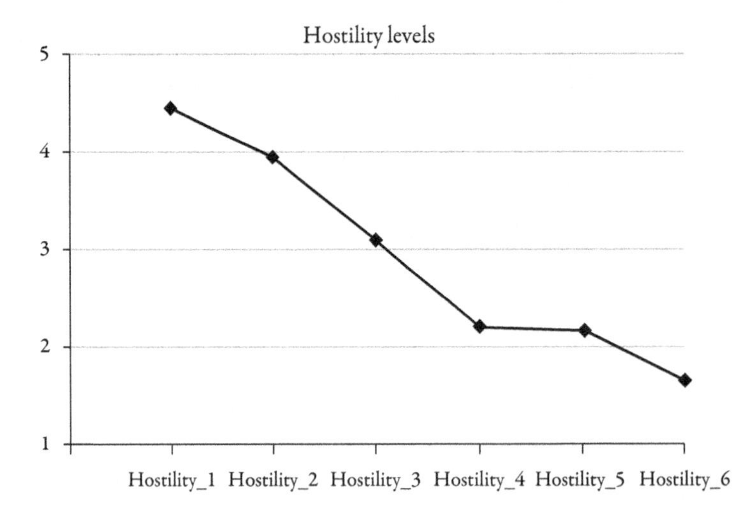

Figure 63.3 Levels of perceived hostility from the avatar.

These preliminary results from in-session measures support the hypothesis that AVATAR therapy could work as an exposure intervention (i.e., graded exposure to the feared stimuli: the voice): the continuous presentation of the experience of seeing and hearing the voice over therapy sessions, along with the modification of the relationship with the voice, may be contributing to the reduction of associated distress and to the disconfirmation of maladaptive beliefs about the voice. However, this hypothesis needs to be further evaluated in relation to final therapy outcomes and reduction of distress associated to the "real" voices.

POTENTIAL MECHANISMS OF ACTION

Analyses of moderators and mediators of therapy are underway, so what follows is preliminary and theoretical. One of the first surprises was that the benefit of therapy was present regardless of the number of voices the person reported, the derogatory nature of the content, or even the extent to which the participant believed the therapy would be helpful. For mediation, we hypothesized that addressing omnipotence would be a key step, as indeed it is, but other aspects, such as the extent to which the individual can get on with their life despite voices are also important.

The exposure and desensitization aspect of early sessions may well play a very important therapeutic role both as a direct consequence of exposure and also by reducing the cognitive avoidance of the fearful information (i.e., the voice and its content). The mechanism of this may be anxiety related: the therapy may be reducing cognitive avoidance of fear-relevant information.[43] Relistening to MP3 recordings of each dialogue between sessions may facilitate this exposure process.

Many participants had initial difficulty looking at and challenging the avatar and displayed a range of safety behaviors including wearing a talisman or special clothing to protect against evil. Knowing that they had created the avatar and that the interaction was under the control of the therapist supported the person take risks in standing up to the avatar in a way that they found impossible with their voices. In several instances, participants experienced their voices speaking over the avatar and threatening dire consequences if the participant continued with therapy but also experienced a sense of having won a victory when the fearful threats did not materialize. The role of the therapist in this phase of therapy was to enact the avatar so that it was a realistic representation of the person's voices but not so hostile that it was overwhelming. This was facilitated by a video link that allowed the therapist to monitor the participant's nonverbal behavior, intervening with advice to avoid safety behaviors or managing the anxiety through different techniques, such as grounding or practising controlled breathing. The therapist was also careful to manage each of these early sessions to ensure that each ended on a "win" for the participant, the avatar conceding power and recognizing a change in the person's ability to fight back.

Beyond this relatively straightforward element of exposure, anxiety management and assertiveness training in the early phase, therapeutic strategies were much more variable. For example, where the therapist thought the voices reflected low self-esteem stemming from earlier trauma, the target might be to help the participant see a link between what they thought about themselves and what the voices typically said and redress the balance by sharing the participant's good qualities with the avatar that could express surprise and appreciation of understanding a positive aspect of the individual. Other scenarios included voices that seemed to reflect earlier experiences of being unfairly treated or having been made to feel socially inferior (typically on grounds of race, education, and sexuality) and here the task was harder and involved challenging the avatar as though it were or represented the source of past injustice. While the original insult might not be addressed, voices echoing the experience could be challenged and brought to apologize for their "behavior." In some cases it was possible to work toward a more explicit understanding of the voices representing things the person thought about themselves. Finally, while the target of many participants was to get rid of the voices, some endpoints were closer to an acceptance and greater tolerance of their presence. A devil is a devil and so unlikely ever to change to be a source of kindness and support but could change in meaning for the individual with a shift from the deeply personal to a background chatter of less personally directed insults, innuendo, and expletives.

The avatar's change and transition of the dialogue was always based on the initial formulation of the participant's presentation and voice function, and discussed throughout the sessions in clinical supervision with the other expert therapists involved in the clinical trial. This is a crucial aspect of AVATAR therapy as we consider that unrealistic or too early changes from the avatar can result in an unsuccessful therapy outcome. Vital emotional meanings of the relationship with the voice, which can only be observed when this relationship between client and voice is enacted in real time, are a key target of the AVATAR therapy that can potentially result in powerful shifts in meaning (e.g., control back to the person) and distress associated with voices.

FUTURE DIRECTIONS OF AVATAR THERAPY

AVATAR therapy is in its infancy. The two clinical trials to date have demonstrated short-term effectiveness but rather less clear evidence for benefit in the longer term. The recent trial showed that even though the initial benefit of avatar endured to the 6-month follow-up, there is no statistically significant difference between AVATAR therapy and supportive counseling on any measure at this point because the supportive counseling group had begun to catch up with AVATAR. It may be that this is just the result of the somewhat greater attrition of follow-up assessments from the supportive counseling arm or that the delayed improvement of supportive counseling reflects better continuing use of the MP3 player recording of therapy. Indeed even if not down to such prosaic explanations, the fact that supportive counseling is effective is not that surprising as there are many studies showing a weak positive effect.[44] Whether comparisons of AVATAR and supportive counseling with treatment as usual would favor AVATAR remains an empirical question that will need to be addressed in later studies. For these, one clear methodological challenge is the quality of the measures used to assess auditory hallucinations. The primary outcome in both studies to date was the PSYRATS-AH, which enquires about voice hearing in the past week—a very short window on problems that clients reported having endured for years. It could be that future studies should supplement cross-sectional measures with momentary assessments of voice hearing and quality (for example, one of the systems that use mobile telephones to record multiple daily snapshots of experiences across a week or more).

Of considerable importance for the wider clinical implementation of this therapy is the recognition that it is focused on just one aspect of the psychotic experience. For many people suffering with psychoses, voice hearing is just one manifestation of a wider experience of persecution. As one of our participants pointed out, just because they no longer heard the voice did not mean that the entity had left the scene. Indeed it was perhaps ever more concerning that there was a silent presence lurking in the sideline, waiting for its time to strike again. Perhaps this could have been addressed by more or longer sessions, or by providing booster "top ups." But it could also be that AVATAR therapy is best seen as just one component of a wider approach, perhaps something to be embedded within more comprehensive cognitive-behavioral therapy for psychosis or as one of a series of focused interventions tackling, for example, ideas of persecution as well as cognitive impairments with or without the use of digital technology.

Another concern for the future is whether the therapy would be enhanced by more immersive VR, such as usage of head-mounted display. Although we do not hypothesize that this would impact therapy outcomes, using more advanced technology could allow the therapist to introduce new techniques or modalities of intervention: e.g., seeing the avatar but also the therapist in the environment or the possibility of changing certain visual aspects of the avatar, such as height or size, to increase sense of control over the voice. In the only such approach that we are aware of to date, a pilot study ($n = 19$) of an immersive VR therapy reported substantial reductions in AVH severity and depressive symptoms and improvements in quality of life that persisted to a 3-month follow-up.[45]

One of the most important future challenges of AVATAR therapy is its establishment in public and private mental health services. Up to this point, the therapy has been delivered only by skilled therapists with substantial expertise in the psychological treatment of psychosis, which limits the implementation of the therapy to other centers or to be delivered by wider mental health workforce. Future therapists delivering AVATAR therapy will require specific training and continuous clinical supervision, as well as the acquisition of the software specifically designed to carry out this therapy.

REFERENCES

1. Rus-Calafell M, Garety P, Sason E, Craig TJK, Valmaggia LR. Virtual reality in the assessment and treatment of psychosis: a systematic review of its utility, acceptability and effectiveness. *Psychol Med.* 2017:1–30.
2. Josman N, Schenirderman AE, Klinger E, Shevil E. Using virtual reality to evaluate executive functioning among persons with schizophrenia: a validity study. *Schizophr Res.* 2009;115(2– 3):270–277.
3. Greenwood KE, Morris R, Smith V, Jones A, Pearman D, Wykes T. Virtual shopping: a viable alternative to direct assessment of real life function? *Schizophr Res.* 2016;172:206–210.
4. Kim K, Kim J-J, Kim J, et al. Characteristics of social perception assessed in schizophrenia using virtual reality. *Cyberpsychol Behav.* 2007;10(2):215–219.
5. Park SH, Ku J, Kim J, et al. Increased personal space of patients with schizophrenia in a virtual social environment. *Psychiatry Research.* 2009;169(3):197–202.
6. Freeman D, Evans N, Lister R, Antley A, Dunn G, Slater M. Height, social comparison, and paranoia: an immersive virtual reality experimental study. *Psychiatry Research.* 2014;218(3):348–352.
7. Freeman D, Pugh K, Vorontsova N, Antley A, Slater M. Testing the continuum of delusional beliefs: an experimental study using virtual reality. *J Abnorm Psychol.* 2010;119:83–92.
8. Stinson K, Valmaggia LR, Antley A, Slater M, Freeman D. Cognitive triggers of auditory hallucinations: an experimental investigation. *J Behav Ther Exp Psychiatry.* 2010;41(3):179–184.
9. Chan CLF, Ngai EKY, Leung PKH, Wong S. Effect of the adapted virtual reality cognitive training program among Chinese older adults with chronic schizophrenia: a pilot study. *International Journal of Geriatric Psychiatry.* 2010;25(6):643–649.

10. Tsang MMY, Man DWK. A virtual reality-based vocational training system (VRVTS) for people with schizophrenia in vocational rehabilitation. *Schizophr Res.* 2013;144(1–3):51–62.
11. Smith MJ, Fleming MF, Wright MA, et al. Virtual reality job interview training and 6-month employment outcomes for individuals with schizophrenia seeking employment. *Schizophrenia Research.* 2015;166(1–3):86–91.
12. Park K, Ku J, Choi S, et al. A virtual reality application in role-plays of social skills training for schizophrenia: a randomized, controlled trial. *Psychiatry Research.* 2011;189(2):166–172.
13. Rus-Calafell M, Gutierrez-Maldonado J, Ribas-Sabate J. A brief cognitive-behavioural social skills training for stabilised outpatients with schizophrenia: a preliminary study. *Schizophr Res.* 2014;143(2):327–336.
14. Freeman D, Bradley J, Antley A, et al. Virtual reality in the treatment of persecutory delusions: randomised controlled experimental study testing how to reduce delusional conviction. *British Journal of Psychiatry.* 2016;209(1):62–67.
15. Gega L, White R, Clarke T, Turner R, Fowler D. Virtual environments using video capture for social phobia with psychosis. *Cyberpsychology, Behavior, and Social Networking.* 2013;16(6):473–479.
16. Leff J. Computer-assisted therapy for medication-resistant auditory hallucinations (vol 202, pg 428, 2013). *British Journal of Psychiatry.* 2013;203(4):428–33.
17. Leff J, Williams G, Huckvale M, Arbuthnot M, Leff AP. Avatar therapy for persecutory auditory hallucinations: what is it and how does it work? *Psychosis.* 2014;6(2):166–176.
18. Mawson A, Cohen K, Berry K. Reviewing evidence for the cognitive model of auditory hallucinations: the relationship between cognitive voice appraisals and distress during psychosis. *Clinical Psychology Review.* 2010;30(2):248–258.
19. McCarthy-Jones S, Resnick PJ. Listening to voice: the use of phenomenology to differentiate malingered from genuine auditory verbal hallucinations. *International Journal of Law and Psychiatry.* 2014;37:183–189.
20. Waters F, Allen P, Aleman A, et al. Auditory hallucinations in schizophrenia and nonschizophrenia populations: a review and integrated model of cognitive mechanisms. *Schizophrenia Bulletin.* 2012;38(4):683–693.
21. Beavan V. Towards a definition of "hearing voices": a phenomenological approach. *Psychosis.* 2011;3(1):63–73.
22. Morrison A, Beck AT, Glentworth D, Dunn H, Reid GS, Larkin W. Imagery and psychotic symptoms: a preliminary investigation. *Behaviour Research and Therapy.* 2002;40:1063–1072.
23. Morrison A. The use of imagery in cognitive therapy for psychosis: a case example. *Memory.* 2010;12:517–524.
24. Woods A, Jones N, Alderson-Day B, Callard F, Fernyhough C. Experiences of hearing voices: analysis of a novel phenomenological survey. *Lancet Psychiatry.* 2016;2:323–331.
25. Nayani TH, David AS. The auditory hallucination: a phenomenological survey. *Psychol Med.* 1996;26(1):177–189.
26. Beck AT, Rector NA. A cognitive model of hallucinations. *Cognitive Therapy and Research.* 2003;27:19–52.
27. Bell V. A community of one: social cognition and auditory verbal hallucinations. *Plos One Biology.* 2013;11(12).
28. Wilkinson S, Bell V. The representation of agents in auditory verbal hallucinations. *Mind Lang.* 2016;31(1):104–126.
29. Birchwood M, Gilbert P, Gilbert J, et al. Interpersonal and role-related schema influence the relationship with the dominant "voice" in schizophrenia: a comparison of three models. *Psychol Med.* 2004;34:1571–1580.
30. Birchwood M, Meaden A, Trower P, Gilbert P, Plaistow J. The power and omnipotence of voices: subordination and entrapment by voices and significant others. *Psychol Med.* 2000;30(2):337–344.
31. Paulik G. The role of social schema in the experience of auditory hallucinations: a systematic review and a proposal for the inclusion of social schema in a cognitive behavioural model of voice hearing. *Clin Psychol Psychot.* 2012;19(6):459–472.
32. Thomas N, Hayward M, Peters E, et al. Psychological therapies for auditory hallucinations (voices): current status and key directions for future research. *Schizophrenia Bulletin.* 2014;40:S202–S212.
33. Corstens D, Longden E, May R. Talking with voices: exploring what is expressed by the voices people hear. *Psychosis.* 2012;4(2):95–104.
34. Hayward M, Strauss C, Bogen-Johnston L. Relating therapy for voices (the R2V study): study protocol for a pilot randomized controlled trial. *Trials.* 2014;15.
35. Hayward M, Jones AM, Bogen-Johnston L, Thomas N, Strauss C. Relating therapy for distressing auditory hallucinations: a pilot randomized controlled trial. *Schizophr Res.* 2017;183:137–142.
36. Hayward M, Overton J, Dorey T, Denney J. Relating therapy for people who hear voices: a case series. *Clin Psychol Psychot.* 2009;16(3):216–227.
37. Birtchnell J. *How humans relate: A new interpersonal theory.* Hove, UK: Psychology Press; 1996.
38. Chadwick P, Lees S, Birchwood M. The revised Beliefs About Voices Questionnaire (BAVQ-R). *Brit J Psychiat.* 2000;177:229–232.
39. Craig TK, Rus-Calafell M, Ward T, et al. AVATAR therapy for auditory verbal hallucinations in people with psychosis: a single-blind, randomised controlled trial. *Lancet Psychiat.* 2018;5(1):31–40.
40. Shawyer F, Ratcliff K, Mackinnon A, Farhall J, Hayes SC, Copolov D. The voices acceptance and action scale (VAAS): pilot data. *J Clin Psychol.* 2007;63(6):593–606.
41. van der Gaag M, Valmaggia LR, Smit F. The effects of individually tailored formulation-based cognitive behavioural therapy in auditory hallucinations and delusions: a meta-analysis. *Schizophr Res.* 2014;156(1):30–37.
42. Slater M, Usoh M, Steed A. Depth of presence in virtual environments. *Presence: Teleoperators and Virtual Environments.* 1994;3:130–144.
43. Foa E, Kozak MJ. Emotional processing of fear: exposure to corrective information. *Psychological Bulleting.* 1986;99:20–35.
44. Lewis SW, Tarrier N, Haddock G, et al. The Socrates trial: a multicentre, randomised, controlled trial of cognitive-behaviour therapy in early schizophrenia. *Schizophr Res.* 2000;41(1):s91–s97.
45. Percie du Sert O, Potvin S. Lipp O, et al. Virtual reality therapy for refractory auditory hallucinations in schizophrenia: a pilot clinical trial. *Schizophrenia Research* 2018;197:176–181.

64

HEALTH IN A CONNECTED WORLD

Philippe Delespaul and Catherine van Zelst

HEALTH: A CHANGING CONCEPT

HEALTH

The WHO in 1948 defined health as "a state of complete physical, mental and social well-being and not merely the absence of disease or infirmity" (WHO, 2006). Huber and colleagues (2011) argued that the concept overemphasizes the abolition of illness (primarily infectious diseases, e.g., malaria) and does not recognize that many people feel healthy despite having an illness or handicap. In search for a better match with experienced health, Huber et al. explored various definitions and proposed a reformulation of health as "positive health": "the ability to adapt and to self-manage" with respect to the domains of mental, physical, and social health challenges.

The theory of salutogenesis, proposed by the sociologist Antonovsky (1987), offers a generic understanding of the development of integrated well-being or health. The central concept is that individuals pursue a "sense of coherence" (SOC) to explore pathways that lead to successful coping and health. Compared to traditional pathogenic oriented theories, salutogenesis proposes a broader perspective on health. For Antonovsky, health is not dichotomous (present or absent) but is best conceptualized as a continuum. Individuals evolve over this dimension and strive toward the health end and thus to increase their SOC and promote coping. The focus shifts to the personal story of the individual rather than to the diagnosis. In this frame of reference the person is understood as an open system in active interaction with the environment (both external and internal conditions) (Langeland, Wahl, Kristoffersen, & Hanestad, 2007). In the pursuit of health individuals are "actors" who strive to well-being, not passive recipients of care who submit to interventions of professionals who "fix" the illness. In "salutogenesis" and "positive health" adaptive strategies within meaningful environments become increasingly important.

NEW MENTAL HEALTH MOVEMENT

The European New Mental Health Movement (NMHM), which originated in the Netherlands, reflects the changing cultural environment of mental healthcare (Delespaul, Milo, Schalken, Boevink, & Van Os, 2016). The movement challenges the underlying assumptions of the organization of services and how professionals provide care for people with mental disorders. The initiative reframes the concept of "mental suffering" from the personal perspectives of actors (suffering patient or relatives) and other stakeholders (neighborhoods, general practitioners, and society at large). It normalizes mental health vulnerabilities as human variation, and alters fundamentally how people relate to individuals with mental health problems. Consequently, managing this human variation becomes a public health challenge. It results in a changed role for society in general and healthcare professionals in particular.

The NMHM believes that often mental health vulnerabilities cannot be cured, but successful adaptive strategies can develop in phases over time. Relevant information to plan interventions can be identified using a few simple questions (van Os, 2014):

1. What happened to you?
2. What are your vulnerabilities and strengths?
3. What are your goals in life?
4. What do you need?

These questions are transparent and have meaning for the individual and their proxies. They contrast with the professional expert language of categorical diagnoses that is often mystifying. They allow a more balanced communication and set the stage for a collaborative relation that facilitates engagement in care.

THE ACTOR IN A MEANINGFUL CONTEXT

GOALS OF THE SERVICE USER

Mental health professionals often formulate goals about what they think a client "should be doing" (e.g., stay out of the hospital, take medications, improve daily living skills) but that may not be relevant for the person. Goals from the mental health professional's point of reference often lack contextual embeddedness and have poor relevance for what the person actually aspires to (Rapp & Goscha, 2006).

Recovery-oriented and person-centered care is, at its core, about getting past the "us/them" dynamic to truly partner with recovering people in their efforts to attain their personally defined and valued goals (Tondora, Miller, Slade, & Davidson, 2014).

HEALTH IN AN ERA OF DIGITAL TRANSFORMATION

eHealth refers to the use of information and communication technologies to improve mental healthcare. Interestingly, this definition is not restricted to the means (technology), but focuses on the goal: improving health of individuals and the way we provide health care (Delespaul et al., 2016).

Virtual reality (VR) is popular within and outside healthcare. In contrast to classic eHealth, VR dramatically changes the way we learn or alter emotions. It is an important first step in the direction of in situ therapy and allows for contextualized learning. In VR, environments are safe, can be modified to the subject's personalized vulnerabilities, and can be presented repeatedly to consolidate learning. Most importantly, VR changes the focus of effective reactivity from cognition (as in cognitive-behavioral therapy [CBT]) to the whole body and the subject's inner environment including expressed emotions and proprioceptive reactions. Learning of mental resilience in the combination of a varying external and internal context, dramatically improves.

mHealth (using mobile technology in healthcare) is a subset of eHealth. The enthusiasm for and the promises of mobile phones and other handheld devices for healthcare initiatives resulted in the emergence of this novel interdisciplinary field (Ben-Zeev et al., 2015). mHealth adopts mobile devices, such as personal digital assistants or smartphones, for in situ assessment and the delivery of treatments in the daily life of patients, while they are engaging in their daily life activities (Myin-Germeys, Klippel, Steinhart, & Reininghaus, 2016).

Often, interventions are multimodal and combine different strategies. *Blended care* points to the combination of face-to-face and eHealth treatment. The wide availability of smartphones and inexpensive applications ("apps") offers unprecedented potential to change our traditional approach to improve mental health in different domains. The Apple App Store and Google Play Store offer a wide range of tools to monitor personal physiological, behavioral, and environmental indicators and to access tailored decision support for collaborative intervention choices. This offers inexpensive solutions, improves dissemination, and crosses the borders of professionally owned technology by making it accessible to everyone. It empowers individuals to facilitate making their own choices for their life. Unfortunately, often technological implementation goes ahead of sound scientific evaluation.

eHealth technology is rapidly evolving, and develops into a challenging and exciting field for innovative assessment and health change solutions (Bartholomew Eldredge et al., 2016). Mobile devices offer a large collection of applications and services that, together, provide valuable resources for both patients and health care providers (Garcia & Repique, 2014).

In a review of the literature, Naslund et al. (2015) study the evidence for using emerging mHealth and eHealth technologies among people with severe mental illness. They grouped interventions into four categories:

1. illness self-management and relapse prevention
2. promoting adherence to medications and/or treatment
3. psychoeducation, supporting recovery, and promoting health and wellness
4. symptom monitoring

The interventions were consistently found to be both feasible and acceptable. Clinical outcomes were variable but offered insight regarding potential effectiveness.

Naslund et al. conclude that mHealth and eHealth interventions are feasible and acceptable among people with SMI, but that it is too early to conclude on effectiveness. Further research is needed to establish effectiveness and cost benefit in this population.

EHEALTH FOR ASSESSMENT

eHealth can potentially transform healthcare delivery for patients with psychosis. It can, for instance, enable early symptom recognition and assessment (Treisman et al., 2016). It can be used to disseminate information (psychoeducation) to patients and family and personalize it to the individual's own profile. eHealth assessments can focus personalized integrating care interventions and facilitate service accessibility.

mHealth assessments can go beyond traditional questionnaires and clinical interviews that focus on symptoms. Smartphones are stuffed with sensors that can be harvested to offer psychiatry a wealth of real-time data regarding patient behavior and physiology (Torous, Staples, & Onnela, 2015). Most of these sensors will not yet generate the relevant clinical indicators. But they may be supplemented with self-reported symptoms, affect variability, and contextual labeling, to enrich personalized assessments and increase the validity. Sensor data, in contrast to self-report, can be collected continually and nonintrusively. They may help clinicians and patients to bridge the gaps that remains even with intensive self-reports.

Virtual reality also has potential in supporting assessment. It was applied to assess paranoia by presenting different situations that match the individual's personal sensitivities. Contrasting neutral social situations with threats enables assessment of unfolding paranoia and to explore the person's individual sensitivities. Virtual reality can safely assess psychotic experiences in psychosis and related diagnoses (Freeman et al., 2017).

Computer-supported assessment can incorporate algorithms that translate state of the art guidelines into customized advice. Such interactive assessment models with custom feedback can facilitate patient engagement by offering real treatment options and explore anticipated outcomes. This results in extensively informed patients, potentially matching the expertise of clinicians. Information technology allows shared decision-making, boosts motivation, and facilitates self-management and rehabilitation (Treisman et al., 2016).

EHEALTH, RECOVERY, AND TREATMENT

Various emerging digital technologies reshape the current healthcare landscape and shift how mental health consumers engage in their own treatment and recovery. eHealth gives us opportunities in the domains of information dispersal; integrating care and extending access; medication adherence, patient engagement, and self-management as well as rehabilitation (Treisman et al., 2016).

Technology can free people from the need for the presence of professionals in their vicinity to "survive" a panic attack, depressive despair, or deep melancholy, or the exacerbation of psychotic symptoms. Professional resources are not available in the daily lives of patients outside of the mental health service. Having alternative prostheses improves autonomy with preservation of safety nets and can facilitate pathways to recovery. The cultural change that normalizes the use of technology as an expansion of personal skills for everyone, makes the adoption of technology for subjects in need more acceptable and reduces stigmatization. Google glass and Apple Augmented Reality (AR) are developed for the entire population, not for vulnerable individuals. But they will offer vulnerable individuals a nonstigmatizing way to develop resilience. These innovative technologies are a tool to enhance one's capacity, whether people have a traditional vulnerability, or not. The broad-scale application, and consequent reduction of costs bringing it into reach of everyone, will revolutionize the lives of people in dependencies due to their vulnerability. While a rollator or a mobility scooter can be the only way to realize mobility for some people, when the technology of self-driving cars will become available for everyone, people in need who have no alternative (blind or with concentration problems) will finally be able to go from A to B with no burden of stigma. Comparably, the increasing convenience of mental healthcare delivery through the use of mHealth technology can be an effective digital strategy to engage mental health service consumers and help them navigate the path of daily life challenges toward continued wellness and recovery (Garcia & Repique, 2014).

EMBEDDED EHEALTH IN DAILY LIFE

With no ambition to be exhaustive, we present some eHealth interventions that focus on the person's daily life and have a good research backing. These examples demonstrate that information technology–related changes in mental health in general will also affect how care is provided to the most ill patients. Two aspects are important:

- for patients suffering from severe mental illness including psychosis, optimal treatment no longer means organizing avoidance and living marginalized lives; optimal treatment in contrast, means that subjects should be coached to engage in exposure to stressful situations and learn to have psychotic experiences without panicking, herewith developing resilience;
- ecological validity is increasingly important. The locus of care shifts to the real ambulant environment, not an office in the city, but the daily lives of people. This requires different assessment strategies and embedded interventions, potentially resulting in better, more generalized well-being.

VIRTUAL REALITY

Difficulties to interact in a meaningful world are at the heart of mental health problems in psychosis and beyond. Recovery is related to thinking, reacting, and behaving differently in these situations. With VR, individuals can engage in simulations of the difficult situations and be coached in the appropriate responses (Freeman et al., 2017).

Freeman et al. (2016) used VR to alter paranoid thoughts in patients suffering from psychosis. They hypothesized that persecutory delusions are the consequence of assumed menaces that are never challenged due to avoidance behaviors. They developed virtual social environments to safely challenge these threats and prevent escape to facilitate new learning. In most VR environments, the subject is an observer (as when watching a 3D movie from a couch) but Freeman used an "immersive" system. The subject physically moves and interacts with objects and virtual people (avatars) in a three-dimensional computer-generated world. The emotions that are generated in VR are similar to those that would occur in the real situation, because they are not generated by cognition alone (as in classic psychotherapy), but also holistically by concurrent behavior and proprioceptive feedback. The impact of graded exposure to real-life situations or VR is comparable for anxiety and expected to be similar in psychosis. The combined VR cognitive therapy was more effective than VR exposure alone.

Virtual reality treatments are conceivable for most aspects of psychotic experiences. Already, interventions for delusions, hallucinations, neurocognition, social cognition, and social skills are under investigation. Virtually reality makes it possible to investigate interactions between individual and social environment in great detail.

The required soft and hardware for VR is quickly becoming less expensive. Early problems such as VR "sickness" due to delays between proprioceptive and visual feedback, are almost gone, thanks to the graphic rendering power of modern (mobile) computers. The current technological packages for VR allow user-level customized set-ups, matching the individual's personal vulnerabilities. Virtual scenarios can be developed in which storylines change in response to the patient's actions, and the level of difficulty is automatically adapted to error rates or arousal (Veling, Moritz, & van der Gaag, 2014). This reduces the risk of escape and ascertains a positive learning environment. One might expect these methods will soon be introduced in nonacademic clinical practices.

TEMSTEM

Temstem (Dutch for "taming voices') was developed as an in situ coping aid for people who hear voices. A student of the technical university of Delft was assigned for one day to live the life of a patient with severe mental illness. He was struck by the disruption and invalidation of the voices that the patient experienced continually and that were related to traumatic moments of familiar people who neglected them or abused them physically or sexually. This left the person withdrawn and marginalized. Temstem is an iPhone and Android-App developed to counteract this detrimental vicious circle. Temstem enables people to keep participating in common daily activities and offers exercises that encourage them to actively develop resilience in their relationship to the voices. The working mechanism is distraction, realized by playing two different mind-absorbing games. The games are language driven and therefore difficult to translate (the original app is in Dutch). In one game (language-tapping) words are displayed and the subjects tap the number of syllabi. Correct answers are counted under time restraints, and subjects are challenged to repeatedly improve their highest score (gamification). Another game (word-link) offers two sets of words, and the user is prompted to form correct combination words. The game has different levels and subjects should find all possible combinations in the shortest possible time. The Temstem working principles are: activation of the language-producing brain regions (and suppress hallucination production); working memory overload, suppressing other memories; and boosting confidence.

SLOMO THERAPY

People with psychosis can worry that others may intentionally cause them harm. Paranoia is one of the most common symptoms in persons with SMI problems and often results in marked distress and disruption. Paranoia is related to the habit to think "fast." Fast thinking can be helpful in some situations but may also contribute to increased stress. SlowMo therapy[1] works by supporting people to notice their upsetting worries and fast thinking habits. The app provides tips to help slow down, scan for new information and explore safer thoughts. SlowMo consists of eight individual face-to-face sessions, assisted by a website with interactive stories and games. People find out how fast thinking habits can contribute to upsetting thoughts. It proposes tips, and subjects can explore what helps for them to slow down their thinking and cope with worries. A mobile app supports the applicability of these strategies in daily life.

PSYMATE

The experience sampling method (ESM, also EMA—ecological momentary assessment) is a structured diary method that uses a random time-sampling strategy to collect mental states in context. The repeated self-observations are conducted and embedded within the natural flow of daily life experiences (Delespaul, 1995). ESM is applied in psychiatry, as well as in patients with somatic illness. The method focuses on symptoms (ill health), as well as on adaptive functioning (well-being), aiming to map (normal) daily life functioning (Verhagen, Hasmi, Drukker, van Os, & Delespaul, 2016). ESM has been used for 30 years in research, and modern IT technology allows it to be implemented in regular clinical practice.

The PsyMate (www.psymate.eu) is an iOS and Android app that implements the PEAS guidelines (Psycho-Ecological-Assessment System) into a development platform. It allows the rapid development of customized daily life assessment strategies and optimized individual ecological treatment. A supplementary reporting module is designed to facilitate the functional analysis of subjects' successful and maladaptive coping. The person's own repertoire is the reference, and this information is provided to incorporate in collaborative decision-making. Daily life assessment strategies can inform patients and clinicians how to improve resilience by maximizing strengths in patients (van Os et al., 2017).

The PsyMate platform has been used to evaluate the daily life impact of virtual reality and Temstem outcome studies among others. It has also been used in collaboration and cocreation with patients to design apps that help develop resilience to stigma experiences, tapering down medication levels, implement acceptance and commitment therapy (ACT), and increase social interactions (Smart4U). Involving service users and other intended end-users in the design is recommended (Hill et al., 2017).

ECOMMUNITIES

One of the principal problems or challenges of social psychiatric interventions is to find substitutes for the relevant communities of our subjects. Traditional communities, such as families, neighborhoods, or school/work environments were exclusively face-to-face and unescapable. Often, these communities were also a primary source of stress. Consequently, they are not the best candidates for support and alternatives are needed that are inexpensive, have a low threshold for access, are continually available, offer more matching options, and, possibly, allow people to explore relations gradually and safely. This is important to overcome past trauma. The Internet might offer that alternative. eCommunities are online platforms where patients, relatives, and professionals can meet without thresholds, to share personal experiences and knowledge and offer help and advice.

The worldwide net has revolutionized knowledge and expertise. Twenty years ago, professionals were the only persons with the knowledge that patients and families require. Now patients and relatives have access to research papers and can be aware of new interventions or a publication that describes undocumented side effects. Progressively, the information available on the Internet is improving in quality. eCommunities are self-correcting, and this results in less biased and

1 https://www.rca.ac.uk/research-innovation/helen-hamlyn-centre/research-projects/2015-projects/slowmo/

often even unbiased information. Modern eCommunities are less silo-like than before, and involve not only patients or families but also clinicians, researchers, and experts. It brings, for instance, a second opinion, potentially from a national expert, within a one-click reach.

This challenges the expertise of the local clinician and fundamentally changes the therapeutic relationship from an expert consultation to shared decision-making. Ideally, all partners in this decision-making process—patients, families, and professionals—can contribute valuable information for assessment and intervention strategies. There is no exclusive expert reference anymore—everyone is expert in its personal right. Moreover, roles diffuse. A professional is also a parent. He or she might have been depressed or even psychotic for a period in their life. They can share this vulnerability without fear that it will hamper their ability to help, since expert knowledge is also present in other persons in the reciprocal communication. Disclosure improves authenticity and empathy and can strengthen the nature of the contribution. Future care will rely on multi-expert meetings of patients, families, and caretakers as well as professionals. These relational changes are not restricted to eCommunities but will spill over to face-to-face contacts.

eCommunities can also be accessed by individuals who do not want to disclose their identity. It is often stated that people with (first-onset) psychosis lack insight. But most psychotic patients we encountered were well aware of the fact that their experiences were different from the experiences of their friends. But they were not willing to accept the solutions proposed by the mental health professionals and consequently were reluctant to engage in mental health services. We can be aware that we drink too much once in a while. But this does not mean that we will agree when someone confronts you with this. We might consider the fact that we should reduce the alcohol intake. But only as a personal choice, a choice that is easily postponed when someone, even with the best intentions, forces us to engage in behavioral change. An anonymous exploration of choices made by peers, as is possible on the Internet, can make us more open for alternative strategies toward change.

What will people search for in eCommunities? The access through the Google search engine might be triggered by diagnostic tags such as schizophrenia, autism, or bipolar disorders. But more often people will access eCommunities using transdiagnostic concepts: "mental health problems," "stress," "hearing voices," "not sleeping well," "loneliness," or "feeling exhausted." Because the focus will shift from the ill part of the individual (diagnosis and symptoms) to the well-being part (adaptational strategies and resilience) eCommunities should develop a language that facilitates the sharing and integration of distributed expertise (of persons with needs, peers, family, and acquaintances, up to the level of generalist and specialist professionals). In these communities, shared knowledge about effects and side effects of psychopharmacological compounds is not more important than insight in how to solve loneliness or reconnect to family after a traumatizing and disruptive episode. eCommunities should therefore not be limited to specific diagnostic domains but focus on the challenges of daily life. This normalization lowers access thresholds and can reduce stigma and burden.

Proud2Bme[2] is a (Dutch) website and eCommunity for young women. It evolves around topics of eating disorders but basically explores the challenges within daily life of a broader population. It covers all life aspects addressed by girls and young woman such as clothing, beauty, dieting, not from an illness perspective, but from a well-being perspective (therefore: "proud to be me"). It is a good example of how to share knowledge and facilitate change without overemphasizing the problems or illness aspects (in this case eating disorders). Psychosenet.nl pursues the same aims in psychosis. Both sites have millions of visitors each year and were created by mental health users, together with mental health professionals. It is expected that eCommunities in the years to come will form the basis of a public mental health movement, guiding people not only to contact and recovery solutions but also to an open market of eHealth and mHealth solutions, rated by consumers.

Bit-to-bit communities are, of course, not the same as face-to-face relations. They can supplement the absence of face-to-face contact, provide safe training environments, and a real relief of stress, or the sense of belonging. But face-to-face contact should not be neglected and eCommunities should also facilitate access. This, of course, requires a "coming-out" from the safe anonymity that can be available on the net. The "outside" world should be open (not restricted to options available within mental health system) because personal interests and hobbies are shared between citizens with and without mental health vulnerabilities. Using these intrinsic affinities is more motivating. People should be able to connect to all people in the neighborhood to start a walking or cooking club, whatever best suits them. The personal affinity, not the diagnosis, should be leading.

DISCUSSION

For years, "e"- and mHealth was a promise. And it was surrounded by hype. Mental health would be revolutionized by computing technology. This started more than 30 years ago. Soon, we came to realize that the reality was sobering. In the early days, the introduction of computers in psychiatry was primarily related to improve diagnostic logistics. The coding of the MMPI scales and the computation of reference table scores were facilitated using the new technology. However, the same classic diagnostic instruments were used and assessment did not fundamentally change clinical practice. The promises remained unfulfilled.

It would be unfair to argue that computers did not alter diagnostic practice. For example, neuropsychological assessments greatly improved using computer-supported standardization, controlled stimulus presentation, and time-controlled outcome measurements. It allowed reliable assessments of newly discovered phenotypes such as reaction

2 www.proud2Bme.nl

time, concentration, visuospatial orientation, and memory. Also, nonintrusive brain-imaging and DNA-mapping techniques supported by smart algorithms improved neuroscience. Centralized large databases allowed us to gain insight in patient illness histories. All these examples are a result of IT innovation in mental healthcare. However, they did not improve clinical practice, nor did they alleviate overall mental suffering.

Even now, more than 30 years after the revolution started, the hype remains. IT technology moved into the therapy room. But recent analyses of the state of the art of eHealth implementation are still disappointing. Not much has changed: we still replicate the same strategy. Traditional diagnostic tools and interventions are translated to include information technology. They are presented as innovations, but mental healthcare remains unchanged. We did not realize that implementing computers is not the ultimate goal. The goal should be to change mental health practice from a conceptual point of view and use computer technology to make it possible to implement and disseminate innovative assessments, interventions, and the way we provide care. To do this, computers should not be the hype or the aim to pursue. They should only be the tools to realize goals that could not be realized without information technology. New information technology can collect terabytes of data, but as a wise man said, "Not all that can be measured, is relevant; not all that's relevant, can be measured."[3] Good use of computers is not different from the use of pencils. We are not searching for applications for pencils but use them to realize goals. In mental health, however, policy makers often set goals to implement more eHealth and in doing so, once again, focus on the tool. The focus should shift more to the relevant clinical and personal challenges in the ecological reality of daily life that are related to the adaptation strategies that guide the process of personal recovery.

We are far away from devices that can read the person's mind and detect psychosis, mania, or depression remotely. Sensors in our phone could possibly be harvested to generate this data in the future. But more importantly, what would be the point? Do we really want an assertive outreach team that remotely monitors 200 psychotic patients living at home or an early detection team that harvests big data and detects all youngsters in a population who are at risk of developing a psychosis? In our modern society people differ in their sensitivity regarding privacy intrusions. Respecting these differences is important, and mental health services cannot assume that privacy-related sensitivities will resolve over time. We have seen over the years how ethical norms may dramatically change. The way Google and Facebook harvest all kinds of data and sell these for profit would have been unthinkable 10 years ago. There is no escape from this connected world, and most of us have accepted this situation because of the gains that smartphones and IT technology have brought. The real answer to the question whether remote monitoring of at-risk mental patients should be developed, is a practical one. Do we have powerful interventions that could be applied successfully, irrespective of the person's motivation? Can we do something that matters, when we reach out because someone in the control room has detected a red flag? The honest answer is that the most successful interventions in psychiatry are collaborative. Therefore, we have to invest in strategies that motivate and engage patients and their networks to develop resilience. Consequently, these monitoring tools should not be in the hands of controlling clinicians but should be used as prosthetic add-ons that patients and relatives use in collaboration to develop resilience. The professional as well as peers, can be "on call" and represent an important safety net. But casting them in the role of controllers will undermine the therapeutic potential. This insight does not come from technology, but is part of understanding human nature.

Mental health IT innovations will not change culture. But culture will change how we will help vulnerable people in the future. Meanwhile, patients still engage in mHealth supported interventions with varying familiarity with mobile technology. Some users are skilled to use the mobile phone features that are necessary to participate (e.g., touchscreens, mobile applications, text messaging, Wi-Fi access). But often participants are unfamiliar with the specific device/model, operating system, or app. This will probably change in the future. For now, however, user demographic (e.g., age, socioeconomic status, education), illness related (e.g., hearing/vision impairments, dexterity and fine motor abilities, cognitive functioning), and environmental factors (e.g., access to electrical outlets for charging, regional wireless infrastructure) impact on the capacity of users to engage with a particular mHealth resource (Ben-Zeev et al., 2015). This is a challenge in research and even more in clinical practice.

CONCLUSION

We live in a quickly changing world. Modern technology already expands our human possibilities on a daily basis. Thanks to TomTom and Garmin, fewer people get lost because they cannot orient themselves. Anxiety can disappear because people can always reach someone on the cellphone. Good technology can empower people. Empowerment is multidimensional and can be facilitated at the individual, organizational, and community level. Environmental conditions that enable and demonstrate empowerment include collective problem-solving, shared leadership and decision-making, accessible government, media and resources, and collective efficacy. Methods by which community members become empowered include participation in decision-making, enactive mastery experiences and feedback, and modeling (Bartholomew Eldredge et al., 2016). These issues are not different when it comes to mental health in general and psychosis in particular.

Individual and network resilience, even for people with severe mental illness and psychosis, will evolve with the cultural development and broad adaptation of technological tools that facilitate empowerment. This requires personal engagement, shared decision-making, a sense of ownership,

3 Attributed to Albert Einstein (http://quoteinvestigator.com/2010/05/26/everything-counts-einstein/).

and belonging. These concepts are crucial, and technology is only relevant when it facilitates this process. Through a collaborative arrangement, all parties are likely to invest in making sustainable and scalable products (Hill et al., 2017).

Mental well-being is contextualized. You do not feel good or bad without a context. Social network and social support theories address the environmental condition of communities that provide emotional, instrumental, informational, and appraisal support as appropriate for the context. Change agents include knowledge, beliefs and attitudes, self-efficacy and skills, facilitation of policies and culture, availability of self-help groups, and network characteristics of reciprocity, intensity, complexity, density, and homogeneity. Strategies to change these behaviors include enhancement of existing networks by skills training to providing and mobilizing support, use of lay health workers, teaching of participatory problem solving, and linking of members to new networks (e.g., mentor programs, buddy systems, and self-help groups) (see Bartholomew Eldredge et al., 2016).

The promise of eHealth and mHealth to alter mental health for people suffering from psychosis remains unfulfilled. But the future looks bright.

REFERENCES

Antonovsky, A. (1987). *Unraveling the mystery of health: how people manage stress and stay well.* San Francisco: Jossey-Bass.

Bartholomew Eldredge, L. K., Markham, C. M., Ruiter, R. A. C., Fernandez, M. E., Kok, G., & Parcel, G. S. (2016). *Planning health promotion programs: an intervention mapping approach.* San Francisco: John Wiley & Sons.

Ben-Zeev, D., Schueller, S. M., Begale, M., Duffecy, J., Kane, J. M., & Mohr, D. C. (2015). Strategies for mHealth research: lessons from 3 mobile intervention studies. *Adm Policy Ment Health, 42*(2), 157–167. doi:10.1007/s10488-014-0556-2

Delespaul, P., Milo, M., Schalken, F., Boevink, W., & Van Os, J. (2016). *GOEDE GGZ! Nieuwe concepten, aangepaste taal, verbeterde organisatie.* Amsterdam: Diagnosis Uitgevers.

Delespaul, P. A. E. G. (1995). *Assessing schizophrenia in daily life: the experience sampling method.* Maastricht: Datawyse/Universitaire Pers Maastricht.

Freeman, D., Bradley, J., Antley, A., Bourke, E., DeWeever, N., Evans, N., Clark, D. M. (2016). Virtual reality in the treatment of persecutory delusions: randomised controlled experimental study testing how to reduce delusional conviction. *Br J Psychiatry, 209*(1), 62–67. doi:10.1192/bjp.bp.115.176438

Freeman, D., Reeve, S., Robinson, A., Ehlers, A., Clark, D., Spanlang, B., & Slater, M. (2017). Virtual reality in the assessment, understanding, and treatment of mental health disorders. *Psychol Med*, 1–8. doi:10.1017/S003329171700040X

Garcia, E. J., & Repique, R. J. (2014). Harnessing mobile health technology to digitally engage mental health consumers in recovery. *J Am Psychiatr Nurses Assoc, 20*(5), 345–346. doi:10.1177/1078390314550788

Hill, C., Martin, J. L., Thomson, S., Scott-Ram, N., Penfold, H., & Creswell, C. (2017). Navigating the challenges of digital health innovation: considerations and solutions in developing online and smartphone-application-based interventions for mental health disorders. *Br J Psychiatry, 211*(2), 65–69. doi:10.1192/bjp.bp.115.180372

Huber, M., Knottnerus, J. A., Green, L., van der Horst, H., Jadad, A. R., Kromhout, D., Smid, H. (2011). How should we define health? *BMJ, 343*, d4163. doi:10.1136/bmj.d4163

Langeland, E., Wahl, A. K., Kristoffersen, K., & Hanestad, B. R. (2007). Promoting coping: salutogenesis among people with mental health problems. *Issues Ment Health Nurs, 28*(3), 275–295. doi:10.1080/01612840601172627

Myin-Germeys, I., Klippel, A., Steinhart, H., & Reininghaus, U. (2016). Ecological momentary interventions in psychiatry. *Curr Opin Psychiatry, 29*(4), 258–263. doi:10.1097/YCO.0000000000000255

Naslund, J. A., Marsch, L. A., McHugo, G. J., & Bartels, S. J. (2015). Emerging mHealth and eHealth interventions for serious mental illness: a review of the literature. *J Ment Health, 24*(5), 321–332. doi:10.3109/09638237.2015.1019054

Rapp, C. A., & Goscha, R. J. (2006). *The strengths model: case management with people with psychiatric disabilities.* New York: Oxford University Press.

Tondora, J., Miller, R., Slade, M., & Davidson, L. (2014). *Partnering for recovery in mental health: a practical guide to person-centered planning.* UK: John Wiley & Sons.

Torous, J., Staples, P., & Onnela, J. P. (2015). Realizing the potential of mobile mental health: new methods for new data in psychiatry. *Curr Psychiatry Rep, 17*(8), 602. doi:10.1007/s11920-015-0602-0

Treisman, G. J., Jayaram, G., Margolis, R. L., Pearlson, G. D., Schmidt, C. W., Mihelish, G. L., Misiuta, I. E. (2016). Perspectives on the use of eHealth in the Management of patients with schizophrenia. *J Nerv Ment Dis, 204*(8), 620–629. doi:10.1097/NMD.0000000000000471

van Os, J. (2014). *De DSM-5 voorbij! Persoonlijke diagnostiek in een nieuwe GGZ.* Leusden: Diagnosis Uitgevers.

van Os, J., Verhagen, S., Marsman, A., Peeters, F., Bak, M., Marcelis, M., Delespaul, P. (2017). The experience sampling method as an mHealth tool to support self-monitoring, self-insight, and personalized health care in clinical practice. *Depression and Anxiety, 34*(6), 481–493. doi:10.1002/da.22647

Veling, W., Moritz, S., & van der Gaag, M. (2014). Brave new worlds—review and update on virtual reality assessment and treatment in psychosis. *Schizophr Bull, 40*(6), 1194–1197. doi:10.1093/schbul/sbu125

WHO. (2006). Constitution of the World Health Organization. Retrieved from http://www.who.int/governance/eb/who_constitution_en.pdf

COGNITIVE REMEDIATION AND OTHER APPROACHES

65

RECOVERY-ORIENTED SERVICES

Mike Slade, Eleanor Longden, Julie Repper, and Samson Tse

INTRODUCTION

The notion of recovery has a long history in mental health services, traditionally understood as synonymous with a "return to normal." A typical definition is that recovery involves full symptom remission, ful-l or part-time work or education, independent living without supervision by informal carers, and having friends with whom activities can be shared, all sustained for a period of at least 2 years.[1] However, in the past few decades a new understanding has emerged, which challenges the view that recovery involves a return to symptom-free "normality." People personally affected by mental illness have become increasingly vocal in communicating what helps in moving beyond the role of "patient." The most widely cited definition is that recovery is "*a deeply personal, unique process of changing one's attitudes, values, feelings, goals, skills, and/or roles*" and "*a way of living a satisfying, hopeful, and contributing life even within the limitations caused by illness*."[2]

This emphasis on experiential knowledge and subjective experience has underpinned significant new developments. In the scientific world, it has led to new approaches to incorporation of knowledge from other academic disciplines into clinical practice, e.g., pedagogy[3] and positive psychology.[4] A new academic discipline of Mad Studies[5] has emerged. The policy context has also altered, with a "recovery orientation" central to national mental health policy in almost all high-income countries and also transnationally.[6] The economic case for a recovery orientation is increasingly compelling.[7] Finally, associated clinical developments have gained traction in mental health systems, e.g., new employment roles (peer support workers, peer trainers, peer researchers), structures (e.g., Recovery Colleges), interventions (Housing First, Individual Placement and Support, No Force First) and working practices (codesign, coproduction, shared decision-making, joint crisis planning).

In this chapter we outline some of the research underpinning these developments, and their implications for services working with people living with psychosis.

RECOVERY IN CONTEXT

Health and illness are rather distinct states that can exist concurrently. Recovery is the bridge between the two that builds on the strengths of health to address the sufferings of illness.[8] The concept of recovery has its roots in the addiction movements[9] of the mid-1930s. Recovery removed the blame for the disease from the individual and empowered people to take control of their own health and well-being. Since the 1980s, the notion of recovery has gradually moved into the mental health field.[8] The two most popular definitions of recovery are the existential, subjective one and the objective one focusing more on the controlling of symptoms such as hallucination and delusion, improved life skills and remediation of cognitive dysfunction.[10]

In the United States, the Substance Abuse and Mental Health Services Administration defines recovery as "a journey of healing and transformation that enables a person to live a meaningful life in a community of his or her choice while striving to achieve maximum human potential."[11] An internationally renowned recovery activist and clinical psychologist, Patricia Deegan, suggests that the goal of recovery is not to become normal but to embrace the journey of becoming more deeply, more fully human.[12] A recovery orientation involves refocusing on a subjectively defined process rather than a clinician-defined outcome.[13] Criticisms have also emerged. Examples of contentious issues include the overuse of the term "recovery," exaggerated claims of its occurrence, and the lack of recognition of the tension between the paternalism of mental health professions and the autonomy and rights of individuals recovering from severe mental illness such as psychosis.[14] The relevance of recovery to individuals in acute psychosis or mental health inpatient services has been questioned: "Clinicians have the uncomfortable experience of competing priorities leading to role tensions, yet advocates raise concerns that recovery is being "commandeered" to individualize social problems, to depoliticize individual experience and to remain focussed on deficit amelioration"[13] (p.148). Rigorous sociological research is needed to better understand the sociocultural meaning and implications of recovery. The mental health recovery movement has to articulate: "recovery from what, recovery of what and recovery to what?"[15] (p.8).

Although it is accepted that the recovery experience is unique to each person, certain common features can be discerned from the last 30 years of research and personal accounts of the recovery process, including, for example, renewing hope and commitment, redefining self, and being supported by others.[16] There is striking overlap between the course of psychotic disorders and the concept of recovery. The imagery associated with psychosis matches well with the description of

Table 65.1 PROPOSED RECOVERY FRAMEWORK

SUBJECTIVE-OBJECTIVE CONTINUUM:	RECOVERY AS A PROCESS:	RECOVERY AS AN OUTCOME:
Subjective: Inside the person, hidden transformation	e.g., accepting diagnosis of psychosis and changing the mindset to embark on the journey of making changes	e.g., sense of self-efficacy, sense of belonging, satisfied with quality of life
Objective: Mental health services, outward changes/achievements	e.g., being in a recovery-enhancing environment	e.g., equal membership and meaningful participation at work, school, and neighborhood

recovery: "Recovery is a process, a way of life, an attitude and a way of approaching the day's challenges. It is not a perfectly linear process. At times, our course is *erratic and we falter, slide back, regroup, and start again*"[12] (p.15, emphasis added). Thus, recovery is also described as a process through which people recover or reclaim their sense of self.[17] Recovery from psychosis (or mental illness in general) over time has been classified into three broad categories:[18] "Syndromal recovery" is defined as no longer meeting diagnostic criteria for an acute psychotic episode; "Symptomatic recovery" as indicating near-absence of auditory hallucination or idea of reference, the classic symptoms of psychosis; "Functional recovery" as regaining premorbid levels of psychosocial status. Some service users find the notion of recovery as an outcome to be unsatisfactory because it implies an evaluation component, where "the patient is only a person if he or she meets some arbitrary and externally imposed criterion"[19] (p.5).

Models of recovery are diverse,[15] representing sets of values and principles, theories of practice based on accumulated research, and debates from over two decades.[20] One example is an integrated sociological model of recovery,[21] identifying how sociodemographic factors (e.g., age, gender) and one's socioeconomic-political positions can lead to stress and symptoms of mental illness, and being able to recognize the symptoms and cope with stigma will have consequences for the person's self-concept and general well-being. Synthesizing the existing literature on the concept of recovery, we propose a conceptual framework of recovery (Table 65.1).

The link between recovery "process" and "outcome" does not operate in a simple, linear fashion. Additionally, recovery as a "process" or an "outcome" should not be seen as being at two opposite ends; rather they interact bidirectionally as the person's recovery journey unfolds. The boundary between "subjective" and "objective" is somewhat arbitrary.[20] Further studies are badly needed to identify the relevant logic and the pathways among the vast array of variables within a recovery model.

To conclude, broadly speaking, among international practitioners and research communities, the development of the recovery approach in mental health has been through three generations: (1) focusing on understanding the narratives and lived experience of individuals with mental illness, (2) development of instruments and their psychometric properties to measure the recovery of individuals and the critical elements embedded in healthcare services that facilitate or hinder a person's recovery; and (3) defining different kinds of recovery-oriented services and technologies (e.g., peer support services, recovery college, strengths model case management, supported employment, wellness recovery action plan) and examining the effectiveness of interventions. We argue that we are now entering Generation 4.0, which concerns not only establishing the evidence for recovery-oriented services but also scaling up interventions for the wider community and informing and shaping policy planning in the mental health sector.

EMPIRICAL EVIDENCE ABOUT RECOVERY

Discourse around psychosis outcomes often maintain a pessimistic emphasis on concepts like "permanence," "impairment," and "psychosocial disability." However, while acknowledging the considerable distress these problems can cause, testimony from those with lived experience of psychosis, as well as academic and clinical literature, demonstrates that such negative expectations are by no means evidence-based. In this respect, a review of empirical research about recovery and mental health (primarily in relation to schizophrenia) has proposed seven evidence-based messages that contest many embedded assumptions surrounding the prognosis for psychotic disorders.[22]

RECOVERY IS SUBJECTIVE

As discussed previously, an emphasis on "clinical recovery"—a dichotomous outcome that can be objectively rated by experts—has been challenged in recent years. As such the concept of "personal recovery" is being increasingly embedded within international mental health policy, wherein recovery is located as a subjective experience rather than an observable state and which is seen as most appropriately defined by the person themselves. Rather than observer-rated or symptom-based outcomes, personal recovery therefore emphasizes factors like hope, identity, meaning, and personal responsibility.[23]

RECOVERY IS NOT UNCOMMON

There is currently an inconclusive evidence-base about the relationship between clinical and personal recovery.[24] However, several longitudinal studies show more than half of people diagnosed with schizophrenia experience clinical recovery,[25] although problems with sampling strategies, follow-up periods, and outcome assessments make it highly likely that the true proportion is underestimated.[22] Combined with the increase in people sharing their stories of living well with psychosis, recovery is clearly emerging as more common than previously acknowledged.[26]

WELLNESS IS NOT THE SAME AS REMISSION

The reasoning bias that, having once been diagnosed with psychosis, no longer being diagnosable indicates an individual is in remission rather than well reflects assumptions of chronicity and deterioration that are toxic in a mental health context. In contrast, it is intuitive that "[t]he more a person can develop a rich and layered identity as a person in recovery, rather than a thin identity as a "patient," the more they will develop resilience and the ability to meet the challenges of life."[22]

DIAGNOSIS IS NOT A ROBUST FOUNDATION

While applying terms like "schizophrenia" to characterize groups and predict outcomes is defensible from a clinical perspective, it is important to acknowledge the contested and controversial nature of psychiatric labels, e.g., limited reliability and validity, and lack of consensus on etiology.[27] While some people find diagnosis helpful others do not, and an uncritical application of clinical labels (i.e., the assumption that they refer to meaningful, invariant individual-level diseases) can have adverse consequences, including diverting attention away from psychosocial factors that may be causing or maintaining distress. In this respect alternative explanations for one's experiences of psychosis, such as a response to abuse or trauma rather than a biological illness, is not only scientifically justified[28] but for some can be a turning point in their recovery journey.[29]

TREATMENT IS ONE ROUTE AMONG MANY TO RECOVERY

Although clinical treatment is often framed as the most likely means of recovery, this position is not empirically justified. On the contrary, factors like connectedness, hope, positive identity, meaning, and empowerment are known to be of importance;[23] they are processes that can, and frequently do, occur outside statutory services. It is of course reasonable to expect the provision of established pharmacological and psychosocial interventions and we do not advocate reducing the allocations of resources for psychiatric care. Nevertheless, while clinical models premise professional intervention as instrumental for improving outcomes, a personal recovery perspective acknowledges that individuals experiencing psychosis can work towards recovery through a range of routes.

SOME PEOPLE CHOOSE NOT TO USE PSYCHIATRIC SERVICES

Barriers to accessing services are sometimes characterized by structural factors like availability, financing, and patient awareness.[30] However, it is also the case that people may decline psychiatric input for a range of reasons, including stigma, recourse to other support networks, or not identifying with medical/psychological models. While it is important to understand the reasons that influence an individual's choice (and the resulting impact of this choice on their experiences) the implication that accessing services should be a default option is not justified; of those who don't elect to use them, some would likely benefit from statutory care whereas others live well outside of services.

THE IMPACT OF MENTAL HEALTH PROBLEMS IS MIXED

While it is important not to trivialize the distress psychosis may cause, assuming the aftermath of such experiences is inevitably negative is both unwarranted and unduly pessimistic. Recovery involves the realization that aspects of mental health problems can provide growth and positive gain,[31] and substantial survivor testimony demonstrates that enduring challenges of this type may ultimately offer a potential for empowerment and psychological growth (e.g., through an increased capacity for self-knowledge, compassion, creativity, and survivor activism[32]).

RECOVERY AT THE INDIVIDUAL LEVEL: RECONSTRUCTING SCHIZOPHRENIA

An inevitable question when considering healing and restitution in mental health is *"recovery from what?"* While some people find clinical/disease frameworks helpful, it is increasingly recognized that theorizing and responding to psychiatric distress using more psychosocial models is also a robustly defensible position. Specifically, the idea of mental health problems as "socially and psychologically intelligible" responses[33] (p.16) has gained considerable traction in the past decade, with an extensive literature demonstrating that a range of environmental factors (particularly, although by no means exclusively, childhood abuse) have a central etiological role. In the case of psychosis (in which biological causes tend to be particularly emphasized), this includes indications of a dose-response relationship (i.e., that risk increases relative to the degree of adversity exposure[34] for reviews) as well as an association that remains statistically significant when controlling for such confounders as substance use, educational level, ethnicity, gender, age, and family history of mental illness. In turn, evidence attests that many structural and functional neurological abnormalities in psychosis patients—typically deemed as signs of a disease process—can be understood as secondary effects to the physiology of stress exposure,[35] in addition to the well-documented finding that adverse life events are often represented in the content of paradigmatic symptoms like delusions and hallucinations.[36]

In their review of this literature, Longden and Read[28] note that:

> [M]ental health professionals have an ethical responsibility to engage with suffering, and to bear witness to the experiences of loss, stress, adversity, abuse and disempowerment that have shaped their patients" lives .... This is not a formulation that prohibits the use of

medication, but rather sees it as one of many elements of an integrated intervention; one that seeks to address an individual's psychological, social, and emotional wounds in healing and restorative ways.

Examples of such ways of working already exist in clinical services (e.g., Soteria, the Open Dialogue family and network approach, the Sanctuary Model). In addition, survivor-led initiatives also exist which have taken an even broader approach to the ways in which subjectively defining the cause of one's mental health challenges can be used in the service of recovery. A notable example is the work of the international Hearing Voices Movement (HVM), an influential collective of experts by experience (voice hearers, their friends and family members) and experts by profession (clinicians, academics) who promote the value of accepting and making sense of voice hearing, as well as providing recovery literature, peer-support groups, and working to challenge stigma. However, while the HVM emphasizes the role of psychosocial adversity in voice hearing, there is no expectation for its members to conform with this assumption; on the contrary, every individual is considered the "owner" and "expert" of their experiences, with all explanatory frameworks (e.g., psychological, biomedical, spiritual) treated as equally worthy of respect and consideration. In doing so, the Movement endeavors to support its members to explore what their experiences mean to them and to provide input to working toward coping and learning from one's voices (and other unusual experiences, such as visions and nonshared beliefs) in one's own way. The popularity of the approach can be seen in its global dissemination in the past 20 years, as well as the growing influence of its principles and philosophies on mental health research and service provision.[37]

Through its work in deconstructing disease models, the HVM has also become a symbol for taking pride in one's identity as a voice hearer: that voices can be understood as a complex, unique and subjectively meaningful experience which, with the right support, can be lived with profitably and peacefully. In this respect, many of its most prominent advocates are former service-users who provide accounts of how initially catastrophic experiences of adversity and extreme mental distress ultimately became a source of emotional growth and empowerment—and how initially terrorizing voices transformed into an important and valued part of this process (e.g., [38]). In this respect, survivor testimony from a range of sources demonstrates how serious mental health difficulties can often be reconstructed into an impetus for self-discovery, inspiration, and personal and interpersonal development.[32,38-41]

RECOVERY AT THE TREATMENT LEVEL: NEW APPROACHES

So what does this new understanding of psychosis mean for clinical services? It has proved challenging to develop a recovery orientation in mental health services that gives primacy to the individual's experience. Indeed, some commentators suggest that recovery has been "hijacked"[42] by professionals, and misuses of the term "recovery" are being identified.[26]

What types of interventions should be available in a recovery-oriented mental health service? A systematic review identified key recovery processes as connectedness, hope and optimism, identity, meaning and purpose, and empowerment (the CHIME framework).[23] These recovery processes differ from traditional clinical outcome targets, so interventions targeted at these processes are needed. Ten empirically supported approaches that are oriented toward supporting people to live as well as possible are now outlined.

APPROACH 1: PEER SUPPORT WORKERS

Peer support emerged from the user/survivor movement, and originally developed outside the mainstream mental health system. Peer support workers are individuals with mental illness who identify themselves as such, and who use their lived experience to support others to recover.[43] A Cochrane review identified eleven randomized trials involving 2,796 people in three countries (Australia, United Kingdom, United States), showing equivalent outcomes from peer support workers compared with professionals employed in similar roles.[44]

APPROACH 2: ADVANCE DIRECTIVES

People with mental illness are almost by definition vulnerable to experiencing emotional crisis. An advance directive involves specifying actions to be taken for the person's health if capacity is lost. Advance directives have strong empirical support,[45] as do a variant increasingly used in mental health called joint crisis plans.[46]

APPROACH 3: WELLNESS RECOVERY ACTION PLANNING

The wellness recovery action planning (WRAP) tools and processes support self-management with a specific focus on recovery-oriented mental health services. WRAP is used to create recovery plans, by guiding individuals and groups of people to reflect on what has assisted them to stay well in the past, and to consider strategies that assisted others with their recovery. RCT evaluation of outcomes for participants (n = 519) at eight outpatient community mental health centers in an 8-week peer-led intervention, compared with usual care and wait-list for WRAP, showed benefits in symptom profile, hope, and quality of life.[47]

APPROACH 4: ILLNESS MANAGEMENT AND RECOVERY

The illness management and recovery program (IMR) is an empirically supported standardized intervention to teach illness self-management strategies to people with a severe mental illness.[48] RCT evaluations indicate IMR can significantly improve symptomatology, functioning, knowledge, and progress toward goals.[49]

APPROACH 5: REFOCUS

The REFOCUS intervention increases the recovery orientation of community adult mental health teams.[50] Mental health staff are trained to use coaching and to implement three working practices (understanding values and treatment preferences, assessing and amplifying strengths, supporting goal-striving) in clinical work. Two multisite trials have shown that the intervention was effective at improving recovery when adequately implemented.[51,52]

APPROACH 6: STRENGTHS MODEL

The strengths model of case management aims to help people with mental health problems to attain goals they set themselves by identifying, securing, and sustaining the range of environmental and personal resources that are needed to live, play, and work in a normally interdependent way in the community.[53] RCT evaluations demonstrate improved outcomes including hospitalization rates, employment and educational attainment, and intrapersonal outcomes such as self-efficacy and sense of hope.[54]

APPROACH 7: RECOVERY COLLEGES OR RECOVERY EDUCATION PROGRAMS

Recovery colleges or recovery education programs are an educational approach to supporting the recovery and reintegration of people with mental health problems. Many are based on a briefing paper[55] by ImROC (discussed later in the chapter), and they involve adult education courses that are coproduced and codelivered by people with lived and professional experience.

APPROACH 8: INDIVIDUAL PLACEMENT AND SUPPORT

People with psychosis who cannot work should have easy access to welfare, and positive incentives to return to work. Individual placement and support is an intervention that provides this support.[56] A Cochrane review synthesized 14 RCTs (n = 2,265), and concluded "supported employment is effective in improving a number of vocational outcomes relevant to people with severe mental illness."[57]

APPROACH 9: HOUSING FIRST

Safe and secure permanent housing can act as a base from which people with psychosis can achieve numerous recovery goals and improve quality of life.[58] The Housing First intervention involves rapid rehousing in suitable independent accommodation, based on personal preference, followed by provision of supportive services to the person once they are housed. A definitive RCT across Canada showed that homeless people living with mental illness (n = 1,198) who received the Housing First intervention had increased housing stability over 24 months compared to those not receiving the intervention.[59]

APPROACH 10: MENTAL HEALTH TRIALOGUES

The active involvement of mental health service users, relatives, and friends is essential for the development of recovery-oriented mental health practice and research.[60] Trialogue groups (also known as psychosis seminars) address this challenge. A mental health trialogue meeting is a community forum where service users, carers, friends, mental health workers, and others with an interest in mental health participate in an open dialogue. In German-speaking countries, well over 100 trialogue groups are regularly attended by 5,000 people.[61]

Other emerging approaches include the Hearing Voices Movement,[62] Open Dialogue,[63] Photovoice,[64] peer-led initiatives such as CommonGround[65] and TREE,[66] service user-led organizations,[67] and family peer support.[68]

RECOVERY AT THE WORKFORCE LEVEL: PEER SUPPORT

Peer support workers (PSWs) are individuals who have progressed in their own recovery from mental illness, identify themselves as such, and have a strong desire to help others with similar conditions by referring to their own lived experience.[69] PSWs can either hold paid positions or support their peers in an informal capacity. To share and engage with their peers, they rely on the knowledge gained through their personal lived experiences with mental illness. Research on the effectiveness of peer-delivered services is growing, and the evidence for its impact originates mostly from the west: Canada, United States, United Kingdom, Australia, and New Zealand.[70] There are indications that peer support services are effective in raising people's level of hope, empowerment, and quality of life; engaging people into care; reducing the use of emergency rooms and hospitals; and reducing substance use among persons with concurrent substance use disorders; and in some cases, the PSWs gain a sense of hope and skills applicable to their own situation.[71]

"Within mental health, peer providers signify a subpopulation engaged in positive change and growth processes"[69] (p.273). PSWs are able to effect positive change at the workforce level, supporting individuals in recovery from psychosis in three ways. First, peer supporters can work with mental health services or systems to create a vision which is "an ideal and unique image of the future"[72] (p.95). Individuals in paid positions with lived experience of mental health issues can play the unique role of cocreating a shared vision for the service that protects and promotes organizational integrity while inspiring greatness in service users and encouraging their empowerment, learning, and adaptation, in turn facilitating workplace change and transformation.[73] Second, PSWs collectively can exert effective role modeling in the form of a visible personal example of recovery from psychosis in the contemporary mental health workforce.[74] Modeling demonstrates PSWs' commitment to an ethos and a set of values (e.g., recovery is not only plausible but also achievable by adopting certain wellness strategies) they seek to institutionalize by their own deeds.[75] Third, PSWs who want to create an impact at the workforce level are able to teach by describing their experience of recovery from psychosis and reflecting on that process.[76] PSWs help shape and alter the motives, values, and goals of the members of the mental health team through the vital teaching role of what recovery is all about. Peer supporters can be Socratic teachers by asking important questions to achieve deeper understanding of recovery from psychosis by providing appropriate examples and personal stories.

RECOVERY AT THE SERVICE LEVEL: TRANSFORMING MENTAL HEALTH SYSTEMS

While it is relatively clear what recovery means for people on their own journey of recovery, and the implications for mental health workers have been explored, this is confounded if their service as a whole is not prioritizing recovery. This requires radical changes in the role and power of people who use services; the stated purpose and vision of services; and the role of communities in supporting all of their members.

ImROC (Implementing Recovery through Organisational Change) uses a number of tools to support and enable organizations to develop a more recovery focused approach:

Recovery focused leadership—The ImROC team comprises a small number of people with extensive experience of driving recovery forward as people who use services, practitioners, managers, and recognized leaders in the field.

Coproduction—The ImROC team always model coproduction with equal value placed on lived and learned expertise and experience.

ImROC uses "ten organizational challenges" (see www.imroc.org) as a basis for assessing the stage of recovery-focused development that an organization has achieved, and uses the results of this to agree to SMART targets. These challenges are:

1. Changing the nature of day-to-day interactions
2. Coproduced learning and development opportunities are available for staff, people using services, and their families to learn together
3. A recovery college is established to provide learning opportunities forward in partnerships with community resources
4. Recovery-focused leadership at every level and a culture of recovery
5. Increasing personalization and choice
6. Reducing restrictive practice; changing conceptions of risk as something to be avoided toward working together to improve safety
7. User involvement is replaced by fully resourced coproduction
8. Transforming the workforce so that lived experience of mental health problems is demonstrably valued in the workforce as a whole and peer workers are employed to explicitly draw on their experience to provide support.
9. Supporting staff to cope effectively with the stressors that are inevitable in working in mental health services
10. Prioritization of life goals (full citizenship and community integration) in all care-planning processes.

Organizational engagement begins with a large meeting with up to 50 people who represent different stakeholders in the organization (service providers working at different levels and in different parts of service; people who use the service and their family members; voluntary sector groups and partner organizations). All focus on each of the 10 challenges in turn, carefully considering which have been developed, to what extent they have progressed, whether this is consistent across the whole service, what accounts for differences, what the experiences of people using the service are, and what further information is needed to be able to provide evidence of progress.

Quality Improvement—The PDSA (Plan, Do, Study, Act) approach is used to frame service development. This is a well tried and tested approach often used in quality improvement. It begins with the organizational assessment of the 10 challenges (earlier) which informs the identification of goals and a SMART action plan to achieve these goals. Once this action has been implemented then the situation is reviewed and further action is planned and implemented in a cyclical manner.

Bespoke organizational support—ImROC supports individual organizations to implement planned action through bespoke consultancy including organizational development approaches such as leadership development, coaching, training, workshops, supervision, and evaluation.

National learning sets—Where a number of organizations have set similar targets ImROC runs national learning sets for teams to hear about best practice, share their own experiences—successes and failures—and solve problems together.

Briefing papers—ImROC publishes its learning in papers that offer accessible up-to-date information relevant to the 10 challenges. These explain the organizational development methodology in detail and provide more detailed accounts of learning.

The ImROC team has learned many lessons through working with many varied organizations. First, it is increasingly apparent that recovery-focused organizations are driven by a genuine understanding of what recovery means. The leadership team must recognize the benefits of working in this way and strive to provide a recovery focused culture. The expertise of practitioners needs to be recognized and built on, and distributed leadership of recovery works well—with recovery champions taking a lead at team level alongside peer leaders.

Leadership is key to the facilitation of recovery. A recovery lead employed to focus solely on recovery, to support teams, "provide air cover," coordinate the development and implementation of a recovery strategy and ensure that recovery becomes "everybody's business" is critical.

The most significant change that the ImROC team has engaged with over the past 7 years is the development of partnerships between mental health services and communities. This is a positive development: many of the challenges faced by people with mental health problems arise from discrimination, lack of opportunity, and isolation/exclusion within their communities. The more visible and constructive their role within communities then the less discrimination they will face, which offers greater opportunity for recovery.

RECOVERY AT THE COMMUNITY LEVEL: SOCIAL AND POLITICAL IMPLICATIONS

The generalizability of the concept of recovery remains a concern. Assumptions embedded in recovery may be "monocultural," and broader concepts of community and cultural resilience and well-being may be needed. For example, an important issue is the collectivist versus individualist value paradigm.[77] In collectivist cultures, such as Maori (the indigenous people of New Zealand) and Chinese ones, emphasis is placed on interdependence among family members and relatives over and above the independence that is often promoted in Western cultures.[78] By investigating factors that facilitate or hinder recovery for individuals from diverse backgrounds, more culturally applicable recovery concepts can be developed which will better address service users' needs and rights.

An understanding of how to transform services is emerging. A synthesis of international guidance on supporting recovery identifies four levels of practice: supporting personally defined recovery (what interventions are offered), working relationship (how interventions are offered), organizational commitment (what is the "core business" of the mental health system?), and promoting citizenship (supporting the experience of wider entitlements of citizenship).[79] Most interventions reviewed in this chapter address the first three of these levels. The final frontier is reducing and removing the barriers that prevent individuals from experiencing full entitlements of citizenship. For mental health systems, this will involve transformation away from a "treat-and-recover" worldview, in which priority is given to the provision of treatments with the aim that the person will then become ready to re-engage with their life. Empirical investigations of the concepts of "work-readiness" (in individual placement and support) and "housing-readiness" (in Housing First) have found them to be inadvertently toxic concepts, which reduce hope and limit expectations. This change of emphasis applies more widely than just support for employment and housing,[80] and includes for example addressing issues of poverty[81] (through personal budgets[82]) and social networks[83] (through asset mapping[84]). However, the broadest—and most important—challenge is societal change, which will involve professionals and people who live with psychosis becoming partners[60] and social activists,[85] to challenge stigmatizing assumptions that people with mental illness cannot, or should not, have the same citizenship entitlements as anyone else in their community.

REFERENCES

1. Libermann RP, Kopelowicz A. Recovery from schizophrenia: a challenge for the 21st Century. *Int Rev Psychiatry* 2002; **14**: 242–255.
2. Anthony WA. Recovery from mental illness: the guiding vision of the mental health system in the 1990s. *Psychosoc Rehabil J* 1993; **16**: 11–23.
3. Oh H. The pedagogy of recovery colleges: clarifying theory. *Ment Health Rev J* 2013; **18**: 240.
4. Slade M, Brownell T, Rashid T, Schrank B. Positive psychotherapy for psychosis. Hove: Routledge; 2017.
5. Russo J, Sweeney, A., editor. Searching for a rose garden: challenging psychiatry, fostering mad studies. Monmouth: PCCS; 2016.
6. World Health Organization. Mental health action plan 2013–2020. Geneva: WHO; 2013.
7. Slade M, McDaid D, Shepherd G, Williams S, Repper J. ImROC Briefing Paper 14. Recovery: the business case. Nottingham: ImROC; 2017.
8. Davidson L, Rakfeldt J, Strauss J. The roots of the recovery movement in psychiatry. Chichester: Wiley-Blackwell; 2010.
9. Sterling EW, von Esenwein SA, Tucker S, Fricks L, Druss BG. Integrating wellness, recovery, and self-management for mental health consumers. *Community Ment Health J* 2010; **46**: 130–138.
10. Liberman RP. Recovery from disability: manual of psychiatric rehabilitation. Washington, DC: American Psychiatric Publishing; 2008.
11. Substance Abuse and Mental Health Services Administration. National Consensus Conference on Mental Health Recovery and Systems Transformation. Rockville, MD: Department of Health and Human Services; 2005.
12. Deegan P. Recovery: the lived experience of rehabilitation. *Psychosoc Rehabil J* 1988; **11**: 11–19.
13. Slade M. Implementing shared decision making in routine mental health care. *World Psychiatry* 2017; **16**: 146–153.
14. Lam MM, Pearson V, Ng RM, Chiu CP, Law CW, Chen EY. What does recovery from psychosis mean? Perceptions of young first-episode patients. *Int J Social Psychiatry* 2011; **57**(6): 580–587.
15. Henderson AR. A substantive theory of recovery from the effects of severe persistent mental illness. *Int J Soc Psychiatry* 2010; **57**: 564–573.
16. Davidson L, O'Connell M, Tondora J, Evans AC. Recovery in serious mental illness: a new wine or just a new bottle? *Professional Psychology* 2005; **36** (5): 480–487.
17. Davidson L, Tondora, J, Ridgway, P. Life is not an "outcome": reflections on recovery as an outcome and as a process. *American Journal of Psychiatric Rehabilitation* 2010; **13**: 1–8.
18. Braslow J. The manufacture of recovery. *Annual Review of Clinical Psychology* 2013; **9**: 781–809.
19. Ralph RO, Corrigan PW. Recovery in mental illness: broadening our understanding of wellness. Washington, DC: American Psychological Association; 2005.

20. Leonhardt B, Huling K, Hamm J, et al. Recovery and serious mental illness: a review of current clinical and research paradigms and future directions. *Expert Rev Neurother* 2017; **17**: 1117–1130.
21. Starnino VR. An integral approach to mental health recovery: implications for social work. *Journal of Human Behavior in the Social Environment* 2009; (19).
22. Slade M, Longden E. Empirical evidence about mental health and recovery. *BMC Psychiatry* 2015; **15**: 285.
23. Leamy M, Bird V, Le Boutillier C, Williams J, Slade M. A conceptual framework for personal recovery in mental health: systematic review and narrative synthesis. *Br J Psychiatry* 2011; **199**: 445–452.
24. van Eck R, Jan Burger, T, Vellinga, A, Schirmbeck, F, de Haan, L. The relationship between clinical and personal recovery in patients with schizophrenia spectrum disorders: a systematic review and meta-analysis. *Schizophr Bull* 2018; **6**: 631–642.
25. Slade M, Amering M, Oades L. Recovery: an international perspective. *Epidemiol Psichiatr Soc* 2008; **17**: 128–137.
26. Slade M, Amering M, Farkas M, et al. Uses and abuses of recovery: implementing recovery-oriented practices in mental health systems. *World Psychiatry* 2014; **13**: 12–20.
27. Boyle M, Johnstone, L. Alternatives to psychiatric diagnosis. *Lancet Psychiatry* 2014; **1**: 409–411.
28. Longden E, Read, J. Social adversity in the etiology of psychosis: a review of the evidence. *Am J Psychother* 2016; **70**: 5–33.
29. Romme M, Escher S, Dillon J, Corstens D, Morris M. Living with voices: 50 stories of recovery. Ross-on-Wye: PCCS; 2009.
30. Nicholas A, Reifels, L., King, K., Pollock, S. Mental health and the NDIS: a literature review. Melbourne: Centre for Mental Health, University of Melbourne; 2014.
31. Repper J. Recovery: a journey of discovery. In: Tee S, Brown J, Carpenter D, ed. Handbook of mental health nursing. London: Hodder & Stoughton; 2012: 100–120.
32. Coleman R. Recovery—an alien concept. Hansell; 1999.
33. Boyle M. The persistence of medicalisation: is the presentation of alternatives part of the problem? In: Coles S, Keenan S, Diamond B, ed. Madness contested: power and practice. Ross-on-Wye, England: PCCS Books; 2013: 3–22.
34. Varese F, Smeets F, Drukker M, et al. Childhood adversities increase the risk of psychosis: a meta-analysis of patient-control, prospective- and cross-sectional cohort studies. *Schizophr Bull* 2012; **38**(4): 661–671.
35. Read J, Fosse, R, Moskowitz, A, Perry, B. The traumagenic neurodevelopmental model of psychosis revisited. *Neuropsychiatry* 2014; **4**: 65–79.
36. Corstens D, Longden, E. The origins of voices: links between voice hearing and life history in a survey of 100 cases. *Psychosis* 2013; **5**: 270–285.
37. Corstens D, Longden, E, McCarthy-Jones, S, Waddingham, R, Thomas, N. Emerging perspectives from the Hearing Voices Movement: implications for research and practice. *Schizophr Bull* 2014; **40**(S4): S285–S94.
38. Dillon J. The personal *is* the political. In: Rapley M, Moncrieff J, Dillon J, ed. De-medicalizing misery: psychiatry, psychology and the human condition. Basingstoke: Palgrave Macmillan; 2011: 141–157.
39. Longden E. Learning from the voices in my head. New York: TED Books; 2013.
40. Comans K. Beyond psychiatry: understanding my own human experience. *Psychosis* 2011; **3**: 242–247.
41. Lampshire D. Living the dream. *Psychosis* 2012; **4**: 172–178.
42. Mental Health "Recovery" Study Working Group. Mental health "recovery": users and refusers. Toronto: Wellesley Institute; 2009.
43. Repper J. ImROC Briefing Paper 7. Peer support workers: a practical guide to implementation. London: Centre for Mental Health; 2013.
44. Pitt V, Lowe D, Hill S, et al. Consumer-providers of care for adult clients of statutory mental health services. *Cochrane Database of Systematic Reviews* 2013; (3).
45. Campbell LA, Kisely SR. Advance treatment directives for people with severe mental illness. *Cochrane Database Syst Rev* 2009; (1): CD005963.
46. Henderson C, Flood C, Leese M, Thornicroft G, Sutherby K, Szmukler G. Effect of joint crisis plans on use of compulsory treatment in psychiatry: single blind randomised controlled trial. *BMJ* 2004; **329**: 136–140.
47. Cook JA, Copeland ME, Jonikas JA, et al. Results of a randomized controlled trial of mental illness self-management using Wellness Recovery Action Planning. *Schizophr Bull* 2012; **38**: 881–891.
48. Gingerich S, Mueser KT. Illness management and recovery. In: Drake RE, Merrens MR, Lynde DW, eds. Evidence-based mental health practice: a textbook. New York: Norton; 2005.
49. Hasson-Ohayon I, Roe, D, Kravetz, S. A randomized controlled trial of the effectiveness of the illness management and recovery program. *Psychiatr Serv* 2007; **58**: 1461–1466.
50. Bird V, Leamy M, Le Boutillier C, Williams J, Slade M. REFOCUS (2nd edition): promoting recovery in mental health services. London: Rethink Mental Illness; 2014.
51. Slade M, Bird V, Clarke E, et al. Supporting recovery in patients with psychosis using adult mental health teams (REFOCUS): a multisite cluster randomised controlled trial. *Lancet Psychiatry* 2015; **2**: 503–514.
52. Meadows G, Brophy L, Shawyer F, et al. REFOCUS-PULSAR recovery-oriented practice training in specialist mental health care: a stepped-wedge cluster randomised controlled trial. *Lancet Psychiatry* 2019; **6**: 103–114.
53. Rapp C, Goscha R. The strengths model: case management with people with psychiatric disabilities. 3rd ed. New York: Oxford University Press; 2012.
54. Tse S, Tsoi E, Hamilton B, O'Hagan M, Shepherd G, Slade M, Whitley R, Petrakis M. Uses of strength-based interventions for people with serious mental illness: A critical review. *Int J Soc Psychiatry* 2016; **62**: 281–291.
55. Perkins R, Repper J, Rinaldi M, Brown H. ImROC 1. Recovery colleges. London: Centre for Mental Health; 2012.
56. Grove B, Locket H, Shepherd G, Bacon J, Rinaldi M. Doing what works—individual placement and support into employment. London: Sainsbury Centre for Mental Health; 2009.
57. Kinoshita Y, Furukawa T, Kinoshita K, Honyashiki M, Omori I, Marshall M, Bond G, Huxley P, Amano N, Kingdon D. Supported employment for adults with severe mental illness. *Cochrane Database of Systematic Reviews* 2013; **9**: CD008297.
58. Boardman J. More than shelter: supported accommodation and mental health. London: Centre for Mental Health; 2016.
59. Stergiopoulos V, Hwang S, Gozdzik A, Nisenbaum R, Latimer E, Rabouin D, Adair C, Bourque J, Connelly J, Frankish J, Katz L, Mason K, Misir V, O'Brien K, Sareen J, Schutz C, Singer A, Streiner D, Vasiliadis H, Goering P. Effect of scattered-site housing using rent supplements and intensive case management on housing stability among homeless adults with mental illness. a randomized trial. *Journal of the American Medical Association* 2015; **313**(9): A905.
60. Wallcraft J, Amering M, Freidin J, Davar B, Froggatt D, Jafri H, Javed A, Katontoka S, Raja S, Rataemane S, Steffen S, Tyano S, Underhill C, Wahlberg H, Warner R, Herrman H. Partnerships for better mental health worldwide: WPA recommendations on best practices in working with service users and family carers. *World Psychiatry* 2011; **10**: 229–236.
61. Bock T, Priebe S. Psychosis seminars: an unconventional approach. *Psychiatr Serv* 2005; **56**: 1441–1443.
62. Jones N, Marino C, Hansen M. The Hearing Voices Movement in the United States: findings from a national survey of group facilitators. *Psychosis* 2016; **8**: 106–117.
63. Seikkula J, Alakare B, Aaltonen J. The comprehensive open-dialogue approach in western Lapland: II. Long-term stability of acute psychosis outcomes in advanced community care. *Psychosis* 2011; **3**: 192–204.
64. Petros R, Solomon P, Linz S, DeCesaris M, Hanrahan M. Autovideography: the lived experience of recovery for adults with serious mental illness. *Psychiatric Quart* 2016; **87**: 417–426.
65. Stein BD, Kogan JN, Mihalyo MJ, et al. Use of a computerized medication shared decision making tool in community mental health

settings: impact on psychotropic medication adherence. *Community Ment Health J* 2013; **49**: 185–192.
66. Boevink W, Kroon H, van Vugt M, Delespaul P, van Os J. A user-developed, user run recovery programme for people with severe mental illness: a randomised control trial. *Psychosis* 2016; **8**: 287–300.
67. Rose D, MacDonald D, Wilson A, Crawford M, Barnes M, Omeni E. Service user led organisations in mental health today. *Journal of Mental Health* 2016; **25**: 254–259.
68. Chien WT, Thompson DR. An RCT with three-year follow-up of peer support groups for Chinese families of persons with schizophrenia. *Psychiatr Serv* 2013; **64**(10): 997–1005.
69. Moran G, Russo-Netzer P. Understanding universal elements in mental health recovery: A cross-examination of peer providers and a non-clinical sample. *Qual Health Res* 2016; **26**: 273–287.
70. Chinman M, George P, Dougherty R, Daniels A, Ghose S, Swift A, Delphin-Rittmon M. Peer support services for individuals with serious mental illnesses: assessing the evidence. *Psychiatr Serv* 2014; **65**: 429–441.
71. Cabassa L, Camacho D, Vélez-Grau C, Stefancic A. Peer-based health interventions for people with serious mental illness: a systematic literature review. *J Psychiatr Res* 2017; **84**: 80–89.
72. Kouzes JM, Posner B.Z. The leadership challenge. Chichester: John Wiley & Sons; 2006.
73. Hurley J, Cashin A, Mills J, Hutchinson M, Graham I. A critical discussion of peer workers: implications for the mental health nursing workforce. *J Psychiatr Ment Health Nurs* 2016; **23**: 129–135.
74. Vandewalle J, Debyser B, Beeckman D, Vandecasteele T, Deproost E, Van Hecke A, Verhaeghe S. Constructing a positive identity: a qualitative study of the driving forces of peer workers in mental healthcare systems. *Int J Ment Health Nurs* 2018; **27**: 378–389.
75. Gillard SG, Edwards C, Gibson SL, Owen K, Wright C. Introducing peer worker roles into UK mental health service teams: A qualitative analysis of the organisational benefits and challenges. *BMC Health Serv Res* 2013; **13**: 188.
76. Gumber S, Stein CH. Consumer perspectives and mental health reform movements in the United States: 30 years of first-person accounts. *Psychiatric Rehab J* 2013; **36**: 187–194.
77. Papadopoulos C, Foster J, Caldwell K. Community Mental Health Journal 2012. "Individualism-Collectivism" as an explanatory devise for mental illness stigma. *Community Ment Health J* 2013; **49**: 270–280.
78. Tse S, Ng R. Applying a mental health recovery approach for people from diverse backgrounds: the case of collectivism and individualism paradigms. *J Psychosoc Rehabil Ment Health* 2014; **1**: 7–13.
79. Le Boutillier C, Leamy M, Bird VJ, Davidson L, Williams J, Slade M. What does recovery mean in practice? A qualitative analysis of international recovery-oriented practice guidance. *Psychiatr Serv* 2011; **62**: 1470–1476.
80. Slade M. Everyday solutions for everyday problems: how mental health systems can support recovery. *Psychiatr Serv* 2012; **63**: 702–704.
81. Psychologists for Social Change. Universal basic income: a psychological impact assessment. London: PAA; 2017.
82. Hamilton S, Tew J, Szymczynska P, Clewett N, Manthorpe J, Larsen J, Pinfold V. Power, choice and control: how do personal budgets affect the experiences of people with mental health problems and their relationships with social workers and other practitioners? *British Journal of Social Work* 2016; **46**: 719–736.
83. Pinfold V, Sweet D, Porter I, Quinn C, Byng R, Griffiths C, Billsborough J, Enki D, Chandler R, Webber M, Larsen J, Carpenter J, Huxley P. Improving community health networks for people with severe mental illness: a case study investigation. *Health Services and Delivery Research* 2015; **3**: 1–236.
84. Pinfold V, Sweet D. Wellbeing networks and asset mapping. London: McPin Foundation; 2015.
85. Slade M. Mental illness and well-being: the central importance of positive psychology and recovery approaches. *BMC Health Services Research* 2010; **10**: 26.

66

NEUROSCIENCE-INFORMED COGNITIVE TRAINING FOR PSYCHOTIC SPECTRUM ILLNESSES

Sophia Vinogradov, Rana Elmaghraby, and Laura Pientka

INTRODUCTION AND OVERVIEW

As discussed in earlier chapters, psychotic illnesses can best be conceptualized as neurocognitive disorders characterized by dysfunction in key information-processing systems of the brain. Neuropsychological domains of impairment include motor skills, attention, working memory, visuospatial abilities, linguistic ability, memory, executive functions, and social cognition, with the greatest deficits observed in verbal memory (reviewed in [1]).

Together, the information-processing deficits of schizophrenia significantly affect patients' abilities to participate in everyday activities and have a negative impact on their overall quality of life. These neuropsychological impairments may reflect an underlying "generalized" or "global" cognitive deficit, suggesting a common neurobiological source (reviewed in [2]). While the exact nature of the underlying "source" is debated and likely to be heterogeneous in origin and nature, individuals with psychotic illness show widespread changes in behavioral and neuroimaging indices of neural system functioning associated with both elemental and complex aspects of perception, cognition, social/emotional information processing, and motivated behavior. Given their functional significance, our field aims to design treatments to target these cognitive deficits.

Decades of research explicitly focused on treatments to target various neuropsychological impairments in schizophrenia, leading to the development of the field of "cognitive remediation," which was formally defined in 2010 by an expert consensus group as "a behavioral training-based intervention that aims to improve cognitive processes (attention, memory, executive function, social cognition, or metacognition) with the goal of durability and generalization."[3] Cognitive remediation in schizophrenia stemmed initially from prior work in the field of traumatic brain injury, was derived from a broad "neurorehabilitation" approach, and was based on a neuropsychological perspective of brain function.[4] Several meta-analyses have demonstrated the effectiveness of such interventions, leading to practical understanding of how to help patients translate neuropsychological gains into functional improvements (reviewed in the following chapter by Wykes et al.).

In this chapter, we offer a complementary neurologically informed "experimental medicine" approach to understanding and treating impaired cognition in psychotic illnesses (see Table 66.1 for a comparison of the two approaches). As systems neurophysiologists, we seek to understand how key dysfunctions in distributed neural systems ultimately lead to widespread cognitive problems in psychosis.[4,5] Such an approach examines behavioral and neuroimaging findings that demonstrate target engagement—reflecting neural system plasticity—and details the cognitive improvements that result from target engagement. We focus on learning-dependent neuroplasticity mechanisms that are explicitly harnessed through focused cognitive training exercises designed to directly target the identified neural system impairments.

In this light, we consider cognitive training to be a neurological treatment rather than a psychological or psychosocial treatment,[6] and we posit that optimal clinical benefit will likely be achieved via integration with novel plasticity-enhancing pharmacologic and/or neuromodulation interventions.[7,8] We also assert that such neuroscience-informed cognitive training methods serve as only one component of a comprehensive treatment program that provides psychoeducation, family support, functional remediation, therapies for meta-cognition, social skills and resiliency, and psychological management of psychosis.

At the heart of this perspective are three axioms: (1) The brain consists of neural systems that support cognitive and affective processing; (2) These systems are inherently plastic and facilitate adaptation to meaningful experiences throughout the life span (learning-induced neuroplasticity); and (3) At a systems level, plasticity implies that the brain represents the relevant sensory and cognitive/affective inputs and action outputs with larger and more coordinated populations of neurons that are distributed throughout the relevant brain regions.[4] Interestingly, in infancy and early childhood, brain plasticity is continuously engaged; in the older individual, brain change is triggered via context and outcome-dependent neuromodulator release from the subcortical limbic system (reviewed by [9]). Thus, harnessing plasticity in a targeted, adaptive, and efficient manner in adults requires precise engineering of the learning experience or learning trials—what we term "targeted cognitive training."

Table 66.1 TWO APPROACHES TO INTERVENTIONS TO ADDRESS COGNITIVE DEFICITS IN SCHIZOPHRENIA
Adapted from[10]

BROAD NEUROPSYCHOLOGICAL REHABILITATION APPROACH	TARGETED COGNITIVE TRAINING APPROACH
Considered a psychological or psychosocial intervention	Considered a neurological intervention
Focused on remediation of a broad array of deficits on the basis of several decades of cognitive psychology and neuropsychology research; sometimes referred to as a "top-down" approach.	Focused on inducing neuroplasticity in specific distributed neural systems on the basis of integrative neuroscience research; sometimes referred to as a "bottom-up" approach.
Dominated by a treatment development paradigm.	Dominated by a mechanistic experimental medicine paradigm.
Multiple active ingredients often embedded together in real-world settings in an effort to maximize behavioral change.	Critical drivers of cognitive change isolated to identify underlying pathophysiology as well as mechanisms and biomarkers of response to intervention.
Treatment may include computerized exercises providing graduated repetitive practice across a wide range of higher-order cognitive domains, usually using the visual modality.	Training consists of computerized exercises that are individually adaptive to drive learners to their threshold at a high reward schedule; exercises focus intensively, in a targeted manner, on relevant early perceptual processes or component cognitive operations along with domain-specific attention and working-memory operations. Most well-studied in the auditory domain.
Training often delivered at a rate of 2–3 sessions per week for a total of approximately 20–30 h, though both lower and higher numbers of sessions over different time spans have been studied.	Training generally aims to be delivered at a rate of 4–5 sessions per week for a total of 20–40 h in a given cognitive domain. However, little is known about the optimal "dose" of training from a neurological point of view.
Examples of available programs include CogPack, CogRehab, RehaCom, Happy Neuron	Examples of available programs include Posit Science and Happy Neuron auditory, visual, and social cognition exercises

Additionally, evidence suggests that broad neuropsychological rehabilitation methods in people with schizophrenia and other brain illnesses may hit a ceiling unless underlying perceptual processing limitations are addressed.[4] Thus, development of cognitive training approaches for psychosis will require strategies that specifically target underlying perceptual processing deficits in relevant individuals, in order to yield robust gains, generalization to broad new learning patterns, and association with improvements in efficiency of distributed neural systems. Indeed, in order for the brain to undergo significant plastic changes, a well-defined specific skill must be successfully practiced at a sufficient level of difficulty for a sufficient amount of time with a robust reward schedule. While this is true for a healthy brain, it appears to be even more so for an impaired brain, such as in schizophrenia and other psychotic illnesses.

DEFINING THE TRAINING TARGETS: PSYCHOTIC ILLNESSES AS DISORDERS OF DISTRIBUTED NEURAL SYSTEMS

As discussed in earlier chapters, a robust body of work supports the concept that psychotic illnesses are disorders of information processing in key neural circuits, with abnormal functional connectivity, neural oscillatory activity, and processing efficiency across cortical and subcortical sectors. Furthermore, these neural system impairments demonstrate consistent associations with behavioral impairments in perception, cognition, emotion, and motivation.

Notably, neural systems are dynamic entities that involve a complex choreography across time, space, and neural oscillatory frequencies among key computational nodes in the brain. Oscillatory activity spans the millisecond cycles of an interneuron-pyramidal neuron microcircuit, the columnar mesocircuit activity that occurs over hundreds of milliseconds, and the coordination of long-range brain macrocircuit interactions that support higher order cognition over seconds, corresponding to high gamma (100 Hz), delta to low gamma (10 Hz), and infraslow (<1 Hz) frequency ranges, respectively.[11,12] This oscillatory activity supports information processing activities of the brain, relying on a neuronal-glial scaffold that has intrinsic plasticity. Harnessing such plasticity in an informed and adaptive manner is the key to improving brain information processing abnormalities in pathological conditions. A key concept here is that the choreography of these oscillatory frequencies involves a circuit with *both* bottom-up/feed-forward mechanisms from sensory processing regions to prefrontal cognitive control operations *and* top-down/feedback mechanisms from prefrontal sectors to sensory inputs.

In one example of such a distributed neural system, Herman and colleagues presented the spatiotemporal characterization of the brain network that supports auditory working memory (Figure 66.1).[13] These studies suggest that treatment strategies aimed at enhancing function of this distributed neural system should focus on the connectivity and timing of interactions between prefrontal and posterior sensory areas to target both the feed-forward and feed-back components of the auditory working memory circuit. The goal is to improve the processing of task-relevant sensory information with a higher degree of fidelity, accuracy, and efficiency, operating in conjunction with prefrontal predictive coding processes that can more efficiently and adaptively anticipate, modulate, and operate on task-relevant sensory information.

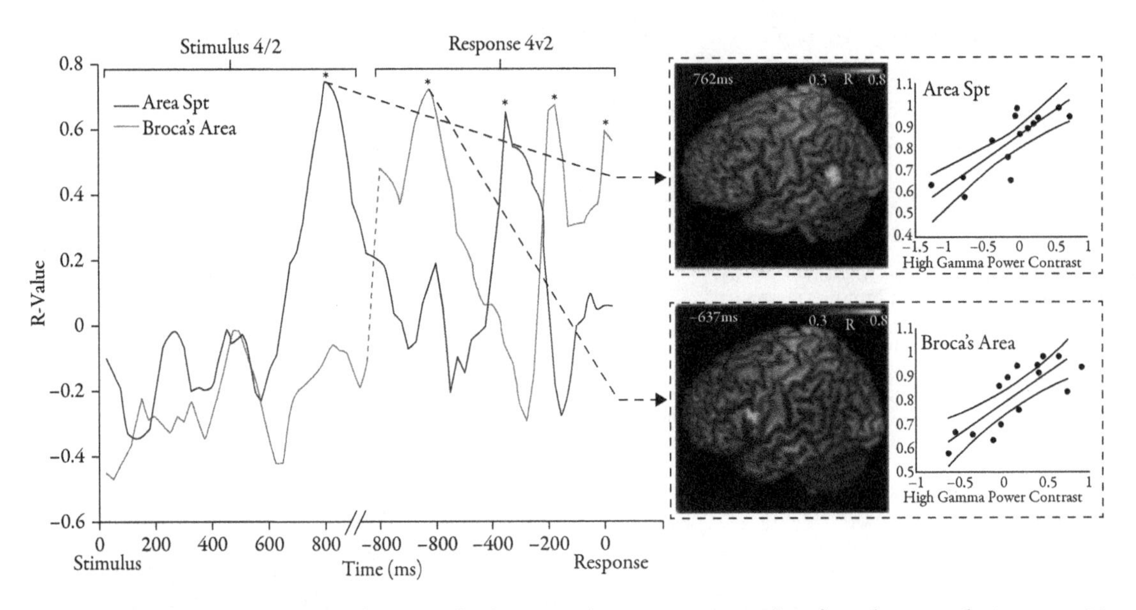

Figure 66.1 **Example of the information-processing choreography that occurs between sensory and prefrontal cortices during a cognitive operation.** In healthy individuals performing a syllable-repetition task that requires auditory working memory, high-gamma oscillatory activity in Broca's area and Area Spt (illustrated in panel on right) alternates temporally in explaining variation in task performance (illustrated in graphs on left). Impairments or inefficiencies in this high-gamma choreography in the prefrontal-sensory network contribute to auditory working-memory deficits in schizophrenia and are amenable to targeted cognitive training. Adapted from Herman et al. 2013.[13] See color insert 5 in front of book for full colorized version.

In this chapter, we describe two examples of cognitive processes—perception and social cognition—that are impaired in psychotic disorders and have high functional significance. Evidence suggests that these cognitive processes and their underlying neural systems can serve as meaningful neuroscience-informed cognitive training targets, driving improvements in both lower-level representational processes and higher-order cognitive and meta-cognitive functions.

Although a robust and venerable body of literature delineates impairments in attention, working memory, and executive function in psychotic illness as well, these impairments are often inextricably entwined with perceptual deficits.[4,14] At present, neuroscience-informed targeted cognitive training approaches are somewhat less well-developed for these higher-order domains, and studies commonly focus on several higher-order functions simultaneously. Therefore, we refer the interested reader to additional sources on working memory training.[15–19]

PERCEPTUAL PROCESSING AS A TREATMENT TARGET

Perceptual processing includes the earliest stages of psychophysical processing of sensory representations in the brain as well as later stages of allocation of attentional resources and associative memory processes to perceptual information. While we only provide a cursory description of abnormalities in perceptual processing in psychosis, there is a deep and rich literature (e.g., [20–22]).

One of the best-studied areas has been in the domain of auditory perceptual dysfunction in schizophrenia and its underlying neurobiological mechanisms. Tone-matching tasks and electrophysiological measures, especially the mismatch negativity (MMN) event-related potential, have yielded valuable insights into auditory cortical neural system dysfunction that contributes to schizophrenia clinical symptoms, including impairments in social cognition and social interaction.[21,23,24] MMN reflects an auditory cortical operation whereby potentially important changes in background auditory information are registered in the brain. In people with schizophrenia, these operations are impaired, leading to a reduction in preattentive sensitivity to changes in background auditory events. These impairments are associated with poorer functioning. Furthermore, the ability to detect changes in tone and rhythm is crucial for detecting the emotional content of voice and for "sounding out" words during reading. Impairments in these processes in schizophrenia are associated with problems in auditory emotion recognition and phonological processing, leading to difficulties with social interactions and reading ability.[21]

A wide range of research on visual perception, including backward masking studies and surround suppression studies, have also demonstrated significant impairments in people with psychotic illnesses.[22] Similar to auditory processing deficits, impairments in early visual system functioning have been associated with poorer real-world outcomes. Butler et al.[25] found that deficits in evoked potential signals predicted impaired contrast sensitivity, which was related to deficits in complex visual processing in people with schizophrenia. Both the evoked potential and the contrast sensitivity measures significantly predicted community functioning. Sergi et al.[26] showed that deficits in early visual processing as assessed by a visual backward masking task were linked to poorer functional status in people with schizophrenia and that this relationship was mediated by social perception performance.

Taken together, these data indicate that impairments in early auditory and visual perceptual operations are associated with a range of higher-order cognitive deficits and poorer psychosocial functioning. Thus, treatments that target the accuracy, fidelity, and "choreography" of information transfer between sensory cortical areas and prefrontal-mediated functions (including attention, working memory, cognitive control, and predictive coding) have a high likelihood of providing significant benefit to patients.

SOCIAL COGNITION AS A TREATMENT TARGET

Social cognition refers to the key neural operations that allow an individual to accurately and efficiently process components of socially relevant information, such as eye gaze, facial expression, facial and vocal emotion, and cognitive theory of mind (TOM). While neuroleptic medication has been instrumental in reducing patients' positive symptoms, many people are left with debilitating negative symptoms, such as social withdrawal, anhedonia, apathy, and blunted emotional response. These symptoms are closely related to impaired reward processing, decreased motivation, and poor functional outcome.[27–29]

Social cognitive deficits have been categorized into four general domains: social cue identification (e.g., eye gaze detection, facial affect identification, and vocal emotional prosody), mentalizing or TOM, experience sharing, and emotion regulation.[30] Given that patients with psychosis often demonstrate difficulty processing elementary components of socially relevant information, social cognitive training exercises have generally focused on identifying social cues, such as recognition of facial emotions, to improve the interpretation of intentions and perspectives of others, although a few have focused on mentalizing or TOM.[30,31]

It is important to think of social cognition treatment targets in terms of their underlying neural systems. Converging evidence indicates that these functionally important behavioral deficits arise from neural system dysfunction along multiple nodes in what is sometimes called the "social brain." This includes visual association areas in the occipital cortex and the superior temporal sulcus, important for processing eye gaze and facial expressions,[32,33] as well as the fusiform gyrus, important for facial recognition. Additional neural system targets that play a role in social cognition include the medial prefrontal cortex and anterior cingulate, which are key for processing of social information and rewards.[34] Dysfunction in the ventral striatum, a key neural system for appropriate reward responsivity, may lead individuals with psychosis to be less responsive to positive reinforcement, including the positive reinforcement of social cues, potentially leading to lack of motivation and social withdrawal.[35] Not surprisingly, the prefrontal cortex has also been implicated in TOM processing, as well as in the working memory and attention demands of social cognition.[36] Finally, the limbic system, specifically the amygdala, functions in the appropriate processing of facial emotion signals and appropriate responses to other emotion cues in the environment.

Our current understanding of the importance of social cognition for community functioning and long-term outcome for individuals with psychosis, along with our understanding of the underlying neural system dysfunction that contribute to the deficits in these processes, provide a strong rationale for the development of social cognition training to drive behavioral improvement in tandem with neural system plasticity.

CURRENT RESEARCH IN TARGETED COGNITIVE TRAINING FOR PSYCHOSIS

In this section, we present a brief summary of research focused on using principles of targeted cognitive training for perceptual processing and for social cognition in individuals with psychosis. Such training is delivered via interactive software, permitting precise control over exercise specifications and delivery. As noted earlier, the goal is to improve the accuracy, fidelity, and "choreography" of information transfer between sensory cortical areas and prefrontally mediated functions.[9] This training differs from conventional cognitive remediation approaches in the following ways:

1. Hundreds to thousands of trials are performed for each exercise.
2. Attention and reward mechanisms are tightly controlled on a trial-by-trial basis.
3. Stimuli are sharply emphasized and designed to drive strong, coherent cortical responses during the earliest phases of perceptual processing, as such responses are critical for effective feed-forward engagement of associative memory processes.[37]
4. Difficulty adjusts on a trial-by-trial basis to maintain 80% correct performance to optimize learning and plasticity via engagement of neuromodulatory release related to salience and reward.
5. Individual exercises are designed to address specific components within a general cognitive domain. For example, the components of auditory syllable identification and auditory working memory are trained to improve verbal memory, and component operations of eye gaze detection and face identification are trained to improve social cognition.

TARGETED COGNITIVE TRAINING OF PERCEPTUAL PROCESSING

BEHAVIORAL FINDINGS

Here, we review research on targeted cognitive training of auditory perceptual processing (for studies of visual perceptual training, see [38]). In a double-blind randomized controlled trial (RCT), auditory training participants showed significant gains in speed of processing, verbal learning, verbal memory, and global cognition relative to computer game control participants.[39] At a 6-month, no-contact follow-up, improvements in cognition were significantly associated with improved functional outcome.[40] In a second double-blind multisite RCT, identical auditory training of 40 hours over 8 weeks in young patients with recent-onset schizophrenia led to significant improvement in verbal memory, problem-solving, and global cognition relative to computer game control patients.[41] Improvement in auditory processing speed after 20 hours of training was significantly associated with improvement in global cognition at the end of 40 hours of training, indicating that successful engagement of improvement in perceptual processing efficiency generalized to improvement in higher-order cognitive functions.

Indeed, in a study of individuals with schizophrenia and schizoaffective disorder who underwent 40–50 hours of targeted cognitive training of auditory processing, Biagianti et al. found a steep improvement in auditory processing speed after 20 hours of training, followed by a plateau in this measure at subsequent time-points.[42] Subjects, who achieved and sustained the fastest auditory processing speed plateau, greatest transfer effects to untrained higher-order cognitive domains. This indicates that plasticity in auditory psychophysical efficiency is related to treatment response.

Popov et al. tested the effects of 20 hours of auditory training over 4 weeks compared to 60- to 90-minute sessions of general cognition exercises 3 times per week for 4 weeks among inpatients in Germany. The auditory training group showed significantly greater gains in verbal working memory and verbal learning.[43]

Not all studies, however, have found significant neuropsychological improvement after auditory system training in schizophrenia. In a multisite feasibility study, outpatients were randomized to auditory training or control computer games.[44] After 20 sessions, in an intent-to-treat-analysis, the auditory training group showed significant gains in verbal learning, APS, and global cognition relative to controls. At study endpoint, however, the effects on verbal learning and global cognition did not reach significance, possibly due to study power and participant attrition. In a single-arm, multisite trial, Murthy et al. found that 8–10 weeks of training resulted in processing speed improvement but no improvement on the Cogstate computerized cognitive assessment battery. Interestingly, in a post hoc analysis, a subgroup of subjects who showed improvement in cognitive performance across the three baseline time points (called "learners") also showed significant gains in processing speed and significant improvement in the cognitive composite score at the end of training.[45] Subject-specific moderators—such as the ability to engage the training target and medication regimen[42,46]—may influence the response to training. Additionally, in a multisite RCT of auditory training vs. computer games combined with open-label lurasidone, Kantrowitz et al. observed no between-group differences in the primary cognitive outcome measure (MATRICS Consensus Cognitive Battery [MCCB] score).[47] In contrast to studies discussed earlier, this study performed a modified intent-to-treat analysis on all subjects who engaged in at least one cognitive training session; however, only 62% of the participants completed >25 cognitive training sessions over a 6-month period, representing a dosing intensity half of that used in prior studies. Interestingly, small-to-moderate effect sizes were seen following auditory training in speed of processing, visual learning, and working memory after the first 20 hours of training, presumably when training was occurring with reasonable intensity.[48] Thus, it appears that auditory training needs to be delivered with a relatively high intensity over a dense training schedule.

NEUROIMAGING FINDINGS

In the last few years, several studies have examined neural system plasticity using magnetoencephalography (MEG) and magnetic resonance imaging (MRI) after targeted cognitive training focused on auditory processing and auditory/verbal working memory dysfunction in schizophrenia. MEG records magnetic fields (measured by sensors outside the head) that are produced by electrical activity within the cortex. It provides a direct measure of electrical activity within the brain with a very high temporal resolution (milliseconds), and is an excellent technique for assessing early neural operations as individuals encode and process information.

Using MEG, Popov et al. studied measures of auditory sensory gating to explore the effects of targeted auditory training vs. CogPack cognitive remediation exercises.[43] Auditory sensory gating is felt to reflect aspects of sensory filtering. Approximately 20 hours of auditory training normalized sensory gating deficits in patients with schizophrenia, while this effect was not evident in subjects who completed CogPack. In a second MEG study, these investigators examined oscillatory high gamma power.[48] After training, subjects showed increased gamma power in response to auditory stimuli compared to those in the control group, and these changes were associated with improvements in cognition and functioning.

In another set of studies, participants with schizophrenia had hemispheric asymmetry abnormalities during auditory processing, but after 50 hours of auditory training, they showed normalization of the asymmetry that was associated with gains in verbal learning.[49] In a second study, Dale et al. found that, after training, participants showed a significant increase in high gamma activity within left dorsolateral prefrontal cortex, plus increased broadband responses in primary auditory cortex[50] (Figure 66.2).

These changes in MEG signal were found to correlate with improvements in executive functioning after training (Figure 66.2, right panel), suggesting that training may improve cognition both by improving local cortical high gamma power and by restoring the choreography between neural activity in prefrontal attention and language regions and activity in auditory cortex. In other words, by targeting auditory perceptual processing and improving the fidelity and power of the auditory cortical signal, this type of training appears to evoke prefrontal plasticity through enhancement of a behaviorally salient feed-forward signal. This, in turn, appears to result in

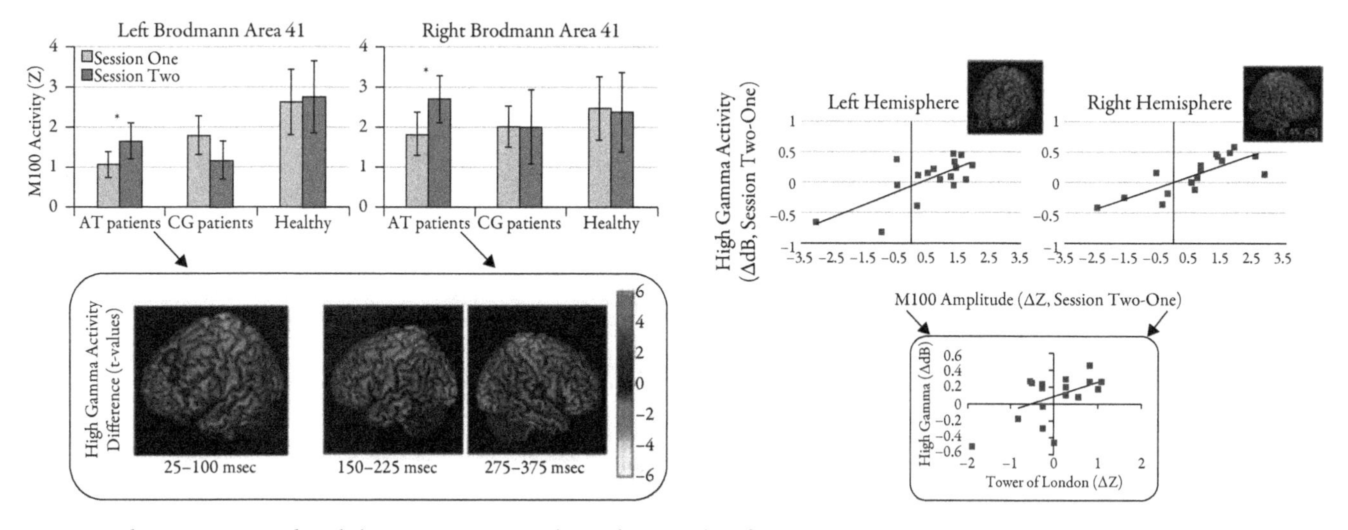

Figure 66.2 Auditory training-induced changes in MEG signal in auditory and prefrontal cortices and relation to cognitive gains.

Top left panel: Broadband M100 auditory response within a priori primary auditory cortex locations is increased in the auditory training patient group. Both healthy participants (imaged at the two time points) and the computer games patient group showed no effect on M100 response.

Bottom left panel: Changes in high gamma activity occurred in the auditory training group in both left dorsolateral prefrontal cortex and inferior frontal gyrus, followed by increased activity in temporal regions.

Right panel: Changes in M100 amplitude were significantly related to subsequent changes in the activity of corresponding left and right hemisphere regions of prefrontal cortex. Late prefrontal changes were also correlated with changes in performance on the untrained Tower of London problem-solving task.

improved preparatory attention processes, which itself may promote improved processing in primary auditory cortex via feed-back mechanisms, potentially translating to improved performance on a higher-order (untrained) executive function task. Thus, the ensemble of MEG data suggests that carefully designed intensive training of auditory processes can increase the fidelity, precision, and signal-to-noise ratio of auditory representations in the brain with accompanying beneficial effects on higher-order cognitive operations.

Finally, Ramsay et al.[51] recently reported an MRI study of thalamic volume changes in response to auditory training in young individuals with recent-onset schizophrenia. They found a significant positive correlation between change in cognition and change in left thalamus volume in the auditory training group and a negative trend in the computer games control group. Greater symptom severity at baseline reduced the likelihood of response to training both with respect to improved cognition and change in thalamic volume, again indicating the importance of individual characteristics in terms of treatment response.

TARGETED COGNITIVE TRAINING OF SOCIAL COGNITION

BEHAVIORAL FINDINGS

In the past 15 years, numerous studies have focused on improving social cognition and social interactions in schizophrenia patients using multiple intervention methods. While the chapter on social cognition delves deeper into the subject, we briefly review targeted cognitive training approaches used to improve basic component processes of social cognition in schizophrenia, with a focus on facial emotion and emotional prosody identification.

In studies of Training of Affect Regulation (TAR), which consists of 12 sessions of computerized and noncomputerized tasks that help subjects identify and verbalize facial features and associate them with basic emotions—subjects' performance, which was impaired before training, fell within the normal range after the intervention.[52,53] TAR induced greater improvements in prosodic affect recognition and global social functioning as well as TOM, social competence, and improved social functioning than CogPack-only training, highlighting the functional impact of facial affect regulation training programs.

Applying principles of targeted "neuroplasticity-informed" training, Nahum et al. developed a set of online exercises focused on component processes of visual and auditory social cognition, such as eye-gaze detection, facial emotion recognition, and vocal emotional prosody identification.[54] Participants with psychotic disorders demonstrated significant improvement on the exercises, approaching the initial performance of healthy controls. They also exhibited improvements on distal measures of facial memory, prosody identification, social functioning, motivation, and reward sensitivity. Fisher et al.[55] recently compared the effects of targeted cognitive training of auditory and visual processing alone and combined with social cognition training. Outpatient participants who received the combined training programs demonstrated greater prosody identification and reward processing than those receiving targeted cognitive training alone. Interestingly, deficits in prosody identification and reward processing had been associated with negative symptoms and poorer functional outcome at baseline, suggesting that improvement in these domains may have longer-term functional benefits.

NEUROIMAGING FINDINGS

To date, studies of the neurological effects of targeted social cognition training have focused mainly on changes in brain activation patterns on functional MRI (fMRI). Hooker et al. found that participants with schizophrenia exhibited a greater increase in postcentral gyrus activity during emotion recognition of both positive and negative emotions after receiving auditory cognitive training plus targeted SCT than those who received auditory cognitive training alone.[56] They also showed neural activation increases in bilateral amygdala, right putamen, and right medial prefrontal cortex, and these regional activation increases, especially in the right amygdala, predicted improvement on a behavioral measure of perceiving emotions.[57]

Reality monitoring—the ability to distinguish self-generated from externally presented information—is impaired in patients with schizophrenia. Unlike healthy subjects, patients with schizophrenia do not show task-associated activation of the medial prefrontal cortex, a region important for successful reality monitoring. Extensive targeted auditory, visual, and social cognition training resulted in improved reality-monitoring performance in patients with schizophrenia compared to those receiving computer game training. This improvement was associated with increased task-related activation of the medial prefrontal cortex similar to baseline levels in healthy subjects.[58] Remarkably, patients who showed larger training-induced increases in medial prefrontal cortex activation during the reality monitoring task also exhibited better real-world social functioning when assessed 6 months later, suggesting that intensive targeted cognitive training of fairly basic cognitive processes in schizophrenia can normalize brain-behavior associations during a nontrained metacognitive process in association with better longer-term real-world functioning.

In a study of TAR by Habel et al., participants exhibited increased activation in the left middle and superior occipital lobe, right inferior and superior parietal cortex, and inferior frontal cortex bilaterally during an emotion recognition task.[59] Furthermore, the fMRI activation changes observed in the TAR group correlated with functional and behavioral improvements, possibly reflecting more efficient use of attentional, perceptual, or cognitive strategies during the processing of social-emotional stimuli.

Several studies have used event-related potential (ERP)-based EEG or MEG methods to evaluate the neural processing changes associated with the effects of social cognition training. Luckhaus et al. administered TAR to male schizophrenia patients who had been committed because of violent offenses, which showed a very high treatment effect that persisted for 2 months post-treatment. While ERPs to a facial affect task were unchanged following treatment, low-resolution brain electromagnetic tomography (LORETA) revealed activation decreases in left-hemispheric parietal-temporal-occipital regions and activation increases in right dorsolateral prefrontal cortex and anterior cingulate, suggesting increased efficiency in structural face decoding and a shift toward a more reflective mode of emotional face decoding that relies on increased frontal brain activity.[60] Stroth et al. further showed significant ERP increases to a facial affect task in parietal and occipital regions in the schizophrenia group after TAR compared to controls, as well as evidence of neuroplasticity changes in the inferior and superior temporal lobe and in the precuneus.[61] In a study by Popova et al., schizophrenia patients were randomly assigned to computer-based facial affect training that focused on affect discrimination and working memory in 20 daily 1-hour sessions; similarly intense, targeted cognitive training on auditory-verbal discrimination and working memory; or treatment as usual.[62] Alpha power increase during the facial affect recognition task was larger following affect training than after treatment as usual. A recent review of 11 articles published between 2000 and 2015 summarized the findings observed in imaging studies of social cognition training in patients with schizophrenia;[63] a reasonably consistent pattern of functional changes was observed in early visual processing areas, the limbic system, and the prefrontal regions. Together, the data indicate that the neural processes supporting social cognitive operations, are modifiable and that social cognition training is associated with neuroplastic changes in the brain that are associated with consequent improvements in social cognitive performance.

FUTURE DIRECTIONS

While emerging research in cognitive training for psychosis shows promise as a targeted "neuroplasticity treatment," some cautionary notes are in order. First, almost by definition, in order to generate meaningful change in neural system functioning, the individual will need to participate in *effortful* training—very similar to going to the gym on a regular basis in order to lose weight. Although some "gamification" can be incorporated in the design of cognitive training exercises, the treatment still requires a high degree of commitment, engagement, and motivation—all functions that are impaired in psychosis. Because of this, adherence can be low, and appropriate supports will need to be in place in real-world clinical settings. Second, we need to understand the exact neural mechanisms that must be harnessed in order to induce robust, meaningful alterations in neural system functioning and ultimately in patient behavior. We must identify the target and how to best engage that target, and we must determine how these targets vary between individuals according to their baseline pathology. Third, we must acknowledge that cognitive gains may be achieved in different individuals via quite different mechanisms of action. As discussed in Fisher et al. ,[10] improvement in global cognition following various types of training may occur via (1) higher fidelity of task-relevant perceptual representations, (2) greater attentional control, (3) more efficient psychomotor processing speed due to repetitive practice, (4) better inhibition of task-irrelevant responses by improvement of distractor suppression, (5) improved motivation and reward processing from increased therapeutic contact, and (6) improved environmental enrichment. Thus, different forms of training may yield changes in any combinations of these factors, but not all mechanisms may be involved in the underlying pathophysiology or important for yielding robust and enduring neurobehavioral improvements in a specific individual.

Another issue is the heterogeneity of responses observed across studies, suggesting the presence of some type of unmeasured individual "plasticity potential" or "learning capacity" that may constrain the response to training. Previous studies in older adults suggest that this may be related to the capacity for sustained attention and that directly training sustained attention can improve learning in an auditory processing task and executive functioning.[64] Interestingly, we have shown that serum anticholinergic burden induced by medications in schizophrenia is negatively associated with cognitive gains after targeted training, consistent with the fact that anticholinergic drugs exert deleterious effects on attentional capacity in patients.[46]

Just as there is much work to be done to understand the science and the neurology of successful cognitive training for people with psychotic illness, there is also much work to be done to develop effective real-world treatments based on these principles. For example, it is likely that enhancement of targeted cognitive training using neuromodulation or pharmacologic adjuncts will prove to be beneficial. Additionally, clinical cognitive training programs are currently limited by time and travel constraints of patients to visit therapists or coaches and the lack of insurance coverage for these approaches. We and others have begun to investigate the use of remote cognitive training in conjunction with adjunctive remote group therapy for social skills and remote motivational coaching for people with psychosis.[65–67] Translation of clinical cognitive training advances to mobile applications or online training exercises may pave the way for greater accessibility to treatment. The future of this field is tantalizing: we seek to develop rigorous, evidence-based personalized cognitive training approaches that combine a deep understanding of how to drive adaptive neurological and behavioral changes with engaging, scalable, and cost-effective digital tools to ultimately promote cognitive recovery, autonomy, and well-being for individuals with psychosis.

REFERENCES

1. Best MW, Bowie CR. A review of cognitive remediation approaches for schizophrenia: from top-down to bottom-up, brain training to psychotherapy. *Expert review of neurotherapeutics*. 2017;17(7): 713–723.
2. Barch DM, Sheffield JM. Cognitive impairments in psychotic disorders: common mechanisms and measurement. *World psychiatry*. 2014;13(3):224–232.
3. Wykes T, Huddy V, Cellard C, McGurk SR, Czobor P. A meta-analysis of cognitive remediation for schizophrenia: methodology and effect sizes. *American journal of psychiatry*. 2011;168(5):472–485.
4. Vinogradov S, Fisher M, de Villers-Sidani E. Cognitive training for impaired neural systems in neuropsychiatric illness. *Neuropsychopharmacology*. 2012;37(1):43–76.
5. MacDonald AW, 3rd, Carter CS. Cognitive experimental approaches to investigating impaired cognition in schizophrenia: a paradigm shift. *Journal of clinical and experimental neuropsychology*. 2002;24(7):873–882.
6. Vinogradov S, Fisher M, Nagarajan S. Cognitive training in schizophrenia: golden age or wild west? *Biological psychiatry*. 2013;73(10):935–937.
7. Swerdlow NR. Beyond antipsychotics: pharmacologically-augmented cognitive therapies (PACTs) for schizophrenia. *Neuropsychopharmacology*. 2012;37(1):310–311.
8. Kantrowitz JT, Epstein ML, Beggel O, et al. Neurophysiological mechanisms of cortical plasticity impairments in schizophrenia and modulation by the NMDA receptor agonist D-serine. *Brain*. 2016;139(Pt 12):3281–3295.
9. Merzenich MM, Van Vleet TM, Nahum M. Brain plasticity-based therapeutics. *Frontiers in human neuroscience*. 2014;8:385.
10. Fisher M, Herman A, Stephens DB, Vinogradov S. Neuroscience-informed computer-assisted cognitive training in schizophrenia. *Annals of the New York Academy of Sciences*. 2016;1366(1):90–114.
11. Mathalon DH, Sohal VS. Neural oscillations and synchrony in brain dysfunction and neuropsychiatric disorders: it's about time. *JAMA psychiatry*. 2015;72(8):840–844.
12. Vinogradov S, Herman A. Psychiatric illnesses as oscillatory connectomopathies. *Neuropsychopharmacology*. 2016;41(1):387–388.
13. Herman AB, Houde JF, Vinogradov S, Nagarajan SS. Parsing the phonological loop: activation timing in the dorsal speech stream determines accuracy in speech reproduction. *Journal of neuroscience*. 2013;33(13):5439–5453.
14. Haenschel C, Linden D. Exploring intermediate phenotypes with EEG: working memory dysfunction in schizophrenia. *Behavioural brain research*. 2011;216(2):481–495.
15. Bor J, Brunelin J, d'Amato T, et al. How can cognitive remediation therapy modulate brain activations in schizophrenia? An fMRI study. *Psychiatry research*. 2011;192(3):160–166.
16. Haut KM, Lim KO, MacDonald A, 3rd. Prefrontal cortical changes following cognitive training in patients with chronic schizophrenia: effects of practice, generalization, and specificity. *Neuropsychopharmacology*. 2010;35(9):1850–1859.
17. Penades R, Gonzalez-Rodriguez A, Catalan R, Segura B, Bernardo M, Junque C. Neuroimaging studies of cognitive remediation in schizophrenia: a systematic and critical review. *World journal of psychiatry*. 2017;7(1):34–43.
18. Ramsay IS, Nienow TM, MacDonald AW, 3rd. Increases in intrinsic thalamocortical connectivity and overall cognition following cognitive remediation in chronic schizophrenia. *Biological psychiatry: Cognitive neuroscience and neuroimaging*. 2017;2(4):355–362.
19. Ramsay IS, Nienow TM, Marggraf MP, MacDonald AW. Neuroplastic changes in patients with schizophrenia undergoing cognitive remediation: triple-blind trial. *British journal of psychiatry*. 2017;210(3):216–222.
20. Javitt DC. Sensory processing in schizophrenia: neither simple nor intact. *Schizophrenia bulletin*. 2009;35(6):1059–1064.
21. Javitt DC, Sweet RA. Auditory dysfunction in schizophrenia: integrating clinical and basic features. *Nature reviews: Neuroscience*. 2015;16(9):535–550.
22. Butler PD, Silverstein SM, Dakin SC. Visual perception and its impairment in schizophrenia. *Biological psychiatry*. 2008;64(1):40–47.
23. Light GA, Braff DL. Mismatch negativity deficits are associated with poor functioning in schizophrenia patients. *Archives of general psychiatry*. 2005;62(2):127–136.
24. Kantrowitz JT, Hoptman MJ, Leitman DI, et al. Neural substrates of auditory emotion recognition deficits in schizophrenia. *Journal of neuroscience*. 2015;35(44):14909–14921.
25. Butler PD, Zemon V, Schechter I, et al. Early-stage visual processing and cortical amplification deficits in schizophrenia. *Archives of general psychiatry*. 2005;62(5):495–504.
26. Sergi MJ, Rassovsky Y, Nuechterlein KH, Green MF. Social perception as a mediator of the influence of early visual processing on functional status in schizophrenia. *American journal of psychiatry*. 2006;163(3):448–454.
27. Gard DE, Fisher M, Garrett C, Genevsky A, Vinogradov S. Motivation and its relationship to neurocognition, social cognition, and functional outcome in schizophrenia. *Schizophrenia research*. 2009;115(1):74–81.
28. Green MF. Impact of cognitive and social cognitive impairment on functional outcomes in patients with schizophrenia. *Journal of clinical psychiatry*. 2016;77 Suppl 2:8–11.

29. Fett AK, Viechtbauer W, Dominguez MD, Penn DL, van Os J, Krabbendam L. The relationship between neurocognition and social cognition with functional outcomes in schizophrenia: a meta-analysis. *Neuroscience and biobehavioral reviews.* 2011;35(3):573–588.
30. Horan WP, Green MF. Treatment of social cognition in schizophrenia: current status and future directions. *Schizophrenia research.* 2017.
31. Kurtz MM, Mueser KT, Thime WR, Corbera S, Wexler BE. Social skills training and computer-assisted cognitive remediation in schizophrenia. *Schizophrenia research.* 2015;162(1–3):35–41.
32. Gobbini MI, Haxby JV. Neural systems for recognition of familiar faces. *Neuropsychologia.* 2007;45(1):32–41.
33. Fusar-Poli P, Placentino A, Carletti F, et al. Functional atlas of emotional faces processing: a voxel-based meta-analysis of 105 functional magnetic resonance imaging studies. *Journal of psychiatry and neuroscience: JPN.* 2009;34(6):418–432.
34. Green MF, Horan WP, Lee J. Social cognition in schizophrenia. *Nature reviews: Neuroscience.* 2015;16(10):620–631.
35. Murray GK, Cheng F, Clark L, et al. Reinforcement and reversal learning in first-episode psychosis. *Schizophrenia bulletin.* 2008;34(5):848–855.
36. Van Overwalle F. Social cognition and the brain: a meta-analysis. *Human brain mapping.* 2009;30(3):829–858.
37. Nahum M, Lee H, Merzenich MM. Principles of neuroplasticity-based rehabilitation. *Progress in brain research.* 2013;207: 141–171.
38. Surti TS, Corbera S, Bell MD, Wexler BE. Successful computer-based visual training specifically predicts visual memory enhancement over verbal memory improvement in schizophrenia. *Schizophrenia research.* 2011;132(2–3):131–134.
39. Fisher M, Holland C, Merzenich MM, Vinogradov S. Using neuroplasticity-based auditory training to improve verbal memory in schizophrenia. *American journal of psychiatry.* 2009;166(7):805–811.
40. Fisher M, Holland C, Subramaniam K, Vinogradov S. Neuroplasticity-based cognitive training in schizophrenia: an interim report on the effects 6 months later. *Schizophrenia bulletin.* 2010;36(4):869–879.
41. Fisher M, Loewy R, Carter C, et al. Neuroplasticity-based auditory training via laptop computer improves cognition in young individuals with recent onset schizophrenia. *Schizophrenia bulletin.* 2015;41(1):250–258.
42. Biagianti B, Roach BJ, Fisher M, et al. Trait aspects of auditory mismatch negativity predict response to auditory training in individuals with early illness schizophrenia. *Neuropsychiatric electrophysiology.* 2017;3.
43. Popov T, Jordanov T, Rockstroh B, Elbert T, Merzenich MM, Miller GA. Specific cognitive training normalizes auditory sensory gating in schizophrenia: a randomized trial. *Biological psychiatry.* 2011;69(5):465–471.
44. Keefe RS, Vinogradov S, Medalia A, et al. Feasibility and pilot efficacy results from the multisite Cognitive Remediation in the Schizophrenia Trials Network (CRSTN) randomized controlled trial. *Journal of clinical psychiatry.* 2012;73(7):1016–1022.
45. Murthy NV, Mahncke H, Wexler BE, et al. Computerized cognitive remediation training for schizophrenia: an open label, multi-site, multinational methodology study. *Schizophrenia research.* 2012;139(1-3):87–91.
46. Vinogradov S, Fisher M, Warm H, Holland C, Kirshner MA, Pollock BG. The cognitive cost of anticholinergic burden: decreased response to cognitive training in schizophrenia. *American journal of psychiatry.* 2009;166(9):1055–1062.
47. Kantrowitz JT, Sharif Z, Medalia A, et al. A multicenter, rater-blinded, randomized controlled study of auditory processing-focused cognitive remediation combined with open-label lurasidone in patients with schizophrenia and schizoaffective disorder. *Journal of clinical psychiatry.* 2016;77(6):799–806.
48. Popov TG, Carolus A, Schubring D, Popova P, Miller GA, Rockstroh BS. Targeted training modifies oscillatory brain activity in schizophrenia patients. *NeuroImage Clinical.* 2015;7:807–814.
49. Adcock RA, Dale C, Fisher M, et al. When top-down meets bottom-up: auditory training enhances verbal memory in schizophrenia. *Schizophrenia bulletin.* 2009;35(6):1132–1141.
50. Dale CL, Brown EG, Fisher M, et al. Auditory cortical plasticity drives training-induced cognitive changes in schizophrenia. *Schizophrenia bulletin.* 2016;42(1):220–228.
51. Ramsay IS, Fryer S, Boos A, et al. Response to targeted cognitive training correlates with change in thalamic volume in a randomized trial for early schizophrenia. *Neuropsychopharmacologyy.* 2017.
52. Frommann N, Streit M, Wolwer W. Remediation of facial affect recognition impairments in patients with schizophrenia: a new training program. *Psychiatry research.* 2003;117(3):281–284.
53. Wolwer W, Frommann N. Social-cognitive remediation in schizophrenia: generalization of effects of the Training of Affect Recognition (TAR). *Schizophrenia bulletin.* 2011;37 Suppl 2:S63–70.
54. Nahum M, Fisher M, Loewy R, et al. A novel, online social cognitive training program for young adults with schizophrenia: A pilot study. *Schizophrenia research: Cognition.* 2014;1(1):e11–e19.
55. Fisher M, Nahum M, Howard E, et al. Supplementing intensive targeted computerized cognitive training with social cognitive exercises for people with schizophrenia: an interim report. *Psychiatric rehabilitation journal.* 2017;40(1):21–32.
56. Hooker CI, Bruce L, Fisher M, Verosky SC, Miyakawa A, Vinogradov S. Neural activity during emotion recognition after combined cognitive plus social cognitive training in schizophrenia. *Schizophrenia research.* 2012;139(1-3):53–59.
57. Hooker CI, Bruce L, Fisher M, et al. The influence of combined cognitive plus social-cognitive training on amygdala response during face emotion recognition in schizophrenia. *Psychiatry research.* 2013;213(2):99–107.
58. Subramaniam K, Luks TL, Fisher M, Simpson GV, Nagarajan S, Vinogradov S. Computerized cognitive training restores neural activity within the reality monitoring network in schizophrenia. *Neuron.* 2012;73(4):842–853.
59. Habel U, Koch K, Kellermann T, et al. Training of affect recognition in schizophrenia: neurobiological correlates. *Social neuroscience.* 2010;5(1):92–104.
60. Luckhaus C, Frommann N, Stroth S, Brinkmeyer J, Wolwer W. Training of affect recognition in schizophrenia patients with violent offences: behavioral treatment effects and electrophysiological correlates. *Social neuroscience.* 2013;8(5):505–514.
61. Stroth S, Kamp D, Drusch K, Frommann N, Wolwer W. Training of affect recognition impacts electrophysiological correlates of facial affect recognition in schizophrenia: analyses of fixation-locked potentials. *World journal of biological psychiatry.* 2015:1–11.
62. Popova P, Popov TG, Wienbruch C, Carolus AM, Miller GA, Rockstroh BS. Changing facial affect recognition in schizophrenia: effects of training on brain dynamics. *NeuroImage Clinical.* 2014;6:156–165.
63. Campos C, Santos S, Gagen E, et al. Neuroplastic changes following social cognition training in schizophrenia: a systematic review. *Neuropsychology review.* 2016;26(3):310–328.
64. Van Vleet TM, DeGutis JM, Merzenich MM, Simpson GV, Zomet A, Dabit S. Targeting alertness to improve cognition in older adults: a preliminary report of benefits in executive function and skill acquisition. *Cortex.* 2016;82:100–118.
65. Biagianti B, Schlosser D, Nahum M, Woolley J, Vinogradov S. Creating Live Interactions to Mitigate Barriers (CLIMB): a mobile intervention to improve social functioning in people with chronic psychotic disorders. *JMIR mental health.* 2016;3(4):e52.
66. Golas AC, Kalache SM, Tsoutsoulas C, Mulsant BH, Bowie CR, Rajji TK. Cognitive remediation for older community-dwelling individuals with schizophrenia: a pilot and feasibility study. *International journal of geriatric psychiatry.* 2015;30(11):1129–1134.
67. Schlosser D, Campellone T, Kim D, et al. Feasibility of PRIME: a cognitive neuroscience-informed mobile app intervention to enhance motivated behavior and improve quality of life in recent onset schizophrenia. *JMIR research protocols.* 2016;5(2):e77.

67

COGNITIVE REMEDIATION

THEORY, META-ANALYTIC EVIDENCE, AND PRACTICE

Til Wykes and Adam Crowther

INTRODUCTION

Individuals with a diagnosis of schizophrenia exhibit a wide range of cognitive difficulties both before the onset of the disorder and afterward even when the acute symptoms of the disorder are absent.[1–4] Even if cognition lies within the normal range it is still below that expected from siblings and/or their parents.[5] Yet we still have few targeted treatments. This chapter presents the psychological theories underlying current treatment as well as the best efficacy evidence.

As described in other chapters in this text describing recovery services, cognitive problems interfere with the attainment of life goals even when the very best recovery-oriented services are provided. But there are also other factors that affect the attainment of life goals, often in concert with cognitive difficulties. Regardless of when the disorder starts, society limits the opportunities of people with severe mental health problems through discrimination and stigma, which prevents the individual from testing or practicing life skills. These societal limitations are being addressed by campaigns around the world such as the Time to Change program in the United Kingdom, which has shown positive effects.[6] But schizophrenia also starts early, thus restricting the chances of learning or practicing skills that are important for future achievement.

Some limitations to achieving life goals do, however, lie with the individuals themselves and their competence, capability, and confidence in performing the tasks and skills that would enable success in a job or social relationships. Cognition is just one of these interlinked factors that need to be addressed in a comprehensive approach to rehabilitation. In clinical practice, changes in cognitive performance achieved through therapy have an effect not only on the individual but also on those who provide them with support, usually making them more optimistic about potential recovery. This increases the number of opportunities offered and raises expectations so that functioning increases slowly over time. However, we have also found opposite effects—a reduction in therapy-induced improvements when the individual loses access to stimulating opportunities provided during therapy. Cognitive remediation (CR) to aid functioning takes place within a care system and a society, and this context will have an impact on the transfer of therapy gains into recovery goals. This context needs to be considered when evaluating efficacy evidence as well as in developments of the underlying theory.

THEORY

Although cognitive difficulties in schizophrenia have been known about and discussed for more than 100 years, therapy development has only taken place over the last 20 years. This is mainly because the theoretical frameworks were pessimistic. Cross-sectional and longitudinal studies showed little change in cognition from the acute to the more chronic forms of the disorder.[7–9] Many theories try to link these cognitive problems to the hallucinations and delusions that are the classic symptoms of an acute phase of schizophrenia (e.g., [10,11]). The aim here was to link a single deficit, such as in self-awareness or theory of mind, to many of the problems experienced by people with schizophrenia. Other theoretical advances, such as the Nuechterlein and Dawson model,[12] suggested that cognitive difficulties were one of the "enduring vulnerability-linked characteristics" (p.301) of individuals who are prone to schizophrenia. Although the two types of model are not contradictory, neither suggested that cognitive difficulty was tractable. Furthermore, emerging findings demonstrated that cognitive difficulties occur many years prior to the onset of schizophrenia, which continued support for these models.[3,13] What spurred the revolution in thinking about the treatment of cognitive problems was a series of studies that put the intractability assumption to the test. This was begun in a well-controlled experiment by Terry Goldberg and colleagues.[14] They tested whether it was possible to teach individuals with chronic schizophrenia how to carry out the Wisconsin Card Sorting Test (WCST)—a well-known test of cognitive flexibility. The study consisted of two training groups and a control group. Both training groups improved their scores but only when they were provided with the highest level of support—card-by-card instruction. When this support was removed performance returned to baseline levels. This study therefore supported the contention of intractable difficulties. In fact, the authors said, "we were confronted by a deficit not amenable to simple and direct instruction" (p.1013). What followed was a number of studies that tested this hypothesis. Most replicated the Goldberg results, but occasionally studies reported gains in learning. The zeitgeist was of intractability, so negative studies were published and skepticism continued to be writ large in a series of articles such as "Cognitive Remediation: Proceed . . . with Caution!"[15] The development of targeted therapy for cognition not only was supported by increasing evidence of tractability but also was stimulated by research demonstrating that cognitive problems were a barrier to living independently and therefore prevented the closure of institutions.

THE BASIC MODEL OF COGNITIVE REMEDIATION

The basis of CR is that if poor cognitive skill predicts poor functioning, then improving cognition should improve functioning (see Figure 67.1). The definition built on by the Cognitive Remediation Experts Workshop (CREW) is that CR "is an intervention targeting cognitive deficit using scientific principles of learning with the ultimate goal of improving functional outcomes. Its effectiveness is enhanced when provided in a context (formal or informal) that provides support and opportunity for extending everyday functioning."

The individual cognitive targets for remediation were originally chosen on the basis of a correlation between performance on the target domain and functional outcomes. However, in some studies change in these cognitive targets did not affect functioning but noncorrelated cognitive factors did (e.g., [16,17]). There are also data demonstrating that functional change can be achieved following CR without a noticeable improvement on cognitive tests (e.g., [18]). Wykes and Huddy[19] highlighted these problems and suggested that the simple model needed to be extended to include moderating or mediating factors not only on cognition change but also on how cognitive change influences functional change. Potential moderating or mediating effects (as opposed to the direct effects of CR on outcome) took a while to be tested. For instance, Wykes and colleagues[20] demonstrated that improved executive functioning had a direct impact on work quality but that this relationship accounted for only 15% of the variance of functional change, leaving 85% still to be explained (see Figure 67.2).

Other variables might also have an influence, such as motivation, and recently Fiszdon and colleagues[21] (personal communication) discovered that motivation prior to remediation did not correlate with maths skills following training but that change in motivation over the training period did have an effect. This suggests that therapy may improve a factor that might help to drive cognitive improvements which then affect functioning. Some variance in outcome must be accounted for by these direct therapy effects on motivation, coping skills generally and self-esteem, which may have an

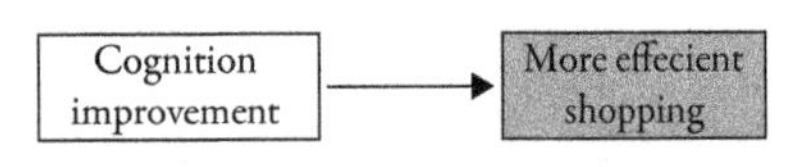

Figure 67.1 How are cognitive gains related to functioning.

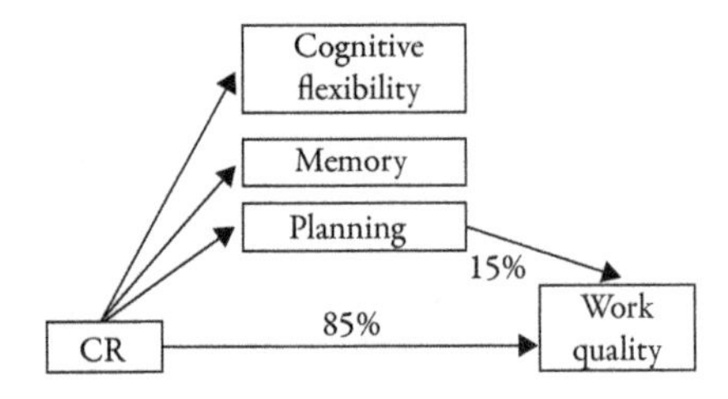

Figure 67.2 A model of how cognitive remediation moderates functional outcome.

impact on self-efficacy. But Wykes and Reeder[22] argue that this unexplained variance can also result from the effects of another factor, metacognition.

Which cognitive skill is essential for functional improvement has been difficult to define, as this depends on the concept of transfer. For CR, near transfer is usually defined as the translation of therapy benefits into improvements on independent cognitive tests and far transfer is the translation of benefits into functional domains. The weight of evidence supports the need to change executive functioning.[16,17,20,23–26] This is a type of metacognition (see later definitions) because it regulates other cognitive skills.

Others consider that CR essentially affects neural plasticity at a low level (sometimes called a "bottom-up" approach) and use evidence from animal models. This remediation model provides low-level tasks (e.g., auditory process tasks) given over a lengthy period to "drive" the neural system. This approach is described in more detail in other chapters. Other bottom-up approaches introduce simple tasks testing attention first and then progress to complex tasks requiring working memory and executive functioning. The underlying mechanisms for both approaches are unclear. Although, for instance, brain-derived neurotrophic factor (BDNF) has been used as a surrogate for increasing neural plasticity, therapy-induced changes[27] have not yet been replicated.[28] Despite differing theoretical approaches, there are few differences in the practical applications for these bottom-up approaches and, as we discuss later, there is little evidence that either theory differentially affects cognitive outcomes. Adopting a metacognitive approach, however, does have different implications for the type of therapy provided and the potential outcomes and so is described in more detail.

METACOGNITION AND SCHIZOPHRENIA

The components of metacognition are monitoring (evaluation of cognitive functioning), control or regulation (directing and evaluating cognitive and behavioral performance), and knowledge (understanding task difficulty and the resources required). Awareness of cognitive problems can be thought of as a form of metacognitive knowledge that can guide the deployment of cognitive resources and their performance at maximal efficiency.[29]

Planning how to remember a shopping list and checking you have received the right change are forms of metacognition called metacognitive regulation. In contrast, metacognitive knowledge refers to what an individual knows (e.g., the most efficient strategies for remembering a list), including what someone knows about their abilities, such as how good their memory is. In order to change behavior, a person needs to understand what they find difficult and what can be improved, and consider strategies that will improve performance, as well as implement, monitor, and evaluate those strategies. All of these components are linked to metacognition and not to basic cognition.

Although there is plenty of evidence that people with a diagnosis of schizophrenia have poor awareness of cognitive abilities (i.e., poor metacognition) other factors moderate this awareness. Cella and colleagues[30] discovered that, after controlling for two potential moderators (self-esteem and negative symptoms), awareness of cognitive problems was actually aligned with the results of objective cognitive tests. This suggests that improvements in self-esteem and opportunities created by reductions in negative symptoms (e.g., medication, other treatments) might aid metacognition—particularly metacognitive knowledge. Poor metacognitive regulation in schizophrenia has also been noted.[31,32]

These metacognitive problems may even be more potent predictors of functioning than cognitive predictors alone.[33–35] In a recent study of 80 people with a diagnosis of schizophrenia, Davies and colleagues[36] demonstrated that metacognition partially mediates the relationship between cognition and functional capacity, and fully mediates the relationship between functional capacity and social and occupational functioning. This suggests that metacognition is vital if the goal of CR is to improve functional outcomes.

If metacognition is a key to transfer, then CR programs need to generate metacognitive improvements in addition to cognitive improvements. For such a development we need to turn to pedagogical knowledge. Metacognitive regulation is essential for learning anything—either a simple test of cognitive performance or a complex functional skill. This is because cognitive control in the learning situation (e.g., the allocation of study time to different parts of a learning task) and effective monitoring are essential for the self-management of learning.[37,38]

A review of meta-analyses of teaching metacognition and strategies in schools demonstrated 7 to 9 months' additional progress on average in pupils' abilities. Importantly for CR, this approach seems to provide the most benefit to lower-achieving students and older people. One of the key issues stressed in the 2008 review[39] is that it is important to ensure that pupils are not supported too much; they should receive scaffolded learning in which the supports are slowly removed to be sure learning is self-supported.

The key to metacognition teaching is to guide students by the explicit teaching of strategies for learning and use of knowledge (i.e., transfer to new situations, teaching metacognitive knowledge). In addition, there needs to be encouragement of self-reflection on learning. There are many reviews of strategic learning[40] but the key principles are to keep success high to maximize motivation, to provide scaffolded learning and to teach strategies explicitly rather than by trial and error. The involvement of metacognition brings a new phase to CR as it suggests further ingredients to target in a tailored approach to functional benefit.

Although theoretical approaches have affected clinical practice in the past, there is agreement across the field that behavioral change is likely to be accompanied by measurable changes in the brain, either through peripheral markers, such as BDNF, or in more complex activation.[41,42,43] In fact longer term beneficial brain changes in the prevention of gray matter loss have been found following CR.[24] This was achieved with a metacognitive approach in addition to social cognition oriented activities. The importance of this result is that CR may be a tool to increase resilience.

META-ANALYTIC AND TRIAL EVIDENCE FOR EFFICACY OF COGNITIVE REMEDIATION

There has been a burgeoning industry in trials, meta-analyses, and systematic reviews concentrating on CR. Although meta-analyses are the gold standard for treatment guidance they can be confusing as they use data in different ways, have different inclusion criteria and test subtle or not so subtle changes following CR. So summarizing their output is complex. Meta-analyses of CR have taken both a narrow and wider scope, with some concentrating on stage of illness or sustainability of effects. Outcomes too differ and can be focused on changes in cognition, functioning, or symptoms. In the 17 meta-analyses published since 2001, few use the CREW definition of CR, which affects the types of studies included. Because meta-analyses are summaries of published data they depend on the quality of the included studies. However, few report trial quality using measures such as the Clinical Trial Assessment Measure (CTAM[44]), Cochrane risk of bias tool, or SASQI (Scale to Assess Scientific Quality of Investigations[45,46]), and when quality ratings are carried out many of the included studies are not highly rated. This may be because early studies were based on feasibility and pilot studies rather than definitive randomized controlled trials (RCTs), and later studies were small extensions to the applicability of CR. Depending on the definition for inclusion, the number of participants ranges from 170 to 3,295. This may or may not affect their conclusions, as small numbers of participants may come from a few high-quality trials and ones including large numbers of participants may be based on many small and inconclusive studies.

The earliest meta-analyses had few studies and either failed to find any trials that fitted the criteria (e.g., Cochrane review[47]—1st version) or included only a few, very disparate types of study. For instance, one[48] included studies lasting only a few days in addition to studies providing therapy over many weeks and failed to include studies that all experts would consider as CR.[49] One influential meta-analysis that had restrictive inclusion criteria (no mixed cognitive/vocational remediation trials, at least 10 subjects per arm, and follow-up data) was published as part of the development of treatment guidance.[50] This meta-analysis only contained 20 RCTs (1,084 participants) and although the entry criteria were restricted, it still managed to include studies not considered as CR. Nevertheless, this meta-analysis did provide some limited evidence that CR had long-lasting effects on cognition and social functioning, but both these effects were driven by large studies—i.e., the meta-analysis team thought that the meta-analytic results were unsafe.

The most diverse meta-analysis that also tested study quality[51] included different types of CR (computerized vs. paper and pencil), different therapy provision (groups, one-to-one with a therapist, or independently) and studies where CR was provided with other recovery programs. This meta-analysis included 2,104 participants in 40 studies. Cognitive remediation had an overall significant positive effect of 0.45 (95% confidence interval [CI] = 0.31–0.59) and most cognitive domains demonstrated a significant—and durable—benefit. All programs seem to have similar effects on cognitive performance, but there were differences on functional outcomes. Studies providing CR on a background of vocational or social rehabilitation showed larger functional change compared to remediation alone (effect size 0.59 vs. 0.28). In addition, programs using a metacognitive approach (e.g., providing strategic teaching) produced more benefit on functioning (effect size 0.47 vs. 0.34). When strategy training was provided in the context of rehabilitation there was a large functioning effect size (0.8 vs. 0.3), but there were only eight studies in this last comparison. The boosting of outcomes through comprehensive rehabilitation has been supported by several recent trials. Bowie[52] discovered that adding CR to skills training boosted the effects of functional adaptation skills training so that improvements not only were larger than with skills training alone but also were durable. Perhaps most importantly, only three people needed to be treated to achieve these effects. In addition, studies by McGurk and colleagues[53,54] demonstrated that CR could help individuals who had previously failed to get a job following a supported work program. Adding CR to supported employment improved outcomes in this previously failing group so that they had consistently better competitive employment outcomes during the follow-up period, including jobs obtained (60% vs. 36%), weeks worked (23.9 vs. 9.2), and wages earned ($3,421 vs. $1,728).

Most other meta-analyses produced similar effect sizes albeit usually with fewer studies. Those concentrating on computerized CR showed improvements in cognition (0.38) and employment (20% higher employment rate and US$959 more in annual earnings).[55,56] It has been suggested that there are advantages to specifically targeting certain problems, but Grynszpan and colleagues[55] found no advantage in their meta-analysis. A further trial[57] investigated specific tailoring of remediation to the problem cognitive domains for individuals. They demonstrated no benefit for the more tailored over the general approach.

So far we have concentrated on the effects of CR on cognition and functioning, which are the two elements mentioned in the CREW definition, but there is also the potential for cognitive improvements to have effects on the experience of positive symptoms. The early theories link cognition and symptoms.[58,59] The Wykes et al. meta-analysis[51] found a small improvement in total symptoms immediately following the end of therapy, but this effect disappeared at follow-up. Symptoms were not the targets for these earlier studies although effects are seen even when they were not a primary outcome. For example, following reports by participants that they had stopped hearing voices since they now had "so much more to think about," Wykes and colleagues further analyzed their data and discovered that those with improvements in memory who had initially heard voices did show reductions in hallucinations.[60]

There is now evidence of general CR benefits for cognition irrespective of the method used, but the evidence on functioning is less clear. Cognitive remediation that includes explicit strategy teaching within a comprehensive rehabilitation approach seems to have the largest effects. But this conclusion was based on only a few studies.[51] However, this result does fit with our own clinical focus and experience, which indicates that engagement with a community that provides opportunities and reinforcement for the transfer of strategies is vital to generalize and maintain therapeutic gains. This result is reinforced in two further meta-analyses. The first[61] compares several psychological treatments and established the effect of CR compared to other studies. Cognitive remediation in high-quality studies demonstrated a significant effect of 0.2 (Hedges' *g*) on the total symptoms score. It was also the most effective treatment compared to all others pooled (social skills training, cognitive behavioral treatment, etc.). But any effects were not durable. More recently Cella and colleagues[62] investigated the effects of CR on negative symptoms in a network meta-analysis. They found a significant small to moderate reduction in negative symptoms in 45 RCTs with 2,511 participants. This benefit was unaffected by whether the comparison was between CR and treatment as usual or an active control condition. The effect at follow-up was larger than the post-treatment effect and was unaffected by the methodological rigor of the studies. They also point out that some studies showed a large effect size and that these were generally studies that included some other rehabilitation program in addition to CR. These additional background programs provide individuals with further opportunities for practicing skills and may therefore have boosted the effects of remediation and could also account for the increased effects at follow-up.

TAILORING TREATMENT

Tailoring treatments to individuals can boost effects, but predictive factors could also allow current treatments to be provided to those who would gain the most and so support the efficient use scarce healthcare resources. For the remaining individuals the treatment would need further development to provide similar effects. The beginning of this approach was to measure simple participant characteristics such as cognitive profiles or demographic variables such as age. The potential for examining age in meta-analyses is low, as they only have access to the summary scores and the majority of studies recruit individuals whose average age is about 32 years old. Two studies that have included older participants[63,64] have shown no improvements with their versions of CR. However, in a recent meta-analysis of stage of illness, and concentrating on early psychosis, Revell and colleagues[65] found few cognitive improvements although there was a small but significant effect of functioning. These authors suggest that part of the difficulty in measuring an effect is baseline functioning. They refer to a study by Bowie and colleagues[66] that compared early course and chronic participants in a CR trial. These researchers found that the early psychosis group had a larger benefit, but that the benefits for this group were greatest when their levels of cognitive problems at baseline were similar to those of chronic participants. So we do not have a clear answer on whether or how age moderates potential benefit.

Other variables that hold some promise in small studies include the gene COMT, where the Met allele has been associated with the largest improvements in verbal and visual learning as well as attention following CR.[67] Similar results were reported by Bosia and colleagues,[68] but there have been some negative findings.[69]

Individual trials provide some insight for the sensible allocation of health resources. For instance, Bell and colleagues[70] demonstrated that at the 2-year follow-up only those with poor functioning at baseline significantly benefited from cognitive enhancement plus supported employment. Those performing above the median showed little or no benefit.

WHAT HAPPENS IN PRACTICE?

Cognitive remediation is like any other psychological therapy; it requires a therapeutic context. This includes having a formulation of the problems and negotiating a shared understanding with the client about their goals and how therapy may help to achieve them. Our bias is therefore to understand the individual's strengths and weaknesses, the pathway to recovery goals and to provide a supportive therapeutic relationship. These are all nonspecific aspects for all psychological treatments, but they are no less important in CR. Not all CR programs offer this approach. Some depend much more on independent working or limited therapist contact such as an occasional phone call. In some programs where clients visit a clinic to carry out the tasks there is also little contact other than helping with technical difficulties.[64] These programs do not have a formalized process for considering strengths and weaknesses but depend on a global approach where all individuals receive the same computerized program, which adapts to the client's performance. Engagement here can depend on paying individuals for each session completed, so this may substitute for the working alliance developed with a therapist. For the client, being able to carry out treatment at home means they have flexibility with appointments. These clients also may feel less stigma than those visiting the clinic. For health services, there is potentially an advantage in not having to pay for extensive personal contact from a trained therapist or coach and also not having to pay for clinic space. The balance of the personal engagement process versus the flexible, less potentially stigmatizing service has yet to be tested on treatment commitment and the number needed to treat for the same benefit.

In addition to therapeutic engagement and alliance there are some key aspects of therapy that need to be considered. More detailed information is provided elsewhere[22,71] but most CR therapies take into account three main factors—errorless learning, scaffolded learning, and strategy skills teaching. The meta-analytic evidence suggests that strategic skills are important only for the transfer of cognitive benefits to functional improvement but other variables will affect the final benefits.

MOTIVATION

Motivation problems are prevalent in schizophrenia,[72] and individuals with this diagnosis have multiple experiences of failure. Ensuring successful task completion keeps positive reinforcement high, and complexity should only be increased following consistent successful performance. Most programs use an 80% success rate. High success levels mean that there is a clear distinction between successful and unsuccessful behaviors. This is sometimes referred to as "errorless learning." Success is also an integral part of motivating individuals to continue to put in effort and engage in therapy to build self-esteem and confidence. Some programs rely solely on the computer program's reward system and some have developed more complex systems transferred from the computer games industry. These reward systems certainly have face validity but again there has not been an explicit test to discover whether these systems would engage individuals in the absence of sessional payment. Individuals can also benefit from further social reward from the therapist or from therapist enquiry so that there is some ownership of success. This can also build metacognitive awareness by requiring self-reflection either through personal or computer responses.

SCAFFOLDED LEARNING

Scaffolded learning is needed so that individuals become aware of their responsibility for successful learning. In our clinic this means that therapists take control of tasks in the early stages and then remove those controls so that the trainee becomes more and more clearly responsible for task success. What helps with this process is the emphasis on similarities between different tasks so that near transfer within the program can be achieved. But it is also vital to provide some support so that clients notice the similarities between remediation tasks and real-world tasks. Some therapies achieve this additional support through bridging groups in which individuals discuss how the skills learned in therapy might transfer. Others incorporate transfer by engaging clients in their environment (e.g., at work, at home), e.g., Thinking Skills at Work program.[54]

STRATEGIC SKILLS LEARNING

There is plenty of evidence, some described earlier, that people with a diagnosis of schizophrenia benefit from strategic support. Akdogan, Izaute, and Bacon[73] demonstrated improved memory performance after providing a guide to memory search and retrieval that reduced or eliminated the difference between people with schizophrenia and controls. Apart from the potential use in transfer of cognitive benefits, these strategies also contribute to task success rates and so support self-efficacy. Education evidence suggests that both didactic teaching and modeling are useful, rather than depending on chance learning or thinking style biases (e.g., continuing to use inefficient strategies).

All of these components—errorless learning, scaffolding, and strategic skills learning—are involved in most CR programs. However, further adaptations need to be made if the remediation program is to improve metacognition. Below is a brief description of some of the activities that are essential and how they might be incorporated into software or therapeutic activities.

A METACOGNITIVE APPROACH TO COGNITIVE REMEDIATION

Our current CR software—Computerized Interactive Remediation of Cognition and Thinking Skills, or CIRCuiTS—fosters the transfer of new cognitive skills to everyday life within the program itself, rather than relying on bridging sessions or adjunctive rehabilitation. This is underpinned by a metacognition model[22,74,75] incorporating two components: (1) metacognitive knowledge about one's own strengths and difficulties and how thinking skills and strategies can affect behavior in general and (2) metacognitive regulation of one's own behavior.

Tasks are presented in the context of virtual situations such as shopping or cooking, but the individual's social milieu is also used to support transfer, and the therapist (and the computer program) cues the potential uses of different strategies and the underlying similarities between tasks and everyday life. Both of these approaches support metacognitive awareness and knowledge and increase the potential for metacognitive regulation, and are based on previous research.[52,54,76,77] Transfer is therefore explicitly targeted and incorporated into CIRCuiTS using bridging by therapists[78] in the delivery of the software tasks themselves rather than being an adjunct to the program.

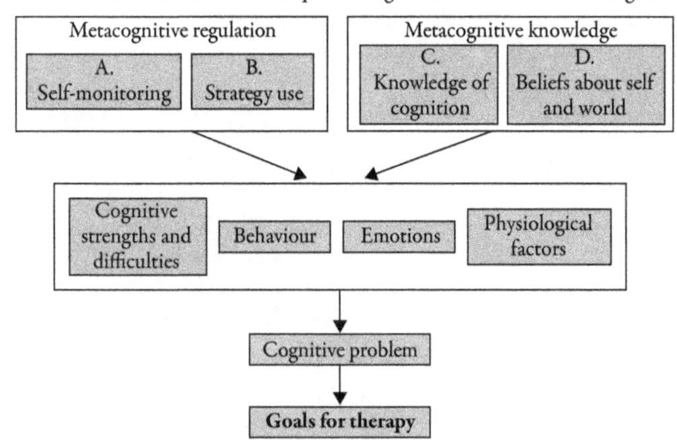

Figure 67.3 A model of how CIRCuiTS develops metacognition to achieve functional goals.

The CIRCuiTS "metacognitive journey" emphasizes the reasons for therapy, how personal goals might be met, the development of metacognition and the transfer process, and monitors achievements. It includes the client's "Cog-SMART" goals, which are real-world goals that link specifically to cognition and the strategies that can be used in real-world situations to achieve those goals (see Figure 67.3).

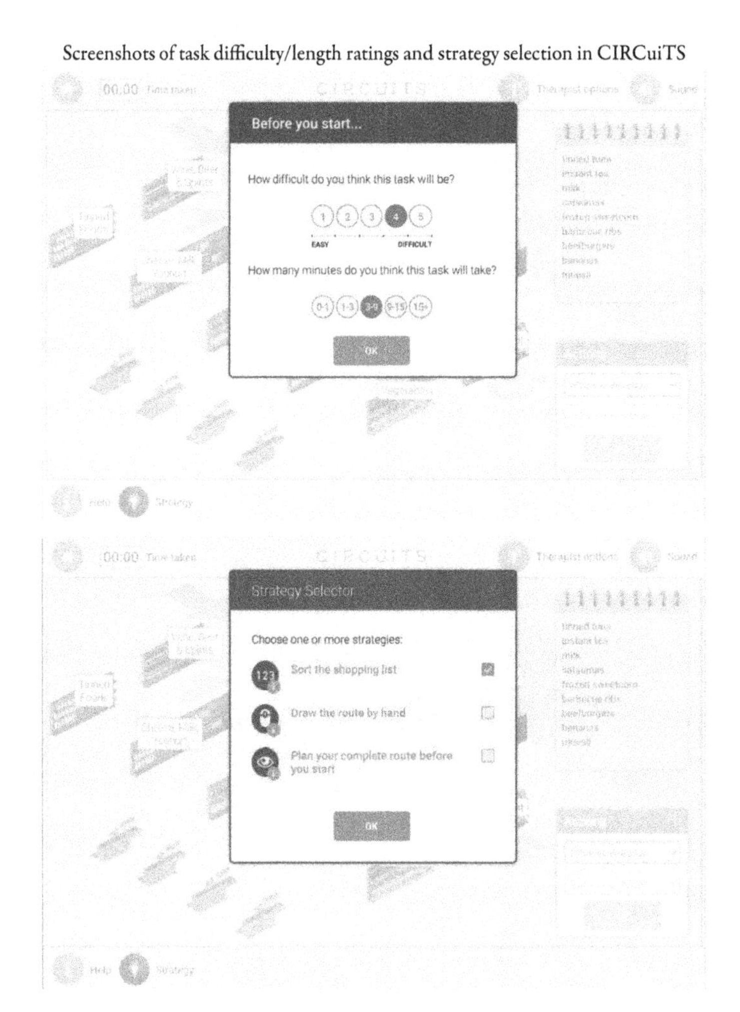

Figure 67.4 Screenshots of task difficulty/length ratings and strategy selection in CIRCuiTS.

METACOGNITIVE REGULATION

Self-monitoring is part of metacognitive regulation and affects the choice of strategies. An initial evaluation is therefore required of the extent of self-monitoring in everyday life and whether this breaks down in particular situations.

Strategy use needs initial assessment to understand which are currently used. Suggested strategies need to link to the clients' strengths.

In beginning a CIRCuiTS task the client needs to decide how much time it will take and how difficult it might be. Then they select strategies to help them complete the task (see Figure 67.4). This encourages a systematic approach but also encourages self-reflection on what the task will entail. At the end of a task the client is provided with feedback on the time it has taken and their accuracy. They are then asked to provide further ratings of the usefulness of the strategies they had chosen. This builds self-awareness as this review allows for both error correction and learning by experience.

METACOGNITIVE KNOWLEDGE

Awareness of one's cognitive strengths and weaknesses and the factors that influence cognitive functioning can have wide-ranging effects on a client's performance. Client ratings are held in their "Library" so they can track their most useful and frequently used strategies (see Figure 67.5). The contrast between helpful or less efficient strategies allows for metacognitive knowledge to be instantiated.

THE FUTURE OF COGNITIVE REMEDIATION

Remediation programs continue to be developed. Some harness underlying plasticity by including aerobic exercise, which is thought to improve cognitive functioning, similarly to CRT, through associated changes in brain structure and increases in BDNF.[79,80] These studies are showing an enhancement of the cognitive benefits following both exercise and remediation, which bode well for the future.

But we have yet to take advantage of already extant data on CR in our trials. Individual variation in outcomes within and across programs may be affected by differing cognitive, genetic, and functioning profiles. The NIMH Database of Cognitive Training and Remediation Studies (DoCTRS), which contains individual trial participant data will allow us to use the variation and so overcome the problems of summarized data in meta-analyses. One paper has already used this potential[81] to demonstrate that improvement in working memory does not impact negative symptoms but that there seems to be a direct CR effect on behavioral negative symptoms but not on expressive ones. These sorts of analyses can hone our efforts to understand not only the ingredients of successful treatment

Figure 67.5 Screenshots of monitoring strategy use in CIRCuiTS.

but also which individuals might benefit from which type of treatment.

CONCLUSION

The CREW definition of CR is that it aims to improve cognition with the target of improved functioning. The results of meta-analyses now suggest that remediation is best provided in the context of formal, rather than informal, opportunities to practice new skills and that a program that incorporates strategy-teaching benefits transfer of skills to functional goals the most. But we do not know how much cognitive benefit is required for that transfer to occur nor how to measure the transfer process. The argument in this chapter is that metacognition is essential for the transfer and that, without metacognition enhancement, even large cognitive benefits are unlikely to be realized in improved life chances.

REFERENCES

1. Cannon TD, Mednick SA, Parnas J. Antecedents of predominantly negative-symptom and predominantly positive-symptom schizophrenia in a high-risk population. *Arch Gen Psychiat.* 1990;47(7):622–632.
2. Seidman LJ, Cherkerzian S, Goldstein JM, Agnew-Blais J, Tsuang MT, Buka SL. Neuropsychological performance and family history in children at age 7 who develop adult schizophrenia or bipolar psychosis in the New England Family Studies. *Psychol Med.* 2013;43(1):119–131.
3. Gur RC, Calkins ME, Satterthwaite TD, et al. Neurocognitive growth charting in psychosis spectrum youths. *Jama Psychiat.* 2014;71(4):366–374.
4. Rund BR. A review of longitudinal studies of cognitive functions in schizophrenia patients. *Schizophr Bull.* 1998;24(3):425–435.
5. Keefe RS, Eesley CE, Poe MP. Defining a cognitive function decrement in schizophrenia. *Biol Psychiatry.* 2005;57(6):688–691.
6. Henderson C, Thornicroft G. Evaluation of the time to change programme in England 2008–2011. *Brit J Psychiat.* 2013;202:S45–S48.
7. McCleery A, Ventura J, Kern RS, et al. Cognitive functioning in first-episode schizophrenia: MATRICS Consensus Cognitive Battery (MCCB) profile of impairment. *Schizophr Research.* 2014;157(1-3):33–39.
8. Barder HE, Sundet K, Rund BR, et al. Ten year neurocognitive trajectories in first-episode psychosis. *Front Hum Neurosci.* 2013;7.
9. Hoff AL, Svetina C, Shields G, Stewart J, DeLisi LE. Ten year longitudinal study of neuropsychological functioning subsequent to a first episode of schizophrenia. *Schizophr Research.* 2005;78(1):27–34.
10. Frith CD. *The cognitive neuropsychology of schizophrenia.* Hove: Erlbaum; 1992.
11. Frith CD, Corcoran R. Exploring "theory of mind" in people with schizophrenia. *Psychol Med.* 1996;26(3):521–530.
12. Nuechterlein KH, Dawson ME. A heuristic vulnerability stress model of schizophrenic episodes. *Schizophr Bull.* 1984;10(2):300–312.
13. Agnew-Blais J, Seidman LJ, Fitzmaurice GM, Smoller JW, Goldstein JM, Buka SL. The interplay of childhood behavior problems and IQ in the development of later schizophrenia and affective psychoses. *Schizophr Research.* 2017;184:45–51.
14. Goldberg TE, Weinberger DR, Berman KF, Pliskin NH, Podd MH. Further evidence for dementia of the prefrontal type in schizophrenia—a controlled-study of teaching the Wisconsin Card Sorting Test. *Arch Gen Psychiat.* 1987;44(11):1008–1014.
15. Hogarty GE, Flesher S. Cognitive remediation in schizophrenia—proceed—with caution! *Schizophr Bull.* 1992;18(1):51–57.
16. Reeder C, Newton E, Frangou S, Wykes T. Which executive skills should we target to affect social functioning and symptom change? A study of a cognitive remediation therapy program. *Schizophr Bull.* 2004;30(1):87–100.
17. Reeder C, Smedley N, Butt K, Bogner D, Wykes T. Cognitive predictors of social functioning improvements following cognitive remediation for schizophrenia. *Schizophr Bull.* 2006;32:S123–S131.
18. Silverstein SM, Hatashita-Wong M, Solak BA, et al. Effectiveness of a two-phase cognitive rehabilitation intervention for severely impaired schizophrenia patients. *Psychol Med.* 2005;35(6):829–837.
19. Wykes T, Huddy V. Cognitive remediation for schizophrenia: it is even more complicated. *Curr Opin Psychiatr.* 2009;22(2):161–167.
20. Wykes T, Reeder C, Huddy V, et al. Developing models of how cognitive improvements change functioning: mediation, moderation and moderated mediation. *Schizophr Research.* 2012;138(1):88–93.
21. Fiszdon JM, Kurtz MM, Choi J, Bell MD, Martino S. Motivational interviewing to increase cognitive rehabilitation adherence in schizophrenia. *Schizophr Bull.* 2016;42(2):327–334.
22. Wykes T, Reeder C. *Cognitive remediation therapy for schizophrenia: theory and practice.* London, UK: Brunner Routledge; 2005.
23. Eack SM, Greenwald DP, Hogarty SS, Keshavan MS. One-year durability of the effects of cognitive enhancement therapy on functional outcome in early schizophrenia. *Schizophr Res.* 2010a;120(1-3):210–216.
24. Eack SM, Hogarty GE, Cho RY, et al. Neuroprotective effects of cognitive enhancement therapy against gray matter loss in early schizophrenia results from a 2-year randomized controlled trial. *Arch Gen Psychiat.* 2010b;67(7):674–682.
25. Penades R, Catalan R, Puig O, et al. Executive function needs to be targeted to improve social functioning with cognitive remediation therapy (CRT) in schizophrenia. *Psychiat Res.* 2010;177(1–2):41–45.
26. Reeder C, Harris V, Pickles A, Patel A, Cella M, Wykes T. Does change in cognitive function predict change in costs of care for people with a schizophrenia diagnosis following cognitive remediation therapy? *Schizophr Bull.* 2014;40(6):1472–1481.
27. Vinogradov S, Fisher M, Holland C, Shelly W, Wolkowitz O, Mellon SH. Is serum brain-derived neurotrophic factor a biomarker for cognitive enhancement in schizophrenia? *Biol Psychiat.* 2009;66(6):549–553.
28. Penades R. *BDNF as a mechanism of cognitive recovery: cognitive remediation in psychiatry.* New York City. 2016.
29. Flavell JH. Meta-cognition and cognitive monitoring—new area of cognitive-developmental inquiry. *Am Psychol.* 1979;34(10):906–911.
30. Cella M, Swan S, Medin E, Reeder C, Wykes T. Metacognitive awareness of cognitive problems in schizophrenia: exploring the role of symptoms and self-esteem. *Psychol Med.* 2014;44(3):469–476.
31. Koren D, Harvey P. Closing the gap between cognitive performance and real-world functional outcome in schizophrenia: the importance of metacognition. *Current Psychiatry Review.* 2006a;2(2):189–198.
32. Koren D, Seidman LJ, Goldsmith M, Harvey PD. Real-world cognitive—and metacognitive—dysfunction in schizophrenia: a new approach for measuring (and remediating) more "right stuff." *Schizophr Bull.* 2006b;32(2):310–326.
33. Hamm JA, Renard SB, Fogley RL, et al. Metacognition and social cognition in schizophrenia: stability and relationship to concurrent and prospective symptom assessments. *J Clin Psychol.* 2012;68(12):1303–1312.
34. Lysaker PH, Gumley A, Luedtke B, et al. Social cognition and metacognition in schizophrenia: evidence of their independence and linkage with outcomes. *Acta Psychiat Scand.* 2013;127(3):239–247.
35. Stratta P, Daneluzzo E, Riccardi I, Bustini M, Rossi A. Metacognitive ability and social functioning are related in persons with schizophrenic disorder. *Schizophr Research.* 2009;108(1-3):301–302.
36. Davies G, Fowler D, Greenwood K. Metacognition as a mediating variable between neurocognition and functional outcome in first episode psychosis. *Schizophr Bull.* 2017;43(4):824–832.
37. Tullis JG, Benjamin AS. On the effectiveness of self-paced learning. *J Mem Lang.* 2011;64(2):109–118.

38. Thiede KW, Anderson MCM, Therriault D. Accuracy of metacognitive monitoring affects learning of texts. *J Educ Psychol.* 2003;95(1):66–73.
39. Dignath C, Buettner G, Langfeldt HP. How can primary school students learn self-regulated learning strategies most effectively? A meta-analysis on self-regulation training programmes. *Educ Res Rev-Neth.* 2008;3(2):101–129.
40. Simpson ML, Nist SL. An update on strategic learning: it's more than textbook reading strategies. *J Adolesc Adult Lit.* 2000;43(6):528–541.
41. Niendam TA, Mathalon DH, Taylor SF, et al. Multi-site fMRI study of cognitive control-related brain activation in early schizophrenia and clinical-high-risk youth. *Schizophr Bull.* 2011;37:223.
42. Wykes T. What are we changing with neurocognitive rehabilitation? Illustrations from two single cases of changes in neuropsychological performance and brain systems as measured by SPECT. *Schizophr Research.* 1998;34(1–2):77–86.
43. Penades R, Pujol N, Catalan R, et al. Brain effects of cognitive remediation therapy in schizophrenia: a structural and functional neuroimaging study. *Biol Psychiat.* 2013;73(10):1015–1023.
44. Wykes T, Steel C, Everitt B, Tarrier N. Cognitive behavior therapy for schizophrenia: effect sizes, clinical models, and methodological rigor. *Schizophr Bull.* 2008;34(3):523–537.
45. Higgins JP, Altman DG, Gotzsche PC, et al. The Cochrane Collaboration's tool for assessing risk of bias in randomised trials. *BMJ.* 2011;343:d5928.
46. Jeste DV, Dunn LB, Folsom DP, Zisook D. Multimedia educational aids for improving consumer knowledge about illness management and treatment decisions: a review of randomized controlled trials. *J Psychiatr Res.* 2008;42(1):1–21.
47. Hayes RL, McGrath JJ. Cognitive rehabilitation for people with schizophrenia and related conditions. *Cochrane Database Syst Rev.* 2000(3):CD000968.
48. Pilling S, Bebbington P, Kuipers E, et al. Psychological treatments in schizophrenia: II. Meta-analyses of randomized controlled trials of social skills training and cognitive remediation. *Psychol Med.* 2002;32(5):783–791.
49. Spaulding WD, Reed D, Sullivan M, Richardson C, Weiler M. Effects of cognitive treatment in psychiatric rehabilitation. *Schizophr Bull.* 1999;25(4):657–676.
50. National Institute for Health and Clinical Excellence. *Schizophrenia: Core interventions in the treatment and management of schizophrenia in adults in primary and secondary care (update): Final version.* https://www.ncbi.nlm.nih.gov/books/NBK11688/#ch8.s612009.
51. Wykes T, Huddy V, Cellard C, McGurk SR, Czobor P. A meta-analysis of cognitive remediation for schizophrenia: methodology and effect sizes. *Am J Psychiat.* 2011;168(5):472–485.
52. Bowie CR, McGurk SR, Mausbach B, Patterson TL, Harvey PD. Combined cognitive remediation and functional skills training for schizophrenia: effects on cognition, functional competence, and real-world behavior. *Am J Psychiat.* 2012;169(7):710–718.
53. McGurk SR, Mueser KT, Pascaris A. Cognitive training and supported employment for persons with severe mental illness: one-year results from a randomized controlled trial. *Schizophr Bull.* 2005;31(4):898–909.
54. McGurk SR, Mueser KT, Xie HY, et al. Cognitive enhancement treatment for people with mental illness who do not respond to supported employment: a randomized controlled trial. *Am J Psychiat.* 2015;172(9):852–861.
55. Grynszpan O, Perbal S, Pelissolo A, et al. Efficacy and specificity of computer-assisted cognitive remediation in schizophrenia: a meta-analytical study. *Psychol Med.* 2011;41(1):163–173.
56. Chan JY, Hirai HW, Tsoi KK. Can computer-assisted cognitive remediation improve employment and productivity outcomes of patients with severe mental illness? A meta-analysis of prospective controlled trials. *J Psychiatr Res.* 2015;68:293–300.
57. Franck N, Duboc C, Sundby C, et al. Specific vs general cognitive remediation for executive functioning in schizophrenia: a multicenter randomized trial. *Schizophr Research.* 2013;147(1):68–74.
58. Frith CD. The cognitive abnormalities underlying the symptomatology and the disability of patients with schizophrenia. *Int Clin Psychopharm.* 1995;10:87–98.
59. Frith CD. The positive and negative symptoms of schizophrenia reflect impairments in the perception and initiation of action. *Psychol Med.* 1987;17(3):631–648.
60. Wykes T, Reeder C, Williams C, Corner J, Rice C, Everitt B. Are the effects of cognitive remediation therapy (CRT) durable? Results from an exploratory trial in schizophrenia. *Schizophr Research.* 2003;61(2–3):163–174.
61. Turner DT, van der Gaag M, Karyotaki E, Cuijpers P. Psychological interventions for psychosis: a meta-analysis of comparative outcome studies. *Am J Psychiat.* 2014;171(5):523–538.
62. Cella M, Preti A, Edwards C, Dow T, Wykes T. Cognitive remediation for negative symptoms of schizophrenia: a network meta-analysis. *Clin Psychol Rev.* 2017;52:43–51.
63. Dickinson D, Tenhula W, Morris S, et al. A randomized, controlled trial of computer-assisted cognitive remediation for schizophrenia. *Am J Psychiat.* 2010;167(2):170–180.
64. Gomar JJ, Valls E, Radua J, et al. A multisite, randomized controlled clinical trial of computerized cognitive remediation therapy for schizophrenia. *Schizophr Bull.* 2015;41(6):1387–1396.
65. Revell ER, Neill JC, Harte M, Khan Z, Drake RJ. A systematic review and meta-analysis of cognitive remediation in early schizophrenia. *Schizophr Research.* 2015;168(1–2):213–222.
66. Bowie CR, Grossman M, Gupta M, Oyewumi LK, Harvey PD. Cognitive remediation in schizophrenia: efficacy and effectiveness in patients with early versus long-term course of illness. *Early Interv Psychia.* 2014;8(1):32–38.
67. Lindenmayer JP, Khan A, Lachman H, et al. COMT genotype and response to cognitive remediation in schizophrenia. *Schizophr Research.* 2015;168(1–2):279–284.
68. Bosia M, Bechi M, Pirovano A, et al. COMT and 5-HT1A-receptor genotypes potentially affect executive functions improvement after cognitive remediation in schizophrenia. *Health Psychol Behav Med.* 2014;2(1):509–516.
69. Greenwood K, Hung CF, Tropeano M, McGuffin P, Wykes T. No association between the Catechol-O-Methyltransferase (COMT) val158met polymorphism and cognitive improvement following cognitive remediation therapy (CRT) in schizophrenia (vol 496, pg 65, 2011). *Neurosci Lett.* 2011;499(1):57.
70. Bell MD, Choi KH, Dyer C, Wexler BE. Benefits of cognitive remediation and supported employment for schizophrenia patients with poor community functioning. *Psychiat Serv.* 2014;65(4):469–475.
71. Wykes T, Crowther A, C. R. A metacognitive approach to cognitive remediation: why we need to attend to it to produce functional outcomes. In: Medalia A, Bowie C, eds. *Cognitive remediation to improve functional outcomes.* New York: Oxford University Press; 2016.
72. Choi J, Medalia A. Intrinsic motivation and learning in a schizophrenia spectrum sample. *Schizophr Res.* 2010;118(1–3):12–19.
73. Akdogan E, Izaute M, Bacon E. Preserved strategic grain-size regulation in memory reporting in patients with schizophrenia. *Biol Psychiat.* 2014;76(2):154–159.
74. Reeder C, Pile V, Crawford P, et al. The feasibility and acceptability to service users of CIRCuiTS, a computerized cognitive remediation therapy programme for schizophrenia. *Behav Cogn Psychoth.* 2016;44(3):288–305.
75. Reeder C, Huddy V, Cella M, et al. A new generation computerised metacognitive cognitive remediation programme for schizophrenia (CIRCuiTS): a randomised controlled trial. *Psychol Med.* 2017;47(15):2720–2730.
76. Eack SM, Hogarty GE, Greenwald DP, Hogarty SS, Keshavan MS. Effects of cognitive enhancement therapy on employment outcomes in early schizophrenia: results from a 2-year randomized trial. *Res Social Work Prac.* 2011;21(1):32–42.
77. Bell M, Bryson G, Greig T, Corcoran C, Wexler BE. Neurocognitive enhancement therapy with work therapy—Effects on neuropsychological test performance. *Arch Gen Psychiat.* 2001;58(8):763–768.

78. Medalia A, Revheim N, Herlands T. *Cognitive remediation for psychological disorders: therapist guide.* New York: Oxford University Press; 2009.
79. Nuechterlein KH, Ventura J, McEwen SC, Gretchen-Doorly D, Vinogradov S, Subotnik KL. Enhancing cognitive training through aerobic exercise after a first schizophrenia episode: theoretical conception and pilot study. *Schizophr Bull.* 2016;42:S44–S52.
80. Malchow B, Keller K, Hasan A, et al. Effects of endurance training combined with cognitive remediation on everyday functioning, symptoms, and cognition in multiepisode schizophrenia patients. *Schizophr Bull.* 2015;41(4):847–858.
81. Cella M, Stahl D, Morris S, Keefe R, Bell M, Wykes T. Effects of cognitive remediation on negative symptoms dimensions: exploring the role of working memory. *Psychol Med.* In press.

68

NONINVASIVE BRAIN STIMULATION TECHNIQUES IN PSYCHOSIS

Marine Mondino, Frédéric Haesebaert, and Jérôme Brunelin

INTRODUCTION

In recent years, a great effort has been made to develop novel therapeutic approaches for psychotic symptoms that resist to the available treatments. One of the proposed nonpharmacological approaches consists in safely modulating brain activity and connectivity in vivo using noninvasive brain stimulation techniques (NIBS). Two NIBS, repetitive transcranial magnetic stimulation (rTMS) and transcranial direct current stimulation (tDCS), have been shown to have potential in reducing symptoms in several psychiatric conditions, including psychotic disorders. The rationale for using NIBS in these indications draws from the concept that symptoms are linked to abnormalities in brain activity and connectivity that can be modulated by NIBS in order to restore normal functioning.

In this chapter, we provide an overview of the use of NIBS, in particular rTMS and tDCS, to reduce symptoms in psychotic disorders. To date, the majority of studies investigating the effects of NIBS on psychotic disorders included patients with schizophrenia and focused on two main categories of treatment-resistant symptoms: auditory verbal hallucinations (AVH) and negative symptoms (NS). More precisely, NIBS were proposed to reduce treatment-resistant AVH by targeting the left temporoparietal junction (TPJ), which have been reported as hyperactive during auditory hallucination occurrence.[1] In order to reduce NS, NIBS were applied in regard to the left dorsolateral prefrontal cortex (DLPFC), a brain region that showed structural and functional abnormalities in patients with NS.[2,3]

THE USE OF REPETITIVE TRANSCRANIAL MAGNETIC STIMULATION IN PSYCHOSIS

PRINCIPLE OF REPETITIVE TRANSCRANIAL MAGNETIC STIMULATION

Transcranial magnetic stimulation consists in applying a high-intensity brief current pulse through an electromagnetic coil placed on the scalp of a subject. According to the principle of electromagnetic induction, it generates a magnetic field, which in turn induces a weak electrical current in the targeted brain area situated under the coil. If the intensity of the TMS pulse is sufficient, it can depolarize neurons. For instance, when a TMS pulse is delivered over the motor cortex at a sufficient intensity, i.e., above the motor threshold (MT) of the subject, it can induce a contraction of a connected muscle that can be measured by a motor-evoked potential (MEP). The MT is commonly assessed at rest (RMT) and is defined as the minimum intensity needed to elicit a MEP of > 50 μV in at least 5 out of 10 consecutive TMS pulses. The RMT depends on each subject and is used to adapt TMS intensity at the individual excitability level in TMS protocols.[4]

When trains of TMS pulses are applied repeatedly over a short period of time, which is called repetitive TMS (rTMS), they can decrease or increase cortical excitability depending on the parameters of stimulation. Neurophysiological studies showed that when applied at high frequency (HF; > 5 Hz), rTMS can have excitatory effects on the targeted cortical excitability, whereas when applied at low frequency (LF; ≤ 1 Hz) rTMS can have inhibitory effects.[5] These effects can outlast the stimulation period. The mechanisms of the rTMS-induced changes are not entirely elucidated but seem to involve long-term potentiation and long-term depression-like changes in synaptic efficacy.

rTMS is a safe NIBS technique with rare adverse effects that are mild and transient, such as headache or discomfort. The feared serious adverse effect is the induction of seizure, however this effect was rarely described when parameters of stimulation followed the international safety guidelines.[6]

EFFECTS OF RTMS ON AUDITORY VERBAL HALLUCINATIONS

The rationale for using rTMS to reduce AVH is based on imaging studies reporting that AVH are associated with hyperactivity of the left TPJ[1] and on neurophysiological studies showing that LF rTMS can reduce cortical excitability. It was thus proposed that LF rTMS applied over the left TPJ would reduce AVH (Figure 68.1). Most of the works published to date focused on AVH in patients with schizophrenia.

In the first works conducted in 1999, Hoffman et al.[7] compared active and sham 1 Hz rTMS (2,400 pulses delivered in 4 days) in 3 patients with schizophrenia using a crossover design. They reported greater reduction of AVH severity

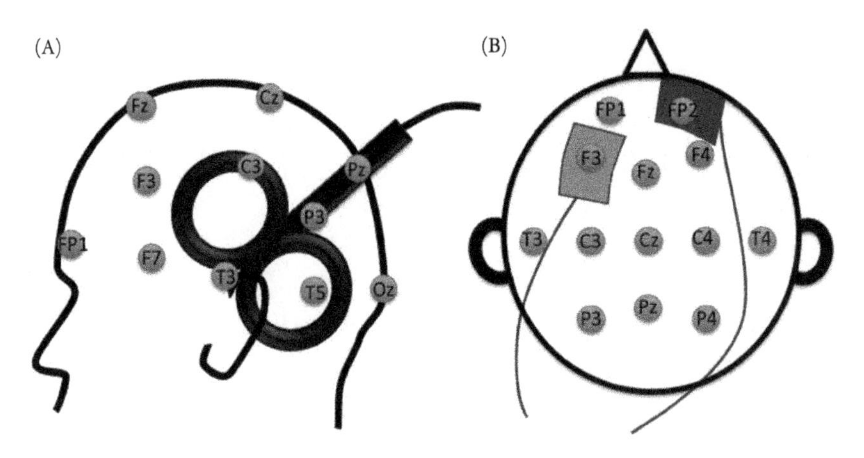

Figure 68.1 **Examples of transcranial magnetic stimulation (TMS) coil location and transcranial direct current Stimulation (tDCS) electrode montage for psychosis.**

Panel A: Example of TMS location with a figure-8 coil applied over the left temporoparietal junction (TPJ) to alleviate auditory verbal hallucinations. The coil is placed midway between T3 and P3 according to the 10–20 EEG international system.

Panel B: Example of tDCS electrodes montage with the anode (gray) applied over the left dorsolateral prefrontal cortex (F3) and the cathode (black) applied over the right supraorbital region (FP2) to alleviate negative symptoms.

following active compared to sham rTMS. Since then, numerous case-reports and open-label studies as well as crossover and parallel, randomized sham-controlled trials have replicated these significant beneficial effects of LF rTMS (mostly 1 Hz) applied over the left TPJ on AVH in patients with schizophrenia. However, there are also several studies that failed to report a significant effect of LF rTMS applied to the left TPJ on AVH (for reviews see[8,9]). Meta-analyses showed that LF rTMS to the left TPJ may be effective to reduce AVH in schizophrenia, although the effect size was moderate[8–10] (Cohen's d ≈ 0.4 depending on meta-analyses) and the level of evidence was graded as low.[10] Remarkably, studies differed on several rTMS parameters such as stimulation intensity (from 80% to 110% of the RMT), number of rTMS sessions (from 1 to 20, with a number of pulses from 1,600 to 23,040) and stimulation site location (using T3–P3 location based on the 10/20 international EEG system or using neuronavigation systems to target the TPJ either anatomically or based on fMRI results). However, using a meta-regression analysis, He et al.[10] failed to demonstrate that either intensity, stimulation site, or total number of pulses were a source of heterogeneity between studies. Interestingly, the beneficial effects of 1 Hz rTMS applied to the left TPJ to reduce AVH was also reported in some specific populations of patients with schizophrenia, such as adolescents with childhood-onset schizophrenia[11,12] or patients with late-onset schizophrenia.[13]

Some studies have tried stimulation sites other than the left TPJ and rTMS frequencies other than 1 Hz to reduce AVH in patients with schizophrenia. For instance, studies found beneficial effects of LF rTMS over the right TPJ or of bilateral stimulation over the left and right TPJ on AVH[14] (but see also[15] for negative results). Regarding frequency, Montagne-Larmurier et al.[16] reported a 41.4% decrease of AVH after 4 sessions of 20Hz rTMS (9600 total pulses) delivered to the left temporal sulcus in an open-label study including 11 patients. In another study, Kim et al.[17] compared the effects of bilateral rTMS at 20Hz or 1Hz frequencies applied on the left and right TPJ or the left and right inferior frontal gyrus. They did not report any significant differences between either stimulation site and frequency and sham. Furthermore, some studies have highlighted the efficacy of bilateral and left-sided unilateral continuous theta burst stimulation (cTBS) to reduce AVH.[18,19] cTBS is a stimulation paradigm in which 3 pulses of rTMS were delivered at 30 or 50 Hz and repeated at intervals of 200 milliseconds (corresponding to theta frequency). Theta burst stimulation is known to have more long-lasting properties on the motor cortex than classical rTMS protocols and has the considerable advantage to be of short duration (less than 1 minute in the case of cTBS).[20] It has been described that cTBS has similar effects to 1 Hz rTMS on AVH.[19] However, in a recent randomized sham-controlled study including 64 patients with psychotic disorder, Koops et al.[21] reported that even if cTBS over the left TPJ decreased AVH, this effect did not differ from sham treatment.

Only few studies investigated the effects of rTMS on AVH outside the frame of schizophrenia diagnosis. Interestingly, 1Hz rTMS applied to the left TPJ was shown to reduce AVH in a patient with Alice in Wonderland syndrome[22] and in a patient with psychotic major depression.[23] Moreover, one case-study reported a reduction of musical auditory hallucinosis in a patient with traumatic brain injury after 10 sessions of 1 Hz rTMS applied over the right posterior temporal cortex.[24] One case study reported a reduction of treatment-resistant AVH in a patient with dementia after 15 sessions of 10 Hz rTMS over the left TPJ.[25]

In sum, studies published to date showed promising results regarding the efficacy on AVH of at least 10 sessions (with at least 1,000 stimulations per session, intensity at least at 90% of RMT) of LF rTMS (1 Hz) over the left TPJ. However, effect size is moderate and there is still a large heterogeneity among studies regarding rTMS parameters, stimulation site, outcomes used to measure AVH, study design, and sham paradigms (e.g., sham coil, active coil unplugged or angulated). Indeed, regarding sham paradigms, the placebo effect in rTMS studies was recently assessed in a meta-analysis[26] and results showed that placebo should be considered as a major source of bias in the assessment of rTMS efficacy. Thus, more sham-controlled trials with larger sample size and better sham paradigms are needed.

EFFECTS OF RTMS ON OTHER PSYCHOTIC SYMPTOMS

It is interesting to note that the effect of 1Hz rTMS on AVH seems to be symptom-specific. Indeed, rTMS to the TPJ seems to have few effects on general and other psychotic symptoms of schizophrenia.[8] Some studies used different rTMS protocols to target other specific psychotic symptoms than AVH. For instance, Jardri et al.[27] showed that 10 daily sessions of 1 Hz rTMS applied to the somatosensory cortex can reduce coenesthetic hallucinations in a patient with schizoaffective disorder. Three case studies reported beneficial effects of 1Hz rTMS on visual hallucinations when targeting the occipital visual cortex, one in a patient with schizophrenia,[28] one in a patient with a partial loss of vision following a myocardial infarction,[29] and one in a blind patient with an equivalent of the Charles Bonnet syndrome.[30] Finally, a case study reported the effects of 1 Hz rTMS applied over the occipito-temporal sulcus on complex hallucinations (involving auditory, visual, and olfactory modalities) in a patient with schizophrenia.

EFFECTS OF RTMS ON NEGATIVE SYMPTOMS

Negative symptoms of schizophrenia are represented by an array of symptoms whose description, delineation, and grouping into more reliable clusters is still under debate. Even if the NIMH-MATRICS consensus statement on NS group proposed to include at least 5 domains (blunted affect, alogia, avolition, anhedonia, asociality),[31] the rTMS field as an adjunctive treatment of NS grew, considering them as a single pathological entity. Moreover first open-label studies investigating the impact of rTMS on NS were modeled on those targeting depressive symptoms (using HF rTMS applied on the left DLPFC) and were focused on a more global effect encompassing NS but not assessing them precisely.[32] Then, studies moved to more precise and hypothesis-driven approaches addressing NS as a result of hypofrontality.[33] Nowadays, some imaging data[3] confirmed a leftward hypofrontality in link with NS and it was showed that HF rTMS targeting the left DLPFC induced dopamine release in subcortical area.[34] Such data gave more pathophysiological support to the therapeutic applications of rTMS on NS. Sham-controlled studies also investigated the impact of rTMS, most of them targeting the left DLPFC with HF rTMS (mainly 10 Hz or even 15 or 20 Hz). These studies involved small sample sizes and NS were assessed with heterogeneous assessment methods, either using specific items of general scales (mostly the Positive and Negative Symptoms scale) or specific scales dedicated to NS (e.g., the Scale for the Assessment of Negative Symptoms). Protocols were heterogeneous regarding stimulation parameters (intensity ranging from 80% to 110% of the individual RMT, number of sessions from 10 to 20, frequency of stimulation from 10 to 20 Hz) and showed a mitigated improvement of NS. Given that, first meta-analyses failed to show a significant improvement and the impact of rTMS was initially considered to be a small.[35] With a growing accumulation of evidence, 5 years later, a further meta-analysis showed a moderated but significant effect size.[36]

Recently, two sham-controlled studies included large samples of patients and reported opposite results. On the one hand, Quan et al.[37] showed a significant 20% reduction of NS in patients who received 20 sessions of active rTMS applied at 10 Hz on the left DLPFC at 80% RMT intensity ($n = 78$), as compared to sham ($n = 39$). The effect lasted 24 weeks. On the other hand, Wobrock et al.[38] included 157 patients and failed to demonstrate the superiority of active rTMS over sham in a design with 15 sessions of 10 Hz rTMS on the left DLPFC at 110% RMT intensity. Including these last results, He et al.[10] performed a meta-analysis focusing on the 10 Hz studies and found that improvement of NS with rTMS did not differ significantly from sham.

When taken overall, the literature of rTMS on NS is promising, but there are numerous unsolved issues regarding stimulation parameters including intensity, total number of sessions, or number of sessions per day. An rTMS protocol for NS is roughly a left DLPFC targeted 10 Hz protocol, lasting a minimum of 3 weeks, including from 10 to 20 sessions, with intensities ranging from 80% to 110% of the RMT (see for reviews [10,36]) with no clear evidence for the superiority to some of these parameters among others.

Besides the 10 Hz protocols targeting the left DLPFC, some authors tested other stimulation parameters or locations to improve treatment impact. Most of them are still at the development phase, with a lack of evidence to affirm their superiority or inferiority toward the 10 Hz protocols. Jin et al.[39] tested an individualized frequency of stimulation based on the subjects EEG alpha rhythm (usually between 8 and 13 Hz) on 11 patients compared to sham, 3 Hz, and 20 Hz rTMS. Alpha rTMS was superior to other conditions but was not compared to a standard 10 Hz rTMS protocol. Other authors tested specific patterns of frequencies such as intermittent TBS (iTBS).[20] In a study included 96 patients randomly assigned to 4 conditions (iTBS, 10 Hz, 20 Hz, and sham), Zhao et al.[40] found that iTBS ($n = 24$) was more effective in decreasing NS than other parameters. In an open-label study, other authors reported the efficacy of iTBS applied over the vermis of the cerebellum to reduce NS.[41] In a recent randomized sham-controlled trial, Dlabac-de Lange et al.[42] reported the efficacy of active bilateral 10 Hz DLPFC stimulation (15 sessions) to decrease NS in a sample of 32 patients. The beneficial effects lasted up to 3 months. Others used another type of coils with deeper effects (called H coil) on the prefrontal region but failed to show a significant superiority to sham stimulation.[43]

Finally, in the late 2010s the field of rTMS in NS is still questionable, with some promising results, but a lack of strong evidence for its efficacy. Authors are still seeking optimal rTMS parameters to improve patients with refractory NS without any consensus. Beyond this we also expect new opportunities with the use of next generation tools for assessment of NS (namely the Brief Negative Symptom Scale and the Clinical Assessment Interview for Negative Symptoms) who offer a refined delineation of NS (showing a two-factor structure with an expressivity and a motivation factor) and should give new insights on the effects of treatments.[44]

THE USE OF TRANSCRANIAL DIRECT CURRENT STIMULATION IN PSYCHOSIS

PRINCIPLE OF TRANSCRANIAL DIRECT CURRENT STIMULATION

Transcranial direct current stimulation consists in applying a low-intensity electric current, commonly 1 or 2 mA, through the brain using two electrodes placed over the scalp. The current delivered by a battery-driven constant current stimulator flows through the brain from the anode (positive electrode) to the cathode (negative electrode). tDCS has been described to induce bidirectional polarity-dependent changes in cortical excitability, namely excitatory and inhibitory effects (for reviews on tDCS effects and mechanisms, see [45,46]). For instance, tDCS applied with the anode over the primary motor cortex can increase MEP, whereas tDCS applied with the cathode over the primary motor cortex can reduce them. The mechanisms by which tDCS modifies cortical excitability are not completely known but seem to be mediated by changes in resting neuronal membrane potential (namely a shift toward depolarization or hyperpolarization). Interestingly, when the stimulation is applied during several minutes, classically 10 to 30 min, the changes in cortical excitability can outlast the stimulation period by an hour or longer. This tDCS-induced long-term plasticity seems linked to glutamatergic NMDA-receptor dependent activity and GABAergic activity.

Compared with other NIBS, tDCS provides considerable advantages: the device is portable, low-cost, and easy to use. Moreover, tDCS induces minor side effects, the mostly frequently reported being itching, tingling, and burning sensations.[47,48] Furthermore, the existence of automated devices delivering active and sham protocols without the awareness of the tDCS operator makes it possible to conduct clinical trials with sham-controlled condition while preserving the integrity of the blinding. Sham protocols commonly consist of delivering active stimulation during a brief period at the beginning of the session (commonly 30 to 60 seconds) followed by no current stimulation during the remaining time of the tDCS session.

EFFECTS OF TDCS ON AUDITORY VERBAL HALLUCINATIONS

In the last decade, several studies have investigated the effects of tDCS on treatment-resistant AVH, mainly in patients with a diagnosis of schizophrenia or schizoaffective disorder. One tDCS protocol is mostly used across studies to reduce treatment-resistant AVH. It consists in a frontotemporal electrode montage, with the anode placed over the left DLPFC and the cathode placed over the left TPJ. The rationale for using this frontotemporal montage is that AVH are associated to hyperactivity within the TPJ[1] and to frontotemporal dysconnectivity.[49] Brunelin et al.[50] conducted a parallel randomized sham-controlled trial using this frontotemporal montage. In this study, 30 patients with a DSM-IV diagnosis of schizophrenia were randomly allocated to receive 10 sessions of 20 minutes of either active (2mA) or sham tDCS delivered twice daily on 5 consecutive days. Electrodes were placed on the scalp based on the 10/20 international EEG system with the center of the anode placed between F3 and FP1 (left DLPFC) and the center of the cathode placed between T3 and P3 (left TPJ). The authors reported a significant 31% decrease of treatment-resistant AVH severity after active tDCS that lasted for at least 3 months. Using the same tDCS protocol, in two randomized sham-controlled trials including samples of patients with schizophrenia that partially overlapped with the initial sample, the same group of authors reported similar findings of AVH reduction.[51,52] Furthermore, promising effects of frontotemporal tDCS (20 min, 2mA) on AVH were also reported in several open-label studies and case-reports (for a review see[53]). Of note, four case-reports showed a complete remission of AVH after tDCS[54–57] and two highlighted the efficacy and safety of maintenance tDCS sessions for 1 and 3 years.[56,58]

Although open-label studies and case-reports should be interpreted cautiously, they provided some relevant findings that might pave the way for the use of tDCS as a treatment in patients with AVH. For instance, some case studies highlighted the potential of frontotemporal tDCS to reduce AVH when used as a monotherapy[54,55,59] or as an add-on treatment to ECT and clozapine.[60] Two case-studies reported the efficacy and safety of frontotemporal tDCS to reduce AVH during pregnancy as a monotherapy[59] or with concomitant antipsychotic medication.[61] In an open-label study, Agarwal et al.[62] showed that tDCS has a lower impact on AVH in patients treated with antipsychotics with high affinity for the dopamine D2 receptor as compared to patients on low affinity antipsychotics or a mixture of the two. In an open-label study including 16 patients, Brunelin et al.[50] reported that tDCS have no effect on AVH in patients with schizophrenia and tobacco use disorder as compared with nonsmokers. Altogether, these findings highlighted the need to take into account the nature of concomitant drug intake when combining with tDCS. Case-reports also provided some insight into the interest of using individualized tDCS parameters, namely regarding stimulation intensity, duration, or electrode montage. In regard to current intensity and duration, Andrade[58] reported the safety and efficacy of increasing the stimulation intensity from 1 to 3 mA and the duration from 20 then 30 minutes. Bose et al.[63] recently reported the efficacy of using a right-sided frontotemporal electrode montage (anode over the right DLPFC and cathode over the right TPJ) for reducing AVH when the commonly used left frontotemporal tDCS failed to induce an effect.

On the other hand, two randomized sham-controlled trials reported no clinical effects of active frontotemporal tDCS as compared to sham tDCS on AVH. It is interesting to note that these two studies have used once-daily sessions of tDCS instead of twice-daily sessions. Fitzgerald et al.[64] failed to report a significant AVH reduction after 15 sessions of either unilateral (with the anode over F3 and the cathode over the T3-P3) or bilateral (with two anodes over F3 and F4 and two cathodes over T3-P3 and T4-P4) tDCS in a crossover randomized sham-controlled trial including 24 patients with a diagnosis of schizophrenia or schizoaffective disorder. Fröhlich et al.[65] conducted a parallel randomized sham-controlled trial including 26 patients with a diagnosis of schizophrenia or schizoaffective disorder. They showed no difference in AVH reduction between active and sham groups after 5 sessions of tDCS delivered with two tDCS devices and three electrodes: one placed over the left DLPFC between F3 and FP1 (the anode of the first tDCS device, with current set at +2 mA), one placed over the left TPJ between T3 and P3 (the cathode of the second tDCS device, with current set at -2 mA) and a common return electrode placed over Cz (posterior midline).

Finally, only one case study showed the effects of tDCS on AVH outside the frame of schizophrenia or schizoaffective disorder. Goyal et al.[66] reported the case of a complete remission of AVH for at least 2 months after 10 sessions of frontotemporal tDCS (2 mA, 20 minutes for the first 5 sessions and 30 minutes for the last 5 sessions) in a patient who suffered from treatment-resistant AVH following discontinuation of alcohol and maintenance of abstinence.

In sum, although results are encouraging, it is still premature to conclude regarding the efficacy of tDCS to alleviate AVH in psychotic disorders. Several factors may account for discrepancies between results. Indeed, studies investigating the effects of frontotemporal tDCS on AVH have used variable parameters in regard to current intensity (from 1 to 3 mA), tDCS duration (from 15 to 30 minutes), electrode size (mostly 35 cm^2 but few studies used 25 cm^2 electrodes), number of sessions (5 to 20), and number of sessions by day (once or twice daily). Moreover, the choice of outcomes (from standardized multidimensional scales to single-item measures) may have influenced the findings, since they may not have the same sensitivity to capture changes in AVH. Further sham-controlled studies with larger samples of subjects are needed to investigate the acute and long-term effects of frontotemporal tDCS on AVH, and more data are needed to determine optimal parameters of tDCS in this indication. Up to now, collected evidence suggests that 10 sessions of tDCS of 20-minute duration and at a 2 mA intensity delivered twice per day may reduce AVH.

EFFECTS OF TDCS OTHER PSYCHOTIC SYMPTOMS

Remarkably, among studies reporting a reduction of AVH in patients with schizophrenia following tDCS, some also observed a decrease in general symptoms of schizophrenia,[50,58,67,68] positive symptoms,[59,69] and delusional thoughts.[60,69] A reduction of NS[51,69,70] and insight into the illness[54,57,71] was also observed after frontotemporal tDCS. For instance, in an open-label study including 21 patients, Bose et al.[71] reported a correlation between improvement of insight into the illness and decrease of AVH in patients with schizophrenia following 10 sessions of frontotemporal tDCS.

Some case reports have specifically investigated the effects of tDCS in reducing other psychotic symptoms than AVH such as hallucinations in other modalities. Two case studies showed the interest of fronto-occipital tDCS (anode over F3 and the cathode over Oz) to reduce treatment-resistant visual hallucinations, one in a patient with schizophrenia[69] and another in a patient with major depression. [72]Moreover, Schwippel et al.[73] described the case of a patient with schizophrenia who reported a considerable alleviation of his treatment-resistant multimodal hallucinations (with auditory, visual, and tactile modalities) after tDCS delivered with the anode over the left TPJ and cathode over the right DLPFC. Since the beneficial effect was transitory and did not last after stimulation, tDCS were pursued at home for 18 months at a frequency from three times a week to daily sessions (approximately 400 tDCS sessions).

Furthermore, one case-report showed that tDCS delivered with the anode over the left DLPFC (F3) and the cathode over the right DLPFC (F4) can completely alleviate catatonia in a patient with medication- and ECT-resistant catatonic schizophrenia.[74]Another case-report showed that tDCS delivered with the cathode placed over the right DLPFC (F4) and the anode over the contralateral deltoid muscle can reduce manic symptoms by approximately 50% in a patient with schizoaffective disorder who suffers from a treatment-resistant manic episode.[75]

EFFECTS OF TDCS ON NEGATIVE SYMPTOMS OF SCHIZOPHRENIA

Besides the effects on AVH, one growing application for the use of tDCS in psychotic disorders is for reducing treatment-resistant NS of schizophrenia. In this indication, a frontal electrode montage was used to target the left DLPFC. The anode was placed over the left DLPFC and the cathode either over the supra orbital region, the right DLPFC, or extracephalically. Palm et al. were the first to report that 10 sessions of tDCS delivered once daily with the anode placed over the left DLPFC (F3) and the cathode electrode placed over the right supra orbital region (FP2, see Figure 68.1) reduced treatment-resistant NS in a patient with schizophrenia[76] and then in a randomized sham-controlled trial including 20 patients with NS.[77] In a second randomized sham-controlled study, Gomes et al.[78] reported a significant 20% reduction in NS after patients received 10 daily sessions of active tDCS with the anode placed over the left DLPFC and the cathode placed over the right DLPFC, as compared to sham. A significant 15% reduction in general symptoms of schizophrenia was also reported after active tDCS. A beneficial effect of tDCS on NS was also reported more recently in an open-label study including 9 patients with schizophrenia who received 10 daily sessions of tDCS with the anode placed over the left DLPFC and the cathode placed over the right deltoid muscle.[79]

OTHER STIMULATION PARAMETERS

Beside the effects of tDCS, new stimulation parameters of transcranial electrical current stimulation can also be useful in patients with psychosis. Among them, two new stimulation strategies were investigated in patients with treatment-resistant NS and AVH. Two case studies reported the beneficial effect of transcranial random noise stimulation (tRNS) to reduce AVH when applied with a frontotemporal montage and to reduce NS when applied with a bifrontal montage (Figure 68.1).[80,81] tRNS consists in delivering HF (100–640 Hz) oscillatory direct current stimulation between 2 electrodes at low intensity (2 mA). Last, in a case series, Kallel et al.[82] reported the clinical interest of 20 sessions of theta-transcranial alternating current stimulation (4 Hz-tACS, 2mA) applied over the left and right DLPFCs to reduce NS and improve insight into the illness in three patients with clozapine-resistant schizophrenia.

CONCLUSION

In this chapter, we reviewed and discussed studies investigating the usefulness of NIBS to reduce symptoms of patients with psychotic disorders. To date, NIBS are used in two main indications: to reduce AVH by targeting the left TPJ with rTMS and/or the left fronto-temporal network with tDCS, and to reduce NS by targeting the DLPFC with rTMS and tDCS. Although results are promising in these indications, the lack of sham-controlled studies conducted with large sample of patients limits the interest for the use of these techniques in clinical routine. Further randomized controlled trials with large samples are thus needed. Moreover, further investigations have to determine optimal stimulation parameters to use for a better impact on symptoms of psychotic disorders.

REFERENCES

1. Jardri R, Pouchet A, Pins D, Thomas P. Cortical activations during auditory verbal hallucinations in schizophrenia: a coordinate-based meta-analysis. *Am J Psychiatry*. 2011;168(1):73–81.
2. Sanfilipo M, Lafargue T, Rusinek H, et al. Volumetric measure of the frontal and temporal lobe regions in schizophrenia: relationship to negative symptoms. *Arch Gen Psychiatry*. 2000;57(5):471–480.
3. Wolf DH, Turetsky BI, Loughead J, et al. Auditory oddball fMRI in schizophrenia: association of negative symptoms with regional hypoactivation to novel distractors. *Brain Imaging Behav*. 2008;2(2):132–145. doi:10.1007/s11682-008-9022-7.
4. Rossini PM, Burke D, Chen R, et al. Non-invasive electrical and magnetic stimulation of the brain, spinal cord, roots and peripheral nerves: basic principles and procedures for routine clinical and research application. An updated report from an I.F.C.N. Committee. *Clin Neurophysiol*. 126(6):1071–1107. doi:10.1016/j.clinph.2015.02.001.
5. Pascual-Leone A, Valls-Solé J, Wassermann EM, Hallett M. Responses to rapid-rate transcranial magnetic stimulation of the human motor cortex. *Brain J Neurol*. 1994;117(Pt 4):847–858.
6. Rossi S, Hallett M, Rossini PM, Pascual-Leone A. Safety, ethical considerations, and application guidelines for the use of transcranial magnetic stimulation in clinical practice and research. *Clin Neurophysiol*. 2009;120(12):2008–2039. doi:10.1016/j.clinph.2009.08.016.
7. Hoffman RE, Boutros NN, Berman RM, et al. Transcranial magnetic stimulation of left temporoparietal cortex in three patients reporting hallucinated "voices." *Biol Psychiatry*. 1999;46(1):130–132.
8. Slotema CW, Blom JD, van Lutterveld R, Hoek HW, Sommer IEC. Review of the efficacy of transcranial magnetic stimulation for auditory verbal hallucinations. *Biol Psychiatry*. 2014;76(2):101–110. doi:10.1016/j.biopsych.2013.09.038.
9. Otani VHO, Shiozawa P, Cordeiro Q, Uchida RR. A systematic review and meta-analysis of the use of repetitive transcranial magnetic stimulation for auditory hallucinations treatment in refractory schizophrenic patients. *Int J Psychiatry Clin Pract*. 2015;19(4):228–232. doi:10.3109/13651501.2014.980830.
10. He H, Lu J, Yang L, et al. Repetitive transcranial magnetic stimulation for treating the symptoms of schizophrenia: a PRISMA compliant meta-analysis. *Clin Neurophysiol Off J Int Fed Clin Neurophysiol*. 2017;128(5):716–724. doi:10.1016/j.clinph.2017.02.007.
10. Jardri R, Bubrovszky M, Demeulemeester M, et al. Repetitive transcranial magnetic stimulation to treat early-onset auditory hallucinations. *J Am Acad Child Adolesc Psychiatry*. 2012;51(9):947–949. doi:10.1016/j.jaac.2012.06.010.
12. Jardri R, Lucas B, Delevoye-Turrell Y, et al. An 11-year-old boy with drug-resistant schizophrenia treated with temporo-parietal rTMS. *Mol Psychiatry*. 2007;12(4):320. doi:10.1038/sj.mp.4001968.
13. Poulet E, Brunelin J, Kallel L, D'Amato T, Saoud M. Maintenance treatment with transcranial magnetic stimulation in a patient with late-onset schizophrenia. *Am J Psychiatry*. 2008;165(4):537–538. doi:10.1176/appi.ajp.2007.07060868.
14. Hoffman RE, Wu K, Pittman B, et al. Transcranial magnetic stimulation of Wernicke's and right homologous sites to curtail "voices": a randomized trial. *Biol Psychiatry*. 2013;73(10):1008–1014. doi:10.1016/j.biopsych.2013.01.016.
15. Jandl M, Steyer J, Weber M, et al. Treating auditory hallucinations by transcranial magnetic stimulation: a randomized controlled cross-over trial. *Neuropsychobiology*. 2006;53(2):63–69. doi:10.1159/000091721.
16. Montagne-Larmurier A, Etard O, Razafimandimby A, Morello R, Dollfus S. Two-day treatment of auditory hallucinations by high frequency rTMS guided by cerebral imaging: a 6 month follow-up pilot study. *Schizophr Res*. 2009;113(1):77–83. doi:10.1016/j.schres.2009.05.006.
17. Kim E-J, Yeo S, Hwang I, et al. Bilateral repetitive transcranial magnetic stimulation for auditory hallucinations in patients with schizophrenia: a randomized controlled, cross-over study. *Clin Psychopharmacol Neurosci*. 2014;12(3):222–228. doi:10.9758/cpn.2014.12.3.222.
18. Plewnia C, Zwissler B, Wasserka B, Fallgatter AJ, Klingberg S. Treatment of auditory hallucinations with bilateral theta burst stimulation: a randomized controlled pilot trial. *Brain Stimulat*. 2014;7(2):340–341. doi:10.1016/j.brs.2014.01.001.
19. Kindler J, Homan P, Flury R, Strik W, Dierks T, Hubl D. Theta burst transcranial magnetic stimulation for the treatment of auditory verbal hallucinations: results of a randomized controlled study. *Psychiatry Res*. 2013;209(1):114–117. doi:10.1016/j.psychres.2013.03.029.
20. Huang Y-Z, Edwards MJ, Rounis E, Bhatia KP, Rothwell JC. Theta burst stimulation of the human motor cortex. *Neuron*. 2005;45(2):201–206. doi:10.1016/j.neuron.2004.12.033.
21. Koops S, van Dellen E, Schutte MJL, Nieuwdorp W, Neggers SFW, Sommer IEC. Theta burst transcranial magnetic stimulation for auditory verbal hallucinations: negative findings from a double-blind-randomized trial. *Schizophr Bull*. 2016;42(1):250–257. doi:10.1093/schbul/sbv100.
22. Blom JD, Looijestijn J, Goekoop R, et al. Treatment of Alice in Wonderland syndrome and verbal auditory hallucinations using repetitive transcranial magnetic stimulation: a case report with fMRI findings. *Psychopathology*. 2011;44(5):337–344. doi:10.1159/000325102.
23. Freitas C, Pearlman C, Pascual-Leone A. Treatment of auditory verbal hallucinations with transcranial magnetic stimulation in a patient with psychotic major depression: one-year follow-up. *Neurocase*. 2012;18(1):57–65. doi:10.1080/13554794.2010.547864.
24. Cosentino G, Giglia G, Palermo A, et al. A case of post-traumatic complex auditory hallucinosis treated with rTMS. *Neurocase*. 2010;16(3):267–272. doi:10.1080/13554790903456191.
25. Marras A, Pallanti S. Transcranial magnetic stimulation for the treatment of pharmacoresistant nondelusional auditory verbal hallucinations in dementia. *Case Rep Psychiatry*. 2013;2013:930304. doi:10.1155/2013/930304.

26. Dollfus S, Lecardeur L, Morello R, Etard O. Placebo response in repetitive transcranial magnetic stimulation trials of treatment of auditory hallucinations in schizophrenia: a meta-analysis. *Schizophr Bull.* 2016;42(2):301–308. doi:10.1093/schbul/sbv076.
27. Jardri R, Pins D, Thomas P. A case of fMRI-guided rTMS treatment of coenesthetic hallucinations. *Am J Psychiatry.* 2008;165(11):1490–1491. doi:10.1176/appi.ajp.2008.08040504.
28. Ghanbari Jolfaei A, Naji B, Nasr Esfehani M. Repetitive transcranial magnetic stimulation in resistant visual hallucinations in a woman with schizophrenia: a case report. *Iran J Psychiatry Behav Sci.* 2016;10(1):e3561. doi:10.17795/ijpbs-3561.
29. Merabet LB, Kobayashi M, Barton J, Pascual-Leone A. Suppression of complex visual hallucinatory experiences by occipital transcranial magnetic stimulation: a case report. *Neurocase.* 2003;9(5):436–440. doi:10.1076/neur.9.5.436.16557.
30. Meppelink AM, de Jong BM, van der Hoeven JH, van Laar T. Lasting visual hallucinations in visual deprivation; fMRI correlates and the influence of rTMS. *J Neurol Neurosurg Psychiatry.* 2010;81(11):1295–1296. doi:10.1136/jnnp.2009.183087.
31. Kirkpatrick B, Fenton WS, Carpenter WTJ, Marder SR. The NIMH-MATRICS consensus statement on negative symptoms. *Schizophr Bull.* 2006;32(2):214–219. doi:10.1093/schbul/sbj053.
32. Geller V, Grisaru N, Abarbanel JM, Lemberg T, Belmaker RH. Slow magnetic stimulation of prefrontal cortex in depression and schizophrenia. *Prog Neuropsychopharmacol Biol Psychiatry.* 1997;21(1):105–110.
33. Cohen E, Bernardo M, Masana J, et al. Repetitive transcranial magnetic stimulation in the treatment of chronic negative schizophrenia: a pilot study. *J Neurol Neurosurg Psychiatry.* 1999;67(1):129–130.
34. Strafella AP, Paus T, Barrett J, Dagher A. Repetitive transcranial magnetic stimulation of the human prefrontal cortex induces dopamine release in the caudate nucleus. *J Neurosci.* 2001;21(15):RC157.
35. Freitas C, Fregni F, Pascual-Leone A. Meta-analysis of the effects of repetitive transcranial magnetic stimulation (rTMS) on negative and positive symptoms in schizophrenia. *Schizophr Res.* 2009;108:11–24. doi:10.1016/j.schres.2008.11.027.
36. Shi C, Yu X, Cheung EFC, Shum DHK, Chan RCK. Revisiting the therapeutic effect of rTMS on negative symptoms in schizophrenia: a meta-analysis. *Psychiatry Res.* 2014;215(3):505–513. doi:10.1016/j.psychres.2013.12.019.
37. Quan WX, Zhu XL, Qiao H, et al. The effects of high-frequency repetitive transcranial magnetic stimulation (rTMS) on negative symptoms of schizophrenia and the follow-up study. *Neurosci Lett.* 2015;584:197–201. doi:10.1016/j.neulet.2014.10.029.
38. Wobrock T, Guse B, Cordes J, et al. Left prefrontal high-frequency repetitive transcranial magnetic stimulation for the treatment of schizophrenia with predominant negative symptoms: a sham-controlled, randomized multicenter trial. *Biol Psychiatry.* 2015;77(11):979–988. doi:10.1016/j.biopsych.2014.10.009.
39. Jin Y, Potkin SG, Kemp AS, et al. Therapeutic effects of individualized alpha frequency transcranial magnetic stimulation (alphaTMS) on the negative symptoms of schizophrenia. *Schizophr Bull.* 2006;32(3):556–561. doi:10.1093/schbul/sbj020.
40. Zhao S, Kong J, Li S, Tong Z, Yang C, Zhong H. Randomized controlled trial of four protocols of repetitive transcranial magnetic stimulation for treating the negative symptoms of schizophrenia. *Shanghai Arch Psychiatry.* 2014;26(1):15–21. doi:10.3969/j.issn.1002-0829.2014.01.003.
41. Demirtas-Tatlidede A, Freitas C, Cromer JR, et al. Safety and proof of principle study of cerebellar vermal theta burst stimulation in refractory schizophrenia. *Schizophr Res.* 2010;124(1–3):91–100. doi:10.1016/j.schres.2010.08.015.
42. Dlabac-de Lange JJ, Bais L, van Es FD, et al. Efficacy of bilateral repetitive transcranial magnetic stimulation for negative symptoms of schizophrenia: results of a multicenter double-blind randomized controlled trial. *Psychol Med.* 2015;45(6):1263–1275. doi:10.1017/S0033291714002360.
43. Rabany L, Deutsch L, Levkovitz Y. Double-blind, randomized sham controlled study of deep-TMS add-on treatment for negative symptoms and cognitive deficits in schizophrenia. *J Psychopharmacol Oxf Engl.* 2014;28(7):686–690. doi:10.1177/0269881114533600.
44. Strauss GP, Gold JM. A psychometric comparison of the clinical assessment interview for negative symptoms and the brief negative symptom scale. *Schizophr Bull.* 2016;42(6):1384–1394. doi:10.1093/schbul/sbw046.
45. Brunoni AR, Nitsche MA, Bolognini N, et al. Clinical research with transcranial direct current stimulation (tDCS): Challenges and future directions. *Brain Stimulat.* 2012;5(3):175–195. doi:10.1016/j.brs.2011.03.002.
46. Stagg CJ, Nitsche MA. Physiological basis of transcranial direct current stimulation. *The Neuroscientist.* 2011;17(1):37–53. doi:10.1177/1073858410386614.
47. Brunoni AR, Amadera J, Berbel B, Volz MS, Rizzerio BG, Fregni F. A systematic review on reporting and assessment of adverse effects associated with transcranial direct current stimulation. *Int J Neuropsychopharmacol.* 2011;14(8):1133–1145. doi:10.1017/S1461145710001690.
48. Kessler SK, Turkeltaub PE, Benson JG, Hamilton RH. Differences in the experience of active and sham transcranial direct current stimulation. *Brain Stimulat.* 2012;5(2):155–162. doi:10.1016/j.brs.2011.02.007.
49. Lawrie SM, Buechel C, Whalley HC, Frith CD, Friston KJ, Johnstone EC. Reduced frontotemporal functional connectivity in schizophrenia associated with auditory hallucinations. *Biol Psychiatry.* 2002;51(12):1008–1011.
50. Brunelin J, Mondino M, Gassab L, et al. Examining transcranial direct-current stimulation (tDCS) as a treatment for hallucinations in schizophrenia. *Am J Psychiatry.* 2012;169(7):719–724. doi:10.1176/appi.ajp.2012.11071091.
51. Mondino M, Jardri R, Suaud-Chagny M-F, Saoud M, Poulet E, Brunelin J. Effects of fronto-temporal transcranial direct current stimulation on auditory verbal hallucinations and resting-state functional connectivity of the left temporo-parietal junction in patients with schizophrenia. *Schizophr Bull.* 2016;42(2):318–326. doi:10.1093/schbul/sbv114.
52. Mondino M, Haesebaert F, Poulet E, Suaud-Chagny M-F, Brunelin J. Fronto-temporal transcranial Direct Current Stimulation (tDCS) reduces source-monitoring deficits and auditory hallucinations in patients with schizophrenia. *Schizophr Res.* 2015;161(2–3):515–516. doi:10.1016/j.schres.2014.10.054.
53. Mondino M, Brunelin J, Palm U, Brunoni AR, Poulet E, Fecteau S. Transcranial direct current stimulation for the treatment of refractory symptoms of schizophrenia. current evidence and future directions. *Curr Pharm Des.* 2015;21(23):3373–3383.
54. Rakesh G, Shivakumar V, Subramaniam A, et al. Monotherapy with tDCS for schizophrenia: a case report. *Brain Stimulat.* 2013;6(4):708–709. doi:10.1016/j.brs.2013.01.007.
55. Goyal P, Kataria L, Andrade C. Transcranial direct current stimulation as monotherapy attenuates auditory hallucinations in treatment-naïve first-episode schizophrenia. *J ECT.* 2016;32(3):e15–16. doi:10.1097/YCT.0000000000000324.
56. Shivakumar V, Narayanaswamy JC, Agarwal SM, Bose A, Subramaniam A, Venkatasubramanian G. Targeted, intermittent booster tDCS: a novel add-on application for maintenance treatment in a schizophrenia patient with refractory auditory verbal hallucinations. *Asian J Psychiatry.* 2014;11:79–80. doi:10.1016/j.ajp.2014.05.012.
57. Shivakumar V, Bose A, Rakesh G, et al. Rapid improvement of auditory verbal hallucinations in schizophrenia after add-on treatment with transcranial direct-current stimulation. *J ECT.* 2013;29(3):e43–44. doi:10.1097/YCT.0b013e318290fa4d.
58. Andrade C. Once- to twice-daily, 3-year domiciliary maintenance transcranial direct current stimulation for severe, disabling, clozapine-refractory continuous auditory hallucinations in schizophrenia. *J ECT.* 2013;29(3):239–242. doi:10.1097/YCT.0b013e3182843866.
59. Strube W, Kirsch B, Padberg F, Hasan A, Palm U. Transcranial direct current stimulation as monotherapy for the treatment of auditory

hallucinations during pregnancy: a case report. *J Clin Psychopharmacol.* 2016;36(5):534–535. doi:10.1097/JCP.0000000000000554.
60. Jacks S, Kalivas B, Mittendorf A, Kindt C, Short EB. Transcranial direct-current stimulation as an adjunct to electroconvulsive therapy and clozapine for refractory psychosis. *Prim Care Companion CNS Disord.* 2014;16(3). doi:10.4088/PCC.14l01635.
61. Shenoy S, Bose A, Chhabra H, et al. Transcranial direct current stimulation (tDCS) for auditory verbal hallucinations in schizophrenia during pregnancy: a case report. *Brain Stimulat.* 2015;8(1):163–164. doi:10.1016/j.brs.2014.10.013.
62. Agarwal SM, Bose A, Shivakumar V, et al. Impact of antipsychotic medication on transcranial direct current stimulation (tDCS) effects in schizophrenia patients. *Psychiatry Res.* 2016;235:97–103. doi:10.1016/j.psychres.2015.11.042.
63. Bose A, Sowmya S, Shenoy S, et al. Clinical utility of attentional salience in treatment of auditory verbal hallucinations in schizophrenia using transcranial direct current stimulation (tDCS). *Schizophr Res.* 2015;164(1–3):279–280. doi:10.1016/j.schres.2015.01.040.
64. Fitzgerald PB, McQueen S, Daskalakis ZJ, Hoy KE. A negative pilot study of daily bimodal transcranial direct current stimulation in schizophrenia. *Brain Stimulat.* 2014;7(6):813–816. doi:10.1016/j.brs.2014.08.002.
65. Fröhlich F, Burrello TN, Mellin JM, et al. Exploratory study of once-daily transcranial direct current stimulation (tDCS) as a treatment for auditory hallucinations in schizophrenia. *Eur Psychiatry J Assoc Eur Psychiatr.* 2016;33:54–60. doi:10.1016/j.eurpsy.2015.11.005.
66. Goyal P, Kataria L, Andrade C. Transcranial direct current stimulation for chronic continuous antipsychotic-refractory auditory hallucinations in alcoholic hallucinosis. *Brain Stimulat.* 2016;9(1):159–160. doi:10.1016/j.brs.2015.10.011.
67. Homan P, Kindler J, Federspiel A, et al. Muting the voice: a case of arterial spin labeling-monitored transcranial direct current stimulation treatment of auditory verbal hallucinations. *Am J Psychiatry.* 2011;168(8):853–854. doi:10.1176/appi.ajp.2011.11030496.
68. Brunelin J, Mondino M, Haesebaert F, Saoud M, Suaud-Chagny MF, Poulet E. Efficacy and safety of bifocal tDCS as an interventional treatment for refractory schizophrenia. *Brain Stimulat.* 2012;5(3):431–432. doi:http://dx.doi.org/10.1016/j.brs.2011.03.010.
69. Shiozawa P, da Silva ME, Cordeiro Q, Fregni F, Brunoni AR. Transcranial direct current stimulation (tDCS) for the treatment of persistent visual and auditory hallucinations in schizophrenia: a case study. *Brain Stimulat.* 2013;6(5):831–833. doi:10.1016/j.brs.2013.03.003.
70. Narayanaswamy JC, Shivakumar V, Bose A, Agarwal SM, Venkatasubramanian G, Gangadhar BN. Sustained improvement of negative symptoms in schizophrenia with add-on tDCS: a case report. *Clin Schizophr Relat Psychoses.* 2014;(aop):1–7. doi:10.3371/CSRP.JNVS.061314.
71. Bose A, Shivakumar V, Narayanaswamy JC, et al. Insight facilitation with add-on tDCS in schizophrenia. *Schizophr Res.* 2014;156(1):63–65. doi:10.1016/j.schres.2014.03.029.
72. Koops S, Sommer IEC. Transcranial direct current stimulation (tDCS) as a treatment for visual hallucinations: a case study. *Psychiatry Res.* April 2017. doi:10.1016/j.psychres.2017.03.054.
73. Schwippel T, Wasserka B, Fallgatter AJ, Plewnia C. Safety and efficacy of long-term home treatment with transcranial direct current stimulation (tDCS) in a case of multimodal hallucinations. *Brain Stimulat.* April 2017. doi:10.1016/j.brs.2017.04.124.
74. Shiozawa P, da Silva ME, Cordeiro Q, Fregni F, Brunoni AR. Transcranial direct current stimulation (tDCS) for catatonic schizophrenia: a case study. *Schizophr Res.* 2013;146(1-3):374–375. doi:10.1016/j.schres.2013.01.030.
75. Sayar GH, Salcini C, Özten E, Gül IG, Eryılmaz G. Transcranial direct current stimulation in a patient with schizoaffective disorder manic episode. *Neuromodulation.* 2014;17(8):743–745. doi:10.1111/ner.12129.
76. Palm U, Keeser D, Blautzik J, et al. Prefrontal transcranial direct current stimulation (tDCS) changes negative symptoms and functional connectivity MRI (fcMRI) in a single case of treatment-resistant schizophrenia. *Schizophr Res.* 2013;150(2–3):583–585. doi:10.1016/j.schres.2013.08.043.
77. Palm U, Keeser D, Hasan A, et al. Prefrontal transcranial direct current stimulation for treatment of schizophrenia with predominant negative symptoms: a double-blind, sham-controlled proof-of-concept study. *Schizophr Bull.* 2016;42(5):1253–1261. doi:10.1093/schbul/sbw041.
78. Gomes JS, Shiozawa P, Dias ÁM, et al. Left dorsolateral prefrontal cortex anodal tDCS effects on negative symptoms in schizophrenia. *Brain Stimulat.* 2015;8(5):989–991. doi:10.1016/j.brs.2015.07.033.
79. Kurimori M, Shiozawa P, Bikson M, Aboseria M, Cordeiro Q. Targeting negative symptoms in schizophrenia: results from a proof-of-concept trial assessing prefrontal anodic tDCS protocol. *Schizophr Res.* 2015;166(1–3):362–363. doi:10.1016/j.schres.2015.05.029.
80. Palm U, Hasan A, Keeser D, Falkai P, Padberg F. Transcranial random noise stimulation for the treatment of negative symptoms in schizophrenia. *Schizophr Res.* 2013;146(1–3):372–373. doi:10.1016/j.schres.2013.03.003.
81. Haesebaert F, Mondino M, Saoud M, Poulet E, Brunelin J. Efficacy and safety of fronto-temporal transcranial random noise stimulation (tRNS) in drug-free patients with schizophrenia: a case study. *Schizophr Res.* 2014;159:251–252.
82. Kallel L, Mondino M, Brunelin J. Effects of theta-rhythm transcranial alternating current stimulation (4.5 Hz-tACS) in patients with clozapine-resistant negative symptoms of schizophrenia: a case series. *J Neural Transm Vienna Austria 1996.* 2016;123(10):1213–1217. doi:10.1007/s00702-016-1574-x.

EARLY INTERVENTIONS

69

TREATMENT APPROACHES IN THE PSYCHOSIS PRODROME

Andrea M. Auther and Barbara A. Cornblatt

INTRODUCTION

The impetus for early intervention for schizophrenia spectrum disorders is based largely on the detrimental outcomes associated with these disorders once established. Although advances in pharmacological treatments have made living with psychotic illness less debilitating in terms of symptom burden, functional recovery (e.g., achieving financial independence through competitive employment and maintaining close personal relationships) has been more elusive. Social and role functioning achievements are particularly vulnerable to derailment, given that illness onset is often in early adulthood when individuals are completing their education and entering the workforce as well as striving to develop intimate relationships. It is believed that both psychotic symptoms and functional deficits may be more amenable to treatment in the period prior to the onset of illness, when symptoms are less severe and social and role functioning is beginning to decline. Thus, intervening in the "high-risk phase" has garnered attention for its potential to prevent or delay the onset of psychosis and improve outcomes by increasing acceptance of treatment and providing support to individuals and their family members.[1] The promise of early intervention has been bolstered by advances in reliably identifying persons at high risk for developing psychosis, the development of new treatment options, and a thriving international research community.

IDENTIFICATION OF HIGH-RISK INDIVIDUALS

Although the "prodromal" or "high-risk" period was recognized over half a century ago,[2] the potential for intervening during the prepsychotic phase of illness was spearheaded in the mid-1990s with insights from Hafner and contemporaries,[3] who recognized that a detectible yet subtle prodromal phase of nonspecific symptoms often preceded psychosis onset and appeared to set the limit on functional attainment. This work, based on retrospective reports of already ill patients, offered the hope that the pre-illness phase could be detectable, and with early intervention, psychosis could be prevented or adverse effects mitigated.

The subsequent development of high-risk criteria and assessment instruments made it possible to prospectively assess high-risk mental states and sparked significant interest in early intervention. The field was galvanized by the early work of Yung and McGorry who made the first attempt to define the high-risk construct based on clinical observations of early referrals to their Personal Assistance Crisis Evaluation (PACE) treatment program in Australia.[4,5] Yung and colleagues introduced the first formal criteria for defining the prodromal state by outlining three specific "at-risk mental states" (ARMS) thought to place a person at enhanced risk (also termed "ultra-high risk") of developing psychosis: (1) Vulnerability group—reflecting a combination of state (present symptoms) and trait (long-standing) factors defined by the presence of a psychotic disorder in a first-degree relative or schizotypal personality disorder accompanied by a significant decline in mental state or functioning; (2) Attenuated psychosis group—defined by the presence of reoccurring attenuated (or subthreshold) symptoms of psychosis as measured by standard psychiatric rating scales; and (3) Brief, limited intermittent psychosis group—referring to the presence of spontaneously remitting psychotic level delusions, hallucinations, or thought disorder. These criteria became the basis of the Comprehensive Assessment of At-Risk Mental States (CAARMS[6]), a commonly used psychosis-risk assessment instrument in Australia and Europe.

Soon after the introduction of ARMS criteria, McGlashan and colleagues in the United States operationally defined these criteria in their own measure, the Structured Interview for Prodromal Syndromes (SIPS) and companion Scale of Prodromal Symptoms (SOPS[7]), which became the standard high-risk instrument in North America and parts of Europe. The SIPS defines risk categories that are similar to the ARMS: Genetic Risk and Deterioration (GRD) syndrome, Attenuated Positive Symptom (APS) syndrome, and Brief Intermittent Psychosis Syndrome (BIPS), and adds specific requirements in terms of onset and frequency of symptoms. Another early group, also based in the United States, the Recognition and Prevention (RAP) program in New York, applied a clinical high-risk strategy based on their neurodevelopmental model of psychosis.[8] This model emphasized vulnerability factors including cognitive, affective, and social/role functioning problems that are thought to precede and underlie psychosis. The RAP developmental high-risk approach served to broaden the concept of the prodrome, in turn providing new targets for intervention beyond attenuated positive symptoms.

Concurrently, but working from a different framework based on "basic symptoms," were German psychiatrists who outlined self-perceived changes in thinking and perception thought to characterize all stages of psychotic illness including the prodromal period. These subjective changes, not typically observable until they impact functioning, include impairments in motivation, affect, thinking and concentration, verbal and motor communications, and perceptions.[9] These perceptual and cognitive concepts were operationally defined by Klosterkotter and colleagues in the Bonn Scale for the Assessment of Basic Symptoms (BSABS[10]). These symptoms were incorporated into the Schizophrenia Proneness Interview-Adult version (SPI-A[11]) and in the comprehensive empirically based Early Recognition Inventory for the Retrospective Assessment of the Onset of Schizophrenia (ERIraos[12]), all of which are used in Germany and throughout Europe.

PREDICTIVE ABILITY OF HIGH-RISK CRITERIA

Regardless of which assessment instrument is used, the ultimate test is whether the high-risk criteria reliably predict psychosis conversion. Numerous papers have been published on the validity of the criteria in individual samples using the measure most typical for that part of the world. An early paper by Klosterkotter and colleagues[10] reassessed 160 psychiatric outpatients after 9 years and found that those meeting "basic symptom" criteria at baseline had a 49.4% rate of conversion to psychosis. Yung and colleagues reported a 40.8% conversion rate at 12 months in 49 subjects meeting ARMS criteria.[13] In the United States, conversion rates based on SIPS high-risk criteria were 28% at 3 years in the RAP program[14] and 35% at 2½ years for the North American Prodrome Longitudinal Study (NAPLS) .[15] However, other studies found relatively lower rates[16] using similar criteria, and published conversion rates have been declining over time.[17] To summarize worldwide rates, a meta-analysis by Fusar-Poli and colleagues[18] incorporated 27 studies from 13 countries with a combined total of 2,502 high-risk youth and found a mean conversion rate of 29% (18% at 6 months, 22% at 1 year, and 36% at 3 years), confirming that the criteria are valid for predicting psychosis.

INTERVENTIONS FOR HIGH-RISK PATIENTS

The development of standardized and reliable assessment measures with evidence of predictive validity brought the field closer to its stated goal of intervening early to possibly prevent or delay the onset of psychosis. This, coupled with the advent of new pharmacologic agents (i.e., second-generation antipsychotics) with improved side-effect profiles, spurred the first worldwide treatment efforts directed toward clinical high-risk adolescents and young adults. Early clinical trials[19–21] demonstrated that treatment-seeking high-risk subjects could be ascertained and that the high-risk criteria were indeed capturing a population of individuals at heightened risk for psychosis. Following the partial success of the first trials, treatment efforts proliferated. However, although high-risk research gained in popularity, early intervention was not without controversy given concerns, for example, about high false positive rates and exposure to antipsychotic medication.[22] There have been a number of published treatment trials for high-risk youth including randomized and naturalistic studies of pharmacologic agents, psychotherapy interventions, family interventions, and cognitive training, either alone or in combination with each other. We present here a brief summary of the key studies and the results of meta-analyses evaluating the effectiveness of these studies on psychosis and other outcomes (see also Table 69.1).

PHARMACOLOGICAL INTERVENTIONS

The first clinical trials, both involving antipsychotic medication, were based on the theory that excess dopamine in the mesolimbic area of the brain was the cause of psychotic symptoms. The use of these agents for the first time in the high-risk phase sparked interest and controversy in the field. The first trial, conducted by McGorry and colleagues in Australia between 1996 and 1999,[19] randomized high-risk subjects to receive low-dose risperidone plus modified cognitive-behavioral therapy (CBT) and needs-based intervention (NBI) or to NBI alone. Subjects receiving the combination treatment were significantly less likely than control subjects to develop psychosis by the end of treatment (10% vs. 36%, respectively), although significance was not sustained after treatment ended (or at long-term follow up[23]). Concurrent with the McGorry trial, McGlashan and colleagues in the United States conducted the first double blind, placebo-controlled study of an atypical antipsychotic, olanzapine, in high-risk participants.[20] Five subjects (16.1%) receiving olanzapine and 11 subjects (37.9%) receiving placebo developed psychosis at the end of treatment, a nonsignificant difference. Secondary outcomes, quality of life, and functioning measures showed little differential improvement, if any at all. The most striking findings from these two early trials were the high rates of nonadherence and attrition. In the McGorry trial, although adherence to the psychosocial interventions was good, medication adherence was variable and impacted the outcome. In the McGlashan sample, although there were no differences in adherence reported, there was a high rate of attrition in the study overall (45% at 1 year and 80% at 2 years). Additionally, significant side effects were reported in the olanzapine group including fatigue and substantial weight gain. Although the McGorry trial offered support for the notion that psychosis could be delayed, at least in the short term, the other trial was largely inconclusive due to difficulty with recruitment and the high drop-out rate resulting in low power. Both studies also suggested that pharmacologic treatments were not highly

Table 69.1 SUMMARY OF HIGH-RISK RANDOMIZED CONTROLLED TRIALS

STUDY	TREATMENT(TX)	COMPARATOR(C)	DURATION OF INTERVENTION	TX EFFECT[A] FOR CONVERSION	TX EFFECT FOR HIGH-RISK SYMPTOMS	TX EFFECT FOR OTHER SYMPTOMS	TX EFFECT FOR FUNCTIONING
Medication Trials							
McGorry, 2002 Phillips, 2007	Risperidone + CBT + NBI N = 31	NBI N = 28	6 months; followed by 6 months NBI; 4 year follow-up	Yes 6 months only	No	No	No
McGlashan, 2006	Olanzapine N = 31	Placebo N = 29	12 months; 12 month follow-up	No	Yes ↓positive symptoms	No	No
Ruhrmann, 2007	Amisulpride + NFI N = 65	NFI N = 59	Up to 2 years; 12 week follow-up	—	Yes ↓basic, positive, negative, general symptoms	No	Yes ↑GAF
McGorry, 2013	CBT + Risperidone N = 43 CBT + Placebo N = 44 ST + Placebo N = 28	Monitoring N = 78	12 months; 12 month follow-up	No	No	No	No
Psychosocial Trials							
Morrison, 2004 Morrison, 2007	CT N = 35	Monitoring N = 23	6 months; 6 month & 3 year follow-up	Yes 12mos only	Yes ↓positive symptoms	—	No
Addington, 2011	CBT N = 27	ST N = 24	6 months; 12 month follow-up	No	No	No	No
van der Gaag, 2012 Ising, 2016	CBT + TAU N = 95	TAU N = 101	6 months; 6, 12, 18 month & 4 year follow-up	Yes 18mos & 4yrs	Yes ↓positive symptoms	Yes ↓distress	No
Morrison, 2012	CT + Monitoring N = 144	Monitoring N = 144	6 months; 12 & 24 month follow-up	No	Yes ↓positive symptoms	No	No
Stain, 2012	CBT N = 30	Non-Directive Reflective Listening N = 27	6 months; 12 month follow-up	No	No	No	No
Miklowitz, 2014	Family-Focused Treatment N = 66	Enhanced Care N = 63	6 months	No	Yes ↓positive symptoms	—	No
Integrated Treatment Trials							
Bechdolf, 2012	IPI N = 63	ST N = 65	12 months; 24 month follow-up	Yes 24mos	Yes ↓positive symptoms	—	—
Nordentoft, 2006	Integrated Treatment N = 42	Standard Treatment N = 37	2 years; 12 & 24 month follow-up	Yes 24mos	Yes ↓negative symptoms	—	—

(*continued*)

Table 69.1 CONTINUED

STUDY	TREATMENT(TX)	COMPARATOR(C)	DURATION OF INTERVENTION	TX EFFECT[a] FOR CONVERSION	TX EFFECT FOR HIGH-RISK SYMPTOMS	TX EFFECT FOR OTHER SYMPTOMS	TX EFFECT FOR FUNCTIONING
Cognitive Remediation Trials							
Piskulic, 2015	Computerized training (N = 18)	Computer Games (N = 14)	40 hours/12 weeks	—	No	No	Yes ↑Social Functioning
Loewy, 2016	Auditory Training (N = 50)	Computer Games (N = 33)	20-40 hours/8 weeks	—	No	Yes ↑Verbal memory	No
Choi, 2017	Processing Speed Training N = 30	Active Control (N = 32)	30 hours/ 8 weeks; 4 month follow-up	—	—	Yes ↑Processing Speed	Yes ↑Social Adjustment
Omega-3 Trials							
Amminger, 2010 Amminger, 2015	Omega-3 (N = 41)	Placebo (N = 40)	12 weeks; 7 year follow-up	Yes 12wks & 7yrs	Yes ↓positive, negative, general symptoms	No	Yes ↑GAF
McGorry, 2017	Omega-3 (N = 153)	Placebo (N = 151)	6 months; 12 month follow-up	No	No	No	No

[a] Treatment effect indicates that the specific treatment (Tx) intervention was superior to the comparator (C) intervention.

RCT = Randomized Controlled Trial; NBI = Needs-Based Intervention; CT = Cognitive Therapy; CBT = Cognitive-Behavioral Therapy; ST = Supportive Therapy; NFI = Needs-Focused Intervention; GAF = Global Assessment of Functioning; TAU = Treatment As Usual; IPI = Integrated Psychological Intervention

desired by participants and could lead to a substantial side-effect burden.

Following these pivotal studies, several other controlled and uncontrolled trials with antipsychotics were conducted. The German Research Network on Schizophrenia[24] randomized high-risk subjects to receive open label amisulpride (AMI) and needs focused intervention (NFI) or NFI alone. After 12 weeks of treatment, subjects in the AMI + NFI group were significantly more likely to have remitted from high-risk criteria (47%) than the control group (21%). Additionally, subjects in the treatment group evidenced significantly greater reduction in positive prodromal and basic symptoms than controls, although both groups improved. However, AMI caused a significant increase in prolactin levels, especially in women who were also taking antidepressants. This study has received relatively little attention in North America, possibly due to reported side effects and the fact that the drug was not approved for use in the United States.

Following up on their original trial, McGorry and colleagues[25] randomly assigned high-risk subjects to receive either CBT with low-dose risperidone, CBT with placebo, or supportive therapy (ST) with placebo, with these groups being compared to a group that refused randomization. Analyses at 12 months did not reveal any significant differences in rates of conversion between the four groups. All groups improved on positive, negative, depressive symptoms, quality of life, and functioning scores, but not differentially so. There were minimal dropouts and few side effects reported; however, adherence to medication/placebo was "poor" for 60%–80% of the subjects in each group. Although the study was underpowered, there was little evidence that specific interventions, especially antipsychotic medication, are necessary to treat many high-risk youth.

Published naturalistic medication prescribing practices in high-risk clinics have also been informative. Medication choice was not restricted in terms of class, agent, dose, or duration, making these study results more generalizable to clinical practice, yet with the obvious limitations of nonrandom assignment. Cornblatt and colleagues[26] in the Recognition and Prevention (RAP) Program in New York found evidence that antidepressants were better tolerated than antipsychotics and were associated with lower rates of conversion, potentially mediated by adherence patterns. Similarly, Fusar-Poli et al.[27] reported on 258 high-risk patients treated in the OASIS clinic in the United Kingdom and found the lowest risk of conversion in those who received therapy and antidepressants and the highest risk in patients who had therapy and antipsychotics. Although randomized trials are needed to confirm the findings, these studies suggest the effectiveness of a nonantipsychotic agent in the treatment of prodromal symptoms. In addition, antidepressants may be useful in treating comorbid conditions, as well as potentially reducing false positives, with less stigma and fewer side effects.

To summarize, antipsychotic medication trials to date have shown minimal effectiveness at reducing conversion rates in the short-term, and even less efficacy after treatment has ended, leading many in the field to conclude that antipsychotic medication should not be a "first-line" treatment for high-risk symptoms.[25] There are still unanswered questions regarding the length of treatment and follow-up that may be necessary to adequately assess efficacy, but given the side effects of some medications and the high drop-out rates seen in several studies, longer-term, sizable replications may prove difficult to conduct. There is also evidence from two naturalistic studies that antidepressants may provide benefits in addition to being better tolerated and accepted, although this requires confirmation in a controlled trial. Although improvements in high-risk symptoms and other psychopathology have been noted in the trials discussed thus far, the medication trials to date have not had an impact on functioning.

PSYCHOSOCIAL INTERVENTIONS

COGNITIVE (BEHAVIORAL) TRIALS

Given concerns about exposure to medication, lack of efficacy and tolerability of medication, and the fact that many high-risk patients present with a myriad of comorbid conditions such as depression and anxiety,[28,29] psychosocial interventions have been explored. These intervention efforts were spearheaded by French and Morrison, who conducted the first randomized controlled trial of psychotherapy in the United Kingdom between late 1999 and 2002.[21] The manualized cognitive therapy (CT) used in this original study[30] was compared to clinical monitoring and both were offered for 6 months. At one year, 2 subjects (6%) in the CT group had transitioned to psychosis compared to 5 subjects (22%) in the monitoring group. Subjects in the CT group also had significantly lower positive symptoms than controls at the end of treatment, but no differences were found for general distress or functioning. This study was well tolerated and had very few dropouts (14%), in contrast to some of the aforementioned medication trials. However, there were issues with therapist adherence, nonblinded ratings, post hoc removal of two subjects, and atypical conversion criteria used. A later report on outcome at 3 years found minimal lasting benefit once treatment terminated.[31]

The relative success of the French and Morrison[30] model led to its modification and use in several subsequent psychotherapy treatment trials. Addington et al.[32] conducted a small single-blind randomized control trial of CBT versus ST in Canada. There were no conversions in the CBT condition and three in the ST condition. Significant improvements in attenuated positive symptoms and comorbid clinical symptoms were seen for both groups, but not for negative symptoms and social functioning. The Dutch Early Detection and Intervention Evaluation (EDIE-NL) program in the Netherlands[29] randomized subjects to Treatment as Usual (TAU), targeting depression and anxiety, or to TAU plus CBT, to address cognitive biases thought to underlie high-risk symptoms. Conversion rates were significantly lower in the combined treatment condition at the 18 month follow up (10% CBT + TAU vs. 21% TAU). There were no significant differences in secondary outcomes measuring depression, anxiety, quality of life, and social functioning at follow up, although both groups improved in these areas. Unlike other trials, a reduction in risk of conversion favoring the CBT + TAU condition remained significant at long-term (4-year) follow-up.[33] The improvements seen in the control conditions in both studies highlight the benefit of supportive interventions in treating affective disorders which may, in turn, impact other outcomes such as attenuated positive symptoms and functioning.

Morrison and colleagues, who conducted the first randomized controlled trial of CT in the prodrome, published results of a second, more methodologically rigorous randomized trial of CT plus monitoring or clinical monitoring alone in 2012.[34] By 24 months, 10 CT subjects and 13 from the monitoring group converted to psychosis (nonsignificant). Aside from an improvement in CAARMS symptoms at 12 months in favor of CT, there were no treatment group differences for distress, depression, anxiety, functioning, or quality of life. However, subjects in both groups improved in all areas over the 24-month study period. While largely an unsuccessful trial, this study, the largest to date, highlighted the power of control treatments based on the observation (from many studies) that high-risk subjects generally show improvements over time, even with minimal intervention. This point is further illustrated by a more recent smaller study that compared CBT to nondirective reflective listening[35] and found low conversion rates (5% CBT, 0% control) after 6 months of treatment and no significant differences on high-risk symptoms or functioning.

CBT-based family therapy interventions have also been employed to treat high-risk youth, the majority of whom still live at home and rely on their families for care and support. Miklowitz and colleagues[36] published the first treatment trial in this area which randomized high-risk youth/families to Family-Focused Treatment (FFT) or psychoeducation (Enhanced Care [EC]). The FFT group showed significantly greater improvement in positive symptoms at 6 months compared to the EC group. Both groups also showed improvements in negative symptoms, Global Assessment of Functioning (GAF) and social and role functioning, but with no treatment effect. Conversion rates were low: 1/55 = 1.8% FFT and 5/47 = 10.6% EC. Although results were promising, the use of concomitant therapy and antipsychotic medication was high and could account for some of the results.

In summary, CBT and family therapy interventions have been proven to be highly feasible and well tolerated in contrast to the medication trials that had significant problems with dropouts and side effects. However, in terms of efficacy, psychosocial treatment trials to date have been disappointingly unsuccessful at reducing rates of conversion to psychosis, with a few exceptions.[21,29] Methodological issues such as small sample sizes, varying length, "dose," and targets of treatment, and low conversion rates have been implicated. However, these same studies have shown significant improvements in secondary outcomes including reduction of attenuated positive symptoms, depression, and anxiety symptoms. This raises the issue of the specificity of the treatments, as comparator treatments (e.g., ST) are often just as powerful in reducing distress related to high-risk and affective symptoms. The one exception to this is that functioning and quality of life measures (which are often proxy measures for functioning) tend to be relatively stable and low in high-risk participants[37] which suggests specific treatments for functioning may be necessary.

INTEGRATED TREATMENT TRIALS

There have been three integrated treatment trials that vary by high-risk population targeted and combinations of interventions. The German Research Network on Schizophrenia[38] designed an integrated intervention to target subtle "basic symptoms" that are thought to precede positive symptoms, and thus, may respond to less intensive (i.e., nonpharmacological) interventions. The manualized Integrated Psychological Intervention (IPI), consisting of individual CBT, group social skills training, computer-based cognitive remediation, and Psychoeducational Multi-Family Group (PMFG) sessions, was superior to ST in reducing the risk of developing "subthreshold psychosis" (i.e., attenuated psychosis symptoms) and full psychosis at 2 years (3.2% IPI, 15.4% ST). The IPI intervention was well tolerated, but the relative merits of each piece of the intervention could not be disentangled.

The OPUS trial in Denmark,[39] focusing on individuals with schizotypal personality disorder (SZT), provided similar psychosocial interventions in their Integrated Treatment (IT; modified assertive community treatment, individual/group social skills training, and PMFG therapy), but also allowed for antipsychotic medication when clinically indicated for both IT and standard care subjects. The integrated treatment significantly reduced risk of conversion at 12 months (8.1% IT, 33% ST) and 24 months (25% IT, 48% ST), although over 60% of both groups were also taking antipsychotic medication at follow-up. This study is in need of replication given the unusually high conversion rate for SZT patients, who often have a low rate compared to traditional high-risk samples. Other issues included nonblinded assessments and nonstandardized treatment components.

One final integrated treatment study, the Early Detection and Intervention for the Prevention of Psychosis Program (EDIPPP[40]), involved PMFG therapy, supported education and employment services, and low-dose antipsychotic medication that was tested at 6 demographically diverse sites in the United States. Subjects were divided into "high-risk" and "low-risk" groups determined by SIPS total positive symptom score, with only high-risk subjects receiving the intervention. The high-risk group evidenced significantly lower SIPS positive symptoms at outcome and better "global outcomes" (i.e., composite score of 10 clinical and functional variables) compared to the low-risk group, with trend differences for negative symptoms and functioning measures. A low number of conversions and nonrandom assignment are clear limitations of this study. However, this study was unique in that it included high-risk research centers as well as early intervention clinical service centers and demonstrated the potential of incorporating research-based interventions into existing clinical services.

Integrated treatment approaches hold broad appeal because they address virtually all of the presenting issues of high-risk patients and their families and several randomized studies have shown promise in delaying psychosis onset.[38,39] However, when successful, it is difficult to know the relative contributions of each treatment component or whether all pieces are needed to impact outcomes. Integrated treatment programs also involve large treatment teams and require multiple appointments per week, which can be difficult on a systems level as well as for families. Of note is an innovative adaptive and sequential treatment approach currently underway in Australia[41] that provides brief supportive strategies, followed by sequential randomized trials of CBT and antidepressant medication, depending on response to the previous intervention. This stepped-care approach has many advantages and can inform on the usefulness of each piece of integrated care.

OMEGA-3 TREATMENT TRIALS

Based on early findings that suggested the benefit of omega-3 poly unsaturated fatty-acids (PUFAs) for symptoms of schizophrenia[42,43] and associated hypotheses on neuroprotective mechanisms of action, this substance was explored in the high-risk stage as a potential novel treatment.

Amminger and colleagues[44] conducted the first randomized, double-blind, placebo-controlled trial of omega-3 PUFAs in high-risk patients referred to a psychosis detection clinic in Vienna. Eighty-one patients received omega-3 PUFAs or placebo for 12 weeks. For the primary outcome, transition to psychosis was significantly reduced in the omega-3 group (2 of 41, 4.9%) compared to the placebo condition (11 of 40, 27.5%), with most conversions occurring by the 6-month follow-up. Additionally, the omega-3 group had significantly lower positive, negative, general, and total scores on the Positive and Negative Syndrome Scale (PANSS) and significantly higher functioning on the GAF at end of treatment and at follow-up compared to the control group. This study was well controlled and had a low attrition rate demonstrating that omega-3's were well tolerated. Furthermore, the lack of significant side effects and the lasting benefits of omega-3 after treatment cessation support the potential of this approach compared to antipsychotic medication.

These authors reassessed the original sample approximately 7 years after baseline[45] (87.7% retention) and found comparable results (cumulative risk of conversion: 4/41 = 9.8% omega-3 and 16/40 = 40% placebo). Additionally, high rates of functional and symptom improvement were seen in the nonconverting omega-3 subjects. The authors hypothesize that adolescence may represent a critical period during which omega-3 supplementation may impact the brain to change adult outcomes, although the particular mechanisms are not well elucidated.

A large-scale replication of these findings was attempted in the NEURAPRO multisite study consisting of 10 early psychosis programs in Asia, Europe, and Australia.[46] In this double blind, randomized, placebo controlled study, supplementation with omega-3 PUFAs or placebo was combined with CBT/case management and provided for 6 months. Approximately one-quarter of the subjects discontinued early or was lost to follow-up. The psychosis transition rates did not significantly differ between the omega-3 and placebo groups at 6 months (6.7% vs. 5.1%) and at 12 months (11.5% vs. 11.2%). Both groups improved on secondary outcomes over time, but not differentially so. Some adverse events were noted (e.g., gastrointestinal effects). Rates of good adherence to study medication (~42%), number of CBT sessions (~10), and concomitant antidepressant medication (~62%) were similar in each group and did not impact results. Thus, this trial failed to replicate the protective effects of omega-3 PUFAs found in the single site trial by Amminger[44] and did not show beneficial effects on secondary symptom and functioning measures either, leading the authors to conclude that omega-3's are likely not effective. One potential caveat is the high rates of antidepressants and cognitive-behavioral strategies, which may have reduced conversion rates and impacted outcomes.

In summary, until a successful replication is conducted, omega-3 supplementation should not be recommended. To this end, results of a second, large multisite replication trial, conducted by the NAPLS group in the United States, are currently pending and have the potential to clarify the inconsistent findings to date.

COGNITIVE REMEDIATION STRATEGIES

Cognitive deficits, long recognized as a core feature of psychotic disorders, have been found to be present prior to illness onset in numerous studies. A meta-analysis of these studies[47] confirmed that high-risk participants show deficits in general intelligence, memory, attention, and executive functioning compared to age-matched healthy controls. Additionally, these cognitive deficits, particularly in the areas of verbal fluency, visual and verbal memory, and working memory, are associated with conversion to psychosis. Given ample evidence that cognitive deficits impact negatively on social and role functioning,[48] efforts to remediate these deficits have been attempted with patients across the psychosis spectrum, with noted benefit seen for patients with established illness.[49] As with other treatment targets (e.g., symptoms), intervening early may have particular benefits as the deficits are not as severe at this stage. According to a recent review, there have been 6 studies of computer based "drill and practice" cognitive remediation interventions with high-risk participants.[50] These studies differ in terms of the remediation programs used and the targets of the interventions and include four randomized controlled trials.

The first randomized trial conducted by the German Research Network on Schizophrenia (Bechdolf and colleagues[38]) included cognitive remediation in an combined treatment package, and the results of this study have already been described in the integrated treatment trial section earlier. Although benefits were seen overall in terms of reduced rates of conversion in the group that received the integrated treatments, the differential impact of the various treatment components could not be determined.

The second study, by Piskulic et al.,[51] did not find a significant impact of cognitive remediation on cognition or symptoms, although there were trend-level improvements on a few measures. The intervention group showed improved social functioning, but this was not related to cognition. It should be noted that this pilot study was small and attrition was very high (48%) in the cognitive remediation group.

Researchers at the University of California, San Francisco, targeted the auditory information-processing systems with computerized training.[52] High-risk youth were randomly assigned to complete 20–40 hours of cognitive training or computer games over 8 weeks. Subjects completed the training/games on laptops at their home and received weekly phone and in-person coaching. Despite this encouragement and payment for their effort, a substantial proportion of the sample did not complete the trial (38% auditory training and 49% computer games). At end of treatment, a significant improvement in verbal memory was seen in the auditory training condition while the computer games condition declined, with a medium effect size of change between groups. Both groups showed improvements in SIPS total and positive symptoms as well as GAF levels, although many subjects were receiving additional treatments. In addition to limiting power in the study, the high attrition rate highlights the limitation of using this training in adolescent samples.

The most recent randomized trial conducted by Choi et al.[53] directly targeted processing speed and incorporated pupillometry, a novel biofeedback system that automatically regulates the cognitive load of the drill and practice tasks. The tablet-based Processing Speed Training (PST), designed to improve the speed of information processing that is thought to be relevant for social interactions, was compared to a matched control group (arcade games). Both groups completed approximately 3½ hours of training/games per week in small groups in the clinic and the 10% attrition rate for PST was substantially lower than other remediation studies. At 2 months, the PST group showed improvements in processing speed, and results were maintained at the 4-month follow-up. Additionally, the PST group had higher self-reported social adjustment at 4 months and this improvement was correlated with change in processing speed scores. The effects sizes ranged from medium to high. While promising, this novel design and treatment will need replication.

To summarize, cognitive remediation interventions, while successful with patients with schizophrenia, have not shown the same promise for high-risk adolescents and young adults, notably in terms of feasibility. The high attrition rates in some studies are likely due to the demands on adolescents to devote time and energy to school and other activities. As role functioning declines, parents may be less inclined to enroll patients in randomized trials that require multiple hours per week of computer games or training. Adolescents may also be less interested in the tedious training and basic design of remediation games, given that many play more exciting and appealing video games in their leisure time. Additionally, high-risk youth are relatively less impaired cognitively, and many are not at true risk of conversion, thus, the intrinsic motivation to devote energy to remediation training may be limited. Novel approaches such as the study by Choi et al.,[53] which involved an automatic adjustment of difficulty to increase intrinsic motivation through successful practice (as opposed to external motivators), resulted in the lowest attrition rate and may be needed to bolster participation. Future research will also need to parse out which subjects are more likely to benefit from the different types of remediation interventions and whether these interventions translate to improvements in real-world functioning. Furthermore, the notion that drill and practice training leads to lasting changes in overall cognition and associated behaviors has not been established, which diminishes our confidence in the findings to date.

META-ANALYTIC RESULTS

There have been several meta-analyses as the results of randomized control trials were published and interest in the field grew. These meta-analyses have largely concluded that specific treatment interventions are superior to control treatments based on the statistical relative risk ratio (RR) of transition to psychosis and the number needed to be treated with a specific intervention for one person to avoid transition (number needed to treat, NNT). While they draw similar conclusions, the results vary given that each meta-analysis includes varying numbers of studies and different treatment approaches. A few of the large meta-analyses are presented here.

Hutton and Taylor[54] reviewed 6 CBT trials with a total of 800 participants. A significant effect on transition favoring CBT was seen at 6 months (RR 0.47, $p = 0.008$), although this result relied heavily on the positive results of one large study.[29] CBT was also associated with a significantly reduced rate of conversion at 12 months (RR 0.45, $p = .001$) with a number needed to treat (NNT) to prevent one conversion of 11 and at 18–24 months (RR 0.41, $p = .002$) with an NNT of 8. The RR values indicate that the risk of conversion was reduced by more than 50% (inverse of RR) at each timepoint. There was little evidence of benefits of CBT on symptom severity reduction and no evidence of impact on functioning (GAF, SOFAS) and quality of life measures at follow-up.

Stafford[55] published results of a meta-analysis the same year and included 11 studies of multiple approaches assessed separately. They came to similar conclusions regarding CBT interventions and their effects on conversion (specifically at 12 months and 24 months) and secondary outcomes. They also examined medication trials and concluded that there was little benefit of medication on reducing conversion rates. The one exception to this was the Omega-3 trial,[44] which at this point was the only study of its kind.

The Fusar-Poli[56] meta-analysis combined 7 treatment trials incorporating different types of interventions and found a RR of 0.34 ($p < .001$) and a NNT of 7 at 1 year. This was followed by a paper by van der Gaag and colleagues[57] that reported on 10 randomized, controlled treatment trials many of which have been described earlier. Three of the studies were deemed poor in quality based on methodological issues. Combining studies of different treatment approaches revealed a statistically significant pooled RR of .46 ($p < .001$) indicating a 54% reduction in risk of conversion at 12 months for participants who received specific treatments compared to controls and a NNT of 9. For the 5 studies with longer term outcomes, the result was significant but less stable. Findings were not significant for impacting functioning in the 6 studies that reported on this.

Thus, when combined in meta-analysis, varied interventions appear to reduce rates of conversion, particularly in the short term (12 months), but more research is needed before any recommendations on specific treatments can be made. Additionally, it is clear that more targeted interventions for other outcomes, namely comorbid diagnoses and functional problems, are needed.

CONCLUSIONS

There is mixed evidence to show that intervening in the high-risk phase can be feasible and generally well tolerated. Compared to other treatments, psychosocial interventions have the appeal of addressing many presenting problems and comorbidities at once with a seemingly more benign side-effect profile. It is less true for antipsychotic medication, where side effects are burdensome and dropouts high, leading to the conclusion that this should not be a first-line intervention. Cognitive remediation has suffered from similar feasible problems as well.

In terms of efficacy, the results are even more inconsistent. Only a few studies have shown clear benefit in terms of reducing psychosis conversions, although a significant reduction has been noted when studies are examined collectively in meta-analyses. Some interventions have additionally been found to impact clinical outcomes such as attenuated psychosis symptoms, depression, and anxiety disorders. These are important treatment targets in their own right, as they are often presenting issues for patients. Interestingly, control treatments such as ST and monitoring have also been shown to improve outcomes, suggesting that only minimal intervention is necessary, which has important implications for the design of future studies. In contrast, it is clear that very few interventions to date have improved social and role functioning, which highlights the need to broaden the focus of early intervention to include these important outcomes, regardless of conversion.

Future studies will also have to address some of the methodological issues present in the studies reviewed in this chapter. Some are easily remedied, such as rater blinding and monitoring of therapist adherence to the interventions. Other issues are more difficult to contain, such as the high rate of potentially confounding concomitant treatments in the trials.

Even when attempts are made to control, for example, antipsychotic medication use, up to 25% of subjects in some centers arrive on these medications[58,59] and exclusion could limit recruitment of an already hard to recruit population. Doing so would also mean that included subjects may not be representative of high-risk subjects in general. Other significant challenges include low adherence to treatment and high attrition rates seen for some interventions that are likely impacted by many factors such as side effects, patient time/availability, and patient choice. Using telemedicine or technology to deliver some services in the home and monitor responses remotely may remedy some of these concerns.

Finally, the low conversion rates seen in many studies limit the ability to draw conclusions from potentially underpowered studies. This issue may be partially related to subject ascertainment strategies. Screening all patients seeking services in one general mental health clinic had good results,[29] but this is too labor intensive to be feasible in all settings. Direct connections to community mental health providers or a sufficiently large system/referral base may be preferable.[8] The pressure to obtain subjects for research studies is heightened by the declining conversion rates, which suggest the need for even larger samples. Additional enrichment strategies could be employed, such as the NAPLS risk calculator,[60] which includes social decline and neurocognitive deficits in addition to selected clinical symptoms as selection criteria. An alternative strategy applies a lead-in period with psychoeducation/support and/or treatment of depression and anxiety, as featured in stepped care and other staging approaches.[41,61] Either of these strategies may reduce false positives and may be more cost effective. Last, as the field advances it will be necessary to demonstrate that treatment strategies can be implemented in community centers (i.e., outside of traditional research settings).

REFERENCES

1. McGlashan TH. Early detection and intervention in psychosis: An ethical paradigm shift. *British Journal of Psychiatry*. 2005;187.
2. Meares A. The diagnosis of prepsychotic schizophrenia. *Lancet*. 1959;1(7063):55–58.
3. Häfner H, Nowotny B. Epidemiology of early-onset schizophrenia. *Eur Arch Psychiatry Clin Neurosci*. 1995;245(2):80–92.
4. Yung AR, McGorry PD, McFarlane CA, Jackson HJ, Patton GC, Rakkar A. Monitoring and care of young people at incipient risk of psychosis. *Schizophr Bull*. 1996;22(2):283–303.
5. Yung AR, Phillips LJ, McGorry PD, McFarlane CA, Francey S, Harrigan S, Patton GC JH. Prediction of psychosis. A step towards indicated prevention of schizophrenia. *Br J Psychiatry Suppl*. 1998;172(33):14–20.
6. Yung AR, Yuen HP, Phillips LJ, Francey S, McGorry PD. Mapping the onset of psychosis: The comprehensive assessment of at risk mental states (CAARMS). *Schizophr Res*. 2005;60(1):30–31.
7. McGlashan TH, Miller TJ, Woods SW. Pre-onset detection and intervention research in schizophrenia psychoses: Current estimates of benefit and risk. *Schizophr Bull*. 2001;27(4):563–570.
8. Cornblatt BA, Lencz T, Smith CW, Correll CU, Auther AM, Nakayama E. The schizophrenia prodrome revisited: A neurodevelopmental perspective. *Schizophr Bull*. 2003;29(4):633–651.
9. Huber G, Gross G. The concept of basic symptoms in schizophrenic and schizoaffective psychoses. *Recent Prog Med*. 1989;80(12):646–652.
10. Klosterkötter J, Hellmich M, Steinmeyer EM S-LF. Diagnosing schizophrenia in the initial prodromal phase. *Arch Gen Psychiatry*. 2001;58(2):158–164.
11. Schultze-Lutter F, Addington J, Ruhrmann S, Klosterkötter J. Schizophrenia proneness interview-adult. 2007.
12. Maurer K, Zink M, Rausch F, Häfner H. The early recognition inventory ERIraos assesses the entire spectrum of symptoms through the course of an at-risk mental state. *Early Interv Psychiatry*. 2016. doi:10.1111/eip.12305
13. Yung AR, Phillips LJ, Yuen HP, Francey SM, McFarlane CA, Hallgren M, McGorry PD. Psychosis prediction: 12-month follow up of a high-risk prodromal group. *Schizophr Res*. 2003;60(1):21–32.
14. Cornblatt BA, Carrión RE, Auther A, et al. Psychosis prevention: A modified clinical high risk perspective from the recognition and prevention (RAP) program. *Am J Psychiatry*. 2015;172(10):986–994.
15. Cannon TD, Cadenhead K, Cornblatt B, et al. Prediction of psychosis in youth at high clinical risk: A multisite longitudinal study in North America. *Arch Gen Psychiatry*. 2008;65(1):28–37.
16. Ruhrmann S, Schultze-Lutter F, Salokangas RK, et al. Prediction of psychosis in adolescents and young adults at high risk: results from the prospective European prediction of psychosis study. *Arch Gen Psychiatry*. 2010;67(3):241–251.
17. Nelson B, Yuen HP, Lin A, et al. Further examination of the reducing transition rate in ultra high risk for psychosis samples: The possible role of earlier intervention. *Schizophr Res*. 2016;174(1-3):43–49.
18. Fusar-Poli P, Bonoldi I, Yung AR, Borgwardt S, Kempton MJ, Valmaggia L, Barale F, Caverzasi E MP. Predicting psychosis: Meta-analysis of transition outcomes in individuals at high clinical risk. *Arch Gen Psychiatry*. 2012;69(3):220–229.
19. McGorry PD, Yung AR, Phillips LJ, et al. Randomized controlled trial of interventions designed to reduce the risk of progression to first-episode psychosis in a clinical sample with subthreshold symptoms. *Arch Gen Psychiatry*. 2002;59(10):921.
20. McGlashan TH, Zipursky RB, Perkins D, et al. Randomized, double-blind trial of olanzapine versus placebo in patients prodromally symptomatic for psychosis. *Am J Psychiatry*. 2006;163(5):790–799.
21. Morrison AP, French P, Walford L, et al. Cognitive therapy for the prevention of psychosis in people at ultra-high risk—randomised controlled trial. *Br J Psychiatry*. 2004;185:291–297.
22. Filaković P, Degmecić D, Koić E BD. Ethics of the early intervention in the treatment of schizophrenia. *Psychiatr Danub*. 2002;19(3):209–215.
23. Phillips LJ, McGorry PD, Yuen HP, et al. Medium term follow-up of a randomized controlled trial of interventions for young people at ultra high risk of psychosis. *Schizophr Res*. 2007;96(1–3):25–33.
24. Ruhrmann S, Bechdolf A, Kühn KU, Wagner M, Schultze-Lutter F, Janssen B, Maurer K, Häfner H, Gaebel W, Möller HJ, Maier W KJL study group. Acute effects of treatment for prodromal symptoms for people putatively in a late initial prodromal state of psychosis. *Br J Psychiatry*. 2007;191:88–95.
25. McGorry PD, Nelson B, Phillips LJ, et al. Randomized controlled trial of interventions for young people at ultra-high risk of psychosis: Twelve-month outcome. *J Clin Psychiatry*. 2013;74(4):349–356.
26. Cornblatt BA, Lencz T, Smith CW, et al. Can antidepressants be used to treat the schizophrenia prodrome? Results of a prospective, naturalistic treatment study of adolescents. *J Clin Psychiatry*. 2007;68(4):546–557.
27. Fusar-Poli P, Frascarelli M, Valmaggia L, et al. Antidepressant, antipsychotic and psychological interventions in subjects at high clinical risk for psychosis: OASIS 6-year naturalistic study. *Psychol Med*. 2015;45(6):1327–1339.
28. Addington J, Cornblatt BA, Cadenhead KS, et al. At clinical high risk for psychosis: outcome for nonconverters. *Am J Psychiatry*. 2011;168(8):800–805.
29. Van Der Gaag M, Nieman DH, Rietdijk J, et al. Cognitive behavioral therapy for subjects at ultrahigh risk for developing psychosis: A randomized controlled clinical trial. *Schizophr Bull*. 2012;38(6):1180–1188.

30. French, P, Morrison A. *Early detection and cognitive therapy for people at high risk of developing psychosis : A treatment approach.* London: John Wiley & Sons; 2004.
31. Morrison AP, French P, Parker S, et al. Three-year follow-up of a randomized controlled trial of cognitive therapy for the prevention of psychosis in people at ultrahigh risk. *Schizophr Bull.* 2007;33(3):682–687.
32. Addington J, Epstein I, Liu L, French P, Boydell K, Zipursky R. A randomized controlled trial of cognitive behavioral therapy for individuals at clinical high risk of psychosis. *Schizophr Res.* 2011;125(1):54–61.
33. Ising HK, Kraan TC, Rietdijk J, et al. Four-year follow-up of cognitive behavioral therapy in persons at ultra-high risk for developing psychosis: The Dutch Early Detection Intervention Evaluation (EDIE-NL) trial. *Schizophr Bull.* 2016;42(5):1243–1252.
34. Morrison AP, French P, Stewart SLK, et al. Early detection and intervention evaluation for people at risk of psychosis: multisite randomised controlled trial. *Br Med J.* 2012;344(April):e2233. doi:10.1136/bmj.e2233
35. Stain HJ, Bucci S, Baker AL, et al. A randomised controlled trial of cognitive behaviour therapy versus non-directive reflective listening for young people at ultra high risk of developing psychosis: The detection and evaluation of psychological therapy (DEPTh) trial. *Schizophr Res.* 2016;176(2–3):212–219.
36. Miklowitz DJ, O'Brien MP, Schlosser DA, et al. Family-focused treatment for adolescents and young adults at high risk for psychosis: Results of a randomized trial. *J Am Acad Child Adolesc Psychiatry.* 2014;53(8):848–858.
37. Fusar-Poli P, Rocchetti M, Sardella A, et al. Disorder, not just state of risk: Meta-analysis of functioning and quality of life in people at high risk of psychosis. *Br J Psychiatry.* 2015;207(3):198–206.
38. Bechdolf A, Wagner M, Ruhrmann S, et al. Preventing progression to first-episode psychosis in early initial prodromal states. *Br J Psychiatry.* 2012;200(1):22–29.
39. Nordentoft M, Thorup A, Petersen L, Ohlenschlaeger J, Melau M, Christensen TØ, Krarup G, Jørgensen P JP. Transition rates from schizotypal disorder to psychotic disorder for first-contact patients included in the OPUS trial. A randomized clinical trial of integrated treatment and standard treatment. *Schizophr Res.* 2006;83(1):29–40.
40. McFarlane WR, Levin B, Travis L, et al. Clinical and functional outcomes after 2 years in the early detection and intervention for the prevention of psychosis multisite effectiveness trial. *Schizophr Bull.* 2015;41(1):30–43.
41. Nelson B, Amminger GP, Yuen HP, Wallis N, Kerr M, Dixon L, Carter C, Loewy R, Niendam TA, Shumway M, Morris S, Blasioli J MP. Staged treatment in early psychosis: A sequential multiple assignment randomised trial of interventions for ultra high risk of psychosis patients. *Early Interv Psychiatry.* 2017;(July18).
42. Emsley R, Myburgh C, Oosthuizen P, Van Rensburg SJ. Randomized, placebo-controlled study of ethyl-eicosapentaenoic acid as supplemental treatment in schizophrenia. *Am J Psychiatry.* 2002;159(9):1596–1598.
43. Peet M, Brind J, Ramchand CN, Shah S, Vankar GK. Two double-blind placebo-controlled pilot studies of eicosapentaenoic acid in the treatment of schizophrenia. *Schizophr Res.* 2001;49(3):243–251.
44. Amminger GP, Schäfer MR, Papageorgiou K, et al. Long-chain omega-3 fatty acids for indicated prevention of psychotic disorders. *Arch Gen Psychiatry.* 2010;67(2):146–154.
45. Amminger GP, Schäfer MR, Schlögelhofer M, Klier CM, McGorry PD. Longer-term outcome in the prevention of psychotic disorders by the Vienna omega-3 study. *Nat Commun.* 2015;6:1–7.
46. McGorry PD, Nelson B, Markulev C, et al. Effect of ω-3 polyunsaturated fatty acids in young people at ultrahigh risk for psychotic disorders: The NEURAPRO randomized clinical trial. *JAMA Psychiatry.* 2017;74(1):19–27.
47. Fusar-Poli P, Deste G, Smieskova R, et al. Cognitive functioning in prodromal psychosis. *Arch Gen Psychiatry.* 2012;69(6).
48. Bowie CR, Leung WW, Reichenberg A, McClure MM, Patterson TL, Heaton RK, Harvey PD. Predicting schizophrenia patients' real world behavior with specific neuropsychological and functional capacity measures. *Biol Psychiatry.* 2008;63:505–511.
49. Wykes T, Huddy V, Cellard C, McGurk S CP. A meta-analysis of cognitive remediation for schizophrenia: Methodology and effect sizes. *Am J Psychiatry.* 2011;168:472–485.
50. Glenthøj LB, Hjorthøj C, Kristensen TD, Davidson CA, Nordentoft M. The effect of cognitive remediation in individuals at ultra-high risk for psychosis: A systematic review. *NPJ Schizophr.* 2017;3(1):20.
51. Piskulic D, Barbato M, Liu L, Addington J. Pilot study of cognitive remediation therapy on cognition in young people at clinical high risk of psychosis. *Psychiatry Res.* 2015;225(1–2):14–21.
52. Loewy R, Fisher M, Schlosser DA, et al. Intensive auditory cognitive training improves verbal memory in adolescents and young adults at clinical high risk for psychosis. *Schizophr Bull.* 2016;42:S118–S126.
53. Choi J, Corcoran CM, Fiszdon JM, et al. Pupillometer-based neurofeedback cognitive training to improve processing speed and social functioning in individuals at clinical high risk for psychosis. *Psychiatr Rehabil J.* 2017;40(1):33–42.
54. Hutton P, Taylor PJ. Cognitive behavioural therapy for psychosis prevention: A systematic review and meta-analysis. *Psychol Med.* 2014;44(3):449–468.
55. Stafford MR, Jackson H, Mayo-Wilson E, Morrison AP, Kendall T. Early interventions to prevent psychosis: systematic review and meta-analysis. *BMJ.* 2013;346:f185.
56. Fusar-Poli P, Borgwardt S, Bechdolf A, et al. The psychosis high-risk state: A comprehensive state-of-the-art review. *Arch Gen Psychiatry.* 2013;70(1):107–120.
57. Van Der Gaag M, Smit F, Bechdolf A, et al. Preventing a first episode of psychosis: Meta-analysis of randomized controlled prevention trials of 12 month and longer-term follow-ups. *Schizophr Res.* 2013;149(1–3):56–62.
58. Bechdolf A, Muller H, Stutzer H, et al. Rationale and baseline characteristics of PREVENT: A second-generation intervention trial in subjects at-risk (Prodromal) of developing first-episode psychosis evaluating cognitive behavior therapy, aripiprazole, and placebo for the prevention of psychosi. *Schizophr Bull.* 2011;37(SUPPL. 2).
59. Woods SW, Addington J, Bearden CE, et al. Psychotropic medication use in youth at high risk for psychosis: Comparison of baseline data from two research cohorts 1998–2005 and 2008–2011. *Schizophr Res.* 2013;148(1–3):99–104.
60. Cannon TD, Yu C, Addington J, et al. An individualized risk calculator for research in prodromal psychosis. *Am J Psychiatry.* 2016;173(10):980–988.
61. McGorry PD, Hickie IB, Yung AR, Pantelis C, Jackson HJ. Clinical staging of psychiatric disorders: A heuristic framework for choosing earlier, safer and more effective interventions. *Aust N Z J Psychiatry.* 2006;40(8):616–622.

70

FROM EARLY INTERVENTION IN PSYCHOSIS TO TRANSFORMATION OF YOUTH MENTAL HEALTH REFORM

Ashok Malla and Patrick McGorry

INTRODUCTION

The introduction of the concept of early intervention (EI) in psychotic disorders, the growth in research in first-episode psychosis (FEP) and clinical high-risk states, and the eventual research-supported transformation of services may turn out to be the most significant development in the field of mental health service delivery since deinstitutionalization began in the second half of the 20th century. The benefits of early intervention in psychoses have been well documented and cover multiple domains of clinical, functional, social, and economic outcomes. Although evidence supporting EI services in psychosis has influenced scaling-up of these specialized models of care in many jurisdictions, the level of scaled-up services remains piecemeal, incompletely resourced and implemented, and confined to a few high-income countries with publicly funded healthcare systems. Yet, it has served as an antidote to pessimism and as a prototype for more radical reform and investment in cost-effective mental healthcare. Given that all major mental disorders have their onset in the same age range (under 25 years), early psychosis is a prototype for a much wider reform directed at all mental disorders. That such a level of transformation of services can be achieved for one group of serious mental disorders, namely schizophrenia and other forms of psychosis, it is logical to think that the same could be achieved for all mental disorders, ranging from mild to severe.

Research generated from EI services and FEP has facilitated a re-examination of our conceptual framework of the nature of individual mental disorders as discrete entities, appearing in their final form as disease states at the time of seeking care. It is now crystal clear that the need for care may long precede the threshold for most of our traditional diagnostic categories. This means we need to introduce the concept of staging to facilitate earlier treatment and align it in a proportional, yet preventive, manner with the needs and prognostic risk of the patient. With such a unique possibility of combining service and conceptual transformations, we are in a position not only to extend more effective service delivery to all forms of mental disorders but also to create unique platforms for investigating the nature and causes of mental disorders and shaping their treatment and outcomes. In this chapter we first briefly review how developments in early intervention in psychotic disorders have influenced transformation of youth mental health services in general, examine the nature, distribution, opportunities, and challenges of various mental disorders at onset during adolescence and emerging adulthood, and explore opportunities that may exist in influencing outcome for all mental disorders when treated closer to their real onset. We also provide a brief overview of current international efforts at both service transformation and generation of new knowledge in the expanded field of youth mental health.

EARLY INTERVENTION IN PSYCHOTIC DISORDERS: LESSONS LEARNED

In this section we connect the lessons learned from more than two decades of work in early intervention in psychosis to the current efforts at transforming services for youth with all severity of mental disorders. Historically the concept of early intervention in psychotic disorders can be traced to H. S. Sullivan (1927), almost a century ago, when he observed that a "psychiatrist sees too many end states and deals professionally with too few of the pre-psychotic."[1] Since then the "period before" (prodrome), comprising nonpsychotic as well as subthreshold positive and negative symptoms of psychosis, has been observed to often precede a full-fledged psychosis and last for several months to years.[2]

The potential to intervene in the period preceding a relapse[3] has predated the more recent effort to prevent the first onset of psychosis.[4] Following several parallel pieces of evidence, observations, and a heuristic belief of providing comprehensive and integrated treatment following onset of psychosis, service models for early intervention began to emerge in the 1990s.[5] A seminal observation[6] that patients for whom antipsychotic therapy had been delayed had worse outcomes than those who received such therapy earlier, began to draw attention to the issue of the duration of untreated psychosis (DUP), a concept that has now been repeatedly and assuredly associated with poor clinical and social outcomes.[7] DUP has also become central to the idea of early intervention.[8] However, the significance of DUP for outcome has not always been appreciated nor incorporated in the process of developing EI services.

A quest to understand sources of delay has encouraged further examination through studies of pathways to care for patients with FEP.[9] This body of work has necessitated a careful examination of and changes to the way mental healthcare systems are organized and how and why it takes so long for young people with FEP to engage in treatment.[10]

While associating delay in treatment to poor outcome may confirm what would appear to be intuitively correct, influencing sources of delay has turned out to be more complex. Several controlled studies have attempted to reduce such delays in treatment (DUP) with mixed results,[11–13] with some having clearly demonstrated the feasibility of reducing such delays. Transforming a system of care may be necessary to reduce such noxious delays in treatment by ensuring rapid and direct access to the population at large, without depending on the prevailing system of care; the latter often operates in silos of child-adolescent vs. adult or primary vs. secondary and tertiary services, with each silo being a barrier in the path of help-seekers. The lesson, however, is obvious that if we want to improve outcome in psychotic disorders, we must find a way to get individuals with FEP into treatment rapidly, with a relatively narrow window of delay in treatment (DUP of probably less than 3 to 6 months) to get the most benefit from EI services[14] This knowledge can now be transferred to youth mental health in general so that the new configurations of services do not make the same error.

EI SERVICE AS INTEGRATED TREATMENT AND REDUCTION OF DELAY IN TREATMENT

Following almost two decades of experience with and studies of EI services, a core set of interventions as well as attempts to reduce delay in treatment are now recognized as forming the essential ingredients of an effective EI service.[15] This involves integrating evidence-based biological (medication) and psychosocial interventions (e.g., family intervention, cognitive behavioral therapy, case management, and vocational interventions), providing this package of interventions in a manner that is acceptable to a younger treatment naïve patient population. This approach, pioneered in Australia in the early 1990s,[5] has been adopted by many others in several high-income countries with publicly funded healthcare systems.

While the integration of evidence-based intervention has been adopted by almost all models of EI services, reduction of delay in treatment through early case identification and fundamental transformation of the system of access to the new EI services have not been implemented by all. This may reflect either the difficult challenges inherent in achieving this level of transformation or the expectation that the system will eventually adapt to the new models of care.[16,17] It has been observed that individuals with a FEP entering many EI services continue to present with long DUPs[18] compared to services that created from the outset an open system of access that did not rely on the referral practices of the existent system[5,12] and engaged in active case identification strategies.[12]

It can, therefore, be concluded that an effective EI service for psychotic disorders must combine an integrated evidence-based treatment program; a client and family centered recovery-oriented culture; raising mental health literacy regarding psychosis and early intervention among potential sources of referral that include health, education, and social programs as well as the local community; and a transformation of access to such care that promotes simplification of the pathways to care and provides rapid assessment and treatment.

An equally, if not more, promising development in the EI field has been an elaboration of the concept of "prodrome" and subsequently clinical high risk state, and the use of the latter for "indicated" prevention. In recent years, there has been considerable success in significantly reducing the rates of conversion to psychosis by 50% from a CHR state.[19–21] This subject matter is covered in a different chapter. Suffice it to say that the lessons learned from EI research and service innovation for psychosis are important for designing any future youth mental health service.

NEW ISSUES ARISING FROM EARLY INTERVENTION IN PSYCHOSIS OF RELEVANCE TO YOUTH MENTAL HEALTH

The success of attempts at indicated prevention of psychosis has raised several interesting issues of relevance to youth mental health. First, it has confirmed that psychotic disorders are often preceded by other, milder forms of symptoms and behaviors and the latter may be amenable to treatment interventions with a better risk benefit ratio than notably antipsychotic medications, thus either preventing or postponing a FEP. Incorporating identification, assessment, and nonpharmacological interventions for those seeking help during CHR states in EI services has been suggested, although most remain research initiatives with some exceptions.[22,23] Scaling-up of these indicated prevention services will require a clear demonstration that the majority of individuals presenting with a FEP have previously experienced a state of CHR.[24] Preliminary evidence suggests the proportion to be between 50% and 60%.[24]

Additional observations from the CHR research and examination of community samples for evidence of psychotic experiences of relevance to youth mental health, include the following:

1. While most patients in a CHR state for psychosis do not develop a psychotic disorder, they almost invariably present with significant distress and functional disability, poor social supports and self-esteem,[23] and relatively poor generic outcomes. While 50% no longer meet criteria for CHR at the end of 2 years,[25] a significant proportion develop or present with mental disorders other than psychosis (e.g., depression, anxiety, substance abuse, eating disorders, and emotional dysregulation, as harbingers of incipient personality disorders);[25] and most require generic and specific interventions targeting their specific presenting problems.[26]

2. A relatively high proportion of the general population report subclinical psychotic experiences without presenting with a psychotic disorder[27] and these are particularly common among adolescents.[27] These experiences tend to persist in one-fifth of adolescents, often start early and are associated with a sevenfold increase in risk of developing not only a psychotic disorder but also of transforming into other forms of mental disorders (e.g., anxiety, depression).[20] Such co-occurrence of nonpsychotic and psychotic experiences early during adolescence and their nonspecific trajectories may reflect a transdiagnostic pluripotential phenotype that may eventually express as one of a variety of mental disorders.[28]

These observations are particularly relevant for understanding all mental disorders presenting in youth and for planning services for them.

EVIDENCE OF EFFECTIVENESS OF EI AND ITS RELEVANCE FOR YOUTH MENTAL HEALTH

Several controlled and naturalistic studies and provide evidence of greater effectiveness of EI compared to regular care on several dimensions such as, remission, relapse, substance abuse, functioning, housing, and residual positive and negative symptoms.[29] These studies mostly tested the model of a specialized EI service incorporating principles and elements described previously. The largest of the randomized controlled trials (RCTs) is the one conducted in Denmark (OPUS) with 547 schizophrenia-spectrum patients randomized to the specialized EI service or regular care for 2 years of treatment and follow-up.[30] Evidence for effectiveness of various modifications of the dominant specialized-care model of EI service has recently emerged from a variety of jurisdictions, such as the United States and Hong Kong.[32] These benefits of EI service, provided in different systems of care are also matched by strong evidence of cost-effectiveness,[33] based on a number of studies that varied from estimating only direct costs to use of sophisticated computer-assisted decision analytic models incorporating direct and indirect costs.

The message from EI services research is clear: integrated services, designed for FEP, produce better outcomes for patients with FEP than does regular care and they are cost-effective. Whether the outcomes have a dose-response relationship to the intensity and breadth of interventions remains largely unexplored. There is some doubt whether such effects persist when the specialized EI service is withdrawn and patients return to regular care.[30] Extension of even low-level intensity of an EI service may help retain such benefits over an extended period of 5 years.[34] More recently, somewhat disparate results from controlled studies conducted in Hong Kong,[35] Canada,[36] and Denmark[37] suggest that intensive EI services for the initial 2 years are likely to produce lasting benefits for some patients while others may need continued support at a fairly high level of intensity beyond the 2-year period to achieve this goal. Further findings of Danish and Canadian studies suggest that benefits of extended EI service may be moderated by length of DUP such that those with shorter DUP (less than 3 months) are the ones to benefit from an extension of the service.[37,38] We can be very confident that for many if not most patients, the benefits of EI will be put at risk if patients are discharged after 2 years to poor quality standard adult mental healthcare.

To summarize, studies of first-episode psychosis and early intervention have revealed a number of insights that are of relevance to and can now be used to guide transformations of service delivery for all mental health problems affecting youth. The lessons learned from EI and FEP include a reconceptualization of mental disorders as being relatively fluid across categories of form, content, and severity, allowing for interventions at lesser levels of severity to prevent greater and more noxious impacts. Key principles that must be incorporated into any effective model of service delivery include quick and unencumbered access, simplicity of pathways to care and treatment based on evidence, a stigma-free culture of collaborative care and continuous evaluations of outcome. The experience has also revealed that research focused on improving care for young people with psychosis is beneficial to advancing scaling-up of these services through convincing policy and funding agencies. While evidence and science have been crucial in achieving this, the role of advocacy on the part of service users and, especially, their families as well as clinicians and scientists interested and involved in this field has been invaluable in convincing politicians and healthcare policy makers to invest in early intervention. This is a lesson that is now being carried through to youth mental health at large (see discussion that follows).

MENTAL HEALTH PROBLEMS OF YOUNG PEOPLE (FROM EARLY ADOLESCENCE THROUGH EARLY ADULTHOOD)

SOCIETAL PERSPECTIVE

It is now well established that mental health problems are a major contributor to loss of gross domestic product (GDP), at par with cardiovascular and other medical disorders,[39] and that an investment in prevention and early intervention makes great economic sense. Most (75%) mental disorders have their first onset during childhood, early adolescence, and young adulthood (before the age of 25), unlike physical health problems, which often peak much later in life.[40] As a substantial proportion of mental disorders appearing in youth tend to surge during the transition between childhood and adulthood, they have a long-lasting social and economic impact not only for the individuals and their families but also for society at large.[41]

The problem for high-income countries is that future productivity of a relatively small proportion of the population under 25 years of age will be necessary to sustain an increasingly aging population. Thus healthy (mental health being the major concern during this period) young people are essential for the survival and future of society. On the other hand, in low- and middle-income countries (LMICs) the relatively high proportion of population under 25 years old

(40%–50%) poses different and equally, if not more, important challenges.[42] Attending to the physical, mental, and social development of children and youth is vital to the survival of LMICs.[43]

SCIENTIFIC PERSPECTIVE

The nature of mental health problems affecting youth needs to be viewed from the perspective of incidence, prevalence, distribution, and burden of disease followed by the current status of care available and its impact on outcomes. As suggested by Jones (2013), two principal areas of inquiry that have provided us relatively reliable data on age at onset of different disorders are: retrospective clinical studies of different disorders (e.g., schizophrenia, depression, anxiety, and conduct disorders) and population-based studies, including birth cohort longitudinal studies and cross-sectional population surveys.[44] Based on the results of these studies schizophrenia and related psychotic disorders first appear in postpubertal period with a peak around 22 years for men and 27 for women,[45] while depression and anxiety are well known to occur during late childhood and early adolescence often continuing during young adulthood and later in life.[46] Anxiety and impulse control disorders, in particular, reach their peak incidence relatively early (age 11), with 75% of cases having appeared prior to the age of 21 and 15, respectively.[47] While the median age at onset for depression is 30 years, 25% of cases appear before the age of 18.[48]

The age at onset of disorders cannot be estimated precisely, partly because most individuals would have experienced some symptoms and behavioral changes much earlier than the first "episode" at a syndromal level. These data are, however, very informative, especially when juxtaposed to the timing of treatment of various disorders and the proportion who actually receive any professional attention.

The National Community Survey (NCS-R) data reveal a very troubling picture of delays and missed opportunities for intervention. While the majority of those with a diagnosable mental disorder eventually make contact with services, these rates vary from 88% to 94% for mood disorders and much lower for anxiety disorders (27%–95%), impulse control disorders (34%–52%), and substance abuse (53%–77%). Most often these contacts are made after prolonged delays that range from 1 or 2 years for psychosis to 6 to 8 years for mood disorders and 9 to 23 years for anxiety disorders.[49] Early intervention services that engage in active case identification and open and rapid access to their services report significantly lower delays ranging from a median of 4 weeks (TIPS)[11] to 11.4 weeks (PEPP-Montréal),[36] compared to those which do not incorporate this component in their treatment program (e.g., OPUS median DUP is 1 year).[37] While such long delays have an impact on outcome when and if eventually treated, there are many dire consequences that are difficult to quantify. These include suicide, traffic accidents, missed employment opportunities, impact on the justice system, physical health, and more.

The state of youth mental health problems, reviewed very briefly earlier, demands a very different conceptual, scientific, and social perspective and service response than has been the case until now. This will require alignment of investigations at the clinical, epidemiological, and neurodevelopmental levels and entertaining very different models of illness than the prevailing categories applied to adult mental disorders. The three first decades of life, during which brain development and maturation occur through changes in structure and connectivity,[50] provide the back drop of vulnerability during which environmental changes can disrupt behavioral adjustments and result in incidents of what we understand as mental disorders. While many of the latter may be transitory, a substantial proportion persist in their original form (e.g., anxiety, mood disorders, impulse disorders) or transition into even more serious disorders such as psychosis and bipolar disorders (often from nonpsychotic conditions) and antisocial personality disorders (often from conduct disorders).

CONNECTING DEVELOPMENTS IN EARLY INTERVENTION IN PSYCHOSIS AND YOUTH MENTAL HEALTH

One of the major accomplishments of research in early intervention in psychosis has been the identification of earlier states of symptoms and behaviors that would not meet criteria for diagnosis of psychosis. As discussed in an earlier section, the identification of mental states and reduced levels of functioning prior to dating the onset of psychosis, has led to a reconceptualization of how we think of mental disorders in general. It has led to identification of the possibility that mental disorders develop in stages[51] that are identifiable in help-seeking individuals and that earlier stages may be milder than subsequent ones and, therefore, amenable to relatively more generic, less noxious interventions. This idea of staging models of mental disorders is heuristically appealing and is supported by some evidence. Further, an earlier state may transform itself into a different category of disorder over time, raising the possibility that there may be basic states of disturbance of mental phenomena that have potential to develop into a variety of disorders ranging from mood disorders to psychosis.[51,52]

HEALTH SYSTEM RESPONSES TO MENTAL HEALTH PROBLEMS OF YOUTH

The ubiquitous nature of current configuration of child-adolescent and adult mental health services functioning as two separate systems, often funded from different sources and situated within different organizational structures, has been shown to be a major impediment to responding adequately and comprehensively to demands of mental health problems prevalent in early adolescence and young adulthood.[53] The current system not only presents enormous difficulties for those transitioning from child-adolescent to adult services but also, even more importantly, is totally inadequate in responding to very large numbers of new cases of mental disorders that often present with a large variation of severity ranging from full-fledged diagnosable disorders to subthreshold symptoms and behaviors. While these presentations lack differentiation into specific disorders, they often require immediate attention for their distress and reduced functioning. The specialized mental healthcare is thwarted in its response to problems of youth by its small-scale divisions into silos and by its prevailing culture of pessimism linked to having to a focus on longstanding mental disorders and disabilities. Primary healthcare, on the other hand, is designed to cater for physical illness, and mainly for young children and older adults while being woefully inadequately structured or resourced to respond to mental health problems of the youth.

In addition to lessons learned from the field of early intervention in psychosis, reviewed earlier, we have also learned that early case identification interventions involving improvement of mental health literacy and the need for early help seeking as well as referral, often lead to a large number of individuals being identified by generic health, social, and educational services, families, and individuals themselves, who present with an array of psychiatric disorders of varying severity, including those with subthreshold symptoms and protean mental states. Such a broader net cast for identification of all cases of mental disorders directed primarily at the age group with the highest incidence (12–25 years) of mental disorders could lead us to address the service gaps for all youth mental health problems. However, this will also lead to very large numbers of "new" cases, given the size of the problem in this age group, requiring an adequate service response.[54] The latter would need to be situated either in a system that is easily accessible without referrals. This could be achieved either in a transformed or enhanced form of primary care given the relatively undifferentiated, mild to moderate severity of majority of cases identified and difficulty in sorting out trajectories of outcomes in psychopathology or functioning at such early stage.[55] While the EI services for psychosis require a horizontal integration with the rest of the system, services for the entire spectrum of youth mental health problems need to be connected both vertically and horizontally to other components of the health (especially primary care) and social services systems.

RECENT DEVELOPMENTS IN YOUTH MENTAL HEALTH SERVICE TRANSFORMATION

In recent years several countries have responded to these challenges through a variation of service and research strategies some examples of which are summarized in this section.

THE AUSTRALIAN RESPONSE

In 2006, the Australian Government established "Headspace," the National Youth Mental Health Foundation, which was tasked with building a national youth mental health service stream designed to provide highly accessible, youth-friendly centers that promote and support early intervention for mental and substance use disorders in young people. Each center, operated by an independent local consortium and overseen by the Headspace national office, provides four core service streams: mental health, drug and alcohol services, primary care (general health, sexual health), and vocational/educational assistance. The therapeutic approach centers on brief psychosocial interventions, used as first-line therapy with the aim of preventing the development of sustained illness, with use of medication as an additive therapy only if the young person does not respond to initial psychosocial interventions, or presents at the outset with more severe symptoms or risk. This stepped-care model ensures that care is linked to the stage of illness, and offers a proactive and preventive approach to therapeutic intervention.

In addition, Headspace also runs a nationwide online support service (eHeadspace; www.eheadspace.org.au), where young people can chat with a mental health professional either online or by telephone and access assessment and therapeutic care, and headspace school support, a suicide postvention program for schools affected by the suicide of a student. A subset of the Headspace centers are supported by "enhanced Headspace" services, which provide primarily EI services for psychotic disorders, in the manner discussed earlier in this chapter. Two separate evaluations have been carried out, one in-house[56] and the other an independent evaluation.[57] The former has revealed that more than 21,000 youth were assessed across 55 Headspace centers, majority (63.7%) of these were female and almost equal proportion were young adolescents (12–18 years old) and young adults (18–25). Most common reasons for presentation were problems with mood, anxiety, and relationships, while work-school, physical health, and alcohol and drug problems were significantly less common presentations. The vast majority (70%) presented with high or very high levels of distress, and the level of distress increased with age, as did the proportion with a diagnosable disorder and functional decline.[56,57] The independent evaluation[57] also reported significant reduction in psychological distress and improvement in suicidal ideation and self-harm compared with two matched groups over time. There are now 110 headspace centers across Australia, which can provide access for over 100,000 young people per annum, also backed up over extended hours by the online e-headspace service. These centers are beginning to be slowly augmented by resources for more complex disorders. This is a very effective scaling-up process with enormous community-wide support and popular with political leaders and policy makers alike.

RESPONSE FROM THE REPUBLIC OF IRELAND

In Ireland, the youth mental health movement can be traced back to high levels of public concern about high rates of youth suicide (4th highest in Europe for those 15–25 years old),[58] mental distress among Irish youth, direct involvement of leaders from the fields of health and allied health and epidemiological research,[58,59] and availability of philanthropic funding.

Concurrently, Jigsaw, a service delivery program with 10 sites that aimed to establish youth-friendly, community-based mental health support structures for young people, were developed across Ireland based on principles, similar to those of Headspace, supported by Headstrong, a dedicated youth mental health organization (www.headstrong.ie), and an online youth mental health organization (ReachOut.com). Jigsaw provides youth-friendly primary mental healthcare and support to young people between the ages of 12 and 25 years, having served over 10,500 Irish youth since 2008.[60]

Evaluations of these services have revealed results similar but not identical to those of Headspace.[56] Youth seeking help in Jigsaw sites are more often female and younger than 18 years old, most often referred by informal sources (family and self), and present with relatively high levels of distress, problems of anxiety, depression, and self-harm, and, for the employable age group, have a high rate of unemployment.[61] Four typologies, namely, developmental, comorbid, anxious, and externalizing[62] have been identified. These categories may be more reflective of presenting problems and amenable to generic interventions than using established diagnostic categories.

CANADIAN RESPONSE TO THE CHALLENGE OF YOUTH MENTAL HEALTH

In Canada, the transformation of youth mental health services, beginning later than that in Australia and Ireland, has taken a somewhat different route through the development of a strategic initiative of Transformational Research in Adolescent (youth 11–25 years) Mental (TRAM) health as part of the new Strategies for Patient Oriented Research (SPOR) program of the Canadian Institutes of Health Research (CIHR), funded jointly by the latter and a philanthropic family foundation (Graham Boeckh Foundation). Launched in October 2012 as a competitive process, the explicit purpose was to establish a national network project that would demonstrate transformation of youth (11–25 years) mental health services and provide evidence of its effectiveness over a period of 5 years. The intention is to bridge the science–practice divide by applying existing evidence to transform the delivery of mental healthcare and to produce better outcomes. The ultimate goal is to scale-up a transformed model of service delivery across the country.

Adolescent/young adult Connections to Community-driven Early Strengths-based and Stigma-free services (ACCESS Open Minds [OM]–Esprits ouverts) has emerged as a service and research network that includes youth, family/career, community organization, service provider, researcher, and policy- and decision-maker stakeholder groups from across the country.[63,64] Influenced largely by evidence from EI services in psychosis and related research and aided by a substantial youth mental health advocacy, ACCESS-OM is currently implementing and evaluating a transformation of youth mental health at a number of sites, representing variation in geography, culture, and service availability in majority of provinces and a territory as well as in multiple indigenous sites. The purpose of this approach is to generate evidence on multiple dimensions to inform development of youth mental health services to benefit all youth in the country. The transformation is designed around five objectives: (1) early case identification; (2) quick access (within 72 hours) for a first assessment; (3) continuous service across the age spectrum of 11–25, if required; (4) connection to specialized services depending on availability; and (5) engagement of youth and their families at all levels of service delivery (for details, see www.accessopenminds.ca). Unlike Headspace and Jigsaw, the new transformed services are derived from a combination of existing resources in the system with addition of new resources in the form of trained personnel, training and building capacity for evaluation, with the transformed services being situated in a range of service environments ranging from enhancement of existing primary care to independent service hubs. Much like Headspace, the transformation is supported by e-mental-health strategies and youth-friendly physical spaces as portals for help-seeking and youth peer-support activities. Access to specialist services varies from embedding specialists within the enhanced primary care to direct links with specialized programs (e.g., early psychosis service, eating disorder program) to provision of specialized services through remote technology. Given the nature of the project, a mixed-methods research and evaluation approach is central to its development and implementation and will inform the process and outcome evidence generated. This should then inform development of various youth mental health services across different provinces and territories as well as indigenous communities.

Some jurisdictions within Canada have already started developing youth mental health services informed by a combination of models such as, Headspace and locally driven initiatives (e.g., Foundry in British Columbia), while others have begun to implement services based on principles and processes articulated in the ACCESS-OM framework and incorporating an evaluation protocol from ACCESS-OM.

One of the major challenges of any transformation of youth mental health services is to match the new improved initial access to services to an equally quick and high-quality specialized services when needed. While the latter are better developed for early psychosis, at least in some parts of Canada, services for other major mental disorders are woefully inadequate, do not operate on early intervention principles, tend to be physician focused, and lack a new culture of care that could address the special needs of youth and their families. Therefore, we face the danger of transforming the front end of the service, perhaps improving care for those with mild and moderately severe mental disorders and leave a large proportion of those requiring specialized care behind. However, this strategy has the huge advantage of making the level of unmet need for care of severe and complex mental illness overt and no longer hidden, since these young people have a stigma-free entry portal for the first time.

OTHER YOUTH MENTAL HEALTH INITIATIVES

In addition to the national initiatives described briefly previously, youth mental health initiatives have also been reported from France, other parts of Europe, and the United Kingdom.[65] The former has established Maisons des Adolescents (N = 104) across different regions of the country with the principles of using multidisciplinary approach and serving multiple needs of adolescents, in a manner that may in some ways be similar to the principles of Headspace, albeit restricted to adolescents. While there are descriptions of these services available there are no formal evaluations reported.

Denmark has established 9 Headspace centers across the nation, and Israel has two as well.

In the United Kingdom, while there is no national initiative, Birmingham, the second-largest city in the country, with a relatively high proportion of youth and multiethnic population, has initiated a new program, Youthspace, under the auspices of the joint Birmingham and Solihull Mental Health Foundation Trust and the Child Adolescent Mental Health Services. The service involves strategic partnership with nonhealth youth agencies, which allows focus on employment and social inclusion along with mental health. Youth access teams developed for this purpose provide quick assessment to referring family physicians, and care is provided in multiple settings with low risk for stigma. Improving transitions across adolescent and adult services is a major focus, as is connection to specialized service teams (e.g., early psychosis, ADHD). No evaluations of this model of service transformation have been published thus far. Finally, Headspace services and other versions of integrated youth mental healthcare are being created in several other countries including the United States and the Netherlands.

CONCLUSION

The worldwide movement for EI in psychotic disorders, supported by strong evidence, for two decades has led the transformation of mental health services for young people presenting with psychotic disorders. Such a transformation has been adopted by several, mostly high-income countries, as part of their public mental health policy. While this is indeed significant progress, much remains to be done in many other high-income countries and all LMICs in order to make a real impact on the lives of millions of young people suffering from psychotic disorders. Part of the benefit of research in EI and FEP has been the successful reconceptualization of mental disorders from their onset, examining them as a process of gradual development in stages and the reality that a substantial proportion of more severe disorders may develop from milder forms appearing earlier. The EI research and service development as well as greater social awareness of the nature and extent of mental health problems, which appear mostly during childhood, but especially early adolescence and young adulthood, have led to development of a variety of responses to service gaps in a few high-income countries and also, in the process, stimulated further research in understanding the origin and development of all mental disorders. In future, this work is likely, via a paradigmatic change, to lead to a more comprehensive understanding of causes, risk factors, course, and outcome of all mental disorders and provide opportunities for prevention and early intervention for a substantial proportion of cases of most mental disorders which currently incur enormous personal, social, and economic costs globally.

ACKNOWLEDGMENT

We acknowledge the significant assistance provided by Bilal Issaoui-Mansouri in preparation of this chapter.

REFERENCES

1. Sullivan HS. The onset of schizophrenia. *American journal of psychiatry.* 1927;84(1):105–134.
2. Yung AR, McGorry PD, McFarlane CA, Jackson HJ, Patton GC, Rakkar A. Monitoring and care of young people at incipient risk of psychosis. *Schizophrenia bulletin.* 1996;22(2):283–303.
3. Malla AK, Norman R. Prodromal symptoms in schizophrenia. *British Journal of Psychiatry.* 1994;164(4):487–493.
4. Falloon IR, Kydd RR, Coverdale JH, Laidlaw TM. Early detection and intervention for initial episodes of schizophrenia. *Schizophrenia bulletin.* 1996;22(2):271–282.
5. McGorry PD, Edwards J, Mihalopoulos C, Harrigan SM, Jackson HJ. EPPIC: an evolving system of early detection and optimal management. *Schizophrenia bulletin.* 1996;22(2):305–326.
6. Wyatt RJ. Early intervention with neuroleptics may decrease the long-term morbidity of schizophrenia. *Schizophrenia research.* 1991;5(3):201–202.
7. Marshall M, Lewis S, Lockwood A, Drake R, Jones P, Croudace T. Association between duration of untreated psychosis and outcome in cohorts of first-episode patients: a systematic review. *Archives of general psychiatry.* 2005;62(9):975–983.
8. Norman RM, Malla AK. Duration of untreated psychosis: a critical examination of the concept and its importance. *Psychological medicine.* 2001;31(3):381–400.
9. Anderson KK, Fuhrer R, Malla AK. The pathways to mental health care of first-episode psychosis patients: a systematic review. *Psychological medicine.* 2010;40(10):1585–1597.
10. Bhui K, Ullrich S, Coid JW. Which pathways to psychiatric care lead to earlier treatment and a shorter duration of first-episode psychosis? *BMC psychiatry.* 2014;14:72.
11. Melle I, Larsen TK, Haahr U, et al. Reducing the duration of untreated first-episode psychosis: effects on clinical presentation. *Archives of general psychiatry.* 2004;61(2):143–150.
12. Malla A, Norman R, Scholten D, Manchanda R, McLean T. A community intervention for early identification of first episode psychosis: impact on duration of untreated psychosis (DUP) and patient characteristics. *Social psychiatry and psychiatric epidemiology.* 2005;40(5):337–344.
13. Srihari VH, Tek C, Pollard J, et al. Reducing the duration of untreated psychosis and its impact in the U.S.: the STEP-ED study. *BMC psychiatry.* 2014;14:335.
14. Larsen TK, Melle I, Auestad B, et al. Early detection of psychosis: positive effects on 5-year outcome. *Psychological medicine.* 2011;41(7):1461–1469.
15. Addington DE, McKenzie E, Norman R, Wang J, Bond GR. Essential evidence-based components of first-episode psychosis services. *Psychiatric services (Washington, D.C.).* 2013;64(5):452–457.
16. Kane JM, Schooler NR, Marcy P, et al. The RAISE early treatment program for first-episode psychosis: background, rationale, and study design. *Journal of clinical psychiatry.* 2015;76(3):240–246.
17. Nordentoft M, Melau M, Iversen T, et al. From research to practice: how OPUS treatment was accepted and implemented throughout Denmark. *Early intervention in psychiatry.* 2015;9(2):156–162.
18. Birchwood M, Connor C, Lester H, et al. Reducing duration of untreated psychosis: care pathways to early intervention in psychosis services. *British journal of psychiatry.* 2013;203(1):58–64.
19. van der Gaag M, Smit F, Bechdolf A, et al. Preventing a first episode of psychosis: meta-analysis of randomized controlled prevention trials of 12 month and longer-term follow-ups. *Schizophrenia research.* 2013;149(1–3):56–62.
20. Fusar-Poli P, Nelson B, Valmaggia L, Yung AR, McGuire PK. Comorbid depressive and anxiety disorders in 509 individuals with an at-risk mental state: impact on psychopathology and transition to psychosis. *Schizophrenia bulletin.* 2014;40(1):120–131.
21. McGorry PD, Nelson B, Markulev C, et al. Effect of omega-3 polyunsaturated fatty acids in young people at ultrahigh risk for psychotic disorders: the NEURAPRO randomized clinical trial. *JAMA psychiatry.* 2017;74(1):19–27.

22. Phillips LJ, Leicester SB, O'Dwyer LE, et al. The PACE clinic: identification and management of young people at "ultra" high risk of psychosis. *Journal of psychiatric practice.* 2002;8(5):255–269.
23. Pruessner M, Faridi K, Shah J, et al. The Clinic for Assessment of Youth at Risk (CAYR): 10 years of service delivery and research targeting the prevention of psychosis in Montreal, Canada. *Early intervention in psychiatry.* 2017;11(2):177–184.
24. Shah JL, Crawford A, Mustafa SS, Iyer SN, Joober R, Malla AK. Is the clinical high-risk state a valid concept? Retrospective examination in a first-episode psychosis sample. *Psychiatric services (Washington, D.C.).* 2017:pp. 1046–1052.
25. Simon AE, Borgwardt S, Riecher-Rossler A, Velthorst E, de Haan L, Fusar-Poli P. Moving beyond transition outcomes: meta-analysis of remission rates in individuals at high clinical risk for psychosis. *Psychiatry research.* 2013;209(3):266–272.
26. Lin A, Wood SJ, Nelson B, Beavan A, McGorry P, Yung AR. Outcomes of nontransitioned cases in a sample at ultra-high risk for psychosis. *American journal of psychiatry.* 2015;172(3):249–258.
27. van Os J, Reininghaus U. Psychosis as a transdiagnostic and extended phenotype in the general population. *World psychiatry).* 2016;15(2):118–124.
28. McGorry P, van Os J. Redeeming diagnosis in psychiatry: timing versus specificity. *Lancet (London, England).* 2013;381(9863):343–345.
29. Fusar-Poli P, McGorry PD, Kane JM. Improving outcomes of first-episode psychosis: an overview. *World psychiatry.* 2017;16(3):251–265.
30. Bertelsen M, Jeppesen P, Petersen L, et al. Five-year follow-up of a randomized multicenter trial of intensive early intervention vs standard treatment for patients with a first episode of psychotic illness: the OPUS trial. *Archives of general psychiatry.* 2008;65(7): 762–771.
31. Kane JM, Robinson DG, Schooler NR, et al. Comprehensive versus usual community care for first-episode psychosis: 2-year outcomes from the NIMH RAISE early treatment program. *American journal of psychiatry.* 2016;173(4):362–372.
32. Chan SKW, So HC, Hui CLM, et al. 10-year outcome study of an early intervention program for psychosis compared with standard care service. *Psychological medicine.* 2015;45(6):1181–1193.
33. Mihalopoulos C, McCrone P, Knapp M, Johannessen JO, Malla A, McGorry P. The costs of early intervention in psychosis: restoring the balance. *Australian and New Zealand journal of psychiatry.* 2012;46(9):808–811.
34. Norman RM, Manchanda R, Malla AK, Windell D, Harricharan R, Northcott S. Symptom and functional outcomes for a 5year early intervention program for psychoses. *Schizophrenia research.* 2011;129(2):111–115.
35. Chang WC, Chan GHK, Jim OTT, et al. Optimal duration of an early intervention programme for first-episode psychosis: randomised controlled trial. *British journal of psychiatry: the journal of mental science.* 2015;206(6):492–500.
36. Malla A, Joober R, Iyer S, et al. Comparing three-year extension of early intervention service to regular care following two years of early intervention service in first-episode psychosis: a randomized single blind clinical trial. *World psychiatry: official journal of the World Psychiatric Association (WPA).* 2017;16(3):278–286.
37. Albert N, Melau M, Jensen H, et al. Five years of specialised early intervention versus two years of specialised early intervention followed by three years of standard treatment for patients with a first episode psychosis: randomised, superiority, parallel group trial in Denmark (OPUS II). *BMJ.* 2017;356:i6681.
38. Malla A, Shah J, Dama M, et al. The effect of duration of untreated psychosis on sustaining symptoms remission in an extended early intervention service for first episode psychosis patients. *Manuscript submitted for publication.*
39. Bloom D, Cafiero E, Jané-Llopis E, et al. *The global economic burden of noncommunicable diseases.* Program on the Global Demography of Aging;2012.
40. Kessler RC, Amminger GP, Aguilar-Gaxiola S, Alonso J, Lee S, Ustun TB. Age of onset of mental disorders: a review of recent literature. *Current opinion in psychiatry.* 2007;20(4):359–364.
41. Patel V, Flisher AJ, Hetrick S, McGorry P. Mental health of young people: a global public-health challenge. *Lancet (London, England).* 2007;369(9569):1302–1313.
42. Bongaarts J. Human population growth and the demographic transition. *Philosophical Transactions of the Royal Society B: Biological Sciences.* 2009;364(1532):2985–2990.
43. Patton GC, Sawyer SM, Santelli JS, et al. Our future: a Lancet commission on adolescent health and wellbeing. *Lancet (London, England).* 2016;387(10036):2423–2478.
44. Jones P. Adult mental health disorders and their age at onset. *British Journal of Psychiatry.* 2013;202(s54):s5–s10.
45. Kirkbride JB, Fearon P, Morgan C, et al. Heterogeneity in incidence rates of schizophrenia and other psychotic syndromes: findings from the 3-center AeSOP study. *Archives of general psychiatry.* 2006;63(3):250–258.
46. Fombonne E, Wostear G, Cooper V, Harrington R, Rutter M. The Maudsley long-term follow-up of child and adolescent depression. 1. Psychiatric outcomes in adulthood. *British journal of psychiatry.* 2001;179:210–217.
47. Kessler RC, Berglund P, Demler O, et al. The epidemiology of major depressive disorder: results from the National Comorbidity Survey Replication (NCS-R). *Jama.* 2003;289(23):3095–3105.
48. Kessler RC, Berglund P, Demler O, Jin R, Merikangas KR, Walters EE. Lifetime prevalence and age-of-onset distributions of DSM-IV disorders in the National Comorbidity Survey Replication. *Archives of general psychiatry.* 2005;62(6):593–602.
49. Wang PS, Berglund P, Olfson M, Pincus HA, Wells KB, Kessler RC. Failure and delay in initial treatment contact after first onset of mental disorders in the National Comorbidity Survey Replication. *Archives of general psychiatry.* 2005;62(6):603–613.
50. Sowell ER, Peterson BS, Thompson PM, Welcome SE, Henkenius AL, Toga AW. Mapping cortical change across the human life span. *Nat Neurosci.* 2003;6(3):309–315.
51. Cosci F, Fava GA. Staging of mental disorders: systematic review. *Psychotherapy and psychosomatics.* 2013;82(1):20–34.
52. Hickie IB, Hermens DF, Naismith SL, et al. Evaluating differential developmental trajectories to adolescent-onset mood and psychotic disorders. *BMC psychiatry.* 2013;13:303.
53. Singh SP, Paul M, Ford T, et al. Process, outcome and experience of transition from child to adult mental healthcare: multiperspective study. *British Journal of Psychiatry.* 2010;197(4):305–312.
54. Copeland W, Shanahan L, Costello EJ, Angold A. Cumulative prevalence of psychiatric disorders by young adulthood: a prospective cohort analysis from the Great Smoky Mountains Study. *Journal of the American Academy of Child and Adolescent Psychiatry.* 2011;50(3):252–261.
55. Cullen W, Broderick N, Connolly D, Meagher D. What is the role of general practice in addressing youth mental health? A discussion paper. *Irish journal of medical science.* 2012;181(2):189–197.
56. Rickwood DJ, Telford NR, Parker AG, Tanti CJ, McGorry PD. Headspace—Australia's innovation in youth mental health: who are the clients and why are they presenting? *Medical journal of Australia.* 2014;200(2):108–111.
57. Hilferty F, Cassells R, Muir K, et al. Is Headspace making a difference to young people's lives? Final Report of the independent evaluation of the Headspace program. 2015.
58. Coughlan H, Tiedt L, Clarke M, et al. Prevalence of DSM-IV mental disorders, deliberate self-harm and suicidal ideation in early adolescence: An Irish population-based study. *Journal of adolescence.* 2014;37(1):1–9.
59. Harley M, Connor D, Clarke M, et al. Prevalence of mental disorder among young adults in Ireland: a population based study. *Irish Journal of Psychological Medicine.* 2015;32(01):79–91.
60. O'Keeffe L, O'Reilly A, O'Brien G, Buckley R, Illback R. Description and outcome evaluation of Jigsaw: an emergent Irish mental health early intervention programme for young people. *Irish Journal of Psychological Medicine.* 2015;32(1):71–77.
61. O'Reilly A, Illback R, Peiper N, O'Keeffe L, Clayton R. Youth engagement with an emerging Irish mental health early intervention

programme (Jigsaw): participant characteristics and implications for service delivery. *Journal of mental health (Abingdon, England).* 2015;24(5):283–288.

62. Peiper N, Clayton R, Wilson R, et al. Empirically derived subtypes of serious emotional disturbance in a large adolescent sample. *Social psychiatry and psychiatric epidemiology.* 2015;50(6):983–994.
63. Malla A, Iyer S, McGorry P, et al. From early intervention in psychosis to youth mental health reform: a review of the evolution and transformation of mental health services for young people. *Social psychiatry and psychiatric epidemiology.* 2016;51(3):319–326.
64. Iyer S, Boksa P, Lal S, et al. Transforming youth mental health: a Canadian perspective. *Irish Journal of Psychological Medicine.* 2015;32(1):51–60.
65. Hetrick SE, Bailey AP, Smith KE, et al. Integrated (one-stop shop) youth health care: best available evidence and future directions. *Medical journal of Australia.* 2017;207(10):5–18.

SECTION 6

FUTURE DIRECTIONS AND OPPORTUNITIES

71

FUTURE DIRECTIONS

MAKING A START TOWARD THE PRIMARY PREVENTION OF PSYCHOSIS

Robin M. Murray, Olesya Ajnakina, and Marta Di Forti

Prevention has a long history in medicine, with early successes such as the clinical trial of lemons by James Lind in 1747 to prevent scurvy in the British Navy, and John Snow's famous removal of the handle of the Broad Street water pump in 1854 to prevent the spread of cholera in London. In this chapter, before considering the prospects for similar interventions in psychosis, we briefly review the approaches to prevention commonly adopted in general medicine.

> **Secondary prevention**
> Secondary prevention seeks to lower the rate of established cases of a disorder in the population through early detection and treatment. For example, mass population screening for tuberculosis by chest x-ray proved highly successful when TB was endemic in Western countries.
>
> **Selective prevention**
> Here the strategy is to target people at high risk, e.g., because of their age or gender, or to identify those who have minimal but detectable signs or symptoms foreshadowing the disorder, e.g., screening by mammography for breast cancer. Selective screening may also target those who manifest a biological marker for the disorder; thus, the identification of individuals positive for HIV long before the development of AIDS brought major benefits. Of course, such approaches depend on the availability of a cost-effective intervention.
>
> **Primary prevention**
> Universal primary prevention targets the whole population and often involves advice to the public to avoid exposure to an important risk factor. This has been successfully employed for tobacco smoking with large, though belated, reductions in smoking rates and smoking-related diseases. Major campaigns, unfortunately as yet less successful, have also been initiated to influence lifestyle in an attempt to reduce cardiovascular disease. As Stewart et al. (2017) point out, "Strong consensus exists . . . regarding the necessity of smoking cessation, weight optimization and the importance of exercise." Again, in relation to asthma, measures have been taken to improve lung health, such as reducing tobacco smoking and reducing indoor and outdoor air pollution and occupational exposures; currently there is much concern, though relatively little action, to reduce diesel particulates in the air (Beasley et al., 2015).

THE SITUATION IN PSYCHOSIS

Thus, preventive approaches are commonplace in medicine. Sadly, they are rarely attempted in modern psychiatry, though in the past they proved valuable in dramatically reducing some neuropsychiatric disorders such as neurosyphilis. However, recently there have been the stirrings of more interest (Arango et al., 2018). This chapter addresses the question of whether it is now time to consider developing strategies for the prevention of psychosis.

SECONDARY PREVENTION

Almost without knowing it, modern psychiatry has been attempting secondary prevention through programs of early detection and treatment. This "early intervention" (EI) approach was initiated by McGorry and colleagues (1996) in Australia, and since 2001 it is mandatory in England that every local psychiatric service should provide EI (Joseph and Birchwood, 2005). The approach spread to Europe and Canada (Malla et al., 2016), and some two decades later it is being discovered in the United States. Evidence for effectiveness of EI services has been demonstrated in the LEO (England) and OPUS (Denmark) studies both in terms of better outcome and in cost-effectiveness for as long as the comprehensive approach is maintained (Craig et al., 2004; Secher et al., 2015). However, when patients leave the early intervention teams 2 or 3 years after illness onset, most of the gains start to be lost. This is not an argument against the value of EI, but rather an indication that the quality of the later services must be improved. Nevertheless, it cannot be said that the EI approach has yet made any major inroads into the prevalence of psychosis.

SELECTIVE PREVENTION

Clinics for those with the "at risk mental state"

During the last two decades, "prodromal" clinics have been developed for young people who are considered to be at high risk of developing psychosis. Such individuals are deemed to show the "at-risk mental state" (ARMS), or the attenuated psychosis syndrome. Initially it was claimed that up to 40% of individuals manifesting the ARMS might develop frank psychosis within 18 months to 2 years (Yung et al., 2005), but recent reports suggest much lower transition rates (Van Os and Guloksuz, 2017). A variety of predominantly psychosocial interventions have been developed, and detection and treatment of young people with the ARMS has become a popular strategy. Thus, Outreach Clinics have been developed in Australia, the United Kingdom, Europe, and North America.

As yet there is little solid evidence that these approaches are cost-effective. Furthermore, it has not been clear what proportion of those who will develop psychosis can be captured by this approach. Therefore, we set out to identify the proportion of patients with a first episode of psychosis (FEP) who had initially presented to prodromal services with the ARMS before making the transition to FEP.

South London has among the largest and most researched prodromal clinics in the world, and these map onto the same geographic area as the services for patients with an FEP. We therefore reviewed all patients who presented to Mental Health Services in South London over a 2-year period with an FEP. We examined those 37 years old and below, an age chosen to be consistent with the age limit for patients attending the ARMS clinics (35 years) plus 2 years to allow them to present with a FEP (Ajnakina et al., 2017).

A total of 16.3% of all those with FEP had previously attended the prodromal services. However, 12.1% were already psychotic, demonstrating that such clinics can provide a useful additional avenue for young people already psychotic to reach psychiatric services. However, only 4.1% of all FEP cases had attended the prodromal service and manifested the ARMS. Of course, it is possible that some such people were prevented from transitioning to frank psychosis through interventions by the prodromal services. But even taking this into consideration, the figures indicate that such clinics are never going to have a major effect on incidence of the disorder.

Interestingly, patients attending the prodromal services were not typical of the majority of those who presented with FEP directly. They were more likely to have a slowly developing rather than acute illness, to be white British, and to be help-seeking and cooperative. It is of course unrealistic to expect prodromal services to see patients who have a sudden acute onset of psychosis or do not accept that they are ill. So although such clinics have undoubtedly have been very valuable for research, it should not be thought that those who attend accurately reflect the characteristics of all those with an FEP.

McGorry, Yung, and their colleagues appear to have accepted the limitations of prodromal clinics, and have now adopted a more comprehensive approach by providing youth mental health services to provide care for those with a range of psychiatric disorders (Rickwood et al., 2014). This approach is more logical and has advantages over clinics with the narrower clientele, though an evaluation of such services in Australia concluded that the early results were mixed (Hilferty et al., 2015).

Earlier identification of those at high risk

Critics of the "at-risk mental state" approach have pointed out that such individuals are already well on their way to developing psychosis (Van Os and Guloksuz, 2017). Might it be possible to identify individuals who are at increased risk of psychosis at a much earlier point? There have been numerous studies identifying behavioral antecedents of psychosis in general or schizophrenia in particular. Thus, those who show delayed milestones, solitary play, or poorer cognitive function than their siblings are at increased risk of later psychosis, as are those who at age 11 report psychotic-like experiences. However, although these characteristics statistically put individuals at risk, the small magnitude of the effect size makes identification impossible. One possibility might be to combine several characteristics. Thus, Laurens et al. (2007) suggested that particular attention should be paid to 9- to 12-year-old children with the triad of speech or developmental delay; social, behavioral, or emotional problem; and psychosis-like experiences. However, to our knowledge, such an approach has not been implemented on a large scale.

Other researchers have sought biological markers for the disorder. One of the great successes of primary prevention in medicine has been the Guthrie test, in which all newborns have a heel stab which allows screening for several congenital disorders including phenylketonuria. The adoption of a low phenylalanine diet has a profoundly beneficial effect on the neurodevelopment of those affected.

Could a similar approach be initiated for psychosis? Copy number variants (CNVs) have been found in excess in schizophrenia, and in theory could be detected at birth. However, CNVs account for only about 3% of cases of schizophrenia, and most of those with a CNV will not develop psychosis. Furthermore, even if those carrying a risk increasing CNV could be identified, there are as yet no known useful interventions.

UNIVERSAL PRIMARY PREVENTION OF PSYCHOSIS?

Variations in the incidence of schizophrenia

In medicine, strategies involving universal primary prevention generally depend on identifying exposures that increase risk of the disorder in question and then publicly advocating avoidance of that risk factor. What do we know about exposures relevant for schizophrenia? If schizophrenia were found to vary in incidence, then this might provide some clues as to the exposures of interest. Unfortunately, for some time, research into this possibility was accidentally discouraged by a WHO study of psychosis in nine countries (Jablensky et al., 1992) which was wrongly interpreted as showing no variation in the incidence of psychosis; thus, Crow (1995) hypothesized that this meant that schizophrenia was intrinsic to the human condition. If that were the case, clearly it would leave little opportunity for prevention.

However, it subsequently became clear that the incidence of schizophrenia, and of psychosis in general, varies considerably. The systematic review by McGrath et al. (2004) suggested a fivefold variation in the incidence of schizophrenia, while the recent EU-GEI study of 16 sites reported that the incidence of psychosis was more than five times higher in certain of the northern European cities studied than that in parts of Italy and Spain (Jongsma et al., 2018).

It is implausible to think that the latter differences are due to genetic differences between northern and southern Europeans. Differential exposure to environmental risk factors is much more likely. But which risk factors are associated with increased risk of psychosis? These have been carefully reviewed by Stilo and Murray (2010); the best replicated findings concern obstetric events, childhood adversity, urban birth and upbringing, migration, adverse life events, and cannabis use. It appears that some, e.g., childhood adversity, are relatively nonspecific whereas others such as being born or brought up in an urban setting, migration, or using cannabis appear to have greater effects on risk of psychosis than on other psychiatric disorders. Here we consider whether these latter three offer prospects for primary prevention.

The example of urbanicity

Although being born or brought up in a city rather than a rural setting is a highly replicated risk factor, urbanicity must be a proxy for one or more specific component causes. High population density, greater exposure to stress, drug abuse crime, and pollution have all been suggested.

Another possibility, which Engemann and her colleagues (2018) have recently addressed is lack of green space. These researchers linked an individual's place of birth and upbringing with a diagnosis of schizophrenia, using the Danish National Registers. Then they used GPS satellite imagery to examine whether being surrounded by more green vegetation at place of childhood residence was related to schizophrenia risk. Growing up surrounded by the lowest amount of green space was associated with a 1.5-fold increased risk of developing schizophrenia compared to that of people living at the highest level of green space in childhood.

Is it possible that cities could be designed so that their poor areas are less toxic? Certainly, planning cities with more green space could benefit not only those at risk of schizophrenia but also children from poor families more widely. However, at present there is insufficient evidence concerning the exact mechanism for us to advocate particular changes in town planning that would diminish the psychotogenicity of cities. Nevertheless, there is an urgent need for the careful evaluation of the effects of urban planning on mental health.

The example of migration/ethnic minority status

The EU-GEI study, discussed earlier, shows a striking difference between the incidence of psychosis in northern European cities and sites in Italy and Spain. One factor that contributed significantly to this was the substantial population of people from ethnic minorities in northern cities, mostly migrants and their children. When migrants/ethnic minorities were removed from the analysis, the incidence of psychosis in South London decreased from 61/100,000 per annum to 45/100,000 per annum. Unfortunately, at present this information is of little use for prevention, because we don't know the mechanism underlying the increased risk in migrants and their children (Jongsma et al., 2018).

The example of cannabis use

Use of a number of drugs has been associated with increased risk of schizophrenia, in particular amphetamines, cocaine, ketamine, and cannabis. Most evidence concerns cannabis. There are now some 14 prospective studies showing that use of cannabis is associated with increased later risk of psychotic symptoms and/or psychotic disorder; in 11 the increase was statistically significant (Murray et al., 2016). The risk increases with the frequency and length of use, and the potency of the cannabis used (Marconi et al., 2016; Di Forti et al., 2014). Di Forti et al. (2015) showed that the population attributable fraction (PAF) for use of high-potency cannabis on FEP in South London was 24%; i.e., 24% of cases would have been prevented had no-one used high-potency cannabis. Di Forti and her colleagues (2019) have recently completed a much larger study examining the effect of cannabis use on the incidence of psychosis across the six countries of the EU-GEI study. The PAFs calculated indicated that if high-potency cannabis were no longer available, 12.2% (95% CI 3.0–16.1) of cases of first-episode psychosis could be prevented across the 11 sites, rising to 30.3% (15.2–40.0) in London and 50.3% (27.4–66.0) in Amsterdam.

Is the evidence sufficiently persuasive to merit preventative action? Gage et al. (2016), who exhaustively scrutinized the epidemiological literature for possible confounding, bias, misclassification, reverse causation, and other explanations for the association between cannabis use and later psychosis, concluded that "epidemiologic studies provide strong enough evidence to warrant a public health message that cannabis use can increase the risk of psychotic disorders." We concur and would remind readers that it took many years for cigarette smoking to be accepted as a cause of lung cancer. Furthermore, 4 decades passed before serious attempts were made to persuade people to stop smoking tobacco. Given the lack of an animal model for psychosis and of the equivalent of painting tobacco tar on mice to demonstrate its carcinogenicity, it is not sensible to wait for absolute proof that cannabis is a component cause of psychosis.

CONCLUSIONS

In the 1940s and 1950s, American psychoanalysts enthusiastically advocated the application of Freudian theory to many aspects of life—from how to bring up children to relations between races, international diplomacy, and the causes of war

(Murray, 1979). However, as the influence of psychoanalysis waned and evidence-based approaches took over, psychiatrists became more cautious about pontifying publicly, including about ways to prevent mental illness.

In the 21st century, as we have seen, much effort has gone into the establishment of "prodromal" clinics for those with the ARMS. However, the evidence we have reviewed suggests that such an approach will never reach more than a small fraction of those about to develop psychosis. Consequently, our thoughts need to turn again to the primary prevention of psychosis, but this time armed with replicable data rather than speculative notions.

Our knowledge concerning the effects of exposures to a range of risk factors on psychosis is mounting, but the evidence is not yet sufficient to initiate universal preventative campaigns, with one exception, the use of cannabis. In our view, there is already sufficient evidence to warrant public education regarding the risks of heavy use of cannabis, particularly of high-potency varieties. Such a campaign has the potential to significantly reduce the incidence of psychosis in countries where high-potency cannabis is readily available. Of course, any such initiative should not exaggerate the risks and take care to ensure that public education does not get confused with the highly charged debate for and against legalization. Indeed, the legalization of cannabis in parts of North America will provide "natural" experiments concerning population exposure to cannabis. It is important that researchers take the opportunity to monitor such changes and their effects on psychosis.

REFERENCES

Ajnakina O, Morgan C, Gayer-Anderson C, Oduola S, Bourque F, Bramley S, et al. Only a small proportion of patients with first episode psychosis come via prodromal services: a retrospective survey of a large UK mental health programme. BMC Psychiatry. 2017;17:308. Published online Aug 25, 2017. doi: 10.1186/s12888-017-1468

Arango C, Díaz-Caneja CM, McGorry PM, Rapoport J, Sommer IE, Vorstman JA, et al. Preventive strategies for mental health. Lancet Psychiatry. 2018. Published online May 14, 2018. http://dx.doi.org/10.1016/S2215-0366(18)30057-9

Beasley R, Semprini A, Mitchell AE. Risk factors for asthma: is prevention possible? Lancet. 2015;386:1075–1085.

Craig TK, Garety P, Power P, Rahaman N, Colbert S, Fornells-Ambrojo M, et al. The Lambeth Early Onset (LEO) team: randomised controlled trial of the effectiveness of specialised care for early psychosis. BMJ. 2004;329(7474):1067. Published online Oct 14, 2004.

Crow TJ. A continuum of psychosis: one gene and not much else. Schizophr Res. 1995;17;135–145.

Di Forti M, Sallis H, Allegri F, Trotta A, Ferraro L, Stilo SA, et al. Daily use, especially of high-potency cannabis, drives the earlier onset of psychosis in cannabis users. Schizophr Bull. 2014;40:1509–1517.

Di Forti M, Marconi A, Carra E, Fraietta S, Trotta A, Bonomo M. et al. Proportion of patients in south London with first-episode psychosis attributable to use of high potency cannabis: a case-control study. Lancet Psychiatry. 2015;2:233–238.

Di Forti M, et al. The contribution of cannabis use to variation in the incidence of psychotic disorder across Europe (EU-GEI): a multicentre case-control study. Lancet Psychiatry. 2019, in press.

Engemann K, Pedersen CB, Arge L, Tsirogiannis C, Mortensen PB, Svenning J-C. Childhood exposure to green space—a novel risk-decreasing mechanism for schizophrenia? Schizophr Res. 2018;online.

Gage SH, Hickman M, Zammit S. Association between cannabis and psychosis: epidemiologic evidence. Biol Psychiatry. 2016;79:549–556.

Hilferty F, Cassells R, Muir K, Duncan A, Christensen D, Mitrou F, et al. (2015). Is Headspace making a difference to young people's lives? Final report of the independent evaluation of the Headspace program. (SPRC Report 08/2015). Sydney: Social Policy Research Centre, UNSW Australia.

Jablensky A, Sartorius N, Ernberg G, et al., Schizophrenia: manifestations, incidence and course in different cultures. WHO ten country study. Psychol Med. 1992;Supplement 20:1–97.

Jongsma HE, Gayer-Anderson C, Lasalvia A, Quattrone D, Mulè A, Szöke A, et al. Treated incidence of psychotic disorders in the multinational EU-GEI study. JAMA Psychiatry. 2018;75(1):36–46.

Joseph R, Birchwood M. The national policy reforms for mental health services and the story of early intervention services in the United Kingdom. J Psychiatry Neurosci. 2005;30:62–365. PMC 1197282. PMID 16151542

Laurens KR, Hodgins S, Maughan B, Murray RM, Rutter ML, Taylor EA. Community screening for psychotic-like experiences and other putative antecedents of schizophrenia in children aged 9-12 years. Schizophr Res. 2007;90(1–3):130–146. Published online Jan 5, 2007.

Lederbogen F, Kirsch P, Haddad L, Streit S, Tost H, Schuch P, et al. City living and urban upbringing affect neural social stress processing in humans. Nature. 2011;474:498–501.

Malla A, Iyer S, McGorry P, Cannon M, Coughlan H, Singh S, Jones P, Joober R. From early intervention in psychosis to youth mental health reform: a review of the evolution and transformation of mental health services for young people. Soc Psychiatry Psychiatr Epidemiol. 2016, 51:319–326.

Marconi A, Di Forti M, Lewis CM, Murray RM, Vassos E. Meta-analysis of the association between the level of cannabis use and risk of psychosis. Schizophr Bull. 2016.

McGorry PD, Edwards J, Mihalopoulos C,Harrigan SM, Jackson HJ. EPPIC: an evolving system of early detection and optimal management. Schizophr Bull. 1996;22(2):305–326.

McGrath J, Saha S, Welham J, El Saadi O, MacCauley C, Chant D. A systematic review of the incidence of schizophrenia. BMC Med. 2004;2:13. Published online 2004 Apr 28. doi: 10.1186/1741-7015-2-13

Murray RM. A reappraisal of American psychiatry. Lancet. 1979;i:255–258.

Murray RM, Quigley H, Quattrone S, Englund A, Di Forti M. Traditional marijuana, high-potency cannabis and synthetic cannabinoids: increasing risk for psychosis. World Psychiatry. 2016;15(3):195–204. Published online 2016 Sep 22. doi: 10.1002/wps.20341

Rickwood DJ, Telford NR, Parker AG, Tanti CJ, McGorry PD Headspace—Australia's innovation in youth mental health: who are the clients and why are they presenting. Med J Aust. 2014;200(2):1–4.

Secher RG, Hjorthøj CR, Austin SF, Thorup A, Jeppesen P, Mors O, Nordentoft M. Ten-year follow-up of the OPUS specialized early intervention trial for patients with a first episode of psychosis. Schizophr Bull. 2015;41(3):617–626. doi: 10.1093/schbul/sbu155. Published online Nov 7, 2014.

Stewart J, Manmathan G, Wilkinson. Primary prevention of cardiovascular disease: a review of contemporary guidance and literature. JRSM Cardiovasc Dis. 2017;6:2048004016687211. Published online 2017 Jan 1. doi: 10.1177/2048004016687211

Stilo SA, Murray RM. The epidemiology of schizophrenia. Dial Clin Neurosci. 2010;12:305–315.

van Os J, Guloksuz S. A critique of the "ultra-high risk" and "transition" paradigm. World Psychiatry. 2017;16:200–206. doi:10.1002/wps.20423

Vassos E, Pedersen CB, Murray RM, et al. Meta-analysis of the association of urbanicity with schizophrenia. Schizophr Bull. 2012;38:1118–1123.

Yung AR, Yuen HP, McGorry PD, Phillips LJ, Kelly D, Dell'Olio M, Francey SM, Cosgrave EM, Killackey E, Stanford C, et al. Mapping the onset of psychosis: the comprehensive assessment of at-risk mental states. Aust N Z J Psychiatry. 2005;39(11–12):964–971. doi: 10.1080/j.1440-1614.2005.01714.x

72

A GLIMPSE FORWARD REGARDING PSYCHOPATHOLOGY OF PSYCHOTIC DISORDERS

William T. Carpenter

Psychotic experiences are common in humans and, when associated with distress, dysfunction, and disability, anchor a spectrum of mental disorders. Persons afflicted with these disorders manifest many biological, social, and psychological differences from persons not afflicted, but a point of rarity on a continuum is evasive. Psychotic experiences even in the non-ill population are associated with decrements in function.[1] In-depth reviews by experts are presented in earlier chapters relating to brain and human functioning. Here I briefly provide a perspective on psychotic disorders and note likely concepts applicable in the near future. The field is rich in hypotheses regarding "how" the brain functions in relation to aspects of psychopathology and will soon be in a better position to more definitively test these hypotheses. The "why" questions are more challenging. Dare we hope for clinically actionable advances? Issues related to (1) early detection and intervention, (2) primary prevention, and (3) reconceptualizing therapeutic targets are considered in the following text.

First, an extremely brief tour of concepts addressing issues in nosology related to psychotic disorders followed by an outline of areas that will affect scientific inquiry and clinical care in the near future. The reader should be aware that my views have been shaped by, and possibly limited by, an early career view that schizophrenia is a poor target for etiology, pathophysiology, and therapeutic discovery.[2] In that paper we considered that syndrome deconstruction was essential; we identified six psychopathology domains for study and borrowed the terms "positive" and "negative" from Hughlings Jackson (for a history of the term originating with J. R. Reynolds, see[3]) to designate separable components of schizophrenia-related psychopathology.[2]

In the mid to late 19th century the most severe mental disorders were known as madness, and afflicted persons were housed in insane asylums. A critical step in forming meaningful diagnostic categories was the identification of the spirochete that caused a psychotic disorder—tertiary syphilis. It was then clear that a specific etiology could result in a great variety of clinical manifestations, presumably depending on which brain areas/networks were infected. But it was also clear that other patients shared the same range of symptoms without the same etiology. In this context Kraepelin provided a fundamental advance in psychiatric nosology by identifying dementia praecox and manic-depressive psychoses as separable illnesses, based on prognosis and specific clinical features. This conceptualization remains influential today with substantial validation. But extensive heterogeneity within each diagnostic class and overlap in clinical features, risk factors, and associated features between diagnostic classes result in critical limitations when used to characterize the individual person in clinical care or participant in clinical research.

Skipping over the issue of diagnostic criteria and reliability, limitations with profound significance include the following:

1. Most currently defined psychotic disorders are clinical syndromes, not disease entities;
2. Within syndromes individual patient differences are substantial;
3. Most symptoms manifested by a person with a specific psychotic disorder are also common in some or many other disorders;
4. Observed psychopathology is on a severity continuum with the general population without a point of rarity distinguishing the ill from the non-ill;
5. Most markers, bio- or otherwise, may separate disorder cohorts from non-ill controls but are not applicable at the individual level, are not established for between psychotic disorders distinction, and are also observed in the general population (there are exceptions such as *22Q11* deletion syndrome, but even here risk is not restricted to schizophrenia);
6. Risk factors associated with psychotic disorders, schizophrenia for example, are not disorder-specific and are also common in the general population;
7. Many genetic and environmental risk factors, active at different times of development, have been identified. But etiological pathways are not established, they may be pleiotropic, and pathophysiological mechanisms may vary between cases; and
8. Whether an individual with risk factors develops a psychotic illness also depends on many normal behavioral (e.g., temperament), environmental (e.g., urban), and genetic factors (e.g., genes and epigenetic changes), not to mention resiliency and compensatory protective factors.

Limitations of current diagnostic classes as defined in DSM-5 are clear, but substantial validity and epidemiological and clinical application is associated with diagnoses such as bipolar disorder or schizophrenia. Consider the role of lithium in bipolar disorder or the contrast between pressured and grandiose speech and behavior in bipolar compared with alogia and apathy in schizophrenia. But the person with bipolar disorder may be depressed with alogia and apathy and the person with schizophrenia may have grandiose delusions. While important similarities and differences exist between cohorts at the diagnostic level, investigations at this level will not reveal whether the same mechanisms are involved in depressive apathy and avolitional apathy. Schizophrenia is an inadequate phenotype for genetic studies, as witnessed by the large sample sizes needed in genome-wide association studies (GWAS) and small effects of genes despite evidence for very substantial contribution of genetic risk to developing a psychotic disorder. And, of course, the genetics of auditory hallucinations may be very different that the genetics of negative symptoms. And even here genetics may differ between subcomponents of the negative symptom construct.

The DSM and ICD diagnostic manuals are sometimes viewed as the source of problems mentioned thus far. This is not the case. Diagnostic manuals are not textbooks of psychopathology or a conceptual framework for understanding mental illness. Rather, DSM/ICD provides a basis for indexing cases, supporting some aspects of statistics and epidemiology, and increasing reliability and similar use of terms on a global basis. Clinical diagnosticians are trained in psychopathology, conceptualize prototypes of major disorder classes, and may use criteria such as contained in DSM-5 as a check/balance on overall evaluation. But the clinician knows, and the scientist should appreciate, that diagnostic manuals are not the solution to the aforementioned problems. Phenomenology of self-experience may distinguish between diagnostic categories or may identify psychopathology as a therapeutic target, but is not represented in diagnostic criteria. Similarly, methods most often used in clinical research fail to address specific psychopathology. Consider the use of total Positive and Negative Syndrome Scale (PANSS) scores in regulatory clinical trials that fail to identify what aspect of illness is responsive to treatment. We have more than 60 years of drugs approved for schizophrenia, but which are widely prescribed in other psychotic and nonpsychotic disorders. But these drugs lack efficacy for apathy, avolition, cognition, motor abnormalities, self-agency, or interpersonal pathology. Failure to deconstruct outcomes resulted in decades' delay in recognizing unmet therapeutic needs.

PARADIGMS FOR DISCOVERY

Organization of scientific investigation at the diagnostic category level does not address heterogeneity. What current paradigms will accelerate acquisition of knowledge? Examples are:

1. Syndrome deconstruction at the symptom level. Develop animal models for low social drive or impaired aspects of cognition or computational models for delusions or thought disorder. Work out mechanisms for the specific psychopathology. Do therapeutic discovery for specific pathology. Then determine whether the mechanism and/or therapeutic discovery is applicable to other symptoms in the same disorder or to similar symptoms in other disorders. The eight dimensions for the psychosis chapter of DSM-5 placed in section 3 are examples of immediate clinical relevance,[4] but many other symptom types can and should be addressed (e.g., motor pathology).
2. Ignore the substantial differences across currently defined classes of psychoses and use an extended psychosis phenotype defined as the symptomatic experience of psychosis.[5] This paradigm mandates that all involved psychopathology be viewed separately and dimensionally. To be worked out is whether each type of hallucination, for example, is a dimension or whether all hallucinations and/or all delusions are grouped as a single reality distortion dimension. This approach allows for early determination of pan-diagnosis relevance of findings. The downside is that it maximizes heterogeneity at the category level, and the assumption that any given aspect of psychopathology has the same pathophysiology in a broad phenotype is speculative.
3. Apply current statistical approaches (see later discussion) to determine most valid construct for psychopathology and identify components in each symptom complex. This approach is based in hierarchical taxonomy and integrates behavioral traits with manifest psychopathology.[6] The HiTOP system elucidates psychopathology at different levels ranging from behavioral constructs such as internalizing, externalizing, and thought disorder/psychoticism to more defined clinical manifestations. New views of the organization of psychopathology at the clinical illness level may emerge and the early stage behavioral platforms may open new approaches to prevention relevant to wide ranges of disorders.[7]
4. Develop hypotheses for new nosology based on endophenotypes using cluster analysis and other techniques to determine naturally occurring groups using extensive clinical information sometimes supplemented with nonclinical behavioral and physiological data. This provides opportunity for a fresh consideration of categories and is based on statistical approaches that maximize between-group differences.[8]
5. Shift focus of investigation from manifest illness to developmental pathways. Study when and how developmental impairments elevate risk for a later psychotic disorder. A risk construct encourages primary prevention research. Alternatively the developmental pathology may be viewed as early manifestations of an illness that later manifest with psychosis. Here early detection may lead to secondary prevention of psychosis.
6. Phenotypes (or endophenotypes, if heritable) may be more proximal to pathophysiology than manifest

symptoms and hence may be a more robust paradigm for discovery.[9] Shift starting point for psychosis studies to a methodology that involves psychopathology or behavior with a known basis in neurobiology. The NIMH Research Domains Criteria (RDoC) is explicit in providing a paradigm that starts with a behavioral construct/neural (anatomy or circuit) platform that is hypothetically linked to a psychopathology dimension not necessarily limited to a single disorder. Application of this paradigm may enhance acquisition of knowledge at the level of molecules, genes, cells, and mechanism pathways that may then be tested in relation to psychopathology hypothetically related to the behavioral construct (https://www.nimh.nih.gov/research-priorities/rdoc/index.shtml). The RDoC approach encourages dimensional concepts and across-diagnostic-boundary research. A motor domain was added to the initial five domains in 2018. An assessment challenge remains with the evaluation of behavioral constructs early in neurodevelopment.

7. Quasi-agnostic investigations based on accumulation of extensive data in large groups and applying machine learning, clustering, and graph and network statistical techniques in effort to define groups related to psychopathology as an alternative to traditional classes of psychoses. While statistical approaches are used in all the previous examples, this approach has advantages over traditional approaches in being able to develop a new range of information on individuals based on technology advances such as mobile devices accessing a range of in-the-moment activity: thought, emotion, context, social, etc. This approach will include information not necessarily hypothesized as relevant to psychopathology and thereby widen the opportunity for novel classification with new hypotheses regarding the organization of psychopathology/nosology.
8. Preclinical investigations related to mechanisms. This includes the range of animal model research but also human models and computational approaches.

The first eight paradigms take different approaches to dependent and independent aspects of study design, but all both guide and rely on preclinical models for fundamental mechanism studies. Each of the first eight approaches may provide insights that enhance the validity of "map-on" phenotypes used in translational mechanism research. All approaches are attractive, each with a different profile of strengths and weaknesses. Overcoming the weakness of clinical syndrome research will facilitate acquisition of fundamental knowledge. For example, anhedonia is observed in some patients in many disorders. Is anhedonia in depression the same pathophysiology as anhedonia in schizophrenia? Will a treatment effect on anhedonia predict a therapeutic effect on other aspects of illness? Is the efficacy of such a treatment the same in schizophrenia as in bipolar? Will a regulatory agency grant an indication for anhedonia; and, if so, in which disorders? What is the translational value of a mechanism for anhedonia identified in a rodent model? And to which disorders does it apply? Chlorpromazine was introduced for schizophrenia in 1952, and minimal progress has been made with "schizophrenia" as the indication. There has been limited exploration of other mechanisms or action. There are a number of unmet therapeutic needs in schizophrenia with implications across many disorders. The paradigms for deconstruction offer new targets for discovery and new investigative approaches. Preclinical models may have a more valid or predictive relationship to psychopathology and, if so, will be crucial in developing novel therapeutics.

ADVANCING KNOWLEDGE WITH HYPOTHESIS TESTING

Long ago, Platt (1964), following Popper (1959), noted that advance in knowledge was more rapid in scientific fields that focused experiments on meaningful hypotheses where results forced an advance in theory by falsifying a hypothesis, forcing change in concept and alternative hypotheses. This is illustrated in a test of the disease entity model of schizophrenia.[10] In the study of psychoses interesting results are often reported that neither confirm or reject a primary hypothesis nor distinguish between alternatives. Often confounds including medication status undermine interpretation, and heterogeneity of subjects weakens results and often is the cause of a failed study. Or the investigative tools are simply not yet available for a definitive study. The human brain is an extraordinarily protected organ. What follows are several suggestions/hopes for the near future:

1. Formulate hypotheses for discreet, deconstructed aspects of psychopathology.
2. Design experiments to test alternative hypotheses. For example, if a neural circuit is confirmed with imaging then show whether the relationship is unique, and which alternative circuits are rejected.
3. Test mechanism hypotheses in human disorders with neuromodulation. For example, if safety issues are resolved, MRI-guided ultrasound can provide precise neuromodulation in deeper parts of the brain to determine whether an aspect of psychopathology can be created, increased, or decreased (if present) with stimulation or inhibition in the hypothesized circuit. Current neuromodulation technology is promising, but target precision is challenging, especially with deeper brain targets.
4. Develop genetic risk information for specific psychopathology phenotypes without the heterogeneity of clinical syndromes. Schizophrenia is not a robust target for genetic interrogation.
5. Change the conceptual approach to therapeutic targets to facilitate novel mechanism of action discovery and new therapies for unmet clinical needs. Examples include:
    a. Motor system abnormalities are common in schizophrenia. Neural pathways are identified and provide molecular targets for potential novel drug discovery.

Motor signs in neurodevelopment may support prevention research, and motor signs during psychosis may provide therapeutic targets. Efficacy on pathophysiology of motor symptoms may benefit patients and may generalize to symptoms in other systems where pathophysiology may involve the same mechanisms.

b. Compensatory and resiliency mechanisms influence vulnerability to illness and course of illness. Therapeutic discovery is accomplished with some psychosocial approaches but has not been mined for pharmacological discovery.

IMPLICATIONS OF DECONSTRUCTION FOR THERAPEUTIC DISCOVERY

Advancing the pharmacotherapy of schizophrenia based on a clinical syndrome and regulatory emphasis on diagnosis has resulted in over five decades of antipsychotic drugs initiating action through the dopamine D2 receptor with a wide array of adverse effects, but only clozapine, itself an old drug, shows strong evidence for a meaningful clinical advance. Diminishing psychotic experience/behavior and reducing psychotic relapse rates is a major accomplishment. What is missing is progress on cognition impairment, avolitional and expressivity pathology of negative symptoms, motor abnormalities except for the treatment of co-occurring catatonia, emotional dysregulation and aggression, a range of functional impairments, and other pathologies afflicting some but not all persons with schizophrenia. Major pharmaceutical companies have withdrawn from any attempt at novel discovery in the absence of known pathophysiology and questionable predictive validity of animal models. Cost of treatment grew exponentially in the absence of novel mechanisms or substantial advance in efficacy.

Syndrome deconstruction has the potential to radically advance therapeutic discovery. Considerations include:

1. Multiple clinical therapeutic targets;
2. Advancing mechanistic understanding of each dimension;
3. More compelling relationship between preclinical models and clinical pathology;
4. Increased chance of identifying biomarkers for specific dimension applicable both for testing efficacy and identifying patients for treatment (experimental or personalized medicine);
5. Opportunity to use biomarkers or behavioral markers to establish across-diagnostic-class application of a treatment; and
6. From an industry perspective, much better information on which to base go/no-go decisions in the drug development process.

An additional consideration is to give emphasis to concepts for therapeutic development not based on reversing pathophysiology. Rather, expand concepts of compensatory mechanism, resilience, secondary prevention of progression, and, one hopes, primary prevention. In this context, it is worth noting that social and psychological treatment development has been more focused on deconstruction. Cognitive-behavioral therapy is different for hallucinations than for negative symptoms, and supported employment has a substantial effect size in controlled clinical trials with a specified functional outcome.

Most of the preceding commentary relates to approaches that are currently available and being further developed. Omitted from the closing discussion is the hope that methods associated with computational psychiatry and the analysis of large datasets with subject information far broader than typical clinical studies will produce new insights that will advance psychopathology concepts and provide mechanism information for a more informed approach to therapeutic discovery.

The vast majority of scientific reports related to psychopathology and mechanisms in psychotic disorders suffer from a host of confounds that distinguish schizophrenia from non-ill controls. Antipsychotic drug confounds undermine most studies. Three approaches, one new and two old, offer solutions that are routinely ignored. These are:

1. Compare subgroups of schizophrenia where the hypothesis relates to a variable present in some and absent in other schizophrenia participants. This was illustrated in the Strong Inference report.[10] A biomarker for anhedonia, for example, could be tested in schizophrenia with and without anhedonia increasing the similarity between groups on confounds including AP drugs. This addresses the common problem of whether a schizophrenia finding is related to pathology or to confounds.
2. Rarely used in this century, off-medication protocols address the problem of drug confounds. The ethical framework for such studies has been detailed. Critical issues include exclusion criteria to address high risk and an intervention approach with targeted medication at first sign of exacerbation.[11]
3. The current emphasis on psychopathology observed across diagnostic boundaries opens a door to explore symptomatically similar symptoms in multiple disorders simultaneously where the nature of confounds will be dissimilar across disorders. AP drugs will not be ubiquitous.

Best immediate opportunities for making a life course difference in the treatment of persons with schizophrenia:

1. Current awareness that treatment is most effective in first episode and that the duration of untreated psychosis can be reduced. Opportunity for full remission is greatest at this stage.[12]
2. Integrative therapeutics may have best opportunity for long-term outcomes.[13]

3. Clinical high risk/attenuated psychosis syndrome construct is a basis for identifying young people with symptoms and behaviors that merit clinical attention and, for the minority who are on a path to full psychosis, offers potential secondary prevention. If a transition to full psychosis is observed, the duration of untreated psychosis will be brief. Evidence on effectiveness of this construct is shallow,[14] but a clinical staging approach to development of psychotic illnesses has a solid basis.
4. In the clinical high-risk paradigm, a majority of patients will not be on a path to full psychosis. Nonetheless, early intervention may improve symptomatic and functional outcomes.

Perhaps the most important opportunity for the near future involves primary prevention of psychosis and related pathologies. Much is known regarding risk factors associated with gestational and early-life effects on brain development and early developmental signs of risk for future illness. The clinical high-risk studies in later adolescence and young adults have provided detection and treatment information for secondary prevention of psychosis. It may be possible to identify individuals before they enter a clinical high-risk state. Mental illness research has been hesitant to explore primary prevention. In part because the large-scale longitudinal studies necessary to establish effectiveness in reducing incidence of illness present a formidable barrier. And in part because risk factors identified to date are common in the history of persons who do not become mentally ill. Recent calls for prevention research and clinical care[15] suggest that much is feasible in this area, and illustrative examples of gestational intervention are compelling examples.[16]

These are speculations that, if true, will greatly increase the plausibility of primary prevention. If behavioral traits at the initial steps in the HiTOP approach are substantially linked to future psychopathology and assessment with relative positive predictive power is achieved, a new door opens on primary prevention. A prevention approach at this behavioral level may be benign, focused on resiliency. Early life risk factors could be addressed without waiting decades to determine whether the incidence of schizophrenia has been reduced. Rather, outcomes at a younger age and related to a number of mental illnesses would greatly increase power to detect. Even more so if the P-factor hypothesized by Caspi and Moffett (2018)[7] is confirmed as substantially related to development of a wide range of mental illnesses.

REFERENCES

1. Nuevo R, Chatterji S, Verdes E, Naidoo N, Arango C, Ayuso-Mateos JL. The continuum of psychotic symptoms in the general population: a cross-national study. *Schizophr Bull* 2012;38:475–485.
2. Strauss JS, Carpenter WT, Jr., Bartko JJ. The diagnosis and understanding of schizophrenia. Part III. Speculations on the processes that underlie schizophrenic symptoms and signs. *Schizophr Bull* 1974;61–69.
3. Berrios GE. Positive and negative symptoms and Jackson: a conceptual history. *Arch Gen Psychiatry* 1985;42:95–97.
4. American Psychiatric Association. *Diagnostic and Statistical Manual of Mental Disorders (DSM-V)*. 5th ed. Washington, DC: American Psychiatric Association, 2000.
5. van OJ, Reininghaus U. Psychosis as a transdiagnostic and extended phenotype in the general population. *World Psychiatry* 2016;15:118–124.
6. Kotov R, Krueger RF, Watson D et al. The Hierarchical Taxonomy of Psychopathology (HiTOP): a dimensional alternative to traditional nosologies. *J Abnorm Psychol* 2017;126:454–477.
7. Caspi A, Moffitt TE. All for one and one for all: mental disorders in one dimension. *Am J Psychiatry* 2018;175:831–844.
8. Clementz BA, Sweeney JA, Hamm JP et al. Identification of distinct psychosis biotypes using brain-based biomarkers. *Am J Psychiatry* 2016;173:373–384.
9. Braff DL, Tamminga CA. Endophenotypes, epigenetics, polygenicity and more: Irv Gottesman's dynamic legacy. *Schizophr Bull* 2017;43:10–16.
10. Carpenter WT, Jr., Buchanan RW, Kirkpatrick B, Tamminga C, Wood F. Strong inference, theory testing, and the neuroanatomy of schizophrenia. *Arch Gen Psychiatry* 1993;50:825–831.
11. Carpenter WT, Jr., Appelbaum PS, Levine RJ. The declaration of Helsinki and clinical trials: a focus on placebo-controlled trials in schizophrenia. *Am J Psychiatry* 2003;160:356–362.
12. Fusar-Poli P, McGorry PD, Kane JM. Improving outcomes of first-episode psychosis: an overview. *World Psychiatry* 2017;16:251–265.
13. Wunderink L, Nieboer RM, Wiersma D, Sytema S, Nienhuis FJ. Recovery in remitted first-episode psychosis at 7 years of follow-up of an early dose reduction/discontinuation or maintenance treatment strategy: long-term follow-up of a 2-year randomized clinical trial. *JAMA Psychiatry* 2013;70:913–920.
14. Davies C, Cipriani A, Ioannidis JPA et al. Lack of evidence to favor specific preventive interventions in psychosis: a network meta-analysis. *World Psychiatry* 2018;17:196–209.
15. Arango C, Diaz-Caneja CM, McGorry PD et al. Preventive strategies for mental health. *Lancet Psychiatry* 2018;5:591–604.
16. Ross RG, Hunter SK, McCarthy L et al. Perinatal choline effects on neonatal pathophysiology related to later schizophrenia risk. *Am J Psychiatry* 2013;170:290–298.

73

TIME FOR CHANGE IN PSYCHOSIS RESEARCH

Brett A. Clementz

I walked to the office the morning after Prince Rogers Nelson's death, listening to *Purple Rain*. The sidewalks were thick with students slogging to early morning classes. At the bottom of Baxter Hill, waiting at a stoplight to enter campus, was an unexpected sight. Despite the University of Georgia's (UGA) colors being red and black, a good quarter of the throng was wearing purple. Purple at UGA! I turned up the volume.

Did these students know Prince should be remembered for more than his music? He was not boisterously political, but The Purple One vigorously championed artists' rights. There were multiple reasons for his battles with the corporate arm of the music industry, but, in essence, the struggles were over control. Or, as Neo says in his confrontation with The Architect in the denouement of *The Matrix Reloaded*: "Choice. The problem is choice."

Class biases and corporate influences affect the academy in multiple ways (1), but I have not obviously felt their impacts. I have wonderful colleagues (as part of the Bipolar-Schizophrenia Network for Intermediate Phenotypes, or B-SNIP) with whom I immensely enjoy collaborating, and have students of high caliber (the "secret sauce" of any successful laboratory operation, to quote Bonnie Basler from her TED presentation on quorum sensing in bacteria). Prince's death, though, and what he stood for, stimulated consideration of whether my main scholarly interest (neural deviation variations in psychosis) had been subject to biasing influences. It was the least I could do to honor my departed age contemporary.

In the current *zeitgeist*, an obvious target of such an inquiry is the Diagnostic and Statistical Manual of Mental Disorders (DSM) project, which was recently modified (DSM-5). What does the scientific public think of the latest iteration? There are multiple published opinions, but Craddock (2) captured widely held sentiments:

> I do not see it as . . . helpful. . . . It is tinkering at the edges. What is needed for psychiatry is a game-changer: a truly new approach to diagnostic classification that better reflects the underlying functions and dysfunctions of the brain . . . DSM-5 . . . exemplifies the shortcomings of the . . . descriptive method and highlights the need for different approaches.

There have been stronger reactions to reliance on DSM-type diagnostic systems. Steve Hyman, speaking with the gravitas earned through a combination of sustained excellence and experience, has been an incisive critic:

> The problematic effects of diagnostic reification were revealed repeatedly in genetic studies, imaging studies, clinical trials . . . where the rigid, operationalized criteria . . . defined the goals of the investigation despite the fact that they appeared to be poor mirrors of nature. (3)
>
> It is critical that scientists are freed from the epistemic blinders and administrative strictures . . . that are imposed by widely accepted but fictive diagnostic categories such as those in the DSM. The price for freedom . . . is that scientists will have to describe the nature and logic of their sampling criteria with great precision and clarity so that their work is replicable. (2)

The preceding quotes describe the sentiments of a nontrivial proportion of clinicians and researchers, and capture four issues of relevance to the future of psychosis research (outlined by the following section headings). Addressing these matters, coming to terms with and resolving them, will hasten enrichment of basic and multilayered understanding of psychosis. The type of understanding that can undergird development of molecular/physiologically informed treatments, perhaps even the discovery of disease entities (4, 5) that have psychosis as a prominent symptom manifestation.

RELIANCE ON CLINICAL DESCRIPTIVE DIAGNOSES

For DSM, psychosis syndromes are variably defined by presence of hallucinations, delusions, disorganized thinking, abnormal motor behavior, and/or deviations in conation, including negative symptoms. Patients are diagnosed following rules prescribed by a committee of experts. Molecular and brain physiology information are absent from the criteria. The committee consults the neuroscience literature, but how such data inform the final diagnostic rules is frequently uncertain. In mature medical classifications diagnoses incorporate etiology and pathophysiology. The DSM diagnostic model, however, is an expected approach at the initial stage of medical syndrome formulation:

> The typical progression of knowledge starts with the identification of the clinical manifestations (the syndrome) and the deviance from the "norm"; understanding of the pathology and aetiology usually comes much later. (4)

Classification of psychosis syndromes started off well, or at least typically. Whether formulation of syndromes with psychosis as a prominent manifestation is still in early phase development is worth consideration. Counting from Kraepelin, the undertaking is elderly; even counting from Feighner (1972), RDC (1978), or DSM-III (1980) places the effort squarely in middle age. There are alternative viewpoints, in this case from Kendler, on the maturity of classifications in psychiatry:

> I am quite worried about whether we can sustain the level of historical discipline to keep the iterative process on track. We are still a young and rather immature science. We are prone to trends and, dare I say, fashions. (6)

Psychosis researchers may overvalue novelty. Perhaps we feel a bit desperate for significant discoveries. The field may be immature, but it is not young. A more insidious problem, one that could short-circuit the iterative process, may be slow recognition that certain theories are false—or at least should be framed differently (7). As David Lykken (8) said: "Theories in psychology are like old soldiers: They are not refuted or replaced . . . they only fade away." Hesitating to cull obsolete theories can delay progress in any field.

Multiple syndromes are defined by presence of a psychosis indicator with one (schizophrenia) defined by chronicity, another as a time-limited phenomenon (brief reactive psychosis), and another as intermediate between these duration extremes (schizophreniform disorder). Creating an at least soft dimensional structure was an explicit desire for disorders listed in the schizophrenia spectrum section of DSM-5 (9). Some diagnoses with psychosis symptoms (e.g., substance-induced psychosis, psychosis due to another medical condition) have a highly probable cause, even though the physiological mechanism for psychosis manifestations are still unknown, others have complex definitions that are blends with other syndromes (e.g., bipolar disorder with psychosis, depression with psychotic features, schizoaffective disorder), and others have more circumscribed manifestations (e.g., delusional disorder).

Fischer and Carpenter (10), in a paper titled, "Will the Kraepelinian dichotomy survive DSM-V?" made an interesting observation:

> the brain generates hallucinations and delusions in so many conditions that it is difficult to understand how [they] have maintained primacy in . . . diagnosis. . . . Psychotic experience is to the diagnosis of mental illness as fever is to the diagnosis of infection—important, but non-decisive.

The questionable differential diagnostic power of psychosis has been known for some time. William Carpenter (11) addressed this issue eloquently while struggling with changing *Schizophrenia Bulletin*'s name to reflect current knowledge; the solution was reasonable, given existing data and practical considerations: *Schizophrenia Bulletin: The Journal of Psychoses and Related Disorders*. Professor Carpenter's concerns with the utility of DSM-like schizophrenia definitions are not new, at least to him (12). Assen Jablensky (6) noted that two large World Health Organization studies highlighted shortcomings of the classical (affective versus nonaffective psychoses) system: (i) psychosis symptoms failed to separate affective and nonaffective psychoses; (ii) more relaxed versus more restrictive clinical definitions failed to create more homogeneous groups, defined by course, outcome, and symptom stability; and (iii) deterioration was not a defining feature of schizophrenia. Adding course and outcome to the differential was not of obvious assistance, a real blow to the classical psychosis nosological system.

As part of the B-SNIP project (13), a multisite study using deep phenotyping to parse psychosis using a combination of clinical and neurobiological features, we formed two symptomatic dimensions that best discriminated DSM schizophrenia, schizoaffective disorder, and bipolar disorder with psychosis. One dimension was labeled '"General Psychosis," given correlates with hallucinations, delusions, blunted affect, unusual thought content, difficulty in abstract thinking, lack of judgment and insight, suspiciousness/persecution, social withdrawal, and lack of spontaneity. Schizophrenia cases scored highest, and bipolar psychosis cases scored lowest on this dimension. The other dimension was labeled "Affective Instability," given correlates with depression/sadness, somatic concern, lassitude, inner tension, reduced sleep, suicidal thought, concentration difficulty, irritability, reduced appetite, and anxiety. Affective instability differentiated the schizoaffective disorder cases from schizophrenia and bipolar psychosis cases (who surprisingly did not differ).

A psychosis severity dimension is hardly novel; it quantifies that DSM is, at least partially and for the major psychoses, a severity scale. The affective instability dimension offers novelty, given its differentiation of schizoaffective disorder from schizophrenia and bipolar psychosis. Schizoaffective disorder is a curious entity. Sometimes this group seems most similar to schizophrenia (14), sometimes it falls between schizophrenia and bipolar psychosis (15, 16), and sometimes schizoaffective disorder seems to deserve unique consideration (17). Neurobiologically, schizoaffective disorder cases are more distributed in their biomarker profiles than either of the other two major psychoses (18), which seems inconsistent with the rationale for inclusion of schizoaffective disorder in the schizophrenia section, that "There is growing evidence that schizoaffective disorder does not represent a distinct nosological category separate from schizophrenia" (19).

Let's consider Fischer and Carpenter's psychosis-fever analogy. For patients presenting with fever (quantifiable with an instrument), the clinical evaluation will assess severity, duration, continuity, and onset, and query accompanying features. Save an instrument for quantification, the clinical evaluation for a psychosis presentation is similar. Hypotheses will be generated, and laboratory tests can assist diagnosis. For pyrexia, differential diagnosis (largely) determines differential treatment. There will be cases where diagnosis is uncertain (pyrexia of unknown origin, or PUO), but diligent follow-up will often reveal the proper diagnosis. For psychosis, the situation is similar on the path to differential diagnosis, but differential treatment options are limited. For the overwhelming majority of psychosis presentations, the outcome of the evaluation will lead to a PUO (psychosis of unknown origin; my term, not DSM's). Follow-up may reveal a diagnosis (20), but this is the exception. Psychiatry, at least for psychosis, is largely defined by the nonspecific treatment of PUOs.

There is also the possible conflation of a symptom (psychosis) with disease. A major advantage for pyrexia is knowledge of a specific physiological mechanism for fever generation (thermoregulatory neurons in anterior hypothalamus), independent of the disease processes. Similar knowledge for psychosis is lacking. Given Fischer and Carpenter's reasonable hypothesis, psychosis indicates a disease process causing dysfunction in a specific brain circuitry: psychosis would be one sign of the primary pathology. For the time being though, Charles Burlingame, former superintendent of the Institute of Living (professional home of my B-SNIP collaborator Godfrey Pearlson) gets the last word (from his 1937 annual report):

> "Psychoses can hardly be called disease entities, even now, but are regarded as manifestations of a disease process, concerning the real sources of which we can do little more than speculate at the present." (as quoted in [21])

PSYCHOSIS DIAGNOSES AND THE REIFICATION PROBLEM

Reification (giving form to an abstraction, idea, concept, or construct) can be a handy device. For the study of psychosis, the problem comes with belief in favorable press coverage, treating a theoretical entity (DSM diagnosis of schizophrenia) as a truth rather than a hypothesis to be tested. With every DSM revision, the form of the schizophrenia construct is similar, with changes reflecting little more than "tinkering at the edges." A successful construct is embedded in a network of findings that establish its superiority over competitors by establishing relationships to other variables via criterion-related, discriminant, and predictive validities (22). Based on the evidence, Hyman and others (e.g., [23–25]) conclude that DSM psychosis diagnoses are failing as theoretical constructs.

There are tendencies for schizophrenia, schizoaffective disorder, and bipolar psychosis to differ in predictable directions on many laboratory and clinical measures. These tendencies are more impressive when considering only schizophrenia versus healthy individuals, and even schizophrenia versus bipolar psychosis cases, although the schizophrenia construct does teeter a bit under the latter comparison. B-SNIP aimed to investigate the neurobiological uniqueness of DSM psychoses using an extensive biomarker panel. With large and representative clinically defined samples there was confidence that comparisons would reveal enduring and meaningful group differentiations (13, 18).

Comparisons across groups on nine biomarker composite variables are shown in Figure 73.1. These plots have three noteworthy features. First, for three cognitive assessments (*antisaccade errors*, *BACS*, *stop signal*), perhaps *N100 ERP*, and *P300 ERP*, there is a severity continuum, with schizophrenia cases being most impaired, bipolar psychosis cases being least impaired. Second, the other biomarkers do not markedly differentiate the psychosis groups from each other and/or from healthy persons. Third, relatives show the same general pattern of effects (but attenuated) as the probands to whom they are related. Only *antisaccade errors* had effect sizes indicating possible usefulness for family/genetic investigations.

DSM psychosis diagnoses had considerable overlap on measures ranging from basic sensory processing through sensorimotor performance, inhibitory control, target detection, memory functions, general executive control abilities, and structural anatomical brain features. This outcome was inconsistent with the thesis that DSM psychosis diagnoses capture neurobiologically distinct syndromes. B-SNIP might have used suboptimal measures (certainly possible), although they were selected for their success in psychosis research. If comparisons were schizophrenia versus healthy persons, schizophrenia would look decent as a neurobiological entity, but that would yield an incomplete picture, especially with regard to the challenge of differential diagnosis for investigating etiologies, pathophysiologies, and targeted treatment plans.

There are other complications. Schizophrenia and bipolar psychosis do not "breed true," which one of my B-SNIP collaborators, Elliot Gershon, observed in the 1980s (26). Smoller (27) highlighted that: (i) persons with serious psychiatric diagnoses can have offspring with any number of psychiatric syndromes; (ii) symptoms and signs not peculiar to any one syndrome (such as psychosis) are heritable independent of their syndromal relationships; (iii) the genetic correlation between schizophrenia and bipolar disorder is high (≈.65; [28]); and (iv) the Psychiatric Genomics Consortium showed substantial overlap of polygenic risk between schizophrenia and bipolar disorder ([29]; see their Figure 3). Sharing of genetic risk does not prove schizophrenia and bipolar psychosis are variations on the same underlying (as yet undiscovered) disease, but it is hard to use available genetic data to support a clear distinction between the syndromes.

Across a range of data types, the distinctions between different syndromes defined by psychosis are fuzzy at best. Nevertheless, leaders of the most recent DSM effort concluded:

> Realistically... DSM users in general are not yet ready for a drastic overhaul of DSM's organization. As such, DSM-V revision experts are examining whether specific indicators can inform and validate the grouping of disorders while maintaining much of the existing categorical framework. (30)

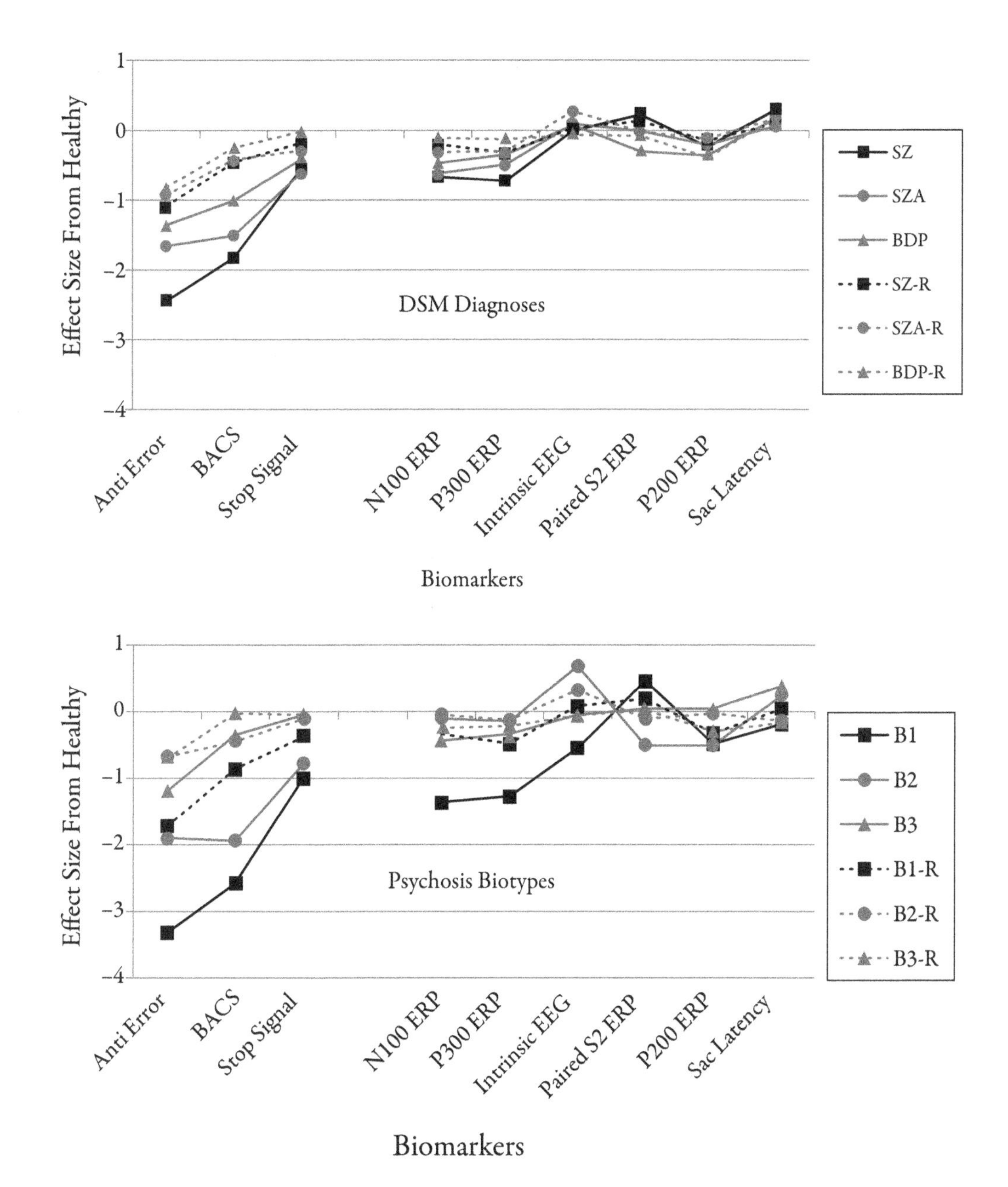

Figure 73.1 Effect size separations from nine biomarker measures from healthy persons ($n = 278$) for psychosis cases ($n = 711$) and their first-degree relatives ($n = 883$) categorized by DSM diagnoses (upper panel) or Biotypes (lower panel). See Clementz et al. (18) for full descriptive details. See color insert 3 in front of book for full colorized version.

McHugh (31) presented a stark comparison:

> all is not well with psychiatry under this new dispensation.... difficulties... are serious enough to challenge the usefulness of any revision of DSM that does not make a significant move to resolve them.... It does not speak to the nature of mental disorders or distinguish them by anything more essential than their clinical appearance. Not a gesture does it make toward... the promise that rational, effective, cause-attacking therapies will eventually replace symptom-focused, palliating ones.

This is a tricky spot. Framers of the DSM project, along with "DSM revision experts," have not maintained a defective system for classification of psychosis syndromes while holding the good system in reserve. Perhaps a partial explanation for maintenance of the DSM project can be abstracted from the conversation between Neo and The Architect in *The Matrix Reloaded*. The Architect informs him that Neo is an anomaly; a result of escalating discontent that if left unchecked can threaten The Architect's beautifully constructed (if fatally flawed) system. Because of such anomalies, The Architect has had to reinitiate the process leading to Neo on five previous occasions; Neo is the sixth iteration. He threatens that if Neo does not comply with the previously established process, allowing for another reinitiation, there will be catastrophe.

E. O. Wilson (32) once wrote, "To be highly successful the scientist must be confident enough to steer for blue water, abandoning sight of land for a while." There have been multiple justifications, eventual catastrophe among them, for delaying what

Professor Kendler (6) calls a system "re-boot." Framers of the DSM project imply the impediment to developing a classification with etiopathological foundations is the lack of supporting data, a legitimate argument only if its circularity is ignored. The gold standards for determining the usefulness of neurobiological measures are DSM-type definitions. If such definitions are orthogonal to neurobiological reality, the ability to identify etiological distinctiveness of psychosis subtypes will be severely limited, the essence of the argument made by Craddock and Owen, Cuthbert and Insel, Hyman, van Os, and many others.

A system of questionable validity is problematic if users perceive there are no options, if they are not free to abandon the system and test alternatives. This has been part of the argument for the slow progress of psychosis research; the DSM project has controlled the narrative and forced use of deficient definitions to obtain reimbursement, secure grant funding, and publish papers (5, 24, 25). It seems unfathomable that a creative proposal to form a promising etiology-based alternative to the DSM, in the context of an innovative overall scientific plan, would be rejected outright by a review committee. But the committee certainly would require the inclusion of DSM diagnoses, and, depending on its composition, may question the pressing need. Many of us have been part of this process; perhaps we have all stayed a bit too close to shore.

PSYCHOSIS MECHANISMS: UNDERLYING FUNCTIONS/ DYSFUNCTIONS OF THE BRAIN

In 1987, Nancy Andreasen (33) proposed that psychosis researchers needed to "move freely between the biotype and the phenomenotype . . . to understand fully the clinical picture and ultimately the cause of schizophrenia." Using laboratory investigation "places our most important goal, the identification of pathophysiology and etiology, at the forefront of investigation." A quarter century later, the B-SNIP consortium accepted Professor Andreasen's challenge. Multiple factors led B-SNIP to question whether DSM psychosis diagnoses were neurobiologically distinct, including the biomarker data presented earlier (see also [34]). Rather than doubting the veracity of our laboratory data, we questioned the construct validity of the DSM diagnoses (18).

Our primary innovations were threefold. First, recruit cases across the psychosis spectrum and exploit variance in the biomarker panel to capture neurobiologically distinctive psychosis subgroups independent of DSM definitions, and then verify the outcome against external validating measures. Similar approaches had been attempted previously (35, 36), with interesting outcomes, but with many fewer biomarkers, many fewer validators, and many fewer observations. Second, parse variance within psychosis only, and apply the obtained solutions to other groups (first-degree relatives, healthy persons) as a validation check on the statistical derivation. The important issue was identifying more homogeneous neurobiological cases than DSM, if they existed, not differentiating psychosis from health (this was already known). Third, rigorously estimate the number of neurobiologically distinct psychosis subgroups given the B-SNIP biomarker data (the number of subgroups was three).

Given the apparent neurobiological distinctiveness of the subgroups, we called them psychosis *Biotypes* (biologically distinctive types). The effect size separations from healthy persons by psychosis Biotype membership are shown in Figure 73.1. There are three noteworthy features of these data. First, the effect size separations of the Biotypes are larger than they were for DSM groups (they should have been, but one never knows). Second, there are three different patterns of abnormalities: (i) on *antisaccade error*, *BACS*, *stop signal*, and (modestly) *saccade latency*, there is a continuum of severity from Biotype-1 (most abnormal) through Biotype-3 (least abnormal) probands; (ii) on *N100* and *P300* there is a different severity continuum from Biotype-1 (most abnormal) through Biotype-2 (least abnormal) probands; and (iii) on *intrinsic EEG*, Biotype-2 shows accentuated activity while on *paired S2 ERP* Biotype-1 shows an accentuated response. The *paired S2 ERP* effect looks consistent with a traditional view that some psychosis cases have difficulties with stimulus filtering (they have modesty elevated values), but a more nuanced consideration reveals this effect is also a function of prestimulus brain activity (25). There was not a single variable on which the Biotypes were indistinguishable. Third, the relatives show the same, but attenuated, pattern of effects as the probands to whom they are related. The cognition measures may have promise as endophenotypes, but perhaps only for Biotype-1. Additionally, external validating measures (social functioning, measures of brain volume from structural magnetic resonance images, clinical diagnoses of the first-degree relatives; [18]), and other testing since the initial Biotypes formulation ([37]; see what follows), demonstrated the superiority of Biotypes versus DSM diagnoses for capturing neurobiological homology. There were more schizophrenia cases in Biotype-1 and more bipolar psychosis cases in Biotype-3, but DSM psychosis diagnoses were spread across Biotypes.

These psychosis Biotypes are a statistical phenomenon, there was no neurobiological model supporting the analyses. It is dustbowl empiricism. But given the outcome, the data provide proof of concept for theory development. One striking feature is that cognition-related biomarkers look like promising endophenotypes primarily for Biotype-1 (about a quarter of the B-SNIP sample). Despite the breadth of the biomarker panel, B-SNIP largely missed biomarkers sensitive to Biotype-3 pathophysiology (about 40% of the B-SNIP sample) as few deviations were seen in this subgroup. Those cases still have disruption in neural circuitry leading to psychosis. Perhaps those will be markers of environmental factors (e.g., [38]). Who knows at this point? That would not make those cases any less neurobiological in their origin. We might call some of them phenocopies, even though some or all still could be genetic (but not familial).

Mitchell (39) drew a parallel between "syndromes" like mental retardation and schizophrenia. Mental retardation is initially defined by scoring beyond an arbitrary cutoff on an IQ test, and when considered as a syndrome looks like a complex disorder. But there are multiple distinct and identifiable factors (many of them genetic, not necessarily familial) that cause an intellectual ability score to fall beyond the arbitrary cut line. If we used only IQ as the phenotype of interest, Mitchell (39) suggests that the mental retardation phenotype would look polygenic. But this conclusion would be false for cases arising secondary to a highly penetrant mutation. The parallels to schizophrenia—psychosis generally, for that matter—are worth considering. The clustering of familial cognition deviations in a small subgroup of B-SNIP cases, despite a continuum of severity across all psychosis, might provide us with an important clue.

The biomarkers identified so far, and most commonly used in psychosis research, have depended on the DSM gold standard. It is easy to suppose that the field has yet to discover critical psychosis-related physiological mechanisms if the target is "a poor mirror of nature." At present the physiological modes of psychosis generation, not to mention the specific deviations that cause psychosis, are "a riddle, wrapped in a mystery, inside an enigma" (to quote Churchill, talking about Russia, of all things). This does not mean we lack interesting clues that inspire testable hypotheses. And when we have promising information on a particular point, we should be keen for its implementation.

Measures of cognition again provide an excellent example. They work well for quantifying brain dysfunction probably because they rely on multiple processes that recruit multiple brain circuitries for their successful performance. There are many paths to dysfunction, so many brain disorders can have phenotypic similarity on measures of complex cognition. Conditioning on presence of psychosis, B-SNIP illustrated that a cognition phenotype differentiates psychosis subtypes and correlates well with widespread structural brain deviations (18). In their statement on the exclusion of cognition from DSM psychosis diagnostic criteria, Barch et al. (19) stated:

> We were concerned that cognitive dysfunction is not a differential diagnostic marker for schizophrenia . . . the profile of cognitive impairments is similar across the non-affective and affective psychoses . . . the preponderance of data suggests that this separation is not sufficient to justify inclusion of cognition as a Criterion "A" symptom of schizophrenia.

There was probably sufficient information in 2013, there is considerably more now, to support inclusion of cognition as an objective laboratory measure of the psychosis severity continuum, and assist the DSM differential diagnosis given the presence of psychosis. Given what we know, an alternative version of the preceding quote, which would have required something closer to McHugh's (31) desired reconceptualization, could look like this:

> We were concerned that [*psychosis symptoms are*] not a differential diagnostic marker for schizophrenia . . . the profile of [*psychosis symptoms*] is similar across the non-affective and affective psychoses . . . the preponderance of data suggests that this separation is not sufficient to justify inclusion of [*psychosis symptoms*] as a Criterion "A" symptom of schizophrenia.

The importance of cognition is obvious regardless of classification scheme (DSM, Biotypes). Other possible biomarkers are not so fortunate. For instance, deviation in level of intrinsic neural activity (present in ongoing signals recorded from humans with EEG/MEG) is observed in psychosis (e.g., [40–43]). Neurophysiological models of psychosis have leaned on this physiological indicator as a possible genetically mediated core deviation (43). Translational models of intrinsic activity deviations hold promise for identifying multiple distinct physiological mechanisms for psychosis manifestation (e.g., [43, 44]). Intrinsic activity deviations may cause diminished signal-to-noise ratios (e.g., [37, 41]), and contribute to, or at least set the stage for, psychosis-related problems like identifying stimulus salience ([40, 45]; see also [46]).

B-SNIP labeled a biomarker composite *intrinsic EEG*, although the contributing variables were diverse. Additional analyses support the thesis that intrinsic activity is important for capturing neural deviation differences between Biotypes (37). So why do we not hear more about intrinsic activity as a core biomarker for any psychosis variation? Let me provide a possible explanation. We analyzed ongoing EEG activity during 150 intervals of 10 sec duration among B-SNIP subjects (the ITI of an auditory task used in Biotypes construction, although these ITI data were not used). A crude illustration of the outcome is presented in Figure 73.2, which plots

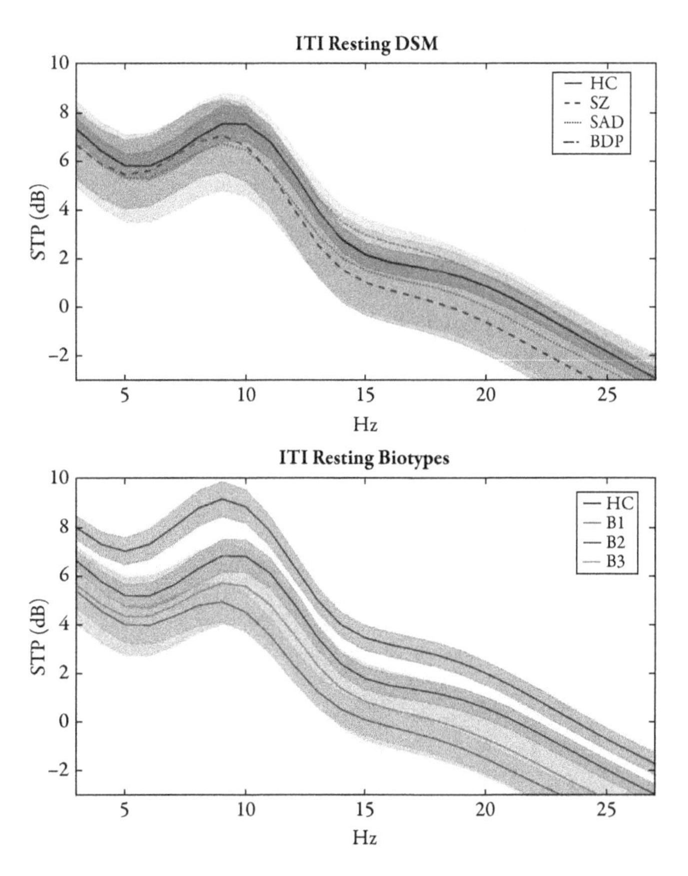

Figure 73.2 EEG power by frequency averaged for single trials from the 10 sec ITI during 150 trials of a paired auditory stimuli task for healthy persons ($n = 278$) and psychosis cases ($n = 711$) categorized by DSM diagnoses (upper panel) or Biotypes (lower panel). Shading illustrates the 95% confidence intervals.

intrinsic activity over all segments as a function of frequency. When plotted by DSM diagnoses, there is not much to consider; bipolar psychosis is a bit high and schizophrenia is a bit low, but the 95% confidence intervals mostly capture all group means. When considered by psychosis Biotypes, however, differences are obvious. In comparison to healthy persons, Biotype-2 probands are notably high, and Biotype-1 probands are notably low. Group separations on this crude metric were better than those obtained with the original *intrinsic EEG* measure used to construct Biotypes (bringing to mind Meehl's Super Bootstraps Theorem; see [47]). Intrinsic neural activity deviation is a biomarker with promise for translational research programs aimed at differential treatment development, but using DSM psychosis diagnoses would obscure its importance for understanding psychosis.

THE GAME CHANGER

George Harrison was posthumously inducted into the Rock and Roll Hall of Fame in 2004. An all-star band played Harrison's "While My Guitar Gently Weeps" as the show's closing number. That same year, Prince was also inducted into the Hall, so he was naturally invited to join the tribute. There are many articles written about that night; I recommend Finn Cohen's piece in the *New York Times* (April 28, 2016, 'The Day Prince's Guitar Wept the Loudest"). I also recommend watching the YouTube video of the performance. Marc Mann, who had been involved in previous Harrison tributes, took the first leads and played Clapton's well-known guitar solo. About 3 minutes in, Prince stepped forward and everything changed. His performance was stunning. A classic song with an iconic solo had been reimagined.

Prince was a game changer. He was on a level few people reach in their profession. This is not just my opinion:

> The greatest performer I have ever seen. A true genius. Musically way ahead of any of us.
>
> —*Sir Elton John*

> I think Prince is stunning. . . . I could never be what he is; he's an absolute genius.
>
> —*Robert Plant*

> Prince . . . was a true genius . . . he was like a light in the darkness.
>
> —*Eric Clapton*

> I never met Mozart, I never met Duke Ellington or Charlie Parker. I never met Elvis. But I met Prince.
>
> —*Bono*

Psychosis research can claim many wonderful scientists of impressive intellect. But, so far, has anyone reimagined the field? Has there been a stunning insight? Meehl's schizotaxia paper (48), which he updated (49), was in the neighborhood of this level of spectacular. Irv Gottesman (50) described the theory as "a product of P.E. Meehl's speculative daring." Yes indeed. Meehl was a genius. So was Gottesman. Both of them, University of Minnesota products (Prince was also a Minnesota man), had outsized influences on psychosis research. We need scientists like them today, modern versions, comprehensively trained in basic and clinical neuroscience. They were experts on the epidemiology, genetics, quantitative, and theoretical ends. A young Meehl paired with a young Gottesman, devoting their joint effort to solving the psychosis puzzle, would be a game changer.

But, alas, Meehl and Gottesman are not here. What can we do? Solving the psychosis puzzle is going to take open minds and elbow grease, and willingness to steer for blue water while being organized in attack but not limited by administrative or conceptual strictures. It is a bit disheartening that Kraepelin (from 1887) sounds contemporary:

> Our science has not arrived at a consensus on even its most fundamental principles. . . . today the different directions of psychiatric research are able only very marginally to complement and support each other . . . [the] countless attempts at classification in the history of our science has involved some intellectual manipulation and violation of the bare empirical evidence . . . in the not too distant future our science, the more it is able to escape the influence of theoretical speculation and fight its way towards sober observation and registration of the facts, will really be able to produce a "clinical science of mental disorders." (cited in [51])

But we can conquer these impediments. The following prescription has six main points, and is partially a compendium of previous suggestions (there might be a few novel proposals). The most important point is that we need to have open (but critical) minds. Reading the heart-wrenching account of a mother's struggle with her child's florid psychosis onset (Tanya Frank, "Unmoored by a Psychotic Break," *New York Times*, April 21, 2017) or remembering interviews with people who have lived with chronic psychosis makes it easier to give these patients' urgent needs our unbiased and best effort.

First, stop blaming Kraepelin (any historical figure really), and stop claiming we have rejected his dichotomy. Historical figures are not the impediment to progress; continued reliance on surface features, failure to implement useful laboratory tests, and denial of the DSM anomaly are not Kraepelin's fault. Kendler (52) summarized how conceptualizations of schizophrenia are remarkably different between historical experts and DSM criteria. Fischer and Carpenter (10) maintain that we cannot reject Kraepelin if we don't use his methodology, which is a pretty fair point. B-SNIP symptom data are illustrative. As we did for DSM diagnoses, we created combinations of psychosis-related clinical features that best differentiated Biotypes, thus building differentiating clinical characteristics based on neurobiology. One dimension, which identified Biotype-1 cases (those with the most profound neurobiological deficits), was labeled "Deficient Conation," given its symptom correlates of difficulty in abstract thinking, blunted affect, conceptual disorganization, poor attention, disorientation, hallucinations, and lack of spontaneity. The other dimension, which identified Biotype-2 cases (the most physiologically dysregulated), was labeled "Neuro-Cognitive Dysregulation," given its symptom correlates of decreased need for sleep, unusual thought content, stereotyped thinking, thought disorder, mannerisms and posturing, concentration difficulty, and anxiety. The differentiating features bear little resemblance to DSM conceptions. One could argue that the historical expert was closer to neurobiological reality.

Second, stop using structured interviews that are organized around diagnostic criteria and predefine the relevance of specific clinical characteristics. Even diagnosticians of high quality are subject to cognitive biases, confirmation bias in this case. We can learn little about the symptomatic structure of psychosis independent of DSM by continued use of the SCID. The Diagnostic Interview for Psychosis, or something like it, may be a good alternative.

Third, it is time for a system reboot; DSM psychosis categories need to be abandoned. This is not a statement on the relative merit of categorical versus dimensional models of psychosis syndromes. There are almost certainly specific etiologies underlying psychosis syndromes (it is not all multifactorial polygenic), but we should stop advocating such distinctions until they are unequivocally demonstrated. Kendell and Jablensky (5) offered excellent advice: (i) surface features are insufficient for deciding the presence of a category with a specific biological mechanism; and (ii) a category should be considered valid only if there is clear bimodality on a physiological measure and/or qualitative differentiation between categories on the physiological measure.

It might be argued that abandoning DSM diagnoses will create havoc for clinicians and researchers. Catastrophe is not the most likely outcome of a system reboot, but the desire to fill a diagnostic vacuum certainly would be motivating. To meet Hyman's challenge of increased descriptive accuracy we take two steps: (i) have a diagnostic code for "Psychosis Syndrome" with criteria broad enough to include something like attenuated psychosis syndrome ([53]; see also [24]). This code could be the organizing influence for treatment planning, insurance reimbursement, and scientific investigation, although final treatment planning, reimbursement, and sample descriptions in scientific papers must be based on (ii) the dimensional assessment scheme provided in DSM-5 and illustrated by Heckers et al.'s (9) psychosis symptom wheels, an excellent descriptive tool. There are eight main ratings: delusions, hallucinations, disorganized speech, abnormal psychomotor behavior, negative symptoms, impaired cognition, depression, and mania. Adding social/occupational functioning might be worthwhile. Each of these ratings could be subdivided (e.g., types of hallucinations, variations on negative symptoms such as emotional blunting, apathy, avolition, subdomains of cognition, etc.), but scientific inquiry should determine their enhancement of knowledge over the broad domains. Perhaps delusions, hallucinations, and disorganized speech are all caused by the same physiological mechanism, but this should be determined by the science.

This leads to the fourth point, the need to study mechanisms for psychosis manifestations, which is not the same as searching for specific disease etiologies. The RDoC (24) is a framework within which to organize the science, perhaps integrated with DSM-free approaches to behavioral manifestations of psychopathology (54). Unfortunately, from the beginning RDoC did not obviously emphasize any association to psychosis research, but that can be rectified. An important aspect of RDoC is formalization of appreciation for levels of analysis. The logical distance from genes to symptoms is considerable. Theories need to be integrated from genes to molecules to circuits to physiology to behavior; E. O. Wilson (32) indicates this is the method of knowledge acquisition in biology. Work at only the symptomatic level independent of physiology, or vice versa, may lead to incomplete, perhaps incorrect, answers to our pressing questions.

Even though RDoC provides a useful framework, it should be modified for the science to work optimally. To wit, specific behavioral, self-report, and paradigm ties to the systems, constructs, and levels of analyses must be eliminated. Including these columns in the matrix is the equivalent of legislating tasks, paradigms, variables, and methods of quantification. Hyman (2) might call such links "fictive." Associations with surface level phenomena must be determined by scientific creativity and inquiry. Who is to say, for instance, that delayed discounting is not an assessment for the neural circuitries involved in cognitive control, goal maintenance, or working memory? Gottesman and McGue (55) had this right: "The relationships between and among endophenotypes will emerge from empirical research that is not fettered by invalid assumptions about what actually goes with what."

Fifth, RDoC findings of relevance to psychosis should be updated at least annually in a publicly available, searchable format. NIMH-funded investigators must provide annual updates on their specific work in virtual interactive conferences organized by a new Psychosis Research Branch. This is a benefit of NIMH funding; we all get smarter by acting as a large consortium working toward a common goal. There is plenty of glory for everyone. Non-NIMH-funded investigators also could be invited to participate. We need to visualize areas of information gain without fear or favor. We need to be lenient, especially in evaluations of new theories, but still fair and critical of the current state of knowledge (56). We must be willing to cull obviously degenerating and/or diminishing-returns research programs (7) and not let our theories be like old soldiers. When areas of promise and/or important gaps in knowledge are identified, the Psychosis Research Branch should evaluate the benefit of issuing RFAs specifically targeting these areas. This would change psychosis research from individuals building their own bungalows to an organized crew constructing a palace (8).

Sixth, someone has to organize this effort. Decisions need to be made on issues like: What part of the matrix should be targeted for optimal effect? Do we have sufficient evidence to claim identification of a physiological mechanism for psychosis? Have we identified a specific disease etiology that causes disruption of that mechanism? What counts as a degenerating research program? This will take people of formidable intellect, broad knowledge, unimpeachable fairness, and unquestioned respect of the scientific community, working within the framework of a Select Committee on Psychosis. In his *New York Times* piece "Diagnosing the D.S.M" (May 11, 2012), Allen Frances, former DSM revision leader, concluded the American Psychiatric Association is "no longer capable of being sole fiduciary" of this task, so the Select Committee should be under some other agency. Dr. Frances makes multiple suggestions; I like the National Academy of Medicine option, in collaboration with the NIMH, who funds the overwhelming majority of the relevant research, and WHO, responsible for maintaining diagnostic consistency around the world.

The Committee will have considerable evidence to evaluate, will need to arrange hearings with investigators of all stripes, and will have to make some tough decisions. It must be responsible for ensuring the publicly searchable RDoC psychosis database provides a veridical representation of the current state of knowledge. I did say elbow grease would be required. Someone has to lead the Committee. The heads of the Academy of Medicine, NIMH, and WHO, should make this call after extensive consultations. I would recommend co-chairs who have been uninvolved in previous DSM revision efforts (makes for a clean slate). There is plenty of talent, both nationally and internationally, from which to draw. Then the Committee membership and their support staff will need to be selected for expertise that covers levels of analysis represented in the RDoC matrix. I am guessing that whoever is selected for committee membership, they will be hoping that before construction begins on a shiny new rich and balanced psychosis knowledge palace, we find ourselves a game changer. Or, more likely, a game changer finds psychosis research.

I know times are changing
It's time we all reach out
For something new that means you too

—Prince (from *Purple Rain*)

REFERENCES

1. Holder CD. Coping with class in science. Science. 2017;355:658.
2. Casey BJ, Craddock N, Cuthbert BN, Hyman SE, Lee FS, Ressler KJ. DSM-5 and RDoC: progress in psychiatry research? Nat Rev Neurosci. 2013;14:810–814.
3. Hyman SE. Can neuroscience be integrated into the DSM-V? Nat Rev Neurosci. 2007;8:725–732.
4. Jablensky A. The disease entity in psychiatry: fact or fiction? Epidemiol Psychiatr Sci. 2012;21:255–264.
5. Kendell R, Jablensky A. Distinguishing between the validity and utility of psychiatric diagnoses. Am J Psychiatry. 2003;160:4–12.
6. Kendler KS, Parnas J: Philosophical issues in psychiatry. II, Nosology. 1st ed. Oxford, Oxford University Press; 2012.
7. Carpenter WT, Jr., Buchanan RW, Kirkpatrick B, Tamminga C, Wood F. Strong inference, theory testing, and the neuroanatomy of schizophrenia. Arch Gen Psychiatry. 1993;50:825–831.
8. Cicchetti D, Grove WM: Thinking clearly about psychology: essays in honor of Paul E. Meehl. Minneapolis, University of Minnesota Press; 1991.
9. Heckers S, Barch DM, Bustillo J, Gaebel W, Gur R, Malaspina D et al. Structure of the psychotic disorders classification in DSM-5. Schizophr Res. 2013;150:11–14.
10. Fischer BA, Carpenter WT, Jr. Will the Kraepelinian dichotomy survive DSM-V? Neuropsychopharmacology. 2009;34:2081–2087.
11. Carpenter WT. Shifting paradigms and the term schizophrenia. Schizophr Bull. 2016;42:863–864.
12. Strauss JS, Carpenter WT, Jr., Bartko JJ. The diagnosis and understanding of schizophrenia: summary and conclusions. Schizophr Bull. 1974:70–80.
13. Tamminga CA, Ivleva EI, Keshavan MS, Pearlson GD, Clementz BA, Witte B et al. Clinical phenotypes of psychosis in the Bipolar-Schizophrenia Network on Intermediate Phenotypes (B-SNIP). Am J Psychiatry. 2013;170:1263–1274.
14. Ivleva EI, Clementz BA, Dutcher AM, Arnold SJM, Jeon-Slaughter H, Aslan S et al. Brain structure biomarkers in the psychosis biotypes: findings from the bipolar-schizophrenia network for intermediate phenotypes. Biol Psychiatry. 2017;82:26–39.
15. Hill SK, Reilly JL, Keefe RS, Gold JM, Bishop JR, Gershon ES et al. Neuropsychological impairments in schizophrenia and psychotic bipolar disorder: findings from the Bipolar-Schizophrenia Network on Intermediate Phenotypes (B-SNIP) study. Am J Psychiatry. 2013;170:1275–1284.
16. Reilly JL, Frankovich K, Hill S, Gershon ES, Keefe RS, Keshavan MS et al. Elevated antisaccade error rate as an intermediate phenotype for psychosis across diagnostic categories. Schizophr Bull. 2014;40:1011–1021.
17. Cardno AG, Owen MJ. Genetic relationships between schizophrenia, bipolar disorder, and schizoaffective disorder. Schizophr Bull. 2014;40:504–515.
18. Clementz BA, Sweeney JA, Hamm JP, Ivleva EI, Ethridge LE, Pearlson GD et al. Identification of distinct psychosis biotypes using brain-based biomarkers. Am J Psychiatry. 2016;173: 373–384.
19. Barch DM, Bustillo J, Gaebel W, Gur R, Heckers S, Malaspina D et al. Logic and justification for dimensional assessment of symptoms and related clinical phenomena in psychosis: relevance to DSM-5. Schizophr Res. 2013;150:15–20.
20. Nelson MA, Lochhead JD, Maguire GA. Anti-NMDA receptor encephalitis presenting as a primary psychotic disorder. Ann Clin Psychiatry. 2017;29:205–206.
21. Dittrich L: Patient H.M.: a story of memory, madness and family secrets. First edition. New York, Random House; 2016.
22. Cronbach LJ, Meehl PE. Construct validity in psychological tests. Psychol Bull. 1955;52:281–302.
23. Craddock N, Owen MJ. The Kraepelinian dichotomy—going, going . . . but still not gone. Br J Psychiatry. 2010;196:92–95.
24. Cuthbert BN, Insel TR. Toward new approaches to psychotic disorders: the NIMH Research Domain Criteria project. Schizophr Bull. 2010;36:1061–1062.
25. Guloksuz S, van Os J. The slow death of the concept of schizophrenia and the painful birth of the psychosis spectrum. Psychol Med. 2017:1–16.
26. Gershon ES. Genetic studies of affective disorders and schizophrenia. Prog Clin Biol Res. 1982;103 Pt A:417–432.
27. Smoller JW. Disorders and borders: psychiatric genetics and nosology. Am J Med Genet B Neuropsychiatr Genet. 2013;162B:559–578.
28. Lichtenstein P, Yip BH, Bjork C, Pawitan Y, Cannon TD, Sullivan PF, Hultman CM. Common genetic determinants of schizophrenia and bipolar disorder in Swedish families: a population-based study. Lancet. 2009;373:234–239.
29. Cross-Disorder Group of the Psychiatric Genomics Consortium. Genetic relationship between five psychiatric disorders estimated from genome-wide SNPs. Nat Genet. 2013;45:984–994.
30. Kupfer DJ, Kuhl EA, Narrow WE, Regier DA. On the road to DSM-V. Cerebrum. 2009.
31. McHugh PR. Psychiatry at a stalemate. Cerebrum. 2009.
32. Wilson EO: Consilience: the unity of knowledge. 1st ed. New York, Knopf: Distributed by Random House; 1998.
33. Andreasen NC. The diagnosis of schizophrenia. Schizophr Bull. 1987;13:9–22.

34. Keshavan MS, Clementz BA, Pearlson GD, Sweeney JA, Tamminga CA. Reimagining psychoses: an agnostic approach to diagnosis. Schizophr Res. 2013;146:10–16.
35. Hall MH, Smoller JW, Cook NR, Schulze K, Hyoun Lee P, Taylor G et al. Patterns of deficits in brain function in bipolar disorder and schizophrenia: a cluster analytic study. Psychiatry Res. 2012;200:272–280.
36. Sponheim SR, Iacono WG, Thuras PD, Beiser M. Using biological indices to classify schizophrenia and other psychotic patients. Schizophr Res. 2001;50:139–150.
37. Hudgens-Haney ME, Ethridge LE, Knight JB, McDowell JE, Keedy SK, Pearlson GD et al. Intrinsic neural activity differences among psychotic illnesses. Psychophysiology. 2017;54:1223–1238.
38. Murray RM. Mistakes I Have Made in My Research Career. Schizophr Bull. 2017;43:253–256.
39. Mitchell KJ. What is complex about complex disorders? Genome Biol. 2012;13:237.
40. Clementz BA, Wang J, Keil A. Normal electrocortical facilitation but abnormal target identification during visual sustained attention in schizophrenia. J Neurosci. 2008;28:13411–13418.
41. Ethridge L, Moratti S, Gao Y, Keil A, Clementz BA. Sustained versus transient brain responses in schizophrenia: the role of intrinsic neural activity. Schizophr Res. 2011;133:106–111.
42. Krishnan GP, Vohs JL, Hetrick WP, Carroll CA, Shekhar A, Bockbrader MA, O'Donnell BF. Steady state visual evoked potential abnormalities in schizophrenia. Clin Neurophysiol. 2005;116:614–624.
43. Rolls ET, Loh M, Deco G, Winterer G. Computational models of schizophrenia and dopamine modulation in the prefrontal cortex. Nat Rev Neurosci. 2008;9:696–709.
44. Hamm JP, Peterka DS, Gogos JA, Yuste R. Altered cortical ensembles in mouse models of schizophrenia. Neuron. 2017;94:153–167 e158.
45. Wang J, Brown R, Dobkins KR, McDowell JE, Clementz BA. Diminished parietal cortex activity associated with poor motion direction discrimination performance in schizophrenia. Cereb Cortex. 2010;20:1749–1755.
46. Hamm JP, Yuste R. Somatostatin interneurons control a key component of mismatch negativity in mouse visual cortex. Cell Rep. 2016;16:597–604.
47. Meehl PE. Factors and taxa, traits and types, differences of degree and differences in kind. J Pers. 1992;60:117–174.
48. Meehl PE. Schizotaxia, schizotypy, schizophrenia. Am Psychol. 1962;17:827–838.
49. Meehl PE. Schizotaxia revisited. Arch Gen Psychiatry. 1989;46:935–944.
50. Gottesman II, Shields J, Hanson DR: Schizophrenia, the epigenetic puzzle. Cambridge; New York, Cambridge University Press; 1982.
51. Kendler KS, Jablensky A. Kraepelin's concept of psychiatric illness. Psychol Med. 2011;41:1119–1126.
52. Kendler KS. Phenomenology of schizophrenia and the representativeness of modern diagnostic criteria. JAMA Psychiatry. 2016;73:1082–1092.
53. Fusar-Poli P, Carpenter WT, Woods SW, McGlashan TH. Attenuated psychosis syndrome: ready for DSM-5.1? Annu Rev Clin Psychol. 2014;10:155–192.
54. Kotov R, Krueger RF, Watson D, Achenbach TM, Althoff RR, Bagby RM et al. The Hierarchical Taxonomy of Psychopathology (HiTOP): a dimensional alternative to traditional nosologies. J Abnorm Psychol. 2017;126:454–477.
55. Cautin RL, Lilienfeld SO: The encyclopedia of clinical psychology. Chichester, West Sussex; Malden, MA, John Wiley and Sons; 2015.
56. Lakatos I: The methodology of scientific research programmes. Cambridge; New York, Cambridge University Press; 1978.

INDEX

For the benefit of digital users, indexed terms that span two pages (e.g., 52–53) may, on occasion, appear on only one of those pages.
Figures, tables, and boxes are indicated by *f*, *t*, and *b* following the page number.

M